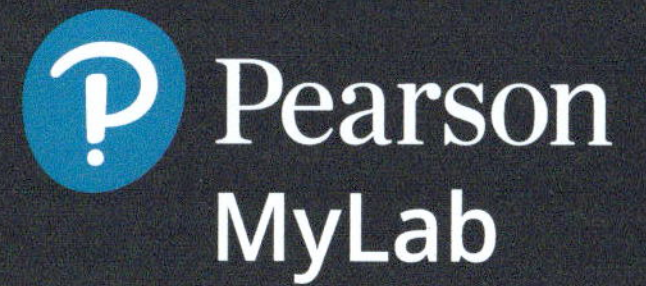

Enrich assignments with question libraries

MyLab Statistics includes a number of question libraries providing additional opportunities for students to practice statistical thinking.

- StatCrunch Projects provide students with opportunities to analyze and interpret data. Each project consists of a series of questions about a large data set in StatCrunch.

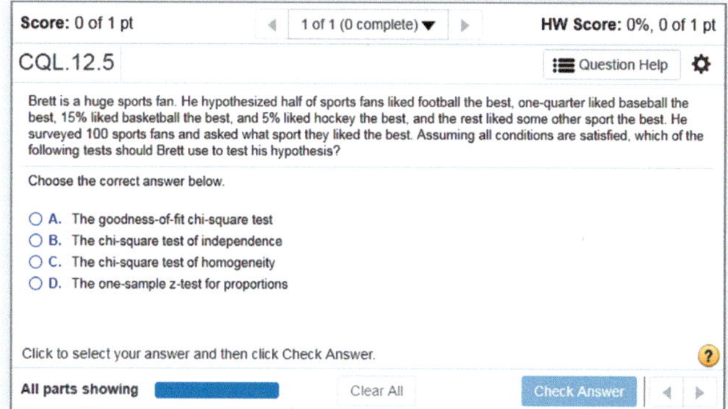

- The Conceptual Question Library offers 1,000 conceptual-based questions to help students internalize concepts, make interpretations, and think critically about statistics.

- The Getting Ready for Statistics Library contains more than 450 exercises on prerequisite topics. Assign these questions to students who may need a little extra practice on their prerequisite skills to be successful in your course.

- The StatTalk Video Library is based on a series of 24 videos, hosted by fun-loving statistician Andrew Vickers, that demonstrate important statistical concepts through interesting stories and real-life events.

Incorporate additional author-created resources in class

Authors infuse their own voice, approach, and experiences teaching statistics into additional text-specific resources, such as interactive applets, technology manuals, workbooks, and more. Check out the Preface to learn more about what's available for this specific title.

FIFTH EDITION

Stats: Data and Models

GLOBAL EDITION

Richard D. De Veaux
Williams College

Paul F. Velleman
Cornell University (Emeritus)

David E. Bock
Cornell University
Ithaca High School (Retired)

Director, Portfolio Management: *Deirdre Lynch*
Courseware Portfolio Manager: *Patrick Barbera*
Courseware Portfolio Management Assistant: *Morgan Danna*
Content Producer: *Tara Corpuz*
Managing Producer: *Scott Disanno*
Producer: *Shannon Bushee*
Manager, Courseware QA: *Mary Durnwald*
Manager, Content Development: *Bob Carroll*
Associate Project Editor, Global Edition: *Aman Kumar*
Associate Media Editor, Global Edition: *Aditee Agarwal*
Managing Producer, Media Production, Global Edition: *Vikram Kumar Medepalli*
Product Marketing Manager: *Emily Ockay*
Field Marketing Manager: *Andrew Noble*
Product Marketing Assistant: *Shannon McCormack*
Senior Author Support/Technology Specialist: *Joe Vetere*
Manager, Rights and Permissions: *Gina Cheselka*
Manufacturing Buyer: *Carol Melville, LSC Communications*
Senior Manufacturing Controller, Global Edition: *Caterina Pellegrino*
Text and Cover Design, Illustrations, Production Coordination, Composition: *SPi Global*
Cover Image: *Jacob Lund/Shutterstock*

Pearson Education Limited

KAO Two
KAO Park
Hockham Way
Harlow
CM17 9SR
United Kingdom

and Associated Companies throughout the world

Visit us on the World Wide Web at: www.pearsonglobaleditions.com

Authorized adaptation from the United States edition, entitled Stats: Data and Models, *5th Edition, ISBN 978-0-13-516382-5 by Richard D. De Veaux, Paul F. Velleman, and David E. Bock, published by Pearson Education © 2020.*

ISBN 10: 1-292-36221-9
ISBN 13: 978-1-292-36221-2

British Library Cataloguing-in-Publication Data
A catalogue record for this book is available from the British Library

1 21

To Sylvia, who has helped me in more ways than she'll ever know,
and to Nicholas, Scyrine, Frederick, and Alexandra,
who make me so proud in everything that they are and do

—Dick

To my sons, David and Zev, from whom I've learned so much,
and to my wife, Sue, for taking a chance on me

—Paul

To Greg and Becca, great fun as kids and great friends as adults,
and especially to my wife and best friend, Joanna, for her
understanding, encouragement, and love

—Dave

MEET THE AUTHORS

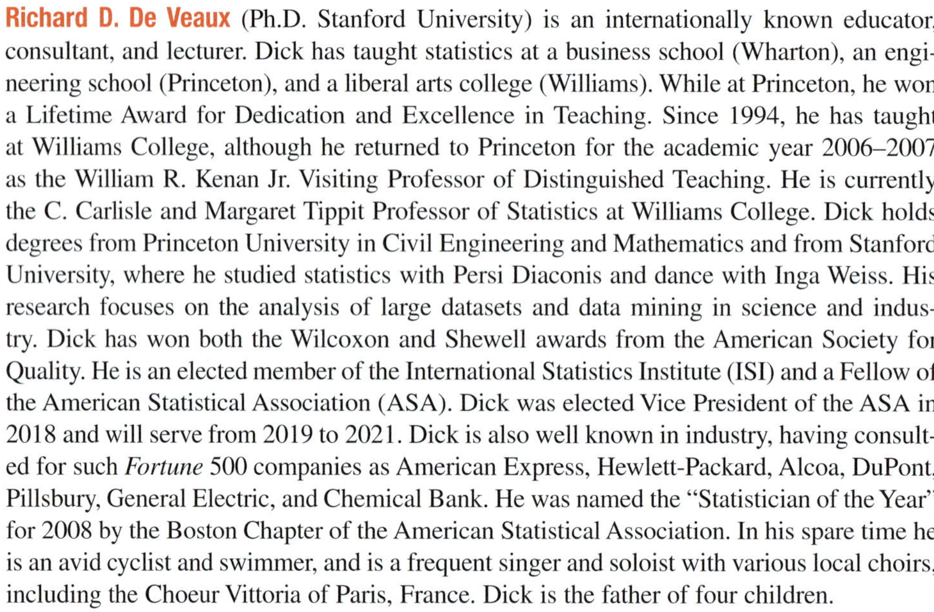

Richard D. De Veaux (Ph.D. Stanford University) is an internationally known educator, consultant, and lecturer. Dick has taught statistics at a business school (Wharton), an engineering school (Princeton), and a liberal arts college (Williams). While at Princeton, he won a Lifetime Award for Dedication and Excellence in Teaching. Since 1994, he has taught at Williams College, although he returned to Princeton for the academic year 2006–2007 as the William R. Kenan Jr. Visiting Professor of Distinguished Teaching. He is currently the C. Carlisle and Margaret Tippit Professor of Statistics at Williams College. Dick holds degrees from Princeton University in Civil Engineering and Mathematics and from Stanford University, where he studied statistics with Persi Diaconis and dance with Inga Weiss. His research focuses on the analysis of large datasets and data mining in science and industry. Dick has won both the Wilcoxon and Shewell awards from the American Society for Quality. He is an elected member of the International Statistics Institute (ISI) and a Fellow of the American Statistical Association (ASA). Dick was elected Vice President of the ASA in 2018 and will serve from 2019 to 2021. Dick is also well known in industry, having consulted for such *Fortune* 500 companies as American Express, Hewlett-Packard, Alcoa, DuPont, Pillsbury, General Electric, and Chemical Bank. He was named the "Statistician of the Year" for 2008 by the Boston Chapter of the American Statistical Association. In his spare time he is an avid cyclist and swimmer, and is a frequent singer and soloist with various local choirs, including the Choeur Vittoria of Paris, France. Dick is the father of four children.

Paul F. Velleman (Ph.D. Princeton University) has an international reputation for innovative statistics education. He designed the Data Desk® software package and is also the author and designer of the award-winning ActivStats® multimedia software, for which he received the EDUCOM Medal for innovative uses of computers in teaching statistics and the ICTCM Award for Innovation in Using Technology in College Mathematics. He is the founder and CEO of Data Description, Inc. (**www.datadesk.com**), which supports both of these programs. Data Description also developed and maintains the Internet site *Data and Story Library* (DASL; **dasl.datadescription.com**), which provides many of the datasets used in this text as well as many others useful for teaching statistics, and the statistics conceptual tools at **astools.datadesk.com**. Paul coauthored (with David Hoaglin) the book *ABCs of Exploratory Data Analysis*. Paul is Emeritus Professor of Statistical Sciences at Cornell University, where he was awarded the MacIntyre Prize for Exemplary Teaching. Paul earned his M.S. and Ph.D. from Princeton University, where he studied with John Tukey. His research often focuses on statistical graphics and data analysis methods. Paul is a Fellow of the American Statistical Association and of the American Association for the Advancement of Science. He was a member of the working group that developed the GAISE 2016 guidelines for teaching statistics.

David E. Bock taught mathematics at Ithaca High School for 35 years. He has taught Statistics at Ithaca High School, Tompkins-Cortland Community College, Ithaca College, and Cornell University. Dave has won numerous teaching awards, including the MAA's Edyth May Sliffe Award for Distinguished High School Mathematics Teaching (twice), Cornell University's Outstanding Educator Award (three times), and has been a finalist for New York State Teacher of the Year.

Dave holds degrees from the University at Albany in Mathematics (B.A.) and Statistics/Education (M.S.). Dave has been a reader and table leader for the AP Statistics exam and a Statistics consultant to the College Board, leading workshops and institutes for AP Statistics teachers. His understanding of how students learn informs much of this book's approach.

Richard De Veaux, Paul Velleman, and David Bock have authored several successful books in the introductory college and AP High School market including *Intro Stats*, Fifth Edition (Pearson, 2018) and *Stats: Modeling the World*, Fifth Edition (Pearson, 2019).

TABLE OF CONTENTS

*Indicates optional sections.

PART V Inference for One Parameter

PART VI Inference for Relationships

PREFACE

Stats: Data and Models, fifth edition, has been especially exciting to develop. The book you hold steps beyond our previous editions in several important ways. Of course, we've kept our conversational style and anecdotes,[1] but we've enriched that material with tools for teaching about randomness, sampling distribution models, and inference throughout the book. And we've expanded discussions of models for data to introduce models with more than two variables earlier in the text. We've taken our inspiration both from our experience in the classroom and from the 2016 revision of the Guidelines for Assessment and Instruction in Statistics Education (GAISE) report adopted by the American Statistical Association. As a result, we increased the text's innovative uses of technology to encourage more statistical thinking, while maintaining its traditional core concepts and coverage. You'll notice that, to expand our attention beyond just one or two variables, we've adjusted the order of some topics.

Innovations

Technology

One of the new GAISE guidelines states: *Use technology to explore concepts and analyze data.* We think a modern statistics text should recognize from the start that statistics is practiced with technology. And so should our students. You won't find tedious calculations worked by hand. You *will* find equation forms that favor intuition over calculation. You'll find extensive use of real data—even large datasets. Throughout, you'll find a focus on statistical thinking rather than calculation. The question that motivates each of our hundreds of examples is not "How do you calculate the answer?" but "How do you think about the answer?"

For this edition of *Stats: Data and Models*, we've taken this principle still further. We have harnessed technology to improve the learning of two of the most difficult concepts in the introductory course: the idea of a sampling distribution and the reasoning of statistical inference.

Multivariable Thinking and Multiple Regression

GAISE's first guideline is to give students experience with multivariable thinking. The world is not univariate, and relationships are not limited to two variables. This edition of *Stats: Data and Models* introduces a third variable as early as Chapter 3's discussion of contingency tables and mosaic plots. Then, following the discussion of correlation and regression as a tool (that is, without inference) in Chapters 6, 7, and 8, we introduce multiple regression in Chapter 9.

Multiple regression may be the most widely used statistical method, and it is certainly one that students need to understand. It is easy to perform multiple regressions with any statistics program, and the exercise of thinking about more than two variables early in the course is worth the effort. We've added new material about interpreting what regression models say. The effectiveness of multiple regression is immediately obvious and makes the reach and power of statistics clear. The use of real data underscores the universal applicability of these methods.

When we return to regression in Chapters 23 and 24 to discuss inference, we can deal with both simple and multiple regression models together. There is nothing different to discuss. (For this reason we set aside the *F*-test until the chapter on ANOVA.)

[1] And footnotes

Innovative ways to teach the logic of statistical inference have received increasing attention. Among these are greater use of computer-based simulations and resampling methods (randomization tests and bootstrapping) to teach concepts of inference.

Bootstrap

The introduction to the new GAISE guidelines explicitly mentions the bootstrap method. The bootstrap is not as widely available or as widely understood as multiple regression. But it follows our presentation naturally. In this edition, we introduce a new feature, **Random Matters**. Random Matters elements in early chapters draw small samples repeatedly from large populations to illustrate how the randomness introduced by sampling leads to both sampling distributions and statistical reasoning for inference. But what can we do when we have only a sample? The bootstrap provides a way to continue this line of thought, now by resampling from the sample at hand.

Bootstrapping provides an elegant way to simulate sampling distributions that we might not otherwise be able to see. And it does not require the assumption of Normality expected by Student's *t*-based methods. However, these methods are not as widely available or widely used in other disciplines, so they should not be the only—or even the principal—methods taught. They may be able to enhance student understanding, but instructors may wish to downplay them if that seems best for a class. We've placed these sections strategically so that instructors can choose the level that they are comfortable with and that works best with their course.

Real Data

GAISE recommends that instructors integrate real data with a context and purpose. More and more high school math teachers are using examples from statistics to demonstrate intuitively how a little bit of math can help us say a lot about the world. So our readers expect statistics to be about real-world insights. *Stats: Data and Models* keeps readers engaged and interested because we show statistics in action right from the start. The exercises pose problems of the kind likely to be encountered in real life and propose ways to think about making inferences almost immediately—and, of course, always with real, up-to-date data.

Let us be clear. *Stats: Data and Models* comes with an archive of more than 500 datasets used in more than 700 applications throughout the book. The datasets are available online at pearsonglobaleditions.com and in MyLab Statistics. Examples that use these datasets cite them in the text. Exercises are marked when they use one of them; exercise names usually indicate the name of the dataset. We encourage students to get the datasets and reproduce our examples using their statistics software, and some of the exercises require that.

Streamlined Content

Following the GAISE recommendations, we've streamlined several parts of the course: Introductory material is covered more rapidly. Today's students have seen a lot of statistics in their K–12 math courses and in their daily contact with online and news sources. We still cover the topics to establish consistent terminology (such as the difference between a histogram and a bar chart). Chapter 2 does most of the work that previously took two chapters.

The Random Matters features show students that statistics vary from sample to sample, show them (empirical) sampling distributions, note the effect of sample size on the shape and variation of the sampling distribution of the mean, and suggest that it looks Normal. As a result, the discussion of the Central Limit Theorem is transformed from the most difficult one in the course to a relatively short discussion ("What you think is true about means really is true; there's this theorem.") that can lead directly to the reasoning of confidence intervals.

Finally, introducing multiple regression doesn't really add much to the lesson on inference for multiple regression because little is new.

GAISE 2016

As we've said, all of these enhancements follow the new Guidelines for Assessment and Instruction in Statistics Education (GAISE) 2016 report adopted by the American Statistical Association:

1. Teach statistical thinking.
   - Teach statistics as an investigative process of problem solving and decision making.
   - Give students experience with multivariable thinking.
2. Focus on conceptual understanding.
3. Integrate real data with a context and purpose.
4. Foster active learning.
5. Use technology to explore concepts and analyze data.
6. Use assessments to improve and evaluate student learning.

The result is a course that is more aligned with the skills needed in the 21st century, one that focuses even more on statistical thinking and makes use of technology in innovative ways, while retaining core principles and topic coverage.

The challenge has been to use this modern point of view to improve learning without discarding what is valuable in the traditional introductory course. Many first statistics courses serve wide audiences of students who need these skills for their own work in disciplines where traditional statistical methods are, well, traditional. So we have not reduced our emphasis on the concepts and methods you expect to find in our texts.

Chapter Order

We've streamlined the presentation of basic topics that most students have already seen. Pie charts, bar charts, histograms, and summary statistics all appear in Chapter 2. Chapter 3 introduces contingency tables, and Chapter 4 discusses comparing distributions. Chapter 5 introduces the Normal model and the 68–95–99.7 Rule. The four chapters of Part II then explore linear relationships among quantitative variables—but here we introduce only the models and how they help us understand relationships. We leave the inference questions until later in the book. Part III discusses how data are gathered by survey and experiment.

Part IV provides background material on probability, random variables, and probability models. In Part V, Chapter 16 introduces confidence intervals for proportions as soon as we've reassured students that their intuition about the sampling distribution of proportions is correct. Chapter 17 formalizes the Central Limit Theorem and introduces Student's *t*-models. Chapter 18 is then about testing hypotheses, and Chapter 19 elaborates further, discussing alpha levels, Type I and Type II errors, power, and effect size. The chapters in Part VI deal with comparing groups (both with proportions and with means), paired samples, chi-square. Finally, Part VII discusses inferences for regression models (both simple and multiple), intelligent uses of multiple regression, and Analysis of Variance, both one- and two-way. A final chapter on data mining looks to the future.

We've found that one of the challenges students face is how to know what technique to use when. In the real world, questions don't come at the ends of the chapters. So, as always, we've provided summaries at the end of each part along with a series of exercises designed to stretch student understanding. These Part Reviews are a mix of questions from all the chapters in that part. The final set are "book-level" review problems that ask students to integrate what they've learned from the entire course. The questions range from simple questions about what method to use in various situations to a more complete data analyses from real data. We hope that these will provide a useful way for students to organize their understanding at the end of the course.

Our Approach

We've discussed how this book is different, but there are some things we haven't changed.

- ***Readability.*** This book doesn't read like other statistics texts. Our style is both colloquial and informative, engaging students to actually read the book to see what it says.
- ***Humor.*** You will find quips and wry comments throughout the narrative, in margin notes, and in footnotes.
- ***Informality.*** Our informal diction doesn't mean that we treat the subject matter lightly or informally. We try to be precise and, wherever possible, we offer deeper explanations and justifications than those found in most introductory texts.
- ***Focused lessons.*** The chapters are shorter than in most other texts so that instructors and students can focus on one topic at a time.
- ***Consistency.*** We try to avoid the "do what we say, not what we do" trap. Having taught the importance of plotting data and checking assumptions and conditions, we model that behavior through the rest of the book. (Check out the exercises in Chapter 24.)
- ***The need to read.*** Statistics is a consistent story about how to understand the world when we have data. The story can't be told piecemeal. This is a book that needs to be read, so we've tried to make the reading experience enjoyable. Students who start with the exercises and then search back for a worked example that looks the same but with different numbers will find that our presentation doesn't support that approach.

Mathematics

Mathematics can make discussions of statistics concepts, probability, and inference clear and concise. We don't shy away from using math where it can clarify without intimidating. But we know that some students are discouraged by equations, so we always provide a verbal description and a numerical example as well.

Nor do we slide in the opposite direction and concentrate on calculation. Although statistics calculations are generally straightforward, they are also usually tedious. And, more to the point, today, virtually all statistics are calculated with technology. We have selected the equations that focus on illuminating concepts and methods rather than for hand calculation. We sometimes give an alternative formula, better suited for hand calculation, for those who find that following the calculation process is a better way to learn about the result.

Technology and Data

We assume that computers and appropriate software are available—at least for demonstration purposes. We hope that students have access to computers and statistics software for their analyses.

We discuss generic computer output at the end of most chapters, but we don't adopt any particular statistics software. The **Tech Support** sections at the ends of chapters offer guidance for seven common software platforms: Data Desk, Excel, JMP, Minitab, SPSS, StatCrunch, and R. We also offer some advice for TI-83/84 Plus graphing calculators, although we hope that those who use them will also have some access to computers and statistics software.

We don't limit ourselves to small, artificial datasets, but base most examples and exercises on real data with a moderate number of cases. Machine-readable versions of the data are available at the book's website at pearsonglobaleditions.com.

Features

Enhancing Understanding

Where Are We Going? Each chapter starts with a paragraph that raises the kinds of questions we deal with in the chapter. A chapter outline organizes the major topics and sections.

New! Random Matters. This new feature travels along a progressive path of understanding randomness and our data. The first Random Matters element begins our thinking about drawing inferences from data. Subsequent Random Matters draw histograms of sample means, introduce the thinking involved in permutation tests, and encourage judgment about how likely the observed statistic seems when viewed against the simulated sampling distribution of the null hypothesis (without, of course, using those terms).

Margin and in-text boxed notes. Throughout each chapter, boxed margin and in-text notes enhance and enrich the text.

Reality Check. We regularly remind students that statistics is about understanding the world with data. Results that make no sense are probably wrong, no matter how carefully we think we did the calculations. Mistakes are often easy to spot with a little thought, so we ask students to stop for a reality check before interpreting their result.

Notation Alert. Throughout this book, we emphasize the importance of clear communication, and proper notation is part of the vocabulary of statistics. We've found that it helps students when we are clear about the letters and symbols statisticians use to mean very specific things, so we've included Notation Alerts whenever we introduce a special notation that students will see again.

Each chapter ends with several elements to help students study and consolidate what they've seen in the chapter.

- **What Can Go Wrong?** sections highlight the most common errors that people make and the misconceptions they have about statistics. One of our goals is to arm students with the tools to detect statistical errors and to offer practice in debunking misuses of statistics, whether intentional or not.
- **Connections** specifically ties the new topics to those learned in previous chapters.
- The **Chapter Review** summarizes the story told by the chapter and provides a bullet list of the major concepts and principles covered.
- A **Review of Terms** is a glossary of all of the special terms introduced in the chapter. In the text, these are printed in **bold** and underlined. The Review provides page references, so students can easily turn back to a full discussion of the term if the brief definition isn't sufficient.

The **Tech Support** section provides the commands in each of the supported statistics packages that deal with the topic covered by the chapter. These are not full documentation, but should be enough to get a student started in the right direction.

Learning by Example

Step-by-Step Examples. We have expanded and updated the examples in our innovative Step-by-Step feature. Each one provides a longer, worked example that guides students through the process of analyzing a problem. The examples follow our three-step Think, Show, Tell organization for approaching a statistics task. They are organized with general explanations of each step on the left and a worked-out solution on the right. The right side of the grid models what would be an "A" level solution to the problem. Step-by-Steps illustrate the importance of thinking about a statistics question (What do we know? What do we hope to learn? Are the assumptions and conditions satisfied?) and reporting our findings (the Tell step). The Show step contains the mechanics of calculating results and conveys our belief that it is only one part of the process. Our emphasis is on statistical thinking, and the pedagogical result is a better understanding of the concept, not just number crunching.

Examples. As we introduce each important concept, we provide a focused example that applies it—usually with real, up-to-the-minute data. Many examples carry the discussion through the chapter, picking up the story and moving it forward as students learn more about the topic.

Just Checking. Just Checking questions are quick checks throughout the chapter; most involve very little calculation. These questions encourage students to pause and think about what they've just read. The Just Checking answers are at the end of the exercise sets in each chapter so students can easily check themselves.

Assessing Understanding

Our **Exercises** have some special features worth noting. First, you'll find relatively simple, focused exercises organized by chapter section. After that come more extensive exercises that may deal with topics from several parts of the chapter or even from previous chapters as they combine with the topics of the chapter at hand. All exercises appear in pairs. The odd-numbered exercises have answers in the back of student texts. Each even-numbered exercise hits the same topic (although not in exactly the same way) as the previous odd exercise. But the even-numbered answers are not provided. If a student is stuck on an even exercise, looking at the previous odd one (and its answer) can often provide the help needed.

More than 600 of our exercises have a T tag next to them to indicate that the dataset referenced in the exercise is available electronically. The exercise title or a note provides the dataset title. Some exercises have a tag to indicate that they call for the student to generate random samples or use randomization methods such as the bootstrap. Although we hope students will have access to computers, we provide ample exercises with full computer output for students to read, interpret, and explain.

We place all the exercises—including section-level exercises—at the end of the chapter. Our writing style is colloquial and encourages reading. We are telling a story about how to understand the world when you have data. Interrupting that story with exercises every few pages would encourage a focus on the calculations rather than the concepts.

Part Reviews. The book is partitioned into seven conceptual parts; each ends with a Part Review. The part review discusses the concepts in that part of the text, tying them together and summarizing the story thus far. Then there are more exercises. These exercises have the advantage (for study purposes) of not being tied to a chapter, so they lack the hints of what to do that would come from that identification. That makes them more like potential exam questions and a good tool for review. Unlike the chapter exercises, these are not paired.

Parts I-VII Cumulative Review Exercises. Cumulative Review exercises are longer and cover concepts from the book as a whole.

Additional Resources Online

Most of the supporting materials can be found online:

- At the book's website at **pearsonglobaleditions.com**
- Within the MyLab Statistics course at **pearsonmylabandmastering.com**

New tools that provide interactive versions of the distribution tables at the back of the book and tools for randomization inference methods such as the bootstrap and for repeated sampling from larger populations can be found online at **astools.datadesk.com**.

MyLab Statistics for *Stats: Data & Models, 5e*

(access code required)

MyLab Statistics is available to accompany Pearson's market-leading text offerings. To give students a consistent tone, voice, and teaching method, each text's flavor and approach are tightly integrated throughout the accompanying MyLab course, making learning the material as seamless as possible.

NEW! StatCrunch Projects

StatCrunch Projects provide opportunities for students to explore data beyond the classroom. In each project, students analyze a data set in StatCrunch® and answer assignable MyLab questions for immediate feedback. StatCrunch Projects span the entire curriculum or focus on certain key concepts. Questions from each project can also be assigned individually.

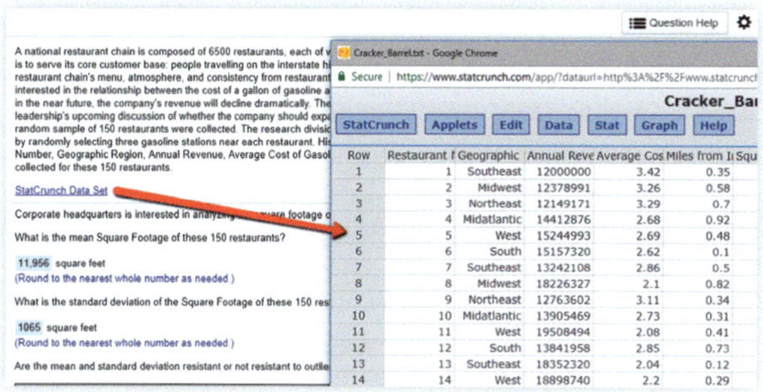

Row	Restaurant I	Geographic	Annual Reve	Average Cos	Miles from I	Squ
1	1	Southeast	12000000	3.42	0.35	
2	2	Midwest	12378991	3.26	0.58	
3	3	Northeast	12149171	3.29	0.7	
4	4	Midatlantic	14412876	2.68	0.92	
5	5	West	15244993	2.69	0.48	
6	6	South	15157320	2.62	0.1	
7	7	Southeast	13242108	2.86	0.5	
8	8	Midwest	18226327	2.1	0.82	
9	9	Northeast	12763602	3.11	0.34	
10	10	Midatlantic	13905469	2.73	0.31	
11	11	West	19508494	2.08	0.41	
12	12	South	13841958	2.85	0.73	
13	13	Southeast	18352320	2.04	0.12	
14	14	West	18898740	2.2	0.29	

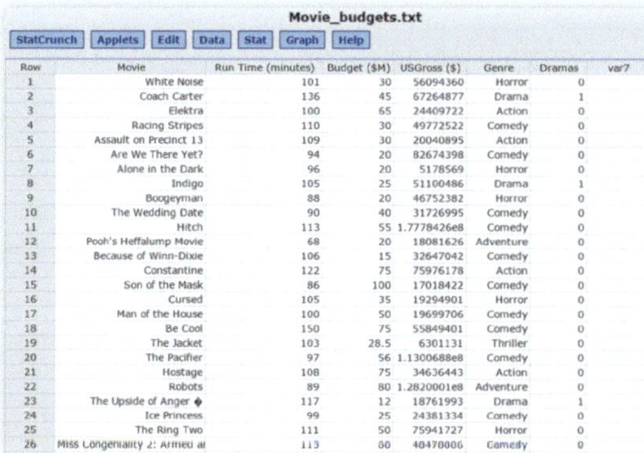

Row	Movie	Run Time (minutes)	Budget ($M)	USGross ($)	Genre	Dramas	var7
1	White Noise	101	30	56094360	Horror	0	
2	Coach Carter	136	45	67264877	Drama	1	
3	Elektra	100	65	24409722	Action	0	
4	Racing Stripes	110	30	49772522	Comedy	0	
5	Assault on Precinct 13	109	30	20040895	Action	0	
6	Are We There Yet?	94	20	82674398	Comedy	0	
7	Alone in the Dark	96	20	5178569	Horror	0	
8	Indigo	105	25	51100486	Drama	1	
9	Boogeyman	88	20	46752382	Horror	0	
10	The Wedding Date	90	40	31726995	Comedy	0	
11	Hitch	113	55	1.7778426e8	Comedy	0	
12	Pooh's Heffalump Movie	68	20	18081626	Adventure	0	
13	Because of Winn-Dixie	106	15	32647042	Comedy	0	
14	Constantine	122	75	75976178	Action	0	
15	Son of the Mask	86	100	17018422	Comedy	0	
16	Cursed	105	35	19294901	Horror	0	
17	Man of the House	100	50	19699706	Comedy	0	
18	Be Cool	150	75	55849401	Comedy	0	
19	The Jacket	103	28.5	6301131	Thriller	0	
20	The Pacifier	97	56	1.1300688e8	Comedy	0	
21	Hostage	108	75	34636443	Action	0	
22	Robots	89	80	1.2820001e8	Adventure	0	
23	The Upside of Anger ◆	117	12	18761993	Drama	1	
24	Ice Princess	99	25	24381334	Comedy	0	
25	The Ring Two	111	50	75941727	Horror	0	
26	Miss Congeniality 2: Armed ar	113	60	48478006	Comedy	0	

UPDATED! Real-World Data

Statistical concepts are applied to everyday life through the extensive current, real-world data examples and exercises provided throughout the text.

Resources for Success

Instructor Resources

Instructor's Solutions Manual (Download Only)

This manual contains detailed solutions to all of the exercises. These files can be downloaded from within MyLab Statistics or from **www.pearsonglobaleditions.com**.

TestGen

TestGen® enables instructors to build, edit, print, and administer tests using a computerized bank of questions developed to cover all the objectives of the text. TestGen is algorithmically based, allowing instructors to create multiple but equivalent versions of the same question or test with the click of a button. Instructors can also modify test bank questions or add new questions. The software and test bank are available for download from Pearson's online catalog, **www.pearsonglobaleditions.com**. The questions are also assignable in MyLab Statistics.

PowerPoint Lecture Slides

PowerPoint Lecture Slides provide an overview of each chapter, stressing important definitions and offering additional examples. They can be downloaded from MyLab Statistics or from **www.pearsonglobaleditions.com**.

Learning Catalytics

Now included in all MyLab Statistics courses, this student response tool uses students' smartphones, tablets, or laptops to engage them in more interactive tasks and thinking during lecture. Learning Catalytics™ fosters student engagement and peer-to-peer learning with real-time analytics. Access pre-built exercises created specifically for statistics.

Question Libraries

In addition to Statcrunch Projects, MyLab Statistics also includes a Getting Ready for Statistics library that contains more than 450 exercises on prerequisite topics and a Conceptual Question Library with 1000 questions that assess conceptual understanding.

Accessibility

Pearson works continuously to ensure our products are as accessible as possible to all students. We are working toward achieving WCAG 2.0 Level AA and Section 508 standards, as expressed in the Pearson Guidelines for Accessible Educational Web Media, **www.pearson.com/mylab/statistics/accessibility**.

Student Resources

Video Resources

Step-by-Step Example videos guide students through the process of analyzing a problem using the "Think, Show, and Tell" strategy from the textbook. StatTalk Videos, hosted by fun-loving statistician Andrew Vickers, demonstrates important statistical concepts through interesting stories and real-life events. StatTalk videos come with accompanying MyLab assessment questions.

StatCrunch

StatCrunch® is powerful web-based statistical software that allows users to collect, crunch, and communicate with data. The vibrant online community offers tens of thousands of shared datasets for students and instructors to analyze, in addition to all of the datasets in the text or online homework. StatCrunch is integrated directly into MyLab Statistics.

Datasets Available Online

Datasets can be found at the book's webpage on **pearsonglobaleditions.com**. Datasets can be easily transferred to any statistics program.

Statistical Software Support

Instructors and students can copy datasets from the text and MyLab exercises directly into software such as StatCrunch, Data desk, or Excel®. Students can also access instructional support tools including tutorial videos, Study Cards, and manuals for a variety of statistical software programs including StatCrunch, Excel, Minitab®, JMP®, R, SPSS, and TI 83/84 calculators.

ACKNOWLEDGMENTS

Many people have contributed to this book throughout all of its editions. This edition never would have seen the light of day without the assistance of the incredible team at Pearson. Director, Portfolio Management Deirdre Lynch was central to the genesis, development, and realization of this project from day one. Our Portfolio Manager, Patrick Barbera, has been invaluable in his support of this edition. Tara Corpuz, Content Producer, kept the cogs from getting into the wheels, where they often wanted to wander. Product Marketing Manager Emily Ockay and Field Marketing Manager Andrew Noble made sure the word got out. Morgan Danna, Portfolio Management Assistant, and Shannon McCormack, Marketing Assistant, were essential in managing all of the behind-the-scenes work. Associate Producer Shannon Bushee put together a top-notch media package for this book. Bob Carroll, Manager of Content Development, for working to ensure digital content meets the mark." Senior Project Manager Chere Bemelmans of the Content Services Centre led us expertly through every stage of production. Manufacturing Buyer Carol Melville, LSC Communications, worked miracles to get this book in your hands.

We would like to draw attention to three people who provided substantial help and support on this edition. First, we would like to acknowledge Nick Horton of Amherst College for his in-depth discussions, guidance, and insights. Nick was invaluable in helping us find the balance between the poles of Normal-based inference and resampling methods. Second, we would like to thank Corey Andreasen of Qatar Academy Doha, Doha, Qatar, and Jared Derksen of Rancho Cucamonga High School for their help with updating the exercises, answers, and datasets.

We'd also like to thank our accuracy checker, Dirk Tempelaar, whose monumental task was to make sure we said what we thought we were saying.

We extend our sincere thanks for the suggestions and contributions made by the following reviewers of this edition:

Ann Cannon
Cornell College

Susan Chimiak
University of Maryland

Lynda Hollingsworth
Northwest Missouri State University

Jeff Kollath
Oregon State University

Cindy Leary
University of Montana

Sheldon Lee
Viterbo University

Pam Omer
Western New England University

Sarah Quesen
West Virginia University

Karin Reinhold
SUNY Albany

Laura Shick
Clemson University

Dirk Tempelaar
Maastricht University

Carol Weideman
St. Petersburg College

Ming Wang
University of Kansas

Lisa Wellinghoff
Wright State

Cathy Zucco-Teveloff
Rider University

We also extend our sincere thanks for the suggestions and contributions made by the following reviewers of the previous editions:

Mary Kay Abbey
Montgomery College

Froozan Pourboghnaf Afiat
Community College of Southern Nevada

Mehdi Afiat
Community College of Southern Nevada

Nazanin Azarnia
Santa Fe Community College

Sanjib Basu
Northern Illinois University

Carl D. Bodenschatz
University of Pittsburgh

Steven Bogart
Shoreline Community College

Ann Cannon
Cornell College

Robert L. Carson
Hagerstown Community College

Jerry Chen
Suffolk County Community College

Rick Denman
Southwestern University

Jeffrey Eldridge
Edmonds Community College

Karen Estes
St. Petersburg Junior College

Richard Friary
Kim (Robinson) Gilbert
Clayton College & State University

Ken Grace
Anoka-Ramsey Community College

Jonathan Graham
University of Montana

Nancy Heckman
University of British Columbia

James Helreich
Marist College

Susan Herring
Sonoma State University

Mary R. Hudachek-Buswell
Clayton State University

Patricia Humphrey
Georgia Southern University

Becky Hurley
Rockingham Community College

Debra Ingram
Arkansas State University

Joseph Kupresanin
Cecil College

Kelly Jackson
Camden County College

Martin Jones
College of Charleston

Rebecka Jornsten
Rutgers University

Michael Kinter
Cuesta College

Kathleen Kone
Community College of Allegheny County

Michael Lichter
State University of New York–Buffalo

Susan Loch
University of Minnesota

Pamela Lockwood
Western Texas A & M University

Wei-Yin Loh
University of Wisconsin–Madison

Steve Marsden
Glendale College

Catherine Matos
Clayton College & State University

Elaine McDonald
Sonoma State University

Jackie Miller
The Ohio State University

Hari Mukerjee
Wichita State University

Helen Noble
San Diego State University

Monica Oabos
Santa Barbara City College

Linda Obeid
Reedley College

Charles C. Okeke
Community College of Southern Nevada

Pamela Omer
Western New England College

Mavis Pararai
Indiana University of Pennsylvania

Gina Reed
Gainesville College

Juana Sanchez
UCLA

Gerald Schoultz
Grand Valley State University

Jim Smart
Tallahassee Community College

Chamont Wang
The College of New Jersey

Edward Welsh
Westfield State College

Heydar Zahedani
California State University, San Marcos

Cathy Zucco-Teveloff
Rider University

Dottie Walton
Cuyahoga Community College

Jay Xu
Williams College

Acknowledgments for the Global Edition

Pearson would like to acknowledge and thank the following for the Global Edition:

Contributors

Lynn Farah
ESCP Europe

Dharini Pathmanathan
University of Malaya

Monique Sciortino
University of Malta

David Suda
Lancaster University

Reviewers

Ruben Garcia Berasategui
Jakarta International College

Margarita Medina
University of Canberra

Kiran Paul
Institute of Bioinformatics and Applied Biotechnology

INDEX OF APPLICATIONS

Accounting

Advertising

Agriculture

Banking

Business (General)

Company Names

Consumers

Demographics

Distribution and Operations Management

E-Commerce

Economics

Education

Energy

Environment

Famous People

Finance and Investments

Food/Drink

Games

Government, Labor, and Law

Human Resource Management/Personnel

Insurance

Manufacturing

Marketing

Media and Entertainment

Pharmaceuticals, Medicine, and Health

Politics and Popular Culture

Quality Control

Real Estate

Salary and Benefits

Sales and Retail

Science

Service Industries and Social Issues

Sports

Surveys and Opinion Polls

Technology

Transportation

Stats Starts Here[1]

WHERE ARE WE GOING?

Statistics gets no respect. People say things like "You can prove anything with statistics." People will write off a claim based on data as "just a statistical trick." And statistics courses don't have the reputation of being students' first choice for a fun elective.

But statistics *is* fun. That's probably not what you heard on the street, but it's true. Statistics is the science of learning from data. A little practice thinking statistically is all it takes to start seeing the world more clearly and accurately.

This is a text about understanding the world by using data. So we'd better start by understanding data. There's more to that than you might have thought.

1.1 What Is Statistics?

People around the world have one thing in common—they all want to figure out what's going on. You'd think with the amount of information available to everyone today this would be an easy task, but actually, as the amount of information grows, so does our need to understand what it can tell us.

At the base of all this information, on the Internet and all around us, are data. We'll talk about data in more detail in the next section, but for now, think of **data** as any collection of numbers, characters, images, or other items that provide information about something. What sense can we make of all this data? You certainly can't make a coherent picture from random pieces of information. Whenever there are data and a need for understanding the world, you'll find statistics.

This text will help you develop the skills you need to understand and communicate the knowledge that can be learned from data. By thinking clearly about the question you're trying to answer and learning the statistical tools to show what the data are saying, you'll acquire the skills to tell clearly what it all means. Our job is to help you make sense of the concepts and methods of statistics and to turn it into a powerful, effective approach to understanding the world through data.

> "But where shall I begin?" asked Alice. "Begin at the beginning," the King said gravely, "and go on till you come to the end: then stop."
>
> —*Lewis Carroll,* Alice's Adventures In Wonderland

[1]We were thinking of calling this chapter "Introduction" but nobody reads the introduction, and we wanted you to read this. We feel safe admitting this down here in the footnotes because nobody reads footnotes either.

> "Data is king at Amazon. Clickstream and purchase data are the crown jewels at Amazon. They help us build features to personalize the Web site experience."
>
> *—Ronny Kohavi, former Director of Data Mining and Personalization, Amazon.com*

Q: What is statistics?
A: Statistics is a way of reasoning, along with a collection of tools and methods, designed to help us understand the world.

Q: What are statistics?
A: Statistics (plural) are particular calculations made from data.

Q: So what is data?
A: You mean "what *are* data?" Data is the plural form. The singular is datum.

Q: OK, OK, so what are data?
A: Data are values along with their context.

The ads say, "Don't drink and drive; you don't want to be a statistic." But you can't be a statistic.

We say, "Don't be a datum."

Data vary. Ask different people the same question and you'll get a variety of answers. Statistics helps us to make sense of the world described by our data by seeing past the underlying variation to find patterns and relationships. This text will teach you skills to help with this task and ways of thinking about variation that are the foundation of sound reasoning about data.

Consider the following:

- If you have a Facebook account, you have probably noticed that the ads you see online tend to match your interests and activities. Coincidence? Hardly. According to *The Wall Street Journal* (10/18/2010),[2] much of your personal information has probably been sold to marketing or tracking companies. Why would Facebook give you a free account and let you upload as much as you want to its site? Because your data are valuable! Using your Facebook profile, a company might build a profile of your interests and activities: what movies and sports you like; your age, sex, education level, and hobbies; where you live; and, of course, who your friends are and what *they* like. From Facebook's point of view, your data are a potential gold mine. Gold ore in the ground is neither very useful nor pretty. But with skill, it can be turned into something both beautiful and valuable. What we're going to talk about is how you can mine your own data and learn valuable insights about the world.
- Americans spend an average of 4.9 hours per day on their smartphones. Trillions of text messages are sent each year.[3] Some of these messages are sent or read while the sender or the receiver is driving. How dangerous is texting while driving?

  How can we study the effect of texting while driving? One way is to measure reaction times of drivers faced with an unexpected event while driving and texting. Researchers at the University of Utah tested drivers on simulators that could present emergency situations. They compared reaction times of sober drivers, drunk drivers, and texting drivers.[4] The results were striking. The texting drivers actually responded more slowly and were more dangerous than drivers who were above the legal limit for alcohol.

In this text, you'll learn how to design and analyze experiments like this. You'll learn how to interpret data and to communicate the message you see to others. You'll also learn how to spot deficiencies and weaknesses in conclusions drawn by others that you see in newspapers and on the Internet every day. Statistics can help you become a more informed citizen by giving you the tools to understand, question, and interpret data.

1.2 Data

STATISTICS IS ABOUT...

- Variation: Data vary because we don't see everything, and even what we do see, we measure imperfectly.
- Learning from data: We hope to learn about the world as best we can from the limited, imperfect data we have.
- Making intelligent decisions: The better we understand the world, the wiser our decisions will be.

Amazon.com opened for business in July 1995, billing itself as "Earth's Biggest Bookstore." By 1997, Amazon had a catalog of more than 2.5 million book titles and had sold books to more than 1.5 million customers in 150 countries. In 2017, the company's sales reached almost $178 billion (more than 30% over the previous year). Amazon has sold a wide variety of merchandise, including a $400,000 necklace, yak cheese from Tibet, and the largest book in the world. How did Amazon become so successful and how can it keep track of so many customers and such a wide variety of products? The answer to both questions is *data*.

But what are data? Think about it for a minute. What exactly *do* we mean by "data"? You might think that data have to be numbers, but data can be text, pictures, web pages,

[2]blogs.wsj.com/digits/2010/10/18/referers-how-facebook-apps-leak-user-ids/

[3]informatemi.com/blog/?p=133

[4]"Text Messaging During Simulated Driving," Drews, F. A., et al., Human Factors: hfs.sagepub.com/content/51/5/762

and even audio and video. If you can sense it, you can measure it. Data are now being collected automatically at such a rate that IBM estimates that "90% of the data in the world today has been created in the last two years alone."[5]

Let's look at some hypothetical values that Amazon might collect:

B0000010AA	0.99	Chris G.	902	105-2686834-3759466	1.99	0.99	Illinois
Los Angeles	Samuel R.	Ohio	N	B000068ZVQ	Amsterdam	New York, New York	Katherine H.
Katherine H.	002-1663369-6638649	Beverly Hills	N	N	103-2628345-9238664	0.99	Massachusetts
312	Monique D.	105-9318443-4200264	413	B00000I5Y6	440	B000002BK9	0.99
Canada	Detroit	440	105-1372500-0198646	N	B002MXA7Q0	Ohio	Y

Try to guess what they represent. Why is that hard? Because there is no *context*. If we don't know what values are measured and what is measured about them, the values are meaningless. We can make the meaning clear if we organize the values into a **data table** such as this one:

Order Number	Name	State/Country	Price	Area Code	Download	Gift?	ASIN	Artist
105-2686834-3759466	Katherine H.	Ohio	0.99	440	Amsterdam	N	B0000015Y6	Cold Play
105-9318443-4200264	Samuel R.	Illinois	1.99	312	Detroit	Y	B000002BK9	Red Hot Chili Peppers
105-1372500-0198646	Chris G.	Massachusetts	0.99	413	New York, New York	N	B000068ZVQ	Frank Sinatra
103-2628345-9238664	Monique D.	Canada	0.99	902	Los Angeles	N	B0000010AA	Blink 182
002-1663369-6638649	Katherine H.	Ohio	0.99	440	Beverly Hills	N	B002MXA7Q0	Weezer

Now we can see that these are purchase records for album download orders from Amazon. The column titles tell what has been recorded. Each row is about a particular purchase.

What information would provide a **context**? Newspaper journalists know that the lead paragraph of a good story should establish the "Five W's": *who, what, when, where,* and (if possible) *why*. Often, we add *how* to the list as well. The answers to the first two questions are essential. If we don't know *what* values are measured and *who* those values are measured on, the values are meaningless.

Who and What

In general, the rows of a data table correspond to individual **cases** about *whom* (or about which, if they're not people) we record some characteristics. Cases go by different names, depending on the situation.

- Individuals who answer a survey are called **respondents**.
- People on whom we experiment are **subjects** or (in an attempt to acknowledge the importance of their role in the experiment) **participants**.

[5]www-01.ibm.com/software/data/bigdata/what-is-big-data.html

DATA BEATS INTUITION

Amazon monitors and updates its website to better serve customers and maximize sales. To decide which changes to make, analysts experiment with new designs, offers, recommendations, and links. Statisticians want to know how long you'll spend browsing the site and whether you'll follow the links or purchase the suggested items. As Ronny Kohavi, former director of Data Mining and Personalization for Amazon, said, "Data trumps intuition. Instead of using our intuition, we experiment on the live site and let our customers tell us what works for them."

- Animals, plants, websites, and other inanimate subjects are often called **experimental units**.
- Often we simply call cases what they are: for example, *customers, economic quarters*, or *companies*.
- In a database, rows are called **records**—in this example, purchase records. Perhaps the most generic term is *cases*, but in any event the rows represent the *Who* of the data.

Look at all the columns to see exactly what each row refers to. Here the cases are different purchase records. You might have thought that each customer was a case, but notice that, for example, Katherine H. appears twice, in both the first and the last rows. A common place to find out exactly what each row refers to is the leftmost column. That value often identifies the cases, in this example, it's the order number. If you collect the data yourself, you'll know what the cases are. But, often, you'll be looking at data that someone else collected and you'll have to ask or figure that out yourself.

Often the cases are a **sample** from some larger **population** that we'd like to understand. Amazon doesn't care about just these customers; it wants to understand the buying patterns of *all* its customers, and, generalizing further, it wants to know how to attract other Internet users who may not have made a purchase from Amazon's site. To be able to generalize from the sample of cases to the larger population, we'll want the sample to be *representative* of that population—a kind of snapshot image of the larger world.

We must know *who* and *what* to analyze data. Without knowing these two, we don't have enough information to start. Of course, we'd always like to know more. The more we know about the data, the more we'll understand about the world. If possible, we'd like to know the *when* and *where* of data as well. Values recorded in 1803 may mean something different than similar values recorded last year. Values measured in Tanzania may differ in meaning from similar measurements made in Mexico. And knowing *why* the data were collected can tell us much about its reliability and quality.

How the Data Are Collected

How the data are collected can make the difference between insight and nonsense. As we'll see later, data that come from a voluntary survey on the Internet are almost always worthless. One primary concern of statistics, to be discussed in Part III, is the design of sound methods for collecting data. Throughout this text, whenever we introduce data, we'll provide a margin note listing the W's (and H) of the data. Identifying the W's is a habit we recommend.

The first step of any data analysis is to know what you are trying to accomplish and what you want to know. To help you use statistics to understand the world and make decisions, we'll lead you through the entire process of *thinking* about the problem, *showing* what you've found, and *telling* others what you've learned. Every guided example in this text is broken into these three steps: *Think*, *Show*, and *Tell*. Identifying the problem and the *who* and *what* of the data is a key part of the *Think* step of any analysis. Make sure you know these before you proceed to *Show* or *Tell* anything about the data.

EXAMPLE 1.1 Identifying the *Who*

In 2015, *Consumer Reports* published an evaluation of 126 computer tablets from a variety of manufacturers.

QUESTION: Describe the population of interest, the sample, and the *Who* of the study.

ANSWER: The magazine is interested in the performance of tablets currently offered for sale. It tested a sample of 126 tablets, which are the *Who* for these data. Each tablet selected represents all similar tablets offered by that manufacturer.

1.3 Variables

The characteristics recorded about each individual are called **variables**. They are usually found as the columns of a data table with a name in the header that identifies what has been recorded. In the Amazon data table we find the variables *Order Number, Name, State/Country, Price*, and so on.

Categorical Variables

Some variables just tell us what group or category each individual belongs to. Are you male or female? Pierced or not? We call variables like these **categorical**, or **qualitative**, **variables**. (You may also see them called **nominal variables** because they name categories.) Some variables are clearly categorical, like the variable *State/Country*. Its values are text and those values tell us what category the particular case falls into. But numerals are often used to label categories, so categorical variable values can also be numerals. For example, Amazon collects telephone area codes that *categorize* each phone number into a geographical region. So area code is considered a categorical variable even though it has numeric values. (But see the story in the following box.)

> **AREA CODES—NUMBERS OR CATEGORIES?**
>
> The *What* and *Why* of area codes are not as simple as they may first seem. When area codes were first introduced, AT&T was still the source of all telephone equipment, and phones had dials.
>
> To reduce wear and tear on the dials, the area codes with the lowest digits (for which the dial would have to spin least) were assigned to the most populous regions—those with the most phone numbers and thus the area codes most likely to be dialed. New York City was assigned 212, Chicago 312, and Los Angeles 213, but rural upstate New York was given 607, Joliet was 815, and San Diego 619. For that reason, at one time the numerical value of an area code could be used to guess something about the population of its region. Since the advent of push-button phones, area codes have finally become just categories.

"Far too many scientists have only a shaky grasp of the statistical techniques they are using. They employ them as an amateur chef employs a cookbook, believing the recipes will work without understanding why. A more *cordon bleu* attitude . . . might lead to fewer statistical soufflés failing to rise."

—*The Economist, June 3, 2004, "Sloppy stats shame science"*

Descriptive responses to questions are often categories. For example, the responses to the questions "Who is your cell phone provider?" and "What is your marital status?" yield categorical values. When Amazon considers a special offer of free shipping to customers, it might first analyze how purchases have been shipped in the recent past. Amazon might start by counting the number of purchases shipped in each category: ground transportation, second-day air, and next-day air. Counting is a natural way to summarize a categorical variable such as *Shipping Method*. Chapter 2 discusses summaries and displays of categorical variables more fully.

Quantitative Variables

When a variable contains measured numerical values with measurement *units*, we call it a **quantitative variable**. Quantitative variables typically record an amount or degree of something. For quantitative variables, its measurement **units** provide a meaning for the numbers. Even more important, units such as yen, cubits, carats, angstroms, nanoseconds, miles per hour, or degrees Celsius tell us the *scale* of measurement, so we know how far apart two values are. Without units, the values of a measured variable have no meaning. It does little good to be promised a raise of 5000 a year if you don't know whether it will be paid in euros, dollars, pennies, yen, or Mauritanian Ouguiya (MRU).[6] We'll see how to display and summarize quantitative variables in Chapter 2.

Some quantitative variables don't have obvious units. The Dow Jones Industrial "Average" has units (points?) but no one talks about them. Percentages are ratios of two quantities and so the units "cancel out." But they are still percentages of something. So although it isn't imperative that a quantitative variable have explicit units, when they are not explicit, be careful to think about whether adding their values, averaging them, or otherwise treating them as numerical makes sense.

Sometimes a variable with numerical values can be treated as either categorical or quantitative depending on what we want to know from it. Amazon could record your *Age* in years. That seems quantitative, and it would be if the company wanted to know the average age of those customers who visit their site after 3 AM. But suppose Amazon wants to decide which album to feature on its site when you visit. Then thinking of your age in one of the categories Child, Teen, Adult, or Senior might be more useful. So, sometimes whether a variable is treated as categorical or quantitative is more about the question we want to ask rather than an intrinsic property of the variable itself.

Identifiers

For a categorical variable like *Survived*, each individual is assigned one of two possible values, say *Alive* or *Dead.*[7] But for a variable with ID numbers, such as a *student ID*, each individual receives a unique value. We call a variable like this, which has exactly as many values as cases, an **identifier variable**. Identifiers are useful, but not typically for analysis.

Amazon wants to know who you are when you sign in again and doesn't want to confuse you with some other customer. So it assigns you a unique identifier. Amazon also wants to send you the right product, so it assigns a unique Amazon Standard Identification Number (ASIN) to each item it carries. You'll want to recognize when a variable is playing the role of an identifier so you aren't tempted to analyze it.

Identifier variables themselves don't tell us anything useful about their categories because we know there is exactly one individual in each. Identifiers are part of what's called **metadata**, or data about the data. Metadata are crucial in this era of large datasets because by uniquely identifying the cases, they make it possible to combine data from different sources, protect (or violate) privacy, and provide unique labels.[8] Many large databases are *relational* databases. In a relational database, different data tables link to one another by matching identifiers. In the Amazon example, the *Customer Number*, *ASIN*, and *Transaction Number* are all identifiers. The IP (Internet Protocol) address of your computer is another identifier, needed so that the electronic messages sent to you can find you.

> **PRIVACY AND THE INTERNET**
>
> You have many identifiers: a Social Security number, a student ID number, possibly a passport number, a health insurance number, and probably a Facebook account name. Privacy experts are worried that Internet thieves may match your identity in these different areas of your life, allowing, for example, your health, education, and financial records to be merged. Even online companies such as Facebook and Google are able to link your online behavior to some of these identifiers, which carries with it both advantages and dangers. The National Strategy for Trusted Identities in Cyberspace (www.wired.com/images_blogs/threatlevel/2011/04/NSTICstrategy_041511.pdf) proposes ways that we may address this challenge in the near future.

Ordinal Variables

A typical course evaluation survey asks, "How valuable do you think this course will be to you?" 1 = Worthless; 2 = Slightly; 3 = Middling; 4 = Reasonably; 5 = Invaluable. Is *Educational Value* categorical or quantitative? Often the best way to tell is to look to the *Why* of the study. A teacher might just count the number of students who gave each response for her course, treating *Educational Value* as a categorical variable. When she wants to see whether the course is improving, she might treat the responses as the *amount* of perceived value—in effect, treating the variable as quantitative.

But what are the units? There is certainly an *order* of perceived worth: Higher numbers indicate higher perceived worth. A course that averages 4.5 seems more valuable than one that averages 2, but we should be careful about treating *Educational Value* as purely quantitative. To treat it as quantitative, she'll have to imagine that it has "educational value units" or some similar arbitrary construct. Because there are no natural units, she

[6] As of 9/7/2018 $1 = 35.95 MRU.

[7] Well, maybe three values if you include Zombies.

[8] The National Security Agency (NSA) made the term "metadata" famous in 2014 by insisting that they only collected metadata on U.S. citizens' phone calls and text messages, not the calls and messages themselves. They later admitted to the bulk collection of actual data.

should be cautious. Variables that report order without natural units are often called **ordinal variables**. But saying "that's an ordinal variable" doesn't get you off the hook. You must still look to the *Why* of your study and understand what you want to learn from the variable to decide whether to treat it as categorical or quantitative.

EXAMPLE 1.2 Identifying the *What* and *Why* of Tablets

RECAP: A *Consumer Reports* article about 126 tablets lists each tablet's manufacturer, price, battery life (hrs.), the operating system (Android, iOS, or Windows), an overall quality score (0–100), and whether or not it has a memory card reader.

QUESTION: Are these variables categorical or quantitative? Include units where appropriate, and describe the *Why* of this investigation.

ANSWER: The variables are

- manufacturer (categorical)
- price (quantitative, $)
- battery life (quantitative, hrs.)
- operating system (categorical)
- quality score (quantitative, no units)
- memory card reader (categorical)

The magazine hopes to provide consumers with the information to choose a good tablet.

JUST CHECKING

In the 2004 Tour de France, Lance Armstrong made history by winning the race for an unprecedented sixth time. In 2005, he became the only 7-time winner and set a new record for the fastest average speed—41.65 kilometers per hour. In 2012, he was banned for life for doping offenses, stripped of all of his titles and his records expunged. You can find data on all the Tour de France races in the dataset **Tour de France 2017**. Here are the first three and last seven lines of the dataset. Keep in mind that the entire dataset has over 100 entries.

1. List as many of the W's as you can for this dataset.
2. Classify each variable as categorical or quantitative; if quantitative, identify the units.

Year	Winner	Country of Origin	Age	Team	Total Time (hours)	Avg. Speed (km/h)	Stages	Total Distance Ridden (km)	Starting Riders	Finishing Riders
1903	Maurice Garin	France	32	La Française	94.55	25.7	6	2428	60	21
1904	Henri Cornet	France	20	Cycles JC	96.10	25.3	6	2428	88	23
1905	Louis Trousseller	France	24	Peugeot	110.45	27.1	11	2994	60	24
...					....					
2011	Cadel Evans	Australia	34	BMC	86.21	39.79	21	3430	198	167
2012	Bradley Wiggins	Great Britain	32	Sky	87.58	39.83	20	3488	198	153
2013	Christopher Froome	Great Britain	28	Sky	94.55	40.55	21	3404	198	169
2014	Vincenzo Nibali	Italy	29	Astana	89.93	40.74	21	3663.5	198	164
2015	Christopher Froome	Great Britain	30	Sky	84.77	39.64	21	3660.3	198	160
2016	Christopher Froome	Great Britain	31	Sky	89.08	39.62	21	3529	198	174
2017	Christopher Froome	Great Britain	32	Sky	86.34	40.997	21	3540	198	167
2018	Geraint Thomas	Great Britain	32	Sky	83.28	40.210	21	3349	176	145

THERE'S A WORLD OF DATA ON THE INTERNET

These days, one of the richest sources of data is the Internet. With a bit of practice, you can learn to find data on almost any subject. Many of the datasets we use in this text were found in this way. The Internet has both advantages and disadvantages as a source of data. Among the advantages are the fact that often you'll be able to find even more current data than those we present. The disadvantage is that references to Internet addresses can "break" as sites evolve, move, and die.

Our solution to these challenges is to offer the best advice we can to help you search for the data, wherever they may be residing. We usually point you to a website. We'll sometimes suggest search terms and offer other guidance.

Some words of caution, though: Data found on Internet sites may not be formatted in the best way for use in statistics software. Although you may see a data table in standard form, an attempt to copy the data may leave you with a single column of values. You may have to work in your favorite statistics or spreadsheet program to reformat the data into variables. You will also probably want to remove commas from large numbers and extra symbols such as money indicators ($, ¥, £); few statistics packages can handle these.

1.4 Models

What is a **model** for data? Models are summaries and simplifications of data that help our understanding in many ways. We'll encounter all sorts of models throughout the text. A model is a simplification of reality that gives us information that we can learn from and use, even though it doesn't represent reality exactly. A model of an airplane in a wind tunnel can give insights about the aerodynamics and flight performance of the plane even though it doesn't show every rivet.[9] In fact, it's precisely because a model is a simplification that we learn from it. Without making models for how data vary, we'd be limited to reporting only what the data we have at hand says. To have an impact on science and society we'll have to generalize those findings to the world at large.

Kepler's laws describing the motion of planets are a great example of a model for data. Using astronomical observations of Tycho Brahe, Kepler saw through the small anomalies in the measurements and came up with three simple "laws"—or models for how the planets move. Here are Brahe's observations on the declination (angle of tilt to the sun) of Mars over a twenty-year period just before 1600:

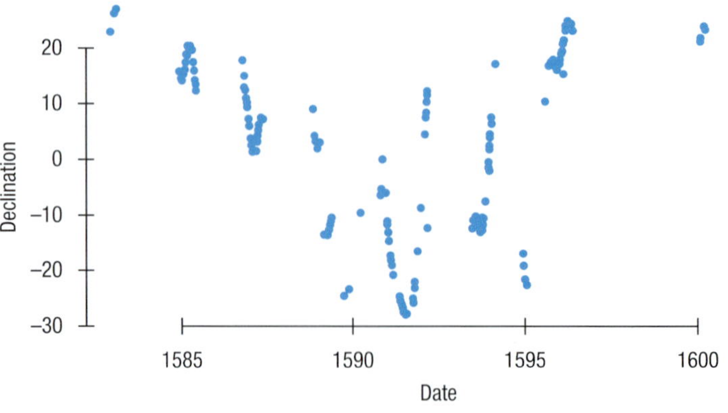

Figure 1.1
A plot of declination against time shows some patterns. There are many missing observations. Can you see the model that Kepler came up with from these data?

[9]Or tell you what movies you might see on the flight.

Here, using modern statistical methods is a plot of the model predictions from the data:

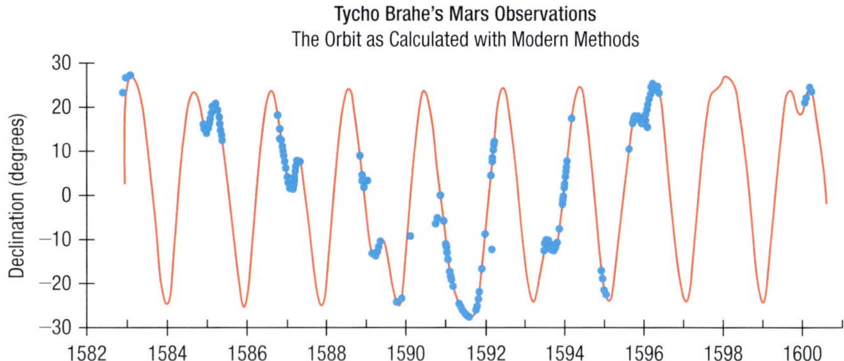

Figure 1.2
The model that Kepler proposed filled in many of the missing points and made the pattern much clearer.

Later, after Newton laid out the physics of gravity, it could be shown that the laws follow from other principles, but Kepler derived the models from data. We may not be able to come up with models as profound as Kepler's, but we'll use models throughout the text. We'll see examples of models as early as Chapter 5 and then put them to use more thoroughly later in the text when we discuss inference.

WHAT CAN GO WRONG?

- **Don't label a variable as categorical or quantitative without thinking about the data and what they represent.** The same variable can sometimes take on different roles.
- **Don't assume that a variable is quantitative just because its values are numbers.** Categories are often given numerical labels. Don't let that fool you into thinking they have quantitative meaning. Look at the context.
- **Always be skeptical.** One reason to analyze data is to discover the truth. Even when you are told a context for the data, it may turn out that the truth is a bit (or even a lot) different. The context colors our interpretation of the data, so those who want to influence what you think may slant the context. A survey that seems to be about all students may in fact report just the opinions of those who visited a fan website. The question that respondents answered may be posed in a way that influences responses.

CHAPTER REVIEW

Understand that data are values, whether numerical or labels, together with their context.

- *Who, what, why, where, when* (and *how*)—the W's—help nail down the context of the data.
- We must know *who, what,* and *why* to be able to say anything useful based on the data. The *Who* are the cases. The *What* are the variables. A variable gives information about each of the cases. The *Why* helps us decide which way to treat the variables.
- Stop and identify the W's whenever you have data, and be sure you can identify the cases and the variables.

Consider the source of your data and the reasons the data were collected. That can help you understand what you might be able to learn from the data.

Identify whether a variable is being used as categorical or quantitative.

- Categorical variables identify a category for each case. Usually we think about the counts of cases that fall in each category. (An exception is an identifier variable that just names each case.)
- Quantitative variables record measurements or amounts of something. They typically have units or are ratios of quantities that have units.
- Sometimes we may treat the same variable as categorical or quantitative depending on what we want to learn from it, which means some variables can't be pigeonholed as one type or the other.

REVIEW OF TERMS

The key terms are in chapter order so you can use this list to review the material in the chapter.

Data Recorded values, whether numbers or labels, together with their context (p. 27).

Data table An arrangement of data in which each row represents a case and each column represents a variable (p. 29).

Context The context ideally tells *who* was measured, *what* was measured, *how* the data were collected, *where* the data were collected, and *when* and *why* the study was performed (p. 29).

Case An individual about whom or which we have data (p. 29).

Respondent Someone who answers, or responds to, a survey (p. 29).

Subject A human experimental unit. Also called a participant (p. 29).

Participant A human experimental unit. Also called a subject (p. 29).

Experimental unit An individual in a study for which or for whom data values are recorded. Human experimental units are usually called subjects or participants (p. 30).

Record Information about an individual in a database (p. 30).

Sample A subset of a population, examined in hope of learning about the population (p. 30).

Population The entire group of individuals or instances about whom we hope to learn (p. 30).

Variable A variable holds information about the same characteristic for many cases (p. 31).

Categorical (or qualitative) variable A variable that names categories with words or numerals (p. 31).

Nominal variable The term "nominal" can be applied to a variable whose values are used only to name categories (p. 31).

Quantitative variable A variable in which the numbers are values of measured quantities (p. 31).

Unit A quantity or amount adopted as a standard of measurement, such as dollars, hours, or grams (p. 31).

Identifier variable A categorical variable that records a unique value for each case, used to name or identify it (p. 32).

Metadata Data about the data. Metadata can provide information to uniquely identify cases, making it possible to combine data from different sources, protect (or violate) privacy, and label cases uniquely (p. 32).

Ordinal variable The term "ordinal" can be applied to a variable whose categorical values possess some kind of order (p. 33).

Model A description or representation, in mathematical and statistical terms, of the behavior of a phenomenon based on data (p. 34).

TECH SUPPORT

Entering Data

These days, nobody does statistics by hand. We use technology: a programmable calculator or a statistics program on a computer. Professionals all use a *statistics package* designed for the purpose. We will provide many examples of results from a statistics package throughout the text. Rather than choosing one in particular, we'll offer generic results that look like those produced by all the major statistics packages but don't exactly match any of them. Then, in the Tech Support section at the end of each chapter, we'll provide hints for getting started on several of the major packages.

If you understand what the computer needs to know to do what you want and what it needs to show you in return, you can figure out the specific details of most packages pretty easily.

For example, to get your data into a computer statistics package, you need to tell the computer:

- Where to find the data. This usually means directing the computer to a file stored on your computer's disk or to data on a database. Or it might just mean that you have copied the data from a spreadsheet program or Internet site and it is currently on your computer's clipboard. Usually, the data should be in the form of a data table with cases in the rows and variables in the columns. Most computer statistics packages prefer the *delimiter* that marks the division between elements of a data table to be a tab character (comma is another common delimiter) and the delimiter that marks the end of a case to be a *return* character. The data used in this text can be found on the text's website at www.pearsonglobaleditions.com.
- Where to put the data. (Usually this is handled automatically.)
- What to call the variables. Some data tables have variable names as the first row of the data, and often statistics packages can take the variable names from the first row automatically.
- Excel is often used to help organize, manipulate, and prepare data for other software packages. Many of the other packages take Excel files as inputs. Alternatively, you can copy a data table from Excel and paste it into many packages, or export Excel spreadsheets as tab delimited (.txt) or comma delimited files (.csv), which can be easily shared and imported into other programs. All data files provided with this text are in tab-delimited text (.txt) format.

EXCEL

To open a file containing data in Excel:

- Choose **File** > **Open**.
- Browse to find the file to open. Excel supports many file formats.
- Other programs can import data from a variety of file formats, but all can read both tab delimited (.txt) and comma delimited (.csv) text files.
- You can also copy tables of data from other sources, such as Internet sites, and paste them into an Excel spreadsheet. Excel can recognize the format of many tables copied this way, but this method may not work for some tables.
- Excel may not recognize the format of the data. If data include dates or other special formats ($, €, ¥, etc.), identify the desired format. Select the cells or columns to reformat and choose **Format** > **Cell**. Often, the General format is the best option for data you plan to move to a statistics package.

DATA DESK

To read data into Data Desk:

- Click the **Open File** icon or choose **File** > **Open**. The dialog lets you specify variable names (or take them from the first row of the data), the delimiter, or how to read formatted data.
- **File** > **Import** works the same way, but instead of starting a new data file, it adds the data in the file to the current data file. Data Desk can work with multiple data tables in the same file.
- If the data are already in another program, such as, for example, a spreadsheet, **Copy** the data table (including the column headings). In Data Desk choose **Edit** > **Paste** variables. There is no need to create variables first; Data Desk does that automatically. You'll see the same dialog as for Open and Import.

JMP

To import a text file:

- Choose **File** > **Open** and select the file from the dialog. At the bottom of the dialog screen you'll see **Open As:**—be sure to change to **Data (Using Preview)**. This will allow you to specify the delimiter and make sure the variable names are correct. (**JMP** also allows various formats to be imported directly, including .xls files.)

You can also paste a dataset in directly (with or without variable names) by selecting:

- **File** > **New** > **New Data Table** and then **Edit** > **Paste** (or **Paste with Column Names** if you copied the names of the variables as well).

Finally, you can import a dataset from a URL directly by selecting:

- **File** > **Internet Open** and pasting in the address of the website. JMP will attempt to find data on the page. It may take a few tries and some edits to get the dataset in correctly.

MINITAB

To import a text or Excel file:

- Choose **File** > **Open Worksheet**. From **Files of type**, choose **Text (*.txt)** or **Excel (*.xls; *xlsx)**.
- Browse to find and select the file.
- In the lower right corner of the dialog, choose **Open** to open the data file alone, or **Merge** to add the data to an existing worksheet.
- Click **Open**.

R

R can import many types of files, but text files (tab or comma delimited) are easiest. If the file is tab delimited and contains the variable names in the first row, then:

> **mydata = read.delim(file.choose())**

will give a dialog where you can pick the file you want to import. It will then be in a data frame called mydata. If the file is comma delimited, use:

> **mydata = read.csv(file.choose())**

COMMENTS

RStudio provides an interactive dialog that may be easier to use. For other options, including the case that the file does not contain variable names, consult **R** help.

SPSS

To import a text file:

- Choose **File** > **Open** > **Data**. Under "Files of type," choose **Text (*.txt,*.dat)**. Select the file you want to import. Click **Open**.
- A window will open called **Text Import Wizard**. Follow the steps, depending on the type of file you want to import.

STATCRUNCH

StatCrunch offers several ways to enter data. Click **MyStatCrunch** > **My Data**. Click a dataset to analyze the data or edit its properties.

Click a dataset link to analyze the data or edit its properties to import a new dataset.

- Choose **Select a file on my computer**,
- Enter the URL of a file,
- Paste data into a form, or
- Type or paste data into a blank data table.

For the "select a file on my computer" option, StatCrunch offers a choice of space, comma, tab, or semicolon delimiters. You may also choose to use the first line as the names of the variables.

After making your choices, select the **Load File** button at the bottom of the screen.

StatCrunch has direct access to the datasets on the text's website.

EXERCISES

SECTION 1.1

1. Grocery shopping Many grocery store chains offer customers a card they can scan when they check out and offer discounts to people who do so. To get the card, customers must give information, including a mailing address and e-mail address. The actual purpose is not to reward loyal customers but to gather data. What data do these cards allow stores to gather, and why would they want that data?

2. Online shopping Online retailers such as Amazon.com keep data on products that customers buy, and even products they look at. What does Amazon hope to gain from such information?

3. Parking lots Sensors in parking lots are able to detect and communicate when spaces are filled in a large covered parking garage next to an urban shopping mall. How might the owners of the parking garage use this information both to attract customers and to help the store owners in the mall make business plans?

4. Satellites and global climate change Satellites send back nearly continuous data on the earth's land masses, oceans, and atmosphere from space. How might researchers use this information in both the short and long terms to help study changes in the earth's climate?

SECTION 1.2

5. Lottery Every year, the administration of a lottery competition performs statistical analysis of previous records. They devise a list of the numbers drawn at each draw, the amount of money won at each draw, the number of times each number has been drawn during that particular year, and the number of draws since each number was last drawn. Identify the *Who* in this list.

6. Nobel laureates The website www.nobelprize.org allows you to look up all the Nobel prizes awarded in any year. The data are not listed in a table. Rather you drag a slider to the year and see a list of the awardees for that year. Describe the *Who* in this scenario.

7. Family growth The National Center for Health Statistics of a country conducts a national survey on family growth, which consists of over 5,000 interviews in each interviewing year. The sample chosen is representative of men and women aged 15–49 living in households. This extensive survey gathers information on "family life, marriage and divorce, pregnancy, infertility, use of contraception, and men's and women's health." Describe the sample, the population, the *Who,* and the *What* of this survey.

8. Facebook. Facebook uploads more than 350 million photos every day onto its servers. For this collection, describe the *Who* and the *What*.

SECTION 1.3

9. Grade levels A person's grade in school is generally identified by a number.

a) Give an example of a *Why* in which grade level is treated as categorical.
b) Give an example of a *Why* in which grade level is treated as quantitative.

10. License plate numbers Passenger car license plates in a city are in the format 1ABC123.

a) In what sense are the car license plate numbers categorical?
b) Is there any ordinal sense to the car license plate numbers? In other words, does a license plate tell you anything about the date when the car was registered?

11. Referendum In a questionnaire administered to the members of a particular gym, the members were asked to specify how many hours they spend training at the gym every week. What kind of variable is the response?

12. Tablet dissolution Manufacturers in a pharmaceutical manufacturing company employ the technique of tablet dissolution to ensure that each drug is delivered properly to patients. This technique is used to measure the rate at which a drug releases from a dosage form. What kind of variable is the company studying?

13. Job satisfaction All employees in a company were asked to respond to the following question relating to their job satisfaction: "How satisfied are you in your current job position?" The possible choices were "Very Satisfied," "Satisfied," "Neutral," "Unsatisfied," and "Very Unsatisfied." What kind of variable is the response?

14. Stress A medical researcher measures the increase in heart rate of patients who are taking a stress test. What kind of variable is the researcher studying?

SECTION 1.4

15. Voting and elections Pollsters are interested in predicting the outcome of elections. Give an example of how they might model whether someone is likely to vote.

16. Weather Meteorologists utilize sophisticated models to predict the weather up to ten days in advance. Give an example of how they might assess their models.

17. The news Find a newspaper or magazine article in which some data are reported. For the data discussed in the article, identify as many of the W's as you can. Include a copy of the article with your report.

18. The Internet Find an Internet source that reports on a study and describes the data. Print out the description and identify as many of the W's as you can.

(Exercises 19–26) *For each description of data, identify Who and What were investigated and the population of interest.*

19. Brain age Researchers have found that women's brains are nearly four years younger than those of men of the same chronological age. To explore the subject, the researchers used a brain scanning technique called positron emission tomography to measure the flow of oxygen and glucose in the brains of 121 women and 84 men aged 20–82. Through the scans, they could notice how sugar was being turned into energy in the volunteers' brains. The researchers also used a computer algorithm to predict the ages of the volunteers based on brain metabolism as measured by the scans. The findings of the study suggest that "changes in how the brain uses energy over a person's lifetime proceed more gradually in women than they do in men." (www.theguardian.com/science/2019/feb/04/womens-brains-are-four-years-younger-than-mens-study-finds)

20. Hula-hoops The hula-hoop, a popular children's toy in the 1950s, has gained popularity as an exercise in recent years. But does it work? To answer this question, the American Council on Exercise conducted a study to evaluate the cardio and calorie-burning benefits of "hooping." Researchers recorded heart rate and oxygen consumption of participants, as well as their individual ratings of perceived exertion, at regular intervals during a 30-minute workout. (www.acefitness.org/certifiednewsarticle/1094/)

21. Bicycle safety Ian Walker, a psychologist at the University of Bath, wondered whether drivers treat bicycle riders differently when they wear helmets. He rigged his bicycle with an ultrasonic sensor that could measure how close each car was that passed him. He then rode on alternating days with and without a helmet. Out of 2500 cars passing him, he found that when he wore his helmet, motorists passed 3.35 inches closer to him, on average, than when his head was bare. (Source: *NY Times*, Dec. 10, 2006)

22. Investments Some companies offer 401(k) retirement plans to employees, permitting them to shift part of their before-tax salaries into investments such as mutual funds. Employers typically match 50% of the employees' contribution up to about 6% of salary. One company, concerned with what it believed was a low employee participation rate in its 401(k) plan, sampled 30 other companies with similar plans and asked for their 401(k) participation rates.

23. Honesty Coffee stations in offices often just ask users to leave money in a tray to pay for their coffee, but many people cheat. Researchers at Newcastle University alternately taped two posters over the coffee station. During one week, it was a picture of flowers; during the other, it was a pair of staring eyes. They found that the average contribution was significantly higher when the eyes poster was up than when the flowers were there. Apparently, the mere feeling of being watched—even by eyes that were not real—was enough to encourage people to behave more honestly. (Source: *NY Times*, Dec. 10, 2006)

24. Blindness A study begun in 2011 examines the use of stem cells in treating two forms of blindness, Stargardt's disease and dry age-related macular degeneration. Each of the 24 patients entered one of two separate trials in which embryonic stem cells were to be used to treat the condition. (www.blindness.org/index.php?view=article&id=2514:stem-cell-clinical-trial-for-stargardt-disease-set-to-begin-&option=com_content&Itemid=122)

25. Not-so-diet soda A look at 474 participants in the San Antonio Longitudinal Study of Aging found that participants who drank two or more diet sodas a day "experienced waist size increases six times greater than those of people who didn't drink diet soda." (*J Am Geriatr Soc.* 2015 Apr;63(4):708–15. doi: 10.1111/jgs.13376. Epub 2015 Mar 17.)

26. Brewery The management of a brewery is adamant on keeping the roasting temperature of barley exactly at the specified value of 232 degrees Celsius. This temperature yields perfectly roasted barley that is full of flavor yet not burnt. The management measured the temperature for 10 randomly selected portions of roasted barley.

(Exercises 27–40) *For each description of data, identify the W's, name the variables, specify for each variable whether its use indicates that it should be treated as categorical or quantitative, and, for any quantitative variable, identify the units in which it was measured (or note that they were not provided).*

27. Weighing bears Because of the difficulty of weighing a bear in the woods, researchers caught and measured 54 bears, recording their weight, neck size, length, and sex. They hoped to find a way to estimate weight from the other, more easily determined quantities.

28. Properties The Property Management Department keeps these records on all properties in a particular district: age, type of property, location, number of residents living in that property, area, number of rooms, and whether the property has a garage.

29. Cocktail menu The owner and manager of a cocktail stall serves several different cocktails. On the stall's menu, the components, the number of calories, and the serving size in milliliters of each served cocktail are specified. This information is useful to assess the nutritional value of the different cocktails.

30. Labor force A representative telephone survey of 2,000 workers was conducted during the first quarter of 2019 for the purpose of gathering information about the labor force. Among the reported results were the respondent's age, gender, region (Northeast, South, etc.), occupation, and yearly average income.

31. Babies Medical researchers at a large city hospital investigating the impact of prenatal care on newborn health collected data from 882 births during 1998–2000. They kept track of the mother's age, the number of weeks the pregnancy lasted, the type of birth (cesarean, induced, natural), the level of prenatal care the mother had (none, minimal, adequate), the birth weight and sex of the baby, and whether the baby exhibited health problems (none, minor, major).

32. Flowers In a study appearing in the journal *Science*, a research team reports that plants in southern England are flowering earlier in the spring. Records of the first flowering dates for 385 species over a period of 47 years show that flowering has advanced an average of 15 days per decade, an indication of climate warming, according to the authors.

33. Herbal medicine Scientists at a major pharmaceutical firm conducted an experiment to study the effectiveness of an herbal compound to treat the common cold. They exposed each patient to a cold virus, then gave them either the herbal compound or a sugar solution known to have no effect on colds. Several days

Place	Name	Official Time	Country
1	Kipchoge, Eliud	02:02:37	KEN
2	Geremew, Mosinet	02:02:55	ETH
3	Wasihun, Mule	02:03:16	ETH
4	Kitata, Tola Shura	02:05:01	ETH
5	Farah, Mo	02:05:39	GBR
6	Tola, Tamirat	02:06:57	ETH
7	Abdi, Bashir	02:07:03	BEL
8	Gebresilasie, Leul	02:07:15	ETH
9	Rachik, Yassine	02:08:05	ITA
10	Hawkins, Callum	02:08:14	GBR

Source: Excerpt from https://results.virginmoneylondonmarathon.com/2019/?pid=leaderboard

later they assessed each patient's condition, using a cold severity scale ranging from 0 to 5. They found no evidence of benefits of the compound.

34. Insurance claims For fraud-detection purposes, an automobile insurance company keeps record of the following information about insurance claims: year of claim, claimant's age, claimant's gender, claimant's marital status, claimant's total economic loss (in thousand dollars), and whether the driver of the claimant's vehicle was uninsured.

35. Streams In performing research for an ecology class, students at a college in upstate New York collect data on streams each year. They record a number of biological, chemical, and physical variables, including the stream name, the substrate of the stream (limestone, shale, or mixed), the acidity of the water (pH), the temperature (°C), and the BCI (a numerical measure of biological diversity).

36. Vehicle registration The Department of Motor Vehicles of a country receives daily applications for motor vehicle registrations. Among the data collected in the applications are the vehicle manufacturer, vehicle type (car, SUV, etc.), year of manufacture, color, engine number, chassis number, and seating capacity.

37. Refrigerators In 2013, *Consumer Reports* published an article evaluating refrigerators. It listed 353 models, giving the brand, cost, size (cu ft), type (such as top freezer), estimated annual energy cost, an overall rating (good, excellent, etc.), and the repair history for that brand (percentage requiring repairs over the past 5 years).

38. Walking in circles People who get lost in the desert, mountains, or woods often seem to wander in circles rather than walk in straight lines. To see whether people naturally walk in circles in the absence of visual clues, researcher Andrea Axtell tested 32 people on a football field. One at a time, they stood at the center of one goal line, were blindfolded, and then tried to walk to the other goal line. She recorded each individual's sex, height, handedness, the number of yards each was able to walk before going out of bounds, and whether each wandered off course to the left or the right. No one made it all the way to the far end of the field without crossing one of the sidelines. (Source: *STATS* No. 39, Winter 2004)

T **39.** London Marathon The London Marathon is an annual marathon held in London, United Kingdom, as part of the World Marathon Majors. It started in 1981 and has been held in the spring of every year since then. The London Marathon is set around the River Thames, beginning around Blackheath and finishing in The Mall alongside St James's Park. This marathon holds the Guinness world record as the largest annual fundraising event in the world. Above are the data for the top ten male finishers aged 18–39 of the 2019 race held on April 28, 2019.

T **40.** Indy 500 2018 The 2.5-mile Indianapolis Motor Speedway has been the home to a race on Memorial Day nearly every year since 1911. Even during the first race, there were controversies. Ralph Mulford was given the checkered flag first but took three extra laps just to make sure he'd completed 500 miles. When he finished, another driver, Ray Harroun, was being presented with the winner's trophy, and Mulford's protests were ignored. Harroun averaged 74.6 mph for the 500 miles. In 2013, the winner, Tony Kanaan, averaged over 187 mph, beating the previous record by over 17 mph!

Here are the data for the first five races and six recent Indianapolis 500 races.

Year	Driver	Time (hr:min:sec)	Speed (mph)
1911	Ray Harroun	6:42:08	74.602
1912	Joe Dawson	6:21:06	78.719
1913	Jules Goux	6:35:05	75.933
1914	René Thomas	6:03:45	82.474
1915	Ralph DePalma	5:33:55.51	89.840
...			
2013	Tony Kanaan	2:40:03.4181	187.433
2014	Ryan Hunter-Reay	2:40:48.2305	186.563
2015	Juan Pablo Montoya	3:05:56.5286	161.341
2016	Alexander Rossi	3:00:02.0872	166.634
2017	Takuma Sato	3:13:3.3584	155.395
2018	Will Power	2:59:42.6365	166.935

41. Formula 1 on the computer Load the Formula 1 data for the ten most recent races in Australia into your preferred statistics package, and answer the following questions:

a) What was the name of the winning driver in 2014?
b) Which country had the highest frequency of winning in the past ten races?
c) What was the winning time in 2011?
d) What is the name of the most recent Finnish winner and in what year did they win?

Year	Driver	Time (hr:min:sec)	Country
2010	Jenson Button	1:33:36.531	UK
2011	Sebastian Vettel	1:29:30.259	Germany
2012	Jenson Button	1:34:09.565	UK
2013	Kimi Räikkönen	1:30:03.225	Finland
2014	Nico Rosberg	1:32:58.710	Germany
2015	Lewis Hamilton	1:31:54.067	UK
2016	Nico Rosberg	1:48:15.565	Germany
2017	Sebastian Vettel	1:24:11.672	Germany
2018	Sebastian Vettel	1:29:33.283	Germany
2019	Valtteri Bottas	1:25:27.325	Finland

Source: Extracted from https://www.formula1.com/en/results.html/2010/races/861/australia/race-result.html

42. Indy 500 2018 on the computer Load the **Indy 500 2018** data into your preferred statistics package and answer the following questions:

a) What was the average speed of the winner in 1920?
b) How many times did Bill Vukovich win the race in the 1950s?
c) How many races took place during the 1940s?

JUST CHECKING

Answers

1. *Who*—Tour de France races; *What*—year, winner, country of origin, age, team, total time, average speed, stages, total distance ridden, starting riders, finishing riders; *How*—official statistics at race; *Where*—France (for the most part); *When*—1903 to 2016; *Why*—not specified (To see progress in speeds of cycling racing?)

2.

Variable	Type	Units
Year	Quantitative or Identifier	Years
Winner	Categorical	
Country of Origin	Categorical	
Age	Quantitative	Years
Team	Categorical	
Total Time	Quantitative	Hours/minutes/seconds
Average Speed	Quantitative	Kilometers per hour
Stages	Quantitative	Counts (stages)
Total Distance	Quantitative	Kilometers
Starting Riders	Quantitative	Counts (riders)
Finishing Riders	Quantitative	Counts (riders)

2 Displaying and Describing Data

WHERE ARE WE GOING?

We can summarize and describe data values in a variety of ways. You'll probably recognize these displays and summaries. This chapter is a fast review of these concepts so we all agree on terms, notation, and methods. We'll be using these displays and descriptions throughout the rest of the text.

What happened on the *Titanic* at 11:40 on the night of April 14, 1912, is well known. Frederick Fleet's cry of "Iceberg, right ahead" and the three accompanying pulls of the crow's nest bell signaled the beginning of a nightmare that has become legend. By 2:15 AM, the *Titanic*, thought by many to be unsinkable, had sunk. Only 712 of the 2208 people on board survived. The others (nearly 1500) met their icy fate in the cold waters of the North Atlantic.

Table 2.1 shows some data about the passengers and crew aboard the *Titanic*. Each case (row) of the data table represents a person on board the ship. The variables are the person's *Name*, *Survival* status (Dead or Alive), *Age* (in years), *Age Category* (Adult or Child), *Sex* (Male or Female), *Price* Paid (in British pounds, £), and ticket *Class* (First, Second, Third, or Crew). Some of these, such as *Age* and *Price*, record numbers. These are called

Table 2.1
Part of a data table showing seven variables for 11 people aboard the *Titanic*.

Name	Survived	Age	Adult/Child	Sex	Price (£)	Class
ABBING, Mr Anthony	Dead	42	Adult	Male	7.55	3
ABBOTT, Mr Ernest Owen	Dead	21	Adult	Male	0	Crew
ABBOTT, Mr Eugene Joseph	Dead	14	Child	Male	20.25	3
ABBOTT, Mr Rossmore Edward	Dead	16	Adult	Male	20.25	3
ABBOTT, Mrs Rhoda Mary "Rosa"	Alive	39	Adult	Female	20.25	3
ABELSETH, Miss Karen Marie	Alive	16	Adult	Female	7.65	3
ABELSETH, Mr Olaus Jörgensen	Alive	25	Adult	Male	7.65	3
ABELSON, Mr Samuel	Dead	30	Adult	Male	24	2
ABELSON, Mrs Hannah	Alive	28	Adult	Female	24	2
ABRAHAMSSON, Mr Abraham August Johannes	Alive	20	Adult	Male	7.93	3
ABRAHIM, Mrs Mary Sophie Halaut	Alive	18	Adult	Female	7.23	3

quantitative variables. Others, like *Survival* and *Class*, place each case in a single category, and are called **categorical** variables. (Data in **Titanic**)

The problem with a data table like this—and in fact with all data tables—is that you can't *see* what's going on. And seeing is just what we want to do. We need ways to show the data so that we can see patterns, relationships, trends, and exceptions.

WHO	People on the *Titanic*
WHAT	Name, survival status, age, adult/child, sex, price paid, ticket class
WHEN	April 14, 1912
WHERE	North Atlantic
HOW	www.encyclopedia-titanica.org
WHY	Historical interest

The Three Rules of Data Analysis

There are three things you should always do first with data:

1. **Make a picture.** A display of your data will reveal things you're not likely to see in a table of numbers and will help you *Think* clearly about the patterns and relationships that may be hiding in your data.
2. **Make a picture.** A well-designed display will *Show* the important features and patterns in your data. It could also show you things you did not expect to see: extraordinary (possibly wrong) data values or unexpected patterns.
3. **Make a picture.** The best way to *Tell* others about your data is with a well-chosen picture.

These are the three rules of data analysis. There are pictures of data throughout the text, and new kinds keep showing up. These days, technology makes drawing pictures of data easy, so there is no reason not to follow the three rules.

We make graphs for two primary reasons: to understand more about data and to show others what we have learned and want them to understand. The first reason calls for simple graphs with little adornment; the second often uses visually appealing additions to draw the viewer's attention. Regardless of their function, graphs should be easy to read and understand and should represent the facts of the data honestly. Axes should be clearly labeled with the names of the variables they display. The intervals set off by "tick marks" should occur at values easy to think about: 5, 10, 15, and 20 are simpler marks than, say, 1.7, 2.3, 2.9, and 3.5. And tick labels that run for several digits are almost never a good idea. Graphs should have a "key" that identifies colors and symbols if those are meaningful in the graph. And all graphs should carry a title or caption that says what the graph displays and suggests what about it is salient or important.

The Area Principle

A bad picture can distort our understanding rather than help it. What impression do you get from Figure 2.1 about who was aboard the ship?

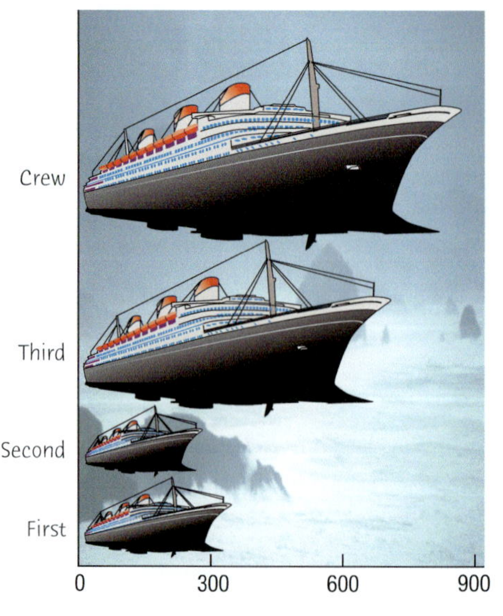

Figure 2.1
How many people were in each class on the *Titanic*? From this display, it looks as though the service must have been great, since most aboard were crew members. Although the length of each ship here corresponds to the correct number, the impression is all wrong. In fact, only about 40% were crew.

The *Titanic* was certainly a luxurious ship, especially for those in first class, but Figure 2.1 gives the mistaken impression that most of the people on the *Titanic* were crew members, with a few passengers along for the ride. What's wrong? The lengths of the ships *do* match the number of people in each ticket class category. However, our eyes tend to be more impressed by the *area* than by other aspects of each ship image. So, even though the *length* of each ship matches up with one of the totals, it's the associated *area* in the image that we notice. There were about 3 times as many crew as second-class passengers, and the ship depicting the number of crew members is about 3 times longer than the ship depicting second-class passengers. The problem is that it occupies about 9 times the area. That just isn't a correct impression.

The best data displays observe a fundamental principle of graphing data called the **area principle**. The area principle says that the area occupied by a part of the graph should correspond to the magnitude of the value it represents. Violations of the area principle are a common way to lie (or, since most mistakes are unintentional, we should say err) with statistics.

2.1 Summarizing and Displaying a Categorical Variable

Frequency Tables

Categorical variables are easy to summarize in a **frequency table** that lists the number of cases in each category along with its name.

For ticket *Class*, the categories are First, Second, Third, and Crew:

Class	Count
First	324
Second	285
Third	710
Crew	889

Table 2.2
A frequency table of the *Titanic* passengers.

Class	Percentage (%)
First	14.67
Second	12.91
Third	32.16
Crew	40.26

Table 2.3
A relative frequency table for the same data.

A **relative frequency table** displays *percentages* (or *proportions*) rather than the counts in each category. Both types of tables show the **distribution** of a categorical variable because they name the possible categories and tell how frequently each occurs. (The percentages should total 100%, although the sum may be a bit too high or low if the individual category percentages have been rounded.)

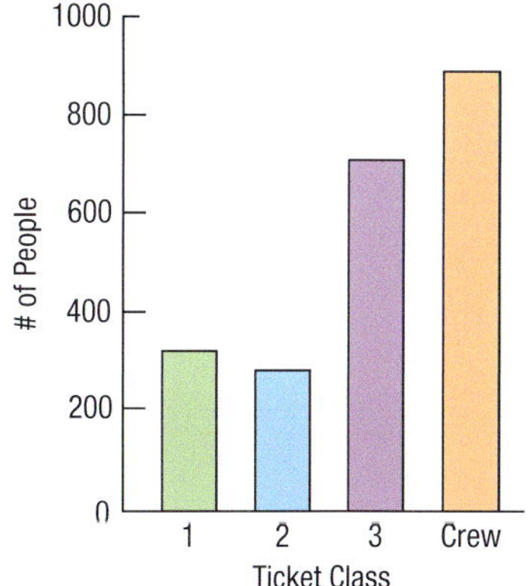

Figure 2.2
People on the Titanic *by Ticket Class.* With the area principle satisfied, we can see the true distribution more clearly.

Bar Charts

Although not as visually entertaining as the ships in Figure 2.1, the **bar chart** in Figure 2.2 gives an *accurate* visual impression of the distribution because it obeys the area principle. Now it's easy to see that the majority of people on board were *not* crew. We can also see that there were about 3 times as many crew members as second-class passengers. And there were more than twice as many third-class passengers as either first- or second-class passengers—something you may have missed in the frequency table. Bar charts make these kinds of comparisons easy and natural.

EXAMPLE 2.1 What Do You Think of Congress?

In December 2017, the Gallup survey asked 1049 people how they viewed a variety of professions. Specifically they asked, "How would you rate the honesty and ethical standards of people in these different fields?" For Members of Congress, the results were

Rating	Percentage (%)
Very high	2
High	9
Average	29
Low	36
Very low	24
No opinion	1

QUESTION: What kind of table is this? What would be an appropriate display?

ANSWER: This is a relative frequency table because the numbers displayed are percentages, not counts. A bar chart would be appropriate:

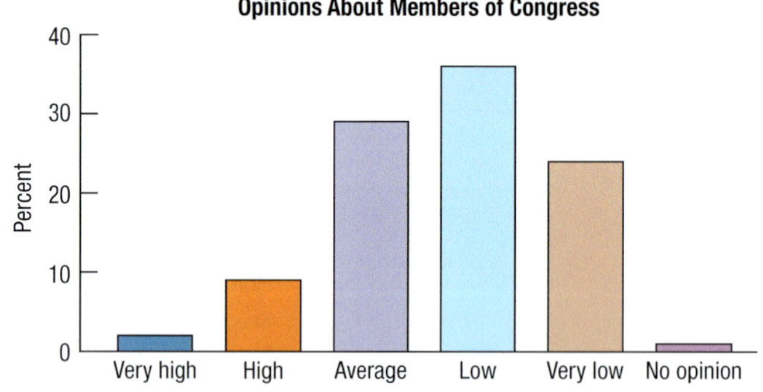

EXAMPLE 2.2 Which Gadgets Do You Use?

In 2014, the Pew Research Organization asked 1005 U.S. adults which of the following electronic items they use: cell phone, smartphone, computer, handheld e-book reader (e.g., Kindle or Nook), or tablet. The results were

Device	Percentage (%) using the device
Cell phone	86.8
Smartphone	54.0
Computer	77.5
E-book reader	32.2
Tablet	41.9

QUESTION: Is this a frequency table, a relative frequency table, or neither? How could you display these data?

ANSWER: This is not a frequency table because the numbers displayed are not counts. Although the numbers are percentages, they do not sum to 100%. A person can use more than one device, so this is not a relative frequency table either. A bar chart might still be appropriate, but the numbers do not sum to 100%.

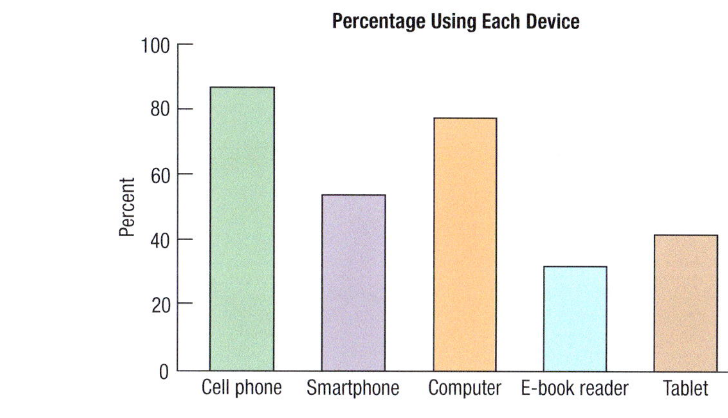

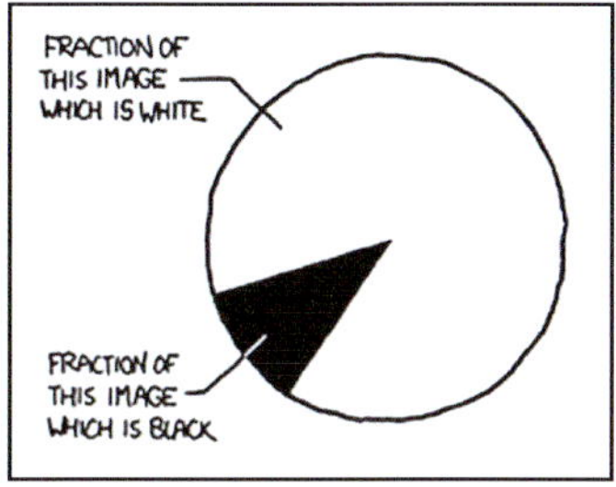

Pie Charts

Pie charts display all the cases as a circle whose slices have areas proportional to each category's fraction of the whole.

Pie charts give a quick impression of the distribution. Because we're used to cutting up pies into 2, 4, or 8 pieces, pie charts are particularly good for seeing relative frequencies near 1/2, 1/4, or 1/8.

Bar charts are almost always better than pie charts for comparing the relative frequencies of categories. Pie charts are widely understood and colorful, and they often appear in reports, but Figure 2.3 shows why statisticians prefer bar charts.

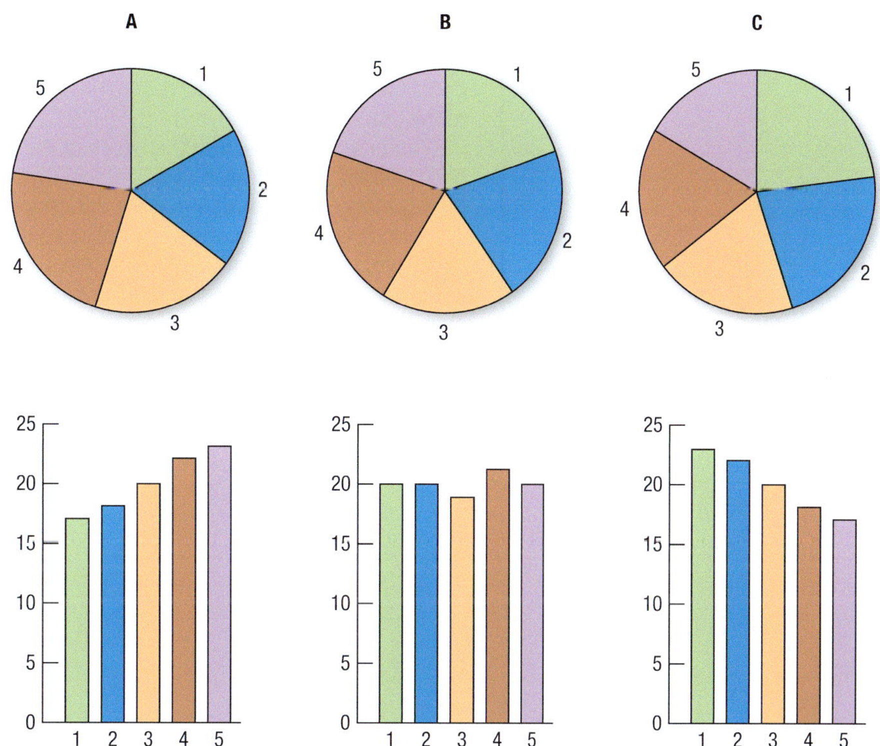

Figure 2.3
Pie charts may be attractive, but it can be hard to see patterns in them. Can you discern the differences in distributions depicted by these pie charts?

Figure 2.4
Bar charts of the same values as shown in Figure 2.3 make it much easier to compare frequencies in groups.

Ring Charts

A ring (or donut) chart is a modified form of pie chart that displays only the "crust" of the pie—a ring that is partitioned into regions proportional in area to each value. You can think of the ring as the bars of a bar chart stuck end to end and wrapped around the circle. Ring charts are somewhere between bar charts and pie charts. They may be easier to read (or not). Judge for yourself:

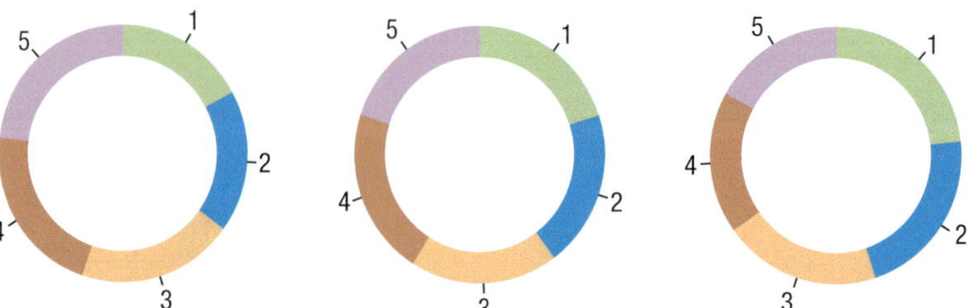

Figure 2.5

Ring charts compromise between pie and bar charts. These ring charts show the same values as the pie charts in Figure 2.3. Do you find it easier to see the patterns?

RANDOM MATTERS

Is it random, or is something systematic going on? Separating the *signal* (the systematic) from the *noise* (the random) is a fundamental skill of statistics.

A geoscientist notices that global temperatures have increased steadily during the past 50 years. Could the pattern be random, or is the earth warming?

An analyst notices that the stock market seems to go up more often on Tuesday afternoons when it rains in Chicago. Is that something she should bank on?

One of the challenges to answering questions like these is that we have only one earth and one stock market history. What if we had two? Or many? Sometimes we can use a computer to *simulate* other situations, to pretend that we have more than one realization of a phenomenon. In these *Random Matters* sections, we'll use the computer as our lab to test what might happen if we could repeat our data collection many times.

You probably know that the "rules of the sea" were enforced on the *Titanic*—women and children were allowed to board the *Titanic* lifeboats before the men. Did ticket class (first, second, or third) also make a difference? Suppose the 712 survivors were chosen at random, giving everyone an equal chance to get into a lifeboat. Would the distribution have been different? Let's look. We selected 712 people at random from the list of those aboard the *Titanic* and made a pie chart of ticket class. We repeated the random selection 24 times, making a new pie chart of each selected group of 712 passengers. Among these pie charts in Figure 2.6 we've "hidden" the actual distribution of survivors. Can you pick out the real distribution? If so, then that might convince you that the lifeboats weren't filled randomly, with everyone getting an equal chance.

Figure 2.6

The distribution of ticket class in 24 simulated lifeboats and the actual distribution of survivors. Can you find the real one? If so, this suggests that people didn't all have an equal chance to survive.[1]

In this example, the difference is pretty obvious. There were more survivors from first and second class and fewer from third class than there would have been were everyone given an equal chance. In other situations, the differences may not be as obvious, so we'll need to develop more sophisticated tools to help distinguish signals from noise.

2.2 Displaying a Quantitative Variable

Histograms

How can we make a bar chart for a quantitative variable? We can't, because quantitative variables don't have categories. Instead, we make a **histogram**.

Histograms and bar charts both use bars, but they are fundamentally different. The bars of a bar chart display the count for each category, so they could be arranged in any order[2]

[1]Wait. Didn't we just say we prefer bar charts? Well, sometimes pie charts are actually a good choice. They are compact, colorful, and—most important—they satisfy the area principle. A figure with 25 bar charts would look much more confusing.

[2]Many statistics programs choose alphabetical order, which is rarely the most useful one.

(and should be displayed with a space between them). The horizontal axis of a bar chart just names the categories. The horizontal axis of a histogram shows the values of the variable in order. A histogram slices up that axis into equal-width bins, and the bars show the counts for each bin. Now **gaps** are meaningful; they show regions with no observations.

Figure 2.7 shows a histogram of the ages of those aboard the *Titanic*. In this histogram, each bin has a width of 5 years, so, for example, the height of the tallest bar shows that the most populous age group was the 20- to 24-year-olds, with over 400 people.[3] The youngest passengers were infants, and the oldest was more than 70 years old.

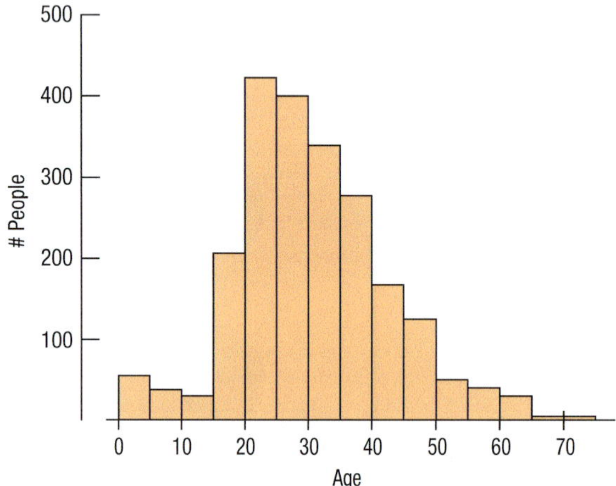

Figure 2.7
A histogram of the distribution of ages of those aboard *Titanic*.

The fact that there are fewer and fewer people in the 5-year bins from age 25 on probably doesn't surprise you either. After all, there are increasingly fewer people of advancing age in the general population as well, and there were no very elderly people on board the *Titanic*. But the bins on the left are a little strange. It looks like there were more infants and toddlers (0–5 years old) than there were preteens.

Does this distribution look plausible? You may not have guessed that fact about the infants and preteens, but it doesn't seem out of the question. It is often a good idea to imagine what the distribution might look like before you make the display. That way you'll be less likely to be fooled by errors in the data or when you accidentally graph the wrong variable.[4]

EXAMPLE 2.3 Earthquakes and Tsunamis

In 2011, the most powerful earthquake ever recorded in Japan created a wall of water that devastated the northeast coast of Japan and left nearly 25,000 people dead or missing. The 2011 tsunami in Japan was caused by a 9.0 magnitude earthquake. It was particularly noted for the damage it caused to the Fukushima Daiichi nuclear power plant, causing a core meltdown and international concern. As disastrous as it was, the Japan tsunami was not nearly as deadly as the 2004 tsunami on the west coast of Sumatra that killed an estimated 227,899 people, making it the

[3]The histogram bar appears to go from 20 to 25, but most statistics programs include values at the lower limit in the bar and put values at the upper limit in the next bin.

[4]You'll notice that we didn't say *if* you graph the wrong variable, but rather *when*. Everyone makes mistakes, and you'll make your share. But if you always think about what your graph or analysis says about the world and judge whether that is reasonable, you can catch many errors before they get away.

most lethal tsunami on record. The earthquake that caused it had magnitude 9.1—more than 25% more powerful than the Japanese earthquake. Were these earthquakes truly extraordinary, or did they just happen at unlucky times and places? The magnitudes (measured or estimated) are available for 999 of the 1116 earthquakes known to have caused tsunamis, dating back to 426 BCE. (Data in **Tsunamis 2018**)

WHO	1116 earthquakes known or suspected to have caused tsunamis for which we have data or good estimates
WHAT	Magnitude (Richter scale), depth (m), date, location, and other variables
WHEN	From 426 BCE to the present
WHERE	All over the earth

QUESTION: What can we learn from these data?

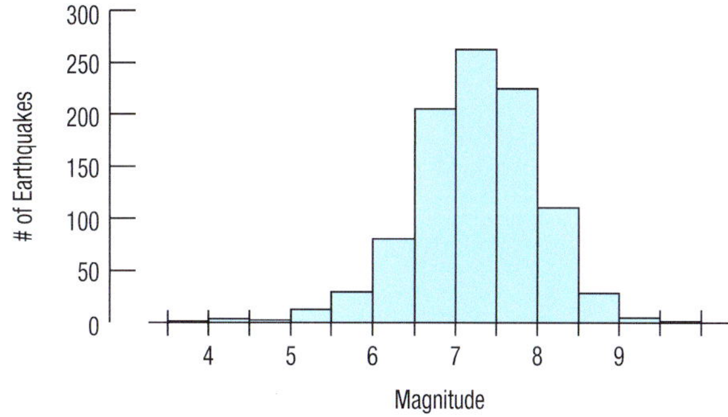

ANSWER: The histogram displays the distribution of earthquake magnitudes on the Richter scale. The height of the tallest bar says that there were about 250 earthquakes with magnitudes between 7.0 and 7.5. We can see that earthquakes typically have magnitudes around 7. Most are between 5.5 and 8.5, but one is less than 4 and a few are 9 or bigger. Relative to the other tsunami-causing earthquakes, the Sumatra and Japan events were extraordinarily powerful.

EXAMPLE 2.4 How Much Do Americans Work?

The Bureau of Labor Statistics (BLS) collects data on many aspects of the U.S. economy. One of the surveys it conducts, the American Time Use Survey (ATUS), asks roughly 11,000 people a year a variety of questions about how they spend their time. For those who are employed, it asks how many hours a week they work. Here is a histogram of the 2270 responses in 2014.

QUESTION: What does the histogram say about how many hours U.S. workers typically work?

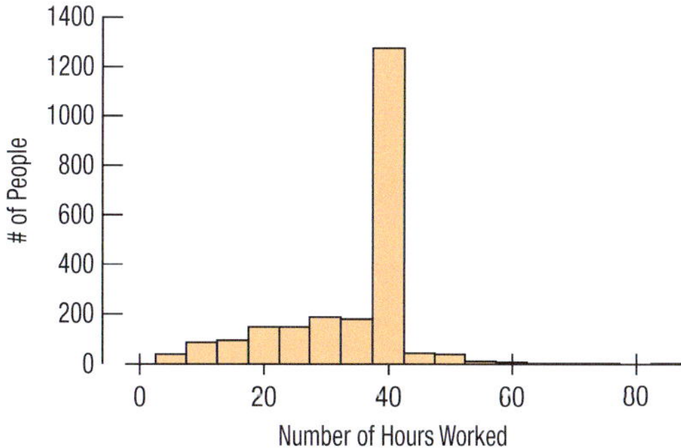

ANSWER: It looks like the vast majority of people (more than 1200 in this study) work right around 40 hours a week. There are some who work less, and a very few who work more.

Stem-and-Leaf Displays

Histograms provide an easy-to-understand summary of the distribution of a quantitative variable, but they don't show the data values themselves. For example, here's a histogram of the pulse rates of 24 women at a health clinic:

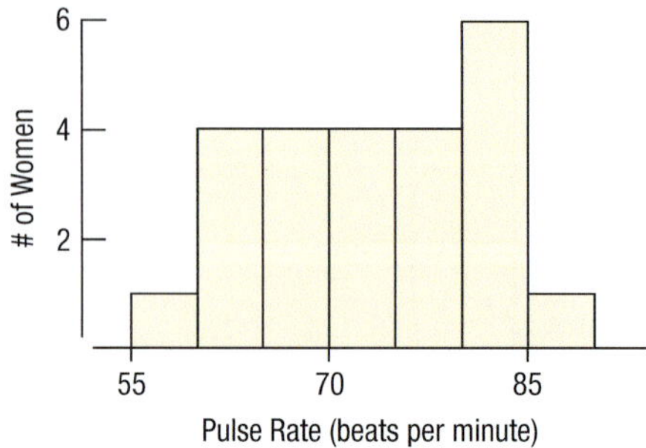

Figure 2.8
A histogram of the pulse rates of 24 women at a health clinic.

Here's a stem-and-leaf display of the same data:

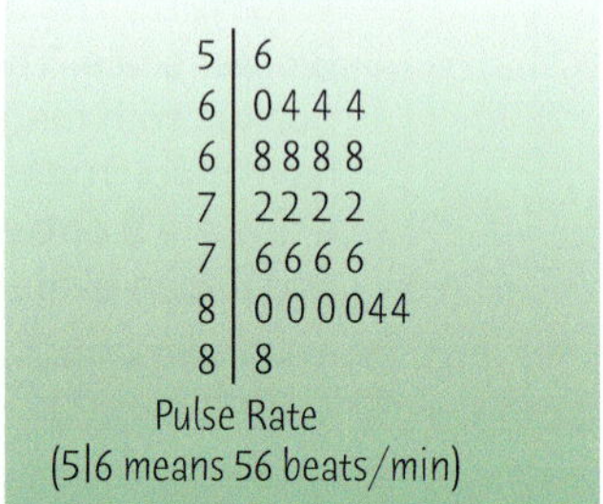

> **STEM-AND-LEAF OR STEMPLOT?**
>
> The stem-and-leaf display was devised by John W. Tukey, one of the greatest statisticians of the 20th century. It is called a "stemplot" in some texts and computer programs.

A **stem-and-leaf display** is like a histogram, but it shows the individual values. It's also easier to make by hand. Turn the stem-and-leaf on its side (or turn your head to the right) and squint at it. It should look roughly like the histogram of the same data. Does it?[5]

The first line of the display, which says 5|6, stands for a pulse of 56 beats per minute (bpm). We've taken the tens place of the number and made that the "stem." Then we sliced off the ones place and made it a "leaf." The next line down is 6|0444, which shows one pulse rate of 60 and three of 64 bpm.

Stem-and-leaf displays are especially useful when you make them by hand for batches of fewer than a few hundred data values. They are a quick way to display—and even to record—numbers. Because the leaves show the individual values, we can sometimes see even more in the data than the distribution's shape. Take another look at all the leaves of the pulse data. See anything unusual? At a glance you can see that they are all even. With a bit more thought you can see that they are all multiples of 4—something you couldn't possibly see from a histogram. How do you think the nurse took these pulses? Counting beats for a full minute, or counting for only 15 seconds and multiplying by 4?

Dotplots

A **dotplot** places a dot along an axis for each case in the data. It's like a stem-and-leaf display, but with dots instead of digits for all the leaves. Dotplots are a great way to display a small dataset. Figure 2.9 shows a dotplot of the time (in seconds) that the winning horse took to win the Kentucky Derby in each race between the first Derby in 1875 and the 2018 Derby. (Data in **Kentucky Derby 2018**)

---

[5] You could make the stem-and-leaf display with the higher values on the top. Putting the lower values at the top matches the histogram; putting the higher values at the top matches the way a vertical axis works in other displays such as dotplots (as we'll see presently).

Dotplots display basic facts about the distribution. We can find the slowest and fastest races by finding the times for the topmost and bottommost dots. It's clear that there are two clusters of points, one just below 160 seconds and the other at about 122 seconds. Something strange happened to the Derby times. Once we know to look for it, we can find out that in 1896 the distance of the Derby race was changed from 1.5 miles to the current 1.25 miles. That explains the two clusters of winning times.

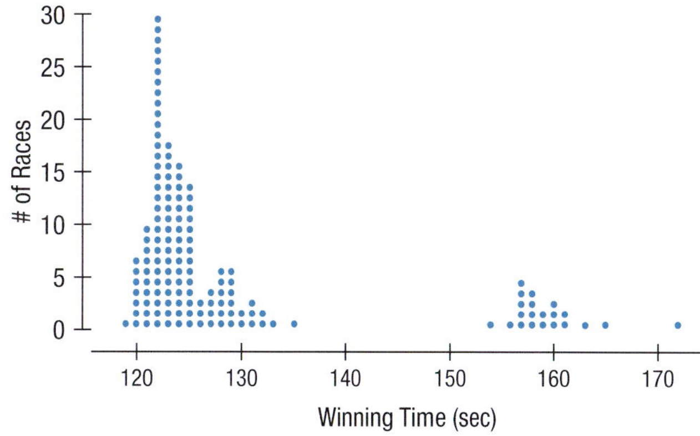

Figure 2.9
A dotplot of Kentucky Derby winning times plots each race as its own dot. We can see two distinct groups corresponding to the two different race distances.

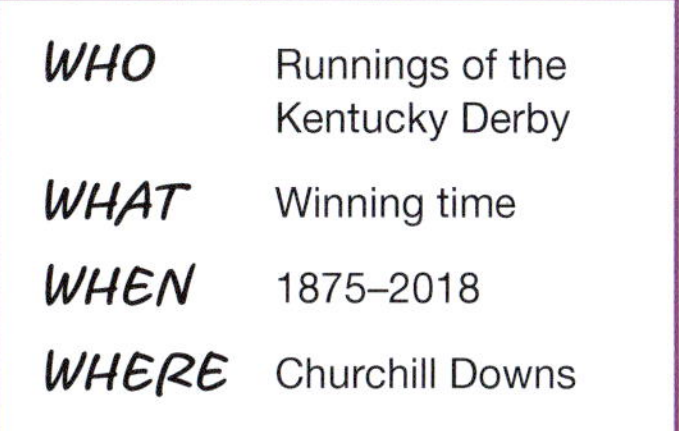

WHO	Runnings of the Kentucky Derby
WHAT	Winning time
WHEN	1875–2018
WHERE	Churchill Downs

*Density Plots

The size of the bins in a histogram can influence its look and our interpretation of the distribution. There is no correct bin size, although recommendations to use between 5 and 20 bins are common. **Density plots** smooth the bins in a histogram to reduce the effect of this choice. How much the bin heights are smoothed is still a choice that affects the shape, but the change in shape is less severe than in a histogram. Here's a density plot of the *Ages* of those on the *Titanic*. Compare it to Figure 2.7.

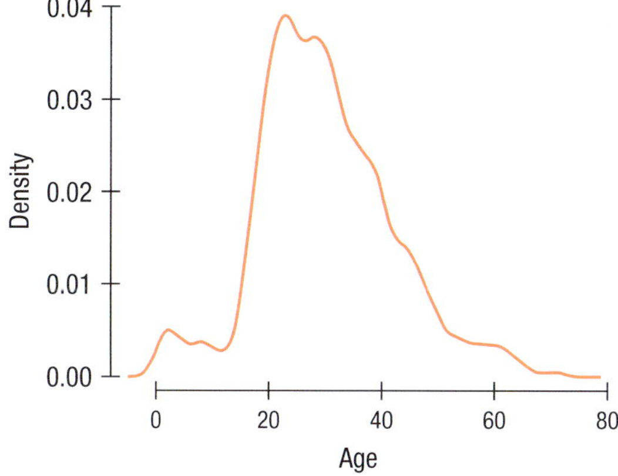

Figure 2.10
A density plot of the *Ages* of those aboard the *Titanic*. We can see, as we did in the histogram, that the most populous age is near 20 and that there are more infants and toddlers than preteens. The density plot does not provide hard cut-offs to the bins, but smooths the distribution over the bins.

Every histogram, stem-and-leaf display, and dotplot tells a story, but you need to develop a vocabulary to help you explain it. Start by talking about three things: its *shape*, *center*, and *spread*.

Think Before You Draw

Before making a pie chart or a bar chart, you should check that you have categorical data. Before making a stem-and-leaf display, a histogram, or a dotplot, you should make sure you are working with quantitative data. Although a bar chart and a histogram may look similar, they're not the same display. You can't display categorical data in a histogram nor quantitative data in a bar chart.

2.3 Shape

We summarize the **shape** of a distribution in terms of three attributes: how many *modes* it has, whether it is *symmetric* or *skewed*, and whether it has any extraordinary cases or *outliers*.

1. *Does the histogram have a single, central hump or several separated humps?* These humps are called **modes**.[6] The mode is sometimes defined as the single value that appears most often. That definition is fine for categorical variables because all we need to do is count the number of cases for each category. For quantitative variables, the mode is more ambiguous. It makes more sense to use the term "mode" to refer to the peak of the histogram rather than as a single summary value. The important feature of the Kentucky Derby races is that there are two distinct modes, representing the two different versions of the race and warning us to consider those two versions separately. The earthquake magnitudes of Example 2.3 have a single mode at just about 7.

   A histogram with one peak, such as the ages (Figure 2.7), is dubbed **unimodal**; histograms with two peaks are **bimodal**, and those with three or more are called **multimodal**.[7]

> **PIE À LA MODE?**
>
> You've heard of pie à la mode. Is there a connection between pie and the mode of a distribution? Actually, there is! The mode of a distribution is a *popular* value near which a lot of the data values gather. And "à la mode" means "in style"—*not* "with ice cream." That just happened to be a *popular* way to have pie in Paris around 1900.

A histogram that doesn't appear to have any mode and in which all the bars are approximately the same height is called **uniform**.

Figure 2.11
In this histogram, the bars are all about the same height. The histogram doesn't appear to have a mode and is called uniform.

2. *Is the histogram* **symmetric***?* Can you fold it along a vertical line through the middle and have the edges match pretty closely, or are more of the values on one side?

Figure 2.12
A symmetric histogram can fold in the middle so that the two sides almost match.

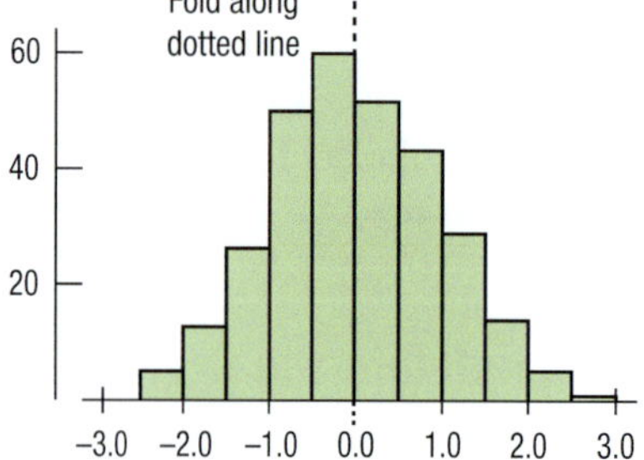

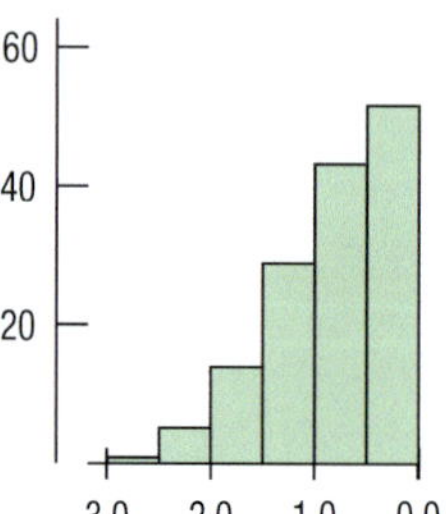

[6]Well, technically, it's the value on the horizontal axis of the histogram that is the mode, but anyone asked to point to the mode would point to the hump.

[7]Apparently, statisticians don't like to count past two.

We generally prefer to work with symmetric distributions because they are easier to summarize, model, and discuss. For example, it is pretty clear where the center of the distribution is located.

The (usually) thinner ends of a distribution are called the **tails**. If one tail stretches out farther than the other, the histogram is said to be **skewed** to the side of the longer tail.

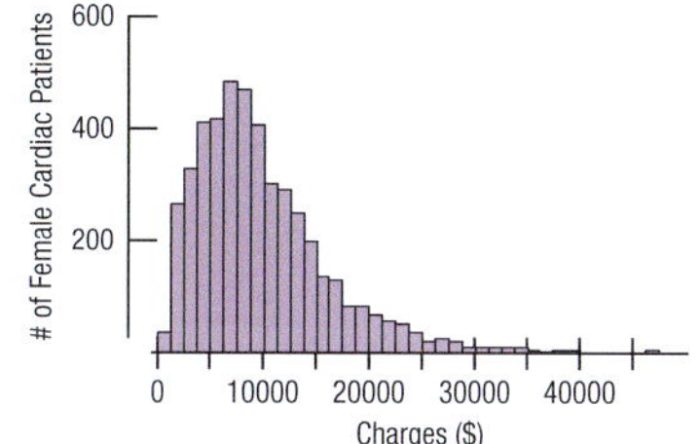

Figure 2.13
Two skewed histograms showing data on two variables for all female heart attack patients in New York State in one year. The blue one (age in years) is skewed to the left. The purple one (charges in $) is skewed to the right. (Data in **Heart attack charges**)

3. *Do any unusual features stick out?* Often such features say something interesting or exciting about the data. Always mention any stragglers, or **outliers**, that stand away from the body of the distribution. If you were collecting data on nose lengths and Pinocchio were in the group, he'd be an outlier, and you'd certainly want to mention it.

   Outliers can affect almost every method we discuss in this course. So we'll always be on the lookout for them. An outlier can be the most informative part of your data. Or it might just be an error. Never throw data away without comment. Treat outliers specially and discuss them when you tell about your data. Or find the error and fix it if you can. Be sure to look for outliers. Always.

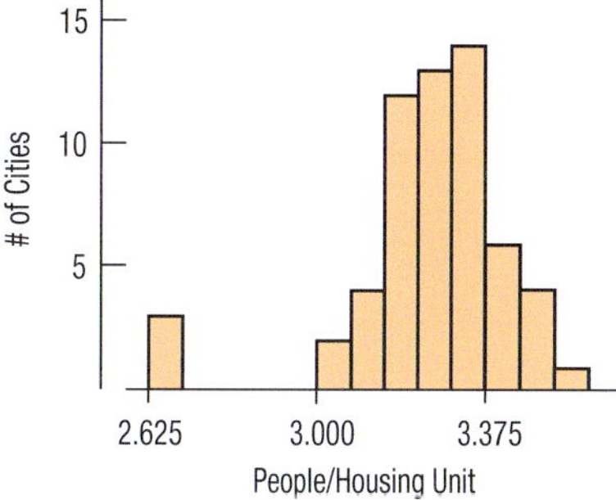

Figure 2.14
A histogram of the number of people per housing unit in a sample of cities. The three cities in the leftmost bar are outliers. We wonder why they are different.

You should also notice any gaps in a distribution, where no data appear at all. Gaps may indicate separate modes or even separate subgroups in your data that you may wish to consider individually. But be careful not to be overzealous when looking for gaps. For small data sets, they may be due to happenstance.

EXAMPLE 2.5 Cost of Living Index

The Cost of Living Index (CLI) summarizes the cost of a representative market basket of goods that includes groceries, restaurants, transportation, utilities, and medical care. Global companies often use the CLI to determine living allowances and salaries for employees. Inflation is often measured by how much the CLI changes from year to year. Relative CLIs can be found for different cities. We have data giving CLI components relative to New York City. For New York City, each index is 100(%). (Data in **Cost_of_living_2018**)

WHO	International cities
WHAT	Cost of Living Index (New York = 100)
WHEN	2018
WHERE	World

QUESTION: A histogram for the Cost of Living Index looks like this:

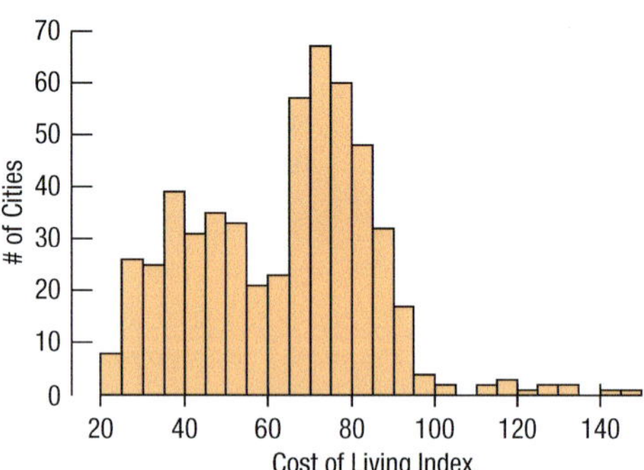

a) If you have access to a statistics program and the data, make a histogram of this variable and show how you did it. (This is an excellent time to fire up your statistics software and start to become familiar with it.) Does your histogram look the same?

b) Describe what the histogram (either the one you made or the one above) says about the data.

ANSWER: Your histogram may look a bit different depending on the plot scale—the number of bars and how wide the bins are. The histogram shows two modes, one just less than 40 and the other near 70. There are also several high values at the right of the distribution, which may deserve attention.

EXAMPLE 2.6 Credit Card Expenditures

A credit card company wants to see how much customers in a particular segment of their market use their credit card. They have provided you with data on the amount spent by 500 selected customers during a 3-month period and have asked you to summarize the expenditures. Of course, you begin by making a histogram. (Data in **Credit card charges**)

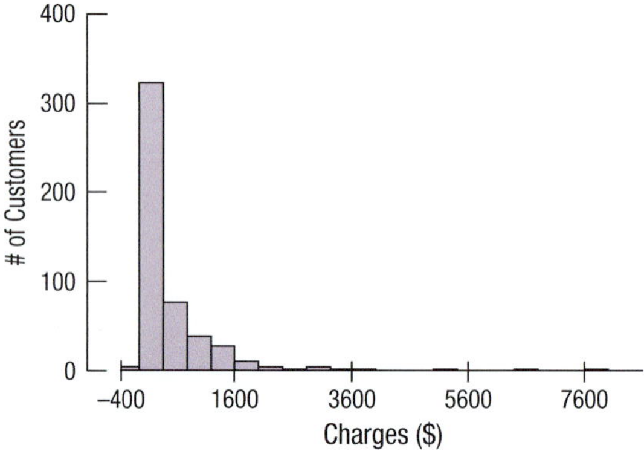

QUESTION: Describe the shape of this distribution.

ANSWER: The distribution of expenditures is unimodal and skewed to the high end. There is an extraordinarily large value at about $7000, and some of the expenditures are negative.

Toto, I've a Feeling We're Not in Math Class Anymore

When Dorothy and her dog Toto land in Oz, everything is more vivid and colorful, but also more dangerous and exciting. Dorothy has new choices to make. She can't always rely on the old definitions, and the yellow brick road has many branches. You may be coming to a similar realization about statistics.

When we summarize data, our goal is usually more than just developing a detailed knowledge of the data we have at hand. We want to know what the data say about the world, so we'd like to know whether the patterns we see in histograms and summary statistics generalize to other individuals and situations.

Because we want to see broad patterns, rather than focus on the details of the dataset we're looking at, many of the most important concepts in statistics are not precisely defined. Whether a histogram is symmetric or skewed, whether it has one or more modes, whether a case is far enough from the rest of the data to be considered an outlier—these are all somewhat vague concepts. They all require judgment.

You may be used to finding a single correct and precise answer in math classes, but in statistics, there may be more than one interpretation. That may make you a little uncomfortable at first, but soon you'll see that leaving room for judgment brings you both power and responsibility. It means that your own knowledge about the world and your judgment matter. You'll use them, along with statistical evidence, to draw conclusions and make decisions about the world.

To recap: We describe the *shape* of a distribution in terms of modes, symmetry, and whether it has any outliers. We usually prefer to work with distributions that are unimodal, symmetric, and have no outliers. We'll refer to this as the USO condition. Now, on to *center* and *spread*.

JUST CHECKING

You may be surprised to find that you already have some intuition about distribution shape. For example, think about what typical values might be and what the largest and smallest values are likely to be. Are the extreme values located symmetrically around the middle? What do you think the distribution of each of the following datasets will look like? Where do you think the center might be? How spread out do you think the values will be?

1. Number of miles run by Saturday morning joggers at a park
2. Hours spent by U.S. adults watching football on Thanksgiving Day
3. Amount of winnings of all people playing a particular state's lottery last week
4. Ages of the faculty members at your school
5. Last digit of students' cell phone numbers on your campus

2.4 Center

How old was a typical crew member on the *Titanic*? The youngest crew member was 14-year-old bellboy William Albert Watson. The captain, Edward John Smith, and the surgeon, Dr. William Francis Norman O'Loughlin, both 62, were the two oldest. But those are the extremes. To answer the question of a typical age, you'd probably look near the **center** of the histogram distribution (Figure 2.15). When the distribution is unimodal and nearly symmetric, the center is easy to find.

The Median

One natural choice of typical value is the **median**—the value that is literally in the middle, with half the values below it and half above it.

The median has the same units as the data. Be sure to include the units whenever you discuss a median. The median age of crew members on the *Titanic* was 30 years.

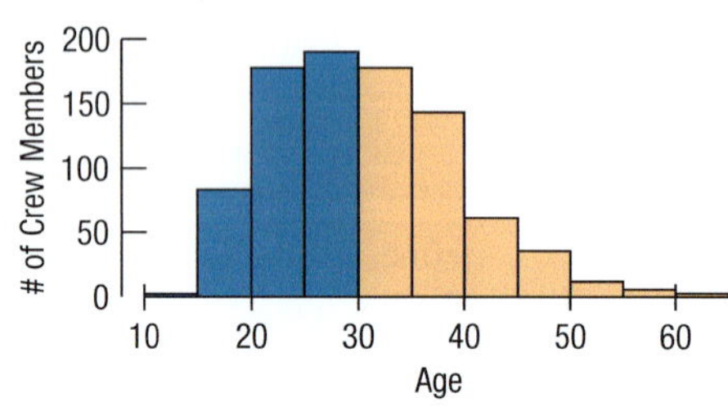

Figure 2.15
The median splits the histogram into two halves of equal area.

BY HAND Finding the Median

Finding the median of a batch of n numbers is easy as long as you remember to order the values first. If n is odd, the median is the middle value. Counting in from the ends, we find this value in the $\frac{n+1}{2}$ position.

When n is even, there are two middle values. So, in this case, the median is the average of the two values in positions $\frac{n}{2}$ and $\frac{n}{2} + 1$.

Here are two examples:

Suppose the batch has these values: 14.1, 3.2, 25.3, 2.8, −17.5, 13.9, 45.8.

First we order the values: −17.5, 2.8, 3.2, 13.9, 14.1, 25.3, 45.8.

Since there are 7 values, the median is the $(7 + 1)/2 = 4$th value, counting from the top or bottom: 13.9. Notice that 3 values are lower, 3 higher.

Suppose we had the same batch with another value at 35.7. Then the ordered values are −17.5, 2.8, 3.2, 13.9, 14.1, 25.3, 35.7, 45.8.

The median is the average of the 8/2, or 4th, and the $(8/2) + 1$, or 5th, values. So the median is $(13.9 + 14.1)/2 = 14.0$. Four data values are lower, and four higher.

The Mean

The most common way to summarize the center of a distribution is with the mean. The mean age of the crew aboard the *Titanic* was 31.1 years—about what we might expect from the histogram in Figure 2.15. You already know how to average values, but this is a good place to introduce notation that we'll use throughout the text. We use the Greek capital letter sigma, Σ, to mean "sum" (sigma is "S" in Greek), and we'll write

$$\bar{y} = \frac{Total}{n} = \frac{\sum y}{n}.$$

The formula says that to find the **mean**, add up all the values of the variable and divide that sum by the number of data values, n—just as you've always done.[8]

NOTATION ALERT

In algebra, you used letters to represent values in a problem, but it didn't matter what letter you picked. You could call the width of a rectangle x or you could call it w (or *Fred*, for that matter). But in statistics, the notation is part of the vocabulary. For example, in statistics n represents the number of data values. Here's another one: Whenever we put a bar over a symbol, it means "find the mean."

[8] You may also see the variable called x and the equation written $\bar{x} = \frac{\sum x}{n}$. Don't let that throw you. You are free to name the variable anything you want. (Well, except n.) We'll generally use y for variables like this that we want to summarize. Later, we'll talk about variables that are used to explain, model, or predict y. We'll call them x. You may also see the summation written with subscripts to indicate counting through the values, but we're summing them all up, so there's no need to clutter the formula that way.

You'd expect the result of averaging to be called the average, but that would be too easy. Instead, the value we calculated is called the mean, $\bar{y}$, and the notation is pronounced "y-bar."

> **DOW JONES INDUSTRIAL MEAN?**
>
> In everyday language, sometimes "average" does mean what we want it to mean. We don't talk about your grade point mean or a baseball player's batting mean or the Dow Jones industrial mean. So we'll continue to say "average" when that seems most natural. When we do, though, you may assume that what we mean is the mean.

The mean feels like the center because it is the point where the histogram balances:

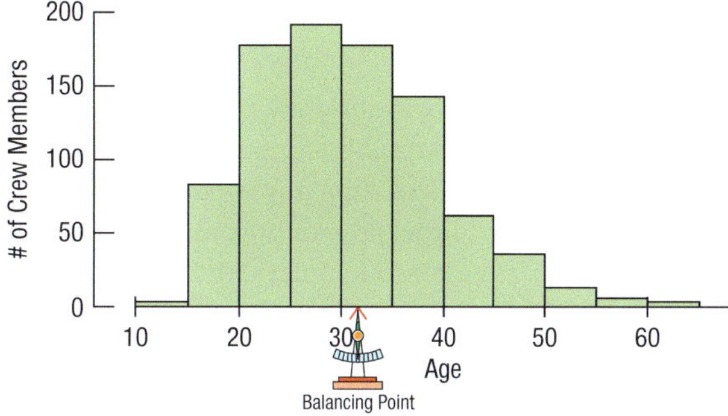

Figure 2.16
The mean is located at the *balancing point* of the histogram.

Mean or Median?

Although the mean and the median both summarize the center of a distribution of data, they do it in different ways. Both can be useful.

The median is less influenced by skewness or outliers than the mean. This makes it a good choice for summarizing skewed distributions such as income or company size. For example, the median ticket price on the *Titanic* might be a better summary for the typical price paid for the voyage. The median price was £14.4.

The mean pays attention to each value in the data. That's good news and bad. Its sensitivity makes the mean appropriate for overall summaries that need to take all of the data into account. But this sensitivity means that outliers and skewed distributions can pull the mean off to the side. For example, although the median ticket price is £14.4, the mean is £33.0 because it has been pulled to the right by a number of very high prices paid in first class and that one very large value of £512.3 (nearly $65,000 in today's dollars!).

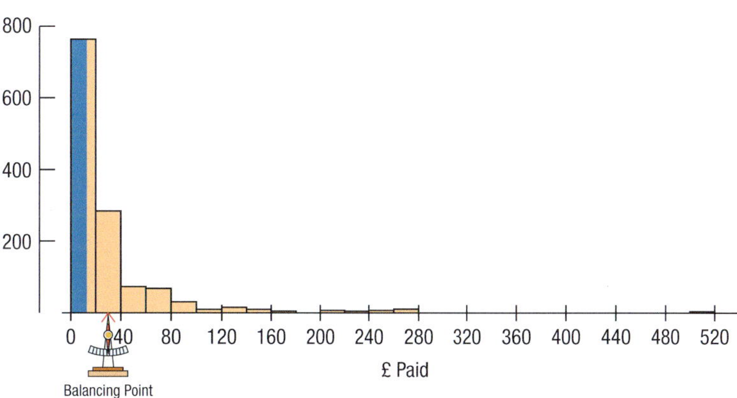

Figure 2.17
The distribution of ticket prices is skewed to the right. The mean (£33.0) is higher than the median (£14.4). The higher prices at the right have pulled the mean toward them and away from the median. (Data in **Titanic**)

Nevertheless, the mean can be a useful summary even for a skewed distribution. For example, the White Star Line, which owned the *Titanic*, might have wanted to know the mean ticket price to get an idea of their revenue per passenger.

So, which summary should you use? Because technology makes it easy to calculate them, it is often a good idea to look at both. If the mean and median differ, you should think about the reasons for that difference. Are there outliers? If so, you should investigate them. Correct them if they are errors; set them aside if they really don't belong with your data. The fact that outliers do not affect the median doesn't get you "off the hook"—don't just use the median and think you've dealt with your outliers. If the mean and median differ because your data distribution is skewed, then you should consider what you want to know about your data. You may end up preferring one or the other—or you may decide to report both summaries. Some application areas may have a standard practice. For example, economists usually summarize income distributions with medians.

2.5 Spread

If every crew member on the *Titanic* had been 30 years old, then the mean and median would both be 30 and we'd know everything about the distribution of ages. The more the data vary, however, the less a measure of center alone can tell us. So we also measure the **spread**—or how much the data values vary around the center. When you describe a distribution numerically, always report a measure of its spread along with its center. Spread, like shape and center, is a general description of the distribution, not a precise quantity.

WHAT WE DON'T KNOW

Statistics pays close attention to what we *don't* know as well as what we do know. When the data we have are only part of the whole—when we are interested in what might happen in the future, when we care about the likely prognosis of patients who haven't yet been treated with a new medicine—we must keep in mind the limits to what we can learn from our data. Understanding how spread out the data are is a first step in understanding what a summary *cannot* tell us about the data. It's the beginning of telling us what we don't know.

The Range

How should we measure the spread? We could simply look at the extent of the data. How far apart are the two extremes? The **range** of the data is defined as the *difference* between the maximum and minimum values:

$$Range = max - min.$$

Notice that the range is a single number, not an interval of values, as you might think from its use in common speech. The maximum *Titanic* crew age is 62 years and the minimum is 14 (the bellboy), so the range is $62 - 14 = 48$ years.

The range has the disadvantage that a single extreme value can make it very large, giving a value that doesn't really represent the data overall. So it isn't a good idea to report the range as your only measure of the spread of the data.

The Interquartile Range

When we summarize a categorical variable, we often look at the percentage of values falling in each category. With a quantitative variable, we talk about the percentage (or fraction) of values found below or above a given value or between two values. And we can turn that around and ask for values that bound a specified percentage of the data. For example, we might seek the value that is just above the lower 25% of the data values. That value is called the lower **quartile**, first quartile, or sometimes just the 25th **percentile**. The upper quartile (or third quartile) is the 75th percentile. And, of course, the median is the 50th percentile—and the second quartile (though we don't usually call it that). We denote the first quartile Q1 and the third quartile Q3.

To describe the spread of a variable, we can find the range of just the middle half of the data—the distance between the quartiles. The difference between the quartiles is called the **interquartile range**. It's commonly abbreviated IQR (and pronounced "eye-cue-are"):

$$IQR = upper\ quartile - lower\ quartile.$$

Why the Quartiles?

Could we use other percentiles besides the quartiles to measure the spread? Sure, we could, but the IQR is the most commonly used percentile difference and the one you're most likely to see in practice.

For the *Titanic* crew ages, the median is 30 years. If we order the 889 ages, the 445th age is 30. If we take the lower 445 ages, their median, 24, gives us Q1. And in the upper half, we find their median, Q3, to be 37 years. The difference between the two quartiles gives the IQR:

$$IQR = 37 - 24 = 13 \text{ years.}$$

Now we know that the range is 48 and that the middle half covers 13 years. Because it describes the range of the middle half, the IQR gives a reasonable summary of the spread of the distribution, as we can see from this histogram:

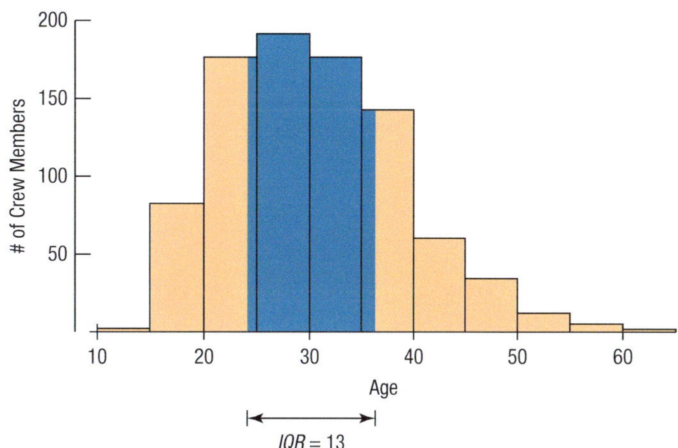

Figure 2.18
The quartiles bound the middle 50% of the values of the distribution, so the median and quartiles divide the data into four intervals of (roughly) equal count. (Data in **Titanic**)

EXAMPLE 2.7 Spread and Credit Card Expenditures

RECAP: The histogram in Example 2.6 showed that the distribution of credit card expenditures is skewed. The quartiles are $73.84 and $624.80.

QUESTION: What is the IQR and why is it a suitable measure of spread?

ANSWER: For these data, the interquartile range (IQR) is $624.80 − $73.84 = $550.96. Like the median, the IQR is not affected by outlying values or by the skewness of the distribution, so it is an appropriate measure of spread for the given expenditures.

NOTATION ALERT

Q1 and Q3 are reserved to refer to the quartiles. But you probably wouldn't have used them for anything else.

So, What Is a Quartile Anyway?

Finding the quartiles sounds easy, but surprisingly, the quartiles are not well defined. It's not always clear how to find a value such that exactly one quarter of the data lies above or below that value. Here's a simple rule for finding quartiles: Find the median of each half of the data split by the median. When n is odd, we include the median with each of the halves. Some other texts omit the median from each half before finding the quartiles. Both methods are commonly used. If you are willing to do a bit more calculating, there are several other methods that locate a quartile somewhere between adjacent data values. We know of at least six different rules for finding quartiles. Remarkably, each one is in use in some software package or calculator.

So don't worry too much about getting the "exact" value for a quartile. All of the methods agree pretty closely when the dataset is large. When the dataset is small, different rules will disagree more, but in that case there's little need to summarize the data anyway.

Remember, statistics is about understanding the world, not about calculating the right number. The "answer" to a statistical question is a sentence about the issue raised in the question.

The Standard Deviation

The IQR is almost always a reasonable summary of spread, but because it uses only the two quartiles of the data, it ignores much of the information about how individual values vary. By contrast, the standard deviation takes into account how far each value is from the mean.

We usually use the mean rather than the median as our central value because the mean has a special property. It turns out that if we sum up the *squares* of all the differences of the data values from any central value, then using the mean makes the sum as small as it could possibly be. This property is called the **least squares property**—something we'll see again. The differences of the values from the mean are called **residuals**.

When we add up these squared residuals and divide by $n - 1$, the result is the **variance**, denoted s^2.

$$s^2 = \frac{\sum (y - \bar{y})^2}{n - 1}$$

Why divide by $n - 1$ rather than n? As we said, the sum of the squared residuals in the numerator is as small as it possibly can be, and that turns out to be just a bit *too* small. Dividing $n - 1$ by fixes it. We'll say more to explain this remarkable fact later.

The variance plays an important role in statistics, but it has a problem as a measure of spread. Whatever the units of the original data, the variance is in *squared* units. We usually want measures of spread to have the same units as the data. So we don't have to talk about squared dollars, squared ages, or mpg^2. So, to get back to the original units, we take the square root. The result, s, is the **standard deviation**:

$$s = \sqrt{\frac{\sum (y - \bar{y})^2}{n - 1}}.$$

NOTATION ALERT

s^2 always denotes a variance, and s always denotes a standard deviation. Sometimes we add subscripts to make it plain what this is a variance or standard deviation of.

We almost always use technology to calculate the variance and standard deviation.

The variance of the *Titanic* crew ages is 73.08 years^2. The standard deviation is 8.55 years. For a roughly symmetric distribution with no outliers, the standard deviation will typically be about $3/4$ as big as the IQR. If the standard deviation is much larger than that, you should look at your histogram again to check for outliers, skewness, or some other unexpected shape.

Thinking About Variation

Statistics is about variation. Measures of variation help us be precise about what we don't know.[9] If many data values are scattered far from the center, the IQR and the standard deviation will be large. If the data values are close to the center, then these measures of spread will be small. If all our data values were exactly the same, we'd have no question about summarizing the center, and all measures of spread would be zero—and we wouldn't need statistics. You might think this would be a big plus, but it would make for a boring world. Fortunately (at least for statistics), data do vary.

One caution: If the distribution has two or more modes, measures of spread are apt to tell us how different the modes are rather than how variable the data values are. Usually we'll learn more by summarizing the center and spread of the data around each mode individually—especially if we understand the reason for the separate modes. For example, the standard deviation of all the Kentucky Derby race times says almost nothing about the speed of race horses. The standard deviations of times for the early, longer races and the standard deviation for the recent shorter races would be far more informative.

Measures of spread tell how well other summaries describe the data. That's why we always (always!) report a spread along with any summary of the center.

JUST CHECKING

6. The U.S. Census Bureau reports the median family income in its summary of census data. Why do you suppose they use the median instead of the mean? What might be the disadvantages of reporting the mean?

7. You've just bought a new car that claims to have a highway fuel efficiency of 31 miles per gallon. Of course, your mileage will "vary." If you had to guess, would you expect the IQR of gas mileage attained by all cars like yours to be 30 mpg, 3 mpg, or 0.3 mpg? Why?

8. A company selling a new MP3 player advertises that the player has a mean lifetime of 5 years. If you were in charge of quality control at the factory, what else would you want to know?

What to Tell About a Quantitative Variable

What should you *Tell* about a quantitative variable?

- Start by making a histogram or stem-and-leaf display, and discuss the shape of the distribution. Note particularly whether it is unimodal, symmetric, and free of outliers.
- Next, discuss the center and spread.
  - Pair the median with the IQR and the mean with the standard deviation. It's more useful to report both a center and spread than to report just one. Reporting a center without a spread can be dangerous. You may appear to know more than you do about the distribution. Reporting only the spread leaves your reader wondering where we are.
  - If the shape is skewed, be sure to report the median and IQR. You may want to include the mean and standard deviation as well and point out why the mean and median differ.
  - If the shape is symmetric, report the mean and standard deviation and possibly the median and IQR as well. For unimodal symmetric data, the IQR is usually a bit larger than the standard deviation. If that's not true of your dataset, look again to check for skewness and outliers.

[9]Wait. What? Yes, this is just what we mean to say. Statisticians spend much of their time measuring and reporting precisely how much we *don't* know. That helps to keep scientists and social scientists from making claims that the data don't support. When we analyze data statistically, that's an important part of the job.

> **HOW "ACCURATE" SHOULD WE BE?**
> Don't report means and standard deviations to a zillion decimal places; such implied accuracy is really meaningless. Although there is no ironclad rule, statisticians commonly report summary statistics to one or two decimal places more than the original data have.

- Discuss any unusual features.
  - If there are multiple modes, try to understand why. If you can identify a reason for separate modes (for example, women and men typically have heart attacks at different ages), it may be a good idea to split the data into separate groups.
  - If there are any clear outliers, point them out. If you are reporting the mean and standard deviation, report them with the outliers present and with the outliers omitted. The differences may be revealing. (Of course, the median and IQR won't be affected very much by the outliers.)

EXAMPLE 2.8 Choosing Summary Statistics

RECAP: You have provided the credit card company's board of directors with a histogram of customer expenditures, and you have summarized the center and spread with the median and IQR. Knowing a little statistics, the directors now insist on having the mean and standard deviation as summaries of the spending data.

QUESTION: You calculate that the mean is \$478.19 and the standard deviation is \$741.87. But you know from the histogram (Example 2.6) that there is an outlier at \$7000. Explain to the directors why the mean and standard deviation should be interpreted differently than the median and IQR. What would you give as reasons?

ANSWER: The high outlier at \$7000 pulls the mean up substantially and inflates the standard deviation. Locating the mean value on the histogram shows that it is not a typical value at all, and the standard deviation suggests that expenditures vary much more than they do. The median and IQR are more resistant to the presence of skewness and outliers, giving more realistic descriptions of center and spread.

STEP-BY-STEP EXAMPLE

Summarizing a Distribution

One of the authors owned a Nissan Maxima for 8 years. Being a statistician, he recorded the car's fuel efficiency (in mpg) each time he filled the tank. He wanted to know what fuel efficiency to expect as "ordinary" for his car. (Hey, he's a statistician. What would you expect?[10]) Knowing this, he was able to predict when he'd need to fill the tank again and to notice if the fuel efficiency suddenly got worse, which could be a sign of trouble. (Data in **Nissan**)

QUESTION: How would you describe the distribution of *Fuel efficiency* for this car?

THINK **PLAN** State what you want to find out.

I want to summarize the distribution of Nissan Maxima fuel efficiency.

VARIABLES Identify the variable and report the W's.

The data are the fuel efficiency values in miles per gallon for the first 100 fill-ups of a Nissan Maxima.

Be sure to check that the data are appropriate for your analysis.

The fuel efficiencies are quantitative with units of miles per gallon. A histogram is an appropriate way to display the distribution. Numerical summaries are appropriate as well.

[10]He also recorded the time of day, temperature, price of gas, and phase of the moon. (OK, maybe not phase of the moon.)

SHOW **MECHANICS** Make a histogram. Based on the shape, choose appropriate numerical summaries.

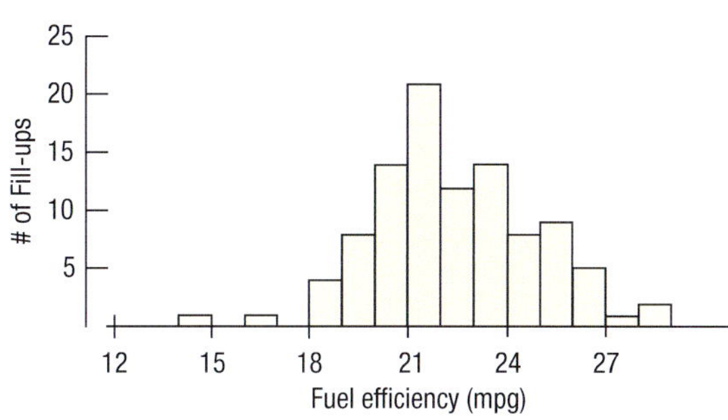

REALITY CHECK A value of 22 mpg seems reasonable for such a car. The spread is reasonable, although the range looks a bit large.

A histogram of the data shows a fairly symmetric distribution with a low outlier.

Count	100
Mean	22.4 mpg
StdDev	2.45
Q1	20.8
Median	22.0
Q3	24.0
IQR	3.2

TELL **CONCLUSION** Summarize and interpret your findings in context. Be sure to discuss the distribution's shape, center, spread, and unusual features (if any).

The mean and median are close, so the outlier doesn't seem to be a problem. I can use the mean and standard deviation.

The distribution of *Fuel efficiency* is unimodal and roughly symmetric, with a mean of 22.4 mpg. There is a low outlier that should be investigated, but it does not influence the mean very much. The standard deviation suggests that from tankful to tankful, I can expect the car's fuel efficiency to differ from the mean by an average of about 2.45 mpg.

I Got a Different Answer: Did I Mess Up?

When you calculate a mean, the computation is clear: You sum all the values and divide by the sample size. You may round your answer less or more than someone else (we recommend one more decimal place than the data), but all books and technologies agree on how to find the mean. Some statistics, however, are more problematic. For example, we've already pointed out that methods of finding quartiles differ.

Differences in numerical results can also arise from decisions made in the middle of calculations. For example, if you are working by hand and you round off your value for the mean before you calculate the sum of squared deviations in the numerator, your standard deviation probably won't agree with a computer program that calculates without rounding. (We do recommend that you use as many digits as you can during the calculation and round only when you are done.)

Don't be overly concerned with these discrepancies, especially if the differences are small. They don't mean that your answer is "wrong," and they usually won't change any conclusion you might draw about the data. Sometimes (in footnotes and in the answer appendix) we'll note alternative results, but we could never list all the possible values, so we'll rely on your common sense to focus on the meaning rather than on the digits. Remember: Answers are sentences—not single numbers!

RANDOM MATTERS

A large financial institution in New York City has about 5000 people working at the Wall Street location. Real estate is expensive in the neighborhood, so some employees live a considerable distance away from the office. The Human Resources Department recently chose 100 employees and interviewed them about their work experience. One of the questions they asked was how long it took to get to work that morning. Here is a histogram of the 100 responses: (Data in **Commute times sample 100**)

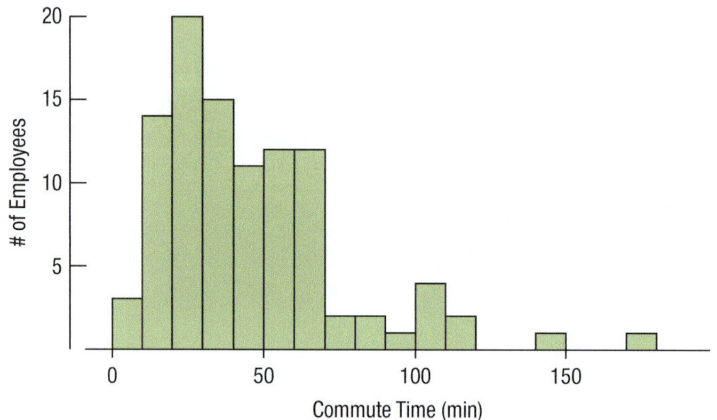

Figure 2.19

The commute times (in minutes) of 100 employees. Mean: 44.98 min, Median: 37.5 min, SD: 30.19 min, IQR: 37 min.

The surveyed employees had a mean commute time of 44.98 minutes with a median time of 37.5 minutes. That's great, but what the human resources folks really wanted to know is the mean commute time for *all* 5000 employees. (You could argue that they should be looking at the median as well.) We often encounter situations like this. By asking a few hundred voters, survey companies try to estimate election outcomes. Government agencies try to estimate hiring by asking a small fraction of businesses. The challenge they all face is that each time a different group is surveyed, the results will vary.

MEANS VARY

How much does a sample mean change from sample to sample? We can answer that question by simulating. We have the data on all 5000 employees,[11] so we can take many samples of size 100 at random and find the mean of each sample. Figure 2.20 shows those means.

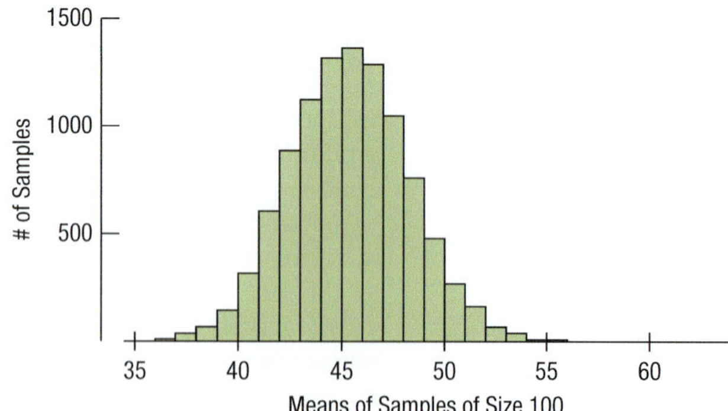

Figure 2.20

The mean commute times of 10,000 different random samples of size 100.

[11]Because of privacy concerns, we have constructed a dataset similar to the actual data that contains the characteristics of the entire 5000 times but protects identities. (Data in **Population Commute Times**)

Some samples had means as small as 40 minutes and others had means over 60 minutes. But most had means in the middle—around 45 minutes. That's not surprising. The mean of all 5000 commute times is in fact 45.4 minutes.

- Statistics vary from sample to sample. Even though an exercise or exam question asks you to find the mean, don't forget that when we only have data for a sample, another sample would have a different mean, so we don't know the mean of the overall population. That's the bad news.
- But most of the sample means are close to each other and do a good job of characterizing commute times. The majority are within 5 minutes on either side of the time for all 5000 employees. Asking everyone in the building isn't practical. Often, we can do almost as well with a sample. That's the good news.

WHAT CAN GO WRONG?

- **Don't violate the area principle.** This is probably the most common mistake in a graphical display. It is often made in the cause of artistic presentation. Here, for example, are two displays of the pie chart of the *Titanic* passengers by class:

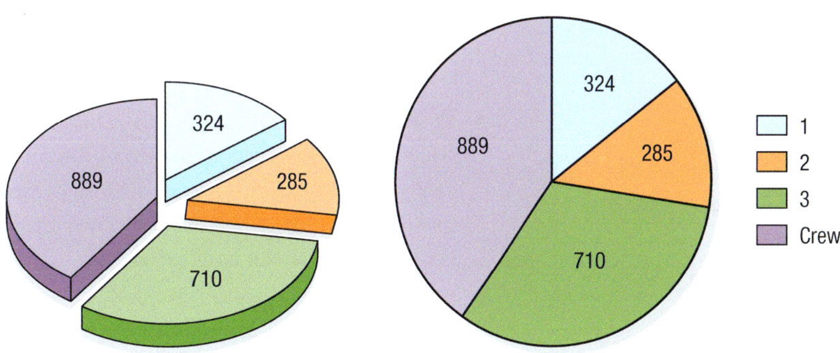

The one on the left looks pretty, doesn't it? But showing the pie on a slant violates the area principle and makes it much more difficult to compare fractions of the whole made up of each class—the main feature that a pie chart ought to show. For example, in the plot on the left, it is hard to see that there were more crew than third-class passengers.

- **Keep it honest.** Here's a pie chart that displays data on the percentage of high school students who engage in specified dangerous behaviors as reported by the Centers for Disease Control and Prevention. What's wrong with this plot?

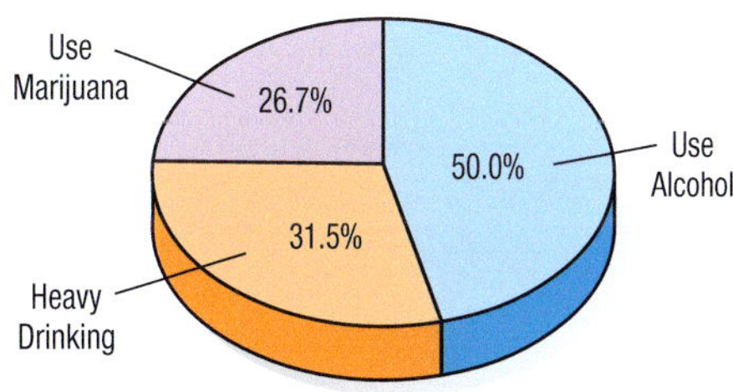

Try adding up the percentages. Or look at the 50% slice. Does it look right? Then think: What are these percentages of? Is there a "whole" that has been sliced up? In a pie chart, the proportions shown by the slices of the pie must add up to 100% and each individual must fall into only one category. Of course, showing the pie on a slant makes it even harder to detect the error.

- **Don't make a histogram of a categorical variable.** Just because the variable contains numbers doesn't mean that it's quantitative. Here's a histogram of the insurance policy numbers of some workers:

Figure 2.21
It's not appropriate to display these data with a histogram.

It's not very informative because the policy numbers are just identifiers. Similarly, a histogram or stem-and-leaf display of a categorical variable makes no sense. A bar chart or pie chart would be more appropriate.

- **Don't look for the shape, center, and spread of a bar chart.** A bar chart that shows the number of values in each category does display the distribution of a categorical variable, but the bars could be arranged in any order from left to right. Concepts like symmetry, center, and spread make sense only for quantitative variables.

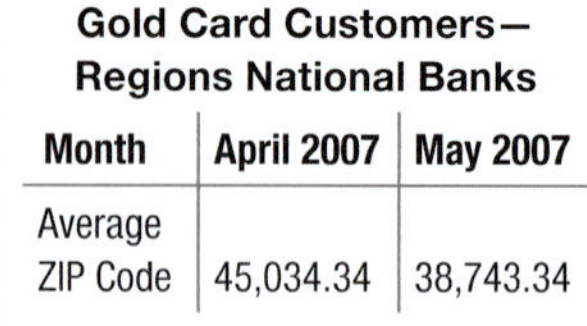

Gold Card Customers—Regions National Banks

Month	April 2007	May 2007
Average ZIP Code	45,034.34	38,743.34

- **Don't compute numerical summaries of a categorical variable.** Neither the mean ZIP code nor the standard deviation of Social Security numbers is meaningful. If the variable is categorical, report summaries such as percentages of individuals in each category. It is easy to make this mistake when using technology to do the summaries for you. After all, the computer doesn't care what the numbers mean.

- **Don't use bars in every display—save them for histograms and bar charts.** The bars in a bar chart indicate how many cases of a categorical variable are in each category. The bars in a histogram indicate the number of cases in each interval of a quantitative variable. In both bar charts and histograms, the bars display counts of data values. Some people create other displays that use bars to represent individual data values. Beware: Such graphs are neither bar charts nor histograms. For example, a student made the following display of the number of juvenile eagles sighted during 13 weeks in the winter at a site in Rock Island, Illinois:

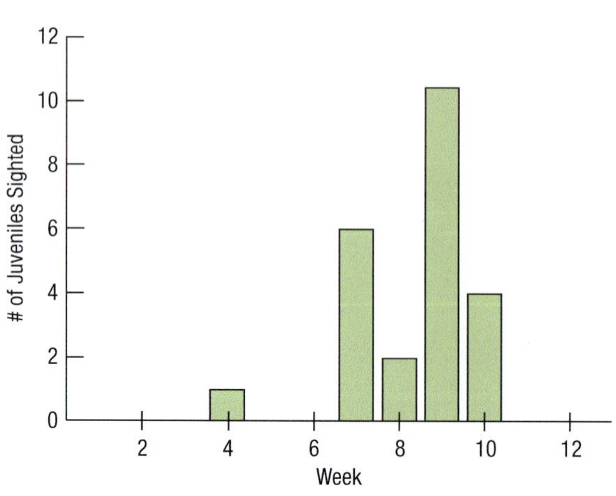

Figure 2.22
A bad graph of the number of juvenile eagles sighted each week of a study.

Unfortunately, the student didn't first think about *why* she was making the display. There are no categories to compare, so a bar chart is not appropriate. If she intended to show the week-to-week pattern, a timeplot (see Chapter 6) would have been appropriate. If she wanted to display the *distribution* of sightings, a histogram would be appropriate, but this isn't a histogram because it doesn't use the x-axis to show numbers of sightings, which is what was measured.[12] But bars suggest to the reader either a bar chart or a histogram and Figure 2.22 is neither. A correct histogram should have a tall bar at "0" to show there were many weeks when no eagles were seen, like this:

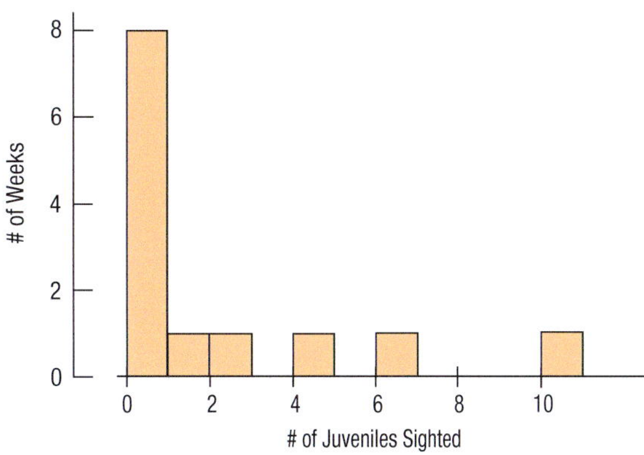

Figure 2.23
A histogram of the eagle-sighting data shows the number of weeks in which different counts of eagles occurred. This display shows the distribution of juvenile-eagle sightings.

- **Choose a bin width appropriate to the data.** Computer programs usually do a pretty good job of choosing histogram bin widths. Often there's an easy way to adjust the width, sometimes interactively. Here are the *Titanic* ages with two (rather extreme) choices for the bin size:

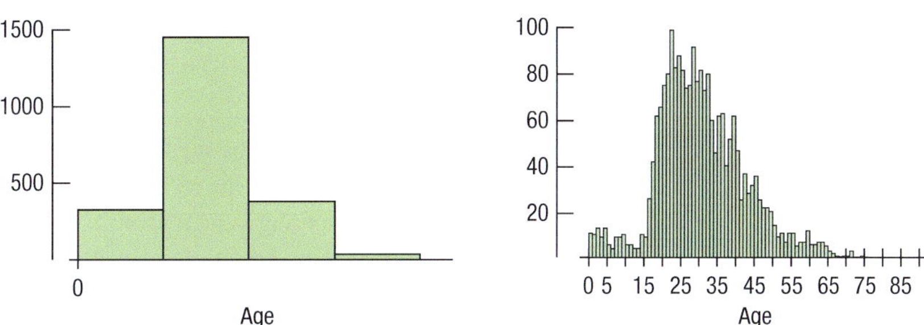

Figure 2.24
Too many or too few bars make histograms less useful. The histogram on the left doesn't show the small mode of children; the one on the right suggests many minor "modes" that are really of no importance at all.

For the usual amounts of data, between 5 and 30 bins show the features of a distribution best. Too few bins tend to hide too much, and too many bins make unimportant features too prominent.

- **Do a reality check.** Don't let the computer or calculator do your thinking for you. Make sure the calculated summaries make sense. For example, does the mean look like it is in the center of the histogram? Think about the spread: An IQR of 50 mpg would clearly be wrong for gas mileage. And no measure of spread can be negative. The standard deviation can take the value 0, but only in the very unusual case that all the data values equal the same number. If you see an IQR or standard deviation equal to 0, it probably means that something's wrong with the data.
- **Don't forget to sort the values before finding the median or percentiles.** It seems obvious, but if you work by hand, it's easy to forget to sort the data first before counting in to find

[12]Edward Tufte, in his book *The Visual Display of Quantitative Information*, proposes that graphs should have a high *data-to-ink ratio*. That is, don't waste a lot of ink to display a single number when a dot would do the job.

medians, quartiles, or other percentiles. Don't report that the median of the five values 194, 5, 1, 17, and 893 is 1 just because 1 is the middle number.

- **Don't worry about small differences due to different methods or rounding.** Finding the 10th percentile or the lower quartile in a dataset sounds easy enough, but it turns out that the definitions are not exactly clear. If you compare different statistics packages or calculators, you may find that they give slightly different answers for the same data. Different programs and calculators may round results in different ways, leading to slightly different answers. These differences, though, are unlikely to be important in interpreting the data, the quartiles, the IQR, or the standard deviation, so don't let them worry you.
- **Don't report too many decimal places.** Statistics programs and calculators often report a ridiculous number of digits. A general rule for numerical summaries is to report one or two more digits than the number of digits in the data. For example, earlier we saw a dotplot of the Kentucky Derby race times. The mean and standard deviation of those times could be reported as

$$\bar{y} = 129.3367361 \text{ sec} \qquad s = 12.94689221 \text{ sec}$$

But we knew the race times only to the nearest quarter-second, so the extra digits are meaningless.

- **Don't round in the middle of a calculation.** Don't report too many decimal places, but it's best not to do any rounding until the end of your calculations. Even though you might report the mean of the earthquake magnitudes as 7.05, it's really 7.05281. Use the more precise number in your calculations if you're finding the standard deviation by hand—or be prepared to see small differences in your final result.
- **Watch out for multiple modes.** The summaries of the Kentucky Derby times are meaningless for another reason. As we saw in the dotplot, the Derby was initially a longer race. It would make much more sense to report that the old 1.5-mile Derby had a mean time of 158.6 seconds, while the current Derby has a mean time of 124.3 seconds. If the distribution has multiple modes, consider separating the data into different groups and summarizing each group separately. But don't be overly impressed with local bumps in a histogram, which may be an artifact of the scale of the histogram. Statistics programs usually permit the re-scaling of histograms to have more (narrower) or fewer (wider) bars. A truly bimodal distribution will appear bimodal in different, reasonable scales.

- **Beware of outliers.** The median and IQR are resistant to outliers, but the mean and standard deviation are not. When the mean and median differ, it is your obligation to figure out why. If the cause is some outlying values, identify and try to understand why they are extreme. If you can't figure out why they are extreme, a safe strategy is to perform the analysis both with and without the outlier(s) and then report the differences. Of course, if you find out that they are errors, either correct or omit them.
- **Beware of inappropriate summaries.** Summarizing a variable when you have not looked at a histogram or dotplot to check for multiple modes, outliers, or skewness invites disaster. Don't accept summary statistics blindly without some evidence that the variables they summarize make sense. The Centerville sign says it all.

CONNECTIONS

The distribution of a variable shows the possible values and the relative frequencies of values or intervals of values. The shape of the distribution of a quantitative variable is an important concept in most of the subsequent chapters. We will be especially interested in distributions that are unimodal and symmetric.

We will most often summarize a quantitative variable with its mean and standard deviation. We will often use standard deviations as rulers to measure differences and distances. Watch for that.

CHAPTER REVIEW

Make and interpret a frequency table for a categorical variable.

- Summarize categorical data by counting the number of cases in each category, sometimes expressing the resulting distribution as percentages.

Make and interpret a bar chart or pie chart.

- Display categorical data using the area principle in either a bar chart or a pie chart.
- Display the distribution of a quantitative variable with a histogram, stem-and-leaf display, or dotplot.
- Understand distributions in terms of their shape, center, and spread.

Describe the shape of a distribution.

- A symmetric distribution has roughly the same shape reflected around the center.
- A skewed distribution extends farther on one side than on the other.
- A unimodal distribution has a single major hump or mode; a bimodal distribution has two; multimodal distributions have more.
- Outliers are values that lie far from the rest of the data.
- Report any other unusual feature of the distribution such as gaps.

Summarize the center of a distribution by computing the mean and median.

- The mean is the sum of the values divided by the count.
- The median is the middle value; half the values are above and half are below the median.
- The mean and median may differ. Understand why and use that information to make intelligent choices of summaries for your data.

Compute the standard deviation and interquartile range (IQR), and know when it is best to use each to summarize the spread.

- The standard deviation is roughly the square root of the "average" squared difference between each data value and the mean. It is the summary of choice for the spread of unimodal, symmetric variables.
- The IQR is the difference between the third and first quartiles. It is the preferred summary of spread for skewed distributions or data with outliers.
- Report the median and IQR when the distribution is skewed. If it's symmetric, summarize the distribution with the mean and standard deviation (and possibly the median and IQR as well). Always pair the median with the IQR and the mean with the standard deviation.

REVIEW OF TERMS

The key terms are in chapter order so you can use this list to review the material in the chapter.

Area principle In a statistical display, each data value should be represented by the same amount of area (p. 45).

Frequency table (Relative frequency table) A frequency table lists the categories in a categorical variable and gives the count (or percentage) of observations for each category (p. 45).

Distribution The distribution of a categorical variable gives

- the possible values of the variable and
- the relative frequency of each value.

The distribution of a quantitative variable slices up all the possible values of the variable into equal-width bins and gives the number of values (or counts) falling into each bin (p. 45).

Bar chart Bar charts show a bar whose area represents the count (or percentage) of observations for each category of a categorical variable (p. 45).

Pie chart Pie charts show how a "whole" is divided into categories. The area of each wedge of the circle corresponds to the proportion in each category (p. 47).

Histogram A histogram uses adjacent bars to show the distribution of a quantitative variable. Each bar represents the frequency (or relative frequency) of values falling in each bin (p. 49).

Gap A region of the distribution where there are no values (p. 50).

Stem-and-leaf display A display that shows quantitative data values in a way that sketches the distribution of the data. It's best described in detail by example (p. 52).

Dotplot A dotplot graphs a dot for each case along a single axis (p. 52).

Density plot A density plot shows the shape of a variable's distribution by "smoothing out" its histogram to make a gentle curve. It may be easier to understand the distribution in this way (p. 53).

Shape To describe the shape of a distribution, look for

- single vs. multiple modes,
- symmetry vs. skewness, and
- outliers and gaps (p. 54).

Mode A hump or local high point in the distribution of a variable. The apparent location of modes can change as the scale of a histogram is changed (p. 54).

Unimodal (Bimodal, Multimodal) Unimodal means having one mode. This is a useful term for describing the shape of a histogram when it's generally mound-shaped. Distributions with two modes are called bimodal. Those with more than two are multimodal (p. 54).

Uniform A distribution that doesn't appear to have any mode and in which all the bars of its histogram are approximately the same height (p. 54).

Symmetric A distribution is symmetric if the two halves on either side of the center look approximately like mirror images of each other (p. 54).

Tails The parts of a distribution that trail off on either side. Distributions can be characterized as having long tails (if they straggle off for some distance) or short tails (if they don't) (p. 55).

Skewed A distribution is skewed if it's not symmetric and one tail stretches out farther than the other. Distributions are said to be skewed left when the longer tail stretches to the left, and skewed right when it goes to the right (p. 55).

Outlier Outliers are extreme values that don't appear to belong with the rest of the data. They may be unusual values that deserve further investigation, or they may be just mistakes; there's no obvious way to tell. Don't delete outliers automatically—you have to think about them. Outliers can affect many statistical analyses, so always be alert for them (p. 55).

Center The place in the distribution of a variable that you'd point to if you wanted to attempt the impossible by summarizing the entire distribution with a single number. Measures of center include the mean and median (p. 57).

Median The median is the middle value, with half of the data above and half below it. If n is even, it is the average of the two middle values. It is usually paired with the IQR (p. 58).

Mean The mean is found by adding up all the data values dividing the sum by the count:

$$\bar{y} = \frac{Total}{n} = \frac{\sum y}{n}.$$

It is usually paired with the standard deviation (p. 58).

Spread A numerical summary of how tightly the values are clustered around the center. Measures of spread include the IQR and standard deviation (p. 60).

Range The difference between the lowest and highest values in a dataset:

$$Range = maximum - minimum \text{ (p. 60).}$$

Quartile The lower quartile (Q1) is the value with a quarter of the data below it. The upper quartile (Q3) has three quarters of the data below it. The median and quartiles divide data into the four parts with approximately equal numbers of data values (p. 61).

Percentile The *i*th percentile is the number that falls above *i*% of the data (p. 61).

Interquartile range (IQR) The IQR is the difference between the first and third quartiles: $IQR = Q3 - Q1$. It is usually reported along with the median (p. 61).

Least squares property The property of a statistic that the sum of the squared deviations of data values from data summaries due to that statistic is as small as it could be for any statistic is called the least squares property. The mean satisfies this property (p. 62).

Residuals A residual is the difference between an observed data value and some summary or model for that value. In this chapter, a residual can be the difference between a data value and the mean, $y_i - \bar{y}$ (p. 62).

Variance The variance is the sum of squared deviations from the mean, divided by the count minus 1:

$$s^2 = \frac{\sum(y - \bar{y})^2}{n - 1} \text{ (p. 62).}$$

Standard deviation The standard deviation is the square root of the variance:

$$s = \sqrt{\frac{\sum(y - \bar{y})^2}{n - 1}} \text{ (p. 62).}$$

TECH SUPPORT

Displaying and Summarizing a Variable

Although every package makes a slightly different bar chart, they all have similar features:

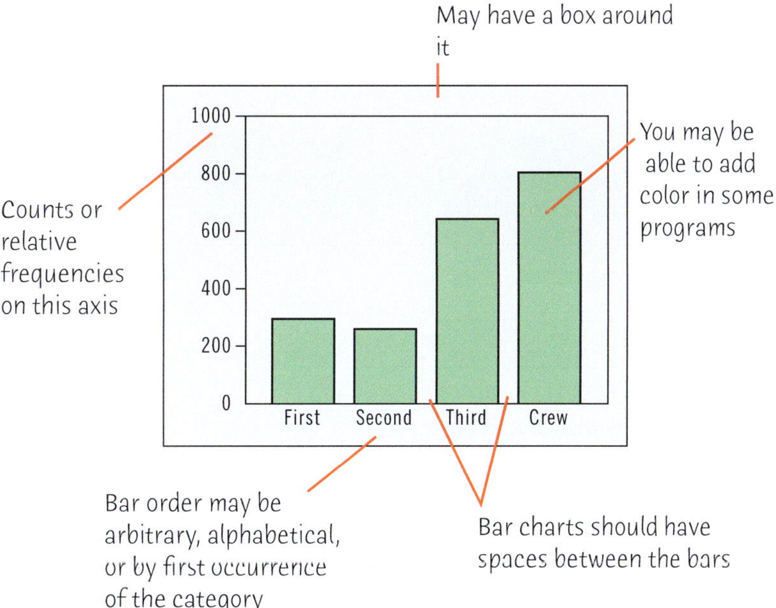

Sometimes the count or a percentage is printed above or on top of each bar to give some additional information. You may find that your statistics package sorts category names in annoying orders by default. For example, many packages sort categories alphabetically or by the order the categories are seen in the dataset. Often, neither of these is the best choice.

Almost any program that displays data can make a histogram, but some do a better job of determining where the bars should start and how they should partition the span of the data.

The vertical scale may be counts or proportions. Sometimes it isn't clear which. But the shape of the histogram is the same either way.

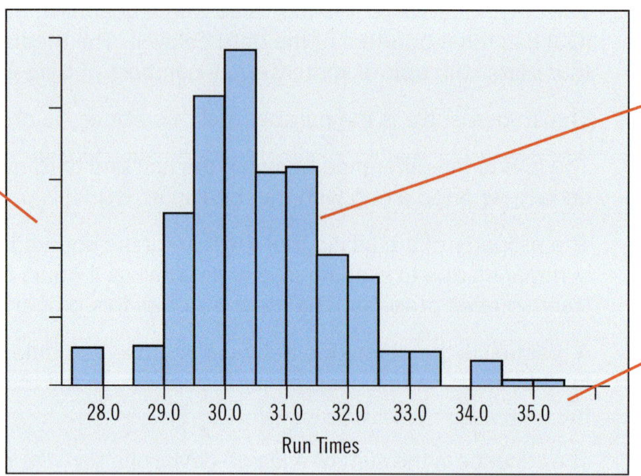

Most packages choose the number of bars for you automatically. Often you can adjust that choice.

The axis should be clearly labeled so you can tell what "pile" each bar represents. You should be able to tell the lower and upper bounds of each bar.

Many statistics packages offer a prepackaged collection of summary measures. The result might look like this:

Variable: Weight
N = 234
Mean = 143.3 Median = 139
St. Dev = 11.1 IQR = 14

Alternatively, a package might make a table for several variables and summary measures:

Variable	N	mean	median	stdev	IQR
Weight	234	143.3	139	11.1	14
Height	234	68.3	68.1	4.3	5
Score	234	86	88	9	5

Packages often provide many more summary statistics than you need. Of course, some of these may not be appropriate when the data are skewed or have outliers. It is your responsibility to check a histogram or stem-and-leaf display and decide which summary statistics to use.

It is common for packages to report summary statistics to many decimal places of "accuracy." But reporting values to six or seven digits beyond the decimal point doesn't mean that those digits have any meaning. It is rare data that have such accuracy in the original measurements, so the extra digits are random "noise." Generally it's a good idea to round values, allowing perhaps one more digit of precision than was given in the original data.

DATA DESK

To make a bar chart, pie chart, histogram, or dotplot:

- In the Toolbar, click on the display you want to make. Data Desk will open a blank window and invite you to drag in a variable.
- Drag the variable to plot into the display window.

To make a frequency table or calculate summary statistics:

- In the Toolbar click on the **Freq** button. Data Desk will open a blank window and invite you to drag in a variable.
- Drag the variable to summarize into the resulting window.
- Click the **HyperView** menu under the triangle in the upper left corner of the window to open a dialog with options of what to calculate. Data Desk will re-compute the window immediately if you select new options.

COMMENTS

Commands such as pie chart and bar chart treat data as categorical even if they are numerals. If you select a quantitative variable by mistake, you may see a large table listing all the individual values or an error message warning of too many categories. If you have data in the form of summary counts for each category and names of the categories, then the **Replicate Y by X** command will generate the corresponding variables with a value for each case. These are the variables that are suitable for these commands.

Commands such as **Histogram**, **Dotplot**, **Frequencies**, and **Summaries** treat variables as quantitative. If you drag in a categorical variable, you'll see an error message warning that there is no quantitative data.

EXCEL

To make a bar chart:

- First make a pivot table (Excel's name for a frequency table). From the Data menu, choose **Pivot Table** and **Pivot Chart Report**.
- When you reach the Layout window, drag your variable to the row area and drag your variable again to the data area. This tells Excel to count the occurrences of each category. Once you have an Excel pivot table, you can construct bar charts and pie charts.
- Click inside the Pivot Table.
- Click the **Pivot Table Chart Wizard** button. Excel creates a bar chart.
- A longer path leads to a pie chart; see your Excel documentation.
- An even longer path leads to a histogram. You must determine the bar boundaries yourself and tell Excel to eliminate the spaces between the bars.

COMMENTS

Excel uses the pivot table to specify the category names and find counts within each category. If you already have that information, you can proceed directly to the Chart Wizard.

In Excel, there is another way to find some of the standard summary statistics. For example, to compute the mean:

- Click on an empty cell.
- Go to the Formulas tab in the Ribbon. Click on the drop-down arrow next to Auto-Sum and choose **Average**.
- Enter the data range in the formula displayed in the empty box you selected earlier.
- Press **Enter**. This computes the mean for the values in that range.

To compute the standard deviation:

- Click on an empty cell.
- Go to the Formulas tab in the Ribbon and click the drop-down arrow next to Auto-Sum and select **More functions** . . .
- In the dialog window that opens, select **STDEV** from the list of functions and click **OK**. A new dialog window opens. Enter a range of fields into the text fields and click **OK**.

Excel computes the standard deviation for the values in that range and places it in the specified cell of the spreadsheet.

JMP

JMP makes a bar chart and frequency table together:

- From the Analyze menu, choose **Distribution**.
- In the Distribution dialog, drag the name of the variable into the empty variable window beside the label Y, Columns; click **OK**.

To make a pie chart:

- Choose **Chart** from the Graph menu.
- In the Chart dialog, select the variable name from the Columns list.
- Click on the button labeled **Statistics**, and select **N** from the drop-down menu.
- Click the **Categories, X, Levels** button to assign the same variable name to the *x*-axis.
- Under Options, click on the second button—labeled **Bar Chart**—and select **Pie** from the drop-down menu.

To make a histogram and find summary statistics:

- Choose **Distribution** from the Analyze menu.
- In the Distribution dialog, drag the name of the variable that you wish to analyze into the empty window beside the label Y, Columns.
- Click **OK**. JMP computes standard summary statistics along with displays of the variables.

COMMENTS

JMP requires you to specify first whether a variable is categorical or quantitative. See the JMP documentation for details.

MINITAB

To make a bar chart:

- Choose **Bar Chart** from the Graph menu.
- Sclcct **Counts of unique values** in the first menu, and select **Simple** for the type of graph. Click **OK**.
- In the Chart dialog, enter the name of the variable that you wish to display in the box labeled Categorical variables.
- Click **OK**.

To make a histogram:

- Choose **Histogram** from the Graph menu.
- Select **Simple** for the type of graph and click **OK**.
- Enter the name of the quantitative variable you wish to display in the box labeled Graph variables. Click **OK**.

To calculate summary statistics:

- Choose **Basic Statistics** from the Stat menu. From the Basic Statistics submenu, choose **Display Descriptive Statistics**.
- Assign variables from the variable list box to the Variables box. Minitab makes a Descriptive Statistics table.

R

To make a bar chart or pie chart in R, you first need to create the frequency table for the desired variable. First run "require(mosaic)"

- tally(~X) will give a frequency table for a single variable X.
- barplot(tally(~X)) will give a bar chart for X.
- Similarly, pie(tally(~X)) will give a pie chart.

For a quantitative variable X:

- favstats(~X) calculates the min, max, quartiles, median, mean, and number of observations.
- mean(~X) gives the mean and sd(~X) gives the standard deviation.
- histogram(~X) produces a histogram. The width and center options can be used to change the default breaks.

COMMENTS

Your variables X and Y may be variables in a data frame. If so, and DATA is the name of the data frame, then you will need to specify the "data=DATA" option for each of these commands. Many other summaries are available, including min(), max(), median(), and quantile(X, prob=p), where p is a probability between 0 and 1.

SPSS

To make a bar chart:

- Open the Chart Builder from the Graphs menu.
- Click the **Gallery** tab.
- Choose **Bar Chart** from the list of chart types.
- Drag the appropriate bar chart onto the canvas.
- Drag a categorical variable onto the *x*-axis drop zone.
- Click **OK**.

A similar path makes a pie chart by choosing **Pie chart** from the list of chart types. To make a histogram in SPSS, open the Chart Builder from the Graphs menu:

- Click the **Gallery** tab.
- Choose **Histogram** from the list of chart types.
- Drag the histogram onto the canvas.
- Drag a scale variable to the *y*-axis drop zone.
- Click **OK**.

To calculate summary statistics:

- Choose **Explore** from the Descriptive Statistics submenu of the Analyze menu. In the Explore dialog, assign one or more variables from the source list to the Dependent List and click the **OK** button.

STATCRUNCH

To make a bar chart or pie chart of a categorical variable:

- Click on **Graph**.
- Choose the type of plot (**Bar Plot** or **Pie Chart**).
- Choose between raw data (**With Data**) or summaries of data (**With Summary**).
- Choose the variable name from the list of Columns; if using summaries, also choose the counts.

For bar graphs, select Type:

- **Frequency, Relative frequency, Percent, Relative frequency (within category)**, or **Percent (within category)**.
- Note that you may elect to group categories under a specified percentage as Other.
- Click on **Compute!**

For a quantitative variable, to calculate summaries:

- Click on **Stat**.
- Choose **Summary Stats > Columns**.
- Choose the variable name from the list of Columns.
- Click On **Compute!**

For a histogram, click on **Graph**.

- Choose **Histogram**.
- Choose the variable name from the list of Columns.
- Click on **Compute!**

COMMENTS

- You may need to hold down the Ctrl or Command key to choose more than one variable to summarize.

TI-83/84 PLUS

The TI-83 won't do displays for categorical variables.

To make a histogram:

- Turn STATPLOT on.
- Choose the histogram icon and specify the List where the data are stored.
- ZoomStat, then adjust the WINDOW appropriately.

To calculate summary statistics:

- Choose **1-VarStats** from the STAT CALC menu and specify the List where the data are stored.

COMMENTS

If the data are stored as a frequency table (say, with data values in L1 and frequencies in L2), set up the Plot with Xlist: L1 and Freq: L2.

EXERCISES

SECTION 2.1

1. Labor status The table below gives the numbers of Maltese persons aged 15 and over during the first quarter of 2019 by labor status.

Labor Status	Frequency
Employed	241,468
Unemployed	8,853
Inactive	165,235

Source: https://nso.gov.mt/en/News_Releases/View_by_Unit/Unit_C2/Labour_Market_Statistics/Documents/2019/News2019_101.pdf

Convert this table to a relative frequency table.

2. Employment by age The distribution of employed persons with a main job by age group is shown in the following bar chart. Change this to a bar chart of relative frequencies.

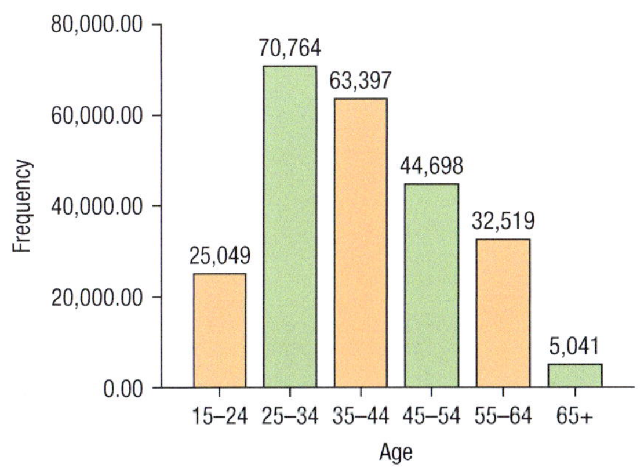

3. Movie genres The pie chart summarizes the genres of the 891 MPAA rated movies released in 2014 and 2015. (Data extracted from **Movies 06-15**)

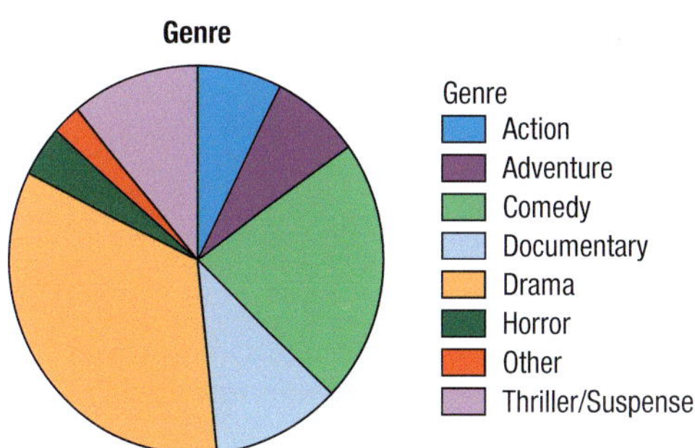

a) Is this an appropriate display for the genres? Why or why not?
b) Which category was least common?

4. Movie ratings The Motion Picture Association of America (MPAA) rates each film to designate the appropriate audience. The ratings are G, PG, PG-13, and R. The pie chart shows the MPAA ratings of the same 891 movies released in 2014 and 2015 as in Exercise 3. (Data extracted from **Movies 06-15**)

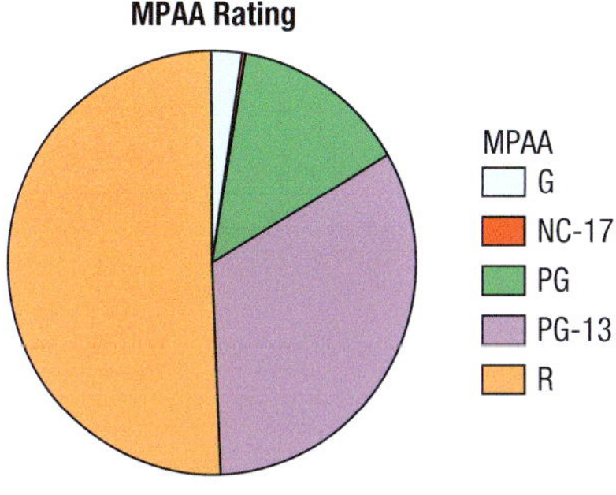

a) Is this an appropriate display for the ratings? Why or why not?
b) Which was the most common rating?

5. Movie ratings The ratings of the 20 top-grossing movies in the years 2006, 2008, 2012, and 2014 are shown in the following bar charts. The pie charts show the same data but are unlabeled. Match each pie chart with the correct year. (Data extracted from **Movies 06-15**)

i)

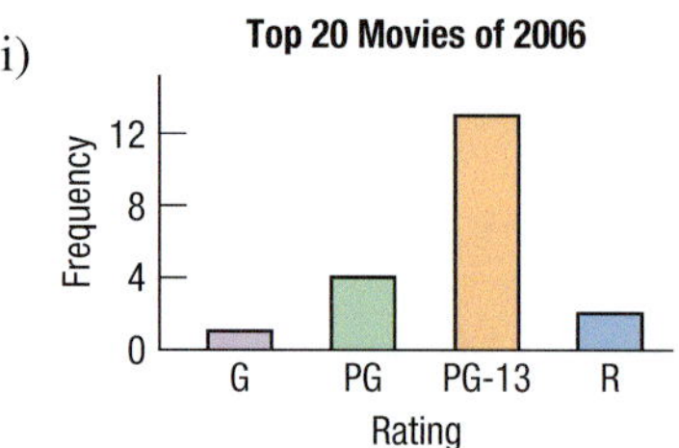

ii)

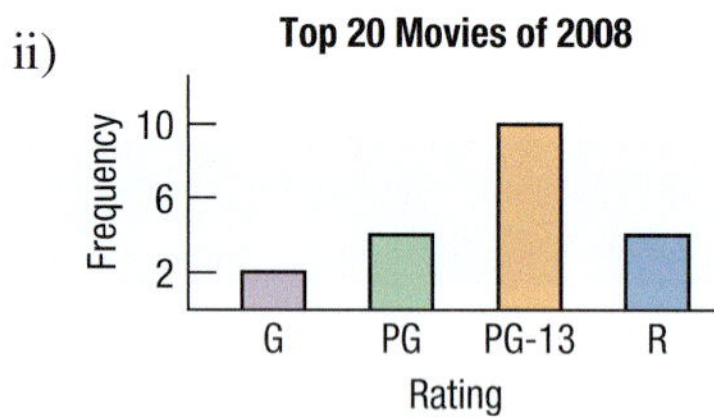

iii)

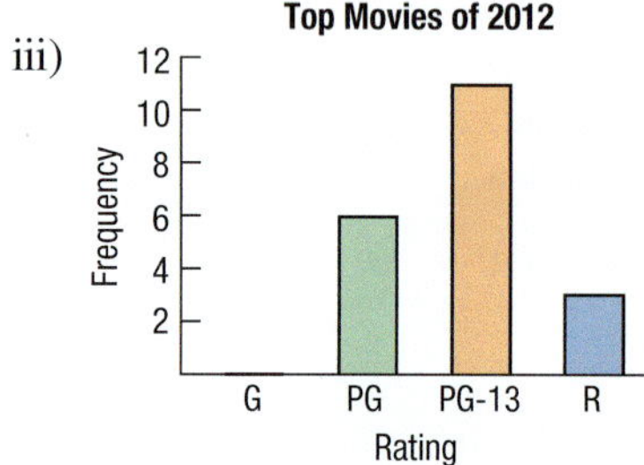

iv)

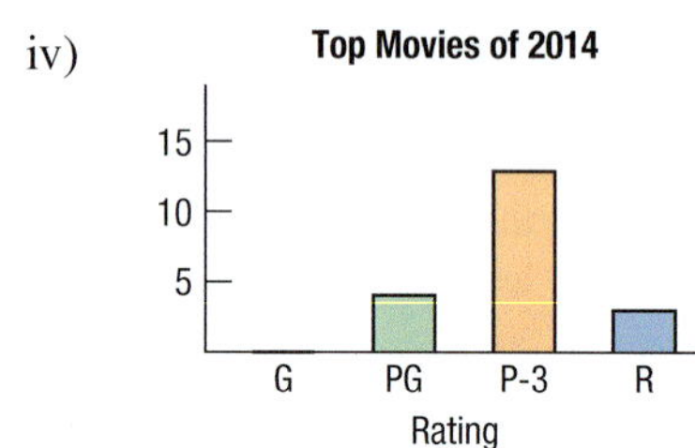

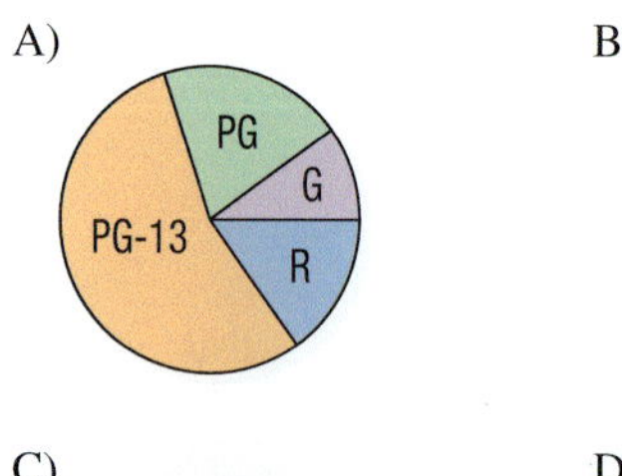

B)

C)

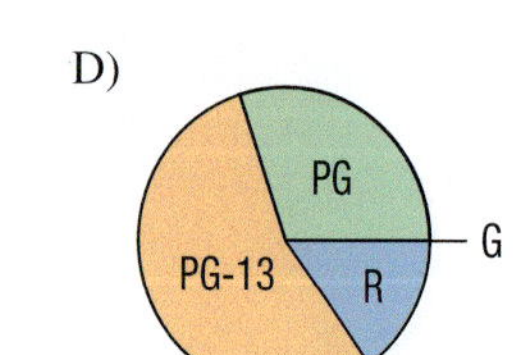

6. Marriage in decline Changing attitudes about marriage and families prompted Pew Research to ask how people felt about particular recent trends. (pewresearch.org/pubs/1802/decline-marriage-rise-new-families) For each trend, participants were asked whether the trend "is a good thing," "is a bad thing," or "makes no difference." Some participants said they didn't know or chose not to respond. The following bar charts show the number of each response for each trend. The pie charts show the same data without the trends identified. Match each pie chart with the correct trend and bar chart.

i)

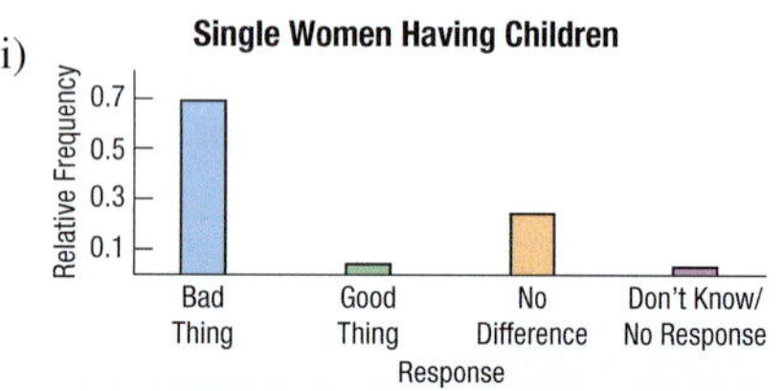

ii)

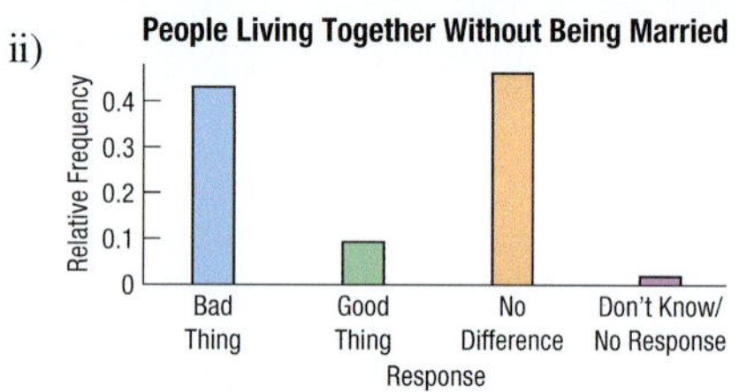

iii)

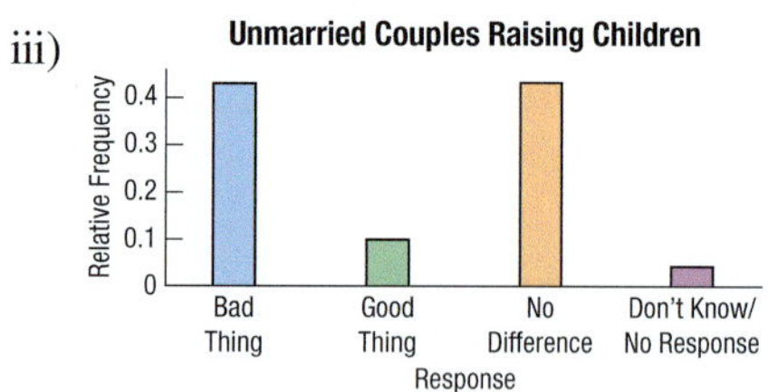

iv)

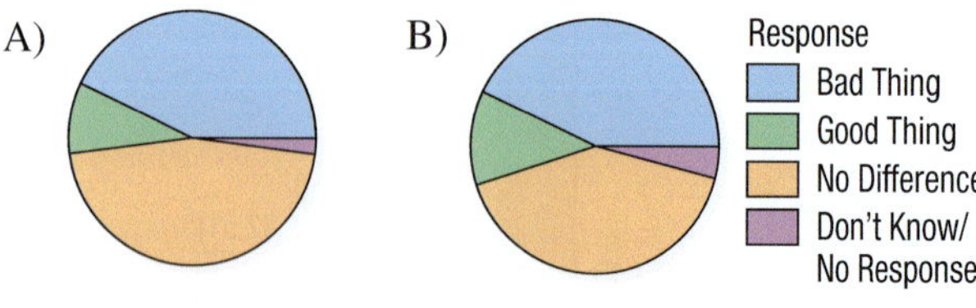

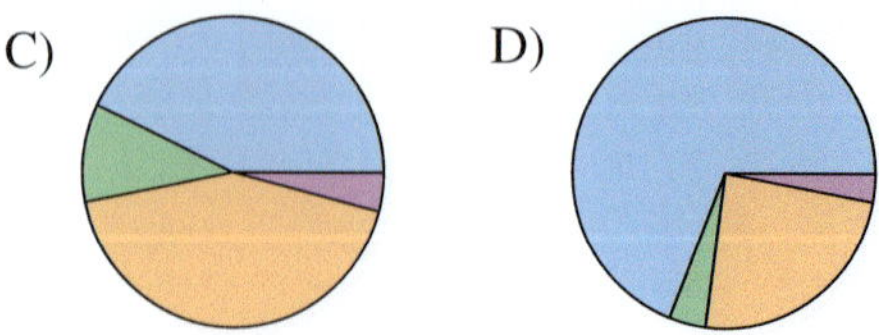

SECTION 2.2

T **7.** Active volcanoes Here are two histograms showing the annual number of active volcanoes (Source: https://volcano.si.edu/faq/index.cfm?question=eruptionsbyyear). One plots the years, and the other plots the number of active volcanoes.

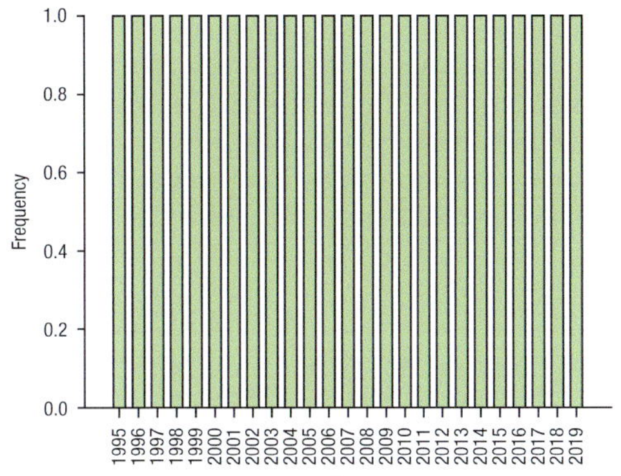

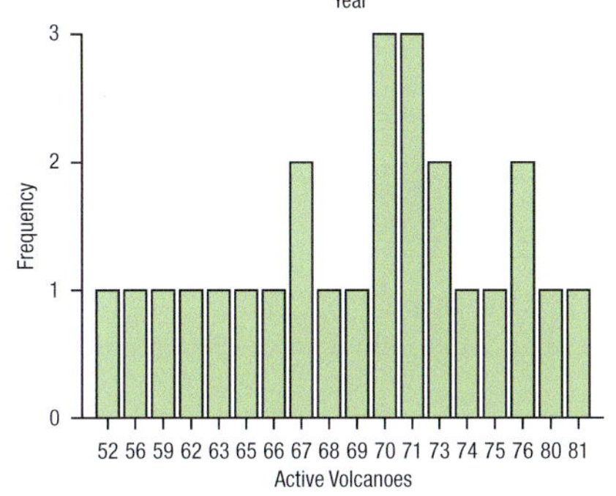

a) For which years is data about active volcanoes available?
b) Why do all the bars in the Year histogram have a height of one unit?
c) Explain what the Active Volcanoes histogram says about active volcanoes.

T **8.** Active volcanoes II Here are the same histograms as in Exercise 7. Note that one bar in the active volcanoes histogram has been highlighted together with the corresponding bars in the Year histogram.

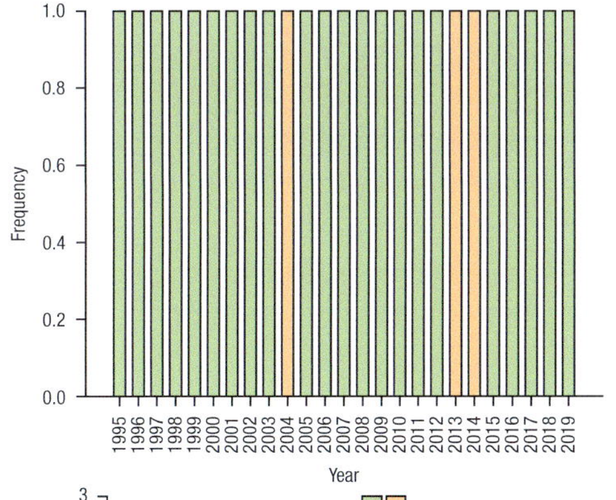

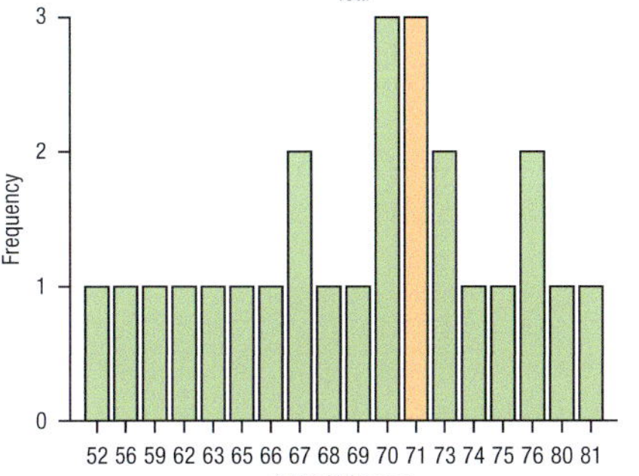

a) Which years are included in the highlighted bar in the Active Volcanoes histogram?
b) How many active volcanoes were there in these years?
c) What do these histograms, taken together, say about active volcanoes?

T **9.** How big is your bicep? A study of body fat on 250 men collected measurements of 12 body parts as well as the percentage of body fat that the men carried. Here is a dotplot of their bicep circumferences (in inches). What does the dotplot say about the distribution of the size of men's biceps? (Data in **Bodyfat**)

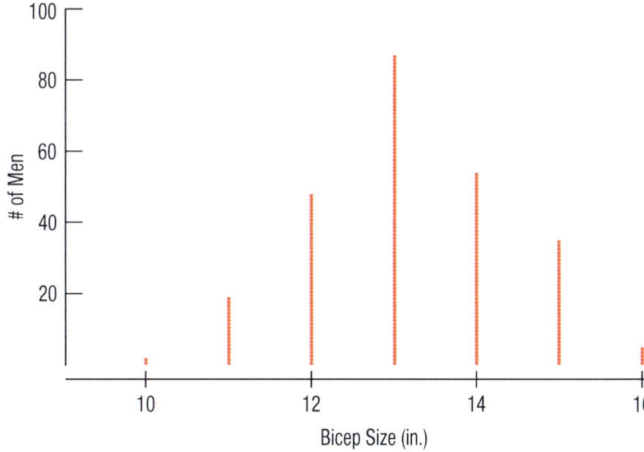

T **10.** How big is your bicep in cm? The study in Exercise 9 actually measured the bicep circumference in centimeters. The dotplot in Exercise 9 was formed by dividing each measurement by 2.54 to convert it to inches. Here is the dotplot of the original values in cm. Do the two dotplots look different? What might account for that?

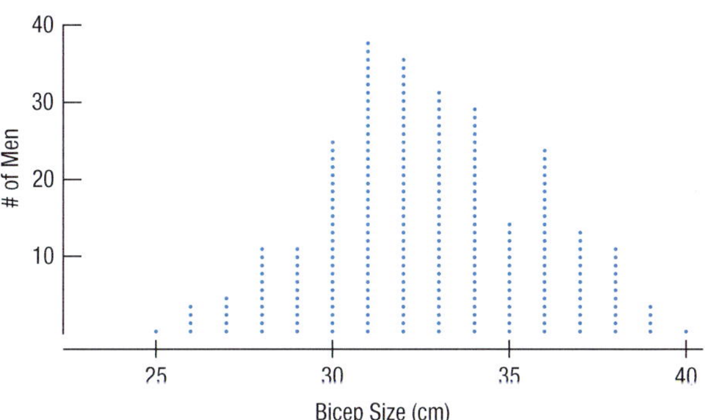

T **11.** Patient calls During a particular month, a gastroenterology provider kept record of the number of telephone calls received by his patients. The following histogram shows the distribution of telephone calls received.

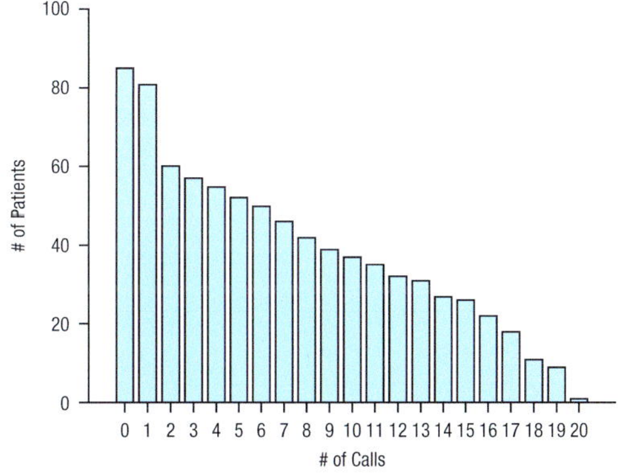

What does the above histogram say about the distribution of patient calls received by the provider?

T **12. Live births** The first histogram below shows the number of live births, in 1,000 births, in the United Kingdom from 2000 to 2018. The second histogram shows the population size, in million inhabitants, in the United Kingdom during the same period.

(Sources: https://www.statista.com/statistics/281981/live-births-in-the-united-kingdom-uk/, https://www.statista.com/statistics/281296/uk-population/)

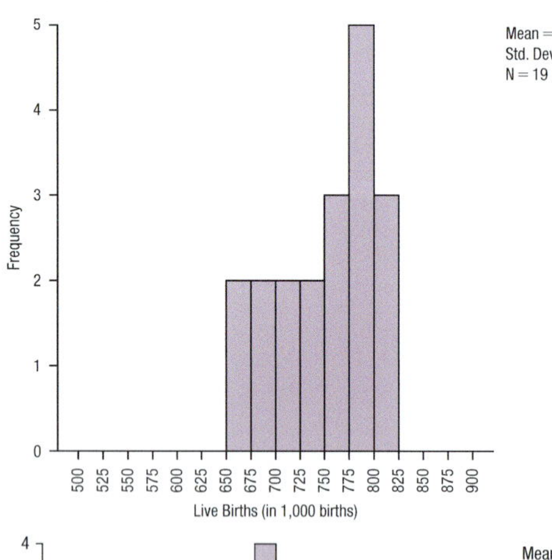

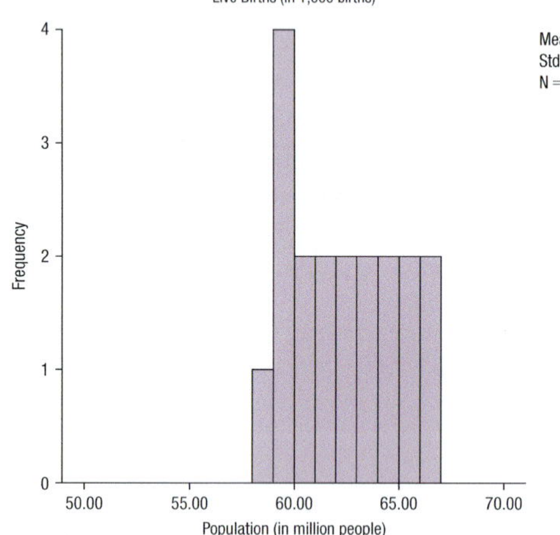

a) What do the histograms say about the distributions?
b) How do you expect the number of live births to be related to the population size?
c) What might be a better way to express the number of live births?

SECTION 2.3

T **13. Biceps revisited** Describe the shape of the distribution of bicep circumferences (in inches) for the 250 men in the study of Exercise 10.

T **14. Patient calls II** For the distribution of patient calls in Exercise 11, describe the shape.

T **15. Life expectancy** Here are the life expectancies at birth in 190 countries (2014) as collected by the World Health Organization.

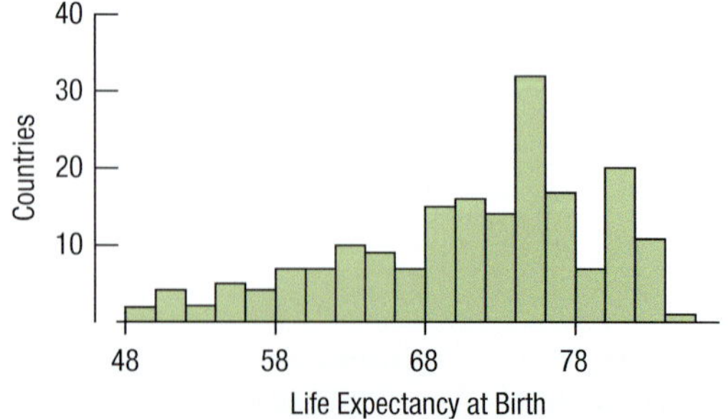

a) Describe the shape.
b) How many modes do you see?

T **16. Shoe sizes** A last is a form, traditionally made of wood, in the shape of the human foot. Lasts of various sizes are used by shoemakers to make shoes. In the United States, shoe sizes are defined differently for men and women:
U.S. men's shoe size = (last size in inches $\times$ 3) $-$ 24
U.S. women's shoe size = (last size in inches $\times$ 3) $-$ 22.5
But in Europe, they are both: Euro size = last size in cm $\times$ 3/2

Here is a histogram of the European shoe sizes of 269 college students (converted from their reported U.S. shoe sizes):

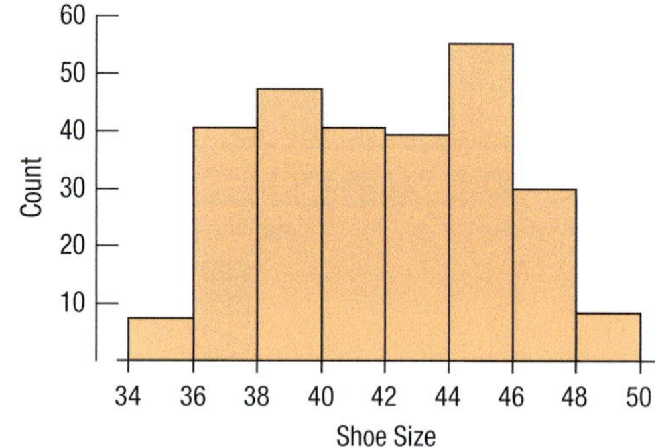

a) Describe the shape.
b) How many modes do you see? What might explain that?

SECTION 2.4

T **17. Life expectancy II** For the 146 life expectancies in Exercise 15,

a) Which would you expect to be larger: the median or the mean? Explain briefly.
b) Which would you report: the mean or the median? Explain briefly.

T **18. Live births II** For the number of live births in Exercise 12, answer the following:

a) Which would you expect to be larger: the median or the mean? Explain.
b) Which would you rather report: the mean or the median? Explain.

T **19. How big is your bicep II?** For the bicep measurements in Exercise 10, would you report the mean, the median, or both? Explain briefly.

T **20. Shoe sizes II** For the shoe sizes in Exercise 16, what might be the problem with either the mean or the median as a measure of center?

SECTION 2.5

T **21. Life expectancy III** For the 190 life expectancies in Exercise 15,

a) Would you report the standard deviation or the IQR?
b) Justify your answer briefly.

T **22. Live births III** For the number of live births in Exercise 12, what is the best measure for reporting the spread of the data? Justify your answer.

T **23. How big is your bicep III?** For the bicep measurements in Exercise 10, would you report the standard deviation or the IQR? Explain briefly.

24. Shoe sizes III For the shoe sizes in Exercise 16, what might be the problem with either the standard deviation or the IQR as a measure of spread?

CHAPTER EXERCISES

25. Graphs in the news Find a bar chart of categorical data from a newspaper, a magazine, or the Internet.

a) Is the graph clearly labeled?
b) Does it violate the area principle?
c) Does the accompanying article tell the W's of the variable?
d) Do you think the article correctly interprets the data? Explain.

26. Graphs in the news II Find a pie chart of categorical data from a newspaper, a magazine, or the Internet.

a) Is the graph clearly labeled?
b) Does it violate the area principle?
c) Does the accompanying article tell the W's of the variable?
d) Do you think the article correctly interprets the data? Explain.

27. Table interpretation Find a frequency table of categorical data from a research paper or an article that interests you online.

a) What is the title of the table? Does it explain the table well?
b) Does the table display counts, percentages, or both?
c) Can you identify the W's of the categorical variable in question?
d) Interpret the data displayed in the frequency table. Compare your interpretation with the one in the paper or article.

28. Table interpretation II Using the frequency table of Exercise 27, do the following:

a) Convert the table to a relative frequency table.
b) Display the data of the frequency table in a bar chart.
c) Discuss what the bar chart reveals about the distribution of the categorical variable.
d) Are there any outliers in the distribution?

29. Histogram Find a histogram that shows the distribution of a variable in a newspaper, a magazine, or the Internet.

a) Does the article identify the W's?
b) Discuss whether the display is appropriate.
c) Discuss what the display reveals about the variable and its distribution.
d) Does the article accurately describe and interpret the data? Explain.

30. Not a histogram Find a graph other than a histogram that shows the distribution of a quantitative variable in a newspaper, a magazine, or the Internet.

a) Does the article identify the W's?
b) Discuss whether the display is appropriate for the data.
c) Discuss what the display reveals about the variable and its distribution.
d) Does the article accurately describe and interpret the data? Explain.

31. EU wages As reported by Eurostat, the average annual net earnings of full-time single workers in the European Union (comprising 28 countries and not inclusive of all countries in the continent of Europe) expressed in euros was 17,749.11 in 2018. (Source: https://appsso.eurostat.ec.europa.eu/nui/show.do?dataset=earn_nt_net&lang=en)

Answer the following:

a) Identify the W's for the variable.
b) What are the units of the variable?
c) Do you think that the average used is the median or the mean? Explain.

32. EU wages II With reference to the data provided in Exercise 31, the EU country having the least annual net earnings in 2018 was Bulgaria with €3,693.70, while the EU country having the highest annual net earnings in 2018 was Luxembourg with €31,620.13.

a) What measure of spread can you deduce from the above data?
b) Do you think that this measure of spread is appropriate for the situation? Explain.
c) Is there any better measure of spread that should be considered?

33. Thinking about shape Would you expect distributions of these variables to be uniform, unimodal, or bimodal? Symmetric or skewed? Explain why.

a) The number of speeding tickets each student in the senior class of a college has ever had
b) Players' scores (number of strokes) at the U.S. Open golf tournament in a given year
c) Weights of female babies born in a particular hospital over the course of a year
d) The length of the average hair on the heads of students in a large class

34. More shapes Would you expect distributions of these variables to be uniform, unimodal, or bimodal? Symmetric or skewed? Explain why.

a) Ages of people at a Little League game
b) Number of siblings of people in your class
c) Pulse rates of college-age males
d) Number of times each face of a die shows in 100 tosses

35. Movie genres again Here is a bar chart summarizing the movie genres from the 891 movies in Exercise 3. (Data extracted from **Movies 06-15**)

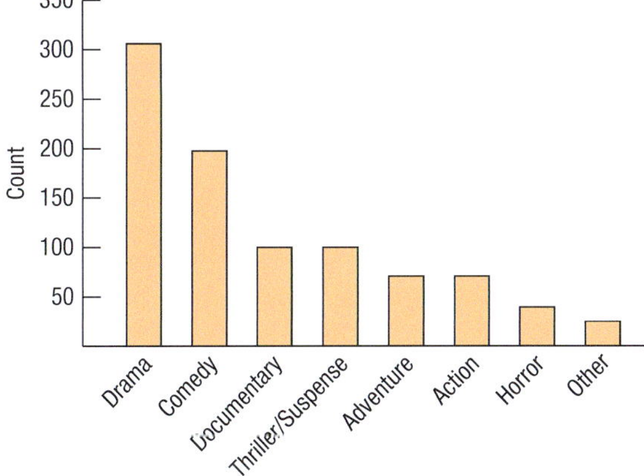

a) Were Thriller/Suspense or Adventure films more common?
b) Is it easier to answer the question from the bar chart or from the pie chart in Exercise 3? Explain.

36. Movie ratings again Here is a bar chart summarizing the movie ratings from the 891 movies shown in Exercise 4.

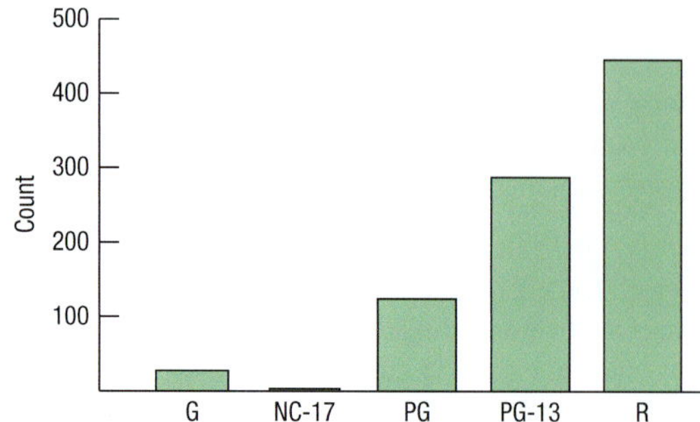

a) Which was the least common rating?
b) Is it easier to answer the question from the bar chart or from the pie chart in Exercise 4? Explain.

37. Judicial activity The following are some statistics concerning the judicial activity of the General Court to the Court of Justice (Source: https://curia.europa.eu/jcms/upload/docs/application/pdf/2012-06/ra2011_statistiques_tribunal_en.pdf). In 2011, there were 722 new cases, 714 completed cases, and 1,308 pending cases. Find the relative frequency distribution of the judicial activity of the General Court, and write a sentence describing it.

38. Judicial activity II The article about the judicial activity of the General Court described in Exercise 37 also indicated that the nature of proceedings of the 722 new cases were as follows: 67 state aid, 39 competition, 219 intellectual property, 264 other direct actions, 44 appeals, 1 appeal concerning interim measures or interventions, and 88 special forms of procedure. Summarize the relative frequency distribution of the proceedings' nature.

39. EU causes of death 2015 In 2015, 5,217,376 persons died in the European Union. Around 1.9 million people died from diseases of the circulatory system, while 1.3 million died from cancer. The third main cause of death in the EU was diseases of the respiratory system, which killed 442,100 persons. A significant share of deaths was also due to accidents and other external causes of deaths (230,000 deaths), diseases of the digestive system (219,000 deaths), mental and behavioral diseases (214,500 deaths), and diseases of the nervous system (213,000 deaths).

(Source: https://ec.europa.eu/eurostat/web/products-eurostat-news/-/DDN-20180314-1?inheritRedirect=true)

a) Convert the above count data into relative frequencies.
b) What percentage of deaths were from causes not listed above?
c) Create an appropriate display for the above data.

40. Plane crashes An investigator compiled information about recent nonmilitary plane crashes. The causes, to the extent that they could be determined, are summarized in the table.

Causes of Fatal Accidents

Cause	Percent
Pilot error	46%
Other human error	8
Weather	9
Mechanical failure	28
Sabotage	9
Other causes	1

Source: www.planecrashinfo.com/cause.htm

a) Is it reasonable to conclude that the weather or mechanical failures caused only about 37% of recent plane crashes?
b) Why do the numbers in the table add to 101%?
c) Create an appropriate display for these data.

T 41. Movie genres once more The movie genres listed in Exercise 35 were originally listed as these:

Genre	Frequency
Action	66
Adventure	69
Black Comedy	16
Comedy	150
Concert/Performance	1
Documentary	99
Drama	303
Horror	38
Multiple Genres	1
Musical	10
Reality	2
Romantic Comedy	30
Thriller/Suspense	98
Western	8

a) What problem would you encounter in trying to make a display of these data?
b) How did the creators of the bar chart in Exercise 35 solve this problem?

T 42. Summer Olympics 2016 Fifty-nine countries won gold medals in the 2016 Summer Olympics. The table lists them, along with the total number of gold medals each won.

Country	Medals	Country	Medals
United States	46	South Africa	2
Great Britain	27	Ukraine	2
China	26	Serbia	2
Russia	19	Poland	2
Germany	17	North Korea	2
Japan	12	Belgium	2
France	10	Thailand	2
South Korea	9	Slovakia	2
Italy	8	Georgia	2
Australia	8	Azerbaijan	1
Netherlands	8	Belarus	1
Hungary	8	Turkey	1
Brazil	7	Armenia	1
Spain	7	Czech Republic	1
Kenya	6	Ethiopia	1
Jamaica	6	Slovenia	1
Croatia	5	Indonesia	1
Cuba	5	Romania	1
New Zealand	4	Bahrain	1
Canada	4	Vietnam	1
Uzbekistan	4	Chinese Taipei	1
Kazakhstan	3	Bahamas	1
Colombia	3	Côte d'Ivoire	1
Switzerland	3	Fiji	1
Iran	3	Jordan	1
Greece	3	Kosovo	1
Argentina	3	Puerto Rico	1
Denmark	2	Singapore	1
Sweden	2	Tajikistan	1

a) Try to make a display of these data. What problems do you encounter?
b) Organize the data so that the graph is more successful.

43. Global warming The Yale Program on Climate Change Communication surveyed 1263 American adults in March 2015 and asked them about their attitudes on global climate change. Here's a display of the percentages of respondents choosing each of the major alternatives offered. List the errors in this display.

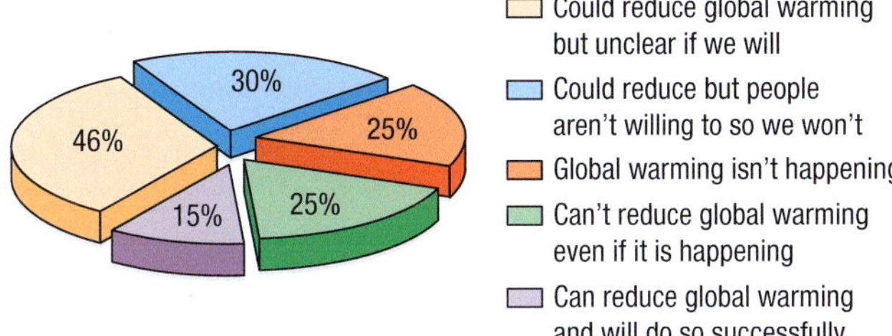

44. Modalities A survey of athletic trainers asked what modalities (treatment methods such as ice, whirlpool, ultrasound, or exercise) they commonly use to treat injuries. Respondents were each asked to list three modalities. The article included the following figure reporting the modalities used:

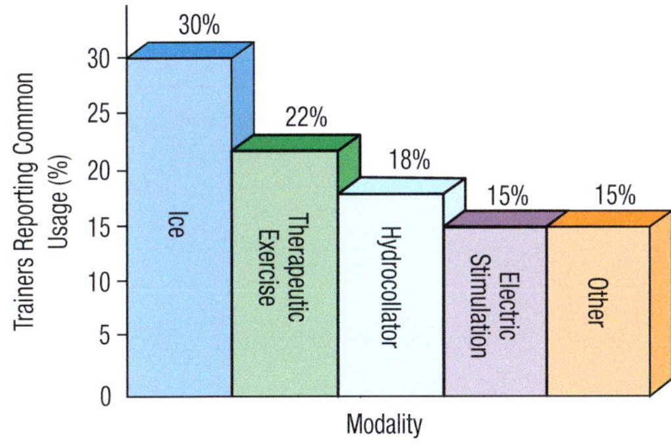

Source: Scott F. Nadler, Michael Prybicien, Gerard A. Malanga, and Dan Sicher, "Complications from Therapeutic Modalities: Results of a National Survey of Athletic Trainers." *Archives of Physical Medical Rehabilitation* 84 (June 2003).

a) What problems do you see with the graph?
b) Consider the percentages for the named modalities. Do you see anything odd about them?

T 45. Cereals The histogram shows the carbohydrate content of 77 breakfast cereals (in grams).

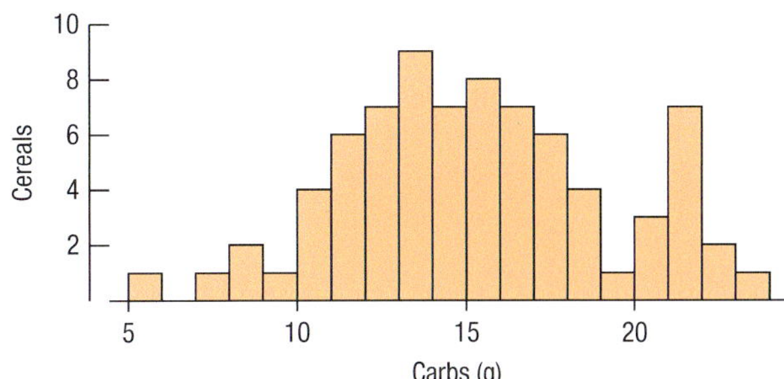

a) Describe this distribution.
b) If you can, open the dataset and identify the cereals with the highest carbohydrate content.

T 46. Run times One of the authors collected the times (in minutes) it took him to run 4 miles on various courses during a 10-year period. Here is a histogram of the times:

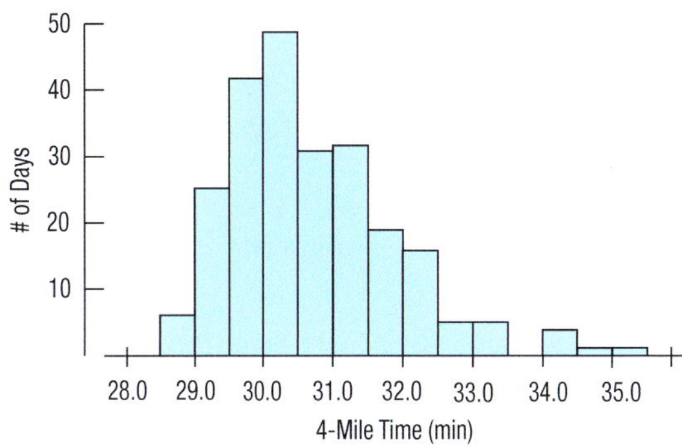

Describe the distribution and summarize the important features. What is it about running that might account for the shape you see?

T 47. Purchasers The histogram below shows the frequencies pertaining to the ages of a sample of 100 female purchasers of a particular household good.

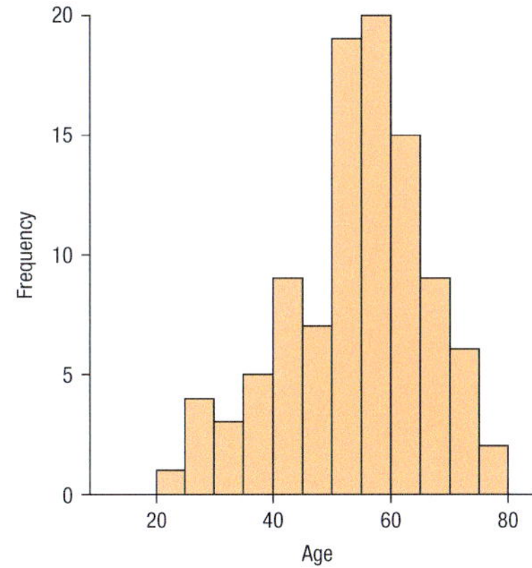

a) From the histogram, would you expect the mean or median to be higher? Explain.
b) Describe the distribution of the ages of the female purchasers.
c) Which summary statistics would you choose to explain the center and spread in the data?

T 48. Bird species 2013 The Cornell Lab of Ornithology holds an annual Christmas Bird Count (www.birdsource.org), in which bird watchers at various locations around the country see how many different species of birds they can spot. Here are the number of species counted from the 20 sites with the most species in 2013:

184	98	101	126	150
166	82	136	124	118
133	83	86	101	105
97	88	131	128	106

a) Create a stem-and-leaf display of these data.
b) Write a brief description of the distribution. Be sure to discuss the overall shape as well as any unusual features.

49. Super Bowl points 2016 How many points do football teams score in the Super Bowl? Here are the total numbers of points scored by both teams in each of the first 50 Super Bowl games. (Data in **Super Bowl 2016**)

45, 47, 23, 30, 29, 27, 21, 31, 22, 38, 46, 37, 66, 50, 37, 47, 44, 47, 54, 56, 59, 52, 36, 65, 39, 61, 69, 43, 75, 44, 56, 55, 53, 39, 41, 37, 69, 61, 45, 31, 46, 31, 50, 48, 56, 38, 65, 51, 52, 34

a) Find the median.
b) Find the quartiles.
c) Make a histogram and write a brief description of the distribution.

50. Super Bowl edge 2016 In the Super Bowl, by how many points does the winning team outscore the losers? Here are the winning margins for the first 50 Super Bowl games. (Data in **Super Bowl 2016**)

25, 19, 9, 16, 3, 21, 7, 17, 10, 4, 18, 17, 4, 12, 17, 5, 10, 29, 22, 36, 19, 32, 4, 45, 1, 13, 35, 17, 23, 10, 14, 7, 15, 7, 27, 3, 27, 3, 3, 11, 12, 3, 4, 14, 6, 4, 3, 35, 4, 14

a) Find the median.
b) Find the quartiles.
c) Make a histogram and write a brief description of the distribution.

51. Test scores, large class Test scores from a large calculus class of 400 are shown in the histogram below.

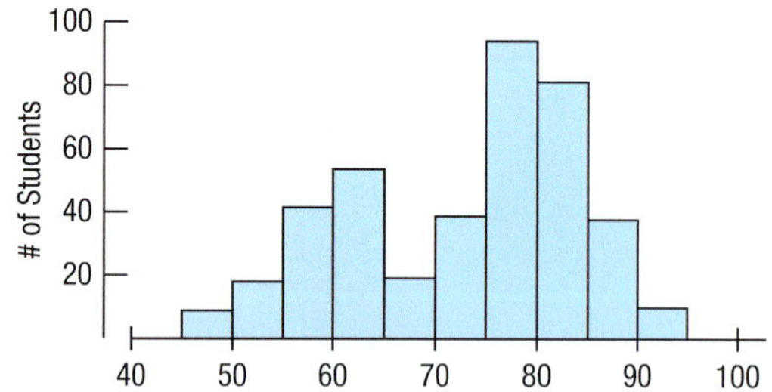

a) Describe the distribution of scores. What might account for this shape?
b) Why might both the mean and median score be misleading as a summary of the center?

52. Test scores, small class Test scores from a calculus section of 40 students are shown in the histogram below.

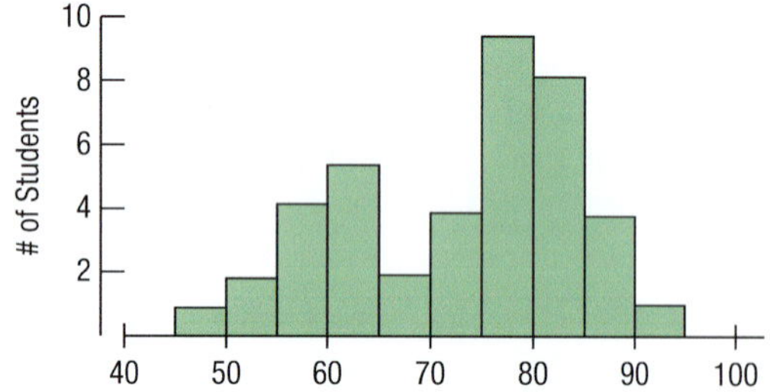

Describe the distribution of scores. Why might you be less sure of the description of the shape as compared to the histogram in Exercise 51?

53. Mistake A clerk entering salary data into a company spreadsheet accidentally put an extra "0" in the boss's salary, listing it as $2,000,000 instead of $200,000. Explain how this error will affect these summary statistics for the company payroll:

a) measures of center: median and mean
b) measures of spread: range, IQR, and standard deviation

54. Sick days During contract negotiations, a company seeks to change the number of sick days employees may take, saying that the annual "average" is 7 days of absence per employee. The union negotiators counter that the "average" employee misses only 3 days of work each year.

a) Explain how both sides might be correct, identifying the measure of center you think each side is using and why the difference might exist.
b) Do their choices of summaries make sense?

55. Casualities The following are the annual numbers of casualties due to terrorism worldwide between 2006 and 2017:

20487, 22719, 15708, 15310, 13186, 12533, 11098, 18066, 32763, 29424, 25722, 18753. Find these statistics:

a) mean
b) median and quartiles
c) range and IQR

(Source: https://www.statista.com/statistics/202871/number-of-fatalities-by-terrorist-attacks-worldwide/)

56. Kidnappings The following are the annual numbers of worldwide kidnappings due to terrorism from 2007 to 2017:

4980, 4680, 10749, 6051, 5554, 1283, 3137, 9461, 12264, 15664, 8937. Find these statistics:

a) mean
b) median and quartiles
c) range and IQR

(Source: https://www.statista.com/statistics/250557/number-of-kidnappings-due-to-terrorism/)

57. Casualties II Using the data from Exercise 55, describe the distribution of the number of casualties due to terrorism worldwide during this time period.

58. Kidnappings II Using the data from Exercise 56, describe the distribution of the number of kidnappings due to terrorism worldwide during this time period.

59. Insurance premiums The histogram below shows the distribution of the sizes of 150 automobile insurance premiums (in Euros).

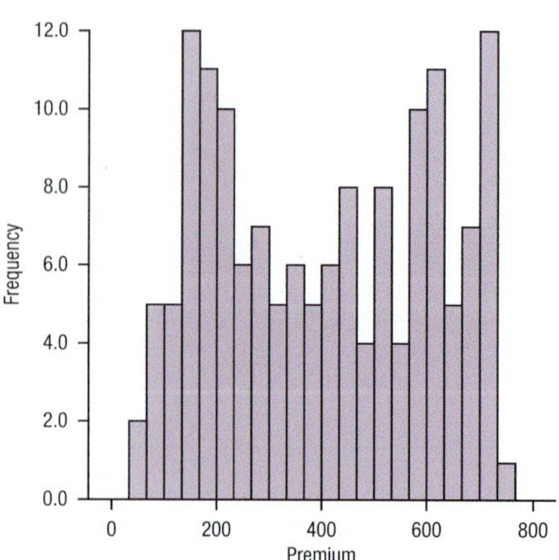

Which summary statistics would you choose to describe the center and spread of the data? Why?

T **60.** Neck size The histogram shows the neck sizes (in inches) of the 250 men recruited for the health study in Utah from Exercise 9. (Data in **Bodyfat.** The neck sizes in **Bodyfat** are in cm. Divide by 2.54 for inches.)

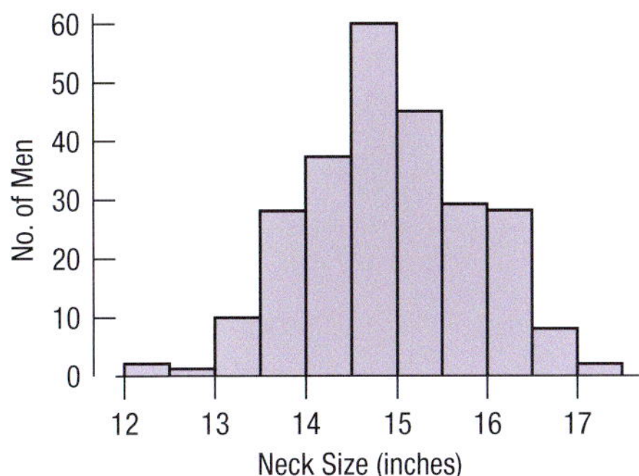

Which summary statistics would you choose to summarize the center and spread in these data? Why?

T **61.** Insurance premiums II Using the histogram of the automobile insurance premiums in Exercise 59, answer the following questions:

a) What is an approximate value of the mean?
b) What is an approximate value of the standard deviation? Explain.

T **62.** Neck sizes again Look again at the histogram of men's neck sizes in Exercise 60.

a) Is the mean closer to 14, 15, or 16 inches? Why?
b) Is the standard deviation closer to 1 inch, 3 inches, or 5 inches? Explain.

T **63.** Movie lengths 2010 The histogram shows the running times in minutes of the 150 top-grossing feature films released in 2010.

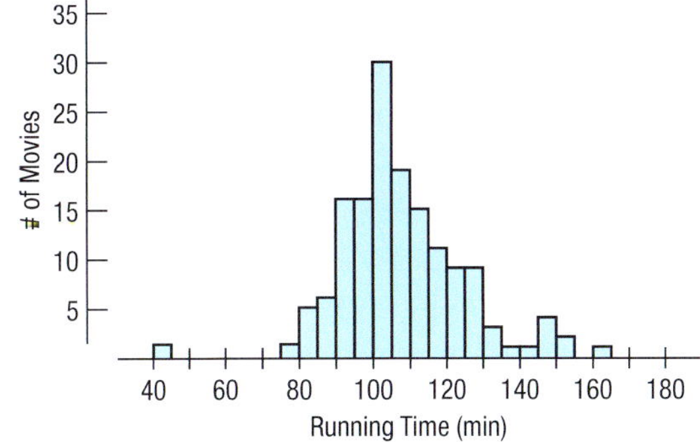

a) You plan to see a movie this weekend. Based on these movies, how long do you expect a typical movie to run?
b) Would you be surprised to find that your movie ran for $2\frac{1}{2}$ hours (150 minutes)?
c) Which would you expect to be higher: the mean or the median run time for all movies? Why?

T **64.** Golf drives 2015 The display shows the average drive distance (in yards) for 199 professional golfers during a week on the men's PGA tour in 2015.

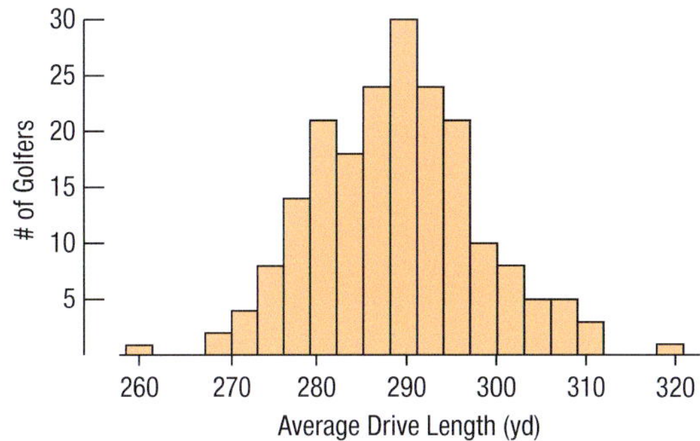

a) Describe this distribution.
b) Approximately what proportion of professional male golfers drive, on average, 280 yards or less?
c) Estimate the mean by examining the histogram.
d) Do you expect the mean to be smaller than, approximately equal to, or larger than the median? Why?

T **65.** Movie lengths II 2010 Exercise 63 looked at the running times of movies released in 2010. The standard deviation of these running times is 16.6 minutes, and the quartiles are $Q1 = 98$ minutes and $Q3 = 116$ minutes.

a) Write a sentence or two describing the spread in running times based on
   i. the quartiles.
   ii. the standard deviation.
b) Do you have any concerns about using either of these descriptions of spread? Explain.

T **66.** Golf drives II 2015 Exercise 64 looked at distances PGA golfers can hit the ball. The standard deviation of these average drive distances is 11.2 yards, and the quartiles are $Q1 = 282.05$ yards and $Q3 = 294.5$ yards.

a) Write a sentence or two describing the spread in distances based on
   i. the quartiles.
   ii. the standard deviation.
b) Do you have any concerns about using either of these descriptions of spread? Explain.

T **67.** Movie earnings 2015 The histogram shows total gross earnings (in millions of dollars) of the top 200 major release movies in 2015. (Data extracted from **Movies_06-15**)

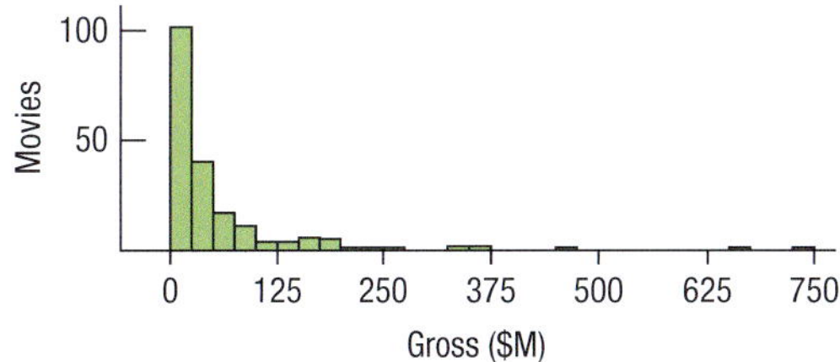

An industry publication reports that the typical movie makes $23.2 million, but a watchdog group concerned with rising ticket prices says that the average earnings are closer to $55.7 million. What statistic do you think each group is using? Explain.

68. Cold weather A meteorologist preparing a talk about global warming compiled a list of weekly low temperatures (in degrees Fahrenheit) he observed at his southern Florida home last year. The coldest temperature for any week was 36°F, but he inadvertently recorded the Celsius value of 2°. Assuming that he correctly listed all the other temperatures, explain how this error will affect these summary statistics:

a) measures of center: mean and median
b) measures of spread: range, IQR, and standard deviation

69. Gasoline 2014 In January 2014, 16 gas stations in eastern Wisconsin posted these prices for a gallon of regular gasoline:

3.23	3.26	3.23	3.39
3.45	3.43	3.42	3.45
3.46	3.11	3.15	3.22
3.21	3.26	3.27	3.28

a) Make a stem-and-leaf display of these gas prices. Use split stems; for example, use two 3.2 stems—one for prices between \$3.20 and \$3.24 and the other for prices from \$3.25 to \$3.29.
b) Describe the shape, center, and spread of this distribution.
c) What unusual feature do you see?

70. The great one During his 20 seasons in the NHL, Wayne Gretzky scored 50% more points than anyone who ever played professional hockey. He accomplished this amazing feat while playing in 280 fewer games than Gordie Howe, the previous record holder. Here are the number of games Gretzky played during each season:

79, 80, 80, 80, 74, 80, 80, 79, 64, 78, 73, 78, 74, 45, 81, 48, 80, 82, 82, 70

a) Create a stem-and-leaf display for these data, using split stems.
b) Describe the shape of the distribution.
c) Describe the center and spread of this distribution.
d) What unusual feature do you see? What might explain this?

71. States The stem-and-leaf display shows populations of the 50 states, in millions of people, according to the 2010 census.

```
0 | 111111111112222333333334444
0 | 555556666667789
1 | 000233
1 | 99
2 |
2 | 5
3 |
3 | 7
(2|5 means 25)
```

a) From the stem-and-leaf display, find the median and the interquartile range.
b) Write a few sentences describing this distribution.

72. Wayne Gretzky In Exercise 70, you examined the number of games played by hockey great Wayne Gretzky during his 20-year career in the NHL.

a) Would you use the median or the mean to describe the center of this distribution? Why?
b) Find the median.
c) Without actually finding the mean, would you expect it to be higher or lower than the median? Explain.

T **73. A-Rod 2016** Alex Rodriguez (known to fans as A-Rod) was the youngest player ever to hit 500 home runs. Here is a stem-and-leaf display of the number of home runs hit by A-Rod during the 1994–2016 seasons. Describe the distribution, mentioning its shape and any unusual features. (Data are given in the stem-and-leaf display.)

```
0 | 00579
1 | 68
2 | 3
3 | 0035566
4 | 12278
5 | 247
(5|2 means 52)
```

T **74. Aircraft accidents** The following data give the numbers of aircraft accidents (rounded to the nearest 10) per year from 1970 to 2017 according to ACRO and Bureau of Aircraft Accident Archives (Source: https://web.archive.org/web/20170728174219/http://www.baaa-acro.com/general-statistics/crashs-rate-per-year/):

300, 270, 340, 330, 270, 320, 280, 340, 360, 330, 330, 270, 250, 240, 230, 260, 250, 280, 250, 270, 260, 240, 270, 280, 230, 270, 250, 230, 230, 220, 200, 210, 200, 200, 180, 190, 190, 170, 190, 160, 160, 150, 150, 140, 120, 120, 100, 50.

a) Create a dotplot of the above data.
b) Describe the distribution of aircraft accidents.

T **75. A-Rod again 2016** Students were asked to make a histogram of the number of home runs Alex Rodriguez hit from 1994 to 2016 (see Exercise 73). One student submitted the following display:

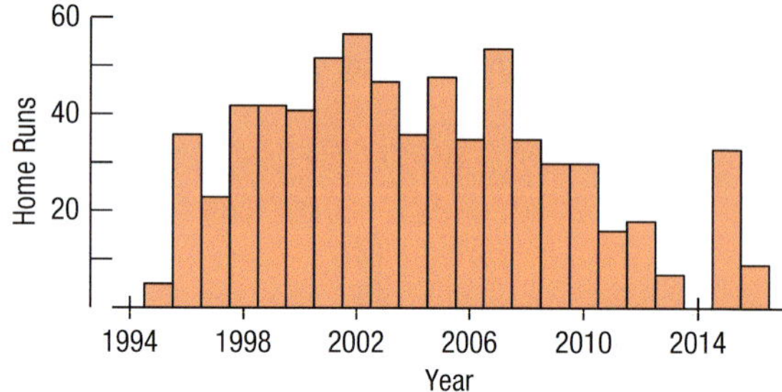

a) Comment on this graph.
b) Create your own histogram of the data.

T **76. Return of the birds 2013** Students were given the assignment to make a histogram of the data on bird counts reported in Exercise 48. One student submitted the following display:

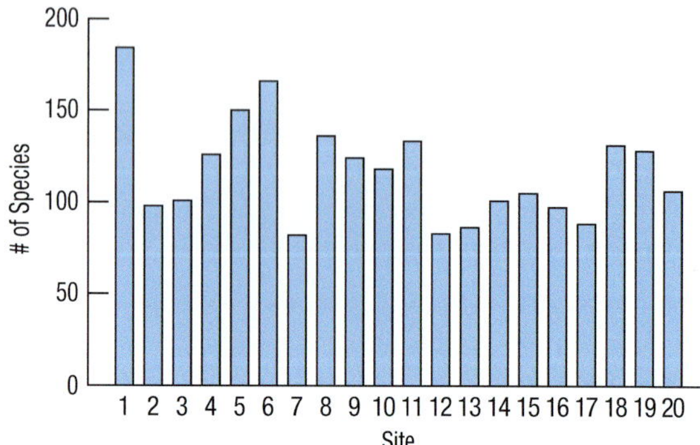

a) Comment on this graph.
b) Create your own histogram of the data.

77. Acid rain Two researchers measured the pH (a scale on which a value of 7 is neutral and values below 7 are acidic) of water collected from rain and snow over a 6-month period in Allegheny County, Pennsylvania:

4.57 5.62 4.12 5.29 4.64 4.31 4.30 4.39 4.45
5.67 4.39 4.52 4.26 4.26 4.40 5.78 4.73 4.56 5.08
4.41 4.12 5.51 4.82 4.63 4.29 4.60

a) Describe their data with a graph and a few sentences.
b) Open the data file to find the dates. Consider the dates corresponding to the least acidic (highest pH) days. Do they have anything in common? (You may need to consult a calendar.)

78. Unemployment The following table displays the unemployment rates as of July 2019 for 30 European countries (Source: https://ec.europa.eu/eurostat/statistics-explained/index.php/Unemployment_statistics). Create an appropriate graph of the below data, and describe the distribution.

Country	Percentage	Country	Percentage
Czechia	2.1	Ireland	5.3
Germany	3.0	Slovakia	5.3
Poland	3.3	Belgium	5.7
Iceland	3.4	Luxembourg	5.7
Malta	3.4	Lithuania	6.4
Netherlands	3.4	Latvia	6.5
Hungary	3.5	Portugal	6.5
Norway	3.6	Finland	6.7
United Kingdom	3.8	Sweden	6.8
Romania	3.9	Cyprus	7.0
Austria	4.4	Croatia	7.1
Bulgaria	4.5	France	8.5
Slovenia	4.5	Italy	9.9
Denmark	4.6	Spain	13.9
Estonia	4.6	Greece	17.2

79. Final grades A professor (of something other than statistics!) distributed the following histogram to show the distribution of grades on his 200-point final exam. Comment on the display.

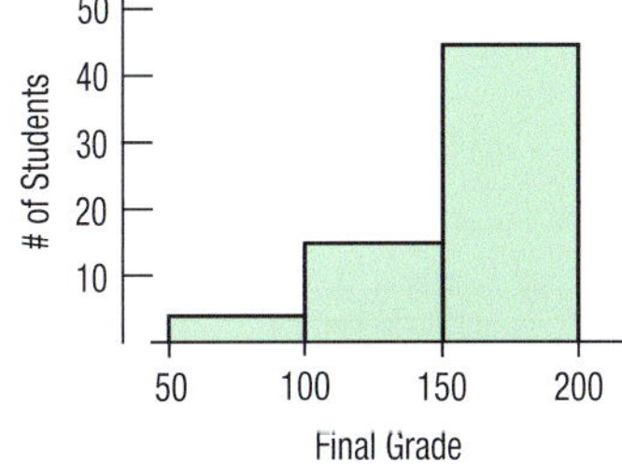

80. Final grades revisited After receiving many complaints about his final-grade histogram from students currently taking a statistics course, the professor from Exercise 79 distributed the following revised histogram:

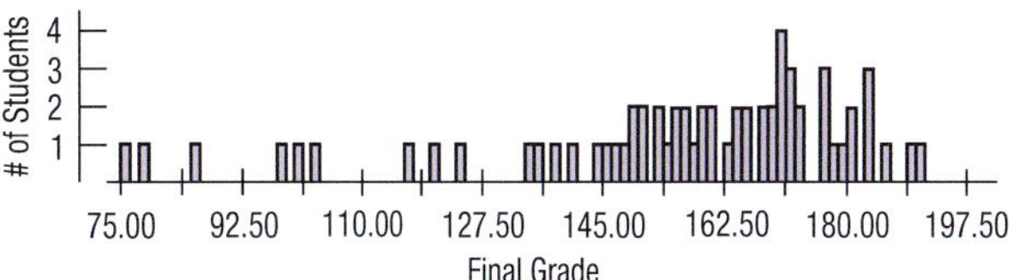

a) Comment on this display.
b) Describe the distribution of grades.

81. ZIP codes Holes-R-Us, an Internet company that sells piercing jewelry, keeps transaction records on its sales. At a recent sales meeting, one of the staff presented a histogram of the ZIP codes of the last 500 customers, so that the staff might understand where sales are coming from. Comment on the usefulness and appropriateness of the display.

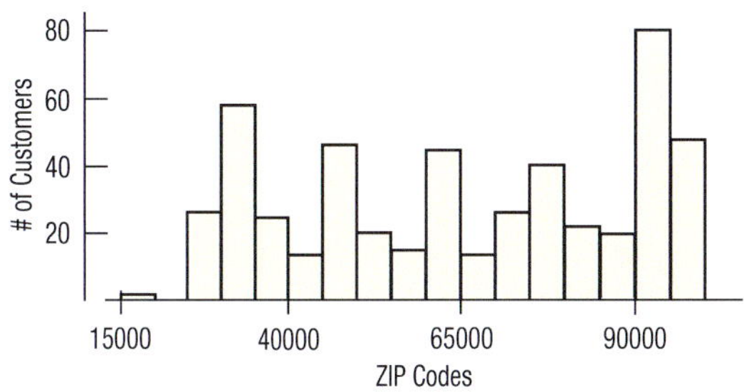

82. ZIP codes revisited Here are some summary statistics to go with the histogram of the ZIP codes of 500 customers from the Holes-R-Us Internet Jewelry Salon that we saw in Exercise 81:

Count	500
Mean	64,970.0
StdDev	23,523.0
Median	64,871
IQR	44,183
Q1	46,050
Q3	90,233

What can these statistics tell you about the company's sales?

T **83. Math scores 2013** The National Center for Education Statistics reports average mathematics achievement scores for eighth graders in all 50 states (nces.ed.gov/nationsreportcard/):

State	Score	State	Score
Alabama	269	Montana	289
Alaska	282	Nebraska	285
Arizona	280	Nevada	278
Arkansas	278	New Hampshire	296
California	276	New Jersey	296
Colorado	290	New Mexico	273
Connecticut	285	New York	282
Delaware	282	North Carolina	286
Florida	281	North Dakota	291
Georgia	279	Ohio	290
Hawaii	281	Oklahoma	276
Idaho	286	Oregon	284
Illinois	285	Pennsylvania	290
Indiana	288	Rhode Island	284
Iowa	285	South Carolina	280
Kansas	290	South Dakota	287
Kentucky	281	Tennessee	278
Louisiana	273	Texas	288
Maine	289	Utah	284
Maryland	287	Vermont	295
Massachusetts	301	Virginia	288
Michigan	280	Washington	290
Minnesota	295	West Virginia	274
Mississippi	271	Wisconsin	289
Missouri	283	Wyoming	288

a) Using technology and the provided data file, find the median, IQR, mean, and standard deviation of these state averages.
b) Which summary statistics would you report for these data? Why?
c) Write a brief summary of the performance of eighth graders nationwide.

T **84. Boomtowns 2015** In 2015, the website NewGeography.com listed its ranking of the best cities for job growth in the United States. The magazine's top 20 large cities, along with their weighted job rating indices, are given in the table. The full dataset contains 70 cities. (newgeography.com/content/004941-large-cities-rankings-2015-best-cities-job-growth)

a) Using technology and the provided data file, make a suitable display of the weighted growth indices for all 70 cities.
b) Summarize the typical growth index among these cities with a median and mean. Why do they differ?
c) Given what you know about the distribution, which of the measures in part b does the better job of summarizing the growth indices? Why?
d) Summarize the spread of the growth index distribution with a standard deviation and with an IQR.
e) Given what you know about the distribution, which of the measures in part d does the better job of summarizing the growth rates? Why?
f) Suppose we subtract from each of the preceding growth rates the average U.S. large-city growth index of 49.23%, so that we can look at how much these indices exceed the U.S. rate. How would this change the values of the summary statistics you calculated above? (*Hint:* You need not recompute any of the summary statistics from scratch.)
g) If we were to omit Austin-Round Rock-San Marcos, TX, from the data, how would you expect the mean, median, standard deviation, and IQR to change? Explain your expectations for each.
h) Write a brief report about all of these growth indices.

City	Index
San Francisco-Redwood City-South San Francisco, CA Metro Div.	97.5
San Jose-Sunnyvale-Santa Clare, CA	97.2
Dallas-Plano-Irving, TX Metro Div.	91.4
Austin-Round Rock, TX	90.9
Nashville-Davidson–Murfreesboro–Franklin, TN	90.8
Houston-The Woodlands-Sugar Land, TX	90.2
Denver-Aurora-Lakewood, CO	89.6
Orlando-Kissimmee-Sanford, FL	88.8
Charlotte-Concord-Gastonia, NC-SC	86.2
San Antonio-New Braunfels, TX	84.9
Riverside-San Bernardino-Ontario, CA	83.9
Atlanta-Sandy Springs-Roswell, GA	83.2
Fort Worth-Arlington, TX Metro Div.	82.8
Seattle-Bellevue-Everett, WA Metro Div.	82.6
Raleigh, NC	82.5
Miami-Miami Beach-Kendall, FL Metro Div.	81.5
New York City, NY	80.9
West Palm Beach-Boca Raton-Delray Beach, FL Metro Div.	80.8
Salt Lake City, UT	78.7
Fort Lauderdale-Pompano Beach-Deerfield Beach, FL Metro Div.	78.2

T **85. Population growth 2010** The following data show the percentage change in population for the 50 states and the District of Columbia from the 2000 census to the 2010 census. Using appropriate graphical displays and summary statistics, write a report on the percentage change in population by state.

State	Change	State	Change
Alabama	7.5	Montana	9.7
Alaska	13.3	Nebraska	6.7
Arizona	24.6	Nevada	35.1
Arkansas	9.1	New Hampshire	6.5
California	10.0	New Jersey	4.5
Colorado	16.9	New Mexico	13.2
Connecticut	4.9	New York	2.1
Delaware	14.6	North Carolina	18.5
District of Columbia	5.2	North Dakota	4.7
Florida	17.6	Ohio	1.6
Georgia	18.3	Oklahoma	8.7
Hawaii	12.3	Oregon	12.0
Idaho	21.1	Pennsylvania	3.4
Illinois	3.3	Rhode Island	0.4
Indiana	6.6	South Carolina	15.3
Iowa	4.1	South Dakota	7.9
Kansas	6.1	Tennessee	11.5
Kentucky	7.4	Texas	20.6
Louisiana	1.4	Utah	23.8
Maine	4.2	Vermont	2.8
Maryland	9.0	Virginia	13.0
Massachusetts	3.1	Washington	14.1
Michigan	−0.6	West Virginia	2.5
Minnesota	7.8	Wisconsin	6.0
Mississippi	4.3	Wyoming	14.1
Missouri	7.0		

Source: www.census.gov/compendia/statab/rankings.html

T **86. Student survey** The dataset **Student survey** contains 299 responses to a student survey from a statistics project. The questions asked included:

- How would you rate yourself politically? (1 = Far left, 9 = Far right)
- What is your gender?
- Do you believe in God?
- Pick a random number from 1 to 10.
- What is your height (in inches)?
- Which is your dominant hand?
- How many dates have you had in the past 6 months?
- How many friends do you have on Facebook?
- What is your weight (in pounds)?
- How many drinks did you have last night?
- Are you a varsity athlete?
- How many songs are on your MP3 player?
- How would you describe your diet?

Using technology and appropriate graphical displays and summary statistics, write a report on several of the variables.

JUST CHECKING

Answers

1. Unimodal, skewed to the high end. Answers on center and spread may vary.
2. Could be symmetric. Center around 3 hours?
3. Very skewed high. Many zeroes.
4. Skewed to low end.
5. Uniform.
6. Incomes are probably skewed to the right and not symmetric, making the median the more appropriate measure of center. The mean will be influenced by high family incomes and may not reflect the "typical" family income as well as the median does. It will give the impression that the typical income is higher than it really is.
7. An IQR of 30 mpg would mean that only 50% of the cars get gas mileages in an interval 30 mpg wide. Fuel economy doesn't vary that much. 3 mpg is reasonable. It seems plausible that 50% of the cars will be within about 3 mpg of each other. An IQR of 0.3 mpg would mean that the gas mileage of half the cars varies little from the estimate. It's unlikely that cars, drivers, and driving conditions are that consistent.
8. I'd like to see the standard deviation and the shape of the distribution.

3

Relationships Between Categorical Variables—Contingency Tables

WHERE ARE WE GOING?

We looked at individual categorical variables in Chapter 2. But things get more exciting when we look at how categorical variables work together. Are men or women more likely to be Democrats? Are people with blue eyes more likely to be left-handed? Is there a relationship between Internet use and race? Questions like these involving two (and more) categorical variables are explored in this chapter.

OkCupid, one of the largest online dating sites in the world, processes about 7.3 million messages per day. The company got into big trouble when it was revealed that they were experimenting on their members. Christian Rudder, one of the founders, was unapologetic, declaring "we experiment on human beings!"

One of the experiments, "Love Is Blind Day," involved suppressing members' photos and measuring whether people responded to messages as often when they could not see what the other person looks like. (They didn't.) Another one, "The Power of Suggestion," involved intentionally giving people matches that didn't correspond to OkCupid's matching algorithm, but telling them they were good fits, to see whether the power of suggestion was sufficient. (It wasn't.)

Rudder has analyzed data about the site's participants to find a number of interesting relationships. He suggests that people are more honest about many things on a site like OkCupid (although perhaps not about their weight or height) than they might be in person.

3.1 Contingency Tables

WHO	Respondents to OkCupid questions
WHAT	Gender (M/F) and pet ownership
WHEN	2015(?)
WHERE	OkCupid.com

Who Likes Dogs—Who Likes Cats?

In Chapter 2 we looked at one variable at a time, but things get more interesting when more variables get into the act. For example, while it might be nice to know whether dogs or cats are more popular, it's even more interesting to know whether the choice is different for men and women. Summary data from OkCupid may help answer this question.

We can arrange the counts of how many pet-owning OkCupid participants have dogs, cats, or both, and the counts of how many participants indentify as male or female in a **contingency table** such as this one: (Data in **OKCupid CatsDogs**)

Table 3.1
Contingency table of *Pets* by *Gender*.

		Gender		
		Female	Male	Total
Pets	Has cats	3412	2388	5800
	Has dogs	3431	3587	7018
	Has both	897	577	1474
	Total	**7740**	**6552**	**14,292**

Each cell of Table 3.1 gives the count for a combination of values of the two variables. The margins of the table, both on the right and at the bottom, give totals. The bottom line of the table is the frequency distribution of *Gender*. The right column is the frequency distribution of the variable *Pets*. When presented like this, in the margins of a contingency table, the frequency distribution of one of the variables is called a **marginal distribution**.

Counts can be interesting, but thinking about percentages is easier and usually more informative. For example, suppose you're a woman who loves pets but is allergic to cats. If you'd like to find a guy who has a dog but not a cat, then knowing that there are 3587 guys with dogs but not cats on OkCupid might be encouraging, but knowing the *proportion* of guys that the dog owners represent is probably more relevant.

Table 3.2
Column percents show the proportions of men and women who own each kind of pet

		Gender		
		Female	Male	Total
Pets	Has cats	44.1%	36.4%	40.6%
	Has dogs	44.3%	54.8%	49.1%
	Has both	11.6%	8.8%	10.3%
	Total	**100%**	**100%**	**100%**

Table 3.2 shows the percents in each column of the table. Looking at these **column percents**, you can see that 54.8% of the pet-owning men have dogs but not cats. The value 54.8% in the Male and Has dogs cell is the number of dog(only)-owning men (3587 according to Table 3.1) divided by the number of pet-owning men (6552). The easy way to tell that these percents divide by the column totals is to notice that their sums in the column margins are 100%. The table shows that although cats and dogs seem to be equally popular among women, men seem to favor dogs over cats.

It seems from Table 3.2 that women may be more likely than men to own both a cat and a dog, but these aren't the right percents to show that properly. To see what proportion of those who have both dogs and cats are men and what proportion are women, look at **row percents**.

Table 3.3

Row percentages show the proportion of each pet ownership category that is male or female.

Pets		Gender: Female	Male	Total
	Has cats	58.8%	41.2%	100%
	Has dogs	48.9%	51.1%	100%
	Has both	60.9%	39.1%	100%
	Total	**54.2%**	**45.8%**	**100%**

Table 3.3 shows that 60.9% of the dual pet owners are women—a much larger proportion than we might have guessed from Table 3.2.

There is a third way to find percents in the entire table. What proportion of all pet owners are women who own both a dog and a cat? Now we are asking for a **table percent** and the denominator of the proportion is the total number of pet owners in the table, 14,292.

Table 3.4

Table percents show the proportion of the total number of respondents in the table falling in each cell.

Pets		Gender: Female	Male	Total
	Has cats	23.9%	16.7%	40.6%
	Has dogs	24.0%	25.1%	49.1%
	Has both	6.3%	4.0%	10.3%
	Total	**54.2%**	**45.8%**	**100%**

Table 3.4 shows, for example, that only 6.3% of the OkCupid pet owners are women who have both a dog and a cat.

EXAMPLE 3.1 Exploring Marginal Distributions

A recent Gallup Poll asked 1008 Americans age 18 and over whether they planned to watch the upcoming Super Bowl. The pollster also asked those who planned to watch whether they were looking forward more to seeing the football game or the commercials. The results are summarized in the table: (Data in **Watch the Super Bowl**)

Watch?		Gender: Male	Female	Total
	Game	279	200	**479**
	Commercials	81	156	**237**
	Won't Watch	132	160	**292**
	Total	**492**	**516**	**1008**

QUESTION: What's the marginal distribution of what they'll *Watch*?

ANSWER: To determine the percentages for the three responses, divide the count for each response by the total number of people polled:

$$\frac{479}{1008} = 47.5\% \quad \frac{273}{1008} = 23.5\% \quad \frac{292}{1008} = 29.0\%.$$

According to the poll, 47.5% of American adults were looking forward to watching the Super Bowl game, 23.5% were looking forward to watching the commercials, and 29% didn't plan to watch at all.

PERCENTAGE ERRORS IN THE NEWS

Former Presidential Press Secretary Sean Spicer cited a study (subsequently called into question) that claimed that 14% of noncitizens were registered to vote. But at a press briefing Spicer concluded that, based on this study, "14 percent of people who voted were non-citizens." (Nicholas Fandosjan, "Trump Won't Back Down from His Voting Fraud Lie. Here Are the Facts," *New York Times*, January 24, 2017)

Percent of What?

The English language can be tricky. If you're asked "What percent of men own only a dog?" it's pretty clear that the question is only about men. So you should restrict your attention only to the men and look at the number of dog owners among the men—in other words, the column percent.

But if you're asked "What percent of pet owners are men who own only a dog?" you have a different question. Be careful; now you need to consider everyone in the table, so 14,292 should be the denominator, and the answer is the table percent.

And if you're asked "What percent of dog owners are men?" you have a third question. Now the denominator is the 7018 dog owners and the answer is the row percent.

Always be sure to ask "percent of what?" That will help you know *Who* we're talking about and whether we want row, column, or table percents. When you come across a table of percents, look at the margins. If the column margins are 100%, the table values are column percents (as in Table 3.2). If the row margins show 100%, the values are row percents (Table 3.3). And if only the grand total in the lower right is 100%, the values are table percents (Table 3.4). It is important to pay attention to these differences; they affect the question that you're asking, and they'll come back again later in this text.

EXAMPLE 3.2

Exploring Percentages: Children and First-Class Ticket Holders First?

Only 31% of those aboard the *Titanic* survived. Was that survival rate the same for all ticket classes? Some accounts of the sinking of the *Titanic* suggest there might have been a relationship between the kind of ticket a passenger held and the passenger's chances of making it into a lifeboat. Here are three contingency tables classifying the 2208 people aboard the *Titanic* according to their ticket class (or whether they were a crew member) and whether they survived. (Data in **Titanic**)

Table percents

Class		Survival		
		Alive	Dead	Total
	1	9.1%	5.6%	**14.7%**
	2	5.4%	7.5%	**12.9%**
	3	8.2%	24.0%	**32.2%**
	Cr	9.6%	30.7%	**40.3%**
	Total	**32.2%**	**67.8%**	**100.0%**

Row percents

Class		Survival		
		Alive	Dead	Total
	1	62.0%	38.0%	**100.0%**
	2	41.8%	58.2%	**100.0%**
	3	25.4%	74.6%	**100.0%**
	Cr	23.8%	76.2%	**100.0%**
	Total	**32.2%**	**67.8%**	**100.0%**

Column percents

Class	Survival: Alive	Dead	Total
1	28.2%	8.2%	**14.7%**
2	16.7%	11.1%	**12.9%**
3	25.3%	35.4%	**32.2%**
Cr	29.8%	45.3%	**40.3%**
Total	**100.0%**	**100.0%**	**100.0%**

QUESTION: Was the chance of survival the same regardless of a passenger's ticket class or whether they were a crew member? Which table is the appropriate one to use to answer that question?

ANSWER: No. From the table of row percents, we see, for example, that 62.0% of first-class ticket holders survived, while only 25.4% of third-class ticket holders survived.

QUESTION: Among the survivors, what group was most common? Which table is appropriate to answer this question?

ANSWER: Now you're asked to focus on the survivors only, so look at the column percents table. Crew members were the most frequent with 29.8% among the survivors. (This makes sense because there were more crew members on board the *Titanic* than passengers of any class, and some were in each lifeboat to handle the boats.)

3.2 Conditional Distributions

To find the percentage of men who own dogs, we focused only on the men. In general, questions that restrict our attention to just one condition of a variable and ask about the distribution of another variable are asking about the **conditional distribution** of that variable. Conditional distributions show the distribution of one variable for only those cases that satisfy a condition on another variable. The pie charts in Figure 3.1 show the distribution of pet ownership choices conditional on the gender of the owner.

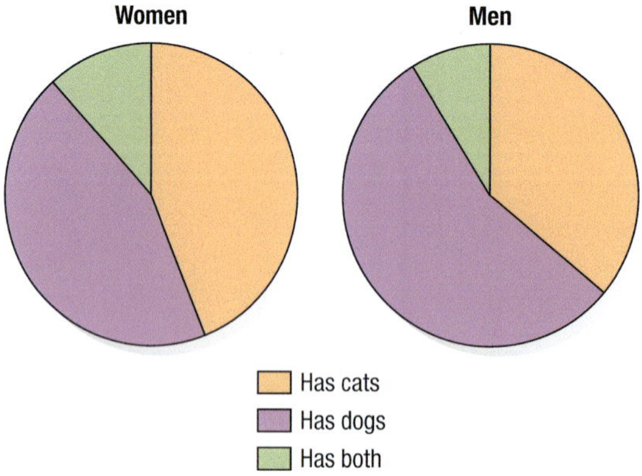

Figure 3.1
Pie charts of the conditional distributions of pet ownership for women (left) and men (right).

Do the distributions in Figure 3.1 appear to be the same? We're primarily concerned with percentages here, so pie charts are a reasonable choice. But pie charts make it difficult to be precise about comparisons. Perhaps a better way to compare conditional distributions is with side-by-side bar charts.

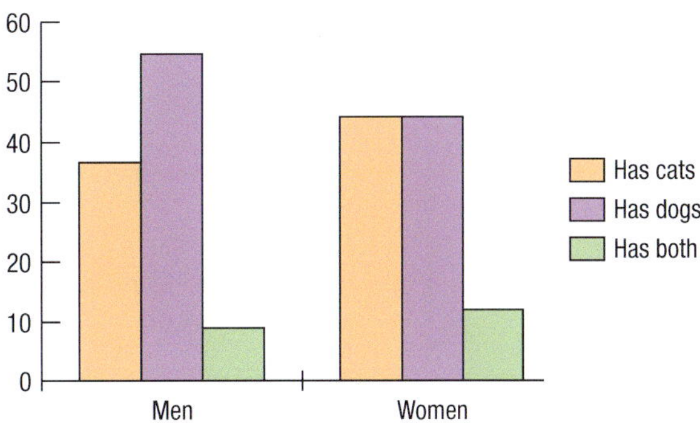

Figure 3.2
Side-by-side bar charts showing the conditional distribution of *Pets* for each *Gender.*

EXAMPLE 3.3 Finding Conditional Distributions: Watching the Super Bowl

RECAP: The table shows results of a poll asking adults whether they were looking forward to the Super Bowl game, looking forward to the commercials, or didn't plan to watch. (Data in **Watch the Super Bowl**)

Watch?	Gender		
	Male	**Female**	**Total**
Game	279	200	**479**
Commercials	81	156	**237**
Won't Watch	132	160	**292**
Total	**492**	**516**	**1008**

QUESTION: How do the conditional distributions of *Gender* differ among the responses to the question about commercials?

ANSWER: Look at the group of people who responded "Commercials" and determine what percent of them were male and female:

$$\frac{81}{237} = 34.2\% \quad \frac{156}{237} = 65.8\%.$$

Women make up a sizable majority of the adult Americans who look forward to seeing Super Bowl commercials more than the game itself. Nearly 66% of people who voiced a preference for the commercials were women; only 34% were men.

Variables can be associated in many ways and to different degrees. The best way to tell whether two variables are associated is to ask whether they are *not*.[1] In a contingency table, when the distribution of *one* variable is the same for all categories of another, the variables are **independent**. There's no association between these variables. We'll see a way to check for independence formally later in the text. For now, we'll just compare the distributions.

[1]This kind of "backward" reasoning shows up surprisingly often in science—and in statistics. We'll see it again.

EXAMPLE 3.4

Looking for Associations Between Variables: Still Watching the Super Bowl

RECAP: The table shows the results of a poll asking 1008 adults whether they were looking forward to the Super Bowl game, looking forward to the commercials, or didn't plan to watch.

Watch?	Gender: Male	Female	Total
Game	279	200	**479**
Commercials	81	156	**237**
Won't Watch	132	160	**292**
Total	**492**	**516**	**1008**

QUESTION: Does it seem that there's an association between choice of what to watch and a person's gender?

ANSWER: First find the distribution of the three responses for the men (the column percentages):

$$\frac{279}{492} = 56.7\% \quad \frac{81}{492} = 16.5\% \quad \frac{132}{492} = 26.8\%.$$

Then do the same for the women who were polled, and display the two distributions with a side-by-side bar chart.

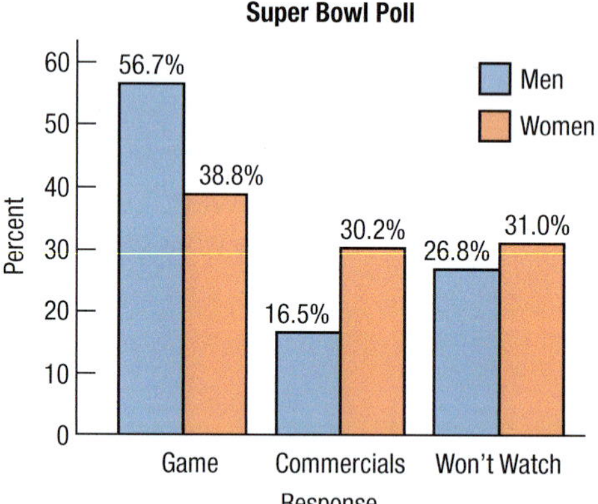

Based on this poll, it appears that women were only slightly more likely than men to say they won't watch the Super Bowl: 31% of the women said they didn't plan to watch, compared to just under 27% of men. Among those who planned to watch, however, there appears to be an association between the viewer's gender and what the viewer is most looking forward to. While more women are interested in the game (38.8%) than the commercials (30.2%), the margin among men is much wider: 56.7% of men said they were looking forward to seeing the game, compared to only 16.5% who cited the commercials.

JUST CHECKING

A statistics class reports the following data on *Gender* and *Eye Color* for students in the class:

		Eye Color			
		Blue	**Brown**	**Green/Hazel/Other**	**Total**
Gender	**Male**	6	20	6	**32**
	Female	4	16	12	**32**
	Total	**10**	**36**	**18**	**64**

1. What percent of females are brown-eyed?
2. What percent of brown-eyed students are female?
3. What percent of students are brown-eyed females?
4. What's the distribution of *Eye Color*?
5. What's the conditional distribution of *Eye Color* for the males?
6. Compare the percent who are female among the blue-eyed students to the percent of all students who are female.
7. Does it seem that *Eye Color* and *Gender* are independent? Explain.

STEP-BY-STEP EXAMPLE

Examining Contingency Tables

We asked for a picture of a man eating fish. This is what we got.

Medical researchers followed 6272 Swedish men for 30 years to see whether there was any association between the amount of fish in their diet and prostate cancer.[2] Their results are summarized in this table: (Data in **Fish diet**)

		Prostate Cancer	
		No	**Yes**
Fish Consumption	**Never/Seldom**	110	14
	Small Part of Diet	2420	201
	Moderate Part	2769	209
	Large Part	507	42

QUESTION: Is there an association between fish consumption and prostate cancer?

THINK **PLAN** State what the problem is about.

I want to know if there is an association between fish consumption and prostate cancer.

VARIABLES Identify the variables and report the W's.

The individuals are 6272 Swedish men followed by medical researchers for 30 years. The variables record their fish consumption and whether or not they were diagnosed with prostate cancer.

Be sure to check the appropriate condition.

✓ **Categorical Data Condition:** I have counts for both fish consumption and cancer diagnosis. The categories of diet do not overlap, and the diagnoses do not overlap. It's okay to draw pie charts or bar charts.

[2] "Fatty fish consumption and risk of prostate cancer," *Lancet*, June 2001. The original study actually used pairs of twins, which enabled the researchers to discern that the risk of cancer for those who never ate fish actually *was* substantially greater. Using pairs is a special way of gathering data. We'll discuss such study design issues and how to analyze the data in later chapters.

SHOW **MECHANICS** Check the marginal distributions first before looking at the two variables together.

Fish Consumption	Prostate Cancer: No	Prostate Cancer: Yes	Total
Never/Seldom	110	14	124 (2.0%)
Small Part of Diet	2420	201	2621 (41.8%)
Moderate Part	2769	209	2978 (47.5%)
Large Part	507	42	549 (8.8%)
Total	5806 (92.6%)	466 (7.4%)	6272 (100%)

Two categories of the diet are quite small, with only 2.0% Never/seldom eating fish and 8.8% in the "Large part" category. Overall, 7.4% of the men in this study had prostate cancer.

Then, make appropriate displays to see whether there is a difference in the relative proportions. These pie charts compare fish consumption for men who have prostate cancer to fish consumption for men who don't.

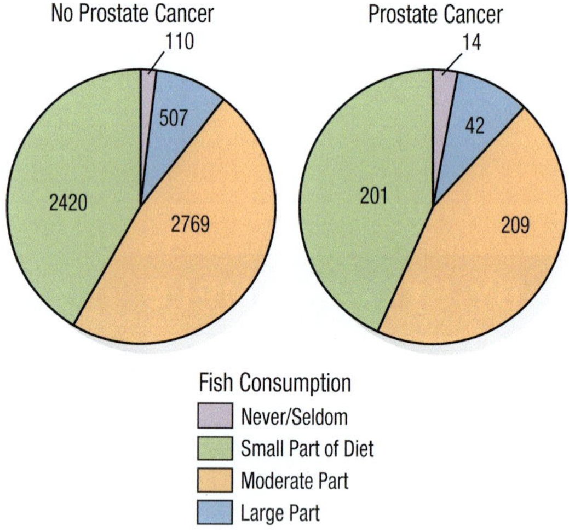

It's hard to see much difference in the pie charts. So, I made a display of the row percentages. Because there are only two alternatives, I chose to display the risk of prostate cancer for each group:

Both pie charts and bar charts can be used to compare conditional distributions. Here we compare prostate cancer rates based on differences in fish consumption.

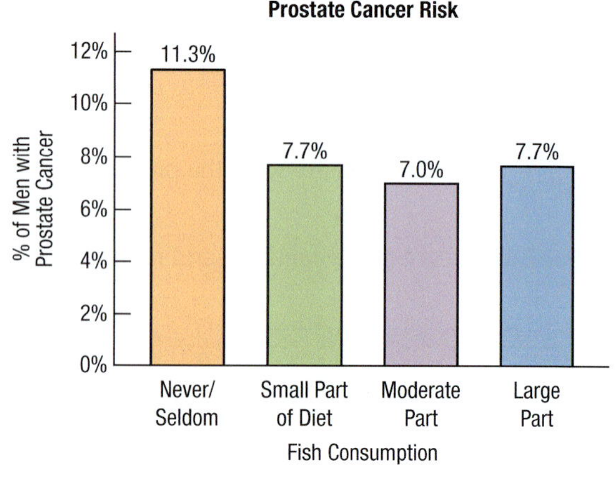

TELL **CONCLUSION** Interpret the patterns in the table and displays in context. If you can, discuss possible real-world consequences. Be careful not to overstate what you see. The results may not generalize to other situations.

Overall, there is a 7.4% rate of prostate cancer among men in this study. Most of the men (89.3%) ate fish either as a moderate or small part of their diet. From the pie charts, it's hard to see a difference in cancer rates among the groups. But in the bar chart, it looks like the cancer rate for those who never/seldom ate fish may be somewhat higher.

However, only 124 of the 6272 men in the study fell into this category, and only 14 of them developed prostate cancer. More study would probably be needed before we would recommend that men change their diets.

This study is an example of looking at a sample of data to learn something about a larger population, one of the main goals of this text. Scientists care about more than these particular 6272 Swedish men. They hope that the men's experiences will say something about the value of eating fish in general. That raises several questions. What population might this sample represent? All Swedish men? All men? How do we know that other factors besides the amount of fish they ate weren't associated with prostate cancer? Perhaps men who eat fish often have other habits that distinguish them from the others and maybe those other habits are what actually kept their cancer rates lower.

Observational studies, like this one, often lead to contradictory results because they can't control all the other factors. In fact, a later paper, published in 2011, based on data from a cancer prevention trial on 3400 men from 1994 to 2003, showed that some fatty acids may actually increase the risk of prostate cancer.[3] We'll discuss the pros and cons of observational studies and experiments where we can control the factors in Chapter 11.

RANDOM MATTERS

Nightmares

Do you sleep on your side? If so, do you favor your right or left side? And can you usually recall your dreams? Researchers interviewed people who could remember their dreams and recall which side they slept on.

They found 63 participants, of whom 41 were right-side sleepers and 22 slept on their left side. Then they interviewed the sleepers about their dreams. Of the 41 right-side sleepers, only 6 reported often having nightmares. But of the 22 left-side sleepers, 9 reported nightmares.[4] (Data in **Nightmares**)

[3] "Serum phospholipid fatty acids and prostate cancer risk: Results from the Prostate Cancer Prevention Trial," *American Journal of Epidemiology*, June 15, 2011.

[4] "Sleeping Position, Dream Emotions, and Subjective Sleep Quality," Mehmet Yucel Agargun, M.D., Murat Boysan, M.A., Lutfu Hanoglu, M.D., www.sleepandhypnosis.org/ing/Pdf/9cf6fb582430425193687f6ac8889038.pdf

The question the researchers asked is a typical one in science: "Can we see a difference?" That is, does the side you sleep on really affect your dreams, or was the observed difference just random? We can organize the data into a contingency table like this:

		Side		
		Right	Left	Totals
Dreams	Nightmares	6	9	15
	Sweet Dreams	35	13	48
	Totals	41	22	63

Now, suppose having nightmares has nothing to do with which side you sleep on. Then we could distribute the 15 nightmares randomly among the sleepers and obtain a new table. For example, a table with randomly distributed nightmares might look like this:

		Side		
		Right	Left	Totals
Dreams	Nightmares	11	4	15
	Sweet Dreams	30	18	48
	Totals	41	22	63

Notice that, with the same 41 right-sleepers and 22 left-sleepers and the same 15 nightmares distributed randomly, the margins of the table stay the same. And if the margins are determined, then if you know just one cell of the table—say, the number of left-sleepers who have nightmares—you can fill in the rest of the table.

The authors did a random scrambling of nightmares 1000 times and looked to see how many left-sleepers had nightmares. Figure 3.3 shows a histogram of what we found.

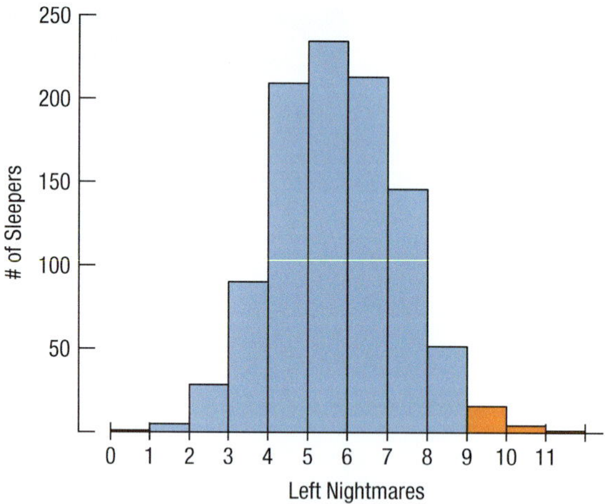

Figure 3.3
The number of left-sleepers with nightmares in 1000 trials in which nightmares were allocated randomly to sleepers. Would 9 or more be unexpected?

Out of 1000 random tables, in only 21 did 9 or more left-sleepers have nightmares. That might be rare enough to convince you of the authors' claim about the risk of sleeping on your left side, and that the researchers' results are not just a random scattering of nightmares. In other words, that we can see a difference.

This may seem convincing. Does it apply to you? The subjects in this study were people who reported not only that they fell asleep and woke up on the same side consistently, but who usually remembered their dreams. Would you be a candidate for this study? Is this group representative enough to draw conclusions about the effect of sleep positions?

3.3 Displaying Contingency Tables

Here's a contingency table of the 2208 people on board the *Titanic* by *Survival* and ticket *Class*:

Table 3.5

A contingency table of *Class* by *Survival* with only counts and column percentages. Each column represents the conditional distribution of *Survival* for a given category of ticket *Class*.

Survival		Class: First	Second	Third	Crew	Total
Alive	Count	201	119	180	212	**712**
	% of Column	62.0%	41.8%	25.4%	23.8%	**32.3%**
Dead	Count	123	166	530	677	**1496**
	% of Column	38.0%	58.2%	74.6%	76.2%	**67.7%**
Total	Count	**324**	**285**	**710**	**889**	**2208**
		100%	**100%**	**100%**	**100%**	**100%**

We could display the *Titanic* information in this table using side-by-side bar charts as in Figure 3.4. Now it's easy to compare the risks. Among first-class passengers, 38% perished, compared to 58.2% for second-class ticket holders, 74.6% for those in third class, and 76.2% for crew members.

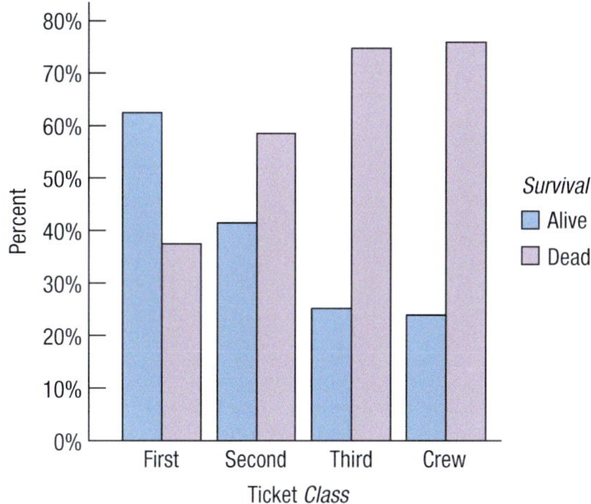

Figure 3.4

Side-by-side bar chart showing the conditional distribution of *Survival* for each category of ticket *Class*. The corresponding pie charts would have only two categories in each of four pies, so bar charts seem the better alternative.

We could also display the *Titanic* information by dividing up bars in proportion to the relative frequencies. The resulting **segmented** (or **stacked**) **bar chart** treats each bar as the "whole" and divides it proportionally into segments corresponding to the

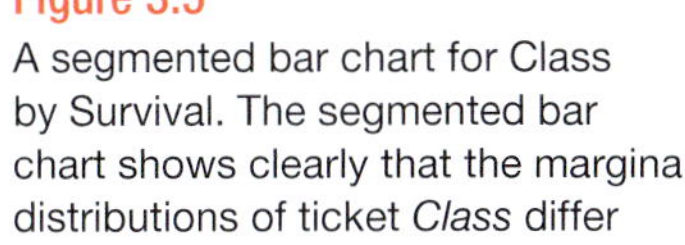

Figure 3.5

A segmented bar chart for Class by Survival. The segmented bar chart shows clearly that the marginal distributions of ticket *Class* differ between survivors and nonsurvivors.

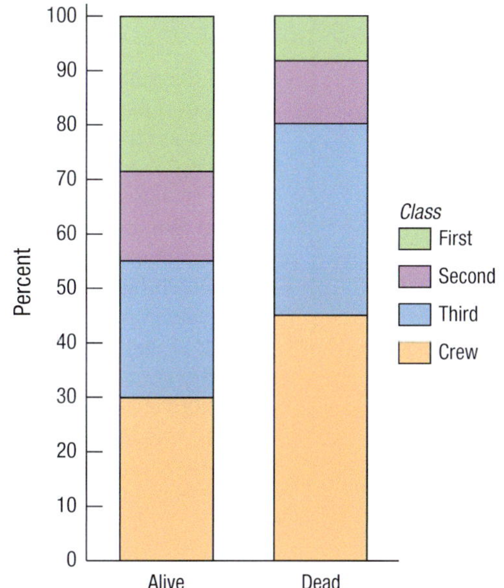

percentage in each group. Now we see that the distributions of ticket *Class* are clearly different, indicating that *Survival* was not independent of ticket *Class*. Notice that although the totals for survivors and nonsurvivors are quite different, the bars are the same height because we have converted the numbers to percentages. Compare this display with the side-by-side bar charts of the same data in Figure 3.4. Which graph do you find easier to understand?

A variant of the segmented bar chart, a **mosaic plot** looks like a segmented bar chart, but obeys the area principle better. Not only is each vertical bar proportional to the number of cases in its group, but each rectangle in the plot is proportional to the number of cases corresponding to it. We can easily see now that only about a third of the people on the *Titanic* survived—something that we can't see in the segmented bar charts. Mosaic plots are increasingly popular for displaying contingency tables and are found in many software packages. They are especially useful when there are many categories, as the next two examples show.

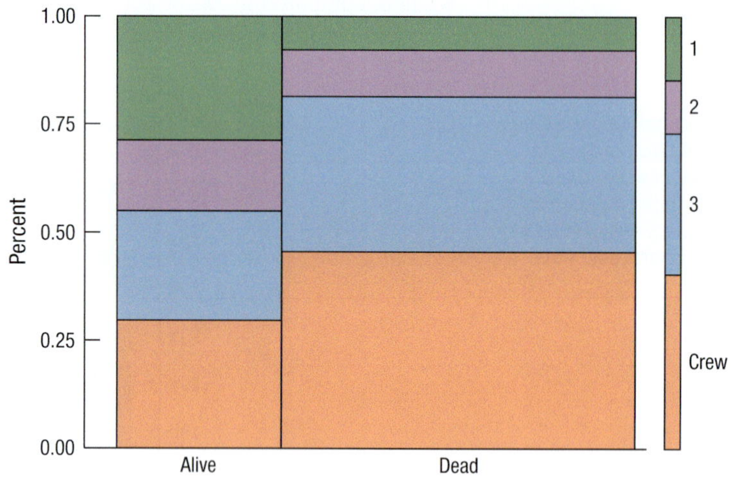

Figure 3.6
A mosaic plot for *Class* by *Survival*. The plot resembles the segmented bar chart in Figure 3.5, but now all areas are proportional to the number of cases corresponding to them. We can see the difference in the marginal distributions of ticket *Class* as before, but now we also see that only about a third of those on board survived.

EXAMPLE 3.5

Looking for Associations in Large Datasets: Text Messages

One of the authors has his four children on his family cell phone plan. He thought he had a good plan, one that covered everyone for 1500 text messages a month, until he received a monthly bill for $497.65, which somewhat exceeded the $149.95 he was expecting. From the Verizon website, he soon realized that the excess charges were due to his daughter who was away at college for the first time. The Verizon website allowed him to download the "metadata" on all calls and text messages. Here is a mosaic plot of the 3277 outgoing text messages by *Date* (on the *x*-axis) and proportion to each number (on the *y*-axis).

For each day (from 9/11 to 10/23), the width of each bar is proportional to the number of text messages sent. The height of each color within the bar represents the proportion of texts going to each of the most frequently texted numbers on that day. The colors show the most frequently texted phone numbers.The daughter explained that she was texting a lot when she first got to campus because she didn't know anyone and was texting her friends at other colleges. But she claimed that the number of texts had really "calmed down" recently.

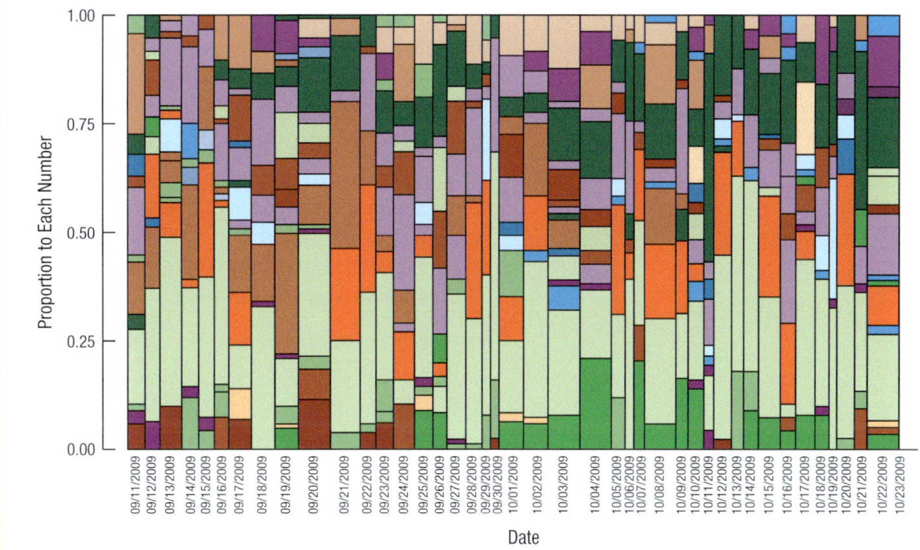

QUESTION: Do you see any evidence that the number of texts decreased over this period?

ANSWER: No, if the number of texts were decreasing we should see wide bars on the left and narrower bars on the right. No such pattern is evident.

The father in Example 3.5 realized that the number of texts was, in fact, not decreasing, and decided to see what other information about the patterns of text sending he could discover.

EXAMPLE 3.6

Looking for Associations in Large Datasets: Text Messages Revisited

RECAP: The Verizon data from Example 3.5 also includes the time of day. Here is a mosaic plot of the same 3277 outgoing texts by hour of the day (0 = midnight through 23 = 11 PM). The order of the hours in this plot goes from 3 AM (3) to 2 AM (2).

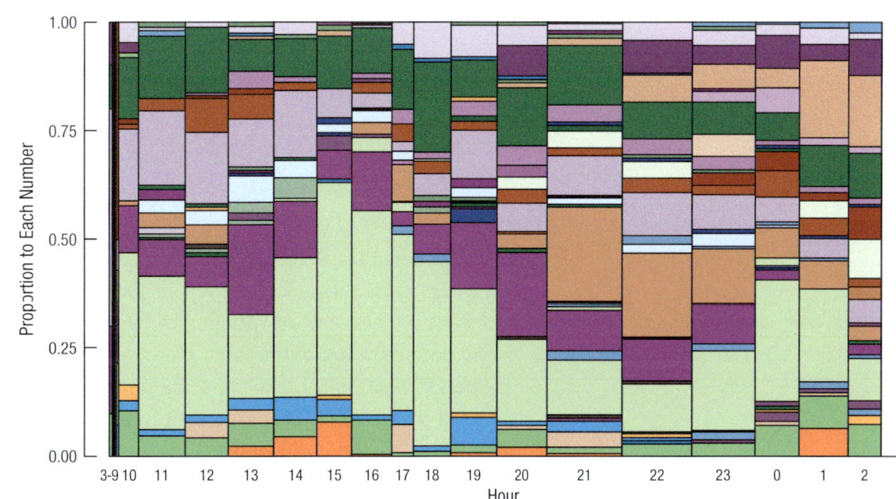

QUESTION: What can you learn about the daily habits of this student from the patterns in her texting? She admitted that she had classes at 11 AM and 2 PM and that her peak study hours were 9–11 PM. How did that information make her dad feel? Note: The first bar on the left contains the hours from 3 AM to 9 AM inclusive.

ANSWER: There is almost no activity between 3 and 9 AM. At 10 AM the texting starts (somewhat slowly as evidenced by the narrow bar at 10), then is fairly active and constant until 5 PM (17). When questioned, the daughter admitted that she often went to the gym at around 5. The texting then increases, with a peak around 9 PM (21), tapering off at bit at midnight (0) to 2 AM. Being a teacher himself, the dad wasn't all that surprised by the patterns.

3.4 Three Categorical Variables

There's no reason to be limited to the relationship between two variables. Considering a third variable may give you a more realistic understanding of your data. For example, recall the allergic OkCupid user searching for a man who has dogs but no cats. A friend points out that OkCupid also asks about drug use, and that's a real turnoff. Is her chance of finding a dog-owning guy hurt if she looks at only non-drug users? Here are two contingency tables like Table 3.2, but now restricted to those who tell OkCupid that they don't use drugs and those who admit to at least some drug use.

Table 3.6 shows that by limiting the search to guys who don't use drugs her chance of finding a guy with dogs but no cats actually increases. 58.5% of non-drug-using male pet owners have dogs but no cats, compared to only 46.1% of drug-using male pet owners.

Table 3.6
Column percents of pet owners differ according to drug use.

Drugs = "No"		Gender		
		Female	Male	Total
Pets	Has cats	40.8%	32.3%	37.1%
	Has dogs	47.0%	58.5%	52.0%
	Has both	12.2%	9.18%	10.9%
	Total	**100%**	**100%**	**100%**

Drugs = "Yes"		Gender		
		Female	Male	Total
Pets	Has cats	51.7%	44.2%	47.7%
	Has dogs	36.3%	46.1%	41.5%
	Has both	12.1%	9.70%	10.8%
	Total	**100%**	**100%**	**100%**

When a third variable is involved, you should be cautious about looking at the overall percentages without considering the third variable. Summaries for a single variable that don't take account of others can be dangerous. Sometimes looking at only two variables can give paradoxical results. Here's a made-up example to illustrate: Moe argues that he's a better pilot than Jill because he managed to land 83% of his last 120 flights on time compared with Jill's 78%. But simple percentages can be misleading. Let's look at the data again, this time including a third variable: what time of day they flew.

EXAMPLE 3.7 Looking for Associations Among Three Variables at Once

How did survival on the *Titanic* depend on the sex and ticket class of individuals aboard? That's a question about three variables: *Sex, Class*, and *Survived*. A mosaic plot can help us see these relationships.

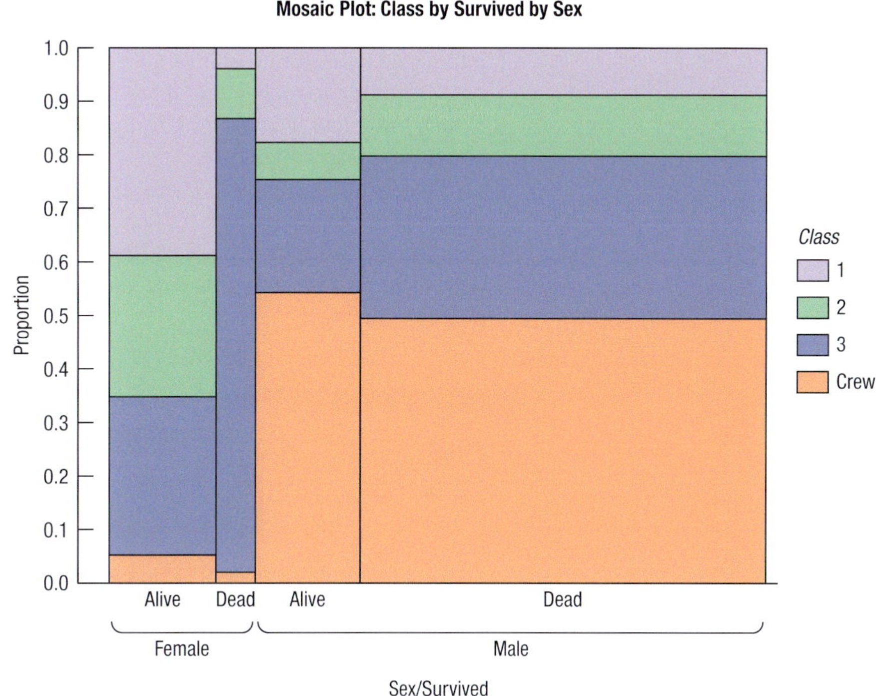

QUESTION: How did survival depend on the sex and ticket class of those aboard the *Titanic*?

ANSWER: With a display like this we can examine three variables at once. First let's look at the individual variables. The widths of the boxes show that there were more men than women aboard. The heights show, for example, that most of the crew were men. Comparing the men to the women, we see that a greater fraction of the women survived. Now look at the *Class* distributions among the men. They are quite similar for survivors and those who died. But among the women, those who died were overwhelmingly third-class passengers. When we looked at *Survival* by *Class* we missed this part of the story. It wasn't just that third-class passengers had a lower chance of survival than first-class passengers; the real difference was among the women. Women were less likely to die than men. And most of those who did were holding third-class tickets.

Table 3.7

On-time flights by *Time of Day* and *Pilot*. Look at the percentages within each *Time of Day* category. Who has a better on-time record during the day? At night? Whose record is better overall?

		Time of Day		
		Day	**Night**	**Overall**
Pilot	**Moe**	90 out of 100 90%	10 out of 20 50%	100 out of 120 83%
	Jill	19 out of 20 95%	75 out of 100 75%	94 out of 120 78%

Moe claims he's better overall, but for day flights, Jill had a 95% on-time rate and Moe only a 90% rate. And at night, Jill was on time 75% of the time and Moe only 50%. So Moe is better "overall," but Jill is better both during the day and at night. How can this be?

What's going on here is that combining results across a third variable (*Time of Day*, in this example) isn't appropriate because Moe and Jill fly different numbers of night and day flights. Jill has mostly night flights, which are more difficult, so her *overall average* is heavily influenced by her nighttime average. Moe, on the other hand, benefits from flying mostly during the day, with its high on-time percentage. With their very different patterns of flying conditions, taking an overall percentage is misleading. It's not a fair comparison.

When this kind of surprising result shows up, it is known as **Simpson's paradox**, named for the statistician who described it in 1951. There have been several well-publicized cases (see Example 3.8). In fact, this result is actually not a paradox at all, but a case of failing to realize that there is an important third variable that was not considered. When you know what to look for, you can explain this kind of result while others worry about the apparent paradox.

The world is complex. There are always other variables that you could include, but it is almost never obvious which they are. Variables that are not included, but that affect the relationship, are called **lurking variables**. You'll need to use your understanding of the world to seek them out. Including lurking variables in your analysis will give you a more accurate picture of the world.

EXAMPLE 3.8

Simpson's Paradox: Gender Discrimination?

One famous example of Simpson's paradox arose during an investigation of admission rates for men and women at the University of California at Berkeley's graduate schools. As reported in an article in *Science*,[5] about 45% of male applicants were admitted, but only about 30% of female applicants got in. A simple comparison of the admittance rates seemed to show clear discrimination:

	Admit	Reject	%Admit
Men	1158	1493	43.7%
Women	557	1278	30.4%

QUESTION: Does this table prove that women were being discriminated against in admissions at Berkeley?

ANSWER: It certainly seemed to be a clear-cut case of gender discrimination, until a statistician showed the table broken down by the school that the candidate applied to. When the data were considered by school (Engineering, Law, Medicine, etc.), it turned out that, within each school, women were admitted at nearly the same or, in some cases, much higher rates than men:

School	Male Admits	Female Admits	Male%	Female%
A	512	89	62.1%	82.4%
B	313	17	60.2%	68.0%
C	120	202	36.9%	34.1%
D	138	131	33.1%	34.9%
E	53	94	27.7%	23.9%
F	22	24	5.9%	7.0%

[5]P. J. Bickel, E. A. Hammel, and J. W. O'Connell (1975), "Sex bias in graduate admissions: Data from Berkeley," *Science* **187** (4175): 398–404.

The lurking variable was the school applied to. Women applied in larger numbers to schools with lower admission rates, such as Law and Medicine. Men tended to apply to Engineering and Science, which had overall admission rates well above 50%. When the average was taken, the women had a lower overall admission rate, but that average missed the important third variable: the *School* applied to.

WHAT CAN GO WRONG?

- **Don't confuse similar-sounding percentages.** These percentages sound similar but are different:

			Class				
			First	Second	Third	Crew	Total
Survival	Alive	Count	201	119	180	212	**712**
	Dead	Count	123	166	530	677	**1496**
	Total	Count	**324**	**285**	**710**	**889**	**2208**

  - The percent of the passengers who were both in first class and survived: This is 201/2208, or 9.1%.
  - The percent of the first-class passengers who survived: This is 201/324, or 62.0%.
  - The percent of the survivors who were in first class: This is 201/712, or 28.2%.

  In each instance, pay attention to the *Who* implicitly defined by the phrase. Often there is a restriction to a smaller group (all aboard the *Titanic*, those in first class, and those who survived, respectively) before a percentage is found. Your discussion of results must make these differences clear.

- **Don't forget to look at the variables separately, too.** When you make a contingency table or display a conditional distribution, be sure you also examine the marginal distributions. It's important to know how many cases are in each category.
- **Be sure to use enough individuals.** When you consider percentages, take care that they are based on a large enough number of individuals. Take care not to make a report such as this one: *We found that 66.67% of the rats improved their performance with training. The other rat died.*
- **Don't overstate your case.** Independence is an important concept, but it is rare for two variables to be *entirely* independent. We can't conclude that one variable has no effect whatsoever on another. Usually, all we know is that little effect was observed in our study. Other studies of other groups under other circumstances could find different results.
- **Watch out for lurking variables.** Combining results across the categories of a third variable may be dangerous. It's always better to compare percentages within each level of the other variable. The overall percent may be misleading. Simpson's paradox is one extreme example.

CONNECTIONS

Focusing on the table, row, or column percents in a contingency table changes *Who* you are interested in. Restricting the *Who* of the study in different ways answers different questions. The concept of conditioning on one variable to look at others will come back as an important concept for thinking about probabilities.

CHAPTER REVIEW

Make and interpret a contingency table.

- When we want to see how two categorical variables are related, put the counts (and/or percents) in a two-way table called a **contingency table**.

Make and interpret bar charts and pie charts of marginal distributions.

- Look at the **marginal distribution** of each variable (found in the margins of the table). Also look at the **conditional distribution** of a variable within each category of the other variable.
- Comparing conditional distributions of one variable across categories of another tells us about the association between variables. If the conditional distributions of one variable are (roughly) the same for every category of the other, the variables are **independent**.
- Consider a third variable whenever it is appropriate, and be able to describe the relationships among the three variables.

REVIEW OF TERMS

The key terms are in chapter order so you can use this list to review the material in the chapter.

Contingency table A contingency table displays counts and, sometimes, percentages of individuals falling into named categories on two or more variables. The table categorizes the individuals on all variables at once to reveal possible patterns in one variable that may be contingent on the category of the other (p. 91).

Marginal distribution In a contingency table, the distribution of either variable alone is called the marginal distribution. The counts or percentages are the totals found in the margins (last row or column) of the table (p. 91).

row percents
column percents
table percents When a cell of a contingency table holds percents, these can be percents of the total in the row or column of that cell, or the total in the entire table. These are row, column, and table percents, respectively (pp. 91–92).

Conditional distribution The distribution of a variable when the *Who* is restricted to consider only a smaller group of individuals is called a conditional distribution (p. 94).

Independence Variables are said to be independent if the conditional distribution of one variable is the same for each category of the other. We'll show how to check for independence in a later chapter (p. 95).

Segmented bar chart A segmented bar chart displays the conditional distribution of a categorical variable within each category of another variable (p. 101).

Mosaic plot A mosaic plot is a graphical representation of a (usually two-way) contingency table. The plot is divided into rectangles so that the area of each rectangle is proportional to the number of cases in the corresponding cell (p. 102).

Simpson's paradox When averages are taken across different groups, they can appear to contradict the overall averages. This is known as "Simpson's paradox" (p. 106).

Lurking variable A lurking variable is one that is not immediately evident in an analysis, but changes the apparent relationships among the variables being studied. (p. 106)

TECH SUPPORT Contingency Tables

Programs have different ways to specify and work with contingency tables and displays of multiple categorical variables.

DATA DESK

- Choose **Contingency Table** from the toolbar or Calc menu.
- Drag your variables into the resulting window according to whether you want them to specify columns or rows of the table.
- From the table's HyperView menu choose **Table Options**.
- In the dialog, select counts and/or row, column, or table percentages.

COMMENTS

Data Desk automatically treats variables selected for this command as categorical variables even if their elements are numerals. Click on any cell of the table to select those cases in all plots and editing windows. Click on margin values to select cases in the corresponding row or column of the table. To examine three-way relationships, click on "No Selector" and either specify a selector or choose "Hot Selector", which will restrict the table to cases selected, for example, in a plot of a third variable. Either update the table or set Automatic Update to see it change instantly.

EXCEL

Excel calls contingency tables "Pivot Tables".

- To make a pivot table, from the Data menu, choose **Pivot Table**.
- In the Layout window, drag your variables to the row area, the column area, and drag your variable again to the data area. This tells Excel to count the occurrences of each category. Once you have an Excel pivot table, you can construct bar charts and pie charts.

JMP

- From the Analyze menu, select **Fit Y by X**.
- Choose one variable as the Y, response variable, and the other as the X, factor variable. Both selected variables must be Nominal or Ordinal.
- JMP will make a plot and a contingency table.

MINITAB

- From the Stat menu, choose the **Tables** submenu.
- In the dialog, identify the columns that make up the table. Minitab will display the table.
- Alternatively, select the **Cross Tabulation** . . . command to see more options for the table, including expected counts and standardized residuals.

R

Using the function xtabs, you can create a contingency table from two variables x and y in a data frame called mydata with the command.

- con.table = xtabs(~x+y,data=mydata)

SPSS

- From the Analyze menu, choose the **Descriptive Statistics** submenu.
- From that submenu, choose **Crosstabs**. . . .
- In the Crosstabs dialog, assign the row and column variables from the variable list. Both variables must be categorical.
- Click the **Cells** button to specify **what** should be displayed.

COMMENTS

SPSS offers only variables that it knows to be categorical in the variable list for the Crosstabs dialog. If the variables you want are missing, check that they have the right type.

STATCRUNCH

To make a contingency table:

- Click on **Stat > Tables > Contingency > With Data** (or **With Summary**).
- Select Row variable and Column Variable (there are options for grouping and type of display).
- Click on **Compute!**

To make a stacked or side-by-side bar chart:

- Click on **Graph > Bar Plot > With Data** (or **With Summary**).
- Select **Column** variable.
- Select **Group by** for second variable.
- Select **Grouping option**. The default is **Split bars**—side-by-side—or select **Stack bars** or **Separate graph for each group**.
- Click on **Compute!**

COMMENTS

There are options for various percentages in the bar charts.

EXERCISES

SECTION 3.1

1. Regulation amendments The administration of a particular University asked 4,000 of its students to express their opinion as to whether they agree or disagree with a set of amendments to the regulations. Students had to choose from "Strongly Agree," "Agree," "Not Sure," "Disagree," and "Strongly Disagree."

	Strongly Agree	Agree	Not Sure	Disagree	Strongly Disagree
Males	499	616	47	395	343
Females	689	571	92	436	312

a) What percent of male students strongly agree with the regulation amendments?

b) What percent of male students disagree or strongly disagree with the regulation amendments? What is the comparable percentage of female students?

c) Compare the distribution of opinions between male and female students.

d) Is it reasonable to conclude that 30% of all University students strongly agree with the regulation amendments? Why or why not?

2. Cyber comparison shopping It has become common for shoppers to comparison shop using the Internet. Respondents to a Pew survey in 2013 who owned cell phones were asked whether they had, in the past 30 days, looked up the price of a product while they were in a store to see if they could get a better price somewhere else. Here is a table of responses by income level:

	<$30K	$30K–$49.9K	$50K–$74.9K	>$75K
Yes	207	115	134	204
No	625	406	260	417

Source: www.pewinternet.org/2012/01/30/the-rise-of-in-store-mobile-commerce/

a) What percent of those earning less than $30K cyber comparison shop? What percent of those earning more than $75K do?

b) What percent of those who cyber comparison shop earn less than $30K?

c) Is it reasonable to assume that 28% of all shoppers use their phones to cyber comparison shop? Why or why not?

SECTION 3.2

3. Regulation amendments again Consider the sample data in Exercise 1.

a) What is the conditional distribution (in percentages) of the opinions of female students on the regulation amendments?

b) Find the conditional distribution of the opinions of male students when the categories are combined into Positive (Strongly Agree or Agree), Neutral (Not Sure), and Negative (Disagree or Strongly Disagree).

4. Cyber comparison shopping again

a) Find the conditional distribution (in percents) of income distribution for those who do not cyber compare prices.

b) Find the conditional distribution (in percents) of incomes for shoppers who do cyber compare prices.

SECTION 3.3

T **5. Diet and politics** The survey of 299 undergraduate students from Exercise 86 in Chapter 2 (data in **Student survey**) asked about respondents' diet preference (Carnivore, Omnivore, Vegetarian) and political alignment (Liberal, Moderate, Conservative). Here is a stacked bar chart of the 285 responses:

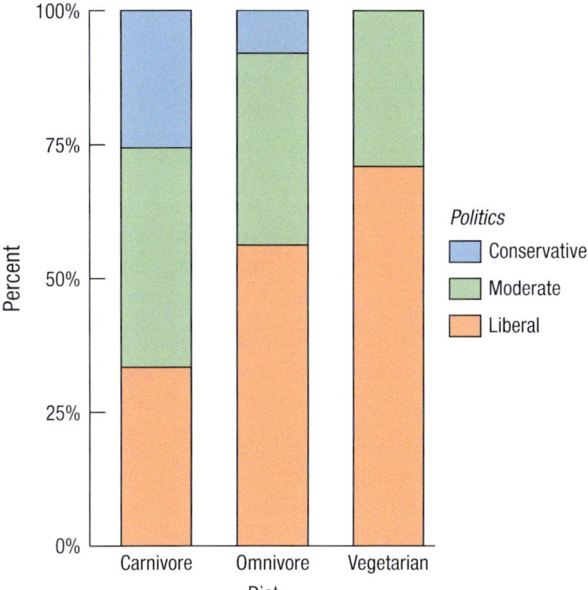

a) Describe what this plot shows using the concept of a conditional distribution.
b) Do you think the differences here are real? Explain.

T **6. Diet and politics revisited** Here are the same data as in Exercise 5 but displayed differently:

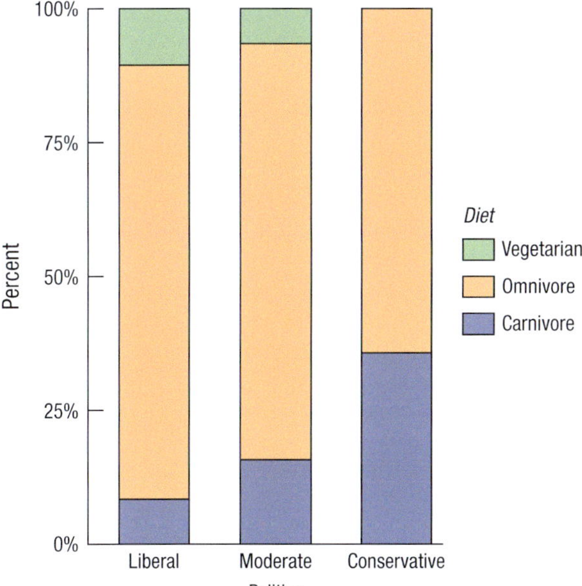

a) Describe what this plot shows using the concept of a conditional distribution.
b) Do you think the differences here are real? Explain.

7. Fish and prostate cancer revisited Here is a mosaic plot of the data on *Fish consumption* and *Prostate cancer* from the Step-by-Step Example on page 97.

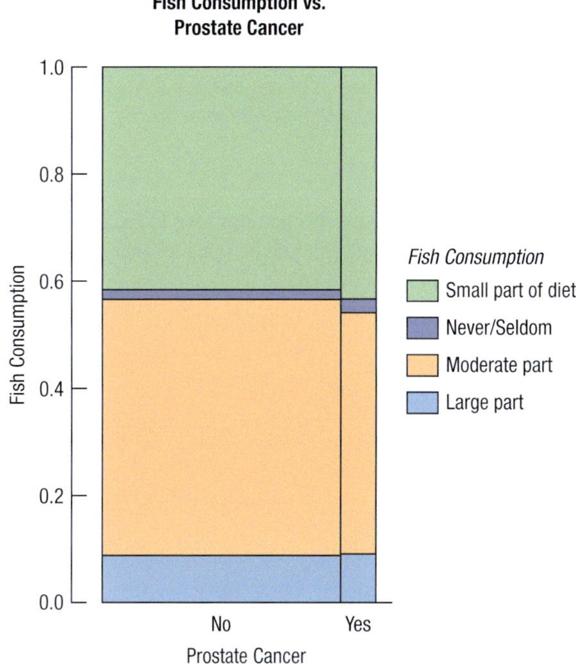

a) From the mosaic plot, about what percent of all men in this survey were diagnosed with prostate cancer?
b) Are there more men who had cancer and never/seldom ate fish, or more who didn't have cancer and never/seldom ate fish?
c) Which is higher: the percent of men with cancer who never/seldom ate fish, or the percent of men without cancer who never/seldom ate fish?

8. Regulation amendments revisited Here is a stacked bar chart of the sample data from Exercise 1 on whether University students agree with a set of regulation amendments from a survey of male and female students:

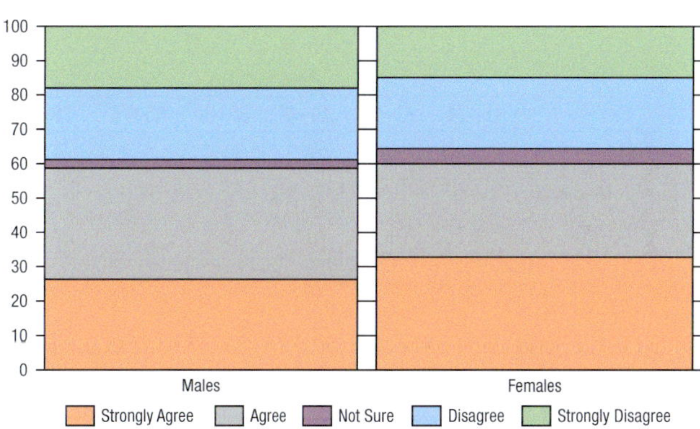

a) From the bar chart, is it easy to see whether more male students said that they strongly disagree with the amendments or more female students said that they strongly disagree? Explain briefly.
b) From the bar chart, is it easy to see how the percentage of male students who said that they strongly agree with the amendments compares with the percentage of female students who said that? Explain briefly.
c) From the bar chart, about what percentage of the male respondents agree with the regulation amendments?
d) From the bar chart, do you think that there is an association between the distribution of responses and whether the respondent is male or female? Explain briefly.

SECTION 3.4

T **9. Diet and politics III** Are the patterns seen in Exercises 5 and 6 relating diet to political opinion the same for men and women? Here are two contingency tables:

Men

	Carnivore	Omnivore	Vegetarian
Liberal	9	74	5
Moderate	12	54	1
Conservative	9	14	0

Women

	Carnivore	Omnivore	Vegetarian
Liberal	4	53	12
Moderate	4	27	6
Conservative	1	4	0

a) Are women or men more likely to be conservative carnivores?
b) Are liberal vegetarians more likely to be women or men?

T **10. Life satisfaction** As part of a survey on well-being of people in the European Union, Eurostat gathered the following data (in percentages) on life satisfaction by age group. (Data from https://ec.europa.eu/eurostat/statistics-explained/index.php?title=Quality_of_life_in_Europe_-_facts_and_views_-_overall_life_satisfaction&oldid=400088).

Life Satisfaction	Age Group					
	16–24	25–34	35–49	50–64	65–74	75+
Low	12.3	16.2	20.3	24.4	22.4	27.4
Medium	58.0	60.1	58.8	56.7	56.0	52.5
High	29.7	23.7	21.0	18.9	21.7	20.1

a) What percentage of people aged 16–24 have medium or high life satisfaction?
b) How does that compare with the same percentage of people aged 65–74?
c) Write a few sentences describing the association between the two extreme age groups and their level of life satisfaction.
d) From this table, can you determine what percent of all surveyed individuals have a medium level of life satisfaction?

CHAPTER EXERCISES

T **11. Movie genres and ratings** The following table summarizes 1529 films from 2014 and 2015 that have been classified into a genre and have a MPAA rating. (Data from **Movies 06-15**)

	G	NC-17	Not Rated	PG	PG-13	R	Total
Action	0	0	35	1	35	30	101
Adventure	2	0	11	38	25	4	80
Comedy	3	0	116	18	51	124	312
Documentary	18	0	189	27	27	27	288
Drama	0	1	226	29	113	160	529
Horror	0	0	13	1	8	29	51
Thriller/Suspense	0	0	34	0	29	69	132
Other	0	0	14	10	3	9	36
Total	23	1	638	124	291	452	1529

a) What percent of these films were rated R?
b) What percent of these films were R-rated comedies?
c) What percent of R-rated films were comedies?
d) What percent of comedies were rated R?

12. Household size The following table displays the distribution of households in the European Union by size for the years 2013–2018. (Percentages may not add to 100% due to rounding.) (Data from https://ec.europa.eu/eurostat/statistics-explained/index.php/Household_composition_statistics#Household_size).

Household Size	Year					
	2013	2014	2015	2016	2017	2018
One person	32.3	32.7	33.3	33.1	33.6	33.9
Two persons	31.7	31.6	31.5	31.7	31.6	31.8
Three persons	16.3	16.1	15.9	15.9	15.8	15.5
Four persons	13.7	13.5	13.4	13.4	13.2	13.1
Five persons or more	6.1	6.0	5.8	5.8	5.8	5.6

a) What percentage of households had a size of two persons in 2015?
b) What percentage of households had a size of at most three persons in 2016?
c) What percentage of households had a size of more than three persons in 2017?
d) Write a few sentences comparing the distributions of household sizes in 2013 and 2018.

13. Body image and gender The following contingency table of categorical data displays an exploration, by the University of Florida, of how body image is related to gender (Data from https://bolt.mph.ufl.edu/6050-6052/unit-1/case-c-c/).

Gender	Body Image			
	About Right	Overweight	Underweight	TOTAL
Female	560	163	37	760
Male	295	72	73	440
TOTAL	855	235	110	1200

a) What is the total sample size?
b) What is the percentage of female participants in the sample?

c) What is the percentage of overweight participants in the sample?
d) Compare the percentage of underweight female participants with that of underweight male participants.

14. Graphs in the news Find a bar graph or pie chart of categorical data from a newspaper, a magazine, or the Internet.

a) Is the graph clearly labeled?
b) Does the graph violate the area principle?
c) Does the accompanying article tell the W's of the variable?
d) Do you think the article correctly interprets the data? Explain.

15. Instagram usage The contingency table below provides data (in percentages) on the distribution of Instagram users worldwide as of July 2019 by age group and gender. (Data from https://www.statista.com/statistics/248769/age-distribution-of-worldwide-instagram-users/). Based on these data, do you think there is an association between Instagram usage and gender, or between Instagram usage and age? Explain.

Gender	Age Group						
	13–17	18–24	25–34	35–44	45–54	55–64	65+
Female	3.2%	15%	17%	8.8%	4.4%	1.8%	1%
Male	3%	16%	17%	7%	3%	1.1%	0.8%

16. Moviegoers and ethnicity The Motion Picture Association of America studies the ethnicity of moviegoers to understand changes in the demographics of moviegoers over time. Here are the numbers of moviegoers (in millions) classified as Hispanic, African American, Caucasian, and Other for the year 2010. Also included are the numbers for the general U.S. population and the number of tickets sold.

	Caucasian	Hispanic	African-American	Other	Total
Population	204.6	49.6	37.2	18.6	310
Moviegoers	88.8	26.8	16.9	8.5	141
Tickets	728	338	143	91	1300
Total	1021.4	414.4	197.1	118.1	1751

a) Compare the conditional distribution of ethnicity for all three groups: the entire population, moviegoers, and ticket holders.
b) Write a brief description of the association between ethnicity and moviegoers.

17. Death from the sky A recent article in *Geophysical Research Letters* (*Asteroid impact effects and their immediate hazards for human populations*, 10.1002/2017GL073191) simulated the consequences of an earth impact by an asteroid of 400 m in diameter. They estimate that for a land impact, 60% of deaths would be caused by wind and shockwave damage, 30% by heat, 1% by flying debris, 0.2 % by cratering, and 0.17% by earthquake.

a) What percent of estimated deaths would they attribute to causes not listed here?
b) Try to create a display for these data. Why is that difficult to do?

18. Cartoons You'll find a number of cartoons throughout this text. Are they simply entertaining, or will they help with learning? Lawrence M. Lesser, Dennis K. Pearl, and John J. Weber III ("Assessing fun items' effectiveness in increasing learning of college introductory statistics students: Results of a randomized experiment," *Journal of Statistics Education*, **24** (2), 2016) showed cartoons and other "fun items" to a group of 80 students and compared their performance on tests with 88 students taught without the fun items. The table gives the number of students scoring correctly on the test.

Topic	Without fun item ($n = 88$)	With fun item ($n = 80$)
Drawing a sample	54	39
What does "random" mean	52	42
Mean versus median	54	44
Categorical versus quantitative variables	15	16
Inadequacy of mean without standard deviation	20	22

a) Find the relative frequency distributions and explain why they are a more appropriate way to compare the two columns.
b) Is there evidence in these data that cartoons help with understanding?

19. Smoking The Centers for Disease Control and Prevention provide data on smoking rates by year and for men and women separately. Here is a table with some of that information:

	Men				Women			
	1974	1994	2004	2014	1974	1994	2004	2014
18–24	42.1%	29.8	25.6	18.5	34.0	28.5	22.9	16.5
25–34	50.5	31.4	26.1	23.7	38.6	30.2	22.6	18.6
35–44	51.0	33.2	26.5	22.0	39.3	27.1	22.7	18.0
45–54	46.8	30.8	26.7	19.9	34.9	24.9	21.4	19.9
55–64	37.7	24.7	22.7	18.8	30.6	20.8	18.4	15.3
≥65	24.8	13.2	9.8	9.8	12.3	11.1	8.2	7.6

a) What was the smoking rate among 18–24-year-old men in 1974?
b) How has the smoking rate among 18–24-year-old men changed from 1974 to 2014?
c) Men who were 18–24 in 1974 were 20 years older—in the 35–44 group—in 1994, and so on diagonally across the table. How has the smoking rate in that cohort of men changed over the 30 years covered by these data?

20. Smoking women Look again at the table of smoking prevalence in Exercise 19.

a) Compare the smoking rate among 18–24-year-old women to that of men during the time covered by this table.
b) Relatively few women over the age of 65 smoke. What other variables might affect this percentage?

T **21. Smartphone ownership** The PEW Research Centre gathered the following data on the percentages of adults in all six sub-Saharan countries who reported that they own a smartphone in 2013 and 2017. (Data from https://www.pewresearch.org/global/2018/10/09/majorities-in-sub-saharan-africa-own-mobile-phones-but-smartphone-adoption-is-modest/).

Sub-Saharan Countries	Smartphone Ownership 2013	2017
South Africa	33%	51%
Ghana	15%	35%
Senegal	13%	34%
Nigeria	19%	32%
Kenya	19%	30%
Tanzania	8%	13%

a) Which sub-Saharan countries have seen an increase in the level of smartphone ownership from 2013 to 2017?
b) Compute the percentage differences from 2013 to 2017 in the level of smartphone ownership for all six sub-Saharan countries.
c) In a few sentences, comment on the percentages of smartphone ownership in Tanzania.

T **22. Education and employment patterns** The following contingency table displays the education and employment patterns of young people (aged between 15 and 29 years) in the European Union for the year 2016. The data is categorized by five-year age groups and the possible selections for the education and employment patterns are the following: (i) exclusively in education, (ii) in education and employment, (iii) exclusively in employment, and (iv) not in employment, nor in education or training. (Data from https://ec.europa.eu/eurostat/en/web/products-eurostat-news/product/-/asset_publisher/VWJkHuaYvLIN/content/EDN-20170811-1/pop_up?_101_INSTANCE_VWJkHuaYvLIN_viewMode=print&_101_INSTANCE_VWJkHuaYvLIN_languageId=en_GB).

Education and Employment Pattern	Age Group 15–19	20–24	25–29
Exclusively in Education	78.1%	32.6%	7.9%
In Education and Employment	11.9%	17.2%	13.6%
Exclusively in Employment	3.8%	33.4%	59.5%
Not in Employment, nor in Education or Training	6.1%	16.7%	18.8%

a) In which of the education and employment pattern categories is the difference between the three age groups greatest?
b) What percentage of individuals aged 25–29 are still in education?
c) What percentage of individuals aged 15–19 are employed? How does this compare with the same percentage of the other two age groups?

23. Communication among teenagers In Pew Research Centre's Teens Relationships Survey of 2014/2015, 1,009 teenagers aged 13 to 17 who have a close friend were asked to identify the three most common ways through which they get in touch with their closest friend. The table below shows the responses (in percentages) categorized by gender. Write a few sentences describing the similarities and differences in communication preferences among male and female teenagers aged 13 to 17 who have a close friend. (Data from https://www.pewinternet.org/2015/08/06/chapter-2-how-teens-hang-out-and-stay-in-touch-with-their-closest-friends/).

Communication Preference	Gender: Males	Females
Text Messaging	75%	84%
Phone Calls	62%	75%
Social Media Site	60%	72%
Video Sharing Site	9%	10%
Blogging Site	2%	6%
Game Site or Server	39%	3%
Discussion Site	5%	3%

24. Being successful revisited Here is a mosaic plot of the data on being successful from Exercise 10:

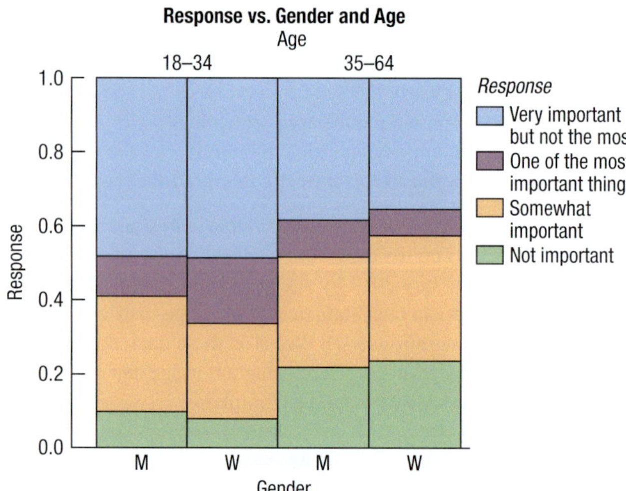

a) Are the differences in sample sizes in the four groups very large? Explain briefly.
b) Which factor seems more important in determining how someone responded: *Age* or *Gender*? Explain briefly.
c) Judging by the top two categories of importance, which of the four groups thinks being successful is most important?

25. Diet and politics IV Here is a mosaic plot of the data on *Diet* and *Politics* from Exercise 5 combined with data on *Gender*.

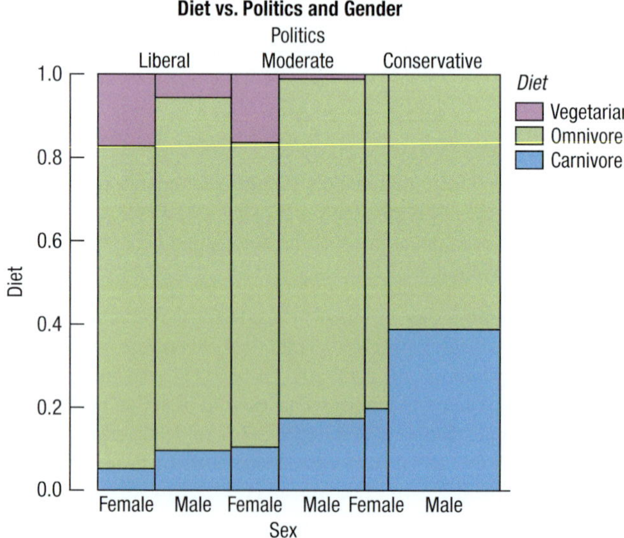

a) Are there more men or women in the survey? Explain briefly.
b) Does there appear to be an association between *Politics* and *Gender*? Explain briefly.
c) Does there appear to be an association between *Politics* and *Diet*? Explain briefly.
d) Does the association between *Politics* and *Diet* seem to differ between men and women? Explain briefly.

26. Assessment performance A report by the Department of Education, United Kingdom, shows the data in the following contingency table, which explores whether gender has an impact in performance in the Year 6 National Assessments. The data, which corresponds to the year 2016, shows the percentages of boys and girls who reached an expected standard in different subjects. (Data from https://www.alcentres.co.uk/does-gender-make-a-difference-in-ks2-sats-performance/). Write a few sentences comparing the two distributions.

Subject	Gender: Boys	Girls
Reading	62%	70%
Grammar, Punctuation, and Spelling	68%	78%
Mathematics	70%	70%
Writing	68%	81%

27. Satisfaction with financial situation The following contingency table contains data gathered by Eurostat from EU-28 in 2013. It illustrates the relationship between income level (measured through the income tercile that the person belongs to on the basis of the distribution at the country level) and satisfaction with the financial situation. (Data from Figure 9, page 27 of https://ec.europa.eu/eurostat/documents/3217494/6856423/KS-05-14-073-EN-N/).

Income Tercile	Satisfaction: Low	Medium	High
Lowest Income Tercile	55.4%	37.8%	6.8%
Second Income Tercile	35.5%	53.1%	11.4%
Highest Income Tercile	20.2%	57.9%	21.9%

a) Is this a table of row percentages, column percentages, or table percentages? How can you tell?

b) Which of the following can you tell from this table? If you can, then give the value specified.

i. The percentage of those respondents who have a low satisfaction level and have an income that falls under the second income tercile.

ii. The percentage of those respondents who have an income that falls under the highest income tercile who have a medium satisfaction level.

iii. The percentage of those respondents who have a high satisfaction level and have an income that falls under the highest income tercile.

iv. The percentage of those respondents who have an income that falls under the lowest income tercile who have a high satisfaction level.

28. Higher education student enrolments The following contingency table contains data gathered by the Higher Education Statistics Agency (HESA) for the academic year 2016/17 on higher education science student enrolments in the United Kingdom by subject area and sex. (Data from https://www.hesa.ac.uk/data-and-analysis/sfr247/figure-14).

Science Subject Areas	Gender: Females	Males
Medicine & dentistry	6.82%	5.36%
Subjects allied to medicine	41.92%	11.78%
Biological sciences	26.00%	16.23%
Veterinary science	1.01%	0.32%
Agriculture & related subjects	2.15%	1.33%
Physical sciences	7.15%	10.84%
Mathematical sciences	2.97%	5.35%
Computer science	3.17%	16.23%
Engineering & technology	5.29%	26.38%
Architecture, building & planning	3.53%	6.19%

a) Is this a table of row percentages, column percentages, or table percentages? How can you tell?

b) Which of the following can you tell from this table? If you can, then give the value specified.

i. The percentage of male science students who studied mathematical sciences.

ii. The percentage of female science students who studied architecture, building, and planning.

iii. The percentage of science students who studied medicine and dentistry.

iv. The percentage of science students who studied agriculture and related subjects.

29. Seniors Prior to graduation, a high school class was surveyed about its plans. The following table displays the results for white and minority students (the "Minority" group included African American, Asian, Hispanic, and Native American students):

Plans	Seniors: White	Minority
4-Year College	198	44
2-Year College	36	6
Military	4	1
Employment	14	3
Other	16	3

a) What percent of the seniors are white?

b) What percent of the seniors are planning to attend a 2-year college?

c) What percent of the seniors are white and planning to attend a 2-year college?

d) What percent of the white seniors are planning to attend a 2-year college?

e) What percent of the seniors planning to attend a 2-year college are white?

30. Politics Students in an Intro Stats course were asked to describe their politics as "Liberal," "Moderate," or "Conservative." Here are the results:

Sex	Politics: L	M	C	Total
Female	35	36	6	77
Male	50	44	21	115
Total	85	80	27	192

a) What percent of the class is male?
b) What percent of the class considers themselves to be "Conservative"?
c) What percent of the males in the class consider themselves to be "Conservative"?
d) What percent of all students in the class are males who consider themselves to be "Conservative"?

31. Movies 06-15 How have movies changed during the decade from 2006 to 2015? Here is a contingency table showing the proportion of movies with each of the MPAA categories in each year:

MPAA2	Year 2006	2015	Total
G	3.61	2.46	3.03
PG	14.9	13.4	14.2
PG-13	31.6	33.1	32.4
R	49.9	51.0	50.4
Total	100	100	100

a) Are these column percents or row percents? How can you tell?
b) Does it look like things have changed much between 2006 and 2015?
Is this pattern the same for each genre of film? Here are the same tables for Dramas and Comedies.
c) What differences do you see between Dramas and Comedies?

Dramas

MPAA2	Year 2006	2015	Total
G	0.595	0	0.303
PG	15.5	9.26	12.4
PG-13	25.6	37.7	31.5
R	58.3	53.1	55.8
Total	100	100	100

Comedies

MPAA2	Year 2006	2015	Total
G	2.42	1.06	1.83
PG	16.1	8.51	12.8
PG-13	38.7	27.7	33.9
R	42.7	62.8	51.4
Total	100	100	100

32. Glasgow's population The following data, gathered by the National Records of Scotland, gives the population size of Glasgow, Scotland in the year 2017 by age group and sex. (Data from https://www.understandingglasgow.com/indicators/population/trends/trend_by_sex_and_age_group).

Age Group	Sex: Males	Females
0–15	50,703	48,434
16–64	216,399	221,512
65+	35,711	48,261

a) What percentage of the males were aged 16–64? Compare the same percentage for the females.
b) Compare the number of males and females in the population at each age group. Write a couple of sentences summarizing what you see.

33. More about seniors Look again at the table of post-graduation plans for the senior class in Exercise 29.

a) Find the conditional distributions (percentages) of plans for the white students.
b) Find the conditional distributions (percentages) of plans for the minority students.
c) Create a graph comparing the plans of white and minority students.
d) Do you see any important differences in the post-graduation plans of white and minority students? Write a brief summary of what these data show, including comparisons of conditional distributions.

34. Politics revisited Look again at the table of political views for the Intro Stats students in Exercise 30.

a) Find the conditional distributions (percentages) of political views for the females.
b) Find the conditional distributions (percentages) of political views for the males.
c) Make a graphical display that compares the two distributions.
d) Do the variables *Politics* and *Sex* appear to be independent? Explain.

35. Student commencements The following contingency table contains Department of Education and Training international student enrolment data. These data are derived from the Provider Registration and International Students Management System (PRISMS) database and counts commencements of students studying in Australia on a student visa in the years 2018 and 2019. (Data from https://internationaleducation.gov.au/research/International-Student-Data/Pages/InternationalStudentData2019.aspx#Pivot_Table).

Sector	Commencements 2018	2019	Total
Higher Education	1,352,347	603,309	1,955,656
VET	1,035,722	447,646	1,483,368
Schools	121,480	54,381	175,861
ELICOS	778,922	280,650	1,059,572
Non-award	299,802	113,974	413,776
Total	3,588,273	1,499,960	5,088,233

a) What percentage of 2018 commencements were VET?
b) What percentage of 2019 commencements were ELICOS?
c) What percentage of higher education commencements happened in the year 2018?
d) What percentage of all commencements in both years happened in the year 2019?

36. More politics Look once more at the table summarizing the political views of Intro Stats students in Exercise 30.

a) Produce a graphical display comparing the conditional distributions of males and females among the three categories of politics.
b) Comment briefly on what you see from the display in part a.

37. Student commencements revisited Examine the table about commencements of students studying in Australia on a student visa, shown in Exercise 35. Compare the distributions of commencements across the different sectors for the years 2018 and 2019.

38. Parking lots A survey of autos parked in student and staff lots at a large university classified the brands by country of origin, as seen in the table.

Origin	Driver: Student	Staff
American	107	105
European	33	12
Asian	55	47

a) What percent of all the cars surveyed were foreign?
b) What percent of the American cars were owned by students?
c) What percent of the students owned American cars?
d) What is the marginal distribution of origin?
e) What are the conditional distributions of origin by driver classification?
f) Do you think that the origin of the car is independent of the type of driver? Explain.

T **39.** Test answers A student had to undergo a test which consisted of 50 Yes/No questions. The following contingency table compares the student's answers with the actual correct answers:

Student's Answers	Actual Answers: Yes	No
Yes	18	9
No	6	17

a) For what percentage of questions was the actual answer a No?
b) For what percentage of questions did the student obtain a correct answer?
c) For what percentage of questions did the student answer an incorrect Yes answer?
d) Do you see evidence of an association between the type of answer and the ability of the student to choose the correct answer? Write a brief explanation, including an appropriate graph.

T **40.** Twin births In 2000, the *Journal of the American Medical Association (JAMA)* published a study that examined pregnancies that resulted in the birth of twins. Births were classified as preterm with intervention (induced labor or cesarean), preterm without procedures, or term/post-term. Researchers also classified the pregnancies by the level of prenatal medical care the mother received (inadequate, adequate, or intensive). The data, from the years 1995–1997, are summarized in the table below. Figures are in thousands of births. (*JAMA* 284 [2000]:335–341)

Twin Births 1995–1997 (in thousands)

Level of Prenatal Care	Preterm (induced or cesarean)	Preterm (without procedures)	Term or Post-Term	Total
Intensive	18	15	28	**61**
Adequate	46	43	65	**154**
Inadequate	12	13	38	**63**
Total	**76**	**71**	**131**	**278**

a) What percent of these mothers received inadequate medical care during their pregnancies?
b) What percent of all twin births were preterm?
c) Among the mothers who received inadequate medical care, what percent of the twin births were preterm?
d) Create an appropriate graph comparing the outcomes of these pregnancies by the level of medical care the mother received.
e) Write a few sentences describing the association between these two variables.

T **41.** Blood pressure A company held a blood pressure screening clinic for its employees. The results are summarized in the table below by *Age* and *Blood pressure*:

Blood Pressure	Age: Under 30	30–49	Over 50
Low	27	37	31
Normal	48	91	93
High	23	51	73

a) Find the marginal distribution of blood pressure level.
b) Find the conditional distribution of blood pressure level within each age group.
c) Compare these distributions with a segmented bar graph.
d) Write a brief description of the association between age and blood pressure among these employees.
e) Does this prove that people's blood pressure increases as they age? Explain.

T **42.** Obesity and exercise The Centers for Disease Control and Prevention (CDC) has estimated that 19.8% of Americans over 15 years old are obese. The CDC conducts a survey on obesity and various behaviors. Here is a table on self-reported exercise classified by body mass index (BMI):

Physical Activity	Body Mass Index: Normal (%)	Overweight (%)	Obese (%)
Inactive	23.8	26.0	35.6
Irregularly Active	27.8	28.7	28.1
Regular, Not Intense	31.6	31.1	27.2
Regular, Intense	16.8	14.2	9.1

a) Are these percentages column percentages, row percentages, or table percentages?
b) Use graphical displays to show different percentages of physical activities for the three BMI groups.
c) Do these data prove that lack of exercise causes obesity? Explain.

43. Anorexia Hearing anecdotal reports that some patients undergoing treatment for the eating disorder anorexia seemed to be responding positively to the antidepressant Prozac, medical researchers conducted an experiment to investigate. They found 93 women being treated for anorexia who volunteered to participate. For one year, 49 randomly selected patients were treated with Prozac and the other 44 were given an inert substance called a placebo. At the end of the year, patients were diagnosed as healthy or relapsed, as summarized in the table:

	Prozac	Placebo	Total
Healthy	35	32	67
Relapse	14	12	26
Total	49	44	93

Do these results provide evidence that Prozac might be helpful in treating anorexia? Explain.

44. Antidepressants and bone fractures For a period of five years, physicians at McGill University Health Center followed more than 5000 adults over the age of 50. The researchers were investigating whether people taking a certain class of antidepressants (SSRIs) might be at greater risk of bone fractures. Their observations are summarized in the table:

	Taking SSRI	No SSRI	Total
Experienced Fractures	14	244	258
No Fractures	123	4627	4750
Total	137	4871	5008

Do these results suggest there's an association between taking SSRI antidepressants and experiencing bone fractures? Explain.

T **45. Drivers' licenses 2014** The table below shows the number of licensed U.S. drivers (in millions) by age and by sex in 2014. (www.fhwa.dot.gov/policyinformation/statistics.cfm)

Age	Male Drivers (Millions)	Female Drivers (Millions)	Total (Millions)
≤19	4.3	4.2	8.5
20–24	8.9	8.7	17.6
25–29	9.3	9.4	18.7
30–34	9.2	9.4	18.7
35–39	8.7	8.8	17.5
40–44	9.1	9.2	18.4
45–49	9.4	9.5	18.9
50–54	10.2	10.4	20.6
55–59	9.7	10.0	19.8
60–64	8.4	8.7	17.1
65–69	6.9	7.1	14.0
70–74	4.8	5.0	9.8
75–79	3.2	3.4	6.6
80–84	2.1	2.3	4.4
≥85	1.6	1.9	3.6
Total	105.9	108.2	214.1

a) What percent of total drivers are under 20?
b) What percent of total drivers are male?
c) Write a few sentences comparing the number of male and female licensed drivers in each age group.
d) Do a driver's age and sex appear to be independent? Explain.

T **46. Tattoos** A study by the University of Texas Southwestern Medical Center examined 626 people to see if an increased risk of contracting hepatitis C was associated with having a tattoo. If the subject had a tattoo, researchers asked whether it had been done in a commercial tattoo parlor or elsewhere. Write a brief description of the association between tattooing and hepatitis C, including an appropriate graphical display.

	Tattoo Done in Commercial Parlor	Tattoo Done Elsewhere	No Tattoo
Has Hepatitis C	17	8	22
No Hepatitis C	35	53	491

T **47. Diet and politics shuffled** In Exercise 5 you were asked whether you thought the differences in political identification across diet preferences were real. To examine that question further, we randomly scrambled the students' politics preferences and created 8 new bar charts of these scrambled distributions (see below).

a) Can you spot the original data among the 8 new charts?
b) How does this support your answer to Exercise 5b? Explain briefly.

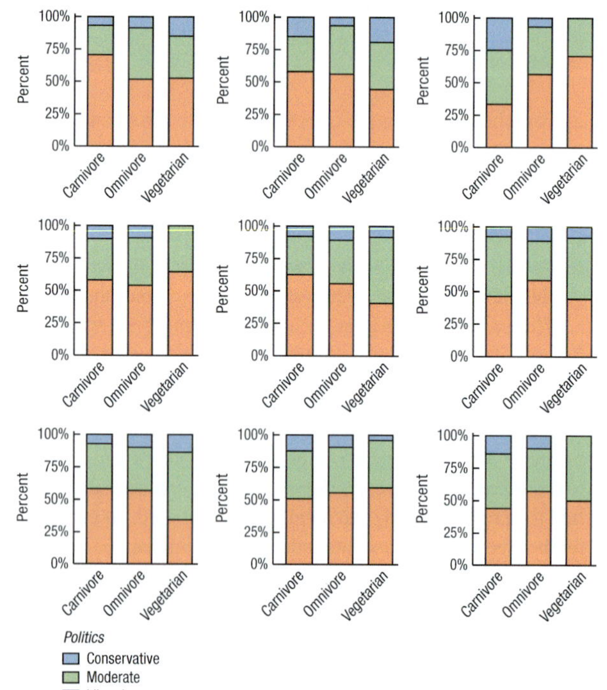

T **48. College values shuffled** In Exercise 8 you were asked to compare the responses of college presidents and all U.S. adults to the question about the value of a college education. To examine that question further, we randomly scrambled the responses and created 15 new mosaic plots of these scrambled distributions (see below).

a) Can you spot the original data among the 15 new plots?
b) How does this support your answer to Exercise 8c? Explain briefly.

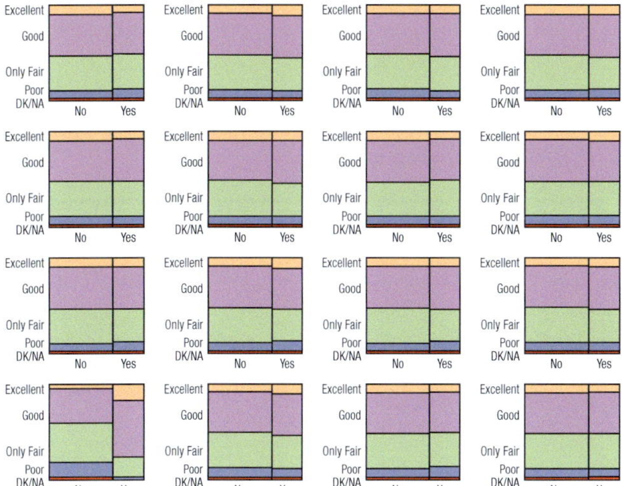

49. Hospitals Most patients who undergo surgery make routine recoveries and are discharged as planned. Others suffer excessive bleeding, infection, or other postsurgical complications and have their discharges from the hospital delayed. Suppose your city has a large hospital and a small hospital, each performing major and minor surgeries. You collect data to see how many surgical patients have their discharges delayed by postsurgical complications, and you find the results shown in the following table:

Discharge Delayed		
	Large Hospital	**Small Hospital**
Major Surgery	120 of 800	10 of 50
Minor Surgery	10 of 200	20 of 250

a) Overall, for what percent of patients was discharge delayed?
b) Were the percentages different for major and minor surgery?
c) Overall, what were the discharge delay rates at each hospital?
d) What were the delay rates at each hospital for each kind of surgery?
e) The small hospital advertises that it has a lower rate of postsurgical complications. Do you agree?
f) Explain, in your own words, why this confusion occurs.

50. Delivery service A company must decide which of two delivery services it will contract with. During a recent trial period, the company shipped numerous packages with each service and kept track of how often deliveries did not arrive on time. Here are the data:

Delivery Service	Type of Service	Number of Deliveries	Number of Late Packages
Pack Rats	Regular	400	12
	Overnight	100	16
Boxes R Us	Regular	100	2
	Overnight	400	28

a) Compare the two services' overall percentages of late deliveries.
b) On the basis of the results in part a, the company has decided to hire Pack Rats. Do you agree that Pack Rats delivers on time more often? Explain.
c) The results here are an instance of what phenomenon?

51. Graduate admissions A 1975 article in the magazine *Science* examined the graduate admissions process at Berkeley for evidence of sex discrimination. The table below shows the number of applicants accepted to each of four graduate programs:

		Males Accepted (of applicants)	Females Accepted (of applicants)
Program	1	511 of 825	89 of 108
	2	352 of 560	17 of 25
	3	137 of 407	132 of 375
	4	22 of 373	24 of 341
	Total	**1022 of 2165**	**262 of 849**

a) What percent of total applicants were admitted?
b) Overall, was a higher percentage of males or females admitted?
c) Compare the percentages of males and females admitted in each program.
d) Which of the comparisons you made do you consider to be the most valid? Why?

52. Be a Simpson Can you design a Simpson's paradox? Two companies are vying for a city's "Best Local Employer" award to be given to the company most committed to hiring local residents. Although both employers hired 300 new people in the past year, Company A brags that it deserves the award because 70% of its new jobs went to local residents, compared to only 60% for Company B. Company B concedes that those percentages are correct, but points out that most of its new jobs were full-time, while most of Company A's were part-time. Not only that, says Company B, but a higher percentage of its full-time jobs went to local residents than did Company A's, and the same was true for part-time jobs. Thus, Company B argues, it's a better local employer than Company A.

Show how it's possible for Company B to fill a higher percentage of both full-time and part-time jobs with local residents, even though Company A hired more local residents overall.

JUST CHECKING

Answers

1. 50.0%
2. 44.4%
3. 25.0%
4. 15.6% Blue, 56.3% Brown, 28.1% Green/Hazel/Other
5. 18.8% Blue, 62.5% Brown, 18.8% Green/Hazel/Other
6. 40% of the blue-eyed students are female, while 50% of all students are female.
7. Since blue-eyed students appear less likely to be female, it seems that *Gender* and *Eye Color* may not be independent. (But the numbers are small.)

4 Understanding and Comparing Distributions

WHERE ARE WE GOING?

Are heart attack rates the same for men and women? Is that expensive beverage container really worth the price? Are wind patterns the same throughout the year? We can answer much more interesting questions about variables when we compare distributions for different groups. These are the kinds of questions where statistics can really help. Some simple graphical displays and summaries can start us thinking about patterns, trends, and models—something we'll do throughout the rest of this text.

WHO	Days during 2011
WHAT	Average daily wind speed (mph), average barometric pressure (mb), average daily temperature (deg Celsius)
WHEN	2011
WHERE	Hopkins Memorial Forest, in western Massachusetts
WHY	Long-term observations to study ecology and climate (Data in **Hopkins Forest**)

The Hopkins Memorial Forest is a 2500-acre reserve in Massachusetts, New York, and Vermont managed by the Williams College Center for Environmental Studies (CES). As part of its mission, the CES monitors forest resources and conditions over the long term.

One of the variables measured in the forest is wind speed. For each day, remote sensors record the average, minimum, and maximum wind speeds. Wind is caused as air flows from areas of high pressure to areas of low pressure. Centers of low pressure often accompany storms, so both high winds and low pressure are associated with some of the fiercest storms. Wind speeds can vary greatly both during a day and from day to day, but if we step back a bit further, we can see patterns. By modeling these patterns, we can understand things about *Average Wind Speed* and the insights it provides about weather that we may not have known.

In Chapter 3, we looked at the association between two categorical variables using contingency tables and displays. Here, we'll look at different ways to examine the relationship between two variables when one is quantitative, and the other indicates groups to compare. We are given wind speed averages for each day of the year. We can gather the days together into groups and compare the wind speeds among the groups. When we change the size of the groups, from the entire year, to seasons, to months, and finally to days, different patterns emerge that together increase our understanding of the entire pattern.

Let's start with the "big picture." Figure 4.1 is a histogram of the *Average Wind Speed* for every day in 2011 along with the median and quartiles. (You'll see why we picked 2011 in a minute.) You can see that the distribution of *Average Wind Speed* is unimodal and skewed to the right. Because of the skewness, we'll report the median and IQR. Median daily wind speed is about 1.12 mph, and on the middle half of the days, the average

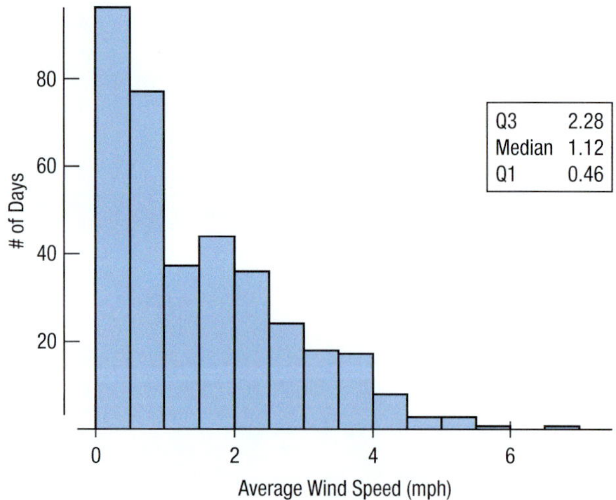

Figure 4.1 A histogram of daily *Average Wind Speed* for 2011. It is unimodal and skewed to the right.

wind speed is between 0.46 and 2.28 mph. We also see several possible **outliers**—values that seem extraordinarily small or (in this case) large—and a rather windy 6.73-mph day. Were these unusual weather events or just the windiest days of the year? To answer that, we'll need to work with the summaries a bit more.

4.1 Displays for Comparing Groups

It is almost always more interesting to compare groups than to summarize data for a single group. For example, it may be interesting to know that the median wind speed in the Hopkins Forest was 1.12 mph in 2011, but that probably isn't something you'd write home about. But what if we ask whether it is windier in the winter or the summer? That could be interesting to sailors (who hope it is windier in the summer) or skiers (who agree with the sailors). Are any months particularly windy? Are weekends a special problem?

Histograms

Let's split the year into two groups: April through September (Spring/Summer) and October through March (Fall/Winter). To compare the groups, we'll create two histograms, being careful to use the same scale. Here are displays of the average daily wind speeds for Spring/Summer (on the left) and Fall/Winter (on the right):

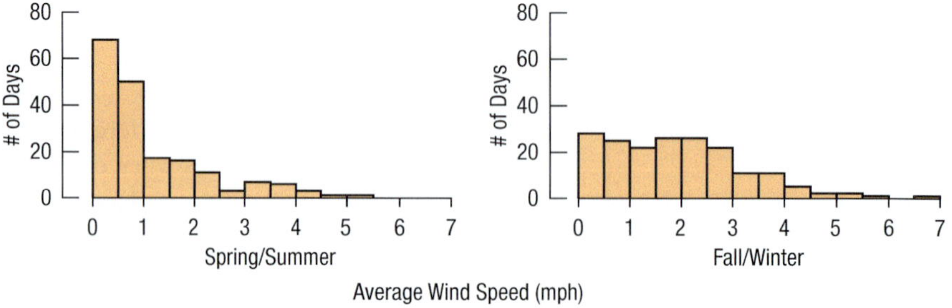

Figure 4.2 Histograms of *Average Wind Speed* for days in Spring/Summer (left) and Fall/Winter (right) show very different patterns.

The distribution of wind speeds for the Spring/Summer months is unimodal and skewed to the right. By contrast, the distribution for the Fall/Winter months is less strongly skewed and more nearly uniform. During the Spring/Summer, a typical day has a mean wind speed of less than 1 mph. In Fall/Winter, the typical wind speed is higher and there is at least one noticeably high value. Wind speeds also vary more in winter (IQR = 1.82 mph) than in summer (IQR = 1.27 mph).

Summaries for *Average Wind Speed* by Season

Season	Mean	StdDev	Median	IQR
Summer	1.11	1.10	0.71	1.27
Winter	1.90	1.29	1.72	1.82

EXAMPLE 4.1

Comparing Groups with Stem-and-Leaf

The Nest Egg Index, devised by the investment firm A.G. Edwards, is a measure of saving and investment performance for each of the 50 states, based on 12 economic factors, including participation in retirement savings plans, personal debt levels, and home ownership. The average index is 100, and the numbers indicate values above or below that average. We could make two histograms, but there are only 50 values, so a back-to-back stem-and-leaf plot is an effective display. Here's one comparing the values of the Nest Egg Index in the Northeast and Midwest states to those in the South and West. In this display, the stems run down the middle of the plot, with the leaves for the two regions to the left or right. Be careful when you read the values on the left: 5 | 8 means a Nest Egg Index of 85%. (Data in **Nest egg index**)

South and West		Northeast and Midwest
3	8	
5778	8	
122444	9	334
7777888899	9	75
003344	10	01223334
6	10	56779
	11	122444

(4 | 9 | 3 means 94% for a South/West state and 93% for a Northeast/Midwest state)

QUESTION: How do Nest Egg indices compare for these regions?

ANSWER: Nest Egg indices were generally higher for states in the Northeast and Midwest than in the South and West. The distribution for the Northeast and Midwest states has most of its values above 100, varying from a low of 93% to a high of 114%. Nine Northeast and Midwest states have higher indices than any states in the South and West.

In Chapter 2, we saw several ways to summarize a variable. At that time, we noted that the median and quartiles are suitable even for data that may be skewed or have outliers and are usually used together. Along with these three values, we can report the maximum and minimum values. These five values together make up the **5-number summary** of the data: the median, quartiles, and extremes (maximum and minimum). It is a useful, concise summary because it gives a good idea of the center (median), spread (IQR), and extent (range) of the values. For example, the 5-number summary for the daily *Average Wind Speed* for 2011 looks like this:

Max	6.73
Q3	2.28
Median	1.12
Q1	0.46
Min	0.00

1.

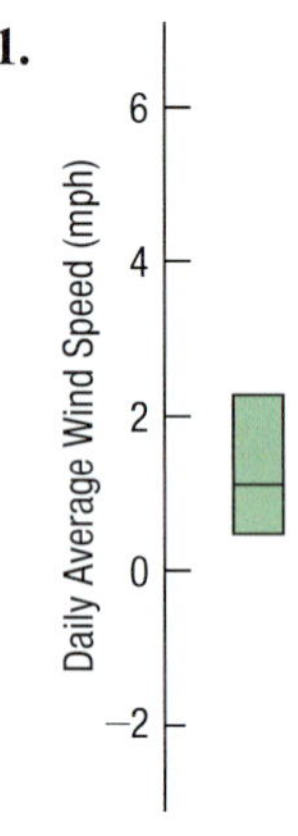

It's a good idea to report the number of data values and the identity of the cases (the *Who*). Here, each case is a day of 2011.

You can see how the 5-number summary provides a good overview of the distribution of the wind speeds. For a start, the median speed is 1.12 mph. The quartiles show that the middle half of days had average wind speeds between 0.46 and 2.28 mph. The IQR is $2.28 - 0.46 = 1.82$ mph. One quarter of the days had average speeds less than 0.46, one quarter of the days had average speeds more than 2.28, and one extraordinary day had an average of 6.73 mph.

Boxplots

Once we have a 5-number summary of a (quantitative) variable, we can display that information in a **boxplot**. To make a boxplot, follow these steps:

2.

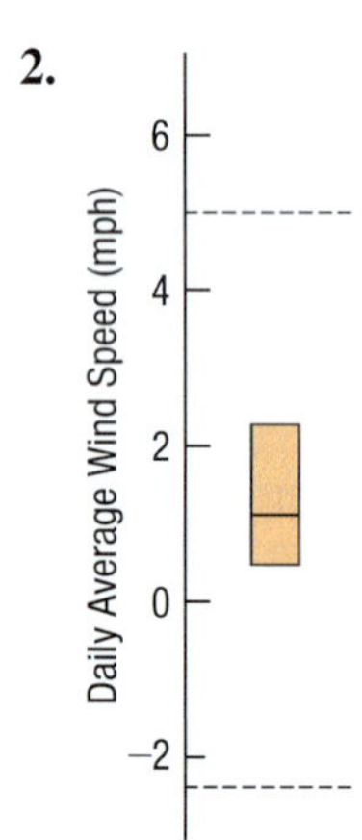

1. Draw a single vertical axis spanning the extent of the data.[1] Draw short horizontal lines at the lower and upper quartiles and at the median. Then connect them with vertical lines to form a box. The box can have any width that looks OK.[2]
2. To construct the boxplot, erect "fences" around the main part of the data. These fences are not part of the final display. If you use technology to make boxplots, you'll never see the fences. They are just outlier "cutoff" values. Place the upper fence 1.5 IQRs above the upper quartile and the lower fence 1.5 IQRs below the lower quartile. For the average wind speed data, we compute

$$\textit{Upper fence} = Q3 + 1.5\,IQR = 2.28 + 1.5(1.82) = 5.01 \text{ mph}$$

and

$$\textit{Lower fence} = Q1 - 1.5\,IQR = 0.46 - 1.5(1.82) = -2.27 \text{ mph}$$

We show the fences here with dotted lines for illustration. You should never include them in your boxplot.

3.

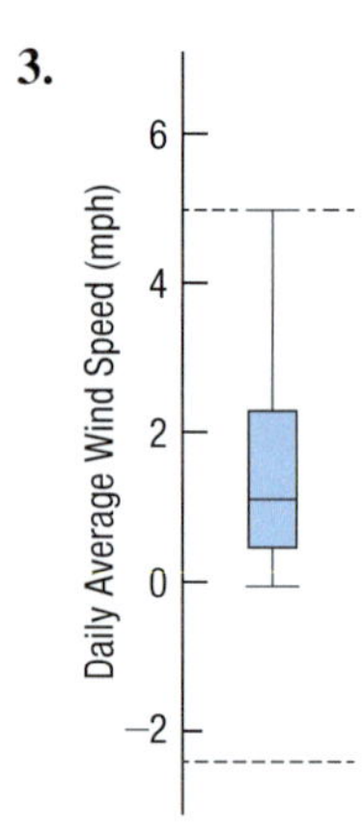

3. We use the fences to grow "whiskers." Draw lines from the ends of the box up and down to *the most extreme data values found within the fences.* If a data value falls outside one of the fences, do *not* connect it with a whisker.
4. Finally, we add the outliers by displaying any data values beyond the fences with special symbols. (We often use a different symbol for **far outliers**—data values farther than 3 IQRs from the quartiles.)

4.

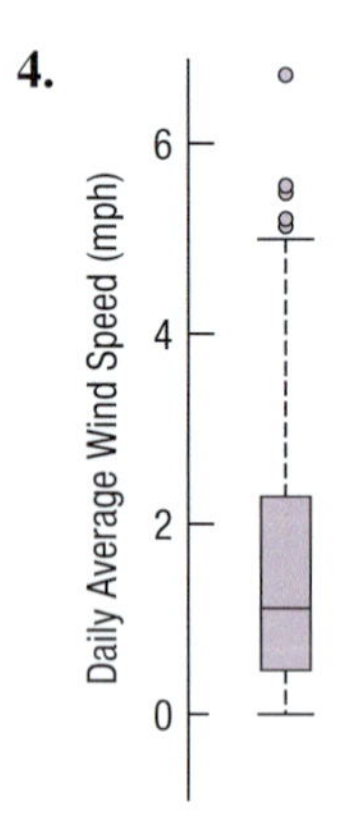

A boxplot highlights several features of the distribution. The central box shows the middle half of the data, between the quartiles. The height of the box is equal to the IQR. If the median is roughly centered between the quartiles, then the middle half of the data is roughly symmetric. If the median is not centered, the distribution is skewed. The whiskers show skewness as well if they are not roughly the same length. Possible outliers are displayed individually, both to keep them out of the way for judging skewness and to encourage you to give them special attention. They may be mistakes, or they may be the most interesting cases in your data. This outlier nomination rule should not be taken as a definition of "outlier." Rather, it is a general guideline to nominate values for special attention because you might decide to treat them as outliers.

For the wind speed data, the central box contains all the days whose speeds are between 0.46 and 2.28 mph. From the shape of the box, it looks like the central part of the distribution is roughly symmetric, but the longer upper whisker indicates that the distribution stretches out at the upper end. We also see the few very windy days we've discussed, including the one with an average wind speed of 6.73 mph. Boxplots are particularly good at pointing out outliers.

Are some months windier than others? Even residents may not have a good idea which parts of the year are the windiest. (Do you know for your hometown?) And what about the spreads? Are wind speeds equally variable from month to month, or do some months show more variation?

[1]The axis could also run horizontally.

[2]Some computer programs draw wider boxes for groups with more cases. That can be useful when comparing those groups.

> ***WHY 1.5 IQRs?***
>
> One of the authors asked the prominent statistician John W. Tukey, the originator of the boxplot, why the outlier nomination rule cut at 1.5 IQRs beyond each quartile. He answered that the reason was that 1 IQR would be too small and 2 IQRs would be too large. That works for us.

Histograms or stem-and-leaf displays are a fine way to look at one distribution or two. But it would be hard to see patterns by comparing 12 histograms. Boxplots offer an ideal balance of information and simplicity, hiding the details while displaying the overall summary information. So we often plot them side by side for groups or categories we wish to compare.

By placing boxplots side by side, you can easily see which groups have higher medians, which have the greater IQRs, where the central 50% of the data is located in each group, and which have the greater overall range. And, when the boxes are in an order, you can get a general idea of patterns in both the centers and the spreads in that order. Equally important, you can see past any outliers in making these comparisons because boxplots display them separately.

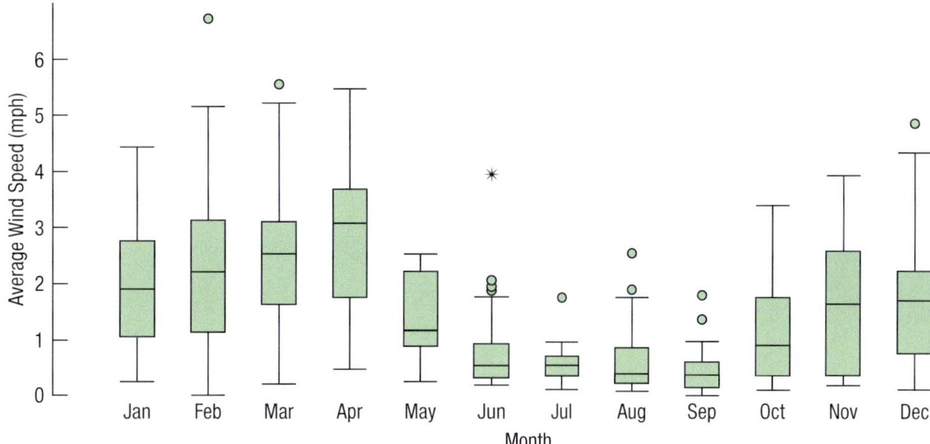

Figure 4.3

Boxplots of the *Average*[3] *Daily Wind Speed* plotted for each *Month* show seasonal patterns in both the centers and spreads. New outliers appear because they are now judged relative to the *Month* in which they occurred.

The boxplots in Figure 4.3 show that wind speeds tend to decrease in the summer. The months in which the winds are both strongest and most variable are October through April.

The boxplot of wind speeds for the entire year nominated only 5 cases as possible outliers. But the monthly boxplots show different outliers than before because days that seemed ordinary when placed against the entire year's data may look like outliers for the month that they're in. Outliers are context dependent. Those windy days in August certainly wouldn't stand out in November or December, but for August, they were remarkable—as we'll soon see.

EXAMPLE 4.2 Comparing Groups with Boxplots

Roller coaster riders want a coaster that goes fast.[4] There are two main types of roller coasters: those with wooden tracks and those with steel tracks. Do they typically run at different speeds? Here are boxplots. (Data in **Coasters 2015**)

---

[3] Here's another warning that the word "average" is used to mean different things. Many "daily average wind speeds" are computed as the midrange—the average of the highest and lowest speeds seen during the day. But the Hopkins Forest values are actually measured continuously all day and averaged over all those observations. When you see the word "average" it is wise to check for the details.

[4] See the Roller Coaster Data Base at www.rcdb.com.

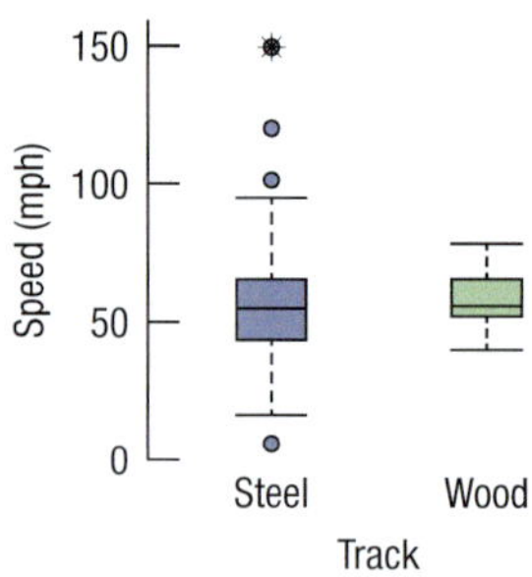

QUESTION: Compare the speeds of wood and steel roller coasters.

ANSWER: The median speed for the wood track coasters is a bit higher than for the steel coasters, but overall their typical speeds are about the same. However, the speeds of the steel track coasters vary much more and include both high and low outliers.

STEP-BY-STEP EXAMPLE

Comparing Groups

Most scientific studies compare two or more groups. It is almost always a good idea to start an analysis of data from such studies by comparing boxplots for the groups. Here's an example:

For her class project, a student compared the efficiency of various coffee cups. She compared four different cups, testing each of them eight different times. Each time, she heated water to 180°F, poured it into a cup, and sealed it. After 30 minutes, she measured the temperature again and recorded the difference in temperature. Because these are temperature differences, smaller differences mean that the liquid stayed hot—just what we would want in a coffee cup. (Data in **Cups**)

QUESTION: What can we say about the effectiveness of these four cups?

THINK **PLAN** State what you want to find out.

I want to compare the effectiveness of four different brands of cups in maintaining temperature. I have eight measurements of Temperature Change for a single example of each of the brands.

VARIABLES Identify the *variables* and report the W's.

Be sure to check the appropriate condition.

✔ **Quantitative Data Condition**: The Temperature Changes are quantitative with units of °F. Boxplots and summary statistics are appropriate for comparing the groups.

SHOW **MECHANICS** Report the 5-number summaries of the four groups. Include the IQR as a check.

	Min	Q1	Median	Q3	Max	IQR
CUPPS	6°F	6	8.25	14.25	18.50	8.25
Nissan	0	1	2	4.50	7	3.50
SIGG	9	11.50	14.25	21.75	24.50	10.25
Starbucks	6	6.50	8.50	14.25	17.50	7.75

Make a picture. Because we want to compare the distributions for four groups, boxplots are an appropriate choice.

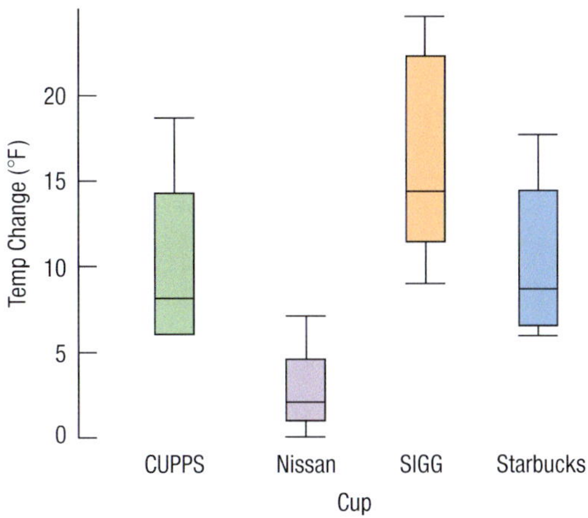

TELL

CONCLUSION Interpret what the boxplots and summaries say about the ability of these cups to retain heat. Compare the shapes, centers, and spreads, and note any outliers.

The individual distributions of temperature changes are all slightly skewed to the high end. The Nissan cup does the best job of keeping liquids hot, with a median loss of only 2°F, and the SIGG cup does the worst, typically losing 14°F. The difference is large enough to be important: A coffee drinker would be likely to notice a 14° drop in temperature. And the cups are clearly different: 75% of the Nissan tests showed less heat loss than any of the other cups in the study. The IQR of results for the Nissan cup is also the smallest of these test cups, indicating that it is a consistent performer.

JUST CHECKING

The Bureau of Transportation Statistics of the U.S. Department of Transportation collects and publishes statistics on airline travel (www.transtats.bts.gov). Here are three displays of the % of flights arriving on time each month from 1994 through June 2016. (Data in **Flights on time 2016**)

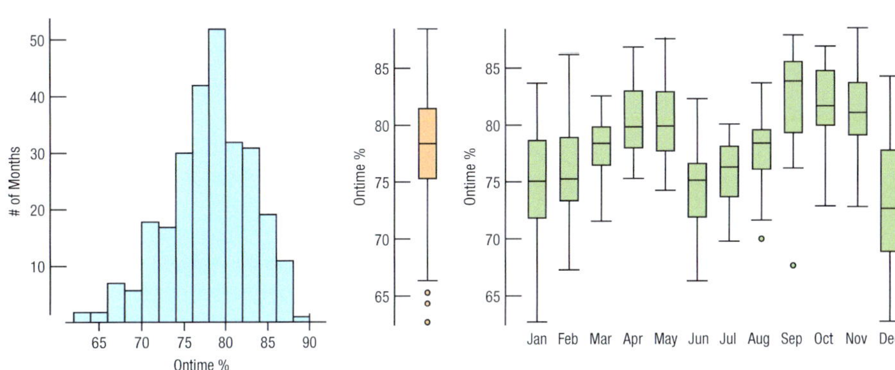

1. Describe what the histogram says about on-time arrivals.
2. What does the boxplot of on-time arrivals suggest that you can't see in the histogram?
3. Describe the patterns shown in the boxplots by month. At what time of year are flights most likely to be on time? Can you suggest reasons for this pattern?

RANDOM MATTERS

John Beale of Stanford, California, was convinced that cars were speeding on his street where the posted speed limit was 20 mph. Being an engineer, he recorded the speed, direction, and acceleration of every car passing his house for several months. We have a sample of 500 speeds and their directions (Up or Down the street) in **Car Speeds**.

Here are side-by-side boxplots of the speeds in the two directions:

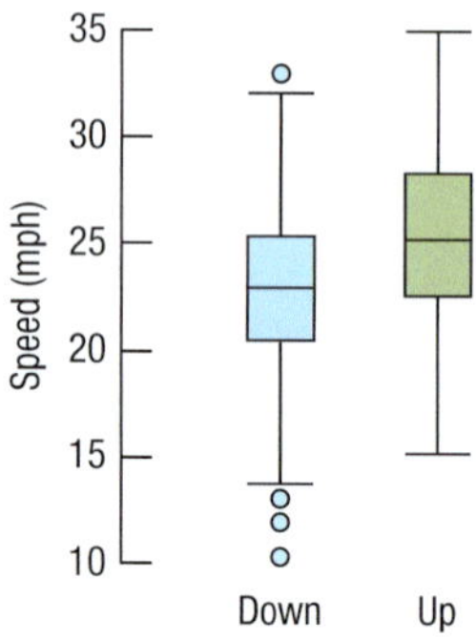

Figure 4.4
Side-by-side boxplots show that cars may generally be going faster up the street than down.

John thought that cars seemed to be going faster up the street and he wondered if that was true. The boxplots show a slight difference: The mean of the 250 speeds going up is 25.25 mph compared to 22.72 mph for the 250 speeds going down. That difference of about 2.53 mph is not that much, but maybe enough to be noticeable.

But, not all cars travel at the same speed. Some of the cars are going over 30 mph, while others are going under 15 mph. Could the apparent difference in the mean speeds be due just to chance, and not to the direction?

How big a difference should we expect to see between the mean speed of cars going up the street and those going down the street? Well, suppose we didn't know what direction each car was driving and just shuffled the directions and assigned a direction at random to each car.[5] There would still probably be some difference in speeds between the "up-street" cars and the "down-street" cars. Another shuffling of the directions would give another difference in mean speeds. And we could shuffle the directions many times to get an idea of the likely size of those differences.

Well, we did just that 10,000 times using a computer program such as the one at astools.datadesk.com. Figure 4.5 on the next page shows a histogram of the differences in speeds for the randomly shuffled groups. Now the question is whether the difference of 2.53 mph in the data seems large compared to what would happen at random if direction didn't matter.

The histogram makes it clear that random groups generally produce differences in mean speeds of 1 mph or less, in either direction. A difference of 2.53 mph would be very unlikely, so this provides strong evidence that the observed difference was not due to chance and that the mean speeds in the two directions are really different.

This sort of logic will be used throughout the text. The amazing fact, which we'll see later, is that we could have predicted how the distribution of differences looks without actually doing the simulation. That's one of the really cool things about statistics. By using mathematics, in many cases we can predict how the simulation will come out and draw conclusions without having to actually do the simulation. Both types of reasoning, with simulation and with mathematics, are important in the chapters that follow.

[5]Here "random" just means that each car has the same chance to be one of those headed in either direction.

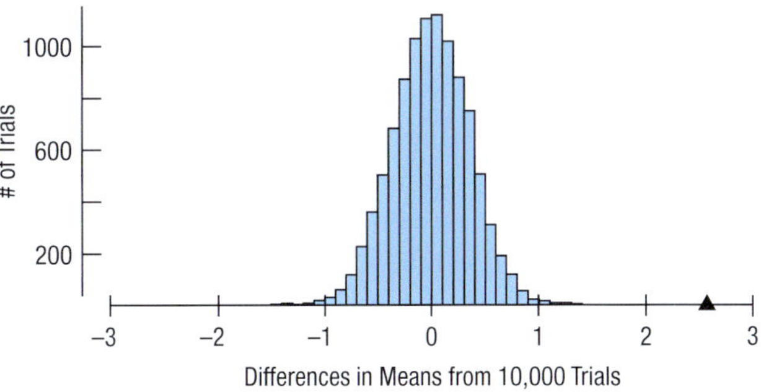

Figure 4.5
From the distribution of differences in means of randomly selected groups of cars we can see that a difference of 2.53 mph is highly unlikely. It appears that the direction was responsible for the observed difference.

4.2 Outliers

In the boxplots for the *Average Wind Speed* by *Month*, several days are nominated by the boxplots as possible outliers. Cases that stand out from the rest of the data almost always deserve attention. An outlier is a value that doesn't fit with the rest of the data, but exactly how different it should be to receive special treatment is a judgment call. Boxplots provide a rule of thumb to highlight these unusual cases, but that rule is only a guide, and it doesn't tell you what to do with them.

Outliers arise for many reasons. They *may* be the most important values in the dataset, pointing out an exceptional case or illuminating a pattern by being the exception to the rule. They may be values that just happen to lie above the limits suggested by the boxplot outlier nomination rule of thumb. Or they may be errors. A decimal point may have been misplaced, digits transposed, or digits repeated or omitted. Some outliers are obviously wrong. For example, if a class survey includes a student who claims to be 170 inches tall (about 14 feet, or 4.3 meters), you can be sure that's an error. But, maybe the units were wrong. (Did the student mean 170 centimeters—about 65 inches?) There are many ways errors can creep into datasets and such errors occur remarkably often. If you can identify the correct value, then you should certainly use it. One important reason to look into outliers is to correct errors in your data.

14-YEAR-OLD WIDOWERS?

Two researchers, Ansley Coale and Fred Stephan, looking at data from the 1950 census, noticed that the number of widowed 14-year-old boys had increased from 85 in 1940 to a whopping 1600 in 1950. The number of divorced 14-year-old boys had increased, too, from 85 to 1240. Oddly, the number of teenaged widowers and divorcees *decreased* for every age group after 14, from 15 to 19. When Coale and Stephan also noticed a large increase in the number of young Native Americans in the Northeast United States, they began to look for the cause of these problems. Data in the 1950 census were recorded on computer cards. Cards are hard to read and mistakes are easy to make. It turned out that data punches had been shifted to the right by one column on hundreds of cards. Because each card column meant something different, the shift turned 43-year-old widowed males into 14-year-olds, 42-year-old divorcees into 14-year-olds, and children of white parents into Native Americans. Not all outliers have such a colorful (or famous) story, but it is always worthwhile to investigate them. And, as in this case, the explanation is often surprising. (A. Coale and F. Stephan, "The case of the Indians and the teen-age widows," *J. Am. Stat. Assoc.* 57 [Jun 1962]: 338–347.)

Values that are large (or small) in one context may not be remarkable in another. The boxplots of *Average Wind Speed* by *Month* show possible outliers in several of the months. The windiest day in February was an outlier not only in that windy month, but for the entire year as well. The windiest day in June was a "far" outlier—lying more than 3 IQRs from the upper quartile for that month, but it wouldn't have been unusual for a winter day. And, the windiest day in August seemed much windier than the days around it even though it was relatively calm.

Many outliers are not wrong; they're just different. And most repay the effort to understand them. You can sometimes learn more from the extraordinary cases than from summaries of the entire dataset.

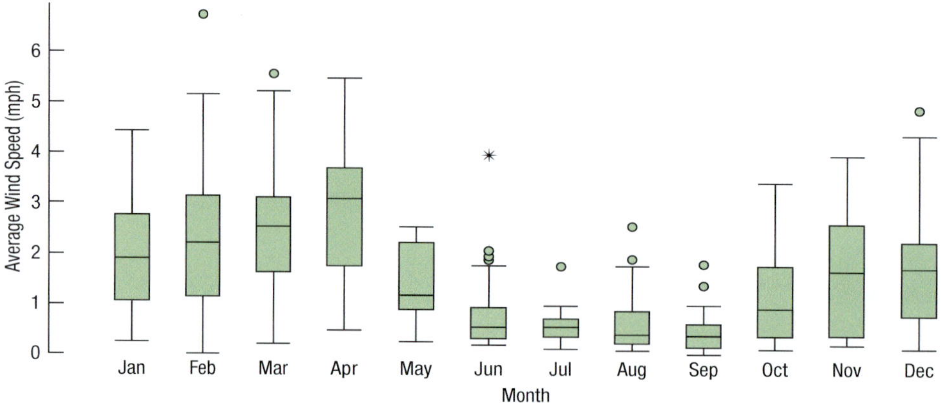

What about those outliers in the Hopkins wind data? That blustery day in February turned out to be a blast that brought four days of subzero temperatures $(-17°\text{F})$ to the region. John Quinlan of the National Weather Service predicted at the time: "The Berkshires have definitely had some of the highest gusts with this wind event. This is definitely the worst we'll be seeing."

What about that extraordinary June day? It was June 2. A search for weather events on the Internet for June 2, 2011, finds a rare tornado in western Massachusetts. The *Berkshire Eagle* reported:

> It probably never occurred to most Berkshire residents horrified by the photos and footage of recent tornado damage in Alabama and Missouri that tornadoes would soon strike close to home. They did, of course, on Wednesday, sowing destruction and in some cases costing lives in Springfield, West Springfield, Monson and other towns near enough to us that the destruction had a sobering effect.

Finally, let's look at the more extreme outlier in August. It wouldn't be remarkable in most other months. It turns out that that was Hurricane Irene, whose eye passed right over the Hopkins Forest. According to *The New York Times*, "It was probably the greatest number of people ever threatened by a single storm in the United States." In fact, the average wind speed that day, 2.53 mph, was much lower than the predictions for Irene had warned.

In the aftermath of Hurricane Irene, the Hoosic River in western Massachusetts rose more than 10 feet over its banks, swallowing portions of Williams College, including the soccer and baseball fields.

Not all outliers are as dramatic as Hurricane Irene, but all deserve attention. If you can correct an error, you should do so (and note the correction). If you can't correct it, or if you confirm that it is correct, you can simply note its existence and leave it in the dataset. But the safest path is to report summaries and analyses with *and* without the outlier so that a reader can judge the influence of the outlier for him- or herself.

There are two things you should *never* do with outliers. You should not leave an outlier in place and proceed as if nothing were unusual. Analyses of data with outliers are very likely to be wrong. The other is to omit an outlier from the analysis without comment. If you want to exclude an outlier, you must announce your decision and, to the extent you can, justify it. But a case lying just over the fence suggested by the boxplot may just be the largest (or smallest) value at the end of a stretched-out tail. A histogram is often a better way to see more detail about how the outlier fits in (or doesn't) with the rest of the data, and how large the gap is between it and the rest of the data.

EXAMPLE 4.3

Checking Out the Outliers

RECAP: We've looked at the speeds of roller coasters and found a difference between steel- and wooden-track coasters. We also noticed an extraordinarily high value.

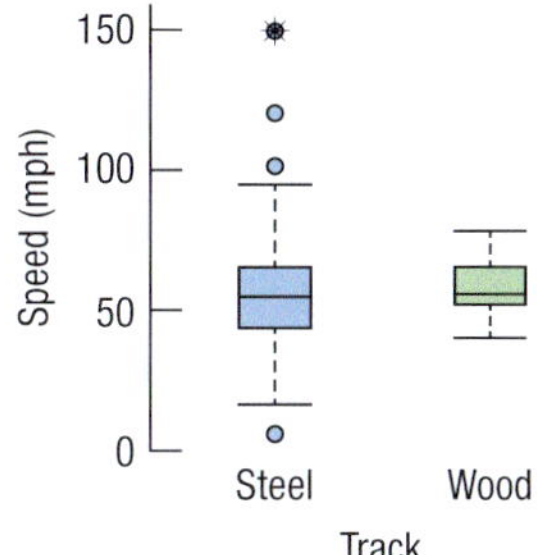

QUESTION: The fastest coaster in this collection turns out to be "Formula Rossa" at Ferrari World in Abu Dhabi. What might make this roller coaster unusual? You'll have to do some research, but that's often what happens with outliers.

ANSWER: Formula Rossa is easy to find in an Internet search. We learn that it is a "hydraulic launch" coaster. That is, it doesn't get its remarkable speed just from gravity, but rather from a kick-start by a hydraulic piston. That could make it different from the other roller coasters. It accelerates to 149 miles per hour (240 km/h) in 4.8 seconds. Riders must wear safety goggles.

4.3 Timeplots: Order, Please!

The Hopkins Forest wind speeds are reported as daily averages. Previously, we grouped the days into months or seasons, but we could look at the wind speed values day by day. Whenever we have data measured over time, it is a good idea to look for patterns by plotting the data in time order. Here are the daily average wind speeds plotted over time:

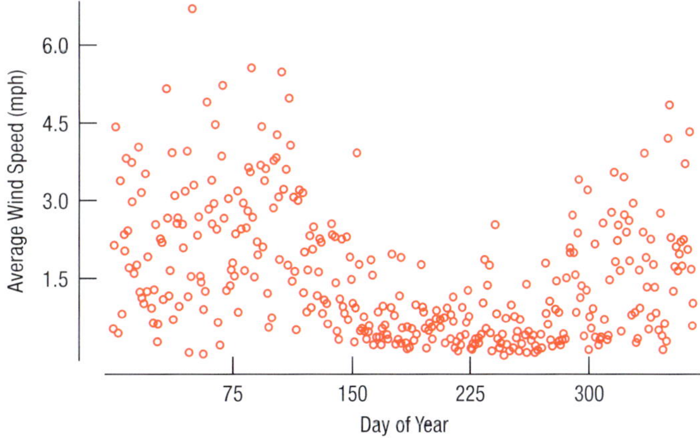

Figure 4.6
A timeplot of *Average WInd Speed* shows the overall pattern and changes in variation.

A display of values against time is sometimes called a **timeplot.** This timeplot reflects the pattern that we saw when we plotted the wind speeds by month. But without the arbitrary divisions between months, we can see a calm period during the summer, starting around day 150 (the beginning of June), when the wind is relatively mild and doesn't vary greatly from day to day. We can also see that the wind becomes both more variable and stronger during the late fall and winter months.

*Smoothing Timeplots

Timeplots often show a great deal of point-to-point variation, as this one does. We usually want to see past this variation to understand any underlying smooth trends, and then also think about how the values vary around that trend—the timeplot version of center and spread. You could take your pencil, squint at the plot a bit to blur things, and sketch a smooth curve. And you'd probably do a pretty good job. But we can also ask a computer to do it for us.

There are many ways for computers to run a smooth trace through a timeplot. Some provide an equation that gives a typical value for any given time point; others just offer a smooth trace. For our purposes here, it doesn't matter all that much. You can try the methods offered by your favorite program and pick the one that looks best for the timeplot you are looking at and the questions you'd like to answer.

You'll often see timeplots drawn with all the points connected, especially in financial publications. This trace through the data falls at one extreme, following every bump. For the wind speeds, the daily fluctuations are so large that connecting the points doesn't help us a great deal:

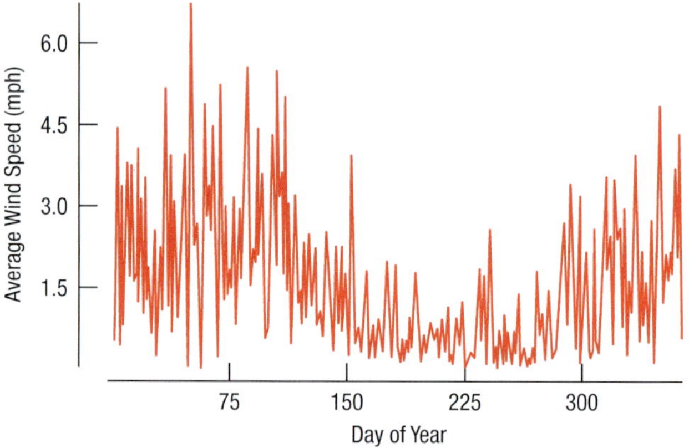

Figure 4.7
The *Average Wind Speeds* of Figure 4.6, drawn by connecting all the points. Sometimes this can help in seeing the underlying pattern, but here there is too much daily variation for this to be very useful.

A smooth trace can highlight long-term patterns and help us see them through the more local variation. Figure 4.8 shows the daily average wind speed values with a smooth trace found by a method called *lowess*, available in many statistics programs.

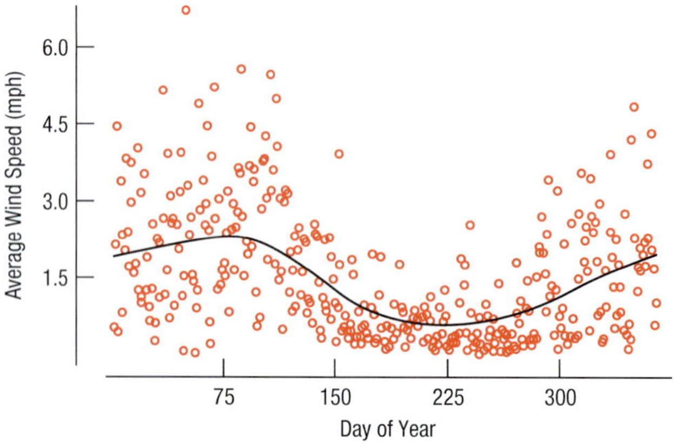

Figure 4.8
The *Average Wind Speeds* with a smooth trace added to help your eye see the annual pattern.

With the smooth trace, it's a bit easier to see a pattern. The trace helps our eye follow the main trend and alerts us to points that don't fit the overall pattern.

EXAMPLE 4.4 Smoothing Timeplots

Many devices now make collecting personal fitness data easy and automatic, but one of the authors has been collecting data on his swimming, cycling, running, and other activities for years. A histogram of his *Weight* over the past three years shows a unimodal, fairly symmetric distribution with a mean of 171.8 lb and a standard deviation of 2.95 lb.

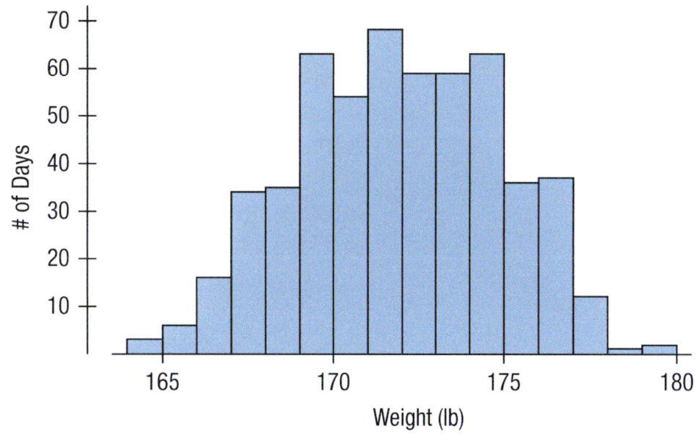

Here's a timeplot of the same data with a smooth trace through it:

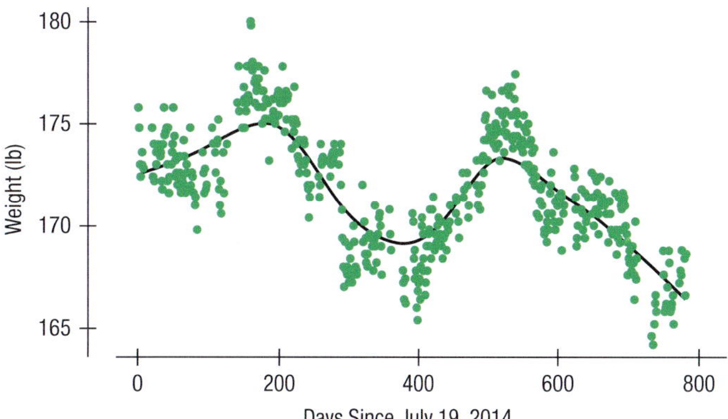

QUESTION: What does the timeplot say about the pattern of weight throughout the year?

ANSWER: There are clear patterns throughout the year. The weight increases during the winter, peaking around Christmas time, and then decreasing in the middle of the summer. Displaying data by month makes the pattern clear:

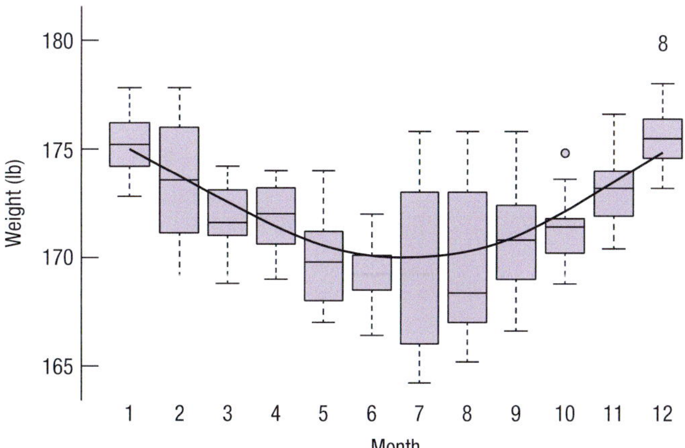

HOW DO SMOOTHERS WORK?

If you planned to go camping in the Hopkins Forest around July 15 and wanted to know how windy it was likely to be, you'd probably look at a typical value for mid-July. Maybe you'd consider all the wind speeds for July, or maybe only those from July 10th to July 20th. That's just the kind of thing we do when we smooth a timeplot.

One simple way to smooth is with a moving average. To find a smooth value for a particular time, we average the values around that point in an interval called the "window." To find the value for the next point, we *move* the window by one point in time and take the new *average*. The size of the window you choose affects how smooth the resulting trace will be. For the Hopkins Forest winds, we might use a 5-day moving average. Stock analysts often use a 50- or 200-day moving average to help them (attempt) to see the underlying pattern in stock price movements.

Can we use smoothing to predict the future? We have only the values of a stock in the past, but (unfortunately) none of the future values. We could use the recent past as our window and take the simple average of those values. A more sophisticated method, exponential smoothing, gives more weight to the recent past values and less and less weight to values as they recede into the past.

Looking Into the Future

People have always wanted to peer into the future, and we predict that they will continue to do so. Sometimes that makes sense. Most likely, the Hopkins Forest climate follows regular seasonal patterns. It's probably safe to predict a less windy June next year and a windier December. But we certainly wouldn't predict another hurricane on August 28.

Other patterns are riskier to extend into the future. If a stock has been rising, will it continue to go up? No stock has ever increased in value indefinitely, and no stock analyst has consistently been able to forecast when a stock's value will turn around. Stock prices, unemployment rates, and other economic, social, or psychological concepts are much harder to predict than physical quantities. The path a ball will follow when thrown from a certain height at a given speed and direction is well understood. Even some dogs can predict the flight of a frisbee well enough to run to where it will be within reach and catch it. The path interest rates will take is much less clear. Unless you have strong (nonstatistical) reasons for doing otherwise, you should resist the temptation to think that any trend you see will continue, even into the near future.

Statistical models often tempt those who use them to think beyond the data. We'll pay close attention later in this text to understanding when, how, and how much we can justify doing that.

4.4 Re-Expressing Data: A First Look

Re-Expressing to Improve Symmetry

When data are skewed, it can be hard to summarize them simply with a center and spread, and hard to decide whether the most extreme values are outliers or just part of the stretched-out tail. How can we say anything useful about such data? The secret is to *re-express* the data by applying a simple function to each value.

Many relationships and "laws" in the sciences and social sciences include functions such as logarithms, square roots, and reciprocals. Similar relationships often show up in data.

In 1980, large companies' chief executive officers (CEOs) made, on average, about 42 times what workers earned. In the next two decades, CEO compensation soared compared to that of the average worker. By 2008, that multiple had jumped to 344. What does the distribution of the compensation of Forbes 500 companies' CEOs look like? Here are a histogram and boxplot for 2014 compensation.[6] (Data in **CEO Compensation 2014**)

[6] www.glassdoor.com/research/ceo-pay-ratio/

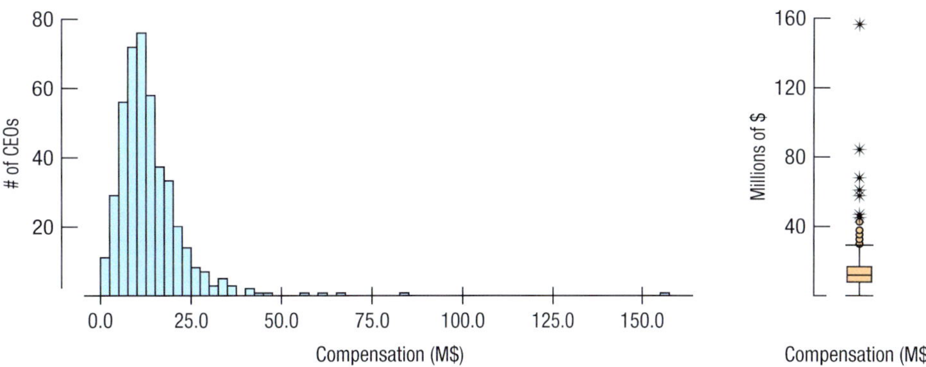

Figure 4.9
CEOs' compensation for the Forbes 500 companies in 2014.

We have data for 434 CEOs and nearly 60 possible histogram bins, about half of which are empty—and several of which contain only one or two CEOs. The boxplot indicates that some CEOs received extraordinarily high pay, while the majority received relatively "little." But look at the values of the bins. The first bin covers compensations of \$0 to \$2,500,000. Imagine receiving a salary survey with these categories:

What is your income?

a) \$0 to \$2,500,000

b) \$2,500,001 to \$5,000,000

c) \$5,000,001 to \$7,500,000

d) More than \$7,500,000

What we *can* see from this histogram and boxplot is that this distribution is highly skewed to the right.

It can be hard to decide what we mean by the "center" of a skewed distribution, so it's hard to pick a typical value to summarize the distribution. What would you say was a typical CEO total compensation? The mean value is \$13,903,006, while the median is "only" \$11,841,179. Each tells us something different about the data.

One approach is to **re-express**, or **transform**, the data by applying a simple function to make the skewed distribution more symmetric. For example, we could take the square root or logarithm of each pay value. Taking logs works pretty well for the CEO compensations, as you can see:[7]

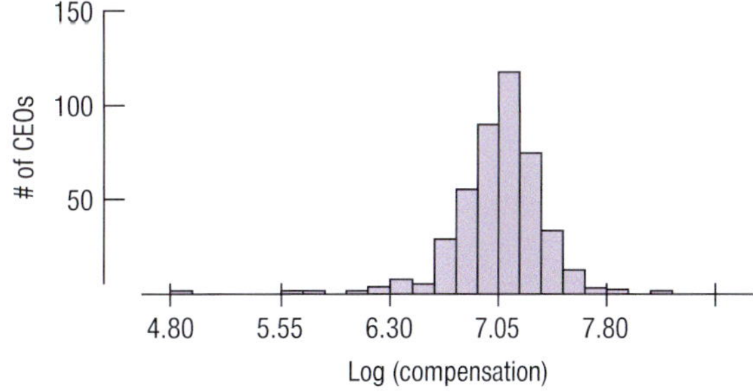

Figure 4.10
The logarithms of 2014 CEO compensations are much more nearly symmetric.

The histogram of the logs of the total CEO compensations is much more nearly symmetric, so we can see that the log compensations are between 5.6, which corresponds to \$610,000, and 8.19, corresponding to about \$156,000,000. And it's easier to talk about a

[7]We've set aside Larry Page (Google), Richard Kinder (Kinder Morgan), and Kosta Kartsotis (Fosil), who had official compensations of \$1, \$1, and \$0, respectively. Richard Hayne of Urban Outfitters is then the low outlier with compensation of \$68,487 and a ratio of only 3.4 times typical worker pay.

typical value for the logs. Both the mean and median log compensations are about 7.0. (That's about $10,000,000 but who's counting?)

Against the background of a generally symmetric main body of data, it's easier to decide whether the largest or smallest pay values are outliers. In fact, the four most highly compensated CEOs are identified as outliers by the boxplot outlier nomination rule of thumb even after this re-expression. It's impressive to be an outlier CEO in annual compensation. It's even more impressive to be an outlier in the log scale!

Variables that are skewed to the right often benefit from a re-expression by square roots, logs, or reciprocals. Those skewed to the left may benefit from squaring the data. Because computers and calculators can do the calculations, re-expressing data is quite easy. Consider re-expression as a helpful tool whenever you have skewed data.

Dealing with Logarithms

You have probably learned about logs in math courses and seen them in psychology or science classes. In this text, we use them only for making data behave better. Base 10 logs are the easiest to understand, but natural logs are often used as well. (Either one is fine.) You can think of base 10 logs as roughly one less than the number of digits you need to write the number. So 100, which is the smallest number to require three digits, has a $\log_{10}$ of 2. And 1000 has a $\log_{10}$ of 3. The $\log_{10}$ of 500 is between 2 and 3, but you'd need a calculator to find that it's approximately 2.7. All salaries of "six figures" have $\log_{10}$ between 5 and 6. Logs are incredibly useful for making skewed data more symmetric. But don't worry—nobody does logs without technology and neither should you. Often, remaking a histogram or other display of the re-expressed data is as easy as pushing another button.

Re-Expressing to Equalize Spread Across Groups

Researchers measured the concentration (nanograms per milliliter) of cotinine in the blood of three groups of people: nonsmokers who have not been exposed to smoke, nonsmokers who have been Exposed To Smoke (ETS), and smokers. Cotinine is left in the blood when the body metabolizes nicotine, so its value is a direct measurement of the effect of passive smoke exposure. The boxplots of the cotinine levels of the three groups tell us that the smokers have higher cotinine levels, but if we want to compare the levels of the passive smokers to those of the nonsmokers, we're in trouble, because on this scale, the cotinine levels for both nonsmoking groups are too low to be seen.

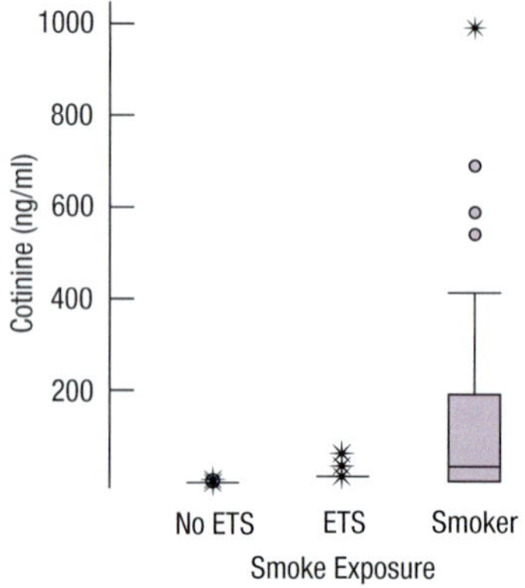

Figure 4.11
Cotinine levels (nanograms per milliliter) for three groups with different exposures to tobacco smoke. Can you compare the ETS (exposed to smoke) and No ETS groups? (Data in **Passive smoke**)

Re-expressing can help alleviate the problem of comparing groups that have very different spreads. For measurements like the cotinine data, whose values can't be negative and whose distributions are skewed to the high end, a good first guess at a re-expression is the logarithm.

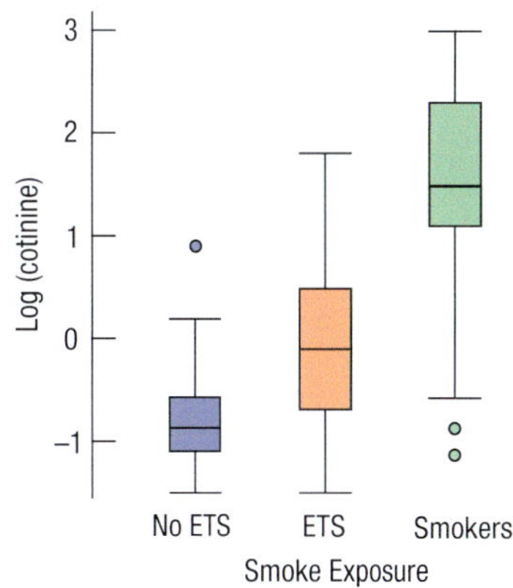

Figure 4.12
Blood cotinine levels after taking logs. What a difference a log makes!

After taking logs, it is easy to see that the nonsmokers exposed to environmental smoke (the ETS group) do show increased levels of (log) cotinine, although not the high levels found in the blood of smokers.

Notice that the same re-expression has also improved the symmetry of the cotinine distribution for smokers and pulled in most of the apparent outliers in all of the groups. It is not unusual for a re-expression that improves one aspect of data to improve others as well. We'll talk about other ways to re-express data as the need arises throughout the text.

WHAT CAN GO WRONG?

- **Avoid inconsistent scales.** Parts of displays should be mutually consistent—no fair changing scales in the middle or plotting two variables on different scales but on the same display. When comparing two groups, be sure to compare them on the same scale.
- **Label clearly.** Variables should be identified clearly and axes labeled so a reader knows what the plot displays.

Here's a remarkable example of a plot gone wrong. It illustrated a news story about rising college costs. It uses timeplots, but it gives a misleading impression. First think about the story you're being told by this display. Then try to figure out what has gone wrong.

What's wrong? Just about everything.

- The horizontal scales are inconsistent. Both lines show trends over time, but exactly for what years? The tuition sequence starts in 1965, but rankings are graphed from 1989. Plotting them on the same (invisible) scale makes it seem that they're for the same years.
- The vertical axis isn't labeled. That hides the fact that it's inconsistent. Does it graph dollars (of tuition) or ranking (of Cornell University)?

  This display violates three of the rules. And it's even worse than that: It violates a rule that we didn't even bother to mention.
- The two inconsistent scales for the vertical axis don't point in the same direction! The line for Cornell's rank shows that it has "plummeted" from 15th place to 6th place in academic rank. Most of us think that's an *improvement,* but that's not the message of this graph.
- **Beware of outliers.** If the data have outliers and you can correct them, you should do so. If they are clearly wrong or impossible, you should remove them and report on them. Otherwise, consider summarizing the data both with and without the outliers.
- **Don't miss an opportunity to re-express.** When the data are severely skewed, or one group has values much different than another, re-expressing a variable can often make an analysis or a display much simpler and clearer.

CONNECTIONS

We discussed the value of summarizing a distribution with shape, center, and spread in Chapter 2, and we developed several ways to measure these attributes. Now we've seen the value of comparing distributions for different groups. Although it can be interesting to summarize a single variable for a single group, it is almost always more interesting to compare groups and look for patterns across several groups. We'll continue to make comparisons like these throughout the rest of our work.

CHAPTER REVIEW

Choose the right tool for comparing distributions.

- Compare the distributions of two or three groups with histograms.
- Compare several groups with boxplots, which make it easy to compare centers and spreads and spot outliers, but hide much of the detail of distribution shape.

Treat outliers with attention and care.

- Outliers are nominated by the boxplot rule, but you must decide what to do with them.
- Track down the background for outliers—it may be informative.

Re-express data to make them easier to work with.

- Re-expression can make skewed distributions more nearly symmetric.
- Re-expression can make the spreads of different groups more nearly comparable.

REVIEW OF TERMS

Outlier Values that are large or small compared to most of the other values in a variable. Whether they are outliers is a judgment call that depends on the context. A boxplot displays values more than 1.5 IQRs beyond the nearest quartile as potential outliers, but that is not a definition of outlier that can be used anywhere. (p. 122)

5-number summary A summary of a variable's distribution that consists of the extremes, the quartiles, and the median (p. 123).

Boxplot A display consisting of a box between the quartiles and "whiskers" extending to the highest and lowest values not nominated as outliers (p. 124).

Far outlier In a boxplot a value more than 3 IQRs beyond the nearest quartile. Such values deserve special attention (p. 124).

Timeplot A timeplot displays data that change over time. Often, successive values are connected with lines to show trends more clearly. Sometimes a smooth curve is added to the plot to help show long-term patterns and trends. (p. 131)

Re-express (Transform) The structure of data may be improved by working with a simple function of the data. The logarithm, square root, and reciprocal are the most common re-expression functions (p. 135).

TECH SUPPORT

Comparing Distributions

Most programs for displaying and analyzing data can display plots to compare the distributions of different groups. Typically, these are boxplots displayed side by side.

Side-by-side boxplots should be on the same y-axis scale so they can be compared.

Some programs offer a graphical way to assess how much the medians differ by drawing a band around the median or by "notching" the boxes.

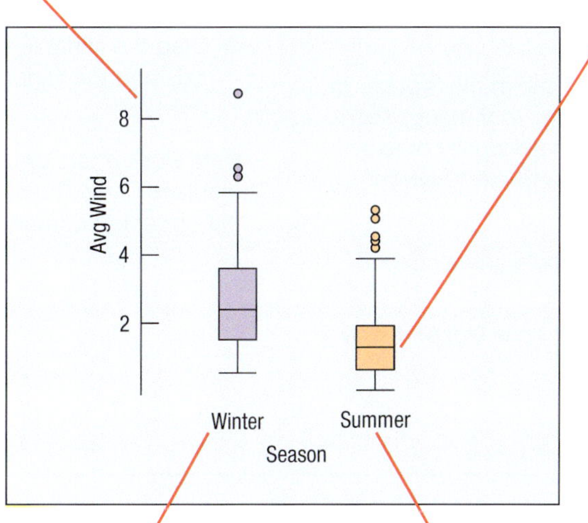

Boxes are typically labeled with a group name. Often they are placed in alphabetical order by group name—not the most useful order.

There are two ways to organize data when we want to compare groups. Each group can be in its own variable (or list, on a calculator). In this form, the experiment comparing cups would have four variables, one for each type of cup:

CUPPS	SIGG	Nissan	Starbucks
6	2	12	13
6	1.5	16	7
6	2	9	7
18.5	3	23	17.5
10	0	11	10
17.5	7	20.5	15.5
11	0.5	12.5	6
6.5	6	24.5	6

But there's another way to think about and organize the data. What is the variable of interest (the *What*) in this experiment? It's the number of degrees lost by the water in each cup. And the *Who* is each time she tested a cup. We could gather all of the temperature values into one variable and put the names of the cups in a second variable listing the individual results, one on each row. Now, the *Who* is clearer—it's an experimental run, one row of the table. Most statistics packages prefer data on groups organized in this way.

Cup	Temperature Difference	Cup	Temperature Difference
CUPPS	6	SIGG	12
CUPPS	6	SIGG	16
CUPPS	6	SIGG	9
CUPPS	18.5	SIGG	23
CUPPS	10	SIGG	11
CUPPS	17.5	SIGG	20.5
CUPPS	11	SIGG	12.5
CUPPS	6.5	SIGG	24.5
Nissan	2	Starbucks	13
Nissan	1.5	Starbucks	7
Nissan	2	Starbucks	7
Nissan	3	Starbucks	17.5
Nissan	0	Starbucks	10
Nissan	7	Starbucks	15.5
Nissan	0.5	Starbucks	6
Nissan	6	Starbucks	6

That's actually the way we've thought about the wind speed data in this chapter, treating wind speeds as one variable and the groups (whether seasons, months, or days) as a second variable.

DATA DESK

If the data are in separate variables:

- ▶ Click on the Single Variable Boxplot tab in the tool bar or choose **Boxplot side by side** from the Plot menu. Then drag variables into the plot window individually or as a group. The boxes will appear in the order in which the variables were selected.

If the data are a single quantitative variable and a second variable holding group names:

- ▶ Click the By Group Boxplot tab or choose **Boxplot y by x** from the Plot menu.
- ▶ Drag the variable that names the groups to the horizontal axis and the variable that holds the values to the vertical axis.

Data Desk offers options for assessing whether any pair of medians differ. Check the Boxplot Options under the menu in the upper left corner of the boxplot window.

EXCEL

Excel cannot make boxplots.

JMP

- ▶ Choose **Fit y by x**.
- ▶ Assign a continuous response variable to Y, Response and a nominal group variable holding the group names to X, Factor, and click **OK**. JMP will offer (among other things) dotplots of the data.
- ▶ Click the red triangle and, under Display Options, select **Boxplots**.

Note: If the variables are of the wrong type, the display options might not offer boxplots.

MINITAB

- ▶ Choose **Boxplot . . .** from the Graph menu.

If your data are in the form of one quantitative variable and one group variable:

- ▶ Choose **One Y** and **with Groups**.

If your data are in separate columns of the worksheet:

- ▶ Choose **Multiple Y's**.

R

If the data for each group are in separate variables, Y1, Y2, Y3, . . .:

- **boxplot(Y1, Y2, Y3 . . .)** will produce side-by-side boxplots.

If the quantitative values are in variable Y and the grouping values are in categorical variable X:

- **boxplot(Y ~ X)** will produce side-by-side boxplots—one for each level of X.

SPSS

To make a boxplot in SPSS, open the **Chart Builder** from the Graphs menu:

- Click the **Gallery** tab.
- Choose **Boxplot** from the list of chart types.
- Drag a single or 2-D (side-by-side) boxplot onto the canvas.
- Drag a scale variable to the *y*-axis drop zone.
- To make side-by-side boxplots, drag a categorical variable to the *x*-axis drop zone.
- Click **OK**.

STATCRUNCH

To make a boxplot:

- Click on **Graph**.
- Choose **Boxplot**.
- Choose the variable name from the list of Columns.
- Click on **Compute!**

TI-83/84 PLUS

To make a boxplot, set up a STAT PLOT by using the boxplot icon.

To compare groups with boxplots, enter the data in lists, for example, in L1 and L2.

Set up STATPLOT's Plot1 to make a boxplot of the L1 data:

- Turn the plot On.
- Choose the first boxplot icon (so the plot will indicate outliers).
- Specify Xlist: L1 and Freq: 1, and select the Mark you want the calculator to use for displaying any outliers.
- Use ZoomStat to display the boxplot for L1. You can now TRACE to see the statistics in the 5-number summary.
- Then set up Plot2 to display the L2 data. This time when you use ZoomStat with both plots turned on, the display shows the boxplots in parallel.

EXERCISES

SECTION 4.1

T **1. Load factors 2017** The Research and Innovative Technology Administration of the Bureau of Transportation Statistics (www.TranStats.bts.gov/Data_Elements.aspx?Data=2) reports load factors (passenger-miles as a percentage of available seat-miles) for commercial airlines for every month from October 2002 through 2017. Here are histograms and summary statistics for the domestic and international load factors for this time period:

Compare and contrast the distributions.

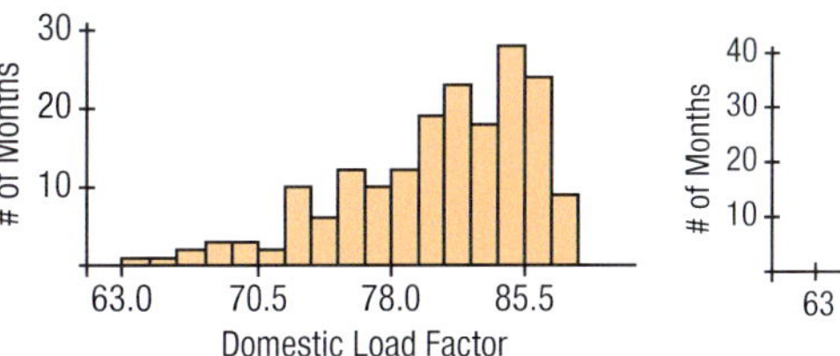

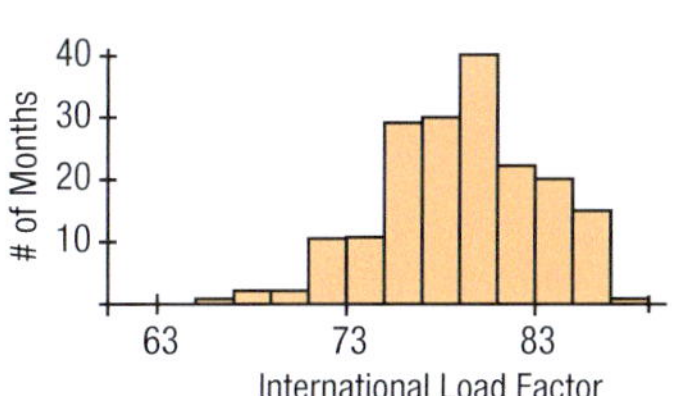

Variable	Count	Mean	Median	StdDev	IQR
Domestic Load Factor	183	80.6441	81.54	5.23262	7.3775
International Load Factor	183	79.2046	79.27	4.22962	5.7900

2. Load factors, 2017 by season We can also compare domestic load factors for September through March versus those for April through August:

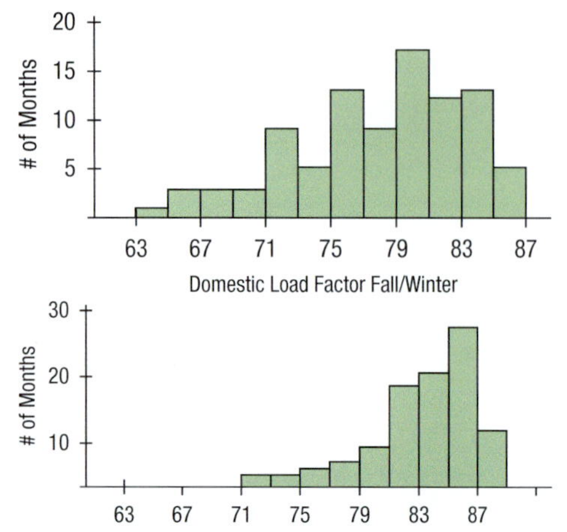

3.

Compare and contrast what these histograms show.

3. Fuel economy The boxplot shows the fuel economy ratings for 67 model year 2012 subcompact cars. Some summary statistics are also provided. The extreme outlier is the Mitsubishi i-MiEV, an electric car whose electricity usage is *equivalent* to 112 miles per gallon.

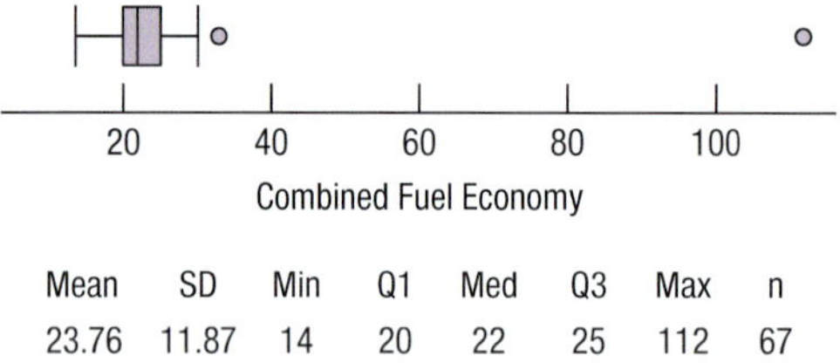

Mean	SD	Min	Q1	Med	Q3	Max	n
23.76	11.87	14	20	22	25	112	67

If that electric car is removed from the dataset, how will the standard deviation be affected? The IQR?

4. Prisons 2014 In a way, boxplots are the opposite of histograms. A histogram divides the number line into equal intervals and displays the number of data values in each interval. A boxplot divides the *data* into equal parts and displays the portion of the number line each part covers. These two plots display the number of incarcerated prisoners in each state at the end of 2014:

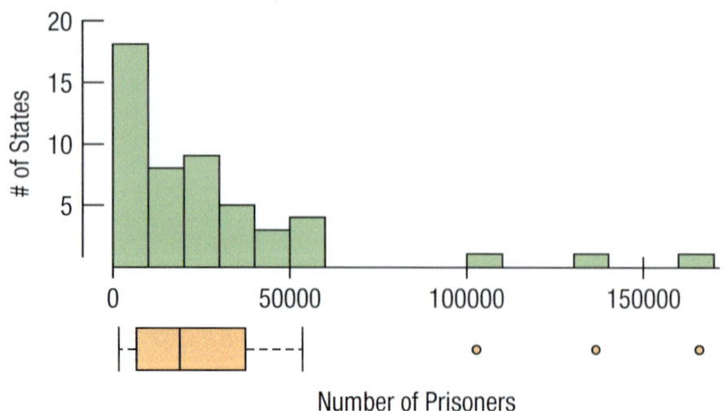

a) Explain how you could tell, by looking at a boxplot, where the tallest bars on the histogram would be located.
b) Explain how both the boxplot and the histogram can indicate a skewed distribution.
c) Identify one feature of the distribution that the histogram shows but the boxplot does not.
d) Identify one feature of the distribution that the boxplot shows but the histogram does not.

5. Load factors 2017 by month Here is a display of the international load factors by month for the period from 2002 to 2017:

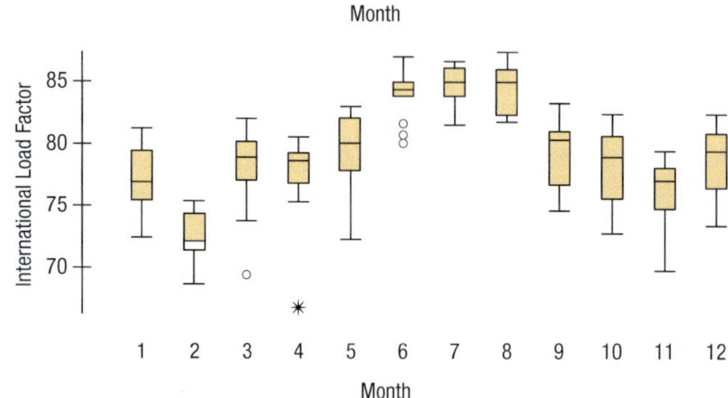

Discuss the patterns you see in this display.

6. Load factors 2017 by year Here is a display of the domestic load factors by year through 2017:

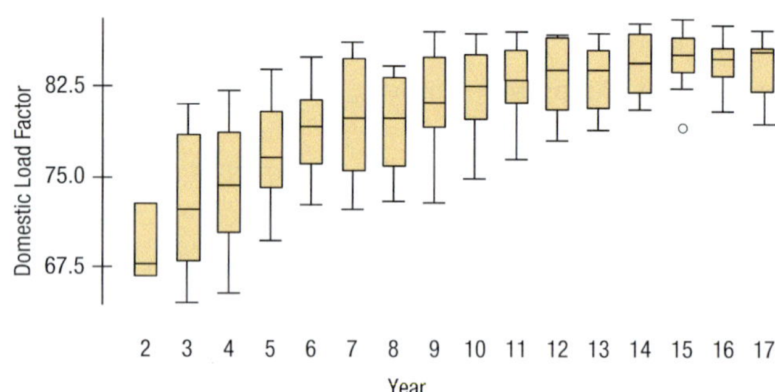

Discuss the patterns you see in this display.

SECTION 4.2

7. Extraordinary months (Data in **Load factors 2017**) Look at the boxplots by month in Exercise 5. The low outlier in April is for April of 2003. At that time, there was concern over the deadly SARS virus, which may have depressed international travel. Do you think the data for April 2003 should be set aside in any overall analysis of load factors? Explain.

8. Load factors 2017 by year again The boxplots of exercise 6 show a single possible outlier. But that month in 2015 would not be an outlier in any of the other months in the display. Do you think it should still be considered an outlier? Explain your reasoning.

SECTION 4.3

T **9. Load factors 2017 over time** Here is a timeplot of each monthly load factor for domestic flights for 2000 to 2013 along with a lowess smooth.

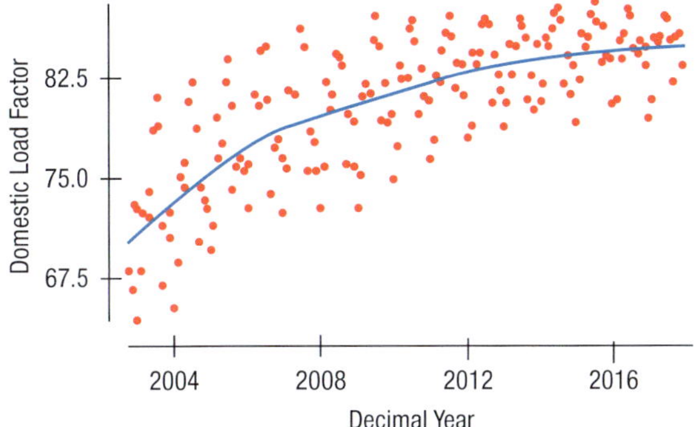

a) Describe the patterns you see in this plot.
b) Do you expect the overall pattern to continue for another decade? Why or why not?

T **10. Load factors 2017 over time, a second look** Here is a timeplot of the same domestic load factors as in Exercise 9, but this time with a different smoother (called a median smoother) applied.

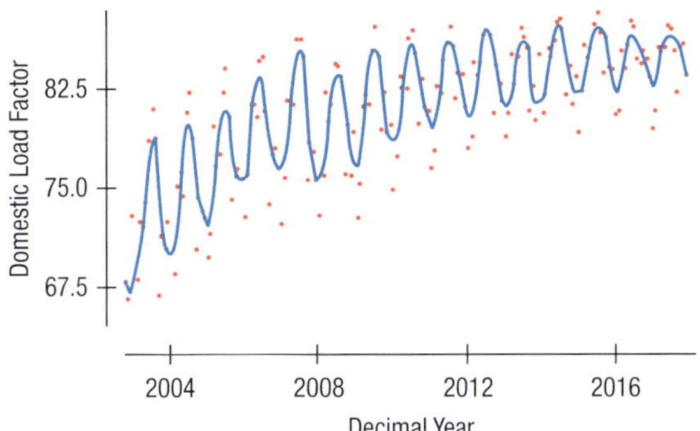

a) What can you see in the time plot with the smoother that you might have missed in the original time plot of exercise 9?
b) Do you expect the patterns shown here to persist into the future?

SECTION 4.4

11. Exoplanets Discoveries of planets beyond our solar system have grown rapidly. Here is a histogram showing the distance (in light-years) from earth to stars that have known planets (as of 2016):

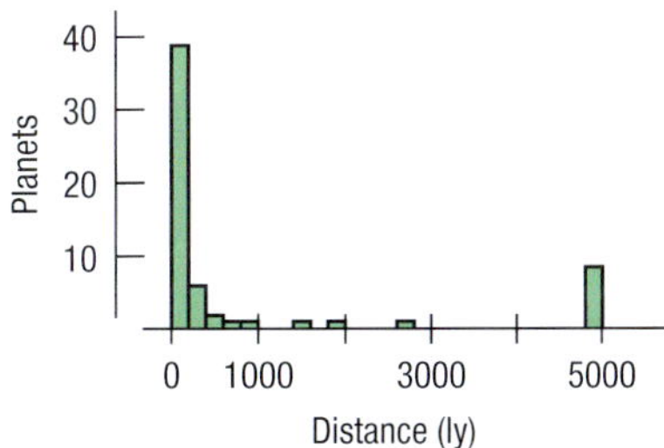

Explain why it might be beneficial to re-express these distances.

12. Exoplanets re-expressed Here are the exoplanet distances of Exercise 11, re-expressed to the log scale:

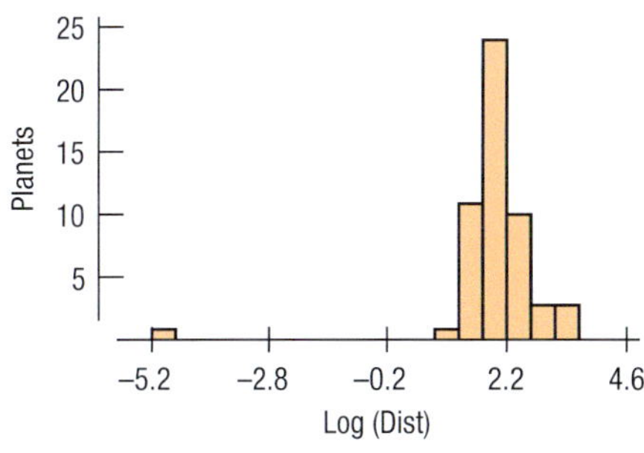

a) Is this a better scale to understand these distances?
b) The low outlier is "sol"– that is, it is the distance of the sun from the earth. Do you think it belongs with these data?

CHAPTER EXERCISES

13. Car values A group of 25 friends compared their car values (in Euro) at the end of 2019 and derived the following statistics.

Median	Minimum	Maximum	Q1	Q3
19685	1475	57183	12507.50	27365.00

a) Which W's of the data can you identify?
b) What kind of display is appropriate to make the data more easily readable?
c) Are there any outliers present in the data? Explain.

14. House prices The following boxplot displays the prices (in British pounds) of 35 houses, 19 of which have a garage and 16 of which do not. Compare the two groups and interpret what you find.

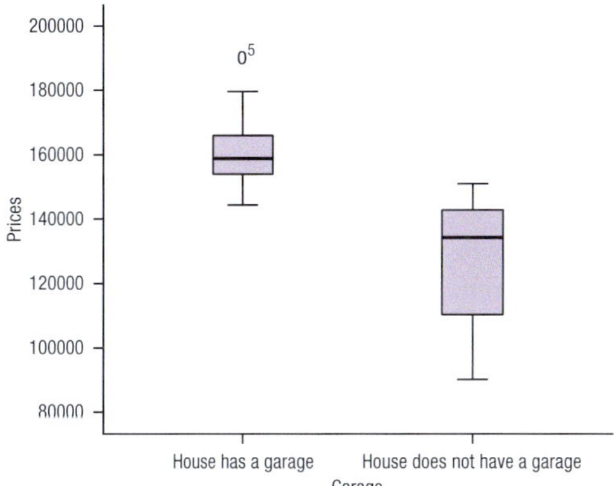

T **15. Cost of living 2016, selected cities** To help travelers know what to expect, researchers collected the prices of commodities in 16 cities throughout the world. Here are boxplots comparing the average prices of a bottle of water, a dozen eggs, and a cappuccino in the 16 cities (prices are all in US$ as of August 2016). (Data in **COL 2016**)

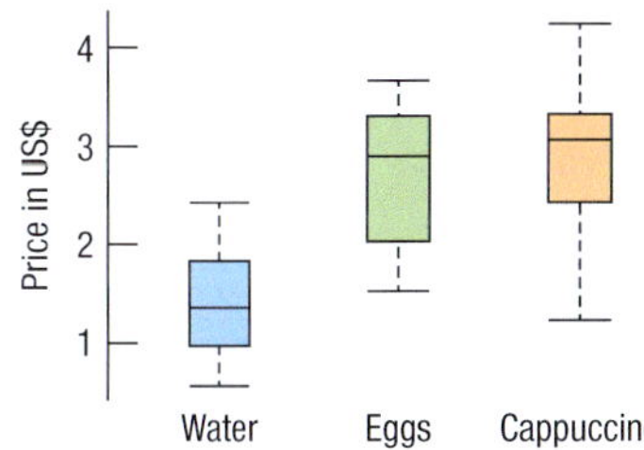

a) In general, which commodity is the most expensive?
b) These are prices for the same 16 cities. But can you tell whether cappuccino is more expensive than eggs in all cities? Explain.

16. Cost of living 2016, more cities Here are the same three prices as in Exercise 15 but for 576 cities around the world. (Prices are all in US$ as of August 2016; data in **COLall 2016**.)

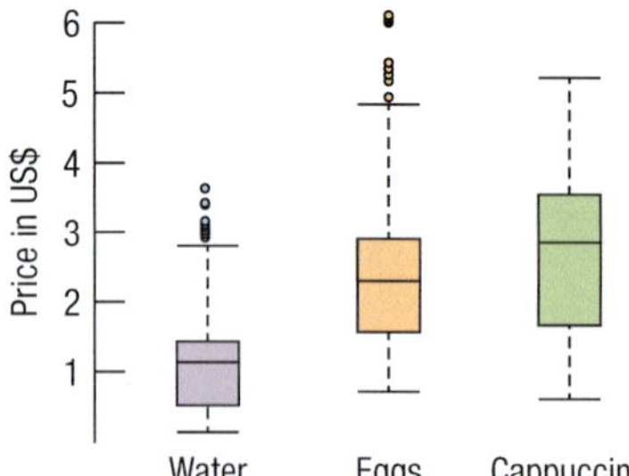

a) In general, which commodity is the most expensive?
b) Is a carton of eggs ever more expensive than a cappuccino? Explain.

17. Rock concert deaths Crowd Management Strategies (www.crowdsafe.com) monitors accidents at rock concerts. In their database, they list the names and other variables of victims whose deaths were attributed to "crowd crush" at rock concerts. Here are the histogram and boxplot of the victims' ages for data from a one-year period:

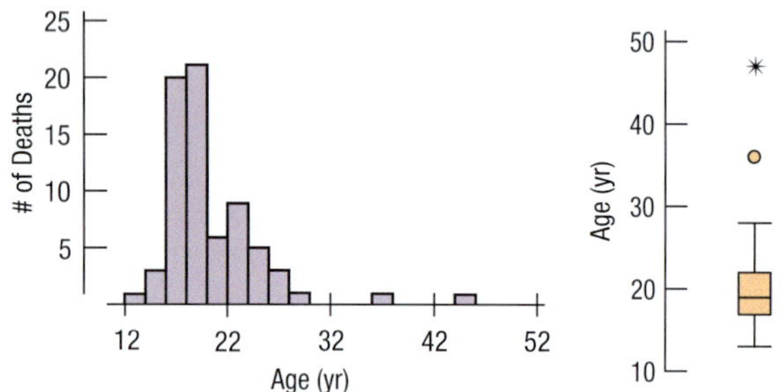

a) What features of the distribution can you see in both the histogram and the boxplot?
b) What features of the distribution can you see in the histogram that you cannot see in the boxplot?
c) What summary statistic would you choose to summarize the center of this distribution? Why?
d) What summary statistic would you choose to summarize the spread of this distribution? Why?

18. Slalom times 2018 The Men's Giant Slalom skiing event consists of two runs whose times are added together for a final score. Two displays of the giant slalom times in the 2018 Winter Olympics at PyeongChang are shown below.

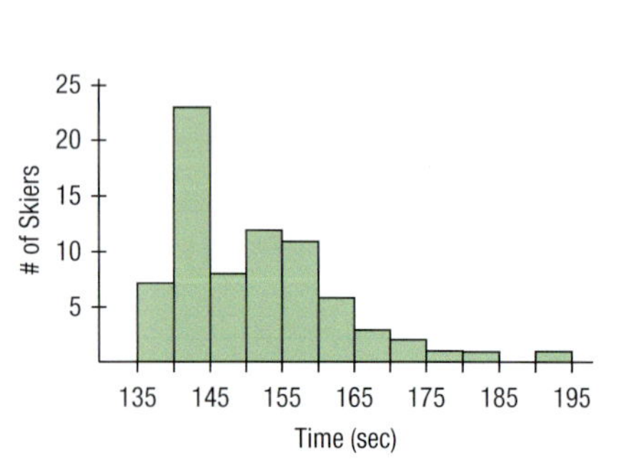

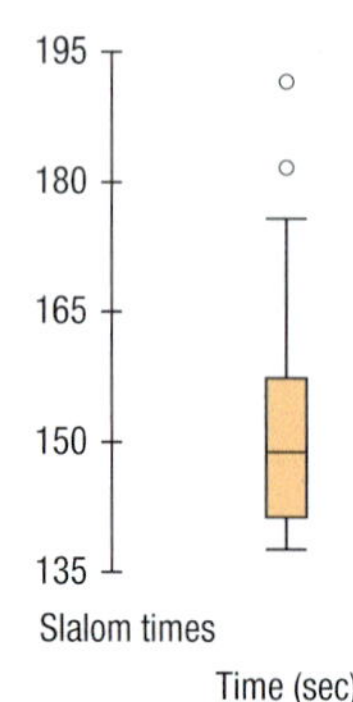

a) What features of the distribution can you see in both the histogram and the boxplot?
b) What summary statistic would you choose to summarize the center of this distribution? Why?
c) What summary statistic would you choose to summarize the spread of this distribution? Why?

19. Sugar in cereals Sugar is a major ingredient in many breakfast cereals. The histogram displays the sugar content as a percentage of weight for 49 brands of cereal. The boxplots compare sugar content for adult and children's cereals.

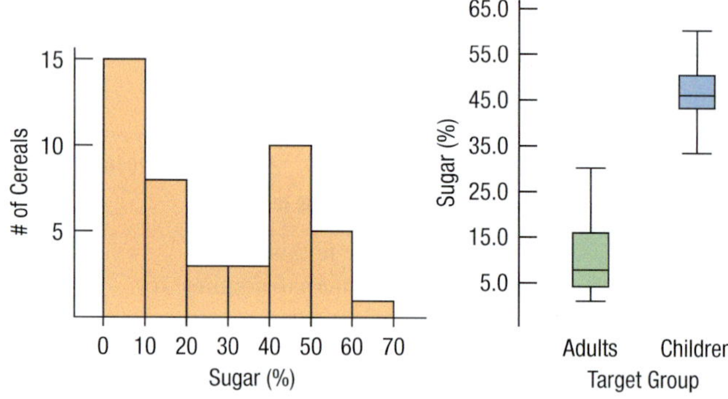

a) What is the range of the sugar contents of these cereals?
b) Describe the shape of the distribution.
c) What aspect of breakfast cereals might account for this shape?
d) Are all children's cereals higher in sugar than adult cereals?
e) Which group of cereals varies more in sugar content? Explain.

20. Tendon transfers People with spinal cord injuries may lose function in some, but not all, of their muscles. The ability to push oneself up is particularly important for shifting position when seated and for transferring into and out of wheelchairs. Surgeons compared two operations to restore the ability to push up in children. The histogram shows scores rating pushing strength two years after surgery, and the boxplots compare results for the two surgical methods. (M. J. Mulcahey, C. Lutz, S. H. Kozen, R. R. Betz, "Prospective evaluation of biceps to triceps and deltoid to triceps for elbow extension in tetraplegia," *Journal of Hand Surgery*, 28, 6, 2003)

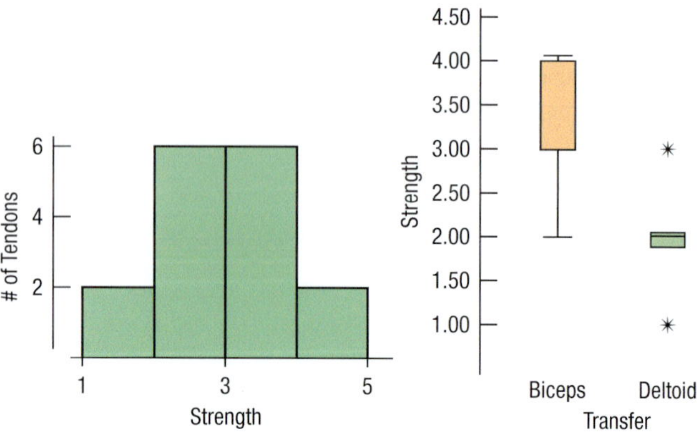

a) Describe the shape of this distribution.
b) What is the range of the strength scores?
c) What fact about the results of the two procedures is hidden in the histogram?
d) Which method had the higher (better) median score?
e) Was that method always better?
f) Which method produced more consistent results? Explain.

T **21.** Population growth 2010, by region (Data in **Population growth 2010**) Below is a "back-to-back" stem-and-leaf display that shows two datasets at once—one going to the left, one to the right. The display compares the percent change in population for two regions of the United States (based on census figures for 2000 and 2010). The fastest growing state was Nevada at 35%. To show the distributions better, this display breaks each stem into two lines, putting leaves 0–4 on one stem and leaves 5–9 on the other.

NE/MW States		S/W States
1	-0	
4433333220	0	145
98777776655	0	8899
32	1	00223344
5	1	57889
	2	11
	2	45
	3	
	3	5

Population Growth rate
(2|1|0 means 12% for a NE/NW state and 10% for a S/W state)

a) Use the data displayed in the stem-and-leaf display to construct comparative boxplots.
b) Write a few sentences describing the difference in growth rates for the two regions of the United States.

T **22.** Shown below are the histogram and summary statistics for the daily number of patients visiting a clinic during a given month.

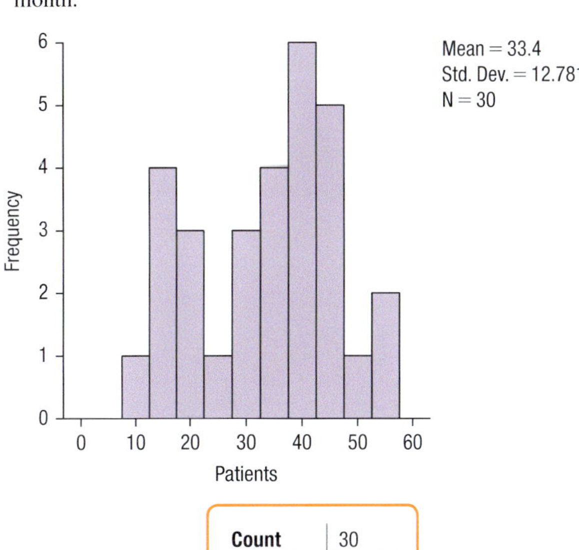

Count	30
Mean	33.40
StdDev	12.781
Median	37
Minimum	10
Maximum	54
Q1	19
Q3	43

a) Which statistics would you use to identify the center and spread of this distribution? Why?
b) Are there any outliers? Explain.
c) Create a boxplot for these data.
d) Write a few sentences describing the distribution.

23. Body temperatures Data from a sample of 65 female patients and 65 male patients on their body temperature (in Fahrenheit) on the last morning during their stay in a short-term hospital are displayed in the following two histograms:

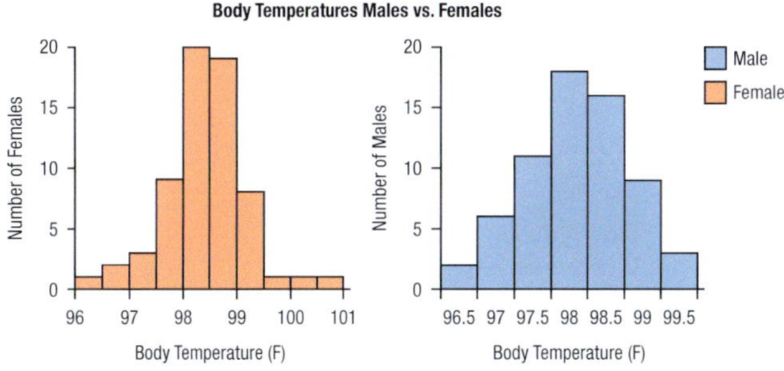

a) What would you change about these histograms to make them easier to compare?
b) Describe these distributions by writing a few sentences comparing the body temperatures of female patients versus those of male patients.
c) Would the shape of the histograms be affected if the body temperatures had to be measured in Celsius rather than Fahrenheit?

24. Deaths 2014 A National Vital Statistics Report (www.cdc.gov/nchs/) provides information on deaths by age, sex, and race. Below are displays of the distributions of ages at death for White and Black males:

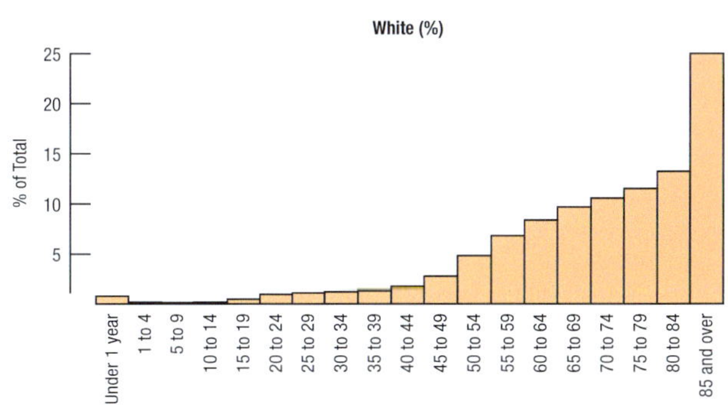

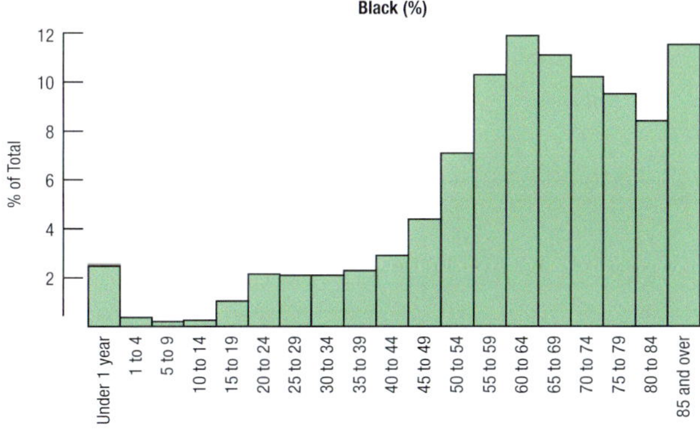

a) Describe the overall shapes of these distributions.
b) How do the distributions differ?
c) Look carefully at the bar definitions. Where do these plots violate the rules for statistical graphs?

25. Women's basketball Here are boxplots of the points scored during the first 10 games of the season for both Scyrine and Alexandra:

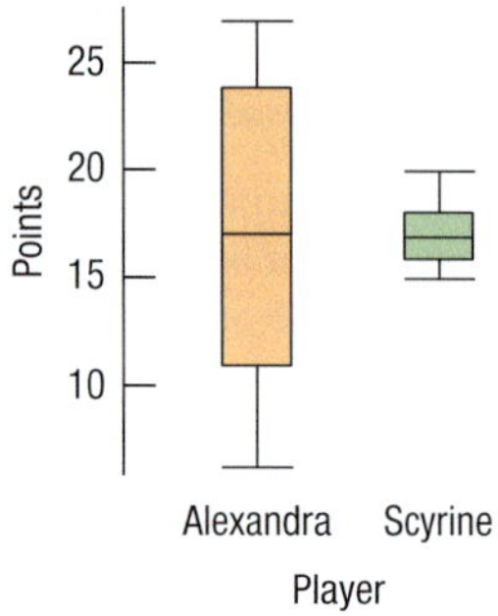

a) Summarize the similarities and differences in their performance so far.

b) The coach can take only one player to the state championship. Which one should she take? Why?

26. Financial service providers The following display shows the distribution of financial service providers (FSP) per inhabitant in 2015 for the countries Austria (AT), Switzerland (CH), Germany (DE), Spain (ES), the Netherlands (NL), and the United Kingdom (UK).

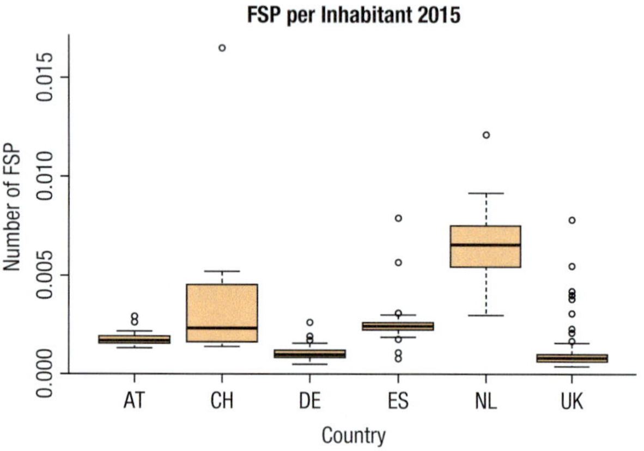

Source: Eurostat 2018a/b/c, Instituto Nacional de Estadística 2018, Centraal Bureau voor de Statistiek 2018, Office of National Statistics 2016, STATcube 2018, Statistik der Unternehmensstruktur 2017, Statistisches Bundesamt 2018, own calculation and presentation

a) Which of the six FSP distributions have outliers?

b) Compare the FSP distribution across the six countries.

27. Market amounts The following two boxplots show the amount (in Euro) spent at a local market by 23 individuals, 11 of whom are males and 12 of whom are females. Write a brief report discussing what these data show.

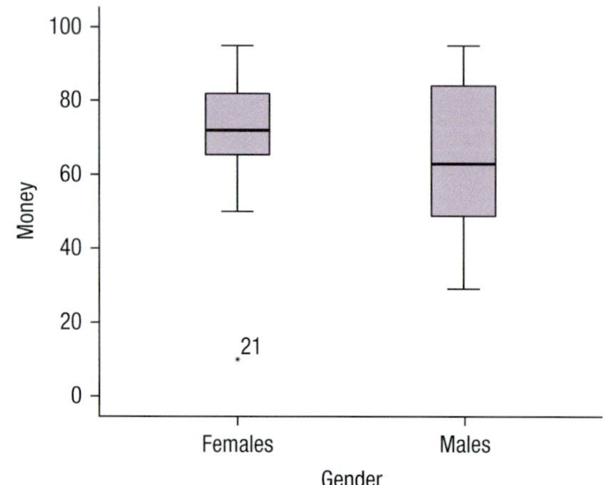

28. Fuel economy and cylinders Describe what these boxplots tell you about the relationship between the number of cylinders a car's engine has and the car's fuel economy (mpg).

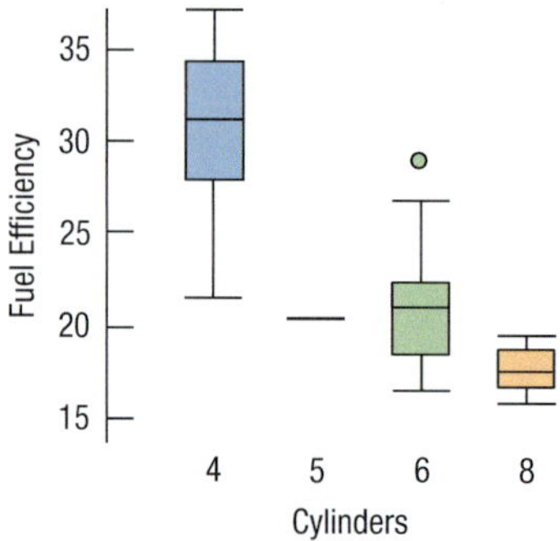

29. Employment rates The following graph displays the EU-28 trends in employment rates for individuals aged 20–64 by gender.

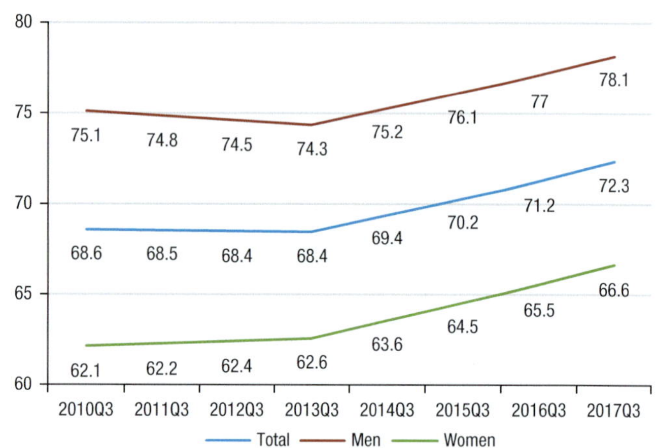

Source: Eurostat, Labour Force Survey

a) Comment on the type of data displayed in the figure.

b) Describe all patterns you see in the figure.

c) What was the highest gender gap in employment rate and in which year did this happen?

30. Ozone Ozone levels (in parts per billion, ppb) were recorded at sites in New Jersey monthly between 1926 and 1971. Here are boxplots of the data for each month (over the 46 years), lined up in order (January = 1):

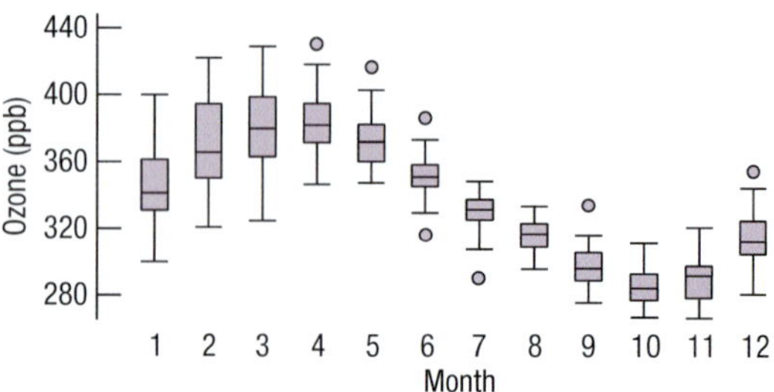

a) In what month was the highest ozone level ever recorded?

b) Which month has the largest IQR?

c) Which month has the smallest range?

d) Write a brief comparison of the ozone levels in January and June.

e) Write a report on the annual patterns you see in the ozone levels.

31. Test scores Three statistics classes all took the same test. Histograms and boxplots of the scores for each class are shown below. Match each class with the corresponding boxplot.

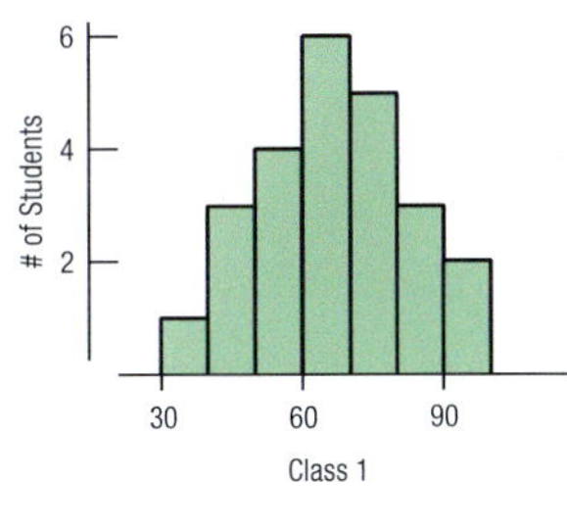

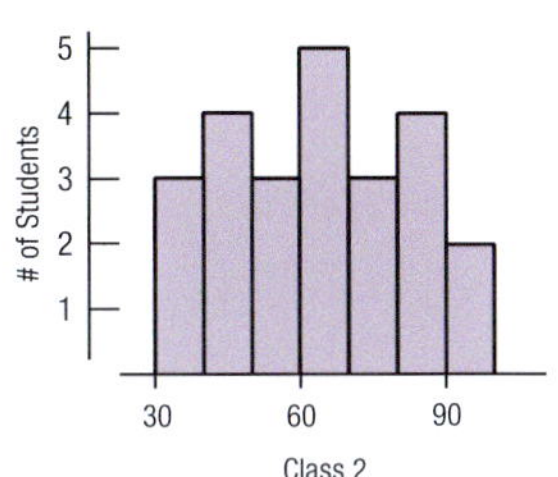

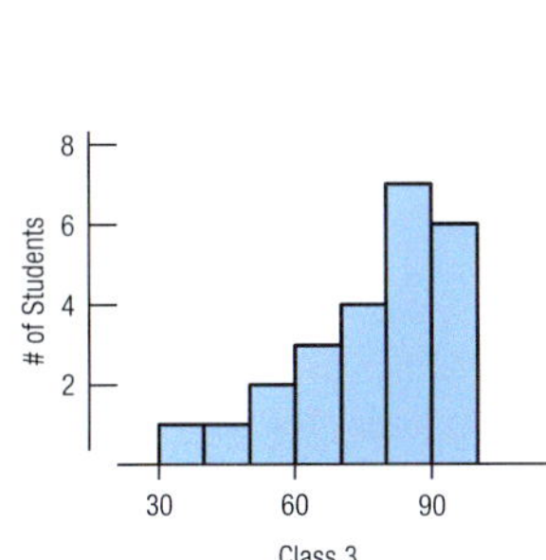

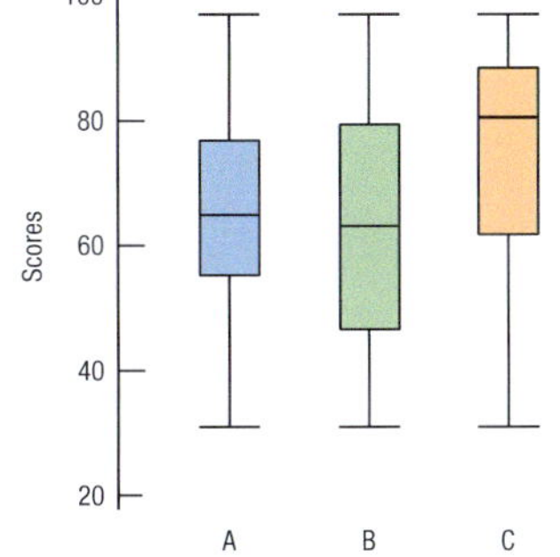

32. Eye and hair color A survey of 1021 school-age children was conducted by randomly selecting children from several large urban elementary schools. Two of the questions concerned eye and hair color. In the survey, the following codes were used:

Hair Color	Eye Color
1 = Blond	1 = Blue
2 = Brown	2 = Green
3 = Black	3 = Brown
4 = Red	4 = Grey
5 = Other	5 = Other

The statistics students analyzing the data were asked to study the relationship between eye and hair color. They produced this plot:

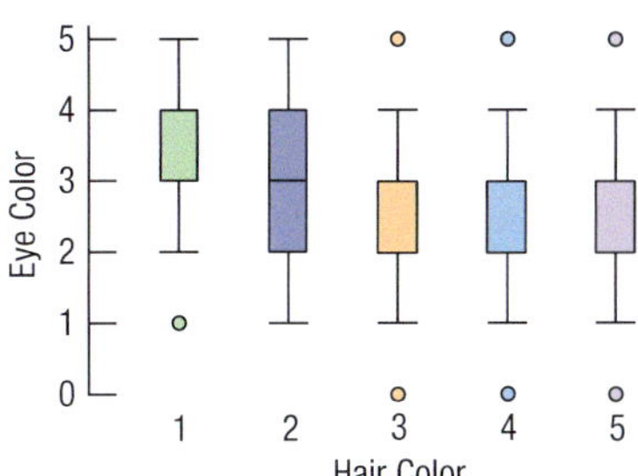

Is their graph appropriate? If so, summarize the findings. If not, explain why not.

33. Graduation? A survey of major universities asked what percentage of incoming freshmen usually graduate "on time" in 4 years. Use the summary statistics given to answer the questions that follow.

	% on Time
Count	48
Mean	68.35
Median	69.90
StdDev	10.20
Min	43.20
Max	87.40
Range	44.20
25th %tile	59.15
75th %tile	74.75

a) Would you describe this distribution as symmetric or skewed? Explain.
b) Are there any outliers? Explain.
c) Create a boxplot of these data.
d) Write a few sentences about the graduation rates.

T **34. Heart beats** Here are summary statistics for the number of heart beats (per minute) of a group of adults:

Count	12
Mean	74.17
Standard Deviation	11.59
Median	71.50
Minimum	55
Maximum	95
Q1	65.25
Q3	83.25

Suppose you didn't have access to the data. Answer the following questions from the summary statistics alone:

a) Would you describe this distribution as symmetric or skewed? Explain.
b) Are there any outliers? Explain.
c) Create a boxplot of these data.
d) Write a few sentences about the heart beats distribution of this group of adults.

35. Caffeine A student study of the effects of caffeine asked volunteers to take a memory test 2 hours after drinking soda. Some drank caffeine-free cola, some drank regular cola (with caffeine), and others drank a mixture of the two (getting a half-dose of caffeine). Here are the 5-number summaries for each group's scores (number of items recalled correctly) on the memory test:

	n	Min	Q1	Median	Q3	Max
No Caffeine	15	16	20	21	24	26
Low Caffeine	15	16	18	21	24	27
High Caffeine	15	12	17	19	22	24

a) Describe the W's for these data.
b) Name the variables and classify each as categorical or quantitative.
c) Create side-by-side boxplots to display these results as best you can with this information.
d) Write a few sentences comparing the performances of the three groups.

36. Anxiety scores Prior to an exam, 30 students in a class took an anxiety test, and the following are the summary statistics for the anxiety scores obtained by gender:

	Count	Mean	Median	SD	Min	Max	Q1	Q3
Male	13	5.08	5.00	3.121	1	10	1.50	7.50
Female	17	6.00	6.00	3.000	1	10	3.50	8.50

a) Create side-by-side boxplots comparing the anxiety scores of male and female students from the information given.
b) Write a brief report on these results. Be sure to discuss the shape, center, and spread of the scores.

T **37. Derby speeds 2017** (Data in **Kentucky Derby 2017**) How fast do horses run? Kentucky Derby winners run well over 30 miles per hour, as shown in this graph. The graph shows the percentage of Derby winners that have run *slower* than each given speed. Note that few have won running less than 33 miles per hour, but about 86% of the winning horses have run less than 37 miles per hour. (A cumulative frequency graph like this is called an "ogive.")

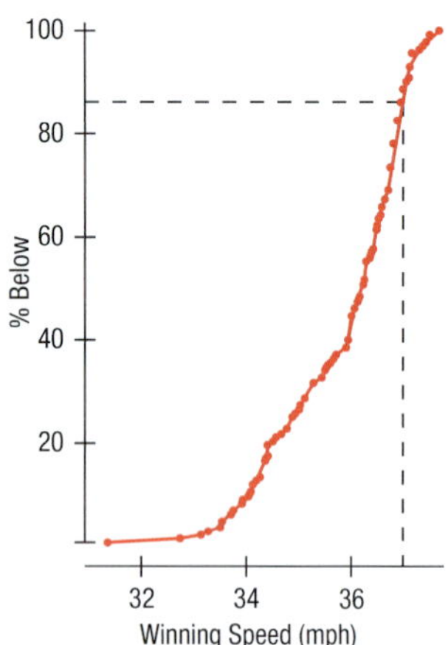

Suppose you had only the ogive (and no access to the data).

a) Estimate the median winning speed.
b) Estimate the quartiles.
c) Estimate the range and the IQR.
d) Create a boxplot of these speeds.
e) Write a few sentences about the speeds of the Kentucky Derby winners.

T **38. Framingham cholesterol** The Framingham Heart Study recorded the cholesterol levels of more than 1400 participants. (Data in **Framingham**) Here is an ogive of the distribution of these cholesterol measures. (An ogive shows the percentage of cases at or below a certain value.) Construct a boxplot for these data, and write a few sentences describing the distribution.

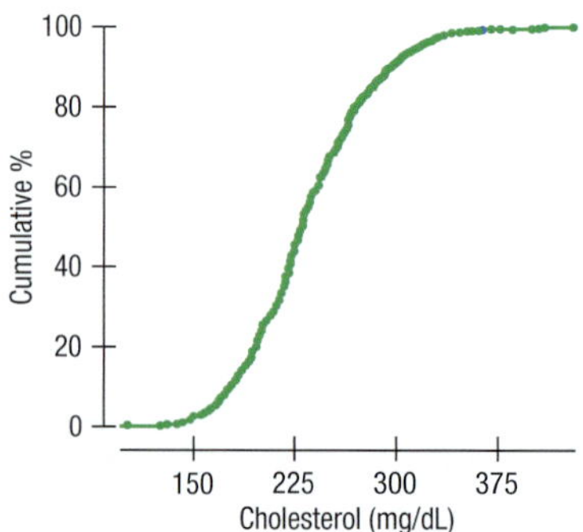

39. Test marks A class of fourth-grade Mathematics students took a November test. The five-number summaries for the 16 girls and 12 boys in the class are shown in the table below:

Girls	70	77	82	85	97
Boys	69	78	84	87	95

a) Which group had the highest mark?
b) Which group had the greater range?
c) Which group had the greater interquartile range?
d) Which group generally did better on the test? Explain.
e) If the mean test mark for girls is 81 and the mean test mark for boys is 83, what is the overall mean for the class?

T **40. Cloud seeding** In an experiment to determine whether seeding clouds with silver iodide increases rainfall, 52 clouds were randomly assigned to be seeded or not. The amount of rain they generated was then measured (in acre-feet). Here are the summary statistics:

	n	Mean	Median	SD	IQR	Q1	Q3
Unseeded	26	164.59	44.20	278.43	138.60	24.40	163
Seeded	26	441.98	221.60	650.79	337.60	92.40	430

a) Which of the summary statistics are most appropriate for describing these distributions. Why?
b) Do you see any evidence that seeding clouds may be effective? Explain.

T **41. Industrial experiment** Engineers at a computer production plant tested two methods for accuracy in drilling holes into a PC board. They tested how fast they could set the drilling machine by running 10 boards at each of two different speeds. To assess the results, they measured the distance (in inches) from the center of a target on the board to the center of the hole. The data and summary statistics are shown in the table:

	Distance (in.)	Speed	Distance (in.)	Speed
	0.000101	Fast	0.000098	Slow
	0.000102	Fast	0.000096	Slow
	0.000100	Fast	0.000097	Slow
	0.000102	Fast	0.000095	Slow
	0.000101	Fast	0.000094	Slow
	0.000103	Fast	0.000098	Slow
	0.000104	Fast	0.000096	Slow
	0.000102	Fast	0.975600	Slow
	0.000102	Fast	0.000097	Slow
	0.000100	Fast	0.000096	Slow
Mean	0.000102		0.097647	
StdDev	0.000001		0.308481	

Write a report summarizing the findings of the experiment. Include appropriate visual and verbal displays of the distributions, and make a recommendation to the engineers if they are most interested in the accuracy of the method.

T **42. Cholesterol and smoking** A study that examined the health risks of smoking measured the cholesterol levels of people who had smoked for at least 25 years and people of similar ages who had smoked for no more than 5 years and then stopped. Create appropriate graphical displays for both groups, and write a brief report comparing their cholesterol levels. Here are the data:

Smokers				Ex-Smokers		
225	211	209	284	250	134	300
258	216	196	288	249	213	310
250	200	209	280	175	174	328
225	256	243	200	160	188	321
213	246	225	237	213	257	292
232	267	232	216	200	271	227
216	243	200	155	238	163	263
216	271	230	309	192	242	249
183	280	217	305	242	267	243
287	217	246	351	217	267	218
200	280	209		217	183	228

T **43. MPG** A consumer organization wants to compare gas mileage figures for several models of cars made in the United States with autos manufactured in other countries. The data for a random sample of cars classified as "midsize" are found in the file **MPG 2016**.

a) Create graphical displays for the three groups.
b) Write a few sentences comparing the distributions.

T **44. Baseball 2016** American League baseball teams play their games with the designated hitter rule, meaning that pitchers do not bat. The league believes that replacing the pitcher, typically a weak hitter, with another player in the batting order produces more runs and generates more interest among fans. Following are the average number of runs scored by each team in the 2016 season:

American League		National League	
Team	**Runs per Game**	**Team**	**Runs per Game**
Baltimore Orioles	4.59	Arizona Diamondbacks	4.64
Boston Red Sox	5.42	Atlanta Braves	4.03
Chicago White Sox	4.23	Chicago Cubs	4.99
Cleveland Indians	4.83	Cincinnati Reds	4.42
Detroit Tigers	4.66	Colorado Rockies	5.22
Houston Astros	4.47	Los Angeles Dodgers	4.48
Kansas City Royals	4.17	Miami Marlins	4.07
Los Angeles Angels	4.43	Milwaukee Brewers	4.14
Minnesota Twins	4.46	NY Mets	4.14
NY Yankees	4.20	Philadelphia Phillies	3.77
Oakland Athletics	4.03	Pittsburgh Pirates	4.50
Seattle Mariners	4.74	San Diego Padres	4.23
Tampa Bay Rays	4.15	San Francisco Giants	4.41
Texas Rangers	4.72	St Louis Cardinals	4.81
Toronto Blue Jays	4.69	Washington Nationals	4.71

a) Create an appropriate graphical display of these data.
b) Write a few sentences comparing the average number of runs scored per game in the two leagues. (Remember: shape, center, spread, unusual features!)
c) The runs per game leaders were the Red Sox and the Rockies in the American and National League, respectively. Did either of those teams score an unusually large number of runs per game? Explain briefly.
d) Is the actual difference in mean runs per game between the leagues in 2016 different from what you might expect by chance if there really were no difference? Using software, shuffle the league labels 1000 times and create a histogram of the differences. What do you conclude?

T **45. Assets** Here is a histogram of the assets (in millions of dollars) of 79 companies chosen from the *Forbes* list of the nation's top corporations: (Data in **Companies**)

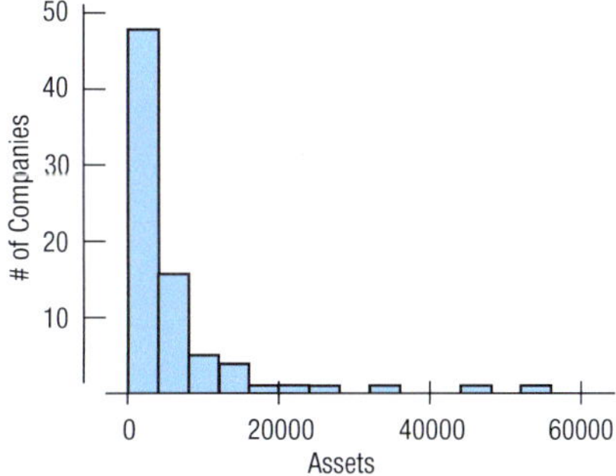

a) What aspect of this distribution makes it difficult to summarize, or to discuss, center and spread?
b) What would you suggest doing with these data if we want to understand them better?

46. Music library Students were asked how many songs they had in their digital music libraries. Here's a display of the responses:

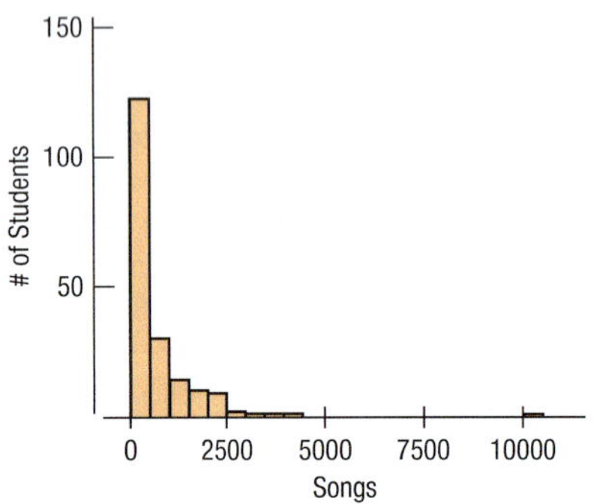

a) What aspect of this distribution makes it difficult to summarize, or to discuss, center and spread?
b) What would you suggest doing with these data if we want to understand them better?

47. Assets again Here are the same data you saw in Exercise 45 after re-expressions as the square root of assets (in $M) and the logarithm of assets (in $M):

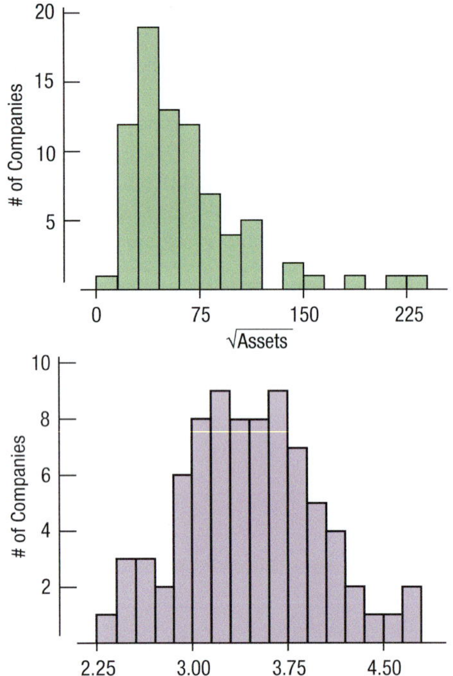

a) Which re-expression do you prefer? Why?
b) In the square root re-expression, what does the value 50 actually indicate about the company's assets?
c) In the logarithm re-expression, what does the value 3 actually indicate about the company's assets?

48. Rainmakers The table lists the amounts of rainfall (in acre-feet) from the 26 clouds seeded with silver iodide discussed in Exercise 40. (Data in **Cloud seeding**)

2745	703	302	242	119	40	7
1697	489	274	200	118	32	4
1656	430	274	198	115	31	
978	334	255	129	92	17	

a) Why is acre-feet a good way to measure the amount of precipitation produced by cloud seeding?
b) Plot these data, and describe the distribution.
c) Create a re-expression of these data that produces a more advantageous distribution.
d) Explain what your re-expressed scale means.

49. Stereograms Stereograms appear to be composed entirely of random dots. However, they contain separate images that a viewer can "fuse" into a three-dimensional (3D) image by staring at the dots while defocusing the eyes. An experiment was performed to determine whether knowledge of the embedded image affected the time required for subjects to fuse the images. One group of subjects (group NV) received no information or just verbal information about the shape of the embedded object. A second group (group VV) received both verbal information and visual information (specifically, a drawing of the object). The experimenters measured how many seconds it took for the subject to report that he or she saw the 3D image.

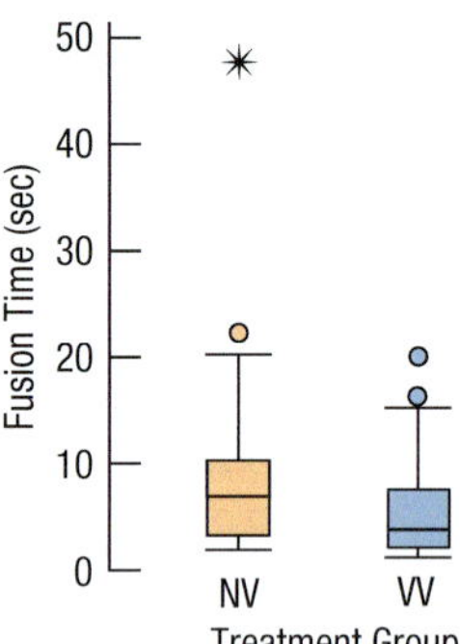

a) What two variables are discussed in this description?
b) For each variable, is it quantitative or categorical? If quantitative, what are the units?
c) The boxplots compare the fusion times for the two treatment groups. Write a few sentences comparing these distributions. What does the experiment show?

50. Stereograms, revisited Because of the skewness of the distributions of fusion times described in Exercise 49, we might consider a re-expression. Here are the boxplots of the *log* of fusion times. Is it better to analyze the original fusion times or the log fusion times? Explain.

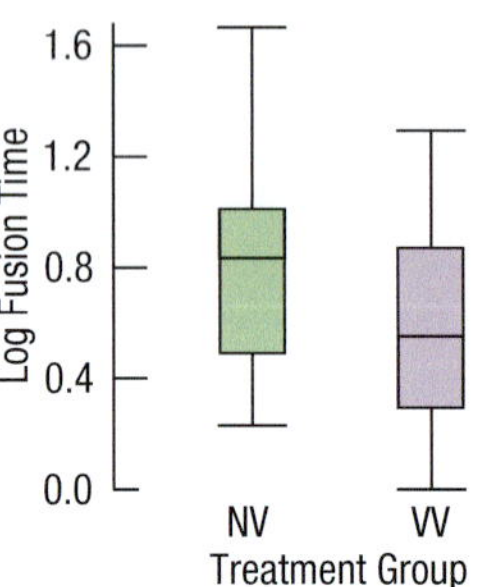

51. Cost of living sampled From Exercise 16 it appeared that the mean cost of a cappuccino was slightly higher than the mean cost of a dozen eggs. Given the variation among the prices, could that difference be due just to chance? To examine that further, we took 1000 random samples of 100 cities and computed the difference between the mean price of a cappuccino and a dozen eggs. The histogram below shows the 1000 mean differences.

a) If there were no real difference between the mean prices, where would you expect the center of the histogram to be?
b) Given the histogram, what do you conclude? Explain briefly.

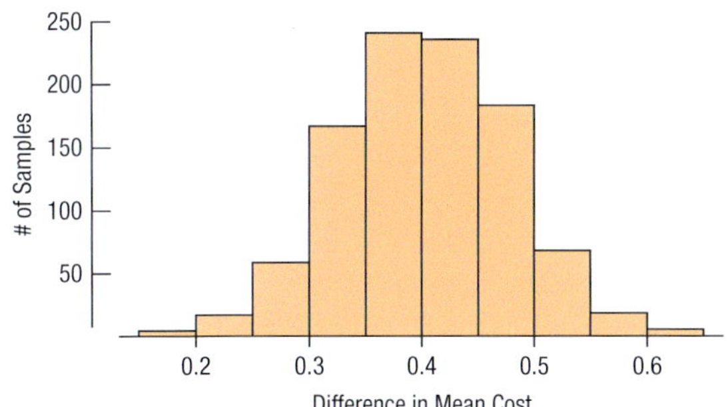

52. Stereograms shuffled From Exercise 50 it appeared that the mean log fusion time for the VV group was greater than that of the NV group. Could the difference be due to chance? The histogram below shows the difference in the means of the log fusion times by randomly selecting two groups (ignoring the group labels) 1000 times. The actual difference was smaller (in absolute value) than the randomly generated ones 16 times out of 1000. What do you conclude? Explain briefly.

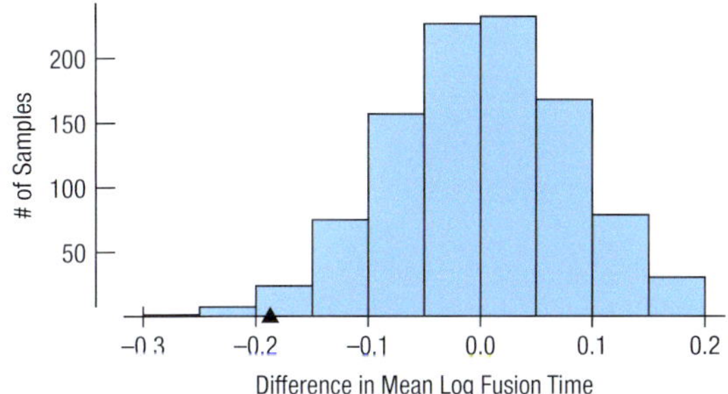

53. Pizza prices by city (Data in **Pizza prices**) The company that sells frozen pizza to stores in four markets in the United States (Denver, Baltimore, Dallas, and Chicago) wants to examine the prices that the stores charge for pizza slices. Here are boxplots comparing data from a sample of stores in each market:

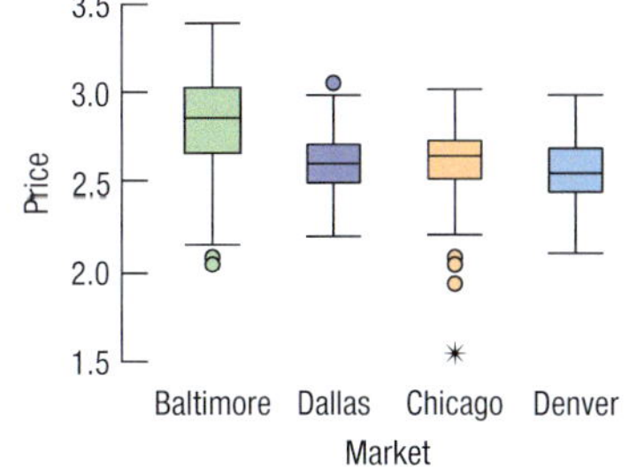

a) Do prices appear to be the same in the four markets? Explain.
b) Does the presence of any outliers affect your overall conclusions about prices in the four markets?

54. Pizza price differences The mean price of pizza in Baltimore was \$2.85, \$0.23 higher than the mean price of \$2.62 in Dallas. To see if that difference was real, or due to chance, we took the 156 prices from Baltimore and Dallas and mixed those 312 prices together. Then we randomly chose 2 groups of 156 prices 10,000 times, and computed the difference in the mean price each time. The histogram shows the distribution of those 10,000 differences. (Data in **Pizza prices**)

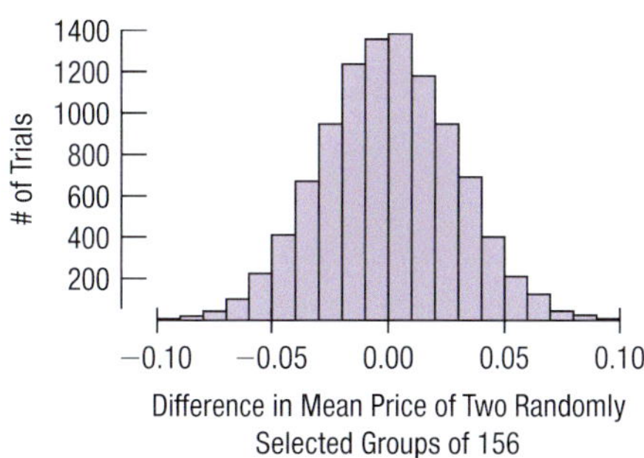

a) Given this histogram, what do you conclude about the actual difference of \$0.23 between the mean prices of Baltimore and Dallas?
b) Do you think the presence of the outliers in the boxplots of Exercise 53 affects your conclusion?
c) Perform a similar analysis using shuffling to compare prices in Chicago and Denver. Is the actual difference in mean prices different from what you might expect by chance? Use any software you choose. Both StatCrunch and the apps at astools.datadesk.com can handle this.

JUST CHECKING

Answers

1. The % on-time arrivals have a unimodal, symmetric distribution centered at about 79%. In most months, between 70% and 86% of the flights arrived on time.
2. The boxplot of % on-time arrivals nominates several months as low outliers.
3. The boxplots by month show a strong seasonal pattern. Flights are more likely to be on time in the spring and fall and less likely to be on time in the winter and summer. One
4. possible reason for the pattern is snowstorms in the winter and thunderstorms in the summer.

5

The Standard Deviation as a Ruler and the Normal Model

WHERE ARE WE GOING?

A college admissions officer is looking at the files of two candidates, one with a total SAT score of 1500, another with an ACT score of 21. Which candidate scored better? How do we compare things when they're measured on different scales?

To answer a question like this, we need to standardize the results. To do that, we need to know two things. First, we need a base value for comparison—we'll often use the mean for that. Next, we need to know how far away we are from the mean. So, we'll need some sort of ruler. Fortunately, the standard deviation is just the thing we need.

The idea of measuring distances from means by counting standard deviations shows up throughout statistics, and it starts right here.

WHO	Olympic competitors
WHAT	Women's Heptathlon event results
WHEN	2016
WHERE	2016 Olympics
WHY	General interest

The women's heptathlon in the Olympics consists of seven track-and-field events: the 200 m and 800 m runs, 100 m high hurdles, shot put, javelin, high jump, and long jump. In the 2016 Olympics, Nafissatou Thiam of Belgium won the long jump with a jump of 6.58 meters, and Katarina Johnson-Thompson of Great Britain won the 200 m run with a time of 23.26 seconds.

Each contestant is awarded points for each event based on her performance. So, which performance deserves more points? It's not clear how to compare them. They aren't measured in the same units, or even in the same *direction* (longer jumps are better but shorter times are better).

To see which value is more extraordinary, we need a way to judge them against the background of data they come from. Chapter 2 discussed the standard deviation, but really didn't tell you what it's for. Statisticians use the standard deviation as a ruler, to judge how far a value is from the center of the distribution. Knowing that a value is 5 units higher than the mean doesn't tell us much unless we know the context, but knowing it's 5 standard deviations above the mean tells us that it's truly extraordinary.

```
200 m Run                    Long Jump
Stem   Leaf                  Stem   Leaf
  23 | 24                      6F | 44555
  23 | 799                     6T | 222233333
  24 | 011123334               6O | 0000111111
  24 | 66677999                5* | 889
  25 | 012334                  5S | 7
  25 |                         5F | 5
  26 | 3
  23|2=23.2 seconds            6F|5=6.5 meters
```

Figure 5.1
Stem-and-leaf displays for the 200 m race and the long jump in the 2016 Olympic heptathlon. Katarina Johnson-Thompson (green scores) won the 200 m, and Nafissatou Thiam (red scores) won the long jump. The stems for the 200 m race run from faster to slower and the stems for the long jump from longer to shorter so that the best scores are at the top of each display. (Data in **Womens heptathalon 2016**).
Note: The stem-and-leaf display on the left uses two stems for each ten values (leaves 0–4 on one and 5–9 on the other). The other display uses five stems for each decade (0/1 labeled "O", 2/3 labeled "T", 4/5 "F", 6/7 "S", and "*" for the last two). These alternatives, along with the standard form, offer three alternatives for stem-and-leaf displays.

Which of the two winning scores is the better one? Thiam's winning 6.58 m long jump was 1.66 SDs better (longer) than the mean. Johnson-Thompson's winning time of 23.26 sec. was 2.02 SDs better (faster) than the mean time. That's more impressive.

We could turn each of the seven results into a score in this way, and add them up to determine the winner. Olympic judges actually use a point system based on similar calculations (but based on performances from many competitions and a larger pool of contestants).

5.1 Using the Standard Deviation to Standardize Values

> *NOTATION ALERT*
> We always use the letter z to denote values that have been standardized with the mean and standard deviation.

Expressing a distance from the mean in standard deviations *standardizes* the performances. To **standardize** a value, we subtract the mean and then divide this difference by the standard deviation:

$$z = \frac{y - \bar{y}}{s}.$$

The values are called **standardized values**, and are commonly denoted with the letter z. Usually we just call them **z-scores**.

z-scores measure the distance of a value from the mean in standard deviations. A z-score of 2 says that a data value is 2 standard deviations above the mean. It doesn't matter whether the original variable was measured in fathoms, dollars, or carats; those units don't apply to z-scores. Data values below the mean have negative z-scores, so a z-score of -1.6 means that the data value was 1.6 standard deviations below the mean. Of course, regardless of the direction, the farther a data value is from the mean, the more unusual it is, so a z-score of -1.3 is more extraordinary than a z-score of 1.2.

Here are the calculations for the heptathlon. For each event, first subtract the mean of all participants from the individual's score. Then divide by the standard deviation to get the z-score. For example, Nafissatou's 6.58 m long jump is compared to the mean of 6.17 m for all 29 heptathletes: $(6.58 - 6.17 = 0.41 \text{ m})$. Dividing this difference by the standard deviation gives her a z-score of 1.66.

		Event	
		Long Jump	**200 m Run**
	Mean	6.17 m	24.58 s
	SD	0.247 m	0.654 s
Thiam	**Performance**	6.58 m	25.10 s
	***z*-score**	$(6.58 \quad 6.17)/0.247 - 1.66$	$(25.10 - 24.58)/0.654 = 0.795$
	Total for two events	$1.66 - 0.795 = 0.865$	
Johnson-Thompson	**Performance**	6.51 m	$(23.26 - 24.58)/0.654 = -2.02$
	***z*-score**	$(6.51 - 6.17)/0.247 = 1.38$	
	Total for two events	$1.38 + 2.02 = 3.40$	

When we combine the two events, we change the sign of the run because faster is *better*. After two events, Thiam's two scores total only 0.865 compared with Johnson-Thompson's 3.40. But Thiam went on to take the gold once all seven events were counted, placing first in the high jump and the shot put as well.

EXAMPLE 5.1 Standardizing Skiing Times

The men's super combined skiing event debuted in the 2010 Winter Olympics in Vancouver. It consists of two races: a downhill and a slalom. Times for the two events are added together, and the skier with the lowest total time wins. At Vancouver, the mean slalom time was 52.67 seconds with a standard deviation of 1.614 seconds. The mean downhill time was 116.26 seconds with a standard deviation of 1.914 seconds. Bode Miller of the United States, who won the gold medal with a combined time of 164.92 seconds, skied the slalom in 51.01 seconds and the downhill in 113.91 seconds.

QUESTION: On which race did he do better compared to the competition?

ANSWER: $z_{slalom} = \frac{y - \bar{y}}{s} = \frac{51.01 - 52.67}{1.614} = -1.03$

$$z_{downhill} = \frac{113.91 - 116.26}{1.914} = -1.23$$

Keeping in mind that faster times are *below* the mean, Miller's downhill time of 1.23 SDs below the mean is even more remarkable than his slalom time, which was 1.03 SDs below the mean.

EXAMPLE 5.2 Combining *z*-Scores

At the 2010 Vancouver Winter Olympics, Bode Miller had earlier earned a bronze medal in the downhill and a silver medal in the super-G slalom before winning the gold in the men's super combined. Ted Ligety, the winner of the gold medal at the 2006 games, had super combined times in 2010 of 50.76 seconds in the slalom (fastest of the 35 finalists) and 115.06 seconds in the downhill for a total of 165.82 seconds, almost a second behind Miller. But he finished in fifth place in 2010. He beat Miller in the slalom, but Miller beat him in the downhill. The downhill is longer than the slalom and so counts for more in the total.

QUESTION: Would the placement have changed if each event had been treated equally by standardizing each and adding the standardized scores?

ANSWER: We've seen that Miller's *z*-scores for slalom and downhill were -1.03 and -1.23, respectively. That's a total of -2.26. Ligety's *z*-scores were

$$z_{slalom} = \frac{y - \bar{y}}{s} = \frac{50.76 - 52.67}{1.614} = -1.18$$

$$z_{downhill} = \frac{115.06 - 116.26}{1.914} = -0.63$$

So his total *z*-score was -1.81. That's not as good as Miller's total of -2.26. So, although Ligety beat Miller in the slalom, Miller beat him by *more* in the downhill. Using the standardized scores would not have put Ligety ahead of Miller.

JUST CHECKING

1. Your statistics teacher has announced that the lower of your two tests will be dropped. You got a 90 on test 1 and an 80 on test 2. You're all set to drop the 80 until she announces that she grades "on a curve." She standardized the scores to decide which is the lower one. If the mean on the first test was 88 with a standard deviation of 4 and the mean on the second was 75 with a standard deviation of 5,
   a) Which one will be dropped?
   b) Does this seem "fair"?
2. A distribution of incomes shows a strongly right skewed distribution. Will the maximum or the minimum income be closer to the mean? Explain.

5.2 Shifting and Scaling

WHO	80 male participants of the NHANES survey between the ages of 19 and 24 who measured between 68 and 70 inches tall
WHAT	Their weights
UNIT	Kilograms
WHEN	2001–2002
WHERE	United States
WHY	To study nutrition, and health issues and trends
HOW	National survey

There are two steps to finding a z-score. First, the data are *shifted* by subtracting the mean. Then, they are *rescaled* by dividing by the standard deviation. How do these operations work?

Shifting to Adjust the Center

Since the 1960s, the Centers for Disease Control and Prevention's National Center for Health Statistics has been collecting health and nutritional information on people of all ages and backgrounds. The National Health and Nutrition Examination Survey (NHANES) of 2001–2002[1] measured a wide variety of variables, including body measurements, cardiovascular fitness, blood chemistry, and demographic information on more than 11,000 individuals. We have data on 80 men between 19 and 24 years old of average height (between 5′8″ and 5′10″ tall). Here are a histogram and boxplot of their weights: (Data in **Mens Weights**)

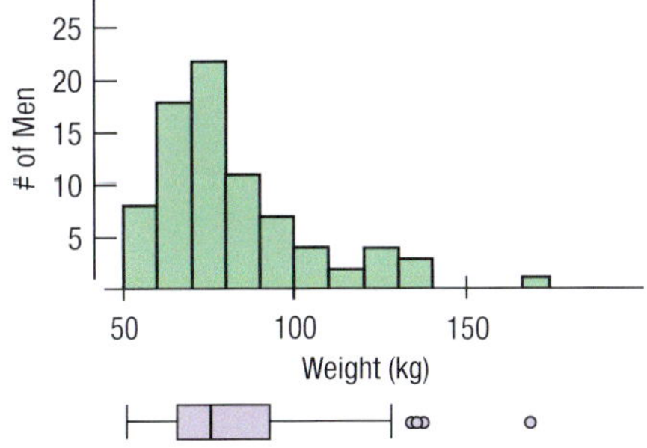

Figure 5.2
Histogram and boxplot for the men's weights. The shape is skewed to the right with several high outliers.

Their mean weight is 82.36 kg. For this age and height group, the National Institutes of Health recommends a maximum healthy weight of 74 kg, but we can see that some of the men are heavier than the recommended weight. To compare their weights to the

[1] www.cdc.gov/nchs/nhanes.htm will take you to the most current version of the NHANES studies.

recommended maximum, we could subtract 74 kg from each of their weights. What would that do to the center, shape, and spread of the histogram? Here's the picture:

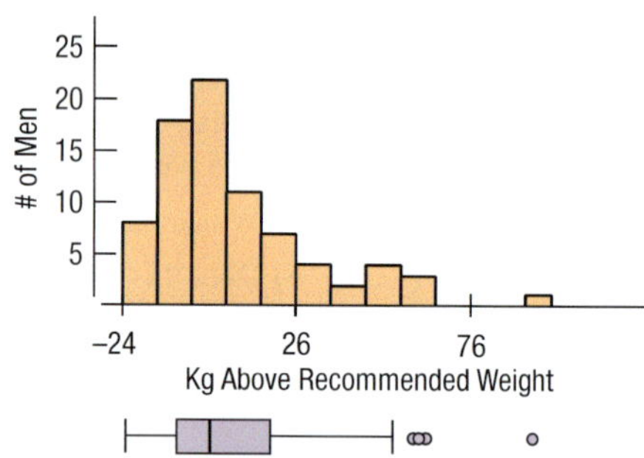

Figure 5.3
Subtracting 74 kilograms shifts the entire histogram down but leaves the spread and the shape exactly the same.

On average, they weigh 82.36 kg, so on average they're 8.36 kg overweight. And, after subtracting 74 from each weight, the mean of the new distribution is $82.36 - 74 = 8.36$ kg. In fact, when we **shift** the data by adding (or subtracting) a constant to each value, all measures of position (center, percentiles, min, max) will increase (or decrease) by the same constant.

What about the spread? What does adding or subtracting a constant value do to the spread of the distribution? Look at the two histograms again. Adding or subtracting a constant changes each data value equally, so the entire distribution just shifts. Its shape doesn't change and neither does the spread. None of the measures of spread we've discussed—not the range, not the IQR, not the standard deviation—changes.

Adding (or subtracting) a constant to every data value adds (or subtracts) the same constant to measures of position, but leaves measures of spread unchanged.

> **SHIFTING IDEAL HEIGHTS**
>
> Doctors' height and weight charts sometimes give ideal weights for various heights that include 2-inch heels. If the mean height of adult women is 66 inches including 2-inch heels, what is the mean height of women without shoes? Each woman is shorter by 2 inches when barefoot, so the mean is decreased by 2 inches, to 64 inches.

Rescaling to Adjust the Scale

Not everyone thinks naturally in metric units. Suppose we want to look at the weights in pounds instead. We'd have to rescale the data. Because there are about 2.2 pounds in every kilogram, we'd convert the weights by multiplying each value by 2.2. Multiplying or dividing each value by a constant—**rescaling** the data—changes the measurement units. Here are histograms of the two weight distributions, plotted on the same scale, so you can see the effect of multiplying:

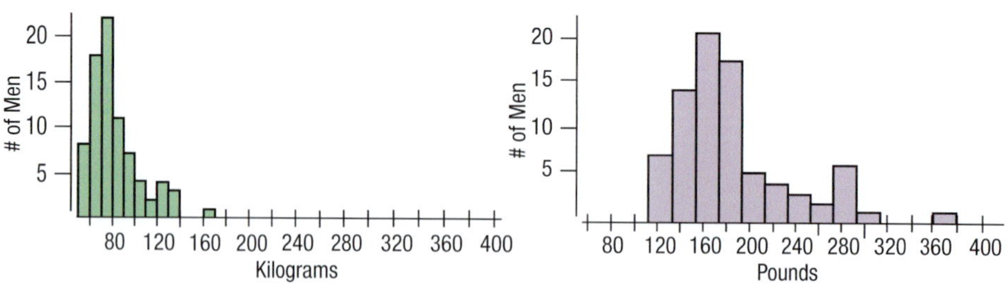

Figure 5.4
Men's weights in both kilograms and pounds. How do the distributions and numerical summaries change?

What happens to the shape of the distribution? Although the histograms don't look exactly alike, we see that the shape really hasn't changed: Both are unimodal and skewed to the right.

What happens to the mean? Not too surprisingly, it gets multiplied by 2.2 as well. The men weigh 82.36 kg on average, which is 181.19 pounds. As the boxplots and 5-number summaries show, all measures of position act the same way. They all get multiplied by this same constant.

What happens to the spread? Take a look at the boxplots in Figure 5.5. The spread in pounds (on the right) is larger. How much larger? If you guessed 2.2 times, you've figured out how measures of spread get rescaled.

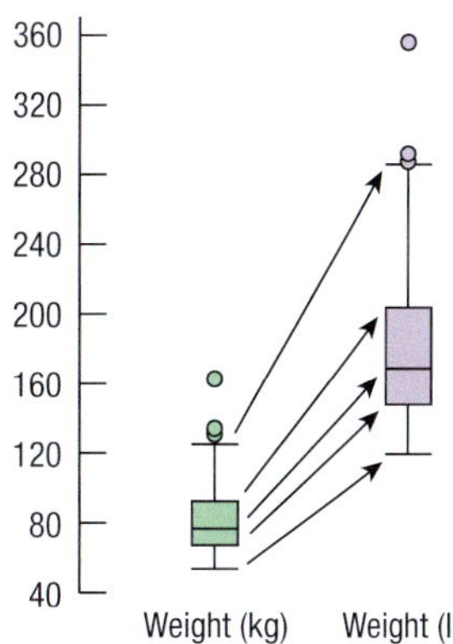

	Weight (kg)	Weight (lb)
Min	54.3	119.46
Q1	67.3	148.06
Median	76.85	169.07
Q3	92.3	203.06
Max	161.50	355.30
IQR	25	55
SD	22.27	48.99

Figure 5.5
The boxplots (drawn on the same scale) show the weights measured in kilograms (on the left) and pounds (on the right). Because 1 kg is about 2.2 lb, all the points in the right box are 2.2 times larger than the corresponding points in the left box. So each measure of position and spread is 2.2 times as large when measured in pounds rather than kilograms.

When we multiply (or divide) all the data values by any constant, all measures of position (such as the mean, median, and percentiles) and measures of spread (such as the range, the IQR, and the standard deviation) are multiplied (or divided) by that same constant.

EXAMPLE 5.3 Rescaling the Men's Combined Times

RECAP: The times in the men's combined event at the winter Olympics are reported in minutes and seconds. The mean and standard deviation of the 34 final super combined times at the 2010 Olympics were 168.93 seconds and 2.90 seconds, respectively.

QUESTION: Suppose instead that we had reported the times in minutes—that is, that each individual time was divided by 60. What would the resulting mean and standard deviation be?

ANSWER: Dividing all the times by 60 would divide both the mean and the standard deviation by 60:

$$\text{Mean} = 168.93/60 = 2.816 \text{ minutes}; \qquad \text{SD} = 2.90/60 = 0.048 \text{ minute.}$$

JUST CHECKING

3. In 1995, the Educational Testing Service (ETS) adjusted the scores of SAT tests. Before ETS recentered the SAT Verbal test, the mean of all test scores was 450.
   a) How would adding 50 points to each score affect the mean?
   b) The standard deviation was 100 points. What would the standard deviation be after adding 50 points?
   c) Suppose we drew boxplots of test takers' scores a year before and a year after the recentering. How would the boxplots of the two years differ?

4. A company manufactures wheels for in-line skates. The diameters of the wheels have a mean of 3 inches and a standard deviation of 0.1 inch. Because so many of their customers use the metric system, the company decided to report their production statistics in millimeters (1 inch = 25.4 mm). They report that the standard deviation is now 2.54 mm. A corporate executive is worried about this increase in variation. Should he be concerned? Explain.

Shifting, Scaling, and *z*-Scores

Standardizing data into *z*-scores is just shifting them by the mean and rescaling them by the standard deviation. Now we can see how standardizing affects the distribution. When we subtract the mean of the data from every data value, we shift the mean to zero. As we have seen, such a shift doesn't change the standard deviation.

When we *divide* each of these shifted values by s, however, the standard deviation should be divided by s as well. Since the standard deviation was s to start with, the new standard deviation becomes 1.

How, then, does standardizing affect the distribution of a variable? Let's consider the three aspects of a distribution: the shape, center, and spread.

z-scores have mean 0 and standard deviation 1.

- *Standardizing into z-scores does not change the **shape** of the distribution of a variable.*
- *Standardizing into z-scores changes the **center** by making the mean 0.*
- *Standardizing into z-scores changes the **spread** by making the standard deviation 1.*

STEP-BY-STEP EXAMPLE

Working with Standardized Variables

Many colleges and universities require applicants to submit scores on standardized tests such as the SAT Writing, Math, and Critical Reading (Verbal) tests. The college your little sister wants to apply to says that while there is no minimum score required, the middle 50% of their students have combined SAT scores between 1530 and 1850. You'd feel confident if you knew her score was in their top 25%, but unfortunately she took the ACT test, an alternative standardized test.

QUESTION: How high does her ACT need to be to make it into the top quarter of equivalent SAT scores?

To answer that question you'll have to standardize all the scores, so you'll need to know the mean and standard deviations of scores for some group on both tests. The college doesn't report the mean or standard deviation for their applicants on either test, so we'll use the group of all test takers nationally. For college-bound seniors, the average combined SAT score is about 1500 and the standard deviation is about 250 points. For the same group, the ACT average is 20.8 with a standard deviation of 4.8.

THINK

PLAN State what you want to find out.

I want to know what ACT score corresponds to the upper-quartile SAT score. I know the mean and standard deviation for both the SAT and ACT scores based on all test takers, but I have no individual data values.

VARIABLES Identify the variables and report the W's (if known).

Check the appropriate conditions.

✓ **Quantitative Data Condition**: Scores for both tests are quantitative but have no meaningful units other than points.

SHOW

MECHANICS Standardize the variables.

The middle 50% of SAT scores at this college fall between 1530 and 1850 points. To be in the top quarter, my sister would have to have a score of at least 1850. That's a z-score of

$$z = \frac{(1850 - 1500)}{250} = 1.40.$$

So an SAT score of 1850 is 1.40 standard deviations above the mean of all test takers.

The ACT value we seek is 1.40 standard deviations above the mean.

For the ACT, 1.40 standard deviations above the mean is $20.8 + 1.40(4.8) = 27.52$.

TELL

CONCLUSION Interpret your results in context.

To be in the top quarter of applicants in terms of combined SAT score, she'd need to have an ACT score of at least 27.52.

5.3 Normal Models

IS THE NORMAL NORMAL?

Don't be misled. The name "Normal" doesn't mean that these are the *usual* shapes for histograms. The name follows a tradition of positive thinking in mathematics and statistics in which functions, equations, and relationships that are easy to work with or have other nice properties are called "normal," "common," "regular," "natural," or similar terms. It's as if by calling them ordinary, we could make them actually occur more often and simplify our lives.

A z-score gives an indication of how unusual a value is because it tells how far it is from the mean. If the data value sits right at the mean, it's not very far at all and its z-score is 0. A z-score of 1 tells us that the data value is 1 standard deviation above the mean, while a z-score of -1 tells us that the value is 1 standard deviation below the mean. How far from 0 does a z-score have to be to be interesting or unusual?

The 68–95–99.7 Rule

In 1733, the French mathematician, Abraham de Moivre, showed that for many unimodal and symmetric distributions, about 68% of the values fall within one standard deviation of the mean, about 95% of the values are found within two standard deviations of the mean, and about 99.7%—virtually all of the values—will be within three standard deviations of the mean.

We call this rule the **68–95–99.7 Rule**.[2]

In 1809, Gauss figured out the formula for the model that accounts for de Moivre's observation. That model is called the **Normal or Gaussian model**.

Gauss's result is remarkable. And it illustrates one of the most important uses of the standard deviation. As we've said earlier, the standard deviation is the statistician's ruler. This model for unimodal, symmetric data gives us even more information because it tells us how likely it is to have z-scores between -1 and 1, between -2 and 2, and between -3 and 3. Now we can also see just how extraordinary a z-score of 5 or 6 would be.

NOTATION ALERT

$N(\mu, \sigma)$ always denotes a Normal model. The μ, pronounced "mew," is the Greek letter for "m" and always represents the mean in a model. The σ, sigma, is the lowercase Greek letter for "s" and always represents the standard deviation in a model.

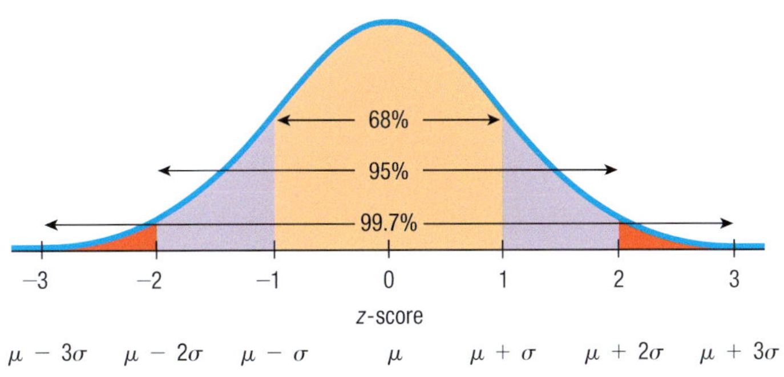

Figure 5.6

Reaching out one, two, and three standard deviations (or z-scores) on a Normal model gives the 68–95–99.7 Rule, seen as proportions of the area under the curve. The top line of numbers shows the z-scores. The bottom line shows the same distances in terms of the mean and standard deviation.

ONE IN A MILLION

These magic 68, 95, 99.7 values come from the Normal model. As a model, it can give us corresponding values for any z-score. For example, it tells us that fewer than 1 out of a million values have z-scores smaller than -5.0 or larger than $+5.0$. So if someone tells you you're "one in a million," they might be talking about your z-score.

Figure 5.6 shows the Normal model in terms of z-scores, but we could just as easily put back in the mean and the standard deviation. When we do, we write $N(\mu, \sigma)$ to represent a Normal model with a mean of μ and a standard deviation of σ. Why the Greek? Well, this mean and standard deviation are not numerical summaries of data. They are characteristics of the model called **parameters**. Parameters are the values we choose that completely specify a model.

We don't want to confuse the parameters with summaries of the data such as $\bar{y}$ and s, so we use special symbols. In statistics, we almost always use Greek letters for parameters. By contrast, summaries of data, like the sample mean, median, or standard deviation, are called **statistics** and are usually written with Latin letters.

[2]This rule is also called the "Empirical Rule" because it originally came from observation. The rule was first published by Abraham de Moivre in 1733, 75 years before the Normal model was discovered. Maybe it should be called "de Moivre's Rule," but those names wouldn't help us remember the important numbers: 68, 95, and 99.7.

If we model data with a Normal model and standardize them using the corresponding μ and σ, we still call the standardized value a **z-score**, and we write

$$z = \frac{y - \mu}{\sigma}.$$

IS THE STANDARD NORMAL A STANDARD?

Yes. We call it the "Standard Normal" because it models standardized values. It is also a "standard" because this is the particular Normal model that we almost always use.

As Figure 5.6 shows, we can think of the Normal in z-scores (numbers of standard deviations above or below the mean) or show the actual units. Usually it's easier to standardize data using the mean and standard deviation first. Then we need only the model $N(0, 1)$ with mean 0 and standard deviation 1. This Normal model is called the **standard Normal model** (or the standard Normal distribution).

EXAMPLE 5.4 Using the 68–95–99.7 Rule

A study of men's health measured 14 body characteristics of 250 men, one of which was wrist circumference (in cm). The mean wrist circumference was 18.22 cm and the standard deviation was 0.91 cm. A histogram of all 250 values looks like this: (Data in **Bodyfat**)

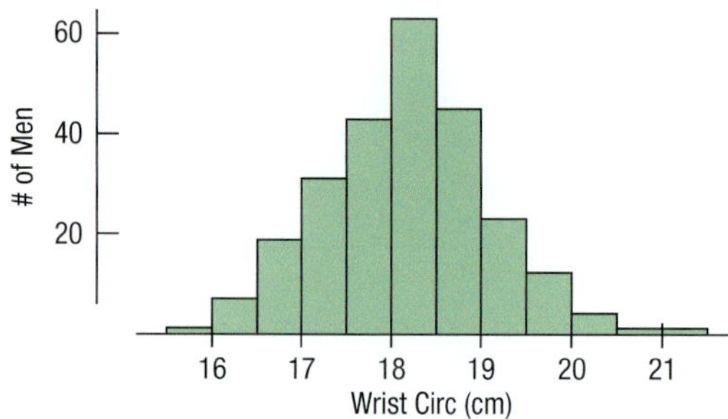

QUESTION: Will the 68–95–99.7 Rule provide useful percentages from these data? About how many of these men does the rule predict to have a wrist circumference larger than 20.04 cm?

ANSWER: Yes, the distribution is unimodal and symmetric and so the 68–95–99.7 Rule should be applicable to these data. 20.04 cm is about 2 standard deviations above the mean. We expect 5% of the data to be either above or below two standard deviations, so 2.5% will be above. That's about 6 men (which, it turns out, is the number of men in the dataset who have wrists larger than 20.04 cm).

Notice how well the 68–95–99.7 Rule works when the distribution is unimodal and symmetric. But, be careful. You shouldn't use a Normal model for just any dataset. Remember that standardizing won't change the shape of the distribution. If the distribution is not unimodal and symmetric to begin with, standardizing won't make it Normal.

EXAMPLE 5.5 Using the 68–95–99.7 Rule II

Beginning in 2017, public companies will be required to disclose the ratio of CEO pay to median worker pay. The Glassdoor Economic Research Blog has published the data for 2014. Here is part of that data: the total compensation (in $M) of CEOs of 434 top public companies. (Data in **CEO Compensation 2014**)

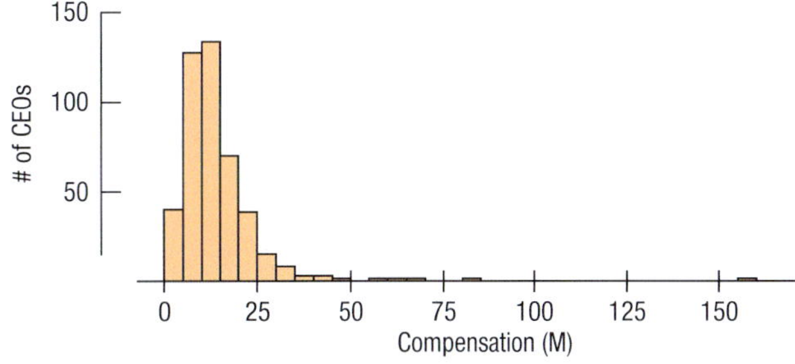

QUESTION: Will the 68–95–99.7 Rule provide useful percentages from these data?

Explain briefly. Using the rule, about how many CEOs should have compensations within one SD of the mean (between \$2.77M and \$25.40M)?

ANSWER: No, the distribution is strongly skewed to the right. We shouldn't expect the 68–95–99.7 Rule to be useful. According to the rule, about 68% of the CEO compensations should be within 1 standard deviation of the mean. One SD below the mean is \$2.77M. But the rule would say that roughly 68% will have compensation greater than \$2.77M and less than \$25.40M. In fact, from the dataset, 391 CEOs (about 90%) had compensations in that interval. The 68–95–99.7% Rule is not appropriate for strongly skewed data such as these.

> "All models are wrong—but some are useful."
>
> —George Box, famous statistician

All models make **assumptions**. Whenever we model—and we'll do that often—we'll be careful to point out the assumptions that we're making. And, what's even more important, we'll check the associated **conditions** in the data to make sure that those assumptions are reasonable.

Notice how poorly the Normal model worked for the CEO compensations in Example 5.5. When we use the Normal model, we assume that the distribution of the data is, well, Normal. Practically speaking, though, there's no way to check whether this Normality Assumption is true. Real data don't behave like mathematical models. Models are idealized; real data are real. The good news, however, is that to use a Normal model, it's sufficient to check that the shape of the data's distribution is unimodal and symmetric. A histogram (or a Normal probability plot—see Section 5.5) is a good way to check. Don't model data with a Normal model without checking whether the following condition is satisfied:

Nearly Normal Condition. The shape of the data's distribution is unimodal and symmetric and there are no obvious outliers. Check this by making a histogram (or a Normal probability plot, which we'll explain in Section 5.5).

JUST CHECKING

5. As a group, the Dutch are among the tallest people in the world. The average Dutch man is 184 cm tall—just over 6 feet (and the average Dutch woman is 170.8 cm tall—just over 5′7″). If a Normal model is appropriate and the standard deviation for men is about 8 cm, what percentage of all Dutch men will be over 2 meters (6′6″) tall?

6. Suppose it takes 20 minutes, on average, to drive to school, with a standard deviation of 2 minutes. Suppose a Normal model is appropriate for the distributions of driving times.
   a) How often will you arrive at school in less than 22 minutes?
   b) How often will it take you more than 24 minutes?
   c) Do you think the distribution of your driving times is unimodal and symmetric?
   d) What does this say about the accuracy of your predictions? Explain.

The First Three Rules for Working with Normal Models

1. Make a picture.
2. Make a picture.
3. Make a picture.

Although we're thinking about models, not histograms of data, the three rules don't change. To help you think clearly, a simple hand-drawn sketch is all you need. Even experienced statisticians sketch pictures to help them think about Normal models. You should too.

Of course, when we have data, we'll also need to make a histogram to check the Nearly Normal Condition to be sure we can use the Normal model to model the data's distribution. Other times, we may be told that a Normal model is appropriate based on prior knowledge of the situation or on theoretical considerations.

How to Sketch a Normal Curve That Looks Normal

To sketch a good Normal curve, you need to remember only three things:

- The Normal curve is bell-shaped and symmetric around its mean. Start at the middle, and sketch to the right and left from there.
- Even though the Normal model extends forever on either side, you need to draw it only for 3 standard deviations. After that, there's so little left that it isn't worth sketching.
- The place where the bell shape changes from curving downward to curving back up—the *inflection point*—is exactly one standard deviation away from the mean.

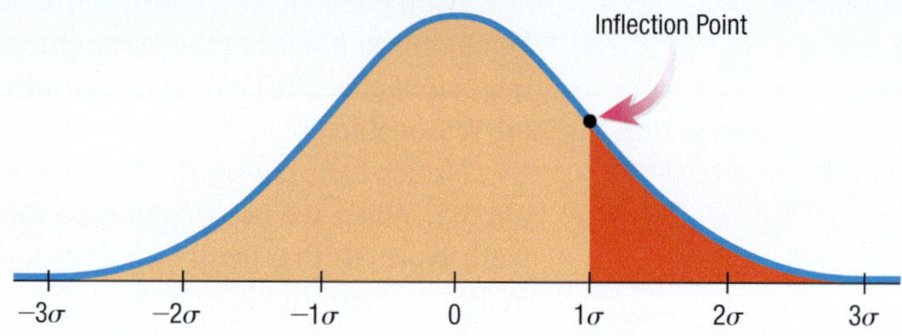

STEP-BY-STEP EXAMPLE Working with the 68–95–99.7 Rule

Each part of the SAT Reasoning Test has a distribution that is roughly unimodal and symmetric and is designed to have an overall mean of about 500 and a standard deviation of 100 for all test takers. In any one year, the mean and standard deviation may differ from these target values by a small amount, but they are a good overall approximation.

QUESTION: Suppose you earned a 600 on one part of your SAT. Where do you stand among all students who took that test?

You could calculate your *z*-score and find out that it's $z = (600 - 500)/100 = 1.0$, but what does that tell you about your percentile? You'll need the Normal model and the 68–95–99.7 Rule to answer that question.

THINK **PLAN** State what you want to know.

I want to see how my SAT score compares with the scores of all other students. To do that, I'll need to model the distribution.

VARIABLES Identify the variable and report the W's.

Let y = my SAT score. Scores are quantitative but have no meaningful units other than points.

Be sure to check the appropriate conditions.

✓ **Nearly Normal Condition:** If I had data, I would check the histogram. I have no data, but I am told that the SAT scores are roughly unimodal and symmetric.

Specify the parameters of your model.

I will model SAT score with a $N(500, 100)$ model.

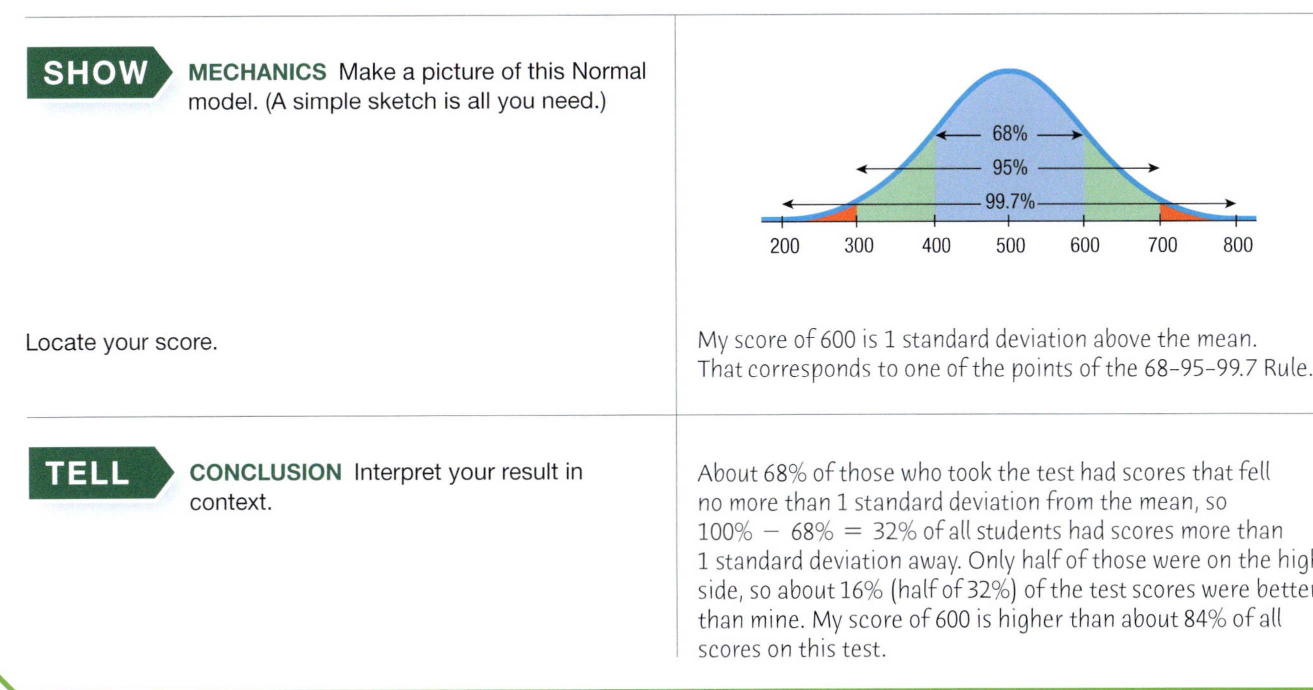

SHOW **MECHANICS** Make a picture of this Normal model. (A simple sketch is all you need.)

Locate your score.

My score of 600 is 1 standard deviation above the mean. That corresponds to one of the points of the 68–95–99.7 Rule.

TELL **CONCLUSION** Interpret your result in context.

About 68% of those who took the test had scores that fell no more than 1 standard deviation from the mean, so 100% − 68% = 32% of all students had scores more than 1 standard deviation away. Only half of those were on the high side, so about 16% (half of 32%) of the test scores were better than mine. My score of 600 is higher than about 84% of all scores on this test.

The bounds of SAT scoring at 200 and 800 can also be explained by the 68–95–99.7 Rule. Because 200 and 800 are three standard deviations from 500, it hardly pays to extend the scoring any farther on either side. We'd get more information only on $100 - 99.7 = 0.3\%$ of students.

RANDOM MATTERS

In Chapter 2, we examined the commuting times of a sample of 100 employees taken from a company with a total of 5000 employees. Here is the distribution of the entire population of 5000 commuting times. The times have a right skewed distribution with a mean time of 45.4 minutes and a standard deviation of 29.5 minutes (median: 38 minutes, IQR: 34 minutes). (Data in **Population commute times**)

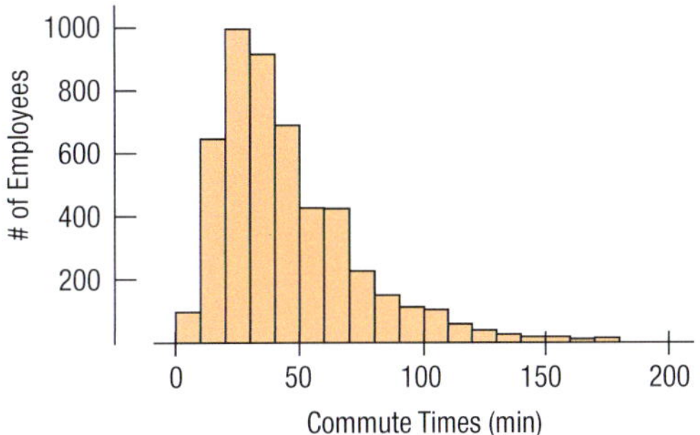

Figure 5.7

The distribution of all the commute times is skewed to the right with a mean of 45.4 minutes and a median of 38 minutes.

We then took 10,000 samples of size 100. The distribution of those means was unimodal and nearly symmetric (Figure 2.20 repeated here).

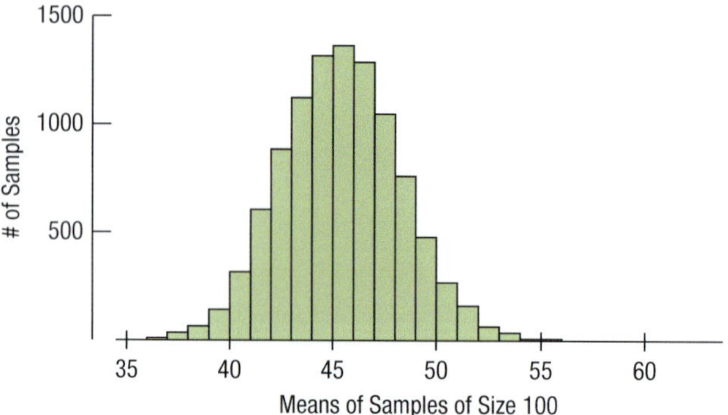

This distribution of sample means is centered at the mean of all 5000 times, but the standard deviation is much smaller than the standard deviation of all the values—it's only 2.87 minutes. Means vary less than observations. Using the 68–95–99.7 Rule on this distribution, we'd estimate that 95% of the samples have means between 39.65 and 51.13 minutes. In fact, in our simulation, 9534 out of 10,000, or 95.3%, were within two standard deviations of the mean. Figure 5.8 shows a histogram with the 95% between those limits shown in blue and the remaining 5% highlighted in red.

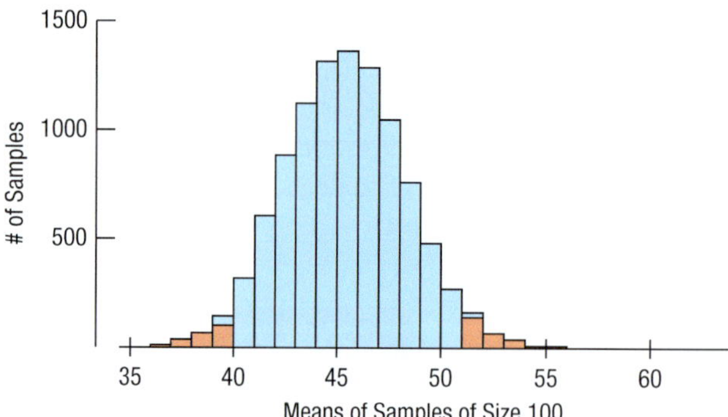

Figure 5.8

About 95% of the samples of commute times shown in Figure 5.7 have means between 39.65 and 51.13 minutes.

There are two striking features of this distribution. First, the distribution of simulated means is much more like a Normal distribution than the original skewed distribution of commuting times. Second, the standard deviation of these means is much smaller than the standard deviation of the commuting times themselves. The standard deviation of these means of samples of 100 is only about 1/10 of the standard deviation of the observations. Hmm . . . 1/10 is $1/\sqrt{100}$. Is that just a coincidence? No. We'll see and use this relationship soon. With these two facts we'll be able to say much more about the population than you might think even when we have only one sample.

The distribution that shows how much the means of samples vary is called the **sampling distribution**. Figure 5.8 shows a simulation of it. We can't show the actual sampling distribution because we'd need to have the sample mean of every possible sample of size 100 from the 5000 commute times. That's about 10^{120} samples (more than a googol). But we can simulate the sampling distribution like we did here. Sampling distributions will be fundamental to our discussion of statistical inference in later chapters.

JUST CHECKING

7. The mean of our original sample of 100 commute times was 55.1 minutes. Looking at the distribution of means in Figure 5.8, was our sample typical? Unusual? How far from the mean of 50.5 minutes would a sample mean have to be before you would consider it unusual?

8. Our original sample of 100 commute times had a standard deviation of 26.74 and a mean of 55.1 minutes.
   a) Looking back at Figure 2.19, would an employee commute time of 65 minutes be unusual?
   b) Would the 68–95–99.7 Rule be a good way to find the percentages of commute times? Why or why not?
   c) Would a sample mean of 65 minutes be unusual? Explain briefly.
   d) Why are your answers for parts a and c so different?

5.4 Working with Normal Percentiles

An SAT score of 600 is easy to assess because we can think of it as one standard deviation above the mean. If your score was 680, though, where do you stand among the rest of the people tested? Your SAT score of 680 has a z-score of $(680 - 500)/100 = 1.80$, so you're somewhere between one and two standard deviations above the mean. We figured out that no more than 16% of people score better than 600. By the same logic, no more than 2.5% of people score better than 700. Can we be more specific than "between 16% and 2.5%"?

When the value doesn't fall exactly one, two, or three standard deviations from the mean, we need help in finding the percentiles. Mathematically, the percentage of values falling between two z-scores is the area under the standard Normal model between those values. To find that area, use technology. In the early 20th century, before computing technology was widely available, statisticians made tables of z-scores and the associated **Normal percentiles**—the percentage of values in a standard Normal distribution found at that z-score or below. These days, even calculators and many smartphone apps can find the percentiles easily.

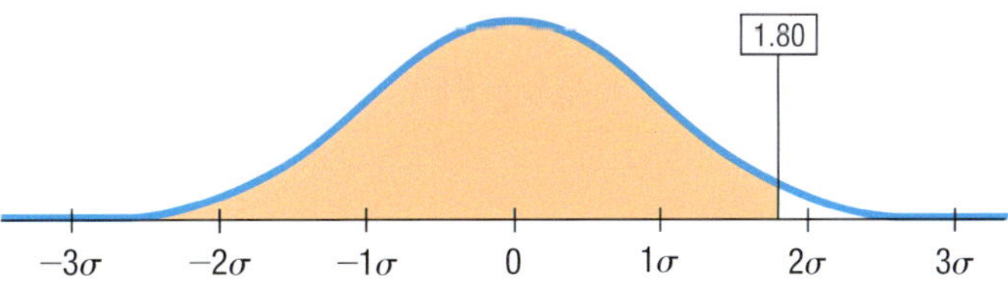

Figure 5.9
An app like the one at astools.datadesk.com/normal_table.html shows the percentage of individuals in a standard Normal distribution falling below any specified z-score value.

Of course, the Normal is not the only model for data. There are models for skewed data (watch out for χ^2 later in the text). But the Normal keeps coming back in different ways, and we'll see it again.

STEP-BY-STEP EXAMPLE

Working with Normal Models Part I

The Normal model is our first model for data. It's the first in a series of modeling situations where we step away from the data at hand to make more general statements about the world. We'll become more practiced in thinking about and learning the details of models as we progress through the text. To give you some practice in thinking about the Normal model, here are several problems that ask you to find percentiles in detail.

QUESTION: What proportion of SAT Reading scores fall between 450 and 600?

THINK

PLAN State the problem.

I want to know the proportion of SAT Reading scores between 450 and 600.

VARIABLES Name the variable.

Let y = SAT Reading score.

Check the appropriate conditions and specify which Normal model to use.

✓ **Nearly Normal Condition:** We are told that SAT Reading scores are nearly Normal.

I'll model SAT Reading scores with a $N(500, 100)$ model, using the mean and standard deviation specified for them.

SHOW

MECHANICS Make a picture of this Normal model. Locate the desired values and shade the region of interest.

Find z-scores for the cut points 450 and 600. Use technology to find the desired proportions, represented by the area under the curve.

Standardizing the two scores, I find that

$$z = \frac{(y - \mu)}{\sigma} = \frac{(600 - 500)}{100} = 1.00$$

and

$$z = \frac{(450 - 500)}{100} = -0.50.$$

So,

$$\text{Area}(450 < y < 600) = \text{Area}(-0.5 < z < 1.0)$$
$$= 0.5328.$$

(If you use a table, then you need to subtract the two areas to find the area *between* the cut points.)

TELL

CONCLUSION Interpret your result in context.

The normal model estimates that about 53.3% of SAT Reading scores fall between 450 and 600.

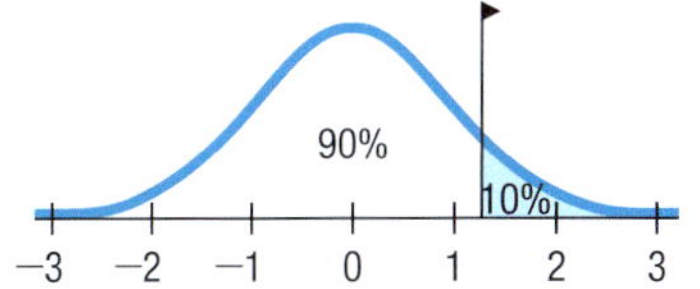

From Percentiles to Scores: *z* in Reverse

Finding areas from *z*-scores is the simplest way to work with the Normal model. But sometimes we start with areas and are asked to work backward to find the corresponding *z*-score or even the original data value. For instance, what *z*-score cuts off the top 10% in a Normal model?

Make a picture like the one shown, shading the rightmost 10% of the area. Notice that this is the 90th percentile, which corresponds to a *z*-score of $z = 1.28$.

STEP-BY-STEP EXAMPLE

Working with Normal Models Part II

QUESTION: Suppose a college says it admits only people with SAT Reading test scores among the top 10%. How high a score does it take to be eligible?

THINK **PLAN** State the problem.

How high an SAT Reading score do I need to be in the top 10% of all test takers?

VARIABLE Define the variable.

Let y = my SAT Reading score.

Check to see if a Normal model is appropriate, and specify which Normal model to use.

✓ **Nearly Normal Condition:** I am told that SAT Reading scores are nearly Normal. I'll model them with $N(500, 100)$.

SHOW **MECHANICS** Make a picture of this Normal model. Locate the desired percentile approximately by shading the rightmost 10% of the area.

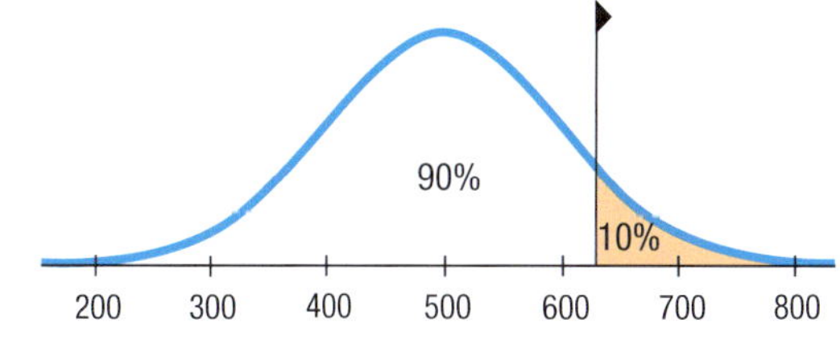

The college takes the top 10%, so its cutoff score is the 90th percentile. Find the corresponding *z*-score using technology or Table Z in the Appendix.

The cut point is $z = 1.28$.

Convert the *z*-score back to the original units.

A *z*-score of 1.28 is 1.28 standard deviations above the mean. Since the SD is 100, that's 128 SAT points. The cutoff is 128 points above the mean of 500, or 628.

TELL **CONCLUSION** Interpret your results in the proper context.

Because the school wants SAT Reading scores in the top 10%, the cutoff is 628. (Actually, since SAT scores are reported only in multiples of 10, I'd have to score at least a 630.)

STEP-BY-STEP EXAMPLE

More Working with Normal Models

Working with Normal percentiles can be a little tricky, depending on how the problem is stated. Here are a few more worked examples of the kind you're likely to see.

A cereal manufacturer has a machine that fills the boxes. Boxes are labeled "16 ounces," so the company wants to have that much cereal in each box, but since no packaging process is perfect, there will be minor variations. If the machine is set at exactly 16 ounces and the Normal model applies (or at least the distribution is roughly symmetric), then about half of the boxes will be underweight, making consumers unhappy and exposing the company to bad publicity and possible lawsuits. To prevent underweight boxes, the manufacturer has to set the mean a little higher than 16.0 ounces.

Based on their experience with the packaging machine, the company believes that the amount of cereal in the boxes fits a Normal model with a standard deviation of 0.2 ounce. The manufacturer decides to set the machine to put an average of 16.3 ounces in each box. Let's use that model to answer a series of questions about these cereal boxes.

QUESTION 1: What fraction of the boxes will be underweight?

THINK **PLAN** State the problem.

What proportion of boxes weigh less than 16 ounces?

VARIABLE Name the variable.

Let y = weight of cereal in a box.

Check to see if a Normal model is appropriate.

✓ **Nearly Normal Condition:** I have no data, so I cannot make a histogram, but I am told that the company believes the distribution of weights from the machine is Normal.

Specify which Normal model to use.

I'll use a $N(16.3, 0.2)$ model.

SHOW **MECHANICS** Make a picture of this Normal model. Locate the value you're interested in on the picture, label it, and shade the appropriate region.

REALITY CHECK Estimate from the picture the percentage of boxes that are underweight. (This will be useful later to check that your answer makes sense.) It looks like a low percentage. Less than 20% for sure.

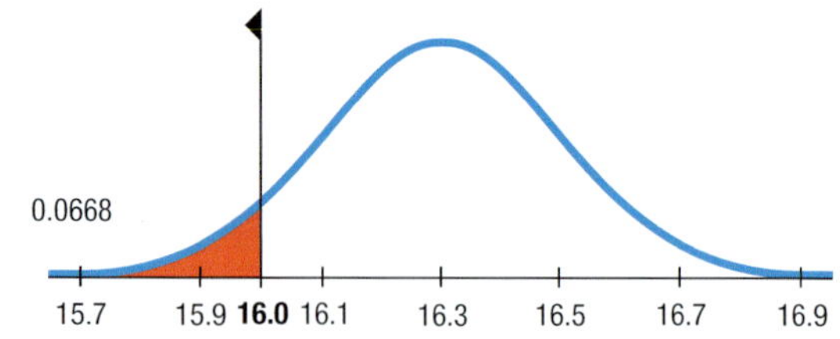

I want to know what fraction of the boxes will weigh less than 16 ounces.

Convert your cutoff value into a z-score.

$$z = \frac{y - \mu}{\sigma} = \frac{16 - 16.3}{0.2} = -1.50$$

Find the area with your calculator (or use the Normal table).

$$Area(y < 16) = Area(z < -1.50) = 0.0668$$

TELL **CONCLUSION** State your conclusion, and check that it's consistent with your earlier guess. It's below 20%—seems okay.

I estimate that approximately 6.7% of the boxes will contain less than 16 ounces of cereal.

QUESTION 2: The company's lawyers say that 6.7% is too high. They insist that no more than 4% of the boxes can be underweight. So the company needs to set the machine to put a little more cereal in each box. What mean setting do they need?

THINK

PLAN State the problem.

What mean weight will reduce the proportion of underweight boxes to 4%?

VARIABLE Name the variable.

Let y = weight of cereal in a box.

Check to see if a Normal model is appropriate.

✓ **Nearly Normal Condition:** I am told that a Normal model applies.

Specify which Normal model to use. This time you are not given a value for the mean!

I don't know μ, the mean amount of cereal. The standard deviation for this machine is 0.2 ounce. The model is $N(\mu, 0.2)$.

REALITY CHECK We found out earlier that setting the machine to $\mu = 16.3$ ounces made 6.7% of the boxes too light. We'll need to raise the mean a bit to reduce this fraction.

No more than 4% of the boxes can be below 16 ounces.

SHOW

MECHANICS Make a picture of this Normal model. Center it at μ (since you don't know the mean), and shade the region below 16 ounces.

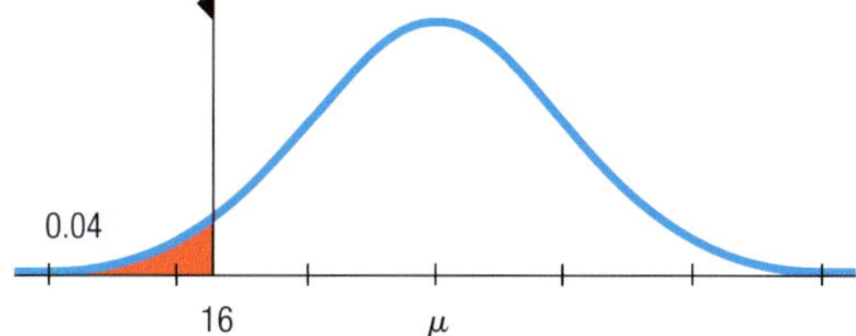

Using your calculator (or the Normal table), find the z-score that cuts off the lowest 4%.

The z-score that has 0.04 area to the left of it is $z = -1.75$.

Use this information to find μ. It's located 1.75 standard deviations to the right of 16. Since σ is 0.2, that's 1.75×0.2, or 0.35 ounce more than 16.

For 16 to be 1.75 standard deviations below the mean, the mean must be

$$16 + 1.75(0.2) = 16.35 \text{ ounces.}$$

CONCLUSION Interpret your result in context.

The company must set the machine to average 16.35 ounces of cereal per box.

(This makes sense; we knew it would have to be just a bit higher than 16.3.)

QUESTION 3: The company president vetoes that plan, saying the company should give away less free cereal, not more. Her goal is to set the machine no higher than 16.2 ounces and still have only 4% underweight boxes. The only way to accomplish this is to reduce the standard deviation. What standard deviation must the company achieve, and what does that mean about the machine?

THINK **PLAN** State the problem.

What standard deviation will allow the mean to be 16.2 ounces and still have only 4% of boxes underweight?

VARIABLE Name the variable.

Let y = weight of cereal in a box.

Check conditions to be sure that a Normal model is appropriate.

✓ **Nearly Normal Condition:** The company believes that the weights are described by a Normal model.

Specify which Normal model to use. This time you don't know σ.

I know the mean, but not the standard deviation, so my model is $N(16.2, \sigma)$.

REALITY CHECK We know the new standard deviation must be less than 0.2 ounce.

SHOW **MECHANICS** Make a picture of this Normal model. Center it at 16.2, and shade the area you're interested in. We want 4% of the area to the left of 16 ounces.

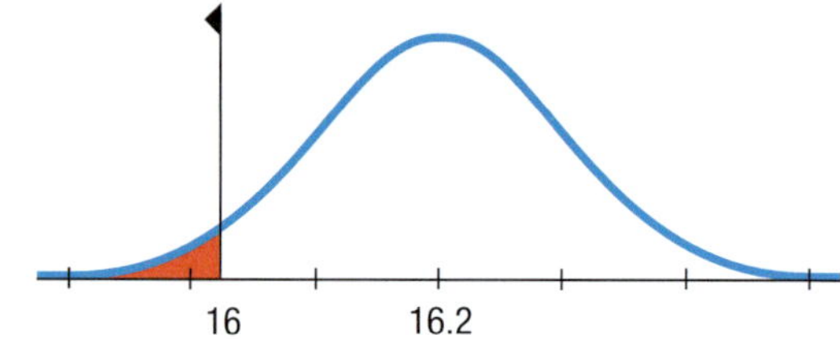

Find the z-score that cuts off the lowest 4%.

I know that the z-score with 4% below it is $z = -1.75$.

Solve for σ. (We need 16 to be 1.75 σ's below 16.2, so 1.75 σ must be 0.2 ounce. You could just start with that equation.)

$$z = \frac{y - \mu}{\sigma}$$

$$-1.75 = \frac{16 - 16.2}{\sigma}$$

$$1.75\sigma = 0.2$$

$$\sigma = 0.114$$

TELL **CONCLUSION** Interpret your result in context.

As we expected, the standard deviation is lower than before—actually, quite a bit lower.

The company must get the machine to box cereal with a standard deviation of only 0.114 ounce. This means the machine must be more consistent (by nearly a factor of 2) in filling the boxes.

5.5 Normal Probability Plots

In the examples we've worked through, we've assumed that the underlying data distribution was roughly unimodal and symmetric so that using a Normal model makes sense. When you have data, you must *check* to see whether a Normal model is reasonable. How? Make a picture, of course! Drawing a histogram of the data and looking at the shape is one good way to see whether a Normal model might work.

There's a more specialized graphical display that can help you to decide whether a Normal model is appropriate: the **Normal probability plot**. If the distribution of the data is roughly Normal, the plot will be roughly a diagonal straight line. Deviations from a straight line indicate that the distribution is not Normal. This plot is usually able to show deviations from Normality more clearly than the corresponding histogram, but it's usually easier to understand *how* a distribution fails to be Normal by looking at its histogram.

Some data on a car's fuel efficiency provide an example of data that are nearly Normal. The overall pattern of the Normal probability plot is straight. The two trailing low values correspond to the values in the histogram that trail off the low end. They're not quite in line with the rest of the dataset. The Normal probability plot shows us that they're a bit lower than we'd expect of the lowest two values in a Normal model.

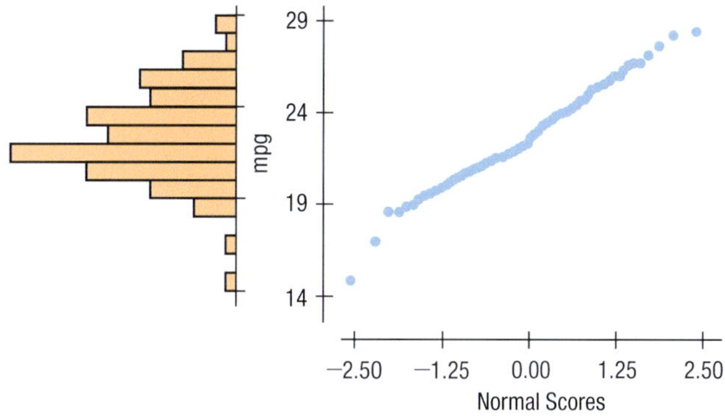

Figure 5.10
Histogram and Normal probability plot for gas mileage (mpg) recorded by one of the authors over the 8 years he owned a Nissan Maxima. (Data in **Nissan**) The vertical axes are the same, so each dot on the probability plot would fall into the bar on the histogram immediately to its left.

By contrast, the Normal probability plot of the men's *Weights* from the NHANES Study is far from straight. The weights are skewed to the high end, and the plot is curved. We'd conclude from these pictures that approximations using the 68–95–99.7 Rule for these data would not be very accurate.

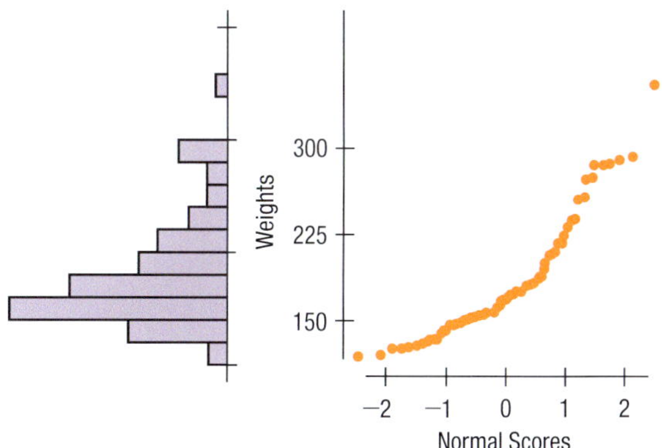

Figure 5.11
Histogram and Normal probability plot for men's weights. (Data in **Mens weights**) Note how a skewed distribution corresponds to a bent probability plot.

How Does a Normal Probability Plot Work?

Why does the Normal probability plot work like that? We looked at 100 fuel efficiency measures for the author's car. The smallest of these has a *z*-score of −3.16. The Normal model can tell us what value to expect for the smallest *z*-score in a batch of 100 if a Normal model were appropriate. That turns out to be −2.58. So our first data value is smaller than we would expect from the Normal.

We can continue this and ask a similar question for each value. For example, the 14th-smallest fuel efficiency has a z-score of almost exactly -1, and that's just what we should expect (well, -1.1 to be exact). A Normal probability plot takes each data value and plots it against the z-score you'd expect that point to have if the distribution were perfectly Normal.[3]

When the values match up well, the line is straight. If one or two points are surprising from the Normal's point of view, they don't line up. When the entire distribution is skewed or different from the Normal in some other way, the values don't match up very well at all and the plot bends.

It turns out to be tricky to find the values we expect. They're called *Normal scores,* but you can't easily look them up in the tables. That's why probability plots are best made with technology and not by hand.

The best advice on using Normal probability plots is to see whether they are straight. If so, then your data look like data from a Normal model. If not, make a histogram to understand how they differ from the model.

WHAT CAN GO WRONG?

- **Don't use a Normal model when the distribution is not unimodal and symmetric.** Normal models are so easy and useful that it is tempting to use them even when they don't describe the data very well. That can lead to wrong conclusions. Don't use a Normal model without first checking the **Nearly Normal Condition**. Look at a picture of the data to check that it is unimodal and symmetric. A histogram, or a Normal probability plot, can help you tell whether a Normal model is appropriate.

  The CEOs (Example 5.5) had a mean total compensation of \$13.90M and a standard deviation of \$11.36M. Using the Normal model rule, we should expect about 95% of the CEOs to have compensations between $-$\$8.82M and \$36.62M. In fact, more than 97% of the CEOs have annual compensations in this range. What went wrong? The distribution is skewed, not symmetric. Using the 68–95–99.7 Rule for data like these will lead to silly results.

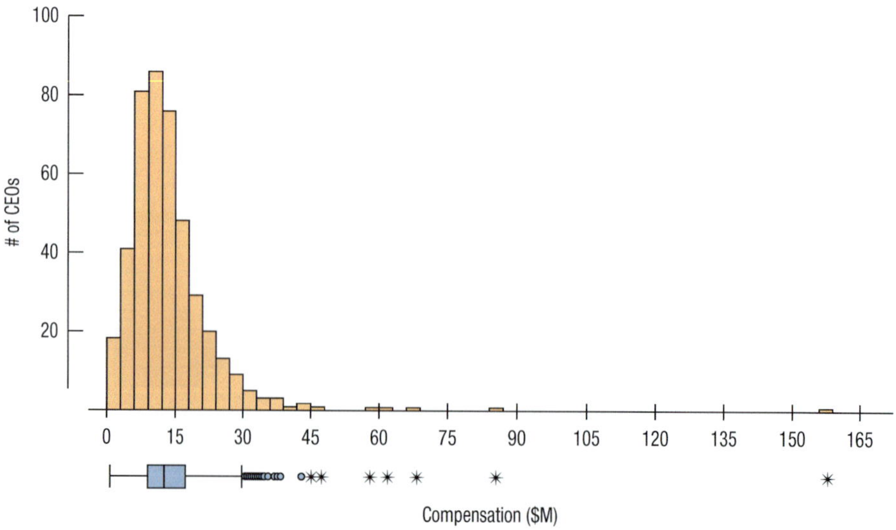

- **Don't use the mean and standard deviation when outliers are present.** Both means and standard deviations can be distorted by outliers, and no model based on distorted values will do a

[3]Some programs make a Normal probability plot with the two axes switched, putting the data on the x-axis and the z-scores on the y-axis.

good job. A z-score calculated from a distribution with outliers may be misleading. It's always a good idea to check for outliers. How? Make a picture.

- **Don't round your results in the middle of a calculation.** We reported the mean of the heptathletes' 200 m run as 24.58 seconds. More precisely, it was 24.582069 seconds.

  You should use all the precision available in the data for all the intermediate steps of a calculation. Using the more precise value for the mean (and also carrying 8 digits for the SD), the z-score calculation for Thiam's run comes out to

$$z = \frac{25.10 - 24.582069}{0.65449751} = 0.79134149.$$

  We'd likely report that as 0.791, as opposed to the rounded-off value of 0.795 we got earlier.

- **Do what we say, not what we do.** When we showed the z-score calculations for Thiam, we rounded the mean to 25.10 s and the SD to 0.654 s. Then to make the story clearer we used *those values* in the displayed calculation.

  We'll continue to show simplified calculations in the text to make the story simpler. When you calculate with full precision, your results may differ slightly from ours. So, we also advise . . .

- **Don't worry about minor differences in results.** Because various calculators and programs may carry different precision in calculations, your answers may differ slightly from those in the text and in the Step-by-Steps, or even from the values given in the answers in the back of the text. Those differences aren't anything to worry about. They're not the main story statistics tries to tell.

CONNECTIONS

Changing the center and spread of a variable is equivalent to changing its *units*. Indeed, the only part of the data's context changed by standardizing is the units. All other aspects of the context do not depend on the choice or modification of measurement units. This fact points out an important distinction between the numbers the data provide for calculation and the meaning of the variables and the relationships among them. Standardizing can make the numbers easier to work with, but it does not alter the meaning.

Another way to look at this is to note that standardizing may change the center and spread values, but it does not affect the *shape* of a distribution. A histogram or boxplot of standardized values looks just the same as the histogram or boxplot of the original values except, perhaps, for the numbers on the axes.

When we summarized *shape, center*, and *spread* for histograms, we compared them to unimodal, symmetric shapes. You couldn't ask for a nicer example than the Normal model. And if the shape *is* like a Normal, we'll use the mean and standard deviation to standardize the values.

CHAPTER REVIEW

Understand how z-scores facilitate comparisons by standardizing variables to have zero mean and unit standard deviation. Recognize normally distributed data by making a histogram and checking whether it is unimodal, symmetric, and bell-shaped, or by making a Normal probability plot using technology and checking whether the plot is roughly a straight line.

- The Normal model is a distribution that will be important for much of the rest of this course.
- Before using a Normal model, we should check that our data are plausibly from a normally distributed population.
- A Normal probability plot provides evidence that the data are Normally distributed if it is linear.

Understand how to use the Normal model to judge whether a value is extreme.

- Standardize values to make z-scores and obtain a standard scale. Then refer to a standard Normal distribution.
- Use the 68–95–99.7 Rule as a rule-of-thumb to judge whether a value is extreme.

Know how to refer to tables or technology to find the probability of a value randomly selected from a Normal model falling in any interval.

- Know how to perform calculations about Normally distributed values and probabilities.

REVIEW OF TERMS

Standardize We standardize to eliminate units. Standardized values can be compared and combined even if the original variables had different units and magnitudes (p. 153).

Standardized value A value found by subtracting the mean and dividing by the standard deviation (p. 153).

z-score A z-score tells how many standard deviations a value is from the mean; z-scores have a mean of 0 and a standard deviation of 1 (pp. 153, 160).

When working with data, use the statistics $\bar{y}$ and s:

$$z = \frac{y - \bar{y}}{s}.$$

When working with models, use the parameters μ and σ:

$$z = \frac{y - \mu}{\sigma}.$$

Shifting Adding a constant to each data value adds the same constant to the mean, the median, and the quartiles, but does not change the standard deviation or IQR (p. 156).

Rescaling Multiplying each data value by a constant multiplies both the measures of position (mean, median, and quartiles) and the measures of spread (standard deviation and IQR) by that constant (p. 156).

68–95–99.7 Rule In a Normal model, about 68% of values fall within 1 standard deviation of the mean, about 95% fall within 2 standard deviations of the mean, and about 99.7% fall within 3 standard deviations of the mean (p. 159).

Normal (or Gaussian) model A useful family of models for unimodal, symmetric distributions (p. 159).

Parameter A numerically valued attribute of a model. For example, the values of μ and σ in a $N(\mu, \sigma)$ model are parameters (p. 159).

Statistic A value calculated from data to summarize aspects of the data. For example, the mean, $\bar{y}$, and standard deviation, s, are statistics (p. 159).

Standard Normal model A Normal model, $N(\mu, \sigma)$ with mean $\mu = 0$ and standard deviation $\sigma = 1$. Also called the standard Normal distribution (p. 160).

Nearly Normal Condition A distribution is nearly Normal if it is unimodal and symmetric. We can check by looking at a histogram or a Normal probability plot (p. 161).

Sampling distribution The sampling distribution of a statistic is the distribution of all values of the statistic if it was computed for every possible sample (of a given size). We can approximate it by drawing many samples and making a histogram of the resulting statistic values (p. 164).

Normal percentile The Normal percentile corresponding to a z-score gives the percentage of values in a standard Normal distribution found at that z-score or below (p. 165).

Normal probability plot A display to help assess whether a distribution of data is approximately Normal. If the plot is nearly straight, the data satisfy the Nearly Normal Condition (p. 171).

TECH SUPPORT

Normal Probability Plots

The best way to tell whether your data can be modeled well by a Normal model is to make a picture or two. We've already talked about making histograms. Normal probability plots are almost never made by hand because the values of the Normal scores are tricky to find. But most statistics software make Normal plots, though various packages call the same plot by different names and array the information differently.

DATA DESK

To make a Normal probability plot in Data Desk:

- Select the Variable.
- Choose **Normal Prob Plot** from the Plot menu.

COMMENTS

Data Desk places the ordered data values on the vertical axis and the Normal scores on the horizontal axis.

You can also find Normal Probability Plot commands in the HyperView menus of variable names on the axis of a histogram and in other plots and analyses we'll see in later chapters.

EXCEL

Excel offers a Normal probability plot as part of the Regression command in the Data Analysis extension, but (as of this writing) it is not a correct Normal probability plot and should not be used.

JMP

To make a Normal quantile plot in JMP:

- Make a histogram using **Distributions** from the Analyze menu.
- Click on the drop-down menu next to the variable name.
- Choose **Normal Quantile Plot** from the drop-down menu.
- JMP opens the plot next to the histogram.

COMMENTS

JMP places the ordered data on the vertical axis and the Normal scores on the horizontal axis. The vertical axis aligns with the histogram's axis, a useful feature.

MINITAB

To make a Normal probability plot in MINITAB:

- Choose **Probability Plot** from the Graph menu.
- Select **Single** for the type of plot. Click **OK**.
- Enter the name of the variable in the Graph variables box. Click **OK**.

COMMENTS

MINITAB places the ordered data on the horizontal axis and the Normal scores on the vertical axis.

R

To make a Normal probability (Q-Q) plot for Y:

- **qqnorm(Y)** will produce the plot.

To standardize a variable Y:

- **Z = (Y − mean(Y))/sd(Y)** will create a standardized variable Z.

COMMENTS

By default, R places the ordered data on the vertical axis and the Normal scores on the horizontal axis, but that can be reversed by setting **datax = TRUE** inside qqnorm.

SPSS

To make a Normal P-P plot in SPSS:

- Choose **P-P** from the Graphs menu.
- Select the variable to be displayed in the source list.
- Click the arrow button to move the variable into the target list.
- Click the **OK** button.

COMMENTS

SPSS places the ordered data on the horizontal axis and the Normal scores on the vertical axis. You may safely ignore the options in the P-P dialog.

STATCRUNCH

To make a Normal probability plot:

- Click on **Graph**.
- Choose **QQ Plot**.
- Choose the variable name from the list of Columns.
- Click on **Compute!**.

To work with Normal percentiles:

- Click on **Stat**.
- Choose **Calculators » Normal**.
- Choose a lower tail ($\le$) or upper tail ($\ge$) region.
- Enter the *z*-score cutoff, and then click on **Compute** to find the probability.

OR

Enter the desired probability, and then click on **Compute** to find the *z*-score cutoff.

TI-83/84 PLUS

To create a Normal percentile plot on the TI-83:

- Set up a **STAT PLOT** using the last of the Types.
- Specify your datalist, and the axis you choose to represent the data.
- Although most people wouldn't open a statistics package just to find a Normal model value they could find in a table, you *would* use a calculator for that function.
- So . . . to find what percent of a Normal model lies between two *z*-scores, choose **normalcdf** from the DISTRibutions menu and enter the command **normalcdf(zLeft, zRight)**.
- To find the *z*-score that corresponds to a given percentile in a Normal model, choose **invNorm** from the DISTRibutions menu and enter the command **invNorm(percentile)**.

COMMENTS

We often want to find Normal percentages from a certain *z*-score to infinity. On the calculator, indicate "infinity" as a very large *z*-score, say, 99. For example, the percentage of a Normal model over 2 standard deviations above the mean can be evaluated with **normalcdf(2, 99)**.

To make a Normal Probability plot:

- Turn a STATPLOT On.
- Tell it to make a Normal probability plot by choosing the last of the icons.
- Specify your datalist and which axis you want the data on. (Use Y to make the plot look like those here.)
- Specify the Mark you want the plot to use.
- Now ZoomStat does the rest.

EXERCISES

SECTION 5.1

1. **Stats test** The mean score on the Stats exam was 75 points with a standard deviation of 5 points, and Gregor's *z*-score was −2. How many points did he score?

2. **Mensa** People with *z*-scores above 2.5 on an IQ test are sometimes classified as geniuses. If IQ scores have a mean of 100 and a standard deviation of 15 points, what IQ score do you need to be considered a genius?

3. **Temperatures** A town's January high temperatures average 36°F with a standard deviation of 10°, while in July the mean high temperature is 74° and the standard deviation is 8°. In which month is it more unusual to have a day with a high temperature of 55°? Explain.

4. Placement exams Clara took her college's placement exams in Physics and Statistics. In Physics, she scored 85 and in Statistics 88. The overall results on the Physics exam had a mean of 74 and a standard deviation of 6, while the mean Statistics score was 80, with a standard deviation of 7. On which exam did she do better compared to the other students?

SECTION 5.2

5. Food packages A food company that ships food packages worldwide claims that its food packages have a median weight of 3 kilograms and an IQR of 1.9 kilograms.

a) The company plans to include a new advertisement sheet weighing 100 grams in each package. What will be the new median and IQR?
b) If the company recorded the shipping weights of these new packages in pounds instead of kilograms, what would be the median and IQR? $(0.4536 \text{ kg} = 1 \text{ lb})$

6. Hotline A company's customer service hotline handles many calls relating to orders, refunds, and other issues. The company's records indicate that the median length of calls to the hotline is 4.4 minutes with an IQR of 2.3 minutes.

a) If the company were to describe the duration of these calls in seconds instead of minutes, what would the median and IQR be?
b) In an effort to speed up the customer service process, the company decides to streamline the series of pushbutton menus customers must navigate, cutting the time by 24 seconds. What will the median and IQR of the length of hotline calls become?

7. Men's shoe sizes In Chapter 2 (Exercise 16) we saw data on shoe sizes of students, reported in European sizes. For the men, the mean size was 44.65 with a standard deviation of 2.03. To convert euro shoe sizes to U.S. sizes for men, use the equation

$$\text{USsize} = \text{EuroSize} \times 0.7865 - 24.$$

a) What is the mean men's shoe size for these respondents in U.S. units?
b) What is the standard deviation in U.S. units?

8. Women's shoe sizes The shoe size data for women has a mean of 38.46 and a standard deviation of 1.84. To convert to U.S. sizes, use

$$\text{USsize} = \text{EuroSize} \times 0.7865 - 22.5.$$

a) What is the mean women's shoe size for these respondents in U.S. units?
b) What is the standard deviation in U.S. units?

SECTION 5.3

9. Newborn baby weights A hospital keeps record of the weights of newborn babies and finds that these are normally distributed with a mean weight of 3.5 kg and a standard deviation of 0.5 kg.

a) Draw the normal model for newborn baby weights. Clearly label it, showing what the 68–95–99.7 rule predicts.
b) In what weight interval would you expect the central 68% of newborn babies to be found?
c) About what percent of newborn babies should weigh more than 4.5 kilograms?
d) About what percent of newborn babies should weigh between 2.5 and 4 kilograms?
e) About what percent of newborn babies should weigh less than 2.5 kilograms?

10. IQ Some IQ tests are standardized to a Normal model, with a mean of 100 and a standard deviation of 15.

a) Draw the model for these IQ scores. Clearly label it, showing what the 68–95–99.7 Rule predicts.
b) In what interval would you expect the central 95% of IQ scores to be found?
c) About what percent of people should have IQ scores above 115?
d) About what percent of people should have IQ scores between 70 and 85?
e) About what percent of people should have IQ scores above 130?

11. Checkup One of the authors has an adopted grandson whose birth family members are very short. After examining him at his 2-year checkup, the boy's pediatrician said that the z-score for his height relative to American 2-year-olds was -1.88. Write a sentence explaining what that means.

12. Stats test Suppose your statistics professor reports test grades as z-scores, and you got a score of 2.20 on an exam.

a) Write a sentence explaining what that means.
b) Your friend got a z-score of -1. If the grades satisfy the Nearly Normal Condition, about what percent of the class scored lower than your friend?

SECTION 5.4

13. Park visitors' heights At an amusement park, visitors have to be taller than 1.5 meters to ride on the Rock n' Roll coaster. From past records, the management of the amusement park knows that the height of park visitors is normally distributed with mean 1.6 meters and standard deviation of 0.4 meters. What percent of park visitors have a height that is

a) above 1.7 meters?
b) below 1.4 meters?
c) between 1.35 and 1.75 meters?

14. IQs revisited Based on the Normal model $N(100, 15)$ describing IQ scores, what percent of people's IQs would you expect to be

a) over 80?
b) under 90?
c) between 112 and 132?

15. ACT scores The Mathematics section of the ACT test had a mean of 20.9 and an SD of 5.3 for the years 2013–2015. If these are well modeled by a Normal distribution, about what percent of students scored

a) over 31?
b) under 18?
c) between 18 and 31?

16. Annual incomes The mean annual income of a group of assistant chefs is €20,500, while the median income is €24,595. The standard deviation is €1,800.

a) If a Normal model is used for these incomes, what would be the annual income of the top 1%?
b) How confident are you in the answer of part (a)?
c) Do you think that the Normal model is appropriate for these incomes? Explain.

SECTION 5.5

17. Music library Corey has 4929 songs in his computer's music library. The lengths of the songs have a mean of 242.4 seconds and standard deviation of 114.51 seconds. A Normal probability plot of the song lengths looks like this:

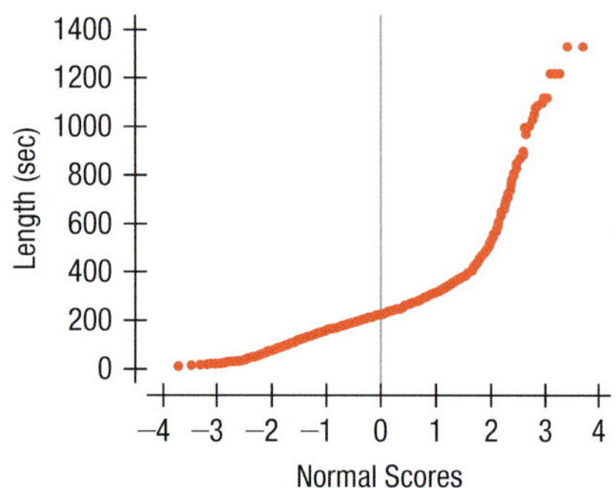

a) Do you think the distribution is Normal? Explain.
b) If it isn't Normal, how does it differ from a Normal model?

T 18. Wisconsin ACT 2015 The histogram shows the distribution of mean ACT composite scores for all Wisconsin public schools in 2015. 80.1% of the data points fall between one standard deviation below the mean and one standard deviation above the mean.

a) Give two reasons that a Normal model is not appropriate for these data.

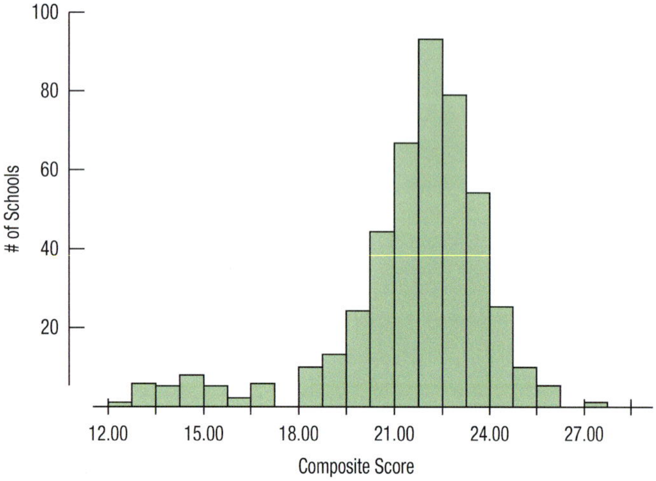

b) The Normal probability plot on the left shows the distribution of these scores. The plot on the right shows the same data with the Milwaukee area schools (mostly in the low mode) removed. What do these plots tell you about the shape of the distributions?

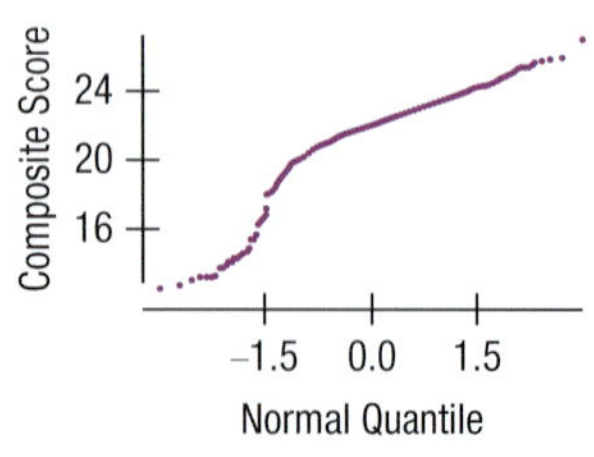

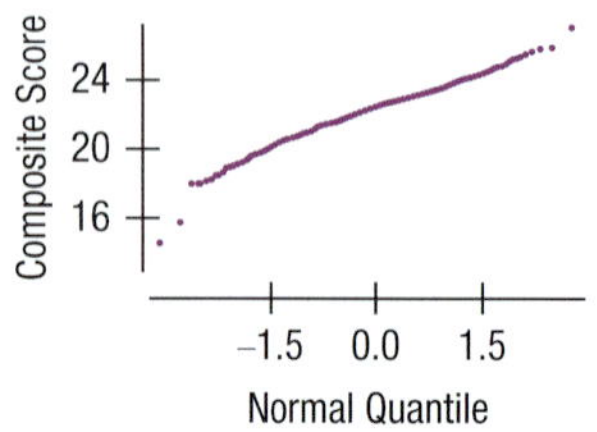

19. Payroll Here are the summary statistics for the weekly payroll of a small company: lowest salary = \$300, mean salary = \$700, median = \$500, range = \$1200, IQR = \$600, first quartile = \$350, standard deviation = \$400.

a) Do you think the distribution of salaries is symmetric, skewed to the left, or skewed to the right? Explain why.
b) Between what two values are the middle 50% of the salaries found?
c) Suppose business has been good and the company gives every employee a \$50 raise. Tell the new value of each of the summary statistics.
d) Instead, suppose the company gives each employee a 10% raise. Tell the new value of each of the summary statistics.

20. Parcel deliveries The parcels delivered by an online clothing company during the year 2019 had weights varying between 21 and 106 ounces, with a mean weight of 48 ounces, and standard deviation of 18 ounces. The quartiles and median weights were 33, 65, and 87 ounces, respectively.

a) Find the range and the IQR of the weights.
b) Do you think the distribution of the weights is symmetric or skewed? Why? If skewed, which way?
c) If these weights were expressed in kilograms (35.274 ounces = 1 kg), what would the mean, standard deviation, quartiles, median, IQR, and range be?
d) When the company ships these parcels, the packing materials add 15 ounces. What are the mean, standard deviation, quartiles, median, IQR, and range of weights of boxes shipped (in ounces)?
e) For the mean of the parcel with packing materials to be at most 60 ounces, what should be the maximum weight of the packing materials? For this maximum weight, what are the standard deviation, quartiles, IQR, and range of weights of boxes shipped (in ounces)?

21. SAT or ACT? Each year thousands of high school students take either the SAT or the ACT, standardized tests used in the college admissions process. Combined SAT Math and Verbal scores go as high as 1600, while the maximum ACT composite score is 36. Since the two exams use very different scales, comparisons of performance are difficult. A convenient rule of thumb is $SAT = 40 \times ACT + 150$; that is, multiply an ACT score by 40 and add 150 points to estimate the equivalent SAT score. An admissions officer reported the following statistics about the ACT scores of 2355 students who applied to her college one year. Find the summaries of equivalent SAT scores.

Lowest score = 19 Mean = 27 Standard deviation = 3

Q3 = 30 Median = 28 IQR = 6

22. Cold U? A high school senior uses the Internet to get information on February temperatures in the town where he'll be going to college. He finds a website with some statistics, but they are given in degrees Celsius. The conversion formula is $°F = 9/5°C + 32$. Determine the Fahrenheit equivalents for the summary information below.

Maximum temperature = 11°C Range = 33°

Mean = 1° Standard deviation = 7°

Median = 2° IQR = 16°

23. Music library again Corey has 4929 songs in his computer's music library. The songs have a mean duration of 242.4 seconds with a standard deviation of 114.51 seconds. *On the Nickel*, by Tom Waits, is 380 seconds long. What is its *z*-score?

24. Windy In the last chapter, we looked at three outliers arising from a plot of *Average Wind Speed* by *Month* in the Hopkins Forest. Each was associated with an unusually strong storm, but which was the most remarkable for its month?

Here are the summary statistics for each of those three months:

	February	June	August
Mean	2.324	0.857	0.63
SD	1.577	0.795	0.597

The outliers had values of 6.73 mph, 3.93 mph, and 2.53 mph, respectively.

a) What are their z-scores?
b) Which was the most extraordinary wind event?

25. Combining test scores The first Stats exam had a mean of 65 and a standard deviation of 10 points; the second had a mean of 80 and a standard deviation of 5 points. Derrick scored an 80 on both tests. Julie scored a 70 on the first test and a 90 on the second. They both totaled 160 points on the two exams, but Julie claims that her total is better. Explain.

26. Combining scores again The first Stat exam had a mean of 80 and a standard deviation of 4 points; the second had a mean of 70 and a standard deviation of 15 points. Reginald scored an 80 on the first test and an 85 on the second. Sara scored an 88 on the first but only a 65 on the second. Although Reginald's total score is higher, Sara feels she should get the higher grade. Explain her point of view.

27. Exam performance Two classmates, Pearl and Faye, wanted to compare their performance on the Mathematics and Accounting exams. Pearl scored 81 on the Mathematics exam and 84 on the Accounting exam, while Faye scored 77 and 90 on the Mathematics and Accounting exams, respectively. The scores of the whole class for the Mathematics exam had a mean of 80 and a standard deviation of 4, and a mean of 82 and a standard deviation of 9 for Accounting.

a) By calculating the average of the Mathematics and Accounting scores for both classmates, indicate which student had the better mean score in the two exams.
b) Which student's overall performance was better?

28. MP3s Two companies market new batteries targeted at owners of personal music players. DuraTunes claims a mean battery life of 11 hours, while RockReady advertises 12 hours.

a) Explain why you would also like to know the standard deviations of the battery lifespans before deciding which brand to buy.
b) Suppose those standard deviations are 2 hours for DuraTunes and 1.5 hours for RockReady. You are headed for 8 hours at the beach. Which battery is most likely to last all day? Explain.
c) If your beach trip is all weekend, and you probably will have the music on for 16 hours, which battery is most likely to last? Explain.

29. Park visitors' heights, part II Using $N(1.6, 0.4)$, the Normal model for heights of park visitors in Exercise 13, answer the following questions:

a) How many standard deviations from the mean would a park visitor with height 1.55 meters be?
b) Which would be more unusual, a park visitor's height being 1.53 meters or 1.71 meters?

T **30. Car speeds 100** John Beale of Stanford, California, recorded the speeds of cars driving past his house, where the speed limit read 20 mph. The mean of 100 readings was 23.84 mph, with a standard deviation of 3.56 mph. (He actually recorded every car for a two-month period. These are 100 representative readings.)

a) How many standard deviations from the mean would a car going under the speed limit be?
b) Which would be more unusual, a car traveling 34 mph or one going 10 mph?

31. Metal plates The thickness of a set of metal plates can be described by a Normal model with a mean of 4.3 mm and a standard deviation of 0.12 mm.

a) Suppose that the company manufacturing these metal plates decides to increase the thickness of each metal plate by 0.2 mm. What would be the new mean and standard deviation?
b) The metal plates sell at €9.95 per mm thickness. Find the mean and standard deviation of the sale prices for all metal plates with the increased thickness.

T **32. Car speeds 100 again** For the car speed data in Exercise 30, recall that the mean speed recorded was 23.84 mph, with a standard deviation of 3.56 mph. To see how many cars are speeding, John subtracts 20 mph from all speeds.

a) What is the mean speed now? What is the new standard deviation?
b) His friend in Berlin wants to study the speeds, so John converts all the original miles-per-hour readings to kilometers per hour by multiplying all speeds by 1.609 (km per mile). What is the mean now? What is the new standard deviation?

33. Metal plates, part II Suppose the company in Exercise 31 sold a set of metal plates whose minimum thickness was 4.1 mm, median 4.42 mm, standard deviation 0.098 mm, and IQR 1.34 mm. They sold for €9.95 per mm thickness, while total production cost was €17.56. Find the minimum, median, standard deviation, and IQR of the net sale prices.

34. Caught speeding Suppose police set up radar surveillance on the Stanford street described in Exercise 30. They handed out a large number of tickets to speeders going a mean of 28 mph, with a standard deviation of 2.4 mph, a maximum of 33 mph, and an IQR of 3.2 mph. Local law prescribes fines of \$100, plus \$10 per mile per hour over the 20 mph speed limit. For example, a driver convicted of going 25 mph would be fined $100 + 10(5) = \$150$. Find the mean, maximum, standard deviation, and IQR of all the potential fines.

35. Middle-aged individuals A researcher has told a colleague about a study he is conducting among middle-aged individuals. He has told her the mean and the standard deviation of the ages of the respondents. The colleague remembers that the mean age is 48; however, she cannot recall whether the standard deviation is 1 year, 2 years, or 5 years. Tell her which one it must have been, and why.

36. Club attendance The management of a club recorded the attendance of people entering the club every Saturday for a whole year. The management reported that the mean attendance was 483 individuals. While the management did not report the standard deviation, which of these values do you think is most likely to be correct and why: 10, 200, or 400 individuals?

37. Short park visitors In Exercise 29, we suggested the model $N(1.6, 0.4)$ for heights of park visitors. What height would be unusual to record for a park visitor? Explain.

38. High IQ Exercise 10 proposes modeling IQ scores with $N(100, 15)$. What IQ would you consider to be unusually high? Explain.

39. Trees A forester measured 27 of the trees in a large woods that is up for sale. He found a mean diameter of 10.4 inches and a standard deviation of 4.7 inches. Suppose that these trees provide an accurate description of the whole forest and that a Normal model applies.

a) Draw the Normal model for tree diameters.
b) What size would you expect the central 95% of all trees to be?
c) About what percent of the trees should be less than an inch in diameter?
d) About what percent of the trees should be between 5.8 and 10.4 inches in diameter?
e) About what percent of the trees should be over 15 inches in diameter?

40. Rivets A company that manufactures rivets believes the shear strength (in pounds) is modeled by $N(800, 50)$.

a) Draw and label the Normal model.
b) Would it be safe to use these rivets in a situation requiring a shear strength of 750 pounds? Explain.
c) About what percent of these rivets would you expect to fall below 900 pounds?
d) Rivets are used in a variety of applications with varying shear strength requirements. What is the maximum shear strength for which you would feel comfortable approving this company's rivets? Explain your reasoning.

41. Trees, part II Later on, the forester in Exercise 39 shows you a histogram of the tree diameters he used in analyzing the woods that was for sale. Do you think he was justified in using a Normal model? Explain, citing some specific concerns.

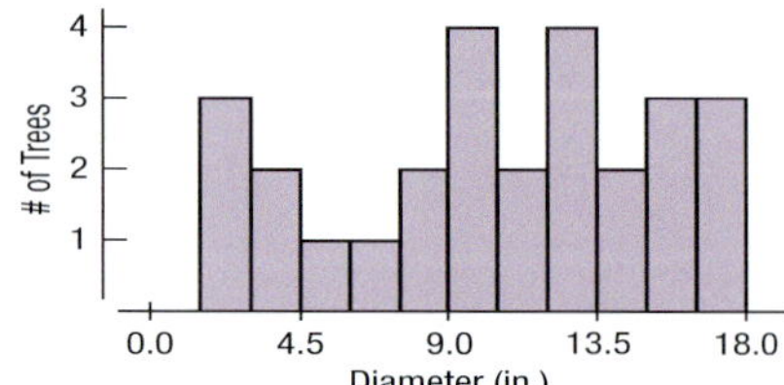

T **42. Car speeds 100, the picture** For the car speed data in Exercise 30, here are the histogram, boxplot, and Normal probability plot of the 100 readings. Do you think it is appropriate to apply a Normal model here? Explain.

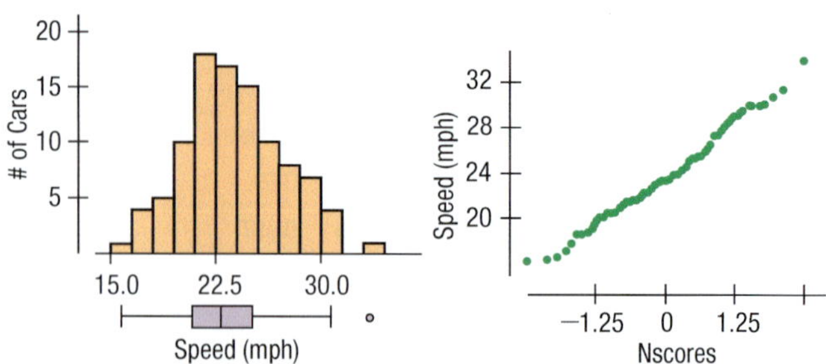

T **43. Winter Olympics 2018** Fifty-three men completed the men's alpine downhill at the 2018 winter Olympics in PyeongChang. The gold medal winner finished in 100.25 seconds. Here are the times (in seconds) for all competitors.

100.25	100.37	100.43	100.79	101.03	101.08
101.19	101.39	101.46	101.62	101.75	101.76
101.86	101.89	102.18	102.22	102.23	102.39
102.53	102.59	102.82	102.84	102.96	102.96
102.98	103.99	103.03	103.19	103.61	103.72
103.78	103.80	103.88	104.02	104.37	104.48
104.65	104.79	105.01	105.21	105.36	105.42
105.61	105.86	106.42	106.60	107.87	107.99
108.57	108.81	109.50	109.98	111.72	

a) The mean time was 103.883 seconds, with a standard deviation of 2.645 seconds. If the Normal model is appropriate, what percent of times will be less than 101.233 seconds?
b) What is the actual percent of times less than 101.233 seconds?
c) Why do you think the two percentages don't agree?
d) Make a histogram of these times. What do you see?

T **44. Check the model** The mean of the 100 car speeds in Exercise 30 was 23.84 mph, with a standard deviation of 3.56 mph.

a) Using a Normal model, what values should border the middle 95% of all car speeds?
b) Here are some summary statistics.

Percentile		Speed
100%	**Max**	34.060
97.5%		30.976
90.0%		28.978
75.0%	**Q3**	25.785
50.0%	**Median**	23.525
25.0%	**Q1**	21.547
10.0%		19.163
2.5%		16.638
0.0%	**Min**	16.270

From your answer in part a, how well does the model do in predicting those percentiles? Are you surprised? Explain.

45. Receivers 2015 NFL data from the 2015 football season reported the number of yards gained by each of the league's 488 receivers:

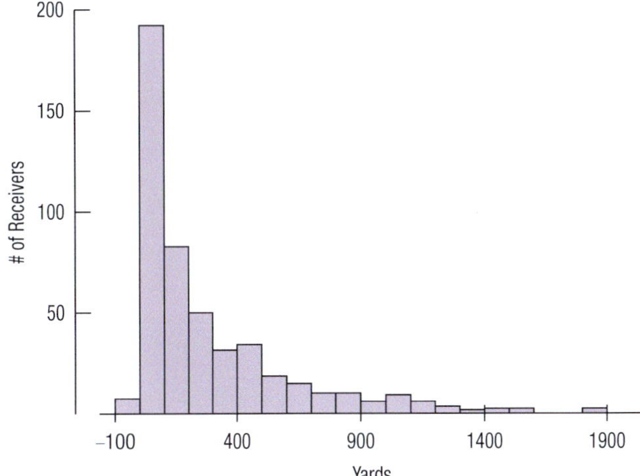

The mean is 274.73 yards, with a standard deviation of 327.32 yards.

a) According to the Normal model, what percent of receivers would you expect to gain more yards than 2 standard deviations above the mean number of yards?
b) For these data, what does that mean?
c) Explain the problem in using a Normal model here.

46. Customer database A large philanthropic organization keeps records on the people who have contributed. In addition to keeping records of past giving, the organization buys demographic data on neighborhoods from the U.S. Census Bureau. Eighteen of these variables concern ethnicity. Here are a histogram and summary statistics for the percentage of whites in the neighborhoods of 500 donors:

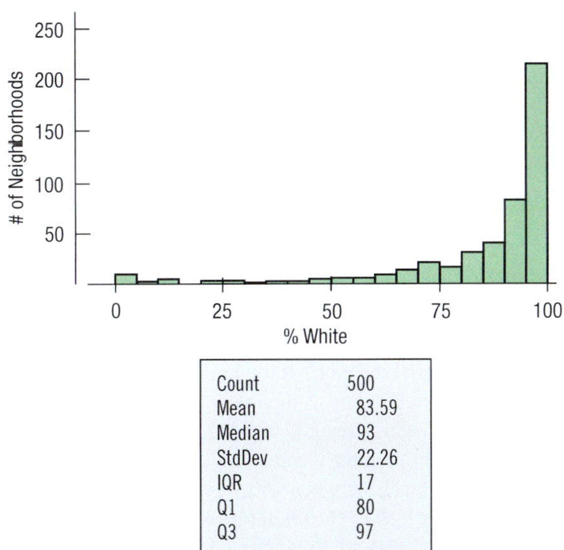

Count	500
Mean	83.59
Median	93
StdDev	22.26
IQR	17
Q1	80
Q3	97

a) Which is a better summary of the percentage of white residents in the neighborhoods, the mean or the median? Explain.
b) Which is a better summary of the spread, the IQR or the standard deviation? Explain.
c) From a Normal model, about what percentage of neighborhoods should have a percent white within one standard deviation of the mean?
d) What percentage of neighborhoods actually have a percent white within one standard deviation of the mean?
e) Explain the discrepancy between parts c and d.

47. CEO compensation 2014 sampled The Glassdoor Economic Research Blog published the compensation (in millions of dollars) for the CEOs of large companies. The distribution looks like this:

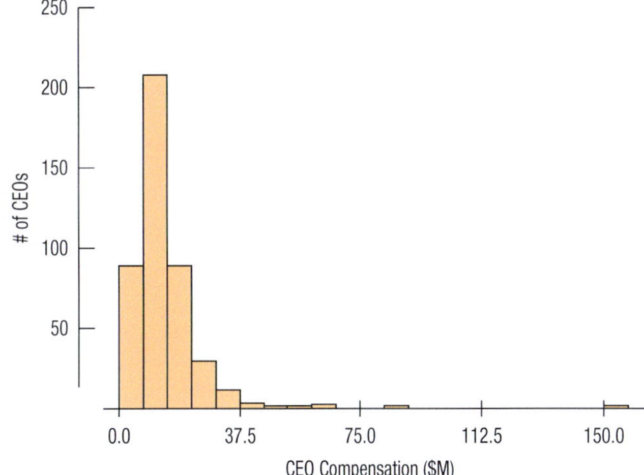

The mean CEO compensation is $14.1M and the standard deviation is $11.32M.

a) According to the Normal model, what percent of CEOs would you expect to earn more than 2 standard deviations above the mean compensation?
b) Is that percentage appropriate for these data?

Use the Simple Sample tool at astools.datadesk.com or sample with any statistics program. Draw 1000 samples of size 30 and make a histogram of the means of these samples. How would you expect the means to vary? Here is the histogram we got when we performed the experiment. Your results, of course, will vary.

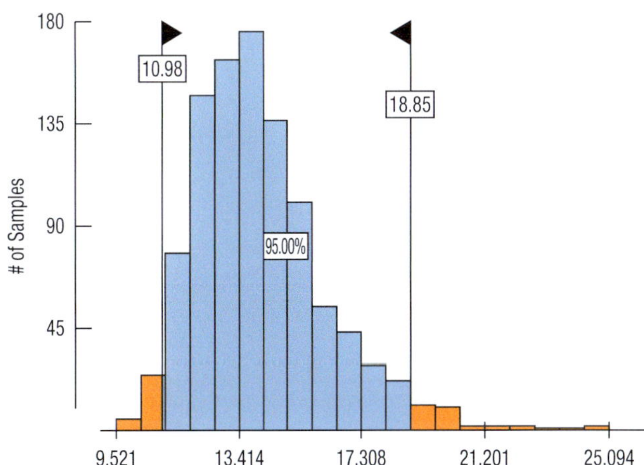

c) The standard deviation of these means is 2.0. The middle 95% of the means is colored in the histogram. Do you think the 68–95–99.7 Rule applies?

Now repeat the experiment, drawing samples of 100 instead. Now our histogram of means looks like this:

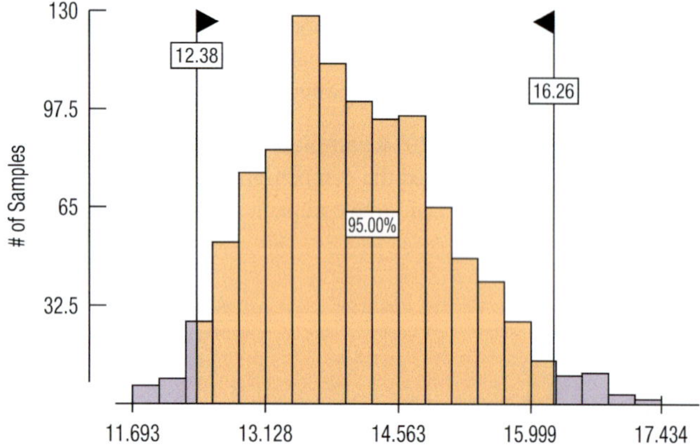

d) Do you think the 68–95–99.7 Rule applies?

Once more, but this time with samples of 200:

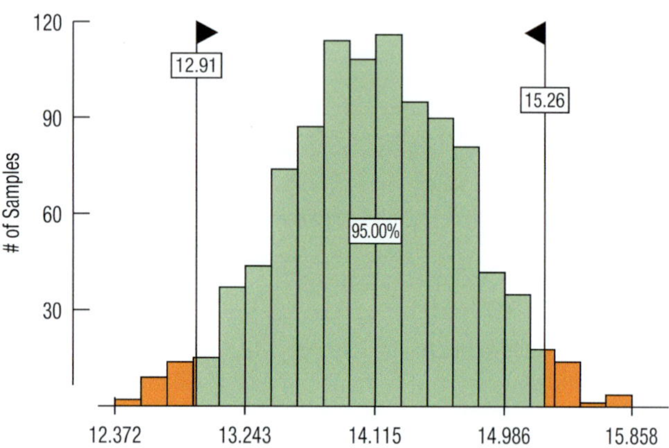

e) Does the 68–95–99.7 Rule give a good idea of the middle 95% of the distribution?

T **48.** CEO compensation logged and sampled Suppose we take logarithms of the CEO compensations in Exercise 47. The histogram of log Compensation looks like this:

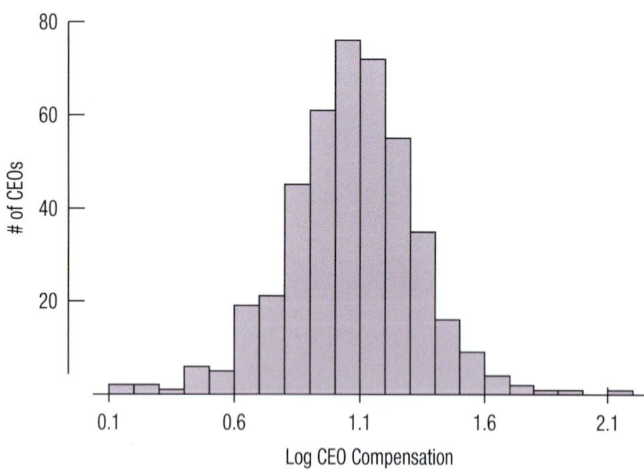

with a mean of 1.07 and a standard deviation of 0.26.

a) According to the Normal model, what percent of CEOs would you expect to earn more than 2 standard deviations above the mean compensation?

b) Is that percentage appropriate for these data?

Now let's draw samples of 30 CEOs from the logged data. We drew 1000 samples and found their means. The distribution of means looks like this:

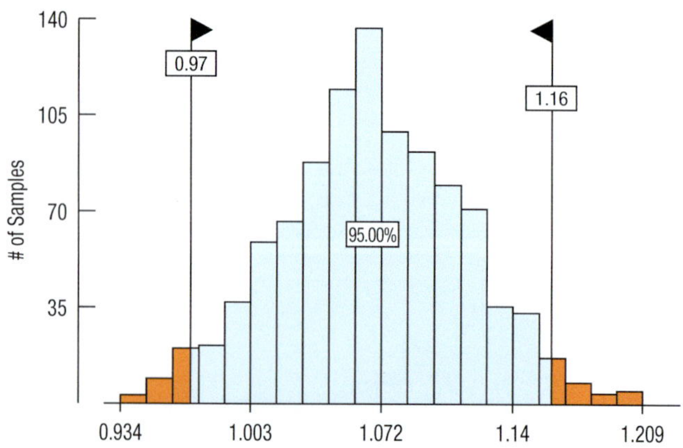

with a standard deviation of 0.05.

c) Do you think the 68–95–99.7 Rule applies to these means?

49. Park visitors' heights, part III Based on the model $N(1.6,0.4)$ describing park visitors' heights from Exercise 29, what are the cutoff values for

a) the highest 10% of the heights?
b) the lowest 10% of the heights?
c) the middle 30% of the heights?

50. More IQs In the Normal model $N(100, 15)$ from Exercise 10, what cutoff value bounds

a) the highest 5% of all IQs?
b) the lowest 30% of the IQs?
c) the middle 80% of the IQs?

51. Cattle, finis Consider the Angus weights model $N(1152, 84)$ one last time.

a) What weight represents the 40th percentile?
b) What weight represents the 99th percentile?
c) What's the IQR of the weights of these Angus steers?

52. IQ, finis Consider the IQ model $N(100, 15)$ one last time.

a) What IQ represents the 15th percentile?
b) What IQ represents the 98th percentile?
c) What's the IQR of the IQs?

53. Cholesterol Assume the cholesterol levels of adult American women can be described by a Normal model with a mean of 188 mg/dL and a standard deviation of 24.

a) Draw and label the Normal model.
b) What percent of adult women do you expect to have cholesterol levels over 200 mg/dL?
c) What percent of adult women do you expect to have cholesterol levels between 150 and 170 mg/dL?
d) Estimate the IQR of the cholesterol levels.
e) Above what value are the highest 15% of women's cholesterol levels?

54. Tires A tire manufacturer believes that the treadlife of its snow tires can be described by a Normal model with a mean of 32,000 miles and standard deviation of 2500 miles.

a) If you buy one of these tires, would it be reasonable for you to hope it will last 40,000 miles? Explain.

b) Approximately what fraction of these tires can be expected to last less than 30,000 miles?
c) Approximately what fraction of these tires can be expected to last between 30,000 and 35,000 miles?
d) Estimate the IQR of the treadlives.
e) In planning a marketing strategy, a local tire dealer wants to offer a refund to any customer whose tires fail to last a certain number of miles. However, the dealer does not want to take too big a risk. If the dealer is willing to give refunds to no more than 1 of every 25 customers, for what mileage can he guarantee these tires to last?

55. Diastolic blood pressure Suppose the diastolic blood pressure of a group of patients can be described by a Normal model with a mean of 85 mmHg and a standard deviation of 10 mmHg.

a) What fraction of patients are expected to have diastolic blood pressure less than 80 mmHg?
b) In what diastolic blood pressure interval are we expected to find the middle 80% of patients?
c) At least what diastolic blood pressure do the highest 5% of patients have?

56. Body temperatures Most people think that the "normal" adult body temperature is 98.6°F. That figure, based on a 19th-century study, has recently been challenged. In a 1992 article in the *Journal of the American Medical Association,* researchers reported that a more accurate figure may be 98.2°F. Furthermore, the standard deviation appeared to be around 0.7°F. Assume that a Normal model is appropriate.

a) In what interval would you expect most people's body temperatures to be? Explain.
b) What fraction of people would be expected to have body temperatures above 98.6°F?
c) Below what body temperature are the coolest 20% of all people?

57. BMI values The ideal BMI for a female is between 18.5 and 24.9. A year ago, a group of female friends had an average BMI of 25.4, and 54% of them exceeded the ideal maximum BMI of 24.9.

a) If a Normal model is appropriate, what was the standard deviation of the BMI values of the females a year ago?
b) The females who had an overweight BMI last year decided to go on a diet. This year, the group has an average BMI of 23.5 and only 28% of them exceed the ideal maximum BMI. What is this year's standard deviation for the appropriate Normal model?
c) Which BMI values are more consistent, last year's or this year's? Explain.

58. Tomatoes Agricultural scientists are working on developing an improved variety of Roma tomatoes. Marketing research indicates that customers are likely to bypass Romas that weigh less than 70 grams. The current variety of Roma plants produces fruit that averages 74 grams, but 11% of the tomatoes are too small. It is reasonable to assume that a Normal model applies.

a) What is the standard deviation of the weights of Romas now being grown?
b) Scientists hope to reduce the frequency of undersized tomatoes to no more than 4%. One way to accomplish this is to raise the average size of the fruit. If the standard deviation remains the same, what target mean should they have as a goal?
c) The researchers produce a new variety with a mean weight of 75 grams, which meets the 4% goal. What is the standard deviation of the weights of these new Romas?
d) Based on their standard deviations, compare the tomatoes produced by the two varieties.

JUST CHECKING

Answers

1. a) On the first test, the mean is 88 and the SD is 4, so $z = (90 - 88)/4 = 0.5$. On the second test, the mean is 75 and the SD is 5, so $z = (80 - 75)/5 = 1.0$. The first test has the lower z-score, so it is the one that will be dropped.
   b) You may not think so, but the second test is 1 standard deviation above the mean, farther away than the first test, so it's the better score relative to the class.
2. With a right skewed distribution the higher values are generally farther from the mean, so the maximum is likely to be farther from the mean than the minimum.
3. a) The mean would increase to 500.
   b) The standard deviation is still 100 points.
   c) The two boxplots would look nearly identical (the shape of the distribution would remain the same), but the later one would be shifted 50 points higher.
4. The standard deviation is now 2.54 millimeters, which is the same as 0.1 inch. Nothing has changed. The standard deviation has "increased" only because we're reporting it in millimeters now, not inches.
5. The mean is 184 centimeters, with a standard deviation of 8 centimeters. 2 meters is 200 centimeters, which is 2 standard deviations above the mean. We expect 5% of the men to be more than 2 standard deviations below or above the mean, so half of those, 2.5%, are likely to be above 2 meters.
6. a) We know that 68% of the time we'll be within 1 standard deviation (2 min) of 20. So 32% of the time we'll arrive in less than 18 or more than 22 minutes. Half of those times (16%) will be greater than 22 minutes, so 84% will be less than 22 minutes.
   b) 24 minutes is 2 standard deviations above the mean. Because of the 95% rule, we know 2.5% of the times will be more than 24 minutes.
   c) Traffic incidents may occasionally increase the time it takes to get to school, so the driving times may be skewed to the right, and there may be outliers.
   d) If so, the Normal model would not be appropriate and the percentages we predict would not be accurate.
7. Not particularly unusual. Answers may vary, but typically we'd think of values 2 SD or more from the mean as unusual.
8. a) No, plenty of commute times were this long.
   b) No, the distribution is not unimodal and symmetric.
   c) Yes, we don't expect sample means to vary that much, for example, in Figure 2.20.
   d) Part a is about a sample value. Part c is about a sample mean. Sample means vary less than individual values.

Review of Part I

EXPLORING AND UNDERSTANDING DATA

Quick Review

It's time to put it all together. Real data don't come tagged with instructions for use. So let's step back and look at how the key concepts and skills we've seen work together. This brief list and the review exercises that follow should help you check your understanding of statistics so far.

- We treat data two ways: as categorical and as quantitative.
- To describe categorical data:
  - Make a picture. Bar graphs work well for comparing counts in categories.
  - Summarize the distribution with a table of counts or relative frequencies (percents) in each category.
  - Pie charts and segmented bar charts display divisions of a whole.
  - Compare distributions with plots side by side.
  - Look for associations between variables by comparing marginal and conditional distributions.
- To describe quantitative data:
  - Make a picture. Use histograms, boxplots, stem-and-leaf displays, density plots, or dotplots. Stem-and-leafs are great when working by hand and good for small datasets. Histograms are a good way to see the distribution. Boxplots are best for comparing several distributions.
  - Describe distributions in terms of their shape, center, and spread, and note any unusual features such as gaps or outliers.
  - The shape of most distributions you'll see will likely be uniform, unimodal, or bimodal. A distribution may be multimodal. If it is unimodal, then it may be symmetric or skewed.
  - A 5-number summary makes a good numerical description of a distribution: min, Q1, median, Q3, and max.
  - If the distribution is skewed, be sure to include the median and interquartile range (IQR) when you describe its center and spread.
  - A distribution that is severely skewed may benefit from re-expressing the data. If it is skewed to the high end, taking logs often works well.
  - If the distribution is unimodal and symmetric, describe its center and spread with the mean and standard deviation.
  - Summary statistics vary from sample to sample. We can draw many random samples from a population and collect all the means (for example) to see how they are distributed.
  - Use the standard deviation as a ruler to tell how unusual an observed value may be, or to compare or combine measurements made on different scales.
  - Shifting a distribution by adding or subtracting a constant affects measures of position but not measures of spread. Rescaling by multiplying or dividing by a constant affects both.
  - When a distribution is roughly unimodal and symmetric, a Normal model may be useful. For Normal models, the 68–95–99.7 Rule is a good rule of thumb.
  - If the Normal model fits well (check a histogram or Normal probability plot), then Normal percentile tables or functions found in most statistics technology can provide more detailed values.

Need more help with some of this? It never hurts to reread sections of the chapters! And in the following pages we offer you more opportunities[1] to review these concepts and skills.

The exercises that follow use the concepts and skills you've learned in the first five chapters. To be more realistic and more useful for your review, they don't tell you which of the concepts or methods you need. But neither will the exam.

[1]If you doubted that we are teachers, this should convince you. Only a teacher would call additional homework exercises "opportunities."

REVIEW EXERCISES

R1.1. Bananas Here are the prices (in cents per pound) of bananas reported from 15 markets surveyed by the U.S. Department of Agriculture.

51	52	45
48	53	52
50	49	52
48	43	46
45	42	50

a) Display these data with an appropriate graph.
b) Report appropriate summary statistics.
c) Write a few sentences about this distribution.

R1.2. Prenatal care Results of a 1996 American Medical Association report about the infant mortality rate for twins carried for the full term of a normal pregnancy are shown on the next page, broken down by the level of prenatal care the mother had received.

Full-Term Pregnancies, Level of Prenatal Care	Infant Mortality Rate Among Twins (deaths per thousand live births)
Intensive	5.4
Adequate	3.9
Inadequate	6.1
Overall	**5.1**

a) Is the overall rate the average of the other three rates? Should it be? Explain.
b) Do these results indicate that adequate prenatal care is important for pregnant women? Explain.
c) Do these results suggest that a woman pregnant with twins should be wary of seeking too much medical care? Explain.

R1.3. Singers by parts The boxplots display the heights (in inches) of 130 members of a choir by the part they sing.

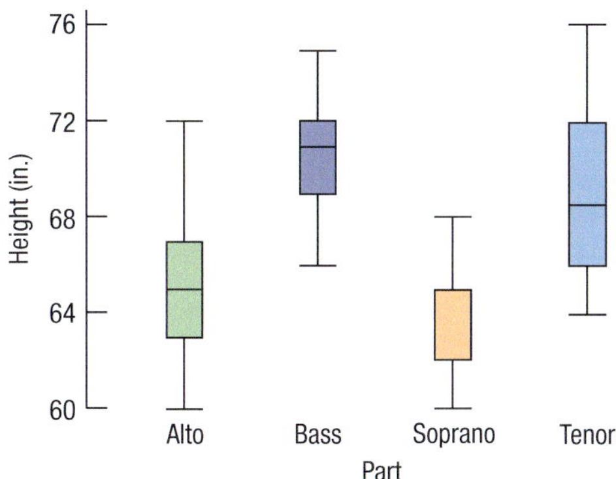

a) It appears that the median height for sopranos is missing, but actually the median and the upper quartile are equal. How could that happen?
b) Write a few sentences describing what you see. These boxplots are arranged in alphabetical order.
c) Suggest a possibly more interesting order.

R1.4. Dialysis In a study of dialysis, researchers found that "of the three patients who were currently on dialysis, 67% had developed blindness and 33% had their toes amputated." What kind of display might be appropriate for these data? Explain.

R1.5. Beanstalks Beanstalk Clubs are social clubs for very tall people. To join, a man must be over 6′2″ tall, and a woman over 5′10″. The National Health Survey suggests that heights of adults may be Normally distributed, with mean heights of 69.1″ for men and 64.0″ for women. The respective standard deviations are 2.8″ and 2.5″.

a) You are probably not surprised to learn that men are generally taller than women, but what does the greater standard deviation for men's heights indicate?
b) Who are more likely to qualify for Beanstalk membership, men or women? Explain.

R1.6. Bread Clarksburg Bakery is trying to predict how many loaves to bake. In the past 100 days, they have sold between 95 and 140 loaves per day. Here is a histogram of the number of loaves they sold for the past 100 days.

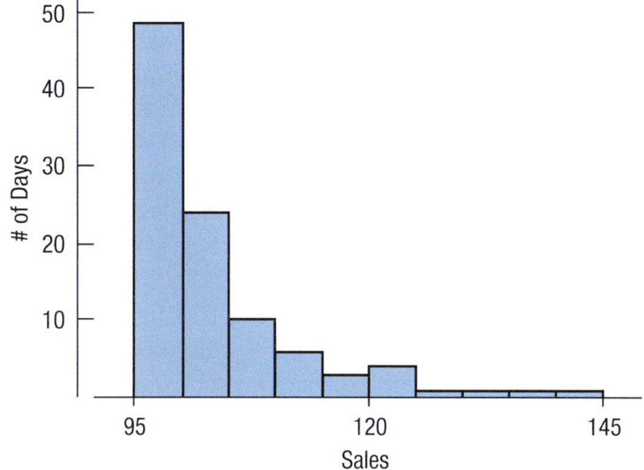

a) Describe the distribution.
b) Which should be larger, the mean number of sales or the median? Explain.
c) Here are the summary statistics for Clarksburg Bakery's bread sales. Use these statistics and the histogram above to create a boxplot. You may approximate the values of any outliers.

Summary of Sales	
Median	100
Min	95
Max	140
25th %tile	97
75th %tile	105.5

d) For these data, the mean was 103 loaves sold per day, with a standard deviation of 9 loaves. Do these statistics suggest that Clarksburg Bakery should expect to sell between 94 and 112 loaves on about 68% of the days? Explain.

R1.7. State University Public relations staff members at State U phoned 850 local residents. After identifying themselves, the callers asked the survey participants their ages, whether they had attended college, and whether they had a favorable opinion of the university. The official report to the university's directors claimed that, in general, people had very favorable opinions about the university.

a) Identify the W's of these data.
b) Identify the variables, classify each as categorical or quantitative, and specify units if relevant.
c) Are you confident about the report's conclusion? Explain.

R1.8. Shenandoah rain Based on long-term investigation, researchers have suggested that the acidity (pH) of rainfall in the Shenandoah Mountains can be described by the Normal model $N(4.9, 0.6)$.

a) Draw and carefully label the model.
b) What percent of storms produce rainfall with pH over 6?
c) What percent of storms produce rainfall with pH under 4?

d) The lower the pH, the more acidic the rain. What is the pH level for the most acidic 20% of all storms?
e) What is the pH level for the least acidic 5% of all storms?
f) What is the IQR for the pH of rainfall?

R1.9. Fraud detection A credit card bank is investigating the incidence of fraudulent card use. The bank suspects that the type of product bought may provide clues to the fraud. To examine this situation, the bank looks at the North American Industry Classification System (NAICS) of the business related to the transaction. This is a code that is used by the U.S. Census Bureau and Statistics Canada to identify the type of every registered business in North America.[2] For example, 1141 designates Fishing, 3121 is Beverage Manufacturing, and 6113 is Colleges, Universities, and Professional schools. Longer codes designate subcategories

A company intern produces the following histogram of the NAICS 4-digit-level codes for 1536 transactions:

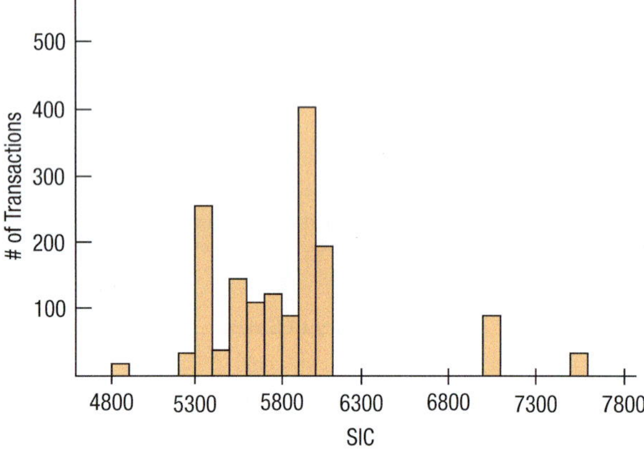

He also reports that the mean NAICS is 5823.13 with a standard deviation of 488.17.

a) Comment on any problems you see with the use of the mean and standard deviation as summary statistics.
b) How well do you think the Normal model will work on these data? Explain.

T **R1.10. Streams** As part of the course work, a class at an upstate NY college collects data on streams each year. Students record a number of biological, chemical, and physical variables, including the stream name, the substrate of the stream (*limestone (L), shale (S),* or *mixed (M)*), the pH, the temperature (°C), and the BCI, a measure of biological diversity.

Substrate	Count	%
Limestone	77	44.8
Mixed	26	15.1
Shale	69	40.1

a) Name each variable, indicating whether it is categorical or quantitative, and give the units if available.
b) These streams have been classified according to their substrate—the composition of soil and rock over which they flow—as summarized in the table. What kind of graph might be used to display these data?

[2]www.census.gov/eos/www/naics/

T **R1.11. Cramming** One Thursday, researchers gave students enrolled in a section of basic Spanish a set of 50 new vocabulary words to memorize. On Friday, the students took a vocabulary test. When they returned to class the following Monday, they were retested—without advance warning. Both sets of test scores for the 25 students are shown below.

Fri	Mon	Fri	Mon
42	36	50	47
44	44	34	34
45	46	38	31
48	38	43	40
44	40	39	41
43	38	46	32
41	37	37	36
35	31	40	31
43	32	41	32
48	37	48	39
43	41	37	31
45	32	36	41
47	44		

a) Create a graphical display to compare the two distributions of scores.
b) Write a few sentences about the scores reported on Friday and Monday.
c) Create a graphical display showing the distribution of the *changes* in student scores.
d) Describe the distribution of changes.

R1.12. e-Books A study by the Pew Internet & American Life Project found that 78% of U.S. residents over 16 years old read a book in the past 12 months. They also found that 21% had read an e-book using a reader or computer during that period. A newspaper reporting on these findings concluded that 99% of U.S. adult residents had read a book in some fashion in the past year. (libraries.pewinternet.org/2012/04/04/the-rise-of-e-reading/) Do you agree? Explain.

R1.13. Let's play cards You pick a card from a standard deck and record its denomination (7, say) and its suit (maybe spades).

a) Is the variable *suit* categorical or quantitative?
b) Name a game you might be playing for which you would consider the variable *denomination* to be categorical. Explain.
c) Name a game you might be playing for which you would consider the variable *denomination* to be quantitative. Explain.

T **R1.14. Accidents** Progressive Insurance asked customers who had been involved in auto accidents how far they were from home when the accident happened. The data are summarized in the table.

Miles from Home	% of Accidents
Less than 1	23
1 to 5	29
6 to 10	17
11 to 15	8
16 to 20	6
Over 20	17

a) Create an appropriate graph of these data.
b) Do these data indicate that driving near home is particularly dangerous? Explain.

R1.15. Hard water In an investigation of environmental causes of disease, data were collected on the annual mortality rate (deaths per 100,000) for males in 61 large towns in England and Wales. In addition, the water hardness was recorded as the calcium concentration (parts per million, ppm) in the drinking water.

a) What are the variables in this study? For each, indicate whether it is quantitative or categorical and what the units are.
b) Here are histograms of calcium concentration and mortality. Describe the distributions of the two variables.

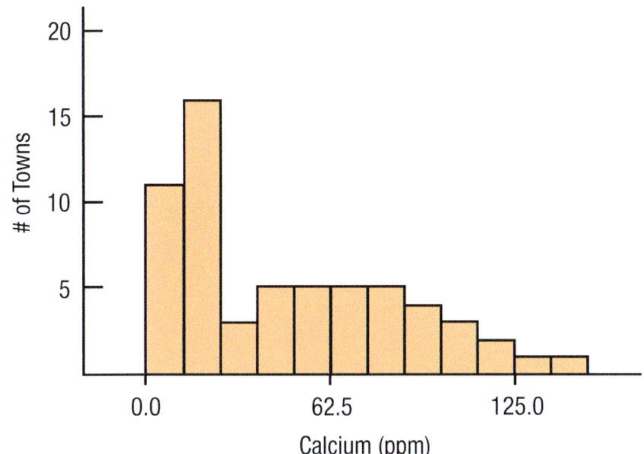

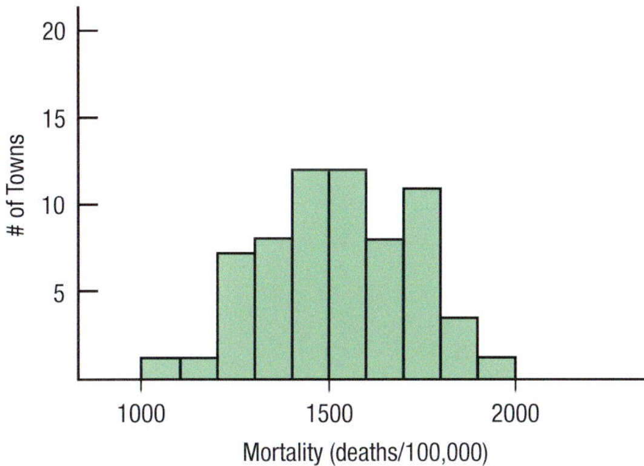

R1.16. Hard water II The dataset from England and Wales also notes for each town whether it was south or north of Derby. Here are some summary statistics and a comparative boxplot for the two regions.

Summary of Mortality

Region	Count	Mean	Median	StdDev
North	34	1631.59	1631	138.470
South	27	1388.85	1369	151.114

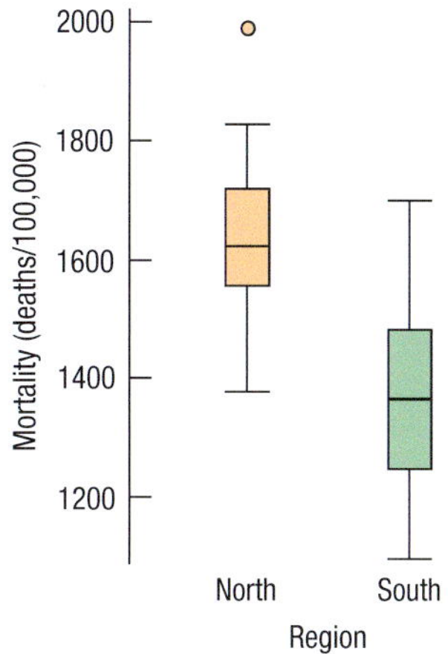

a) What is the overall mean mortality rate for the two regions?
b) Do you see evidence of a difference in mortality rates? Explain.

R1.17. Seasons Average daily temperatures in January and July for 60 large U.S. cities are graphed in the histograms below. (Data in **City climate**)

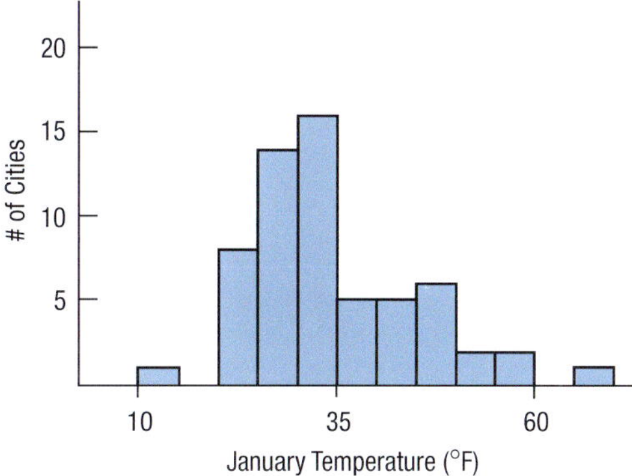

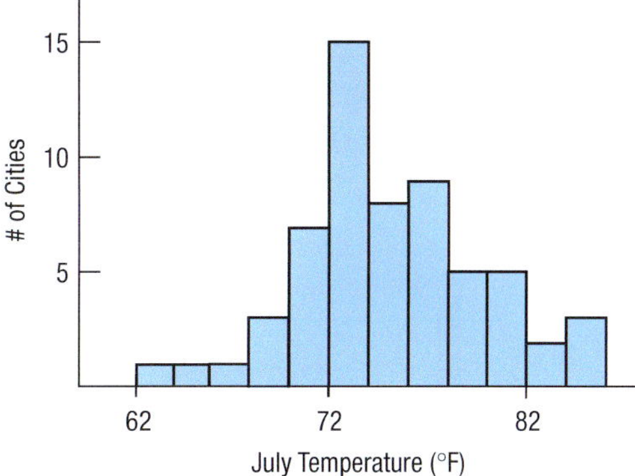

a) What aspect of these histograms makes it difficult to compare the distributions?
b) What differences do you see between the distributions of January and July average temperatures?

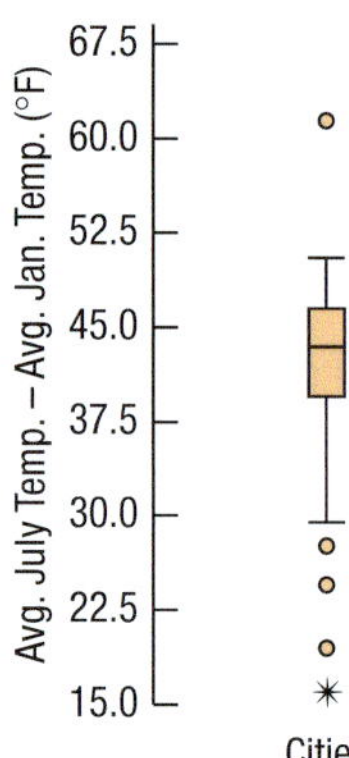

c) Differences in temperatures (July − January) for each of the cities are displayed in the boxplot above. Write a few sentences describing what you see.

T **R1.18.** Old Faithful It is a common belief that Yellowstone's most famous geyser erupts once an hour at very predictable intervals. The histogram below shows the time gaps (in minutes) between 222 successive eruptions. Describe this distribution.

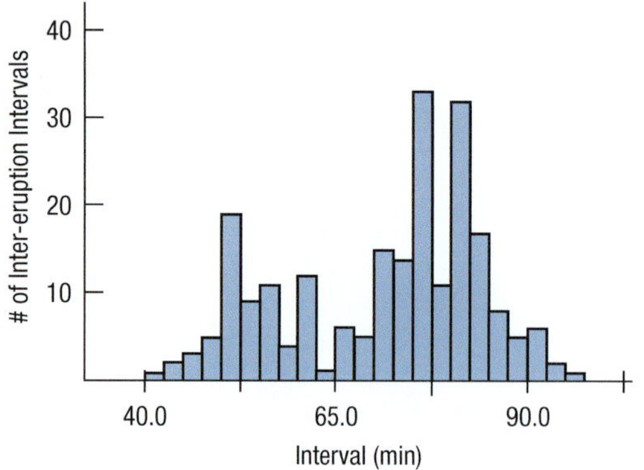

T **R1.19.** Old Faithful? Does the duration of an eruption have an effect on the length of time that elapses before the next eruption?

a) The histogram below shows the duration (in minutes) of those 222 eruptions. Describe this distribution.

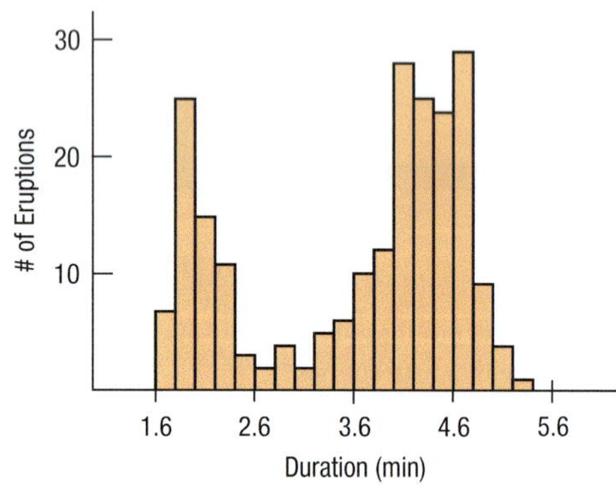

b) Explain why it is not appropriate to find summary statistics for this distribution.

c) Let's classify the eruptions as "long" or "short," depending on whether they last more than 3 minutes or not. Describe what you see in the comparative boxplots.

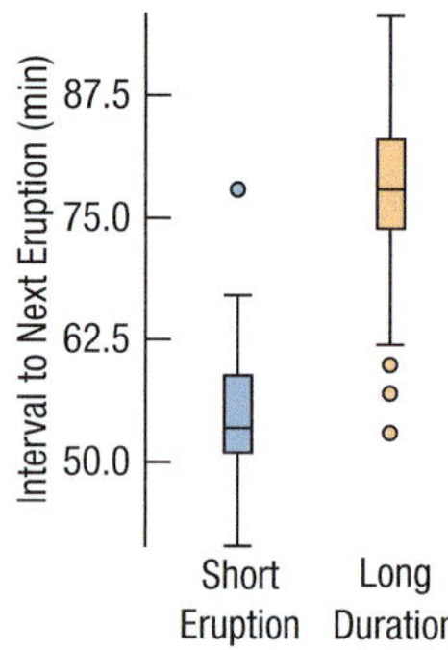

R1.20. Teen drivers 2013 The National Highway Traffic Safety Administration reported that there were 3206 fatal accidents involving drivers between the ages of 15 and 19 years old the previous year, of which 65.5% involved male drivers. Of the male drivers, 18.4% involved drinking, while of the female drivers, 10.8% involved drinking. Assuming roughly equal numbers of male and female drivers, use these statistics to explain the concept of independence.

T **R1.21.** Liberty's nose Is the Statue of Liberty's nose too long? Her nose measures 4′6″, but she is a large statue, after all. Her arm is 42 feet long. That means her arm is $42/4.5 = 9.3$ times as long as her nose. Is that a reasonable ratio? Shown in the table are arm and nose lengths of 18 girls in a statistics class, and the ratio of arm-to-nose length for each.

Arm (cm)	Nose (cm)	Arm/Nose Ratio
73.8	5.0	14.8
74.0	4.5	16.4
69.5	4.5	15.4
62.5	4.7	13.3
68.6	4.4	15.6
64.5	4.8	13.4
68.2	4.8	14.2
63.5	4.4	14.4
63.5	5.4	11.8
67.0	4.6	14.6
67.4	4.4	15.3
70.7	4.3	16.4
69.4	4.1	16.9
71.7	4.5	15.9
69.0	4.4	15.7
69.8	4.5	15.5
71.0	4.8	14.8
71.3	4.7	15.2

a) Make an appropriate plot and describe the distribution of the ratios.

b) Summarize the ratios numerically, choosing appropriate measures of center and spread.

c) Is the ratio of 9.3 for the Statue of Liberty unrealistically low? Explain.

R1.22. Women's Short Track 2018 The times of skaters in the qualifying heats for the women's short track race at the 2018 Olympics in PyeongChang are given in the table below.

a) The mean finishing time was 45.075 seconds, with a standard deviation of 4.50 seconds. If the Normal model is appropriate, what percent of the times should be within 4 seconds of 45.075?
b) What percent of the times actually fall within this interval?
c) Explain the discrepancy between parts a and b.

Name	Country	Time
Kim Boutin	Canada	43.634
Natalia Maliszewska	Poland	43.725
Lana Gehring	United States	43.825
Lucia Peretti	Italy	43.994
Arianna Fontana	Italy	43.214
Andrea Keszler	Hungary	43.274
Kathryn Thomson	Great Britain	61.029
Sofia Prosvirnova	Athletes from Russia	43.376
Yara van Kerkhof	Netherlands	43.430
Bianca Walter	Germany	43.541
Sumire Kikuchi	Japan	44.838
Elise Christie	Great Britain	42.872
Qu Chunyu	China	42.971
Shim Suk-hee	South Korea	43.048
Veronique Pierron	France	43.148
Fan Kexin	China	43.350
Maame Biney	United States	43.665
Kim A-lang	South Korea	43.724
Anastassiya Krestova	Kazakhstan	43.821
Martina Valcepina	Italy	43.698
Han Yutong	China	43.719
Tifany Huot-Marchand	France	44.659
Marianne St-Gelais	Canada	43.437
Anna Seidel	Germany	43.742
Lara van Ruijven	Netherlands	43.771
Magdalena Warakomska	Poland	44.311
Choi Min-jeong	South Korea	42.870
Petra Jaszapati	Hungary	55.670
Emina Malagich	Athletes from Russia	56.830

R1.23. Sample A study in South Africa focusing on the impact of health insurance identified 1590 children at birth and then sought to conduct follow-up health studies 5 years later. Only 416 of the original group participated in the 5-year follow-up study. This made researchers concerned that the follow-up group might not accurately resemble the total group in terms of health insurance. The following table summarizes the two groups by race and by presence of medical insurance when the child was born. Carefully explain how this study demonstrates Simpson's paradox. (*Birth to Ten Study*, Medical Research Council, South Africa)

		Number (%) Insured	
		Follow-Up	**Not Traced**
Race	**Black**	36 of 404 (8.9%)	91 of 1048 (8.7%)
	White	10 of 12 (83.3%)	104 of 126 (82.5%)
	Overall	46 of 416 (11.1%)	195 of 1174 (16.6%)

R1.24. Sluggers Babe Ruth was the first great "slugger" in baseball. His record of 60 home runs in one season held for 34 years until Roger Maris hit 61 in 1961. Mark McGwire (with the aid of steroids) set a new standard of 70 in 1998. Listed below are the home run totals for each season McGwire played. Also listed are Babe Ruth's home run totals.

McGwire: 3*, 49, 32, 33, 39, 22, 42, 9*, 9*, 39, 52, 58, 70, 65, 32*, 29*

Ruth: 54, 59, 35, 41, 46, 25, 47, 60, 54, 46, 49, 46, 41, 34, 22

a) Find the 5-number summary for McGwire's career.
b) Do any of his seasons appear to be outliers? Explain.
c) McGwire played in only 18 games at the end of his first big league season, and missed major portions of some other seasons because of injuries to his back and knees. Those seasons might not be representative of his abilities. They are marked with asterisks in the list above. Omit these values and make parallel boxplots comparing McGwire's career to Babe Ruth's.
d) Write a few sentences comparing the two sluggers.
e) Create side-by-side stem-and-leaf displays comparing the careers of the two players.
f) What aspects of the distributions are apparent in the stem-and-leaf displays that did not clearly show in the boxplots?

R1.25. Be quick! Avoiding an accident when driving can depend on reaction time. That time, measured from the moment the driver first sees the danger until he or she steps on the brake pedal, is thought to follow a Normal model with a mean of 1.5 seconds and a standard deviation of 0.18 second.

a) Use the 68–95–99.7 Rule to draw the Normal model.
b) Write a few sentences describing driver reaction times.
c) What percent of drivers have a reaction time less than 1.25 seconds?
d) What percent of drivers have reaction times between 1.6 and 1.8 seconds?
e) What is the interquartile range of reaction times?
f) Describe the reaction times of the slowest 1/3 of all drivers.

R1.26. Music and memory Is it a good idea to listen to music when studying for a big test? In a study conducted by some statistics students, 62 people were randomly assigned to listen to rap music, Mozart, or no music while attempting to memorize objects pictured on a page. They were then asked to list all the objects they could remember. Here are the 5-number summaries for each group:

	n	Min	Q1	Median	Q3	Max
Rap	29	5	8	10	12	25
Mozart	20	4	7	10	12	27
None	13	8	9.5	13	17	24

a) Describe the W's for these data: *Who, What, Where, Why, When, How.*
b) Name the variables and classify each as categorical or quantitative.
c) Create parallel boxplots as best you can from these summary statistics to display these results.
d) Write a few sentences comparing the performances of the three groups.

T **R1.27.** Mail Here are the number of pieces of mail received at a school office for 36 days.

123	70	90	151	115	97
80	78	72	100	128	130
52	103	138	66	135	76
112	92	93	143	100	88
118	118	106	110	75	60
95	131	59	115	105	85

a) Plot these data.
b) Find appropriate summary statistics.
c) Write a brief description of the school's mail deliveries.
d) What percent of the days actually lie within one standard deviation of the mean? Comment.

T **R1.28.** Birth order Is your birth order related to your choice of major? A statistics professor at a large university polled his students to find out what their majors were and what position they held in the family birth order. The results are summarized in the table.

a) What percent of these students are oldest or only children?
b) What percent of Humanities majors are oldest children?
c) What percent of oldest children are Humanities students?
d) What percent of the students are oldest children majoring in the Humanities?

		Birth Order*				
		1	2	3	4+	Total
Major	Math/Science	34	14	6	3	57
	Agriculture	52	27	5	9	93
	Humanities	15	17	8	3	43
	Other	12	11	1	6	30
	Total	113	69	20	21	223

*1 = oldest or only child

R1.29. Herbal medicine Researchers for the Herbal Medicine Council collected information on people's experiences with a new herbal remedy for colds. They went to a store that sold natural health products. There they asked 100 customers whether they had taken the cold remedy and, if so, to rate its effectiveness (on a scale from 1 to 10) in curing their symptoms. The Council concluded that this product was highly effective in treating the common cold.

a) Identify the W's of these data.
b) Identify the variables, classify each as categorical or quantitative, and specify units if relevant.
c) Are you confident about the Council's conclusion? Explain.

T **R1.30.** Birth order revisited Consider again the data on birth order and college majors in Exercise R1.28.

a) What is the marginal distribution of majors?
b) What is the conditional distribution of majors for the oldest children?
c) What is the conditional distribution of majors for the children born second?
d) Do you think that college major appears to be independent of birth order? Explain.

R1.31. Engines One measure of the size of an automobile engine is its "displacement," the total volume (in liters or cubic inches) of its cylinders. Summary statistics for several models of new cars are shown. These displacements were measured in cubic inches.

Summary of Displacement	
Count	38
Mean	177.29
Median	148.5
StdDev	88.88
Range	275
25th %tile	105
75th %tile	231

a) How many cars were measured?
b) Why might the mean be so much larger than the median?
c) Describe the center and spread of this distribution with appropriate statistics.
d) Your neighbor is bragging about the 227-cubic-inch engine he bought in his new car. Is that engine unusually large? Explain.
e) Are there any engines in this dataset that you would consider to be outliers? Explain.
f) Is it reasonable to expect that about 68% of car engines measure between 88 and 266 cubic inches? (That's 177.289 ± 88.8767.) Explain.
g) We can convert all the data from cubic inches to cubic centimeters (cc) by multiplying by 16.4. For example, a 200-cubic-inch engine has a displacement of 3280 cc. How would such a conversion affect each of the summary statistics?

R1.32. Engines, again Horsepower is another measure commonly used to describe auto engines. Here are the summary statistics and histogram displaying horsepowers of the same group of 38 cars discussed in Exercise R1.31.

Summary of Horsepower	
Count	38
Mean	101.7
Median	100
StdDev	26.4
Range	90
25th %tile	78
75th %tile	125

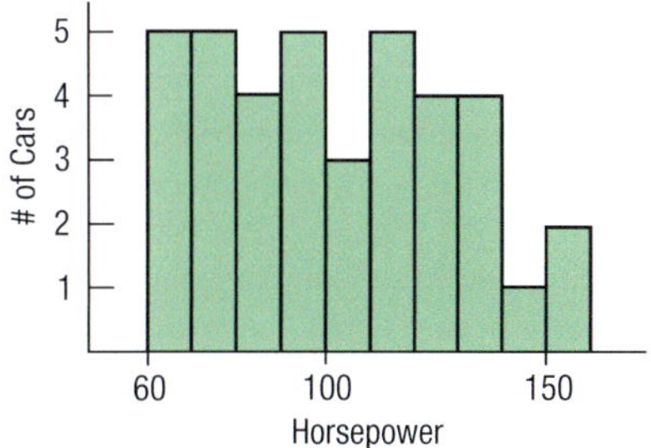

a) Describe the shape, center, and spread of this distribution.
b) What is the interquartile range?
c) Are any of these engines outliers in terms of horsepower? Explain.
d) Do you think the 68–95–99.7 Rule applies to the horsepower of auto engines? Explain.
e) From the histogram, make a rough estimate of the percentage of these engines whose horsepower is within one standard deviation of the mean.
f) A fuel additive boasts in its advertising that it can "add 10 horsepower to any car." Assuming that is true, what would happen to each of these summary statistics if this additive were used in all the cars?

R1.33. Age and party 2011 The Pew Research Center conducts surveys regularly asking respondents which political party they identify with or lean toward. Among their results is the following table relating preferred political party and age.

		Party			
		Republican/ Lean Rep.	Democrat/ Lean Dem.	Neither	Total
Age	18–29	318	424	73	**815**
	30–49	991	1058	203	**2252**
	50–64	1260	1407	264	**2931**
	65 +	1136	1087	193	**2416**
	Total	**3705**	**3976**	**733**	**8414**

a) What percent of people surveyed were Republicans or leaned Republican?
b) Do you think this might be a reasonable estimate of the percentage of all voters who are Republicans or lean Republican? Explain.
c) What percent of people surveyed were under 30 or over 65?
d) What percent of people were classified as "Neither" and under the age of 30?
e) What percent of the people classified as "Neither" were under 30?
f) What percent of people under 30 were classified as "Neither"?

R1.34. Pay According to the Bureau of Labor Statistics, the mean hourly wage for Chief Executives in 2009 was $80.43 and the median hourly wage was $77.27. By contrast, for General and Operations Managers, the mean hourly wage was $53.15 and the median was $44.55. Are these wage distributions likely to be symmetric, skewed left, or skewed right? Explain.

R1.35. Age and party 2011 II Consider again the Pew Research Center results on age and political party in Exercise R1.33.

a) What is the marginal distribution of party affiliation?
b) Create segmented bar graphs displaying the conditional distribution of party affiliation for each age group.
c) Summarize these poll results in a few sentences that might appear in a newspaper article about party affiliation in the United States.
d) Do you think party affiliation is independent of the voter's age? Explain.

R1.36. Bike safety 2015 The Bicycle Helmet Safety Institute website includes a report on the number of bicycle fatalities per year in the United States. The table below shows the counts for the years 1994–2015.

Year	Bicycle Fatalities
1994	802
1995	833
1996	765
1997	814
1998	760
1999	754
2000	693
2001	732
2002	665
2003	629
2004	727
2005	784
2006	769
2007	699
2008	716
2009	628
2010	616
2011	675
2012	726
2013	743
2014	726
2015	811

a) What are the W's for these data?
b) Display the data in a stem-and-leaf display.
c) Display the data in a timeplot.
d) What is apparent in the stem-and-leaf display that is hard to see in the timeplot?
e) What is apparent in the timeplot that is hard to see in the stem-and-leaf display?
f) Write a few sentences about bicycle fatalities in the United States.

R1.37. Some assembly required A company that markets build-it-yourself furniture sells a computer desk that is advertised with the claim "less than an hour to assemble." However, through postpurchase surveys the company has learned that only 25% of its customers succeeded in building the desk in under an hour. The mean time was 1.29 hours. The company assumes that consumer assembly time follows a Normal model.

a) Find the standard deviation of the assembly time model.
b) One way the company could solve this problem would be to change the advertising claim. What assembly time should the company quote in order that 60% of customers succeed in finishing the desk by then?
c) Wishing to maintain the "less than an hour" claim, the company hopes that revising the instructions and labeling the parts more clearly can improve the 1-hour success rate to 60%. If the standard deviation stays the same, what new lower mean time does the company need to achieve?
d) Months later, another postpurchase survey shows that new instructions and part labeling did lower the mean assembly time, but only to 55 minutes. Nonetheless, the company did achieve the 60%-in-an-hour goal, too. How was that possible?

T **R1.38.** Global500 2014 Here is a stem-and-leaf display showing profits (in $M) for 30 of the 500 largest global corporations (as measured by revenue). The stems are split; each stem represents a span of 5000 ($M), from a profit of 43,000 ($M) to a loss of 7000 ($M). Use the stem-and-leaf to answer the questions.

Stem	Leaf	Count
4	3	1
3	67	2
3	3	1
2	7	1
2	13	2
1	66899	5
1	1123	4
0	55677899	8
0	13334	5
−0		
−0	7	1

Profits ($M)

−0|7 means −7000 ($M)

a) Find the 5-number summary.
b) Draw a boxplot for these data.
c) Find the mean and standard deviation.
d) Describe the distribution of profits for these corporations.

R1.39. Shelves shuffled (Data in **Cereals**) Consumer groups are concerned that cereals with a high sugar content (usually designed for children) are placed just where kids are most likely to see them—in the middle shelf of the supermarket. The variable *Middle* indicates whether the cereal is located on shelf 2, the middle shelf.

a) Compare the sugar content of the cereals on the middle shelf versus those on shelves 1 and 3 with displays and summary statistics.
b) By shuffling the variable *Middle* 1000 times, investigate whether the mean sugar content difference between the two groups of cereals could have arisen by chance. What do you conclude?

R1.40. Salty fries? (Data in **Burger King items**) Is the mean amount of salt higher in menu items that contain meat?

a) Compare the sodium content of the meat and non-meat items with displays and summary statistics.
b) By shuffling the variable *Meat* 1000 times, investigate whether the mean sodium content difference between the two groups of items could have arisen by chance. What do you conclude?

T **R1.41.** Hopkins Forest investigation The **Hopkins Forest** dataset includes all 24 weather variables reported by the researchers. Many of the variables (e.g., temperature, relative humidity, solar radiation, wind) are reported as daily averages, minima and maxima. Using any of these variables, compare the distributions of the daily minima and maxima in a few sentences. Use summary statistics and appropriate graphical displays.

T **R1.42.** *Titanic* investigation The **Titanic** dataset includes more variables than just those discussed in Chapter 2. Others include the crew's job and where each person boarded the ship. Stories, biographies, and pictures can be found on this site: www.encyclopedia-titanica.org/. Using the dataset, investigate some of the variables and their associations. Write a short report on what you discover. Be sure to include summary statistics, tables, and graphical displays.

T **R1.43.** Student survey investigation The **Student survey** dataset introduced in the Chapter 3 exercises includes responses to 13 questions. Investigate the associations among the variables that you find interesting. Write a short report on what you discover. Be sure to include summary statistics, tables, and graphical displays.

T **R1.44.** Movies investigation The dataset **Movies 06-15** introduced in the Chapter 3 exercises includes the distributor, number of tickets sold, and gross revenue in addition to the MPAA rating and the genre for each of the 10 years 2006 to 2015. Investigate the associations among the variables that you find interesting. Write a short report on what you discover. Be sure to include summary statistics, tables, and graphical displays.

6 Scatterplots, Association, and Correlation

WHERE ARE WE GOING?

Is the price of sneakers related to how long they last? Is your alertness in class related to how much (or little) sleep you got the night before?

In this chapter, we'll look at relationships between two quantitative variables. We'll start by looking at scatterplots and describing the essence of what we see—the direction, form, and strength of the association. Then, as we did for histograms, we'll find a quantitative summary of what we learned from the display. We'll use the correlation to measure the strength of the association we see in the scatterplot.

WHO	Years 1970–2016
WHAT	Year and Mean error in the position of Atlantic hurricanes as predicted 72 hours ahead by the NHC
UNITS	Years and Nautical miles
WHEN	1970–2016
WHERE	Atlantic Ocean, the Gulf of Mexico and the Caribbean
WHY	NHC wants to improve prediction models

The typhoon that hit the Philippines in late 2013 was the strongest cyclone ever recorded both in terms of wind speed at landfall and low pressure. It devastated the coast of that island country and killed over 6000 people. In 2015, Hurricane Patricia recorded sustained winds of 200 miles per hour (mph): the strongest ever recorded at sea. Earlier that year, Typhoon Soudelor made landfall with gusts up to 145 mph in Japan and 50 inches of rain in Taiwan. In 2011, Hurricane Irene was smaller and less powerful, but it still caused 7 billion dollars in damage and killed 33 people. Irene's impact was high for a smaller storm because of its track, along the heavily populated East Coast, directly over New York City, and into New England.

Where will a hurricane go? The impact of a hurricane depends on both its path and strength, so the National Hurricane Center (NHC) of the National Oceanic and Atmospheric Administration (NOAA) tries to predict the path each hurricane will take. But hurricanes tend to wander around aimlessly and are pushed by fronts and other weather phenomena in their area, so they are notoriously difficult to predict. Even relatively small changes in a hurricane's track can make big differences in the damage it causes.

To improve hurricane prediction, NOAA[1] relies on sophisticated computer models and has been working for decades to improve them. How well are they doing? Have predictions improved in recent years? Has the improvement been consistent? Figure 6.1 shows the mean error in nautical miles of the NHC's 72-hour predictions of Atlantic hurricanes for each year plotted against the year. NOAA refers to these errors as the Forecast error or the Prediction error and reports annual results.

[1] www.nhc.noaa.gov (Data in **Tracking Hurricanes 2016**)

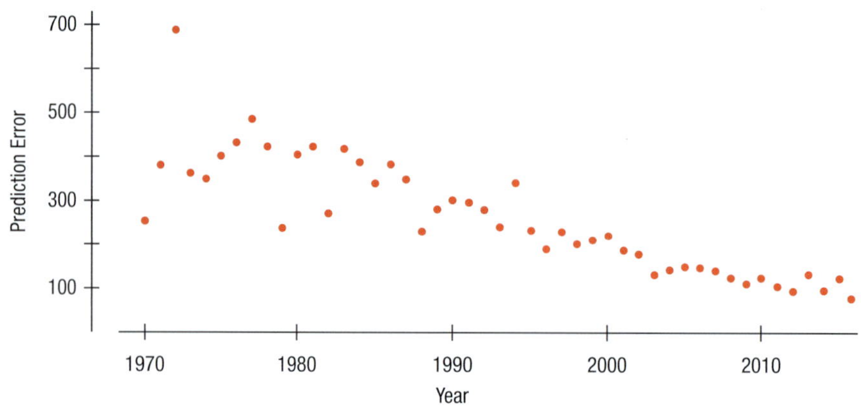

Figure 6.1
A scatterplot of the average tracking error in nautical miles of the predicted position of Atlantic hurricanes for predictions made by the National Hurricane Center of NOAA, plotted against the *Year* in which the predictions were made.

Where Did the Origin Go?

Scatterplots usually don't—and shouldn't—show the origin (the place where both *x* and *y* are zero) because often neither variable has values near 0. The display should focus on the part of the coordinate plane that actually contains the data. In our example about hurricanes, none of the prediction errors or years were anywhere near 0, so the computer drew the scatterplot with axes that don't meet.

Figure 6.1 is an example of a general kind of display called a **scatterplot**. Because they show the relationship between two quantitative variables, scatterplots may be the most common displays for data. By just looking at them, you can see patterns, trends, relationships, and even the occasional extraordinary value sitting apart from the others. As the great philosopher Yogi Berra[2] once said, "You can observe a lot just by watching."[3] Scatterplots are the best way to start observing the relationship between two *quantitative* variables. When the *x*-variable is *Time,* as it is here, the plot is often referred to as a **timeplot**.

From Figure 6.1, it's clear that predictions have improved. The plot shows a fairly steady decline in the average tracking error, from almost 500 nautical miles in the late 1970s to less than 100 nautical miles in 2014. We can also see a few years when predictions were unusually good and that 1972 was a really bad year for predicting hurricane tracks.

Relationships between variables are often at the heart of what we'd like to learn from data:

- Are grades higher now than they used to be?
- Do people tend to reach puberty at a younger age than in previous generations?
- Does applying magnets to parts of the body relieve pain? If so, are stronger magnets more effective?
- Do students learn better when they get more sleep?

Questions such as these relate two quantitative variables and ask whether there is an **association** between them. Scatterplots are the ideal way to *picture* such associations.

6.1 Scatterplots

How would you describe the association of hurricane *Prediction Error* and *Year*? Everyone looks at scatterplots. But, if asked, many people would find it hard to say what to look for in a scatterplot. What do *you* see? Try to describe the scatterplot of *Prediction Error* against *Year.*

You might say that the **direction** of the association is important. Over time, the NHC's prediction errors have decreased. A pattern like this that runs from the upper left to the lower right is said to be **negative**. A pattern running the other way is called **positive**.

Look for **Direction**: What's my sign—positive, negative, or neither?

The second thing to look for in a scatterplot is its **form**. A plot that appears as a cloud or swarm of points stretched out in a generally consistent, straight form is called linear. For example, the scatterplot of *Prediction Error* vs. *Year* has such an underlying linear form, although some points stray away from it.

[2]Hall of Fame catcher, outfielder, and manager of the New York Mets and Yankees.

[3]But then he also said, "I really didn't say everything I said." So we can't really be sure.

Look for **Form**: straight, curved, something exotic, or no pattern at all?

If the relationship isn't straight, but curves gently while still increasing or decreasing steadily 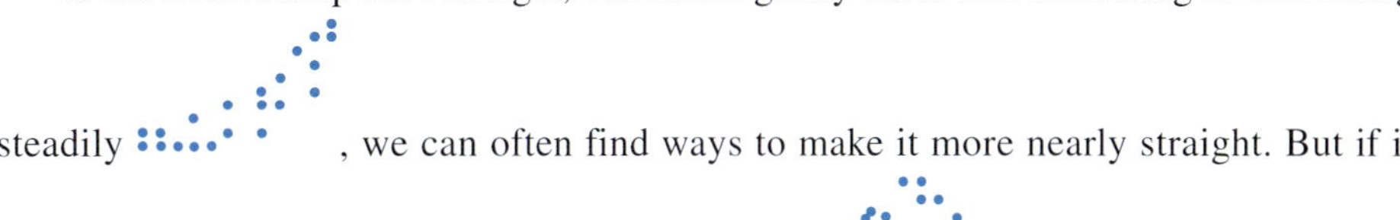, we can often find ways to make it more nearly straight. But if it curves sharply up and down, for example like this: , there is much less we can say about it with the methods of this text.

Look for **Strength**: how much scatter?

The third feature to look for in a scatterplot is the **strength** of the relationship. At one extreme, do the points appear tightly clustered in a single stream (whether straight, curved, or bending all over the place)? Or, at the other extreme, does the swarm of points seem to form a vague cloud through which we can barely discern any trend or pattern? The *Prediction error* vs. *Year* plot (Figure 6.1) shows moderate scatter around a generally straight form. This indicates that the linear trend of improving prediction is pretty consistent and moderately strong.

Look for **Unusual Features**: are there outliers or subgroups?

Finally, always look for the unexpected. Often the most interesting thing to see in a scatterplot is something you never thought to look for. One example of such a surprise is an **outlier** standing away from the overall pattern of the scatterplot. Such a point is almost always interesting and always deserves special attention. In the scatterplot of prediction errors, the year 1972 stands out as a year with very high prediction errors. An Internet search shows that it was a relatively quiet hurricane season. However, it included the very unusual—and deadly—Hurricane Agnes, which combined with another low-pressure center to ravage the northeastern United States, killing 122 and causing 1.3 billion 1972 dollars in damage. Possibly, Agnes was also unusually difficult to predict.

You should also look for clusters or subgroups that stand away from the rest of the plot or that show a trend in a different direction. Deviating groups should raise questions about why they are different. They may be a clue that you should split the data into subgroups instead of looking at them all together.

EXAMPLE 6.1 Comparing Prices Worldwide

If you travel overseas, you know that what's really important is not the amount in your wallet but the amount it can buy. UBS (one of the largest banks in the world) prepared a report comparing prices, wages, and other economic conditions in cities around the world for its international clients. Some of the variables it measured in 73 cities are *Cost of Living, Food Costs*, *Average Hourly Wage*, average number of *Working Hours* per Year, average number of *Vacation Days*, hours of work (at the average wage) needed to buy an *iPhone*, minutes of work needed to buy a *Big Mac*, and *Women's Clothing Costs*.[4] For your burger fix, you might want to live in Tokyo where it takes only about 9 minutes of work to afford a Big Mac. In Nairobi, you'd have to work almost an hour and a half.

Of course, these variables are associated, but do they consistently reflect costs of living? Plotting pairs of variables can reveal how and even if they are associated.

[4]Detail of the methodology can be found in the report *Prices and Earning: A comparison of purchasing power around the globe/2012 edition*, www.economist.com/node/14288808?story_id=14288808. (Data in **Prices and Earnings**)

The variety of these associations illustrates different directions and kinds of association patterns you might see in other scatterplots.

QUESTION: Describe the patterns shown by each of these plots.

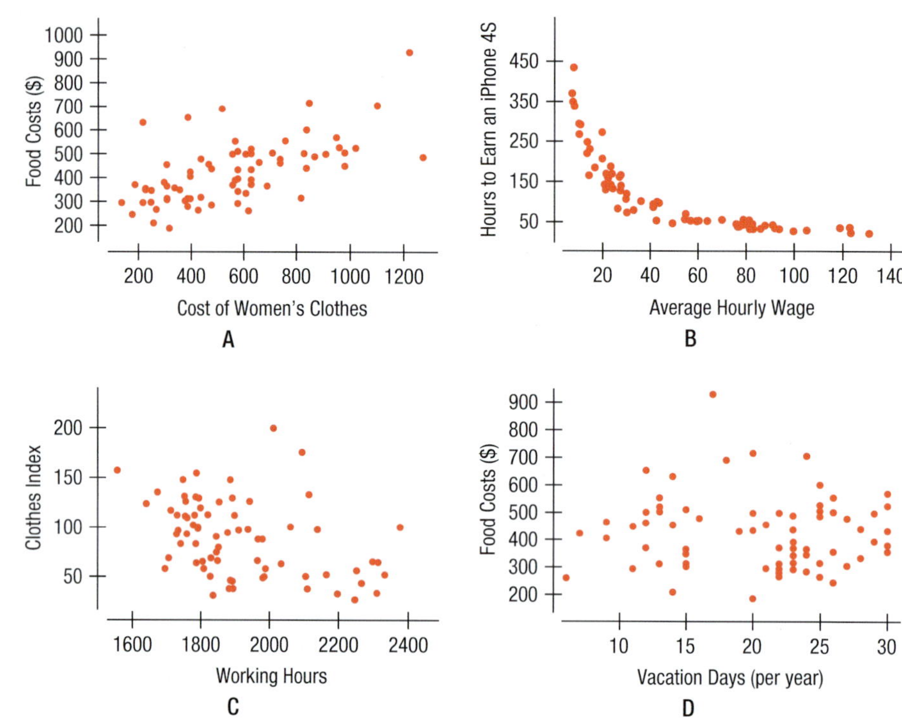

ANSWER: In Plot A, the association between *Food Costs* and *Cost of Women's Clothes* is positive and straight with a few high outliers, and moderately strong. In Plot B, the association between *Hours to Earn an iPhone4S* and *Average Hourly Wage* is negative and strong, but the form is not straight. In Plot C, the association between the *Clothes Index* and *Working Hours* is weak, negative, and generally straight with one or two possible high outliers. In Plot D, there does not appear to be any association between *Food Costs* and *Vacation Days*.

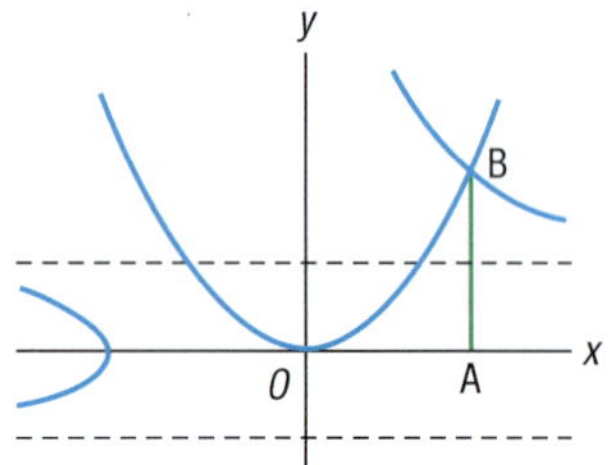

A figure based on Newton's *Enumeration of lines of the third order*[5] labels the axes in the modern way.

SCATTERPLOT HISTORY

The idea of using two axes at right angles to define a field on which to display values can be traced back to the philosopher, scientist, and mathematician René Descartes (1596–1650). The playing field he defined in this way is formally called a Cartesian plane in his honor. Sir Isaac Newton (1642–1727) may have been the first to use the now-standard convention of denoting the vertical axis *y* and the horizontal axis *x*.

Both Descartes and Newton used the Cartesian plane to plot functions, as you have probably done yourself. Plotting individual data values works in a similar way. Each case is displayed as a dot or symbol on a scatterplot at a position that corresponds to its values on two variables—one measured along the *x*-axis and the other along the *y*-axis. When you work with data you should label the axes with the names of the variables rather than just using *y* and *x*.

[5]Sir Isaac Newton's *Enumeration of lines of the third order, generation of curves by shadows, organic description of curves, and construction of equations by curves*, p. 86. Translated from the Latin. With notes and examples, by C.R.M. Talbot. Available at quod.lib.umich.edu/u/umhistmath/ABQ9451.0001.001/81?rgn=full+text;view=pdf

Roles for Variables

Which variable should go on the x-axis and which on the y-axis? What we want to know about the relationship can tell us how to make the plot. We often have questions such as

- Do baseball teams that score more runs sell more tickets to their games?
- Do older houses sell for less than newer ones of comparable size and quality?
- Do students who score higher on their SAT tests have higher grade point averages in college?
- Can we estimate a person's percent body fat accurately by measuring their waist or wrist size?

NOTATION ALERT

In statistics, the assignment of variables to the x- and y-axes (and the choice of notation for them in formulas) often conveys information about their roles as predictor or response variable. So x and y are reserved letters as well, but not just for labeling the axes of a scatterplot.

In these examples, the two variables play different roles. We'll call the variable of interest the **response variable** and the other the **explanatory** or **predictor variable**.[6] We'll continue our practice of naming the variable of interest y. Naturally we'll plot it on the y-axis and place the explanatory variable on the x-axis. Sometimes, we'll call them the **x- and y-variables**. For example, we wanted to know how accurate hurricane prediction is given the year, so prediction error was on the y-axis and year on the x-axis. When you make a scatterplot, you can assume that those who view it will think this way, so choose which variables to assign to which axes carefully.

The roles that we choose for variables are more about how we *think* about them than about the variables themselves. Just placing a variable on the x-axis doesn't necessarily mean that it explains or predicts *anything*. And the variable on the y-axis may not respond to it in any way. We plotted prediction error on the y-axis against year on the x-axis because the National Hurricane Center is interested in how their predictions have changed over time. Could we have plotted them the other way? In this case, it's hard to imagine reversing the roles—knowing the prediction error and wanting to guess in what year it happened. But for some scatterplots, it can make sense to use either choice, so you have to think about how the choice of role helps to answer the question you have.

6.2 Correlation

WHO	Students
WHAT	Height (inches), weight (pounds)
WHERE	Ithaca, NY
WHY	Data for class
HOW	Survey

Data collected from students in statistics classes included their *Height* (in inches) and *Weight* (in pounds). It's no great surprise that there is a positive association between the two. As you might suspect, taller students tend to weigh more. (If we had reversed the roles and chosen weight as the explanatory variable, we might say that heavier students trend to be taller.)[7] And the form of the scatterplot is fairly straight as well, although there appears to bc a high outlier.

[6]The x- and y-variables are sometimes referred to as the *independent* and *dependent* variables, respectively. The idea was that the y-variable depended on the x-variable and the x-variable acted independently. These names, however, conflict with other uses of the same terms in statistics, so we prefer not to use them.

[7]The young son of one of the authors, when told (as he often was) that he was tall for his age, used to point out that, actually, he was young for his height.

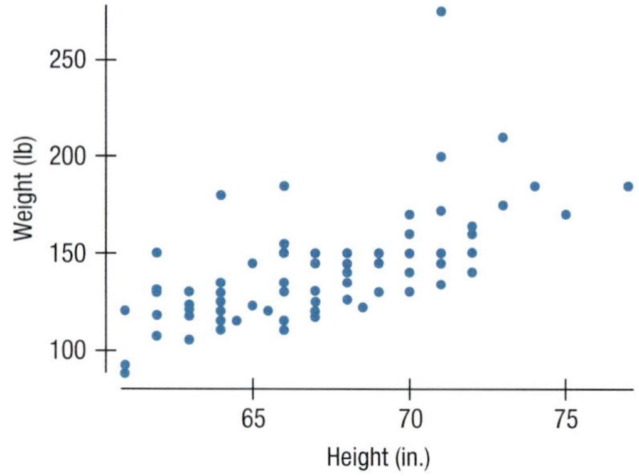

Figure 6.2
Weight (lb) versus *Height* (in.) from a statistics class. (Data in **Heights and weights**)

The pattern in the scatterplots looks straight, and the positive association is clear, but how strong is that association? If you had to put a number (say, between 0 and 1) on the strength, what would you say? The strength of the association shouldn't depend on the units we choose for *x* or *y*. For example, if we had measured weight in *kg* instead of *lb* and height in *cm* instead of *inches*, the plot (Figure 6.3) would look the same.

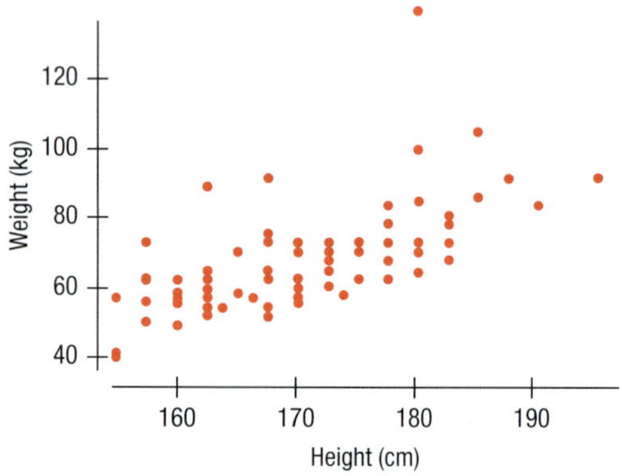

Figure 6.3
Changing the units (from pounds to kilograms for *Weight* and from inches to centimeters for *Height*) did not change the direction, form, or strength of the association between the variables.

If we subtract the mean from each variable, that shouldn't change the strength of the association either. We would just move the means of both variables to zero. (See Figure 6.4.) This makes it easier to see what each point contributes to the strength of the association because we can tell which points are above and which are below the means of each variable. Figure 6.4 colors them to make this even clearer. The green points in the upper right (first quadrant) and lower left (third quadrant) are consistent with a positive association while the red points in the other two quadrants are consistent with a negative association. Points lying on the *x*- or *y*-axis don't really add any information; they are colored blue.

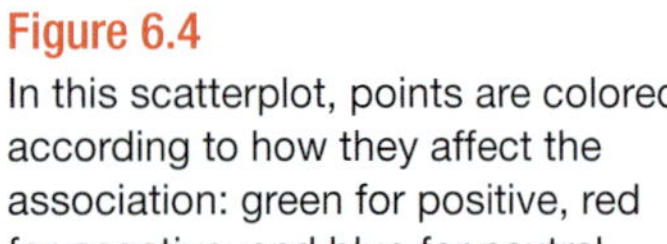

Figure 6.4
In this scatterplot, points are colored according to how they affect the association: green for positive, red for negative, and blue for neutral.

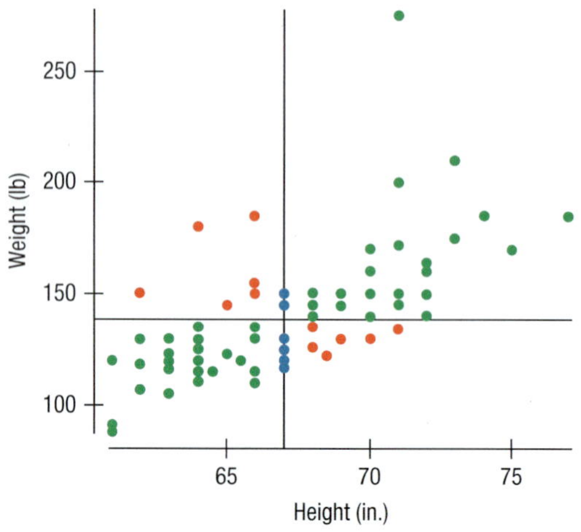

We said that the units shouldn't affect our measure of strength. Now that we've subtracted means, a natural way to remove the units is to standardize each variable and work instead with the z-scores. Recall that to find z-scores for any variable, we subtract its mean from each value and divide by its standard deviation. So instead of plotting (x, y), we plot:

$$(z_x, z_y) = \left(\frac{x - \bar{x}}{s_x}, \frac{y - \bar{y}}{s_y}\right).$$

z-Scores do just what we want. They center the plotted points around the origin and they remove the original units, replacing them with standard deviations. As we've seen, measuring differences in standard deviations is a fundamental idea in statistics.

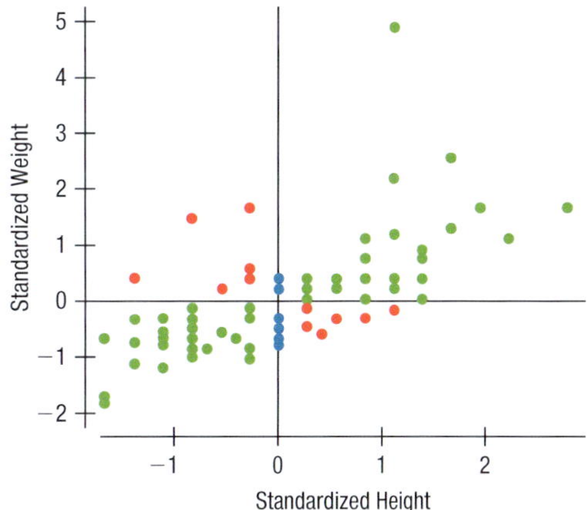

Figure 6.5
Standardized *Weight* and *Height*. We use the standard deviation to judge the distance of each point from the center of the plot and the influence it should have.

For the green points in Figure 6.5, both z-scores are positive, or both are negative. In either case, their product, $z_x z_y$, is positive. But the red points have opposite signs, so their products are negative. We can measure the strength of the association by adding up these products: $\sum z_x z_y$. That summarizes both the direction and strength of the association. Points farther from the origin have larger z-scores, so they'll contribute more to the sum. But the sum can keep growing as we consider more points. To keep that from happening, the natural (for statisticians, anyway) thing to do is to divide the sum by $n - 1$.[8]

The result is the famous **correlation coefficient**:

$$r = \frac{\sum z_x z_y}{n - 1}.$$

> *NOTATION ALERT*
> The letter r is always used for correlation, so you can't use it for anything else in statistics. Whenever you see an r, it's safe to assume it's a correlation.

Dividing the sum by $n - 1$ serves two purposes. It adjusts the strength for the number of points and it makes the correlation lie between values of -1 and $+1$. For the students' heights and weights, the correlation is 0.644. Because it is based on z-scores, which have no units, the correlation has no units either. It will stay the same if you change from inches to centimeters, fathoms, or Ångstroms.

There are a number of alternative formulas for the correlation coefficient, using x and y in their original units, that you may encounter. Here are two of the most common:

$$r = \frac{\sum(x - \bar{x})(y - \bar{y})}{\sqrt{\sum(x - \bar{x})^2(y - \bar{y})^2}} = \frac{\sum(x - \bar{x})(y - \bar{y})}{(n - 1)s_x s_y}.$$

These formulas are convenient for computing correlation by hand, but the z-score form is best for understanding what correlation means.

[8]Yes, the same $n - 1$ as in the standard deviation calculation. And we offer the same promise to explain it later.

BY HAND Finding the correlation coefficient

To find the correlation coefficient by hand, start with the summary statistics for both variables: $\bar{x}, \bar{y}, s_x$, and s_y. For each data pair, find the deviations from the means for both variables: $(x - \bar{x})$ and $(y - \bar{y})$, and multiply these deviations together: $(x - \bar{x})(y - \bar{y})$. Add up these products for all data pairs.

To get the correlation, r, divide this sum by the product $(n - 1)s_x s_y$.

Here's an example:

Suppose the data pairs are

x	6	10	14	19	21
y	5	3	7	8	12

Then $\bar{x} = 14, \bar{y} = 7, s_x = 6.20$, and $s_y = 3.39$

Deviations from $\bar{x}$	Deviations from $\bar{y}$	Product $(x - \bar{x})(y - \bar{y})$
6 − 14 = −8	5 − 7 = −2	−8 × −2 = 16
10 − 14 = −4	3 − 7 = −4	16
14 − 14 = 0	7 − 7 = 0	0
19 − 14 = 5	8 − 7 = 1	5
21 − 14 = 7	12 − 7 = 5	35

Add the products: $16 + 16 + 0 + 5 + 35 = 72$
Finally, divide by $(n - 1) \times s_x \times s_y = (5 - 1) \times 6.20 \times 3.39 = 84.07$
The ratio is the correlation coefficient:

$$r = 72/84.07 = 0.856$$

Assumptions and Conditions for Correlation

Correlation measures the strength of *linear* association between two *quantitative* variables. To interpret a correlation, we must assume that there is a true underlying linear relationship. Of course, in general, we can't know that. But we *can* check whether that assumption is plausible by looking at the data we do have. To distinguish what we *assume* about variables from what we can check by looking at the data, we'll call the things we check *conditions*. Checking the conditions almost always means either thinking about how the data were collected or looking at a graph—but if you follow the first three rules, you'll have already made the plot, so that takes no extra effort.

There are three conditions to check before you use a correlation:

- **Quantitative Variables Condition** Don't make the common error of calling an association involving a categorical variable a correlation. Correlation is only about quantitative variables.
- **Straight Enough Condition** The best check for the assumption that the variables are truly linearly related is to look at the scatterplot to see whether it looks reasonably straight. That's a judgment call, but not a difficult one.
- **No Outliers Condition** Outliers can distort the correlation dramatically, making a weak association look strong or a strong one look weak. Outliers can even change the sign of the correlation. But it's easy to see outliers in the scatterplot, so to check this condition, just look.

Correlations are often reported without supporting data or plots. Nevertheless, you can still think about the conditions. Be cautious in interpreting (or accepting others' interpretations of) a correlation when you can't check the conditions.

EXAMPLE 6.2 Correlations for Scatterplot Patterns

Look back at the scatterplots of the economic variables in cities around the world. The correlations for those plots are (A to D) 0.614, −0.791, −0.388, and −0.040, respectively.

QUESTION: Check the conditions for using correlation. If you feel they are satisfied, interpret the correlation.

ANSWER: All of the variables examined are quantitative and none of the plots shows an outlier. However, the relationship between *Hours to Earn an iPhone* and *Average Wage* is not straight, so the correlation coefficient isn't an appropriate summary. For the others:

A correlation of 0.614 between *Food Costs* and *Women's Clothing Costs* indicates a moderately strong positive association.

A correlation of −0.388 between *Clothes Index* and *Working Hours* indicates a moderately weak negative association.

The small correlation value of −0.040 between *Food Costs* and *Vacation Days* suggests that there may be no linear association between them.

JUST CHECKING

Your statistics instructor tells you that the correlation between the scores (points out of 50) on Exam 1 and Exam 2 was 0.75.

1. Before answering any questions about the correlation, what would you like to see? Why?
2. If she adds 10 points to each Exam 1 score, how will this change the correlation?
3. If she standardizes scores on each exam, how will this affect the correlation?
4. In general, if someone did poorly on Exam 1, are they likely to have done poorly or well on Exam 2? Explain.
5. If someone did poorly on Exam 1, can you be sure that they did poorly on Exam 2 as well? Explain.

STEP-BY-STEP EXAMPLE Looking at Association

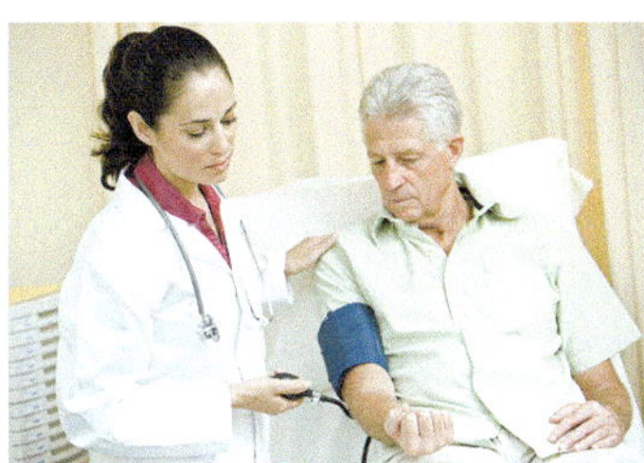

When blood pressure is measured, it is reported as two values: systolic blood pressure and diastolic blood pressure.

QUESTIONS: How are these variables related to each other? Do they tend to be both hlgh or both low? How strongly assoclated are they?

THINK **PLAN** State what you are trying to investigate.

I'll examine the relationship between two measures of blood pressure. (Data in **Framingham**)

VARIABLES Identify the two quantitative variables whose relationship you wish to examine. Report the W's, and be sure both variables are recorded for the same individuals.

The variables are systolic and diastolic blood pressure (*SBP* and *DBP*), recorded in millimeters of mercury (mm Hg) for each of 1406 participants in the Framingham Heart Study, a long-running health study in Framingham, Massachusetts.[9]

PLOT Make the scatterplot. Use a computer program or graphing calculator if possible.

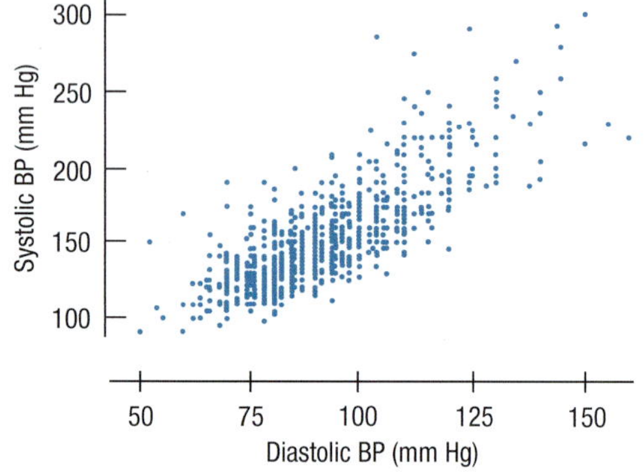

Check the conditions.

✓ **Quantitative Variables Condition**: Both variables are quantitative with units of mm Hg.

✓ **Straight Enough Condition**: The scatterplot is quite straight.

✓ **Outliers?** There are no extreme outliers.

REALITY CHECK Looks like a strong positive linear association. We shouldn't be surprised if the correlation coefficient is positive and fairly large.

I have two quantitative variables that satisfy the conditions, so correlation is a suitable measure of association.

SHOW **MECHANICS** We usually calculate correlations with technology. Here there are 1406 cases, so we'd never try it by hand.

The correlation coefficient is $r = 0.792$.

TELL **CONCLUSION** Describe the direction, form, and strength you see in the plot, along with any unusual points or features. Be sure to state your interpretations in the proper context.

The scatterplot shows a positive direction, with higher *SBP* going with higher *DBP*. The plot is generally straight, with a moderate amount of scatter. The correlation of 0.792 is consistent with what I saw in the scatterplot.

Correlation Properties

Here's a useful list of facts about the correlation coefficient:

- The sign of a correlation coefficient gives the direction of the association.
- Correlation is always between -1 and $+1$. Correlation *can* be exactly equal to -1.0 or $+1.0$, but these values are unusual in real data because they mean that all the data points fall *exactly* on a single straight line.
- Correlation treats x and y symmetrically. The correlation of x with y is the same as the correlation of y with x.

[9]www.framinghamheartstudy.org

- Correlation has no units. This fact can be especially appropriate when the data's units are somewhat vague to begin with (IQ score, personality index, socialization, and so on). Correlation is sometimes given as a percentage, but you probably shouldn't do that because it suggests a percentage of *something*—and correlation, lacking units, has no "something" of which to be a percentage.
- Correlation is not affected by changes in the center or scale of either variable. Changing the units or baseline of either variable has no effect on the correlation coefficient. Correlation depends only on the *z*-scores, and they are unaffected by changes in center or scale.
- Correlation measures the strength of the *linear* association between the two variables. Variables can be strongly associated but still have a small correlation if the association isn't linear.
- Correlation is sensitive to outliers. A single outlying value can make a small correlation large or make a large one small.

HOW STRONG IS STRONG?

You'll often see correlations characterized as "weak," "moderate," or "strong," but be careful. There's no agreement on what those terms mean. The same numerical correlation might be strong in one context and weak in another. You might be thrilled to discover a correlation of 0.7 between the new summary of the economy you've come up with and stock market prices, but you'd consider it a design failure if you found a correlation of "only" 0.7 between two tests intended to measure the same skill. Deliberately vague terms like "weak," "moderate," or "strong" that describe a linear association can be useful additions to the numerical summary that correlation provides. But be sure to include the correlation and show a scatterplot so others can judge for themselves.

EXAMPLE 6.3 Changing Scales

RECAP: Several measures of prices and wages in cities around the world show a variety of relationships, some of which we can summarize with correlations.

QUESTION: Suppose that, instead of measuring prices in U.S. dollars and recording work time in hours, we had used euros and minutes. How would those changes affect the conditions, the values of correlation, or our interpretation of the relationships involving those variables?

ANSWER: Not at all. Correlation is based on standardized values (*z*-scores), so the conditions, value of *r*, and interpretation are all unaffected by changes in units.

RANDOM MATTERS Correlations Vary

Earlier we saw that means vary from sample to sample. The shape of their sampling distribution was unimodal and symmetric, even when the observations were drawn from a distribution that wasn't. It shouldn't be a surprise that correlations vary from sample to sample too. But is their sampling distribution symmetric as well? Before we do a simulation, think about it. If the two variables were really uncorrelated, it would be just as likely for a random sample to have a small positive correlation as a small negative one. If the two variables were strongly related—say, with a correlation of 0.98—a random sample might show a correlation of 0.9. But no sample can have a correlation greater than 1, so more samples would have a smaller correlation and very few could show a larger one.[10] In fact, unless the true correlation is 0, the sampling distribution has to be skewed toward zero. Let's run a simulation to see.

We'll take samples of size 50 from the population of all live births in the United States in 1998. There were 3,945,192 live births. The mother's age was recorded for all of them, but the father's age was recorded for "only" 3,377,721 of them. We'll treat those complete pairs of ages as our population. A sample of 50 of these parents shows that the relationship between the age of the mother and the age of the father is positive, straight, and strong (Figure 6.6). The correlation of ages in all the pairs in the population is 0.757.

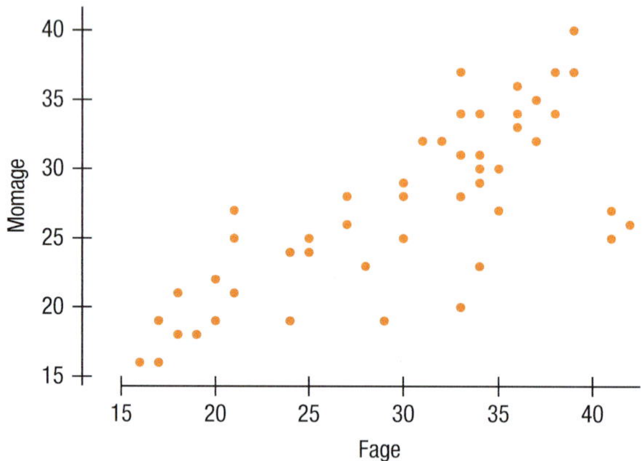

Figure 6.6
Mother's age vs. father's age for a sample of 50 live births.

To examine the sampling distribution of the correlation we drew 10,000 samples of 50 pairs of parents and found the correlation of mother's and father's ages from each sample. Figure 6.7 shows a histogram of the correlations.

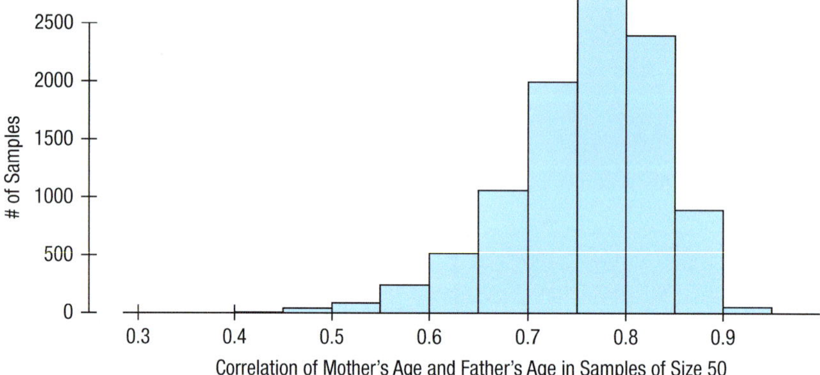

Figure 6.7
Correlations of 10,000 samples of 50 parent pairs.

As we suspected, this distribution is skewed to the left. It has its mode near the population correlation of 0.757. And there are more samples with correlations less than 0.657 than there are with correlations greater than 0.857.

We can learn two things from this experiment. First, correlations do vary. Some in our experiment are as small as 0.4, although most are between 0.7 and 0.8. Second, the distribution of the correlations is skewed toward zero. That tells us that as correlations vary from sample to sample, it is more likely that we'll see a correlation that is somewhat smaller than we might have seen in a different sample. That's probably a good thing because it makes it less likely that we'll be impressed with a large correlation that happened due to sample-to-sample variation.

[10]If the correlation were large and negative, it would bump up against the bound at −1, and random correlations would tend to be closer to zero.

Sir Maurice Kendall (1907–1983) was a prominent statistician, writer, and teacher. He made significant contributions to the study of randomness (developing tests for randomness that are still used), the theory of statistics, and rank-based statistics. He also served as director of the World Fertility Study, for which the United Nations awarded him their Peace Medal.

*Measuring Trend: Kendall's Tau

Survey researchers often ask people to rank their opinions or attitudes on a scale that lacks units. (Common scales run from 1 to 5, 1 to 7, or 1 to 10.) For example, we might ask:

"How would you assess the pace of your statistics course so far?"

1 = Way too slow	*2 = A little too slow*	*3 = About right*
4 = A little too fast	*5 = Way too fast*	

Scales of this sort that attempt to measure attitudes numerically are called Likert scales, after their developer, Rensis Likert. Likert scales have order. The higher the number, the faster you think the course has gone, but a 4 doesn't necessarily mean *twice* as fast as 2. So we may want to be careful when applying methods such as correlation that require the Straight Enough Condition. That may be too strong a requirement when all we may want to ask of the data is whether there is a consistent trend.

Suppose we collected data from lots of courses and tried to relate perceived pace to the number of students in the courses. We might wonder if students perceive the course as being faster in larger courses, but it isn't necessarily clear what it would mean for *pace* to be related to *class size* in a linear way. So the correlation coefficient might not be the appropriate measure. Still, we might want to know if the speed of the course and its size are associated. To answer this question, we can use an alternative measure of association, Kendall's tau.

Pronunciation Note:

Many people pronounce Likert's name with a long "i" as in "like." But he actually pronounced it with a short "i" as in "lick."

Kendall's tau is a statistic designed to assess how close the relationship between two variables is to being *monotone*. A monotone relationship is one that consistently increases or decreases, but not necessarily in a linear fashion. That is often all we are interested in. And, of course, if we only care about a consistent trend, we don't have to worry that one or both of our variables might not be on a perfectly quantitative scale.

Kendall's tau measures monotonicity directly. For each pair of points in a scatterplot, it records only whether the slope of a line between those two points is positive, negative, or zero. (If the points have the same x-value, the slope between them is ignored.) In a monotone plot, these pairwise slopes would all be positive or all be negative. In a non-monotone plot, the counts of positive and negative slopes will tend to balance out. Tau (often written with the Greek letter τ) is the difference between the number of positive slopes and the number of negative slopes divided by the total number of slopes between pairs.

> NOTATION ALERT
>
> The Greek letter tau is most often used for Kendall's tau, but it is sometimes used for other things in statistics. The context is usually clear.

Obviously, tau can take values between −1.0 and +1.0. If every pairwise slope is negative, then tau will equal −1.0, and the plot will show a single stream of points trending consistently downward. If every pairwise slope is positive, then tau will equal +1.0, and the plot will show a single stream of points trending consistently upward. In a generally horizontal plot or one that goes up and then down, the counts of positive and negative slopes will balance out and tau will be near zero.

A Monotonous Relationship?

You probably think of monotonous as *boring*. A monotone relationship shares the same root, but it's only the *sameness* that's important. A monotone relationship is one that increases or decreases consistently (but it doesn't have to be boring).

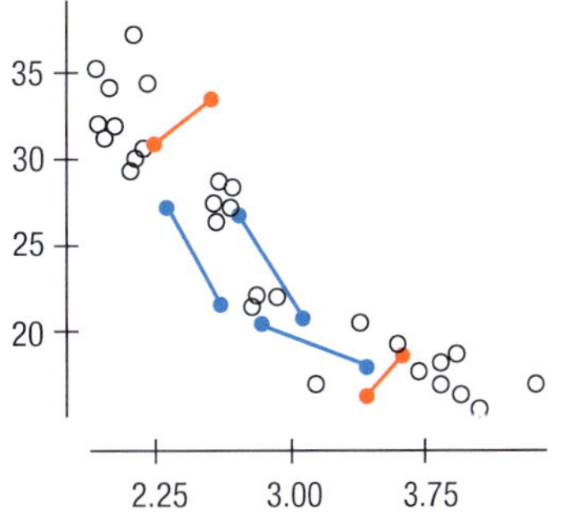

Figure 6.8

For each pair of points, Kendall's tau records whether the slope between them is positive (red), negative (blue), or zero.

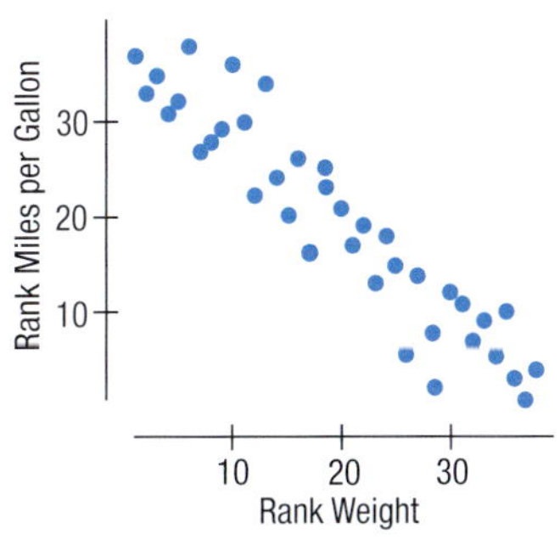

Figure 6.9

Spearman's rho finds the correlation of the *ranks* of the two variables.

NOTATION ALERT

The Greek letter rho has some other uses in statistics, but the context usually makes the meaning clear.

*Nonparametric Association: Spearman's Rho

One of the problems we have seen with the correlation coefficient is that it is very sensitive to violations of the Straight Enough Condition. Both outliers and bends in the data make it impossible to interpret correlation. **Spearman's rho** (often denoted by the Greek letter ρ) can deal with both of these problems. Rho replaces the original data values with their *ranks* within each variable. That is, it replaces the lowest value in x by the number 1, the next lowest value by 2, and so on, until the highest value is assigned the value n. The same ranking method is applied to the y-variable. A scatterplot of the y-ranks against the x-ranks shows the same general *trend* as the original data, going up whenever the data went up and down whenever the data went down. If the original scatterplot shows a consistent but bent trend, however, the scatterplot of the ranks is likely to be more nearly linear. And if either variable has an extreme outlying value, the ranking process keeps it from being extreme by just counting it as the highest or lowest value, ignoring how extreme it really is.

Charles Edward Spearman, FRS (1863–1945), was an English psychologist known for work in statistics, as a pioneer of factor analysis, and for Spearman's rank correlation coefficient. He also did fundamental work on measures of human intelligence.

Spearman's rho is the correlation of the two rank variables. Because this is a correlation coefficient, it must be between -1.0 and 1.0.

Both Kendall's tau and Spearman's rho have advantages over the correlation coefficient r. For one, they can be used even when we know only the ranks. Also, they measure the consistency of the trend between the variables without insisting that the trend be linear.

They have the added advantage that they are not much affected by outliers. Spearman's rho limits the outlier to the value of its rank, and Kendall's tau cares only about the sign of each slope between points, not how *steep* it might be. Neither tau nor rho is changed at all by re-expressing variables by functions that don't alter the order of values (that is, by just the sort of functions we might use to straighten a scatterplot). But, unlike the correlation coefficient, neither statistic can be used as a base for more advanced or complex methods, so they tend to be specialized methods used when we care primarily about consistent trend between two variables.

Both of these measures are examples of what are called *nonparametric* or *distribution-free* methods. We first discussed parameters in Chapter 5. They are the constants in a model. The correlation coefficient attempts to estimate a particular parameter in the Normal model for two quantitative variables. Kendall's tau and Spearman's rho are less specific. They measure association, but there is no parameter that they are tied to and no specific model they require.

We'll run into more examples of nonparametric methods throughout the book. Because they tend to be somewhat specialized methods, we'll continue to put them in starred sections in the chapters where they appear.

6.3 Warning: Correlation ≠ Causation

Whenever we have a strong correlation, it's tempting to try to explain it by imagining that the predictor variable has *caused* the response to change. Humans are like that; we tend to see causes and effects in everything.

Sometimes this tendency can be amusing. A scatterplot of the human population (y) of Oldenburg, Germany, in the beginning of the 1930s plotted against the number of storks nesting in the town (x) shows a tempting pattern.

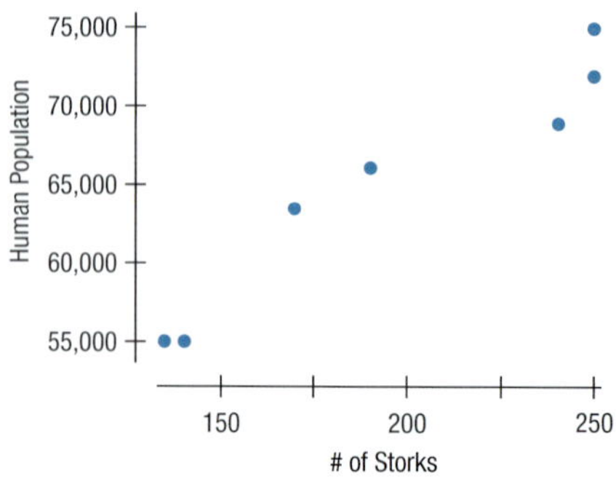

Figure 6.10
The number of storks in Oldenburg, Germany, plotted against the population of the town for 7 years in the 1930s. The association is clear. How about the causation? (*Statistics for Experimenters,* Box, Hunter and Hunter. Originally from *Ornithologische Monatsberichte*, 44, no. 2) (Data in **Storks**)

Anyone who has seen the beginning of the movie *Dumbo* remembers Mrs. Jumbo anxiously waiting for the stork to bring her new baby. Even though you know it's silly, you can't help but think for a minute that this plot shows that storks are the culprits. The two variables are obviously related to each other (the correlation is 0.97!), but that doesn't prove that storks bring babies.

It turns out that storks nest on house chimneys. More people means more babies, but also more houses, more nesting sites, and so more storks. Storks are "caused" if at all, by families—who also cause babies, but you can't tell from the scatterplot or correlation. You need additional information—not just the data—to determine the real mechanism.

A scatterplot of the damage (in dollars) caused to houses by fires would show a strong correlation with the number of firefighters at the scene. Surely the damage doesn't cause firefighters. And firefighters do seem to cause damage, spraying water all around and chopping holes. Does that mean we shouldn't call the fire department? Of course not. There is an underlying variable that leads to both more damage and more firefighters: the size of the blaze.

We call a hidden variable, like the fire size or the number of families, that stands behind a relationship and determines it by simultaneously affecting the other two variables a **lurking variable**. You can often debunk claims made about data by finding a lurking variable behind the scenes.

However, whether you find a lurking variable behind a relationship or not, remember: Scatterplots and correlation coefficients *never* prove causation. That's one reason it took so long for the U.S. Surgeon General to get warning labels on cigarettes. Although there was plenty of evidence that increased smoking was *associated* with increased levels of lung cancer, it took years to provide evidence that smoking actually *causes* lung cancer.

DOES CANCER CAUSE SMOKING?

Even if the correlation of two variables is due to a causal relationship, the correlation itself cannot tell us what causes what.

Sir Ronald Aylmer Fisher (1890–1962) was one of the greatest statisticians of the 20th century. Fisher testified in court (in testimony paid for by the tobacco companies) that a causal relationship might underlie the correlation of smoking and cancer:

> "Is it possible, then, that lung cancer . . . is one of the causes of smoking cigarettes? I don't think it can be excluded . . . the pre-cancerous condition is one involving a certain amount of slight chronic inflammation. . . .
>
> A slight cause of irritation . . . is commonly accompanied by pulling out a cigarette, and getting a little compensation for life's minor ills in that way. And . . . is not unlikely to be associated with smoking more frequently."

Ironically, the proof that smoking indeed is the cause of many cancers came from experiments conducted following the principles of experiment design and analysis that Fisher himself developed—and that we'll see in Chapter 11.

Correlation Tables

It is common in some fields to compute the correlations between every pair of variables in a collection of variables and arrange these correlations in a table. The row and column headings name the variables, and the cells hold the correlations.

Correlation tables are compact and give a lot of summary information at a glance. They can be an efficient way to start to look at a large dataset, but a dangerous one. By presenting all of these correlations without any checks for linearity and outliers, the correlation table risks showing correlations that by random chance are larger than the underlying relationship would warrant, truly small correlations that have been inflated by outliers, truly large correlations that are hidden by outliers, and correlations of any size that may be meaningless because the underlying form is not linear.

The diagonal cells of a correlation table always show correlations of exactly 1. (Can you see why?) Correlation tables are commonly offered by statistics packages on computers. These same packages often offer simple ways to make all the scatterplots that go with these correlations.

	Assets	Sales	Market Value	Profits	Cash Flow	Employees
Assets	1.000					
Sales	0.746	1.000				
Market Value	0.682	0.879	1.000			
Profits	0.602	0.814	0.968	1.000		
Cash Flow	0.641	0.855	0.970	0.989	1.000	
Employees	0.594	0.924	0.818	0.762	0.787	1.000

Table 6.1

A correlation table of data reported by *Forbes* magazine on several financial measures for large companies. From this table, can you be sure that the variables are linearly associated and free from outliers?

*6.4 Straightening Scatterplots

Correlation is a suitable measure of strength for straight relationships only. When a scatterplot shows a bent form that has an increasing or decreasing trend, we can often straighten the form of the plot by re-expressing one or both variables.

Some camera lenses have an adjustable aperture, the hole that lets the light in. The size of the aperture is expressed in a mysterious number called the f/stop. Each increase of one f/stop number corresponds to a halving of the light that is allowed to come through. The f/stops of one digital camera are: (Data in **F-stops**)

f/stop: 2.8 4 5.6 8 11 16 22 32.

When you halve the shutter speed, you cut down the light, so you have to open the aperture one notch. We could experiment to find the best f/stop value for each shutter speed. A table of recommended shutter speeds and f/stops for a camera lists the relationship like this:

Shutter speed:	1/1000	1/500	1/250	1/125	1/60	1/30	1/15	1/8
f/stop:	2.8	4	5.6	8	11	16	22	32

The correlation of these shutter speeds and f/stops is 0.979. That sounds pretty high. But a high correlation doesn't necessarily mean a strong linear relationship. When we check the scatterplot (we *always* check the scatterplot), it shows that something is not quite right:

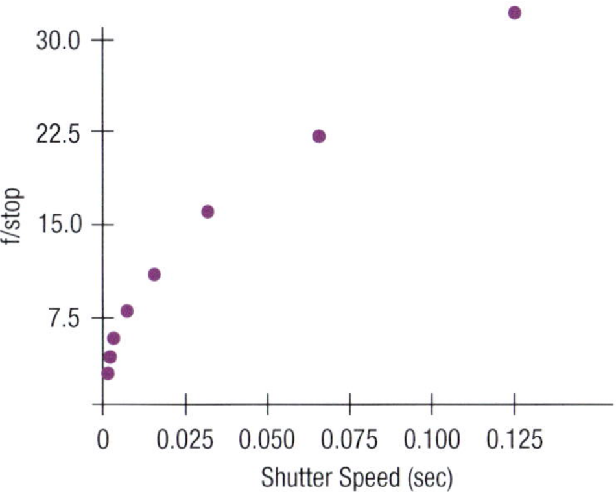

Figure 6.11
A scatterplot of *f/stop* vs. *Shutter Speed* shows a bent relationship.

We can see that the f/stop is not *linearly* related to the shutter speed. Can we find a transformation of *f/stop* that straightens out the line?

In Chapter 4, we saw that **re-expressing** data by raising each value to a power or taking the logarithm could make the shape of the histogram of the data more nearly symmetric and could make the spreads of different groups more nearly equal. Now we hope to make a curved relationship straighter. Remarkably, these three goals are closely related, so we can use the same methods to achieve all of them.

The Ladder of Powers

The secret is to select a re-expression from a simple collection of functions—the powers and the logarithm. We can raise each data value in a quantitative variable to the same power. It is important to recall a few facts about powers:

- The ½ power is the same as taking the square root.
- The -1 power is the reciprocal: $y^{-1} = 1/y$
- Putting those together, $y^{-1/2} = 1/\sqrt{y}$

In addition, we'll use the logarithm in place of the "0" power. Although it doesn't matter for this purpose, the base-10 logarithm is usually easier to think about.[11]

> **RE-EXPRESSIONS IN SCIENTIFIC LAWS**
> Scientific laws often include simple re-expressions. For example, in psychology, Fechner's Law states that sensation increases as the logarithm of stimulus intensity ($S = k \log R$).

[11] You may have learned facts about calculating with logarithms. Don't worry; you won't need them here.

Where to start? It turns out that certain kinds of data are more likely to be helped by particular re-expressions. Knowing that gives you a good place to start your search for a re-expression.

Power	Name	Comment
2	The square of the data values, y^2.	Try this for unimodal distributions that are skewed to the left or to re-express *y* in a scatterplot that bends downward.
1	The raw data—no change at all. This is "home base." The farther you step from here up or down the ladder, the greater the effect.	Data that can take on both positive and negative values with no bounds are less likely to benefit from re-expression.
1/2	The square root of the data values, $\sqrt{y}$.	Counts often benefit from a square root re-expression. For counted data, start here.
"0"	Although mathematicians define the "0-th" power differently,[12] for us the place is held by the logarithm. You may feel uneasy about logarithms. Don't worry; the computer or calculator does the work.[13]	Measurements that cannot be negative, and especially values that grow by percentage increases such as salaries or populations, often benefit from a log re-expression. When in doubt, start here. If your data have zeros, try adding a small constant to each value before finding the logs.
−1/2	The (negative) reciprocal square root, $-1/\sqrt{y}$.	An uncommon re-expression, but sometimes useful. Changing the sign to take the *negative* of the reciprocal square root preserves the direction of relationships, making things a bit simpler.
−1	The (negative) reciprocal, $-1/y$.	Ratios of two quantities (miles per hour, for example) often benefit from a reciprocal. (You have about a 50-50 chance that the original ratio was taken in the "wrong" order for simple statistical analysis and would benefit from re-expression.) Often, the reciprocal will have simple units (hours per mile). Change the sign if you want to preserve the direction of relationships. If your data have zeros, try adding a small constant to all values before finding the reciprocal.

The powers and logarithm have an important advantage as re-expressions: They are *invertible*. That is, you can get back to the original values easily. For example, if we re-express *y* by the square root, we can get back to the original values by squaring the re-expressed values. To undo a logarithm, raise 10 to the power of each value.[14]

The positive powers and the logarithm preserve the order of values—the larger of two values is still the larger one after re-expression. Negative powers reverse the order $(4 > 2$ but $\frac{1}{4} < \frac{1}{2})$ so we sometimes negate the values $(-\frac{1}{4} > -\frac{1}{2})$ to preserve the order.

Most important, the *effect* that any of these re-expressions has on straightening a scatterplot is ordered. Choose a power. If it doesn't straighten the plot by enough, then try the next power. If it overshoots so the plot bends the other way, then step back. That insight makes it easy to find an effective re-expression, and it leads to referring to these functions as the **ladder of powers**.

But don't forget—when you take a negative power, the direction of the relationship will change. That's OK. You can always change the sign of the response variable if you want to keep the same direction. With modern technology, finding a suitable re-expression is no harder than the push of a button.

---

[12]You may remember that for any nonzero number *y*, $y^0 = 1$. This is not a very exciting re-expression for data; every data value would be the same. We use the logarithm in its place.

[13]Your calculator or software package probably gives you a choice between "base 10" logarithms and "natural (base *e*)" logarithms. Don't worry about that. It doesn't matter at all which you use; they have exactly the same effect on the data. If you want to choose, base-10 logarithms can be a bit easier to interpret.

[14]Your computer or calculator knows how.

f/Stops Again

Following the advice in the table, we can try taking the logarithm of the f/stop values.

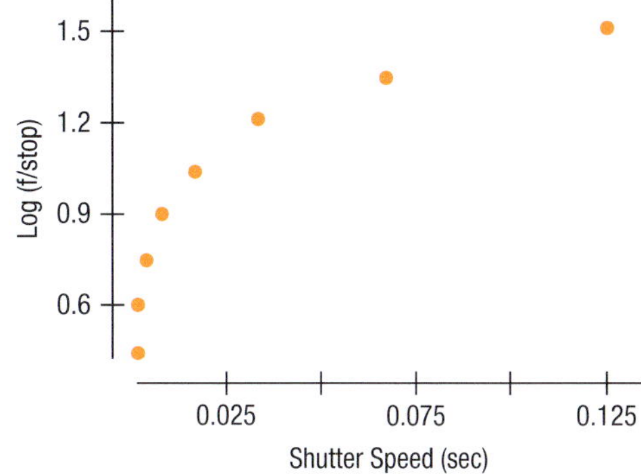

Figure 6.12
Log(f/stop) values vs. *Shutter Speed*.

But, as Figure 6.12 shows, the logarithm of f/stop values is even *more* curved vs. *Shutter Speed*. The ladder of powers says that we'd do better if we move in the opposite direction. That would take us to the square.

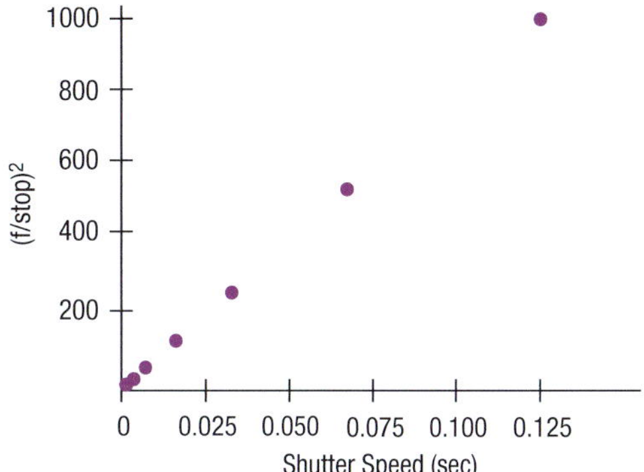

Figure 6.13
(f/stop)^2 vs. *Shutter Speed*.

That scatterplot looks much straighter.[15] It's appropriate to summarize the strength of this relationship with a correlation, which turns out to be 0.998.[16] You can find more discussion of finding appropriate re-expressions in Chapter 8.

WHAT CAN GO WRONG?

- **Don't say "correlation" when you mean "association."** How often have you heard the word "correlation"? Chances are pretty good that when you've heard the term, it's been misused. It's one of the most widely misused statistics terms, and given how often statistics are misused, that's saying a lot. One of the problems is that many people use the specific term *correlation* when they really mean the more general term *association*. "Association" is a deliberately vague term describing the relationship between two variables. "Correlation" is a precise term that measures the strength and direction of the linear relationship between quantitative variables.

---

[15]Sometimes, we can do a "reality check" on our choice of re-expression. In this case, a bit of research reveals that f/stops are related to the diameter of the open shutter. Since the amount of light that enters is determined by the *area* of the open shutter, which is related to the diameter by squaring, the square re-expression seems reasonable. Not all re-expressions have such nice explanations, but it's a good idea to think about them.

[16]That's a high correlation, but it does not confirm that we have the best re-expression. After all, the correlation of the original relationship was 0.979—high enough to satisfy almost anyone. To judge a re-expression look at the scatterplot. Your eye is a better judge of straightness than any statistic.

- **Don't correlate categorical variables.** Did you know that there's a strong correlation between playing an instrument and drinking coffee? No? One reason might be that the statement doesn't make sense. People who misuse the term "correlation" to mean "association" often fail to notice whether the variables they discuss are quantitative. But correlation is only valid for quantitative variables. Be sure to check that the variables are quantitative.
- **Don't confuse correlation with causation.** One of the most common mistakes people make in interpreting statistics occurs when they observe a high correlation between two variables and jump to the perhaps tempting conclusion that one thing must be causing the other. Scatterplots and correlations *never* demonstrate causation. At best, these statistical tools can only reveal an association between variables, and that's a far cry from establishing cause and effect.
- **Make sure the association is linear.** Not all associations between quantitative variables are linear. Correlation can miss even a strong nonlinear association. A student project evaluating the quality of brownies baked at different temperatures reports a correlation of -0.05 between judges' scores and baking temperature. That seems to say there is no relationship—until we look at the scatterplot.

The relationship between brownie taste *Score* and *Baking Temperature* is strong, but not at all linear.

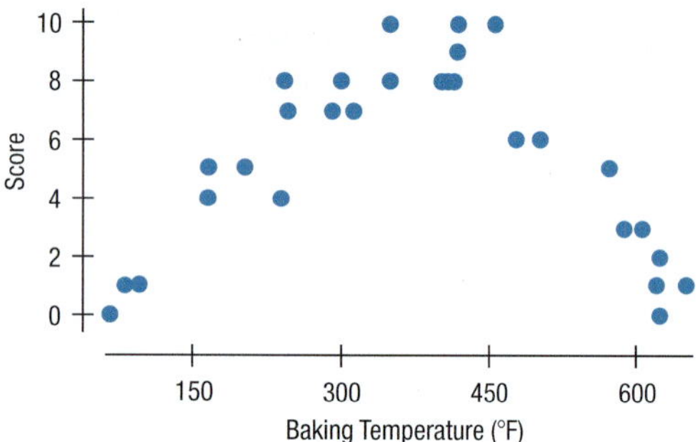

There is a strong association, but the relationship is not linear. Don't forget to check the Straight Enough Condition.

- **Don't assume the relationship is linear just because the correlation coefficient is high.** The correlation of the *Time* of the winner to complete the Indianapolis 500 race (for those races that actually lasted for 500 miles) with the *Year* of the race (for years from the beginning of the race until 1980) is 97.4%, but the scatterplot shows a bend. The relationship of *Speed* (essentially 1/*Time*) with *Year* is straighter. A high correlation is no guarantee of straightness. Nor is it safe to use correlation to judge the best re-expression. It's always important to look at the scatterplot.

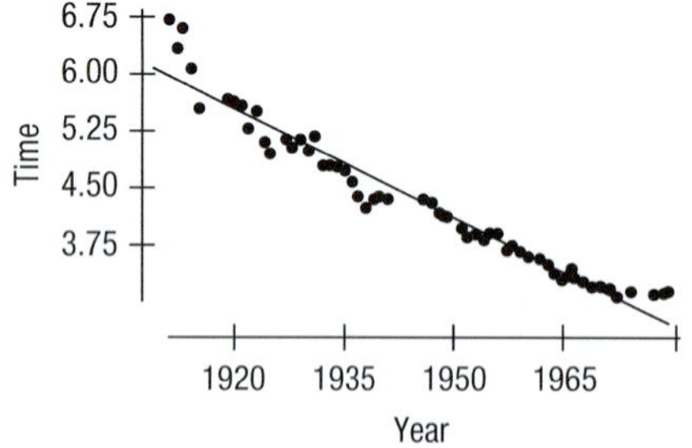

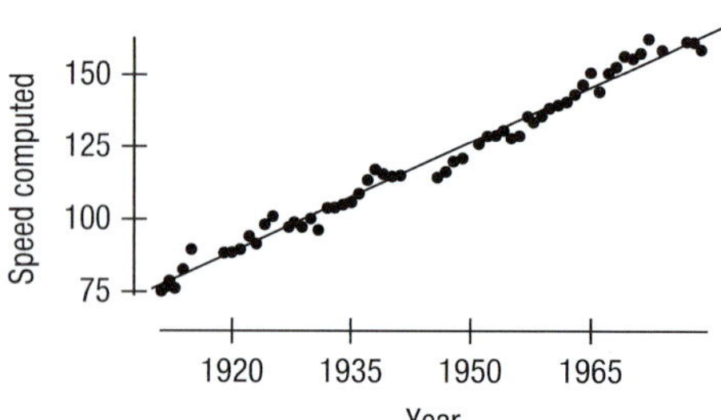

- **Beware of outliers.** You can't interpret a correlation coefficient safely without a background check for outliers. Here's a silly example:

  The relationship between IQ and shoe size among comedians shows a surprisingly strong positive correlation of 0.50. To check assumptions, we look at the scatterplot. What is the relationship?

The correlation is 0.50, but it is all due to the outlier (the green x) in the upper right corner. Who does that point represent?

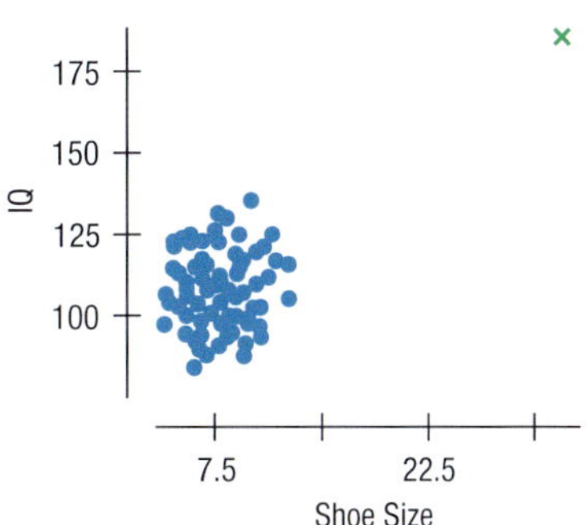

The outlier is Bozo the Clown, known for his large shoes, and widely acknowledged to be a comic "genius." Without Bozo, the correlation is near zero.

Even a single outlier can dominate the correlation value because its z-scores will be large and their product even larger. That's why you need to check the scatterplot for outliers.

CONNECTIONS

Scatterplots are the basic tool for examining the relationship between two quantitative variables. We start with a picture when we want to understand the distribution of a single variable, and we always make a scatterplot to begin to understand the relationship between two quantitative variables. Always.

We used z-scores as a way to measure the statistical distance of data values from their means. Now we've seen the z-scores of x and y working together to build the correlation coefficient. Correlation is a summary statistic like the mean and standard deviation—only it summarizes the strength of a linear relationship. And we interpret it as we did z-scores, using the standard deviations as our rulers in both x and y.

CHAPTER REVIEW

Make a scatterplot to display the relationship between two quantitative variables.

- Look at the direction, form, and strength of the relationship, and any outliers that stand away from the overall pattern.

Provided the form of the relationship is straight, summarize its strength with a correlation, r.

- The sign of the correlation gives the direction of the relationship.
- $-1 \le r \le 1$; A correlation of 1 or -1 is a perfect linear relationship. A correlation of 0 indicates that there is no linear relationship.
- Correlation has no units, so shifting or scaling the data, standardizing, or even swapping the variables has no effect on the numerical value.

A large correlation is not a sign of a causal relationship.

Correlations, like other statistics, vary from sample to sample. The distribution of correlations is unimodal, but it is only symmetric if the underlying correlation is 0.5.

Recognize nonlinearity, and straighten plots.

- If a scatterplot of y vs. x isn't straight, correlation isn't appropriate.
- Re-expressing one or both variables can often improve the straightness of the relationship.
- The powers, roots, and the logarithm provide an ordered collection of re-expressions so you can search up and down the "ladder of powers" to find an appropriate one.

REVIEW OF TERMS

Scatterplots A scatterplot shows the relationship between two quantitative variables measured on the same cases (p. 194).

Timeplot A timeplot displays data that change over time. Often, successive values are connected with lines to show trends more clearly. Sometimes a smooth curve is added to the plot to help show long-term patterns and trends (p. 194).

Association

- **Direction:** A positive direction or association means that, in general, as one variable increases, so does the other. When increases in one variable generally correspond to decreases in the other, the association is negative (p. 194).
- **Form:** The form we care about most is straight, but you should certainly describe other patterns you see in scatterplots (p. 194).
- **Strength:** A scatterplot is said to show a strong association if there is little scatter around the underlying relationship (p. 195).

Outlier A point that does not fit the overall pattern seen in the scatterplot (p. 195).

Response variable, Explanatory variable, *x*-variable, or predictor *y*-variable In a scatterplot, you must choose a role for each variable. Assign to the *y*-axis the response variable that you hope to predict or explain. Assign to the *x*-axis the explanatory or predictor variable that accounts for, explains, predicts, or is otherwise responsible for the *y*-variable (p. 197).

Correlation coefficient The correlation coefficient is a numerical measure of the direction and strength of a linear association (p. 199):

$$r = \frac{\sum z_x z_y}{n - 1}.$$

Kendall's tau A nonparametric measure of association that measures monotonicity by finding the slopes between all pairs of points in the data. It is computed as (p. 205):

$$\frac{\text{positive slopes} - \text{negative slopes}}{\frac{1}{2}n(n - 1)}$$

Spearman's rho A nonparametric measure of association found by first converting both variable to ranks and then finding the correlation between the ranks. (p. 206)

Lurking variable A variable other than *x* and *y* that simultaneously affects both variables, accounting for the correlation between the two (p. 207).

Re-expression We re-express data by taking the logarithm, the square root, the reciprocal, or some other mathematical operation of all values of a variable (p. 209).

Ladder of powers The ladder of powers places in order the effects that many re-expressions have on the data (p. 210).

TECH SUPPORT

Scatterplots and Correlation

Statistics packages generally make it easy to look at a scatterplot to check whether the correlation is an appropriate summary of the relationship. Some packages make this easier than others.

Many packages allow you to modify or enhance a scatterplot, altering the axis labels, the axis numbering, the plot symbols, or the colors used. Some options, such as color and symbol choice, can be used to display additional information on the scatterplot.

DATA DESK

To make a scatterplot of two variables:

- Click the scatterplot tool in the tool bar or choose **Scatterplot** from the Calc menu. Data Desk opens a plot window.
- Drag variables to the Y and X axes and drop them there.
- You can find the correlation by choosing **Correlation** from the scatterplot's HyperView menu.

Alternatively, click the **Correlation tool** in the toolbar and drag variables into the Correlation window it opens.

COMMENTS

We prefer that you look at the scatterplot first and then find the correlation. But if you've found the correlation first, click on the correlation value to drop down a menu that offers to make the scatterplot.

EXCEL

To make a scatterplot in Excel:

- Select the columns of data to use in the scatterplot. You can select more than one column by holding down the control key while clicking.
- In the Insert tab, click on the **Scatter** button and select the **Scatter with only Markers** chart from the menu.

Unfortunately, the plot this creates is often statistically useless. To make the plot useful, we need to change the display:

- With the chart selected click on the **Gridlines** button in the Layout tab to cause the Chart Tools tab to appear.
- Within Primary Horizontal Gridlines, select **None**. This will remove the gridlines from the scatterplot.
- To change the axis scaling, click on the numbers of each axis of the chart, and click on the **Format Selection** button in the Layout tab.
- Select the **Fixed** option instead of the Auto option, and type a value more suited for the scatterplot. You can use the pop-up dialog window as a straightedge to approximate the appropriate values.

Excel automatically places the leftmost of the two columns you select on the *x*-axis, and the rightmost one on the *y*-axis. If that's not what you'd prefer for your plot, you'll want to switch them.

To switch the X- and Y-variables:

- Click the chart to access the **Chart Tools** tabs.
- Click on the **Select Data** button in the Design tab.
- In the pop-up window's Legend Entries box, click on **Edit**.
- Highlight and delete everything in the Series X Values line, and select new data from the spreadsheet. (Note that selecting the column would inadvertently select the title of the column, which would not work well here.)
- Do the same with the Series Y Values line.
- Press **OK**, then press **OK** again.

COMMENTS

The CORREL(array1, array2) function computes the correlation coefficient in a cell.

- You can also find correlations with the Analysis Toolpack.

JMP

To make a scatterplot and compute correlation:

- Choose **Fit Y by X** from the Analyze menu.
- In the Fit Y by X dialog, drag the Y variable into the Y, Response box, and drag the X variable into the X, Factor box.
- Click the **OK** button.

Once JMP has made the scatterplot, click on the red triangle next to the plot title to reveal a menu of options:

- Select **Density Ellipse** and select 0.95. JMP draws an ellipse around the data and reveals the Correlation tab.
- Click the blue triangle next to Correlation to reveal a table containing the correlation coefficient.

MINITAB

To make a scatterplot:

- Choose **Scatterplot** from the Graph menu.
- Choose **Simple** for the type of graph. Click **OK**.
- Enter variable names for the Y variable and X variable into the table. Click **OK**.

To compute a correlation coefficient:

- Choose **Basic Statistics** from the Stat menu.
- From the Basic Statistics submenu, choose **Correlation**. Specify the names of at least two quantitative variables in the Variables box.
- Click **OK** to compute the correlation table.

R

To make a scatterplot of two variables, X and Y:

- **plot(X, Y)** or equivalently **plot(Y~X)** will produce the plot.
- **cor(X, Y)** finds the correlation.
  In the mosaic library, the syntax is cor (Y~X, data=DATA) even though correlation is symmetric with respect to Y and X.

COMMENTS

Your variables X and Y may be variables in a data frame. If so, and DATA is the name of the data frame, then you will need to attach the data frame, or use

with(DATA,plot(Y~X)) or plot (Y~X,data=DATA).

SPSS

To make a scatterplot in SPSS:

- Open the Chart Builder from the Graphs menu. Then
- Click the **Gallery** tab.
- Choose **Scatterplot** from the list of chart types.
- Drag the scatterplot onto the canvas.
- Drag a scale variable you want as the response variable to the *y*-axis drop zone.
- Drag a scale variable you want as the factor or predictor to the *x*-axis drop zone.
- Click **OK**.

To compute a correlation coefficient:

- Choose **Correlate** from the Analyze menu.
- From the Correlate submenu, choose **Bivariate**.
- In the Bivariate Correlations dialog, use the arrow button to move variables between the source and target lists.

Make sure the **Pearson** option is selected in the Correlation Coefficients field.

STATCRUNCH

To make a scatterplot:

- Click on **Graph**.
- Choose **Scatter Plot**.
- Choose **X** and **Y** variable names from the list of **Columns**.
- Click on **Compute**.

To find a correlation:

- Click on **Stat**.
- Choose **Summary Stats** » **Correlation**.
- Choose two variable names from the list of **Columns**. (You may need to hold down the ctrl or command key to choose the second one.)
- Click on **Compute**.

TI-83/84 PLUS

To create a scatterplot:

- Set up the **STAT PLOT** by choosing the **scatterplot** icon (the first option). Specify the lists where the data are stored as Xlist and Ylist.
- Set the graphing **WINDOW** to the appropriate scale and **GRAPH** (or take the easy way out and just ZoomStat!).

To find the correlation:

- Go to **STAT CALC** menu and select **8: LinReg(a + bx).**
- Then specify the lists where the data are stored. The final command you will enter should look like **LinReg(a + bx)L1, L2.**

COMMENTS

Notice that if you **TRACE** the scatterplot, the calculator will tell you the x- and y-value at each point.

If the calculator does not tell you the correlation after you enter a LinReg command, try this: Hit **2nd CATALOG**. You now see a list of everything the calculator knows how to do. Scroll down until you find **DiagnosticOn**. Hit **ENTER** twice. (It should say Done.) Now and forevermore (or until you change batteries), you can find a correlation by using your calculator.

EXERCISES

SECTION 6.1

1. Association Suppose you were to collect data for each pair of variables. You want to make a scatterplot. Which variable would you use as the explanatory variable and which as the response variable? Why? What would you expect to see in the scatterplot? Discuss the likely direction, form, and strength.

a) Apples: weight in grams, weight in ounces
b) Apples: circumference (inches), weight (ounces)
c) College freshmen: shoe size, grade point average
d) Gasoline: number of miles you drove since filling up, gallons remaining in your tank

2. Association II Suppose you were to collect data for each pair of variables. You want to make a scatterplot. Which variable would you use as the explanatory variable and which as the response variable? Why? What would you expect to see in the scatterplot? Discuss the likely direction, form, and strength.

a) T-shirts at a store: price each, number sold
b) Scuba diving: depth, water pressure
c) Scuba diving: depth, visibility
d) All elementary school students: weight, score on a reading test

3. Bookstore sales Consider the following data from a small bookstore:

Number of Sales People Working	Sales (in $1000)
4	15
10	15
12	17
14	20
14	20
16	21
17	23
17	24
18	25
20	25

a) Prepare a scatterplot of *Sales* against *Number of sales people working*.
b) What can you say about the direction of the association?
c) What can you say about the form of the relationship?
d) What can you say about the strength of the relationship?
e) Does the scatterplot show any outliers?

T **4. Disk drives 2016** Disk drives have been getting larger. Their capacity is now often given in *terabytes* (TB), where 1 TB is roughly equal to 1000 gigabytes, or about a trillion bytes. A search of prices for external disk drives found the following data:

Capacity (in TB)	Price (in $)
0.5	64.11
1	85.11
2	107.97
3	105.11
4	145.34
6	420.97
8	619.97
12	1065.97
32	4453.00

a) Prepare a scatterplot of *Price* against *Capacity*.
b) What can you say about the direction of the association?
c) What can you say about the form of the relationship?
d) What can you say about the strength of the relationship?
e) Does the scatterplot show any outliers?

SECTION 6.2

5. Correlation facts If we assume that the conditions for correlation are met, which of the following are true? If false, explain briefly.

a) A correlation of -0.98 indicates a strong, negative association.
b) Multiplying every value of x by 2 will double the correlation.
c) The units of the correlation are the same as the units of y.

6. Correlation facts II If we assume that the conditions for correlation are met, which of the following are true? If false, explain briefly.

a) A correlation of 0.02 indicates a strong, positive association.
b) Standardizing the variables will make the correlation 0.
c) Adding an outlier can dramatically change the correlation.

SECTION 6.3

7. Bookstore sales again A larger firm is considering acquiring the bookstore of Exercise 3. An analyst for the firm, noting the relationship seen in Exercise 3, suggests that when they acquire the store they should hire more people because that will drive higher sales. Is his conclusion justified? What alternative explanations can you offer? Use appropriate statistics terminology.

8. Blizzards A study finds that during blizzards, online sales are highly associated with the number of snow plows on the road; the more plows, the more online purchases. The director of an association of online merchants suggests that the organization should encourage municipalities to send out more plows whenever it snows because, he says, that will increase business. Comment.

SECTION 6.4

***9.** Salaries and logs For an analysis of the salaries of your company, you plot the salaries of all employees against the number of years they have worked for the company. You find that plotting the base-10 logarithm of salary makes the plot much straighter. A part-time shipping clerk who has worked at the company for one year earns \$10,000. A manager earns \$150,000 after 15 years with the firm. The CEO, who founded the company 30 years ago, receives \$1,250,000. What are the values you will plot? Will the plot of these three points be straight enough?

***10.** Dexterity scores Scores on a test of dexterity are recorded by timing a subject who is inverting pegs by picking them up with one hand, manipulating them to turn them over, and then placing them back in a frame. A typical 3-year-old needs about 6.0 seconds to invert a peg, but a 6-year-old takes 3.1 seconds and an 11-year-old can do it in 2.0 seconds. Plot these three points. Now try the reciprocal re-expression $(1/y)$ vs. age. Which version is straighter? What interpretation would you give to the reciprocal?

CHAPTER EXERCISES

11. Association III Suppose you were to collect data for each pair of variables. You want to make a scatterplot. Which variable would you use as the explanatory variable and which as the response variable? Why? What would you expect to see in the scatterplot? Discuss the likely direction, form, and strength.

a) When climbing mountains: altitude, temperature
b) For each week: ice cream cone sales, air-conditioner sales
c) People: age, grip strength
d) Drivers: blood alcohol level, reaction time

12. Association IV Suppose you were to collect data for each pair of variables. You want to make a scatterplot. Which variable would you use as the explanatory variable and which as the response variable? Why? What would you expect to see in the scatterplot? Discuss the likely direction, form, and strength.

a) Legal consultation time, cost
b) Lightning strikes: distance from lightning, time delay of the thunder
c) A streetlight: its apparent brightness, your distance from it
d) Cars: weight of car, age of owner

13. Scatterplots Which of these scatterplots show

a) little or no association?
b) a negative association?
c) a linear association?
d) a moderately strong association?
e) a very strong association?

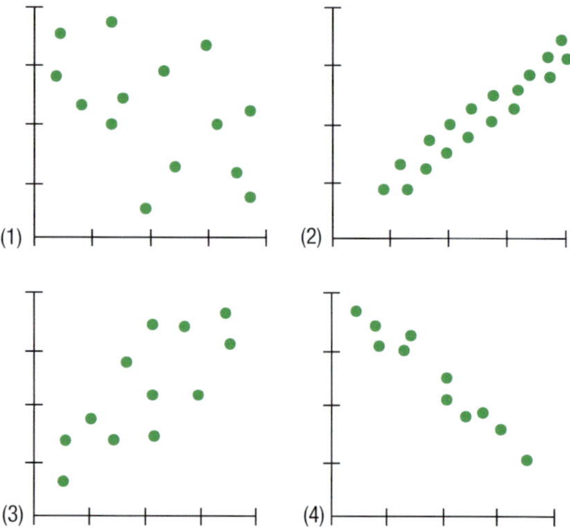

14. Scatterplots II Which of these scatterplots show

a) little or no association?
b) a negative association?
c) a linear association?
d) a moderately strong association?
e) a very strong association?

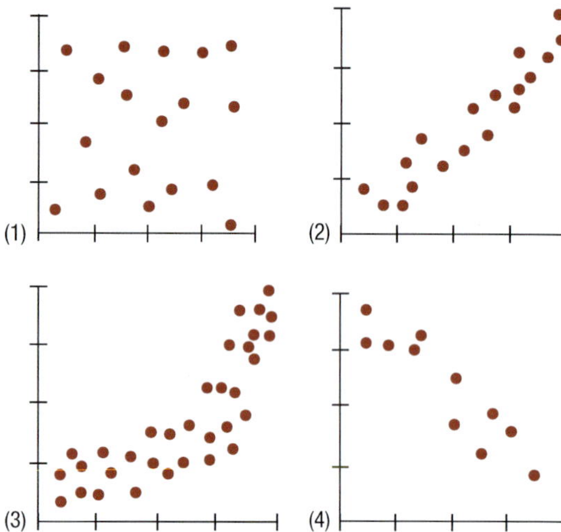

T **15.** Performance IQ scores vs. brain size. A study examined brain size (measured as pixels counted in a digitized magnetic resonance image [MRI] of a cross section of the brain) and IQ (4 performance scales of the Wechsler IQ test) for college students. The scatterplot shows the *Performance IQ* scores vs. *Brain Size*. Comment on the association between *Brain Size* and *Performance IQ* as seen in the scatterplot. (Data in **IQ brain**)

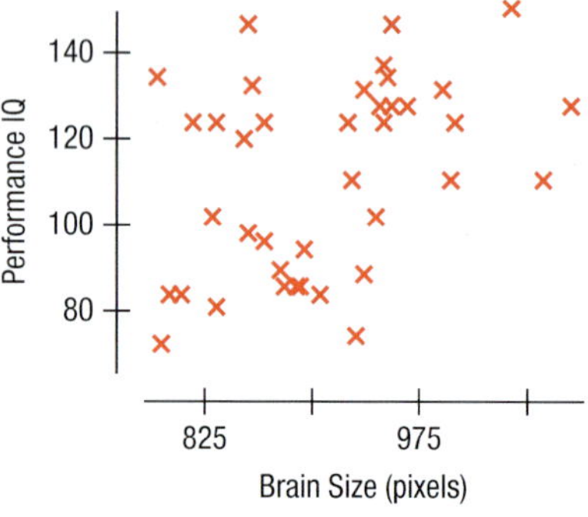

T **16. Kentucky Derby 2017** The fastest horse in Kentucky Derby history was Secretariat in 1973. The scatterplot shows speed (in miles per hour) of the winning horses each year.

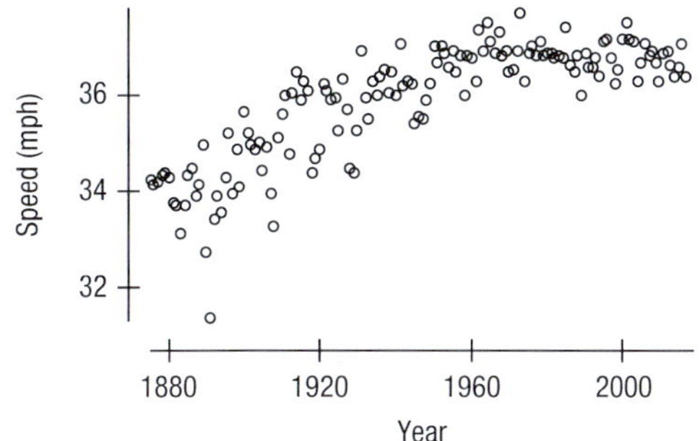

What do you see? In most sporting events, performances have improved and continue to improve, so surely we anticipate a positive direction. But what of the form? Has the performance increased at the same rate throughout the past 140 years?

17. Firing pottery A ceramics factory can fire eight large batches of pottery a day. Sometimes a few of the pieces break in the process. In order to understand the problem better, the factory records the number of broken pieces in each batch for three days and then creates the scatterplot shown.

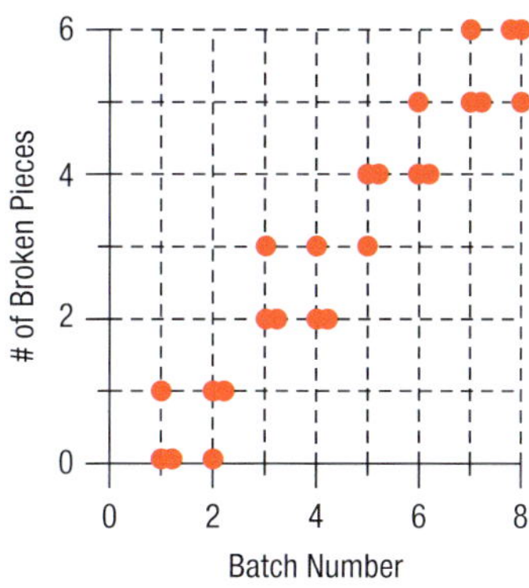

a) Make a histogram showing the distribution of the number of broken pieces in the 24 batches of pottery examined.
b) Describe the distribution as shown in the histogram. What feature of the problem is more apparent in the histogram than in the scatterplot?
c) What aspect of the company's problem is more apparent in the scatterplot?

18. Coffee sales Owners of a new coffee shop tracked sales for the first 20 days and displayed the data in a scatterplot (by day).

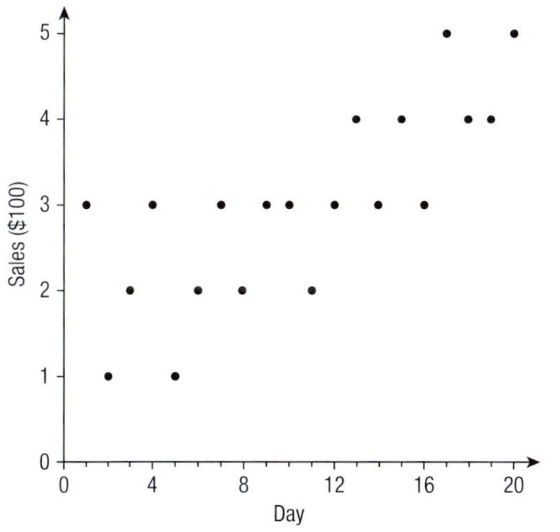

a) Make a histogram of the daily sales since the shop has been in business.
b) State one fact that is obvious from the scatterplot, but not from the histogram.
c) State one fact that is obvious from the histogram, but not from the scatterplot.

19. Matching Here are several scatterplots. The calculated correlations are 0.006, −0.977, −0.487, and 0.777. Which is which?

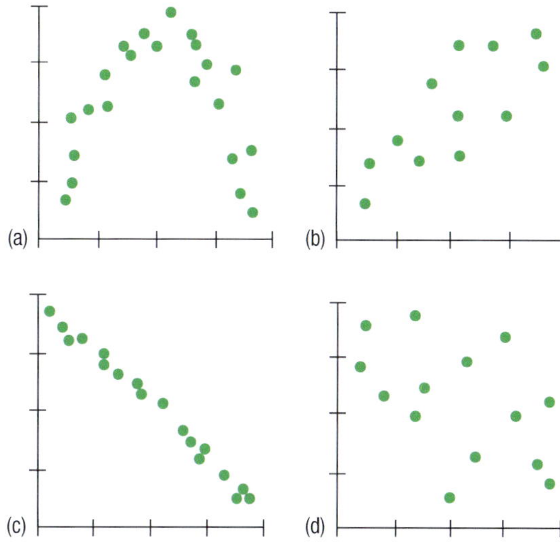

20. Matching II Here are several scatterplots. The calculated correlations are −0.977, −0.021, 0.736, and 0.951. Which is which?

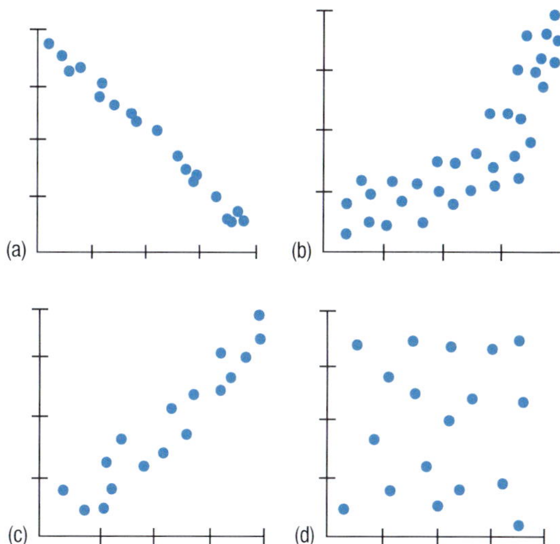

21. Politics A candidate for office claims that "there is a correlation between television watching and crime." Criticize this statement on statistical grounds.

22. Car thefts The National Insurance Crime Bureau reports that Honda Accords, Honda Civics, and Toyota Camrys are the cars most frequently reported stolen, while Ford Tauruses, Pontiac Vibes, and Buick LeSabres are stolen least often. Is it reasonable to say that there's a correlation between the type of car you own and the risk that it will be stolen?

T **23. Coasters 2015** Most roller coasters get their speed by dropping down a steep initial incline, so it makes sense that the height of that drop might be related to the speed of the coaster. Here's a scatterplot of top *Speed* and largest *Drop* for 75 roller coasters around the world.

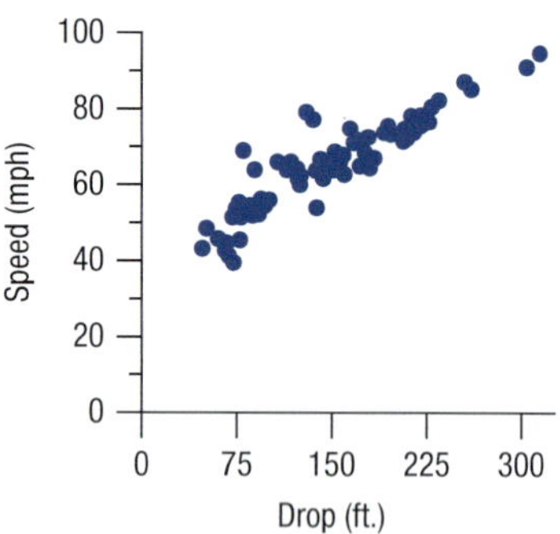

a) Does the scatterplot indicate that it is appropriate to calculate the correlation? Explain.

b) The correlation of *Speed* and *Drop* is 0.91. Describe the association.

T **24. Antidepressants** A study compared the effectiveness of several antidepressants by examining the experiments in which they had passed the FDA requirements. Each of those experiments compared the active drug with a placebo, an inert pill given to some of the subjects. In each experiment, some patients treated with the placebo had improved, a phenomenon called the *placebo effect*. Patients' depression levels were evaluated on the Hamilton Depression Rating Scale, where larger numbers indicate greater improvement. (The Hamilton scale is a widely accepted standard that was used in each of the independently run studies.) The following scatterplot compares mean improvement levels for the antidepressants and placebos for several experiments.

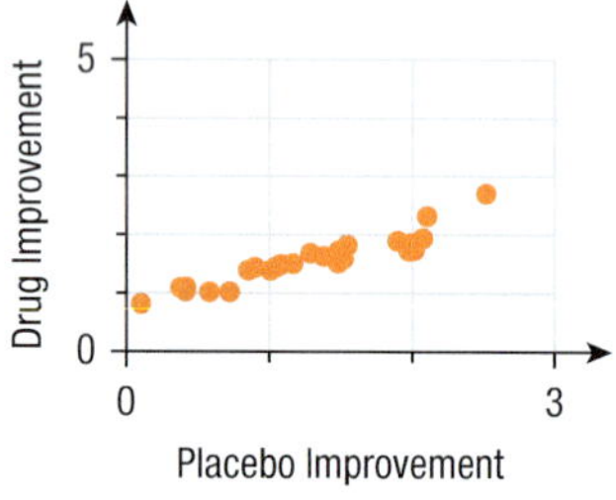

a) Is it appropriate to calculate the correlation? Explain.

b) The correlation is 0.895. Explain what we have learned about the results of these experiments.

T **25. Streams and hard water** In a study of streams in the Adirondack Mountains, the following relationship was found between the water's pH and its hardness (measured in grains):

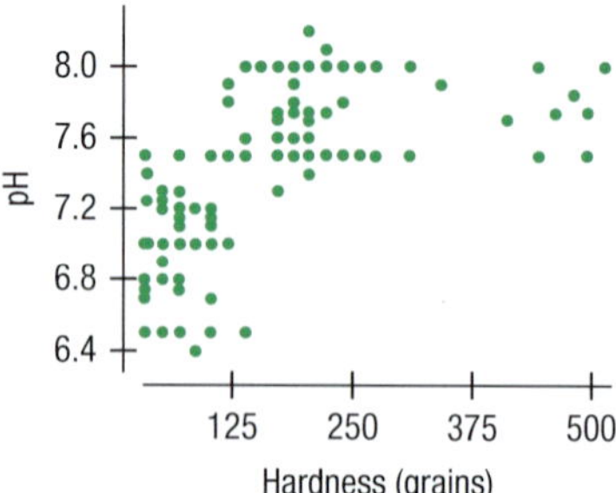

Is it appropriate to summarize the strength of association with a correlation? Explain. (Data in **Streams**)

26. Traffic headaches A study of traffic delays in 68 cities found the following relationship between *Total Delay* (in total hours lost) and *Mean Highway Speed*:

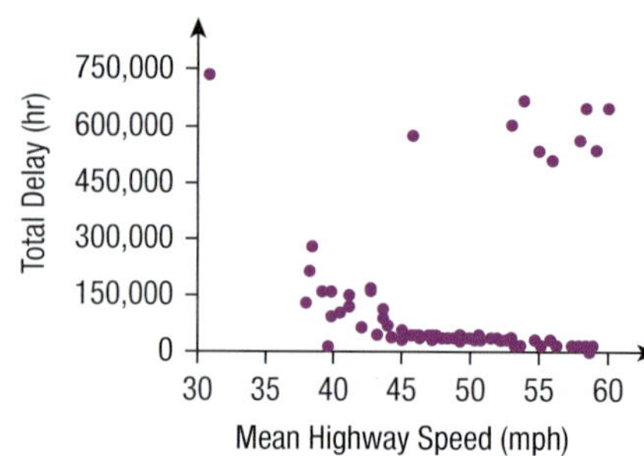

Is it appropriate to summarize the strength of association with a correlation? Explain.

27. Cold nights Is there an association between time of year and the nighttime temperature in North Dakota? A researcher assigned the numbers 1–365 to the days January 1–December 31 and recorded the temperature at 2:00 A.M. for each. What might you expect the correlation between *Day Number* and *Temperature* to be? Explain.

28. Association V A researcher investigating the association between two variables collected some data and was surprised when he calculated the correlation. He had expected to find a fairly strong association, yet the correlation was near 0. Discouraged, he didn't bother making a scatterplot. Explain to him how the scatterplot could still reveal the strong association he anticipated.

29. Prediction units The errors in predicting hurricane tracks (examined in this chapter) were given in nautical miles. A statutory mile is 0.86898 nautical mile. Most people living on the Gulf Coast of the United States would prefer to know the prediction errors in statutory miles rather than nautical miles. Explain why converting the errors to statutory miles would not change the correlation between *Prediction Error* and *Year.*

30. More predictions Hurricane Katrina's hurricane force winds extended 120 miles from its center. Katrina was a big storm, and that affects how we think about the prediction errors. Suppose we add 120 miles to each error to get an idea of how far from the predicted track we might still find damaging winds. Explain what would happen to the correlation between *Prediction Error* and *Year*, and why.

31. Correlation errors Your economics instructor assigns your class to investigate factors associated with the gross domestic product (*GDP*) of nations. Each student examines a different factor (such as *Life Expectancy*, *Literacy Rate*, etc.) for a few countries and reports to the class. Apparently, some of your classmates do not understand statistics very well because you know several of their conclusions are incorrect. Explain the mistakes in their statements:

a) "My very low correlation of −0.772 shows that there is almost no association between *GDP* and *Infant Mortality Rate.*"

b) "There was a correlation of 0.44 between *GDP* and *Continent.*"

32. More correlation errors Students in the economics class discussed in Exercise 31 also wrote these conclusions. Explain the mistakes they made.

a) "There was a very strong correlation of 1.22 between *Life Expectancy* and *GDP*."
b) "The correlation between *Literacy Rate* and *GDP* was 0.83. This shows that countries wanting to increase their standard of living should invest heavily in education."

33. Height and reading A researcher studies children in elementary school and finds a strong positive linear association between height and reading scores.

a) Does this mean that taller children are generally better readers?
b) What might explain the strong correlation?

34. Smartphones and life expectancy A survey of the world's nations in 2014 shows a strong positive correlation between percentage of the country using smartphones and life expectancy in years at birth.

a) Does this mean that smartphones are good for your health?
b) What might explain the strong correlation?

35. Correlation conclusions I The correlation between *Age* and *Income* as measured on 120 people is $r = 0.65$. Explain whether or not each of these possible conclusions is justified:

a) When *Age* increases, *Income* increases as well.
b) The form of the relationship between *Age* and *Income* is straight.
c) There are no outliers in the scatterplot of *Income* vs. *Age*.
d) Whether we measure *Age* in months instead of years, the correlation will decrease.

36. Correlation conclusions II The correlation between *Fuel Efficiency* (as measured by miles per gallon) and *Price* of 150 cars at a large dealership is $r = -0.34$. Explain whether or not each of these possible conclusions is justified:

a) The more you pay, the lower the fuel efficiency of your car will be.
b) The form of the relationship between *Fuel Efficiency* and *Price* is moderately straight.
c) There are several outliers that explain the low correlation.
d) If we measure *Fuel Efficiency* in kilometers per liter instead of miles per gallon, the correlation will increase.

37. Baldness and heart disease Medical researchers followed 1435 middle-aged men for a period of 5 years, measuring the amount of *Baldness* present (none = 1, little = 2, some = 3, much = 4, extreme = 5) and presence of *Heart Disease* (No = 0, Yes = 1). They found a correlation of 0.089 between the two variables. Comment on their conclusion that this shows that baldness is not a possible cause of heart disease.

38. Sample survey A polling organization is checking its database to see if the data sources it used sampled the same ZIP codes. The variable *Datasource* = 1 if the data source is MetroMedia, 2 if the data source is DataQwest, and 3 if it's RollingPoll. The organization finds that the correlation between five-digit ZIP code and *Datasource* is −0.0229. It concludes that the correlation is low enough to state that there is no dependency between *ZIP Code* and *Source of Data*. Comment.

T **39.** Income and housing Here is a scatterplot of the *Housing Cost Index* vs. the *Median Family Income* for 10 regions in a country. The correlation is 0.66.

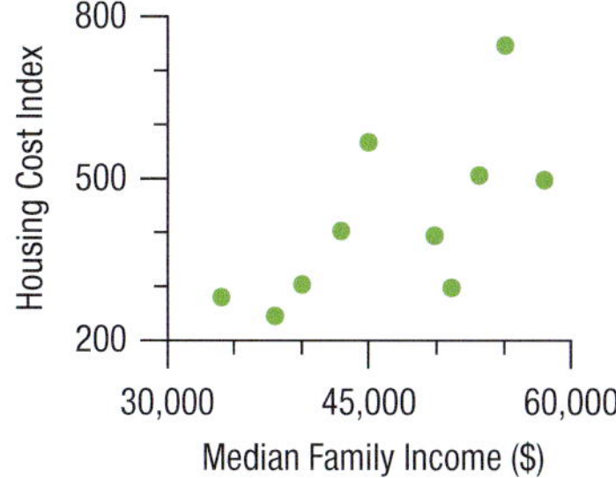

a) Describe the relationship between the *Housing Cost Index* and the *Median Family Income* by region.
b) If we standardized both variables, what would the correlation coefficient between the standardized variables be?
c) If we had measured *Median Family Income* in thousands of dollars instead of dollars, how would the correlation change?
d) One region of the country has a housing cost index of 225 and a median income of about \$45,000. If we were to include this region in the dataset, how would that affect the correlation coefficient?
e) Do these data provide proof that by raising the median family income in a state, the housing cost index will rise as a result? Explain.
*f) For these data, Kendall's tau is 0.58. Does that provide proof that by raising the median income in a region the Housing Cost Index will rise as a result? Explain what Kendall's tau says and does not say.

T **40.** Interest rates and mortgages 2015 Since 1980, average mortgage interest rates have fluctuated from a low of nearly 3% to a high of over 14%. Is there a relationship between the amount of money people borrow and the interest rate that's offered? Here is a scatterplot of *Mortgage Loan Amount* in a country (in trillions of dollars) vs. yearly *Interest Rate* since 1980. The correlation is −0.68.

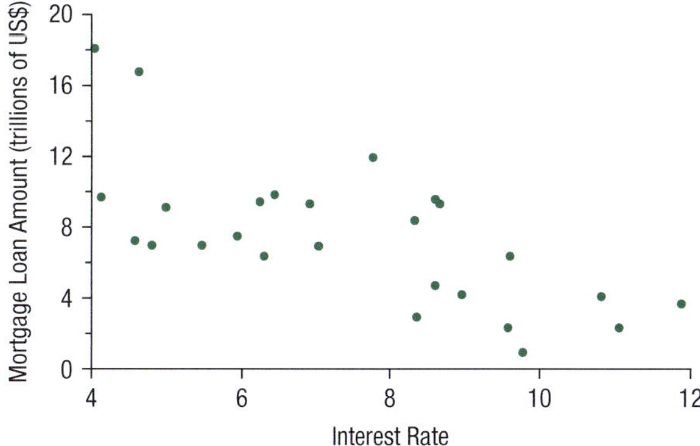

a) Describe the relationship between *Mortgage Loan Amount* and *Interest Rate*.
b) If we standardized both variables, what would the correlation coefficient between the standardized variables be?
c) If we were to measure *Mortgage Loan Amount* in billions of dollars instead of trillions of dollars, how would the correlation coefficient change?

d) Suppose that next year, interest rates were 6% and mortgages totaled 35 trillion. How would including that year with these data affect the correlation coefficient?
e) Do these data provide proof that if mortgage rates are lowered, people will take out larger mortgages? Explain.
f) For these data, Kendall's tau is −0.65. Does that provide proof that if mortgage rates are lowered, people will take out more mortgages? Explain what Kendall's tau says and does not say.

T **41. Fuel economy 2016** Here are engine size (displacement, in liters) and gas mileage (estimated combined city and highway) for a random sample of 10 model cars (taken from **Fuel economy 2016** and identified in the data with *Sample* = "Yes")

Make	Model	Displacement	Comb MPG
BMW	535d	3	30
Porsche	Boxster GTS	3.4	25
Jaguar	F-TYPE S Convertible Manual	3	18
Cadillac	ATS	2	23
BMW	328d × Drive	2	34
Porsche	911 Carrera 4 Cabriolet	3.4	22
Volkswagen	Beetle	2	26
Ford	Focus FWD FFV	2	30
Kia	Forte Koup	1.6	25
Mercedes-Benz	AMG SL 63	5.5	19

a) Make a scatterplot for these data.
b) Describe the direction, form, and strength of the plot.
c) Find the correlation between engine size and miles per gallon.
d) Write a few sentences telling what the plot says about fuel economy.

T **42. Drug abuse** A survey was conducted in the United States and 10 countries of Western Europe to determine the percentage of teenagers who had used marijuana and other drugs. The results are summarized in the following table.

	Percent Who Have Used	
Country	**Marijuana**	**Other Drugs**
Czech Rep.	22	4
Denmark	17	3
England	40	21
Finland	5	1
Ireland	37	16
Italy	19	8
No. Ireland	23	14
Norway	6	3
Portugal	7	3
Scotland	53	31
United States	34	24

a) Create a scatterplot.
b) What is the correlation between the percent of teens who have used marijuana and the percent who have used other drugs?
c) Write a brief description of the association.
d) Do these results confirm that marijuana is a "gateway drug," that is, that marijuana use leads to the use of other drugs? Explain.

T **43. Burgers** Fast food is often considered unhealthy because much of it is high in both fat and sodium. But are the two related? Here are the fat and sodium contents of several brands of burgers.

Fat (g)	21	32	34	36	40	40	44
Sodium (mg)	914	1501	1300	862	1178	1259	942

a) Analyze the association between fat content and sodium using correlation and scatterplots.
*b) Find Spearman's rho for these data. Compare it with the Pearson correlation. Comment.

T **44. Burgers again** In the previous exercise, you analyzed the association between the amounts of fat and sodium in fast food hamburgers. What about fat and calories? Here are data for the same burgers:

Fat (g)	24	29	33	38	40	40	43
Calories	422	579	573	588	645	685	664

a) Analyze the association between fat content and calories using correlation and scatterplots.
*b) Find Spearman's rho for these data. Compare it with the Pearson correlation. Comment.

T **45. Attendance 2016** American League baseball games are played under the designated hitter rule, meaning that pitchers, often weak hitters, do not come to bat. Baseball owners believe that the designated hitter rule means more runs scored, which in turn means higher attendance. Is there evidence that more fans attend games if the teams score more runs? Data collected from a particular baseball season indicate a correlation of 0.387 between runs scored and the average number of people at the home games.

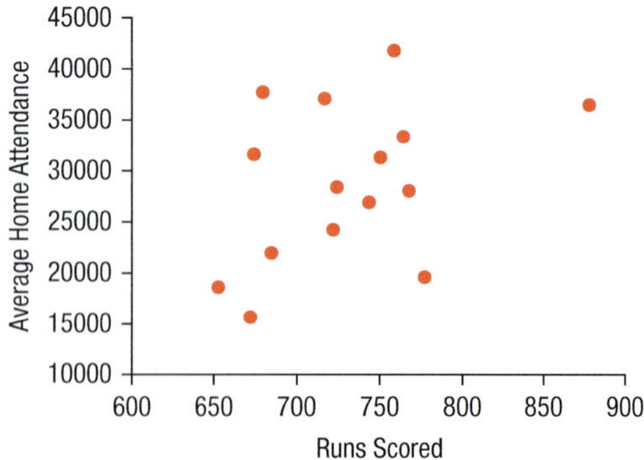

a) Does the scatterplot indicate that it's appropriate to calculate a correlation? Explain.
b) Describe the association between attendance and runs scored.
c) Does this association prove that the owners are right that more fans will come to games if the teams score more runs?

46. Second inning 2016 Perhaps fans are just more interested in teams that win. The displays below are based on American League teams for the 2016 season. (Data in **Attendance 2016**)

	Wins	Runs	Avg Home Att
Wins	1.000		
Runs	0.646	1.000	
Avg Home Att	0.457	0.431	1.000

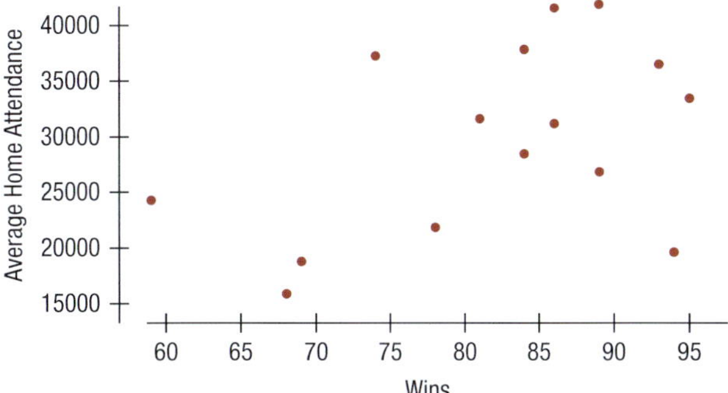

a) Do winning teams generally enjoy greater attendance at their home games? Describe the association.
b) Is attendance more strongly associated with winning or scoring runs? Explain.
c) How strongly is scoring runs associated with winning games?

47. Coasters 2015 sampled Return to the data on roller coasters seen in Exercise 23. Here is the distribution of the lengths of the rides of 241 coasters.

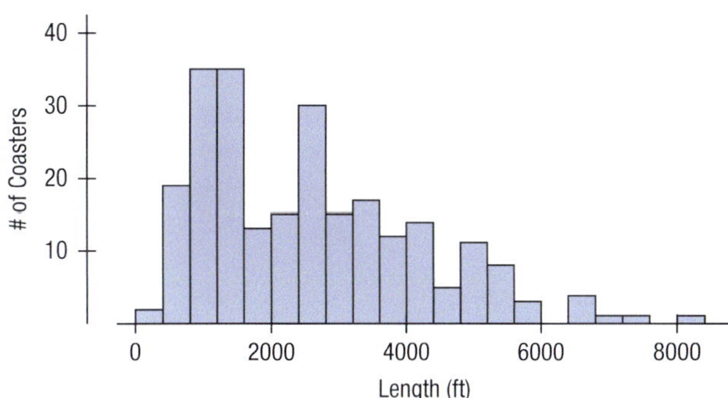

We drew samples of 60 coasters from the full set of 241. We then repeated this 1000 times. For each sample, we found the mean and median. Here are histograms of the 1000 means and 1000 medians:

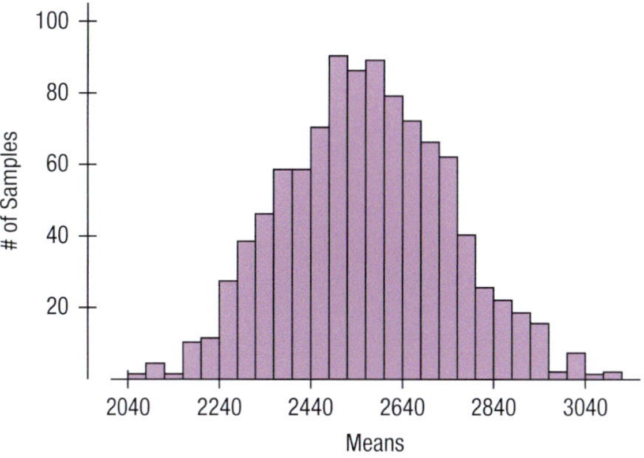

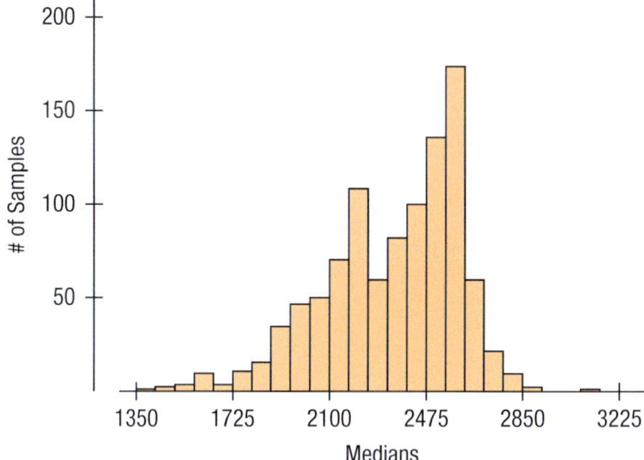

For which of these distributions would you use the 68–95–99.7 Rule? Explain.

48. Housing price sampled In a dataset of 1057 New York homes offered for sale, a histogram of the ages looks like this:

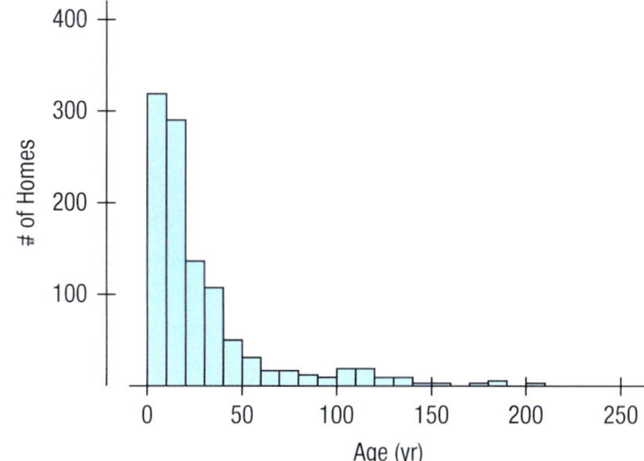

We drew 1000 samples of 105 homes from this dataset and found the IQR and the range of each sample. Below are histograms of the 1000 IQRs and 1000 ranges. For which of these distributions would you use the 68–95–99.7 Rule? Explain.

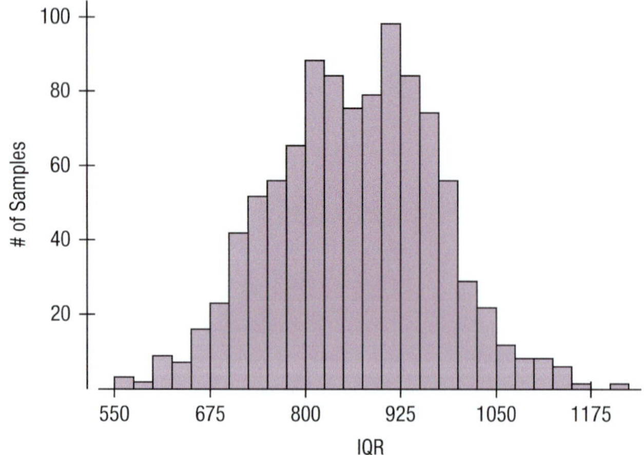

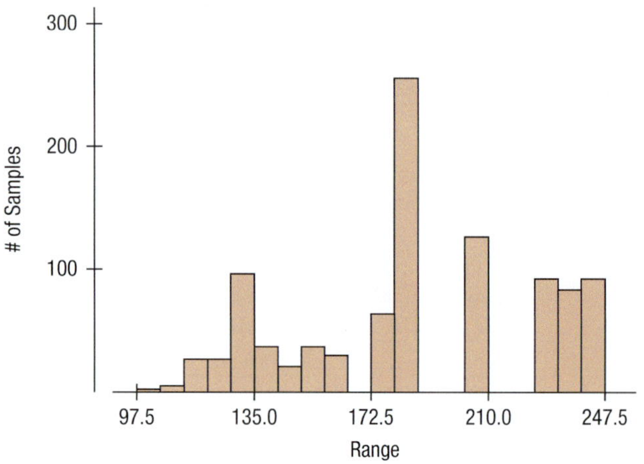

T **49. Thrills 2013** In a poll, a roller-coaster ride called Bizarro dropped to 4th after earning the top steel coaster rank for six straight years. Data on the top 11 steel coasters from this poll are at the bottom of the page. What do these data indicate about the *Length* of the track and the *Duration* of the ride you can expect?

T **50. Thrills II** For the roller coaster data in Exercise 49:

a) Examine the relationship between *Initial Drop* and *Speed*.
b) Examine the relationship between *Initial Drop* and *Height*.
c) What conclusions can you safely draw about the initial drop of a roller coaster? Is *Initial Drop* strongly correlated with other variables as well?

T **51. Thrills III** For the roller coaster data in Exercise 49:

a) Explain why in looking for a variable that explains rank, you will be hoping for a negative correlation.
b) Do any of the provided variables provide a strong predictor for roller coaster rank?

T **52. Vehicle weights** The department of transportation hoped that they could measure the weights of big trucks without actually stopping the vehicles by using a newly developed "weight-in-motion" scale. To see if the new device was accurate, they conducted a calibration test. They weighed several stopped trucks (*Static Weight*) and assumed that this weight was correct. Then they weighed the trucks again while they were moving to see how well the new scale could estimate the actual weight. Their data are given in the table at the top of the next page.

Rank	Roller Coaster	Park	Initial Drop (ft)	Duration (sec)	Height (ft)	Max Speed (mph)	Max Vert Angle (degrees)	Length (ft)
1	Expedition GeForce	Holiday Park	184	75	188	74.6	82	4003
2	New Texas Giant	Six Flags	147		153	65	79	4200
3	Intimidator 305	Kings Dominion	300	98	305	93	85	5100
4	Bizarro	SF New England	221	155	208	77	72	5400
5	Skyrush	Hersheypark	212	60	200	76	85	3600
7	Kawasemi	Tobu Zoo	98.4	60	102	53	101.7	2477
9	Nemisis	Alton Towers	104	80	43	50	40	2349
10	Iron Rattler	Six Flags	171		179	70	81	3266
11	Katun	Mirabilandia	148	142	169	65		3937
13	Leviathan	Medieval Fair	306	208	306	92	80	5486
14	Millennium Force	Cedar Point	300	140	310	93	80	6595

Weights (1000s of lbs)	
Weight-in-Motion	**Static Weight**
29.6	29.4
34.1	33.8
25.7	26.9
26.5	27.4
36.2	35.9
31.3	30.8
39.2	36.8
31.4	31.6
27.6	29.2
40.5	34.4

a) Make a scatterplot for these data.
b) Describe the direction, form, and strength of the plot.
c) Write a few sentences telling what the plot says about the data. (*Note*: The sentences should be about weighing trucks, not about scatterplots.)
d) Find the correlation.
e) If the trucks were weighed in kilograms, how would this change the correlation? (1 kilogram = 2.2 pounds)
f) Do any points deviate from the overall pattern? What does the plot say about a possible recalibration of the weight-in-motion scale?

T **53. Planets (more or less)** Is there any pattern to the locations of the planets in a distant solar system? The table shows the average distance of each of the planets from the star.

Planet	Position Number	Distance from Sun (million miles)
A	1	36
B	2	72
C	3	90
D	4	145
E	5	489
F	6	990
G	7	1798
H	8	2796
J	9	3673

a) Make a scatterplot, and describe the association. (Remember: direction, form, and strength!)
b) Why would you not want to talk about the correlation between a planet's *Position Number* and *Distance* from the sun?
c) Make a scatterplot showing the logarithm of *Distance* vs. *Position Number*. What is better about this scatterplot?

T **54. Flights 2016** Here are the number of domestic flights flown in each year from 2000 to 2016

Year	Flights
2000	7,908,038
2001	7,678,978
2002	8,023,994
2003	9,512,088
2004	9,948,102
2005	10,196,781
2006	9,883,088
2007	9,883,849
2008	9,544,573
2009	8,689,374
2010	8,601,762
2011	8,493,859
2012	8,592,730
2013	8,380,145
2014	8,010,515
2015	7,941,104
2016	4,028,372

a) Find the correlation of *Flights* with *Year*.
b) Make a scatterplot, and describe the trend.
c) Why is the correlation you found in part a not a suitable summary of the strength of the association?
d) It turns out that the value reported for 2016 was only for the period January to June. What should we have done with that point?
*e) Find Kendall's tau for these data. Would that be an appropriate measure for summarizing their relationship? Explain.

JUST CHECKING

Answers

1. We know the scores are quantitative. We should check to see whether the Straight Enough Condition and the Outlier Condition are satisfied by looking at a scatterplot of the two scores.
2. It won't change.
3. It won't change.
4. They are likely to have done poorly. The positive correlation means that low scores on Exam 1 are associated with low scores on Exam 2 (and similarly for high scores).
5. No. The general association is positive, but individual performances may vary.

7 Linear Regression

WHERE ARE WE GOING?

It's often hard to measure wind speeds of a hurricane directly. But lower pressure in storms is associated with higher wind speeds. In fact, for hurricanes, the correlation is −0.88. If we know that a hurricane has a central pressure of 860 millibars (mb), how high would we expect the wind speed to be? The correlation alone won't tell us the answer.

We need a model to be able to use one variable to predict another. Using linear regression, we can understand the relationship between two quantitative variables better and make predictions.

WHO	Items on the Burger King menu
WHAT	Protein content and total fat content
UNITS	Grams of protein Grams of fat
HOW	Supplied by BK on request or at their website

The Whopper™ has been Burger King's signature sandwich since 1957. One Triple Whopper with cheese provides 71 grams of protein—all the protein you need in a day. It also supplies 1230 calories and 82 grams of fat. The Daily Value (based on a 2000-calorie diet) for fat is 65 grams. So after a Triple Whopper you might want the rest of your calories that day to be fat-free.[1]

Of course, the Whopper isn't the only item Burger King sells. How are fat and protein related on the entire BK menu? The scatterplot of the *Fat* (in grams) vs. the *Protein* (in grams) for foods sold at Burger King shows a positive, moderately strong, linear relationship.

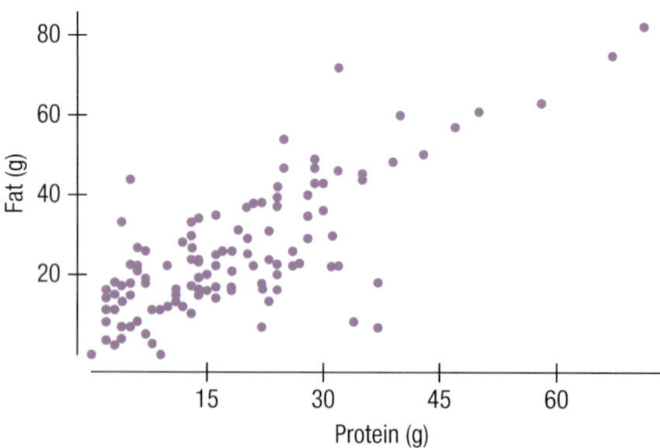

Figure 7.1
Total *Fat* vs. *Protein* for 122 items on the BK menu. The Triple Whopper is in the upper right corner. (Data in **Burger King items**)

[1] Sorry about the fries.

7.1 Least Squares: The Line of "Best Fit"

If you want 25 grams of protein in your lunch, how much fat should you expect to consume at Burger King? The correlation between *Fat* and *Protein* is 0.76, which tells us that the linear association seen in the scatterplot is fairly strong. But *strength* of the relationship is only part of the picture. The correlation says, "The linear association between these two variables is fairly strong," but it doesn't tell us *what the line is.*

Of course, the points in the scatterplot don't all line up. But anyone looking at the plot can see that the points move from the lower left to the upper right in a consistent way. You could probably sketch a straight line that summarizes that relationship pretty well. We'll take that one step further and find a **linear model** that gives an equation of a straight line through the data. This model can predict the fat content of any Burger King food, given its protein.

> "Statisticians, like artists, have the bad habit of falling in love with their models."
>
> —George Box, famous statistician

Of course, no line can go through all the points, but a linear model can summarize the general pattern with only a couple of parameters. Like all **models** of the real world, the line will be wrong—wrong in the sense that it can't match reality *exactly*. But it can help us understand how the variables are associated. Without models, our understanding of the world is limited to only the data we have at hand.

Not only can't we draw a line through all the points, the best line might not even hit *any* of the points. Then how can it be the "best" line? We want to find the line that somehow comes *closer* to all the points than any other line. Some of the points will be above the line and some below. For example, the line might suggest that a BK Tendercrisp® chicken sandwich (without mayonnaise) with 31 grams of protein[2] should have 36.6 grams of fat when, in fact, it actually has only 22. We call the estimate made from a model the **predicted value** and write it as $\hat{y}$ (called *y-hat*) to distinguish it from the observed value, y. The difference between the observed value and its predicted value is called its **residual**. The residual value tells how well the model predicted the observed value at that point. The BK Tendercrisp chicken residual is $y - \hat{y} = 22 - 36.6 = -14.6$ g of fat.

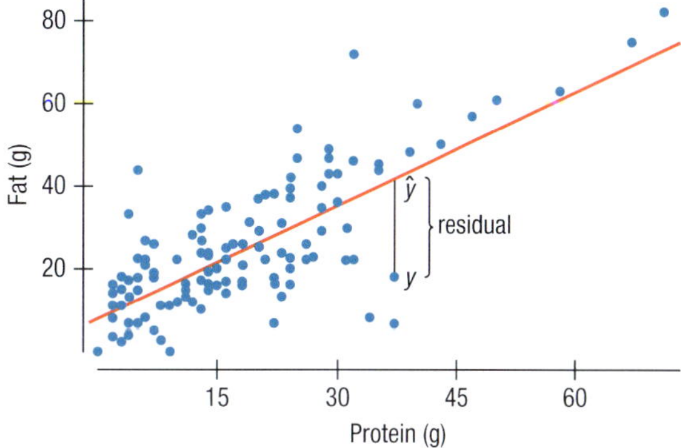

Figure 7.2 The residual is the difference between the observed y-value and the predicted $\hat{y}$-value. So it is a difference in the y direction.

To find the residuals, we always subtract the predicted value from the observed one:

$$\text{Residual} = \text{Observed value} - \text{Predicted value}.$$

The negative residual tells us that the observed fat content of the BK Tendercrisp chicken is about 14.6 grams *less* than the model predicts for a typical Burger King menu item with 31 grams of protein. We often use residuals in this way. The line tells us the average fat content we'd expect for a given amount of protein. Items with negative residuals have less fat than we'd expect, while those with positive residuals have more. In fact, the model we'll find balances the overestimates with the underestimates so that the mean residual is zero.

Residuals

A **positive** residual means the observed value is larger than the predicted value—an underestimate. A **negative** residual shows that the model made an overestimate.

[2]The sandwich comes with mayo unless you ask for it without. That adds an extra 24 grams of fat, which is more than the original sandwich contained.

Our challenge now is how to find the right line.

The *size* of the residuals tells us how well our line fits the data. A line that fits well will have small residuals. The best line will be the one with the smallest spread of data around the line. We measure that spread with the standard deviation of the residuals. The standard deviation sums the squares of deviations, so we should minimize the sum of the squares of the residuals. The line of best fit is the line for which the sum of the squared residuals is smallest, and is called the **least squares** line.

You might think that finding this line would be pretty hard. Surprisingly, it's not. However, you're not going to want to do it by hand.

> *WHO'S ON FIRST*
>
> In 1805, the French mathematician Adrien-Marie Legendre was the first to publish the "least squares" solution to the problem of fitting a line to data when the points don't all fall exactly on the line. The main challenge was how to distribute the errors "fairly." After considerable thought, he decided to minimize the sum of the squares of what we now call the residuals. When Legendre published his paper, though, the German mathematician Carl Friedrich Gauss claimed he had been using the method since 1795. Gauss later referred to the "least squares" solution as "*our* method" (*principium nostrum*), which certainly didn't help his relationship with Legendre.

7.2 The Linear Model

> *NOTATION ALERT*
>
> "Putting a hat on it" is standard statistics notation to indicate that something has been predicted by a model. Whenever you see a hat over a variable name or symbol, you can assume it is the predicted version of that variable or symbol (and look around for the model).
>
> In a linear model, we use b_1 for the slope and b_0 for the y-intercept.

You may remember from algebra that a straight line can be written as

$$y = mx + b.$$

We'll use this form for our linear model, but in statistics we use slightly different notation:

$$\hat{y} = b_0 + b_1 x.$$

We write $\hat{y}$ (y-hat) to emphasize that the points that satisfy this equation are just our *predicted* values, not the actual data values (which scatter around the line). If the model is a good one, the data values will scatter closely around it.

We write b_1 and b_0 for the slope and intercept of the line. The b's are called the *coefficients* of the linear model. The coefficient b_1 is the **slope**, which tells how rapidly $\hat{y}$ changes with respect to x. The coefficient b_0 is the **intercept**, which tells where the line hits (intercepts) the y-axis.[3]

For the Burger King menu items, the line of best fit is

$$\widehat{Fat} = 8.4 + 0.91\ Protein.$$

What does this mean? The slope, 0.91, says that a Burger King item with one more gram of protein can be expected, on average, to have 0.91 more grams of fat. Less formally, we might say that, on average, Burger King foods pack about 0.91 grams of fat per gram of protein. Slopes are always expressed in y-units per x-unit. They tell how the y-variable changes (in its units) for a one-unit change in the x-variable. When you see a phrase like "students per teacher" or "kilobytes per second," think slope.

What about the intercept, 8.4? Algebraically, that's the value the line takes when x is zero. Here, our model predicts that even a BK item with no protein would have, on average, about 8.4 grams of fat. Is that reasonable? Well, there are two items on the

[3]We change from $mx + b$ to $b_0 + b_1x$ for a reason: Soon we'll want to add more x's to the model to make it more realistic, and we don't want to use up the entire alphabet. What would we use for the next coefficient after m? The next letter is n and that one's already taken. o? See our point? Sometimes subscripts are the best approach.

menu with 0 grams of protein (apple fries and applesauce). Neither has any fat either (they are essentially pure carbohydrate), but we could imagine a protein-less item with 6.8 grams of fat. But often 0 is not a plausible value for x (the year 0, a baby born weighing 0 grams, . . .). Then the intercept serves only as a starting value for our predictions, and we don't interpret it as a meaningful predicted value.

EXAMPLE 7.1

A Linear Model for Hurricanes

The barometric pressure at the center of a hurricane is often used to measure the strength of the hurricane because it can predict the maximum wind speed of the storm. A scatterplot shows that the relationship is straight, strong, and negative. It has a correlation of −0.898. (Data in **Hurricanes 2015**)

Using technology to fit the straight line, we find

$$\widehat{MaxWindSpeed} = 1031.24 - 0.9748\ CentralPressure.$$

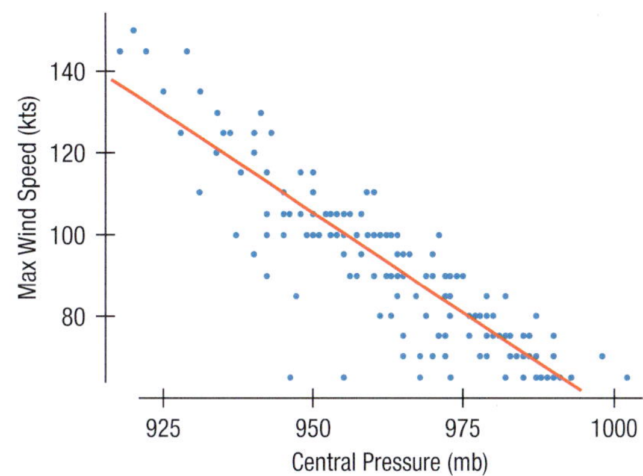

QUESTION: Interpret this model. What does the slope mean in this context? Does the intercept have a meaningful interpretation?

ANSWER: The negative slope says that lower *CentralPressure* is associated, on average, with higher *MaxWindSpeed*. This is consistent with the way hurricanes work: From physics, we know that low central pressure pulls in moist air, driving the rotation and the resulting destructive winds. The slope's value says that, on average, the maximum wind speed is about 0.975 knots higher for every 1-millibar drop in central pressure. Of course, we can't conclude from the regression that pressure causes wind speed—and, in fact, higher wind speeds also reduce the central pressure. The regression only estimates the slope of the relationship, not its underlying mechanism.

And it's not meaningful to interpret the intercept as the wind speed predicted for a central pressure of zero—that would be a vacuum. Instead, it is merely a starting value for the model.

7.3 Finding the Least Squares Line

Slope

$$b_1 = r\frac{s_y}{s_x}$$

How do we find the actual values of slope and intercept for the least squares line? The formulas are built from the summary statistics you already know: the correlation (to tell us the strength of the linear association), the standard deviations (to give us the units), and the means (to tell us where to put the line). Remarkably, that's all you need to find the value of the slope:

$$b_1 = r\frac{s_y}{s_x}.$$

Units of y per Unit of x

Get into the habit of identifying the units by writing down "y-units per x-unit," with the unit names put in place. You'll find it'll really help you to Tell about the line in context.

We've already seen that the correlation tells us the *sign* as well as the strength of the relationship, so it should be no surprise that the slope inherits this sign as well.

Correlations don't have units, but slopes do. How x and y are measured—what units they have—doesn't affect their correlation, but can change their slope. If children grow an average of 3 inches per year, that's 0.21 millimeters per day. For the slope, it matters whether you express age in days or years and whether you measure height in inches or millimeters. Changing the units of x and y affects their standard deviations directly. And that's how the slope gets its units—from the ratio of the two standard deviations. So, the units of the slope are a ratio too. The units of the slope are always the units of y *per* unit of x.

What about the intercept? If you had to predict the y-value for a data point whose x-value was the average, $\bar{x}$, you'd probably pick $\bar{y}$, and that's just what the least squares line does. Putting that into our equation, we find

$$\bar{y} = b_0 + b_1\bar{x},$$

or, rearranging the terms,

$$b_0 = \bar{y} - b_1\bar{x}.$$

What to Say About a Regression

A regression summarizes the relationship between two variables from the data at hand using a straight line. The slope summarizes the linear relationship—how y changes for different values of x. Slopes are the ratios of the y-change to the x-change, summarized conveniently with the phrase "y-units *per* x-unit" (of course, with the actual units named). For the regression of fat on protein in Burger King food items, we found

$$\widehat{Fat} = 8.4 + 0.91\ Protein.$$

What does this say? It tells us how much fat we expect to find, on average, in an item with a given amount of protein. The slope 0.91 says that we expect (or predict) items containing one more gram of protein, on average, to have an extra 0.91 grams of fat. You might say that an additional gram of protein is associated, on average, with an extra 0.91 grams of fat. Because the regression is based only on the data we have, it says nothing about what "might happen" or why it happens. Do not say anything that suggests that changing x will result in a change in y. Don't say: "Adding one more gram of protein will add 0.91 grams of fat." That's just not true. Even including words like "on average" or "expect" doesn't rescue this description from saying more than you know.

In this case, we know from science that protein and fat are only associated. Items with more protein *tend* to have more fat, but adding protein itself doesn't "cause" fat to increase. But often we don't know why x and y are associated. A simple linear regression can't tell us. Don't assume anything more.

Even when we suspect that a change in x will cause a change in y, a regression can't tell us that it does. In Chapter 6 we warned that correlation does not prove causation. The same is true for regression.

EXAMPLE 7.2 Finding the Regression Equation

Let's try out the slope and intercept formulas on the "BK data" and see how to find the line. We checked the conditions when we calculated the correlation. The summary statistics are as follows:

Protein	Fat
$\bar{x} = 18.0$ g	$\bar{y} = 24.8$ g
$s_x = 13.5$ g	$s_y = 16.2$ g
$r = 0.76$	

So, the slope is

$$b_1 = r\frac{s_y}{s_x} = 0.76 \times \frac{16.2 \text{ g fat}}{13.5 \text{ g protein}} = 0.91 \text{ grams of fat per gram of protein.}$$

The intercept is

$$b_0 = \bar{y} - b_1\bar{x} = 24.8 \text{ g fat} - 0.91 \frac{\text{g fat}}{\text{g protein}} \times 18.0 \text{ g protein} = 8.4 \text{ g fat.}$$

Putting these results back into the equation gives

$$\widehat{Fat} = 8.4 \text{ g fat} + 0.91 \frac{\text{g fat}}{\text{g protein}} Protein,$$

or more simply,

$$\widehat{Fat} = 8.4 + 0.91\, Protein.$$

In summary, the predicted fat of a Burger King food item in grams is $8.4 + 0.91 \times$ the protein content (in grams).

With an estimated linear model, it's easy to predict fat content for any menu item we want. For example, for the BK Tendercrisp chicken sandwich with 31 grams of protein, we can plug in 31 grams for the amount of protein and see that the predicted fat content is $8.4 + 0.91(31) = 36.6$ grams of fat. (Actually, rounding has caught up with us. Using full precision not shown here, the predicted value to one decimal place should be 36.7. Always use full precision until you report your results). Because the BK Tendercrisp chicken sandwich actually has 22 grams of fat, its residual is

$$Fat - \widehat{Fat} = 22 - 36.7 = -14.7 \text{ g.}$$

Least squares lines are commonly called regression lines. In a few pages, we'll see that this name is an accident of history. For now, you just need to know that "regression" almost always means "the linear model fit by least squares."

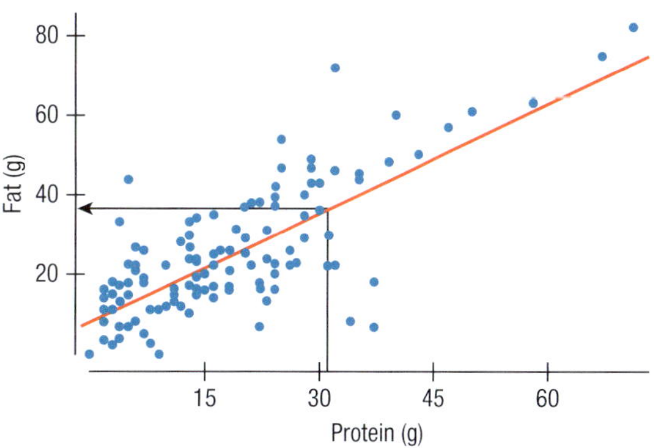

Figure 7.3
Burger King menu items with the regression line. The regression model lets us make predictions. The predicted fat value for an item with 31 grams of protein is 36.6 g.

To use a regression model, we should check the same **conditions** for regression as we did for correlation, making sure that the variables are **Quantitative**, that the relationship is **Straight Enough**, and that there are **No Outliers**.

STEP-BY-STEP EXAMPLE

Calculating a Regression Equation

During the evening rush hour of August 1, 2007, an eight-lane steel truss bridge spanning the Mississippi River in Minneapolis, Minnesota, collapsed without warning, sending cars plummeting into the river, killing 13 and injuring 145. Although similar events had brought attention to our aging infrastructure, this disaster put the spotlight on the problem and raised the awareness of the general public.

How can we tell which bridges are safe?

Most states conduct regular safety checks, giving a bridge a structural deficiency score on various scales. The New York State Department of Transportation uses a scale that runs from 1 to 7, with a score of 5 or less indicating "deficient." Many factors contribute to the deterioration of a bridge, including amount of traffic, material used in the bridge, weather, and bridge design.

New York has more than 17,000 bridges. We have available data on the 193 bridges of Tompkins County.[4] (Data in **Tompkins County Bridges 2016**)

One natural concern is the age of a bridge. A model that relates age to safety score might help the DOT focus inspectors' efforts where they are most needed.

QUESTION: What is the relationship between the age of a bridge and its safety rating?

THINK **PLAN** State the problem.

I want to model the relationship between the age of a bridge in Tompkins County, New York, and its safety rating.

VARIABLES Identify the variables and report the W's.

I have data giving the Safety Score and Age at time of inspection for 193 bridges constructed or replaced since 1900.

Just as we did for correlation, check the conditions for a regression by making a picture. Never fit a regression without looking at the scatterplot first.

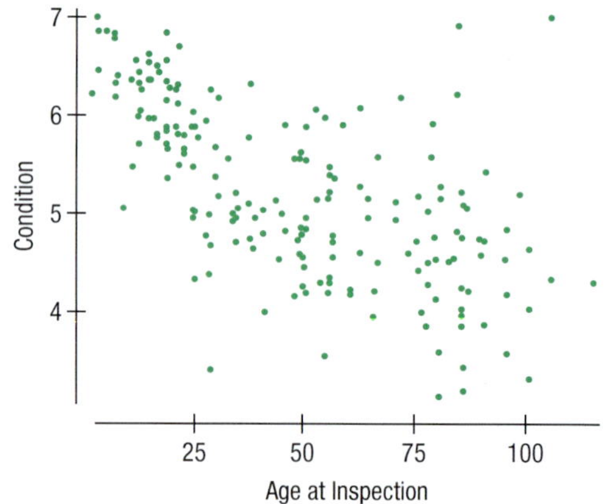

Conditions:

✔ **Quantitative Variables**: Yes, although safety rating has no units. Age is in years.

✔ **Straight Enough**: Scatterplot looks straight.

✔ **No Outliers**: Two points in upper right may deserve attention.

It is OK to use a linear regression to model this relationship.

[4] Coincidentally, the home county of two of the authors.

SHOW **MECHANICS** Find the equation of the regression line. Summary statistics give the building blocks of the calculation (though we will generally use statistical software to calculate regression slopes and intercepts.)

(We generally report summary statistics to one more digit of accuracy than the data. We do the same for intercept and predicted values, but for slopes we usually report an additional digit. Remember, though, not to round until you finish computing an answer.)[5]

Age

$\bar{x} = 47.86$

$s_x = 28.70$

Safety Score (points from 1 to 7)

$\bar{y} = 5.24$

$s_y = 0.887$

Correlation

$r = -0.634$

Find the slope, b_1.

$$b_1 = r\frac{s_y}{s_x}$$

Find the intercept, b_0.

$$= -0.634\frac{0.887}{28.70}$$

$= -0.0196$ points per year

Write the equation of the model, using meaningful variable names.

$$b_0 = \bar{y} - b_1\bar{x} = 5.24 - (-0.0196)47.46$$

$$= 6.18$$

The least squares line is

$$\hat{y} = 6.18 - 0.0196x$$

or

$$\widehat{Rating} = 6.18 - 0.0196\,Age$$

TELL **CONCLUSION** Interpret what you have found in the context of the question. Discuss in terms of the variables and their units.

The condition of the bridges in Tompkins County, New York, is generally lower for older bridges. Specifically, we expect bridges that are a year older to have safety ratings that are, on average, 0.02 points lower, on a scale of 1 to 7. The model uses a base of 6.2, which is quite reasonable because a new bridge (0 years of age) should have a safety score near the maximum of 7.

Because I have data only from one county, I can't tell from these data whether this model would apply to bridges in other counties of New York or in other locations.

RANDOM MATTERS

We've looked at data for the 193 bridges in one county of New York and found a slope of −0.0196 for the relationship of safety rating to the year the bridge was built. We predict that a bridge will lose one rating point, on average, about every 46 years. There are more than 17,000 rated bridges in the state. If we drew a random sample of the same size as Tompkins County, would we get the same slope? We know that other samples of 193 bridges from the population of New York bridges will be different, but how different will the estimated slopes of the regressions relating safety rating to age be? We can simulate doing exactly that to find out. (Data in **New York bridges 2016**)

We drew a sample of 193 bridges from the population of 17,493 highway bridges in New York State and found the slope of the regression of the safety *Rating* on the

[5]We warned you that we'll round in the intermediate steps of a calculation to show the steps more clearly, and we've done that here. If you repeat these calculations yourself on a calculator or statistics program, you may get somewhat different results.

Year in which the bridge was built. We then repeated that 1000 times, recording the slopes for each random sample. Here's a histogram of those slopes:

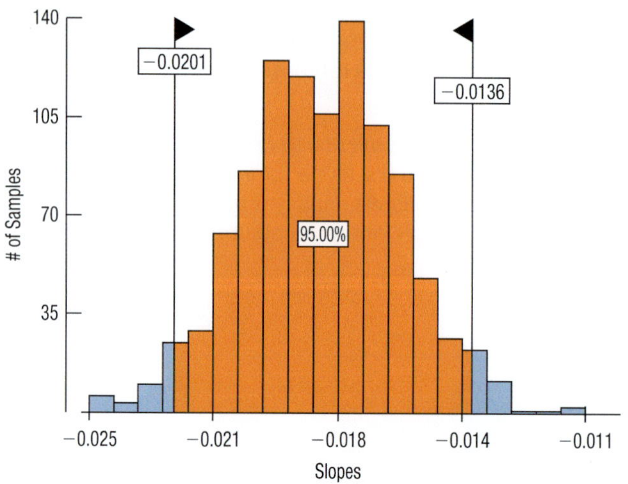

The slopes vary from sample to sample, but most (in fact 95% of them) are between −0.0201 and −0.0136. Tompkins Country does not appear to be unusual.

This reminds us to keep in mind that, although you can find a regression model for any sample, another (random) sample would typically give a (slightly) different slope.

7.4 Regression to the Mean

Suppose you were told that a new male student was about to join the class, and you were asked to guess his height in inches. What would be your guess? Since you have no other information, a good guess would be the mean height of all male students. (If you thought about this in z-scores, you'd say $\hat{z}_{Height} = 0$.) Now suppose you are also told that this student has a grade point average (*GPA*) of 3.9—about 2 SDs above the mean *GPA*. Would that change your guess? Probably not. The correlation between *GPA* and *Height* is near 0; knowing something about *GPA* wouldn't tell you anything about *Height*.

On the other hand, if you were told that, measured in centimeters, the student's *Height* was 2 SDs above the mean, you'd know his *Height* in inches exactly. There's a perfect correlation between *Height* in inches and *Height* in centimeters, so you'd be able to predict perfectly. In fact, you'd predict that he's 2 SDs above the mean *Height* in inches as well. Or, writing this as z-scores, you could say: $\hat{z}_{Htin} = z_{Htcm}$.

What if you're told that the student is 2 SDs above the mean in *Shoe Size*? Would you still guess that he's of average *Height*? You might guess that he's taller than average, since there's a positive correlation between *Height* and *Shoe Size*. But would you guess that he's 2 SDs above the mean? When there was no correlation, we didn't move away from the mean at all. With a perfect correlation, we moved our guess the full 2 SDs. Any correlation between these extremes should lead us to guess somewhere between 0 and 2 SDs above the mean. If we think in z-scores, the relationship is remarkably simple. When the correlation between x and y is r, and x has a z-score of z_x, the predicted value of y is just $\hat{z}_y = rz_x$. For example, if our student was 2 SDs above the mean *Shoe Size*, the formula tells us to guess $r \times 2$ standard deviations above the mean for his *Height*.

How did we get that? We know from the least squares line that $\hat{y} = b_0 + b_1x$. We also know that $b_0 = \bar{y} - b_1\bar{x}$. If we standardize x and y, both their means will be 0, and so for standardized variables, $b_0 = 0$. And the slope $b_1 = r\frac{s_y}{s_x}$. For standardized variables, both s_y and s_x are 1, so $b_1 = r$. In other words, $\hat{z}_y = rz_x$.

WHY IS CORRELATION "r"?

In his original paper, Galton standardized the variables and so the slope was the same as the correlation. He used r to stand for this coefficient from the (standardized) regression.

Even more important than deriving the formula is understanding what it says about regression. While we rarely standardize variables to do a regression, we can think about what the z-score regression means. It says that when the correlation is r, and an x-value lies k standard deviations from its mean, our prediction for y is $r \times k$ standard deviations from its mean as we can see in the following plot.

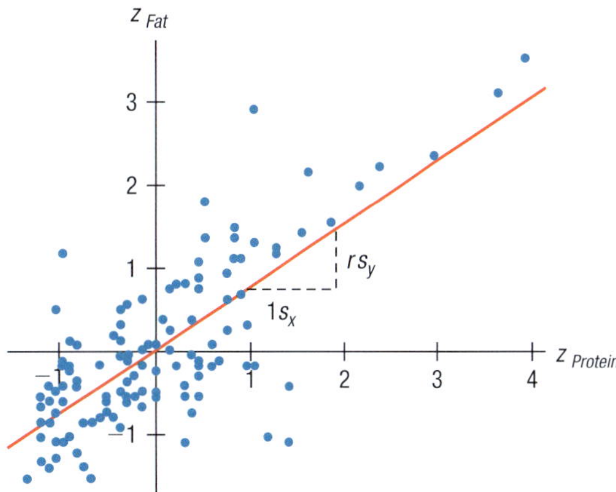

Figure 7.4
Standardized Fat vs. *Standardized Protein* with the regression line. Foods that are one standard deviation away from the mean in *Protein* are expected to be about r standard deviations (in the same direction) away from the mean in *Fat*.

The z-score equation has another profound consequence. It says that for linear relationships, you can never predict that y will be *farther* from its mean than x was from its mean. That's because r is always between -1 and 1. So, each predicted y tends to be closer to its mean (in standard deviations) than its corresponding x was. Sir Francis Galton discovered this property when trying to predict heights of offspring from their parents. He found that although tall parents tend to have tall children, their children are closer to the mean of all offspring than their parents were. He called this phenomenon "regression to mediocrity." He had hoped that tall parents would have children whose heights were even more extraordinary, but that's not what he found. Often when people relate essentially the same variable in two different groups, or at two different times, they see this same phenomenon—the tendency of the response variable to be closer to the mean than the predicted value. Many people have made the error of trying to interpret this by thinking that the performance of those far from the mean is deteriorating, but it's just a mathematical fact about the correlation. So, today we try to be less judgmental about this phenomenon and we call it **regression to the mean**. We managed to get rid of the term "mediocrity," but the name regression stuck as a name for the whole least squares fitting procedure—and that's where we get the term **regression line**.

JUST CHECKING

A scatterplot of house *Price* (in dollars) vs. house *Size* (in square feet) for houses sold recently in Saratoga, New York, shows a relationship that is straight, with only moderate scatter and no outliers. The correlation between house *Price* and house *Size* is 0.76. The least squares line is

$$\widehat{Price} = 8400 + 88.67\ Size.$$

1. You go to an open house and find that the house is 1 SD above the mean *Size*. How many SDs above the mean *Price* would you predict it to cost?
2. You read an ad for a house that's 2 SDs below the mean *Size*. What would you guess about its price?
3. A friend tells you about a house whose *Size* in square meters (he's European) is 1.5 standard deviations above the mean. What would you guess about its *Size* in square feet?
4. Suppose the standard deviation of the house *Prices* is \$77,000. How much more expensive than the mean *Price* would you predict for the house in Part 1 (that's one SD above the mean in *Size*)?

5. Suppose the mean *Price* of all these houses is \$168,000. What is the predicted *Price* for the house in Part 4 that's one SD above the mean in *Size*?

6. The mean *Size* of the houses is 1800 sq ft and the standard deviation is 660 sq ft. Predict the *Price* of a 2460 sq ft house (1 SD above the mean) using the least squares equation in the original units to verify that you get the same prediction as in Part 5.

Math Box

Where does the equation of the line of best fit come from? And what does it really mean?

We can write the equation of any line if we know a point on the line and the slope. Finding a point on the line isn't hard. For the BK sandwiches, it's logical to predict that a sandwich with average protein will have average fat. In fact, it's fairly easy to show that the best fit line *always* passes through the point $(\bar{x}, \bar{y})$.

To find the slope, we'll again convert both x and y to z-scores. We need to keep a few facts in mind:

1. The mean of any set of z-scores is 0. Since we know the line passes through $(\bar{x}, \bar{y})$ this tells us that the line of best fit for z-scores passes through the origin (0, 0).
2. The standard deviation of a set of z-scores is 1, and so, of course, are the variances. This means, in particular, that

$$\frac{\sum(z_y - \bar{z}_y)^2}{(n-1)} = \frac{\sum(z_y - 0)^2}{(n-1)} = \frac{\sum z_y^{\,2}}{(n-1)} = 1,$$

a fact that will be important soon.

3. Remember from Chapter 6, the correlation is $r = \frac{\sum z_x z_y}{(n-1)}$, also important soon.

We want to find the slope of the best fit line. Because it passes through the origin, its equation will be of the simple form $\hat{z}_y = mz_x$. We want to find the value of m that minimizes the sum of the squared residuals $\sum(z_y - \hat{z}_y)^2$. Because it actually makes things simpler, we'll divide that sum by $(n-1)$ and minimize this quantity instead.

Minimize: $\dfrac{\sum(z_y - \hat{z}_y)^2}{(n-1)}$

Since $\hat{z}_y = mz_x$, we are minimizing $\dfrac{\sum(z_y - mz_x)^2}{(n-1)}$

We can expand this and write it as: $\dfrac{\sum z_y^{\,2}}{(n-1)} - 2m\dfrac{\sum z_x z_y}{(n-1)} + m^2\dfrac{\sum z_x^{\,2}}{(n-1)}$

4. Notice that the first term is 1 (because it's the variance of the z_y), the second term is $-2mr$, and the last term is m^2 (because the variance of z_x is 1 as well). So to minimize the sum of squared residuals we only have to minimize: $1 - 2mr + m^2$. This last expression is quadratic, which, as you may remember, describes a parabola. And a parabola of the form $y = ax^2 + bx + c$ reaches its minimum at its turning point, which occurs when $x = -\frac{b}{2a}$. So, for this particular quadratic, we can minimize the sum of the squared residuals by choosing $m = -\frac{-(2r)}{2(1)} = r$.

This is really an amazing fact and explains to some extent the importance of the correlation as a summary of the strength of the linear relationship between *x* and *y*. To say it again: The slope of the best fit line for the z-scores (the standardized values) of any two variables *x* and *y* is their correlation, *r*.

We've converted to z-scores to make the mathematics easier, but what does this result mean for any two variables *x* and *y*?

- A slope of *r* for the best fit line for z-scores means that if we move 1 standard deviation in z_x, we expect to move r standard deviations in z_y. So a BK item that is one standard deviation above the mean in protein would be expected to be *r* standard deviations above the mean in fat. And a sandwich that is 2 standard deviations higher than the mean in protein would be expected to be $2 \times r$ standard deviations higher than the mean in fat. If the correlation were 1, we'd expect it to be a full 2 standard deviations higher in fat. On the other hand, if the correlation were 0, we'd predict it to have the mean fat content. For the Burger King items, the correlation actually is 0.83, so for a sandwich 2 standard deviations higher than the mean in protein, we'd predict its fat content to be 1.66 standard deviations higher than the average fat content.

  Rather than just saying that we'd predict a BK sandwich to be *r* standard deviations above the mean in fat for each standard deviation above the mean in protein, we can write the relationship in the original units by multiplying by the ratio of the standard deviations.

  The slope of the regression line is then $b = r\frac{s_y}{s_x}$.

- We know that choosing $m = r$ minimizes the sum of the squared residuals, but how small does that sum get? Equation 4 told us that the sum of the squared residuals divided by $(n - 1)$ is $1 - 2mr + m^2$. When $m = r$, this quantity is $1 - 2r \times r + r^2 = 1 - r^2$. Because it's a sum of squares it has to be positive, so $1 - r^2 \geq 0$. That means that $r^2 \leq 1$, and therefore, $-1 \leq r \leq 1$. In other words, it shows that the correlation is always between -1 and $+1$.

The quantity r^2 is also important in its own right. Look back at the expression

$$\frac{\sum(z_y - \hat{z}_y)^2}{(n - 1)}.$$

Compare it to the variance of the z_y: $Var(z_y) = \frac{\sum(z_y - \bar{z}_y)^2}{(n - 1)}$. Almost the same, aren't they? The variance of z_y measures its variation from the mean. The sum of the residuals measures variation from the predicted values, from the regression line. The variance of z_y is 1. Now that we've minimized the variation around the regression line it's only $1 - r^2$. In other words, the regression line has removed (or accounted for) r^2 of the original variance in *y*. Knowing how much variance the regression removes helps us assess the strength of our model. We'll use r^2 to help us do just that.

7.5 Examining the Residuals

> **NOTATION ALERT**
>
> Galton used *r* for correlation, so we use *e* for "error." But it's not a mistake. Statisticians often refer to variability not explained by a model as error.

The residuals are defined as

$$\text{Residual} = \text{Observed value} - \text{Predicted value},$$

or in symbols,

$$e = y - \hat{y}.$$

When we want to know how well the model fits, we can ask instead what the model missed. No model is perfect, so it's important to know how and where it fails. To see that, we look at the residuals.

EXAMPLE 7.3

Katrina's Residual

RECAP: The linear model relating hurricanes' wind speeds to their central pressures was

$$\widehat{MaxWindSpeed} = 1031.24 - 0.9748\ CentralPressure.$$

Let's use this model to make predictions and see how those predictions do.

QUESTION: Hurricane Katrina had a central pressure measured at 920 millibars. What does our regression model predict for her maximum wind speed? How good is that prediction, given that Katrina's actual wind speed was measured at 150 knots?

ANSWER: Substituting 920 for the central pressure in the regression model equation gives

$$\widehat{MaxWindSpeed} = 1031.24 - 0.9748(920) = 134.4.$$

The regression model predicts a maximum wind speed of 134.4 knots for Hurricane Katrina.

The residual for this prediction is the observed value minus the predicted value:

$$150 - 134.4 = 15.6\ kts.$$

In the case of Hurricane Katrina, the model predicts a wind speed 15.6 knots lower than was actually observed.

Residuals help us see whether the model makes sense. When a regression model is appropriate, it should model the underlying relationship. Nothing interesting should be left behind. So after we fit a regression model, we usually plot the residuals in the hope of finding . . . nothing interesting.

A scatterplot of the residuals vs. the x-values should be the most boring scatterplot you've ever seen. It shouldn't have any interesting features, like a direction or shape. It should stretch horizontally, with about the same amount of scatter throughout. It should show no bends, and it should have no outliers. If you see any of these features, investigate them if possible, and find out what the regression model missed.

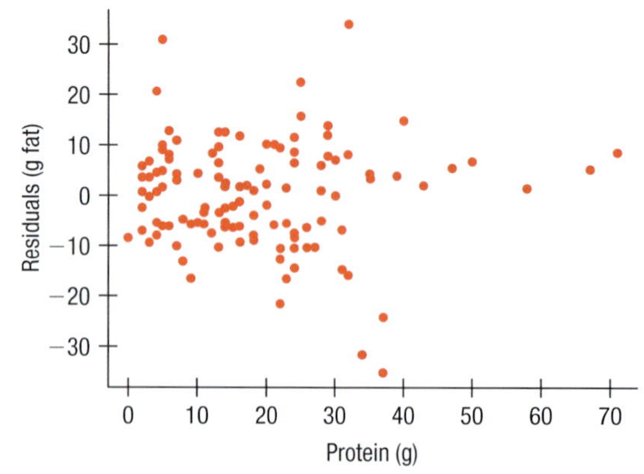

Figure 7.5
The residuals for the BK menu regression look appropriately boring. There are no obvious patterns, although there are a few points with large residuals. The negative ones turn out to be grilled chicken items, which are among the lowest fat (per protein) items on the menu. The two high outliers contain hash browns, which are high fat per protein items.

Most computer statistics packages plot the residuals against the predicted values $\hat{y}$, rather than against x.[6] When the slope is negative, the two versions are mirror images. When the slope is positive, they're virtually identical except for the axis labels. Since all we care about is the patterns (or, better, lack of patterns) in the plot, it really doesn't matter which way we plot the residuals.

The Residual Standard Deviation

Why $n - 2$?

Why $n - 2$ rather than $n - 1$? We used $n - 1$ for s when we estimated the mean. Now we're estimating both a slope and an intercept. Looks like a pattern—and it is. We subtract one more for each parameter we estimate.

If the residuals show no interesting pattern when we plot them against x, we can look at how big they are. After all, we're trying to make them as small as possible. Since their mean is always zero, though, it's only sensible to look at how much they vary. The standard deviation of the residuals, s_e, gives us a measure of how much the points spread around the regression line. Of course, for this summary to make sense, the residuals should all share the same underlying spread. That's why we check to make sure that the residual plot has about the same amount of scatter throughout.

This gives us a new assumption: the **Equal Variance Assumption**. The associated condition to check is the **Does the Plot Thicken? Condition**. We ask if the plot thickens because the usual violation to the assumption is that the spread increases as x or the predicted values increase. However, to be thorough, we check to make sure that the spread is about the same throughout. We can check that either in the original scatterplot of y against x or in the scatterplot of residuals.

We estimate the **standard deviation of the residuals** in almost the way you'd expect:

$$s_e = \sqrt{\frac{\sum e^2}{n - 2}}.$$

We don't need to subtract the mean because the mean of the residuals $\bar{e} = 0$. The $n - 2$ is an adjustment, like the $n - 1$ in the denominator of the variance, which we'll discuss in a later chapter.

For the Burger King foods, the standard deviation of the residuals is 10.6 grams of fat. That looks about right in the scatterplot of residuals. The residual for the BK Tendercrisp chicken was −14.7 grams, just under 1.5 standard deviations.

It's a good idea to make a histogram of the residuals. If we see a unimodal, symmetric histogram, then we can apply the 68–95–99.7 Rule to see how well the regression model describes the data. In particular, we know that 95% of the residuals should be no farther from 0 than $2s_e$. The Burger King residuals look like this:

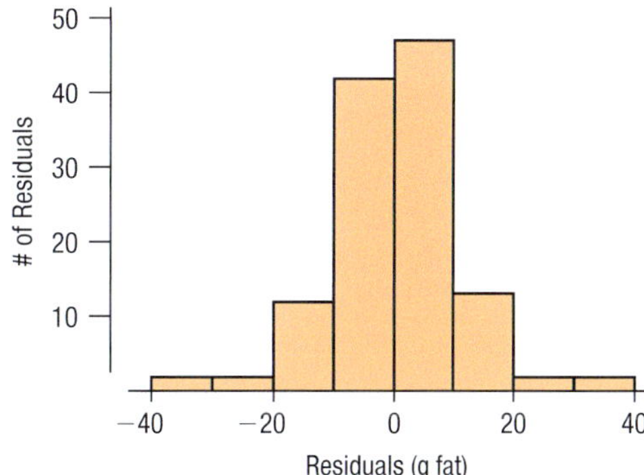

Figure 7.6
The histogram of residuals is symmetric and unimodal, centered at 0, with a standard deviation of 10.6. Only a few values lie outside of 2 standard deviations. The low ones are the chicken items mentioned before. The high ones contain hash browns.

[6] They have a good reason for this choice. When regression models use more than one x-variable, there can be many x's to plot against, but there's still only one $\hat{y}$.

Sure enough, most are within 2(10.6), or 21.2, grams of fat from 0.

The size of s_e provides our first indication of how useful the regression might be. Because roughly 95% of the residuals will be within $2s_e$ from the line of best fit, we can't expect our predictions to be more accurate than that. For x-values far from the mean, we shouldn't be even that optimistic. For the Burger King items, that means we need to accept prediction errors of at least plus or minus 21.2 grams of fat. If that range is too wide, the regression might not be a useful tool for prediction.

7.6 R^2—The Variation Accounted for by the Model

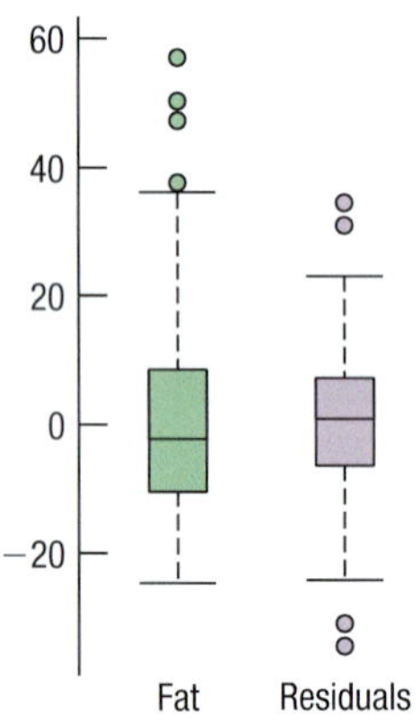

Figure 7.7
Compare the variability of total *Fat* with the residuals from the regression. The means have been subtracted to make it easier to compare spreads. The variation left in the residuals is unaccounted for by the model, but it's less than the variation in the original data.

The variation in the residuals is the key to assessing how well the model fits the data. Let's compare the variation of the response variable with the variation of the residuals. The total *Fat* has a standard deviation of 16.2 grams. The standard deviation of the residuals is 10.6 grams. If the correlation were 1.0 and the model predicted the *Fat* values perfectly, the residuals would all be zero and have no variation. We couldn't possibly do any better than that.

On the other hand, if the correlation were zero, the model would simply predict 24.8 grams of *Fat* (the mean) for all menu items. The residuals from that prediction would just be the observed *Fat* values minus their mean. These residuals would have the same variability as the original data because, as we know, just subtracting the mean doesn't change the spread.

How well does the BK regression model do? Look at the boxplots. The variation in the residuals is smaller than in the data, but certainly bigger than zero. That's nice to know, but how much of the variation is still left in the residuals? If you had to put a number between 0% and 100% on the fraction of the variation left in the residuals, what would you say?

All regression models fall somewhere between the two extremes of zero correlation and perfect correlation. We'd like to gauge where our model falls. Can we use the correlation to do that? Well, a regression model with correlation −0.5 is doing as well as one with correlation +0.5. They just have different directions. But if we *square* the correlation coefficient, we'll get a value between 0 and 1, and the direction won't matter. All regression analyses include this statistic, although by tradition, it is written with a capital letter, $\boldsymbol{R^2}$, and pronounced "R-squared." The squared correlation, r^2, gives the fraction of the data's variation accounted for by the model, and $1 - r^2$ is the fraction of the original variation left in the residuals. For the Burger King model, $r^2 = 0.76^2 = 0.58$, and $1 - r^2$ is 0.42, so 42% of the variability in total *Fat* has been left in the residuals. How close was that to your guess?

An R^2 of 0 means that none of the variance in the data is in the model; all of it is still in the residuals. It would be hard to imagine using that model for anything. Because R^2 is a fraction of a whole, it is often given as a percentage.[7] For the Burger King data, R^2 is 58%.

When interpreting a regression model, you need to be able to explain what R^2 means. According to our linear model, 58% of the variability in the fat content of Burger King items is accounted for by variation in the protein content.

Twice as Strong?

Is a correlation of 0.80 twice as strong as a correlation of 0.40? Not if you think in terms of R^2. A correlation of 0.80 means an R^2 of $0.80^2 = 64\%$. A correlation of 0.40 means an R^2 of $0.40^2 = 16\%$—only a quarter as much of the variability accounted for. A correlation of 0.80 gives an R^2 *four* times as strong as a correlation of 0.40 and accounts for four times as much of the variability.

How Can We See That R^2 Is Really the Fraction of Variance Accounted for by the Model?

It's a simple calculation. The variance of the fat content of the Burger King foods is $16.2^2 = 262.44$. If we treat the residuals as data, the variance of the residuals is 110.53.[8] As a fraction, that's $110.53/262.44 = 0.42$ *or* 42%. That's the fraction of the variance that is *not* accounted for by the model. The fraction that is accounted for is $100\% - 42\% = 58\%$, just the value we got for R^2.

[7] By contrast, we usually give correlation coefficients as decimal values between −1.0 and 1.0.

[8] This isn't quite the same as squaring the s_e that we discussed on the previous pages, but it's very close. We'll deal with the distinction in Chapter 23.

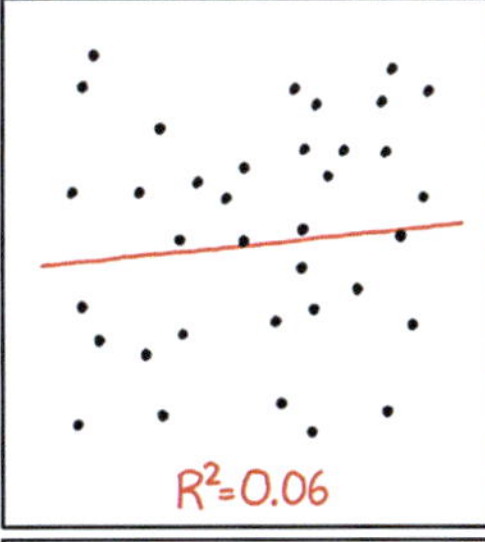

I DON'T TRUST LINEAR REGRESSIONS WHEN IT'S HARDER TO GUESS THE DIRECTION OF THE CORRELATION FROM THE SCATTER PLOT THAN TO FIND NEW CONSTELLATIONS ON IT.

EXAMPLE 7.4 Interpreting R^2

RECAP: Our regression model that predicts maximum wind speed in hurricanes based on the storm's central pressure has $R^2 = 80.6\%$.

QUESTION: What does that say about our regression model?

ANSWER: An R^2 of 80.6% indicates that 80.6% of the variation in maximum wind speed can be accounted for by the hurricane's central pressure. Other factors, such as temperature and whether the storm is over water or land, may account for some of the remaining variation.

JUST CHECKING

Back to our regression of house *Price* ($) on house *Size* (square feet):

$$\widehat{Price} = 8400 + 88.67\, Size.$$

The R^2 value is reported as 57.8%, and the standard deviation of the residuals is $53,790.

7. What does the R^2 value mean about the relationship of price and size?
8. Is the correlation of price and size positive or negative? How do you know?
9. You find that your house is worth $50,000 more than the regression model predicts. You are undoubtedly pleased, but is this actually a surprisingly large residual?

SOME EXTREME TALES

One major company developed a method to differentiate between proteins. To do so, they had to distinguish between regressions with R^2 of 99.99% and 99.98%. For this application, 99.98% was not high enough.

On the other hand, the head of a hedge fund reports that although his regressions give R^2 below 2%, they are highly successful because those used by his competition are even lower.

How Big Should R^2 Be?

R^2 is always between 0% and 100%. But what's a "good" R^2 value? The answer depends on the kind of data you are analyzing and on what you want to do with it. Just as with correlation, there is no value for R^2 that automatically determines that the regression is "good." Data from scientific experiments often have R^2 in the 80% to 90% range and even higher. Data from observational studies and surveys, though, often show relatively weak associations because it's so difficult to measure responses reliably. An R^2 of 50% to 30% or even lower might be taken as evidence of a useful regression. The standard deviation of the residuals can give us more information about the usefulness of the regression by telling us how much scatter there is around the line.

As we've seen, an R^2 of 100% is a perfect fit, with no scatter around the line. The s_e would be zero. All of the variance is accounted for by the model and none is left in the residuals at all. This sounds great, but it's too good to be true for real data.[9]

Along with the slope and intercept for a regression, you should always report R^2 and s_e so that readers can judge for themselves how successful the regression is at fitting the data. Statistics is about variation, and R^2 measures the success of the regression model in terms of the fraction of the variation of y accounted for by the regression. The residual standard deviation, s_e, tells us how far the points are likely to be from the fitted line. R^2 is the first part of a regression that many people look at because, along with the scatterplot, it tells whether the regression model is even worth thinking about.

[9]If you see an R^2 of 100%, it's a good idea to figure out what happened. You may have discovered a new law of physics, but it's much more likely that you accidentally regressed two variables that measure the same thing.

Predicting in the Other Direction—A Tale of Two Regressions

Regression slopes may not behave exactly the way you'd expect at first. Our regression model for the Burger King sandwiches was $\widehat{Fat} = 8.4 + 0.91\ Protein$. That equation allowed us to estimate that a sandwich with 31 grams of protein would have 36.6 grams of fat. Suppose, though, that we knew the fat content and wanted to predict the amount of protein. It might seem natural to think that by solving the equation for *Protein* we'd get a model for predicting *Protein* from *Fat*. But that doesn't work.

Our original model is $\hat{y} = b_0 + b_1 x$, but the new one needs to evaluate an $\hat{x}$ based on a value of y. We don't have y in our original model, only $\hat{y}$, and that makes all the difference. Our model doesn't fit the BK data values perfectly, and the least squares criterion focuses on the *vertical* (y) errors the model makes in using x to model y—not on *horizontal* errors related to x.

A quick look at the equations reveals why. Simply solving for x would give a new line whose slope is the reciprocal of ours. To model y in terms of x, our slope is $b_1 = r\frac{s_y}{s_x}$. To model x in terms of y, we'd need to use the slope $b_1 = r\frac{s_x}{s_y}$. That's *not* the reciprocal of ours.

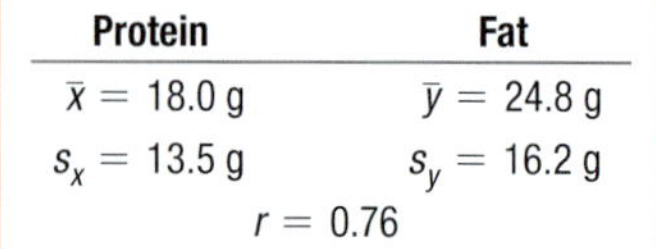

Protein	Fat
$\bar{x} = 18.0$ g	$\bar{y} = 24.8$ g
$s_x = 13.5$ g	$s_y = 16.2$ g
$r = 0.76$	

Sure, if the correlation, r, were 1.0 or -1.0 the slopes *would* be reciprocals, but that would happen only if we had a perfect fit. Real data don't follow perfect straight lines, so in the real world y and $\hat{y}$ aren't the same, r is a fraction, and the slopes of the two models are not simple reciprocals of one another. Also, if the standard deviations were equal—for example, if we standardize both variables—the two slopes would be *the same*. Far from being reciprocals, both would be equal to the correlation—but we already knew that the correlation of x with y is the same as the correlation of y with x.

Otherwise, slopes of the two lines will not be reciprocals, so we can't derive one equation from the other. If we want to predict x from y, we need to create that model from the data. For example, to predict *Protein* from *Fat* we can't just invert the model we found before. Instead we find the slope, $b_1 = 0.76\frac{13.5}{16.2} = 0.63$ grams of protein per gram of fat. The regression model is then $\widehat{Protein} = 2.29 + 0.63\ Fat$.

Moral of the story: Decide which variable you want to use (x) to predict values for the other (y). Then find the model that does that. If, later, you want to make predictions in the other direction, start over and create the other model from scratch.

7.7 Regression Assumptions and Conditions

The linear regression model may be the most widely used model in all of statistics. It has everything we could want in a model: two easily estimated parameters, a meaningful measure of how well the model fits the data, and the ability to predict new values. It even provides an easy way to see violations of conditions in plots of the residuals.

Like all models, though, linear models are only appropriate if some assumptions are true. We can't confirm assumptions, but we often can check related conditions.

First, be sure that both variables are quantitative. It makes no sense to perform a regression on categorical variables. After all, what could the slope possibly mean? Always check the **Quantitative Variables Condition**.

The linear model only makes sense if the relationship is linear. It is easy to check the associated **Straight Enough Condition**. Just look at the scatterplot of y vs. x. You don't need a *perfectly* straight plot, but it must be straight enough for the linear model to make sense. If you try to model a curved relationship with a straight line, you'll usually get just what you deserve. If the scatterplot is not straight enough, stop here. You can't use a linear model for *any* two variables, even if they are related. They must have a *linear* association, or the model won't mean a thing.

Make a Picture (or Two)

You can't check the conditions just by checking boxes. You need to examine both the original scatterplot of *y* against *x* before you fit the model, and the plot of residuals afterward. These plots can save you from making embarrassing errors and losing points on the exam.

For the standard deviation of the residuals to summarize the scatter of all the residuals, the residuals must share the same spread. That's an assumption. But if the scatterplot of *y* vs. *x* looks equally spread out everywhere and (often more vividly) if the *residual plot* of residuals vs. predicted values also has a consistent spread, then the assumption is reasonable. The most common violation of that Equal Variance Assumption is for the residuals to spread out more for *larger* values of *x*, so a good mnemonic for this check is the **Does the Plot Thicken? Condition**.

Outlying points can dramatically change a regression model. They can even change the sign of the slope, which would give a very different impression of the relationship between the variables if you only look at the regression model. So check the **Outlier Condition**. Check both the scatterplot of *y* against *x*, and the residual plot to be sure there are no outliers. The residual plot often shows violations more clearly and may reveal other unexpected patterns or interesting quirks in the data. Of course, any outliers are likely to be interesting and informative, so be sure to look into why they are unusual.

To summarize:

Before starting, be sure to check the

- **Quantitative Variable Condition** If either *y* or *x* is categorical, you can't make a scatterplot and you can't perform a regression. Stop.

From the scatterplot of *y* against *x*, check the

- **Straight Enough Condition** Is the relationship between *y* and *x* straight enough to proceed with a linear regression model?
- **Outlier Condition** Are there any outliers that might dramatically influence the fit of the least squares line?
- **Does the Plot Thicken? Condition** Does the spread of the data around the generally straight relationship seem to be consistent for all values of *x*?

After fitting the regression model, make a plot of residuals against the predicted values and look for

- Any bends that would violate the **Straight Enough Condition**,
- Any outliers that weren't clear before, and
- Any change in the spread of the residuals from one part of the plot to another.

STEP-BY-STEP EXAMPLE Regression

If you plan to hit the fast-food joints for lunch, you should have a good breakfast. Nutritionists, concerned about "empty calories" in breakfast cereals, recorded facts about 77 cereals, including their *Calories* per serving and *Sugar* content (in grams). (Data in **Cereals**)

QUESTION: How are calories and sugar content related in breakfast cereals?

THINK **PLAN** State the problem.

I am interested in the relationship between sugar content and calories in cereals.

VARIABLES Name the variables and report the W's.

I have two quantitative variables, *Calories* and *Sugar* content per serving, measured on 77 breakfast cereals. The units of measurement are calories and grams of sugar, respectively.

✓ **Quantitative Variables**: Both variables are quantitative. The units of measurement for *Calories* and *Sugar* are calories and grams, respectively.

I'll check the remaining conditions from the scatterplot:

Check the conditions for a regression by making a picture. Never fit a regression without looking at the scatterplot first. Calories are reported to the nearest 10 calories and sugar content is reported to the nearest gram, so many cereals have the same values for both variables. In order to show them better, we have added a little random noise to both variables in the scatterplot, a process called *jittering*. Of course, we use the actual values for calculation.

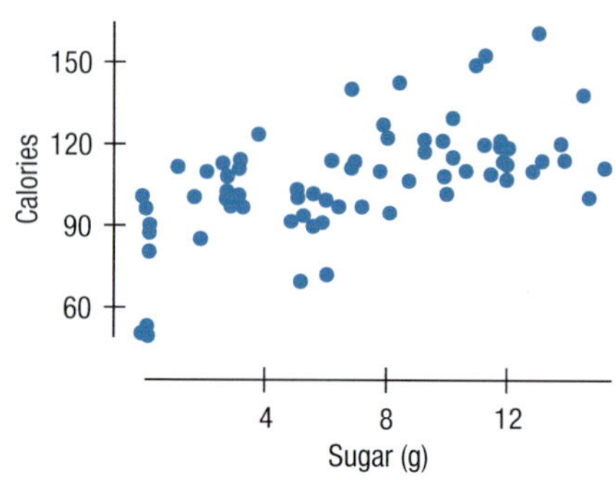

✓ **Outlier Condition**: There are no obvious outliers or groups.

✓ **Straight Enough**: The scatterplot looks straight to me.

✓ **Does the Plot Thicken?** The spread around the line looks about the same throughout, but I'll check it again in the residuals.

I will fit a regression model to these data.

SHOW **MECHANICS** If there are no clear violations of the conditions, fit a straight line model of the form $\hat{y} = b_0 + b_1 x$ to the data. Summary statistics give the building blocks of the calculation (though one would usually use software to calculate these quantities).

Calories

$\bar{y} = 106.88$ *calories*

$s_y = 19.48$ *calories*

Sugar

$\bar{x} = 6.94$ *grams*

$s_x = 4.42$ *grams*

Correlation

$r = 0.564$

Find the slope.

$$b_1 = r\frac{s_y}{s_x} = 0.564\frac{19.48}{4.42}$$

$= 2.49$ *calories per gram of sugar*

Find the intercept.

$$b_0 = \bar{y} - b_1\bar{x} = 106.88 - 2.49(6.94)$$

$= 89.6$ *calories*

Write the equation, using meaningful variable names.

So the least squares line is

$\hat{y} = 89.6 + 2.49\,x$,

or $\widehat{Calories} = 89.6 + 2.49\,Sugar$.

State the value of R^2.

Squaring the correlation gives

$R^2 = 0.564^2 = 0.318$ or 31.8%.

TELL **CONCLUSION** Describe what the model says in words and numbers. Be sure to use the names of the variables and their units.

The scatterplot shows a positive, linear relationship and no outliers. The least squares regression line fit through these data has the equation

$$\widehat{Calories} = 89.6 + 2.49\ Sugar.$$

The key to interpreting a regression model is to start with the phrase "b_1 y-units per x-unit," substituting the estimated value of the slope for b_1 and the names of the respective units.

The slope says that we expect cereals to have, on average, about 2.49 calories per gram of sugar.

The intercept is then a starting or base value. It may (as in this example) be meaningful, or (when $x = 0$ is not realistic) it may just be a starting value.

The intercept predicts that a serving of a sugar-free cereal would average about 89.6 calories.

R^2 gives the fraction of the variability of y accounted for by the linear regression model.

The R^2 says that 31.8% of the variability in *Calories* is accounted for by variation in *Sugar* content.

Find the standard deviation of the residuals, s_e, and compare it to the original, s_y.

$s_e = 16.2$ calories. That's smaller than the original SD of 19.5, but still fairly large. A prediction error of plus or minus 32.4 calories may too large for the regression model to be useful.

THINK AGAIN **CHECK AGAIN** Even though we looked at the scatterplot *before* fitting a regression model, a plot of the residuals is essential to any regression analysis because it is the best check for additional patterns and interesting quirks in the data. (As we did earlier, the points have been jittered to see the pattern more clearly.)

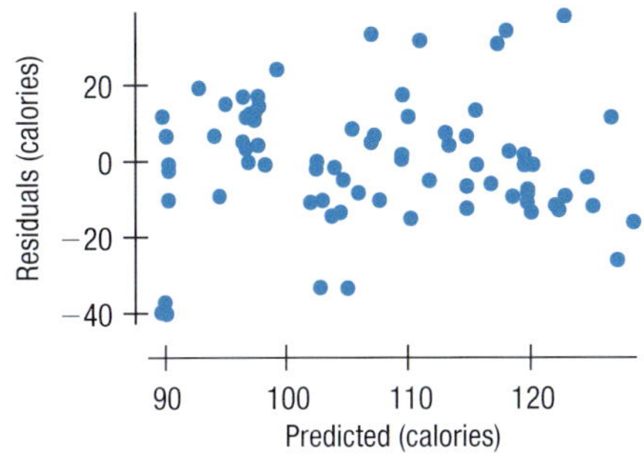

The residuals show a horizontal direction, a shapeless form, and roughly equal scatter for all predicted values. The linear model appears to be appropriate.

Reality Check: Is the Regression Reasonable?

> **REGRESSION: ADJECTIVE, NOUN, OR VERB?**
> You may see the term *regression* used in different ways. There are many ways to fit a line to data, but the term "regression line" or "regression" without any other qualifiers always means least squares. People also use *regression* as a verb when they speak of *regressing* a y-variable on an x-variable to mean fitting a linear model.

Statistics don't come out of nowhere. They are based on data. The results of a statistical analysis should reinforce your common sense, not fly in its face. If the results are surprising, then either you've learned something new about the world or your analysis is wrong.

Whenever you perform a regression, think about the coefficients and ask whether they make sense. Is a slope of 2.5 calories per gram of sugar reasonable? That's hard to say right off. We know from the summary statistics that a typical cereal has about 100 calories and 7 grams of sugar per serving. A gram of sugar contributes some calories (actually, 4, but you don't need to know that), so calories should go up with increasing sugar. The direction of the slope seems right.

To see if the *size* of the slope is reasonable, a useful trick is to consider its order of magnitude. Start by asking if deflating the slope by a factor of 10 seems reasonable. Is 0.25 calories per gram of sugar enough? The 7 grams of sugar found in the average cereal would contribute less than 2 calories. That seems too small.

Then try inflating the slope by a factor of 10. Is 25 calories per gram reasonable? The average cereal would have 175 calories from sugar alone. The average cereal has only 100 calories per serving, though, so that slope seems too big.

We have tried inflating the slope by a factor of 10 and deflating it by 10 and found both to be unreasonable. So, like Goldilocks, we're left with the value in the middle that's just right. And an increase of 2.5 calories per gram of sugar is certainly *plausible.*

The small effort of asking yourself whether the regression equation is plausible is repaid whenever you catch errors or avoid saying something silly or absurd about the data. It's too easy to take something that comes out of a computer at face value and assume that it makes sense.

Always be skeptical and ask yourself if the answer is reasonable.

EXAMPLE 7.5 Causation and Regression

RECAP: Our regression model predicting hurricane wind speeds from the central pressure was reasonably successful. The negative slope indicates that, in general, storms with lower central pressures have stronger winds.

QUESTION: Can we conclude that lower central barometric pressure *causes* the higher wind speeds in hurricanes?

ANSWER: No. While it may be true that lower pressure causes higher winds, a regression model for observed data such as these cannot demonstrate causation. Perhaps higher wind speeds reduce the barometric pressure, or perhaps both pressure and wind speed are driven by some other variable we have not observed.

(As it happens, in hurricanes it is reasonable to say that the low central pressure at the eye is responsible for the high winds because it draws moist, warm air into the center of the storm, where it swirls around, generating the winds. But as is often the case, things aren't quite that simple. The winds themselves contribute to lowering the pressure at the center of the storm as it becomes organized into a hurricane. The lesson is that to understand causation in hurricanes, we must do more than just model the relationship of two variables; we must study the mechanism itself.)

WHAT CAN GO WRONG?

There are many ways in which data that appear at first to be good candidates for regression analysis may be unsuitable. And there are ways that people use regression that can lead them astray. Here's an overview of the most common problems. We'll discuss them at length in the next chapter.

- **Don't fit a straight line to a nonlinear relationship.** Linear regression is suited only to relationships that are, well, *linear.* Fortunately, we can often improve the linearity easily by using re-expression. We'll come back to that topic in Chapter 8.
- **Don't ignore outliers.** Outliers can have a serious impact on the fitted model. You should identify them and think about why they are extraordinary. If they turn out not to be obvious errors, read the next chapter for advice.
- **Don't invert the regression.** The BK regression model was $\widehat{Fat} = 8.4 + 0.91\, Protein$. Knowing protein content, we can predict the amount of fat. But that doesn't let us switch the regression around. We can't use this model to predict protein values from fat values. To model y from x, the least squares slope is $b_1 = r\frac{s_y}{s_x}$. To model x in terms of y, we'd find $b_1 = r\frac{s_x}{s_y}$.

That's not the reciprocal of the first slope (unless the correlation is 1.0). To swap the predictor–response roles of the variables in a regression (which can sometimes make sense), we must fit a new regression equation (it turns out to be $\widehat{Protein} = 2.29 + 0.63\ Fat$).

- **Don't claim causation.** Even when it's tempting, don't say that x causes y. Don't say that increasing x by one unit results in (adds, causes, etc.) a corresponding change in y. The regression model describes the current data, and the association between x and y. It says nothing about what happens to y if x changes.

CONNECTIONS

The linear model is one of the most important models in statistics. Chapter 6 talked about the assignment of variables to the y- and x-axes. That didn't matter to correlation, but it does matter to regression because y is predicted by x in the regression model.

The connection of R^2 to correlation is obvious, although it may not be immediately clear that just by squaring the correlation we can learn the fraction of the variability of y accounted for by a regression on x. We'll return to this in subsequent chapters.

We made a big fuss about knowing the units of your quantitative variables. We didn't need units for correlation, but without the units we can't define the slope of a regression. A regression makes no sense if you don't know the *Who*, the *What*, and the *Units* of both your variables.

We've summed squared deviations before when we computed the standard deviation and variance. That's not coincidental. They are closely connected to regression.

CHAPTER REVIEW

Model linear relationships with a linear model.

- Examine residuals to assess how well a model fits the data.

The best fit model is the one that minimizes the sum of the squared residuals—the Least Squares Model.

Find the coefficients of the least squares line from the correlation and the summary statistics of each variable: $b_1 = r\frac{s_y}{s_x}$.

$$b_0 = \bar{y} - b_1\bar{x}.$$

Understand the correlation coefficient as the number of standard deviations by which one variable is expected to change for a one standard deviation change in the other.

- Recognize the sometimes-surprising phenomenon of regression to the mean that can occur when two variables measuring the same, or similar, things are correlated.

Always examine the residuals to check for violations of assumptions and conditions and to identify any outliers.

Always report and interpret R^2, which reports the fraction of the variation of y accounted for by the linear model.

- R^2 is the square of the correlation coefficient.

Be sure to check the conditions for regression before reporting or interpreting a regression model. Check that:

- The relationship is Straight Enough.
- A scatterplot does not "thicken" in any region.
- There are no outliers.

Interpret a regression slope as y-units per x-unit.

- Regression slopes vary from sample to sample. A different random sample from the same population will likely have a different slope. We can gather the slopes from many samples to see how they are distributed.

REVIEW OF TERMS

Linear model An equation of the form

$$\hat{y} = b_0 + b_1x.$$

To interpret a linear model, we need to know the variables (along with their W's) and their units (p. 227).

Model An equation or formula that simplifies and represents reality (p. 227).

Predicted value The value of $\hat{y}$ found for a given x-value in the data. A predicted value is found by substituting the x-value in the regression equation. The predicted values are the values on the fitted line; the points $(x, \hat{y})$ all lie exactly on the fitted line (p. 227).

The predicted values are found from the linear model that we fit:

$$\hat{y} = b_0 + b_1x.$$

Residuals The differences between data values and the corresponding values predicted by the regression model—or, more generally, values predicted by any model (p. 227):

$$\text{Residual} = \text{Observed value} - \text{Predicted value} = y - \hat{y}.$$

Least squares The least squares criterion specifies the unique line (sometimes called the line of best fit) that minimizes the variance of the residuals or, equivalently, the sum of the squared residuals (p. 228).

Slope The slope, b_1, gives a value in "y-units *per* x-unit." Changes of one unit in x are associated with changes of b_1 units in predicted values of y (p. 228).

The slope can be found by

$$b_1 = r\frac{s_y}{s_x}.$$

Intercept The intercept, b_0, gives a starting value in y-units. It's the $\hat{y}$-value when x is 0. You can find the intercept from $b_0 = \bar{y} - b_1\bar{x}$ (p. 228).

Regression to the mean Because the correlation is always less than 1.0 in magnitude, each predicted $\hat{y}$ tends to be fewer standard deviations from its mean than its corresponding x was from its mean (p. 235).

Regression line (line of best fit) The particular linear equation

$$\hat{y} = b_0 + b_1x$$

that satisfies the least squares criterion is called the least squares regression line. Casually, we often just call it the regression line, or the line of best fit (p. 235).

Standard deviation of the residuals (s_e) The standard deviation of the residuals is found by $s_e = \sqrt{\frac{\sum e^2}{n-2}}$. When the assumptions and conditions are met, the residuals can be well described by using this standard deviation and the 68–95–99.7 Rule (p. 239).

R^2 The square of the correlation between y and x (p. 240).

- R^2 gives the fraction of the variability of y accounted for by the least squares linear regression on x.
- R^2 is an overall measure of how successful the regression is in linearly relating y to x.

TECH SUPPORT

Regression

All statistics packages make a table of results for a regression. These tables may differ slightly from one package to another, but all are essentially the same—and all include much more than we need to know for now. Every computer regression table includes a section that looks something like this:

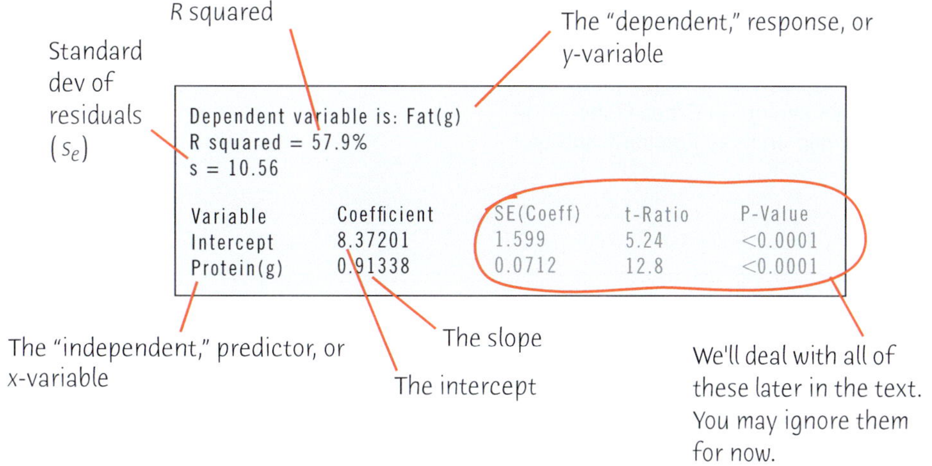

Dependent variable is: Fat(g)
R squared = 57.9%
s = 10.56

Variable	Coefficient	SE(Coeff)	t-Ratio	P-Value
Intercept	8.37201	1.599	5.24	<0.0001
Protein(g)	0.91338	0.0712	12.8	<0.0001

The slope and intercept coefficient are given in a table such as this one. Usually the slope is labeled with the name of the *x*-variable, and the intercept is labeled "Intercept" or "Constant." So the regression equation shown here is

$$\widehat{Fat} = 8.37201 + 0.91338\ Protein.$$

It is not unusual for statistics packages to give many more digits of the estimated slope and intercept than could possibly be estimated from the data. (The original data were reported to the nearest gram.) Ordinarily, you should round most of the reported numbers to one digit more than the precision of the data, and the slope to two. We will learn about the other numbers in the regression table later in the text. For now, all you need to be able to do is find the coefficients, the s_e, and the R^2 value.

DATA DESK

- Select the **Regression** button in the control bar. Data Desk will make a Regression window.
- Drag y and x variables into the appropriate parts of the table.
- Or in a scatterplot HyperView menu, choose **Regression** to compute the regression.
- To display the line, from a scatterplot HyperView menu, choose **Add Regression Line**.
- To plot the residuals, click on the **HyperView** menu in the upper left corner of the **Regression** output table and select the residual plot you want from the menu that drops down.

COMMENTS

Alternatively,

- Select the **y-variable** and the **x-variable** icons.
- In the Calc menu, choose **Regression**.

EXCEL

- Click on a blank cell in the spreadsheet.
- Go to the **Formulas** tab in the Ribbon and click **More Functions » Statistical**.
- The data analysis add-in includes a Regression command.
- In the dialog that pops up, enter the range of one of the variables in the space provided.
- Enter the range of the other variable in the space provided.
- Click **OK**.
- Excel offers alternatives. The LINEST function performs regressions. It is an "array function" in Excel. Consult the Excel documentation for help on array functions and on LINEST.
- The SLOPE(y-range, x-range) and INTERCEPT(y-range, x-range) functions compute the slope and intercept, but don't provide the rest of the usual regression results.

COMMENTS

The correlation is computed in the selected cell. Correlations computed this way will update if any of the data values are changed.

JMP

- Choose **Fit Y by X** from the Analyze menu.
- Specify the *y*-variable in the Select Columns box and click the **y, Response** button.
- Specify the *x*-variable and click the **X, Factor** button.
- Click **OK** to make a scatterplot.
- In the scatterplot window, click on the red triangle beside the heading labeled **Bivariate Fit** ... and choose **Fit Line**. JMP draws the least squares regression line on the scatterplot and displays the results of the regression in tables below the plot.

MINITAB

- Choose **Regression** from the Stat menu.
- From the Regression submenu, choose **Fitted Line Plot**.
- In the Fitted Line Plot dialog, click in the **Response Y** box, and assign the *y*-variable from the variable list.
- Click in the **Predictor X** box, and assign the *x*-variable from the Variable list. Make sure that the Type of Regression Model is set to Linear. Click the **OK** button.

R

- **lm(Y~X)** produces the linear model.
- **summary(lm(Y~X))** produces more information about the model.

COMMENTS

Typically, your variables X and Y will be in a data frame. If DATA is the name of the data frame, then

- **lm(Y~X,data=DATA)** is the preferred syntax.

Note: "lm" is short for "linear model".

SPSS

- Choose **Interactive** from the Graphs menu.
- From the interactive Graphs submenu, choose **Scatterplot**.
- In the Create Scatterplot dialog, drag the *y*-variable into the ***y*-axis target**, and the *x*-variable into the ***x*-axis target**.
- Click on the **Fit** tab.
- Choose **Regression** from the Method popup menu. Click the **OK** button.

STATCRUNCH

- Click on **Stat**.
- Choose **Regression » Simple Linear**.
- Choose X and Y variable names from the list of columns.
- Choose the plots you would like to include.
- Click on **Compute!** to see the regression analysis.

COMMENTS

Remember to check the scatterplot to be sure a linear model is appropriate.
Note that before you **Compute!** you can choose to:

- enter an X-value for which you want to find the predicted Y-value;
- save all the fitted values;
- save the residuals;
- ask for a residuals plot.

TI-83/84 PLUS

To find the equation of the regression line (add the line to a scatterplot), choose **LinReg(a+bx)**, tell it the list names, and then add a comma to specify a function name (from **VARS Y-Vars 1:Function**). The final command looks like

LinReg(a+bx)L1, L2, Y1.

- To make a residuals plot, set up a **STATPLOT** as a scatterplot.
- Specify your explanatory data list as Xlist.
- For Ylist, import the name RESID from the **LIST NAMES** menu. **ZoomStat** will now create the residuals plot.

COMMENTS

Each time you execute a **LinReg** command, the calculator automatically computes the residuals and stores them in a data list named RESID. If you want to see them, go to **STAT EDIT**. Space through the names of the lists until you find a blank. Import RESID from the LIST NAMES menu. Now every time you have the calculator compute a regression analysis, it will show you the residuals.

EXERCISES

SECTION 7.1

1. True or false If false, explain briefly.

a) We choose the linear model that passes through the most data points on the scatterplot.
b) The residuals are the observed y-values minus the y-values predicted by the linear model.
c) Least squares means that the square of the largest residual is as small as it could possibly be.

2. True or false II If false, explain briefly.

a) Some of the residuals from a least squares linear model will be positive and some will be negative.
b) Least squares means that some of the squares of the residuals are minimized.
c) We write $\hat{y}$ to denote the predicted values and y to denote the observed values.

SECTION 7.2

3. Least squares interpretations A least squares regression line was calculated to relate the length (cm) of newborn boys to their weight in kg. The line is $\widehat{weight} = -5.36 + 0.1675\ length$. Explain in words what this model means. Should new parents (who tend to worry) be concerned if their newborn's length and weight don't fit this equation?

4. Residual interpretations The newborn grandson of one of the authors was 48 cm long and weighed 3 kg. According to the regression model of Exercise 3, what was his residual? What does that say about him?

SECTION 7.3

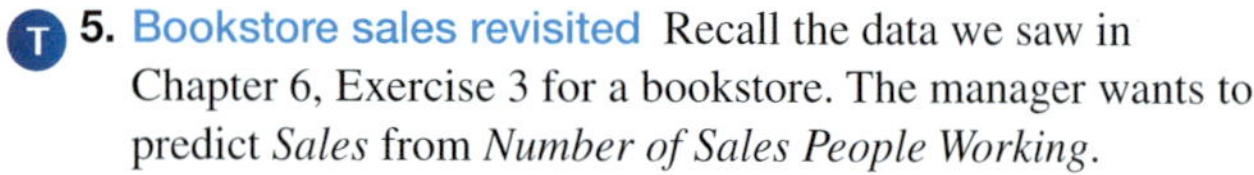

T 5. Bookstore sales revisited Recall the data we saw in Chapter 6, Exercise 3 for a bookstore. The manager wants to predict *Sales* from *Number of Sales People Working*.

Number of Sales People Working	Sales (in $1000)
2	10
3	11
7	13
9	14
10	18
10	20
12	20
15	22
16	22
20	26
$\bar{x} = 10.4$	$\bar{y} = 17.6$
$SD(x) = 5.64$	$SD(y) = 5.34$
$r = 0.965$	

a) Find the slope estimate, b_1.
b) What does it mean, in this context?
c) Find the intercept, b_0.
d) What does it mean, in this context? Is it meaningful?
e) Write down the equation that predicts *Sales* from *Number of Sales People Working*.
f) If 18 people are working, what *Sales* do you predict?
g) If sales are actually $25,000, what is the value of the residual?
h) Have we overestimated or underestimated the sales?

T 6. Disk drives 2016 again Recall the data on disk drives we saw in Chapter 6, Exercise 4. Suppose we want to predict *Price* from *Capacity*.

Capacity (in TB)	Price (in $)
0.50	59.99
1.0	79.99
2.0	111.97
3.0	109.99
4.0	149.99
6.0	423.34
8.0	596.11
12.0	1079.99
32.0	4461.00
$\bar{x} = 7.6111$	$\bar{y} = 785.819$
$SD(x) = 9.855$	$SD(y) = 1418.67$
$r = 0.9876$	

a) Find the slope estimate, b_1.
b) What does it mean, in this context?
c) Find the intercept, b_0.
d) What does it mean, in this context? Is it meaningful?
e) Write down the equation that predicts *Price* from *Capacity*.
f) What would you predict for the price of a 20 TB drive?
g) A 20 TB drive on Amazon.com was listed at $2017.86. According to the model, does this seem like a good buy? How much would you save compared to what you expected to pay?
h) Does the model overestimate or underestimate the price?
i) The correlation is very high. Does this mean that the model is accurate? Explain. (Hint: Revisit the scatterplot from Chapter 6, Exercise 4.)

SECTION 7.4

7. Sophomore slump? A CEO complains that the winners of his "rookie junior executive of the year" award often turn out to have less impressive performance the following year. He wonders whether the award actually encourages them to slack off. Can you offer a better explanation?

8. Sophomore slump again? An online investment blogger advises investing in mutual funds that have performed badly the past year because "regression to the mean tells us that they will do well next year." Is he correct?

SECTION 7.5

T 9. Bookstore sales once more Here are the residuals for a regression of *Sales* on *Number of Sales People Working* for the bookstore of Exercise 5:

Number of Sales People Working	Residual
2	0.07
3	0.16
7	−1.49
9	−2.32
10	0.77
10	2.77
12	0.94
15	0.20
16	−0.72
20	−0.37

a) What are the units of the residuals?
b) Which residual contributes the most to the sum that was minimized according to the least squares criterion to find this regression?
c) Which residual contributes least to that sum?

T 10. Disk drives 2016, residuals Here are the residuals for a regression of *Price* on *Capacity* for the hard drives of Exercise 6 (based on the hand-computed coefficients).

Capacity	Residual
0.50	285.16
1.0	234.07
2.0	123.88
3.0	−20.27
4.0	−122.44
6.0	−133.43
8.0	−245.00
12.0	−329.80
32.0	207.81

a) Which residual contributes the most to the sum that is minimized by the least squares criterion?
b) Five of the residuals are negative. What does that mean about those drives? Be specific and use the correct units.

SECTION 7.6

11. Bookstore sales last time For the regression model for the bookstore of Exercise 5, what is the value of R^2 and what does it mean?

12. Disk drives encore For the hard drive data of Exercise 6, find and interpret the value of R^2.

SECTION 7.7

13. Residual plots Here are residual plots (residuals plotted against predicted values) for three linear regression models. Indicate which condition appears to be violated (Linearity, Outlier, or Equal Spread) in each case.

a)

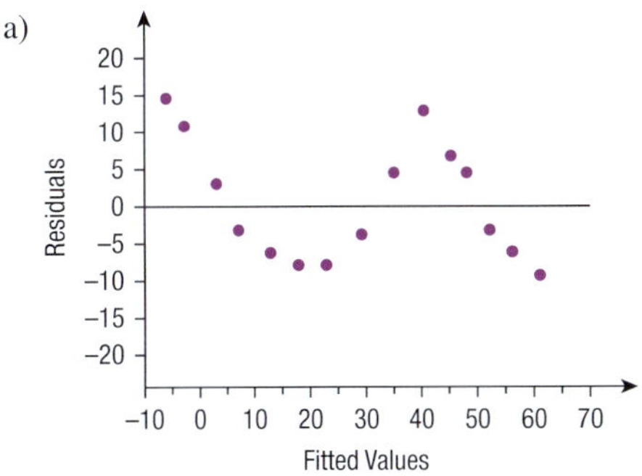

b)

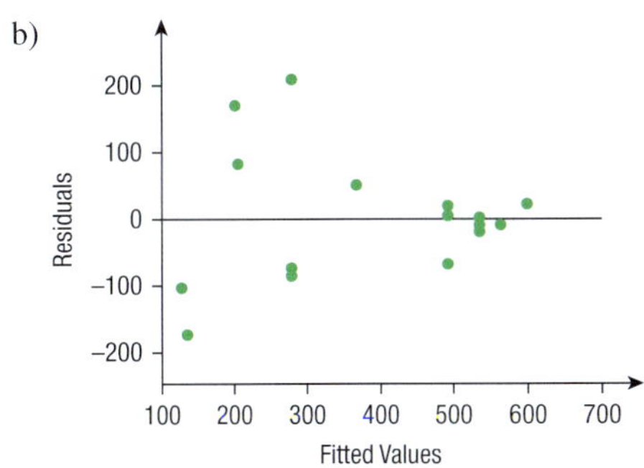

c)

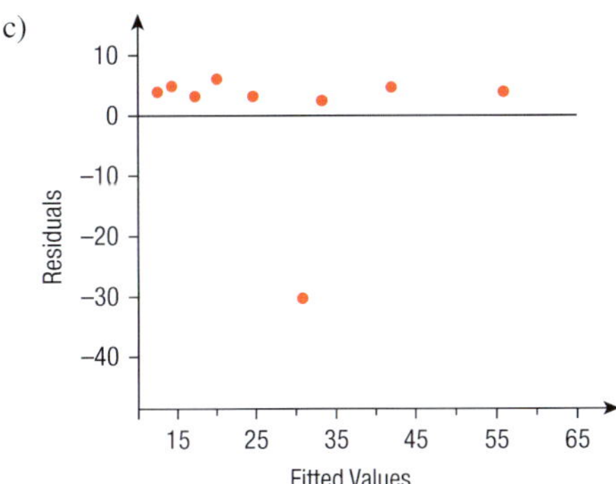

14. Disk drives 2016, residuals again Here is a scatterplot of the residuals from the regression of the hard drive prices on their sizes from Exercise 6.

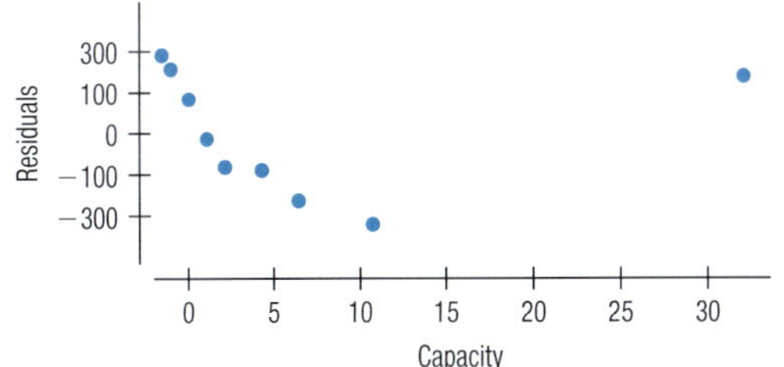

a) Are any assumptions or conditions violated? If so, which ones?
b) What would you recommend about this regression?

CHAPTER EXERCISES

15. Cereals For many people, breakfast cereal is an important source of fiber in their diets. Cereals also contain potassium, a mineral shown to be associated with maintaining a healthy blood pressure. An analysis of the amount of fiber (in grams) and the potassium content (in milligrams) in servings of 77 breakfast cereals produced the regression model $\widehat{Potassium} = 39 + 28\,Fiber$. If your cereal provides 10 grams of fiber per serving, how much potassium does the model estimate you will get?

16. Engine size In Chapter 6, Exercise 41 we examined the relationship between the fuel economy (*CombinedMPG*) and *Displacement* (in liters) for 1211 models of cars. (Data in **Fuel economy 2016**) Further analysis produces the regression model $\widehat{CombinedMPG} = 33.46 - 3.23\,Displacement$. If the car you are thinking of buying has a 4-liter engine, what does this model suggest your gas mileage would be?

17. More cereal Exercise 15 describes a regression model that estimates a cereal's potassium content from the amount of fiber it contains. In this context, what does it mean to say that a cereal has a negative residual?

18. Engine size again Exercise 16 describes a regression model that uses a car's engine displacement to estimate its fuel economy. In this context, what does it mean to say that a certain car has a positive residual?

19. Another bowl In Exercise 15, the regression model $\widehat{Potassium} = 38 + 27\,Fiber$ relates fiber (in grams) and potassium content (in milligrams) in servings of breakfast cereals. Explain what the slope means.

20. More engine size In Exercise 16, the regression model $\widehat{CombinedMPG} = 33.46 - 3.23\,Displacement$ relates cars' engine size to their fuel economy (Combined mpg). Explain what the slope means.

21. Cereal again The correlation between a cereal's fiber and potassium contents is $r = 0.903$. What fraction of the variability in potassium is accounted for by the amount of fiber that servings contain?

22. Another car The correlation between a car's engine size and its fuel economy (in mpg) is $r = -0.774$. What fraction of the variability in fuel economy is accounted for by the engine size?

23. Last bowl! For Exercise 15's regression model predicting potassium content (in milligrams) from the amount of fiber (in grams) in breakfast cereals, $s_e = 30.77$. Explain in this context what that means.

24. Last tank! For Exercise 16's regression model predicting fuel economy (in mpg) from the car's engine size, $s_e = 3.522$. Explain in this context what that means.

25. Regression equations Fill in the missing information in the following table.

	$\bar{x}$	s_x	$\bar{y}$	s_y	r	$\hat{y} = b_0 + b_1x$
a)	49	4.7	8	5.1	0.3	
b)	16	7.7	34	4.5	−0.8	
c)	20	9.2			−0.8	$\hat{y} = 201 - 8x$
d)	40.7	1.9		35.5		$\hat{y} = -164 + 15x$

26. More regression equations Fill in the missing information in the following table.

	$\bar{x}$	s_x	$\bar{y}$	s_y	r	$\hat{y} = b_0 + b_1x$
a)	25	2	24	8	−0.3	
b)	200	18	40	8.4	0.9	
c)		0.8	37	13		$\hat{y} = -15 + 13x$
d)			18	4	−0.7	$\hat{y} = 32 - 2x$

27. Residuals Tell what each of the residual plots below indicates about the appropriateness of the linear model that was fit to the data.

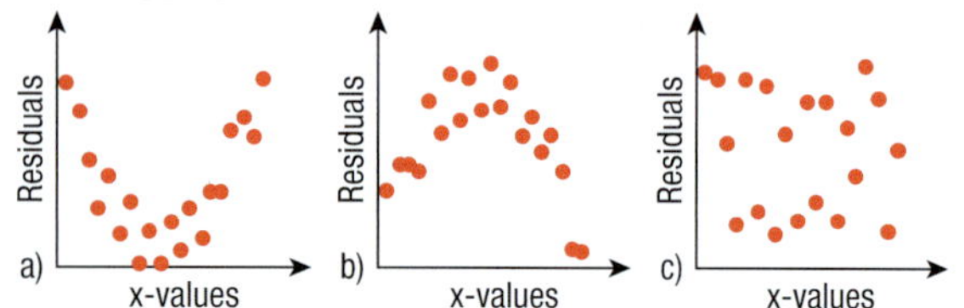

28. Residuals Tell what each of the residual plots below indicates about the appropriateness of the linear model that was fit to the data.

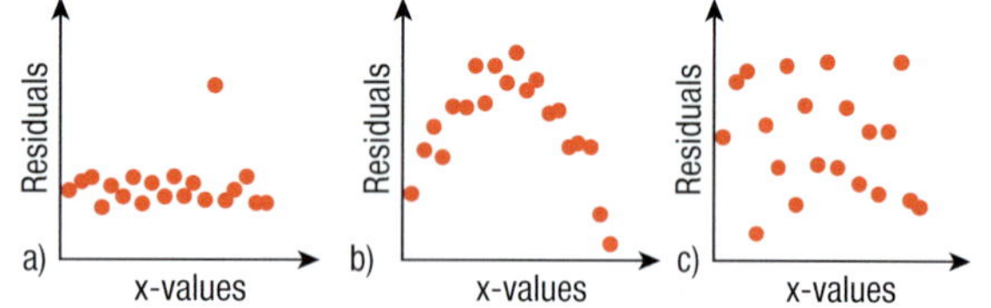

29. Real estate A random sample of records of home sales from Feb. 15 to Apr. 30, 1993, from the files maintained by the Albuquerque Board of Realtors gives the *Price* and *Size* (in square feet) of 117 homes. A regression to predict *Price* (in thousands of dollars) from *Size* has an R^2 of 71.4%. The residuals plot indicated that a linear model is appropriate.

a) What are the variables and units in this regression?
b) What units does the slope have?
c) Do you think the slope is positive or negative? Explain.

T **30. Coasters 2015, revisited** The dataset on roller coasters lists the *Duration* of the ride in seconds in addition to the *Drop* height in feet for some of the coasters. One coaster (the "Tower of Terror") is unusual for having a large drop but a short ride. After setting it aside, a regression to predict *Duration* from *Drop* for the remaining 90 coasters has $R^2 = 29.4\%$.

a) What are the variables and units in this regression?
b) What units does the slope have?
c) Do you think the slope is positive or negative? Explain.

31. What slope? If you create a regression model for predicting the *Weight* of a car (in pounds) from its *Length* (in feet), is the slope most likely to be 3, 30, 300, or 3000? Explain.

32. What slope again? If you create a regression model for estimating the *Height* of a pine tree (in feet) based on the *Circumference* of its trunk (in inches), is the slope most likely to be 0.1, 1, 10, or 100? Explain.

33. Real estate again The regression of *Price* on *Size* of homes in Albuquerque had $R^2 = 71.4\%$, as described in Exercise 29. Write a sentence (in context, of course) summarizing what the R^2 says about this regression.

T **34. Coasters 2015 again** Exercise 30 examined the association between the *Duration* of a roller coaster ride and the height of its initial *Drop*, reporting that $R^2 = 29.4\%$. Write a sentence (in context, of course) summarizing what the R^2 says about this regression.

35. Misinterpretations A biology student who created a regression model to use a bird's *Height* when perched for predicting its *Wingspan* made these two statements. Assuming the calculations were done correctly, explain what is wrong with each interpretation.

a) My R^2 of 93% shows that this linear model is appropriate.
b) A bird 10 inches tall will have a wingspan of 17 inches.

36. More misinterpretations A sociology student investigated the association between a country's *Literacy Rate* and *Life Expectancy*, and then drew the conclusions listed below. Explain why each statement is incorrect. (Assume that all the calculations were done properly.)

a) The R^2 of 64% means that the *Literacy Rate* determines 64% of the *Life Expectancy* for a country.
b) The slope of the line shows that an increase of 5% in *Literacy Rate* will produce a 2-year improvement in *Life Expectancy*.

37. Real estate redux The regression of *Price* on *Size* of homes in Albuquerque had $R^2 = 71.4\%$, as described in Exercise 29.

a) What is the correlation between *Size* and *Price*?
b) What would you predict about the *Price* of a home 1 SD above average in *Size*?
c) What would you predict about the *Price* of a home 2 SDs below average in *Size*?

T **38. Another ride** The regression of *Duration* of a roller coaster ride on the height of its initial *Drop*, described in Exercise 30, had $R^2 = 29.4\%$.

a) What is the correlation between *Drop* and *Duration*?
b) What would you predict about the *Duration* of the ride on a coaster whose initial *Drop* was 1 standard deviation below the mean *Drop*?
c) What would you predict about the *Duration* of the ride on a coaster whose initial *Drop* was 3 standard deviations above the mean *Drop*?

39. ESP People who claim to have "extrasensory perception (ESP)" participate in a screening test in which they have to guess which of several images someone is thinking of. You and a friend both took the test. You scored 2 standard deviations above the mean, and your friend scored 1 standard deviation below the mean. The researchers offer everyone the opportunity to take a retest.

a) Should you choose to take this retest? Explain.
b) Now explain to your friend what his decision should be and why.

40. *SI* jinx Players in any sport who are having great seasons, turning in performances that are much better than anyone might have anticipated, often are pictured on the cover of *Sports Illustrated*. Frequently, their performances then falter somewhat, leading some athletes to believe in a "*Sports Illustrated* jinx." Similarly, it is common for phenomenal rookies to have less stellar second seasons—the so-called "sophomore slump." While fans, athletes, and analysts have proposed many theories about what leads to such declines, a statistician might offer a simpler (statistical) explanation. Explain.

41. More real estate Consider the Albuquerque home sales from Exercise 29 again. The regression analysis gives the model $\widehat{Price} = 47.82 + 0.061\ Size$.

a) Explain what the slope of the line says about housing prices and house size.
b) What price would you predict for a 3000-square-foot house in this market?
c) A real estate agent shows a potential buyer a 1200-square-foot home, saying that the asking price is \$6000 less than what one would expect to pay for a house of this size. What is the asking price, and what is the \$6000 called?

T **42.** Last ride Consider the roller coasters (with the outlier removed) described in Exercise 30 again. The regression analysis gives the model $\widehat{Duration} = 87.22 + 0.389\ Drop$.

a) Explain what the slope of the line says about how long a roller coaster ride may last and the height of the coaster.
b) A new roller coaster advertises an initial drop of 200 feet. How long would you predict the rides last?
c) Another coaster with a 150-foot initial drop advertises a 2 and a half-minute ride. Is this longer or shorter than you'd expect? By how much? What's that called?

T **43.** Cigarettes Is the nicotine content of a cigarette related to the "tar"? A collection of data (in milligrams) on 29 cigarettes produced the scatterplot, residuals plot, and regression analysis shown:

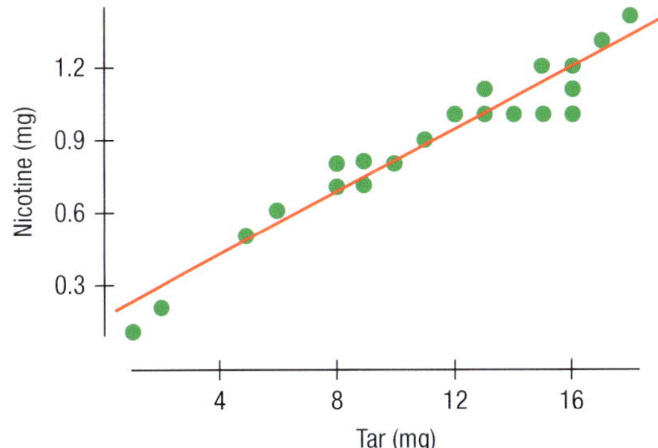

Response variable is: Nicotine
R squared = 92.4%

Variable	Coefficient
Intercept	0.154030
Tar	0.065052

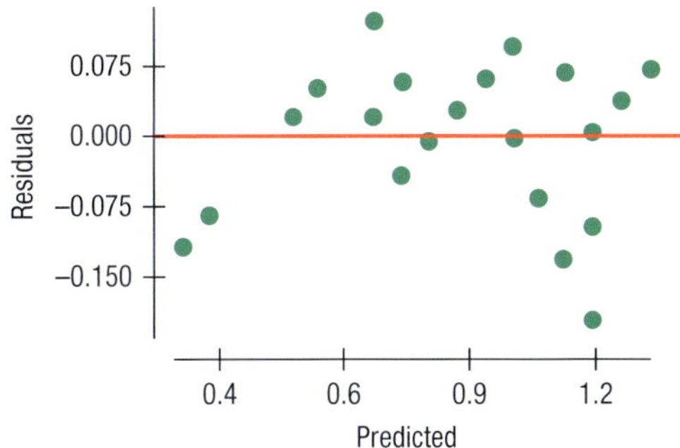

a) Do you think a linear model is appropriate here? Explain.
b) Explain the meaning of R^2 in this context.

T **44.** Attendance 2016, revisited In Chapter 6, Exercise 45 looked at the relationship between the number of runs scored by American League baseball teams and the average attendance at their home games for the 2016 season. Here are the scatterplot, the residuals plot, and part of the regression analysis for *all* major league teams in 2016 (National League teams in red, American League in blue):

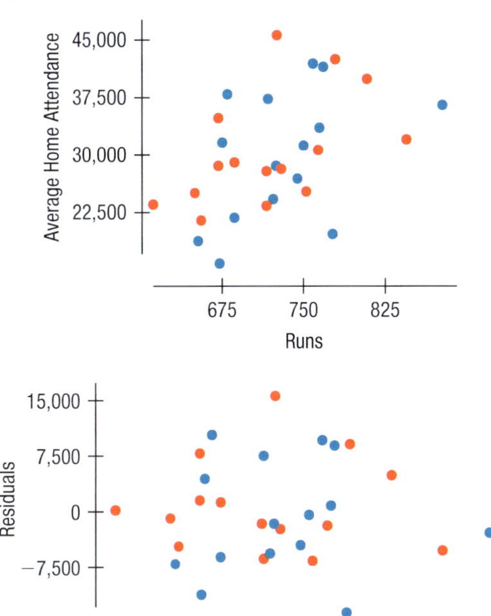

Response variable is: Home Avg Att
R squared = 21.0%
s = 7002

Variable	Coefficient	SE(Coeff)
Intercept	−12882.2	15815
Runs	59.389	21.75

a) Do you think a linear model is appropriate here? Explain.
b) Interpret the meaning of R^2 in this context.
c) Do the residuals show any pattern worth remarking on?
d) The point at the top of the plots is the L.A. Dodgers. What can you say about the residual for the Dodgers?

T **45. Another cigarette** Consider again the regression of *Nicotine* content on *Tar* (both in milligrams) for the cigarettes examined in Exercise 43.

a) What is the correlation between *Tar* and *Nicotine*?
b) What would you predict about the average *Nicotine* content of cigarettes that are 2 standard deviations below average in *Tar* content?
c) If a cigarette is 1 standard deviation above average in *Nicotine* content, what do you suspect is true about its *Tar* content?

T **46. Attendance 2016, revisited** Consider again the regression of *Home Average Attendance* on *Runs* for the baseball teams examined in Exercise 44.

a) What is the correlation between *Runs* and *Home Average Attendance*?
b) What would you predict about the *Home Average Attendance* for a team that is 2 standard deviations above average in *Runs*?
c) If a team is 1 standard deviation below average in attendance, what would you predict about the number of runs the team has scored?

T **47. Last cigarette** Take another look at the regression analysis of tar and nicotine content of the cigarettes in Exercise 43.

a) Write the equation of the regression line.
b) Estimate the *Nicotine* content of cigarettes with 4 milligrams of *Tar.*
c) Interpret the meaning of the slope of the regression line in this context.
d) What does the y-intercept mean?
e) If a new brand of cigarette contains 7 milligrams of tar and a nicotine level whose residual is -0.05 mg, what is the nicotine content?

T **48. Attendance 2016, last inning** Refer again to the regression analysis for home average attendance and games won by baseball teams, seen in Exercise 44.

a) Write the equation of the regression line.
b) Estimate the *Home Average Attendance* for a team with 750 *Runs.*
c) Interpret the meaning of the slope of the regression line in this context.
d) In general, what would a negative residual mean in this context?

T **49. Income and housing revisited** In Chapter 6, Exercise 39, we learned that the Office of Federal Housing Enterprise Oversight (OFHEO) collects data on various aspects of housing costs around the United States. Here's a scatterplot (by state) of the *Housing Cost Index* (HCI) vs. the *Median Family Income* (MFI) for the 50 states. The correlation is $r = 0.624$. The mean HCI is 342.3, with a standard deviation of 119.07. The mean MFI is \$46,210, with a standard deviation of \$7003.55.

a) Is a regression analysis appropriate? Explain.
b) What is the equation that predicts Housing Cost Index from median family income?
c) For a state with MFI = \$44,993, what would be the predicted HCI?
d) Washington, DC, has an MFI of \$44,993 and an HCI of 548.02. How far off is the prediction in part b from the actual HCI?
e) If we standardized both variables, what would be the regression equation that predicts standardized HCI from standardized MFI?
f) If we standardized both variables, what would be the regression equation that predicts standardized MFI from standardized HCI?

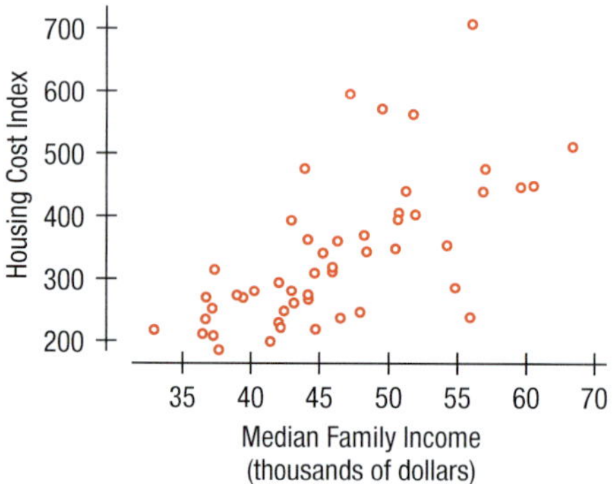

T **50. Interest rates and mortgages 2015 again** In Chapter 6, Exercise 40, we saw a plot of mortgages in the United States (in trillions of 2013 dollars) vs. the interest rate at various times over the past 25 years. The correlation is $r = -0.845$. The mean mortgage amount is \$8.207 T and the mean interest rate is 6.989%. The standard deviations are \$4.527 T for mortgage amounts and 2.139% for the interest rates.

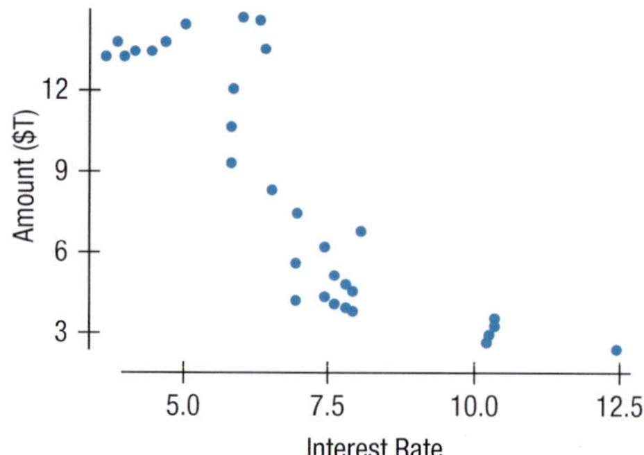

a) Is a regression model appropriate for predicting mortgage amount from interest rates? Explain.
b) Regardless of your answer to part a, find the equation that predicts mortgage amount from interest rates.
c) What would you predict the mortgage amount would be if the interest rates climbed to 13%?
d) Do you have any reservations about your prediction in part c? Explain.
e) If we standardized both variables, what would be the regression equation that predicts standardized mortgage amount from standardized interest rates?
f) If we standardized both variables, what would be the regression equation that predicts standardized interest rates from standardized mortgage amount?

51. Online clothes An online clothing retailer keeps track of its customers' purchases. For those customers who signed up for the company's credit card, the company also has information on the customer's *Age* and *Income*. A random sample of 25 of these customers shows the following scatterplot of *Total Yearly Purchases* by *Age:*

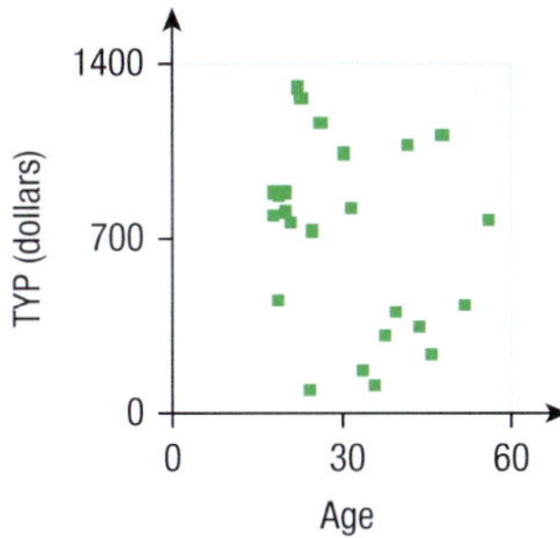

The correlation between *Total Yearly Purchases* and *Age* is $r = 0.028$. Summary statistics for the two variables are:

	Mean	SD
Age	25.572 yr	8.551 yr
Total Yearly Purchase	$567.323	$258.717

a) What is the linear regression equation for predicting *Total Yearly Purchase* from *Age*?
b) Do the assumptions and conditions for regression appear to be met?
c) What is the predicted *Total Yearly Purchase* for a 19-year-old? For a 49-year-old?
d) What percent of the variability in *Total Yearly Purchases* is accounted for by this model?
e) Do you think the regression might be a useful one for the company? Explain.

52. Online clothes II For the online clothing retailer discussed in the previous problem, the scatterplot of *Total Yearly Purchases* by *Income* looks like this:

The correlation between *Total Yearly Purchases* and *Income* is 0.722. Summary statistics for the two variables are:

	Mean	SD
Income	$50,343.40	$16,952.50
Total Yearly Purchase	$572.52	$253.62

a) What is the linear regression equation for predicting *Total Yearly Purchase* from *Income*?
b) Do the assumptions and conditions for regression appear to be met?
c) What is the predicted *Total Yearly Purchase* for someone with a yearly *Income* of $20,000? For someone with an annual *Income* of $80,000?
d) What percent of the variability in *Total Yearly Purchases* is accounted for by this model?
e) Do you think the regression might be a useful one for the company? Comment.

T **53. SAT scores** The SAT is a test often used as part of an application to college. SAT scores are between 200 and 800, but have no units. Tests are given in both Math and Verbal areas. SAT-Math problems require the ability to read and understand the questions, but can a person's verbal score be used to predict the math score? Verbal and Math SAT scores of a high school graduating class are displayed in the scatterplot, with the regression line added.

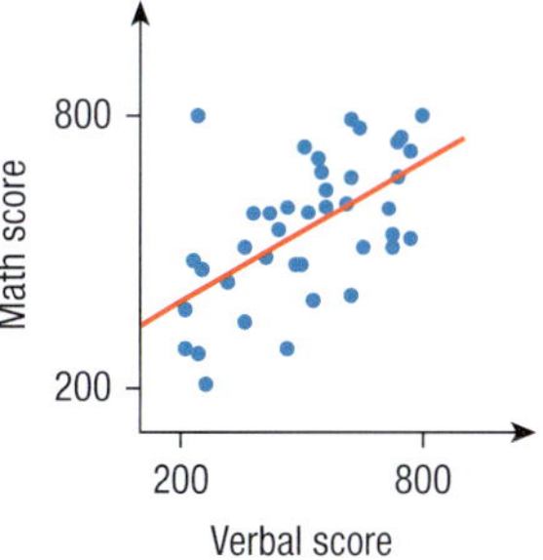

a) Describe the relationship.
b) Are there any students whose scores do not seem to fit the overall pattern?
c) For these data, $r = 0.576$. Interpret this statistic.
d) These verbal scores averaged 513.4, with a standard deviation of 176.8, and the math scores averaged 554.1, with a standard deviation of 157.2. Write the equation of the regression line predicting math scores from verbal scores.
e) Interpret the slope of this line.
f) Predict the math score of a student with a verbal score of 432.
g) Every year, some students score a perfect 1600 (800 on both tests). Based on this model, what would such a student's residual be for her math score?

54. Success in college Colleges use SAT scores in the admissions process because they believe these scores provide some insight into how a high school student will perform at the college level. Suppose the entering freshmen at a certain college have mean combined *SAT scores* of 1247, with a standard deviation of 128. In the first semester, these students attained a mean *GPA* of 2.66, with a standard deviation of 0.59. A scatterplot showed the association to be reasonably linear, and the correlation between *SAT* score and *GPA* was 0.41.

a) Write the equation of the regression line.
b) Explain what the *y*-intercept of the regression line indicates.
c) Interpret the slope of the regression line.
d) Predict the *GPA* of a freshman who scored a combined 1400.
e) Based upon these statistics, how effective do you think SAT scores would be in predicting academic success during the first semester of the freshman year at this college? Explain.
f) As a student, would you rather have a positive or a negative residual in this context? Explain.

55. SAT, take 2 Suppose we wanted to use SAT math scores to estimate verbal scores based on the information in Exercise 53.

a) What is the correlation?
b) Write the equation of the line of regression predicting verbal scores from math scores.
c) In general, what would a positive residual mean in this context?
d) A person tells you her math score was 500. Predict her verbal score.
e) Using that predicted verbal score and the equation you created in Exercise 53, predict her math score.
f) Why doesn't the result in part e come out to 500?

56. Success, part 2 Based on the statistics for college freshmen given in Exercise 54, what SAT score would you predict for a freshmen who attained a first-semester GPA of 3.0?

57. Wildfires 2015 The number of wildfires in a country has been declining. A scatterplot and analysis of the amount of land (in millions of acres) lost to wildfires is shown below.

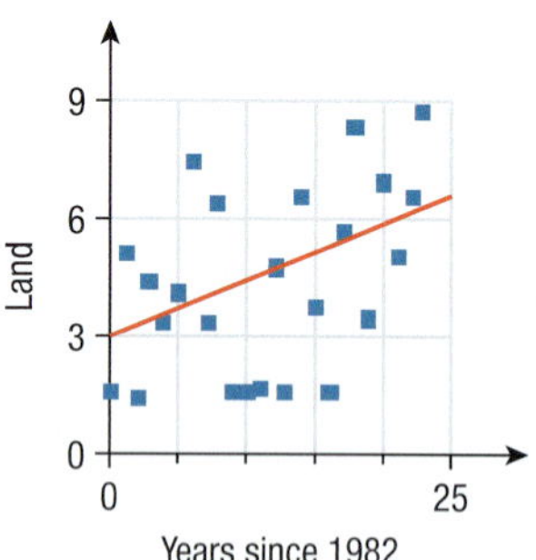

Response variable is: Land Lost to Wildfires
R squared = 22.7%
s = 1.804

Variable	Coefficient
Intercept	2.99035
Years since 1982	0.14445

a) Is a linear model appropriate for these data? Explain.
b) Interpret the slope in this context.
c) Can we interpret the intercept? Why or why not?
d) What does the value of *s* say about the size of the residuals? What does it say about the effectiveness of the model?
e) What does R^2 mean in this context?

58. Wildfires 2015—sizes We saw in Exercise 57 that the number of fires was nearly constant. But has the damage they cause remained constant as well? Here's a regression that examines the trend in *Acres per Fire* (in hundreds of thousands of acres) together with some supporting plots:

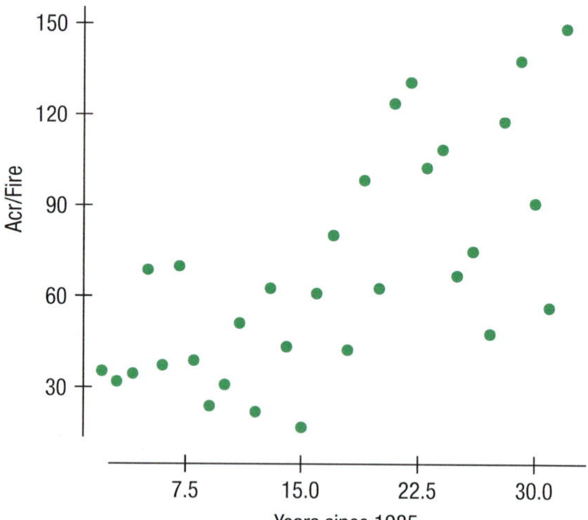

Response variable is: Acres/Fire
R squared = 45.5%
s = 27.72

Variable	Coefficient
Intercept	21.7466
Years since 1985	2.73606

a) Is the regression model appropriate for these data? Explain.
b) Interpret the slope in this context.
c) Can we interpret the intercept? Why or why not?
d) What interpretation (if any) can you give for the R^2 in the regression table?

59. Used cars 2014 A website lists numerous used cars for sale within a 250-mile radius of a town. The table lists the ages of the cars and the advertised prices.

a) Make a scatterplot for these data.
b) Describe the association between *Age* and *Price* of a used car.
c) Do you think a linear model is appropriate?
d) Computer software says that $R^2 = 89.3\%$. What is the correlation between *Age* and *Price*?

e) Explain the meaning of R^2 in this context.
f) Why doesn't this model explain 100% of the variability in the price of a used car?

Age (yr)	Price Advertised ($)
1	12,940
1	10,945
2	10,440
3	10,985
3	10,950
4	6,980
4	7,995
5	8,715
5	6,960
6	5,980
6	4,965
9	3,230
9	2,230
9	3,990
11	2,935
11	2,965
13	1,760

T **60. Drug abuse revisited** Chapter 6, Exercise 42 examines results of a survey conducted in the United States and 10 countries of Western Europe to determine the percentage of teenagers who had used marijuana and other drugs. Below is the scatterplot. Summary statistics showed that the mean percent that had used marijuana was 23.9%, with a standard deviation of 15.6%. An average of 11.6% of teens had used other drugs, with a standard deviation of 10.2%.

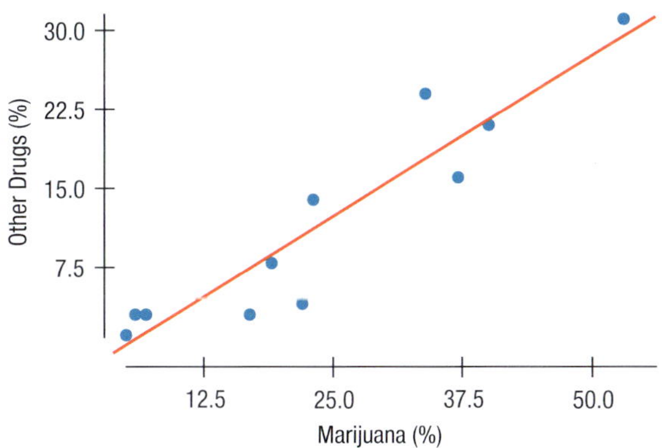

a) Do you think a linear model is appropriate? Explain.
b) For this regression, R^2 is 87.3%. Interpret this statistic in this context.
c) Write the equation you would use to estimate the percentage of teens who use other drugs from the percentage who have used marijuana.
d) Explain in context what the slope of this line means.
e) Do these results confirm that marijuana is a "gateway drug," that is, that marijuana use leads to the use of other drugs?

T **61. More used cars 2014** Use the advertised prices for Toyota Corollas given in Exercise 59 to create a linear model for the relationship between a car's *Age* and its *Price*.

a) Find the equation of the regression line.
b) Explain the meaning of the slope of the line.
c) Explain the meaning of the *y*-intercept of the line.
d) If you want to sell a 7-year-old Corolla, what price seems appropriate?
e) You have a chance to buy one of two cars. They are about the same age and appear to be in equally good condition. Would you rather buy the one with a positive residual or the one with a negative residual? Explain.
f) You see a "For Sale" sign on a 10-year-old Corolla stating the asking price as \$8,500. What is the residual?
g) Would this regression model be useful in establishing a fair price for a 25-year-old car? Explain.

T **62. Veggie burger 2014** Burger King introduced a meat-free burger in 2002. The nutrition label for the 2014 BK Veggie burger (no mayo) is shown here: (Data in **Burger King items**)

Nutrition Facts

Calories	320
Fat	8g*
Sodium	840g
Sugars	9g
Protein	21g
Carbohydrates	42g
Dietary Fiber	4g
Cholesterol	0

*(1 gram of saturated fat)

RECOMMENDED DAILY VALUES (based on a 2,000-calorie/day diet)

Iron	0%
Vitamin A	0%
Vitamin C	0%
Calcium	0%

a) Use the regression model created in this chapter, $\widehat{Fat} = 8.4 + 0.91\ Protein$, to predict the fat content of this burger from its protein content given here.
b) What is its residual? How would you explain the residual?
c) Write a brief report about the *Fat* and *Protein* content of this menu item. Be sure to talk about the variables by name and in the correct units.

T **63. Burgers revisited** In Chapter 6, you examined the association between the amounts of *Fat* and *Calories* in fast-food hamburgers. Here are the data:

Fat (g)	19	31	34	35	39	39	43
Calories	410	580	590	570	640	680	660

a) Create a scatterplot of *Calories* vs. *Fat*.
b) Interpret the value of R^2 in this context.
c) Write the equation of the line of regression.
d) Use the residuals plot to explain whether your linear model is appropriate.
e) Explain the meaning of the *y*-intercept of the line.
f) Explain the meaning of the slope of the line.
g) A new burger containing 28 grams of fat is introduced. According to this model, its residual for calories is +33. How many calories does the burger have?

64. Chicken Chicken sandwiches are often advertised as a healthier alternative to beef because many are lower in fat. Tests on 11 brands of fast-food chicken sandwiches produced the following summary statistics and scatterplot from a graphing calculator:

	Fat(g)	Calories
Mean	20.6	472.7
St. Dev.	9.8	144.2
Correlation	0.947	

a) Do you think a linear model is appropriate in this situation?
b) Describe the strength of this association.
c) Write the equation of the regression line.
d) Explain the meaning of the slope.
e) Explain the meaning of the y-intercept.
f) What does it mean if a certain sandwich has a negative residual?
g) If a chicken sandwich and a burger each advertised 35 grams of fat, which would you expect to have more calories (see Exercise 63)?

T 65. A second helping of burgers In Exercise 63, you created a model that can estimate the number of *Calories* in a burger when the *Fat* content is known.

a) Explain why you cannot use that model to estimate the fat content of a burger with 600 calories.
b) Using an appropriate model, estimate the fat content of a burger with 600 calories.

T 66. Cost of living 2016 A website records the cost of living (COL) for 100 cities around the world. It reports the typical cost of a number of staples. Here are a scatterplot and regression relating the cost of a cappuccino to the cost of a third of a liter of water:

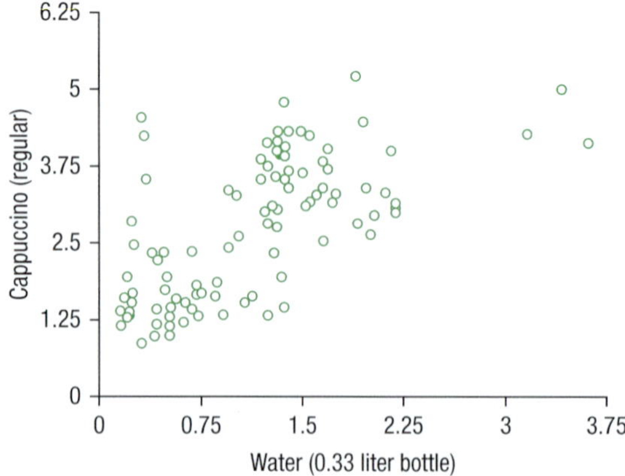

a) Using this information, describe the association between the costs of a cappuccino and a third of a liter of water.
b) The correlation is 0.655. Find and interpret the value of R^2.
c) The regression equation predicting the cost of a cappuccino from the cost of a liter of water is

$$\widehat{Cappuccino\ cost} = 1.522 + 1.0770\ Water\ cost$$

In one city, a third of a liter of water costs \$0.33, and a cappuccino is \$4.22. Calculate and interpret the residual for the city.

T 67. Bridges We saw in this chapter that in a certain town, older bridges were in worse condition than newer ones. The given town is a rural area. Is this relationship true in cities as well? Here are data on the *Condition* (as measured by the state Department of Transportation Condition Index) and *Age at Inspection* for bridges in a particular city.

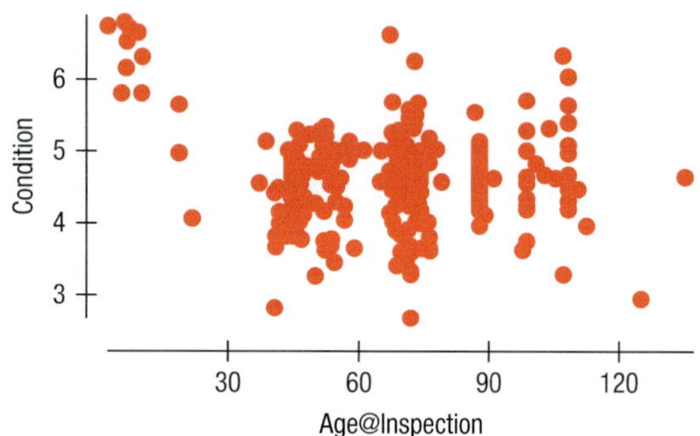

Dependent variable is Condition
R-squared = 2.6%
s = 0.6708

Variable	Coefficient
Intercept	4.95147
Age@Inspection	−0.00481

a) This city defines any bridge with a condition score less than 5 as *deficient*. What does this model predict for the condition scores of bridges in the city?
b) Our earlier model found that the condition of bridges in Tompkins County was decreasing at about 0.025 per year. What does this model say about bridges in this city?
c) How much faith would you place in this model? Explain.

T 68. Birthrates 2015 The table shows the number of live births per 1000 population in the United States, starting in 1965. (National Center for Health Statistics, www.cdc.gov/nchs/)

Year	1965	1970	1975	1980	1985	1990	1995	2000	2005	2010	2015
Rate	19.4	18.4	14.8	15.9	15.6	16.4	14.8	14.4	14.0	13.0	12.4

a) Make a scatterplot and describe the general trend in *Birthrates*. (Enter *Year* as years since 1900: 65, 70, 75, etc.)
b) Find the equation of the regression line.
c) Check to see if the line is an appropriate model. Explain.
d) Interpret the slope of the line.
e) The table gives rates only at intervals. Estimate what the rate was in 1978.
f) In 1978, the birthrate was actually 15.0. How close did your model come?

g) The birthrate in 2020 was not yet available when this was written. Predict the birthrate in 2020 from your model. Comment on your faith in this prediction.
h) Predict the *Birthrate* for 2050. Comment on your faith in this prediction.

69. Climate change 2016 The earth's climate is getting warmer. The most common theory attributes the increase to an increase in atmospheric levels of carbon dioxide (CO_2), a greenhouse gas. The mean annual carbon dioxide in parts per million (ppm) measured at the top of a mountain and the mean annual air temperature over both land and sea across the globe, in degrees Celsius (C).

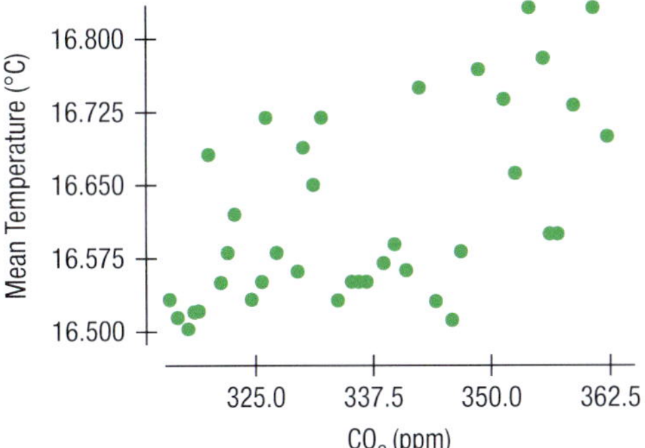

A regression predicting *Mean Temperature* anomaly from CO_2 produces the following output table:

Response variable is: Mean Temperature
R-squared = 33.7%
s = 0.0885

Variable	Coefficient
Intercept	15.308
CO_2 (ppm)	0.004

a) What is the correlation between CO_2 and *Mean Temperature*?
b) Explain the meaning of *R*-squared in this context.
c) Give the regression equation.
d) What is the meaning of the slope in this equation?
e) What is the meaning of the *y*-intercept of this equation?
f) Here is a scatterplot of the residuals vs. predicted values. Does this plot show evidence of the violation of any assumptions behind the regression? If so, which ones?

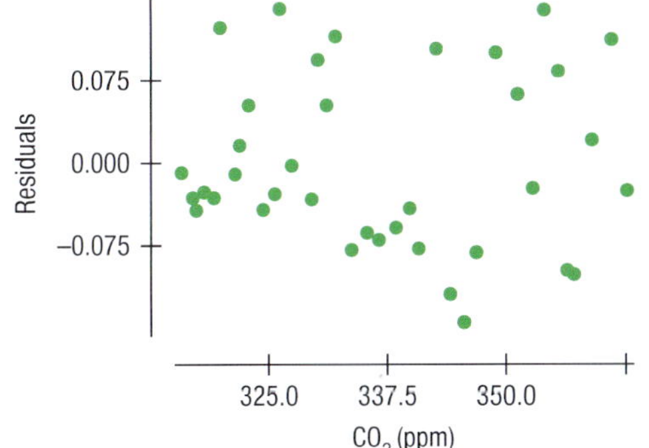

g) CO_2 levels will probably reach 369 ppm by 2050. What mean temperature *anomaly* does the regression predict for that concentration of CO_2?
h) Does the answer in part g mean that when the CO_2 level hits 369 ppm, the mean temperature anomaly will reach the predicted level? Explain briefly.

70. Climate change 2016, revisited In Exercise 69, we saw the relationship between CO_2 measured at Mauna Loa and average global temperature anomaly from 1959 to 2016. Here is a plot of average global temperatures plotted against the yearly final value of the Dow Jones Industrial Average for the same time period.

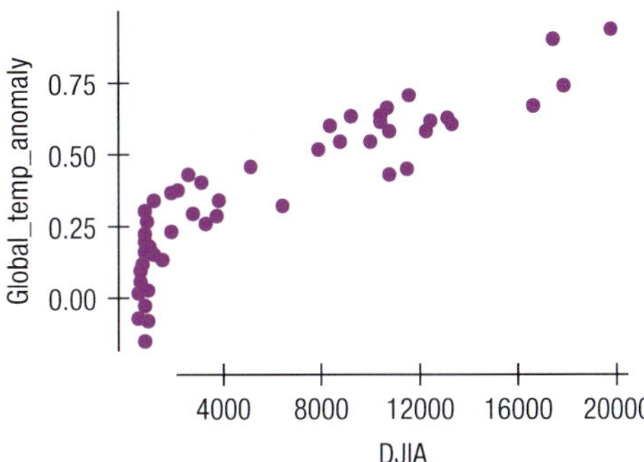

A regression produces the following output table (in part):

Response variable is: Global_temp_anomaly
R-squared = 80.2% s = 0.1229

Variable	Coefficient
Intercept	0.084216
DJIA	4.44440e-5

a) What is the correlation between the *DJIA* and *Temperature*?
b) Explain the meaning of *R*-squared in this context.
c) Give the regression equation.
d) What is the meaning of the slope in this equation?
e) What is the meaning of the intercept of this equation?
f) Here is a scatterplot of the residuals vs. predicted values. Does this plot show evidence of the violation of any assumptions behind the regression? If so, which ones?

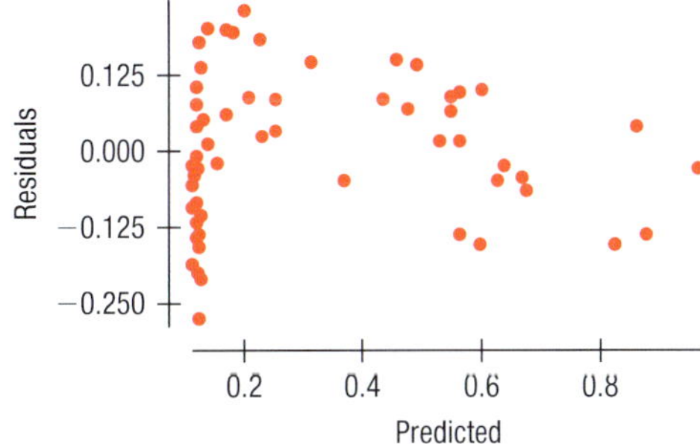

g) Suppose the Dow hits 25,000 in 2020. What mean Temperature does the regression predict for that level?
h) Does the answer in part g mean that when the Dow hits 25,000, the Temperature will reach the predicted level? Explain briefly.

71. Body fat It is difficult to determine a person's body fat percentage accurately without immersing him or her in water. Researchers hoping to find ways to make a good estimate immersed 20 male subjects, then measured their waists and recorded their weights shown in the table at the top of the next column.

a) Create a model to predict *%Body Fat* from *Weight*.
b) Do you think a linear model is appropriate? Explain.
c) Interpret the slope of your model.
d) Is your model likely to make reliable estimates? Explain.
e) What is the residual for a person who weighs 227 pounds and has 27% body fat?

Weight (lb)	Body Fat (%)	Weight (lb)	Body Fat (%)
177	5	187	11
179	21	241	19
202	15	176	23
158	6	170	8
193	23	244	39
193	32	158	11
204	34	217	28
172	22	161	12
185	25	148	9
186	31	218	27

T **72. Body fat again** Would a model that uses the person's *Waist* size be able to predict the *%Body Fat* more accurately than one that uses *Weight*? Using the data in Exercise 71, create and analyze that model.

T **73. Women's heptathlon revisited** We discussed the women's 2016 Olympic heptathlon in Chapter 5. Here are the results from the high jump, 800-meter run, and long jump for the 27 women who successfully completed all three events in the 2016 Olympics:

Name	High Jump	800 m	Long Jump
Evelis Aguilar	1.74	134.32	6.23
Nadine Broersen	1.77	137.55	6.15
Katerina Cachov	1.77	138.95	5.91
Vanessa Chefer	1.68	134.2	6.1
Ivona Dadic	1.77	135.64	6.05
Jessica Ennis-Hill	1.89	129.07	6.34
Alysbeth Felix	1.68	135.32	6.22
Laura Ikauniece-Admidina	1.77	129.43	6.12
Katarina Johnson-Thompson	1.98	130.47	6.51
Akela Jones	1.89	161.12	6.3
Hanna Kasyanova	1.77	136.58	5.88
Eliska Klucinova	1.8	142.81	6.08
Xenia Krizsan	1.77	133.46	6.08
Heather Miller-Koch	1.8	126.82	6.16
Antoinette Nana Djimou Ida	1.77	140.36	6.43
Barbara Nwaba	1.83	131.61	5.81
Jennifer Oeser	1.86	133.82	6.19
Claudia Rath	1.74	127.22	6.55
Yorgelis Rodríguez	1.861	134.65	6.25
Carolin Schafer	1.83	136.52	6.2
Brianne Theisen-Eaton	1.86	129.5	6.48
Nafissatou Thiam	1.98	136.54	6.58
Anouk Vetter	1.77	137.71	6.1
Nadine Visser	1.68	134.47	6.35
Kendell Williams	1.83	136.24	6.31
Sofia Yfantidou	1.65	150.08	5.51
Gyorgyi Zsivoczky-Farkas	1.86	131.76	6.31

Let's examine the association among these events. Perform a regression to predict long jump performance from 800 m run times.

a) What is the regression equation? What does the slope mean?
b) What percent of the variability in long jumps can be accounted for by differences in 800-m times?
c) Do good long jumpers tend to be fast runners? (Be careful—low times are good for running events and high distances are good for jumps.)
d) What does the residuals plot reveal about the model?
e) Do you think this is a useful model? Would you use it to predict long jump performance? (Compare the residual standard deviation to the standard deviation of the long jumps.)

T **74. Heptathlon revisited again** We saw the data for the women's 2016 Olympic heptathlon in Exercise 73. Are the two jumping events associated? Perform a regression of the long-jump results on the high-jump results.

a) What is the regression equation? What does the slope mean?
b) What percentage of the variability in long jumps can be accounted for by high-jump performances?
c) Do good high jumpers tend to be good long jumpers?
d) What does the residuals plot reveal about the model?
e) Do you think this is a useful model? Would you use it to predict long-jump performance? (Compare the residual standard deviation to the standard deviation of the long jumps.)

T **75. Hard water** In an investigation of environmental causes of disease, data were collected on the annual mortality rate (deaths per 100,000) for males in 20 large towns. In addition, the water hardness was recorded as the calcium concentration (parts per million, ppm) in the drinking water. The following display shows the relationship between *Mortality* and *Calcium* concentration for these towns:

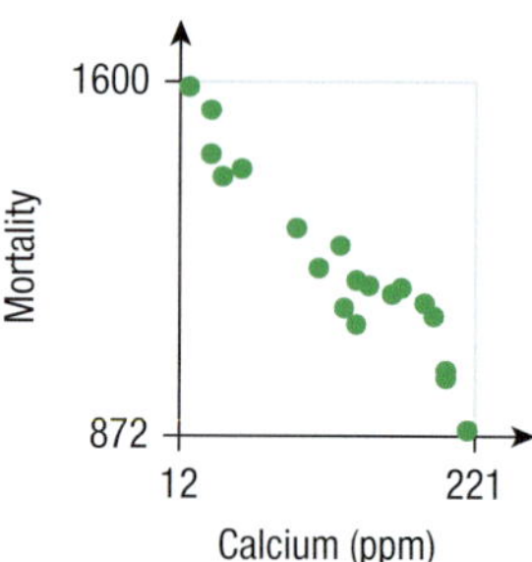

a) Describe what you see in this scatterplot, in context.
b) Here is the regression analysis of *Mortality* and *Calcium* concentration. What is the regression equation?

Dependent variable is Mortality
R-squared = 90.3%
s = 61.217

Variable	Coefficient	SE(Coeff)	t-Ratio	P-Value
Intercept	1579.734	31.53	50.1	<0.0001
Calcium	−2.811	0.22	−12.78	<0.0001

c) Interpret the slope and intercept of the line, in context.
d) The largest residual is 86.1. Explain what this value means.
e) The hardness of one town's municipal water is about 67 ppm of calcium. Use this equation to predict the mortality rate in that town.
f) Explain the meaning of *R*-squared in this situation.

T **76.** Gators Wildlife researchers monitor many wildlife populations by taking aerial photographs. Can they estimate the weights of alligators accurately from the air? Here is a regression analysis of the *Weight* of alligators (in pounds) and their *Length* (in inches) based on data collected about captured alligators.

Dependent variable is Weight
R-squared = 83.6%
s = 54.01

Variable	Coefficient	SE(Coeff)	t-Ratio	P-Value
Intercept	−393.3	47.53	−8.27	<0.0001
Length	5.9	0.5448	10.8	<0.0001

a) Did they choose the correct variable to use as the dependent variable and the predictor? Explain.
b) What is the correlation between an alligator's length and weight?
c) Plot the data. Is this regression appropriate? Explain.
d) Propose an alternative model for these data.
e) Which model would you use to predict alligator weight?
f) Do you think your predictions would be accurate? Explain.

77. Least squares Consider the four points (10,10), (20,50), (40,20), and (50,80). The least squares line is $\hat{y} = 7.0 + 1.1x$. Explain what "least squares" means, using these data as a specific example.

78. Least squares Consider the four points (200,1950), (400,1650), (600,1800), and (800,1600). The least squares line is $\hat{y} = 1975 - 0.45x$. Explain what "least squares" means, using these data as a specific example.

T **79.** Fuel economy 2016 revisited Exercise 41 of Chapter 6 looked at a sample of 35 vehicles to examine the relationship between gas mileage and engine displacement. The full dataset holds data on 1211 cars. How well did our sample of 35 represent the underlying relationship between displacement and fuel efficiency? If you can use the computer tools available, then draw many samples of 35 from the full dataset (without replacement), find regression slopes, and comment on how much the slope values differ. If you don't have access to the computer tools, here are the results we found for one such experiment. (Note: Your results of a new experiment should be different from these.) Here is a plot showing the 1000 slopes we found:

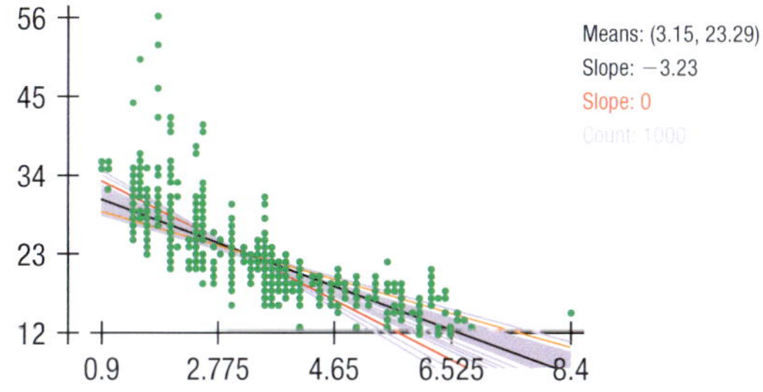

Here is a histogram of the slopes of those lines with the middle 95% of them selected:

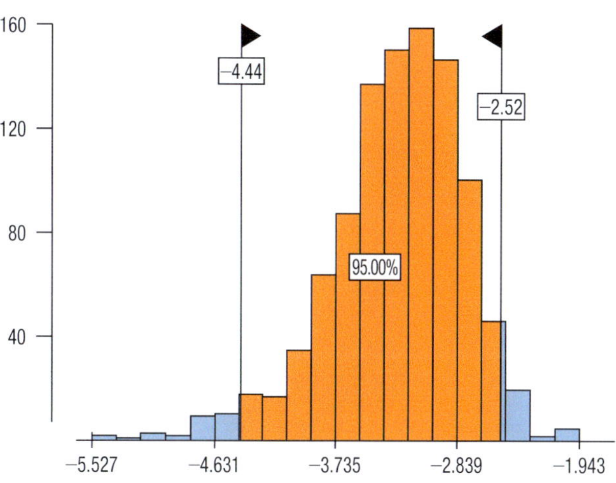

a) Describe how the sample slopes vary.
b) The sample in Chapter 6, Exercise 41 has a slope of −4.36. Is this typical of the slopes you found for such samples?
c) If you performed the experiment for yourself, how did your results differ from those reported here?

T **80.** Receivers 2015 The data file **Receivers 2015** holds information about the 488 NFL players who caught at least one pass during the 2015 football season. A typical 53-man roster has about 13 players who would be expected to catch passes (primarily wide receivers, tight ends, and running backs). We'll examine the relationship between the number of *Yards* gained during the season and the number of *Receptions*. If you have the computer tools available, then draw samples of 13 from the full data file (without replacement) and find the slope of the regression of *Yards* on *Receptions* for each sample. If you don't have access to the computer tools, here are the results we found for one such experiment. (Note: Your results of a new experiment should be different from these.) Here are plots showing the 500 slopes we found:

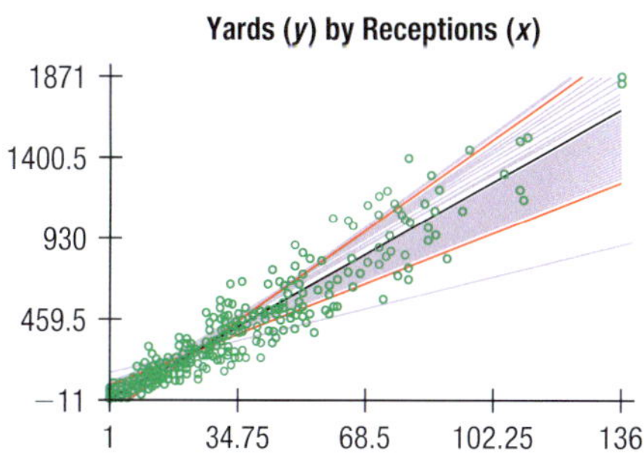

Here is a histogram of the slopes of those lines with the middle 95% of them selected:

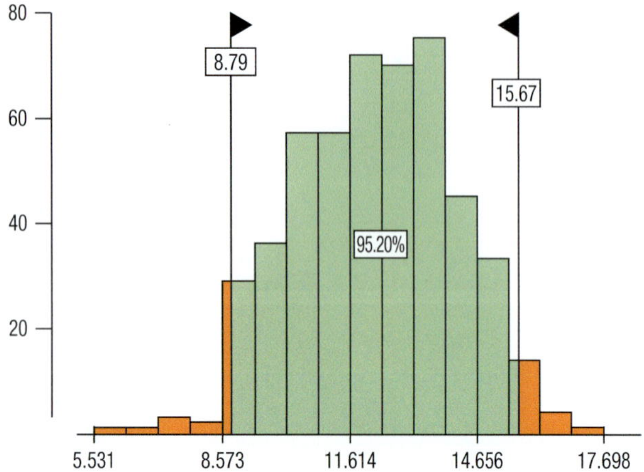

a) Describe how the sample slopes vary.
b) How would you interpret the meaning of a slope in the regression of *Yards* on *Receptions*?
c) Based on your answer in part b, what might you advise an NFL coach to expect from his 13 receivers in the course of the season?
d) If you performed the experiment for yourself, how did your results differ from those reported here?

JUST CHECKING

Answers

1. You should expect the price to be 0.76 standard deviations above the mean.
2. You should expect the price to be $2(0.76) = 1.52$ standard deviations below the mean.
3. The home is 1.5 standard deviations above the mean in size no matter how size is measured.
4. $0.76 \times \$77{,}000 = \$58{,}520$
5. $\$168{,}000 + 58{,}520 = \$226{,}520$
6. $8400 + 88.67 \times 2460 = \$226{,}528$. The difference is due to rounding error when the regression coefficients are kept to only 2 decimal places.
7. Differences in the size of houses account for about 57.8% of the variation in the house prices.
8. It's positive. The correlation and the slope have the same sign.
9. No, the standard deviation of the residuals is $53,790. We shouldn't be surprised by any residual smaller in magnitude than 2 standard deviations, and a residual of $50,000 is less than 1 standard deviation from 0.

8

Regression Wisdom

WHERE ARE WE GOING?

What happens when we fit a regression model to data that aren't straight? How bad will the predictions be? How can we tell if the model is appropriate? Questions like these are as important as fitting the model itself. In this chapter, we'll see how to tell whether a regression model is sensible and what to do if it isn't.

Regression is used every day throughout the world to predict customer loyalty, numbers of admissions at hospitals, sales of automobiles, and many other things. Because regression is so widely used, it's also widely abused and misinterpreted. This chapter presents examples of regressions in which things are not quite as simple as they may have seemed at first and shows how you can still use regression to discover what the data have to say.

8.1 Examining Residuals

Straight Enough Condition

We can't *know* whether the **Linearity Assumption** is true, but we can see if it's *plausible* by checking the **Straight Enough Condition**.

No regression analysis is complete without a display of the residuals to check that the linear model is reasonable. Because the residuals are what is "left over" after the model describes the relationship, they often reveal subtleties that were not clear from a plot of the original data. Sometimes these are additional details that help confirm or refine our understanding. Sometimes they reveal violations of the regression conditions that require our attention.

For example, the fundamental assumption in working with a linear model is that the relationship you are modeling is, in fact, linear. That sounds obvious, but you can't take it for granted. It may be hard to detect nonlinearity from the scatterplot you looked at before you fit the regression model. Sometimes you can see important features such as nonlinearity more readily when you plot the residuals.

Getting the "Bends": When the Residuals Aren't Straight

Jessica Meir and Paul Ponganis study emperor penguins at the Scripps Institution of Oceanography's Center for Marine Biotechnology and Biomedicine at the University of California at San Diego. Says Jessica:

> *Emperor penguins are the most accomplished divers among birds, making routine dives of 5–12 minutes, with the longest recorded dive over 27 minutes. These birds can also dive to depths of over 500 meters! Since air-breathing animals like penguins must hold their breath while submerged, the duration of any given dive depends on how much oxygen is in the bird's body at the beginning of the dive, how quickly that oxygen gets used, and the lowest level of oxygen the bird can tolerate. The rate of oxygen depletion is primarily determined by the penguin's heart rate. Consequently, studies of heart rates during dives can help us understand how these animals regulate their oxygen consumption in order to make such impressive dives.**

The researchers equip emperor penguins with devices that record their heart rates during dives. Here's a scatterplot of the *Dive Heart Rate* (beats per minute) and the *Duration* (minutes) of dives by these high-tech penguins. (Data in **Penguins**)

Figure 8.1
The scatterplot of *Dive Heart Rate* in beats per minute (bpm) vs. *Duration* (minutes) shows a strong, roughly linear, negative association.

The scatterplot has a consistent trend from the upper left to the lower right and a moderately strong negative association ($R^2 = 71.5\%$). The linear regression equation

$$\widehat{DiveHeartRate} = 96.9 - 5.47\ Duration$$

says that for longer dives, the average *Dive Heart Rate* is lower by about 5.47 beats per dive minute, starting from a value of 96.9 beats per minute.

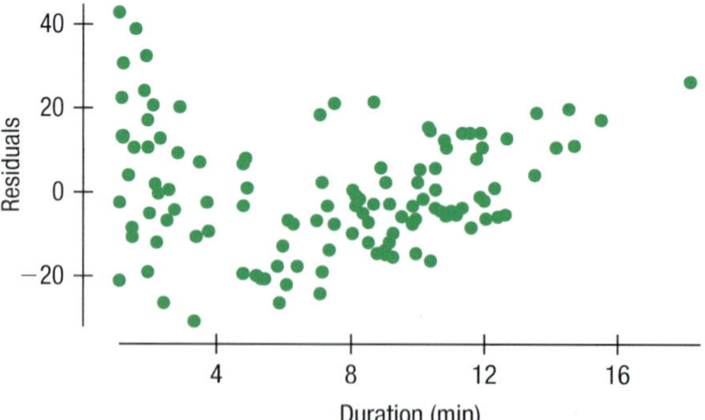

Figure 8.2
Plotting the residuals against *Duration* reveals a bend. It was also in the original scatterplot, but here it's easier to see.

The scatterplot of the residuals against *Duration* makes things clearer. The Linearity Assumption says we should see no pattern here, but the residual plot shows a strong bend, starting high on the left, dropping down in the middle of the plot, and rising again at the right. Graphs of residuals often reveal patterns such as this that were not readily apparent in the original scatterplot.

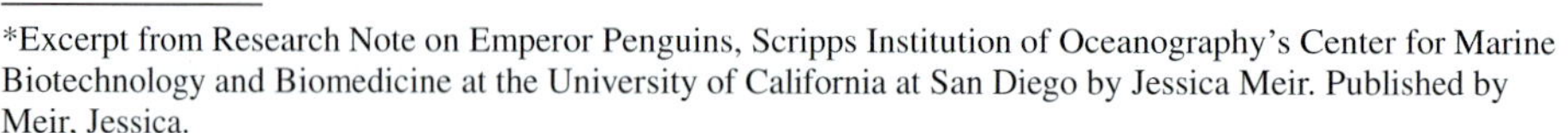

*Excerpt from Research Note on Emperor Penguins, Scripps Institution of Oceanography's Center for Marine Biotechnology and Biomedicine at the University of California at San Diego by Jessica Meir. Published by Meir, Jessica.

Now looking back at the original scatterplot, we can see that the scatter of points isn't really straight. There's a slight bend to that plot, but the bend is much easier to see in the residuals. Even though it means rechecking the Straight Enough Condition *after* you find the regression, it's always a good idea to check your scatterplot of the residuals for bends that you might have overlooked in the original scatterplot. We probably want to consider re-expressing one of the variables (as discussed in Section 6.4) to help straighten out the relationship.

Sifting Residuals for Groups

In the Step-By-Step analysis in Chapter 7, to predict *Calories* from *Sugar* content in breakfast cereals, we examined a scatterplot of the residuals. (Data in **Cereals**) Our first impression was that it had no particular structure. But let's look again.

Figure 8.3
A histogram of the breakfast cereal regression residuals from a regression of from Chapter 7 shows small modes both above and below the central large mode. These may be worth a second look.

Look at the histogram of the residuals. How would you describe its shape? It looks like there might be small modes on both sides of the central body of the data. One group of cereals seems to stand out (on the left) as having large negative residuals, with fewer calories than our regression model predicts, and another stands out (on the right) with large positive residuals. Whenever we suspect multiple modes, we ask whether there might be different subgroups in the data.

Let's look more carefully at those points in the residual plot (see Figure 8.4). Now we can see that those two groups do stand away from the central pattern in the scatterplot. The high-residual cereals are Just Right Fruit & Nut; Muesli Raisins, Dates & Almonds; Peaches & Pecans; Mueslix Crispy Blend; and Nutri-Grain Almond Raisin. Do these cereals seem to have something in common? They all present themselves as "healthy." This might be surprising, but in fact, "healthy" cereals often contain more fat, and therefore more calories, than we might expect from looking at their sugar content alone.

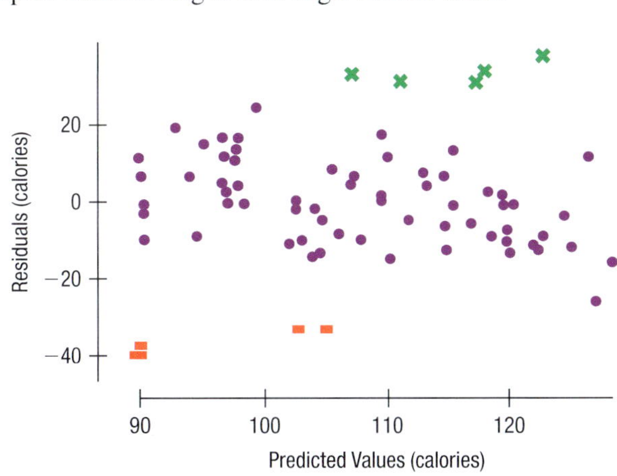

Figure 8.4
A scatterplot of the breakfast cereal regression residuals vs. predicted values (points have been jittered). The green "**x**" points are cereals whose calorie content is higher than the linear model predicts. The red "–" points show cereals with fewer calories than the model predicts. Is there something special about these cereals?

The low-residual cereals are Puffed Rice, Puffed Wheat, three bran cereals, and Golden Crisp. You might not have grouped these cereals together before. What they have in common is a low calorie count *relative to their sugar content*—even though their sugar contents are quite different.

These observations may not lead us to question the overall linear model, but they do help us understand that other factors may be part of the story. An examination of residuals often leads us to discover groups of observations that are different from the rest.

When we discover that there is more than one group in a regression, we may decide to analyze the groups separately, using a different model for each group. Or we can stick with the original model and simply note that there are groups that are a little different. Either way, the model will be wrong, but useful, so it will improve our understanding of the data.

Subsets

Here's an important unstated condition for fitting models: **All the data must come from the same population.**

Cereal manufacturers aim cereals at different segments of the market. Supermarkets and cereal manufacturers try to attract different customers by placing different types of cereals on certain shelves. Cereals for kids tend to be on the "kid's shelf," at their eye level. Toddlers wouldn't be likely to grab a box from this shelf and beg, "Mom, can we please get this All-Bran with Extra Fiber?"

Should we take this extra information into account in our analysis? Figure 8.5 shows a scatterplot of *Calories* and *Sugar*, colored according to the shelf on which the cereals were found and with a separate regression line fit for each. The top shelf is clearly different. We might want to report two regressions, one for the top shelf and one for the bottom two shelves.[1]

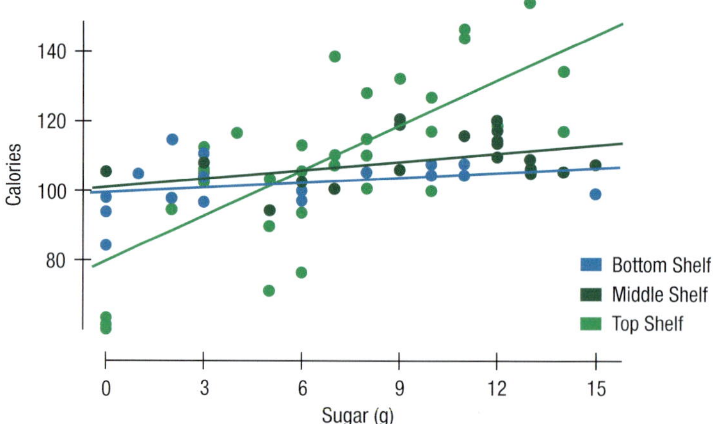

Figure 8.5
Calories and *Sugar* (y-values jittered) colored according to the shelf on which the cereal was found in a supermarket, with regression lines fit for each shelf individually. Do these data appear homogeneous? That is, do all the cereals seem to be from the same population of cereals? Or are there different kinds of cereals that we might want to consider separately?

8.2 Extrapolation: Reaching Beyond the Data

Linear models give a predicted value for each case in the data. Put a new x-value into the equation, and it gives a predicted value, $\hat{y}$, to go with it. But when the new x-value lies far from the data we used to build the regression, how trustworthy is the prediction?

The simple answer is that the farther the new x-value is from $\bar{x}$, the less trust we should place in the predicted value. Once we venture into new x territory, such a prediction is called an **extrapolation**. Extrapolations are dubious because they require the very questionable assumption that nothing about the relationship between x and y changes even at extreme values of x and beyond.

> "Prediction is difficult, especially about the future."
>
> —Niels Bohr, Danish physicist

Extrapolation is a good way to see just where the limits of our model may be. But it requires caution. When the x-variable is *Time*, extrapolation becomes an attempt to peer

[1]More complex models can take into account both sugar content and shelf information. This kind of *multiple regression* model is a natural extension of the model we're using here. We will see these models in Chapter 9.

into the future. People have always wanted to see into the future, and it doesn't take a crystal ball to foresee that they always will. In the past, seers, oracles, and wizards were called on to predict the future. Today, mediums, fortune-tellers, astrologers, and Tarot card readers still find many customers.

Those with a more scientific outlook may use a linear model as their digital crystal ball. Some physical phenomena do exhibit a kind of "inertia" that allows us to guess that current systematic behavior will continue, but be careful in counting on that kind of regularity in phenomena such as stock prices, sales figures, hurricane tracks, or public opinion.

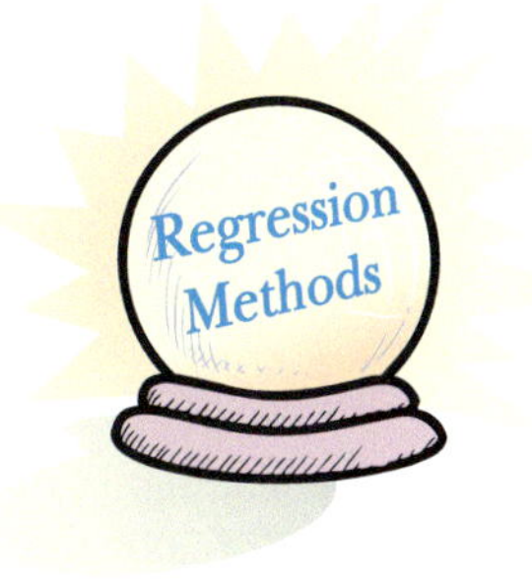

Extrapolating from current trends is so tempting that even professional forecasters sometimes expect too much from their models—and sometimes the errors are striking. In the mid-1970s, oil prices surged and long lines at gas stations were common. In 1970, oil cost about $17 a barrel (in 2016 dollars)—about what it had cost for 20 years or so. But then, within just a few years, the price surged to over $40. In 1975, a survey of 15 top econometric forecasting models (built by groups that included Nobel prize–winning economists) found predictions for 1985 oil prices that ranged from $300 to over $700 a barrel (in 2016 dollars). How close were these forecasts?

Here's a scatterplot (with regression line) of oil prices from 1971 to 1982 (in 2016 dollars). (Data in **Historical oil prices 2016**)

When the Data Are Years . . .

. . . we usually don't enter them as four-digit numbers. Here we used 0 for 1970, 10 for 1980, and so on. Or we may simply enter two digits, using 82 for 1982, for instance. Rescaling years like this often makes calculations easier and equations simpler. We recommend you do it, too. But be careful: If 1982 is 82, then 2004 is 104 (not 4), right?

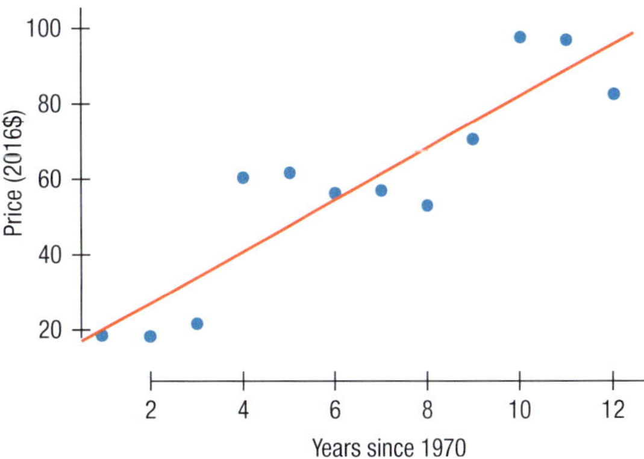

Figure 8.6
The price of a barrel of oil increased by about $6.89 per year from 1971 to 1982.

The regression model

$$\widehat{Price} = 13.11 + 6.89\ Years\ Since\ 1970$$

says that prices had been going up 6.89 dollars per year, or about $68.88 in 10 years. If you assume that they would *keep going up*, it's not hard to imagine almost any price you want. That would result in a forecast of $337 for the year 2017.

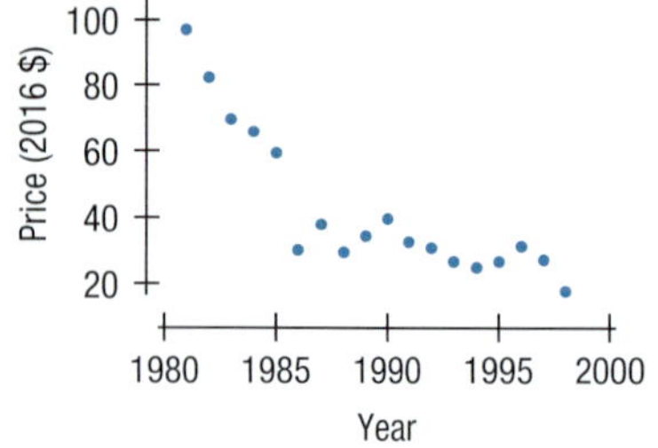

Figure 8.7
Oil prices from 1981 to 1998 *decreased* by about \$3.50 per year.

So, how did the forecasters do? Well, in the period from 1982 to 1998 oil prices didn't exactly continue that steady increase. In fact, they went down so much that by 1998, prices (adjusted for inflation) were the lowest they'd been since before World War II.

Not one of the experts' models predicted that.

Of course, these decreases clearly couldn't continue, or oil would have been free by the year 2000. The Energy Information Administration (EIA) has provided the U.S. Congress with both short- and long-term oil price forecasts every year since 1979. In that year they predicted that 1995 prices would be \$87 (in today's dollars). As we've just seen, they were closer to \$25. So, 20 years later, what did the EIA learn about forecasting? Let's see how well their 1998 and 1999 forecasts for the present have done. Here's a timeplot of the EIA's predictions and the actual prices (in 2016 dollars).

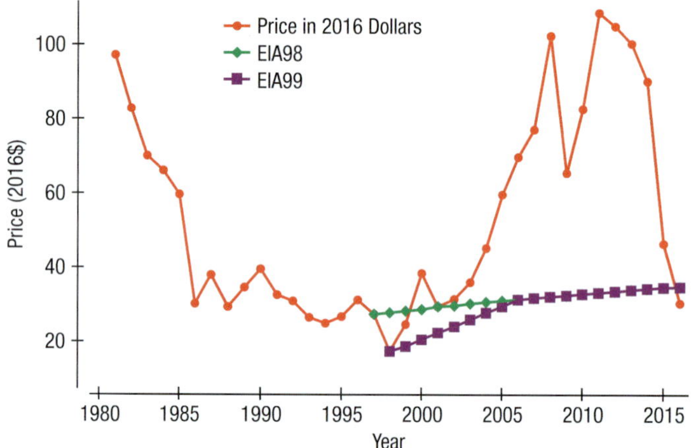

Figure 8.8
EIA forecasts and actual oil prices from 1981 to 2015. Ironically, the EIA forecast for 2016 is almost correct, but forecasters seem to have missed the run-up between 2005 and 2014.

Oops! They seemed to have missed the sharp run-up in oil prices leading up to the financial crisis of 2008, and then the sharp drop in prices at the beginning of 2009 back to about \$64 per barrel and the bounce back up to \$106 after that (and the subsequent drop, and the plunge to \$30, and . . .).

Where do you think oil prices will go in the next decade? *Your* guess may be as good as anyone's!

Of course, knowing that extrapolation requires thought and caution doesn't stop people. The temptation to see into the future is hard to resist. So our more realistic advice is this:

> *If you extrapolate into the future, at least don't believe blindly that the prediction will come true.*

EXAMPLE 8.1 Extrapolation: Reaching Beyond the Data

The U.S. Census Bureau (www.census.gov) reports the median age at first marriage for men and women. (Data in **Marriage age 2017**) Here's a regression of median *Age* (at first marriage) for men against *Year* (since 1890) at every census from 1890 to 1940:

R-squared = 92.6%
s = 0.2417

Variable	**Coefficient**
Intercept	26.07
Year	−0.04

The regression equation is

$$\widehat{Age} = 26.07 - 0.04\ Year.$$

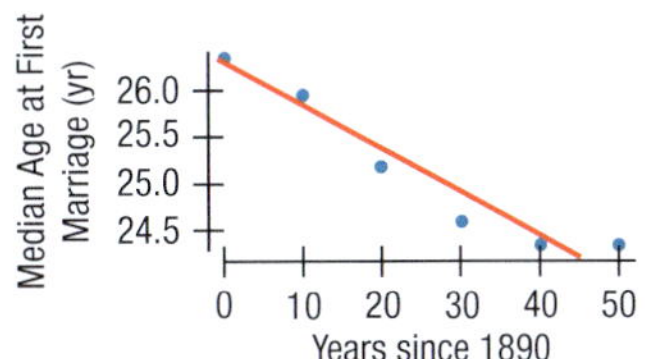

The median age at which men first married fell at the rate of about a year every 25 years from 1890 to 1940.

QUESTION: What would this model predict as the age at first marriage for men in the year 2000?

ANSWER: When *Year* counts from 0 in 1890, the year 2000 is "110." Substituting 110 for *Year*, we find that the model predicts a first marriage *Age* of $26.07 - 0.04 \times 110 = 21.7$ years old.

QUESTION: In the year 2017, the median age at first marriage for men was 29.5 years. What's gone wrong?

ANSWER: It is never safe to extrapolate beyond the data very far. The regression was calculated for years up to 1940. To see how absurd a prediction from that period can be when extrapolated into the present look at a scatterplot of the median *Age* at first marriage for men for all the data from 1890 to 2017:

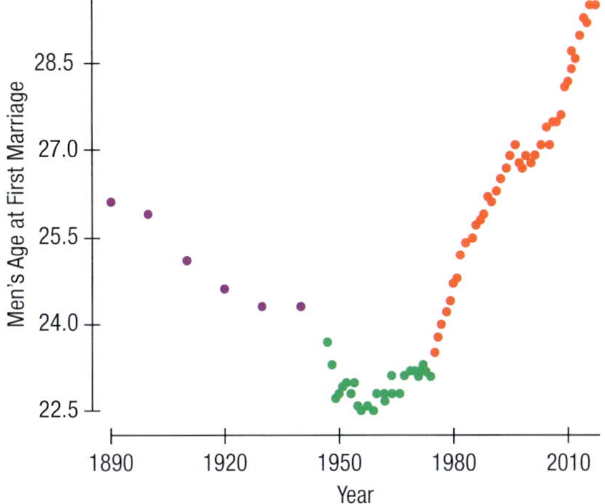

Median *Age* at first marriage (years of age) for men in the United States vs. *Year*. The regression model above was fit only to the first 50 years of the data (shown in purple), which looked nicely linear. But the linear pattern could not have continued, and in fact it changed in direction, steepness, and strength.

Now we can see why the extrapolation failed. Although the trend in *Age* at first marriage was linear and negative for the first part of the century, after World War II, it leveled off for about 30 years. Since 1980 or so, it has risen steadily. To characterize age at first marriage, we should probably treat these three time periods separately.

Predicting Changes

Not only is it incorrect and dangerous to interpret association as causation, but when using regression there's a more subtle danger. Never interpret a regression slope coefficient as predicting how y is likely to change if its x-value in the data were changed. Here's an example: In Chapter 7, we found a regression model relating calories in breakfast cereals to their sugar content as

$$\widehat{Calories} = 89.6 + 2.49\,Sugar.$$

It might be tempting to interpret this slope as implying that adding 1 gram of sugar is expected to *lead to* a change of 2.49 calories. Predicting a change of b_1 in y for a 1 unit change in x might be true in some instances, but it isn't a consequence of the regression. In fact, adding a gram of sugar to a cereal increases its calorie content by 3.90 calories—that's the calorie content of a gram of sugar.

A safer interpretation of the slope is that cereals that have a gram more sugar in them tend to have about 2.49 more calories per serving. That is, the regression model describes how the cereals differ, but does not tell us how they might change if circumstances were different.

We've warned against using a regression model to predict far from the x-values in the data. You should also be cautious about using a regression model to predict what might happen to cases in the regression if they were changed. Regression models describe the data as they are, not as they might be under other circumstances.

8.3 Outliers, Leverage, and Influence

The outcome of the 2000 U.S. presidential election was determined in Florida amid much controversy. Even years later, historians continue to debate who really received the most votes. The main race was between George W. Bush and Al Gore, but two minor candidates played a significant role. To the political right of the major party candidates was Pat Buchanan, while to the political left was Ralph Nader. Generally, Nader earned more votes than Buchanan throughout the state. We would expect counties with larger vote totals to give more votes to each candidate. Here's a regression relating *Buchanan*'s vote totals by county in the state of Florida to *Nader*'s: (Data in **Election 2000**)

Dependent variable is Buchanan
R-squared = 42.8%

Variable	**Coefficient**
Intercept	50.3
Nader	0.14

The regression model:

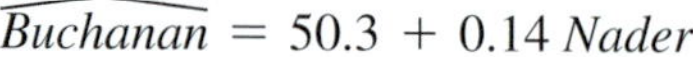

$$\widehat{Buchanan} = 50.3 + 0.14\,Nader$$

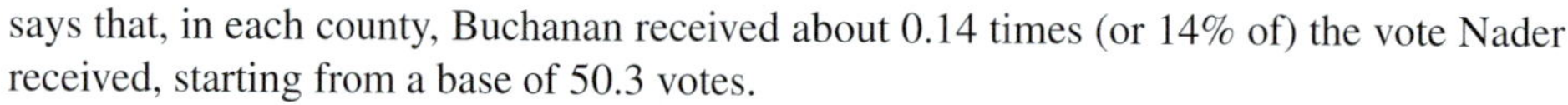

says that, in each county, Buchanan received about 0.14 times (or 14% of) the vote Nader received, starting from a base of 50.3 votes.

This seems like a reasonable regression, with an R^2 of almost 43%. But we've violated all three Rules of Data Analysis by going straight to the regression table without making a picture.

Here's a scatterplot that shows the vote for Buchanan in each county of Florida plotted against the vote for Nader. The striking **outlier** is Palm Beach County.

> Nature is nowhere accustomed more openly to display her secret mysteries than in cases where she shows traces of her workings apart from the beaten path.
>
> —William Harvey (1657)

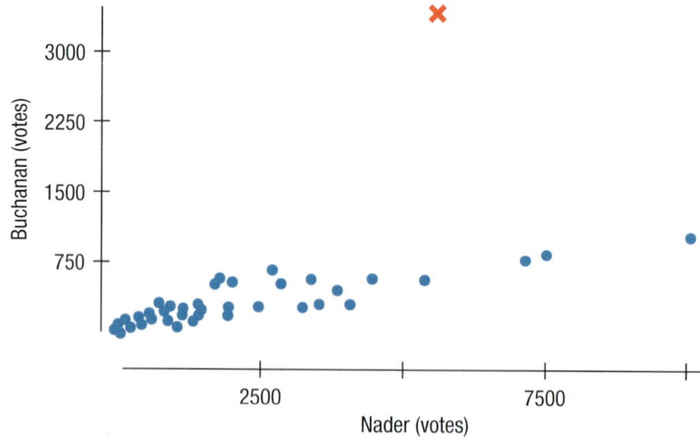

Figure 8.9
Votes received by Buchanan against votes for Nader in all Florida counties in the presidential election of 2000. The red "x" point is Palm Beach County, home of the "butterfly ballot."

The so-called "butterfly ballot," used only in Palm Beach County, was a source of controversy. It has been claimed that the format of this ballot confused voters so that some who intended to vote for the Democrat, Al Gore, punched the wrong hole next to his name and, as a result, voted for Buchanan.

Figure 8.10
The red line shows the effect that one unusual point can have on a regression.

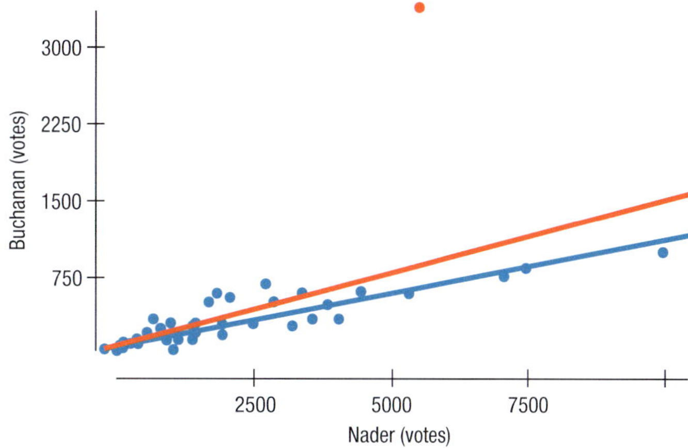

> Give me a place to stand and I will move the Earth.
>
> —Archimedes (287–211 BCE)

> For whoever knows the ways of Nature will more easily notice her deviations; and, on the other hand, whoever knows her deviations will more accurately describe her ways.
>
> —Francis Bacon (1561–1626)

The scatterplot shows a strong, positive, linear association, and one striking point. With Palm Beach removed from the regression, the R^2 jumps from 42.8% to 82.1% and the slope of the line changes to 0.1, suggesting that Buchanan received only about 10% of the vote that Nader received. Palm Beach County now stands out, not as a Buchanan stronghold, but rather as a clear violation of the model that begs for explanation.

One of the great values of models is that they help us to see when and how data values are unusual. In regression, a point can stand out in two different ways. First, a data value can have a large residual, as Palm Beach County does in this example. Because they seem to be different from the other cases, points whose residuals are large always deserve special attention.

A data point can also be unusual if its x-value is far from the mean of the x-values. Such a point is said to have high **leverage**. The physical image of a lever is exactly right. We know the line must pass through $(\bar{x}, \bar{y})$, so you can picture that point as the fulcrum of the lever. Just as sitting farther from the hinge on a see-saw gives you more leverage to pull it your way, points with values far from $\bar{x}$ pull more strongly on the regression line.

A point with high leverage has the potential to change the regression line. But it doesn't always use that potential. If the point lines up with the pattern of the other points, then including it doesn't change our estimate of the line. How can you tell if a high-leverage point actually changes the model? Just fit the linear model twice, both with and without the point in question. We say that a point is **influential** if omitting it from the analysis changes the model enough to make a meaningful difference.[2]

Influence depends on both leverage and residual; a case with high leverage whose y-value sits right on the line fit to the rest of the data is not influential. Removing that case won't change the slope, even if it does affect R^2. A case with modest leverage but a very large residual (such as Palm Beach County) can be influential. Of course, if a point has enough leverage, it can pull the line right to it. Then it's highly influential, but its residual is small. The only way to be sure is to fit both regressions.

Unusual points in a regression often tell us a great deal about the data and the model. We face a challenge: The best way to identify unusual points is against the background of a model, but a model dominated by a single case is unlikely to be useful for identifying unusual cases. (That insight's at least 400 years old. See the quote in the margin.) You can set aside cases and discuss what the model looks like with and without them, but arbitrarily deleting cases can give a false sense of how well the model fits the data. Your goal should be understanding the data, not making R^2 as big as you can.

In 2000, George W. Bush won Florida (and thus the presidency) by only a few hundred votes, so Palm Beach County's residual is big enough to be meaningful. It's the rare unusual point that determines a presidency, but all are worth examining and trying to understand.

[2]How big a change would make a difference depends on the circumstances and is a judgment call. Note that some texts use the term *influential point* for any observation that influences the slope, intercept, or R^2. We'll reserve the term for points that influence the slope.

A point with so much influence that it pulls the regression line close to it can make its residual deceptively small. Influential points like that can have a shocking effect on the regression. Here's a plot of *IQ* against *Shoe Size*, again from the fanciful study of intelligence and foot size in comedians we saw in Chapter 6. The linear regression output shows

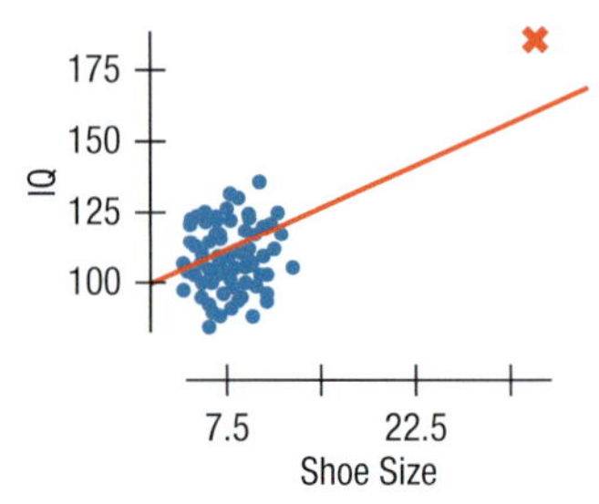

Figure 8.11
Bozo's extraordinarily large shoes give his data point high leverage in the regression. Wherever Bozo's *IQ* falls, the regression line will follow.

Dependent variable is IQ
R-squared = 24.8

Variable	Coefficient
Intercept	93.3265
Shoe size	2.08318

Although this is a silly example, it illustrates an important and common potential problem: Almost all of the variance accounted for ($R^2 = 24.8\%$) is due to *one* point, namely, Bozo. Without Bozo, there is almost no correlation between *Shoe Size* and *IQ*. Look what happens to the regression when we take him out:

Dependent variable is IQ
R-squared 0.7%

Variable	Coefficient
Intercept	105.458
Shoe size	−0.460194

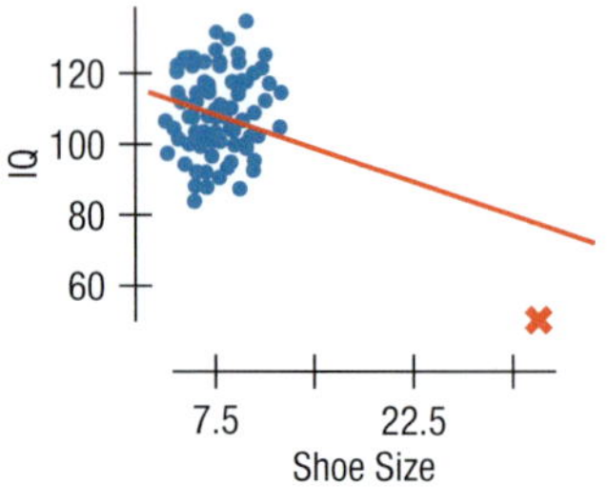

Figure 8.12
If Bozo's IQ were low, the regression slope would change from positive to negative. A single influential point can change a regression model drastically.

The R^2 value is now 0.7%—a very weak linear relationship (as one might expect!). One single point exhibits a great influence on the regression analysis.

What would have happened if Bozo hadn't shown his comic genius on IQ tests? Suppose his measured *IQ* had been only 50. The slope of the line would then drop from 0.96 IQ point/shoe size to −0.69 IQ point/shoe size. No matter where Bozo's *IQ* is, the line follows it because his *Shoe Size*, being so far from the mean *Shoe Size*, makes Bozo a high-leverage point.

This example is far-fetched, but similar situations occur in real life. For example, a regression of sales against floor space for hardware stores that looked primarily at small-town businesses could be dominated in a similar way if Lowe's was included.

Warning

Influential points can hide in plots of residuals. Points with high leverage pull the line close to them, so they often have small residuals. You'll see influential points more easily in scatterplots of the original data or by finding a regression model with and without the points.

JUST CHECKING

Each of these scatterplots shows an unusual point. For each, tell whether the point is a high-leverage point, would have a large residual, or is influential.

1.

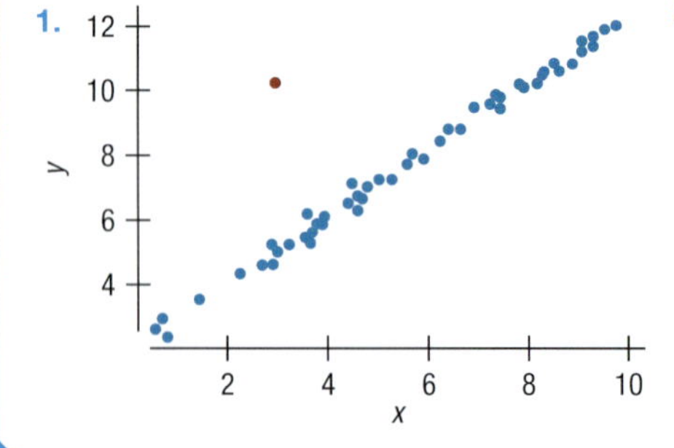

2.

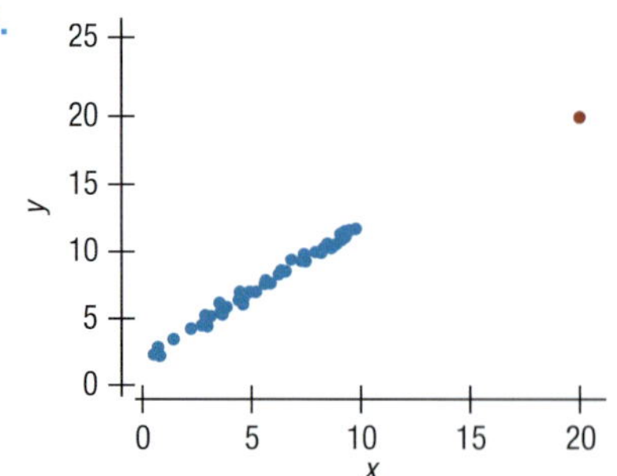

3.

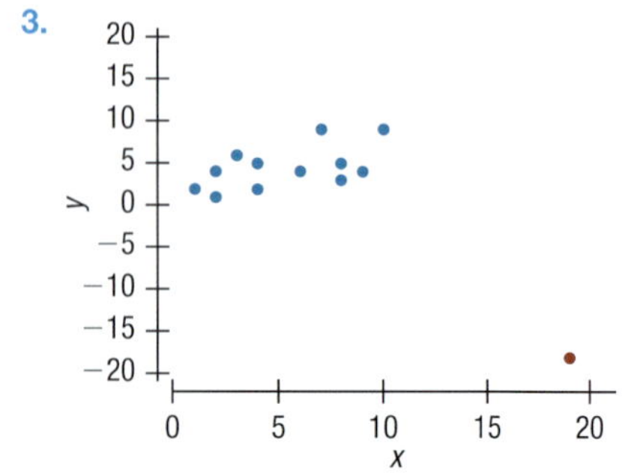

8.4 Lurking Variables and Causation

Causal Interpretations

One common way to interpret a regression slope is to say that "a change of 1 unit in *x* results in a change of b_1 units in *y*." This way of saying things encourages causal thinking. Beware.

In Chapter 6, we tried to make it clear that no matter how strong the correlation is between two variables, it doesn't show that one variable causes the other. Putting a regression line through a cloud of points just increases the temptation to think and to say that the *x*-variable *causes* the *y*-variable. Just to make sure, let's repeat the point again: No matter how strong the association, no matter how large the R^2 value, no matter how straight the line, you cannot conclude from a regression alone that one variable *causes* the other. There's always the possibility that some third variable is driving both of the variables you have observed. With observational data, as opposed to data from a designed experiment, there is no way to be sure that such a **lurking variable** is not responsible for the apparent association.[3]

Here's an example: The scatterplot shows the *Life Expectancy* (average of men and women, in years) for each of 40 countries, plotted against the square root of the number of *Doctors per person* in the country. (The square root helps to make the relationship satisfy the Straight Enough Condition, as we saw back in Chapter 6. Data in **Doctors and life expectancy**)

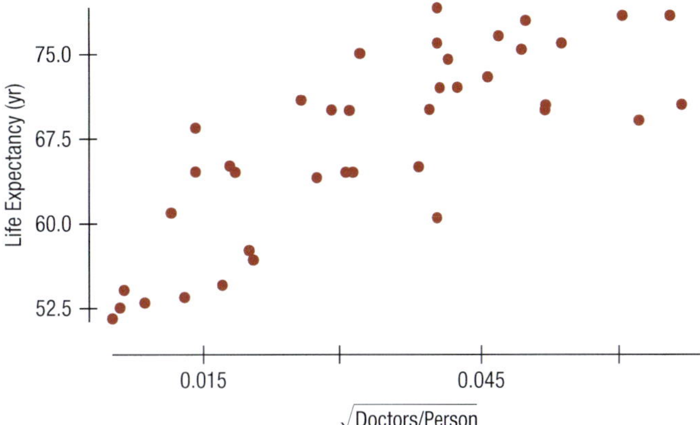

Figure 8.13
The relationship between *Life Expectancy* (years) and availability of *Doctors* (measured as $\sqrt{(Doctors/Person)}$ for countries of the world is strong, positive, and linear.

The strong positive association ($R^2 = 62.9\%$) seems to confirm our expectation that more *Doctors per Person* improves health care, leading to longer lifetimes and a greater *Life Expectancy*. The strength of the association would *seem* to argue that we should send more doctors to developing countries to increase life expectancy.

That conclusion is about the consequences of a change. Would sending more doctors increase life expectancy? Specifically, do doctors *cause* greater life expectancy? Perhaps, but these are observed data, so there may be another explanation for the association.

In Figure 8.14, the similar-looking scatterplot's *x*-variable is the square root of the number of *Televisions per Person* in each country. The positive association in this scatterplot

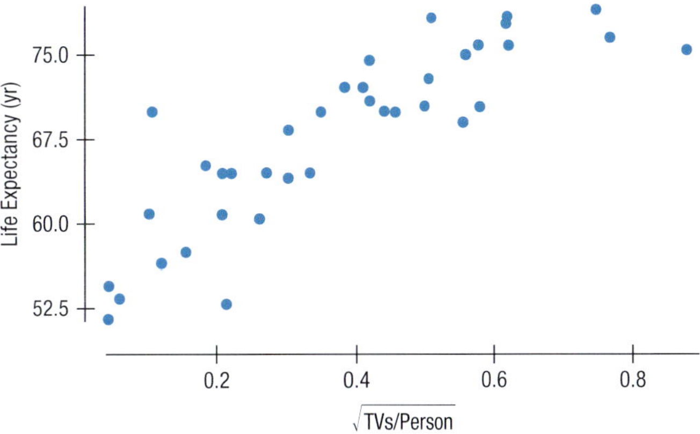

Figure 8.14
To increase life expectancy, don't send doctors, send TVs; they're cheaper and more fun. Or maybe that's not the right interpretation of this scatterplot of *Life Expectancy* against availability of TVs (as $\sqrt{TVs/person}$).

[3]Chapter 11 discusses observational data and experiments at greater length.

is even *stronger* than the association in the previous plot ($R^2 = 72.5\%$). We can fit the linear model, and quite possibly use the number of TVs as a way to predict life expectancy. Should we conclude that increasing the number of TVs actually extends lifetimes? If so, we should send TVs instead of doctors to developing countries. Not only is the correlation with life expectancy higher, but TVs are much cheaper than doctors and possibly more fun.

What's wrong with this reasoning? Maybe we were a bit hasty earlier when we concluded that doctors *cause* longer lives. Maybe there's a lurking variable here. Countries with higher standards of living have both longer life expectancies *and* more doctors (and more TVs). Could higher living standards cause changes in the other variables? If so, then improving living standards might be expected to prolong lives, increase the number of doctors, and increase the number of TVs.

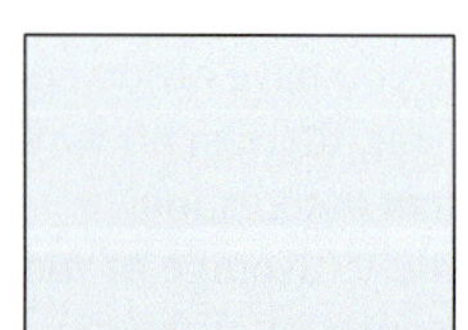

To think about lurking variables you must think "outside the box." What variables are *not* in your data that ought to be?[4]

From this example, you can see how easy it is to fall into the trap of mistakenly inferring causality from a regression. For all we know, doctors (or TVs!) *do* increase life expectancy. But we can't tell that from data like these, no matter how much we'd like to. Resist the temptation to conclude that x causes y from a regression, no matter how obvious that conclusion seems to you.

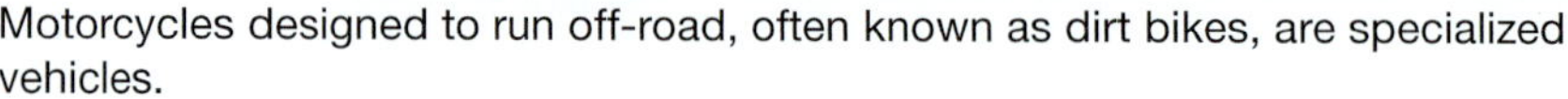

EXAMPLE 8.2 Using Several of These Methods Together

Motorcycles designed to run off-road, often known as dirt bikes, are specialized vehicles.

We have data on 114 dirt bikes. (Data in **Dirt bikes 2014**) Some cost as little as $1399, while others are substantially more expensive. Let's investigate how the size and type of engine contribute to the cost of a dirt bike. As always, we start with a scatterplot.

Here's a scatterplot of the manufacturer's suggested retail price (*MSRP*) in dollars against the engine *Displacement* (in cubic centimeters), along with a regression analysis:

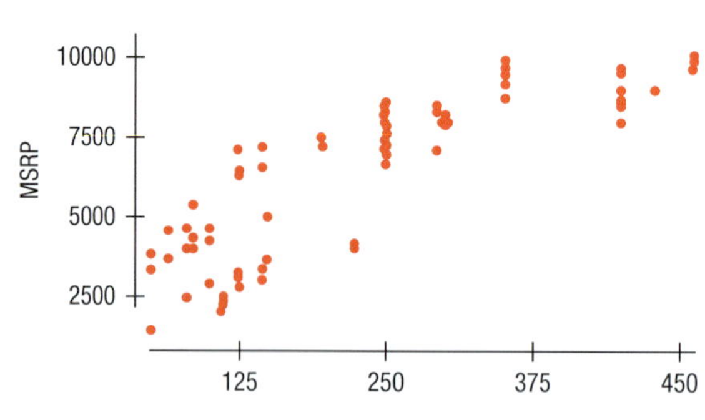

Dependent variable is MSRP
R-squared = 70.4%   s = 1335

Variable	**Coefficient**
Intercept	2856.84
Displacement	15.2567

QUESTION: What do you see in the scatterplot?

ANSWER: There is a strong positive association between the engine displacement of dirt bikes and the manufacturer's suggested retail price. There is some curvature and

[4] If you wonder about the graphic: this is one of the boxes you should think outside of. There are many others. Stay alert for them.

there seems to be a group of points with small displacement whose prices are substantially below the rest but show a similar slope.

Here's a scatterplot of the residuals:

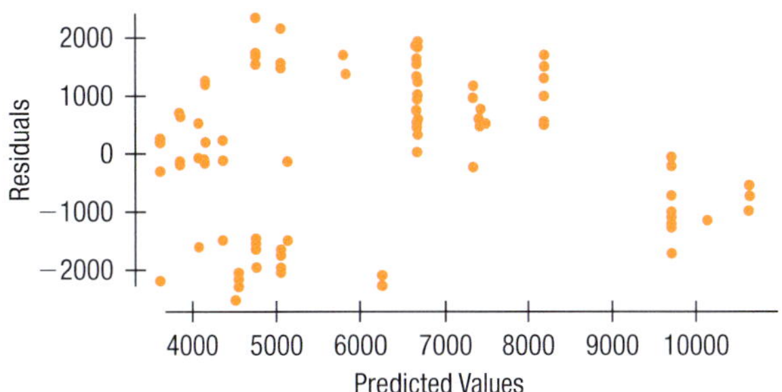

QUESTION: What do you see in the residuals plot?

ANSWER: The points at the far right and the bottom left don't fit well with the other dirt bikes. Overall, there appears to be a bend in the relationship, so a linear model may not be appropriate.

Let's try a re-expression. Here's a scatterplot showing *MSRP* against the cube root of *Displacement* to make the relationship closer to straight. (Since displacement is measured in cubic centimeters, its cube root has the simple units of centimeters.) In addition, we've colored the plot according to the cooling method used in the bike's engine: liquid or air. Each group is shown with its own regression line, as we did for the cereals on different shelves.

QUESTION: What does this plot say about dirt bikes?

ANSWER: There appears to be a positive, linear relationship between *MSRP* and the cube root of *Displacement*. In general, the larger the engine a bike has, the higher the suggested price. Liquid-cooled dirt bikes, however, typically cost more than air-cooled bikes with comparable displacement.

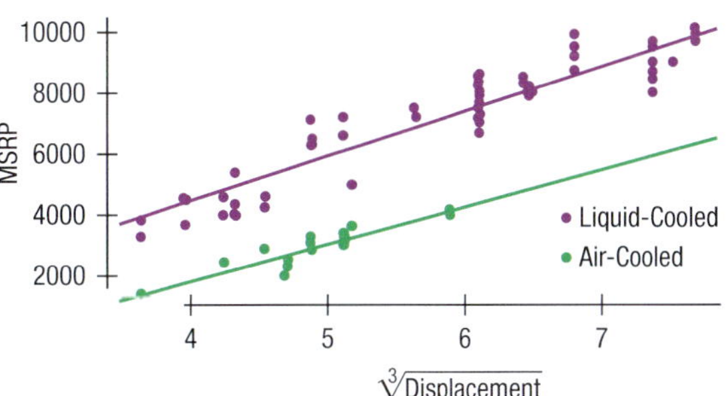

8.5 Working with Summary Values

Scatterplots of statistics summarized over groups tend to show less variability than we would see if we measured the same variable on individuals. This is because the summary statistics themselves vary less than the data on the individuals do—a fact we will make more specific in coming chapters.

In the life expectancy and TVs example, we have no good measure of exposure to doctors or to TV on an individual basis. But if we did, we should expect the scatterplot to show more variability and the corresponding R^2 to be smaller. The bottom line is that you should be a bit suspicious of conclusions based on regressions of summary data. They may look better than they really are.

In Chapter 6, we looked at the heights and weights of individual students. (Data in **Heights and weights**) There, we saw a correlation of 0.644, so R^2 is 41.5%.

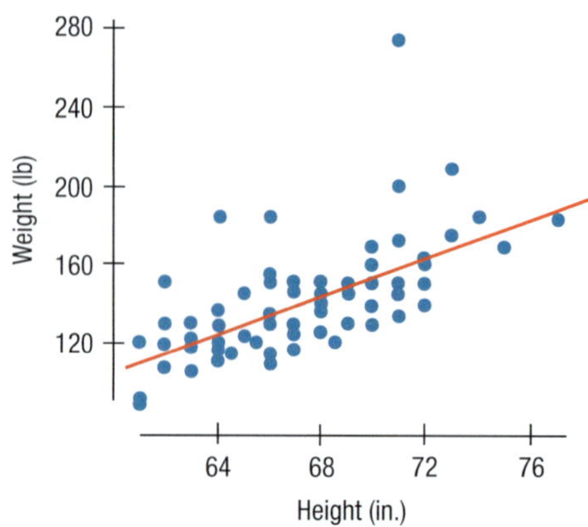

Figure 8.15
Weight (lb) against *Height* (in.) for a sample of men. There's a strong, positive, linear association.

Suppose, instead of data on individuals, we knew only the mean weight for each height value. The scatterplot of mean weight by height would show less scatter. And R^2 would increase to 80.1%.

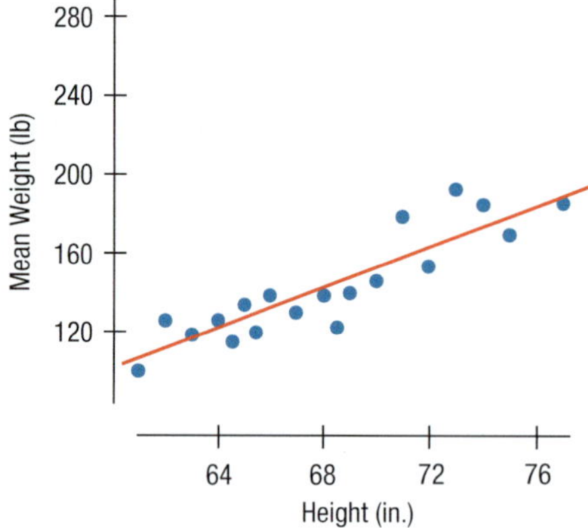

Figure 8.16
Mean Weight (lb) shows a stronger linear association with *Height* than do the weights of individuals. Means vary less than individual values.

Scatterplots of summary statistics show less scatter than the baseline data on individuals and can give a false impression of how well a line summarizes the data. There's no simple correction for this phenomenon. Once we're given summary data, there's no simple way to get the original values back.

*8.6 Straightening Scatterplots—The Three Goals

WHO	38 cars
WHAT	Weight, MPG
HOW	MPG measured on a track. Note: not EPA values
WHY	Evaluation of performance for ratings

We introduced the Ladder of Powers in Chapter 6 as a guide to finding a good re-expression. In this chapter we'll expand the discussion and give examples to show you how to extend your ability to analyze data by using re-expressions.

We know from common sense and from physics that heavier cars need more fuel, but exactly how does a car's weight affect its fuel efficiency? Here are the scatterplot of *Weight* (in pounds) and *Fuel Efficiency* (in miles per gallon) for 38 cars, and the residuals plot. (Data in **Fuel efficiency**)

Hmm… . Even though R^2 is 81.6%, the residuals don't show the random scatter we were hoping for. The shape is clearly bent. Looking back at the first scatterplot, you can probably see the slight bending. Think about the regression line through the points. How heavy would a car have to be to have a predicted gas mileage of 0? It looks like the *Fuel Efficiency* would go negative at about 6000 pounds. The Maybach 62 was a luxury car

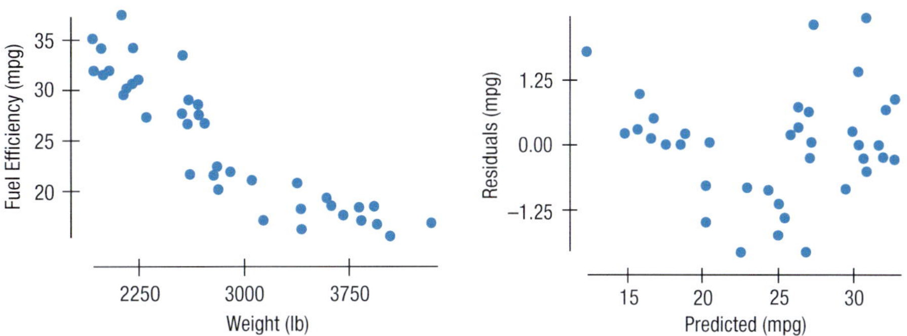

Figure 8.17

Fuel Efficiency (mpg) vs. *Weight* for 38 cars as reported by *Consumer Reports.* The scatterplot shows a negative direction, roughly linear shape, and strong relationship. However, the residuals from a regression of *Fuel Efficiency* on *Weight* reveal a bent shape when plotted against the predicted values. Looking back at the original scatterplot, you may be able to see the bend.

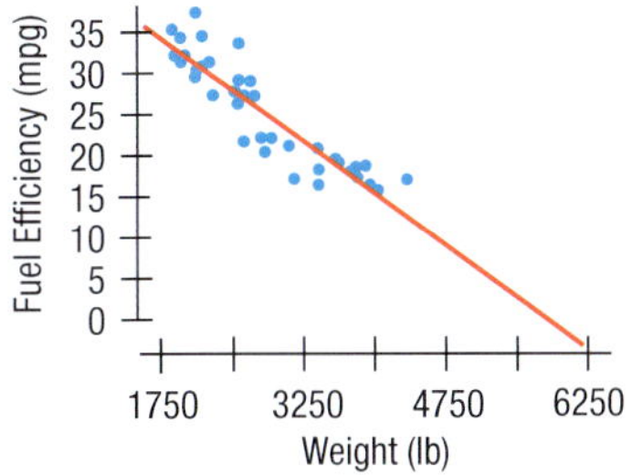

Figure 8.18

Extrapolating the regression line gives an absurd answer for vehicles that weigh as little as 6000 pounds.

produced by Daimler AG to compete with Rolls-Royce and Bentley. It was the heaviest "production car" in the world at 6184 lb. Although not known for its fuel efficiency (rated at 10 mpg city), it did get more than the *minus* 2.6 mpg predicted from the model. Sadly after selling only 2100 of them in 10 years, Daimler stopped production in 2013. The 2012 model sold (without extra features) for $427,700. Extrapolation always requires caution, but it can go dangerously wrong when your model is wrong, because wrong models tend to do even worse the farther you get from the middle of the data.

The bend in the relationship between *Fuel Efficiency* and *Weight* is the kind of failure to satisfy the conditions for an analysis that we can repair by re-expressing the data. Instead of looking at miles per gallon, we could take the reciprocal and work with gallons per hundred miles.[5]

> *"GALLONS PER HUNDRED MILES—WHAT AN ABSURD WAY TO MEASURE FUEL EFFICIENCY! WHO WOULD EVER DO IT THAT WAY?"*
>
> Not all re-expressions are easy to understand, but in this case the answer is "Everyone except U.S. drivers." Most of the world measures fuel efficiency in liters per 100 kilometers (L/100 km). This is the same reciprocal form (fuel amount per distance driven) and differs from gallons per 100 miles only by a constant multiple of about 2.38. It has been suggested that most of the world thinks, "I've got to go 100 km; how much gas do I need?" But Americans think, "I've got 10 gallons in the tank. How far can I drive?" In much the same way, re-expressions "think" about the data differently but don't change what they mean.

The direction of the association is positive now, since we're measuring gas consumption and heavier cars consume more gas per mile. The relationship is much straighter, as we can see from a scatterplot of the regression residuals.

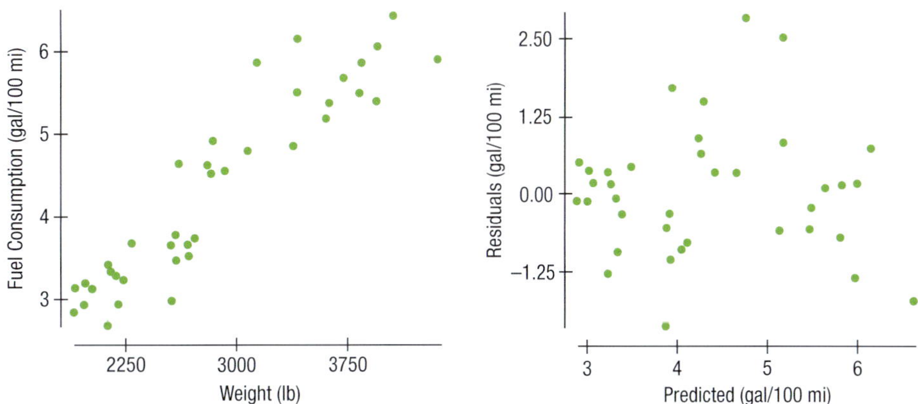

Figure 8.19

The reciprocal $(1/y)$ is measured in gallons per mile. Gallons per 100 miles gives more meaningful numbers. The reciprocal is more nearly linear against *Weight* than the original variable, but the re-expression changes the direction of the relationship. The residuals from the regression of *Fuel Consumption* (gal/100 mi) on *Weight* show less of a pattern than before.

[5]Multiplying by 100 to get gallons per 100 miles simply makes the numbers easier to think about: You might have a good idea of how many gallons your car needs to drive 100 miles, but probably a much poorer sense of how much gas you need to go just 1 mile.

This is more the kind of boring residuals plot (no direction, no particular shape, no outliers, no bends) that we hope to see, so we have reason to think that the Straight Enough Condition is now satisfied. Now here's the payoff: What does the reciprocal model say about the Maybach? The regression line fit to *Fuel Consumption* vs. *Weight* predicts somewhere near 9.26 for a car weighing 6184 pounds. What does this mean? It means the car is predicted to use 9.26 gallons for every 100 miles, or in other words,

$$\frac{100\ miles}{9.26\ gallons} = 10.8\ mpg.$$

That's a much more reasonable prediction and very close to the reported value of 10 miles per gallon (of course, *your* mileage may vary ...).

Goals of Re-Expression for Regression

Re-expressing data for regression can help make our data more suitable for regression in several ways. Here are three goals for re-expressing data for regression.

Goal 1

Make the form of a scatterplot more nearly linear. Scatterplots with a straight form are easier to model. We saw an example of scatterplot straightening in Chapter 6. The greater value of re-expression to straighten a relationship is that we can fit a linear model once the relationship is straight.

Physical therapists measure a patient's manual dexterity with a simple task. The patient picks up small cylinders from a 4 × 4 frame with one hand, flips them over (still with one hand), and replaces them in the frame. The task is timed for all 16 cylinders. Researchers used this tool to study how dexterity improves with age in children.[6] Figure 8.20 shows the relationship. (Data in **Hand dexterity**)

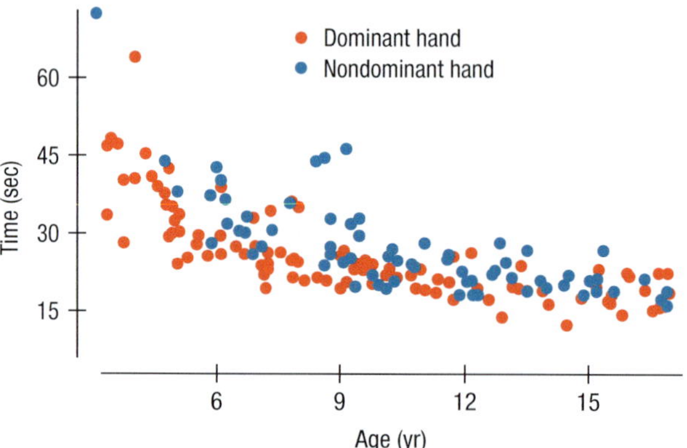

Figure 8.20

The relationship between *Time* and *Age* is curved, and appears to be more variable on the left than on the right. The two colors plot dominant and nondominant hands.

Often data recorded as time to complete a task can benefit from being re-expressed as a reciprocal. Instead of seconds/task, the re-expressed variable is in units of tasks/second—essentially a measure of speed. When *Time* is re-expressed as 1/*Time*, the result is Figure 8.21.

[6]"Hand dexterity in children: Administration and normative values of the Functional Dexterity Test (FDT), Gloria R. Gogola, MD, Paul F. Velleman, PhD, Shuai Xu, BS, MS, Adrianne M. Morse, BA, Barbara Lacy, BS, Dorit Aaron, MA OTR CHT FAOTA, *J Hand Surg Am.* 2013 Dec; 38(12):2426–31. doi: 10.1016/j.jhsa.2013.08.123. Epub 2013 Nov. 1. You can see the task in the photograph at the beginning of Chapter 18.

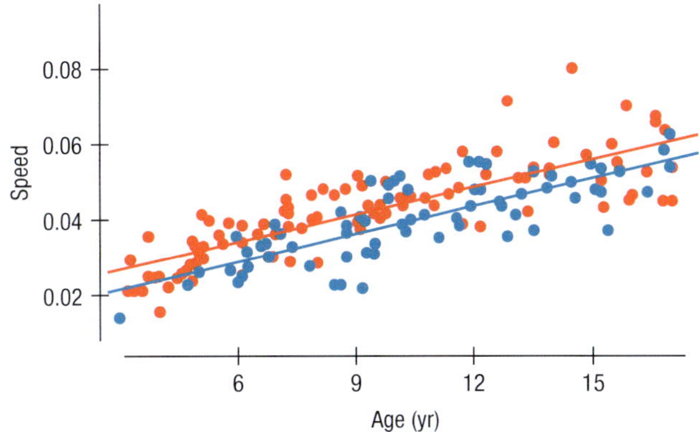

Figure 8.21
Re-expressing *Time* as *Speed* = 1/*Time* gives a plot that is straight and more nearly equal in variability throughout. In addition, the regression lines for dominant and nondominant hands are now parallel.

Goal 2

Make the scatter in a scatterplot spread out evenly rather than thickening at one end. Having an even scatter is a condition of many methods of statistics. As Figure 8.21 shows, a well-chosen re-expression can even out the spread.

Goal 3

Make the distribution of the residuals more nearly Normal. Although not strictly necessary until we talk about inference for regression, a side benefit of a good re-expression is that when we make the plot of y versus x more linear and the spread around the line more even, we'll often see the shape of the histogram of the residuals become more symmetric as well.

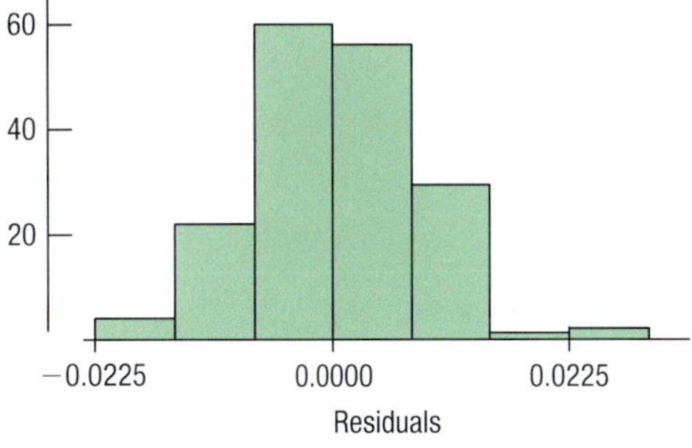

Figure 8.22
The residuals for a regression of *Speed* on *Age* are unimodal and nearly symmetric.

*8.7 Finding a Good Re-Expression

Symmetric distributions are easier to summarize and understand. It is even difficult to say just where the center of a skewed distribution is. And linear relationships are simpler and are easier to model than curved relationships. But just haphazardly trying various **re-expressions** to seek a useful one can be frustrating. You saw the **Ladder of Powers** in Chapter 6. By systematically going up or down the ladder, you can often find a re-expression that will make your analysis simpler. The comments next to each power in the ladder suggest where to start looking for a re-expression for a single variable depending on your situation. Here is the Ladder of Powers again, for convenience. (See Table 8.1 on the next page.)

But, for scatterplots, which direction on the ladder should you go? A simple rule due to John Tukey[7] can help guide you. Look at Figure 8.23. Now look at your scatterplot and see which quarter of the circle the curve of your plot most resembles. For example, the plot of *Time* vs. *Age* in Figure 8.20 looks like quadrant 3 in the lower left. Tukey's rule says to take y down the ladder if it most resembles quadrant 3 or 4. And the reciprocal is the -1 power—down the ladder from the "1" power where we started.

[7]We mentioned John Tukey back in Chapter 2 as the creator of the stem-and-leaf display and in Chapter 4 as the originator of the boxplot. He also coined the term "bit" for a binary digit in a computer and made many important advances in statistics—and taught two of your authors.

Table 8.1
The Ladder of Powers with comments to help guide an initial choice of re-expression for a single variable.

Power	Name	Comment
2	The square of the data values, y^2.	Try this for unimodal distributions that are skewed to the left.
1	The raw data—no change at all. This is "home base." The farther you step from here up or down the ladder, the greater the effect.	Data that can take on both positive and negative values with no bounds are less likely to benefit from re-expression.
1/2	The square root of the data values, $\sqrt{y}$.	Counts often benefit from a square root re-expression. For counted data, start here.
"0"	The logarithm of y. It doesn't matter whether you take the log base 10, the natural log, or any other base.	Measurements that cannot be negative, and especially values that grow by percentage increases such as salaries or populations, often benefit from a log re-expression. When in doubt, start here. If your data have zeros, try adding a small constant to all values before finding the logs.
−1/2	The (negative) reciprocal square root, $-1/\sqrt{y}$.	An uncommon re-expression, but sometimes useful. Changing the sign to take the *negative* of the reciprocal square root preserves the direction of relationships, making things a bit simpler.
−1	The (negative) reciprocal, $-1/y$.	Ratios of two quantities (miles per hour, for example) often benefit from a reciprocal. (You have about a 50–50 chance that the original ratio was taken in the "wrong" order for simple statistical analysis and would benefit from re-expression.) Often, the reciprocal will have simple units (hours per mile). Change the sign if you want to preserve the direction of relationships. If your data have zeros, try adding a small constant to all values before finding the reciprocal.

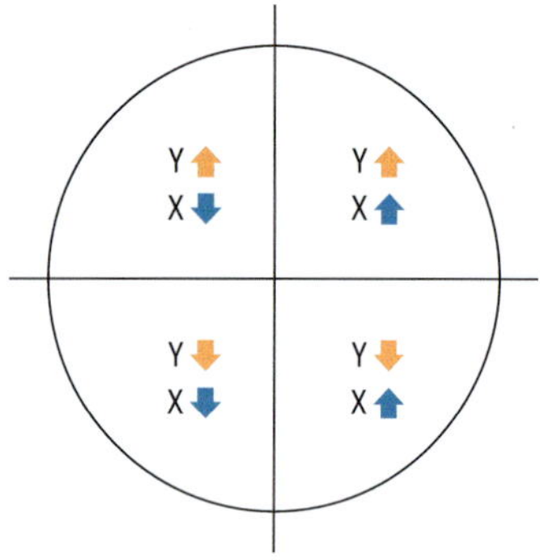

Figure 8.23
Tukey's circle of re-expressions suggests what direction to move on the Ladder of Powers for each variable depending on what quadrant the scatterplot resembles. For example, if the scatterplot is curved upward like quadrant 3 (bottom left), then either y or x should move down the ladder.

One note of caution: If your curve looks like two quadrants at the same time or curves in other ways, you won't be able to straighten the relationship with the methods in this chapter.

You should judge the straightness of a relationship by looking at a scatterplot. A high R^2—even a *very* high R^2—does not guarantee that the relationship is straight. And outliers can influence the correlation, either increasing or decreasing it. It's best to look at the scatterplots and choose the re-expression that makes the relationship straightest, the spread most constant, and the histogram of the response variable the most symmetric. It's often easy to find a single re-expression that does a good job on all three criteria. But no re-expression will be perfect. You will need to use your judgment.

JUST CHECKING

4. You want to model the relationship between the number of birds counted at a nesting site and the temperature (in degrees Celsius). The scatterplot of counts vs. temperature shows an upwardly curving pattern, with more birds spotted at higher temperatures. What transformation (if any) of the bird counts might you start with?
5. You want to model the relationship between prices for various items in Paris and in Hong Kong. The scatterplot of Hong Kong prices vs. Parisian prices shows a generally straight pattern with a small amount of scatter. What transformation (if any) of the Hong Kong prices might you start with?
6. You want to model the population growth of the United States over the past 200 years. The scatterplot shows a strongly upwardly curved pattern. What transformation (if any) of the population might you start with?

STEP-BY-STEP EXAMPLE

Re-Expressing to Straighten a Scatterplot

We've seen curvature in the relationship between emperor penguins' diving heart rates and the duration of the dive. In addition, the histogram of heart rates was skewed and the boxplots of heart rates by individual birds showed skewness to the high end. (Data in **Penguins**)

QUESTIONS: What re-expression might straighten out the relationship?

THINK **PLAN** State the problem.

I want to fit a linear model for the *Diving Heart Rate (DHR)* and *Duration* of a penguin dive.

VARIABLES Identify the variables and report the W's.

I have the *DHR* in beats per minute and the *Duration* in minutes.

PLOT Check that even if there is a curve, the overall pattern does not reach a minimum or maximum and then turn around and go back. An up-and-down curve can't be fixed by re-expression.

The plot shows a negative direction and an association that is curved like the third quadrant of Tukey's circle. The correlation is −0.846.

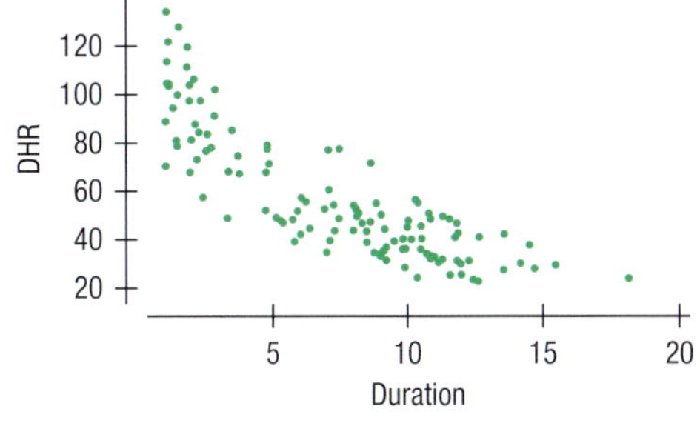

The plot of residuals against predicted values shows both curvature and increasing spread.

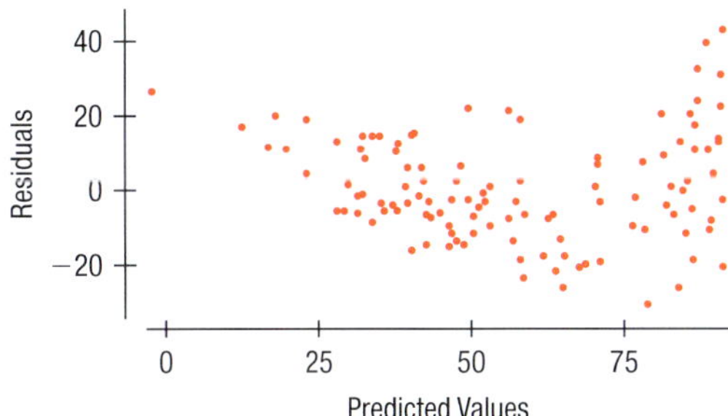

The histogram of **residuals** shows **a slight** right skewness.

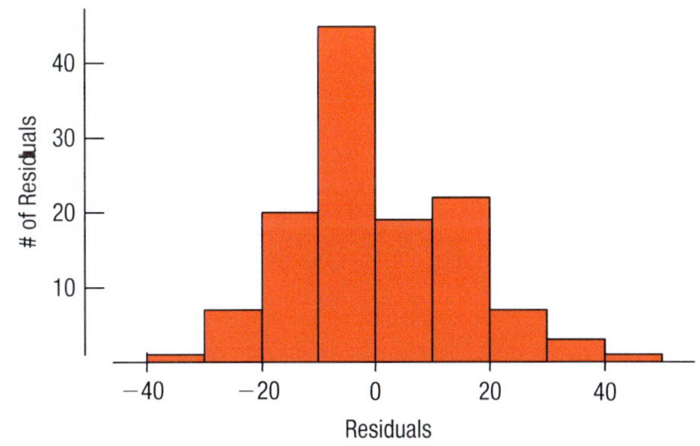

SHOW **MECHANICS** Try a re-expression. See what quadrant the curve most resembles and try moving a re-expression of *y* in that direction. The lesson of the Ladder of Powers is that if we're moving in the right direction but have not had sufficient effect, we should go farther along the ladder.

Tukey's circle suggests going down the ladder. We'll try the square root of *DHR*. Here's a plot of the square root of *DHR* against *Duration*:

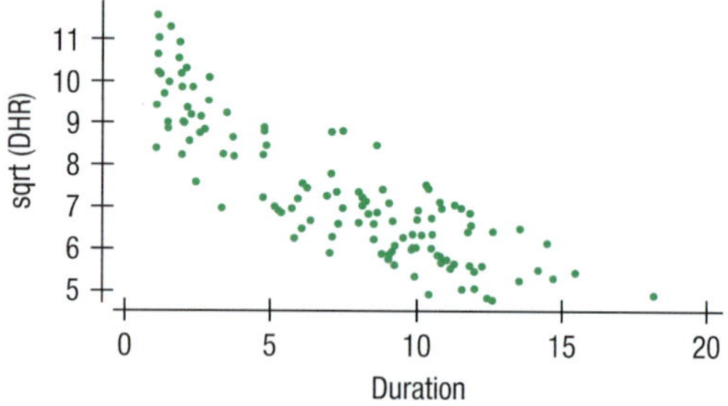

This example shows improvement, but is still not straight. We'll go further down the ladder and try logarithm.

The plot is less bent, but still not straight. Stepping from the 1/2 power to the "0" power, we try the logarithm of *DHR* against *Duration*:

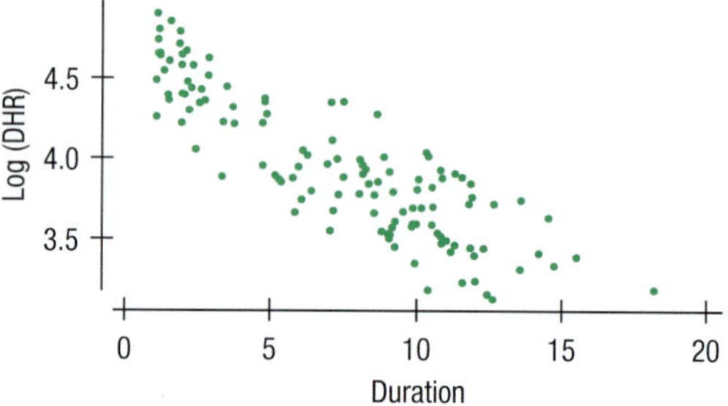

The plot of residuals against predicted values is much straighter and the spread is more even

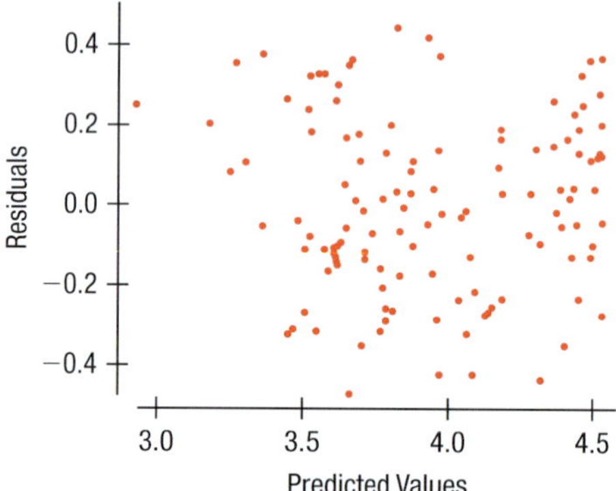

Often a re-expression that improves straightness improves the symmetry of the residuals as well. That's the case here.

Let's look at the histogram of the residuals from this regression:

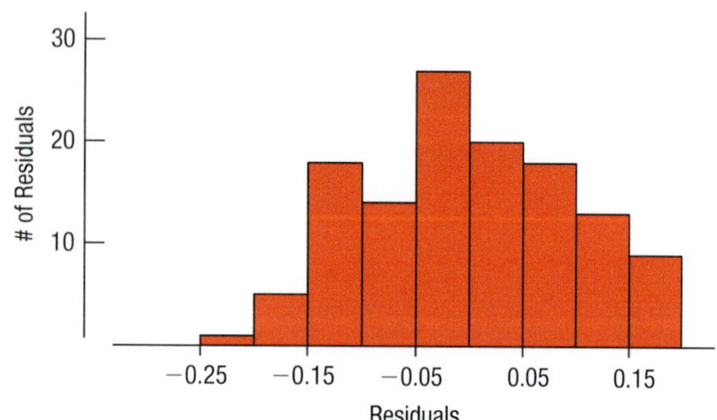

There is no harm in going too far down the ladder as we can always "back up." That way we know we've gone far enough. It looks here like we've gone too far. If anything the curve is now curving downward instead of upward. We used the negative reciprocal to keep the direction the same (negative in this case).

The histogram is more symmetric. This re-expression might work. The correlation is now −0.869. However, I'll go a little farther to make sure I've gone far enough. Here's a plot of the (negative) reciprocal of *DHR* against *Duration:*

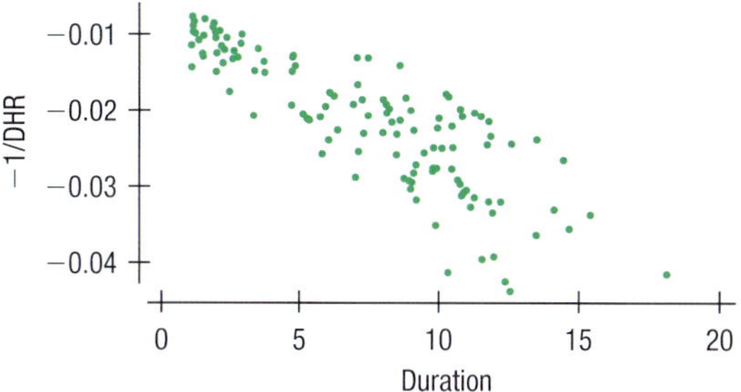

Now the plot seems to curve slightly downward, it spreads out more on the right, the correlation is −0.845, the residual plot shows increasing spread on the left, and my histogram now shows skewness in the opposite direction.

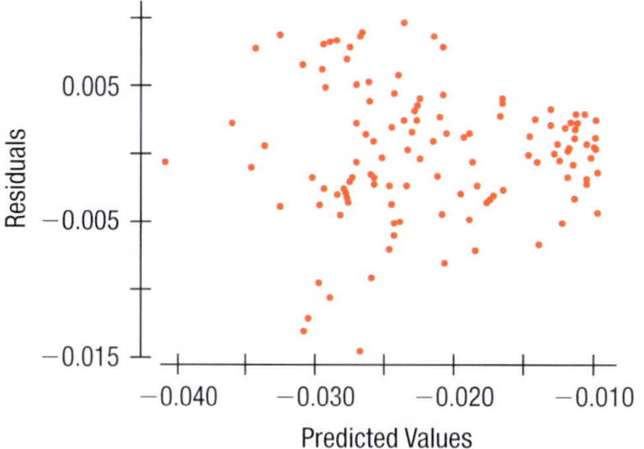

The histogram also reflects that we've gone too far. Now the skewness is in the opposite direction.

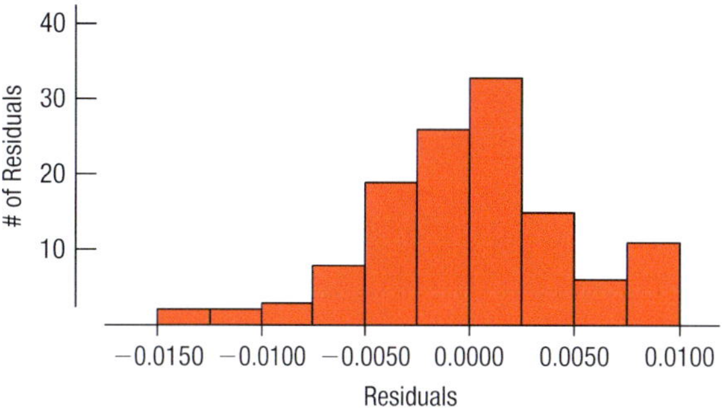

It appears that I've gone too far. I'll back up one "rung" and try −1/sqrt(*DHR*):

We may have to choose between two adjacent re-expressions. For most data analyses, it really doesn't matter which we choose.

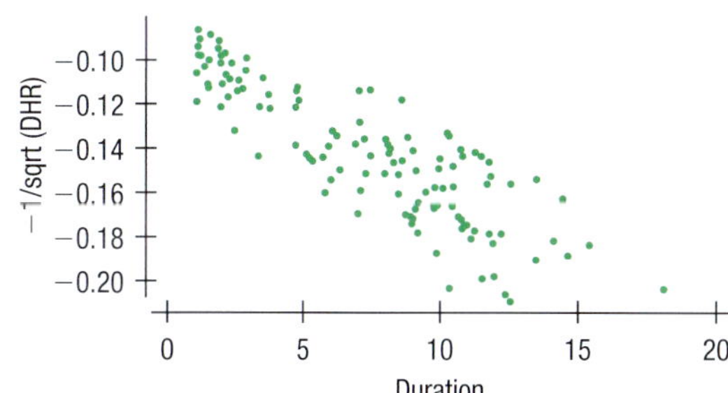

This looks pretty good and the correlation is −0.863. Let's check the histogram of the residuals:

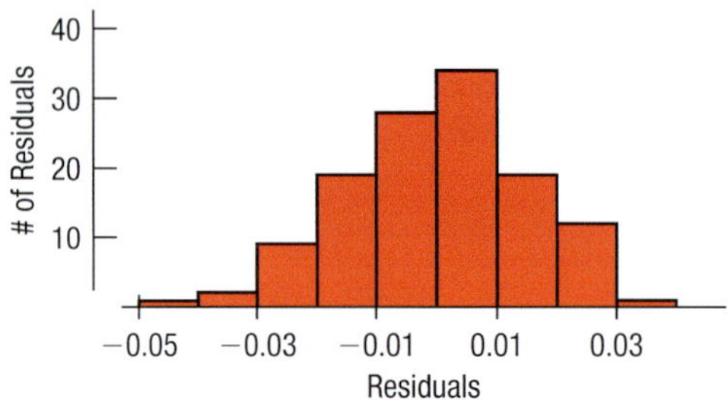

The histogram is symmetric as well.

TELL **CONCLUSION** Specify your choice of re-expression. If there's some natural interpretation (as for gallons per 100 miles), give that.

It is often hard to choose between two re-expressions that produce similar effects. Here, log(DHR) would have done about as well. The choice is often a judgment call.

Now that the re-expressed data satisfy the Straight Enough Condition, we can fit a linear model by least squares. We find that

$$\frac{-1}{\sqrt{\widehat{DHR}}} = -0.0955 - 0.00642\ Duration.$$

We can use this model to predict the heart beats resulting from a dive of, say, 10 minutes:

$$\frac{-1}{\sqrt{\widehat{DHR}}} = -0.0955 - 0.00642 \times 10 = -0.1597.$$

We could leave the result in these units $(-1/\sqrt{beats})$. Sometimes the new units may be as meaningful as the original, but here we want to transform the predicted value back into beats. Fortunately, each of the re-expressions in the Ladder of Powers can be reversed.

To reverse the process, we first take the reciprocal: $\sqrt{\widehat{DHR}} = -1/(-0.1597) = 6.2617$. Then squaring gets us back to the original units:

$$\widehat{DHR} = 6.2617^2 = 39.2\ beats.$$

This may be the most painful part of the re-expression. Getting back to the original units can be a little work. Nevertheless, it's worth the effort to always consider re-expression. Re-expressions extend the reach of all of your statistics tools by helping more data satisfy the conditions they require. Just think how much more useful this course just became!

WHAT CAN GO WRONG?

This entire chapter has held warnings about things that can go wrong in a regression analysis. So let's just recap. When you make a linear model:

- **Make sure the relationship is straight.** Check the Straight Enough Condition. Always examine the residuals for evidence that the Linearity Assumption has failed. It's often easier to see deviations from a straight line in the residuals plot than in the scatterplot of the original data. Pay special attention to the most extreme residuals because they may have something to add to the story told by the linear model.

- **Be on guard for different groups in your regression.** Check for evidence that the data consist of separate subsets. If you find subsets that behave differently, consider fitting a different linear model to each subset.
- **Beware of extrapolating.** Beware of extrapolation beyond the *x*-values that were used to fit the model. Although it's common to use linear models to extrapolate, the practice is dangerous.
- **Beware especially of extrapolating into the future!** Be especially cautious about extrapolating into the future with linear models. To predict the future, you must assume that future changes will continue at the same rate you've observed in the past. Predicting the future is particularly tempting and particularly dangerous.
- **Look for unusual points.** Unusual points always deserve attention and may well reveal more about your data than the rest of the points combined. Always look for them and try to understand why they stand apart. A scatterplot of the data is a good way to see high-leverage and influential points. A scatterplot of the residuals against the predicted values is a good tool for finding points with large residuals.
- **Beware of high-leverage points and especially of those that are influential.** Influential points can alter the regression model a great deal. The resulting model may say more about one or two points than about the overall relationship.
- **Consider comparing two regressions.** To see the impact of outliers on a regression, it's often wise to run two regressions, one with and one without the extraordinary points, and then to discuss the differences.
- **Treat unusual points honestly.** If you remove enough carefully selected points, you will eventually get a regression with a high R^2, but it won't give you much understanding. Some variables are not related in a way that's simple enough for a linear model to fit very well. When that happens, report the failure and stop.
- **Beware of lurking variables.** Think about lurking variables before interpreting a linear model. It's particularly tempting to explain a strong regression by thinking that the *x*-variable *causes* the *y*-variable. A linear model alone can never demonstrate such causation, in part because it cannot eliminate the chance that a lurking variable has caused the variation in both *x* and *y*.
- **Watch out when dealing with data that are summaries.** Be cautious in working with data values that are themselves summaries, such as means or medians. Such statistics are less variable than the data on which they are based, so they tend to inflate the impression of the strength of a relationship.
- **Don't expect your model to be perfect.** In Chapter 5 we quoted statistician George Box: "All models are wrong, but some are useful." Be aware that the real world is a messy place and data can be uncooperative. Don't expect to find one elusive re-expression that magically irons out every kink in your scatterplot and produces perfect residuals. You aren't looking for the Right Model, because that mythical creature doesn't exist. Find a useful model and use it wisely.
- **Don't stray too far from the ladder.** It's wise not to stray too far from the powers that we suggest in the Ladder of Powers. Taking the *y*-values to an extremely high power may artificially inflate R^2, but it won't give a useful or meaningful model, so it doesn't really simplify anything. It's better to stick to powers between 2 and −2. Even in that range, you should prefer the simpler powers in the ladder to those in the cracks. A square root is easier to understand than the 0.413 power. That simplicity may compensate for a slightly less straight relationship.

CONNECTIONS

We should always be alert to things that could go wrong if we were to use statistics without thinking carefully. Regression opens new vistas of potential problems. But each one relates to issues we've thought about before.

It is always important that our data be from a single homogeneous group and not made up of disparate groups. We looked for multiple modes in single variables. Now we check scatterplots for evidence of subgroups in our data. As with modes, it's often best to split the data and analyze the groups separately.

Our concern with unusual points and their potential influence also harkens back to our earlier concern with outliers in histograms and boxplots—and for many of the same reasons. As we've seen here, regression offers such points new scope for mischief.

The risks of interpreting linear models as causal or predictive arose in Chapters 6 and 7. And they're important enough to mention again in later chapters.

We have seen several ways to model or summarize data. Each requires that the data have a particular simple structure. We seek symmetry for summaries of center and spread and to use a Normal model. We seek equal variation across groups when we compare groups with boxplots or want to compare their centers. We seek linear shape in a scatterplot so that we can use correlation to summarize the scatter and regression to fit a linear model.

Data often satisfy the requirements to use statistics methods. But often they do not. Our choice is to stop with just displays, to use much more complex methods, or to re-express the data so that we can use the simpler methods we have developed.

In this fundamental sense, this chapter connects to everything we have done thus far and to all of the methods we will introduce throughout the rest of the text. Re-expression greatly extends the reach and applicability of all of these methods.

CHAPTER REVIEW

Be skeptical of regression models. Always plot and examine the residuals for unexpected behavior. Be alert to a variety of possible violations of the standard regression assumptions and know what to do when you find them.

Be alert for subgroups in the data.

- Often these will turn out to be separate groups that should not be analyzed together in a single analysis.
- Often identifying subgroups can help us understand what is going on in the data.

Be especially cautious about extrapolating beyond the data.

Look out for unusual and extraordinary observations.

- Cases that are extreme in x have high leverage and can affect a regression model strongly.
- Cases that are extreme in y have large residuals and are called outliers.
- Cases that have both high leverage and large residuals are influential. Setting them aside will change the regression model, so you should consider whether their influence on the model is appropriate or desirable.

Interpret regression models appropriately. Don't infer causation from a regression model. Notice when you are working with summary values.

- Summaries vary less than the data they summarize, so they may give the impression of greater certainty than your model deserves.

Diagnose and treat nonlinearity.

- If a scatterplot of y vs. x isn't straight, a linear regression model isn't appropriate.
- Re-expressing one or both variables can often improve the straightness of the relationship.
- The powers, roots, and the logarithm provide an ordered collection of re-expressions so you can search up and down the Ladder of Powers to find an appropriate one.
- Understand and use re-expression to make better models.
- Understand that models aren't perfect, but recognize when a well-chosen re-expression can help you improve and simplify your analysis.
- Understand the value of re-expressing data to improve symmetry, to make the scatter around a line more constant, or to make a scatterplot more linear.
- Recognize when the pattern of the data indicates that no re-expression can improve the structure of the data.

- Know how to re-express data with powers and how to find an effective re-expression for your data using your statistics software or calculator. Understand how to use the Ladder of Powers to find a good re-expression.
- Be able to reverse any of the common re-expressions to put a predicted value or residual back into the original units.
- Be able to describe a summary or display of a re-expressed variable, making clear how it was re-expressed and giving its re-expressed units.
- Be able to describe a regression model fit to re-expressed data in terms of the re-expressed variables.

REVIEW OF TERMS

Extrapolation Although linear models provide an easy way to predict values of *y* for a given value of *x*, it is unsafe to predict for values of *x* far from the ones used to find the linear model equation. Such extrapolation may pretend to see into the future, but the predictions should not be trusted (p. 268).

Outlier Any data point that stands away from the others can be called an outlier. In regression, cases can be extraordinary in two ways: by having a large residual—being an outlier—or by having high leverage (p. 272).

Leverage Data points whose *x*-values are far from the mean of *x* are said to exert leverage on a linear model. High-leverage points pull the line close to them, and so they can have a large effect on the line, sometimes completely determining the slope and intercept. With high enough leverage, their residuals can be deceptively small (p. 273).

Influential point A point that, if omitted from the data, results in a very different regression model (p. 273).

Lurking variable A variable that is not explicitly part of a model but affects the way the variables in the model appear to be related. Because we can never be certain that observational data are not hiding a lurking variable that influences both *x* and *y*, it is never safe to conclude that a linear model demonstrates a causal relationship, no matter how strong the linear association (p. 275).

***Re-expression** We re-express data by taking the logarithm, the square root, the reciprocal, or some other mathematical operation of all values of a variable (p. 281).

***Ladder of Powers** The Ladder of Powers places in order the effects that many re-expressions have on the data (p. 281).

TECH SUPPORT

Regression and Re-expression

Regression Diagnosis

Most statistics technology offers simple ways to check whether your data satisfy the conditions for regression. We have already seen that these programs can make a simple scatterplot. They can also help check the conditions by plotting residuals.

DATA DESK

- Click on the **HyperView** menu on the Regression output table. A menu drops down to offer scatterplots of residuals against predicted values, Normal probability plots of residuals, or just the ability to save the residuals and predicted values.
- Click on the name of a predictor in the regression table to be offered a scatterplot of the residuals against that predictor.

COMMENTS

If you change any of the variables in the regression analysis, Data Desk will offer to update the plots of residuals.

EXCEL

The Data Analysis add-in for Excel includes a Regression command. The dialog box it shows offers to make plots of residuals.

COMMENTS

Do not use the Normal probability plot offered in the regression dialog. It is not what it claims to be and is wrong.

JMP

- From the Analyze menu, choose **Fit Y by X**. Select **Fit Line**.
- Under Linear Fit, select **Plot Residuals**. You can also choose to **Save Residuals**.
- Subsequently, from the Distribution menu, choose **Normal quantile plot** or **histogram** for the residuals.

MINITAB

- From the Stat menu, choose **Regression**.
- From the Regression submenu, select **Regression** again.
- In the Regression dialog, enter the response variable name in the Response box predictor variable name in the Predictor box.
- To specify saved results, in the Regression dialog, click **Storage**.
- Check Residuals and Fits. Click **OK**.
- To specify displays, in the Regression dialog, click **Graphs**.
- Under Residual Plots, select **Individual plots** and check Residuals versus fits.
- Click **OK**. Now back in the Regression dialog, click **OK**. Minitab computes the regression and the requested saved values and graphs.

R

Save the regression model object by giving it a name, such as "myreg":

- myreg = lm(Y~X) or myreg = lm(Y~X, data = DATA) where DATA is the name of the data frame.
- **plot(residuals(myreg)~predict(myreg))** plots the residuals against the predicted values.
- **qqnorm(residuals(myreg))** gives a Normal probability plot of the residuals.
- **plot(myreg)** gives similar plots (but not exactly the same).

SPSS

- From the Analyze menu, choose **Regression**.
- From the Regression submenu, choose **Linear**.
- After assigning variables to their roles in the regression, click the **Plots ...** button.

In the Plots dialog, you can specify a Normal probability plot of residuals and scatterplots of various versions of standardized residuals and predicted values.

COMMENTS

A plot of *ZRESID against *PRED will look most like the residual plots we've discussed. SPSS standardizes the residuals by dividing by their standard deviation. (There's no need to subtract their mean; it must be zero.) The standardization doesn't affect the scatterplot.

STATCRUNCH

To create a residuals plot:

- Click on **Stat**.
- Choose **Regression** » **Simple Linear** and choose **X** and **Y**.
- Indicate which type of residuals plot you want.
- Click on **Compute!**

TI-83/84 PLUS

- To make a residuals plot, set up a STATPLOT as a scatterplot.
- Specify your explanatory data list as **Xlist.**
- For Ylist, import the name RESID from the LIST NAMES menu. ZoomStat will now create the residuals plot.

COMMENTS

Each time you execute a LinReg command, the calculator automatically computes the residuals and stores them in a data list named RESID. If you want to see them, go to STAT EDIT. Space through the names of the lists until you find a blank. Import RESID from the LIST NAMES menu. Now every time you have the calculator compute a regression analysis, it will show you the residuals.

Re-Expression

Computers and calculators make it easy to re-express data. Most statistics packages offer a way to re-express and compute with variables. Some packages permit you to specify the power of a re-expression with a slider or other moveable control, possibly while watching the consequences of the re-expression on a plot or analysis. This, of course, is a very effective way to find a good re-expression.

DATA DESK

To re-express a variable in Data Desk, select the variable and Choose the function to re-express it from the Manip » Transform menu. Square root, log, reciprocal, and reciprocal root are immediately available. You can also make a derived variable and type the function. Data Desk makes a new derived variable that holds the re-expressed values. Any value changed in the original variable will immediately be re-expressed in the derived variable.

COMMENTS

Or choose **Manip** » **Transform** » **Dynamic** » **Box-Cox** to generate a continuously changeable variable and a slider that specifies the power. Drag the new variable into a plot or analysis and set plots to **Automatic Update** in their HyperView menus. Then you can watch them change dynamically as you drag the slider.

EXCEL

To re-express a variable in Excel, use Excel's built-in functions as you would for any calculation. When you change a value in the original column, Excel will change the re-expressed value.

JMP

To re-express a variable in JMP, double-click to the right of the last column of data to create a new column. Name the new column and select it. Choose **Formula** from the Cols menu. In the Formula dialog, choose the transformation and variable that you wish to assign to the new column. Click the **OK** button. JMP places the re-expressed data in the new column.

As of version 11, JMP allows re-expression "on the fly" in most dialogs. For example, in the Fit Y by X platform, right click on either variable to bring up a Transform option that offers most re-expressions.

COMMENTS

The log and square root re-expressions are found in the Transcendental menu of functions in the formula dialog.

MINITAB

To re-express a variable in MINITAB, choose **Calculator** from the Calc menu. In the Calculator dialog, specify a name for the new re-expressed variable. Use the Functions List, the calculator buttons, and the Variables list box to build the expression. Click **OK**.

R

To re-express a variable in R, either create a new variable, as, for example:

```
> Log.y = log(y) #natural log
```

or re-express inside a formula. For example:

```
> lm(log(y) ~ x)
```

SPSS

To re-express a variable in SPSS, choose **Compute** from the Transform menu. Enter a name in the Target Variable field. Use the calculator and Function List to build the expression. Move a variable to be re-expressed from the source list to the Numeric Expression field. Click the **OK** button.

STATCRUNCH

From the Data menu, select **Compute Expression.**

In the Compute Expression box, type the expression and name the new variable.

Click the **Compute!** button to make the new variable.

COMMENTS

The regression dialog also offers simple transformations of both X and Y as an option.

TI-83/84 PLUS

To re-express data stored in a list, perform the re-expression on the whole list and store it in another list. For example, to use the logarithms of the data in L1, enter the command **log(L1) STO L2**.

EXERCISES

SECTION 8.1

1. Credit card spending An analysis of spending by a sample of credit card bank cardholders shows that spending by cardholders in January (*Jan*) is related to their spending in December (*Dec*):

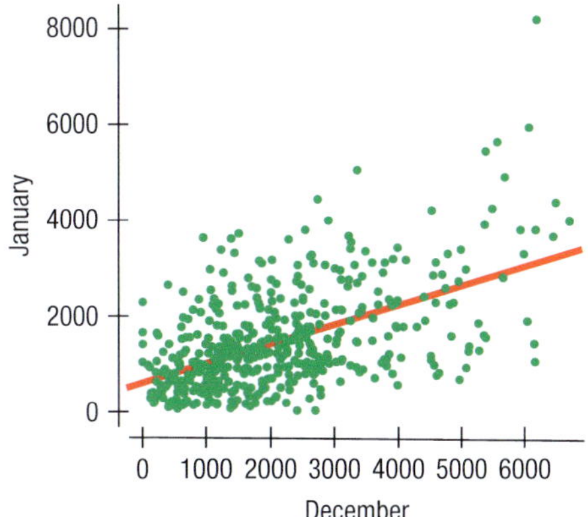

The assumptions and conditions of the linear regression seemed to be satisfied and an analyst was about to predict January spending using the model

$$\widehat{Jan} = \$612.07 + 0.403\ Dec.$$

Another analyst worried that different types of cardholders might behave differently. She examined the spending patterns of the cardholders and placed them into five market *Segments*. When she plotted the data using different colors and symbols for the five different segments, she found the following:

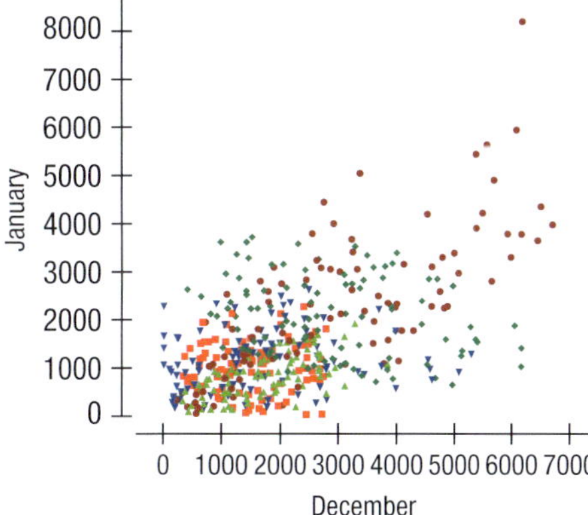

Look at this plot carefully and discuss why she might be worried about the predictions from the model $\widehat{Jan} = \$612.07 + 0.403\ Dec$.

2. Revenue and talent cost A concert production company examined its records. The manager made the scatterplot at the top of the next column. The company places concerts in two venues, a smaller, more intimate theater (plotted with blue circles) and a larger auditorium-style venue (red x's).

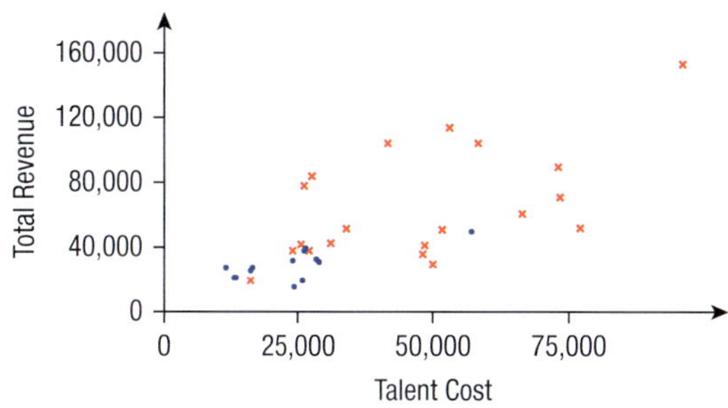

a) Describe the relationship between *Talent Cost* and *Total Revenue*. (Remember: direction, form, strength, and outliers.)
b) How are the results for the two venues similar?
c) How are they different?

3. Market segments The analyst in Exercise 1 tried fitting the regression line to each market segment separately and found the following:

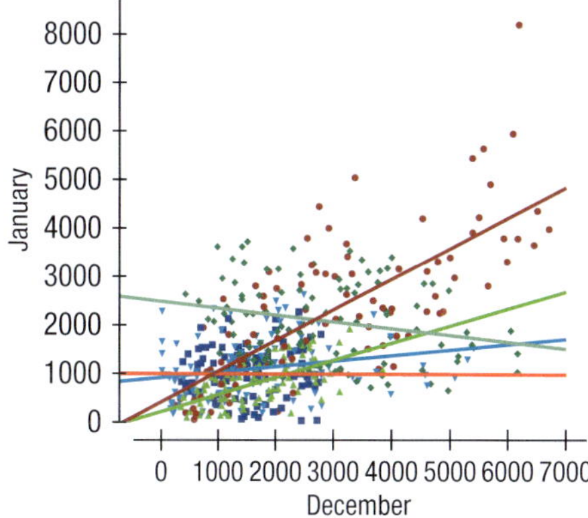

What does this say about her concern in Exercise 1? Was she justified in worrying that the overall model $\widehat{Jan} = \$612.07 + 0.403\ Dec$ might not accurately summarize the relationship? Explain briefly.

4. Revenue and ticket sales The concert production company of Exercise 2 made a second scatterplot, this time relating *Total Revenue* to *Ticket Sales*.

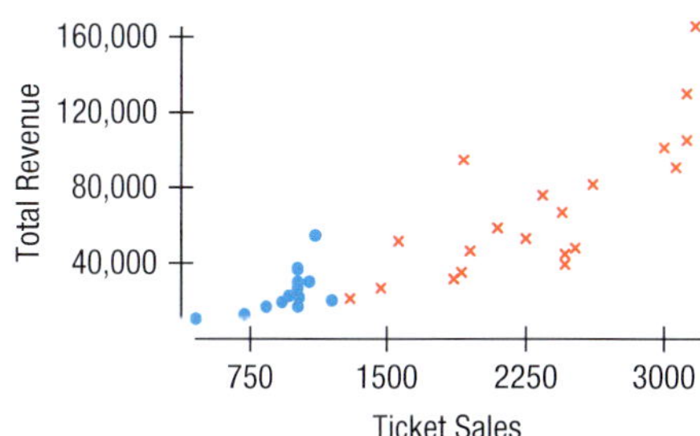

a) Describe the relationship between *Ticket Sales* and *Total Revenue*.
b) How are the results for the two venues similar?
c) How are they different?

SECTION 8.2

5. Cell phone costs Noting a recent study predicting the increase in cell phone costs, a friend remarks that by the time he's a grandfather, no one will be able to afford a cell phone. Explain where his thinking went awry.

6. Stopping times Using data from 20 compact cars, a consumer group develops a model that predicts the stopping time for a vehicle by using its weight. You consider using this model to predict the stopping time for your large SUV. Explain why this is not advisable.

7. Revenue and large venues A regression of *Total Revenue* on *Ticket Sales* by the concert production company of Exercises 2 and 4 finds the model

$$\widehat{Revenue} = -14{,}228 + 36.87\ TicketSales.$$

a) Management is considering adding a stadium-style venue that would seat 10,000. What does this model predict that revenue would be if the new venue were to sell out?
b) Why would it be unwise to assume that this model accurately predicts revenue for this situation?

8. Revenue and advanced sales The production company of Exercise 7 offers advanced sales to "Frequent Buyers" through its website. Here's a relevant scatterplot:

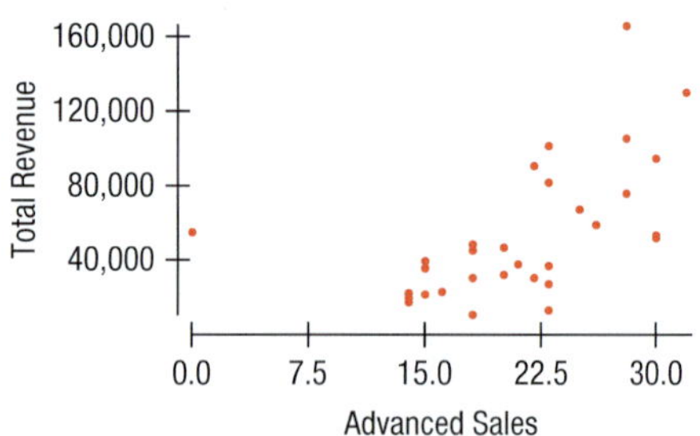

One performer refused to permit advanced sales. What effect has that point had on the regression to model *Total Revenue* from *Advanced Sales*?

SECTION 8.3

9. Abalone Abalones are edible sea snails that include over 100 species. A researcher is working with a model that uses the number of rings in an abalone's shell to predict its age. He finds an observation that he believes has been miscalculated. After deleting this outlier, he redoes the calculation. Does it appear that this outlier was exerting very much influence?

Before

R-squared = 50.6%

Variable	Coefficient
Intercept	1.794
Rings	0.49

After

R-squared = 73.3%

Variable	Coefficient
Intercept	1.45
Rings	1.19

10. Abalone again The researcher in Exercise 9 is content with the second regression. But he has found a number of shells that have large residuals and is considering removing all of them. Is this good practice?

SECTION 8.4

11. Skinned knees There is a strong correlation between the temperature and the number of skinned knees on playgrounds. Does this tell us that warm weather causes children to trip?

12. Cell phones and life expectancy The correlation between cell phone usage and life expectancy is very high. Should we buy cell phones to help people live longer?

SECTION 8.5

13. Grading A team of calculus teachers is analyzing student scores on a final exam compared to the midterm scores. One teacher proposes that they already have every teacher's class averages and they should just work with those averages. Explain why this is problematic.

14. Average GPA An athletic director proudly states that he has used the average GPAs of the university's sports teams and is predicting a high graduation rate for the teams. Why is this method unsafe?

SECTION 8.6

15. Residuals Suppose you have fit a linear model to some data and now take a look at the residuals. For each of the following possible residuals plots, tell whether you would try a re-expression and, if so, why.

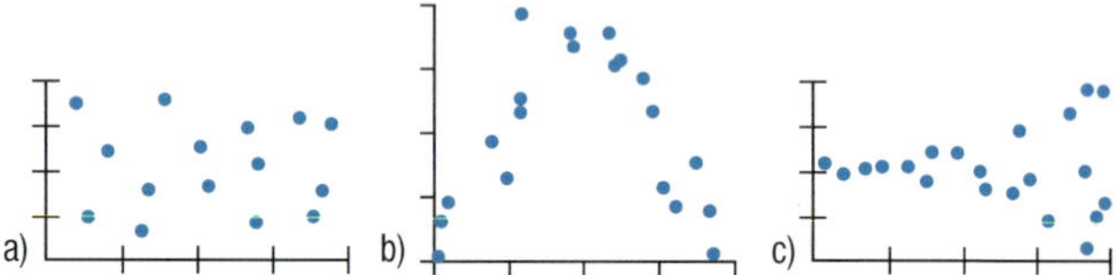

16. More residuals Suppose you have fit a linear model to some data and now take a look at the residuals. For each of the following possible residuals plots, tell whether you would try a re-expression and, if so, why.

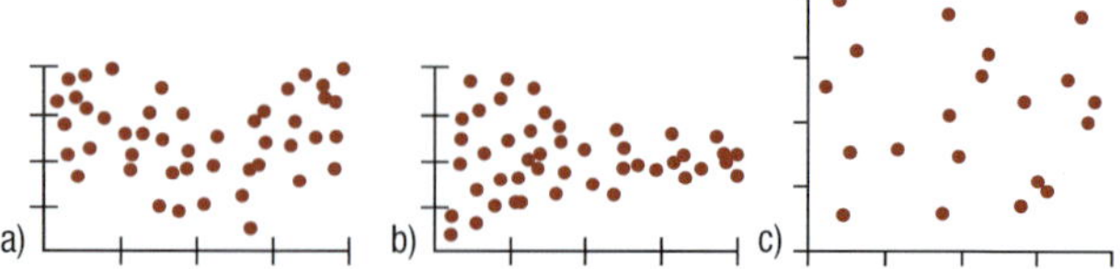

17. BK protein Recall the data about the Burger King menu items in Chapter 7. Here are boxplots of protein content comparing items that contain meat with those that do not. The plot on the right graphs log(*Protein*). Which of the goals of re-expression does this illustrate?

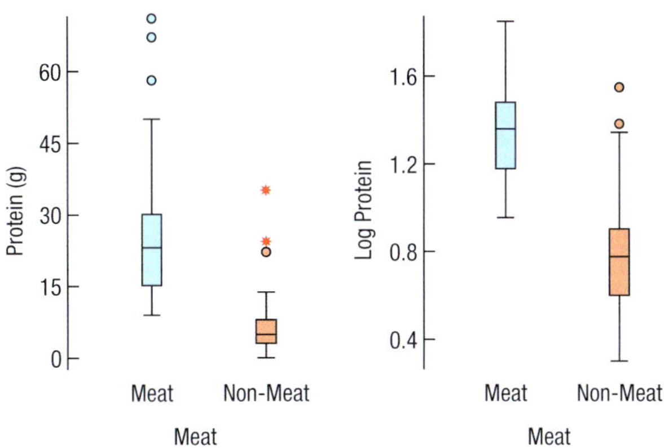

T 18. TVs and life expectancy Recall the example of life expectancy vs. TVs per person in the chapter. In that example, we use the square root of TVs per person. Here are the original data and the re-expressed version. Which of the goals of re-expression does this illustrate? (Data in **Doctors and life expectancy**)

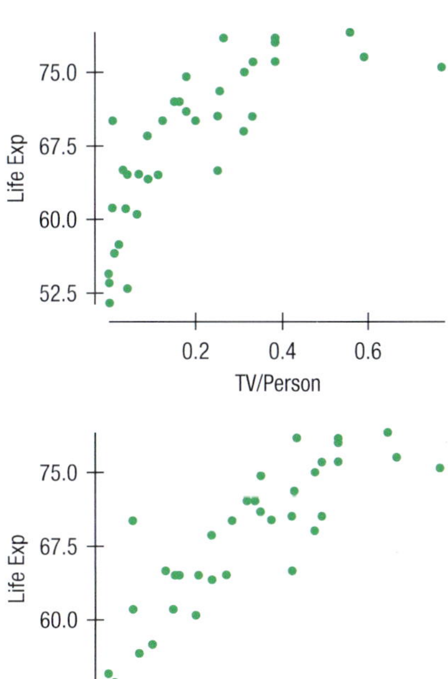

SECTION 8.7

19. BK protein again Exercise 17 looked at the distribution of protein in the Burger King menu items, comparing meat and non-meat items. That exercise offered the logarithm as a re-expression of Protein. Here are two other alternatives, the square root and the reciprocal. Would you still prefer the log? Explain why.

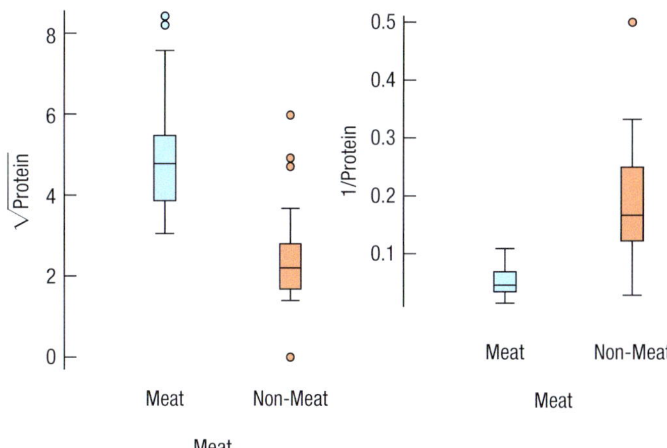

T 20. TVs and life expectancy Exercise 18 revisited the relationship between life expectancy and TVs per capita and saw that re-expression to the square root of TVs per capita made the plot more nearly straight. But was that the best choice of re-expression? Here is a scatterplot of life expectancy vs. the logarithm of TVs per person. How can you tell that this re-expression is too far along the Ladder of Powers?

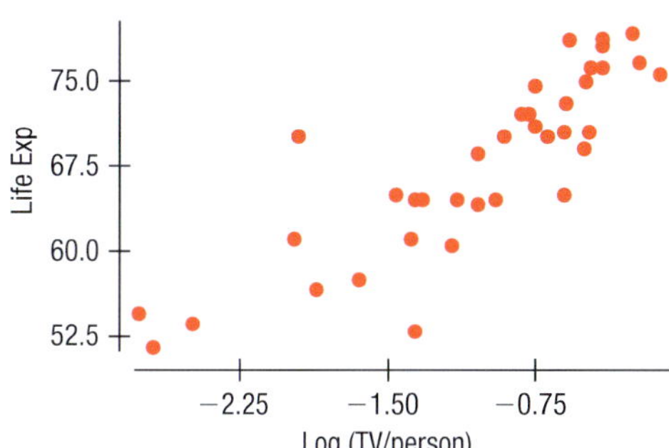

CHAPTER EXERCISES

T 21. Marriage age 2015 Is there evidence that the age at which women get married has changed over the past 100 years? The scatterplot shows the trend in age at first marriage for American women (www.census.gov).

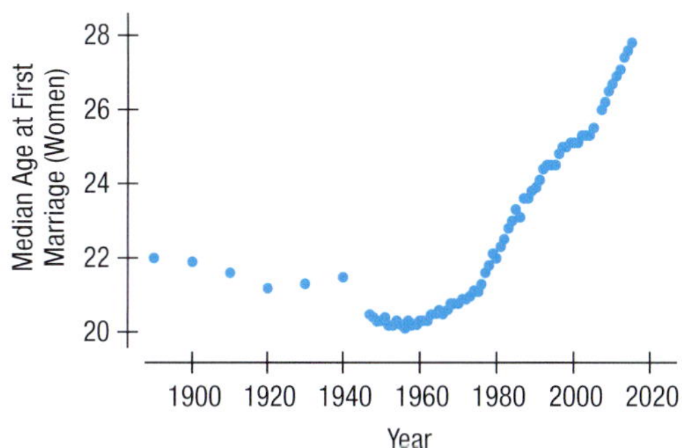

a) Is there a clear pattern? Describe the trend.
b) Is the association strong?
c) Is the correlation high? Explain.
d) Is a linear model appropriate? Explain.

T **22. Smoking 2014** The Centers for Disease Control and Prevention tracks cigarette smoking in the United States (www.cdc.gov/nchs). How has the percentage of people who smoke changed since the danger became clear during the last half of the 20th century? The scatterplot shows percentages of smokers among men 18–24 years of age, as estimated by surveys, from 1965 through 2014.

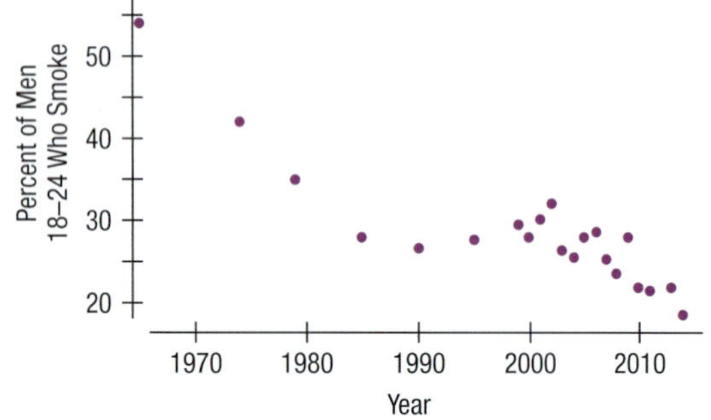

a) Is there a clear pattern? Describe the trend.
b) Is the association strong?
c) Is a linear model appropriate? Explain.

T **23. Human Development Index 2015** The Human Development Index (HDI) is used to summarize in one number the health, education, and economics of a country. The gross domestic product per capita (GDPPC), by contrast, is often used to summarize the overall economic strength of a country. Is the HDI related to the GDPPC? A scatterplot of HDI against GDPPC is shown to the right.

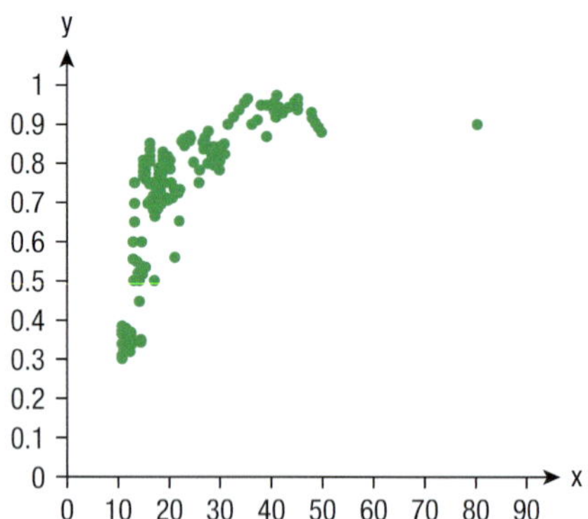

a) Explain why fitting a linear model to these data might be misleading.
b) If you fit a linear model to the data, what do you think a scatterplot of residuals will look like?

T **24. HDI 2015 revisited** As explained in Exercise 23, the Human Development Index (HDI) is a measure that attempts to summarize in one number the progress in health, education, and economics of a country. The percentage of older people (65 and older) in a country is positively associated with its HDI. Can the percentage of older adults be used to predict the HDI? Here is a scatterplot of the two variables.

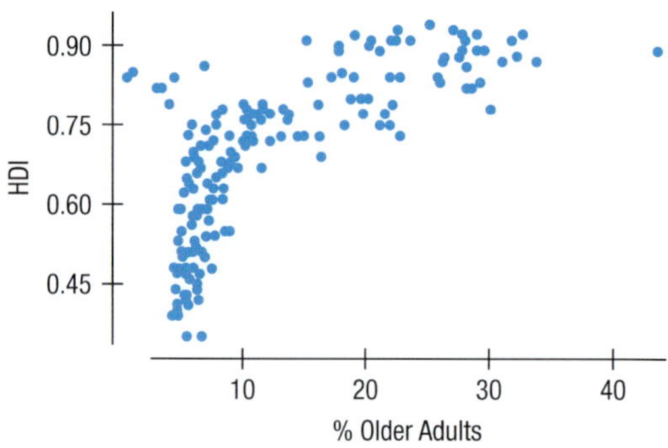

a) Explain why fitting a linear model to these data might be misleading.
b) If you fit a linear model to the data, what do you think a scatterplot of residuals vs. predicted *HDI* will look like?

25. Good model? In justifying his choice of a model, a student wrote, "I know this is the correct model because $R^2 = 99.4\%$."

a) Is this reasoning correct? Explain.
b) Does this model allow the student to make accurate predictions? Explain.

26. Bad model? A student who has created a linear model is disappointed to find that her R^2 value is a very low 13%.

a) Does this mean that a linear model is not appropriate? Explain.
b) Does this model allow the student to make accurate predictions? Explain.

27. Movie dramas Here's a scatterplot of the production budgets (in millions of dollars) vs. the running time (in minutes) for major release movies in 2005. Dramas are plotted as red x's and all other genres are plotted as blue dots. (The re-make of *King Kong* is plotted as a black "-". At the time it was the most expensive movie ever made, and not typical of any genre.) A separate least squares regression line has been fitted to each group. For the following questions, just examine the plot:

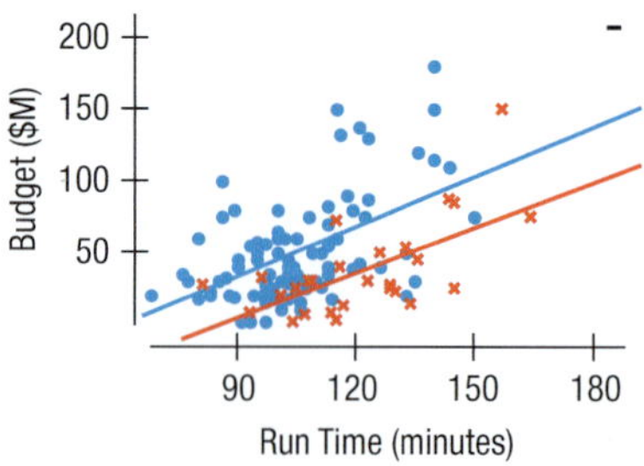

a) What are the units for the slopes of these lines?
b) In what way are dramas and other movies similar with respect to this relationship?
c) In what way are dramas different from other genres of movies with respect to this relationship?

T **28. Smoking 2014, women and men** In Exercise 22, we examined the percentage of men aged 18–24 who smoked from 1965 to 2014 according to the Centers for Disease Control and

Prevention. How about women? Here's a scatterplot showing the corresponding percentages for both men and women along with least squares lines for each.:

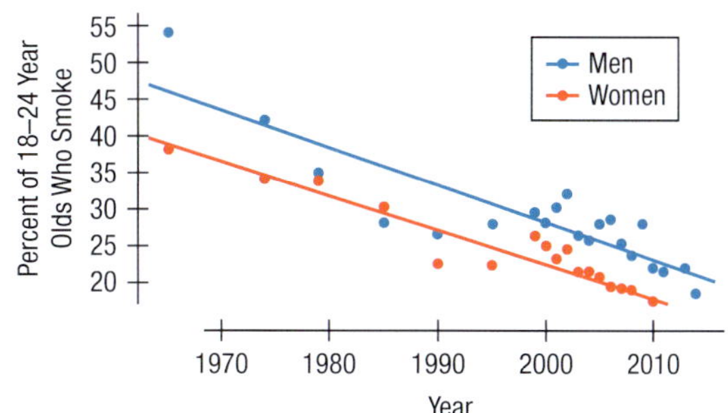

a) In what ways are the trends in smoking behavior similar for men and women?
b) How do the smoking rates for women differ from those for men?
c) Viewed alone, the trend for men may have seemed to violate the Linearity Condition. How about the trend for women? Does the consistency of the two patterns encourage you to think that a linear model for the trend in men might be appropriate? (Note: there is no correct answer to this question; it is raised for you to think about.)

T **29. Oakland passengers 2016** The scatterplot below shows the number of passengers at Oakland (CA) airport month by month since 1997 (oaklandairport.com/news/statistics/passenger-history/).

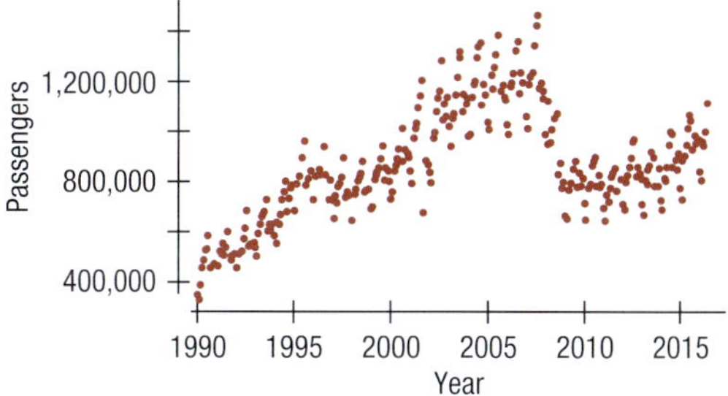

a) Describe the patterns in passengers at Oakland airport that you see in this time plot.
b) Until 2009, analysts got fairly good predictions using a linear model. Why might that not be the case now?
c) If they considered only the data from 2009 to the present, might they get reasonable predictions into the future?

T **30. Tracking hurricanes 2016** In Chapter 6, we saw data on the errors (in nautical miles) made by the National Hurricane Center in predicting the path of hurricanes. The scatterplot below shows the trend in the 24-hour tracking errors since 1970 (www.nhc.noaa.gov).

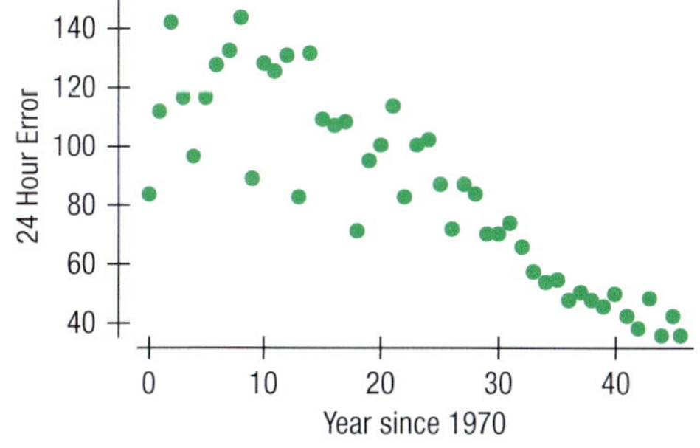

Dependent variable is Error
R-squared = 77.5% s_e = 15.44

Variable	Coefficient
Intercept	133.130
Years - 1970	−2.067

a) Interpret the slope and intercept of the model.
b) Interpret s_e in this context.
c) The Center would like to achieve an average tracking error of 45 nautical miles by 2020. Will they make it? Defend your response.
d) What if their goal were an average tracking error of 25 nautical miles?
e) What cautions would you state about your conclusion?

31. Unusual points Each of these four scatterplots shows a cluster of points and one "stray" point. For each, answer these questions:

1) In what way is the point unusual? Does it have high leverage, a large residual, or both?
2) Do you think that point is an influential point?
3) If that point were removed, would the correlation become stronger or weaker? Explain.
4) If that point were removed, would the slope of the regression line increase or decrease? Explain.

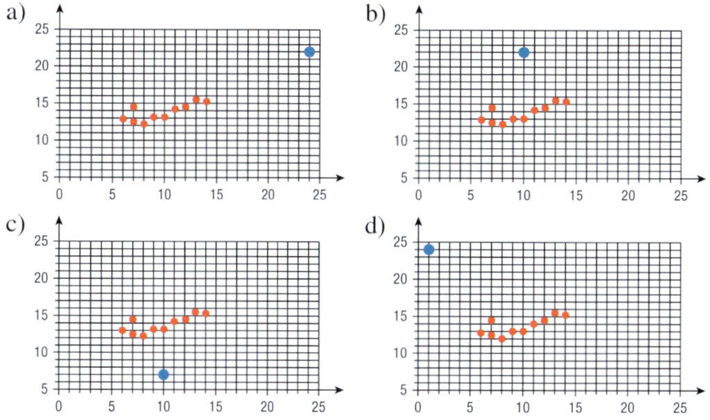

32. More unusual points Each of the following scatterplots shows a cluster of points and one "stray" point. For each, answer these questions:

1) In what way is the point unusual? Does it have high leverage, a large residual, or both?
2) Do you think that point is an influential point?
3) If that point were removed, would the correlation become stronger or weaker? Explain.
4) If that point were removed, would the slope of the regression line increase or decrease? Explain.

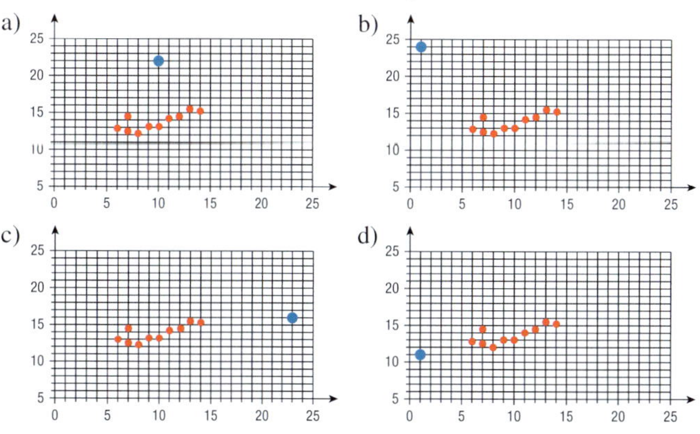

33. The extra point The scatterplot shows five blue data points at the left. Not surprisingly, the correlation for these points is $r = 0$. Suppose *one* additional data point is added at one of the five positions suggested below in red. Match each point (a–e) with the correct new correlation from the list given.

1) 0.75
2) −0.90
3) 0.05
4) 0.00
5) −0.40

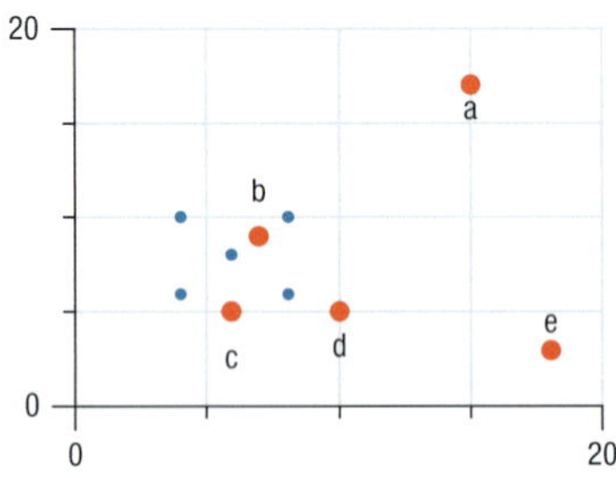

34. The extra point revisited The accompanying scatterplot shows five blue data points. These five points produce a regression line with slope 0. Match each one of the red points (a through e) with the slope of the line after that one point is added.

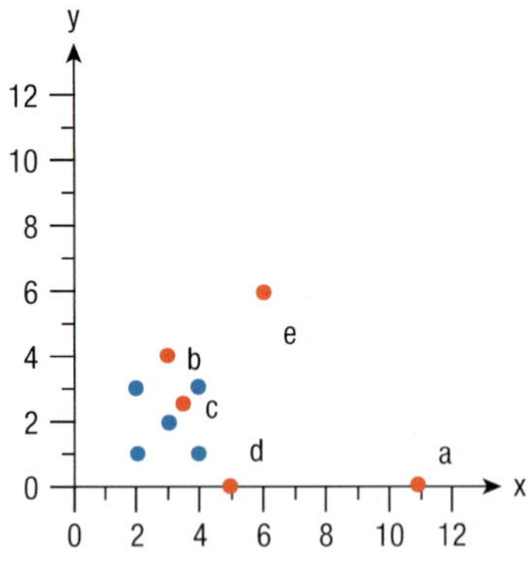

1) 0.87
2) 0.05
3) 0.00
4) −0.45
5) −0.23

35. What's the cause? Suppose a researcher studying health issues measures blood pressure and the percentage of body fat for several adult males and finds a strong positive association. Describe three different possible cause-and-effect relationships that might be present.

36. What's the effect? A researcher studying violent behavior in elementary school children asks the children's parents how much time each child spends playing computer games and has their teachers rate each child on the level of aggressiveness they display while playing with other children. Suppose that the researcher finds a moderately strong positive correlation. Describe three different possible cause-and-effect explanations for this relationship.

37. Reading To measure progress in reading ability, students at an elementary school take a reading comprehension test every year. Scores are measured in "grade-level" units; that is, a score of 4.2 means that a student is reading at slightly above the expected level for a fourth grader. The school principal prepares a report to parents that includes a graph showing the mean reading score for each grade. In his comments, he points out that the strong positive trend demonstrates the success of the school's reading program.

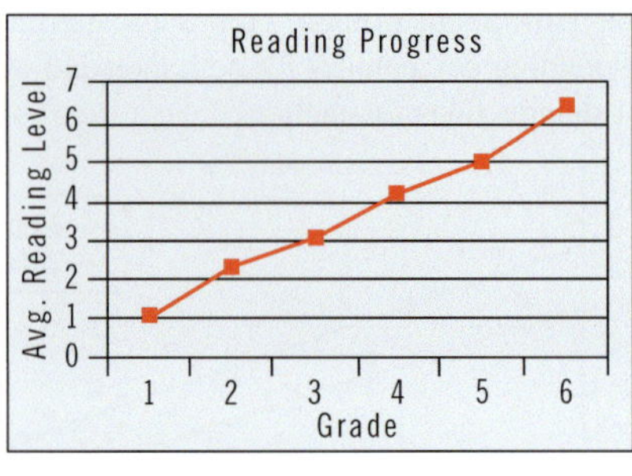

a) Does this graph indicate that students are making satisfactory progress in reading? Explain.
b) What would you estimate the correlation between *Grade* and *Average Reading Level* to be?
c) If, instead of this plot showing average reading levels, the principal had produced a scatterplot of the reading levels of all the individual students, would you expect the correlation to be the same, higher, or lower? Explain.
d) Although the principal did not do a regression analysis, someone as statistically astute as you might do that. (But don't bother.) What value of the slope of that line would you view as demonstrating acceptable progress in reading comprehension? Explain.

38. Grades A college admissions officer, defending the college's use of SAT scores in the admissions process, produced the following graph. It shows the mean GPAs for last year's freshmen, grouped by SAT scores. How strong is the evidence that *SAT Score* is a good predictor of *GPA*? What concerns you about the graph, the statistical methodology or the conclusions reached?

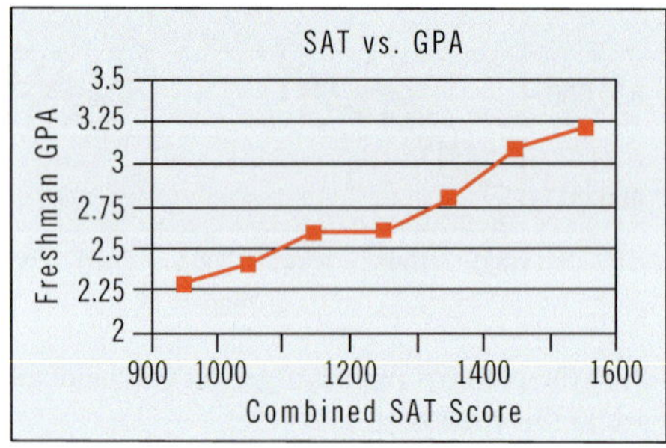

39. Heating After keeping track of his heating expenses for several winters, a homeowner believes he can estimate the monthly cost from the average daily Fahrenheit temperature by using the model $\widehat{Cost} = 101 - 2.16\, Temp$. Here is the residuals plot for his data:

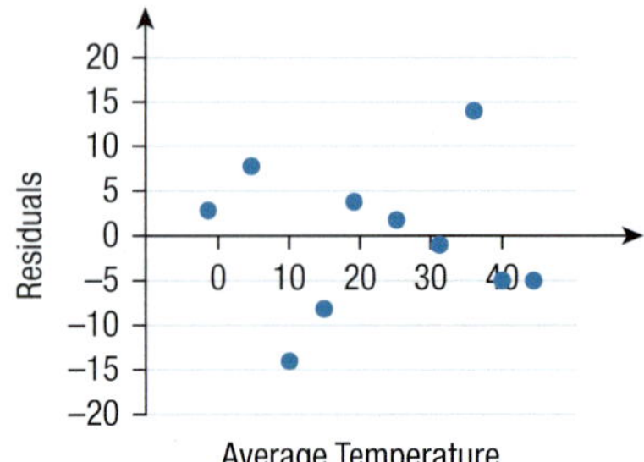

a) Interpret the slope of the line in this context.
b) Interpret the y-intercept of the line in this context.
c) During months when the temperature stays around freezing(32° F), would you expect cost predictions based on this model to be accurate, too low, or too high? Explain.

d) What heating cost does the model predict for a month that averages 10°?
e) During one of the months on which the model was based, the temperature did average 10°. What were the actual heating costs for that month?
f) Should the homeowner use this model? Explain.
g) Would this model be more successful if the temperature were expressed in degrees Celsius? Explain.

40. Speed How does the speed at which you drive affect your fuel economy? To find out, researchers drove a compact car for 200 miles at speeds ranging from 35 to 75 miles per hour. From their data, they created the model *Fuel Efficiency* = 32 − 0.1 *Speed* and created this residual plot:

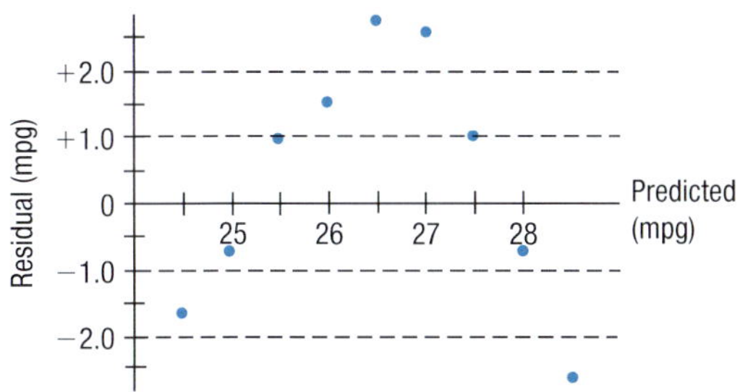

a) Interpret the slope of this line in context.
b) Explain why it's silly to attach any meaning to the y-intercept.
c) When this model predicts high *Fuel Efficiency*, what can you say about those predictions?
d) What *Fuel Efficiency* does the model predict when the car is driven at 50 mph?
e) What was the actual *Fuel Efficiency* when the car was driven at 45 mph?
f) Do you think there appears to be a strong association between *Speed* and *Fuel Efficiency*? Explain.
g) Do you think this is the appropriate model for that association? Explain.

T 41. TBill rates 2016 Here are a hypothetical plot and regression output showing the federal rate on a three-month bond from 1950 to 1980, and a regression model fit to the relationship between the *Rate* (in %) and *Years Since 1950*

a) What is the correlation between *Rate* and *Year*?
b) Interpret the slope and intercept.
c) What does this model predict for the interest rate in the year 2000?
d) Would you expect this prediction to be accurate? Explain.

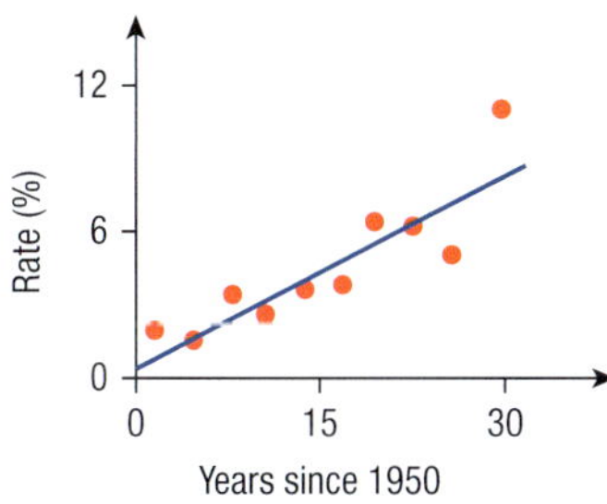

Dependent Variable is: Rate
R-squared = 76.2% s = 1.433

Variable	Coefficient
Intercept	0.431718
Year - 1950	0.261595

T 42. Marriage age, 2015 The graph shows the ages of both men and women at first marriage (www.census.gov).

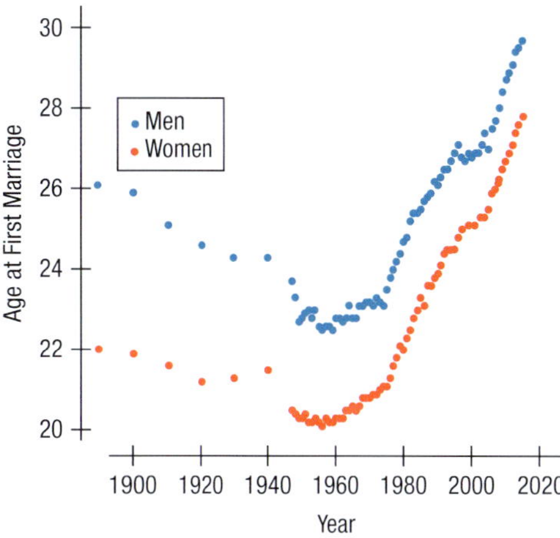

Clearly, the patterns for men and women are similar. But are the two lines getting closer together?

Here are a timeplot showing the *difference* in average age (men's age − women's age) at first marriage, the regression analysis, and the associated residuals plot.

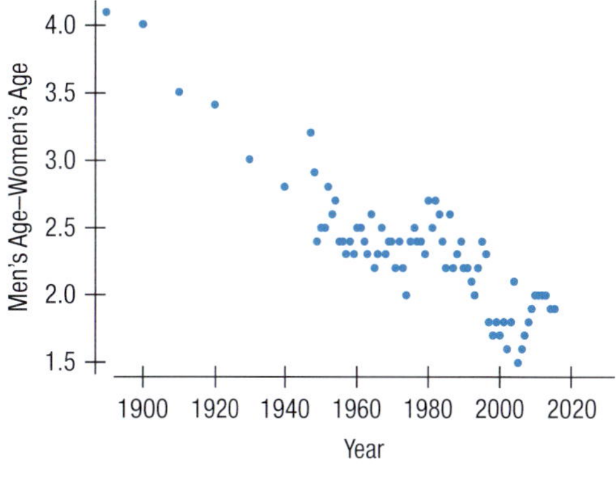

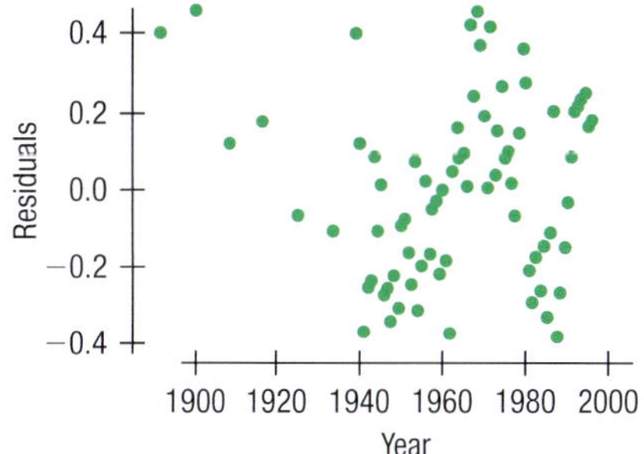

Dependent Variable is Age Difference
R-squared = 77.5% s = 0.2299

Variable	Coefficient
Intercept	33.604
Year	−0.01582

a) What is the correlation between *Age Difference* and *Year*?
b) Interpret the slope of this line.
c) Predict the average age difference in 2020.
d) Describe reasons why you might not place much faith in that prediction.

T **43. TBill rates 2016 revisited** In Exercise 41, you investigated the federal rate on 3-month Treasury bills between 1950 and 1980. The scatterplot below shows that the trend changed dramatically after 1980, so we computed a new regression model for the years 1981 to 2015.

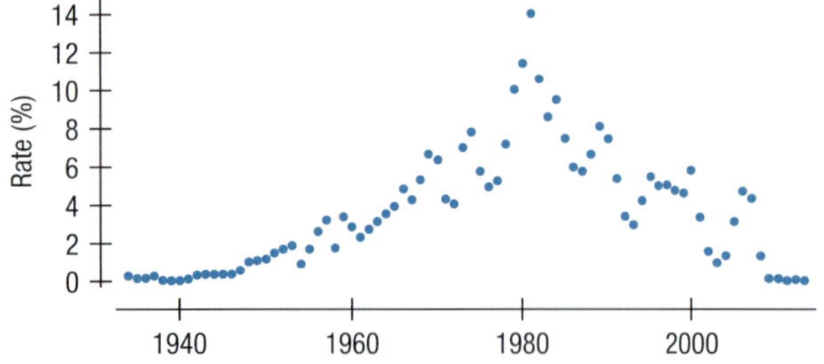

Here's the model for the data from 1981 to 2015 (in years since 1950):

Dependent variable is Rate
R-squared = 79.1% s = 1.584

Variable	Coefficient
Intercept	18.5922
Year - 1950	−0.29646

a) How does this model compare to the one in Exercise 41?
b) What does this model estimate the interest rate will be in 2020? How does this compare to the rate you predicted in Exercise 41?
c) Do you trust this newer predicted value? Explain.
d) Would you use either of these models to predict the TBill rate in the future? Explain.

T **44. Marriage age 2015 again** Has the trend of decreasing difference in age at first marriage seen in Exercise 42 gotten stronger recently? The scatterplot and residual plot for the data from 1980 through 2015, along with a regression for just those years, are below.

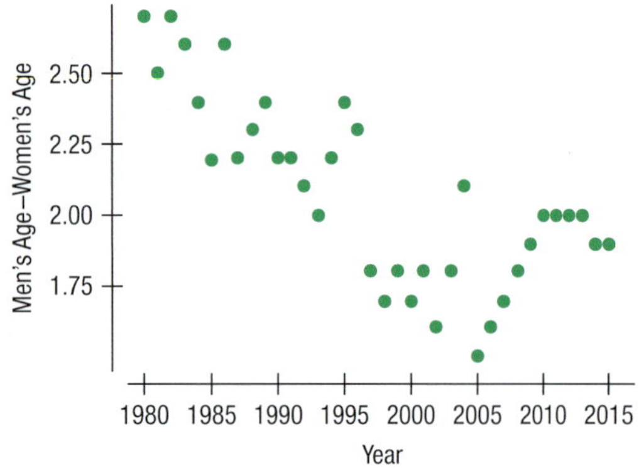

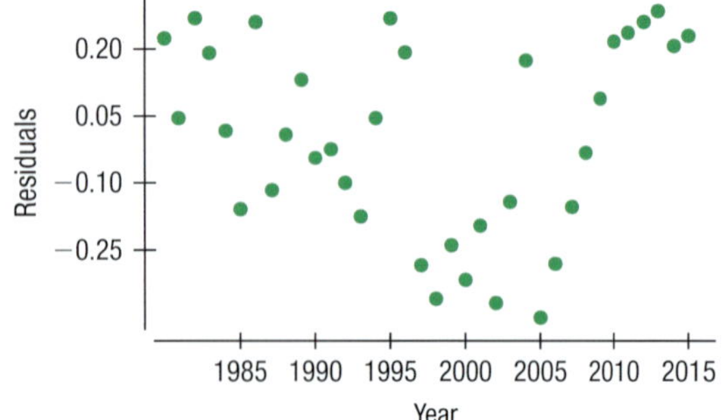

Response variable is Men − Women
R-squared = 55.4% s = 0.2206

Variable	Coefficient
Intercept	48.038
Year	−0.0230

a) Is this linear model appropriate for the post-1980 data? Explain.
b) What does the slope say about marriage ages?
c) Explain why it's not reasonable to interpret the y-intercept.

T **45. Gestation** For humans, pregnancy lasts about 280 days. In other species of animals, the length of time from conception to birth varies. Is there any evidence that the gestation period is related to the animal's life span? The first scatterplot shows *Gestation Period* (in days) vs. *Life Expectancy* (in years) for 18 species of mammals. The highlighted point at the far right represents humans.

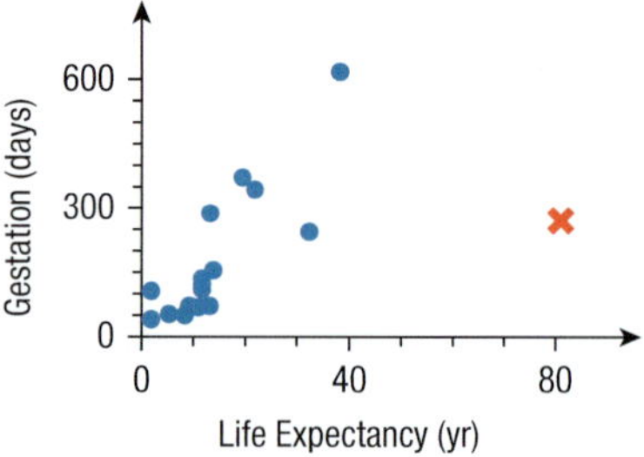

a) For these data, $r = 0.541$, not a very strong relationship. Do you think the association would be stronger or weaker if humans were removed? Explain.
b) Is there reasonable justification for removing humans from the dataset? Explain.
c) Here are the scatterplot and regression analysis for the 17 nonhuman species. Comment on the strength of the association.

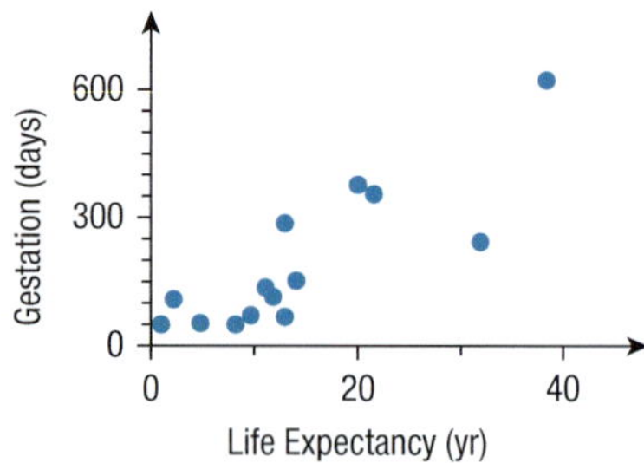

Dependent Variable is: Gestation
R − squared = 86.7%

Variable	Coefficient
Constant	88.1723
LifExp	12.9722

d) Interpret the slope of the line.
e) Some species of monkeys have a life expectancy of about 32 years. Estimate the expected gestation period of one of these monkeys.

46. Swim the lake 2016 People swam across a certain lake (52 km, or about 32.3 mi) 42 times between 1954 and 2016. We might be interested in whether the swimmers are getting any faster or slower. Here are the regression of the crossing *Times* (minutes) against the *Year* since 1954 of the crossing and a plot of the residuals against *Year since 1954:*

Dependent variable is: Time
R-squared = 1.1% s = 495.4

Variable	**Coefficient**
Intercept	−3717.31
Year	2.51068

a) What does the R^2 mean for this regression?
b) Are the swimmers getting faster or slower? Explain.
c) The outlier seen in the residuals plot is a crossing by a woman in 1987 in which she swam a round trip, north to south, and then back again. Clearly, this swim doesn't belong with the others. Would removing it change the model a lot? Explain.

47. Elephants and hippos We removed humans from the scatterplot of the **Gestation** data in Exercise 45 because our species was an outlier in life expectancy. The resulting scatterplot (below) shows two points that now may be of concern. The point in the upper right corner of this scatterplot is for elephants, and the other point at the far right is for hippos.

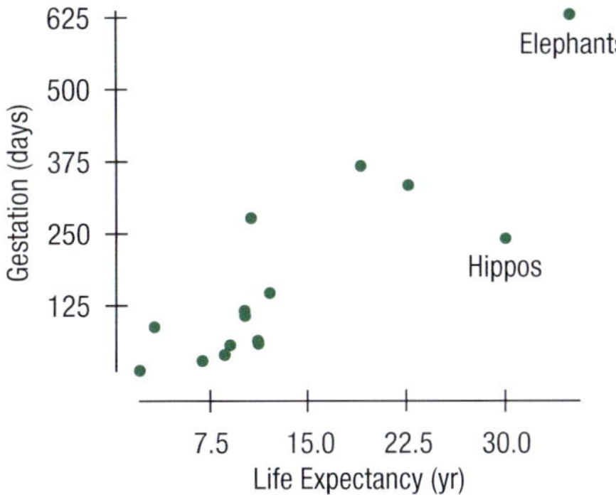

a) By removing one of these points, we could make the association appear to be stronger. Which point? Explain.
b) Would the slope of the line increase or decrease?
c) Should we just keep removing animals to increase the strength of the model? Explain.
d) If we remove elephants from the scatterplot, the slope of the regression line becomes 11.6 days per year. Do you think elephants were an influential point? Explain.

48. Another swim In Exercise 46, we saw in the **Swim the Lake 2016** data that Vicki Keith's round-trip swim of Lake Ontario was an obvious outlier among the other one-way times. Here is the new regression after this unusual point is removed:

Dependent variable is Time
R-Squared = 2.8% s = 302.1

Variable	**Coefficient**
Intercept	1135.99
Year since 1954	3.024

a) In this new model, the value of s_e is smaller. Explain what that means in this context.
b) Are you more convinced (compared to the previous regression) that Ontario swimmers are getting faster (or slower)?

49. Marriage age 2015 predictions Look again at the graph of the age at first marriage for women in Exercise 42. Here is a regression model for the data on women, along with a residuals plot:

Response variable is: Women
R-squared = 61.1%
s = 1.474

Variable	**Coefficient**
Intercept	−112.543
Year	0.068479

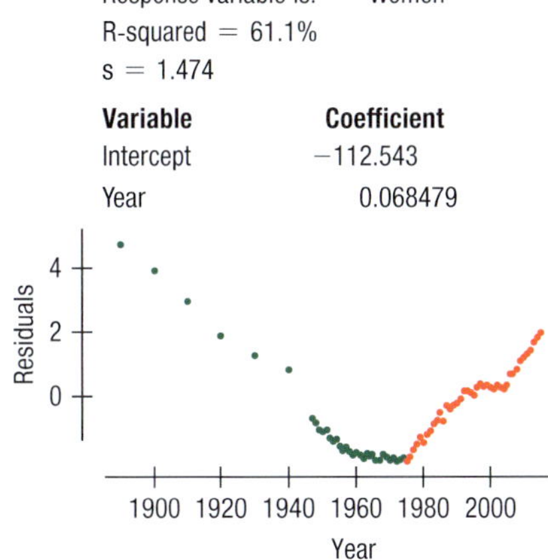

a) Based on this model, what would you predict the marriage age will be for women in 2025?
b) How much faith do you place in this prediction? Explain.
c) Would you use this model to make a prediction about your grandchildren, say, 50 years from now? Explain.

Now, let's restrict our model to the years 1975–2015. Here are the regression and residual plot:

Response variable is: Women
75 total cases of which 34 are missing
R squared = 98.2%
s = 0.2450

Variable	**Coefficient**
Intercept	−274.742
Year	0.149983

d) Based on this model, what would you predict the marriage age will be for women in 2025?
e) How much faith do you place in this prediction? Explain.
f) Would you use this model to make a prediction about your grandchildren, say, 50 years from now? Explain.

T 50. Bridges covered In Chapter 7, we found a relationship between the age of a bridge in Tompkins County, New York, and its condition as found by inspection. (Data in **Tompkins County Bridges 2016**) But we considered only bridges built or replaced since 1900. Tompkins County is the home of the oldest covered bridge in daily use in New York State. Built in 1853, it was recently judged to have a condition of 4.57. Here is a regression of *Condition* on the year in which a bridge was *Built*, computed for bridges built since 1900.

Dependent variable is Condition
R-squared = 40.0% s = 0.6891

Variable	Coefficient
Intercept	−33.4115
Built	0.0196

a) If we use this regression to predict the condition of the covered bridge, what would its residual be?
b) If we add the covered bridge to the data, what would you expect to happen to the regression slope? Explain.

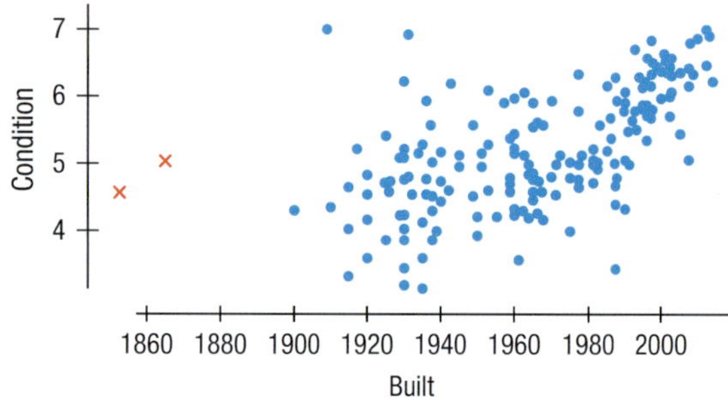

c) If we add the covered bridge to the data, what would you expect to happen to the R^2? Explain.
d) The bridge was extensively restored in 1972. If we use that date instead as the *Built* date, do you find the condition of the bridge remarkable?

T 51. Fertility and life expectancy 2014 A certain institution's data file holds data on the *Fertility* rate (births per woman) and the female *Life Expectancy* at birth (in years) for 200 countries of the world.

Response variable is: Life expectancy
219 total cases of which 19 are missing
R-squared = 72.7%
s = 4.510

Variable	Coefficient
Intercept	81.0372
Fertility	−4.86041

Here is a scatterplot of the data.

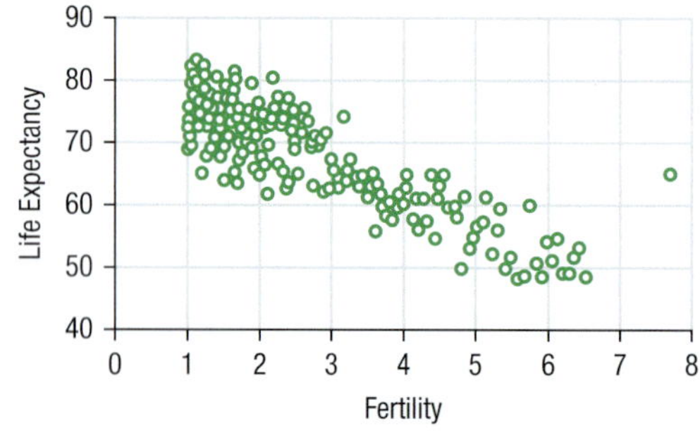

a) Are the conditions for regression satisfied? Discuss. Here is a plot of the residuals.

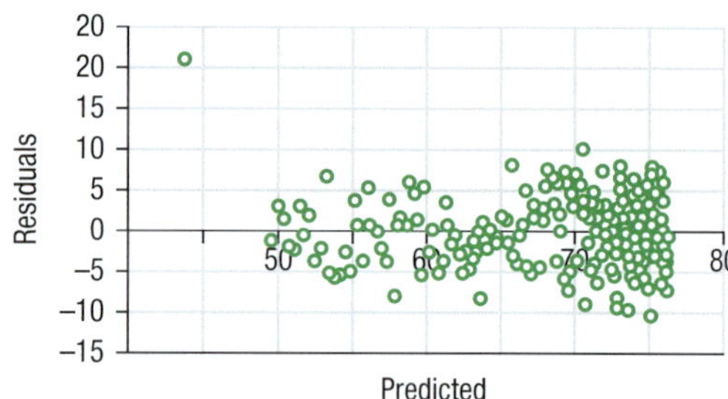

b) Is there an outlier? If so, identify it. Which data value is unusual?
c) Interpret the value of R^2.
d) If government leaders want to increase life expectancy, should they encourage women to have fewer children? Explain.

T 52. Tour de France 2017 We met the Tour de France dataset in Chapter 1 (in Just Checking). One hundred years ago, the fastest rider finished the course at an average speed of about 25.3 kph (around 15.8 mph). By the 21st century, winning riders were averaging over 40 kph (nearly 25 mph).

a) Make a scatterplot of *Avg Speed* against *Year.* Describe the relationship of *Avg Speed* and *Year*, being careful to point out any unusual features in the plot.
b) Find the regression equation of *Avg Speed* on *Year.*
c) Are the conditions for regression met? Comment.

T 53. Inflation 2016 The Consumer Price Index (CPI) tracks the prices of consumer goods in the United States, as shown in the following table. The CPI is reported monthly, but we can look at selected values. The table shows the January CPI at five-year intervals. It indicates, for example, that the average item costing $17.90 in 1926 cost $236.92 in the year 2016.

Year	JanCPI	Year	JanCPI
1916	10.4	1971	39.8
1921	19.0	1976	55.6
1926	17.9	1981	87.0
1931	15.9	1986	109.6
1936	13.8	1991	134.6
1941	14.1	1996	154.4
1946	18.2	2001	175.1
1951	25.4	2006	198.3
1956	26.8	2011	220.223
1961	29.8	2016	236.916
1966	31.8		

a) Make a scatterplot showing the trend in consumer prices. Describe what you see.
b) Be an economic forecaster: Project increases in the cost of living over the next decade. Justify decisions you make in creating your model.

T 54. Second stage 2016 Look once more at the data from Tour de France 2016. In Exercise 52, we looked at the whole history of the race, but now let's consider just the modern era from 1967 on.

a) Make a scatterplot and find the regression of *Avg Speed* by *Year* only for years from 1967 to the present. Are the conditions for regression met?
b) The years 1999–2005 have been disqualified because of doping. How does this help explain the fact that 9 out of 10 residuals in the final 10 years are negative?
c) Would you extrapolate from this model to predict next year's winning speed?

55. Oakland passengers 2016 revisited In Exercise 29, we considered whether a linear model would be appropriate to describe the trend in the number of passengers departing from the Oakland (CA) airport each month since the start of 1997. If we fit a regression model, we obtain this residual plot. We've added lines to show the order of the values in time:

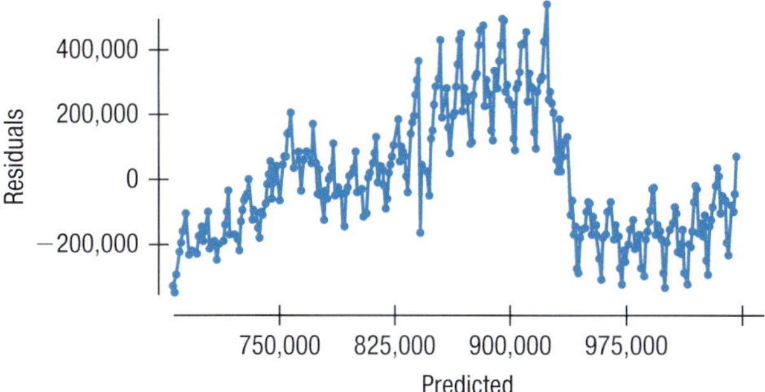

a) Can you account for the pattern shown here?
b) Would a re-expression help us deal with this pattern? Explain.

56. Hopkins winds, revisited In Chapter 4, we examined the wind speeds in the Hopkins Forest over the course of a year. Here's the scatterplot we saw then: (Data in **Hopkins Forest**)

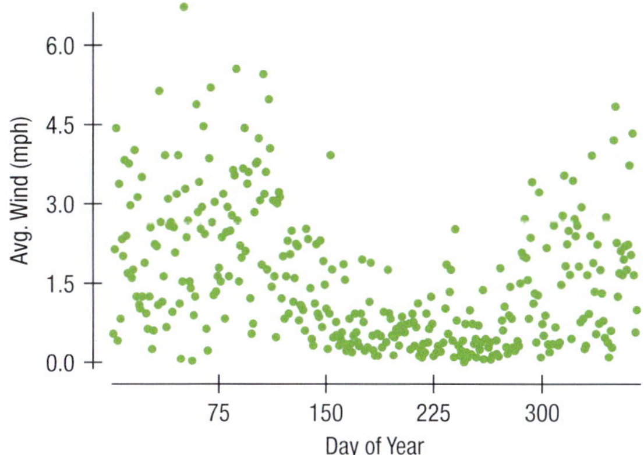

a) Describe the pattern you see here.
b) Should we try re-expressing either variable to make this plot straighter? Explain.

57. Gas mileage As the example in the chapter indicates, one of the important factors determining a car's *Fuel Efficiency* is its *Weight*. Let's examine this relationship again, for 11 cars.

a) Describe the association between these variables shown in the scatterplot at the top of the next column.

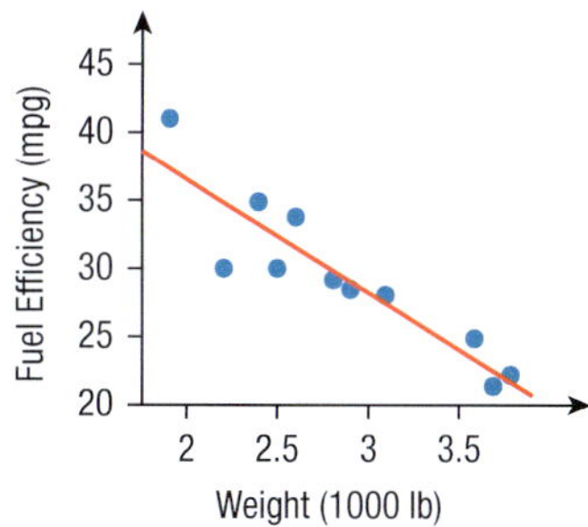

b) Here is the regression analysis for the linear model. What does the slope of the line say about this relationship?

Dependent variable is: Fuel Efficiency
R-squared = 82.9%

Variable	**Coefficient**
Intercept	52.8879
Weight	−8.20243

c) Do you think this linear model is appropriate? Use the residuals plot to explain your decision.

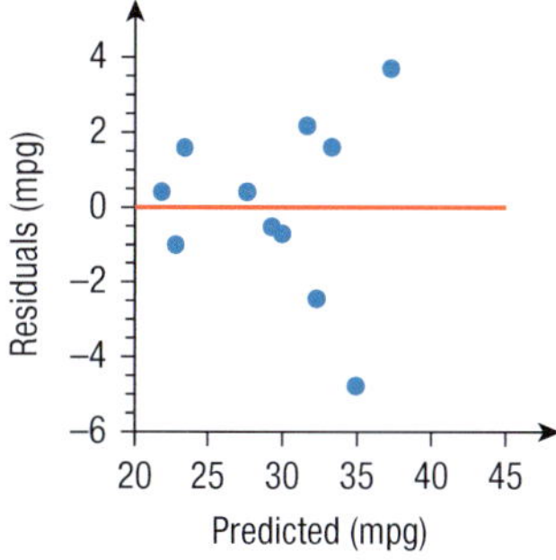

58. Crowdedness In a *Chance* magazine article (Summer 2005), Danielle Vasilescu and Howard Wainer used data from the United Nations Center for Human Settlements to investigate aspects of living conditions for several countries. Among the variables they looked at were the country's per capita gross domestic product (*GDP*, in $) and *Crowdedness*, defined as the average number of persons per room living in homes there. This scatterplot displays these data for 56 countries:

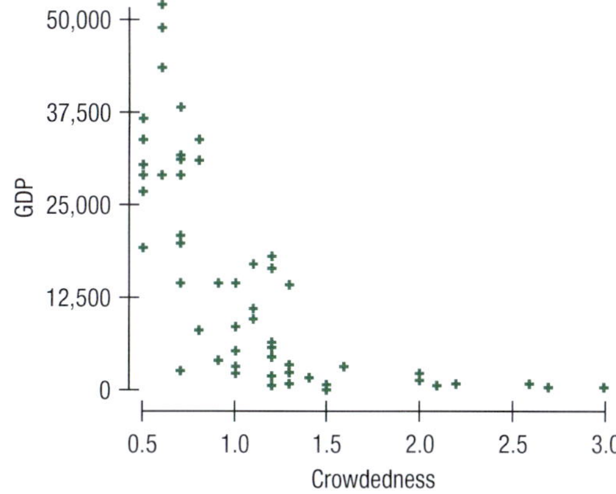

a) Explain why you should re-express these data before trying to fit a model.
b) What re-expression of *GDP* would you try as a starting point?

59. Gas mileage, revisited Let's try the re-expressed variable *Fuel Consumption* (gal/100 mi) to examine the fuel efficiency of the 11 cars in Exercise 57. Here are the revised regression analysis and residuals plot:

Dependent variable is: Fuel Consumption
R-squared = 89.2%

Variable	Coefficient
Intercept	0.624932
Weight	1.17791

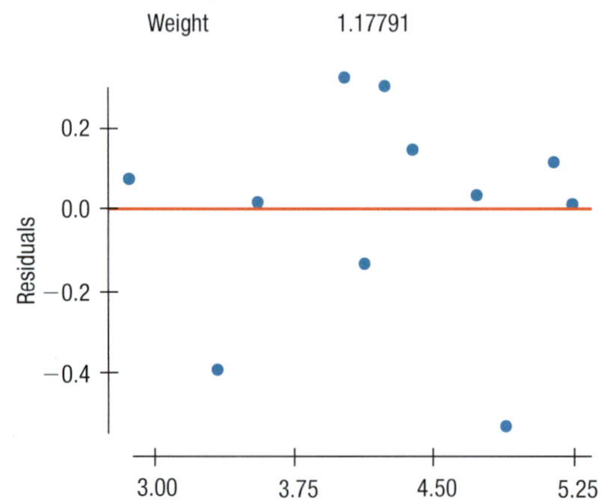

a) Explain why this model appears to be better than the linear model.
b) Using the regression analysis above, write an equation of this model.
c) Interpret the slope of this line.
d) Based on this model, how many miles per gallon would you expect a 3500-pound car to get?

T **60.** Crowdedness again In Exercise 58 we looked at United Nations data about a country's *GDP* and the average number of people per room (*Crowdedness*) in housing there. For a re-expression, a student tried the reciprocal $-10000/GDP$, representing the number of people per \$10,000 of gross domestic product. Here are the results, plotted against *Crowdedness:*

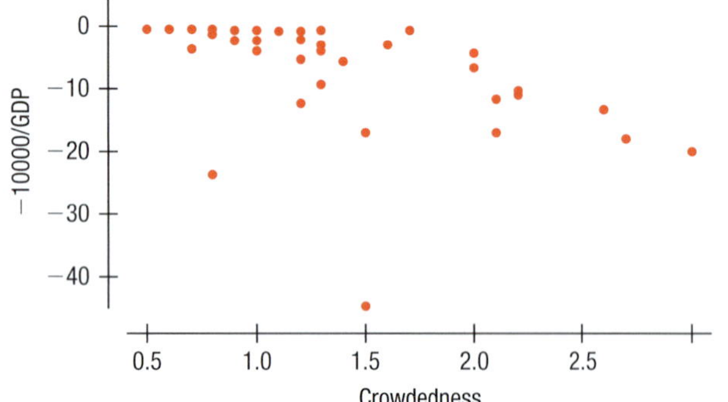

a) Is this a useful re-expression? Explain.
b) The low outliers are Gambia and Guinea. If they were set aside or corrected, would this then be a useful re-expression?
c) Why could you not see these outliers in the plot of the original data in Exercise 58?
d) Look up the GDP of Gambia and correct it in the data. Is it now an outlier (after re-expression)?

T **61.** GDP 2016 The scatterplot shows the gross domestic product (GDP) of a nation in trillions of 2009 dollars plotted against years since 1950.

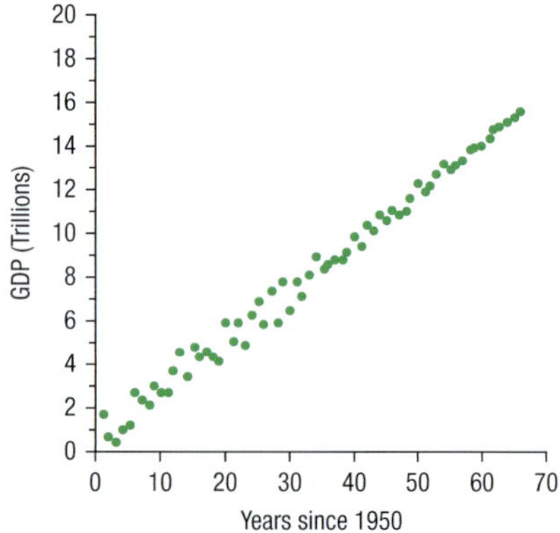

A linear model fit to the relationship looks like this

Dependent variable is GDP (Trillions)
R-squared = 98.9% s = 0.4527

Variable	Coefficient
Intercept	0.803
Years since 1950	0.225

a) Does the value 98.9% suggest that this is a good model? Explain.
b) Here's a scatterplot of the residuals. Now do you think this is a good model for these data? Explain.

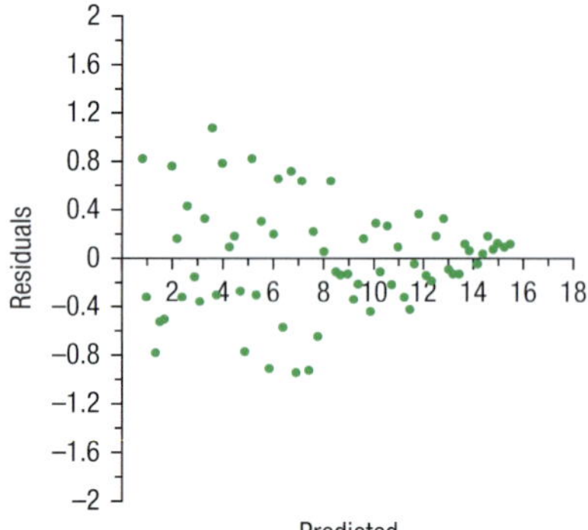

T ***62.** TBill rates 2016, once more The 3-month Treasury bill interest rate is watched by investors and economists. Here's the scatterplot of the 3-month Treasury bill rate since 1934 that we saw in Exercise 43:

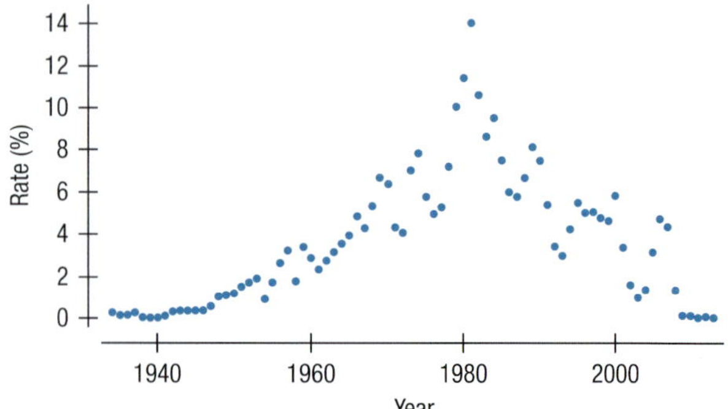

Clearly, the relationship is not linear. Can it be made nearly linear with a re-expression? If so, which one would you suggest? If not, why not?

T **63. Better GDP model?** Consider again the post-1950 trend in U.S. GDP we examined in Exercise 61. Here are regression output and a residual plot when we use the log of GDP in the model. Is this a better model for GDP? Explain. Would you want to consider a different re-expression? If so, which one?

Dependent variable is: LogGDP
R-squared = 99.0% s = 0.0268

Variable	Coefficient
Intercept	0.3918
Years since 1950	0.01358

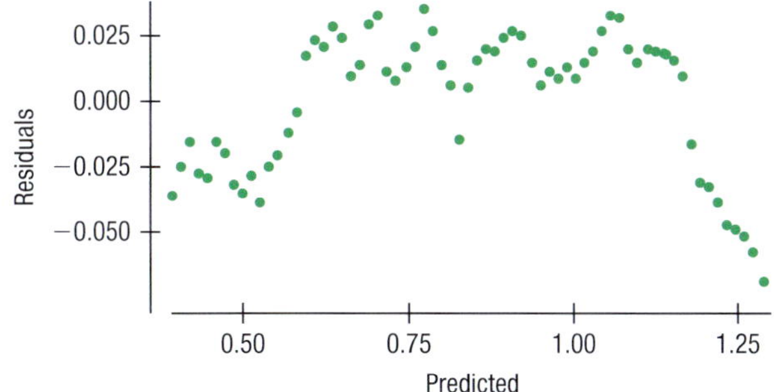

T ***64. Boyle** Scientist Robert Boyle examined the relationship between the volume in which a gas is contained and the pressure in its container. He used a cylindrical container with a moveable top that could be raised or lowered to change the volume. He measured the *Height* in inches by counting equally spaced marks on the cylinder, and measured the *Pressure* in inches of mercury (as in a barometer). Here is Boyle's original table, but you can find the data in the file **Boyle**. Find a suitable re-expression of *Pressure*, fit a regression model, and check whether it is appropriate for modeling the relationship between *Pressure* and *Height*.

A Table of the Condenſation of the Air

A	A	B	C	D	E
48	12	00	Added to $29\frac{1}{8}$ makes	$29\frac{2}{16}$	$29\frac{2}{16}$
46	$11\frac{1}{2}$	$01\frac{7}{16}$		$30\frac{9}{16}$	$30\frac{6}{16}$
44	11	$02\frac{13}{16}$		$31\frac{15}{16}$	$31\frac{12}{16}$
42	$10\frac{1}{2}$	$04\frac{6}{16}$		$33\frac{8}{16}$	$33\frac{1}{7}$
40	10	$06\frac{3}{16}$		$35\frac{5}{16}$	35--
38	$9\frac{1}{2}$	$07\frac{14}{16}$		37--	$36\frac{15}{19}$
36	9	$10\frac{2}{16}$		$39\frac{5}{16}$	$38\frac{7}{8}$
34	$8\frac{1}{2}$	$12\frac{8}{16}$		$41\frac{10}{16}$	$41\frac{2}{17}$
32	8	$15\frac{1}{16}$		$44\frac{3}{16}$	$43\frac{11}{16}$
30	$7\frac{1}{2}$	$17\frac{15}{16}$		$47\frac{1}{16}$	$46\frac{3}{5}$
28	7	$21\frac{3}{16}$		$50\frac{5}{16}$	50--
26	$6\frac{1}{2}$	$25\frac{3}{16}$		$54\frac{5}{16}$	$53\frac{10}{13}$
24	6	$29\frac{11}{16}$		$58\frac{13}{16}$	$58\frac{2}{8}$
23	$5\frac{3}{4}$	$32\frac{3}{16}$		$61\frac{5}{16}$	$60\frac{18}{23}$
22	$5\frac{1}{2}$	$34\frac{15}{16}$		$64\frac{1}{16}$	$63\frac{6}{11}$
21	$5\frac{1}{4}$	$37\frac{15}{16}$		$67\frac{1}{16}$	$66\frac{4}{7}$
20	5	$41\frac{9}{16}$		$70\frac{11}{16}$	70--
19	$4\frac{3}{4}$	45--		$74\frac{2}{16}$	$73\frac{11}{19}$
18	$4\frac{1}{2}$	$48\frac{12}{16}$		$77\frac{14}{16}$	$77\frac{2}{3}$
17	$4\frac{1}{4}$	$53\frac{11}{16}$		$82\frac{12}{16}$	$82\frac{4}{17}$
16	4	$58\frac{2}{16}$		$87\frac{14}{16}$	$87\frac{3}{8}$
15	$3\frac{3}{4}$	$63\frac{15}{16}$		$93\frac{1}{16}$	$93\frac{1}{5}$
14	$3\frac{1}{2}$	$71\frac{5}{16}$		$100\frac{7}{16}$	$99\frac{6}{7}$
13	$3\frac{1}{4}$	$78\frac{11}{16}$		$107\frac{13}{16}$	$107\frac{7}{13}$
12	3	$88\frac{7}{16}$		$117\frac{9}{16}$	$116\frac{4}{8}$

AA. The number of equal ſpaces in the ſhorter leg, that contained the ſame parcel of Air diverſly extended.
B. The height of the Mercurial Cylinder in the longer leg, that compreſs'd the Air into thoſe dimenſions.
C. The height of a Mercurial Cylinder that counterbalanc'd the preſſure of the Atmoſphere.
D. The Aggregate of the two laſt Columns *B* and *C*, exhibiting the preſſure ſuſtained by the included Air.
E. What that preſſure ſhould be according to the *Hypotheſis*, that ſuppoſes the preſſures and expanſions to be in reciprocal proportion.

T ***65. Brakes** The following table shows stopping distances in feet for a car tested 3 times at each of 5 speeds. We hope to create a model that predicts *Stopping Distance* from the *Speed* of the car.

Speed (mph)	Stopping Distances (ft)
20	64, 63, 59
30	113, 119, 103
40	151, 171, 167
50	234, 200, 242
60	322, 316, 275

a) Explain why a linear model is not appropriate.
b) Re-express the data to straighten the scatterplot.
c) Create an appropriate model.
d) Estimate the stopping distance for a car traveling 45 mph.
e) Estimate the stopping distance for a car traveling 70 mph.
f) How much confidence do you place in these predictions? Why?

T ***66. Pendulum** A student experimenting with a pendulum counted the number of full swings the pendulum made in 20 seconds for various lengths of string. Her data are shown here.

Length (in.)	6.5	9	11.5	14.5	18	21	24	27	30	37.5
Number of Swings	22	20	17	16	14	13	13	12	11	10

a) Explain why a linear model is not appropriate for using the *Length* of a pendulum to predict the *Number of Swings* in 20 seconds.
b) Re-express the data to straighten the scatterplot.
c) Create an appropriate model.
d) Estimate the number of swings for a pendulum with a 4-inch string.
e) Estimate the number of swings for a pendulum with a 48-inch string.
f) How much confidence do you place in these predictions? Why?

T ***67. Baseball salaries 2015** Ballplayers have been signing ever-larger contracts. The highest salaries (in millions of dollars per season) for each year since 1874 are in the data file **Baseball salaries 2015**.

a) Make a scatterplot of *Adjusted salary* vs. *Year*. Does it look straight?
b) Find the best re-expression of *Adjusted salary* that you can to straighten out the scatterplot.
c) Fit the regression for the re-expression you found in part b, and write a sentence telling what it says.
d) Plot the residuals of the model in part c against *Year*.
e) Comment on the model you found given the residual plot in part d.

T ***68. Planets** Here is a table of the 9 sun-orbiting objects formerly known as planets.

Planet	Position Number	Distance from Sun (million miles)	Length of Year (Earth years)
Mercury	1	36.254	0.24
Venus	2	66.931	0.62
Earth	3	92.960	1
Mars	4	141.299	1.88
Jupiter	5	483.392	11.86
Saturn	6	886.838	29.46
Uranus	7	1782.97	84.01
Neptune	8	2794.37	164.8
Pluto	9	3671.92	248

a) Plot the *Length* of the year against the *Distance* from the sun. Describe the shape of your plot.
b) Re-express one or both variables to straighten the plot. Use the re-expressed data to create a model describing the length of a planet's year based on its distance from the sun.
c) Comment on how well your model fits the data.

T ***69. Is Pluto a planet?** Let's look again at the pattern in the locations of the planets in our solar system seen in the table in Exercise 68.

a) Re-express the distances to create a model for the *Distance* from the sun based on the planet's *Position*.
b) Based on this model, would you agree with the International Astronomical Union that Pluto is not a planet? Explain.

T ***70. Planets and asteroids** The asteroid belt between Mars and Jupiter may be the remnants of a failed planet. If so, then Jupiter is really in position 6, Saturn is in 7, and so on. Repeat Exercise 69, using this revised method of numbering the positions. Which method seems to work better?

T ***71. Planets and Eris** In July 2005, astronomers Mike Brown, Chad Trujillo, and David Rabinowitz announced the discovery of a sun-orbiting object, since named Eris,[8] that is 5% larger than Pluto. Eris orbits the sun once every 560 earth years at an average distance of about 6300 million miles from the sun. Based on its *Position*, how does Eris's *Distance* from the sun (re-expressed to logs) compare with the prediction made by your model of Exercise 70?

T ***72. Planets, models, and laws** The model you found in Exercise 68 is a relationship noted in the 17th century by Kepler as his Third Law of Planetary Motion. It was subsequently explained as a consequence of Newton's Law of Gravitation. The models for Exercises 69, 70, and 71 relate to what is sometimes called the Titius-Bode "law," a pattern noticed in the 18th century but lacking any scientific explanation.

Compare how well the re-expressed data are described by their respective linear models. What aspect of the model of Exercise 68 suggests that we have found a physical law? In the future, we may learn enough about planetary systems around other stars to tell whether the Titius-Bode pattern applies there. If you discovered that another planetary system followed the same pattern, how would it change your opinion about whether this is a real natural "law"? What would you think if some of the extrasolar planetary systems being discovered do not follow this pattern?

***73. Logs (not logarithms)** The value of a log is based on the number of board feet of lumber the log may contain. (A board foot is the equivalent of a piece of wood 1 inch thick, 12 inches wide, and 1 foot long. For example, a 2″ × 4″ piece that is 12 feet long contains 8 board feet.) To estimate the amount of lumber in a log, buyers measure the diameter inside the bark at the smaller end. Then they look in a table based on the Doyle Log Scale. The table below shows the estimates for logs 16 feet long.

Diameter of Log	8	12	16	20	24	28
Board Feet	169	441	841	1369	2025	2809

a) What model does this scale use?
b) How much lumber would you estimate that a log 25 inches in diameter contains?
c) What does this model suggest about logs 30 inches in diameter?

T ***74. Weightlifting 2016** Listed below are the world record men's weightlifting performances as of 2016.

Weight Class (kg)	Record Holder	Country	Total Weight
56	Long Quingquan	China	307
62	Kim Un-Guk	North Korea	327
69	Galain Boevski	Bulgaria	357
77	Lu Xiaojun	China	379
85	Kianoush Rostami	Iran	396
94	Ilya Ilyin	Kazakhstan	418
105	Andrei Aramnau	Belarus	436
105+	Lasha Tlakhadze	Georgia	473

a) Create a linear model for the *Weight Lifted* in each *Weight Class*, leaving out the 105+ unlimited class.
b) Check the residuals plot. Is your linear model appropriate?
c) Create a better model by re-expressing *Weight Lifted* and explain how you found it.
d) Explain why you think your new model is better.
e) The record holder of the Unlimited weight class weighs 157 kg. Predict how much he can lift with the models from parts a and c.

***75. Life expectancy history** The table gives the *Life Expectancy* for a certain demographic group in a certain region for every decade during the past 120 years (1 = 1900 to 1910, 2 = 1911 to 1920, etc.). Create a model to predict future increases in life expectancy. Hint: Try "Plan B."

Decade	1	2	3	4	5	6	7	8	9	10	11	12
Life exp	46.8	48.4	54.5	60.0	62.3	66.3	67.8	68.5	70.7	72.6	75.2	76.0

[8]Eris is the Greek goddess of warfare and strife who caused a quarrel among the other goddesses that led to the Trojan War. In the astronomical world, Eris stirred up trouble when the question of its proper designation led to the raucous meeting of the IAU in Prague where IAU members voted to demote Pluto and Eris to dwarf-planet status (www.gps.caltech.edu/~mbrown/planetlila/#paper).

T ***76.** Lifting more weight In Exercise 74 you examined the record weightlifting performances for the Olympics. You found a re-expression of *Weight Lifted*.

a) Find a model for *Weight Lifted* by re-expressing *Weight Class* instead of *Weight Lifted*.
b) Compare this model to the one you found in Exercise 74.
c) Predict the *Weight Lifted* by the 157 kg record holder in Exercise 74, part c.
d) Which prediction do you think is better? Explain.
e) The record holder is Lasha Talakhadze, who lifted 473 kg at the 2016 Rio de Janeiro Olympics. Which model predicted it better?

T ***77.** Slower is cheaper? Researchers studying how a car's *Fuel Efficiency* varies with its *Speed* drove a compact car 200 miles at various speeds on a test track. Their data are shown in the table.

Speed (mph)	35	40	45	50	55	60	65	70	75
Miles per gal	15.9	19.4	22.8	24.9	25.9	26.2	25.2	23.9	21.7

Create a linear model for this relationship, and report any concerns you may have about the model.

T ***78.** Oranges The table below shows that as the number of oranges on a tree increases, the fruit tends to get smaller. Create a model for this relationship, and express any concerns you may have.

Number of Oranges/Tree	Average Weight/Fruit (lb)
50	0.61
100	0.59
150	0.57
200	0.56
250	0.53
300	0.52
350	0.51
400	0.49
450	0.48
500	0.46
600	0.45
700	0.43
800	0.41
900	0.40

T ***79.** Years to live, 2016 Insurance companies and other organizations use actuarial tables to estimate the remaining lifespans of their customers. The data file gives life expectancy and estimated additional years of life for black males in a particular country.

Age	10	20	30	40	50	60	70	80	90	100
Years Left	61.2	52.4	40.8	33.7	27.9	19.5	13.3	8.6	4.8	2.5

a) Fit a regression model to predict *Life expectancy* from *Age*. Does it look like a good fit? Now plot the residuals.
b) Find a re-expression to create a better model. Predict the life expectancy of a 21-year-old black man.
c) Are you satisfied that your model has accounted for the relationship between *Life expectancy* and *Age*? Explain.

***80.** Tree growth A 1996 study examined the growth of grapefruit trees in Texas, determining the average trunk *Diameter* (in inches) for trees of varying *Ages:*

Age (yr)	2	4	6	8	10	12	14	16	18	20
Diameter (in.)	2.1	3.9	5.2	6.2	6.9	7.6	8.3	9.1	10.0	11.4

a) Fit a linear model to these data. What concerns do you have about the model?
b) If data had been given for individual trees instead of averages, would you expect the fit to be stronger, less strong, or about the same? Explain.

JUST CHECKING

Answers

1. Not high leverage, not influential, large residual
2. High leverage, not influential, small residual
3. High leverage, influential, not large residual
4. Counts are often best transformed by using the square root.
5. None. The relationship is already straight.
6. Even though, technically, the population values are counts, you should probably try a stronger transformation like log(population) because populations grow in proportion to their size.

9

Multiple Regression

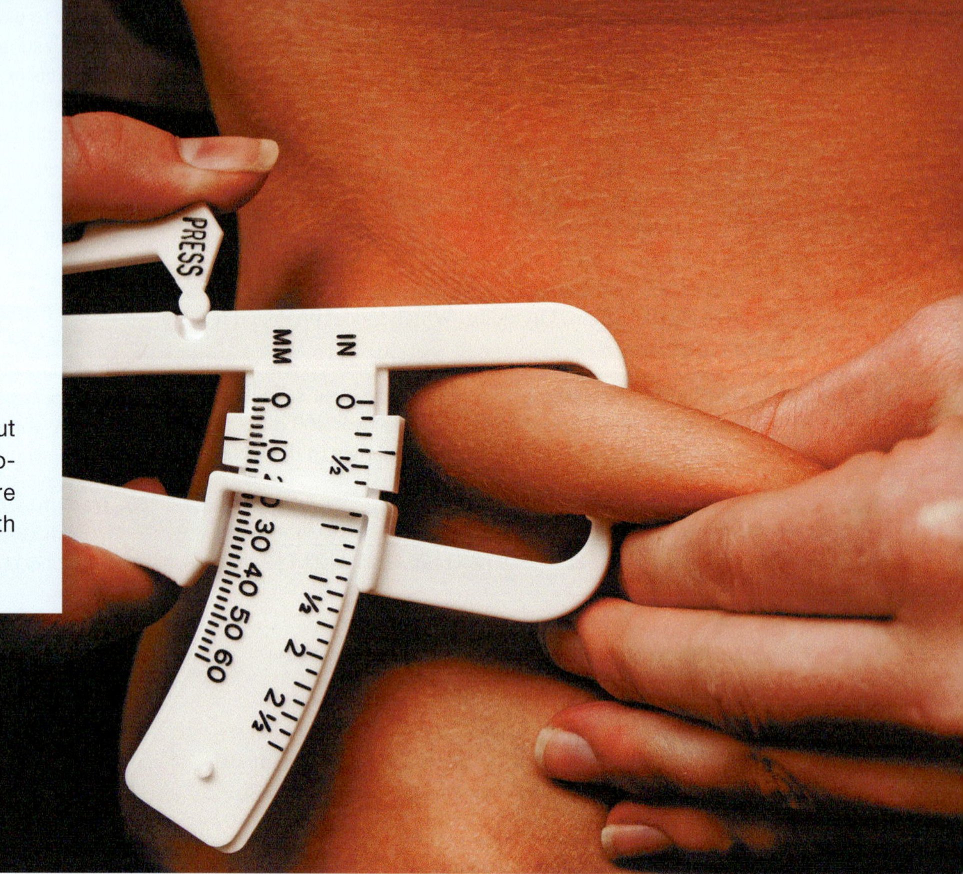

WHERE ARE WE GOING?

Linear regression models are often useful, but the world is usually not so simple that a two-variable model does the trick. For a more realistic understanding, we need models with several variables.

Three percent of a man's body is essential fat. For a woman, the percentage is closer to 12.5%. As the name implies, essential fat is necessary for a normal, healthy body. Fat is stored in small amounts throughout the body. Too much body fat, however, can be dangerous to health. For men between 18 and 39 years old, a healthy percentage of body fat varies from 8% to 19%. For women of the same age, it's 21% to 32%.

Measuring body fat can be tedious and expensive. The "standard reference" measurement is by dual-energy X-ray absorptiometry (DEXA), which involves two low-dose X-ray generators and takes from 10 to 20 minutes.

How close can we get to a usable prediction of body fat from easily measurable variables such as *Height, Weight*, and *Waist* size? Here's a scatterplot of *%Body Fat* plotted against *Waist* size for a sample of 250 men of various ages. (Data in **Bodyfat**)

WHO	250 male subjects
WHAT	Body fat and waist size
UNITS	% Body fat and inches
WHEN	1990s
WHERE	United States
WHY	Scientific research

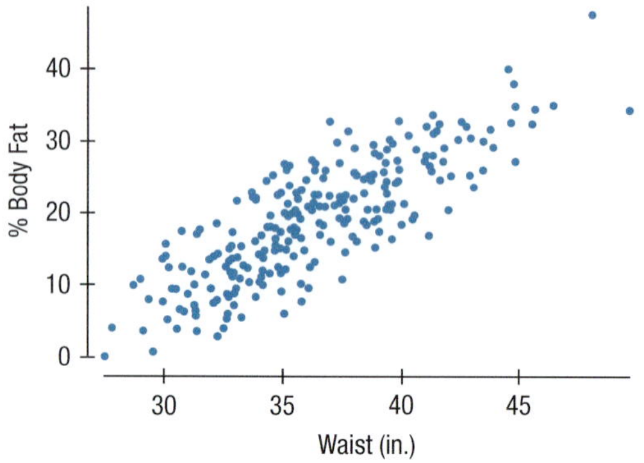

Figure 9.1
The relationship between *%Body Fat* and *Waist* size for 250 men

The plot is clearly straight, so we can find a least squares regression line. The equation of the least squares line for these data is $\widehat{\%Body\ Fat} = -42.7 + 1.7\ Waist$. The slope says that, on average, men who have an additional inch around the waist are expected to have about 1.7% more body fat.

This regression does pretty well. The standard deviation of the residuals is just 4.713 *%Body Fat*. The R^2 of 67.8% says that the regression model accounts for almost 68% of the variability in *%Body Fat* just by knowing *Waist* size.

But that remaining 32% of the variance is bugging us. Couldn't we do a better job of accounting for *%Body Fat* if we weren't limited to a single predictor? In the full dataset there were 15 other measurements on the 250 men. We might be able to use other predictor variables to help us account for the leftover variation that wasn't accounted for by waist size.

9.1 What Is Multiple Regression?

A Note on Terminology

When we have two or more predictors and fit a linear model by least squares, we are formally said to fit a least squares linear multiple regression. Most folks just call it "multiple regression." You may also see the abbreviation OLS used with this kind of analysis. It stands for "Ordinary Least Squares."

Does a regression with two predictors even make sense? It does—and that's fortunate because the world is too complex a place for linear regression to model it with a single predictor. A regression with two or more predictor variables is called a **multiple regression**. (When we need to note the difference, a regression on a single predictor is called a simple regression.) We'd never try to find a regression by hand, and even calculators aren't really up to the task. This is a job for a statistics program on a computer. If you know how to find the regression of *%Body Fat* on *Waist* size with a statistics package, you can usually just add *Height* to the list of predictors without having to think hard about how to do it.

For simple regression, we found the **least squares solution**, the one whose coefficients made the sum of the squared residuals as small as possible. For multiple regression, we'll do the same thing, but this time with more coefficients. Remarkably, we can still solve this problem. Even better, a statistics package can find the coefficients of the least squares model easily.

Here's a typical example of a multiple regression table:

Table 9.1

A multiple regression table provides many values of interest. But for now, we'll just look at the R^2, the standard deviation of the residuals, and the coefficients. You can ignore the grayed-out parts of this table until we come back to the topic later in the text.

Dependent variable is: %Body Fat
R-squared = 71.3,
s = 4.460 with 250 − 3 = 247 degrees of freedom

Variable	Coefficient	SE(Coeff)	t-ratio	P-value
Intercept	−3.10088	7.686	−0.403	0.6870
Waist	1.77309	0.0716	24.8	<0.0001
Height	−0.60154	0.1099	−5.47	<0.0001

We have already seen many of the important numbers in this table. Most of them continue to mean what they did in a simple regression.

R^2 gives the fraction of the variability of *%Body Fat* accounted for by the multiple regression model. The multiple regression model accounts for 71.3% of the variability in *%Body Fat*. That's an improvement over 67.8% with *Waist* alone predicting *%Body Fat*. We shouldn't be surprised that R^2 has gone up. It was the hope of accounting for some of the leftover variability that led us to try a second predictor.

The standard deviation of the residuals is still denoted s (or sometimes s_e to distinguish it from the standard deviation of y). It has gone down from 4.713 to 4.460. As before, we can think of the 68–95–99.7 Rule. The regression model will, of course, make errors—indeed, it is unlikely to fit any of the 250 men perfectly. But about 68% of those errors are likely to be smaller than 4.46 *%Body Fat*, and 95% are likely to be less than twice that.

Other values are usually reported as part of a regression. We've shown them here because you are likely to see them in any computer-generated regression table. But we've made them gray because we aren't going to talk about them yet. You can safely ignore them for now. Don't worry, we'll discuss them all in later chapters.

For each predictor, the table gives a coefficient next to the name of the variable it goes with. Using the coefficients from this table, we can write the regression model:

$$\widehat{\%Body\ Fat} = -3.10 + 1.77\ Waist - 0.60\ Height.$$

As before, we define the residuals as

$$Residuals = \%Body\ Fat - \widehat{\%Body\ Fat}.$$

We've fit this model with the same least squares principle: The sum of the squared residuals is as small as possible for any choice of coefficients.

So what's different and why is understanding multiple regression important? There are several answers to these questions. First—and most important—the meaning of the coefficients in the regression model has changed in a subtle but important way. Because that change is not obvious, multiple regression coefficients are often misinterpreted.

Second, multiple regression is an extraordinarily versatile calculation, underlying many widely used statistics methods. A sound understanding of the multiple regression model will help you to understand these other applications.

Third, multiple regression offers a glimpse into statistical models that use more than two quantitative variables. The real world is complex. Simple models of the kind we've seen so far are a great start, but often they're just not detailed enough to be useful for understanding, predicting, and decision making. Models that use several variables can be a big step toward realistic and useful modeling of complex phenomena and relationships.

EXAMPLE 9.1 Modeling Home Prices

As a class project, students in a large statistics class collected publicly available information on recent home sales in their hometowns. There are 894 properties. These are not a random sample, but they may be representative of home sales during a short period of time, nationwide. (Data in **Real estate**)

Variables available include the price paid, the size of the living area (sq ft), the number of bedrooms, the number of bathrooms, the year of construction, the lot size (acres), and a coding of the location as urban, suburban, or rural made by the student who collected the data.

Here's a regression to model the sale price from the living area (sq ft) and the number of bedrooms.

Dependent variable is: Price
R-squared = 14.6%
s = 266899 with 894 − 3 = 891 degrees of freedom

Variable	**Coefficient**	SE(Coeff)	t-ratio	P-value
Intercept	308100	41148	7.49	<0.0001
Living area	135.089	11.48	11.8	<0.0001
Bedrooms	−43346.8	12844	−3.37	0.0008

QUESTION: How should we interpret the regression output?

ANSWER: The model is

$$\widehat{Price} = 308{,}100 + 135\ Living\ Area - 43{,}347\ Bedrooms.$$

The *R*-squared says that this model accounts for 14.6% of the variation in *Price*. But the value of *s* leads us to doubt that this model would provide very good predictions because the standard deviation of the residuals is more than $266,000. Nevertheless, we may be able to learn about home prices.

9.2 Interpreting Multiple Regression Coefficients

The multiple regression model suggests that height might be important in predicting body fat in men, but it isn't immediately obvious why that should be true. What's the relationship between *%Body Fat* and *Height* in men? We know how to approach this question; we follow the three rules from Chapter 2 and make a picture. Here's the scatterplot:

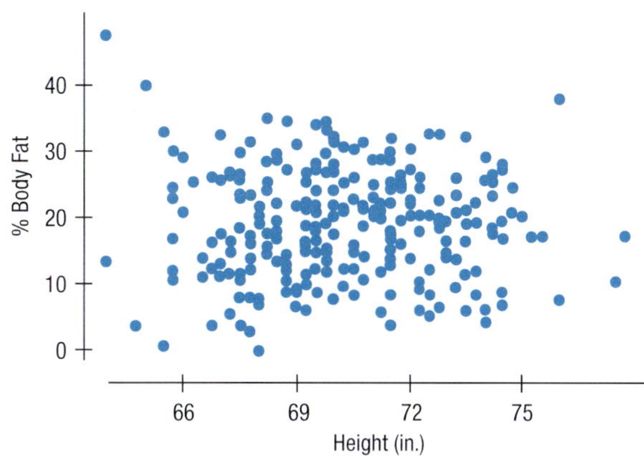

Figure 9.2
The scatterplot of *%Body Fat* against *Height* seems to say that there is little relationship between these variables.

As their name reminds us, residuals are what's left over after we fit a model. That lets us remove the effects of some variables. The residuals are what's left.

That doesn't look very promising. It doesn't look like *Height* tells us much about *%Body Fat* at all. You just can't tell much about a man's *%Body Fat* from his *Height*. But in the multiple regression it certainly seemed that *Height did* contribute to the multiple regression model. How could that be?

The answer is that the multiple regression coefficient of *Height* takes account of the other predictor, *Waist* size, in the regression model. In a multiple regression, each coefficient must be interpreted as the relationship between *y* and that *x after allowing for the linear effects of the other x's on both variables*. So the coefficient for *Height* is about the relationship between *%Body Fat* and *Height* after we allow for *Waist* size. But what does that mean?

Think about all men whose waist size is about 37 inches—right in the middle of our data. If we think only about these men, what do we expect the relationship between *Height* and *%Body Fat* to be? Now a negative association makes sense because taller men probably have less body fat than shorter men who have the same waist size. Let's look at the plot. Figure 9.3 shows the men with waist sizes between 36 and 38 inches as blue dots.

What about the coefficient of *Waist*? Well, it no longer means what it did in the simple regression. Multiple regression treats all the predictors alike, so the coefficient of *Waist* now tells about how the body fat% of men of the same height tends to vary with their waist size. In the simple regression, the coefficient of *Waist* was 1.7. It hasn't changed much in the multiple regression. (It is now 1.77.) But its meaning *has* changed.

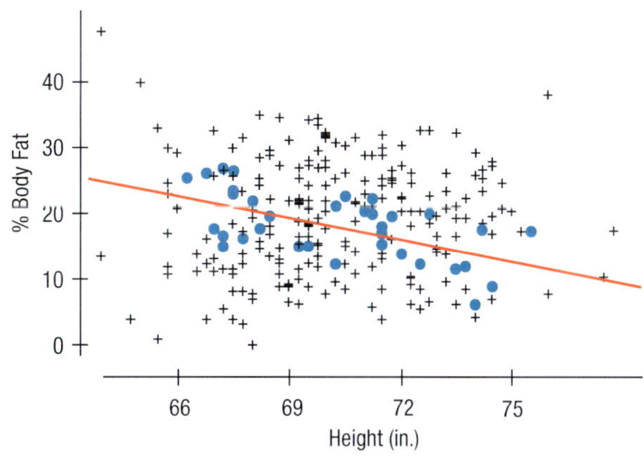

Figure 9.3
When we restrict our attention to men with waist sizes between 36 and 38 inches (points in blue), we can see a relationship between *%Body Fat* and *Height*.

Overall, there's little relationship between *%Body Fat* and *Height*, as you can see from the full set of points. But when we focus on particular waist sizes, there is a relationship between body fat and height. This relationship is *conditional* because we've restricted our attention to those men with a given waist size. Among men with that waist size, those who are an inch taller generally have about 0.60% less body fat. If that relationship is consistent for each waist size, then the multiple regression coefficient will estimate it. The simple regression coefficient simply couldn't see it.

The presence of several predictors in a multiple regression makes the interpretation of a multiple regression slope less straightforward than it is in simple regression. One way to think about the coefficient b_j for a predictor x_j is the slope of y on x_j after the effects of all the other predictors have been taken out of both y and x_j. When it makes sense, b_j can also be interpreted as the effect of x_j on y for fixed values of all the other predictors.

When predictor variables are highly correlated, as they are in many real situations, the interpretation is more challenging. It may not be possible to hold one x constant while changing another. And removing the effects of one predictor from another predictor with which it is correlated can leave behind a variable that doesn't look very much like the original x_j, changing the meaning of b_j.

In general, we can't assume that a variable's coefficient will stay roughly the same when new predictors are included in the model. Often coefficients change in unexpected ways—even changing sign. You should be alert for possible changes in value, and for certain changes in meaning.

EXAMPLE 9.2 Interpreting Coefficients

RECAP: We looked at a multiple regression to predict the price of a house from its living area and the number of bedrooms. We found the model

$$\textit{Price} = 308{,}100 + 135\ \textit{Living Area} - 43{,}347\ \textit{Bedrooms}.$$

However, common sense says that houses with more bedrooms are usually worth more. And, in fact, the simple regression of *Price* on *Bedrooms* finds the model

$$\textit{Price} = 338{,}975 + 40{,}234\ \textit{Bedrooms}.$$

QUESTION: How should we understand the coefficient of *Bedrooms* in the multiple regression?

ANSWER: The coefficient of *Bedrooms* in the multiple regression does not mean that houses with more bedrooms are generally worth less. It must be interpreted taking account of the other predictor (*Living area*) in the regression. If we consider houses with a given amount of living area, those that devote more of that area to bedrooms must have either smaller bedrooms or less living area for other parts of the house. Those differences could result in reducing the home's value.

Multiple Regression Coefficients Can Be Surprising

The multiple regression model looks simple and straightforward:

$$\hat{y} = b_0 + b_1x_1 + \cdots + b_kx_k.$$

It looks like each b_j tells us the effect of its associated predictor, x_j, on the response variable, y. But that is not so. This is, without a doubt, the most common error that people make with multiple regression.

- It is possible for there to be no simple relationship between y and x_j, and yet for b_j in a multiple regression to be quite different from 0. We saw this happen for the coefficient of *Height* in our example.
- It is also possible for there to be a strong two-variable relationship between y and x_j, and yet for b_j in a multiple regression to be almost 0. If we're trying to model the horsepower of a car, using both its weight and its engine size, it may turn out that the coefficient for *Engine Size* is nearly 0. That doesn't mean that engine size isn't important for understanding horsepower. It simply means that after allowing for the weight of the car, the engine size doesn't give much additional information.
- It is even possible for there to be a strong linear relationship between y and x_j in one direction, and yet b_j can be of the *opposite* sign in a multiple regression. More expensive cars tend to be bigger, and since bigger cars have worse fuel efficiency, the price of a car has a slightly negative association with fuel efficiency. But in a multiple regression of *Fuel Efficiency* on *Weight* and *Price*, the coefficient of *Price* may be positive. If so, it means that among cars of the same weight, more expensive cars have better fuel efficiency. The simple regression on *Price*, though, has the opposite direction because, overall, more expensive cars are bigger. This switch in sign may seem a little strange at first, but it's not really a contradiction at all. It's due to the change in the meaning of the coefficient of *Price* when it is in a multiple regression rather than a simple regression.

So we'll say it once more: The coefficient of x_j in a multiple regression depends as much on the other predictors as it does on x_j. Remember that when you interpret a multiple regression model.

9.3 The Multiple Regression Model—Assumptions and Conditions

Check the Residual Plot (Part 1)

The residuals should appear to have no pattern with respect to the predicted values.

We can write an estimated multiple regression model, numbering the predictors arbitrarily (we don't care which one is x_1) and writing b's for the regression coefficients, like this:

$$\hat{y} = b_0 + b_1x_1 + \cdots + b_kx_k.$$

We then find the residuals as

$$e = y - \hat{y}.$$

(Recall that we use e for the residuals, representing the "errors" that the regression model makes because it can't fit a line through all the points in a typical scatterplot. And besides, r was already taken by correlation.)

Of course, the multiple regression model is not limited to two predictor variables, and regression model equations can involve any number (a typical letter to use is k) of predictors. That doesn't really change anything, so we'll often discuss the two-predictor version just for simplicity.

The assumptions and conditions for the multiple regression model are nearly the same as for simple regression, but with more variables in the model, we'll have to make a few changes.

Check the Residual Plot (Part 2)

The residuals should appear to be randomly scattered and show no patterns or clumps when plotted against the predicted values.

Linearity Assumption

We are fitting a linear model.[1] For that to be the right kind of model, we need an underlying linear relationship. But now we're thinking about several predictors. To see whether the assumption is reasonable, we'll check the Straight Enough Condition for *each* of the predictors.

[1]By *linear* we mean that each x appears simply multiplied by its coefficient and added to the model. No x appears in an exponent or in some other more complicated function. That means that as we consider cases whose x-values are higher, the corresponding predictions for y will be different by a constant rate if nothing else changes.

Straight Enough Condition: Scatterplots of y against each of the predictors are reasonably straight. As we have seen with *Height* in the body fat example, the scatterplots need not show a strong (or any!) slope; we just check that there isn't a bend or other nonlinearity. For the body fat data, the scatterplot is beautifully linear for *Waist*, as we saw in Figure 9.1. For *Height*, we saw no relationship at all, but at least there was no bend.

As we did in simple regression, it's a good idea to check the residuals for linearity after we fit the model. It's good practice to plot the residuals against the predicted values and check for patterns, especially bends or other nonlinearities.

Check the Residual Plot (Part 3)

The spread of the residuals should be uniform when plotted against any of the x's or against the predicted values.

If we're willing to assume that the multiple regression model is reasonable, we can fit the regression model by least squares. That is, we'll find the coefficient values that make the sum of the squared residuals as small as possible.

Regression methods are often applied to data that were "found" on an Internet site, in a database, or in a data warehouse, and were not originally intended for statistical analysis. Regression models fit to such data may still do a good job of modeling the data at hand, but without some reason to believe that the data are representative of a larger population, you might reasonably be reluctant to believe that the model generalizes to other situations.

You should also check displays of the regression residuals for evidence of patterns, trends, or clumping, any of which suggest that there's more to be learned about the data beyond the regression model. As one example, if one of the x-variables is related to time, be sure to check whether the residuals have a pattern when plotted against the data sequence or against the variable *Time*. The body fat data were collected on a sample of men. The men were not related in any way, so we don't expect to find a pattern in the residuals.

Equal Variance Assumption

The variability of the errors should be about the same for all values of each predictor. Once again, we look at scatterplots.

Does the Plot Thicken? Condition: Scatterplots of the regression residuals against each x or against the predicted values, $\hat{y}$, offer a visual check.[2] The spread around the line should be nearly constant. Be alert for a "fan" shape or other tendency for the variability to grow or shrink in one part of the scatterplot.

Here are the residuals from our multiple regression for *%Body Fat* plotted against the residuals. The plot shows no patterns that might indicate a problem.

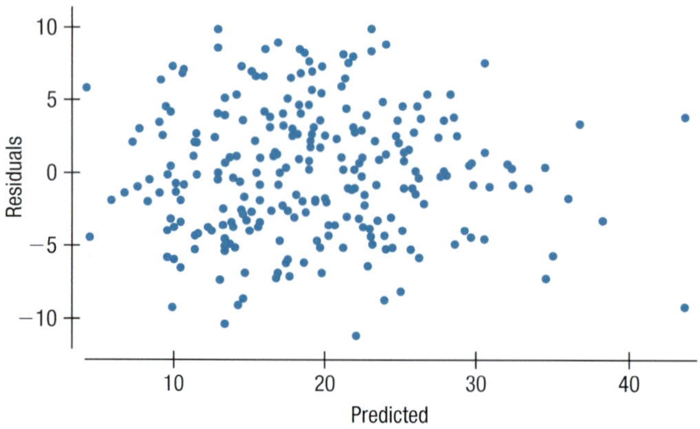

Figure 9.4
Residuals from the *%Body Fat* regression plotted against predicted values show no pattern. That's a good indication that the Straight Enough Condition has been satisfied.

[2] For a simple regression, we plotted residuals against x or $\hat{y}$, but now there is more than one x. So a reasonable alternative is to plot them against a simple (linear) combination of the x's. And that's just what $\hat{y}$ is. (Do you read the footnotes? Are you feeling a sense of déjà vu? If not yet, you may soon.)

If the plot of residuals doesn't thicken, then the residuals are about as variable in one part of the data as in another. If that's so, it makes sense to compute a single estimate of the standard deviation of the residuals and interpret it. The formula for the residual standard deviation is very much like the standard deviation for the residuals we found for a simple regression, but with one minor change:

$$s_e = \sqrt{\frac{\sum e^2}{n - 3}}$$

There's no need to subtract a mean in the numerator because the mean of the residuals must always be zero. And now we are dividing by $n - 3$. Why 3? Well, for the same reason we subtracted 1 in the standard deviation of y and 2 in the residual standard deviation for a simple regression. For the standard deviation of y we estimated one mean; for a simple regression, we found two coefficients; and now we have estimated three. (This is a pattern that we'll explain further in a later chapter.)

If a histogram of the residuals is unimodal and symmetric, then the 68–95–99.7 Rule applies. We can say that about 95% of the errors made by the regression model are smaller than $2s_e$. If residual plots show no pattern and if the plots don't thicken, we can feel good about interpreting the regression model.

Check the Residuals

We assume that the errors around the regression model at each value of the x-variables have a distribution that is unimodal, symmetric, and without outliers. Look at a histogram or Normal probability plot of the residuals. The histogram of residuals in the *%Body Fat* regression certainly looks unimodal and symmetric, and the Normal probability plot is fairly straight. And, as we have said before, this assumption becomes less important as the sample size grows.

Let's summarize all the checks of conditions that we've made and the order in which we've made them:

1. Check the Straight Enough Condition with scatterplots of the y-variable against each x-variable.
2. If the scatterplots are straight enough (that is, if it looks like the regression model is plausible), fit a multiple regression model to the data. (Otherwise, either stop or consider re-expressing an x- or the y-variable.)
3. Find the residuals and predicted values.
4. Make a scatterplot of the residuals against the predicted values. This plot should look patternless. Check in particular for any bend (which would suggest that the data weren't all that straight after all) and for any thickening. If there's a bend and especially if the plot thickens, consider re-expressing the y-variable and starting over.
5. Think about how the data were collected. Was suitable randomization used? Are the data representative of some identifiable population? If the data are measured over time, check for evidence of patterns by plotting the residuals against time to look for patterns.
6. Check a histogram of the residuals to see that the distribution is unimodal, symmetric, and without outliers. Deal with any outliers, correcting or omitting them. If the distribution of the residuals is not unimodal, consider whether there may be subgroups in the data that should be modeled individually. If the residuals are skewed, consider re-expressing the y-variable.
7. If the conditions check out, feel free to interpret the regression model and use it for prediction.

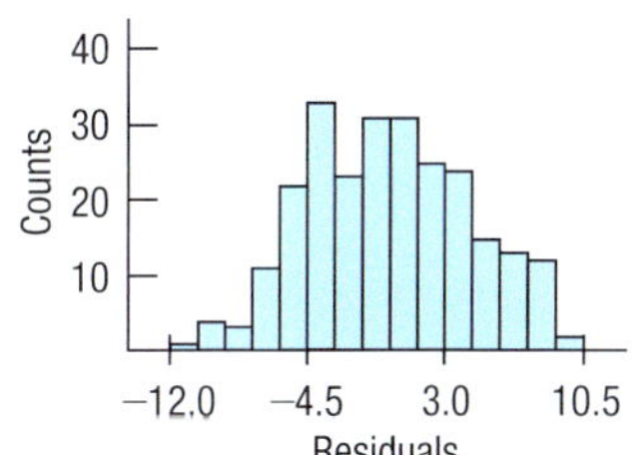

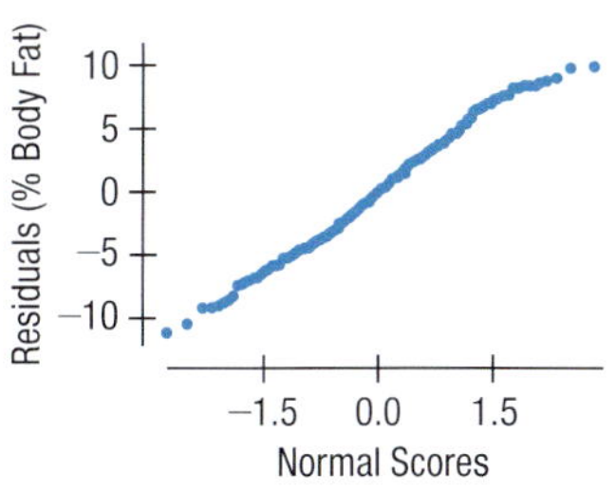

Figure 9.5

Check a histogram of the residuals. The distribution of the residuals should be unimodal and symmetric. Or check a Normal probability plot to see whether it is straight.

STEP-BY-STEP EXAMPLE

Multiple Regression

QUESTION: Zillow.com attracts millions of users each month who are interested in how much their house is worth. Maybe a multiple regression model could help. (Data in **Housing prices**)

VARIABLES Name the variables, report the W's, and specify the questions of interest.

We want to build a regression model to predict house prices for a region of upstate New York. We have data on *Price* ($), *Living Area* (sq ft), *Age* (years), and number of *Bedrooms*.

PLAN Think about the assumptions and check the conditions.

✓ **Straight Enough Condition:** The plot of *Price* against *Living Area* is reasonably straight, as is the plot of *Price* against *Bedrooms*. The plot of *Price* against *Age*, however, is not straight.

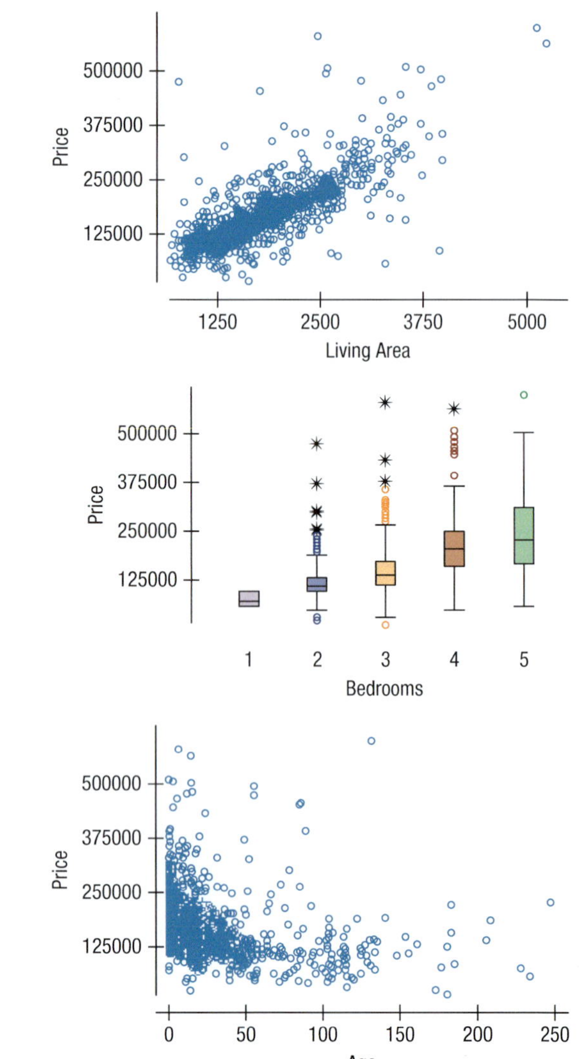

See Chapter 8 for a discussion of re-expressing variables to straighten scatterplots.

The alternative of using the logarithm of *Age* looks better. [Because there are some new houses with $Age = 0$ and we can't find the log of 0, we'll add 1 and find $log(Age + 1)$.]

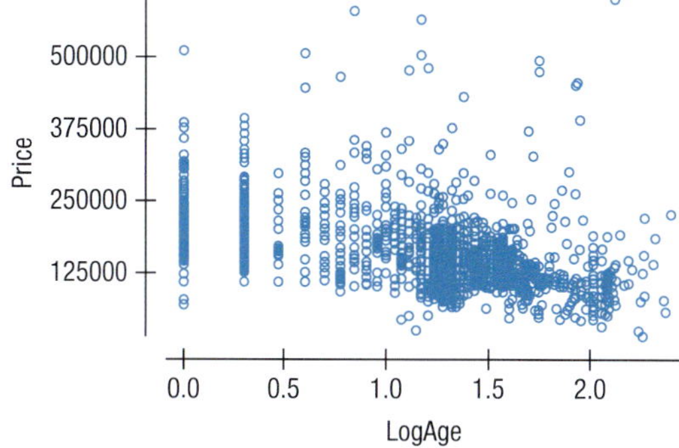

Now we can find the regression and examine the residuals.

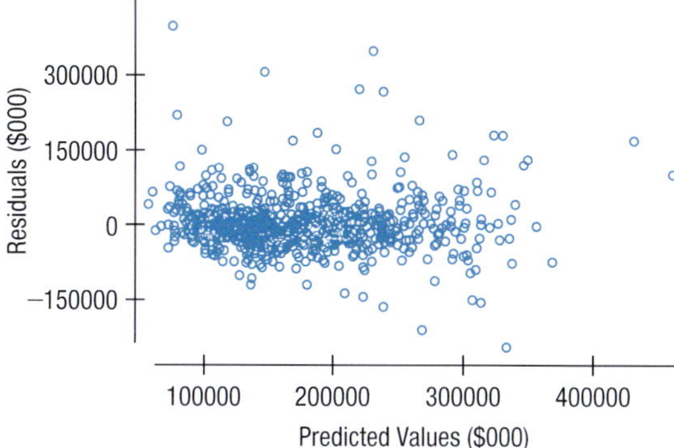

The plot of residuals against predicted values is straight, but some residuals seem unusually large or small and could deserve attention.

SHOW **MECHANICS** Linearity is all we need to fit a regression model. We always find multiple regression models with computer software. A table like this one isn't identical to what any particular statistics package produces, but it is enough like all of them to look familiar.

Here is the computer output for the regression:

Dependent variable is: Price
s = 49620 with 1057 − 4 = 1053 degrees of freedom

Variable	Coefficient	SE(Coeff)	t-ratio	p-value
Intercept	44797.2	8357	5.36	<0.0001
Living Area	87.2598	3.365	25.9	<0.0001
LogAge	−14439.1	2991	−4.83	<0.0001
Bedrooms	−5902.76	2774	−2.13	0.0336

The regression equation is

$$\widehat{Price} = 44{,}797.20 + 87.26\ LivingArea - 14{,}439\ LogAge - 5902.76\ Bedrooms.$$

TELL **INTERPRETATION**

With an R^2 of 58.8% this regression accounts for only a little more than half the variation in price of a house. The standard deviation of the residuals is $49,620. By our rule of thumb, the model could make errors of as much as twice that, or almost $100,000. That probably would not be good enough for most uses of such a model.

However, we may learn something from interpreting the coefficients. The coefficient of *Living Area* says that, after allowing for the other two predictors, houses cost an estimated $87.26 per square foot. That may not be a bad working estimate.

The negative coefficient of *LogAge* tells us that older houses cost less, but before we get too attached to the signs of the coefficients, consider the coefficient of *Bedrooms*, which seems to indicate that each bedroom reduces the price of a house—hardly what we expect from common understanding of house prices. The key here is the required qualifying phrase "after allowing for the effects of the other variables." Generally, houses with more bedrooms are larger overall. But for houses of a fixed size, it may be that devoting too much space to bedrooms at the expense of other living areas can make a house less attractive. This is a good reminder not to interpret multiple regression coefficients as if they estimate the two-variable relationship between predictor and response, ignoring the other predictors.

9.4 Partial Regression Plots

Each coefficient in a multiple regression has a corresponding plot that shows just what we mean by the relationship of its predictor with the response *after removing the linear effects of the other predictors from both y and x*. This kind of plot can be looked at in just the way you would use a simple scatterplot to visualize the relationship between two variables modeled by a simple regression slope.

Many statistics programs offer to make these plots, either as an option when you request a regression or as an addition after it has been computed. They are usually called **partial regression plots**, but you may see them called partial residual plots, added variable plots, or component-plus-residual plots. Figure 9.6 shows a partial regression plot for the coefficient of *Height* in the multiple regression predicting *%Body Fat*. Partial regression plots have several nice features:

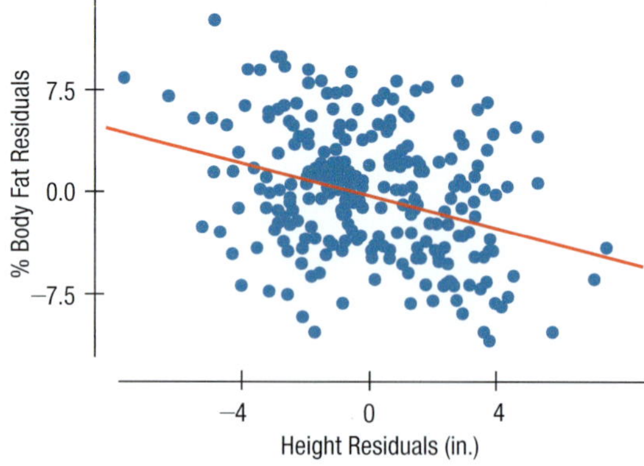

Figure 9.6
A partial regression plot for the coefficient of *Height* in the multiple regression model has a slope equal to the coefficient value in the multiple regression model and residuals equal to the final regression residuals.

- A partial regression plot has a direction that corresponds to the *sign* of its multiple regression coefficient.
- You can look at the plot to judge whether its *form* is straight enough for the corresponding variable to belong safely in the multiple regression model.
- A least squares line fit to the plot will have a *slope* equal to the coefficient the plot illustrates.
- The *residuals* around that line are the same as the final residuals of the multiple regression, allowing you to judge the strength of the estimation of the plot's coefficient.

- The influence of individual data values on the coefficient can be judged just as you would judge the influence of individual points on a simple least squares regression. *Outliers* and *high-leverage* points can be seen in just the way they would appear in a simple scatterplot.

How Do Partial Regression Plots Work?

It isn't necessary to know how a statistics program makes a partial regression plot, and you'd never want to do the calculations by hand. But it can be informative to know how it is done. The steps are simple and use things we already know about. To keep things simple, let's consider the partial regression plot for b_1:

1. Compute the (possibly multiple) regression of y on all the *other* x's except x_1.
2. Calculate the residuals from that regression. Call them $e_{y.[1]}$ (Here the dot notation means "residuals after regression on" and the [ ] notation means "all but". Thus the subscript says "residuals after regression on all but x_1."
3. Compute the (possibly multiple) regression of x_1 on all the other x's except (of course) x_1.
4. Calculate the residuals from that regression. Call them $e_{1.[1]}$.
5. Plot $e_{y.[1]}$ vs. $e_{1.[1]}$. This is the partial regression plot.

The interpretation of a multiple regression coefficient requires careful thought. The corresponding partial regression plot simplifies things and returns us to the simple relationship between a scatterplot and its regression slope.

A partial regression plot accomplishes this by displaying the *interpretation* of its coefficient. It "removes the effects of the other predictors from both y and x" in the regression, leaving only the part of the model accounted for by the coefficient being displayed.

If the partial regression plot looks straight and has a strong trend (with little scatter around the line), then that coefficient is likely to be estimated reliably. If there is much scatter or a nonlinear pattern, you should interpret the corresponding coefficient cautiously. If you see an outlier or high-leverage point, then that case could be influential on that particular coefficient in the multiple regression.

It is a good idea to make and examine a partial regression plot for any coefficient in a multiple regression whose meaning you care about in your analysis. That may not be every coefficient because some may be there to account for background effects that you are not particularly concerned with.

JUST CHECKING

Recall the regression example in Chapter 7 to predict hurricane maximum wind speed from central barometric pressure. Another researcher, interested in the possibility that global warming was causing hurricanes to become stronger, added the variable *Year* as a predictor and obtained the following regression: (Data in **Hurricanes 2015**)

Dependent variable is: Max. Winds (kn)
R-squared = 80.6 s = 8.13

Variable	**Coefficient**
Intercept	1032.01
Central Pressure	−0.975
Year	−0.00031

1. Interpret the R^2 of this regression.
2. Interpret the coefficient of *Central Pressure*.
3. The researcher concluded that "There has been no change over time in the strength of Atlantic hurricanes." Is this conclusion a sound interpretation of the regression model?

*9.5 Indicator Variables

Roller coasters are an old thrill that continues to grow in popularity. Engineers and designers compete to make them bigger and faster. For a two-minute ride on the best roller coasters, fans will wait hours. Can we learn what makes a roller coaster fast? Or how long the ride will last? Table 9.2 shows data on some of the fastest roller coasters in the world.

Name	Park	Country	Type	Duration (sec)	Speed (mph)	Height (ft)	Drop (ft)	Length (ft)	Inversion?
New Mexico Rattler	Cliff's Amusement Park	USA	Wood	75	47	80	75	2750	No
Fujiyama	Fuji-Q Highlands	Japan	Steel	216	80.8	259.2	229.7	6708.67	No
Goliath	Six Flags Magic Mountain	USA	Steel	180	85	235	255	4500	No
Great American Scream Machine	Six Flags Great Adventure	USA	Steel	140	68	173	155	3800	Yes
Hangman	Wild Adventures	USA	Steel	125	55	115	95	2170	Yes
Hayabusa	Tokyo SummerLand	Japan	Steel	108	60.3	137.8	124.67	2559.1	No
Hercules	Dorney Park	USA	Wood	135	65	95	151	4000	No
Hurricane	Myrtle Beach Pavilion	USA	Wood	120	55	101.5	100	3800	No

Table 9.2

A small selection of coasters from the larger dataset available in **Coasters** 2015.

WHO	Roller coasters
WHAT	See Table 9.2. (For multiple regression we have to know "What" and the units for each variable.)
WHERE	Worldwide
WHEN	All were in operation in 2015.
SOURCE	The Roller Coaster database, www.rcdb.com

Here are the variables and their units:

- *Type* indicates what kind of track the roller coaster has. The possible values are "wood" and "steel." (The frame usually is of the same construction as the track, but doesn't have to be.)
- *Duration* is the duration of the ride in seconds.
- *Speed* is top speed in miles per hour.
- *Height* is maximum height above ground level in feet.
- *Drop* is greatest drop in feet.
- *Length* is total length of the track in feet.
- *Inversion* reports whether riders are turned upside down during the ride. It has the values "yes" and "no."

Let's consider how the ride's *Duration* is related to the size of that initial stomach-wrenching drop that starts a roller coaster ride. We have data on both those variables for only 89 of the 241 coasters in our dataset, but there's no reason to believe that the data are missing in any patterned way so we'll look at those 89 coasters.[3]

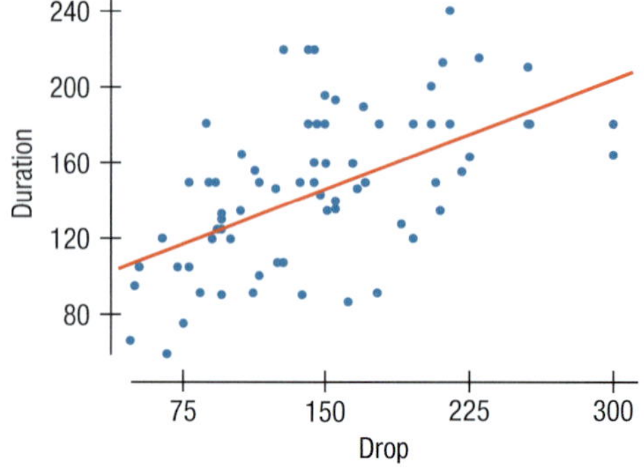

Figure 9.7

Duration (sec) of the ride appears to be linearly related to the maximum *Drop* (ft) of the ride. The residuals show no special patterns.

[3]Well, to be honest, we set aside two coasters: the Tower of Terror, which has been discontinued and dismantled, and was just a sheer drop and not really a coaster, and the Xcelerator because it uses a different method for acceleration so its largest drop is not the source of speed.

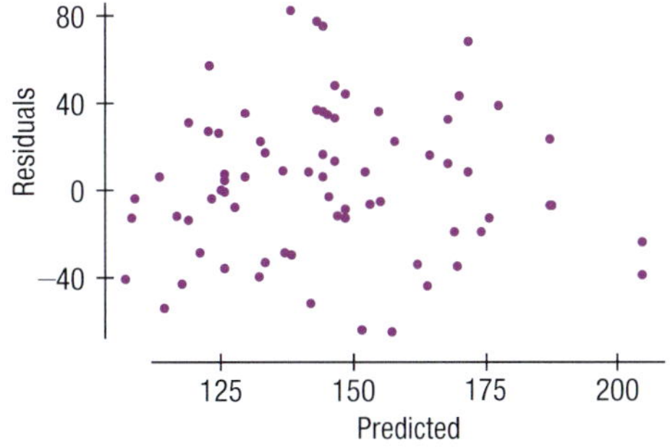

Response variable is: Duration
R-squared = 30.3%   s = 33.27

Variable	Coefficient
Intercept	88.487
Drop	0.386

The regression conditions seem to be met. From a base of about 88.5 seconds, the duration of the ride increases by about 0.386 seconds per foot of drop—coasters with bigger drops have rides that last longer.

Of course, there's more to these data. One interesting variable might not be one you'd naturally think of. Many modern coasters have "inversions." That's a nice way of saying that they turn riders upside down, with loops, corkscrews, or other devices. These inversions add excitement, but they must be carefully engineered, and that enforces some speed limits on that portion of the ride. We'd like to add the information of whether the roller coaster has an inversion to our model. Until now, all our predictor variables have been quantitative. Whether or not a roller coaster has any inversions is a categorical variable ("yes" or "no"). Can we introduce the categorical variable *Inversions* as a predictor in our regression model? What would it mean if we did?

Let's start with a plot. Figure 9.8 shows the same scatterplot of duration against drop, but now with the roller coasters that have inversions shown as red x's and a separate regression line drawn for each type of roller coaster. It's easy to see that, for a given drop, the roller coasters with inversions take a bit longer, and that for each type of roller coaster, the slopes of the relationship between duration and drop are nearly the same.

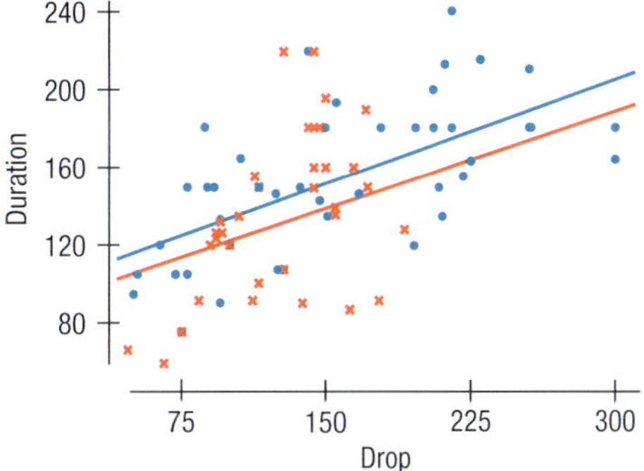

Figure 9.8
The two lines fit to coasters with inversions and without are roughly parallel.

We could split the data into two groups—the 44 coasters in these data without inversions and the 45 coasters with inversions—and compute the regression for each group. That would look like this:

Response variable is: Duration
cases selected according to No Inversions
R-squared = 40.6%   s = 29.86

Variable	Coefficient
Intercept	98.64
Drop	0.358

Response variable is: Duration
cases selected according to Inversions
R-squared = 10.0%   s = 36.00

Variable	Coefficient
Intercept	89.2007
Drop	0.336

As the scatterplot showed, the slopes are very similar, but the intercepts are different.

When we have a situation like this with roughly parallel regressions for each group, there's an easy way to add the group information to a single regression model. We make up a special variable that *indicates* what type of roller coaster we have, giving it the value 1 for roller coasters that have inversions and the value 0 for those that don't. (We could have reversed the coding; it's an arbitrary choice.[4]) Such variables are called **indicator variables** or *indicators* because they indicate which of two categories each case is in.[5] The only condition that must be satisfied for using an indicator variable is that the slopes for each of the groups must be roughly parallel. That's easy to check in the scatterplot.[6]

When we add our new indicator, *Inversions*, to the regression model, the model looks like this:

Response variable is: Duration
R-squared = 32.6% s = 32.92

Variable	Coefficient
Intercept	99.39
Drop	0.353
Inversions	−12.35

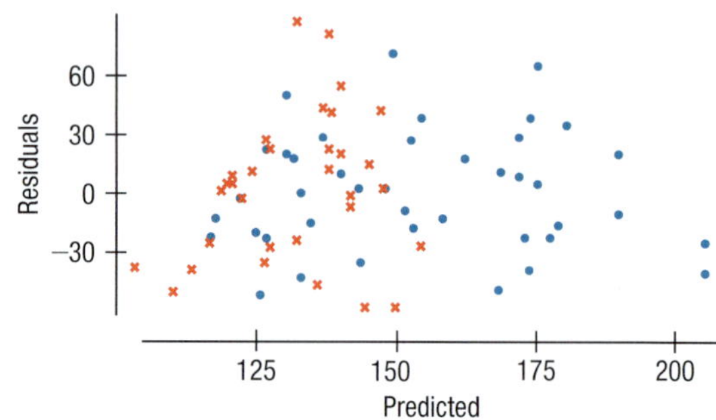

This looks like a better model than the simple regression for all the data. The R^2 is larger; the standard deviation of the residuals, *s*, is smaller; and the residuals look reasonable. And the coefficient of *Drop* is unchanged. But what does the coefficient for *Inversions* mean?

Let's see how an indicator variable works when we calculate predicted values for two of the roller coasters listed in Table 9.2:

Name	Park	Country	Type	Duration	Speed	Height	Drop	Length	Inversion?
Hangman	Wild Adventures	USA	Steel	125	55	115	95	2170	Yes
Hayabusa	Tokyo SummerLand	Japan	Steel	108	60.3	137.8	124.67	2559.1	No

The model says that for all coasters, the predicted *Duration* is

$$99.39 + 0.353 \times Drop - 12.35 \times Inversions.$$

For *Hayabusa*, the drop is 124.67 ft and the value of *Inversions* is 0, so the model predicts a duration of[7]

$$99.39 + 0.353 \times 55 - 12.35 \times 0 = 118.805 \text{ seconds.}$$

That's not far from the actual duration of 108 seconds.

[4]Some implementations of indicator variables use −1 and 1 for the levels of the categories.

[5]They are also commonly called *dummies* or *dummy variables*. But this sounds like an insult, so the more politically correct term is indicator variable.

[6]The fact that the individual regression lines are nearly parallel is really a part of the Straight Enough Condition. You should check that the lines are nearly parallel before using this method. Or read on to see what to do if they are not parallel enough.

[7]We round coefficient values when we write the model but calculate with the full precision, rounding at the end of the calculation.

For the *Hangman*, the drop is 95 ft. It has an inversion, so the value of *Inversions* is 1, and the model predicts a duration of

$$99.39 + 0.353 \times 95 - 12.35 \times 1 = 120.575 \text{ seconds.}$$

That compares well with the actual duration of 125 seconds.

Notice how the indicator works in the model. When there is an inversion (as for *Hangman*), the value 1 for the indicator causes the amount of the indicator's coefficient, −12.35, to be added to the prediction. When there is no inversion (as in *Hayabusa*), the indicator is zero, so nothing is added. Looking back at the scatterplot, we can see that this is exactly what we need. The difference between the two lines is a vertical shift of about 12 seconds.

Be careful interpreting the coefficients of indicator variables. We usually think of the coefficients in a multiple regression as slopes. But coefficients of indicator variables act differently. They're vertical shifts that keep the slopes for the other variables apart.

EXAMPLE 9.3 Using Indicator Variables

In Example 8.2 of the previous chapter, we saw that the manufacturer's suggested retail price (MSRP) of "dirt bikes" could be modeled well by knowing their displacement. After re-expressing *Displacement* to make the relationship straighter, we noted that two distinct groups of bikes could be seen in a scatterplot, each with roughly the same slope relating *MSRP and* $\sqrt[3]{Displacement}$. (Data in **Dirt bikes 2014**)

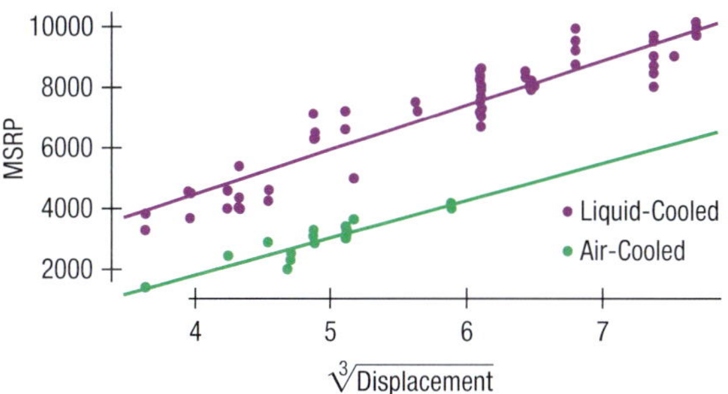

At each value of *Displacement*, bikes with liquid-cooled engines were more expensive than bikes with air-cooled engines. Because, as you can see in the scatterplot, the slopes of the relationship were very similar, this is a relationship that can be modeled well by introducing an indicator variable. Here is a regression model with the variable *Engine Cooling* equal to 1 for liquid-cooled engines and 0 for air-cooled engines:

Response variable is: MSRP
R-squared = 94.2%   s = 602.7

Variable	Coefficient
Intercept	−3814.94
Disp^1/3	1341.41
Engine cooling	2908.13

QUESTION: How should the coefficient of *Engine cooling* be interpreted in this model?

ANSWER: At each displacement size, bikes with liquid-cooled engines cost, on average, about $2908 more than comparably sized air-cooled engines. Overall, taking this difference into account, an engine's displacement accounts for more than 94% of the variation in MSRP. (In simple terms, when buying a dirt bike, you are paying for the size and cooling type of the engine. Other features make little price difference.)

Adjusting for Different Slopes

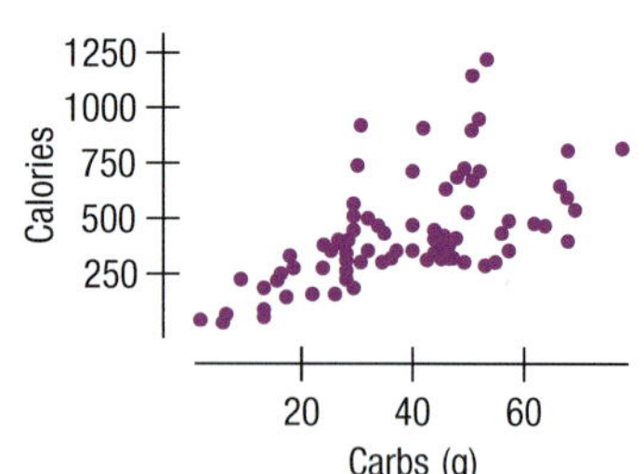

Figure 9.9

Calories of Burger King foods plotted against *Carbs* seems to fan out.

What if the lines aren't parallel? An indicator variable that is 0 or 1 can only shift the line up and down. It can't change the slope, so it works only when we have lines with the same slope and different intercepts.

Let's return to the Burger King data we looked at in Chapter 7 and look at how *Calories* are related to *Carbohydrates* (*Carbs* for short) for BK menu items other than breakfast items. Figure 9.9 shows the scatterplot.

It's not surprising to see that more *Carbs* goes with more *Calories*, but the plot seems to thicken as we move from left to right. Could there be something else going on?[8]

Burger King foods can be divided into two groups: those with meat (including chicken and fish) and those without. When we color the plot (red for meat, blue for non-meat) and look at the regressions for each group, we see a different picture. (Breakfast items often are a mix, with bacon on the side, for example. So we don't include them here.)

Clearly, meat-based dishes have more calories for each gram of carbohydrate than do other Burger King foods. But the regression model can't account for the kind of difference we see here by just including an indicator variable. It isn't just the height of the lines that is different; they have entirely different slopes. How can we deal with that in our regression model?

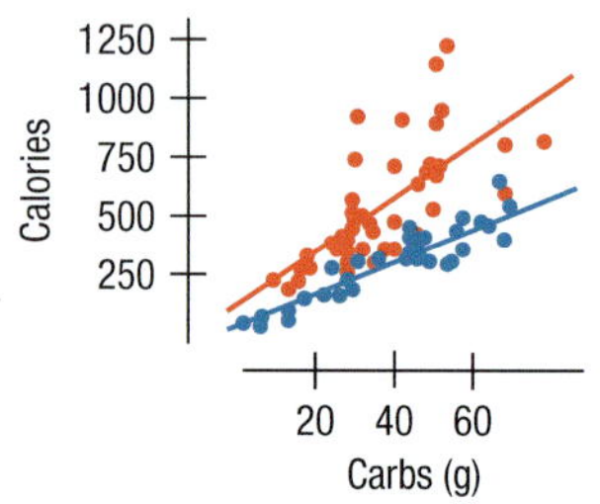

Figure 9.10

Plotting the meat-based and non-meat items separately, we see two distinct linear patterns.

The trick is to adjust the slopes with another constructed variable. This one is the *product* of an indicator and the predictor variable. The coefficient of this constructed **interaction term** in a multiple regression gives an adjustment to the slope, b_1, to be made for the individuals in the indicated group. Here we have the indicator variable *Meat*, which is 1 for meat-containing foods and 0 for the others. We then construct an interaction variable, *Carbs*Meat,* which is just the product of those two variables. That's right; just multiply them. The resulting variable has the value of *Carbs* for foods containing meat (those coded 1 in the *Meat* indicator) and the value 0 for the others. By including the interaction variable in the model, we can adjust the slope of the line fit to the meat-containing foods. Here's the resulting analysis:

Dependent variable is: Calories
R-squared = 60.7%   s = 146.5

Variable	**Coefficient**
Intercept	83.5330
Carbs(g)	6.25484
Meat	120.220
Carbs*Meat	2.14516

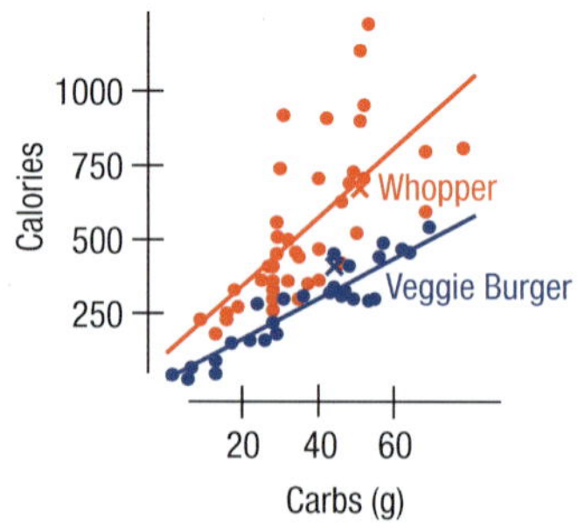

Figure 9.11

The Whopper and Veggie Burger belong to different groups.

What does the coefficient for the indicator *Meat* mean? It provides a different intercept to separate the meat and non-meat items at the origin (where *Carbs* = 0).

The coefficient of the interaction term, *Carbs*Meat*, says that the slope relating calories to carbohydrates is steeper by 2.145 calories per carbohydrate gram for meat-containing foods than for meat-free foods.

$$83.53 + 6.25\,Carbs + 120.22\,Meat + 2.15\,Carbs{*}Meat.$$

Let's see how these adjustments work. A BK Whopper has 51 g of *Carbohydrates* and is a meat dish. The model predicts its *Calories* as

$$83.53 + 6.25 \times 51 + 120.22 \times 1 + 2.15 \times 51 \times 1 = 632.15,$$

not far from the measured calorie count of 670. By contrast, the Veggie Burger, with 44 g of *Carbohydrates*, is predicted to have

$$83.53 + 6.25 \times 44 + 120.22 \times 0 + 2.15 \times 0 \times 44 = 358.7$$

not far from the 410 measured officially. The last two terms in the equation for the Veggie Burger are just zero because the indicator for *Meat* is 0 for the Veggie Burger.

[8]Would we even ask if there weren't?

One, Two, Many

Chapter 8 examined data on breakfast cereals and noted that the data (in **Cereals**) includes the shelf of the market on which the cereal was found. It seems natural to represent *Shelf* as an indicator variable, but it has three levels, not two, so the usual 0/1 coding just isn't going to work.

We can still use indicators for categorical predictors with multiple levels. The secret is to define a separate indicator for each level—and then to choose one level as your "base" and leave its indicator out. (Leave one out? Why? Well, recall that indicators only adjust the level and not the slopes of a regression, so they modify the intercept term in the model. For a simple two-level indicator, the intercept is what it should be for the "0" level of the indicator and changes for the group coded "1". For multiple levels, the intercept fits the right level for the "base" group, and the others are then adjusted relative to that level.)

For the cereals, we can define indicator variables named *shelf2* and *shelf3*, which are, respectively, 1 for cereals on shelf 2 (and 0 otherwise) and 1 for cereals on shelf 3 (and 0 otherwise). We'll make shelf 1 our base shelf. Here's a regression relating the sugar (g) and sodium (g) contents of these cereals and including the indicators. As Figure 9.12 shows, the three groups have roughly parallel slopes, so adjusting just the levels makes sense:

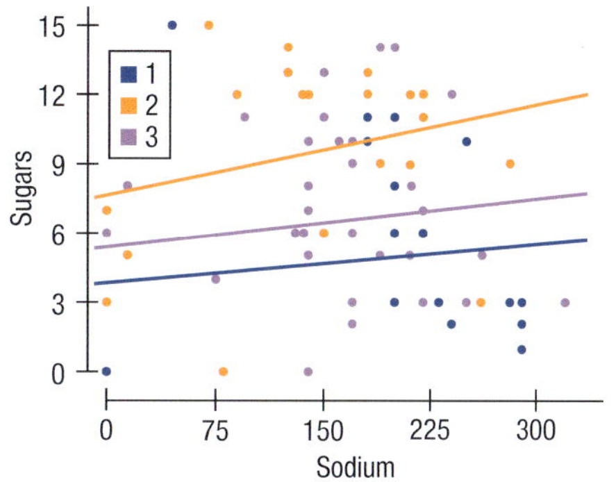

Figure 9.12
Sugars vs. *Sodium* for breakfast cereals, with a regression line for those on each of three shelves showing a roughly parallel relationship across shelves.

Response variable is: sugars
R-squared = 18.7% s = 4.070

Variable	Coefficient
Intercept	3.44674
sodium	0.00796
shelf_2	5.01217
shelf_3	1.81821

EXAMPLE 9.4

Indicators for Variables with Several Levels

Diamonds are priced by the four C's (carat size, color, clarity, and cut). Color is rated in levels. In this dataset the diamonds range from D (colorless) to K (yellow). Adding indicators for color to a model of (square root) *Price* by *Carat Size* allows for different intercepts for each color level.

QUESTION: How can we incorporate information such as this in a multiple regression model?

ANSWER: We can make Color D the base color and include indicators for the 7 other colors. Now we have 8 parallel lines for (square root) *Price* by *Carat Size* (depending on the color).

Response variable is: $\sqrt{\text{Price}}$
R-squared = 85.8% s = 7.218

Variable	Coefficient	SE(Coeff)	t-ratio	p-value
Intercept	13.1946	0.5488	24.043	<2e-16
Carat.Size	61.2491	0.5032	121.722	<2e-16
ColorE	−2.1027	0.5399	−3.895	0.000101
ColorF	−2.8640	0.5576	−5.136	3.00e-07
ColorG	−3.6320	0.5769	−6.296	3.57e-10
ColorH	−7.8948	0.5858	−13.477	<2e-16
ColorI	−11.8542	0.6261	−18.932	<2e-16
ColorJ	−16.6404	0.6637	−25.071	<2e-16
ColorK	−21.3577	0.8282	−25.787	<2e-16

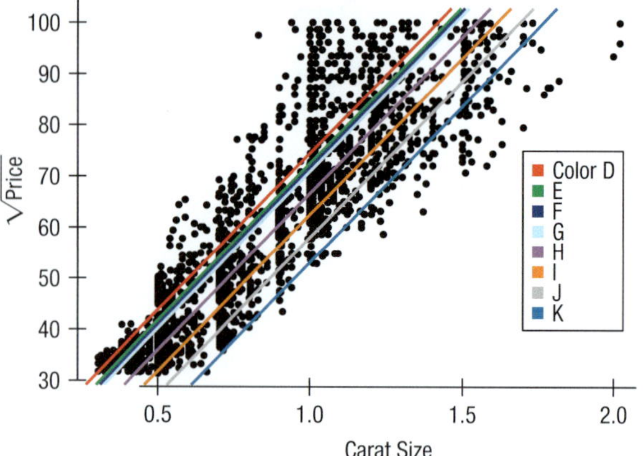

QUESTION: How can we make a model allowing for different slopes?

ANSWER: We use Color D as the base color again and add 7 interaction terms as well as the 7 indicator variables.

Response variable is: $\sqrt{\text{Price}}$
R-squared = 86.5% s = 7.058

Variable	Coefficient	SE(Coeff)	t-ratio	p-value
Intercept	9.3239	1.2142	7.679	2.23E-14
Carat.Size	67.0408	1.7025	39.379	<2e-16
ColorE	−0.5392	1.5075	−0.358	0.72063
ColorF	−2.3716	1.5627	−1.518	0.12922
ColorG	−2.6709	1.6643	−1.605	0.10867
ColorH	−3.9177	1.8248	−2.147	0.03189
ColorI	−2.5481	1.9301	−1.32	0.18689
ColorJ	−5.4176	2.0716	−2.615	0.00897
ColorK	0.5976	2.7815	0.215	0.82991
Carat.Size*ColorE	−2.4007	2.0999	−1.143	0.25305
Carat.Size*ColorF	−1.3211	2.0954	−0.63	0.52843
Carat.Size*ColorG	−2.5457	2.0868	−1.22	0.2226
Carat.Size*ColorH	−5.9017	2.1774	−2.71	0.00676
Carat.Size*ColorI	−10.9139	2.1812	−5.004	5.99E-07
Carat.Size*ColorJ	−12.4948	2.2531	−5.546	3.22E-08
Carat.Size*ColorK	−21.4477	2.6978	−7.95	2.72E-15

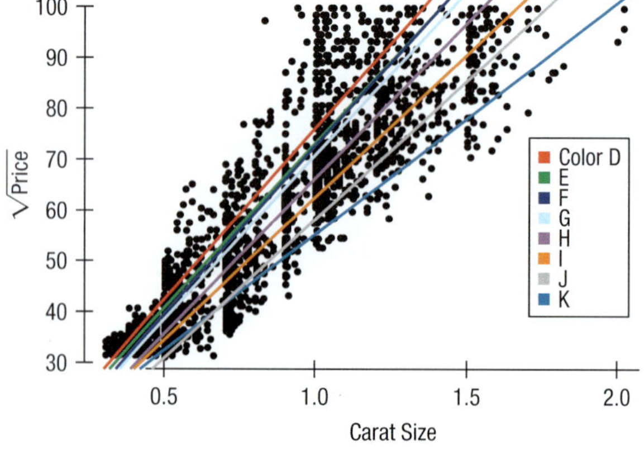

WHAT CAN GO WRONG?

Interpreting Coefficients

- **Don't claim to "hold everything else constant" for a single individual.** It's often meaningless to say that a regression coefficient says what we expect to happen if all variables but one were held constant for an individual and the predictor in question changed. While it's mathematically correct, it often just doesn't make any sense. We can't gain a year of experience or have another child without getting a year older. Instead, we can think about all those who fit given criteria on some predictors and ask about the conditional relationship between y and one x for those individuals. The coefficient -0.60 of *Height* for predicting *%Body Fat* says that among men of the same *Waist* size, those who are one inch taller in *Height* tend to be, on average, 0.60% lower in *%Body Fat.* The multiple regression coefficient measures that average conditional relationship.
- **Don't interpret regression causally.** Regressions are usually applied to observational data. Without a properly designed experiment (see Chapter 11), we can't draw conclusions about causes and effects. We can never be certain that there are no variables lurking in the background, causing everything we've seen. Don't interpret b_1, the coefficient of x_1 in the multiple regression, by saying, "If we were to change an individual's x_1 by 1 unit (holding the other x's constant) it would change his y by b_1 units." We don't know from the regression what applying a change to an individual would do; the regression models the data as they are, not as they might be after a change.
- **Be cautious about interpreting a regression model as predictive.** Yes, we do call the x's predictors, and you can certainly plug in values for each of the x's and find a corresponding predicted value, $\hat{y}$. But the term "prediction" suggests extrapolation into the future or beyond the data, and we know that we can get into trouble when we use models to estimate $\hat{y}$-values for x's not in the range of the data. Be careful not to extrapolate very far from the span of your data. In simple regression it was easy to tell when you extrapolated. With many predictor variables, it's often harder to know when you are outside the bounds of your original data.[9] We usually think of fitting models to the data more as modeling than as prediction, so that's often a more appropriate term.
- **Don't think that the sign of a coefficient is special.** Sometimes our primary interest in a predictor is whether it has a positive or negative association with y. As we have seen, though, the sign of the coefficient also depends on the other predictors in the model. Don't look at the sign in isolation and conclude that "the direction of the relationship is positive (or negative)." Just like the value of the coefficient, the sign is about the relationship after allowing for the linear effects of the other predictors. The sign of a variable can change depending on which other predictors are in or out of the model. This is one of the most common misinterpretations of regression coefficients. Be alert for it.

And, as always...

- **Don't fit a linear regression to data that aren't straight.** Linearity is the most fundamental regression assumption. If the relationship between y and the x's isn't approximately linear, there's no sense in fitting a linear model to it. What we mean by "linear" is a model of the form we have been writing for the regression. When we have two predictors, this is the equation of a plane, which is linear in the sense of being flat in all directions. With more predictors, the geometry is harder to visualize, but the simple structure of the model is consistent; the predicted values change consistently with equal size changes in any predictor.

  Usually we're satisfied when plots of y against each of the x's are straight enough. We'll also check a scatterplot of the residuals against the predicted values for evidence of nonlinearity.

[9]With several predictors it is easy to wander beyond the data. Even for values that are individually ordinary, a combination may be extraordinary. For example, both 28-inch waists and 76-inch heights can be found in men in the body fat study, but a single individual with both these measurements would not be at all typical, so a prediction for such a tall, thin man would be an extrapolation beyond the data.

- **Watch out for the plot thickening.** We always calculate the standard deviation of the residuals. But that estimate assumes that the standard deviation is the same throughout the range of the x's so that we can combine all the residuals when we estimate it. If s_e changes with any x, these estimates won't make sense. The most common check is a plot of the residuals against the predicted values. If plots of residuals against several of the predictors all show a thickening, and especially if they also show a bend, then consider re-expressing y. If the scatterplot against only one predictor shows thickening, consider re-expressing that predictor.
- **The errors should be unimodal, symmetric, and without outliers.** A histogram of the residuals should be unimodal and symmetric.
- **Watch out for high-leverage points and outliers.** We always have to be on the lookout for a few points that have undue influence on our model, and regression is certainly no exception. Partial regression plots are a good place to look for influential points and to understand how they affect each of the coefficients.

CONNECTIONS

We would never consider a regression analysis without first making scatterplots. The aspects of scatterplots that we always look for—their direction, form, and strength—relate directly to regression and to partial regression plots, and we assess the Unimodal, Symmetric, and without Outliers (USO) condition by examining the shape of a residual histogram or a Normal probability plot.

CHAPTER REVIEW

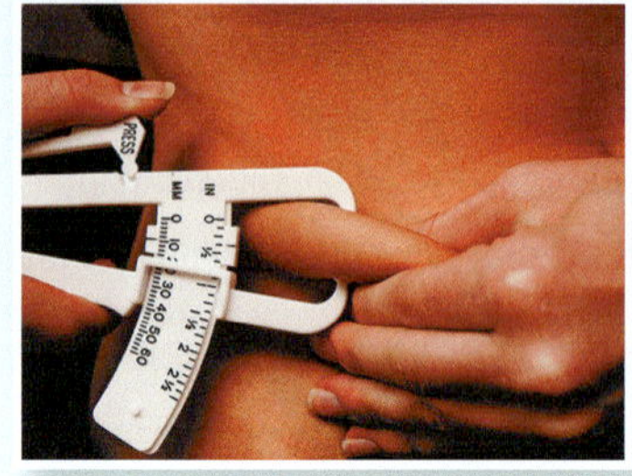

In the previous chapters we saw how to find and interpret a least squares linear regression model. Now we've seen that much of what we know about those models is also true for multiple regression:

- The presence of several predictors in a multiple regression makes the interpretation of a multiple regression slope less straightforward than it is in simple regression. One way to think about the coefficient b_j for a predictor x_j is the slope of y on x_j after the effects of all the other predictors have been taken out of both y and x_j. This is what a partial regression plot shows. When it makes sense, b_j can also be interpreted as the effect of x_j on y for fixed values of all the other predictors.
- When predictor variables are highly correlated, as they are in many real situations, the interpretation is more challenging. It may not be possible to hold one x constant while changing another. And removing the effects of one predictor from a correlated predictor can leave behind a variable that doesn't look very much like the original x_j, changing the meaning of b_j. Use caution when trying to interpret the slopes of highly correlated predictors.
- The assumptions and conditions are the same: linearity (checked now with scatterplots of y against each x), independence (think about it), constant variance (checked with the scatterplot of residuals against predicted values), and residuals that have a unimodal and symmetric distribution (checked with a histogram or probability plot).
- R^2 is still the fraction of the variation in y accounted for by the regression model.
- s_e is still the standard deviation of the residuals—a good indication of the precision of the model.
- The regression table produced by any statistics package shows a row for each coefficient, giving its estimate, and other statistics we'll learn about later in the course.
- A partial regression plot shows the relationship described by its corresponding multiple regression coefficient, but can be interpreted just like a simple scatterplot to understand the effects of individual cases.
- Just as we saw with simple regressions, other random samples of data from the same population will yield slightly different multiple regression analyses. Don't think that the one you have is the only one possible.

- Indicator variables can be used to include categorical predictors in a regression model, using one indicator for each category and leaving out one as the "base" category.

Finally, we've learned that multiple regression models extend our ability to model the world to many more situations, but with that power comes the responsibility to interpret the coefficients carefully.

REVIEW OF TERMS

Multiple regression A linear regression with two or more predictors whose coefficients are found to minimize the sum of the squared residuals is a least squares linear multiple regression. But it is usually just called a multiple regression. When the distinction is needed, a least squares linear regression with a single predictor is called a simple regression. The multiple regression model is

$$y = \beta_0 + \beta_1 x_1 + \cdots + \beta_k x_k + \varepsilon \text{ (p. 309).}$$

Least squares We still fit multiple regression models by choosing the coefficients that make the sum of the squared residuals as small as possible, the method of least squares (p. 309).

Partial regression plot The partial regression plot for a specified coefficient is a display that helps us understand the meaning of that coefficient in a multiple regression. It has a slope equal to the coefficient value and shows the influences of each case on that value. Partial regression plots display the residuals when y is regressed on the other predictors against the residuals when the specified x is regressed on the other predictors (p. 318).

Indicator variable An indicator variable is one constructed to indicate in which of two groups each case belongs. Indicators are typically assigned the value 1 for the group of interest and the value 0 for others. Including an indicator in a multiple regression adjusts the intercept for the group coded 1 (p. 322).

Interaction term A constructed variable that can account for different slopes for different groups in a multiple regression (p. 324).

TECH SUPPORT

Regression Analysis

All statistics packages make a table of results for a regression. If you can read a package's regression output table for simple regression, then you can read its table for a multiple regression.

Most packages offer to plot residuals against predicted values. Some will also plot residuals against the x's. With some packages you must request plots of the residuals when you request the regression. Others let you find the regression first and then analyze the residuals afterward. Either way, your analysis is not complete if you don't check the residuals with a histogram or Normal probability plot and a scatterplot of the residuals against the x's or the predicted values.

One good way to check assumptions before embarking on a multiple regression analysis is with a scatterplot matrix. This is sometimes abbreviated SPLOM in commands.

Multiple regressions are always found with a computer or programmable calculator. Before computers were available, a full multiple regression analysis could take months or even years of work.

DATA DESK

- From the toolbar or Calc menu, choose **Regression**.
- Data Desk opens a regression window. Drag variables into position in the regression table: response variable to the top of the table, predictors to the bottom part.
- Data Desk displays the regression table.
- Select plots of residuals from the Regression table's HyperView menu. To make a partial regression plot, click on the coefficient of interest and choose from the HyperView menu.

COMMENTS

You can change the regression by dragging the icon of another variable over either the Y- or an X-variable name in the table and dropping it there. You can add a predictor to the model by dragging its icon into that part of the table. The regression, predicted values, residuals, and diagnostic statistics will recompute automatically.

EXCEL

- Select **Data Analysis** from the Analysis Group on the Data Tab.
- Select **Regression** from the Analysis Tools list.
- Click the **OK** button.
- Enter the data range holding the *Y*-variable in the box labeled *Y*-range.
- Enter the range of cells holding the *X*-variables in the box labeled *X*-range.
- Select the **New Worksheet Ply** option.
- Select **Residuals** options. Click the **OK** button.

Alternatively, the LINEST function can compute a multiple regression. LINEST is an array function. Consult your Excel documentation about array functions and LINEST.

COMMENTS

The Y and X ranges do not need to be in the same rows of the spreadsheet, although they must cover the same number of cells. But it is a good idea to arrange your data in parallel columns as in a data table. The X-variables must be in adjacent columns. No cells in the data range may hold non-numeric values.

JMP

- From the Analyze menu select **Fit Model.**
- Specify the response, *Y*. Assign the predictors, *X*, in the Construct Model Effects dialog box.
- Click on **Run Model.**

COMMENTS

JMP chooses a regression analysis when the response variable is "Continuous." The predictors can be any combination of quantitative or categorical. If you get a different analysis, check the variable types.

MINITAB

- Choose **Regression** from the Stat menu.
- Choose **Regression...** from the Regression submenu.
- In the Regression dialog, assign the *Y*-variable to the Response box and assign the *X*-variables to the Predictors box.
- Click the **Graphs** button.
- In the Regression-Graphs dialog, select **Standardized residuals,** and check **Normal plot of residuals** and **Residuals versus fits.**
- Click the **OK** button to return to the Regression dialog.
- Click the **OK** button to compute the regression.

R

Suppose the response variable *y* and predictor variables $x_1, \ldots, x_k$ are in a data frame called mydata. To fit a multiple regression of *y* on x_1 and x_2:

- mylm = lm(y ~ x_1 + x_2, data = mydata)
- summary(mylm) # gives the details of the fit, including the ANOVA table
- plot(mylm) # gives a variety of plots

To fit the model with *all* the predictors in the data frame,

- mylm = lm(y ~., data = mydata) # The period means use all other variables

COMMENTS

To get confidence or prediction intervals use:

- predict(mylm, interval = "confidence")

or

- predict(mylm, interval = "prediction")

SPSS

- Choose **Regression** from the Analyze menu.
- Choose **Linear** from the Regression submenu.
- When the Linear Regression dialog appears, select the *Y*-variable and move it to the dependent target. Then move the *X*-variables to the independent target.
- Click the **Plots** button.
- In the Linear Regression Plots dialog, choose to plot the *SRESIDs against the *ZPRED values.
- For partial regression plots, click **Next** to find the options.
- Click the **Continue** button to return to the Linear Regression dialog.
- Click the **OK** button to compute the regression.

STATCRUNCH

To find a multiple regression of a response variable *y* on two factors, x_1 and x_2, from the Stat menu, choose **Regression » Multiple Linear**.

In the dialog, specify the response (*Y*) variable. Specify the predictors (*X* variables) and Interactions (optional).

Click **Compute!** to perform the regression.

COMMENTS

You can save residuals and predicted values among other quantities with the save options.

TI-83/84 PLUS

COMMENTS

You need a special program to compute a multiple regression on the TI-83.

EXERCISES

SECTION 9.1

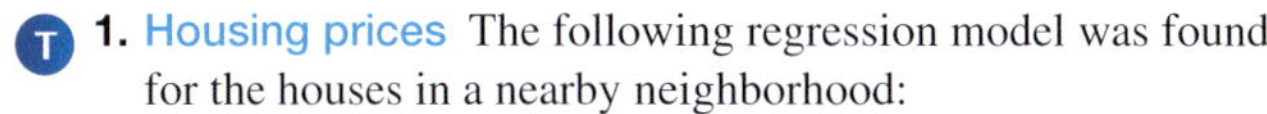

1. Housing prices The following regression model was found for the houses in a nearby neighborhood:

$$\widehat{Price} = 21{,}037.97 - 8321.13\,Bedrooms + 84.28\,Living\,Area.$$

a) Find the predicted price of a 2-bedroom, 1000-sq-ft house from this model.

b) The house just sold for \$184,000. Find the residual corresponding to this house.

c) What does that residual say about this transaction?

2. Candy sales A candy maker surveyed chocolate bars available in a local supermarket and found the following least squares regression model:

$$\widehat{Calories} = 26.4 + 11.34\,Fat(g) + 3.88\,Sugar(g).$$

a) The hand-crafted chocolate she makes has 10 g of fat and 17 g of sugar. How many calories does the model predict for a serving?

b) In fact, a laboratory test shows that her candy has 195 calories per serving. Find the residual corresponding to this candy. (Be sure to include the units.)

c) What does that residual say about her candy?

SECTION 9.2

3. Movie profits What can predict how much a motion picture will make? We have data on 593 recent releases that includes the *Gross* (in \$M), the *Budget* (\$M), the *Run Time* (minutes), and the score given by the critics on an aggregator website. The first several entries in the data table look like this:

Movie	Gross (\$M)	Budget (\$M)	Run Time (minutes)	Critics Score
Django Unchained	162.81	100	165	88
Parental Guidance	77.26	25	105	18
The Impossible	19.02	40	114	81
Jack Reacher	80.07	60	130	61
This is 40	67.54	35	134	51
The Guilt Trip	37.13	40	92	38
The Hobbit: An Unexpected Journey	303.00	250	169	64
Playing for Keeps	13.10	35	105	4
Killing Them Softly	15.03	15	97	75
Red Dawn	44.81	65	93	12
Silver Linings Playbook	132.09	21	127	87
Life of Pi	124.99	120	127	87
Rise of the Guardians	103.41	145	97	74
Twilight Saga: Breaking Dawn	281.29	136.2	115	49
Skyfall	304.36	200	143	92

We want a regression model to predict *Gross*. Parts of the regression output computed in Excel look like this:

Dependent variable is: Gross (\$M)
R-squared = 77.7%
s = 2.41 with 593 − 4 = 589 degrees of freedom

Variable	Coefficient	SE(Coeff)	*t*-ratio	P-value
Intercept	−45.8877	15.5896	−2.94	0.0034
Budget (\$M)	0.9895	0.0491	20.15	<0.0001
Run Time (min)	0.3855	0.1538	2.51	0.0123
Critics Score	0.6308	0.0925	6.82	<0.0001

a) Write the multiple regression equation.

b) What is the interpretation of the coefficient of *Budget* in this regression model?

T 4. Movie profits again A middle manager at an entertainment company, upon seeing the analysis of Exercise 3, concludes that longer movies make more money. He argues that his company's films should all be padded by 30 minutes to improve their gross. Explain the flaw in his interpretation of this model.

SECTION 9.3

T 5. More movies profits For the movies examined in Exercise 3, here is a scatterplot of *USGross* vs. *Budget*:

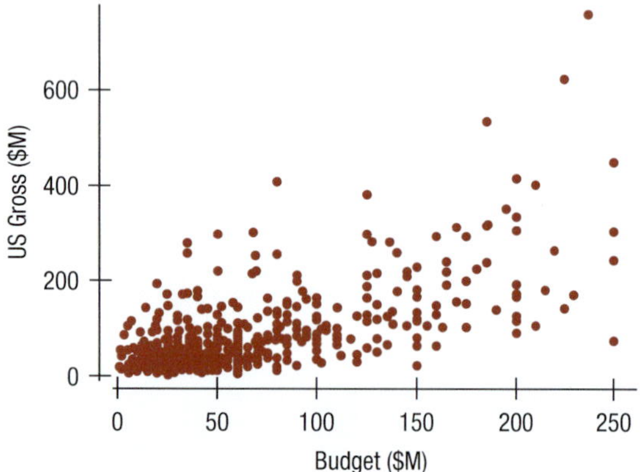

What (if anything) does this scatterplot tell us about the following assumptions and conditions for the regression?

a) Linearity Condition
b) Equal Spread Condition
c) Normality Assumption

T 6. Movie residuals For the movies regression in Exercise 3, here is a histogram of the residuals. What does it tell us about the assumptions and conditions below?

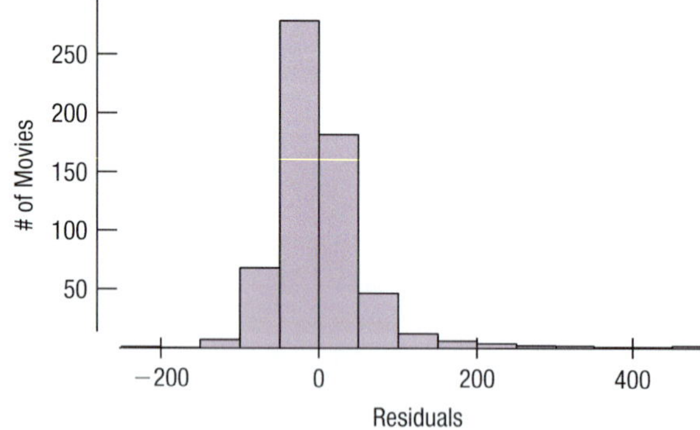

a) Linearity Condition
b) Nearly Normal Condition
c) Equal Spread Condition

SECTION 9.4

T 7. Movie profits once more Look back at the regression in Exercise 3. Here is the partial regression plot for the coefficient of *Budget*.

a) What is the slope of the least squares regression line in the partial regression plot?

b) The point plotted with a red "x" is the movie *Avatar*, which had an extraordinarily large budget. If that movie were removed from the data, how would that affect the coefficient of *Budget* in the multiple regression?
 i. It would get larger (more positive).
 ii. It would get smaller.
 iii. It would become zero.
 iv. We can't tell from this graph.

T 8. Hopkins Forest wind In Chapters 4 and 6 we've seen data from the Hopkins Forest. Here's a regression that models the maximum daily wind speed in terms of the average temperature and precipitation:

Response variable is: Max wind (mph)
R-squared = 9.0% s = 6.815

Variable	Coefficient
Intercept	22.1975
Avg.Temp (F)	−0.113310
Precip	2.01300

a) Write the regression model.
b) What is the interpretation of the coefficient of *Precip*?

Here is the partial regression plot for the coefficient of *Precip* in the regression:

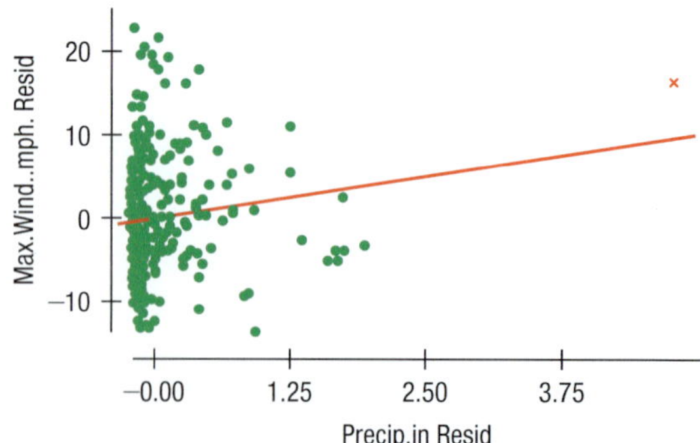

c) What is the slope of the least squares regression line in the partial regression plot?
d) The point plotted with an "x" is for 8/28/2011—the date of Hurricane Irene. If we were to remove it from the data, what would happen to the coefficient of *Precip*?

SECTION 9.5

9. Indicators For each of these potential predictor variables, say whether they should be represented in a regression model by indicator variables. If so, then suggest what specific indicators should be used (that is, what values they would have).

a) In a regression to predict income, the sex of respondents in a survey
b) In a regression to predict the square footage available for rent, the number of stories in a commercial building
c) In a regression to predict the amount an individual's medical insurance would pay for an operation, whether the individual was over 65 (and eligible for Medicare)

10. More indicators For each of these potential predictor variables say whether they should be represented in a regression model by indicator variables. If so, then suggest what specific indicators should be used (that is, what values they would have).

a) In a regression to predict income, the age of respondents in a survey
b) In a regression to predict the square footage available for rent, whether a commercial building has an elevator or not
c) In a regression to predict annual medical expenses, whether a person was a child (in pediatric care), an adult, or a senior (over 65 years old)

11. Interpretations A regression performed to predict the selling price of houses found the equation

$$\widehat{Price} = 169{,}328 + 35.3\,Area + 0.718\,Lotsize - 6543\,Age$$

where *Price* is in dollars, *Area* is in square feet, *Lotsize* is in square feet, and *Age* is in years. The R^2 is 92%. One of the interpretations below is correct. Which is it? Explain what's wrong with the others.

a) Each year a house *Ages* it is worth \$6543 less.
b) Every extra square foot of *Area* is associated with an additional \$35.30 in average price, for houses with a given *Lotsize* and *Age*.
c) Every dollar in price means *Lotsize* increases 0.718 square feet.
d) This model fits 92% of the data points exactly.

12. More interpretations A household appliance manufacturer wants to analyze the relationship between total sales and the company's three primary means of advertising (television, magazines, and radio). All values were in millions of dollars. They found the regression equation

$$\widehat{Sales} = 250 + 6.75\,TV + 3.5\,Radio + 2.3\,Magazines.$$

One of the interpretations below is correct. Which is it? Explain what's wrong with the others.

a) If they did no advertising, their income would be \$250 million.
b) Every million dollars spent on radio makes sales increase \$3.5 million, all other things being equal.
c) Every million dollars spent on magazines increases TV spending \$2.3 million.
d) Sales increase on average about \$6.75 million for each million spent on TV, after allowing for the effects of the other kinds of advertising.

13. Predicting final exams How well do exams given during the semester predict performance on the final? One class had three tests during the semester. Computer output of the regression gives

Dependent variable is: Final
s = 13.46 R-Sq = 77.7%

Predictor	Coeff
Intercept	−6.74
Test1	0.2570
Test2	0.3917
Test3	0.9011

a) Write the equation of the regression model.
b) How much of the variation in final exam scores is accounted for by the regression model?
c) Explain in context what the coefficient of *Test3* scores means.
d) A student argues that the first exam doesn't help to predict final performance because the R^2 is almost as good without it in the model. She suggests that this exam not be given at all. Does *Test1* have no effect on the final exam score? Can you tell from this model? (*Hint*: Do you think test scores are related to each other?)

T **14.** Scottish hill races Hill running—races up and down hills—has a written history in Scotland dating back to the year 1040. Races are held throughout the year at different locations around Scotland. A recent compilation of information for 90 races (for which full information was available and omitting two unusual races) includes the *Distance* (km), the *Climb* (m), and the *Record Time* (minutes). A regression to predict the men's records as of 2008 looks like this:

Dependent variable is: Men's Time (mins)
R-squared = 98.0%
s = 6.623 with 90 − 3 = 87 degrees of freedom

Variable	Coefficient	SE(Coeff)	t-ratio	P-value
Intercept	−10.3723	1.245	−8.33	<0.0001
Climb (m)	0.034227	0.0022	15.7	<0.0001
Distance (km)	4.04204	0.1448	27.9	<0.0001

a) Write the regression equation. Give a brief report on what it says about men's record times in hill races.
b) Interpret the value of R^2 in this regression.
c) What does the coefficient of *Climb* mean in this regression?

15. Attendance 2016 Several exercises in Chapter 7 showed that attendance at American League baseball games increased with the number of runs scored. But fans may respond more to winning teams than to high-scoring games. Here is a regression of average attendance on both the average runs per game and the number of games won during the 2016 season:

Response variable is: Total Attendance
cases selected according to American League
R-squared = 24.0% s = 642476 with 15 − 3 = 12 degrees of freedom

Variable	Coefficient
Intercept	−1272076
Won	19711.9
Runs	2825.73

a) The regression using only *Runs* as a predictor has a slope coefficient of 5180.47. Can you account for the difference in coefficients when *Won* is included in the model?

Here is a plot of the residuals:

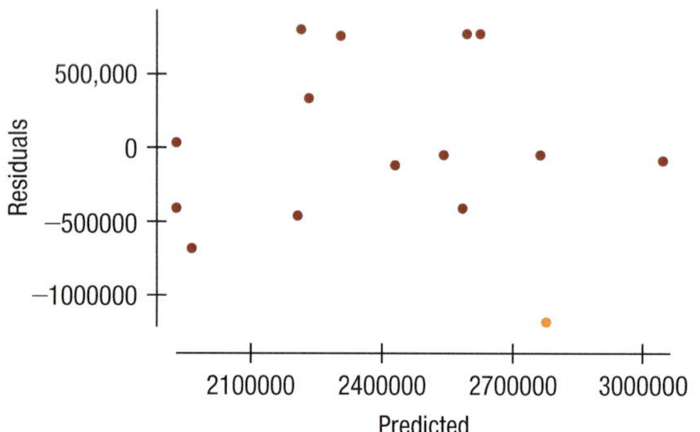

b) The point plotted in red is the Cleveland Indians, who won the AL pennant that year. What does the residual say about attendance at Indian games?

16. Candy bars: calories A student collected nutrition data about candy bars by reading the labels in a supermarket. Because candy bars have different "serving sizes," the data are given as values per serving. Here is a regression predicting calories from the sugar (g/serving). (Five sugar-free candy bars have been omitted from this analysis.)

Response variable is: Calories/serving
R-squared = 37.4% s = 0.3443

Variable	Coefficient
Intercept	6.92032
Sugar/serving	−3.91491

a) What is a proper interpretation of the coefficient of *Sugar/serving*?

Here is another regression, now with protein (g/serving) as a second predictor:

Response variable is: Calories/serving
R-squared = 44.5% s = 0.3312

Variable	Coefficient
Intercept	5.61689
Sugar/serving	−2.18057
Protein (g/serving)	6.04437

b) What is a proper interpretation of the coefficient of *Sugar/serving* in this regression?

c) How can you account for the change in *Sugar* coefficient values?

17. Home prices Many variables have an impact on determining the price of a house. Among these are *Living Area* of the house (square feet) and number of *Bathrooms*. Information for a random sample of homes for sale in a certain area was obtained from the Internet. Regression output modeling the asking *Price* with *Living Area* and number of *Bathrooms* gave the following result:

Dependent variable is: Price
s = 66053 R-Sq = 70.2%

Predictor	Coeff
Intercept	−150765
Bathrooms	9306
LivingArea	141.74

a) Write the regression equation.

b) Explain in context what the coefficient of *Living Area* means.

c) The owner of a construction firm, upon seeing this model, says that the slope for bathrooms is too small. He says that when *he* adds another bathroom, it increases the value by more than $9530. Can you think of an explanation for his question?

18. More hill races Here is the regression for the women's records for the same Scottish hill races we considered in Exercise 14:

Dependent variable is: Women's Time (mins)
R-squared = 96.7% s = 10.06

Variable	Coefficient
Intercept	−11.6545
Climb (m)	0.045195
Distance	4.43427

a) Compare the regression model for the women's records with that found for the men's records in Exercise 14.

Here's a scatterplot of the residuals for this regression:

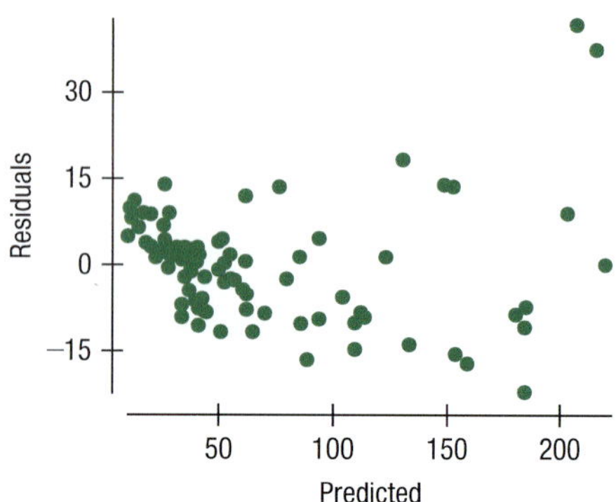

b) Discuss the residuals and what they say about the assumptions and conditions for this regression.

19. Predicting finals II Here are some diagnostic plots for the final exam data from Exercise 13. These were generated by a computer package and may look different from the plots generated by the packages you use. (In particular, note that the axes of the Normal probability plot are swapped relative to the plots we've made in the text. We only care about the pattern of this plot, so it shouldn't affect your interpretation.) Examine these plots and discuss whether the assumptions and conditions for the multiple regression seem reasonable.

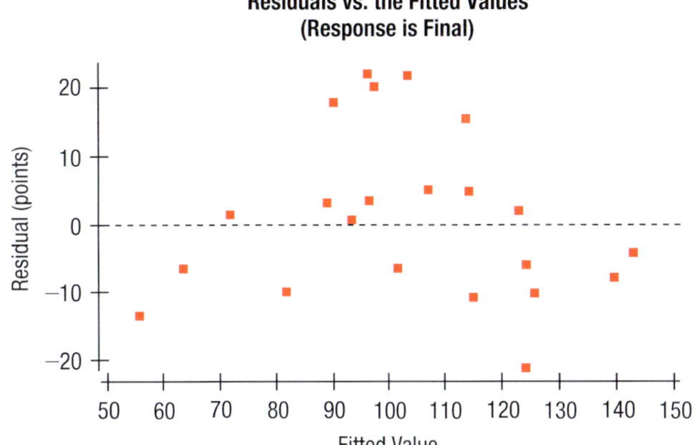

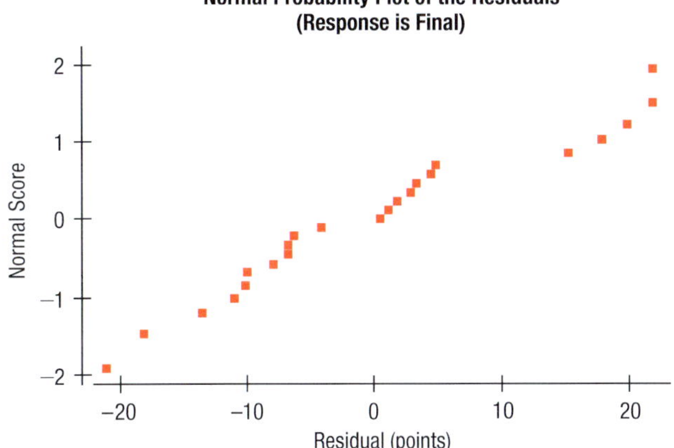

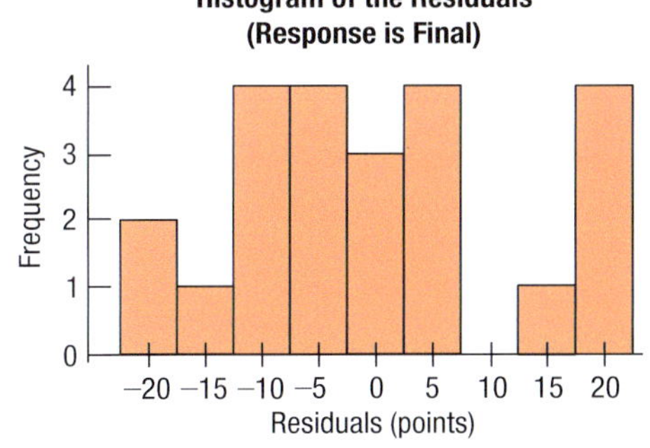

20. Home prices II Here are some diagnostic plots for the home prices data from Exercise 17. These were generated by a computer package and may look different from the plots generated by the packages you use. (In particular, note that the axes of the Normal probability plot are swapped relative to the plots we've made in the text. We only care about the pattern of this plot, so it shouldn't affect your interpretation.) Examine these plots and discuss whether the assumptions and conditions for the multiple regression seem reasonable.

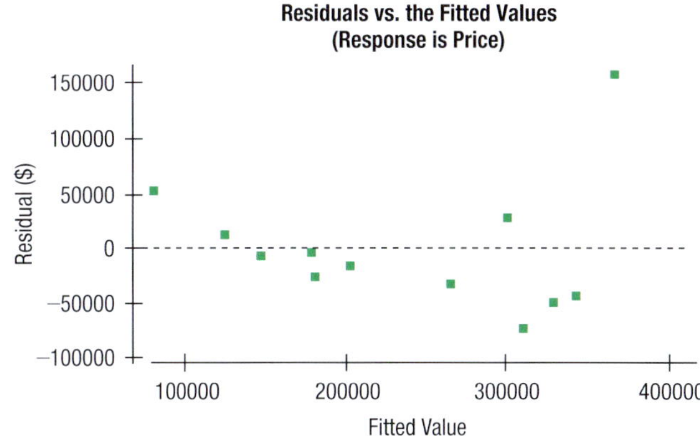

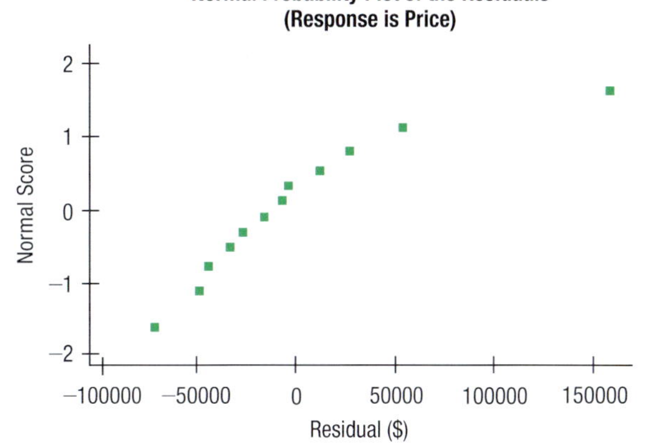

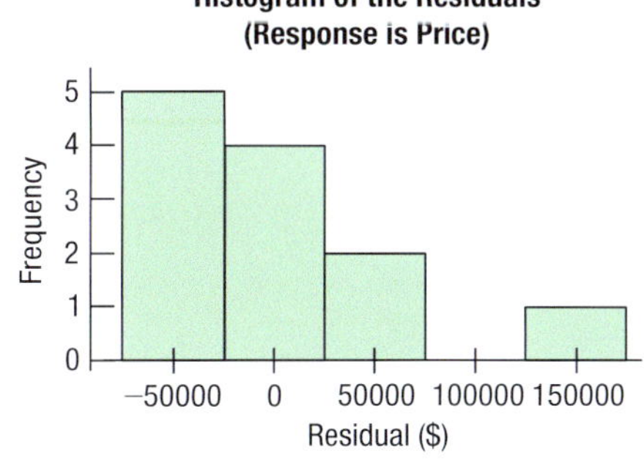

21. Admin performance An organization has undertaken a study of the yearly salaries (in thousands of dollars) of 30 administrative assistants. The organization wants to predict salaries from several other variables.

The variables considered to be potential predictors of salary are:

$X1$ = months of service
$X2$ = years of education
$X3$ = score on standardized test
$X4$ = words per minute (wpm) typing speed
$X5$ = ability to take dictation in words per minute

A multiple regression model with all five variables was run on a computer package, resulting in the following output:

Variable	Coefficient
Intercept	9.876
X1	0.111
X2	0.068
X3	0.098
X4	0.017
X5	0.065
$s = 0.383$	$R^2 = 0.863$

Assume that the residual plots show no violations of the conditions for using a linear regression model.

a) What is the regression equation?
b) From this model, what is the predicted *Salary* (in thousands of dollars) of a secretary with 12 years (144 months) of experience, 9th grade education (9 years of education), a 53 on the standardized test, 61 wpm typing speed, and the ability to take 30 wpm dictation?
c) A correlation of Age with *Salary* finds $r = 0.709$, and the scatterplot shows a moderately strong positive linear association. However, if $X6 = Age$ is added to the multiple regression, the estimated coefficient of *Age* turns out to be $b_6 = -0.122$. Explain some possible causes for this apparent change of direction in the relationship between age and salary.

22. GPA and SATs A large section of Stat 101 was asked to fill out a survey on grade point average and SAT scores. A regression was run to find out how well Math and Verbal SAT scores could predict academic performance as measured by GPA. The regression was run on a computer package with the following output:

Response: GPA

Variable	Coefficient	SE(Coeff)	t-ratio	P-value
Intercept	0.574908	0.253859	2.26	0.0298
SAT Verbal	0.001311	0.000502	2.61	0.0130
SAT Math	0.001905	0.000551	3.46	0.0014

a) What is the regression equation?
b) From this model, what is the predicted GPA of a student with a SAT Verbal score of 520 and a SAT Math score of 650?
c) What else would you want to know about this regression before writing a report about the relationship between SAT scores and grade point averages? Why would these be important to know?

T **23. Body fat revisited** The dataset on body fat contains 15 body measurements on 250 men from 22 to 81 years old. Is average *%Body Fat* related to *Weight*? Here's a scatterplot:

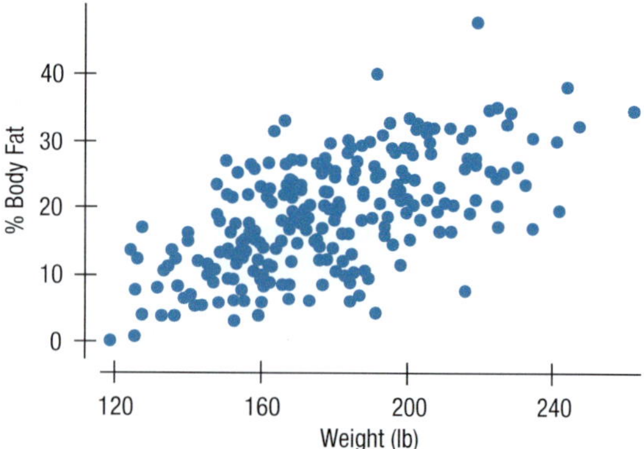

And here's the simple regression:

Dependent variable is: Pct BF
R-squared = 38.1% s = 6.538

Variable	Coefficient	SE(Coeff)	t-ratio	P-value
Intercept	−14.6931	2.760	−5.32	<0.0001
Weight	0.18937	0.0153	12.4	<0.0001

a) What does the slope coefficient mean in this regression?

What happens if we add *Height* and *Waist* to that regression? Recall that we've already checked the assumptions and conditions for regression on *Waist* size and *Height* in the chapter. Here is the output from a regression on all three variables:

Dependent variable is: Pct BF
R-squared = 72.5% s = 4.376

Variable	Coefficient	SE(Coeff)	t-ratio	P-value
Intercept	−31.4830	11.54	−2.73	0.0068
Waist	2.31848	0.1820	12.7	<0.0001
Height	−0.224932	0.1583	−1.42	0.1567
Weight	−0.100572	0.0310	−3.25	0.0013

b) Interpret the slope for *Weight*. How can the coefficient for *Weight* in this model be negative when its coefficient was positive in the simple regression model?
c) What else would you like to see? Explain briefly.

T **24. Cereals** We saw in Chapter 7 that the calorie content of a breakfast cereal is linearly associated with its sugar content. Is that the whole story? Here's the output of a regression model that regresses *Calories* per serving on each serving's *Protein(g)*, *Fat(g)*, *Fiber(g)*, *Carbohydrate(g)*, and *Sugars(g)* content.

Dependent variable is: Calories
R-squared = 93.6% s = 5.113

Variable	Coefficient	SE(Coeff)	t-ratio	P-value
Intercept	−0.879994	4.383	−0.201	0.8414
Protein	3.60495	0.6977	5.17	<0.0001
Fat	8.56877	0.6625	12.9	<0.0001
Fiber	0.309180	0.3337	0.927	3.572
Carbo	4.13996	0.2049	20.2	<0.0001
Sugars	4.00677	0.1719	23.3	<0.0001

a) What is the regression equation?
b) To check the conditions, what plots of the data might you want to examine?
c) What does the coefficient of *Fat* mean in this model?

25. Cereals again We saw a model in Exercise 24 for the calorie count of a breakfast cereal. Can we predict the calories of a serving from its vitamin and mineral content? Here's a multiple regression model of *Calories* per serving on its *Sodium (mg)*, *Potassium (mg)*, and *Sugars (g)*:

Dependent variable is: Calories
R-squared = 38.4% s = 15.60

Variable	Coefficient	SE(Coeff)	t-ratio	P-value
Intercept	83.0469	5.198	16.0	<0.0001
Sodium	0.05721	0.0215	2.67	0.0094
Potass	−0.01933	0.0251	−0.769	0.4441
Sugars	2.38757	0.4066	5.87	<0.0001

Assuming that the conditions for multiple regression are met,

a) What is the regression equation?
b) To check the conditions, what plots of the data might you want to examine?
c) Will adding *Potassium* to a breakfast cereal lower its *Calories*? Explain briefly.

26. Grades The dataset **Grades** shows the five scores from an Introductory statistics course. Find a model for final exam score by trying all possible models with two predictor variables. Which model would you choose? Be sure to check the conditions for multiple regression.

***27. Hand dexterity** Researchers studied the dexterity of children using a timed test called the Functional Dexterity Test (FDT). Of course, they needed to allow for differences in dexterity between their subjects' dominant and non-dominant hands. Here is a scatterplot of the speed at which subjects completed the task vs. their age, and lines for dominant and nondominant hands:

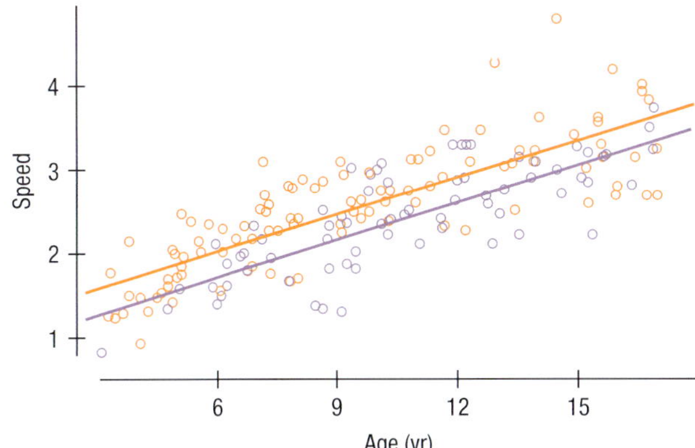

Here is a corresponding regression. The variable *Dominant* is 1 for trials in which a child used his or her dominant hand and 0 for trials with the non-dominant hand. Speed is recorded in tasks/second.

Response variable is: Speed
R-squared = 64.2% s = 0.4171

Variable	Coefficient	SE(Coeff)	t-ratio	P-value
Intercept	0.866000	0.1023	8.47	<0.0001
Age(yr)	0.146239	0.0084	17.4	<0.0001
Dominant	0.304262	0.0654	4.65	<0.0001

a) What is the proper interpretation of the predictor *Dominant* in this regression?
b) Are the conditions required for using an indicator variable met for these data? Explain.

***28. Candy bars with nuts** The data on candy bars per serving from Exercise 16 also has information on whether the candy bar has nuts. The variable *Nuts* is coded 0 for candy bars without nuts and 1 for those with nuts. Here's the regression model of Exercise 16 with *Nuts* added:

Response variable is: Calories
R-squared = 46.7% s = 0.3318

Variable	Coefficient
Intercept	5.74529
Sugar	−2.44376
Protein	7.13429
Nuts	−0.155754

a) What is a proper interpretation of the coefficient of *Nuts* in this regression?
b) Did adding *Nuts* improve the regression model? Explain.

***29. Scottish hill races, men and women** The Scottish hill races considered in Exercise 14 are run by both men and women. We can combine the data to fit a single model for both. Here is one such model. The variable *Dist*Sex* is the product of the variable *Distance* and the indicator variable *Sex*, which is 0 for women and 1 for men.

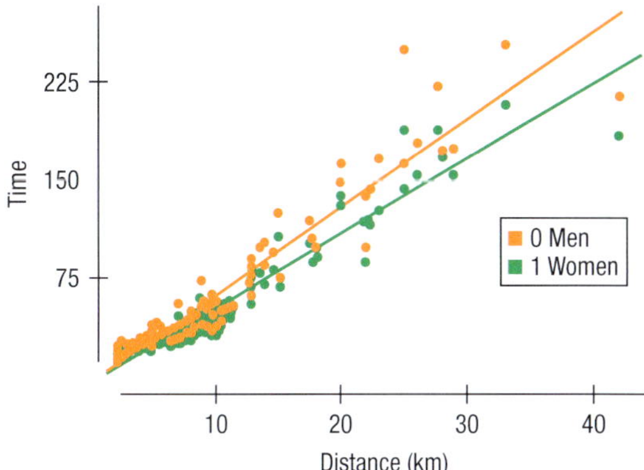

Response variable is: Time(mins)
R-squared = 90.9% s = 15.49

Variable	Coefficient	SE(Coeff)	t-ratio	P-value
Intercept	−7.23696	2.029	−3.57	0.0005
Distance(km)	6.76479	0.1734	39.0	<0.0001
Dist*Sex	−0.944957	0.1715	−5.51	<0.0001

a) What is the purpose of including the variable *Dist*Sex* in this model?
b) What does the coefficient of that variable mean in this model?

T *30. Scottish hill races, men and women climbing The Scottish hill races considered in Exercises 14 and 29 are run by both men and women. We can combine the data to fit a single model for both. Here is one such model. The variable *Climb**Sex* is the product of the variable *Climb* and the indicator variable *Sex*, which is 0 for women and 1 for men.

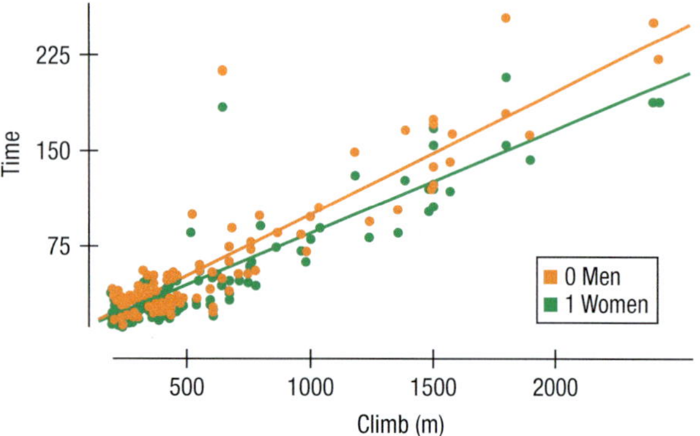

Response variable is: Time(mins)
R-squared = 81.2% s = 22.29

Variable	Coefficient	SE(Coeff)	t-ratio	P-value
Intercept	2.89214	2.738	1.06	0.2923
Climb (m)	0.096929	0.0038	25.4	<0.0001
Climb*Sex	−0.015548	0.0040	−3.93	0.0001

a) What is the purpose of including the variable *Climb**Sex* in this model?

b) What does the coefficient of that variable mean in this model?

JUST CHECKING

Answers

1. 80.6% of the variation in *Maximum Wind Speed* can be accounted for by this multiple regression on *Central Pressure* and *Year*.
2. The maximum wind speed of hurricanes is about 0.975 knotlower per millibar of central pressure, after allowing for changes over time. Put another way, lower central pressure at the eye of a hurricane leads to higher winds.
3. No. The coefficient of *Year* in this regression is quite small and possibly not really different from zero, but this can't prove that there have been no changes during these years. That isn't what this coefficient means in the multiple regression. Suppose, for example, that hurricanes have actually become stronger, on average, over time but that the mean central pressure has declined and the central wind speed has increased correspondingly.

Review of Part II

EXPLORING RELATIONSHIPS BETWEEN VARIABLES

Quick Review

You have now survived your second major unit of statistics. Here's a brief summary of the key concepts and skills:

- To explore relationships in categorical data, review Chapter 3.
- To explore relationships in quantitative data:
  - Make a picture. Use a scatterplot. Put the explanatory variable on the x-axis and the response variable on the y-axis.
  - Describe the association between two quantitative variables in terms of direction, form, and strength.
  - The amount of scatter determines the strength of the association.
  - If, as one variable increases so does the other, the association is positive. If one increases as the other decreases, it's negative.
  - If the form of the association is linear, calculate a correlation to measure its strength numerically, and do a regression analysis to model it.
  - Correlations closer to -1 or $+1$ indicate stronger linear associations. Correlations near 0 indicate weak linear relationships, but other forms of association may still be present.
  - The line of best fit is also called the least squares regression line because it minimizes the sum of the squared residuals.
  - The regression line predicts values of the response variable from values of the explanatory variable.
  - A residual is the difference between the true value of the response variable and the value predicted by the regression model.
  - The slope of the line is a rate of change, best described in "y-units" per "x-unit."
  - A multiple regression fits a linear model that predicts a response variable from a quantitative variable and at least one quantitative or indicator variable.
  - A multiple regression coefficient gives the slope of the relationship between y and a particular predictor *after accounting for the effects of the other predictors.*
  - To display what it means to account for the effects of the other predictors, make a partial regression plot.
  - R^2 gives the fraction of the variation in the response variable that is accounted for by the model.
  - The standard deviation of the residuals measures the amount of scatter around the line.
  - Outliers and influential points can distort any of our models.
  - If you see a pattern (a curve) in the residuals plot, your chosen model is not appropriate; use a different model. You may, for example, straighten the relationship by re-expressing one of the variables.
  - To straighten bent relationships, re-express the data using logarithms or a power (squares, square roots, reciprocals, etc.).
  - Always remember that an association is not necessarily an indication that one of the variables causes the other.

Need more help with some of this? Try rereading some sections of Chapters 6 through 9. And see below for more opportunities to review these concepts and skills.

> "One must learn by doing the thing; though you think you know it, you have no certainty until you try."
>
> —Sophocles (495–406 B.C.E.)

REVIEW EXERCISES

R2.1. College Every year, *US News and World Report* publishes a special issue on many U.S. colleges and universities. The scatterplots have *Student/Faculty Ratio* (number of students per faculty member) for the colleges and universities on the y-axes plotted against 4 other variables. The correct correlations for these scatterplots appear in this list. Match them.

$-0.98 \quad -0.71 \quad -0.51 \quad 0.09 \quad 0.23 \quad 0.69$

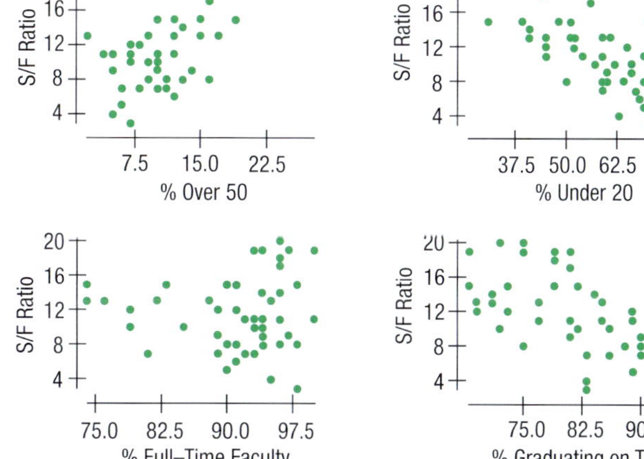

R2.2. Togetherness Are good grades in high school associated with family togetherness? A random sample of 142 high school students was asked how many meals per week their families ate together. Their responses produced a mean of 3.78 meals per week, with a standard deviation of 2.2. Researchers then matched these responses against the students' grade point averages (GPAs). The scatterplot appeared to be reasonably linear, so they created a line of regression. No apparent pattern emerged in the residuals plot. The equation of the line was $\widehat{GPA} = 2.73 + 0.11\ Meals$.

a) Interpret the y-intercept in this context.
b) Interpret the slope in this context.
c) What was the mean GPA for these students?
d) If a student in this study had a negative residual, what did that mean?
e) Upon hearing of this study, a counselor recommended that parents who want to improve the grades their children get should get the family to eat together more often. Do you agree with this interpretation? Explain.

T **R2.3.** Vineyards Here are the scatterplot and regression analysis for *Case Prices* of 36 wines from vineyards in the Finger Lakes region of New York State and the *Ages* of the vineyards. (Data in **Vineyards full**)

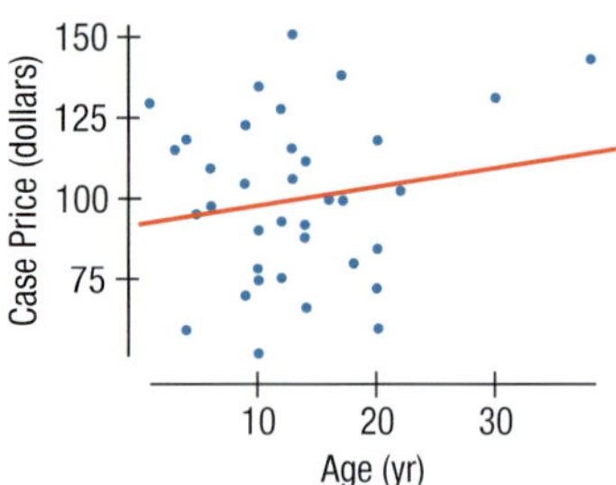

Dependent variable is: Case Price
R-squared = 2.7%

Variable	Coefficient
Constant	92.7650
Age	0.567284

a) Does it appear that vineyards in business longer get higher prices for their wines? Explain.
b) What does this analysis tell us about vineyards in the rest of the world?
c) Write the regression equation.
d) Explain why that equation is essentially useless.

T **R2.4.** Vineyards again Instead of *Age*, perhaps the *Size* of the vineyard (in acres) is associated with the price of the wines. Look at the scatterplot:

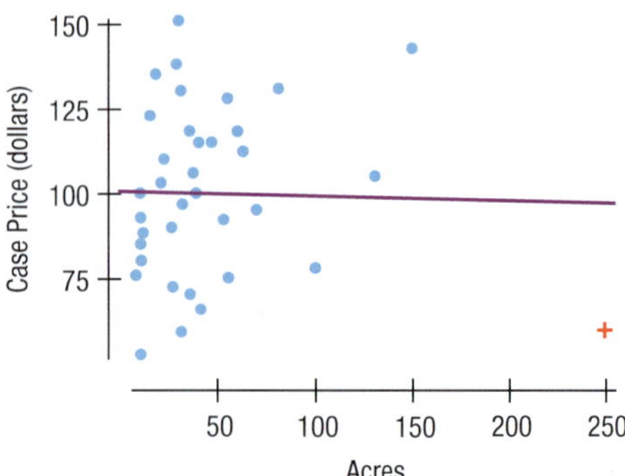

a) Do you see any evidence of an association?
b) What concern do you have about this scatterplot?
c) If the red "+" data point is removed, would the correlation become stronger or weaker? Explain.
d) If the red "+" data point is removed, would the slope of the line increase or decrease? Explain.

T **R2.5.** Twins by year 2014 In January 2012, the *New York Times* published a story called *Twin Births in the U.S., Like Never Before*, in which they reported a 76 percent increase in the rate of twin births from 1980 to 2009. Here are the number of twin births each year (per 1000 live births). (www.cdc.gov/nchs/births.htm)

Year	Twin Birth (per 1000 live births)	Year	Twin Birth (per 1000 live births)
1980	18.92	1998	28.08
1981	19.21	1999	28.87
1982	19.46	2000	29.30
1983	19.86	2001	30.12
1984	19.88	2002	31.11
1985	20.50	2003	31.46
1986	21.30	2004	32.15
1987	21.36	2005	32.17
1988	21.80	2006	32.11
1989	22.41	2007	32.19
1990	22.46	2008	32.62
1991	23.05	2009	33.22
1992	23.35	2010	33.15
1993	23.88	2011	33.20
1994	24.39	2012	33.15
1995	24.86	2013	33.7
1996	25.84	2014	33.9
1997	26.83		

a) Using the data only up to 2009 (as in the *NYTimes* article), find the equation of the regression line for predicting the number of twin births by *Years since 1980*.
b) Explain in this context what the slope means.
c) Predict the number of twin births in the United States for the year 2014. Then compare your prediction to the actual value.
d) Fit a new regression model to the entire data sequence. Comment on the fit. Now plot the residuals. Are you satisfied with the model? Explain.

R2.6. Dow Jones 2015 The Dow Jones stock index measures the performance of the stocks of America's largest companies (finance.yahoo.com). A regression of the Dow prices on years 1972–2015 looks like this:

Dependent variable is: Dow Index
R-squared = 88.4% s = 1790

Variable	Coefficient
Intercept	−2891.43
Year since 1970	379.917

a) What is the correlation between *Dow Index* and *Year*?
b) Write the regression equation.
c) Explain in this context what the equation says.
d) Here's a scatterplot of the residuals. Comment on the appropriateness of the model.

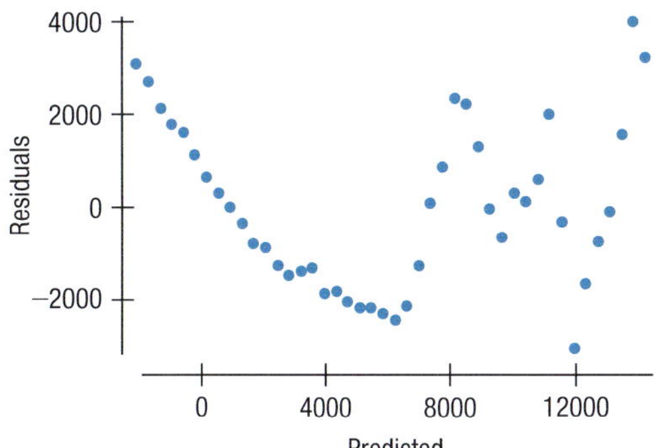

R2.7. Streams Biologists studying the effects of acid rain on wildlife collected data from 163 streams in the Adirondack Mountains. They recorded the *pH* (acidity) of the water and the *BCI*, a measure of biological diversity, and they calculated $R^2 = 27\%$. Here's a scatterplot of *BCI* against *pH*:

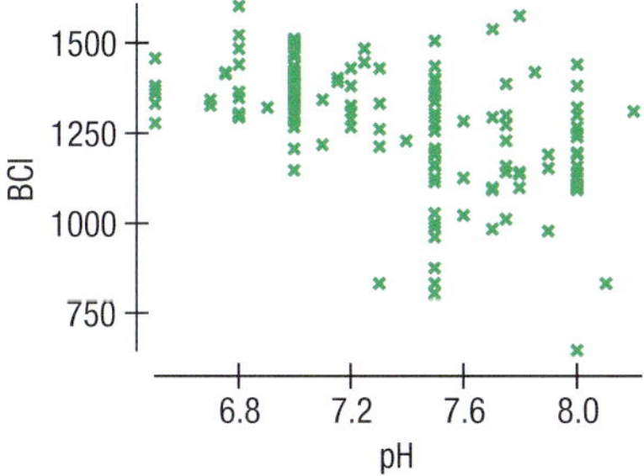

a) What is the correlation between *pH* and *BCI*?
b) Describe the association between these two variables.
c) If a stream has average *pH*, what would you predict about the *BCI*?
d) In a stream where the *pH* is 3 standard deviations above average, what would you predict about the *BCI*?

R2.8. Manatees 2015 Marine biologists warn that the growing number of powerboats registered in Florida threatens the existence of manatees. The data in the table come from the Florida Fish and Wildlife Conservation Commission (myfwc.com/research/manatee/) and the U.S. Coast Guard Office of Auxiliary and Boating Safety (www.uscgboating.org/library/accident-statistics/Recreational-Boating-Statistics-2015.pdf).

Year	Manatees Killed	Powerboat Registrations (in 1000s)
1982	13	447
1983	21	460
1984	24	481
1985	16	498
1986	24	513
1987	20	512
1988	15	527
1989	34	559
1990	33	585
1992	33	614
1993	39	646
1994	43	675
1995	50	711
1996	47	719
1997	53	716
1998	38	716
1999	35	716
2000	49	735
2001	81	860
2002	95	923
2003	73	940
2004	69	946
2005	79	974
2006	92	988
2007	73	992
2008	90	932
2009	97	949
2010	83	914
2011	88	890
2012	81	870
2013	73	871
2014	69	873.5
2015	86	889

a) In this context, which is the explanatory variable?
b) Make a scatterplot of these data and describe the association you see.
c) Find the correlation between *Boat Registrations* and *Manatee Deaths*.
d) Interpret the value of R^2.
e) Does your analysis prove that powerboats are killing manatees?

R2.9. Streams II Exercise R2.7 examined the correlation between *BCI* and *pH* in streams sampled in the Adirondack Mountains. Here is the corresponding regression model:

Response variable is: BCI
R squared = 27.1% s = 140.4

Variable	Coefficient
Intercept	2733.37
pH	−197.694

a) Write the regression model.
b) What is the interpretation of the coefficient of *pH*?
c) What would you predict the *BCI* would be for a stream with a *pH* of 8.2?

R2.10. A manatee model 2015 Continue your analysis of the manatee situation from Exercise R2.8.

a) Create a linear model of the association between *Manatee Deaths* and *Powerboat Registrations*.
b) Interpret the slope of your model.
c) Interpret the *y*-intercept of your model.
d) Which is better for the manatees, positive residuals or negative residuals? Explain.
e) Create a regression model just for the years from 2001 to 2015. Interpret the slope of your model.
f) What do these models suggest about the future for the manatee?

R2.11 Streams III Exercise R2.9 fit a regression model to the relationship between *BCI* and *pH* in streams sampled in the Adirondack Mountains. More variables are available. For example, scientists also recorded the water hardness. Here's a new model:

Response variable is: BCI
R squared = 29.6% s = 138.4

Variable	Coefficient
Intercept	2342.95
pH	−137.833
Hard	−0.337210

a) Write the regression model.
b) Is this a more successful model (for example, will it provide better predictions) than the model of Exercise R2.9?
c) What would you predict the *BCI* would be for a stream with a *pH* of 8.2 and a hardness value of 205?
d) There is such a stream in the data. Its *BCI* is 1309. What is the difference between your prediction of part c and the observed value? What is that called?
e) What is the interpretation of the coefficient of *pH* in this regression?

R2.12. Final exam A statistics instructor created a linear regression equation to predict students' final exam scores from their midterm exam scores. The regression equation was $\widehat{Fin} = 10 + 0.9\,Mid$.

a) If Susan scored a 70 on the midterm, what did the instructor predict for her score on the final?
b) Susan got an 80 on the final. How big is her residual?
c) If the standard deviation of the final was 12 points and the standard deviation of the midterm was 10 points, what is the correlation between the two tests?
d) How many points would someone need to score on the midterm to have a predicted final score of 100?
e) Suppose someone scored 100 on the final. Explain why you can't estimate this student's midterm score from the information given.
f) One of the students in the class scored 100 on the midterm but got overconfident, slacked off, and scored only 15 on the final exam. What is the residual for this student?
g) No other student in the class "achieved" such a dramatic turnaround. If the instructor decides not to include this student's scores when constructing a new regression model, will the R^2 value of the regression increase, decrease, or remain the same? Explain.
h) Will the slope of the new line increase or decrease?

R2.13. Traffic Highway planners investigated the relationship between traffic *Density* (number of automobiles per mile) and the average *Speed* of the traffic on a moderately large city thoroughfare. The data were collected at the same location at 10 different times over a span of 3 months. They found a mean traffic *Density* of 68.6 cars per mile (cpm) with standard deviation of 27.07 cpm. Overall, the cars' average *Speed* was 26.38 mph, with standard deviation of 9.68 mph. These researchers found the regression line for these data to be $\widehat{Speed} = 50.55 - 0.352\,Density$.

a) What is the value of the correlation coefficient between *Speed* and *Density*?
b) What percent of the variation in average *Speed* is explained by traffic *Density*?
c) Predict the average *Speed* of traffic on the thoroughfare when the traffic *Density* is 50 cpm.
d) What is the value of the residual for a traffic *Density* of 56 cpm with an observed *Speed* of 32.5 mph?
e) The dataset initially included the point *Density* = 125 cpm, *Speed* = 55 mph. This point was considered an outlier and was not included in the analysis. Will the slope increase, decrease, or remain the same if we redo the analysis and include this point?
f) Will the correlation become stronger, weaker, or remain the same if we redo the analysis and include this point (125, 55)?
g) A European member of the research team measured the *Speed* of the cars in kilometers per hour (1 km ≈ 0.62 miles) and the traffic *Density* in cars per kilometer. Find the value of his calculated correlation between speed and density.

R2.14. Cramming One Thursday, researchers gave students enrolled in a section of basic Spanish a set of 50 new vocabulary words to memorize. On Friday, the students took a vocabulary test. When they returned to class the following Monday, they were retested—without advance warning. Here are the test scores for the 25 students.

Fri.	Mon.	Fri.	Mon.	Fri.	Mon.
42	36	48	37	39	41
44	44	43	41	46	32
45	46	45	32	37	36
48	38	47	44	40	31
44	40	50	47	41	32
43	38	34	34	48	39
41	37	38	31	37	31
35	31	43	40	36	41
43	32				

a) What is the correlation between *Friday* and *Monday* scores?
b) What does a scatterplot show about the association between the scores?
c) What does it mean for a student to have a positive residual?
d) What would you predict about a student whose *Friday* score was one standard deviation below average?
e) Write the equation of the regression line.
f) Predict the *Monday* score of a student who earned a 40 on Friday.

R2.15. Cars, correlations What factor most explains differences in *Fuel Efficiency* among cars? Below is a correlation matrix exploring that relationship for the car's *Weight* (1000 lb), *Horsepower*, *Displacement*, and number of *Cylinders*. (Data in **Cars**)

	MPG	Weight	Horsepower	Displacement	Cylinders
MPG	1.000				
Weight	−0.903	1.000			
Horsepower	−0.871	0.917	1.000		
Displacement	−0.786	0.951	0.872	1.000	
Cylinders	−0.806	0.917	0.864	0.940	1.000

a) Which factor seems most strongly associated with *Fuel Efficiency*?
b) What does the negative correlation indicate?
c) Explain the meaning of R^2 for that relationship.

R2.16. Cars, associations Look again at the correlation table for cars in the previous exercise.

a) Which two variables in the table exhibit the strongest association?
b) Is that strong association necessarily cause-and-effect? Offer at least two explanations why that association might be so strong.
c) Engine displacements for U.S.-made cars are often measured in cubic inches. For many foreign cars, the units are either cubic centimeters or liters. How would changing from cubic inches to liters affect the calculated correlations involving *Displacement*?
d) What would you predict about the *Fuel Efficiency* of a car whose engine *Displacement* is one standard deviation above the mean?

R2.17. Cars, horsepower Can we predict the *Horsepower* of the engine that manufacturers will put in a car by knowing the *Weight* of the car? Here are the regression analysis and residuals plot:

Dependent variable is: Horsepower
R-squared = 84.1%

Variable	Coefficient
Intercept	3.49834
Weight	34.3144

a) Write the equation of the regression line.
b) Do you think the car's *Weight* is measured in pounds or thousands of pounds? Explain.
c) Do you think this linear model is appropriate? Explain.
d) The highest point in the residuals plot, representing a residual of 22.5 horsepower, is for a Chevy weighing 2595 pounds. How much horsepower does this car have?

R2.18. Cars, fuel efficiency Consider a regression to predict the fuel efficiency (as miles per gallon, *MPG*) of the cars in the Cars data file. Here is one regression model using the *Weight* and the *Drive Ratio*:

Response variable is: MPG
R squared = 89.5% s = 2.186

Variable	Coefficient
Intercept	70.9191
Weight	−10.8315
Drive Ratio	−4.89716

a) What is the interpretation of the coefficient of *Weight* in this model?
b) Do you think this is a successful model for predicting the fuel efficiency of a car?
c) These cars were measured more than two decades ago. Do you think this model should apply to more modern cars? What exceptions would you suggest?

R2.19. Cars, more efficient? Here is a scatterplot of the residuals from the regression in Exercise R2.18:

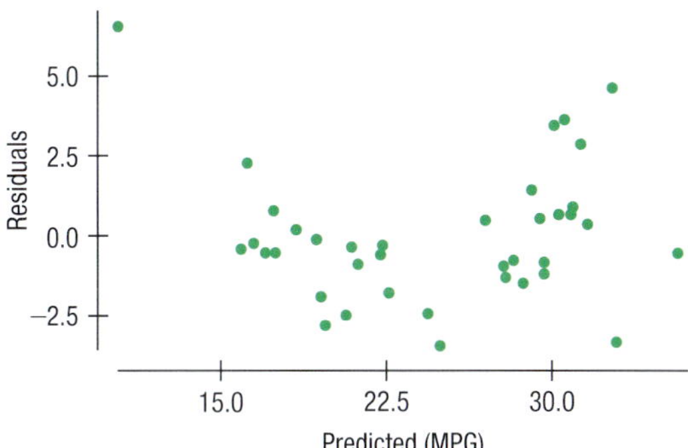

a) Does the residual plot suggest that the regression conditions were satisfied? Explain.

In the United States, fuel efficiency is usually measured as we did here, in miles per gallon. But in the rest of the world, the common measure is the *reciprocal*. There the measure is liters/100 km. The equivalent in U.S. terms is gallons per 100 miles. Here are a new regression model using *Gallons/100Miles* as the response variable and the corresponding residual plot:

Response variable is: 100/MPG
R squared = 92.0% s = 0.3355

Variable	Coefficient
Intercept	−3.49083
Weight	1.90186
Drive Ratio	0.768277

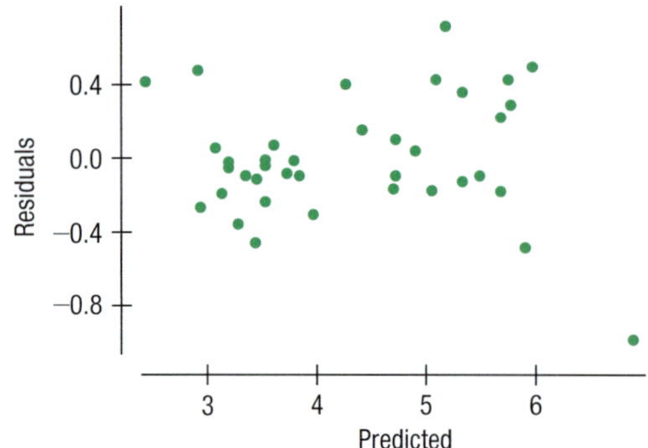

b) Would you prefer this model to the one in Exercise R2.18? Explain.

R2.20. Colorblind Although some women are colorblind, this condition is found primarily in men. Why is it wrong to say there's a strong correlation between *Sex* and *Colorblindness*?

T **R2.21.** Old Faithful again There is evidence that eruptions of Old Faithful can best be predicted by knowing the duration of the previous eruption.

a) Describe what you see in the scatterplot of *Intervals* between eruptions vs. *Duration* of the previous eruption.

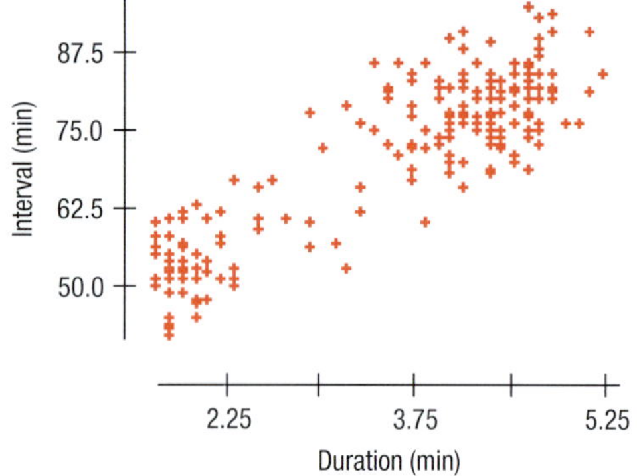

b) Write the equation of the line of best fit. Here's the regression analysis:

Dependent variable is: Interval
R-squared = 77.0% s = 6.16 min

Variable	Coefficient
Intercept	33.9668
Duration	10.3582

c) Carefully explain what the slope of the line means in this context.
d) How accurate do you expect predictions based on this model to be? Cite statistical evidence.
e) If you just witnessed an eruption that lasted 4 minutes, how long do you predict you'll have to wait to see the next eruption?
f) So you waited, and the next eruption came in 79 minutes. Use this as an example to define a residual.

T **R2.22.** Crocodile lengths The ranges inhabited by the Indian gharial crocodile and the Australian saltwater crocodile overlap in Bangladesh. Suppose a very large crocodile skeleton is found there, and we wish to determine the species of the animal. Wildlife scientists have measured the lengths of the heads and the complete bodies of several crocs (in centimeters) of each species, creating the regression analyses below:

Indian Crocodile
Dependent variable is IBody
R-squared = 97.2%

Variable	Coefficient
Intercept	−69.3693
IHead	7.40004

Australian Crocodile
Dependent variable is ABody
R-squared = 98.1%

Variable	Coefficient
Intercept	−21.3429
AHead	7.82761

a) Do the associations between the sizes of the heads and bodies of the two species appear to be strong? Explain.
b) In what ways are the two relationships similar? Explain.
c) What is different about the two models? What does that mean?
d) The crocodile skeleton found had a head length of 62 cm and a body length of 380 cm. Which species do you think it was? Explain why.

T **R2.23.** How old is that tree? One can determine how old a tree is by counting its rings, but that requires either cutting the tree down or extracting a sample from the tree's core. Can we estimate the tree's age simply from its diameter? A forester measured 27 trees of the same species that had been cut down, and counted the rings to determine the ages of the trees.

Diameter (in.)	Age (yr)	Diameter (in.)	Age (yr)
1.8	4	10.3	23
1.8	5	14.3	25
2.2	8	13.2	28
4.4	8	9.9	29
6.6	8	13.2	30
4.4	10	15.4	30
7.7	10	17.6	33
10.8	12	14.3	34
7.7	13	15.4	35
5.5	14	11.0	38
9.9	16	15.4	38
10.1	18	16.5	40
12.1	20	16.5	42
12.8	22		

a) Find the correlation between *Diameter* and *Age*. Does this suggest that a linear model may be appropriate? Explain.
b) Create a scatterplot and describe the association.
c) Create the linear model.
d) Check the residuals. Explain why a linear model is probably not appropriate.
e) If you used this model, would it generally overestimate or underestimate the ages of very large trees? Explain.

T **R2.24.** Improving trees In the last exercise, you saw that the linear model had some deficiencies. Let's create a better model.

a) Perhaps the cross-sectional area of a tree would be a better predictor of its age. Since area is measured in square units, try re-expressing the data by squaring the diameters. Does the scatterplot look better?
b) Create a model that predicts *Age* from the square of the *Diameter*.

c) Check the residuals plot for this new model. Is this model more appropriate? Why?
d) Estimate the age of a tree 18 inches in diameter.

R2.25. Big screen An electronics website collects data on the size of new HD flat-panel televisions (measuring the diagonal of the screen in inches) to predict the cost (in hundreds of dollars). Which of these is most likely to be the slope of the regression line: 0.03, 0.3, 3, 30? Explain.

T **R2.26.** Smoking and pregnancy 2011 The Child Trends Data Bank monitors issues related to children. The table shows a 50-state average of the percent of expectant mothers who smoked cigarettes during their pregnancies.

Year	% Smoking While Pregnant	Year	% Smoking While Pregnant
1990	19.2	2001	13.8
1991	18.7	2002	13.3
1992	17.9	2003	12.7
1993	16.8	2004	10.9
1994	16.0	2005	10.1
1995	15.4	2006	10.0
1996	15.3	2007	10.4
1997	14.9	2008	9.7
1998	14.8	2009	9.3
1999	14.1	2010	9.2
2000	14.0	2011	9.0

a) Create a scatterplot and describe the trend you see.
b) Find the correlation.
c) How is the value of the correlation affected by the fact that the data are averages rather than percentages for each of the 50 states?
d) Write a linear model and interpret the slope in context.

T **R2.27.** No smoking? The downward trend in smoking you saw in the last exercise is good news for the health of babies, but will it ever stop?

a) Explain why you can't use the linear model you created in Exercise R2.26 to see when smoking during pregnancy will cease altogether.
b) Create a model that could estimate the year in which the level of smoking would be 0%.
c) Comment on the reliability of such a prediction.

R2.28. Tips It's commonly believed that people use tips to reward good service. A researcher for the hospitality industry examined tips and ratings of service quality from 2645 dining parties at 21 different restaurants. The correlation between ratings of service and tip percentages was 0.11. (M. Lynn and M. McCall, "Gratitude and Gratuity," *Journal of Socio-Economics* 29: 203–214)

a) Describe the relationship between *Quality of Service* and *Tip Size*.
b) Find and interpret the value of R^2 in this context.

R2.29. U.S. cities Data from 50 large U.S. cities show the mean *January Temperature* and the *Latitude*. Describe what you see in the scatterplot.

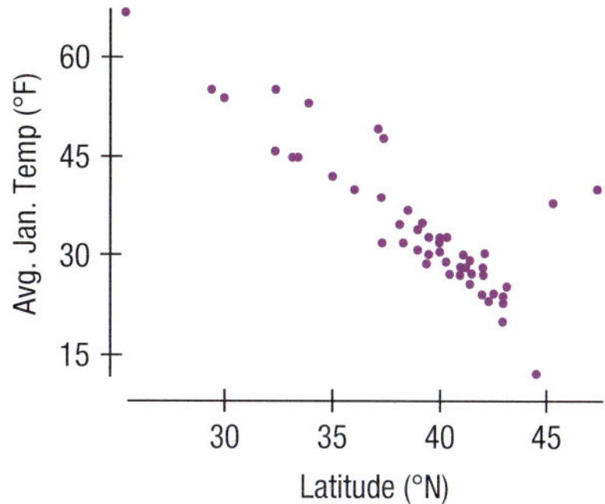

R2.30. Correlations The study of U.S. cities in Exercise R2.29 found the mean *January Temperature* (degrees Fahrenheit), *Altitude* (feet above sea level), and *Latitude* (degrees north of the equator) for 55 cities. Here's the correlation matrix:

	Jan. Temp	Latitude	Altitude
Jan. Temp	1.000		
Latitude	−0.848	1.000	
Altitude	−0.369	0.184	1.000

a) Which seems to be more useful in predicting *January Temperature: Altitude* or *Latitude*? Explain.
b) If the *Temperature* were measured in degrees Celsius, what would be the correlation between *Temperature* and *Latitude*?
c) If the *Temperature* were measured in degrees Celsius and the *Altitude* in meters, what would be the correlation? Explain.
d) What would you predict about the *January Temperature* in a city whose *Altitude* is two standard deviations higher than the average *Altitude*?

R2.31. Winter in the city Summary statistics for the data relating the *Latitude* and average *January temperature* for 55 large U.S. cities are given below.

Variable	Mean	StdDev
Latitude	39.02	5.42
JanTemp	26.44	13.49

Correlation = −0.848

a) What percent of the variation in *January Temperature* can be explained by variation in *Latitude*?
b) What is indicated by the fact that the correlation is negative?
c) Write the equation of the line of regression for predicting average *January Temperature* from *Latitude*.
d) Explain what the slope of the line means.
e) Do you think the y-intercept is meaningful? Explain.
f) The latitude of Denver is 40°N. Predict the mean January temperature there.
g) What does it mean if the residual for a city is positive?

R2.32. Depression and the Internet The September 1998 issue of the *American Psychologist* published an article by Kraut et al. that reported on an experiment examining "the social and psychological impact of the Internet on 169 people in 73 households during their first 1 to 2 years online." In the experiment, 73 households were offered free Internet access for 1 or 2 years in return for allowing their time and activity online to be tracked. The members of the households who participated in the study were also given a battery of tests at the beginning and again at the end of the study. The conclusion of the study made news headlines: Those who spent more time online tended to be more depressed at the end of the experiment. Although the paper reports a more complex model, the basic result can be summarized in the following regression of *Depression* (at the end of the study, in "depression scale units") vs. *Internet Use* (in mean hours per week):

Dependent variable is: Depression
R-squared = 4.6%   s = 0.4563

Variable	Coefficient
Intercept	0.5655
Internet use	0.0199

The news reports about this study clearly concluded that using the Internet causes depression. Discuss whether such a conclusion can be drawn from this regression. If so, discuss the supporting evidence. If not, say why not.

R2.33. Olympic jumps 2016 How are Olympic performances in various events related? The plot shows winning long-jump and high-jump distances, in meters, for the Summer Olympics from 1912 through 2016:

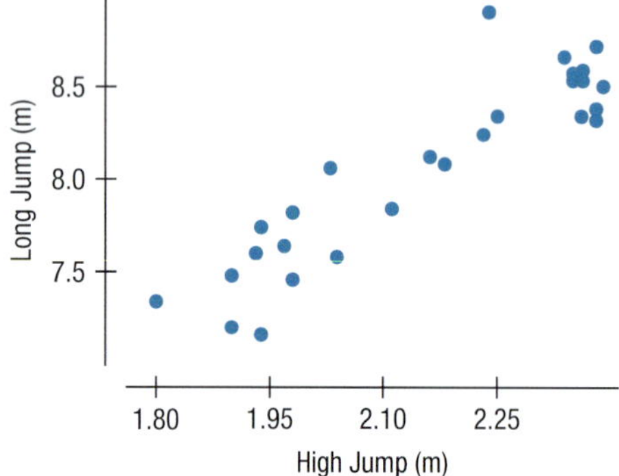

a) Describe the association.
b) Do long-jump performances somehow influence the high-jumpers? How do you account for the relationship you see?
c) The correlation for the given scatterplot is 0.910. If we converted the jump lengths to centimeters by multiplying by 100, would that make the actual correlation higher or lower?
d) What would you predict about the long jump in a year when the high-jumper jumped one standard deviation better than the average high jump?

R2.34. Modeling jumps 2016 Here are the summary statistics for the Olympic jumps displayed in the previous exercise.

Event	Mean	StdDev
High Jump	2.15667	0.195271
Long Jump	8.06222	0.507606

Correlation = 0.90992

a) Write the equation of the line of regression for estimating *High Jump* from *Long Jump*.
b) Interpret the slope of the line.
c) In a year when the long jump is 8.9 m, what high jump would you predict?
d) Why can't you use this line to estimate the long jump for a year when you know the high jump was 2.2 m?
e) Write the equation of the line you need to make that prediction.

R2.35. French Consider the association between a student's score on a French vocabulary test and the weight of the student. What direction and strength of correlation would you expect in each of the following situations? Explain.

a) The students are all in third grade.
b) The students are in third through twelfth grades in the same school district.
c) The students are in tenth grade in France.
d) The students are in third through twelfth grades in France.

R2.36. Twins Twins are often born at less than 9 months gestation. The graph from the *Journal of the American Medical Association (JAMA)* shows the rate of preterm twin births in the United States over the past 20 years. In this study, *JAMA* categorized mothers by the level of prenatal medical care they received: inadequate, adequate, or intensive.

a) Describe the overall trend in preterm twin births.
b) Describe any differences you see in this trend, depending on the level of prenatal medical care the mother received.
c) Should expectant mothers be advised to cut back on the level of medical care they seek in the hope of avoiding preterm births? Explain.

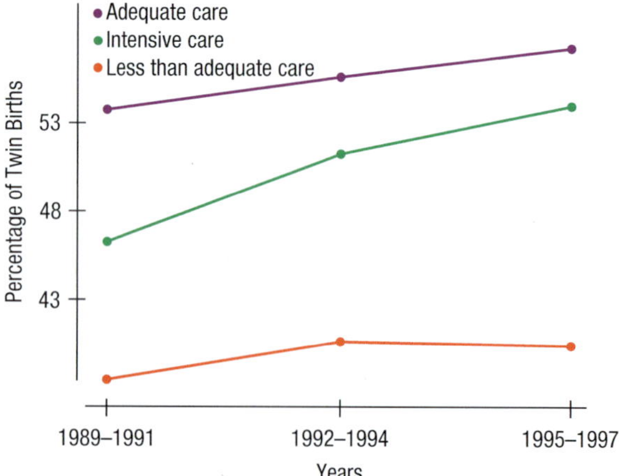

"Preterm Birth Rate per 100 live twin births among U.S. twins by intensive, adequate, and less than adequate prenatal care utilization, 1981–1997." (Source: *JAMA* 284[2000]: 335–341)

R2.37. Lunchtime Does how long toddlers sit at the lunch table help predict how much they eat? The table and graph show the number of minutes the kids stayed at the table and the number of calories they consumed. Create and interpret a model for these data.

Calories	Time	Calories	Time
472	21.4	450	42.4
498	30.8	410	43.1
465	37.7	504	29.2
456	33.5	437	31.3
423	32.8	489	28.6
437	39.5	436	32.9
508	22.8	480	30.6
431	34.1	439	35.1
479	33.9	444	33.0
454	43.8	408	43.7

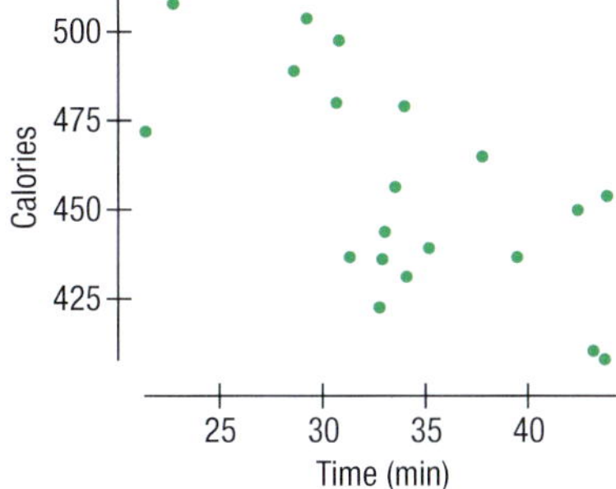

R2.38. Gasoline Since clean-air regulations have dictated the use of unleaded gasoline, the supply of leaded gas in New York state has diminished. The following table was given on the August 2001 New York State Math B exam, a statewide achievement test for high school students.

Year	1984	1988	1992	1996	2000
Gallons (1000's)	150	124	104	76	50

a) Create a linear model and predict the number of gallons available in 2015. Comment.
b) The exam then asked students to estimate the year when leaded gasoline will first become unavailable, expecting them to use the model from part a to answer the question. Explain why that method is incorrect.
c) Create a model that *would* be appropriate for that task, and make the estimate.
d) The "wrong" answer from the other model is fairly accurate in this case. Why?

R2.39. Tobacco and alcohol Are people who use tobacco products more likely to consume alcohol? Here are data on household spending (in pounds) taken by the British government on 11 regions in Great Britain. Do tobacco and alcohol spending appear to be related? What questions do you have about these data? What conclusions can you draw?

Region	Alcohol	Tobacco
North	6.47	4.03
Yorkshire	6.13	3.76
Northeast	6.19	3.77
East Midlands	4.89	3.34
West Midlands	5.63	3.47
East Anglia	4.52	2.92
Southeast	5.89	3.20
Southwest	4.79	2.71
Wales	5.27	3.53
Scotland	6.08	4.51
Northern Ireland	4.02	4.56

R2.40. Williams football The Sears Cup was established in 1993 to honor institutions that maintain a broad-based athletic program, achieving success in many sports, both men's and women's. In the years following its Division III inception in 1995, the cup was won by Williams College 15 of 17 years. Why did the football team win so much? Was it because they were heavier than their opponents? The table shows the average team weights for selected years from 1973 to 1993.

Year	Weight (lb)	Year	Weight (lb)
1973	185.5	1983	192.0
1975	182.4	1987	196.9
1977	182.1	1989	202.9
1979	191.1	1991	206.0
1981	189.4	1993	198.7

a) Fit a straight line to the relationship between *Weight* and *Year*.
b) Does a straight line seem reasonable?
c) Predict the average weight of the team for the year 2015. Does this seem reasonable?
d) What about the prediction for the year 2103? Explain.
e) What about the prediction for the year 3003? Explain.

R2.41. Models Find the predicted value of y, using each model for $x = 10$.

a) $\hat{y} = 2 + 0.8 \ln x$ b) $\log \hat{y} = 5 - 0.23x$

c) $\frac{1}{\sqrt{\hat{y}}} = 17.1 - 1.66x$

R2.42. Williams vs. Texas Here are the average weights of the football team for the University of Texas for various years in the 20th century.

Year	1905	1919	1932	1945	1955	1965
Weight (lb)	164	163	181	192	195	199

a) Fit a straight line to the relationship of *Weight* by *Year* for Texas football players.
b) According to these models, in what year will the predicted weight of the Williams College team from Exercise R2.40 first be more than the weight of the University of Texas team?
c) Do you believe this? Explain.

T R2.43. Vehicle weights The Minnesota Department of Transportation hoped that they could measure the weights of big trucks without actually stopping the vehicles by using a newly developed "weigh-in-motion" scale. After installation of the scale, a study was conducted to find out whether the scale's readings correspond to the true weights of the trucks being monitored. In Chapter 6, Exercise 52, you examined the scatterplot for the data they collected, finding the association to be approximately linear with $R^2 = 93\%$. Their regression equation is $\widehat{Wt} = 10.85 + 0.64\,Scale$, where both the scale reading and the predicted weight of the truck are measured in thousands of pounds.

a) Estimate the weight of a truck if this scale read 31,200 pounds.
b) If that truck actually weighed 32,120 pounds, what was the residual?
c) If the scale reads 35,590 pounds, and the truck has a residual of −2440 pounds, how much does it actually weigh?
d) If the police plan to use this scale to issue tickets to trucks that appear to be overloaded, will negative or positive residuals be a greater problem? Explain.

T R2.44. Companies How are a company's profits related to its sales? Let's examine data from 71 large U.S. corporations. All amounts are in millions of dollars.

a) Histograms of *Profits* and *Sales* and histograms of the logarithms of *Profits* and *Sales* are seen below. Why are the re-expressed data better for regression?

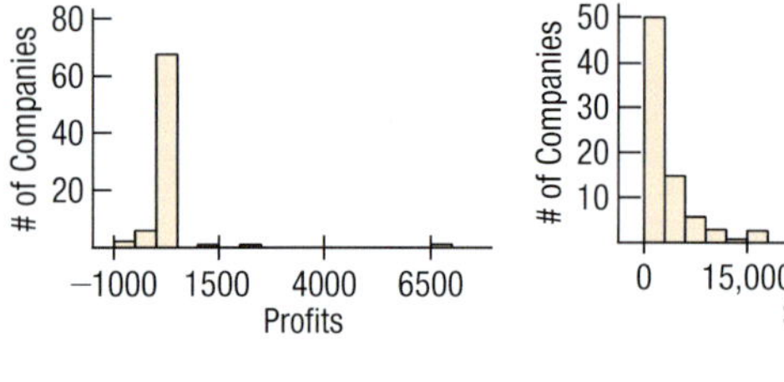

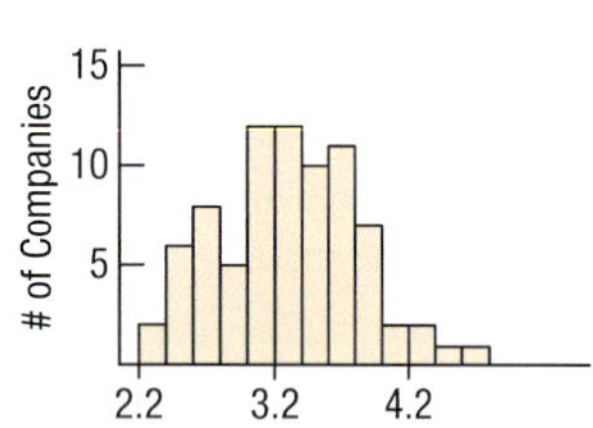

b) Here are the scatterplot and residuals plot for the regression of logarithm of *Profits* vs. log of *Sales*. Do you think this model is appropriate? Explain.

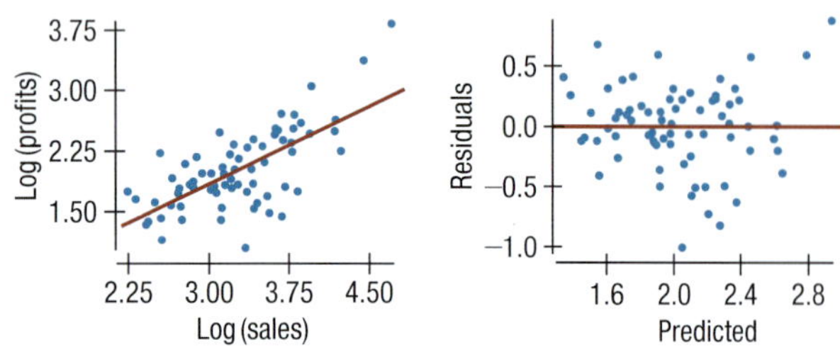

c) Here's the regression analysis. Write the equation.

Dependent variable is: Log Profit
R-squared = 48.1%

Variable	Coefficient
Intercept	−0.106259
LogSales	0.647798

d) Use your equation to estimate profits earned by a company with sales of 2.5 billion dollars. (That's 2500 million.)

T R2.45. Down the drain Most water tanks have a drain plug so that the tank may be emptied when it's to be moved or repaired. How long it takes a certain size of tank to drain depends on the size of the plug, as shown in the table. Create a model.

Plug Dia (in.)	$\frac{3}{8}$	$\frac{1}{2}$	$\frac{3}{4}$	1	$1\frac{1}{4}$	$1\frac{1}{2}$	2
Drain Time (min)	140	80	35	20	13	10	5

T R2.46. Chips A start-up company has developed an improved electronic chip for use in laboratory equipment. The company needs to project the manufacturing cost, so it develops a spreadsheet model that takes into account the purchase of production equipment, overhead, raw materials, depreciation, maintenance, and other business costs. The spreadsheet estimates the cost of producing 10,000 to 200,000 chips per year, as seen in the table. Develop a regression model to predict *Costs* based on the *Level* of production.

Chips Produced (1000s)	Cost per Chip ($)	Chips Produced (1000s)	Cost per Chip ($)
10	146.10	90	47.22
20	105.80	100	44.31
30	85.75	120	42.88
40	77.02	140	39.05
50	66.10	160	37.47
60	63.92	180	35.09
70	58.80	200	34.04
80	50.91		

T *R2.47. Companies assets and sales How are the assets and sales of major corporations related? The dataset is the same as that for Exercise 2.44. The variable *Banks* in the dataset is 1 for the banks and 0 for the other companies. As Exercise 2.44 showed, it is best to take the logarithm of most of these variables. Here are a scatterplot and a regression relating the logarithm of assets and the logarithm of sales. The banks are shown in red:

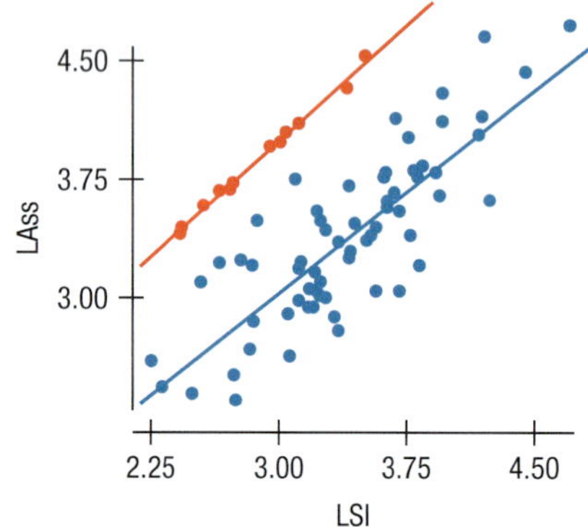

Response variable is: LogAssets
R squared = 73.7% s = 0.2741

Variable	Coefficient
Intercept	0.422563
LogSales	0.868488
Banks	0.945675

a) What is an appropriate interpretation of the coefficient of *LogSales*?
b) What is an appropriate interpretation of the coefficient of *Banks*?
c) Is the use of a variable such as *Banks* appropriate for these data? Explain.

T * **R2.48 Real estate** As a class project, students in a large statistics class collected publicly available information on recent home sales in their hometowns. There are 894 properties. Among the variables available is an indication of whether the home was in an urban, suburban, or rural setting. Here is a scatterplot relating the selling *Price* of homes to their *Living area* (sq ft), colored according to whether or not they were in a suburban setting.

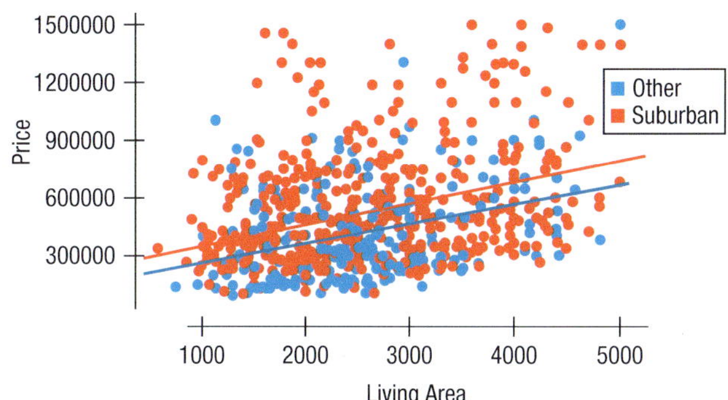

Response variable is: Price
R squared = 16.3% s = 264220

Variable	Coefficient
Intercept	136512
Living area	110.318
Suburb	104610

a) What is an appropriate interpretation of the coefficient of *Living area*?
b) What is an appropriate interpretation of the coefficient of *Suburb*?
c) Is it appropriate to use a variable such as *Suburb* in this model? Explain.

T **R2.49. Real estate, bathrooms** As a class project, students in a large statistics class collected publicly available information on recent home sales in their hometowns. There are 894 properties. Important predictors of the price of a home are its living area (sq ft) and the number of bathrooms. In fact, the correlation of *Price* with *Bathrooms* is 0.378. Here is a regression:

Response variable is: Price
R squared = 16.6% s = 263970

Variable	Coefficient
Intercept	126832
Living area	64.6077
Bathrooms	75020.3

a) What is a correct interpretation of the coefficient of *Bathrooms*?

Here is a partial regression plot for the coefficient of *Bathrooms* along with a least squares regression line:

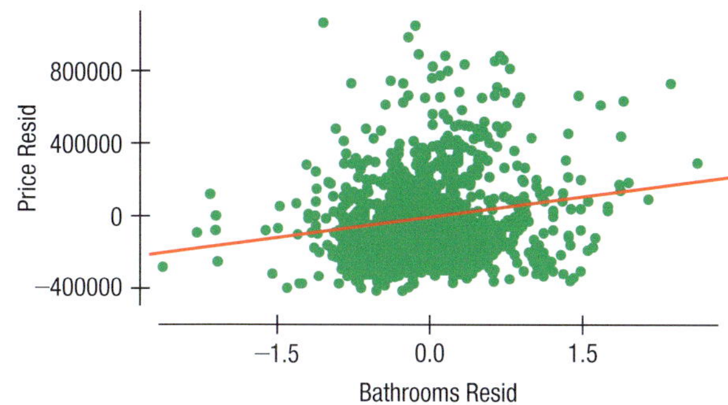

b) What is the slope of the regression line in this plot?

T **R2.50. Real estate, bedrooms** For the real estate data of the previous exercise, consider the value of the number of bedrooms in modeling the price of a home. The correlation between *Price* and *Bedrooms* is 0.116. Here is a regression model:

Response variable is: Price
R squared = 14.6% s = 266899

Variable	Coefficient
Intercept	308100
Living area	135.089
Bedrooms	−43346.8

a) What is a correct interpretation of the coefficient of *Bedrooms*?
b) The correlation of *Price* and *Bedrooms* is positive. How can the regression coefficient of *Bedrooms* be negative?

Here is a partial regression plot for the coefficient of *Bedrooms* along with a least squares regression line:

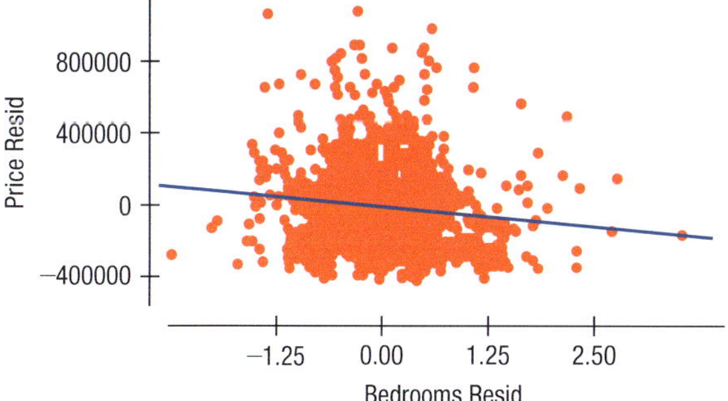

c) What is the slope of the regression line in the partial regression plot?
d) A homeowner proposes to convert his living room to a bedroom, thinking that more bedrooms would make his home more valuable. What do you think? Explain.

R2.51 Penguins again In Chapter 8 we learned about the extraordinary depth and duration of the dives taken by penguins. In that chapter we modeled a re-expression of *Heart rate* with the *Duration* (min) of dives. The data also include the depth of each dive. Here is an analysis that incorporates both variables:

Response variable is: LogHeartRate
R squared = 79.5% s = 0.0869

Variable	Coefficient
Intercept	1.99213
Duration(min)	−0.04510
Depth(m)	0.000887

a) What is an appropriate interpretation of the coefficient of *Duration* in this regression?

Here is a partial regression plot for the coefficient of *Duration*:

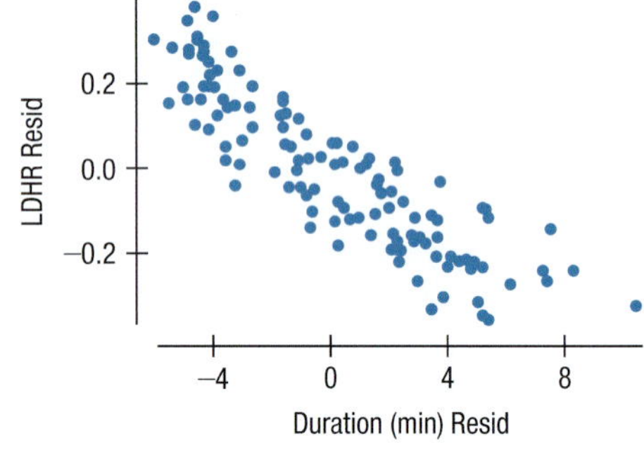

b) Based on this plot, do you think the coefficient of *Duration* is a good estimate of the relationship of *Duration* to (log) *Heart rate* in these penguins? Explain.

R2.52. Doctors and life expectancy and TVs Chapter 8 looked at the data on life expectancy in different countries as they related to the (square root of the) number of doctors and to the (square root of the) number of TVs. Here's a regression using both variables to predict life expectancy:

Response variable is: Life expectancy
R squared = 76.6% s = 3.877

Variable	Coefficient
Intercept	54.3108
$\sqrt{\text{TV/person}}$	22.9032
$\sqrt{\text{Doctors/person}}$	136.196

a) The slope coefficient for the regression of *Life expectancy* on $\sqrt{Doctors/person}$ is 369.249. Why is the coefficient for $\sqrt{Doctors/person}$ so different in this regression?

Here is a partial regression plot for the coefficient of $\sqrt{TV/person}$:

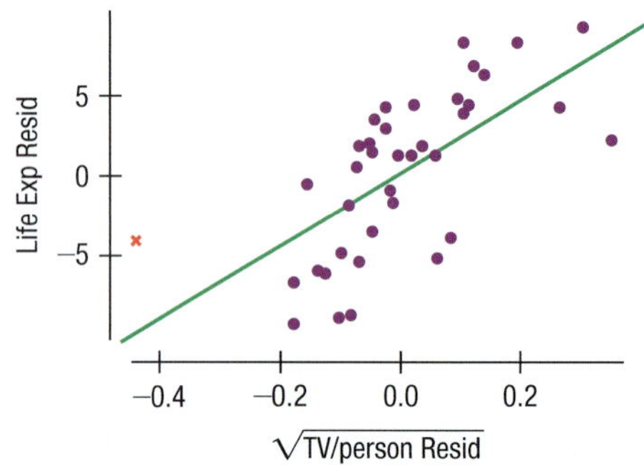

The country corresponding to the "x" in this plot is North Korea.

b) What is the slope of the least square line shown in the plot?

c) What effect is North Korea having on the estimation of the coefficient of $\sqrt{TV/person}$?

d) Suppose we decide that North Korea is sufficiently unusual that it should not be included in this analysis. What would you expect its omission to do to the coefficient of $\sqrt{TV/person}$?

10 Sample Surveys

WHERE ARE WE GOING?

We see surveys all the time. How can asking just a thousand people tell us much about a national election? And how *do* pollsters select the respondents? It turns out that there are many ways to select a good sample. But there are just three main ideas to understand.

Donald J. Trump began his presidency as the least popular new president in the history of modern polling. But exactly how unpopular was he? The Gallup Poll tracks presidential job approval. They found that at his inauguration, 45% of a sample of 1500 U.S. adults approved of President Trump's job performance and 45% disapproved. (The previous highest disapproval at inauguration was for George W. Bush when 25% disapproved.)

Gallup found that 45% approved of Trump, but at the same time, a poll of 1190 American voters by the Quinnipiac University Poll found that only 36% of respondents approved of the way President Trump was handling his job.

Both Gallup and Quinnipiac claim that these estimates are close to the true percentages that they would have found if they had asked all U.S. adults. But, their numbers are clearly very different. How we understand and account for these differences is a central topic of statistics and will concern us for most of the rest of this text. To make business decisions, to do science, to choose wise investments, or to understand how voters think they'll vote in the next election, we need to stretch beyond the data we have at hand to the world at large. That step from a small sample to the entire population is made possible by the methods you'll learn in this course.

10.1 The Three Big Ideas of Sampling

To make that stretch, we need three ideas. You'll find the first one natural. The second may be more surprising. The third is one of the strange but true facts that often confuse those who don't know statistics.

Idea 1: Examine a Part of the Whole

The first idea is to draw a sample. We'd like to know about an entire group of individuals—a **population**—but examining all of them is usually impractical, if not impossible. So we settle for examining a smaller group of individuals—a **sample**—selected from the population.

You do this every day. For example, suppose you wonder how the vegetable soup you're cooking for dinner tonight is going to go over with your friends. To decide whether it meets your standards, you only need to try a small amount. You might taste just a spoonful or two. You trust that the taste will *represent* the flavor of the entire pot. You know that a small sample, if selected properly, can represent the entire population.

It's hard to go a day without hearing about the latest opinion poll. These polls are examples of **sample surveys**, designed to ask questions of a small group of people in the hope of learning something about the entire population. How can the pollsters claim that a sample is representative of the entire population? Professional pollsters work quite hard to ensure that the "taste"—the sample that they take—represents the population.

The W's and Sampling

The population we are interested in is usually determined by *Why* we are doing the study. The individuals on whom we collect data are the sample.

Bias Selecting a sample to represent the population fairly is more difficult than it sounds. Polls or surveys may overlook subgroups that are harder to find (such as the homeless) or favor others (such as Internet users who like to respond to online surveys). Sampling methods that, by their nature, tend to over- or underemphasize some characteristics of the population are said to be **biased**. Bias is the bane of sampling—the one thing above all to avoid. Conclusions based on samples drawn with biased methods are inherently flawed. There is usually no way to fix bias after the sample is drawn and no way to salvage useful information from it.

Here's a famous example of a really dismal failure. By the beginning of the 20th century, it was common for newspapers to ask readers to return "straw" ballots on a variety of topics. (Today's Internet surveys are the same idea, gone electronic.)

During the period from 1916 to 1936, the magazine *Literary Digest* regularly surveyed public opinion and forecast election results correctly. During the 1936 presidential campaign between Alf Landon and Franklin Delano Roosevelt, it mailed more than 10 million ballots and got back an astonishing 2.4 million. (Polls were still a relatively novel idea, and many people thought it was important to send back their opinions.) The results were clear: Alf Landon would be the next president by a landslide, 57% to 43%. You remember President Landon? No? In fact, Landon carried only two states. Roosevelt won, 62% to 37%, and, perhaps coincidentally, the *Digest* went bankrupt soon afterward.

What went wrong? The *Digest* used the phone book, as many surveys do.[1] But in 1936, at the height of the Great Depression, telephones were a luxury, so the *Digest under*-sampled poor voters. The campaign of 1936 focused on the economy, and those who were less well off were more likely to vote for the Democrat. So the *Digest*'s sample was hopelessly biased.

How do modern polls get their samples to *represent* the entire population? You might think that they'd handpick individuals to sample with care and precision. But in fact, they do something quite different: They select individuals to sample *at random*.

Idea 2: Randomize

Think back to the soup sample. Suppose you add some salt to the pot. If you sample it from the top before stirring, you'll get the misleading idea that the whole pot is salty. If you sample from the bottom, you'll get an equally misleading idea that the whole pot is bland. By stirring, you *randomize* the amount of salt throughout the pot, making each taste more typical of the whole pot.

[1] Today, phone numbers are computer-generated to make sure that unlisted numbers are included. But special methods must be used for the growing part of the population that do not have landline access.

In 1936, a young pollster named George Gallup used a subsample of only 3000 of the 2.4 million responses that the *Literary Digest* received to reproduce the wrong prediction of Landon's victory over Roosevelt. He then used an entirely different sample of 50,000 and predicted that Roosevelt would get 56% of the vote to Landon's 44%. His sample was apparently much more *representative* of the actual voting populace. The Gallup Organization went on to become one of the leading polling companies.

Not only does randomization protect you against factors that you know are in the data, but it can also help protect against factors that you didn't know were there. Suppose that while you weren't looking, a friend added a handful of peas to the soup. If they're down at the bottom of the pot, and you don't randomize the soup by stirring, your test spoonful won't have any peas. By stirring in the salt, you *also* randomize the peas throughout the pot, making your sample taste more typical of the overall pot *even though you didn't know the peas were there.* So randomizing protects us even in this case.

How do we "stir" people in a survey? We select them at random. **Randomizing** protects us from the influences of *all* the features of our population by making sure that, *on average*, the sample looks like the rest of the population. The importance of deliberately using randomness is one of the great insights of statistics.

Why Not Match the Sample to the Population?

Rather than randomizing, we could try to design our sample so that the people we choose are typical in terms of every characteristic we can think of. We might want the income levels of those we sample to match the population. How about age? Political affiliation? Marital status? Having children? Living in the suburbs? We can't possibly think of all the things that might be important. Even if we could, we wouldn't be able to match our sample to the population for all these characteristics.

EXAMPLE 10.1 Is a Random Sample Representative?

Here are summary statistics comparing two samples of 8000 drawn at random from a company's database of 3.5 million customers:

	Mean Age (yr)	White (%)	Female (%)	Mean # of Children	Mean Income Bracket (1–7)	Mean Wealth Bracket (1–9)	Homeowner? (% Yes)
Sample 1	61.4	85.12	56.2	1.54	3.91	5.29	71.36
Sample 2	61.2	84.44	56.4	1.51	3.88	5.33	72.30

QUESTION: Do you think these samples are representative of the population of the company's customers? Explain.

ANSWER: The two samples look very similar with respect to these seven variables. It appears that randomizing has automatically matched them pretty closely. We can reasonably assume that since the two samples don't differ too much from each other, they don't differ much from the rest of the population either.

Idea 3: It's the Sample Size

How large a random sample do we need for the sample to be reasonably representative of the population? Most people think that we need a large fraction of the population, but it turns out that all that matters is the *number of individuals in the sample.* A random sample of 100 students in a university represents the student body just about as well as a random sample of 100 voters represents the entire electorate of the United States. This is the *third* idea and probably the most surprising one in designing surveys.

How can it be that only the size of the sample, and not the population, matters? Well, let's return one last time to that pot of soup. If you're cooking for a banquet rather than just for a few people, your pot will be bigger, but do you need a bigger spoon to decide how the soup tastes? Of course not. The same-size spoonful is probably enough to make a decision about the entire pot, no matter how large the pot. The *fraction* of the population that you've sampled doesn't matter.[2] It's the **sample size**—or the number of individuals in the sample—that's important.

How big a sample do you need? That depends on what you're estimating. To get an idea of what's really in the soup, you'll need a large enough taste to get a *representative* sample from the pot. For a survey that tries to find the proportion of the population falling into a category, you'll usually need several hundred respondents to say anything precise enough to be useful.[3]

WHAT DO THE POLLSTERS DO?

How do professional polling agencies do their work? The most common polling method today is to contact respondents by telephone. Computers generate random telephone numbers, so pollsters can even call some people with unlisted phone numbers. The interviewer may then ask to speak with the adult who is home who had the most recent birthday or use some other essentially random way to select among those available. In phrasing questions, pollsters often list alternative responses (such as candidates' names) in different orders to avoid biases that might favor the first name on the list.

Do these methods work? The Pew Research Center for the People and the Press, reporting on one survey, says that

> Across five days of interviewing, surveys today are able to make some kind of contact with the vast majority of households (76%). But because of busy schedules, skepticism and outright refusals, interviews were completed in just 38% of households that were reached using standard polling procedures.

Nevertheless, studies indicate that those actually sampled can give a good snapshot of larger populations from which the surveyed households were drawn.

LARGER SAMPLES?

A friend who knows that you are taking statistics asks your advice on her study. What can you possibly say that will be helpful? Just say, "If you could just get a larger sample, it would probably improve your study." Even though a larger sample might not be worth the cost, it will almost always make the results more precise.

Does a Census Make Sense? Why bother determining the right sample size? Wouldn't it be better to just include everyone and "sample" the entire population? Such a special sample is called a **census**. Although a census would appear to provide the best possible information about the population, there are a number of reasons why it might not.

First, it can be difficult to complete a census. Some individuals in the population will be hard (and expensive) to locate. Or a census might just be impractical. If you were a taste tester for the Hostess™ Company, you probably wouldn't want to census *all* the Twinkies on the production line. Not only might this be life-endangering, but you wouldn't have any left to sell.

Second, populations rarely stand still. In populations of people, babies are born and folks die or leave the country. In opinion surveys, events may cause a shift in opinion during the survey. A census takes longer to complete and the population changes while you work. A sample surveyed in just a few days may give more accurate information.

[2]Well, that's not exactly true. If the population is small enough and the sample is more than 10% of the whole population, it *can* matter. It doesn't matter whenever, as usual, our sample is a very small fraction of the population.

[3]Chapter 13 gives the details behind this statement and shows how to decide on a sample size for a survey.

Third, taking a census can be more complex than sampling. For example, the U.S. Census records too many college students. Many are counted once with their families and are then counted a second time in a report filed by their schools.

THE UNDERCOUNT

It's particularly difficult to compile a complete census of a population as large, complex, and spread out as the U.S. population. The U.S. Census is known to miss some residents. On occasion, the undercount has been striking. For example, there have been blocks in inner cities in which the number of residents recorded by the Census was smaller than the number of electric meters for which bills were being paid. What makes the problem particularly important is that some groups have a higher probability of being missed than others—undocumented immigrants, the homeless, the poor. The Census Bureau proposed the use of random sampling to estimate the number of residents missed by the ordinary census. Unfortunately, the resulting debate has become more political than statistical.

10.2 Populations and Parameters

Statistics and Parameters I

Any quantity that we calculate from data could be called a "statistic." But in practice, we usually use a statistic to estimate a population parameter.

Statistics and Parameters II

Remember: Population parameters are not just unknown—they are almost always *unknowable*. The best we can do is to estimate them with sample statistics.

A study found that teens were less likely to "buckle up." The National Center for Chronic Disease Prevention and Health Promotion reports that 21.7% of U.S. teens never or rarely wear seat belts. We're sure they didn't take a census, so what *does* the 21.7% mean? We can't know what percentage of teenagers wear seat belts. Reality is just too complex. But we can simplify the question by building a model.

Models use mathematics to represent reality. Parameters are the key numbers in those models. A parameter used in a model for a population is sometimes called (redundantly) a **population parameter**.

We use summaries of the data to estimate the population parameters. Any summary found from the data is a **statistic**. Sometimes you'll see the (also redundant) term **sample statistic**.[4]

We've already met two parameters in Chapter 5: the mean, μ, and the standard deviation, σ. We'll try to keep denoting population model parameters with Greek letters and the corresponding statistics with Latin letters. Usually, but not always, the letter used for the statistic and the parameter correspond in a natural way. So the standard deviation of the data is s, and the corresponding parameter is σ (Greek for s). In Chapter 6, we used r to denote the sample correlation. The corresponding correlation in a model for the population would be called ρ (rho). In Chapters 7 and 8, b_1 represented the slope of a linear regression estimated from the data. But when we think about a (linear) *model* for the population, we denote the slope parameter β_1 (beta).

Get the pattern? Good. Now it breaks down. We denote the mean of a population model with μ (because μ is the Greek letter for m). It might make sense to denote the sample mean with m, but long-standing convention is to put a bar over anything when we average it, so we write $\bar{y}$. What about proportions? Suppose we want to talk about the proportion of teens who don't wear seat belts. If we use p to denote the proportion from the data, what is the corresponding model parameter? By all rights, it should be π. But statements like $\pi = 0.25$ might be confusing because π has been equal to 3.1415926 . . . for so long, and it's worked so *well*. So, once again we violate the rule. We'll use p for the population model parameter and $\hat{p}$ for the proportion from the data (since, like $\hat{y}$ in regression, it's an estimated value).

[4]Where else besides a sample *could* a statistic come from?

Here's a table summarizing the notation:

NOTATION ALERT

This entire table is a notation alert.

Name	Statistic	Parameter
Mean	$\bar{y}$	μ (mu, pronounced "meeoo," not "moo")
Standard Deviation	s	σ (sigma)
Correlation	r	ρ (rho, pronounced like "row")
Regression Coefficient	b	β (beta, pronounced "baytah"[5])
Proportion	$\hat{p}$	p (pronounced "pee"[6])

We draw samples because we can't work with the entire population, but we want the statistics we compute from a sample to reflect the corresponding parameters accurately. A sample that does this is said to be **representative**. A biased sampling methodology tends to over- or underestimate the parameter of interest.

JUST CHECKING

1. Various claims are often made for surveys. Why is each of the following claims not correct?
   a) It is always better to take a census than to draw a sample.
   b) Stopping students on their way out of the cafeteria is a good way to sample if we want to know about the quality of the food there.
   c) We drew a sample of 100 from the 3000 students in a school. To get the same level of precision for a town of 30,000 residents, we'll need a sample of 1000.
   d) A poll taken at a statistics support website garnered 12,357 responses. The majority said they enjoy doing statistics homework. With a sample size that large, we can be pretty sure that most statistics students feel this way too.
   e) The true percentage of all statistics students who enjoy the homework is called a "population statistic."

10.3 Simple Random Samples

How would you select a representative sample? Most people would say that every individual in the population should have an equal chance to be selected, and certainly that seems fair. But it's not sufficient. There are many ways to give everyone an equal chance that still wouldn't give a representative sample. Consider, for example, a school that has equal numbers of males and females. We could sample like this: Flip a coin. If it comes up heads, select 100 female students at random. If it comes up tails, select 100 males at random. Everyone has an equal chance of selection, but every sample is of only a single sex—hardly representative.

We need to do better. Suppose we insist that every possible *sample* of the size we plan to draw have an equal chance to be selected. This ensures that situations like the one just described are not likely to occur and still guarantees that each person has an equal chance of being selected. What's different is that with this method, each *combination* of people has an equal chance of being selected as well. A sample drawn in this way is called a **Simple Random Sample**, usually abbreviated **SRS**. An SRS is the standard against which we measure other sampling methods, and the sampling method on which the theory of working with sampled data is based.

To select a sample at random, first define where the sample will come from. The **sampling frame** is a list of individuals from which the sample is drawn. For example,

[5] If you're from the United States. If you're British or Canadian, it's "beetah."

[6] Just in case you weren't sure.

Error Okay; Bias No Way!

Sampling variability is sometimes referred to as *sampling error*, making it sound like it's some kind of mistake. It's not. We understand that samples will vary, so "sampling error" is to be expected. It's *bias* we must strive to avoid. Bias means our sampling method distorts our view of the population, and that will surely lead to mistakes.

a random sample of students at a college might be selected from a list of all registered full-time students. In defining the sampling frame, you must deal with the details of defining the population. Are part-time students included? How about those who are attending school elsewhere and transferring credits back to the college?

There are several ways to draw an SRS from a sampling frame. Two simple ideas are:

- Assign a random number with several digits (say, from 0 to 10,000) to each individual. Then sort the random numbers into numerical order, keeping each name with its number. The first n names are then a random sample of that size.
- Assign each individual a single random digit, 0 to 9. Then those with a specific random digit (say, 5) are a 10% SRS.

Samples drawn at random generally differ one from another. Each draw of random numbers selects *different* people for our sample. As we have seen, different random samples yield different values for the statistics we compute. We call these sample-to-sample differences **sampling variability**. Surprisingly, sampling variability isn't a problem; it's an opportunity. In subsequent chapters we'll investigate what the variation in a sample can tell us about its population.

EXAMPLE 10.2 Using Random Numbers to Get an SRS

There are 80 students enrolled in an introductory statistics class; you are to select a sample of 5.

QUESTION: How can you select an SRS of 5 students using these random digits found on the Internet: 05166 29305 77482?

ANSWER: First, I'll number the students from 00 to 79. Taking the random numbers two digits at a time gives me 05, 16, 62, 93, 05, 77, and 48. I'll ignore 93 because the students were numbered only up to 79. And, so as not to pick the same person twice, I'll skip the repeated number 05. My simple random sample consists of students with the numbers 05, 16, 62, 77, and 48.

RANDOM MATTERS

How well do random samples represent a population? That depends in part on how large a sample you draw. Figure 10.1 gives a histogram of the birthweights of all 3,940,552 babies born in the United States in 1998 for whom we have recorded birthweights. (Data in **All births1998**) It shows a smooth, unimodal distribution, slightly skewed to the low end.

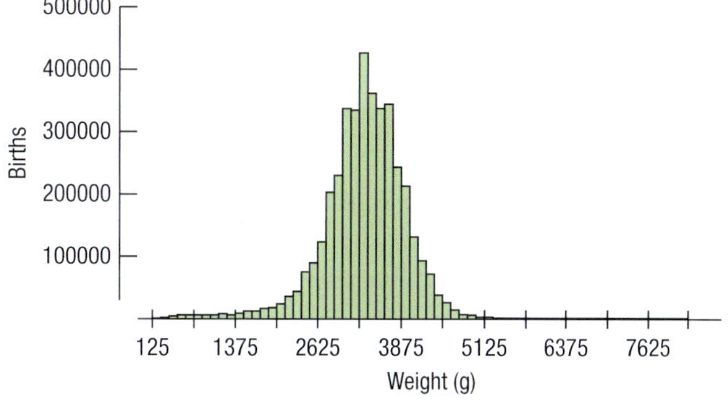

Figure 10.1
Birthweights of 3,940,552 babies born in the United States in 1998.

Figure 10.2 shows histograms of simple random samples drawn from this population of babies. The top row shows histograms of samples of 100, the second row shows samples of 250, and the bottom row shows histograms for samples of 1000.

We saw in Chapter 2 that means vary from sample to sample; now we can see why. The samples in the top row resemble the population distribution, but they vary across the row quite a bit. Not only their means, but their shape and spread vary as well.

By looking down the columns, you can see that larger samples look more and more like the population. The histograms in the bottom row look both more like the population distribution and more like each other.

The fact that larger samples are more consistent and look more like the population from which they're drawn is key to many; of the calculations that we'll see later in the text.

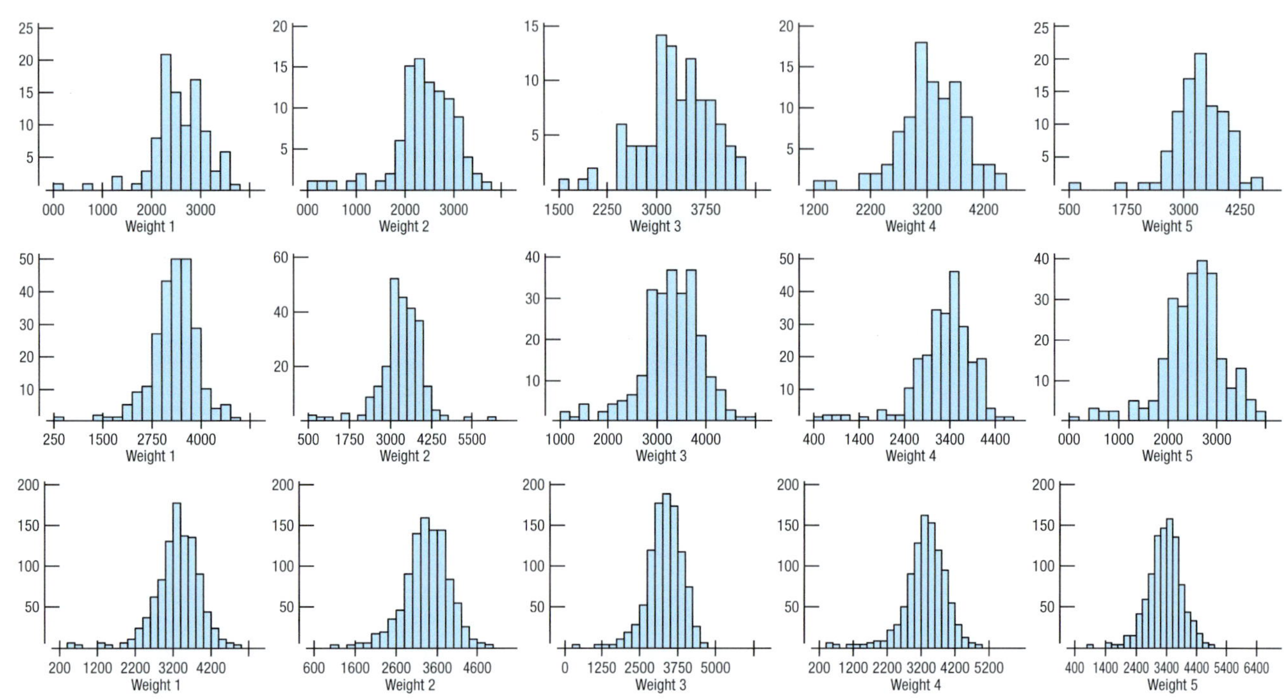

Figure 10.2

Samples from the population of all 1998 births. Samples of 100 in the top row, of 250 in the middle row, and of 1000 in the bottom row.

10.4 Other Sampling Designs

Simple random sampling is not the only fair way to sample. More complicated designs may save time or money or help avoid potential bias and improve sample accuracy. All statistical sampling designs have in common the idea that chance, rather than human choice, is used to select the sample.

Stratified Sampling

Sometimes the population is first sliced into homogeneous groups, called *strata*, before the sample is selected. Then simple random sampling is used within each stratum and the results are combined. This common sampling design is called **stratified random sampling**.

For example, suppose we wanted to learn how students feel about building a new \$100 million stadium for the football team at a large university at which 6% of the students are varsity athletes. Simple random samples of 50 students could end up with too few athletes. In fact, nearly 15% of such random samples will have no athletes at all.

If the opinion of varsity athletes differed from the rest of the student body, we could easily reach the wrong conclusion about the support. A stratified sample could ensure the proportions are correct by selecting 3 athletes at random and 47 non-athletes at random. This would guarantee that the proportions of athletes in our samples match the proportions in the population, and that should help the survey more accurately represent the opinion of the student body.

You can imagine the importance of stratifying by race, income, age, and other characteristics, depending on the questions in the survey. Samples taken within a stratum vary less, so estimates can be more precise. This reduced sampling variability is the most important benefit of stratifying. Stratified sampling can also help us notice important differences among groups. But be careful. As we saw in Chapter 3, if we unthinkingly combine grouped data, we risk reaching the wrong conclusion, becoming victims of Simpson's paradox.

EXAMPLE 10.3 Stratifying the Sample

Suppose you are trying to find out what first-year students think of the food served on campus. Food Services believes that men and women typically have different opinions about the importance of the salad bar.

QUESTION: How should you adjust your sampling strategy to allow for this difference?

ANSWER: I will stratify my sample by drawing an SRS of men and a separate SRS of women—assuming that the data from the registrar include information about each person's sex.

Cluster and Multistage Sampling

Suppose that the administration of a university wants to advertise the average starting salary offered to its graduates. They know, for example, that graduating law students are likely to earn more (at least at first) than philosophy majors. They could collect a sample stratifying by major, but it might be easier to select a few dorms to survey. If each dorm is representative of the distribution of majors in the university, then each dorm would be a *cluster*, a smaller representative. Selecting one or a few clusters at random, and then drawing a random sample or a complete census within each cluster, can make the survey more practical. If each cluster represents the full population fairly, cluster sampling will be unbiased.

EXAMPLE 10.4 Cluster Sampling

RECAP: In trying to find out what first-years think about the food served on campus, you've considered both an SRS and a stratified sample. Now you have run into a problem: It's simply too difficult and time consuming to track down the individuals whose names were chosen for your sample. Fortunately, first-years at your school are all housed in 10 freshman dorms.

QUESTIONS: How could you use this fact to draw a **cluster sample**? How might that alleviate the problem? What concerns do you have?

ANSWER: To draw a cluster sample, I would select a few freshman dorms at random and then try to contact everyone in each selected dorm. I could save time by simply knocking on doors on a given evening and interviewing people. I'd have to assume that first-years are assigned to dorms pretty much at random and that the people I'm able to contact are representative of everyone in the dorm.

What's the difference between cluster sampling and stratified sampling? Their purposes are different. We stratify to ensure that our sample represents different groups in the population. Strata are internally homogeneous, but differ from one another. By contrast, we select clusters to make sampling more practical or affordable. Clusters can be heterogeneous; we want our sample of clusters to provide a representative sample.

Sampling schemes that combine several methods are called **multistage samples**. The example of using dorms as clusters is actually a multistage sample in which dorms are first selected as clusters and then another sampling scheme is applied within each dorm.

EXAMPLE 10.5 Multistage Sampling

RECAP: Learning that first-years are housed in separate dorms allowed you to sample their attitudes about the campus food by going to dorms chosen at random, but you're still concerned about possible differences in opinions between men and women. It turns out that these freshman dorms house men and women on alternate floors.

QUESTION: How can you design a sampling plan that uses this fact to your advantage?

ANSWER: Now I can stratify my sample by sex. I would first choose one or two dorms at random and then select the same number of dorm floors at random from among those that house men and, separately, from among those that house women. I could then draw an SRS of residents on the selected floors.

Systematic Samples

Some samples select individuals systematically. For example, you might survey every 10th person on an alphabetical list of students. To make it random, you still must start the systematic selection from a randomly selected individual. When the order of the list is not associated in any way with the responses sought, **systematic sampling** can give a representative sample.

Systematic sampling can be much less expensive than true random sampling. When you use a systematic sample, you should justify the assumption that the systematic method is not associated with any of the measured variables. For example, if you decided to sample students in their dorm rooms by knocking on every other door, a dorm in which male and female rooms were alternated could result in a sample that was all male or all female.

JUST CHECKING

2. We need to survey a random sample of the 300 passengers on a flight from San Francisco to Tokyo. Name each sampling method described below.
   a) Pick every 10th passenger as people board the plane.
   b) From the boarding list, randomly choose 5 people flying first class and 25 of the other passengers.
   c) Randomly generate 30 seat numbers and survey the passengers who sit there.
   d) Randomly select a seat position (right window, right center, right aisle, etc.) and survey all the passengers sitting in those seats.

STEP-BY-STEP EXAMPLE

Sampling

The assignment says, "Conduct your own sample survey to find out how many hours per week students at your school spend watching TV during the school year." Let's see how we might do this step by step. (Remember, though—actually collecting the data from your sample can be difficult and time consuming.)

QUESTION: How would you design this survey?

THINK

PLAN State what you want to know.

I wanted to design a study to find out how many hours of TV students at my school watch.

POPULATION AND PARAMETER Identify the W's of the study. The *Why* determines the population and the associated sampling frame. The *What* identifies the parameter of interest and the variables measured. The *Who* is the sample we actually draw. The *How*, *When*, and *Where* are given by the sampling plan.

Often, thinking about the *Why* will help you see whether the sampling frame and plan are adequate to learn about the population.

The population studied was students at our school. I obtained a list of all students currently enrolled and used it as the sampling frame. The parameter of interest was the number of TV hours watched per week during the school year, which I attempted to measure by asking students how much TV they watched during the previous week.

SAMPLING PLAN Specify the sampling method and the sample size, n. Specify how the sample was actually drawn. What is the sampling frame? How was the randomization performed?

A good description should be complete enough to allow someone to replicate the procedure, drawing another sample from the same population in the same manner.

I decided against stratifying by class or sex because I didn't think TV watching would differ much between males and females or across classes. I selected a simple random sample of students from the list. I obtained an alphabetical list of students, assigned each a random digit between 0 and 9, and then selected all students who were assigned a "4." This method generated a sample of 212 students from the population of 2133 students.

SHOW

SAMPLING PRACTICE Specify *When, Where*, and *How* the sampling was performed. Specify any other details of your survey, such as how respondents were contacted, what incentives were offered to encourage them to respond, how nonrespondents were treated, and so on.

The survey was taken over the period Oct. 15 to Oct. 25. Surveys were sent to selected students by e-mail, with the request that they respond by e-mail as well. Students who could not be reached by e-mail were handed the survey in person.

TELL

SUMMARY AND CONCLUSION This report should include a discussion of all the elements. In addition, it's good practice to discuss any special circumstances. Professional polling organizations report the *When* of their samples but will also note, for example, any important news that might have changed respondents' opinions during the sampling process. In this survey, perhaps, a major news story or sporting event might change students' TV viewing behavior.

The question you ask also matters. It's better to be specific ("How many hours did you watch TV last week?") than to ask a general question ("How many hours of TV do you usually watch in a week?").

During the period Oct. 15 to Oct. 25, 212 students were randomly selected, using a simple random sample from a list of all students currently enrolled. The survey they received asked the following question: "How many hours did you spend watching television last week?"

Of the 212 students surveyed, 110 responded. It's possible that the nonrespondents differ in the number of TV hours watched from those who responded, but I was unable to follow up on them due to limited time and funds. The 110 respondents reported an average 3.62 hours of TV watching per week. The median was only 2 hours per week. A histogram of the data shows that the distribution is highly right-skewed, indicating that the median might be a more appropriate summary of the typical TV watching of the students.

The report should show a display of the data, provide and interpret the statistics from the sample, and state the conclusions that you reached about the population.

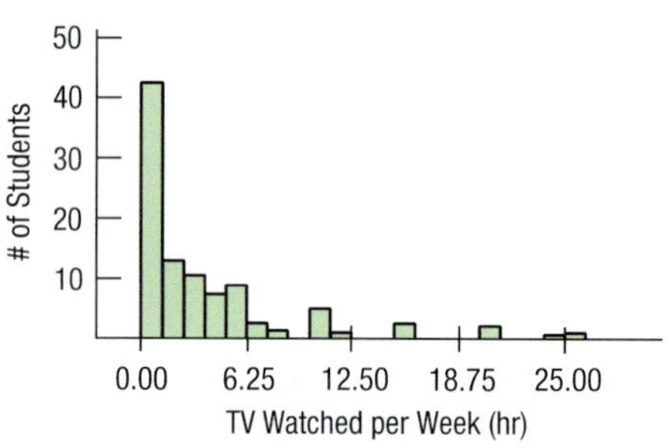

Most of the students (90%) watch between 0 and 10 hours per week, while 30% reported watching less than 1 hour per week. A few watch much more. About 3% reported watching more than 20 hours per week.

10.5 From the Population to the Sample: You Can't Always Get What You Want[7]

What's the Sample?

The population we want to study is determined by asking *why*. When we design a survey, we use the term "sample" to refer to the individuals selected, from whom we hope to obtain responses. Unfortunately, the real sample is just those we can reach to obtain responses—the *who* of the study. These are slightly different uses of the same term *sample*. The context usually makes clear which we mean, but it's important to realize that the difference between the two samples could undermine even a well-designed study.

Before you start a survey, think first about the population you want to study. For example, if a pollster wants to predict the outcome of the next presidential election, what's the correct population? All adults? All citizens? All registered voters? The pollster would really like to sample from all those who will vote in the next election—a population that is impossible to identify before Election Day. The list of those in the population that you can actually identify is called the sampling frame.

The pollster might use a list of registered voters as the sampling frame. Usually, the sampling frame is not the group you *really* want to know about (all those registered to vote are not all likely to show up).

Nonresponse is a problem in many surveys. You sample from the sampling frame. These are the individuals for whom you *intend* to measure responses—for example, the people the pollster telephones. But you're not likely to get responses from all of them. ("I know it's dinnertime, but I'm sure you wouldn't mind answering a few questions. It'll only take 20 minutes or so. Oh, you're busy?") The actual respondents (your sample) are the individuals about whom you do get data and can draw conclusions.

Unfortunately, your sample might not be representative of the sampling frame or the population.

The real world is complex. Most professional polls spend enormous resources constructing a sampling frame that is close to the intended population. Then they use a variety of multistage sampling to attempt to ensure a representative sample. Advanced courses in sample surveys focus on both the methods and the consequences of these designs. Most common statistics methods are based on the assumption that our samples are simple random samples from a population. Even though this is rarely true, these methods work quite well.

[7]But if you try sometimes you just might find you get what you need.—The Rolling Stones

10.6 The Valid Survey

It isn't sufficient to just draw a sample and start asking questions. You'll want your survey to be *valid*. A valid survey yields the information you are seeking about the population you are interested in. Before setting out to survey, ask yourself

- What do I want to know?
- Am I asking the right respondents?
- Am I asking the right questions?
- What would I do with the answers if I had them; would they address the things I want to know?

These questions may sound obvious, but there are a number of pitfalls to avoid.

Know what you want to know. Before considering a survey, understand what you hope to learn and about whom you hope to learn it.

Use the right sampling frame. A valid survey obtains responses from the appropriate respondents. Be sure you have a suitable *sampling frame.* Have you identified the population of interest and sampled from it appropriately? A company might survey customers who returned warranty registration cards—a readily available sampling frame. But if the company wants to make their product more attractive, the most important population is the customers who rejected their product in favor of one from a competitor.

Tune your instrument. You may be tempted to ask questions you don't really need, but longer survey instruments yield fewer responses and thus a greater chance of nonresponse bias.

Ask specific rather than general questions. People are not very good at estimating their typical behavior, so it is better to ask "How many hours did you sleep last night?" than "How much do you usually sleep?" Sure, some responses will include some unusual events (My dog was sick; I was up all night.), but overall you'll get better data.

Ask for quantitative results when possible. "How many magazines did you read last week?" is better than "How much do you read: A lot, A moderate amount, A little, or Not at all?"

Be careful in phrasing questions. A respondent may not understand the question—or may understand the question differently than the researcher intended it. ("Does anyone in your family ride a motorcycle?" Do you mean just me, my spouse, and my children? Or does "family" include my father, my siblings, and my second cousin once removed? And does a motor scooter count?) Respondents are unlikely (or may not have the opportunity) to ask for clarification. A question like "Do you approve of the recent actions of the Ambassador to Mexico?" is likely not to measure what you want if many respondents don't know who the Ambassador is or what he or she recently did.

Respondents may even lie or shade their responses if they feel embarrassed by the question ("Did you have too much to drink last night?"), are intimidated or insulted by the question ("Could you understand our new *Instructions for Dummies* manual, or was it too difficult for you?"), or if they want to avoid offending the interviewer ("Would you hire a man with a tattoo?" asked by a tattooed interviewer). Also, be careful to avoid phrases that have double or regional meanings. "How often do you go to town?" might be interpreted differently by different people and cultures.

> **A SHORT SURVEY**
>
> Given that the *New York Times* reports[8] that statisticians can earn $125,000 at top companies their first year on the job, do you think this course will be valuable to *you*?

> **PILOT TESTING A SURVEY QUESTION**
>
> A researcher distributed a survey to an organization before some economizing changes were made. She asked how people felt about a proposed cutback in secretarial and administrative support on a seven-point scale from Very Happy to Very Unhappy.
>
> But virtually all respondents were very unhappy about the cutbacks, so the results weren't particularly useful. If she had pretested the question, she might have chosen a scale that ran from unhappy to outraged.

Even subtle differences in phrasing can make a difference. In January 2006, the *New York Times* asked half of the 1229 U.S. adults in their sample the following question:

> *After 9/11, President Bush authorized government wiretaps on some phone calls in the U.S. without getting court warrants, saying this was necessary to reduce the threat of terrorism. Do you approve or disapprove of this?*[9]

They found that 53% of respondents approved. But when they asked the other half of their sample a question with different phrasing,

> *After 9/11, George W. Bush authorized government wiretaps on some phone calls in the U.S. without getting court warrants. Do you approve or disapprove of this?*[10]

only 46% approved.

Be careful in phrasing answers. It's often a good idea to offer choices rather than inviting a free response. Open-ended answers can be difficult to analyze. "How did you like the movie?" may start an interesting debate, but it may be better to give a range of possible responses. Be sure to phrase answers in a neutral way. When asking "Do you support higher school taxes?" positive responses could be worded "Yes," "Yes, it is important for our children," or "Yes, our future depends on it." But those are not equivalent answers.

The best way to protect a survey from unanticipated measurement errors is to perform a pilot survey. A **pilot** is a trial run of the survey you eventually plan to give to a larger group, using a draft of your survey questions administered to a small sample drawn from the same sampling frame you intend to use. By analyzing the results from this smaller survey, you can often discover ways to improve your survey instrument.

10.7 Common Sampling Mistakes, or How to Sample Badly

Bad sample designs yield worthless data. Many of the most convenient forms of sampling can be seriously biased. And there is no way to correct for the bias from a bad sample. So it's wise to pay attention to sample design—and to beware of reports based on poor samples.

Mistake 1: Sample Volunteers

One of the most common dangerous sampling methods is a voluntary response sample. In a **voluntary response sample**, a large group of individuals is invited to respond, and those who choose to respond are counted. The respondents, rather than the researcher, decide who will be in the sample. This method is used by call-in shows, 900 numbers, Internet polls, and letters written to members of Congress. Voluntary response samples are almost always biased, so conclusions drawn from them are almost always wrong.

Voluntary response samples are often biased toward those with strong opinions or those who are strongly motivated. People with very negative opinions tend to respond more often than those with equally strong positive opinions. The sample is not representative, even though every individual in the population may have been offered the chance to respond. The resulting **voluntary response bias** invalidates the survey.

[8] www.nytimes.com/2009/08/06/technology/06stats.html

[9] Excerpt from "The New York Times/CBS News Poll, January 20–25, 2006". Published in *The New York Times*, © 2006.

[10] Ibid.

IF YOU HAD IT TO DO OVER AGAIN, WOULD YOU HAVE CHILDREN?

Ann Landers, the advice columnist, asked parents this question. The overwhelming majority—70% of the more than 10,000 people who wrote in—said no, kids weren't worth it. A more carefully designed survey later showed that about 90% of parents actually are happy with their decision to have children. What accounts for the striking difference in these two results? What parents do you think are most likely to respond to the original question?

EXAMPLE 10.6 Voluntary Response Sample

RECAP: You're trying to find out what freshmen think of the food served on campus, and have thought of a variety of sampling methods, all time consuming. A friend suggests that you set up a "Tell Us What You Think" website and invite freshmen to visit the site to complete a questionnaire.

QUESTION: What's wrong with this idea?

ANSWER: Letting each freshman decide whether to participate makes this a voluntary response survey. Students who were dissatisfied might be more likely to go to the website to record their complaints, and this could give me a biased view of the opinions of all freshmen.

Mistake 2: Sample Conveniently

Do you use the Internet?
Click here ◯ for yes
Click here ◯ for no

Another sampling method that doesn't work is convenience sampling. As the name suggests, in **convenience sampling** we simply include the individuals who are convenient for us to sample. Unfortunately, this group may not be representative of the population. Here's an amusing example. Back in 2001, when computer use in the home was not as common as it is today, a survey of 437 potential home buyers in Orange County, California, reached the surprising conclusion that

> *All but 2 percent of the buyers have at least one computer at home, and 62 percent have two or more. Of those with a computer, 99 percent are connected to the Internet (Jennifer Hieger, "Portrait of Homebuyer Household: 2 Kids and a PC,"* Orange County Register, *27 July 2001).*[11]

How was the survey conducted? On the Internet!

Internet Surveys

Internet convenience surveys are worthless. As voluntary response surveys, they have no well-defined sampling frame (all those who use the Internet and visit their site?) and thus report no useful information. Do not believe them.

EXAMPLE 10.7 Convenience Sample

RECAP: To try to gauge freshman opinion about the food served on campus, Food Services suggests that you just stand outside a school cafeteria at lunchtime and stop people to ask them questions.

QUESTION: What's wrong with this sampling strategy?

ANSWER: This would be a convenience sample, and it's likely to be biased. I would miss people who use the cafeteria for dinner, but not for lunch, and I'd never hear from anyone who hates the food so much that they have stopped coming to the school cafeteria.

[11] Excerpt from "Portrait of Home buyer Household: 2 Kids and a PC" by Jennifer Hieger. Published by *Orange County Register*, © 27 July 2001.

Mistake 3: Use a Bad Sampling Frame

An SRS from an incomplete sampling frame introduces bias because the individuals included may differ from the ones not in the frame. People in prison, homeless people, students, and long-term travelers are all likely to be missed. Professional polling companies now need special procedures to be sure they include in their sampling frame people who can be reached only by cell phone.

Mistake 4: Undercoverage

Many survey designs suffer from **undercoverage**, in which some portion of the population is not sampled at all or has a smaller representation in the sample than it has in the population. Undercoverage can arise for a number of reasons, but it's always a potential source of bias.

Telephone surveys are usually conducted when you are likely to be home, such as dinnertime. If you eat out often, you may be less likely to be surveyed, a possible source of undercoverage.

Nonresponse Bias

A common and serious potential source of bias for most surveys is **nonresponse bias**. No survey succeeds in getting responses from everyone. The problem is that those who don't respond may differ from those who do. And they may differ on just the variables we care about. Rather than sending out a large number of surveys for which the response rate will be low, it is often better to design a smaller randomized survey for which you have the resources to ensure a high response rate. One of the problems with nonresponse bias is that it's usually impossible to tell what the nonrespondents might have said.

It turns out that the *Literary Digest* survey was wrong on two counts. First, their list of 10 million people was not representative. There was a selection bias in their sampling frame. There was also a nonresponse bias. We know this because the *Digest* also surveyed a *systematic* sample in Chicago, sending the same question used in the larger survey to every third registered voter. They *still* got a result in favor of Landon, even though Chicago voted overwhelmingly for Roosevelt in the election. This suggests that the Roosevelt supporters were less likely to respond to the *Digest* survey. There's a modern version of this problem: It's been suggested that those who screen their calls with caller ID or an answering machine, and so might not talk to a pollster, may differ in wealth or political views from those who just answer the phone.

Response Bias

Response bias[12] refers to anything in the survey design that influences the responses. Response biases include the tendency of respondents to tailor their responses to try to please the interviewer, the natural unwillingness of respondents to reveal personal facts or admit to illegal or unapproved behavior, and the ways in which the wording of the questions can influence responses.

How to Think About Biases

- **Look for biases in any survey you encounter.** If you design one of your own, ask someone else to help look for biases that may not be obvious to you. And do this *before* you collect your data. There's no way to recover from a biased sampling method or a survey that asks biased questions. Sorry, it just can't be done. A bigger sample size for a biased study just gives you a bigger useless study. A really big sample gives you a really big useless study. (Think of the 2.4 million *Literary Digest* responses.)

[12]Response bias is not the opposite of nonresponse bias. (We don't make these terms up; we just try to explain them.)

- **Spend your time and resources reducing biases.** No other use of resources is as worthwhile as reducing the biases.
- **Think about the members of the population who could have been excluded from your study.** Be careful not to claim that you have learned anything about them.
- **If you can, pilot-test your survey.** Administer the survey in the exact form that you intend to use it to a small sample drawn from the population you intend to sample. Look for misunderstandings, misinterpretation, confusion, or other possible biases. Then refine your survey instrument.
- **Always report your sampling methods in detail.** Others may be able to detect biases where you did not expect to find them.

WHAT CAN GO WRONG?

- The principal thing that can go wrong in sampling is that the sample can fail to represent the population. Unfortunately, this can happen in many different ways and for many different reasons. We've considered many of them in the chapter.
- It is also an error to draw too small a sample for your needs regardless of how well you draw the sample. However, generally it is more worthwhile to devote effort on improving the quality of your sample than on expanding its size.

CONNECTIONS

With this chapter, we take our first formal steps to relate our sample data to a larger population. Some of these ideas have been lurking in the background as we sought patterns and summaries for data. Even when we only worked with the data at hand, we often thought about implications for a larger population of individuals.

Notice the ongoing central importance of models. We've seen models in several ways in previous chapters. Here, we recognize the value of a model for a population. The parameters of such a model are values we will often want to estimate using statistics such as those we've been calculating. The connections to summary statistics for center, spread, correlation, and slope are obvious.

We now have a specific application for random numbers. We have used randomization to understand how data represents the world and we shall use it in subsequent chapters as the basis for drawing inferences. Now we need randomization to get good-quality data from the real world.

CHAPTER REVIEW

Know the three ideas of sampling.

- Examine a part of the whole: A sample can give information about the population.
- Randomize to make the sample representative.
- The sample size is what matters. It's the size of the sample—and not its fraction of the larger population—that determines the precision of the statistics it yields.

Be able to draw a simple random sample (SRS) using a table of random digits or a list of random numbers from technology or an Internet site (such as the Simple Sample tool at astools.datadesk.com).

- In a **simple random sample** (SRS), every possible group of *n* individuals has an equal chance of being our sample.

- Understand that repeated samples can each represent a larger population, but differ from each other. Know that larger samples vary from one sample to another less than smaller samples.

Understand the advantages and disadvantages of other sampling methods:

- **Stratified samples** can reduce sampling variability by identifying homogeneous subgroups and then randomly sampling within each.
- **Cluster samples** randomly select among heterogeneous subgroups, making our sampling tasks more manageable.
- **Systematic samples** can work in some situations and are often the least expensive method of sampling. But we still want to start them randomly.
- **Multistage samples** combine several random sampling methods.

Identify and avoid causes of bias.

- **Voluntary response samples** are almost always biased and should be avoided and distrusted.
- **Convenience samples** are likely to be flawed for similar reasons.
- **Bad sampling frames** can lead to samples that don't represent the population of interest.
- **Undercoverage** occurs when individuals from a subgroup of the population are selected less often than they should be.
- **Nonresponse bias** can arise when sampled individuals will not or cannot respond.
- **Response bias** arises when respondents' answers might be affected by external influences, such as question wording or interviewer behavior.
- Use best practices when designing a sample survey to improve the chance that your results will be valid.

REVIEW OF TERMS

Population The entire group of individuals or instances about whom we hope to learn (p. 352).

Sample A (representative) subset of a population, examined in the hope of learning about the population (p. 352).

Sample survey A study that asks questions of a sample drawn from some population in the hope of learning something about the entire population. Polls taken to assess voter preferences are common sample surveys (p. 352).

Bias Any systematic failure of a sampling method to represent its population is bias. Biased sampling methods tend to over- or underestimate parameters. It is almost impossible to recover from bias, so efforts to avoid it are well spent (p. 352).

Common errors include

- relying on voluntary response.
- undercoverage of the population.
- nonresponse bias.
- response bias.

Randomization The best defense against bias is randomization, in which each individual is given a fair, random chance of selection (p. 353).

Sample size The number of individuals in a sample. The sample size determines how well the sample represents the population, not the fraction of the population sampled (p. 354).

Census A sample that consists of the entire population (p. 354).

Population parameter A numerically valued attribute of a model for a population. We rarely expect to know the true value of a population parameter, but we do hope to estimate it from sampled data. For example, the mean income of all employed people in the country is a population parameter (p. 355).

Statistic, sample statistic Statistics are values calculated for sampled data. Those that correspond to, and thus estimate, a population parameter are of particular interest. For example, the mean income of all employed people in a representative sample can provide a good estimate of the corresponding population parameter. The term "sample statistic" is sometimes used, usually to parallel the corresponding term "population parameter" (p. 355).

Representative A sample is said to be representative if the statistics computed from it accurately reflect the corresponding population parameters (p. 356).

Simple Random Sample (SRS) A Simple Random Sample of sample size *n* is a sample in which each set of *n* elements in the population has an equal chance of selection (p. 356).

Sampling frame A list of individuals from whom the sample is drawn is called the sampling frame. Individuals who may be in the population of interest, but who are not in the sampling frame, cannot be included in any sample (p. 356).

Sampling variability The natural tendency of randomly drawn samples to differ, one from another. Sometimes, unfortunately, called *sampling error*, sampling variability is no error at all, but just the natural result of random sampling (p. 357).

Stratified random sample A sampling design in which the population is divided into several subpopulations, or strata, and random samples are then drawn from each stratum. If the strata are homogeneous, but are different from each other, a stratified sample may yield more consistent results than an SRS (p. 358).

Cluster sample A sampling design in which entire groups, or **clusters**, are chosen at random. Cluster sampling is usually selected as a matter of convenience, practicality, or cost. Clusters are heterogeneous. A sample of clusters should be representative of the population (p. 359).

Multistage sample Sampling schemes that combine several sampling methods. For example, a national polling service may stratify the country by geographical regions, select a random sample of cities from each region, and then interview a cluster of residents in each city (p. 360).

Systematic sample A sample drawn by selecting individuals systematically from a sampling frame. When there is no relationship between the order of the sampling frame and the variables of interest, a systematic sample can be representative (p. 360).

Pilot study A small trial run of a survey to check whether questions are clear. A pilot study can reduce errors due to ambiguous questions (p. 364).

Voluntary response sample A sample in which individuals can choose on their own whether to participate. Samples based on voluntary response are always invalid and cannot be recovered, no matter how large the sample size (p. 364).

Voluntary response bias Bias introduced to a sample when individuals can choose on their own whether to participate in the sample. Samples based on voluntary response are always invalid and cannot be recovered, no matter how large the sample size (p. 364).

Convenience sample A convenience sample consists of the individuals who are conveniently available. Convenience samples often fail to be representative because every individual in the population is not equally convenient to sample (p. 365).

Undercoverage A sampling scheme that biases the sample in a way that gives a part of the population less representation in the sample than it has in the population suffers from undercoverage (p. 366).

Nonresponse bias Bias introduced when a large fraction of those sampled fails to respond. Those who do respond are likely to not represent the entire population. Voluntary response bias is a form of nonresponse bias, but nonresponse may occur for other reasons. For example, those who are at work during the day won't respond to a telephone survey conducted only during working hours (p. 366).

Response bias Anything in a survey design that influences responses falls under the heading of response bias. One typical response bias arises from the wording of questions, which may suggest a favored response. Voters, for example, are more likely to express support of "the president" than support of the particular person holding that office at the moment (p. 366).

TECH SUPPORT

Sampling

Computer-generated pseudorandom numbers are usually good enough for drawing random samples. But there is little reason not to use the truly random values available on the Internet.

Here's a convenient way to draw an SRS of a specified size using a computer-based sampling frame. The sampling frame can be a list of names or of identification numbers arrayed, for example, as a column in a spreadsheet, statistics program, or database:

1. Generate random numbers of enough digits so that each exceeds the size of the sampling frame list by several digits. This makes duplication unlikely.
2. Assign the random numbers arbitrarily to individuals in the sampling frame list. For example, put them in an adjacent column.
3. Sort the list of random numbers, carrying along the sampling frame list.
4. Now the first *n* values in the sorted sampling frame column are an SRS of *n* values from the entire sampling frame.

EXERCISES

SECTION 10.1

1. University education Administrators at a university were interested in estimating the percentage of students who are the first in their family to go to college. The student body has about 50,000 members. How might the administrators answer their question by applying the three Big Ideas?

2. Satisfied workers The managers of a large company wished to know the percentage of employees who feel "extremely satisfied" to work there. The company has roughly 33,000 employees. They contacted a random sample of employees and asked them about their job satisfaction, obtaining 418 completed responses. How does their study deal with the three Big Ideas of sampling?

SECTION 10.2

3. A&M again The president of the university plans a speech to an alumni group. He plans to talk about the proportion of students who responded in the survey that they are the first in their family to attend college, but the first draft of his speech treats that proportion as the actual proportion of current A&M students who are the first in their families to attend college. Explain to the president the difference between the proportion of respondents who are first attenders and the proportion of the entire student body that are first attenders. Use appropriate statistics terminology.

4. Satisfied respondents The company's annual report states, "Our survey shows that 87.34% of our employees are 'very happy' working here." Comment on that claim. Use appropriate statistics terminology.

SECTION 10.3

5. Sampling students A professor teaching a large lecture class of 450 students samples her class by rolling a die. Then, starting with the row number on the die (1 to 6), she passes out a survey to every fourth row of the large lecture hall. She says that this is a simple random sample because everyone had an equal opportunity to sit in any seat and because she randomized the choice of rows. What do you think? Be specific.

6. Sampling satisfaction A company hoping to assess employee satisfaction surveys employees by assigning computer-generated random numbers to each employee on a list of all employees and then contacting all those whose assigned random number is divisible by 6. Is this a simple random sample?

SECTION 10.4

7. Sampling A&M students For each scenario, identify the kind of sample used by the university administrators from Exercise 1:

a) Select several dormitories at random and contact everyone living in the selected dorms.
b) Using a computer-based list of registered students, contact 200 freshmen, 200 sophomores, 200 juniors, and 200 seniors selected at random from each class.
c) Using a computer-based alphabetical list of registered students, select one of the first 25 on the list by random and then contact the student whose name is 50 names later, and then every 50 names beyond that.

8. Satisfactory satisfaction samples For each scenario, determine the sampling method used by the managers from Exercise 2.

a) Use the company e-mail directory to contact 150 employees from among those employed for less than 5 years, 150 from among those employed for 5–10 years, and 150 from among those employed for more than 10 years.
b) Use the company e-mail directory to contact every 50th employee on the list.
c) Select several divisions of the company at random. Within each division, draw an SRS of employees to contact.

SECTION 10.6

9. Survey students What problems do you see with asking the following question of students? "Are you the first member of your family to seek higher education?"

10. Happy employees The company plans to have the head of each corporate division hold a meeting of their employees to ask whether they are happy on their jobs. They will ask people to raise their hands to indicate whether they are happy. What problems do you see with this plan?

SECTION 10.7

11. Student samples The university administration of Exercise 1 is considering a variety of ways to sample students for a survey. For each of these proposed survey designs, identify the problem.

a) Publish an advertisement inviting students to visit a website and answer questions.
b) Set up a table in the student union and ask students to stop and answer a survey.

12. Surveying employees The company of Exercise 2 is considering ways to survey their employees. For each of these proposed designs, identify the problem.

a) Leave a stack of surveys out in the employee cafeteria so people can pick them up and return them.
b) Stuff a questionnaire in the mailbox of each employee with the request that they fill it out and return it.

CHAPTER EXERCISES

13. Roper Through their *Roper Reports Worldwide*, GfK Roper conducts a global consumer survey to help multinational companies understand different consumer attitudes throughout the world. Within 30 countries, the researchers interview 1000 people aged 13–65. Their samples are designed so that they get 500 males and 500 females in each country. (www.gfkamerica.com)

a) Are they using a simple random sample? Explain.
b) What kind of design do you think they are using?

14. Student center survey For their class project, a group of statistics students decide to survey the student body to assess opinions about the proposed new student center. Their sample of 220 contained 55 first-year students, 55 sophomores, 55 juniors, and 55 seniors.

a) Do you think the group was using an SRS? Why?
b) What sampling design do you think they used?

15. Drug tests Major League Baseball tests players to see whether they are using performance-enhancing drugs. Officials select a team at random, and a drug-testing crew shows up unannounced to test all 33 players on the team. Each testing day can be considered a study of drug use in Major League Baseball.

a) What kind of sample is this?
b) Is that choice appropriate?

16. Gallup At its website (www.gallup.com), the Gallup Poll publishes results of a new survey each day. Scroll down to the end, and you'll find a statement that includes words such as these:

Results are based on telephone interviews with 1016 national adults, aged 18 and older, conducted March 7–10, 2014. . . .

In addition to sampling error, question wording and practical difficulties in conducting surveys can introduce error or bias into the findings of public opinion polls.

a) For this survey, identify the population of interest.
b) Gallup performs its surveys by phoning numbers generated at random by a computer program. What is the sampling frame?
c) What problems, if any, would you be concerned about in matching the sampling frame with the population?

In Exercises 17 to 23, for the reports about statistical studies, identify the following items (if possible). If you can't tell, then say so—this often happens when we read about a survey.

a) The population
b) The population parameter of interest
c) The sampling frame
d) The sample
e) The sampling method, including whether or not randomization was employed
f) Who (if anyone) was left out of the study
g) Any potential sources of bias you can detect and any problems you see in generalizing to the population of interest

17. Medical treatments Consumers Union asked all subscribers whether they had used alternative medical treatments and, if so, whether they had benefited from them. For almost all of the treatments, approximately 15% of those responding reported cures or substantial improvement in their condition.

18. Social life A question posted on the gamefaqs.com website asked visitors to the site, "Do you have an active social life outside the Internet?" 22% of the 55,581 respondents said "No" or "Not really, most of my personal contact is online."

19. Mayoral race Hoping to learn what issues may resonate with voters in the coming election, the campaign director for a mayoral candidate selects four blocks from each of the city's election districts. Staff members go there and interview all the adult residents they can find.

20. Soil samples The Environmental Protection Agency took soil samples at 20 locations near a former industrial waste dump and checked each for evidence of toxic chemicals. They found no elevated levels of any harmful substances.

21. Roadblock State police set up a roadblock to estimate the percentage of cars with up-to-date registration, insurance, and safety inspection stickers. They usually find problems with about 15% of the cars they stop.

22. Snack foods A company packaging snack foods maintains quality control by randomly selecting 12 cases from each day's production and weighing the bags. Then they open one bag from each case and inspect the contents.

23. Orange samples Food-safety inspectors visit produce farms unannounced and take samples of a day's worth of oranges to test for contamination. If the samples are found to contain dirt, antibiotics unsuited for human consumption, or other foreign matter, the oranges will be destroyed. The farm will be regularly reinspected until purity is restored.

24. Mistaken poll A local TV station conducted a "PulsePoll" about the upcoming mayoral election. Evening news viewers were invited to text in their votes, with the results to be announced on the late-night news. Based on the texts, the station predicted that Amabo would win the election with 52% of the vote. They were wrong: Amabo lost, getting only 46% of the vote. Do you think the station's faulty prediction is more likely to be a result of bias or sampling error? Explain.

25. Another mistaken poll Prior to the mayoral election discussed in Exercise 24, the newspaper also conducted a poll. The paper surveyed a random sample of registered voters stratified by political party, age, sex, and area of residence. This poll predicted that Amabo would win the election with 52% of the vote. The newspaper was wrong: Amabo lost, getting only 46% of the vote. Do you think the newspaper's faulty prediction is more likely to be a result of bias or sampling error? Explain.

26. Parent opinion, part 1 In a large city school system with 50 elementary schools, the school board is considering the adoption of a new policy that would require elementary students to pass a test in order to be promoted to the next grade. The PTA wants to find out whether parents agree with this plan. Listed below are some of the ideas proposed for gathering data. For each, indicate what kind of sampling strategy is involved, and what (if any) biases might result.

a) Put a big ad in the newspaper asking people to log their opinions on the PTA website.
b) Randomly select one of the elementary schools and contact every parent by phone.
c) Send a survey home with every student, and ask parents to fill it out and return it the next day.
d) Randomly select 10 parents from each elementary school. Send them a survey, and follow up with a phone call if they do not return the survey within a week.

27. Parent opinion, part 2 Let's revisit the school system described in Exercise 26. Four new sampling strategies have been proposed to help the PTA determine whether parents favor requiring elementary students to pass a test in order to be promoted to the next grade. For each, indicate what kind of sampling strategy is involved and what (if any) biases might result.

a) Run a poll on the local TV news, asking people to text in whether they favor or oppose the plan.
b) Hold a PTA meeting at each of the 20 elementary schools, and tally the opinions expressed by those who attend the meetings.
c) Randomly select one class at each elementary school and contact each of those parents.
d) Go through the district's enrollment records, selecting every 40th parent. PTA volunteers will go to those homes to interview the people chosen.

28. Churches For your political science class, you'd like to take a survey from a sample of all the Catholic church members in your city. A list of churches shows 16 Catholic churches within the city limits. Rather than try to obtain a list of all members of all these churches, you decide to pick 3 churches at random. For those churches, you'll ask to get a list of all current members and contact 100 members at random.

a) What kind of design have you used?
b) What could go wrong with your design?

29. Playground Some people have been complaining that the children's playground at a municipal park is too small and is in need of repair. Managers of the park decide to survey city residents to see if they believe the playground should be rebuilt. They hand out questionnaires to parents who bring children to the park. Describe possible biases in this sample.

30. Roller coasters An amusement park has opened a new roller coaster. It is so popular that people are waiting for up to 3 hours for a 2-minute ride. Concerned about how patrons (who paid a large amount to enter the park and ride on the rides) feel about this, they survey every person in line for the roller coaster, starting from a randomly selected individual.

a) What kind of sample is this?
b) What is the sampling frame?
c) Is it likely to be representative?
d) What members of the population of interest are omitted?

31. Playground, act two The survey described in Exercise 29 asked: *Many people believe this playground is too small and in need of repair. Do you think the playground should be repaired and expanded even if that means raising the entrance fee to the park?*

Describe two ways this question may lead to response bias.

32. Wording the survey Two members of the PTA committee in Exercises 26 and 27 have proposed different questions to ask in seeking parents' opinions.

Question 1: *Should elementary school–age children have to pass high-stakes tests in order to remain with their classmates?*
Question 2: *Should schools and students be held accountable for meeting yearly learning goals by testing students before they advance to the next grade?*

a) Do you think responses to these two questions might differ? How? What kind of bias is this?
b) Propose a question with more neutral wording that might better assess parental opinion.

33. Banning ephedra An online poll on a website asked:

A nationwide ban of the diet supplement ephedra went into effect recently. The herbal stimulant has been linked to 155 deaths and many more heart attacks and strokes. Ephedra manufacturer NVE Pharmaceuticals, claiming that the FDA lacked proof that ephedra is dangerous if used as directed, was denied a temporary restraining order on the ban by a federal judge. Do you think that ephedra should continue to be banned nationwide?

65% of 17,303 respondents said "yes." Comment on each of the following statements about this poll:

a) With a sample size that large, we can be pretty certain we know the true proportion of Americans who think ephedra should be banned.
b) The wording of the question is clearly very biased.
c) The sampling frame is all Internet users.
d) Results of this voluntary response survey can't be reliably generalized to any population of interest.

34. Survey questions Examine each of the following questions for possible bias. If you think the question is biased, indicate how and propose a better question.

a) Should companies that pollute the environment be compelled to pay the costs of cleanup?
b) Given that 16-year-olds are old enough to drive in the United States and to serve in the military, is it fair to set the voting age at 18?

35. More survey questions Examine each of the following questions for possible bias. If you think the question is biased, indicate how and propose a better question.

a) Do you think high school students should be required to wear uniforms?
b) Given humanity's great tradition of exploration, do you favor continued funding for space flights?

36. Phone surveys Any time we conduct a survey, we must take care to avoid undercoverage. Suppose we plan to select 500 names from the city phone book, call their homes between noon and 4 PM, and interview whoever answers, anticipating contacts with at least 200 people.

a) Why is it difficult to use a simple random sample here?
b) Describe a more convenient, but still random, sampling strategy.
c) What kinds of households are likely to be included in the eventual sample of opinion? Excluded?
d) Suppose, instead, that we continue calling each number, perhaps in the morning or evening, until an adult is contacted and interviewed. How does this improve the sampling design?
e) Random-digit dialing machines can generate the phone calls for us. How would this improve our design? Is anyone still excluded?

37. Cell phone survey What about drawing a random sample only from cell phone exchanges? Discuss the advantages and disadvantages of such a sampling method compared with surveying randomly generated telephone numbers from non–cell phone exchanges. Do you think these advantages and disadvantages have changed over time? How do you expect they'll change in the future?

38. Arm length How long is your arm compared with your hand size? Put your right thumb at your left shoulder bone, stretch your hand open wide, and extend your hand down your arm. Put your thumb at the place where your little finger is, and extend down the arm again. Repeat this a third time. Now your little finger will probably have reached the back of your left hand. If your arm is less than four hand widths, turn your hand sideways and count finger widths until you reach the end of your middle finger.

a) How many hand and finger widths is your arm?
b) Suppose you repeat your measurement 15 times and average your results. What parameter would this average estimate? What is the population?
c) Suppose you now collect arm lengths measured in this way from 6 friends and average these 6 measurements. What is the population now? What parameter would this average estimate?
d) Do you think these 6 arm lengths are likely to be representative of the population of arm lengths in your community? In the country? Why or why not?

39. Fuel economy Occasionally, when I fill my car with gas, I figure out how many miles per gallon my car got. I wrote down those results after six fill-ups in the past few months. Overall, it appears my car gets 28.1 miles per gallon.

a) What statistic have I calculated?
b) What is the parameter I'm trying to estimate?
c) How might my results be biased?
d) When the Environmental Protection Agency (EPA) checks a car like mine to predict its fuel economy, what parameter is it trying to estimate?

40. Accounting Between quarterly audits, a company likes to check on its accounting procedures to address any problems before they become serious. The accounting staff processes payments on about 120 orders each day. The next day, the supervisor rechecks 10 of the transactions to be sure they were processed properly.

a) Propose a sampling strategy for the supervisor.
b) How would you modify that strategy if the company makes both wholesale and retail sales, requiring different bookkeeping procedures?

41. Happy workers? A manufacturing company employs 14 project managers, 370 supervisors, and 388 laborers. In an effort to keep informed about any possible sources of employee discontent, management wants to conduct job satisfaction interviews with a sample of employees every month.

a) Do you see any potential danger in the company's plan? Explain.
b) Propose a sampling strategy that uses a simple random sample.
c) Why do you think a simple random sample might not provide the representative opinion the company seeks?
d) Propose a better sampling strategy.
e) Listed below are the last names of the project managers. Use random numbers to select two people to be interviewed. Explain your method carefully.

Barrett	Bowman	Chen
DeLara	DeRoos	Grigorov
Maceli	Mulvaney	Pagliarulo
Rosica	Smithson	Tadros
William	Yamamoto	

42. Quality control Sammy's Salsa, a small local company, produces 20 cases of salsa a day. Each case contains 12 jars and is imprinted with a code indicating the date and batch number. To help maintain consistency, at the end of each day, Sammy selects three jars of salsa, weighs the contents, and tastes the product. Help Sammy select the sample jars. Today's cases are coded 07N61 through 07N80.

a) Carefully explain your sampling strategy.
b) Show how to use random numbers to pick 3 jars.
c) Did you use a simple random sample? Explain.

43. A fish story Concerned about reports of discolored scales on fish caught downstream from a newly sited chemical plant, scientists set up a field station in a shoreline public park. For one week, they asked fishermen there to bring any fish they caught to the field station for a brief inspection. At the end of the week, the scientists said that 19% of the 332 fish that were submitted for inspection displayed the discoloration. From this information, can the researchers estimate what proportion of fish in the river have discolored scales? Explain.

44. Another fish story In recent years, beaches around the world have seen fish washing up on shore in large numbers. One group of scientists thought that leakage from the damaged Fukushima nuclear power plant in Japan might be a cause. They propose to measure the mean amount of radiation in a sample of dead fish on a certain beach. Will this give them a good estimate of the mean amount of radiation that fish received from Fukushima?

45. Sampling methods Consider each of these situations. Do you think the proposed sampling method is appropriate? Explain.

a) We want to know what percentage of local doctors accept patients who avail state-sponsored medical assistance. We call the offices of 75 doctors randomly selected from local Yellow Pages listings.
b) We want to know what percentage of local businesses anticipate hiring additional employees in the upcoming quarter. We randomly select a page in the Yellow Pages and call every business listed there.

46. More sampling methods Consider each of these situations. Do you think the proposed sampling method is appropriate? Explain.

a) We want to know if there is neighborhood support to turn a vacant lot into a playground. We spend a Saturday afternoon going door-to-door in the neighborhood, asking people to sign a petition.
b) We want to know if students at our college are satisfied with the selection of food available on campus. We go to the largest cafeteria and interview every 10th person in line.

JUST CHECKING

Answers

1. a) It can be hard to reach all members of a population, and it can take so long that circumstances change, affecting the responses. A well-designed sample is often a better choice.
   b) This sample is probably biased—students who didn't like the food at the cafeteria might not choose to eat there.
   c) No, only the sample size matters, not the fraction of the overall population.
   d) Students who frequent this website might be more enthusiastic about statistics than the overall population of statistics students. A large sample cannot compensate for bias.
   e) It's the population "parameter." "Statistics" describe samples.
2. a) systematic
   b) stratified
   c) simple
   d) cluster

11

Experiments and Observational Studies

WHERE ARE WE GOING?

Experiments are the "Gold Standard" of data collection. No drug comes to market without at least one FDA-approved experiment to demonstrate its safety and effectiveness. Much of what we know in science and social science comes from carefully designed experiments.

The Four Principles of Experimental Design (Control what you can, Randomize for the rest, Replicate the trials, and, when appropriate, Block to remove identifiable variation) describe what makes a sound experiment and how to understand the results.

Who gets good grades? And, more important, why? Is there something schools and parents could do to help weaker students improve their grades? Some people think they have an answer: music! No, not your iPod, but an instrument. In a study conducted at Mission Viejo High School, in California, researchers compared the scholastic performance of music students with that of non-music students. The music students had a much higher overall grade point average than the non-music students, 3.59 to 2.91. And a whopping 16% of the music students had all A's compared with only 5% of the non-music students.

As a result of this study, many parent groups and educators pressed for expanded music programs in the nation's schools. They argued that the work ethic, discipline, and feeling of accomplishment fostered by learning to play an instrument also enhance a person's ability to succeed in school. They thought that involving more students in music would raise academic performance. What do you think? Does this study provide solid evidence? Or are there other possible explanations for the difference in grades? Is there any way to really prove such a conjecture?

11.1 Observational Studies

This research tried to show an association between music education and grades. But it wasn't a survey. Nor did it assign students to get music education. Instead, it simply observed students "in the wild," recording the choices they made and the outcome. Such studies are called **observational studies**. In observational studies, researchers don't *assign* choices; they simply observe them. In addition, this was a **retrospective study**, because researchers first identified subjects who studied music and then collected data on their past grades.

What's wrong with concluding that music education causes good grades? The claim that music study *caused* higher grades depends on there being *no other differences* between the groups that could account for the differences in grades. But there are lots of variables that might cause the groups to perform differently. Students who study music may have better work habits to start with. Music students may have more parental support and that support may have enhanced their academic performance too. Maybe they came from wealthier homes and had other advantages. Or it could be that smarter kids just like to play musical instruments.

RETROSPECTIVE STUDIES CAN GIVE VALUABLE CLUES

For rare illnesses, it's not practical to draw a large enough sample to see many ill respondents, so the only option remaining is to develop retrospective data. For example, researchers can interview those who have become ill. The likely causes of both Legionnaires' disease and HIV were identified from such retrospective studies of the small populations who were first infected. But to confirm the causes, researchers needed laboratory-based experiments.

Observational studies are used widely in public health and marketing. Those that study rare outcomes, such as specific diseases, are often retrospective. They first identify people with the disease and then look into their history and heritage in search of things that may be related to their condition. But retrospective studies have a restricted view of the world because they are usually limited to a small part of the entire population. And because retrospective records are based on historical data and memories, they can have errors. (Do you recall *exactly* what you ate even yesterday? How about last Wednesday?)

If we identify subjects in advance and collect data as events unfold, we have a **prospective study**. That's a more reliable approach. For example, we might select a group of infants to study, follow whether they later take music lessons, and track their academic performance.

Although an observational study may identify important variables related to an outcome of interest, there is no guarantee that we have found the right or the most important related variables. Students who choose to study an instrument might still differ from the others in some important way that we failed to observe. It may be this difference—whether we know what it is or not—rather than music itself that leads to better grades. It's just not possible for observational studies, whether prospective or retrospective, to demonstrate a causal relationship.

EXAMPLE 11.1

Designing an Observational Study

In early 2007, a larger-than-usual number of cats and dogs developed kidney failure; many died. Initially, researchers didn't know why, so they used an observational study to investigate.

QUESTION: Suppose you were called on to plan a study seeking the cause of this problem. Would your design be retrospective or prospective? Explain why.

ANSWER: I would use a retrospective observational study. Even though the incidence of disease was higher than usual, it was still rare. Surveying all pets would have been impractical. Instead, it makes sense to locate some who were sick and ask about their diets, exposure to toxins, and other possible causes.

11.2 Randomized, Comparative Experiments

"He that leaves nothing to chance will do few things ill, but he will do very few things."

—Lord Halifax (1633–1695)

Is it *ever* possible to get convincing evidence of a cause-and-effect relationship? Well, yes it is, but we have to take a different approach. We could take a group of third graders, randomly assign half to take music lessons, and forbid the other half to do so. Then we could compare their grades several years later. This kind of study design is an experiment.

An experiment requires a **random assignment** of subjects to treatments. Only an experiment can justify a claim like "Music lessons cause higher grades." Questions such as "Does taking vitamin C reduce the chance of getting a cold?" and "Does working with computers improve performance in statistics class?" and "Is this drug a safe and effective treatment for that disease?" require a designed experiment to establish cause and effect.

Experiments study the relationship between two or more variables. An experimenter must identify at least one explanatory variable, called a **factor**, to manipulate and at least

Experimental design was advanced in the 19th century by work in psychophysics by Gustav Fechner (1801–1887), the founder of experimental psychology. Fechner designed ingenious experiments that exhibited many of the features of modern designed experiments. Fechner was careful to control for the effects of factors that might affect his results. For example, in his 1860 book *Elemente der Psychophysik* he cautioned readers to group experiment trials together to minimize the possible effects of time of day and fatigue.

one **response variable** to measure. What distinguishes an **experiment** from other types of investigation is that the experimenter actively and deliberately manipulates the factors to control the details of the possible treatments, and assigns the subjects to those treatments *at random*. The experimenter then *compares* responses for different groups of subjects who have been treated differently. For example, we might design an experiment to see whether the amount of sleep and exercise you get affects your grades.

The individuals on whom or which we experiment are known by a variety of terms. When they are human, we usually call them **subjects or participants**. Other individuals (rats, days, petri dishes of bacteria) are commonly referred to by the more generic term **experimental unit**. When we recruit subjects for our sleep deprivation experiment, we'll probably have better luck inviting participants than advertising for experimental units.

The specific values that the experimenter chooses for a factor are called the **levels** of the factor. We might assign our participants to sleep for 4, 6, or 8 hours. Often there are several factors at a variety of levels. (Our participants will also be assigned to a treadmill for 0 or 30 minutes.) The combination of specific levels from all the factors that an experimental unit receives is known as its **treatment**. (Our participants could have any one of six different treatments—three sleep levels, each at two exercise levels.)

How should we assign our participants to these treatments? Some students prefer 4 hours of sleep, while others need 8. Some exercise regularly; others are couch potatoes. Should we let the students choose the treatments they'd prefer? No. That would not be a good idea. Nor should we assign them according to what we feel might be best. To have any hope of drawing a fair conclusion, we must assign our participants to their treatments *at random*.

It may be obvious to you that we shouldn't let the students choose the treatment they'd prefer, but the need for random assignment is a lesson that was once hard for some to accept. For example, physicians naturally prefer to assign patients to the therapy that they think best rather than have a random element such as a coin flip determine the treatment.[1] But we've known for more than a century that for the results of an experiment to be valid, we must use deliberate randomization.

In Summary—An Experiment

Manipulates the factor levels to create treatments. *Randomly assigns* subjects to these treatment levels. *Compares* the responses of the subject groups across treatment levels.

THE WOMEN'S HEALTH INITIATIVE

In 2010, the National Center for Biotechnology Information released data from the Women's Health Initiative, a major 15-year research program funded by the National Institutes of Health.[2] It consists of both an observational study with more than 93,000 participants and several randomized comparative experiments. The goals of this study included

- giving reliable estimates of the extent to which known risk factors predict heart disease, cancers, and fractures;
- identifying "new" risk factors for these and other diseases in women;
- comparing risk factors, presence of disease at the start of the study, and new occurrences of disease during the study across all study components; and
- creating a future resource to identify biological indicators of disease, especially substances and factors found in blood.

That is, the study sought to identify possible risk factors and assess how serious they might be and to build up data that might be checked retrospectively as the women in the study continue to be followed. There would be no way to find out these things with an experiment because the task includes identifying new risk factors. If we don't know those risk factors, we could never control them as factors in an experiment.

By contrast, one of the clinical trials (randomized experiments) that received much press attention randomly assigned postmenopausal women to take either hormone replacement therapy or an inactive pill. The results concluded that hormone replacement with estrogen carried increased risks of stroke.

[1] If we (or the physician) knew which was the better treatment, we wouldn't be performing the experiment.

[2] www.nhlbi.nih.gov/whi/background.htm

THE FDA

No drug can be sold in the United States without first showing, in a suitably designed experiment approved by the Food and Drug Administration (FDA), that it's safe and effective. The small print on the booklet that comes with many prescription drugs usually describes the outcomes of that experiment.

EXAMPLE 11.2

Determining the Treatments and Response Variable

RECAP: In 2007, deaths of a large number of pet dogs and cats were ultimately traced to contamination of some brands of pet food. The manufacturer now claims that the food is safe, but before it can be released, it must be tested.

QUESTION: In an experiment to test whether the food is now safe for dogs to eat,[3] what would be the treatments and what would be the response variable?

ANSWER: The treatments would be ordinary-size portions of two dog foods: the new one from the company (the *test food*) and one that I was certain was safe (perhaps prepared in my kitchen or laboratory). The response would be a veterinarian's assessment of the health of the test animals.

11.3 The Four Principles of Experimental Design

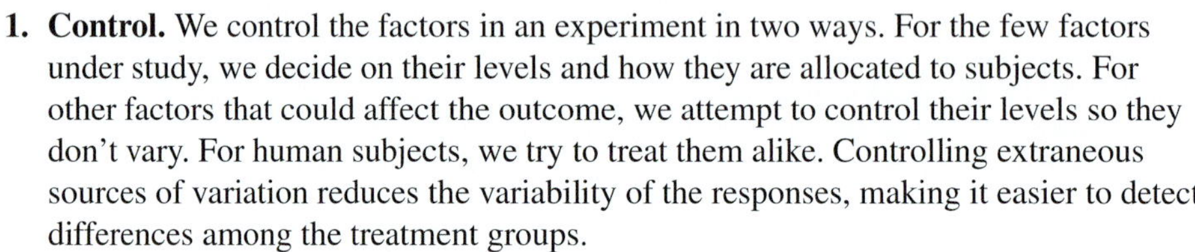

1. **Control.** We control the factors in an experiment in two ways. For the few factors under study, we decide on their levels and how they are allocated to subjects. For other factors that could affect the outcome, we attempt to control their levels so they don't vary. For human subjects, we try to treat them alike. Controlling extraneous sources of variation reduces the variability of the responses, making it easier to detect differences among the treatment groups.

   Although we **control** both experimental factors and other sources of variation, we think of them and deal with them very differently. We control a factor by assigning subjects to different factor levels so we can see how the response changes at those different levels. But we control other sources of variation to *prevent* them from changing and affecting the response.

The deep insight that experiments should use random assignment is quite an old one. It can be attributed to the American philosopher and scientist C. S. Peirce in his experiments with J. Jastrow, published in 1885.

2. **Randomize.** As in sample surveys, randomization allows us to equalize the effects of unknown or uncontrollable sources of variation. It does not eliminate the effects of these sources, but by distributing them equally (on average) across the treatment levels, it makes comparisons among the treatments fair. Assigning experimental units to treatments at random allows us to use the powerful methods of statistics to draw conclusions from an experiment. When we assign subjects to treatments at random, we reduce bias due to uncontrolled sources of variation.

   Experimenters often control factors that are easy or inexpensive to control. They **randomize** to protect against the effects of other factors, even factors they haven't thought about. How to choose between the two strategies is best summed up in the old adage that says "control what you can, and randomize the rest."

3. **Replicate.** Drawing conclusions about the world is impossible unless we repeat, or **replicate**, our results. Two kinds of replication show up in comparative experiments. First, we should apply each treatment to a number of subjects. Only with such replication can we estimate the variability of responses. If we have not assessed the variation, the experiment is not complete. The outcome of an experiment on a single subject is an anecdote, not data.

[3] It may disturb you (as it does us) to think of deliberately putting dogs at risk in this experiment, but in fact that is what is done. The risk is borne by a small number of dogs so that the far larger population of dogs can be kept safe.

A second kind of **replication** shows up when the entire experiment is repeated on a different population of experimental units. What is true of the students in Psych 101 who volunteered for the sleep experiment may be is true of all humans, but we'll feel more confident if our results for the experiment are *replicated* in another part of the country, with people of different ages, and at different times of the year. Replication of an entire experiment with the controlled sources of variation at different levels is an essential step in science.

4. **Block.** Suppose the participants available for a study of balance include two members of the varsity girls gymnastics team and 10 other students. Randomizing may place both gymnasts in the same treatment group. In the long run, if we could perform the experiment over and over, it would all equalize. But wouldn't it be better to assign one gymnast to each group and five of the other students to each group (at random)? Doing this improves fairness in the *short* run.

The differences in the students' gymnastics training is likely to affect what we want to measure. Whenever identifiable differences among the experimental units might affect the outcome being studied, it is wise to collect the units into homogeneous groups or **blocks**. Here, the variable "Training" is a blocking variable, and the levels of "Experience" are called blocks.

Grouping similar individuals into blocks can help us account for unwanted variation among the subjects, allowing us to see differences in the treatments that might otherwise be obscured. Blocking is an important addition to the principles of randomization, control, and replication. However, unlike the first three, blocking is not required in an experimental design.

EXAMPLE 11.3 Control, Randomize, and Replicate

RECAP: We're planning an experiment to see whether the new pet food is safe for dogs to eat. We'll feed some animals the new food and others a food known to be safe, comparing their health after a period of time.

QUESTION: In this experiment, how will you implement the principles of control, randomization, and replication?

ANSWER: I'd control the portion sizes eaten by the dogs. To reduce possible variability from factors other than the food, I'd standardize other aspects of their environments—housing the dogs in similar pens and ensuring that each got the same amount of water, exercise, play, and sleep time, for example. I might restrict the experiment to a single breed of dog and to adult dogs to further minimize variation.

To equalize traits, pre-existing conditions, and other unknown influences, I would assign dogs to the two feed treatments randomly.

I would replicate by assigning more than one dog to each treatment to allow for variability among individual dogs. If I had the time and funding, I might replicate the entire experiment using the same breed of dog, or use several different breeds to further generalize my results.

Diagrams

An experiment is carried out over time with specific actions occurring in a specified order. A diagram of the procedure can help in thinking about experiments.[4]

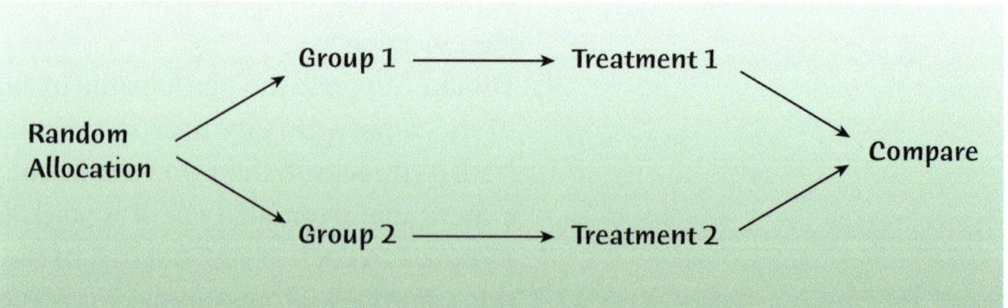

The diagram emphasizes the random allocation of subjects to treatment groups, the separate treatments applied to these groups, and the ultimate comparison of results. It's best to specify the responses that will be compared. A good way to start comparing results for the treatment groups is with boxplots.

STEP-BY-STEP EXAMPLE

Designing an Experiment

A student, preparing for a triathlon, suspected that the 45 minutes each day she spent training in a chlorinated pool was damaging her nail polish. She wished to investigate whether the color of the nail polish might make a difference. How could she design an experiment to investigate? (Data in **Nail polish**)

Let's work through the design, step by step. We'll design a completely randomized experiment in one factor. The statements in the right column are the kinds of things you would need to say in proposing an experiment. You'd include them in the "methods" section of a report once the experiment is run.

A **completely randomized experiment** is the ideal simple design, just as a *simple random sample* is the ideal simple sample—and for many of the same reasons.

QUESTION: How would you design an experiment to test whether two nail colors resisted chipping differently when challenged by soaks in chlorine water?

THINK **PLAN** State what you want to know.	I want to know whether two colors of nail polish resist chipping differently when challenged by soaks in chlorine water.
RESPONSE Specify the response variable.	I can measure the percentage of the nail polish chipped away by examining scanned images of the nails with a computer program.
TREATMENTS Specify the factor levels and the treatments.	The factor is nail polish *Color*. The factor levels are *Red* and *Nude*, two common nail polish colors.

[4]Diagrams of this sort were introduced by David Moore in his textbooks and are still widely used.

EXPERIMENTAL UNITS Specify the experimental units.

To control the surface colored by the polish, reducing variability that might be found with natural nails, I'll obtain 30 acrylic nails and glue each one to a chopstick.

EXPERIMENT DESIGN Observe the principles of design:

Control any sources of variability you know of and can control.

I will use the same brand of acrylic nails and the same brand of nail polish. All nails will be painted at the same time and soaked at the same time in the same chlorine solution for the same amount of time.

Replicate results by placing more than one nail in each treatment group.

I'll use 15 nails in each group.

Randomly assign experimental units to treatments to equalize the effects of unknown or uncontrollable sources of variation.

I will randomly divide the nails into two groups for painting with each color.

Describe how the randomization will be accomplished.

I will number the nails from 1 to 30. I will obtain 30 random numbers from an Internet site and sort them, carrying along the nail ID numbers. Then I'll assign the first 15 nails to one color and the remaining ones to the other.

MAKE A PICTURE A diagram of your design can help you think about it clearly.

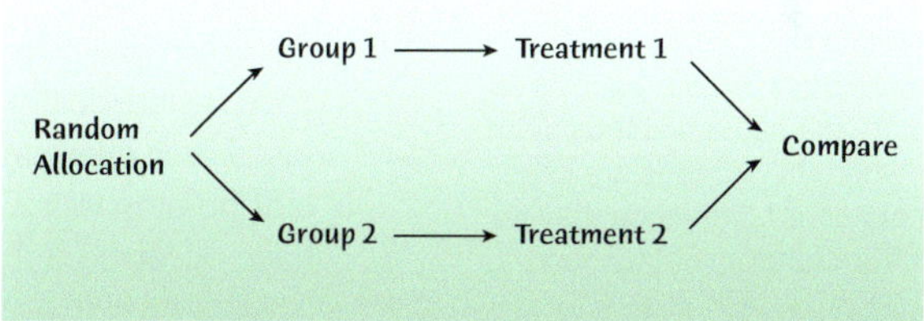

Specify any other experiment details. Give enough details so that another experimenter could exactly replicate your experiment. It's generally better to include details that might seem irrelevant than to leave out matters that could turn out to make a difference.

All nails will be soaked together for 45 minutes in a chlorine solution that matches typical swimming pool levels of chlorine. After the nails dry, I will tap each one 100 times on a computer keyboard to simulate daily stresses. I will repeat this treatment for 5 days and measure the result at the end of that time.

Specify how to measure the response.

I will scan each nail and use a digitized image of the nail to measure the fraction of the polish that has chipped off by comparing the before and after areas of nail polish using Adobe Illustrator.

SHOW Once you collect the data, you'll need to display them and compare the results for the treatment groups.

I will display the results with side-by-side boxplots to compare the treatment groups.

I will then compare the means of the groups.

TELL To answer the initial question, we ask whether the differences observed in the means of the groups are meaningful.

Because this is a randomized experiment, we can attribute significant differences to the treatments. To do this properly, we'll need methods from "statistical inference," the subject of the rest of this text.

If the difference in mean percent chipped between the groups is greater than I would expect from the variation I observe among the nails within each treatment group, I may be able to conclude that this difference can be attributed to the differences in nail polish color.

Does the Difference Make a Difference?

A new Kickstarter project claims it will produce running shoes "guaranteed to make you faster." Your friend knows one of the developers and has snagged a pair to test. The two of you take them down to the track. Her standard time in the 400 m is 55 sec. On the first run, she runs it 2 seconds faster. Are you convinced? What if over the course of the next two days her average of 10 runs with the new shoes is 2 sec faster than with her standard shoes. Are you convinced now? What if the average had been 10 sec faster?

You want to know if the faster time is really due to the new shoes, or to random variation. If the shoes only have a small **effect**, it will be hard to tell the difference. If your friend's performance is inconsistent, it will be hard to tell the difference as well. And, if she runs the course only a couple of times, it will be very hard to tell the difference. These are the three things that affect your ability to draw conclusions.

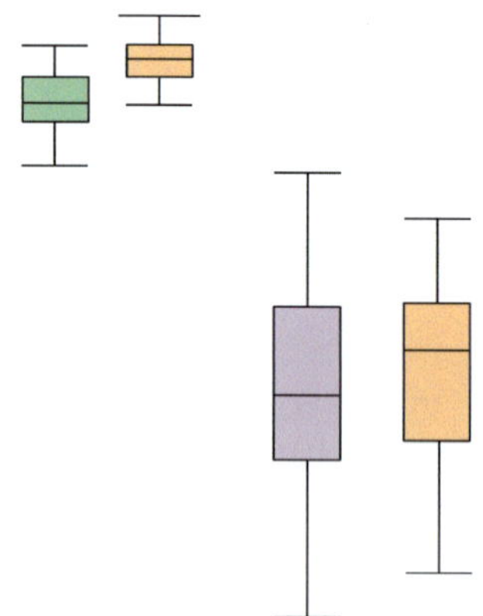

Figure 11.1
The boxplots in both pairs have centers the same distance apart, but when the spreads are large, the observed difference may be just from random fluctuation.

Observations vary. What if, in fact, your friend knows that the standard deviation of her times in the past month is about 1 second. Now, her single run is 2 standard deviations faster with the new shoes. From the 68–95–99.7% Rule, we know that's unusual. But you'd be more convinced if the average of 10 runs is 2 standard deviations faster. When the observed difference is large enough in standard deviations to be convincing, we say it's **statistically significant**.

Boxplots are a great way to visualize this difference. Figure 11.1 shows the same running shoe effect on two different runners, a consistent runner on the left and much more variable runner on the right. In the left pair the variation is quite small within treatment groups, making the effect clear. In the right pair, that same effect looks less impressive. We'd say the difference is statistically significant in the left pair and not statistically significant in the right pair.

Later chapters show how statistical tests quantify this intuition. For now, the important point is that a difference is statistically significant if we don't believe that it's likely to have occurred only by chance.

JUST CHECKING

1. At one time, a method called "gastric freezing" was used to treat people with peptic ulcers. An inflatable bladder was inserted down the esophagus and into the stomach, and then a cold liquid was pumped into the bladder. Now you can find the following notice on the Internet site of a major insurance company:

   [Our company] does not cover gastric freezing (intragastric hypothermia) for chronic peptic ulcer disease. . . .

   Gastric freezing for chronic peptic ulcer disease is a non-surgical treatment which was popular about 20 years ago but now is seldom performed. It has been abandoned due to a high complication rate, only temporary improvement experienced by patients, and a lack of effectiveness when tested by double-blind, controlled clinical trials.

   What did that "controlled clinical trial" (experiment) probably look like? (Don't worry about "double-blind"; we'll get to that soon.)

   a) What was the factor in this experiment?
   b) What was the response variable?
   c) What were the treatments?
   d) How did researchers decide which subjects received which treatment?
   e) Were the results statistically significant?

Experiments and Samples

Both experiments and sample surveys use randomization to get unbiased data. But they do so in different ways and for different purposes. Sample surveys try to estimate population parameters, so the sample needs to be as representative of the population as possible. By contrast, experiments try to assess the effects of treatments. Experimental units are not generally drawn randomly from the population. For example, a medical experiment may deal only with local patients who have the disease under study. The randomization is in the assignment of their therapy. A sample should exhibit the diversity and variability of the population, but for an experiment the more homogeneous the subjects the easier it will be to spot differences in the effects of the treatments.

Unless the experimental units are chosen from the population at random, you should be cautious about generalizing experiment results to larger populations until the experiment has been repeated under different circumstances. Results become more persuasive if they remain the same in completely different settings, such as in a different season, in a different country, or for a different species.

Nevertheless, experiments can draw stronger conclusions than surveys. Only by actively manipulating the factors, can an experimenter conclude that observed changes are due to the factors themselves. This is the fundamental insight of the scientific method.

A Random Sample? Probably Not

Experiments are rarely performed on random samples from a population. Don't describe the subjects in an experiment as a random sample unless they really are. More likely, the randomization was in assigning subjects to treatments.

11.4 Control Groups

The runner wanted to be sure that the effect she saw wasn't just due to her increased training. So, before each run, she flipped a coin to see whether she'd use the new shoes or her usual ones. That provided her with a baseline for her to compare the new results. Such a baseline is called a **control treatment**, and the experimental units (in this case runs) assigned to it are called a **control group**.

This is an entirely different use of the word "control" than we saw in the previous section. There, we *controlled* extraneous sources of variation by keeping them constant. Here, we use a control treatment as another *level* of the factor included so we can compare the treatment results to a situation in which "nothing happens."

Blinding

Humans are notoriously susceptible to errors in judgment.[5] All of us. When we know what treatment was assigned, it's difficult not to let that knowledge influence our assessment of the response, even when we try to be careful.

Suppose you were trying to investigate which brand of cola students really prefer. You set up an experiment to see which of the three competing brands students prefer (or whether they can tell the difference at all). But people have brand loyalties. Your experiment subjects probably prefer one brand already. If they knew which brand they were tasting, it might influence their rating. To avoid this problem, it would be better to disguise the brands as much as possible. This strategy is called **blinding** the participants to the treatment.[6]

But it isn't just the subjects who should be blind. Experimenters themselves often subconsciously behave in ways that favor what they believe. Even technicians may treat plants or test animals differently if, for example, they expect them to die. An animal that

[5]For example, here we are in Chapter 11 and you're still reading the footnotes.

[6]C. S. Peirce, in the same 1885 work in which he introduced randomization, also recommended blinding.

starts doing a little better than others by showing an increased appetite may get fed a bit more than the experimental protocol specifies.

People are so good at picking up subtle cues about treatments that the best (in fact, the *only*) defense against such biases in experiments on human subjects is to keep *anyone* who could affect the outcome or the measurement of the response from knowing which subjects have been assigned to which treatments. So, not only should your cola-tasting subjects be blinded, but also *you*, as the experimenter, shouldn't know which drink is which, either—at least until you're ready to analyze the results.

There are two main classes of individuals who can affect the outcome of the experiment:

- those who could influence the results (the subjects, treatment administrators, or technicians)
- those who evaluate the results (judges, treating physicians, etc.)

When all the individuals in either one of these classes are blinded, an experiment is said to be **single-blind**. When everyone in *both* classes is blinded, we call the experiment **double-blind**. Even if several individuals in one class are blinded—for example, both the patients and the technicians who administer the treatment—the study would still be just single-blind. If only some of the individuals in a class are blind—for example, if subjects are not told of their treatment, but the administering technician is not blind—there is a substantial risk that subjects can discern their treatment from subtle cues in the technician's behavior or that the technician might inadvertently treat subjects differently. Such experiments cannot be considered truly blind.

To improve the nail polish experiment design, the experimenter could have the amount of chipping measured by an independent judge who was shown only monochrome images and didn't know how the nails had been treated. That would make the experiment single-blind. To make the experiment double-blind, we could put colored glasses on the person painting the nails, soaking them in the chlorine water, and tapping them on the keyboard so she couldn't see the colors of the nails.

BLINDING BY MISLEADING

Social science experiments can sometimes blind subjects by misleading them about the purpose of a study. One of the authors participated as an undergraduate in a (now infamous) psychology experiment using such a blinding method. The subjects were told that the experiment was about three-dimensional spatial perception and were assigned to draw a horse. While they were drawing, they heard a loud noise and groaning from the next room. The *real* purpose of the experiment was to see whether the social pressure of being in groups made people react to disaster differently. Subjects had been randomly assigned to draw either in groups or alone; that was the treatment. The experimenter was not interested in the drawings, but the subjects were blinded to the treatment because they were misled.

EXAMPLE 11.4 Blinding

RECAP: In our experiment to see if the new pet food is now safe, we're feeding one group of dogs the new food and another group a food we know to be safe. Our response variable is the health of the animals as assessed by a veterinarian.

QUESTIONS: Should the vet be blinded? Why or why not? How would you do this? (Extra credit: Can this experiment be double-blind? Would that mean that the test animals wouldn't know what they were eating?)

ANSWERS: Whenever the response variable involves judgment, it is a good idea to blind the evaluator to the treatments. The veterinarian should not be told which dogs ate which foods.

Extra credit: There is a need for double-blinding. In this case, the workers who care for and feed the animals should not be aware of which dogs are receiving which food. We'll need to make the "safe" food look as much like the "test" food as possible.

Placebos

Often, simply applying *any* treatment can induce an improvement. Every parent knows the medicinal value of a kiss to make a toddler's scrape stop hurting. Some of the improvement seen with a treatment—even an effective treatment—can be due simply to the act of treating. In addition, regression to the mean may account for some patients' improvement.[7] To separate the effect of the treatment from these other effects we can use a control treatment that mimics the treatment itself.

A "fake" treatment that looks just like the treatments being tested is called a **placebo**. Placebos are the best way to blind human participants from knowing whether they are receiving the treatment or not. One common version of a placebo in drug testing is a "sugar pill." The fact is that subjects treated with a placebo sometimes improve. It's not unusual for 20% or more of subjects given a placebo to report reduction in pain, improved movement, or greater alertness, or even to demonstrate improved health or performance. This **placebo effect** highlights both the importance of effective blinding and the importance of comparing treatments with a control. Placebo controls are so effective that you should use them as an essential tool for blinding whenever possible.

In summary, the best experiments are usually

- randomized.
- comparative.
- double-blind.
- placebo-controlled.

ACTIVE PLACEBOS

The placebo effect is stronger when placebo treatments are administered with authority or by a figure who appears to be an authority. "Doctors" in white coats generate a stronger effect than salespeople in polyester suits. But the placebo effect is not reduced much even when subjects know that the effect exists. People often suspect that they've gotten the placebo if nothing at all happens. So, recently, drug manufacturers have gone so far in making placebos realistic that they cause the same side effects as the drug being tested! Such "active placebos" usually induce a stronger placebo effect. When those side effects include loss of appetite or hair, the practice may raise ethical questions.

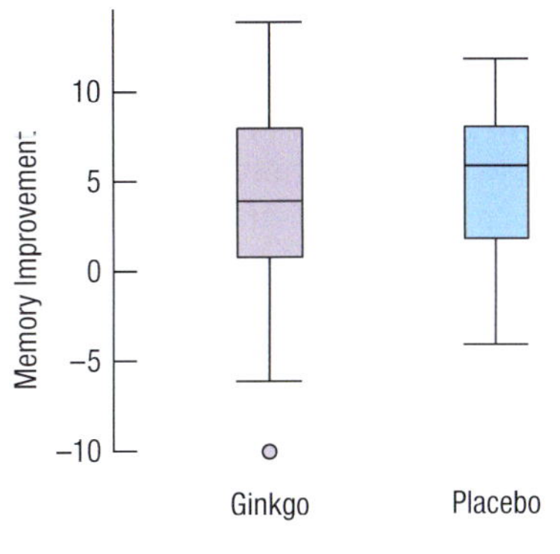

REAL-WORLD EXAMPLE: DOES GINKGO BILOBA IMPROVE MEMORY?

Researchers investigated the purported memory-enhancing effect of ginkgo biloba tree extract.[8] In a randomized, comparative, double-blind, placebo-controlled study, they administered treatments to 230 elderly community members. One group received Ginkoba™ according to the manufacturer's instructions. The other received a similar-looking placebo. Thirteen different tests of memory were administered before and after treatment. The placebo group showed greater improvement on 7 of the tests, the treatment group on the other 6. None showed any significant differences. In the margin are boxplots of one measure.

[7] See Section 7.4 for a refresher on regression to the mean.

[8] P. R. Solomon, F. Adams, A. Silver, J. Zimmer, and R. De Veaux, "Ginkgo for Memory Enhancement. A Randomized Controlled Trial." *JAMA* 288 (2002): 835–840.

JUST CHECKING

2. Recall the experiment about gastric freezing, an old method for treating peptic ulcers that you read about in the first Just Checking. Doctors would insert an inflatable bladder down the patient's esophagus and into the stomach and then pump in a cold liquid. A major insurance company now states that it doesn't cover this treatment because "double-blind, controlled clinical trials" failed to demonstrate that gastric freezing was effective.

   a) What does it mean that the experiment was double-blind?
   b) Why would you recommend a placebo control?
   c) Suppose that researchers suspected that the effectiveness of the gastric freezing treatment might depend on whether a patient had recently developed the peptic ulcer or had been suffering from the condition for a long time. How might the researchers have designed the experiment?

11.5 Blocking

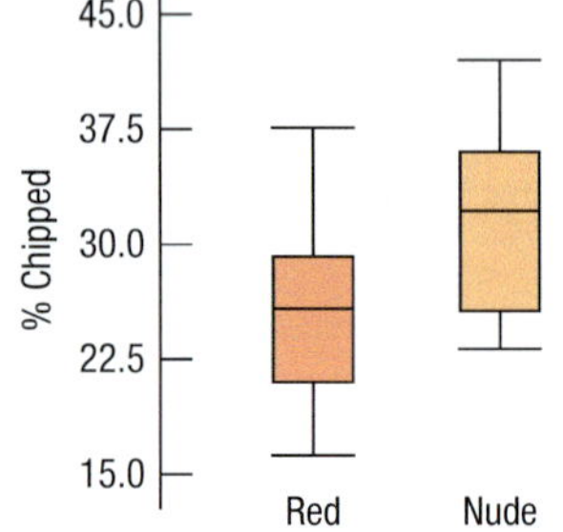

The nail polish experiment showed a difference in the durability of the two colors (see the boxplots in the margin). But the student wondered whether her artificial treatments in fact reflected the experience of swimming and dealing with the everyday stresses on her nails. So she designed a follow-up study. Her plan was to paint her own nails and record chipping during several days as she swam each day and went about her usual work schedule. In this experiment, her experimental units would be her own fingernails. One possible design would have been to paint her left hand with one color and her right hand with the other. But she is right-handed and thought that the stresses on her nails might be different between the two hands.

How could she design the experiment so that the differences in hands don't mess up her attempts to see differences in the nail polish colors?

She could define each hand to be a block. She could then randomly assign half the fingers of a hand to one color and half to the other and randomly do the same for the other hand. Each treatment (color) is still assigned to the same number of fingers. The randomization occurs only within the blocks. The design is a **randomized block design**.

In effect, she can conduct a separate experiment for the fingers of each hand and then combine the results. If her right hand is harder on nail polish because she is right-handed, she can still compare the two polish colors within that hand. The picture depicts the design.

In an observational study, we sometimes pair participants that are similar in ways *not* under study. By **matching** participants in this way, we can reduce variation in much the

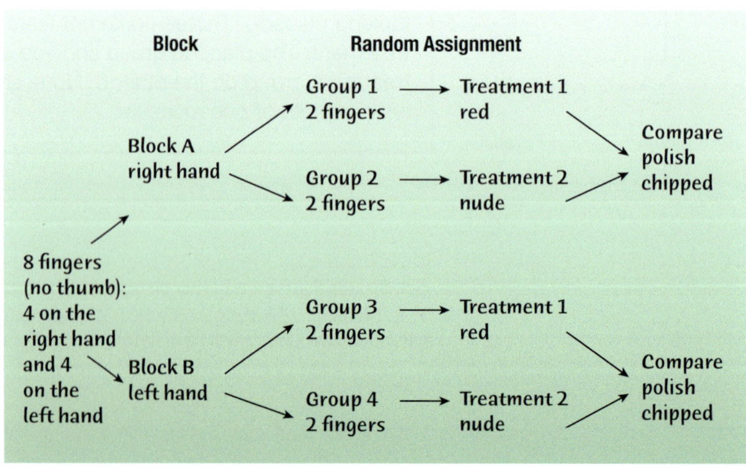

same way as blocking. For example, a retrospective study of music education and grades might match each student who studies an instrument with someone of the same sex who is similar in family income but didn't study an instrument. We could then compare grades of music students with those of non-music students. The matching would reduce the variation due to income and sex differences.

Blocking for experiments is the same idea as stratifying is for sampling. Both methods group together subjects that are similar and randomize within those groups as a way to remove unwanted variation. (But be careful to keep the terms straight. Don't say that we "stratify" an experiment or "block" a sample.) We use blocks to reduce variability so we can see the effects of the factors; we're not usually interested in studying the effects of the blocks themselves.

EXAMPLE 11.5

Blocking

RECAP: In 2007, pet food contamination put cats at risk, as well as dogs. Our experiment should probably test the safety of the new food on both animals.

QUESTIONS: Why shouldn't we randomly assign a mix of cats and dogs to the two treatment groups? What would you recommend instead?

ANSWERS: Dogs and cats might respond differently to the foods, and that variability could obscure my results. Blocking by species can remove that superfluous variation. I'd randomize cats to the two treatments (test food and safe food) separately from the dogs. I'd measure their responses separately and look at the results afterward.

11.6 Confounding

Professor Stephen Ceci of Cornell University performed an experiment to investigate the effect of a teacher's classroom style on student evaluations. He taught a class in developmental psychology during two successive terms to a total of 472 students in two very similar classes. He kept everything about his teaching identical (same text, same syllabus, same office hours, etc.) and modified only his style in class. During the fall term, he maintained a subdued demeanor. During the spring term, he used expansive gestures and lectured with more enthusiasm, varying his vocal pitch and using more hand gestures. He administered a standard student evaluation form at the end of each term.

The students in the fall term class rated him only an average teacher. Those in the spring term class rated him an excellent teacher, praising his knowledge and accessibility, and even the quality of the textbook. On the question "How much did you learn in the course?," the average response changed from 2.93 to 4.05 on a 5-point scale.[9]

How much of the difference he observed was due to his difference in manner, and how much might have been due to the season of the year? Fall term in Ithaca, New York (home of Cornell University), starts out colorful and pleasantly warm but ends cold and bleak. Spring term starts out bitter and snowy and ends with blooming flowers and singing birds. Might students' overall happiness have been affected by the season and reflected in their evaluations?

Unfortunately, there's no way to tell. Nothing in the data enables us to tease apart these two effects, because all the students who experienced the subdued manner did so during the fall term and all who experienced the expansive manner did so during the spring. When the levels of one factor are associated with the levels of another factor, we say that these two factors are **confounded**.

[9]But the two classes performed almost identically well on the final exam.

In some experiments, such as this one, and some observational studies as well, it's just not possible to avoid some confounding. Professor Ceci could have randomly assigned students to one of two classes during the same term, but then we might question whether mornings or afternoons were better, or whether he really delivered the same class the second time (after practicing on the first class). Or he could have had another professor deliver the second class, but that would have raised more serious issues about differences in the two professors and concern over more serious confounding.

EXAMPLE 11.6 Confounding

RECAP: After many dogs and cats suffered health problems caused by contaminated foods, we're trying to find out whether a newly formulated pet food is safe. Our experiment will feed some animals the new food and others a food known to be safe, and a veterinarian will check the response.

QUESTION: Why would it be a bad design to feed the test food to some dogs and the safe food to cats?

ANSWER: This would create confounding. We would not be able to tell whether any differences in animals' health were attributable to the food they had eaten or to differences in how the two species responded.

A Two-Factor Example

Confounding can also arise from a badly designed multifactor experiment. Here's a classic. A credit card bank wanted to test the sensitivity of the market to two factors: the annual fee charged for a card and the annual percentage rate charged. Not wanting to scrimp on sample size, the bank selected 100,000 people at random from a mailing list. It sent out 50,000 offers with a low rate and no fee and 50,000 offers with a higher rate and a $50 annual fee. Not surprising, people preferred the low-rate, no-fee card. In fact, they signed up for that card at over twice the rate as the other offer. And because of the large sample size, the bank was able to estimate the difference precisely. But the question the bank really wanted to answer was "how much of the change was due to the rate, and how much was due to the fee?" Unfortunately, there's simply no way to separate out the two effects. If the bank had tested all four treatments—low rate with no fee, low rate with $50 fee, high rate with no fee, and high rate with $50 fee—each to 25,000 people, it could have learned about both factors and could have also seen what happens when the two factors occur in combination.

Lurking and Confounding?

A lurking variable creates an association between two other variables that tempts us to think that one may cause the other. Recall from the example in Chapter 8 that people in countries with more TV sets per capita tend to have longer lives. You shouldn't conclude it's the TVs "causing" longer life. It's more likely that a generally higher standard of living allows people to afford more TVs and get better health care, too. Our data revealed an association between TVs and life expectancy, but economic conditions were a likely lurking variable. A lurking variable, then, is usually thought of as a variable associated with both y and x that makes it appear that x may be causing y.

Confounding and lurking variables are very similar. Imagine an observational study hoping to understand the relationship between herbal supplements and patient health finds that patients who take the supplements report fewer colds. However, if they find from their survey that the patients who take the herbal supplements also tend to take larger doses of Vitamin C, we would say that taking Vitamin C is a confounder of herbal supplements. Had we not asked the question at all, and we later found that taking Vitamin C was more effective in preventing colds than the herbal supplement, we might call Vitamin C a lurking variable in the original study.

Both confounding and lurking variables are outside influences that make it harder to understand the relationship we are modeling. It's important to realize that in any observational study or even in a carefully designed experiment, there may be variables that influence the relationship between that variable and the response other than the ones being studied. You should always be alert for the possible effects of other variables on the coefficients you care about. Be especially wary of variables that you might not have considered.

WHAT CAN GO WRONG?

- **Don't give up just because you can't run an experiment.** Sometimes we can't run an experiment because we can't identify or control the factors. Sometimes it would simply be unethical to run the experiment. (Consider randomly assigning students to take—and be graded in—a statistics course deliberately taught to be boring and difficult or one that had an unlimited budget to use multimedia, real-world examples, and field trips.) If we can't perform an experiment, an observational study may be a good choice.
- **Beware of confounding.** Be aware of variables that may be confounded. In a prospective study, it may be possible to stratify the subjects by levels of one variable. In an experiment, unmeasured confounders will be balanced (on average) by randomization. To include a variable that may be a confounder, it is a good idea to block by the potential confounder to ensure that the levels are balanced. And always think about possible lurking variables that may be influencing the response that aren't in your study as well.
- **Bad things can happen even to good experiments.** Protect yourself by recording additional information. An experiment in which the air conditioning failed for 2 weeks, affecting the results, was saved by recording the temperature (although that was not originally one of the factors) and estimating the effect the higher temperature had on the response.[10]

  It's generally good practice to collect as much information as possible about your experimental units and the circumstances of the experiment. For example, in the nail polish experiment, it would be wise to record details (temperature, humidity) that might affect the durability of the polish on the acrylic nails. Sometimes we can use this extra information during the analysis to reduce biases.
- **Don't spend your entire budget on the first run.** Just as it's a good idea to pretest a survey, it's always wise to try a small pilot experiment before running the full-scale experiment. You may learn, for example, how to choose factor levels more effectively, about effects you forgot to control, and about unanticipated confoundings.

[10]R. D. De Veaux and M. Szelewski, "Optimizing Automatic Splitless Injection Parameters for Gas Chromatographic Environmental Analysis." *Journal of Chromatographic Science* 27, no. 9 (1989): 513–518.

CONNECTIONS

The fundamental role of randomization in experiments clearly points back to our discussions of randomization, to our experiments with simulations, and to our use of randomization in sampling. The similarities and differences between experiments and samples are important to keep in mind and can make each concept clearer.

If you think that blocking in an experiment resembles stratifying in a sample, you're quite right. Both are ways of removing variation we can identify to help us see past the variation in the data.

Experiments compare groups of subjects that have been treated differently. Graphics such as boxplots that help us compare groups are closely related to these ideas. Think about what we look for in a boxplot to tell whether two groups look really different, and you'll be thinking about the same issues as experiment designers.

Generally, we're going to consider how different the mean responses are for different treatment groups. And we're going to judge whether those differences are large by using standard deviations as rulers. (That's why we needed to replicate results for each treatment; we need to be able to estimate those standard deviations.) The discussion in Chapter 2 introduced this fundamental statistical thought, and it's going to keep coming back over and over again. Statistics is about variation.

We'll see a number of ways to analyze results from experiments in subsequent chapters.

CHAPTER REVIEW

Recognize observational studies.

- A retrospective study looks at an outcome in the present and looks for facts in the past that relate to it.
- A prospective study selects subjects and follows them as events unfold.

Know the elements of a designed randomized experiment.

- *Experimental units* (sometimes called *subjects* or *participants*) are assigned at random to *treatments*.
  - The experimenter manipulates *factors*, setting them to specified *levels* to establish the treatments.
- A quantitative *response variable* is measured or observed for each experimental unit.
- We can attribute differences in the response to the differences among the treatments.

State and apply the Four Principles of Experimental Design.

- *Control* sources of variation other than the factors being tested. Make the conditions as similar as possible for all treatment groups except for differences among the treatments.
- *Randomize* the assignment of participants to treatments.
- *Replicate* by applying each treatment to more than one participant.
- *Block* the experiment by grouping together participants who are similar in important ways that you cannot control.

Work with *blinding* and *control groups*.

- A *single-blind* study is one in which *either* all those who can affect the results *or* all those who evaluate the results are kept ignorant of which subjects receive which treatments.
- A *double-blind* study is one in which *both* all who can affect the results *and* all who evaluate the results are ignorant of which subjects receive which treatments.

- A *control group* is assigned to a null treatment or to the best available alternative treatment.
  - Control participants are often administered a *placebo* or null treatment that mimics the treatment being studied but is known to be inactive. This is one way to blind participants.
- Understand the differences between experiments and surveys.
  - Surveys try to estimate facts (parameter) about a population, so they require a representative random sample from that population.
  - Experiments try to estimate the differences in the effects of treatments. They randomize a group of experimental units to treatments, but there is no need for the experimental units to be a representative sample from the population.
- Be alert for possible confounding due to a variable that is not under control affecting the responses differentially.

REVIEW OF TERMS

Observational study A study based on data in which no manipulation of factors has been employed (p. 375).

Retrospective study An observational study in which subjects are selected and then their previous conditions or behaviors are determined. Retrospective studies need not be based on random samples and they usually focus on estimating differences between groups or associations between variables (p. 375).

Prospective study An observational study in which subjects are followed to observe future outcomes. Because no treatments are deliberately applied, a prospective study is not an experiment. Nevertheless, prospective studies typically focus on estimating differences among groups that might appear as the groups are followed during the course of the study (p. 376).

Random assignment To be valid, an experiment must assign experimental units to treatment groups using some form of randomization (p. 376).

Factor A variable whose levels are manipulated by the experimenter. Experiments attempt to discover the effects that differences in factor levels may have on the responses of the experimental units (p. 376).

Response variable A variable whose values are compared across different treatments. In a randomized experiment, large response differences can be attributed to the effect of differences in treatment level (p. 377).

Experiment An experiment *manipulates* factor levels to create treatments, *randomly assigns* subjects to these treatment levels, and then *compares* the responses of the subject groups across treatment levels (p. 377).

Subjects or participants The individuals who participate in an experiment, especially when they are human. A more general term is experimental unit (p. 377).

Experimental units Individuals on whom an experiment is performed. Usually called **subjects** or **participants** when they are human (p. 377).

Levels The specific values that the experimenter chooses for a factor (p. 377).

Treatment The process, intervention, or other controlled circumstance applied to randomly assigned experimental units. Treatments are the different levels of a single factor or are made up of combinations of levels of two or more factors (p. 377).

Principles of experimental design

- **Control** aspects of the experiment that we know may have an effect on the response, but that are not the factors being studied (p. 378).
- **Randomize** subjects to treatments to even out effects that we cannot control (p. 378).
- **Replicate** over as many subjects as possible. Results for a single subject are just anecdotes. If, as often happens, the subjects of the experiment are not a representative sample from the population of interest, replicate the entire study with a different group of subjects, preferably from a different part of the population (pp. 378–379).
- **Block** to reduce the effects of identifiable attributes of the subjects that cannot be controlled (p. 379).

Block When groups of experimental units are similar in a way that is not a factor under study, it is often a good idea to gather them together into blocks and then randomize the assignment of treatments within each block. By blocking in this way, we isolate the variability attributable to the differences between the blocks so that we can see the differences caused by the treatments more clearly (p. 379).

Completely randomized experiment In a completely randomized experiment, all experimental units have an equal chance of receiving any treatment (p. 380).

Effect size The effect of a treatment is the magnitude of the difference in responses. When the effect size is larger it is easier to discern a difference in the treatment groups. (p. 382)

Statistically significant When an observed difference is too large for us to believe that it is likely to have occurred naturally, we consider the difference to be statistically significant. Subsequent chapters will show specific calculations and give rules, but the principle remains the same (p. 382).

Control group
Control treatment The experimental units assigned to a baseline treatment level (called the "control treatment"), typically either the default treatment, which is well understood, or a null, placebo treatment. Their responses provide a basis for comparison (p. 383).

Blinding Any individual associated with an experiment who is not aware of how subjects have been allocated to treatment groups is said to be blinded (p. 383).

Single-blind and double-blind There are two main classes of individuals who can affect the outcome of an experiment:

- Those who could *influence the results* (the subjects, treatment administrators, or technicians).
- Those who *evaluate the results* (judges, treating physicians, etc.).

When every individual in *either* of these classes is blinded, an experiment is said to be single-blind (p. 384). When everyone in *both* classes is blinded, we call the experiment double-blind (p. 384).

Placebo A treatment known to have no effect, administered so that all groups experience the same conditions. Many subjects respond to such a treatment (a response known as a placebo effect). Only by comparing with a placebo can we be sure that the observed effect of a treatment is not due simply to the placebo effect (p. 385).

Placebo effect The tendency of many human subjects (often 20% or more of experiment subjects) to show a response even when administered a placebo (p. 385).

Randomized block design An experiment design in which participants are randomly assigned to treatments within each block (p. 386).

Matching In a retrospective or prospective study, participants who are similar in ways not under study may be matched and then compared with each other on the variables of interest. Matching, like blocking, reduces unwanted variation (p. 386).

Confounding When the levels of one factor are associated with the levels of another factor in such a way that their effects on a response are difficult to disentangle, we say that these two factors are confounded (p. 387).

TECH SUPPORT

Experiments

Most experiments are analyzed with a statistics package. You should almost always display the results of a comparative experiment with side-by-side boxplots. You may also want to display the means and standard deviations of the treatment groups in a table.

The analyses offered by statistics packages for comparative randomized experiments fall under the general heading of Analysis of Variance, usually abbreviated ANOVA.

EXERCISES

SECTION 11.1

1. Steroids The 1990s and early 2000s could be considered the steroids era in Major League Baseball, as many players have admitted to using the drug to increase performance on the field. If a sports writer wanted to compare home run totals from the steroids era to an earlier decade, say the 1960s, explain why this would be an observational study. Could the writer conclude that it was the steroids that caused the increase in home runs? Why or why not?

2. E-commerce A business student conjectures that the Internet caused companies to become more profitable, since many transactions previously handled "face-to-face" could now be completed online. The student compares earnings from a sample of companies from the 1980s to a sample from the 2010s. Explain why this is an observational study. If indeed profitability increased, can she conclude the Internet was the cause? Why or why not?

SECTION 11.2

3. Tips A pizza delivery driver, always trying to increase tips, runs an experiment on his next 40 deliveries. He flips a coin to decide whether or not to call a customer from his mobile phone when he is five minutes away, hoping this slight bump in customer service will lead to a slight bump in tips. After 40 deliveries, he will compare the average tip percentage between the customers he called and those he did not. What are the experimental units and how did he randomize treatments?

4. Tomatoes You want to compare the tastiness and juiciness of tomatoes grown with three amounts of a new fertilizer: none, half the recommended amount, and the full recommended amount. You allocate 24 tomato plants to receive each amount of fertilizer, assigning them at random. What are the experimental units? What is the response variable?

5. Tips II For the experiment described in Exercise 3, list the factor, the levels, and the response variable.

6. Tomatoes II For the experiment described in Exercise 4, name the factor and its levels. How might the response be measured?

SECTION 11.3

7. Tips again For the experiment of Exercise 3, name some variables the driver did or should have controlled. Was the experiment randomized and replicated?

8. Tomatoes again For the experiment of Exercise 4, discuss variables that could be controlled or that could not be controlled. Is the experiment randomized and replicated?

SECTION 11.4

9. More tips Is the experiment of Exercise 3 blind? Can it be double-blind? Explain.

10. More tomatoes If the tomato taster doesn't know how the tomatoes have been treated, is the experiment single- or double-blind? How might the blinding be improved further?

SECTION 11.5

11. Block that tip The driver of Exercise 3 wants to know about tipping in general. So he recruits several other drivers to participate in the experiment. Each driver randomly decides whether to phone customers before delivery and records the tip percentage. Is this experiment blocked? Is that a good idea?

12. Blocking tomatoes To obtain enough plants for the tomato experiment of Exercise 4, experimenters have to purchase plants from two different garden centers. They then randomly assign the plants from each garden center to all three fertilizer treatments. Is the experiment blocked? Is that a good idea?

SECTION 11.6

13. Confounded tips For the experiment of Exercise 3, name some confounding variables that might influence the experiment's results.

14. Tomatoes finis What factors might confound the results of the experiment in Exercise 4?

CHAPTER EXERCISES

15. Standardized test scores For his statistics class experiment, researcher J. Gilbert decided to study how parents' income affects children's performance on standardized tests like the SAT. He proposed to collect information from a random sample of test takers and examine the relationship between parental income and SAT score.

a) Is this an experiment? If not, what kind of study is it?
b) If there is a relationship between parental income and SAT score, why can't we conclude that differences in score are caused by differences in parental income?

16. Heart attacks and height Researchers who examined health records of thousands of males found that men who died of myocardial infarction (heart attack) tended to be shorter than men who did not.

a) Is this an experiment? If not, what kind of study is it?
b) Is it correct to conclude that shorter men are at higher risk of dying from a heart attack? Explain.

17. MS and vitamin D Multiple sclerosis (MS) is an autoimmune disease that strikes more often the farther people live from the equator. Could vitamin D—which most people get from the sun's ultraviolet rays—be a factor? Researchers compared vitamin D levels in blood samples from 190 exchange students who have developed MS with blood samples of nearly 280 who have not. The samples were taken, on average, five years before the disease was diagnosed. Those with the highest blood vitamin D levels had a 67% lower risk of MS than those with the lowest levels.

a) What kind of study was this?
b) Is that an appropriate choice for investigating this problem? Explain.
c) Who were the subjects?
d) What were the variables?

18. Super Bowl commercials When spending large amounts to purchase advertising time, companies want to know what audience they'll reach. In January 2011, a poll asked 1008 adults whether they planned to watch the upcoming Super Bowl. Men and women were asked separately whether they were looking forward more to the football game or to watching the commercials. Among the men, 70% were planning to watch and were looking forward primarily to the game. Among women, 60% were looking forward primarily to the game.

a) Was this a stratified sample or a blocked experiment? Explain.
b) Was the design of the study appropriate for the advertisers' questions?

19. Menopause Researchers studied the herb black cohosh as a treatment for hot flashes caused by menopause. They randomly assigned 393 women aged 45 to 55 who reported at least two hot flashes a day to one of five groups: (1) black cohosh, (2) a multiherb supplement with black cohosh, (3) the multiherb supplement plus advice to consume more soy foods, (4) estrogen replacement therapy, or (5) a placebo. After a year, only the women given estrogen replacement therapy had symptoms different from those of the placebo group.

a) What kind of study was this?
b) Is that an appropriate choice for this problem?
c) Who were the subjects?
d) Identify the treatment and response variables.

20. Honesty Coffee stations in offices often just ask users to leave money in a tray to pay for their coffee, but many people cheat. Researchers at Newcastle University replaced the picture of flowers on the wall behind the coffee station with a picture of staring eyes. They found that the average contribution increased significantly above the well-established standard when people felt they were being watched, even though the eyes were patently not real. (*The New York Times* 12/10/06)

a) Was this a survey, an observational study, or an experiment? How can we tell?
b) Identify the variables.
c) What does "increased significantly" mean in a statistical sense?

21–34. What's the design? *Read each brief report of statistical research, and identify*

a) whether it was an observational study or an experiment.

If it was an observational study, identify (if possible)

b) whether it was retrospective or prospective.
c) the subjects studied, and how they were selected.
d) the parameter of interest.
e) the nature and scope of the conclusion the study can reach.

If it was an experiment, identify (if possible)

b) the subjects studied.
c) the factor(s) in the experiment and the number of levels for each.
d) the number of treatments.
e) the response variable measured.
f) the design (completely randomized, blocked, or matched).
g) whether it was blind (or double-blind).
h) the nature and scope of the conclusion the experiment can reach.

21. Over a five-month period, among 72 people with bipolar disorder, patients who were given a 626 mg/day dose of omega-3 fats from fish oil improved more than those given a placebo.

22. Among a group of disabled women aged 60 and older who were tracked for several years, those who had a vitamin B_{12} deficiency were twice as likely to suffer severe depression as those who did not.

23. In a test of roughly 560 widows, those with moderately high blood pressure (averaging 164/89 mm Hg) did worse on tests of memory and reaction time than those with normal blood pressure.

24. Is diet or exercise effective in combating insomnia? Some believe that cutting out desserts can help alleviate the problem, while others recommend exercise. Forty volunteers suffering from insomnia agreed to participate in a month-long test. Half were randomly assigned to a special no-desserts diet; the others continued desserts as usual. Half of the people in each of these groups were randomly assigned to an exercise program, while the others did not exercise. Those who ate no desserts and engaged in exercise showed the most improvement.

25. There is some concern that if women over the age of 57 who take supplemental hormones also drink alcohol, their hormone levels will drop too low. Twelve volunteers who were receiving supplemental estrogen were randomly divided into two groups, as were 13 other volunteers not on estrogen. In each case, one group drank an alcoholic beverage, the other a nonalcoholic beverage. An hour later, everyone's estrogen level was checked. Only those on supplemental hormones who drank alcohol showed a marked decrease.

26. Researchers have linked an increase in the incidence of breast cancer in Italy to dioxin released by an industrial accident in 1976. The study identified 981 women who lived near the site of the accident and were under age 40 at the time. Fifteen of the women had developed breast cancer at an unusually young average age of 45. Medical records showed that they had heightened concentrations of dioxin in their blood and that each tenfold increase in dioxin level was associated with a doubling of the risk of breast cancer. (*Science News*, Aug. 3, 2002)

27. A popular science magazine reported that a study of women in Finland indicated that having sons shortened the life spans of mothers by about 25 weeks per son, but that daughters helped to lengthen the mothers' lives. The data came from church records from the period 1623 to 1724.

28. Scientists at a major pharmaceutical firm investigated the effectiveness of a new medicine to treat the measles. They exposed each subject to a measles virus, then gave him or her either the new medicine or a sugar solution known to have no effect on measles. Several days later, they assessed the patient's condition, using a severity scale ranging from 0 to 5. They found no evidence of benefits associated with the medicine.

29. A science journal reported that, contrary to popular belief, depressed individuals cry no more often in response to sad situations than nondepressed people. Researchers studied 65 women and 32 men with major depression and 71 men and 35 women with no depression. They showed the subjects a sad film about a boy whose father has died, noting whether or not the subjects cried. Women cried more often than men, but there were no significant differences between the depressed and nondepressed groups.

30. Some people who race greyhounds give the dogs large doses of vitamin C in the belief that the dogs will run faster. Investigators at the University of Florida tried three different diets in random order on each of five racing greyhounds. They were surprised to find that when the dogs ate high amounts of vitamin C they ran more slowly. (*Science News*, July 20, 2002)

31. Some people claim they can get relief from sharp neck pain by drinking a large glass of ice-cold water. Researchers plan to enlist several people who suffer from sharp neck pain in a test. When a participant experiences a sharp neck pain, he or she will take a pill that may be a standard pain reliever or a placebo. Half of each group will also drink ice-cold water whereas the other half will drink regular water. Each participant, equally likely to get either treatment, will then report the level of pain relief they experience.

32. A pet supply company wants to compare a new lower-calorie food with their standard horse food to see if it's effective in helping inactive horses lower blood pressure. They have found several horse owners willing to participate in the trial. The horses have been classified as small (<600 lbs), medium (600 to 1000 lbs), or large (>1000 lbs) breeds, and the company will supply some owners of each size of horse with one of the two foods. The owners have agreed not to feed their horses anything else for a period of 2 months, after which the horses' blood pressure will be checked.

33. Athletes who had suffered hamstring injuries were randomly assigned to one of two exercise programs. Those who engaged in static stretching returned to sports activity in a mean of 50.2 days faster than those assigned to a program of agility and trunk stabilization exercises.

34. Pew Research compared respondents to an ordinary 5-day telephone survey with respondents to a 4-month-long rigorous survey designed to generate the highest possible response rate. They were especially interested in identifying any variables for which those who responded to the ordinary survey were different from those who could be reached only by the rigorous survey.

35. Omega-3 Exercise 21 describes an experiment that showed that high doses of omega-3 fats might be of benefit to people with bipolar disorder. The experiment involved a control group of subjects who received a placebo. Why didn't the experimenters just give everyone the omega-3 fats to see if they improved?

36. Insomnia Exercise 24 describes an experiment showing that exercise helped people sleep better. The experiment involved other groups of subjects who didn't exercise. Why didn't the experimenters just have everyone exercise and see if their ability to sleep improved?

37. Omega-3, revisited Exercises 21 and 35 describe an experiment investigating a dietary approach to treating bipolar disorder. Researchers randomly assigned 30 subjects to two treatment groups, one group taking a high dose of omega-3 fats and the other a placebo.

a) Why was it important to randomize in assigning the subjects to the two groups?
b) What would be the advantages and disadvantages of using 100 subjects instead of 30?

38. Insomnia, again Exercises 24 and 36 describe an experiment investigating the effectiveness of exercise in combating insomnia. Researchers randomly assigned half of the 40 volunteers to an exercise program.

a) Why was it important to randomize in deciding who would exercise?
b) What would be the advantages and disadvantages of using 100 subjects instead of 40?

39. Omega-3, finis Exercises 21, 35, and 37 describe an experiment investigating the effectiveness of omega-3 fats in treating bipolar disorder. Suppose some of the 30 subjects were very active people who walked a lot or got vigorous exercise several times a week, while others tended to be more sedentary, working office jobs and watching a lot of TV. Why might researchers choose to block the subjects by activity level before randomly assigning them to the omega-3 and placebo groups?

40. Insomnia, at last Exercises 24, 36, and 38 describe an experiment investigating the effectiveness of exercise in combating insomnia. Suppose some of the 40 subjects had maintained a healthy weight, but others were quite overweight. Why might researchers choose to block the subjects by weight level before randomly assigning some of each group to the exercise program?

41. Injuries Exercise 33 describes an experiment that studies hamstring injuries. Describe a strategy to randomly assign injured athletes to the two exercise programs. What should be done if one of the athletes says he'd prefer the other program?

42. Tomatoes II Describe a strategy to randomly split 24 tomato plants into the three groups for the completely randomized single-factor experiment of Exercise 4.

43. Shoes A running-shoe manufacturer wants to test the effect of its new sprinting shoe on 100-meter dash times. The company sponsors 5 athletes who are running the 100-meter dash in the 2012 Summer Olympic games. To test the shoe, it has all 5 runners run the 100-meter dash with a competitor's shoe and then again with their new shoe. The company uses the difference in times as the response variable.

a) Suggest some improvements to the design.
b) Why might the shoe manufacturer not be able to generalize the results they find to all runners?

44. Swimsuits A swimsuit manufacturer wants to test the speed of their newly designed suit. The company designs an experiment by having 13 randomly selected Olympic swimmers swim as fast as they can with their old swimsuit first and then swim the same event again with the new, expensive swimsuit. The company will use the difference in times as the response variable. Criticize the experiment and point out some of the problems with generalizing the results.

45. Hamstrings Exercise 33 discussed an experiment to see if the time it took athletes with hamstring injuries to be able to return to sports was different depending on which of two exercise programs they engaged in.

a) Explain why it was important to assign the athletes to the two different treatments randomly.
b) There was no control group consisting of athletes who did not participate in a special exercise program. Explain the advantage of including such a group.
c) How might blinding have been used?
d) One group returned to sports activity in a mean of 37.4 days ($SD = 27.6$ days) and the other in a mean of 22.2 days ($SD = 8.3$ days). Do you think this difference is statistically significant? Explain.

46. Diet and blood pressure An experiment showed that subjects fed the Paleo diet were able to lower their blood pressure by an average of 5.7 points compared to a group fed a "control diet." All meals were prepared by dieticians.

a) Why were the subjects randomly assigned to the diets instead of letting people pick what they wanted to eat?
b) Why were the meals prepared by dieticians?
c) Why did the researchers need the control group? If the Paleo diet group's blood pressure was lower at the end of the experiment than at the beginning, wouldn't that prove the effectiveness of that diet?
d) What additional information would you want to know in order to decide whether an average reduction in blood pressure of 5.7 points was statistically significant?

47. Mozart Will listening to a Mozart piano sonata make you smarter? In a study published in the journal *Psychological Science*, Rauscher, Shaw, and Ky reported that when students were given a spatial reasoning section of a standard IQ test, those who listened to Mozart for 10 minutes improved their scores more than those who simply sat quietly.

a) These researchers said the differences were statistically significant. Explain what that means in context.
b) Steele, Bass, and Crook tried to replicate the original study. In their study, also published in *Psychological Science*, the subjects were 125 college students who participated in the experiment for course credit. Subjects first took the test. Then they were assigned to one of three groups: listening to a Mozart piano sonata, listening to music by Philip Glass, and sitting for 10 minutes in silence. Three days after the treatments, they were retested. Draw a diagram displaying the design of this experiment.
c) These boxplots show the differences in score before and after treatment for the three groups. Did the Mozart group show improvement?

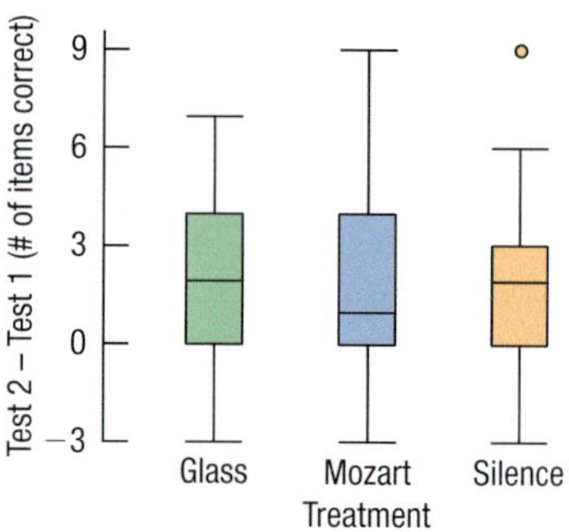

d) Do you think the results prove that listening to Mozart is beneficial? Explain.

48. Contrast baths Contrast bath treatments use the immersion of an injured limb alternately in water of two contrasting temperatures. Those who use the method claim that it can reduce swelling. Researchers compared three treatments: (1) contrast baths and exercise, (2) contrast baths alone, and (3) exercise alone. (R. G. Janssen, D. A. Schwartz, and P. F. Velleman, "A Randomized Controlled Study of Contrast Baths on Patients with Carpal Tunnel Syndrome," *Journal of Hand Therapy*, 2009) They report the following boxplots comparing the change in hand volume after treatment:

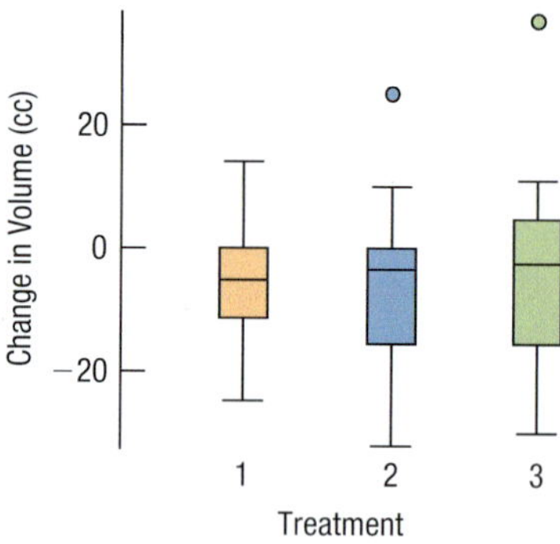

a) The researchers conclude that the differences were not statistically significant. Explain what that means in context.
b) The title says that the study was randomized and controlled. Explain what that probably means for this study.
c) The study did not use a placebo treatment. What was done instead? Do you think that was an appropriate choice? Explain.

49. Wine A study published in certain medical journals casts significant doubt on suggestions that adults who drink wine have higher levels of "good" cholesterol and fewer heart attacks. These researchers followed a group of individuals born at a hospital between 1936 and 1955 for 15 years. Their study found that in this group, the adults who drank wine were richer and better educated than those who did not.

a) What kind of study was this?
b) It is generally true that people with high levels of education and high socioeconomic status are healthier than others. How does this call into question the supposed health benefits of wine?
c) Can studies such as these prove causation (that wine helps prevent heart attacks, that drinking wine makes one richer, that being rich helps prevent heart attacks, etc.)? Explain.

50. Swimming Recently, a group of adults who swim regularly for exercise were evaluated for depression. It turned out that these swimmers were less likely to be depressed than the general population. The researchers said the difference was statistically significant.

a) What does "statistically significant" mean in this context?
b) Is this an experiment or an observational study? Explain.
c) News reports claimed this study proved that swimming can prevent depression. Explain why this conclusion is not justified by the study. Include an example of a possible lurking variable.
d) But perhaps it is true. We wonder if exercise can ward off depression, and whether anaerobic exercise (like weight training) is as effective as aerobic exercise (like swimming). We find 120 volunteers not currently engaged in a regular program of exercise. Design an appropriate experiment.

51. Dowsing Before drilling for water, many rural homeowners hire a dowser (a person who claims to be able to sense the presence of underground water using a forked stick). Suppose we wish to set up an experiment to test one dowser's ability. We get 28 identical containers, fill half of them with water, and ask him to tell which ones they are.

a) How will we randomize this procedure?
b) The dowser correctly identifies the contents of 16 out of 28 containers. Do you think this level of success is statistically significant? Explain.
c) How many correct identifications (out of 28) would the dowser have to make to convince you that the forked-stick trick works? Explain.

52. Healing A medical researcher suspects that giving post-surgical patients large doses of vitamin E will speed their recovery times by helping their incisions heal more quickly. Design an experiment to test this conjecture. Be sure to identify the factors, levels, treatments, response variable, and the role of randomization.

53. Reading Some schools teach reading using phonics (the sounds made by letters) and others using whole language (word recognition). Suppose a school district wants to know which method works better. Suggest a design for an appropriate experiment.

54. Gas mileage Do cars get better gas mileage with premium instead of regular unleaded gasoline? It might be possible to test some engines in a laboratory, but we'd rather use real cars and real drivers in real day-to-day driving, so we get 34 volunteers. Design the experiment.

55. Weekend deaths A study suggests that it's dangerous to enter a hospital on a weekend. During a six-year period, researchers tracked over 1 million emergency admissions to hospitals in a certain region. Their findings revealed that patients admitted on weekends had a much higher risk of death than those who went on weekdays.

a) The researchers said the difference in death rates was "statistically significant." Explain in this context what that means.
b) What kind of study was this? Explain.
c) If you think you're quite ill on a Saturday, should you wait until Monday to seek medical help? Explain.
d) Suggest some possible explanations for this troubling finding.

56. Shingles A research doctor has discovered a new ointment that she believes will be more effective than the current medication used to decrease the severity of shingles and the pain level associated with this skin rash. Eight patients have volunteered to participate in the initial trials of this ointment. You are the statistician hired as a consultant to help design an experiment.

a) Describe how you will conduct this experiment.
b) Suppose the eight patients' last names start with the letters A to H. Using the random numbers listed below, show which patients you will assign to each treatment. Explain your randomization procedure clearly.

41098 18329 78458 31685 55259

c) Can you make this experiment double-blind? How?
d) The initial experiment revealed that males and females may respond differently to the ointment. Further testing of the drug's effectiveness is now planned, and many patients have volunteered. What changes in your first design, if any, would you make for this second stage of testing?

57. Beetles Hoping to learn how to control crop damage by a certain species of beetle, a researcher plans to test two different pesticides in small plots of corn. A few days after application of the chemicals, he'll check the number of beetle larvae found on each plant. The researcher wants to know whether either pesticide works and whether there is a significant difference in effectiveness between them. Design an appropriate experiment.

58. SAT prep Can special study courses actually help raise SAT scores? One organization says that the 50 students they tutored achieved an average gain of 60 points when they retook the test.

a) Explain why this does not necessarily prove that the special course caused the scores to go up.
b) Propose a design for an experiment that could test the effectiveness of the tutorial course.
c) Suppose you suspect that the tutorial course might be more helpful for students whose initial scores were particularly low. How would this affect your proposed design?

59. Safety switch An industrial machine requires an emergency shutoff switch that must be designed so that it can be easily operated with either hand. Design an experiment to find out whether workers will be able to deactivate the machine as quickly with their left hands as with their right hands. Be sure to explain the role of randomization in your design.

60. Washing clothes A consumer group wants to test the effectiveness of a new "organic" laundry detergent and make recommendations to customers about how to best use the product. They intentionally stain 30 white T-shirts with grass in order to see how well the detergent will clean them. They want to try the detergent in cold water and in hot water on both the "regular" and "delicates" wash cycles. Design an appropriate experiment, indicating the number of factors, levels, and treatments. Explain the role of randomization in your experiment.

61. Skydiving, anyone? A humor piece published in the *British Medical Journal* ("Parachute Use to Prevent Death and Major Trauma Related to Gravitational Challenge: Systematic Review of Randomized Control Trials," Gordon, Smith, and Pell, *BMJ*, 2003:327) notes that we can't tell for sure whether parachutes are safe and effective because there has never been a properly randomized, double-blind, placebo-controlled study of parachute effectiveness in skydiving. (Yes, this is the sort of thing statisticians find funny. . . .) Suppose you were designing such a study.

a) What is the factor in this experiment?
b) What experimental units would you propose?[11]
c) What would serve as a placebo for this study?
d) What would the treatments be?
e) What would the response variable be?
f) What sources of variability would you control?
g) How would you randomize this "experiment"?
h) How would you make the experiment double-blind?

JUST CHECKING

Answers

1. a) The factor was type of treatment for peptic ulcer.
   b) The response variable could be a measure of relief from gastric ulcer pain or an evaluation by a physician of the state of the disease.
   c) Treatments would be gastric freezing and some alternative control treatment.
   d) Treatments should be assigned randomly.
   e) No. The website reports "lack of effectiveness," indicating that no large differences in patient healing were noted.
2. a) Neither the patients who received the treatment nor the doctor who evaluated them to see if they had improved knew what treatment they had received.
   b) The placebo is needed to accomplish blinding. The best alternative would be using body-temperature liquid rather than the freezing liquid.
   c) The researchers should block the subjects by the length of time they had had the ulcer, then randomly assign subjects in each block to the freezing and placebo groups.

[11]Don't include your statistics instructor!

Review of Part III

GATHERING DATA

Quick Review

Before you can make a boxplot, calculate a mean, describe a distribution, or fit a line, you must have meaningful data to work with. Getting good data is essential to any investigation. No amount of clever analysis can make up for badly collected data. Here's a brief summary of the key concepts and skills:

- The way you gather data depends both on what you want to discover and on what is practical.
- To answer questions about a target population, collect information from a sample with a **survey** or poll.
  - Choose the sample randomly. Random sampling designs include simple, stratified, systematic, cluster, and multistage.
  - A simple random sample draws without restriction from the entire target population.
  - When there are subgroups within the population that may respond differently, use a stratified sample.
  - Avoid bias, a systematic distortion of the results. Sample designs that allow undercoverage or response bias and designs such as voluntary response or convenience samples don't faithfully represent the population.
  - Samples will naturally vary one from another. This sample-to-sample variation is called sampling error. Each sample only approximates the target population.
- **Observational studies** collect information from a sample drawn from a target population.
  - Retrospective studies examine existing data. Prospective studies identify subjects in advance, then follow them to collect data as the data are created, perhaps over many years.
  - Observational studies can spot associations between variables but cannot establish cause and effect. It's impossible to eliminate the possibility of lurking or confounding variables.
- To see how different treatments influence a response variable, design an **experiment**.
  - Assign subjects to treatments randomly. If you don't assign treatments randomly, your experiment is not likely to yield valid results.
  - Control known sources of variation as much as possible. Reduce variation that cannot be controlled by using blocking, if possible.
  - Replicate the experiment, assigning several subjects to each treatment level.
  - If possible, replicate the entire experiment with an entirely different collection of subjects.
  - A well-designed experiment can provide evidence that changes in the factors cause changes in the response variable.

Now for more opportunities to review these concepts and skills . . .

REVIEW EXERCISES

R3.1–R3.18. What design? *Analyze the design of each research example reported. Is it a sample survey, an observational study, or an experiment? If a sample, what are the population, the parameter of interest, and the sampling procedure? If an observational study, was it retrospective or prospective? If an experiment, describe the factors, treatments, randomization, response variable, and any blocking, matching, or blinding that may be present. In each, what kind of conclusions can be reached?*

R3.1. Researchers identified 242 children in the Cleveland area who had been born prematurely (at about 29 weeks). They examined these children at age 8 and again at age 20, comparing them to another group of 233 children not born prematurely. Their report, published in the *New England Journal of Medicine*, said the "preemies" engaged in significantly less risky behavior than the others. Differences showed up in the use of alcohol and marijuana, conviction of crimes, and teenage pregnancy.

R3.2. The journal *Circulation* reported that among 1900 people who had heart attacks, those who drank an average of 19 cups of tea a week were 44% more likely than nondrinkers to survive at least 3 years after the attack.

R3.3. Researchers at the Purina Pet Institute studied Labrador retrievers for evidence of a relationship between diet and longevity. At 8 weeks of age, 2 puppies of the same sex and weight were randomly assigned to one of two groups—a total of 48 dogs in all. One group was allowed to eat all they wanted, while the other group was fed a diet about 25% lower in calories. The median life span of dogs fed the restricted diet was 22 months longer than that of other dogs. (*Science News* 161, no. 19)

R3.4. The radioactive gas radon, found in some homes, poses a health risk to residents. To assess the level of contamination in their area, a county health department wants to test a few homes. If the risk seems high, they will publicize the results to emphasize the need for home testing. Officials plan to use the local property tax list to randomly choose 25 homes from various areas of the county.

R3.5. Data were collected over a decade from 1021 men and women with a recent history of precancerous colon polyps. Participants were randomly assigned to receive folic acid (a B vitamin) or a placebo, and the study concluded that those receiving the folic acid may actually increase their risk of developing additional precancerous growths. Previous studies suggested that taking folic acid may help to prevent colorectal cancer. (*JAMA* 2007, 297)

R3.6. In the journal *Science*, a research team reported that plants in southern England are flowering earlier in the spring. Records of the first flowering dates for 385 species over a period of 47 years indicate that flowering has advanced an average of 15 days per decade, an indication of climate warming, according to the authors.

R3.7. Fireworks manufacturers face a dilemma. They must be sure that the rockets work properly, but test-firing a rocket essentially destroys it. On the other hand, not testing the product leaves open the danger that they sell a bunch of duds, leading to unhappy customers and loss of future sales. The solution, of course, is to test a few of the rockets produced each day, assuming that if those tested work properly, the others are ready for sale.

R3.8. People who read the last page of a mystery novel first generally like stories better. Researchers recruited 819 college students to read short stories, and for one story, they were given a spoiler paragraph beforehand. On the second and third story, the spoiler was incorporated as the opening paragraph or not given at all. Overall, participants liked the stories best after first reading spoilers. (*Psychological Science*, August 12, 2011)

R3.9. Does keeping a child's lunch in an insulated bag, even with ice packs, protect the food from warming to temperatures where germs can proliferate? Researchers used an electric temperature gun on 235 lunches at preschools 90 minutes before they were to be eaten. Of the lunches with ice packs, over 90% of them were at unsafe temperatures. The study was of particular interest because preschoolers develop up to four times as many foodborne infections as do adults. (*Science News*, August 9, 2011)

R3.10. Some doctors have expressed concern that men who have vasectomies seemed more likely to develop prostate cancer. Medical researchers used a national cancer registry to identify 923 men who had had prostate cancer and 1224 men of similar ages who had not. Roughly one quarter of the men in each group had undergone a vasectomy, many more than 25 years before the study. The study's authors concluded that there is strong evidence that having the operation presents no long-term risk for developing prostate cancer. (*Science News*, July 20, 2002)

R3.11. Widely used antidepressants may reduce ominous brain plaques associated with Alzheimer's disease. In the study, mice genetically engineered to have large amounts of brain plaque were given a class of antidepressants that boost serotonin in the brain. After a single dose, the plaque levels dropped, and after four months, the mice had about half the brain plaques as the mice that didn't take the drug. (*Proceedings of the National Academy of Sciences*, August 22, 2011)

R3.12. An artisan wants to create pottery that has the appearance of age. He prepares several samples of clay with four different glazes and test fires them in a kiln at three different temperature settings.

R3.13. Tests of gene therapy on laboratory rats have raised hopes of stopping the degeneration of tissue that characterizes chronic heart failure. Researchers at the University of California, San Diego, used hamsters with cardiac disease, randomly assigning 30 to receive the gene therapy and leaving the other 28 untreated. Five weeks after treatment the gene therapy group's heart muscles stabilized, while those of the untreated hamsters continued to weaken. (*Science News*, July 27, 2002)

R3.14. People aged 50 to 71 were initially contacted in the mid-1990s to participate in a study about smoking and bladder cancer. Data were collected from more than 280,000 men and 186,000 women from eight states who answered questions about their health, smoking history, alcohol intake, diet, physical activity, and other lifestyle factors. When the study ended in 2006, about half the bladder cancer cases in adults age 50 and older were traceable to smoking. (*Journal of the American Medical Association*, August 17, 2011)

R3.15. An orange-juice processing plant will accept a shipment of fruit only after several hundred oranges selected from various locations within the truck are carefully inspected. If too many show signs of unsuitability for juice (bruised, rotten, unripe, etc.), the whole truckload is rejected.

R3.16. A soft-drink manufacturer must be sure the bottle caps on the soda are fully sealed and will not come off easily. Inspectors pull a few bottles off the production line at regular intervals and test the caps. If they detect any problems, they will stop the bottling process to adjust or repair the machine that caps the bottles.

R3.17. Older Americans with a college education are significantly more likely to be emotionally well-off than are people in this age group with less education. Among those aged 65 and older, 35% scored 90 or above on the Emotional Health Index, but for those with a college degree, the percentage rose to 43% (post-graduate degree, 46%). The results are based on phone interviews conducted between January 2010 and July 2011. (gallup.com, August 19, 2011)

R3.18. Does the use of computer software in introductory statistics classes lead to better understanding of the concepts? A professor teaching two sections of statistics decides to investigate. She teaches both sections using the same lectures and assignments, but gives one class statistics software to help them with their homework. The classes take the same final exam, and graders do not know which students used computers during the semester. The professor is also concerned that students who have had calculus may perform differently from those who have not, so she plans to compare software vs. no-software scores separately for these two groups of students.

R3.19. Commuter sample The data file **Commuter sample** holds a sample drawn from the **Population commute times** data set. Using your statistics program make histograms of the sample and the population. Discuss how they are similar and how they differ.

R3.20. Samples of bridges Use the statistics package of your choice or the simple sample tool at astools.datadesk.com to draw samples of the conditions from the **New York bridges 2016** data file. Draw a sample of 50, a sample of 100, a sample of 200, and a sample of 500. Compare the histograms of the samples to a histogram of the full data. Write a brief report on what you conclude. What attributes of the full data distribution can you see in the samples? Which ones are harder to see?

R3.21. Alternate-day fasting A paper published in 2017 in *JAMA Internal Medicine* (jamanetwork.com/journals/jamainternalmedicine/fullarticle/2623528) reported on a study of alternate-day fasting as a weight-loss method. One hundred obese persons were assigned at random to one of three groups: an alternate-day fasting group, a calorie restrictive group, and a control. The alternate-day fasting group alternately consumed 25% of their usual caloric intake during lunch on fasting days and 125% on the alternating days. The calorie restrictive group consumed 75% of baseline energy over three meals each day. The control group ate as usual. The study reports that there was essentially no difference in weight loss between the alternate-day fasting group and the calorie-restricted group, both losing an average of 6.8% of their weight. From this description, identify:

a) The participants.
b) The treatments.
c) The response.
d) Was the study blinded? If not, should it have been blinded?
e) The participants were not a random sample from the population. Is that a problem for this study?

R3.22. Cell phone risks Researchers at the Washington University School of Medicine randomly placed 480 rats into one of three chambers containing radio antennas. One group was exposed to digital cell phone radio waves, the second to analog cell phone waves, and the third group to no radio waves. Two years later, the rats were examined for signs of brain tumors. In June 2002, the scientists said that differences among the three groups were not statistically significant.

a) Is this a study or an experiment? Explain.
b) Explain in this context what "not statistically significant" means.
c) Comment on the fact that this research was funded by Motorola, a manufacturer of cell phones.

R3.23. Tips In restaurants, servers rely on tips as a major source of income. Does serving candy after the meal produce larger tips? To find out, two waiters determined randomly whether or not to give candy to 92 dining parties. They recorded the sizes of the tips and reported that guests getting candy tipped an average of 17.8% of the bill, compared with an average tip of only 15.1% from those who got no candy. ("Sweetening the Till: The Use of Candy to Increase Restaurant Tipping," *Journal of Applied Social Psychology* 32, no. 2 [2002]: 300–309)

a) Was this an experiment or an observational study? Explain.
b) Is it reasonable to conclude that the candy caused guests to tip more? Explain.
c) The researchers said the difference was statistically significant. Explain in this context what that means.

R3.24. Tips, take 2 In another experiment to see if getting candy after a meal would induce customers to leave a bigger tip, a waitress randomly decided what to do with 80 dining parties. Some parties received no candy, some just one piece, and some two pieces. Others initially got just one piece of candy, and then the waitress suggested that they take another piece. She recorded the tips received, finding that, in general, the more candy, the higher the tip, but the highest tips (23%) came from the parties who got one piece and then were offered more. ("Sweetening the Till: The Use of Candy to Increase Restaurant Tipping," *Journal of Applied Social Psychology* 32, no. 2 [2002]: 300–309)

a) Diagram this experiment.
b) How many factors are there? How many levels?
c) How many treatments are there?
d) What is the response variable?
e) Did this experiment involve blinding? Double-blinding?
f) In what way might the waitress, perhaps unintentionally, have biased the results?

R3.25. Timing In August 2011, a Sodahead.com voluntary response poll asked site visitors, "Obama is on Vacation Again: Does He Have the Worst Timing Ever?" 56% of the 629 votes were for "Yes." During the week of the poll, a 5.8 earthquake struck near Washington, D.C., and Hurricane Irene made its way up the East Coast. What types of bias may be present in the results of the poll?

R3.26. Laundry An experiment to test a new laundry detergent, SparkleKleen, is being conducted by a consumer advocate group. They would like to compare its performance with that of a laboratory standard detergent they have used in previous experiments. They can stain 16 swatches of cloth with 2 tsp of a common staining compound and then use a well-calibrated optical scanner to detect the amount of the stain left after washing. To save time in the experiment, several suggestions have been made. Comment on the possible merits and drawbacks of each one.

a) Since data for the laboratory standard detergent are already available from previous experiments, for this experiment wash all 16 swatches with SparkleKleen, and compare the results with the previous data.
b) Use both detergents with eight separate runs each, but to save time, use only a 10-second wash time with very hot water.
c) To ease bookkeeping, first run all of the standard detergent washes on eight swatches, then run all of the SparkleKleen washes on the other eight swatches.
d) Rather than run the experiment, use data from the company that produced SparkleKleen, and compare them with past data from the standard detergent.

R3.27. How long is 30 seconds? Sofie, Ryan, and Alessandra wanted to design an experiment to find out how distraction affects our ability to judge time. The experiment consisted of starting a clock (out of view of the subject) and then asking the subject to tell them when they thought 30 seconds had passed. For each of four subjects, they repeated the experiment eight times under different treatment conditions: eyes open or closed, having music on or off, and sitting or moving around.

a) Identify the factors and the factor levels.
b) What kind of variable is *subject*?
c) How many runs did they perform?
d) They suspect that subjects may do slightly better the more times they perform the task. Ryan argues that randomizing the run order but having all four subjects use the same run order is a good idea. Sofie insists that they should randomize the run order for all four subjects. What do you think? Explain briefly.

R3.28. Cookies Mary Beth, Nigel, and Molly want to design an experiment to find the recipe for the best chocolate chip cookies. They will try to keep the size of the cookies the same, but use cooking times of 10 and 15 minutes. They will use three different temperatures: 325° F, 375° F and 425° F and use either 5 or 10 chips in each cookie. Six of their friends will taste the cookies and rank them in order.

a) What are the factors and levels?
b) What does blinding mean in the context of this experiment? Is it double-blinded?
c) How many different cookies will each judge taste?
d) What kind of variable is the response variable?
e) What are some of the challenges of carrying out the experiment?

R3.29. Homecoming A college statistics class conducted a survey concerning community attitudes about the college's large homecoming celebration. That survey drew its sample in the following manner: Telephone numbers were generated at random by selecting one of the local telephone exchanges (first three digits) at random and then generating a random four-digit number to follow the exchange. If a person answered the phone and the call was to a residence, then that person was taken to be the subject for interview. (Undergraduate students and those under voting age were excluded, as was anyone who could not speak English.) Calls were placed until a sample of 200 eligible respondents had been reached.

a) Did every telephone number that could occur in that community have an equal chance of being generated?
b) Did this method of generating telephone numbers result in a Simple Random Sample (SRS) of local residences? Explain.
c) Did this method generate an SRS of local voters? Explain.
d) Is this method unbiased in generating samples of households? Explain.

R3.30. Youthful appearance *Readers' Digest* (April 2002, p. 152) reported results of several surveys that asked graduate students to examine photographs of men and women and try to guess their ages. Researchers compared these guesses with the number of times the people in the pictures reported having sexual intercourse. It turned out that those who had been more sexually active were judged as looking younger, and that the difference was described as "statistically significant." Psychologist David Weeks, who compiled the research, speculated that lovemaking boosts hormones that "reduce fatty tissue and increase lean muscle, giving a more youthful appearance."

a) What does "statistically significant" mean in this context?
b) Explain in statistical terms why you might be skeptical about Dr. Weeks's conclusion. Propose an alternative explanation for these results.

R3.31. Smoking and Alzheimer's Medical studies indicate that smokers are less likely to develop Alzheimer's disease than people who never smoked.

a) Does this prove that smoking may offer some protection against Alzheimer's? Explain.
b) Offer an alternative explanation for this association.
c) How would you conduct a study to investigate this?

R3.32. Antacids A researcher wants to compare the performance of three types of antacid in volunteers suffering from acid reflux disease. Because men and women may react differently to this medication, the subjects are split into two groups, by sex. Subjects in each group are randomly assigned to take one of the antacids or to take a sugar pill made to look the same. The subjects will rate their level of discomfort 30 minutes after eating.

a) What kind of design is this?
b) The experiment uses volunteers rather than a random sample of all people suffering from acid reflux disease. Does this make the results invalid? Explain.
c) How may the use of the placebo confound this experiment? Explain.

R3.33. Sex and violence Does the content of a television program affect viewers' memory of the products advertised in commercials? Design an experiment to compare the ability of viewers to recall brand names of items featured in commercials during programs with violent content, sexual content, or neutral content.

R3.34. Pubs In England, a Leeds University researcher said that the local watering hole's welcoming atmosphere helps men get rid of the stresses of modern life and is vital for their psychological well-being. Author of the report, Dr. Colin Gill, said rather than complain, women should encourage men to "pop out for a swift half." "Pub-time allows men to bond with friends and colleagues," he said. "Men need break-out time as much as women and are mentally healthier for it." Gill added that men might feel unfulfilled or empty if they had not been to the pub for a week. The report, commissioned by alcohol-free beer brand Kaliber, surveyed 900 men on their reasons for going to the pub. More than 40% said they went for the conversation, with relaxation and a friendly atmosphere being the other most common reasons. Only 1 in 10 listed alcohol as the overriding reason.

Let's examine this news story from a statistical perspective.

a) What are the W's: *Who, What, When, Where, Why, How*?
b) What population does the researcher think the study applies to?

c) What is the most important thing about the selection process that the article does *not* tell us?
d) How do *you* think the 900 respondents were selected? (Name a method of drawing a sample that is likely to have been used.)
e) Do you think the report that only 10% of respondents listed alcohol as an important reason for going to the pub might be a biased result? Why?

R3.35. Age and party 2008 The Pew Research Center collected data from national exits polls conducted by *NBC News* after the 2008 presidential election. The following table shows information regarding voter age and party preference:

	Republican	Democrat	Other	Total
18–29	260	390	351	1001
30–44	320	379	300	999
45–64	329	369	300	998
65+	361	392	251	1004
Total	1270	1530	1202	4002

a) What sampling strategy do you think the pollsters used? Explain.
b) What percentage of the people surveyed were Democrats?
c) Do you think this is a good estimate of the percentage of voters in the United States who are registered Democrats? Why or why not?
d) In creating this sample design, what question do you think the pollsters were trying to answer?

R3.36. Bias? Political analyst Michael Barone has written that "conservatives are more likely than others to refuse to respond to polls, particularly those polls taken by media outlets that conservatives consider biased" (*The Weekly Standard*, March 10, 1997). The Pew Research Foundation tested this assertion by asking the same questions in a national survey run by standard methods and in a more rigorous survey that was a true SRS with careful follow-up to encourage participation. The response rate in the "standard survey" was 42%. The response rate in the "rigorous survey" was 71%.

a) What kind of bias does Barone claim may exist in polls?
b) What is the population for these surveys?
c) On the question of political position, the Pew researchers report the following table:

	Standard Survey	Rigorous Survey
Conservative	37%	35%
Moderate	40%	41%
Liberal	19%	20%

What makes you think these results are incomplete?
d) The Pew researchers report that differences between opinions expressed on the two surveys were not statistically significant. Explain what "not statistically significant" means in this context.

R3.37. Save the grapes Vineyard owners have problems with birds that like to eat the ripening grapes. Some vineyards use scarecrows to try to keep birds away. Others use netting that covers the plants. Owners really would like to know if either method works and, if so, which one is better. One owner has offered to let you use his vineyard this year for an experiment. Propose a design. Carefully indicate how you would set up the experiment, specifying the factor(s) and response variable.

R3.38. Bats It's generally believed that baseball players can hit the ball farther with aluminum bats than with the traditional wooden ones. Is that true? And, if so, how much farther? Players on your local high school baseball team have agreed to help you find out. Design an appropriate experiment.

R3.39. Acupuncture Research reported in 2008 brings to light the effectiveness of treating chronic lower back pain with different methods. One-third of nearly 1200 volunteers were administered conventional treatment (drugs, physical therapy, and exercise). The remaining patients got 30-minute acupuncture sessions. Half of these patients were punctured at sites suspected of being useful and half received needles at other spots on their bodies. Comparable shares of each acupuncture group, roughly 45%, reported decreased back pain for at least six months after their sessions ended. This was almost twice as high as those receiving the conventional therapy, leading the researchers to conclude that results were statistically significant.

a) Why did the researchers feel it was necessary to have some of the patients undergo a "fake" acupuncture?
b) Because patients had to consent to participate in this experiment, the subjects were essentially self-selected—a kind of voluntary response group. Explain why that does not invalidate the findings of the experiment.
c) What does "statistically significant" mean in the context of this experiment?

R3.40. Fuel efficiency Wayne Collier designed an experiment to measure the fuel efficiency of his family car under different tire pressures. For each run, he set the tire pressure to either 28 or 32 psi and then measured the miles driven on a highway (I-95 between Mills River and Pisgah Forest, NC) until he ran out of fuel using 2-liters of fuel each time. He also used two different types of gasoline (regular and premium). To run the experiment he made some alterations to the normal flow of gasoline to the engine. In Wayne's words, "I inserted a T-junction into the fuel line just before the fuel filter, and a line into the passenger compartment of my car, where it joined with graduated 2-liter Rubbermaid bottle that I mounted in a box where the passenger seat is normally fastened. Then I sealed off the fuel-return line, which under normal operation sends excess fuel from the fuel pump back to the fuel tank."

a) Identify the factors and the levels in the experiment.
b) How many times would he need to drive to make sure all treatments are represented and each treatment combination is replicated?
c) For simplicity, he wants to run all the regular gasoline runs first. Explain why this might not be a good idea. What would you suggest?
d) If he wanted to enlist a friend to use his car to double the number or runs by replicating all treatments, what kind of new factor would he be introducing?

R3.41. Security There are 20 first-class passengers and 120 coach passengers scheduled on a flight. In addition to the usual security screening, 10% of the passengers will be subjected to a more complete search.

a) Describe a sampling strategy to randomly select those to be searched.

b) Here is the first-class passenger list and a set of random digits. Select two passengers to be searched, carefully demonstrating your process.

65436 71127 04879 41516 20451 02227 94769 23593

Bergman	Cox	Fontana	Perl
Bowman	DeLara	Forester	Rabkin
Burkhauser	Delli-Bovi	Frongillo	Roufaiel
Castillo	Dugan	Furnas	Swafford
Clancy	Febo	LePage	Testut

c) Explain how you would use a random number table to select the coach passengers to be searched.

R3.42. Internet speed Carsten, Matt, and Rainer designed an experiment to see how different environments affect the Internet speed around campus. They used their own Mac computer and a PC belonging to the school and tested each in two different libraries, the main and the science library. Other factors included Time of Week (weekday or weekend), Time of Day (before or after 5 PM), and how busy the computer was (running other jobs or not). The measured the time it took to download a 50 Mb file from the school course management system.

a) Name the factors and levels.

b) How many runs will they need to run all treatment levels?

c) In what order should they perform all the runs?

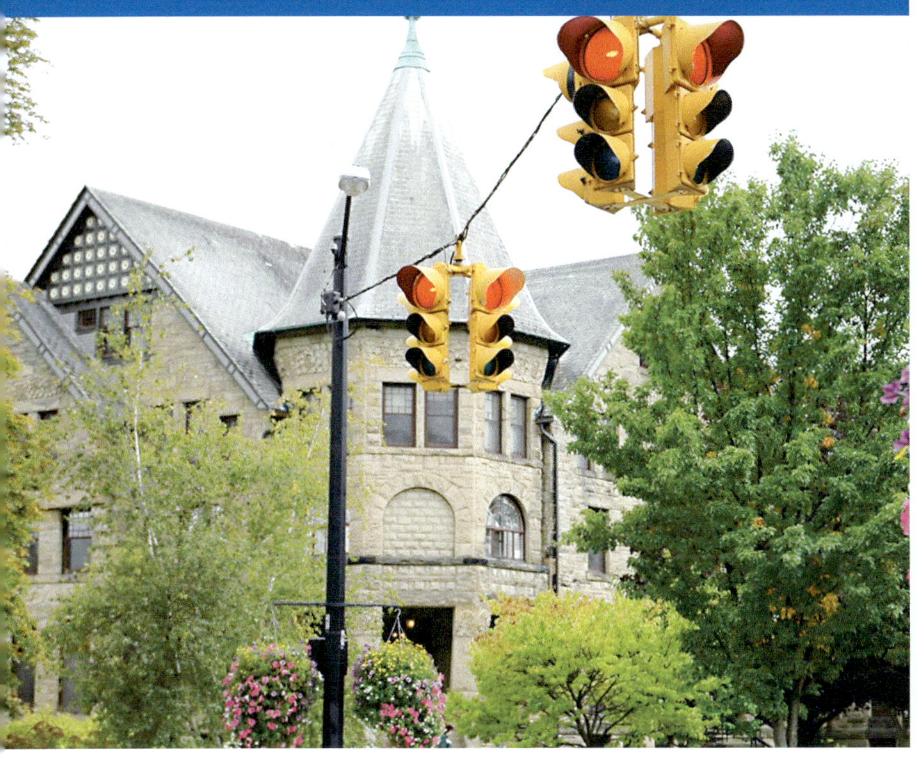

12

From Randomness to Probability

WHERE ARE WE GOING?

Flip a coin. Can you predict the outcome? It's hard to guess the outcome of just one flip because the outcome is random. If you flip it over and over, though, you can predict the *proportion* of heads you're likely to see in the long run.

It's this long-term predictability of randomness that we'll use throughout the rest of the text. To do that, we'll need to talk about the probability of different outcomes and learn some rules for dealing with them.

Early humans saw a world filled with random events. To help them make sense of the chaos around them, they sought out seers, consulted oracles, and read tea leaves. As science developed, we learned to recognize some events as predictable. We can now forecast the change of seasons, tell when eclipses will occur precisely, and even make a reasonably good guess at how warm it will be tomorrow. But many other events are still essentially random. Will the stock market go up or down today? When will the next car pass this corner?

But we have also learned to understand randomness. The surprising fact is that in the long run, many truly random phenomena settle down in a way that's consistent and predictable. It's this property of random phenomena that makes the next steps we're about to take in statistics possible. The previous two chapters showed that randomness plays a critical role in gathering data. That fact alone makes it important to understand how random events behave. From here on, randomness will be fundamental to how we think about data.

12.1 Random Phenomena

Every day you drive through the intersection at College and Main. Even though it may seem that the light is never green when you get there, you know this can't really be true. In fact, if you try really hard, you can recall just sailing through the green light once in a while.

What's random here? The light itself is governed by a timer. Its pattern isn't haphazard. In fact, the light may even be red at precisely the same times each day. It's the time you arrive at the light that is *random*. It isn't that your driving is erratic. Even if you try to leave your house at exactly the same time every day, whether the light is red or green as *you* reach the intersection is a random phenomenon.[1] For us, a **random phenomenon** is a

[1]If you somehow managed to leave your house at *precisely* the same time every day and there was *no* variation in the time it took you to get to the light, then there wouldn't be any randomness, but that's not very realistic.

situation in which we know what outcomes can possibly occur, but we don't know which particular outcome will happen. Even though the color of the light is random[2] as you approach it, some *fraction* of the time, the light will be green. How can you figure out what that fraction is?

You might record what happens at the intersection each day and graph the *accumulated percentage* of green lights like this:

Figure 12.1
The accumulated percentage of times the light is green settles down as you see more outcomes.

Day	Light	% Green
1	Green	100
2	Red	50
3	Green	66.7
4	Green	75
5	Red	60
6	Red	50
⋮	⋮	⋮

The first day you recorded the light, it was green. Then on the next five days, it was red, then green again, then green, red, and red. When you plot the percentage of green lights against days, the graph starts at 100% (because the first time, the light was green, so 1 out of 1, for 100%). Then the next day it was red, so the accumulated percentage drops to 50% (1 out of 2). The third day it was green again (2 out of 3, or 67% green), then green (3 out of 4, or 75%), then red twice in a row (3 out of 5, for 60% green, and then 3 out of 6, for 50%), and so on. As you collect a new data value for each day, each new outcome becomes a smaller and smaller fraction of the accumulated experience, so, in the long run, the graph settles down. As it settles down, you can see that, in fact, the light is green about 35% of the time.

Trials, Outcomes, and Events

A phenomenon consists of trials. Each trial has an outcome. Outcomes combine to make events.

In general, each occasion upon which we observe a random phenomenon is called a **trial**. At each trial, we note the value of the random phenomenon, and call that the trial's **outcome**. (If this language reminds you of Chapter 10, that's *not* unintentional.)

For the traffic light, there are really three possible outcomes: red, yellow, or green. Often we're more interested in a combination of outcomes rather than in the individual ones. When you see the light turn yellow, what do *you* do? If you race through the intersection, then you treat the yellow more like a green light. If you step on the brakes, you treat it more like a red light. Either way, you might want to group the yellow with one or the other. When we combine outcomes like that, the resulting combination is an **event**.[3] We call the collection of *all possible outcomes* the **sample space**.[4] We'll denote the sample space **S**. (Some books are even fancier and use the Greek letter Ω.) For the traffic light, **S** = {red, green, yellow}. If you flip a coin once, the sample space is very simple: **S** = {H, T}. If you flip two coins, it's more complicated because now there are four outcomes, so S = {HH, HT, TH, TT}. If *ABC News* takes a sample of 1023 randomly chosen U.S. adults for a poll, the sample space is incomprehensibly enormous because it would list every combination of 1023 adults you could take from the approximately 250 million adults in the United States.

The Sample Space

For a random phenomenon, the sample space, S, is the set of all possible outcomes of each trial.

[2]Even though the randomness here comes from the uncertainly in our arrival time, we can think of the light itself as showing a color at random.

[3]Each individual outcome is also an event.

[4]Mathematicians like to use the term "space" as a fancy name for a set. Sort of like referring to that closet colleges call a dorm room as "living space." But it's really just the set of all outcomes.

> "For even the most stupid of men . . . is convinced that the more observations have been made, the less danger there is of wandering from one's goal."
>
> —Jacob Bernoulli, 1713, first to prove the LLN

Although he could have said it much more gently, Bernoulli is trying to tell us how intuitive the LLN actually is, even though it wasn't formally proved until the 18th century. Most of us would guess that the law is true from our everyday experiences.

The Law of Large Numbers

What's the *probability* of a green light at College and Main? Based on the graph, it looks like the relative frequency of green lights settles down to about 35%, so saying that the probability is about 0.35 seems like a reasonable answer. But do random phenomena always behave well enough for this to make sense? Might the relative frequency of an event bounce back and forth between two values forever, never settling on just one number?

Fortunately, a principle called the **Law of Large Numbers** (LLN) gives us the guarantee we need. The LLN says that as we repeat a random process over and over, the proportion of times that an event occurs does settle down to one number. We call this number the **probability** of the event. But the Law of Large Numbers requires two key assumptions. First, the random phenomenon we're studying must not change—the outcomes must have the same probabilities for each trial. And, the events must be **independent**.[5] Informally, independence means that the outcome of one trial doesn't affect the outcomes of the others. (We'll see a formal definition of independent events in the next chapter.) The LLN says that as the number of independent trials increases, the long-run *relative frequency* of repeated events gets closer and closer to a single value.

Because the LLN guarantees that relative frequencies settle down in the long run, we can give a name to the value that they approach. We call it the probability of the event. If the relative frequency of green lights at that intersection settles down to 35% in the long run, we say that the probability of encountering a green light is 0.35, and we write $P(\text{green}) = 0.35$. Because this definition is based on repeatedly observing the event's outcome, this definition of probability is often called **empirical probability**.

Empirical Probability

For any event A,

$$P(\mathbf{A}) = \frac{\#\text{ times } \mathbf{A} \text{ occurs}}{\text{total } \#\text{ of trials}}$$

in the long run.

The Nonexistent Law of Averages

Even though the LLN seems natural, it is often misunderstood because the idea of the *long run* is hard to grasp. Many people believe, for example, that an outcome of a random event that hasn't occurred in many trials is "due" to occur. Many gamblers bet on numbers that haven't been seen for a while, mistakenly believing that they're likely to come up sooner. A common term for this is the "Law of Averages." After all, we know that in the long run, the relative frequency will settle down to the probability of that outcome, so now we have some "catching up" to do, right?

Wrong. The Law of Large Numbers says nothing about short-run behavior. Relative frequencies even out *only in the long run*. And, according to the LLN, the long run is *really* long (*infinitely* long, in fact).

The so-called Law of Averages doesn't exist at all. But you'll hear people talk about it as if it does. Is a good hitter in baseball who has struck out the last six times *due* for a hit his next time up? If you've been doing particularly well in weekly quizzes in statistics class, are you *due* for a bad grade? No. This isn't the way random phenomena work. There is *no* Law of Averages for short runs.

The Law of Averages

Don't let yourself think that there's a Law of Averages that promises short-term compensation for recent deviations from expected behavior. A belief in such a "Law" can lead to money lost in gambling and to poor business decisions.

EXAMPLE 12.1 Coins and The Law of Averages

You've just flipped a fair coin and seen six heads in a row.

QUESTION: Does the coin "owe" you some tails? Suppose you spend that coin and your friend gets it in change. When she starts flipping the coin, should she expect a run of tails?

ANSWER: Of course not. Each flip is a new event. The coin can't "remember" what it did in the past, so it can't "owe" any particular outcomes in the future.

---

[5] There are stronger forms of the Law that don't require independence, but for our purposes, this form is general enough.

THE LAW OF AVERAGES IN EVERYDAY LIFE

"Dear Abby: My husband and I just had our eighth child. Another girl, and I am really one disappointed woman. I suppose I should thank God she was healthy, but, Abby, this one was supposed to have been a boy. Even the doctor told me that the law of averages was in our favor 100 to one." (Excerpt from "The Law of Averages" by Abigail Van Buren from The Nature of Mathematics, by Karl J. Smith. Published by Brooks/Cole, Cengage Learning, © 1991.)

Just to see how this works in practice, the authors ran a simulation of 100,000 flips of a fair coin. We collected 100,000 random numbers, letting the numbers 0 to 4 represent heads and the numbers 5 to 9 represent tails. In our 100,000 "flips," there were 2981 streaks of at least 5 heads. The "Law of Averages" suggests that the next flip after a run of 5 heads should be tails more often to even things out. Actually, the next flip was heads more often than tails: 1550 times to 1431 times. That's 52% heads. You can perform a similar simulation easily on a computer. Try it!

Of course, sometimes an apparent drift from what we expect means that the probabilities are, in fact, *not* what we thought. If you get 10 heads in a row, maybe the coin has heads on both sides!

The lesson of the LLN is that sequences of random events don't compensate in the *short* run and don't need to do so to get back to the right long-run probability. If the probability of an outcome doesn't change and the events are independent, the probability of any outcome in another trial is *always* what it was, no matter what has happened in other trials.

"Slump? I ain't in no slump. I just ain't hittin'."

—Yogi Berra

BEAT THE CASINO

Keno is a simple casino game in which numbers from 1 to 80 are chosen. The numbers, as in most lottery games, are supposed to be equally likely. Payoffs are made depending on how many of those numbers you match on your card. A group of graduate students from a statistics department decided to take a field trip to Reno. They (*very* discreetly) wrote down the outcomes of the games for a couple of days, then drove back to test whether the numbers were, in fact, equally likely. It turned out that some numbers were *more likely* to come up than others. Rather than bet on the Law of Averages and put their money on the numbers that were "due," the students put their faith in the LLN—and all their (and their friends') money on the numbers that had come up before. After they pocketed more than $50,000, they were escorted off the premises and invited never to show their faces in that casino again.

JUST CHECKING

1. One common proposal for beating the lottery is to note which numbers have come up lately, eliminate those from consideration, and bet on numbers that have not come up for a long time. Proponents of this method argue that in the long run, every number should be selected equally often, so those that haven't come up are due. Explain why this is faulty reasoning.

12.2 Modeling Probability

Probability was first studied extensively by a group of French mathematicians who were interested in games of chance.[6] Rather than *experiment* with the games (and risk losing their money), they developed mathematical models. When the probability comes from a mathematical model and not from observation, it is called **theoretical probability**. To make things simple (as we usually do when we build models), they started by looking at games in which the different outcomes were equally likely. Fortunately, many games of chance are like that. Any of 52 cards is equally likely to be the next one dealt from a well-shuffled deck. Each face of a die is equally likely to land up (or at least it *should be*).

[6] OK, gambling.

It's easy to find probabilities for events that are made up of several *equally likely* outcomes. We just count all the outcomes that the event contains. The probability of the event is the number of outcomes in the event divided by the total number of possible outcomes. We can write

$$P(\mathbf{A}) = \frac{\#\text{ outcomes in } \mathbf{A}}{\#\text{ of possible outcomes}}.$$

For example, the probability of drawing a face card (JQK) from a deck is

$$P(\text{face card}) = \frac{\#\text{ face cards}}{\#\text{ cards}} = \frac{12}{52} = \frac{3}{13}.$$

NOTATION ALERT

We often use capital letters—and usually from the beginning of the alphabet—to denote events. We *always* use P to denote
probability. So,

$$P(\mathbf{A}) = 0.35$$

means "the probability of the event **A** is 0.35."

When being formal, use decimals (or fractions) for the probability values, but sometimes, especially when talking more informally, it's easier to use percentages.

HOW HARD CAN COUNTING BE?

Finding the probability of any event when the outcomes are equally likely is straightforward, but not necessarily easy. It gets hard when the number of outcomes in the event (and in the sample space) gets big. Think about flipping two coins. The sample space is $\mathbf{S} = \{HH, HT, TH, TT\}$ and each outcome is equally likely. So, what's the probability of getting exactly one head and one tail? Let's call that event **A**. Well, there are two outcomes in the event $\mathbf{A} = \{HT, TH\}$ out of the 4 possible equally likely ones in S, so $P(\mathbf{A}) = \frac{2}{4}$, or $\frac{1}{2}$.

OK, now flip 100 coins. What's the probability of exactly 67 heads? Well, first, how many outcomes are in the sample space? $\mathbf{S} = \{HHHHHHHHHHH \ldots H, HH \ldots T, \ldots\}$ Um . . . . a lot. In fact, there are 1,267,650,600,228,229,401,496,703, 205,376 different outcomes possible when flipping 100 coins. And that's just the denominator of the probability! To answer the question, we'd still have to figure out how many ways there are to get 67 heads. We'll see how in Chapter 15; stay tuned!

Don't get trapped into thinking that random events are always equally likely. The chance of winning a lottery—especially lotteries with very large payoffs—is small. Regardless, people continue to buy tickets. In an attempt to understand why, an interviewer asked someone who had just purchased a lottery ticket, "What do you think your chances are of winning the lottery?" The reply was, "Oh, about 50– 50." The shocked interviewer asked, "How do you get that?" to which the response was, "Well, the way I figure it, either I win or I don't!"

The moral of this story is that events are *not* always equally likely.

Personal Probability

What's the probability that your grade in this statistics course will be an A? You may be able to come up with a number that seems reasonable. Of course, no matter how confident or depressed you feel about your chance of success, your probability should be between 0 and 1. How did you come up with this probability? It can't be an empirical probability. For that, you'd have to take the course over and over (and over . . .), and forget everything after each time so the probability of getting an A would stay the same. But people use probability in a third sense as well.

We use the language of probability in everyday speech to express a degree of uncertainty *without* basing it on long-run relative frequencies or mathematical models. Your personal assessment of your chances of getting an A expresses your uncertainty about the outcome. That uncertainty may be based on how comfortable you're feeling in the course or on your midterm grade, but it can't be based on long-run behavior. We call this third kind of probability a **subjective or personal probability**.

Although personal probabilities may be based on experience, they're not based either on long-run relative frequencies or on equally likely events. So they don't display the kind of consistency that we'll need probabilities to have. For that reason, we'll stick to formally defined probabilities. You should be alert to the difference.

John Venn (1834– 1923) created the Venn diagram. His book on probability, *The Logic of Chance*, was "strikingly original and considerably influenced the development of the theory of Statistics," according to John Maynard Keynes, one of the luminaries of Economics.

WHICH KIND OF PROBABILITY?

The line between personal probability and the other two probabilities can be a fuzzy one. When a weather forecaster predicts a 40% probability of rain, is this a personal probability or a relative frequency probability? The claim may be that 40% of the time, when the map looks like this, it has rained (over some period of time). Or the forecaster may be stating a personal opinion that is based on years of experience and reflects a sense of what has happened in the past in similar situations. When you hear a probability stated, try to ascertain what kind of probability is intended.

The First Three Rules for Working with Probability

1. Make a picture.
2. Make a picture.
3. Make a picture.

We're dealing with probabilities now, not data, but the three rules don't change. The most common kind of picture to make is called a Venn diagram. We'll use Venn diagrams throughout the rest of this chapter. Even experienced statisticians make Venn diagrams to help them think about probabilities of compound and overlapping events. You should, too.

12.3 Formal Probability

For some people, the phrase "50/50" means something vague like "I don't know" or "whatever." But when we discuss probabilities of outcomes, it takes on the precise meaning of *equally likely*. Speaking vaguely about probabilities will get us into trouble, so whenever we talk about probabilities, we'll need to be precise.[7] And to do that, we'll need to develop some formal rules[8] about how probability works.

Rule 1. If the probability is 0, the event never occurs, and likewise if it has probability 1, it *always* occurs. Even if you think an event is very unlikely, its probability can't be negative, and even if you're sure it will happen, its probability can't be greater than 1. (Think about relative frequencies.) So we require that

A probability is a number between 0 and 1.
For any event A, $0 \leq P(\text{A}) \leq 1$.

SURPRISING PROBABILITIES

We've been careful to discuss probabilities only for situations in which the outcomes were finite, or even countably infinite. But if the outcomes can take on *any* numerical value at all (we say they are *continuous*), things can get surprising. For example, what is the probability that a randomly selected child will be *exactly* 3 feet tall? Well, if we mean 3.00000 . . . feet, the answer is zero. No randomly selected child—even one whose height would be recorded as 3 feet, will be *exactly* 3 feet tall (to an infinite number of decimal places). But, if you've grown taller than 3 feet, there must have been a time in your life when you actually *were* exactly 3 feet tall, even if only for a second. So this is an outcome with probability 0 that not only has happened—it has happened to *you*.

[7]And to be precise, we will be talking only about sample spaces where we can enumerate all the outcomes. Mathematicians call this a countable number of outcomes.

[8]Actually, in mathematical terms, these are axioms—statements that we assume to be true of probability. We'll derive other rules from these in the next chapter.

Rule 2. If a random phenomenon has only one possible outcome, it's not very interesting (or very random). So we need to distribute the probabilities among all the outcomes a trial can have. How can we do that so that it makes sense? For example, consider what you're doing as you read this text. The possible outcomes might be

A: You read to the end of this chapter before stopping.
B: You finish this section but stop reading before the end of the chapter.
C: You bail out before the end of this section.

When we assign probabilities to these outcomes, the first thing to be sure of is that we distribute all of the available probability. If we look at all the events in the entire sample space, the probability of that collection of events has to be 1. So the probability of the entire sample space is 1.

Making this more formal gives the **Probability Assignment Rule:**

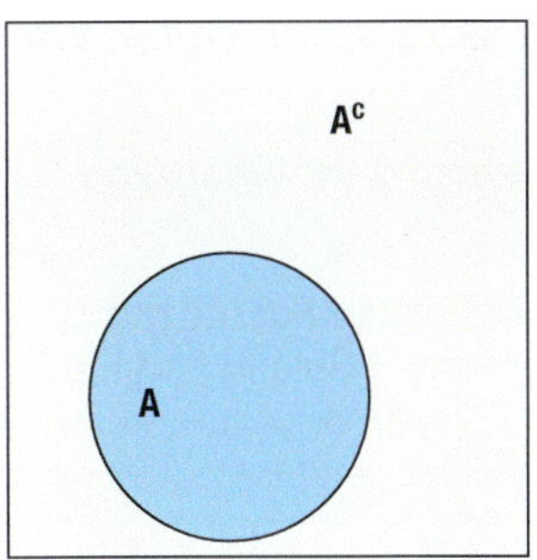

The set **A** and its complement **A^C**. Together, they make up the entire sample space **S**.

The set of all possible outcomes of a trial must have probability 1.

$$P(S) = 1$$

Rule 3. Suppose the probability that you get to class on time is 0.8. What's the probability that you don't get to class on time? Yes, it's 0.2. The set of outcomes that are *not* in the event **A** is called the complement of **A**, and is denoted $\mathbf{A}^C$. This leads to the **Complement Rule**:

The probability of an event not occurring is 1 minus the probability that it does occur.

$$P(\mathbf{A}^C) = 1 - P(\mathbf{A})$$

EXAMPLE 12.2 Applying The Complement Rule

RECAP: We opened the chapter by looking at the traffic light at the corner of College and Main, observing that when we arrive at that intersection, the light is green about 35% of the time.

QUESTION: If $P(\text{green}) = 0.35$, what's the probability the light isn't green when you get to College and Main?

ANSWER: "Not green" is the complement of "green," so $P(\text{not green}) = 1 - P(\text{green})$
$= 1 - 0.35 = 0.65$

There's a 65% chance I won't have a green light.

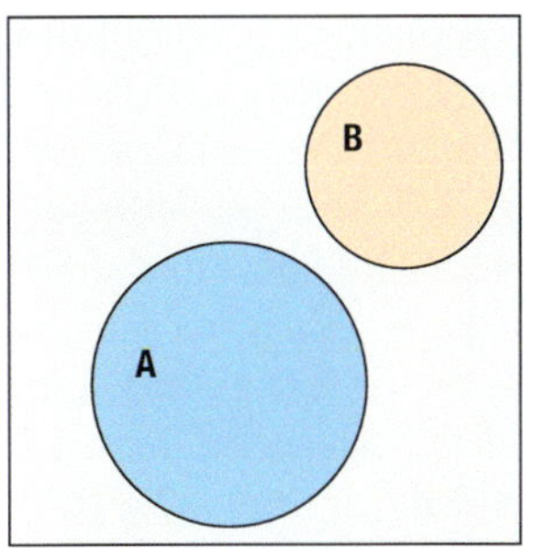

A simple Venn diagram of two disjoint sets, **A** and **B**.

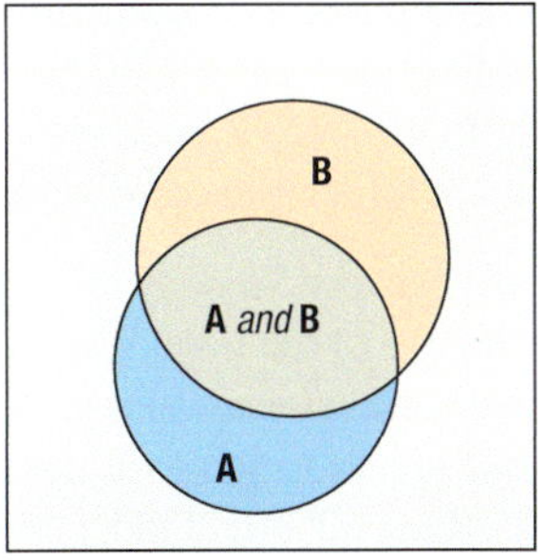

Two sets **A** and **B** that are not disjoint. The event (**A** *and* **B**) is their intersection.

Rule 4. Suppose the probability that a randomly selected student is a sophomore (**A**) is 0.20, and the probability that he or she is a junior (**B**) is 0.30. What is the probability that the student is *either* a sophomore *or* a junior, written $P(\mathbf{A} \text{ or } \mathbf{B})$? If you guessed 0.50, you've deduced the **Addition Rule,** which says that you can add the probabilities of events that are disjoint.[9] To see whether two events are disjoint, we take them apart into their component outcomes and check whether they have any outcomes in common. **Disjoint** (or mutually exclusive) events have no outcomes in common. The **Addition Rule** states,

For two disjoint events A and B, the probability that one *or* the other occurs is the sum of the probabilities of the two events.

$$P(\mathbf{A} \text{ or } \mathbf{B}) = P(\mathbf{A}) + P(\mathbf{B}), \text{ provided that A and B are disjoint.}$$

[9]You may see $P(\mathbf{A} \text{ or } \mathbf{B})$ written as $P(\mathbf{A} \cup \mathbf{B})$. The symbol $\cup$ means "union," representing outcomes that are in event **A** *or* event **B** (or both). The symbol $\cap$ means "intersection," representing outcomes that are in both event **A** *and* event **B**. You may sometimes see $P(\mathbf{A} \text{ and } \mathbf{B})$ written as $P(\mathbf{A} \cap \mathbf{B})$.

EXAMPLE 12.3 Applying The Addition Rule

RECAP: When you get to the light at College and Main, it's red, green, or yellow. We know that $P(\text{green}) = 0.35$.

QUESTION: Suppose we find out that P(yellow) yellow) is about 0.04. What's the probability the light is red?

ANSWER: The light must be red, green, or yellow, so if we can figure out the probability that the light is green or yellow, we can use the Complement Rule to find the probability that it is red. To find the probability that the light is green or yellow, I can use the Addition Rule because these are disjoint events: The light can't be both green and yellow at the same time.

$$P(\text{green OR yellow}) = 0.35 + 0.04 = 0.39$$

Red is the only remaining alternative, and the probabilities must add up to 1, so

$$\begin{aligned} P(\text{red}) &= P(\text{not green OR yellow}) \\ &= 1 - P(\text{green OR yellow}) \\ &= 1 - 0.39 = 0.61 \end{aligned}$$

> "Baseball is 90% mental. The other half is physical."
>
> —Yogi Berra

The Addition Rule can be extended to any number of disjoint events, and that's helpful for checking probability assignments. Because individual sample space outcomes are always disjoint, we have an easy way to check whether the probabilities we've assigned to the possible outcomes are legitimate. The Probability Assignment Rule tells us that to be a **legitimate assignment of probabilities**, the sum of the probabilities of all possible outcomes must be exactly 1. No more, no less. For example, if we were told that the probabilities of selecting at random a freshman, sophomore, junior, or senior from all the undergraduates at a school were 0.25, 0.23, 0.22, and 0.20, respectively, we would know that something was wrong. These "probabilities" sum to only 0.90, so this is not a legitimate probability assignment. Either a value is wrong, or we just missed some possible outcomes, like "pre-freshman" or "postgraduate" categories that soak up the remaining 0.10. Similarly, a claim that the probabilities were 0.26, 0.27, 0.29, and 0.30 would be wrong because these "probabilities" sum to more than 1.

But be careful: The Addition Rule doesn't work for events that aren't disjoint. If the probability of owning a smartphone is 0.50 and the probability of owning a computer is 0.90, the probability of owning either a smartphone or a computer may be pretty high, but it is *not* 1.40! Why can't you add probabilities like this? Because these events are not disjoint. You *can* own both. In the next chapter, we'll see how to add probabilities for events like these, but we'll need another rule.

Rule 5. Suppose your job requires you to fly from Atlanta to Houston every Monday morning. The airline's website reports that this flight is on time 85% of the time. What's the chance that it will be on time two weeks in a row? That's the same as asking for the probability that your flight is on time this week *and* it's on time again next week. For independent events, the answer is very simple. Remember that independence means that the outcome of one event doesn't influence the outcome of the other. What happens with your flight this week doesn't influence whether it will be on time next week, so it's reasonable to assume that those events are independent. The **Multiplication Rule** says that for independent events, to find the probability that both events occur, we just multiply the probabilities together. Formally,

For two independent events A and B, the probability that both A *and* B occur is the product of the probabilities of the two events.

$$P(\text{A } and \text{ B}) = P(\text{A}) \times P(\text{B}), \text{ provided that A and B are independent.}$$

This rule can be extended to more than two independent events. What's the chance of your flight being on time for a month—four Mondays in a row? We can multiply the probabilities of it happening each week:

$$0.85 \times 0.85 \times 0.85 \times 0.85 = 0.522$$

or just over 50– 50. Of course, to calculate this probability, we have used the assumption that the four events are independent.

Many statistics methods require an **Independence Assumption,** but *assuming* independence doesn't make it true. Always *Think* about whether that assumption is reasonable before using the Multiplication Rule.

EXAMPLE 12.4 Applying The Multiplication Rule (and Others)

RECAP: We've determined that the probability that we encounter a green light at the corner of College and Main is 0.35, a yellow light 0.04, and a red light 0.61. Let's think about how many times during your morning commute in the week ahead you might hit a red light there.

QUESTION: What's the probability you find the light red both Monday and Tuesday?

ANSWER: Because the color of the light I see on Monday doesn't influence the color I'll see on Tuesday, these are independent events; I can use the Multiplication Rule:

$$\begin{aligned} P(\text{red Monday AND red Tuesday}) &= P(\text{red}) \times P(\text{red}) \\ &= (0.61)(0.61) \\ &= 0.3721 \end{aligned}$$

There's about a 37% chance I'll hit red lights both Monday and Tuesday mornings.

QUESTION: What's the probability you don't encounter a red light until Wednesday?

ANSWER: For that to happen, I'd have to see green or yellow on Monday, green or yellow on Tuesday, and then red on Wednesday. I can simplify this by thinking of it as not red on Monday, not red on Tuesday, and then red on Wednesday.

$$P(\text{not red}) = 1 - P(\text{red}) = 1 - 0.61 = 0.39, \text{ so}$$

$$\begin{aligned} &P(\text{not red Monday AND not red Tuesday AND red Wednesday}) \\ &= P(\text{not red}) \times P(\text{not red}) \times P(\text{red}) \\ &= (0.39)(0.39)(0.61) \\ &= 0.092781 \end{aligned}$$

There's about a 9% chance that this week I'll hit my first red light there on Wednesday morning.

QUESTION: What's the probability that you'll have to stop *at least once* during the week?

ANSWER: Having to stop at least once means that I have to stop for the light 1, 2, 3, 4, or 5 times next week. It's easier to think about the complement: never having to stop at a red light. Having to stop at least once means that I didn't make it through the week with no red lights.

P (having to stop at the light at least once in 5 days)

$$\begin{aligned} &= 1 - P(\text{no red lights for 5 days in a row}) \\ &= 1 - P(\text{not red AND not red AND not red AND not red AND not red}) \\ &= 1 - (0.39)(0.39)(0.39)(0.39)(0.39) \\ &= 1 - (0.39)^5 \\ &= 1 - 0.0090 \\ &= 0.991 \end{aligned}$$

I'm not likely to make it through the intersection of College and Main without having to stop sometime this week. There's over a 99% chance I'll hit at least one red light there.

SOME = AT LEAST ONE

In informal English, you may see "some" used to mean "at least one." "What's the probability that some of the eggs in that carton are broken?" means at least one.

AT LEAST

Note that the phrase "at least" is often a tip-off to think about the complement. Something that happens *at least once* does happen. Happening at least once is the complement of not happening at all, and that's easier to find.

JUST CHECKING

1. Opinion polling organizations contact their respondents by telephone. Random telephone numbers are generated, and interviewers try to contact those households. According to the Pew Research Center for the People and the Press, in 2012, the contact rate was about 62%. We can reasonably assume each household's response to be independent of the others. What is the probability that
   a) the interviewer successfully contacts the next household on her list?
   b) the interviewer successfully contacts both of the next two households on her list?
   c) the interviewer's first successful contact is the third household on the list?
   d) the interviewer makes at least one successful contact among the next five households on the list?

STEP-BY-STEP EXAMPLE Probability

The five rules we've seen can be used in a number of different combinations to answer a surprising number of questions. Let's try one to see how we might go about it.

M&M's® Milk Chocolate candies now come in 7 colors, but they've changed over time. In 1995, Americans voted to change tan M&M's (which had replaced violet in 1949) to blue. In 2002, Mars™, the parent company of M&M's, used the Internet to solicit global opinion for a seventh color. To decide which color to add, Mars surveyed kids in nearly every country of the world and asked them to vote among purple, pink, and teal. The global winner was purple! In the United States, 42% of those who voted said purple, 37% said teal, and only 19% said pink. But in Japan, the percentages were 38% pink, 36% teal, and only 16% purple. Let's use Japan's percentages to ask some questions:

1. What's the probability that a Japanese M&M's survey respondent selected at random chose either pink or teal?
2. If we pick two respondents at random, what's the probability that both chose purple?
3. If we pick three respondents at random, what's the probability that *at least one* chose purple?

THINK The probability of an event is its long-term relative frequency. It can be determined in several ways: by looking at many replications of an event, by deducing it from equally likely events, or by using some other information. Here, we are told the relative frequencies of the three responses.

Make sure the probabilities are legitimate. Here, they're not. Either there was a mistake, or the other voters must have chosen a color other than the three given. A check of the reports from other countries shows a similar deficit, so probably we're missing those who had no preference or who wrote in another color.

According to the M&M's website:

$$P(\text{pink}) = 0.38$$
$$P(\text{teal}) = 0.36$$
$$P(\text{purple}) = 0.16$$

Each is between 0 and 1, but they don't add up to 1. The remaining 10% must have not expressed a preference or written in another color. I'll put them together into "other" and add $P(\text{other}) = 0.10$.

Now I have a legitimate assignment of probabilities.

QUESTION 1. What's the probability that a Japanese M&M's survey respondent selected at random chose either pink or teal?

THINK	**PLAN** Plan Decide which rules to use and check the conditions they require.	The events "pink" and "teal" are disjoint (because one respondent can't choose both). I can apply the Addition Rule.
SHOW	**MECHANICS** Show your work.	$P(\text{pink or teal}) = P(\text{pink}) + P(\text{teal})$ $= 0.38 + 0.36 = 0.74$
TELL	**CONCLUSION** Interpret your results in the proper context.	The probability that the respondent chose pink or teal is 0.74.

QUESTION 2. If we pick two respondents at random, what's the probability that they both said purple?

THINK	**PLAN** The word "both" suggests we want *P*(**A** *and* **B**), which calls for the Multiplication Rule. Think about the assumption.	✓ **Independence Assumption**: The choice made by one respondent does not affect the choice of the other, so the events are independent. I can use the Multiplication Rule.
SHOW	**MECHANICS** Show your work. For both respondents to choose purple, each one has to choose purple.	P (both purple) $= P(\text{first purple and second purple})$ $= P(\text{first purple}) \times P(\text{second purple})$ $= 0.16 \times 0.16 = 0.0256$
TELL	**CONCLUSION** Interpret your results in the proper context.	The probability that both chose purple is 0.0256.

QUESTION 3. If we pick three respondents at random, what's the probability that at least one chose purple?

THINK	**PLAN** The phrase "at least . . . " often flags a question best answered by looking at the complement, and that's the best approach here. The complement of "At least one preferred purple" is "None of them preferred purple." Think about the assumption.	$P(\text{at least one purple})$ $= P(\{\text{none purple}\}^C)$ $= 1 - P(\text{none purple}).$ $P(\text{none purple}) = P(\text{not purple } \textit{and} \text{ not purple } \textit{and} \text{ not purple}).$ ✓ **Independence Assumption**: These are independent events because they are choices by three random respondents. I can use the Multiplication Rule.
SHOW	**MECHANICS** We calculate *P*(none purple) by using the Multiplication Rule.	$P(\text{none purple}) = P(\text{first not purple}) \times$ $P(\text{second not purple}) \times P(\text{third not purple})$ $= [P(\textit{not purple})]^3.$ $P(\text{not purple}) = 1 - P(\text{purple})$ $= 1 - 0.16 = 0.84.$ So $P(\text{none purple}) = (0.84)^3 = 0.5927.$

Then we can use the Complement Rule to get the probability we want.	$P(\text{at least 1 purple}) = 1 - P(\text{none purple})$ $= 1 - 0.5927 = 0.4073.$
TELL **CONCLUSION** Interpret your results in the proper context.	There's about a 40.7% chance that at least one of the respondents chose purple.

WHAT CAN GO WRONG?

- **Beware of probabilities that don't add up to 1.** To be a legitimate probability assignment, the sum of the probabilities for all possible outcomes must total 1. If the sum is less than 1, you may need to add another category ("other") and assign the remaining probability to that outcome. If the sum is more than 1, check that the outcomes are disjoint. If they're not, then you can't assign probabilities by just counting relative frequencies.
- **Don't add probabilities of events if they're not disjoint.** Events must be disjoint to use the Addition Rule. The probability of being younger than 80 *or* a female is not the probability of being younger than 80 *plus* the probability of being female. That sum may be more than 1.
- **Don't multiply probabilities of events if they're not independent.** The probability of selecting a student at random who is over 6′10″ tall *and* on the basketball team is *not* the probability the student is over 6′10″ tall *times* the probability he's on the basketball team. Knowing that the student is over 6′10″ changes the probability of his being on the basketball team. You can't multiply these probabilities. The multiplication of probabilities of events that are not independent is one of the most common errors people make in dealing with probabilities.
- **Don't confuse disjoint and independent.** Disjoint events *can't* be independent. If **A** = {you get an A in this class} and **B** = {you get a B in this class}, **A** and **B** are disjoint. Are they independent? If you find out that **A** is true, does that change the probability of **B**? You bet it does! So they can't be independent. We'll return to this issue in the next chapter.

CONNECTIONS

We began thinking about independence back in Chapter 2 when we looked at contingency tables and asked whether the distribution of one variable was the same for each category of another. Then, in Chapter 10, we saw that independence was fundamental to drawing a Simple Random Sample. For computing compound probabilities, we again ask about independence. And we'll continue to think about independence throughout the rest of the text.

Our interest in probability extends back to the start of the text. We've talked about "relative frequencies" often. But—let's be honest—that's just a casual term for probability. For example, you can now rephrase the 68– 95– 99.7 Rule we saw in Chapter 5 to talk about the *probability* that a random value selected from a Normal model will fall within 1, 2, or 3 standard deviations of the mean.

Why not just say "probability" from the start? Well, we didn't need any of the formal rules of this chapter (or the next one), so there was no point to weighing down the discussion with those rules. And "relative frequency" is the right intuitive way to think about probability in this course, so you've been thinking right all along.

Keep it up.

CHAPTER REVIEW

Apply the facts about probability to determine whether an assignment of probabilities is legitimate.

- Probability is long-run relative frequency.
- Individual probabilities must be between 0 and 1.
- The sum of probabilities assigned to all outcomes must be 1.

Understand the Law of Large Numbers and that the common understanding of the "Law of Averages" is false.

Know the rules of probability and how to apply them.

- **Rule 1:** A probability is a number between 0 and 1.
- For any event A, $0 \le P(\mathbf{A}) \le 1$.
- **Rule 2: Probability Assignment Rule**:

$$P(S) = 1.$$

- **Rule 3: The Complement Rule** says that $P(\textit{not } \mathbf{A}) = P(\mathbf{A}^C) = 1 - P(\mathbf{A})$.
- **Rule 4:** The **Addition Rule** for disjoint events says that $P(\mathbf{A} \textit{ or } \mathbf{B}) = P(\mathbf{A}) + P(\mathbf{B})$ provided events A and B are disjoint.
- **Rule 5:** The **Multiplication Rule** for independent events says that $P(\mathbf{A} \textit{ and } \mathbf{B}) = P(\mathbf{A}) \times P(\mathbf{B})$ provided events **A** and **B** are independent.

REVIEW OF TERMS

Random phenomenon A phenomenon is random if we know what outcomes could happen, but not which particular values will happen (p. 405).

Trial A single attempt or realization of a random phenomenon (p. 406).

Outcome The value measured, observed, or reported for an individual instance of trial (p. 406).

Event A collection of outcomes. Usually, we identify events so that we can attach probabilities to them. We denote events with bold capital letters such as A, B, or C (p. 406).

Sample space The collection of all possible outcome values. The collection of values in the sample space has a probability of 1. We denote the sample space with a boldface capital S (p. 406).

Law of Large Numbers This law states that the long-run *relative frequency* of an event's occurrence gets closer and closer to the *true* relative frequency as the number of trials increases (p. 407).

Independence (informally) Two events are *independent* if learning that one event occurs does not change the probability that the other event occurs (p. 407).

Probability The probability of an event is a number between 0 and 1 that reports the likelihood of that event's occurrence. We write *P*(A) for the probability of the event A (p. 407).

Empirical probability When the probability comes from the long-run relative frequency of the event's occurrence, it is an empirical probability (p. 407).

Theoretical probability When the probability comes from a model (such as equally likely outcomes), it is a theoretical probability (p. 408).

Personal (or subjective) Probability When the probability is subjective and represents your personal degree of belief, it is a probability personal probability (p. 409).

Probability Assignment Rule The probability of the entire sample space must be 1. $P(\mathbf{S}) = 1$ (p. 411).

Complement Rule The probability of an event not occurring is 1 minus the probability that it does occur (p. 411).

$$P(\mathbf{A}^C) = 1 - P(\mathbf{A})$$

Addition Rule If A and B are disjoint events, then the probability of A *or* B is (p. 411)

$$P(\mathbf{A} \text{ or } \mathbf{B}) = P(\mathbf{A}) + P(\mathbf{B}).$$

Disjoint (mutually exclusive) Two events are disjoint if they share no outcomes in common. If A and B are disjoint, then knowing that A occurs tells us that B cannot occur. Disjoint events are also called "mutually exclusive" (p. 411).

Legitimate assignment of probabilities An assignment of probabilities to outcomes is legitimate if (p. 412)

- each probability is between 0 and 1 (inclusive).
- the sum of the probabilities is 1.

Multiplication Rule If A and B are independent events, then the probability of A *and* B is (p. 412)

$$P(\mathbf{A} \text{ and } \mathbf{B}) = P(\mathbf{A}) \times P(\mathbf{B}).$$

EXERCISES

SECTION 12.1

1. Flipping a coin Flipping a fair coin is said to randomly generate heads and tails with equal probability. Explain what random means in this context.

2. Dice Rolling a fair six-sided die is supposed to randomly generate the numbers 1 through 6. Explain what random means in this context.

3. Flipping a coin II Your friend says: "I flipped five heads in a row! The next one has to be tails!" Explain why this thinking is incorrect.

4. Dice II After rolling doubles on a pair of dice three times in a row, your friend exclaims, "I can't get doubles four times in a row!" Explain why this thinking is incorrect.

SECTION 12.2

5. Wardrobe In your dresser are five blue shirts, three red shirts, and two black shirts.

a) What is the probability of randomly selecting a red shirt?
b) What is the probability that a randomly selected shirt is not black?

6. Playlists Your list of favorite songs contains 10 rock songs, 7 rap songs, and 3 country songs.

a) What is the probability that a randomly played song is a rap song?
b) What is the probability that a randomly played song is not country?

SECTION 12.3

7. Cell phones and surveys A 2018 study conducted by the Australian Communications and Media Authority found that 31% of Australian households had no fixed landline at home. This raises concerns about the accuracy of certain surveys, as they depend on random-digit dialing to households via landlines. Suppose four Australian households are picked at random:

a) What is the probability that all four of them have a landline?
b) What is the probability that at least one of them does not have a landline?
c) What is the probability that at least one of them has a landline?

(Source: https://www.smh.com.au/technology/the-long-inexorable-decline-of-the-landline-in-australia-20161221-gtfjsp.html)

8. Cell phones and surveys II A 2017 report by the National Health Interview Survey claims that 53.3% of adults in American homes used only wireless service (and no landline). We randomly select six adults:

a) What is the probability that all of these adults had a landline?
b) What is the probability that all of these adults used only wireless service?
c) What is the probability that at least one of these adults had a landline?

(Source: https://www.cdc.gov/nchs/data/nhis/earlyrelease/wireless201806.pdf)

CHAPTER EXERCISES

9. Sample spaces For each of the following, list the sample space and tell whether you think the events are equally likely:

a) Toss 2 coins; record the order of heads and tails.
b) A family has 3 children; record the number of boys.
c) Flip a coin until you get a head or 3 consecutive tails; record each flip.
d) Roll two dice; record the larger number (or simply the number in case of a tie).

10. Sample spaces For each of the following, list the sample space and tell whether you think the events are equally likely:

a) Roll two dice; record the sum of the numbers.
b) A family has 3 children; record each child's sex in order of birth.
c) Toss four coins; record the number of tails.
d) Toss a coin 10 times; record the length of the longest run of heads.

11. Roulette A casino claims that its roulette wheel is truly random. What should that claim mean?

12. Rain The weather reporter on TV makes predictions such as a 25% chance of rain. What do you think is the meaning of such a phrase?

13. Winter Comment on the following quotation:
"What I think is our best determination is it will be a colder than normal winter," said Pamela Naber Knox, a Wisconsin state climatologist. "I'm basing that on a couple of different things. First, in looking at the past few winters, there has been a lack of really cold weather. Even though we are not supposed to use the law of averages, we are due." (Source: Excerpt from "*The Law of Averages*" by Naber Knox from Activity-Based Statistics: Student Guide by Richard L. Scheaffer, Jeffrey Witmer, Ann Watkins, Mrudulla Gnanadesikan. Published by Springer/Sci-Tech/Trade, © 1996.)

14. Snow After an unusually dry autumn, a radio announcer is heard to say, "Watch out! We'll pay for these sunny days later on this winter." Explain what he's trying to say, and comment on the validity of his reasoning.

15. Cold streak A tennis player loses 4 out of the first 5 games in a set against an equally skilled opponent, where the odds give the tennis player a 50% chance of both winning and losing. However, he is extremely confident that he will win his next game. Comment on his reasoning.

16. Crash Commercial airplanes have an excellent safety record. Nevertheless, there are crashes occasionally, with the loss of many lives. In the weeks following a crash, airlines often report a drop in the number of passengers, probably because people are afraid to risk flying.

a) A travel agent suggests that since the Law of Averages makes it highly unlikely to have two plane crashes within a few weeks of each other, flying soon after a crash is the safest time. What do you think?
b) If the airline industry proudly announces that it has set a new record for the longest period of safe flights, would you be reluctant to fly? Are the airlines due to have a crash?

17. Auto insurance Insurance companies collect annual payments from drivers in exchange for paying for the cost of accidents.

a) Why should you be reluctant to accept a $1500 payment from your neighbor to cover his automobile accidents in the next year?
b) Why can the insurance company make that offer?

18. Jackpot On February 11, 2009, the AP news wire released the following story:

(*LAS VEGAS, Nev.*)—A man in town to watch the NCAA basketball tournament hit a $38.7 million jackpot on Friday, the biggest slot machine payout ever. The 25-year-old software engineer from Los Angeles, whose name was not released at his request, won after putting three $1 coins in a machine at the Excalibur hotel-casino, said Rick Sorensen, a spokesman for slot machine maker International Game Technology.

a) How can the Excalibur afford to give away millions of dollars on a $3 bet?
b) Why was the maker willing to make a statement? Wouldn't most businesses want to keep such a huge loss quiet?

19. Spinner The plastic arrow on a spinner for a child's game stops rotating to point at a color that will determine what happens next. Which of the following probability assignments are legitimate?

	Probabilities of . . .			
	Red	**Yellow**	**Green**	**Blue**
a)	0.25	0.25	0.25	0.25
b)	0.10	0.20	0.30	0.40
c)	0.20	0.30	0.40	0.50
d)	0	0	1.00	0
e)	0.10	0.20	1.20	−1.50

20. Scratch off Many stores run "secret sales": Shoppers receive cards that determine how large a discount they get, but the percentage is revealed by scratching off that black stuff (what *is* that?) only after the purchase has been totaled at the cash register. The store is required to reveal (in the fine print) the distribution of discounts available. Which of these probability assignments are legitimate?

	Probabilities of . . .			
	10% Off	**20% Off**	**30% Off**	**50% Off**
a)	0.20	0.20	0.20	0.20
b)	0.50	0.30	0.20	0.10
c)	0.80	0.10	0.05	0.05
d)	0.75	0.25	0.25	−0.25
e)	1.00	0	0	0

21. Electronics Suppose that 46% of families living in a certain county own a computer and 18% own an HDTV. The Addition Rule might suggest, then, that 64% of families own either a computer or an HDTV. What's wrong with that reasoning?

22. Homes Funding for many schools comes from taxes based on assessed values of local properties. People's homes are assessed higher if they have extra features such as garages and swimming pools. Assessment records in a certain school district indicate that 37% of the homes have garages and 3% have swimming pools. The Addition Rule might suggest, then, that 40% of residences have a garage or a pool. What's wrong with that reasoning?

23. Speeders Traffic checks on a certain section of highway suggest that 60% of drivers are speeding there. Since $0.6 \times 0.6 = 0.36$, the Multiplication Rule might suggest that there's a 36% chance that two vehicles in a row are both speeding. What's wrong with that reasoning?

24. Lefties Although it's hard to be definitive in classifying people as right- or left-handed, some studies suggest that about 14% of people are left-handed. Since $0.14 \times 0.14 = 0.0196$, the Multiplication Rule might suggest that there's about a 2% chance that a brother and a sister are both lefties. What's wrong with that reasoning?

25. University admissions For the incoming class of 2023, university admissions for the most selective universities were extremely competitive. Three of these universities accepted 4.5%, 5.91%, and 5.77% of their applicants, respectively.

Noah applied to all three. Assuming that he is a typical applicant, he figures that his chances of getting into all three is less than 0.02%.

a) How has he arrived to this conclusion?
b) What additional assumption is he making?
c) Do you agree with his conclusion?

(Source: https://www.crimsoneducation.org/ca-fr/blog/2023-Ivy-League-admissions-results-are-in)

26. University admissions In Exercise 25, we see that for the incoming class of 2023, three of the top universities accepted 4.5%, 5.91%, and 5.77% of their applicants, respectively. Noah applied to all three. He figures that his chances of getting into at least one of them must be about 16.18%.

a) How has he arrived at this conclusion?
b) What additional assumption is he making?
c) Do you agree with his conclusion?

27. Car repairs A consumer organization estimates that over a 1-year period, 20% of cars will need to be repaired only once, 5% will need repairs exactly twice, and 2% of cars will require three or more repairs. What is the probability that a car chosen at random will need

a) no repairs?
b) no more than one repair?
c) two or more repairs?

28. Stats projects In a large introductory statistics lecture hall, the professor reports that 55% of the students enrolled have never taken a calculus course, 32% have taken only one semester of calculus, and the rest have taken two or more semesters of calculus. The professor randomly assigns students to groups of three to work on a project for the course. What is the probability that the first groupmate you meet has studied

a) two or more semesters of calculus?
b) some calculus?
c) no more than one semester of calculus?

29. More repairs Consider again the auto repair rates described in Exercise 27. If you own three cars, use both multiplication rule and addition rule to find the probability that:

a) none will need repair.
b) all will need repair.
c) two or more cars will need repair.

30. Another project You are assigned to be part of a group of three students from the intro stats class described in Exercise 28. What is the probability that of your other two groupmates,

a) neither has studied calculus?
b) both have studied at least one semester of calculus?
c) at least one has had more than one semester of calculus?

31. Repairs again You use the multiplication rule and, sometimes, the addition rule to calculate repair probabilities in Exercise 29.

a) What must be true about your cars in order to make use of the multiplication rule valid?
b) Do you think this assumption is reasonable? Explain.

32. Final project You used the Multiplication Rule to calculate probabilities about the calculus background of your statistics groupmates in Exercise 30.

a) What must be true about the groups in order to make that approach valid?
b) Do you think this assumption is reasonable? Explain.

33. Shootings A poll in December 2015 asked the respondents of a country whether they were worried if someone in their family would become a victim of a mass shooting. The poll asked 1028 adults to classify their worry between "very worried," "somewhat worried," "not too worried," and "not worried at all." The following are the results:

Responses	Number
Very Worried	105
Somewhat Worried	301
Not Too Worried	353
Not Worried at All	269
Total	**1028**

If we select a person at random from this sample of 1028 adults,

a) what is the probability that the person responded that they are not too worried?
b) what is the probability that the person responded that they are somewhat or very worried?

(Source: https://news.gallup.com/poll/220634/four-americans-fear-victim-mass-shooting.aspx?&g_source=link_newsv9&g_campaign=item_234800&g_medium=copy)

34. GM foods A research poll in 2018 with 2537 adult participants asked whether GM foods are better or worse for one's health. Here is how they responded:

Responses	Number
Worse	1249
Neither Better nor Worse	1123
Better	134
No Answer	31
Total	**2537**

If we select a respondent at random from this sample of 2537 adults,

a) what is the probability that the selected person thinks GM foods are worse for one's health?
b) what is the probability that the selected person answers the question?

(Source: https://www.pewresearch.org/science/2018/11/19/public-perspectives-on-food-risks/)

35. More Shootings Exercise 33 shows the results of a poll about shootings. Suppose we select four people at random from this sample.

a) What is the probability that all four responded that they are very or somewhat worried?
b) What is the probability that none responded that they are very or somewhat worried?
c) What assumption did you make in computing these probabilities?
d) Explain why you think this assumption is reasonable.

36. GM foods revisited, revisited Consider again the results of the poll about GM foods discussed in Exercise 34. If we select two people at random from this sample,

a) what is the probability that they think GM foods are worse?
b) what is the probability that neither thinks that GM foods are worse?
c) what is the probability that the first thinks GM foods are worse, and the second thinks GM foods are better?
d) what assumption did you make in computing these probabilities?
e) explain why you think this assumption is reasonable.

37. Polling As mentioned in the chapter, opinion-polling organizations contact their respondents by sampling random telephone numbers. Although interviewers can reach about 62% of U.S. households, the percentage of those contacted who agree to cooperate with the survey fell from 43% in 1997 to 14% in 2012 (Source: Pew Research Center for the People and the Press). Each household, of course, is independent of the others. Using the cooperation rate from 2012,

a) what is the probability that the next household on the list will be contacted but will refuse to cooperate?
b) what is the probability of failing to contact a household or of contacting the household but not getting them to agree to the interview?
c) show another way to calculate the probability in part b.

38. Polling, part II According to Pew Research, the contact rate (probability of contacting a selected household) was 90% in 1997 and 62% in 2012. However, the cooperation rate (probability of someone at the contacted household agreeing to be interviewed) was 43% in 1997 and dropped to 14% in 2012.

a) What is the probability (in 2012) of obtaining an interview with the next household on the sample list? (To obtain an interview, an interviewer must both contact the household and then get agreement for the interview.)
b) Was it more likely to obtain an interview from a randomly selected household in 1997 or in 2012?

39. M&M's The Mars company says that before the introduction of purple, yellow candies made up 20% of their plain M&M's, red another 20%, and orange, blue, and green each made up 10%. The rest were brown.

a) If you pick an M&M at random, what is the probability that
   1. it is brown?
   2. it is yellow or orange?
   3. it is not green?
   4. it is striped?
b) If you pick three M&M's in a row, what is the probability ability that
   1. they are all brown?
   2. the third one is the first one that's red?
   3. none are yellow?
   4. at least one is green?

40. Blood The American Red Cross says that about 51% of African-Americans have Type O blood, 26% Type A, 19% Type B, and the rest Type AB.

a) An African-American volunteers to give blood. What is the probability that this donor
   i. has Type AB blood?
   ii. has Type A or Type B blood?
   iii. is not Type O?
b) Among four potential African-American donors, what is the probability that
   i. all are Type O?
   ii. no one is Type O?
   iii. they are not all Type B?
   iv. at least one person is Type A?

(Source: https://www.livescience.com/36559-common-blood-type-donation.html)

41. Disjoint or independent? In Exercise 39, you calculated probabilities of getting various M&M's. Some of your answers depended on the assumption that the outcomes described were *disjoint*; that is, they could not both happen at the same time. Other answers depended on the assumption that the events were *independent*; that is, the occurrence of one of them doesn't affect the probability of the other. Do you understand the difference between disjoint and independent?

a) If you draw one M&M, are the events of getting a red one and getting an orange one disjoint, independent, or neither?
b) If you draw two M&M's one after the other, are the events of getting a red on the first and a red on the second disjoint, independent, or neither?
c) Can disjoint events ever be independent? Explain.

42. Blood In Exercise 40, you calculated probabilities involving various blood types. Some of your answers depended on the assumption that the outcomes described were *disjoint*; that is, they could not both happen at the same time. Other answers depended on the assumption that the events were *independent*; that is, the occurrence of one of the does not affect the probability of the other. Do you understand the difference between disjoint and independent?

a) If you examine one African-American donor, are the events that the person is Type O and that same person is Type AB disjoint, independent, or neither?
b) If you examine two African-American donors, are the events that the first is Type O and the second is Type AB disjoint, independent, or neither?
c) Can disjoint events ever be independent? Explain.

43. Dice You roll a fair die three times. What is the probability that

a) you roll all 6's?
b) you roll all odd numbers?
c) none of your rolls gets a number divisible by 3?
d) you roll at least one 5?
e) the numbers you roll are not all 5's?

44. Slot machine A slot machine has three wheels that spin independently. Each has 10 equally likely symbols: 4 bars, 3 lemons, 2 cherries, and a bell. If you play, what is the probability that

a) you get 3 lemons?
b) you get no fruit symbols?
c) you get 3 bells (the jackpot)?
d) you get no bells?
e) you get at least one bar (an automatic loser)?

45. Striker A certain football/soccer striker scores 14.6% of his shots. If his shots are independent, what is the probability that he

a) makes three consecutive shots without a goal?
b) scores his first goal on the second shot?
c) scores at least once in the first two shots?
d) scores all his first four shots?

46. The train To get to work, a commuter must cross train tracks. The time the train arrives varies slightly from day to day, but the commuter estimates he'll get stopped on about 15% of work days. During a certain 5-day work week, what is the probability that he

a) gets stopped on Monday and again on Tuesday?
b) gets stopped for the first time on Thursday?
c) gets stopped every day?
d) gets stopped at least once during the week?

47. Union membership Suppose that in a particular city, 25% of workers have obligatory union membership and 30% of workers have voluntary union membership, while 45% of members do not belong to a workers' union. You are conducting a poll by selecting workers from a database at random. In your first three calls, what is the probability that you talk to

a) all workers belonging to a union?
b) no obligatory union member?
c) at least one voluntary union member?

48. Religion Census reports for a city indicate that 62% of residents classify themselves as Christian, 12% as Jewish, and 16% as members of other religions (Muslims, Buddhists, etc.). The remaining residents classify themselves as nonreligious. A polling organization seeking information about public opinions wants to be sure to talk with people holding a variety of religious views, and makes random phone calls. Among the first four people they call, what is the probability they reach

a) all Christians?
b) no Jews?
c) at least one person who is nonreligious?

49. Lights You purchased a five-pack of new light bulbs that were recalled because 6% of the lights did not work. What is the probability that at least one of your lights is defective?

50. Pepsi For a sales promotion, the manufacturer places winning symbols under the caps of 10% of all Pepsi bottles. You buy a six-pack. What is the probability that you win something?

51. 9/11? On September 11, 2002, the first anniversary of the terrorist attack on the World Trade Center, the New York State Lottery's daily number came up 9– 1– 1. An interesting coincidence or a cosmic sign?

a) What is the probability that the winning three numbers match the date on any given day?
b) What is the probability that a whole year passes without this happening?
c) What is the probability that the date and winning lottery number match at least once during any year?
d) If every one of the 50 states has a three-digit lottery, what is the probability that at least one of them will come up 9– 1– 1 on September 11?

52. Red cards You shuffle a deck of cards and then start turning them over one at a time. The first one is red. So is the second. And the third. In fact, you are surprised to get 10 red cards in a row. You start thinking, "The next one is due to be black!"

a) Are you correct in thinking that there's a higher probability that the next card will be black than red? Explain.
b) Is this an example of the Law of Large Numbers? Explain.

JUST CHECKING

ANSWERS

1. The LLN works only in the long run, not in the short run. The random methods for selecting lottery numbers have no memory of previous picks, so there is no change in the probability that a certain number will come up.
2. a) 0.62
   b) $(0.62)(0.62) = 0.3844$
   c) $(1 - 0.62)^2(0.62) = 0.089528$
   d) $1 - (1 - 0.62)^5 = 0.992076$

13 Probability Rules!

WHERE ARE WE GOING?

Is the probability of a car accident the same for all age groups? Insurance companies want to know so they can set their rates. Is everyone at equal risk for getting the flu next winter? Medical researchers use factors such as age, sex, lifestyle, and family history to estimate the probability for each individual. These are examples of **conditional** probabilities. We'll see how the probability of an event can change when we learn more about the situation at hand. Sometimes the results are surprising.

Pull a bill from your wallet or pocket without looking at it. An outcome of this trial is the bill you select. The sample space is all the bills in circulation: **S** = {\$1 bill, \$2 bill, \$5 bill, \$10 bill, \$20 bill, \$50 bill, \$100 bill}.[1] These are *all* the possible outcomes. (In spite of what you may have seen in bank robbery movies, there are no \$500 or \$1000 bills.)

We can combine the outcomes in different ways to make many different events. For example, the event **A** = {\$1, \$5, \$10} represents selecting a \$1, \$5, or \$10 bill. The event **B** = {a bill that does not have a president on it} is the collection of outcomes (Don't look! Can you name them?): {\$10 (Hamilton), \$100 (Franklin)}. The event **C** = {enough money to pay for a \$12 meal with one bill} is the set of outcomes {\$20, \$50, \$}.

Notice that these outcomes are not equally likely. You'd no doubt be more surprised (and pleased) to pull out a \$100 bill than a \$1 bill—it's not as likely, though. You probably carry many more \$1 than \$100 bills, but without information about the probability of each outcome, we can't calculate the probability of an event.

The probability of the event **C** (getting a bill worth more than \$12) is *not* 3/7. There are 7 possible outcomes, and 3 of them exceed \$12, but they are *not equally* likely.

13.1 The General Addition Rule

Now look at the bill in your hand. There are images of famous buildings in the center of the backs of all but two bills in circulation. The \$1 bill has the word ONE in the center, and the \$2 bill shows the signing of the Declaration of Independence.

[1]Well, technically, the sample space is all the bills in your pocket. You may be quite sure there isn't a \$100 bill in there, but *we* don't know that, so humor us that it's at least *possible* that any legal bill could be there.

What's the probability of randomly selecting $\mathbf{A}$ = {a bill with an odd-numbered value} *or* $\mathbf{B}$ = {a bill with a building on the reverse}? We know $\mathbf{A} = \{\$1, \$5\}$ and $\mathbf{B}$ = {\$5, \$10, \$20, \$50, \$100}. But $P(\mathbf{A}\ or\ \mathbf{B})$ is not simply the sum $P(\mathbf{A}) + P(\mathbf{B})$, because the events **A** and **B** are not disjoint (the \$5 bill is in both sets). So what can we do? We'll need a new probability rule.

As the diagrams show, we can't use the Addition Rule and add the two probabilities because the events are not disjoint; they overlap. There's an outcome (the \$5 bill) in the *intersection* of **A** and **B**. The Venn diagram represents the sample space. Notice that the \$2 bill has neither a building nor an odd denomination, so it sits outside both circles.

The \$5 bill plays a crucial role here because it is both odd *and* has a building on the reverse. It's in both **A** and **B**, which places it in the *intersection* of the two circles. The reason we can't simply add the probabilities of **A** and **B** is that we'd count the \$5 bill twice.

If we did add the two probabilities, we could compensate by *subtracting* the probability of that \$5 bill. So,

$$
\begin{aligned}
&P(\text{odd number value or building})\\
&\quad = P(\text{odd number value}) + P(\text{building}) - P(\text{odd number value } \textit{and} \text{ building})\\
&\quad = P(\$1, \$5) + P(\$5, \$10, \$20, \$50, \$100) - P(\$5).
\end{aligned}
$$

This method works in general. We add the probabilities of two events and then subtract out the probability of their intersection. This approach gives us the **General Addition Rule**, which does not require disjoint events:

$$P(\mathbf{A}\ or\ \mathbf{B}) = P(\mathbf{A}) + P(\mathbf{B}) - P(\mathbf{A}\ and\ \mathbf{B}).$$

Drawing a picture using Venn diagrams often makes the rule easier to see and understand.

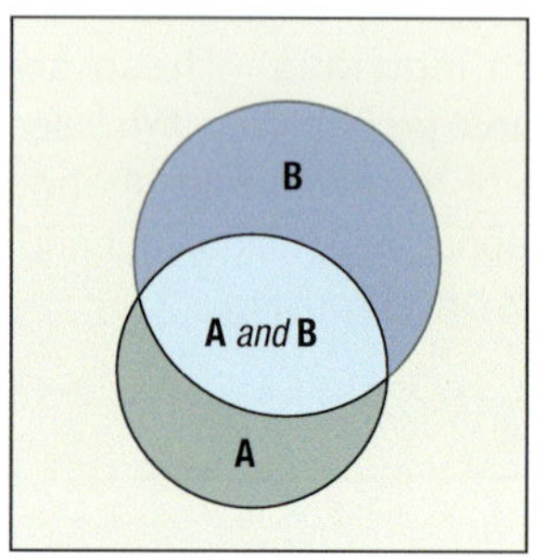

Events **A** and **B** and their intersection.

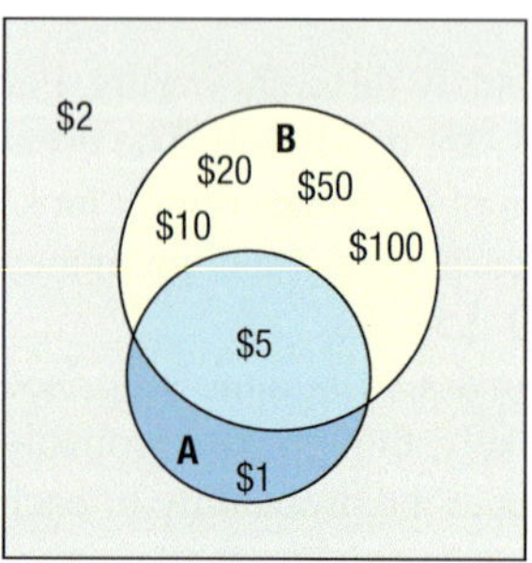

Denominations of bills that are odd (**A**) or that have a building on the reverse side (**B**). The two sets both include the \$5 bill, and both exclude the \$2 bill.

EXAMPLE 13.1 Using the General Addition Rule

A survey of college students asked the questions: "Are you currently in a relationship?" and "Are you involved in intercollegiate or club sports?" The survey found that 33% were currently in a relationship, and 25% were involved in sports. Eleven percent responded "yes" to both.

QUESTION: What's the probability that a randomly selected student either is in a relationship or is involved in athletics?

ANSWER: Let **R** = {student in a relationship}and **S** = {student involved in sports}.

$$
\begin{aligned}
P(\text{a student either is in a relationship or involved in sports}) &= P(\mathbf{R}\ or\ \mathbf{S})\\
&= P(\mathbf{R}) + P(\mathbf{S}) - P(\mathbf{R}\ and\ \mathbf{S})\\
&= 0.33 + 0.25 - 0.11\\
&= 0.47
\end{aligned}
$$

There's a 47% chance that a randomly selected college student either is in a relationship or is involved in sports. Notice that this use of the word "or" includes "or both." That's the usual meaning of *or* in English. If we wanted to exclude those students who were involved in sports and in a relationship we'd have to ask, "What's the probability that a selected student either is in a relationship or is involved in sports, but not both?"

WOULD YOU LIKE DESSERT OR COFFEE?

Natural language can be ambiguous. In this question, what are your choices? Can you have only one of the two, or could you simply answer "yes" and have both dessert and coffee?

In statistics, when we say "or" we always mean the inclusive version. In other words, the probability of **A** or **B** means the probability of either **A** or **B** or both. But in everyday language, "or" is usually the exclusive version meaning one or the other but not both. For example, if you are asked "Would you like the steak or the vegetarian entrée?" you can't get both. This difference can be confusing for students seeing the statistical language for the first time.

The General Addition Rule subtracts the probability of the outcomes in **A** *and* **B** because we've counted those outcomes *twice*. But they're still there. On the other hand, if we really want the probability of **A** or **B**, but NOT both, we have to get rid of the outcomes in {**A** *and* **B**}. So $P(\mathbf{A}$ *or* $\mathbf{B}$, but *not* both) $= P(\mathbf{A} \text{ or } \mathbf{B}) - P(\mathbf{A} \text{ and } \mathbf{B}) - P(\mathbf{A}) + P(\mathbf{B}) - 2 \times P(\mathbf{A} \text{ and } \mathbf{B})$. Now we've subtracted P(**A** *and* **B**) twice—once because we don't want to double-count these events and a second time because we really didn't want to count them at all. Confused? *Make a picture.*

EXAMPLE 13.2 Using Venn Diagrams

RECAP: We return to our survey of college students: 33% are in a relationship, 25% are involved in sports, and 11% are in both.

QUESTIONS: Based on a Venn diagram, what is the probability that a randomly selected student

a. is not in a relationship and is not involved in sports?

b. is in a relationship but not involved in sports?

c. either is in a relationship or is involved in sports, but not both?

ANSWER: Let **R** = {student is in a relationship} and **S** = {student is involved in sports}. In the Venn diagram, the intersection of the circles is $P(\mathbf{R} \text{ and } \mathbf{S}) = 0.11$. Since $P(\mathbf{R}) = 0.33$, $P(\mathbf{R} \text{ and } \mathbf{S^C}) = 0.33 - 0.11 = 0.22$. Also, $P(\mathbf{R^C} \text{ and } \mathbf{S}) = 0.25 - 0.11 = 0.14$. Now, $0.22 + 0.11 + 0.14 = 0.47$, leaving $1 - 0.47 = 0.53$ for the region outside both circles.

Now, P(is not in a relationship and is not involved in sports)
$= P(\mathbf{R^C} \text{ and } \mathbf{S^C}) = 0.53$

P(is in a relationship but not involved in sports) $= P(\mathbf{R} \text{ and } \mathbf{S^C}) = 0.22$

P(is in a relationship or involved in sports, but not both)
$= P(\mathbf{R} \text{ and } \mathbf{S^C}) + P(\mathbf{R^C} \text{ and } \mathbf{S}) = 0.22 + 0.14 = 0.36.$

Or

$P(\mathbf{R}) + P(\mathbf{S}) - 2P(\mathbf{R} \text{ and } \mathbf{S}) = 0.33 + 0.25 - 2(0.11) = 0.36.$

JUST CHECKING

1. Back in Chapter 1, we suggested that you sample some pages of this text at random to see whether they held some sort of data display or an equation. We actually did just that. We drew a representative sample and found the following:

48% of pages had some kind of data display,

27% of pages had an equation, and

7% of pages had both a data display and an equation.

a) Display these results in a Venn diagram.

b) What is the probability that a randomly selected sample page had neither a data display nor an equation?

c) What is the probability that a randomly selected sample page had a data display but no equation?

STEP-BY-STEP EXAMPLE

Using the General Addition Rule

You almost certainly qualify as an "online adult"—an adult who uses the Internet. Pew Research tracks the behavior of online adults. In 2013 they found that 71% were Facebook users, 18% used Twitter, and 15% used both. www.pewinternet.org/fact-sheet/social-media/has the latest data.

QUESTIONS: For a randomly selected online adult, what is the probability that the person

1. uses either Facebook or Twitter?
2. uses either Facebook or Twitter, but not both?
3. doesn't use Facebook or Twitter?

THINK

PLAN Define the events we're interested in. There are no conditions to check; the General Addition Rule works for any events!

Let $\mathbf{A} = \{\text{respondent uses Facebook}\}$.

Let $\mathbf{B} = \{\text{respondent uses Twitter}\}$.

PLOT Make a picture, and use the given probabilities to find the probability for each region.

I know that

$$P(\mathbf{A}) = 0.71$$
$$P(\mathbf{B}) = 0.18$$
$$P(\mathbf{A} \textit{ and } \mathbf{B}) = 0.15$$

The blue region represents **A** but not **B**. The green intersection region represents **A** *and* **B**. Note that since $P(\mathbf{A}) = 0.71$ and $P(\mathbf{A} \textit{ and } \mathbf{B}) = 0.15$, the probability of **A** but not **B** must be $0.71 - 0.15 = 0.56$.

So

$$P(\mathbf{A} \text{ and } \mathbf{B}^{\mathbf{C}}) = 0.71 - 0.15 = 0.56$$

The yellow region is **B** but not **A**.

$$P(\mathbf{B} \text{ and } \mathbf{A}^{\mathbf{C}}) = 0.18 - 0.15 = 0.03$$

The gray region outside both circles represents the outcome neither **A** nor **B**. All the probabilities must total 1, so you can determine the probability of that region by subtraction.

$$P(\mathbf{A}^{\mathbf{C}} \text{ and } \mathbf{B}^{\mathbf{C}}) = 1 - (0.56 + 0.15 + 0.03) = 0.26$$

Now, figure out what you want to know. The probabilities can come from the diagram or a formula. Sometimes translating the words to equations is the trickiest step.

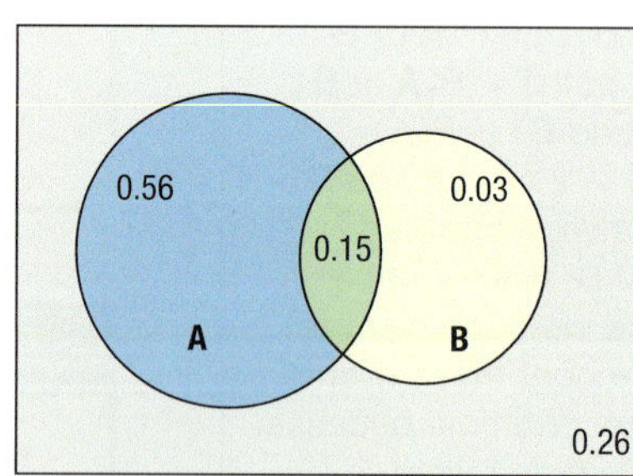

QUESTION 1. What is the probability that the person uses either Facebook or Twitter?

SHOW

MECHANICS The probability that the respondent uses Facebook or Twitter is $P(\mathbf{A}$ *or* $\mathbf{B})$. We can use the General Addition Rule, or we can add the probabilities seen in the diagram.

$$P(\mathbf{A} \textit{ or } \mathbf{B}) = P(\mathbf{A}) + P(\mathbf{B}) - P(\mathbf{A} \textit{ and } \mathbf{B}) = 0.71 + 0.18 - 0.15 = 0.74$$

OR

$$P(\mathbf{A} \textit{ or } \mathbf{B}) = 0.56 + 0.15 + 0.03 = 0.74$$

TELL **CONCLUSION** Don't forget to interpret your result in context.	76% of online adults use Facebook or Twitter.

QUESTION 2. What is the probability that the person uses either Facebook or Twitter, but not both?

SHOW **MECHANICS** We can use the rule, or just add the appropriate probabilities seen in the Venn diagram.	$P(\mathbf{A} \text{ or } \mathbf{B} \text{ but NOT both}) = P(\mathbf{A} \text{ or } \mathbf{B}) - P(\mathbf{A} \text{ and } \mathbf{B})$ $= 0.74 - 0.15 = 0.59$ *OR* $P(\mathbf{A} \text{ or } \mathbf{B} \text{ but NOT both})$ $= P(\mathbf{A} \text{ and } \mathbf{B}^C) + P(\mathbf{B} \text{ and } \mathbf{A}^C)$ $= 0.56 + 0.03 = 0.59$
TELL **CONCLUSION** Interpret your result in context.	59% of online adults use Facebook or Twitter, but not both.

QUESTION 3. What is the probability that the person doesn't use Facebook or Twitter?

SHOW **MECHANICS** Not using Facebook or Twitter is the complement of using one or the other. Use the Complement Rule or just notice that "not using Facebook or Twitter" is represented by the region outside both circles.	$P(\text{not Facebook or Twitter}) =$ $1 - P(\text{either Facebook or Twitter})$ $= 1 - P(\mathbf{A} \text{ or } \mathbf{B})$ $= 1 - 0.74 = 0.26$ *OR* $P(\mathbf{A}^C \text{ and } \mathbf{B}^C) = 0.26$
TELL **CONCLUSION** Interpret your result in context.	Only 26% of online adults use neither Facebook nor Twitter.

13.2 Conditional Probability and the General Multiplication Rule

Two psychologists surveyed 478 children in grades 4, 5, and 6 in elementary schools in Michigan. They stratified their sample, drawing roughly 1/3 from rural, 1/3 from suburban, and 1/3 from urban schools. Among other questions, they asked the students whether their primary goal was to get good grades, to be popular, or to be good at sports. One question of interest was whether boys and girls at this age had similar goals.

Here's a *contingency table* giving counts of the students by their goals and sex:

Table 13.1
The distribution of goals for boys and girls.

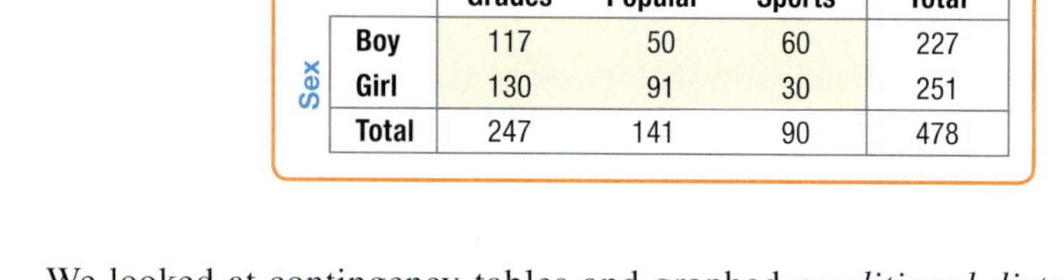

Sex	Goals: Grades	Popular	Sports	Total
Boy	117	50	60	227
Girl	130	91	30	251
Total	247	141	90	478

Girls

Boys

Grades
Popular
Sports

Figure 13.1
The distribution of goals for boys and girls.

We looked at contingency tables and graphed *conditional distributions* back in Chapter 2. The pie charts show the *relative frequencies* with which boys and girls named the three goals. It's only a short step from these relative frequencies to probabilities.

Let's focus on this study and make the sample space just the set of these 478 students. If we select a student at random from this study, the probability we select a girl is just the corresponding relative frequency (since we're equally likely to select any of the 478 students). There are 251 girls in the data out of a total of 478, giving a probability of

$$P(\text{girl}) = 251/478 = 0.525$$

The same method works for more complicated events like intersections. For example, what's the probability of selecting a girl whose goal is to be popular? Well, 91 girls named popularity as their goal, so the probability is

$$P(\text{girl } \textit{and} \text{ popular}) = 91/478 = 0.190$$

The probability of selecting a student whose goal is to excel at sports is

$$P(\text{sports}) = 90/478 = 0.188$$

What if we are given the information that the selected student is a girl? Would that change the probability that the selected student's goal is sports? You bet it would! The pie charts show that girls are much less likely to say their goal is to excel at sports than are boys. When we restrict our focus to girls, we look only at the girls' row of the table. Of the 251 girls, only 30 of them said their goal was to excel at sports.

We write the probability that a selected student wants to excel at sports *given that we have selected a girl* as

$$P(\text{sports} \mid \text{girl}) = 30/251 = 0.120$$

For boys, we look at the conditional distribution of goals given "boy" shown in the top row of the table. There, of the 227 boys, 60 said their goal was to excel at sports. So, $P(\text{sports} \mid \text{boy}) = 60/227 = 0.264$, more than twice the girls' probability.

In general, when we want the probability of an event from a *conditional* distribution, we write $P(\mathbf{B} \mid \mathbf{A})$ and pronounce it "the probability of **B** *given* **A**." A probability that takes into account a given *condition* such as this is called a **Conditional Probability**.

Let's look at what we did. We worked with the counts, but we could work with the probabilities just as well. There were 30 students who both were girls and had sports as their goal, and there are 251 girls. So we found the probability to be 30/251. To find the probability of the event **B** *given* the event **A**, we restrict our attention to the outcomes in **A**. We then find in what fraction of *those* outcomes **B** also occurred. Formally, we write

NOTATION ALERT
$P(\mathbf{B} \mid \mathbf{A})$ is the conditional probability of **B** *given* **A**.

$$P(\mathbf{B} \mid \mathbf{A}) = \frac{P(\mathbf{A} \textit{ and } \mathbf{B})}{P(\mathbf{A})}.$$

Thinking this through, we can see that it's just what we've been doing, but now with probabilities rather than with counts. Look back at the girls for whom sports was the goal. How did we calculate $P(\text{sports} \mid \text{girl})$?

The rule says to use probabilities. It says to find $P(\mathbf{A}\ \textit{and}\ \mathbf{B})/P(\mathbf{A})$. The result is the same whether we use counts or probabilities because the total number in the sample cancels out:

$$\frac{P(\text{sports } \textit{and} \text{ girl})}{P(\text{girl})} = \frac{30/478}{251/478} = \frac{30}{251} = 0.12.$$

To use the formula for conditional probability, we're supposed to insist on one restriction. The formula doesn't work if $P(\mathbf{A})$ is 0. After all, we can't be "given" the fact that **A** was true if the probability of **A** is 0!

Let's take our rule out for a spin. What's the probability that we have selected a girl *given* that the selected student's goal is popularity? Applying the rule, we get

$$P(\text{girl} \mid \text{popular}) = \frac{P(\text{girl } \textit{and} \text{ popular})}{P(\text{popular})}$$
$$= \frac{91/478}{141/478} = \frac{91}{141} = 0.65.$$

EXAMPLE 13.3

Finding a Conditional Probability

RECAP: Our survey found that 33% of college students are in a relationship, 25% are involved in sports, and 11% are both.

QUESTION: While dining in a campus facility open only to students, you meet a student who is on a sports team. What is the probability that your new acquaintance is in a relationship?

ANSWER: Let $\mathbf{R} = \{\text{student in a relationship}\}$ and $\mathbf{S} = \{\text{student is involved in sports}\}$.

$$P(\text{student is in a relationship given that the student is involved in sports}) = P(\mathbf{R} \mid \mathbf{S})$$
$$= \frac{P(\mathbf{R} \textit{ and } \mathbf{S})}{P(\mathbf{S})}$$
$$= \frac{0.11}{0.25}$$
$$\approx 0.44$$

There's a probability of about 0.44 that a student involved with sports is in a relationship. Notice that this is higher than the probability for all students.

The General Multiplication Rule

Remember the Multiplication Rule for the probability of **A** *and* **B**? It said

$$P(\mathbf{A} \textit{ and } \mathbf{B}) = P(\mathbf{A}) \times P(\mathbf{B}) \text{ when } \mathbf{A} \text{ and } \mathbf{B} \text{ are independent.}$$

Now we can write a more general rule that doesn't require independence. In fact, we've *already* written it down. We just need to rearrange the equation a bit.

The equation in the definition for conditional probability contains the probability of **A** *and* **B**. Rewriting the equation gives

$$P(\mathbf{A} \textit{ and } \mathbf{B}) = P(\mathbf{A}) \times P(\mathbf{B} \mid \mathbf{A}).$$

This is a **General Multiplication Rule** for compound events that does not require the events to be independent. Better than that, it even makes sense. The probability that two events, **A** and **B**, *both* occur is the probability that event **A** occurs multiplied by the probability that event **B** *also* occurs—that is, by the probability that event **B** occurs *given* that event **A** occurs.

Of course, there's nothing special about which set we call **A** and which one we call **B**. We should be able to state this the other way around. And indeed we can. It is equally true that

$$P(\mathbf{A} \textit{ and } \mathbf{B}) = P(\mathbf{B}) \times P(\mathbf{A} \mid \mathbf{B}).$$

13.3 Independence

Independence

If we had to pick one idea in this chapter that you should understand and remember, it's the definition and meaning of independence. We'll need this idea in every one of the chapters that follow.

Let's return to the question of just what it means for events to be independent. We've said informally that what we mean by independence is that the outcome of one event does not influence the probability of the other. With our new notation for conditional probabilities, we can write a formal definition: Events **A** and **B** are **independent** if (and *only* if)

$$P(\mathbf{B} \mid \mathbf{A}) = P(\mathbf{B}).$$

From now on, use the formal rule whenever you can. The Multiplication Rule for independent events that we saw in Chapter 12 is just a special case of the General Multiplication Rule. The general rule says

$$P(\mathbf{A} \textit{ and } \mathbf{B}) = P(\mathbf{A}) \times P(\mathbf{B} \mid \mathrm{A}),$$

whether the events are independent or not. But when events **A** and **B** are independent, we can write $P(\mathbf{B})$ for $P(\mathbf{B} \mid \mathbf{A})$ and we get back our simple rule:

$$P(\mathbf{A} \textit{ and } \mathbf{B}) = P(\mathbf{A}) \times P(\mathbf{B}).$$

Sometimes people use this statement as the definition of independent events, but we find the other definition more intuitive. Either way, the idea is that for independent events, the probability of one doesn't change when the other occurs.

Is the probability of having good grades as a goal independent of the sex of the responding student? Looks like it might be. We need to check whether

$$P(\text{grades} \mid \text{girl}) = P(\text{grades})$$

$$\frac{130}{251} = 0.52 \stackrel{?}{=} \frac{247}{478} = 0.52$$

To two decimal place accuracy, it looks like we can consider choosing good grades as a goal to be independent of sex.

On the other hand, $P(\text{sports})$ is 90/478, or about 18.8%, but $P(\text{sports} \mid \text{boy})$ is $60/227 = 26.4\%$. Because these probabilities aren't equal, we can be pretty sure that choosing success in sports as a goal is not independent of the student's sex.

EXAMPLE 13.4 Checking for Independence

RECAP: Our survey told us that 33% of college students are in a relationship, 25% are involved in sports, and 11% are both.

QUESTION: Are being in a relationship and being involved in sports independent? Are they disjoint?

ANSWER: Let $\mathbf{R} = \{\text{student is in a relationship}\}$ and $\mathbf{S} = \{\text{student is involved in sports}\}$. If these events are independent, then knowing that a student does sports doesn't affect the probability that he or she is in a relationship.

But we saw that $P(\mathbf{R} \mid \mathbf{S}) = P(\mathbf{R} \text{ and } \mathbf{S})/P(\mathbf{S}) = 0.44$ and we know that $P(\mathbf{R}) = 0.33$, so the events are not independent. Students who are involved in sports are more likely to be in a relationship on this campus.

These events are not disjoint either. In fact, 11% of all students on this campus are both involved in sports and in a relationship.

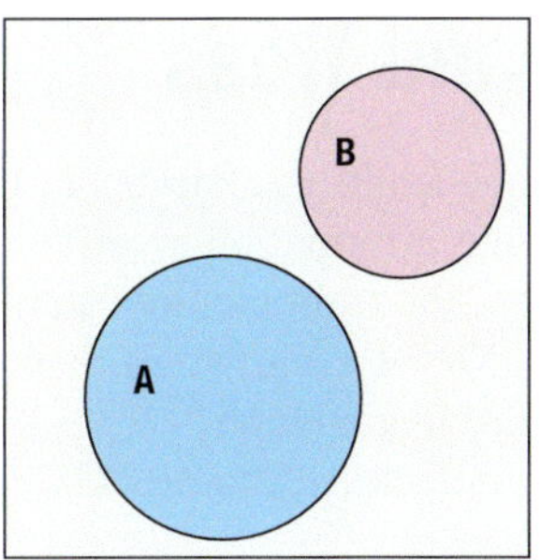

Figure 13.2

Because events **A** and **B** are mutually exclusive, learning that **A** happened tells us that **B** didn't. The probability of **B** has changed from whatever it was to zero. So disjoint events **A** and **B** are not independent.

Independent ≠ Disjoint

Are disjoint events independent? Both disjoint and independent seem to imply separation and distinctness, but in fact disjoint events *cannot* be independent.[2] Let's see why. Consider the two disjoint events {you get an A in this course} and {you get a B in this course}. They're disjoint because they have no outcomes in common. Suppose you learn that you *did* get an A in the course. Now what is the probability that you got a B? You can't get both grades, so it must be 0.

Think about what that means. Knowing that the first event (getting an A) occurred changed your probability for the second event (down to 0). So these events aren't independent.

Mutually exclusive events can't be independent. They have no outcomes in common, so if one occurs, the other doesn't. A common error is to treat disjoint events as if they were independent, and apply the Multiplication Rule for independent events. Don't make that mistake.

JUST CHECKING

2. The American Association for Public Opinion Research (AAPOR) is an association of about 1600 individuals who share an interest in public opinion and survey research. They report that typically as few as 10% of random phone calls result in a completed interview. Reasons are varied, but some of the most common include no answer, refusal to cooperate, and failure to complete the call.

   Which of the following events are independent, which are disjoint, and which are neither independent nor disjoint?

   a) **A** = Your telephone number is randomly selected. **B** = You're not at home at dinnertime when they call.

   b) **A** = As a selected subject, you complete the interview. **B** = As a selected subject, you refuse to cooperate.

   c) **A** = You are not at home when they call at 11 a.m. **B** = You are employed full-time.

Depending on Independence

It's much easier to think about independent events than to deal with conditional probabilities, but treating all events as independent can lead to wrong, and possibly costly, decisions.

For example, insurance companies that insure against natural disasters like hurricanes or earthquakes might want to avoid insuring too many buildings in the same city. For an event like the Japan earthquake, the probability of two buildings suffering damage is not independent if they're in the same city. Even though the probability of a building being destroyed by an earthquake may be small, the probability that two neighboring buildings are both destroyed by an earthquake is not the product of the two. In fact, it may be close to the probability that a single one is destroyed. The probabilities can't be multiplied unless the events are independent. For that reason, it's safer for a company to hold policies on buildings that are geographically spread out because then the events are closer to being independent.

Whenever you see probabilities multiplied together, stop and ask whether you think they are really independent.

[2]Well, technically two disjoint events *can* be independent, but only if the probability of one of the events is 0. For practical purposes, though, we can ignore this case. After all, as statisticians we don't anticipate having data about things that never happen.

13.4 Picturing Probability: Tables, Venn Diagrams, and Trees

In the Facebook–Twitter example, we found that 71% of online adults use Facebook, 18% use Twitter, and 15% use both. That may not look like enough information to answer all the questions you[3] might have about the problem. But often, the right picture can help. Surprisingly there are three very different ways to do this. We've seen two of them already: Venn diagrams and probability tables. We'll meet a third in a minute.

Let's try to put what we know into a table. Translating percentages to probabilities, what we know looks like this:

		Use Facebook		
		Yes	No	Total
Use Twitter	Yes	0.15		0.18
	No			
	Total	0.71		1.00

Notice that the 0.71 and 0.18 are *marginal* probabilities and so they go into the *margins*. The 0.15 is the probability of using both sites—Facebook and Twitter—so that's a *joint* probability. Those belong in the interior of the table.

Because the cells of the table show disjoint events, the probabilities always add to the marginal totals going across rows or down columns. So, filling in the rest of the table is quick:

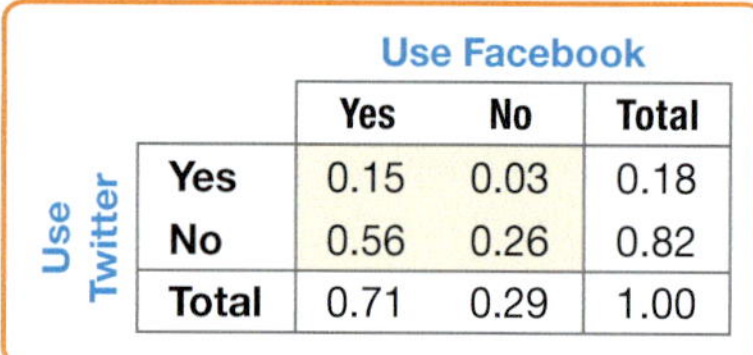

		Use Facebook		
		Yes	No	Total
Use Twitter	Yes	0.15	0.03	0.18
	No	0.56	0.26	0.82
	Total	0.71	0.29	1.00

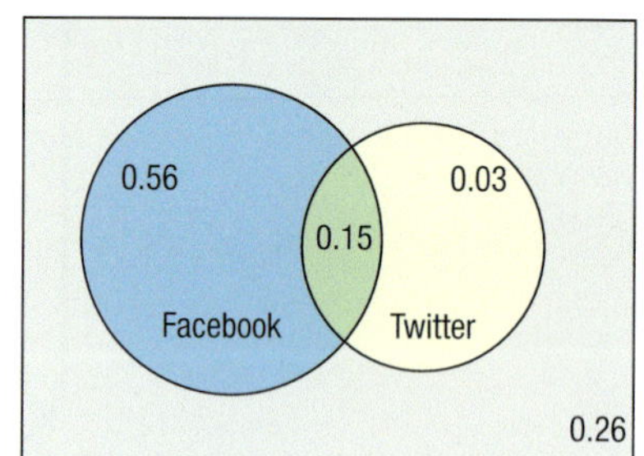

Compare this with the Venn diagram. Notice which entries in the table match up with the sets in this diagram. Whether a Venn diagram or a table is better to use will depend on what you are given and the questions you're being asked. Try both.

STEP-BY-STEP EXAMPLE

Are the Events Disjoint? Independent?

Let's take another look at the social networking survey. Researchers report that 71% of online adults use Facebook, 18% use Twitter, and 15% use both.

QUESTIONS: 1. Are using Facebook and using Twitter mutually exclusive?
2. Are Facebooking and tweeting independent?

[3]or your instructor.

THINK **PLAN** Define the events we're interested in.

State the given probabilities.

Let $\mathbf{A}$ = {respondent uses Facebook}.

Let $\mathbf{B}$ = {respondent uses Twitter}.

I know that $P(\mathbf{A}) = 0.71$

$P(\mathbf{B}) = 0.18$

$P(\mathbf{A} \text{ and } \mathbf{B}) = 0.15$

QUESTION 1. Are using Facebook and using Twitter mutually exclusive?

SHOW **MECHANICS** Disjoint events cannot *both* happen at the same time, so check to see if $P(\mathbf{A} \text{ and } \mathbf{B}) = 0$.

$P(\mathbf{A} \text{ and } \mathbf{B}) = 0.15$. Since some adults use both Facebook and Twitter, $P(\mathbf{A} \text{ and } \mathbf{B}) \neq 0$. The events are not mutually exclusive.

TELL **CONCLUSION** State your conclusion in context.

15% of online adults use both Facebook and Twitter, so these are not disjoint events.

QUESTION 2. Are Facebooking and tweeting independent?

THINK **PLAN** Make a table.

		Use Facebook		
		Yes	**No**	**Total**
Use Twitter	**Yes**	0.15	0.03	**0.18**
	No	0.56	0.26	**0.82**
	Total	**0.71**	**0.29**	**1.00**

SHOW **MECHANICS** Does using Facebook change the probability of using twitter? That is, does $P(\mathbf{B} \mid \mathbf{A}) = P(\mathbf{B})$?
Because the two probabilities are *not* the same, the events are not independent.

$$P(\mathbf{B} \mid \mathbf{A}) = \frac{P(\mathbf{A} \text{ and } \mathbf{B})}{P(\mathbf{A})} = \frac{0.15}{0.71} \approx 0.21$$

$P(\mathbf{B}) = 0.62$

$P(\mathbf{B} \mid \mathbf{A}) \neq P(\mathbf{B})$

TELL **CONCLUSION** Interpret your results in context.

Overall, 18% of online adults use Twitter, but 21% of Facebook users use Twitter. Since respondents who use Facebook are more likely to tweet, the two events are not independent.

JUST CHECKING

3. Remember our sample of pages in this text from the earlier Just Checking . . .?

48% of pages had a data display,

27% of pages had an equation, and

7% of pages had both a data display and an equation.

a) Make a contingency table for the variables *display* and *equation*.

b) What is the probability that a randomly selected sample page with an equation also had a data display?

c) Are having an equation and having a data display disjoint events?

d) Are having an equation and having a data display independent events?

"Why," said the Dodo, "the best way to explain it is to do it."

—*Lewis Carroll*

Tree Diagrams

For men, binge drinking is defined as having five or more drinks in a row, and for women as having four or more drinks in a row. (The difference is because of the average difference in weight.) According to a study by the Harvard School of Public Health, 44% of college students engage in binge drinking, 37% drink moderately, and 19% abstain entirely. Another study, published in the *American Journal of Health Behavior*, finds that among binge drinkers aged 21 to 34, 17% have been involved in an alcohol-related automobile accident, while among non-bingers of the same age, only 9% have been involved in such accidents.

What's the probability that a randomly selected college student will be a binge drinker and has had an alcohol-related car accident?

To start, we see that the probability of selecting a binge drinker is about 44%. To find the probability of selecting someone who is both a binge drinker and a driver with an alcohol-related accident, we would need to pull out the General Multiplication Rule and multiply the probability of one of the events by the conditional probability of the other given the first.

Or we *could* make a picture. Which would you prefer?

We thought so.

Neither tables nor Venn diagrams can handle conditional probabilities. The kind of picture that helps us look at conditional probabilities is called a **tree diagram** because it shows sequences of events as paths that look like branches of a tree. It is a good idea to make a tree diagram almost any time you plan to use the General Multiplication Rule. The number of different paths we can take can get large, so we usually draw the tree starting from the left and growing vine-like across the page, although sometimes you'll see them drawn from the bottom up or top down.

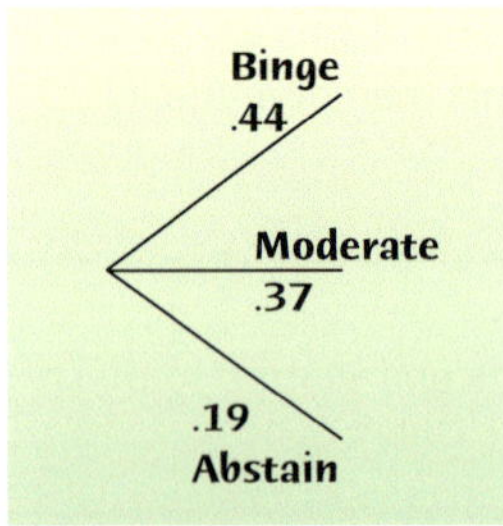

Figure 13.3
We can diagram the three outcomes of drinking and indicate their respective probabilities with a simple tree diagram.

The first branch of our tree (Figure 13.3) separates students according to their drinking habits. We label each branch of the tree with a possible outcome and its corresponding probability.

Notice that we cover all possible outcomes with the branches. The probabilities add up to one. But we're also interested in car accidents. The probability of having an alcohol-related accident *depends* on one's drinking behavior. Because the probabilities are *conditional,* we draw the alternatives separately on each branch of the tree.

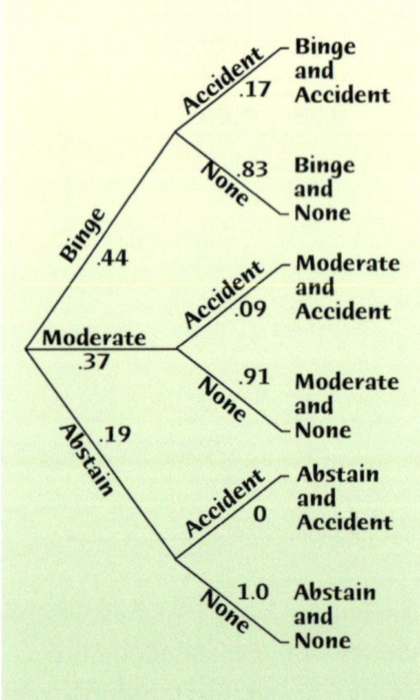

Figure 13.4
Extending the tree diagram, we can show both drinking and accident outcomes. The accident probabilities are conditional on the drinking outcomes, and they change depending on which branch we follow. Because we are concerned only with *alcohol-related* accidents, the conditional probability $P(\text{accident}|\text{abstinence})$ must be 0.

On each of the second set of branches, we write the possible outcomes associated with having an alcohol-related car accident (having an accident or not) and the associated probability (Figure 13.4). These probabilities are different because they are *conditional* depending on the student's drinking behavior. (It shouldn't be too surprising that those who binge drink have a higher probability of alcohol-related accidents.) The probabilities add up to one, because given the outcome on the first branch, these outcomes cover all the possibilities. Looking back at the General Multiplication Rule, we can see how the tree depicts the calculation. To find the probability that a randomly selected student will be a binge drinker who has had an alcohol-related car accident, we follow the top branches. The probability of selecting a binger is 0.44. The conditional probability of an accident *given* binge drinking is 0.17. The General Multiplication Rule tells us that to find the *joint* probability of being a binge drinker and having an accident, we multiply these two probabilities together:

$$\begin{aligned} P(\text{binge } \textit{and} \text{ accident}) &= P(\text{binge}) \times P(\text{accident}\,|\,\text{binge}) \\ &= 0.44 \times 0.17 = 0.075 \end{aligned}$$

And we can do the same for each combination of outcomes:

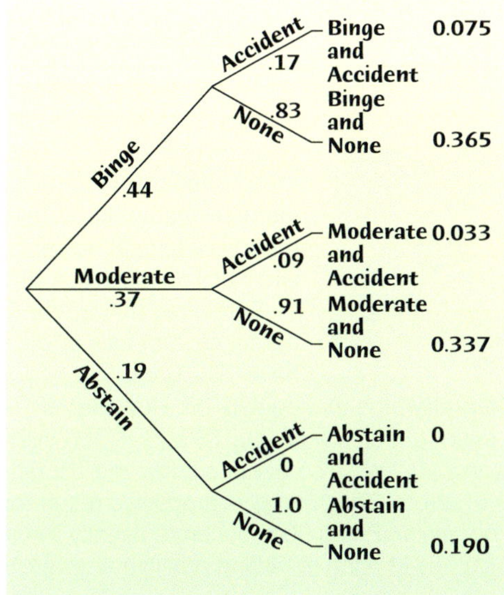

Figure 13.5
We can find the probabilities of compound events by multiplying the probabilities along the branch of the tree that leads to the event, just the way the General Multiplication Rule specifies. The probability of abstaining and having an alcohol-related accident is, of course, zero.

All the outcomes at the far right are disjoint because at each branch of the tree we chose between disjoint alternatives. And they are *all* the possibilities, so the probabilities on the far right must add up to one.

Because the final outcomes are disjoint, we can add up their probabilities to get probabilities for compound events. For example, what's the probability that a selected student has had an alcohol-related car accident? We simply find *all* the outcomes on the far right in which an accident has happened. There are three and we can add their probabilities: $0.075 + 0.033 + 0 = 0.108$—almost an 11% chance.

13.5 Reversing the Conditioning and Bayes' Rule

If we know a student has had an alcohol-related accident, what's the probability that the student is a binge drinker? That's an interesting question, but we can't just read it from the tree. The tree gives us $P(\text{accident} \mid \text{binge})$, but we want $P(\text{binge} \mid \text{accident})$—conditioning in the other direction. The two probabilities are definitely *not* the same. We have reversed the conditioning.

We may not have the conditional probability we want, but we do know everything we need to know to find it. To find a conditional probability, we need the probability that both events happen divided by the probability that the given event occurs. We have already found the probability of an alcohol-related accident: $0.075 + 0.033 + 0 = 0.108$.

The joint probability that a student is both a binge drinker and someone who's had an alcohol-related accident is found at the top branch: 0.075. We've restricted the *Who* of the problem to the students with alcohol-related accidents, so we divide the two to find the conditional probability:

$$P(\text{binge} \mid \text{accident}) = \frac{P(\text{binge } \textit{and} \text{ accident})}{P(\text{accident})}$$

$$= \frac{0.075}{0.108} = 0.694$$

The chance that a student who has an alcohol-related car accident is a binge drinker is more than 69%! As we said, reversing the conditioning is rarely intuitive, but tree diagrams help us keep track of the calculation when there aren't too many alternatives to consider.

STEP-BY-STEP EXAMPLE

Reversing the Conditioning

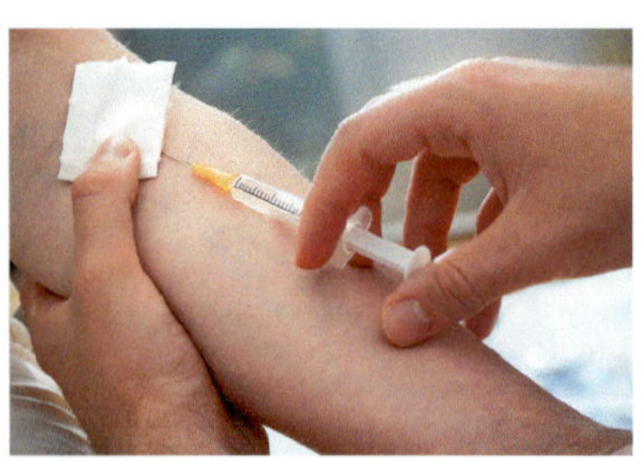

When the authors were in college, there were only three requirements for graduation that were the same for all students: You had to be able to tread water for 2 minutes, you had to learn a foreign language, and you had to be free of tuberculosis. For the last requirement, all freshmen had to take a TB screening test that consisted of a nurse jabbing what looked like a corncob holder into your forearm. You were then expected to report back in 48 hours to have it checked. If you were healthy and TB-free, your arm was supposed to look as though you'd never had the test.

Sometime during the 48 hours, one of us had a reaction. When he finally saw the nurse, his arm was about 50% bigger than normal and a very unhealthy red. Did he have TB? The nurse had said that the test was about 99% effective, so it seemed that the chances must be pretty high that he had TB. How high do you think the chances were? Go ahead and guess. Guess low.

We'll call **TB** the event of actually having TB and + the event of testing positive. To start a tree, we need to know $P(\mathbf{TB})$, the probability of having TB.[4] We also need to know the conditional probabilities $P(+\mid\mathbf{TB})$ and $P(+\mid\mathbf{TB^C})$. Diagnostic tests can make two kinds of errors. They can give a positive result for a healthy person (a *false positive*) or a negative result for a sick person (a *false negative*). Being 99% accurate usually means a false-positive rate of 1%. That is, someone who doesn't have the disease has a 1% chance of testing positive anyway. We can write $P(+\mid\mathbf{TB^C}) = 0.01$.

Since a false negative is more serious (because a sick person might not get treatment), tests are usually constructed to have a lower false-negative rate. We don't know exactly, but let's assume a 0.1% false-negative rate. So only 0.1% of sick people test negative. We can write $P(-\mid\mathbf{TB}) = 0.001$.

THINK **PLAN** Define the events we're interested in and their probabilities.

Let $\mathbf{TB} = \{\text{having } \mathbf{TB}\}$ and $\mathbf{TB^C} = \{\text{no TB}\}$

$+ = \{\text{testing positive}\}$ and

$- = \{\text{testing negative}\}$

Figure out what you want to know in terms of the events. Use the notation of conditional probability to write the event whose probability you want to find.

I know that $P(+\mid\mathbf{TB^C}) = 0.01$ and $P(-\mid\mathbf{TB}) = 0.001$. I also know that $P(\mathbf{TB}) = 0.00005$.

I'm interested in the probability that the author had TB given that he tested positive: $P(\mathbf{TB}\mid +)$.

SHOW **PLOT** Draw the tree diagram. When probabilities are very small like these are, be careful to keep all the significant digits.

To finish the tree we need $P(\mathbf{TB^C})$, $P(-\mid\mathbf{TB^C})$, and $P(+\mid\mathbf{TB})$. We can find each of these from the Complement Rule:

$$P(\mathbf{TB^C}) = 1 - P(\mathbf{TB}) = 0.99995$$

$$P(-\mid\mathbf{TB^C}) = 1 - P(+\mid\mathbf{TB^C}) = 1 - 0.01 = 0.99 \text{ and}$$

$$P(+\mid\mathbf{TB}) = 1 - P(-\mid\mathbf{TB}) = 1 - 0.001 = 0.999$$

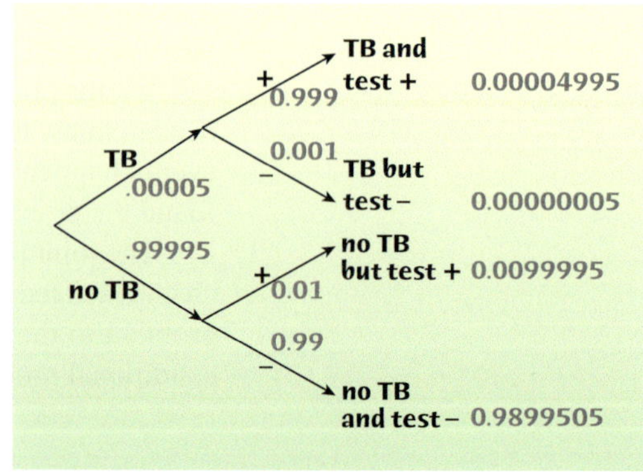

MECHANICS Multiply along the branches to find the probabilities of the four possible outcomes. Check your work by seeing if they total 1.

(Check: 0.00004995 + 0.00000005 + 0.0099995 + 0.98995050 = 1)

[4]This isn't given, so we looked it up. Although TB is a matter of serious concern to public health officials, it is a fairly uncommon disease, with an incidence of about 5 cases per 100,000 in the United States (see www.cdc.gov/tb/default.htm).

Add up the probabilities corresponding to the condition of interest—in this case, testing positive. We can add the probabilities from the tree twigs that correspond to testing positive because the tree shows disjoint events.

$$P(+) = P(\mathbf{TB}\ and\ +) + P(\mathbf{TB^C}\ and\ +)$$
$$P(+) = 0.00004995 + 0.0099995 = 0.01004945$$

Divide the probability of both events occurring (here, having TB and a positive test) by the probability of satisfying the condition (testing positive).

$$P(\mathbf{TB} \mid +) = \frac{P(\mathbf{TB}\ and\ +)}{P(+)} = \frac{0.00004995}{0.01004945} = 0.00497$$

TELL **CONCLUSION** Interpret your result in context.

The chance of having TB after a positive test is less than 0.5%.

When we reverse the order of conditioning, we change the *Who* we are concerned with. With events of low probability, the result can be surprising. That's the reason patients who test positive for HIV, for example, are always told to seek medical counseling. They may have only a small chance of actually being infected. That's why global drug or disease testing can have unexpected consequences if people interpret *testing* positive as *being* positive.

The Reverend Thomas Bayes is credited posthumously with the rule that is the foundation of Bayesian Statistics.

Bayes' Rule

When we have $P(\mathbf{A} \mid \mathbf{B})$ but want the *reverse* probability $P(\mathbf{B} \mid \mathbf{A})$, we need to find $P(\mathbf{A}\ and\ \mathbf{B})$ and $P(\mathbf{A})$. A tree is often a convenient way of finding these probabilities. It can work even when we have more than two possible events, as we saw in the binge-drinking example. Instead of using the tree, we *could* write the calculation algebraically, showing exactly how we found the quantities that we needed: $P(\mathbf{A}\ and\ \mathbf{B})$ and $P(\mathbf{A})$. The result is a formula known as Bayes' Rule, after the Reverend Thomas Bayes (1702?–1761), who was credited with the rule after his death, when he could no longer defend himself. Bayes' Rule is quite important in statistics and is the foundation of an approach to statistical analysis known as Bayesian Statistics. Although the simple rule deals with two alternative outcomes, the rule can be extended to the situation in which there are more than two branches to the first split of the tree. The principle remains the same (although the math gets more difficult). Bayes' Rule is just a formula for reversing the probability from the conditional probability that you're originally given. Bayes' Rule for two events says that $P(\mathbf{B} \mid \mathrm{A}) = \dfrac{P(\mathbf{A} \mid \mathbf{B})P(\mathbf{B})}{P(\mathbf{A} \mid \mathbf{B})P(\mathbf{B}) + P(\mathbf{A} \mid \mathbf{B}^C)P(\mathbf{B}^C)}$. Masochists may wish to try it with the TB testing probabilities. (It's easier to just draw the tree, isn't it?)

EXAMPLE 13.5 **Reversing the Conditioning**

A recent Maryland highway safety study found that in 77% of all accidents the driver was wearing a seatbelt. Accident reports indicated that 92% of those drivers escaped serious injury (defined as hospitalization or death), but only 63% of the nonbelted drivers were so fortunate.

QUESTION: What's the probability that a driver who was seriously injured wasn't wearing a seatbelt?

ANSWER: Let B = the driver was wearing a seatbelt, and NB = no belt.

Let I = serious injury or death, and OK = not seriously injured.

I know $P(\mathrm{B}) = 0.77$, so $P(\mathrm{NB}) = 1 - 0.77 = 0.23$.

Also, $P(\mathrm{OK} \mid \mathrm{B}) = 0.92$, so $P(\mathrm{I} \mid \mathrm{B}) = 0.08$
and $P(\mathrm{OK} \mid \mathrm{NB}) = 0.63$, so $P(\mathrm{I} \mid \mathrm{NB}) = 0.37$

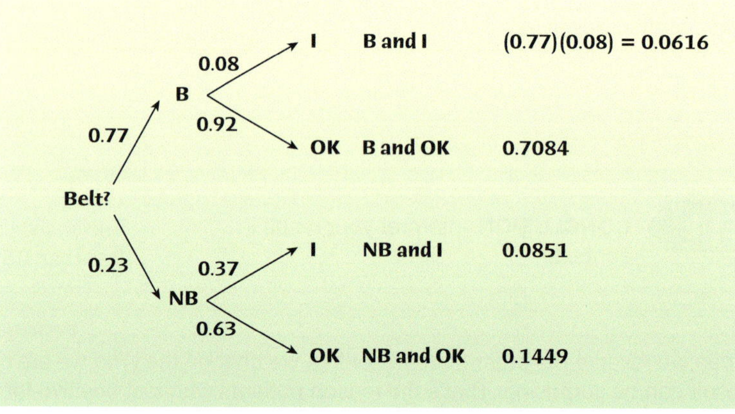

$$P(\mathrm{NB} \mid \mathrm{I}) = \frac{P(\mathrm{NB}\ and\ \mathrm{I})}{P(\mathrm{I})} = \frac{0.0851}{0.0616 + 0.0851} = 0.58$$

Even though only 23% of drivers weren't wearing seatbelts, they accounted for 58% of all the deaths and serious injuries.

Just some advice from your friends, the authors: *Please buckle up!* (We want you to finish this course.)

Nicholas Saunderson (1682–1739) was a blind English mathematician who invented a tactile board to help other blind people do mathematics. And he may have been the true originator of "Bayes' Rule."

WHO DISCOVERED BAYES' RULE?

Stigler's "Law of Eponymy" states that discoveries named for someone (eponyms) are usually named for the wrong person. Steven Stigler, who admits he didn't originate the law, is an expert on the history of statistics, and he suspected that the law might apply to Bayes' Rule. He looked at the possibility that another candidate–one Nicholas Saunderson–was the real discoverer, not the Reverend Bayes. He assembled historical evidence and compared probabilities that the historical events would have happened *given* that Bayes was the discoverer of the rule, with the corresponding probabilities *given* that Saunderson was the discoverer. Of course, what he really wanted to know were the probabilities that Bayes or Saunderson was the discoverer *given* the historical events. How did he *reverse* the conditional probabilities? He used Bayes' Rule and concluded that, actually, it's more likely that Saunderson is the real originator of the rule.

But that doesn't change our tradition of naming the rule for Bayes and calling the branch of statistics arising from this approach Bayesian Statistics.

WHAT CAN GO WRONG?

- **Don't use a simple probability rule where a general rule is appropriate.** Don't assume independence without reason to believe it. Don't assume that outcomes are disjoint without checking that they are. Remember that the general rules always apply, even when outcomes are in fact independent or disjoint.

- **Don't reverse conditioning naïvely.** As we have seen, the probability of **A** *given* **B** may not, and, in general does not, resemble the probability of **B** *given* **A**. The true probability may be counterintuitive.
- **Don't confuse "disjoint" with "independent."** Disjoint events *cannot* happen at the same time. When one happens, you know the other did not, so $P(\mathbf{B} \mid \mathbf{A}) = 0$. Independent events *must* be able to happen at the same time. When one happens, you know it has no effect on the other, so $P(\mathbf{B} \mid \mathbf{A}) = P(\mathbf{B})$.

CONNECTIONS

This chapter shows the unintuitive side of probability. If you've been thinking, "My mind doesn't work this way," you're probably right. Humans don't seem to find conditional and compound probabilities natural and often have trouble with them. But, by being careful to follow the rules outlined in this chapter you can avoid the common mistakes.

Our central connection is to the guiding principle that statistics is about understanding the world. The events discussed in this chapter are close to the kinds of real-world situations in which understanding probabilities matters. The methods and concepts of this chapter are the tools you need to understand the part of the real world that deals with the outcomes of complex, uncertain events.

CHAPTER REVIEW

Know the general rules of probability and how to apply them.

- The General Addition Rule says that $P(\mathbf{A} \text{ or } \mathbf{B}) = P(\mathbf{A}) + P(\mathbf{B}) - P(\mathbf{A} \text{ and } \mathbf{B})$.
- The General Multiplication Rule says that $P(\mathbf{A} \text{ and } \mathbf{B}) = P(\mathbf{A}) \times P(\mathbf{B} \mid \mathbf{A})$.

Know that the conditional probability of an event **B** given the event **A** is $P(\mathbf{B} \mid \mathbf{A}) = P(\mathbf{A} \text{ and } \mathbf{B})/P(\mathbf{A})$.

Know how to define and use independence.

- Events **A** and **B** are independent if $P(\mathbf{A} \mid \mathbf{B}) = P(\mathbf{A})$.

Know how to make and use tables, Venn diagrams, and tree diagrams to solve problems involving probability.

Know how to use tree diagrams and Bayes' Rule to reverse the conditioning.

REVIEW OF TERMS

General Addition Rule For any two events, **A** and **B**, the probability of **A** *or* **B** is (p. 424)

$$P(\mathbf{A} \text{ or } \mathbf{B}) = P(\mathbf{A}) + P(\mathbf{B}) - P(\mathbf{A} \text{ and } \mathbf{B}).$$

Conditional probability The conditional probability of the event **B** given the event **A** has occurred is:

$$P(\mathbf{B} \mid \mathbf{A}) = \frac{P(\mathbf{A} \text{ and } \mathbf{B})}{P(\mathbf{A})}.$$

$P(\mathbf{B} \mid \mathbf{A})$ is read "the probability of **B** *given* **A**" (p. 428).

General Multiplication Rule For any two events, **A** and **B**, the probability of **A** and **B** is (p. 429)

$$P(\mathbf{A} \text{ and } \mathbf{B}) = P(\mathbf{A}) \times P(\mathbf{B} \mid \mathbf{A}).$$

Independence (used formally) Events **A** and **B** are independent when $P(\mathbf{B} \mid \mathbf{A}) = P(\mathbf{B})$ (p. 430).

Tree diagram A display of conditional events or probabilities that is helpful in thinking through conditioning (p. 434).

EXERCISES

SECTION 13.1

1. Pet ownership Suppose that 25% of people have a dog, 29% of people have a cat, and 12% of people own both. What is the probability that someone owns a dog or a cat?

2. Internship For a course program, the chance of being offered an internship at company A is 35%, the chance of being offered an internship at company B is 30%, while the chance of being offered an internship at both is 20%. What is the probability that a randomly selected applicant gets selected by at least one company?

SECTION 13.2

3. Sports What is the probability that a person likes to watch football, given that she also likes to watch basketball?

	Football	No Football
Basketball	27	13
No Basketball	38	22

4. Sports again From Exercise 3, if someone doesn't like to watch basketball, what is the probability that she will be a football fan?

5. Late to the train A student figures that he has a 30% chance of being let out of class late. If he leaves class late, there is a 45% chance that he will miss his train. What is the probability that it will cause him to miss the train?

6. Field goals A nervous kicker usually makes 70% of his first field goal attempts. If he makes his first attempt, his success rate rises to 90%. What is the probability that he makes his first two kicks?

SECTION 13.3

7. Titanic On the *Titanic*, the probability of survival was 0.323. Among first-class passengers, it was 0.625. Were *survival* and *ticket class* independent? Explain.

8. Births If the sex of a child is independent of all other births, is the probability of a woman giving birth to a girl after having four boys greater than it was on her first birth? Explain.

SECTION 13.4

9. Internet usage The Asian continent contains 55% of the world population and 57% of the world's population uses the Internet. Suppose that 29% of the world's population does not live on the Asian continent and uses the Internet. Make a probability table. Why is a table better than a tree here?

(Source: https://www.internetworldstats.com/stats3.htm)

10. Online banking A national survey indicated that 30% of adults conduct their banking online. It also found that 40% are younger than 50, and that 25% are younger than 50 and conduct their banking online. Make a probability table. Why is a table better than a tree here?

11. Internet again Suppose that the information in Exercise 9 has been presented in the following way. 55% of the world population lives on the Asian continent. 51% of the Asian continent's population use the Internet. On the other hand, 64% of the rest of the world's population uses the Internet. Draw a tree for this situation. Why is the tree better than the table in this case? Where are the joint probabilities found?

12. Online banking again Suppose that the information in Exercise 10 had been presented in the following way. A national survey of bank customers finds that 40% are younger than 50. Of those younger than 50, 5 of 8 conduct their banking online. Of those older than 50, only 1 of 12 banks online. Draw a tree for this situation. Why is the tree better than the table in this case? Where are the joint probabilities found?

SECTION 13.5

13. Internet final Given the probabilities in Exercise 11, what is the probability that a person is from the Asian continent and that they use the Internet? Has the probability that they are from the Asian continent increased or decreased with additional information?

14. Online banking last time Given the probabilities in Exercise 12, what is the probability that a person is younger than 50 given that she uses online banking? Has the probability that she is younger than 50 increased or decreased with the additional information?

CHAPTER EXERCISES

15. Smartphone Recent research suggests that 50% of people in the 30–49 age bracket chose "smartphone only" as their preferred means to access the Internet, 20% chose "desktop/laptop/tablet only," while 27% chose "both." What is the probability that a person in the 30–49 age bracket:

a) uses the Internet?
b) does not use the Internet?
c) uses a smartphone to access the Internet?

16. School In a particular school, 45% of pupils have another sibling in the same school, and 28% of pupils have another relative who is not a sibling. Furthermore, 15% of pupils have both siblings and other relatives. What is the probability that a pupil chosen at random has

a) a sibling in the school but not another relative?
b) have either a sibling in the school or another relative?
c) neither have a sibling in the school nor another relative?

17. Amenities A check of dorm rooms on a large college campus revealed that 38% had refrigerators, 52% had TVs, and 21% had both a TV and a refrigerator. What's the probability that a randomly selected dorm room has

a) a TV but no refrigerator?
b) a TV or a refrigerator, but not both?
c) neither a TV nor a refrigerator?

18. Workers Employment data at a large company reveal that 72% of the workers are married, that 44% are college graduates, and

that half of the college grads are married. What's the probability that a randomly chosen worker

a) is neither married nor a college graduate?
b) is married but not a college graduate?
c) is married or a college graduate?

19. Global survey The marketing research organization GfK Roper conducts a yearly survey on consumer attitudes worldwide. They collect demographic information on the roughly 1500 respondents from each country that they survey. Here is a table showing the number of people with various levels of education in five countries:

Educational Level by Country

	Post-Graduate	College	Some High School	Primary or Less	No Answer	Total
China	7	315	671	506	3	1502
France	69	388	766	309	7	1539
India	161	514	622	227	11	1535
U.K.	58	207	1240	32	20	1557
U.S.	84	486	896	87	4	1557
Total	**379**	**1910**	**4195**	**1161**	**45**	**7690**

If we select someone at random from this survey,

a) what is the probability that the person is from the United States?
b) what is the probability that the person completed his or her education before college?
c) what is the probability that the person is from France *or* did some post-graduate study?
d) what is the probability that the person is from France *and* finished only primary school or less?

20. Birth order A survey of students in a large Introductory Statistics class asked about their birth order ($1 =$ oldest or only child) and which college of the university they were enrolled in. Here are the data:

College	Birth Order: 1 or Only	2 or More	Total
Arts & Sciences	34	23	57
Agriculture	52	41	93
Human Ecology	15	28	43
Other	12	18	30
Total	**113**	**110**	**223**

Suppose we select a student at random from this class. What is the probability that the person is

a) a Human Ecology student?
b) a firstborn student?
c) firstborn *and* a Human Ecology student?
d) firstborn *or* a Human Ecology student?

21. Cards You draw a card at random from a standard deck of 52 cards. Find each of the following conditional probabilities:

a) The card is a heart, given that it is red.
b) The card is red, given that it is a heart.
c) The card is an ace, given that it is red.
d) The card is a queen, given that it is a face card.

22. Pets In its monthly report, the local animal shelter states that it currently has 24 dogs and 18 cats available for adoption. Eight of the dogs and 6 of the cats are male. Find each of the following conditional probabilities if an animal is selected at random:

a) The pet is male, given that it is a cat.
b) The pet is a cat, given that it is female.
c) The pet is female, given that it is a dog.

23. Job application The probabilities that an applicant of a particular job has the required qualifications and/or the required experience are shown in the table.

Qualifications	Experience: Yes	No
Yes	0.66	0.06
No	0.13	0.15

What's the probability that

a) an applicant has both required qualifications and experience?
b) an applicant has required qualifications?
c) an applicant with the required experience has the required qualifications?
d) an applicant with the required qualifications does not have the required experience?

24. Polls The table shows the political affiliations of British voters with clear voting intentions according to gender in a YouGov poll in April 2017.

Political Affiliation

Gender	Conservative	Labour	Liberal Democrat	Other Parties	Will Not Vote
Male	0.18	0.1	0.05	0.1	0.06
Female	0.19	0.12	0.05	0.08	0.07

a) What is the probability that
   i) a randomly chosen voter will vote Conservative?
   ii) a female voter votes Labour?
   iii) a voter who will vote for a party other than Conservative, Labour, or Liberal Democrat is male?
b) A new party thinks it stands a good chance of gaining the votes of Conservative or male voters. What proportion of voters is that?

(Source: https://d25d2506sfb94s.cloudfront.net/cumulus_uploads/document/pvsh4yddit/InternalResults_170420_Demographics_W.pdf)

25. Global survey, take 2 Look again at the table summarizing the Roper survey in Exercise 19.

a) If we select a respondent at random, what's the probability we choose a person from the United States who has done post-graduate study?
b) Among the respondents who have done post-graduate study, what's the probability the person is from the United States?
c) What's the probability that a respondent from the United States has done post-graduate study?
d) What's the probability that a respondent from China has only a primary-level education?
e) What's the probability that a respondent with only a primary-level education is from China?

26. Birth order, take 2 Look again at the data about birth order of Intro Stats students and their choices of colleges shown in Exercise 20.

a) If we select a student at random, what's the probability the person is an Arts and Sciences student who is a second child (or more)?
b) Among the Arts and Sciences students, what's the probability a student was a second child (or more)?
c) Among second children (or more), what's the probability the student is enrolled in Arts and Sciences?
d) What's the probability that a first or only child is enrolled in the Agriculture College?
e) What is the probability that an Agriculture student is a first or only child?

27. Sick kids Seventy percent of kids who visit a doctor have a fever, and 30% of kids with a fever also have sore throats. What's the probability that a kid who goes to the doctor has a fever and a sore throat?

28. Sick cars Twenty percent of cars that are inspected have faulty pollution control systems. The cost of repairing a pollution control system exceeds $100 about 40% of the time. When a driver takes her car in for inspection, what's the probability that she will end up paying more than $100 to repair the pollution control system?

29. Cards You are dealt a hand of three cards, one at a time. Find the probability of each of the following.

a) The first heart you get is the third card dealt.
b) Your cards are all red (that is, all diamonds or hearts).
c) You get no spades.
d) You have at least one ace.

30. Another hand You pick three cards at random from a deck. Find the probability of each event described below.

a) You get no aces.
b) You get all hearts.
c) The third card is your first red card.
d) You have at least one diamond.

31. Batteries A junk box in your room contains a dozen old batteries, five of which are totally dead. You start picking batteries one at a time and testing them. Find the probability of each outcome.

a) The first two you choose are both good.
b) At least one of the first three works.
c) The first four you pick all work.
d) You have to pick five batteries to find one that works.

32. Shirts The soccer team's shirts have arrived in a big box, and people just start grabbing them, looking for the right size. The box contains 4 medium, 10 large, and 6 extra-large shirts. You want a medium for you and one for your sister. Find the probability of each event described.

a) The first two you grab are the wrong sizes.
b) The first medium shirt you find is the third one you check.
c) The first four shirts you pick are all extra-large.
d) At least one of the first four shirts you check is a medium.

33. Eligibility A university requires its biology majors to take a course called BioResearch. The prerequisite for this course is that students must have taken either a statistics course or a computer course. By the time they are juniors, 52% of the biology majors have taken statistics, 23% have had a computer course, and 7% have done both.

a) What percent of the junior biology majors are ineligible for BioResearch?
b) What's the probability that a junior biology major who has taken statistics has also taken a computer course?
c) Are taking these two courses disjoint events? Explain.
d) Are taking these two courses independent events? Explain.

34. In a particular city, 53% of workers use their car to go to work, 45% use public transport, and 18% use both means of transport. We select a worker at random.

a) What is the probability that the worker uses neither a car nor public transport to go to work?
b) What is the probability that the worker uses also public transport if they use a car?
c) Are the use of a car and the use of public transport independent events? Explain.
d) Are the use of a car and the use of public transport mutually exclusive events? Explain.

35. Public health A public health department knows that the chance that a country resident contracts some form of cancer in his life is 30%, the chance that a resident has private health insurance is 25%, and the chance that a resident being treated for cancer has private health insurance is 7.5%. For a randomly selected resident:

a) What is the probability of the resident neither contracting some form of cancer nor having private health insurance?
b) What is the probability that the resident has private health insurance, given that they contract some form of cancer?
c) Are the event of contracting cancer in one's lifetime and the event of having private health insurance independent? Explain.
d) Are the two events mentioned in (c) mutually exclusive? Explain.

36. School again According to Exercise 16, in a particular school, 45% of pupils have another sibling in the same school, 28% of pupils have another relative who is not a sibling, and 15% of pupils have both siblings and other relatives.

a) What is the probability that a pupil who has a sibling in the same school also has an another relative who is not a sibling?
b) Are having a sibling in the same school and having another relative who is not a sibling disjoint events? Explain.
c) Are having a sibling in the same school and having another relative who is not a sibling independent events? Explain.

37. Cards If you draw a card at random from a well-shuffled deck, is getting an ace independent of the suit? Explain.

38. Pets, again The local animal shelter in Exercise 22 reported that it currently has 24 dogs and 18 cats available for adoption; 8 of the dogs and 6 of the cats are male. Are the species and sex of the animals independent? Explain.

39. Unsafe food Early in 2010, *Consumer Reports* published the results of an extensive investigation of broiler chickens purchased from food stores in 23 states. Tests for bacteria in the meat showed that 62% of the chickens were contaminated with campylobacter, 14% with salmonella, and 9% with both.

a) What's the probability that a tested chicken was not contaminated with either kind of bacteria?
b) Are contamination with the two kinds of bacteria disjoint? Explain.
c) Are contamination with the two kinds of bacteria independent? Explain.

40. Birth order, finis In Exercises 20 and 26, we looked at the birth orders and college choices of some Intro Stats students. For these students:

a) Are enrolling in Agriculture and Human Ecology disjoint? Explain.
b) Are enrolling in Agriculture and Human Ecology independent? Explain.
c) Are being firstborn and enrolling in Human Ecology disjoint? Explain.
d) Are being firstborn and enrolling in Human Ecology independent? Explain.

41. Men's health again Given the table of probabilities from Exercise 23, are high blood pressure and high cholesterol independent? Explain.

Cholesterol	Blood Pressure: High	Blood Pressure: OK
High	0.11	0.21
OK	0.16	0.52

42. Given the table of probabilities from Exercise 24, are gender and party affiliation independent? Explain.

Gender	Political Affiliation: Conservative	Labour	Liberal Democrat	Other Parties	Will Not Vote	Total
Male	0.18	0.1	0.05	0.1	0.06	0.49
Female	0.19	0.12	0.05	0.08	0.07	0.51
Total	0.37	0.22	0.1	0.18	0.13	1

43. Phone service According to estimates from the federal government's 2010 National Health Interview Survey, based on face-to-face interviews in 16,676 households, approximately 63.6% of U.S. adults have both a landline in their residence and a cell phone, 25.4% have only cell phone service but no landline, and 1.8% have no telephone service at all.

a) Polling agencies won't phone cell phone numbers because customers object to paying for such calls. What proportion of U.S. households can be reached by a landline call?
b) Are having a cell phone and having a landline independent? Explain.

44. Snoring After surveying 995 adults, 81.5% of whom were over 30, the National Sleep Foundation reported that 36.8% of all the adults snored. 32% of the respondents were snorers over the age of 30.

a) What percent of the respondents were 30 or less and did not snore?
b) Is snoring independent of age? Explain.

45. Gender A poll conducted by Gallup classified respondents by sex and political party, as shown in the table. Is party affiliation independent of the respondents' sex? Explain.

	Democrat	Republican	Independent
Male	32	28	34
Female	41	25	26

46. Cars A random survey of autos parked in student and staff lots at a large university classified the brands by country of origin, as seen in the table. Is country of origin independent of type of driver?

Origin	Driver: Student	Driver: Staff
American	107	105
European	33	12
Asian	55	47

47. Television A television station conducting market research on its viewership finds that 21% of households enjoy its music programs. If a household enjoys its music programs, the chance that it also enjoys its sports programs is 71%, but if not, the chance that it also enjoys its sports programs is 16%.

a) Are the event that the household enjoys music programs and the event that the household enjoys sports programs independent? Explain.
b) What is the probability that the selected household neither enjoys music programs nor sports programs?

48. Graduation A private college report contains these statistics:

70% of incoming freshmen attended public schools.

75% of public school students who enroll as freshmen eventually graduate.

90% of other freshmen eventually graduate.

a) Is there any evidence that a freshman's chances to graduate may depend upon what kind of high school the student attended? Explain.
b) What percent of freshmen eventually graduate?

49. Television again Recall Exercise 47. Suppose that the market researcher asks the household whether they enjoy the station's sports programs first and gets yes for an answer. What is the probability that the household also enjoys the music programs?

50. Graduation, part II What percent of students who graduate from the college in Exercise 48 attended a public high school?

51. Absenteeism A company's records indicate that on any given day about 1% of their day-shift employees and 2% of the night-shift employees will miss work. Sixty percent of the employees work the day shift.

a) Is absenteeism independent of shift worked? Explain.
b) What percent of employees are absent on any given day?

52. E-readers PewInternet reported in January of 2014 that 32% of U.S. adults own at least one e-reader, and that 28% of U.S. adults read at least one e-book in the previous year (and thus, presumably, owned an e-reader). Overall, 76% of U.S. adults read at least one book (electronic or otherwise) in the previous year. (http://pewinternet.org/Reports/2014/E-Reading-Update/A-Snapshot-of-Reading-in-America-in-2013/Whos-reading-and-how-a-demographic-portrait.aspx).

a) Explain how these statistics indicate that owning an e-reader and reading at least one book in the previous year are not independent.
b) What's the probability that a randomly selected U.S. adult has an e-reader but didn't use it to read an e-book in the previous year?

53. Absenteeism, part II At the company described in Exercise 51, what percent of the absent employees are on the night shift?

54. E-readers II Given the e-reader data presented in Exercise 52:

a) If a randomly selected U.S. adult has an e-reader, what is the probability that he or she hasn't read an e-book in the past year?
b) Is it more or less likely that a randomly selected U.S. adult who does not own an e-reader would have read no books in the past year?

55. Drunks Police often set up sobriety checkpoints—roadblocks where drivers are asked a few brief questions to allow the officer to judge whether or not the person may have been drinking. If the officer does not suspect a problem, drivers are released to go on their way. Otherwise, drivers are detained for a Breathalyzer test that will determine whether or not they will be arrested. The police say that based on the brief initial stop, trained officers can make the right decision 80% of the time. Suppose the police operate a sobriety checkpoint after 9:00 P.M. on a Saturday night, a time when national traffic safety experts suspect that about 12% of drivers have been drinking.

a) You are stopped at the checkpoint and, of course, have not been drinking. What's the probability that you are detained for further testing?
b) What's the probability that any given driver will be detained?
c) What's the probability that a driver who is detained has actually been drinking?
d) What's the probability that a driver who was released had actually been drinking?

56. No-shows An airline offers discounted "advance-purchase" fares to customers who buy tickets more than 30 days before travel and charges "regular" fares for tickets purchased during those last 30 days. The company has noticed that 60% of its customers take advantage of the advance-purchase fares. The "no-show" rate among people who paid regular fares is 30%, but only 5% of customers with advance-purchase tickets are no-shows.

a) What percent of all ticket holders are no-shows?
b) What's the probability that a customer who didn't show had an advance-purchase ticket?
c) Is being a no-show independent of the type of ticket a passenger holds? Explain.

57. Dishwashers Dan's Diner employs three dishwashers. Al washes 40% of the dishes and breaks only 1% of those he handles. Betty and Chuck each wash 30% of the dishes, and Betty breaks only 1% of hers, but Chuck breaks 3% of the dishes he washes. (He, of course, will need a new job soon. . . .) You go to Dan's for supper one night and hear a dish break at the sink. What's the probability that Chuck is on the job?

58. Parts A company manufacturing electronic components for home entertainment systems buys electrical connectors from three suppliers. The company prefers to use supplier A because only 1% of those connectors prove to be defective, but supplier A can deliver only 70% of the connectors needed. The company must also purchase connectors from two other suppliers, 20% from supplier B and the rest from supplier C. The rates of defective connectors from B and C are 2% and 4%, respectively. You buy one of these components, and when you try to use it you find that the connector is defective. What's the probability that your component came from supplier A?

59. HIV testing In July 2005, the journal *Annals of Internal Medicine* published a report on the reliability of HIV testing. Results of a large study suggested that among people with HIV, 99.7% of tests conducted were (correctly) positive, while for people without HIV 98.5% of the tests were (correctly) negative. A clinic serving an at-risk population offers free HIV testing, believing that 15% of the patients may actually carry HIV. What's the probability that a patient testing negative is truly free of HIV?

60. Polygraphs Lie detectors are controversial instruments, barred from use as evidence in many courts. Nonetheless, many employers use lie detector screening as part of their hiring process in the hope that they can avoid hiring people who might be dishonest. There has been some research, but no agreement, about the reliability of polygraph tests. Based on this research, suppose that a polygraph can detect 65% of lies, but incorrectly identifies 15% of true statements as lies.

61. A certain company believes that 95% of its job applicants are trustworthy. The company gives everyone a polygraph test, asking, "Have you ever stolen anything from your place of work?" Naturally, all the applicants answer "No," but the polygraph identifies some of those answers as lies, making the person ineligible for a job. What's the probability that a job applicant rejected under suspicion of dishonesty was actually trustworthy?

JUST CHECKING

Answers

1. a)

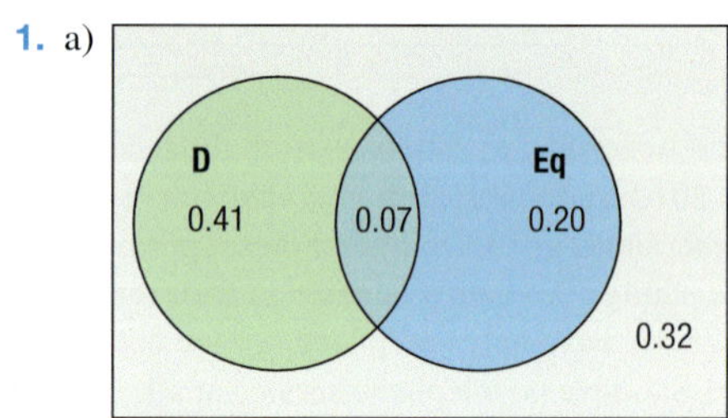

b) 0.32
c) 0.41

2. a) Independent
b) Disjoint
c) Neither

3. a)

		Equation		
		Yes	No	Total
Display	Yes	0.07	0.41	**0.48**
	No	0.20	0.32	**0.52**
	Total	**0.27**	**0.73**	**1.00**

b) $P(\mathbf{D}|\mathbf{Eq}) = P(\mathbf{D}\ and\ \mathbf{Eq})/P(\mathbf{Eq}) = 0.07/0.27 = 0.259$
c) No, pages can (and 7% do) have both.
d) To be independent, we'd need $P(\mathbf{D}|\mathbf{Eq}) = P(\mathbf{D})$. $P(\mathbf{D}/\mathbf{Eq}) = 0.259$, but $P(\mathbf{D}) = 0.48$. Overall, 48% of pages have data displays, but only about 26% of pages with equations do. They do not appear to be independent.

14

Random Variables

WHERE ARE WE GOING?

How long do products last? Should you expect your computer to die just after the warranty runs out? How can you reduce your risk for developing hepatitis? Businesses, medical researchers, and other scientists all use probability to determine risk factors to help answer questions like these. To do that, they model the probability of outcomes using a special kind of variable— a *random* variable. Using random variables can help us talk about and predict random behavior.

Insurance companies make bets. They bet that you're going to live a long life. You bet that you're going to die sooner. Both you and the insurance company want the company to stay in business, so it's important to find a "fair price" for your bet. Of course, the right price for *you* depends on many factors, and nobody can predict exactly how long you'll live. But when the company averages over enough customers, it can make reasonably accurate estimates of the amount it can expect to collect on a policy before it has to pay its benefit.

Here's a simple example. An insurance company offers a "death and disability" policy that pays $10,000 when you die or $5000 if you are permanently disabled. It charges a premium of only $50 a year for this benefit. Is the company likely to make a profit selling such a plan? To answer this question, the company needs to know the *probability* that its clients will die or be disabled in any year. From actuarial information like this, the company can calculate the expected value of this policy.

> **WHAT IS AN ACTUARY?**
>
> Actuaries are the daring people who put a price on risk, estimating the likelihood and costs of rare events, so they can be insured. That takes financial, statistical, and business skills. It also makes them invaluable to many businesses. Actuaries are rather rare themselves; only about 19,000 work in North America. Perhaps because of this, they are well paid. If you're enjoying this course, you may want to look into a career as an actuary. Contact the Society of Actuaries or the Casualty Actuarial Society (who, despite what you may think, did not pay for this blurb).

14.1 Center: The Expected Value

We'll want to build a probability model in order to answer the questions about the insurance company's risk. First we need to define a few terms. The amount the company pays out on an individual policy is called a **random variable** because its numeric value is based on the outcome of a random event. We use a capital letter, like X, to denote a random

> **NOTATION ALERT**
> The most common letters for random variables are X, Y, and Z. But be cautious. If you see any capital letter, it just might denote a random variable.

variable. We'll denote a particular value that it can have by the corresponding lowercase letter, in this case x. For the insurance company, x can be \$10,000 (if you die that year), \$5000 (if you are disabled), or \$0 (if neither occurs). Because we can list all the outcomes, we might formally call this random variable a **discrete** random variable. Otherwise, we'd call it a **continuous** random variable. (We will say more about the properties of continuous random variables later in this chapter.) The collection of all the possible values and the probabilities that they occur is called the **probability model** for a discrete random variable.

Suppose, for example, that the death rate in any year is 1 out of every 1000 people, and that another 2 out of 1000 suffer some kind of disability. Then we can display the probability model for this insurance policy in a table like this:

Policyholder Outcome	Payout x	Probability $P(X = x)$
Death	10,000	$\frac{1}{1000}$
Disability	5000	$\frac{2}{1000}$
Neither	0	$\frac{997}{1000}$

To see what the insurance company can expect, imagine that it insures exactly 1000 people. Further imagine that, in perfect accordance with the probabilities, 1 of the policyholders dies, 2 are disabled, and the remaining 997 survive the year unscathed. The company would pay \$10,000 to one client and \$5000 to each of 2 clients. That's a total of \$20,000, or an average of $20000/1000 = \$20$ per policy. Since it is charging people \$50 for the policy, the company expects to make a profit of \$30 per customer. Not bad!

> **NOTATION ALERT**
> The expected value (or mean) of a random variable is written $E(X)$ or μ.

We can't predict what *will* happen during any given year, but we can say what we *expect* to happen. To do this, we (or, rather, the insurance company) need the probability model. The expected value of a policy is a parameter of this model. In fact, it's the mean. We'll signify this with the notation μ (for population mean) or $E(X)$ for expected value. This isn't an average of some data values, so we won't estimate it. Instead, we assume that the probabilities are known and simply calculate the expected value from them.

How did we come up with \$20 as the expected value of a policy payout? Here's the calculation. As we've seen, it often simplifies probability calculations to think about some (convenient) number of outcomes. For example, we could imagine that we have exactly 1000 clients. Of those, exactly 1 died and 2 were disabled, corresponding to what the probabilities would say.

$$\mu = E(X) = \frac{10{,}000(1) + 5000(2) + 0(997)}{1000}$$

So our total payout comes to \$20,000, or \$20 per policy.

Instead of writing the expected value as one big fraction, we can rewrite it as separate terms with a common denominator of 1000.

$$\begin{aligned}\mu &= E(X)\\ &= \$10{,}000\left(\frac{1}{1000}\right) + \$5000\left(\frac{2}{1000}\right) + \$0\left(\frac{997}{1000}\right)\\ &= \$20\end{aligned}$$

How convenient! See the probabilities? For each policy, there's a 1/1000 chance that we'll have to pay \$10,000 for a death and a 2/1000 chance that we'll have to pay \$5000 for a disability. Of course, there's a 997/1000 chance that we won't have to pay anything.

Take a good look at the expression now. It's easy to calculate the **expected value** of a (discrete) random variable—just multiply each possible value by the probability that it occurs, and find the sum:

$$\mu = E(X) = \sum xP(x).$$

Be sure that every possible outcome is included in the sum. And verify that you have a valid probability model to start with—the probabilities should each be between 0 and 1 and should sum to one.

EXAMPLE 14.1 Love and Expected Value

On Valentine's Day the *Quiet Nook* restaurant offers a *Lucky Lovers Special* that could save couples money on their romantic dinners. When the waiter brings the check, he'll also bring the four aces from a deck of cards. He'll shuffle them and lay them out face down on the table. The couple will then get to turn one card over. If it's a black ace, they'll owe the full amount, but if it's the ace of hearts, the waiter will give them a $20 Lucky Lovers discount. If they first turn over the ace of diamonds (hey—at least it's red!), they'll then get to turn over one of the remaining cards, earning a $10 discount for finding the ace of hearts this time.

QUESTION: Based on a probability model for the size of the Lucky Lovers discounts the restaurant will award, what's the expected discount for a couple?

ANSWER: Let X = the Lucky Lovers discount. The probabilities of the three outcomes are:

$$P(X = 20) = P(A\heartsuit) = \frac{1}{4}$$

$$P(X = 10) = P(A\diamondsuit, \text{ then } A\heartsuit) = P(A\diamondsuit) \times P(A\heartsuit \mid A\diamondsuit) = \frac{1}{4} \times \frac{1}{3} = \frac{1}{12}$$

$$P(X = 0) = P(X \neq 20 \text{ or } 10) = 1 - \left(\frac{1}{4} + \frac{1}{12}\right) = \frac{2}{3}.$$

My probability model is:

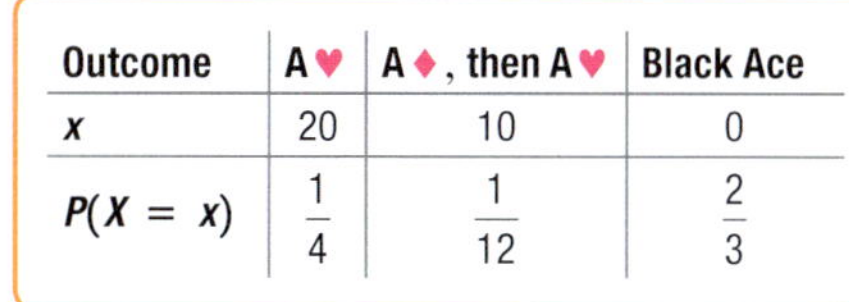

Outcome	A♥	A♦, then A♥	Black Ace
x	20	10	0
$P(X = x)$	$\frac{1}{4}$	$\frac{1}{12}$	$\frac{2}{3}$

$$E(X) = 20 \times \frac{1}{4} + 10 \times \frac{1}{12} + 0 \times \frac{2}{3} = \frac{70}{12} \approx 5.83$$

Couples dining at the Quiet Nook can expect an average discount of $5.83.

JUST CHECKING

1. One of the authors took his minivan in for repair recently because the air conditioner was cutting out intermittently. The mechanic identified the problem as dirt in a control unit. He said that in about 75% of such cases, drawing down and then recharging the coolant a couple of times cleans up the problem—and costs only $60. If that fails, then the control unit must be replaced at an additional cost of $100 for parts and $40 for labor.

   a) Define the random variable and construct the probability model.

   b) What is the expected value of the cost of this repair?

   c) What does that mean in this context?

   Oh—in case you were wondering—the $60 fix worked!

14.2 Spread: The Standard Deviation

Of course, this expected value (or mean) is not what actually happens to any *particular* policyholder. No individual policy actually costs the company $20. We are dealing with random events, so some policyholders receive big payouts, others nothing. Because the insurance company must anticipate this variability, it needs to know the *standard deviation* of the random variable.

For data, we calculated the **standard deviation** by first computing the deviation from the mean and squaring it. We do that with (discrete) random variables as well. First, we find the deviation of each payout from the mean (expected value):

Policyholder Outcome	Payout x	Probability $P(X = x)$	Deviation $(x - \mu)$
Death	10,000	$\frac{1}{1000}$	$(10{,}000 - 20) = 9980$
Disability	5000	$\frac{2}{1000}$	$(5000 - 20) = 4980$
Neither	0	$\frac{997}{1000}$	$(0 - 20) = -20$

Next, we square each deviation. The **variance** is the expected value of those squared deviations, so we multiply each by the appropriate probability and sum those products. That gives us the variance of X. Here's what it looks like:

$$Var(X) = 9980^2\left(\frac{1}{1000}\right) + 4980^2\left(\frac{2}{1000}\right) + (-20)^2\left(\frac{997}{1000}\right) = 149{,}600.$$

Finally, we take the square root to get the standard deviation:

$$SD(X) = \sqrt{149{,}600} \approx \$386.78.$$

The insurance company can expect an average payout of \$20 per policy, with a standard deviation of \$386.78.

Think about that. The company charges \$50 for each policy and expects to pay out \$20 per policy. Sounds like an easy way to make \$30. In fact, most of the time (probability 997/1000) the company pockets the entire \$50. But would you consider selling your roommate such a policy? The problem is that occasionally the company loses big. With probability 1/1000, it will pay out \$10,000, and with probability 2/1000, it will pay out \$5000. That may be more risk than you're willing to take on. The standard deviation of \$386.78 gives an indication that it's no sure thing. That's a pretty big spread (and risk) for an average profit of \$30.

Here are the formulas for what we just did. Because these are parameters of our probability model, the variance and standard deviation can also be written as σ^2 and σ. You should recognize both kinds of notation.

$$\sigma^2 = Var(X) = \sum (x - \mu)^2 P(x)$$

$$\sigma = SD(X) = \sqrt{Var(X)}$$

EXAMPLE 14.2 Finding the Standard Deviation

RECAP: The probability model for the Lucky Lovers restaurant discount is

Outcome	A♥	A♦, then A♥	Black Ace
x	20	10	0
$P(X = x)$	$\frac{1}{4}$	$\frac{1}{12}$	$\frac{2}{3}$

We found that couples can expect an average discount of $\mu = \$5.83$.

QUESTION: What's the standard deviation of the discounts?

ANSWER: First find the variance:

$$Var(X) = \sum (x - \mu)^2 \times P(x) = (20 - 5.83)^2 \times \frac{1}{4} + (10 - 5.83)^2 \times \frac{1}{12} + (0 - 5.83)^2 \times \frac{2}{3} \approx 74.306.$$

$$\text{So, } SD(X) = \sqrt{74.306} \approx \$8.62$$

Couples can expect the Lucky Lovers discounts to average \$5.83, with a standard deviation of \$8.62.

STEP-BY-STEP EXAMPLE

Expected Values and Standard Deviations for Discrete Random Variables

As the head of inventory for Knowway Computer Company, you were thrilled that you had managed to ship 2 computers to your biggest client the day the order arrived. You are horrified, though, to find out that someone had restocked refurbished computers in with the new computers in your storeroom. The shipped computers were selected randomly from the 15 computers in stock, but 4 of those were actually refurbished.

If your client gets 2 new computers, things are fine. If the client gets one refurbished computer, it will be sent back at your expense—\$100—and you can replace it. However, if both computers are refurbished, the client will cancel the order this month and you'll lose a total of \$1000.

QUESTION: What's the expected value and the standard deviation of the company's loss?

THINK **PLAN** State the problem.

I want to find the company's expected loss for shipping refurbished computers and the standard deviation.

VARIABLE Define the random variable.

Let X = amount of loss.

PLOT Make a picture. This is another job for tree diagrams.

If you prefer calculation to drawing, find $P(\mathbf{NN})$ and $P(\mathbf{RR})$, then use the Complement Rule to find $P(\mathbf{NR}$ *or* $\mathbf{RN})$.

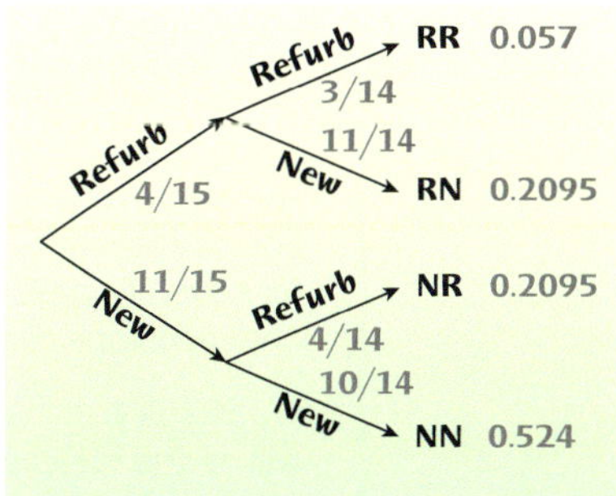

MODEL List the possible values of the random variable, and determine the probability model.

Outcome	x	$P(X = x)$
Two Refurbs	1000	$P(\mathbf{RR}) = 0.057$
One Refurb	100	$P(\mathbf{NR} \text{ or } \mathbf{RN}) = 0.2095 + 0.2095 = 0.419$
New/New	0	$P(\mathbf{NN}) = 0.524$

SHOW **MECHANICS** Find the expected value.

$$E(X) = 0(0.524) + 100(0.419) + 1000(0.057) = \$98.90$$

Find the variance.

$$Var(X) = (0 - 98.90)^2(0.524) + (100 - 98.90)^2(0.419) + (1000 - 98.90)^2(0.057) = 51{,}408.79$$

Find the standard deviation.

$$SD(X) = \sqrt{51{,}408.79} = \$226.735$$

TELL **CONCLUSION** Interpret your results in context.

REALITY CHECK Both numbers seem reasonable. The expected value of \$98.90 is between the extremes of \$0 and \$1000, and there's great variability in the outcome values.

I expect this mistake to cost the firm \$98.90, with a standard deviation of \$226.74. The large standard deviation reflects the fact that there's a pretty large range of possible losses.

14.3 Shifting and Combining Random Variables

Our insurance company expected to pay out an average of \$20 per policy, with a standard deviation of about \$387. If we take the \$50 premium into account, we see the company makes a profit of $50 - 20 = \$30$ per policy. Suppose the company lowers the premium by \$5 to \$45. It's pretty clear that the expected profit also drops an average of \$5 per policy, to $45 - 20 = \$25$.

What about the standard deviation? We know that adding or subtracting a constant from data shifts the mean but doesn't change the variance or standard deviation. The same is true of random variables.[1]

$$E(X \pm c) = E(X) \pm c \qquad Var(X \pm c) = Var(X)$$

EXAMPLE 14.3 Adding a Constant

RECAP: We've determined that couples dining at the *Quiet Nook* can expect Lucky Lovers discounts averaging \$5.83 with a standard deviation of \$8.62. Suppose that for several weeks the restaurant has also been distributing coupons worth \$5 off any one meal (one discount per table).

QUESTION: If every couple dining there on Valentine's Day brings a coupon, what will be the mean and standard deviation of the total discounts they'll receive?

ANSWER: Let D = total discount (Lucky Lovers plus the coupon); then $D = X + 5$.

$$E(D) = E(X + 5) = E(X) + 5 = 5.83 + 5 = \$10.83$$
$$Var(D) = Var(X + 5) = Var(X) = 8.62^2$$
$$SD(D) = \sqrt{Var(X)} = \$8.62$$

Couples with the coupon can expect total discounts averaging \$10.83. The standard deviation is still \$8.62.

[1] The rules in this section are true for both discrete *and* continuous random variables.

Back to insurance … What if the company decides to double all the payouts—that is, pay \$20,000 for death and \$10,000 for disability? This would double the average payout per policy and also increase the variability in payouts. We have seen that multiplying or dividing all data values by a constant changes both the mean and the standard deviation by the same factor. Variance, being the square of standard deviation, changes by the square of the constant. The same is true of random variables. In general, multiplying each value of a random variable by a constant multiplies the mean by that constant and the variance by the *square* of the constant.

$$E(aX) = aE(X) \qquad Var(aX) = a^2Var(X)$$

EXAMPLE 14.4

Double the Love

RECAP: On Valentine's Day at the *Quiet Nook*, couples may get a Lucky Lovers discount averaging \$5.83 with a standard deviation of \$8.62. When two couples dine together on a single check, the restaurant doubles the discount offer—\$40 for the ace of hearts on the first card and \$20 on the second.

QUESTION: What are the mean and standard deviation of discounts for such foursomes?

$$\begin{aligned} E(2X) &= 2E(X) = 2(5.83) = \$11.66 \\ Var(2X) &= 2^2Var(X) = 2^2 \cdot 8.62^2 = 297.2176 \\ SD(2X) &= \sqrt{297.2176} = \$17.24 \end{aligned}$$

ANSWER: If the restaurant doubles the discount offer, two couples dining together can expect to save an average of \$11.66 with a standard deviation of \$17.24.

This insurance company sells policies to more than just one person. How can we figure means and variances for a collection of customers? For example, how can the company find the total expected value (and standard deviation) of policies taken over all policyholders? Consider a simple case: just two customers, Mr. Ecks and Ms. Wye. With an expected payout of \$20 on each policy, we might predict a total of \$20 + \$20 = \$40 to be paid out on the two policies. Nothing surprising there. The expected value of the sum is the sum of the expected values.

$$E(X + Y) = E(X) + E(Y)$$

The variability is another matter. Is the risk of insuring two people the same as the risk of insuring one person for twice as much? We wouldn't expect both clients to die or become disabled in the same year. Because we've spread the risk, the standard deviation should be smaller. Indeed, this is the fundamental principle behind insurance. By spreading the risk among many policies, a company can keep the standard deviation quite small and predict costs more accurately.

But how much smaller is the standard deviation of the sum? It turns out that, if the random variables are independent, there is a simple Addition Rule for variances: *The variance of the sum of two independent random variables is the sum of their individual variances.*

For Mr. Ecks and Ms. Wye, the insurance company can expect their outcomes to be independent, so (using X for Mr. Ecks's payout and Y for Ms. Wye's)

$$\begin{aligned} Var(X + Y) &= Var(X) + Var(Y) \\ &= 149{,}600 + 149{,}600 \\ &= 299{,}200. \end{aligned}$$

If they had insured only Mr. Ecks for twice as much, there would only be one outcome rather than two *independent* outcomes, so the variance would have been

$$Var(2X) = 2^2 Var(X) = 4 \times 149{,}600 = 598{,}400, \text{ or}$$

twice as big as with two independent policies.

Of course, variances are in squared units. The company would prefer to know standard deviations, which are in dollars. The standard deviation of the payout for two independent policies is $\sqrt{299{,}200} = \$546.99$. But the standard deviation of the payout for a single policy of twice the size is $\sqrt{598{,}400} = \$773.56$, or about 40% more.

If the company has two customers, then, it will have an expected annual total payout of \$40 with a standard deviation of about \$547.

EXAMPLE 14.5

Adding the Discounts

RECAP: The Valentine's Day Lucky Lovers discount for couples averages \$5.83 with a standard deviation of \$8.62. We've seen that if the restaurant doubles the discount offer for two couples dining together on a single check, they can expect to save \$11.66 with a standard deviation of \$17.24. Some couples decide instead to get separate checks and pool their two discounts.

QUESTION: You and your amour go to this restaurant with another couple and agree to share any benefit from this promotion. Does it matter whether you pay separately or together?

ANSWER: Let X_1 and X_2 represent the two separate discounts, and T the total; then $T = X_1 + X_2$.

$$E(T) = E(X_1 + X_2) = E(X_1) + E(X_2) = 5.83 + 5.83 = \$11.66,$$

so the expected saving is the same either way.

The cards are reshuffled for each couple's turn, so the discounts couples receive are independent. It's okay to add the variances:

$$Var(T) = Var(X_1 + X_2) = Var(X_1) + Var(X_2) = 8.62^2 + 8.62^2 = 148.6088$$

$$SD(T) = \sqrt{148.6088} = \$12.19$$

When two couples get separate checks, there's less variation in their total discount. The standard deviation is \$12.19, compared to \$17.24 for couples who play for the double discount on a single check.

Pythagorean Theorem of Statistics

We often use the standard deviation to measure variability, but when we add independent random variables, we use their variances. Think of the Pythagorean Theorem. In a right triangle (only), the *square* of the length of the hypotenuse is the sum of the *squares* of the lengths of the other two sides:

$$c^2 = a^2 + b^2.$$

For independent random variables (only), the *square* of the standard deviation of their sum is the sum of the *squares* of their standard deviations:

$$SD^2(X+Y) = SD^2(X) + SD^2(Y).$$

It's simpler to write this with *variances*:

For independent random variables, X and Y,

$$Var(X+Y) = Var(X) + Var(Y).$$

In general,

- *The mean of the sum of two random variables is the sum of the means.*
- *The mean of the difference of two random variables is the difference of the means.*
- *If the random variables are independent, the variance of their sum or difference is always the sum of the variances.*

$$E(X \pm Y) = E(X) \pm E(Y) \quad Var(X \pm Y) = Var(X) + Var(Y)$$

Wait a minute! Is that third part correct? Do we always *add* variances? Yes. Think about the two insurance policies. Suppose we want to know the mean and standard deviation of the *difference* in payouts to the two clients. Since each policy has an expected payout of \$20, the expected difference is $20 - 20 = \$0$. If we also subtract variances, we get \$0, too, and that surely doesn't make sense. Note that if the outcomes for the two clients are independent, the difference in payouts could range from $\$10{,}000 - \$0 = \$10{,}000$ to $\$0 - \$10{,}000 = -\$10{,}000$, a spread of \$20,000. The variability in differences increases as much as the variability in sums. If the company has two customers, the difference in payouts has a mean of \$0 and a standard deviation of about \$547 (again).

EXAMPLE 14.6 Working with Differences

RECAP: The Lucky Lovers discount at the *Quiet Nook* averages \$5.83 with a standard deviation of \$8.62. Just up the street, the *Wise Fool* restaurant has a competing Lottery of Love promotion. There a couple can select a specially prepared chocolate from a large bowl and unwrap it to learn the size of their discount. The restaurant's manager says the discounts vary with an average of \$10.00 and a standard deviation of \$15.00.

QUESTION: How much more can you expect to save at the *Wise Fool*? With what standard deviation?

ANSWER: Let W = discount at the *Wise Fool*, X = the discount at the *Quiet Nook*, and D = the difference: $D = W - X$. These are different promotions at separate restaurants, so the outcomes are independent.

$$E(W - X) = E(W) - E(X) = 10.00 - 5.83 = \$4.17$$

$$\begin{aligned} SD(W - X) &= \sqrt{Var(W - X)} \\ &= \sqrt{Var(W) + Var(X)} \\ &= \sqrt{15^2 + 8.62^2} \\ &\approx \$17.30 \end{aligned}$$

Discounts at the *Wise Fool* will average \$4.17 more than at the *Quiet Nook*, with a standard deviation of the difference of \$17.30.

For Random Variables, Does $X + X + X = 3X$?

Maybe, but be careful. As we've just seen, insuring one person for \$30,000 is not the same risk as insuring three people for \$10,000 each. When each instance represents a different outcome for the same random variable, it's easy to fall into the trap of writing all of them with the same symbol. Don't make this common mistake. Make sure you write each instance as a *different* random variable. Just because each random variable describes a similar situation doesn't mean that each random outcome will be the same.

These are *random* variables, not the variables you saw in algebra. Being random, they take on different values each time they're evaluated. So what you really mean is $X_1 + X_2 + X_3$. Written this way, it's clear that the sum shouldn't necessarily equal 3 times *anything*.

EXAMPLE 14.7 Summing a Series of Outcomes

RECAP: The *Quiet Nook*'s Lucky Lovers promotion offers couples discounts averaging \$5.83 with a standard deviation of \$8.62. The restaurant owner is planning on serving 40 couples on Valentine's Day.

QUESTION: What's the expected total of the discounts the owner will give? With what standard deviation?

ANSWER: Let $X_1, X_2, X_3, \ldots, X_{40}$ represent the discounts to the 40 couples, and T the total of all the discounts. Then:

$$T = X_1 + X_2 + X_3 + \cdots + X_{40}$$

$$E(T) = E(X_1 + X_2 + X_3 + \cdots + X_{40})$$

$$= E(X_1) + E(X_2) + E(X_3) + \cdots + E(X_{40})$$
$$= 5.83 + 5.83 + 5.83 + \cdots + 5.83$$
$$= \$233.20$$

Reshuffling cards between couples makes the discounts independent, so:

$$SD(T) = \sqrt{Var(X_1 + X_2 + X_3 + \cdots + X_{40})}$$
$$= \sqrt{Var(X_1) + Var(X_2) + Var(X_3) + \cdots + Var(X_{40})}$$
$$= \sqrt{8.62^2 + 8.62^2 + 8.62^2 + \cdots + 8.62^2}$$
$$\approx \$54.52$$

The restaurant owner can expect the 40 couples to win discounts totaling \$233.20, with a standard deviation of \$54.52.

JUST CHECKING

2. Suppose the time it takes a customer to get and pay for seats at the ticket window of a baseball park is a random variable with a mean of 100 seconds and a standard deviation of 50 seconds. When you get there, you find only two people in line in front of you.
   a) How long do you expect to wait for your turn to get tickets?
   b) What's the standard deviation of your wait time?
   c) What assumption did you make about the two customers in finding the standard deviation?

STEP-BY-STEP EXAMPLE

Hitting the Road: Combining Means and Variances

You're planning to spend next year wandering through the mountains of Kyrgyzstan. You plan to sell your used SUV so you can purchase an off-road Honda motor scooter when you get there. Used SUVs of the year and mileage of yours are selling for a mean of \$6940 with a standard deviation of \$250. Your research shows that scooters in Kyrgyzstan are going for about 65,000 Kyrgyzstan som with a standard deviation of 500 som. One U.S. dollar is worth about 54.5 Kyrgyzstan som (54 som and 50 tylyn).

QUESTION: How much cash can you expect to pocket after you sell your SUV and buy the scooter?

THINK **PLAN** State the problem.	I want to model how much money I'd have (in som) after selling my SUV and buying the scooter.
VARIABLES Define the random variables.	Let A = sale price of my SUV (in dollars), B = price of a scooter (in som), and D = profit (in som)
Write an appropriate equation. Think about the assumptions.	$D = 54.5A - B$ ✓ **Independence Assumption**: The prices are independent.

SHOW **MECHANICS** Find the expected value, using the appropriate rules.	$E(D) = E(54.5A - B)$ $= 54.5E(A) - E(B)$ $= 54.5(6{,}940) - (65{,}000)$ $E(D) = 313{,}230$ som
Find the variance, using the appropriate rules. Be sure to check the assumptions first!	Since sale and purchase prices are independent, $Var(D) = Var(54.5A - B)$ $= Var(54.5A) + Var(B)$ $= (54.5)^2 Var(A) + Var(B)$ $= 2970.25(250)^2 + (500)^2$ $Var(D) = 185{,}890{,}625$
Find the standard deviation.	$SD(D) = \sqrt{185{,}890{,}625} = 13{,}634.17$ som
TELL **CONCLUSION** Interpret your results in context. (Here that means talking about dollars.)	I can expect to clear about 313,230 som (\$5747) with a standard deviation of 13,634.17 som (\$250).
REALITY CHECK Given the initial cost estimates, the mean and standard deviation seem reasonable.	

*Correlation and Covariance

In Chapter 6 we saw that the association of two variables could be measured with their correlation. What about random variables? We can talk about the correlation between random variables, too. But it's easier to start with a related concept called covariance.

If X is a random variable with expected value $E(X) = \mu$ and Y is a random variable with expected value $E(Y) = \nu$, then the **covariance** of X and Y is defined as

$$Cov(X, Y) = E((X - \mu)(Y - \nu)).$$

The covariance, like the correlation, measures how X and Y vary together (co = *together*). Think about a scatterplot of two positively correlated variables. When X is above its mean, Y tends to be, too. And when X is below its mean, Y tends to be below its mean. So the product in the covariance formula is positive.

Covariance has a few properties worth remembering:

- $Cov(X, Y) = Cov(Y, X)$
- $Cov(X, X) = Var(X)$
- $Cov(cX, dY) = cdCov(X, Y)$, for any constants c and d
- $Cov(X, Y) = E(XY) - \mu\nu$
- If X and Y are independent, then $Cov(X, Y) = 0$—but be careful not to assume that the converse is true; it is possible for two variables to have zero covariance yet not be independent.

The covariance gives us the extra information we need to find the variance of the sum or difference of two random variables when they are not independent:

$$Var(X \pm Y) = Var(X) + Var(Y) \pm 2Cov(X, Y).$$

When X and Y are independent, their covariance is zero, so we have the Pythagorean Theorem of Statistics, as we saw earlier.[2]

[2]If you want to follow the geometry one step further, the more general form for correlated variables is related to the Law of Cosines.

Covariance, unlike correlation, doesn't have to be between -1 and 1. If X and Y have large values, the covariance will be large as well. And that's its problem. If we know the covariance between X and Y is -64.5, it doesn't give us a real sense of how negatively related they are. To fix the "problem," we can divide the covariance by each of the standard deviations to get the **correlation**:

$$Corr(X, Y) = \frac{Cov(X, Y)}{\sigma_X \sigma_Y}.$$

Correlation of random variables is usually denoted as ρ. Correlation of random variables has many of the properties we saw in Chapter 6 for the correlation coefficient, r.

- The sign of a correlation coefficient gives the direction of the association between random variables.
- Correlation is always between -1 and $+1$.
- Correlation treats X and Y symmetrically. The correlation of X with Y is the same as the correlation of Y with X.
- Unlike covariance, correlation has no units.
- Correlation is not affected by changes in the center or scale of either variable.

14.4 Continuous Random Variables

A Normal random variable differs from the discrete random variables that we have been discussing because a Normal random variable can take on any value. It's an example of a continuous random variable. You remember from Chapter 5 that for a Normal, we find the probability that it lies between two values by finding the area under the Normal curve. The curve we use to work with the Normal distribution is called the Normal model.[3]

For any continuous random variable, this curve (or model) doesn't give the probability directly as the probability models for discrete random variables did. Instead, to find the probability that the random variable is in any interval, you find the area below the curve over that interval as we did for the Normal model in Chapter 5. For the standard Normal,

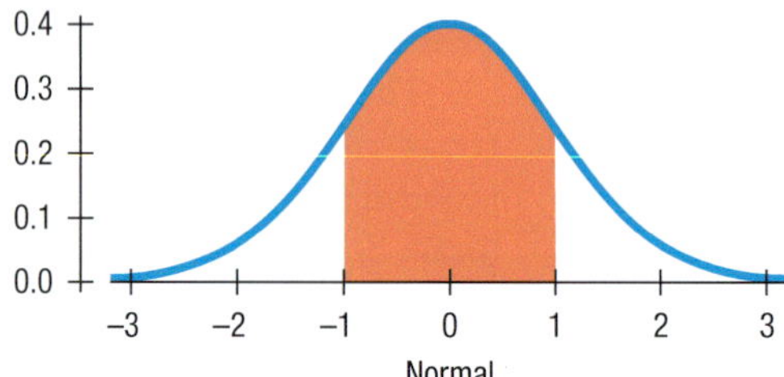

Figure 14.1
The standard Normal curve (with mean 0 and standard deviation 1). The probability of finding a z-score in any interval is the area over that interval under the curve. For example, the probability that the z-score falls between -1 and 1 is about 68%, which can be seen approximately from the density function or found more precisely from a table or technology.

shown in Figure 14.1, we can see that the area below the curve between -1 and 1 is about 68%, which is where the 68–95–99.7 Rule comes from.

It's important to remember that, for a continuous random variable, X, $P(X = x)$ is 0 for every value of x! That may seem strange at first, but since the probability is the area under the curve over an interval, as the interval gets smaller and smaller, the probability does too. Finally, when the interval is just a point, there is no area—and no probability. (See the box on next page.)

Some continuous random variables have Normal models; others may be skewed, uniform, or bimodal. Regardless of shape, all continuous random variables have means

[3]For continuous random variables, the curve is also called the probability density function (pdf) of the random variable.

How Can *Every* Value Have Probability 0?

We can find a probability for any interval of z-scores. But the probability for a single z-score is zero. How can that be? Let's look at the standard Normal random variable, Z. We could find (from a table, website, or computer program) that the probability that Z lies between 0 and 1 is 0.3413.

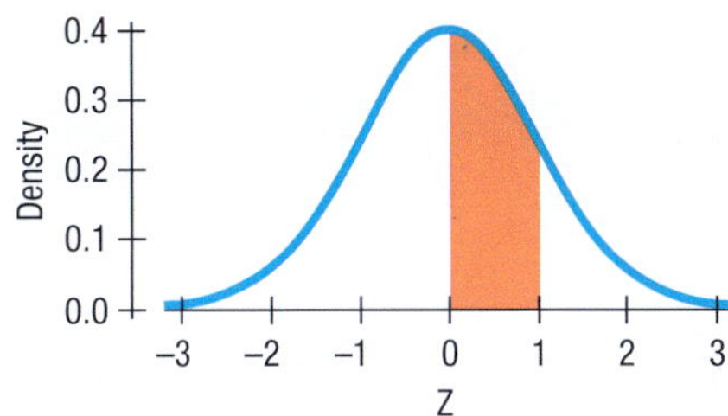

That's the area under the Normal pdf (in red) between the values 0 and 1.

So, what's the probability that Z is between 0 and 1/10?

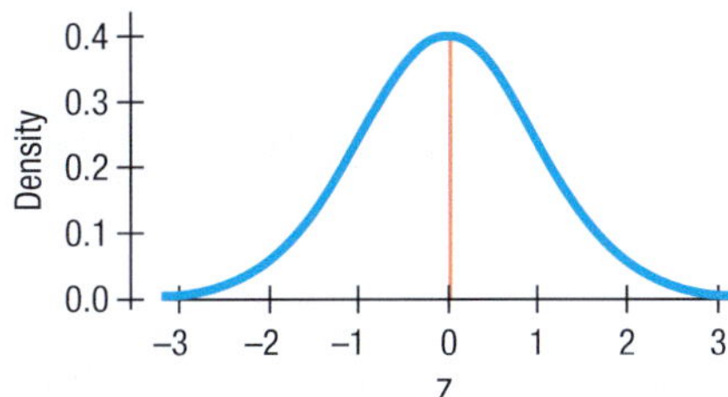

That area is only 0.0398. What is the chance then that Z will fall between 0 and 1/100? There's not much area—the probability is only 0.0040. If we kept going, the probability would keep getting smaller. The probability that Z is between 0 and 1/100,000 is less than 0.00001.

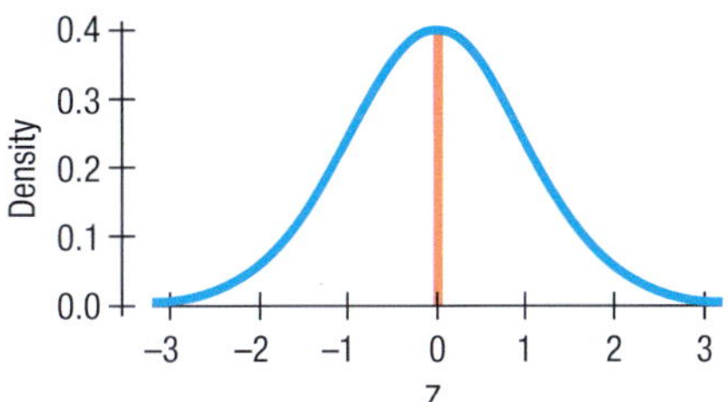

So, what's the probability that Z is *exactly* 0? Well, there's *no* area under the curve right at $x = 0$, so the probability is 0. It's only intervals that have positive probability, but that's OK. In real life we never mean exactly 0.0000000000 or any other value. If you say "exactly 164 pounds," you might really mean between 163.5 and 164.5 pounds or even between 163.99 and 164.01 pounds, but realistically not 164.000000000 . . . pounds.

(which we also call *expected values*) and variances. In this text we won't worry about how to calculate them, but we can still work with models for continuous random variables when we're given these parameters. Chapter 15 gives some examples of other continuous random variables.

We know that the expected value of the sum of *any* two random variables is the sum of their expected values. We know how to compute the standard deviation of the sum as well. But, we also know that when we add two random variables together, the sum generally has a different probability model than the one we started with. Amazingly, independent Normal random variables behave better. The probability model for the sum of any number of independent Normal random variables is still Normal.

This is just one of the things that is special about the Normal. However, if this mathematical good behavior alone doesn't impress you, here's a real-world example of how we can use this fact.

STEP-BY-STEP EXAMPLE

Packaging Stereos

A company manufactures small stereo systems. At the end of the production line, the stereos are packaged and prepared for shipping. Stage 1 of this process is called "packing." Workers must collect all the system components (a main unit, two speakers, a power cord, an antenna, and some wires), put each in plastic bags, and then place everything inside a protective styrofoam form. The packed form then moves on to Stage 2, called "boxing." There, workers place the form and a packet of instructions in a cardboard box, close it, then seal and label the box for shipping.

The company says that times required for the packing stage can be described by a Normal model with a mean of 9 minutes and standard deviation of 1.5 minutes. The times for the boxing stage can also be modeled as Normal, with a mean of 6 minutes and standard deviation of 1 minute.

QUESTION:

1. What is the probability that packing two consecutive systems takes over 20 minutes?
2. What percentage of the stereo systems take longer to pack than to box?

QUESTION 1: What is the probability that packing two consecutive systems takes over 20 minutes?

THINK

PLAN State the problem.

I want to estimate the probability that packing two consecutive systems takes over 20 minutes.

VARIABLES Define your random variables.

Let P_1 = time for packing the first system
P_2 = time for packing the second
T = total time to pack two systems

Write an appropriate equation.

$T = P_1 + P_2$

Think about the assumptions. Sums of independent Normal random variables follow a Normal model. Such simplicity isn't true in general.

✓ **Normal Model Assumption**: We are told that both random variables follow Normal models.

✓ **Independence Assumption**: We can reasonably assume that the two packing times are independent.

SHOW

MECHANICS Find the expected value.

$$E(T) = E(P_1 + P_2) = E(P_1) + E(P_2) = 9 + 9 = 18 \text{ minutes}$$

For sums of independent random variables, variances add. (We don't need the variables to be Normal for this to be true—just independent.)

Since the times are independent,

$$Var(T) = Var(P_1 + P_2) = Var(P_1) + Var(P_2) = 1.5^2 + 1.5^2$$

$$Var(T) = 4.50$$

Find the standard deviation.

$$SD(T) = \sqrt{4.50} \approx 2.12 \text{ minutes}$$

Now we use the fact that both random variables follow Normal models to say that their sum is also Normal.

Sketch a picture of the Normal model for the total time, shading the region representing over 20 minutes.

Find the z-score for 20 minutes.

Use technology or Table Z to find the probability.

I'll model T with $N(18, 2.12)$.

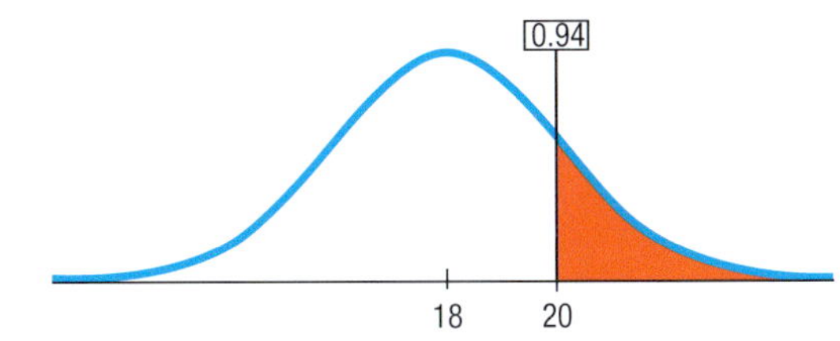

$$z = \frac{20 - 18}{2.12} = 0.94$$

$$P(T > 20) = P(z > 0.94) = 0.1736$$

CONCLUSION Interpret your result in context.

There's a little more than a 17% chance that it will take a total of over 20 minutes to pack two consecutive stereo systems.

QUESTION 2: What percentage of the stereo systems take longer to pack than to box?

PLAN State the question.

I want to estimate the percentage of the stereo systems that take longer to pack than to box.

VARIABLES Define your random variables.

Let P = time for packing a system
B = time for boxing a system
D = difference in times to pack and box a system

Write an appropriate equation.

$D = P - B$

What are we trying to find? Notice that we can tell which of two quantities is greater by subtracting and asking whether the difference is positive or negative.

The probability that it takes longer to pack than to box a system is the probability that the difference $P - B$ is greater than zero.

Don't forget to think about the assumptions.

✓ **Normal Model Assumption:** We are told that both random variables follow Normal models.

✓ **Independence Assumption:** We can assume that the times it takes to pack and to box a system are independent.

SHOW

MECHANICS Find the expected value.

$$E(D) = E(P - B) = E(P) - E(B) = 9 - 6 = 3 \text{ minutes}$$

For the difference of independent random variables, variances add.

Since the times are independent,

$$Var(D) = Var(P - B) = Var(P) + Var(B) = 1.5^2 + 1^2$$

$$Var(D) = 3.25$$

Find the standard deviation.

$$SD(D) = \sqrt{3.25} \approx 1.80 \text{ minutes}$$

State what model you will use.

I'll model D with $N(3, 1.80)$

Sketch a picture of the Normal model for the difference in times, and shade the region representing a difference greater than zero.

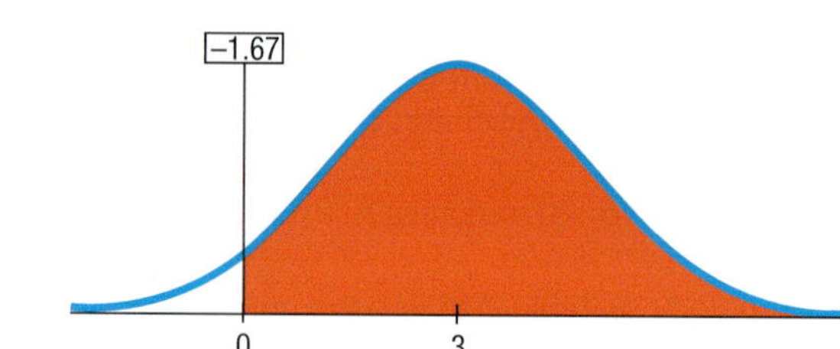

Find the z-score for 0 minutes, then use Table Z or technology to find the probability.

$$z = \frac{0 - 3}{1.80} = -1.67$$

$$P(D > 0) = P(z > -1.67) = 0.9525$$

TELL **CONCLUSION** Interpret your result in context.

About 95% of all the stereo systems will require more time for packing than for boxing.

WHAT CAN GO WRONG?

- **Probability models are still just models.** Models can be useful, but they are not reality. Think about the assumptions behind your models. Are your dice really perfectly fair? (They are probably pretty close.) But when you hear that the probability of a nuclear accident is 1/10,000,000 per year, is that likely to be a precise value? Question probabilities as you would data.
- **If the model is wrong, so is everything else.** Before you try to find the mean or standard deviation of a random variable, check to make sure the probability model is reasonable. As a start, the probabilities in your model should add up to 1. If not, you may have calculated a probability incorrectly or left out a value of the random variable. For instance, in the insurance example, the description mentions only death and disability. Good health is by far the most likely outcome, not to mention the best for both you and the insurance company (who gets to keep your money). Don't overlook that.
- **Don't assume everything's Normal.** Just because a random variable is continuous or you happen to know a mean and standard deviation doesn't mean that a Normal model will be useful. You must *Think* about whether the **Normality Assumption** is justified. Using a Normal model when it really does not apply will lead to wrong answers and misleading conclusions.
- To find the expected value of the sum or difference of random variables, we simply add or subtract means. Center is easy; spread is trickier. Watch out for some common traps.
- **Watch out for variables that aren't independent.** You can add expected values of *any* two random variables, but you can only add variances of independent random variables. Suppose a survey includes questions about the number of hours of sleep people get each night and also the number of hours they are awake each day. From their answers, we find the mean and standard deviation of hours asleep and hours awake. The expected total must be 24 hours; after all, people are either asleep or awake.[4] The means still add just fine. Since all the totals are exactly 24 hours, however, the standard deviation of the total will be 0. We can't add variances here because the number of hours you're awake depends on the number of hours you're asleep. Be sure to check for independence before adding variances.

[4]Although some students do manage to attain a state of consciousness somewhere between sleeping and wakefulness during statistics class.

- **Don't forget: Variances of independent random variables add. Standard deviations don't.**
- **Don't forget: Variances of independent random variables add, even when you're looking at the difference between them.**
- **Don't write independent instances of a random variable with notation that looks like they are the same variables.** Make sure you write each instance as a different random variable. Just because each random variable describes a similar situation doesn't mean that each random outcome will be the same. These are *random* variables, not the variables you saw in algebra. Write $X_1 + X_2 + X_3$ rather than $X + X + X$.

CONNECTIONS

We've seen means, variances, and standard deviations of data. We know that they estimate parameters of models for these data. Now we're looking at the probability models directly. We have only parameters because there are no data to summarize.

It should be no surprise that expected values and standard deviations adjust to shifts and changes of units in the same way as the corresponding data summaries. The fact that we can add variances of independent random quantities is fundamental and will explain why a number of statistical methods work the way they do.

CHAPTER REVIEW

We've learned to work with random variables. We can use the probability model for a discrete random variable to find its expected value and its standard deviation.

We've learned that the mean of the sum or difference of two random variables, discrete or continuous, is just the sum or difference of their means. And we've learned the Pythagorean Theorem of Statistics: *For independent random variables,* the variance of their sum or difference is always the *sum* of their variances.

Finally, we've learned that Normal models are once again special. Sums or differences of Normally distributed random variables also follow Normal models.

REVIEW OF TERMS

Random variable A random variable assumes any of several different numeric values as a result of some random event. Random variables are denoted by a capital letter such as X (p. 445).

Discrete random variable A random variable that can take one of a finite number[5] of distinct outcomes is called a discrete random variable (p. 446).

Continuous random variable A random variable that can take any numeric value within a range of values is called a continuous random variable. The range may be infinite or bounded at either or both ends (p. 446).

Probability model The probability model is a function that associates a probability P with each value of a discrete random variable X, denoted $P(X = x)$, or with any interval of values of a continuous random variable (p. 446).

[5]Technically, there could be an infinite number of outcomes, as long as they're *countable*. Essentially that means we can imagine listing them all in order, like the counting numbers 1, 2, 3, 4, 5, …

Expected value The expected value of a random variable is its theoretical long-run average value, the center of its model. Denoted μ or $E(X)$, it is found (if the random variable is discrete) by summing the products of variable values and probabilities (p. 446):

$$\mu = E(X) = \sum xP(x).$$

Standard deviation The standard deviation of a random variable describes the spread in the model, and is the square root of the variance (p. 448):

$$\sigma = SD(X) = \sqrt{Var(X)}.$$

Variance The variance of a random variable is the expected value of the squared deviation from the mean. For discrete random variables, it can be calculated as (p. 448):

$$\sigma^2 = Var(X) = \sum (x - \mu)^2 P(x).$$

Changing a random variable by a constant

$E(X \pm c) = E(X) \pm c$ $\quad Var(X \pm c) = Var(X)$ (p. 450)

$E(aX) = aE(X)$ $\quad Var(aX) = a^2 Var(X)$ (p. 451)

Adding or subtracting random variables

$E(X \pm Y) = E(X) \pm E(Y)$ (p. 452)

$Var(X \pm Y) = Var(X) + Var(Y)$ *If X and Y are independent*, (p. 452)

(The Pythagorean Theorem of Statistics) (p. 452)

***Covariance** The covariance of random variables X and Y is $Cov(X, Y) = E((X - \mu)(Y - \nu))$ where $\mu = E(X)$ and $\nu = E(Y)$. In general (no need to assume independence) $Var(X \pm Y) = Var(X) + Var(Y) \pm 2Cov(X, Y)$ (p. 455).

TECH SUPPORT

Random Variables

Statistics packages deal with data, not with random variables. Nevertheless, the calculations needed to find means and standard deviations of random variables are little more than weighted means. Most packages can manage that, but then they are just being overblown calculators. For technological assistance with these calculations, we recommend you use your computer or smart phone as a calculator, or just pull out your calculator.

TI-83/84 PLUS

To calculate the mean and standard deviation of a discrete random variable, enter the probability model in two lists:

- In one list (say, L1) enter the x-values of the variable.
- In a second list (say, L2) enter the associated probabilities $P(X = x)$.
- From the **STAT CALC** menu select **1-VarStats** and specify the two lists. You'll see the mean and standard deviation.

COMMENTS

You can enter the probabilities as fractions; the calculator will change them to decimals for you.

Notice that the calculator knows enough to call the standard deviation σ, but mistakenly uses $\bar{x}$ when it should say μ. Make sure you don't make that mistake!

EXERCISES

SECTION 14.1

1. Pets In a particular city, the following are the probabilities of number of pets in a household (the probability of having 4 or more pets is negligible, therefore ignored). What is the average number of pets per household?

# Pets	0	1	2	3
Probability	0.2	0.3	0.25	0.25

2. Ratings On an online marketplace, a product has the following ratings, with the corresponding probabilities. What is the average rating score?

Rating	1	2	3	4	5
Probability	0.01	0.03	0.06	0.2	0.7

SECTION 14.2

3. Pets again What is the standard deviation for Exercise 1?

4. Ratings again What is the standard deviation for Exercise 2?

SECTION 14.3

5. Salary An employer pays a mean salary for a 5-day workweek of \$1250 with a standard deviation of \$129. On the weekends, his salary expenses have a mean of \$450 with a standard deviation of \$57. What is the mean and standard deviation of his total weekly salaries?

6. Golf scores A golfer keeps track of his score for playing nine holes of golf (half a normal golf round). His mean score is 85 with a standard deviation of 11. Assuming that the second 9 has the same mean and standard deviation, what is the mean and standard deviation of his total score if he plays a full 18 holes?

SECTION 14.4

7. Battery The life span of a calculator battery is Normally distributed with a mean of 45 hours and a standard deviation of 5 hours. What is the probability that a battery lasts more than 53 hours?

8. Ketchup An automatic filling machine in a factory fills bottles of ketchup with a mean of 16.1 oz and a standard deviation of 0.05 oz with a distribution that can be well modeled by a Normal model. What is the probability that your bottle of ketchup contains less than 16 oz?

CHAPTER EXERCISES

9. Expected value Find the expected value of each random variable:

a)

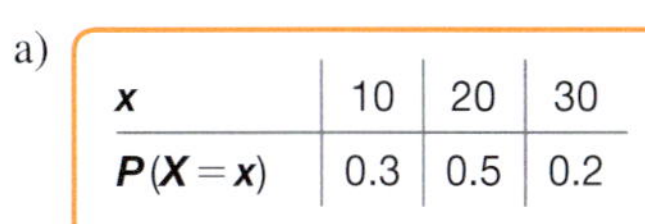

x	10	20	30
$P(X = x)$	0.3	0.5	0.2

b)

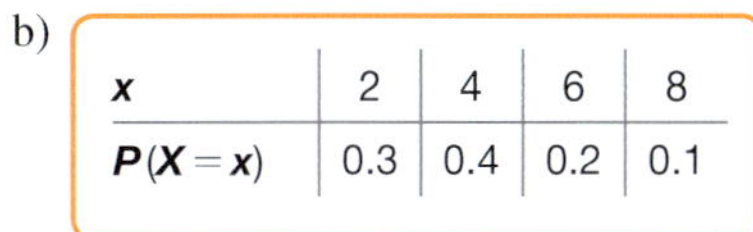

x	2	4	6	8
$P(X = x)$	0.3	0.4	0.2	0.1

10. Expected value Find the expected value of each random variable:

a)

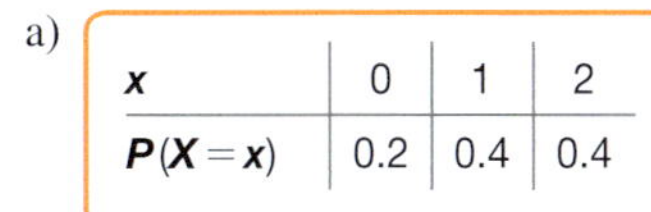

x	0	1	2
$P(X = x)$	0.2	0.4	0.4

b)

x	100	200	300	400
$P(X = x)$	0.1	0.2	0.5	0.2

11. Shares For a particular share on the stock market with value of \$1 (per share), there is a 40% chance of retaining the same value the next day, while outcomes \$0.8, \$0.9, and \$1.2 are equally likely.

a) Create a probability model for gains/losses on the value of one share.
b) Find the expected gains/losses.
c) Is it worth buying these shares?

12. Asset The price of a particular asset today is 100. There is a 50/50 chance that its price the next day will either go up or down by 10.

a) Create a probability model for the price of the asset in two days.
b) Find the expected price of the model in two days.
c) Do you expect to gain, lose, or stay the same, on average?

13. Driving license In a particular country, the chances of obtaining a driver's license on any single attempt is 0.9; however, the law does not allow for any one person to sit for a driver's license exam more than three times.

a) Create a probability model for the number of attempts for a person trying to obtain a driver's license.
b) Find the expected number of attempts.
c) Find the expected number of unsuccessful attempts.

14. Game show In a game show on television, a participant has up to 3 attempts to find the box that contains 100 points from among 4 boxes. Each attempt costs him 10 points.

a) Create a probability model for the net points scored.
b) Find the expected number of attempts.
c) Find the expected net points scored.

15. Software A small software company bids on two contracts and knows it can only get one of them. It anticipates a profit of \$50,000 if it gets the larger contract and a profit of \$20,000 on the smaller contract. The company estimates there's a 30% chance it will get the larger contract and a 60% chance it will get the smaller contract. Assuming the contracts will be awarded independently, what's the expected profit?

16. Racehorse A man buys a racehorse for \$20,000 and enters it in two races. He plans to sell the horse afterward, hoping to make a profit. If the horse wins both races, its value will jump to \$100,000. If it wins one of the races, it will be worth \$50,000. If it loses both races, it will be worth only \$10,000. The man believes there's a 20% chance that the horse will win the first race and a 30% chance it will win the second one. Assuming that the two races are independent events, find the man's expected profit.

17. Variation 1 Find the standard deviations of the random variables in Exercise 9.

18. Variation 2 Find the standard deviations of the random variables in Exercise 10.

19. Shares again Find the standard deviation of the gains/losses in Exercise 11.

20. Assets, again Find the standard deviation of the price of the asset in Exercise 12.

21. Driving license, again Find the standard deviation of the number of attempts for a person trying to obtain a driver's license in Exercise 13.

22. Game show, again Find the standard deviation of the net points scored in Exercise 14.

23. Reservation The probability model below describes the number of people in table reservation in a restaurant.

Number of people	1	2	3	4
Probability	0.1	0.2	0.2	0.5

a) How many people should the restaurant expect on the next reservation?
b) What is the standard deviation?

24. Bus A bus must pass through five stations before reaching its destination, and if there are any people at the station, it will have to stop. Given past history, the probability model for the number of stops the bus has to make during its journey is shown below.

X = # of stops	0	1	2	3	4	5
$P(X = x)$	0.05	0.1	0.1	0.5	0.15	0.1

a) How many stops should be expected on any bus journey?
b) What is the standard deviation?

25. Defects A consumer organization inspecting new cars found that many had appearance defects (dents, scratches, paint chips, etc.). While none had more than three of these defects, 7% had three, 11% two, and 21% one defect. Find the expected number of appearance defects in a new car and the standard deviation.

26. Insurance An insurance policy costs \$100 and will pay policyholders \$10,000 if they suffer a major injury (resulting in hospitalization) or \$3000 if they suffer a minor injury (resulting in lost time from work). The company estimates that each year 1 in every 2000 policyholders may have a major injury, and 1 in 500 a minor injury only.

a) Create a probability model for the profit on a policy.
b) What's the company's expected profit on this policy?
c) What's the standard deviation?

27. Cancelled flights Mary is deciding whether to book the cheaper flight home from college after her final exams, but she's unsure when her last exam will be. She thinks there is only a 20% chance that the exam will be scheduled after the last day she can get a seat on the cheaper flight. If it is and she has to cancel the flight, she will lose \$150. If she can take the cheaper flight, she will save \$100.

a) If she books the cheaper flight, what can she expect to gain, on average?
b) What is the standard deviation?

28. Day trading An option to buy a stock is priced at \$200. If the stock closes above 30 on May 15, the option will be worth \$1000. If it closes below 20, the option will be worth nothing, and if it closes between 20 and 30 (inclusively), the option will be worth \$200. A trader thinks there is a 50% chance that the stock will close in the 20–30 range, a 20% chance that it will close above 30, and a 30% chance that it will fall below 20 on May 15.

a) Should she buy the stock option?
b) How much does she expect to gain?
c) What is the standard deviation of her gain?

29. Contest You play two games against the same opponent. The probability you win the first game is 0.4. If you win the first game, the probability you also win the second is 0.2. If you lose the first game, the probability that you win the second is 0.3.

a) Are the two games independent? Explain.
b) What's the probability you lose both games?
c) What's the probability you win both games?
d) Let random variable X be the number of games you win. Find the probability model for X.
e) What are the expected value and standard deviation?

30. Contracts Your company bids for two contracts. You believe the probability you get contract #1 is 0.8. If you get contract #1, the probability you also get contract #2 will be 0.2, and if you do not get #1, the probability you get #2 will be 0.3.

a) Are the two contracts independent? Explain.
b) Find the probability you get both contracts.
c) Find the probability you get no contract.
d) Let X be the number of contracts you get. Find the probability model for X.
e) Find the expected value and standard deviation.

31. Batteries In a group of 10 batteries, 3 are dead. You choose 2 batteries at random.

a) Create a probability model for the number of good batteries you get.
b) What's the expected number of good ones you get?
c) What's the standard deviation?

32. Kittens In a litter of seven kittens, three are female. You pick two kittens at random.

a) Create a probability model for the number of male kittens you get.
b) What's the expected number of males?
c) What's the standard deviation?

33. Random variables Given independent random variables with means and standard deviations as shown, find the mean and standard deviation of:

a) $3X$
b) $Y + 6$
c) $X + Y$
d) $X - Y$
e) $X_1 + X_2$

	Mean	SD
X	10	2
Y	20	5

34. Random variables Given independent random variables with means and standard deviations as shown, find the mean and standard deviation of:

a) $X - 20$
b) $0.5Y$
c) $X + Y$
d) $X - Y$
e) $Y_1 + Y_2$

	Mean	SD
X	80	12
Y	12	3

35. Random variables Given independent random variables with means and standard deviations as shown, find the mean and standard deviation of:

a) $0.8Y$
b) $2X - 100$
c) $X + 2Y$
d) $3X - Y$
e) $Y_1 + Y_2$

	Mean	SD
X	120	12
Y	300	16

36. Random variables Given independent random variables with means and standard deviations as shown, find the mean and standard deviation of:

a) $2Y + 20$
b) $3X$
c) $0.25X + Y$
d) $X - 5Y$
e) $X_1 + X_2 + X_3$

	Mean	SD
X	80	12
Y	12	3

37. Eggs A grocery supplier believes that in a dozen eggs, the mean number of broken ones is 0.6 with a standard deviation of 0.5 eggs. You buy 3 dozen eggs without checking them.

a) How many broken eggs do you expect to get?
b) What's the standard deviation?
c) What assumptions did you have to make about the eggs in order to answer this question?

38. Garden A company selling vegetable seeds in packets of 20 estimates that the mean number of seeds that will actually grow is 18, with a standard deviation of 1.2 seeds. You buy 5 different seed packets.

a) How many bad (non-growing) seeds do you expect to get?
b) What's the standard deviation?
c) What assumptions did you make about the seeds? Do you think that assumption is warranted? Explain.

39. Reservation again Find the mean and standard deviation of the number of people the restaurant in Exercise 23 should expect in the next ten reservations.

40. Bus again Find the mean and standard deviation of the number of stops the bus in Exercise 24 is expected to make in the next 20 identical journeys.

41. Tickets A delivery company's trucks occasionally get parking tickets, and based on past experience, the company plans that each truck will average 1.3 tickets a month, with a standard deviation of 0.7 tickets.

a) If they have 18 trucks, what are the mean and standard deviation of the total number of parking tickets the company will have to pay this month?
b) What assumption did you make in answering?

42. Donations Organizers of a televised fundraiser know from past experience that most people donate small amounts ($10–$25), some donate larger amounts ($50–$100), and a few people make very generous donations of $250, $500, or more. Historically, pledges average about $32 with a standard deviation of $54.

a) If 120 people call in pledges, what are the mean and standard deviation of the total amount raised?
b) What assumption did you make in answering this question?

43. Fire! An insurance company estimates that it should make an annual profit of $150 on each homeowner's policy written, with a standard deviation of $6000.

a) Why is the standard deviation so large?
b) If it writes only two of these policies, what are the mean and standard deviation of the annual profit?
c) If it writes 10,000 of these policies, what are the mean and standard deviation of the annual profit?
d) Is the company likely to be profitable? Explain.
e) What assumptions underlie your analysis? Can you think of circumstances under which those assumptions might be violated? Explain.

44. Insurance For a particular insurance policy, individual claims are, on average, 100, while the standard deviation is 1000. Policy holders with zero claims are also included in calculating this.

a) Why is the standard deviation so large?
b) If 5 policy holders are picked at random, what are the mean and standard deviation of the sum of their claims?
c) If there are 10,000 policy holders in total, what are the mean and standard deviation of the sum of the claims?
d) If the insurance policy charges a 130 premium, is it likely to be profitable? Explain.

45. Cereal The amount of cereal that can be poured into a small bowl varies with a mean of 1.5 ounces and a standard deviation of 0.3 ounces. A large bowl holds a mean of 2.5 ounces with a standard deviation of 0.4 ounces. You open a new box of cereal and pour one large and one small bowl.

a) How much more cereal do you expect to be in the large bowl?
b) What's the standard deviation of this difference?
c) If the difference follows a Normal model, what's the probability the small bowl contains more cereal than the large one?
d) What are the mean and standard deviation of the total amount of cereal in the two bowls?
e) If the total follows a Normal model, what's the probability you poured out more than 4.5 ounces of cereal in the two bowls together?
f) The amount of cereal the manufacturer puts in the boxes is a random variable with a mean of 16.3 ounces and a standard deviation of 0.2 ounces. Find the expected amount of cereal left in the box and the standard deviation.

46. Homework A survey in a particular school indicates that girls dedicate, on average, 4.1 hours a week on homework, with a standard deviation of 2.37, while boys dedicate, on average, 3.68 hours a week, with a standard deviation of 2.54.

a) What's the expected difference in hours per week on homework between girls and boys?
b) What is the standard deviation of that difference?
c) If hours can be described by normal models, what's the probability that hours a week dedicated on homework are higher for a randomly chosen boy than for a randomly chosen girl?
d) What concerns do you have?

47. More cereal In Exercise 45 we poured a large and a small bowl of cereal from a box. Suppose the amount of cereal that the manufacturer puts in the boxes is a random variable with mean 16.2 ounces and standard deviation 0.1 ounces.

a) Find the expected amount of cereal left in the box.
b) What's the standard deviation?
c) If the weight of the remaining cereal can be described by a Normal model, what's the probability that the box still contains more than 13 ounces?

48. More pets You're thinking about getting two dogs and a cat. Assume that annual veterinary expenses are independent and have a Normal model with the means and standard deviations described in Exercise 46.

a) Define appropriate variables and express the total annual veterinary costs you may have.
b) Describe the model for this total cost. Be sure to specify its name, expected value, and standard deviation.
c) What's the probability that your total expenses will exceed $400?

49. Medley In the 4 × 100 medley relay event, four swimmers swim 100 yards, each using a different stroke. A college team preparing for the conference championship looks at the times their swimmers have posted and creates a model based on the following assumptions:

- The swimmers' performances are independent.
- Each swimmer's times follow a Normal model.
- The means and standard deviations of the times (in seconds) are as shown:

Swimmer	Mean	SD
1 (backstroke)	50.72	0.24
2 (breaststroke)	55.51	0.22
3 (butterfly)	49.43	0.25
4 (freestyle)	44.91	0.21

a) What are the mean and standard deviation for the relay team's total time in this event?
b) The team's best time so far this season was 3:19.48. (That's 199.48 seconds.) Do you think the team is likely to swim faster than this at the conference championship? Explain.

50. Bikes Bicycles arrive at a bike shop in boxes. Before they can be sold, they must be unpacked, assembled, and tuned (lubricated, adjusted, etc.). Based on past experience, the shop manager makes the following assumptions about how long this may take:

- The times for each setup phase are independent.
- The times for each phase follow a Normal model.
- The means and standard deviations of the times (in minutes) are as shown:

Phase	Mean	SD
Unpacking	3.5	0.7
Assembly	21.8	2.4
Tuning	12.3	2.7

a) What are the mean and standard deviation for the total bicycle setup time?
b) A customer decides to buy a bike like one of the display models but wants a different color. The shop has one, still in the box. The manager says they can have it ready in half an

hour. Do you think the bike will be set up and ready to go as promised? Explain.

51. Farmers' market A farmer has 100 lb of apples and 50 lb of potatoes for sale. The market price for apples (per pound) each day is a random variable with a mean of 0.5 dollars and a standard deviation of 0.2 dollars. Similarly, for a pound of potatoes, the mean price is 0.3 dollars and the standard deviation is 0.1 dollars. It also costs him 2 dollars to bring all the apples and potatoes to the market. The market is busy with eager shoppers, so we can assume that he'll be able to sell all of each type of produce at that day's price.

a) Define your random variables, and use them to express the farmer's net income.
b) Find the mean.
c) Find the standard deviation of the net income.
d) Do you need to make any assumptions in calculating the mean? How about the standard deviation?

52. Bike sale The bicycle shop in Exercise 50 will be offering 2 specially priced children's models at a sidewalk sale. The basic model will sell for $120 and the deluxe model for $150. Past experience indicates that sales of the basic model will have a mean of 5.4 bikes with a standard deviation of 1.2, and sales of the deluxe model will have a mean of 3.2 bikes with a standard deviation of 0.8 bikes. The cost of setting up for the sidewalk sale is $200.

a) Define random variables and use them to express the bicycle shop's net income.
b) What's the mean of the net income?
c) What's the standard deviation of the net income?
d) Do you need to make any assumptions in calculating the mean? How about the standard deviation?

53. Lunch deals At a certain food outlet, all customers buy a lunch deal, and some also get dessert. The shop owner believes that the number of lunch deals he sells each day is normally distributed with mean 200 and standard deviation 15. He also believes that the number of desserts he sells each day is independent of the lunch deal sales and is normally distributed with a mean of 120 and a standard deviation of 10.

a) The shop is open 5 days a week on weekdays. Assuming day-to-day sales are independent, what is the probability that he will sell more than 1100 lunch deals in a week?
b) If the shop owner makes a profit of 1.5 units of currency on each lunch deal and of 0.5 units of currency on each dessert, can he reasonably expect to have a day's profit of over 500? Explain.
c) What is the probability that on any given day he'll sell dessert to more than half of his lunch deal customers?

54. Weightlifting The Atlas BodyBuilding Company (ABC) sells "starter sets" of barbells that consist of one bar, two 20-pound weights, and four 5-pound weights. The bars weigh an average of 10 pounds with a standard deviation of 0.25 pounds. The weights average the specified amounts, but the standard deviations are 0.2 pounds for the 20-pounders and 0.1 pounds for the 5-pounders. We can assume that all the weights are Normally distributed.

a) ABC ships these starter sets to customers in two boxes: The bar goes in one box and the six weights go in another. What's the probability that the total weight in that second box exceeds 60.5 pounds? Define your variables clearly and state any assumptions you make.
b) It costs ABC $0.40 per pound to ship the box containing the weights. Because it's an odd-shaped package, though, shipping the bar costs $0.50 a pound plus a $6.00 surcharge. Find the mean and standard deviation of the company's total cost for shipping a starter set.
c) Suppose a customer puts a 20-pound weight at one end of the bar and the four 5-pound weights at the other end. Although he expects the two ends to weigh the same, they might differ slightly. What's the probability the difference is more than a quarter of a pound?

JUST CHECKING

Answers

1. a)

Outcome	X = Cost	Probability
Recharging works	$60	0.75
Replace control unit	$200	0.25

b) $60(0.75) + 200(0.25) = \$95$

c) Car owners with this problem will spend an average of $95 to get it fixed.

2. a) $100 + 100 = 200$ seconds

b) $\sqrt{50^2 + 50^2} = 70.7$ seconds

c) The times for the two customers are independent.

15 Probability Models

WHERE ARE WE GOING?

How many frogs do you have to kiss before you meet Prince Charming? Is finding 10 leukemia cases in a neighborhood adjacent to an EPA superfund site grounds for a lawsuit? We could simulate to answer almost any question involving probabilities. But in some situations, we can do better with a simple probability model. We'll meet a few of the most common models in this chapter.

Suppose a cereal manufacturer puts pictures of famous athletes on cards in boxes of cereal, in the hope of increasing sales. The manufacturer announces that 20% of the boxes contain a picture of Hope Solo, 30% a picture of Danica Patrick, and the rest a picture of Blake Griffin.

15.1 Bernoulli Trials

You're a huge Hope Solo fan. You don't care about completing the whole sports card collection, but you've just *got* to have the Hope Solo picture. How many boxes do you expect you'll have to open before you find her? This isn't the same question that we asked before, but this situation is simple enough for a probability model.

We'll assume that pictures are distributed at random and we'll trust the manufacturer's claim that 20% of the cards are Hope. So, when you open the box, the probability that you succeed in finding Hope is 0.20. We'll call the act of opening *each* box a trial, and note that:

- There are only two possible outcomes (called *success* and *failure*) on each trial. Either you get Hope's picture (success), or you don't (failure).
- The probability of success, denoted p, is the same on every trial. Here $p = 0.20$.
- The trials are independent. Finding Hope in the first box does not change what might happen when you reach for the next box.

Situations like this occur often and are called **Bernoulli trials**. Common examples of Bernoulli trials include tossing a coin, looking for defective products rolling off an assembly

Daniel Bernoulli (1700–1782) was the nephew of Jacob, whom you saw in Chapter 12. He was the first to work out the mathematics for what we now call Bernoulli trials.

line, or even shooting free throws in a basketball game. We can use Bernoulli trials to build a wide variety of useful probability models.

Back to Hope. We want to know how many boxes we'll need to open to find her card. Let's call this random variable $Y = \#$ boxes and build a probability model for it. What's the probability you find her picture in the first box of cereal? It's 20%, of course. We could write $P(Y = 1) = 0.2$.

How about the probability that you don't find Hope until the second box? Well, that means you fail on the first trial and then succeed on the second. With the probability of success 0.2, the probability of failure, denoted q, is $1 - 0.2 = 0.8$. Since the trials are independent, the probability of getting your first success on the second trial is $P(Y = 2) = (0.8)(0.2) = 0.16$.

Of course, you could have a run of bad luck. Maybe you won't find Hope until the fifth box of cereal. What are the chances of that? You'd have to fail 4 straight times and then succeed, so $P(Y = 5) = (0.8)^4(0.2) = 0.08192$.

How many boxes might you expect to have to open? We could reason that since Hope's picture is in 20% of the boxes, or 1 in 5, we expect to find her picture, on average, in the fifth box; that is, $E(Y) = \frac{1}{0.2} = 5$ boxes. That's correct, but not easy to prove.

15.2 The Geometric Model

We want to model how long it will take to achieve the first success in a series of Bernoulli trials. The model that tells us this probability is called the **Geometric probability model**. Geometric models are completely specified by one parameter, p, the probability of success, and are denoted Geom(p). Since achieving the first success on trial number x requires first experiencing $x - 1$ failures, the probabilities are easily expressed by a formula.

NOTATION ALERT

Now we have two more reserved letters. Whenever we deal with Bernoulli trials, p represents the probability of success and q the probability of failure. (Of course, $q = 1 - p$.)

Geometric Probability Model for Bernoulli Trials: Geom(p)

p = probability of success (and $q = 1 - p$ = probability of failure)
X = number of trials until the first success occurs

$$P(X = x) = q^{x-1}p$$

Expected value: $E(X) = \mu = \frac{1}{p}$

Standard deviation: $\sigma = \sqrt{\frac{q}{p^2}}$

EXAMPLE 15.1 Spam and the Geometric Model

Postini is a global company specializing in communications security. The company monitors over 1 billion Internet messages per day and recently reported that 91% of e-mails are spam!

Let's assume that your e-mail is typical—91% spam. We'll also assume you aren't using a spam filter, so every message gets dumped in your inbox. And, since spam comes from many different sources, we'll consider your messages to be independent.

QUESTIONS: Overnight your inbox collects e-mail. When you first check your e-mail in the morning, about how many spam e-mails should you expect to have to wade through and discard before you find a real message? What's the probability that the fourth message in your inbox is the first one that isn't spam?

ANSWER: There are two outcomes: a real message (success) and spam (failure). Since 91% of e-mails are spam, the probability of success $p = 1 - 0.91 = 0.09$.

Let $X =$ the number of e-mails I'll check until I find a real message. I assume that the messages arrive independently and in a random order. I can use the model Geom(0.09).

$$E(X) = \frac{1}{p} = \frac{1}{0.09} = 11.1$$

$$P(X = 4) = (0.91)^3(0.09) = 0.0678$$

On average, I expect to have to check just over 11 e-mails before I find a real message. There's slightly less than a 7% chance that my first real message will be the fourth one I check.

Math Box

We want to find the mean (expected value) of random variable X, using a Geometric model with probability of success p.

First, write the probabilities:

x	1	2	3	4	...
$P(X = x)$	p	qp	q^2p	q^3p	...

The expected value is: $E(X) = 1p + 2qp + 3q^2p + 4q^3p + \cdots$

Let $p = 1 - q$: $= (1 - q) + 2q(1 - q) + 3q^2(1 - q) + 4q^3(1 - q) + \cdots$

Simplify: $= 1 - q + 2q - 2q^2 + 3q^2 - 3q^3 + 4q^3 - 4q^4 + \cdots$

That's an infinite geometric series, with first term 1 and common ratio q: $= 1 + q + q^2 + q^3 + \cdots$

$$= \frac{1}{1 - q}$$

So, finally . . . $E(X) = \frac{1}{p}$.

Independence

One of the important requirements for Bernoulli trials is that the trials be independent. Sometimes that's a reasonable assumption—when tossing a coin or rolling a die, for example. But that becomes a problem when (often!) we're looking at situations involving samples chosen without replacement. We said that whether we find a Hope Solo card in one box has no effect on the probabilities in other boxes. This is *almost* true. Technically, if exactly 20% of the boxes have Hope Solo cards, then when you find one, you've reduced the number of remaining Hope Solo cards. With a few million boxes of cereal, though, the difference is hardly worth mentioning.

But if you knew there were 2 Hope Solo cards hiding in the 10 boxes of cereal on the market shelf, then finding one in the first box you try would clearly change your chances of finding Hope in the next box.

If we had an infinite number of boxes, there wouldn't be a problem. It's selecting from a finite population that causes the probabilities to change, making the trials not independent. Obviously, taking 2 out of 10 boxes changes the probability. Taking even a few hundred out of millions, though, makes very little difference. Fortunately, we have a rule of thumb for the in-between cases. It turns out that if we look at less than 10% of the population, we can pretend that the trials are independent and still calculate probabilities that are quite accurate.

The 10% Condition: Bernoulli trials must be independent. If that assumption is violated, it is still okay to proceed as long as the sample is smaller than 10% of the population.

STEP-BY-STEP EXAMPLE

Working with a Geometric Model

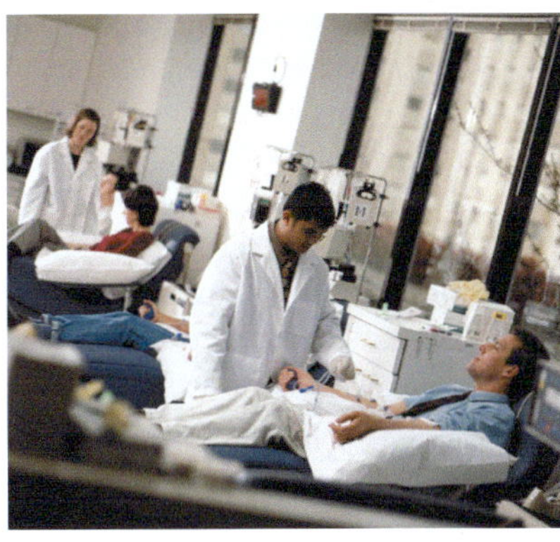

People with O-negative blood are called "universal donors" because O-negative blood can be given to anyone else, regardless of the recipient's blood type. Only about 6% of people have O-negative blood.

QUESTIONS:

1. If donors line up at random for a blood drive, how many do you expect to examine before you find someone who has O-negative blood?
2. What's the probability that the first O-negative donor found is one of the first four people in line?

THINK

PLAN State the questions.

I want to estimate how many people I'll need to check to find an O-negative donor, and the probability that 1 of the first 4 people is O-negative.

Check to see that these are Bernoulli trials.

✓ There are two outcomes:

success = O-negative

failure = other blood types

✓ The probability of success for each person is $p = 0.06$, because they lined up randomly.

✓ ***10% Condition***: Trials aren't independent because the population is finite, but the donors lined up are fewer than 10% of all possible donors.

VARIABLE Define the random variable.

Let X = number of donors until one is O negative.

MODEL Specify the model.

I can model X with Geom(0.06).

SHOW

MECHANICS Find the mean.

Calculate the probability of success on one of the first four trials. That's the probability that $X = 1, 2, 3,$ *or* 4.

$$E(X) = \frac{1}{0.06} \approx 16.7$$

$$\begin{aligned} P(X \leq 4) &= P(X = 1) + P(X = 2) + P(X = 3) + P(X = 4) \\ &= (0.06) + (0.94)(0.06) + (0.94)^2(0.06) + (0.94)^3(0.06) \\ &\approx 0.2193 \end{aligned}$$

TELL

CONCLUSION Interpret your results in context.

Blood drives such as this one expect to examine an average of 16.7 people to find a universal donor. About 22% of the time there will be one within the first 4 people in line.

15.3 The Binomial Model

We can use the Bernoulli trials to answer other questions. Suppose you buy 5 boxes of cereal. What's the probability you get *exactly* 2 pictures of Hope Solo? Before, we asked how long it would take until our first success. Now we want to find the probability of getting 2 successes among the 5 trials. We are still talking about Bernoulli trials, but we're asking a different question.

This time we're interested in the *number of successes* in the 5 trials, so we'll call it X = number of successes. We want to find $P(X = 2)$. This is an example of a **Binomial probability**. It takes two parameters to define this **Binomial model**: the number of trials, n, and the probability of success, p. We denote this model Binom(n, p). Here, $n = 5$ trials, and $p = 0.2$, the probability of finding a Hope Solo card in any trial.

Exactly 2 successes in 5 trials means 2 successes and 3 failures. It seems logical that the probability should be $(0.2)^2(0.8)^3$. Too bad! It's not that easy. That calculation would give you the probability of finding Hope in the first 2 boxes and not in the next 3—*in that order*. But you could find Hope in the third and fifth boxes and still have 2 successes. The probability of those outcomes in that particular order is (0.8) (0.8) (0.2) (0.8) (0.2). That's also $(0.2)^2(0.8)^3$. In fact, the probability will always be the same, no matter what order the successes and failures occur in. Anytime we get 2 successes in 5 trials, no matter what the order, the probability will be $(0.2)^2(0.8)^3$. We just need to take account of all the possible orders in which the outcomes can occur.

Fortunately, these possible orders are *disjoint*. (For example, if your two successes came on the first two trials, they couldn't come on the last two.) So we can use the Addition Rule and add up the probabilities for all the possible orderings. Since the probabilities are all the same, we only need to know how many orders are possible. For small numbers, we can just make a tree diagram and count the branches. For larger numbers this isn't practical, so we let the computer or calculator do the work.

Each different order in which we can have k successes in n trials is called a "combination." The total number of ways that can happen is written $\binom{n}{k}$ or ${}_nC_k$ and pronounced "n choose k."

> **NOTATION ALERT**
> Now punctuation! Throughout mathematics $n!$, pronounced "n factorial," is the product of all the integers from 1 to n. For example, $4! = 4 \cdot 3 \cdot 2 \cdot 1 = 24$.

$${}_nC_k = \frac{n!}{k!(n-k)!} \text{ where } n! \text{ (pronounced "}n\text{ factorial") } = n \times (n-1) \times \cdots \times 1$$

For 2 successes in 5 trials,

$${}_5C_2 = \frac{5!}{2!(5-2)!} = \frac{5 \times 4 \times 3 \times 2 \times 1}{2 \times 1 \times 3 \times 2 \times 1} = \frac{5 \times 4}{2 \times 1} = 10.$$

So there are 10 ways to get 2 Hope pictures in 5 boxes, and the probability of each is $(0.2)^2(0.8)^3$. Now we can find what we wanted:

$$P(\#\text{success} = 2) = 10(0.2)^2(0.8)^3 = 0.2048$$

In general, the probability of exactly k successes in n trials is ${}_nC_k\, p^k q^{n-k}$.

Using this formula, we could find the expected value by adding up $xP(X = x)$ for all values, but it would be a long, hard way to get an answer that you already know intuitively. What's the expected value? If we have 5 boxes, and Hope's picture is in 20% of them, then we would expect to have $5(0.2) = 1$ success. If we had 100 trials with probability of success 0.2, how many successes would you expect? Can you think of any reason not to say 20? It seems so simple that most people wouldn't even stop to think about it. You just multiply the probability of success by n. In other words, $E(X) = np$.

The standard deviation is less obvious; you can't just rely on your intuition. Fortunately, the formula for the standard deviation also boils down to something simple: $SD(X) = \sqrt{npq}$. (If you're curious about where that comes from, it's in the Math Box, too!) In 100 boxes of cereal, we expect to find 20 Hope Solo cards, with a standard deviation of $\sqrt{100 \times 0.8 \times 0.2} = 4$ boxes.

Time to summarize. A Binomial probability model describes the number of successes in a specified number of trials. It takes two parameters to specify this model: the number of trials n and the probability of success p.

Binomial Probability Model For Bernoulli Trials: Binom(n, p)

n = number of trials
p = probability of success (and $q = 1 - p$ = probability of failure)
X = number of successes in n trials

$$P(X = x) = {}_nC_x\, p^x q^{n-x}, \text{ where } {}_nC_x = \frac{n!}{x!(n-x)!}$$

Mean: $\mu = np$
Standard Deviation: $\sigma = \sqrt{npq}$

Math Box

To derive the formulas for the mean and standard deviation of a Binomial model we start with the most basic situation.

Consider a single Bernoulli trial with probability of success p. Let's find the mean and variance of the number of successes.

Here's the probability model for the number of successes:

x	0	1
$P(X = x)$	q	p

Find the expected value:

$$E(X) = 0q + 1p$$
$$E(X) = p$$

And now the variance:

$$\begin{aligned} Var(X) &= (0 - p)^2 q + (1 - p)^2 p \\ &= p^2 q + q^2 p \\ &= pq(p + q) \\ &= pq(1) \\ Var(X) &= pq \end{aligned}$$

What happens when there is more than one trial, though? A Binomial model simply counts the number of successes in a series of n independent Bernoulli trials. That makes it easy to find the mean and standard deviation of a Binomial random variable, Y.

$$\begin{aligned} \text{Let } Y &= X_1 + X_2 + X_3 + \cdots + X_n \\ E(Y) &= E(X_1 + X_2 + X_3 + \cdots + X_n) \\ &= E(X_1) + E(X_2) + E(X_3) + \cdots + E(X_n) \\ &= p + p + p + \cdots + p \text{ (There are } n \text{ terms.)} \end{aligned}$$

So, as we thought, the mean is $E(Y) = np$.

And since the trials are independent, the variances add:

$$\begin{aligned} Var(Y) &= Var(X_1 + X_2 + X_3 + \cdots + X_n) \\ &= Var(X_1) + Var(X_2) + Var(X_3) + \cdots + Var(X_n) \\ &= pq + pq + pq + \cdots + pq \text{ (Again, } n \text{ terms.)} \\ Var(Y) &= npq \end{aligned}$$

Voilà! The standard deviation is $SD(Y) = \sqrt{npq}$.

EXAMPLE 15.2 Spam and the Binomial Model

RECAP: The communications monitoring company *Postini* has reported that 91% of e-mail messages are spam. Suppose your inbox contains 25 messages.

QUESTIONS: What are the mean and standard deviation of the number of real messages you should expect to find in your inbox? What's the probability that you'll find only 1 or 2 real messages?

ANSWER: I assume that messages arrive independently and at random, with the probability of success (a real message) $p = 1 - 0.91 = 0.09$. Let $X =$ the number of real messages among 25. I can use the model Binom(25, 0.09).

$$E(X) = np = 25(0.09) = 2.25$$

$$SD(X) = \sqrt{npq} = \sqrt{25(0.09)(0.91)} = 1.43$$

$$\begin{aligned} P(X = 1 \text{ or } 2) &= P(X = 1) + P(X = 2) \\ &= {}_{25}C_1(0.09)^1(0.91)^{24} + {}_{25}C_2(0.09)^2(0.91)^{23} \\ &= 0.2340 + 0.2777 \\ &= 0.5117 \end{aligned}$$

Among 25 e-mail messages, I expect to find an average of 2.25 that aren't spam, with a standard deviation of 1.43 messages. There's just over a 50% chance that 1 or 2 of my 25 e-mails will be real messages.

STEP-BY-STEP EXAMPLE Working with a Binomial Model

Suppose 20 donors come to a blood drive. Recall that 6% of people are "universal donors."

QUESTIONS:

1. What are the mean and standard deviation of the number of universal donors among them?
2. What is the probability that there are 2 or 3 universal donors?

THINK

PLAN State the question.

I want to know the mean and standard deviation of the number of universal donors among 20 people, and the probability that there are 2 or 3 of them.

Check to see that these are Bernoulli trials.

✓ There are two outcomes:

success = O-negative

failure = other blood types

✓ $p = 0.06$, because people have lined up at random.

✓ *10% Condition*: Trials are not independent, because the population is finite, but fewer than 10% of all possible donors are lined up.

VARIABLE Define the random variable.

Let $X =$ number of O-negative donors among $n = 20$ people.

MODEL Specify the model.

I can model X with Binom(20, 0.06).

SHOW	**MECHANICS** Find the expected value and standard deviation.	$E(X) = np = 20(0.06) = 1.2$ $SD(X) = \sqrt{npq} = \sqrt{20(0.06)(0.94)} \approx 1.06$ $P(X = 2 \text{ or } 3) = P(X = 2) + P(X = 3)$ $= {}_{20}C_2(0.06)^2(0.94)^{18} + {}_{20}C_3(0.06)^3(0.94)^{17}$ $\approx 0.2246 + 0.0860$ $= 0.3106$
TELL	**CONCLUSION** Interpret your results in context.	In groups of 20 randomly selected blood donors, I expect to find an average of 1.2 universal donors, with a standard deviation of 1.06. About 31% of the time, I'd find 2 or 3 universal donors among the 20 people.

15.4 Approximating the Binomial with a Normal Model

Suppose the Tennessee Red Cross anticipates the need for at least 1850 units of O-negative blood this year. It estimates that it will collect blood from 32,000 donors. How great is the risk that the Tennessee Red Cross will fall short of meeting its need? We've just learned how to calculate such probabilities. We can use the Binomial model with $n = 32{,}000$ and $p = 0.06$. The probability of getting *exactly* 1850 units of O-negative blood from 32,000 donors is ${}_{32000}C_{1850} \times 0.06^{1850} \times 0.94^{30150}$. No calculator on earth can calculate that first term (it has more than 100,000 digits).[1] And that's just the beginning. The problem said *at least* 1850, so we have to do it again for 1851, for 1852, and all the way up to 32,000. No thanks.

When we're dealing with a large number of trials like this, making direct calculations of the probabilities becomes tedious (or outright impossible). Here an old friend—the Normal model—comes to the rescue.

The Binomial model has mean $np = 1920$ and standard deviation $\sqrt{npq} \approx 42.48$. We could try approximating its distribution with a Normal model, using the same mean and standard deviation. Remarkably enough, that turns out to be a very good approximation. (We'll see why presently.) With that approximation, we can find the *probability:*

$$P(X < 1850) = P\left(z < \frac{1850 - 1920}{42.48}\right) \approx P(z < -1.65) \approx 0.05$$

There seems to be about a 5% chance that this Red Cross chapter will run short of O-negative blood.

Can we always use a Normal model to make estimates of Binomial probabilities? No. Consider the Hope Solo situation—pictures in 20% of the cereal boxes. If we buy five boxes, the actual Binomial probabilities that we get 0, 1, 2, 3, 4, or 5 pictures of Hope are 33%, 41%, 20%, 5%, 1%, and 0.03%, respectively. The first histogram shows that this

[1] If your calculator *can* find Binom(32000, 0.06), then it's smart enough to use an approximation. Read on to see how you can, too.

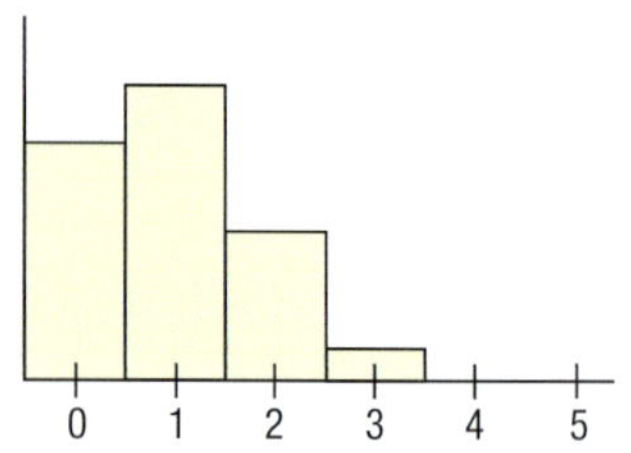

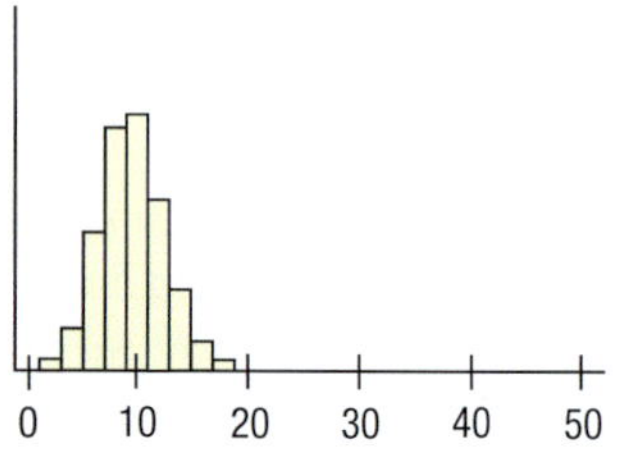

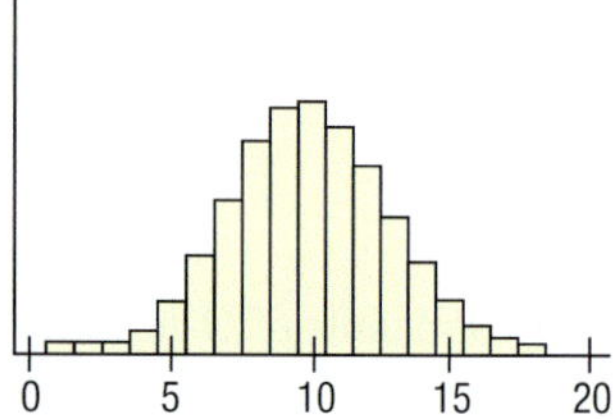

probability model is skewed. That makes it clear that we should not try to estimate these probabilities by using a Normal model.

Now suppose we open 50 boxes of this cereal and count the number of Hope Solo pictures we find. The second histogram shows this probability model. It is centered at $np = 50(0.2) = 10$ pictures, as expected, and it appears to be fairly symmetric around that center. Let's have a closer look.

The third histogram again shows Binom(50, 0.2), this time magnified somewhat and centered at the expected value of 10 pictures of Hope. It looks close to Normal, for sure. With this larger sample size, it appears that a Normal model might be a useful approximation.

A Normal model, then, is a close enough approximation only for a large enough number of trials. And what we mean by "large enough" depends on the probability of success. We'd need a larger sample if the probability of success were very low (or very high). It turns out that a Normal model works pretty well if we expect to see at least 10 successes and 10 failures. That is, we check the **Success/Failure Condition**.

The Success/Failure Condition: A Binomial model is approximately Normal if we expect at least 10 successes and 10 failures:

$$np \geq 10 \text{ and } nq \geq 10.$$

Math Box

It's easy to see where the magic number 10 comes from. You just need to remember how Normal models work. The problem is that a Normal model extends infinitely in both directions. But a Binomial model must have between 0 and *n* successes, so if we use a Normal to approximate a Binomial, we have to cut off its tails. That's not very important if the center of the Normal model is so far from 0 and *n* that the lost tails have only a negligible area. More than three standard deviations should do it, because a Normal model has little probability past that.

So the mean needs to be at least 3 standard deviations from 0 and at least 3 standard deviations from *n*. Let's look at the 0 end.

We require:	$\mu - 3\sigma > 0$
Or in other words:	$\mu > 3\sigma$
For a Binomial, that's:	$np > 3\sqrt{npq}$
Squaring yields:	$n^2p^2 > 9npq$
Now simplify:	$np > 9q$
Since $q \leq 1$, we can require:	$np > 9$

For simplicity, we usually require that *np* (and *nq* for the other tail) be at least 10 to use the Normal approximation, the Success/Failure Condition.[2]

[2]Looking at the final step, we see that we need $np > 9$ in the worst case, when q (or p) is near 1, making the Binomial model quite skewed. When q and p are near 0.5—say between 0.4 and 0.6—the Binomial model is nearly symmetric and $np > 5$ ought to be safe enough. Although we'll always check for 10 expected successes and failures, keep in mind that for values of p near 0.5, we can be somewhat more forgiving.

EXAMPLE 15.3 Spam And The Normal Approximation To The Binomial

RECAP: The communications monitoring company *Postini* has reported that 91% of e-mail messages are spam. Recently, you installed a spam filter. You observe that over the past week it okayed only 151 of 1422 e-mails you received, classifying the rest as junk. Should you worry that the filtering is too aggressive?

QUESTIONS: What's the probability that no more than 151 of 1422 e-mails is a real message?

ANSWER: I assume that messages arrive randomly and independently, with a probability of success (a real message) $p = 0.09$. The model Binom(1422, 0.09) applies, but will be hard to work with. Checking conditions for the Normal approximation, I see that:

✓ These messages represent less than 10% of all e-mail traffic.

✓ I expect $np = (1422)(0.09) = 127.98$ real messages and $nq = (1422)(0.91) = 1294.02$ spam messages, both far greater than 10.

It's okay to approximate this Binomial probability by using a Normal model.

$$\mu = np = 1422(0.09) = 127.98$$

$$\sigma = \sqrt{npq} = \sqrt{1422(0.09)(0.91)} \approx 10.79$$

$$P(x \leq 151) = P\left(z \leq \frac{151 - 127.98}{10.79}\right)$$
$$= P(z \leq 2.13)$$
$$= 0.9834$$

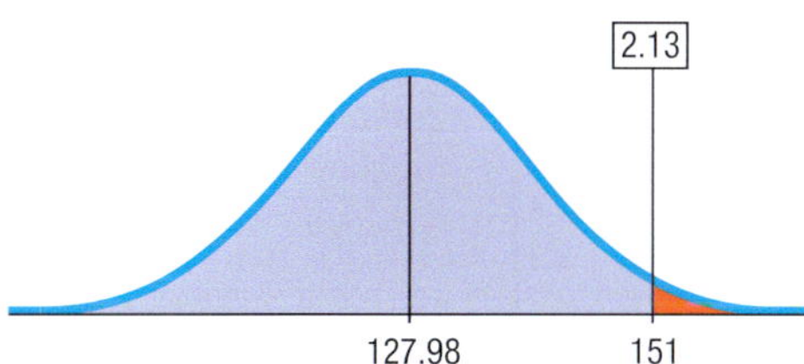

Among my 1422 e-mails, there's over a 98% chance that no more than 151 of them were real messages, so the filter may be working properly.

RANDOM MATTERS Could the Normal Really Be a Good Model for the Binomial?

The Binomial model is just counting successes in a number of trials where the outcomes are only success or failure. Each individual Bernoulli trial is about as discrete as an outcome can be: just 0 or 1. The Normal, on the other hand, models a continuous random variable. How can it be a good model for the sum? Let's see how this can work with a simulation.

We'll start with a large population of just 0's and 1's—for example, we drew a sample of 39,452 from a large dataset of all live births in the United States in 1998. One variable available there is the sex of the child, coded as 0 for Female and 1 for Male.

If 30 babies were born in the past week in a small town, what's the probability that we'd find at least half girls listed in the town paper? It is known that boy babies are slightly more likely at birth—a probability of about 0.5115. So we could calculate the Binomial probability of, for example, 15 girls in a sample of 30 births as

$$P(X = 15) = {}_{30}C_{15} \times 0.5115^{15} \times (1 - 0.5115)^{15}.$$

And if we wanted to know the probability of 15 or more girls in the sample, we'd have to add up 15 such probabilities. That makes using the Normal model instead seem tempting.

But can it work?

We can simulate a sample of size 30 of 0's and 1's using $P(X = 1) = 0.4885$. Here's one such sample:

1 1 0 1 1 1 0 1 0 0 0 1 0 1 0 0 0 0 0 0 0 1 1 0 0 0 0 0 0 0

This sample has 10 girls.

Repeating this 10,000 times, keeping track of the number of girls in each sample of 30, produced the following histogram for the number of girls:

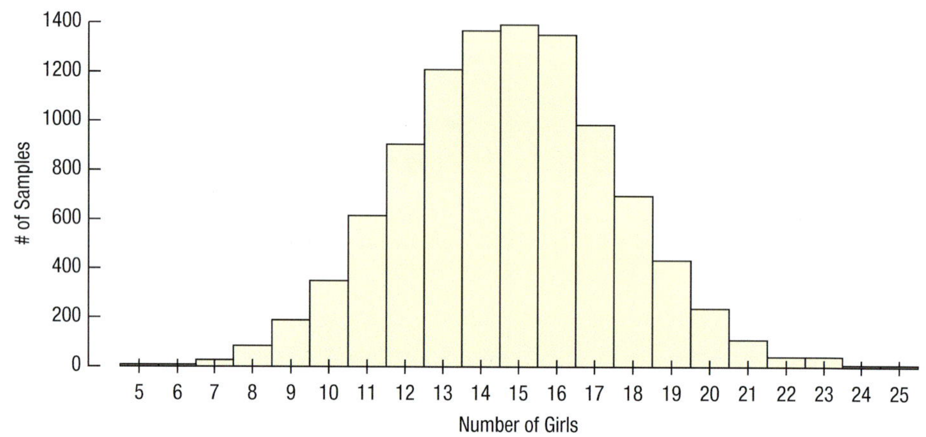

If we squint, it certainly looks like a Normal model might work. Let's put a Normal curve with the correct mean and standard deviation on top of the histogram. The mean should be $np = 30*0.4885 = 14.66$ and the standard deviation should be $\sqrt{npq} = \sqrt{30(0.4885)(0.5115)} = 2.74$:

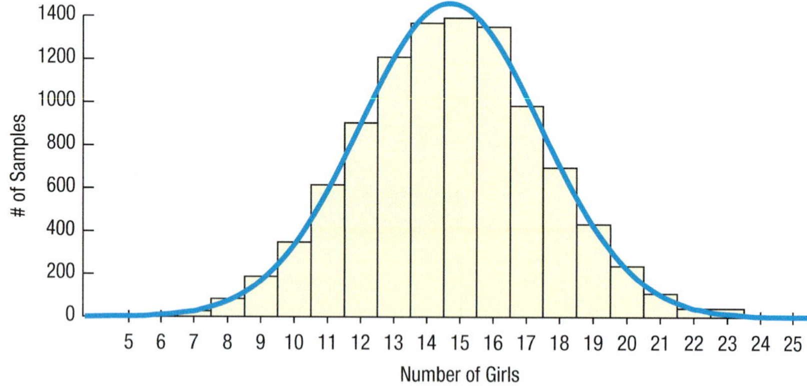

Now, let's find the area under the red curve that corresponds to 15 or more girls in a sample. Notice that to include the entire bar of 15, we should start the area at $x = 14.5$ (the left boundary of the bar that represents 15). (Sliding over 0.5 to include the entire bar is called the "continuity correction." See the next section for details.) So, we want $P(X>= 14.5)$ and not $P(X>=15)$, where our X is Normal(14.66, 2.74). Looking this up (from a table, computer program, or the Internet), we find: $P(X>=14.5) = 0.5226$.

How well does this match the simulation?

$\text{Sum}(\text{Counts}>=15) = 5253$, a match to two decimal places.

When does this work? Well, if we had taken samples of size 3, instead of 30, we would get a histogram that looks something like:

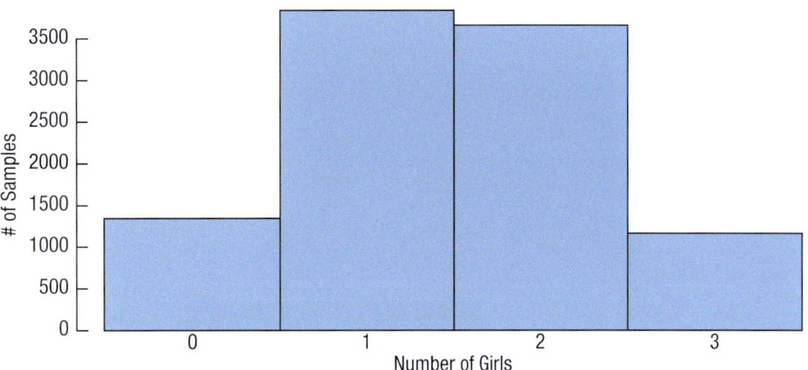

So, a Normal model isn't likely to work well here. The general rule for when the Normal is a good model for the Binomial is exactly the Success/Failure Condition: np and nq >=10.

*15.5 The Continuity Correction

When we used the Normal model to approximate the Binomial probabilities in the last section, we actually pulled a fast one on you. The Binomial is discrete, giving probabilities for specific counts, but the Normal is a **continuous random variable** that can take on *any value*. As we've seen, the probability that a Normal takes on any *specific* value is 0. So, for a Binomial with $n = 50$ and $p = 0.2$, the probability of exactly 10 successes is $P(X = 10) = 0.1398$. But for the Normal, $P(X = 10) = 0$. So there's a problem. The box below describes a solution, called the **continuity correction**.

*The Continuity Correction

When we use a continuous model to model a set of discrete events, we may need to make an adjustment called the **continuity correction**. We approximated the Binomial distribution (50, 0.2) with a Normal distribution. But what does the Normal distribution say about the probability that $X = 10$? Every specific value in the Normal probability model has probability 0. That's not the answer we want.

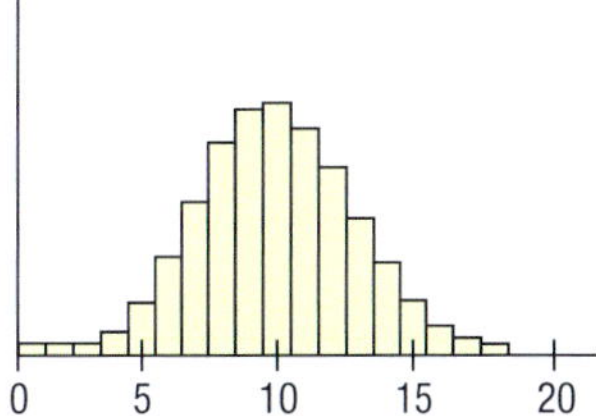

Because X is really discrete, it takes on the exact values 0, 1, 2, . . . , 50, each with positive probability. The histogram holds the secret to the correction. Look at the bin corresponding to $X = 10$ in the histogram. It goes from 9.5 to 10.5. What we really want is to find the area under the Normal curve *between* 9.5 and 10.5.

So when we use the Normal distribution to approximate discrete events, we go halfway to the next value on the left and/or the right. We approximate $P(X = 10)$ by finding $P(9.5 \leq X \leq 10.5)$. For a Binomial (50, 0.2), $\mu = 10$ and $\sigma = 2.83$.

$$\text{So } P(9.5 \leq X \leq 10.5) \approx P\left(\frac{9.5 - 10}{2.83} \leq z \leq \frac{10.5 - 10}{2.83}\right)$$
$$= P(-0.177 \leq z \leq 0.177)$$
$$= 0.1405$$

By comparison, the *exact* Binomial probability is 0.1398.

EXAMPLE 15.4 Using the Normal Distribution

Some LCD panels have stuck or "dead" pixels that have defective transistors and are permanently unlit. If a panel has too many dead pixels, it must be rejected. A manufacturer knows that, when the production line is working correctly, the probability of rejecting a panel is 0.07.

QUESTIONS:

1. How many screens do they expect to reject in a day's production run of 500 screens? What is the standard deviation?
2. If they reject 40 screens today, is that a large enough number that they should be concerned that something may have gone wrong with the production line?
3. In the past week of 5 days of production, they've rejected 200 screens—an average of 40 per day. Should that raise concerns?

ANSWERS:

1. $\mu = 0.07 \times 500 = 35$ is the expected number of rejects

   $\sigma = \sqrt{npq} = \sqrt{500 \times 0.07 \times 0.93} = 5.7$
2. $P(X \geq 40) = P\left(z \geq \frac{40 - 35}{5.7}\right) = P(z \geq 0.877) \approx 0.19$, not an extraordinarily large number of rejects
3. Using the Normal approximation:

   $\mu = 0.07 \times 2500 = 175$

   $\sigma = \sqrt{2500 \times 0.07 \times 0.93} = 12.757$

   $P(X \geq 200) = P\left(z \geq \frac{200 - 175}{12.757}\right) = P(z \geq 1.96) \approx 0.025$

Yes, this seems to be a number of rejects that would occur by chance rarely if nothing were wrong.

15.6 The Poisson Model

In the early 1990s, a leukemia cluster was identified in the Massachusetts town of Woburn. Many more cases of leukemia, a cancer that originates in a cell in the marrow of bone, appeared in this small town than would be predicted. Was it evidence of a problem in the town, or was it chance? That question led to a famous trial in which the families of eight leukemia victims sued, and became grist for the book and movie *A Civil Action*. Following an 80-day trial, the judge called for a retrial after dismissing the jury's contradictory and confusing findings. Shortly thereafter, the chemical companies and the families settled.

When rare events occur together or in clusters, people often want to know if that happened just by chance or whether something else is going on. If we assume that the events occur independently, we can use a Binomial model to find the probability that a cluster of events like this occurs. For rare events, p will be quite small, and when n is large, it may be difficult to compute the exact probability that a certain size cluster occurs. To see why, let's try to compute the probability that a cluster of cases of size x occurs in Woburn. We'll use the *national average* of leukemia incidence to get a value for p, and the population of Woburn as our value for n. In the United States in the early 1990s, there were about 30,800 new cases of leukemia each year and about 280,000,000 people, giving a value for p of about 0.00011. The population of Woburn was about $n = 35{,}000$. We'd expect $np = 3.85$ new cases of leukemia in Woburn. How unlikely would 8 or more cases be? To answer that, we'll need to calculate the complement, adding the probabilities of no cases, exactly 1 case, etc., up to 7 cases. Each of those probabilities would have the form ${}_{35000}C_x p^x q^{35000-x}$. To find ${}_{35000}C_x$, we'll need 35,000!—a daunting number even for computers. In the 18th century, when people first were interested in computing such probabilities, it was impossible. We can easily go beyond the capabilities of today's computers by making n large enough and p small enough. We could use the Normal model to help us approximate the probabilities as long as np is at least 10, but for rare events such as leukemia in Woburn, when np falls much below 10, the Normal probabilities won't be accurate.

Simeon Denis Poisson[3] was a French mathematician interested in events with very small probability. He originally derived his model to approximate the Binomial model when the probability of a success, p, is very small and the number of trials, n, is very large. Poisson's contribution was providing a simple approximation to find that probability. When you see the formula, however, you won't necessarily see the connection to the Binomial. The Poisson's parameter is often denoted by λ, the mean of the distribution. To use the **Poisson model** to approximate the Binomial, we'll make their means match, so we set $\lambda = np$.

NOTATION ALERT

We almost always use λ (the Greek letter lambda) for the mean of a Poisson model.

Poisson Probability Model for Successes: *Poisson* (λ)

λ = mean number of successes.

X = number of successes.

$$P(X = x) = \frac{e^{-\lambda}\lambda^x}{x!}$$

Expected value: $E(X) = \lambda$

Standard deviation: $SD(X) = \sqrt{\lambda}$

The Poisson model is a reasonably good approximation of the Binomial when $n \geq 20$ with $p \leq 0.05$ or $n \geq 100$ with $p \leq 0.10$.

Using Poisson's model, we can easily find the probabilities of a given size cluster of leukemia cases. Using $\lambda = np = 35000 \times 0.00011 = 3.85$ new cases a year, we find that the probability of seeing exactly x cases in a year is $P(X = x) = \frac{e^{-3.85}3.85^x}{x!}$. By adding up these probabilities for $x = 0, 1, \ldots 7$, we'd find that the probability of 8 or more cases with $\lambda = 3.85$ is about 0.043. That's small but not terribly unusual.

[3] *Poisson* is a French name (meaning "fish"), properly pronounced "pwa sohn" and not with an "oy" sound as in poison.

WHERE DOES E COME FROM?

You might know that $e = 2.7182818\ldots$ (to 7 decimal places), but does it have any real meaning? Yes! In fact, one of the places e originally turned up was in calculating how much more money you'd earn if you could get interest compounded more often. Suppose you receive an offer to earn 100% per year simple interest. (Take it!) At the end of the year, you'd have twice as much money as when you started. But suppose that you have the option of having the interest calculated and paid at the end of every month instead. Then each month you'll earn (100/12)% interest. At the end of the year, you'll wind up with $(1 + 1/12)^{12} = 2.613$ times as much instead of 2. If you could get the interest paid every day, you'd get $(1 + 1/365)^{365} = 2.715$ times as much. If you could convince the bank to compound and pay the interest every second, you'd get $(1 + 1/31536000)^{31536000} = 2.7182818$ times as much at year's end. This is where e shows up. In the limit, if you could get the interest compounded continually, you'd get e times as much. In other words, as *n* gets large, the limit of $(1 + 1/n)^n = e$. Who discovered this fact? Jacob Bernoulli, in 1683!

POISSON IN HISTORY

In his book *Gravity's Rainbow,* Thomas Pynchon describes using the Poisson to model the bombs dropping over London during World War II. "The rockets *are* distributing about London just as Poisson's equation in the textbook predicts. As the data keep coming in, Roger looks more and more like a prophet . . . all I'm doing is plugging numbers into a well-known equation, you can look it up in the book and do it yourself. . . . "

In spite of its origins as an approximation to the Binomial, the Poisson model is also used directly to model the probability of the occurrence of events for a variety of phenomena. It's a good model to consider whenever your data consist of counts of occurrences. It requires only that the events be independent and that the mean number of occurrences stay constant for the duration of the data collection (and beyond, if we hope to make predictions).

One nice feature of the Poisson model is that it scales according to the sample size.[4] For example, if we know that the average number of occurrences in a town the size of Woburn, with 35,000 people, is 3.85, we know that the average number of occurrences in a town of only 3500 residents is 0.385. We can use that new value of λ to calculate the probabilities for the smaller town. Using the Poisson formula, we'd find that the probability of 0 occurrences for a town of 3500 is about 0.68.

One of the consequences of the Poisson model is that, as long as the mean rate of occurrences stays constant, the occurrence of past events doesn't change the probability of future events. That is, even though events that occur according to the Poisson model *appear* to cluster, the probability of *another* event occurring is still the same. This is counterintuitive and may be one reason why so many people believe in the form of the "Law of Averages." When the Poisson model was used to model bombs dropping over London in World War II, the model said that even though several bombs had hit a particular sector, the probability of *another* hit was still the same. So, there was no point in moving people from one sector to another, even after several hits. You can imagine how difficult it must have been trying to convince people of that! This same phenomenon leads many people to think they see patterns in which an athlete, slot machine, or financial market gets "hot." But the occurrence of several successes in a row is not unusual and is rarely evidence that the mean has changed. Careful studies have been made of "hot streaks," but none has ever demonstrated that they actually exist.

JUST CHECKING

As we noted a few chapters ago, the Pew Research Center (www.pewresearch.org) reports that they are actually able to contact only 76% of the randomly selected households drawn for a telephone survey.

1. Explain why these phone calls can be considered Bernoulli trials.
2. Which of the models of this chapter (Geometric, Binomial, Normal) would you use to model the number of successful contacts from a list of 1000 sampled households? Explain.

[4]Because *Poisson* means "fish" in French, you can remember this fact by the bilingual pun "*Poisson* scales"!

3. Pew further reports that even after they contacted a household, only 38% agree to be interviewed, so the probability of getting a completed interview for a randomly selected household is only 0.29. Which of the models of this chapter would you use to model the number of households Pew has to call before they get the first completed interview?

4. Suppose that in the course of a year, legitimate survey organizations (not folks pretending to take a survey but actually trying to sell you something or change your vote) sample 70,000 of the approximately 107,000,000 households in the United States. You wonder if people from your school are represented in these polls. Counting all the living alumni from your school, there are 10,000 households. How would you model the number of these alumni households you'd expect will be contacted next year by a legitimate poll?

15.7 Other Continuous Random Variables: The Uniform and the Exponential

Many phenomena in business can be modeled by continuous random variables. The Normal model is important, but it is only one of many different models. Entire courses are devoted to studying which models work well in different situations, but we'll introduce just two others that are commonly used: the Uniform and the Exponential.

The Uniform Distribution

We've already seen the discrete version of the **Uniform probability model**. It has the property that all events are equally likely: With n possible outcomes, $P(X = x) = 1/n$ for all possible values x. For the continuous Uniform, we'd want all *intervals* of the same length to have the same probability. So, not surprisingly the probability model for the continuous Uniform random variable looks flat. It can be defined by the formula

$$f(x) = \begin{cases} \dfrac{1}{b - a} & \text{if } a \le x \le b \\ 0 & \text{otherwise} \end{cases}$$

From Figure 15.1, it's easy to see that the probability that X lies in any interval between a and b is the same as any other interval of the same length. In fact, the probability is just the ratio of the length of the interval to the total length: $b - a$. In other words:

For values c and d ($c \le d$) both within the interval $[a, b]$:

$$P(c \le X \le d) = \frac{(d - c)}{(b - a)}$$

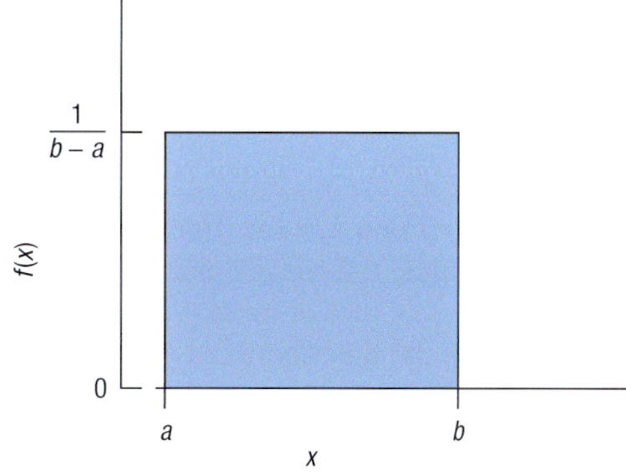

Figure 15.1
The probability model of a continuous Uniform random variable on the interval from a to b.

As an example, suppose you arrive at a bus stop and want to model how long you'll wait for the next bus. The sign says that busses arrive about every 20 minutes, but no other information is given. You might assume that the arrival is equally likely to be anywhere in the next 20 minutes, and so the probability model would be

$$f(x) = \begin{cases} \frac{1}{20} & \text{if} \quad 0 \le x \le 20 \\ 0 & \text{otherwise} \end{cases}$$

and would look as shown in Figure 15.2.

Figure 15.2
The probability model for a continuous Uniform random variable on the interval [0,20]. Notice that the mean (the balancing point) of the distribution is at 10 minutes and that the area of the box is 1.

Just as the mean of a data distribution is the balancing point of a histogram, the mean of any continuous random variable is the balancing point of the density function. Looking at Figure 15.2, we can see that the balancing point is halfway between the end points at 10 minutes. In general, the expected value is:

$$E(X) = \frac{a + b}{2}$$

for a Uniform distribution on the interval (a, b). With $a = 0$ and $b = 20$, the expected value would be 10 minutes.

The variance and standard deviation are less intuitive:

$$Var(X) = \frac{(b - a)^2}{12}; SD(X) = \sqrt{\frac{(b - a)^2}{12}}.$$

Using these formulas, our bus wait will have an expected value of 10 minutes with a standard deviation of $\sqrt{\frac{(20 - 0)^2}{12}} = 5.77$ minutes.

The Exponential Model

We've seen that the Poisson model is a good model for the arrival, or occurrence, of events. Suppose, for example, that we use the Poisson to model the probability of x visits to our website within the next minute. Then, the **Exponential probability model** with parameter λ can be used to model the time *between* those events. Because *time* is a continuous quantity, it has a continuous model with the form:

$$f(x) = \lambda e^{-\lambda x} \quad \text{for } x \ge 0 \text{ and } \lambda > 0$$

The use of the parameter λ again is not coincidental. It highlights the relationship between the Exponential and the Poisson.

If a discrete random variable can be modeled by a Poisson model with rate λ, then the times between events can be modeled by the continuous Exponential model with the same parameter λ. The mean of the Poisson random variable is λ and the mean of the

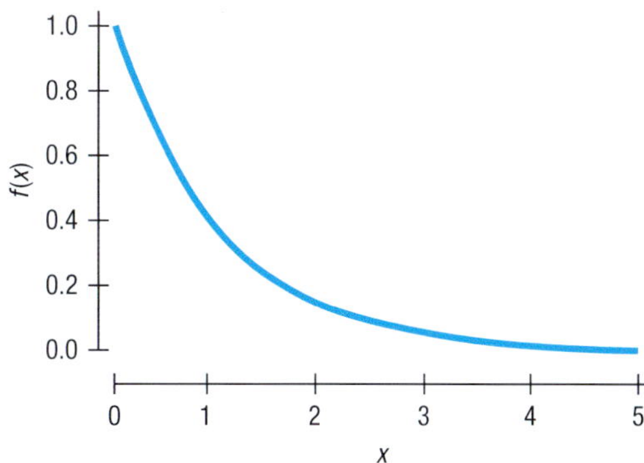

The Exponential probability model (with $\lambda = 1$). The probability that x lies between any two values corresponds to the area under the curve between the two values.

Exponential is $1/\lambda$. The inverse relationship between the two means makes intuitive sense. If λ increases and we expect *more* hits per minute, then the expected time between hits should go down. The standard deviation of the Exponential random variable is $1/\lambda$.

Probabilities of an Exponential random variable can be found through the probability model. Fortunately, the probability that x lies between any two values, s and t ($s \leq t$), has a particularly easy form:

$$P(s \leq X \leq t) = e^{-\lambda s} - e^{-\lambda t}.$$

In particular, by setting s to be 0, we can find the probability that the waiting time will be less than t:[5]

$$P(X \leq t) = P(0 \leq X \leq t) = e^{-\lambda 0} - e^{-\lambda t} = 1 - e^{-\lambda t}.$$

If arrivals of hits to our website can be well modeled by a Poisson with $\lambda = 4/\text{minute}$, then the probability that we'll have to wait less than 20 seconds (1/3 of a minute) is $F(1/3) = P(0 \leq X \leq 1/3) = 1 - e^{-4/3} = 0.736$. That seems about right. Arrivals are coming about every 15 seconds on average, so we shouldn't be surprised that nearly 75% of the time we won't have to wait more than 20 seconds for the next hit.

WHAT CAN GO WRONG?

- **Be sure you have Bernoulli trials.** Be sure to check the requirements first: two possible outcomes per trial ("success" and "failure"), a constant probability of success, and independence. Remember to check the 10% Condition when sampling without replacement.
- **Don't confuse Geometric and Binomial models.** Both involve Bernoulli trials, but the issues are different. If you are repeating trials until your first success, that's a Geometric probability. You don't know in advance how many trials you'll need—theoretically, it could take forever. If you are counting the number of successes in a specified number of trials, that's a Binomial probability.
- **Don't use the Normal approximation with small *n*.** To use a Normal approximation in place of a Binomial model, there must be at least 10 expected successes and 10 expected failures.

[5]The function $P(X \leq t) = F(t)$ is called the cumulative distribution function (cdf) of the random variable X. The probability model is formally called the density.

◆ **Don't assume everything's Normal.** Just because a random variable is continuous or you happen to know a mean and standard deviation doesn't mean that a Normal model will be useful. You must think about whether the **Normality Assumption** is justified. Using a Normal model when it really does not apply will lead to wrong answers and misleading conclusions.

A sample of CEOs has a mean total compensation of \$10,307,311.87 with a standard deviation of \$17,964,615.16. Using the Normal model rule, we should expect about 68% of the CEOs to have compensations between −\$7,657,303.29 and \$28,271,927.03. In fact, more than 90% of the CEOs have annual compensations in this range. What went wrong? The distribution is skewed, not symmetric. Using the 68–95–99.7 Rule for data like these will lead to silly results.

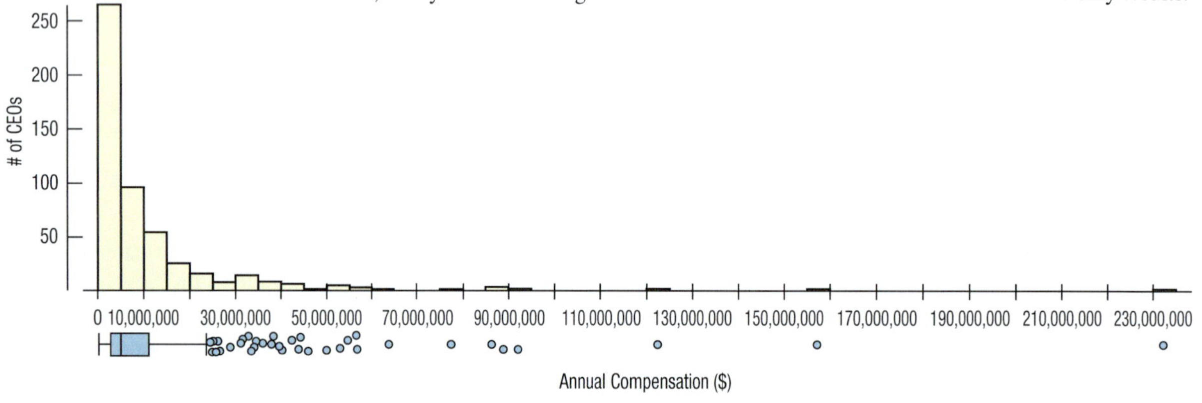

CONNECTIONS

This chapter builds on what we know about random variables. We now have more probability models to join the Normal model.

There are a number of "forward" connections from this chapter. We'll see the **10% Condition** and the **Success/Failure Condition** often. And the facts about the Binomial distribution can help explain how proportions behave as we'll see in the next chapter.

CHAPTER REVIEW

We've learned that Bernoulli trials show up in lots of places. Depending on the random variable of interest, we can use one of three models to estimate probabilities for Bernoulli trials:

- a Geometric model when we're interested in the number of Bernoulli trials until the next success;
- a Binomial model when we're interested in the number of successes in a certain number of Bernoulli trials;
- a Normal model to approximate a Binomial model when we expect at least 10 successes and 10 failures;

We can use a Poisson model to approximate a Binomial model when the probability of success, p, is very small and the number of trials, n, is very large;

- an Exponential model can model the time between two random events.

REVIEW OF TERMS

Bernoulli trials, if . . .

1. there are two possible outcomes (p. 468).
2. the probability of success is constant.
3. the trials are independent.

Geometric probability model A Geometric model is appropriate for a random variable that counts the number of Bernoulli trials until the first success (p. 469).

Binomial probability model A Binomial model is appropriate for a random variable that counts the number of successes in a fixed number of Bernoulli trials (p. 472).

Success/Failure Condition For a Normal model to be a good approximation of a Binomial model, we must expect at least 10 successes and 10 failures. That is, $np \geq 10$ and $nq \geq 10$ (p. 476).

Continuous random variable A continuous random variable is one that can take on any value in an interval of values (p. 479).

Continuity correction An adjustment made when approximating a discrete distribution (e.g., the Binomial) by a continuous model (e.g., the Normal; p. 479).

Poisson probability model A Poisson model can be used to approximate a Binomial model when *p* is small and *n* is large. Use $\lambda = np$ as the mean for the Poisson. The Poisson can also be used to model counts of a wide variety of phenomena (p. 481).

Uniform probability model A Uniform model gives equal probability to intervals of the same size (p. 483).

Exponential probability model An Exponential model can be used to estimate the time between two random events that occur following a Poisson probability model (p. 484).

TECH SUPPORT

Probability Models

Most statistics packages offer functions that compute Binomial probabilities, and many offer functions for Geometric and Poisson probabilities as well. Some technology solutions automatically use the Normal approximation for the Binomial when the exact calculations become unmanageable. And many websites offer Normal probability calculations.

The only important differences among these functions are in what they are named and the order of their arguments.

In these functions, pdf stands for "probability density function"—what we've been calling a probability model. For a discrete random variables (e.g., Geometric, Binomial, Poisson) the pdf(x) gives the probability $P(X = x)$. For a continuous random variable (e.g., Normal, Exponential), pdf(x) gives the value of the curve under which the probability lies. To find the probability, use the cdf(x), the "cumulative distribution function", that accumulates the probability and gives $P(X \leq x)$.

Generically, the functions are as follows:

Geometric pdf(*prob*, *x*)	Finds the individual Geometric probability of getting the first success on trial *x* when the probability of success is *prob*. Note: Some Geometric models model the number of failures before the first success.	For example, the probability of finding the first Hope Solo picture in the fifth cereal box is Geometric pdf(0.2, 5)
Geometric cdf(*prob*, *x*)	Finds the cumulative probability of getting the first success on or before trial *x*, when the probability of success is *prob*.	For example, the total probability of finding Hope's picture in one of the first 4 boxes is Geometric cdf(0.2, 4)
Binomial pdf(*n*, *prob*, *x*)	Finds the probability of getting *x* successes in *n* trials when the probability of success is *prob*.	For example, Binomial pdf(5, 0.2, 2) is the probability of finding Hope's picture exactly twice among 5 boxes of cereal.
Binomial cdf(*n*, *prob*, *x*)	The cdf of the Binomial	
Poisson pdf(λ, *x*)	Finds the probability of *x* successes when the mean number of successes is $\lambda = np$.	Best for large *n* and small probability of success.
Poisson cdf(λ, *x*)	The cdf of the Poisson	
Norm (x, mu, sigma)	The pdf or cdf of the Normal.	
Exp(*x*, λ)	The pdf or cdf of the Exponential	

DATA DESK

BinomDistr(*x, n, prob*) (pdf)
CumBinomDistr(*x, n, prob*) (cdf)
PoisDistr(*x, mean*) (pdf)
CumPoisDistr(*x, mean*) (cdf)
ZDistr(*x*) (pdf)
CumZDistr(*x*) (cdf)

COMMENTS

Data Desk does not compute Geometric probabilities. These functions work in derived variables or in scratchpads.

EXCEL

Binomdist(*x, n, prob, cumulative*)
Norm.dist(*x, mean, standard_dev, cumulative*)

COMMENTS

Set cumulative = *true* for cdf, *false* for pdf.

Excel's function fails when *x* or *n* is large.

Possibly, it does not use the Normal approximation.

Excel does not compute Geometric or Poisson probabilities.

JMP

Binomial Probability (*prob, n, x*) (pdf)
Binomial Distribution (*prob, n, x*) (cdf)
Poisson Probability (*mean, k*) (pdf)
Poisson Distribution (*mean, k*) (cdf)
Normal Distribution(*x, mean* = 0, *sd* = 1) (cdf)

COMMENTS

JMP does not compute Geometric probabilities.

MINITAB

Choose **Probability Distributions** from the **Calc** menu.

- dgeom(x,p) Gives $P(X = x)$ for the Geometric with probability of success p
- pgeom(x,p) Gives the cumulative probability $P(X \leq x)$ for the Geometric

Choose **Binomial** from the Probability Distributions submenu.

To calculate the probability of getting *x* successes in *n* trials, choose **Probability**.

To calculate the probability of getting *x* or fewer successes among *n* trials, choose **Cumulative Probability**.

For Poisson, choose **Poisson** from the Probability Distribution submenu.

For Geometric, choose **Geometric** from the Probability Distribution submenu.

R

The standard library stats contains pdf's and cdf's of many common models:

- **dgeom(x,p)** # Gives $P(X = x)$ for the Geometric with probability *p*
- **pgeom(x,p)** # Gives the cdf $P(X \leq x)$ for the Geometric with probability *p*
- **dbinom(x,n,p)** # Gives $P(X = x)$ for the Binomial with *n* and *p*
- **pbinom(x,n,p)** # Gives the cdf $P(X \leq x)$ for the Binomial with *n* and *p*
- **dpois(x,lambda)** # Gives $P(X = x)$ for the Poisson with mean lambda
- **ppois(x,lambda)** # Gives the cdf $P(X \leq x)$ for the Poisson with mean lambda
- **dnorm(x,mu,sig)** # Gives the pdf for the Normal model
- **pnorm(x,mu,sig)** # Gives the cdf for the Normal model
- **dexp(x,rate)** # Gives the pdf for the Exponential with rate = λ
- **pexp(x,rate)** # Gives the cdf for the Exponential with rate = λ

SPSS

PDF.GEOM(*x, prob*)
CDF.GEOM(*x, prob*)
PDF.Poisson(*x, mean*)
CDF.Poisson(*x, mean*)
PDF.BINOM(*x, n, prob*)
CDF.BINOM(*x, n, prob*)
CDF.NORMAL(*quant, mean, stddev*)
PDF.NORMAL(*quant, mean, stddev*)

TI-83/84 PLUS

geometpdf(*prob, x*)
geometcdf(*prob, x*)
binompdf(*n, prob, x*)
binomcdf(*n, prob, x*)
poissonpdf(*mean, x*)
poissoncdf(*mean, x*)

COMMENTS

Find these commands in the 2nd DISTR menu (the calculator refers to models as "distributions").

EXERCISES

SECTION 15.1

1. Bernoulli Do these situations involve Bernoulli trials? Explain.

a) We roll 50 dice to find the distribution of the number of spots on the faces.

b) How likely is it that in a group of 120 the majority may have Type A blood, given that Type A is found in 43% of the population?

c) We deal 7 cards from a deck and get all hearts. How likely is that?

d) We wish to predict the outcome of a vote on the school budget, and poll 500 of the 3000 likely voters to see how many favor the proposed budget.

e) A company realizes that about 10% of its packages are not being sealed properly. In a case of 24, is it likely that more than 3 are unsealed?

2. Bernoulli 2 Do these situations involve Bernoulli trials? Explain.

a) You are rolling 5 dice and need to get at least two 6's to win the game.

b) We record the distribution of eye colors found in a group of 500 people.

c) A manufacturer recalls a doll because about 3% have buttons that are not properly attached. Customers return 37 of these dolls to the local toy store. Is the manufacturer likely to find any dangerous buttons?

d) A city council of 11 Republicans and 8 Democrats picks a committee of 4 at random. What's the probability they choose all Democrats?

e) A 2002 Rutgers University study found that 74% of high school students have cheated on a test at least once. Your local high-school principal conducts a survey in homerooms and gets responses that admit to cheating from 322 of the 481 students.

SECTION 15.2

3. Toasters A manufacturer ships toasters in cartons of 20. In each carton, they estimate a 5% chance that one of the toasters will need to be sent back for minor repairs. What is the probability that in a carton, there will be exactly 3 toasters that need repair?

4. Soccer A soccer team estimates that they will score on 8% of the corner kicks. In next week's game, the team hopes to kick 15 corner kicks. What are the chances that they will score on 2 of those opportunities?

SECTION 15.3

5. Toasters again In a batch of 10,000 toasters, what are the chances that fewer than 450 need to be returned?

6. Soccer again If this team has 200 corner kicks over the season, what are the chances that they score more than 22 times?

SECTION 15.4

7. Sell! A car dealership sells an average of 5 cars in a day. Using the Poisson model, what is the probability that the dealer sells 3 cars tomorrow?

8. Passing on A large hospital has an average of 7 fatalities in a week. Using the Poisson model, what is the probability that this week it has 10 fatalities?

SECTION 15.7

9. Telephone numbers A cable provider wants to contact customers in a particular telephone exchange to see how satisfied they are with the new digital TV service the company has provided. All numbers are in the 452 exchange, so there are 10,000 possible numbers from 452-0000 to 452-9999. If they select the numbers with equal probability:

a) What distribution would they use to model the selection?

b) The new business "incubator" was assigned the 200 numbers between 452-2500 and 452-2699, but these businesses don't subscribe to digital TV. What is the probability that the randomly selected number will be for an incubator business?

c) Numbers above 9000 were only released for domestic use last year, so they went to newly constructed residences. What is the probability that a randomly selected number will be one of these?

10. Serial numbers In an effort to check the quality of their cell phones, a manufacturing manager decides to take a random sample of 10 cell phones from yesterday's production run, which produced cell phones with serial numbers ranging (according to when they were produced) from 43005000 to 43005999. If each of the 1000 phones is equally likely to be selected:

a) What distribution would they use to model the selection?

b) What is the probability that a randomly selected cell phone will be one of the last 100 to be produced?

c) What is the probability that the first cell phone selected is either from the last 200 to be produced or from the first 50 to be produced?

11. Component lifetimes Lifetimes of electronic components can often be modeled by an Exponential model. Suppose quality control engineers want to model the lifetime of a hard drive to have a mean lifetime of 3 years.

a) What value of λ should they use?

b) With this model, what would the probability be that a hard drive lasts 5 years or less?

12. Website sales Suppose occurrences of sales on a small company's website are well modeled by a Poisson model with $\lambda = 5/\text{hour}$.

a) If a sale just occurred, what is the expected waiting time until the next sale?

b) What is the probability that the next sale will happen in the next 6 minutes?

CHAPTER EXERCISES

13. Simulating the model Think about the Hope Solo picture search again. You are opening boxes of cereal one at a time looking for her picture, which is in 20% of the boxes. You want to know how many boxes you might have to open in order to find Hope.

a) Describe how you would simulate the search for Hope using random numbers.

b) Run at least 30 trials.

c) Based on your simulation, estimate the probabilities that you might find your first picture of Hope in the first box, the second, etc.

d) Calculate the actual probability model.

e) Compare the distribution of outcomes in your simulation to the probability model.

14. Simulation II You are one space short of winning a child's board game and must roll a 1 on a die to claim victory. You want to know how many rolls it might take.

a) Describe how you would simulate rolling the die until you get a 1.

b) Run at least 30 trials.

c) Based on your simulation, estimate the probabilities that you might win on the first roll, the second, the third, etc.

d) Calculate the actual probability model.

e) Compare the distribution of outcomes in your simulation to the probability model.

15. Hope, again Let's take one last look at the Hope Solo picture search. You know her picture is in 20% of the cereal boxes. You buy five boxes to see how many pictures of Hope you might get.

a) Describe how you would simulate the number of pictures of Hope you might find in five boxes of cereal.

b) Run at least 30 trials.

c) Based on your simulation, estimate the probabilities that you get no pictures of Hope, 1 picture, 2 pictures, etc.

d) Find the actual probability model.

e) Compare the distribution of outcomes in your simulation to the probability model.

16. Seatbelts Suppose 75% of all drivers always wear their seatbelts. Let's investigate how many of the drivers might be belted among five cars waiting at a traffic light.

a) Describe how you would simulate the number of seatbelt-wearing drivers among the five cars.

b) Run at least 30 trials.

c) Based on your simulation, estimate the probabilities there are no belted drivers, exactly one, two, etc.

d) Find the actual probability model.

e) Compare the distribution of outcomes in your simulation to the probability model.

17. On time A Department of Transportation report about air travel found that, nationwide, 76% of all flights are on time. Suppose you are at the airport and your flight is one of 50 scheduled to take off in the next two hours. Can you consider these departures to be Bernoulli trials? Explain.

18. Lost luggage A Department of Transportation report about air travel found that airlines misplace about 5 bags per

1000 passengers. Suppose you are traveling with a group of people who have checked 22 pieces of luggage on your flight. Can you consider the fate of these bags to be Bernoulli trials? Explain.

19. Ping A computer sends a periodical ping to a server, and the server is available 90%. Assuming each ping trial is independent from the others, find the probability that, on any particular day,

a) the computer finds the server unavailable on its fourth ping.

b) the computer finds the server available on its third ping.

c) it takes the computer more than two pings to find the server available.

20. Appliances Suppose 5% stock of a particular appliance of a particular brand sold at a large department store will have a fault of some kind.

a) What is the probability that the sixth appliance sold is the first faulty one?

b) What is the probability that it takes more than twelve sales to sell a faulty one?

21. More pings For the computer in Exercise 19, what is the expected number of pings until a server is found unavailable?

22. Appliances ahoy For the appliances described in Exercise 20, how many does the department store expect to sell before selling a faulty one?

23. Customer center operator Raaj works at the customer service call center of a major credit card bank. Cardholders call for a variety of reasons, but regardless of their reason for calling, if they hold a platinum card, Raaj is instructed to offer them a double-miles promotion. About 10% of all cardholders hold platinum cards, and about 50% of those will take the double-miles promotion. On average, how many calls will Raaj have to take before finding the first cardholder to take the double-miles promotion?

24. Cold calls Justine works for an organization committed to raising money for Alzheimer's research. From past experience, the organization knows that about 20% of all potential donors will agree to give something if contacted by phone. They also know that of all people donating, about 5% will give \$100 or more. On average, how many potential donors will she have to contact until she gets her first \$100 donor?

25. Blood Only 4% of people have Type AB blood.

a) On average, how many donors must be checked to find someone with Type AB blood?

b) What's the probability that there is a Type AB donor among the first 5 people checked?

c) What's the probability that the first Type AB donor will be found among the first 6 people?

d) What's the probability that we won't find a Type AB donor before the 10th person?

26. Color blindness About 8% of males are color-blind. A researcher needs some color-blind subjects for an experiment and begins checking potential subjects.

a) On average, how many men should the researcher expect to check to find one who is color-blind?

b) What's the probability that she won't find anyone color-blind among the first 4 men she checks?

c) What's the probability that the first color-blind man found will be the sixth person checked?

d) What's the probability that she finds someone who is color-blind before checking the 10th man?

27. Coins and intuition If you flip a fair coin 100 times,

a) Intuitively, how many heads do you expect?

b) Use the formula for expected value to verify your intuition.

28. Roulette and intuition An American roulette wheel has 38 slots, of which 18 are red, 18 are black, and 2 are green (0 and 00). If you spin the wheel 38 times,

a) Intuitively, how many times would you expect the ball to wind up in a green slot?

b) Use the formula for expected value to verify your intuition.

29. Cold calling A sales department knows that 30% of cold calls for a particular product end up in a sale. If 6 potential customers are selected at random, find the probability of each of the following outcomes.

a) The first customer to buy is the third one.

b) At least one sale happens.

c) The first sale happens with the second or third customer.

d) There are exactly two sales from all contacted persons.

e) There are no more than two sales from all contacted persons.

f) There are at least three sales from all contacted persons.

30. Shooting range A marksman in a shooting range is able to hit a plate 70% of the time. Assume each shot is independent of the others. If the marksman shoots 5 times, what is the probability of each of the following results?

a) The marksman's first hit comes on the second shot.

b) The marksman misses at least one plate.

c) The marksman's first hit comes on the third or fourth shot.

d) The marksman hits exactly three times.

e) The marksman hits at least three times.

f) The marksman hits at most three times.

31. Cold calling, redux Consider our group of 6 people in Exercise 29.

a) How many sales do you expect?

b) What is the standard deviation on the number of sales?

c) If we keep calling until we get a sale, how many calls do we expect to make?

32. More shooting Consider our marksman from Exercise 30.

a) How many plates is the marksman expected to hit?

b) With what standard deviation?

c) If the marksman keeps shooting until she hits a plate, how many tries do you expect it will take?

33. More cold calling Suppose we choose 10 people instead of 6 in Exercise 29.

a) Find the mean and standard deviation of the number of calls that don't lead to sales.

b) What's the probability that not all calls end without a sale?

c) What is the probability that not more than 3 calls end without a sale?

d) What is the probability that 5 calls end without a sale?

e) What is the probability that the majority of the calls end without a sale?

34. Still more shooting Suppose the marksman from Exercise 30 shoots 8 times.

a) Find the mean and standard deviation of the number of hits.

b) What is the probability that the marksman never misses?

c) What is the probability that there are no more than 6 hits?

d) What is the probability that there are exactly 6 hits?

e) What is the probability that there are more hits than misses?

35. Vision It is generally believed that nearsightedness affects about 12% of all children. A school district tests the vision of 169 incoming kindergarten children. How many would you expect to be nearsighted? With what standard deviation?

36. International students At a certain college, 6% of all students come from outside the United States. Incoming students there are assigned at random to freshman dorms, where students live in residential clusters of 40 freshmen sharing a common lounge area. How many international students would you expect to find in a typical cluster? With what standard deviation?

37. Electric scooters In a company, 90% of electric scooters produced pass first inspection. Assume that the quality of each electric scooter is independent of the other. If 10 electric scooters are made in one day, what is the probability that:

a) all electric scooters pass first inspection?

b) exactly 8 electric scooters pass first inspection?

c) at least 8 electric scooters pass first inspection?

d) no more than 8 electric scooters pass first inspection?

38. Frogs A wildlife biologist examines frogs for a genetic trait he suspects may be linked to sensitivity to industrial toxins in the environment. Previous research had established that this trait is usually found in 1 of every 8 frogs. He collects and examines a dozen frogs. If the frequency of the trait has not changed, what's the probability he finds the trait in

a) none of the 12 frogs?

b) at least 2 frogs?

c) 3 or 4 frogs?

d) no more than 4 frogs?

39. More electric scooters Suppose the company in Exercise 37 expands and now produces 120 scooters in one day.

a) What are the mean and standard deviation of the number of electric scooters that pass first inspection?

b) Verify that you can use a normal model to approximate the distribution of the number of electric scooters passing first inspection.

c) Use the 68-95-99.7 Rule to describe this distribution.

d) What is the probability that at least 100 electric scooters pass first inspection?

40. More shooting The marksman in Exercise 30 will be shooting 100 plates in a competition.

a) What are the mean and standard deviation of the number of times they hit?

b) Is a normal model appropriate here? Explain.

c) Use the 68-95-99.7 Rule to describe the distribution of the number of hits the marksman may get.

d) Would you be surprised if the marksman hits at most 60 times?

41. Apples An orchard owner knows that he'll have to use about 6% of the apples he harvests for cider because they will have bruises or blemishes. He expects a tree to produce about 300 apples.

a) Describe an appropriate model for the number of cider apples that may come from that tree. Justify your model.

b) Find the probability there will be no more than a dozen cider apples.

c) Is it likely there will be more than 50 cider apples? Explain.

42. Frogs, part II Based on concerns raised by his preliminary research, the biologist in Exercise 38 decides to collect and examine 150 frogs.

a) Assuming the frequency of the trait is still 1 in 8, determine the mean and standard deviation of the number of frogs with the trait he should expect to find in his sample.

b) Verify that he can use a Normal model to approximate the distribution of the number of frogs with the trait.

c) He found the trait in 22 of his frogs. Do you think this proves that the trait has become more common? Explain.

43. Cold calling, again The sales department intends to do 1000 cold calls in a week and stocks 300 items. On average, 900 cold calls are answered, and we can assume that 30% of these will result in a sale. What's the probability that there are more sales than what is available in stock?

44. No-shows An airline, believing that 5% of passengers fail to show up for flights, overbooks (sells more tickets than there are seats). Suppose a plane will hold 265 passengers, and the airline sells 275 tickets. What's the probability the airline will not have enough seats, so someone gets bumped?

45. Annoying phone calls A newly hired telemarketer is told he will probably make a sale on about 12% of his phone calls. The first week he called 200 people, but only made 10 sales. Should he suspect he was misled about the true success rate? Explain.

46. The euro Shortly after the introduction of the euro coin in Belgium, newspapers around the world published articles claiming the coin is biased. The stories were based on reports that someone had spun the coin 250 times and gotten 140 heads—that's 56% heads. Do you think this is evidence that spinning a euro is unfair? Explain.

47. Hurricanes, redux We have looked at various facts about hurricanes in previous chapters. Suppose we find that the arrivals of hurricanes can be modeled by a Poisson distribution with mean 2.45.

a) What's the probability of no hurricanes next year?

b) What's the probability that during the next two years, there's exactly 1 hurricane?

48. Bank tellers I am the only bank teller on duty at my local bank. I need to run out for 10 minutes, but I don't want to miss any customers. Suppose the arrival of customers can be modeled by a Poisson distribution with mean 2 customers per hour.

a) What's the probability that no one will arrive in the next 10 minutes?

b) What's the probability that 2 or more people arrive in the next 10 minutes?

c) You've just served 2 customers who came in one after the other. Is this a better time to run out?

49. TB Worldwide, tuberculosis is the number-one cause of death from an infectious disease. But the majority of cases are in developing countries. In the U.S., the probability of contracting TB is small, with p about 0.0005 for a new case in a given year. In a town of 8000 people:

a) What's the expected number of new cases?

b) Use the Poisson model to approximate the probability that there will be at least one new case of TB next year.

50. Earthquakes Suppose the probability of a major earthquake on a given day is 1 out of 10,000.

a) What's the expected number of major earthquakes in the next 1000 days?

b) Use the Poisson model to approximate the probability that there will be at least one major earthquake in the next 1000 days.

51. Seatbelts II Police estimate that 80% of drivers now wear their seatbelts. They set up a safety roadblock, stopping cars to check for seatbelt use.

a) How many cars do they expect to stop before finding a driver whose seatbelt is not buckled?

b) What's the probability that the first unbelted driver is in the sixth car stopped?

c) What's the probability that the first 10 drivers are all wearing their seatbelts?

d) If they stop 30 cars during the first hour, find the mean and standard deviation of the number of drivers expected to be wearing seatbelts.

e) If they stop 120 cars during this safety check, what's the probability they find at least 20 drivers not wearing their seatbelts?

52. Rickets Vitamin D is essential for strong, healthy bones. Our bodies produce vitamin D naturally when sunlight falls upon the skin, or it can be taken as a dietary supplement. Although the bone disease rickets was largely eliminated in England during the 1950s, some people there are concerned that this generation of children is at increased risk because they are more likely to watch TV or play computer games than spend time outdoors. Recent research indicated that about 20% of British children are deficient in vitamin D. Suppose doctors test a group of elementary school children.

a) What's the probability that the first vitamin D–deficient child is the eighth one tested?

b) What's the probability that the first 10 children tested are all okay?

c) How many kids do they expect to test before finding one who has this vitamin deficiency?

d) They will test 50 students at the third-grade level. Find the mean and standard deviation of the number who may be deficient in vitamin D.

e) If they test 320 children at this school, what's the probability that no more than 50 of them have the vitamin deficiency?

53. ESP Scientists wish to test the mind-reading ability of a person who claims to have ESP. They use five cards with different and distinctive symbols (square, circle, triangle, line, squiggle). Someone picks a card at random and thinks about the symbol. The "mind reader" must correctly identify which symbol was on the card. If the test consists of 100 trials, how many would this person need to get right in order to convince you that ESP may actually exist? Explain.

54. True–false A true–false test consists of 50 questions. How many does a student have to get right to convince you that he is not merely guessing? Explain.

55. Ice hockey An ice hockey player usually scores 33% of his shots on goal. He has scored 3 in a row. Is this evidence that he is having a good game tonight? In other words, is this streak so unusual that it means the probability that he scores a shot on goal has changed? Explain.

56. Different gun The marksman in Exercise 30 uses a different gun, hoping that it will improve the success rate to more than 70%. The marksman is delighted when, in their first test, they hit 10 consecutive plates. Do you think this is compelling evidence that the new gun is better? In other words, is a streak like this unusual for the marksman? Explain.

57. Ice hockey, again The ice hockey player in Exercise 55 has new blades, which he thinks will improve his game. Off his past 60 shots, he has scored 30, much better than the 33% he usually scores. Do you think his chances of making a shot really increased? In other words, is making at least 30 out of 60 shots really unusual for him? Do you think his new blades are responsible for this?

58. New gun, again The marksman in Exercise 56 continues shooting, ending up with 90 hits out of a 100. Are you now convinced that the new gun is better? Explain.

59. Web visitors A website manager has noticed that during the evening hours, about 3 people per minute check out from their shopping cart and make an online purchase. She believes that each purchase is independent of the others and wants to model the number of purchases per minute.

a) What model might you suggest to model the number of purchases per minute?

b) What is the probability that in any one minute at least one purchase is made?

c) What is the probability that no one makes a purchase in the next two minutes?

60. Quality control In an effort to improve the quality of their cell phones, a manufacturing manager records the number of faulty phones in each day's production run. The manager notices that the number of faulty cell phones in a production run of cell phones is usually small and that the quality of one day's run seems to have no bearing on the next day.

a) What model might you use to model the number of faulty cell phones produced in one day?

b) If the mean number of faulty cell phones is 2 per day, what is the probability that no faulty cell phones will be produced tomorrow?

c) If the mean number of faulty cell phones is 2 per day, what is the probability that 3 or more faulty cell phones were produced in today's run?

***61.** Web visitors, part 2 The website manager in Exercise 59 wants to model the time between purchases. Recall that the mean number of purchases in the evening is 3 per minute.

a) What model would you use to model the time between events?

b) What is the mean time between purchases?

c) What is the probability that the time to the next purchase will be between 1 and 2 minutes?

***62.** Quality control, part 2 The cell phone manufacturer in Exercise 60 wants to model the time between events. The mean number of defective cell phones is 2 per day.

a) What model would you use to model the time between events?

b) What would the probability be that the time to the next failure is 1 day or less?

c) What is the mean time between failures?

JUST CHECKING

Answers

1. There are two outcomes (contact, no contact), the probability of contact is 0.76, and random calls should be independent.
2. Binomial, with $n = 1000$ and $p = 0.76$. For actual calculations, we could approximate using a Normal model with $\mu = np = 1000(0.76) = 760$ and $\sigma = \sqrt{npq} = \sqrt{1000(0.76)(0.24)} \approx 13.5$.
3. Geometric, with $p = 0.29$.
4. Poisson, with $\lambda = np = 10000\left(\frac{70000}{107000000}\right) = 6.54$.

Review of Part IV

RANDOMNESS AND PROBABILITY

Quick Review

Here's a brief summary of the key concepts and skills in probability and probability modeling:

- The Law of Large Numbers says that the more times we try something, the closer the results will come to theoretical perfection.
  - Don't mistakenly misinterpret the Law of Large Numbers as the "Law of Averages." There's no such thing.
- Basic rules of probability can handle most situations:
  - To find the probability that an event OR another event happens, add their probabilities and subtract the probability that both happen.
  - To find the probability that an event AND another independent event both happen, multiply probabilities.
  - Conditional probabilities tell you how likely one event is to happen, knowing that another event has happened.
  - Mutually exclusive events (also called "disjoint") cannot both happen at the same time.
  - Two events are independent if the occurrence of one doesn't change the probability that the other happens.
- A probability model for a random variable describes the theoretical distribution of outcomes.
  - The mean of a random variable is its expected value.
  - For sums or differences of independent random variables, variances add.
  - To estimate probabilities involving quantitative variables, you may be able to use a Normal model—but only if the distribution of the variable is unimodal and symmetric.
  - To estimate the probability you'll get your first success on a certain trial, use a Geometric model.
  - To estimate the probability you'll get a certain number of successes in a specified number of independent trials, use a Binomial model.
  - To estimate the probability of the number of occurrences of a relatively rare phenomenon, consider using a Poisson model.
  - To estimate the interval between two occurrences of a random phenomenon, use the Exponential model.

Ready? Here are some opportunities to check your understanding of these ideas.

REVIEW EXERCISES

R4.1. Quality control A consumer organization estimates that 29% of new cars have a cosmetic defect, such as a scratch or a dent, when they are delivered to car dealers. This same organization believes that 7% have a functional defect—something that does not work properly—and that 2% of new cars have both kinds of problems.

a) If you buy a new car, what's the probability that it has some kind of defect?
b) What's the probability it has a cosmetic defect but no functional defect?
c) If you notice a dent on a new car, what's the probability it has a functional defect?
d) Are the two kinds of defects disjoint events? Explain.
e) Do you think the two kinds of defects are independent events? Explain.

R4.2. Workers A company's human resources officer reports a breakdown of employees by job type and sex shown in the table.

		Sex	
		Male	Female
Job Type	Management	7	6
	Supervision	8	12
	Production	45	72

a) What's the probability that a worker selected at random is
   i) female?
   ii) female or a production worker?
   iii) female, if the person works in production?
   iv) a production worker, if the person is female?
b) Do these data suggest that job type is independent of being male or female? Explain.

R4.3. Airfares Each year a company must send 3 officials to a meeting in China and 5 officials to a meeting in France. Airline ticket prices vary from time to time, but the company purchases all tickets for a country at the same price. Past experience has shown that tickets to China have a mean price of \$1000, with a standard deviation of \$150, while the mean airfare to France is \$500, with a standard deviation of \$100.

a) Define random variables and use them to express the total amount the company will have to spend to send these delegations to the two meetings.
b) Find the mean and standard deviation of this total cost.
c) Find the mean and standard deviation of the difference in price of a ticket to China and a ticket to France.
d) Do you need to make any assumptions in calculating these means? How about the standard deviations?

R4.4. Bipolar Psychiatrists estimate that about 1 in 100 adults suffers from bipolar disorder. What's the probability that in a city of 10,000 there are more than 200 people with this condition? Be sure to verify that a Normal model can be used here.

R4.5. A game To play a game, you must pay \$5 for each play. There is a 10% chance you will win \$5, a 40% chance you will win \$7, and a 50% chance you will win only \$3.

a) What are the mean and standard deviation of your net winnings?
b) You play twice. Assuming the plays are independent events, what are the mean and standard deviation of your total winnings?

R4.6. Emergency switch Safety engineers must determine whether industrial workers can operate a machine's emergency shutoff device. Among a group of test subjects, 66% were successful with their left hands, 82% with their right hands, and 51% with both hands.

a) What percent of these workers could not operate the switch with either hand?
b) Are success with right and left hands independent events? Explain.
c) Are success with right and left hands mutually exclusive? Explain.

R4.7. Twins In the United States, the probability of having twins (usually about 1 in 90 births) rises to about 1 in 10 for women who have been taking the fertility drug Clomid. Among a group of 10 pregnant women, what's the probability that

a) at least one will have twins if none were taking a fertility drug?
b) at least one will have twins if all were taking Clomid?
c) at least one will have twins if half were taking Clomid?

R4.8. Deductible A car owner may buy insurance that will pay the full price of repairing the car after an at-fault accident, or save \$12 a year by getting a policy with a \$500 deductible. Her insurance company says that about 0.5% of drivers in her area have an at-fault auto accident during any given year. Based on this information, should she buy the policy with the deductible or not? How does the value of her car influence this decision?

R4.9. More twins A group of 5 women became pregnant while undergoing fertility treatments with the drug Clomid, discussed in Exercise 7. What's the probability that

a) none will have twins?
b) exactly 1 will have twins?
c) at least 3 will have twins?

R4.10. At fault The car insurance company in Exercise 8 believes that about 0.5% of drivers have an at-fault accident during a given year. Suppose the company insures 1355 drivers in that city.

a) What are the mean and standard deviation of the number who may have at-fault accidents?
b) Can you describe the distribution of these accidents with a Normal model? Explain.

R4.11. Twins, part III At a large fertility clinic, 152 women became pregnant while taking Clomid. (See Exercise 7.)

a) What are the mean and standard deviation of the number of twin births we might expect?
b) Can we use a Normal model in this situation? Explain.
c) What's the probability that no more than 10 of the women have twins?

R4.12. Child's play In a board game you determine the number of spaces you may move by spinning a spinner and rolling a die. The spinner has three regions: Half of the spinner is marked "5," and the other half is equally divided between "10" and "20." The six faces of the die show 0, 0, 1, 2, 3, and 4 spots. When it's your turn, you spin and roll, adding the numbers together to determine how far you may move.

a) Create a probability model for the outcome on the spinner.
b) Find the mean and standard deviation of the spinner results.
c) Create a probability model for the outcome on the die.
d) Find the mean and standard deviation of the die results.
e) Find the mean and standard deviation of the number of spaces you get to move.

R4.13. Language Neurological research has shown that in about 80% of people, language abilities reside in the brain's left side. Another 10% display right-brain language centers, and the remaining 10% have two-sided language control. (The latter two groups are mainly left-handers; *Science News*, 161 no. 24 [2002].)

a) Assume that a freshman composition class contains 25 randomly selected people. What's the probability that no more than 15 of them have left-brain language control?
b) In a randomly assigned group of 5 of these students, what's the probability that no one has two-sided language control?
c) In the entire freshman class of 1200 students, how many would you expect to find of each type?
d) What are the mean and standard deviation of the number of these freshmen who might be right-brained in language abilities?
e) If an assumption of Normality is justified, use the 68–95–99.7 Rule to describe how many students in the freshman class might have right-brain language control.

R4.14. Play, again If you land in a "penalty zone" on the game board described in Exercise 12, your move will be determined by subtracting the roll of the die from the result on the spinner. Now what are the mean and standard deviation of the number of spots you may move?

R4.15. Beanstalks In some cities tall people who want to meet and socialize with other tall people can join Beanstalk Clubs. To qualify, a man must be over 6′2″ tall, and a woman over 5′10″. According to the National Health Survey, heights of adults may have a Normal model with mean heights of 69.1″ for men and 64.0″ for women. The respective standard deviations are 2.8″ and 2.5″.

a) You're probably not surprised to learn that men are generally taller than women, but what does the greater standard deviation for men's heights indicate?
b) Are men or women more likely to qualify for Beanstalk membership?

c) Beanstalk members believe that height is an important factor when people select their spouses. To investigate, we select at random a married man and, independently, a married woman. Define two random variables, and use them to express how many inches taller the man is than the woman.
d) What's the mean of this difference?
e) What's the standard deviation of this difference?
f) What's the probability that the man is taller than the woman (that the difference in heights is greater than 0)?
g) Suppose a survey of married couples reveals that 92% of the husbands were taller than their wives. Based on your answer to part f, do you believe that people's choice of spouses is independent of height? Explain.

R4.16. Stocks Since the stock market began in 1872, stock prices have risen in about 73% of the years. Assuming that market performance is independent from year to year, what's the probability that

a) the market will rise for 3 consecutive years?
b) the market will rise 3 years out of the next 5?
c) the market will fall during at least 1 of the next 5 years?
d) the market will rise during a majority of years over the next decade?

R4.17. Multiple choice A multiple choice test has 50 questions, with 4 answer choices each. You must get at least 30 correct to pass the test, and the questions are very difficult.

a) Are you likely to be able to pass by guessing on every question? Explain.
b) Suppose, after studying for a while, you believe you have raised your chances of getting each question right to 70%. How likely are you to pass now?
c) Assuming you are operating at the 70% level and the instructor arranges questions randomly, what's the probability that the third question is the first one you get right?

R4.18. Stock strategy Many investment advisors argue that after stocks have declined in value for 2 consecutive years, people should invest heavily because the market rarely declines 3 years in a row.

a) Since the stock market began in 1872, there have been two consecutive losing years eight times. In six of those cases, the market rose during the following year. Does this confirm the advice?
b) Overall, stocks have risen in value during 95 of the 130 years since the market began in 1872. How is this fact relevant in assessing the statistical reasoning of the advisors?

R4.19. Insurance A 65-year-old woman takes out a $100,000 term life insurance policy. The company charges an annual premium of $520. Estimate the company's expected profit on such policies if mortality tables indicate that only 2.6% of women age 65 die within a year.

R4.20. Teen smoking The Centers for Disease Control and Prevention say that about 18% of high-school students smoke tobacco (down from a high of 38% in 1997). Suppose you randomly select high-school students to survey them on their attitudes toward scenes of smoking in the movies. What's the probability that

a) none of the first 4 students you interview is a smoker?
b) the first smoker is the sixth person you choose?
c) there are no more than 2 smokers among 10 people you choose?

R4.21. Passing stats Molly's college offers two sections of Statistics 101. From what she has heard about the two professors listed, Molly estimates that her chances of passing the course are 0.80 if she gets Professor Scedastic and 0.60 if she gets Professor Kurtosis. The registrar uses a lottery to randomly assign the 120 enrolled students based on the number of available seats in each class. There are 70 seats in Professor Scedastic's class and 50 in Professor Kurtosis's class.

a) What's the probability that Molly will pass Statistics?
b) At the end of the semester, we find out that Molly failed. What's the probability that she got Professor Kurtosis?

R4.22. Teen smoking II Suppose that, as reported by the Centers for Disease Control, about 18% of high school students smoke tobacco. You randomly select 120 high school students to survey them on their attitudes toward scenes of smoking in the movies.

a) What's the expected number of smokers?
b) What's the standard deviation of the number of smokers?
c) The number of smokers among 120 randomly selected students will vary from group to group. Explain why that number can be described with a Normal model.
d) Using the 68–95–99.7 Rule, create and interpret a model for the number of smokers among your group of 120 students.

R4.23. Random variables Given independent random variables with means and standard deviations as shown, find the mean and standard deviation of each of these variables:

a) $X + 50$
b) $10Y$
c) $X + 0.5Y$
d) $X - Y$
e) $X + Y$

	Mean	SD
X	50	8
Y	100	6

R4.24. Merger Explain why the facts you know about variances of independent random variables might encourage two small insurance companies to merge. (*Hint:* Think about the expected amount and potential variability in payouts for the separate and the merged companies.)

R4.25. Youth survey According to a recent Gallup survey, 93% of teens use the Internet, but there are differences in how teen boys and girls say they use computers. The telephone poll found that 77% of boys had played computer games in the past week, compared with 65% of girls. On the other hand, 76% of girls said they had e-mailed friends in the past week, compared with only 65% of boys.

a) For boys, the cited percentages are 77% playing computer games and 65% using e-mail. That total is 142%, so there is obviously a mistake in the report. No? Explain.
b) Based on these results, do you think playing games and using e-mail are mutually exclusive? Explain.
c) Do you think whether a child e-mails friends is independent of being a boy or a girl? Explain.
d) Suppose that in fact 93% of the teens in your area do use the Internet. You want to interview a few who do not, so

you start contacting teenagers at random. What is the probability that it takes you 5 interviews until you find the first person who does not use the Internet?

R4.26. Meals A college student on a seven-day meal plan reports that the amount of money he spends daily on food varies with a mean of $13.50 and a standard deviation of $7.

a) What are the mean and standard deviation of the amount he might spend in two consecutive days?
b) What assumption did you make in order to find that standard deviation? Are there any reasons you might question that assumption?
c) Estimate his average weekly food costs, and the standard deviation.
d) Do you think it likely he might spend less than $50 in a week? Explain, including any assumptions you make in your analysis.

R4.27. Travel to Kyrgyzstan Your pocket copy of *Kyrgyzstan on 4237 ± 360 Som a Day* claims that you can expect to spend about 4237 som each day with a standard deviation of 360 som. How well can you estimate your expenses for the trip?

a) Your budget allows you to spend 90,000 som. To the nearest day, how long can you afford to stay in Kyrgyzstan, on average?
b) What's the standard deviation of your expenses for a trip of that duration?
c) You doubt that your total expenses will exceed your expectations by more than two standard deviations. How much extra money should you bring? On average, how much of a "cushion" will you have per day?

R4.28. Picking melons Two stores sell watermelons. At the first store the melons weigh an average of 22 pounds, with a standard deviation of 2.5 pounds. At the second store the melons are smaller, with a mean of 18 pounds and a standard deviation of 2 pounds. You select a melon at random at each store.

a) What's the mean difference in weights of the melons?
b) What's the standard deviation of the difference in weights?
c) If a Normal model can be used to describe the difference in weights, what's the probability that the melon you got at the first store is heavier?

R4.29. Home, sweet home According to the 2000 Census, 66% of U.S. households own the home they live in. A mayoral candidate conducts a survey of 820 randomly selected homes in your city and finds only 523 owned by the current residents. The candidate then attacks the incumbent mayor, saying that there is an unusually low level of homeownership in the city. Do you agree? Explain.

R4.30. Buying melons The first store in Exercise 28 sells watermelons for 32 cents a pound. The second store is having a sale on watermelons—only 25 cents a pound. Find the mean and standard deviation of the difference in the price you may pay for melons randomly selected at each store.

R4.31. Who's the boss? The 2013 State of Women-Owned Businesses Report commissioned by American Express (www.womenable.com/content/userfiles/2013_State_of_Women-Owned_Businesses_Report_FINAL.pdf) says that, excluding large, publicly traded firms, women-owned firms make up 30% of the privately held firm population, a number that has increased by 59% since 1997. You call some firms doing business locally, assuming that the national percentage is true in your area.

a) What's the probability that the first 3 you call are all owned by women?
b) What's the probability that none of your first 4 calls finds a firm that is owned by a woman?
c) Suppose none of your first 5 calls found a firm owned by a woman. What's the probability that your next call does?

R4.32. Jerseys A Statistics professor comes home to find that all four of his children got white team shirts from soccer camp this year. He concludes that this year, unlike other years, the camp must not be using a variety of colors. But then he finds out that in each child's age group there are 4 teams, only 1 of which wears white shirts. Each child just happened to get on the white team at random.

a) Why was he so surprised? If each age group uses the same 4 colors, what's the probability that all four kids would get the same-color shirt?
b) What's the probability that all 4 would get white shirts?
c) We lied. Actually, in the oldest child's group there are 6 teams instead of the 4 teams in each of the other three groups. How does this change the probability you calculated in part b?

R4.33. When to stop In Exercise 27 of the Review Exercises for Part III, we posed this question:

You play a game that involves rolling a die. You can roll as many times as you want, and your score is the total for all the rolls. But ... if you roll a 6, your score is 0 and your turn is over. What might be a good strategy for a game like this?

You attempted to devise a good strategy by simulating several plays to see what might happen. Let's try calculating a strategy.

a) On what roll would you expect to get a 6 for the first time?
b) So, roll *one time less* than that. Assuming all those rolls were not 6's, what's your expected score?
c) What's the probability that you can roll that many times without getting a 6?

R4.34. Plan B Here's another attempt at developing a good strategy for the dice game in Exercise 33. Instead of stopping after a certain number of rolls, you could decide to stop when your score reaches a certain number of points.

a) How many points would you expect a roll to *add* to your score?
b) In terms of your current score, how many points would you expect a roll to *subtract* from your score?
c) Based on your answers in parts a and b, at what score will another roll "break even"?
d) Describe the strategy this result suggests.

R4.35. Technology on campus Every 5 years the Conference Board of the Mathematical Sciences surveys college math departments. In 2000 the board reported that 51% of all undergraduates taking Calculus I were in classes that used graphing calculators and 31% were in classes that used computer assignments. Suppose that 16% used both calculators and computers.

a) What percent used neither kind of technology?
b) What percent used calculators but not computers?
c) What percent of the calculator users had computer assignments?
d) Based on this survey, do calculator and computer use appear to be independent events? Explain.

R4.36. Dogs A census by the county dog control officer found that 18% of homes kept one dog as a pet, 4% had two dogs, and 1% had three or more. If a salesman visits two homes selected at random, what's the probability he encounters

a) no dogs?
b) some dogs?
c) dogs in each home?
d) more than one dog in each home?

R4.37. O-rings Failures of O-rings on the space shuttle are fairly rare, but often disastrous, events. If we are testing O-rings, suppose that the probability of a failure of any one O-ring is 0.01. Let X be the number of failures in the next 10 O-rings tested.

a) What model might you use to model X?
b) What is the mean number of failures in the next 10 O-rings?
c) What is the probability that there is exactly one failure in the next 10 O-rings?
d) What is the probability that there is at least one failure in the next 10 O-rings?

R4.38. Volcanoes Almost every year, there is some incidence of volcanic activity on the island of Japan. In 2005 there were 5 volcanic episodes, defined as either eruptions or sizable seismic activity. Suppose the mean number of episodes is 2.4 per year. Let X be the number of episodes in the 2-year period 2010–2011.

a) What model might you use to model X?
b) What is the mean number of episodes in this period?
c) What is the probability that there will be no episodes in this period?
d) What is the probability that there are more than three episodes in this period?

R4.39. Socks In your sock drawer you have 4 blue socks, 5 gray socks, and 3 black ones. Half asleep one morning, you grab 2 socks at random and put them on. Find the probability you end up wearing

a) 2 blue socks.
b) no gray socks.
c) at least 1 black sock.
d) a green sock.
e) matching socks.

R4.40. Coins A coin is to be tossed 36 times.

a) What are the mean and standard deviation of the number of heads?
b) Suppose the resulting number of heads is unusual, two standard deviations above the mean. How many "extra" heads were observed?
c) If the coin were tossed 100 times, would you still consider the same number of extra heads unusual? Explain.
d) In the 100 tosses, how many extra heads would you need to observe in order to say the results were unusual?
e) Explain how these results refute the "Law of Averages" but confirm the Law of Large Numbers.

R4.41. The Drake equation In 1961 astronomer Frank Drake developed an equation to try to estimate the number of extraterrestrial civilizations in our galaxy that might be able to communicate with us via radio transmissions. Now largely accepted by the scientific community, the Drake equation has helped spur efforts by radio astronomers to search for extraterrestrial intelligence.

Here is the equation:

$$N_C = N \cdot f_p \cdot n_e \cdot f_l \cdot f_i \cdot f_c \cdot f_L$$

OK, it looks a little messy, but here's what it means:

Factor	What It Represents	Possible Value
N	Number of stars in the Milky Way Galaxy	200–400 billion
f_p	Probability that a star has planets	20%–50%
n_e	Number of planets in a solar system capable of sustaining earth-type life	1? 2?
f_l	Probability that life develops on a planet with a suitable environment	1%–100%
f_i	Probability that life evolves intelligence	50%?
f_c	Probability that intelligent life develops radio communication	10%–20%
f_L	Fraction of the planet's life for which the civilization survives	$\frac{1}{1{,}000{,}000}$?
N_c	Number of extraterrestrial civilizations in our galaxy with which we could communicate	?

So, how many ETs are out there? That depends; values chosen for the many factors in the equation depend on ever-evolving scientific knowledge and one's personal guesses. But now, some questions.

a) What quantity is calculated by the first product, $N \cdot f_p$?
b) What quantity is calculated by the product, $N \cdot f_p \cdot n_e \cdot f_l$?

c) What probability is calculated by the product $f_l \cdot f_i$?
d) Which of the factors in the formula are conditional probabilities? Restate each in a way that makes the condition clear.

Note: A quick Internet search will find you a site where you can play with the Drake equation yourself.

R4.42. Recalls In a car rental company's fleet, 70% of the cars are American brands, 20% are Japanese, and the rest are German. The company notes that manufacturers' recalls seem to affect 2% of the American cars, but only 1% of the others.

a) What's the probability that a randomly chosen car is recalled?
b) What's the probability that a recalled car is American?

R4.43. Pregnant? Suppose that 70% of the women who suspect they may be pregnant and purchase an in-home pregnancy test are actually pregnant. Further suppose that the test is 98% accurate. What's the probability that a woman whose test indicates that she is pregnant actually is?

R4.44. Door prize You are among 100 people attending a charity fundraiser at which a large-screen TV will be given away as a door prize. To determine who wins, 99 white balls and 1 red ball have been placed in a box and thoroughly mixed. The guests will line up and, one at a time, pick a ball from the box. Whoever gets the red ball wins the TV, but if the ball is white, it is returned to the box. If none of the 100 guests gets the red ball, the TV will be auctioned off for additional benefit of the charity.

a) What's the probability that the first person in line wins the TV?
b) You are the third person in line. What's the probability that you win the TV?
c) What's the probability that the charity gets to sell the TV because no one wins?
d) Suppose you get to pick your spot in line. Where would you want to be in order to maximize your chances of winning?
e) After hearing some protest about the plan, the organizers decide to award the prize by not returning the white balls to the box, thus ensuring that 1 of the 100 people will draw the red ball and win the TV. Now what position in line would you choose in order to maximize your chances?

16

Sampling Distribution Models and Confidence Intervals for Proportions

WHERE ARE WE GOING?

A poll based on a random sample of U.S. adults reports that 29% of those asked admit that they don't always drive as safely as they should because they are on their cell phone. How much can we trust this statistic? After all, they didn't ask *everyone*. Maybe the true proportion is 33% or 25%. How reliable *are* statistics based on random samples? We'll see that we can be surprisingly precise about how much we expect statistics from random samples to vary. This will enable us to start generalizing from samples we have at hand to the population at large.

The National Center for Health Statistics published data on all 3,945,192 U.S. live births in 1998 (www.cdc.gov/nchs/data_access/vitalstatsonline.htm), recording information on the babies (such as birthweight and Apgar score) and on the mother. They may have missed a few, but we'll treat this as the population of 1998 babies. Babies born before the 37th week of gestation are consider pre-term and, according to the Mayo Clinic (mayoclinic.org), are at risk for both short and long-term health problems. In 1998, 11.6% of births in the United States were premature (preemies). Figure 16.1 shows the distribution of gestation times

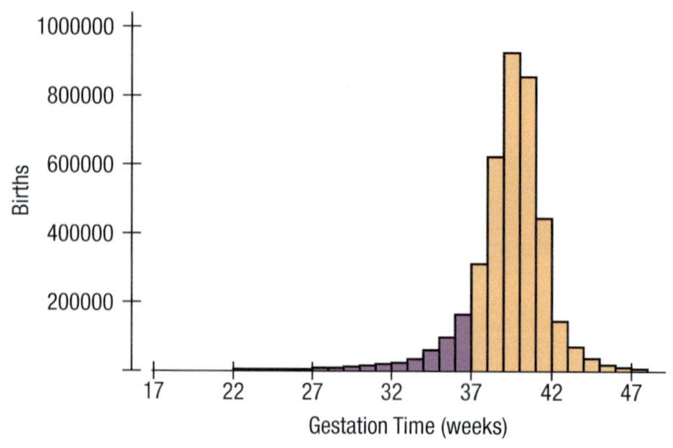

Figure 16.1
Gestation times (in weeks) of 3,904,759 babies born in the United States in 1998. The distribution is slightly skewed to the left. About 11.6% of the babies are preemies (with gestation times less than 37 weeks), highlighted in purple. (Note: Values for 40,433 births were not available.)

WHO	U.S. babies
WHAT	Gestation times
WHEN	1998
WHERE	United States
WHY	Public health

16.1 The Sampling Distribution Model for a Proportion

From any one sample, the best guess of the true proportion is the proportion in the sample. But since each random sample is different, our challenge—and the challenge faced by anyone who has only a sample—is not only to guess the true proportion in the entire population, but also to guess how good that guess is. Quantifying the uncertainty in our guesses is the core of statistical thinking.

You've already seen how variable statistics can be from the simulation exercises of previous chapters. You can use that variation to judge how close your guess is to the true answer. Think of it like this: Imagine throwing a dart at the bull's-eye of a dartboard after putting a plain sheet of paper in front of the target to record where the dart hits. Now, take the paper off, show it to a friend, and ask her to guess the location of the bull's-eye. Her best guess has to be where the dart hit.

But that's not likely to be exactly right. Instead, let her draw a circle that she thinks will contain the bull's-eye. How large a circle should she draw? She has no idea how close your throw was to the bull's-eye, so she'll find it hard to judge the right size for her circle.

What if you'd thrown several darts? She'd probably guess that the bull's-eye is in the middle of the cluster of holes, and by looking at your consistency (or lack of), she now has a much better idea of how big a circle to draw.

It is unusual to have an entire population to work with. Usually, we draw a sample at random. But, as we have seen in several of the Random Matters simulations of previous chapters, samples vary from one to the next. In Chapter 5, we called the distribution of values found for a statistic across many random samples the *sampling distribution* of that statistic. Now we can make a more formal definition. Suppose we could examine *all* possible samples of size 100 from the population of 4,000,000. The distribution of those proportions from all possible samples would be the **sampling distribution** of the proportion.[1] Unfortunately, the number of all possible samples of size 100 drawn from a population of 3,945,192 is about 4×10^{501}, a number so big we can't even imagine it.[2] But we've seen in previous chapters that we only need to generate a few thousand random samples from a population to get an approximation of the sampling distribution of a statistic.

[1]A word of caution. Until now, we've been plotting the *distribution of the sample*, a display of the actual data that were collected in that one sample. But now we've plotted the *sampling distribution*; a display of summary statistics ($\hat{p}$'s, for example) for many different samples. "Sample distribution" and "sampling distribution" sound a lot alike, but they refer to very different things. (Sorry about that—we didn't make up the terms.) And the distinction is critical. Whenever you read or write something about one of these, choose your terms carefully.

[2]And neither can you, probably. The number of atoms in the universe is "only" about 4×10^{81}.

RANDOM MATTERS A Sampling Distribution for a Proportion

To simulate the sampling distribution of a proportion, we draw independent random samples from the population and calculate the proportion for each sample. Figure 16.2 shows the distribution of the proportions of preemies from a simulation drawing 1000 random samples of size 100 from the 1998 babies. To understand the sampling distribution, consider the facts about this histogram as we did back in Chapter 2.

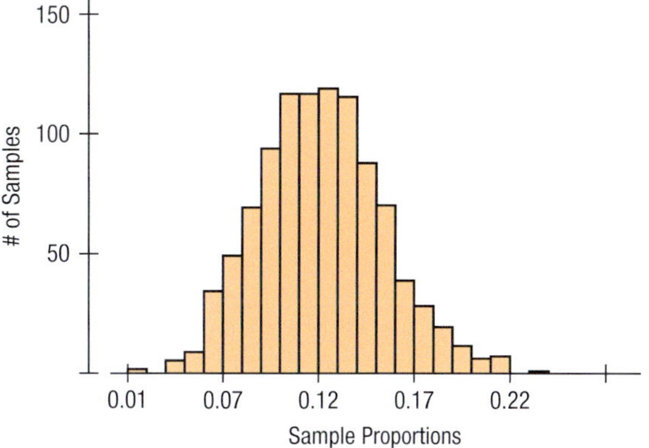

Figure 16.2

A histogram of the proportions of preemies in 1000 independent samples of 100 births drawn at random from the population of babies shows a distribution that is centered at the population proportion, 11.6%, with a standard deviation of 3.2%.

1. The shape of the histogram is unimodal and symmetric even though the population distribution was skewed to the left. In fact, it looks like a Normal model might fit.
2. The sample proportions fall symmetrically around the population proportion (which we know is 11.6%). (In fact, the mean of all these proportions agrees with the population proportion to at least one decimal place.)
3. The standard deviation of the sampling distribution is 3.2%.

The facts we can see from this simulated histogram tell us what to expect for the sampling distribution of any proportion. The sampling distribution will be unimodal and symmetric, centered at the true proportion, and with a standard deviation that, as we will soon see, we can estimate. These facts can be proven mathematically, but they are much more vivid and easier to understand from a simulation such as this one.

The Normal Model(!)

Abraham de Moivre (1667–1754)

It should be no surprise that the samples don't all have the same proportion. Each $\hat{p}$ comes from a different simulated sample. But, it's an amazing and fortunate fact that, as the simulation suggests, a Normal model is just the right one for the histogram of sample proportions. In fact, this was proved mathematically by Abraham de Moivre over 300 years ago. And because we know this mathematically, *we don't have to draw all those samples*.

There is no reason you should guess that the Normal model would be the one we need here,[3] and, indeed, the importance of de Moivre's result was not immediately understood by his contemporaries in 1738. But (unlike them) we know how useful the Normal model can be.

Modeling how sample statistics, such as proportions or means, vary from sample to sample is one of the most powerful ideas we'll see in this course. A sampling distribution model for how a statistic from a sample varies from sample to sample allows us to quantify that variation and to make statements about where we think the corresponding population parameter is.

The fact that the sampling distribution of a sample proportion follows a Normal model is the important idea in this chapter. Once we can make a model for random behavior, we

[3]Well, the fact that we spent much of Chapter 5 on the Normal model might have been a hint.

NOTATION ALERT

The letter p is our choice for the *parameter* of the model for proportions. It violates our "Greek letters for parameters" rule, but if we stuck to that, our natural choice would be π, which could be confusing.

So, we'll use p for the model parameter (the probability of a success) and $\hat{p}$ for the observed proportion in a sample. We'll also use q for the probability of a failure ($q = 1 - p$) and $\hat{q}$ for its observed value.

In Chapter 7, we introduced $\hat{y}$ as the predicted value for y. The "hat" here plays a similar role. It indicates that $\hat{p}$—the observed proportion in our data—is our *estimate* of the parameter p.

But be careful. We've already used capital P for a general probability. And we'll soon see another use of P in the next chapter! There are a lot of p's in this course; you'll need to think clearly about the context to keep them straight.

can understand and use that model. The rest of this chapter explores some of those uses, and the rest of the text applies the concept of sampling distribution models to the statistics we've been using to understand our data.

Which Normal?

To use a Normal model, we need to specify two parameters: its mean and standard deviation. The center of the histogram is naturally at p, so that's what we'll use for the mean of the Normal model.

What about the standard deviation? For quantitative data, the mean gives us no information about the standard deviation. If we told you that a batch of bike helmets had a mean diameter of 26 centimeters and asked what the standard deviation was, you'd probably have no idea. And there's no reason why you should. There's no information about σ from knowing the value of μ for quantitative data.

But there's a special fact about proportions that makes the standard deviation easy to find. Once we know the mean, p, (and write $q = 1 - p$) we automatically also know the standard deviation of $\hat{p}$. De Moivre showed (as we saw in Chapter 15) that for a Binomial model, the standard deviation of the *number* of successes is $\sqrt{npq}$. Now we want the standard deviation of the *proportion* of successes, $\hat{p}$. The proportion is just the number of successes divided by the number of trials, n, (and dividing by a constant doesn't change the distribution shape) so the standard deviation is also divided by n:

$$\sigma(\hat{p}) = SD(\hat{p}) = \frac{\sqrt{npq}}{n} = \sqrt{\frac{pq}{n}}.$$

When we draw Simple Random Samples of n individuals, the proportions we find will vary from sample to sample. As long as n is reasonably large,[4] we can model the distribution of these sample proportions with a probability model that is

$$N\left(p, \sqrt{\frac{pq}{n}}\right).$$

Be careful! What is this standard deviation? It's not the standard deviation of the data. It's the standard deviation of the proportions of all possible samples of n values from the population. Suppose we have a sample of 100 births from the population of all births in a year. What should we expect the sampling distribution of the proportion of premature births to be? Because we have the population, we know that the true proportion of premature births is $p = 11.6\%$ preemies. The formula then says that the standard deviation of proportions from samples of size 100 is[5]

$$\sqrt{\frac{pq}{n}} = \sqrt{\frac{(0.116)(0.884)}{100}} = 0.032.$$

Figure 16.3 puts together the facts that the sampling distribution is Normal, centered at p, and has standard deviation $\sqrt{\frac{pq}{n}}$.

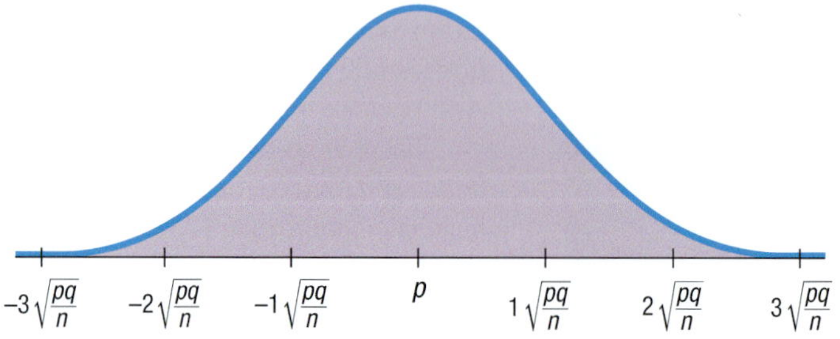

Figure 16.3
The sampling distribution model of a proportion is Normal with mean p and standard deviation $\sqrt{\frac{pq}{n}}$.

De Moivre, of course, had no Normal tables but he actually published a result that is what we call the 68–95–99.7 rule based on his best approximations to the Normal model.

[4]For smaller n, we could use a Binomial model (see Chapter 15).

[5]The standard deviation is 0.032 or 3.2%. Remember that the standard deviation always has the same units as the data. Here our units are %, or "percentage points." The standard deviation isn't 3.2% of anything, it is just 3.2 percentage points. If that's confusing, try writing the units out as "percentage points" instead of using the symbol %. Many polling agencies now do that too.

Here's the pay-off. Because we have a Normal model, we can use the 68–95–99.7 Rule or look up exact probabilities using a table or technology. For example, 95% of Normally distributed values are within roughly two standard deviations of the mean. So about 95% of samples will have sample proportion within two standard deviations of the true population proportion. Knowing that, we should not be surprised if various polls that may appear to ask the same question report a variety of results, but we *would* be surprised if the variety was too large because we now know how much variation to expect from sample to sample. Such sample-to-sample variation is sometimes called **sampling error**. It's not really an *error* at all, but just *variability* you'd expect to see from one sample to another. A better term would be **sampling variability**.[6]

16.2 When Does the Normal Model Work? Assumptions and Conditions

De Moivre claimed that the sampling distribution of a Binomial—and thus, of a proportion, can be modeled well by a Normal model. But, does it always work? Well, no. For example, if we drew samples of size 2, the only possible proportion values would be 0, 0.5, and 1. There's no way a histogram consisting only of those values could look like a Normal model.

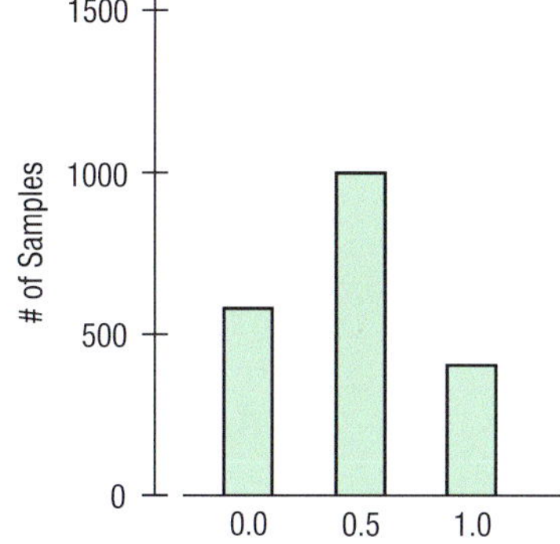

Figure 16.4
Proportions from samples of size 2 can take on only three possible values. A Normal model does not work well.

In fact, de Moivre's claim is only *approximately* true. (But that's OK. After all, models are supposed to be only approximately true.) The Normal model becomes a better and better representation of the sampling distribution as the size of the samples gets bigger.[7] Samples of size 1 or 2 won't work at all, but the distributions of sample proportions of larger samples are modeled well by the Normal model as long as p isn't too close to 0 or 1.

Populations with a true proportion, p, close to 0 or 1 can be a problem. Suppose a basketball coach surveys students to see how many male high school seniors are over 6′6″. What will the proportions in samples of size 1000 look like? If the true proportion of students that tall is 0.001, then the coach is likely to get only a few seniors over 6′6″ in any random sample of 1000. Most samples will have proportions of 0/1000, 1/1000, 2/1000,

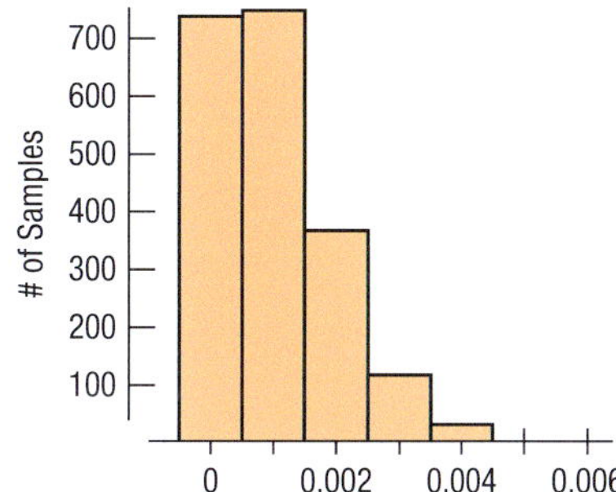

Figure 16.5
The distribution of sample proportions for 2000 samples of size 1000 with $p = 0.001$. Because the true proportion is so small, the sampling distribution is skewed to the right and the Normal model won't work well.

[6]Of course, polls differ for many reasons in addition to sampling variability. These reasons can include minor differences in question wording and in the ways that the random samples are selected. But polling agencies report their "margin of error" based on sampling variability. We'll see how they find that later in the chapter.

[7]Formally, we say the claim is true in the limit as n grows.

3/1000, 4/1000 with only a very few samples having a higher proportion. Figure 16.5 shows a simulation of 2000 samples of size 1000 with $p = 0.001$:

The distribution is skewed to the right because p is so close to 0. (Had p been very close to 1, it would have been skewed to the left.) So, even though n is large, p is too small, and so the Normal model still won't work well.

When should you use the Normal model with mean p and standard deviation $\sqrt{\frac{pq}{n}}$ as a model for the sampling distribution of a sample proportion? Just check these assumptions and conditions:

The Independence Assumption: The individuals in the samples (whose proportion we are finding) must be independent of each other. If not, the standard deviation be distorted. You can't know if an assumption is true or not, but you *can* check to see whether the data were collected in a way that makes this assumption plausible.

If your data come from an experiment in which subjects were randomly assigned to treatments or from a survey based on a Simple Random Sample, then the data satisfy the **Randomization Condition** and are plausibly independent. If not, you'll need to think carefully about whether it is honest to consider your cases independent with respect to the variables of interest and whether the cases you have are representative of the population you care about.

10% Condition: It can also be a problem to sample too much of the population. Once you've sampled more than about 10% of the population, the remaining individuals have a higher probability of being selected (at random) next, and the standard deviation will be smaller than the model predicts. There are special formulas to adjust the standard deviation, but, as the saying goes, they are "beyond the scope of this course." You should report when you're sampling more than 10%, but it shouldn't stop your work.

Success/Failure Condition: The sample size, n, has to be large enough, and p (and $q = 1 - p$) can't be too close to 0 or 1. Putting those two concerns together, check for at least 10 "successes" and 10 "failures." That is, make sure that $n\hat{p}$ and $n\hat{q}$ are both at least 10.

The Success/Failure Condition encourages a large sample size; the 10% Condition warns against sampling too much of a limited population. But this apparent conflict is rarely a problem. Most populations are safely large enough. Usually you should devote your effort and expense to gathering a larger sample or more experimental units.

Successes and Failures

The terms "success" and "failure" for the outcomes that have probability p and q are common in statistics. But they are completely arbitrary labels. When we say that a disease occurs with probability p, we certainly don't mean that getting sick is a "success" in the ordinary sense of the word.

In summary, if the assumptions and conditions are met, then you can use the Normal model to model the sampling distribution of the sample proportion. In other words:

The Sampling Distribution Model for a Proportion

Provided that the sampled values are independent and the sample size is large enough, the sampling distribution of $\hat{p}$ is modeled by a Normal model with mean p and standard deviation $SD(\hat{p}) = \sqrt{\frac{pq}{n}}$.

Without sampling distribution models, the rest of statistics just wouldn't exist and this text would stop here.[8] Sampling distribution models are what make statistics work. They inform us about the amount of variation we should expect when we sample. With a sampling distribution model, we can make informed decisions about how precise our estimate of the true value of a parameter might be. That's exactly what we'll be doing for the rest of this text.

Sampling distribution models enable us to say something about the population when all we have is data from a sample. By imagining what *might* happen if we were to draw many, many samples from the same population, we can learn a lot about how close the statistics

[8] Don't get your hopes up.

computed from our one particular sample may be to the corresponding population parameters they estimate. That's the path to the *margin of error* you hear about in polls and surveys. We'll see how to determine that later in this chapter.

EXAMPLE 16.1 Using the Sampling Distribution Model for Proportions

The Centers for Disease Control and Prevention reports that 22% of 18-year-old women in the United States have a body mass index (BMI)[9] of 30 or more—a value considered by the National Heart, Lung, and Blood Institute to be associated with increased health risk.

As part of a routine health check at a large college, the physical education department usually requires students to come in to be measured and weighed. This year, the department decided to try out a self-report system. It asked 200 randomly selected female students to report their heights and weights (from which their BMIs could be calculated). Only 31 of these students had BMIs greater than 30.

QUESTION: Is this proportion of high-BMI students unusually small?

ANSWER: First, check:

✓ Randomization Condition: The department drew a random sample, so the respondents should be independent.

✓ 10% Condition: If the college class was only a few hundred students, we'd need those special formulas. For a "large college," we're probably OK as it is.

✓ Success/Failure Condition: There were 31 "successes"—students with BMIs greater than 30—and 169 "failures." Both are at least 10.

It's OK to use a Normal model to describe the sampling distribution of the proportion of respondents with BMIs above 30.

The phys ed department observed $\hat{p} = \frac{31}{200} = 0.155$.

The department expected $p = 0.22$, with $SD(\hat{p}) = \sqrt{\frac{pq}{n}} = \sqrt{\frac{(0.22)(0.78)}{200}} = 0.029$,

so $z = \frac{\hat{p} - p}{SD(\hat{p})} = \frac{0.155 - 0.22}{0.029} = -2.24$.

By the 68–95–99.7 Rule, I know that values more than 2 standard deviations below the mean of a Normal model show up less than 2.5% of the time. Perhaps women at this college differ from the general population, or maybe self-reporting doesn't provide accurate heights and weights.

JUST CHECKING

1. You want to poll a random sample of 100 students on campus to see if they are in favor of the proposed location for the new student center. Of course, you'll get just one number, your sample proportion, $\hat{p}$. But if you imagined all the possible samples of 100 students you could draw and imagined the histogram of all the sample proportions from these samples, what shape would it have?
2. Where would the center of that histogram be?
3. If you think that about half the students are in favor of the plan, what would the standard deviation of the sample proportions be?

[9] $BMI = \text{weight in kg}/(\text{height in m})^2$

STEP-BY-STEP EXAMPLE

Working with Sampling Distribution Models for Proportions

Suppose that about 13% of the population is left-handed.[10] A 200-seat school auditorium has been built with 15 "lefty seats," seats that have the built-in desk on the left rather than the right arm of the chair. (For the right-handed readers among you, have you ever tried to take notes in a chair with the desk on the left side?)

QUESTION: In a class of 90 students, what's the probability that there will not be enough seats for the left-handed students?

THINK

PLAN State what we want to know.

I want to find the probability that in a group of 90 students, more than 15 will be left-handed. Since 15 out of 90 is 16.7%, I need the probability of finding more than 16.7% left-handed students out of a sample of 90 if the proportion of lefties is 13%.

MODEL Think about the assumptions and check the conditions.

You might be able to think of cases where the **Independence Assumption** is not plausible—for example, if the students are all related, or if they were selected for being left- or right-handed. But for students in a class, the assumption of independence with respect to handedness seems reasonable.

✓ **Independence Assumption**: The 90 students in the class are not a random sample, but it is reasonable to assume that the probability that one student is left-handed is not changed by the fact that another student is right- or left-handed. It is also reasonable to assume that the students in the class are representative of students in general with respect to handedness.

✓ **10% Condition**: 90 is surely less than 10% of the population of all students so there is no need for special formulas. (Even if the school itself is small, I'm thinking of the population of all *possible* students who could have gone to the school.)

✓ **Success/Failure Condition**:

$$np = 90(0.13) = 11.7 \geq 10$$
$$nq = 90(0.87) = 78.3 \geq 10$$

State the parameters and the sampling distribution model.

The population proportion is $p = 0.13$. The conditions are satisfied, so I'll model the sampling distribution of $\hat{p}$ with a Normal model with mean 0.13 and a standard deviation of

$$SD(\hat{p}) = \sqrt{\frac{pq}{n}} = \sqrt{\frac{(0.13)(0.87)}{90}} \approx 0.035.$$

My model for $\hat{p}$ is $N(0.13, 0.035)$.

SHOW

PLOT Make a picture. Sketch the model and shade the area we're interested in, in this case the area to the right of 16.7%.

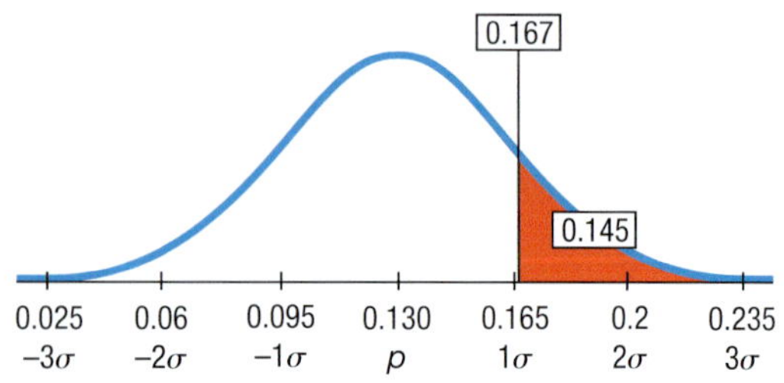

[10]Actually, there is little agreement among experts about the proportion of lefties in the population—or even how to define "left-handedness." Estimates range from 8% to 15%.

MECHANICS Use the standard deviation as a ruler to find the *z*-score of the cutoff proportion. We see that 16.7% lefties would be just over one standard deviation above the mean.

$$z = \frac{\hat{p} - p}{SD(\hat{p})} = \frac{0.167 - 0.13}{0.035} = 1.06$$

Find the resulting probability from a table of Normal probabilities, a computer program, or a calculator.

$$P(\hat{p} > 0.167) = P(z > 1.06) = 0.1446$$

TELL

CONCLUSION Interpret the probability in the context of the question.

There is about a 14.5% chance that there will not be enough seats for the left-handed students in the class.

RANDOM MATTERS Does the Normal Model Always Work? Sampling Distributions for Other Statistics

The sampling distribution for the proportion is especially useful because the Normal model provides such a good approximation. And it would certainly be useful to know the sampling distribution for *any* statistic that we can calculate, not just the sample proportion. We saw in the Random Matters of Chapter 6 that the sampling distribution of the correlation coefficient was skewed, and was only symmetric (and, in fact, Normal looking) when the population correlation was 0.5. How about other statistics? Would you expect the sampling distribution of the minimum, the maximum, or the variance of a sample to be Normally distributed? What about the median?

In Chapter 9, we looked at a study of body measurements of 250 men. (Data in **Bodyfat**) The median of all 250 men's weights is 176.125 lb and the variance is 730.9 lb. We can treat these 250 men as a population, draw repeated random samples of 10, and compute the median, the variance, and the minimum for each sample.

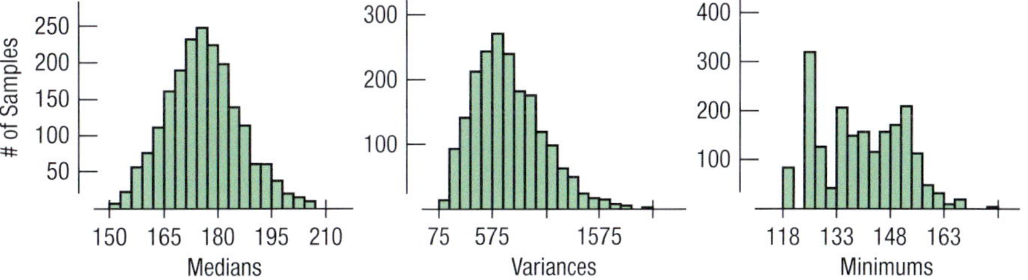

Each of these histograms depicts the sampling distribution of its respective statistic. And it is easy to see that they aren't all the same. The sampling distribution of the median is unimodal and symmetric. The sampling distribution of the variance is skewed to the right. And the sampling distribution of the minimum is, well, messy.

We can simulate to find the sampling distribution of *any* statistic we like: the maximum, the IQR, the 37th percentile, anything. But it is more useful to have a mathematical result that tells us the sampling distribution model. Both the proportion and the mean have sampling distributions that we know from mathematics to be well approximated by a Normal model. And that is a fundamental fact about the sampling distribution models for the two summary statistics that we use most often.

16.3 A Confidence Interval for a Proportion

Let's think about the proportion of preemies in a random sample of 100 U.S. births. We know the sampling distribution model for this proportion is approximately Normal (under certain assumptions, which we should be careful to check) and that its mean is 0.116, the true proportion of preemies in 1998. And (because we know the true proportion, p) we can calculate the standard deviation:

$$SD(\hat{p}) = \sqrt{\frac{pq}{n}} = \sqrt{\frac{0.116 \times (1 - 0.116)}{100}} = 0.032$$

Now we know that the sampling model for $\hat{p}$ should look like this:

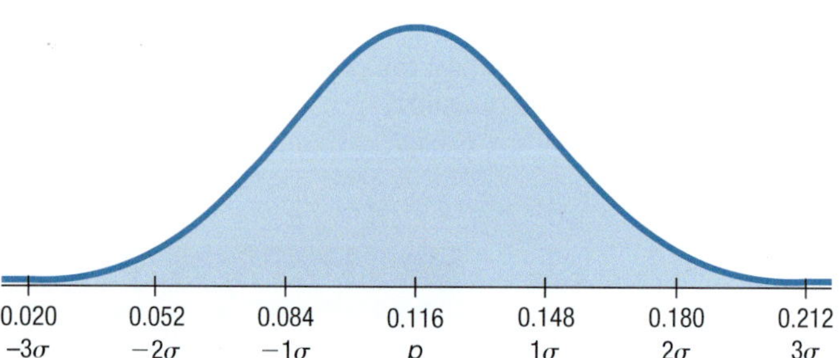

Figure 16.6
Using 0.116 for the mean and 0.032 for the standard deviation gives this Normal model for Figure 16.2's histogram of the proportions of premature babies in many independent samples of 100 births. Compare it to Figure 16.2.

That's great, but how does this help us when we have only one random sample? That's the usual situation. Suppose we draw a random sample of 100 births and find 14 preemies. From the sample, we'd guess that the true proportion is 0.14, and of course, we'd be wrong. Can we tell how wrong we are? It may seem like we're back throwing only one dart, but that's the magic of de Moivre's theorem. He told us that the standard deviation is

$$SD(\hat{p}) = \sqrt{\frac{pq}{n}}.$$

but that doesn't solve our problem, because the sample only provides the sample proportion, $\hat{p}$, not the true proportion, p. What if we use that $\hat{p}$? We can *estimate* the standard deviation by:

$$SE(\hat{p}) = \sqrt{\frac{\hat{p}\hat{q}}{n}}.$$

> **NOTATION ALERT**
> Remember that $\hat{p}$ is our sample-based estimate of the true proportion p. Recall also that q is just shorthand for $1 - p$, and $\hat{q} = 1 - \hat{p}$.
> When we use $\hat{p}$ to estimate the standard deviation of the sampling distribution model, we call that the standard error and write
> $$SE(\hat{p}) = \sqrt{\frac{\hat{p}\hat{q}}{n}}.$$

This seems logical. What else could we do? To keep track of what we're doing, when we estimate the standard deviation of the sampling distribution using $\hat{p}$, we give it a special name. We call it the **standard error** of $\hat{p}$ and we write $SE(\hat{p})$.[11] For our example that's 0.035. So, based on the sample of 100, our best guess is that $p = 0.14$ and that the standard deviation of the sampling distribution is 0.035. How can we use this model?

We know how to relate values in a Normal model to probabilities. Sample proportions are random, varying from sample to sample, but we can use probabilities to understand what values are most likely. We know, for example, that in a Normal model 95% of values are no more than about 2 standard deviations away from the mean. We have a sampling distribution that is approximately Normal and an estimate of its mean and standard deviation, so we can apply that understanding.

For example, we can say that for about 95% of random samples, the sample proportion, $\hat{p}$ will be no more than 2 *SE*s away from the true population proportion p. So let's look at this from $\hat{p}$'s point of view. If I'm $\hat{p}$, there's a 95% chance that p is no more than 2 *SE*s away from me. If I reach out 2 *SE*s, or 2×0.035, away from me on both sides, I'm 95% sure that p will be within my grasp. Now I've got him! Probably. Of course, even if my interval does catch p, I still don't know its true value. The best I can do is an interval that probably includes it, and even then I can't be certain that it contains p.

[11]Some statisticians (and books) refer to the standard deviation of the sampling distribution as the (theoretical) standard error, and refer to our estimate as the (estimated) standard error. But your statistics software (whatever software you are using) just calls the estimated version the "standard error." We think that's easier, so we will too.

What we have just constructed is called a **confidence interval**. It is an interval of plausible values for a parameter along with a probability that the interval actually contains the true parameter value. We can express the interval as $0.14 \pm 2 \times 0.035$, or we can calculate the values and write $(0.07. 0.21)$ and say that we are 95% *confident* that the interval contains the true proportion of preemies in the population. Our degree of confidence is sometimes called the **level** of the confidence interval.

Critical Values

> **NOTATION ALERT**
> We'll put an asterisk on a letter to indicate a critical value, so z^* is always a critical value from a Normal model.

We used 2*SE* to give us a 95% confidence interval. To change the confidence level, we'd need to change the *number* of *SE*s. And there's no reason to stick to the 68-95-99.7 rule; we have a Normal model and know how to use any z-score value. For any desired confidence level, the corresponding multiplier of the *SE* is called the **critical value**. Because it's based on the Normal model for this interval, it is denoted z^*. For any confidence level, we can find the corresponding critical value from a computer, a calculator, or a Normal probability table, such as Table Z.

Of course, as we know from Chapter 5, the precise critical value for a 95% confidence interval is $z^* = 1.96$. That is, 95% of a Normal model is found within ± 1.96 standard deviations of the mean. We've been using $z^* = 2$ from the 68–95–99.7 Rule because it's easy to remember.

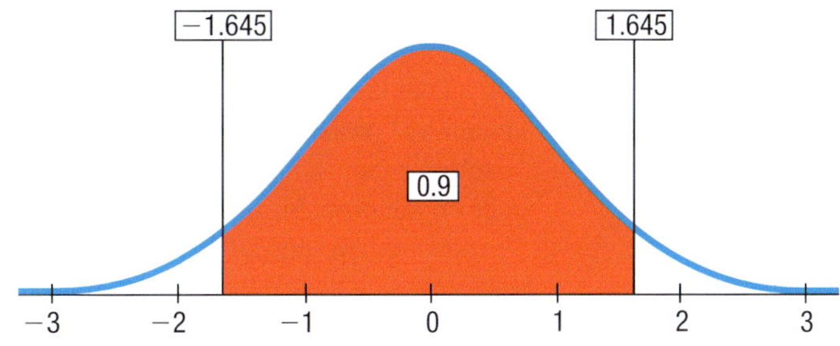

Figure 16.7
For a 90% confidence interval, the critical value is 1.645, because, for a Normal model, 90% of the values are within 1.645 standard deviations from the mean.

> **ONE-PROPORTION Z-INTERVAL**
> When the conditions are met, we are ready to find a level C confidence interval for the population proportion, p. The confidence interval is $\hat{p} \pm z^* \times SE(\hat{p})$ where the standard deviation of the proportion is estimated by $SE(\hat{p}) = \sqrt{\frac{\hat{p}\hat{q}}{n}}$ and the critical value, z^*, specifies the number of *SE*s needed for C% of random samples to yield confidence intervals that capture the true parameter value.

Figure 16.8
Reaching out 2 *SE*s on either side of $\hat{p}$ makes us 95% confident that we'll trap the true proportion, p.

So what can we really say about p from a sample? Without knowing the population, here's a list of things we'd like to be able to say, in order of strongest to weakest and the reasons we can't say most of them:

1. **"14% of *all* babies born in 1998 were born prematurely."** No. It would be nice to be able to make absolute statements about population values with certainty, but usually we just don't have enough information to do that. There's no way to be sure that the population proportion is the same as the sample proportion; in fact, it almost certainly isn't. Observations vary. Another sample would yield a different sample proportion.
2. **"It is *probably* true that 14% of all babies born in 1998 were born prematurely."** No again. In fact, even if we didn't know the true proportion, we'd know that it's probably not 14%. So the statement is not true.
3. **"We don't know exactly what proportion of babies born in 1998 were born prematurely, but we *know* that it's within the interval 14% $\pm$ 2 $\times$ 3.5%."** That is, in the interval between 7% and 21%. This is getting closer, but it's still wrong. We just can't be certain. We can't know *for sure* that the true proportion is in this interval—or in any particular interval.
4. **"We don't know exactly what proportion of babies born in 1998 were born prematurely, but the interval from 7% to 21% *probably* contains the true proportion."** Now we're getting closer. We've admitted that we're unsure about two things. First, we need an interval, not just a value, to try to capture the true proportion. Second, we aren't even certain that the true proportion is in that interval, but we're "pretty sure" that it is.

> "Far better an approximate answer to the right question, . . . than an exact answer to the wrong question."
>
> —John W. Tukey

That last statement may be true, but we can tighten it up by quantifying what we mean by "probably." We saw that 95% of the time when we reach out 2 *SE*s from $\hat{p}$, we capture p, *so we can be 95% confident that this is one of those times*. After putting a number on the probability that this interval covers the true proportion, we've given our best guess of where the parameter is and how certain we are that it's within some range.

5. **"We are 95% confident that between 7% and 21% of babies born in 1998 were premature."** Statements like these are called *confidence intervals*. They're the best we can do.

Each confidence interval discussed in the text has a name. You'll see many different kinds of confidence intervals in the following chapters. Some will be about more than *one* sample, some will be about statistics other than *proportions*, and some will use models other than the Normal. The interval calculated and interpreted here is sometimes called a **one-proportion *z*-interval**.[12]

JUST CHECKING

A Pew Research study regarding cell phones asked about cell phone experience. One growing concern is unsolicited advertising in the form of text messages. Pew asked cell phone owners, "Have you ever received unsolicited text messages on your cell phone from advertisers?" and 17% reported that they had. Pew estimates a 95% confidence interval to be 0.17 $\pm$ 0.04, or between 13% and 21%.

Are the following statements about people who have cell phones correct interpretations of that confidence interval? Explain.

4. In Pew's sample, somewhere between 13% and 21% of respondents reported that they had received unsolicited advertising text messages.
5. We can be 95% confident that 17% of U.S. cell phone owners have received unsolicited advertising text messages.

[12]In fact, this confidence interval is so standard for a single proportion that you may see it simply called a "confidence interval for the proportion."

6. We are 95% confident that between 13% and 21% of all U.S. cell phone owners have received unsolicited advertising text messages.
7. We know that between 13% and 21% of all U.S. cell phone owners have received unsolicited advertising text messages.
8. Ninety-five percent of all U.S. cell phone owners have received unsolicited advertising text messages.

16.4 Interpreting Confidence Intervals: What Does 95% Confidence Really Mean?

What do we mean when we say we have 95% confidence that our interval contains the true proportion? Formally, what we mean is that "95% of samples of this size drawn at random from this population will produce confidence intervals that capture the true proportion." This is correct, but a little long winded, so we sometimes say, "We are 95% confident that the true proportion lies in our interval." Our uncertainty is about whether the particular sample we have at hand is one of the successful ones or one of the 5% that fail to produce an interval that captures the true value.

We know that proportions vary from sample to sample. If other researchers select their own samples of 1998 babies, the sample proportions will almost certainly differ. When each researcher tries to estimate the true rate of premature births in the population, they'll center *their* confidence intervals at the proportions they observed in their own samples. Each researcher will end up with a different interval.

Our interval guessed the true proportion of preemies to be between about 7% and 21%. Another sample might find 17 of 100 to be preemies and get a confidence interval from 9.5% to 24.5%. Still another might find only 10 of 100 for a confidence interval from 4% to 16%. And so on. We don't expect all of these intervals to contain the true proportion of preemies. Some will be duds, missing the population proportion entirely. But we expect, on average, that 95% of them will be winners and capture the true proportion.

RANDOM MATTERS Confidence Intervals

We can visualize what a confidence interval means by simulation. We draw 20 samples from the population of the births data, find the proportion of preemies, and calculate the confidence intervals using the formulas from de Moivre's theorem. Figure 16.9 shows the confidence intervals. The red dots plot the proportions of

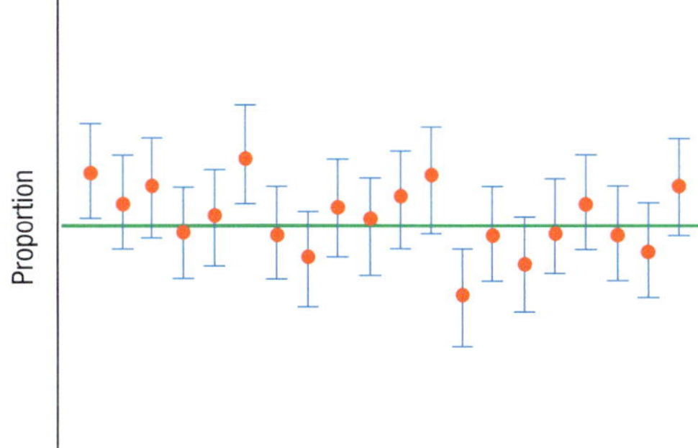

Figure 16.9
The horizontal green line shows the true percentage of preemies born in 1998 (11.6%). Most of the 20 simulated samples produced confidence intervals that captured the true value, but a few missed.

preemies in each sample. The green line is the true proportion of preemies in the population (11.6%). Most of the intervals include the true value, but a few missed. The important thing to notice is that it is the intervals that vary from sample to sample; the green line doesn't move.

There's a huge number of possible samples that *could* be drawn, each with its own sample proportion. These are just some of them. Each sample proportion can be used to make a confidence interval. That's a large pile of possible confidence intervals, and when we find a confidence interval from data, it is just one of those in the pile. Did *our* confidence interval "work"? When we work with a sample, we can't be sure, because in general we can't know the true proportion.[13] However, de Moivre assures us that (in the long run) 95% of the intervals are winners, covering the true value, and only 5% are duds. *That's* why we're 95% confident that our interval is a winner! (Even though we may never know.)

EXAMPLE 16.2

Polls and Confidence Intervals

In January and February 2016, Pew Research polled 4654 U.S. adults, asking them, "How often do you read any newspapers in print?" For the first time, a majority of adults responded "Hardly ever" or "Never."

QUESTION: It is standard among pollsters to use a 95% confidence level unless otherwise stated. Given that, what do these researchers mean by their confidence interval in this context?

ANSWER: If this polling were done repeatedly, 95% of all random samples would yield confidence intervals that contain the true proportion of all U.S. adults who hardly ever or never read a newspaper in print.

So, What *Can* I Say?

Confidence intervals are based on random samples, so the interval is random, too. De Moivre's theorem tells us that 95% of the random samples will yield intervals that capture the true value. That's what we mean by being 95% confident.

A Chapter 13 Step-by-Step Example considered the results of a Pew Research survey of 1520 U.S. adults that found that 68% of all U.S. adults (79% of those on line) use Facebook. The facts about the sampling distribution of proportions say to estimate that the true proportion of U.S. adults who use Facebook is about 0.68, and to estimate the standard deviation of the sampling distribution for that estimate as

$$SE(\hat{p}) = \sqrt{\frac{0.68 \times (1 - 0.68)}{1520}} = 0.012.$$

So a 95% confidence interval would be $0.68 \pm 2 \times 0.012$, or from 0.656 to 0.704.

Technically, we should say, "I am 95% confident that the interval from 65.6% to 70.4% captures the true proportion of U.S. adults." That formal phrasing emphasizes that

[13]Of course, in this example, we *do* know the true proportion because we have the entire population of births. That's a very unusual situation, but useful here to show how confidence intervals work.

our confidence (and our uncertainty) is about the interval, not the true proportion. But you may choose a more casual phrasing like "I am 95% confident that between 65.6% and 70.4% of U.S. adults use Facebook." Because you've made it clear that the uncertainty is yours and you didn't suggest that the randomness is in the true proportion, this is OK. Keep in mind that it's the interval that's random and is the focus of both our confidence and doubt.

16.5 Margin of Error: Certainty vs. Precision

We've just claimed that with a certain confidence we've captured the true proportion of all U.S. adults who are Facebook users. Our confidence interval had the form

$$\hat{p} \pm 2\,SE(\hat{p}).$$

The extent of the interval on either side of $\hat{p}$ is called the **margin of error** (*ME*). We'll want to use the same approach for many other situations besides estimating proportions. In fact, almost any population parameter—a proportion, a mean, or a regression slope, for example—can be estimated with some margin of error. The margin of error is a way to describe our uncertainty in estimating the population value. We'll see how to find a margin of error for each of these values and for others.

For all of those statistics, regardless of how we calculate the margin of error, we'll be able to construct a confidence interval that looks like this:

$$Estimate \pm ME.$$

The margin of error for our 95% confidence interval was 2 *SE*. What if we wanted to be more confident? To be more confident, we'd need to capture *p* more often, and to do that we'll need to make the interval wider. For example, if we want to be 99.7% confident, the margin of error will have to be 3 *SE*.

Figure 16.10
Reaching out 3 *SE*s on either side of $\hat{p}$ makes us 99.7% confident we'll trap the true proportion *p*. Compare with Figure 16.8.

The more confident we want to be, the larger the margin of error must be. We can be 100% confident that the proportion who are Facebook users is between 0% and 100%, but this isn't very useful. On the other hand, we could give a confidence interval from 67.8% to 68.2%, but we can't be very confident about a precise statement like this. Every confidence interval is a balance between certainty and precision.

The tension between certainty and precision is always there. Fortunately, in most cases we can be both sufficiently certain and sufficiently precise to make useful statements. There is no simple answer to the conflict. You must choose a confidence level yourself. The data can't do it for you. The choice of confidence level is somewhat arbitrary. The most commonly chosen confidence levels are 90%, 95%, and 99%, but, in theory, any percentage can be used. (In practice, though, using something like 92.9% or 97.2% is likely to make people think you're up to something.)

EXAMPLE 16.3

Finding the Margin of Error (Take 1)

RECAP: The Pew Research poll asking "How often do you read any newspapers in print?" reported a margin of error of 1.5%. It is a convention among pollsters to use a 95% confidence level and to report the "worst case" margin of error, based on $p = 0.5$.

QUESTION: How did the researchers calculate their margin of error?

ANSWER: Assuming $p = 0.5$, for random samples of $n = 4654$,

$$SD(\hat{p}) = \sqrt{\frac{pq}{n}} = \sqrt{\frac{(0.5)(0.5)}{4654}} = 0.0073.$$

For a 95% confidence level, $ME = 2(0.0073) = 0.0146$, so their margin of error is just a bit under 1.5%.

EXAMPLE 16.4

Finding the Margin of Error (Take 2)

RECAP: In the poll on reading newspapers in print, Pew Research found that 52% of adults reported "Hardly ever" or "Never." They reported a 95% confidence interval with a margin of error of 1.5%.

QUESTIONS: Using the critical value of z and the standard error based on the observed proportion, what would be the margin of error for a 90% confidence interval? What's good and bad about this change?

$$\text{With } n = 4654 \text{ and } \hat{p} = 0.52,\ SE(\hat{p}) = \sqrt{\frac{\hat{p}\hat{q}}{n}} = \sqrt{\frac{(0.52)(0.48)}{4654}} = 0.0073.$$

For a 90% confidence level, $z^* = 1.645$, so $ME = 1.645(0.0073) = 0.0012$.

ANSWER: Now the margin of error is only about 1.2%, producing a narrower interval. What's good about the change is that we now have a smaller interval, but what's bad is that we are less certain that the interval actually contains the true proportion of adults who never or hardly ever read a newspaper in print.

JUST CHECKING

Think some more about the 95% confidence interval originally created for the proportion of U.S. adults who never or hardly ever read a newspaper in print.

9. If the researchers wanted to be 98% confident, would their confidence interval need to be wider or narrower?

10. The study's margin of error was about 1.5%. If the researchers wanted to reduce it to 1%, would their level of confidence be higher or lower?

11. If the researchers had polled more people, would the interval's margin of error have been larger or smaller?

We've just made some statements about newspaper readership that might alarm media publishers. Those statements were possible because we used a Normal model for the sampling distribution. But is that model appropriate?

We saw the assumptions and conditions for using the Normal to model the sampling distribution for a proportion earlier in this chapter. The assumptions and conditions for a confidence interval are the same because they are based on the same sampling distribution. You should think about the assumptions to see if they are appropriate and check the corresponding conditions. Violations of them can render the confidence interval invalid, so you should check them whenever you see a confidence interval for a proportion. Briefly, be sure to check:

Independence Assumption and the associated **Randomization Condition**.

10% Condition: Not a concern unless your population is very small.

Success/Failure Condition.

STEP-BY-STEP EXAMPLE

A Confidence Interval for a Proportion

WHO	Adults in the United States
WHAT	Response to a question about the death penalty
WHEN	October 2015
WHERE	United States
HOW	1015 adults were randomly sampled and asked by the Gallup Poll
WHY	Public opinion research

In October 2015, the Gallup Poll[14] asked 1015 randomly sampled adults the question "Generally speaking, do you believe the death penalty is applied fairly or unfairly in this country today?" Of these, 53% answered "Fairly," 41% said "Unfairly," and 6% said they didn't know.

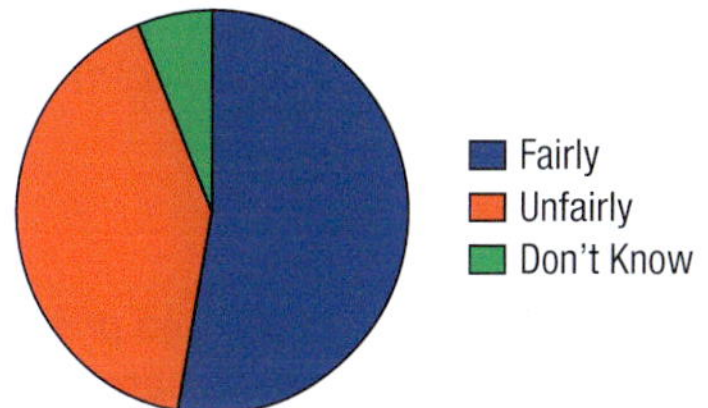

QUESTION: From this survey, what can we conclude about the opinions of *all* adults?

To answer this question, we'll build a confidence interval for the proportion of all U.S. adults who believe the death penalty is applied fairly. There are four steps to building a confidence interval for proportions: Plan, Model, Mechanics, and Conclusion.

THINK

PLAN State the problem and the W's.
Identify the *parameter* you wish to estimate.
Identify the *population* about which you wish to make statements.

I want to find an interval that is likely, with 95% confidence, to contain the true proportion, p, of U.S. adults who think the death penalty is applied fairly. I have a random sample of **1015** U.S. adults.

[14] news.gallup.com/poll/186218/solid-majority-continue-support-death-penalty.aspx

Choose and state a confidence level. **MODEL** Think about the assumptions and check the conditions.	✓ **Randomization Condition:** Gallup drew a random sample from all U.S. adults. I can be confident that the respondents are independent. ✓ **10% Condition:** The sample is certainly less than 10% of the population. ✓ **Success/Failure Condition:** $n\hat{p} = \mathbf{1015}(53\%) = \mathbf{538} \geq 10$ and $n\hat{q} = \mathbf{1015}(47\%) = \mathbf{477} \geq 10$, so the sample appears to be large enough to use the Normal model.
State the sampling distribution model for the statistic. Choose your method.	The conditions are satisfied, so I can use a Normal model to find a one-proportion z-interval.
SHOW **MECHANICS** Construct the confidence interval. First find the standard error. (Remember: It's called the "standard error" because we don't know p and have to use $\hat{p}$ instead.)	$n = \mathbf{1015}, \hat{p} = 0.53$, so $SE(\hat{p}) = \sqrt{\frac{\hat{p}\hat{q}}{n}} = \sqrt{\frac{(0.53)(0.47)}{\mathbf{1015}}} = 0.0\mathbf{157}.$ Because the sampling model is Normal, for a 95% confidence interval, the critical value $z^* = 1.96$.
Next find the margin of error. We could informally use 2 for our critical value, but 1.96 (found from a table or technology) is more accurate.	The margin of error is $ME = z^* \times SE(\hat{p}) = 1.96(0.0\mathbf{157}) = 0.031.$
Write the confidence interval (CI). **REALITY CHECK** The CI is centered at the sample proportion and about as wide as we might expect for a sample of about 1000.	So the 95% confidence interval is 0.53 ± 0.031 or $(0.\mathbf{499}, 0.\mathbf{561})$.
TELL **CONCLUSION** Interpret the confidence interval in the proper context. We're 95% confident that our interval captured the true proportion.	I am 95% confident that between **49.9**% and **56.1**% of all U.S. adults think that the death penalty is applied fairly.

*16.6 Choosing the Sample Size

The question of how large a sample to take is an important step in planning any study. We weren't ready to make that calculation when we first looked at study design in Chapter 11, but now we can—and we always should.

Suppose a candidate is planning a poll and wants to estimate voter support within 3% with 95% confidence. How large a sample does she need?

Let's look at the margin of error:

$$ME = z^* \sqrt{\frac{\hat{p}\hat{q}}{n}}$$

$$0.03 = 1.96\sqrt{\frac{\hat{p}\hat{q}}{n}}.$$

We want to find n, the sample size. To find n we need a value for $\hat{p}$. We don't know $\hat{p}$ because we don't have a sample yet, but we can probably guess a value. To be safe, we can take the value that makes $\hat{p}\hat{q}$ (and therefore n) largest. That value is $\hat{p} = \hat{q} = 0.50$, so if we use that value for $\hat{p}$, we'll certainly be safe. Our candidate probably expects to be near 50% anyway.

Our equation, then, is

$$0.03 = 1.96\sqrt{\frac{(0.5)(0.5)}{n}}.$$

To solve for n, we first multiply both sides of the equation by $\sqrt{n}$ and then divide by 0.03:

$$0.03\sqrt{n} = 1.96\sqrt{(0.5)(0.5)}$$

$$\sqrt{n} = \frac{1.96\sqrt{(0.5)(0.5)}}{0.03} \approx 32.67.$$

Notice that evaluating this expression tells us the *square root* of the sample size. We need to square that result to find n:

$$n \approx (32.67)^2 \approx 1067.1.$$

When computing sample sizes, we always round up (to ensure the confidence level we wanted) and conclude that we need at least 1068 respondents to keep the margin of error as small as 3% with a confidence level of 95%.

What Do I Use Instead of $\hat{p}$?

Often we have an estimate of the population proportion based on experience or perhaps a previous study. If so, use that value as $\hat{p}$ in calculating what size sample you need. If not, the cautious approach is to use $\hat{p} = 0.5$ in the sample size calculation; that will determine the largest sample necessary regardless of the true proportion.

EXAMPLE 16.5 Choosing a Sample Size

In a March 2016 survey by Gallup of 1019 U.S. adults, 65% believe that global warming is due more to human activity than natural changes in the environment. The poll reports a margin of error of 4%.

Suppose an environmental group planning a follow-up survey of voters' opinions on global warming wants to determine a 95% confidence interval with a margin of error of no more than ±2%.

QUESTION: How large a sample do they need? (You could take $p = 0.5$, but we have data that indicate $p = 0.65$, so we can use that.)

ANSWER:

$$ME = z^*\sqrt{\frac{\hat{p}\hat{q}}{n}}$$

$$0.02 = 1.96\sqrt{\frac{(0.65)(0.35)}{n}}$$

$$\sqrt{n} = 1.96\frac{\sqrt{(0.65)(0.35)}}{0.02} \approx 46.74$$

$$n = 46.74^2 = 2184.91$$

The environmental group's survey will need at least 2185 respondents.

Unfortunately, bigger samples cost more money and more effort. Because the standard error declines only with the *square root* of the sample size, to cut the standard error (and thus the *ME*) in half, we must roughly *quadruple* the sample size.

Generally, a margin of error of 5% or less is acceptable, but different circumstances call for different standards. For a pilot study, a margin of error of 10% may be fine, so a sample of 100 will do quite well. In a close election, a polling organization might want to

> **HOW BIG SHOULD A MARGIN OF ERROR BE?**
>
> Public opinion polls often sample about 1000 people, which gives an *ME* of 3% when $\hat{p} = 0.5$. But businesses and nonprofit organizations typically use much larger samples to estimate the proportion who will accept a direct mail offer. Why? Because that proportion is very low—often far below 5%. An *ME* of 3% wouldn't be precise enough. An *ME* like 0.1% would be more useful, and that requires a very large sample size.

get the margin of error down to 2%. Drawing a large sample to get a smaller *ME*, however, can run into trouble. It takes time to survey 2400 people, and a survey that extends over a week or more may be trying to hit a target that moves during the time of the survey. An important event can change public opinion in the middle of the survey process. Professional polling organizations often report both an instantaneous estimate and a rolling average of several recent polls in order to increase the sample size.

The sample size for a survey is the number of respondents, not the number of people to whom questionnaires were sent or whose phone numbers were dialed. And keep in mind that a low response rate turns any study essentially into a voluntary response study, which is of little value for inferring population values. It's almost always better to spend resources on increasing the response rate than on surveying a larger group. A full or nearly full response by a modest-size sample can yield more useful results.

Surveys are not the only place where proportions pop up. Banks sample huge mailing lists to estimate what proportion of people will accept a credit card offer. Even pilot studies may mail offers to over 50,000 customers. Most don't respond; but, in this case, that doesn't make the sample smaller—they simply said, "No thanks." Those who do respond want the card. To the bank, the response rate[15] is $\hat{p}$. With a typical success rate around 0.5%, the bank needs a very small margin of error—often as low as 0.1%—to make a sound business decision. That calls for a large sample, and the bank must take care in estimating the size needed. For our election poll calculation, we used $\hat{p} = 0.5$, both because it's safe and because we honestly believed the true p to be near 0.5. If the bank used that value, they'd get an absurd answer. Instead, they base their calculation on a proportion closer to the one they expect to find.

EXAMPLE 16.6 Sample Size Revisited

A credit card company is about to send out a mailing to test the market for a new credit card. From that sample, they want to estimate the true proportion of people who will sign up for the card nationwide. A pilot study suggests that about 0.5% of the people receiving the offer will accept it.

QUESTION: To be within a tenth of a percentage point (0.001) of the true rate with 95% confidence, how big does the test mailing have to be?

ANSWER: Using the estimate $\hat{p} = 0.5\%$:

$$ME = 0.001 = z^*\sqrt{\frac{\hat{p}\hat{q}}{n}} = 1.96\sqrt{\frac{(0.005)(0.995)}{n}}$$

$$(0.001)^2 = 1.96^2\,\frac{(0.005)(0.995)}{n} \Rightarrow n = \frac{1.96^2(0.005)(0.995)}{(0.001)^2}$$

$$= 19{,}111.96 \text{ or } 19{,}112$$

That's a lot, but it's actually a reasonable size for a trial mailing such as this. Note, however, that if they had assumed 0.50 for the value of p, they would have found

$$ME = 0.001 = z^*\sqrt{\frac{pq}{n}} = 1.96\sqrt{\frac{(0.5)(0.5)}{n}}$$

$$(0.001)^2 = 1.96^2\,\frac{(0.5)(0.5)}{n} \Rightarrow n = \frac{1.96^2(0.5)(0.5)}{(0.001)^2} = 960{,}400.$$

Quite a different (and unreasonable) result.

[15] In marketing studies, every mailing yields a response—"yes" or "no"—and "response rate" means the proportion of customers who accept an offer. That's not the way we use the term for survey responses.

WHAT CAN GO WRONG?

Confidence intervals are powerful tools. Not only do they tell what we know about the parameter value, but—more important—they also tell what we *don't* know. To use confidence intervals effectively, you must be clear about what you say about them. A sample of 200 births from the 1998 U.S. births finds 13% of the babies were born prematurely (gestation weeks <37) with a 95% confidence interval from 8.3% to 17.7%. (Data in **Babysamp 98**)

Don't Misstate What the Interval Means

- **Don't claim that other samples will agree with yours.** Keep in mind that the confidence interval makes a statement about the true population proportion. An interpretation such as "In 95% of samples of 200 babies born in 1998, the proportion born premature is between 8.3% and 17.7%" is just wrong. The interval isn't about sample proportions but about the population proportion.
- **Don't be certain about the parameter.** Saying "Between 8.3% and 17.7% of babies born in 1998 were premature" asserts that the population proportion cannot be outside that interval. Of course, we can't be absolutely certain of that unless we have the entire population. (But we can be reasonably sure.)
- **Don't forget: It's about the parameter.** Don't say, "I'm 95% confident that $\hat{p}$ is between 8.3% and 17.7%." Of course you are—in fact, we calculated that $\hat{p} = 13.0\%$ so we know the sample proportion is in the interval! The confidence interval is about the (unknown) population parameter, p.
- **Don't claim to know too much.** Don't say, "I'm 95% confident that between 8.3% and 17.7% of all babies ever born are premature." This sample was from babies born in the U.S. in 1998.
- **Do take responsibility.** Confidence intervals are about *un*certainty. *You* are the one who is uncertain, not the parameter. You have to accept the responsibility and consequences of the fact that not all the intervals you compute will capture the true value. In fact, about 5% of the time a 95% confidence interval such as this one will fail to capture the true value of the parameter. You *can* say "I am 95% confident that between 8.3% and 17.7% of babies born in the U.S. in 1998 were premature"[16]
- **Don't suggest that the parameter varies.** A statement like "There is a 95% chance that the true proportion is between 8.3% and 17.7%" sounds as though you think the population proportion wanders around and sometimes happens to fall between 8.3% and 17.7%. When you interpret a confidence interval, make it clear that *you* know that the population parameter is fixed and that it is the interval that varies from sample to sample.
- **Do treat the whole interval equally.** Although a confidence interval is a set of plausible values for the parameter, don't think that the values in the middle of a confidence interval are somehow "more plausible" than the values near the edges. Your interval provides no information about where in your current interval (if at all) the parameter value is most likely to be hiding.

Other Confidence Interval Cautions

- **Beware of margins of error that are too large to be useful.** A confidence interval that says that the percentage of premature babies is between 10% and 90% wouldn't be of much use. Most likely, you have some sense of how large a margin of error you can tolerate. What can you do?

  One way to make the margin of error smaller is to reduce your level of confidence. But that may not be a useful solution. It's a rare study that reports confidence levels lower than 80%. Levels of 95% or 99% are more common.

  The time to think about whether your margin of error is small enough to be useful is when you design your study. Don't wait until you compute your confidence interval. To get a narrower interval without giving up confidence, you need to have less variability in your sample proportion. How can you do that? Choose a larger sample.

[16]When we are being very careful we say, "95% of samples of this size will produce confidence intervals that capture the true proportion of babies born premature in the U.S. in 1998."

Confidence intervals and margins of error are often reported along with poll results and other analyses. But it's easy to misuse them and wise to be aware of the ways things can go wrong.

- **Watch out for biased sampling.** Don't forget about the potential sources of bias in surveys that we discussed in Chapter 10. Just because we have more statistical machinery now doesn't mean we can forget what we've already learned. A questionnaire that finds that most people enjoy filling out surveys suffers from nonresponse bias if only 10% of the surveys are returned. Even though now we're able to put confidence intervals around this (biased) estimate, the nonresponse has made it useless.
- **Think about independence.** The assumption that the values in our sample are mutually independent is one that we usually cannot check. It always pays to think about it, though. For example, if we sampled Instagram users and all of the friends in their network, we might get a biased estimate because friends in the same network may behave similarly. That would violate the Independence Assumption and could severely affect our sample proportion. To avoid this, the researchers should be careful in their sample design to ensure randomization.
- **Don't assume the Normal always works.** The sampling distribution of a proportion is well modeled by the Normal as long as the assumptions and conditions hold. We'll see in the next chapter that a similar result is true for means as well. But not all summary statistics have sampling distributions that are Normal.

CONNECTIONS

Now we can see a practical application of sampling distributions. To find a confidence interval, we lay out an interval measured in standard deviations. We're using the standard deviation as a ruler again. But now the standard deviation we need is the standard deviation of the sampling distribution. That's the one that tells how much the proportion varies. (And when we estimate it from the data, we call it a standard error.)

CHAPTER REVIEW

Construct a confidence interval for a proportion, p, as the statistic, $\hat{p}$, plus and minus a **margin of error**.

- The margin of error consists of a **critical value** based on the sampling model times a **standard error** based on the sample.
- The critical value is found from the Normal model.
- The standard error of a sample proportion is calculated as $\sqrt{\frac{\hat{p}\hat{q}}{n}}$.

Interpret a confidence interval correctly.

- You can claim to have the specified level of confidence that the interval you have computed actually covers the true value.

Understand the relationship of the sample size, n, to both the certainty (confidence level) and precision (margin of error).

- For the same sample size and true population proportion, more certainty means less precision (wider interval) and more precision (narrower interval) implies less certainty.

Know and check the assumptions and conditions for finding and interpreting confidence intervals.

- Independence Assumption or Randomization Condition
- 10% Condition
- Success/Failure Condition

Be able to invert the calculation of the margin of error to find the sample size required, given a proportion, a confidence level, and a desired margin of error.

REVIEW OF TERMS

Sampling distribution model
Different random samples give different values for a statistic. The distribution of the statistics over all possible samples is called the sampling distribution. The sampling distribution model shows the behavior of the statistic over all the possible samples for the same size *n* (p. 502).

Sampling variability (sampling error)
The variability we expect to see from one random sample to another. It is sometimes called sampling error, but sampling variability is the better term (p. 505).

Standard error
When we estimate the standard deviation of a sampling distribution using statistics found from the data, the estimate is called a standard error (p. 510):

$$SE(\hat{p}) = \sqrt{\frac{\hat{p}\hat{q}}{n}}.$$

Confidence interval
A level C confidence interval for a model parameter is an interval of values usually of the form

Estimate $\pm$ *margin of error*

found from data in such a way that *C*% of all random samples will yield intervals that capture the true parameter value (p. 511).

Critical value
The number of standard errors to move away from the mean of the sampling distribution to correspond to the specified level of confidence. The critical value, for a Normal sampling distribution, denoted z^*, is usually found from a table or with technology (p. 511).

One-proportion *z*-interval
A confidence interval for the true value of a proportion. The confidence interval is

$$\hat{p} \pm z^* SE(\hat{p}),$$

where z^* is a critical value from the standard Normal model corresponding to the specified confidence level (p. 512).

Margin of error
In a confidence interval, the extent of the interval on either side of the observed statistic value is called the margin of error. A margin of error is typically the product of a critical value from the sampling distribution and a standard error from the data. A small margin of error corresponds to a confidence interval that pins down the parameter precisely. A large margin of error corresponds to a confidence interval that gives relatively little information about the estimated parameter. For a proportion (p. 515):

$$ME = z^* \sqrt{\frac{\hat{p}\hat{q}}{n}}.$$

TECH SUPPORT

Confidence Intervals for Proportions

Confidence intervals for proportions are so easy and natural that many statistics packages don't offer special commands for them. Most statistics programs want the "raw data" for computations. For proportions, the raw data are the "success" and "failure" status for each case. Usually, these are given as 1 or 0, but they might be category names like "yes" and "no." Often, we just know the proportion of successes, $\hat{p}$, and the total count, *n*. Computer packages don't usually deal with summary data like this easily, but the statistics routines found on many graphing calculators allow you to create confidence intervals from summaries of the data—usually all you need to enter are the number of successes and the sample size.

In some programs, you can reconstruct variables of 0's and 1's with the given proportions. But even when you have (or can reconstruct) the raw data values, you may not get *exactly* the same margin of error from a computer package as you would find working by hand. The reason is that some packages make approximations or use other methods. The result is very close but not exactly the same. Fortunately, statistics means never having to say you're certain, so the approximate result is good enough.

DATA DESK

Data Desk does not offer built-in methods for inference with proportions.

COMMENTS

For summarized data, open a Scratchpad to compute the standard deviation and margin of error by typing the calculation. Then use **z-interval for individual μs**.

EXCEL

Inference methods for proportions are not part of the standard Excel tool set.

COMMENTS

For summarized data, type the calculation into any cell and evaluate it.

JMP

For a categorical variable that holds category labels, the Distribution platform includes tests and intervals for proportions. For summarized data, put the category names in one variable and the frequencies in an adjacent variable. Designate the frequency column to have the role of frequency. Then use the Distribution platform.

COMMENTS

JMP uses slightly different methods for proportion inferences than those discussed in this text. Your answers are likely to be slightly different, especially for small samples.

MINITAB

- Choose **Basic Statistics** from the Stat menu.
- Choose **1Proportion** from the Basic Statistics submenu.
- If the data are category names in a variable, assign the variable from the variable list box to the Samples in columns box. If you have summarized data, click the **Summarized Data** button and fill in the number of trials and the number of successes.
- Click the **Options** button and specify the remaining details.
- If you have a large sample, check **Use test and interval based on normal distribution**.

Click the **OK** button.

COMMENTS

When working from a variable that names categories, MINITAB treats the last category as the "success" category. You can specify how the categories should be ordered.

R

The standard libraries in R do not contain a function for the confidence of a proportion, but a simple function can be written to do so. For example:

```
pconfint=function(phat,n,conf=.95)
  {
    se = sqrt(phat*(1-phat)/n)
    al2 = 1-(1-conf)/2
    zstar = qnorm(al2)
    ul = phat+zstar*se
    ll = phat·zstar*se
    return(c(ll,ul))
  }
```

For example, pconfint(0.3,500) will give a 95% confidence interval for *p* based on 150 successes out of 500.

SPSS

SPSS does not find confidence intervals for proportions.

STATCRUNCH

To create a confidence interval for a proportion using summaries:

- Click on **Stat**.
- Choose **Proportion Stats** » **One sample** » **With summary**.
- Enter the **Number of successes** (x) and **Number of observations** (n).
- Indicate **Confidence Interval** (Standard-Wald), and then enter the **Level** of confidence.
- Click on **Compute!**

To create a confidence interval for a proportion using data:

- Click on **Stat**.
- Choose **Proportion Stats** » **One sample** » **With data**.
- Choose the variable **Column** listing the Outcomes.
- Enter the outcome to be considered a Success.
- Indicate **Confidence Interval**, and then enter the **Level** of confidence.
- Click on **Compute!**

TI-83/84 PLUS

To calculate a confidence interval for a population proportion:

- Go to the STATSTESTS menu and select **A:1-PropZInt**.
- Enter the number of successes observed and the sample size.
- Specify a confidence level.
- Calculate the interval.

COMMENTS

Beware: When you enter the value of x, you need the count, not the percentage. The count must be a whole number. If the number of successes is given as a percentage, you must first multiply *np* and round the result.

EXERCISES

SECTION 16.1

1. Website An investment company is planning to upgrade the mobile access to their website, but they'd like to know the proportion of their customers who access it from their smartphones. They draw a random sample of 200 from customers who recently logged in and check their IP address. Suppose that the true proportion of smartphone users is 36%.

a) What would you expect the shape of the sampling distribution for the sample proportion to be?
b) What would be the mean of this sampling distribution?
c) What would be the standard deviation of the sampling distribution?

2. Marketing The proportion of adult women in China was approximately 49% in 2016. A marketing survey telephones 400 people at random.

a) What proportion of the sample of 400 would you expect to be women?
b) What would the standard deviation of the sampling distribution be?
c) How many women, on average, would you expect to find in a sample of that size?

(Source: https://tradingeconomics.com/china/population-female-percent-of-total-wb-data.html)

3. Send money When they send out their fundraising letters, a philanthropic organization typically gets a return from about 5% of the people on their mailing list. To see what the response rate might be for future appeals, they did a simulation using samples of size 20, 50, 100, and 200. For each sample size, they simulated 1000 mailings with success rate $p = 0.05$ and constructed the histogram of the 1000 sample proportions, shown below. Explain what these histograms show about the sampling distribution model for sample proportions. Be sure to talk about shape, center, and spread.

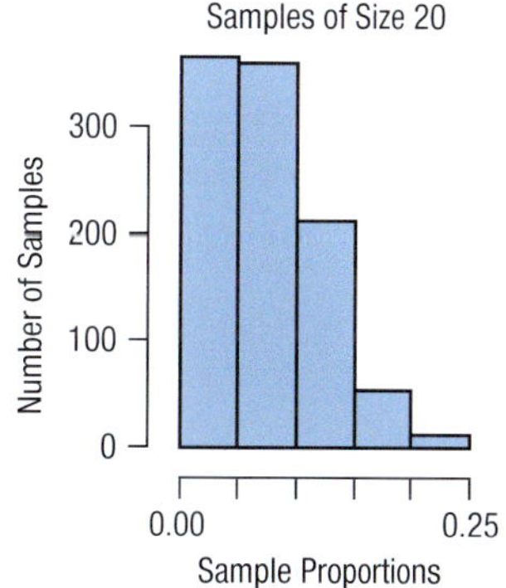

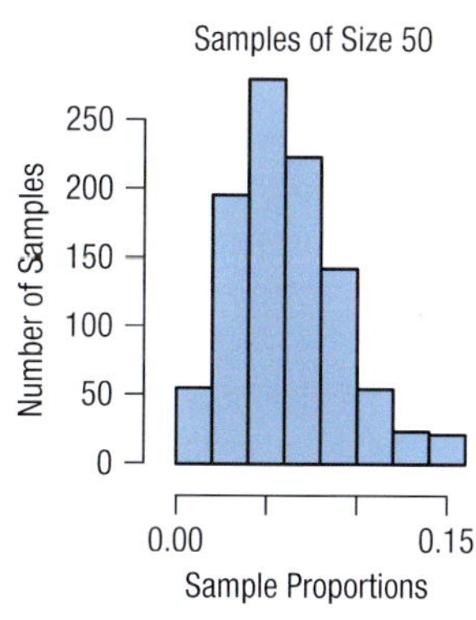

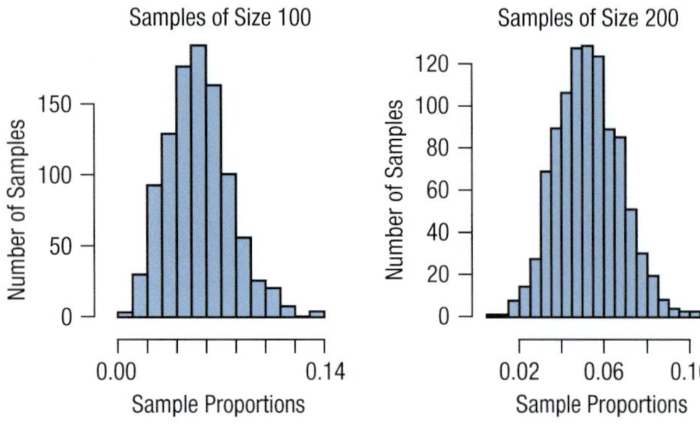

4. Character recognition An automatic character recognition device can successfully read about 85% of handwritten credit card applications. To estimate what might happen when this device reads a stack of applications, the company did a simulation using samples of size 20, 50, 75, and 100. For each sample size, they simulated 1000 samples with success rate $p = 0.85$ and constructed the histogram of the 1000 sample proportions, shown here. Explain what these histograms show about the sampling distribution model for sample proportions. Be sure to talk about shape, center, and spread.

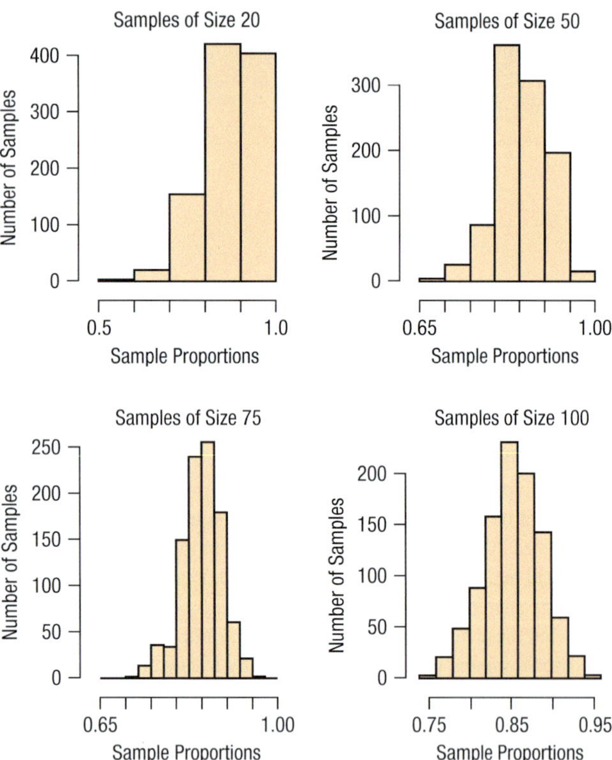

5. Living online Digital Clarity, in 2014, polled a random sample of 1300 UK young adults (ages 18–25) about Internet use. 16% of them reported going online several times a day—a fact of great interest to advertisers.

a) Explain the meaning of $\hat{p} = 0.16$ in the context of this situation.
b) Calculate the standard error of $\hat{p}$.
c) Explain what this standard error means in the context of this situation.

(Source: https://www.bbc.com/news/technology-29627896)

6. How's life? Gallup regularly conducts a poll using a "Cantril scale," which asks respondents to imagine a ladder with 10 rungs. Rung 0 represents the worst possible life, and rung 10 represents the best possible life. Respondents are asked what rung they would say they are on. Responses are classified as "Thriving" (standing on rung 7 or higher, and expecting to be on rung 8 or higher five years from now), "Suffering" (standing on rung 4 or lower and expecting to be on rung 4 or lower five years from now), or "Struggling" (not thriving or suffering). In the first half of 2016, Gallup found that the index had reached a new high of 55.7% thriving with a sample size of 105,000. (www.gallup.com/poll/194816/americans-life-evaluations-improve-during-obama-era.aspx)

a) Explain the meaning of $\hat{p} = 0.557$ in the context of this situation.
b) Calculate the standard error of $\hat{p}$.
c) Explain what this standard error means in the context of this situation.

SECTION 16.2

7. Marriage According to a Japanese National Fertility Survey in 2015, 12% of Japanese men aged 18–34 do not intend to ever marry. That is based on a random sample of 2705 Japanese men. Is it reasonable for them to use a normal model for the sampling distribution of the sample proportion? Why or why not?

(Source: http://www.ipss.go.jp/ps-doukou/e/doukou15/Nfs15R_summary_eng.pdf)

8. Campus sample For her final project, Stacy plans on surveying a random sample of 50 students on whether they plan to go to Paris for spring break. From past years, she guesses that about 10% of the class goes. Is it reasonable for her to use a Normal model for the sampling distribution of the sample proportion? Why or why not?

9. Send more money The philanthropic organization in Exercise 3 expects about a 5% success rate when they send fundraising letters to the people on their mailing list. In Exercise 3, you looked at the histograms showing distributions of sample proportions from 1000 simulated mailings for samples of size 20, 50, 100, and 200. Here are both the theoretical and observed statistics for these simulated samples:

n	Observed Mean	Theoretical Mean	Observed st. dev	Theoretical st. dev
20	0.0497	0.05	0.0479	0.0487
50	0.0516	0.05	0.0309	0.0308
100	0.0497	0.05	0.0215	0.0218
200	0.0501	0.05	0.0152	0.0154

a) Looking at the histograms of Exercise 3, at what sample size would you be comfortable using the Normal model as an approximation for the sampling distribution?
b) What does the Success/Failure Condition say about the choice you made in part a?

10. Character recognition, again The automatic character recognition device discussed in Exercise 4 successfully reads about 85% of handwritten credit card applications. In Exercise 4, you looked at the histograms showing distributions of sample proportions from 1000 simulated samples of size 20, 50, 75, and 100. The sample statistics and theoretical values for each simulation are as follows:

n	Observed Mean	Theoretical Mean	Observed st. dev.	Theoretical st. dev.
20	0.8481	0.85	0.0803	0.0799
50	0.8507	0.85	0.0509	0.0505
75	0.8481	0.85	0.0406	0.0412
100	0.8488	0.85	0.0354	0.0357

a) Looking at the histograms in Exercise 4, at what sample size would you be comfortable using the Normal model as an approximation for the sampling distribution?
b) What does the Success/Failure Condition say about the choice you made in part a?

11. Sample maximum The distribution of scores on a statistics test for a particular class is skewed to the left. The professor wants to predict the maximum score and so wants to understand the distribution of the sample maximum. She simulates the distribution of the maximum of the test for 30 different tests (with $n = 5$). The histogram below shows a simulated sampling distribution of the sample maximum from these tests.

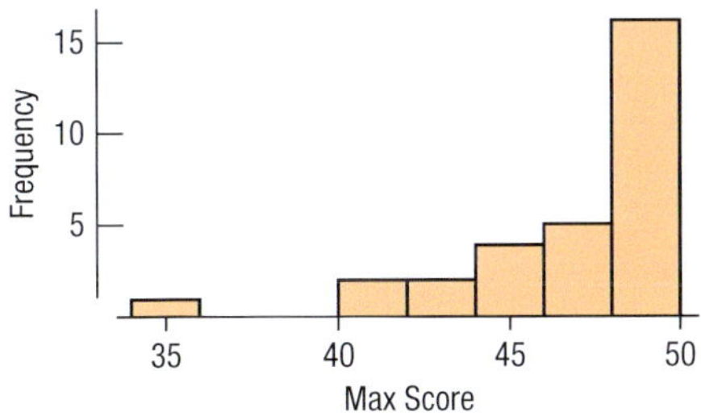

a) Would a Normal model be a useful model for this sampling distribution? Explain.
b) The mean of this distribution is 46.3 and the SD is 3.5. Would you expect about 95% of the samples to have their maximums within 7 of 46.3? Why or why not?

12. Soup A machine is supposed to fill cans with 16 oz of soup. Of course, there will be some variation in the amount actually dispensed, and measurement errors are often approximately Normally distributed. The manager would like to understand the variability of the variances of the samples, so he collects information from the last 250 batches of size 10 and plots a histogram of the variances:

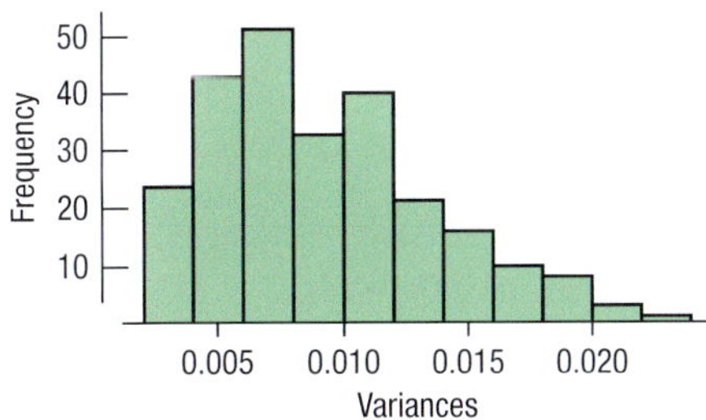

a) Would a Normal model be a useful model for this sampling distribution? Explain.
b) The mean of this distribution is 0.009 and the SD is 0.004. Would you expect about 95% of the samples to have their variances within 0.008 of 0.009? Why or why not?

SECTION 16.3

13. Still living online The 95% confidence interval for the number of young adults in Exercise 5 who reported that they went online several times daily is from 14% to 18%.

a) Interpret the interval in this context.
b) Explain the meaning of "95% confident" in this context.

14. Corporal punishment In a 2019 YouGov research study on corporal punishment in schools (https://www.businessinsider.my/almost-50-of-malaysian-parents-think-physical-punishment-should-be-allowed-in-schools-yougov/), 47% of randomly selected parents from a country agreed to allow corporal punishment in schools. The 95% confidence interval is from 43.1% to 50.9% ($n = 619$).

a) Interpret the interval in this context.
b) Explain the meaning of "95% confident" in this context.

SECTION 16.4

15. Brexit A YouGov poll of 2066 UK adults aged 18 and above in July 2019 asked a standard polling question of whether Britain was "Right" or "Wrong" to vote to leave the European Union. 46% said that the decision is right vs. 54% who said that it was a wrong decision.

a) Calculate the margin of error for the proportion of all UK adults who think that it was a "wrong" decision for 90% confidence.
b) Explain in a simple sentence what your margin of error means.

(Source: https://whatukthinks.org/eu/questions/in-highsight-do-you-think-britain-was-right-or-wrong-to-vote-to-leave-the-eu/?removed)

16. Smoking The Gallup Poll described in Exercise 6 also asked about smoking. Only 18% of those polled reported that they smoked.

a) Calculate the margin of error for the proportion of all American adults who smoke with 99% confidence.
b) Explain in a simple sentence what your margin of error means.

SECTION 16.5

17. Brexit again Consider the poll of Exercise 15.

a) Are the assumptions and conditions met?
b) Would the margin of error be larger or smaller for 95% confidence? Explain.

18. More corporal punishment In Exercise 14, we saw that 47% of surveyed parents agreed to corporal punishment in schools.

a) Are the conditions for constructing a confidence interval met?
b) Would the margin of error be larger or smaller for 90% confidence? Explain.

*SECTION 16.6

19. Graduation It's believed that as many as 30% of adults over 50 never graduated from high school. We wish to see if this percentage is the same among the 25 to 30 age group.

a) How many of this younger age group must we survey in order to estimate the proportion of non-grads to within 6% with 90% confidence?
b) Suppose we want to cut the margin of error to 4%. What's the necessary sample size?
c) What sample size would produce a margin of error of 3%?

20. Hiring In preparing a report on the economy, we need to estimate the percentage of businesses that plan to hire additional employees in the next 60 days.

a) How many randomly selected employers must we contact in order to create an estimate in which we are 98% confident with a margin of error of 5%?
b) Suppose we want to reduce the margin of error to 3%. What sample size will suffice?
c) Why might it not be worth the effort to try to get an interval with a margin of error of only 1%?

CHAPTER EXERCISES

21. Margin of error A TV newscaster reports the results of a poll of voters and then says, "The margin of error is plus or minus 3%." Explain carefully what that means.

22. Another margin of error A medical researcher estimates the percentage of children exposed to lead-based paint, adding that he believes his estimate has a margin of error of about 4%. Explain what the margin of error means.

23. Conditions For each situation described below, identify the population and the sample, explain what p and $\hat{p}$ represent, and tell whether the methods of this chapter can be used to create a confidence interval.

a) Police set up an auto checkpoint at which drivers are stopped and their cars inspected for safety problems. They find that 14 of the 134 cars stopped have at least one safety violation. They want to estimate the percentage of all cars that may be unsafe.
b) A TV talk show asks viewers to register their opinions on prayer in schools by logging on to a website. Of the 602 people who voted, 488 favored prayer in schools. We want to estimate the level of support among the general public.
c) A school is considering requiring students to wear uniforms. The PTA surveys parent opinion by sending a questionnaire home with all 1245 students; 380 surveys are returned, with 228 families in favor of the change.
d) A college admits 1632 freshmen one year, and four years later, 1388 of them graduate on time. The college wants to estimate the percentage of all their freshman enrollees who graduate on time.

24. More conditions Consider each situation described. Identify the population and the sample, explain what p and $\hat{p}$ represent, and tell whether the methods of this chapter can be used to create a confidence interval.

a) A consumer group hoping to assess customer experiences with auto dealers surveys 167 people who recently bought new cars; 3% of them expressed dissatisfaction with the salesperson.
b) What percent of college students have cell phones? 2883 students were asked as they entered a football stadium, and 2430 said they had phones with them.
c) Two hundred forty potato plants in a field in Maine are randomly checked, and only 7 show signs of blight. How severe is the blight problem for the U.S. potato industry?
d) Twelve of the 309 employees of a small company suffered an injury on the job last year. What can the company expect in future years?

25. Conclusions A catalog sales company promises to deliver orders placed on the Internet within 3 days. Follow-up calls to a few randomly selected customers show that a 95% confidence interval for the proportion of all orders that arrive on time is 88% ± 6%. What does this mean? Are these conclusions correct? Explain.

a) Between 82% and 94% of all orders arrive on time.
b) Ninety-five percent of all random samples of customers will show that 88% of orders arrive on time.
c) Ninety-five percent of all random samples of customers will show that 82% to 94% of orders arrive on time.
d) We are 95% sure that between 82% and 94% of the orders placed by the sampled customers arrived on time.
e) On 95% of the days, between 82% and 94% of the orders will arrive on time.

26. More conclusions In January 2002, two students made worldwide headlines by spinning a Belgian euro 250 times and getting 140 heads—that's 56%. That makes the 90% confidence interval (51%, 61%). What does this mean? Are these conclusions correct? Explain.

a) Between 51% and 61% of all euros are unfair.
b) We are 90% sure that in this experiment this euro landed heads on between 51% and 61% of the spins.
c) We are 90% sure that spun euros will land heads between 51% and 61% of the time.
d) If you spin a euro many times, you can be 90% sure of getting between 51% and 61% heads.
e) Ninety percent of all spun euros will land heads between 51% and 61% of the time.

27. Confidence intervals Several factors are involved in the creation of a confidence interval. Among them are the sample size, the level of confidence, and the margin of error. Which statements are true?

a) For a given sample size, higher confidence means a smaller margin of error.
b) For a specified confidence level, larger samples provide smaller margins of error.
c) For a fixed margin of error, larger samples provide greater confidence.
d) For a given confidence level, halving the margin of error requires a sample twice as large.

28. Confidence intervals, again Several factors are involved in the creation of a confidence interval. Among them are the sample size, the level of confidence, and the margin of error. Which statements are true?

a) For a given sample size, reducing the margin of error will mean lower confidence.
b) For a certain confidence level, you can get a smaller margin of error by selecting a bigger sample.
c) For a fixed margin of error, smaller samples will mean lower confidence.
d) For a given confidence level, a sample 9 times as large will make a margin of error one third as big.

29. Cars What fraction of cars made in Japan? The computer output below summarizes the results of a random sample of 50 autos. Explain carefully what it tells you.

z-Interval for proportion
With 90.00% confidence,
0.29938661 < P(japan) < 0.46984416

30. Parole A study of 902 decisions (to grant parole or not) made by the Nebraska Board of Parole produced the following computer output. Assuming these cases are representative of all cases that may come before the Board, what can you conclude?

z-Interval for proportion
With 95.00% confidence,
0.56100658 < P(parole) < 0.62524619

31. Mislabeled fish In 2010, biologists from University College Dublin (https://www.ucd.ie/news/2010/04APR10/230410_cod.html) analyzed 156 samples of fish sold as "cod" or "haddock" in Ireland using DNA barcoding to genetically identify the species of the fish. Laboratory results indicated that 25% of the fish were mislabeled according to the European Union (EU) regulations.

a) Construct a 95% confidence interval for the proportion of all fish sold as "cod" or "haddock" in Ireland that are mislabeled.
b) Explain what your confidence interval says about "cod" or "haddock" sold in Ireland.
c) A 2010 journal article by Miller and Mariani said that the strict EU policies in place to regulate seafood labeling have not been adequately implemented and enforced. Suppose an official said, "That's only 1215 packages out of the millions of pieces of fish sold in a year. With the small number tested, I don't know that one would want to change one's buying habits." (An official was quoted similarly in a different but similar context). Is this argument valid? Explain.

(Source: https://esajournals.onlinelibrary.wiley.com/doi/10.1890/090212)

32. Mislabeled fish, second course A study similar to the one described in Exercise 31 found that 12 of 30 "red snapper" packages tested were a different kind of fish.

a) Are the conditions for creating a confidence interval satisfied? Explain.
b) Construct a 95% confidence interval.
c) Explain what your confidence interval says about "red snapper" sold.

33. Baseball fans In a poll taken in December 2012, Gallup asked 1006 national adults whether they were baseball fans; 48% said they were. Almost five years earlier, in February 2008, only 35% of a similar-size sample had reported being baseball fans.

a) Find the margin of error for the 2012 poll if we want 90% confidence in our estimate of the percent of national adults who are baseball fans.
b) Explain what that margin of error means.
c) If we wanted to be 99% confident, would the margin of error be larger or smaller? Explain.
d) Find that margin of error.
e) In general, if all other aspects of the situation remain the same, will smaller margins of error produce greater or less confidence in the interval?

34. Still living online The Digital Clarity poll described in Exercise 5 found that 16% of a sample of 1300 UK young adults (ages 18–25) go online several times a day. (Treat this as a Simple Random Sample.)

a) Find the margin of error for this poll if we want 95% confidence in our estimate of the percent of UK young adults who go online several times a day.
b) Explain what that margin of error means.
c) If we only need to be 90% confident, will the margin of error be larger or smaller? Explain.
d) Find that margin of error.
e) In general, if all other aspects of the situation remain the same, would smaller samples produce smaller or larger margins of error?

35. Contributions, please The Care Association for the Blind is a philanthropic organization that relies on contributions. They send free mailing labels and greeting cards to potential donors on their list and ask for a voluntary contribution. To test a new campaign, they recently sent letters to a random sample of 50,000 potential donors and received 2380 donations.

a) Give a 95% confidence interval for the true proportion of their entire mailing list who may donate.
b) A staff member thinks that the true rate is 5.5%. Given the confidence interval you found, do you find that percentage plausible?

36. Take the offer Top One, a major credit card company, is planning a new offer for their current cardholders. The offer will give double airline miles on purchases for the next 6 months if the cardholder goes online and registers for the offer. To test the effectiveness of the campaign, Top One recently sent out offers to a random sample of 100,000 cardholders. Of those, 2468 registered.

a) Give a 95% confidence interval for the true proportion of those cardholders who will register for the offer.
b) If the acceptance rate is only 2% or less, the campaign won't be worth the expense. Given the confidence interval you found, what would you say?

37. Teenage drivers An insurance company checks police records on 582 accidents selected at random and notes that teenagers were at the wheel in 91 of them.

a) Create a 95% confidence interval for the percentage of all auto accidents that involve teenage drivers.
b) Explain what your interval means.
c) Explain what "95% confidence" means.
d) A politician urging tighter restrictions on drivers' licenses issued to teens says, "In one of every five auto accidents, a teenager is behind the wheel." Does your confidence interval support or contradict this statement? Explain.

38. Junk mail Direct mail advertisers send solicitations (a.k.a. "junk mail") to thousands of potential customers in the hope that some will buy the company's product. The acceptance rate is usually quite low. Suppose a company wants to test the response to a new flyer, and sends it to 1000 people randomly selected from their mailing list of over 200,000 people. They get orders from 123 of the recipients.

a) Create a 90% confidence interval for the percentage of people the company contacts who may buy something.
b) Explain what this interval means.
c) Explain what "90% confidence" means.
d) The company must decide whether to now do a mass mailing. The mailing won't be cost-effective unless it produces at least a 5% return. What does your confidence interval suggest? Explain.

39. Safe food Some food retailers propose subjecting food to a low level of radiation in order to improve safety, but the sale of such "irradiated" food is opposed by many people. Suppose a grocer wants to find out what his customers think. He has cashiers distribute surveys at checkout and ask customers to fill them out and drop them in a box near the front door. He gets responses from 156 customers, of whom 100 oppose the radiation treatments. What can the grocer conclude about the opinions of all his customers?

40. Local news The mayor of a small city has suggested that the state locate a new prison there, arguing that the construction project and resulting jobs will be good for the local economy. A total of 183 residents show up for a public hearing on the proposal, and a show of hands finds only 31 in favor of the prison project. What can the city council conclude about public support for the mayor's initiative?

41. Death penalty, again In the survey on the death penalty you read about in the Step-by-Step Example, the Gallup Poll actually split the sample at random, asking 510 respondents the question quoted earlier, "Generally speaking, do you believe the death penalty is applied fairly or unfairly in this country today?" The other 510 were asked, "Generally speaking, do you believe the death penalty is applied unfairly or fairly in this country today?" Seems like the same question, but sometimes the order of the choices matters. Suppose that for the second way of phrasing it, 64% said they thought the death penalty was fairly applied. (Recall that 53% of the original 510 thought the same thing.)

a) What kind of bias may be present here?
b) If we combine them, considering the overall group to be one larger random sample of 1020 respondents, what is a 95% confidence interval for the proportion of the general public that thinks the death penalty is being fairly applied?
c) How does the margin of error based on this pooled sample compare with the margins of error from the separate groups? Why?

42. Gambling A city ballot includes a local initiative that would legalize gambling. The issue is hotly contested, and two groups decide to conduct polls to predict the outcome. The local newspaper finds that 53% of 1200 randomly selected voters plan to vote "yes," while a college statistics class finds 54% of 450 randomly selected voters in support. Both groups will create 95% confidence intervals.

a) Without finding the confidence intervals, explain which one will have the larger margin of error.
b) Find both confidence intervals.
c) Which group concludes that the outcome is too close to call? Why?

43. Rickets Vitamin D, whether ingested as a dietary supplement or produced naturally when sunlight falls on the skin, is essential for strong, healthy bones. The bone disease rickets was largely eliminated in England during the 1950s, but now there is concern that a generation of children more likely to watch TV or play computer games than spend time outdoors is at increased risk. A recent study of 2700 children randomly selected from all parts of England found 20% of them deficient in vitamin D.

a) Find a 98% confidence interval.
b) Explain carefully what your interval means.
c) Explain what "98% confidence" means.

44. Private tuition A 2018 Sutton Trust survey found that 27% of young people in England and Wales attended private tuition. This poll was based on a random sample of 2381 young people aged between 11 and 16.

(Source: https://www.suttontrust.com/newsarchive/27-of-pupils-pay-to-have-private-tuition/)

a) Find a 90% confidence interval for the proportion of young people in England and Wales who attended private tuition.
b) Interpret your interval in this context.
c) Explain what "90% confidence" means.
d) Do these data refute a pundit's claim that 1/5 of the English and Welsh attended private tuition.

45. Privacy or security? In January 2014 AP-GfK polled 1060 U.S. adults to find if people were more concerned with privacy or security. Privacy concerns outweighed concerns about being safe from terrorists for 646 out of the 1060 polled. Of the 1060 adults, about 180 are 65 and older and their concerns may be different from the rest of the population.

a) Do you expect the 95% confidence interval for the true proportion of those 65 and over who are more concerned with privacy to be wider or narrower than the 95% confidence interval for the true proportion of all U.S. adults? Explain.
b) If 13% of the 1060 polled were 18–24 years old, would you expect the margin of error for this group to be larger or smaller than for the 65 and older group? Explain.

46. Back to campus ACT, Inc. reported that 74% of 1644 randomly selected college freshmen returned to college the next year. The study was stratified by type of college—public or private. The retention rates were 71.9% among 505 students enrolled in public colleges and 74.9% among 1139 students enrolled in private colleges.

a) Will the 95% confidence interval for the true national retention rate in private colleges be wider or narrower than the 95% confidence interval for the retention rate in public colleges? Explain.
b) Do you expect the margin of error for the overall retention rate to be larger or smaller? Explain.

47. Deer ticks Wildlife biologists inspect 153 deer taken by hunters and find 32 of them carrying ticks that test positive for Lyme disease.

a) Create a 90% confidence interval for the percentage of deer that may carry such ticks.
b) If the scientists want to cut the margin of error in half, how many deer must they inspect?
c) What concerns do you have about this sample?

48. Back to campus II Suppose ACT, Inc. wants to update their information from Exercise 46 on the percentage of freshmen that return for a second year of college.

a) They want to cut the stated margin of error in half. How many college freshmen must be surveyed?
b) Do you have any concerns about this sample? Explain.

***49. Graduation, again** As in Exercise 19, we hope to estimate the percentage of adults aged 25 to 30 who never graduated from high school. What sample size would allow us to increase our confidence level to 95% while reducing the margin of error to only 2%?

***50. Better hiring info** Editors of the business report in Exercise 20 are willing to accept a margin of error of 4% but want 99% confidence. How many randomly selected employers will they need to contact?

***51. Pilot study** A state's environmental agency worries that many cars may be violating clean air emissions standards. The agency hopes to check a sample of vehicles in order to estimate that percentage with a margin of error of 4% and 95% confidence. To gauge the size of the problem, the agency first picks 50 cars and finds 8 with faulty emissions systems. How many should be sampled for a full investigation?

***52. Another pilot study** During routine screening, a doctor notices that 31% of her adult patients show higher than normal levels of glucose in their blood—a possible warning signal for diabetes. Hearing this, some medical researchers decide to conduct a large-scale study, hoping to estimate the proportion to within 5% with 98% confidence. How many randomly selected adults must they test?

53. Approval rating A newspaper reports that the governor's approval rating stands at 65%. The article adds that the poll is based on a random sample of 972 adults and has a margin of error of 2.5%. What level of confidence did the pollsters use?

54. Amendment A TV news reporter says that a proposed constitutional amendment is likely to win approval in the upcoming election because a poll of 1505 likely voters indicated that 52% would vote in favor. The reporter goes on to say that the margin of error for this poll was 3%.

a) Explain why the poll is actually inconclusive.
b) What confidence level did the pollsters use?

JUST CHECKING

Answers

1. Normal
2. It would be centered at the true proportion of students who are in favor.
3. 0.05
4. No. We know that in the sample 17% said "yes"; there's no need for a margin of error.
5. No, we are 95% confident that the percentage falls in some interval, not exactly on a particular value.
6. Yes. That's what the confidence interval means.
7. No. We don't know for sure that's true; we are only 95% confident.
8. No. That's our level of confidence, not the proportion of people receiving unsolicited text messages. The sample suggests the proportion is much lower.
9. Wider.
10. Lower.
11. Smaller.

17

Confidence Intervals for Means

WHERE ARE WE GOING?

We've learned how to generalize from the data at hand to the world at large for proportions. But not all data are as simple as Yes or No. In this chapter, we'll learn how to make confidence intervals for the mean of a quantitative variable.

Premature babies are at risk for a variety of reasons. One of the main problems is their low birthweight. Doctors monitor a baby's birthweight to anticipate possible health issues and appropriate intervention. Birthweight is one of the variables in the data provided by the National Center for Health Statistics, which we looked at in Chapter 16. Figure 17.1 shows the distribution of birthweights for the entire population of U.S. babies born in 1998.

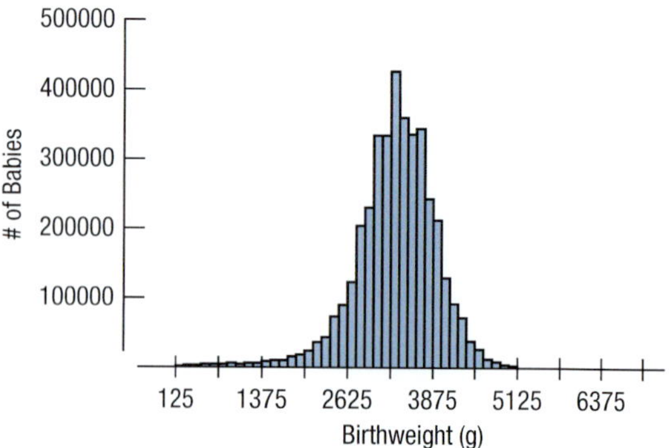

Figure 17.1
Birthweights of all 3,945,192 babies born in the United States in 1998 (4640 were missing birthweight values). The distribution is skewed slightly to the left with very long tails on both sides. It is hard to see any babies whose weights exceeded 6000 grams (6 kg) in this histogram, but in fact there were 144 whose weights were between 6 and 8 kg.

This is an unusual situation; we have the entire population. Typically, we'll only have a sample to work with. From that sample we'll want to estimate the population mean and construct a confidence interval for it. Chapter 16 showed how to make a confidence interval for a proportion, but birthweights are quantitative. What's different? Will the same ideas that we used in the last chapter work here?

17.1 The Central Limit Theorem

The babies' weights differ from proportions in one important way. Proportions are summaries of individual responses that had two possible values such as "yes" and "no," "male" and "female," or "1" and "0." Quantitative variables report a quantitative value for each case. The first thing to do with a quantitative variable is to make a picture—typically a histogram—to get an idea of how it is distributed.

To understand the sampling distribution of a mean, we'll start as we did in the last chapter by simulating the sampling distribution. We'll take several thousand random samples of size $n = 100$, calculate the mean of each sample, and then display the means in a histogram (see Figure 17.2).

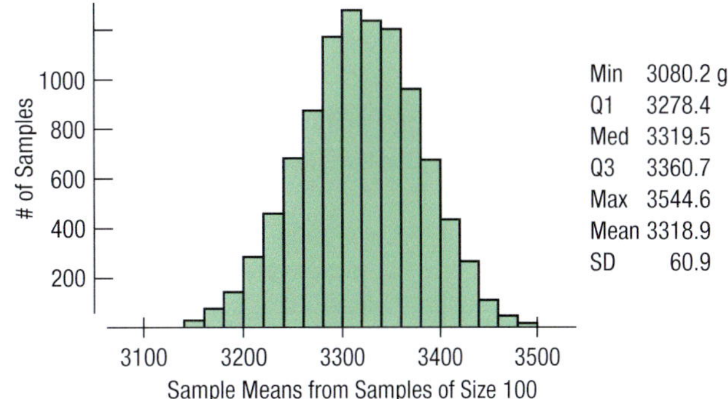

Figure 17.2
A histogram of the means of 10,000 samples drawn at random from the population of babies' birthweights has a distribution that is centered at the population mean (3318.9 g) with a standard deviation much smaller than the population standard deviation of the birthweights (which was 609.5).

The histogram has three important features:

1. Its shape is symmetric even though the population distribution was slightly skewed to the left.
2. Its center (specifically, its mean, 3318.9 g) is very close to the population mean (which we know is 3318.9 g).
3. Its spread (measured by its standard deviation) is 60.9 g—much smaller than the standard deviation of the population (which is 609.5). In fact, it's about 1/10 of the population value. (Note for now that $10 = \sqrt{100}$, the size of the samples.)

So it seems that the means of many independent samples from a population have a distribution that looks Normal. Could de Moivre's result for proportions work for means as well? Several of the leading mathematicians of the 18th century worked on generalizing de Moivre's result. And in 1812, Pierre-Simon Laplace was able to prove that what we suspect from our simulations—that the distribution of means tends to follow a Normal model centered at the true population mean—is, in fact, true. Laplace's result is called the **Central Limit Theorem** (CLT), and is considered to be the fundamental theorem of statistics.[1]

What do we mean by "tends to follow"? Well, Laplace's theorem is a *limit* theorem. That is, it is more and more nearly true as the sample size, *n*, gets larger and larger. A sample size of 100 is often large enough to make it work well. The remarkable thing is that it is true *regardless of the distribution of the population.*

Let's say that again: The sampling distribution of *any* mean becomes more nearly Normal as the sample size grows. All we need is for the observations to be independent. We don't even care about the shape of the population distribution![2] This surprising fact

[1]The word "central" in the name of the theorem means "fundamental." It doesn't refer to the center of a distribution.

[2]OK, one technical condition. The data must come from a population with a finite variance. You probably can't imagine a population with an infinite variance, but statisticians can construct such things, so we have to discuss them in footnotes like this. It really makes no difference in how you think about the important stuff, so feel free to forget we mentioned it.

> "I know of scarcely anything so apt to impress the imagination as the wonderful form of cosmic order expressed by the "Law of Frequency of Error." The law would have been personified by the Greeks and deified, if they had known of it."
>
> —Sir Francis Galton (for whom the "Law of Frequency of Error" was the term for what is now called the Central Limit Theorem)

caused quite a stir (at least in mathematics circles) because the fact that it works no matter how the data are distributed seems remarkable, but its great value to statistical analysis started with the observation by the mathematician Carl Friedrich Gauss that it meant that the least squares criterion, which we have seen for regression, is in some sense the "best" way to find a good description of data in the face of random errors.

How large a sample do we need for the CLT to be approximately correct? Well, if our data are Normally distributed to begin with, then if we draw samples of size 1 (whose means are, of course, just the data values themselves) and make a histogram, it will (of course) be Normal. So for Normally distributed data, the CLT is true even for samples of one. In general, when the distribution of the data is unimodal and symmetric, the CLT works well even for small sample sizes. But the really remarkable part of Laplace's theorem is that it works even if we start with a distribution that's strongly skewed. We won't prove the CLT here, but you can see it work for yourself with simulations such as those we've done in previous chapters.

RANDOM MATTERS The Central Limit Theorem

Let's consider the CEO compensations we saw back in Chapter 4. (Data in **CEO compensation 2014**) We urge you to try these simulations for yourself (e.g., with the Simple Sample tool at https://astools.datadesk.com). Of course, your results will differ at random from those here, but we are confident that the CLT is true, so you will see the same patterns, if not the same exact results.

Here is a histogram of the CEO compensations (in thousands of dollars):

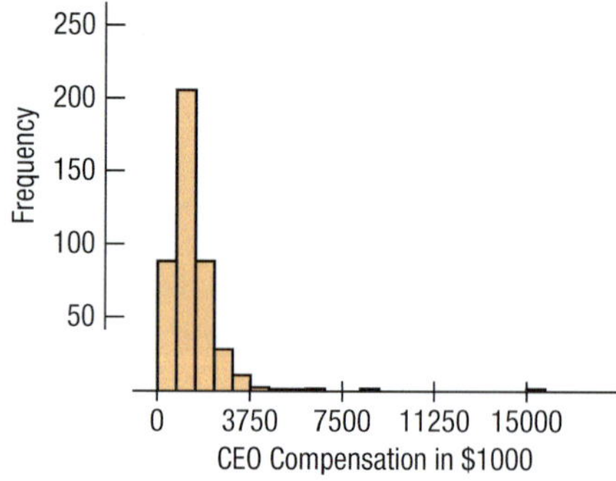

Figure 17.3
The distribution of CEO compensations is highly right skewed.

If we sample from this distribution, there's a good chance that we'll get an exceptionally large value. So some samples will have sample means much larger than others. Here is the sampling distribution of the means from 1000 samples of size 10:

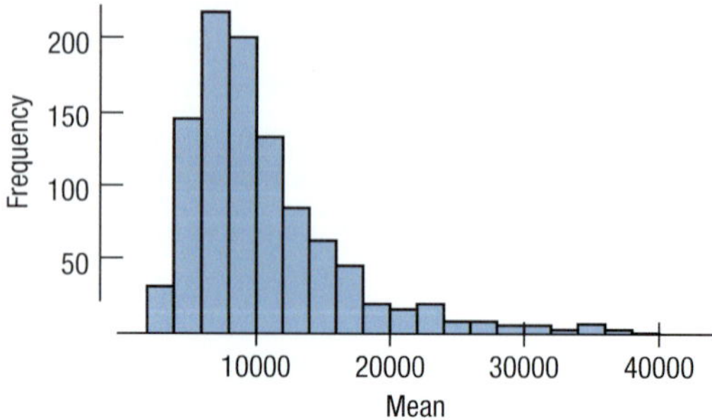

Figure 17.4
Samples of size 10 from the CEOs have means whose distribution is still right skewed, but less so than the distribution of the individual values shown in Figure 17.3.

This distribution is not *as* skewed as the original distribution, but still strongly right skewed. We wouldn't ordinarily find a mean of such a skewed distribution, but bear with us for a few steps. The mean of these values is around 10,000 (in $1000s), which actually matches the mean of all 434 CEOs in our population. What happens if we take a larger sample? Here is the sampling distribution of means from samples of size 50:

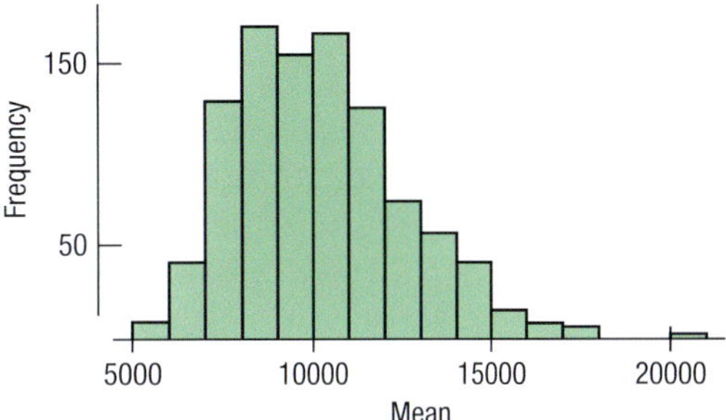

Figure 17.5
Means of samples of size 50 have a distribution that is only moderately right skewed.

This distribution is less skewed than the corresponding distribution from smaller samples and its mean is again near 10,000. Will this continue as we increase the sample size? Let's try samples of size 100:

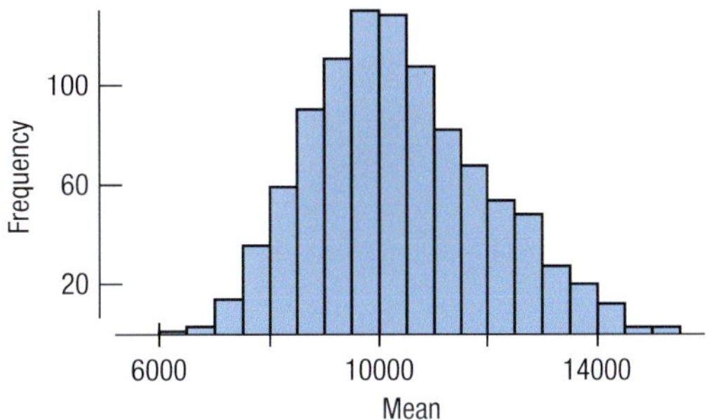

Figure 17.6
The means of samples of size 100 are nearly symmetric.

As we take larger samples, the distribution of means becomes more and more symmetric. By the time we get to samples of size 200, the distribution is quite symmetric and, of course, has a mean quite close to 10,000.

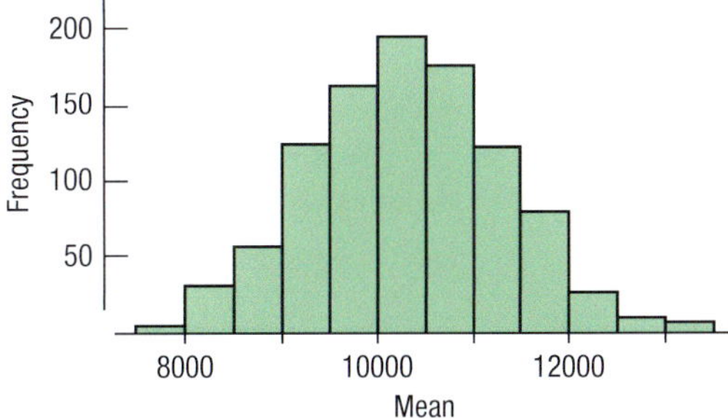

Figure 17.7
When the sample size is 200, even though we started with a highly skewed distribution, the sampling distribution of means is now almost perfectly symmetric.

IS 30 A MAGIC NUMBER?

There is folklore around that a sample size of 30 is large enough to ensure that the sampling distribution of the mean will be Normal enough to base confidence intervals on it. But—that's just wrong. It depends on the skewness of the distribution of the data you start with. If your data have a fairly symmetric distribution, a sample of only a few will produce a sampling distribution of the mean that's nearly Normal, but if, like the CEO salaries, the data are strongly skewed, you'll need a larger sample size before the skewness in the mean's sampling distribution calms down. Some historians have suggested that the 30 rule came from the number of lines of the *t*-table (see the next section) that could fit on one page.

What about a really bimodal population, one that consists of only 0's and 1's? The Central Limit Theorem says that even means of samples from this population will follow a Normal sampling distribution model. But wait. Suppose we have a categorical variable and we assign a 1 to each individual in the category and a 0 to each individual not in the category. And then we find the mean of these 0's and 1's. That's the same as counting the number of individuals who are in the category and dividing by n. That mean will be . . . the *sample proportion*, $\hat{p}$, of individuals who are in the category (a "success"). So maybe it wasn't so surprising after all that proportions, like means, have Normal sampling distribution models; in fact, de Moivre's theorem about proportions is just a special case of Laplace's. Of course, for such an extremely bimodal population, we'd need a reasonably large sample size—and that's where the sample size condition for proportions comes in.

There's just one more fact we need. In Figure 17.2, the means vary with a standard deviation of 60.9 g. That's just about 1/10 as big as the standard deviation of all the birthweights in Figure 17.1, which was 609.5 g. Coincidence? Hardly! You know that means vary less than observations.[3] (Would you rather have your homework grade depend on one exercise in this text, or on the average of many?) In fact, if the population has standard deviation σ, means of a sample of size n will vary with a standard deviation of $\sigma/\sqrt{n}$.

The Central Limit Theorem

When a random sample is drawn from any population with mean μ and standard deviation σ, its sample mean, $\bar{y}$, has a sampling distribution with the same *mean* μ but whose *standard deviation* is $\frac{\sigma}{\sqrt{n}}$ $\left(\text{and we write } \sigma(\bar{y}) = SD(\bar{y}) = \frac{\sigma}{\sqrt{n}}\right)$.

No matter what population the random sample comes from, the *shape* of the sampling distribution is approximately Normal as long as the sample size is large enough. The larger the sample used, the more closely the Normal approximates the sampling distribution for the mean.

It seems we have all we need to construct a confidence interval for the mean. The CLT tells us that the sampling distribution is Normal and centered at the true population mean μ with a standard deviation of $\sigma/\sqrt{n}$.

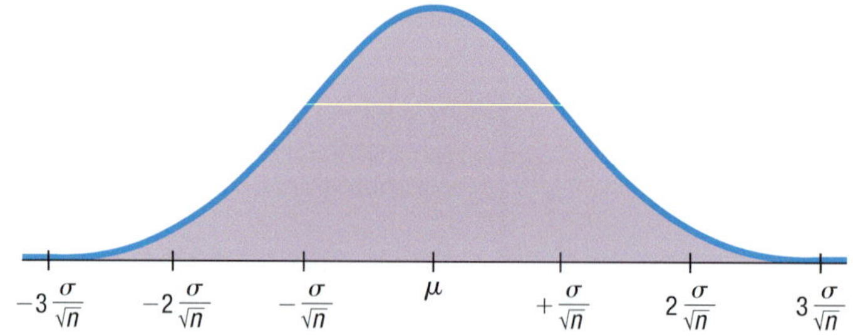

Figure 17.8
The sampling distribution of a mean tends toward one like this—a Normal model centered at μ with standard deviation $\sigma/\sqrt{n}$.

Standard Error

Because we estimate the standard deviation of the sampling distribution model from the data, it's a *standard error*. So we use the $SE(\bar{y})$ notation. Remember, though, that it's just the estimated standard deviation of the sampling distribution model for means.

We know that *our* sample mean $\bar{y}$ is somewhere in this picture. But where? 95% of samples will have means within 1.96 standard deviations of μ. So, if we made μ traps for each sample going out $1.96 \times \frac{\sigma}{\sqrt{n}}$ on either side of $\bar{y}$, 95% of those traps would capture μ.

But we don't know σ. The natural thing to do would be to use s from our sample in its place. That would estimate the standard deviation of $\bar{y}$, $SD(\bar{y}) = \frac{\sigma}{\sqrt{n}}$, with the *standard error* of $\bar{y}$, $SE(\bar{y}) = s/\sqrt{n}$.

Early in the 20th century, people used this standard error with the Normal model, assuming it would work. And for large sample sizes it *did* work pretty well. But they began to notice problems with smaller samples. The sample standard deviation, s, like any

[3]If you use the Simple Sample tool, you can drag the bounds to see how the variability of the means compares with the original variation in the data.

other statistic, varies from sample to sample. And this extra variation in the standard error was messing up the margins of error.

William S. Gosset is the man who investigated this fact. He realized that not only do we need to allow for the extra variation with larger margins of error, but we even need a new sampling distribution model. In fact, we need a whole *family* of models, depending on the sample size, n. These models are unimodal, symmetric, bell-shaped models, but the smaller our sample, the more we must stretch out the tails. Gosset's work transformed statistics, but most people who use his work don't even know his name.

17.2 A Confidence Interval for the Mean

W. S. Gosset (1876–1937) is thought by many to be the founder of modern statistics because of his paper, which showed how to perform inference for a mean when the standard deviation of the population was not known. To find the sampling distribution of $\frac{\bar{y} - \mu}{s/\sqrt{n}}$, Gosset simulated it *by hand*.

Gosset had a job that might make you envious. He was the chief Experimental Brewer for the Guinness Brewery in Dublin, Ireland. The brewery was a pioneer in scientific brewing and Gosset's job was to meet the demands of the brewery's many discerning customers by developing the best stout (a thick, dark beer) possible.

Gosset's experiments often required as much as a day to make the necessary chemical measurements or a full year to grow a new crop of hops. For these reasons, not to mention his health, his samples sizes were small—often as small as 3 or 4.

When he calculated means of these small samples and used the standard error formula, it didn't seem to work right. Guinness granted Gosset time off to earn a graduate degree in the emerging field of statistics, and naturally he chose this problem to work on. He figured out that when he used the standard error, $\frac{s}{\sqrt{n}}$, as an estimate of the standard deviation of the mean, the shape of the sampling model changed. He even figured out what the new model should be.

The Guinness Company may have been ahead of its time in using statistical methods to manage quality, but they also had a policy that forbade their employees to publish. Gosset pleaded that his results were of no specific value to brewers and was allowed to publish under the pseudonym "Student," chosen for him by Guinness's managing director. Accounts differ about whether the use of a pseudonym was to avoid ill feelings within the company or to hide from competing brewers the fact that Guinness was using statistical methods. In fact, Guinness was alone in applying Gosset's results in their quality assurance operations.

It was a number of years before the true value of "Student's" results was recognized. By then, statisticians knew Gosset well, as he continued to contribute to the young field of statistics. But this important result is still widely known as **Student's *t***.

The sampling distribution models that Gosset found are always bell-shaped, but the details change with the sample size. When the sample size is very large, the model is nearly Normal, but for small samples the tails of the distribution are much fatter than the Normal, so values far from the mean are more common (see Figure 17.9). These Student's t-models form a whole *family* of related distributions that depend on the sample size. Statisticians use the term **degrees of freedom** to label families of distributions.[4] We denote degrees of freedom as df and the model as t_{df}, with the degrees of freedom as a subscript. For the t-model, degrees of freedom are simple: $df = n - 1$.

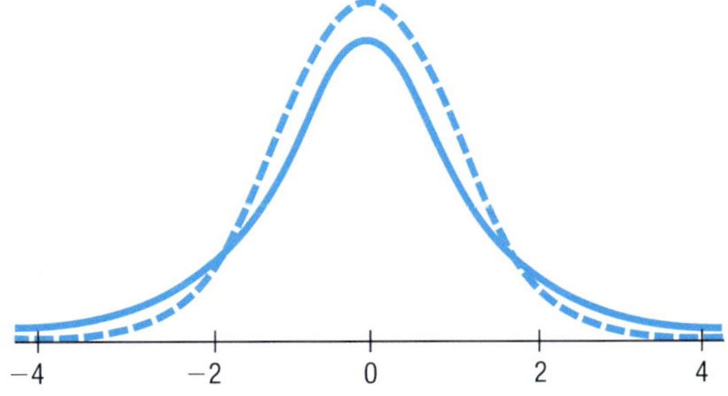

Figure 17.9
The t-model (solid curve) on 2 degrees of freedom has fatter tails than the Normal model (dashed curve). So the 68–95–99.7 Rule doesn't work for t-models with only a few degrees of freedom. It may not look like a big difference, but a t with 2 df is more than 4 times as likely to have a value greater than 2 than a standard Normal.

[4]Degrees of freedom were first introduced as a concept by Gauss as early as 1821. Gosset used the idea in describing the t family in 1908. Fisher first used the term degrees of freedom in 1922 when talking about another family, the chi-square (see Chapter 22).

A Practical Sampling Distribution Model

When certain assumptions and conditions[5] are met, the standardized sample mean,

$$t = \frac{\bar{y} - \mu}{SE(\bar{y})},$$

follows a Student's t-model with $n - 1$ degrees of freedom. We estimate the standard deviation with

$$SE(\bar{y}) = \frac{s}{\sqrt{n}}.$$

Degrees of Freedom

When we introduced the formula for the standard deviation, we promised to explain why we divide by $n - 1$ rather than by n. The reason is closely tied to the concept of degrees of freedom.

If only we knew the true population mean, μ, we would use it in our formula for the sample standard deviation:

$$s = \sqrt{\frac{\sum(y - \mu)^2}{n}}. \qquad \text{(Equation 1)}^6$$

But we don't know μ, so we naturally use $\bar{y}$ in its place. And that causes a small problem. For any sample, the data values will generally be closer to their own sample mean, $\bar{y}$, than to the true population mean, μ. Why is that? Imagine that we take a random sample of 10 high school seniors. The mean SAT Verbal score is 500 in the United States. But the sample mean, $\bar{y}$, for *these* 10 seniors won't be exactly 500. Are the 10 seniors' scores closer to 500 or $\bar{y}$? They'll always be closer to their own average $\bar{y}$. So, when we calculate s using $\sum(y - \bar{y})^2$ instead of $\sum(y - \mu)^2$ in Equation 1, our standard deviation estimate is too small. The amazing mathematical fact is that we can fix it by dividing by $n - 1$ instead of by n. This difference is much more important when n is small than when the sample size is large. The t-distribution inherits this same number, and we call $n - 1$ the degrees of freedom.

What Did Gosset Do?

Amazingly, Gosset used simulation to find the shape of the sampling distribution. In 1908? We've been using computers to draw many samples from a population. Gosset did it by hand. He first copied 3000 values of criminals' finger lengths from a journal article[7] onto cards. He shuffled the cards into 750 groups of 4, and calculated the mean and the standard deviation for each group, by hand (!). He then made a histogram of the standardized means, subtracting the population mean (which he knew) from each sample mean, but dividing by each sample's own standard error. His histogram was unimodal, symmetric, and bell-shaped, but it had fatter tails than a Normal model, which led him to find the t-distribution.

You could reproduce what Gosset did a lot easier—and a lot faster with a statistics package. (He spent over a year on the project.) (Data in **Fingers and heights**)

Constructing a Confidence Interval for the Mean

We construct a confidence interval for a mean in much the same way as we made one for a proportion in Chapter 16. There we wrote

$$\hat{p} \pm z^* \times SE(\hat{p}).$$

[5]You can probably guess what they are. See the next section.

[6]Statistics textbooks usually have equation numbers so they can talk about equations by name. We haven't needed equation numbers yet, but we admit it's useful here, so this is our first.

[7]In those days, journal articles appeared in print, so cut and paste really meant it.

We centered it at the sample proportion and reached out a number of standard errors on each side, the number depending on the level of confidence and on the Normal model (hence the z^*). Now we do the same for means, using Gosset's t:

$$\bar{y} \pm t^*_{n-1} \times SE(\bar{y}).$$

When Gosset corrected the model to take account of the extra uncertainty, the margin of error got bigger because the t-distribution has fatter tails. When you use Gosset's model instead of the Normal model, your confidence intervals will be just a bit wider. That's the correction you need. By using the t-model, you've compensated for the extra variability in precisely the right way.

NOTATION ALERT

Ever since Gosset, t has been reserved in statistics for his distribution.

NOTATION ALERT

When we found critical values from a Normal model, we called them z^*. When we use a Student's t-model, we'll denote the critical values with t^*. You can find them using the app at astools.datadesk.com or in Table T at the back of the text.

One-Sample t-Interval for the Mean

When the assumptions and conditions[8] are met, we are ready to find the confidence interval for the population mean, μ. The confidence interval is

$$\bar{y} \pm t^*_{n-1} \times SE(\bar{y}),$$

where the standard error of the mean is $SE(\bar{y}) = \frac{s}{\sqrt{n}}$.

The critical value t^*_{n-1} depends on the particular confidence level, C, that you specify and on the number of degrees of freedom, $n - 1$, which we get from the sample size.

EXAMPLE 17.1 A One-Sample t-Interval for the Mean

In 2004, a team of researchers published a study of contaminants in farmed salmon.[9] Fish from many sources were analyzed for 14 organic contaminants. The study expressed concerns about the level of contaminants found. One of those was the insecticide mirex, which has been shown to be carcinogenic and is suspected to be toxic to the liver, kidneys, and endocrine system. One farm in particular produced salmon with very high levels of mirex. After those outliers are removed, summaries for the mirex concentrations (in parts per million) in the rest of the farmed salmon are: (Data in **Farmed salmon**)

$$n = 150 \qquad \bar{y} = 0.0913 \text{ ppm} \qquad s = 0.0495 \text{ ppm}.$$

QUESTION: What does a 95% confidence interval say about mirex?

ANSWER: $df = 150 - 1 = 149$ $\quad t^*_{149} \approx 1.977$ (from Table T, using 140 df) (actually, $t^*_{149} \approx 1.976$ from technology)

$$SE(\bar{y}) = \frac{s}{\sqrt{n}} = \frac{0.0495}{\sqrt{150}} = 0.0040$$

So the confidence interval for μ is
$$\bar{y} \pm t^*_{149} \times SE(\bar{y}) = 0.0913 \pm 1.977(0.0040) = 0.0913 \pm 0.0079 = (0.0834, 0.0992)$$

I'm 95% confident that the mean level of mirex concentration in farm-raised salmon is between 0.0834 and 0.0992 parts per million.

---

[8]Yes, the same ones.

[9]Ronald A. Hites, Jeffery A. Foran, David O. Carpenter, M. Coreen Hamilton, Barbara A. Knuth, and Steven J. Schwager, "Global Assessment of Organic Contaminants in Farmed Salmon," *Science* 9 January 2004: Vol. 303, no. 5655, pp. 226–229.

Student's t-models are unimodal, symmetric, and bell-shaped, just like the Normal. But t-models with only a few degrees of freedom have longer tails and a larger standard deviation than the Normal. (That's what makes the margin of error bigger.) As the degrees of freedom increase, the t-models look more and more like the standard Normal. In fact, the t-model with infinite degrees of freedom is exactly Normal.[10] This is great news if you happen to have an infinite number of data values, but that's not likely. Fortunately, above a few hundred degrees of freedom it's very hard to tell the difference. Of course, in the rare situation that we *know* σ, it would be foolish not to use that information. And if we don't have to estimate σ, we can use the Normal model.

z or t?

If you know σ, use z. (That's rare!)
Whenever you use s to estimate σ, use t.

WHEN σ IS KNOWN

Administrators of a hospital in Nashville were concerned about the prenatal care given to mothers in their part of the city. To study this, they examined the gestation times of babies born there. They drew a sample of 70 babies born in their hospital in the previous 6 months. Human gestation times for healthy pregnancies are known to be well modeled by a Normal with a mean of 280 days and a standard deviation of about 14 days. The hospital administrators wanted to construct a confidence interval for the mean gestation time of their hospital to compare against the known standard. If they use the established value for the standard deviation, 14 days, rather than estimating it from their sample, they can base their interval on the Normal model and use z rather than Student's t.

Sir Ronald Fisher (1890–1962) was one of the founders of modern statistics.

Assumptions and Conditions

After Gosset found the t-model by simulation, he made some assumptions so the math would work out. Years later, when Sir Ronald A. Fisher confirmed Gosset's model, he needed the same assumptions. These are the assumptions we need to use the Student's t-models.

Independence Assumption The data values should be mutually independent. There's really no way to check independence of the data by looking at the sample, but you should think about whether the assumption is reasonable.

Randomization Condition: This condition is satisfied if the data arise from a random sample or suitably randomized experiment. Randomly sampled data—and especially data from a Simple Random Sample—are almost surely independent. If the data don't satisfy the **Randomization Condition** then you should think about whether the values are likely to be independent for the variables you are concerned with and whether the sample you have is likely to be representative of the population you wish to learn about.

In the rare case that you have a sample that is more than 10% of the population, you may want to consider using special formulas that adjust for that. But that's not a common concern for means.

When the Assumptions Fail

When you check conditions you usually hope to make a meaningful analysis of your data. The conditions serve as *disqualifiers*—keep going unless there's a serious problem. If you find minor issues, note them and express caution about your results. If the sample is not a Simple Random Sample (SRS), but you believe it's representative of some population, limit your conclusions accordingly. If there are outliers, perform the analysis both with and without them. If the sample looks bimodal, try to analyze subgroups separately. Only when there's major trouble—like a strongly skewed small sample or an obviously nonrepresentative sample—are you unable to proceed at all.

[10] Formally, in the limit as n goes to infinity.

Normal Population Assumption Student's t-models won't work for small samples that are badly skewed. How skewed is too skewed? Well, formally, we assume that the data are from a population that follows a Normal model. Practically speaking, we can't be sure this is true.

And it's almost certainly *not* true. Models are idealized; real data are, well, real—*never* exactly Normal. The good news, however, is that even for small samples, it's sufficient to check the . . .

Nearly Normal Condition: The data come from a distribution that is unimodal and symmetric.

Check this condition by making a histogram or Normal probability plot. Normality is less important for larger sample sizes. Just our luck: It matters most when it's hardest to check.[11]

For very small samples ($n < 15$ or so), the data should follow a Normal model pretty closely. Of course, with so little data, it can be hard to tell. But if you do find outliers or strong skewness, don't use these methods.

For moderate sample sizes (n between 15 and 40 or so), the t methods will work well as long as the data are unimodal and reasonably symmetric. Make a histogram.

When the sample size is larger than 40 or 50, the t methods are safe to use unless the data are extremely skewed. Be sure to make a histogram. If you find outliers in the data, it's always a good idea to perform the analysis twice, once with and once without the outliers, even for large samples. Outliers may well hold additional information about the data, but you may decide to give them individual attention and then summarize the rest of the data. If you find multiple modes, you may well have different groups that should be analyzed and understood separately.

EXAMPLE 17.2

Checking Assumptions and Conditions for Student's t

RECAP: Researchers purchased whole farmed salmon from 51 farms in eight regions in six countries. The histogram shows the concentrations of the insecticide mirex in 150 farmed salmon.

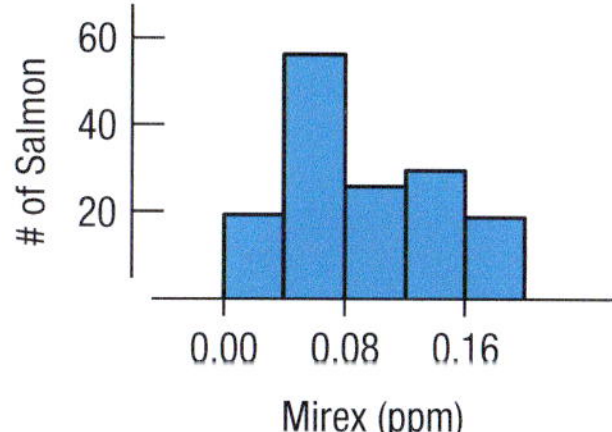

QUESTION: Are the assumptions and conditions for using Student's t satisfied?

ANSWER:

✓ **Independence Assumption:** The fish were not a random sample because no simple population existed to sample from. But they were raised in many different places, and samples were purchased independently from several sources, so they were likely to be independent and to represent the population of farmed salmon worldwide.

✓ **Nearly Normal Condition:** The histogram of the data is unimodal. Although it may be somewhat skewed to the right, this is not a concern with a sample size of 150.

It's okay to use these data for inference about farm-raised salmon.

---

[11]There are formal tests of Normality, but they don't really help. When we have a small sample—just when we really care about checking Normality—these tests are not very effective. So it doesn't make much sense to use them in deciding whether to perform a t-test. We don't recommend that you use them.

JUST CHECKING

Every 10 years, the United States takes a census. The census tries to count every resident. There have been two forms, known as the "short form," answered by most people, and the "long form," slogged through by about one in six or seven households chosen at random. (For the 2010 Census, the long form was replaced by the American Community Survey.) According to the Census Bureau (www.census.gov), ". . . each estimate based on the long form responses has an associated confidence interval."

1. Why does the Census Bureau need a confidence interval for long-form information but not for the questions that appear on both the long and short forms?
2. Why must the Census Bureau base these confidence intervals on *t*-models?

The Census Bureau goes on to say, "These confidence intervals are wider . . . for geographic areas with smaller populations and for characteristics that occur less frequently in the area being examined (such as the proportion of people in poverty in a middle-income neighborhood)."

3. Why is this so? For example, why should a confidence interval for the mean amount families spend monthly on housing be wider for a sparsely populated area of farms in the Midwest than for a densely populated area of an urban center? How does the formula show this will happen?

To deal with this problem, the Census Bureau reports long-form data only for ". . . geographic areas from which about two hundred or more long forms were completed—which are large enough to produce good quality estimates. If smaller weighting areas had been used, the confidence intervals around the estimates would have been significantly wider, rendering many estimates less useful. . . ."

4. Suppose the Census Bureau decided to report on areas from which only 50 long forms were completed. What effect would that have on a 95% confidence interval for, say, the mean cost of housing? Specifically, which values used in the formula for the margin of error would change? Which would change a lot and which would change only slightly?
5. Approximately how much wider would that confidence interval based on 50 forms be than the one based on 200 forms?

STEP-BY-STEP EXAMPLE

A One-Sample *t*-Interval for the Mean

Let's build a 90% confidence interval for the mean birthweight of a U.S. baby born in 1998. We'll draw a Simple Random Sample of 30 babies from our population.

QUESTION: What can we say about the mean birthweight of all babies from just a sample?

THINK

PLAN State what we want to know. Identify the parameter of interest.

Identify the variables and review the W's.

Make a picture. Check the distribution shape and look for skewness, multiple modes, and outliers.

REALITY CHECK The histogram center is at about 3500 grams, and the data lie between 2495 and 4485 grams.

I want to find a 90% confidence interval for the mean birthweight, μ.

I have a random sample of 30 babies.

Here's a histogram of the 30 observed birthweights.

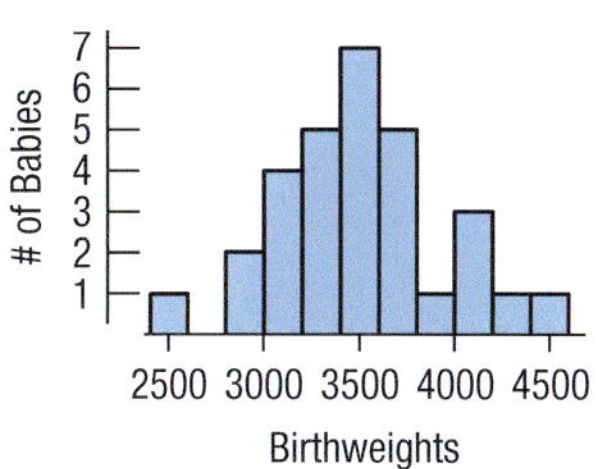

MODEL Think about the assumptions and check the conditions.

Because this was a randomized survey, we check the Randomization Condition.

State the sampling distribution model for the statistic.

Choose your method.

✓ **Randomization Condition**: These are data from a **Simple Random Sample**, so **values** are likely to be independent.

✓ **Nearly Normal Condition**: The histogram of the **birthweights** is **unimodal and symmetric**.

The conditions are satisfied, so I will use a Student's t-model with

$$n - 1 = 29 \text{ degrees of freedom}$$

and find a **one-sample t-interval for the mean**.

SHOW

MECHANICS Construct the confidence interval.

Be sure to include the units along with the statistics.

You can find the critical value from a Student's t-table, a computer program, or a calculator. We have $30 - 1 = 29$ degrees of freedom. The selected confidence level says that we want 90% of the probability to be caught in the middle, so we exclude 5% in *each* tail, for a total of 10%. The degrees of freedom and 5% tail probability are all we need to know to find the critical value.

REALITY CHECK The result looks plausible and in line with what we thought.

Calculating from the data:

$n = 30$ babies

$\bar{y} = 3498.7$ grams

$s = 434.2$ grams

The standard error of $\bar{y}$ is

$$SE(\bar{y}) = \frac{s}{\sqrt{n}} = \frac{434.2}{\sqrt{30}} = 79.27 \text{ grams.}$$

The 90% critical value is $t^*_{29} = 1.699$, so the margin of error is

$$\begin{aligned} ME &= t^*_{29} \times SE(\bar{y}) \\ &= 1.699(79.27) = 134.68. \end{aligned}$$

The 90% confidence interval for the mean birthweight is $3498.7 \pm 134.68 = (3364.0, 3633.4)$ grams.

CONCLUSION Interpret the confidence interval in the proper context.

When we construct confidence intervals in this way, we expect 90% of them to cover the true mean and 10% to miss the true value. That's what "90% confident" means.

I am 90% confident that the interval from 3364.0 to 3633.4 grams contains the true mean birthweight of U.S. babies born in 1998.

17.3 Interpreting Confidence Intervals

Statisticians make mistakes. And they're not ashamed of them. They admit that they don't know the answer, but they try to be precise about how wrong they are. They even use the term "margin of error" in making a confidence interval.

So, be careful when you interpret a confidence interval. You cannot be certain that the interval covers the true mean; you can only express a degree of confidence—and you might be wrong. Look back at the interval we just found in the Step-by-Step Example. You know the population mean—it is 3318.9—*and our interval failed to cover it.*[12] When you make a 90% confidence interval, you must allow for the 10% chance that it won't actually include the population mean. So you can't be certain—only confident to a certain degree.

Here are some of the most common mistakes people make when talking about confidence intervals:

- ***Don't say***, "*90% of all* babies weigh between 3364.0 and 3633.4 grams at birth." The confidence interval is about the *mean* birthweight, not about the *individual* babies. Babies' birthweights vary much more than that (as you can see from Figure 17.1). Confidence intervals for means are narrower.
- ***Don't say***, "We are 90% confident that *a randomly selected baby* will weigh between 3364.0 and 3633.4 grams." That's the same mistake. Birthweights are much more spread out. We are 90% confident that the *mean* birthweight is between 3364.0 and 3633.4 grams.
- ***Don't say***, "The mean birthweight is 3498.7 grams 90% of the time." That's about means, but still wrong. That might sound "statistical," but what could it even mean? And what's time got to do with it?
- Finally, ***don't say***, "90% *of all samples* will have a mean birthweight between 3364.0 and 3633.4 grams." That statement suggests that *this* interval somehow sets a standard for every other interval. In fact, this interval is no more (or less) likely to be correct than any other. You could say that 90% of all possible samples will produce intervals that actually do contain the true mean birthweight. (The problem is that, because we'll never know where the true mean birthweight really is, we can't know if our sample was one of those 90%.)
- ***Do say***, "I am 90% confident that the true mean birthweight is between 3364.0 and 3633.4 grams." Technically, what this means is "90% of all random samples will produce intervals that cover the true value." But that's a bit too technical to be useful to most people.

So What *Should* We Say?

Since 90% of random samples yield an interval that captures the true mean, we *should* say, "I am 90% confident that the interval from 3364.0 to 3633.4 grams contains the mean birthweight of U.S. babies." It's also okay to say something less formal: "I am 90% confident that the mean birthweight is between 3364.0 and 3633.4 grams." Remember: *Our uncertainty is about the interval, not the true mean*. The interval varies randomly. The true mean **birthweight** is neither variable nor random—just (**usually**) unknown.

JUST CHECKING

In discussing estimates based on the long-form samples, the Census Bureau notes, "The disadvantage . . . is that . . . estimates of characteristics that are also reported on the short form will not match the [long-form estimates]."

The short-form estimates are values from a complete census, so they are the "true" values—something we don't usually have when we do inference.

6. Suppose we use long-form data to make 95% confidence intervals for the mean age of residents for each of 100 of the Census-defined areas. How many of these 100 intervals should we expect will fail to include the true mean age (as determined from the complete short-form Census data)?

[12] If you noticed that, you're really paying attention and not just reading the footnotes.

*17.4 Picking Our Interval up by Our Bootstraps

When we looked at the population of all babies born in 1998 in the United States, we simulated the sampling distribution of the mean birthweights by drawing 10,000 samples of size 100 from the population, and we got this distribution (see Figure 17.2, repeated here):

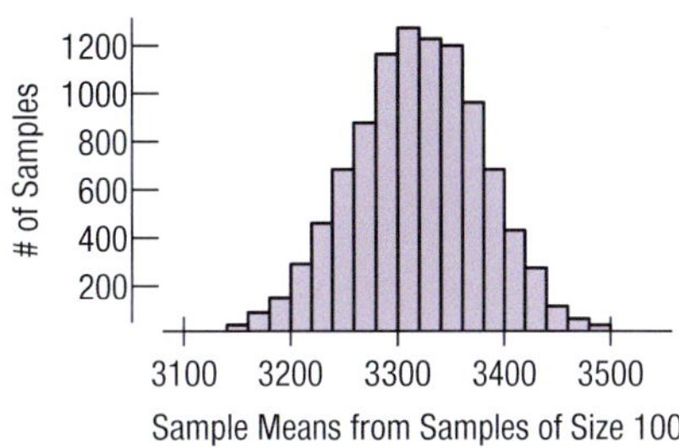

But usually we don't have a population. (If we did, we wouldn't need to sample.) When we had only a sample, we relied on a powerful theorem to tell us about the sampling distribution, for reasonable size samples. We used that theorem to build a confidence interval for the mean.

That's great, but we need to keep in mind three things:

1. The confidence interval (unlike the sampling distribution) is centered at $\bar{y}$ rather than at μ.
2. We need to know how far to reach out from $\bar{y}$, so we need to estimate the population standard deviation. Estimating σ means we need to refer to Student's *t*-models.
3. Using Student's *t* requires the assumption that the underlying data follow a Normal model. Practically, we need to check that the data distribution is at least unimodal and reasonably symmetric, with no outliers.

It turns out that there's a way to approach this problem that relies on the power of modern computers to simplify the process.

Here's the idea. We don't have the population, but we do have a random sample. That sample is supposed to be representative. So, what if we make lots of copies of this sample, build a *pseudo-population*, and sample repeatedly from this population? As before, we'll find the means of our samples and collect those means in a histogram. From the histogram we'll see how the sample means are distributed and—most important—how much they vary.[13]

Figure 17.10 shows what happened when we drew a random sample of birthweights, built a large pseudo-population from it, and drew samples of size 100 from it 10,000 times.

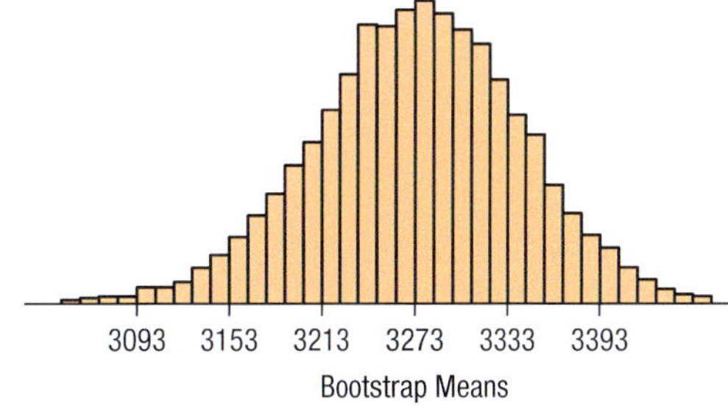

Figure 17.10
A histogram of the means of 10,000 samples drawn from our pseudo-population of birthweights looks a lot like the histogram of 10,000 samples drawn from the real population in Figure 17.2.

Figure 17.10 looks a lot like the histogram in Figure 17.2 where we took samples from the real population. But how can that be? It seems impossible that we could get an entire

[13]There's a computing trick that makes this easier to do with a statistics program. We work only with the original sample and select the values at random but *with replacement*. That is, after we select a value, we "throw it back in the pool" so it can be selected again.

distribution of means starting from only one sample. It's as if we pulled ourselves up by our bootstraps.[14] The discoverer of the method thought it seemed like magic as well, and named the method bootstrapping. We call each of the samples a **bootstrap sample** and refer to sampling from the sample in this way as **resampling**.

How can we use this idea to construct a confidence interval? Well, this histogram is centered at $\bar{y}$ and is unimodal and symmetric. And, somewhat amazingly, its standard deviation is about the same as the standard deviation of the distribution of means drawn from the true population. So, we have all we need to make a confidence interval for μ!

> *BOOTSTRAP BACKGROUND*
>
> The bootstrap method was first described by the prominent statistician Bradley Efron. He based it on previous work by John Tukey, who we've already met as the originator of the boxplot and stem-and-leaf display. Tukey called *his* method the *jackknife* because it was so generally useful. He based his results on work by Maurice Quenouille in the middle of the 20th century.

A Bootstrap Confidence Interval

Look at the bootstrap distribution of sample means in Figure 17.10. It is already centered at $\bar{y}$. And, because it has about the same standard deviation as the sampling distribution, we can take the middle 95% of these means for a confidence interval.

Figure 17.11 shows the histogram of Figure 17.10 with the middle 95% of the values selected. Statistics programs that offer bootstrap confidence intervals for the mean often show a graph much like the one in Figure 17.11.

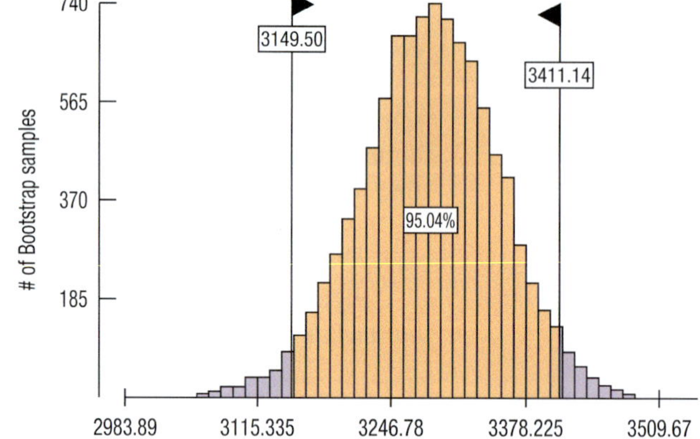

Figure 17.11
A histogram of 10,000 bootstrap sample means based on a sample of 100 birthweights. The middle 95% of the values is highlighted.

In this example, the middle 95% of bootstrap means are between 3149.50 and 3411.14 grams. Those limits propose a 95% **bootstrap confidence interval** for the population mean. We might say that we have 95% confidence that this interval includes the true mean.

Notice that we said "a" 95% confidence interval and not "the" 95% confidence interval. If we drew bootstrap samples again, we'd get different bootstrap means and a (slightly) different interval. You and your classmates will get different bootstrap confidence intervals[15] even when you use the same data.

---

[14]The term is sometimes attributed to a story in Rudolf Erich Raspe's *The Surprising Adventures of Baron Munchausen*, but in that story Baron Munchausen pulls himself (and his horse) out of a swamp by his hair (specifically, his pigtail), not by his bootstraps.

[15]This could drive whoever grades your homework crazy.

Bootstrap methods are newer and not as widely used as methods based on the Central Limit Theorem. And they require a computer to draw all those samples. But even if you don't use them frequently, they offer valuable insight into how confidence intervals work and how to interpret them.

Comparing the Methods

We now have two ways to identify an interval that contains plausible values for the population mean. For a particular sample, these methods won't give exactly the same answer. In fact, the bootstrap method will give a slightly different interval each time we run the resampling program. This time it gave an interval from 3149.5 to 3411 gm and the CLT method gave an interval from 3152 to 3415 gm. The difference in the intervals is a helpful reminder that confidence intervals are *random*. That is, we expect confidence intervals to vary from sample to sample. And for bootstraps, from run to run. But that's OK. All we are claiming is that, in the long run, about 95% of these intervals will include the true mean, and certainly two different, but similar, intervals can each do that.

STEP-BY-STEP EXAMPLE

A Bootstrap Confidence Interval for the Mean

Recall in Chapter 2 that the Human Resources Department of a large financial institution wanted to know how long it took employees to get to work (see p. 40). We had a sample of 100 employees. Let's build a 95% bootstrap confidence interval for the mean commute time.

QUESTION: What can we say about the mean commute time of all employees by bootstrapping a single sample?

THINK

PLAN State what we want to know. Identify the parameter of interest.

I want to find a 95% confidence interval for the mean commute time, μ.

Identify the variables and review the W's.

I have a random sample of 100 employees.

Make a picture. Check the distribution shape and look for skewness, multiple modes, and outliers.

Here's a histogram of a random sample of 100 observed commute times.

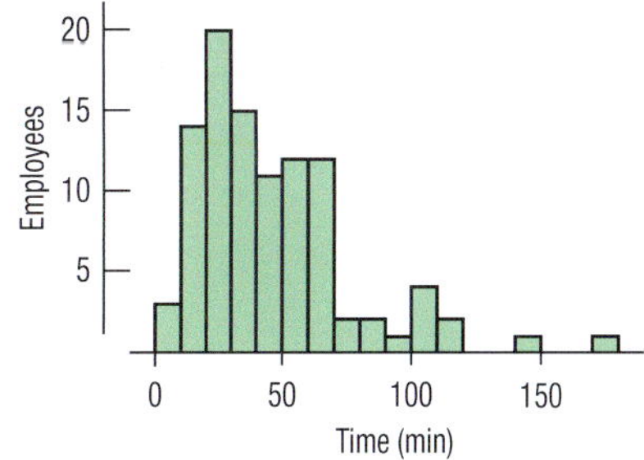

REALITY CHECK Commute times of 40–80 minutes are quite reasonable. The maximum commute time seems long, but possible.

MODEL Think about the assumptions and check the conditions.

✔ **Randomization Condition**: These are data from a Simple Random Sample, so values are likely to be independent.

Because this was a randomized survey, we check the Randomization Condition.

The commute times histogram is skewed to the right. In Chapter 2, we considered whether the median might be a more appropriate summary, but HR has requested information about the mean. Fortunately, the bootstrap method doesn't require the Nearly Normal Condition.

We checked the high values and verified that they were correctly recorded.

SHOW **MECHANICS** Construct the bootstrap confidence interval using appropriate software.

Be sure to include the units along with the statistics.

For our sample,

$\bar{y} = 44.98$ min

$SD = 30.19$ min

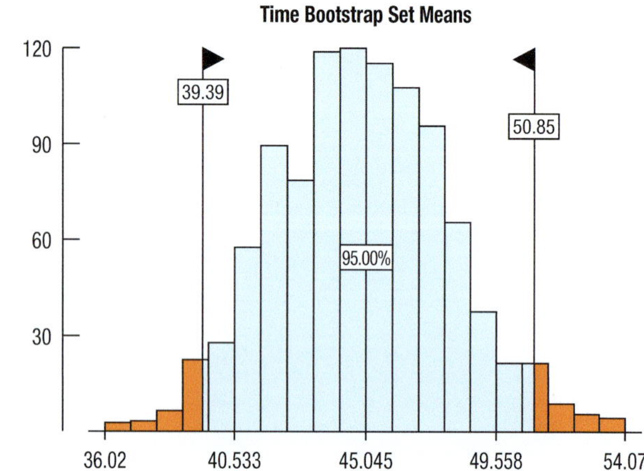

The 95% confidence interval for the mean commute time is (39.39, 50.85) minutes.

REALITY CHECK The result looks plausible and in line with what we thought.

TELL **CONCLUSION** Interpret the confidence interval in the proper context.

When we construct confidence intervals in this way, we expect 95% of them to cover the true mean and 5% to miss the true value. That's what "95% confident" means.

I am 95% confident that the interval from 39.39 minutes to 50.85 minutes contains the true mean commute time of employees in this company.

17.5 Thoughts About Confidence Intervals

The 5000 employees from whom we drew the sample for the Step-by-Step example are a diverse group. Some live virtually next to the office in Lower Manhattan, commuting only a minute or two to work, and others live far out on Long Island or in New Jersey and Pennsylvania, commuting several hours a day. On average, they commute about 45.4 minutes a day.

We have two very different methods for finding a confidence interval for the mean given a random sample. One is based on the Central Limit Theorem, using Gosset's t correction when estimating the standard deviation of the mean. The other, based on the bootstrap, resamples the data to estimate how much the sample mean varies and looks at the histogram of the bootstrap means. It constructs the confidence interval by identifying the central values of this histogram as the plausible values for the true mean.

Does it make any difference which method you use? Well, yes and no. Yes, in the sense that you'll get slightly different answers. In fact, if you repeat the bootstrap, you (or your classmates) will get slightly different answers each time you resample, even with the same data. And of course, if you draw a new sample from the population, you'll get a new confidence interval no matter which method you use. But really the answer is no, it doesn't make much difference to decisions about the real world. A confidence interval for the mean is simply a carefully stated guess about where we think the mean might be, along with an indication of how successful we think that guess might be.

There are always constraints on what we can know about a population from a sample. Some arise from the size of the sample, others from the quality of the data and how the data were gathered. A confidence interval communicates some of those constraints, but you should keep in mind the limits of your data.

For example, the t-interval for the mean commute times for our sample gives (39.0, 51.0) minutes as the 95% confidence interval. Since we know $\mu = 45.43$, we can see that it "worked." One bootstrap interval that we ran gave (39.4, 50.9) minutes. Most software reports intervals like this to ridiculous fake precision. For example:

```
95 percent confidence interval:
38.99045, 50.96955
sample estimates:
mean of x = 44.98
```

The commuters reported their commute time in minutes. (And some may have rounded to the nearest quarter hour.) So five decimal places (about 600 microseconds) is a "precision" that just isn't available from the data, nor one that makes any sense.

Similarly, it makes very little difference whether you choose a 95% confidence level or use 98% or 93%. Most people have little intuition for the differences among those.

The *important point* about confidence intervals is not their precision. A confidence interval is, by nature, random. It changes with the sample and, for the bootstrap, with each simulation as well. But every confidence interval contains both our best guess of the mean and how precise we think that guess is.

In the example, we can see that the mean commute time may be about 5 minutes shorter or longer than the sample mean of 45.0 minutes. That's probably all the HR Department wanted to know.

People often misinterpret confidence intervals. When we asked employees of the company what the 95% interval (39.0, 51.0) minutes meant, most answered "95% of us commute between 39 and 51 minutes to work." That would imply that most of the employees live in this ring around Lower Manhattan.

That's clearly not right. It's the *mean* commute time that we've captured, not the commuters' times themselves. We've seen that people's commute times vary widely, from a few minutes to several hours. A larger sample would make the confidence interval even smaller, but wouldn't make people live nearer to work.

It's important to know how confidence intervals are constructed and to check the assumptions and conditions to be sure that they are valid. Remember that they represent our best *guess* of where we think the mean is—and how confident we are in that guess.

WHAT CAN GO WRONG?

The most fundamental issue you face is knowing when to use Student's t methods.

- **Don't confuse proportions and means.** When you treat your data as categorical, counting successes and summarizing with a sample proportion, make inferences using the Normal model methods you learned about in Chapter 16. When you treat your data as quantitative, summarizing with a sample mean, make your inferences using Student's t methods.

Student's t methods work only when the Normality Assumption is true. Naturally, many of the ways things can go wrong turn out to be different ways that the Normality Assumption can fail. It's always a good idea to look for the most common kinds of failure. It turns out that you can even fix some of them.

- **Beware of multimodality.** The Nearly Normal Condition clearly fails if a histogram of the data has two or more modes. When you see this, look for the possibility that your data come from two groups. If so, your best bet is to try to separate the data into different groups. (Use the variables to help distinguish the modes, if possible. For example, if the modes seem to be composed mostly of men in one and women in the other, split the data according to sex.) Then you could analyze each group separately.

Don't Ignore Outliers

As tempting as it is to get rid of annoying values, you can't just throw away outliers and not discuss them. It isn't appropriate to lop off the highest or lowest values just to improve your results.

- **Beware of skewed data.** Student's t methods assume Normality. Check that with a Normal probability plot and a histogram of the data. If the data are very skewed, you might try re-expressing the variable. Re-expressing may yield a distribution that is unimodal and symmetric, more appropriate for Student's t inference methods for means. Re-expression cannot help if the sample distribution is not unimodal. Some people may object to re-expressing the data, but unless your sample is very large, you just can't use the methods of this chapter on skewed data.
- **Set outliers aside.** Student's t methods are built on the mean and standard deviation, so we should beware of outliers when using them. When you make a histogram to check the Nearly Normal Condition, be sure to check for outliers as well. If you find some, consider doing the analysis twice, both with the outliers excluded and with them included in the data, to get a sense of how much they affect the results.

  The suggestion that you can perform an analysis with outliers removed may be controversial in some disciplines. Setting aside outliers is seen by some as "cheating." But an analysis of data with outliers left in place is *always* wrong. The outliers violate the Nearly Normal Condition and also the implicit assumption of a homogeneous population, so they invalidate inference procedures. An analysis of the nonoutlying points, along with a separate discussion of the outliers, is often much more informative and can reveal important aspects of the data.

  How can you tell whether there are outliers in your data? The "outlier nomination rule" of boxplots can offer some guidance, but it's just a rule of thumb and not an absolute definition. The best practical definition is that a value is an outlier if removing it substantially changes your conclusions about the data. You won't want a single value to determine your understanding of the world unless you are very, very sure that it is correct and similar in nature to the other cases in your data. Of course, when the outliers affect your conclusion, this can lead to the uncomfortable state of not really knowing what to conclude. Such situations call for you to use your knowledge of the real world and your understanding of the data you are working with.[16]

  Of course, Normality issues aren't the only risks you face when doing inferences about means. Remember to *Think* about the usual suspects.
- **Watch out for bias.** Measurements of all kinds can be biased. If your observations differ from the true mean in a systematic way, your confidence interval may not capture the true mean. And there is no sample size that will save you. A bathroom scale that's 5 pounds off will be 5 pounds off even if you weigh yourself 100 times and take the average. We've seen several sources of bias in surveys, and measurements can be biased, too. Be sure to think about possible sources of bias in your measurements.
- **Make sure cases are independent.** Student's t methods also require the sampled values to be mutually independent. We check for random sampling. You should also think hard about whether there are likely violations of independence in the data collection method. If there are, be very cautious about using these methods.
- **Make sure that data are from an appropriately randomized sample.** Ideally, all data that we analyze are drawn from a Simple Random Sample or generated by a randomized experiment. When they're not, be careful about making inferences from them. You may still compute a confidence interval correctly, but this might not save you from making a serious mistake in inference.
- **Interpret your confidence interval correctly.** Many statements that sound tempting are, in fact, misinterpretations of a confidence interval for a mean. You might want to have another look at some of the common mistakes, explained on page 544. Keep in mind that a confidence interval is about the mean of the population, not about the means of samples, individuals in samples, or individuals in the population.

[16]An important reason for *you* to know statistics rather than let someone else analyze your data.

CONNECTIONS

The steps for finding a confidence interval for means are just like the corresponding steps for proportions. Even the form of the calculations is similar. As the z-statistic did for proportions, the t-statistic tells us how many standard errors to go out from the sample mean. For means, though, we have to estimate the standard error separately. This added uncertainty changes the model for the sampling distribution from z to t.

As with all of our inference methods, the randomization applied in drawing a random sample or in randomizing a comparative experiment is what generates the sampling distribution. Randomization is what makes inference in this way possible at all.

The new concept of degrees of freedom connects back to the denominator of the sample standard deviation calculation, as shown earlier.

There's just no escaping histograms and Normal probability plots. The Nearly Normal Condition required to use Student's t can be checked best by making appropriate displays of the data. Back when we first used histograms, we looked at their shape and, in particular, checked whether they were unimodal and symmetric, and whether they showed any outliers. Those are just the features we check for here. The Normal probability plot zeros in on the Normal model a little more precisely.

CHAPTER REVIEW

Know the sampling distribution of the mean.

- To apply the Central Limit Theorem for the mean in practical applications, we must estimate the standard deviation. This *standard error* is

$$SE(\bar{y}) = \frac{s}{\sqrt{n}}.$$

- When we use the *SE*, the sampling distribution that allows for the additional uncertainty is Student's t.

Construct confidence intervals for the true mean, μ.

- A confidence interval for the mean has the form $\bar{y} \pm ME$.
- The margin of error is $ME = t^*_{df} \times SE(\bar{y})$.
- Find critical values by technology or from tables.
- When constructing confidence intervals for means, the correct degrees of freedom is $n - 1$.
- Check the assumptions and conditions before using any sampling distribution for constructing a confidence interval.

Alternatively, consider finding a confidence interval with a bootstrap.

Write clear summaries to interpret a confidence interval.

REVIEW OF TERMS

Central Limit Theorem The Central Limit Theorem (CLT) states that the sampling distribution model of the sample mean (and proportion) from a random sample is approximately Normal for large n, *regardless of the distribution of the population, as long as the observations are independent* (p. 533).

Student's *t* A family of distributions indexed by its degrees of freedom. The t-models are unimodal, symmetric, and bell shaped, but have fatter tails and a narrower center than the Normal model. As the degrees of freedom increase, t-distributions approach the Normal (p. 537).

Degrees of freedom for Student's *t*-distribution For the t-distribution, the degrees of freedom are equal to $n - 1$, where n is the sample size (p. 537).

One-sample t-interval for the mean A one-sample t-interval for the population mean is (p. 539)

$$\bar{y} \pm t^*_{n-1} \times SE(\bar{y}), \text{ where } SE(\bar{y}) = \frac{s}{\sqrt{n}}.$$

The critical value t^*_{n-1} depends on the particular confidence level, C, that you specify and on the number of degrees of freedom, $n - 1$.

Resampling A common way to perform bootstrap calculations is to repeatedly sample from your data *with replacement*, thus re-sampling the data (p. 546).

Bootstrap confidence interval A bootstrap confidence interval is found by identifying the central $C\%$ of a bootstrap distribution based on the available data (p. 546).

TECH SUPPORT

Confidence Intervals for Means

Statistics packages offer convenient ways to make histograms of the data. Even better for assessing near-Normality is a Normal probability plot. When you work on a computer, there is simply no excuse for skipping the step of plotting the data to check that it is nearly Normal. *Beware:* Statistics packages don't agree on whether to place the Normal scores on the x-axis (as we have done) or the y-axis. Read the axis labels.

Any standard statistics package can compute a confidence interval.

The commands to do inference for means on common statistics programs and calculators are not always obvious. (By contrast, the resulting output is usually clearly labeled and easy to read.) The guides for each program can help you start navigating.

DATA DESK

- From the Calc menu, choose **Estimate** for confidence intervals.
- Select the interval from the drop-down menu and make other choices in the dialog.
- Drag variables into the Estimate window.

EXCEL

Specify formulas. Find t^* with the TINV(alpha, df) function.

COMMENTS

Not really automatic. For the examples in this chapter, substitute 0.05 for "alpha" in the TINV command.

JMP

- From the Analyze menu, select **Distribution**.
- For a confidence interval, scroll down to the Moments section to find the interval limits.
- Then fill in the resulting dialog.

COMMENTS

"Moment" is a fancy statistical term for means, standard deviations, and other related statistics.

MINITAB

- From the Stat menu, choose the **Basic Statistics** submenu.
- From that menu, choose **1-sample t. . . .**
- Then fill in the dialog.

COMMENTS

The dialog offers a clear choice between confidence interval and test.

R

To produce a confidence interval (default is 95%), create a vector of data in x and then:

- **t.test**(x, alternative = c("two.sided", "less", "greater"), mu = 0, conf.level = 0.95)

provides the confidence interval for a specified alternative along with additional statistics for performing a hypothesis test (as we'll see in the next chapter).

SPSS

- From the Analyze menu, choose the **Compare Means** submenu.

COMMENTS

The commands suggest neither a single mean nor an interval. But the results provide both a test and an interval.

STATCRUNCH

To do inference for a mean using summaries:

- Click on **Stat**.
- Choose **T Stats** > **One sample** > **with summary**.
- Enter the Sample mean, Sample std dev, and Sample size.
- Indicate **Confidence Interval**, and then enter the **Level** of confidence.
- Click on **Compute!**

To do inference for a mean using data:

- Click on **Stat**.
- Choose **T Stats** > **One sample** > **with data**.
- Choose the variable **Column**.
- Indicate **Confidence Interval**, and then enter the **Level** of confidence.
- Click on **Compute!**

TI-83/84 PLUS

Finding a confidence interval:

- In the STAT TESTS menu, choose **8:TInterval**.
- Specify whether you are using data stored in a list or whether you will enter the mean, standard deviation, and sample size.
- You must also specify the desired level of confidence.

EXERCISES

SECTION 17.1

1. Salmon A specialty food company sells whole King Salmon to various customers. The mean weight of these salmon is 35 pounds with a standard deviation of 2 pounds. The company ships them to restaurants in boxes of 4 salmon, to grocery stores in cartons of 16 salmon, and to discount outlet stores in pallets of 100 salmon. To forecast costs, the shipping department needs to estimate the standard deviation of the mean weight of the salmon in each type of shipment.

a) Find the standard deviations of the mean weight of the salmon in each type of shipment.

b) The distribution of the salmon weights turns out to be skewed to the high end. Would the distribution of shipping weights be better characterized by a Normal model for the boxes or pallets? Explain.

2. LSAT The LSAT (a test taken for law school admission) has a mean score of 151 with a standard deviation of 9 and a unimodal, symmetric distribution of scores. A test preparation organization teaches small classes of 9 students at a time. A larger organization teaches classes of 25 students at a time. Both organizations publish the mean scores of all their classes.

a) What would you expect the sampling distribution of mean class scores to be for each organization?
b) If either organization has a graduating class with a mean score of 160, they'll take out a full-page ad in the local school paper to advertise. Which organization is more likely to have that success? Explain.
c) Both organizations advertise that if any class has an average score below 145, they'll pay for everyone to retake the LSAT. Which organization is at greater risk to have to pay?

3. Tips A waiter believes the distribution of his tips has a model that is slightly skewed to the right, with a mean of \$8.50 and a standard deviation of \$4.30.

a) Explain why you cannot determine the probability that a given party will tip him at least \$18 using the Normal model.
b) Can you estimate the probability that the next 5 parties will tip an average of at least \$13 using the Normal model? Explain.
c) Is it likely that his 10 parties today will tip an average of at least \$13? Explain.

4. Groceries A grocery store's receipts show that Sunday customer purchases have a skewed distribution with a mean of \$28 and a standard deviation of \$20.

a) Explain why you cannot determine the probability that the next Sunday customer will spend at least \$34.
b) Can you estimate the probability that the next 10 Sunday customers will spend an average of at least \$34? Explain.
c) Is it likely that the next 60 Sunday customers will spend an average of at least \$34? Explain.

5. More tips The waiter in Exercise 3 usually waits on about 50 parties over a weekend of work.

a) Estimate the probability that he will earn at least \$550 in tips.
b) How much does he earn on the "worst" 10% of such weekends?

6. More groceries Suppose the store in Exercise 4 had 300 customers this Sunday.

a) Estimate the probability that the store's revenues were at least \$9,000.
b) If, on a typical Sunday, the store serves 300 customers, how much does the store take in on the "best" 10% of such days?

SECTION 17.2

7. *t*-models, part I Using the *t*-tables, software, or a calculator, estimate

a) the critical value of t for a 98% confidence interval with $df = 13$.
b) the critical value of t for a 90% confidence interval with $df = 95$.

8. *t*-models, part II Using the *t*-tables, software, or a calculator, estimate

a) the critical value of t for a 95% confidence interval with $df = 7$.
b) the critical value of t for a 99% confidence interval with $df = 102$.

9. *t*-models, part III Describe how the shape, center, and spread of t-models change as the number of degrees of freedom increases.

10. *t*-models, part IV Describe how the critical value of t for a 95% confidence interval changes as the number of degrees of freedom increases.

11. Home sales The housing market recovered slowly from the economic crisis of 2008. Recently, in one large community, realtors randomly sampled 36 bids from potential buyers to estimate the average loss in home value. The sample showed the average loss from the peak in 2008 was \$9560 with a standard deviation of \$1500.

a) What assumptions and conditions must be checked before finding a confidence interval? How would you check them?
b) Find a 95% confidence interval for the mean loss in value per home.

12. Home sales again In the previous exercise, you found a 95% confidence interval to estimate the average loss in home value.

a) Suppose the standard deviation of the losses had been \$3000 instead of \$1500. What would the larger standard deviation do to the width of the confidence interval (assuming the same level of confidence)?
b) Your classmate suggests that the margin of error in the interval could be reduced if the confidence level were changed to 90% instead of 95%. Do you agree with this statement? Why or why not?
c) Instead of changing the level of confidence, would it be more statistically appropriate to draw a bigger sample?

SECTION 17.3

13. Home sales revisited For the confidence interval you found in Exercise 11, interpret this interval and explain what 95% confidence means in this context.

14. Salaries A survey finds that a 99% confidence interval for the mean salary of a police patrol officer in Paris, France, in 2018 is €33,876 and €58,073. A student is surprised that such few police officers make more than \$57,000. Explain what is wrong with the student's interpretation.

15. Cattle Livestock are given a special feed supplement to see if it will promote weight gain. Researchers report that the 77 cows studied gained an average of 56 pounds, and that a 95% confidence interval for the mean weight gain this supplement produces has a margin of error of ± 11 pounds. Some students wrote the following conclusions. Did anyone interpret the interval correctly? Explain any misinterpretations.

a) 95% of the cows studied gained between 45 and 67 pounds.
b) We're 95% sure that a cow fed this supplement will gain between 45 and 67 pounds.
c) We're 95% sure that the average weight gain among the cows in this study was between 45 and 67 pounds.
d) The average weight gain of cows fed this supplement will be between 45 and 67 pounds 95% of the time.
e) If this supplement is tested on another sample of cows, there is a 95% chance that their average weight gain will be between 45 and 67 pounds.

16. Teachers Software analysis of the salaries of a random sample of 300 London teachers produced the confidence interval shown below. Which conclusion is correct? What's wrong with the others?

t-Interval for μ: with 95% Confidence
$44489 < \mu(\text{TchPay}) < 48244$

a) If we took many random samples of 300 London teachers, about 95 out of 100 of them would produce this confidence interval.
b) If we took many random samples of London teachers, about 95 out of 100 of them would produce a confidence interval that contained the mean salary of all London teachers.
c) About 95 out of 100 London teachers earn between £44,489 and £48,244.
d) About 95 out of 100 of the teachers surveyed earn between £44,489 and £48,244.
e) We are 95% confident that the average teacher salary in the United Kingdom is between £44,489 and £48,244.

*SECTION 17.4

T ***17. Framingham revisited** In Chapter 4, Exercise 38, we saw an ogive of the distribution of cholesterol levels (in mg/dL) of a random sample of 1406 participants (taken in 1948 from Framingham, MA). To find a bootstrap confidence interval for the mean cholesterol, a student took 1000 bootstrap samples, calculated the mean of each, and found the following histogram of the bootstrap means:

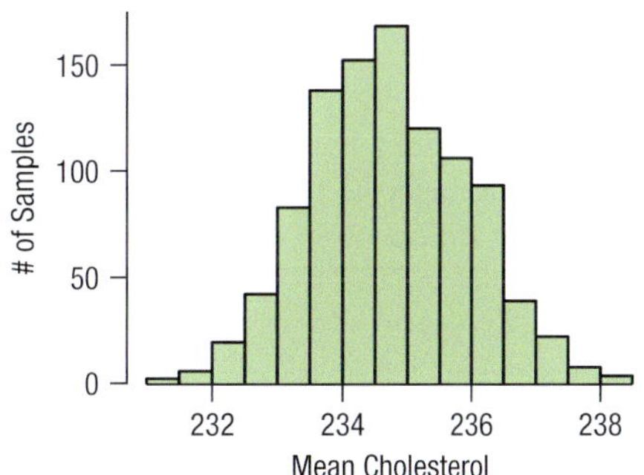

Summary statistics of the means show:

0.5%	1%	2%	2.5%	5%	95%	97.5%	98%	99%	99.5%
231.890	232.130	232.370	232.434	232.769	236.763	237.130	237.281	237.526	237.873

a) Use the data above to construct a 95% confidence interval for the mean cholesterol.
b) Interpret the interval you constructed in part a.
c) What assumptions did you make, if any, in interpreting the interval?

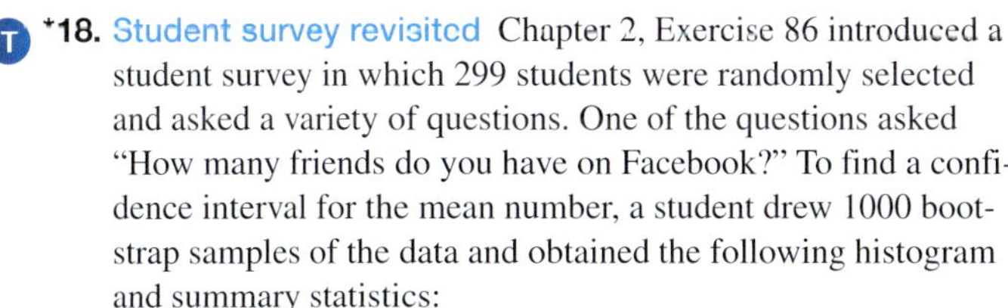

T ***18. Student survey revisited** Chapter 2, Exercise 86 introduced a student survey in which 299 students were randomly selected and asked a variety of questions. One of the questions asked "How many friends do you have on Facebook?" To find a confidence interval for the mean number, a student drew 1000 bootstrap samples of the data and obtained the following histogram and summary statistics:

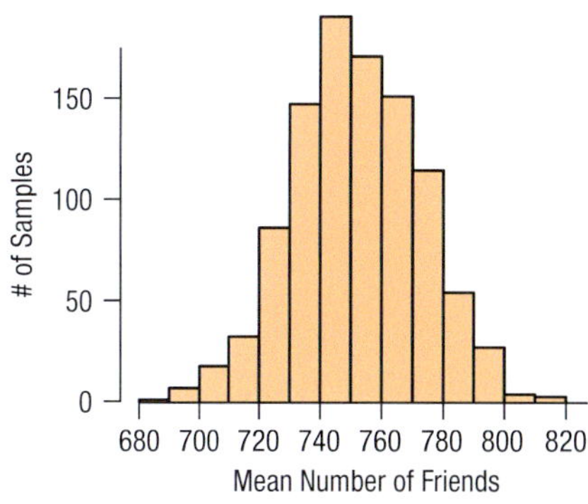

Summary statistics of the means show:

0.5%	1%	2%	2.5%	5%	95%	97.5%	98%	99%	99.5%
698.301	702.498	707.779	709.967	718.279	785.623	792.099	793.411	797.358	800.702

a) Use the data above to construct a 98% confidence interval for the mean number of Facebook friends a student has.
b) Interpret the interval you constructed in part a.
c) What assumptions did you make, if any, in interpreting the interval?

SECTION 17.5

T **19. Shoe sizes revisited** Chapter 2, Exercise 16 showed the histogram of the European shoe sizes from a sample of 269 college students. Looking at the men only, a 95% confidence interval for the mean shoe size shows:

95 percent confidence interval:
44.3071 44.9900

The student knows that European shoes are sized only in whole and half sizes, so is surprised that most men wear size 44.5 shoes. What is wrong with the student's reasoning?

20. Bird counts A biology class conducts a bird count every week during the semester. Using the number of species counted each week, a student finds the following confidence interval for the mean number of species counted:

90 percent confidence interval:
17.22 19.78

Assuming that species have to be whole numbers, the student reports that 90% of the bird counts saw 17, 18, or 19 species. Comment on the student's report.

21. Meal plan After surveying students at Humboldt Universität, a campus organization calculated that a 90% confidence interval for the mean cost of food for one term (of three in the Humboldt trimester calendar) is (€600, €900).

Now the organization is trying to write its report and is considering the following interpretations. Comment on each.

a) 90% of all students pay between €600 and €900 for food.
b) 90% of the sampled students paid between €600 and €900.
c) We're 90% sure that students in this sample averaged between €600 and €900 for food.
d) 90% of all samples of students will have average food costs between €600 and €900.
e) We're 90% sure that the average amount all students pay is between €600 and €900.

22. Snow Based on meteorological data for the past century, a local TV weather forecaster estimates that the region's average winter snowfall is 23", with a margin of error of ± 2 inches. Assuming he used a 95% confidence interval, how should viewers interpret this news? Comment on each of these statements:

a) During 95 of the past 100 winters, the region got between 21" and 25" of snow.
b) There's a 95% chance the region will get between 21" and 25" of snow this winter.
c) There will be between 21" and 25" of snow on the ground for 95% of the winter days.
d) Residents can be 95% sure that the area's average snowfall is between 21" and 25".
e) Residents can be 95% confident that the average snowfall during the past century was between 21" and 25" per winter.

CHAPTER EXERCISES

23. Pulse rates A medical researcher measured the pulse rates (beats per minute) of a sample of randomly selected adults and found the following Student's t-based confidence interval:

With 95.00% Confidence,
$70.887604 < \mu(Pulse) < 74.497011$

a) Explain carefully what the software output means.
b) What's the margin of error for this interval?
c) If the researcher had calculated a 99% confidence interval, would the margin of error be larger or smaller? Explain.

24. Crawling Data collected by child development scientists produced this confidence interval for the average age (in weeks) at which babies begin to crawl:

t-Interval for μ (95.00% Confidence): $30.65 < \mu(\text{age}) < 32.89$

a) Explain carefully what the software output means.
b) What is the margin of error for this interval?
c) If the researcher had calculated a 90% confidence interval, would the margin of error be larger or smaller? Explain.

25. CEO compensation A sample of 20 CEOs from the Forbes 500 shows total annual compensations ranging from a minimum of \$0.1 to \$62.24 million. The average for these 20 CEOs is \$7.946 million. Here's a histogram:

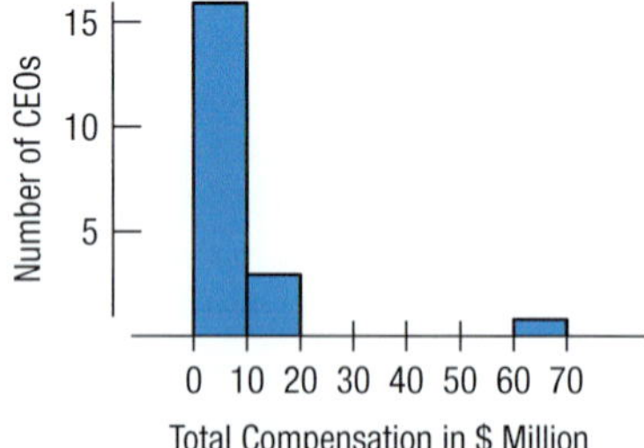

Based on these data, a computer program found that a 95% confidence interval for the mean annual compensation of all Forbes 500 CEOs is (1.69, 14.20) \$ million. Why should you be hesitant to trust this confidence interval?

26. Credit card charges A credit card company takes a random sample of 100 cardholders to see how much they charged on their card last month. Here's a histogram.

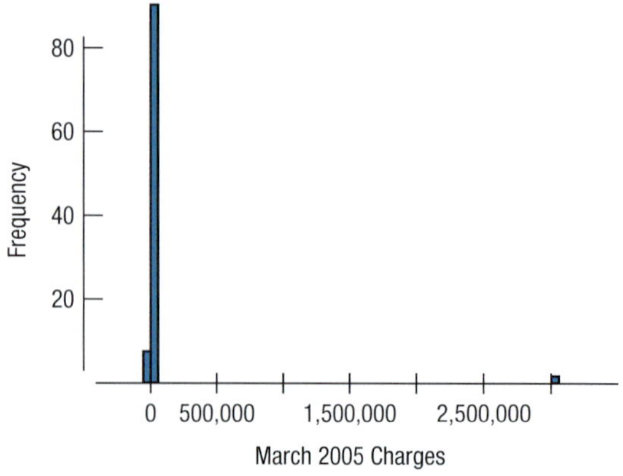

A computer program found that the resulting 95% confidence interval for the mean amount spent in March 2011 is (−\$28,366.84, \$90,691.49). Explain why the analysts didn't find the confidence interval useful, and explain what went wrong.

27. Cholesterol In the Mona Lisa study about cardiovascular risk factors in the French population (Archives of Cardiovascular Disease 2013 - https://www.sciencedirect.com/science/article/pii/S1875213612003105), HDL cholesterol of 4609 French adults averaged 1.48 mmol/L with a standard deviation of 0.36 mmol/L.

a) Can you apply the Central Limit Theorem to describe the distribution of the cholesterol measurements? Why or why not?
b) Can you apply the Central Limit Theorem to describe the sampling distribution model for the sample mean of French adults? Why or why not?
c) Sketch and clearly label the sampling model of the mean cholesterol levels of samples of size 4609 based on the 68-95-99.7 Rule.

28. Systolic Blood Pressure In the Mona Lisa study of cardiovascular risk factors in the French population (Archives of Cardiovascular Disease 2013 - https://www.sciencedirect.com/science/article/pii/S1875213612003105), systolic blood pressure of 4609 French adults averaged 134.4 mmHg with a standard deviation of 20.6 mmHg.

a) Can you apply the Central Limit Theorem to describe the distribution of systolic blood pressure measurements? Why or why not?
b) Can you apply the Central Limit Theorem to describe the sampling distribution model for the sample mean of French adults? Why or why not?
c) Sketch and clearly label the sampling model of the mean systolic blood pressure levels of samples of size 4609 based on the 68-95-99.7 Rule.

29. Normal temperature The researcher described in Exercise 23 also measured the body temperatures of that randomly selected group of adults. Here are summaries of the data he collected. We wish to estimate the average (or "normal") temperature among the adult population.

Summary	Temperature
Count	52
Mean	98.285
Median	98.200
MidRange	98.600
StdDev	0.6824
Range	2.800
IntQRange	1.050

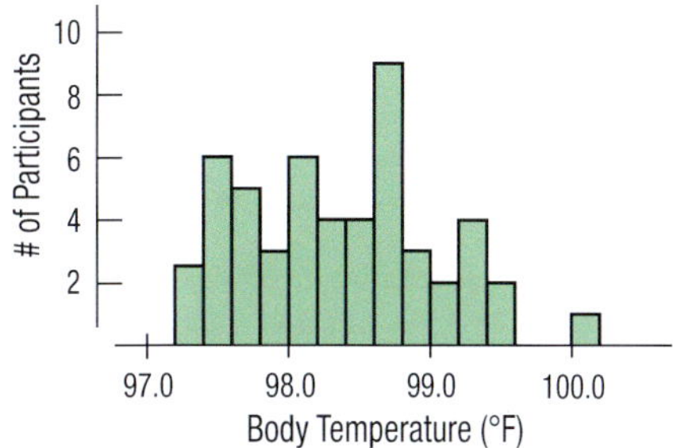

a) Check the conditions for creating a t-interval.
b) Find a 98% confidence interval for mean body temperature.
c) Explain the meaning of that interval.
d) Explain what "98% confidence" means in this context.
e) 98.6°F is commonly assumed to be "normal." Do these data suggest otherwise? Explain.

30. Parking Hoping to lure more shoppers downtown, a city builds a new public parking garage in the central business district. The city plans to pay for the structure through parking fees. During a two-month period (44 weekdays), daily fees collected averaged \$126, with a standard deviation of \$15.

a) What assumptions must you make in order to use these statistics for inference?
b) Write a 90% confidence interval for the mean daily income this parking garage will generate.
c) Interpret this confidence interval in context.
d) Explain what "90% confidence" means in this context.
e) The consultant who advised the city on this project predicted that parking revenues would average \$130 per day. Based on your confidence interval, do you think the consultant was correct? Why?

T **31.** Normal temperature, part II Consider again the statistics about human body temperature in Exercise 29.

a) Would a 90% confidence interval be wider or narrower than the 98% confidence interval you calculated before? Explain. (Don't compute the new interval.)
b) What are the advantages and disadvantages of the 98% confidence interval?
c) If we conduct further research, this time using a sample of 500 adults, how would you expect the 98% confidence interval to change? Explain.

32. Parking II Suppose that, for budget planning purposes, the city in Exercise 30 needs a better estimate of the mean daily income from parking fees.

a) Someone suggests that the city use its data to create a 95% confidence interval instead of the 90% interval first created. How would this interval be better for the city? (You need not actually create the new interval.)
b) How would the 95% interval be worse for the planners?
c) How could they achieve an interval estimate that would better serve their planning needs?

33. Speed of light In 1882, Michelson measured the speed of light (usually denoted c as in Einstein's famous equation $E = mc^2$). His values are in km/sec and have 299,000 subtracted from them. He reported the results of 23 trials with a mean of 756.22 and a standard deviation of 107.12.

a) Find a 95% confidence interval for the true speed of light from these statistics.
b) State in words what this interval means. Keep in mind that the speed of light is a physical constant that, as far as we know, has a value that is true throughout the universe.
c) What assumptions must you make in order to use your method?

T **34.** Michelson After his first attempt to determine the speed of light (described in Exercise 33), Michelson conducted an "improved" experiment. In 1897, he reported results of 100 trials with a mean of 852.4 and a standard deviation of 79.0.

a) What is the standard error of the mean for these data?
b) Without computing it, how would you expect a 95% confidence interval for the second experiment to differ from the confidence interval for the first? Note at least three specific reasons why they might differ, and indicate the ways in which these differences would change the interval.
c) According to Stephen M. Stigler (*The Annals of Statistics* 5:4, 1075 [1977]), the true speed of light is 299,710.5 km/sec, corresponding to a value of 710.5 for Michelson's 1897 measurements. What does this indicate about Michelson's two experiments? Find a new confidence interval and explain using your confidence interval.

T **35.** Flights on time 2016 What are the chances your flight will leave on time? The Bureau of Transportation Statistics of the U.S. Department of Transportation publishes information about airline performance. We saw the data in a Just Checking exercise in Chapter 4. Here are a histogram and summary statistics for the percentage of flights departing on time each month from January 1994 through June 2016. (www.transtats.bts.gov/HomeDrillChart.asp)

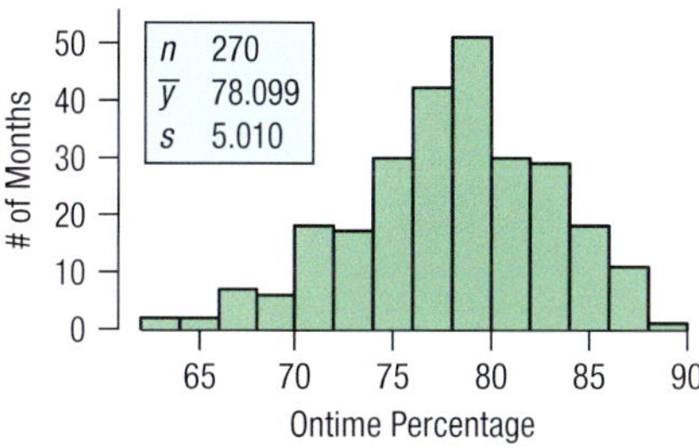

There is no evidence of a trend over time.

a) Check the assumptions and conditions for inference.
b) Find a 90% confidence interval for the true percentage of flights that depart on time.
c) Interpret this interval for a traveler planning to fly.

T **36.** Flights on time 2016 revisited Will your flight get you to your destination on time? The Bureau of Transportation Statistics reported the percentage of flights that were delayed each month from 1994 through June of 2016. Here's a histogram, along with some summary statistics:

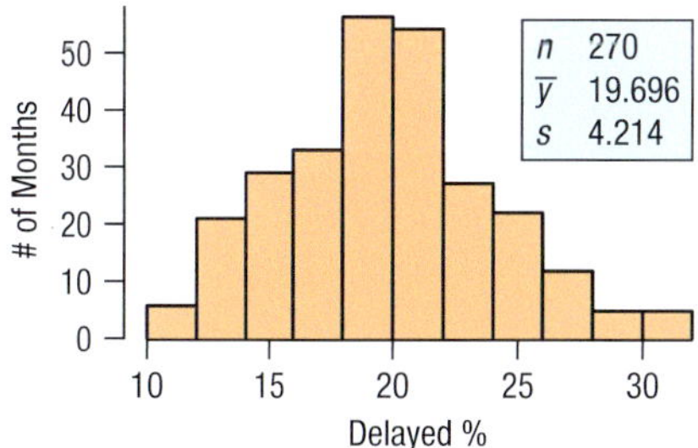

We can consider these data to be a representative sample of all months. There is no evidence of a time trend ($r = 0.07$).

a) Check the assumptions and conditions for inference about the mean.
b) Find a 99% confidence interval for the true percentage of flights that arrive late.
c) Interpret this interval for a traveler planning to fly.

T **37. Farmed salmon, second look** This chapter's Examples 17.1 and 17.2 looked at mirex contamination in farmed salmon. We first found a 95% confidence interval for the mean concentration to be 0.0834 to 0.0992 parts per million (ppm). The EPA sets a limit of 0.08 ppm to be considered safe. What does the confidence interval say about the safety of the farmed salmon?

38. Burgers A nutrition lab tested 50 fast-food chains burgers to see if their mean calorie content was less than the 650-calorie upper limit recommended as intake per lunch or dinner meal. A 95% confidence interval estimated the mean calories content for these burgers at 634.7 to 660.2 calories. Given this, what would you tell the lab about whether the burgers meet the recommendation?

39. Pizza A researcher investigates whether the mean cholesterol level among those who eat frozen pizza exceeds 220 mg/dL, the value considered to indicate a health risk. The confidence interval for the mean is (225.5, 240.8) mg/dL. Explain what this indicates for the health risk.

40. Footballs A sports federation sets specifications for footballs. For example, the mean weight should not be less than 400 grams at the beginning of a game when measured by an apparatus approved by the federation. Suppose a manufacturer introduces a new kind of ball and provides a sample for testing. Based on these data, the federation comes up with a 95% confidence interval for the mean weight from 387.2 to 419.6 grams.

What does this say about the weight of the new ball?

T **41. Fuel economy 2016 revisited** In Chapter 6, Exercise 41, we examined the average fuel economy of 35 2016 model vehicles. (Data in **Fuel economy 2016** with sample = "Yes")

a) Find and interpret a 95% confidence interval for the gas mileage of 2016 vehicles.
b) Do you think that this confidence interval captures the mean gas mileage for all 2016 vehicles?

T **42. Computer lab fees** The technology committee has stated that the average time spent by students per lab visit has increased, and the increase supports the need for increased lab fees. To substantiate this claim, the committee randomly samples 12 student lab visits and notes the amount of time spent using the computer. The times in minutes are as follows:

Time	
52	74
57	53
54	136
76	73
62	8
52	62

a) Plot the data. Are any of the observations outliers? Explain.
b) The previous mean amount of time spent using the lab computer was 55 minutes. Find a 95% confidence interval for the true mean. What do you conclude about the claim? If there are outliers, find intervals with and without the outliers present.

T **43. Waist size** A study measured the *Waist Size* of 250 men, finding a mean of 36.33 inches and a standard deviation of 4.02 inches. Here is a histogram of these measurements: (Data in **Bodyfat**)

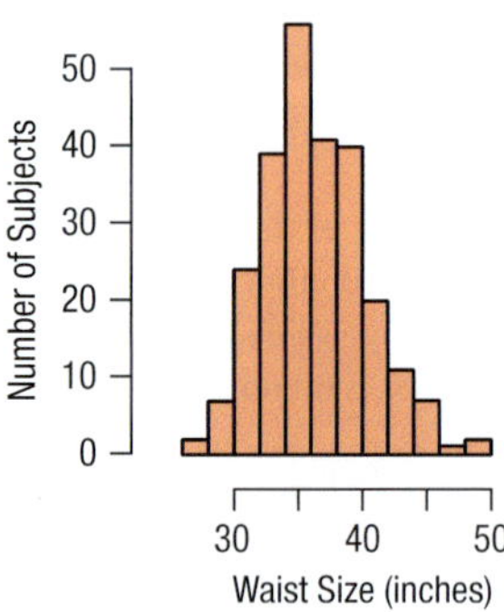

a) Describe the histogram of *Waist Size*.
b) To explore how the mean might vary from sample to sample, they simulated by drawing many samples of size 2, 5, 10, and 20, with replacement, from the 250 measurements. Here are histograms of the sample means for each simulation. Explain how these histograms demonstrate what the Central Limit Theorem says about the sampling distribution model for sample means.

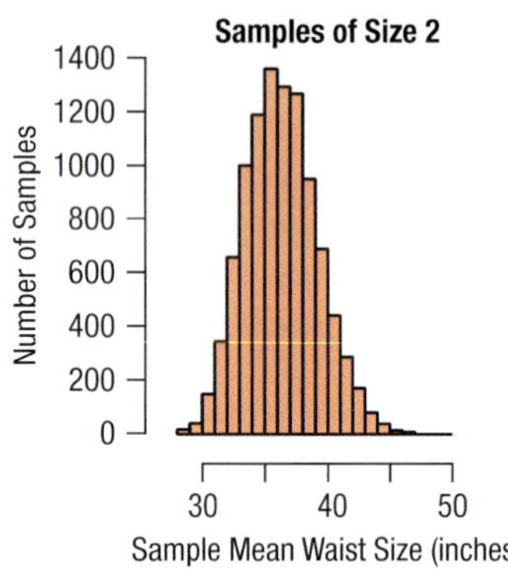

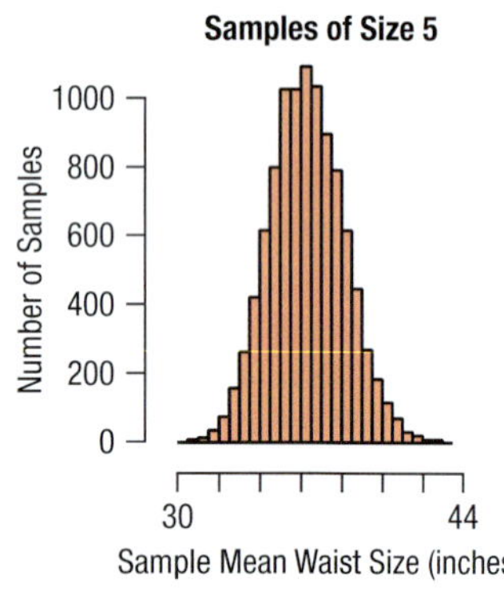

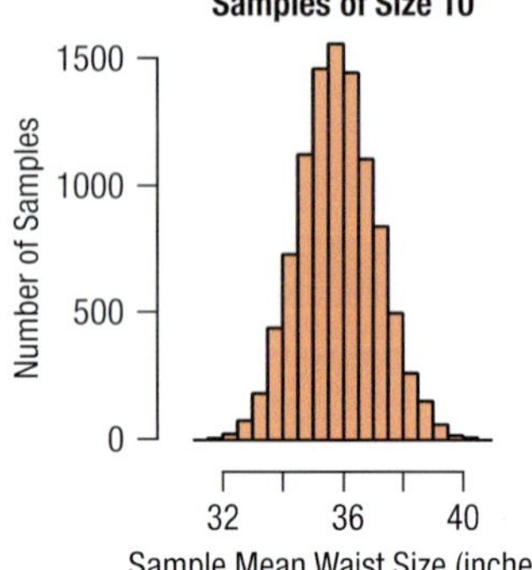

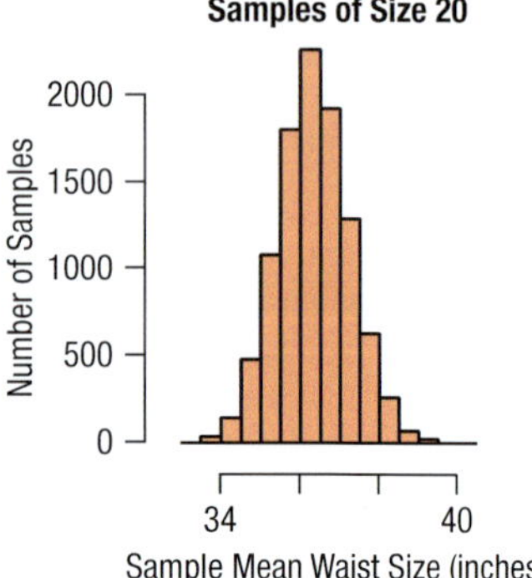

44. CEO compensation The total compensation of the chief executive officers (CEOs) of the 800 largest U.S. companies (the *Fortune* 800) averaged (in thousands of dollars) 10,307.31 with a standard deviation (also in \$1000) of 17,964.62. Here is a histogram of their annual compensations (in \$1000):

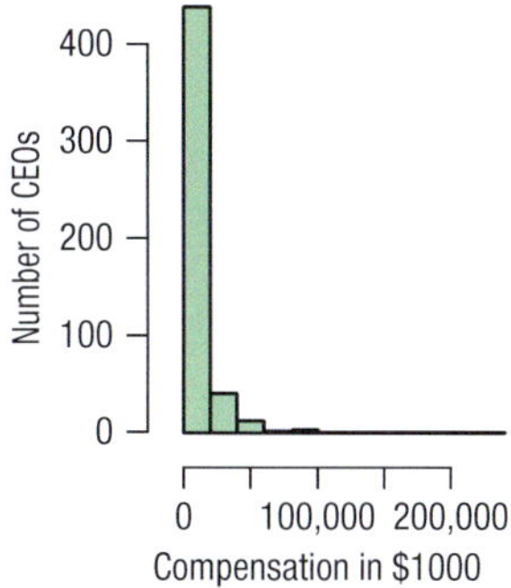

a) Describe the histogram of *Total Compensation*. A research organization simulated sample means by drawing samples of 30, 50, 100, and 200, with replacement, from the 800 CEOs. The histograms show the distributions of means for many samples of each size.

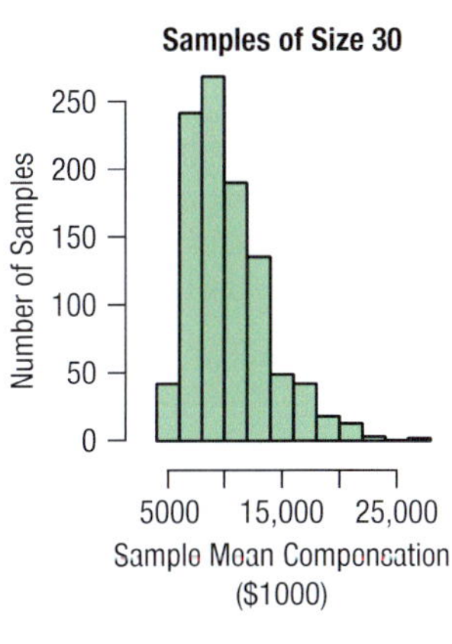

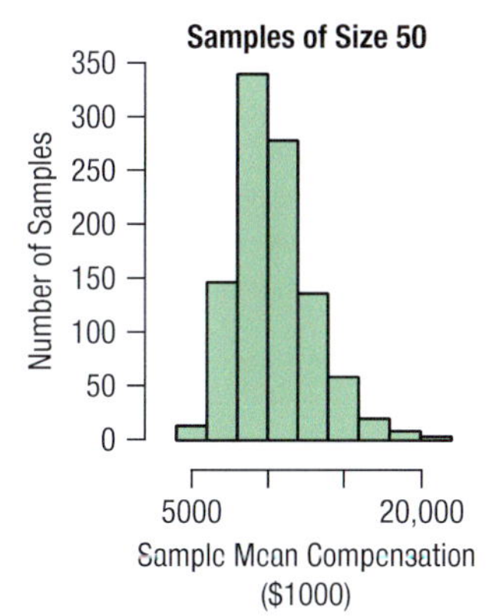

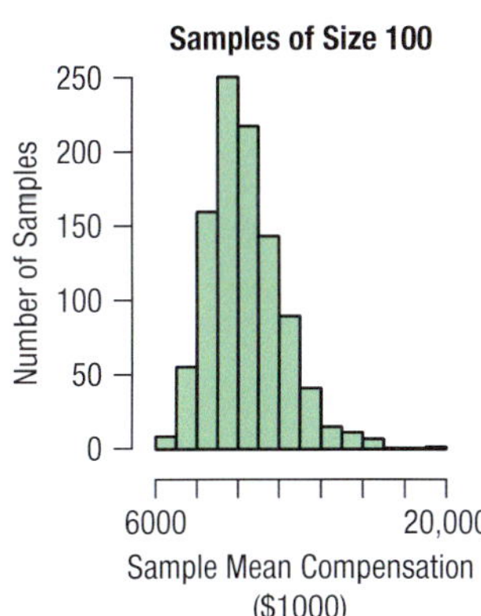

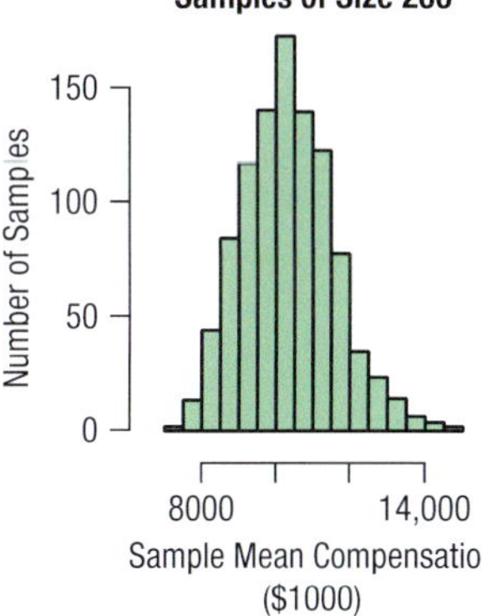

b) Explain how these histograms demonstrate what the Central Limit Theorem says about the sampling distribution model for sample means. Be sure to talk about shape, center, and spread.

c) Comment on the "rule of thumb" that "With a sample size of at least 30, the sampling distribution of the mean is Normal."

T **45. Waist size, revisited** Researchers measured the *Waist Sizes* of 250 men in a study on body fat. The true mean and standard deviation of the *Waist Sizes* for the 250 men are 36.33 inches and 4.019 inches, respectively. In Exercise 43, you looked at the histograms of simulations that drew samples of sizes 2, 5, 10, and 20 (with replacement). The summary statistics for these simulations were as follows:

n	Mean	st. dev.
2	36.314	2.855
5	36.314	1.805
10	36.341	1.276
20	36.339	0.895

a) According to the Central Limit Theorem, what should the theoretical mean and standard deviation be for each of these sample sizes?

b) How close are the theoretical values to what was observed in the simulation?

c) Looking at the histograms in Exercise 43, at what sample size would you be comfortable using the Normal model as an approximation for the sampling distribution?

d) What about the shape of the distribution of *Waist Size* explains your choice of sample size in part c?

T **46. CEOs, revisited** In Exercise 44, you looked at the annual compensation for 800 CEOs, for which the true mean and standard deviation were (in thousands of dollars) 10,307.31 and 17,964.62, respectively. A simulation drew samples of sizes 30, 50, 100, and 200 (with replacement) from the total annual compensations of the *Fortune* 800 CEOs. The summary statistics for these simulations were as follows:

n	Mean	st. dev.
30	10,251.73	3359.64
50	10,343.93	2483.84
100	10,329.94	1779.18
200	10,340.37	1230.79

a) According to the Central Limit Theorem, what should the theoretical mean and standard deviation be for each of these sample sizes?

b) How close are the theoretical values to what was observed from the simulation?

47. GPAs A college's data about the incoming freshmen indicate that the mean of their high school GPAs was 3.4, with a standard deviation of 0.35; the distribution was roughly mound-shaped and only slightly skewed. The students are randomly assigned to freshman writing seminars in groups of 25. What might the mean GPA of one of these seminar groups be? Describe the appropriate sampling distribution model—shape, center, and spread—with attention to assumptions and conditions. Make a sketch using the 68–95–99.7 Rule.

48. Home values Assessment records indicate that the value of homes in a small city is skewed right, with a mean of \$140,000 and standard deviation of \$60,000. To check the accuracy of the assessment data, officials plan to conduct a detailed appraisal of 100 homes selected at random. Using the 68–95–99.7 Rule, draw and label an appropriate sampling model for the mean value of the homes selected.

49. Lucky spot? A reporter working on a story about the New York lottery contacted one of the authors of this text, wanting help analyzing data to see if some ticket sales outlets were more likely to produce winners. His data for each of the 966 New York lottery outlets are graphed below; the scatterplot shows the ratio *TotalPaid/TotalSales* vs. *TotalSales* for the state's "instant winner" games for all of 2007.

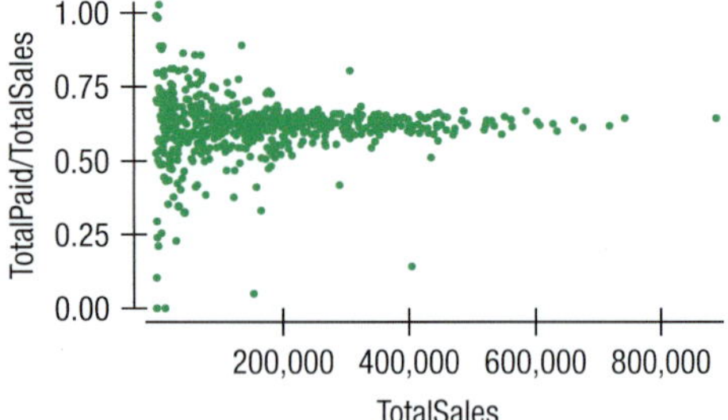

The reporter thinks that by identifying the outlets with the highest fraction of bets paid out, players might be able to increase their chances of winning. Typically—but not always—instant winners are paid immediately (instantly) at the store at which they are purchased. However, the fact that tickets may be scratched off and then cashed in at any outlet may account for some outlets paying out more than they take in. The few with very low payouts may be on interstate highways where players may purchase cards but then leave.

a) Explain why the plot has this funnel shape.
b) Explain why the reporter's idea wouldn't have worked anyway.

50. Safe cities ABIS Group Insurance Company identified the 10 safest and 10 least-safe Mexican cities from among the 250 largest cities in Mexico, based on the mean number of years drivers went between automobile accidents. The cities on both lists were all smaller than the 10 largest cities. Using facts about the sampling distribution model of the mean, explain why this is not statistically unusual.

51. Pregnancy Assume that the duration of human pregnancies can be described by a Normal model with mean 266 days and standard deviation 16 days.

a) What percentage of pregnancies should last between 270 and 280 days?
b) At least how many days should the longest 25% of all pregnancies last?
c) Suppose a certain obstetrician is currently providing prenatal care to 60 pregnant women. Let $\bar{y}$ represent the mean length of their pregnancies. According to the Central Limit Theorem, what's the distribution of this sample mean, $\bar{y}$? Specify the model, mean, and standard deviation.
d) What's the probability that the mean duration of these patients' pregnancies will be less than 260 days?

52. Rainfall Statistics from Cornell's Northeast Regional Climate Center indicate that Ithaca, New York, gets an average of 35.4" of rain each year, with a standard deviation of 4.2". Assume that a Normal model applies.

a) During what percentage of years does Ithaca get more than 40" of rain?
b) Less than how much rain falls in the driest 20% of all years?
c) A Cornell University student is in Ithaca for 4 years. Let $\bar{y}$ represent the mean amount of rain for those 4 years. Describe the sampling distribution model of this sample mean, $\bar{y}$.
d) What's the probability that those 4 years average less than 30" of rain?

53. Pregnant again The duration of human pregnancies may not actually follow the Normal model described in Exercise 51.

a) Explain why it may be somewhat skewed to the left.
b) If the correct model is in fact skewed, does that change your answers to parts a, b, and c of Exercise 51? Explain why or why not for each.

54. At work Some business analysts estimate that the length of time people work at a job has a mean of 6.2 years and a standard deviation of 4.5 years.

a) Explain why you suspect this distribution may be skewed to the right.
b) Explain why you could estimate the probability that 100 people selected at random had worked for their employers an average of 10 years or more, but you could not estimate the probability that an individual had done so.

T **55. Ruffles** Students investigating the packaging of potato chips purchased 6 bags of Lay's Ruffles marked with a net weight of 28.3 grams. They carefully weighed the contents of each bag, recording the following weights (in grams): 29.3, 28.2, 29.1, 28.7, 28.9, 28.5.

a) Do these data satisfy the assumptions for inference? Explain.
b) Find the mean and standard deviation of the weights.
c) Create a 95% confidence interval for the mean weight of such bags of chips.
d) Explain in context what your interval means.
e) Comment on the company's stated net weight of 28.3 grams.
*f) Why might finding a bootstrap confidence interval not be a good idea for these data?

T **56. Doritos** Some students checked 6 bags of Doritos marked with a net weight of 28.3 grams. They carefully weighed the contents of each bag, recording the following weights (in grams): 29.2, 28.5, 28.7, 28.9, 29.1, 29.5.

a) Do these data satisfy the assumptions for inference? Explain.
b) Find the mean and standard deviation of the weights.
c) Create a 95% confidence interval for the mean weight of such bags of chips.

d) Explain in context what your interval means.
e) Comment on the company's stated net weight of 28.3 grams.
*f) Why might finding a bootstrap confidence interval not be a good idea for these data?

T 57. Popcorn Yvon Hopps ran an experiment to determine optimum power and time settings for microwave popcorn. His goal was to find a combination of power and time that would deliver high-quality popcorn with less than 10% of the kernels left unpopped, on average. After experimenting with several bags, he determined that power 9 at 4 minutes was the best combination. To be sure that the method was successful, he popped 8 more bags of popcorn (selected at random) at this setting. All were of high quality, with the following percentages of uncooked popcorn: 7, 13.2, 10, 6, 7.8, 2.8, 2.2, 5.2. Does the 95% confidence interval suggest that he met his goal of an average of no more than 10% uncooked kernels? Explain.

T 58. Ski wax Bjork Larsen was trying to decide whether to use a new racing wax for cross-country skis. He decided that the wax would be worth the price if he could average less than 55 seconds on a course he knew well, so he planned to study the wax by racing on the course 8 times. His 8 race times were 56.3, 65.9, 50.5, 52.4, 46.5, 57.8, 52.2, and 43.2 seconds. Should he buy the wax? Explain by using a confidence interval

T 59. Chips Ahoy! In 1998, as an advertising campaign, the Nabisco Company announced a "1000 Chips Challenge," claiming that every 18-ounce bag of their Chips Ahoy! cookies contained at least 1000 chocolate chips. Dedicated statistics students at the Air Force Academy (no kidding) purchased some randomly selected bags of cookies and counted the chocolate chips. Some of their data are given below.

1219	1214	1087	1200	1419	1121	1325	1345
1244	1258	1356	1132	1191	1270	1295	1135

a) Check the assumptions and conditions for inference. Comment on any concerns you have.
b) Create a 95% confidence interval for the average number of chips in bags of Chips Ahoy! cookies.
c) What does this confidence interval say about Nabisco's claim?

T 60. Yogurt *Consumer Reports* tested 11 brands of vanilla yogurt and found these numbers of calories per serving:

130	160	150	120	120	110	170	160	110	130	90

a) Check the assumptions and conditions.
b) Create a 95% confidence interval for the average calorie content of vanilla yogurt.
c) A diet guide claims that you will get an average of 120 calories from a serving of vanilla yogurt. What does this confidence interval indicate about this claim?

T 61. Maze Psychology experiments sometimes involve testing the ability of rats to navigate mazes. The mazes are classified according to difficulty, as measured by the mean length of time it takes rats to find the food at the end. One researcher needs a maze that will take rats an average of about one minute to solve. He tests one maze on several rats, collecting the data shown.

Time (sec)	
38.4	57.6
46.2	55.5
62.5	49.5
38.0	40.9
62.8	44.3
33.9	93.8
50.4	47.9
35.0	69.2
52.8	46.2
60.1	56.3
55.1	

a) Find a 95% confidence interval for the mean time to find food.
b) Plot the data. Do you think the conditions are satisfied? Explain.
c) Eliminate the outlier and find the confidence interval again.
d) Do you think this maze meets the requirement that the maze takes at least a minute to complete on average? Explain.

62. Stopping distance 60 A tire manufacturer is considering a newly designed tread pattern for its all-weather tires. Tests have indicated that these tires will provide better gas mileage and longer tread life. The last remaining test is for braking effectiveness. The company hopes the tire will allow a car traveling at 60 mph to come to a complete stop within an average of 125 feet after the brakes are applied. The distances (in feet) for 10 stops on a test track were 129, 128, 130, 132, 135, 123, 102, 125, 128, and 130. Should the company adopt the new tread pattern? Using a confidence interval as evidence, what would you recommend to the company on whether the tire performs as they hope? Explain how you dealt with the outlier and why you made the recommendation you did.

T 63. Golf drives 2015 The Professional Golfers Association reported the average distance that 199 professional golfers drove the ball (in yd) during a week in 2015. Here is a histogram of those drives. Assume this is a representative sample of all professional golfers.

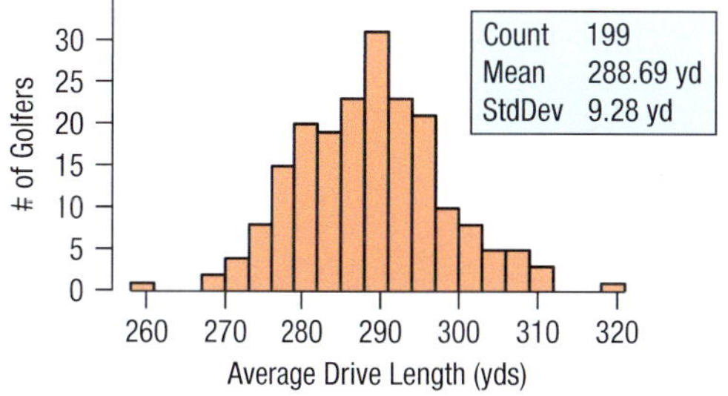

a) Find a 95% confidence interval for the mean of these drive distances.
b) Do you think that this interval captures the mean of golf drives for all golfers in 2015? Discuss.
c) These data are the *mean* drive distances for each golfer during a week. Should that be a concern in interpreting the interval? (*Hint*: Review the What Can Go Wrong warnings of Chapter 8. Chapter 8?! Yes, Chapter 8.)

T **64. Wind power** Should you generate electricity with your own personal wind turbine? That depends on whether you have enough wind on your site. To produce enough energy, your site should have an annual average wind speed above 8 miles per hour, according to the Wind Energy Association. One candidate site was monitored for a year, with wind speeds recorded every 6 hours. A total of 1114 readings of wind speed averaged 8.019 mph with a standard deviation of 3.813 mph. You've been asked to make a statistical report to help the landowner decide whether to place a wind turbine at this site.

a) Discuss the assumptions and conditions for constructing a confidence interval with these data. Here are some plots that may help you decide.

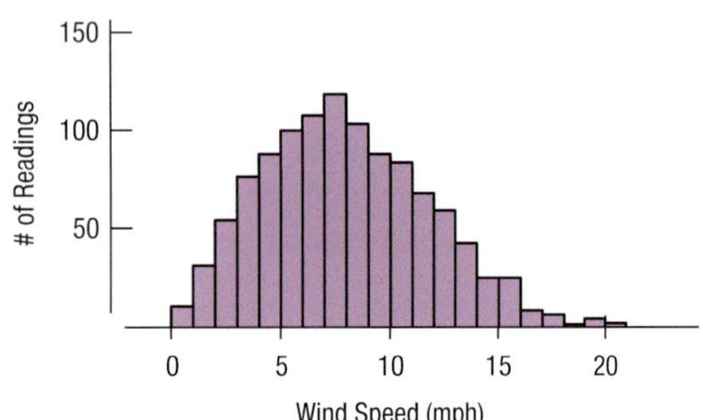

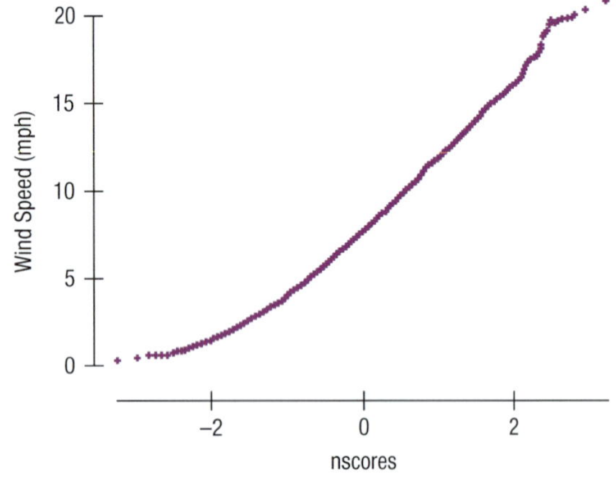

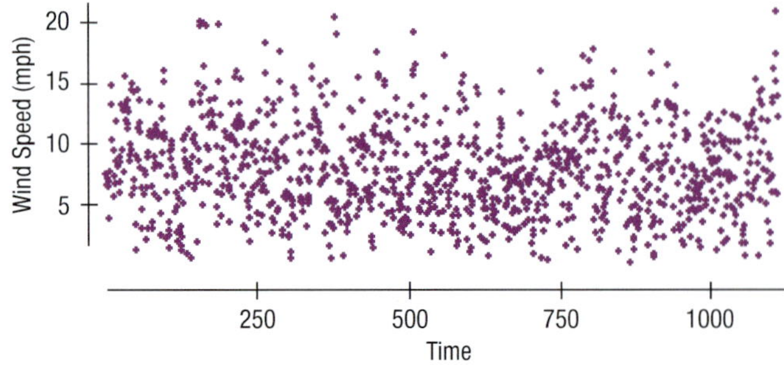

b) Based on your confidence interval, what would you tell the landowner about whether this site is suitable for a small wind turbine? Explain.

JUST CHECKING

Answers

1. Questions on the short form are answered by everyone in the population. This is a census, so means or proportions *are* the true population values. The long forms are given just to a sample of the population. When we estimate parameters from a sample, we use a confidence interval to take sample-to-sample variability into account.
2. They don't know the population standard deviation, so they must use the sample standard deviation as an estimate. The additional uncertainty is taken into account by t-models.
3. The margin of error for a confidence interval for a mean depends, in part, on the standard error,

$$SE(\bar{y}) = \frac{s}{\sqrt{n}}.$$

Since n is in the denominator, smaller sample sizes lead to larger *SE*s and correspondingly wider intervals. Long forms returned by one in every six or seven households in a less populous area will be a smaller sample.
4. The critical values for t with fewer degrees of freedom would be slightly larger. The $\sqrt{n}$ part of the standard error changes a lot, making the *SE* much larger. Both would increase the margin of error.
5. The smaller sample is one fourth as large, so the confidence interval would be roughly twice as wide.
6. We expect 95% of such intervals to cover the true value, so we would expect about 5 of the 100 intervals to miss.

18

Testing Hypotheses

WHERE ARE WE GOING?

Do people ages 18–24 really prefer Pepsi to Coke? Does this new allergy medication really reduce symptoms more than a placebo? There are times when we want to make a *decision.* To do that, we'll propose a model for the situation at hand and test a hypothesis about that model. The result will help us answer the real-world question.

"Half the money I spend on advertising is wasted; the trouble is I don't know which half."

—John Wanamaker (attributed)

Ingots are huge pieces of metal, sometimes weighing more than 20,000 pounds, made in a giant mold. They must be cast in one large piece for use in fabricating large structural parts for cars and planes. As the liquid metal cools, cracks can develop on the surface of the ingot, which can propagate into the zone required for the part, compromising its integrity. Airplane manufacturers insist that metal for their planes be defect-free, so the ingot must be made over if any cracking is detected.

Even though the metal from the cracked ingot is recycled, the cost runs into the tens of thousands of dollars to recast an ingot, not to mention the energy waste. About $2/3$ of all aluminum ingots produced in the United States use a process called the "direct chill" method designed to reduce recasting. Metal manufacturers would like to avoid cracking if at all possible. But the casting process is complicated and not everything is completely under control. It's estimated that about 5% of aluminum ingots need to be recast because of cracking. That rate depends on the size of the ingot cast. In one plant that specializes in very large (over 30,000 lb) ingots designed for the airplane industry, about 20% of the ingots have had some kind of surface crack. In an attempt to reduce the cracking proportion, the plant engineers and chemists recently tried out some changes in the casting process. Since then, 400 ingots have been cast and only 68 (17%) of them have cracked. Has the new method worked? Has the cracking rate really decreased, or was 17% just due to luck? Is the 17% cracking rate merely a result of natural sampling variability, or is this enough evidence to justify a change to the new method?

People want to make informed decisions like this all the time. Does the new website design increase our click-through rate? Has the "click-it or ticket" campaign increased compliance with seat belt laws? Did the Super Bowl ad we bought actually increase sales? To answer such questions so that we can make intelligent decisions, we test *hypotheses*.

18.1 Hypotheses

> *HYPOTHESIS*
>
> n.; pl. {Hypotheses}. A supposition; a proposition or principle which is supposed or taken for granted, in order to draw a conclusion or inference for proof of the point in question; something not proved, but assumed for the purpose of argument.
>
> —*Webster's Unabridged Dictionary, 1913*

> *NOTATION ALERT*
>
> Capital H is the standard letter for hypotheses. H_0 always labels the null hypothesis, and H_A labels the alternative hypothesis.

If the changes they made lowered the cracking rate from 20%, management will need to decide whether the costs of the new method warrant the changes. Managers are naturally cautious, and because humans are natural skeptics, they assume the new method makes no difference—but they hope the data can convince them otherwise. The starting hypothesis to be tested is called the **null hypothesis**—null because it assumes that nothing has changed. We denote it H_0. It specifies a parameter—here the proportion of cracked ingots—and a value—that the cracking rate is 20%. We usually write this in the form H_0: *parameter* = *hypothesized value*. So, for the ingots we would write H_0: $p = 0.20$.

The **alternative hypothesis**, which we denote H_A, is not a single value, but contains all the other values of the parameter. We can write H_A: $p \neq 0.20$.

What would convince you that the cracking rate had actually changed? If the rate dropped from 20% to 19.8%, would that convince you? After all, observed proportions do vary, so even if the changes had no effect, you'd expect some difference. What if it dropped to 1%? Would random fluctuation account for a change that big? That's the crucial question in a hypothesis test. As usual in statistics, when thinking about the size of a change, we *naturally* think of using its standard deviation to measure that change.[1] We ask how many standard deviations the observed value is from the hypothesized value, and we know how to find the standard deviation of a proportion:

$$SD(\hat{p}) = \sqrt{\frac{pq}{n}} = \sqrt{\frac{(0.20)(0.80)}{400}} = 0.02.$$

Why Is This a Standard Deviation and Not a Standard Error?

Remember that we reserve the term "standard error" for the *estimate* of the standard deviation of the sampling distribution. But we're not estimating here—we have a value of p from our null hypothesis model. To remind us that the parameter value comes from the null hypothesis, it is sometimes written as p_0 and the standard deviation as $SD(\hat{p}) = \sqrt{p_0 q_0 / n}$. That's different than when we found a confidence interval for p. In that case we couldn't assume that we knew its value, so we estimated the standard deviation from the sample value $\hat{p}$.

If the changes have no effect, then the true cracking rate is still 0.20, and for samples of 400, the standard deviation of $\hat{p}$ is 0.02. We know from the sampling distribution of $\hat{p}$ that in 95% of samples of this size, the engineers will see a cracking rate within 0.04 of 0.20 just by chance. In other words, they expect to see between 64 and 96 cracked ingots (see Figure 18.1). The engineers saw 68 cracked ingots out of 400 (a rate of 0.17). Given an assumed rate of 0.20, is that a surprising number? Would you say that the cracking rate has changed?

Figure 18.1
A simulation of 10,000 samples of 400 ingots with a cracking rate of 20% shows how we should expect the number of cracked ingots to vary.

[1] It's Chapter 18. Did you?

Now that we have the Central Limit Theorem we don't need to rely on simulation. The CLT tells us that we can use the Normal model to find the probability instead. The engineers observed a cracking rate of 0.17—a difference of 0.03 from the standard (null hypothesis) value of 0.20. Using the fact that the standard deviation is 0.02, we can find the area in the two tails of the Normal model that lie more than 0.03 away from 0.20 (see Figure 18.2). That tells us how rare our observed rate is.

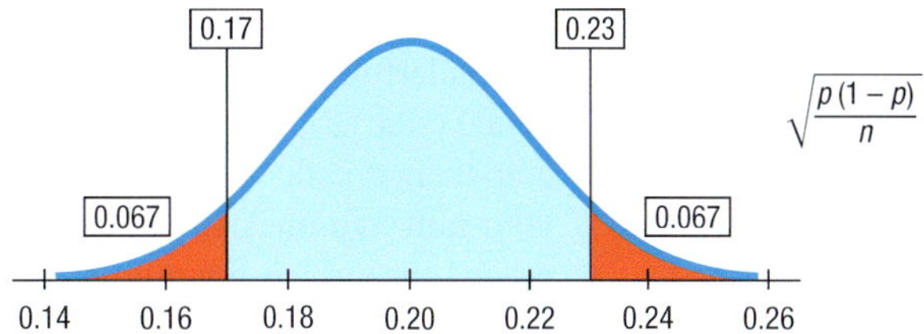

Figure 18.2
How likely is it to have a sample proportion less than 17% or more than 23% when the true proportion is 20%? Each red area is 6.7% of the total area, and they sum to 0.13. That's not very unlikely.

You know how to find that probability using either technology, such as the app at www.astools.datadesk.com, or Table Z at the back of the book. (See Chapter 5.) As you can see in Figure 18.2, the probability comes to about 0.13. In other words, a sample of 400 ingots with a cracking rate this far from 0.20 would happen about 13% of the time just by chance. That doesn't seem very unusual, so the observed proportion of 0.17, even though it's lower, doesn't provide evidence that the new method changed the cracking rate.

EXAMPLE 18.1 Framing Hypotheses

Summit Projects was a full-service interactive agency, based in Hood River, OR, that offered companies a variety of website services. One of Summit's clients, SmartWool®, produces and sells wool apparel, including the famous SmartWool socks. After Summit redesigned SmartWool's apparel website, analysts at SmartWool wondered whether traffic has changed since the new website went live. In particular, an analyst might want to know if the proportion of visits resulting in a sale has increased, decreased, or stayed pretty much the same since the new site went online. She might also wonder if the average sale amount has changed.

QUESTION: If the old site's proportion was 15%, what are appropriate null and alternative hypotheses for the proportion?

ANSWER: Let p = proportion of visits that result in a sale. Then

$$H_0: p = 0.15 \text{ vs. } H_A: p \neq 0.15.$$

A Trial as a Hypothesis Test

Management would be really interested to learn that the engineers' changes changed the cracking rate. But to test it, they assumed that the rate had *not* changed. Does this reasoning seem backward? That could be because we usually prefer to think about getting things right rather than getting them wrong. But, you've seen this reasoning before in a different context. This is the logic of jury trials.

Let's suppose a defendant has been accused of robbery. In British common law and those systems derived from it (including U.S. law), the null hypothesis is that the defendant is innocent. Instructions to juries are quite explicit about this.

How is the null hypothesis tested? The prosecution first collects evidence. ("If the defendant were innocent, wouldn't it be remarkable that the police found him at the scene of the crime with a bag full of money in his hand, a mask on his face, and a getaway car parked outside?") For us, the data are the evidence.

The next step is to judge the evidence. Evaluating the evidence is the responsibility of the jury in a trial, but it falls on your shoulders in hypothesis testing. The jury considers the evidence in light of the *presumption* of innocence and judges whether the evidence against the defendant would be plausible *if the defendant were in fact innocent*.

Like the jury, you ask, "Could these data plausibly have happened by chance if the null hypothesis were true?" If they are very unlikely to have occurred, then the evidence raises a reasonable doubt about the null hypothesis.

Ultimately, you must make a decision. The standard of "beyond a reasonable doubt" is wonderfully ambiguous because it leaves the jury to decide the degree to which the evidence contradicts the hypothesis of innocence. Juries don't explicitly use probability to help them decide whether to reject that hypothesis. But when you ask the same question of your null hypothesis, you have the advantage of being able to quantify exactly how surprising the evidence would be if the null hypothesis were true.

How unlikely is unlikely? Some people set rigid standards, like 1 time out of $20(0.05)$ or 1 time out of $100(0.01)$.[2] But if *you* have to make the decision, you must judge for yourself in each situation whether the probability of observing your data is small enough to constitute "reasonable doubt."

18.2 P-Values

NOTATION ALERT

We have many P's to keep straight. We use an uppercase P for probabilities, as in $P(\mathbf{A})$, and for the special probability we care about in hypothesis testing, the P-value.

We use lowercase p to denote our model's underlying proportion parameter and $\hat{p}$ to denote our observed proportion statistic.

To test a hypothesis we must answer the question "Are the data surprising, given the null hypothesis?" So we need *probability*—specifically, the probability of seeing data like these (or something even less likely) *given* the null hypothesis. In the ingots example, this came to 0.13. This probability is the value on which we base our decision, so statisticians give it a special name: the **P-value**.[3] Usually, you'll use a sampling distribution model or a simulation to find P-values. Either way, the computer will do the heavy lifting.

When a P-value is very low, there are only two possibilities. Either the null hypothesis is correct and we've just seen something remarkable, or the null hypothesis is wrong (and the reason for a low P-value is that the model was wrong). Now we have a choice. Should we decide that a rare event has happened to us, or should we trust that the data were not unusual and that our null model was wrong? We don't believe in rare events,[4] so a low enough P-value leads us to reject the null hypothesis. There is no hard and fast rule about how low the P-value has to be. In fact, it depends on the consequences of our decision and the size of the change we've observed.

> "If the People fail to satisfy their burden of proof, you must find the defendant not guilty."
>
> —NY state jury instructions

When the P-value is high, we haven't seen anything unlikely or surprising at all. The data are consistent with the model from the null hypothesis, and we have no reason to reject it. Does that mean we've proved it? No. Many other models could be consistent with the data we've seen, so *we haven't proven anything*. The most we can say is that the null model doesn't appear to be false. Formally, we "fail to reject" the null hypothesis. That's a pretty weak conclusion, but it's all we can do with a high P-value.

BEYOND A REASONABLE DOUBT

We ask whether the data were unlikely beyond a reasonable doubt. We've just calculated that probability. The probability that the observed statistic value (or an even more extreme value) could occur if the null model were true—in this case, 0.13—is the P-value.

EXAMPLE 18.2 Using P-Values to Make Decisions

QUESTION: The SmartWool analyst in Example 18.1 collects a representative sample of visits since the new website has gone online and finds that the P-value for the test of whether the proportion of visits resulting in a sale has changed is 0.0015 and the P-value for the test of whether the mean sale amount has changed is 0.3740. What conclusions can she draw?

---

[2] See Section 18.6 for warnings about bright line decisions like this.

[3] You'd think if it were that special it would have a better name, but "P-value" is about as creative as statisticians get.

[4] Or at least we think that they don't happen to us.

ANSWER: Even though the low P-value for the proportion of visits resulting in sales suggests that the proportion has changed, there is little evidence that the amount of the mean sale has changed. To make a decision, the analyst would need to do further analysis.

> "The null hypothesis is never proved or established, but is possibly disproved, in the course of experimentation. Every experiment may be said to exist only in order to give the facts a chance of disproving the null hypothesis."
>
> —Sir Ronald Fisher, The Design of Experiments

What to Do with an "Innocent" Defendant

Back to the jury trial. The jury assumes the defendant is innocent, but when the evidence is not strong enough to reject that hypothesis, they say "not guilty." They do not claim that the null hypothesis is true and say that the defendant is innocent. All they say is that they have not seen sufficient evidence to convict. The defendant may, in fact, be innocent, but the jury has no way to be sure.

In the same way, when the P-value is large, the most we can do is to "fail to reject" our null hypothesis. We never declare the null hypothesis to be true (or "accept" the null), because we simply do not know whether it's true or not. (But, unlike a jury trial, there is no "double indemnity" in science. More data may become available in the future.)

Don't "Accept" the Null Hypothesis

Think about the null hypothesis H_0: All swans are white. Does collecting a sample of 100 white swans prove the null hypothesis? The data are *consistent* with this hypothesis and seem to lend support to it, but they don't *prove* it. In fact, all we can do is disprove the null hypothesis—for example, by finding just one non-white swan.

JUST CHECKING

1. A research team wants to know if aspirin helps to thin blood. The null hypothesis says that it doesn't. They test 12 patients, observe the proportion with thinner blood, and get a P-value of 0.32. They proclaim that aspirin doesn't work. What would you say?
2. An allergy drug has been tested and found to give relief to 75% of the patients in a large clinical trial. Now the scientists want to see if the new, improved version works even better. What would the null hypothesis be?
3. The new drug is tested and the P-value is 0.0001. What would you conclude about the new drug?

Alternative Alternatives

Tests on the ingot data can be viewed in two different ways. We know the old cracking rate is 20%, so the null hypothesis is

$$H_0: p = 0.20.$$

But we have a choice of alternative hypotheses. A metallurgist working for the company might be interested in *any* change in the cracking rate due to the new process. Even if the rate got worse, she might learn something useful from it. In that case, she's interested in possible changes on both sides of the null hypothesis. So she would write her alternative hypothesis as

$$H_A: p \neq 0.20.$$

An alternative hypothesis such as this is known as a **two-sided alternative** because we are equally interested in deviations on either side of the null hypothesis value. For two-sided alternatives, the P-value is the probability of deviating in *either* direction from the null hypothesis value.

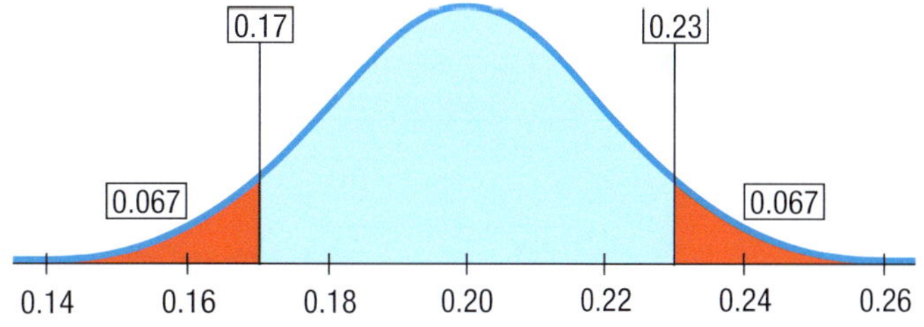

But management wants to know only if the cracking rate has *decreased* to below 20%. Knowing how to *increase* the cracking rate may not interest them. To make that explicit, they could write their alternative hypothesis as

$$H_A: p < 0.20.$$

An alternative hypothesis that focuses on deviations from the null hypothesis value in only one direction is called a **one-sided alternative**.[5]

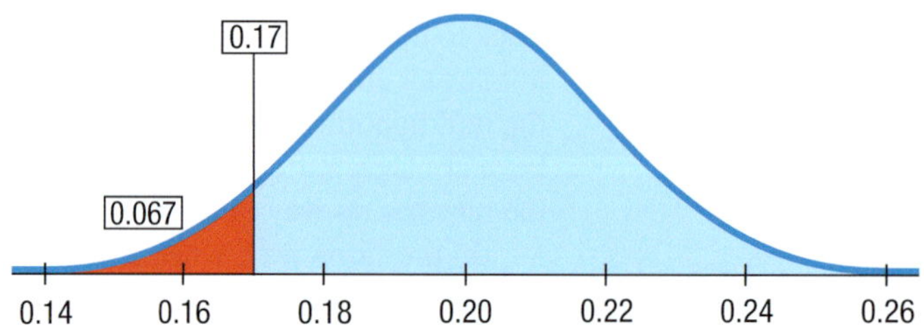

For a hypothesis test with a one-sided alternative, the P-value is the probability of deviating *only in the direction of the alternative* away from the null hypothesis value. For the same data, the one-sided P-value is half the two-sided P-value. So, a one-sided test will reject the null hypothesis more often. This is good and bad. It's great that it rejects the null hypothesis more often when it's false, but it also rejects it more often when it's true. We discuss this issue in detail in the next chapter. A two-sided test is always more conservative. Because its P-value is twice as big as either one-sided alternative, it will reject the null hypothesis less often. Unless you can justify the choice of a one-sided test, it's usually better to assume the alternative is two-sided. One advantage of a two-sided test is that the results are easily matched to the confidence interval. See Section 18.5.

18.3 The Reasoning of Hypothesis Testing

As the subsequent chapters will show, the reasoning of hypothesis testing is essentially the same no matter what we test. For example, we'll test proportions, means, differences between groups, and even regression coefficients, but the logic stays the same. Here's a path to follow.

1. Hypotheses

First, we state the null hypothesis. That's usually the skeptical claim that nothing's different. Are we considering a (New! Improved!) possibly better method? The null hypothesis says, "Oh yeah? Convince me!" To convert a skeptic, we must pile up enough evidence against the null hypothesis that we can reasonably reject it.

In statistical hypothesis testing, hypotheses are almost always about model parameters. To assess how unlikely our data may be, we need a null model. The null hypothesis specifies a particular parameter value to use in our model. In the usual shorthand, we write H_0: *parameter* = *hypothesized value*. The alternative hypothesis, H_A, contains the values of the parameter we consider plausible when we reject the null.

How to Say It

You might think that the 0 in H_0 should be pronounced as "zero" or "0," but it's actually pronounced "naught" as in "all is for naught." It is a label for "nothing new," not the first in a sequence of hypotheses.

---

[5] These are also called a **two- and one-tailed alternatives**, because the probabilities we care about are found in the tails of the sampling distribution.

EXAMPLE 18.3

Writing Hypotheses

A large city's Department of Motor Vehicles claimed that 80% of candidates pass driving tests, but a newspaper reporter's survey of 90 randomly selected local teens who had taken the test found only 61 who passed.

QUESTION: Does this finding suggest that the passing rate for teenagers is lower than the DMV reported? Write appropriate hypotheses.

ANSWER: I'll assume that the passing rate for teenagers is the same as the DMV's overall rate of 80%, unless there's strong evidence that it's lower.

$$H_0: p = 0.80$$

$$H_A: p < 0.80$$

2. Model

To plan a statistical hypothesis test, specify the *model* you will use to test the null hypothesis and the parameter of interest. Of course, all models require assumptions, so you will need to state them and check any corresponding conditions.

Your Model step should end with a statement such as

Because the conditions are satisfied, I can model the sampling distribution of the proportion with a Normal model.

Watch out, though. Your Model step could end with

Because the conditions are not satisfied, I can't proceed with the test.

If that's the case, stop and reconsider.

Because the test about a single proportion is based on the Normal model, it is called a **one-proportion *z*-test**.[6] Each test in the text has a name that you should include in your report. Some tests will be about more than one sample, some will involve statistics other than proportions or means, and some will use models other than the Normal. For each test, be sure to check the appropriate assumptions and conditions. This will usually require a plot of your data.

When the Conditions Fail . . .

You might proceed with caution, explicitly stating your concerns. Or you may need to do the analysis with and without an outlier, or on different subgroups, or after re-expressing the response variable. Or you may not be able to proceed at all.

One-Proportion *z*-Test

The conditions for the one-proportion *z*-test are the same as for the one-proportion *z*-interval. We test the hypothesis $H_0: p = p_0$ using the statistic $z = \frac{(\hat{p} - p_0)}{SD(\hat{p})}$. We use the hypothesized proportion to find the standard deviation, $SD(\hat{p}) = \sqrt{\frac{p_0 q_0}{n}}$.

When the conditions are met and the null hypothesis is true, this statistic follows the standard Normal model, so we can use that model to obtain a P-value.

[6] It's also called the "one-sample test for a proportion."

EXAMPLE 18.4 Checking the Conditions

RECAP: A large city's DMV claimed that 80% of candidates pass driving tests. A reporter has results from a survey of 90 randomly selected local teens who had taken the test.

QUESTION: Are the conditions for inference satisfied?

✓ **Randomization Condition:** The 90 teens surveyed were a random sample of local teenage driving candidates.

✓ **10% Condition:** 90 is fewer than 10% of the teenagers who take driving tests in a large city.

✓ **Success/Failure Condition:** We expect $np_0 = 90(0.80) = 72$ successes and $nq_0 = 90(0.20) = 18$ failures. Both are at least 10.

ANSWER: The conditions are satisfied, so it's okay to use a Normal model and perform a one-proportion *z*-test.

3. Mechanics

The "mechanics" are the calculation of our test statistic and P-value. Different tests will have different formulas, different test statistics, and different sampling distributions. Usually, the mechanics are handled by a statistics program.

Conditional Probability

Did you notice that a P-value is a conditional probability? It's the probability that the observed results could have happened *if (or given that) the null hypothesis were true.*

EXAMPLE 18.5 Finding a P-Value

RECAP: A large city's DMV claimed that 80% of candidates pass driving tests, but a survey of 90 randomly selected local teens who had taken the test found only 61 who passed.

QUESTION: What's the P-value for the one-proportion *z*-test?

ANSWER: I have $n = 90$, $x = 61$, and a hypothesized $p = 0.80$.

$$\hat{p} = \frac{61}{90} \approx 0.678$$

$$SD(\hat{p}) = \sqrt{\frac{p_0 q_0}{n}} = \sqrt{\frac{(0.8)(0.2)}{90}} \approx 0.042$$

$$z = \frac{\hat{p} - p_0}{SD(\hat{p})} = \frac{0.678 - 0.800}{0.042} \approx -2.90$$

$$\text{P-value} = P(z < -2.90) = 0.002$$

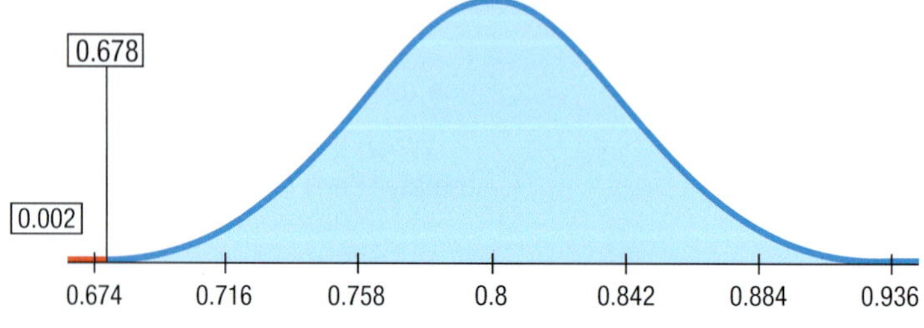

How Much Does It Cost?

Formal tests of a null hypothesis base the decision of whether to reject the null hypothesis solely on the size of the P-value. But in real life, we want to evaluate the costs of our decisions as well. How much would you be willing to pay for a faster computer? Shouldn't your decision depend on how much faster? And on how much more it costs? Costs are not just monetary either. Would you use the same standard of proof for testing the safety of an airplane as for the speed of your new computer?

"... They make things admirably plain, But one hard question will remain: If one hypothesis you lose, Another in its place you choose ..."

—James Russell Lowell, Credidimus Jovem Regnare

4. Conclusion

The conclusion in a hypothesis test is always a statement about the null hypothesis. The conclusion must state either that we reject or that we fail to reject the null hypothesis. And, as always, the conclusion should be stated in context.

Your conclusion about the null hypothesis should never be the end of a testing procedure. Usually there are actions to take or policies to change. In our ingot example, management must decide whether to continue the changes proposed by the engineers. The decision always includes the practical consideration of whether the new method is worth the cost. Suppose management decides to reject the null hypothesis of 20% cracking in favor of the alternative that the percentage has changed. They must still evaluate how much the new method changes the cracking rate and how much it would cost to accomplish that change. The *size of the effect* is always a concern when we test hypotheses. A good way to look at the **effect size** is to examine a confidence interval.

EXAMPLE 18.6 Stating the Conclusion

RECAP: A large city's DMV claimed that 80% of candidates pass driving tests. Data from a reporter's survey of randomly selected local teens who had taken the test produced a P-value of 0.002.

QUESTION: What can the reporter conclude? And how might the reporter explain what the P-value means for the newspaper story?

ANSWER: Because the P-value of 0.002 is very small, I reject the null hypothesis. These survey data provide strong evidence that the passing rate for teenagers taking the driving test is lower than 80%.

If the passing rate for teenage driving candidates were actually 80%, we'd expect to see success rates this low in only about 1 in 500 (0.2%) samples of this size. This seems quite unlikely, casting doubt that the DMV's stated success rate applies to teens.

STEP-BY-STEP EXAMPLE Testing a Hypothesis

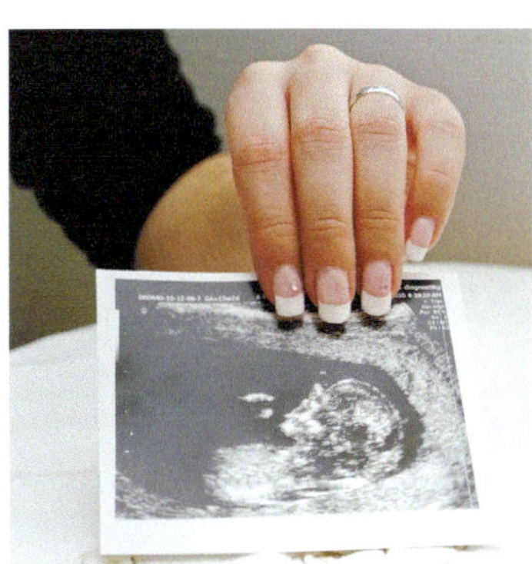

Advances in medical care such as prenatal ultrasound examination now make it possible to determine a child's sex early in a pregnancy. There is a fear that in some cultures some parents may use this technology to select the sex of their children. A study from Punjab, India,[7] reports that, in one hospital, 56.9% of the 550 live births that year were boys. It's a medical fact that male babies are slightly more common than female babies. The study's authors report a baseline for this region of 51.7% male live births.

QUESTION: Is there evidence that the proportion of male births is different for this hospital?

[7] E. E. Booth, M. Verma, and R. S. Beri, "Fetal Sex Determination in Infants in Punjab, India: Correlations and Implications," *BMJ* 309: 1259–1261.

THINK **PLAN** State what we want to know.

Define the variables and discuss the W's.

I want to know whether the proportion of male births in this hospital is different from the established baseline of 51.7%. The data are the recorded sexes of the 550 live births from a hospital in Punjab, India, in 1993, collected for a study on fetal sex determination. The parameter of interest, p, is the proportion of male births:

HYPOTHESES The null hypothesis makes the claim of no difference from the baseline.

Before seeing the data, we were interested in any change in male births, so the alternative hypothesis is two-sided.

$$H_0: p = 0.517$$

$$H_A: p \neq 0.517$$

MODEL Think about the assumptions and check the appropriate conditions.

For testing proportions, the conditions are the same ones we had for making confidence intervals, except that we check the **Success/Failure Condition** with the *hypothesized* proportions rather than with the *observed* proportions.

✓ **Independence Assumption**: There is no reason to think that the sex of one baby can affect the sex of other babies, so births can reasonably be assumed to be independent with regard to the sex of the child.

✓ **Randomization Condition**: The 550 live births are not a random sample, so I must be cautious about any general conclusions. I hope that this is a representative year, and I think that the births at this hospital may be typical of this area of India.

✓ **10% Condition**: I would like to be able to make statements about births at similar hospitals in India. These 550 births are fewer than 10% of all of those births.

✓ **Success/Failure Condition**: Both $np_0 = 550(0.517) = 284.35$ and $nq_0 = 550(0.483) = 265.65$ are greater than 10; I expect the births of at least 10 boys and at least 10 girls, so the sample is large enough.

Specify the sampling distribution model.

Tell what test you plan to use.

The conditions are satisfied, so I can use a Normal model and perform a **one-proportion z-test**.

SHOW **MECHANICS** The null model gives us the mean, and (because we are working with proportions) the mean gives us the standard deviation.

The null model is a Normal distribution with a mean of 0.517 and a standard deviation of

$$SD(\hat{p}) = \sqrt{\frac{p_0 q_0}{n}} = \sqrt{\frac{(0.517)(1 - 0.517)}{550}}$$

$$= 0.0213.$$

We find the z-score for the observed proportion to find out how many standard deviations it is from the hypothesized proportion.

The observed proportion, $\hat{p}$, is 0.569, so

$$z = \frac{\hat{p} - p_0}{SD(\hat{p})} = \frac{0.569 - 0.517}{0.0213} = 2.44.$$

The sample proportion lies 2.44 standard deviations above the mean.

Make a picture. Sketch a Normal model centered at $p_0 = 0.517$. Shade the region to the right of the observed proportion, and because this is a two-tailed test, also shade the corresponding region in the other tail.

From the z-score, we can find the P-value, which tells us the probability of observing a value that extreme (or more). Use technology or a table (see below).

Because this is a two-tailed test, the P-value is the probability of observing an outcome more than 2.44 standard deviations from the mean of a Normal model *in either direction.* We must therefore *double* the probability we find in the upper tail.

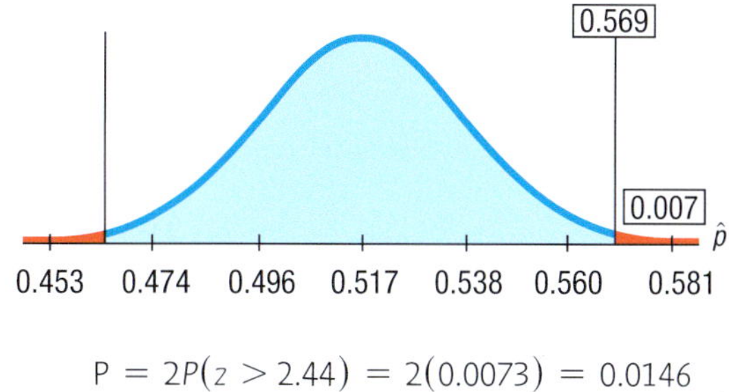

$$P = 2P(z > 2.44) = 2(0.0073) = 0.0146$$

TELL **CONCLUSION** State your conclusion in context.

This P-value is roughly 1 time in 70. That's clearly significant, but don't jump to other conclusions. We can't be sure how this deviation came about. For instance, we don't know whether this hospital is typical, or whether the time period studied was selected at random.

The P-value of 0.0146 says that if the true proportion of male babies were still at 51.7%, then an observed proportion as different as 56.9% male babies would occur at random only about 15 times in 1000. With a P-value this small, I reject H_0. This is strong evidence that the proportion of boys is not equal to the baseline for the region. It appears that the proportion of boys may be **artificially inflated**.

18.4 A Hypothesis Test for the Mean

A hospital in Nashville is considering changes to the prenatal care they offer. They collected the gestation times of 70 pregnancies that ended in live births. The established human gestation time is 266 days. Were their mean gestation times different? (Data in **Nashville**)

The test called for here is based on Student's *t* because it is a test about the mean and we don't know the standard deviation. It is called a **one-sample *t*-test for the mean**. The rubric for testing says to first state the hypotheses.

1. **Hypotheses.** The parameter of interest is the mean gestation time, and the null value is given by general medical knowledge. We can write

$$H_0: \mu = 266 \text{ vs. } H_A: \mu \neq 266.$$

2. **Model.** In Chapter 17, we used Student's *t* to build confidence intervals for the mean. That's what we'll do for hypothesis tests as well—and for the same reason we had then: We don't know the standard deviation, σ. We don't have a random sample, but we think it is representative. The values are almost surely independent. A histogram of the data is unimodal and nearly symmetric with no outliers. The sample mean, $\bar{y}$, is 260.31 and the standard deviation, s, is 18.26 days.

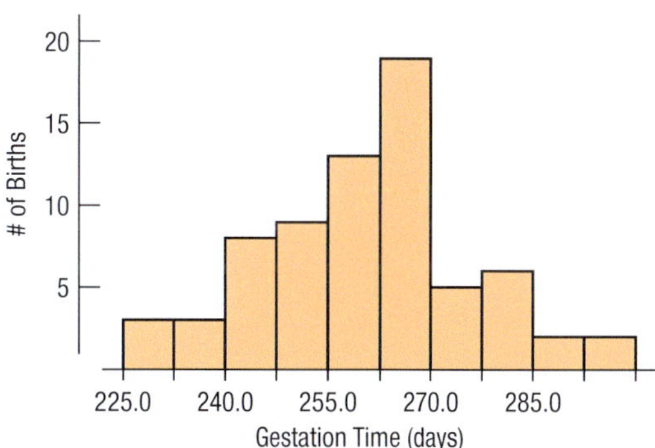

3. **Mechanics.** The calculations are similar to those for a confidence interval:

$$t = \frac{(\bar{y} - \mu_0)}{s/\sqrt{n}} = \frac{260.31 - 266}{15.2577/\sqrt{70}} = -3.118.$$

The t distribution on $n - 1 = 69\ df$ looks like this:

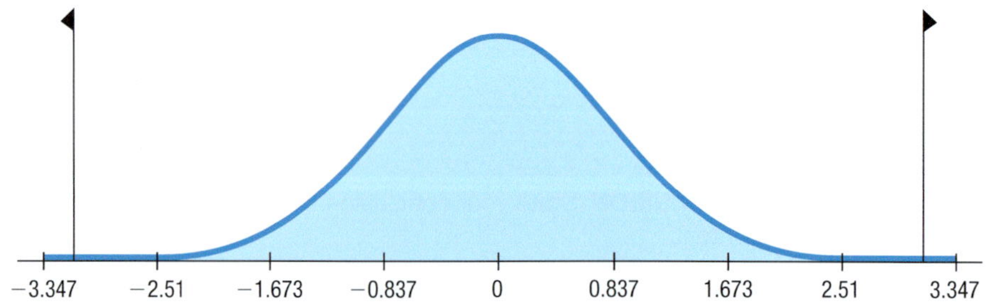

which shows the P-value to be 0.0027.

4. **Conclusion.** With a P-value this small, we reject the null hypothesis. In fact, a 95% confidence interval is (256.7, 264.0) days, so plausible values for the mean gestation time are below the standard mean of 266. However, all of the plausible values are within about a week of normal gestation, so the hospital administrators may not see this as a reason for alarm.

One-Sample *t*-Test for the Mean

The assumptions and conditions for the one-sample *t*-test for the mean are the same as for the one-sample *t*-interval (see Chapter 17). We test the hypothesis H_0: $\mu = \mu_0$ using the statistic

$$t_{n-1} = \frac{\bar{y} - \mu_0}{SE(\bar{y})}.$$

The standard error of $\bar{y}$ is $SE(\bar{y}) = \frac{s}{\sqrt{n}}$.

When the conditions are met and the null hypothesis is true, this statistic follows a Student's *t*-model with $n - 1$ degrees of freedom. We use that model to obtain a P-value.

EXAMPLE 18.7 A One-Sample *t*-Test for the Mean

Researchers tested 150 farm-raised salmon for organic contaminants. They found the mean concentration of the carcinogenic insecticide mirex to be 0.0913 parts per million, with standard deviation 0.0495 ppm. As a safety recommendation to recreational fishers, the Environmental Protection Agency's (EPA) recommended "screening value" for mirex is 0.08 ppm (Data in **Farmed salmon**).

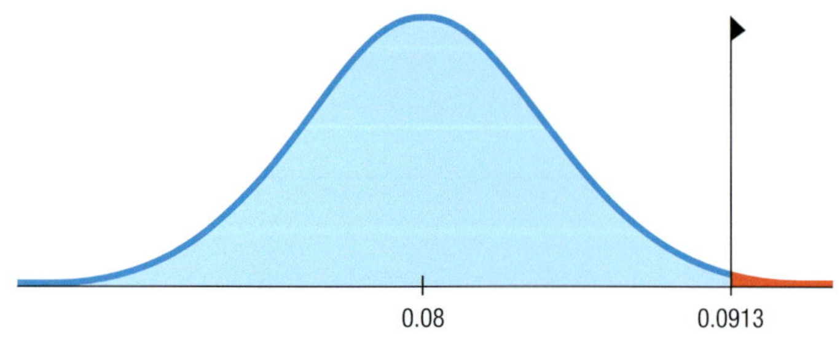

QUESTION: Are farmed salmon contaminated beyond the level permitted by the EPA?

ANSWER: (We've already checked the conditions; see page 541.)

$$H_0: \mu = 0.08$$
$$H_A: \mu \neq 0.08$$

These data satisfy the conditions for inference; I'll do a one-sample *t*-test for the mean:

$$n = 150, df = 149$$
$$\bar{y} = 0.0913, s = 0.0495$$
$$SE(\bar{y}) = \frac{0.0495}{\sqrt{150}} = 0.0040$$
$$t_{149} = \frac{0.0913 - 0.08}{0.0040} = 2.825$$
$$\text{P-value} = 0.0054 \text{ (from technology)}$$

With a P-value that low, I reject the null hypothesis and conclude that, in farm-raised salmon, the mirex contamination level does not conform to the EPA screening value. In fact, we already knew this. The confidence interval we found in Chapter 17 showed that the plausible values for the mean mirex contamination are in the interval (0.0834, 0.0992) ppm—above the EPA screening value.

WHEN σ IS KNOWN

If the hospital administrators use the established value of 14 days for the standard deviation of healthy gestation times, as they did for the confidence interval, they can use a *z*-test instead of a *t*-test. Practically, this will make little difference, but, when the science is solid, it removes the variability that results from estimating the standard deviation from samples. In this case, instead of a $t_{69} = -3.118$ they would find

$$z = \frac{260.31 - 266}{14/\sqrt{70}} = -3.40$$

which would lead to the same conclusion (that the mean gestation time of the hospital is not 266 days) with a slightly smaller P-value.

JUST CHECKING

4. The research team that wants to know if aspirin helps to thin blood also measures the plasma viscosity (PV) of the 12 patients before and after taking aspirin. The mean of the 12 differences (after − before) is 0.8 standard errors below 0 for a one-sided P-value of 0.22. They claim that on the basis of this trial, aspirin does not reduce PV. What would you say?

5. A marketing team wants to test whether their Facebook page increases revenue. They find that people who "like" their page spend more, on average, than their other customers by about $5.06 a month. The P-value is 0.049. They tell management that they have strong evidence that the Facebook page is working and is worth the time and effort to maintain it. What would you say?

Psychologists Jim Maas and Rebecca Robbins, in their book *Sleep for Success!*, say that

> In general, high school and college students are the most pathologically sleep-deprived segment of the population. Their alertness during the day is on a par with that of untreated narcoleptics and those with untreated sleep apnea. Not surprisingly, teens are also 71 percent more likely to drive drowsy and/or fall asleep at the wheel compared to other age groups. (Males under the age of twenty-six are particularly at risk.)

They report that adults require between 7 and 9 hours of sleep each night and claim that college students require 9.25 hours of sleep to be fully alert.[8] They note that "There is a 19 percent memory deficit in sleep-deprived individuals" (p. 35).

STEP-BY-STEP EXAMPLE

A One-Sample *t*-Test for the Mean

The Sleep Foundation (www.sleepfoundation.org) says that adults should get at least 7 hours of sleep each night. That's less than Maas and Robbins recommend. A survey of students at a small school in the northeast United States asked, among other things, "How much did you sleep last night?" Let's apply the one-sample *t*-test to see if the students are (on average) achieving the minimum. (Data in **Sleep**)

QUESTION: Is the mean amount that college students sleep at least as much as the 7-hour minimum recommended for adults?

THINK **PLAN** State what we want to test. Make clear what the population and parameter are.

Identify the variables and review the W's.

I want to know whether the mean amount that college students sleep equals the recommended minimum of 7 hours per night. I have a random sample of 25 student responses of their sleep amounts. I'll assume they are typical of college students.

HYPOTHESES The null hypothesis is that the true mean sleep time is equal to the minimum recommended.

H_0: Mean sleep, $\mu = 7$ hours

H_A: Mean sleep, $\mu < 7$ hours

Make a picture. Check the distribution for skewness, multiple modes, and outliers.

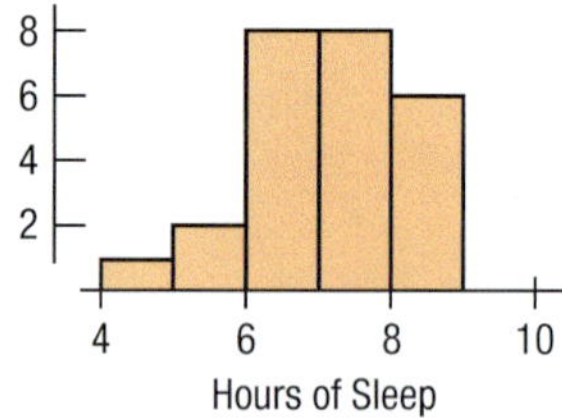

REALITY CHECK The histogram of the observed hours of sleep is clustered around a value less than 7. But is this enough evidence to suggest that the mean for all college students is different from 7?

MODEL Think about the assumptions and check the conditions.

✓ **Randomization Condition**: The students were sampled in a randomized survey, so the amounts they sleep are likely to be mutually independent.

✓ **Nearly Normal Condition**: The histogram of the speeds is unimodal and reasonably symmetric.

State the sampling distribution model. (Be sure to include the degrees of freedom.)

Choose your method.

The conditions are satisfied, so I'll use a Student's *t*-model with $(n - 1) = 24$ degrees of freedom to do a **one-sample *t*-test for the mean**.

SHOW **MECHANICS** Be sure to include the units when you write down what you know from the data.

From the data,

$$n = 25 \text{ students}$$
$$\bar{y} = 6.64 \text{ hours}$$
$$s = 1.075 \text{ hours}$$

$$SE(\bar{y}) = \frac{s}{\sqrt{n}} = \frac{1.075}{\sqrt{25}} = 0.215 \text{ hour.}$$

We use the null model to find the P-value. Make a picture of the *t*-model with 24 degrees of freedom.

The *t*-statistic calculation is just a standardized value, like *z*. We subtract the hypothesized mean and divide by the standard error.

[8] When was the last time you got 9.25 hours of sleep?

The P-value is the probability of observing a sample mean as small as 6.64 (or smaller) *if* the true mean were 7, as the null hypothesis states. We can find this P-value from a table, calculator, or computer program.

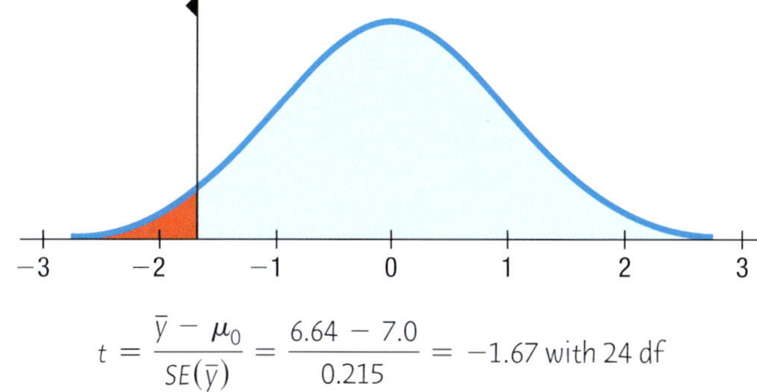

$$t = \frac{\bar{y} - \mu_0}{SE(\bar{y})} = \frac{6.64 - 7.0}{0.215} = -1.67 \text{ with } 24 \text{ df}$$

REALITY CHECK The *t*-statistic is negative because the observed mean is below the hypothesized value. That makes sense because we suspect students get too little sleep, not too much.

The observed mean is 1.67 standard errors below the hypothesized value. Using technology, I find

P-value = 0.0536.

TELL **CONCLUSION** Link the P-value to your decision about H_0, and state your conclusion in context.

The P-value of 0.0536 says that our sample of 25 students does not allow us to reject the null hypothesis that the mean amount of sleep is 7 hours. The 95% confidence interval is (6.20, 7.08) hours, and it does include 7. On the other hand, the interval extends as low as 6.2 hours. If a value that low would result in negative health impacts to students, it might be worth collecting more data to reduce the margin of error and obtain evidence on which to base some action.

18.5 Intervals and Tests

Carl Wunderlich (1815–1877), the father of clinical thermometry.

We've just seen two examples in which the confidence interval answered the question about the null hypothesis. The exercises in Chapter 17 provided many examples as well. The confidence interval contains all the plausible values of the parameter. If it doesn't contain the hypothesized value, then that value isn't plausible, and we should reject the null hypothesis. Confidence intervals and significance tests are built from the same calculations. In fact, they are really just two ways of looking at the same question.

Confidence intervals and hypothesis tests look at the same problem from two different perspectives. The confidence interval is *data-centric*. Its center is the statistic computed from the data. It then finds an interval of plausible values for the parameter by extending around that statistic. When testing a hypothesis using a confidence interval, we ask if the proposed parameter value is consistent with our interval and we reject the null hypothesis if that proposed value is not in the interval. By contrast, a hypothesis test is *model-centric*. It starts with a model centered at the *proposed null parameter value*, and asks if the *data* are consistent with that model. If the data are too unusual (as measured by the model), then we reject the hypothesis. So, they are both doing the same thing, but from different points of view.

How is the confidence level related to the P-value? To be precise, a level C confidence interval contains *all* of the plausible null hypothesis values that would *not* be rejected if you use a P-value of $(1 - C)$ as the cutoff for deciding to reject H_0.

When you've performed a hypothesis test, the corresponding confidence interval can provide additional information. By providing the plausible values for the parameter, it can help you judge the importance of your result. If your hypothesized value was far from the observed statistic (and you therefore rejected your null hypothesis), the null value has no information about your data, so a confidence interval provides the *only* information about plausible values.

Summary	Temperature
Count	52
Mean	98.285
Median	98.200
MidRange	98.600
StdDev	0.6824
Range	2.800
IntQRange	1.050

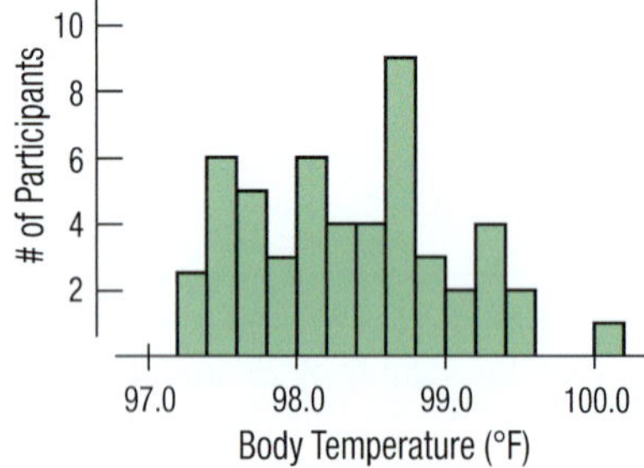

Here's an example. In Chapter 17, Exercise 29, we looked at 52 temperature measurements of healthy adults. (Data in **Normal temperature**) Although we all "know" that "normal" body temperature is 98.6°F (or 37°C), the mean of these 52 people is only 98.285°F. The 98.6 number comes from the work of Carl Wunderlich, a German medical professor of the mid-19th century who measured the temperatures of tens of thousands of patients over an 18-year period using a foot-long thermometer that took 20 minutes to get a stable reading. Could 98.6 and Wunderlich be wrong? In fact, some research published nearly 30 years ago in the *Journal of the American Medical Association*[9] asked the same question.

We'll examine Wunderlich's hypothesis in two ways: by constructing a confidence interval for the true mean and by performing a hypothesis test of $\mu = 98.6$. We'll use 99% as our confidence level, which equates to a level of significance of 0.01. The confidence interval is

$$\bar{y} \pm t_{51,0.005}\frac{s}{\sqrt{n}} = 98.285 \pm 2.676\frac{0.6824}{\sqrt{52}} = (98.032, 98.538).$$

As we can see, the upper end misses 98.6 by nearly a half a degree.

What does a hypothesis test say about it?

$$t = \frac{98.285 - 98.6}{0.6824/\sqrt{52}} = -3.3387$$

$$df = 51, \text{P-value} = 0.0016$$

The hypothesis test rejects the null hypothesis of 98.6 with a P-value just over 0.001, strong evidence that the mean is not 98.6. We'll return to this example in the *Random Matters* section.

The Special Case of Proportions

The relationship between confidence intervals and hypothesis tests works for almost every inference procedure we'll see in this text. We use the standard error of the statistic as a ruler for both. For the confidence interval, we reach out a number (often near 2) of standard errors on each side of the statistic. For a hypothesis test, we measure how many standard errors away our data lie from the hypothesized value and compute a P-value. But proportions are a little special. When we test a hypothesis about a proportion, not only do we use the hypothesized null value as the center of our null distribution, but we use it to compute the spread of that distribution as well. In this case, we're not estimating anything, so we call it a standard deviation, not a standard error, and write it as

$$SD(\hat{p}) = \sqrt{\frac{pq}{n}}.$$

But when we construct a confidence interval, there's no null hypothesis value. So we use the observed proportion, $\hat{p}$, and calculate its standard error: $SE(\hat{p}) = \sqrt{\frac{\hat{p}\hat{q}}{n}}$.

Does this make a difference? Usually no. When the observed proportion is near the hypothesized value, the *SE* and *SD* are very similar, so the usual relationship between the test and the interval works reasonably well. But if the hypothesized value is quite far from the observed value, the relationship between test and interval breaks down.

Here's an example: Suppose you want to test whether the coin that is flipped to determine who kicks off in your school's football games is fair. The natural null hypothesis is H_0: $p = 0.50$. Suppose you flip the coin 100 times, and get only 3 Heads. There's no way that could happen with a fair coin. That's 9.4 standard deviations below 50%, with a P-value of about 10^{-25}. You would obviously claim that the coin is rigged. With only

[9]Philip A. Mackowiak, MD; Steven S. Wasserman, PhD; and Myron M. Levine, MD; *Journal of the American Medical Association* 268: 1578–1580 (1992).

3 "successes" you can't make a confidence interval using the methods of this chapter. But that's no reason not to test the hypothesis—and that's why the Success/Failure Condition for a hypothesis test of a proportion uses np_0 and not $n\hat{p}$. When testing a hypothesis, you should always use the null value of p when calculating $SD(\hat{p})$.

*RANDOM MATTERS Bootstrap Hypothesis Tests and Intervals

Section 17.4 in Chapter 17 showed how to construct a bootstrap confidence interval. Applying that to the body temperatures, we create new "samples" by resampling the 52 temperatures (with replacement). For each sample we compute the mean. For a 99% confidence interval, we'd find the central 99% of the values. Here's a histogram of the means of 10,000 resamples of the temperatures.

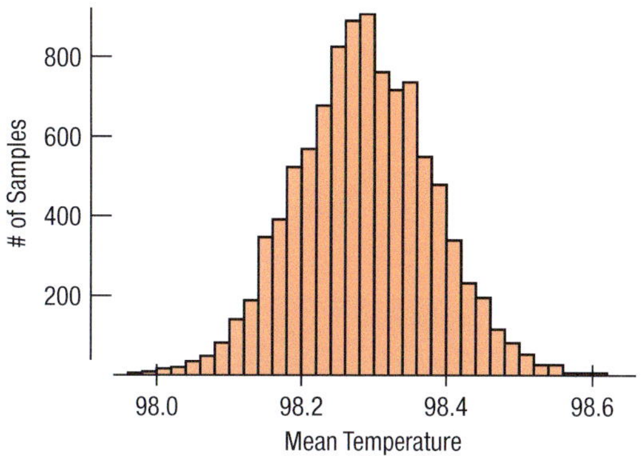

From the 0.5th and 99.5th percentiles, a 99% confidence interval finds

0.5%	99.5%
98.04231	98.53078

very close to our *t*-based interval.

Because 98.6 is not in the confidence interval, we can reject the null hypothesis. And because the confidence level is 0.99, we know that the P-value is less than 0.01, but we can't pin it down more precisely than that.

If we want more information about the P-value, we'll need to shift our distribution to center it at the null hypothesis, making it a *model-centric* distribution instead of the *data-centric* one we just created. The one we used to create the confidence interval was centered at the mean of our sample, 98.285°F. To adjust the distribution to our null hypothesis mean of 98.6°F, we can simply add 0.315 to each of our original temperatures before we resample—or, equivalently, we can add 0.315 to each of the bootstrap means. That will shift the distribution without changing its standard deviation (as we learned back in Chapter 2). The shifted bootstrap distribution looks like this:

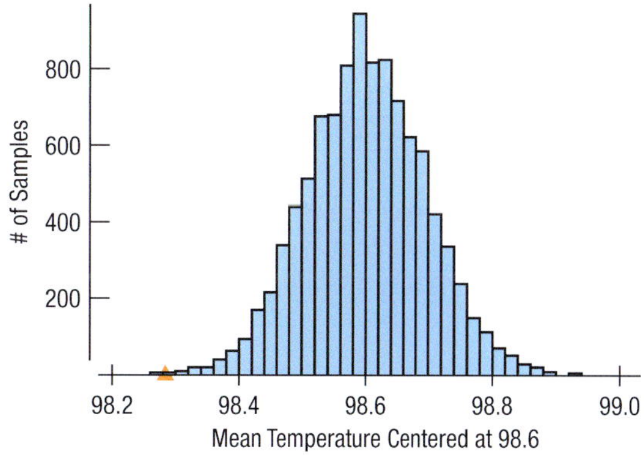

The P-value is the chance of seeing a sufficiently extreme value for this shifted distribution. So we ask, "How many times was there a value as far from 98.6 as the sample mean we did observe?" That value is shown in red on the histogram. In this example, the answer is 9 times out of 10,000, which gives a bootstrap P-value of 0.0009. The *t*-based P-value was 0.0016, so the two methods draw essentially the same conclusion:[10] We reject the hypothesis that the true mean temperature of healthy adults is 98.6°F. In fact, modern research has concluded that the mean human body temperature is about 98.2°F for temperatures taken orally (as Wunderlich did).

Here are the steps to follow to use the bootstrap to test a hypothesis $H_0: \mu = \mu_0$.

1. Find the sample mean, $\bar{y}$, from your data. Add $(\mu_0 - \bar{y})$ to each observation in your sample to shift the center of this sample to μ_0. Bootstrap this shifted sample *k* times, recording the mean each time. (Equivalently, bootstrap the original sample *k* times, record the means, and then add $(\mu_0 - \bar{y})$ to each mean.) *Note:* You get to choose *k*. Typical values for *k* are at least 1000 or even 10,000.
2. You have produced a simulated sampling distribution centered at the null hypothesis value of μ_0. Find the proportion of times the bootstrapped sample means fall as far or farther from the mean of this sampling distribution as the original observed sample mean, $\bar{y}$. That proportion is a P-value. *Note:* It is considered bad form to report a P-value of zero even if no resampled means are that extreme. After all, there is always the possibility that with more trials you'd see a sample mean that extraordinary.

STEP-BY-STEP EXAMPLE

Tests and Intervals

Anyone who plays or watches sports has heard of the "home field advantage." Tournaments in many sports are designed to try to neutralize the advantage of the home team or player. Most people believe that teams tend to win more often when they play at home. But do they?

If there were no home field advantage, the home teams would win about half of all games played. To test this, we'll use the games in the Major League Baseball 2013 season. That year, there were 2431 regular-season games. It turns out that the home team won 1308 of the 2431 games, or 53.81% of the time.

QUESTION: Could this deviation from 50% be explained just from natural sampling variability, or is it evidence to suggest that there really is a home field advantage, at least in professional baseball?

THINK **PLAN** State what we want to know.

Define the variables and discuss the W's.

I want to know whether the home team in professional baseball is more likely to win. The data are all 2431 games from the 2013 Major League Baseball season. The variable is whether or not the home team won. The parameter of interest is the proportion of home team wins. If there's no advantage, I'd expect that proportion to be 0.50.

HYPOTHESES The null hypothesis makes the claim of no difference from the baseline. Here, that means no home field advantage. The alternative hypothesis is that there is in fact a home field advantage.

$$H_0: p = 0.50$$

$$H_A: p > 0.50$$

[10] With only 52 observations, we couldn't draw meaningful differences between these P-values even if we wanted to.

MODEL Think about the assumptions and check the appropriate conditions. This is not a random sample. If we wanted to talk only about this season there would be no inference. So, we view the 2431 games here not as a random sample, but as a representative collection of games. Our inference is about all years of Major League Baseball.

✓ **Independence Assumption**: Generally, the outcome of one game has no effect on the outcome of another game. But this may not be strictly true. For example, if a key player is injured, the probability that the team will win in the next couple of games may decrease slightly, but independence is still roughly true. The data come from one entire season, but I expect other seasons to be similar.

I'm not just interested in this particular year, and those games, while not randomly selected, should be a reasonable representative sample of all Major League Baseball games in the recent past and near future.

✓ **10% Condition**: We are interested in home field advantage for Major League Baseball for all seasons. While not a random sample, these 2431 games are fewer than 10% of all games played over the years.

✓ **Success/Failure Condition**: Both $np_0 = 2431(0.50) = 1215.5$ and $nq_0 = 2431(0.50) = 1215.5$ are at least 10.

Specify the sampling distribution model.

State what test you plan to use.

Because the conditions are satisfied, I'll use a Normal model for the sampling distribution of the proportion and do a **one-proportion z-test**.

SHOW **MECHANICS** The null model gives us the mean, and (because we are working with proportions) the mean gives us the standard deviation.

The null model is a Normal distribution with a mean of 0.50 and a standard deviation of

$$SD(\hat{p}) = \sqrt{\frac{p_0 q_0}{n}} = \sqrt{\frac{(0.5)(1-0.5)}{2431}}$$

$$= 0.010141.$$

Next, we find the z-score for the observed proportion, to find out how many standard deviations it is from the hypothesized proportion.

The observed proportion, $\hat{p}$, is 0.53805.

So the z-value is

$$z = \frac{0.53805 - 0.5}{0.010141} = 3.75.$$

From the z-score, we can find the P-value, which tells us the probability of observing a value that extreme (or more).

The sample proportion lies 3.75 standard deviations above the mean.

The probability of observing a value 3.75 or more standard deviations higher than the mean of a Normal model can be found to be less than 0.0001.

The corresponding P-value is less than 0.0001.

TELL **CONCLUSION** State your conclusion about the parameter—in context, of course!

The P-value of 0.0001 says that if the true proportion of home team wins were 0.50, then an observed value of 0.53805 (or **more extreme**) would occur less than one time in 1000. With a P-value so small, I reject H_0. I have reasonable evidence that the true proportion of home team wins is $> 50\%$, and that therefore there is a home field advantage.

QUESTION: OK, but how big a difference are we talking about? Just knowing that there is an effect is only part of the answer. Let's find a confidence interval for the home field advantage.

MODEL Think about the assumptions and check the conditions.

✓ **Success/Failure Condition**: There were 1308 home team wins and 1123 losses, both at least 10.

The conditions are identical to those for the hypothesis test, with one difference: Now we are not given a hypothesized proportion, p_0, so we must instead work with the observed results.

Specify the sampling distribution model.

Tell what method you plan to use.

The conditions are satisfied, so I can model the sampling distribution of the proportion with a Normal model and find a **one-proportion z-interval**.

SHOW **MECHANICS** We can't find the sampling model standard deviation from the null model proportion. (In fact, we've just rejected it.) Instead, we find the standard error of $\hat{p}$ from the *observed* proportions. Other than that substitution, the calculation looks the same as for the hypothesis test.

$$SE(\hat{p}) = \sqrt{\frac{\hat{p}\hat{q}}{n}} = \sqrt{\frac{(0.53805)(1 - 0.53805)}{2431}}$$

$$= 0.01011$$

The sampling model is Normal, so for a 95% confidence interval, the critical value $z^* = 1.96$.

With this large a sample size, the difference is negligible, but in smaller samples, it could make a bigger difference.

The margin of error is

$$ME = z^* \times SE(\hat{p}) = 1.96 \times 0.01011 = 0.0198.$$

So the 95% confidence interval is

$$0.53805 \pm 0.0198 \text{ or } (0.5182, 0.5579).$$

TELL **CONCLUSION** Confidence intervals help us think about the size of the effect. Here we can see that the home field advantage may affect enough games to make a real difference.

I am 95% confident that, in professional baseball, home teams win between 51.82% and 55.79% of the games.

In a season of 162 games, the low end of this interval, 51.82% of the 81 home games, would mean nearly one and a half extra home victories, on average. The upper end, 55.79%, would mean more than 4 extra wins.

18.6 P-Values and Decisions: What to Tell About a Hypothesis Test

Hypothesis tests are particularly useful when we must make a decision. Is the defendant guilty or not? Should we choose print advertising or television? The absolute nature of the hypothesis test decision, however, makes some people (including the authors) uneasy. Whenever possible, it's a good idea to report a confidence interval for the parameter of interest as well.

How small should the P-value be to reject the null hypothesis? A jury needs enough evidence to show the defendant guilty "beyond a reasonable doubt." How does that translate to P-values? The answer is that there is no good, universal answer. How small the P-value has to be to reject the null hypothesis is highly context-dependent. When we're

> "An extraordinary claim requires extraordinary proof."
> —Marcello Truzzi

This saying is often quoted by scientists without attributing it to Truzzi. But he appears to have published it first (in On the Extraordinary: An Attempt at Clarification, *Zetetic Scholar*, Vol. 1, No. 1, p. 11, 1978).

DON'T WE WANT TO REJECT THE NULL?

Often the folks who collect the data or perform the experiment hope to reject the null. (They hope the new drug is better than the placebo, or the new ad campaign is better than the old one.) But when we practice statistics, we can't allow that hope to affect our decision. The essential attitude for a hypothesis tester is skepticism. Until we become convinced otherwise, we cling to the null's assertion that there's nothing unusual, no effect, no difference, etc. As in a jury trial, the burden of proof rests with the alternative hypothesis—innocent until proven guilty. When you test a hypothesis, you must act as judge and jury, but you are not the prosecutor.

screening for a disease and want to be sure we treat all those who are sick, we may be willing to reject the null hypothesis of no disease with a P-value as large as 0.10. That would mean that 10% of the healthy people would be treated as sick and subjected to further testing. We might rather treat (or recommend further testing for) the occasional healthy person than fail to treat someone who was really sick. But a long-standing hypothesis, believed by many to be true, needs stronger evidence (and a correspondingly small P-value) to reject it.

See if you require the same P-value to reject each of the following null hypotheses:

- A renowned musicologist claims that she can distinguish between the works of Mozart and Haydn simply by hearing a randomly selected 20 seconds of music from any work by either composer. What's the null hypothesis? If she's just guessing, she'll get 50% of the pieces correct, on average. So our null hypothesis is that p equals 50%. If she's for real, she'll get more than 50% correct. Now, we present her with 10 pieces of Mozart or Haydn chosen at random. She gets 9 out of 10 correct. It turns out that the P-value associated with that result is 0.011. (In other words, if you tried to just guess, you'd get at least 9 out of 10 correct only about 1% of the time.) What would *you* conclude? Most people would probably reject the null hypothesis and be convinced that she has some ability to do as she claims. Why? Because the P-value is small and we don't have any particular reason to doubt the alternative.
- On the other hand, imagine a student who bets that he can make a flipped coin land the way he wants just by thinking hard. To test him, we flip a fair coin 10 times. Suppose he gets 9 out of 10 right. This also has a P-value of 0.011. Are you willing now to reject this null hypothesis? Are you convinced that he's not just lucky? What amount of evidence *would* convince you? We require more evidence if rejecting the null hypothesis would contradict long-standing beliefs or other scientific results. Of course, with sufficient evidence we would revise our opinions (and scientific theories). That's how science makes progress.

Another factor in choosing a P-value is the importance of the issue being tested. Consider the following two tests:

- A researcher claims that the proportion of college students who hold part-time jobs now is higher than the proportion known to hold such jobs a decade ago. You might be willing to believe the claim (and reject the null hypothesis of no change) with a P-value of 0.05.
- An engineer claims that even though there were several problems with the rivets holding the wing on an airplane in their fleet, they've retested the proportion of faulty rivets and now the P-value is small enough to reject the null hypothesis that the proportion is the same. What P-value would be small enough to get you to fly on that plane?

Your conclusion about any null hypothesis should always be accompanied by the P-value of the test and, ideally, a confidence interval. Don't just declare the null hypothesis rejected or not rejected. Report the P-value to show the strength of the evidence against the hypothesis and a confidence interval to show the effect size. This will let each reader decide whether or not to reject the null hypothesis and whether or not to consider the result important.

When you reject a null hypothesis you conclude that the parameter value lies in the alternative. But the alternative is absurdly large—usually every possible value except the null. A confidence interval is based on the observed data and provides a much more useful set of plausible values. In fact, it's likely that you didn't believe the null value anyway. (Is the coin *exactly* fair? Is $P(\text{head}) = 0.5000000\ldots$ and not 0.500000001?) So what are hypothesis tests good for? Well, sometimes we need to make a decision. Setting an arbitrary threshold for the P-value provides a "bright line" decision rule. And the P-value provides useful information about how far the data are from the null value.

P-values have become controversial because some people base decisions solely on their P-values without regard to assumptions, conditions, effect size, or cost. The American

Statistical Association recently published a statement about P-values.[11] They recommend six principles underlying the proper interpretation of P-values:

1. P-values can indicate how incompatible the data are with a specified statistical model.
2. P-values do not measure the probability that the studied hypothesis is true, or the probability that the data were produced by random chance alone.
3. Scientific conclusions and business or policy decisions should not be based only on whether a P-value passes a specific threshold.
4. Proper inference requires full reporting and transparency.
5. A P-value, or statistical significance, does not measure the size of an effect or the importance of a result.
6. By itself, a P-value does not provide a good measure of evidence regarding a model or hypothesis.

JUST CHECKING

6. A bank is testing a new method for getting delinquent customers to pay their past-due credit card bills. The standard way was to send a letter (costing about $0.40) asking the customer to pay. That worked 30% of the time. They want to test a new method that involves sending a DVD to customers encouraging them to contact the bank and set up a payment plan. Developing and sending the video costs about $10.00 per customer. What is the parameter of interest? What are the null and alternative hypotheses?
7. The bank sets up an experiment to test the effectiveness of the DVD. They mail it out to several randomly selected delinquent customers and keep track of how many actually do contact the bank to arrange payments. The bank's statistician calculates a P-value of 0.003. What does this P-value suggest about the DVD?
8. The statistician tells the bank's management that the results are clear and that they should switch to the DVD method. Do you agree? What else might you want to know?

WHAT CAN GO WRONG?

Hypothesis tests are so widely used—and so widely misused—that we've devoted all of Chapter 19 to discussing the pitfalls involved, but there are a few issues that we can talk about already.

- **Don't base your null hypothesis on what you see in the data.** You are not allowed to look at the data first and then adjust your null hypothesis so that it will be rejected. When your sample value turns out to be $\hat{p} = 51.8\%$, with a standard deviation of 1%, don't form a null hypothesis like H_0: $p = 49.8\%$, knowing that you can reject it. You should always *Think* about the situation you are investigating and make your null hypothesis describe the "nothing interesting" or "nothing has changed" scenario. No peeking at the data!
- **Don't make your null hypothesis what you want to show to be true.** Remember, the null hypothesis is the status quo, the nothing-is-strange-here position a skeptic would take. You wonder whether the data cast doubt on that. You can reject the null hypothesis, but you can never "accept" or "prove" the null.

[11]You have seen these issues discussed already in the chapter. For more details, see the paper at www.amstat.org/asa/files/pdfs/P-ValueStatement.pdf.

- **Don't forget to check the conditions.** The reasoning of inference depends on randomization. No amount of care in calculating a test result can recover from biased sampling. The probabilities we compute depend on the Independence Assumption. And the sample must be large enough to justify the use of the null model.
- **Don't "accept" the null hypothesis.** You may not have found enough evidence to reject it, but you surely have *not* proven it's true!
- **If you fail to reject the null hypothesis, don't think that a bigger sample would be more likely to lead to rejection.** If the results you looked at were "almost" significant, it's enticing to think that because you would have rejected the null had these same observations come from a larger sample, then a larger sample would surely lead to rejection. Don't be misled. Remember, each sample is different, and a larger sample won't necessarily duplicate your current observations. Indeed, the Central Limit Theorem tells us that statistics will vary *less* in larger samples. We should therefore expect such results to be less extreme. Maybe they'd be statistically significant but maybe (perhaps even probably) not. Even if you fail to reject the null hypothesis, it's a good idea to examine a confidence interval. If none of the plausible parameter values in the interval would matter to you (for example, because none would be *practically* significant), then even a larger study with a correspondingly smaller standard error is unlikely to be worthwhile.

CONNECTIONS

Hypothesis tests and confidence intervals share many of the same concepts. Both rely on sampling distribution models, and because the models are the same and require the same assumptions, both check the same conditions. They also calculate many of the same statistics. Like confidence intervals, hypothesis tests use the standard deviation of the sampling distribution as a ruler, as we first saw in Chapter 5.

P-values give the probability of observing the result we have seen (or one even more extreme) *given* that the null hypothesis is true.

For testing, we find ourselves looking at z- and t-scores, and we compute the P-value by finding the distance of our test statistic from the center of the null model.

As with all of our inference methods, the randomization applied in drawing a random sample or in randomizing a comparative experiment is what generates the sampling distribution. Randomization is what makes inference in this way possible at all.

The new concept of degrees of freedom connects back to the denominator of the sample standard deviation calculation, in Chapter 2.

There's just no escaping histograms and Normal probability plots. The Nearly Normal Condition required to use Student's t can be checked best by making appropriate displays of the data. Back when we first used histograms, we looked at their shape and, in particular, checked whether they were unimodal and symmetric, and whether they showed any outliers. Those are just the features we check for here. The Normal probability plot zeros in on the Normal model a little more precisely.

CHAPTER REVIEW

Know how to formulate a null and an alternative hypothesis for a question of interest.

- The null hypothesis specifies a parameter and a (null) value for that parameter.
- The alternative hypothesis specifies a range of plausible values should we reject the null.

Be able to perform a hypothesis test for a proportion.

- The null hypothesis has the form $H_0: p = p_0$.
- We find the standard deviation of the sampling distribution of the sample proportion by assuming that the null hypothesis is true:

$$SD(\hat{p}) = \sqrt{\frac{p_0 q_0}{n}}.$$

- We refer the statistic $z = \frac{\hat{p} - p_0}{SD(\hat{p})}$ to the standard Normal model.

Be able to perform a hypothesis test for a mean.

- To apply the Central Limit Theorem for the mean in practical applications, we must estimate the standard deviation. This *standard error* is

$$SE(\bar{y}) = \frac{s}{\sqrt{n}}.$$

- When we use the *SE*, the sampling distribution that allows for the additional uncertainty is Student's *t*-model on $n - 1$ degrees of freedom.
- Find critical values by technology or from tables.
- Check the assumptions and conditions before using any sampling distribution for inference.

Write clear summaries to interpret a confidence interval or state a hypothesis test's conclusion.

Understand P-values.

- A P-value is the estimated probability of observing a statistic value at least as far from the (null) hypothesized value as the one we have actually observed.
- A small P-value indicates that the statistic we have observed would be unlikely were the null hypothesis true. That leads us to doubt the null.
- A large P-value just tells us that we have insufficient evidence to doubt the null hypothesis. In particular, it does not prove the null to be true.

Know the reasoning of hypothesis testing.

- State the **hypotheses**.
- Determine (and check assumptions for) the sampling distribution **model**.
- Calculate the test statistic—the **mechanics**.
- State your **conclusions and decisions**.

Be able to decide on a two-sided or one-sided alternative hypothesis, and justify your decision.

Know that confidence intervals and hypothesis tests go hand in hand in helping us think about models.

- A hypothesis test makes a yes/no decision about the plausibility of the value of a parameter value.
- A confidence interval shows us the range of plausible values for the parameter.

REVIEW OF TERMS

Hypothesis A model or proposition that we adopt in order to test (p. 564).

Null hypothesis The claim being assessed in a hypothesis test that states "no change from the traditional value," "no effect," "no difference," or "no relationship." For a claim to be a testable null hypothesis, it must specify a value for some population parameter that can form the basis for assuming a sampling distribution for a test statistic (p. 564).

Alternative hypothesis The alternative hypothesis proposes what we should conclude if we reject the null hypothesis (p. 564).

P-value The probability of observing a value for a test statistic at least as far from the hypothesized value as the statistic value actually observed if the null hypothesis is true. A small P-value indicates either that the observation is improbable or that the probability calculation was based on incorrect assumptions. The assumed truth of the null hypothesis is the assumption under suspicion (p. 566).

Two-sided alternative (Two-tailed alternative) An alternative hypothesis is two-sided (H_A: $p \neq p_0$) when we are interested in deviations in *either* direction away from the hypothesized parameter value (p. 567).

One-sided alternative (One-tailed alternative) An alternative hypothesis is one-sided (e.g., H_A: $p > p_0$ or H_A: $p < p_0$) when we are interested in deviations in *only one* direction away from the hypothesized parameter value (p. 568).

One-proportion *z*-test A test of the null hypothesis that the proportion of a single sample equals a specified value (H_0: $p = p_0$) by referring the statistic $z = \frac{\hat{p} - p_0}{SD(\hat{p})}$ to a standard Normal model (p. 569).

Effect size The difference between the null hypothesis value and the true value of a model parameter (p. 571).

One-sample *t*-test for the mean The one-sample *t*-test for the mean tests the hypothesis H_0: $\mu = \mu_0$ using the statistic (p. 573)

$$t_{n-1} = \frac{\bar{y} - \mu_0}{SE(\bar{y})}.$$

The standard error of $\bar{y}$ is

$$SE(\bar{y}) = \frac{s}{\sqrt{n}}.$$

TECH SUPPORT

Hypothesis Tests

Hypothesis tests for proportions are so easy and natural that many statistics packages don't offer special commands for them. Most statistics programs want to know the "success" and "failure" status for each case. Usually these are given as 1 or 0, but they might be category names like "yes" and "no." Often you just know the proportion of successes, $\hat{p}$, and the total count, n. Computer packages don't usually deal naturally with summary data like these, but the statistics routines found on many graphing calculators do. These calculators allow you to test hypotheses from summaries of the data—usually, all you need to enter are the number of successes and the sample size.

In some programs you can reconstruct the original values. But even when you have reconstructed (or can reconstruct) the raw data values, often you won't get exactly the same test statistic from a computer package as you would find working by hand. The reason is that when the packages treat the proportion as a mean, they make some approximations. The result is very close, but not exactly the same.

For quantitative data, statistics packages offer convenient ways to make histograms of the data. Even better for assessing near-Normality is a Normal probability plot. When you work on a computer, there is simply no excuse for skipping the step of plotting the data to check that it is nearly Normal. *Beware:* Statistics packages don't agree on whether to place the Normal scores on the *x*-axis (as we have done) or the *y*-axis. Read the axis labels.

Any standard statistics package can compute a hypothesis test for a mean. Here's what the package output might look like in general (although no package we know gives the results in exactly this form):[12]

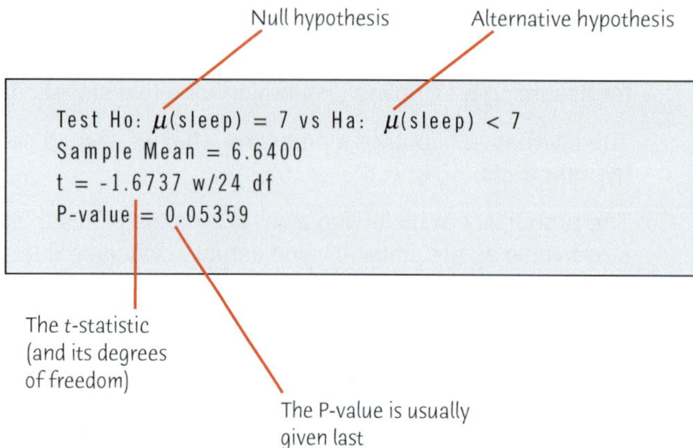

The package computes the sample mean and sample standard deviation of the variable and finds the P-value from the *t*-distribution based on the appropriate number of degrees of freedom. All modern statistics packages report P-values. The package may also provide additional information such as the sample mean, sample standard deviation, *t*-statistic value, and degrees of freedom. These are useful for interpreting the resulting P-value and telling the difference between a meaningful result and one that is merely statistically significant. Statistics packages that report the estimated standard deviation of the sampling distribution usually label it "standard error" or "*SE*."

Inference results are also sometimes reported in a table. You may have to read carefully to find the values you need. Often, test results and the corresponding confidence interval bounds are given together. And often you must read carefully to find the alternative hypotheses. Here's an example of that kind of output:

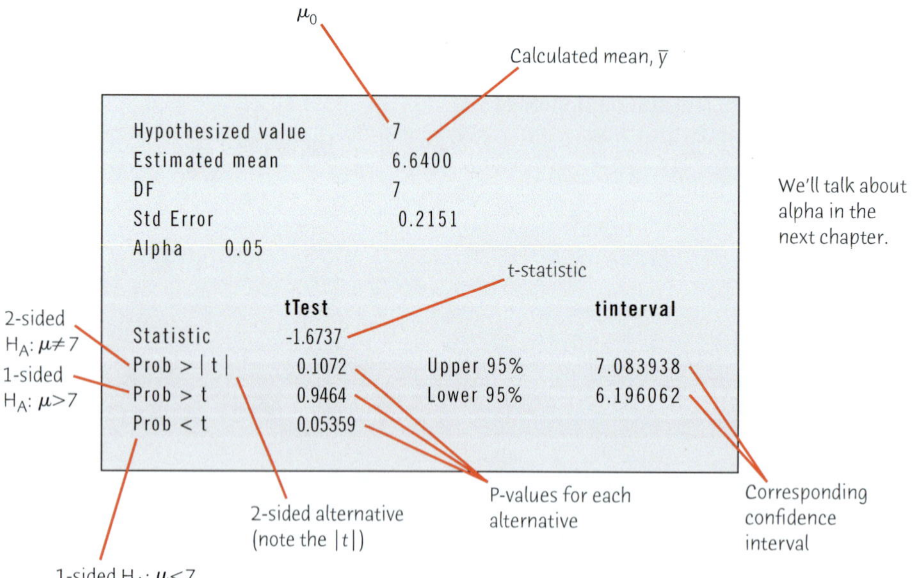

[12]Many statistics packages keep as many as 16 digits for all intermediate calculations. If we had kept as many, our results in the Step-by-Step Example would have been closer to these.

DATA DESK

Data Desk does not offer built-in methods for inference with proportions. The Replicate Y by X command in the Manip menu will "reconstruct" summarized count data so that you can display it.

For means:

- Select variables.
- From the Calc menu, choose **Test** for hypothesis tests.
- Select the test from the drop-down menu and make other choices in the dialog.

COMMENTS

For summarized data, open a Scratchpad to compute the standard deviation and margin of error by typing the calculation. Then perform the test with the z-test for individual μ's found in the Test command.

EXCEL

Inference methods for proportions are not part of the standard Excel tool set.

For means, specify formulas. Find t^* with the TINV(alpha, df) function.

COMMENTS

For summarized data, type the calculation into any cell and evaluate it.

Hypothesis tests are not really automatic. There's no easy way to find P-values in Excel. For the examples in this chapter, substitute 0.05 for "alpha" in the TINV command.

JMP

For a categorical variable that holds category labels, the Distribution platform includes tests of proportions. For summarized data:

- Put the category names in one variable and the frequencies in an adjacent variable.
- Designate the frequency column to have the role of frequency. Then use the Distribution platform.

For quantitative variables:

- From the Analyze menu, select **Distribution**.
- For a hypothesis test, click the red triangle next to the variable's name and choose **Test Mean** from the menu.
- Then fill in the resulting dialog.

COMMENTS

JMP uses slightly different methods for proportion inferences than those discussed in this text. Your answers are likely to be slightly different.

MINITAB

For proportions:

- Choose **Basic Statistics** from the Stat menu.
- Choose **1Proportion** from the Basic Statistics submenu.
- If the data are category names in a variable, assign the variable from the variable list box to the Samples in columns box.
- If you have summarized data, click the **Summarized Data** button and fill in the number of trials and the number of successes.
- Click the **Options** button and specify the remaining details.
- If you have a large sample, check Use test and interval based on Normal distribution.
- Click the **OK** button.

COMMENTS

When working from a variable that names categories, Minitab treats the last category as the "success" category. You can specify how the categories should be ordered.

For means:

- From the Stat menu, choose the **Basic Statistics** submenu.
- From that menu, choose **1-sample t. . . .**
- Then fill in the dialog.

COMMENTS

The dialog offers a clear choice between confidence interval and test.

R

For proportions, in library(stats):

- prop.test(X, n, p = NULL, alternative = c("two.sided," "less," "greater"), conf.level = 0.95, correct=FALSE)

will test the hypothesis that $p = p_0$ against various alternatives. For example with 260 successes out of 500, to test that $p = 0.5$ vs. $p \neq 0.5$, use:

prop.test(260,500,0.5, "two.sided," correct=FALSE)

For means, to test the hypothesis that μ = mu (default is mu = 0) against an alternative (default is two-sided) and to produce a confidence interval (default is 95%), create a vector of data in x and then:

- **t.test**(x, alternative = c("two.sided", "less", "greater"), mu = 0, conf.level = 0.95)

provides the *t*-statistic, P-value, degrees of freedom, and the confidence interval for a specified alternative.

SPSS

SPSS does not offer hypothesis tests for proportions.

For means:

- From the Analyze menu, choose the **Compare Means** submenu.
- From that, choose the **One-Sample t-test** command.

COMMENTS

The commands suggest neither a single mean nor an interval. But the results provide both a test and an interval.

STATCRUNCH

To test a hypothesis for a proportion using summaries:

- Click on **Stat**.
- Choose **Proportion Stats** > **One Sample** > **With Summary**.
- Enter the **# of successes** (x) and **# of observations** (n).
- Under **Perform:** indicate **Hypothesis Test** for p, then enter the hypothesized Null proportion, and choose the alternative hypothesis.
- Click on **Compute!**

To test a hypothesis for a proportion using data:

- Click on **Stat**.
- Choose **Proportions Stats** > **One Sample** > **With Data**.
- Choose the variable from **Values in:**
- Enter the outcome to be considered a Success.
- Under **Perform:** indicate **Hypothesis Test** for p, then enter the hypothesized Null proportion, and choose the alternative hypothesis.
- Click on **Compute!**

To do inference for a mean using summaries:

- Click on **Stat**.
- Choose **T Stats** > **One Sample** > **With Summary**.
- Enter the Sample mean, Sample std dev, and Sample size.
- Under **Perform:** indicate **Hypothesis Test** for μ, then enter the hypothesized Null mean, and choose the alternative hypothesis.

  OR

  Indicate **Confidence Interval**, and then enter the Level of confidence.
- Click on **Compute!**

To do inference for a mean using data:

- Click on **Stat**.
- Choose **T Stats** > **One Sample** > **With Data**.
- Choose the variable of interest under **Select Column(s).**
- Under **Perform:** indicate **Hypothesis Test,** then enter the hypothesized Null mean, and choose the alternative hypothesis.

  OR

  Indicate **Confidence Interval**, and then enter the Level of confidence.
- Click on **Compute!**

TI-83/84 PLUS

To do the mechanics of a hypothesis test for a proportion:

- Select **5:1-PropZTest** from the STAT TESTS menu.
- Specify the hypothesized proportion.
- Enter the observed value of *x*.
- Specify the sample size.
- Indicate what kind of test you want: one-tailed lower tail, two-tailed, or one-tailed upper tail.
- Calculate the result.

Testing a hypothesis about a mean:

- In the STAT TESTS menu, choose **2:T-Test**. You may specify that you are using data stored in a list, or you may enter the mean, standard deviation, and size of your sample. You must also specify the hypothesized model mean and whether the test is to be two-tailed, lower-tailed, or upper-tailed.

COMMENTS

Beware: When you enter the value of x, you need the *count*, not the percentage. The count must be a whole number. If the number of successes is given as a percent, you must first multiply *np* and round the result to obtain x.

EXERCISES

SECTION 18.1

1. Better than aspirin? A very large study showed that aspirin reduced the rate of first heart attacks by 42%. A pharmaceutical company thinks they have a drug that will be more effective than aspirin and plans to do a randomized clinical trial to test the new drug. What is the null hypothesis the company will use?

2. Psychic Your friend claims to be psychic. You are skeptical. To test this, you take a deck of 52 playing cards and have your friend try to identify the color of the card (black or red), without looking, of course! State the null hypothesis for your experiment.

3. Parameters and hypotheses For each of the following situations, define the parameter (proportion, probability, or mean), and write the null and alternative hypotheses in terms of parameter values. Example: We want to know if the proportion of "up days" in the stock market is 50%. Answer: Let p = the proportion of "up days." H_0: $p = 0.5$ vs. H_A: $p \neq 0.5$.

a) A casino wants to know if their slot machine really delivers the 2 in 100 win rate that it claims.
b) Last year, customers spent an average of $54.81 per visit to a company's website. Based on a random sample of purchases this year, the company wants to know if the mean this year has changed.
c) A pharmaceutical company wonders if their new drug has a cure rate different from the 27% reported by a placebo.
d) A bank wants to know if the percentage of customers using their website has changed from the 35% that used it before their system crashed last week.

4. Hypotheses and parameters As in Exercise 3, for each of the following situations, define the parameter and write the null and alternative hypotheses in terms of parameter values.

a) Seat-belt compliance among university students in Kuala Lumpur was 40% in 1993. The city wants to know if it has changed. (https://www.researchgate.net/publication/319363427_Seat_Belt_Compliance_and_University_Students)
b) Last year, a survey found that 53% of the employees were willing to pay for on-site day care. The company wants to know if that has changed.
c) Regular card customers have a default rate of 2.3%. A credit card bank wants to know if that rate is different for their Gold card customers.
d) Regular card customers have been with the company for an average of 15.2 months. The credit card bank wants to know if their Gold card customers have been with the company on average the same amount of time.

SECTION 18.2

5. Better than aspirin again? Referring to the study of Exercise 1:

a) Is the alternative to the null hypothesis more naturally one-sided or two-sided? Explain.
b) The P-value from a clinical trial testing the hypothesis is 0.0031. What do you conclude?
c) What would you have concluded if the P-value had been 0.31?

6. GRE performance A test preparation company claims that more than 50% of the students who take their GRE prep course improve their scores by at least 10 points.

a) Is the alternative to the null hypothesis more naturally one-sided or two-sided? Explain.
b) A test run with randomly selected participants gives a P-value of 0.981. What do you conclude?
c) What would you have concluded if the P-value had been 0.019?

SECTION 18.3

7. Brexit According to the EU Referendum results in 2016, 51.9% of the United Kingdom has voted to leave the European Union. Before the official results were disclosed, the Wealden District Council had predicted that they will have a higher proportion of people in favor of leaving the United Kingdom. The statement was made based on the most recent survey data, which was a random sample of 1300 district residents where 744 people voted to leave the United Kingdom. (https://www.bbc.com/news/politics/eu_referendum/results)

a) State the hypotheses.
b) Name the model, and check appropriate conditions for a hypothesis test.
c) Draw and label a sketch, and then calculate the test statistic and P-value.
d) State your conclusion.

8. Seasonal rentals According to a recent study conducted by the Paris city hall, 7% of apartments in the first four districts are rented on a seasonal basis. The mayor of the 10th district wonders if the rate in the district is different from this. He randomly selects 800 apartments in his district and finds that 32 of them are used for seasonal rentals. (https://immobilier.lefigaro.fr/article/6-chiffres-pour-savoir-ce-que-pesent-airbnb-et-la-location-touristique-a-paris_858dd228-d269-11e6-a9e9-ec8855ae56b6/)

a) State the hypotheses.
b) Name the model, and check appropriate conditions for a hypothesis test.
c) Draw and label a sketch, and then calculate the test statistic and P-value.
d) State your conclusion.

SECTION 18.4

9. GRE performance again Instead of advertising the percentage of customers who improve by at least 10 points, a manager suggests testing whether the mean score improves at all. For each customer they record the difference in score before and after taking the course (After − Before).

a) State the null and alternative hypotheses.
b) The P-value from the test is 0.65. Does this provide any evidence that their course works?
c) From part b, what can you tell, if anything, about the mean difference in the sample scores?

10. Marriage In 1970, a European study indicated that the age at which women first married in Europe had a mean of 22.7 years. It is widely suspected that young people today are waiting longer to get married. We want to find out if the mean age of first marriage has increased since then. (https://ec.europa.eu/eurostat/statistics-explained/index.php?title=Main_Page & https://soc.kuleuven.be/ceso/historischedemografie/resources/pdf/WOG%20working%20paper15.pdf)

a) Write appropriate hypotheses.
b) We plan to test our hypothesis by selecting a random sample of 60 women who married for the first time last year. Do you think the necessary assumptions for inference are satisfied? Explain.
c) Describe the approximate sampling distribution model for the mean age in such samples.
d) The women in our sample married at an average age of 23.6 years with a standard deviation of 4.2 years. That results in a t-statistic of 1.66. What is the P-value for this?
e) Explain what this P-value means in this context.
f) What's your conclusion?

11. Pizza A researcher tests whether the mean cholesterol level among those who eat frozen pizza exceeds the value considered to indicate a health risk. She gets a P-value of 0.07. Explain in this context what the "7%" represents.

12. Golf balls The United States Golf Association (USGA) sets performance standards for golf balls. For example, the initial velocity of the ball may not exceed 250 feet per second when measured by an apparatus approved by the USGA. Suppose a manufacturer introduces a new kind of ball and provides a sample for testing. Based on the mean speed in the test, the USGA comes up with a P-value of 0.34. Explain in this context what the "34%" represents.

SECTION 18.5

13. Bad medicine Occasionally, a report comes out that a drug that cures some disease turns out to have a nasty side effect. For example, some antidepressant drugs may cause suicidal thoughts in younger patients. A researcher wants to study such a drug and look for evidence of a side effect.

a) If the test yields a low P-value and the researcher rejects the null hypothesis, but there is actually no ill side effect of the drug, what are the consequences of such an error?
b) If the test yields a high P-value and the researcher fails to reject the null hypothesis, but there *is* a bad side effect of the drug, what are the consequences of such an error?

14. Expensive medicine Developing a new drug can be an expensive process, resulting in high costs to patients. A pharmaceutical company has developed a new drug to reduce cholesterol, and it will conduct a clinical trial to compare the effectiveness to the most widely used current treatment. The results will be analyzed using a hypothesis test.

a) If the test yields a low P-value and the researcher rejects the null hypothesis that the new drug is not more effective, but it actually is not better, what are the consequences of such an error?
b) If the test yields a high P-value and the researcher fails to reject the null hypothesis, but the new drug *is* more effective, what are the consequences of such an error?

CHAPTER EXERCISES

15. Hypotheses Write the null and alternative hypotheses you would use to test each of the following situations:

a) A governor is concerned about his "negatives"—the percentage of state residents who express disapproval of his job performance. His political committee pays for a series of TV ads, hoping that they can keep the negatives below 30%. They will use follow-up polling to assess the ads' effectiveness.
b) Is a coin fair?
c) Only about 20% of people who try to quit smoking succeed. Sellers of a motivational tape claim that listening to the recorded messages can help people quit.

16. More hypotheses Write the null and alternative hypotheses you would use to test each situation.

a) In the 1950s, only about 40% of high school graduates went on to college. Has the percentage changed?
b) Twenty percent of cars of a certain model have needed costly transmission work after being driven between 50,000 and

100,000 miles. The manufacturer hopes that a redesign of a transmission component has solved this problem.

c) We field-test a new-flavor soft drink, planning to market it only if we are sure that over 60% of the people like the flavor.

17. Negatives After the political ad campaign described in Exercise 15, part a, pollsters check the governor's negatives. They test the hypothesis that the ads produced no change against the alternative that the negatives are now below 30% and find a P-value of 0.22. Which conclusion is appropriate? Explain.

a) There's a 22% chance that the ads worked.
b) There's a 78% chance that the ads worked.
c) There's a 22% chance that their poll is correct.
d) There's a 22% chance that natural sampling variation could produce poll results like these if there's really no change in public opinion.

18. Dice The seller of a loaded die claims that it will favor the outcome 6. We don't believe that claim, and roll the die 200 times to test an appropriate hypothesis. Our P-value turns out to be 0.03. Which conclusion is appropriate? Explain.

a) There's a 3% chance that the die is fair.
b) There's a 97% chance that the die is fair.
c) There's a 3% chance that a loaded die could randomly produce the results we observed, so it's reasonable to conclude that the die is fair.
d) There's a 3% chance that a fair die could randomly produce the results we observed, so it's reasonable to conclude that the die is loaded.

19. Relief A company's old antacid formula provided relief for 70% of the people who used it. The company tests a new formula to see if it is better and gets a P-value of 0.27. Is it reasonable to conclude that the new formula and the old one are equally effective? Explain.

20. Cars A survey investigating whether the proportion of today's high school seniors who own their own cars is higher than it was a decade ago finds a P-value of 0.017. Is it reasonable to conclude that more high schoolers have cars? Explain.

21. He cheats? A friend of yours claims that when he tosses a coin he can control the outcome. You are skeptical and want him to prove it. He tosses the coin, and you call heads; it's tails. You try again and lose again.

a) Do two losses in a row convince you that he really can control the toss? Explain.
b) You try a third time, and again you lose. What's the probability of losing three tosses in a row if the process is fair?
c) Would three losses in a row convince you that your friend controls the outcome? Explain.
d) How many times in a row would you have to lose to be pretty sure that this friend really can control the toss? Justify your answer by calculating a probability and explaining what it means.

22. Candy Someone hands you a box of a dozen chocolate-covered candies, telling you that half are vanilla creams and the other half peanut butter. You pick candies at random and discover the first three you eat are all vanilla.

a) If there really were 6 vanilla and 6 peanut butter candies in the box, what is the probability that you would have picked three vanillas in a row?
b) Do you think there really might have been 6 of each? Explain.
c) Would you continue to believe that half are vanilla if the fourth one you try is also vanilla? Explain.

23. Smartphones Many people have trouble setting up all the features of their smartphones, so a company has developed what it hopes will be easier instructions. The goal is to have at least 96% of customers succeed. The company tests the new system on 200 people, of whom 188 were successful. Is this strong evidence that the new system fails to meet the company's goal? A student's test of this hypothesis is shown. How many mistakes can you find?

$H_0: \hat{p} = 0.96$

$H_A: \hat{p} \neq 0.96$

SRS, $0.96(200) > 10$

$\frac{188}{200} = 0.94; \quad SD(\hat{p}) = \sqrt{\frac{(0.94)(0.06)}{200}} = 0.017$

$z = \frac{0.96 - 0.94}{0.017} = 1.18$

$P = P(z > 1.18) = 0.12$

There is strong evidence the new instructions don't work.

24. Obesity 2018 In 2018, the World Health Organization (WHO) reported that 17% of European adults are obese. A health service planning a new awareness campaign in Hungary, where the rate is the highest in Europe, polls a random sample of 1000 adults living there. In this sample, 212 people were found to be obese based on their answers to a health questionnaire.

Do these responses provide strong evidence that the 17% figure is not an accurate proportion? Correct the mistakes you find in a student's attempt to test an appropriate hypothesis.

(https://www.touteleurope.eu/actualite/epidemie-d-obesite-en-europe-un-enjeu-de-sante-publique.html, https://www.who.int/news-room/fact-sheets/detail/obesity-and-overweight, https://www.who.int/fr/news-room/fact-sheets/detail/obesity-and-overweight)

$H_0: \hat{p} = 0.17$

$H_A: \hat{p} > 0.17$

SRS, $1000 \geq 10$

$\frac{212}{1000} = 0.212; \quad SD(\hat{p}) = \sqrt{\frac{0.212 \times 0.799}{1000}} \approx 0.0129$

$z = \frac{0.212 - 0.17}{0.0129} = 3.256$

$P - value - P(z < 3.256) - 0.9998$

There is more than 99.98% chance that the stated percentage is correct.

25. Dowsing In a rural area, only about 30% of the wells that are drilled find adequate water at a depth of 100 feet or less. A local man claims to be able to find water by "dowsing"—using a forked stick to indicate where the well should be drilled. You check with 80 of his customers and find that 27 have wells less than 100 feet deep. What do you conclude about his claim?

a) Write appropriate hypotheses.
b) Check the necessary assumptions and conditions.
c) Perform the mechanics of the test. What is the P-value?
d) Explain carefully what the P-value means in context.
e) What's your conclusion?

26. Abnormalities According to the Eurocat registries, congenital abnormalities affected about 2.16% of births in Europe in the early 2000s. Some people believe that the increase in the number of chemicals in the environment has led to an increase in the incidence of abnormalities. A recent study examined

400 European children and found that 20 of them showed signs of an abnormality. Is this strong evidence that the risk has increased?

(https://www.ncbi.nlm.nih.gov/pmc/articles/PMC5886482/)

a) Write appropriate hypotheses.
b) Check the necessary assumptions and conditions.
c) Perform the mechanics of the test. What is the P-value?
d) Explain carefully what the P-value means in context.
e) What's your conclusion?

27. Absentees A certain Ministry of Education report states that 35% of students in a large region had not been absent from school even once during the previous month. In a survey conducted in a given year, responses from 7500 students in the region showed that this figure had increased to 36.1%. Officials would, of course, be less concerned if student attendance were indeed increasing. Do these figures give evidence of a significant change in student attendance?

a) Write appropriate hypotheses.
b) Check the assumptions and conditions.
c) Perform the test, and find the P-value.
d) State your conclusion.
e) Do you think this difference is meaningful? Explain.

28. Educated mothers According to a Ministry of Education report for a large region, 42% of students reported that their mothers had graduated from college. In a survey conducted in a given year, responses from 7500 students in the region found that this figure had grown to 43%. Is this evidence of an increase in education level among mothers?

a) Write appropriate hypotheses.
b) Check the assumptions and conditions.
c) Perform the test and find the P-value.
d) State your conclusion.
e) Do you think this difference is meaningful? Explain.

29. Contributions, please II We learned in Chapter 16, Exercise 35 that the Care Association for the Blind recently sent letters to a random sample of 50,000 potential donors and received 2380 donations. They've had a contribution rate of 5.5% in past campaigns, but a staff member worries that the rate is lower now that they've redesigned their letter. Is there evidence that the 4.76% they received is evidence of a real drop in the contribution rate?

a) What are the hypotheses?
b) Are the assumptions and conditions for inference met?
c) Do you think the rate would drop? Explain.

30. Take the offer II We saw in Chapter 16, Exercise 36 that Top One tested the effectiveness of a double-miles campaign by recently sending out offers to a random sample of 100,000 cardholders. Of those, 2468 registered for the promotion. Even though this is nearly a 2.5% rate, a staff member suspects that the success rate for the full campaign will be no different than the standard 2.2% rate that they are used to seeing in similar campaigns. What do you predict?

a) What are the hypotheses?
b) Are the assumptions and conditions for inference met?
c) Do you think the rate would change if they use this fundraising campaign? Explain.

31. Pollution A company with a fleet of 150 cars found that the emissions systems of 7 out of the 22 they tested failed to meet pollution control guidelines. Is this strong evidence that more than 20% of the fleet might be out of compliance? Test an appropriate hypothesis and state your conclusion. Be sure the appropriate assumptions and conditions are satisfied before you proceed.

32. Scratch and dent An appliance manufacturer stockpiles washers and dryers in a large warehouse for shipment to retail stores. Sometimes in handling them the appliances get damaged. Even though the damage may be minor, the company must sell those machines at drastically reduced prices. The company goal is to keep the level of damaged machines below 2%. One day an inspector randomly checks 60 washers and finds that 5 of them have scratches or dents. Is this strong evidence that the warehouse is failing to meet the company goal? Test an appropriate hypothesis and state your conclusion. Be sure the appropriate assumptions and conditions are satisfied before you proceed.

33. Twins A national vital statistics report indicated that about 3% of all births produced twins. Is the rate of twin births the same among very young mothers? Data from a large city hospital found that only 7 sets of twins were born to 469 teenage girls. Test an appropriate hypothesis and state your conclusion. Be sure the appropriate assumptions and conditions are satisfied before you proceed.

34. Football 2016 During the first 15 weeks of the 2016 season, the home team won 137 of the 238 regular-season National Football League games. Is this strong evidence of a home field advantage in professional football? Test an appropriate hypothesis and state your conclusion. Be sure the appropriate assumptions and conditions are satisfied before you proceed.

35. WebZine A magazine is considering the launch of an online edition. The magazine plans to go ahead only if it's convinced that more than 25% of current readers would subscribe. The magazine contacted a Simple Random Sample of 500 current subscribers, and 137 of those surveyed expressed interest. What should the company do? Test an appropriate hypothesis and state your conclusion. Be sure the appropriate assumptions and conditions are satisfied before you proceed.

36. Seeds A garden center wants to store leftover packets of vegetable seeds for sale the following spring, but the center is concerned that the seeds may not germinate at the same rate a year later. The manager finds a packet of last year's green bean seeds and plants them as a test. Although the packet claims a germination rate of 92%, only 171 of 200 test seeds sprout. Is this evidence that the seeds have lost viability during a year in storage? Test an appropriate hypothesis and state your conclusion. Be sure the appropriate assumptions and conditions are satisfied before you proceed.

37. Women executives A company is criticized because only 13 of 43 people in executive-level positions are women. The company explains that although this proportion is lower than it might wish, it's not a surprising value given that only 40% of all its employees are women. What do you think? Test an appropriate hypothesis and state your conclusion. Be sure the appropriate assumptions and conditions are satisfied before you proceed.

38. Jury Census data for a certain county show that 19% of the adult residents are Hispanic. Suppose 72 people are called for jury duty and only 9 of them are Hispanic. Does this apparent underrepresentation of Hispanics call into question the fairness of the jury selection system? Explain.

39. Dropouts 2013 According to a report, the overall dropout average rate, which includes all students who do not complete upper secondary education, for a large country was around 15%. Subsequent studies show that the trend in the country is toward a decrease in the percentage of young people abandoning education, except for some regions where the dropout rates are among the highest. Data shows that 171 among 1024 high school students in such a region dropped out in a given year. Is this evidence that their dropout rate may be increasing? Explain.

40. Acid rain A study of the effects of acid rain on trees in a certain forest shows that 46 of 150 trees sampled exhibited some sort of damage from acid rain. This rate seemed to be higher than the 17% quoted in an ecosystem study report on the average proportion of damaged trees in the region. Does the sample suggest that trees in the given forest are more susceptible than trees in the rest of the region? Comment, and write your own conclusions based on an appropriate confidence interval as well as a hypothesis test. Include any assumptions you made about the data.

41. Lost luggage An airline's public relations department says that the airline rarely loses passengers' luggage. It further claims that on those occasions when luggage is lost, 90% is recovered and delivered to its owner within 24 hours. A consumer group that surveyed a large number of air travelers found that only 103 of 122 people who lost luggage on that airline were reunited with the missing items by the next day. Does this cast doubt on the airline's claim? Explain.

42. TV ads A start-up company is about to market a new computer printer. It decides to gamble by running commercials during the Super Bowl. The company hopes that name recognition will be worth the high cost of the ads. The goal of the company is that over 40% of the public recognize its brand name and associate it with computer equipment. The day after the game, a pollster contacts 420 randomly chosen adults and finds that 181 of them know that this company manufactures printers. Would you recommend that the company continue to advertise during Super Bowls? Explain.

43. John Wayne Like a lot of other Americans, John Wayne died of cancer. But is there more to this story? In 1955, Wayne was in Utah shooting the film *The Conqueror*. Across the state line, in Nevada, the United States military was testing atomic bombs. Radioactive fallout from those tests drifted across the filming location. A total of 46 of the 220 people working on the film eventually died of cancer. Cancer experts estimate that one would expect only about 30 cancer deaths in a group this size.

a) Is the death rate among the movie crew unusually high?
b) Does this prove that exposure to radiation increases the risk of cancer?

44. AP Stats 2016 The College Board reported that 60.3% of all students who took the 2016 AP Statistics exam earned scores of 3 or higher. One teacher wondered if the performance of her school was better. She believed that year's students to be typical of those who will take AP Stats at that school and was pleased when 34 of her 54 students achieved scores of 3 or better.

a) How many standard deviations above the national rate did her students score? Does that seem like a lot? Explain.
b) Can she claim that her school is better? Explain.

T **45. Normal temperature again** From the measurements of body temperature in Chapter 17, Exercise 29, you created a confidence interval for the true mean body temperature of healthy adults.

Summary	Temperature
Count	52
Mean	98.285
Median	98.200
MidRange	98.600
StdDev	0.6824
Range	2.800
IntQRange	1.050

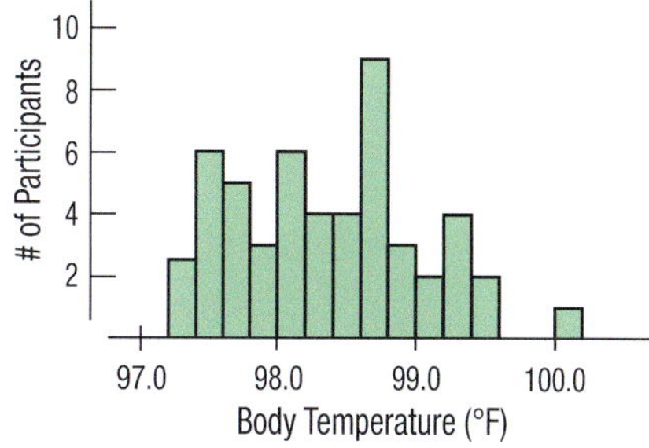

a) 98.6°F is commonly assumed to be "normal." Set up the null and alternative hypotheses for testing this.
b) Check the conditions for performing the test.
c) What do the data say about the test in part a?

46. Burgers again The nutrition lab in Chapter 17, Exercise 38 tested 50 burgers to see if their mean calorie content was less than the 650-calorie upper limit recommended as intake per lunch or dinner meal. The mean calorie content for the sample was 647.45 with a standard deviation of 20 calories. Assume that the assumptions and conditions for the test are met.

a) Test the hypothesis that the mean calorie content meets the regulation.
b) Will a larger sample size ensure that the regulations are met?

47. More pizza The researcher from Chapter 17, Exercise 39 tests whether the mean cholesterol level among those who eat frozen pizza exceeds the value considered to indicate a health risk. She gets a P-value of 0.07. Explain in this context what the "7%" represents.

T **48. Computer lab fees again** The technology committee of Chapter 17, Exercise 42 wants to perform a test to see if the mean amount of time students are spending in the lab has increased from 55 minutes. Here are the data from a random sample of 12 students.

Time	
52	74
57	53
54	136
76	73
62	8
52	62

a) Plot the data. Are any of the observations outliers? Explain.
b) What do you conclude about the claim? If there are outliers, perform the test with and without the outliers present.

T **49. More Ruffles** Recall from Chapter 17, Exercise 55 that students investigated the packaging of potato chips. They purchased 6 bags of Lay's Ruffles marked with a net weight of 28.3 grams. They carefully weighed the contents of each bag, recording the following weights (in grams): 29.3, 28.2, 29.1, 28.7, 28.9, 28.5.

a) Do these data satisfy the assumptions for inference? Explain.
b) Find the mean and standard deviation of the weights.
c) Test the hypothesis that the net weight is as claimed.

T **50. More Doritos** We saw in Chapter 17, Exercise 56 that some students checked 6 bags of Doritos marked with a net weight of 28.3 grams. They carefully weighed the contents of each bag, recording the following weights (in grams): 29.2, 28.5, 28.7, 28.9, 29.1, 29.5.

a) Do these data satisfy the assumptions for inference? Explain.
b) Find the mean and standard deviation of the weights.
c) Test the hypothesis that the net weight is as claimed.

T **51. More popcorn** In Chapter 17, Exercise 57 we saw that Yvon Hopps ran an experiment to determine optimum power and time settings for microwave popcorn. His goal was to find a combination of power and time that would deliver high-quality popcorn with less than 10% of the kernels left unpopped, on average. After experimenting with several bags, he determined that power 9 at 4 minutes was the best combination. To be sure that the method was successful, he popped 8 more bags of popcorn (selected at random) at this setting. All were of high quality, with the following percentages of uncooked popcorn: 7, 13.2, 10, 6, 7.8, 2.8, 2.2, 5.2. Use a test of hypothesis to decide if Yvon has met his goal.

T **52. More ski wax** From Chapter 17, Exercise 58, Bjork Larsen was trying to decide whether to use a new racing wax for cross-country skis. He decided that the wax would be worth the price if he could average less than 55 seconds on a course he knew well, so he planned to study the wax by racing on the course 8 times. His 8 race times were 56.3, 65.9, 50.5, 52.4, 46.5, 57.8, 52.2, and 43.2 seconds. Should he buy the wax? Explain by performing an appropriate hypothesis test.

T **53. Chips Ahoy! again** As we learned in Chapter 17, Exercise 59, in 1998, as an advertising campaign, the Nabisco Company announced a "1000 Chips Challenge," claiming that every 18-ounce bag of their Chips Ahoy! cookies contained at least 1000 chocolate chips. Dedicated statistics students at the Air Force Academy purchased some randomly selected bags of cookies and counted the chocolate chips. Some of their data are given below.

1219 1214 1087 1200 1419 1121 1325 1345
1244 1258 1356 1132 1191 1270 1295 1135

a) Check the assumptions and conditions for inference. Comment on any concerns you have.
b) Test their claim by performing an appropriate hypothesis test.

T **54. More yogurt** As we saw in Chapter 17, Exercise 60, *Consumer Reports* tested 11 brands of vanilla yogurt and found these numbers of calories per serving:

130 160 150 120 120 110 170 160 110 130 90

a) Check the assumptions and conditions.
b) A diet guide claims that you will get an average of 120 calories from a serving of vanilla yogurt. Use an appropriate hypothesis test to comment on their claim.

T **55. Maze** Here are the data from the researcher studying the reaction times of rats from Chapter 17, Exercise 61. Recall that he has a requirement that the maze take about a minute to complete on average.

Time (sec)	
38.4	57.6
46.2	55.5
62.5	49.5
38.0	40.9
62.8	44.3
33.9	93.8
50.4	47.9
35.0	69.2
52.8	46.2
60.1	56.3
55.1	

a) Plot the data. Do you think the conditions are satisfied? Explain.
b) Do you think this maze meets the requirement that the maze takes at most a minute to complete, on average? Perform the test with and without the outlier and write a couple of sentences to the experimenter about whether the maze meets the requirement.

T **56. Facebook friends** According to www.marketingcharts.com/, the average 18–24-year old has 649 Facebook friends. The student who collected the survey data in **Student survey** wanted to test if the mean number is higher at his school. Using his data, test an appropriate hypothesis and write a couple of sentences summarizing what you discover.

T ***57. Maze revisited** A student resampled the **Maze** times from Exercise 55 1000 times. The histogram shows the distributions of the means, and a summary of the quantiles is shown below it.

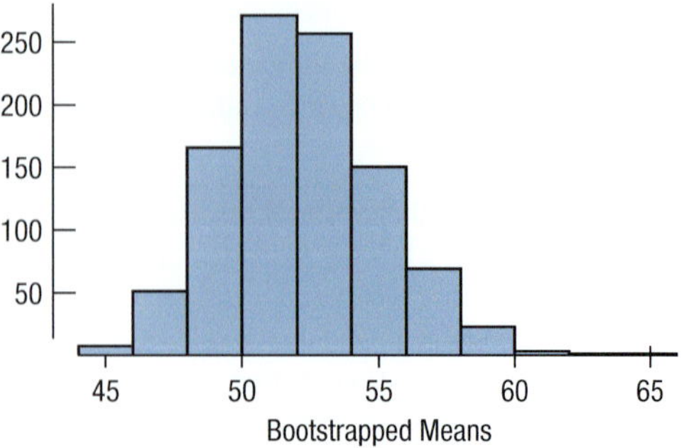

0.05%	0.1%	0.5%	1%	5%	50%	95%	99%	99.5%	99.9%	99.95%
44.657	44.985	45.271	46.119	47.642	52.021	57.220	59.363	59.939	62.088	63.132

a) Find a 99% bootstrap confidence interval for the true mean time it takes to complete the maze.
b) Why is the sampling distribution slightly skewed to the right?
c) How does the skewness affect the confidence interval?
d) What does the confidence interval say about the hypothesis that it takes a minute on average?

T ***58. Facebook friends again** A bootstrap test of the hypothesis in Exercise 56 produced the following distribution (shifted to center the histogram at the hypothesized mean of 649): (Data in **Student survey**)

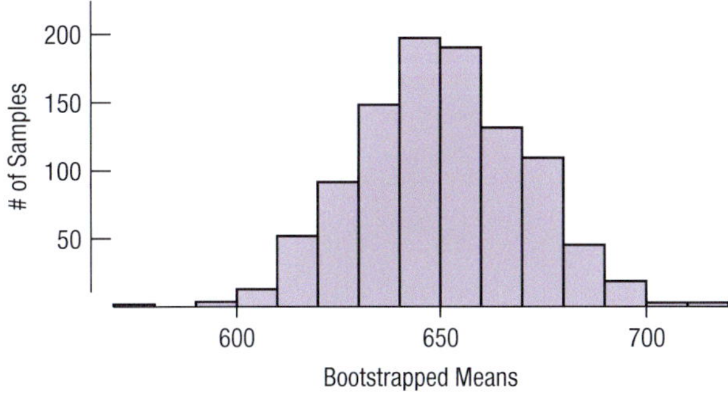

a) What is the P-value of this test?
b) Write a sentence or two with your conclusion.

***59. Cholesterol** According to the National Institutes of Health (NIH) adults should have a total cholesterol level of below 240 mg/dL. The dataset **Framingham** contains the total cholesterol levels of 1406 participants in the Framingham Heart Study.

a) Find the mean total cholesterol of these participants.
b) Find a bootstrapped confidence interval for the mean total cholesterol of American adults, assuming this is a random sample, or representative of U.S. adults.
c) Find the P-value of a bootstrapped test for the hypothesis that the mean cholesterol is 240 mg/dL.
d) Compare the answers in parts b and c and comment.

***60. BMI** According to the Centers for Disease Control (CDC), a body mass index (BMI) of more than 24.9 is considered "overweight." BMI is calculated by the formula:

$$BMI = 703 \times Weight\ (\text{lb})/Height^2\ (\text{in.})$$

Using the data in **Bodyfat**, calculate the BMI of the 250 men in the dataset.

a) Find the mean BMI of the men.
b) Find a bootstrapped confidence interval for the mean BMI of American men, assuming this is a random sample of men.
c) Find the P-value of a bootstrapped test for the hypothesis that the mean BMI is 24.9.
d) Compare the answers in parts b and c and comment.

JUST CHECKING

Answers

1. You can't conclude that the null hypothesis is true. You can conclude only that the experiment was unable to reject the null hypothesis. They were unable, on the basis of 12 patients, to show that aspirin was effective.
2. The null hypothesis is H_0: $p = 0.75$.
3. With a P-value of 0.0001, this is very strong evidence against the null hypothesis. We can reject H_0 and conclude that the improved version of the drug gives relief to a higher proportion of patients.
4. You can't conclude that the null hypothesis is true. You can conclude only that the experiment was unable to reject the null hypothesis. They were unable, on the basis of 12 patients, to show that aspirin was effective in lowering the mean PV level.
5. The P-value shows some (weak) evidence that the mean might be higher. However, this is not a random sample of customers. People who spend more at their website might be more inclined to "like" their page. There is no evidence that the presence of the Facebook page is causing the spending to increase, or that even if it is increasing that it's worth the investment.
6. The parameter of interest is the proportion, p, of all delinquent customers who will pay their bills. H_0: $p = 0.30$ and H_A: $p > 0.30$.
7. The very low P-value leads us to reject the null hypothesis. There is strong evidence that the DVD is more effective in getting people to start paying their debts than just sending a letter had been.
8. All we know is that there is strong evidence to suggest that $p > 0.30$. We don't know how much higher than 30% the new proportion is. We'd like to see a confidence interval to see if the new method is worth the cost.

19

More About Tests and Intervals

WHERE ARE WE GOING?

A news headline reports a "statistically significant" increase in global temperatures. Another says that studies have found no "statistically significant" benefits of taking vitamin C to prevent colds. What does significance really mean? Can failing to reject the null hypothesis sometimes be as important as rejecting it? Knowing what the hypotheses were and how the researchers arrived at their conclusion can help you decide whether you agree with the findings of their studies. We'll look at hypothesis testing in more depth in this chapter.

WHO	Florida motorcycle riders aged 20 and younger involved in motorcycle accidents
WHAT	% wearing helmets
WHEN	2001–2003
WHERE	Florida
WHY	Assessment of injury rates commissioned by the National Highway Traffic Safety Administration (NHTSA)

All motorcycle riders were once required to wear helmets in Florida. In 2000, the law was changed to allow riders 21 and older to ride helmetless. Even though the law did not apply to those under 21, a report by the Preusser Group (one.nhtsa.gov/people/injury/pedbimot/motorcycle/FlaMCReport/pages/Index.htm) suggests that helmet use may have declined in this group as well.

It isn't very practical to survey young motorcycle riders. (How could you construct a sampling frame? If you contacted licensed riders, would they admit to riding illegally without a helmet?) To avoid these problems, the researchers adopted a different strategy. They looked only at police reports of motorcycle accidents, which record whether the rider wore a helmet and give the rider's age.

Before the change in the helmet law, 60% of youths involved in a motorcycle accident had been wearing their helmets. During the three years following the law change, 396 of 781 young riders who were involved in accidents were wearing helmets. That's only 50.7%. Is this evidence of a decline in helmet-wearing, or just the natural fluctuation of such statistics?

EXAMPLE 19.1 Writing Hypotheses

The diabetes drug Avandia® was approved to treat Type 2 diabetes in 1999. But an article in the *New England Journal of Medicine* (*NEJM*)[1] raised concerns that the drug might carry an increased risk of heart attack. This study combined results from a

[1] Steven E. Nissen, M.D., and Kathy Wolski, M.P.H., "Effect of Rosiglitazone on the Risk of Myocardial Infarction and Death from Cardiovascular Causes," *NEJM* 2007; 356.

number of other separate studies to obtain an overall sample of 4485 diabetes patients taking Avandia. People with Type 2 diabetes are known to have about a 20.2% chance of suffering a heart attack within a seven-year period. According to the article's author, Dr. Steven E. Nissen,[2] the risk found in the *NEJM* study was equivalent to a 28.9% chance of heart attack over seven years. The FDA is the government agency responsible for relabeling Avandia to warn of the risk if it is judged to be unsafe. Although the statistical methods they use are more sophisticated, we can get an idea of their reasoning with the tools we have learned.

QUESTION: What hypotheses about the seven-year heart attack risk would you test? Explain.

ANSWER:

$$H_0: p = 0.202$$
$$H_A: p \neq 0.202$$

The parameter of interest is the proportion of diabetes patients suffering a heart attack in seven years. The FDA is interested in any deviation from the expected rate of heart attacks.

19.1 Interpreting P-Values

Which Conditional?

Suppose that as a political science major you are offered the chance to be a White House intern. There would be a very high probability that next summer you'd be in Washington, D.C. That is, $P(\text{Washington} \mid \text{Intern})$ would be high. But if we find a student in Washington, D.C., is it likely that he's a White House intern? Almost surely not; $P(\text{Intern} \mid \text{Washington})$ is low. You can't switch around conditional probabilities. The P-value is $P(\text{data} \mid H_0)$. We might wish we could report $P(H_0 \mid \text{data})$, but these two quantities are NOT the same.

A P-value is a conditional probability. It tells us the probability of getting results at least as unusual as the observed statistic, *given* that the null hypothesis is true. We can write P-value $= P(\text{observed statistic value [or even more extreme]} \mid H_0)$.

Writing the P-value this way helps to make clear that the P-value is *not* the probability that the null hypothesis is true. It is a probability about the data. Because this point is so often confused, let's say that again:

The P-value is not the probability that the null hypothesis is true.[3]

The P-value is not even the conditional probability that the null hypothesis is true given the data.

We can find the P-value, $P(\text{observed statistic value} \mid H_0)$, because we are assuming H_0 and it gives the parameter values we need to calculate the required probability. But there's no direct way to reverse the conditioning and find $P(H_0 \mid \text{observed statistic value})$.[4] As tempting as it may be to say that a P-value of 0.03 means there's a 3% chance that the null hypothesis is true, that just isn't right. All we can say is that, *given the null hypothesis*, there's a 3% chance of observing the statistic value that we have actually observed (or one more unlike the null value).

What to Do with a Low P-Value

How small the P-value has to be for you to reject the null hypothesis depends on a lot of things, not all of which can be precisely quantified. The P-value should serve as a measure of the strength of the evidence against the null hypothesis, but should never serve as a hard and fast rule for decisions. You have to take that responsibility on yourself.

[2]Interview reported in the *New York Times* [May 26, 2007].

[3]And yes, this will be on the exam!

[4]The approach to statistical inference known as Bayesian Statistics addresses the question in just this way, but it requires more advanced mathematics and more assumptions. See p. 411 for more about the founding father of this approach.

STEP-BY-STEP EXAMPLE

Interpreting a Low P-Value

QUESTION: Has helmet use in Florida changed among riders under the age of 21 subsequent to the change in the helmet laws?

THINK

PLAN State the problem and discuss the variables and the W's.

I want to know whether the rate of helmet wearing among Florida's motorcycle riders under the age of 21 changed after the law changed to allow older riders to go without helmets. The proportion before the law was passed was 60% so I'll use that as my null hypothesis value. I have data from accident records showing 396 of 781 young riders were wearing helmets.

HYPOTHESES The null hypothesis is established by the rate set before the change in the law. The study was concerned with safety, so they'll want to know of any change in helmet use, making this a lower-tail test.

$$H_0: p = 0.60$$
$$H_A: p \neq 0.60$$

SHOW

MODEL Check the conditions.

✓ **Independence Assumption**: The data are for riders involved in accidents during a three-year period. Individuals are independent of one another.

✓ **Randomization Condition**: Helmet use data were collected as part of a complex, multistage sample, but it is randomized and great effort is taken to make it representative. It is safe to treat it as though it were a random sample.

✓ **Success/Failure Condition**: We'd expect $np = 781(0.6) = 468.6$ helmeted riders and $nq = 781(0.4) = 312.4$ non-helmeted. Both are at least 10.

Specify the sampling distribution model and name the test.

The conditions are satisfied, so I can use a Normal model and perform a **one-proportion z-test**.

SHOW

MECHANICS Find the standard deviation of the sampling model using the hypothesized proportion.

Find the z-score for the observed proportion.

There were 396 helmet wearers among the 781 accident victims.

$$\hat{p} = \frac{396}{781} = 0.507$$

$$SD(\hat{p}) = \sqrt{\frac{p_0 q_0}{n}} = \sqrt{\frac{(0.60)(0.40)}{781}} = 0.0175$$

$$z = \frac{\hat{p} - p_0}{SD(\hat{p})} = \frac{0.507 - 0.60}{0.0175} = -5.31$$

Make a picture. Sketch a Normal model centered at the hypothesized helmet rate of 60%. This is a lower-tail test, so shade the region to the left of the observed rate.

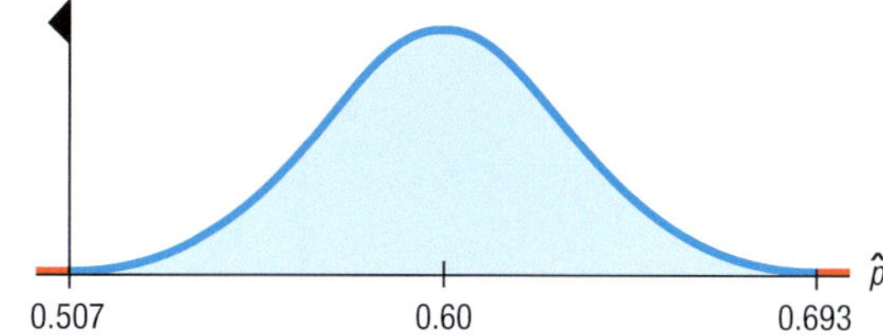

Given this z-score, the P-value is obviously very low.

The observed helmet rate is 5.31 standard deviations below the former rate. The corresponding P-value is less than 0.001.

TELL **CONCLUSION** Link the P-value to your decision about the null hypothesis, and then state your conclusion in context.

The very small P-value says that if the true rate of helmet-wearing among riders under 21 were still 60%, the probability of observing a rate this different from 60% in a sample like this is less than 1 chance in 1000, so I reject the null hypothesis. There is strong evidence that there has been a change in helmet use among riders under 21.

The P-value in the helmet example is quite small—less than 0.001. There is strong evidence that the rate is no longer 60%, but the small P-value by itself says nothing about how much it changed (or in what direction). To answer that question, construct a confidence interval:

$$\hat{p} \pm z^*\sqrt{\frac{\hat{p}\hat{q}}{n}} = 0.507 \pm 1.96(0.0175) = (0.473, 0.541)$$

(using 95% confidence).

The confidence interval provides the additional information. Now we see that even in the best case, the rate is below 55%. Whether a change that large makes an important difference in safety is a judgment that depends on the situation, but not on the P-value. In fact, Florida made the decision to require a motorcycle "endorsement" for all riders. For riders under 21, that requires a motorcycle safety course. The percentage of unendorsed riders involved in crashes dropped considerably afterward.[5]

> "The wise man proportions his belief to the evidence."
>
> —David Hume, "Enquiry Concerning Human Understanding," 1748

How Guilty Is the Suspect?

The smaller the P-value is, the more confident we can be in rejecting the null hypothesis. However, it does not make the null hypothesis any more false. Think again about the jury trial. Our null hypothesis is that the defendant is innocent. But, the bank's security camera showed the robber was male and about the same size as the defendant. We're starting to question his innocence a little. Then witnesses add that the robber wore a blue jacket just like the one the police found in a garbage can behind the defendant's house. Well, if he's innocent, then that doesn't seem very likely, does it? If he's really innocent, the probability that all of these could have happened is getting pretty low. As the evidence rolls in, our P-value may become small enough to be called "beyond a reasonable doubt" and lead to a conviction. Each new piece of evidence strains our belief in the null a bit more. The more compelling the evidence—the more unlikely it would be were he innocent—the more convinced we become that he's guilty.

But even though it may make *us* more confident in declaring him guilty, additional evidence does not make *him* any guiltier. Either he robbed the bank or he didn't. Additional evidence just makes us more confident that we did the right thing when we convicted him. The lower the P-value, the more comfortable we feel about our decision to reject the null hypothesis, but the null hypothesis doesn't get any more false.

> "You're so guilty now."
>
> —Rearview Mirror

[5] www.ridesmartflorida.com

EXAMPLE 19.2

Thinking About the P-Value

RECAP: A *New England Journal of Medicine* paper reported that the seven-year risk of heart attack in diabetes patients taking the drug Avandia was increased from the baseline of 20.2% to an estimated risk of 28.9% and said the P-value was 0.03.[6]

QUESTION: How should the P-value be interpreted in this context?

ANSWER: The P-value is the probability of seeing a high heart attack rate that differs from the expected rate among the people studied if, in fact, taking Avandia really didn't change the risk at all.

What to Do with a High P-Value

Therapeutic touch (TT), taught in many schools of nursing, is a therapy in which the practitioner moves her hands near, but does not touch, a patient in an attempt to manipulate a "human energy field." Therapeutic touch practitioners believe that by adjusting this field they can promote healing. However, no instrument has ever detected a human energy field, and no experiment has ever shown that TT practitioners can detect such a field.

In 1998, the *Journal of the American Medical Association* published a paper reporting work by a then nine-year-old girl.[7] She had performed a simple experiment in which she challenged 15 TT practitioners to detect whether her unseen hand was hovering over their left or right hand (selected by the flip of a coin).

The practitioners "warmed up" with a period during which they could see the experimenter's hand, and each said that they could detect the girl's human energy field. Then a screen was placed so that the practitioners could not see the girl's hand, and they attempted 10 trials each. Overall, of 150 trials, the TT practitioners were successful only 70 times—a success proportion of 46.7%.

The null hypothesis here is that the TT practitioners were just guessing. If that were the case, since the hand was chosen using a coin flip, the practitioners would guess correctly 50% of the time. So the null hypothesis is that $p = 0.5$.

When we find $SD(\hat{p}) = 0.041$ (or 4.1%) we can see that 46.7% is about 1 SD *below* the hypothesized proportion.

$$SD(\hat{p}) = \sqrt{\frac{p_0 q_0}{n}} = \sqrt{\frac{(0.5)(0.5)}{150}} \approx 0.041$$

The observed proportion, $\hat{p}$, is 0.467.

$$z = \frac{\hat{p} - p_0}{SD(\hat{p})} = \frac{0.467 - 0.5}{0.041} = -0.805$$

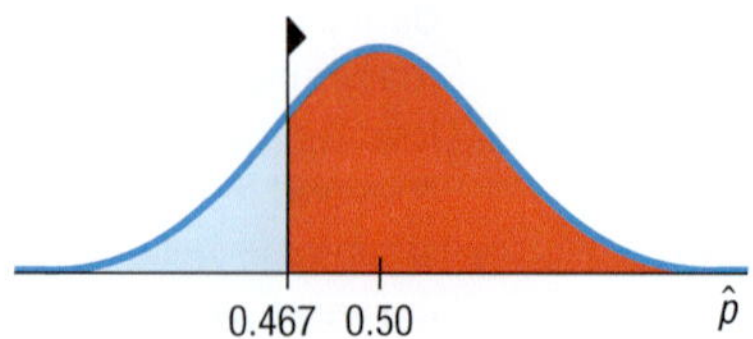

The corresponding confidence interval is (38.7%, 54.5%) correct guesses. The observed success rate is 0.805 standard deviations below the hypothesized mean with a P-value of 0.4208.

[6]The methods in the article are more sophisticated than those we have seen, but the interpretation of the P-value is essentially the same.

[7]L. Rosa, E. Rosa, L. Sarner, and S. Barrett, "A Close Look at Therapeutic Touch," *JAMA* 279(13) [1 April 1998]: 1005–1010.

CLICKBAIT-CORRECTED P-VALUE:

$$P_{CL} = P_{TRADITIONAL} \cdot \frac{CLICK(H_1)}{CLICK(H_0)}$$

H_0: NULL HYPOTHESIS ("CHOCOLATE HAS NO EFFECT ON ATHLETIC PERFORMANCE")

H_1: ALTERNATIVE HYPOTHESIS ("CHOCOLATE BOOSTS ATHLETIC PERFORMANCE")

CLICK(H): FRACTION OF TEST SUBJECTS WHO CLICK ON A HEADLINE ANNOUNCING THAT H IS TRUE

If the practitioners had been highly successful, we would have seen a low P-value. In that case, we would then have concluded that they could actually detect a human energy field.

But that's not what happened. What we observed was a $\hat{p} = 0.467$ success rate. The P-value is very large. Deviations of less than one standard error happen all the time. Not only that, but the practitioners want to show that the method *works*. This deviation is on the "wrong" side of the null hypothesis value. For an insurance company to justify covering this therapy, the practitioners should be doing better than guessing, not worse!

Large P-values just mean that what we've observed isn't surprising given the null model. A big P-value doesn't prove that the null hypothesis is true, but it certainly offers no evidence that it's *not* true. When we see a large P-value, all we can say is that we "don't reject the null hypothesis."

EXAMPLE 19.3

More About P-Values

RECAP: Before the *New England Journal of Medicine* study of diabetics who took Avandia claimed an increased risk of heart attack with a P-value of 0.03, an earlier study (the ADOPT study) had estimated the seven-year risk to be 26.9% and reported a P-value of 0.27.

QUESTION: Why did the researchers in the ADOPT study not express alarm about the increased risk they had seen?

ANSWER: A P-value of 0.27 means that a heart attack rate such as the one they observed could be expected in 27% of similar experiments even if, in fact, there were no increased risk from taking Avandia. That's not remarkable enough to reject the null hypothesis. In other words, the earlier ADOPT study wasn't convincing.

19.2 Alpha Levels and Critical Values

NOTATION ALERT

The first Greek letter, α, is used in statistics for the threshold value of a hypothesis test. You'll hear it referred to as the alpha level. Common values are 0.10, 0.05, 0.01, and 0.001.

Up to now, we have avoided the question of how small the P-value needs to be in order to reject the null. One strategy is to set an arbitrary threshold before collecting the data. Then, if the P-value falls below that "bright-line," reject the null hypothesis and call the result **statistically significant**. The threshold is called an **alpha level** and labeled with the Greek letter α. Common α-levels are 0.10, 0.05, 0.01, and 0.001. There's nothing special or memorable about the letter α, but this is the standard terminology and notation, so we're stuck with it.

It can be hard to justify a particular choice of α, though, so 0.05 is often used. But you have to select the alpha level *before* you look at the data. Otherwise you can be accused of cheating by tuning your alpha level to suit the data.

WHERE DID THE VALUE 0.05 COME FROM?

In 1931, in a famous book called *The Design of Experiments*, Sir Ronald Fisher (see p. 583) discussed the amount of evidence needed to reject a null hypothesis. He said that it was *situation dependent*, but remarked, somewhat casually, that for many scientific applications, 1 out of 20 *might be* a reasonable value. Since then, some people—indeed some entire disciplines—have treated the number 0.05 as sacrosanct.

The alpha level is also called the **significance level**. When we reject the null hypothesis, we say that the test is "significant at that level." For example, we might say that we reject the null hypothesis "at the 5% level of significance."

The automatic nature of the reject/fail-to-reject decision when we use an alpha level may make you uncomfortable. If your P-value falls just slightly above your alpha level, you're not allowed to reject the null. Yet a P-value just barely below the alpha level leads to rejection. If this bothers you, you're in good company. Many statisticians think it better to report the P-value than to base a decision on an arbitrary alpha level.

IT COULD HAPPEN TO YOU!

Of course, if the null hypothesis *is* true, no matter what alpha level you choose, you still have a probability of rejecting the null hypothesis by mistake. This is the rare event we want to protect ourselves against. When we do reject the null hypothesis, no one ever thinks that *this* is one of those rare times. As statistician Stu Hunter notes, "*The statistician says 'rare events do happen—but not to me!'*"

It's in the Stars

Some disciplines carry the idea further and code P-values by their size. In this scheme, a P-value between 0.05 and 0.01 gets highlighted by *. A P-value between 0.01 and 0.001 gets **, and a P-value less than 0.001 gets ***. This can be a convenient summary of the weight of evidence against the null hypothesis if it's not taken too literally. But we warn you against taking the distinctions too seriously and against making a black-and-white decision near the boundaries. The boundaries are a matter of tradition, not science; there is nothing special about 0.05. A P-value of 0.051 should be looked at very seriously and not casually thrown away just because it's larger than 0.05, and one that's 0.009 is not very different from one that's 0.01.

When you decide to declare a verdict, it's always a good idea to report the P-value as an indication of the strength of the evidence. Sometimes it's best to report that the conclusion is not yet clear and to suggest that more data be gathered. (In a trial, a jury may "hang" and be unable to return a verdict.) In these cases, the P-value is the best summary we have of what the data say or fail to say about the null hypothesis.

For every null model, an α-level specifies an area under the curve. The cutpoint for that area is called a **critical value**. For the Normal and t-models, the critical values are denoted z^* and t^*. Before computers were common, P-values were hard to find. It was easier to select a few common alpha levels (0.05, 0.01, 0.001, for example) and find the corresponding critical values for the Normal or t-models. (Of course, for the t-models, you need a different set of critical values for each value of the degrees of freedom.)

In Chapter 5 we used critical values of 1, 2, and 3 when we talked about the 68–95–99.7 Rule. In testing hypotheses, people are more likely to choose alpha levels like 0.1, 0.05, 0.01, or 0.001 with corresponding z^* critical values of 1.645, 1.96, 2.576, and 3.29. (For a one-sided alternative, divide the α-level in half.)

α	z^* Critical Value (two-sided)
0.10	1.645
0.05	1.96
0.01	2.576
0.001	3.29

Table 19.1
The Normal model critical values for common α-levels. For a one-sided test, simply divide the α-level in half.

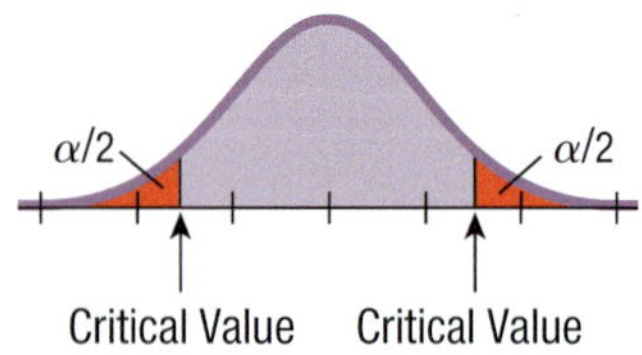

Figure 19.1
For a two-sided alternative, the critical value splits α equally into two tails.

1.96 OR 2

If you want to make a decision on the fly without technology, remember "2." That's our old friend from the 68–95–99.7 Rule. It's roughly the critical value for testing a hypothesis against a two-sided alternative at $\alpha = 0.05$.

A more exact critical value for z is 1.96. For t, the value is 2.000 at 60 *df*. As the number of degrees of freedom increases, this critical value approaches 1.96. If the number of degrees of freedom is less than 60, the critical value is larger. If you are performing a hypothesis test for a mean with only three values, the 0.05 critical value for t is 4.303. But by 28 *df*, it rounds to 2.0, and even for samples bigger than 15, using "2" for the 0.05 critical value is good enough for most decisions.

JUST CHECKING

1. An experiment to test the fairness of a roulette wheel gives a P-value of 0.967. What would you conclude?
2. A town is voting on a change to zoning that will allow a casino in the south part of town, near the mall. A pollster asks 500 people how they will vote and tests whether the proportion of yes votes is 0.50. He reports that the one-sample z-statistic is 4.56. Is the vote close?
3. If the true proportion of yes votes is 50.5%, do you think a sample of 500 people would likely lead to a z-statistic that large? What if the true proportion is 80%?

19.3 Practical vs. Statistical Significance

STATISTICALLY SIGNIFICANT? YES, BUT IS IT IMPORTANT?

A large insurance company mined its data and found a statistically significant ($P = 0.04$) difference between the mean value of policies sold in 2013 and 2014. The difference in the mean values was $9.83. Even though it was statistically significant, management did not see this as an important difference when a typical policy sold for more than $1000. On the other hand, even a clinically important improvement of 10% in cure rate with a new treatment is not likely to be statistically significant in a study of fewer than 225 patients. A small clinical trial would probably not be conclusive.

You've probably heard or read discussions in which a statistical result is called "significant." Sounds important and scientific, doesn't it? But really, all that means is that the test statistic had a P-value lower than the specified alpha level. And if you weren't told that alpha level, it's hard to know exactly what "significant" means. Sometimes the term is used to suggest that the result is meaningful or important. But think about it. A test with a small P-value may be surprising (if you believe the null value), but a small P-value says nothing about the size of the effect—and that's what determines whether the result actually makes a difference.[8] And, as we noted earlier, the null hypothesis may have been an absurd value (e.g., no ingot cracking or every rider wearing a helmet). Don't be lulled into thinking that statistical significance carries with it any sense of practical importance or impact.

When we reject a null hypothesis, what we really care about is the actual change or difference in the data. The difference between the value you see in your data and the null value is called the **effect size**. For large samples, even a small, unimportant effect size can be statistically significant. On the other hand, if the sample is not large enough, even a large financially or scientifically important effect may not be statistically significant.

It's good practice to report a confidence interval for the parameter along with the P-value to indicate the range of plausible values for the parameter. The confidence interval is centered on the observed effect, and puts bounds on how big or small the effect size may actually be.

EXAMPLE 19.4 Are They Speeding?

College Terrace is a neighborhood of Palo Alto, California, on the edge of Stanford University. The town has implemented a series of traffic-calming measures over the past decade including building speed tables, center median islands, and traffic circles and reducing the number of through streets. The posted speed limit in College Terrace is 25 mph. The town reports that speeds and traffic volumes in the neighborhood have been "noticeably reduced," so they are seeking approval to make the changes permanent.

A Palo Alto resident, convinced that cars were still speeding by his house, measured the speeds of every car going by his house by using two infrared photoelectric gates connected to a microcontroller. We can examine 250 of the speeds randomly selected from more than 2000 cars that he recorded.

QUESTION: Is the mean speed of the cars different than 25 mph?

[8]Academic research journals often require statistically significant results before they'll consider a paper for publication. Such a result might then be "significant" to the researcher, whose career depends on publishing. But that may not make the result important to anyone else.

ANSWER: The hypotheses are

$$H_0: \mu = 25 \text{ mph}$$
$$H_A: \mu \neq 25 \text{ mph}$$

Independence Assumption: The data were a census of cars over several days. We have drawn a random sample of about 10% of the data. That choice makes it reasonable to assume that the car speeds in our sample are independent, although cars driving near each other may not have been. That random selection makes independence a safe assumption.

Nearly Normal Condition: The histogram of the data is unimodal and fairly symmetric. With a sample size of 250, the sampling model for the mean will not be affected much by the slight skew to the high side. There are no outliers on either side.

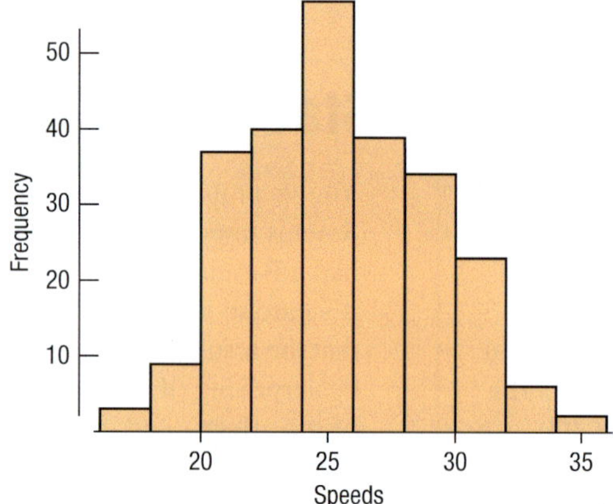

The calculations show:

$$n = 250, \quad df = 249$$
$$\bar{y} = 25.55; \quad s = 3.618$$
$$SE(\bar{y}) = \frac{3.618}{\sqrt{250}} = 0.2288$$
$$t_{249} = \frac{25.55 - 25}{0.2288} = 2.404$$

resulting in a P-value = 0.0162.

With a P-value that small, I reject the null hypothesis and conclude that the mean speed is not 25 mph.

A 95% confidence interval for the true mean speed is (25.099, 26.001), which shows that on average, the cars are, in fact, speeding.

But wait. Yes, the hypothesis test leads us to conclude that the true mean speed is greater than 25 mph. The difference is "statistically significant," but would anyone care? The confidence interval shows that cars are speeding in College Terrace by about 1 mph, on average. Would the town of Palo Alto decide to spend money on traffic-slowing measures? The true mean speed is probably no more than 26 mph. So, although the results are highly statistically significant, they are not *practically significant*. Be careful not to confuse the two.

EXAMPLE 19.5 Making a Decision Based on a Confidence Interval

RECAP: The baseline seven-year risk of heart attacks for diabetics is 20.2%. A *NEJM* study reported a 95% confidence interval equivalent to 20.8% to 40.0% for the risk among patients taking the diabetes drug Avandia.

QUESTION: What did this confidence interval suggest to the FDA about the safety of the drug?

ANSWER: The FDA could be 95% confident that the interval from 20.8% to 40.0% included the true risk of heart attack for diabetes patients taking Avandia. Because the interval doesn't cover the baseline risk of 20.2%, there is evidence that Avandia alters the risk of heart attacks for diabetics. Because the interval is above the baseline, the evidence is that Avandia increases that risk.

The confidence interval from Example 19.5 for the risk of heart attacks for diabetics showed that the risk may have been only 20.8% compared to 20.2%. That doesn't seem practically significant. But the upper end of the interval was 40% and that seemed large enough to cause concern. In fact, the FDA later concluded that the drug may have been responsible for up to 100,000 extra deaths. In 2012 Smith-Kline, the manufacturer of the drug, pled guilty to failing to report clinical data and agreed to pay over $3 billion to settle thousands of lawsuits.

JUST CHECKING

4. In Chapter 18, we encountered a bank that wondered if it could get more customers to make payments on delinquent balances by sending them a DVD urging them to set up a payment plan. Well, the bank just got back the results on their test of this strategy. A 90% confidence interval for the success rate is (0.29, 0.45). Their old send-a-letter method had worked 30% of the time. Can you reject the null hypothesis that the proportion is still 30% at $\alpha = 0.05$? Explain.
5. Given the confidence interval the bank found in their trial of DVDs, what would you recommend that they do? Should they scrap the DVD strategy?
6. A researcher reports that teens are spending more time on average online now than they were 10 years ago and reports that the P-value for the difference is 0.001. Is the result "statistically significant"? Is it practically significant?
7. If the null hypothesis is, in fact, true, how often will a test give a statistically significant result just by chance if the researchers use an α-level of 0.05?

19.4 Errors

Nobody's perfect. Even with lots of evidence, we can still make the wrong decision. In fact, when we perform a hypothesis test, we can make mistakes in *two* ways:

I. The null hypothesis is true, but we mistakenly reject it.

II. The null hypothesis is false, but we fail to reject it.

These two types of errors are known as **Type I** and **Type II errors**. One way to keep the names straight is to remember that we start by assuming the null hypothesis is true, so a Type I error is the first kind of error we could make.

In medical disease testing, the null hypothesis is usually the assumption that a person is healthy. The alternative is that he or she has the disease we're testing for. So a Type I error is a *false positive:* A healthy person is diagnosed with the disease. A Type II error, in which an infected person is diagnosed as disease free, is a *false negative.* These errors have other names, depending on the particular discipline and context.

Which type of error is more serious depends on the situation. In the jury trial, a Type I error occurs if the jury convicts an innocent person. A Type II error occurs if the jury fails to convict a guilty person. Which seems more serious? In medical diagnosis, a false

> **FALSE-POSITIVE CONSEQUENCES**
>
> Some false-positive results mean no more than an unnecessary chest X-ray. But for a drug test or a disease like AIDS, a false-positive result that is not kept confidential could have serious consequences.

negative could mean that a sick patient goes untreated. A false positive might mean that the person must undergo further tests. In a statistics final exam (with H_0: the student has learned only 60% of the material), a Type I error would be passing a student who in fact learned less than 60% of the material, while a Type II error would be failing a student who knew enough to pass. Which of these errors seems more serious? It depends on the situation, the cost, and your point of view.

Here's an illustration of the situations:

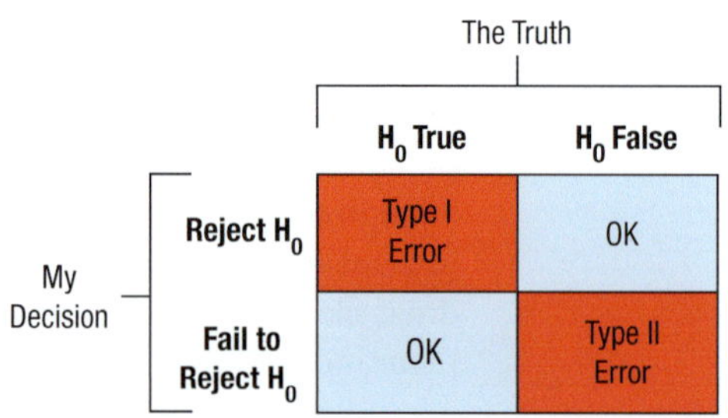

How often will a Type I error occur? It happens when the null hypothesis is true but we've had the bad luck to draw an unusual sample. To reject H_0, the P-value must fall below α. When H_0 is true, that happens *exactly* with probability α. So when you choose level α, you're setting the probability of a Type I error to α.

What if H_0 is not true? Then we can't possibly make a Type I error. You can't get a false positive from a sick person. **A Type I error can happen only when H_0 is true.**

When H_0 is false but we *fail* to reject it, we have made a Type II error. We assign the letter $\boldsymbol{\beta}$ to the probability of this mistake. What's the value of β? That's harder to assess than α because we don't know the true value of the parameter. When H_0 is true, it specifies a single parameter value. But when H_0 is false, we don't know the parameter value and there are many possible values. We can compute the probability β for any parameter value in H_A. But the one we should choose depends on the situation.

Finding β

The null hypothesis specifies a single value for the parameter. So it's easy to calculate the probability of a Type I error. But the alternative gives a whole range of possible values, and we may want to find a β for several of them. Type II errors are often reported on a graph that gives an error rate for different values of β.

We could reduce β for *all* alternative parameter values by increasing α. By making it easier to reject the null, we'd be more likely to reject it whether it's true or not. So we'd reduce β, the chance that we fail to reject a false null—but we'd make more Type I errors. This tension between Type I and Type II errors is inevitable. In the political arena, think of the ongoing debate between those who favor provisions to reduce Type I errors in the courts (supporting defendants' rights, requiring warrants for wiretaps, providing legal representation for those who can't afford it) and those who advocate changes to reduce Type II errors (admitting into evidence confessions made when no lawyer is present, eavesdropping on conferences with lawyers, restricting paths of appeal, etc.).

The only way to reduce *both* types of error is to collect more evidence or, in statistical terms, to collect more data. For example, the Innocence Project (innocenceproject.org) uses DNA testing, which may not have been available or used in some cases, to exonerate the wrongly convicted by introducing additional data to reduce Type I errors in the legal system. Too often, studies fail because their sample sizes are too small to detect the change they are looking for.

Of course, what we really want to do is to detect a false null hypothesis. When H_0 is false and we reject it, we have done the right thing. A test's ability to detect a false null hypothesis is called the power of the test. In a jury trial, power is the ability of the criminal justice system to convict people who are guilty—a good thing!

NOTATION ALERT

In statistics, α is almost always saved for the alpha level. But β has already been used for the parameters of a linear model. Fortunately, it's usually clear whether we're talking about a Type II error probability or the slope or intercept of a regression model.

FINDING THE SAMPLE SIZE

We have seen ways to find a sample size by specifying the margin of error. Choosing the sample size to achieve a specified β (for a particular alternative value) is sometimes more appropriate, but the calculation is more complex and lies beyond the scope of this text.

EXAMPLE 19.6 Thinking About Errors

RECAP: The issue of the *New England Journal of Medicine* (*NEJM*) in which the Avandia study appeared also included an editorial that said, in part, "A few events either way might have changed the findings for myocardial infarction [medical jargon for "heart attack"] or for death from cardiovascular causes. In this setting, the possibility that the findings were due to chance cannot be excluded."

QUESTIONS: What kind of error would the researchers have made if, in fact, their findings were due to chance? What could be the consequences of this error?

ANSWERS: The null hypothesis said the risk didn't change, but the researchers rejected that model and claimed evidence of a higher risk. If these findings were just due to chance, they rejected a true null hypothesis—a Type I error.

If, in fact, Avandia carried no extra risk, then patients might be deprived of its benefits for no good reason.

FISHER AND $\alpha = 0.05$

Why did Sir Ronald Fisher suggest 0.05 as a criterion for testing hypotheses? It turns out that he had in mind small initial studies. Fisher was concerned that they might make too many Type II errors—failing to discover an important effect—if too strict a criterion were used.

The increased risk of Type I errors arising from a generous criterion didn't concern him as much because he expected a significant result to be followed with a replication or a larger study. The probability that two independent studies tested at $\alpha = 0.05$ would both make Type I errors is $0.05 \times 0.05 = 0.0025$, so Fisher didn't think that Type I errors in initial studies were a major concern.

The widespread use of the relatively generous 0.05 criterion even in large studies is most likely not what Fisher had in mind.

Power

When we failed to reject the null hypothesis about TT practitioners, did we prove that they were just guessing? No, it could be that they actually *can* discern a human energy field but we just couldn't tell. The confidence interval was (38.7%, 54.5%), so based on this study, a 54.5% success rate is possible, but so is doing much worse than just guessing.

Whenever a study fails to reject its null hypothesis, it's natural to wonder whether we looked hard enough. Might we have committed a Type II error? Would a narrower confidence interval resolve any remaining doubts? Perhaps β was high because we had too small a sample.

When the null hypothesis actually *is* false, we hope our test is strong enough to reject it. The probability of making that correct decision is called the **power** of the test. That's the probability that it *succeeds* in rejecting the false null hypothesis, so it's just $1 - \beta$.

EXAMPLE 19.7 Errors and Power

RECAP: The study of Avandia published in the *NEJM* actually combined results from 47 different trials—a method called *meta-analysis*. The drug's manufacturer, GlaxoSmithKline (GSK), issued a statement that pointed out, "Each study is designed differently and looks at unique questions: For example, individual studies vary in size and length, in the type of patients who participated, and in the outcomes they investigate."

Nevertheless, by combining data from many studies, meta-analyses can achieve a much larger sample size.

QUESTION: How would increasing the sample size affect the power of the test?

ANSWER: If Avandia really did increase the seven-year heart attack rate, doctors needed to know. To overlook that would have been a Type II error putting patients at greater risk. Increasing the sample size could increase the power of the analysis, because it would make the confidence intervals smaller, making it more likely to detect a true difference.

Effect Size

The effect size is central to how we think about the power of a hypothesis test. It's easier to detect larger effects, so for the same sample size, a larger effect size results in higher power. Small effects are naturally more difficult to detect. They'll result in more Type II errors and therefore lower power. Knowing the effect size and the sample size helps us determine the power. But when we design a study we won't know the effect size, so we can only imagine possible effect sizes and look at their consequences. How can we decide what effect sizes to look at?

One way to think about the effect size is to ask "*How big a difference would matter?*" The answer to this question depends on who is asking it and why. For example, if therapeutic touch practitioners really can detect a human energy field, but only 54.5% of the time, that might not be consistent enough for health insurers to justify covering the cost. But *any* real ability to detect a previously unknown human energy field would be of great interest to scientists.[9]

When designing a study it is important to know what effect size you would consider meaningful and how variable your data are likely to be. Once you know both of those, a bit of algebra gives an estimate of *n*, the sample size you'll need. This is a common calculation to perform when designing a study so you can be sure to gather a large enough sample (and, speaking practically, to gather enough funds to pay for that large a sample).

How can you know how variable your data will be before you run the study? You may need a small "pilot" study or base your calculations on previous similar studies.[10]

Whenever you fail to reject your null hypothesis, you should think about whether your test had sufficient power.

JUST CHECKING

8. Remember the bank that's sending out DVDs to try to get customers to make payments on delinquent loans? It is looking for evidence that the costlier DVD strategy produces a higher success rate than the letters it has been sending. Explain what a Type I error is in this context and what the consequences would be to the bank.

9. What's a Type II error in the bank experiment context, and what would the consequences be?

10. For the bank, which situation has higher power: a strategy that works really well, actually getting 60% of people to pay off their balances, or a strategy that barely increases the payoff rate to 32%? Explain briefly.

[9]And would probably lead to a Nobel Prize.

[10]When estimating a proportion, you're in luck. A good guess of the proportion value will give you an estimate of the standard deviation. And we can always use a "conservative" guess that $p_0 = 0.5$, which gives the largest estimated standard deviation. That's what public opinion pollsters typically do to get an idea of the sample size they need to obtain any margin of error they want.

A Picture Worth $\frac{1}{P(z > 3.09)}$ Words

It makes intuitive sense that the larger the effect size, the easier it should be to see it. Obtaining a larger sample size decreases the probability of a Type II error, so it increases the power. It also makes sense that the more we're willing to accept a Type I error, the less likely we will be to make a Type II error.

> **NOTATION ALERT**
>
> We've attached symbols to many of the p's. Let's keep them straight. p is a true proportion parameter. p_0 is a hypothesized value of p. $\hat{p}$ is an observed proportion. p^* is a critical value of a proportion corresponding to a specified α.

Figure 19.2 shows a good way to visualize the relationships among these concepts. It uses a simple one-sided hypothesis test about a proportion for illustration, but a similar figure could be drawn for other situations.

In testing H_0: $p = p_0$ against the one-sided alternative H_A: $p > p_0$, we'll reject the null if the observed proportion, $\hat{p}$, is big enough. By big enough, we mean $\hat{p} > p^*$ for some critical value p^* (shown as the red region in the right tail of the upper curve). For example, we might be willing to believe the ability of therapeutic touch practitioners if they were successful in 65% of our trials. This is what the upper model shows. It's a picture of the sampling distribution model for the proportion if the null hypothesis were true. We'd make a Type I error whenever the sample gave us $\hat{p} > p^*$, because we would reject the (true) null hypothesis. And unusual samples like that would happen only with probability α.

In reality, though, the null hypothesis is rarely *exactly* true. The lower probability model supposes that H_0 is not true. In particular, it supposes that the true value is p, not p_0. (Perhaps the TT practitioner really can detect the human energy field 72% of the time.)

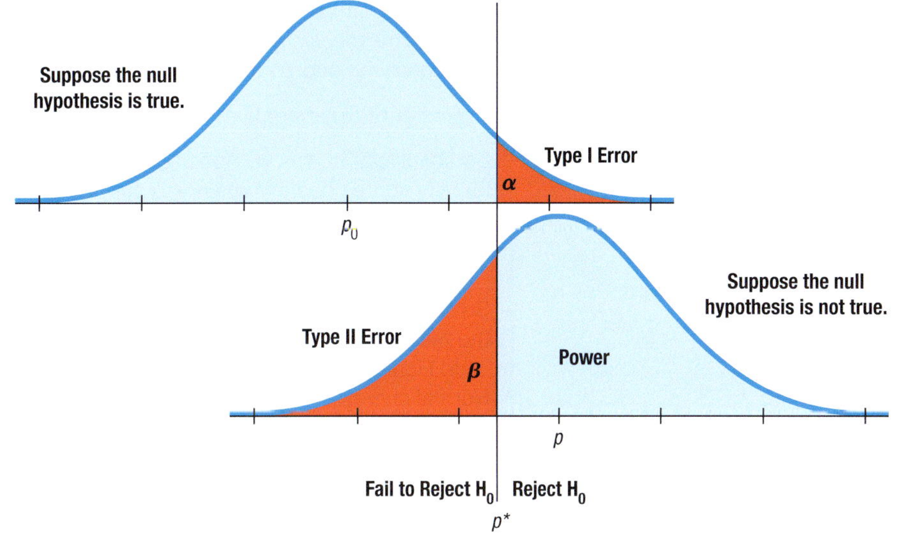

Figure 19.2
The power of a test is the probability that it rejects a false null hypothesis. The upper figure shows the null hypothesis model for a hypothesis test about a proportion. We'd reject the null in a one-sided test if we observed a value of $\hat{p}$ in the red region to the right of the critical value, p^*. The lower figure shows the true model. If the true value of p is greater than p_0, then we're more likely to observe a value that exceeds the critical value and make the correct decision to reject the null hypothesis. The power of the test is the probability represented by the area of the blue region on the right of the lower figure. Of course, even drawing samples whose observed proportions are distributed around p, we'll sometimes get a value in the red region on the left and make a Type II error of failing to reject the null.

The lower figure shows a distribution of possible observed $\hat{p}$ values around this true value. As we saw in the Random Matters section of Chapter 18 (p. 579–580), sample proportions will vary from sample to sample. Because of this sampling variability, sometimes $\hat{p} < p^*$ and we fail to reject the (false) null hypothesis. Suppose a TT practitioner with a true ability level of 72% is actually successful on fewer than 65% of our tests. Then we'd make a Type II error. The area under the curve to the left of p^* in the bottom model represents how often this happens. The probability is β. In this picture, β is less than half, so most of the time we *do* make the right decision. The *power* of the test—the probability that we make the right decision—is shown as the region to the right of p^*. It's $1 - \beta$.

We calculate p^* based on the upper model because p^* depends only on the null model and the alpha level. No matter what the true proportion, no matter whether the practitioners can detect a human energy field 90%, 53%, or 2% of the time, p^* doesn't change. After all, we don't *know* the truth, so we can't use it to determine the critical value. But we always reject H_0 when $\hat{p} > p^*$.

How often we correctly reject H_0 when it's *false* depends on the effect size. We can see from the picture that if the effect size were larger (the true proportion were farther above the hypothesized value), the bottom curve would shift to the right, making the power greater.

We can see several important relationships from this figure:

- Power $= 1 - \beta$.
- Reducing α to lower the chance of committing a Type I error will move the critical value, p^*, to the right (in this example). This will have the effect of increasing β, the probability of a Type II error, and correspondingly reducing the power.
- The larger the real difference between the hypothesized value, p_0, and the true population value, p, the smaller the chance of making a Type II error and the greater the power of the test. If the two proportions are very far apart, the two models will barely overlap, and we will not be likely to make any Type II errors at all—but then, we are unlikely to really need a formal hypothesis-testing procedure to see such an obvious difference. If the TT practitioners were successful almost all the time, we'd be able to see that with even a small experiment.

SENSITIVITY AND SPECIFICITY

The terms *sensitivity* and *specificity* often appear in medical studies. In these studies, the null hypothesis is that the person is healthy and the alternative is that the person is sick. The *specificity* of a test measures its ability to correctly identify only the healthy (the ones who should test negative). It's defined as

number of true negatives/(number of true negatives + number of false positives).

So *specificity* $= 1 - \alpha$. Specificity gives you the probability that the test will correctly identify you as healthy when you are healthy.

The *sensitivity* is the ability to detect the disease. It's defined as

number of true positives/(number of true positives + number of false negatives).

So *sensitivity* $= 1 - \beta =$ power of the test. Sensitivity gives you the probability that the test will correctly identify you as sick when you are sick.

In Figure 19.2, the sensitivity and specificity are the blue areas of the corresponding Normal curves.

Reducing Both Type I and Type II Errors

Figure 19.2 seems to show that if we reduce Type I error, we automatically must increase Type II error. But there is a way to reduce both. Can you think of it?

If we can make both curves narrower, as shown in Figure 19.3, then both the probability of Type I errors and the probability of Type II errors will decrease, and the power of the test will increase.

How can we accomplish that? The only way is to reduce the standard deviations, which we can do by increasing the sample size. Compare the distributions in Figure 19.3 to those in Figure 19.2. You'll see that both sampling distributions are much narrower. The means haven't moved, but if we keep α the same size, the critical value, p^*, moves closer to p_0 and farther from p. That means that larger sample size has increased the power of the test as well, as you can see by the smaller β.

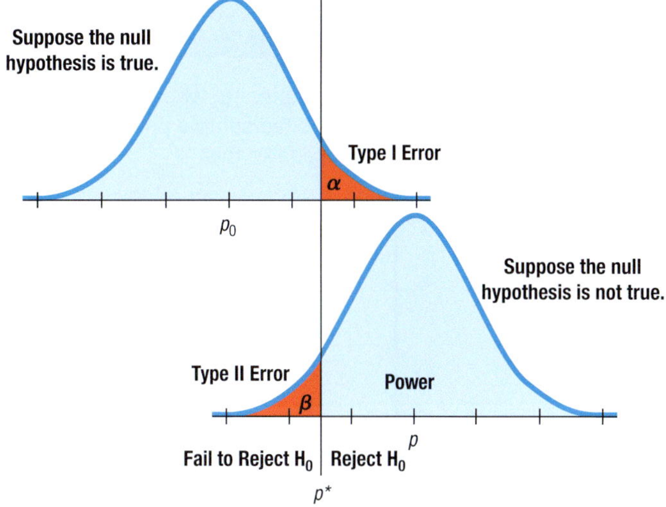

Figure 19.3
Making the standard deviations smaller makes it possible to increase the power while reducing the Type I error rate at the same time.

EXAMPLE 19.8 Sample Size, Errors, and Power

RECAP: The meta-analysis of the risks of heart attacks in patients taking the diabetes drug Avandia combined results from 47 smaller studies. As GlaxoSmithKline (GSK), the drug's manufacturer, pointed out in their rebuttal, "Data from the ADOPT clinical trial did show a small increase in reports of myocardial infarction among the *Avandia*-treated group . . . however, the number of events is too small to reach a reliable conclusion about the role any of the medicines may have played in this finding."

QUESTIONS: Why would this smaller study have been less likely to detect the difference in risk? What are the appropriate statistical concepts for comparing the smaller studies? What are the possible drawbacks of a very large sample?

ANSWERS: Smaller studies are subject to greater sampling variability; that is, the sampling distributions they estimate have a larger standard deviation for the sample proportion. That gives small studies less power: They'd be less able to discern whether an apparently higher risk was merely the result of chance variation or evidence of real danger. The FDA doesn't want to restrict the use of a drug that's safe and effective (Type I error), nor do they want patients to continue taking a medication that puts them at risk (Type II error). Larger sample sizes can reduce the risk of both kinds of error. Greater power (the probability of rejecting a false null hypothesis) means a better chance of spotting a genuinely higher risk of heart attacks. Of course, larger studies are more expensive. Their larger power also makes them able to detect differences that may not be clinically or financially significant. This tension among the power, cost, and sample size lies at the heart of most clinical trial designs.

On September 23, 2010, "The U.S. Food and Drug Administration announced that it will significantly restrict the use of the diabetes drug Avandia (rosiglitazone) to patients with Type 2 diabetes who cannot control their diabetes on other medications. These new restrictions are in response to data that suggest an elevated risk of cardiovascular events, such as heart attack and stroke, in patients treated with Avandia." (www.fda.gov/drugs/DrugSafety/PostmarketDrugSafetyInformationforPatientsand Providers/ucm226956.htm)

RANDOM MATTERS Testing Without Models

In Chapter 18, when we tested whether the ingot cracking rate had changed, we simulated sample proportions based on the hypothesized rate of 20%. The histogram looked like this:

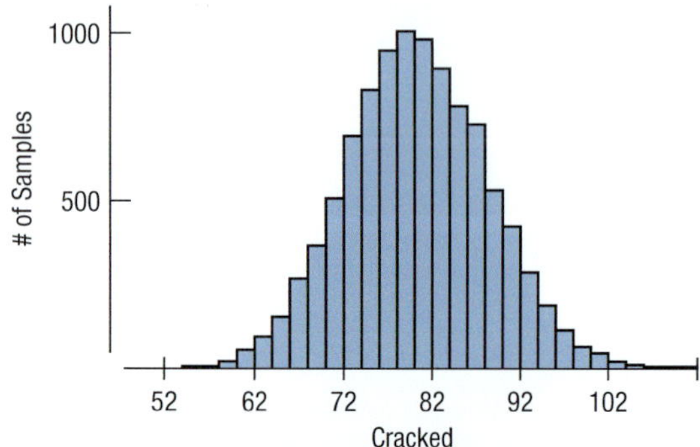

The simulation reported a cracking rate less than 17% (or more than 23%) in about 15% of the trials. This gives us a simulated P-value, since it shows us the probability of a result that extreme or more assuming the null model. (By comparison, the model-based P-value was 0.13.)

Can we do the same thing for means? In Chapter 18 we used a *t*-model and tested to see if the mean gestation times of 70 women differed from the standard duration of 266 days. To test the hypothesis with a simulation, all we need is to generate sample means from a distribution with mean 266 and standard deviation . . . oops! We know the mean, but the standard deviation isn't part of the null hypothesis. We could do a simulation for proportions because there is a special link between proportions and their standard deviations. When a hypothesis specifies the null proportion, that tells us the standard deviation to use for a sample proportion. But that's not true for means.

We can use the bootstrap to make a confidence interval for a mean, as we did in Chapter 17. But here we have a hypothesized value, μ_0. Our hypothesis says, in effect, that our sample was drawn (at random) from a population with that hypothesized mean. One way to test the null hypothesis would be to ask whether the bootstrap interval includes μ_0. If the null value is within the interval, then it is a plausible value and we wouldn't reject the null hypothesis. But this approach doesn't generate a P-value.

We can simulate a sampling distribution by bootstrapping, but the problem is that the bootstrap distribution is centered at the sample mean (260.3 days for the Nashville times). When we test a hypothesis, we *assume* that it is true. And we know the hypothesized mean. So, we shift the bootstrap distribution to be centered at μ_0 adjust by adding $(\mu_0 - \bar{y})$ to each bootstrapped mean.

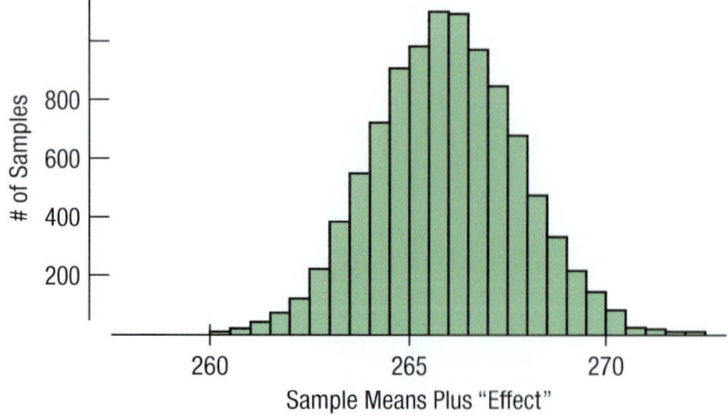

Figure 19.4

Centering the bootstrap distribution at the null hypothesis allows us to see how often a value as extreme as the one we observed would happen. Here it happened 18 times (10 below 260.3 and 8 above 271.7) for a P-value of 0.0018.

A P-value is the probability of seeing the statistic value we actually obtained (here $\bar{y}$) *if* the null hypothesis were true. The shifted bootstrapped distribution depicts a world in which the null is true, so we can ask how many times did this distribution produce a mean as extreme as the sample mean we observed. In this example, how many times did we see a resampled mean less than 260.3 or greater than 271.7? In our trial, the answer was 18 (10 below and 8 above). We ran 10,000 trials, so the proportion of results that extreme was 0.0018, and that's our P-value. When we did a *t*-test, we found a P-value of 0.0026—slightly larger, but fairly close.

Here are three steps to follow to use the bootstrap to test a hypothesis H_0: $\mu = \mu_0$:

1. Shift your data so that they have the mean that was specified by the null hypothesis by adding $(\mu_0 - \bar{y})$ to each value.
2. Perform a bootstrap for the mean of this shifted population, resampling *k* times and recording the means. (You get to choose *k*. Typical values for *k* are at least 1000 or even 10,000.)
3. Find the proportion of times the bootstrap sample means fall as far or farther from the mean of the sampling distribution than the original observed sample mean. That proportion is a P-value. It is considered bad form to report a P-value of zero even if no resampled means are that extreme. After all, there is always the possibility that with more trials you'd see a sample mean that extraordinary.

WHAT CAN GO WRONG?

- **Don't interpret the P-value as the probability that H_0 is true.** The P-value is about the data, not the hypothesis. It's the probability of observing data this unusual, *given* that H_0 is true, not the other way around.
- **Don't believe too strongly in arbitrary alpha levels.** There's not really much difference between a P-value of 0.051 and a P-value of 0.049, but sometimes it's regarded as the difference between night (having to refrain from rejecting H_0) and day (being able to shout to the world that your results are "statistically significant"). It may just be better to report the P-value and a confidence interval and let the world decide along with you.
- **Don't confuse practical and statistical significance.** A large sample size can make it easy to discern even a trivial change from the null hypothesis value. On the other hand, an important difference can be missed if your test lacks sufficient power.
- **Don't forget that in spite of all your care, you might make a wrong decision.** We can never reduce the probability of a Type I error (α) or of a Type II error (β) to zero (but increasing the sample size helps).

CONNECTIONS

All of the hypothesis tests we'll see boil down to the same question: "Is the difference between two quantities large enough to draw conclusions that go beyond this particular sample?" We always measure "how large" by finding a ratio of this difference to the standard deviation of the sampling distribution of the statistic. Using the standard deviation as our ruler for inference is one of the core ideas of statistical thinking.

We've discussed the close relationship between hypothesis tests and confidence intervals. They are two sides of the same coin.

This chapter also has natural links to the discussion of probability, to the Normal model, and to the three previous chapters on inference.

CHAPTER REVIEW

Understand P-values.

- A P-value is the estimated probability of observing a statistic value at least as far from the (null) hypothesized value as the one we have actually observed.
- A small P-value indicates that the statistic we have observed would be unlikely were the null hypothesis true. That leads us to doubt the null.
- A large P-value just tells us that we have insufficient evidence to doubt the null hypothesis. In particular, it does not prove the null to be true.

Know that α is the probability of making a Type I error—rejecting the null hypothesis when it is, in fact, true.

Know that statistical significance simply implies strong evidence that the null hypothesis is false, not that the difference is important.

Know that sometimes P-values are compared to a predetermined α-level, called a critical value, to decide whether to reject the null hypothesis.

Know the value of estimating and reporting the effect size.

- A test may be statistically significant, but practically meaningless if the estimated effect is of trivial importance.

Be aware of the risks of making errors when testing hypotheses.

- A Type I error can occur when rejecting a null hypothesis if that hypothesis is, in fact, true. The probability of this is α.
- A Type II error can occur when failing to reject a null hypothesis if that hypothesis is, in fact, false. The probability of this is β.

Understand the concept of the power of a test.

- We are particularly concerned with power when we fail to reject a null hypothesis.
- The power of a test reports, for a specified effect size, the probability that the test would reject a false null hypothesis.
- Remember that increasing the sample size will generally improve the power of any test.

REVIEW OF TERMS

Statistically significant When the P-value falls below the alpha level, we say that the test is "statistically significant" at that alpha level (p. 603).

Alpha level The threshold P-value that determines when we reject a null hypothesis. If we observe a statistic whose P-value based on the null hypothesis is less than α, we reject that null hypothesis (p. 603).

Significance level The alpha level is also called the significance level, most often in a phrase such as a conclusion that a particular test is "significant at the 5% significance level" (p. 603).

Critical value The value in the sampling distribution model of the statistic whose P-value is equal to the alpha level. The critical value is often denoted with an asterisk, as z^* and t^*, for example (p. 604).

Effect size The difference between the null hypothesis value and the corresponding value that you observe is called the effect size (p. 605).

Type I error The error of rejecting a null hypothesis when in fact it is true (also called a "false positive"). The probability of a Type I error is α (p. 607).

Type II error The error of failing to reject a null hypothesis when in fact it is false (also called a "false negative") (p. 607).

β The probability of a Type II error is commonly denoted β and depends on the effect size (p. 608).

Power The probability that a hypothesis test will correctly reject a false null hypothesis is the power of the test. To find power, we must specify a particular alternative parameter value as the "true" value. For any specific value in the alternative, the power is $1 - \beta$ (p. 609).

EXERCISES

SECTION 19.1

1. True or false Which of the following are true? If false, explain briefly.

a) A P-value of 0.01 means that the null hypothesis is false.
b) A P-value of 0.01 means that the null hypothesis has a 0.01 chance of being true.
c) A P-value of 0.01 is evidence against the null hypothesis.
d) A P-value of 0.01 means we should definitely reject the null hypothesis.

2. False or true Which of the following are true? If false, explain briefly.

a) If the null hypothesis is true, you'll get a high P-value.
b) If the null hypothesis is true, a P-value of 0.01 will occur about 1% of the time.
c) A P-value of 0.90 means that the null hypothesis has a good chance of being true.
d) A P-value of 0.90 is strong evidence that the null hypothesis is true.

3. P-values Which of the following are true? If false, explain briefly.

a) A very high P-value is strong evidence that the null hypothesis is false.
b) A very low P-value proves that the null hypothesis is false.
c) A high P-value shows that the null hypothesis is true.
d) A P-value below 0.05 is always considered sufficient evidence to reject a null hypothesis.

4. More P-values Which of the following are true? If false, explain briefly.

a) A very low P-value provides evidence against the null hypothesis.
b) A high P-value is strong evidence in favor of the null hypothesis.
c) A P-value above 0.10 shows that the null hypothesis is true.
d) If the null hypothesis is true, you can't get a P-value below 0.01.

SECTION 19.2

5. Alpha true and false Which of the following statements are true? If false, explain briefly.

a) Using an alpha level of 0.05, a P-value of 0.04 results in rejecting the null hypothesis.
b) The alpha level depends on the sample size.
c) With an alpha level of 0.01, a P-value of 0.10 results in rejecting the null hypothesis.
d) Using an alpha level of 0.05, a P-value of 0.06 means the null hypothesis is true.

6. Alpha false and true Which of the following statements are true? If false, explain briefly.

a) It is better to use an alpha level of 0.05 than an alpha level of 0.01.
b) If we use an alpha level of 0.01, then a P-value of 0.001 is statistically significant.
c) If we use an alpha level of 0.01, then we reject the null hypothesis if the P-value is 0.001.
d) If the P-value is 0.01, we reject the null hypothesis for any alpha level greater than 0.01.

7. Critical values For each of the following situations, find the critical value(s) for z or t.

a) H_0: $p = 0.5$ vs. H_A: $p \neq 0.5$ at $\alpha = 0.05$.
b) H_0: $p = 0.4$ vs. H_A: $p > 0.4$ at $\alpha = 0.05$.
c) H_0: $\mu = 10$ vs. H_A: $\mu \neq 10$ at $\alpha = 0.05$; $n = 36$.
d) H_0: $p = 0.5$ vs. H_A: $p > 0.5$ at $\alpha = 0.01$; $n = 345$.
e) H_0: $\mu = 20$ vs. H_A: $\mu < 20$ at $\alpha = 0.01$; $n = 1000$.

8. More critical values For each of the following situations, find the critical value for z or t.

a) H_0: $\mu = 105$ vs. H_A: $\mu \neq 105$ at $\alpha = 0.05$; $n = 61$.
b) H_0: $p = 0.05$ vs. H_A: $p > 0.05$ at $\alpha = 0.05$.
c) H_0: $p = 0.6$ vs. H_A: $p \neq 0.6$ at $\alpha = 0.01$.
d) H_0: $p = 0.5$ vs. H_A: $p < 0.5$ at $\alpha = 0.01$; $n = 500$.
e) H_0: $p = 0.2$ vs. H_A: $p < 0.2$ at $\alpha = 0.01$.

SECTION 19.3

9. Significant? Public health officials believe that 98% of children have been vaccinated against measles. A random survey of medical records at many schools across the country found that, among more than 13,000 children, only 97.4% had been vaccinated. A statistician would reject the 98% hypothesis with a P-value of $P < 0.0001$.

a) Explain what the P-value means in this context.
b) The result is statistically significant, but is it important? Comment.

10. Significant again? A new reading program may reduce the number of elementary school students who read below grade level. The company that developed this program supplied materials and teacher training for a large-scale test involving nearly 8500 children in several different school districts. Statistical analysis of the results showed that the percentage of students who did not meet the grade-level goal was reduced from 15.9% to 15.1%. The hypothesis that the new reading program produced no improvement was rejected with a P-value of 0.023.

a) Explain what the P-value means in this context.
b) Even though this reading method has been shown to be significantly better, why might you not recommend that your local school adopt it?

SECTION 19.4

11. Errors For each of the following situations, state whether a Type I, a Type II, or neither error has been made. Explain briefly.

a) A bank wants to know if the enrollment on their website is above 30% based on a small sample of customers. They test H_0: $p = 0.3$ vs. H_A: $p > 0.3$ and reject the null hypothesis. Later they find out that actually 28% of all customers enrolled.
b) A student tests 100 students to determine whether other students on her campus prefer Coke or Pepsi and finds no evidence that preference for Coke is not 0.5. Later, a marketing company tests all students on campus and finds no difference.
c) A human resource analyst wants to know if the applicants this year score, on average, higher on their placement exam than the 52.5 points the candidates averaged last year. She samples 50 recent tests and finds the average to be 54.1 points. She fails to reject the null hypothesis that the mean is 52.5 points. At the end of the year, they find that the candidates this year had a mean of 55.3 points.
d) A pharmaceutical company tests whether a drug lifts the headache relief rate from the 25% achieved by the placebo. They fail to reject the null hypothesis because the P-value is 0.465. Further testing shows that the drug actually relieves headaches in 38% of people.

12. More errors For each of the following situations, state whether a Type I, a Type II, or neither error has been made.

a) A test of H_0: $\mu = 25$ vs. H_A: $\mu > 25$ rejects the null hypothesis. Later it is discovered that $\mu = 24.9$.
b) A test of H_0: $p = 0.8$ vs. H_A: $p < 0.8$ fails to reject the null hypothesis. Later it is discovered that $p = 0.9$.
c) A test of H_0: $p = 0.5$ vs. H_A: $p \neq 0.5$ rejects the null hypothesis. Later it is discovered that $p = 0.65$.
d) A test of H_0: $p = 0.7$ vs. H_A: $p < 0.7$ fails to reject the null hypothesis. Later it is discovered that $p = 0.6$.

CHAPTER EXERCISES

13. P-value A medical researcher tested a new treatment for poison ivy against the traditional ointment. He concluded that the new treatment is more effective. Explain what the P-value of 0.021 means in this context.

14. Another P-value Have harsher penalties and ad campaigns increased seatbelt use among drivers and passengers? Observations of commuter traffic failed to find evidence of a significant change compared with three years ago. Explain what the study's P-value of 0.25 means in this context.

15. Alpha A medical researcher testing new medication for dizziness has concluded that the new tested medicine is significantly more efficient than the current one. She made this decision based on a test using $\alpha = 0.05$. Would she have made the same decision at $\alpha = 0.10$? How about $\alpha = 0.01$? Explain.

16. Alpha, again Environmentalists concerned about the impact of pesticides on human health found that there was no evidence of a higher rate of contamination among farmers using a new organic-based pesticide. They based this conclusion on a test using $\alpha = 0.05$. Would they have made the same decision at $\alpha = 0.10$? How about $\alpha = 0.02$? Explain.

17. Groceries A study conducted by the INSEE about the evolution of grocery shopping habits in France since 1974 shows that among some 5000 respondents (18 years+) surveyed in 2010, 2310 said they go grocery shopping on Sundays.

a) Estimate the percentage of all French adults who go grocery shopping on Sundays. Use a 95% confidence interval. Check the conditions first.
b) A grocery store owner believes that 43% of French adults go grocery shopping on Sundays and is using this as an argument to convince his staff to take shifts on Sundays. Accordingly, he wishes to conduct a hypothesis test to see if the fraction is in fact lower than 43%. What does your confidence interval indicate? Explain.
c) What is the level of significance of this test? Explain.

(www.insee.fr/fr/ffc/ipweb/ip1533/ip1533.pdf)

18. Is the euro fair? Soon after the euro was introduced as currency in Europe, it was widely reported that someone had spun a euro coin 250 times and gotten heads 140 times. We wish to test a hypothesis about the fairness of spinning the coin.

a) Estimate the true proportion of heads. Use a 95% confidence interval. Don't forget to check the conditions.
b) Does your confidence interval provide evidence that the coin is unfair when spun? Explain.
c) What is the significance level of this test? Explain.

19. Mayor approval In April 2015, exactly one year after being elected Mayor of Paris, Anne Hidalgo's approval rating stood at 53% according to a poll organized by IFOP for a French journal surveying 1001 randomly selected Parisians.

a) Make a 95% confidence interval for her approval rating by all Parisians.
b) In March 2016, a new poll also organized by IFOP showed that the rating stood at 42%. Based on the confidence interval, test the null hypothesis that Hidalgo's approval rating remained essentially the same after one year.

(https://www.lejdd.fr/JDD-Paris/Hidalgo-avis-contrastes-sur-son-action-un-an-apres-son-election-selon-un-sondage-Ifop-JDD-726500)

20. Hard times According to a study published by the Guardian in 2018, around 7.7% of workers will work at least two jobs to pay their bills.

We assume this number was calculated based on a random sample of 1000 workers.

a) Estimate the true percentage of workers who will work at least two jobs by constructing a 95% confidence interval.
b) A pundit on a TV news show claimed that only 5.8% of workers work at least two jobs. Use your confidence interval to test whether his claim is plausible given the study findings.

(https://www.theguardian.com/business/grogonomics/2019/mar/18/more-people-than-ever-need-a-second-job-to-help-pay-the-bills)

21. Dogs Canine hip dysplasia is a degenerative disease that causes pain in many dogs. Sometimes advanced warning signs appear in puppies as young as 6 months. A veterinarian checked 42 puppies whose owners brought them to a vaccination clinic, and she found 5 with early hip dysplasia. She considers this group to be a random sample of all puppies.

a) Explain why we cannot use this information to construct a confidence interval for the rate of occurrence of early hip dysplasia among all 6-month-old puppies.
*b) Could you use a bootstrap hypothesis test? Why or why not?

22. Fans A survey of 92 randomly selected people standing in line to enter a tennis game at the Roland-Garros stadium found that 85 of them were fans of Rafael Nadal.

a) Explain why we cannot use this information to construct a confidence interval for the proportion of all people at the game who are fans of Rafael Nadal.
*b) Would a bootstrap confidence interval be a good idea?

23. Loans Before lending someone money, banks must decide whether they believe the applicant will repay the loan. One strategy used is a point system. Loan officers assess information about the applicant, totaling points they award for the person's income level, credit history, current debt burden, and so on. The higher the point total, the more convinced the bank is that it's safe to make the loan. Any applicant with a lower point total than a certain cutoff score is denied a loan.

We can think of this decision as a hypothesis test. Since the bank makes its profit from the interest collected on repaid loans, their null hypothesis is that the applicant will repay the loan and therefore should get the money. Only if the person's score falls below the minimum cutoff will the bank reject the null and deny the loan. This system is reasonably reliable, but, of course, sometimes there are mistakes.

a) When a person defaults on a loan, which type of error did the bank make?
b) Which kind of error is it when the bank misses an opportunity to make a loan to someone who would have repaid it?
c) Suppose the bank decides to lower the cutoff score from 250 points to 200. Is that analogous to choosing a higher or lower value of α for a hypothesis test? Explain.
d) What impact does this change in the cutoff value have on the chance of each type of error?

24. Spam Spam filters try to sort your e-mails, deciding which are real messages and which are unwanted. One method used is a point system. The filter reads each incoming e-mail and assigns points to the sender, the subject, key words in the message, and so on. The higher the point total, the more likely it is that the message is unwanted. The filter has a cutoff value for the point total; any message rated lower than that cutoff passes through to your inbox, and the rest, suspected to be spam, are diverted to the junk mailbox.

We can think of the filter's decision as a hypothesis test. The null hypothesis is that the e-mail is a real message and should go to your inbox. A higher point total provides evidence that the message may be spam; when there's sufficient evidence, the filter rejects the null, classifying the message as junk. This usually works pretty well, but, of course, sometimes the filter makes a mistake.

a) When the filter allows spam to slip through into your inbox, which kind of error is that?
b) Which kind of error is it when a real message gets classified as junk?
c) Some filters allow the user (that's you) to adjust the cutoff. Suppose your filter has a default cutoff of 50 points, but you reset it to 60. Is that analogous to choosing a higher or lower value of α for a hypothesis test? Explain.
d) What impact does this change in the cutoff value have on the chance of each type of error?

25. Second loan Exercise 23 describes the loan score method a bank uses to decide which applicants it will lend money. Only if the total points awarded for various aspects of an applicant's financial condition fail to add up to a minimum cutoff score set by the bank will the loan be denied.

a) In this context, what is meant by the power of the test?
b) What could the bank do to increase the power?
c) What's the disadvantage of doing that?

26. More spam Consider again the points-based spam filter described in Exercise 24. When the points assigned to various components of an e-mail exceed the cutoff value you've set, the filter rejects its null hypothesis (that the message is real) and diverts that e-mail to a junk mailbox.

a) In this context, what is meant by the power of the test?
b) What could you do to increase the filter's power?
c) What's the disadvantage of doing that?

27. Homeowners 2018 According to recent data collected by Eurostat regarding homeownership rates across Europe, in 2018, 42.5% of people in Switzerland own their homes, the lowest rate in Europe. A study conducted in a small town in Switzerland reveals that the ownership rate is much lower. The city council is debating a plan to offer tax breaks to first-time home buyers to encourage people to become homeowners. They decide to adopt the plan on a two-year trial basis and use the data they collect to make a decision about continuing the tax breaks. Since this plan costs the city tax revenues, they will continue to use it only if there is strong evidence that the rate of home ownership is increasing.

a) In words, what will their hypotheses be?
b) What would a Type I error be?
c) What would a Type II error be?
d) For each type of error, tell who would be harmed.
e) What would the power of the test represent in this context?

(https://tradingeconomics.com/european-union/home-ownership-rate; http://appsso.eurostat.ec.europa.eu/nui/show.do?dataset=ilc_lvho02&lang=en)

28. Alzheimer's Testing for Alzheimer's disease can be a long and expensive process, consisting of lengthy tests and medical diagnosis. A group of researchers (Solomon et al., 1998) devised a 7-minute test to serve as a quick screen for the disease for use in the general population of senior citizens. A patient who tested positive would then go through the more expensive battery of tests and medical diagnosis. The authors reported a false-positive rate of 4% and a false-negative rate of 8%.

a) Put this in the context of a hypothesis test. What are the null and alternative hypotheses?
b) What would a Type I error mean?
c) What would a Type II error mean?
d) Which is worse here, a Type I or Type II error? Explain.
e) What is the power of this test?

29. Testing cars A clean air standard requires that vehicle exhaust emissions not exceed specified limits for various pollutants. Many states require that cars be tested annually to be sure they meet these standards. Suppose state regulators double-check a random sample of cars that a suspect repair shop has certified as okay. They will revoke the shop's license if they find significant evidence that the shop is certifying vehicles that do not meet standards.

a) In this context, what is a Type I error?
b) In this context, what is a Type II error?
c) Which type of error would the shop's owner consider more serious?
d) Which type of error might environmentalists consider more serious?

30. Quality control Production managers on an assembly line must monitor the output to be sure that the level of defective products remains small. They periodically inspect a random sample of the items produced. If they find a significant increase in the proportion of items that must be rejected, they will halt the assembly process until the problem can be identified and repaired.

a) In this context, what is a Type I error?
b) In this context, what is a Type II error?
c) Which type of error would the factory owner consider more serious?
d) Which type of error might customers consider more serious?

31. Cars, again As in Exercise 29, state regulators are checking up on repair shops to see if they are certifying vehicles that do not meet pollution standards.

a) In this context, what is meant by the power of the test the regulators are conducting?
b) Will the power be greater if they test 20 or 40 cars? Why?
c) Will the power be greater if they use a 5% or a 10% level of significance? Why?
d) Will the power be greater if the repair shop's inspectors are only a little out of compliance or a lot? Why?

32. Production Consider again the task of the quality control inspectors in Exercise 30.

a) In this context, what is meant by the power of the test the inspectors conduct?
b) They are currently testing 5 items each hour. Someone has proposed that they test 10 instead. What are the advantages and disadvantages of such a change?
c) Their test currently uses a 5% level of significance. What are the advantages and disadvantages of changing to an alpha level of 1%?
d) Suppose that, as a day passes, one of the machines on the assembly line produces more and more items that are defective. How will this affect the power of the test?

33. Equal opportunity? A company is sued for gender discrimination because only 41% of the newly hired candidates were women when 62% of all applicants were women. Is this strong evidence that the company's hiring practices are discriminatory?

a) Is this a one-tailed or a two-tailed test? Why?
b) In this context, what would a Type I error be?
c) In this context, what would a Type II error be?
d) In this context, what is meant by the power of the test?
e) If the hypothesis is tested at the 5% level of significance instead of 1%, how will this affect the power of the test?
f) The lawsuit is based on the hiring of 42 employees. Is the power of the test higher than, lower than, or the same as it would be if it were based on 92 hires?

34. Stop signs Highway safety engineers test new road signs, hoping that increased reflectivity will make them more visible to drivers. Volunteers drive through a test course with several of the new- and old-style signs and rate which kind shows up the best.

a) Is this a one-tailed or a two-tailed test? Why?
b) In this context, what would a Type I error be?
c) In this context, what would a Type II error be?
d) In this context, what is meant by the power of the test?
e) If the hypothesis is tested at the 1% level of significance instead of 5%, how will this affect the power of the test?
f) The engineers hoped to base their decision on the reactions of 50 drivers, but time and budget constraints may force them to cut back to 20. How would this affect the power of the test? Explain.

35. Motivating learning A statistics professor has observed that, for several years, students score an average of 67 points out of 100 on the semester exam.

A colleague suggested implementing regular graded assignments to help the students prepare progressively for the exam and give them a chance to improve their grade. This would require a lot of preparation and follow-up on behalf of the professor.

a) Is this a one-tailed or two-tailed test? Explain.
b) Write the null and alternative hypotheses.
c) In this context, explain what would happen if the professor makes a Type I error.
d) In this context, explain what would happen if the professor makes a Type II error.
e) What is meant by the power of this test?

36. Ads A company is willing to renew its advertising contract with a city's public transportation operator only if the operator can prove that more than 23% of the residents of the city have seen their advertisement posters/billboards during a given week. The transportation operator conducts a survey among 500 randomly selected public transportation users.

a) What are the hypotheses?
b) The operator plans to conduct this test using a 10% level of significance, but the company wants the significance level lowered to 5%. Why?
c) What is meant by the power of this test?
d) For which level of significance will the power of this test be higher? Why?
e) They finally agree to use $\alpha = 0.05$, but the company proposes that the operator surveys 620 people instead of the 500 initially proposed. Will that make the risk of Type II error higher or lower? Explain.

37. Motivating learning, part II 227 students signed up for the Stats course in Exercise 35. They completed regular graded assignments and scored 72 points on the final with a standard deviation of 6.4 points.

a) Should the professor invest time and effort on implementing the assignments system? Support your recommendation with an appropriate test.
b) Does this improvement seem to be practically significant?

38. Testing the ads The company in Exercise 36 contacts 620 people selected at random and only 157 remember the ad.

a) Should the company renew the contract? Support your recommendation with an appropriate test.
b) Explain what your P-value means in this context.

39. TV safety The manufacturer of a metal stand for home TV sets must be sure that its product will not fail under the weight of the TV. Since some larger sets weigh nearly 300 pounds, the company's safety inspectors have set a standard of ensuring that the stands can support an average of over 500 pounds. Their inspectors regularly subject a random sample of the stands to increasing weight until they fail. They test the hypothesis H_0: $\mu = 500$ against H_A: $\mu > 500$, using the level of significance $\alpha = 0.01$. If the sample of stands fails to pass this safety test, the inspectors will not certify the product for sale to the general public.

a) Is this an upper-tail or lower-tail test? In the context of the problem, why do you think this is important?
b) Explain what will happen if the inspectors commit a Type I error.
c) Explain what will happen if the inspectors commit a Type II error.

40. Catheters During an angiogram, heart problems can be examined via a small tube (a catheter) threaded into the heart from a vein in the patient's leg. It's important that the company that manufactures the catheter maintain a diameter of 2.00 mm. (The standard deviation is quite small.) Each day, quality control personnel make several measurements to test H_0: $\mu = 2.00$ against H_A: $\mu \neq 2.00$ at a significance level of $\alpha = 0.05$. If they discover a problem, they will stop the manufacturing process until it is corrected.

a) Is this a one-sided or two-sided test? In the context of the problem, why do you think this is important?
b) Explain in this context what happens if the quality control people commit a Type I error.
c) Explain in this context what happens if the quality control people commit a Type II error.

41. TV safety, revisited The manufacturer of the metal TV stands in Exercise 39 is thinking of revising its safety test.

a) If the company's lawyers are worried about being sued for selling an unsafe product, should they increase or decrease the value of α? Explain.
b) In this context, what is meant by the power of the test?
c) If the company wants to increase the power of the test, what options does it have? Explain the advantages and disadvantages of each option.

42. Catheters, again The catheter company in Exercise 40 is reviewing its testing procedure.

a) Suppose the significance level is changed to $\alpha = 0.01$. Will the probability of a Type II error increase, decrease, or remain the same?
b) What is meant by the power of the test the company conducts?
c) Suppose the manufacturing process is slipping out of proper adjustment. As the actual mean diameter of the catheters produced gets farther and farther above the desired 2.00 mm, will the power of the quality control test increase, decrease, or remain the same?
d) What could they do to improve the power of the test?

43. Two coins In a drawer are two coins. They look the same, but one coin produces heads 90% of the time when spun while the other one produces heads only 30% of the time. You select one of the coins. You are allowed to spin it *once* and then must decide whether the coin is the 90%- or the 30%-head coin. Your null hypothesis is that your coin produces 90% heads.

a) What is the alternative hypothesis?
b) Given that the outcome of your spin is tails, what would you decide? What if it were heads?
c) How large is α in this case?
d) How large is the power of this test? (*Hint:* How many possibilities are in the alternative hypothesis?)
e) How could you lower the probability of a Type I error and increase the power of the test at the same time?

44. Faulty or not? You are in charge of shipping computers to customers. You learn that a faulty chip was put into some of the machines. There's a simple test you can perform, but it's not perfect. All but 4% of the time, a good chip passes the test, but unfortunately, 35% of the bad chips pass the test, too. You have to decide on the basis of one test whether the chip is good or bad. Make this a hypothesis test.

a) What are the null and alternative hypotheses?
b) Given that a computer fails the test, what would you decide? What if it passes the test?
c) How large is α for this test?
d) What is the power of this test? (*Hint:* How many possibilities are in the alternative hypothesis?)

45. Hoops A basketball player with a poor foul-shot record practices intensively during the off-season. He tells the coach that he has raised his proficiency from 65% to 85%. Dubious, the coach asks him to take 10 shots, and is surprised when the player hits 9 out of 10. Did the player prove that he has improved?

a) Suppose the player really is no better than before—still a 65% shooter. What's the probability he can hit at least 9 of 10 shots anyway? (*Hint:* Use a Binomial model.)
b) If that is what happened, now the coach thinks the player has improved when he has not. Which type of error is that?
c) If the player really can hit 85% now, and it takes at least 9 out of 10 successful shots to convince the coach, what's the power of the test?
d) List two ways the coach and player could increase the power to detect any improvement.

46. Pottery An artist experimenting with clay to create pottery with a special texture has been experiencing difficulty with these special pieces. About 40% break in the kiln during firing. Hoping to solve this problem, she buys some more expensive clay from another supplier. She plans to make and fire 10 pieces and will decide to use the new clay if at most one of them breaks.

a) Suppose the new, expensive clay really is no better than her usual clay. What's the probability that this test convinces her to use it anyway? (*Hint:* Use a Binomial model.)
b) If she decides to switch to the new clay and it is no better, what kind of error did she commit?
c) If the new clay really can reduce breakage to only 20%, what's the probability that her test will not detect the improvement?
d) How can she improve the power of her test? Offer at least two suggestions.

47. Chips Ahoy! bootstrapped Exercise 53 of Chapter 18 asked for a Student's t-based test of the hypothesis that every bag of Chips Ahoy! cookies had at least 1000 chips. Here is a histogram of 10,000 bootstrapped means based on the sample of packages in the data file.

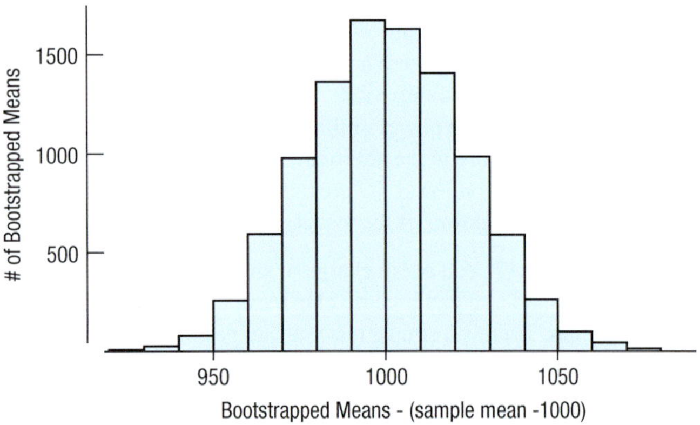

What P-value would you assign to the hypothesis test?

48. Farmed salmon bootstrap In Chapter 18 we saw data from samples of farmed salmon and examined the mirex content. The EPA sets a limit of 0.08 ppm as a maximum safe value. We performed a bootstrap on these data, drawing 10,000 resamples. Here is the resulting histogram of bootstrapped mean mirex levels: (The bootstrapped mean is 0.09134.)

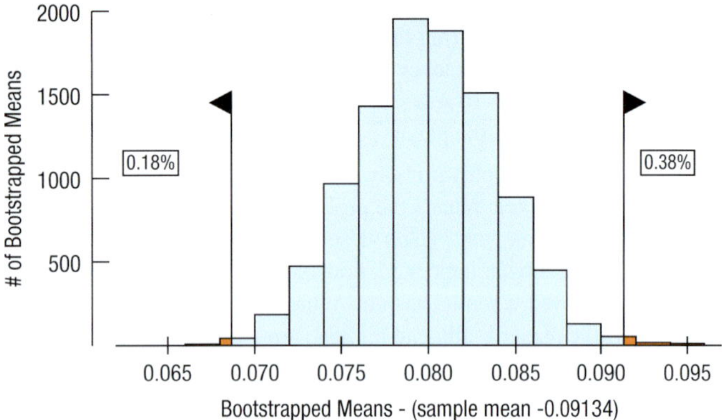

a) What P-value would you assign to the hypothesis test?
b) The EPA is really only concerned with contamination above 0.08 ppm, so perhaps a one-sided test would be more appropriate. What P-value do you now get?

JUST CHECKING

Answers

1. With a p-value of 0.967, you can't reject the null hypothesis. The experiment shows no evidence that the wheel is not fair.
2. No. A z-score of 4.56 says that the observed proportion is more than 4 standard deviations away from the hypothesized value of 50%.
3. No. If the true proportion is 50.5% then the standard deviation of the observed proportion is about 0.022. A z-score of 4.56 would then be more than 10% away from 50%. A true proportion of 80% should yield an even larger z-score than 4.56.
4. The confidence interval is from 29% to 45%. The DVD strategy is more expensive and may not be worth it. We can't distinguish the success rate from 30% given the results of this experiment, but 45% would represent a large improvement. The bank should consider another trial, increasing their sample size to get a narrower confidence interval.
5. At $\alpha = 0.05$, you can't reject the null hypothesis because 0.30 is contained in the 90% confidence interval—it's plausible that sending the DVDs is no more effective than just sending letters.
6. A P-value of 0.001 is "statistically significant" for most reasonable alpha levels. We can't tell whether the difference is practically significant without knowing the actual difference in proportions, which is not given.
7. The alpha level is the probability of making a Type I error, so a test at 5% has a 5% chance of rejecting a true null hypothesis.
8. A Type I error would mean deciding that the DVD success rate is higher than 30% when it really isn't. They would adopt a more expensive method for collecting payments that's no better than the less expensive strategy.
9. A Type II error would mean deciding that there's not enough evidence to say that the DVD strategy works when in fact it does. The bank would fail to discover an effective method for increasing their revenue from delinquent accounts.
10. 60%; the larger the effect size, the greater the power. It's easier to detect an improvement to a 60% success rate than to a 32% rate.

Review of Part V

INFERENCE FOR ONE PARAMETER

Quick Review

Here's a brief summary of the key concepts and skills that you'll need to learn what samples can tell us about the populations from which they are drawn:

- Sampling models describe the variability of sample statistics using a remarkable result called the Central Limit Theorem.
  - When the number of trials is sufficiently large, proportions found in different samples vary according to an approximately Normal model.
  - When samples are sufficiently large, the means of different samples vary, with an approximately Normal model.
  - The variability of sample statistics decreases as sample size increases.
  - Many statistical inference procedures are based on the Central Limit Theorem.
  - Inference can also be performed using randomization methods and the bootstrap.
  - No inference procedure is valid unless the underlying assumptions are true. Always check the conditions before proceeding.
- Statistical inference procedures for means are also based on the Central Limit Theorem, but we don't usually know the population standard deviation. Student's t-models take the additional uncertainty of independently estimating the standard deviation into account.
- A confidence interval uses a sample statistic (such as a proportion or mean) to estimate a range of plausible values for the parameter of a population model.
  - All confidence intervals involve an estimate of the parameter, a margin of error, and a level of confidence.
  - For confidence intervals based on a given sample, the greater the margin of error, the higher the confidence.
  - At a given level of confidence, the larger the sample, the smaller the margin of error.
- A hypothesis test proposes a model for the population, then examines the observed statistics to see if that model is plausible.
  - A null hypothesis suggests a parameter value for the population model. Usually, we assume there is nothing interesting, unusual, or different about the sample results.
  - The alternative hypothesis states what we will believe if the sample results turn out to be inconsistent with our null model.
  - We compare the difference between the statistic and the hypothesized value with the standard deviation of the statistic. It's the sampling distribution of this ratio that gives us a P-value.
  - The P-value of the test is the conditional probability that the null model could produce results at least as extreme as those observed in the sample or the experiment just as a result of sampling error.
  - A low P-value indicates evidence against the null model. If it is sufficiently low, we reject the null model.
  - A high P-value indicates that the sample results are not inconsistent with the null model, so we cannot reject it. However, this does not prove the null model is true.
  - Sometimes we will mistakenly reject the null hypothesis even though it's actually true—that's called a Type I error. If we fail to reject a false null hypothesis, we commit a Type II error.
  - The power of a test measures its ability to detect a false null hypothesis.
  - You can lower the risk of a Type I error by requiring a higher standard of proof (lower P-value) before rejecting the null hypothesis. But this will raise the risk of a Type II error and decrease the power of the test.
  - The only way to increase the power of a test while decreasing the chance of committing either error is to design a study based on a larger sample.

And now for some opportunities to review these concepts and skills . . .

REVIEW EXERCISES

R5.1. Babies The National Perinatal Statistics Unit of the Sydney Children's Hospital reports that the mean birth weight of all babies born in birth centers in Australia in a recent year was 3564 grams—about 7.86 pounds. A Missouri hospital reports that the average weight of 112 babies born there last year was 7.68 pounds, with a standard deviation of 1.31 pounds. We want to see if U.S. babies weigh the same, on average, as Australian babies.

a) Why is this a one-sample test?
b) If we believe the Missouri babies fairly represent American newborns, is there any evidence that U.S. babies and Australian babies do not weigh the same at birth?

R5.2. Archery A champion archer can generally hit the bull's-eye 80% of the time. Suppose she shoots 200 arrows during competition. Let $\hat{p}$ represent the percentage of bull's-eyes she gets (the sample proportion).

a) What are the mean and standard deviation of the sampling distribution model for $\hat{p}$?
b) Is a Normal model appropriate here? Explain.
c) Sketch the sampling model, using the 68–95–99.7 Rule.
d) What's the probability that she gets at least 85% bull's-eyes?

R5.3. Color-blind Medical literature says that about 8% of males are color-blind. A university's introductory psychology course is taught in a large lecture hall. Among the students, there are 325 males. Each semester when the professor discusses visual perception, he shows the class a test for color blindness. The percentage of males who are color-blind varies from semester to semester.

a) Is the sampling distribution model for the sample proportion likely to be Normal? Explain.
b) What are the mean and standard deviation of this sampling distribution model?
c) Sketch the sampling model, using the 68–95–99.7 Rule.
d) Write a few sentences explaining what the model says about this professor's class.

R5.4. Hamsters How large are hamster litters? Among 47 golden hamster litters recorded, there were an average of 7.72 baby hamsters, with a standard deviation of 2.5 hamsters per litter.

a) Create and interpret a 90% confidence interval.
b) Would a 98% confidence interval have a larger or smaller margin of error? Explain.
c) Would you need more or fewer litters to create a 95% confidence interval of the same size as the one in part a)?

R5.5. Polling 2016 The 2016 U.S. presidential election was unusual in several ways. First, the candidate who won the most electoral votes, Donald Trump, did not win the most popular votes. Second, several minor-party candidates received enough votes to possibly affect the outcome. The official results showed that Hillary Clinton received 48.04% of the popular vote, Donald Trump received 45.95%, Gary Johnson got 3.28%, and Jill Stein won 1.06%. After the election, there was much discussion about the polls, which had indicated that Clinton would win. Suppose you had taken a Simple Random Sample of 1000 voters in an exit poll and asked them for whom they had voted.

a) Would you always get 480 votes for Clinton and 459 votes for Trump?
b) In 95% of such polls, your sample proportion of voters for Trump should be between what two values?
c) What might be a problem in finding a 95% confidence interval for the true proportions of Stein voters from this sample?
d) Would you expect the sample proportion of Johnson votes to vary more than, less than, or about the same as the sample proportion of Trump votes? Why?

R5.6. Fake news In a survey of 1002 U.S. adults in December 2016 by Pew Research (www.journalism.org/2016/12/15/many-americans-believe-fake-news-is-sowing-confusion/), 64% of adult respondents say they think that made-up "news" is causing a great deal of confusion about the basic facts of current issues and events.

a) Pew reports a margin of error of $\pm 3.6\%$ for this result. Explain what the margin of error means.
b) Pew reports that 39% of respondents are "very confident" that *they* can recognize fake news. Find a confidence interval for the true proportion who are very confident.
c) Pew reports that only 6% of respondents are not at all confident in their ability to detect fake news. Would a confidence interval for the proportion not confident be wider or narrower than the confidence interval for those who are very confident that you found in part b?

R5.7. Scrabble Using a computer to play many simulated games of Scrabble, researcher Charles Robinove found that the letter "A" occurred in 54% of the hands. This study had a margin of error of $\pm 10\%$. (*Chance*, 15, no. 1 [2002])

a) Explain what the margin of error means in this context.
b) Why might the margin of error be so large?
c) Probability theory predicts that the letter "A" should appear in 63% of the hands. Does this make you concerned that the simulation might be faulty? Explain.

R5.8. Bimodal We are sampling randomly from a distribution known to be bimodal.

a) As our sample size increases, what's the expected shape of the sample's distribution?
b) What's the expected value of our sample's mean? Does the size of the sample matter?
c) How is the variability of sample means related to the standard deviation of the population? Does the size of the sample matter?
d) How is the shape of the sampling distribution model affected by the sample size?

R5.9. Gay marriage In June 2017, Pew Research asked a random sample of 2504 U.S. adults, "Do you strongly favor, favor, oppose, or strongly oppose allowing gays and lesbians to marry legally?" (www.people-press.org/2017/06/26/support-for-same-sex-marriage-grows-even-among-groups-that-had-been-skeptical/). Of those polled, 62% said they favored marriage equality, the highest percentage to date.

a) Create a 95% confidence interval for the percentage of all U.S. adults who support marriage equality.
b) Based on your confidence interval, can you conclude that a majority of U.S. adults support legally recognizing gay marriages? Explain.
c) If pollsters wanted to follow up on this poll with another survey that could determine the level of support for gay marriage to within 2% with 98% confidence, how many people should they poll?

R5.10. Who's the boss? The 2016 State of Women-Owned Businesses Report commissioned by American Express (www.womenable.com/70/the-state-of-women-owned-businesses-in-the-u.s.:-2016) says that, excluding large, publicly traded firms, women-owned firms make up 38% of the privately held firm population, a number that doubled since 1997. You call some firms doing business locally, assuming that the national percentage is true in your area.

a) What's the probability that the first 3 you call are all owned by women?
b) What's the probability that none of your first 4 calls finds a firm that is owned by a woman?
c) Suppose none of your first 5 calls found a firm owned by a woman. What's the probability that your next call does?

R5.11. Living at home According to the U.S. American Community Survey (ACS), by 2014, for the first time in the history of the ACS, more U.S. 18- to 34-year-olds reported living with a parent than were living independently in their own homes. The survey found that 32.1% were living with parents vs. 31.6% heading a household with a spouse or partner. About 14% were living alone, were a single parent, or lived with roommates. The results are based on a survey of 648,118 respondents (www.pewsocialtrends.org/2016/05/24/for-first-time-in-modern-era-living-with-parents-edges-out-other-living-arrangements-for-18-to-34-year-olds/).

a) Create a 95% confidence interval for the proportion of 18- to 34-year-olds who were living with their parents.
b) Interpret the interval in this context.
c) Explain in this context what "95% confidence" means.

R5.12. Polling disclaimer A newspaper article that reported the results of an election poll included the following explanation:

> *The Associated Press poll on the 2016 presidential campaign is based on telephone interviews with 798 randomly selected registered voters from all states except Alaska and Hawaii.*
>
> *The results were weighted to represent the population by demographic factors such as age, sex, region, and education.*
>
> *No more than 1 time in 20 should chance variations in the sample cause the results to vary by more than 4 percentage points from the answers that would be obtained if all Americans were polled.*
>
> *The margin of sampling error is larger for responses of subgroups, such as income categories or those in political parties. There are other sources of potential error in polls, including the wording and order of questions.*

a) Did they describe the 5 W's well?
b) What kind of sampling design could take into account the several demographic factors listed?
c) What was the margin of error of this poll?
d) What was the confidence level?
e) Why is the margin of error larger for subgroups?
f) Which kinds of potential bias did they caution readers about?

R5.13. Enough eggs? One of the important issues for poultry farmers is the production rate—the percentage of days on which a given hen actually lays an egg. Ideally, that would be 100% (an egg every day), but realistically, hens tend to lay eggs on about 3 of every 4 days. ISA Babcock wants to advertise the production rate for the B300 Layer as a 95% confidence interval with a margin of error of $\pm 2\%$. How many hens must they collect data on?

R5.14. Largemouth bass Organizers of a fishing tournament believe that the lake holds a sizable population of largemouth bass. They assume that the weights of these fish have a model that is skewed to the right with a mean of 3.5 pounds and a standard deviation of 2.2 pounds.

a) Explain why a skewed model makes sense here.
b) Explain why you cannot determine the probability that a largemouth bass randomly selected ("caught") from the lake weighs over 3 pounds.
c) Each fisherman in the contest catches 5 fish each day. Can you determine the probability that someone's catch averages over 3 pounds? Explain.
d) The 12 fishermen competing each caught the limit of 5 fish. What's the probability that the total catch of 60 fish averaged more than 3 pounds?

R5.15. Cheating A Rutgers University study found that many high school students cheat on tests. The researchers surveyed a random sample of 4500 high school students nationwide; 74% of them said they had cheated at least once.

a) Create a 90% confidence interval for the level of cheating among high school students. Don't forget to check the appropriate conditions.
b) Interpret your interval.
c) Explain what "90% confidence" means.
d) Would a 95% confidence interval be wider or narrower? Explain without actually calculating the interval.

R5.16. Language Neurological research has shown that in about 80% of people language abilities reside in the brain's left side. Another 10% display right-brain language centers, and the remaining 10% have two-sided language control. (The latter two groups are mainly left-handers.) (*Science News*, 161, no. 24 [2002])

a) We select 60 people at random. Is it reasonable to use a Normal model to describe the possible distribution of the proportion of the group that has left-brain language control? Explain.
b) What's the probability that our group has at least 75% left-brainers?
c) If the group had consisted of 100 people, would that probability be higher, lower, or about the same? Explain why, without actually calculating the probability.
d) How large a group would almost certainly guarantee at least 75% left-brainers? Explain.

R5.17. Religion 2014 The 2014 U.S. Religious Landscape Study interviewed more than 35,000 Americans from all 50 states about their beliefs and the role of religion in their lives. The fastest-growing group is the 22.8% who are "Nones"—those who are not affiliated with any organized religion or have no religious beliefs.

a) Create a 95% confidence interval for the proportion of all American adults who are Nones. (Use 35,000 as the sample size.)
b) Interpret your interval in context.
c) Explain carefully what "95% confidence" means in this context.

R5.18. Teen smoking 2015 The Centers for Disease Control and Prevention reports that 9.3% of surveyed high school students reported in 2015 that they had smoked cigarettes in the past 30 days. A college has 522 students in its freshman class. How likely is it that more than 10% of them are smokers?

R5.19. Alcohol abuse Growing concern about binge drinking among college students has prompted one large state university to conduct a survey to assess the size of the problem on its campus. The university plans to randomly select students and ask how many have been drunk during the past week. If the school hopes to estimate the true proportion among all its students with 90% confidence and a margin of error of $\pm 4\%$, how many students must be surveyed?

R5.20. Errors An auto parts company advertises that its special oil additive will make the engine "run smoother, cleaner, longer, with fewer repairs." An independent laboratory decides to test part of this claim. It arranges to have a taxicab company's fleet of cars use the additive, and follows them to see how many need repairs. At the end of a year the laboratory will compare the percentage of cars that need repairs to the historical repair percentage.

a) What kind of a study is this?
b) Will they do a one-tailed or a two-tailed test?
c) Explain in this context what a Type I error would be.
d) Explain in this context what a Type II error would be.
e) Which type of error would the additive manufacturer consider more serious?
f) If the cabs with the additive do indeed run significantly better, can the company conclude it is an effect of the additive? Can they generalize this result and recommend the additive for all cars? Explain.
g) How might you improve the study?

R5.21. Safety Observers in Texas watched children at play in eight communities. Of the 814 children seen biking, roller skating, or skateboarding, only 14% wore a helmet.

a) Create and interpret an appropriate 95% confidence interval.
b) What concerns do you have about this study that might make your confidence interval unreliable?
c) Suppose we want to do this study again, picking various communities and locations at random, and hope to end up with a 98% confidence interval having a margin of error of $\pm 4\%$. How many children must we observe?

R5.22. Fried PCs A computer company recently experienced a disastrous fire that ruined some of its inventory. Unfortunately, during the panic of the fire, some of the damaged computers were sent to another warehouse, where they were mixed with undamaged computers. The engineer responsible for quality control would like to check out each computer in order to decide whether it's undamaged or damaged. Each computer undergoes a series of 100 tests. The number of tests it fails will be used to make the decision. If it fails more than a certain number, it will be classified as damaged and then scrapped. From past history, the distribution of the number of tests failed is known for both undamaged and damaged computers. The probabilities associated with each outcome are listed in the table below:

Number of Tests Failed	0	1	2	3	4	5	>5
Undamaged (%)	80	13	2	4	1	0	0
Damaged (%)	0	10	70	5	4	1	10

The table indicates, for example, that 80% of the undamaged computers have no failures, while 70% of the damaged computers have 2 failures.

a) To the engineers, this is a hypothesis-testing situation. State the null and alternative hypotheses.
b) Someone suggests classifying a computer as damaged if it fails any of the tests. Discuss the advantages and disadvantages of this test plan.
c) What number of tests would a computer have to fail in order to be classified as damaged if the engineers want to have the probability of a Type I error equal to 5%?
d) What's the power of the test plan in part c?
e) A colleague points out that by increasing α just 2%, the power can be increased substantially. Explain.

R5.23. Power We are replicating an experiment. How will each of the following changes affect the power of our test? Indicate whether it will increase, decrease, or remain the same, assuming that all other aspects of the situation remain unchanged.

a) We increase the number of subjects from 40 to 100.
b) We require a higher standard of proof, changing from $\alpha = 0.05$ to $\alpha = 0.01$.

R5.24. Approval 2016 President Obama was very popular at the end of his eight years in office. A CNN/ORC poll of 1000 U.S. adults conducted in the week before the end of his term found that 63% of Americans said they held a favorable view of the President (elections.huffingtonpost.com/pollster/polls/cnn-27029). Commentators noted that, although popular, Obama was not as popular as Bill Clinton had been at the end of his presidency. Clinton had set a mark of 66% approval. What do you think? Was Obama less popular than the mark set by Clinton?

R5.25. Grade inflation In 1996, 20% of all students at a major university had an overall grade point average of 3.5 or higher (on a scale of 4.0). In 2012, a random sample of 1100 student records found that 25% had a GPA of 3.5 or higher. Is this evidence of grade inflation?

R5.26. Name recognition An advertising agency won't sign an athlete to do product endorsements unless it is sure the person is known to more than 25% of its target audience. The agency always conducts a poll of 500 people to investigate the athlete's name recognition before offering a contract. Then it tests H_0: $p = 0.25$ against H_A: $p > 0.25$ at a 5% level of significance.

a) Why does the company use upper-tail tests in this situation?
b) Explain what Type I and Type II errors would represent in this context, and describe the risk that each error poses to the company.
c) The company is thinking of changing its test to use a 10% level of significance. How would this change the company's exposure to each type of risk?

R5.27. Name recognition, part II The advertising company described in Exercise R5.26 is thinking about signing a WNBA star to an endorsement deal. In its poll, 27% of the respondents could identify her.

a) Fans who never took statistics can't understand why the company did not offer this WNBA player an endorsement contract even though the 27% recognition rate in the poll is above the 25% threshold. Explain it to them.
b) Suppose that further polling reveals that this WNBA star really is known to about 30% of the target audience. Did the company initially commit a Type I or Type II error in not signing her?
c) Would the power of the company's test have been higher or lower if the player were more famous? Explain.

R5.28. Dropouts One study comparing various treatments for the eating disorder anorexia nervosa initially enlisted 198 subjects, but found overall that 105 failed to complete their assigned treatment programs. Construct and interpret an appropriate confidence interval. Discuss any reservations you have about this inference.

R5.29. Women The U.S. Census Bureau reports that 36% of all U.S. businesses are owned by women (www.entrepreneur.com/article/252048). A Colorado consulting firm surveys a random sample of 410 businesses in the Denver area and finds that 164 of them have women owners. Should the firm conclude that its area is unusual? Test an appropriate hypothesis and state your conclusion.

R5.30. Speeding A newspaper report in August 2002 raised the issue of racial bias in the issuance of speeding tickets. The following facts were noted:

- Sixteen percent of drivers registered in New Jersey are black.
- Of the 324 speeding tickets issued in one month on a 65-mph section of the New Jersey Turnpike, 25% went to black drivers.

a) Is the percentage of speeding tickets issued to blacks unusually high compared with the state's registration information?
b) Does this prove that racial profiling was used?
c) What other statistics would you like to know about this situation?

R5.31. Petitions To get a voter initiative on a state ballot, petitions that contain at least 250,000 valid voter signatures must be filed with the Elections Commission. The board then has 60 days to certify the petitions. A group wanting to create a statewide system of universal health insurance has just filed petitions with a total of 304,266 signatures. As a first step in the process, the Board selects an SRS of 2000 signatures and checks them against local voter lists. Only 1772 of them turn out to be valid.

a) What percent of the sample signatures were valid?
b) What percent of the petition signatures submitted must be valid in order to have the initiative certified by the Elections Commission?
c) What will happen if the Elections Commission commits a Type I error?
d) What will happen if the Elections Commission commits a Type II error?
e) Does the sample provide evidence in support of certification? Explain.
f) What could the Elections Commission do to increase the power of the test?

R5.32. Meals A college student is on a "meal program." His budget allows him to spend an average of $10 per day for the semester. He keeps track of his daily food expenses for 2 weeks; the data are given in the table below. Is there strong evidence that he will overspend his food allowance? Explain.

Date	Cost ($)
7/29	15.20
7/30	23.20
7/31	3.20
8/1	9.80
8/2	19.53
8/3	6.25
8/4	0
8/5	8.55
8/6	20.05
8/7	14.95
8/8	23.45
8/9	6.75
8/10	0
8/11	9.01

R5.33. Occupy Wall Street In 2011, the Occupy Wall Street movement protested the concentration of wealth and power in the United States. A 2012 University of Delaware survey asked a random sample of 901 American adults whether they agreed or disagreed with the following statement:

> The Occupy Wall Street protesters offered new insights on social issues.

Of those asked, 59.9% said they strongly or somewhat agreed with this statement. We know that if we could ask the entire population of American adults, we would not find that exactly 59.9% think that Wall Street workers would be willing to break the law to make money. Construct a 95% confidence interval for the true percentage of American adults who agree with the statement.

R5.34. Streams Researchers in the Adirondack Mountains collect data on a random sample of streams each year. One of the variables recorded is the substrate of the stream—the type of soil and rock over which they flow. The researchers found that 69 of the 172 sampled streams had a substrate of shale. Construct a 95% confidence interval for the proportion of Adirondack streams with a shale substrate. Clearly interpret your interval in this context.

R5.35. Skin cancer In February 2012, MedPage Today reported that researchers used vemurafenib to treat metastatic melanoma (skin cancer). Out of 152 patients, 53% had a partial or complete response to vemurafenib.

a) Write a 95% confidence interval for the proportion helped by the treatment, and interpret it in this context.
b) If researchers subsequently hope to produce an estimate (with 95% confidence) of treatment effectiveness for metastatic melanoma that has a margin of error of only 6%, how many patients should they study?

R5.36. Bread Clarksburg Bakery is trying to predict how many loaves to bake. In the past 100 days, the bakery has sold between 95 and 140 loaves per day. Here are a histogram and the summary statistics for the number of loaves sold for the past 100 days.

Summary of Sales	
Mean	103
Median	100
StdDev	9.000
Min	95
Max	140
Lower 25th %tile	97
Upper 25th %tile	105.5

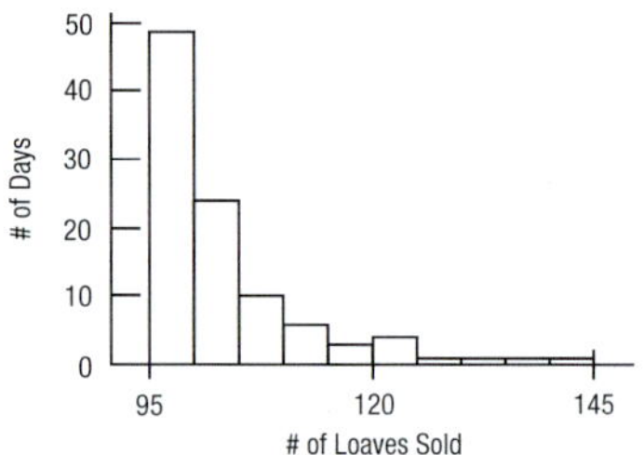

a) Can you use these data to estimate the number of loaves sold on the busiest 10% of all days? Explain.
b) Explain why you can use these data to construct a 95% confidence interval for the mean number of loaves sold per day.
c) Calculate a 95% confidence interval and carefully interpret what that confidence interval means.
d) If the bakery would have been satisfied with a confidence interval whose margin of error was twice as wide, how many days' data could they have used?
e) When the bakery opened, the owners estimated that they would sell an average of 100 loaves per day. Does your confidence interval provide strong evidence that this estimate was incorrect? Explain.

T **R5.37. Fritos®** As a project for an Introductory Statistics course, students checked 6 bags of Fritos marked with a net weight of 35.4 grams. They carefully weighed the contents of each bag, recording the following weights (in grams): 35.5, 35.3, 35.1, 36.4, 35.4, 35.5. Is there evidence that the mean weight of bags of Fritos is less than advertised?

a) Write appropriate hypotheses.
b) Do these data satisfy the assumptions for inference? Explain.
c) Test your hypothesis using all 6 weights.
d) Retest your hypothesis with the one unusually high weight removed.
e) What would you conclude about the stated net weight?

R5.38. And it means? Every statement about a confidence interval contains two parts—the level of confidence and the interval. Suppose that an insurance agent estimating the mean loss claimed by clients after home burglaries created the 95% confidence interval ($1644, $2391).

a) What's the margin of error for this estimate?
b) Carefully explain what the interval means.
c) Carefully explain what the 95% confidence level means.

R5.39. Batteries We work for the "Watchdog for the Consumer" consumer advocacy group. We've been asked to look at a battery company that claims its batteries last an average of 100 hours under normal use. There have been several complaints that the batteries don't last that long, so we decide to test them. To do this we select 16 batteries and run them until they die. They lasted a mean of 97 hours, with a standard deviation of 12 hours.

a) One of the editors of our newsletter (who does not know statistics) says that 97 hours is a lot less than the advertised 100 hours, so we should reject the company's claim. Explain to him the problem with doing that.
b) What are the null and alternative hypotheses?
c) What assumptions must we make in order to proceed with inference?
d) At a 5% level of significance, what do you conclude?
e) Suppose that, in fact, the average life of the company's batteries is only 98 hours. Has an error been made in part d? If so, what kind?

R5.40. Recalls In a car rental company's fleet, 70% of the cars are American brands, 20% are Japanese, and the rest are German. The company notes that manufacturers' recalls seem to affect 2% of the American cars, but only 1% of the others.

a) What's the probability that a randomly chosen car is recalled?
b) What's the probability that a recalled car is American?

20

Comparing Groups

WHERE ARE WE GOING?

Do people really feel better about the economy this month compared to last month, or was that increase just sampling variation? Do students who use technology learn statistics better than those who do not? In practice, it's much more common to compare two groups than to test whether a single proportion or mean is equal to a given number. Comparing two groups is very much like testing one. The standard errors are different, but the concepts are really the same.

Do men take more risks than women? Psychologists have documented that in many situations, men choose riskier behavior than women do. A seat-belt observation study in Massachusetts found that, not surprisingly, male drivers wear seat belts less often than women do.

But the study also found an unexpected pattern. Men's belt-wearing jumped more than 16 percentage points when they were sitting next to a female passenger. Seat-belt use was recorded at 161 locations in Massachusetts, using random-sampling methods developed by the National Highway Traffic Safety Administration (NHTSA). Female drivers wore belts more than 70% of the time, regardless of the sex of their passengers. Of 4208 male drivers with female passengers, 2777 (66.0%) were belted. But among 2763 male drivers with male passengers only, 1363 (49.3%) wore seat belts. This was a random sample, but it suggests there may be a shift in men's risk-taking behavior when women are present.

We can draw conclusions about differences between two groups in the same way that we did for conclusions about a single group. Almost. We just have a few extra things to check and a slightly different calculation. The checks are natural, when you think about them. And the calculations are usually handled by technology.

20.1 A Confidence Interval for the Difference Between Two Proportions

WHO	6971 male drivers
WHAT	Seat-belt use
WHY	Highway safety
WHEN	2007
WHERE	Massachusetts

To find a confidence interval for the difference between two proportions we'll need to know its sampling distribution. In Chapter 16 we saw that for large enough samples, any sample proportion has an approximately Normal sampling distribution. Will the same be true for the difference of two proportions? We can take many samples of the 6971 male drivers. Some will have male passengers and some female passengers, so we can calculate the difference between the proportions of seat-belt users with male and with female passengers. In this way, we can simulate the sampling distribution of the difference. Doing that 10,000 times gives us the distribution in Figure 20.1. Yes, it certainly looks Normal. And it isn't hard to believe that it is centered at the observed difference of proportions. But the *standard deviation* of the difference of two proportions requires a new formula. (For an outline of the derivation, see Section 20.8.)

$$SD(\hat{p}_1 - \hat{p}_2) = \sqrt{\frac{p_1q_1}{n_1} + \frac{p_2q_2}{n_2}}$$

Why Normal?

One of the special properties of the Normal model is that sums and differences of independent Normal random variables also follow a Normal model. That's the reason we use a Normal model for the difference of two independent proportions.

It turns out that what the simulation suggests can be proven mathematically. The sampling distribution model for the difference between two independent proportions is Normal:

The Sampling Distribution Model for the Difference Between Proportions

When certain assumptions and conditions are met, the sampling distribution of $\hat{p}_1 - \hat{p}_2$ is modeled by a Normal distribution centered at $p_1 - p_2$ with standard deviation

$$SD(\hat{p}_1 - \hat{p}_2) = \sqrt{\frac{p_1q_1}{n_1} + \frac{p_2q_2}{n_2}}$$

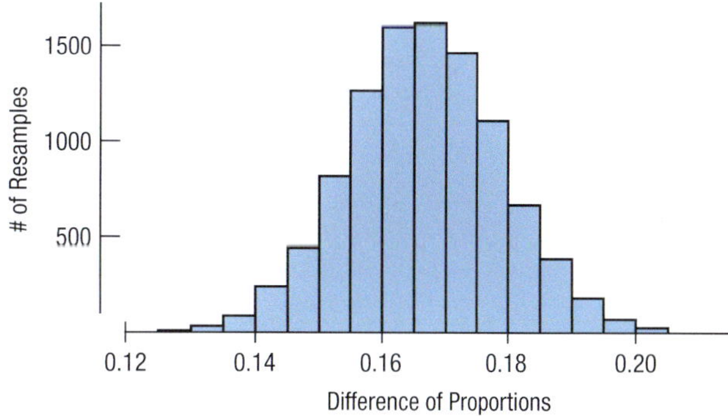

Figure 20.1
The distribution of the simulated difference between the proportions of two groups when the true underlying difference is 16.7% appears to be Normal and centered at 16.7%.

The sampling distribution model and its standard deviation give us all we need to find a margin of error for the difference in proportions—or at least they would if we knew the true proportions, p_1 and p_2. Because we don't know the true values, we'll work with the observed proportions, $\hat{p}_1$ and $\hat{p}_2$, and use $SE(\hat{p}_1 - \hat{p}_2)$ to estimate the standard deviation. The rest is just like a one-proportion z-interval.

$$SE(\hat{p}_1 - \hat{p}_2) = \sqrt{\frac{\hat{p}_1\hat{q}_1}{n_1} + \frac{\hat{p}_2\hat{q}_2}{n_2}}$$

For the seat-belt example $\hat{p}_1 = 0.660$, $\hat{p}_2 = 0.493$, $\hat{q}_1 = 1 - \hat{p}_1 = 0.340$, $\hat{q}_2 = 1 - \hat{p}_2 = 0.507$, and $\sqrt{\frac{\hat{p}_1\hat{q}_1}{4208} + \frac{\hat{p}_2\hat{q}_2}{2763}} = 0.0120$. This standard error agrees with the standard deviation in the simulation of Figure 20.1 to four decimal places.

EXAMPLE 20.1

Finding the Standard Error of a Difference in Proportions

A recent survey of 886 randomly selected teenagers (aged 12–17) found that more than half of them had online profiles.[1] Some researchers and privacy advocates are concerned about the possible access to personal information about teens in public places on the Internet. There appear to be differences between boys and girls in their online behavior. Among teens aged 15–17, 57% of the 248 boys had posted profiles, compared to 70% of the 256 girls. Let's start the process of estimating how large the true gender gap might be.

QUESTION: What's the standard error of the difference in sample proportions?

ANSWER: Because the boys and girls were selected at random, it's reasonable to assume their behaviors are independent, so it's okay to add the variances:

$$SE(\hat{p}_{boys}) = \sqrt{\frac{0.57 \times 0.43}{248}} = 0.0314 \qquad SE(\hat{p}_{girls}) = \sqrt{\frac{0.70 \times 0.30}{256}} = 0.0286$$

$$SE(\hat{p}_{girls} - \hat{p}_{boys}) = \sqrt{0.0314^2 + 0.0286^2} = 0.0425$$

So, as long as certain assumptions and conditions are met (see next section), we have all we need to find a confidence interval for the difference of two proportions.

A Two-Proportion *z*-Interval

When the conditions are met, we are ready to find the confidence interval for the difference of two proportions, $p_1 - p_2$. The confidence interval is

$$(\hat{p}_1 - \hat{p}_2) \pm z^* \times SE(\hat{p}_1 - \hat{p}_2),$$

where we find the standard error of the difference,

$$SE(\hat{p}_1 - \hat{p}_2) = \sqrt{\frac{\hat{p}_1\hat{q}_1}{n_1} + \frac{\hat{p}_2\hat{q}_2}{n_2}},$$

from the observed proportions.

The critical value z^* comes from the Normal model (as signified by the letter "z") and depends on the particular confidence level, C, that we specify.

EXAMPLE 20.2

Finding a Two-Proportion *z*-Interval

RECAP: Among randomly sampled teens aged 15–17, 57% of the 248 boys had posted online profiles, compared to 70% of the 256 girls. We calculated the standard error for the difference in sample proportions to be $SE(\hat{p}_{girls} - \hat{p}_{boys}) = 0.0425$ and found that the assumptions and conditions required for inference checked out.

QUESTION: What does a confidence interval say about the difference in online behavior?

[1] Princeton Survey Research Associates International for the Pew Internet & American Life Project.

ANSWER:

A 95% confidence interval for $p_{girls} - p_{boys}$ is $(\hat{p}_{girls} - \hat{p}_{boys}) \pm z^*SE(\hat{p}_{girls} - \hat{p}_{boys})$:

$$(0.70 - 0.57) \pm 1.96(0.0425)$$
$$0.13 \pm 0.083$$
$$(4.7\%, 21.3\%).$$

We can be 95% confident that among teens aged 15–17, the proportion of girls who post online profiles is between 4.7 and 21.3 percentage points higher than the proportion of boys who do. It seems clear that teen girls are more likely to post profiles than are boys the same age.

20.2 Assumptions and Conditions for Comparing Proportions

For a single proportion, we needed to assume that the observations were independent of each other. If not, the formula for the standard deviation would be wrong. As an extreme, imagine that we conduct a sample by asking the same person whether she or he owns a pet 1000 times. There's not likely to be much variation! The less independent the observations are, the smaller the variance will be. So, for simplicity we assume the observations are independent. When we compare the proportions in two groups, we still need that assumption for *each group*.

If you look more closely at the formula for the standard deviation of the difference, you'll see that each term under the square root is the variance of the sample proportion of one of the groups. The formula for the difference adds those two variances. Mathematically, that requires another strong assumption: We need to assume that the observations in the two groups are independent of each other. We call that the Independent Groups Assumption. Finally, for the Normal model to work well, we need to assume that the sample size in each group is large enough.

Here's a summary of the assumptions and conditions:

Independence

Independence Assumption: *Within each group*, individual responses should be independent of each other. Knowing one response should provide no information about other responses.

Randomization Condition: If the responses are selected with randomization, their independence is likely.

Independent Groups Assumption: The responses in the two groups we're comparing must also be independent *of each other*. Knowing how one group responds should not provide information about the other group. Usually, the independence of the groups from each other is evident from the way the data were collected.

A common violation of the Independent Groups Assumption is for the groups to be *paired*. For example, the opinions of husbands and wives won't always agree, but wives' preferences may provide some information about their husbands' likely responses. We'll deal with this situation in the next chapter.

Sample Size

Each of the groups must be big enough. If you are comparing proportions, you'll need larger groups to estimate proportions that are near 0% or 100%. Check the Success/Failure Condition for each group:

Success/Failure Condition: Both groups are big enough that at least 10 successes and at least 10 failures have been observed in each. For testing hypotheses, we should *expect* at least 10 successes and 10 failures if the null hypothesis is true.

EXAMPLE 20.3 Checking Assumptions and Conditions

RECAP: Among teens aged 15–17, 57% of the 248 boys had posted profiles, compared to 70% of the 256 girls.

QUESTION: Can we use these results to make inferences about all 15–17-year-olds?

ANSWER:

✓ **Randomization Condition:** The sample of boys and the sample of girls were both chosen randomly.

✓ **Independent Groups Assumption:** Because the samples were selected at random, it's reasonable to believe the boys' online behaviors are independent of the girls' online behaviors.

✓ **Success/Failure Condition:** Among the boys, $248(0.57) = 141$ had online profiles and the other $248(0.43) = 107$ did not. For the girls, $256(0.70) = 179$ successes and $256(0.30) = 77$ failures. All counts are at least 10.

Because all the assumptions and conditions are satisfied, it's ok to proceed with inference for the difference in proportions.

(Note that when we find the *observed* counts of successes and failures, we round off to whole numbers. We're using the reported percentages to recover the actual counts.)

STEP-BY-STEP EXAMPLE A Two-Proportion z-Interval

Now we are ready to be more precise about the passenger-based gap in male drivers' seat-belt use. We'll estimate the difference with a confidence interval using a **two-proportion *z*-interval** and follow the four confidence interval steps.

QUESTION: How much difference is there in the proportion of male drivers who wear seat belts when sitting next to a male passenger and the proportion who wear seat belts when sitting next to a female passenger?

THINK **PLAN** State what you want to know. Discuss the variables and the W's.

Identify the parameter you wish to estimate. (It doesn't usually matter mathematically in which direction we subtract, but it is much easier for people to understand the meaning of the interval when we choose the direction with a positive difference.)

I want to estimate the true difference in the population proportion, p_M, of male drivers who wear seat belts when sitting next to a male passenger and p_F, the proportion who wear seat belts when sitting next to a female passenger. The data are from a random sample of drivers in Massachusetts in 2007, observed according to procedures developed by the NHTSA. The parameter of interest is the difference $p_F - p_M$.

Choose and state a confidence level.

I will find a 95% confidence interval for this parameter.

MODEL Think about the assumptions and check the conditions.

✓ **Randomization Condition:** The NHTSA methods are more complex than an SRS, but they result in a suitable random sample so it's reasonable to assume that the driver behavior is independent from car to car.

✓ **Independent Groups Assumption:** It's reasonable to believe that seat-belt use among drivers with male passengers and those with female passengers are independent.

We are working with estimated proportions, so the Success/Failure Condition must hold for each group.

✓ **Success/Failure Condition**: Among male drivers with female passengers, 2777 wore seat belts and 1431 did not; of those driving with male passengers, 1363 wore seat belts and 1400 did not. Each group contained far more than 10 successes and 10 failures.

State the sampling distribution model for the statistic.

Choose your method.

Under these conditions, the sampling distribution of the difference between the sample proportions is approximately Normal, so I'll find a two-proportion z-interval.

SHOW **MECHANICS** Construct the confidence interval.

I know

$$n_F = 4208,\ n_M = 2763.$$

The observed sample proportions are

$$\hat{p}_F = \frac{2777}{4208} = 0.660,\ \hat{p}_M = \frac{1363}{2763} = 0.493.$$

As often happens, the key step in finding the confidence interval is estimating the standard deviation of the sampling distribution model of the statistic. Here the statistic is the difference in the proportions of men who wore seat belts when they had a female passenger and those who had a male passenger. Substitute the data values into the formula.

I'll estimate the SD of the difference with

$$SE(\hat{p}_F - \hat{p}_M) = \sqrt{\frac{\hat{p}_F\hat{q}_F}{n_F} + \frac{\hat{p}_M\hat{q}_M}{n_M}}$$

$$= \sqrt{\frac{(0.660)(0.340)}{4208} + \frac{(0.493)(0.507)}{2763}}$$

$$= 0.012.$$

The sampling distribution is Normal, so the critical value for a 95% confidence interval, z^*, is 1.96. The margin of error is the critical value times the SE.

$$ME = z^* \times SE(\hat{p}_F - \hat{p}_M)$$

$$= 1.96(0.012) = 0.024$$

The confidence interval is the statistic $\pm ME$.

The observed difference in proportions is $\hat{p}_F - \hat{p}_M = 0.660 - 0.493 = 0.167$, so the 95% confidence interval is

$$0.167 \pm 0.024$$

or 14.3% to 19.1%.

TELL **CONCLUSION** Interpret your confidence interval in the proper context. (Remember: We're 95% confident that our interval captured the true difference.)

I am 95% confident that the proportion of male drivers who wear seat belts when driving with a female passenger is between 14.3 and 19.1 percentage points higher than the proportion who wear seat belts when driving with a male passenger.

This is an interesting result—but be careful not to try to say too much! In Massachusetts, overall seat-belt use is lower than the national average, so these results may not generalize to other states. You can probably think of several alternative explanations. For example, perhaps age is a lurking variable: Maybe older men are more likely to wear seat belts and also more likely to be driving with their wives. Or maybe men who don't wear seat belts have trouble attracting women!

JUST CHECKING

A public broadcasting station plans to launch a special appeal for additional contributions from current members. Unsure of the most effective way to contact people, they run an experiment. They randomly select two groups of current members. They send the same request for donations to everyone, but it goes to one group by e-mail and to the other group by regular mail. The station was successful in getting contributions from 26% of the members they e-mailed but only from 15% of those who received the request by regular mail. A 90% confidence interval estimated the difference in donation rates to be 11% $\pm$ 7%.

1. Interpret the confidence interval in this context.
2. Based on this confidence interval, what conclusion would we reach if we tested the hypothesis that there's no difference in the response rates to the two methods of fundraising? Explain.

20.3 The Two-Sample *z*-Test: Testing the Difference Between Proportions

WHO	Randomly sampled adults in different countries
WHAT	Proportion who use their laptops or computers in the hour before sleep
WHEN	2013
WHERE	United States, Japan, Canada, Germany, Mexico, United Kingdom
WHY	To study sleep behaviors

The 2013 International Bedroom Poll[2] surveyed about 250 randomly sampled adult residents in each of six countries asking about their sleep habits. Sleep researchers generally report that:

> . . . people who regularly use their computers or laptops in the hour before trying to go to sleep are *less* likely to report getting a good night's sleep . . . , more likely to be categorized as "sleepy" . . . , and more likely to drive drowsy . . . than their counterparts.[3]

The poll found that 128, or 51%, of 251 U.S. respondents reported using a computer, laptop, or electronic tablet in the hour before trying to go to sleep almost every night. By contrast, 162, or 65%, of 250 Japanese respondents made the same admission. Is this difference of 14% real, or is it likely to be due only to natural fluctuations in the sample?

The question calls for a hypothesis test. Now the parameter of interest is the true *difference* between the (reported) pre-sleep habits of the two groups.

What's the appropriate null hypothesis? That's easy here. We hypothesize that there is no difference in the proportions. This is such a natural null hypothesis that we rarely consider any other. But instead of writing $H_0: p_1 = p_2$, we treat the difference as the parameter that we are testing:

$$H_0: p_1 - p_2 = 0.$$

The standard error of the difference in proportions is

$$SE(\hat{p}_1 - \hat{p}_2) = \sqrt{\frac{\hat{p}_1\hat{q}_1}{n_1} + \frac{\hat{p}_2\hat{q}_2}{n_2}}$$

$$= \sqrt{\frac{0.51 \times (1 - 0.51)}{251} + \frac{0.65 \times (1 - 0.65)}{250}} = 0.0437.$$

[2]sleepfoundation.org/sites/default/files/RPT495a.pdf

[3]Excerpt from "The 2011 Sleep in America Poll" from National Sleep Foundation. Copyright © 2011, published by National Sleep Foundation.

We can divide the difference, 0.14, by this SE to get 0.14/0.0437 = 3.20. We know from the 68–95–99.7 Rule that a z-value this large is not very likely, so we'd reject the null hypothesis. We'd need a table or tool to find an exact P-value, but we don't really need one to draw a conclusion. There is a difference between U.S. and Japanese sleepers.

A Two-Proportion *z*-Test

The conditions for the two-proportion *z*-test are the same as for the two-proportion *z*-interval. We are testing the hypothesis

$$H_0: p_1 - p_2 = 0.$$

We estimate the standard error as

$$SE(\hat{p}_1 - \hat{p}_2) = \sqrt{\frac{\hat{p}_1\hat{q}_1}{n_1} + \frac{\hat{p}_2\hat{q}_2}{n_2}}.$$

Now we find the test statistic:

$$z = \frac{(\hat{p}_1 - \hat{p}_2) - 0}{SE(\hat{p}_1 - \hat{p}_2)}.$$

When the conditions are met and the null hypothesis is true, this statistic follows the standard Normal model, so we can use that model to obtain a P-value.

Rounding

When finding the number of successes, round the values to integers. For example, we were told that 70% of the 293 Gen-Yers reported using the Internet before sleep. But 293 × 0.7 = 205.1. So, we round down to the nearest whole number to find the count that could have yielded the rounded percent we were given.

Sometimes the formula for the SE takes into account the fact that the null hypothesis assumes the two proportions are equal. It then estimates the *overall* proportion of successes in the two groups and uses that for both $\hat{p}_1$ and $\hat{p}_2$ in our formula. This is called pooling the proportions. Pooling is discussed in Section 20.7 below. It is, in fact, theoretically more correct. But pooling for this test is unlikely to make a difference in any practical decision and is not implemented in most software, so we think you can usually ignore it.

STEP-BY-STEP EXAMPLE

A Two-Proportion *z*-Test

The 2011 Sleep in America Poll surveyed 1508 randomly sampled U.S. residents aged 13 to 65 asking about their sleep habits and, in particular, their use of technology around the time they try to go to sleep. The researchers report that: "This research shows that people who regularly use their computers or laptops in the hour before trying to go to sleep are less likely to report getting a good night's sleep . . . , more likely to be categorized as "sleepy" . . . , and more likely to drive drowsy . . . than their counterparts." (The 2011 Sleep in America Poll" from National Sleep Foundation. Copyright © 2011, published by National Sleep Foundation.)

The poll found that 205 of 293 or 70% of those 19–29 years old ("Gen-Y") reported using the Internet in the hour before trying to go to sleep at least a few nights a week. By contrast, 235 of 469 or 50% of those aged 30–45 ("Gen-X") reported Internet use before sleep.

QUESTION: Is this difference of 20% real or is it likely to be due only to natural fluctuations in the sample?

THINK **PLAN** State what you want to know. Discuss the variables and the W's.

I want to know whether pre-sleep Internet surfing rates differ for "Generation Y's" (19- to 29-year-olds) relative to "Generation X's" (30–45). The data are from a random sample of 1508 U.S. adults surveyed in the 2011 Sleep in America Poll. Of these, 293 Gen-Y's and 469 Gen-X's responded to the question about surfing, indicating whether or not they had surfed the Internet at least a few nights a week before trying to go to sleep.

HYPOTHESES The study simply broke down the responses by age, so there is no sense that either alternative was preferred. A two-sided alternative hypothesis is appropriate.

H_0: There is no difference in pre-sleep surfing rates in the two age groups:

$$p_{GenY} - p_{GenX} = 0.$$

H_A: The rates are different: $p_{GenY} - p_{GenX} \neq 0$.

MODEL Think about the assumptions and check the conditions.

✔ **Randomization Condition**: The respondents were randomly selected by telephone number and stratified by sex and region, so they should be independent.

✔ **Independent Groups Assumption**: The two groups are independent of each other because the sample was selected at random.

✔ **Success/Failure Condition**: The observed numbers of both successes and failures are much more than 10 for both groups.

State the null model.

Choose your method.

Because the conditions are satisfied, I'll use a Normal model and perform a two-proportion z-test.

SHOW **MECHANICS** Find the SE to estimate $SD(p_{GenY} - p_{GenX})$.

$$n_{GenY} = 293,\ y_{GenY} = 205,\ \hat{p}_{GenY} = 0.700$$

$$n_{GenX} = 469,\ y_{GenX} = 235,\ \hat{p}_{GenX} = 0.501$$

$$SE(\hat{p}_{GenY} - \hat{p}_{GenX}) = \sqrt{\frac{\hat{p}_{GenY}\hat{q}_{GenY}}{n_{GenY}} + \frac{\hat{p}_{GenX}\hat{q}_{GenX}}{n_{GenX}}} = \sqrt{\frac{0.700 \times 0.300}{293} + \frac{0.501 \times 0.499}{469}} = 0.0354$$

The observed difference in sample proportions is

$$\hat{p}_{GenY} - \hat{p}_{GenX} = 0.700 - 0.501 = 0.199.$$

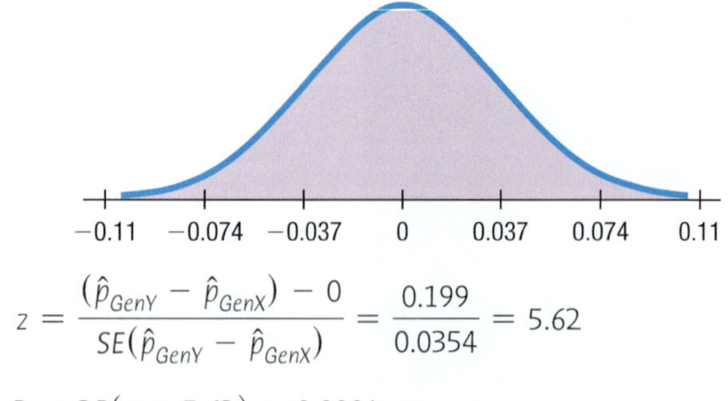

Find the z-score for the observed difference in proportions, 0.199.

$$z = \frac{(\hat{p}_{GenY} - \hat{p}_{GenX}) - 0}{SE(\hat{p}_{GenY} - \hat{p}_{GenX})} = \frac{0.199}{0.0354} = 5.62$$

Find the P-value using Table Z or technology. Because this is a two-tailed test, we must *double* the probability we find in the upper tail.

$$P = 2P(z > 5.62) \leq 0.0001$$

TELL **CONCLUSION** Link the P-value to your decision about the null hypothesis, and state your conclusion in context.

The P-value says that if there really were no difference in surfing rates between the two groups, then the difference observed in this study would be very rare indeed. We can conclude that there is in fact a difference in the rate of surfing between GenY and GenX adults.

JUST CHECKING

3. A February 2014 public opinion poll asked 1000 randomly selected adults whether the United States should decrease the amount of immigration allowed; 36% of those responding said "yes." In June 1995, a random sample of 1000 had found that 65% of adults thought immigration should be curtailed. To see if that percentage has decreased, why can't we just use a one-proportion z-test of H_0: $p = 0.65$ and see what the P-value for $\hat{p} = 0.36$ is?

4. For opinion polls like this, which has more variability: the percentage of respondents answering "yes" in either year or the difference in the percentages between the two years?

EXAMPLE 20.4 Two-Proportion *z*-Test

RECAP: Another concern of the study on teens' online profiles was safety and privacy. In the random sample, girls were less likely than boys to say that they are easy to find online from their profiles. Only 19% (62 girls) of 325 teen girls with profiles say that they are easy to find, while 28% (75 boys) of the 268 boys with profiles say the same.

QUESTION: Are these results evidence of a real difference between boys and girls? Perform a two-proportion *z*-test and discuss what you find.

ANSWER:

$$H_0: p_{boys} - p_{girls} = 0$$

$$H_A: p_{boys} - p_{girls} \neq 0$$

✓ **Randomization Condition:** The sample of boys and the sample of girls were both chosen randomly, so they should be independent.

✓ **Independent Groups Assumption:** Because the samples were selected at random, it's reasonable to believe the boys' perceptions are independent of the girls'.

✓ **Success/Failure Condition:** Among the girls, there were 62 "successes" and 263 failures, and among boys, 75 successes and 193 failures. These counts are at least 10 for each group.

Because all the assumptions and conditions are satisfied, it's ok to do a **two-proportion *z*-test**:

$$SE(\hat{p}_{boys} - \hat{p}_{girls}) = \sqrt{\frac{0.28 \times 0.72}{268} + \frac{0.19 \times 0.81}{325}} = 0.035$$

$$z = \frac{(0.28 - 0.19) - 0}{0.035} = 2.57$$

$$P(z > 2.57) = 0.0051$$

This is a two-tailed test, so the P-value $= 2(0.0051) = 0.010$. Because this P-value is very small, I reject the null hypothesis. This study provides strong evidence that there really is a difference in the proportions of teen girls and boys who say they are easy to find online.

20.4 A Confidence Interval for the Difference Between Two Means

WHO	University students
WHAT	Prices offered for a used camera
UNITS	$
WHY	Study of the effects of friendship on transactions
WHEN	1990s
WHERE	Cornell University

If you bought a used camera in good condition from a friend, would you pay the same as you would if you bought the same camera from a stranger? A researcher at Cornell University[4] wanted to know how friendship might affect simple sales such as this. She randomly divided subjects into two groups and gave each group descriptions of items they might want to buy. One group was told to imagine buying from a friend whom they expected to see again. The other group was told to imagine buying from a stranger. (Data in **Buy from a friend**)

Here are the prices they offered for a used camera in good condition:

Price Offered for a Used Camera ($)

Buying from a Friend	Buying from a Stranger
275	260
300	250
260	175
300	130
255	200
275	225
290	240
300	

Those buying from a friend offered an average of \$281.88 and those buying from a stranger offered an average of \$211.43. But that's just from these 15 subjects. How big a difference should we expect in general?

Comparing two means is just like comparing two proportions. In the camera purchase experiment, for example, the parameter of interest is the difference between the *mean* amount offered to friends and the mean amount offered to strangers, $\mu_1 - \mu_2$.

The rest is essentially the same as before. The statistic of interest is the difference in the two observed means, $\bar{y}_1 - \bar{y}_2$. We'll start with this statistic to build a confidence interval, and we'll use the same standard deviation and sampling distribution model for the hypothesis test.

To find the standard deviation of the sampling distribution of the difference between the two independent sample means, we add their variances and then take a square root (see Section 20.6 for more explanation and intuition):

$$SD(\bar{y}_1 - \bar{y}_2) = \sqrt{Var(\bar{y}_1) + Var(\bar{y}_2)}$$

$$= \sqrt{\frac{\sigma_1^2}{n_1} + \frac{\sigma_2^2}{n_2}}.$$

Of course, we still don't know the true standard deviations of the two groups, σ_1 and σ_2, so as usual, we'll use the estimates, s_1 and s_2. Using the estimates gives us the *standard error*:

$$SE(\bar{y}_1 - \bar{y}_2) = \sqrt{\frac{s_1^2}{n_1} + \frac{s_2^2}{n_2}}.$$

We'll use the standard error to judge how big the difference really is. Because we are working with means and estimating the standard error of their difference using the data, you shouldn't be surprised that the sampling model is a Student's t, as it was for individual means in Chapter 17.

---

[4]J. J. Halpern, "The Transaction Index: A Method for Standardizing Comparisons of Transaction Characteristics Across Different Contexts," *Group Decision and Negotiation,* 6: 557–572, 1997.

Maybe Not

In 2018, questions about Dr. Wansink's data and methodology led several journals to formally withdraw many of his papers, and Cornell suspended him from conducting further research citing "academic misconduct." It appears that he engaged in *P-hacking*, trying out a variety of null hypotheses in search of a test that is statistically significant.

EXAMPLE 20.5

Finding the Standard Error of the Difference in Independent Sample Means

Can you tell how much you are eating from how full you are? Or do you need visual cues? Researchers[5] constructed a table with two ordinary 18 oz soup bowls and two identical-looking bowls that had been modified to slowly, imperceptibly, refill as they were emptied. They assigned experiment participants to the bowls randomly and served them tomato soup. Those eating from the ordinary bowls had their bowls refilled by ladle whenever they were one-quarter full. If people judge their portions by internal cues, they should eat about the same amount. How big a difference was there in the amount of soup consumed? The table summarizes their results.

	Ordinary Bowl	Refilling Bowl
n	27	27
$\bar{y}$	8.5 oz	14.7 oz
s	6.1 oz	8.4 oz

QUESTION: How much variability do we expect in the difference between the two means? Find the standard error.

ANSWER: Participants were randomly assigned to bowls, so the two groups should be independent. It's ok to add variances.

$$SE(\bar{y}_{refill} - \bar{y}_{ordinary}) = \sqrt{\frac{s_r^2}{n_r} + \frac{s_o^2}{n_o}} = \sqrt{\frac{8.4^2}{27} + \frac{6.1^2}{27}} = 2.0 \text{ oz}$$

The confidence interval we build is called a **two-sample *t*-interval** (for the difference in means). The corresponding hypothesis test is called a two-sample *t*-test. The interval looks just like all the others we've seen—the statistic plus or minus an estimated margin of error:

$$(\bar{y}_1 - \bar{y}_2) \pm ME$$

$$\text{where } ME = t^* \times SE(\bar{y}_1 - \bar{y}_2).$$

z or *t*?

If you know σ, use *z*. (That's rare!) Whenever you use *s* to estimate σ, use *t*.

This formula is almost the same as the confidence interval for the difference of two proportions. It's just that here we use a Student's *t*-model instead of a Normal model to find the appropriate critical value corresponding to our chosen confidence level. (So we write it as t^*.)

What are we missing? Only the degrees of freedom for the Student's *t*-model. Unfortunately, *that* formula is strange.

The deep, dark secret is that the sampling model isn't *really* Student's *t*, but only something close. The trick is that by using a special, adjusted degrees-of-freedom value, we can make it so close to a Student's *t*-model that nobody can tell the difference. The adjustment formula is straightforward but doesn't help our understanding much, so we leave it to the computer or calculator. (If you are curious and really want to see the formula, look in the footnote.[6])

---

[5] Brian Wansink, James E. Painter, and Jill North, "Bottomless Bowls: Why Visual Cues of Portion Size May Influence Intake," *Obesity Research*, Vol. 13, No. 1, January 2005.

[6]
$$df = \frac{\left(\frac{s_1^2}{n_1} + \frac{s_2^2}{n_2}\right)^2}{\frac{1}{n_1 - 1}\left(\frac{s_1^2}{n_1}\right)^2 + \frac{1}{n_2 - 1}\left(\frac{s_2^2}{n_2}\right)^2}$$

Are you sorry you looked? This formula usually doesn't even give a whole number. If you are using a table, you'll need a whole number, so round down to be safe. If you are using technology, it's even easier. The approximation formulas that computers and calculators use for the Student's *t*-distribution can deal with fractional degrees of freedom.

A Sampling Distribution Model for the Difference Between Two Means

When the conditions are met, the sampling distribution of the standardized sample difference between the means of two independent groups,

$$t = \frac{(\bar{y}_1 - \bar{y}_2) - (\mu_1 - \mu_2)}{SE(\bar{y}_1 - \bar{y}_2)},$$

can be modeled by a Student's t-model with a number of degrees of freedom found with a special formula. We estimate the standard error with

$$SE(\bar{y}_1 - \bar{y}_2) = \sqrt{\frac{s_1^2}{n_1} + \frac{s_2^2}{n_2}}.$$

Assumptions and Conditions

This test is sometimes called the two *independent samples* t-test because it is only appropriate when the responses for the two groups are independent of each other. No statistical test can verify this assumption. You have to think about how the data were collected. Paired or matched data cannot be analyzed with these methods because paired groups are (kind of obviously) not independent (see Chapter 21).

We need exactly the same assumptions about independence as we had when comparing proportions for two groups:

Independence

Independence Assumption: Within each group, individual responses should be independent of each other. Knowing one response should provide no information about other responses.

Randomization Condition: If the responses are selected with randomization, their independence is likely.

Independent Groups Assumption: The responses in the two groups we're comparing must also be independent *of each other*. Knowing how one group responds should not provide information about the other group. Usually, the independence of the groups from each other is evident from the way the data were collected.

A common violation of the Independent Groups Assumption is for the groups to be *paired*. See the next chapter.

Whenever we have quantitative data, we should check for outliers, skewness, multiple modes, and other surprises. Now that we have two groups, we need to check for each group. The shape of the sampling distribution depends on the assumption that the distribution of values in each group is unimodal and roughly symmetric with no outliers. More formally, the populations should be Normal.

Normal Populations

If you are comparing the means of two groups, then, as we did before with Student's t-models, you must assume that the underlying populations are *each* Normally distributed. Check the . . .

Nearly Normal Condition: Check this for *both* groups; a violation by either one violates the condition. As we saw for single sample means, the Normality Assumption matters most when sample sizes are small. For samples of $n < 15$ in either group, you should not use these methods if the histogram or Normal probability plot shows severe skewness. For n's closer to 40, a mildly skewed histogram is OK, but you should remark on any outliers you find and not work with severely skewed data. When both groups are bigger than 40, the Central Limit Theorem starts to kick in no matter how the data are distributed, so the Nearly Normal Condition for the data matters less. Even in large samples, however, you should still be on the lookout for outliers, extreme skewness, and multiple modes.

EXAMPLE 20.6 Checking Assumptions and Conditions

RECAP: Researchers randomly assigned people to eat soup from one of two bowls: 27 got ordinary bowls that were refilled by ladle, and 27 others got bowls that secretly refilled slowly as the people ate.

QUESTION: Can the researchers use their data to make inferences about the role of visual cues in determining how much people eat?

ANSWER:

✓ **Randomization Condition:** The fact that subjects were randomly assigned to treatments makes independence between subjects plausible.

✓ **Nearly Normal Condition:** The histograms for both groups look unimodal (see the histograms in the margin) but somewhat skewed to the right. I believe both groups are large enough (27) to allow use of *t* methods.

✓ **Independent Groups Assumption:** Randomization to treatment groups guarantees this.

It's ok to construct a two-sample *t*-interval for the difference in means.

Note: When you check the Nearly Normal Condition it's important that you include the graphs you looked at (histograms or Normal probability plots).

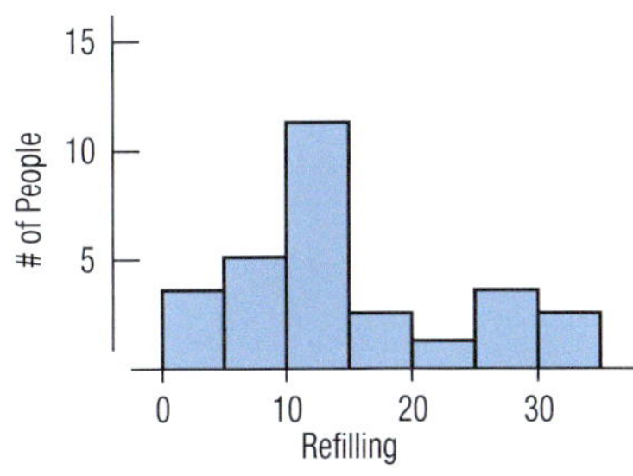

An Easier Rule?

The formula for the degrees of freedom of the sampling distribution of the difference between two means is long, but the number of degrees of freedom is always at *least* the smaller of the two *n*'s, minus 1. Wouldn't it be easier to just use that value? You could, but *that* approximation can be a poor choice because it can give fewer than *half* the degrees of freedom you're entitled to from the correct formula.

A Two-Sample *t*-Interval for the Difference Between Means

When the conditions are met, we are ready to find the confidence interval for the difference between means of two independent groups, $\mu_1 - \mu_2$. The confidence interval is

$$(\bar{y}_1 - \bar{y}_2) \pm t^*_{df} \times SE(\bar{y}_1 - \bar{y}_2),$$

where the standard error of the difference of the means is

$$SE(\bar{y}_1 - \bar{y}_2) = \sqrt{\frac{s_1^2}{n_1} + \frac{s_2^2}{n_2}}.$$

The critical value t^*_{df} depends on the particular confidence level, *C*, that you specify and on the number of degrees of freedom, which we get from the sample sizes and a special formula.

EXAMPLE 20.7 Finding a Confidence Interval for the Difference in Sample Means

RECAP: Researchers studying the role of internal and visual cues in determining how much people eat conducted an experiment in which some people ate soup from bowls that secretly refilled. The results are summarized in the table.

	Ordinary Bowl	Refilling Bowl
n	27	27
$\bar{y}$	8.5 oz	14.7 oz
s	6.1 oz	8.4 oz

We've already checked the assumptions and conditions, and have found the standard error for the difference in means to be $SE(\bar{y}_{refill} - \bar{y}_{ordinary}) = 2.0$ oz.

QUESTION: What does a 95% confidence interval say about the difference in mean amounts eaten?

ANSWER: The observed difference in means is $\bar{y}_{refill} - \bar{y}_{ordinary} = 14.7 - 8.5 = 6.2$ oz.

$$df = 47.46 \quad t^*_{47.46} = 2.011 \text{ (Table T gives } t^*_{45} = 2.014.)$$

$$ME = t^* \times SE(\bar{y}_{refill} - \bar{y}_{ordinary}) = 2.011(2.0) = 4.02 \text{ oz}$$

The 95% confidence interval for $\mu_{refill} - \mu_{ordinary}$ is 6.2 ± 4.02, *or* (2.18 oz, 10.22 oz).

I am 95% confident that people eating from a subtly refilling bowl will eat an average of between 2.18 and 10.22 more ounces of soup than those eating from an ordinary bowl.

JUST CHECKING

Carpal tunnel syndrome (CTS) causes pain and tingling in the hand. It can be bad enough to keep sufferers awake at night and restrict their daily activities. Researchers studied the effectiveness of two alternative surgical treatments for CTS (Mackenzie, Hainer, and Wheatley, *Annals of Plastic Surgery*, 2000). Patients were randomly assigned to have endoscopic or open-incision surgery. Four weeks later, the endoscopic surgery patients demonstrated a mean pinch strength of 9.1 kg compared to 7.6 kg for the open-incision patients.

5. Why is the randomization of the patients into the two treatments important?
6. A 95% confidence interval for the difference in mean strength is about (0.04 kg, 2.96 kg). Explain what this interval means.
7. Why might we want to examine such a confidence interval in deciding between these two surgical procedures?
8. Why might you want to see the data before trusting the confidence interval?

20.5 The Two-Sample *t*-Test: Testing for the Difference Between Two Means

Sometimes we want to test whether the population means of two groups are the same, so we'll need a hypothesis test. Actually, you already know enough to construct this test. The test is properly called the **two-sample *t*-test for the difference between two means**, but it is used so often that it is usually just called the **two-sample *t*-test**. The test statistic looks just like the others we've seen and uses the same standard error as the confidence interval. It finds the difference between the observed group means and compares it with a hypothesized value for that difference.

When we write the null hypothesis, we could write: H_0: $\mu_1 = \mu_2$. But to allow the possibility of testing whether the difference is something other than 0, we test the *difference* between the two means. We'll call that hypothesized difference Δ_0 ("delta naught"). It's so common for that hypothesized difference to be zero that we often just assume $\Delta_0 = 0$ and write: H_0: $\mu_1 - \mu_2 = 0$, but in general we'll write: H_0: $\mu_1 - \mu_2 = \Delta_0$.

Then, as usual, we compare the difference in the means with the standard error of that difference. We already know that, for a difference between independent means, we can find P-values from a Student's *t*-model on that special number of degrees of freedom.

NOTATION ALERT

Δ_0—delta naught—isn't so standard that you can assume everyone will understand it. We use it because Δ is the Greek letter (good for a parameter) "D" for "difference." You should say "delta naught" rather than "delta zero"—that's a standard way to make zero sound fancy for parameters associated with null hypotheses.

A Two-Sample *t*-Test for the Difference Between Means

The conditions for the two-sample *t*-test for the difference between the means of two independent groups are the same as for the two-sample *t*-interval. We test the hypothesis

$$H_0: \mu_1 - \mu_2 = \Delta_0,$$

where the hypothesized difference is almost always 0, using the statistic

$$t = \frac{(\bar{y}_1 - \bar{y}_2) - \Delta_0}{SE(\bar{y}_1 - \bar{y}_2)}.$$

The standard error of $\bar{y}_1 - \bar{y}_2$ is

$$SE(\bar{y}_1 - \bar{y}_2) = \sqrt{\frac{s_1^2}{n_1} + \frac{s_2^2}{n_2}}.$$

When the conditions are met and the null hypothesis is true, this statistic can be closely modeled by a Student's *t*-model with a number of degrees of freedom given by a special formula. We use that model to obtain a P-value.

STEP-BY-STEP EXAMPLE

A Two-Sample *t*-Test for the Difference Between Two Means

RECALL: We want to know if people will pay the same amount, on average, when buying a used camera from a friend or a stranger. The usual null hypothesis is that there's no difference in means. That's just the right null hypothesis for the camera purchase prices.

QUESTION: Is there a difference in the average price people would offer a friend rather than a stranger?

THINK

PLAN State what we want to know.

Identify the *parameter* you wish to estimate. Here our parameter is the difference in the means, not the individual group means.

Identify the variables and check the W's.

I want to know whether people are likely to offer a different amount for a used camera when buying from a friend than when buying from a stranger. I plan to test whether the difference between mean amounts is zero. I have bid prices from 8 subjects buying from a friend and 7 buying from a stranger, found in a randomized experiment.

HYPOTHESES State the null and alternative hypotheses.

The research claim is that friendship changes what people are willing to pay.[7] The natural null hypothesis is that friendship makes no difference.

H_0: The difference in mean price offered to friends and the mean price offered to strangers is zero:

$$\mu_F - \mu_S = 0.$$

[7]This claim is a good example of what is called a "research hypothesis" in many social sciences. The only way to formally test a research hypothesis is to deny that it's true and see whether we can reject that null hypothesis.

We didn't start with any knowledge of whether friendship might increase or decrease the price, so we choose a two-sided alternative.

H_A: The difference in mean prices is not zero:

$$\mu_F - \mu_S \neq 0.$$

Make a picture. Boxplots are the display of choice for comparing groups. We'll also want to check the distribution of each group. Histograms or Normal probability plots do a better job there.

REALITY CHECK Looks like the prices are higher if you buy from a friend, but we can't be sure. You can't tell from looking at the boxplots whether the difference is statistically significant—the plot shows spreads on the same scale as the data, and we know those don't add. You'll need to add the *variances* to get a suitable ruler for comparing the difference.

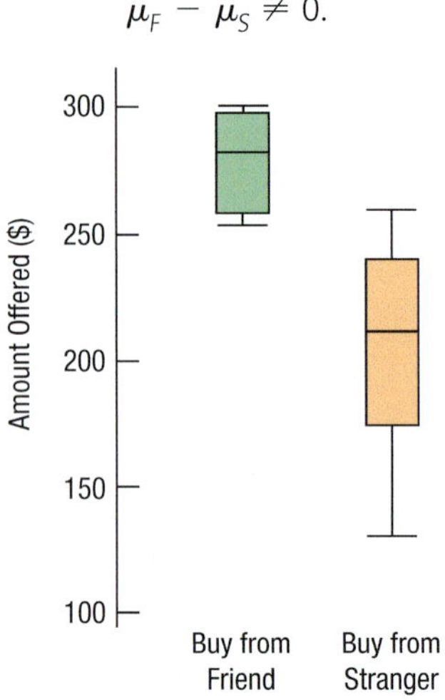

MODEL Think about the assumptions and check the conditions.

✔ **Randomization Condition:** The experiment was randomized. Subjects were assigned to treatment groups at random and so the prices paid should be independent.

✔ **Independent Groups Assumption:** Randomizing the experiment gives independent groups.

✔ **Nearly Normal Condition:** Histograms of the two sets of prices are reasonably unimodal and symmetric given the small sample sizes:

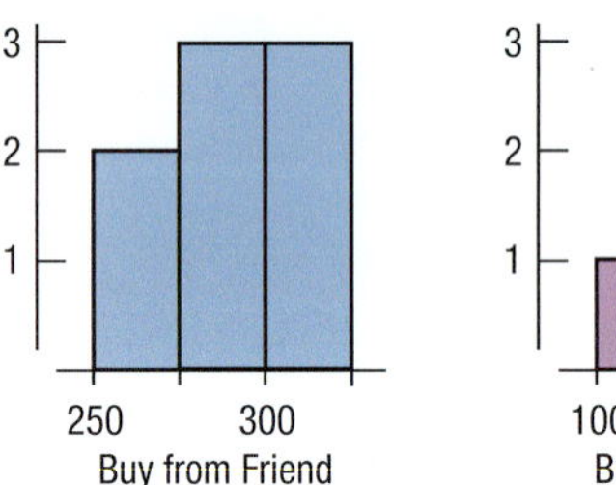

State the sampling distribution model.

Specify your method.

The assumptions are reasonable and the conditions are ok, so I'll use a Student's *t*-model to perform a **two-sample *t*-test**.

SHOW **MECHANICS** List the summary statistics. Be sure to use proper notation.

From the data:

$n_F = 8$ $\quad n_S = 7$

$\bar{y}_F = \$281.88$ $\quad \bar{y}_S = \$211.43$

$s_F = \$18.31$ $\quad s_S = \$46.43$

Use the null model to find the P-value. First determine the standard error of the difference between sample means.

For independent groups,

$$SE(\bar{y}_F - \bar{y}_S) = \sqrt{SE^2(\bar{y}_F) + SE^2(\bar{y}_S)}$$

$$= \sqrt{\frac{s_F^2}{n_F} + \frac{s_S^2}{n_S}}$$

$$= \sqrt{\frac{18.31^2}{8} + \frac{46.43^2}{7}}$$

$$= 18.70.$$

The observed difference is

$$(\bar{y}_F - \bar{y}_S) = 281.88 - 211.43 = \$70.45.$$

Make a picture. Sketch the *t*-model centered at the hypothesized difference of zero. Because this is a two-tailed test, shade the region to the right of the observed difference and the corresponding region in the other tail.

3.77

−4σ −3σ −2σ −1σ 0 1σ 2σ 3σ 4σ

Find the *t*-value.

$$t = \frac{(\bar{y}_F - \bar{y}_S) - 0}{SE(\bar{y}_F - \bar{y}_S)} = \frac{70.45}{18.70} = 3.77$$

A statistics program or graphing calculator can find the P-value using the fractional degrees of freedom from the approximation formula. If you are doing a test like this without technology, you could use the smaller sample size to determine degrees of freedom. In this case, $n_2 - 1 = 6$.

The approximation formula gives 7.62 degrees of freedom.[8]

$$\text{P-value} = 2P(t_{7.62} > 3.77) = 0.006$$

TELL **CONCLUSION** Link the P-value to your decision about the null hypothesis, and state the conclusion in context.

Be cautious about generalizing to items whose prices are outside the range of those in this study.

If there were no difference in the mean prices, a difference this large would occur only 6 times in 1000. That's too rare to believe, so I reject the null hypothesis and conclude that people are likely to offer a friend more than they'd offer a stranger for a used camera (and possibly for other, similar items).

JUST CHECKING

Recall the experiment comparing patients 4 weeks after surgery for carpal tunnel syndrome. The patients who had endoscopic surgery demonstrated a mean pinch strength of 9.1 kg compared to 7.6 kg for the open-incision patients.

9. What hypotheses would you test?
10. The P-value of the test was less than 0.05. State a brief conclusion.
11. The study reports work on 36 "hands," but there were only 26 patients. In fact, 7 of the endoscopic surgery patients had both hands operated on, as did 3 of the open-incision group. Does this alter your thinking about any of the assumptions? Explain.

[8]If you were daring enough to calculate that messy degrees of freedom formula by hand with the values given here, you'd get about 7.74. Computers work with more precision for the standard deviations than we gave in our example. Many computer programs will round the final result down to 7 degrees of freedom. All give about the same result for the P-value, so it doesn't really matter—the conclusion would be the same.

EXAMPLE 20.8 A Two-Sample t-Test

Many office "coffee stations" collect voluntary payments for the food consumed. Researchers at the University of Newcastle upon Tyne performed an experiment to see whether the image of eyes watching would change employee behavior.[9] They alternated pictures of eyes looking at the viewer with pictures of flowers each week on the cupboard behind the "honesty box." They measured the consumption of milk (in liters) to approximate the amount of food consumed and recorded the contributions (in £) each week per liter of milk. The table summarizes their results.

	Eyes	Flowers
n (# weeks)	5	5
$\bar{y}$	0.417£/l	0.151£/l
s	0.1811£/l	0.067£/l

QUESTION: Do these results provide evidence that there really is a difference in honesty even when it's only photographs of eyes that are "watching"?

ANSWER:

$$H_0: \mu_{eyes} - \mu_{flowers} = 0$$

$$H_A: \mu_{eyes} - \mu_{flowers} \neq 0$$

✔ **Independence Assumption:** The amount paid by one person should be independent of the amount paid by others. However, this is an observational study, not a randomized experiment.

✔ **Nearly Normal Condition:** I don't have the data to check, but it seems unlikely there would be outliers in either group. I could be more certain if I could see histograms for both groups.

✔ **Independent Groups Assumption:** The same workers were recorded each week, but week-to-week independence is plausible.

It's ok to do a two-sample t-test for the difference in means:

$$SE(\bar{y}_{eyes} - \bar{y}_{flowers}) = \sqrt{\frac{s^2_{eyes}}{n_{eyes}} + \frac{s^2_{flowers}}{n_{flowers}}} = \sqrt{\frac{0.1811^2}{5} + \frac{0.067^2}{5}} = 0.0864$$

$df = 5.07$

$$t_5 = \frac{(\bar{y}_{eyes} - \bar{y}_{flowers}) - 0}{SE(\bar{y}_{eyes} - \bar{y}_{flowers})} = \frac{0.417 - 0.151}{0.0864} = 3.08$$

$$\text{P-value} = 2 \times P(t_5 > 3.08) = 2 \times 0.0137 = 0.027$$

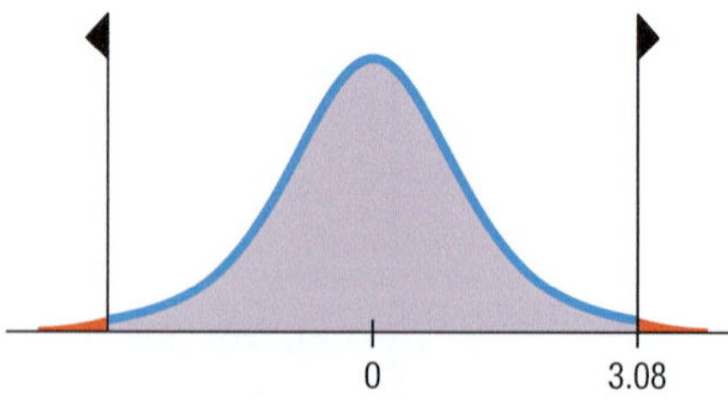

Assuming the data were free of outliers, the very low P-value leads me to reject the null hypothesis. This study provides evidence that people will leave higher average voluntary payments for food if pictures of eyes are "watching."

(Note: In Table T, we can see that at 5 df, $t = 3.08$ lies between the critical values for $P = 0.02$ and $P = 0.05$, so we could report $P < 0.05$.)

[9] Melissa Bateson, Daniel Nettle, and Gilbert Roberts, "Cues of Being Watched Enhance Cooperation in a Real-World Setting," *Biol. Lett.* doi: 10.1098/rsbl.2006.0509.

*Tukey's Quick Test

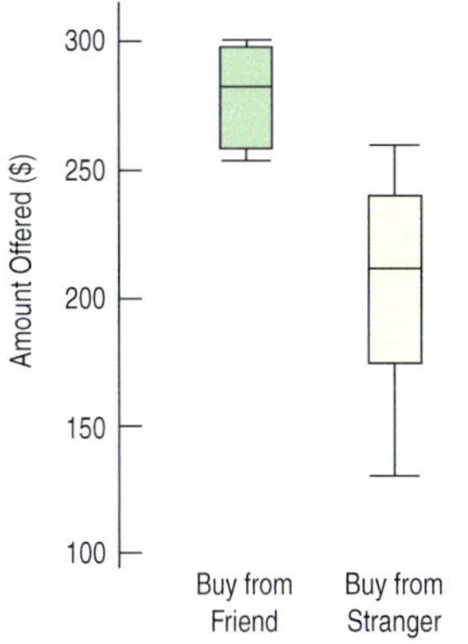

Buying from a Friend	Buying from a Stranger
$275	$260
300	250
260	175
300	130
255	200
275	225
290	240
300	

The famous statistician John Tukey (originator of the stem-and-leaf display and the boxplot) was once challenged to come up with a simpler alternative to the two-sample *t*-test that, like the 68-95-99.7 Rule, had critical values that could be remembered easily. The test he came up with asks you only to count and to remember three numbers: 7, 10, and 13.

When you first looked at the boxplots of the price data, you might have noticed that they didn't overlap very much. That's the basis for Tukey's test.

To use Tukey's test, one group must have the highest value and the other, the lowest. We just count how many values in the high group are higher than *all* the values of the lower group. Add to this the number of values in the low group that are lower than *all* the values of the higher group. (Count ties as 1/2.) Now if this total is 7 or more, we can reject the null hypothesis of equal means at $\alpha = 0.05$. The "critical values" of 10 and 13 give us α's of 0.01 and 0.001.

Let's try it. The "Friend" group has the highest value ($300) and the "Stranger" group has the lowest ($130). Six of the values in the Friend group (shown in red) are higher than the highest value of the Stranger group ($260) and one is a tie. Six of the Stranger values (shown in green) are lower than the lowest value for Friends. That's a total of 12 1/2 values that leak over. That's more than 10, but less than 13. So the P-value is between 0.01 and 0.001—just what we found with the two-sample *t*.

This is a remarkably good test. The only assumption it requires is that the two samples be independent. It's so simple to do that there's no reason not to do one to check your two-sample *t* results. If they disagree, check the assumptions. Tukey's quick test, however, is not as widely known or accepted as the two-sample *t*-test, so you still need to know and use the two-sample *t*.

*A Rank Sum Test

Another distribution-free test of whether two means are equal uses the *ranks* of the data in the two samples. Tukey's test ranks all the data and then counts the excesses on either side. Like Tukey's test, the **Wilcoxon rank sum (or Mann-Whitney) test** first ranks the combined sample from the two groups together from smallest to largest, assigning each observation a rank from 1 to N, where N is $n_1 + n_2$, the total number of observations in both groups. (If there are ties, we use the average rank.) The test statistic, *W*, is the sum of the ranks of the first group. If *W* is too large or too small, it is evidence that the groups do not have equal means (actually, the test is more general and tests whether the two distributions are the same).

The hypothesis test is based on a couple of facts that can be derived mathematically. When the null hypothesis is true, the test statistic, *W*, has mean $\mu_W = \frac{n_1(N+1)}{2}$ and variance $Var(W) = \frac{n_1 n_2(N+1)}{12}$. Now, for any but very small sample sizes (both larger than about 7 should do), we can use a *z*-test with $z = \frac{W - \mu_W}{SD(W)}$ to test the hypothesis.

Let's try it on the buying from friends data. Here are the data ranked from 1 to 15. There are two ties at ranks 8 and 9, two ties at ranks 10 and 11, and three ties at ranks 13, 14, and 15:

Data	130	175	200	225	240	250	255	260	260	275	275	290	300	300	300
Rank	1	2	3	4	5	6	7	8.5	8.5	10.5	10.5	12	14	14	14
Group	S	S	S	S	S	S	F	S	F	F	F	F	F	F	F

Now we'll add up all the ranks of the Friends group:

$$W = 7 + 8.5 + 10.5 + 10.5 + 12 + 14 + 14 + 14 = 90.5$$

The mean of W when the null hypothesis is true is $\mu_W = \dfrac{8(15 + 1)}{2} = 64$ and the SD is

$$SD(W) = \sqrt{Var(W)} = \sqrt{\frac{8 \times 7(15 + 1)}{12}} = 8.64, \text{ so } z = \frac{90.5 - 64}{8.64} = 3.07$$

with a two-sided P-value of 0.0021. As with both other tests we looked at, this P-value is between 0.01 and 0.001. We reach the same conclusion as we did with Tukey's test and with the two-sample t-test. Of course, like Tukey's test, the rank sum test has the advantage that it doesn't depend on the Nearly Normal Condition. And, like Tukey's test, it will be less powerful than the two-sample t-test when that condition *is* satisfied because it doesn't use all the information in the data.

*20.6 Randomization Tests and Confidence Intervals for Two Means

In Section 4.3 we looked at John Beale's data on car speeds going up and down his street in Palo Alto, California. By drawing many random samples, we simulated the sampling distribution of the differences in the mean speeds of the two directions. (Data in **Car speeds**)

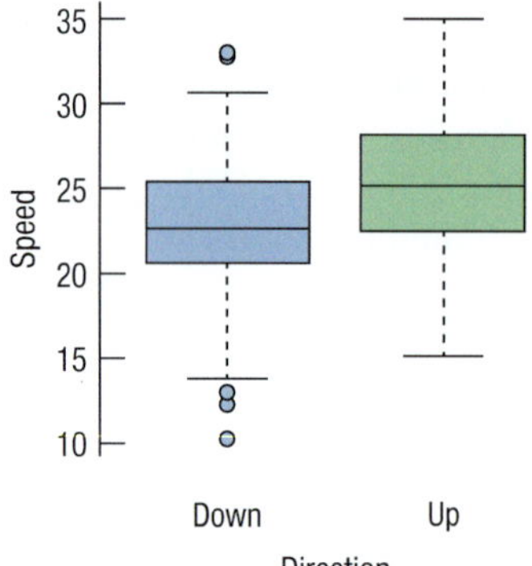

Speeds (mph) Direction	Down	Up
Min	10.27	15.08
Q1	20.47	22.50
Median	22.88	25.16
Mean	22.72	25.25
Q3	25.35	28.16
Max	32.95	34.97

Figure 20.2
(see also Figure 4.4) Side-by-side boxplots show that cars may generally be going faster up the street than down.

We could test whether the mean speed was the same for cars in each direction using a two-sample t-test, but let's try a different approach. As usual, whenever we test a null hypothesis we start by assuming that it is true. Here, we're assuming that the mean speeds are the same in both directions. Let's just select 250 cars at random and call them "Up" and call the other 250 cars "Down." The difference in mean speeds between these groups should be zero. But, of course, if we do this over and over, our actual results will vary. Simulating the random assignment many times and finding the difference in the means will generate a sampling distribution on which we can locate the difference we actually saw. (See Figure 20.3.) If there were no difference, then we should be able to label a car at random as going up or down the street." (Another way to think about selecting the two groups is to imagine that we shuffled the 500 labels "Up" and "Down," distributing them at random.)

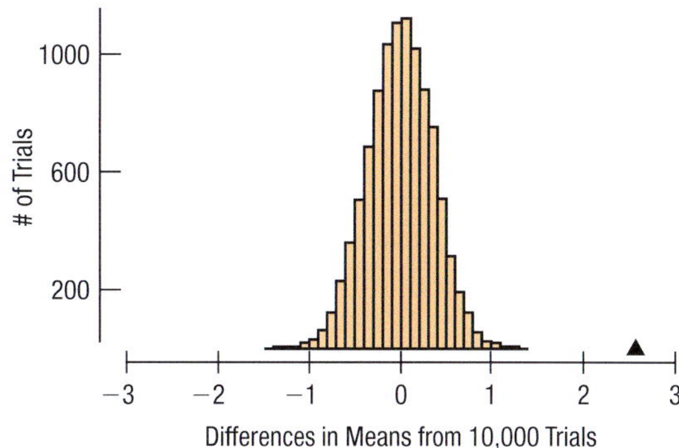

Figure 20.3
From the distribution of differences of means of randomly selected groups of cars we can see that the observed difference of 2.53 mph is highly unlikely. It appears that there is a difference in car speeds in the two directions.

In Chapter 4 we simulated such a random assignment 10,000 times and got the histogram of differences in Figure 20.3:

As we did in the Random Matters section of Chapter 19, we count the number of times our observed value was exceeded by a simulated value (in either tail). In our 10,000 trials we never saw a difference as big as 2.53. But we don't say the P-value is 0. If we simulated 1,000,000 differences we might see one larger than 2.53. All we can say that is that the P-value is < 0.0001, since a value as large as 2.53 didn't occur in 10,000 trials. This test is called a **randomization test for the difference of two means**.

What about a confidence interval for the difference? A confidence interval doesn't depend on the null hypothesis being true. Now the labels may actually matter—we just don't know. The null hypothesis said that they didn't, but the confidence interval shouldn't rely on that. Now we want to center our interval at the actual observed difference, so we resample from the original dataset, keeping the labels and creating many bootstrap samples of 500 cars. For each of these new samples we compute the difference in mean speeds going up and down the street. Repeating that 10,000 times gives us a simulated sampling distribution for the difference in means that doesn't rely on the null hypothesis as shown in Figure 20.4. We construct a 95% confidence interval using the middle 95% of that distribution. Our simulation gave us (1.878, 3.199) mph for the true mean difference in speed of cars going up and down his street.

Figure 20.4
The differences in means of the two groups using 10,000 bootstrapped samples. The simulated sampling distribution is now centered at the observed difference of 2.53 mph and the quantiles give us a confidence interval for the true mean difference. Using the 2.5th and 97.5th percentiles gives a 95% confidence interval of (1.878, 3.199) mph.

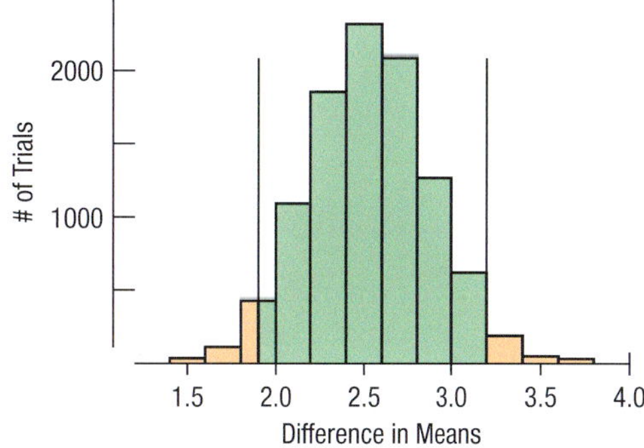

We got a P-value < 0.0001 for the hypothesis of equal means with a 95% confidence interval of $(1.878, 3.199)$ mph. How would the standard t-test and interval compare? The t-test shows $t = -7.575$, $df = 496.06$, P-value < 0.0001, and the confidence interval is $(1.877, 3.192)$ mph.

Normal based methods and resampling methods for tests and confidence intervals don't always match so closely, but here the data are quite unimodal and symmetric in each group and the sample sizes (250) in each group are large. Unfortunately, resampling methods break down just when traditional methods do—when the data are highly skewed, have outliers, or the sample sizes are small. So, even though these resampling methods seem to have fewer assumptions and provide a nice alternative to standard methods, be cautious in applying them without checking the data.

*20.7 Pooling

Hypothesis tests for the difference between two groups compare the observed difference in proportions or means with a standard error of that difference. The better the estimate of the standard error, the more power the tests will have.[10] If you are willing to make the additional assumption that the two groups have the same variance, you can improve the power of your tests.

For proportions, that assumption is natural—it's part of the null hypothesis. If you assume that two proportions are the same, then their variances will be the same as well (as long as the sample sizes are equal). In that case, it is OK to combine or **pool** the data from the two groups when estimating the standard error.

The pooled standard error for the difference in two proportions combines the "successes" from both groups and divides by the total number of cases. You can use this pooled proportion to find the standard error with the usual formula:

$$\hat{p}_{pooled} = \frac{successes_1 + successes_2}{n_1 + n_2}$$

$$\hat{q}_{pooled} = 1 - \hat{p}_{pooled}$$

$$SE_{pooled}(\hat{p}_1 - \hat{p}_2) = \sqrt{\frac{\hat{p}_{pooled} \times \hat{q}_{pooled}}{n_1} + \frac{\hat{p}_{pooled} \times \hat{q}_{pooled}}{n_2}}$$

For means, the assumption that the two groups have the same variance is much less natural. The null hypothesis assumes the means are equal, but usually there's no reason to believe that the variances should be as well. Until the 1950s, the pooled *t-test* was the only test available. The special result that provides the two-sample t was published in 1946.

The pooled estimate finds a weighted average of the two variances using the degrees of freedom of each variance as the weights. You can use that pooled estimate for each group variance in the standard error formula:

$$s^2_{pooled} = \frac{(n_1 - 1)s_1^2 + (n_2 - 1)s_2^2}{(n_1 - 1) + (n_2 - 1)}$$

$$SE_{pooled}(\bar{y}_1 - \bar{y}_2) = \sqrt{\frac{s^2_{pooled}}{n_1} + \frac{s^2_{pooled}}{n_2}} = s_{pooled}\sqrt{\frac{1}{n_1} + \frac{1}{n_2}}.$$

Is the Pool All Wet?

Generally, we don't advise pooling. However, there are times when the assumption of equal variances may make sense.

In the special case of testing whether the proportions of two groups are equal, the special property that links a proportion with its standard deviation kicks in. Recall that when we test a hypothesis we pretend that it is true. And if the proportions were true, then their variances would be equal. So we can defend pooling. Of course, this argument doesn't apply to finding a confidence interval. However, you'll find that few statistics programs bother to pool for testing proportions. For practical decisions the extra effort should make little difference.

For means, the special case that arises most often is for data that have come from a randomized experiment. When subjects are randomly assigned to two treatment groups, it is certainly true that those groups have the same variance *before the treatments are applied.* The common null hypothesis that the group means are equal *after* treatment essentially says that the treatments had no effect—and if that's true, then the variances should still be equal.

As it turns out, there is little advantage to pooling for testing the difference between two means. However, the method used for comparing means of more than two groups, called analysis of variance, does use pooled estimates, so it is good idea to have some understanding of the concept.

[10] Recall from Chapter 19 that the power of a test is a measure of its ability to reject a false null hypothesis.

*20.8 The Standard Deviation of a Difference

You may have noticed that the formula for the standard deviation of the difference of two proportions (and two means) summed two variances. The reason for this is a fact about independent random variables:

> *The variance of the sum or difference of two independent random variables is the sum of their variances.*

This is such an important (and powerful) idea in statistics that it's worth pausing a moment to review the reasoning. Here's some intuition about why variation increases even when we subtract two random quantities.

Let's say everyone in your classroom has a full box of cereal that's labeled 16 ounces. We know that's not exact: there's some variation from box to box. Now everyone pours a 2-ounce bowl of cereal. There'll be some variation in how much each person pours, too. How much variation will there be in the amount of cereal that is left in each box? Even though the amount left in the box is the *difference* between the original amount and the amount that was poured, there will be even *more* variation than there was in either the original amount or the amount poured. This is because the amount left depends on *both* the variation in how much was originally in the box and the variation in the amount poured.

According to the rule, the variance of the amount of cereal left in the box would now be the *sum* of the two *variances.*

For independent random variables, **variances add**.

We want a standard deviation, not a variance, but that's just a square root away. We can write symbolically what we've just said:

$$Var(X - Y) = Var(X) + Var(Y), \text{ so}$$
$$SD(X - Y) = \sqrt{SD^2(X) + SD^2(Y)} = \sqrt{Var(X) + Var(Y)}.$$

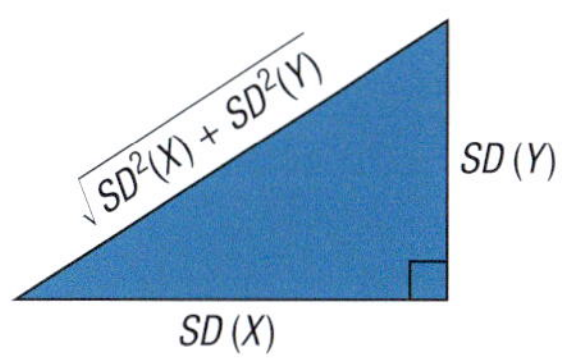

This form should remind you of the Pythagorean Theorem. And this result is sometimes called the *Pythagorean Theorem of Statistics.*

Be careful, though—just as the Pythagorean Theorem works only for right triangles, the formula works only for independent random variables.[11] Always check for independence before using it.

The Standard Deviation and Standard Error of the Difference Between Two Proportions

We'll apply what we've learned about the variance of a difference to proportions first. Fortunately, proportions observed in independent random samples *are* independent, so we can put the two proportions in for X and Y and add their variances. We just need to use careful notation to keep things straight.

When we have two samples, each can have a different size and proportion value, so we keep them straight with subscripts. Often we choose subscripts that remind us of the groups. For our example, we might use "$_M$" and "$_F$" but generically we'll just use "$_1$" and "$_2$." We will represent the two sample proportions as $\hat{p}_1$ and $\hat{p}_2$, and the two sample sizes as n_1 and n_2.

The standard deviations of the sample proportions are $SD(\hat{p}_1) = \sqrt{\frac{p_1q_1}{n_1}}$ and $SD(\hat{p}_2) = \sqrt{\frac{p_2q_2}{n_2}}$, so the variance of the difference in the proportions is

$$Var(\hat{p}_1 - \hat{p}_2) = \left(\sqrt{\frac{p_1q_1}{n_1}}\right)^2 + \left(\sqrt{\frac{p_2q_2}{n_2}}\right)^2 = \frac{p_1q_1}{n_1} + \frac{p_2q_2}{n_2}.$$

The standard deviation is the square root of that variance:

$$SD(\hat{p}_1 - \hat{p}_2) = \sqrt{\frac{p_1q_1}{n_1} + \frac{p_2q_2}{n_2}}.$$

[11] It is not coincidental that Pythagoras needs a right angle and we need independence. The mathematics behind this fact shows that they are the same condition.

We usually don't know the true values of p_1 and p_2. When we have the sample proportions in hand from the data, we use them to estimate the variances. So the standard error is

$$SE(\hat{p}_1 - \hat{p}_2) = \sqrt{\frac{\hat{p}_1\hat{q}_1}{n_1} + \frac{\hat{p}_2\hat{q}_2}{n_2}}.$$

The Standard Deviation and Standard Error of the Difference Between Two Means

Finding the standard deviation of the difference of two means is essentially the same idea. We start with the variance of each group mean, add them, and take the square root:

$$SD(\bar{y}_1 - \bar{y}_2) = \sqrt{Var(\bar{y}_1) + Var(\bar{y}_2)}$$

$$= \sqrt{\frac{\sigma_1^2}{n_1} + \frac{\sigma_2^2}{n_2}}.$$

Of course, we still don't know the true standard deviations of the two groups, σ_1 and σ_2, so as usual, we'll use the estimates, s_1 and s_2. Using the estimates gives us the *standard error*:

$$SE(\bar{y}_1 - \bar{y}_2) = \sqrt{\frac{s_1^2}{n_1} + \frac{s_2^2}{n_2}}.$$

We'll use the standard error to judge how big the difference really is. Because we are working with means and estimating the standard error of their difference using the data, we shouldn't be surprised that the sampling model is a Student's t.

WHAT CAN GO WRONG?

- **Don't use two-sample methods when the samples aren't independent.** These methods give wrong answers when this assumption of independence is violated. Good random sampling is usually the best assurance of independent groups. Make sure there is no relationship between the two groups. Matched-pairs designs arise often and are important, but can't be analyzed with the methods of this chapter.
- **Look at the plots.** The usual cautions about checking for outliers and non-Normal distributions apply, of course. The simple defense is to make and examine boxplots. You may be surprised how often this simple step saves you from the wrong or even absurd conclusions that can be generated by a single undetected outlier. You don't want to conclude that two methods have very different means just because one observation is atypical.
- **Be cautious if you apply inference methods where there was no randomization.** If the data do not come from representative random samples or from a properly randomized experiment, then the inference about the differences between the groups may be wrong.
- **Don't interpret a significant difference in proportions or means causally.** Studies find that people with higher incomes are more likely to snore. Would surgery to increase your snoring be a wise investment? Probably not. It turns out that older people are more likely to snore, and they are also likely to earn more. In a prospective or retrospective study, there is always the danger that other lurking variables not accounted for are the real reason for an observed difference. Be careful not to jump to conclusions about causality.

DO WHAT WE SAY, NOT WHAT WE DO . . .

Precision machines used in industry often have a bewildering number of parameters that have to be set, so experiments are performed in an attempt to try to find the best settings. Such was the case for a hole-punching machine used by a well-known computer manufacturer to make printed circuit boards. The data were analyzed by one of the authors, but because he was in a hurry, he didn't look at the boxplots first and just performed *t*-tests on the experimental factors. When he found extremely small P-values even for factors that made no sense, he plotted the data. Sure enough, there was one observation 1,000,000 times bigger than the others. It turns out that it had been recorded in microns (millionths of an inch), while all the rest were in inches.

CONNECTIONS

We looked at contingency tables for two categorical variables in Chapter 3. Differences in proportions are just 2 × 2 contingency tables. You'll often see data presented in this way. For example, the data on surfing the Internet could be shown as

	Gen-Y	Gen-X	Total
Surf	205	235	**440**
Don't Surf	88	234	**322**
Total	**293**	**469**	**762**

We tested whether the column percentages of Internet surfers were the same for the two age groups.

It should be clear by now that the inference methods we are learning work in a consistent pattern. Although the methods have different standard errors, the step-by-step procedures are almost identical. In particular, we divide a statistic by its standard error. You should feel right at home.

We first compared groups with boxplots in Chapter 4. Back then, we made general statements about the shape, center, and spread of each group. When we compared groups, we asked whether their centers looked different compared to how spread out the distributions were. Now, we've made that kind of thinking precise, with confidence intervals for the difference and tests of whether the means are the same.

We use Student's *t* as we did for single sample means, and for the same reasons: We are using standard errors from the data to estimate the standard deviation of the sample statistic. As before, to work with Student's *t*-models, we need to check the Nearly Normal Condition. Histograms and Normal probability plots are the best methods for such checks.

CHAPTER REVIEW

Know how to construct and interpret confidence intervals for the difference between proportions or means of two independent groups.

- The two-proportion *z*-interval is appropriate for proportions.
- The two-sample *t*-interval is appropriate for means.
- The assumptions and conditions are the same for both, except that for means, we must also assume that the values in each group follow a Normal model.

Be able to perform and interpret a two-sample *t*-test of the difference between the means of two independent groups.

- Understand the relationship between testing and providing a confidence interval.
- The most common null hypothesis is that the means are equal.

Be able to perform and interpret a two-sample z-test of whether proportions in two independent groups are equal.

Recognize that in special cases in which it is reasonable to assume equal variances between the groups, we can also use a pooled standard error calculation for testing the difference between means.

- This may make sense particularly for randomized experiments in which the randomization has produced groups with equal variance to start with and the null hypothesis is that a treatment under study has had no effect.

For all comparisons of two groups, keep in mind that we require both independence of observations within each group and independence between the groups themselves.

- These methods are not appropriate for paired or matched data.

REVIEW OF TERMS

Sampling distribution of the difference between two proportions

The sampling distribution of $\hat{p}_1 - \hat{p}_2$ is, under appropriate assumptions, modeled by a Normal model with mean $\mu = p_1 - p_2$ and standard deviation $SD(\hat{p}_1 - \hat{p}_2) = \sqrt{\frac{p_1q_1}{n_1} + \frac{p_2q_2}{n_2}}$. (p. 631)

Two-proportion z-interval

A two-proportion z-interval gives a confidence interval for the true difference in proportions, $p_1 - p_2$, in two independent groups.

The confidence interval is $(\hat{p}_1 - \hat{p}_2) \pm z^* \times SE(\hat{p}_1 - \hat{p}_2)$, where z^* is a critical value from the standard Normal model corresponding to the specified confidence level. (p. 634)

Two-proportion z-test

Test the null hypothesis H_0: $p_1 - p_2 = 0$ by comparing the statistic

$$z = \frac{(\hat{p}_1 - \hat{p}_2) - 0}{SE(\hat{p}_1 - \hat{p}_2)}$$

to a standard Normal model. (p. 639)

Two-sample t-interval for the difference between means

A confidence interval for the difference between the means of two independent groups found as

$$(\bar{y}_1 - \bar{y}_2) \pm t^*_{df} \times SE(\bar{y}_1 - \bar{y}_2),$$

where

$$SE(\bar{y}_1 - \bar{y}_2) = \sqrt{\frac{s_1^2}{n_1} + \frac{s_2^2}{n_2}}$$

and the number of degrees of freedom is given by a special formula. (p. 641)

Two-sample t-test for the difference between means

A hypothesis test for the difference between the means of two independent groups. It tests the null hypothesis

$$H_0: \mu_1 - \mu_2 = \Delta_0,$$

where the hypothesized difference, Δ_0, is almost always 0, using the statistic

$$t_{df} = \frac{(\bar{y}_1 - \bar{y}_2) - \Delta_0}{SE(\bar{y}_1 - \bar{y}_2)},$$

with the number of degrees of freedom given by the special formula. (p. 644)

Wilcoxon rank sum (or Mann-Whitney) test

A test of the difference between two groups that first ranks the combined sample from the two groups together from smallest to largest, assigning each observation a rank from 1 to $n_1 + n_2$. The test statistic, W, is the sum of the ranks of the first group. (p. 649)

Randomization test for the difference between two means A randomization test that the difference between the means of two groups is zero is based on a simulated sampling distribution of the difference based on the null hypothesis that the difference is zero. The simulation proceeds by randomly selecting two groups from the original data (or shuffling the group identities) and then computing the difference in the means of the resulting randomly assigned groups. Repeating that many times yields a sampling distribution for the difference. A hypothesis test can be based on this sampling distribution. (p. 651)

Pooling Data from two or more populations may sometimes be combined, or *pooled*, to estimate a statistic (typically a pooled variance) when the estimated value is assumed to be the same in both populations. The resulting larger sample size may lead to an estimate with lower sample variance. However, pooled estimates are appropriate only when the required assumptions are true. (p. 652)

TECH SUPPORT

Two-Sample Methods for Proportions

It is so common to test against the null hypothesis of no difference between the two true proportions that most statistics programs simply assume this null hypothesis. And most will automatically use the pooled standard deviation. If you wish to test a different null (say, that the true difference is 0.3), you may have to search for a way to do it.

Many statistics packages don't offer special commands for inference for differences between proportions. As with inference for single proportions, most statistics programs want the "success" and "failure" status for each case. Usually these are given as 1 or 0, but they might be category names like "yes" and "no." Often you just know the proportions of successes, $\hat{p}_1$ and $\hat{p}_2$, and the counts, n_1 and n_2. Computer packages don't usually deal with summary data like these easily. Calculators typically do a better job.

In some programs, you can reconstruct the original values. But even when you have (or can reconstruct) the raw data values, often you won't get *exactly* the same test statistic from a computer package as you would find working by hand. The reason is that when the packages treat the proportion as a mean, they make some approximations. The result is very close, but not exactly the same.

DATA DESK

Data Desk does not offer built-in methods for inference with proportions. Use Replicate Y by X to construct data corresponding to given proportions and totals. The Test and Estimate commands in the Calc menu offer two-sample t methods to test or find confidence intervals for means of two independent groups. Select the two variables first, then go to the menus.

COMMENTS

For inference on proportions with summarized data, open a Scratchpad to compute the standard deviations and margin of error by typing the calculation.

EXCEL

Inference methods for proportions are not part of the standard Excel tool set.

COMMENTS

For summarized data, type the calculation into any cell and evaluate it.

JMP

For a categorical variable that holds category labels, the Distribution platform includes tests and intervals of proportions:

- For summarized data, put the category names in one variable and the frequencies in an adjacent variable.
- Designate the frequency column to have the role of frequency. Then use the Distribution platform.

COMMENTS

JMP uses slightly different methods for proportion inferences than those discussed in this text. Your answers are likely to be slightly different.

MINITAB

To find a hypothesis test for a proportion:

- Choose **Basic Statistics** from the Stat menu.
- Choose **2Proportions** … from the Basic Statistics submenu. If the data are organized as category names in one column and case IDs in another, assign the variables from the variable list box to the Samples in one column box.
- If the data are organized as two separate columns of responses, click on **Samples in different columns:** and assign the variables from the variable list box. If you have summarized data, click the **Summarized Data** button and fill in the number of trials and the number of successes for each group.
- Click the **Options** button and specify the remaining details. Remember to click the Use pooled estimate of *p* for test box when testing the null hypothesis of no difference between proportions.
- Click the **OK** button.

COMMENTS

When working from a variable that names categories, MINITAB treats the last category as the "success" category. You can specify how the categories should be ordered.

R

- Create a vector of successes = c(X1,X2) where X1 is the number of successes in group 1 and X2 is the number of successes in group 2.
- Create a vector of sample.sizes = c(n1,n2) where n1 and n2 are the sample sizes for the two groups, respectively.
- To test the equality of the two proportions:
  - **prop.test(successes,sample.sizes,alternative = "two.sided",conf.level = 0.95,correct = FALSE)**

will test the null hypothesis of equal proportions against the two-sided alternative and give a 95% confidence level.

COMMENTS

R uses the χ^2 statistic instead of the z-statistic for testing. The corresponding z-statistic can be found by taking the square root of the X-squared (χ^2) statistic in the output. The P-values are the same. Other possibilities for alternative are "greater" and "less".

SPSS

SPSS does not perform hypothesis tests for proportions.

STATCRUNCH

To do inference for the difference between two proportions using summaries:

- Click on **Stat**.
- Choose **Proportion Stats » Two sample » with summary**.
- Enter the Number of successes (x) and Number of observations (n) in each group.
- Indicate **Hypothesis Test**, then enter the hypothesized Null proportion difference (usually 0), and choose the **Alternative** hypothesis.

  OR

  Indicate **Confidence Interval**, and then enter the Level of confidence.
- Click on **Compute!**

To do inference for the difference between two proportions using data:

- Click on **Stat**.
- Choose **Proportion Stats » Two sample » with Data**.
- Enter the number of successes (x) and number of observations (n) in each group
- Indicate **Hypothesis Test**, then enter the hypothesized Null proportion difference (usually 0), and choose the **Alternative** hypothesis.

  OR

  Indicate **Confidence Interval**, and then enter the Level of confidence.
- Click on **Compute!**

TI-83/84 PLUS

To calculate a confidence interval for the difference between two population proportions:

- Select **B:2-PropZInt** from the STAT TESTS menu.
- Enter the observed counts and the sample sizes for both samples.
- Specify a confidence level.
- Calculate the interval.

To do the mechanics of a hypothesis test for equality of population proportions:

- Select **6:2-PropZTest** from the STAT TESTS menu.
- Enter the observed counts and sample sizes.
- Indicate what kind of test you want: one-tail upper tail, lower tail, or two-tail.
- Calculate the result.

COMMENTS

Beware: When you enter the value of x, you need the *count*, not the percentage. The count must be a whole number.

Two-Sample Methods for Means

Here's some typical computer package output with comments:

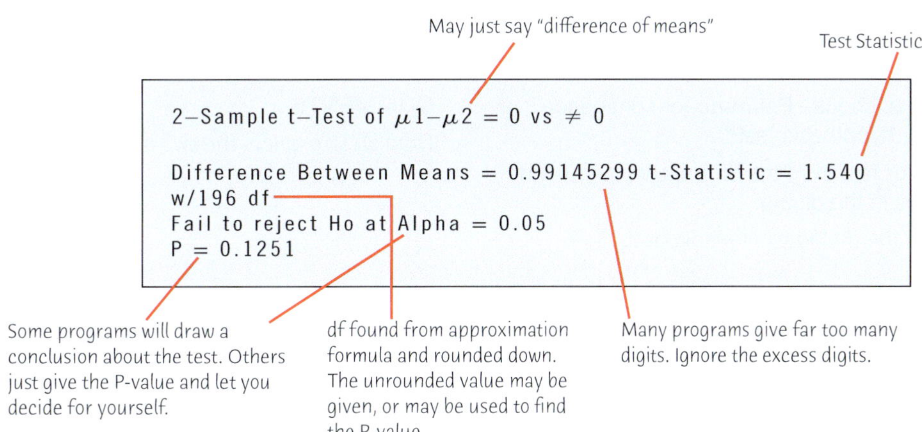

Most statistics packages compute the test statistic for you and report a P-value corresponding to that statistic. And, of course, statistics packages make it easy to examine the boxplots of the two groups, so you have no excuse for skipping this important check.

Some statistics software automatically tries to test whether the variances of the two groups are equal. Some automatically offer both the two-sample-*t* and pooled-*t* results. Ignore the test for the variances; it has little power in any situation in which its results could matter. If the pooled and two-sample methods differ in any important way, you should stick with the two-sample method. Most likely, the Equal Variance Assumption needed for the pooled method has failed.

The degrees of freedom approximation usually gives a fractional value. Most packages seem to round the approximate value down to the next smallest integer (although they may actually compute the P-value with the fractional value, gaining a tiny amount of power).

There are two ways to organize data when we want to compare two independent groups. The data can be in two columns, as in the table for camera prices. Each column can be thought of as a variable. In this method, the variables for the camera example would be *Prices to Friend* and *Prices to Stranger*. Graphing calculators usually prefer this form, and some computer programs can use it as well.

There's another way to think about the data. What is the response variable for the camera experiment? It's the *Price* offered. But the values of this variable are in both columns, and actually there's an experiment factor here, too—namely, whether the *Seller* was a *Friend* or a *Stranger*. So, we could put the data into two different columns, one with the *Prices* in it and one with the *Seller*. Then the data would look like this:

Seller	Price
Friend	275
Friend	300
Friend	260
Friend	300
Friend	255
Friend	275
Friend	290
Friend	300
Stranger	260
Stranger	250
Stranger	175
Stranger	130
Stranger	200
Stranger	225
Stranger	240

This way of organizing the data makes sense as well. Now the factor and the response variables are clearly visible. You'll have to see which method your program requires. Some packages even allow you to structure the data either way.

The commands to do inference for two independent groups on common statistics technology are not always found in obvious places. Here are some starting guidelines.

DATA DESK

- From the Calc menu, choose **Estimate** for confidence intervals or Test for hypothesis tests.
- Select the interval or test from the drop-down menu and make other choices in the dialog.
- Drag variables into the dialog where indicated.

COMMENTS

Data Desk expects the two groups to be in separate variables.

EXCEL

- From the Data Tab, Analysis Group, choose **Data Analysis**.
- Alternatively (if the Data Analysis Tool Pack is not installed), in the Formulas Tab, choose **More functions** » **Statistical** » **TTEST**, and specify **Type = 3** in the resulting dialog.
- Fill in the cell ranges for the two groups, the hypothesized difference, and the alpha level.

COMMENTS

Excel expects the two groups to be in separate cell ranges. Notice that, contrary to Excel's wording, we do not need to assume that the variances are *not* equal; we simply choose not to assume that they *are* equal.

JMP

- From the Analyze menu, select **Fit y by x**.
- Select variables: a **Y, Response** variable that holds the data and an **X, Factor** variable that holds the group names. JMP will make a dotplot.
- Click the **red triangle** in the dotplot title, and choose **Unequal variances**. The *t*-test is at the bottom of the resulting table.
- Find the P-value from the Prob > F section of the table (they are the same).

COMMENTS

JMP expects data in one variable and category names in the other. Don't be misled: There is no need for the variances to be unequal to use two-sample *t* methods.

MINITAB

- From the Stat menu, choose the **Basic Statistics** submenu.
- From that menu, choose **2-sample *t*** . . . . Then fill in the dialog.

COMMENTS

The dialog offers a choice of data in two variables, or data in one variable and category names in the other.

R

To test the hypothesis that $\mu_1 = \mu_2$ against an alternative (default is two-sided):

- Create a vector for the data of each group, say *X* and *Y*, and produce the confidence interval (default is 95%):
- t.test(X,Y, alternative = c("two.sided", "less", "greater"), conf.level = 0.95)

will produce the *t*-statistic, degrees of freedom, P-value, and confidence interval for a specified alternative.

COMMENTS

This is the same function as for the one-sample *t*-test, but with two vectors. In R, the default *t*-test does not assume equal variances in the two groups (default is **var.equal = FALSE**). To get the "pooled" t-test, add **var.equal = TRUE** to the function call.

SPSS

- From the Analyze menu, choose the **Compare Means** submenu.
- From that, choose the **Independent-Samples t-test** command. Specify the data variable and group variable.
- Then type in the labels used in the group variable. SPSS offers both the two-sample and pooled-*t* results in the same table.

COMMENTS

SPSS expects the data in one variable and group names in the other. If there are more than two group names in the group variable, only the two that are named in the dialog box will be compared.

STATCRUNCH

To do inference for the difference between two means using summaries:

- Click on **Stat**.
- Choose **T Stats** » **Two sample** » **with summary**.
- Enter the **Sample mean, Standard deviation**, and sample **Size** for each group.
- Indicate **Hypothesis Test**, then enter the hypothesized Null mean difference (usually 0), and choose the **Alternative** hypothesis.

  OR

  Indicate **Confidence Interval**, and then enter the **Level** of confidence.
- Click on **Compute!**

To do inference for the difference between two means using data:

- Click on **Stat**.
- Choose **T Stats** » **Two sample** » **with data**.
- Choose the variable Column for each group.
- Indicate **Hypothesis Test**, then enter the hypothesized **Null mean** difference (usually 0), and choose the **Alternative** hypothesis.

  OR

  Indicate **Confidence Interval**, and then enter the **Level** of confidence.
- Click on **Compute!**

TI-83/84 PLUS

For a confidence interval:

- In the STAT TESTS menu, choose **0:2-SampTInt**.
- You may specify that you are using data stored in two lists, or you may enter the means, standard deviations, and sizes of both samples.
- You must also indicate whether to pool the variances (when in doubt, say no) and specify the desired level of confidence.

To test a hypothesis:

- In the STAT TESTS menu, choose **4:2-SampT Test**.
- You may specify if you are using data stored in two lists, or you may enter the means, standard deviations, and sizes of both samples.
- You must also indicate whether to pool the variances (when in doubt, say no) and specify whether the test is to be two-tail, lower-tail, or upper-tail.

EXERCISES

SECTION 20.1

1. Canada Suppose an advocacy organization surveys 960 Canadians and 192 of them reported being born in another country (www.unitednorthamerica.org/simdiff.htm). Similarly, 170 out of 1250 U.S. citizens reported being foreign-born. Find the standard error of the difference in sample proportions.

2. Non-profits Do people who work for non-profit organizations differ from those who work at for-profit companies when it comes to personal job satisfaction? Separate random samples were collected by a polling agency to investigate the difference. Data collected from 422 employees at non-profit organizations revealed that 377 of them were "highly satisfied." From the for-profit companies, 431 out 518 employees reported the same level of satisfaction. Find the standard error of the difference in sample proportions.

3. Canada, deux The information in Exercise 1 was used to create a 95% two-proportion confidence interval for the difference between Canadians and U.S. citizens who were born in foreign countries. Interpret this interval with a sentence in context.

$$95\% \text{ confidence interval for } p_{\text{Canadians}} - p_{\text{Americans}} \text{ is } (3.24\%, 9.56\%).$$

4. Non-profits, part 2 The researchers from Exercise 2 created a 95% two-proportion confidence interval for the difference in those who are "highly satisfied" when comparing people who work at non-profits to people who work at for-profit companies. Interpret the interval with a sentence in context.

$$95\% \text{ confidence interval for } p_{\text{non-profits}} - p_{\text{for-profits}} = (1.77\%, 10.50\%).$$

5. Canada, trois For the interval given in Exercise 3, explain what "95% confidence" means.

6. Non-profits, part 3 For the interval given in Exercise 4, explain what "95% confidence" means.

SECTION 20.2

7. Canada, encore If the information in Exercise 1 is to be used to make inferences about the proportion all Canadians and all U.S. citizens born in other countries, what conditions must be met before proceeding? Are they met? Explain.

8. Non-profits, again If the information in Exercise 2 is to be used to make inferences about all people who work at non-profits and for-profit companies, what conditions must be met before proceeding? List them and explain if they are met.

SECTION 20.3

9. Canada, test The researchers from Exercise 1 want to test if the proportions of foreign born are the same in the United States and Canada. What is the appropriate standard error to use for the hypothesis test?

a) What is the difference in the proportions of foreign born residents in both countries?
b) What is the value of the z-statistic?
c) What do you conclude at $\alpha = 0.05$?

10. Non-profits test Complete the analysis begun in Exercise 2.

a) What is the difference in the proportions of the two types of companies?
b) What is the value of the z-statistic?
c) What do you conclude at $\alpha = 0.05$?

SECTION 20.4

11. Cost of shopping Do consumers spend more on a trip to Sainsbury's or Tesco? Suppose researchers interested in this question collected a systematic sample from 82 Sainsbury's customers and 78 Tesco customers by asking customers for their purchase amount as they left the stores. The data collected are summarized in the table below.

To perform inference on these two samples, what conditions must be met? Are they? Explain.

	Sainbury's	Tesco
n	82	78
$\bar{y}$	£43	£51
s	£20	£18

12. Athlete ages A sports reporter suggests that professional baseball players must be, on average, older than professional football players, since football is a contact sport and players are more susceptible to concussions and serious injuries. Using data from sports.yahoo.com, one player was selected at random from each team in both professional baseball (MLB) and professional football (NFL). The data are summarized below.

	MLB	NFL
n	30	32
$\bar{y}$	27.5	26.16
s	3.94	2.78

To perform inference on these two samples, what conditions must be met? Are they? Explain.

13. Cost of shopping, again Using the summary statistics provided in Exercise 11, researchers calculated a 95% confidence interval for the mean difference between Sainsbury's and Tesco purchase amounts. The interval was (−13.93; −2.07). Explain in context what this interval means.

14. Athlete ages, again Using the summary statistics provided in Exercise 12, the sports reporter calculated the following 95% confidence interval for the mean difference between major league baseball players and professional football players. The 95% interval for $\mu_{MLB} - \mu_{NFL}$ was $(-0.41, 3.09)$. Summarize in context what the interval means.

SECTION 20.5

15. Cost of shopping, once more The researchers in Exercise 11 decide to test the hypothesis that the means are equal. The degrees of freedom formula gives 157.525 df. Test the null hypothesis at $\alpha = 0.05$.

16. Athlete ages, ninth inning The researchers in Exercise 12 decide to test the hypothesis. The degrees of freedom formula gives 51.83 df. Test the null hypothesis at $\alpha = 0.05$. Is the alternative one- or two-sided?

*SECTION 20.7

17. Cost of shopping, yet again Repeat the test you did in Exercise 15 but assume that the variances of purchase amounts are the same at Tesco and Sainsbury's. Did your conclusion change? Why do you think that is so?

18. Athlete ages, overtime Repeat the test you did in Exercise 16, but assume that the variances of ages are the same in the two leagues. Did your conclusion change? Why do you think that is?

19. Cost of shopping, once more Looking back at Exercise 11, instead of comparing two very similar stores, suppose the researchers had compared purchases at two clothing shops: one that specializes in casual prêt-à-porter and another that sells fancy designer dresses. Would the pooled t-test be a good choice here? Why or why not? Which test or confidence interval would be appropriate for testing whether the mean purchase amounts at the two shops are the same?

20. Athletes, extra innings Looking back at Exercise 12, instead of comparing the ages of players in Major League Baseball to players in the National Football League, what if they had compared the ages in Major League Baseball to a Little League composed of third graders? Would the pooled t-test be a good choice here? Why or why not? Which test or confidence interval would be appropriate for testing the difference in mean ages in that case?

CHAPTER EXERCISES

21. Online social networking Many reports published through online blogs provide statistics related to social media. According to an online blog, a social media platform reached 60.6% of Internet users worldwide last year. This proportion was closer to 50% in 2010. What does it mean to say that the difference in proportions is "significant"?

22. Science news At the end of 2013, the Pew Project for Excellence in Journalism investigated where people are getting their news. In the study 22% of people 18–29 years old said they still read newspapers as one of their sources of news, while only 18% of people 30–49 said the same. What does it mean to say that the difference is not significant?

23. Revealing information Eight hundred eighty-six randomly sampled teens were asked which of several personal items of information they thought it ok to share with someone they had just met. Forty-four percent said it was ok to share their e-mail addresses, but only 29% said they would give out their cell phone numbers. A researcher claims that a two-proportion z-test could tell whether there was a real difference among all teens. Explain why that test would not be appropriate for these data.

24. Regulating access When a random sample of 935 parents were asked about rules in their homes, 77% said they had rules about the kinds of TV shows their children could watch. Among the 790 of those parents whose teenage children had Internet access, 85% had rules about the kinds of Internet sites their teens could visit. That looks like a difference, but can we tell? Explain why a two-sample z-test would not be appropriate here.

25. Gender gap A presidential candidate fears he has a problem with women voters. His campaign staff plans to run a poll to assess the situation. They'll randomly sample 300 men and 300 women, asking if they have a favorable impression of the candidate. Obviously, the staff can't know this, but suppose the candidate has a positive image with 59% of males but with only 53% of females.

a) What kind of sampling design is his staff planning to use?
b) What difference would you expect the poll to show?
c) Of course, sampling error means the poll won't reflect the difference perfectly. What's the standard deviation for the difference in the proportions?
d) Sketch a sampling model for the size difference in proportions of men and women with favorable impressions of this candidate that might appear in a poll like this.
e) Could the campaign be misled by the poll, concluding that there really is no gender gap? Explain.

26. Buy it again? A consumer magazine plans to poll car owners to see if they are happy enough with their vehicles that they would purchase the same model again. They'll randomly select 450 owners of American-made cars and 450 owners of Japanese models. Obviously, the actual opinions of the entire population couldn't be known, but suppose 76% of owners of American cars and 78% of owners of Japanese cars would purchase another.

a) What kind of sampling design is the magazine planning to use?
b) What difference would you expect their poll to show?
c) Of course, sampling error means the poll won't reflect the difference perfectly. What's the standard deviation for the difference in the proportions?
d) Sketch a sampling model for the difference in proportions that might appear in a poll like this.
e) Could the magazine be misled by the poll, concluding that owners of American cars are much happier with their vehicles than owners of Japanese cars? Explain.

27. Arthritis According to the National Health Survey 2017–2018 results reported by the Australian Bureau of Statistics, for randomly selected Australians aged 65 and older, 399 of 1000 men and 573 of 1000 women suffered from some form of arthritis.

a) Are the assumptions and conditions necessary for inference satisfied? Explain.
b) Create a 95% confidence interval for the difference in the proportions of senior men and women who have this disease.
c) Interpret your interval in this context.
d) Does this confidence interval suggest that arthritis is more likely to afflict women than men? Explain.

(https://www.abs.gov.au/ausstats/abs@.nsf/Lookup/by%20Subject/4364.0.55.001~2017-18~Main%20Features~Arthritis%20and%20osteoporosis~30)

28. Graduation According to a report published by the OECD Better Life Index about Switzerland, 89% of men aged 25–64 have successfully completed upper secondary education compared to 86% of women. We assume that these proportions are computed based on data collected from random samples of 12,000 men and 12,000 women, respectively.

a) Are the assumptions and conditions necessary for inference satisfied? Explain.
b) Create a 95% confidence interval for the difference in graduation rates between males and females.
c) Interpret your confidence interval.
d) Does this provide strong evidence that men are more likely than women to complete upper secondary education? Explain.

(http://www.oecdbetterlifeindex.org/countries/switzerland/)

29. Pets Researchers at a cancer institute released the results of a study that investigated the effect of weed-killing herbicides on house pets. They examined 850 dogs from homes where a herbicide was used on a regular basis, diagnosing malignant lymphoma in 470 of them. Of the 160 dogs from homes where no herbicides were used, only 22 were found to have lymphoma.

a) What's the standard error of the difference in the two proportions?
b) Construct a 95% confidence interval for this difference.
c) State an appropriate conclusion.

30. Carpal tunnel The painful wrist condition called carpal tunnel syndrome can be treated with surgery or, less invasively, with wrist splints. Recently, *Time* magazine reported on a study of 176 patients. Among the half that had surgery, 80% showed improvement after three months, but only 48% of those who used the wrist splints improved.

a) What's the standard error of the difference in the two proportions?
b) Construct a 95% confidence interval for this difference.
c) State an appropriate conclusion.

31. Prostate cancer There has been debate among doctors over whether surgery can prolong life among men suffering from prostate cancer, a type of cancer that typically develops and spreads very slowly. Recently, *The New England Journal of Medicine* published results of some Scandinavian research. Men diagnosed with prostate cancer were randomly assigned to either undergo surgery or not. Among the 347 men who had surgery, 16 eventually died of prostate cancer, compared with 31 of the 348 men who did not have surgery.

a) Was this an experiment or an observational study? Explain.
b) Create a 95% confidence interval for the difference in rates of death for the two groups of men.
c) Based on your confidence interval, is there evidence that surgery may be effective in preventing death from prostate cancer? Explain.

32. Gender and smoking 2016 Data collected in 2016 by the World Bank revealed that in the Pacific state of Nauru, 43% of women smoke as compared to 37% of men. Suppose these proportions were calculated based on samples of 3400 women and 3600 men.

a) Create a 90% confidence interval for the difference in the percentage of smokers between men and women in Nauru.
b) Do these numbers indicate a gender-based difference in smoking in Nauru? Explain, using your confidence interval to test an appropriate hypothesis.
c) What alpha level did your test use?

(https://ourworldindata.org/smoking)

33. Ear infections A new vaccine was recently tested to see if it could prevent the painful and recurrent ear infections that many infants suffer from. *The Lancet*, a medical journal, reported a study in which babies about a year old were randomly divided into two groups. One group received vaccinations; the other did not. During the following year, only 333 of 2455 vaccinated children had ear infections, compared to 499 of 2452 unvaccinated children in the control group.

a) Are the conditions for inference satisfied?
b) Find a 95% confidence interval for the difference in rates of ear infection.
c) Use your confidence interval to explain whether you think the vaccine is effective.

34. Anorexia The *Journal of the American Medical Association* reported on an experiment intended to see if the drug Prozac® could be used as a treatment for the eating disorder anorexia nervosa. The subjects, women being treated for anorexia, were randomly divided into two groups. Of the 49 who received Prozac, 35 were deemed healthy a year later, compared to 32 of the 44 who got the placebo.

a) Are the conditions for inference satisfied?
b) Find a 95% confidence interval for the difference in outcomes.
c) Use your confidence interval to explain whether you think Prozac is effective.

35. Another ear infection In Exercise 33, you used a confidence interval to examine the effectiveness of a vaccine against ear infections in babies. Suppose that instead you had conducted a hypothesis test. (Answer these questions *without* actually doing the test.)

a) What hypotheses would you test?
b) State a conclusion based on your confidence interval.
c) If that conclusion is wrong, which type of error did you make?
d) What would be the consequences of such an error?

36. Anorexia again In Exercise 34, you used a confidence interval to examine the effectiveness of Prozac in treating anorexia nervosa. Suppose that instead you had conducted a hypothesis test. (Answer these questions *without* actually doing the test.)

a) What hypotheses would you test?
b) State a conclusion based on your confidence interval.
c) If that conclusion is wrong, which type of error did you make?
d) What would be the consequences of such an error?

37. Teen smoking A Vermont study published by the American Academy of Pediatrics examined parental influence on teenagers' decisions to smoke. A group of students who had never smoked were questioned about their parents' attitudes toward smoking. These students were questioned again two years later to see if they had started smoking. The researchers found that, among the 284 students who indicated that their parents disapproved of kids smoking, 54 had become established smokers. Among the 41 students who initially said their parents were lenient about smoking, 11 became smokers. Do these data provide strong evidence that parental attitude influences teenagers' decisions about smoking?

a) What kind of design did the researchers use?
b) Write appropriate hypotheses.
c) Are the assumptions and conditions necessary for inference satisfied?
d) Test the hypothesis and state your conclusion.
e) Explain in this context what your P-value means.
f) If it is later found that parental attitudes actually do influence teens' decisions to smoke, which type of error did you commit?
g) Create a 95% CI.
h) Interpret your interval.
i) Carefully explain what "95% confidence" means.

38. Depression A study published in the *Archives of General Psychiatry* examined the impact of depression on a patient's ability to survive cardiac disease. Researchers identified 450 people with cardiac disease, evaluated them for depression, and followed the group for 4 years. Of the 361 patients with no depression, 67 died. Of the 89 patients with minor or major depression, 26 died. Among people who suffer from cardiac disease, are depressed patients more likely to die than non-depressed ones?

a) What kind of design was used to collect these data?
b) Write appropriate hypotheses.
c) Are the assumptions and conditions necessary for inference satisfied?
d) Test the hypothesis and state your conclusion.
e) Explain in this context what your P-value means.
f) If your conclusion is actually incorrect, which type of error did you commit?
g) Create a 95% CI.
h) Interpret your interval.
i) Carefully explain what "95% confidence" means.

39. Birthweight The *Journal of the American Medical Association* reported a study examining the possible impact of air pollution caused by the 9/11 attack on New York's World Trade Center on the weight of babies. Researchers found that 8% of 182 babies born to mothers who were exposed to heavy doses of soot and ash on September 11 were classified as having low birthweight. Only 4% of 2300 babies born in another New York City hospital whose mothers had not been near the site of the disaster were similarly classified. Does this indicate a possibility that air pollution might be linked to a significantly higher proportion of low-weight babies?

a) Test an appropriate hypothesis at $\alpha = 0.10$ and state your conclusion.
b) If you concluded there is a difference, estimate that difference with a confidence interval and interpret that interval in context.

40. Politics and sex One month before the election, a poll of 630 randomly selected voters showed 54% planning to vote for a certain candidate. A week later, it became known that he had had an extramarital affair, and a new poll showed only 51% of 1010 voters supporting him. Do these results indicate a decrease in voter support for his candidacy?

a) Test an appropriate hypothesis and state your conclusion.
b) If you concluded there was a difference, estimate that difference with a confidence interval and interpret your interval in context.

41. Mammograms It's widely believed that regular mammogram screening may detect breast cancer early, resulting in fewer deaths from that disease. One study that investigated this issue over a period of 18 years was published during the 1970s. Among 30,565 women who had never had mammograms, 196 died of breast cancer, while only 153 of 30,131 who had undergone screening died of breast cancer.

a) Do these results suggest that mammograms may be an effective screening tool to reduce breast cancer deaths?
b) If your conclusion is incorrect, what type of error have you committed?

42. Mammograms redux In 2001, the conclusion of the study outlined in Exercise 41 was questioned. A new 9-year study was conducted in Sweden, comparing 21,088 women who had mammograms with 21,195 who did not. Of the women who underwent screening, 63 died of breast cancer, compared to 66 deaths among the control group. (*The New York Times*, Dec. 9, 2001)

a) Do these results support the effectiveness of regular mammograms in preventing deaths from breast cancer?
b) If your conclusion is incorrect, what kind of error have you committed?

43. Pain Researchers comparing the effectiveness of two pain medications randomly selected a group of patients who had been complaining of a certain kind of joint pain. They randomly divided these people into two groups, then administered the pain killers. Of the 112 people in the group who received medication A, 84 said this pain reliever was effective. Of the 108 people in the other group, 66 reported that pain reliever B was effective.

a) Write a 95% confidence interval for the percent of people who may get relief from this kind of joint pain by using medication A. Interpret your interval.
b) Write a 95% confidence interval for the percent of people who may get relief by using medication B. Interpret your interval.
c) Do the intervals for A and B overlap? What do you think this means about the comparative effectiveness of these medications?
d) Find a 95% confidence interval for the difference in the proportions of people who may find these medications effective. Interpret your interval.
e) Does this interval contain zero? What does that mean?
f) Why do the results in parts c and e seem contradictory? If we want to compare the effectiveness of these two pain relievers, which is the correct approach? Why?

44. Gender gap Candidates for political office realize that different levels of support among men and women may be a crucial factor in determining the outcome of an election. One candidate finds that 52% of 473 men polled say they will vote for him, but only 45% of the 522 women in the poll express support.

a) Write a 95% confidence interval for the percent of male voters who may vote for this candidate. Interpret your interval.
b) Write a 95% confidence interval for the percent of female voters who may vote for him. Interpret your interval.
c) Do the intervals for males and females overlap? What do you think this means about the gender gap?
d) Find a 95% confidence interval for the difference in the proportions of males and females who will vote for this candidate. Interpret your interval.
e) Does this interval contain zero? What does that mean?
f) Why do the results in parts c and e seem contradictory? If we want to see if there is a gender gap among voters with respect to this candidate, which is the correct approach? Why?

45. Presidential bounce According to a poll by a political thinktank, the approval rating for a country's president surged 11.2 percentage points, from 23.3% to 34.5%, since a set of bills were passed by their government.

Assume these proportions were obtained through polls randomly sampling 1600 likely voters each time, would you consider that there was a significant "bounce" effect?

46. Stay-at-home dads A *Time* magazine article about a survey of men's attitudes reported that 11 of 161 black respondents and 20 of 358 Latino respondents responded "Yes" to the question "Are you a stay-at-home dad?" How big is the difference in proportions in the two populations?

a) Construct and interpret an appropriate confidence interval.
b) Overall, the survey contacted 1302 men and claims a margin of error of $\pm 2.9\%$. Why is the margin of error different for your confidence interval?

47. Sensitive men In the same article from Exercise 46, *Time* magazine, reporting on a survey of men's attitudes, noted that "Young men are more comfortable than older men talking about their problems." The survey reported that 80 of 129 surveyed 18- to 24-year-old men and 98 of 184 25- to 34-year-old men said they were comfortable. What do you think? Is *Time*'s interpretation justified by these numbers?

48. Carbs Recently, the Gallup International Poll asked 1010 European adults if they actively try to avoid carbohydrates in their diet. That number increased to 25% from 19% in a similar 2002 poll. Is this a statistically significant increase? Explain.

(https://www.gallup-international.com/about/; https://www.gallup-international.com/survey-results/)

49. Food preference GfK Roper Consulting gathers information on consumer preferences around the world to help companies monitor attitudes about health, food, and healthcare products. They asked people in many different cultures how they felt about the following statement:

> *I have a strong preference for regional or traditional products and dishes from where I come from.*

In a random sample of 800 respondents, 417 of 646 people who live in urban environments agreed (either completely or somewhat) with that statement, compared to 78 out of 154 people who live in rural areas.

Based on this sample, is there evidence that the percentage of people agreeing with the statement about regional preferences differs between all urban and rural dwellers?

50. Fast food The global survey we learned about in Exercise 49 also asked respondents how they felt about the statement "I try to avoid eating fast foods." The random sample of 800 included 411 people 35 years old or younger, and of those, 197 agreed (completely or somewhat) with the statement. Of the 389 people over 35 years old, 246 people agreed with the statement. Is there evidence that the percentage of people avoiding fast food is different in the two age groups?

51. Hot dogs In the July 2007 issue, *Consumer Reports* examined the calorie content of two kinds of hot dogs: meat (usually a mixture of pork, turkey, and chicken) and all beef. The researchers purchased samples of several different brands. The meat hot dogs averaged 111.7 calories, compared to 135.4 for the beef hot dogs. A test of the null hypothesis that there's no difference in mean calorie content yields a P-value of 0.124. Would a 95% confidence interval for $\mu_{Meat} - \mu_{Beef}$ include 0? Explain.

52. Washers In the June 2007 issue, *Consumer Reports* also examined the relative merits of top-loading and front-loading washing machines, testing samples of several different brands of each type. Suppose the study tested the null hypothesis that top- and front-loading machines don't differ in their mean costs, and the test had a P-value of 0.32. Would a 95% confidence interval for $\mu_{top} - \mu_{front}$ contain 0? Explain.

53. Hot dogs, second helping The *Consumer Reports* article described in Exercise 51 also listed the fat content (in grams) for samples of beef and meat hot dogs. The resulting 90% confidence interval for $\mu_{Meat} - \mu_{Beef}$ is $(-6.5, -1.4)$.

a) The endpoints of this confidence interval are negative numbers. What does that indicate?

b) What does the fact that the confidence interval does not contain 0 indicate?

c) If we use this confidence interval to test the hypothesis that $\mu_{Meat} - \mu_{Beef} = 0$, what's the corresponding alpha level?

54. Second load of wash The *Consumer Reports* article described in Exercise 52 continued their investigation of washing machines. One of the variables the article reported was "cycle time," the number of minutes it took each machine to wash a load of clothes. Among the machines rated good to excellent, the 98% confidence interval for the difference in mean cycle time $(\mu_{Top} - \mu_{Front})$ is $(-40, -22)$.

a) The endpoints of this confidence interval are negative numbers. What does that indicate?

b) What does the fact that the confidence interval does not contain 0 indicate?

c) If we use this confidence interval to test the hypothesis that $\mu_{Top} - \mu_{Front} = 0$, what's the corresponding alpha level?

55. Hot dogs, last one In Exercise 53, we saw a 90% confidence interval of $(-6.5, -1.4)$ grams for $\mu_{Meat} - \mu_{Beef}$, the difference in mean fat content for meat vs. all-beef hot dogs. Explain why you think each of the following statements is true or false:

a) If I eat a meat hot dog instead of a beef dog, there's a 90% chance I'll consume less fat.

b) 90% of meat hot dogs have between 1.4 and 6.5 grams less fat than a beef hot dog.

c) I'm 90% confident that meat hot dogs average between 1.4 and 6.5 grams less fat than the beef hot dogs.

d) If I were to get more samples of both kinds of hot dogs, 90% of the time the meat hot dogs would average between 1.4 and 6.5 grams less fat than the beef hot dogs.

e) If I tested more samples, I'd expect about 90% of the resulting confidence intervals to include the true difference in mean fat content between the two kinds of hot dogs.

56. Third load of wash In Exercise 54, we saw a 98% confidence interval of $(-40, -22)$ minutes for $\mu_{Top} - \mu_{Front}$, the difference in time it takes top-loading and front-loading washers to do a load of clothes. Explain why you think each of the following statements is true or false:

a) 98% of top loaders are 22 to 40 minutes faster than front loaders.

b) If I choose the laundromat's top loader, there's a 98% chance that my clothes will be done faster than if I had chosen the front loader.

c) If I tried more samples of both kinds of washing machines, in about 98% of these samples I'd expect the top loaders to be an average of 22 to 40 minutes faster.

d) If I tried more samples, I'd expect about 98% of the resulting confidence intervals to include the true difference in mean cycle time for the two types of washing machines.

e) I'm 98% confident that top loaders wash clothes an average of 22 to 40 minutes faster than front-loading machines.

57. Learning math The Core Plus Mathematics Project (CPMP) is an innovative approach to teaching Mathematics that engages students in group investigations and mathematical modeling. After field tests in 36 high schools over a three-year period, researchers compared the performances of CPMP students with those taught using a traditional curriculum. In one test, students had to solve applied algebra problems using calculators. Scores for 320 CPMP students were compared to those of a control group of 273 students in a traditional math program. Computer software was used to create a confidence interval for the difference in mean scores. (*Journal for Research in Mathematics Education*, 31, no. 3)

Conf level: 95% Variable: Mu(CPMP)−Mu(Ctrl)
Interval: (5.573, 11.427)

a) What's the margin of error for this confidence interval?

b) If we had created a 98% CI, would the margin of error be larger or smaller?

c) Explain what the calculated interval means in this context.

d) Does this result suggest that students who learn mathematics with CPMP will have significantly higher mean scores in algebra than those in traditional programs? Explain.

T **58. Stereograms** In Exercises 49 and 50, Chapter 4, we looked at data from an experiment to determine whether visual information about an image helped people "see" the image in 3D.

2-Sample *t*-Interval for $\mu 1 - \mu 2$
Conf level = 90% df = 70
μ(NV) − μ(VV) interval: (0.55, 5.47)

a) Interpret your interval in context.

b) Does it appear that viewing a picture of the image helps people "see" the 3D image in a stereogram?

c) What's the margin of error for this interval?

d) Explain carefully what the 90% confidence level means.

e) Would you expect a 99% confidence interval to be wider or narrower? Explain.

f) Might that change your conclusion in part b? Explain.

59. CPMP, again During the study described in Exercise 57, students in both CPMP and traditional classes took another algebra test that did not allow them to use calculators. The table below shows the results. Are the mean scores of the two groups significantly different?

Math Program	*n*	Mean	SD
CPMP	312	29.0	18.8
Traditional	265	38.4	16.2

a) Write appropriate hypotheses.

b) Do you think the assumptions for inference are satisfied? Explain.

c) Here is computer output for this hypothesis test. Explain what the P-value means in this context.

2-Sample *t*-Test of $\mu 1 - \mu 2 \neq 0$
t-Statistic = −6.451 w/574.8761 df
P < 0.0001

d) State a conclusion about the CPMP program.

60. CPMP and word problems The study of the new CPMP Mathematics methodology described in Exercise 57 also tested students' abilities to solve word problems. This table shows how the CPMP and traditional groups performed. (The df are 590.049.) What do you conclude?

Math Program	*n*	Mean	SD
CPMP	320	57.4	32.1
Traditional	273	53.9	28.5

61. Commuting A man who moves to a new city sees that there are two routes he could take to work. A neighbor who has lived there a long time tells him Route A will average 5 minutes faster than Route B. The man decides to experiment. Each day, he flips a coin to determine which way to go, driving each route 20 days. He finds that Route A takes an average of 40 minutes, with standard deviation 3 minutes, and Route B takes an average of 43 minutes, with standard deviation 2 minutes. Histograms of travel times for the routes are roughly symmetric and show no outliers.

a) Find a 95% confidence interval for the difference in average commuting time for the two routes. (From technology, $df = 33.1$.)
b) Should the man believe the old-timer's claim that he can save an average of 5 minutes a day by always driving Route A? Explain.

62. Pulse rates A researcher wanted to see whether there is a significant difference in resting pulse rates for men and women. The data she collected are displayed in the boxplots and summarized below.

	Sex	
	Male	**Female**
Count	28	24
Mean	72.75	72.625
Median	73	73
StdDev	5.37225	7.69987
Range	20	29
IQR	9	12.5

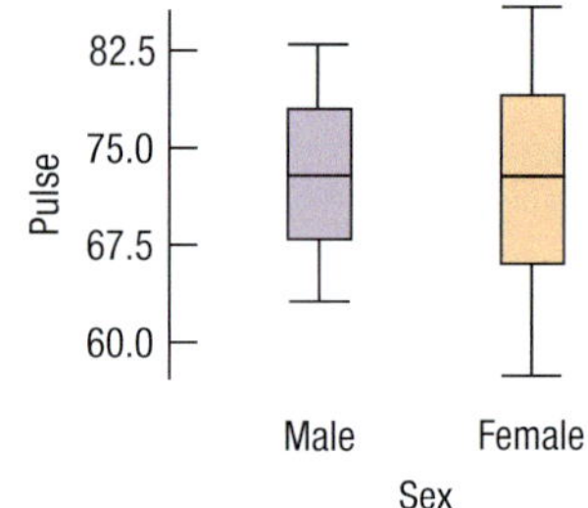

a) What do the boxplots suggest about differences between male and female pulse rates?
b) Is it appropriate to analyze these data using the methods of inference discussed in this chapter? Explain.
c) Create a 90% confidence interval for the difference in mean pulse rates. (From technology, $df = 40.2$.)
d) Does the confidence interval confirm your answer to part a? Explain.

63. View of the water How much extra is having a waterfront property worth? A student took a random sample of 170 recently sold properties in Upstate New York to examine the question. Here are her summaries and boxplots of the two groups of prices:

Non-Waterfront Properties		Waterfront Properties	
n	100	n	70
$\bar{y}$	$219,896.60	$\bar{y}$	$319,906.40
s	$94,627.15	s	$153,303.80

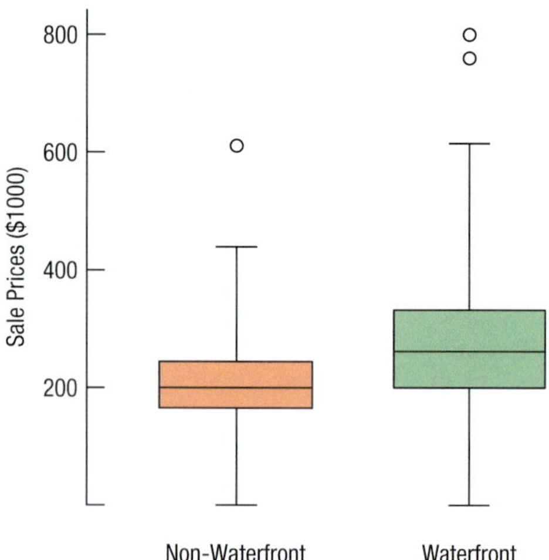

Construct and interpret a 95% confidence interval for the mean additional amount that waterfront property is worth. (From technology, $df = 105.48$.)

64. New construction The house sales we looked at in Exercise 63 also listed whether the home was new construction or not. Find and interpret a 95% confidence interval for how much more an agent can expect to sell a new home for. (From technology, $df = 197.8$.) Here are the summaries and boxplots of the *Sale Prices*:

Old Construction		New Construction	
n	100	n	100
$\bar{y}$	$201,707.50	$\bar{y}$	$267,878.10
s	$96,116.88	s	$93,302.18

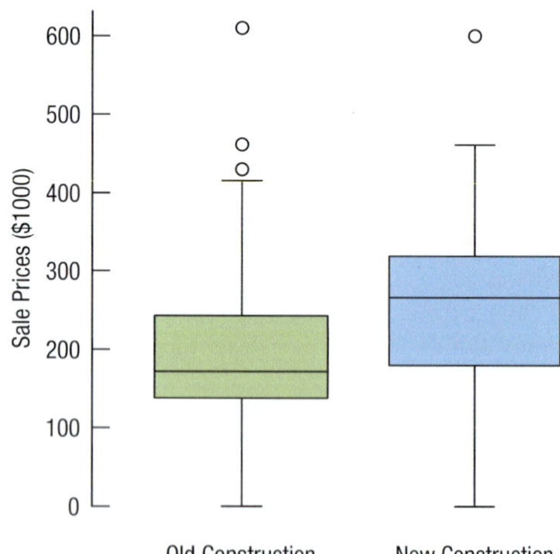

T **65. Sugar in cereal** The data below show the sugar content (as a percentage of weight) of several national brands of children's and adults' cereals. Create and interpret a 95% confidence interval for the difference in mean sugar content. Be sure to check the necessary assumptions and conditions.

Children's cereals: 40.3, 55, 45.7, 43.3, 50.3, 45.9, 53.5, 43, 44.2, 44, 47.4, 44, 33.6, 55.1, 48.8, 50.4, 37.8, 60.3, 46.6

Adults' cereals: 20, 30.2, 2.2, 7.5, 4.4, 22.2, 16.6, 14.5, 21.4, 3.3, 6.6, 7.8, 10.6, 16.2, 14.5, 4.1, 15.8, 4.1, 2.4, 3.5, 8.5, 10, 1, 4.4, 1.3, 8.1, 4.7, 18.4

66. Egyptians Some archaeologists theorize that ancient Egyptians interbred with several different immigrant populations over thousands of years. To see if there is any indication of changes in body structure that might have resulted, they measured 30 skulls of male Egyptians dated from 4000 B.C.E. and 30 others dated from 200 B.C.E. (A. Thomson and R. Randall-Maciver, *Ancient Races of the Thebaid*, Oxford: Oxford University Press, 1905)

a) Are these data appropriate for inference? Explain.
b) Create a 95% confidence interval for the difference in mean skull breadth between these two eras.
c) Do these data provide evidence that the mean breadth of males' skulls changed over this period? Explain.
*d) Perform Tukey's test for the difference. Do your conclusions of part c change?
*e) Perform a rank sum test for the difference. Do your conclusions of part c change?

Maximum Skull Breadth (mm)

4000 B.C.E.	4000 B.C.E.	200 B.C.E.	200 B.C.E.
131	131	141	131
125	135	141	129
131	132	135	136
119	139	133	131
136	132	131	139
138	126	140	144
139	135	139	141
125	134	140	130
131	128	138	133
134	130	132	138
129	138	134	131
134	128	135	136
126	127	133	132
132	131	136	135
141	124	134	141

67. Reading An educator believes that new reading activities for elementary school children will improve reading comprehension scores. She randomly assigns third graders to an eight-week program in which some will use these activities and others will experience traditional teaching methods. At the end of the experiment, both groups take a reading comprehension exam. Their scores are shown in the back-to-back stem-and-leaf display. Do these results suggest that the new activities are better? Test an appropriate hypothesis and state your conclusion.

(*Would Tukey's test be appropriate here? Explain. *Use a rank sum test to test an appropriate hypothesis as well. Compare your conclusions with your previous one.)

New Activities		Control
	1	07
4	2	068
3	3	377
96333	4	12222238
9876432	5	355
721	6	02
1	7	
	8	5

68. Streams In Chapter 6, Exercise 25, we looked at collected samples of water from streams in the Adirondack Mountains to investigate the effects of acid rain. Researchers measured the pH (acidity) of the water and classified the streams with respect to the kind of substrate (type of rock over which they flow).

A lower pH means the water is more acidic. Here is a boxplot of the pH of the streams by substrate (limestone, mixed, or shale):

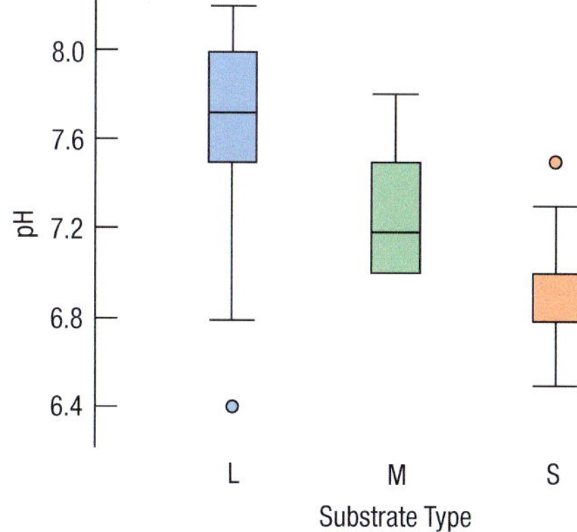

Here are selected parts of a software analysis comparing the pH of streams with limestone and shale substrates:

2-Sample *t*-Test of $\mu_1 - \mu_2$
Difference Between Means = 0.735
t-Statistic = 16.30 w/133 df
p $\leq$ 0.0001

a) State the null and alternative hypotheses for this test.
b) From the information you have, do the assumptions and conditions appear to be met?
c) What conclusion would you draw?

69. Cholesterol and gender Are men or women at higher risk for having high cholesterol? The answer may surprise you. Use the data in **Framingham**.

a) Make boxplots comparing the cholesterol levels of men and women.
b) Does it seem convincing from the boxplots that one gender has higher average levels of cholesterol?
c) Test the hypothesis that the mean levels are equal and report a 95% confidence interval for the mean difference.
d) Does your conclusion match what you said in part b? Explain.
e) Would removing the outliers change your conclusion? Explain briefly.

70. Memory Does ginkgo biloba enhance memory? In an experiment to find out, subjects were assigned randomly to take ginkgo biloba supplements or a placebo. Their memory was tested to see whether it improved. The numbers reported are the number of items recalled before and after taking a supplement. Here are boxplots comparing the two groups. At the top of the next page is some computer output from a two-sample *t*-test computed for the data.

2-Sample t-Test of $\mu_G - \mu_P > 0$
Difference Between Means = −0.9914
t-Statistic = −1.540 w/196 df
P = 0.9374

a) Explain in this context what the P-value means.
b) State your conclusion about the effectiveness of ginkgo biloba.
c) Proponents of ginkgo biloba continue to insist that it works. What type of error do they claim your conclusion makes? Explain.

71. Home runs 2016 American League baseball teams play their games with the designated hitter rule, meaning that pitchers do not bat. The league believes that replacing the pitcher, traditionally a weak hitter, with another player in the batting order produces more runs and generates more interest among fans. Below are the average numbers of home runs hit per game in American League and National League stadiums for the 2016 season. (Data in **Baseball 2016**)

American		National	
1.56	1.23	1.17	1.35
1.28	1.13	0.76	0.99
1.04	1.04	1.24	0.95
1.15	1.38	1.01	1.09
1.31	1.33	1.26	0.80
1.22	1.33	1.17	1.39
0.91	1.36	0.80	1.25
0.96		1.20	

a) Create an appropriate display of these data. What do you see?
b) With a 95% confidence interval, estimate the mean number of home runs hit in American League games.
c) Coors Field, in Denver, stands a mile above sea level, an altitude far greater than that of any other major league ballpark. Some believe that the thinner air makes it harder for pitchers to throw curve balls and easier for batters to hit the ball a long way. Do you think the 1.26 home runs hit per game at Coors is unusual? (Denver is a National League team.) Explain.
d) Explain why you should not use two separate confidence intervals to decide whether the two leagues differ in average number of runs scored.
e) Using a 95% confidence interval, estimate the difference between the mean number of home runs hit in American and National League games.
f) Interpret your interval.
g) Does this interval suggest that the two leagues may differ in average number of home runs hit per game?

72. Hard water by region Recall from Chapter 7, Exercise 75, that data were collected on the annual mortality rate (deaths per 100,000) for males in 61 large towns in England and Wales. In addition, the water hardness was recorded as the calcium concentration (parts per million, ppm) in the drinking water. The dataset also notes, for each town, whether it was south or north of Derby. Is there a significant difference in mortality rates in the two regions? Here are the summary statistics. (Data in **Hard water**)

Summary of: **mortality**
For categories in: **Region**

Group	Count	Mean	Median	StdDev
North	34	1631.59	1631	138.470
South	27	1388.85	1369	151.114

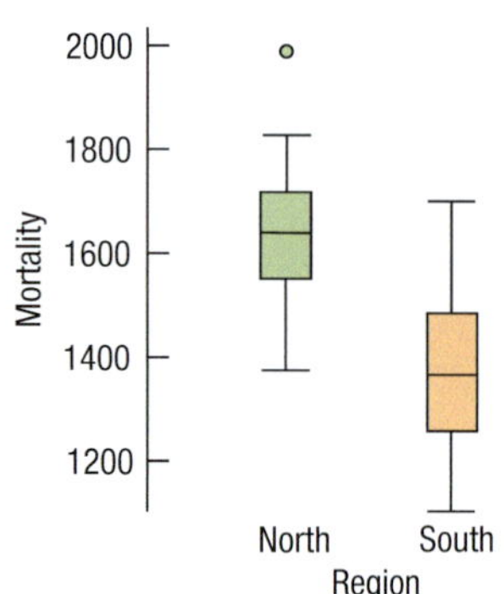

a) Test appropriate hypotheses and state your conclusion.
b) The boxplots of the two distributions show an outlier among the data in the North region. What effect might that have had on your test?

73. Job satisfaction A company institutes an exercise break for its workers to see if this will improve job satisfaction, as measured by a questionnaire that assesses workers' satisfaction. Scores for 10 randomly selected workers before and after implementation of the exercise program are shown. The company wants to assess the effectiveness of the exercise program. Explain why you can't use the methods discussed in this chapter to do that.

Worker Number	Job Satisfaction Index Before	After
1	34	33
2	28	36
3	29	50
4	45	41
5	26	37
6	27	41
7	24	39
8	15	21
9	15	20
10	27	37

T **74.** Summer school Having done poorly on their math final exams in June, six students repeat the course in summer school, then take another exam in August. If we consider these students representative of all students who might attend this summer school in other years, explain why you can't use the methods discussed in this chapter to test whether the class was beneficial.

June	54	49	68	66	62	62
Aug.	50	65	74	64	68	72

75. Color and memory Several studies in psychology show that colors play an important role in memory performance. A well-known experiment consists in showing subjects cardboard sheets of different colors displaying the same objects or words, then asking them to recall the items they saw.

Assume we reproduce that experiment with 300 volunteers randomly assigned to three groups (100 per group) according to the cardboard sheet colors: red, blue, and black. Results are summarized below.

(https://www.academia.edu/35114668/_PDF_EFFECT_OF_COLOUR_ON_MEMORY_DOES_COLOUR_BACKGROUND_AFFECT_THE_RATE_OF_RETENTION?source=swp_share https://scholar.colorado.edu/cgi/viewcontent.cgi?article=2764&context=honr_theses)

	Red	Blue	Black
Mean	2.33	1.89	2.10
SD	1.27	1.34	1.16

a) Do these results indicate that viewer memory may differ depending on cardboard color? A test of the hypothesis that there is no difference in memory performance between red and black cardboards has a P-value of 0.12. State your conclusion.

b) Is there evidence that viewer memory performance differs between red and blue cardboards? Test an appropriate hypothesis, and state your conclusion.

76. Ad campaign You are a consultant to the marketing department of a business preparing to launch an ad campaign for a new product. The company can afford to post ads in the metro and has decided not to use black background posters.

You read the study described in Exercise 75, then use a computer to create a confidence interval for the difference in mean number of items remembered between the groups seeing red and blue cardboards.

TWO-SAMPLE T

95% CI FOR MUred – MUBlue: (0.077, 0.803)

a) At the meeting of the marketing staff, you have to explain what this output means. What will you say?

b) What advice would you give the company about the upcoming ad campaign?

77. Hungry? Researchers investigated how the size of a bowl affects how much ice cream people tend to scoop when serving themselves.[12] At an "ice cream social," people were randomly given either a 17-oz or a 34-oz bowl (both large enough that they would not be filled to capacity). They were then invited to scoop as much ice cream as they liked. Did the bowl size change the selected portion size? Here are the summaries:

Small Bowl		Large Bowl	
n	26	n	22
$\bar{y}$	5.07 oz	$\bar{y}$	6.58 oz
s	1.84 oz	s	2.91 oz

Test an appropriate hypothesis and state your conclusions. For assumptions and conditions that you cannot test, you may assume that they are sufficiently satisfied to proceed.

78. Thirsty? Researchers randomly assigned participants either a tall, thin "highball" glass or a short, wide "tumbler," each of which held 355 ml. Participants were asked to pour a shot (1.5 oz = 44.3 ml) into their glass. Did the shape of the glass make a difference in how much liquid they poured?[13] Here are the summaries:

Highball		Tumbler	
n	99	n	99
$\bar{y}$	42.2 ml	$\bar{y}$	60.9 ml
s	16.2 ml	s	17.9 ml

Test an appropriate hypothesis and state your conclusions. For assumptions and conditions that you cannot test, you may assume that they are sufficiently satisfied to proceed.

T **79.** Swim the lake 2016 revisited As we saw in Chapter 8, Exercise 46, between 1954 and 2016, swimmers have crossed Lake Ontario 62 times. Both women and men have made the crossing. Here are some plots (we've omitted a crossing by Vikki Keith, who swam a round trip—north to south to north—in 3390 minutes):

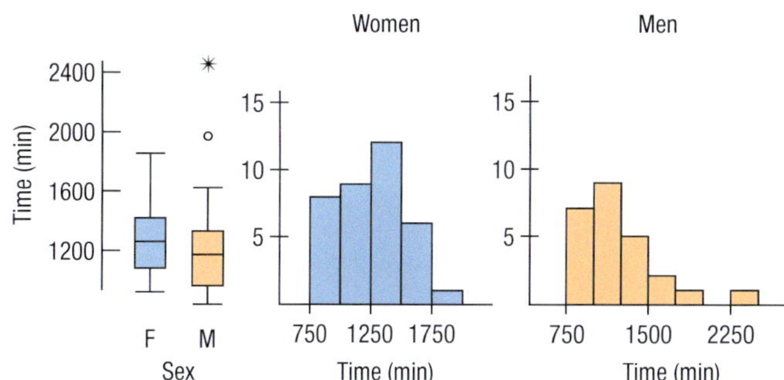

[12]Brian Wansink, Koert van Ittersum, and James E. Painter, "Ice Cream Illusions: Bowls, Spoons, and Self-Served Portion Sizes," *Am. J. Prev. Med.* 2006.

[13]Brian Wansink and Koert van Ittersum, "Shape of Glass and Amount of Alcohol Poured: Comparative Study of Effect of Practice and Concentration," *BMJ* 331: 1512–1514, 2005.

Summary statistics are as follows:

Summary of Time (min)			
Group	Count	Mean	StdDev
F	36	1262.08	254.701
M	25	1226.29	368.022

Is there a difference between the mean amount of time (in minutes) it takes female and male swimmers to swim the lake?

a) Construct and interpret a 95% confidence interval for the difference between female and male crossing times (technology gives 39 df).
b) Find the P-value for the appropriate hypothesis test.
c) Comment on the assumptions and conditions and your conclusions.

T **80. Still swimming** Here's some additional information about the Ontario crossing times presented in Exercise 79. It is generally thought to be harder to swim across the lake from north to south. Indeed, this has been done only 5 times. Every one of those crossings was by a woman. If we omit those 5 crossings, the boxplots look like this:

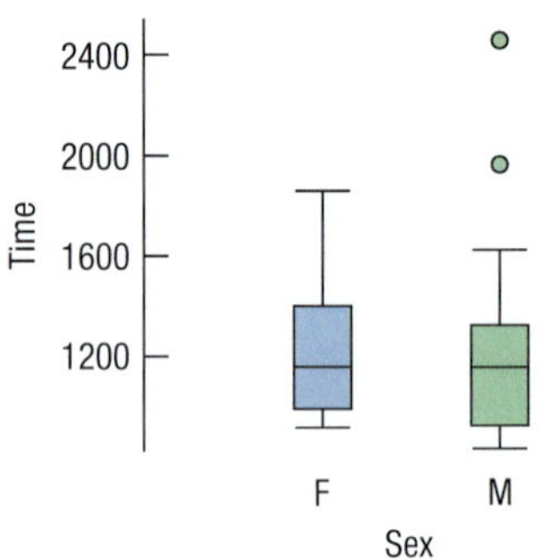

Although not designated as an outlier, the slowest female time belongs to Vikki Keith, who crossed the lake swimming only butterfly stroke. Omitting that extraordinary swim gives the following summary statistics:

Summary of Time (min)			
Sex	Count	Mean	StdDev
F	30	1208.80	215.473
M	25	1226.29	368.022

a) Do women or men appear to be faster at swimming across the lake? Support your answer by interpreting a confidence interval. (Use 37 df.)
b) Some might argue that the men are being unfairly slowed down by two extraordinarily slow swims. Here are the summaries without those two swims. Repeat part a based on these data. Does your conclusion change? (Use 47 df.)

Summary of Time (min)			
SEX	Count	Mean	StdDev
F	30	1208.80	215.47
M	23	1140.57	214.722

c) Vikki Keith was responsible for two of the more remarkable crossings, but she also swam Lake Ontario two other times. In fact, of the 50 crossings in this analysis, 7 were repeat crossings by a swimmer who'd crossed the lake before. How does this fact affect your thoughts about the confidence interval?

T **81. Running heats London** In Olympic running events, preliminary heats are determined by random draw, so we should expect the abilities of runners in the various heats to be about the same, on average. Here are the times (in seconds) for the 400-m women's run in the 2012 Olympics in London for preliminary heats 2 and 5. Is there any evidence that the mean time to finish is different for randomized heats? Explain. Be sure to include a discussion of assumptions and conditions for your analysis. (Note: One runner in heat 2 did not finish and one runner in heat 5 did not start.)

(*Use Tukey's test to compare the mean times in each heat. Do your conclusions change? *Use a rank sum test to compare the mean mean times in each heat. Do your conclusions change? *Use a randomization test to test the difference. Do your conclusions change?)

Country	Name	Time	Heat
BOT	Amantle Montsho	50.4	2
JAM	Christine Day	51.05	2
GBR	Shana Cox	52.01	2
GUY	Aliann Pompey	52.1	2
ANT	Afia Charles	54.25	2
NCA	Ingrid Narvaez	59.55	2
BAH	Shaunae Miller	DNF	2
RUS	Antonina Krivoshapka	50.75	5
UKR	Alina Lohvynenko	52.08	5
GBR	Lee McConnell	52.23	5
SWE	Moa Hjelmer	52.86	5
MAW	Ambwene Simukonda	54.2	5
FIJ	Danielle Alakija	56.77	5
GRN	Kanika Beckles	DNS	5

T **82. Swimming heats London** In Exercise 81, we looked at the times in two different heats for the 400-m women's run from the 2012 Olympics. Unlike track events, swimming heats are *not* determined at random. Instead, swimmers are seeded so that better swimmers are placed in later heats. Here are the times (in seconds) for the women's 400-m freestyle from heats 2 and 5. Do these results suggest that the mean times of seeded heats are not equal? Explain. Include a discussion of assumptions and conditions for your analysis.

(*Use Tukey's test to compare the mean times in each heat. Do your conclusions change? *Use a rank sum test to compare the mean times in each heat. Do your conclusions change?) (*Use randomization test to test the difference. Do your conclusions change?)

Country	Name	Time	Heat
BUL	Nina Rangelova	251.7	2
CHI	Kristel Köbrich	252.0	2
JPN	Aya Takano	252.3	2
LIE	Julia Hassler	253.0	2
MEX	Susana Escobar	254.8	2
THA	Nathanan Junkrajang	256.5	2
SIN	Lynette Lim	258.6	2
KOR	Kim Ga-Eul	283.5	2
FRA	Camille Muffat	243.3	5
USA	Allison Schmitt	243.3	5
NZL	Lauren Boyle	243.6	5
DEN	Lotte Friis	244.2	5
ESP	Melanie Costa	246.8	5
AUS	Bronte Barratt	248.0	5
HUN	Boglarka Kapas	250.0	5
RUS	Elena Sokolova	252.2	5

83. Tees Does it matter what kind of tee a golfer places the ball on? The company that manufactures "Stinger" tees claims that the thinner shaft and smaller head lessen resistance and drag, reducing spin and allowing the ball to travel farther. In August 2003, Golf Laboratories, Inc., compared the distance traveled by golf balls hit off regular wooden tees to those hit off Stinger tees. All the balls were struck by the same golf club using a robotic device set to swing the club head at approximately 95 miles per hour. Summary statistics from the test are shown in the table below. Assume that 6 balls were hit off each tee and that the data were suitable for inference.

		Total Distance (yards)	Ball Velocity (mph)	Club Velocity (mph)
Regular Tee	Avg.	227.17	127.00	96.17
	SD	2.14	0.89	0.41
Stinger Tee	Avg.	241.00	128.83	96.17
	SD	2.76	0.41	0.52

Is there evidence that balls hit off the Stinger tees have a higher initial velocity?

84. Golf again Given the test results on golf tees described in Exercise 83, is there evidence that balls hit off Stinger tees travel farther? Again, assume that 6 balls were hit off each tee and that the data were suitable for inference.

85. Music and memory Is it a good idea to listen to music when studying for a big test? In a study conducted by some statistics students, 62 people were randomly assigned to listen to rap music, music by Mozart, or no music while attempting to memorize objects pictured on a page. They were then asked to list all the objects they could remember. Here are summary statistics for each group:

	Rap	Mozart	No Music
Count	29	20	13
Mean	10.72	10.00	12.77
SD	3.99	3.19	4.73

a) Does it appear that it is better to study while listening to Mozart than to rap music? Test an appropriate hypothesis and state your conclusion.
b) Create a 90% confidence interval for the mean difference in memory score between students who study to Mozart and those who listen to no music at all. Interpret your interval.

86. Rap Using the results of the experiment described in Exercise 85, does it matter whether one listens to rap music while studying, or is it better to study without music at all?

a) Test an appropriate hypothesis and state your conclusion.
b) If you concluded there is a difference, estimate the size of that difference with a confidence interval and explain what your interval means.

T ***87. Attendance 2016 revisited** We have seen data on ballpark attendance in Chapters 6, 7, and 9. Now we find that National League teams drew in, on average, nearly 60,000 more fans per season than American League teams. That translates to over $1,000,000 a year. To see whether that difference is statistically significant:

a) Make a boxplot of the *Home Attendance* by *League*.
b) Look at this histogram of 1000 differences in means of *Home Attendance* by *League* obtained by shuffling the *League* label among the 30 teams. What does it say about whether 60,000 is a statistically significant difference?

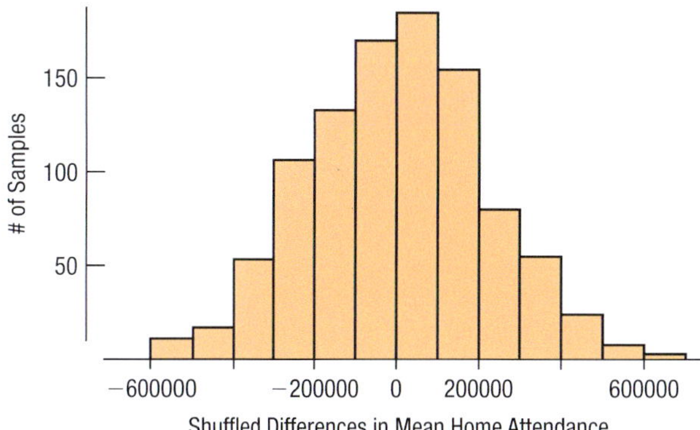

T ***88. Hard water revisited** In Exercise 72, we saw that the towns in the region south of Derby seemed to have fewer deaths than the towns in the north. To see whether that difference is statistically significant, look at this histogram of 1000 differences in mean *Mortality* by *Region*. What does it say about whether a difference of 242.7 is statistically significant?

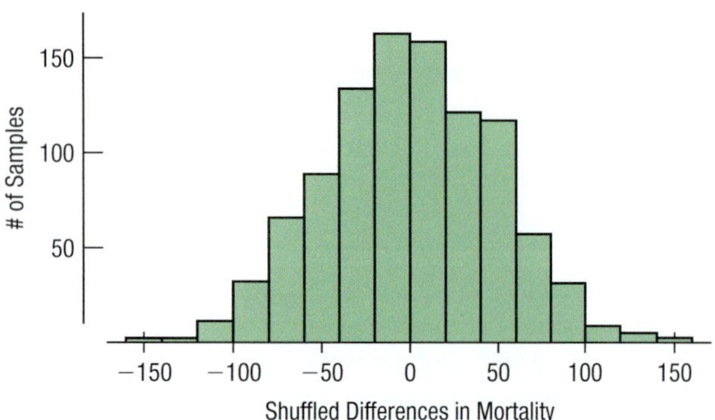

***89. Cholesterol and gender II** Using the data in **Framingham**, repeat the analysis in Exercise 69, this time using a randomization test to see if the difference in mean cholesterol levels is statistically significant. Be sure to include a P-value.

***90. Memory II** Does ginkgo biloba enhance memory? Repeat the analysis in Exercise 70, this time using a randomization test to see if the difference in mean number of items recalled is statistically significant. Be sure to include a P-value.

JUST CHECKING

Answers

1. We're 90% confident that if members are contacted by e-mail, the donation rate will be between 4 and 18 percentage points higher than if they receive regular mail.
2. Since a difference of 0 is not in the confidence interval, we'd reject the null hypothesis. There is evidence that more members will donate if contacted by e-mail.
3. The proportion from the sample in 1995 has variability, too. If we do a one-proportion z-test, we won't take that variability into account and our P-value will be incorrect.
4. The difference in the proportions between the two years has more variability than either individual proportion. The variance of the difference is the sum of the two variances.
5. Randomization should balance unknown sources of variability in the two groups of patients and helps us believe the two groups are independent.
6. We can be 95% confident that after 4 weeks endoscopic surgery patients will have a mean pinch strength between 0.04 kg and 2.96 kg higher than open-incision patients.
7. The lower bound of this interval is close to 0, so the difference may not be great enough that patients could actually notice the difference. We may want to consider other issues such as cost or risk in making a recommendation about the two surgical procedures.
8. Without data, we can't check the Nearly Normal Condition.
9. H_0: Mean pinch strength is the same after both surgeries $(\mu_E - \mu_0 = 0)$.
   H_A: Mean pinch strength is different after the two surgeries $(\mu_E - \mu_0 \neq 0)$.
10. With a P-value this low, we reject the null hypothesis. We can conclude that mean pinch strength differs after 4 weeks in patients who undergo endoscopic surgery vs. patients who have open-incision surgery. Results suggest that the endoscopic surgery patients may be stronger, on average.
11. If some patients contributed two hands to the study, then the groups may not be internally independent. It is reasonable to assume that two hands from the same patient might respond in similar ways to similar treatments.

21

Paired Samples and Blocks

WHERE ARE WE GOING?

How much will an LSAT prep course raise scores? Are boys better at computer games than their sisters? Questions like these look at paired variables. When pairs of observations go together naturally, they can't be independent. In this chapter, you'll see what to do with paired data.

WHO	Children in an experiment
WHAT	Speed of inverting cylinders
UNITS	Cylinders/second
WHEN	2013
WHERE	Israel
WHY	To study development of dexterity

Only about 1% of people are truly ambidextrous. The vast majority of us have a dominant hand. (If you're not sure which is your dominant hand, think about which one you use to hold a soup spoon.) But how unbalanced is your dexterity? One common test, the Functional Dexterity Test (FDT), measures the time it takes to place a series of cylinders into a set of holes. In one experiment researchers in Israel measured the time it took 93 children aged 5 to 18 complete the task using each hand. (We have converted their results, by taking the reciprocal of the time to obtain the children's *Speed* in cylinders per second.)

Here are the data for the first seven of 93 children tested: (Data in the right two columns of **Dexterity**)

	Speed (Cylinders/sec)		
Age (mo)	Dominant Hand	Non-dominant Hand	Gender
117	0.35335689	0.21624544	male
101	0.25719338	0.34349506	male
135	0.53655265	0.49735779	male
119	0.44370494	0.49612403	male
124	0.48280024	0.38806694	female
127	0.52390308	0.4217185	female
101	0.45506257	0.3808617	male

How much faster is the dominant hand, on average? We have two samples, so we can do a t-test to compare the two means. Or can we? The *t-test* assumes that the groups are independent. We have both hands measured for each child. Some children are naturally more dextrous than others, and 18-year-olds are generally much faster than 5-year-olds. In fact, the correlation between the measurements on the two hands is 0.49. These measurements are clearly *not* independent.

21.1 Paired Data

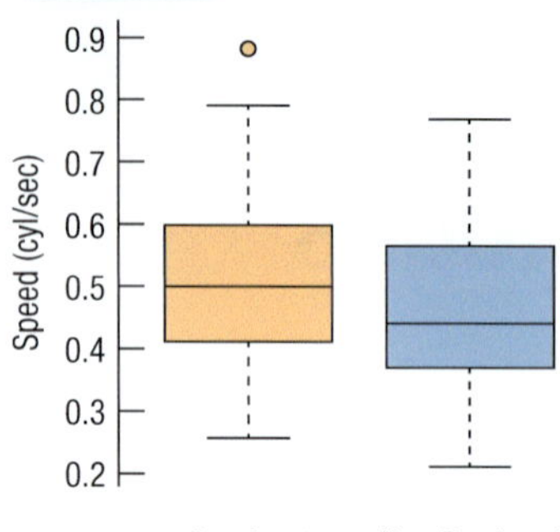

Figure 21.1
Using boxplots to compare dominant and non-dominant hands shows little because it ignores the fact that the measurements are in pairs.

Data such as these are called **paired.** We have the speeds for each child for each hand. We want to compare the mean speed for hands across all the children, so what we're interested in is the *difference* in speeds for each child.

Paired data arise in a number of ways. Perhaps the most common way is to compare subjects with themselves before and after a treatment or, as here, for two similar measurements. When pairs arise from an experiment, the pairing is a type of *blocking*. When they arise from an observational study, it is a form of *matching*.

EXAMPLE 21.1 Identifying Paired Data

Do flexible work schedules reduce the demand for resources? The Lake County, Illinois, Health Department experimented with a flexible four-day workweek. For a year, the department recorded the mileage driven by 11 field workers on an ordinary five-day workweek. Then it changed to a flexible four-day workweek and recorded mileage for another year.[1] The data are shown below and found in **Work week**.

Name	5-Day Mileage	4-Day Mileage
Jeff	2798	2914
Betty	7724	6112
Roger	7505	6177
Tom	838	1102
Aimee	4592	3281
Greg	8107	4997
Larry G.	1228	1695
Tad	8718	6606
Larry M.	1097	1063
Leslie	8089	6392
Lee	3807	3362

QUESTION: Why are these data paired?

ANSWER: The mileage data are paired because each driver's mileage is measured before and after the change in schedule. I'd expect drivers who drove more than others before the schedule change to continue to drive more afterwards, so the two sets of mileages can't be considered independent.

Pairing isn't a problem; it's an opportunity. If you know the data are paired, you can take advantage of that fact—in fact, you *must* take advantage of it. Remember: The two-sample *t-test* requires the two samples to be independent. Paired data aren't. There is no test to determine whether your data are paired. You must determine that from understanding how they were collected and what they mean (check the W's).

[1]Charles S. Catlin, "Four-day Work Week Improves Environment," *Journal of Environmental Health*, Denver, 59:7, 1997.

Once we recognize that the hand dexterity data are matched pairs, it makes sense to consider the difference in speeds. So we look at the *pairwise* differences:

	Speed (cylinders/sec)		
Child	Dominant	Non-dominant	Difference
1	0.3533569	0.21624544	0.13711145
2	0.2571934	0.34349506	−0.08630168
3	0.5365527	0.49735779	0.03919486
4	0.4437049	0.49612403	−0.05241909
5	0.4828002	0.38806694	0.0947333
6	0.5239031	0.4217185	0.10218458
7	0.4550626	0.3808617	0.07420087
8	0.3941858	0.40322581	−0.00904005
9	0.4514673	0.32780168	0.12366559
10	0.5270092	0.27091094	0.25609828
11	0.564573	0.41515309	0.14941995
12	0.6527948	0.29778522	0.35500956
13	0.4207205	0.33741038	0.0833101
14	0.3201921	0.2329984	0.08719372
15	0.3437164	0.24121815	0.10249828
16	0.4280364	0.61185468	−0.1838183
17	0.5563282	0.52066385	0.03566438
18	0.4651163	0.41078306	0.05433322

Because it is the *differences* we care about, we'll treat them as if *they* were the data, ignoring the original two columns. Now that we have only one column of values to consider, we can use a one-sample t-test. Mechanically, a **paired t-test** is just a one-sample t-test for the means of these pairwise differences. The sample size is the number of pairs. (Here there are 93.)

The mechanics of the paired t-test are not new. They're the same as a one-sample t-test using the differences as the data . You've already seen the *Show!*

21.2 The Paired t-Test

The assumptions and conditions for a paired t-test are what you might expect:

Paired Data Condition: The data must be paired. You can't just decide to pair data when in fact the samples are independent. When you have two groups with the same number of observations, it may be tempting to match them up. Don't, unless you are prepared to justify your claim that the data are paired.

On the other hand, be sure to recognize paired data when you have them. Remember that two-sample t methods aren't valid without independent groups, and paired groups aren't independent. Although this is a strictly required assumption, it is one that can be easy to check if you understand how the data were collected.

Independence Assumption: If the data are paired, the *groups* are not independent. For these methods, it's the pairwise *differences* that must be independent of each other. There's no reason to believe that the difference in dexterity speed of one child could affect the difference in speed for another child.

An excellent way to be assured of independence is to generate or sample the data with suitable randomization. With paired data, randomness can arise in many ways. The pairs may be a random sample. In an experiment, the order of the two treatments may be randomly assigned, or the treatments may be randomly assigned to one member of each pair. In a before-and-after study, you may believe that the observed differences are a representative sample from a population of interest. If you have any doubts, you'll need to include a control group to be able to draw conclusions. What you want to know usually focuses your attention on where the randomness should be. In the hand dexterity example, the children tested were a representative (and possibly random) sample of children.

Nearly Normal Condition: We need to assume that the population of *differences* follows a Normal model. We don't need to check the individual groups. This condition can be checked with a histogram or Normal probability plot of the *differences*—but not of the individual groups. As with the one-sample *t*-methods, this assumption matters less the more pairs there are to consider. You may be pleasantly surprised when you check this condition. Even if your original measurements are skewed or bimodal, the *differences* may be nearly Normal. After all, the individual who was way out in the tail on an initial measurement is likely to still be out there on the second one, giving a perfectly ordinary difference.

EXAMPLE 21.2 Checking Assumptions and Conditions

RECAP: Field workers for a health department compared driving mileage on a five-day work schedule with mileage on a new four-day schedule. To see if the new schedule changed the amount of driving they did, we'll look at paired differences in mileages before and after.

Name	5-Day Mileage	4-Day Mileage	Difference
Jeff	2798	2914	−116
Betty	7724	6112	1612
Roger	7505	6177	1328
Tom	838	1102	−264
Aimee	4592	3281	1311
Greg	8107	4997	3110
Larry G.	1228	1695	−467
Tad	8718	6606	2112
Larry M.	1097	1063	34
Leslie	8089	6392	1697
Lee	3807	3362	445

QUESTION: Is it okay to use these data to test whether the new schedule changed the amount of driving?

ANSWER:

✓ **Paired Data Condition:** The data are paired because each value is the mileage driven by the same person before and after a change in work schedule.

✓ **Independence Assumption:** The driving behavior of any individual worker is independent of the others, so the differences are mutually independent.

✓ **Randomization Condition:** The mileages are the sums of many individual trips, each of which experienced random events that arose while driving. Repeating the experiment in two new years would give randomly different values.

✔ **Nearly Normal Condition:** The histogram of the mileage differences is unimodal and symmetric:

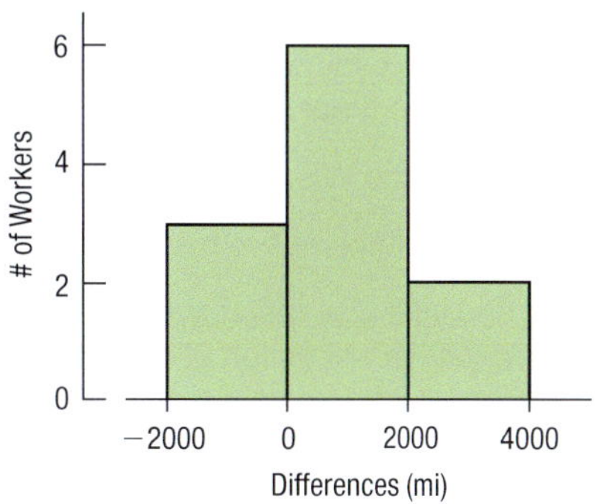

Since the assumptions and conditions are satisfied, it's okay to use paired *t* methods for these data.

The steps in testing a hypothesis for paired differences are very much like the steps for a one-sample *t*-test for a mean.

The Paired *t*-Test

When the conditions are met, we are ready to test whether the mean of paired differences is significantly different from zero. We test the hypothesis

$$H_0: \mu_d = \Delta_0,$$

where the *d*'s are the pairwise differences and Δ_0 is almost always 0.

We use the statistic

$$t_{n-1} = \frac{\bar{d} - \Delta_0}{SE(\bar{d})},$$

where $\bar{d}$ is the mean of the pairwise differences, *n* is the number of *pairs,* and

$$SE(\bar{d}) = \frac{s_d}{\sqrt{n}}.$$

$SE(\bar{d})$ is the ordinary standard error for the mean, applied to the differences.

When the conditions are met and the null hypothesis is true, we can model the sampling distribution of this statistic with a Student's *t*-model with $n - 1$ degrees of freedom, and use that model to obtain a P-value.

STEP-BY-STEP EXAMPLE

A Paired *t*-Test

Speed-skating races are run in pairs. Two skaters start at the same time, one on the inner lane and one on the outer lane. Halfway through the race, they cross over, switching lanes so that each will skate the same distance in each lane. Even though this seems fair, at the 2006 Olympics some fans thought there might have been an advantage to starting on the outside. After all, the winner, Cindy Klassen, started on the outside and skated a remarkable 1.47 seconds faster than the silver medalist.

QUESTION: Was there a difference in speeds between the inner and outer speed-skating lanes at the 2006 Winter Olympics? (Data in **Womens speed skating 2006**)

THINK **PLAN** State what we want to know.

Identify the *parameter* we wish to estimate. Here our parameter is the mean difference in race times.

Identify the variables and check the W's.

I want to know whether there really was a difference in the speeds of the two lanes for speed skating at the 2006 Olympics. I have data for 17 pairs of racers at the women's 1500-m race.

HYPOTHESES State the null and alternative hypotheses.

Although fans suspected one lane was faster, we can't use the data we have to specify the direction of a test. We (and Olympic officials) would be interested in a difference in either direction, so we'd better test a two-sided alternative.

H_0: Neither lane offered an advantage:

$$\mu_d = 0.$$

H_A: The mean difference is different from zero:

$$\mu_d \neq 0.$$

REALITY CHECK The individual differences are all in seconds. We should expect the mean difference to be comparable in magnitude.

MODEL Think about the assumptions and check the conditions.

State why you think the data are paired. Simply having the same number of individuals in each group and displaying them in side-by-side columns doesn't make them paired.

Think about what we hope to learn and where the randomization comes from. Here, the randomization comes from the racer pairings and lane assignments.

Make a picture—just one. Don't plot separate distributions of the two groups—that entirely misses the pairing. For paired data, it's the Normality of the *differences* that we care about. Treat those paired differences as you would a single variable, and check the Nearly Normal Condition with a histogram or a Normal probability plot.

✓ **Paired Data Condition**: The data are paired because racers compete in pairs.

✓ **Independence Assumption**: Each race is independent of the others, so the differences are mutually independent. Skaters are assigned to lanes at random.

✓ **Nearly Normal Condition**: The histogram of the differences is unimodal and symmetric:

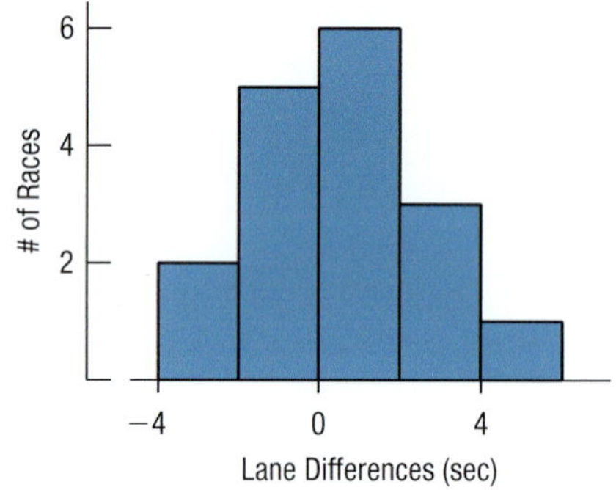

Specify the sampling distribution model.

Choose the method.

The conditions are met, so I'll use a Student's *t*-model with $(n - 1) = 16$ degrees of freedom, and perform a **paired *t*-test**.

SHOW **MECHANICS** *n* is the number of *pairs*—in this case, the number of races.

$\bar{d}$ is the mean difference.

s_d is the standard deviation of the differences.

Find the standard error and the *t*-score of the observed mean difference. There is nothing new in the mechanics of the paired *t* methods. These are the mechanics of the *t*-test for a mean applied to the differences.

The data give

$$n = 17 \text{ pairs}$$
$$\bar{d} = 0.499 \text{ seconds}$$
$$s_d = 2.333 \text{ seconds.}$$

I estimate the standard deviation of $\bar{d}$ using

$$SE(\bar{d}) = \frac{s_d}{\sqrt{n}} = \frac{2.333}{\sqrt{17}} = 0.5658$$

$$\text{So } t_{16} = \frac{\bar{d} - 0}{SE(\bar{d})} = \frac{0.499}{0.5658} = 0.882$$

Make a picture. Sketch a t-model centered at the hypothesized mean of 0. Because this is a two-tailed test, shade both the region to the right of the observed mean difference of 0.499 seconds and the corresponding region in the lower tail.

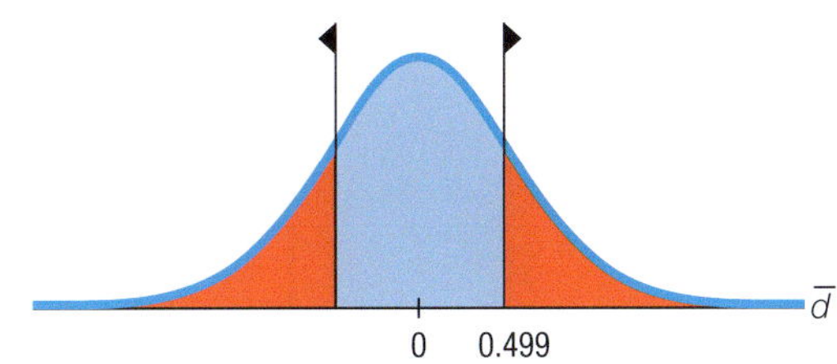

Find the P-value, using technology.

P-value $= 2P(t_{16} > 0.882) = 0.39$

REALITY CHECK The mean difference is 0.499 seconds. That may not seem like much, but a smaller difference determined the Silver and Bronze medals. The standard error is about this big, so a t-value less than 1.0 isn't surprising. Nor is a large P-value.

TELL **CONCLUSION** Link the P-value to your decision about H_0, and state your conclusion in context.

The P-value is large. Events that happen more than a third of the time are not remarkable. So, even though there is an observed difference between the lanes, I can't conclude that it isn't due simply to random chance. It appears the fans may have interpreted a random fluctuation in the data as favoring one lane. There's insufficient evidence to declare any lack of fairness.

EXAMPLE 21.3 Doing a Paired t-Test

RECAP: We want to test whether a change from a five-day workweek to a four-day workweek could change the amount driven by field workers of a health department. We've already confirmed that the assumptions and conditions for a paired t-test are met.

QUESTION: Is there evidence that a four-day workweek would change how many miles workers drive?

ANSWER: H_0: The change in the health department workers' schedules didn't change the mean mileage driven; the mean difference is zero:

$$\mu_d = 0.$$

H_A: The mean difference is different from zero:

$$\mu_d \neq 0.$$

The conditions are met, so I'll use a Student's t-model with $(n - 1) = 10$ degrees of freedom and perform a **paired t-test**.

The data give

$$n = 11 \text{ pairs}$$

$$\bar{d} = 982 \text{ miles}$$

$$s_d = 1139.6 \text{ miles}$$

$$SE(\bar{d}) = \frac{s_d}{\sqrt{n}} = \frac{1139.6}{\sqrt{11}} = 343.6$$

$$\text{So } t_{10} = \frac{\bar{d} - 0}{SE(\bar{d})} = \frac{982.0}{343.6} = 2.86$$

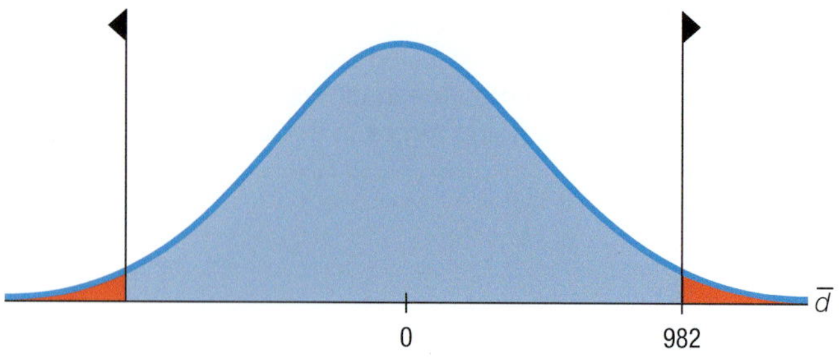

$$\text{P-value} = 2P(t_{10} > 2.86) = 0.017$$

The P-value is small, so we reject the null hypothesis and conclude that the change in workweek may reduce the mileage driven by workers.

Note: We should propose a course of action, but it's hard to tell from the hypothesis test whether the reduction matters. Is the observed difference in mileage (the *effect size*) important in the sense of reducing air pollution or costs, or is it merely statistically significant? To help make that decision, we should look at a confidence interval. If the difference in mileage proves to be large in a practical sense, then we might recommend a change in schedule for the rest of the department.

21.3 Confidence Intervals for Matched Pairs

WHO	170 randomly sampled couples
WHAT	Ages
UNITS	Years
WHEN	Recently
WHERE	Britain

In developed countries, the average age of women is generally higher than that of men. After all, women tend to live longer and male children are more likely to die than females. But if we look at *married couples*, husbands tend to be slightly older than wives. How much older, on average, are husbands? We have data from a random sample of 200 British couples. Only 170 couples provided ages for both husband and wife, so we can work only with that many pairs. Let's form a confidence interval for the mean difference of husband's and wife's ages for these 170 couples. Here are the first 7 pairs. (Full data in **Couples**)

Wife's Age	Husband's Age	Difference (husband − wife)
43	49	6
28	25	−3
30	40	10
57	52	−5
52	58	6
27	32	5
52	43	−9
⋮	⋮	⋮

Clearly, these data are paired. The survey selected *couples* at random, not individuals. We're interested in the mean age difference within couples. To construct a confidence interval for the true mean difference in ages we'll use a **paired *t*-interval**, and we'll construct it just as you should expect.

Paired *t*-Interval

When the conditions are met, we are ready to find the confidence interval for the mean of the paired differences. The confidence interval is

$$\bar{d} \pm t^*_{n-1} \times SE(\bar{d}),$$

where the standard error of the mean difference is $SE(\bar{d}) = \frac{s_d}{\sqrt{n}}$.

The critical value t^* from the Student's t-model depends on the particular confidence level, C, that you specify and on the degrees of freedom, $n - 1$, which is based on the number of pairs, n.

Making confidence intervals for matched pairs follows exactly the steps for a one-sample t-interval.

STEP-BY-STEP EXAMPLE

A Paired t-Interval

Using the Functional Dexterity Test data discussed at the beginning of the chapter, we can try to assess how much more dexterous a typical person is with the dominant hand.

QUESTION: How big a difference is there, on average, between the dexterity of dominant and non-dominant hands?

THINK

PLAN State what we want to know.

I want to estimate the mean difference in dexterity between dominant and non-dominant hands.

Identify the variables and check the W's.

Identify the parameter you wish to estimate. For a paired analysis, the parameter of interest is the mean of the differences. The population of interest is the population of differences.

MODEL Think about the assumptions and check the conditions.

✓ **Paired Data Condition**: The data are paired because they are measured on hands for the same individuals.

✓ **Independence Assumption**: The data are from an experiment. Dexterity of one individual should be independent of dexterity for any other.

Make a picture. We focus on the differences, so a histogram or Normal probability plot is best here.

✓ **Nearly Normal Condition**: The histogram of the dominant − non-dominant differences is unimodal and symmetric:

REALITY CHECK The histogram shows that dominant hands are generally faster at the FDT task by about 0.01 cylinders/second.

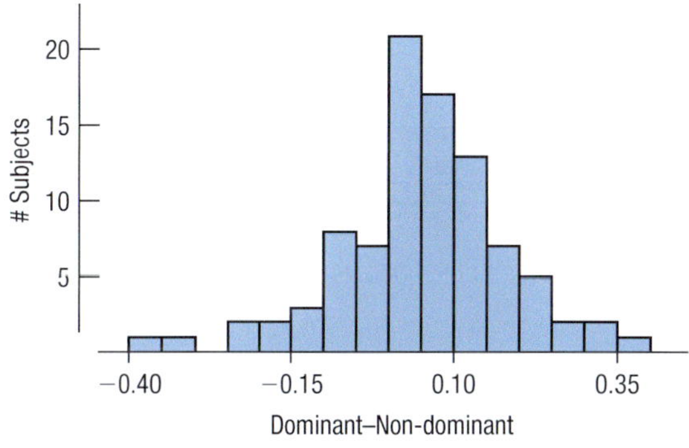

State the sampling distribution model.

Choose your method.

The conditions are met, so I can use a Student's t-model with $(n - 1) = 92$ degrees of freedom and find a **paired t-interval**.

SHOW **MECHANICS** n is the number of *pairs*—here, the number of children tested.

$$n = 93 \text{ participants}$$
$$\bar{d} = 0.054 \text{ cylinders/sec}$$
$$s_d = 0.132 \text{ cylinders/sec}$$

$\bar{d}$ is the mean difference.

I estimate the standard error of $\bar{d}$ as

s_d is the standard deviation of the differences.

$$SE(\bar{d}) = \frac{s_d}{\sqrt{n}} = \frac{0.132}{\sqrt{93}} = 0.0137.$$

Be sure to include the units along with the statistics.

The *df* for the *t*-model is $n - 1 = 92$.

The critical value we need to make a 95% interval comes from Table T, a computer program such as the interactive table at astools.datadesk.com, or a calculator.

The 95% critical value for t_{92} (from technology) is 1.986.

The margin of error is

$ME = t^*_{92} \times SE(\bar{d}) = 1.986(0.0137) = 0.0272$. So the 95% confidence interval is 0.051 ± 0.0272 *years*, or an interval of (0.027, 0.081) cylinders/sec.

REALITY CHECK This result makes sense. Our everyday experience is that we are generally more dexterous with our dominant hand.

TELL **CONCLUSION** Interpret the confidence interval in context.

I am 95% confident that the mean difference in dexterity measured on the FDT between dominant and non-dominant hands is between 0.027 and 0.081 cylinders/sec.

Effect Size

When we examined the speed-skating times, we failed to reject the null hypothesis, so we couldn't be certain whether there really was a difference between the lanes. Maybe there wasn't any difference, or maybe whatever difference there might have been was just too small to matter at all. Were the fans right to be concerned?

We can't tell from the hypothesis test, but using the same summary statistics, we can find that the corresponding 95% confidence interval for the mean difference is $(-0.70 < \mu_d < 1.70)$ seconds.

A confidence interval is a good way to get a sense for the size of the effect we're trying to understand. That gives us a plausible range of values for the true mean difference in lane times. If differences of 1.7 seconds were too small to matter in 1500-m Olympic speed skating, we'd be pretty sure there was no need for concern.

But in fact, except for the gap between the Gold and Silver medal performances, the successive gaps between each skater and the next-faster one were *all* less than the high end of this interval, and most were right around the middle of the interval.

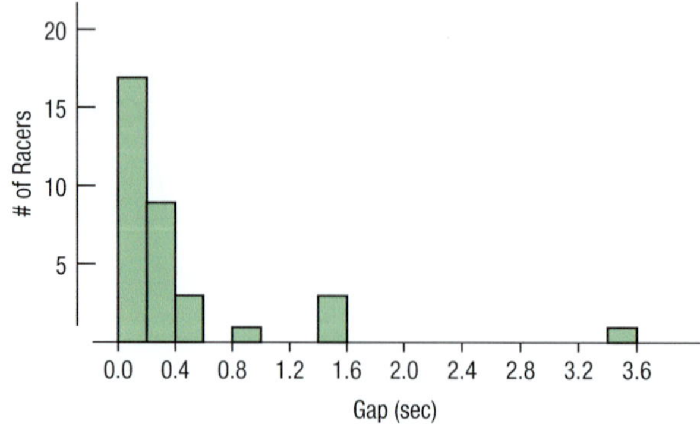

Figure 21.2
The distribution of the differences in skating times for two lanes.

So even though we were unable to discern a real difference, the confidence interval shows that the effects we're considering may be big enough to be important. We may want to continue this investigation by checking out other races on this ice and being alert for possible differences at other venues.

EXAMPLE 21.4

Looking at Effect Size with a Paired *t* Confidence Interval

RECAP: We know that, on average, the switch from a five-day workweek to a four-day workweek reduced the mileage driven by field workers in that Illinois health department. However, finding that there is a significant difference doesn't necessarily mean that difference is meaningful or worthwhile. To assess the size of the effect, we need a confidence interval. We already know the assumptions and conditions are met.

QUESTION: By how much, on average, might a change in workweek schedule reduce the mileage driven by workers?

ANSWER: $\bar{d} = 982$ mi $\quad SE(\bar{d}) = 343.6 \quad t^*_{10} = 2.228$ (*for* 95%)

$$ME = t^*_{10} \times SE(\bar{d}) = 2.228(343.6) = 765.54$$

So the 95% confidence interval for μ_d is 982 ± 765.54 *or* (216.46, 1747.54) fewer miles.

With 95% confidence, I estimate that by switching to a four-day workweek employees would drive an average of between 216 and 1748 fewer miles per year. With high gas prices, this could save a lot of money.

21.4 Blocking

Because the sample of British husbands and wives (data in **Couples**) includes both older and younger couples, there's a lot of variation in the ages of the men and in the ages of the women. In fact, that variation is so great that a boxplot of the two groups would show little difference. But that would be the wrong plot. It's the *difference* we care about. Pairing isolates the extra variation and allows us to focus on the individual differences. Chapter 11 showed how to design an experiment with blocking to isolate the variability between identifiable groups of subjects. Blocking makes it easier to see the variability among treatment groups that is attributable to their responses to the treatment. A paired design is an example of blocking.

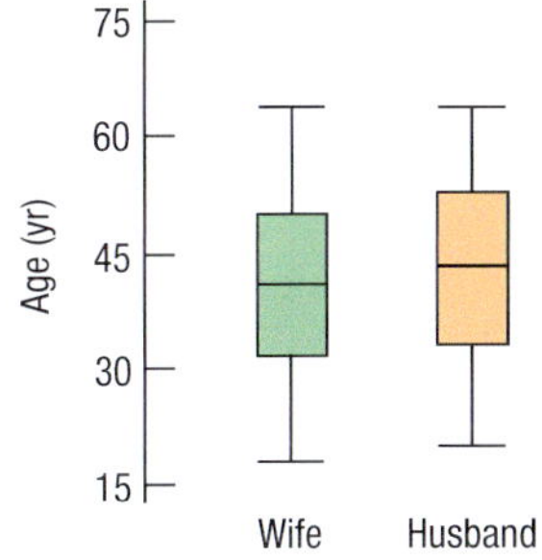

Figure 21.3
This display is worthless. It does no good to compare all the wives as a group with all the husbands. We care about the paired differences.

A paired study has roughly half the degrees of freedom of a two-sample test. You may see discussions that suggest that in "choosing" a paired analysis we "give up" these degrees of freedom. This isn't really true, though. If the data are paired, then there never were additional degrees of freedom, and we have no "choice." Only the paired differences are independent, not the individual observations. The fact of the pairing determines how many degrees of freedom are available.

Matching pairs generally removes so much extra variation that it more than compensates for having only half the degrees of freedom, so it is usually a good choice when you design a study. Of course, inappropriate matching when the groups are in fact independent (say, by matching on the first letter of the last name of subjects) would cost degrees of freedom without the benefit of reducing the variance. When you design a study or experiment, you should consider using a paired design if possible.

What's Independent?

The methods of this chapter require the pairs to be independent of each other. They make no assumptions about the individuals—only about the pairs. By contrast, many other inference methods require independence among all the individuals.

You will sometimes see the paired *t* methods of this chapter referred to as methods for "dependent" samples to differentiate them from the independence we required for one-sample *t* methods. But in fact, the only "dependence" in paired data is the pairing itself. That's not a failure of statistical independence; it's a feature of the design for the data collection.

JUST CHECKING

Think about each of the situations described below.

- Would you use a two-sample *t* or paired *t* method (or neither)? Why? or Why not?
- Would a P-value be useful or would a confidence interval be sufficient to answer the question?

1. Random samples of 50 men and 50 women are asked to imagine buying a birthday present for their best friend. We want to estimate the difference in how much they are willing to spend.
2. Mothers of twins were surveyed and asked how often in the past month strangers had asked whether the twins were identical.
3. Are parents equally strict with boys and girls? In a random sample of families, researchers asked a brother and sister from each family to rate how strict their parents were.
4. Forty-eight overweight subjects are randomly assigned to either aerobic or stretching exercise programs. They are weighed at the beginning and at the end of the experiment to see how much weight they lost.
   a) We want to estimate the mean amount of weight lost by those doing aerobic exercise.
   b) We want to know which program is more effective at reducing weight.
5. Couples at a dance club were separated and each person was asked to rate the band. Do men or women like this band more?

RANDOM MATTERS A Bootstrapped Paired Data Confidence Interval and Hypothesis Test

The bootstrap offers an alternative way to find a confidence interval for paired differences. You still must be sure that your data are paired, but for this method you no longer need the assumption that the differences follow a Normal model. There's really nothing new here. Inferences for paired differences use the same methods we saw for a single sample.

Revisiting the hand dexterity data, we re-sampled the differences 5000 times. The resulting bootstrap sampling distribution of the mean differences from each of the 5000 samples looks like this:

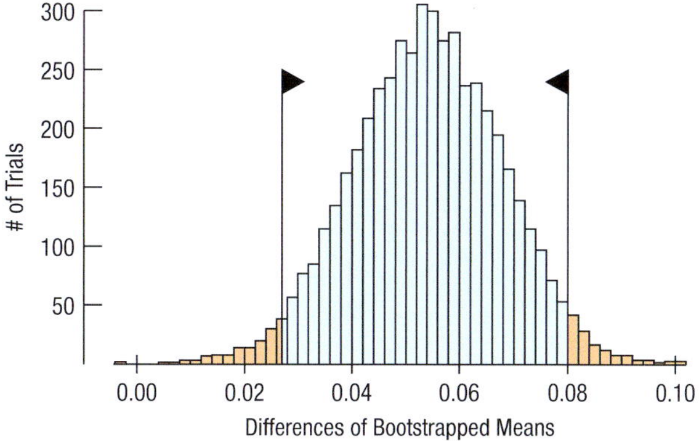

Figure 21.4

The means of 5000 bootstrap samples from the paired differences in hand dexterity speed.

The flags show the 2.5% and 97.5% percentiles of this distribution. They show that the bootstrap 95% confidence interval is from 0.027 to 0.080 (cylinders inverted per second). Using the software, you can slide the flags to see other intervals. For comparison, the *t*-based 95% confidence interval is nearly the same: (0.027, 0.081).

The natural null hypothesis is that the difference of the pairs is, on average, zero. To get a P-value for that hypothesis, we need to center the bootstrap distribution at that null hypothesis value so that we can see how extraordinary our mean difference would be if the hypothesis were true. In Figure 21.4, the bootstrap distribution is centered at the observed mean difference (0.0524). To center it at zero, we subtract that observed value from each of the 5000 bootstrapped values. Then we find the sample value we observed on that re-centered distribution (either side for a two-sided test), shown by the flags below. To calculate the P-value we count how many bootstrap values fell beyond those flags and divide by the number of trials. Here there was one instance (out of 5000), so we estimate the P-value as 1/50,000 = 0.0002. For comparison, the *t*-test gives a P-value of 0.00014—nearly the same.

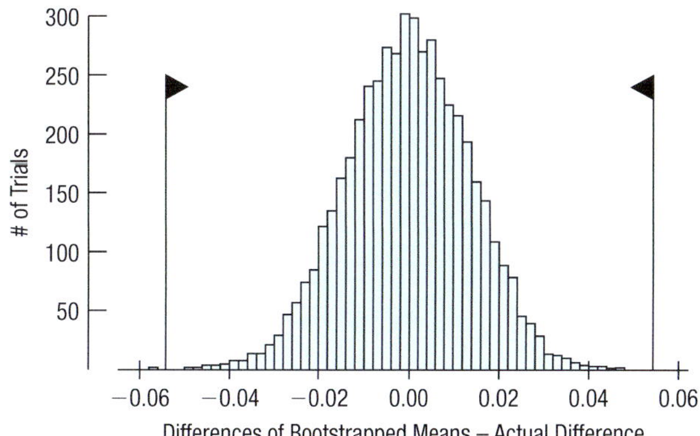

Figure 21.5

The bootstrap sampling distribution of Figure 21.4 shifted to be centered at zero, the null hypothesis value.

WHAT CAN GO WRONG?

- **Don't use a two-sample *t*-test when you have paired data.** See the What Can Go Wrong? discussion in Chapter 20.
- **Don't use a paired *t* method when the samples aren't paired.** Just because two groups have the same number of observations doesn't mean they can be paired, even if they are shown side by side in a table. We might have 25 men and 25 women in our study, but they might be completely independent of one another. If they were siblings or spouses, we might consider them paired. Remember that you cannot *choose* which method to use based on your preferences. If the data are from two independent samples, use two-sample *t* methods. If the data are from an experiment in which observations were paired, you must use a paired method. If the data are from an observational study, you must be able to defend your decision to use matched pairs or independent groups.
- **Don't forget outliers.** The outliers we care about now are in the differences. A subject who is extraordinary both before and after a treatment may still have a perfectly typical difference. But one outlying difference can completely distort your conclusions. Be sure to plot the differences (even if you also plot the data).
- **Don't look for the difference between the means of paired groups with side-by-side boxplots.** The point of the paired analysis is to remove extra variation. The boxplots of each group still contain that variation. Comparing them is likely to be misleading.

CONNECTIONS

The most important connection is to the concept of blocking that we first discussed when we considered designed experiments in Chapter 11. Pairing is a basic and very effective form of blocking.

Of course, the details of the mechanics for paired *t*-tests and intervals are identical to those for the one-sample *t* methods. Everything we know about those methods applies here.

The connection to the two-sample methods of Chapter 20 is that when the data are naturally paired, those methods are not appropriate because paired data fail the required condition of independence.

CHAPTER REVIEW

Recognize when data are paired or matched.

Know how to construct a confidence interval for the mean difference in paired data.

Know how to perform a hypothesis test about the mean difference (usually with a null of zero representing no difference between the groups).

REVIEW OF TERMS

Paired data Data are paired when the observations are collected in pairs or the observations in one group are naturally related to observations in the other. The simplest form of pairing is to measure each subject twice—often before and after a treatment is applied. More sophisticated forms of pairing in experiments are a form of blocking and arise in other contexts. Pairing in observational and survey data is a form of matching (p. 676).

Paired *t*-test A hypothesis test for the mean of the pairwise differences of two groups. It tests the null hypothesis

$$H_0: \mu_d = \Delta_0,$$

where the hypothesized difference is almost always 0, using the statistic

$$t = \frac{\bar{d} - \Delta_0}{SE(\bar{d})}$$

with $n - 1$ degrees of freedom, where $SE(\bar{d}) = \frac{s_d}{\sqrt{n}}$, and n is the number of pairs (p. 677).

Paired *t* confidence interval A confidence interval for the mean of the pairwise differences between paired groups found as

$$\bar{d} \pm t^*_{n-1} \times SE(\bar{d}),\ \text{where } SE(\bar{d}) = \frac{s_d}{\sqrt{n}} \text{ and } n \text{ is the number of pairs (p. 682).}$$

TECH SUPPORT

Paired *t*

Most statistics programs can compute paired *t* analyses. Some may want you to find the differences yourself and use the one-sample *t* methods. Those that perform the entire procedure will need to know the two variables to compare. The computer, of course, cannot verify that the variables are naturally paired. Most programs will check whether the two variables have the same number of observations, but some stop there, and that can cause trouble. Most programs will automatically omit any pair that is missing a value for either variable (as we did with the British couples). You must look carefully to see whether that has happened.

As we've seen with other inference results, some packages pack a lot of information into a simple table, but you must locate what you want for yourself. Here's a generic example with comments:

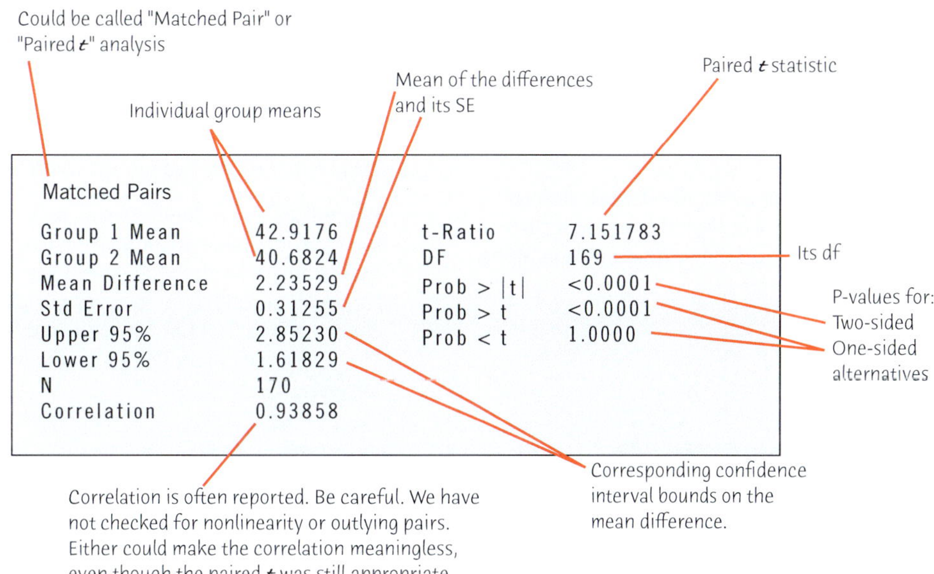

Other packages try to be more descriptive. It may be easier to find the results, but you may get less information from the output table.

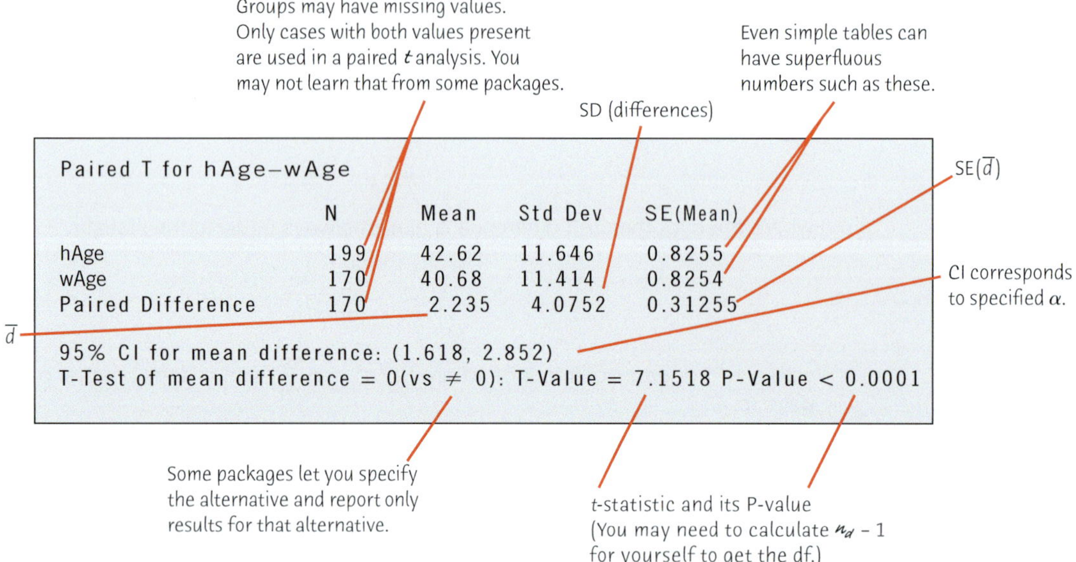

Computers make it easy to examine the boxplots of the two groups and the histogram of the differences—both important steps. Some programs offer a scatterplot of the two variables. That can be helpful. In terms of the scatterplot, a paired *t*-test is about whether the points tend to be above or below the line $y = x$. (Note, though, that pairing says nothing about whether the scatterplot should be straight. That doesn't matter for our *t* methods.)

DATA DESK

- Select variables.
- From the Calc menu, choose **Estimate** for confidence intervals or **Test** for hypothesis tests.
- Select the interval or test from the drop-down menu, and make other choices in the dialog.

COMMENTS

Data Desk expects the two groups to be in separate variables and in the same "Relation"—that is, about the same cases.

EXCEL

- Select **Data Analysis** from the Analysis Group on the Data Tab.
- From the Data Analysis menu, choose ***t*-test: paired two-sample for Means**. Fill in the cell ranges for the two groups, the hypothesized difference, and the alpha level.

COMMENTS

Excel expects the two groups to be in separate cell ranges.

Warning: Do not compute this test in Excel without checking for missing values. If there are any missing values (empty cells), Excel can give a wrong answer. Excel compacts each list, pushing values up to cover the missing cells, and then checks only that it has the same number of values in each list. The result is mismatched pairs and an entirely wrong analysis.

JMP

- From the Analyze menu, select **Matched Pairs**.
- Specify the columns holding the two groups in the Y Paired Response dialog.
- Click **OK**.

MINITAB

- From the Stat menu, choose the **Basic Statistics** submenu.
- From that menu, choose **Paired *t*** . . .
- Then fill in the dialog.

COMMENTS

Minitab takes "First sample" minus "Second sample."

R

To test the hypothesis that $\mu_1 = \mu_2$ for paired data against an alternative (default is two-sided), create a vector for the data of each group, such as x and y, and produce the confidence interval (default is 95%):

- **t.test**(x,y, alternative = c("two.sided", "less", "greater"), paired = TRUE, conf.level = **0.95**)

will produce the *t*-statistic, degrees of freedom, P-value, and confidence interval for a specified alternative.

COMMENTS

This is the same function as for both the one- and two-sample *t*-test, but with paired=True.

SPSS

- From the Analyze menu, choose the **Compare Means** submenu.
- From that, choose the **Paired-Samples *t*-test** command.
- Select pairs of variables to compare, and click the arrow to add them to the selection box.

COMMENTS

You can compare several pairs of variables at once. Options include the choice to exclude cases missing in any pair from all tests.

STATCRUNCH

To do inference for the mean of paired differences:

- Click on **Stat**.
- Choose **T Stats** » **Paired**.
- Choose the **Column** for each variable.
- Check the **Histogram with mean** option in the list of Optional graphs and tables so you can look at a histogram to be sure the Nearly Normal Condition is satisfied.
- Indicate **Hypothesis Test**, then enter the hypothesized **Null mean** difference (usually 0), and choose the **Alternative** hypothesis.

  OR

  Indicate **Confidence Interval**, and then enter the **Level** of confidence.
- Click on **Compute!**

TI-83/84 PLUS

If the data are stored in two lists, say, L1 and L2, create a list of the differences:

L**1** − L**2** → L**3**. (The arrow is the STO button.)

- Since inference for paired differences uses one-sample *t*-procedures, select **2:T-Test** or **8:TInterval** from the **STAT TESTS** menu.
- Specify as the data the list of differences you just created in L3, and apply the procedure.

EXERCISES

SECTION 21.1

1. Which method? Which of the following scenarios should be analyzed as paired data?

a) Students attend an IQ training course. Their "before" and "after" IQ scores are compared.
b) 15 male and 15 female students in a school compete in a quiz. We compare their scores.
c) Body weights of 30 university students were recorded before enrolling for a weight-loss program. Six months later, the same students were weighed again. Did they lose weight after enrolling for the program?

2. Which method II? Which of the following scenarios should be analyzed as paired data?

a) Twins were asked about the number of hours spent studying every day. We want to see if older twins study longer than the younger twins.
b) 60 diabetic patients are given Drug A and 60 are given Drug B. Which subjects show lower sugar levels in 6 months?
c) A group of male staff and a group of female staff are asked about the quality of the office cafeteria. Do staff opinions change in one year?

SECTION 21.2

3. Cars and trucks We have data on the city and highway fuel efficiency of 651 cars and trucks.

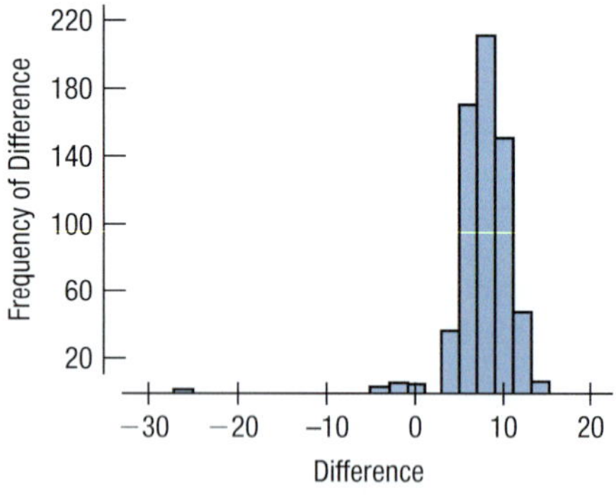

a) Would it be appropriate to use paired t methods to compare the city fuel efficiency of the cars and the trucks?
b) Would it be appropriate to use paired t methods to compare the city and highway fuel efficiencies of these vehicles?
c) A histogram of the differences (highway – city) is shown. Are the conditions for inference satisfied?

T **4. Vehicle weights II** The calibration test for a new weight-in-motion method of weighing trucks was introduced in Chapter 6, exercise 52. Is this method consistent with the traditional method of static weighing? Are the conditions for matched pairs inference satisfied? Weights are in 1000s of pounds.

Weight-in-Motion	Static Weight	Diff (Static − Motion)
26.0	27.9	−1.9
29.9	29.1	0.8
39.5	38.0	1.5
25.1	27.0	−1.9
31.6	30.3	1.3
36.2	34.5	1.7
25.1	27.8	−2.7
31.0	29.6	1.4
35.6	33.1	2.5
40.2	35.5	4.7

SECTION 21.3

5. Cars and trucks again In Exercise 3, after deleting an outlying value of −26.5, the mean difference in fuel efficiencies for the remaining 650 vehicles was 7.25 mpg with a standard deviation of 2.30 mpg. Find a 95% confidence interval for this difference and interpret it in context.

T **6. Vehicle weights III** Find a 98% confidence interval of the weight differences in Exercise 4. Interpret this interval in context.

SECTION 21.4

7. Blocking cars and trucks Thinking about the data on fuel efficiency in Exercise 3, why is the blocking accomplished by a matched pairs analysis particularly important for a sample that has both cars and trucks?

T **8. Vehicle weights IV** Consider the weights from Exercise 4. The side-by-side boxplots below show little difference between the two groups. Should this be sufficient to draw a conclusion about the accuracy of the weigh-in-motion scale?

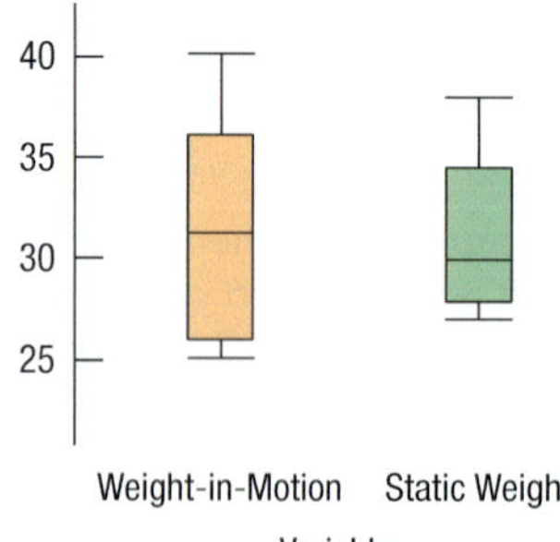

CHAPTER EXERCISES

9. More milk Can supplementing dairy cows with yeast in their feed increase milk production? Researchers want to design an experiment to find out. They have 100 cows available. They have two kinds of feed: the regular feed and the new feed with yeast. They plan to run their experiment for a month, recording the amount of milk (in liters) each cow produces.

a) Design an experiment that will require a two-sample *t*-procedure to analyze the results.
b) Design an experiment that will require a matched-pairs *t*-procedure to analyze the results.
c) Which experiment would you consider the stronger design? Why?

10. Music According to a study, those who listened to music completed their tasks more efficiently than those who didn't. Some researchers were interested to see if people can work as effectively with music as without. The researchers will time some volunteers to see how long it takes them to solve some simple arithmetic problems. During some of the trials, the room will be quiet; during other trials in the same room, subjects will be listening to some light music.

a) Design an experiment that will require a two-sample *t*-procedure to analyze the results.
b) Design an experiment that will require a matched-pairs *t*-procedure to analyze the results.
c) Which experiment would you consider the stronger design? Why?

11. Celebrity images Ads for many products use popular celebrity images to try to attract attention to the product. But do these ads bring people's attention to the item that was being advertised? We want to design an experiment to see if the presence of celebrity images in an advertisement affects people's ability to remember the product.

a) Describe an experiment design requiring a matched-pairs *t*-procedure to analyze the results.
b) Describe an experiment design requiring an independent sample procedure to analyze the results.

12. Freshman 15? Many people believe that students gain weight as freshmen. Suppose we plan to conduct a study to see if this is true.

a) Describe a study design that would require a matched-pairs *t*-procedure to analyze the results.
b) Describe a study design that would require a two-sample *t*-procedure to analyze the results.

13. Women Values for the labor force participation rate of women (LFPR) are published by the U.S. Bureau of Labor Statistics. We are interested in whether there was a difference between female participation in 1968 and 1972, a time of rapid change for women. We check LFPR values for 19 randomly selected cities for 1968 and 1972. Shown below is software output for two possible tests:

Paired *t*-Test of $\mu(1 - 2)$
Test Ho: $\mu(1972-1968) = 0$ vs Ha: $\mu(1972-1968) \neq 0$
Mean of Paired Differences = 0.0337
t-Statistic = 2.458 with 18 df
$p = 0.0244$

2-Sample *t*-Test of $\mu1 - \mu2$
Ho: $\mu1 - \mu2 = 0$ Ha: $\mu1 - \mu2 \neq 0$
Test Ho: $\mu(1972) - \mu(1968) = 0$ vs
Ha: $\mu(1972) - \mu(1968) \neq 0$
Difference Between Means = 0.0337
t-Statistic = 1.496 with 35 df
$p = 0.1434$

a) Which of these tests is appropriate for these data? Explain.
b) Using the test you selected, state your conclusion.

T **14. Cloud seeding** Simpson, Alsen, and Eden (*Technometrics* 1975) report the results of trials in which clouds were seeded and the amount of rainfall recorded. The authors report on 26 seeded and 26 unseeded clouds in order of the amount of rainfall, largest amount first. Here are two possible tests to study the question of whether cloud seeding works.

Paired *t*-Test of $\mu(1 - 2)$
Mean of Paired Differences = −277.39615
t-Statistic = −3.641 with 25 df
$p = 0.0012$

2-Sample *t*-Test of $\mu1 - \mu2$
Difference Between Means = −277.4
t-Statistic = −1.998 with 33 df
$p = 0.0538$

a) Which of these tests is appropriate for these data? Explain.
b) Using the test you selected, state your conclusion.

T **15. Friday the 13th, traffic** The *British Medical Journal* (1993; 307:1584) published an article titled, "Is Friday the 13th Bad for Your Health?" Researchers in Britain examined how Friday the 13th affects human behavior. One question was whether people tend to stay at home more on Friday the 13th. The data below are the number of cars passing Junctions 9 and 10 on the M25 motorway for consecutive Fridays (the 6th and 13th) for five different periods.

Dates	6th	13th	Junction
1990, July	139,246	138,548	9
1990, July	134,012	132,908	10
1991, September	137,055	136,018	9
1991, September	133,732	131,843	10
1991, December	123,552	121,641	9
1991, December	121,139	118,723	10
1992, March	128,293	125,532	9
1992, March	124,631	120,249	10
1992, November	124,609	122,770	9
1992, November	117,584	117,263	10

Here are summaries of two possible analyses:

Paired *t*-Test; Mean of Paired Differences: 1835.8
t-Statistic = 4.936 with 9 df
P = 0.0008

2-Sample *t*-Test
Difference Between Means: 1835.8
t-Statistic = 0.5499 with 17 df
P = 0.5891

a) Which of the tests is appropriate for these data? Explain.
b) Using the test you selected, state your conclusion.
c) Are the assumptions and conditions for inference met?

16. Friday the 13th, accidents The researchers in Exercise 15 also examined the number of people admitted to emergency rooms for vehicular accidents on 12 Friday evenings (6 each on the 6th and 13th).

Year	Month	6th	13th
1989	October	9	13
1990	July	6	12
1991	September	11	14
1991	December	11	10
1992	March	3	4
1992	November	5	12

Based on these data, is there evidence that more people are admitted, on average, on Friday the 13th? Here are two possible analyses of the data:

Paired t-Test of $\mu(1 - 2) = 0$ vs. $\mu(1 - 2) < 0$
Mean of Paired Differences $= -3.333$
t-Statistic $= -2.7116$ with 5 df
P $= 0.0211$

2-Sample t-Test of $\mu1 = \mu2$ vs. $\mu1 < \mu2$
Difference Between Means $= -3.333$
t-Statistic $= -1.6644$ with 9.940 df
P $= 0.0636$

a) Which of these tests is appropriate for these data? Explain.
b) Using the test you selected, state your conclusion.
c) Are the assumptions and conditions for inference met?

17. Online insurance I After seeing countless commercials claiming one can get cheaper car insurance from an online company, a local insurance agent was concerned that he might lose some customers. To investigate, he randomly selected profiles (type of car, coverage, driving record, etc.) for 10 of his clients and checked online price quotes for their policies. The comparisons are shown in the table below. His statistical software produced the following summaries (where $PriceDiff = Local - Online$):

Variable	Count	Mean	StdDev
Local	10	792.900	235.431
Online	10	750.800	254.956
PriceDiff	10	42.100	174.964

Local	Online	PriceDiff
550	388	162
856	593	263
460	497	−37
1248	910	338
580	665	−85
1022	1263	−241
773	703	70
830	789	41
900	1001	−101
710	699	11

At first, the insurance agent wondered whether there was some kind of mistake in this output. He thought the Pythagorean Theorem of Statistics should work for finding the standard deviation of the price differences—in other words, that

$$SD(Local - Online) = \sqrt{SD^2(Local) + SD^2(Online)}$$

But when he checked, he found that $\sqrt{(235.431)^2 + (254.956)^2} = 347.0303$, not 174.964 as given by the software. Tell him where his mistake is.

18. Wind speed, part I To select the site for an electricity generating wind turbine, wind speeds were recorded at several potential sites every 6 hours for a year. Two sites not far from each other looked good. Each had a mean wind speed high enough to qualify, but we should choose the site with a higher average daily wind speed. Because the sites are near each other and the wind speeds were recorded at the same times, we should view the speeds as paired. Here are the summaries of the speeds (in miles per hour):

Variable	Count	Mean	StdDev
site 2	1120	7.405	3.625
site 4	1120	7.169	3.332
site 2 – site 4	1120	0.236	2.405

Is there a mistake in this output? Why doesn't the Pythagorean Theorem of Statistics work here? In other words, shouldn't

$$SD(site\ 2 - site\ 4) = \sqrt{SD^2(site\ 2) + SD^2(site\ 4)}?$$

But $\sqrt{(3.625)^2 + (3.332)^2} = 4.924$, not 2.551 as given by the software. Explain why this happened.

19. Online insurance II In Exercise 17, we saw summary statistics for 10 drivers' car insurance premiums quoted by a local agent and an online company. Here are the displays for each company's quotes and for the difference $(Local - Online)$:

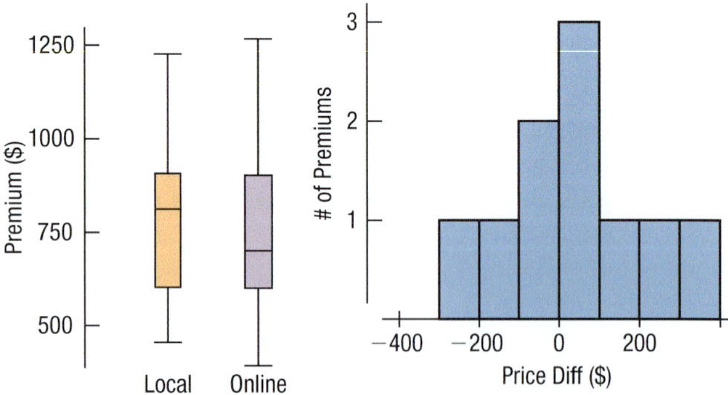

a) Which of the summaries would help you decide whether the online company offers cheaper insurance? Why?
b) The standard deviation of *PriceDiff* is quite a bit smaller than the standard deviation of prices quoted by either the local or online companies. Discuss why.
c) Using the information you have, discuss the assumptions and conditions for inference with these data.

20. Wind speed, part II In Exercise 18, we saw summary statistics for wind speeds at two sites near each other, both being considered as locations for an electricity-generating wind turbine. The data, recorded every 6 hours for a year, showed each of the sites had a mean wind speed high enough to qualify, but how can we tell which site is best? Here are some displays:

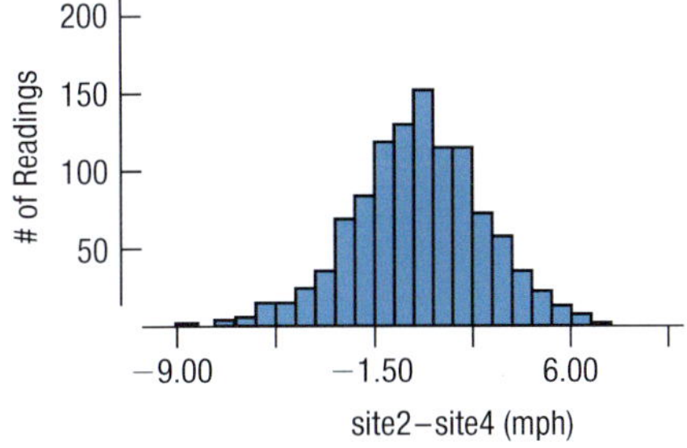

a) The boxplots show outliers for each site, yet the histogram shows none. Discuss why.
b) Which of the summaries would you use to select between these sites? Why?
c) Using the information you have, discuss the assumptions and conditions for paired t inference for these data. (*Hint:* Think hard about the independence assumption in particular.)

21. Online insurance III Exercises 17 and 19 give summaries and displays for car insurance premiums quoted by a local agent and an online company. Test an appropriate hypothesis to see if there is evidence that drivers might save money by switching to the online company.

22. Wind speed, part III Exercises 18 and 20 give summaries and displays for two potential sites for a wind turbine. Test an appropriate hypothesis to see if there is evidence that either of these sites has a higher average wind speed.

23. City temperatures The following table gives the average daily high temperatures in January and July for several cities. Find a 95% confidence interval for the mean temperature difference between summer and winter.

a) Check the assumptions and conditions. If you find a city that doesn't belong with the others, set it aside.
b) Compute your interval for the data (adjusted, if necessary), and explain what it means.
*c) Find a bootstrap confidence interval for the difference. Compare it to your paired t-interval.

Location	Jan.	July	Location	Jan.	July
Amsterdam (Netherlands)	41	69	Madrid (Spain)	50	89
Athens (Greece)	54	90	Montreal (Canada)	22	79
Auckland (New Zealand)	73	56	Moscow (Russia)	21	76
Beijing (China)	35	87	Nassau (Bahamas)	77	88
Belgrade (Yugoslavia)	38	81	Oslo (Norway)	30	73
Berlin (Germany)	35	74	Paris (France)	42	76
Cairo (Egypt)	65	96	Prague (Czech Republic)	34	74
Calcutta (India)	80	89	Quebec (Canada)	19	77
Copenhagen (Denmark)	36	72	Rome (Italy)	54	88
Dublin (Ireland)	47	67	Seoul (Korea)	33	84
Glasgow (Scotland)	43	66	Stockholm (Sweden)	31	70
Hamilton (Bermuda)	68	85	Taipei (Taiwan)	66	92
Helsinki (Finland)	27	71	Tokyo (Japan)	48	84
Hong Kong (China)	67	90	Toronto (Canada)	30	79
Istanbul (Turkey)	48	78	Vancouver (Canada)	42	71
Jerusalem (Israel)	55	87	Vienna (Austria)	34	75
Lisbon (Portugal)	56	79	Zurich (Switzerland)	36	77
London (United Kingdom)	44	73			

24. London Marathon 2018 The table below shows the winning times (in minutes) for men and women in the Virgin Money London Marathon between 1981 and 2018. (www.virginmoneylondonmarathon.com) Assuming that performances in the Virgin Money London Marathon resemble performances elsewhere, we can think of these data as a sample of performances in marathon competitions.

a) Create a 90% confidence interval for the mean difference in winning times for male and female marathon competitors.
*b) Find a 90% bootstrap interval for the mean difference. How does it compare with your paired t-interval?

Year	Men	Women	Year	Men	Women
1981	131.8	150.0	2000	126.6	144.6
1982	129.4	149.7	2001	127.2	144.0
1983	129.7	145.5	2002	125.6	138.9
1984	130.0	144.4	2003	127.9	135.4
1985	128.3	141.1	2004	126.3	142.6
1986	130.0	144.9	2005	127.4	137.7
1987	129.8	142.8	2006	126.7	139.6
1988	130.3	145.7	2007	127.7	140.6
1989	129.1	145.9	2008	125.3	144.2
1990	130.2	146.5	2009	125.2	142.2
1991	129.3	146.2	2010	125.3	142.6
1992	130.0	149.7	2011	124.7	139.3
1993	130.8	147.2	2012	124.7	138.6
1994	128.9	152.6	2013	126.1	140.3
1995	128.5	147.7	2014	124.5	140.4
1996	130.0	147.9	2015	124.7	143.4
1997	127.9	146.9	2016	123.1	143.0
1998	128.0	146.4	2017	125.8	137.0
1999	128.0	143.4	2018	124.3	138.5

T **25. Push-ups** Every year, the students at Gossett High School take a physical fitness test during their gym classes. One component of the test asks them to do as many push-ups as they can. Results for one class are shown below, separately for boys and girls. Assuming that students at Gossett are assigned to gym classes at random, create a 90% confidence interval for how many more push-ups boys can do than girls, on average, at that high school.

Boys	18	28	33	17	24	35	28	24	27	15	12	36
Girls	25	6	12	16	3	12	19	26	12	27	33	9

T **26. Brain waves** An experiment was performed to see whether sensory deprivation over an extended period of time has any effect on the alpha-wave patterns produced by the brain. To determine this, 20 subjects, inmates in a Canadian prison, were randomly split into two groups. Members of one group were placed in solitary confinement. Those in the other group were allowed to remain in their own cells. Seven days later, alpha-wave frequencies were measured for all subjects, as shown in the table below. (P. Gendreau et al., "Changes in EEG Alpha Frequency and Evoked Response Latency During Solitary Confinement," *Journal of Abnormal Psychology* 79 [1972]: 54–59)

Nonconfined	Confined
10.7	9.6
10.7	10.4
10.4	9.7
10.9	10.3
10.5	9.2
10.3	9.3
9.6	9.9
11.1	9.5
11.2	9.0
10.4	10.9

a) What are the null and alternative hypotheses? Be sure to define all the terms and symbols you use.
b) Are the assumptions necessary for inference met?
c) Perform the appropriate test, indicating the formula you used, the calculated value of the test statistic, the df, and the P-value.
d) State your conclusion.

T **27. Job satisfaction** (When you first read about this exercise break plan in Chapter 20, you did not have an inference method that would work. Try again now.) A company institutes an exercise break for its workers to see if it will improve job satisfaction, as measured by a questionnaire that assesses workers' satisfaction. Scores for 10 randomly selected workers before and after the implementation of the exercise program are shown in the table at the top of the next column.

a) Identify the procedure you would use to assess the effectiveness of the exercise program, and check to see if the conditions allow the use of that procedure.
b) Test an appropriate hypothesis and state your conclusion.
c) If your conclusion turns out to be incorrect, what kind of error did you commit?

Worker Number	Job Satisfaction Index Before	After
1	34	33
2	28	36
3	29	50
4	45	41
5	26	37
6	27	41
7	24	39
8	15	21
9	15	20
10	27	37

T **28. Summer school** (When you first read about the summer school issue in Chapter 20, you did not have an inference method that would work. Try again now.) Having done poorly on their math final exams in June, six students repeat the course in summer school and take another exam in August. Here are the exam scores:

June	54	49	68	66	62	62
Aug.	50	65	74	64	68	72

a) If we consider these students to be representative of all students who might attend this summer school in other years, do these results provide evidence that the program is worthwhile?
b) This conclusion, of course, may be incorrect. If so, which type of error was made?

T **29. Yogurt flavors** Is there a significant difference in calories between servings of strawberry and vanilla yogurt? Based on the data shown in the table, test an appropriate hypothesis and state your conclusion. Don't forget to check assumptions and conditions!

	Calories per Serving	
Brand	**Strawberry**	**Vanilla**
America's Choice	210	200
Breyer's Lowfat	220	220
Columbo	220	180
Dannon Light'n Fit	120	120
Dannon Lowfat	210	230
Dannon la Crème	140	140
Great Value	180	80
La Yogurt	170	160
Mountain High	200	170
Stonyfield Farm	100	120
Yoplait Custard	190	190
Yoplait Light	100	100

T **30. Gasoline** Many drivers of cars that can run on regular gas actually buy premium in the belief that they will get better gas mileage. To test that belief, we use 10 cars from a company fleet in which all the cars run on regular gas. Each car is filled first with either regular or premium gasoline, decided by a coin toss, and the mileage for that tankful is recorded. Then the mileage is recorded again for the same cars for a tankful of the other kind of gasoline. We don't let the drivers know about this experiment.

Here are the results (miles per gallon):

Car #	1	2	3	4	5	6	7	8	9	10
Regular	16	20	21	22	23	22	27	25	27	28
Premium	19	22	24	24	25	25	26	26	28	32

a) Is there evidence that cars get significantly better fuel economy with premium gasoline?
b) How big might that difference be? Check a 90% confidence interval.
c) Even if the difference is significant, why might the company choose to stick with regular gasoline?
d) Suppose you had done a "bad thing." (We're sure you didn't.) Suppose you had mistakenly treated these data as two independent samples instead of matched pairs. What would the significance test have found? Carefully explain why the results are so different.

T **31. Stopping distance** A tire manufacturer tested the braking performance of one of its tire models on a test track. The company tried the tires on 10 different cars, recording the stopping distance for each car on both wet and dry pavement. Results are shown in the table

	Stopping Distance (ft)	
Car #	Dry Pavement	Wet Pavement
1	153	205
2	145	215
3	132	144
4	135	188
5	140	191
6	135	175
7	132	199
8	125	178
9	138	196
10	167	216

a) Write a 95% confidence interval for the mean dry pavement stopping distance. Be sure to check the appropriate assumptions and conditions, and explain what your interval means.
b) Write a 95% confidence interval for the mean increase in stopping distance on wet pavement. Be sure to check the appropriate assumptions and conditions, and explain what your interval means.

T **32. Stopping distance** For another test of the tires in Exercise 31, a car made repeated stops from 60 miles per hour. The test was run on both dry and wet pavement, with results as shown in the table. (Note that actual *braking distance*, which takes into account the driver's reaction time, is much longer, typically nearly 300 feet at 60 mph!)

a) Write a 95% confidence interval for the mean dry pavement stopping distance. Be sure to check the appropriate assumptions and conditions, and explain what your interval means.
b) Write a 95% confidence interval for the mean increase in stopping distance on wet pavement. Be sure to check the appropriate assumptions and conditions, and explain what your interval means.

Stopping Distance (ft)	
Dry Pavement	Wet Pavement
145	215
153	192
142	221
144	209
132	199
149	206
127	203
140	176
136	184
133	222

T **33. Tuition 2013** The Reddin Survey is a comprehensive survey of UK university tuition fees. A random sample of 26 public colleges and universities listed at www.thecompleteuniversityguide.co.uk yielded the following data for MBA students from the United Kingdom and overseas.

University	MBA Home (£)	MBA Overseas (£)
Abertay University	8750	11300
Bangor University	10500	14000
Birmingham City University	8500	12650
Buckinghamshire New University	8800	10300
Cardiff Metropolitan University	9000	11500
Cardiff University	11475	18660
Coventry University	8485	12183
Edinburgh Napier University	12340	13330
Leeds Trinity University	4900	10000
London Metropolitan University	11000	12600
Queen Margaret University	9140	12460
The University of Buckingham	9000	13500
The University of Nottingham	21000	22000
University for the Creative Arts	13680	14650
University of Bedfordshire	6750	11250
University of Bolton	7800	11000
University of Brighton	13140	16740
University of Central Lancashire	11500	12000
University of Cumbria	7060	11349
University of Derby	9000	12810
University of Glasgow	14500	21500
University of Greenwich	12500	13350
University of Leeds	19500	22500
University of Leicester	11405	18320
University of Liverpool	12000	17325
University of Northampton	8550	11500

a) Check the assumptions and conditions for inference on the mean difference in tuition.
b) Create a 90% confidence interval for the mean difference in cost taking account of what you found in part a, and interpret your interval in context.
c) A national magazine claims that public institutions charge home residents an average of £3500 less than those from overseas for tuition each year. What does your confidence interval indicate about this assertion?

T **34. Celebrity images, part II** In Exercise 11, you considered the question of whether celebrity images in ads affected people's ability to remember the item being advertised. To investigate, a group of statistics students cut ads out of magazines. They were careful to find two ads for each of 10 similar items, one with a celebrity image and one without. They arranged the ads in random order and had 39 subjects look at them for one minute. Then they asked the subjects to list as many of the products as they could remember. Their data are shown in the table. Is there evidence that the celebrity images mattered?

	Ads Remembered			Ads Remembered	
Subject Number	Celebrity Image	No Celebrity Image	Subject Number	Celebrity Image	No Celebrity Image
1	3	3	21	3	4
2	7	8	22	5	3
3	4	2	23	3	3
4	5	4	24	6	4
5	2	1	25	4	5
6	2	2	26	1	3
7	4	6	27	2	2
8	8	5	28	1	3
9	3	7	29	7	6
10	4	3	30	5	6
11	1	3	31	5	4
12	4	3	32	5	6
13	5	2	33	3	0
14	6	3	34	5	4
15	3	2	35	3	4
16	6	3	36	3	3
17	3	3	37	5	5
18	1	3	38	3	4
19	4	4	39	5	4
20	3	3			

T **35. Strikes** Advertisements for an instructional video claim that the techniques will improve the ability of Little League pitchers to throw strikes and that, after undergoing the training, players will be able to throw strikes on at least 60% of their pitches. To test this claim, we have 20 Little Leaguers throw 50 pitches each, and we record the number of strikes. After the players participate in the training program, we repeat the test. The table (at the top of the next column) shows the number of strikes each player threw before and after the training.

a) Is there evidence that after training players can throw strikes more than 60% of the time?
b) Is there evidence that the training is effective in improving a player's ability to throw strikes?
*c) Use a bootstrap method to test the appropriate hypothesis. Do your conclusions change from those in part a?

Number of Strikes (out of 50)		Number of Strikes (out of 50)	
Before	After	Before	After
28	35	33	33
29	36	33	35
30	32	34	32
32	28	34	30
32	30	34	33
32	31	35	34
32	32	36	37
32	34	36	33
32	35	37	35
33	36	37	32

T **36. Freshman 15, revisited** In Exercise 12, you thought about how to design a study to see if it's true that students tend to gain weight during their first year in college. Well, Cornell Professor of Nutrition David Levitsky did just that. He recruited students from two large sections of an introductory health course. Although they were volunteers, they appeared to match the rest of the freshman class in terms of demographic variables such as sex and ethnicity. The students were weighed during the first week of the semester, then again 12 weeks later. (Weights are in pounds.)

a) Based on Professor Levitsky's data, estimate the mean weight gain in first-semester freshmen and comment on the "freshman 15."
*b) Construct a bootstrap confidence interval for the weight gain and compare it with the paired t method.

Subject Number	Initial Weight	Terminal Weight	Subject Number	Initial Weight	Terminal Weight
1	171	168	25	135	139
2	110	111	26	148	150
3	134	136	27	110	112
4	115	119	28	160	163
5	150	155	29	220	224
6	104	106	30	132	133
7	142	148	31	145	147
8	120	124	32	141	141
9	144	148	33	158	160
10	156	154	34	135	134
11	114	114	35	148	150
12	121	123	36	164	165
13	122	126	37	137	138
14	120	115	38	198	201
15	115	118	39	122	124
16	110	113	40	146	146
17	142	146	41	150	151
18	127	127	42	187	192
19	102	105	43	94	96
20	125	125	44	105	105
21	157	158	45	127	130
22	119	126	46	142	144
23	113	114	47	140	143
24	120	128	48	107	107

Subject Number	Initial Weight	Terminal Weight	Subject Number	Initial Weight	Terminal Weight
49	104	105	59	155	158
50	111	112	60	118	120
51	160	162	61	149	150
52	134	134	62	149	149
53	151	151	63	122	121
54	127	130	64	155	158
55	106	108	65	160	161
56	185	188	66	115	119
57	125	128	67	167	170
58	125	126	68	131	131

T **37.** Wheelchair marathon 2016 The Boston Marathon has had a wheelchair division since 1977. Who do you think is typically faster, the men's marathon winner on foot or the women's wheelchair marathon winner? Because the conditions differ from year to year, and speeds have improved over the years, it seems best to treat these as paired measurements. Here are summary statistics for the pairwise differences in finishing time (in minutes):

Summary of wheelchrF – runM
N = 40
Mean = –7.27
SD = 33.568

a) Comment on the assumptions and conditions.
b) Assuming that these times are representative of such races and the differences appeared acceptable for inference, construct and interpret a 95% confidence interval for the mean difference in finishing times.
c) Would a hypothesis test at $\alpha = 0.05$ reject the null hypothesis of no difference? What conclusion would you draw?

T **38.** Marathon start-up years 2016 When we considered the Boston Marathon in Exercise 37, we were unable to check the Nearly Normal Condition. Here's a histogram of the differences:

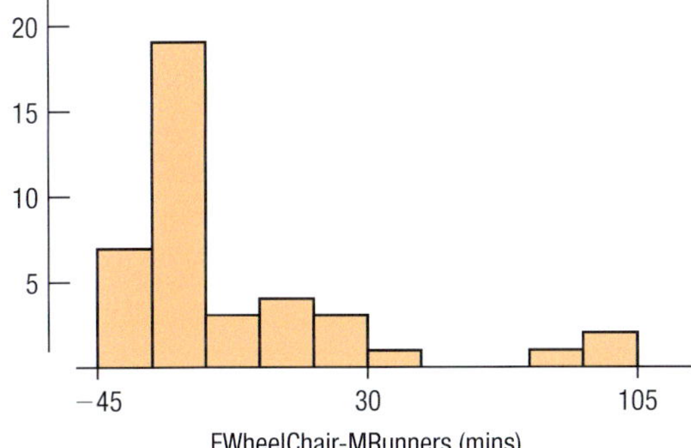

The three largest differences are in the first three years of wheelchair competition: 1977, 1978, and 1979. Often the start-up years of new events are different; later on, more athletes train and compete. If we omit those three years, the summary statistics change as follows:

Summary of wheelchrF – runM
n = 37
Mean = –15.29
SD = 18.14

a) Comment on the assumptions and conditions.
b) Assuming that these times are representative of such races, construct and interpret a 95% confidence interval for the mean difference in finishing time.
c) Would a hypothesis test at $\alpha = 0.05$ reject the null hypothesis of no difference? What conclusion would you draw?

39. BST Many dairy cows now receive injections of BST, a hormone intended to spur greater milk production. After the first injection, a test herd of 70 Ayrshire cows increased their mean daily production from 45 pounds to 65 pounds of milk. The standard deviation of the increases was 5.5 pounds. We want to estimate the mean increase a farmer could expect in his own cows.

a) Check the assumptions and conditions for inference.
b) Write a 95% confidence interval.
c) Explain what your interval means in this context.
d) Given the cost of BST, a farmer believes he cannot afford to use it unless he is sure of attaining at least a 25% increase in milk production. Based on your confidence interval, what advice would you give him?

40. BST II In the experiment about hormone injections in cows described in Exercise 39, a group of 61 Jersey cows increased average milk production from 41 pounds to 53 pounds per day, with a standard deviation of 4.9 pounds. Is this evidence that the hormone may be more effective in one breed than the other? Test an appropriate hypothesis, and state your conclusion. Be sure to discuss any assumptions you make.

JUST CHECKING

Answers

1. These are independent groups sampled at random, so use a two-sample t confidence interval to estimate the size of the difference.
2. There is only one sample. Use a one-sample t-interval.
3. A brother and sister from the same family represent a matched pair. The question calls for a paired t-test.
4. a) A before-and-after study calls for paired t methods. To estimate the loss, find a confidence interval for the before–after differences.

   b) The two treatment groups were assigned randomly, so they are independent. Use a two-sample t-test to assess whether the mean weight losses differ.
5. Sometimes it just isn't clear. Most likely, couples would discuss the band or even decide to go to the club because they both like a particular band. If we think that's likely, then these data are paired. But maybe not. If we asked them their opinions of, say, the decor or furnishings at the club, the fact that they were couples might not affect the independence of their answers.

22 Comparing Counts

WHERE ARE WE GOING?

Is your favorite color related to how much education you've had? A survey found a higher percentage of those naming blue and a lower percentage saying red among adults with only a high school education compared with adults with more education. Could this be just random fluctuation, or is the distribution of color preference different for different education levels? We saw tables of counts and percentages in Chapter 2. In this chapter, we'll see how to test the strength of the patterns we saw in those tables.

WHO	Executives of Fortune 400 companies
WHAT	Zodiac birth sign
WHY	Maybe the researcher was a Gemini and naturally curious?

Does your zodiac sign predict how successful you will be later in life? *Fortune* magazine collected the zodiac signs of 256 heads of the largest 400 companies. The table shows the number of births for each sign. (Data in **Zodiac**)

Births	Sign	Births	Sign
23	Aries	18	Libra
20	Taurus	21	Scorpio
18	Gemini	19	Sagittarius
23	Cancer	22	Capricorn
20	Leo	24	Aquarius
19	Virgo	29	Pisces

We can see some variation in the number of births per sign, and there *are* more Pisces, but is that enough to claim that successful people are more likely to be born under some signs than others?

22.1 Goodness-of-Fit Tests

If these 256 births were distributed uniformly across the year, we would expect about $1/12$ of them to occur under each sign of the zodiac. That suggests $256/12$, or about 21.3 births per sign. How closely do the observed numbers of births per sign fit this simple "null" model?

A hypothesis test to address this question is called a test of **"goodness-of-fit."** The name suggests a certain badness-of-grammar, but it is quite standard. After all, we are asking whether the model that births are uniformly distributed over the signs fits the data good, . . . er, well. Goodness-of-fit involves testing a hypothesis. We have specified a model for the distribution and want to know whether it fits. There is no single parameter to estimate, so a confidence interval wouldn't make much sense.

If the question were about only one astrological sign (for example, "Are executives more likely to be Pisces?"[1]), we could use a one-proportion z-test and ask if the true proportion of executives with that sign is equal to $1/12$. However, here we have 12 hypothesized proportions, one for each sign. We need a test that considers all of them together and gives an overall idea of whether the observed distribution differs from the hypothesized one.

EXAMPLE 22.1 Finding Expected Counts

Birth month may not be related to success as a CEO, but what about on the ball field? It has been proposed by some researchers that children who are the older ones in their class at school naturally perform better in sports and that these children then get more coaching and encouragement. Could that make a difference in who makes it to the professional level in sports?

Month	Ballplayer Count	National Birth %	Month	Ballplayer Count	National Birth %
1	137	8%	7	102	9%
2	121	7%	8	165	9%
3	116	8%	9	134	9%
4	121	8%	10	115	9%
5	126	8%	11	105	8%
6	114	8%	12	122	9%
			Total	**1478**	**100%**

Baseball is a remarkable sport, in part because so much data are available, including the birth date of every player who ever played in a Major League game. Since the effect we're suspecting may be due to relatively recent policies (and to keep the sample size moderate), we'll consider the birth months of 1478 Major League players born since 1975. We can also look up the national demographic statistics to find what percentage of people were born in each month. Let's test whether the observed distribution of ballplayers' birth months shows just random fluctuations or whether it represents a real deviation from the national pattern. (Data in **Ballplayer births**)

QUESTION: How can we find the expected counts?

ANSWER: There are 1478 players in this set of data. I found the national percentage of births in each month. Based on the national birth percentages, I'd expect 8% of players to have been born in January, and $1478(0.08) = 118.24$. I won't round off, because expected "counts" needn't be integers. Multiplying 1478 by each of the birth percentages gives the expected counts shown in the table in the margin.

Month	Expected	Month	Expected
1	118.24	7	133.02
2	103.46	8	133.02
3	118.24	9	133.02
4	118.24	10	133.02
5	118.24	11	118.24
6	118.24	12	133.02

Assumptions and Conditions

These data are organized in tables as we saw in Chapter 3, and the assumptions and conditions reflect that. Rather than having an observation for each individual, we typically work with summary counts in categories. In our example, we don't see the birth signs of each of the 256 executives, only the totals for each sign.

[1]A question someone actually asked us. We suspect he's a Pisces.

Counted Data Condition The data must be *counts* for the categories of a categorical variable. This might seem a simplistic, even silly condition. But many kinds of values can be assigned to categories, and it is unfortunately common to find the methods of this chapter applied incorrectly to proportions, percentages, or measurements just because values happen to be organized in a table. So check to be sure the values in each **cell** really are counts.

Independence Assumption The counts in the cells should be independent of each other. The easiest case is when the individuals who are counted in the cells are sampled independently from some population. That's what we'd like to have if we want to draw conclusions about that population. Randomness can arise in other ways, though. For example, these Fortune 400 executives are not a random sample of company executives, but there is no reason to think that their birth dates should not be randomly distributed throughout the year. If we want to generalize to a larger population, we should check the Randomization Condition.

We can use the methods of this chapter to assess patterns in a table provided the individuals counted are independent. But if we want to generalize our conclusions to a larger population, the individuals who have been counted should be representative of that population. The easiest way to be sure of this is if they are a random sample from the population of interest.

Sample Size Assumption There must be enough data for the methods to work, so we usually check the **Expected Cell Frequency Condition**. We should expect to see at least 5 individuals in each cell.

The Expected Cell Frequency Condition sounds like—and is, in fact, quite similar to—the condition that np and nq be at least 10 when we tested proportions. In the astrology example, assuming equal births in each zodiac sign leads us to expect 21.3 births per sign, so the condition is easily met here.

EXAMPLE 22.2 Checking Assumptions and Conditions

RECAP: Are professional baseball players more likely to be born in some months than in others? We have observed and expected counts for 1478 players born since 1975.

QUESTION: Are the assumptions and conditions met for performing a goodness-of-fit test?

ANSWER:

✓ **Counted Data Condition:** I have month-by-month counts of ballplayer births.

✓ **Independence Assumption:** These births were independent.

Even though these players are not from a random sample, we can think of them as representative of players past and future. I'll proceed with caution.

✓ **Expected Cell Frequency Condition:** The expected counts range from 103.46 to 133.02, all much greater than 5.

It's okay to use these data for a goodness-of-fit test.

Calculations

We have observed a count in each category from the data, and have an expected count for each category from the hypothesized proportions. Are the differences just natural sampling variability, or are they so large that they indicate something important? It's natural to look at the *differences* between these observed and expected counts, denoted $(Obs - Exp)$. We'd like to think about the total of the differences, but just adding them won't work because some differences are positive; others negative. We've been in this predicament before—once when we looked at deviations from the mean and again when we dealt with

NOTATION ALERT

We compare the counts *observed* in each cell with the counts we *expect* to find. The usual notation uses O's and E's or abbreviations such as those we've used here. The method for finding the expected counts depends on the model.

NOTATION ALERT

The only use of the Greek letter χ (chi) in statistics is to represent this statistic and the associated sampling distribution. This is another violation of our "rule" that Greek letters represent population parameters. Here we are using a Greek letter simply to name a family of distribution models and a statistic.

residuals. In fact, these *are* residuals. They're just the differences between the observed data and the counts given by the (null) model. We handle these residuals in essentially the same way we did in regression: We square them. That gives us positive values and focuses attention on any cells with large differences from what we expected. Because the differences between observed and expected counts generally get larger the more data we have, we also need to get an idea of the *relative* sizes of the differences. To do that, we divide each squared difference by the expected count for that cell.

The test statistic, called the **chi-square** (or chi-squared) **statistic**, is found by adding up the squares of the deviations between the observed and expected counts divided by the expected counts:

$$\chi^2 = \sum_{all\ cells} \frac{(Obs - Exp)^2}{Exp}.$$

The chi-square statistic is denoted χ^2, where χ is the Greek letter chi (pronounced "ky" as in "sky"). It refers to a family of sampling distribution models we have not seen before called (remarkably enough) the **chi-square models**.

This family of models, like the Student's t-models, differ only in the number of degrees of freedom. The number of degrees of freedom for a goodness-of-fit test is $n - 1$. Here, however, n is *not* the sample size, but instead is the number of categories. For the zodiac example, we have 12 signs, so our χ^2 statistic has 11 degrees of freedom.

Chi-Square P-Values

The chi-square statistic for tables of counts is used only for testing hypotheses, not for constructing confidence intervals. If the observed counts don't match the expected, the statistic will be large. It can't be "too small." That would just mean that our model *really* fit the data well. So the chi-square test is always one-sided. If the calculated statistic value is large enough, we'll reject the null hypothesis. What could be simpler?

If you don't have technology handy, it's easy to read the χ^2 table (Table X in Appendix D).

A portion of Table X.

Values of χ^2_α

Right-Tail Probability df	0.10	0.05	0.025	0.01	0.005
1	2.706	3.841	5.024	6.635	7.879
2	4.605	5.991	7.378	9.210	10.597
3	6.251	7.815	9.348	11.345	12.838
4	7.779	9.488	11.143	13.277	14.860
5	9.236	11.070	12.833	15.086	16.750
6	10.645	12.592	14.449	16.812	18.548
7	12.017	14.067	16.013	18.475	20.278
8	13.362	15.507	17.535	20.090	21.955
9	14.684	16.919	19.023	21.666	23.589
10	15.987	18.307	20.483	23.209	25.188
11	17.275	19.675	21.920	24.725	26.757
12	18.549	21.026	23.337	26.217	28.300
13	19.812	23.362	24.736	27.688	29.819
14	21.064	23.685	26.119	29.141	31.319

The usual selected P-values are at the top of the columns. As with the t-tables, we have only selected probabilities, so the best we can do is to trap a P-value between two of the values in the table. Just find the row for the correct number of degrees of freedom and read across to find where your calculated χ^2 value falls. Of course, technology can find an exact P-value, and that's usually what we'll see.

Even though its mechanics work like a one-sided test, the interpretation of a chi-square test is in some sense *many*-sided. With more than two proportions, there are many ways the null hypothesis can be wrong. By squaring the differences, we made all the deviations positive, whether our observed counts were higher or lower than expected. There's no direction to the rejection of the null model. All we know is that it doesn't fit.

EXAMPLE 22.3 Doing a Goodness-of-Fit Test

RECAP: The birth months of Major League baseball players are appropriate for performing a χ^2 test.

QUESTIONS: What are the hypotheses, and what does the test show?

ANSWERS: H_0: The distribution of birth months for Major League ballplayers is the same as that for the general population.

H_A: The distribution of birth months for Major League ballplayers differs from that of the rest of the population.

$$df = 12 - 1 = 11$$

$$\chi^2 = \sum \frac{(Obs - Exp)^2}{Exp}$$

$$= \frac{(137 - 118.24)^2}{118.24} + \frac{(121 - 103.46)^2}{103.46} + \cdots$$

$$= 26.48 \text{ (by technology)}$$

$$\text{P-value} = P(\chi^2_{11} \geq 26.48) = 0.0055 \text{ (by technology)}$$

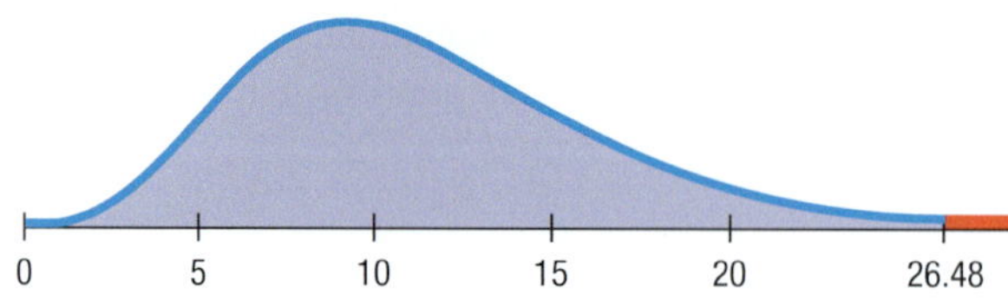

Because of the small P-value, I reject H_0; there's evidence that birth months of Major League ballplayers have a different distribution from the rest of us.

STEP-BY-STEP EXAMPLE A Chi-Square Test for Goodness-of-Fit

We have counts of 256 executives in 12 zodiac sign categories. The natural null hypothesis is that birth dates of executives are divided equally among all the zodiac signs. The test statistic looks at how closely the observed data match this idealized situation.

QUESTION: Are zodiac signs of CEOs distributed uniformly?

THINK **PLAN** State what you want to know.

Identify the variables and check the W's.

I want to know whether births of successful people are uniformly distributed across the signs of the zodiac. I have counts of 256 Fortune 400 executives, categorized by their birth sign.

HYPOTHESES State the null and alternative hypotheses. For χ^2 tests, it's usually easier to do that in words than in symbols.

H_0: Births are uniformly distributed over zodiac signs.[2]

H_A: Births are not uniformly distributed over zodiac signs.

MODEL Make a picture. The null hypothesis is that the frequencies are equal, so a bar chart (with a line at the hypothesized "equal" value) is a good display.

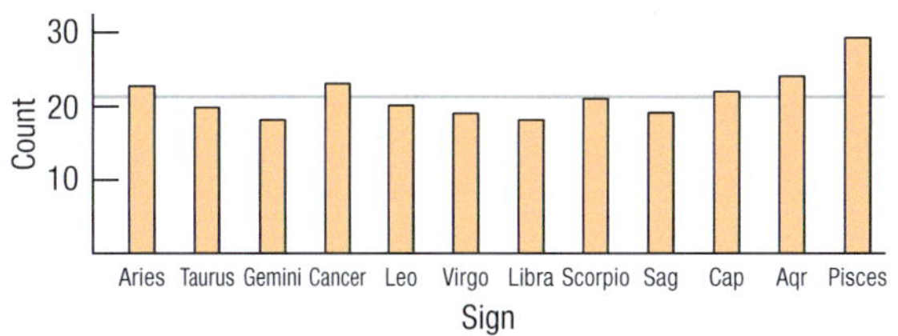

The bar chart shows some variation from sign to sign, and Pisces is the most frequent. But it is hard to tell whether the variation is more than I'd expect from random variation.

Think about the assumptions and check the conditions.

We don't need randomization to assert that birth dates are independent. But we do want our sample to be representative of executives so we can draw inferences about the population of executives. That's a reasonable assumption about this sample.

✓ **Counted Data Condition**: I have counts of the number of executives in 12 categories.

✓ **Independence Assumption**: The birth dates of executives should be independent of each other. This is a convenience sample of executives. It is random, but not a SRS. However, there's no reason to suspect bias.

✓ **Expected Cell Frequency Condition**: The null hypothesis expects that 1/12 of the 256 births, or 21.333, should occur in each sign. These expected values are all at least 5, so the condition is satisfied.

Specify the sampling distribution model.

Name the test you will use.

The conditions are satisfied, so I'll use a χ^2 model with $12 - 1 = 11$ degrees of freedom and do a **chi-square goodness-of-fit test**.

SHOW **MECHANICS** Each cell contributes an $\frac{(Obs - Exp)^2}{Exp}$ value to the chi-square sum. We add up these components for each zodiac sign. If you do it by hand, it can be helpful to arrange the calculation in a table. We show that after this Step-by-Step Example.

The expected value for each zodiac sign is 21.333.

$$\chi^2 = \sum \frac{(Obs - Exp)^2}{Exp} = \frac{(23 - 21.333)^2}{21.333} + \frac{(20 - 21.333)^2}{21.333} + \cdots$$

$$= 5.094 \text{ for all 12 signs.}$$

The P-value is the area in the upper tail of the χ^2 model above the computed χ^2 value.

The χ^2 models are skewed to the high end and change shape depending on the degrees of freedom. The P-value considers only the right tail. Large χ^2 statistic values correspond to small P values, which lead us to reject the null hypothesis.

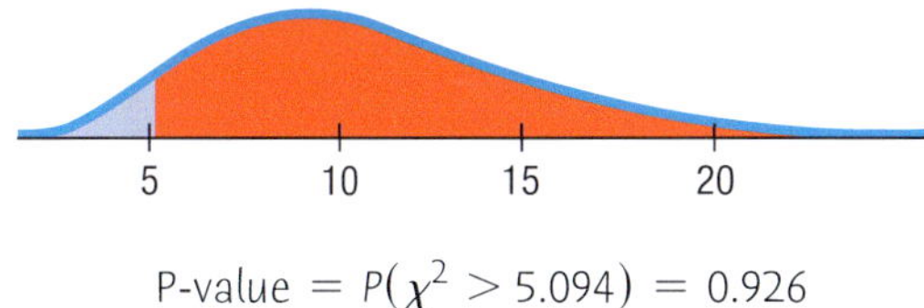

$$\text{P-value} = P(\chi^2 > 5.094) = 0.926$$

[2]It may seem that we have broken our rule of thumb that null hypotheses should specify parameter values. If you want to get formal about it, the null hypothesis is that

$$p_{\text{Aries}} = p_{\text{Taurus}} = \cdots = p_{\text{Pisces}}.$$

That is, we hypothesize that the true proportions of births of CEOs under each sign are equal. The role of the null hypothesis is to specify the model so that we can compute the test statistic. That's what this one does.

TELL **CONCLUSION** Link the P-value to your decision. Remember to state your conclusion in terms of what the data mean, rather than just making a statement about the distribution of counts.

The P-value of 0.926 says that if the zodiac signs of executives were in fact distributed uniformly, an observed chi-square value of 5.09 or higher would occur about 93% of the time. This certainly isn't unusual, so I fail to reject the null hypothesis, and conclude that these data show virtually no evidence of nonuniform distribution of zodiac signs among executives.

The Chi-Square Calculation

Let's make the chi-square procedure very clear. Here are the steps:

1. **Find the expected values.** These come from the null hypothesis model. Every model gives a hypothesized proportion for each cell. The expected value is the product of the total number of observations times this proportion.

   For our example, the null model hypothesizes *equal* proportions. With 12 signs, 1/12 of the 256 executives should be in each category. The expected number for each sign is 21.333.
2. **Compute the residuals.** Once you have expected values for each cell, find the residuals, *Observed* − *Expected.*
3. **Square the residuals.**
4. **Compute the components.** Now find the **component**, $\frac{(Observed - Expected)^2}{Expected}$, for each cell.
5. **Find the sum of the components.** That's the chi-square statistic.
6. **Find the degrees of freedom.** It's equal to the number of cells minus one. For the zodiac signs, that's 12 − 1 = 11 degrees of freedom.
7. **Test the hypothesis.** Large chi-square values mean lots of deviation from the hypothesized model, so they give small P-values. Look up the critical value from a table of chi-square values, or use technology to find the P-value directly.

The steps of the chi-square calculations are often laid out in tables. If you don't have a statistics program handy, these calculations are particularly well suited to being performed on a spreadsheet. Use one row for each category, and columns for observed counts, expected counts, residuals, squared residuals, and the contributions to the chi-square total like this:

Sign	Observed	Expected	Residual = $(Obs - Exp)$	$(Obs - Exp)^2$	Component = $\frac{(Obs - Exp)^2}{Exp}$
Aries	23	21.333	1.667	2.778889	0.130262
Taurus	20	21.333	−1.333	1.776889	0.083293
Gemini	18	21.333	−3.333	11.108889	0.520737
Cancer	23	21.333	1.667	2.778889	0.130262
Leo	20	21.333	−1.333	1.776889	0.083293
Virgo	19	21.333	−2.333	5.442889	0.255139
Libra	18	21.333	−3.333	11.108889	0.520737
Scorpio	21	21.333	−0.333	0.110889	0.005198
Sagittarius	19	21.333	−2.333	5.442889	0.255139
Capricorn	22	21.333	0.667	0.444889	0.020854
Aquarius	24	21.333	2.667	7.112889	0.333422
Pisces	29	21.333	7.667	58.782889	2.755491
					$\sum = 5.094$

HOW BIG IS BIG?

When we calculated χ^2 for the zodiac sign example, we got 5.094. That value would have been big for z or t, leading us to reject the null hypothesis. Not here. Were you surprised that $\chi^2 = 5.094$ had a huge P-value of 0.926? What *is* big for a χ^2 statistic, anyway?

Think about how χ^2 is calculated. In every cell, any deviation from the expected count contributes to the sum. Large deviations generally contribute more, but if there are a lot of cells, even small deviations can add up, making the χ^2 value larger. So the more cells there are, the higher the value of χ^2 has to get before it becomes noteworthy. For χ^2, then, the decision about how big is big depends on the number of degrees of freedom.

Unlike the Normal and t families, χ^2 models are skewed. Curves in the χ^2 family change both shape and center as the number of degrees of freedom grows. Here, for example, are the χ^2 curves for 5 and 9 degrees of freedom.

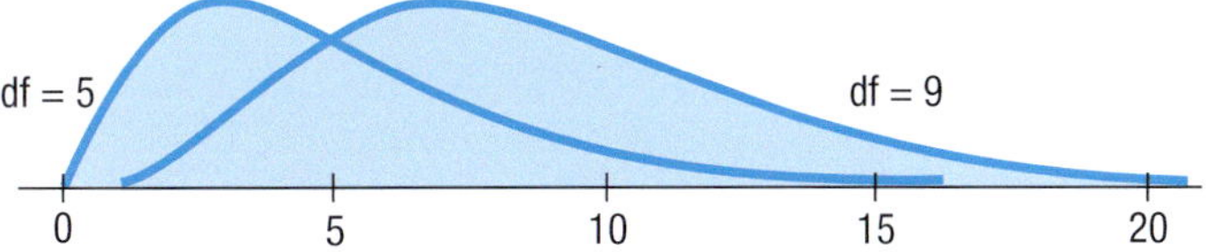

Notice that the value $\chi^2 = 10$ might seem somewhat extreme when there are 5 degrees of freedom, but appears to be rather ordinary for 9 degrees of freedom. Here are two simple facts to help you think about χ^2 models:

- The mode is at $\chi^2 = df - 2$. (Look back at the curves; their peaks are at 3 and 7, see?)
- The expected value (mean) of a χ^2 model is its number of degrees of freedom. That's a bit to the right of the mode—as we would expect for a skewed distribution.

Our test for zodiac birthdays had 11 df, so the relevant χ^2 curve peaks at 9 and has a mean of 11. Knowing that, we might have easily guessed that the calculated χ^2 value of 5.094 wasn't going to be significant.

The Trouble with Goodness-of-Fit Tests: What's the Alternative?

Goodness-of-fit tests are likely to be performed by people who have a theory of what the proportions *should* be in each category and who believe their theory to be true. Unfortunately, the only *null* hypothesis available for a goodness-of-fit test is that the theory is true. And as we know, the hypothesis-testing procedure allows us only to *reject* the null or *fail to reject* it. We can never confirm that a theory is in fact true, which is often what people want to do.

Unfortunately, they're stuck. At best, we can point out that the data are consistent with the proposed theory. But this doesn't *prove* the theory. The data *could* be consistent with the model even if the theory were wrong. In that case, we fail to reject the null hypothesis but can't conclude anything for sure about whether the theory is true.

And we can't fix the problem by turning things around. Suppose we try to make our favored hypothesis the alternative. Then it is impossible to pick a single null. For example, suppose, as a doubter of astrology, you want to prove that the distribution of executive births is uniform. If you choose uniform as the null hypothesis, you can only *fail* to reject it. So you'd like uniformity to be your alternative hypothesis. Which particular violation of equally distributed births would you choose as your null? The problem is that the model can be wrong in many, many ways. There's no way to frame a null hypothesis the other way around. There's just no way to prove that a favored model is true.

Why Can't We Prove the Null?

A biologist wanted to show that her inheritance theory about fruit flies is valid. It says that 10% of the flies should be type 1, 70% type 2, and 20% type 3. After her students collected data on 100 flies, she did a goodness-of-fit test and found a P-value of 0.07. She started celebrating, since her null hypothesis wasn't rejected—that is, until her students collected data on 100 more flies. With 200 flies, the P-value dropped to 0.02. Although she knew the answer was probably no, she asked the statistician somewhat hopefully if she could just ignore half the data and stick with the original 100. By this reasoning we could always "prove the null" just by not collecting much data. With only a little data, the chances are good that they'll be consistent with almost anything. But they also have little chance of disproving anything either. In this case, the test has no power. Don't be lured into this scientist's reasoning. With data, more is always better. But you can't ever prove that your null hypothesis is true.

22.2 Chi-Square Test of Homogeneity

Many universities survey graduating classes to determine the plans of the graduates. We might wonder whether the plans of students are the same at different schools within a university. Here's a **two-way table** for graduates from several schools at one university. Each cell of the table shows how many students from a particular school made a certain choice.

Table 22.1
Post-graduation activities of the class of 2011 for several colleges of a large university. (Note: ILR is the School of Industrial and Labor Relations.)

	Agriculture	Arts & Sciences	Engineering	ILR	Total
Employed	209	198	177	101	**685**
Grad School	104	171	158	33	**466**
Other	135	115	39	16	**305**
Total	**448**	**484**	**374**	**150**	**1456**

Because class sizes are so different, we see differences better by examining the proportions for each class rather than the counts:

Table 22.2
Activities of graduates as a percentage of respondents from each school.

	Agriculture	Arts & Sciences	Engineering	ILR	Total
Employed	46.7%	40.9%	47.3%	67.3%	**47.0%**
Grad School	23.2%	35.3%	42.2%	22.0%	**32.0%**
Other	30.1%	23.8%	10.4%	10.7%	**20.9%**
Total	**100.0%**	**100.0%**	**100.0%**	**100.0%**	**100.0%**

WHO	Graduates from four schools at an upstate New York university
WHAT	Post-graduation activities
WHEN	2011
WHY	Survey for general information

We already know how to test whether *two* proportions are the same. For example, we could use a two-proportion z-test to see whether the proportion of students choosing graduate school is the same for Agriculture students as for Engineering students. But now we have more than two groups. We want to test whether the students' choices are the same across all four schools. The z-test for two proportions generalizes to a **chi-square tests of homogeneity and independence**.

Chi-square again? It turns out that the mechanics of this test are *identical* to the chi-square test for goodness-of-fit that we just saw. (How similar can you get?) Why the different names then? The tests are really quite different. The goodness-of-fit test compared counts with a theoretical model. But here we're asking whether the distribution of choices is the same among different groups, so we find the expected counts for each category directly from the data. As a result, we count the degrees of freedom slightly differently as well.

The term "homogeneity" means that things are the same. Here, we ask whether the post-graduation choices made by students are the *same* for these four schools. The **chi-square test of homogeneity** comes with a built-in null hypothesis: We hypothesize that the distribution does not change from group to group. The test looks for differences large enough to step beyond what we might expect from random sample-to-sample variation. It can reveal a large deviation in a single category or small, but persistent, differences over all the categories—or anything in between.

A test of homogeneity measures one variable and asks if the distributions of that variable are the same across several groups. If instead of groups we have a second variable, then the test becomes a test of independence. In a test of independence, there are two variables being measured and we want to know if the two distributions are independent. The mechanics of the two tests are exactly the same. The only difference is in the inference we make. Section 22.4 discusses tests of independence and gives examples of each so you can learn to recognize the difference.

The assumptions and conditions for both tests are the same as for the chi-square test for goodness-of-fit.

Calculations

The null hypothesis of the homogeneity test says that the distribution of the proportions of graduates choosing each alternative is the same for all four schools, so we can estimate those overall proportions by pooling our data from the four schools together. The variable is *Career Choice* and the groups are the different schools. Within each school, the expected proportion for each choice is just the overall proportion of all students making that choice. The expected counts are those proportions applied to the number of students in each graduating class.

For example, overall, 685, or about 47.0%, of the 1456 students who responded to the survey were employed. If the distributions are homogeneous (as the null hypothesis asserts), then 47% of the 448 Agriculture school graduates (or about 210.76 students) should be employed. Similarly, 47% of the 374 Engineering grads (or about 175.95) should be employed.

Working in this way, we (or, more likely, the computer) can fill in expected values for each cell. Because these are theoretical values, they don't have to be integers. The expected values look like this:

Table 22.3
Expected values for the graduates.

	Agriculture	Arts & Sciences	Engineering	ILR	Total
Employed	210.769	227.706	175.955	70.570	**685**
Grad School	143.385	154.907	119.701	48.008	**466**
Other	93.846	101.387	78.345	31.422	**305**
Total	**448**	**484**	**374**	**150**	**1456**

Following the pattern of the goodness-of-fit test, we compute the component for each cell of the table. For the highlighted cell, employed students graduating from the Ag school, that's

$$\frac{(Obs - Exp)^2}{Exp} = \frac{(209 - 210.769)^2}{210.769} = 0.0148.$$

Summing these components across all cells gives

$$\chi^2 = \sum_{all\ cells} \frac{(Obs - Exp)^2}{Exp} = 93.66.$$

How about the degrees of freedom? We don't really need to calculate all the expected values in the table. We know there is a total of 685 employed students, so once we find the expected values for three of the schools, we can determine the expected number for the fourth by just subtracting. Similarly, we know how many students graduated from each

> *NOTATION ALERT*
> For a contingency table, R represents the number of rows and C the number of columns.

school, so after filling in two rows, we can find the expected values for the remaining row by subtracting. To fill out the table, we need to know the counts in only $R - 1$ rows and $C - 1$ columns. So the table has $(R - 1)(C - 1)$ degrees of freedom.

In our example, we need to calculate only 2 choices in each column and counts for 3 of the 4 colleges, for a total of $2 \times 3 = 6$ degrees of freedom. We'll need the degrees of freedom to find a P-value for the chi-square statistic.

STEP-BY-STEP EXAMPLE

A Chi-Square Test for Homogeneity

We have reports from four schools on the post-graduation activities of their graduating classes.

QUESTION: Are the distributions of students' choices of post-graduation activities the same across all the schools?

THINK **PLAN** State what you want to know.

Identify the variables and check the W's.

I want to test whether post-graduation choices are the same for students from each of four schools. I have a table of counts classifying each school's respondents according to their activities.

HYPOTHESES State the null and alternative hypotheses.

H_0: Students' post-graduation activities are distributed in the same way for all four schools.

H_A: Students' plans do not have the same distribution.

MODEL Make a picture: A side-by-side bar chart shows the four distributions of post-graduation activities. Plot column percents to remove the effect of class size differences. A split bar chart would also be an appropriate choice.

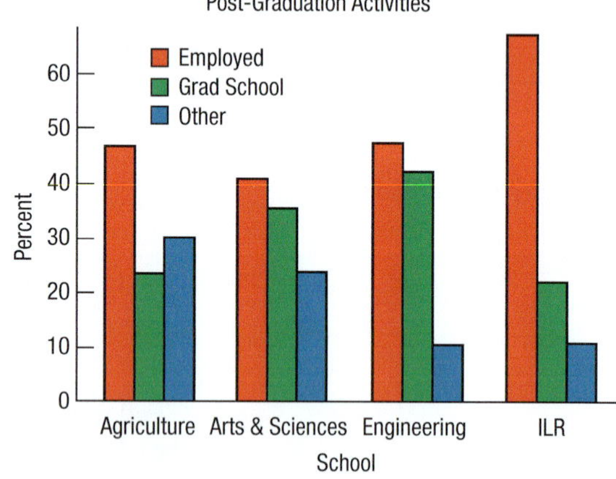

A side-by-side bar chart shows how the distributions of choices differ across the four schools.

Think about the assumptions and check the conditions.

✓ **Counted Data Condition:** I have counts of the number of students in categories.

✓ **Independence Assumption:** Even though this isn't a random sample, student plans should be largely independent of each other. The occasional friends who decide to join Teach for America together or couples who make grad school decisions together are too rare to affect this analysis.

✓ **Expected Cell Frequency Condition:** The expected values (shown below) are all at least 5.

State the sampling distribution model and name the test you will use.

The conditions seem to be met, so I can use a χ^2 model with $(3-1)\times(4-1)=6$ degrees of freedom and do a **chi-square test of homogeneity**.

SHOW **MECHANICS** Show the expected counts for each cell of the data table. You could make separate tables for the observed and expected counts, or put both counts in each cell as shown here. While observed counts must be whole numbers, expected counts rarely are—don't be tempted to round those off.

	Ag	A&S	Eng	ILR
Empl.	209 210.769	198 227.706	177 175.955	101 70.570
Grad School	104 143.385	171 154.907	158 119.701	33 48.008
Other	135 93.846	115 101.387	39 78.345	16 31.422

Calculate χ^2.

$$\chi^2 = \sum_{all\ cells} \frac{(Obs - Exp)^2}{Exp}$$

$$= \frac{(209 - 210.769)^2}{210.769} + \cdots$$

$$= 93.66$$

The shape of a χ^2 model depends on the degrees of freedom. A χ^2 model with 6 df is skewed to the high end.

The P-value considers only the right tail. Here, the calculated value of the χ^2 statistic is off the scale, so the P-value is quite small.

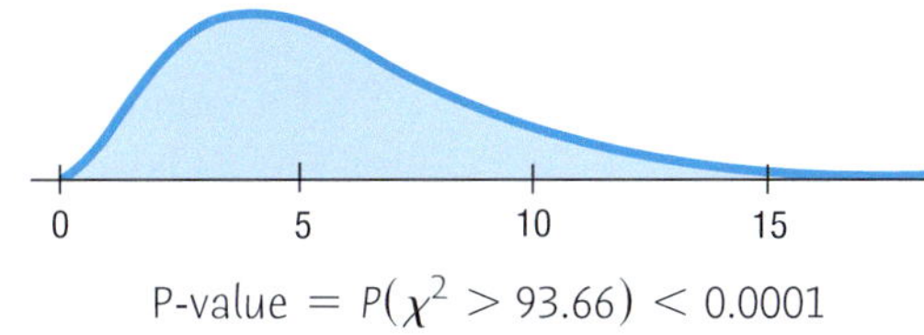

$$\text{P-value} = P(\chi^2 > 93.66) < 0.0001$$

TELL **CONCLUSION** State your conclusion in the context of the data. You should specifically talk about whether the distributions for the groups appear to be different.

The P-value is very small, so I reject the null hypothesis and conclude that there's evidence that the post-graduation activities of students from these four schools don't have the same distribution.

If you find that simply rejecting the hypothesis of homogeneity is a bit unsatisfying, you're in good company. OK, so the post-graduation plans are different. What we'd really like to know is what the differences are, where they're the greatest, and where they're smallest. The test for homogeneity doesn't answer these interesting questions, but it does provide some evidence that can help us.

22.3 Examining the Residuals

Whenever we reject the null hypothesis, it's a good idea to examine residuals. For chi-square tests, we want to compare residuals for cells that may have very different counts. So we're better off standardizing the residuals. We know the mean residual is zero,[3] but we need to know each residual's standard deviation. When we tested proportions, we saw a link between the expected proportion and its standard deviation. For counts, there's

[3]Residual = Observed − Expected. Because the total of the expected values is set to be the same as the observed total, the residuals must sum to zero.

a similar link. To standardize a cell's residual, we just divide by the square root of its expected value:

$$c = \frac{(Obs - Exp)}{\sqrt{Exp}}.$$

Notice that these **standardized residuals** are just the square roots of the **components** we calculated for each cell, and their sign indicates whether we observed more cases than we expected, or fewer.

The standardized residuals give us a chance to think about the underlying patterns and to consider the ways in which the distribution of post-graduation plans may differ from school to school. Now that we've subtracted the mean (the residual, $Obs - Exp$, has mean 0, so it's already subtracted) and divided by their standard deviations, these are z-scores. If the null hypothesis were true, we could even appeal to the Central Limit Theorem, think of the Normal model, and use the 68–95–99.7 Rule to judge how extraordinary the large ones are.

Here are the standardized residuals:

Table 22.4

Standardized residuals can help show how the table differs from the null hypothesis pattern.

	Ag	A&S	Eng	ILR
Employed	−0.121866	−1.96860	0.078805	3.62235
Grad School	−3.28909	1.29304	3.50062	−2.16607
Other	4.24817	1.35192	−4.44511	−2.75117

The column for Engineering students immediately attracts our attention. It holds both the largest positive and the largest negative standardized residuals. It looks like Engineering school graduates are more likely to go on to graduate work and very unlikely to take time off for "volunteering and travel, among other activities" (as the "Other" category is explained). By contrast, Ag school graduates seem to be less likely to pursue graduate work immediately after school.

EXAMPLE 22.4 Looking at χ^2, Residuals

RECAP: Some people suggest that schoolchildren who are the older ones in their class naturally perform better in sports and therefore get more coaching and encouragement. To see if there's any evidence for this, we looked at Major League Baseball players born since 1975. A goodness-of-fit test found their birth months to have a distribution that's significantly different from the rest of us. The table shows the standardized residuals.

Month	Residual	Month	Residual
1	1.73	7	−2.69
2	1.72	8	2.77
3	−0.21	9	0.08
4	0.25	10	−1.56
5	0.71	11	−1.22
6	−0.39	12	−0.96

QUESTION: What's different about the distribution of birth months among Major League ballplayers?

ANSWER: It appears that, compared to the general population, fewer ballplayers than expected were born in July and more than expected in August. Either month would make them the younger kids in their grades in school, so these data don't offer support for the conjecture that being older is an advantage in terms of a career as a professional baseball player.

JUST CHECKING

Tiny black potato flea beetles can damage potato plants in a vegetable garden. These pests chew holes in the leaves, causing the plants to wither or die. They can be killed with an insecticide, but a canola oil spray has been suggested as a non-chemical "natural" method of controlling the beetles. To conduct an experiment to test the effectiveness of the natural spray, we gather 500 beetles and place them in three Plexiglas® containers. Two hundred beetles go in the first container, where we spray them with the canola oil mixture. Another 200 beetles go in the second container; we spray them with the insecticide. The remaining 100 beetles in the last container serve as a control group; we simply spray them with water. Then we wait 6 hours and count the number of surviving beetles in each container.

1. Why do we need the control group?
2. What would our null hypothesis be?
3. After the experiment is over, we could summarize the results in a table as shown. How many degrees of freedom does our χ^2 test have?

	Natural Spray	Insecticide	Water	Total
Survived				
Died				
Total	200	200	100	500

4. Suppose that, all together, 125 beetles survived. (That's the first-row total.) What's the expected count in the first cell—survivors among those sprayed with the natural spray?
5. If it turns out that only 40 of the beetles in the first container survived, what's the calculated component of χ^2 for that cell?
6. If the total calculated value of χ^2 for this table turns out to be around 10, would you expect the P-value of our test to be large or small? Explain.

22.4 Chi-Square Test of Independence

A study from the University of Texas Southwestern Medical Center examined 626 people being treated for non–blood-related diseases to see whether the risk of hepatitis C was related to whether people had tattoos and to where they got their tattoos. Hepatitis C causes about 10,000 deaths each year in the United States but often goes undetected for years after infection. (Data in **Tattoos**)

The data from this study can be summarized in a two-way table, as follows:

Table 22.5
Counts of patients classified by their hepatitis C test status according to whether they had a tattoo from a tattoo parlor or from another source, or had no tattoo.

	Hepatitis C	No Hepatitis C	Total
Tattoo, Parlor	17	35	**52**
Tattoo, Elsewhere	8	53	**61**
None	22	491	**513**
Total	**47**	**579**	**626**

These data differ from the kinds of data we've considered before in this chapter because they categorize subjects from a single group on *two* categorical variables rather than on only one, so now we have a test of independence. The categorical variables here are *Hepatitis C Status* ("Hepatitis C" or "No Hepatitis C") and *Tattoo Status* ("Parlor," "Elsewhere," "None"). We've seen counts classified by two categorical variables displayed like this in Chapter 3, so we know such tables are called contingency tables. **Contingency tables**

WHO	Patients being treated for non–blood-related disorders
WHAT	Tattoo status and hepatitis C status
WHEN	1991, 1992
WHERE	Texas

categorize counts on two (or more) variables so that we can see whether the distribution of counts on one variable is *contingent*—or depends—on the other.

The natural question to ask of these data is whether the chance of having hepatitis C is *independent* of tattoo status. Recall that for events **A** and **B** to be independent, $P(\mathbf{A})$ must equal $P(\mathbf{A} \mid \mathbf{B})$. Here, this means the probability that a randomly selected patient has hepatitis C should be the same regardless of the patient's tattoo status. If *Hepatitis Status* is independent of tattoos, we'd expect the proportion of people testing positive for hepatitis to be the same for the three levels of *Tattoo Status*. Of course, for real data we won't expect them to be exactly the same, so we look to see how close they are. This sounds a lot like the test of homogeneity, but the difference is that now we have two categorical variables measured on a single population. For the homogeneity test, we had a single categorical variable measured on two or more populations. We asked if the distribution of the variable was the same across all the groups. But now we ask a different question: "Are the *variables* independent?" rather than "Are the *groups* homogeneous?" These are subtle differences, but they are important when we state hypotheses and draw conclusions. When we ask whether two variables measured on the same population are independent, we're performing a **chi-square test of independence**.

EXAMPLE 22.5 Which χ^2 Test?

Many states and localities now collect data on traffic stops regarding the race of the driver. The initial concern was that Black drivers were being stopped more often (the "crime" ironically called "Driving While Black"). With more data in hand, attention has turned to other issues. For example, data from 2533 traffic stops in Cincinnati[4] report the race of the driver (Black, White, or Other) and whether the traffic stop resulted in a search of the vehicle.

		Race			
		Black	**White**	**Other**	**Total**
Search	**No**	787	594	27	**1408**
	Yes	813	293	19	**1125**
	Total	**1600**	**887**	**46**	**2533**

QUESTIONS: Which test would be appropriate to examine whether race is a factor in vehicle searches? What are the hypotheses?

ANSWERS: These data represent one group of traffic stops in Cincinnati, categorized on two variables, Race and Search. I'll do a chi-square test of independence.

H_0: Whether or not police search a vehicle is independent of the race of the driver.

H_A: Decisions to search vehicles are not independent of the driver's race.

Assumptions and Conditions

Of course, we still need counts and enough data so that the expected values are at least 5 in each cell.

If we're interested in the independence of variables, we usually want to generalize from the data to some population. In that case, we'll need to check that the data are a representative random sample from that population.

[4]John E. Eck, Lin Liu, and Lisa Growette Bostaph, Police Vehicle Stops in Cincinnati, Oct. 1, 2003, available at www.cincinnati-oh.gov. Data for other localities can be found by searching from archive.is/www.racialprofilinganalysis.neu.edu.

STEP-BY-STEP EXAMPLE

A Chi-Square Test for Independence

We have counts of 626 individuals categorized according to their "tattoo status" and their "hepatitis status."

QUESTION: Are tattoo status and hepatitis status independent?

THINK

PLAN State what you want to know.

Identify the variables and check the W's.

I want to test whether the categorical variables Tattoo Status and Hepatitis Status are statistically independent. I have a contingency table of 626 Texas patients under treatment for a non-blood-related disease.

HYPOTHESES State the null and alternative hypotheses.

We perform a test of independence when we suspect the variables may not be independent. We are on the familiar ground of making a claim (in this case, that knowing *Tattoo Status* will change probabilities for *Hepatitis C Status*) and testing the null hypothesis that it is *not* true.

H_0: *Tattoo Status* and *Hepatitis Status* are independent.[5]

H_A: *Tattoo Status* and *Hepatitis Status* are not independent.

MODEL Make a picture. Because these are only two categories—Hepatitis C and No Hepatitis C—a simple bar chart of the distribution of tattoo sources for Hep C patients shows all the information.

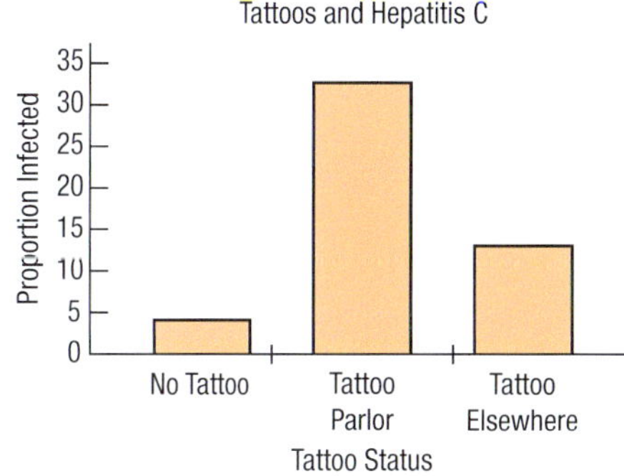

The bar chart suggests strong differences in Hepatitis C risk based on tattoo status.

Think about the assumptions and check the conditions.

✓ **Counted Data Condition**: I have counts of individuals categorized on two variables.

✓ **Independence Assumption**: These data are from a retrospective study of patients being treated for something unrelated to hepatitis. Although they are not an SRS, they are likely to be independent of each other.

[5] Once again, parameters are hard to express. The hypothesis of independence itself tells us how to find expected values for each cell of the contingency table. That's all we need.

✔ **Expected Cell Frequency Condition**: The expected values do not meet the condition that all are at least 5.

This table shows both the observed and expected counts for each cell. The expected counts are calculated exactly as they were for a test of homogeneity; in the first cell, for example, we expect $\frac{52}{626}$ (that's 8.3%) of 47.

	Hepatitis C	No Hepatitis C	Total
Tattoo, Parlor	17 3.904	35 48.096	52
Tattoo, Elsewhere	8 4.580	53 56.420	61
None	22 38.516	491 474.484	513
Total	47	579	626

Warning: Be wary of proceeding when there are small expected counts. If we see expected counts that fall far short of 5, or if many cells violate the condition, we should not use χ^2. (We will soon discuss ways you can fix the problem.) If you do continue, always check the residuals to be sure those cells did not have a major influence on your result.

Specify the model.

Name the test you will use.

Although the Expected Cell Frequency Condition is not satisfied, the values are close to 5. I'll go ahead, but I'll check the residuals carefully. I'll use a χ^2 model with $(3 - 1) \times (2 - 1) = 2$ df and do a **chi-square test of independence**.

SHOW **MECHANICS** Calculate χ^2.

The shape of a chi-square model depends on its degrees of freedom. With 2 df, the model looks quite different, as you can see here. We still care only about the right tail.

$$\chi^2 = \sum_{\text{all cells}} \frac{(Obs - Exp)^2}{Exp}$$

$$= \frac{(17 - 3.094)^2}{3.094} + \cdots = 57.91$$

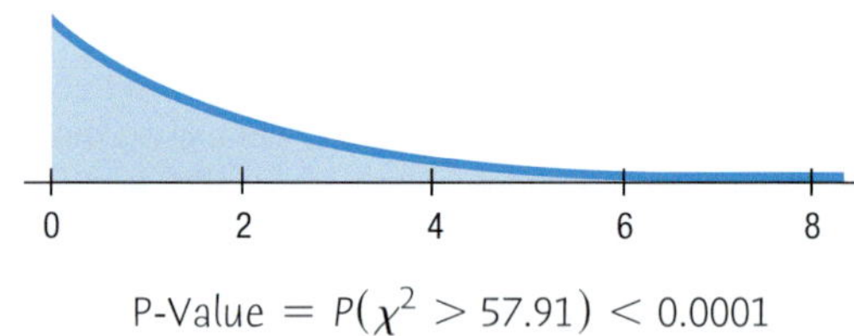

P-Value $= P(\chi^2 > 57.91) < 0.0001$

TELL **CONCLUSION** Link the P-value to your decision. State your conclusion about the independence of the two variables.

(We should be wary of this conclusion because of the small expected counts. A complete solution must include the additional analysis, recalculation, and final conclusion discussed in the following section.)

The P-value is very small, so I reject the null hypothesis and conclude that *Hepatitis Status* is not independent of *Tattoo Status*. Because the Expected Cell Frequency Condition was violated, I need to check that the two cells with small expected counts did not influence this result too greatly.

EXAMPLE 22.6 Chi-Square Mechanics

RECAP: We have data that allow us to investigate whether police searches of vehicles they stop are independent of the driver's race.

QUESTIONS: What are the degrees of freedom for this test? What is the expected frequency of searches for the Black drivers who were stopped? What's that cell's component in the χ^2 computation? And how is the standardized residual for that cell computed?

		Race			
		Black	**White**	**Other**	**Total**
Search	**No**	787	594	27	**1408**
	Yes	813	293	19	**1125**
	Total	**1600**	**887**	**46**	**2533**

ANSWERS: This is a 2 × 3 contingency table, so df = (2 − 1)(3 − 1) = 2.

Overall, 1125 of 2533 vehicles were searched. If searches are conducted independent of race, then I'd expect $\frac{1125}{2533}$ of the 1600 Black drivers to have been searched: $\frac{1125}{2533} \times 1600 \approx 710.62$.

That cell's term in the χ^2 calculation is $\frac{(Obs - Exp)^2}{Exp} = \frac{(813 - 710.62)^2}{710.62} = 14.75$.

The standardized residual for that cell is $\frac{Obs - Exp}{\sqrt{Exp}} = \frac{813 - 710.62}{\sqrt{710.62}} = 3.84$.

Examine the Residuals

Each cell of the contingency table contributes a term to the chi-square sum. As we did earlier, we should examine the residuals because we have rejected the null hypothesis. In this instance, we have an additional concern that the cells with small expected frequencies not be the ones that make the chi-square statistic large.

Our interest in the data arises from the potential for improving public health. If patients with tattoos are more likely to test positive for hepatitis C, perhaps physicians should be advised to suggest blood tests for such patients.

The standardized residuals look like this:

Table 22.6
Standardized residuals for the hepatitis and tattoos data. Are any of them particularly large in magnitude?

	Hepatitis C	No Hepatitis C
Tattoo, Parlor	6.628	−1.888
Tattoo, Elsewhere	1.598	−0.455
None	−2.661	0.758

The chi-square value of 57.91 is the sum of the squares of these six values. The cell for people with tattoos obtained in a tattoo parlor who have hepatitis C is large and positive, indicating there are more people in that cell than the null hypothesis of independence would predict. Maybe tattoo parlors are a source of infection or maybe those who go to tattoo parlors also engage in risky behavior.

The second-largest component is a negative value for those with no tattoos who test positive for hepatitis C. A negative value says that there are fewer people in this cell than independence would expect. That is, those who have no tattoos are less likely to be infected with hepatitis C than we might expect if the two variables were independent.

What about the cells with small expected counts? The formula for the chi-square standardized residuals divides each residual by the square root of the expected frequency. Too small an expected frequency can arbitrarily inflate the residual and lead to an inflated chi-square statistic. Any expected count close to the arbitrary minimum of 5 calls for checking that cell's standardized residual to be sure it is not particularly large. In this case, the standardized residual for the "Hepatitis C and Tattoo, Elsewhere" cell is not particularly large, but the standardized residual for the "Hepatitis C and Tattoo, Parlor" cell is large.

We might choose not to report the results because of concern with the small expected frequency. Alternatively, we could include a warning along with our report of the results. Yet another approach is to combine categories to get a larger category total and correspondingly larger expected frequencies, if there are some categories that can be appropriately combined. Here, we might naturally combine the two rows for tattoos, obtaining a 2×2 table:

Table 22.7
Combining the two tattoo categories gives a table with all expected counts greater than 5.

	Hepatitis C	No Hepatitis C	Total
Tattoo	25	88	113
None	22	491	513
Total	47	579	626

TELL ALL

This table has expected values of at least 5 in every cell, and a chi-square value of 42.42 on 1 degree of freedom. The corresponding P-value is <0.0001.

We conclude that *Tattoo Status* and *Hepatitis C Status* are not independent. The data *suggest* that tattoo parlors may be a particular problem, but we haven't enough data to draw that conclusion.

EXAMPLE 22.7 Writing Conclusions for χ^2 Tests

RECAP: We're looking at Cincinnati traffic stop data to see if police decisions about searching cars show evidence of racial bias. With 2 df, technology calculates $\chi^2 = 73.25$, a P-value less than 0.0001, and these standardized residuals:

		Race		
		Black	White	Other
Search	No	−3.43	4.55	0.28
	Yes	3.84	−5.09	−0.31

QUESTION: What's your conclusion?

ANSWER: The very low P-value leads me to reject the null hypothesis. There's strong evidence that police decisions to search cars at traffic stops are associated with the driver's race.

The largest residuals are for White drivers, who are searched less often than independence would predict. It appears that Black drivers' cars are searched more often.

Chi-Square and Causation

Chi-square tests are common. Tests for independence are especially widespread. Unfortunately, many people interpret a small P-value as proof of causation. We know better. Just as correlation between quantitative variables does not demonstrate causation, a failure of independence between two categorical variables does not show a cause-and-effect relationship between them, nor should we say that one variable *depends* on the other.

The chi-square test for independence treats the two variables symmetrically. There is no way to differentiate the direction of any possible causation from one variable to the other. In our example, it is unlikely that having hepatitis causes one to crave a tattoo, but other examples are not so clear.

In this case, it's easy to imagine that lurking variables are responsible for the observed lack of independence. Perhaps the lifestyles of some people include both tattoos and behaviors that put them at increased risk of hepatitis C, such as body piercings or even drug use. Even a small subpopulation of people with such a lifestyle among those with tattoos might be enough to create the observed result. After all, we observed only 25 patients with both tattoos and hepatitis.

In some sense, a failure of independence between two categorical variables is less impressive than a strong, consistent, linear association between quantitative variables. Two categorical variables can fail the test of independence in many ways, including ways that show no consistent pattern of failure. Examination of the chi-square standardized residuals can help you think about the underlying patterns.

JUST CHECKING

Which of the three chi-square tests—goodness-of-fit, homogeneity, or independence—would you use in each of the following situations?

7. A restaurant manager wonders whether customers who dine on Friday nights have the same preferences among the four "chef's special" entrées as those who dine on Saturday nights. One weekend he has the wait staff record which entrées were ordered each night. Assuming these customers to be typical of all weekend diners, he'll compare the distributions of meals chosen Friday and Saturday.
8. Company policy calls for parking spaces to be assigned to everyone at random, but you suspect that may not be so. There are three lots of equal size: lot A, next to the building; lot B, a bit farther away; and lot C, on the other side of the highway. You gather data about 120 managers to see how many were assigned parking in each lot.
9. Is a student's social life affected by where the student lives? A campus survey asked a random sample of students whether they lived in a dormitory, in off-campus housing, or at home, and whether they had been out on a date 0, 1–2, 3–4, or 5 or more times in the past two weeks.

WHAT CAN GO WRONG?

- **Don't use chi-square methods unless you have counts.** All three of the chi-square tests apply only to counts. Other kinds of data can be arrayed in two-way tables. Just because numbers are in a two-way table doesn't make them suitable for chi-square analysis. Data reported as proportions or percentages can be suitable for chi-square procedures, *but only after they are converted to counts*. If you try to do the calculations without first finding the counts, your results will be wrong.
- **Beware large samples.** Beware *large* samples?! That's not the advice you're used to hearing. The chi-square tests, however, are unusual. Be wary of chi-square tests performed on very large samples. No hypothesized distribution fits perfectly, no two groups are exactly homogeneous, and two variables are rarely perfectly independent. The degrees of freedom for chi-square tests don't grow with the sample size. With a sufficiently large sample size, a chi-square test can always reject the null hypothesis. But we have no measure of how far the data are from the null model. There are no confidence intervals to help us judge the effect size.
- **Don't say that one variable "depends" on the other just because they're not independent.** Dependence suggests a pattern and implies causation, but variables can fail to be independent in many different ways. When variables fail the test for independence, you might just say they are "associated."

CONNECTIONS

Chi-square methods relate naturally to inference methods for proportions. We can think of a test of homogeneity as stepping from a comparison of two proportions to a question of whether three or more proportions are equal. The standard deviations of the residuals in each cell are linked to the expected counts much like the standard deviations we found for proportions.

Independence is, of course, a fundamental concept in statistics. But chi-square tests do not offer a general way to check on independence for all those times when we have had to assume it.

Stacked bar charts or side-by-side pie charts can help us think about patterns in two-way tables. A histogram or boxplot of the standardized residuals can help locate extraordinary values.

CHAPTER REVIEW

Recognize when a chi-square test of goodness-of-fit, homogeneity, or independence is appropriate.

For each test, find the expected cell frequencies.

For each test, check the assumptions and corresponding conditions and know how to complete the test.

- Counted Data Condition.
- Independence Assumption; randomization makes independence more plausible.
- Sample size assumption with the Expected Cell Frequency Condition; expect at least 5 observations in each cell.

Interpret a chi-square test.

- Even though we might believe the model, we cannot prove that the data fit the model with a chi-square test because that would mean confirming the null hypothesis.

Examine the standardized residuals to understand what cells were responsible for rejecting a null hypothesis.

State the null hypothesis for a test of independence and understand how that is different from the null hypothesis for a test of homogeneity.

- Both are computed the same way. You may not find both offered by your technology. You can use either one as long as you interpret your result correctly.

REVIEW OF TERMS

Chi-square goodness-of-fit test A test of whether the distribution of counts in one categorical variable matches the distribution predicted by a model is called a test of goodness-of-fit. In a chi-square goodness-of-fit test, the expected counts come from the predicting model. The test finds a P-value from a chi-square model with $n - 1$ degrees of freedom, where n is the number of categories in the categorical variable (p. 701).

Cell A cell is one element of a table corresponding to a specific row and a specific column. Table cells can hold counts, percentages, or measurements on other variables. Or they can hold several values (p. 702).

Expected Cell Frequency Condition We should expect to see at least 5 individuals in each cell (p. 702).

Chi-square statistic The chi-square statistic can be used to test whether the observed counts in a frequency distribution or contingency table match the counts we would expect according to some model. It is calculated as

$$\chi^2 = \sum_{\text{all cells}} \frac{(Obs - Exp)^2}{Exp}.$$

Chi-square statistics differ in how expected counts are found, depending on the question asked (p. 703).

Chi-square model Chi-square models are skewed to the right. They are parameterized by their degrees of freedom and become less skewed with increasing degrees of freedom (p. 703).

Chi-square component The components of a chi-square calculation are

$$\frac{(Observed - Expected)^2}{Expected},$$

found for each cell of the table (p. 706).

Two-way table Each *cell* of a two-way table shows counts of individuals. One way classifies a sample according to a categorical variable. The other way can classify different groups of individuals according to the same variable or classify the same individuals according to a different categorical variable (p. 708).

Chi-square test of homogeneity A test comparing the distribution of counts for *two or more groups* on the same categorical variable. A chi-square test of homogeneity finds expected counts based on the overall frequencies, adjusted for the totals in each group under the (null hypothesis) assumption that the distributions are the same for each group. We find a P-value from a chi-square distribution with $(\#Rows - 1) \times (\#Cols - 1)$ degrees of freedom, where *#Rows* gives the number of categories and *#Cols* gives the number of independent groups (p. 709).

Standardized residual In each cell of a table of counts, a standardized residual is the square root of the chi-square component for that cell with the sign of the *Observed* − *Expected* difference:

$$\frac{(Obs - Exp)}{\sqrt{Exp}}.$$

When we reject a chi-square test, an examination of the standardized residuals can sometimes reveal more about how the data deviate from the null model (p. 712).

Contingency table A two-way table that classifies individuals according to two categorical variables (p. 713).

Chi-square test of independence A test of whether two categorical variables are independent examines the distribution of counts for *one group of individuals* classified according to both variables. A chi-square test of *independence* finds expected counts by assuming that knowing the marginal totals tells us the cell frequencies, assuming that there is no association between the variables. This turns out to be the same calculation as a test of homogeneity. We find a P-value from a chi-square distribution with $(\#Rows - 1) \times (\#Cols - 1)$ degrees of freedom, where *#Rows* gives the number of categories in one variable and *#Cols* gives the number of categories in the other (p. 714).

TECH SUPPORT

Chi-Square

Most statistics packages associate chi-square tests with contingency tables. Often chi-square is available as an option only when you make a contingency table. This organization can make it hard to locate the chi-square test and may confuse the three different roles that the chi-square test can take. In particular, chi-square tests for goodness-of-fit may be hard to find or missing entirely. Chi-square tests for homogeneity are computationally the same as chi-square tests for independence, so you may have to perform the mechanics as if they were tests of independence and interpret them afterwards as tests of homogeneity.

Most statistics packages work with data on individuals rather than with the summary counts. If the only information you have is the table of counts, you may find it more difficult to get a statistics package to compute chi-square. Some packages offer a way to reconstruct the data from the summary counts so that they can then be passed back through the chi-square calculation, finding the cell counts again. Many packages offer chi-square standardized residuals (although they may be called something else).

DATA DESK

- Select the contingency table icon in the tool bar.
- Drag variables into the table specifying which is the row and which is the column variable.
- From the table's HyperView menu, choose **Table Options**. (Or Choose **Calc** » **Calculation Options** » **Table options**.)
- In the dialog, check the boxes for **Chi Square** and for **Standardized Residuals**. Data Desk will display the chi-square and its P-value below the table, and the standardized residuals within the table.

COMMENTS

Data Desk automatically treats variables selected for this command as categorical variables even if their elements are numerals. The **Compute Counts** command in the table's HyperView menu will make variables that hold the table contents (as selected in the Table Options dialog), including the standardized residuals.

EXCEL

Excel offers the function

CHITEST(actual_range, expected_range), which computes a chi-square value for homogeneity. Both ranges are of the form UpperLeftcell:LowerRightCell, specifying two rectangular tables that must hold counts (although Excel will not check for integer values). The two tables must be of the same size and shape.

COMMENTS

Excel's documentation claims this is a test for independence and labels the input ranges accordingly, but Excel offers no way to find expected counts, so the function is not particularly useful for testing independence. You can use this function only if you already know both tables of counts or are willing to program additional calculations.

JMP

- From the Analyze menu, select **Fit Y by X**.
- Choose one variable as the Y, response variable, and the other as the X, factor variable. Both selected variables must be Nominal or Ordinal.
- JMP will make a plot and a contingency table. Below the contingency table, JMP offers a Tests panel. In that panel, the Chi Square for independence is called a Pearson ChiSquare. The table also offers the P-value.
- Click on the **Contingency Table** title bar to drop down a menu that offers to include a Deviation and Cell Chi square in each cell of the table.

COMMENTS

JMP will choose a chi-square analysis for a Fit Y by X if both variables are nominal or ordinal (marked with an N or O), but not otherwise. Be sure the variables have the right type.

Deviations are the observed − expected differences in counts. Cell chi-squares are the squares of the standardized residuals. Refer to the deviations for the sign of the difference.

Look under Distributions in the Analyze menu to find a chi-square test for goodness-of-fit.

MINITAB

- From the Stat menu, choose the **Tables** submenu.
- From that menu, choose **Chi Square Test** . . . .
- In the dialog, identify the columns that make up the table. Minitab will display the table and print the chi-square value and its P-value.

COMMENTS

Alternatively, select the **Cross Tabulation** . . . command to see more options for the table, including expected counts and standardized residuals.

R

Goodness-of-fit test:

To test the probabilities in a vector prob with observed values in a vector x:

- **chisq.test**(x,p=prob)

Test of independence or homogeneity:

With counts in a contingency table (a matrix called, say, con.table), the test of independence (or homogeneity) is found by

- **chisq.test**(con.table)

COMMENTS

Using the function xtabs you can create a contingency table from two variables x and y in a data frame called mydata by

- con.table = xtabs(~x+y,data=mydata) then
- **chisq.test**(con.table)

SPSS

- From the Analyze menu, choose the **Descriptive Statistics** submenu.
- From that submenu, choose **Crosstabs** . . . .
- In the Crosstabs dialog, assign the row and column variables from the variable list. Both variables must be categorical.
- Click the **Cells** button to specify that standardized residuals should be displayed.
- Click the **Statistics** button to specify a chi-square test.

COMMENTS

SPSS offers only variables that it knows to be categorical in the variable list for the Crosstabs dialog. If the variables you want are missing, check that they have the right type.

STATCRUNCH

To perform a goodness-of-fit test:

- Enter the observed counts in one column of a data table and the expected counts in another.
- Click on **Stat**.
- Choose **Goodness-of-fit » Chi-Square test**.
- Choose the **Observed Column** and the **Expected Column**.
- Click on **Compute!**

COMMENTS

These chi-square tests may also be performed using the actual data table instead of summary counts. See the StatCrunch Help page for details.

To perform a test of homogeneity or independence:

- Create a table (without totals):
- Name the first column as one variable, enter the categories underneath.
- Name the adjacent columns as the categories of the other variable, entering the observed counts underneath.
- Click on **Stat**.
- Choose **Tables » Contingency » with summary**.
- Choose the **Columns** holding counts.
- Choose the **Row labels column**.
- Enter the **Column variable** name.
- Click on **Compute!**

TI-83/84 PLUS

To test goodness-of-fit:

- Enter the observed counts in one list and the expected counts in another list. Expected counts can be entered as n*p and the calculator will compute them for you.
- From the STAT TESTS menu, select χ^2 **GOF-Test**. Specify the lists where you stored the observed and expected counts.
- Enter the number of degrees of freedom, and then hit **Calculate**.

COMMENTS

The cell-by-cell components will be calculated automatically and stored in a list named **CNTRB**.

EXERCISES

SECTION 22.1

1. Human births If there is no seasonal effect on human births, we would expect equal numbers of children to be born in each season (winter, spring, summer, and fall). A student takes a census of her statistics class and finds that of the 120 students in the class, 27 were born in winter, 37 in spring, 31 in summer, and 25 in fall. She wonders if the excess in the spring is an indication that births are not uniform throughout the year.

a) What is the expected number of births in each season if there is no "seasonal effect" on births?
b) Compute the χ^2 statistic.
c) How many degrees of freedom does the χ^2 statistic have?

2. Bank cards At a major credit card bank, the percentages of people who historically apply for the Silver, Gold, and Platinum cards are 60%, 30%, and 10%, respectively. In a recent sample of customers responding to a promotion, of 200 customers, 111 applied for Silver, 59 for Gold, and 30 for Platinum. Is there evidence to suggest that the percentages for this promotion may be different from the historical proportions?

a) What is the expected number of customers applying for each type of card in this sample if the historical proportions are still true?
b) Compute the χ^2 statistic.
c) How many degrees of freedom does the χ^2 statistic have?

3. Human births, again For the births in Exercise 1,

a) If there is no seasonal effect, about how big, on average, would you expect the χ^2 statistic to be (what is the mean of the χ^2 distribution)?
b) Does the statistic you computed in Exercise 1 seem large in comparison to this mean? Explain briefly.
c) What does that say about the null hypothesis?
d) Find the $\alpha = 0.05$ critical value for the χ^2 distribution with the appropriate number of df.
e) Using the critical value, what do you conclude about the null hypothesis at $\alpha = 0.05$?

4. Bank cards, again For the customers in Exercise 2,

a) If the customers apply for the three cards according to the historical proportions, about how big, on average, would you expect the χ^2 statistic to be (what is the mean of the χ^2 distribution)?
b) Does the statistic you computed in Exercise 2 seem large in comparison to this mean? Explain briefly.
c) What does that say about the null hypothesis?
d) Find the $\alpha = 0.05$ critical value for the χ^2 distribution with the appropriate number of df.
e) Using the critical value, what do you conclude about the null hypothesis at $\alpha = 0.05$?

SECTION 22.2

5. Customer ages An analyst at a local bank wonders if the age distribution of customers coming for service at his branch in town is the same as at the branch located near the mall. He selects 100 transactions at random from each branch and researches the age information for the associated customer. Here are the data:

	Age			
	Less Than 30	**30–55**	**56 or Older**	**Total**
In-Town Branch	20	35	45	**100**
Mall Branch	35	55	10	**100**
Total	**55**	**90**	**55**	**200**

a) What is the null hypothesis?
b) What type of test is this?
c) What are the expected numbers for each cell if the null hypothesis is true?
d) Find the χ^2 statistic.
e) How many degrees of freedom does it have?
f) Find the P-value.
g) What do you conclude?

6. Bank cards, once more A market researcher working for the bank in Exercise 2 wants to know if the distribution of applications by card is the same for the past three mailings. She takes a random sample of 200 from each mailing and counts the number applying for Silver, Gold, and Platinum. The data follow:

	Type of Card			
	Silver	**Gold**	**Platinum**	**Total**
Mailing 1	120	50	30	**200**
Mailing 2	115	50	35	**200**
Mailing 3	105	55	40	**200**
Total	**340**	**155**	**105**	**600**

a) What is the null hypothesis?
b) What type of test is this?
c) What are the expected numbers for each cell if the null hypothesis is true?
d) Find the χ^2 statistic.
e) How many degrees of freedom does it have?
f) Find the P-value.
g) What do you conclude?

SECTION 22.3

7. Human births, last time For the data in Exercise 1,

a) Compute the standardized residual for each season.
b) Are any of these particularly large? (Compared to what?)
c) Why should you have anticipated the answer to part b?

8. Bank cards, last time For the data in Exercise 2,

a) Compute the standardized residual for each type of card.
b) Are any of these particularly large? (Compared to what?)
c) What does the answer to part b say about this new group of customers?

SECTION 22.4

9. *Iliad* injuries 800 BCE Homer's *Iliad* is an epic poem, compiled around 800 BCE, that describes several weeks of the last year of the 10-year siege of Troy (Ilion) by the Achaeans. The story centers on the rage of the great warrior Achilles. But it includes many details of injuries and outcomes and is thus the oldest record of Greek medicine. Here is a table of 135 of the recorded injuries for which both injury site and outcome are provided in the *Iliad*. Are some kinds of injuries more lethal than others?

		Lethal?		
		Fatal	Not fatal	Total
Injury Site	Body	56	6	62
	Head/neck	43	1	44
	Limb	10	19	29
	Total	109	26	135

a) Under the null hypothesis, what are the expected values?
b) Compute the χ^2 statistic.
c) How many degrees of freedom does it have?
d) Find the P-value.
e) What do you conclude?

10. *Iliad* weapons The *Iliad* also reports the cause of many injuries. Here is a table summarizing those reports for the 152 injuries for which the *Iliad* provides that information. Is there an association?

		Injury Site			
		Body	Head/Neck	Limb	Total
Weapon	Arrow	5	2	5	12
	Ground/Rock	1	5	5	11
	Sword	61	38	23	122
	Total	67	45	33	145

a) Under the null hypothesis, what are the expected values?
b) Compute the χ^2 statistic.
c) How many degrees of freedom does it have?
d) Find the P-value.
e) What do you conclude?

CHAPTER EXERCISES

11. Which test? For each of the following situations, state whether you'd use a chi-square goodness-of-fit test, a chi-square test of homogeneity, a chi-square test of independence, or some other statistical test:

a) A brokerage firm wants to see whether the type of account a customer has (Silver, Gold, or Platinum) affects the type of trades that customer makes (in person, by phone, or on the Internet). It collects a random sample of trades made for its customers over the past year and performs a test.
b) That brokerage firm also wants to know if the type of account affects the size of the account (in dollars). It performs a test to see if the mean size of the account is the same for the three account types.
c) The academic research office at a large community college wants to see whether the distribution of courses chosen (Humanities, Social Science, or Science) is different for its residential and nonresidential students. It assembles last semester's data and performs a test.

12. Which test, again? For each of the following situations, state whether you'd use a chi-square goodness-of-fit test, a chi-square test of homogeneity, a chi-square test of independence, or some other statistical test:

a) Is the quality of a car affected by what day it was built? A car manufacturer examines a random sample of the warranty claims filed over the past two years to test whether defects are randomly distributed across days of the workweek.
b) A medical researcher wants to know if blood cholesterol level is related to heart disease. She examines a database of 10,000 patients, testing whether the cholesterol level (in milligrams) is related to whether or not a person has heart disease.
c) A student wants to find out whether political leaning (liberal, moderate, or conservative) is related to choice of major. He surveys 500 randomly chosen students and performs a test.

13. Dice After getting trounced by your little brother in a children's game, you suspect the die he gave you to roll may be unfair. To check, you roll it 96 times, recording the number of times each face appears. Do these results cast doubt on the die's fairness?

Face	Count
1	8
2	24
3	24
4	19
5	8
6	13

a) If the die is fair, how many times would you expect each face to show?
b) To see if these results are unusual, will you test goodness-of-fit, homogeneity, or independence?
c) State your hypotheses.
d) Check the conditions.
e) How many degrees of freedom are there?
f) Find χ^2 and the P-value.
g) State your conclusion.

14. A candy company says that until very recently yellow candies made up 20% of its candies, red another 20% and orange, blue, and green 10% each. The rest are brown. On his way home from work the day he was writing these exercises, one of the authors bought a bag of plain candies. He got 27 yellow ones, 21 red, 15 orange, 14 blue, 7 green, and 21 brown. Is this sample consistent with the company's stated proportions? Test an appropriate hypothesis, and state your conclusion.

a) If the candies are packaged in the stated proportions, how many of each color should the author have expected to get in his bag?
b) To see if his bag was unusual, should he test goodness-of-fit, homogeneity, or independence?
c) State the hypotheses.
d) Check the conditions.
e) How many degrees of freedom are there?
f) Find χ^2 and the P-value.
g) State a conclusion.

15. Nuts A company says its premium mixture of nuts contains 12% Brazil nuts, 21% cashews, 18% almonds, and 10% hazelnuts, and the rest are peanuts. You buy a large can and separate the various kinds of nuts. On weighing them, you find there are 110 grams of Brazil nuts, 184 grams of cashews, 207 grams of almonds, 77 grams of hazelnuts, and 448 grams of peanuts. You wonder whether your mix is significantly different from what the company advertises.

a) Explain why the chi-square goodness-of-fit test is not an appropriate way to find out.

b) What might you do instead of weighing the nuts in order to use a χ^2 test?

16. Mileage A salesman who is on the road visiting clients thinks that, on average, he drives the same distance each day of the week. He keeps track of his mileage for several weeks and discovers that he averages 123 miles on Mondays, 196 miles on Tuesdays, 168 miles on Wednesdays, 173 miles on Thursdays, and 113 miles on Fridays. He wonders if this evidence contradicts his belief in a uniform distribution of miles across the days of the week. Explain why it is not appropriate to test his hypothesis using the chi-square goodness-of-fit test.

17. City police and race Census data for a city indicate that 26.7% of the under-18 population is white, 29.7% black, 31.8% Latino, 9.7% Asian, and 2.1% other ethnicities. A civil body in the city points out that, of 26,181 police officers, 64.8% are white, 14.5% black, 19.1% Latino, 1.4% Asian, and the remaining are other ethnicities. Do the police officers reflect the ethnic composition of the city's youth? Test an appropriate hypothesis, and state your conclusion.

18. Violence against women In its study *When Men Murder Women: An Analysis of 2009 Homicide Data*, 2011, the Violence Policy Center (www.vpc.org) reported that 1818 women were murdered by men in 2009. Of these victims, a weapon could be identified for 1654 of them. Of those for whom a weapon could be identified, 861 were killed by guns, 364 by knives or other cutting instruments, 214 by other weapons, and 215 by personal attack (battery, strangulation, etc.). The FBI's Uniform Crime Report says that, among all murders nationwide, the weapon use rates were as follows: guns 63.4%, knives 13.1%, other weapons 16.8%, personal attack 6.7%. Is there evidence that violence against women involves different weapons than other violent attacks in the United States?

19. Fruit flies Offspring of certain fruit flies may have ebony or green bodies and normal wings or short wings. Genetic theory predicts that these traits will appear in the ratio 9:3:3:1 (9 ebony, normal: 3 ebony, short: 3 green, normal: 1 green, short). A researcher checks 100 such flies and finds the distribution of the traits to be 61, 20, 10, and 9, respectively.

a) Are the results this researcher observed consistent with the theoretical distribution predicted by the genetic model?

b) If the researcher had examined 200 flies and counted exactly twice as many in each category—122, 40, 20, 18—what conclusion would he have reached?

c) Why is there a discrepancy between the two conclusions?

T **20. Pi** Many people know the mathematical constant π is approximately 3.14. But that's not exact. To be more precise, here are 20 decimal places: 3.14159265358979323846. Still not exact, though. In fact, the actual value is irrational, a decimal that goes on forever without any repeating pattern. But notice that there are no 0's and only one 7 in the 20 decimal places above. Does that pattern persist, or do all the digits show up with equal frequency? The table shows the number of times each digit appears in the first million digits. Test the hypothesis that the digits 0 through 9 are uniformly distributed in the decimal representation of π.

The First Million Digits of π

Digit	Count
0	99,959
1	99,758
2	100,026
3	100,229
4	100,230
5	100,359
6	99,548
7	99,800
8	99,985
9	100,106

T **21. Hurricane frequencies** The National Hurricane Center provides data that list the numbers of large (category 3, 4, or 5) hurricanes that have struck the United States, by decade since 1851 (www.nhc.noaa.gov/dcmi.shtml). The data are given below.

Decade	Count	Decade	Count
1851–1860	6	1931–1940	8
1861–1870	1	1941–1950	10
1871–1880	7	1951–1960	9
1881–1890	5	1961–1970	6
1891–1900	8	1971–1980	4
1901–1910	4	1981–1990	4
1911–1920	7	1991–2000	5
1921–1930	5	2001–2010	7

Recently, there's been some concern that perhaps the number of large hurricanes has been increasing. The natural null hypothesis would be that the frequency of such hurricanes has remained constant.

a) With 96 large hurricanes observed over the 16 periods, what are the expected value(s) for each cell?

b) What kind of chi-square test would be appropriate?

c) State the null and alternative hypotheses.

d) How many degrees of freedom are there?

e) The value of χ^2 is 12.67. What's the P-value?

f) State your conclusion.

T **22. Lottery numbers** The fairness of the South African lottery was recently challenged by one of the country's political parties. The lottery publishes historical statistics at its website (www.nationallottery.co.za). Here is a table of the number of times each number appeared in the lottery and as the "Powerball" number as of June 2007:

Number	Lotto	Powerball	Number	Lotto	Powerball
1	81	14	26	78	12
2	91	16	27	83	16
3	78	14	28	76	7
4	77	12	29	76	12
5	67	16	30	99	16
6	87	12	31	78	10
7	88	15	32	73	15
8	90	16	33	81	14
9	80	9	34	81	13
10	77	19	35	77	15
11	84	12	36	73	8
12	68	14	37	64	17
13	79	9	38	70	11
14	90	12	39	67	14
15	82	9	40	75	13
16	103	15	41	84	11
17	78	14	42	79	8
18	85	14	43	74	14
19	67	18	44	87	14
20	90	13	45	82	19
21	77	13	46	91	10
22	78	17	47	86	16
23	90	14	48	88	21
24	80	8	49	76	13
25	65	11			

We wonder if all the numbers are equally likely to be the "Powerball."

a) What kind of test should we perform?
b) There are 655 Powerball observations. What are the appropriate expected value(s) for the test?
c) State the null and alternative hypotheses.
d) How many degrees of freedom are there?
e) The value of χ^2 is 34.5. What's the P-value?
f) State your conclusion.

23. Childbirth, part 1 There is some concern that if a woman has an epidural to reduce pain during childbirth, the drug can get into the baby's bloodstream, making the baby sleepier and less willing to breastfeed. The *International Breastfeeding Journal* published results of a study conducted at Sydney University. Researchers followed up on 1178 births, noting whether the mother had an epidural and whether the baby was still nursing after 6 months. Below are their results.

a) What kind of test would be appropriate?
b) State the null and alternative hypotheses.

		Epidural?		
		Yes	No	Total
Breastfeeding at 6 Months?	Yes	206	498	704
	No	190	284	474
	Total	396	782	1178

24. Does your doctor know? A survey[6] of articles from the *New England Journal of Medicine (NEJM)* classified them according to the principal statistics methods used. The articles recorded were all noneditorial articles appearing during the indicated years. Let's just look at whether these articles used statistics at all.

	Publication Year			
	1978–79	1989	2004–05	Total
No Stats	90	14	40	144
Stats	242	101	271	614
Total	332	115	311	758

Has there been a change in the use of statistics?

a) What kind of test would be appropriate?
b) State the null and alternative hypotheses.

25. Childbirth, part 2 In Exercise 23, the table shows results of a study investigating whether aftereffects of epidurals administered during childbirth might interfere with successful breastfeeding. We're planning to do a chi-square test.

a) How many degrees of freedom are there?
b) The smallest expected count will be in the epidural/no breastfeeding cell. What is it?
c) Check the assumptions and conditions for inference.

26. Does your doctor know? (part 2) The table in Exercise 24 shows whether *NEJM* medical articles during various time periods included statistics or not. We're planning to do a chi-square test.

a) How many degrees of freedom are there?
b) The smallest expected count will be in the 1989/No cell. What is it?
c) Check the assumptions and conditions for inference.

27. Childbirth, part 3 In Exercises 23 and 25, we've begun to examine the possible impact of epidurals on successful breastfeeding.

a) Calculate the component of chi-square for the epidural/no breastfeeding cell.
b) For this test, $\chi^2 = 14.87$. What's the P-value?
c) State your conclusion.

28. Does your doctor know? (part 3) In Exercises 24 and 26, we've begun to examine whether the use of statistics in *NEJM* medical articles has changed over time.

a) Calculate the component of chi-square for the 1989/No cell.
b) For this test, $\chi^2 = 25.28$. What's the P-value?
c) State your conclusion.

29. Childbirth, part 4 In Exercises 23, 25, and 27, we've tested a hypothesis about the impact of epidurals on successful breastfeeding. The following table shows the test's residuals.

		Epidural?	
		Yes	No
Breastfeeding at 6 Months?	Yes	−1.99	1.42
	No	2.43	−1.73

a) Show how the residual for the epidural/no breastfeeding cell was calculated.
b) What can you conclude from the standardized residuals?

[6] Suzanne S. Switzer and Nicholas J. Horton, "What Your Doctor Should Know about Statistics (but Perhaps Doesn't)," *Chance*, 20:1, 2007.

30. Does your doctor know? (part 4) In Exercises 24, 26, and 28, we've tested a hypothesis about whether the use of statistics in *NEJM* medical articles has changed over time. The table shows the test's residuals.

	1978–79	1989	2004–05
No Stats	3.39	−1.68	−2.48
Stats	−1.64	0.81	1.20

a) Show how the residual for the 1989/No cell was calculated.
b) What can you conclude from the patterns in the standardized residuals?

31. Childbirth, part 5 In Exercises 23, 25, 27, and 29, we've looked at a study examining epidurals as one factor that might inhibit successful breastfeeding of newborn babies. Suppose a broader study included several additional issues, including whether the mother drank alcohol, whether this was a first child, and whether the parents occasionally supplemented breastfeeding with bottled formula. Why would it not be appropriate to use chi-square methods on the 2 × 8 table with yes/no columns for each potential factor?

32. Does your doctor know? (part 5) In Exercises 24, 26, 28, and 30, we considered data on articles in the *NEJM*. The original study listed 23 different statistics methods. (The list read: *t*-tests, contingency tables, linear regression, ...) Why would it not be appropriate to use a chi-square test on the 23 × 3 table with a row for each method?

T **33. *Titanic*** Here is a table we first saw in Chapter 2 showing who survived the sinking of the *Titanic* based on whether they were crew members, or passengers booked in first-, second-, or third-class staterooms:

	Crew	First	Second	Third	Total
Alive	212	201	119	180	712
Dead	677	123	166	530	1496
Total	889	324	285	710	2208

a) If we draw an individual at random, what's the probability that we will draw a member of the crew?
b) What's the probability of randomly selecting a third-class passenger who survived?
c) What's the probability of a randomly selected passenger surviving, given that the passenger was a first-class passenger?
d) If someone's chances of surviving were the same regardless of their status on the ship, how many members of the crew would you expect to have lived?
e) Are *Survival* and *Ticket Class* related? State the null and alternative hypotheses.
f) Give the degrees of freedom for the test.
g) The chi-square value for the table is 187.6 and the corresponding P-value is barely greater than 0. State your conclusions about the hypotheses.

T **34. NYPD** The table below shows the rank attained by male and female officers in the New York City Police Department (NYPD). Do these data indicate that men and women are equitably represented at all levels of the department?

Rank	Male	Female
Officer	21,900	4,281
Detective	4,058	806
Sergeant	3,898	415
Lieutenant	1,333	89
Captain	359	12
Higher Ranks	218	10

a) What's the probability that a person selected at random from the NYPD is a female?
b) What's the probability that a person selected at random from the NYPD is a detective?
c) Assuming no bias in promotions, how many female detectives would you expect the NYPD to have?
d) To see if there is evidence of differences in ranks attained by males and females, will you test goodness-of-fit, homogeneity, or independence?
e) State the hypotheses.
f) Check the conditions.
g) How many degrees of freedom are there?
h) The chi-square value for the table is 290.1 and the P-value is less than 0.0001. State your conclusion about the hypotheses.

T **35. *Titanic*, again** Examine and comment on this table of the standardized residuals for the chi-square test you looked at in Exercise 33.

	Crew	First	Second	Third
Alive	−4.41	9.44	2.83	−3.24
Dead	3.04	−6.51	−1.95	2.23

T **36. NYPD again** Examine and comment on this table of the standardized residuals for the chi-square test you looked at in Exercise 34.

	Male	Female
Officer	−2.34	5.57
Detective	−1.18	2.80
Sergeant	3.84	−9.14
Lieutenant	3.58	−8.52
Captain	2.46	−5.86
Higher Ranks	1.74	−4.14

T **37. Cranberry juice** It's common folk wisdom that drinking cranberry juice can help prevent urinary tract infections in women. In 2001, the *British Medical Journal* reported the results of a Finnish study in which three groups of 50 women were monitored for these infections over 6 months. One group drank cranberry juice daily, another group drank a lactobacillus drink, and the third drank neither of those beverages, serving as a control group. In the control group, 18 women developed at least one infection, compared to 20 of those who consumed the lactobacillus drink and only 8 of those who drank cranberry juice. Does this study provide supporting evidence for the value of cranberry juice in warding off urinary tract infections?

a) Is this a survey, a retrospective study, a prospective study, or an experiment? Explain.
b) Will you test goodness-of-fit, homogeneity, or independence?

c) State the hypotheses.
d) Check the conditions.
e) How many degrees of freedom are there?
f) Find χ^2 and the P-value.
g) State your conclusion.
h) If you concluded that the groups are not the same, analyze the differences using the standardized residuals of your calculations.

T **38. Car origins** A random survey of autos parked in the student lot and the staff lot at a large university classified the brands by country of origin, as seen in the table. Are there differences in the national origins of cars driven by students and staff?

		Driver	
		Student	Staff
Origin	American	107	105
	European	33	12
	Asian	55	47

a) Is this a test of independence or homogeneity?
b) Write appropriate hypotheses.
c) Check the necessary assumptions and conditions.
d) Find the P-value of your test.
e) State your conclusion and analysis.

T **39. Montana** A poll conducted by the University of Montana classified respondents by whether they were male or female and political party, as shown in the table. We wonder if there is evidence of an association between being male or female and party affiliation.

	Democrat	Republican	Independent
Male	36	45	24
Female	48	33	16

a) Is this a test of homogeneity or independence?
b) Write an appropriate hypothesis.
c) Are the conditions for inference satisfied?
d) Find the P-value for your test.
e) State a complete conclusion.

T **40. Fish diet** Medical researchers followed 6272 Swedish men for 30 years to see if there was any association between the amount of fish in their diet and prostate cancer. ("Fatty Fish Consumption and Risk of Prostate Cancer," *Lancet*, June 2001)

Fish Consumption	No Prostate Cancer	Prostate Cancers
Never/Seldom	110	14
Small Part of Diet	2420	201
Moderate Part	2769	209
Large Part	507	42

a) Is this a survey, a retrospective study, a prospective study, or an experiment? Explain.
b) Is this a test of homogeneity or independence?
c) Do you see evidence of an association between the amount of fish in a man's diet and his risk of developing prostate cancer?
d) Does this study prove that eating fish does not prevent prostate cancer? Explain.

T **41. Montana revisited** The poll described in Exercise 39 also investigated the respondents' party affiliations based on what area of the state they lived in. Test an appropriate hypothesis about this table and state your conclusions. (Data in **Montana revisited**)

	Democrat	Republican	Independent
West	39	17	12
Northeast	15	30	12
Southeast	30	31	16

T **42. Working parents** In April 2009, Gallup published results from data collected from a large sample of adults in the 27 European Union member states. One of the questions asked was, "Which is the most practicable and realistic option for child care, taking into account the need to earn a living?" The counts below are representative of the entire collection of responses.

	Male	Female
Both Parents Work Full Time	161	140
One Works Full Time, Other Part Time	259	308
One Works Full Time, Other Stays Home for Kids	189	161
Both Parents Work Part Time	49	63
No Opinion	42	28

Source: www.gallup.com/poll/117358/Work-Life-Balance-Tilts-Against-Women-Single-Parents.aspx

a) Is this a survey, a retrospective study, a prospective study, or an experiment?
b) Will you test goodness-of-fit, homogeneity, or independence?
c) Based on these results, do you think men and women have differing opinions when it comes to raising children?

43. Grades Two different professors teach an introductory statistics course. The table shows the distribution of final grades they reported. We wonder whether one of these professors is an "easier" grader.

	Prof. Alpha	Prof. Beta
A	3	9
B	11	12
C	14	8
D	9	2
F	3	1

a) Will you test goodness-of-fit, homogeneity, or independence?
b) Write appropriate hypotheses.
c) Find the expected counts for each cell, and explain why the chi-square procedures are not appropriate.

T **44. Full moon** Some people believe that a full moon elicits unusual behavior in people. The table shows the number of arrests made in a small town during weeks of six full moons and six other randomly selected weeks in the same year. We wonder if there is evidence of a difference in the types of illegal activity that take place.

	Full Moon	Not Full
Violent (murder, assault, rape, etc.)	2	3
Property (burglary, vandalism, etc.)	17	21
Drugs/Alcohol	27	19
Domestic Abuse	11	14
Other Offenses	9	6

a) Will you test goodness-of-fit, homogeneity, or independence?
b) Write appropriate hypotheses.
c) Find the expected counts for each cell, and explain why the chi-square procedures are not appropriate.

45. Grades, again In some situations where the expected cell counts are too small, as in the case of the grades given by Professors Alpha and Beta in Exercise 43, we can complete an analysis anyway. We can often proceed after combining cells in some way that makes sense and also produces a table in which the conditions are satisfied. Here, we create a new table displaying the same data, but calling D's and F's "Below C":

	Prof. Alpha	Prof. Beta
A	3	9
B	11	12
C	14	8
Below C	12	3

a) Find the expected counts for each cell in this new table, and explain why a chi-square procedure is now appropriate.
b) With this change in the table, what has happened to the number of degrees of freedom?
c) Test your hypothesis about the two professors, and state an appropriate conclusion.

T **46. Full moon, next phase** In Exercise 44, you found that the expected cell counts failed to satisfy the conditions for inference.

a) Find a sensible way to combine some cells that will make the expected counts acceptable.
b) Test a hypothesis about the full moon and state your conclusion.

T **47. Racial steering** A subtle form of racial discrimination in housing is "racial steering." Racial steering occurs when real estate agents show prospective buyers only homes in neighborhoods already dominated by that family's race. This violates the Fair Housing Act of 1968. According to an article in *Chance* magazine (Vol. 14, no. 2 [2001]), tenants at a large apartment complex filed a lawsuit alleging racial steering. The complex is divided into two parts: Section A and Section B. The plaintiffs claimed that white potential renters were steered to Section A, while African Americans were steered to Section B. The table displays the data that were presented in court to show the locations of recently rented apartments. Do you think there is evidence of racial steering?

New Renters			
	White	Black	Total
Section A	87	8	95
Section B	83	34	117
Total	170	42	212

T **48. Survival on the *Titanic*** Newspaper headlines at the time, and traditional wisdom in the succeeding decades, have held that women and children escaped the *Titanic* in greater proportions than men. Here's a table with the relevant data. Do you think that survival was independent of whether the person was male or female? Explain.

	Female	Male	Total
Alive	359	353	712
Dead	130	1366	1496
Total	489	1719	2208

T **49. Pregnancies** Most pregnancies are full term, but some are preterm (less than 37 weeks). Of those that are preterm, the Centers for Disease Control and Prevention classifies them as early (less than 34 weeks) and late (34 to 36 weeks). A December 2010 National Vital Statistics Report examined those outcomes in the United States broken down by age of the mother. The table shows counts consistent with that report. Is there evidence that the outcomes are not independent of age group?

	Early Preterm	Late Preterm
Under 20	129	270
20 to 29	243	612
30 to 39	165	424
40 or Over	18	39

Source: www.cdc.gov/nchs/data/nvsr/nvsr59/nvsr59_01.pdf

T **50. Education by age** Use the survey results in the table to investigate differences in education level attained among different age groups in the United States.

	Age Group				
Education Level	25–34	35–44	45–54	55–64	≥65
Not HS Grad	27	50	52	71	101
HS	82	19	88	83	59
1–3 Years College	43	56	26	20	20
≥4 Years College	48	75	34	26	20

JUST CHECKING

Answers

1. We need to know how well beetles can survive 6 hours in a Plexiglas® box so that we have a baseline to compare the treatments.
2. There's no difference in survival rate in the three groups.
3. $(2 - 1)(3 - 1) = 2$ df
4. 50
5. 2
6. The mean value for a χ^2 with 2 df is 2, so 10 seems pretty large. The P-value is probably small.
7. This is a test of homogeneity. The clue is that the question asks whether the distributions are alike.
8. This is a test of goodness-of-fit. We want to test the model of equal assignment to all lots against what actually happened.
9. This is a test of independence. We have responses on two variables for the same individuals.

23 Inferences for Regression

WHERE ARE WE GOING?

A scatterplot of IQ vs. brain size shows a mildly positive association. Could this just be due to chance? A hypothesis test is clearly what we need. We can estimate the slope, but how reliable is our estimate? In this chapter we'll apply what we know about tests and confidence intervals for means to regression.

WHO	250 male subjects
WHAT	Body fat and waist size
UNITS	% Body fat and inches
WHEN	1990s
WHERE	Brigham Young University
WHY	Scientific research

In earlier chapters we looked at the problem of estimating body fat from easy to measure variables like *Height* or *Weight*. We were able to find both a simple regression model with one predictor and multiple regression models with two or more predictors. Now that we've seen statistical inference at work, it seems natural to apply it to regression models.

Here's the scatterplot we saw in Chapter 9 of *%Body Fat* plotted against *Waist* size for a sample of 250 males of various ages.? (Data (still) in **Bodyfat**)

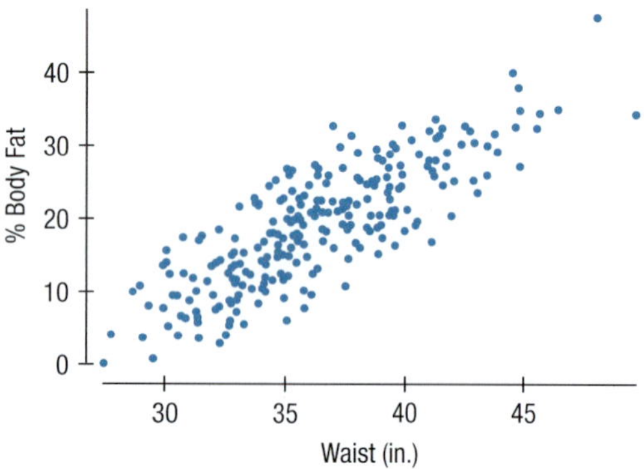

Figure 23.1
Percent Body Fat vs. *Waist* size for 250 men of various ages. The scatterplot shows a strong, positive, linear relationship.

23.1 The Regression Model

Back in Chapter 7, we modeled relationships like this by fitting a least squares line. The plot is clearly straight, so it seems a sensible thing to do. The equation of the least squares line for these data is

$$\widehat{\%Body\ Fat} = -42.7 + 1.7\ Waist.$$

The slope says that *%Body Fat* in men is, on average, greater by 1.7 for each additional inch around the waist.

How useful is this model? When we fit linear models before, we used them only descriptively. We interpreted the slope and intercept as descriptions of the data. But now we'd like to do more. We'd like to understand what the regression model can tell us beyond the 250 men in this study. To do that, we'll want to make confidence intervals and test hypotheses about the slope and intercept of the regression line.

When we found a confidence interval for a mean, we could imagine a single, true underlying value for the mean. When we tested whether two means or two proportions were equal, we imagined a true underlying difference. But what does it mean to do inference for regression? We know better than to think that the data would line up perfectly on a straight line even if we knew every population value. After all, even in our sample, not all men who have 38-inch waists have the same *%Body Fat*. In fact, there's a whole distribution of *%Body Fat* for these men:

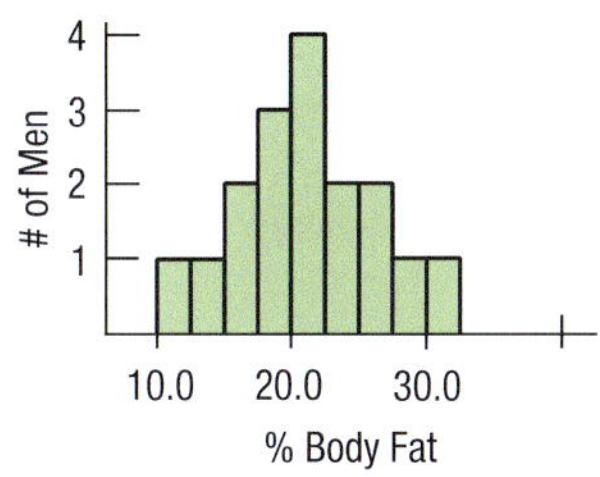

Figure 23.2
The distribution *of %Body Fat* for men with a *Waist* size of 38 inches is unimodal and symmetric.

This is true at each *Waist* size. To model the relationship between *%Body Fat* and *Waist* size for *all* men, we imagine an idealized regression line that goes through the *means* of the distributions of *%Body Fat* for each *Waist* size (see Figure 23.3). Those means fall along the line, even though the individuals are scattered around it. We know that this model is not a perfect description of how the variables are associated, but it may be useful for predicting *%Body Fat* and for understanding how it's related to *Waist* size.

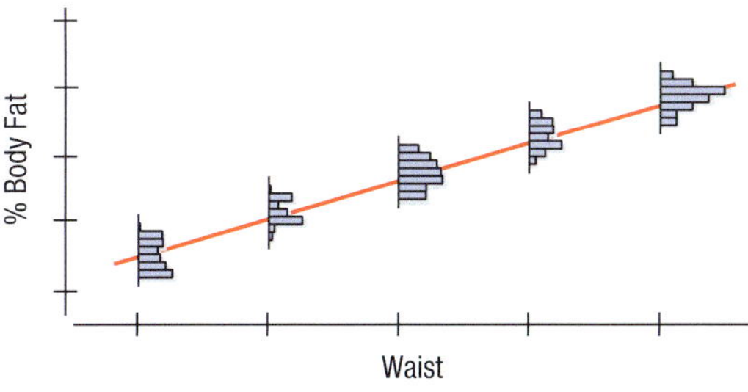

Figure 23.3
There's a distribution of *%Body Fat* for each value of *Waist* size. We'd like the means of these distributions to line up.

NOTATION ALERT

This time we used up only one Greek letter for two things. Lowercase Greek β (beta) is the natural choice to correspond to the b's in the regression equation. We used β before for the probability of a Type II error, but there's little chance of confusion here.

If only we had all the values in the population, we could find the slope and intercept of this *idealized regression line* explicitly by using least squares. Following our usual conventions, we write the idealized line with Greek letters and consider the coefficients (the slope and intercept) to be *parameters:* β_0 is the intercept and β_1 is the slope. Corresponding to our fitted line of $\hat{y} = b_0 + b_1x$, we write $\mu_y = \beta_0 + \beta_1x$. Why μ_y instead of $\hat{y}$? Because this is a model. There is a distribution of *%Body Fat* for each *Waist* size. The model places the *means* of the distributions of *%Body Fat* for each *Waist* size on the same straight line.

Of course, not all the individual y's are at these means. (In fact, the line will miss most—and quite possibly all—of the plotted points.) Some individuals lie above and some below the line, so, like all models, this one makes errors. Lots of them. In fact, one at each point. These errors are random and, of course, can be positive or negative. They are model errors, so we use a Greek letter and denote them by ε.

When we put the errors into the equation, we can account for each individual y-value:

$$y = \beta_0 + \beta_1 x + \varepsilon.$$

This equation is now true for each data point (since there is an ε to soak up the error), so the model gives a (true) value of y for any value of x. (If only we knew the parameters!)

An idealized model such as this provides a summary of the relationship between *%Body Fat* and *Waist* size. Like all models, it simplifies the real situation. Although we know there is more to predicting body fat than waist size alone, this simplification might help us to think about the relationship and assess how well *%Body Fat* can be predicted from simpler measurements.

RANDOM MATTERS Slopes Vary

In Chapter 7 we saw that regression slopes (of bridge safety rating vs. age) vary from sample to sample. And we noted that a histogram of those slopes looked unimodal and symmetric. (Check page 234) Now we know that what the histogram depicted is the sampling distribution of the slopes. We can do the same for the men in the body fat study. For bridges we had the entire population of New York bridges and we took 1000 random samples of 190 bridges from it. Now we don't have measurements on all men, but only a sample of 250 men, so we'll re-sample those 250 as we did for the bootstrap in Chapter 17.

Here is the result of drawing 1000 re-samples of the 250 men and fitting a least squares line to each:

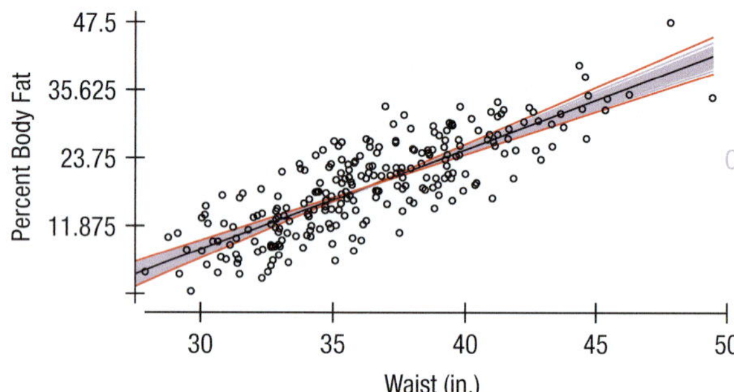

Figure 23.4
1000 re-samples of the 250 men in the body fat study show how the slope of the relationship between *%Body Fat* and size varies from sample to sample.

A histogram of these slopes shows the sampling distribution. Are you surprised to see that (once again) the sampling distribution is unimodal and symmetric and can be modeled well by a Normal?

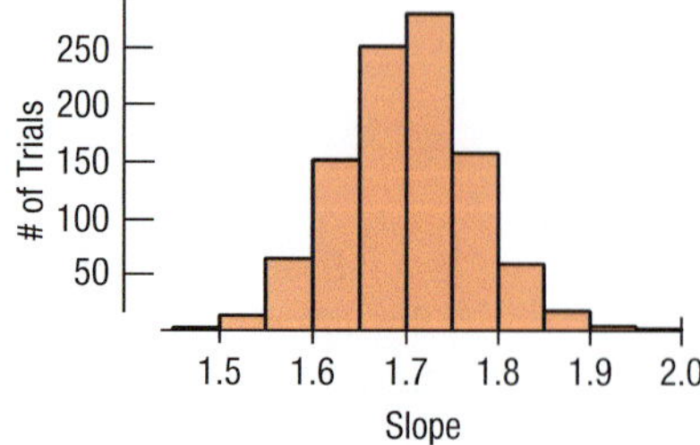

Figure 23.5
The distribution of the slopes of the 1000 re-samples shown in Figure 23.4.

We estimate the β's by finding a regression line, $\hat{y} = b_0 + b_1x$, as we did in Chapter 7. The residuals, $e = y - \hat{y}$, are the sample-based versions of the errors, ε. We'll use them to help us assess the regression model.

We know that slopes and intercepts vary from the true population values. The sample slope and intercept give us a best guess of those values. If the data follow certain assumptions and conditions, we can also estimate the uncertainty of those estimates with confidence intervals and test hypotheses in much the same way we've done for means.

23.2 Assumptions and Conditions

Back in Chapter 7 when we fit lines to data, we needed to check only the Straight Enough Condition. Now, when we want to make inferences about the coefficients of the line, we'll have to make more assumptions. As we've done before with regression, we'll check some basic conditions to make sure it's reasonable to fit the regression and then check other conditions *after* we find the regression equation and residuals.

If our initial assumptions are not true, it makes no sense to check the later ones. So we've numbered the assumptions to keep them in order.

Check the Scatterplot

The shape must be linear or we can't use linear regression at all.

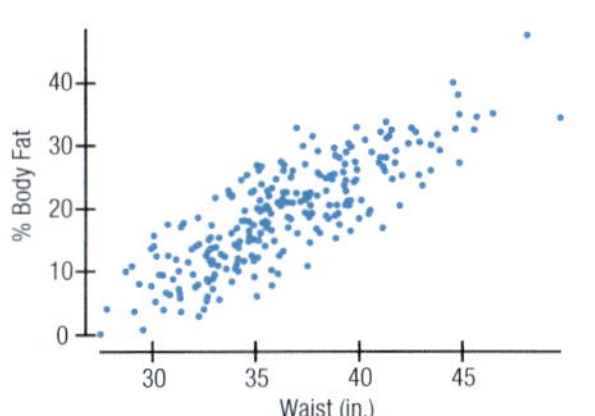

Figure 23.6

The relationship between *%Body Fat* and *Waist* size is straight enough.

1. Linearity Assumption

If the true relationship is far from linear and we use a straight line to fit the data, our entire analysis will be useless, so we always check this first.

The **Straight Enough Condition** is satisfied if a scatterplot looks straight. It's generally not a good idea to draw a line through the scatterplot when checking. That can fool your eyes into seeing the plot as more straight. Sometimes it's easier to see violations of the Straight Enough Condition by looking at a scatterplot of the residuals against x or against the predicted values, $\hat{y}$. That plot will have a horizontal direction and should have no pattern if the condition is satisfied.

If the scatterplot is straight enough, we can go on to some assumptions about the errors. If not, stop here, or consider re-expressing the data to make the scatterplot more nearly linear. (see Chapter 8.) For the *%Body Fat* data, the scatterplot is beautifully linear. Of course, the data must be quantitative for this to make sense. Check the **Quantitative Data Condition**.

Check the Residuals Plot (1)

The residuals should appear to be randomly scattered.

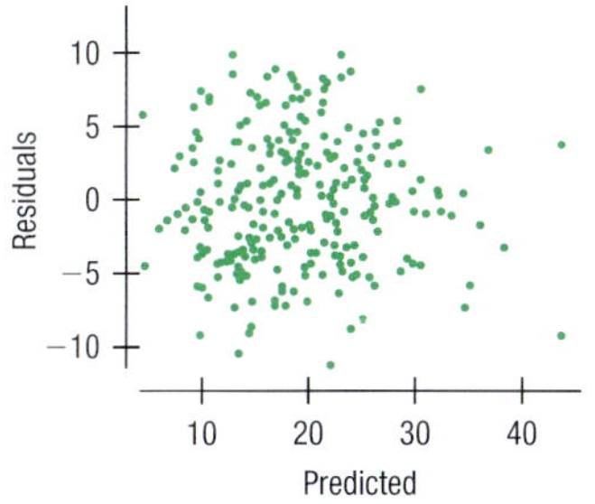

Figure 23.7

The residuals show only random scatter when plotted against the predicted values.

2. Independence Assumption

For inferences to work, the errors in the true underlying regression model (the ε's) must be independent of each other. As usual, there's no way to be sure that the Independence Assumption is true.

Usually when we care about inference for the regression parameters, it's because we think our regression model might apply to a larger population. In such cases, we can check that the individuals are a representative (ideally, random) sample from that population.

We can also check displays of the regression residuals for evidence of patterns, trends, or clumping, any of which would suggest a failure of independence. In the special case when an x-variable is related to time, a common violation of the Independence Assumption is for the errors to be correlated. (The error our model makes today may be similar to the one it made for yesterday.) This violation can be checked by plotting the residuals against the *time* variable and looking for patterns. A further check can be made by plotting the residuals against those same residuals offset by one time position.[1] Neither plot should show patterns.

The *%Body Fat* data were collected on a sample of men taken to be representative. The subjects were not related in any way, so we can be pretty sure that their measurements are independent. The residuals plot in Figure 23.7 shows no pattern.

[1]Such an offset version is said to be "lagged."

Check the Residuals Plot (2)

The vertical spread of the residuals should be roughly the same everywhere.

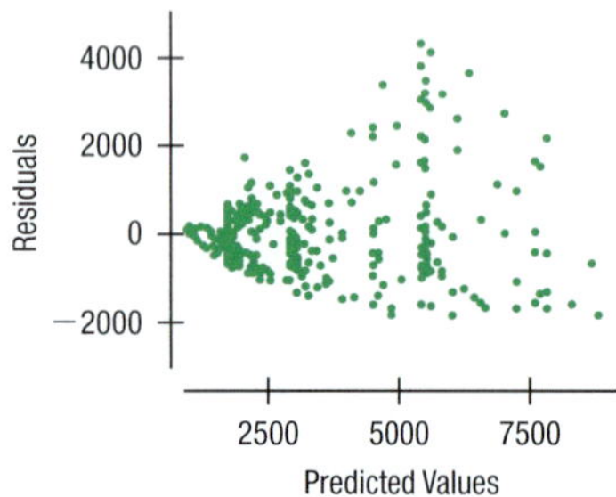

Figure 23.8

A scatterplot of residuals against predicted values can reveal plot thickening. In this plot of the residuals from a regression of diamond prices on carat weight, we see that larger diamonds have more price variation than smaller diamonds. When the Equal Variance Assumption is violated, we can't summarize how the residuals vary with a single number.

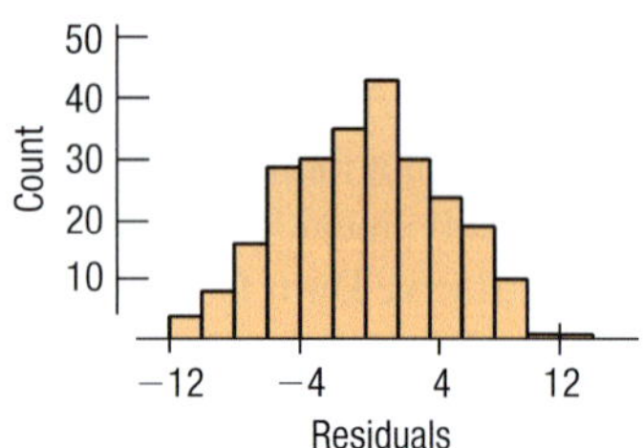

Figure 23.9

A histogram of the residuals is one way to check whether they are nearly Normal. Alternatively, we can look at a Normal probability plot.

Check a Histogram of the Residuals

The distribution of the residuals should be unimodal and symmetric.

3. Equal Variance Assumption

The variability of y should be about the same for all values of the x's. When we first looked at regression in Chapters 7 and 9, we looked at the standard deviation of the residuals (s_e) to measure the size of the scatter. Now we'll need this standard deviation to build confidence intervals and test hypotheses. The standard deviation of the residuals is the building block for the standard errors of all the regression parameters. But it makes sense only if the scatter of the residuals is the same everywhere. In effect, the standard deviation of the residuals "pools" information across all of the individual distributions at each x-value, and pooled estimates are appropriate only when they combine information for groups with the same variance.

Practically, what we can check is the **Does the Plot Thicken? Condition**. A scatterplot of y against the residuals offers a visual check. Fortunately, we've already made one. Make sure the spread across the plot is nearly constant. Be alert for a "fan" shape or other tendency for the variation to grow or shrink in one part of the scatterplot (as in the plot in Figure 23.8).

If the residuals show no special patterns, the data are independent, and the plot doesn't thicken, you can move on to the final assumption.

4. Normal Population Assumption

We assume the errors around the idealized regression line at each value of the x's follow a Normal model. We need this assumption so that we can use a Student's t-model for inference.

As we have at other times when we've used Student's t, we'll settle for the residuals satisfying the **Nearly Normal Condition** and the **Outlier Condition**. Look at a histogram or Normal probability plot of the residuals.[2]

The histogram of residuals in the *%Body Fat* regression certainly looks nearly Normal. As we have noted before, the Normality Assumption becomes less important as the sample size grows, because the model is about means and the Central Limit Theorem takes over.

If all four assumptions were true, the idealized one-predictor regression model would look like this:

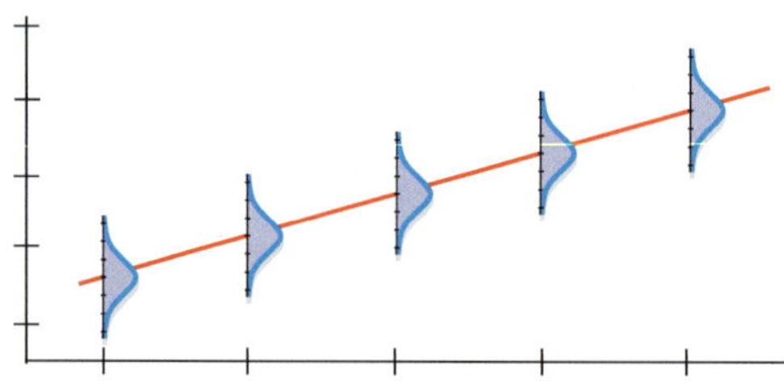

Figure 23.10

The regression model has a distribution of y-values for each x-value. These distributions follow a Normal model with means lined up along the line and with the same standard deviations.

At each value of x, there is a distribution of y-values that follows a Normal model, and each of these Normal models is centered on the line and has the same standard deviation. Of course, we don't expect the assumptions to be exactly true, and we know that all models are wrong, but the linear model is often close enough to be very useful.

[2] *This* is why we have to check the conditions in order. We have to check that the residuals are independent and that the variation is the same for all x's so that we can lump all the residuals together for a single check of the Nearly Normal Condition and so it makes sense to estimate the variance of the residuals.

EXAMPLE 23.1 Checking Assumptions and Conditions

WHO	39 impact craters
WHAT	Diameter and age
UNITS	km and millions of years ago
WHEN	Past 35 million years
WHERE	Worldwide
WHY	Scientific research

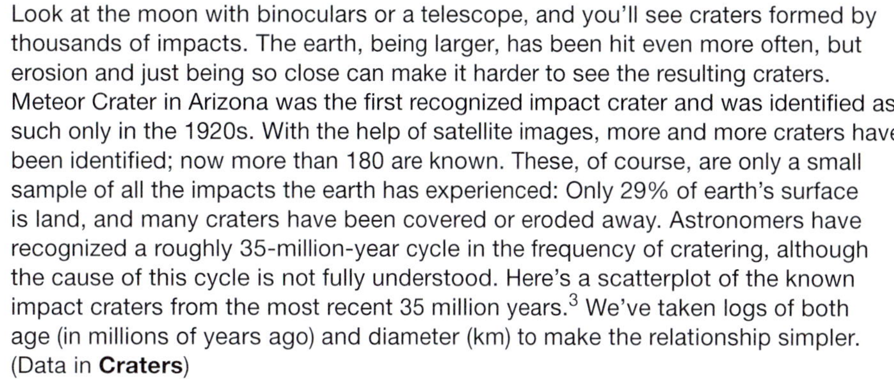

Look at the moon with binoculars or a telescope, and you'll see craters formed by thousands of impacts. The earth, being larger, has been hit even more often, but erosion and just being so close can make it harder to see the resulting craters. Meteor Crater in Arizona was the first recognized impact crater and was identified as such only in the 1920s. With the help of satellite images, more and more craters have been identified; now more than 180 are known. These, of course, are only a small sample of all the impacts the earth has experienced: Only 29% of earth's surface is land, and many craters have been covered or eroded away. Astronomers have recognized a roughly 35-million-year cycle in the frequency of cratering, although the cause of this cycle is not fully understood. Here's a scatterplot of the known impact craters from the most recent 35 million years.[3] We've taken logs of both age (in millions of years ago) and diameter (km) to make the relationship simpler. (Data in **Craters**)

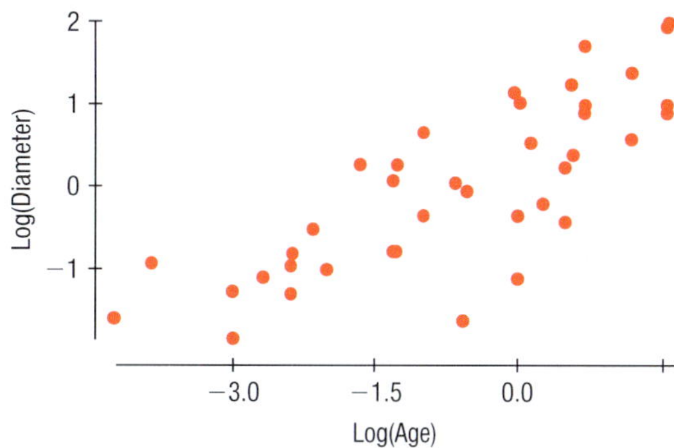

QUESTION: Are the assumptions and conditions satisfied for fitting a linear regression model to these data?

ANSWER:

✓ **Linearity Assumption:** The scatterplot of *y* vs. *x* is straight enough.

✓ **Independence Assumption:** Sizes of impact craters are likely to be generally independent. However, these are the only known craters, and may differ from others that have disappeared or not yet been found. We'll need to be careful not to generalize my conclusions too broadly.

✓ **Does the Plot Thicken? Condition:** After fitting a linear model, we find the residuals shown.

Two points seem to give the impression that the residuals may be more variable for higher predicted values than for lower ones, but this doesn't seem to be a serious violation of the Equal Variance Assumption.

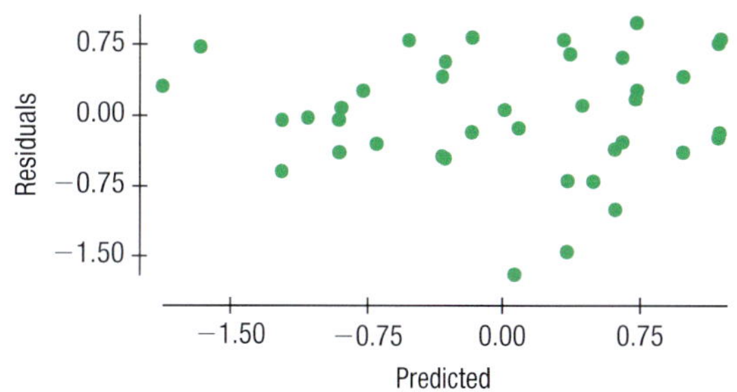

[3]Data, pictures, and much more information at the Earth Impact Database found at www.passc.net/EarthImpactDatabase/index.html.

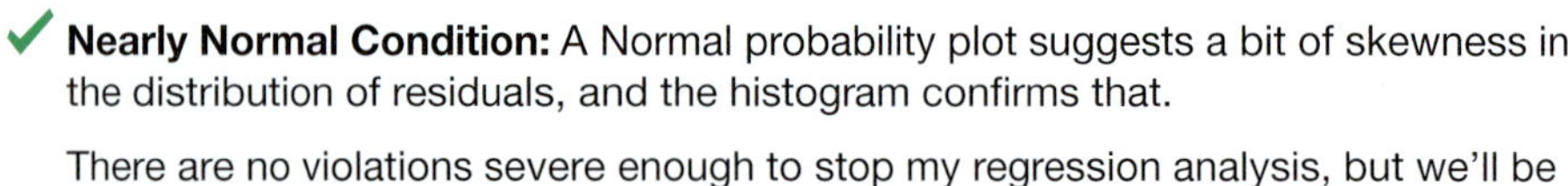

✓ **Nearly Normal Condition:** A Normal probability plot suggests a bit of skewness in the distribution of residuals, and the histogram confirms that.

There are no violations severe enough to stop my regression analysis, but we'll be cautious about our conclusions.

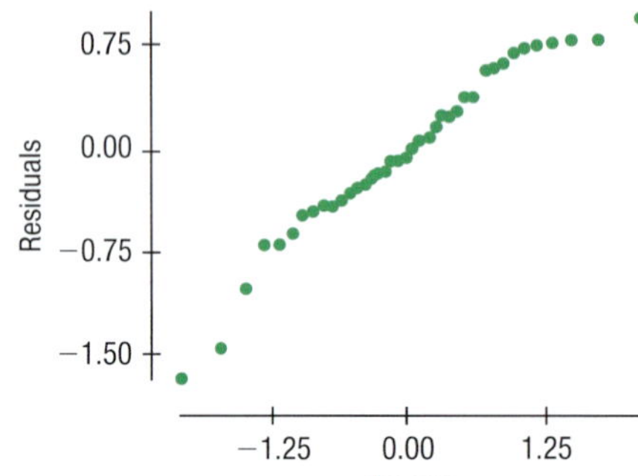

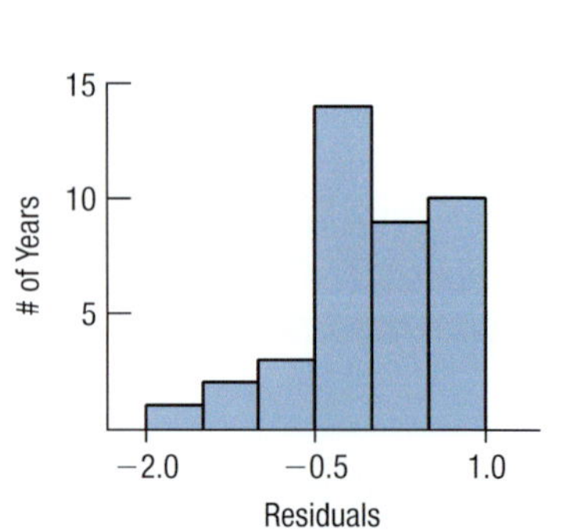

Which Come First: The Conditions or the Residuals?

> "Truth will emerge more readily from error than from confusion."
>
> —Francis Bacon (1561–1626)

The best way to check many of the conditions is with the residuals, but we get the residuals only *after* we compute the regression. Before we compute the regression, however, we should check at least one of the conditions.

So we work in this order:

1. Make a scatterplot of y against at least the most important x's and check the Straight Enough Condition. (If the relationship is curved, try re-expressing the data. Or stop.)
2. If the data are straight enough, fit a regression and find the predicted values, $\hat{y}$, and the residuals, e.
3. Make a scatterplot of the residuals against x or against the predicted values. This plot should have no pattern. Check in particular for any bend (which would suggest that the data weren't all that straight after all), for any thickening (or thinning), and, of course, for any outliers. (If there are outliers, and you can correct them or justify removing them, do so and go back to step 1, or consider performing two regressions—one with and one without the outliers.)
4. If the data are measured over time, plot the residuals against time to check for evidence of patterns that might suggest they are not independent.
5. If the scatterplots look OK, then make a histogram and Normal probability plot of the residuals to check the Nearly Normal Condition.
6. If all the conditions seem to be reasonably satisfied, go ahead with inference.

STEP-BY-STEP EXAMPLE Regression Inference

If our data can jump through all these hoops, we're ready to do regression inference. Let's see how much more we can learn about body fat and waist size from a regression model.

QUESTIONS: What is the relationship between *%Body Fat* and *Waist* size in men? What model best predicts body fat from waist size, and how well does it do the job?

THINK **PLAN** Specify the question of interest.

Name the variables and report the W's.

Identify the parameters you want to estimate.

I have quantitative body measurements on 250 adult males from the BYU Human Performance Research Center. I want to understand the relationship between *%Body Fat* and *Waist* size.

MODEL Think about the assumptions and check the conditions.

Make pictures. For regression inference, you'll need a scatterplot, a residuals plot, and either a histogram or a Normal probability plot of the residuals.

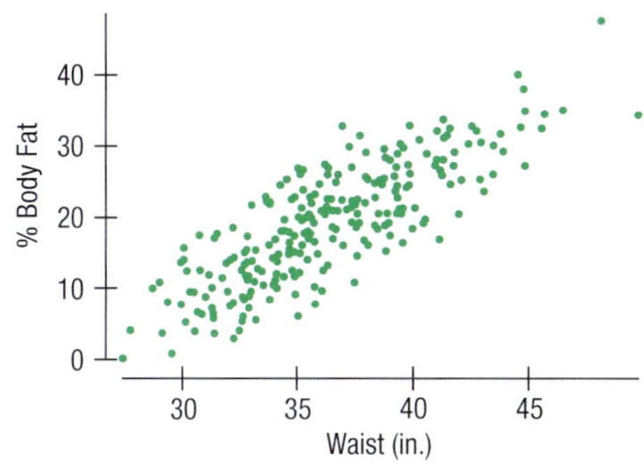

✓ **Straight Enough Condition:** There's no obvious bend in the original scatterplot of the data or in the plot of residuals against predicted values.

✓ **Independence Assumption:** These data are not collected over time, and there's no reason to think that the *%Body Fat* of one man influences the *%Body Fat* of another.

(We've seen plots of the residuals already. See Figure 23.9.)

✓ **Does the Plot Thicken? Condition:** Neither the original scatterplot nor the residual scatterplot shows any changes in the spread about the line.

✓ **Nearly Normal Condition, Outlier Condition:** A histogram of the residuals is unimodal and symmetric. The Normal probability plot of the residuals is quite straight, indicating that the Normal model is reasonable for the errors.

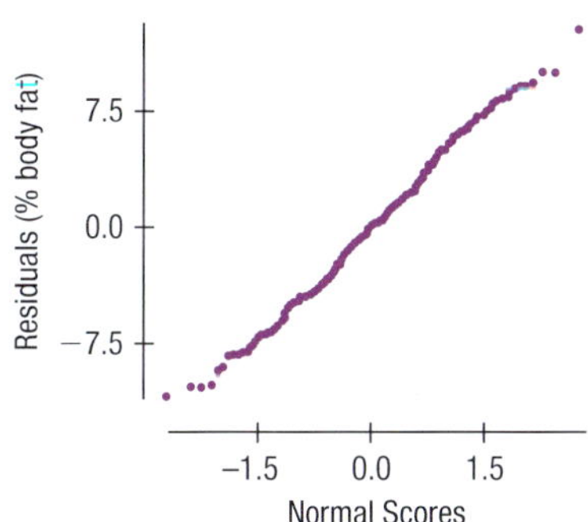

Choose your method.

Under these conditions a **regression model** is appropriate.

SHOW **MECHANICS** Let's just "push the button" and see what the regression looks like.

The formula for the regression equation can be found in Chapter 7, and the standard error formulas will be shown a bit later, but regressions are almost always computed with a computer program or calculator.

Here's the computer output for this regression:

Dependent variable is %BF
R-squared = 67.8%
s = 4.713 with 250 − 2 = 248 degrees of freedom

Variable	Coeff	SE(Coeff)	t-Ratio	P-Value
Intercept	−42.734	2.717	−15.7	<0.0001
Waist	1.70	0.0743	22.9	<0.0001

Write the regression equation.

The estimated regression equation is

$$\widehat{\%Body\ Fat} = -42.73 + 1.70\,Waist.$$

TELL **CONCLUSION** Interpret your results in context.

The R^2 for the regression is 67.8%. *Waist* size seems to account for about 2/3 of the *%Body Fat* variation in men. The slope of the regression says that *%Body Fat* increases by about 1.7 percentage points per inch of *Waist* size, on average.

MORE INTERPRETATION We haven't worked it out in detail yet, but the output gives us numbers labeled as *t*-statistics and corresponding P-values, and we have a general idea of what those mean.

(Now it's time to learn more about regression inference so we can figure out what the rest of the output means.)

The standard error of 0.07 for the slope is much smaller than the slope itself, so it looks like the estimate is reasonably precise. And there are a couple of *t*-ratios and P-values given. Because the P-values are small, it appears that some null hypotheses can be rejected.

23.3 Regression Inference and Intuition

Wait a minute! We just pulled a fast one. We pushed the "regression button" on our computer or calculator and the computer produced a table with a P-value. But what hypothesis is being tested?

For regression, the null hypothesis is so natural that it is rare to see any other considered. The natural null hypothesis is that the slope is zero and the alternative is (almost) always two-sided.

$$H_0: \beta_1 = 0$$
$$H_A: \beta_1 \neq 0$$

The P-values in the table come from tests of this hypothesis for each coefficient. Of course, to find a P-value, we need a sampling distribution. And it isn't sufficient to just say (as we have done) that the sampling distribution of the estimated coefficient is Normal. A Normal model needs a mean and standard deviation. The mean will be 0 (from the null hypothesis), but the standard errors must come from the data. We know that with different samples of 250 men, each sample would have produced its own regression line with slightly different b_0's and b_1's—as Figure 23.4 shows.

In practice, everyone relies on technology to find the standard errors of the coefficients. But the components of that calculation are easy to understand:

- **Spread around the model.** Here are two situations. Which situation would be more likely to yield a consistent slope? That is, if we were to sample over and over from the two underlying populations that these samples come from and compute all the slopes, which group of slopes would vary less?

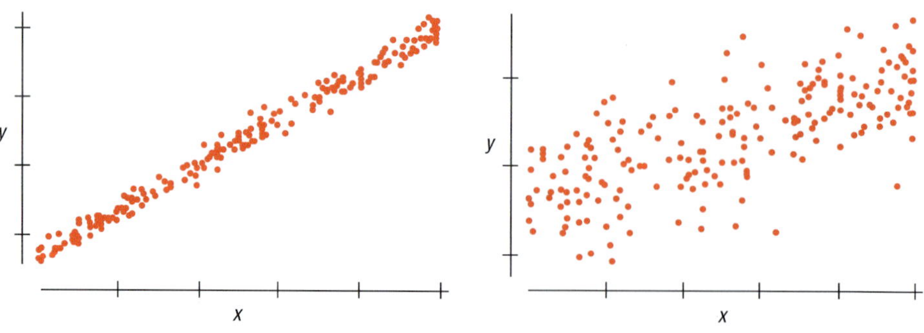

Figure 23.11
Which of these scatterplots shows a situation that would give the more consistent regression slope estimate if we were to sample repeatedly from its underlying population?

Clearly, data like those in the left plot will give more consistent coefficient estimates.

Less scatter around the regression model means the slope will be more consistent from sample to sample. The spread around the line is measured with the **residual standard deviation**, s_e. You can always find s_e in the regression output, often just labeled s. You're not likely to calculate the residual standard deviation by hand. When we saw this formula in Chapter 7, we said that it looks a lot like the standard deviation of y, only subtracting the predicted values rather than the mean and dividing by $n - 2$ instead of $n - 1$.

$$s_e = \sqrt{\frac{\sum (y - \hat{y})^2}{n - 2}}.$$

The less scatter around the line, the smaller the residual standard deviation and the stronger the relationship between x and y.

Some people prefer to assess the strength of a regression by looking at s_e rather than R^2. After all, s_e has the same units as y, and because it's the standard deviation of the errors around the line, it tells you how close the data are to our model. By contrast, R^2 is the proportion of the variation of y accounted for by x. We say, why not look at both?

$n - 2$?

For standard deviation (in Chapter 2), we divided by $n - 1$ because we didn't know the true mean and had to estimate it. Now it's later in the course and there's even more we don't know. Here we don't know *two* things: the slope and the intercept. If we knew them both, we'd divide by n and have n degrees of freedom. When we *estimate* both, however, we adjust by subtracting 2, so we divide by $n - 2$ and (as we will see soon) have 2 fewer degrees of freedom.

◆ **Spread of the x's.** Here are two more situations. Which of these would yield more consistent slope estimates?

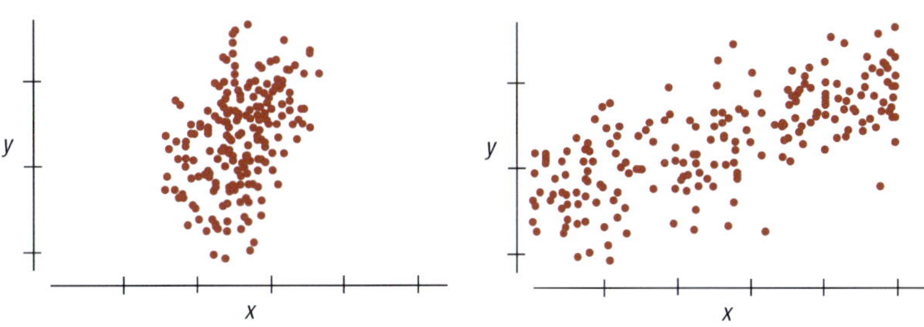

Figure 23.12
Which of these scatterplots shows a situation that would give the more consistent regression slope estimates if we were to sample repeatedly from the underlying population?

A plot like the one on the right has a broader range of x-values, so it gives a more stable base for the slope. We'd expect the slopes of samples from situations like that to vary less from sample to sample. If s_x, the standard deviation of x, is large, it provides a more stable regression.

In a multiple regression, this aspect of the data is a bit more complex. It is the standard deviation of x after allowing for the other x's in the model. Fortunately, the computer takes care of that (and we *always* use a computer to find multiple regression models).

◆ **Sample size.** What about these two?

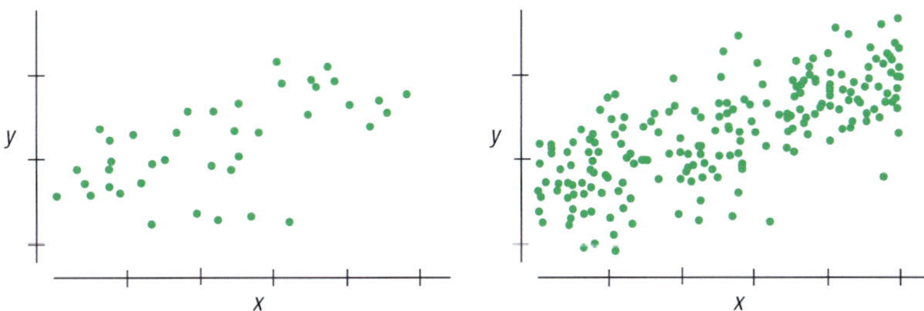

Figure 23.13
Which of these scatterplots shows a situation that would give the more consistent regression slope estimates if we were to sample repeatedly from the underlying population?

It shouldn't be a surprise that having a larger sample size, n, gives more consistent estimates from sample to sample.

> **NOTATION ALERT**
> Don't confuse the standard deviation of the residuals, s_e, with the standard error of a coefficient, $SE(b_j)$. The first measures the scatter around the line, and the second tells us how reliably we can estimate the slope.

Standard Error for the Slope

Three aspects of the scatterplot, then, affect the standard error of the regression slope:

- Spread around the model: s_e
- Variation among the x-values
- Sample size: n

These are in fact the *only* things that affect the standard error of the slope.

We know the b's vary from sample to sample. As you'd expect, the sampling distribution model of each coefficient is centered at its parameter value—its β. We estimate the standard deviations with the SE's found in the regression output table. What about the shape of these sampling distributions? Here the Central Limit Theorem and Gosset come to the rescue again. When we standardize the coefficients by subtracting the model mean (usually zero) and dividing by their estimated standard errors, we get a Student's t-model with $n - 2$ degrees of freedom:

$$\frac{b_1 - \beta_1}{SE(b_1)} \sim t_{n-2}.$$

Calculating the Standard Error of the Slope

For a simple regression, the standard error for the slope is

$$SE(b_1) = \frac{s_e}{\sqrt{n-1}\, s_x}$$

but you usually won't have to calculate it by hand. That's one of the things computers are particularly good at.

A Sampling Distribution for Regression Slopes

When the conditions are met, the standardized estimated regression slope,

$$t = \frac{b_1 - \beta_1}{SE(b_1)},$$

follows a Student's t-model with $n - 2$ degrees of freedom. We estimate the standard error with technology and find it in the regression output table.

What if the Slope Were 0?

If $b_1 = 0$, our prediction is $\hat{y} = b_0 + 0x$. The equation collapses to just $\hat{y} = b_0$. Now x is nowhere in sight, so y doesn't depend on x at all.

And b_0 would turn out to be $\bar{y}$. Why? We know that $b_0 = \bar{y} - b_1\bar{x}$, but when $b_1 = 0$, that becomes simply $b_0 = \bar{y}$. It turns out, then, that when the slope is 0, the equation is just $\hat{y} = \bar{y}$; at every value of x, we always predict the mean value for y.

Confidence Intervals and Hypothesis Tests

Armed with a model for the sampling distribution, we can make confidence intervals and perform hypothesis tests.

Inference for the Regression Slope

When the conditions are met, we can find a confidence interval for the slope as

$$b_1 - SE(b_1) \times t^*_{n-2} < \beta_1 < b_1 + SE(b_1) \times t^*_{n-2}.$$

The standard null hypothesis is that the slope is zero. The test takes the usual form:

$$\frac{b_1 - 0}{SE(b_1)}.$$

The same formulas apply to the intercept term, although its standard error is found a bit differently. Your technology will take care of that for you. It is much less common to test the null hypothesis $H_0{:}\,\beta_0 = 0$, but you can do it if you wish.

Inference for a Regression Model

1. Check the assumptions and conditions. Plot the data. Plot the residuals. If things look OK, then . . .
2. Look at the R^2 and s_e. Does it seem that the model accounts for much of the variation in y? If so, then . . .

3. Use the P-values to test the hypotheses that the coefficients are really zero. If you can reject the null hypothesis, then . . .
4. State and interpret the regression model.

Where should you find the R^2, s_e, coefficients, and P-values? In the regression table. Different statistics packages work a bit differently, but all seem to produce essentially the same regression table. And you'll see versions of this table in research papers that report regression models. So it is wise to learn to find the values you want in the table.

EXAMPLE 23.2 Confidence Interval and Hypothesis Test for a Slope

QUESTION: For the regression of *%Body Fat* on *Waist* size, what's a 95% confidence interval for the slope, and how do we test whether the slope is 0?

The coefficient for *Waist* size is 1.70%/in. with a standard error of 0.074%/in. With $n = 250$, there are $n - 2 = 248$ degrees of freedom and $t^*_{0.025,248} = 1.970$.

ANSWER: Once we have the standard error, we find the confidence interval from

$$b_1 \pm t^*_{0.025,248}\, SE(b_1) = 1.70 \pm 1.970(0.074) = (1.55, 1.85)\%/\text{in.}$$

With 95% confidence we can say that the true slope of *%Body Fat* per inch of *Waist* size is between 1.55 and 1.85% per inch.

To test whether the slope is 0, we first notice that the value 0 is far from the ends of the confidence interval, so 0 is not a plausible value. To find the P-value, we first find how many standard errors the slope is from 0:

$$t_{248} = \frac{1.70 - 0}{0.074} = 22.97.$$

Falling 22.97 standard errors away from 0 by chance is highly unlikely. The P-value is <0.0001. We can be quite confident that the coefficient is not zero. Based on the regression model, we can conclude that, *%Body Fat* increases, on average, with *Waist* size, at a rate between 1.55 and 1.85% per inch.

RANDOM MATTERS Slopes Vary–Revisited

When we re-sampled the body fat data, fitting a regression to 1000 re-samples, we found the histograms of slopes to be unimodal and symmetric (see Figure 23.5). The 2.5th and 97.5th percentiles of the 1000 slopes are

2.5%	97.5%
1.558259	1.845261

which gives a 95% bootstrap confidence interval of (1.56, 1.85) %/inch, nearly exactly what we saw from the Normal-based method. We could test the hypothesis that the slope is 0 simply by checking whether 0 is in the confidence interval. Finding the P-value takes a bit more work. To find a P-value, we subtract b_1 (here, 1.70) from each of the re-sampled slopes to center the distribution at the null hypothesis. Then we count the trials that resulted in slopes more extreme than the one in our sample. In this simulation, none of 1000 trials gave a slope more than 1.70 in either direction from 0. But we can't say that the P-value is zero. The best we can say is that the simulated P-value is <0.001. We can't be more precise about the P-value unless we run more trials.

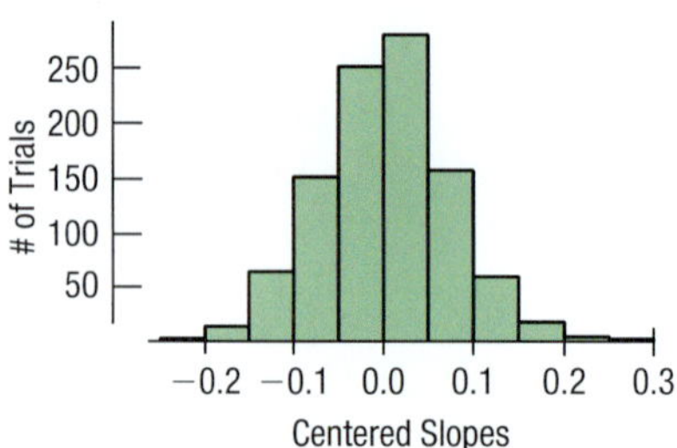

Figure 23.14
To find a P-value for testing the hypothesis that the slope is 0, we subtract the slope of the sample (1.70) from the slope of each trial. The histogram shows that none of these centered slopes falls more than 1.70 units away in either direction, so the P-value is less than 1 out of 1000 (0.001), consistent with what we saw from Normal theory.

23.4 The Regression Table

Here's the regression table we saw in the Step-by-Step Example:

Table 23.1
The regression table from the Step-by-Step Example.

Dependent variable is %BF
R-squared = 67.8%
s = 4.713 with 250 − 2 = 248 degrees of freedom

Variable	Coeff	SE(Coeff)	t-Ratio	P-Value
Intercept	−42.734	2.717	−15.7	<0.0001
Waist	1.70	0.0743	22.9	<0.0001

The coefficients are the best estimates of the parameters in the regression model.

The standard errors are estimated standard deviations of the sampling distribution of the coefficients. We've seen that you can find these by simulation, but most software will base them on theory.

The *t*-ratios in the next column show how many standard errors the coefficients are away from zero (because zero is the assumed null hypothesis value). You already know how to think about them: The 68–95–99.7 Rule is a good rule of thumb. Each *t*-ratio can be used to find a ***t*-test** or **confidence interval for the regression coefficient.** The natural null hypothesis for a test about a coefficient is that it is zero: $H_0: \beta = 0$. That's the truly "null" result—and one people are often interested in testing—because if the coefficient were zero, then the corresponding variable wouldn't contribute to the model at all.

The final column holds P-values just like those we've found for other hypothesis tests. In this case, a small P-value says that if the true parameter were zero, it would be unlikely to see a coefficient this far from zero, so we would reject the null hypothesis.

Other parts of this table are familiar as well. R^2 is still the fraction of the variability of y accounted for by the regression model, and s is still the standard deviation of the regression residuals, which we have called s_e.

EXAMPLE 23.3 Interpreting a Regression Model

RECAP: On a log scale, there seems to be a linear relationship between the diameter and the age of recent terrestrial impact craters. We have regression output from statistics software:

Dependent variable is LogDiam
R-squared = 63.6%
s = 0.6362 with 39 − 2 = 37 degrees of freedom

Variable	Coefficient	Se(coeff)	t-Ratio	P-Value
Intercept	0.358262	0.1106	3.24	0.0025
LogAge	0.526674	0.0655	8.05	≤0.0001

QUESTION: What's the regression model, and what can it tell us?

ANSWER: For terrestrial impact craters younger than 35 million years, the logarithm of *Diameter* grows linearly with the logarithm of *Age:*

$$\widehat{logDiam} = 0.358 + 0.527\ logAge.$$

The P-value for each coefficient's *t*-statistic is very small, so we can be quite confident that neither coefficient is zero. Based on the regression model, we can conclude that, on average, the older a crater is, the larger it tends to be. This model accounts for 63.6% of the variation in *logDiam*.

Although it is possible that impacts (and their craters) are getting smaller, it is more likely that we are seeing the effects of age on craters. Small craters are probably more likely to erode or become buried or otherwise be difficult to find as they age. Larger craters may survive the huge expanses of geologic time more successfully.

23.5 Multiple Regression Inference

In Chapter 9 we saw that a second predictor, *Height*, improved the model for *%Body Fat*. The regression table (Table 9.1) had a section faded out. But now you know everything you need to be able to understand the rest of the regression table. Here is the full table:

Table 23.2
The multiple regression table seen in Table 9.1, now with all values present.

Dependent variable is: %Body Fat
R squared = 71.3%
s = 4.460 with 250 − 3 = 247 degrees of freedom

Variable	Coefficient	SE(Coeff)	t-ratio	P-value
Intercept	−3.10088	7.686	−0.403	0.6870
Waist	1.77309	0.0716	24.8	<0.0001
Height	−0.601539	0.1099	−5.47	<0.0001

Recall that in a multiple regression each coefficient is interpreted as the effect of its associated predictor *after allowing for the effects of the other predictors*. For example, the coefficient of *Height* says that *among men with the same waist size*, taller men tend, on average, to have a *%Body Fat* that is lower by about 0.6 percent per inch of height. This makes sense. A five foot tall man with a 40-inch waist might be seen as chunky, but a 6'5" basketball player with a 40-inch waist might seem skinny.

The other values in this table mean the same thing as they did before. R^2 is still the fraction of the variability of y accounted for by the regression model. Although s is still the standard deviation of the regression residuals, there is a minor adjustment: The number of degrees of freedom in this table is $n - 3$. The rule is that the number of degrees of freedom is n (the sample size) minus one for each estimated coefficient, so in this case $n - 3$.

The standard error, *t*-statistic, and P-values mean the same thing in the multiple regression as they meant in a simple regression. Each coefficient has a Normal sampling distribution. Those sampling distributions are centered at their respective true (i.e., population) coefficient values. And each has a standard deviation. Because we don't know the underlying population standard deviations, we use standard errors and refer to Student's *t*-model to find P-values. The *t*-ratios and corresponding P-values in each row of the table refer to their corresponding coefficients.

The complication in multiple regression is that all of these values are interrelated. Including any new predictor or changing any data value can change any or all of the other numbers in the table. For example, the coefficient of *Height* was −0.09 when it was the sole predictor of *%Body Fat* and is now −0.60. Even more interesting is the change in its P-value, which was 0.64 for *Height* alone, saying that *Height* was not a successful predictor of *%Body Fat*. In the multiple regression the P-value is 0.001, saying that *Height* is indeed an important predictor after we allow for *Waist* size. And we can see from the increased R^2, the added complication of an additional predictor was worthwhile in improving the fit of the regression model.

JUST CHECKING

Researchers in Food Science studied how big people's mouths tend to be. They measured mouth volume by pouring water into the mouths of subjects who lay on their backs. Unless this is your idea of a good time, it would be helpful to have a model to estimate mouth volume more simply. Fortunately, mouth volume is related to height. (Mouth volume is measured in cubic centimeters and height in meters.)

The data were checked and deemed suitable for regression. Take a look at the computer output. (Data in **Mouth_volume**)

1. What does the *t*-ratio of 3.27 for the slope tell about this relationship? How does the P-value help your understanding?
2. Would you say that measuring a person's height could reliably be used as a substitute for the wetter method of determining how big a person's mouth is? What numbers in the output helped you reach that conclusion?
3. What does the value of s_e add to this discussion?

Summary of	Mouth Volume
Mean	60.2704
StdDev	16.8777

Dependent variable is Mouth Volume
R-squared = 15.3%
s = 15.66 with 61 − 2 = 59 degrees of freedom

Variable	Coefficient	SE(coeff)	t-Ratio	P-Value
Intercept	−44.7113	32.16	−1.39	0.1697
Height	61.3787	18.77	3.27	0.0018

Here's a multiple regression with the variable *Age* added to the previous model:

Response variable is: Mouth_Volume
R squared = 19.3%
s = 15.42 with 61 − 3 = 58 degrees of freedom

Variable	Coefficient	SE(Coeff)	t-ratio	P-value
Intercept	−51.0122	31.88	−1.60	0.1151
Height	58.1009	18.58	3.13	0.0028
Age	0.437288	0.2588	1.69	0.0965

4. Is this a better model? Explain.
5. How has the interpretation of the coefficient of *Height* changed?
6. Test the standard null hypothesis on the coefficient of *Age*.

Collinearity

In Chapter 9 we examined data on roller coasters and found that the duration of the ride depended on, among other things, the drop—that initial stomach-turning plunge down the high hill that powers the coaster through its run. (Data in **Coasters 2015;** Xcelerator and Tower of Terror omitted)

Response variable is: Duration
R squared = 30.3%
s = 33.27 with 89 − 2 = 87 degrees of freedom

Variable	Coefficient	SE(Coeff)	t-ratio	P-value
Intercept	88.4869	9.524	9.29	<0.0001
Drop	0.386339	0.0628	6.15	<0.0001

At that time, the *SE*, *t*-ratio, and P-value were skipped, but now we can see that *Drop* is indeed a good predictor of the *Duration* of the ride with a tiny P-value and a large *t*-ratio.

Adding a second predictor should only improve the model, so let's add the maximum *Speed* of the coaster to the model:

Response variable is: Duration
R squared = 36.5%
s = 31.94 with 89 − 3 = 86 degrees of freedom

Variable	Coefficient	SE(Coeff)	t-ratio	P-value
Intercept	−6.39318	34.06	−0.188	0.8515
Drop	−0.139870	0.1917	−0.730	0.4675
Speed	2.70299	0.9346	2.89	0.0048

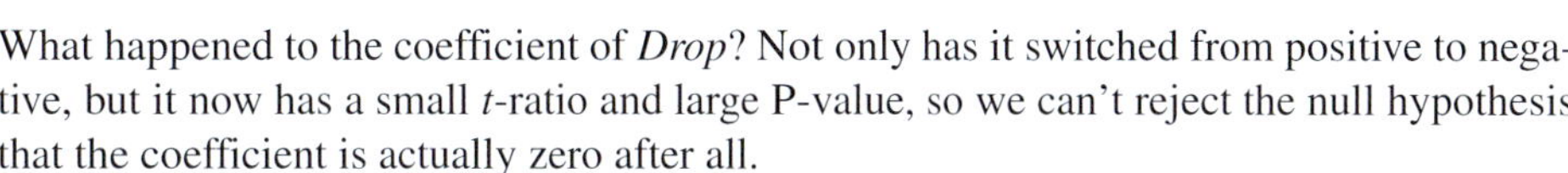

Figure 23.15
Drop and *Speed* are collinear. They line up on a single narrow line (and are thus highly correlated). Imagine balancing a table top on these points. It would be unstable.

What happened to the coefficient of *Drop*? Not only has it switched from positive to negative, but it now has a small *t*-ratio and large P-value, so we can't reject the null hypothesis that the coefficient is actually zero after all.

What we have seen here is a problem known as *collinearity*. Specifically, *Drop* and *Speed* are highly correlated with each other. As a result, the effect of *Drop after allowing for the effect of Speed* is negligible.

Whenever you have several predictors, you must think about how the predictors are related to each other. When predictors are unrelated to each other, each provides new information to help account for more of the variation in *y*. But when there are several predictors, the model will work best if they vary in different ways so that the multiple regression has a stable base. If you wanted to build a deck on the back of your house, you wouldn't build it with supports placed just along one diagonal. Instead, you'd want the supports spread out in different directions as much as possible to make the deck stable. We're in a similar situation with multiple regression. When predictors are highly correlated, they line up together, which makes the regression they support balance precariously. Even small variations can rock it one way or the other. A more stable model can be built when predictors have low correlation and the points are spread out.

Multicollinearity?

You may find this problem referred to as "multicollinearity." But there is no such thing as "unicollinearity"—we need at least two predictors for there to be a linear association between them—so there is no need for the extra two syllables.

When two or more predictors are linearly related, they are said to be **collinear**. The general problem of predictors with close (but perhaps not perfect) linear relationships is called the problem of **collinearity**.

It is not sufficient to just check the pairwise correlations. Collinearity can occur among several predictors in a larger multiple regression. Fortunately, there's an easy way to assess collinearity. To measure how much one predictor is linearly related to the others, just find the regression of that predictor on the others[4] and look at the R^2. That R^2 gives the fraction of the variability of the predictor in question that is accounted for by the other predictors. So $1 - R^2$ is the amount of the predictor's variance that is left after we allow for the effects of the other predictors. That's what the predictor has left to bring to the

[4]The residuals from this regression are plotted as the *x*-axis of the partial regression plot for this variable. (See Section 9.4 for a reminder about partial regression plots.) So if they have a very small variance, you can see it by looking at the *x*-axis labels of the partial regression plot, and get a sense of how precarious a line fit to the partial regression plot—and its corresponding multiple regression coefficient—may be.

Why not Just Look at the Correlations?

Why not just examine the table of correlations of all the predictors to search for collinearity? This would find associations among *pairs* of predictors. But collinearity can—and does—occur among *several* predictors working together. You won't discover that more subtle collinearity with a correlation table.

regression model. And we know that a predictor with little variance can't do a good job of predicting. Another consequence of collinearity is that the variance of the predictor's coefficient is multiplied by the reciprocal of that small $1 - R^2$, making it larger (and the corresponding t-statistic smaller). In fact, that reciprocal, $1/(1 - R^2)$, is called the **Variance Inflation Factor (VIF)**. You'll may see it in analyses to diagnose or further understand a regression model.

Collinearity can hurt a regression analysis in yet another way. We've seen that the variance of a predictor appears in the denominator of the standard error of its associated coefficient. A predictor reduced to small random residuals after "removing the effects of the other predictors" will vary little. And that reduced variance leads to a larger *SE*.

As a final blow, when a predictor is collinear with the other predictors, it's often difficult to figure out what its coefficient means in the multiple regression. We've blithely talked about "removing the effects of the other predictors," but if predictors are collinear, then when we do that, there may not be much left. What does the drop down that roller coaster hill contribute to the ride other than the initial (and highest) speed?

When a predictor is collinear with the other predictors in the model, two things can happen:

1. Its coefficient can be surprising, taking on an unanticipated sign or being unexpectedly large or small.
2. The standard error of its coefficient can be inflated, leading to a smaller t-statistic and correspondingly large P-value—even for a variable that you know is associated with the response variable.

What should you do about a collinear regression model? The simplest cure is to remove some of the predictors. That simplifies the model and usually improves the t-statistics. And, if several predictors provide pretty much the same information, removing some of them won't hurt the model. Which predictors should you remove? Keep those that are most reliably measured, those that are least expensive to find, or even the ones that are politically important.

CHOOSING A SENSIBLE MODEL

The Mathematics Department at a large university built a regression model to help them predict success in graduate study. They were shocked when the coefficient for Mathematics GRE score was not significant. But the Math GRE was collinear with some of the other predictors, such as math course GPA and Verbal GRE, which made its coefficient not significant. They decided to omit some of the other predictors and retain Math GRE as a predictor because that model seemed more appropriate—even though it predicted no better (and no worse) than others without Math GRE.

Much more can be said about multiple regression. Indeed, it is common to find an entire course on the subject. Multiple regression may be the most widely used statistics model-building method and is found in many different disciplines. What we've covered here should give you the basic knowledge you need to read articles in which regression is used. But if you want to perform your own multiple regression analyses, we urge you to study the subject in greater depth.

23.6 Confidence and Prediction Intervals

We know how to compute predicted values of y for any value of x. We first did that in Chapter 7. This predicted value would be our best estimate, but it's still just an informed guess. Now that we have standard errors, we can use those to construct intervals for both the predictions and our parameters to report our uncertainty honestly.

From our model of *%Body Fat* and *Waist* size, we might want to use *Waist* size to get a reasonable estimate of *%Body Fat*, and we want to know how precise that prediction will be. The precision depends on which question we ask:

- Do we want to know the uncertainty in the mean *%Body Fat* for *all* men with a *Waist* size of, say, 38 inches?
- Or do we want to estimate the uncertainty in the *%Body Fat* for a *particular* man with a 38-inch *Waist*?

Mean vs. Individual Predictions

A physiologist might be interested in the general relationship of bodyfat percentage to measurements of body properties. For her, a confidence interval for the predicted mean would be appropriate. But if you want to predict your own bodyfat percentage without the technology needed for precise measurement, you'll probably be disappointed with the larger size of the prediction interval.

What's the difference between the two questions? The predicted *%Body Fat* is the same, but we know that we can estimate means much more precisely than we can predict for an individual. So we can predict the *mean %Body Fat* for *all* men whose *Waist* size is 38 inches with a lot more precision than we can predict the *%Body Fat* of a *particular individual* whose *Waist* size happens to be 38 inches. We call the interval around the mean a **confidence interval for the mean prediction** because the mean can be thought of as a parameter. By contrast, we call the interval around an individual prediction a **prediction interval**.

Both intervals have the form of all our other confidence intervals: We want to be able to predict for any new value of x, so we'll call the x-value "x sub new" and write it x_ν.[5] The regression equation predicts *%Body Fat* as $\hat{y}_\nu = b_0 + b_1 x_\nu$. Using that notation, the intervals have the form:

$$\hat{y}_\nu \pm t^*_{n-2} \times SE.$$

They are both centered at the predicted value and refer to the t-distribution with $n - 2$ degrees of freedom for values of t^*. The difference between them is just in their standard errors. The standard error for the confidence interval will be much smaller than the standard error for the prediction interval. Figure 23.16 shows both intervals. Notice that they are narrower in the middle of the x range. It is easier to predict near the mean than for more extreme values. Extrapolation is, as we know, hazardous to your health.

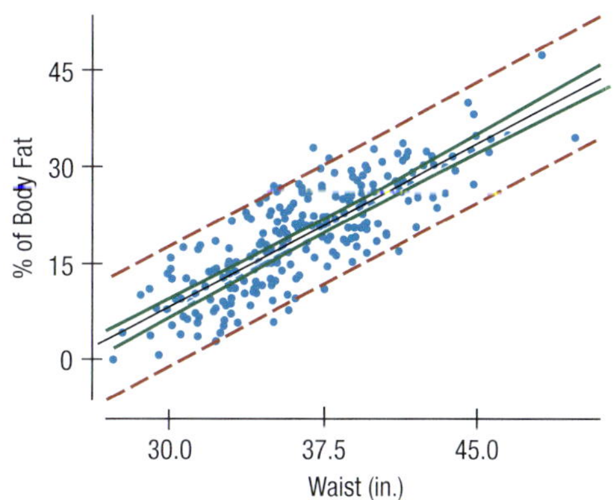

Figure 23.16

A scatterplot of *%Body Fat* vs. *Waist* size with a least squares regression line. The solid green lines near the regression line show the extent of the 95% confidence intervals for mean *%Body Fat* at each *Waist* size. Notice that there is more uncertainty the farther we are from the mean of the x-values. The dashed red lines show the prediction intervals. Most of the data points are contained within the prediction intervals—as we'd expect for predicting individual values. The narrower intervals are for the mean values, which vary less than individual values.

*Formulas for the Standard Errors of Confidence and Prediction Intervals

Much of the intuition we had about slopes varying carries over to confidence and prediction intervals:

1. When there's less spread around the line, it's easier to be precise.
2. If we're more certain of the slope, we'll be more certain of our predictions.
3. If we have more data, our predictions will be more precise.
4. And predictions closer to the mean of x will have less variability than predictions farther from it.

[5] Yes, this is a bilingual pun. The Greek letter ν is called "nu." Don't blame me; my co-author suggested this.

These are the four factors that contribute to the standard errors. Because the factors are independent of each other, we can add their variances to find the total variability.[6] The resulting formula for the standard error of the predicted *mean* value explicitly takes into account each of the factors:

$$SE(\hat{\mu}_\nu) = \sqrt{SE^2(b_1) \times (x_\nu - \bar{x})^2 + \frac{s_e^2}{n}}.$$

What's different about predicting an individual value? We have all this variability and an extra term that accounts for the variability of individuals around their mean. That appears as the extra variance term, s_e^2:

$$SE(\hat{y}_\nu) = \sqrt{SE^2(b_1) \times (x_\nu - \bar{x})^2 + \frac{s_e^2}{n} + s_e^2}.$$

Remember to keep this distinction between the two kinds of standard errors when looking at computer output. The smaller one is for the *confidence interval for the mean value,* and the larger one is for the *prediction interval for an individual value*.

EXAMPLE 23.4 Intervals for Predictions

QUESTION: How well can the regression model predict the mean *%Body Fat* for men with 38-inch *Waists*?

ANSWER: The regression output table (still on page 744) provides most of the numbers we need:

$$s_e = 4.713$$
$$n = 250$$
$$SE(b_1) = 0.074, \text{ and from the data we need to know that}$$
$$\bar{x} = 36.3$$

The regression model gives a predicted value at $x_\nu = 38$ of

$$\hat{y}_\nu = -42.7 + 1.7(38) = 21.9\%.$$

The standard error for the confidence interval is

$$SE(\hat{\mu}_\nu) = \sqrt{0.074^2 \times (38 - 36.3)^2 + \frac{4.713^2}{250}} = 0.32\%.$$

The t^*–value for 95% confidence at 248 df is 1.97. Putting it all together, we find the margin of error as

$$ME = 1.97(0.32) = 0.63\%.$$

So, we are 95% confident that the mean *%Body Fat* for men with 38-inch *Waists* is

$$21.9\% \pm 0.63\%.$$

QUESTION: Predict the *%Body Fat* for an individual man with a 38-inch *Waist.*

ANSWER: We need the larger standard error:

$$SE(\hat{y}_\nu) = \sqrt{0.074^2 \times (38 - 36.3)^2 + \frac{4.713^2}{250} + 4.713^2} = 4.72\%.$$

[6]This is another example of the "Pythagorean Theorem of Statistics" we saw in Section 20.8. It says that when adding or subtracting independent random quantities, their variances can be added to find the variance of the result. There we added only two variances, but the principle applies to any number of mutually independent random quantities.

The corresponding margin of error is

$$ME = 1.97(4.72) = 9.30\%,$$

so the prediction interval is

$$21.9\% \pm 9.30\%.$$

We can be fairly precise about the mean body fat we expect for men with 38-inch waists. Our confidence interval has a margin of error of less than a percent. That may be interesting medically. But the researchers hoped to be able to predict body fat% for an individual man using simple measurements, and our prediction interval goes from 12.6% to 31.2%. Is that useful? Well, the lower end is about what you'd expect for an adult male athlete. But the American Council on Exercise considers anything above 25% as obese. Knowing only a man's waist size clearly isn't going to do the job.

*23.7 Logistic Regression

The Pima Indians of southern Arizona are a unique community. Their ancestors were among the first people to cross over into the Americas some 30,000 years ago. For at least two millennia, they have lived in the Sonoran Desert near the Gila River. Known throughout history as a generous people, they have given of themselves for many years helping researchers at the National Institutes of Health study certain diseases like diabetes and obesity. Young Pima Indians often marry other Pimas, making them an ideal group for genetic researchers to study. Pimas also have an extremely high incidence of diabetes.

Researchers investigating factors for increased risk of diabetes examined data on 768 adult women of Pima Indian heritage.[7] One possible predictor is the body mass index (BMI), calculated as weight/height^2, where weight is measured in kilograms and height in meters. We are interested in the relationship between *BMI* and the incidence of diabetes. We might start by looking at boxplots of *BMI* for each group: (Data in **Pima Indians**)

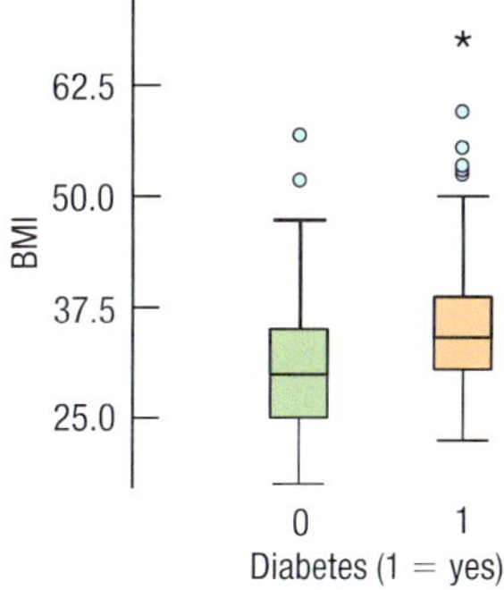

Figure 23.17
Side-by-side boxplots for the two *Diabetes* groups (1 = has diabetes; 0 = doesn't have) show elevated body mass index (BMI) for the women who have diabetes.

From the boxplots, we see that the group with diabetes has a higher mean *BMI*. (A *t*-test would show the difference to be more than 9 *SE*s from 0 with a P-value < 0.0001.) There is clearly a relationship. Here we've displayed *BMI* as the response and *Diabetes* as the predictor. But the researchers are interested in predicting the increased risk of *Diabetes* due to increased *BMI*, not the other way around.

[7]Data from the UCI Machine Learning Repository.

Diabetes is usually treated as a categorical variable. It might seem unwise at first, but coding *Diabetes* as 1 and *No Diabetes* as 0 gives rise to a useful model as we'll soon see. Using this coding, we can reverse roles, treat *Diabetes* as the response, and make this scatterplot:

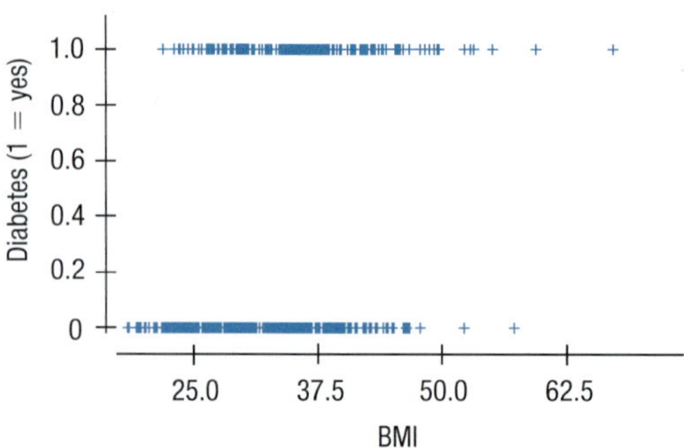

Figure 23.18
A scatterplot of *Diabetes* by *BMI*, using the 0–1 coding for *Diabetes*, shows a shift in *BMI* for the two groups.

We now have the problem of how to model this relationship. Fitting a linear regression to these data has some obvious problems:

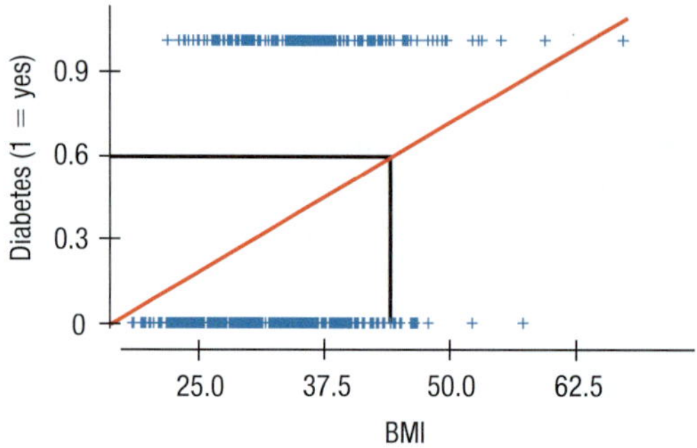

Figure 23.19
A regression of *Diabetes* on *BMI* using the 0–1 coding for *Diabetes*. The linear regression is not appropriate because even if the predicted values are interpreted as the probability of having the disease, this line will predict probabilities greater than 1 and less than 0.

The equation says: $\widehat{Diabetes} = -0.351 + 0.022\,BMI$. Does this make *any* sense? Suppose someone had a *BMI* of 44. The equation predicts 0.62. What would you guess is the chance that she will have diabetes? If you said about 62%, then you're using the line to model the *probability* of having *Diabetes*.

There are some obvious problems with this model, though. What's the probability that someone with a *BMI* of 10 has diabetes? It's low, but the equation predicts −0.13, obviously an impossible probability. And if we imagined someone with a *BMI* of 70, we might suspect that the probability of her having diabetes is pretty high, but not the predicted value of 1.19.

One simple fix to this problem might be to set all negative probabilities to 0 and all probabilities greater than 1 to 1. That would give a model that looked like this:

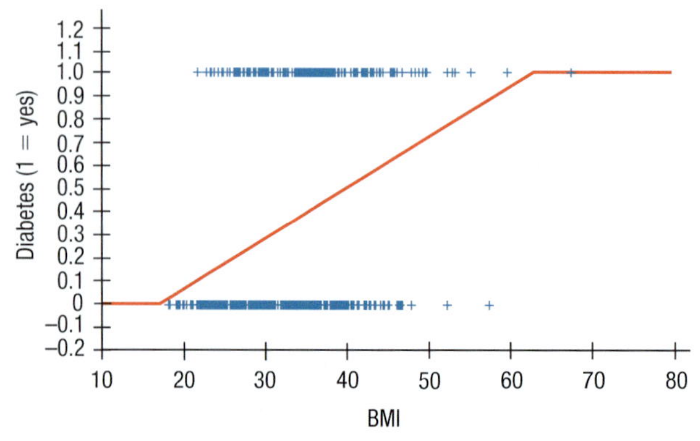

Figure 23.20
This model eliminates the problem of probability values that are negative or that exceed 1, but the corners are neither aesthetically pleasing nor scientifically meaningful.

This avoids one problem, but it can't really be correct. The occurrence of *Diabetes* can't be either certain $(p = 1)$ or impossible $(p = 0)$ based only on *BMI*. And real-world changes are likely to be smooth, so we prefer models with smooth transitions rather than corners. That makes good predictive sense, too. There's no reason to expect sharp changes at certain *BMI* values. So instead, we can use a smooth curve like this to model the probability of having *Diabetes*.

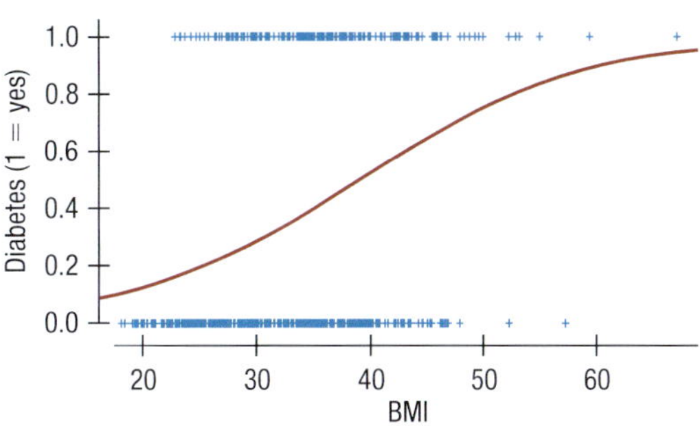

Figure 23.21
The smooth curve models the probability of having *Diabetes* as a function of *BMI* in a sensible way. The logistic curve shown here is just one of a number of choices for the form of the curve, but all are fairly similar in shape and in the resulting predicted probabilities.

This smoother version is a sensible way to model the probability of having *Diabetes* as a function of *BMI*. There are many curves in mathematics with shapes like this that we might use for our model. One of the most common is the *logistic curve*. The regression based on this curve is called **logistic regression**.

The equation for logistic regression can be written like an ordinary regression on a transformed *y*-variable:

$$\ln\left(\frac{\hat{p}}{1 - \hat{p}}\right) = b_0 + b_1 x.$$

The expression on the left-hand side gives the *logistic* curve. The logarithm used is the natural, or base *e*, log, although that doesn't really matter to the shape of the model.

It turns out that the logistic curve has a particularly useful interpretation. Racetrack enthusiasts know that when p is a probability, $\frac{p}{1 - p}$ is the *odds* in favor of a success. For example, when the probability of success $p, = 1/3$, we get the ratio $\frac{1/3}{2/3} = \frac{1}{2}$. We'd say that the odds in favor of success are 1:2 (or we'd probably say the odds *against* it are 2:1). Logistic regression models the *logarithm* of the odds as a linear function of x. In fact, nobody really thinks in terms of the log of the odds ratio. But it's the combination of that ratio and the logarithm that gets us the nice S-curved shape. What is important is that we can work backward from a log odds ratio to get the probability—which is often easier to think about.

Because we're not fitting a straight line to the data directly, we can't use ordinary least squares to estimate the parameters b_0 and b_1.[8] Instead, we use special *nonlinear* methods that require a good deal more computation—but computers take care of that for us. For the Pima Indians dataset, the equation is

$$\ln\left(\frac{\hat{p}}{1 - \hat{p}}\right) = -4.00 + 0.1025\, BMI.$$

[8]This is a tricky point. We've fit regressions with transformed y's before without special methods. If our data consisted of observed *proportions* of people, we could transform the data using log $(p/(1 - p))$ and use linear regression to fit the equation. But all we have are individual 0's and 1's. To be able to fit this equation with our raw data, we need special nonlinear methods.

A computer output will typically provide a table like the following:

Term	Estimate	Std Error	ChiSquare	P-Value
Intercept	−3.9967	0.4288	86.8576	<0.00001
BMI	0.1025	0.0126	66.0872	<0.00001

We usually don't interpret the slope itself (unless you happen to think naturally in log odds), but we can perform a test on whether the slope is 0, similar to the *t*-test we did for linear regression. Unlike linear regression, the ratio of the estimate to its standard error does not have a *t*-distribution, but the square of that ratio has a χ^2 distribution. The P-values for both the slope and the intercept clearly indicate that neither is 0.

Once we have decided that the slope is not 0, we can use the model to predict the probabilities. Solving the equation $\ln\left(\frac{\hat{p}}{1-\hat{p}}\right) = b_0 + b_1 x$ for $\hat{p}$ gives

$$\hat{p} = \frac{1}{1 + e^{-(b_0 + b_1 x)}}$$

and the logistic equation guarantees that the estimate of $\hat{p}$ will be between 0 and 1. Fortunately, technology can provide these probability estimates, producing the curve shown previously and giving an estimate for the probability at any *BMI* value.

Response variables that are dichotomous (having only two possible values) like the variable *Diabetes* are common, so there's a widespread need to model data like these. Phone companies want to predict who will switch carriers, credit card companies want to know which transactions are likely to be fraudulent, and loan companies want to know who is most likely to declare bankruptcy. All of these are potential application areas for logistic regression. It's not surprising that logistic regression has become an important modeling tool in the toolbox of analysts in science and industry in the past decade. By understanding the basics of what logistic regression can do, you can expand your ability to apply regression to many other important applications.

*23.8 More About Regression

Much more can be said about regression than we can discuss in an introductory text. Regression may be the most widely used tool for modeling relationships among variables, and it forms the core of many of the methods that are currently used in advanced data analysis, data mining, and analytics. So your understanding of both simple and multiple regression should provide a good foundation for learning about and understanding many advanced methods.

To point the way, let's consider two parts of regression tables that you may have seen if you have been finding regressions with a statistics program, but that we've omitted from our regression tables.

Adjusted R^2 and Other Measures for Model Selection

We've seen that R^2 tells us how much of the variance of *y* is accounted for by the regression model. There's no cutoff value of R^2 that helps us decide if a model is good enough, but, in general, a higher R^2 is more desirable. But, when comparing alternative models with different numbers of predictors, R^2 has a problem. Adding a new predictor variable to a model can never decrease R^2, so following a strategy of maximizing it will lead you to a model with as many predictors as you can find. Unfortunately, that model is unlikely to predict new values very well, because some of those predictors may be modeling noise (the error) and not the actual variation in *y*, a phenomenon known as **overfitting** the data.

Choosing a good model with many potential predictors can be challenging because, as we've seen, each predictor affects the coefficients of all the other predictors. Often there is no clear "best" regression model; several may do perfectly good jobs modeling the data. In fact, there is nothing wrong with maintaining two or three alternative models while you explore your data, seeking an understanding of how the variables relate and ferreting out influential points, outliers, or errors in the data. Deciding on the best model relies on your judgment, common sense, knowledge of the subject matter being modeled, and the quality of the data. The science of selecting among regression models is one of the hottest areas of current statistics research.

A number of measures are reported with every regression that can help you compare models. You may have noticed a statistic called **adjusted R^2** (often R^2 adjusted). Adjusted R^2, as the name implies, attempts to adjust R^2 by adding a *penalty* for each predictor in the model so that including more predictors doesn't automatically make the value bigger. That makes it useful for comparing alternative multiple regression models for the same data.[9] Between two alternative regression models for the same data, the one with the higher adjusted R^2 may be preferred.

The ANOVA Table

Many computer regression tables include an additional table that looks like this:

Source	Sum of Squares	df	Mean Square	F-ratio
Regression	3.83413	1	3.83413	490
Residual	0.438420	56	0.007829	

These numbers are the building blocks of the regression. For example, the "Sum of Squares" in the row for "Residual" is just that very same sum of squared residuals that the least squares criterion has been minimizing. The degrees of freedom that go with it in the "df" column is the degrees of freedom we've seen for the standard deviation of the residuals: $n -$ (number of predictors $+$ 1). The "Sum of Squares" for "Regression" is found as $\sum(\hat{y}_i - \bar{y})^2$, and the degrees of freedom is the number of predictors in the model.

Each "Mean Square" is a sum of squares divided by its associated df. And the "F-ratio" is the ratio of the mean square for regression divided by the mean square residual.

So what do these numbers tell us about the regression? First, the mean square residual is actually the *variance* of the residuals, and so it is the square of s_e. The F-ratio is a statistic used to test the null hypothesis that all the coefficients (except the intercept) are really zero. That's the null hypothesis that the entire regression is worthless and we should have just used $\bar{y}$ instead of our model to predict y. But don't we already have a test based on the t-statistics to test that? Well, yes, for a simple regression we do. In that simple case with just one predictor, the F-ratio is just the square of the t-statistic for the slope coefficient and the tests have the same P-value. But in a multiple regression, as we've learned, individual coefficients (and their t-statistics) aren't as simple. For multiple regression each predictor's coefficient can be tested individually using a t-test, but the F-statistic tests the null hypothesis that *all* the coefficients are zero.

The ANOVA table points the way to a large collection of related analyses known as the Analysis of Variance (hence "ANOVA"), which we discuss in later chapters. Analysis of variance is typically used to analyze data from experiments where the response variable is quantitative and the predictors (usually called factors) are categorical. We can, of course, represent those categorical factors with indicator variables, as we have seen. If we do that, then the ANOVA table holds many of the statistics we'll want for understanding the effects of the factors on the response.

[9]However, many statisticians think that adjusted R^2 doesn't go far enough. Several alternatives have been proposed with names like Aikaike's Information Criterion (AIC), Bayesian Information Criterion (BIC), and Mallow's C_p statistic. Each of these applies a slightly different penalty for each added predictor and you may run into these, especially when dealing with large datasets.

There is, of course, much more to know about this and many subtleties to the design of experiments. Entire courses are taught on this subject and it is essential in many sciences. But the fundamental idea to understand here is that all of the related analyses are intimately related to least squares multiple regression and everything you've learned about it.

We hope you'll have the interest and opportunity to study some of these fascinating topics and their applications.

WHAT CAN GO WRONG?

In this chapter, we've added inference to the regression explorations that we did in Chapters 7, 8, and 9. Everything covered in those chapters that could go wrong with regression can still go wrong. It's probably a good time to review Chapters 8 and 9. Take your time; we'll wait.

With inference, we've put numbers on our estimates and predictions, but these numbers are only as good as the model. Here are the main things to watch out for:

- **Don't fit a linear regression to data that aren't straight.** The Linearity Assumption is the most fundamental assumption. If the relationship between x and y isn't approximately linear, there's no sense in fitting a straight line to it. In a multiple regression, check partial regression plots for nonlinearity.

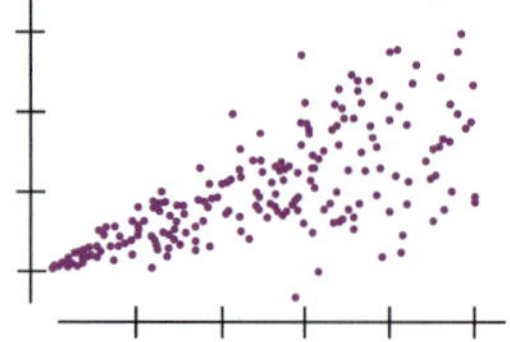

- **Watch out for the plot thickening.** We've warned before about the plot of residuals thickening. Now we can see why that's a problem. All the standard errors we've seen depend on the standard deviation of the residuals, s_e, estimated as s in most regression output tables. But if the spread is not consistent everywhere, a single value of s_e won't pick that up. The standard errors will use the average spread around the line and the *SE*s will be wrong. A re-expression of y is often a good fix for changing spread.
- **Make sure the errors are Normal.** As long as there's a consistent spread around the regression model, a histogram or Normal probability plot of the residuals provides a way to check that their distribution is nearly Normal. The theoretical results for Student's t require that Normality. Without it, we can't trust the t-statistics or their P-values.
- **Beware of collinearity.** When the x's are related to each other, it can cause many kind things to go wrong. Coefficients don't mean what they appear to (because "removing the effects of the other predictors" can leave behind a very different variable), coefficients can change sign or value with the introduction of a new predictor, and standard errors of the coefficients can be inflated (leading to large P-values even for variables that ought to be good predictors). Be alert for such apparent anomalies. (One good hint is a statistically significant F-statistic, rejecting the hypothesis that all the coefficients are zero, but no significant t-statistics; each coefficient appearing to say "not me" because each one "thinks" the others are doing the job of predicting y.) The simplest fix is to use fewer predictors.
- **Watch out for extrapolation.** It's tempting to think that because we have prediction *intervals*, they'll take care of all our uncertainty so we don't have to worry about extrapolating. Wrong. The interval is only as good as the model. The uncertainty our intervals predict is correct only if our model is true. There's no way to adjust for wrong models. That's why it's always dangerous to predict for x-values that lie far from the center of the data.
- **Watch out for influential points and outliers.** We always have to be on the lookout for a few points that have undue influence on our estimated model—and regression is certainly no exception.
- **Be cautious of one-tailed tests.** Tests of hypotheses about regression coefficients are usually two-tailed, because in a multiple regression, it is almost impossible to reliably predict the signs of coefficients ("after removing the effects of the other predictors"). If you insist on a one-tailed test, you'll need to cut the P-value in the regression table in half. And you'll need to actively defend your decision.

CONNECTIONS

Regression inference is connected to almost everything we've done so far. Scatterplots are essential for checking linearity and whether the plot thickens. Histograms and Normal probability plots come into play to check the Nearly Normal Condition. And we're still thinking about the same attributes of the data in these plots as we were back in the first part of the text.

Regression inference is also connected to just about every inference method we have seen for measured data. The assumption that the spread of data about the line is constant is essentially the same as the assumption of equal variances required for the pooled-t methods. Our use of all the residuals together to estimate their standard deviation is a form of pooling.

Inference for regression is closely related to inference for means, so your understanding of means transfers directly to your understanding of regression. Here's a table that displays the similarities:

	Mean	Simple Regression Slope	Multiple Regression Coefficients
Parameter	μ	β_1	$\beta_j\ j = 0, \ldots, k$
Statistic	$\bar{y}$	b_1	$b_j\ j = 0, \ldots, k$
Population Spread Estimate	$s_y = \sqrt{\frac{\sum(y - \bar{y})^2}{n - 1}}$	$s_e = \sqrt{\frac{\sum(y - \hat{y})^2}{n - 2}}$	$s_e = \sqrt{\frac{\sum(y - \hat{y})^2}{n - (k + 1)}}$
Standard Error of the Statistic	$SE(\bar{y}) = \frac{s_y}{\sqrt{n}}$	$SE(b_1) = \frac{s_e}{s_x\sqrt{n - 1}}$	$SE(b_k)$ found by computer
Test Statistic	$\frac{\bar{y} - \mu_0}{SE(\bar{y})} \sim t_{n-1}$	$\frac{b_1 - \beta_1}{SE(b_1)} \sim t_{n-2}$	$\frac{b_k - \beta_k}{SE(b_k)} \sim t_{n-(k+1)}$
Margin of Error	$ME = t^*_{n-1} \times SE(\bar{y})$	$ME = t^*_{n-2} \times SE(b_1)$	$ME = t^*_{n-(k+1)} \times SE(b_k)$

CHAPTER REVIEW

Apply your understanding of inference for means using Student's t to inference about regression coefficients.

Know the assumptions and conditions for inference about regression coefficients and how to check them, in this order:

- **Linearity Assumption**, checked with the Straight Enough Condition by examining a scatterplot of y vs. x or of the residuals plotted against the predicted values.
- **Independence Assumption**, which can't be checked, but is more plausible if the data were collected with appropriate randomization—the Randomization Condition.
- **Equal Variance Assumption**, which requires that the spread around the regression model be the same everywhere. We check it with the Does the Plot Thicken? Condition, assessed with a scatterplot of the residuals versus the predicted values.
- **Normal Population Assumption**, which is required to use Student's t-models unless the sample size is large. Check it with the Nearly Normal Condition by making a histogram or Normal probability plot of the residuals.

Know the components of the standard error of the slope coefficient:

- The standard deviation of the residuals, $s_e = \sqrt{\frac{\sum(y - \hat{y})^2}{n - 2}}$
- The standard deviation of x, $s_x = \sqrt{\frac{\sum(x - \bar{x})^2}{n - 1}}$
- The sample size, n

Understand that standard errors for multiple regression coefficients require more complex calculations and are generally performed by technology.

Be able to find and interpret the standard error of the slope.

- $SE(b_1) = \dfrac{s_e}{s_x\sqrt{n-1}}$
- The standard error of the slope is the estimated standard deviation of the sampling distribution of the slope.

State and test the standard null hypothesis on the slope.

- $H_0: \beta_1 = 0$. This would mean that x and y are not linearly related.
- We test this null hypothesis using the t-statistic: $t = \dfrac{b_1 - 0}{SE(b_1)}$.

Construct and interpret a confidence interval for the predicted mean value corresponding to a specified value, x_ν.

- $\hat{y}_\nu \pm t^*_{n-2} \times SE(\hat{\mu}_\nu)$, where $SE(\hat{\mu}_\nu) = \sqrt{SE^2(b_1) \times (x_\nu - \bar{x})^2 + \dfrac{s_e^2}{n}}$

Construct and interpret a confidence interval for an individual predicted value corresponding to a specified value, x_ν.

- $\hat{y}_\nu \pm t^*_{n-2} \times SE(\hat{y}_\nu)$, where $SE(\hat{y}_\nu) = \sqrt{SE^2(b_1) \times (x_\nu - \bar{x})^2 + \dfrac{s_e^2}{n} + s_e^2}$

REVIEW OF TERMS

Residual standard deviation The spread of the data around the regression line is measured with the residual standard deviation, s_e:

$$s_e = \sqrt{\frac{\Sigma(y-\hat{y})^2}{n-2}} = \sqrt{\frac{\Sigma e^2}{n-2}} \text{ (p. 741).}$$

***t*-test for the regression slope coefficient** When the assumptions are satisfied, we can perform a test for the slope coefficient. We usually test the null hypothesis that the true value of the slope is zero against the alternative that it is not. A zero slope would indicate a complete absence of linear relationship between y and x.

To test $H_0: \beta_1 = 0$, we find

$$t = \frac{b_1 - 0}{SE(b_1)},$$

where

$$SE(b_1) = \frac{s_e}{\sqrt{n-1}\, s_x}, \quad s_e = \sqrt{\frac{\Sigma(y-\hat{y})^2}{n-2}},$$

n is the number of cases, and s_x is the standard deviation of the x-values. We find the P-value from the Student's t-model with $n - 2$ degrees of freedom (p. 744).

Confidence interval for the regression coefficient When the assumptions are satisfied, we can find a confidence interval for a regression coefficient parameter from $b_1 \pm t^*_{n-k} \times SE(b_1)$. The critical value, t^*_{n-k}, depends on the confidence level specified and on Student's t-model with $n - k$ degrees of freedom and k is the number of estimated coefficients (including the intercept) (p. 744).

Collinearity In a multiple regression if any predictor is highly correlated with one or more of the other predictors the regression may suffer from collinearity. The consequences of collinearity are that coefficients must be interpreted carefully, their standard errors may be inflated, and predictors that are in fact highly related to the response variable may have coefficients that are not statistically significant. One measure of collinearity is the Variance Inflation Factor (p. 747).

Variance Inflation Factor (VIF) Find a regression of x_j on the other x's in a regression model and find the R^2 for that regression, R_j^2. The VIF for the coefficient b_j is then

$$VIF = \frac{1}{(1 - R_j^2)} \quad \text{(p. 748)}$$

Confidence interval for a predicted mean value Different samples will give different estimates of the regression model and, so, different predicted values for the same value of x. We find a confidence interval for the mean of these predicted values at a specified x-value, x_ν, as $\hat{y}_\nu \pm t^*_{n-2} \times SE(\hat{\mu}_\nu)$, where

$$SE(\hat{\mu}_\nu) = \sqrt{SE^2(b_1) \times (x_\nu - \bar{x})^2 + \frac{s_e^2}{n}}.$$

The critical value, t^*_{n-2}, depends on the specified confidence level and the Student's t-model with $n - 2$ degrees of freedom (p. 749).

Prediction interval for an individual Different samples will give different estimates of the regression model and, so, different predicted values for the same value of x. We can make a confidence interval to capture a certain percentage of the entire distribution of predicted values. This makes it much wider than the corresponding confidence interval for the mean. The confidence interval takes the form $\hat{y}_\nu \pm t^*_{n-2} \times SE(\hat{y}_\nu)$, where

$$SE(\hat{y}_\nu) = \sqrt{SE^2(b_1) \times (x_\nu - \bar{x})^2 + \frac{s_e^2}{n} + s_e^2}.$$

The critical value, t^*_{n-2}, depends on the specified confidence level and the Student's t-model with $n - 2$ degrees of freedom (p. 749).

***Logistic regression** A regression model suitable for a categorical response variable with two categories (p. 753).

Adjusted R^2 A modification of the R^2 statistic that adjusts for the number of predictors in the regression model. It can be used to compare regression models with different numbers of predictors (p. 755).

TECH SUPPORT

Inference for Regression Analysis

All statistics packages make a table of results for a regression. These tables differ slightly from one package to another, but all are essentially the same. We've seen two examples of such tables already.

All packages offer analyses of the residuals. With some, you must request plots of the residuals as you request the regression. Others let you find the regression first and then analyze the residuals afterward. Either way, your analysis is not complete if you don't check the residuals with a histogram or Normal probability plot and a scatterplot of the residuals against x or the predicted values.

You should, of course, always look at the scatterplot of your two variables before computing a regression.

Regressions are almost always found with a computer or calculator. The calculations are too long to do conveniently by hand for datasets of any reasonable size. No matter how the regression is computed, the results are usually presented in a table that has a standard form. Here's a portion of a typical regression results table, along with annotations showing where the numbers come from:

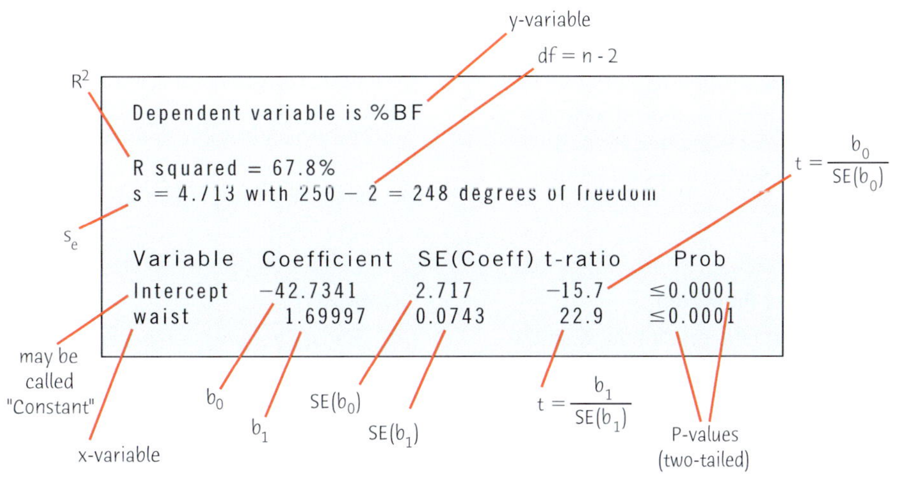

The regression table gives the coefficients (once you find them in the middle of all this other information), so we can see that the regression equation is

$$\widehat{\%BF} = -42.73 + 1.7\ Waist$$

and that the R^2 for the regression is 67.8%.

The column of *t*-ratios gives the test statistics for the respective null hypotheses that the true values of the coefficients are zero. The corresponding P-values are also usually reported.

DATA DESK

- In the tool ribbon or from the Calc menu, choose Regression. Data desk makes a regression table.
- Drag variables into the appropriate parts of the table. Data desk recomputes the regression immediately.
- Select plots of residuals from the Regression table's HyperView menu.
- Click on any coefficient to drop down a HyperView that offers the partial regression plot for that coefficient.

COMMENTS

You can replace any variable in the regression table by dragging the icon of another variable over the variable name in the table and dropping it there. You can add predictors to the model by dragging them into the predictor part of the table and dropping them there. You can remove any variable by click and hold over the variable's name.

EXCEL

- Select **Data Analysis** from the Analysis Group on the Data Tab.
- Select **Regression** from the Analysis Tools list.
- Click the **OK** button.
- Enter the data range holding the Y-variable in the box labeled Y-range.
- Enter the range of cells holding the X-variable in the box labeled X-range.
- Select the **New Worksheet Ply** option.
- Select Residuals options. Click the OK button.

Alternatively,

The LINEST function can compute a multiple regression. LINEST is an array function, producing a table of results that is not identical to, but contains the same results as the regression table shown above. Consult your Excel documentation for details on array functions and the LINEST function.

COMMENTS

The Y and X ranges do not need to be in the same rows of the spreadsheet, although they must cover the same number of cells. But it is a good idea to arrange your data in parallel columns as in a data table.

Excel calls the standard deviation of the residuals the Standard Error. This is a common error. Don't be confused; it is not $SE(y)$, but rather s_e.

JMP

For simple regression:

- From the Analyze menu, select **Fit Y** by **X**.
- Select variables: a Y, Response variable, and an X, Factor variable. Both must be continuous (quantitative).
- JMP makes a scatterplot.
- Click on the red triangle beside the heading labeled **Bivariate Fit** . . . and choose **Fit Line**. JMP draws the least squares regression line on the scatterplot and displays the results of the regression in tables below the plot.
- The portion of the table labeled Parameter Estimates gives the coefficients and their standard errors, *t*-ratios, and P-values.

COMMENTS

JMP chooses a regression analysis when both variables are Continuous. If you get a different analysis, check the variable types.

The Parameter table does not include the residual standard deviation s_e. You can find that as Root Mean Square Error in the Summary of Fit panel of the output.

For multiple regression:

- From the Analyze menu, select **Fit Model.**
- Specify the response, Y. Assign the predictors, X, in the Construct Model Effects dialog box.
- Click on **Run Model.**

COMMENTS

JMP chooses a regression analysis when the response variable is Continuous. The predictors can be any combination of quantitative or categorical. If you get a different analysis, check the variable types.

MINITAB

- Choose **Regression** from the Stat menu.
- Choose **Regression** . . . from the Regression submenu.
- In the Regression dialog, assign the Y-variable to the Response box and assign the X-variable to the Predictors box.
- Click the **Graphs** button.
- In the Regression-Graphs dialog, select **Standardized residuals**, and check **Normal plot of residuals** and **Residuals versus fits**.
- Click the **OK** button to return to the Regression dialog.
- Click the **OK** button to compute the regression.

COMMENTS

In Minitab Express, regression and multiple regression are found in the Regression menu of the toolbar. Minitab offers an output option that flags individual values that may be unusual due to large residual or leverage.

R

Suppose the response variable *y* and predictor variables $x_1, \ldots, x_k$ are in a data frame called mydata. To fit a regression of *y* on x_1 and x_2:

- mylm = lm(y ~ x1 + x2, data = mydata)
- summary(mylm) # gives the details of the fit, including estimates, SEs, and the ANOVA table
- plot(mylm) # gives a variety of plots
- To fit the model with *all* the predictors in the data frame,
- mylm = lm(y ~., data = mydata) # The period means use all other variables

COMMENTS

To get confidence or prediction intervals use:

- predict(mylm,interval = "confidence")

or

- predict(mylm, interval = "prediction")

Predictions on points not found in the original data frame can be found from predict as well. In the new data frame (called, say, mynewdata), there must be a predictor variable with the same name as the original "x" variable. Then predict(mylm, newdata=mynewdata) will produce predictions at all the "x" values of mynewdata.

SPSS

- Choose **Regression** from the Analyze menu.
- Choose **Linear** from the Regression submenu.
- In the Linear Regression dialog that appears, select the Y-variable and move it to the dependent target. Then move the X-variable to the independent target.
- Click the **Plots** button.
- In the Linear Regression Plots dialog, choose to plot the *SRESIDs against the *ZPRED values.
- To make partial regression plots, click **Produce all partial plots.**
- Click the **Continue** button to return to the Linear Regression dialog.
- Click the **OK** button to compute the regression.

STATCRUNCH

- Click on **Stat**.
- Choose **Regression** » **Multiple Linear**.
- Choose X and Y variable names from the list of columns.
- Indicate that you want to see a residuals plot and a histogram of the residuals.
- Click on **Compute!**
- Click on > to see any plots you chose.

COMMENTS

Be sure to check the conditions for regression inference by looking at both the residuals plot and a histogram of the residuals.

TI-83/84 PLUS

You need a special program to compute a multiple regression on the TI-83.

EXERCISES

SECTION 23.1

T **1. Graduate earnings** Does attending college pay back the investment? What factors predict higher earnings for graduates? A magazine surveyed graduates, asking about their point of view of the colleges they had attended. One good predictor of early career earnings ($/year) turned out to be the average SAT score of entering students. Here are the regression model and associated plots. Write the regression model, and explain what the slope coefficient means in this context.

Responsible Variable is Earn
R squared = 31.6%
S = 5647 with 706 − 2 = 704 degrees of freedom

Variable	Coefficient	SE(Coeff)	t-ratio	P-value
Intercept	14318.69	1706	8.39	<0.0001
SAT	27.12	1.506	18.01	<0.0001

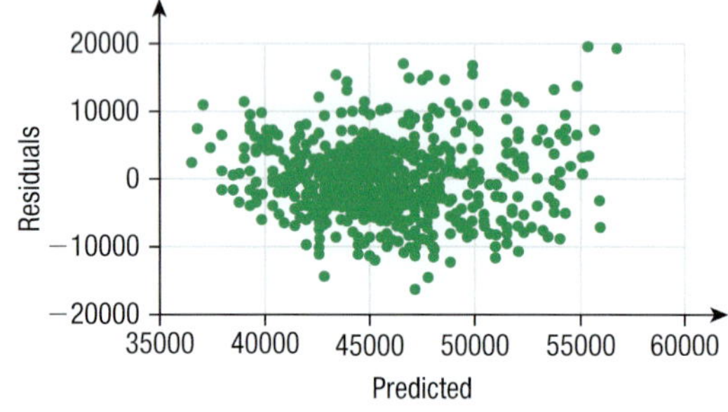

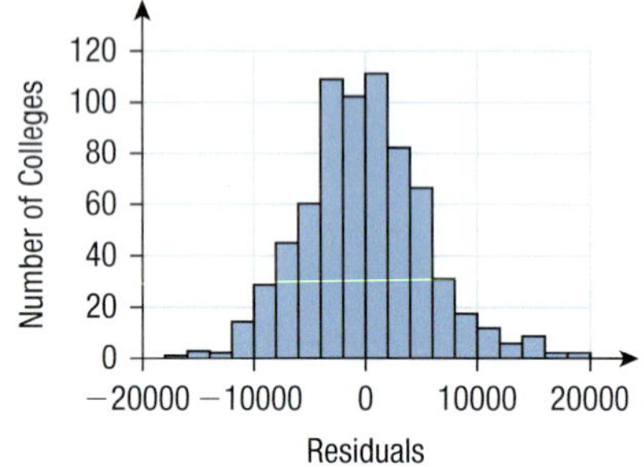

T **2. Shoot to score 2016** A college hockey coach collected data from the 2016–2017 National Hockey League season. He hopes to convince his players that the number of shots taken has an effect on the number of goals scored. The coach performed a preliminary analysis, using the scoring statistics from 65 offensive players who had played at least 44 games by the middle of the season. (If you use the data file, note that it includes defensive players as well. Use the variable *Offense* to select the players in this analysis.) He predicts *Goals* from number of *Shots*. Write the regression model and explain what the slope coefficient means in this context.

Response variable is: Goals
R squared = 49.9%
s = 2.983 with 65 − 2 = 63 degrees of freedom

Variable	Coefficient	SE(Coeff)	t-ratio	P-value
Intercept	1.13495	1.231	0.922	0.3602
Shots	0.099267	0.0125	7.93	<0.0001

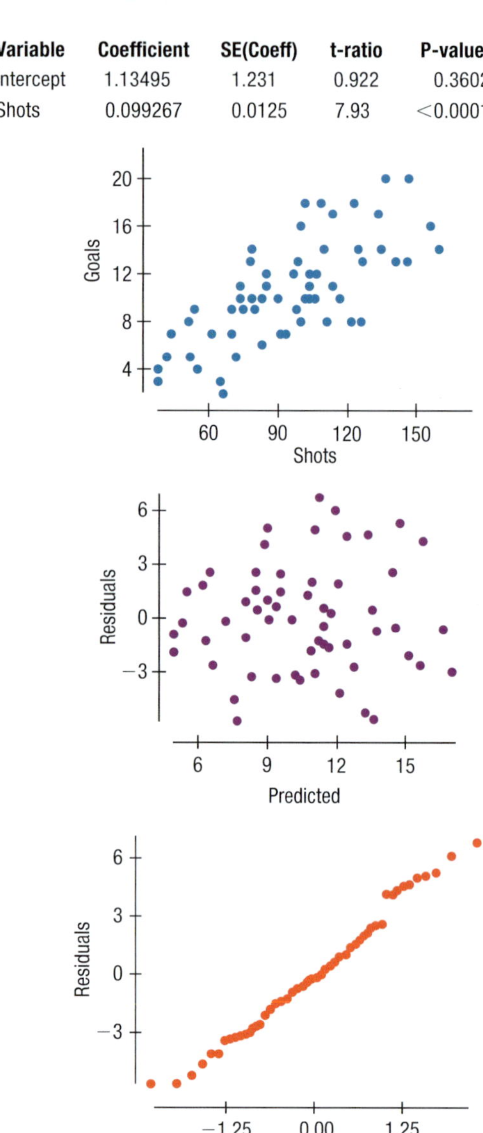

SECTION 23.2

T **3. Earnings II** Discuss the assumptions and conditions necessary for proceeding with the regression analysis in Exercise 1. Do you think the conditions are satisfied?

T **4. Shoot to score II** Discuss the assumptions and conditions necessary for proceeding with the regression analysis in Exercise 2. Do you think the conditions are satisfied?

SECTION 23.3

T **5. Earnings, part III** Using the regression output in Exercise 1, identify the residual standard deviation and explain what it means in the context of the problem.

T **6. Shoot to score, hat trick** Using the regression output from Exercise 2, identify the residual standard deviation and explain its meaning with a sentence in context.

T **7. Earnings, part IV** Continuing with the regression of Exercise 1, write a sentence that explains the meaning of the standard error of the slope of the regression line, $SE(b_1) = 1.545$, and the corresponding P-value.

T **8. Shoot to score, another one** Returning to the results of Exercise 2, write a sentence to explain the meaning of the standard error of the slope of the regression line, $SE(b_1) = 0.0125$, and the corresponding P-value.

SECTION 23.4

T **9. Earnings, part V** Use the information in Exercise 1 to test the hypotheses $H_0: \beta_1 = 0$ vs. $H_A: \beta_1 \neq 0$. What do you conclude about the relationship between earnings and SAT scores?

T **10. Shoot to score, number five** What can the hockey coach in Exercise 2 conclude about shooting and scoring goals from the fact that the P-value < 0.0001 for the slope of the regression line? Write a sentence in context.

T **11. Earnings VI** Construct a 95% confidence interval for the slope of the regression line in Exercise 1. Interpret the meaning of the interval. Be sure to state it in the context of the data and the question about the data.

T **12. Shoot to score, overtime** The coach in Exercise 2 found a 95% confidence interval for the slope of his regression line. Recall that he is trying to understand how the number of goals scored is related to shots taken. Interpret with a sentence the meaning of the interval $0.099267 \pm 2 \times 0.0125$

SECTION 23.5

T **13. Earnings and need** Continuing with the data from Exercise 1, here's a regression with the percent of students who receive merit-based financial aid included in the model:

Response variable is: Earn
R squared = 35.5%
s = 5409 with 687 − 3 = 684 degrees of freedom

Variable	Coefficient	SE(Coeff)	t-ratio	P-value
Intercept	23974.2	2327	10.3	<0.0001
SAT	23.1880	1.658	14.0	<0.0001
%need	−8500.75	1329	−6.40	<0.0001

a) Write the regression model.
b) What is the interpretation of the coefficient of *SAT* in this model? How does it differ from the interpretation in Exercise 1?

T **14. Shoot to score, time on ice** The players on the team in Exercise 2 point out to the coach that they can't shoot if they are not on the ice. They add the variable *TimeOnIce/Game (TOI/G)* (in minutes per game) to the regression: (Reminder: if you are using the full dataset be sure to select only the offence players.)

Response variable is: Goals
R squared = 53.7%
s = 2.891 with 65 − 3 = 62 degrees of freedom

Variable	Coefficient	SE(Coeff)	t-ratio	P-value
Intercept	−3.90580	2.539	−1.54	0.1291
Shots	0.070019	0.0178	3.94	0.0002
TOI/G	0.458088	0.2037	2.25	0.0281

a) Write the regression model.
b) What is the interpretation of the coefficient of *Shots* in this model? How does it differ from the interpretation in Exercise 2?
c) Find and interpret a 95% confidence interval for the coefficient of *TOI/G*.

T **15. Earnings and more** A second predictor in Exercise 13 improved the regression model of Exercise 1, so let's try a third. Here's a model with average ACT score of the entering class included:

Response variable is: Earn
R squared = 36.5%
s = 5372 with 687 − 4 = 683 degrees of freedom

Variable	Coefficient	SE(Coeff)	t-ratio	P-value
Intercept	25162.8	2340	10.8	<0.0001
SAT	10.1117	4.355	2.32	0.0205
%need	−8564.03	1320	−6.49	<0.0001
ACT	551.243	170.0	3.24	0.0012

a) The coefficient of *SAT* in this model is quite different from the *SAT* coefficient in the original model of Exercise 1 or the multiple regression model of Exercise 13. Why the change?
b) Find a 95% confidence interval for the coefficient of *SAT*. How does it compare with the one you found in Exercise 13?
c) The *t*-ratio associated with the *SAT* coefficient is now much smaller and the corresponding P-value much larger. Explain why this has happened.

T **16. Shoot to score, double overtime** Continuing from Exercise 14, the coach responds to the players by claiming that shooting accuracy is more important than time on the ice. He adds *Shoot%* (% of shots on goal) to the model.

Response variable is: Goals
R squared = 95.7%
s = 0.8850 with 65 − 4 = 61 degrees of freedom

Variable	Coefficient	SE(Coeff)	t-ratio	P-value
Intercept	−9.11025	0.8057	−11.3	<0.0001
Shots	0.111264	0.0057	19.5	<0.0001
TOI/G	−0.040008	0.0656	−0.610	0.5441
Shoot%	0.872339	0.0356	24.5	<0.0001

a) The coefficient of *TOI/G* in this model is quite different from the *TOI/G* coefficient in the previous model of Exercise 14. Why the change?
b) Find a 95% confidence interval for the coefficient of *TOI/G*. How does it compare with the one you found in Exercise 14?
c) The *t*-ratio associated with the *TOI/G* coefficient is now much smaller and the corresponding P-value much larger. Explain why this has happened.

SECTION 23.6

17. Earnings, predictions Naturally, you would like to know what you are going to earn in the next few years. Explain why a regression model such as the ones we have found won't do a very good job of such a prediction. (Sorry.)

18. Shoot to score, predictions The coach we've been following wants to predict how many goals each of his players will score this season. Explain why a model like the ones we've made won't be very successful at doing that.

CHAPTER EXERCISES

19. Earnings, planning An SAT preparation course wants to advertise based on the analyses we've seen that raising your SAT scores will increase your eventual earnings. Is that conclusion supported by these analyses?

20. Shoot to score, shootout The coach from Exercise 2 called a team meeting to summarize the results from his study. Would it be a good strategy to tell the players that all they need to do is to shoot more and the goals will follow?

21. Tracking hurricanes 2016 In Chapter 6, we looked at data from the National Oceanic and Atmospheric Administration about their success in predicting hurricane tracks. Here is a scatterplot of the error (in nautical miles) for predicting hurricane locations 24 hours in the future vs. the year in which the prediction (and the hurricane) occurred.

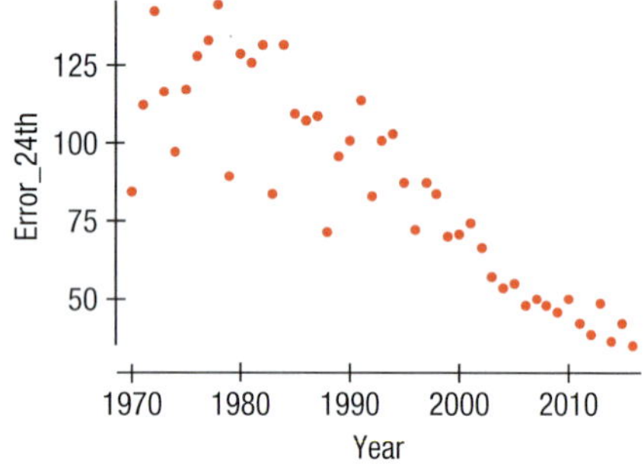

In Chapter 6, we could describe this relationship only in general terms. Now we can learn more. Here is the regression analysis:

Response variable is: Error_24h
R squared = 77.5%   R squared (adjusted) = 77.0%
s = 15.44 with 47 − 2 = 45 degrees of freedom

Variable	Coefficient	SE(Coeff)	t-ratio	P-value
Intercept	4205.28	331.0	12.7	<0.0001
Year	−2.06708	0.1661	−12.4	<0.0001

a) Explain in words and numbers what the regression says.
b) State the hypothesis about the slope (both numerically and in words) that describes how hurricane prediction quality has changed.
c) Assuming that the assumptions for inference are satisfied, perform the hypothesis test and state your conclusion. Be sure to state it in terms of prediction errors and years.
d) Explain what the *R*-squared means in terms of this regression.

22. Drug use 2013 The *2013 World Drug Report* investigated the prevalence of drug use as a percentage of the population aged 15 to 64. Data from 32 European countries are shown in the following scatterplot and regression analysis. (*World Drug Report*, 2013. www.unodc.org/unodc/en/data-and-analysis/WDR-2013.html)

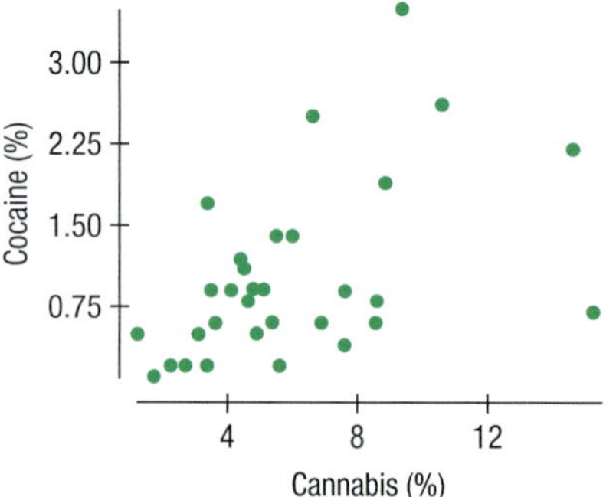

Response variable is Cocaine
R-squared = 25.8%
s = 0.7076 with 32 − 2 = 30 degrees of freedom

Variable	Coefficient	SE(Coeff)	t-Ratio	P-Value
Intercept	0.296899	0.2627	1.13	0.2673
Cannabis	0.123638	0.0382	3.23	0.0030

a) Explain in context what the regression says.
b) State the hypothesis about the slope (both numerically and in words) that describes how use of cannabis is associated with other drugs.
c) Assuming that the assumptions for inference are satisfied, perform the hypothesis test and state your conclusion in context.
d) Explain what *R*-squared means in context.
e) Do these results indicate that cannabis use leads to the use of harder drugs? Explain.

23. Sea ice Climate scientists have been observing the extent of sea ice in the northern Arctic using satellite observations. Many have expressed concern because in recent decades the extent of sea ice has declined precipitously—possibly due to global climate change. Here is an analysis relating the minimum (September) *Extent* of sea ice (km^2) to the *Mean Global Temperature* (°C) for a set of years:

Response Variable is: Extent
R-squared = 88.7%
S = 0.4437 with 37 − 2 = 35 degrees of freedom

Variable	Coefficient	SE(Coeff)	t-ratio	P-value
Intercept	97.3704	5.546	17.56	<0.0001
Temperature	−6.01784	0.3633	−16.56	<0.0001

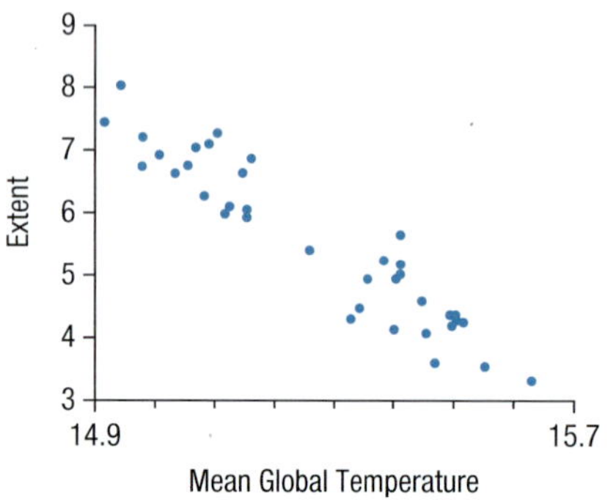

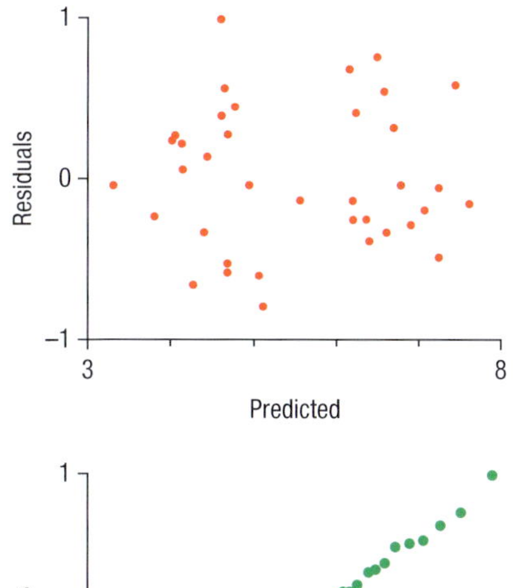

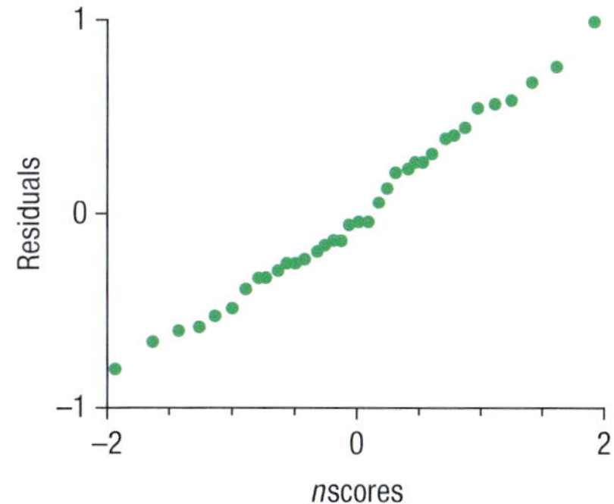

a) Explain in context what the regression says.
b) Check the assumptions and conditions for regression inference.
c) The output reports $s = 0.4437$. Explain what that means in this context.
d) What's the value of the standard error of the slope of the regression line?
e) Explain what that means in this context.
f) Does this analysis prove that global temperature changes are causing sea ice to melt? Explain.

24. How does the price of a house depend on its size? Data from a town on 20 randomly selected houses that had been sold include data on price (\$1000s) and size (1000 ft^2), producing the following graphs and computer output:

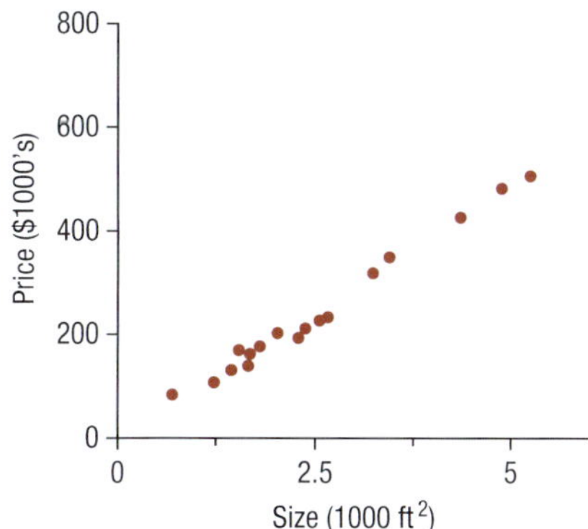

Dependent Variable is: price
R-squared = 98.7%
S = 14.44 with 20 − 2 = 18 degrees of freedom

Variable	Coefficient	SE(Coeff)	t-ratio	P-value
Intercept	−7.99	7.19	−1.11	0.2809
Size	99.12	2.67	37.19	<0.0001

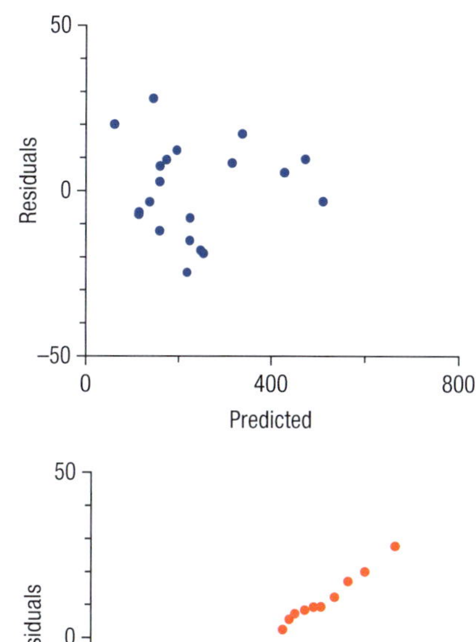

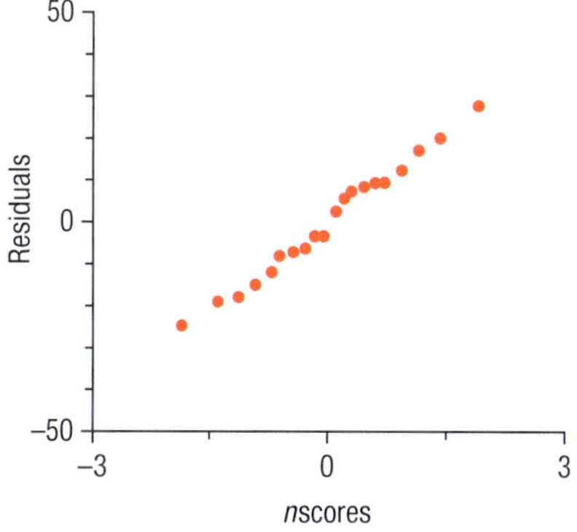

a) Explain in context what the regression says.
b) The intercept is negative. What does this mean? (Hint: Notice the P-value.)
c) The output reports $s = 14.44$. Explain what that means in this context.
d) What's the value of the standard error of the slope of the regression line?
e) Explain what that means in this context.

25. More sea ice Exercise 23 shows computer output examining the association between Arctic sea ice extent and global mean temperature. Find a 95% confidence interval for the slope and interpret it in context.

26. Second home Exercise 24 shows computer output examining the association between the sizes of houses and their sale prices.

a) Check the assumptions and conditions for inference.
b) Find a 95% confidence interval for the slope and interpret it in context.

27. Hot dogs Healthy eating probably doesn't include hot dogs, but if you are going to have one, you'd probably hope it's low in both calories and sodium. A magazine listed the number of calories and sodium content (in milligrams) for 13 brands of hot dogs it tested. Examine the association, assuming that the data satisfy the conditions for inference.

Dependent Variable is Sodium
R-squared = 63.8%
S = 71.21 with 13 − 2 = 11 degrees of freedom

Variable	Coefficient	SE(Coeff)	t-ratio	P-value
Constant	84.7047	78.74	1.08	0.3051
Calories	2.37982	0.5409	4.40	0.0011

a) State the appropriate hypotheses about the slope.
b) Test your hypotheses, and state your conclusion in the proper context.

28. Cholesterol Does a person's cholesterol level tend to change with age? Data collected from 1446 adults aged 45 to 62 produced the regression analysis shown. Assuming that the data satisfy the conditions for inference, examine the association between age and cholesterol level.

Dependent Variable is Cholesterol

Variable	Coefficient	SE(Coeff)	t-ratio	P-value
Intercept	130.322	20.73	6.29	≤0.0001
Age	0.891458	0.2434	3.66	0.0003

a) State the appropriate hypothesis for the slope.
b) Test your hypothesis, and state your conclusion in the proper context.

29. Second frank Look again at Exercise 27's regression output for the calorie and sodium content of hot dogs.

a) The output reports $s = 59.66$. Explain what that means in this context.
b) What's the value of the standard error of the slope of the regression line?
c) Explain what that means in this context.

30. More cholesterol Look again at Exercise 28's regression output for age and cholesterol level. (Data in **Framingham**)

a) The output reports $s = 46.16$. Explain what that means in this context.
b) What's the value of the standard error of the slope of the regression line?
c) Explain what that means in this context.

31. Last dog Based on the regression output seen in Exercise 27, create a 95% confidence interval for the slope of the regression line and interpret your interval in context.

32. Cholesterol, finis Based on the regression output seen in Exercise 28, create a 95% confidence interval for the slope of the regression line and interpret it in context.

33. Marriage age 2017 Chapter 8, Exercises 42, 44, and 49, looked at the how the age at first marriage has changed over time for men and women. One trend was that people have been waiting until they are older to get married.

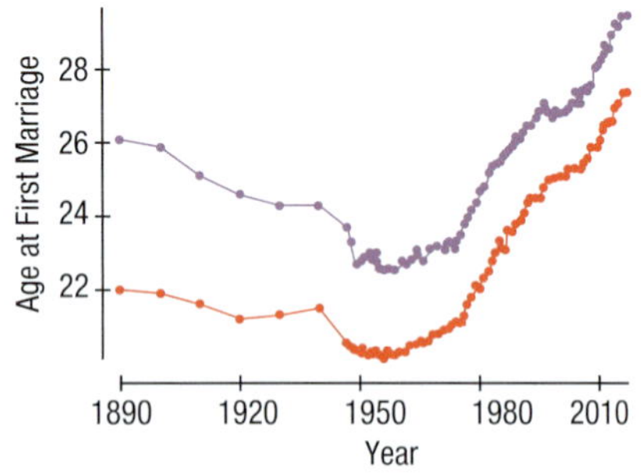

Generally, men are older at their first marriage than are women, but the gap seems to have been closing. Has the difference in age at first marriage between men and women really been declining?

Response variable is: Men − Women
R squared = 69.9%
s = 0.2487 with 78 − 2 = 76 degrees of freedom

Variable	Coefficient	SE(Coeff)	t-ratio	P-value
Intercept	29.6175	2.049	14.5	<0.0001
Year	−0.013782	0.0010	−13.3	<0.0001

a) Write appropriate hypotheses.
b) Here are plots of the residuals. Comment on what they say about the regression.

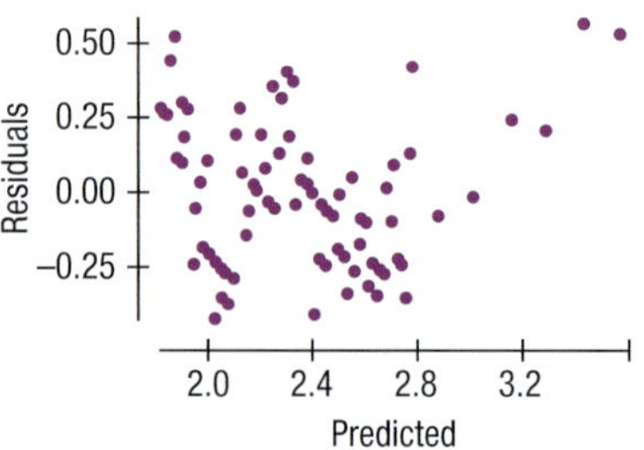

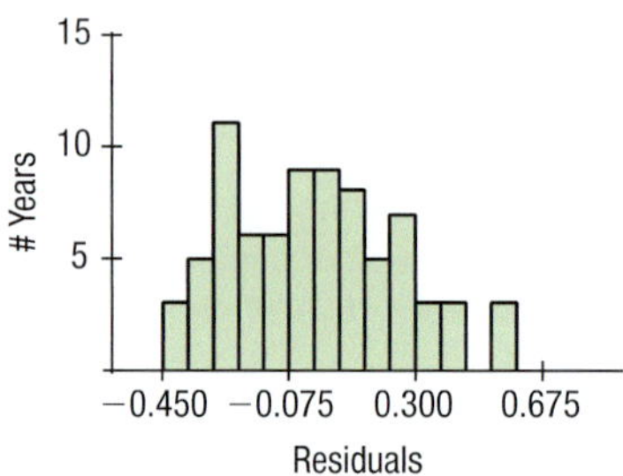

And here is a Normal probability plot of the residuals:

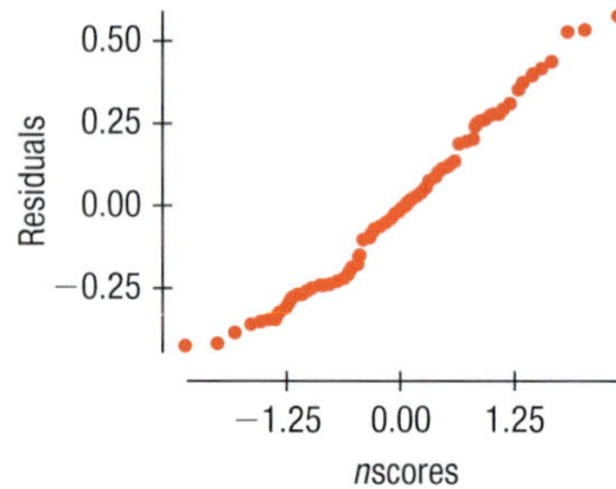

c) Test the hypothesis and state your conclusion about the trend in age at first marriage.

34. Used Civics 2017 On January 22, 2017, www.autotrader.com listed 55 used Honda Civics for sale by owner. Here's a scatterplot of the asking price vs. the number of miles on the odometer (in thousands):

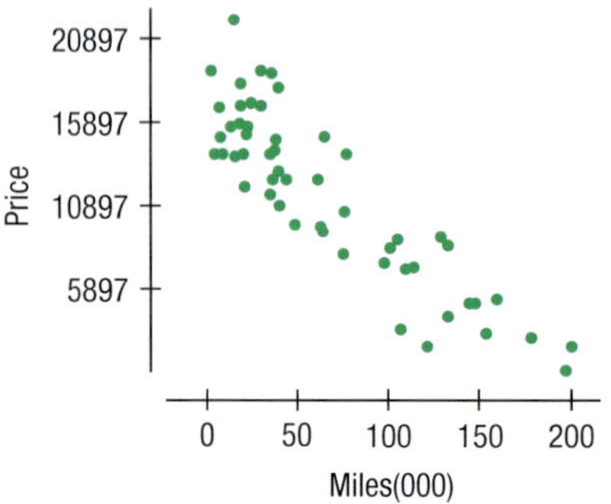

a) Do you think a linear model is appropriate? Explain.

Here is the regression model:

Response variable is: Price
R squared = 79.3%
s = 2363 with 55 − 2 = 53 degrees of freedom

Variable	Coefficient	SE(Coeff)
Intercept	17164.3	500.2
Miles (000)	−84.1570	5.907

b) State the null and alternative hypotheses under investigation.
c) Assuming that the assumptions for regression inference are reasonable, find the *t*- and P-values.
d) State your conclusion.

Here is a plot of the residuals.

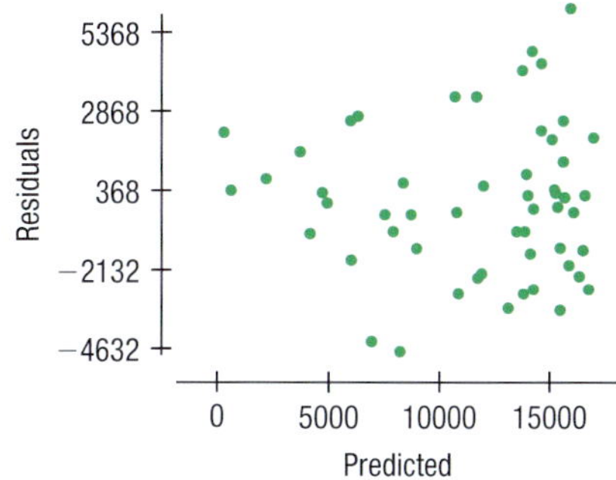

e) Do the assumptions and conditions for regression inference appear to be satisfied? If not, what would you suggest doing to improve the model?

35. Marriage age 2015, again Based on the analysis of marriage ages given in Exercise 33, find a 95% confidence interval for the rate at which the age gap is closing. Explain what your confidence interval means.

36. Used Civics 2017, again Based on the analysis of used car prices in Exercise 34, create a 95% confidence interval for the slope of the regression line and explain what your interval means in context.

37. Streams Biologists studying the effects of acid rain on wildlife collected data from 19 streams. They recorded the *pH* (acidity) of the water and the *BCI*, a measure of biological diversity. Here's a scatterplot of *BCI* against *pH*:

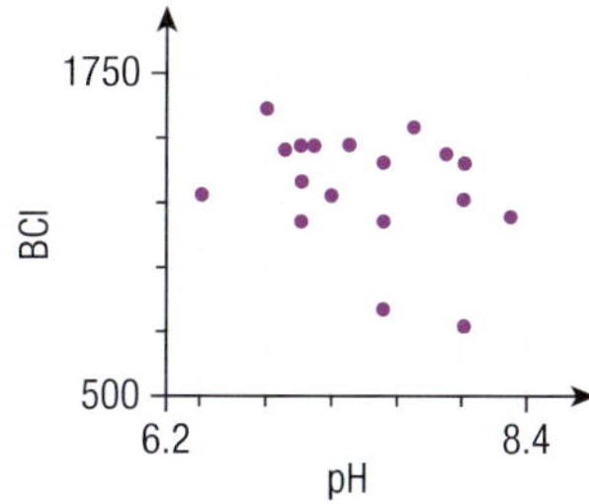

And here is part of the regression analysis:

Dependent Variable is: BCI
R squared = 9.0%
S = 236.4 with 19 − 2 = 17 degrees of freedom

Variable	Coefficient	SE(Coeff)
Intercept	2353.09	826.6
pH	−144.244	111.46

a) State the null and alternative hypotheses under investigation.
b) Assuming that the assumptions for regression inference are reasonable, find the *t*- and P-values.
c) State your conclusion.

38. Civics again The price of a car depends on its age as well as on its mileage. Here is a regression in which the age of the cars (in years) is included in the regression model from Exercise 34:

Response variable is: Price
R squared = 86.1%
s = 1957 with 55 − 3 = 52 degrees of freedom

Variable	Coefficient	SE(Coeff)	t-ratio	P-value
Intercept	17751.4	430.5	41.2	<0.0001
miles(000)	−53.5638	7.809	−6.86	<0.0001
age	−389.373	77.47	−5.03	<0.0001

a) What is the interpretation of the coefficient of *miles(000)* in this regression? Why is it so different from the interpretation in the model of Exercise 34?

Here is a a partial regression plot for the coefficient of *miles(000)*:

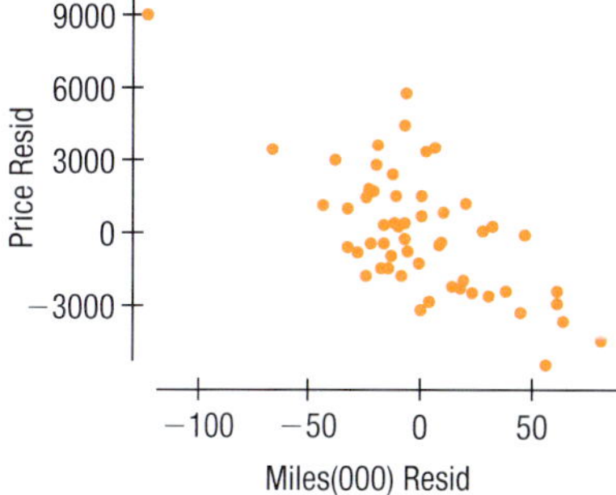

b) What is the slope of this plot? How do you know?
c) What effect does the point in the upper left have on the coefficient and P-value of *miles(000)*?

39. Streams again Here is a regression model in which the variable *Alkali* has been included in the model from Exercise 37:

Response variable is: BCI
R squared = 37.8%
s = 130.1 with 163 − 3 = 160 degrees of freedom

Variable	Coefficient	SE(Coeff)	t-ratio	P-value
Intercept	1641.48	271.1	6.06	<0.0001
pH	−28.9009	39.92	−0.724	0.4701
Alkali	−1.45740	0.2774	−5.25	<0.0001

a) State the null and alternative hypotheses about the coefficient of *pH* in this regression.
b) Assuming that the assumptions for regression inference are reasonable, test the hypothesis and state your conclusion.

c) Account for any differences with the conclusion you drew in Exercise 37.

Here is a partial regression plot for the coefficient of *Alkali*:

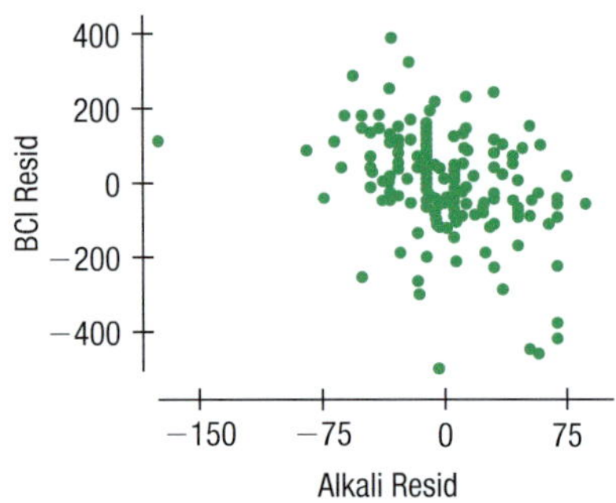

d) What is the least squares slope of the partial regression plot? How do you know?

e) What effect does the point on the far left of the plot have on the coefficient of *Alkali*?

40. Fuel economy A consumer organization has reported test data for 50 car models. We will examine the association between the weight of the car (in thousands of pounds) and the fuel efficiency (in miles per gallon). Here are the scatterplot, summary statistics, and regression analysis:

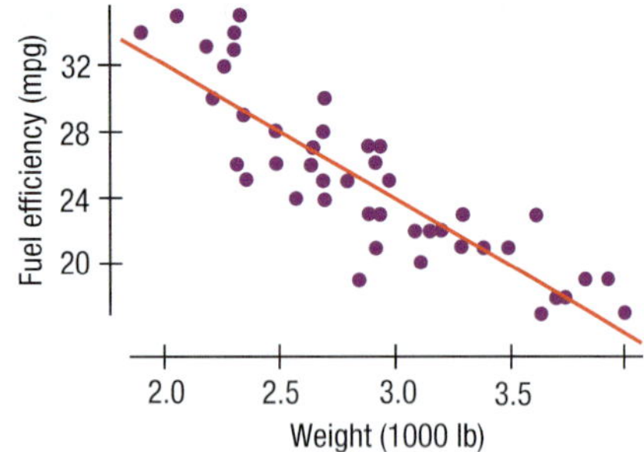

Variable	Count	Mean	StdDev
MPG	50	25.0200	4.83394
Wt(1000's)	50	2.88780	0.511656

Dependent variable is MPG
R-squared = 75.6%
s = 2.413 with 50 − 2 = 48 df

Variable	Coefficient	SE(Coeff)	t-Ratio	P-Value
Intercept	48.7393	1.976	24.7	≤0.0001
Weight	−8.2136	0.674	−12.2	≤0.0001

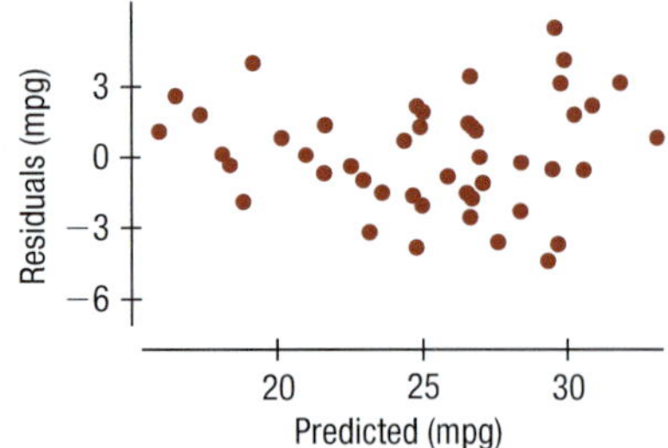

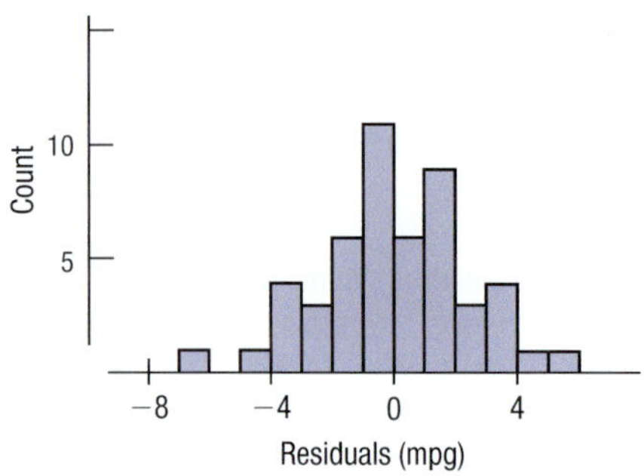

a) Is there strong evidence of an association between the weight of a car and its gas mileage? Write an appropriate hypothesis.

b) Are the assumptions for regression satisfied?

c) Test your hypothesis and state your conclusion.

41. Climate change 2016 Data collected from around the globe show that the earth is getting warmer. The generally accepted explanation relates climate change to an increase in atmospheric levels of carbon dioxide (CO_2) because CO_2 is a greenhouse gas that traps the heat of the sun. A measurement of the mean annual concentration in the atmosphere (parts per million) is taken every year, as well as the global temperature anomaly, or the difference in mean global temperature relative to a base period from 1981 to 2010 in degrees Celsius. The accompanying output shows the scatterplot and regression for the years from 1959 to 2016:

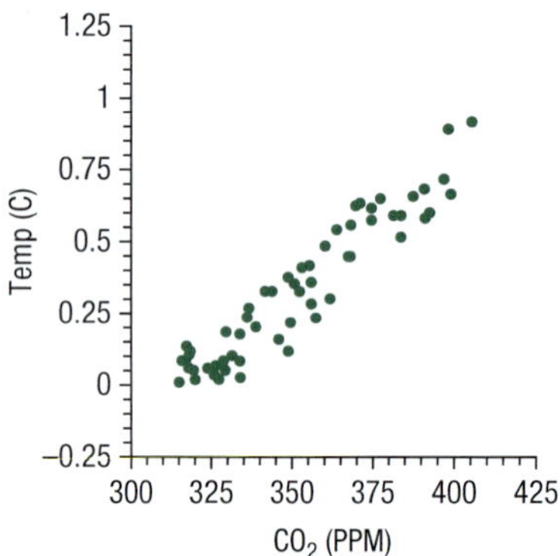

Response Variable is: Global Temperature Anomaly
R squared = 89.6%
S = 0.0810 with 58 − 2 = 56 degrees of freedom

Variable	Coefficient	SE(Coeff)	t-ratio	P-value
Intercept	−2.84396	0.1451	−19.61	<0.0001
CO_2	0.00902	0.0004	21.97	<0.0001

a) Write the equation of the regression line.

b) Is there evidence of an association between CO_2 level and global temperature anomaly?

c) Do you think predictions made by this regression will be very accurate? Explain.

d) Does this regression prove that increasing CO_2 levels are causing global warming? Discuss.

T **42.** Fuel economy, part II Consider again the data in Exercise 40 about the gas mileage and weights of cars.

a) Create a 95% confidence interval for the slope of the regression line.
b) Explain in this context what your confidence interval means.

T **43.** Climate change, part II Consider the CO_2 and global temperature data of Exercise 41.

a) Find a 90% confidence interval for the slope of the true line describing the association between *Temp* and CO_2.
b) Explain in this context what your confidence interval means.

T **44.** Fuel economy, part III Consider again the data in Exercise 40 about the gas mileage and weights of cars.

a) Create a 95% confidence interval for the average fuel efficiency among cars weighing 2500 pounds, and explain what your interval means.
b) Create a 95% prediction interval for the gas mileage you might get driving your new 3450-pound SUV, and explain what that interval means.

T **45.** Climate change again Consider once again the CO_2 and global temperature data of Exercise 41. The mean CO_2 level for these data is 352.566 ppm.

a) Find a 90% confidence interval for the mean global temperature anomaly if the CO_2 level reaches 450 ppm.
b) Find a 90% prediction interval for the mean global temperature anomaly if the CO_2 level reaches 450 ppm.
c) The goal of the Paris Climate agreement is equivalent to a temperature anomaly of about 1.3 °C in our data. If CO_2 levels do reach 450 ppm, would this be a plausible value?

T **46.** Cereals A healthy cereal should be low in both calories and sodium. Data for 96 cereals were examined and judged acceptable for inference. The 96 cereals had between 50 and 160 calories per serving and between 0 and 320 mg of sodium per serving. Here's the regression analysis:

R-squared = 9.7%
S = 83.66 with 96 − 2 = 94 degrees of freedom

Variable	Coefficient	SE(Coeff)	t-ratio	P-value
Intercept	21.0087	39.58	0.531	0.5967
Calories	0.00902	0.6646	2.81	0.0060

a) Is there an association between the number of calories and the sodium content of cereals? Explain.
b) Do you think this association is strong enough to be useful? Explain.

T **47.** Brain size Is your IQ related to the size of your brain? A group of female college students took a test that measured their verbal IQs and also underwent an MRI scan to measure the size of their brains (in 1000s of pixels). The scatterplot and regression analysis are shown, and the assumptions for inference were satisfied.

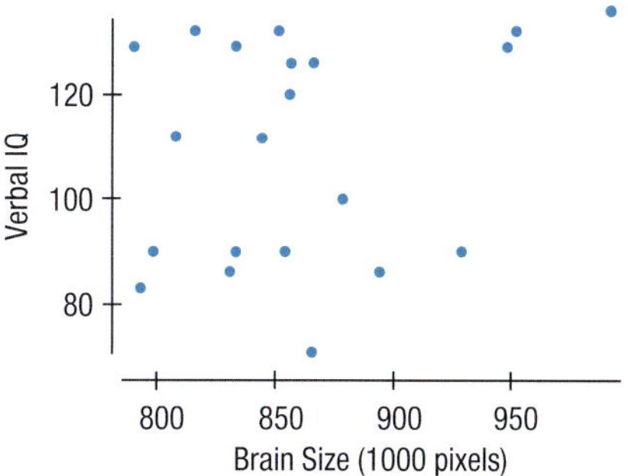

Dependent variable is IQ_Verbal
R-squared = 6.5%

Variable	Coefficient	SE(Coeff)
Intercept	24.1835	76.38
Size	0.0988	0.088

a) Test an appropriate hypothesis about the association between brain size and IQ.
b) State your conclusion about the strength of this association.

T **48.** Cereals, part 2 Further analysis of the data for the breakfast cereals in Exercise 46 looked for an association between *Fiber* content and *Calories* by attempting to construct a linear model. Here are three graphs. Which of the assumptions for inference are violated? Explain.

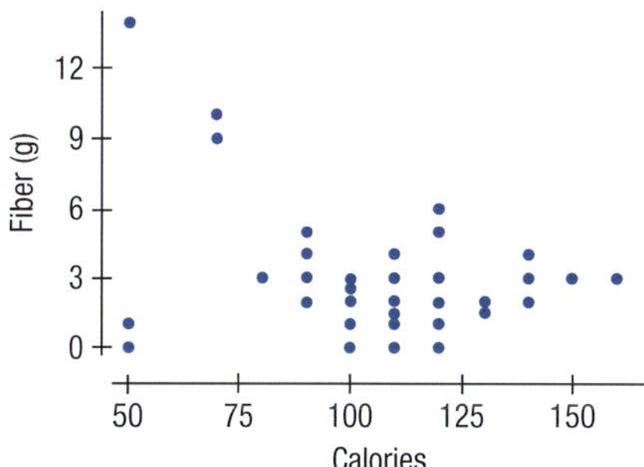

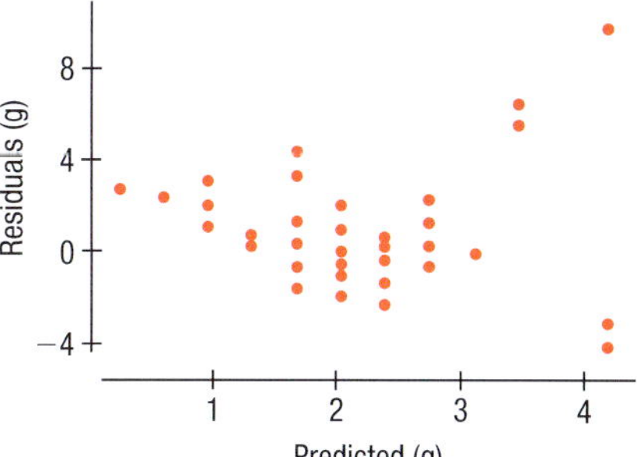

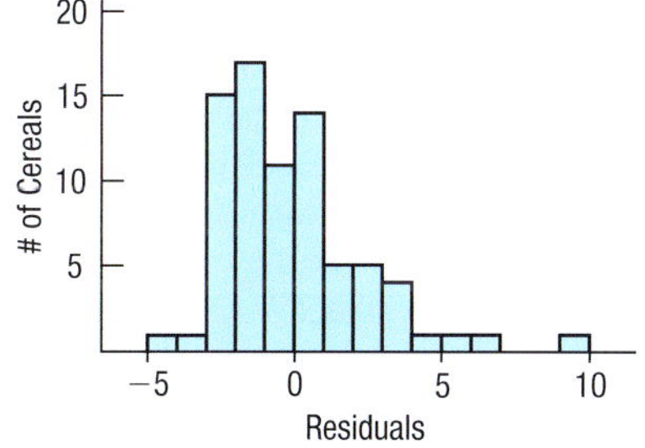

T **49. City climate** The output shows an attempt to model the association between average *January Temperature* (in degrees Fahrenheit) and *Latitude* (in degrees north of the equator) for 59 cities. Which of the assumptions for inference do you think are violated? Explain.

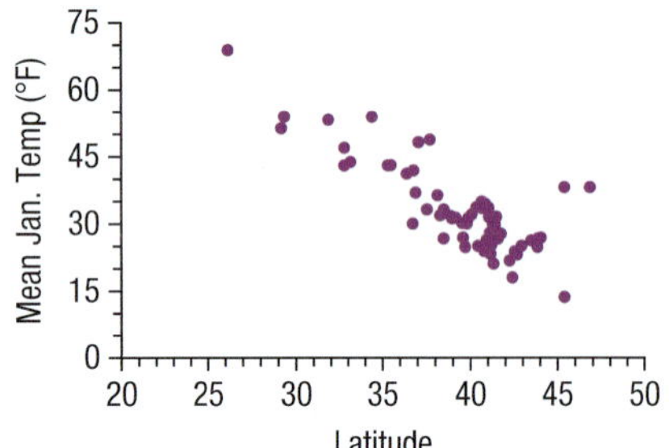

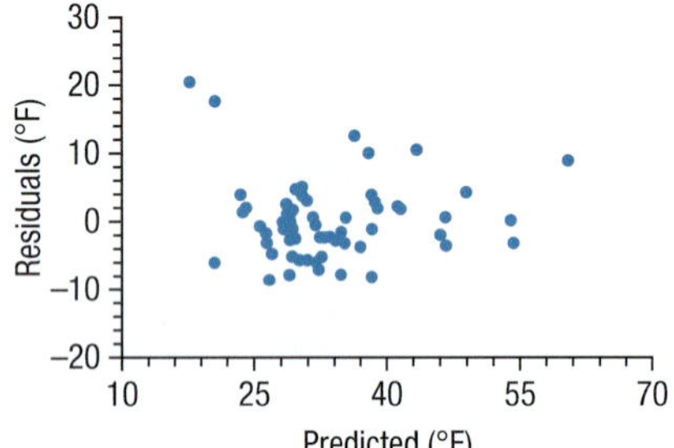

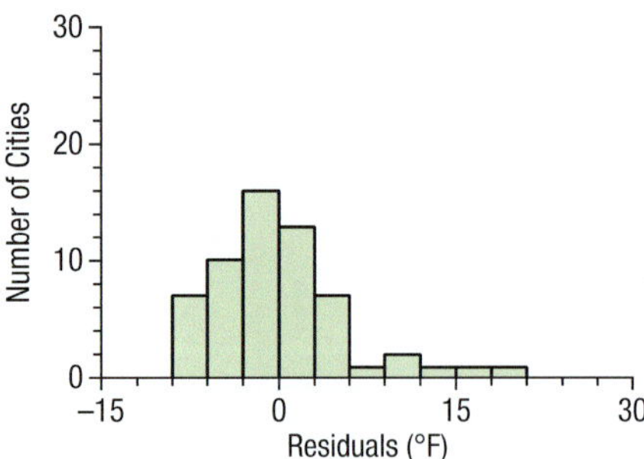

T **50. See ice?** A skeptic suggests that reduced sea ice isn't due to global climate change at all. He offers the following model, including *Year since 1979* as another predictor as an alternative to the model in Exercise 23 (Data in **Sea ice**):

Response variable is: Extent
No Selector
R squared = 74.3%
s = 0.5722 with 37 − 3 = 34 degrees of freedom

Variable	Coefficient	SE(Coeff)	t-ratio	P-value
Intercept	15.6311	16.39	0.954	0.3468
Mean global Temp	−0.515332	1.095	−0.471	0.6409
Year since 1979	−0.078959	0.0198	−3.98	0.0003

a) The coefficient of *Mean global temp* now has a P-value of 0.6409. Interpret that value in the context of this model.
b) Would you conclude that global temperature is not, in fact, related to sea ice extent? Why or why not?
c) What do you think might be the reason that the coefficient and P-value for *Mean global Temp* are so different between this model and the one of Exercise 23?

51. Ozone and population A government environmental agency is examining the relationship between the ozone level (in parts per million) and the population (in millions) of certain cities. Part of the regression analysis is shown.

Dependent Variable is Ozone
R-squared = 80.7%
S = 5.942 with 16 − 2 = 14 degrees of freedom

Variable	Coefficient	SE(Coeff)
Intercept	18.124	2.336
Pop	6.045	1.605

a) We suspect that the ozone level is related to the population of a city. Is the relationship statistically significant? Assuming the conditions for inference are satisfied, test an appropriate hypothesis, and state your conclusion in context.
b) Do you think that the population of a city is a useful predictor of ozone level? Use the values of both R^2 and s in your explanation.

T **52. Sales and profits** A business analyst was interested in the relationship between a company's sales and its profits. She collected data (in millions of dollars) from a random sample of big business companies and created the regression analysis and summary statistics shown. The assumptions for regression inference appeared to be satisfied.

	Profits	Sales
Count	79	79
Mean	205.067	4171.38
Variance	635,026	49,157,000
Std Dev	796.885	7011.205

Dependent Variable is Profits
R-squared = 67.7% S = 474.5
with 79 − 2 = 77 degrees of freedom

Variable	Coefficient	SE(Coeff)
Intercept	−115.908	63.95
Sales	0.076947	0.0077

a) Is there a statistically significant association between sales and profits? Test an appropriate hypothesis, and state your conclusion in context.
b) Do you think that a company's sales serve as a useful predictor of its profits? Use the values of both R^2 and s in your explanation.

53. Ozone, again Consider again the relationship between the population and ozone level of U.S. cities that you analyzed in Exercise 51.

a) Give a 90% confidence interval for the slope of the relationship between ozone level and population.
b) For the cities studied, the mean population was 1.7 million people. The population of Boston is approximately 0.6 million people. Predict the mean ozone level for cities of that size with an interval in which you have 90% confidence.

T **54. More sales and profits** Consider again the relationship between the sales and profits of Fortune 500 companies that you analyzed in Exercise 52.

a) Find a 95% confidence interval for the slope of the regression line. Interpret your interval in context.
b) Last year, the drug manufacturer Eli Lilly, Inc., reported gross sales of \$23 billion (that's \$23,000 million). Create a 95% prediction interval for the company's profits, and interpret your interval in context.

T **55. Tablet computers 2014** A website listed the battery life (in hours) and luminous intensity (i.e., screen brightness, in cd/m^2) for a sample of tablet computers. We want to know if *Battery life* is related to the maximum *Screen Brightness*.

Dependent Variable Video Battery life (in hours)
R-squared = 4.43%
S = 1.946 with 69 − 2 = 67 degrees of freedom

Variable	Coefficient	SE(Coeff)	t-ratio	P-value
Intercept	6.24114	0.97561	6.22	0.000
Max Brightness	0.00442	0.0028	2.03	0.0463

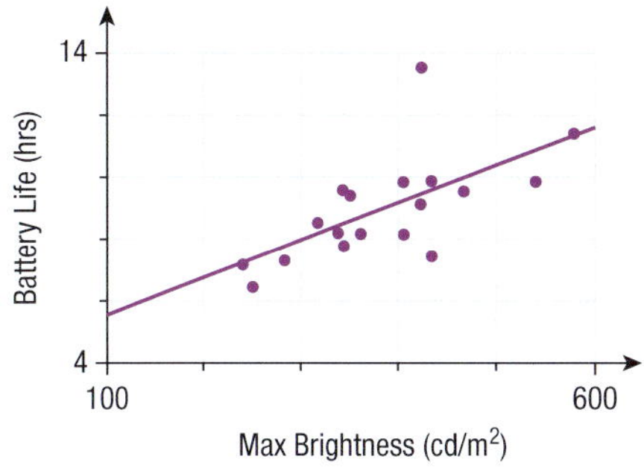

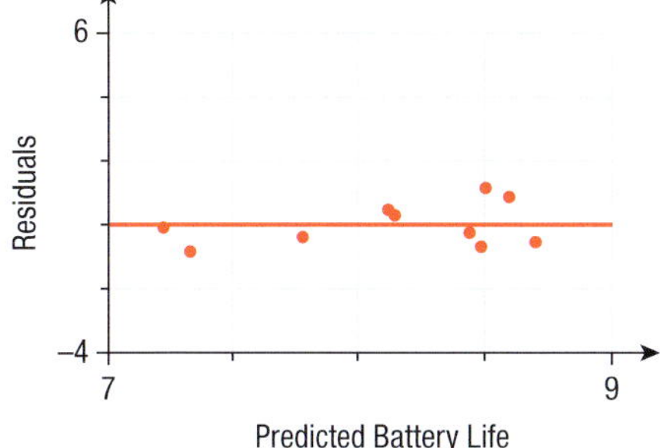

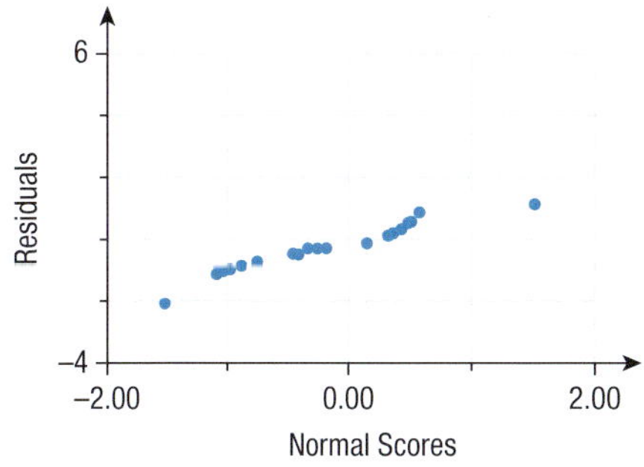

a) How many tablet computers were tested?
b) Are the conditions for inference satisfied? Explain.
c) Is there evidence of an association between maximum brightness of the screen and battery life? Test an appropriate hypothesis, and state your conclusion.
d) Is the association strong? Explain.
e) What is the equation of the regression line?
f) Create a 95% confidence interval for the slope of the true line.
g) Interpret your interval in this context.

T **56. Crawling** University researchers wondered whether temperature might influence the age at which babies learn to crawl. Perhaps the extra clothing that babies wear in cold weather would restrict movement and delay the age at which they started crawling. Data were collected on 208 boys and 206 girls. Parents reported the month of the baby's birth and the age (in weeks) at which their child first crawled. The table gives the average *Temperature* (°F) when the babies were 6 months old and average *Crawling Age* (in weeks) for each month of the year. Make the plots and compute the analyses necessary to answer the following questions.

City	6-Month Temperature	Average Crawling Age
1	43	32.32
2	79	30.84
3	81	30.43
4	49	33.36
5	55	28.56
6	39	33.24
7	53	32.52
8	60	30.89
9	51	32.89
10	55	30.95
11	37	33.83
12	68	32.14
13	53	32.01
14	61	31.37
15	63	32.17
16	56	30.99
17	61	32.64
18	45	32.22
19	50	31.16

a) Would this Association appear to be weaker, stronger, or the same if data had been plotted for individual babies instead of using monthly averages? Explain.
b) Is there evidence of an association between *Temperature* and *Crawling Age*? Test an appropriate hypothesis, and state your conclusion. Don't forget to check the assumptions.
c) Create and interpret a 90% confidence interval for the slope of the true relationship.

T **57. Midterms** The dataset below shows midterm and homework scores from an introductory statistics course.

Student	Midterm 1	Midterm 2
1	49	60
2	51	34
3	58	36
4	60	84
5	62	66
6	64	28
7	70	66
8	70	59
9	72	46
10	75	88
11	78	79
12	81	78
13	80	63
14	85	61
15	86	99
16	88	51

Student	Midterm 1	Midterm 2
17	88	61
18	92	95
19	93	83
20	93	90
21	95	70
22	97	68
23	94	73
24	98	90

a) Fit a model predicting the second midterm score from the first.
b) Comment on the model you found, including a discussion of the assumptions and conditions for regression. Is the coefficient for the slope statistically significant?
c) A student comments that because the P-value for the slope is very small, Midterm 2 is very well predicted from Midterm 1. So, he reasons, next term the professor can give just one midterm. What do you think?

T **58. Midterms?** The professor teaching the introductory statistics class discussed in Exercise 57 wonders whether performance on homework can accurately predict midterm scores.

a) To investigate it, she fits a regression of the sum of the two midterms scores on homework scores. Fit the regression model.
b) Comment on the model including a discussion of the assumptions and conditions for regression. Is the coefficient for the slope "statistically significant"?
c) Do you think she can accurately judge a student's performance without giving the midterms? Explain.

T **59. Strike two** Remember the Little League instructional video discussed in Chapter 21, Exercise 35? Ads claimed it would improve the performances of Little League pitchers. To test this claim, 20 Little Leaguers threw 50 pitches each, and we recorded the number of strikes. After the players participated in the training program, we repeated the test. The table shows the number of strikes each player threw before and after the training. A test of paired differences failed to show that this training improves ability to throw strikes. Is there any evidence that the effectiveness of the video (*After – Before*) depends on the player's initial ability to throw strikes (*Before*)? Test an appropriate hypothesis and state your conclusion. Propose an explanation for what you find. (Data in **Strikes**)

Number of Strikes (out of 50)

Before	After	Before	After
28	35	33	33
29	36	33	35
30	32	34	32
32	28	34	30
32	30	34	33
32	31	35	34
32	32	36	37
32	34	36	33
32	35	37	35
33	36	37	32

T **60. All the efficiency money can buy 2011** A sample of 84 model-2011 cars from an online information service was examined to see how fuel efficiency (as highway mpg) relates to the cost (Manufacturer's Suggested Retail Price in dollars) of cars. Here are displays and computer output:

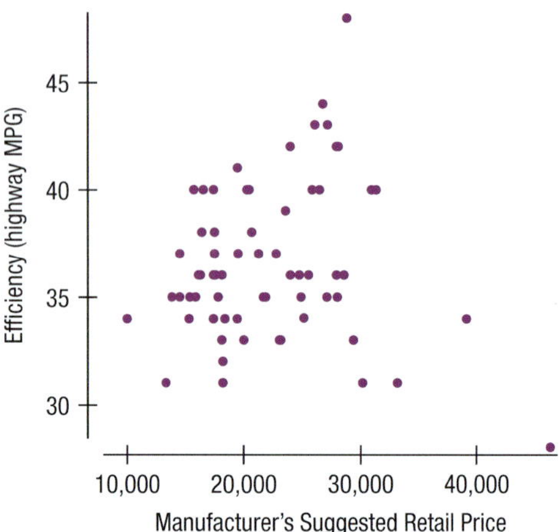

Dependent variable is MPG
R-squared = 0.0216%
s = 3.54

Variable	Coefficient	SE(Coeff)	t-Ratio	P-Value
Intercept	36.514	1.496	24.406	<0.0001
Slope	−8.089E −6	6.439E −5	−0.1256	0.900

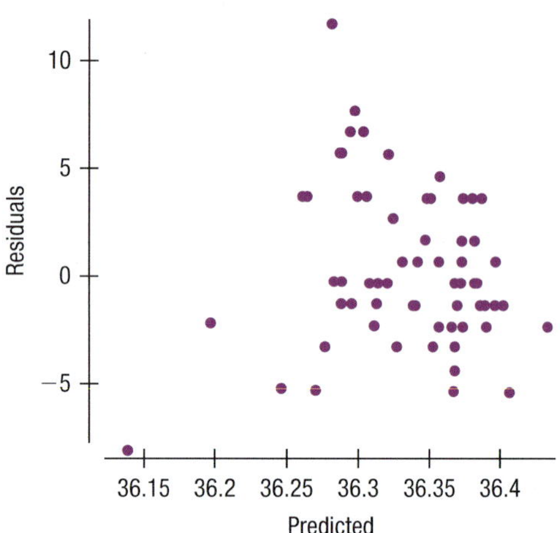

a) State what you want to know, identify the variables, and give the appropriate hypotheses.
b) Check the assumptions and conditions.
c) If the conditions are met, complete the analysis.

T **61. Education and mortality** The following software output is based on the mortality rate (deaths per 100,000 people) and the education level (average number of years in school) for 28 cities.

Variable	Count	Mean	StdDev
Mortality	28	958.1071	47.8054
Education	28	10.852	0.821464

Dependent Variable is: Mortality
R-squared = 42.6%
S = 36.92 with 28 − 2 = 26 degrees of freedom

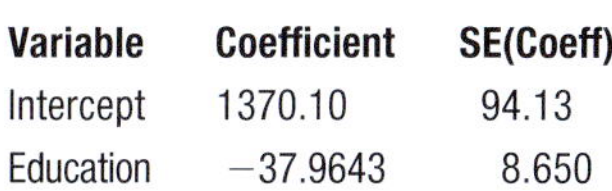

Variable	Coefficient	SE(Coeff)
Intercept	1370.10	94.13
Education	−37.9643	8.650

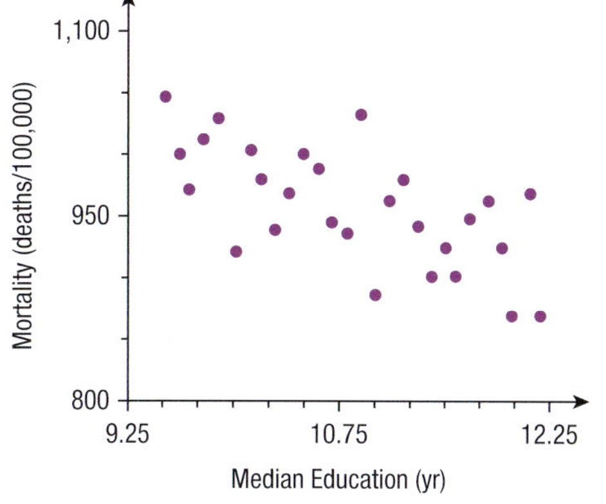

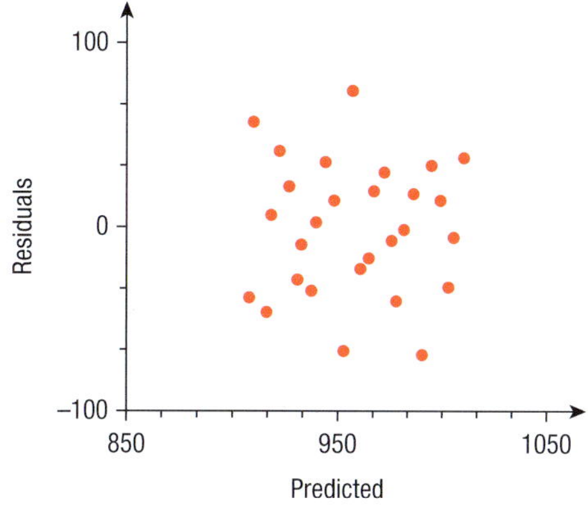

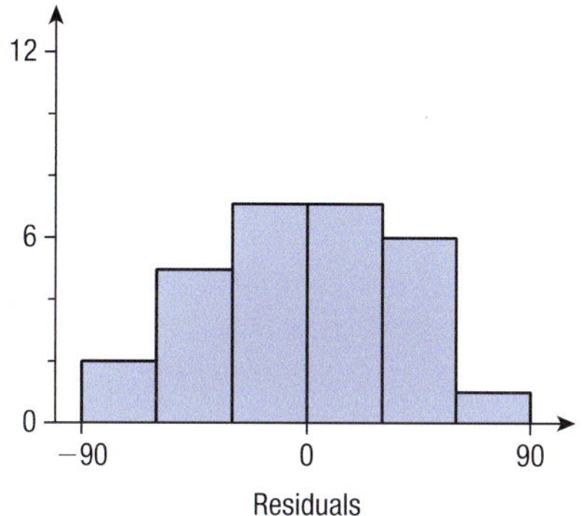

a) Comment on the assumptions for inference.
b) Is there evidence of a strong association between the level of *Education* in a city and the *Mortality* rate? Test an appropriate hypothesis, and state your conclusion.
c) Can we conclude that getting more education is likely (on average) to prolong your life? Why or why not?
d) Find a 95% confidence interval for the slope of the true relationship.
e) Explain what your interval means.
f) Find a 95% confidence interval for the average *Mortality* rate in cities where the adult population completed an average of 12 years of school.

62. Property assessments The following software output provides information about the *Size* (in square feet) of 18 homes in Ithaca, New York, and the city's assessed *Value* of those homes.

Variable	Count	Mean	StdDev	Range
Size	18	2003.39	264.727	890
Value	18	60946.7	5527.62	19710

Dependent variable is Value
R-squared = 32.5%
s = 4682 with 18 − 2 = 16 degrees of freedom

Variable	Coefficient	SE(Coeff)
Intercept	37108.8	8664
Size	11.8987	4.290

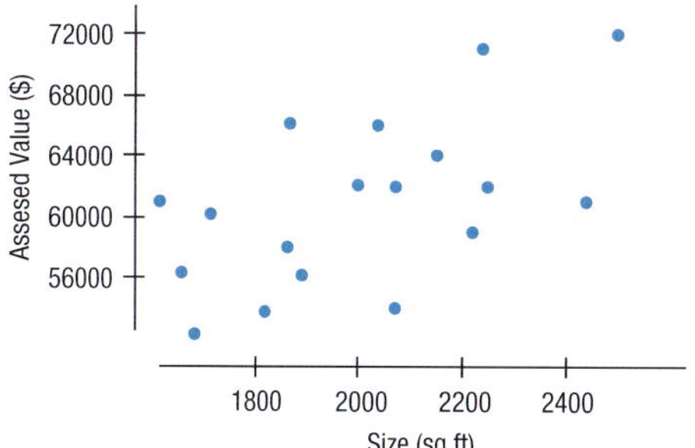

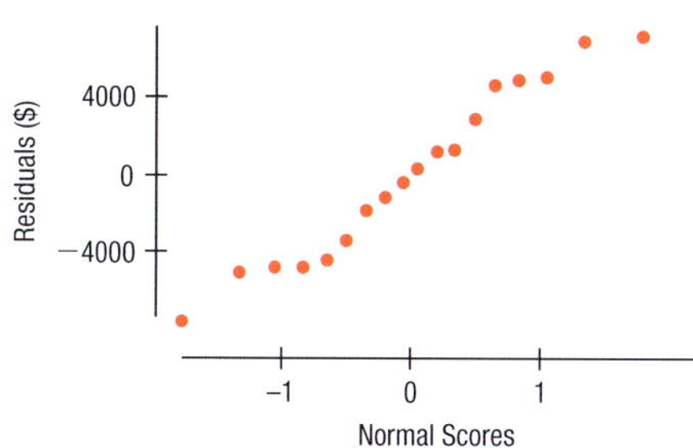

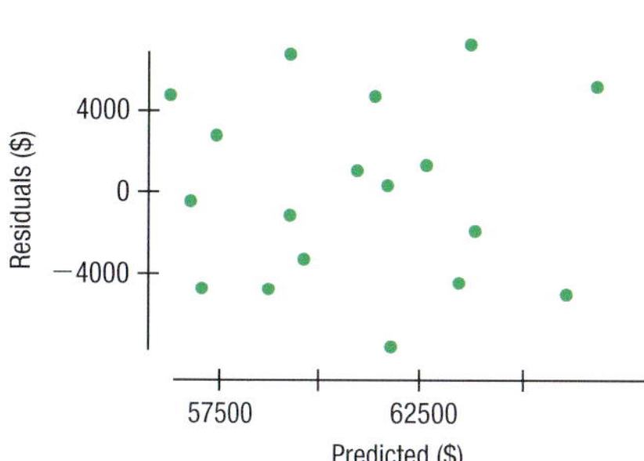

a) Explain why inference for linear regression is appropriate with these data.
b) Is there a significant association between the *Size* of a home and its assessed *Value*? Test an appropriate hypothesis and state your conclusion.
c) What percentage of the variability in assessed *Value* is explained by this regression?
d) Give a 90% confidence interval for the slope of the true regression line, and explain its meaning in the proper context.
e) From this analysis, can we conclude that adding a room to your house will increase its assessed *Value*? Why or why not?
f) The owner of a home measuring 2100 square feet files an appeal, claiming that the $70,200 assessed *Value* is too high. Do you agree? Explain your reasoning.

***63. Embrace or protect?** In January 2017, a research group asked people which statement they identified more with:

1) Our lives are threatened by terrorists, criminals, and immigrants and our priority should be to **protect** ourselves. Or
2) It's a big, beautiful world, mostly full of good people, and we must find a way to **embrace** each other and not allow ourselves to become isolated.

Does a person's age affect how they responded to the embrace/protect question? Here's some output from a logistic regression model (Embrace = 1, Protect = 0):

Logistic Regression Table

Predictor	Coeff	SE(Coeff)	Chisq	P-value
Intercept	0.5651	0.1802	9.83	0.0019
Age	−0.0152	0.000459	11.0	0.0015

a) Write out the estimated regression equation.
b) Is there evidence that age is associated with the response? Explain.
c) On average, is an older person more or less likely to respond "Embrace"? Explain.

***64. Cost of higher education** Are there fundamental differences between liberal arts colleges and universities? In this case, we have information on the top 25 liberal arts colleges and the top 25 universities in the United States. We will consider the type of school as our response variable and use the percent of students who were in the top 10% of their high school class and the amount of money spent per student by the college or university as our explanatory variables. The output from this logistic regression is given below.

Logistic Regression Table

Predictor	Coeff	SE(Coeff)	Chisq	P
Intercept	−13.1461	3.98629	10.89	0.001
Top 10%	0.0845	0.0396345	4.54	0.033
$/Student	0.0003	0.0000860	9.12	0.003

a) Write the estimated regression equation.
b) Is percent of students in the top 10% of their high school class statistically significant in predicting whether or not the school is a university? Explain.
c) Is the amount of money spent per student statistically significant in predicting whether or not the school is a university? Explain.

JUST CHECKING

Answers

1. A high *t*-ratio of 3.27 indicates that the slope is different from zero—that is, that there is a linear relationship between height and mouth size. The small P-value says that a slope this large would be very unlikely to occur by chance if, in fact, there was no linear relationship between the variables.
2. Not really. The R^2 for this regression is only 15.3%, so height doesn't account for very much of the variability in mouth size.
3. The value of s_e tells the standard deviation of the residuals. Mouth sizes have a mean of 60.3 cubic centimeters. A standard deviation of 15.7 in the residuals indicates that the errors made by this regression model can be quite large relative to what we are estimating. Errors of 15 to 30 cubic centimeters would be common.
4. The R^2 is a bit higher, but not very much, and there is a second predictor, which should increase the R^2 automatically.
5. The coefficient of *Height* now estimates the effect of height on mouth volume for individuals of a given age.
6. The P-value for *Age* is 0.0965. This is not statistically significant at the 0.05 alpha level, but is marginally significant at 0.10. I would probably choose not to reject the null hypothesis in this case.

Review of Part VI

INFERENCE FOR RELATIONSHIPS

Quick Review

We now turn our focus to relationships among variables. This is where the effectiveness of statistics becomes clear. Statistical inference about how variables are related is essential to science, social science, health, and business. In this part we have considered comparing groups with means and proportions, comparing groups that have a natural pairing, comparing proportions in several categories and between two categorical variables, and inferential decisions about linear regression. That may seem like a wide variety of ideas, but they are all held together by the central concept that by understanding the ways in which variables and groups relate to each other, we can understand how the world works. Here's a brief summary of the key concepts and skills:

- A confidence interval uses a sample statistic to estimate a range of possible values for the parameter of a population model.
- A hypothesis test proposes a model for the population, then examines the observed statistics to see if the model is plausible.
- Statistical inference procedures for comparing proportions are based on the Central Limit Theorem. We can make inferences about the difference of two proportions using Normal models.
- Statistical inference procedures for comparing means and for estimating regression coefficients are also based on the Central Limit Theorem, but we don't usually know the population standard deviation. Student's *t*-models take the additional uncertainty of independently estimating the standard deviation into account.
  - We can make inferences about the difference of two independent means, the mean of paired differences, or the intercept or slope of a linear regression using *t*-models.
  - No inference procedure is valid unless the underlying assumptions are true. Always think about the assumptions and check the conditions before proceeding. For regression, take care to check them in the right order.
  - Because *t*-models assume that samples are drawn from Normal populations, data in the sample (or, for regression, the residuals) should appear to be nearly Normal. Skewness and outliers are particularly problematic.
  - To identify the appropriate statistic for comparing the means of groups, you must think carefully about how the data were collected. You may use two-sample *t* procedures only if the groups are independent.
  - Unless there is some obvious reason to suspect that two independent populations have the same standard deviation, you should not pool the variances. It is never wrong to use unpooled *t* procedures.
  - If two groups are paired, the data are *not* from independent groups. You must use matched-pairs *t* procedures and test the mean difference rather than the difference in the means.
  - Linear regression inference is only valid if the relationship between the two variables is straight. Examine a scatterplot.
- Not all sampling distributions are unimodal, symmetric, or bell-shaped. Inferences about distributions of counts use chi-square models, which are unimodal but skewed to the high end. Nevertheless, the sampling distribution plays the same role in inference, helping us to translate between probabilities and values based on data.
  - To see if an observed distribution is consistent with a proposed model, use a chi-square goodness-of-fit test.
  - To see if two or more observed distributions could have arisen from populations with the same model, use a test of homogeneity.
  - To see if two categorical variables are independent, perform a chi-square test of independence.
- You can now use statistical inference to answer questions about means, proportions, distributions, associations, and linear regression models.
  - No inference procedure is valid unless the underlying assumptions are true. Always check the conditions before proceeding.
  - You can make inferences about the difference between two proportions using Normal models.
  - You can make inferences about the difference between two independent means, or about the mean of paired differences using *t*-models.
  - You can make inferences about distributions using chi-square models.
  - You can make inferences about association between categorical variables using chi-square models.
  - You can make inferences about the coefficients in a linear regression model using *t*-models.

Now for some opportunities to review these concepts. Be careful. You have a lot of thinking to do. These review exercises mix questions about proportions, means, chi square, and regression. You have to determine which of our inference procedures is appropriate in each situation. Then you have to check the proper assumptions and conditions. Keeping track of those can be difficult, so first we summarize the many procedures with their corresponding assumptions and conditions on the next page. Look them over carefully . . . then, on to the Exercises!

Assumptions for Inference	And the Conditions That Support or Override Them
Proportions (z)	
◆ **One sample**	
1. Individuals are independent.	1. SRS and $n < 10\%$ of the population.
2. Sample is sufficiently large.	2. Successes and failures ≥ 10.
◆ **Two sample**	
1. Samples are independent.	1. (Think about how the data were collected.)
2. Data in each sample are independent.	2. Both are SRSs and $n < 10\%$ of populations OR random allocation.
3. Both samples are sufficiently large.	3. Successes and failures ≥ 10 for both.
Means (t)	
◆ **One sample** ($df = n - 1$)	
1. Individuals are independent.	1. SRS and $n < 10\%$ of the population.
2. Population has a Normal model.	2. Histogram is unimodal and symmetric.*
◆ **Two independent samples** (df from technology)	
1. Samples are independent.	1. (Think about the design.)
2. Data in each sample are independent.	2. SRSs and $n < 10\%$ OR random allocation.
3. Both populations are Normal.	3. Both histograms are unimodal and symmetric.*
◆ **Matched pairs** ($df = n - 1$)	
1. Data are matched; n pairs.	1. (Think about the design.)
2. Individuals are independent.	2. SRSs and $n < 10\%$ OR random allocation.
3. Population of differences is Normal.	3. Histogram of differences is unimodal and symmetric.
Distributions/Association (χ^2)	
◆ **Goodness-of-fit** [df = # of cells −1; one variable, one sample compared with population model]	
1. Data are counts.	1. (Are they?)
2. Data in sample are independent.	2. SRS and $n < 10\%$ of the population.
3. Sample is sufficiently large.	3. All expected counts ≥ 5.
◆ **Homogeneity** [$df = (r - 1)(c - 1)$; samples from many populations compared on one variable]	
1. Data are counts.	1. (Are they?)
2. Data in samples are independent.	2. SRSs and $n < 10\%$ OR random allocation.
3. Groups are sufficiently large.	3. All expected counts ≥ 5.
◆ **Independence** [$df = (r - 1)(c - 1)$; sample from one population classified on two variables]	
1. Data are counts.	1. (Are they?)
2. Data are independent.	2. SRSs and $n < 10\%$ of the population.
3. Group is sufficiently large.	3. All expected counts ≥ 5.
◆ **Regression with k quantitative predictors** (t, $df = n - k - 1$)	
1. Form of relationship is linear.	1. Scatterplot of residuals against predicted values shows no special structure.
2. Errors are independent.	2. No apparent pattern in plot of residuals against predicted values.
3. Variability of errors is constant.	3. Plot of residuals against predicted values has constant spread, doesn't "thicken."
4. Errors follow a Normal model.	4. Histogram of residuals is approximately unimodal and symmetric, or Normal probability plot is reasonably straight.*

*Less critical as n increases

Quick Guide to Inference

Think			Show				Tell?
Inference about?	One sample or two?	Procedure	Model	Parameter	Estimate	SE	Chapter
Proportions	One sample	1-Proportion z-interval	z	p	$\hat{p}$	$\sqrt{\frac{\hat{p}\hat{q}}{n}}$	16
		1-Proportion z-test				$\sqrt{\frac{p_0q_0}{n}}$	18
	Two independent groups	2-Proportion z-interval 2-Proportion z-test	z	$p_1 - p_2$	$\hat{p}_1 - \hat{p}_2$	$\sqrt{\frac{\hat{p}_1\hat{q}_1}{n_1} + \frac{\hat{p}_2\hat{q}_2}{n_2}}$	20
Means	One sample	t-interval t-test	t $\text{df} = n - 1$	μ	$\bar{y}$	$\frac{s}{\sqrt{n}}$	17
	Two independent groups	2-Sample t-test 2-Sample t-interval	t df from technology	$\mu_1 - \mu_2$	$\bar{y}_1 - \bar{y}_2$	$\sqrt{\frac{s_1^2}{n_1} + \frac{s_2^2}{n_2}}$	20
	n Matched pairs	Paired t-test Paired t-interval	t $\text{df} = n - 1$	μ_d	$\bar{d}$	$\frac{s_d}{\sqrt{n}}$	21
Distributions (one categorical variable)	One sample	Goodness-of-fit	χ^2 $\text{df} = cells - 1$			$\sum \frac{(Obs - Exp)^2}{Exp}$	22
	Many independent samples	Homogeneity χ^2 test	χ^2 $\text{df} = (r - 1)(c - 1)$				
Independence (two categorical variables)	One sample	Independence χ^2 test					
Association (two quantitative variables)	One sample	Linear regression t-test or confidence interval for β	t $\text{df} = n - 2$	β_1	b_1	$\frac{s_e}{s_x\sqrt{n-1}}$ (compute with technology)	23
		Confidence interval for μ_ν		μ_ν	$\hat{y}_\nu$	$\sqrt{SE^2(b_1) \times (x_\nu - \bar{x})^2 + \frac{s_e^2}{n}}$	
		Prediction interval for y_ν		y_ν	$\hat{y}_\nu$	$\sqrt{SE^2(b_1) \times (x_\nu - \bar{x})^2 + \frac{s_e^2}{n} + s_e^2}$	
Association (k predictors)	One sample	Multiple regression	t $\text{df} = n - k - 1$	β_j	b_j	Compute with technology	

REVIEW EXERCISES

R6.1. Herbal cancer A report in the *New England Journal of Medicine* notes growing evidence that the herb *Aristolochia fangchi* can cause urinary tract cancer in those who take it. Suppose you are asked to design an experiment to study this claim. Imagine that you have data on urinary tract cancers in subjects who have used this herb and similar subjects who have not used it and that you can measure incidences of cancer and precancerous lesions in these subjects. State the null and alternative hypotheses you would use in your study.

R6.2. Birth days During a 2-month period, 72 babies were born at the Tompkins Community Hospital in upstate New York. The table shows how many babies were born on each day of the week.

Day	Births
Mon.	7
Tues.	17
Wed.	8
Thurs.	12
Fri.	9
Sat.	10
Sun.	9

a) If births are uniformly distributed across all days of the week, how many would you expect on each day?
b) Test the hypothesis that babies are equally likely to be born on any of the days of the week.
c) Given the results of part b, do you think that the 7 births on Monday and 17 births on Tuesday indicate that women might be less likely to give birth on Monday, or more likely to give birth on Tuesday?
d) Can you think of any reasons why births may not occur completely at random?

R6.3. Surgery and germs Joseph Lister (for whom Listerine is named!) was a British physician who was interested in the role of bacteria in human infections. He suspected that germs were involved in transmitting infection, so he tried using carbolic acid as an operating room disinfectant. In 75 amputations, he used carbolic acid 40 times. Of the 40 amputations using carbolic acid, 34 of the patients lived. Of the 35 amputations without carbolic acid, 19 patients lived. The question of interest is whether carbolic acid is effective in increasing the chances of surviving an amputation.

a) What kind of a study was this?
b) What do you conclude? Support your conclusion by testing an appropriate hypothesis.
c) What reservations do you have about the design of the study?

R6.4. Free throws 2017 At the middle of the 2016–2017 NBA season, James Hardin led the league by making 468 of 544 free throws, for a success rate of 86%. But Russell Westbook was close behind with 425 of 517 (82.2%).

a) Find a 95% confidence interval for the difference in their free throw percentages.
b) Based on your confidence interval, is it certain that Hardin is better than Westbrook at making free throws?

R6.5. Twins There is some indication in medical literature that doctors may have become more aggressive in inducing labor or doing preterm cesarean sections when a woman is carrying twins. Records at a large hospital show that, of the 43 sets of twins born in 2000, 20 were delivered before the 37th week of pregnancy. In 2010, 26 of 48 sets of twins were born preterm. Does this indicate an increase in the incidence of early births of twins? Test an appropriate hypothesis and state your conclusion.

R6.6. Eclampsia It's estimated that 50,000 pregnant women worldwide die each year of eclampsia, a condition involving elevated blood pressure and seizures. A research team from 175 hospitals in 33 countries investigated the effectiveness of magnesium sulfate in preventing the occurrence of eclampsia in at-risk patients. Results are summarized below. (*Lancet*, June 1, 2002)

Treatment		Total Subjects	Reported Side Effects	Developed Eclampsia	Deaths
	Magnesium Sulfate	4999	1201	40	11
	Placebo	4993	228	96	20

a) Write a 95% confidence interval for the increase in the proportion of women who may develop side effects from this treatment. Interpret your interval.
b) Is there evidence that the treatment may be effective in preventing the development of eclampsia? Test an appropriate hypothesis and state your conclusion.

R6.7. Eclampsia deaths Refer again to the research summarized in Exercise R6.6. Is there any evidence that when eclampsia does occur, the magnesium sulfide treatment may help prevent the woman's death?

a) Write an appropriate hypothesis.
b) Check the assumptions and conditions.
c) Find the P-value of the test.
d) What do you conclude about the magnesium sulfide treatment?
e) If your conclusion is wrong, which type of error have you made?
f) Name two things you could do to increase the power of this test.
g) What are the advantages and disadvantages of those two options?

R6.8. Perfect pitch A recent study of perfect pitch tested 2700 students in American music conservatories. It found that 7% of non-Asian and 32% of Asian students have perfect pitch. A test of the difference in proportions resulted in a P-value of <0.0001.

a) What are the researchers' null and alternative hypotheses?
b) State your conclusion.
c) Explain in this context what the P-value means.
d) The researchers claimed that the data prove that genetic differences between the two populations cause a difference in the frequency of occurrence of perfect pitch. Do you agree? Why or why not?

R6.9. More errors A corporation with a fleet of vehicles wanted to test the cost-effectiveness of using Motor Silk oil additive. For the study, 6100 delivery and passenger vehicles were tested for the same 3-month period in one year and then again in the subsequent year. In the initial year, the fleet was driven without Motor Silk. Then in the second year, Motor Silk was used according to the standard instructions. The average fuel economy increased from 18.97 mpg to 21.72 mpg.

a) What kind of a study is this?
b) Will they do a one-tailed or a two-tailed test?
c) Explain in this context what a Type I error would be.
d) Explain in this context what a Type II error would be.
e) Which type of error would the additive manufacturer consider more serious?
f) If the vehicles with the additive are indeed statistically significantly better, can the company conclude it is an effect of the additive? Can they generalize this result and recommend the additive for all cars? Explain.

R6.10. Preemies Among 242 Cleveland-area children born prematurely at low birthweights between 1977 and 1979, only 74% graduated from high school. Among a comparison group of 233 children of normal birthweight, 83% were high school graduates. ("Outcomes in Young Adulthood for Very-Low-Birth-Weight Infants," *New England Journal of Medicine*, 346, no. 3)

a) Create a 95% confidence interval for the difference in graduation rates between children of normal and of very low birthweights. Be sure to check the appropriate assumptions and conditions.
b) Does this provide evidence that premature birth may be a risk factor for not finishing high school? Use your confidence interval to test an appropriate hypothesis.
c) Suppose your conclusion is incorrect. Which type of error did you make?

T **R6.11.** Crawling A study found that babies born at different times of the year may develop the ability to crawl at different ages. The authors of the study suggested that these differences may be related to the temperature at the time the infant is 6 months old. (Benson and Janette, "Season of birth and onset of locomotion: Theoretical and methodological implications," *Infant Behavior and Development* 16:1, pp 69–81)

a) The study found that 32 babies born in January crawled at an average age of 29.84 weeks, with a standard deviation of 7.08 weeks. Among 21 July babies, crawling ages averaged 33.64 weeks, with a standard deviation of 6.91 weeks. Is this difference significant?
b) For 26 babies born in April, the mean and standard deviation were 31.84 and 6.21 weeks, while for 44 October babies the mean and standard deviation of crawling ages were 33.35 and 7.29 weeks. Is this difference significant?
c) Are these results consistent with the researcher's claim?

T **R6.12.** Mazes and smells Can pleasant smells improve learning? Researchers timed 21 subjects as they tried to complete paper-and-pencil mazes. Each subject attempted a maze both with and without the presence of a floral aroma. Subjects were randomized with respect to whether they did the scented trial first or second. Some of the data collected are shown in the table. Is there any evidence that the floral scent improved the subjects' ability to complete the mazes? (A. R. Hirsch and L. H. Johnston, "Odors and Learning." Chicago: Smell and Taste Treatment and Research Foundation)

Time to Complete the Maze (sec)

Unscented	Scented	Unscented	Scented
25.7	30.2	61.5	48.4
41.9	56.7	44.6	32.0
51.9	42.4	35.3	48.1
32.2	34.4	37.2	33.7
64.7	44.8	39.4	42.6
31.4	42.9	77.4	54.9
40.1	42.7	52.8	64.5
43.2	24.8	63.6	43.1
33.9	25.1	56.6	52.8
40.4	59.2	58.9	44.3
58.0	42.2		

R6.13. Pottery Archaeologists can use the chemical composition of clay found in pottery artifacts to determine whether different sites were populated by the same ancient people. They collected five samples of Romano-British pottery from each of two sites in Great Britain—the Ashley Rails site and the New Forest site—and measured the percentage of aluminum oxide in each. Based on these data, do you think the same people used these two kiln sites? Base your conclusion on a 95% confidence interval for the difference in aluminum oxide content of pottery made at the sites. (A. Tubb, A. J. Parker, and G. Nickless, "The Analysis of Romano-British Pottery by Atomic Absorption Spectrophotometry," *Archaeometry*, 22:153–171)

Ashley Rails	19.1	14.8	16.7	18.3	17.7
New Forest	20.8	18.0	18.0	15.8	18.3

R6.14. Grant writing Does race matter when applying for National Institutes of Health grants? A study found that of 58,148 applications submitted by white researchers, 15,700 were accepted and funded by the NIH. Additionally, 198 of the 1164 applications submitted by black researchers were funded. (*Science*, August 19, 2011)

a) Is there evidence that the chance for funding is higher for white researchers?
b) What kind of study is this? How does that affect the inference you made in part a?

R6.15. Feeding fish In the midwestern United States, a large aquaculture industry raises largemouth bass. Researchers wanted to know whether the fish would grow better if fed a natural diet of fathead minnows or an artificial diet of food pellets. They stocked six ponds with bass fingerlings weighing about 8 grams. For one year, the fish in three of the ponds were fed minnows, and the others were fed the commercially prepared pellets. The fish were then harvested, weighed, and measured. The bass fed a natural food source had a higher average length (19.6 cm) and weight (95.9 g) than those fed the commercial fish food (17.3 cm and 72.0 g, respectively). The researchers reported P-values for both measurements to be less than 0.001.

a) Explain to someone who has not studied statistics what the P-values mean here.
b) What advice should the researchers give the people who raise largemouth bass?
c) If that advice turns out to be incorrect, what type of error occurred?

T **R6.16.** Seat belts 2015 The National Highway Traffic Safety Administration reported seat belt use and fatalities in car accidents. (*Seat belt use in 2015—use rates in the states and territories*. Report no. DOT HS-812-274) Is the rate of seat belt use different in New England compared to the Mountain states? We hope to create a 95% confidence interval for the difference in seat belt use proportions. Are these data appropriate for inference? If so, create the interval and make a concluding remark. Data are percentages.

New England	85	86	74	70	87	86		
Mountain	87	85	81	77	92	93	87	80

R6.17. Age In a study of how depression may affect one's ability to survive a heart attack, the researchers reported the ages of the two groups they examined. The mean age of 2397 patients without cardiac disease was 69.8 years (SD = 8.7 years), while for the 450 patients with cardiac disease, the mean and standard deviation of the ages were 74.0 and 7.9, respectively.

a) Create a 95% confidence interval for the difference in mean ages of the two groups.
b) How might an age difference confound these research findings about the relationship between depression and ability to survive a heart attack?

R6.18. Smoking In the depression and heart attack research described in Exercise R6.17, 32% of the diseased group were smokers, compared with only 23.7% of those free of heart disease.

a) Create a 95% confidence interval for the difference in the proportions of smokers in the two groups.
b) Is this evidence that the two groups in the study were different? Explain.
c) Could this be a problem in analyzing the results of the study? Explain.

R6.19. Eating disorders A study conducted in the multicultural Spanish city of Ceuta investigated the relationship between religion and the prevalence of eating disorders. Students aged 12–20 were selected from three public schools. In the study, suppose there were 150 Muslim students and 46 were diagnosed with eating disorders. Of the 200 Christian students, 17% were diagnosed the same way.

a) Create a 95% confidence interval for the difference in eating disorder rates when comparing Muslims to Christians.
b) Based on your interval, are you convinced that there is a true difference when comparing rates between the religions? Explain.

R6.20. Cesareans Some people fear that differences in insurance coverage can affect health care decisions. A survey of several randomly selected hospitals found that 16.6% of 223 recent births in Vermont involved cesarean deliveries, compared to 18.8% of 186 births in New Hampshire. Is this evidence that the rates of cesarean births in the two states are different?

R6.21. Teach for America Several programs attempt to address the shortage of qualified teachers by placing uncertified instructors in schools with acute needs—often in inner cities. A study compared students taught by certified teachers to others taught by uncertified teachers in the same schools. Reading scores of the students of certified teachers averaged 35.62 points with standard deviation 9.31. The scores of students instructed by uncertified teachers had mean 32.48 points with standard deviation 9.43 points on the same test. There were 44 students in each group. The appropriate *t*-procedure has 86 degrees of freedom. Is there evidence of lower scores with uncertified teachers? Discuss. (*The Effectiveness of "Teach for America" and Other Under-certified Teachers on Student Academic Achievement: A Case of Harmful Public Policy*. Education Policy Analysis Archives)

T **R6.22.** Legionnaires' disease In 1974, the Bellevue-Stratford Hotel in Philadelphia was the scene of an outbreak of what later became known as Legionnaires' disease. The cause of the disease was finally discovered to be bacteria that thrived in the air-conditioning units of the hotel. Owners of the Rip Van Winkle Motel, hearing about the Bellevue-Stratford, replace their air-conditioning system. The following data are the bacteria counts in the air of eight rooms, before and after a new air-conditioning system was installed (measured in colonies per cubic foot of air). The objective is to find out whether the new system has succeeded in lowering the bacteria count. You are the statistician assigned to report to the hotel whether the strategy has worked.

a) Base your analysis on a confidence interval. Be sure to list all your assumptions, methods, and conclusions.
*b) Repeat part a using a bootstrap confidence interval, and compare your results to part a.

Room Number	Before	After
121	11.8	10.1
163	8.2	7.2
125	7.1	3.8
264	14.0	12.0
233	10.8	8.3
218	10.1	10.5
324	14.6	12.1
325	14.0	13.7

R6.23. Teach for America, part II The study described in Exercise R6.21 also looked at scores in mathematics and language. Here are software outputs for the appropriate tests. Explain what they show.

Mathematics

T-TEST OF Mu(1) – Mu(2) = 0
Mu(Cert) – Mu(NoCert) = 4.53   t(86) = 2.95   p = 0.002

Language

T-TEST OF Mu(1) – Mu(2) = 0
Mu(Cert) – Mu(NoCert) = 2.13   t(84) = 1.71   p = 0.045

T **R6.24.** Fisher's irises In 1936 Sir Ronald Fisher presented data on irises as the example in a famous statistics paper. Ever since, "Fisher's Iris data" have been a feature of statistics texts. We didn't want to be an exception. Can measurements of the petal length of flowers be of value when you need to determine the species of iris? Here are the summary statistics from measurements of the petals of two species of irises. (R. A. Fisher, "The Use of Multiple Measurements in Axonomic Problems," *Annals of Eugenics* 7 [1936]:179–188)

	Species	
	Versicolor	**Virginica**
Count	50	50
Mean	55.52	43.22
Median	55.50	44.00
SD	5.519	5.362
Min	45	30
Max	69	56
Lower Quartile	51	40
Upper Quartile	59	47

a) Make parallel boxplots of petal lengths for the two species.
b) Describe the differences seen in the boxplots.
c) Write a 95% confidence interval for this difference.
d) Explain what your interval means.
e) Based on your confidence interval, is there evidence of a difference in petal length? Explain.

*f) By resampling the data, create a 95% bootstrap confidence interval and compare your results to those above. Did your conclusion in part e change?

R6.25. Insulin and diet A study published in the *Journal of the American Medical Association* examined people to see if they showed any signs of IRS (insulin resistance syndrome) involving major risk factors for Type 2 diabetes and heart disease. Among 102 subjects who consumed dairy products more than 35 times per week, 24 were identified with IRS. In comparison, IRS was identified in 85 of 190 individuals with the lowest dairy consumption, fewer than 10 times per week.

a) Is this strong evidence that IRS risk is different in people who frequently consume dairy products than in those who do not?
b) Does this prove that dairy consumption influences the development of IRS? Explain.

T **R6.26.** Cloud seeding In an experiment to determine whether seeding clouds with silver iodide increases rainfall, 52 clouds were randomly assigned to be seeded or not. The amount of rain they generated was then measured (in acre-feet).

a) Create a 95% confidence interval for the average amount of additional rain created by seeding clouds. Explain what your interval means.
*b) Repeat part a using a bootstrap confidence interval based on shuffling the categories, and compare your results to part a. Discuss why the results might differ. (Hint: Plot the data, plot the data, plot the data.)

	Unseeded Clouds	Seeded Clouds
Count	26	26
Mean	164.588	441.985
Median	44.200	221.600
SD	278.426	650.787
IntQRange	138.600	337.600
25%ile	24.400	92.400
75%ile	163	430

T **R6.27.** Cloud seeding re-expressed

a) Make histograms of the cloud seeding data for both unseeded and seeded clouds. Do you think either of the inference methods used in Exercise R6.26 is appropriate?
b) Find a re-expression that improves the distributions of the cloud seeding rainfall amounts.
c) Repeat the tests of Exercise R6.26. Comment on what you discover.

R6.28. Genetics Two human traits controlled by a single gene are the ability to roll one's tongue and whether one's ear lobes are free or attached to the neck. Genetic theory says that people will have neither, one, or both of these traits in the ratio 1:3:3:9 (1 attached, noncurling; 3 attached, curling; 3 free, noncurling; 9 free, curling). An introductory biology class of 122 students collected the data shown. Are they consistent with the genetic theory? Test an appropriate hypothesis and state your conclusion.

	Trait			
	Attached, Noncurling	**Attached, Curling**	**Free, Noncurling**	**Free, Curling**
Count	10	22	31	59

T **R6.29.** Tableware Nambe Mills manufactures plates, bowls, and other tableware made from an alloy of several metals. Each item must go through several steps, including polishing. To better understand the production process and its impact on pricing, the company checked the polishing time (in minutes) and the retail price (in US$) of these items. The regression analysis is shown below. The scatterplot showed a linear pattern, and residuals were deemed suitable for inference.

Dependent variable is Price
R-squared = 84.5%
s = 20.50 with 59 – 2 = 57 degrees of freedom

Variable	Coefficient	SE(Coeff)
Intercept	−2.89054	5.730
Time	2.49244	0.1416

a) How many different products were included in this analysis?
b) What fraction of the variation in retail price is explained by the polishing time?
c) Create a 95% confidence interval for the slope of this relationship.
d) Interpret your interval in this context.

R6.30. Hard water In an investigation of environmental causes of disease, data were collected on the annual mortality rate (deaths per 100,000) for males in 61 large towns in England and Wales. In addition, the water hardness was recorded as the calcium concentration (parts per million, or ppm) in the drinking water. Here are the scatterplot and regression analysis of the relationship between mortality and calcium concentration.

Dependent variable is Mortality
R-squared = 43%
s = 143.0 with 61 − 2 = 59 degrees of freedom

Variable	Coefficient	SE(Coeff)
Intercept	1676	29.30
Calcium	−3.23	0.48

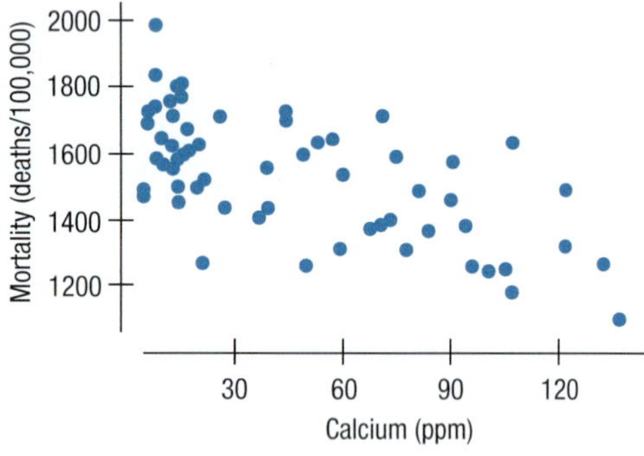

a) Is there an association between the hardness of the water and the mortality rate? Write the appropriate hypothesis.
b) Assuming the assumptions for regression inference are met, what do you conclude?
c) Create a 95% confidence interval for the slope of the true line relating calcium concentration and mortality.
d) Interpret your interval in context.

R6.31. Wealth redistribution 2015 The following table is based on a Gallup Poll of 1015 U.S. adults on April 9–12, 2015. Respondents were classified as high income (over $75,000), middle income ($30k–$75k), or low income (less than $30k). Those polled were asked for their views on redistributing U.S. wealth by heavily taxing the rich. The data are summarized in the table below.

	Should Redistribute Wealth	Should Not	No Opinion
High Income	426	579	10
Middle Income	558	447	10
Low Income	619	355	41

Is there any evidence that income level is associated with feelings toward wealth distribution in the United States? Test an appropriate hypothesis about this table, and state your conclusions.

R6.32. Wild horses Large herds of wild horses can become a problem on some federal lands in the West. Researchers hoping to improve the management of these herds collected data to see if they could predict the number of foals that would be born based on the size of the current herd. Their attempt to model this herd growth is summarized in the output shown.

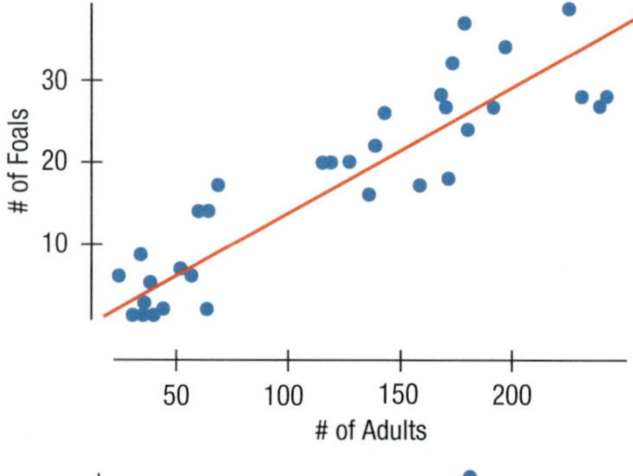

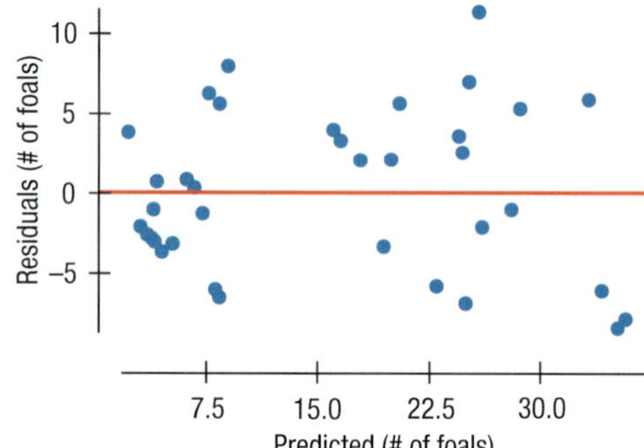

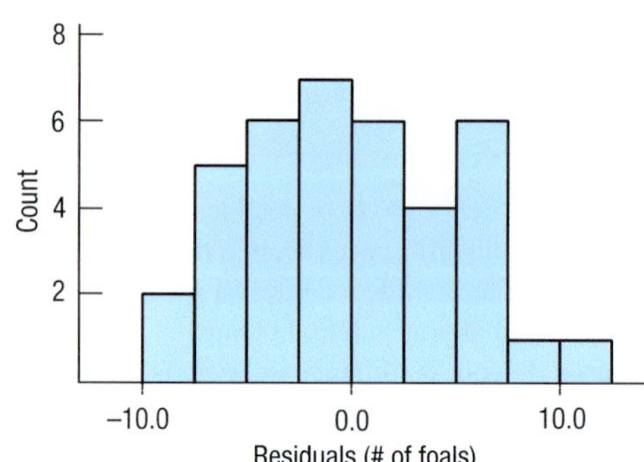

Variable	Count	Mean	StdDev
Adults	38	110.237	71.1809
Foals	38	15.3947	11.9945

Dependent variable is Foals
R-squared = 83.5%
s = 4.941 with 38 − 2 = 36 degrees of freedom

Variable	Coefficient	SE(Coeff)	t-Ratio	P-Value
Intercept	−1.57835	1.492	−1.06	0.2970
Adults	0.153969	0.0114	13.5	≤0.0001

a) How many herds of wild horses were studied?
b) Are the conditions necessary for inference satisfied? Explain.
c) Create a 95% confidence interval for the slope of this relationship.
d) Explain in this context what that slope means.
e) Suppose that a new herd with 80 adult horses is located. Estimate, with a 90% prediction interval, the number of foals that may be born.

R6.33. Lefties and music In an experiment to see if left- and right-handed people have different abilities in music, subjects heard a tone and were then asked to identify which of several other tones matched the first. Of 76 right-handed subjects, 38 were successful in completing this test, compared with 33 of 53 lefties. Is this strong evidence of a difference in musical abilities based on handedness?

R6.34. AP statistics scores 2016 In 2016, about 200,000 statistics students nationwide took the Advanced Placement Examination in statistics. The national distribution of scores and the results at Ithaca High School are shown in the table.

Score	National Distribution	Ithaca High School Counts
5	13.9%	27
4	21.7%	18
3	24.7%	10
2	15.7%	11
1	24.0%	3

Is the distribution of scores at this high school significantly different from the national results? Explain.

T **R6.35. Twin births** In 2000, the *Journal of the American Medical Association* published a study that examined a sample of pregnancies that resulted in the birth of twins. Births were classified as preterm with intervention (induced labor or cesarean), preterm without such procedures, or term or postterm. Researchers also classified the pregnancies by the level of prenatal medical care the mother received (inadequate, adequate, or intensive). The data, from the years 1995–1997, are summarized in the table below. Figures are in thousands of births. (*JAMA* 284 [2000]: 335–341)

Twin Births, 1995–1997 (in thousands)

Level of Prenatal Care	Preterm (induced or cesarean)	Preterm (without procedures)	Term or Postterm	Total
Intensive	18	15	28	**61**
Adequate	46	43	65	**154**
Inadequate	12	13	38	**63**
Total	**76**	**71**	**131**	**278**

Is there evidence of an association between the duration of the pregnancy and the level of care received by the mother?

T **R6.36. Twins by year** Are twin births becoming more common? Part II Review Exercise R2.5 looked at the number of twin births by year from 1980 to 2014. Now we can include some inference:

Response variable is: TwinBirths/1000
R squared = 97.1%
s = 0.9184 with 35 − 2 = 33 degrees of freedom

Variable	Coefficient	SE(Coeff)	t-Ratio	P-Value
Intercept	18.2127	0.3039	59.9	<0.0001
Year since 1980	0.508261	0.0154	33.1	<0.0001

a) State and test the standard null hypothesis for the slope.

An Internet blogger claims that the increase in twin births can be explained by the increase in CO_2 levels in the atmosphere. He posts this regression for the same time period:

Response variable is: TwinBirths/1000
R squared = 97.4
s = 0.8751 with 35 − 3 = 32 degrees of freedom

Variable	Coefficient	SE(Coeff)	t-Ratio	P-Value
Intercept	92.3382	35.58	2.60	0.0142
Year since 1980	0.891661	0.1846	4.83	<0.0001
CO2 (ppm)	−0.220668	0.1059	−2.08	0.0453

b) Do you think the growth in twin births is due to the amount of CO_2 in the atmosphere? Explain your reasoning.

R6.37. Retirement planning According to the 2011 Retirement Confidence Survey, run by the Employee Benefit Research Institute, 37% of men reported being "a lot behind schedule" while 43% of women answered in the same way. Is this evidence that in general, more women are feeling this way when it comes to retirement planning? Assuming the study had 722 men and 701 women, run the appropriate hypothesis test.

T **R6.38. Age and party 2016** The Pew Research Center conducted a representative telephone survey in October of 2016. Among the reported results was the following table concerning the preferred political party affiliation of respondents and their ages for white voters. Is there evidence of age-based differences in party affiliation in the United States for white voters?

	Leaning Republican	Leaning Democrat	Total
18–29	148	248	396
30–49	330	397	727
50–64	375	360	735
65+	284	236	520
Total	1137	1241	2378

a) Will you conduct a test of homogeneity or independence? Why?
b) Test an appropriate hypothesis.
c) State your conclusion, including an analysis of differences you find (if any).

R6.39. Eye and hair color A survey of 1021 school-age children was conducted by randomly selecting children from several large urban elementary schools. Two of the questions concerned eye and hair color. In the survey, the following codes were used:

Hair Color	Eye Color
1 = Blond	1 = Blue
2 = Brown	2 = Green
3 = Black	3 = Brown
4 = Red	4 = Grey
5 = Other	5 = Other

The statistics students analyzing the data were asked to study the relationship between eye and hair color.

a) One group of students produced the output shown below. What kind of analysis is this? What are the null and alternative hypotheses? Is the analysis appropriate? If so, summarize the findings, being sure to include any assumptions you've made and/or limitations to the analysis. If it's not an appropriate analysis, state explicitly why not.

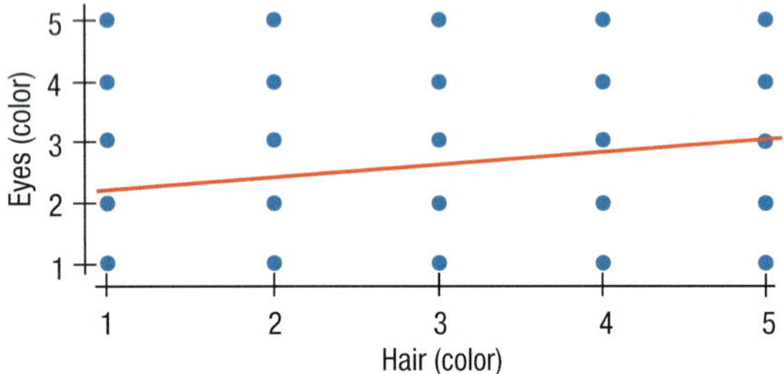

Dependent variable is Eyes
R-squared = 3.7%
s = 1.112 with 1021 − 2 = 1019 degrees of freedom

Variable	Coefficient	SE(Coeff)	t-Ratio	P-Value
Intercept	1.99541	0.08346	23.9	≤ 0.0001
Hair	0.211809	0.03372	6.28	≤ 0.0001

b) A second group of students used the same data to produce the output shown below. The table displays counts and standardized residuals in each cell. What kind of analysis is this? What are the null and alternative hypotheses? Is the analysis appropriate? If so, summarize the findings, being sure to include any assumptions you've made and/or limitations to the analysis. If it's not an appropriate analysis, state explicitly why not.

Hair Color	Eye Color 1	2	3	4	5
1	143 7.67540	30 0.41799	58 −5.88169	15 −0.63925	12 −0.31451
2	90 −2.57141	45 0.29019	215 1.72235	30 0.49189	20 −0.08246
3	28 −5.39425	15 −2.34780	190 6.28154	10 −1.76376	10 −0.80382
4	30 2.06116	15 2.71589	10 −4.05540	10 2.37402	5 0.75993
5	10 −0.52195	5 0.33262	15 −0.94192	5 1.36326	5 2.07578

$$\sum \frac{(Obs - Exp)^2}{Exp} = 223.6 \quad \text{P-value} < 0.00001$$

R6.40. Barbershop music At a barbershop music singing competition, choruses are judged on three scales: *Music* (quality of the arrangement, etc.), *Performance*, and *Singing*. The scales are supposed to be independent of each other, and each is scored by a different judge, but a friend claims that he can predict a chorus's *Singing* score from the other two scores. He offers the following regression based on the scores of all 34 choruses in a recent competition:

Dependent variable is: Singing
R-squared = 90.9% R-squared (adjusted) = 90.3%
s = 6.483 with 34 − 3 = 31 degrees of freedom

Variable	Coefficient	SE(Coeff)	t-ratio	P-value
Intercept	2.08926	7.973	0.262	0.7950
Performance	0.793407	0.0976	8.13	≤0.0001
Music	0.219100	0.1196	1.83	0.0766

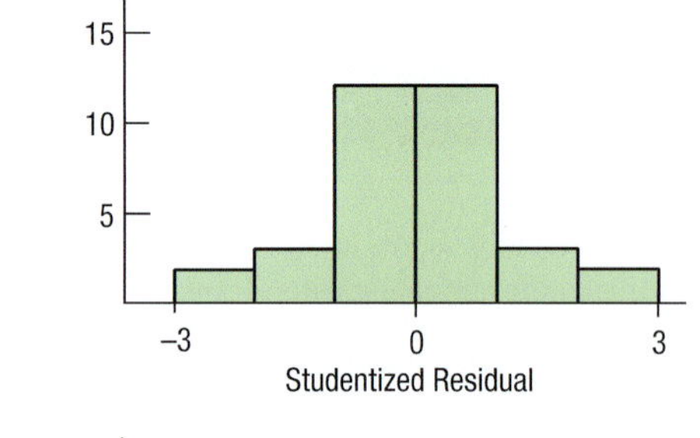

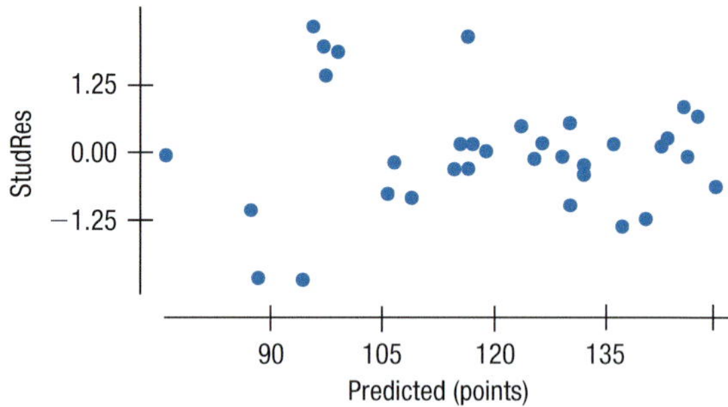

a) What do you think of your friend's claim? Can he predict *Singing* scores? Explain.

b) State the standard null hypothesis for the coefficient of *Performance* and complete the *t*-test at the 5% level. State your conclusion.

c) Complete the analysis. Check assumptions and conditions to the extent you can with the information provided.

R6.41. Cereals and fiber (Data in **Cereals**). A regression of the calories in breakfast cereals on their carbohydrate content (g) looks like this:

Response variable is: calories
R squared = 6.4% s = 18.98 with 77 − 2 = 75 degrees of freedom

Variable	Coefficient	SE(Coeff)	t-Ratio	P-Value
Intercept	88.1521	8.566	10.3	<0.0001
carbo	1.26185	0.5584	2.26	0.0267

A second regression with *fiber* content included gives this model:

Response variable is: calories
R squared = 11.0% s = 18.63 with 77 − 3 = 74 degrees of freedom

Variable	Coefficient	SE(Coeff)	t-Ratio	P-Value
Intercept	98.6677	9.987	9.88	<0.0001
carbo	0.827390	0.5916	1.40	0.1661
fiber	−1.88968	0.9678	−1.95	0.0546

a) Give an interpretation of the coefficient of *carbo* in both regressions.
b) Why did the P-value on *carbo* change from 0.0267 in the first regression to 0.1661 in the second regression?
c) What statistics or graphs would you look at to understand this change?

R6.42. Family planning A 1954 study of 1438 pregnant women examined the association between the woman's education level and the occurrence of unplanned pregnancies, producing these data:

	Education Level		
	<3 Yr HS	3+ Yr HS	Some College
Number of Pregnancies	591	608	239
% Unplanned	66.2%	55.4%	42.7%

Do these data provide evidence of an association between family planning and education level? (*Fertility Planning and Fertility Rates by Socio-Economic Status*, Social and Psychological Factors Affecting Fertility, 1954)

T **R6.43.** Cereals with bran Exercise R6.41 looked at regressions to model calories in breakfast cereals based on their carbohydrates and fiber content. Here is a scatterplot of carbohydrates vs. fiber content for these cereals:

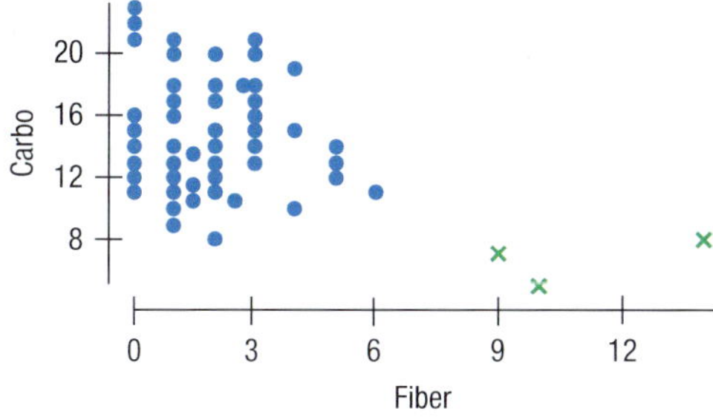

The cereals plotted with green x's are the three bran cereals in the data. How does this plot help explain the change in the coefficient of *carbo* seen in Exercise R6.41?

T **R6.44.** Old Faithful We saw in Part II Review Exercise R2.21 that Old Faithful eruptions do not occur at constant intervals and the intervals may vary greatly depending on the duration of the previous eruption. In that exercise, we fit a regression model, which we can now improve with *t*- statistics and P-values.

Dependent variable is Interval
R-squared = 77.0%
s = 6.159 with 222 − 2 = 220 degrees of freedom

Variable	Coefficient	SE(Coeff)	t-Ratio	P-Value
Intercept	33.9668	1.428	23.8	≤0.0001
Duration	10.3582	0.3822	27.1	≤0.0001

a) State and test the natural null hypothesis for the slope.

Here are summary statistics for the variables:

Variable	Mean	StdDev
Duration	3.57613	1.08395
Interval	71.0090	12.7992

b) Create a 95% confidence interval for the mean length of time that will elapse following a 2-minute eruption.
c) You arrive at Old Faithful just as an eruption ends. Witnesses say it lasted 4 minutes. Create a 95% prediction interval for the length of time you should expect to wait to see the next eruption.

R6.45. Togetherness Are good grades in high school associated with family togetherness? A Simple Random Sample of 142 high-school students was asked how many meals per week their families ate together. Their responses produced a mean of 3.78 meals per week, with a standard deviation of 2.2. Researchers then matched these responses against the students' grade point averages. The scatterplot appeared to be reasonably linear, so they went ahead with the regression analysis, seen below. No apparent pattern emerged in the residuals plot.

Dependent variable is GPA
R-squared = 11.0%
s = 0.6682 with 142 − 2 = 140 df

Variable	Coefficient	SE(Coeff)
Intercept	2.7288	0.1148
Meals/wk	0.1093	0.0263

a) Is there evidence of an association? Test an appropriate hypothesis and state your conclusion.
b) Do you think this association would be useful in predicting a student's grade point average? Explain.
c) Are your answers to parts a and b contradictory? Explain.

R6.46. Learning math Developers of a new math curriculum called "Accelerated Math" compared performances of students taught by their system with control groups of students in the same schools who were taught using traditional instructional methods and materials. Statistics about pretest and posttest scores are shown in the table. (J. Ysseldyke and S. Tardrew, *Differentiating Math Instruction*, Renaissance Learning)

a) Did the groups differ in average math score at the start of this study?
b) Did the group taught using the Accelerated Math program show a significant improvement in test scores?
c) Did the control group show a significant improvement in test scores?
d) Were gains significantly higher for the Accelerated Math group than for the control group?

		Instructional Method	
	Number of Students	Acc. Math 231	Control 245
Pretest	Mean	560.01	549.65
	St. Dev	84.29	71.68
Posttest	Mean	637.55	588.76
	St. Dev	82.9	83.24
Individual Gain	Mean	77.53	39.11
	St. Dev.	78.01	66.25

R6.47. Juvenile offenders According to an article in the *Journal of Consulting and Clinical Psychology*, Charles Borduin pioneered a treatment called Multisystemic Therapy (MST) as a way to prevent serious mental health problems in adolescents. The therapy involves a total support network including family and community, rather than the more common individual therapy (e.g., visits to a therapist). After a 22-year-long study, one notable fact was that while 15.5% of juveniles who received individual therapy were arrested for a violent felony, only 4.3% of the juveniles treated with MST had done so.

a) Suppose the results are based on sample sizes of 125 juveniles in each group. Create a 99% confidence interval for the reduction in violent felony rate when comparing MST to the traditional individual therapy.

b) Using your interval, is there evidence of a true reduction for the whole population? Which population is the study investigating?

T **R6.48. Dairy sales** Peninsula Creameries sells both cottage cheese and ice cream. The CEO recently noticed that in months when the company sells more cottage cheese, it seems to sell more ice cream as well. Two of his aides were assigned to test whether this is true or not. The first aide's plot and analysis of sales data for the past 12 months (in millions of pounds for cottage cheese and for ice cream) appear below.

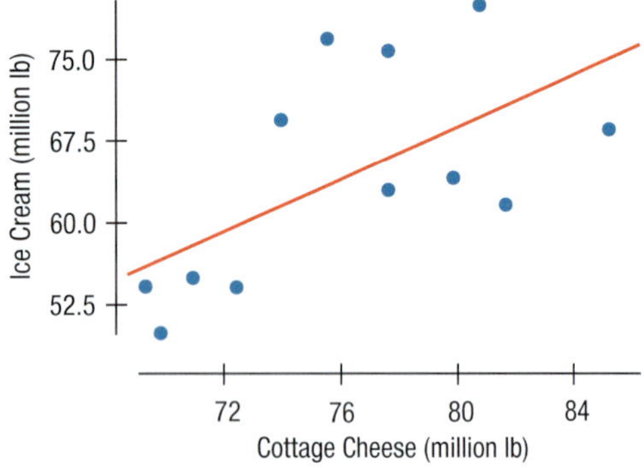

Dependent variable is Ice cream
R-squared = 36.9%
s = 8.320 with 12 − 2 = 10 degrees of freedom

Variable	Coefficient	SE(Coeff)	t-Ratio	P-Value
Intercept	−26.5306	37.68	−0.704	0.4975
Cottage C . . .	1.19334	0.4936	2.42	0.0362

The other aide looked at the differences in sales of ice cream and cottage cheese for each month and created the following output:

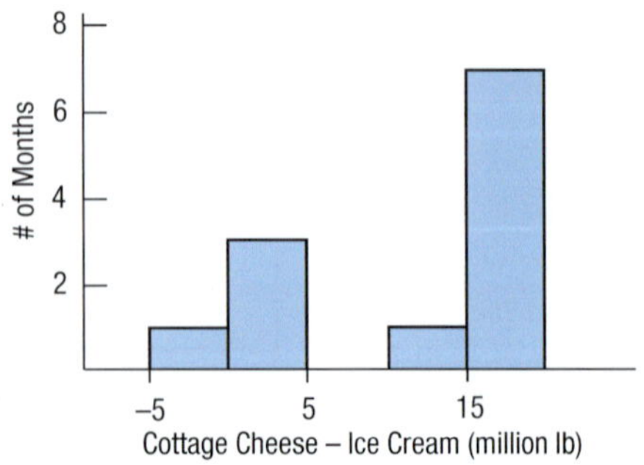

Cottage Cheese − Ice Cream

Count	12
Mean	11.8000
Median	15.3500
StdDev	7.99386
IntQRange	14.3000
25th %tile	3.20000
75th %tile	17.5000

Test H_0: $\mu(CC - IC) = 0$ vs H_a: $\mu(CC - IC) \neq 0$
Sample Mean = 11.800000 t-Statistic = 5.113 w/11 df
Prob = 0.0003
Lower 95% bound = 6.7209429
Upper 95% bound = 16.879057

a) Which analysis would you use to answer the CEO's question? Why?

b) What would you tell the CEO?

c) Which analysis would you use to test whether the company sells more cottage cheese or ice cream in a typical year? Why?

d) What would you tell the CEO about this other result?

e) What assumptions are you making in the analysis you chose in part a? What assumptions are you making in the analysis in part c?

f) Next month's cottage cheese sales are 82 million pounds. Ice cream sales are not yet available. How much ice cream do you predict Peninsula Creameries will sell?

g) Give a 95% confidence interval for the true slope of the regression equation of ice cream sales by cottage cheese sales.

h) Explain what your interval means.

T **R6.49. Diet** Thirteen overweight women volunteered for a study to determine whether eating specially prepared crackers before a meal could help them lose weight. The subjects were randomly assigned to eat crackers with different types of fiber (bran fiber, gum fiber, both, and a control cracker). Unfortunately, some of the women developed uncomfortable bloating and upset stomachs. Researchers suspected that some of the crackers might be at fault. The contingency table of "Cracker" versus "Bloat" shows the relationship between the four different types of crackers and the reported bloating. The study was paid for by the manufacturers of the gum fiber. What would you recommend to them about the prospects for marketing their new diet cracker?

		Bloat	
		Little/None	Moderate/Severe
Cracker	Bran	11	2
	Gum	4	9
	Combo	7	6
	Control	8	4

T **R6.50. Cramming** Students in two basic Spanish classes were required to learn 50 new vocabulary words. One group of 45 students received the list on Monday and studied the words all week. Statistics summarizing this group's scores on Friday's quiz are given. The other group of 25 students did not get the vocabulary list until Thursday. They also took the quiz on Friday, after "cramming" Thursday night.

Then, when they returned to class the following Monday, they were retested—without advance warning. Both sets of test scores for these students are shown.

Group 1
Fri.
Number of students = 45
Mean = 43.2 (of 50)
StDev = 3.4
Students passing (score ≥ 40) = 33%

Group 2			
Fri.	Mon.	Fri.	Mon.
42	36	50	47
44	44	34	34
45	46	38	31
48	38	43	40
44	40	39	41
43	38	46	32
41	37	37	36
35	31	40	31
43	32	41	32
48	37	48	39
43	41	37	31
45	32	36	41
47	44		

a) Did the week-long study group have a mean score significantly higher than that of the overnight crammers?
b) Was there a significant difference in the percentages of students who passed the quiz on Friday?
c) Is there any evidence that when students cram for a test, their "learning" does not last for 3 days?
d) Use a 95% confidence interval to estimate the mean number of words that might be forgotten by crammers.
e) Is there any evidence that how much students forget depends on how much they "learned" to begin with?

T **R6.51.** Hearing Fitting someone for a hearing aid requires assessing the patient's hearing ability. In one method of assessment, the patient listens to a tape of 50 English words. The tape is played at low volume, and the patient is asked to repeat the words. The patient's hearing ability score is the number of words perceived correctly. Four tapes of equivalent difficulty are available so that each ear can be tested with more than one hearing aid. These lists were created to be equally difficult to perceive in silence, but hearing aids must work in the presence of background noise. Researchers had 24 subjects with normal hearing compare two of the tapes when a background noise was present, with the order of the tapes randomized. Is it reasonable to assume that the two lists are still equivalent for purposes of the hearing test when there is background noise? Base your decision on a confidence interval for the mean difference in the number of words people might misunderstand. (Faith Loven, *A Study of the Interlist Equivalency of the CID W-22 Word List Presented in Quiet and in Noise*. University of Iowa [1981])

Subject	List A	List B
1	24	26
2	32	24
3	20	22
4	14	18
5	32	24
6	22	30
7	20	22
8	26	28
9	26	30
10	38	16
11	30	18
12	16	34
13	36	32
14	32	34
15	38	32
16	14	18
17	26	20
18	14	20
19	38	40
20	20	26
21	14	14
22	18	14
23	22	30
24	34	42

T **R6.52.** Newspapers Who reads the newspaper more, men or women? Eurostat, an agency of the European Union (EU), conducts surveys on several aspects of daily life in EU countries. Recently, the agency asked samples of 1000 respondents in each of 14 European countries whether they read the newspaper on a daily basis. Below are the data by country and gender.

	% Reading a Newspaper Daily	
Country	Men	Women
Belgium	56.3	45.5
Denmark	76.8	70.3
Germany	79.9	76.8
Greece	22.5	17.2
Spain	46.2	24.8
Ireland	58.0	54.0
Italy	50.2	29.8
Luxembourg	71.0	67.0
Netherlands	71.3	63.0
Austria	78.2	74.1
Portugal	58.3	24.1
Finland	93.0	90.0
Sweden	89.0	88.0
U.K.	32.6	30.4

a) Examine the differences in the percentages for each country. Which of these countries seem to be outliers? What do they have in common?
b) After eliminating the outliers, is there evidence that in Europe men are more likely than women to read the newspaper?

24 Multiple Regression Wisdom

Burger King publishes many facts about their menu items (Data in **Burger King items**). We may want to understand how the number of calories in a BK item depends on other nutritional aspects. Elsewhere in this text we've looked at simple analyses of these data. Now let's consider more of the variables together. The data in a text like this often seem to be clean and clear. But in the real world, the first step in most data analyses is to format and clean the data. That's a topic that deserves a longer discussion than we have room for here, but let's do a quick overview.

24.1 Cleaning and Formatting Data

The Burger King nutritional data can be found online at bk.com/pdfs/nutrition.pdf, formatted in a neat table. It looks like this:

Table 24.1
A portion of the table of Burger King nutrition information

Nutrition Facts	serving size (g)	Calories	Calories from fat	Total fat (g)	Saturated Fat (g)	Trans Fat (g)	Chol (mg)	Sodium (mg)	Total Carb (g)	Dietary Fiber (g)	Total Sugar (g)	Protein (g)
						WHOPPER® Sandwiches						
WHOPPER® Sandwich	270	660	360	40	12	1.5	90	980	49	2	11	28
WHOPPER® Sandwich with Cheese	292	740	420	46	16	2	115	1340	50	2	11	32
Bacon & Cheese WHOPPER® Sandwich	303	790	460	51	17	2	125	1560	50	2	11	35
BBQ Bacon WHOPPER® Sandwich	303	800	460	51	17	2	125	1540	53	2	14	35
DOUBLE WHOPPER® Sandwich	354	900	520	58	20	3	175	1050	49	2	11	48
DOUBLE WHOPPER® Sandwich with Cheese	377	980	580	64	24	3	195	1410	50	2	11	52
TRIPLE WHOPPER® Sandwich with Cheese	461	1220	740	82	32	4.5	280	1470	50	2	11	71
WHOPPER JR.® Sandwich	134	310	160	18	5	0.5	40	390	27	1	7	13
						FLAME BROILED BURGERS						
BACON KING™ Sandwich	356	1150	710	79	31	3.5	240	2150	49	2	10	61
...												

WHO	Burger King menu items
WHAT	Various nutrition facts
WHEN	2017
WHERE	bk.com/pdfs/nutrition/pdf
WHY	General interest

Even in data as clearly formatted as this, you will need to do some **data cleaning**. First, you'll notice special characters (™, ®) scattered among the names of the menu items. Many statistics programs won't accept case names with special characters. The variable names themselves are images rather than text, so you can't just copy them into your data file. And the table itself is interrupted by banners that say such things as "WHOPPER® Sandwiches," which are not data and should be removed. When you work with data found on the Internet or available from a source other than your own work, you can expect to spend as much or more time simply getting your data ready to use in the statistics software of your choice as you will performing statistical analyses. And if you collect data yourself, you should take great care in recording and formatting it with future uses in mind.

You may need to restructure your data. For example, you may want to **aggregate** cases, summarizing values for cases in the same group. You might, for instance find the average number of calories for entrees vs. desserts. Another option is to **filter** the data to focus on a subset. The dessert items in the BK data are quite different from the main course items, so you may wish to filter them out of your analysis.[1] Sometimes data on the same items can be found in different places. You may need to **merge** or **join** cases together. Merging is a traditional database operation that requires you to have an indexing variable that names the cases with the same names in each data file to enable the matching of cases. If you have additional data measured on different cases but on the same variables, you may want to **append** the two datasets into a single, larger one. One data cleaning function that is often needed is to remove duplicate records, often called **deduping**. It is common, for example, for records of contributors kept by a nonprofit agency to accidentally enter a contributor more than once, possibly with a just slightly different name (for example, including a middle initial or changing the honorific from Ms. to Dr. or from Mr to Mr.) Finally, it is common for entries in a given variable to be inconsistent. *Sex* may be recorded as M or Male. Or there may be a mix of numeric codes (1 = Male, 2 = Female) and text versions. No statistical analysis is possible until the coding of a variable is consistent. Finally, you may want to create new variables by combining some of those in your data. The specialized statistics methods often called Statistical Machine Learning prefer the term **feature extraction** for this operation and the term **feature selection** for the choosing of variables from what might be a very large collection of potential predictors.

Once your data are clean enough to read and calculate with, you should check the variables for outliers, anomalies, errors, blunders, and the need to re-express them to improve their structure or to create new variables from the original ones. One useful tool for identifying variables that need attention in datasets of between 3 and 15 or so variables is the **scatterplot matrix**, sometimes abbreviated **splom**. A scatterplot matrix consists of a scatterplot of each pair of variables arranged in a table so that across each row of the table the same variable is on the y-axis and down each column the same variable is on the x-axis. Figure 24.1 shows an example for variables from the Burger King data. Statistics programs vary slightly in how they make their scatterplot matrix displays. Some fill the lower left triangle of the table, although that just duplicates plots from the upper triangle. Some put a histogram of each variable on the diagonal for its row (as y) and column (as x); others put a Normal probability plot there. Still others just name the variable. A scatterplot matrix provides a broad overview of your variables. It isn't intended to be read in detail, but rather to indicate whether relationships are generally straight (or curved and in need of re-expression), whether there are outliers or subgroups that stick out, and whether the individual variables are roughly unimodal and symmetric. In Figure 24.1, a first glance suggests that the variable *Sugar* may divide the items into two groups—not too surprising, since dessert items are included—and that *Cholesterol* may have a few outliers—something to keep an eye on.

[1] Or, as some prefer, to start with dessert.

Figure 24.1
A scatterplot matrix of Burger King variables.

Leverage

One of the important outcomes of the iterative cycle of data wrangling and data exploration is the discovery of outliers. As we've seen, outliers may be mistakes, such as a decimal point in the wrong place, or just an unusual observation. Deciding what to do with an outlier requires judgment, especially if the outlier has a strong effect on the analysis.

In multiple regression, outliers can be even more problematic. One observation can affect some or all of the coefficients of the model, and even the selection of the variables to go into the model. Because it involves several predictors, observations with high influence are not as easy to spot as they were in simple regression. In a regression of a single predictor and a response, it's easy to see if a value has high leverage, because it's far from the mean of the x-values in a scatterplot. In a multiple regression with k predictor variables, things are more complicated. A point might actually not be far from any of the x means and yet still exert large leverage because it has an unusual *combination* of predictor values. Even a graphics program designed to display points in high dimensional spaces may not make it obvious. Fortunately, we can calculate leverage values for each case. These are standard for most multiple regression programs.

The **leverage** of a case in multiple regression is easy to understand. For any case, add 1 to its y-value. Recompute the regression, and see how much the *predicted* value of the case changes. The amount of the change is the leverage. It can never be greater than 1 or less than 0. A point with zero leverage has no effect at all on the regression slopes, although it does participate in the calculations of the intercept, R^2, s, and the F- and t-statistics. The leverage of the ith point in a dataset is often denoted by h_i.

A point with high leverage may not actually influence the regression coefficients if it follows the pattern of the model set by the other points, but it's worth examining simply because of its *potential* to do so. Looking at leverage values can be an effective way to discover cases that are extraordinary on a combination of *x*-variables. Such cases often deserve special attention.

A case can have large leverage in two different ways:

- The case might be extraordinary in one or more individual variables. For example, the food item with the most fat may stand out. This is a good way to identify errors because they may not conform with the rest of the data.
- The case may be extraordinary in a combination of variables. For example, a special low-fat veggie burger may have a surprisingly high sodium level and that combination of low fat and high salt may be unusual. Looking at leverage values can be a very effective way to discover cases that are extraordinary on a combination of *x*-variables.

There are no tests for whether the leverage of a case is too large. The average leverage value of all cases in a regression is $1/n$, but that doesn't give you much of a guide. One common approach is to examine a histogram or probability plot of the leverages. Any case whose leverage stands out relative to the others probably deserves your attention. You should try to understand why any high-leverage case is extraordinary. You may have found an error or a telling exception.

Because leverages can help to identify cases that are extraordinary in ways not necessarily obvious from simple histograms of individual variables or scatterplots of pairs of variables, it is a good idea to include a look at the leverages of cases on all the variables you are considering for an analysis as part of your data cleaning routines.[2] For the Burger King data, the leverages look like this:

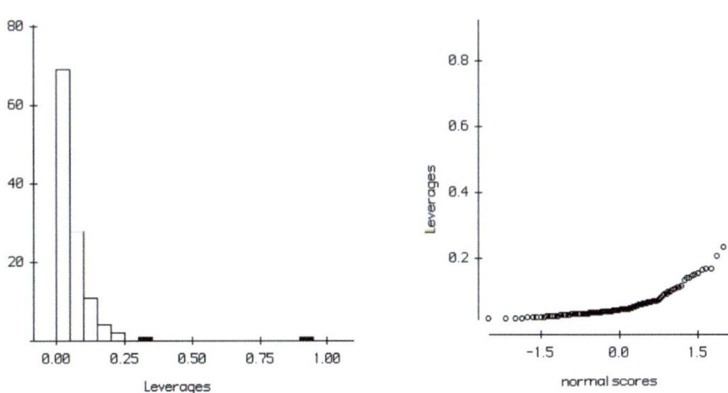

Figure 24.2
Leverages of the Burger King data show some high-leverage points.

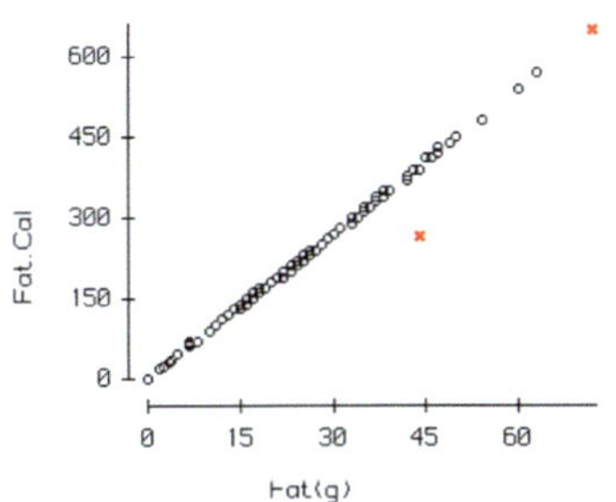

Figure 24.3
A scatterplot of *Fat calories* and *Fat(g)* shows the error.

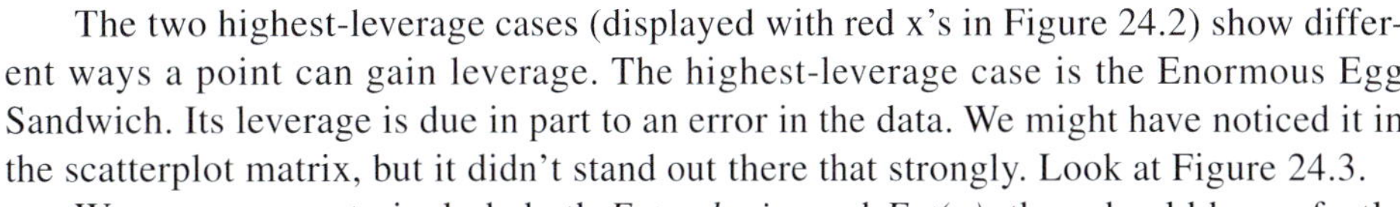

The two highest-leverage cases (displayed with red x's in Figure 24.2) show different ways a point can gain leverage. The highest-leverage case is the Enormous Egg Sandwich. Its leverage is due in part to an error in the data. We might have noticed it in the scatterplot matrix, but it didn't stand out there that strongly. Look at Figure 24.3.

We were wrong to include both *Fat calories* and *Fat(g)*; they should be perfectly related. Except that they aren't in the case of the Enormous Egg Sandwich. At least one of those values must be an error. We'll be careful not to put both variables in a regression model.

[2]Many statistics programs only offer leverages as a calculation for the predictors in a regression analysis. To find the leverages for many variables, just generate a variable full of random numbers or choose some other variable to declare as *y* and regress it on all the variables of interest. Then find the leverages and throw away the regression.

To identify the other high-leverage case—the Ultimate Breakfast Platter—we'll turn to another display. A **parallel coordinate plot** displays several variables side-by-side as vertical dotplots and then connects the points for each case. Figure 24.4 shows the plot with the two high-leverage points, each shown as a red line connecting values for each variable.

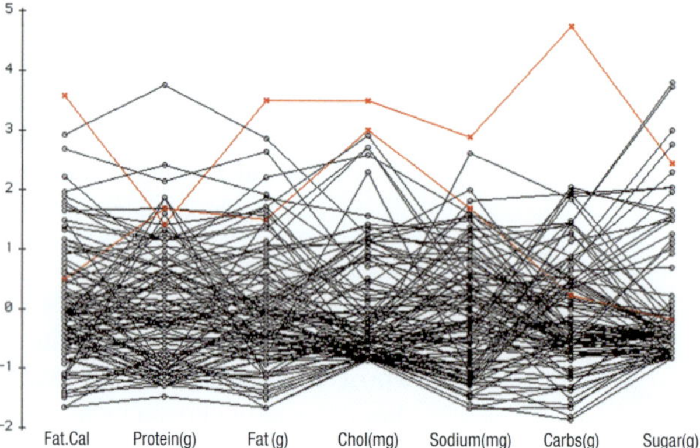

Figure 24.4
A parallel coordinate plot connects the points for each case.

The connected points at the top for *Fat, Cholesterol, Sodium,* and *Carbs* are for the item with the second highest leverage, the Ultimate Breakfast Platter. They show that its leverage is due to being the largest value on each of these variables, although it is not all that extraordinary on any of them (well, except maybe on *Carbs*). This case is not an error in the data.[3]

EXAMPLE 24.1 Looking at Leverages

As a class project, students in a large statistics class collected publicly available information on recent home sales in their hometowns (Data in **Real Estate**). There are 894 properties. These are not a random sample, but they may be representative of home sales during a short period of time, nationwide. Let's consider leverages for these data.

Dependent variable is: Price
R squared = 23.1% R squared (adjusted) = 22.8%
s = 253709 with 893 − 5 = 888 degrees of freedom

Variable	Coefficient	SE(Coeff)	t-ratio	P-value
Intercept	322470	40192	8.02	≤0.0001
Living area	92.6272	13.09	7.08	≤0.0001
Bedrooms	−69720.6	12764	−5.46	≤0.0001
Bathrooms	82577.6	13410	6.16	≤0.0001
Rural	−161575	23313	−6.93	≤0.0001

[3]We'll leave for you the question of whether it would be an error to choose it for breakfast.

QUESTION: What do diagnostic statistics tell us about these data and this model?

A boxplot of the leverage values shows one extraordinarily large leverage:

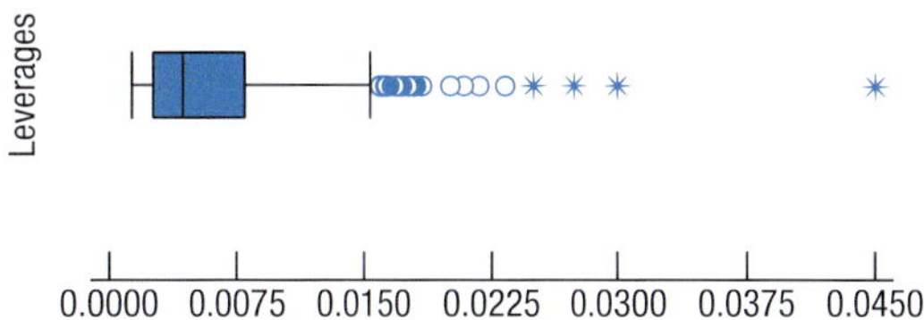

Investigation of that case reveals it to be a home that sold for $489,900 that has 8 bedrooms and only 2.5 baths. This is a particularly unusual combination, especially for a home with that value. If we were pursuing this analysis further, we'd want to check the records for this house to be sure that the numbers of bedrooms and bathrooms were recorded accurately.

If this section has left you with the impression that there is often a great deal of work to do before you can start to use all your statistics skills, that's quite correct. Experienced data analysts often say that the major part of their work is data cleaning.

24.2 Diagnosing Regression Models: Looking at the Cases

We often use regression analyses to try to understand the world. By working with the data and creating models, we can learn a great deal about the relationships among variables. As we saw with simple regression, sometimes we can learn as much from the cases that *don't* fit the model as from the bulk of cases that do. Extraordinary cases often tell us more about the world simply by the ways in which they fail to conform and the reasons we can discover for those deviations.

If a case doesn't conform to the others, we should identify it and, if possible, understand why it is different. As in simple regression, a case can be extraordinary by standing away from the model in the y direction or by having unusual values in an x-variable. In multiple regression it can also be extraordinary by having an unusual *combination* of values in the x-variables. Deviations in the y direction show up in the residuals. Deviations in the x's show up as *leverage*.

EXAMPLE 24.2 Leverage in a Multiple Regression

Let's return to the Burger King items. This time, we'll look at only items that are not for breakfast. Here's a regression to predict *Calories*:

Response variable is: Calories
Cases selected according to Not Breakfast
122 total cases of which 45 are missing
R squared = 99.4% R squared (adjusted) = 99.4%
s = 18.96 with 77 − 5 = 72 degrees of freedom

Variable	Coefficient	SE(Coeff)	t-ratio	P-value
Intercept	0.428142	5.661	0.076	0.9399
Fat(g)	8.98350	0.2808	32.0	<0.0001
Sodium(mg)	0.049603	0.0063	7.87	<0.0001
Carbs(g)	3.77941	0.1506	25.1	<0.0001
Chol(mg)	0.822855	0.0840	9.80	<0.0001

QUESTION: What can we learn about this model with diagnostics?

ANSWER: A Normal probability plot is one way to identify high leverage cases:

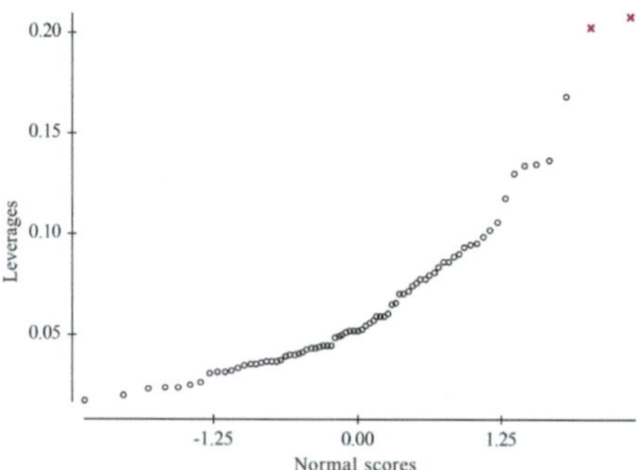

The two highest-leverage items are the Triple Whopper and the Triple Whopper with cheese.

And the parallel coordinate plot tells us that *Fat* and *Cholesterol* are the likely culprits.

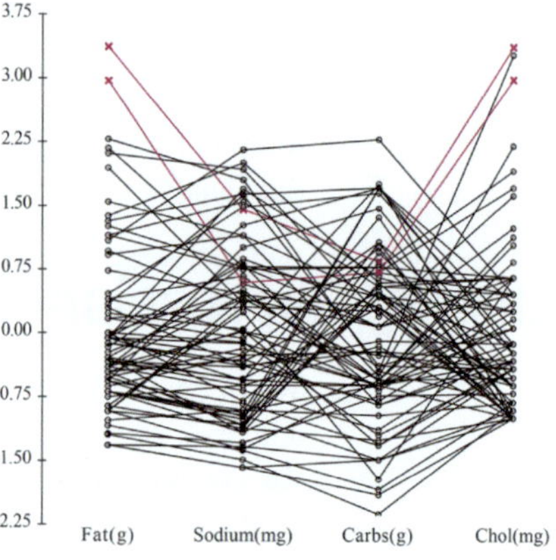

Residuals and Standardized Residuals

Residuals are not all alike. Consider a point with leverage 1.0. That's the highest a leverage can be, and it means that the line follows the point perfectly. So, a point like that must have a zero residual. And since we know the residual exactly, that residual has zero standard deviation. This tendency is true in general: The larger the leverage, the *smaller* the standard deviation of its residual.[4]

When we want to compare values that have differing standard deviations, it's a good idea to standardize them.[5] We can do that with the regression residuals, dividing each one by an estimate of its own standard deviation. When we do that, the resulting values follow a Student's t-distribution. In fact, these standardized residuals are called **Studentized residuals**. It's a good idea to examine the Studentized residuals (rather than the simple residuals) to assess the Nearly Normal Condition and the **Does the Plot Thicken? Condition.** Any Studentized residual that stands out from the others deserves our attention.[6]

It may occur to you that we've always plotted the *unstandardized* residuals when we made regression models. And we've treated them as if they all had the same standard deviation

[4]Technically, $SD(e_i) = \sigma\sqrt{1 - h_i}$ where h_i is the leverage of the i-th case, e_i is its residual, and σ is the standard deviation of the regression model errors.

[5]Be cautious when you encounter the term "standardized residual." It is used in different books and by different statistics packages to mean quite different things. Be sure to check the meaning.

[6]There's more than one way to Studentize residuals according to how you estimate σ. You may find statistics packages referring to *externally Studentized residuals* and *internally Studentized residuals*. It is the *externally* Studentized version that follows a t-distribution, so those are the ones we recommend.

when we checked the **Nearly Normal Condition**. It turns out that this was a simplification. It didn't matter much for simple regression, but for multiple regression models, it's a better idea to use the Studentized residuals when checking the Nearly Normal Condition. (Of course, Student's t isn't exactly Normal either—that's why we say "nearly" Normal.)

Influential Cases

A case that has *both* high leverage and large Studentized residuals is likely to have changed the regression model substantially all by itself. Such a case is said to be **influential**. An influential case cries out for special attention because removing it is likely to give a very different regression model.

The surest way to tell whether a case is influential is to omit it and see how much the regression model changes. You should call a case "influential" if omitting it changes the regression model by enough to matter for *your* purposes. To identify possibly influential cases, check the leverage and Studentized residuals. Two statistics that combine leverage and Studentized residuals into a single measure of influence, Cook's distance and DFFITs, are offered by many statistics programs. If either measure is unusually large for a case, that case should be checked as a possibly influential point.

When a regression analysis has cases that have both high leverage and large Studentized residuals, it would be irresponsible to report only the regression on all the data. You should also compute and discuss the regression found with such cases removed, and discuss the extraordinary cases individually if they offer additional insight. If your interest is to understand the world, the extraordinary cases may well tell you more than the rest of the model. If your only interest is in the model (for example, because you hope to use it for prediction), then you'll want to be certain that the model wasn't determined by only a few influential cases, but instead was built on the broader base of the bulk of your data.

Recent events have focused attention on airport screening of passengers. But screening has a longer history. The *Sourcebook of Criminal Justice Statistics Online* lists the numbers of various violations found by airport screeners for each of several types of violations in each year from 1977 to 2000 (Data in **Airport screening**). Here's a regression of the number of long guns (rifles and the like) found vs. the number of times false information was discovered.

Response variable is: Long guns
R-squared = 3.8%   R-squared (adjusted) = −0.6%
s = 56.34 with 24 − 2 = 22 degrees of freedom

Variable	Coefficient	SE(Coeff)	t-ratio
Intercept	87.0071	19.68	4.42
False information	0.246762	0.2657	0.929

That summary doesn't look like it's a particularly successful regression. The R^2 is only 3.8%, and the P-value for *False Info* is large. But a look at the scatterplot tells us more.

The unusual cases are from 1988 and 2000. In 2000, there were nearly 300 long gun confiscations. But because this point is at a typical value for false information, it doesn't have a strong effect on the slope of the regression line. But in 1988, the number of false information reports jumped over 200. The resulting case has high leverage because it is so far from the x-values of the other points. It's easy to see the influence of that one high-leverage case if we look at the regression lines with and without that case (Figure 24.5).

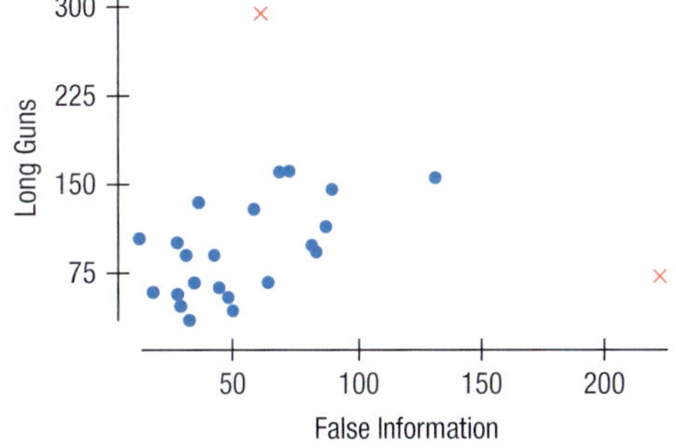

Figure 24.5
A high-leverage point can hide a strong relationship, so that you can't see it in the regression. Make a plot.

For the airport inspections, the leverage of 1988 is 0.63. That's quite high. If there had been even one fewer long gun discovered that year (decreasing the *observed* y-value by 1), the *predicted* y-value for 1988 would have decreased by 0.63, dragging the regression line down still farther. For comparison, the point for 2000 that has an extraordinary value for Long guns only has leverage 0.42. We would consider it an outlier because it is far from the other values in the y (here, Long gun) direction. But because it is near the mean in x (False information) it doesn't have enough leverage to change the slope of the regression line.

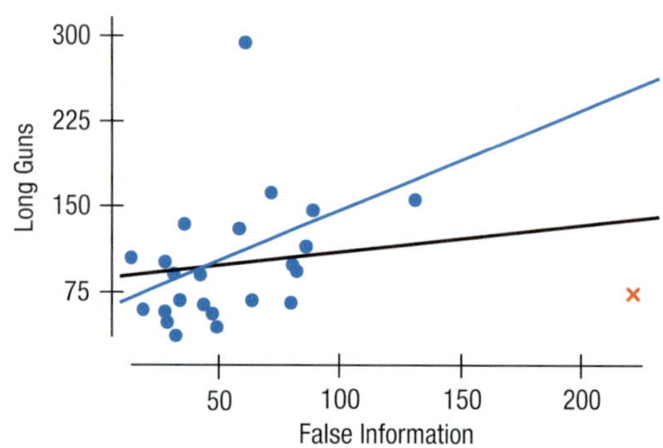

Figure 24.6
A single high-leverage point can change the regression slope quite a bit. The line omitting the point for 1988 is quite different from the line that includes the outlier.

STEP-BY-STEP EXAMPLE

Diagnosing A Multiple Regression

Let's consider what makes a roller coaster fast and then diagnose the model to understand more. Roller coasters get their *Speed* from gravity (the "coaster" part), so we'd naturally look to such variables as the *Height* and largest *Drop* as predictors. Let's make and diagnose that multiple regression. (Data in **Coasters 2015)**

VARIABLES Name the variables, report the W's, and specify the questions of interest.

I have data on 241 roller coasters that give their top *Speed* (mph), maximum *Height*, and largest *Drop* (both in feet).

PLOT

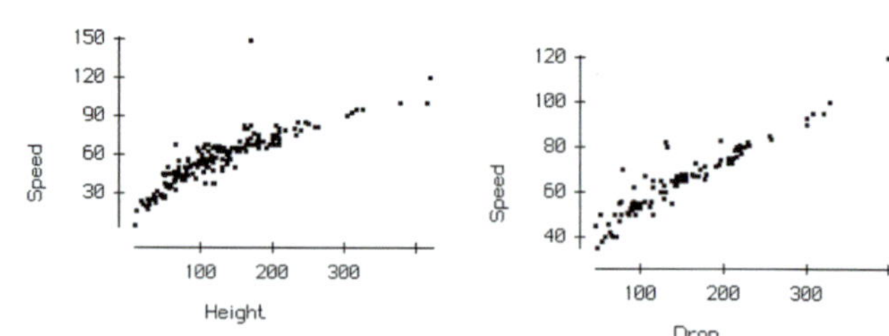

The *Speed* scatterplot shows an outlier. It is the *Formula Rossa* coaster in Abu Dhabi. This is a "launch" coaster that does not depend on the drop from an initial tower for its speed. We'll omit it from our analysis because it is a fundamentally different kind of coaster.

✔ **Straight Enough Condition**: The plots look reasonably straight.

✔ **Independence Assumption**: There are only a few manufacturers of roller coasters worldwide, and coasters made by the same company may be similar in some respects, but each roller coaster in our data is individualized for its site, so the coasters are likely to be independent.

PLAN Think about the assumptions and check the conditions.

Because these conditions are met I computed the regression model and found the Studentized residuals.

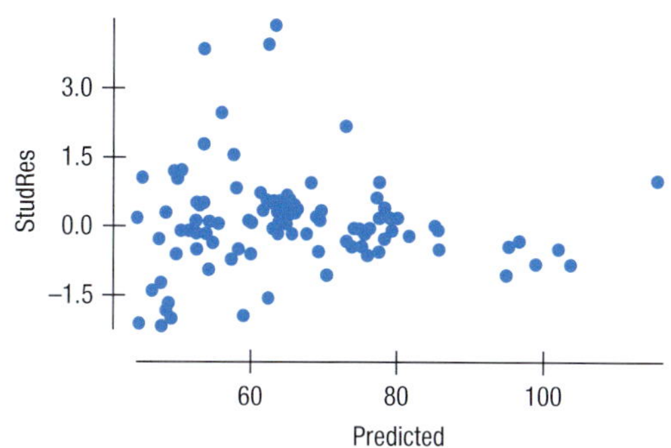

✓ **Straight Enough Condition (2)**: The values for one roller coaster don't seem to affect the values for the others in any systematic fashion. This makes the Independence Assumption more plausible.

✓ **Does the Plot Thicken? Condition**: The scatterplot of Studentized residuals against predicted values shows no obvious changes in the spread about the line. There do seem to be some large residuals that might be outliers.

Actually, we need the Nearly Normal Condition only if we want to do inference, but it's hard not to look at the P-values, so we usually check it out. In a multiple regression, it's best to check the Studentized residuals, although the difference is rarely large enough to change our assessment of the Normality.

✓ **Nearly Normal Condition**: A histogram of the Studentized residuals is unimodal and reasonably symmetric, but has three high outliers.

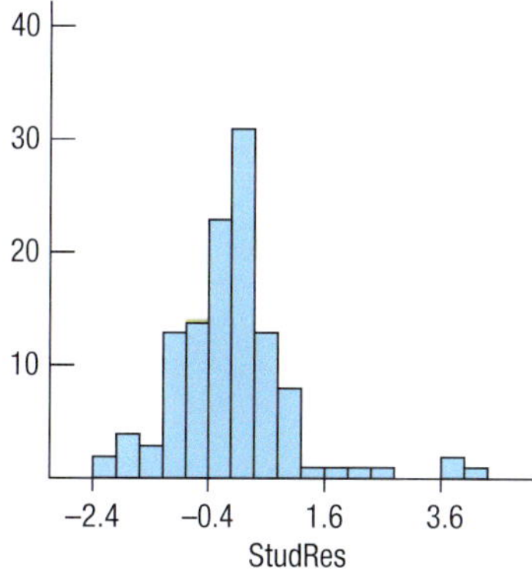

Choose your method.

✗ **Outlier Condition**: The histogram suggests that there may be a few large positive residuals. I'll want to look at those.

Under these conditions, the multiple regression model is appropriate as long as we are cautious about the possible outliers.

SHOW **MECHANICS**

Here is the computer output for the regression:

Response variable is: Speed
R squared = 89.6% R squared (adjusted) = 89.4%
ε = 4.746 with 118 − 3 = 115 degrees of freedom

Variable	Coefficient	SE(Coeff)	t-ratio	P-value
Intercept	34.5913	1.053	32.8	<0.0001
Height	0.037202	0.0164	2.27	0.0253
Drop	0.163203	0.0171	9.52	<0.0001

The estimated regression equation is

$$\widehat{Speed} = 34.6 + 0.037\,Height + 0.163\,Drop$$

INTERPRETATION

The R^2 for the regression is 89.6%. *Height* and *Drop* account for 89.6% of the variation in *Speed* in roller coasters like these. Both *Height* and *Drop* contribute significantly to the *Speed* of a roller coaster.

DIAGNOSIS

A histogram of the leverages shows one roller coaster with a rather high leverage of more than 0.18.

LEVERAGE Most computer regression programs will calculate leverages. There is a leverage value for each case.

It may not be necessary to remove high leverage points from the model, but it's certainly wise to know where they are and, if possible, why they are unusual.

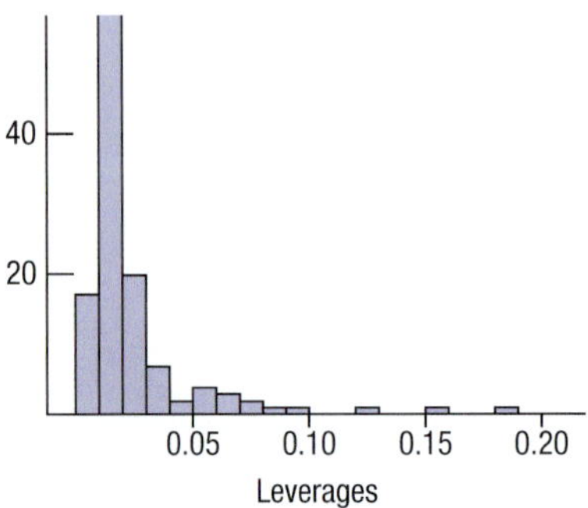

This high-leverage point is the *Oblivion* coaster in Alton, England. Neither its *Height* nor its *Drop* is extraordinary. To see what's going on, I made a scatterplot of *Drop* against *Height* with *Oblivion* shown as a red x.

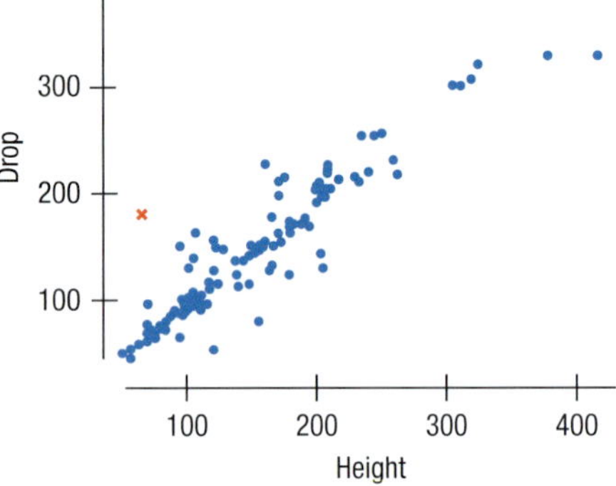

RESIDUALS At this point, we might consider recomputing the regression model after removing these three coasters. That's what we do in the next section.

Although *Oblivion's* maximum height is a modest 65 feet, it has a surprisingly long drop of 180 feet. At first, that seemed like an error, but looking it up, I discovered that the unusual feature of the *Oblivion* coaster is that it plunges riders down a deep hole below the ground.

The histogram of the Studentized residuals (above) also nominates some cases for special attention. The bars on the right of the histogram hold three roller coasters with large positive residuals: the *Xcelerator, Hypersonic XCL,* and *Volcano, the Blast Coaster.* New technologies such as hydraulics or compressed air are used to launch all three roller coasters. These three coasters are different in that their speed doesn't depend only on gravity.

The *Oblivion* roller coaster plunges into a hole in the ground.

Diagnosis Wrap-Up

What have we learned from diagnosing the roller coaster regression? We've discovered four roller coasters that may be influencing the model. And for each of them, we've been able to understand why and how they differed from the others. The oddness of *Oblivion* in plunging into a hole in the ground may lead us to prefer *Drop* as a predictor of speed rather than *Height*.

The three influential cases turned out to be different from the other roller coasters because they are "blast coasters" that don't rely only on gravity for their acceleration. Although we can't count on always discovering why influential cases are special, diagnosing influential cases raises the question of what about them might be different. Understanding influential cases can help us understand our data better.

When there are influential cases, we always want to consider the regression model without them:

Response variable is: Speed
R-squared = 91.5%   R-squared (adjusted) = 91.3%
s = 3.683 with 102 − 3 = 99 degrees of freedom

Variable	Coefficient	SE(Coeff)	t-ratio	P-value
Intercept	34.3993	0.9926	34.7	<0.0001
Drop	0.198348	0.0153	13.0	<0.0001
Height	0.00183	0.0155	0.118	0.9062

Without the three blast coasters, *Height* no longer appears to be important in the model, so we might try omitting it:

Response variable is: Speed
R-squared = 91.5%   R-squared (adjusted) = 91.4%
s = 3.664 with 102 − 2 = 100 degrees of freedom

Variable	Coefficient	SE(Coeff)	t-ratio	P-value
Intercept	34.4285	0.9567	36.0	<0.0001
Drop	0.200004	0.0061	32.7	<0.0001

That looks like a good model. It seems that our diagnosis has led us back to a simple regression.

Indicators for Influence

One good way to examine the effect of an extraordinary case on a regression is to construct a special indicator variable that is zero for all cases *except* the one we want to isolate. Including such an indicator in the regression model has the same effect as removing the case from the data, but it has two special advantages. First, it makes it clear to anyone looking at the regression model that we have treated that case specially. Second, the *t*-statistic for the indicator variable's coefficient can be used as a test of whether the case is influential. If the P-value is small, then that case really didn't fit well with the rest of the data. Typically, we name such an indicator with the identifier of the case we want to remove. Here's the last roller coaster model in which we have removed the influence of the three blast coasters by constructing indicators for them instead of by removing them from the data. Notice that the coefficients for the other predictors are just the same as the ones we found by omitting the cases.

Response variable is: Speed
R-squared = 91.8% R-squared (adjusted) = 91.5%
s = 3.664 with 105 − 5 = 100 degrees of freedom

Variable	Coefficient	SE(Coeff)	t-ratio	P-value
Intercept	34.4285	0.9567	36.0	<0.0001
Drop	0.200004	0.0061	32.7	<0.0001
Xcelerator	21.5711	3.684	5.86	<0.0001
HyperSonic	18.9711	3.683	5.15	<0.0001
Volcano	19.5713	3.704	5.28	<0.0001

The P-values for the three indicator variables confirm that each of these roller coasters doesn't fit with the others.

24.3 Building Multiple Regression Models

> "It is the mark of an educated mind to be able to entertain a thought without accepting it."
>
> —Aristotle

When many possible predictors are available, we will naturally want to select only a few of them for a regression model. But which ones? The first and most important thing to realize is that often there is no such thing as the "best" regression model. (After all, all models are wrong.) Several alternative models may be useful or insightful. The "best" for one purpose may not be best for another. And the one with the highest R^2 may not be best for many purposes. There is nothing wrong with continuing to work with several models without choosing among them.

Multiple regressions are subtle. The coefficients often don't mean what they at first appear to mean. The choice of which predictors to use determines almost everything about the regression.

Predictors interact with each other, which complicates interpretation and understanding. So it is usually best to build a *parsimonious* model, using as few predictors as you can. On the other hand, we don't want to leave out predictors that are theoretically or practically important. Making this trade-off is the heart of the challenge of selecting a good model. The best regression models, in addition to satisfying the assumptions and conditions of multiple regression, have:

- Relatively few predictors to keep the model simple.
- A relatively high R^2 indicating that much of the variability in y is accounted for by the regression model.
- A relatively small value of s, the standard deviation of the residuals, indicating that the magnitude of the errors is small.
- Relatively small P-values for their F- and t-statistics, showing that the overall model is better than a simple summary with the mean, and that the individual coefficients are reliably different from zero.
- No cases with extraordinarily high leverage that might dominate and alter the model.
- No cases with extraordinarily large residuals, and Studentized residuals that appear to be Nearly Normal. Outliers can alter the model and certainly weaken the power of any test statistics. And the Nearly Normal Condition is required for inference.
- Predictors that are reliably measured and relatively unrelated to each other.

The term "relatively" in this list is meant to suggest that you should favor models with these attributes over others that satisfy them less, but, of course, there are many trade-offs and no absolute rules.

Cases with high leverage or large residuals can be dealt with by introducing indicator variables.

In addition to favoring predictors that can be measured reliably, you may want to favor those that are less expensive to measure, especially if your model is intended for prediction with values not yet measured.

Seeking Multiple Regression Models Automatically

How can we find the best multiple regression model? The list of desirable features we just looked at should make it clear that there is no simple definition of the "best" model. A computer can try all combinations of the predictors to find the regression model with the highest R^2, or optimize some other criterion,[7] but models found that way are not best for all purposes, and may not even be particularly good for many purposes.

Another alternative is to have the computer build a regression "stepwise." In a **stepwise regression**, at each step, a predictor is either added to or removed from the model. The predictor chosen to add is the one whose addition increases the R^2 the most (or similarly improves some other measure). The predictor chosen to remove is the one whose removal reduces the R^2 least (or similarly loses the least on some other measure). The hope is that by following this path, the computer can settle on a good model. The model will gain or lose a predictor only if that change in the model makes a big enough improvement in the performance measure. The changes stop when no more changes pass the criterion.

Stepping in the Wrong Direction

Here's an example of how stepwise regression can go astray. We might want to find a regression to model *Horsepower* in a sample of cars from the car's engine size (*Displacement*) and its *Weight*. The simple correlations are as follows:

	HP	Disp	Wt
Horsepower	1.000		
Displacement	0.872	1.000	
Weight	0.917	0.951	1.000

Because *Weight* has a slightly higher correlation with *Horsepower*, stepwise regression will choose it first. Then, because *Weight* and engine size (*Displacement*) are so highly correlated, once *Weight* is in the model, *Displacement* won't be added to the model. But *Weight* is, at best, a lurking variable leading to both the need for more horsepower and a larger engine. Don't try to tell an engineer that the best way to increase horsepower is to add weight to the car and that the engine size isn't important! From an engineering standpoint, *Displacement* is a far more appropriate predictor of *Horsepower*, but stepwise regression for these data doesn't find that model.

[7]This is literally true. Even for many variables and a moderately large number of cases, it is computationally possible to find the "best subset" of predictors that maximizes R^2. Many statistics programs offer this capability.

Stepwise methods can be valuable when there are hundreds or thousands of potential predictors, as can happen in data mining applications. They can build models that are useful for prediction or as starting points in the search for better models. Because they do each step *automatically*, however, stepwise methods are inevitably affected by influential points and nonlinear relationships. A better strategy might be to mimic the stepwise procedure yourself, but more carefully. You could consider adding or removing a variable yourself with a careful look at the assumptions and conditions each time a variable is considered. That kind of guided stepwise method is still not guaranteed to find a good model, but it may be a sensible way to search among the potential candidates.

Building Regression Models Sequentially

You can build a regression model by adding variables to a growing regression. Each time you add a predictor, you hope to account for a little more of the variation in the response. What's left over is the residuals. At each step, consider the predictors still available to you. Those that are most highly correlated with the current residuals are the ones that are most likely to improve the model.

If you see a variable with a high correlation at this stage and it is *not* among those that you thought were important, stop and think about it. Is it correlated with another predictor or with several other predictors? Don't let a variable that doesn't make sense enter the model just because it has a high correlation, but at the same time, don't exclude a predictor just because you didn't initially think it was important. (That would be a good way to make sure that you never learn anything new.) Finding the balance between these two choices underlies the art of successful model building.

Alternatively, you can start with all available predictors in the model and remove those with small t-ratios. At each step make a plot of the residuals to check for outliers, and check the leverages (say, with a histogram of the leverage values) to be sure there are no high-leverage points. Influential cases can strongly affect which variables appear to be good or poor predictors in the model. It's also a good idea to check that a predictor doesn't appear to be unimportant in the model only because it's correlated with other predictors in the model. It may (as is true of *Displacement* in the example of predicting *Horsepower*) actually be a more useful or meaningful predictor than some of those in the model.

In either method, adding or removing a predictor will usually change *all* of the coefficients, sometimes by quite a bit.

STEP-BY-STEP EXAMPLE

Building Multiple Regression Models

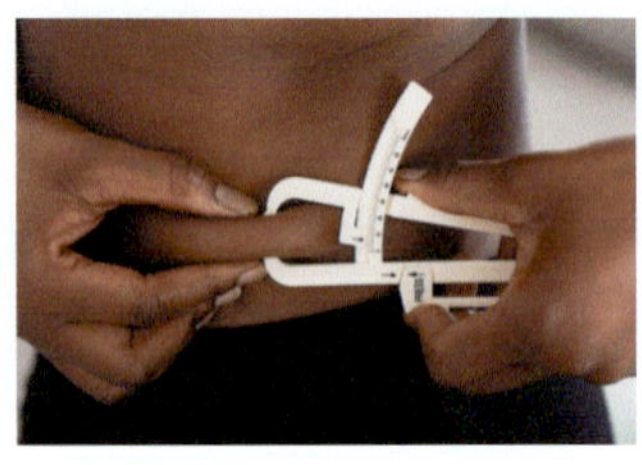

QUESTION: Let's return to the bodyfat data (Data in, um, **Bodyfat**). Percent body fat is an important predictor of patient health. Physicians and researchers would be thrilled to be able to measure % body fat using only easily measured aspects of a patient rather than subject them to immersion in a tank, an x-ray machine, or an expensive MRI. In Chapter 9, we fit a multiple regression model using *Height* and *Weight* as predictors of % body fat. That makes sense, but can we do better? There are 11 other measurements available on each subject including *Age (yr), Neck (cm), Chest (cm), Waist (in.), Hip (cm), Thigh (cm), Knee (cm), Ankle (cm), Bicep (cm), Forearm (cm),* and *Wrist (cm)*. What combination of predictors might lead to a useful model to predict % body fat?

THINK **VARIABLES** Name the variables, report the W's, and specify the questions of interest.

We want to build a regression model to predict % body fat in men of various ages. We have data on Weight (lb), Height (in.), Age (yr), Neck (cm), Chest (cm), Waist (in.), Hip (cm), Thigh (cm), Knee (cm), Ankle (cm), Bicep (cm), Forearm (cm), and Wrist (cm).

✔ **Straight Enough Condition:** A scatterplot matrix reveals that many of the predictor variables are related, a fact that shouldn't surprise us given the nature of the measurements. We also see that % body fat is positively correlated with all the predictors except *Height*, which we saw and explained in Chapter 9. The relationships with *%Bodyfat* all seem fairly straight with no outliers except for *Ankle* size, where two outliers appear far from the center.

PLAN Think about the assumptions and check the conditions.

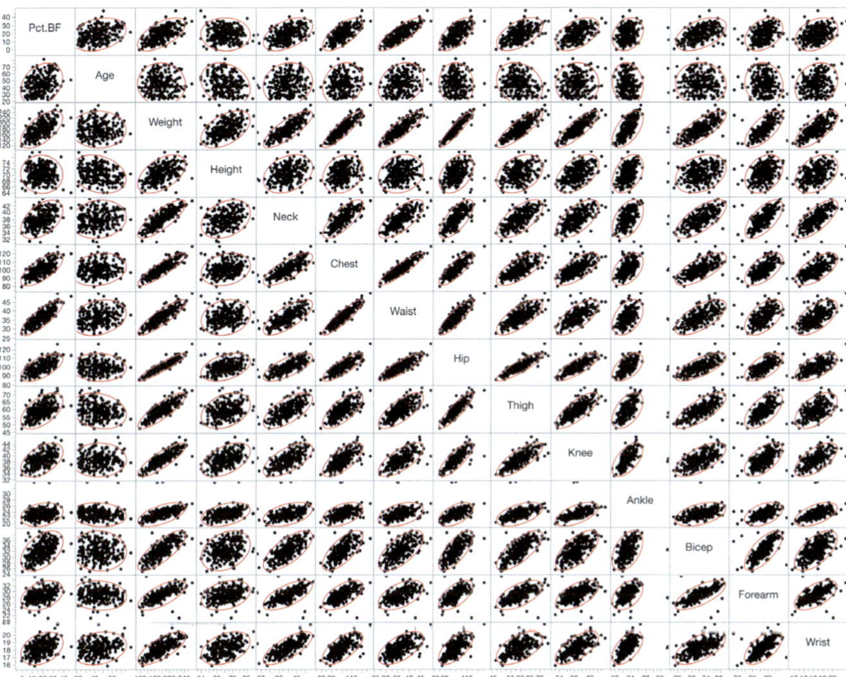

I've computed the leverage values for the 250 cases and found two points with extremely high leverage values:

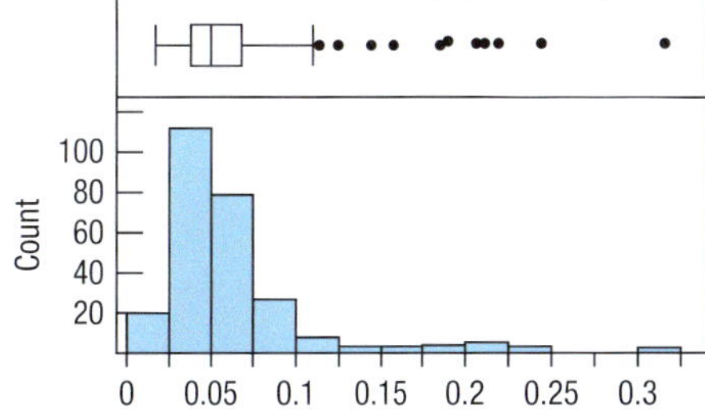

Not surprisingly, the two highest cases at the right in the boxplot are the extreme *Ankle* size cases. A check on the Internet reveals that ankle sizes that large are highly unlikely. I therefore decided to set aside these two cases. Reassessing leverage with the remaining 248 cases shows several high values:

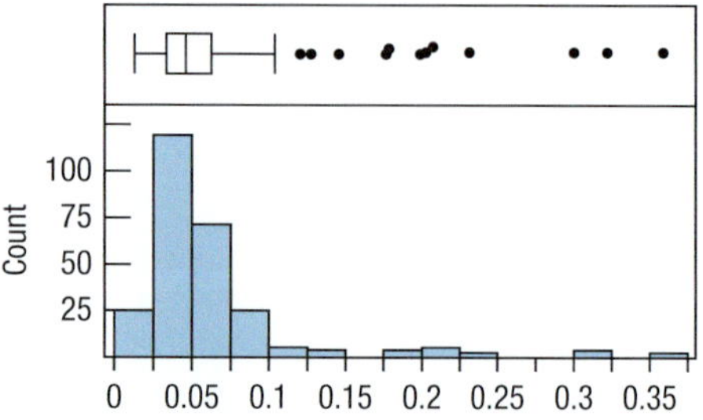

However, none of these points seems likely to exert undue influence on the regression. They are all somewhat large or small values on one of more predictors, but none is particularly extreme:

Now we can find the regression and examine the residuals. We start with a regression containing all the predictor variables.

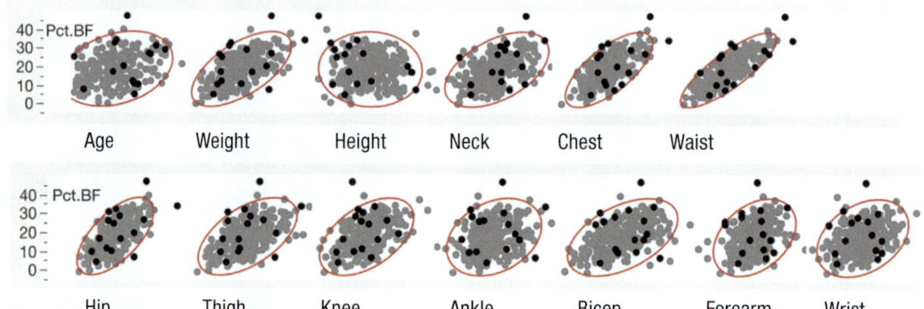

First, I built a model with all predictors.

Parameter Estimates

Term	Estimate	Std Error	t Ratio	P-value
Int	0.5995677	23.84262	0.03	0.9800
Age	0.0625713	0.032862	1.90	0.0581
Weight	−0.019941	0.068275	−0.29	0.7705
Height	−0.223478	0.192945	−1.16	0.2479
Neck	−0.380267	0.23795	−1.60	0.1114
Chest	−0.111463	0.108522	−1.03	0.3054
Waist	2.3048432	0.232738	9.90	<.0001
Hip	−0.163049	0.146762	−1.11	0.2677
Thigh	0.1834224	0.147816	1.24	0.2159
Knee	−0.009386	0.2508	−0.04	0.9702
Ankle	−0.007003	0.361046	−0.02	0.9845
Bicep	0.1592473	0.172653	0.92	0.3573
Forearm	0.2789955	0.207851	1.34	0.1808
Wrist	−1.671603	0.558038	−3.00	0.0030

Many of the coefficients were not significant, but the plot of residuals against predicted values showed no patterns.

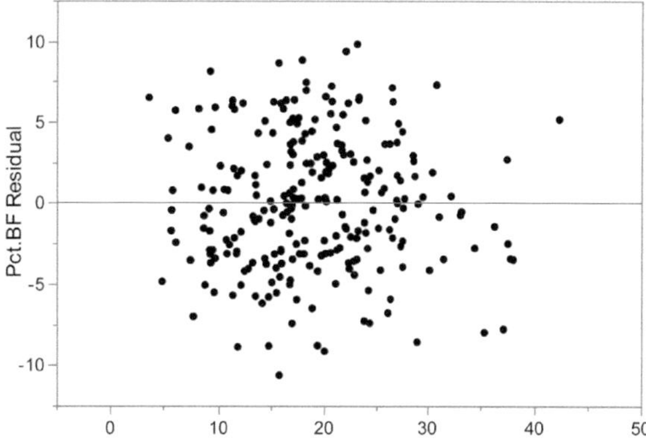

SHOW **MECHANICS** We can use stepwise in either forward or backward mode. If we use forward, we start with nothing in the model, adding variables one at a time until no more predictors are significant. In backward mode we start with all variables in the model and eliminate non-significant predictors. We can use a variety of model selection criteria.

We'll start with forward.

Forward stepwise regression started by adding *Waist*, then *Weight*, *Wrist, and Bicep*. At that point no other predictors were significant at the 0.10 level.

Term	Estimate	Std Error	t Ratio	P-value
Intercept	−31.58422	7.002424	−4.51	<.0001
Weight	−0.116075	0.02747	−4.23	<.0001
Waist	2.4517657	0.139611	17.56	<.0001
Bicep	0.2790974	0.150757	1.85	0.0653
Wrist	−1.471359	0.432999	−3.40	0.0008

R^2 0.741 R^2Adjusted 0.737 Residual Std Error 4.261

The residual plot shows no problems with the model assumptions:

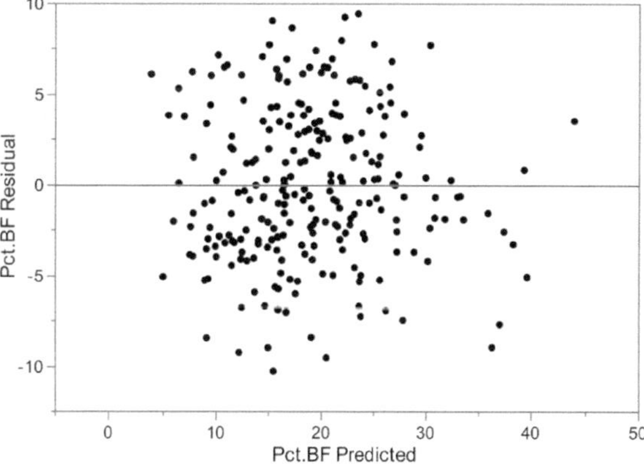

Backward stepwise regression started from a model with all 13 predictors, then removed *Ankle, Knee, Weight, Bicep, Chest, Forearm, Hip, Thigh*, and *Neck*, resulting in the following model:

Parameter Estimates

Term	Estimate	Std Error	t Ratio	P-value
Intercept	2.2683152	8.07915	0.28	0.7791
Age	0.0514829	0.02389	2.15	0.0321
Height	−0.311538	0.12155	−2.56	0.0110
Waist	1.9625448	0.085251	23.02	<.0001
Wrist	−1.919412	0.40848	−4.70	<.0001

R^2 0.740 R^2Adjusted 0.736 Residual Std Error 4.266

It may be interesting to note that while *Weight* is certainly a contributor to the amount of body fat in a human body, the second model predicts body fat using *Age* and *Height* rather than *Weight*. This demonstrates one of the dangers of automatic model selection.

The residuals from this model show no problems with the model assumptions.

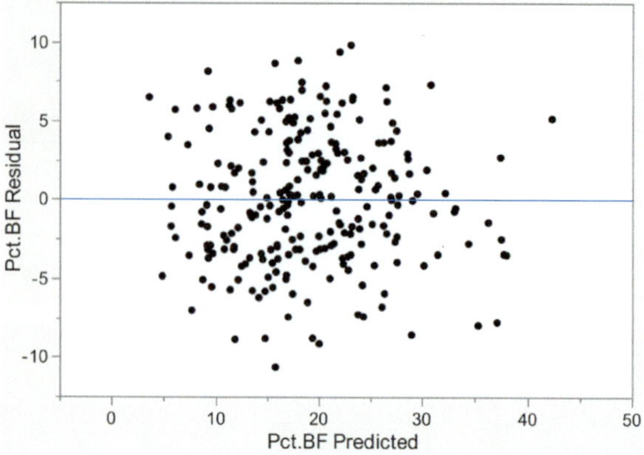

Although they involve different sets of predictors, the two models are very similar. Both use 4 predictors and have an R^2 value about 0.74 and a residual standard error of 4.26 %. Body measurements are highly correlated. I could discuss the models with a physiologist to see which is more useful, but statistically they are essentially indistinguishable.

TELL **INTERPRETATION**

With an R^2 of 74% these regression models account for nearly 3/4 of the variation in body fat. The standard deviation of the residuals is about 4.26%. By our rule of thumb, the model could make errors of as much as twice that, or more than 9.5%. Normal male body fat is around 20%. For a man whose predictors are near the mean of each variable, that would result in a prediction interval wider than 10.5% to 29.5%. The lower value of that interval includes professional athletes. Technically the man would be classified as "very lean." The upper end, 29.5%, is nearly obese, which is defined as a body fat percentage of 30% or more. So, although the models have a respectable R^2 value, they are essentially useless for predicting an individual's body fat. The hope that 13 measurements could be used in models to predict body fat does not seem to be realized.

Model Selection: Training the Model

When we fit a multiple regression model to data, using stepwise or any other model building strategy, we use various criteria like R^2, adjusted R^2, t-statistics, and F-statistics to decide which predictor variables to include. All of these measures (and many others that we didn't discuss) are based on the assumption that the errors are Normally distributed. The reason we don't keep adding variables to a regression model is the possibility that we will fit the data too well. That sounds like a good thing, but with too many variables we run into the danger of fitting the noise in the data as well as the signal that we're trying to see. The result is that the R^2 for the data we used to fit the model will be very high, but when we fit the model to new data, it will drop significantly. This is called **overfitting** the data. We'd like to find a model that works reasonably well for the data we use to fit and also predicts well in new situations.

Rather than wait for new data, we can simulate how well the model works on new data by splitting the data we have randomly into two, using one part for fitting the data and the other part for testing. The first dataset is called the **training set** and the other is called the **test set**. The test set is left alone and is not used for fitting the model. This strategy of using random subsets to select and test the model is called **cross-validation**.

The simplest way to construct the datasets is to choose percentages of the original dataset you want to denote as training, tuning, and test and then randomly select cases according to the percentages. There is no good rule of thumb for knowing what percentages to use, but typically, the majority of the data goes into training. Choices like 75% training, 10% tuning, and 15% test are not uncommon. When datasets are very small, other strategies are needed.

The R^2 on the test set is a more realistic measure of how the model will perform on new datasets, although it's important to remember that future datasets may be more different from the training data than a test dataset that is randomly chosen from the data at hand. Models don't last forever. Environments change, and even if a model predicts well for customers today, there is no guarantee that it will work as well in a week, a month, or a year from now.

An even more thorough way to evaluate models is to split the dataset randomly into training and test repeatedly, cross-validating more than once. Typically 90% of the data are used to fit and 10% used to evaluate in each iteration. When repeated test sets are used, they are often called **hold out samples** or simply **hold outs**. The model fitting is repeated on all the training sets, computing the sum of squared errors (SSE) on each hold out. The total SSE or R^2 on all the hold out sets will be the final measure of how well the model fits.

Choosing a model always requires judgment. It's good to remember George Box's warning that "All models are wrong, but some are useful." Choosing a model based on any one criterion is dangerous. Domain knowledge (one aspect of which is called common sense), performance, and appropriateness (assumptions and conditions) are all important when choosing a model.

Which model is better? That depends on what you want to know. Remember— all models are wrong. But both may offer useful information and insights.

JUST CHECKING

1. Give two ways that we use histograms to support the construction, inference, and understanding of multiple regression models.
2. Give two ways that we use scatterplots to support the construction, inference, and understanding of multiple regression models.
3. What role does the Normal model play in the construction, inference, and understanding of multiple regression models?

Regression Roles

We build regression models for a number of reasons. One reason is to model how variables are related to each other in the hope of understanding the relationships. Another is to build a model that might be used to predict values for a response variable when given values for the predictor variables.

When we hope to understand, we are often particularly interested in simple, straightforward models in which predictors are as unrelated to each other as possible. We are especially happy when the t-statistics are large, indicating that the predictors each contribute to the model. We are likely to want to look at partial regression plots to understand the coefficients and to check that no outliers or influential points are affecting them.

When prediction is our goal, we are more likely to care about the overall R^2. Good prediction occurs when much of the variability in y is accounted for by the model. We might be willing to keep variables in our model that have relatively small t-statistics simply for the stability that having several predictors can provide. We care less whether the predictors are related to each other because we don't intend to interpret the coefficients anyway.

In both roles, we may include some predictors to "get them out of the way." Regression offers a way to approximately control for factors when we have observational data because each coefficient measures effects *after removing the effects* of the other predictors. Of course, it would be better to control for factors in a randomized experiment, but often that's just not possible.

Collinearity

WHO	Books for sale on Amazon.com
WHAT	Various facts, including list price, thickness, and weight
WHEN	2016
WHERE	Data accessed from Amazon.com
WHY	Class project to generate data for building a regression model

As a class project, students collected data on books found at Amazon.com (Data in **Amazon books**). Starting from a book they liked (and excluding their favorite statistics text), they then generated a random number and counted out that many books in the "Customers who bought this item also bought" list to find the next book, repeating the process ten times. For each book they recorded all the facts about the book available at the Amazon site, including its height, width, weight, number of pages, and, of course, its price.

How expensive are books per inch of thickness? Could you estimate the value of your library with a tape measure?

Response variable is: List Price
R squared = 2.3% R squared (adjusted) = 2.0%
s = 13.78 with 323 − 2 = 321 degrees of freedom

Variable	Coefficient	SE(Coeff)	t-ratio	P-value
Intercept	13.1828	2.039	6.47	<0.0001
Thickness	5.75107	2.081	2.76	0.0060

According to this regression, books cost about $5.75 per inch of thickness.

If you are building bookshelves, you'll also want to allow for the weight of your books. Are books more expensive per pound than, say, steak? Let's add *Weight* to the model:

Response variable is: List Price
R squared = 20.5% R squared (adjusted) = 20.0%
s = 11.79 with 315 − 3 = 312 degrees of freedom

Variable	Coefficient	SE(Coeff)	t-ratio	P-value
Intercept	13.1321	1.801	7.29	<0.0001
Thickness	−11.3675	2.558	−4.44	<0.0001
Weight (oz)	1.19767	0.1371	8.73	<0.0001

Both predictors have large t-ratios and small P-values, but the coefficient of *Thickness* has now flipped from strongly positive to even more strongly negative. What happened? The problem is that *Thickness* and *Weight* are themselves closely associated; thicker books weigh more. The coefficient of *Thickness* now reports how the price of a book depends upon its thickness *after allowing for* its *Weight*. Perhaps, for a given weight, books made with thinner paper are generally more expensive.

When we have several predictors, we must think about how the predictors are related to each other. When predictors are unrelated to each other, each provides new information to help account for more of the variation in y. Just as we need a predictor to have a large enough variability to provide a stable base for simple regression, when we have several predictors, we need for them to vary in different directions for the multiple regression to have a stable base. If you wanted to build a deck on the back of your house, you wouldn't build it with supports placed just along one diagonal. Instead, you'd want the supports spread out in different directions as much as possible to make the deck stable. We're in a similar situation with multiple regression. When predictors are highly correlated, they line up together, which makes the regression they support balance precariously. Even small variations can rock it one way or the other. A more stable model can be built when predictors have low correlation and the points are spread out.

MULTI-COLLINEARITY?
You may find this problem referred to as "multicollinearity." But there is no such thing as "unicollinearity"—we need at least two predictors for there to be a linear association between them—so there is no need for the extra two syllables.

When two or more predictors are linearly related, they are said to be **collinear**. The general problem of predictors with close (but perhaps not perfect) linear relationships is called the problem of **collinearity**.

Fortunately, there's an easy way to assess collinearity. To measure how much one predictor is linearly related to the others, just find the regression of that predictor on the others[8] and look at the R^2. That R^2 gives the fraction of the variability of the predictor in question that is accounted for by the other predictors. So $1 - R^2$ is the amount of the predictor's variance that is left after we allow for the effects of the other predictors. That's what the predictor has left to bring to the regression model. And we know that a predictor with little variance can't do a good job of predicting.[9]

Collinearity can hurt our analysis in yet another way. We've seen that the variance of a predictor plays a role in the standard error of its associated coefficient. Small variance leads to a larger SE. In fact, it's exactly this leftover variance that shows up in the formula for the SE of the coefficient.

WHY NOT JUST LOOK AT THE CORRELATIONS?

It's sometimes suggested that we examine the table of correlations of all the predictors to search for collinearity. But this will find only associations among *pairs* of predictors. Collinearity can—and does—occur among *several* predictors working together. You won't find that more subtle collinearity with a correlation table.

As a final blow, when a predictor is collinear with the other predictors, it's often difficult to figure out what its coefficient means in the multiple regression. We've blithely talked about "removing the effects of the other predictors," but now when we do that, there may not be much left. What is left is not likely to be about the original predictor, but more about the fractional part of that predictor not associated with the others. In a regression of horsepower on weight and engine size, once we've removed the effect of weight on horsepower, engine size doesn't tell us anything *more* about horsepower. That's certainly not the same as saying that engine size doesn't tell us anything about horsepower. It's just that most cars with big engines also weigh a lot.

When a predictor is collinear with the other predictors in the model, two things can happen:

1. Its coefficient can be surprising, taking on an unanticipated sign or being unexpectedly large or small.
2. The standard error of its coefficient can be large, leading to a smaller t-statistic and correspondingly large P-value.

One telltale sign of collinearity is the paradoxical situation in which the overall F-test for the multiple regression model is significant, showing that at least one of the coefficients is discernibly different from zero, and yet most or all of the individual coefficients have small t-values, each in effect, denying that *it* is the significant one.

What should you do about a collinear regression model? The simplest cure is to remove some of the predictors. That both simplifies the model and generally improves the t-statistics. And, if several predictors give pretty much the same information, removing some of them won't hurt the model. Which should you remove? Keep the predictors that are most reliably measured, least expensive to find, or even those that are politically important.

Choosing a Sensible Model

The mathematics department at a large university built a regression model to help them predict success in graduate study. They were shocked when the coefficient for Mathematics GRE score was not significant. But the Math GRE was collinear with some of the other predictors, such as math course GPA and Verbal GRE, which made its slope not significant. They decided to omit some of the other predictors and retain Math GRE as a predictor because that model seemed more appropriate—even though it predicted no better (and no worse) than others without Math GRE.

---

[8]The residuals from this regression are plotted as the x-axis of the partial regression plot for this variable. So if they have a very small variance, you can see it by looking at the x-axis labels of the partial regression plot, and get a sense of how precarious a line fit to the partial regression plot—and its corresponding multiple regression coefficient—may be.

[9]The statistic $1 - R^2$ for the R^2 found from the regression of one predictor on the other predictors in the model is also called the *Variance Inflation Factor,* or *VIF,* in some computer programs and books.

WHAT CAN GO WRONG?

In the Oscar-winning movie *The Bridge on the River Kwai* and in the book on which it is based,[10] the character Colonel Green famously says, "As I've told you before, in a job like yours, even when it's finished, there's always one more thing to do." It is wise to keep Colonel Green's advice in mind when building, analyzing, and understanding multiple regression models.

- **Beware of collinearities.** When the predictors are linearly related to each other, they add little to the regression model after allowing for the contributions of the other predictors. Check the R^2s when each predictor is regressed on the others. If these are high, consider omitting some of the predictors.
- **Don't check for collinearity only by looking at pairwise correlations.** Collinearity is a relationship among any number of the predictors. Pairwise correlations can't always show that. (Of course, a high pairwise correlation between two predictors does indicate collinearity of a special kind.)
- **Don't be fooled when high-influence points and collinearity show up together.** A single high-influence point can be the difference between your predictors being collinear and seeming not to be collinear. (Picture that deck supported only along its diagonal and with a single additional post in another corner. Supported in this way, the deck is stable, but the height of that single post completely determines the tilt of the deck, so it's very influential.) Removing a high-influence point may surprise you with unexpected collinearity. Alternatively, a single value that is extreme on several predictors can make them appear to be collinear when in fact they would not be if you removed that point. Removing that point may make apparent collinearities disappear (and would probably result in a more useful regression model).
- **Beware missing data.** Values may be missing or unavailable for any case in any variable. In simple regression, when the cases are missing for reasons that are unrelated to the variable we're trying to predict, that's not a problem. We just analyze the cases for which we have data. But when several variables participate in a multiple regression, any case with data missing on any of the variables will be omitted from the analysis. You can unexpectedly find yourself with a much smaller set of data than you started with. Be especially careful, when comparing regression models with different predictors, that the cases participating in the models are the same.
- **Remember linearity.** The **Linearity Assumption** (and the **Straight Enough Condition**) require linear relationships among the variables in a regression model. As you build and compare regression models, be sure to plot the data to check that it is straight. Violations of this assumption make everything else about a regression model invalid.
- **Watch out for high-influence points and outliers.** We always have to be on the lookout for cases that have an undue influence on our model. Partial regression plots are a good place to look for influential points and to understand how they affect each of the coefficients.

CONNECTIONS

We've generalized the concept of leverage that we first saw in Chapter 8. Everything we said about how to think about these ideas back in Chapters 8 and 10 still applies to the multiple regression model.

Don't forget that the Straight Enough Condition is essential to all of regression. At any stage in developing a model, if the scatterplot that you check is not straight, consider re-expressing the variables to make the relationship straighter. The topics of Chapter 9 will help you with that.

[10]The author of the book, Pierre Boulle, also wrote the book and script for *Planet of the Apes*. The director, David Lean, also directed *Lawrence of Arabia*.

CHAPTER REVIEW

In earlier chapters, we learned that multiple regression is a natural way to extend linear regression models to include several predictors. Now we've learned that multiple regression is both more powerful and more complex than it may appear at first. As with other chapters in this text whose titles spoke of greater "wisdom," this chapter has drawn us deeper into the uses and cautions of multiple regression. In addition, we've seen that statistical analyses often have to start with efforts at cleaning and formatting data.

- Know the basic principles and actions of data cleaning and formatting. Ideally, know how to perform them using your statistics software of choice.
- Beware unusual cases. A single case can have high leverage, allowing it to influence the entire regression. Such cases should be treated specially, possibly by fitting the model both with and without them or by including indicator variables to isolate their influence.
- With complex models, be careful in interpreting the coefficients. Associations among the predictors can change the coefficients to values that can be quite different from the coefficient in the simple regression of a predictor and the response, even changing the sign.
- Understand that building multiple regression models is an art that speaks to the central goal of statistical analysis: understanding the world with data. We've learned that there is no right model. We've seen that the same response variable can be modeled with several alternative models, each showing us different aspects of the data and of the relationships among the variables and nominating different cases as special and deserving of our attention.
- We've also seen that everything we've discussed throughout this text fits together to help us understand the world. The graphical methods are the same ones we learned in the early chapters, and the inference methods are those we originally developed for means. In short, there's been a consistent tale of how we understand data to which we've added more and more detail and richness, but which has been consistent throughout.

REVIEW OF TERMS

Data cleaning A general term that refers to the actions needed to make data ready for analysis (p. 789).

Aggregation Combining information for cases by grouping them together according to values on some identifying variable (p. 789).

Filtering Selecting a subset of cases for analyses according to some criterion based on the data (p. 789).

Merging (Joining) Combining data from two files that have been recorded on the same cases, but on different variables and/or in a different order. Merging requires an *index* variable that identifies cases uniquely so that they may be matched (p. 789).

Appending Joining additional cases measured on the same variables to an existing set of data (p. 789).

Deduping Removing duplicate records or cases so that each case appears only once in the final data (p. 789).

Feature extraction Constructing new variables by combining existing ones with mathematical functions (p. 789).

Feature selection Choosing some variables—typically predictors for a statistical model—from a larger (possibly very large) collection of potential predictors (p. 789).

Scatterplot matrix (Splom) A display in which scatterplots of every pair of variables are arrayed so that each variable appears on the y-axis of plots on one row and on the x-axis of plots in one column. Scatterplot matrices provide a quick overview of the structure and relationships of the variables (p. 789).

Leverage The leverage of a case measures how far its x-values are from the center of the x's and, consequently, how much influence it can exert on the regression model. Points with high leverage can determine a regression model and should, therefore, be examined carefully (p. 790).

Parallel coordinate plot A display in which several variables are plotted as dotplots on parallel, vertical axes and individual cases are connected horizontally with lines. Variables are usually standardized (p. 792).

Studentized residual When a residual is divided by an independent estimate of its standard deviation, the result is a Studentized residual. The type of Studentized residual that has a *t*-distribution is an *externally* Studentized residual (p. 794).

Influential case A case is *influential* on a multiple regression model if, when it is omitted, the model changes by enough to matter for your purposes. (There is no specific amount of change defined to declare a case influential.) Cases with high leverage and large Studentized residual are likely to be influential (p. 795)

Stepwise regression An automated method of building regression models in which predictors are added to or removed from the model one at a time in an attempt to optimize a measure of the success of the regression. Stepwise methods rarely find the best model and are easily influenced by influential cases, but they can be valuable in winnowing down a large collection of candidate predictors (p. 801).

Cross validation A method of evaluating the performance of a model. In one round of cross-validation the model is fit on the training set and evaluated on the hold out (or test) set (p. 806).

Test (or hold out) set The dataset used in a supervised classification or regression problem not used to build the predictive model. The test set is withheld (held out) from the model building stage of the process, and the model's predictions on the test set are used to evaluate the model performance (pp. 806–807).

Training set The dataset used in a supervised classification or regression problem to build the predictive model (p. 806).

Overfitting The problem of fitting a model typically using all the data (instead of using a hold out set), in which the model fits the training data very well, but predicts poorly on data it hasn't used (p. 806).

Collinearity When one (or more) of the predictors can be fit closely by a multiple regression on the other predictors, we have collinearity. When collinear predictors are in a regression model, they may have unexpected coefficients and often have inflated standard errors (and correspondingly small *t*-statistics) (p. 808).

TECH SUPPORT

Regression Diagnosis

Statistics packages differ in how much information they provide to diagnose a multiple regression. Most packages provide leverage values. Many provide far more, including statistics that we have not discussed. But for all, the principle is the same. We hope to discover any cases that don't behave like the others in the context of the regression model and then to understand why they are special.

Many of the ideas in this chapter rely on the concept of examining a regression model and then finding a new one based on your growing understanding of the model and the data. Regression diagnosis is meant to provide steps along that road. A thorough regression analysis may involve finding and diagnosing several models.

DATA DESK

Request diagnostic statistics and graphs from the HyperView menus in the regression output table. Most will update and can be set to update automatically when the model or data change.

COMMENTS

You can add a predictor to the regression by dragging its icon into the table, or replace variables by dragging the icon over their name in the table. Click on a predictor's name to drop down a menu that lets you remove it from the model.

EXCEL

Excel does not offer diagnostic statistics with its regression function.

COMMENTS

Although the dialog offers a Normal probability plot of the residuals, the data analysis add-in does not make a correct probability plot, so don't use this option. The "standardized residuals" are just the residuals divided by their standard deviation (with the wrong df), so they too should be ignored.

JMP

- From the **Analyze** menu select **Fit Model**.
- Specify the response, Y. Assign the predictors, X, in the **Construct Model Effects** dialog box.
- Click on **Run Model**.
- Click on the red triangle in the title of the Model output to find a variety of plots and diagnostics available.

COMMENTS

JMP chooses a regression analysis when the response variable is "Continuous."

MINITAB

- Choose **Regression** from the **Stat** menu.
- Choose **Regression . . .** from the **Regression** submenu.
- In the Regression dialog, assign the Y variable to the Response box and assign the X-variables to the Predictors box.
- Click the **Storage** button.
- In the Regression Storage dialog, you can select a variety of diagnostic statistics. They will be stored in columns of your worksheet.
- Click the **OK** button to return to the Regression dialog.
- To specify displays, click **Graphs,** and check the displays you want.
- Click the **OK** button to return to the Regression dialog.
- Click the **OK** button to compute the regression.

COMMENTS

You will probably want to make displays of the stored diagnostic statistics. Use the usual Minitab methods for creating displays.

R

Suppose the response variable *y* and predictor variables *x1, . . . ,xk* are in a data frame called mydata. After fitting a multiple regression of *y* on *x1* and *x2* via:

- mylm = lm(y~x1 + x2, data = mydata)
- summary(mylm) # gives the details of the fit, including the ANOVA table
- plot(mylm) #gives a variety of plots
- lm.influence(mylm) #gives a variety of regression diagnostic values

To get partial regression plots (called Added Variable plots in **R**), you need the library car:

- library(car)

Then to get the partial regression plots:

- avPlots(mylm) #one plot for each predictor variable–interactions not permitted

STATCRUNCH

StatCrunch offers some of the diagnostic statistics discussed in this chapter in the regression dialog.

SPSS

- Choose **Regression** from the **Analyze** menu.
- Choose **Linear** from the **Regression** submenu.
- When the Linear Regression dialog appears, select the Y-variable and move it to the dependent target. Then move the X-variables to the independent target.
- Click the **Save** button.
- In the Linear Regression Save dialog, choose diagnostic statistics. These will be saved in your worksheet along with your data.
- Click the **Continue** button to return to the Linear Regression dialog.
- Click the **OK** button to compute the regression.

TI-83/84 PLUS

COMMENTS

You need a special program to compute a multiple regression on the TI-83.

EXERCISES

SECTION 24.1

For exercises 24.1 and 24.2, indicate whether the operation described is aggregation, filtering, merging, appending, deduping, or feature extraction.

1. Combining and separating

a) In the Burger King items data, use one of the variables to separate the items containing meat from the items that do not contain meat, and analyze those separately.

b) Combine data about McDonald's menu items using the same variables with the data from Burger King.

2. New cases and variables

a) Look up additional nutrition information about the BK items and combine a file holding that information with the existing BK data.

b) Define the new variable Fat/Carb as the ratio of Fat grams to Carbohydrate grams in each BK item

SECTION 24.2

3. Residual, leverage, influence For each of the following cases, would your primary concern about them be that they had a large residual, large leverage, or likely large influence on the regression model? Explain your thinking.

a) In a regression to predict the construction cost of roller coasters from the length of track, the height of the highest point, and the type of construction (metal or wood), the *Kingda Ka* coaster, which opened in 2005 and, at 456 ft, is currently the tallest.

b) In a regression to predict income of graduates of a college five years after graduation, a graduate who created a high-tech start-up company based on his senior thesis, and sold it for several million dollars.

4. Residual, leverage, influence, 2 For each of the following cases, would your primary concern about them be that they had a large residual, large leverage, or likely large influence on the regression model?

a) In a regression to predict freshman grade point averages as part of the admissions process, a student whose Math SAT was 750, whose Verbal SAT was 585, and who had a 4.0 GPA at the end of her freshman year.

b) In a regression to predict life expectancy in countries of the world from generally available demographic, economic, and health statistics, a country that, due to a high prevalence of HIV, has an unusually low life expectancy.

SECTION 24.3

5. Significant coefficient? In a regression to predict compensation of employees in a large firm, the predictors in the regression were *Years with the Firm*, *Age*, and *Years of Experience*.

The coefficient of *Age* is negative and statistically significantly different from zero. Does this mean that the company pays workers less as they get older? Explain.

6. Better model? Jamal wants to impress his boss. He builds a regression model to predict sales that has 20 predictors and an R^2 of 83%. Stefan builds a competing model with only 5 predictors, but an R^2 of only 77%. Which model is likely to be most useful for understanding the drivers of sales? How could the boss tell? Explain.

CHAPTER EXERCISES

T 7. Climate change 2016 Recent concern with the rise in global temperatures has focused attention on the level of carbon dioxide (CO_2) in the atmosphere. The National Oceanic and Atmospheric Administration (NOAA) records the CO_2 levels in the atmosphere atop the Mauna Loa volcano in Hawaii, far from any industrial contamination, and calculates the annual overall temperature of the atmosphere and the oceans using an established method. (See www.ncdc.noaa.gov/cag/data-info/global.) It is widely acknowledged that global average temperatures have been rising. A politician, arguing that human behavior is not responsible for the rise in temperatures, presents the following regression:

Response variable is: Avg Global temp
cases selected according to years 1975—2016
R squared = 86.0% R squared (adjusted) = 85.3%
s = 0.0876 with 42 − 3 = 39 degrees of freedom

Variable	Coefficient	SE(Coeff)	t-ratio	P-value
Intercept	6.65061	23.94	0.278	0.7826
CO_2(ppm)	0.008656	0.0077	1.12	0.2697
year	0.002271	0.0134	0.169	0.8663

He argues that because the coefficient of CO_2 has a large, nonsignificant P-value, we should conclude that there is no effect of rising CO_2 levels.

You look at the data and find the following correlations:

	G temp	year	CO_2
Global temp	1.000		
year	0.934	1.000	
CO_2	0.947	0.993	1.000

a) What do you think accounts for the small t-ratios in the regression when there clearly is an association between *Average Global temperature* and each of the predictors?
b) What data formatting operation was performed to focus on these years?

T 8. Climate change 2016 again A different regression from the one in Exercise 7, now using all of the available data (years 1959 to 2016), looks like this:

Response variable is: Avg Global temp
R squared = 90.0% R squared (adjusted) = 89.6%
s = 0.0882 with 58 − 3 = 55 degrees of freedom

Variable	Coefficient	SE(Coeff)	t-ratio	P-value
Intercept	22.2985	10.03	2.22	0.0303
CO_2(ppm)	0.014131	0.0037	3.84	0.0003
year	−0.00657	0.0057	−1.15	0.2532

a) Does this regression model prove that increases in CO_2 are responsible for the rise in *Average Global Temperature* (as in Exercise 7)
b) What would make it appropriate to limit the regression model to only some of the years in this series—or do you think it would always be inappropriate? (*Note*: There isn't really a correct answer to this, but it is worth thinking about and recording your opinion.)

9. Healthy breakfast, sick data A regression model for data on breakfast cereals originally looked like this:

Dependent variable is: Calories
R-squared = 84.5%,
R-squared (adjusted) = 83.4%,
s = 7.946 with 77 − 6 = 71 degrees of freedom

Source	Sum of Squares	DF	Mean Square	F-ratio
Regression	24366.8	5	4873.36	77.2
Residual	4483.15	71	63.1430	

Variable	Coefficient	SE(Coeff)	t-ratio	P-value
Intercept	20.2457	5.988	3.38	0.0006
Protein	5.69578	1.075	5.30	≤0.0001
Fat	8.35912	1.034	8.08	≤0.0001
Fiber	−1.02047	0.4853	−2.10	0.0198
Carbs	2.93505	0.2662	11.03	≤0.0001
Sugars	3.31871	0.2597	12.78	≤0.0001

Let's take a closer look at the coefficient for *Fiber*. Here's the partial regression plot for *Fiber* in that regression model:

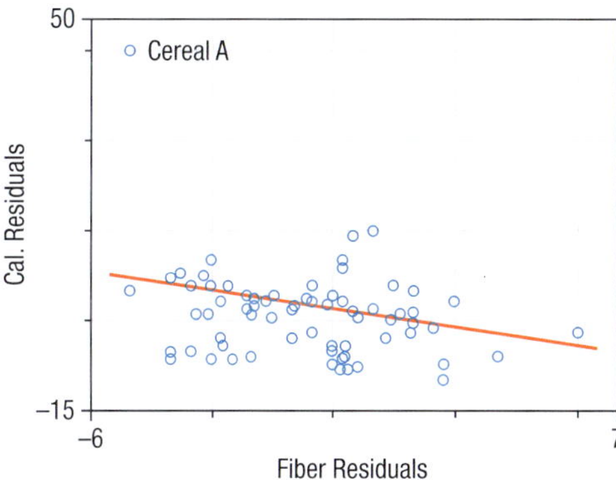

a) The line on the plot is the least squares line fit to this plot. What is its slope? (You may need to look back at the facts about partial regression plots in Chapter 23.)
b) One point is labeled as corresponding to Quaker Oatmeal. What effect does this point have on the slope of the line? (Does it make it larger, smaller, or have no effect at all?)

Here is the same regression with Quaker Oatmeal removed from the data:

Dependent variable is: Calories
77 total cases of which 1 is missing
R-squared = 93.9%, R-squared (adjusted) = 93.5%,
$s = 5.002$ with $76 - 6 = 70$ degrees of freedom

Source	Sum of Squares	DF	Mean Square	F-ratio	P-value
Regression	27052.4	5	5410.49	216	<0.001
Residual	1751.51	70	25.0216		

Variable	Coefficient	SE(Coeff)	t-ratio	P-value
Intercept	−1.25891	4.292	−0.293	0.7701
Protein	3.88601	0.6963	5.58	<0.0001
Fat	8.69834	0.6512	13.4	<0.0001
Fiber	0.250140	0.3277	0.763	0.4478
Carbs	4.14458	0.2005	20.7	<0.0001
Sugars	3.96806	0.1692	23.4	<0.0001

c) Compare this regression with the previous one. In particular, which model is likely to make the best predictions of calories? Which seems to fit the data better?

d) How would you interpret the coefficient of *Fiber* in this model? Does *Fiber* contribute significantly to modeling calories?

(In fact, the data for Quaker Oatmeal was determined to be in error and was corrected for the subsequent analyses seen elsewhere in this book.)

10. Fifty states In previous chapters we have looked at data from the 50 states. Here's an analysis of data from a few years earlier. The *Murder* rate is per 100,000, *HS Graduation* rate is in %, *Income* is per capita income in dollars, *Illiteracy* rate is per 1000, and *Life Expectancy* is in years. We are trying to find a regression model for *Life Expectancy*.
Here's the result of a regression on all the available predictors:

Dependent variable is: Lifeexp
R-squared = 67.0% R-squared (adjusted) = 64.0%
$s = 0.8049$ with $50 - 5 = 45$ degrees of freedom

Source	Sum of Squares	DF	Mean Square	F-ratio
Regression	59.1430	4	14.7858	22.8
Residual	24.1560	45	0.6479	

Variable	Coefficient	SE(Coeff)	t-ratio	P-value
Intercept	69.4833	1.325	52.4	<0.0001
Murder	−0.261940	0.0445	−5.89	<0.0001
HS grad	0.046144	0.0218	2.11	0.0403
Income	1.24948e–4	0.0002	0.516	0.6084
Illiteracy	0.276077	0.3105	0.889	0.3787

Here's a histogram of the leverages and a scatterplot of the externally Studentized residuals against the leverages:

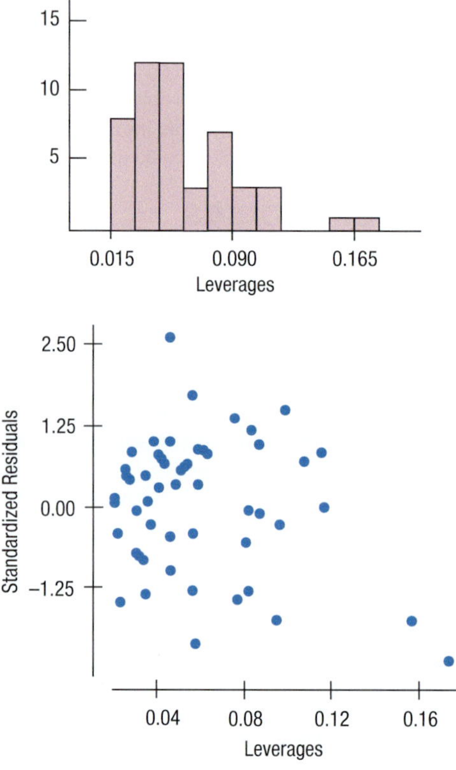

a) The two states with high leverages and large (negative) Studentized residuals are Nevada and Alaska. Do you think they are likely to be influential in the regression? From just the information you have here, why or why not?

Here's the regression with indicator variables for Alaska and Nevada added to the model to remove those states from affecting the model:

Dependent variable is: Lifeexp
R-squared = 74.1% R-squared (adjusted) = 70.4%
$s = 0.7299$ with $50 - 7 = 43$ degrees of freedom

Variable	Coefficient	SE(Coeff)	t-ratio	P-value
Intercept	66.9280	1.442	46.4	<0.0001
Murder	−0.207019	0.0446	−4.64	<0.0001
HS grad	0.065474	0.0206	3.18	0.0027
Income	3.91600e-4	0.0002	1.63	0.1105
Illiteracy	0.302803	0.2984	1.01	0.3159
Alaska	−2.57295	0.9039	−2.85	0.0067
Nevada	−1.95392	0.8355	−2.34	0.0241

b) What evidence do you have that Nevada and Alaska are outliers with respect to this model? Do you think they should continue to be treated specially? Why?

c) Would you consider removing any of the predictors from this model? Why or why not?

11. Cereals In Exercise 25 of Chapter 9, we considered a multiple regression model for predicting calories in breakfast cereals. The regression looked like this:

Dependent variable is: Calories
R-squared = 38.4% R-squared (adjusted) = 35.9%
s = 15.60 with 77 − 4 = 73 degrees of freedom

Source	Sum of Squares	DF	Mean Square	F-ratio	P-value
Regression	11091.8	3	3697.28	15.2	<0.0001
Residual	17760.1	73	243.289		

Variable	Coefficient	SE(Coeff)	t-ratio	P-value
Intercept	83.0469	5.198	16.0	<0.0001
Sodium	0.057211	0.0215	2.67	0.0094
Potassium	−0.019328	0.0251	−0.769	0.4441
Sugars	2.38757	0.4066	5.87	<0.0001

Here's a histogram of the leverages and a partial regression plot for *Potassium* in which the three high-leverage points are plotted with red x's. (They are *All-Bran, 100% Bran,* and *All-Bran with Extra Fiber.*)

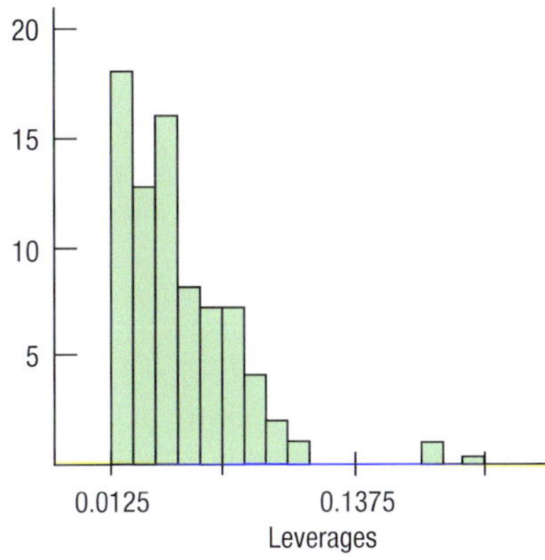

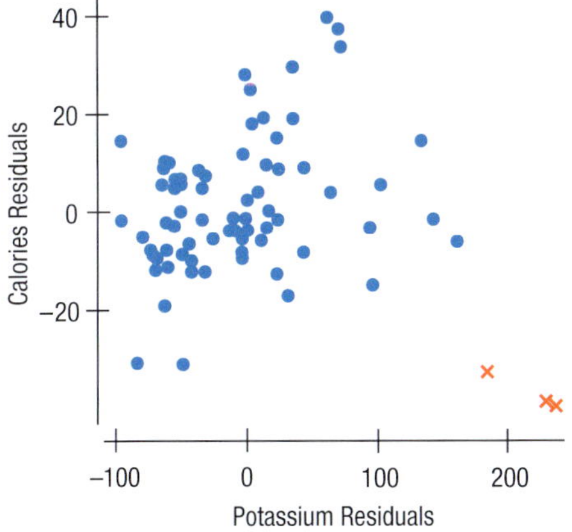

With this additional information, answer the following:

a) How would you interpret the coefficient of *Potassium* in the multiple regression?
b) *Without doing any calculating,* how would you expect the coefficient and t-statistic for *Potassium* to change if we were to omit the three high-leverage points?

Here's a histogram of the externally Studentized residuals. The selected bar, holding the two most negative residuals, holds the two bran cereals that had the largest leverages.

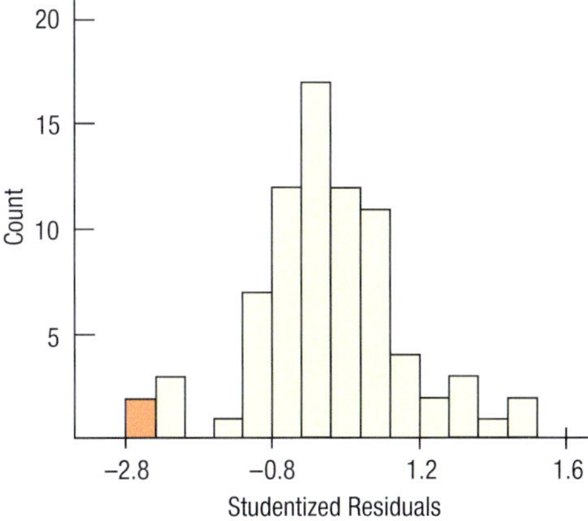

With this additional information, answer the following:

c) What term would you apply to these two cases? Why?
d) Do you think they should be omitted from this analysis? Why or why not? (*Note*: There is no correct choice. What matters is your reasons.)

12. Scottish hill races 2008 In Chapter 9, Exercises 14, 18, 29, and 30, we considered data on hill races in Scotland. These are overland races that climb and descend hills—sometimes several hills in the course of one race. Here is a regression analysis to predict the *Women's Record* times from the *Distance* and total vertical *Climb* of the races:

Dependent variable is: Women's record
R-squared = 96.7% R-squared (adjusted) = 96.7%
s = 10.06 with 90 − 3 = 87 degrees of freedom

Source	Sum of Squares	DF	Mean Square	F-ratio
Regression	261029	2	130515	1288
Residual	8813.02	87	101.299	

Variable	Coefficient	SE(Coeff)	t-ratio	P-value
Intercept	−11.6545	1.891	−6.16	<0.0001
Distance	4.43427	0.2200	20.2	<0.0001
Climb	0.045195	0.0033	13.7	<0.0001

On the next page, you'll see the scatterplot of externally Studentized residuals against predicted values, as well as a histogram of leverages for this regression:

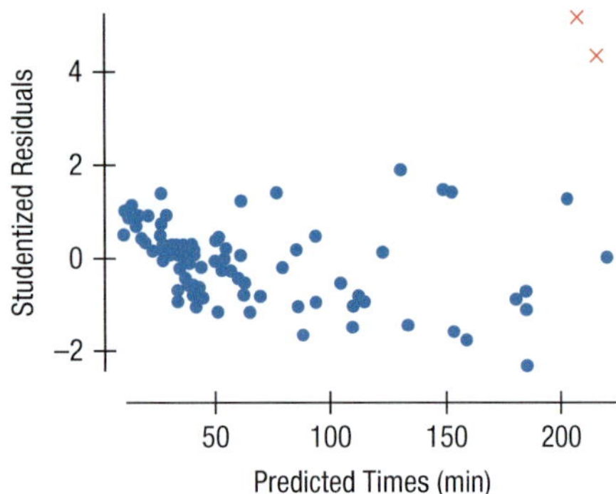

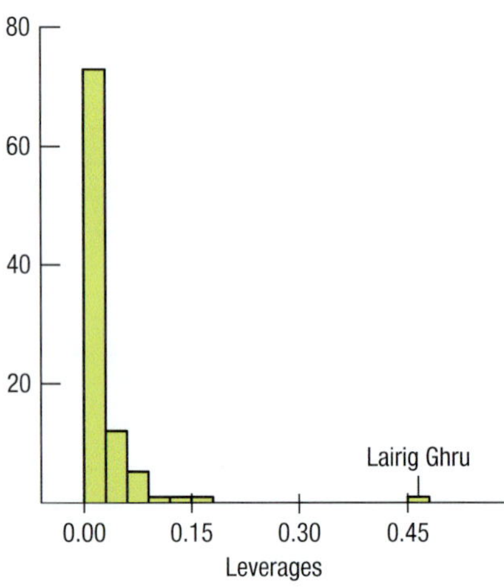

a) Comment on what these diagnostic displays indicate.

b) The two races with the largest Studentized residuals are the Arochar Alps race and the Glenshee 9. Both are relatively new races, having been run only one or two times with relatively few participants. What effects can you be reasonably sure they have had on the regression? What displays would you want to see to investigate other effects? Explain.

c) If you have access to a suitable statistics package, make the diagnostic plots you would like to see and discuss what you find.

13. World data The United States Central Intelligence Agency maintains a public site called the World Factbook at www.cia.gov/library/publications/the-worldfactbook/. There you'll find a wealth of variables about all the countries of the world. Let's examine some of them as possible participants in a regression model.

(*Note*: The following odd-numbered exercises will lead you through an analysis of these data. At each step you are asked to comment. There are no strict right or wrong answers to these questions, but there certainly are things you should notice and be able to comment on. So, taken together, these are an exercise in thinking about and building a multiple regression model.)

Let's consider the following selected variables:

Life expectancy (years)

Population

Birthrate (births/1,000 persons/year at midyear)

Obesity% (% of a country's population with a BMI $\geq$ 30.0)

Education (expenditure as %GDP)

GDP/capita

Imports/capita (US$, exchange rate basis)

Exports/capita

Airports/capita (total number of airports or airfields recognizable from the air)

Cell phones/100 inhabitants

Here are histograms of the variables:

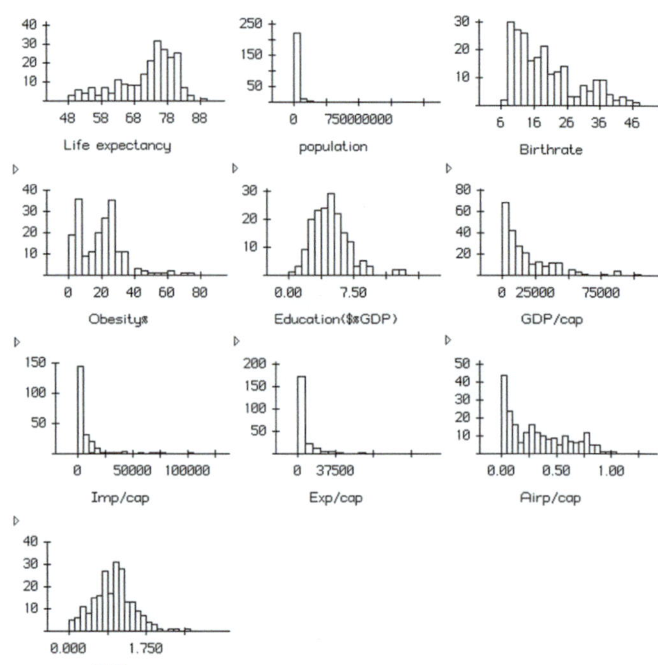

a) Comment on these variables. Which are suitable for use in a regression model? Which would benefit from re-expression (and, if so, why)? Note any other issues.

Here are histograms in which some of the variables have been re-expressed:

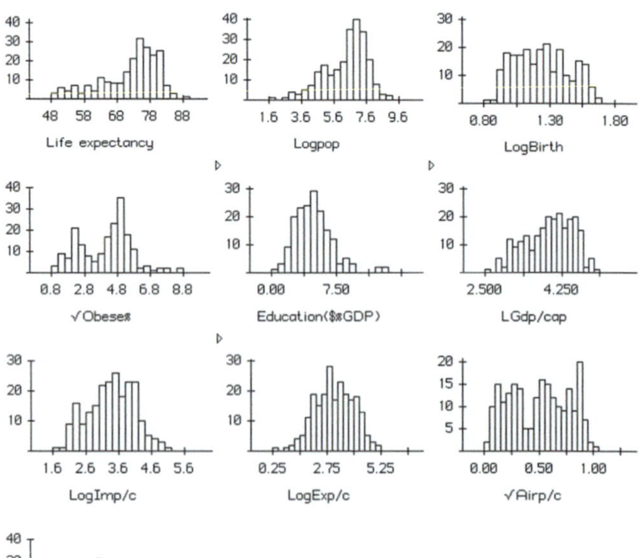

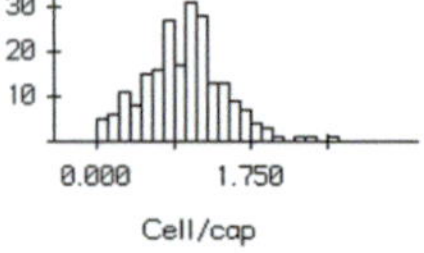

b) Comment: Are these variables now suitable for use in a regression model?

14. Pizza Consumers' Union rated frozen pizzas. Their report includes the number of *Calories, Fat* content, and *Type* (cheese or pepperoni, represented here as an indicator variable that is 1 for cheese and 0 for pepperoni). Here's a regression model to predict the *"Score"* awarded each pizza from these variables:

Dependent variable is: Score
R-squared = 28.7% R-squared (adjusted) = 20.2%
s = 19.79 with 29 − 4 = 25 degrees of freedom

Source	Sum of Squares	DF	Mean Square	F-ratio
Regression	3947.34	3	1315.78	3.36
Residual	9791.35	25	391.654	

Variable	Coefficient	SE(Coeff)	t-ratio	P-value
Intercept	−148.817	77.99	−1.91	0.0679
Calories	0.743023	0.3066	2.42	0.0229
Fat	−3.89135	2.138	−1.82	0.0807
Type	15.6344	8.103	1.93	0.0651

a) What is the interpretation of the coefficient of *Type* in this regression?
b) What displays would you like to see to check assumptions and conditions for this model?

15. World data correlated Here is a scatterplot matrix of the variables as re-expressed in Exercise 13 using a version that places Normal probability plots on the diagonal.

a) Comment on their suitability for a regression model to predict *Life expectancy*. The points are colored according to values of *Life expectancy*.

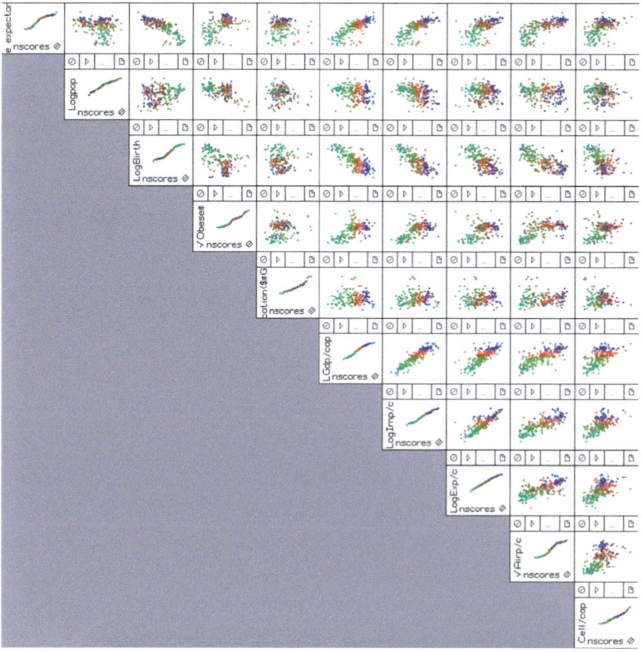

b) Here is a correlation table for the variables graphed in in the scatterplot matrix. Suppose we want to build a regression model to predict *Life expectancy* from some of the other variables. Which predictors might you try? Do you see any potential problems we should be alert for? (Use both the correlations and the scatterplot matrix in your discussion.)

	Life expect	Logpop	LogBirth	√Obese%	Education	LGDP/cap	LogImp/cap	LogExp/cap	√Airp/cap	Cell/cap
Life expect	1.000									
Logpop	−0.240	1.000								
LogBirth	−0.821	0.211	1.000							
√Obese%	0.573	−0.503	−0.510	1.000						
Education	−0.017	−0.116	−0.046	0.225	1.000					
LogGDP/cap	0.794	−0.236	−0.794	0.585	0.068	1.000				
LogImp/cap	0.772	−0.391	−0.793	0.640	0.085	0.919	1.000			
LogExp/cap	0.596	−0.023	−0.649	0.453	0.021	0.801	0.837	1.000		
√Airp/cap	0.758	−0.231	−0.772	0.558	0.069	0.781	0.801	0.687	1.000	
Cell/cap	0.512	−0.088	−0.536	0.431	−0.108	0.639	0.643	0.621	0.579	1.000

16. Gourmet pizza Here's a plot of the Studentized residuals against the predicted values for the regression model found in Exercise 14:

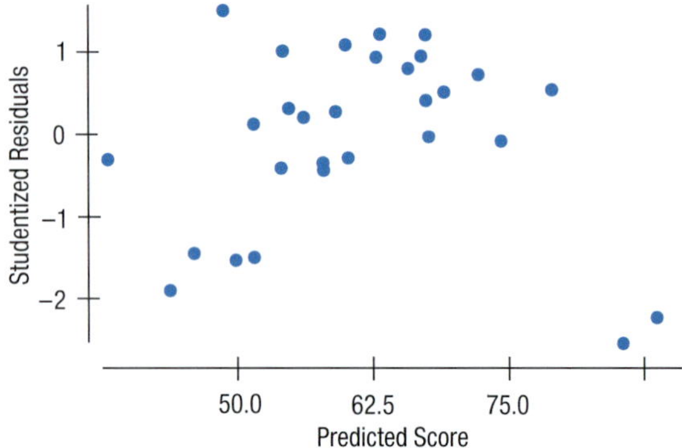

The two extraordinary cases in the plot of residuals are Reggio's and Michelina's, two gourmet pizzas.

a) Interpret these residuals. What do they say about these two brands of frozen pizza? Be specific—that is, talk about the *Scores* they received and might have been expected to receive.

We can create indicator variables to isolate these cases. Adding them to the model results in the following model:

Dependent variable is: Score
R-squared = 65.2% R-squared (adjusted) = 57.7%
s = 14.41 with 29 − 6 = 23 degrees of freedom

Source	Sum of Squares	DF	Mean Square	F-ratio
Regression	8964.13	5	1792.83	8.64
Residual	4774.56	23	207.590	

Variable	Coefficient	SE(Coeff)	t-ratio	P-value
Intercept	−363.109	72.15	−5.03	<0.0001
Calories	1.56772	0.2824	5.55	<0.0001
Fat	−8.82748	1.887	−4.68	0.0001
Type	25.1540	6.214	4.05	0.0005
Reggio's	−67.6401	17.86	−3.79	0.0010
Michelina's	−67.0036	16.62	−4.03	0.0005

b) What does the coefficient of *Michelina's* mean in this regression model? Do you think that Michelina's pizza is an outlier for this model for these data? Explain.

17. World data regression At the top of the next column is a regression analysis to predict *Life expectancy* using the data of Exercise 15 and a plot of the residuals

Response variable is: Life expectancy
240 total cases of which 20 are missing
R squared = 73.2% R squared (adjusted) = 73.0%
s = 4.582 with 220 − 3 = 217 degrees of freedom

Source	Sum of Squares	df	Mean Square	F-ratio
Regression	12462.4	2	6231.18	297
Residual	4555.24	217	20.9919	

Variable	Coefficient	SE(Coeff)	t-ratio	P-value
Intercept	75.5592	6.451	11.7	<0.0001
LogBirthrate	−22.6213	2.478	−9.13	<0.0001
LogGdp/cap	6.11560	0.9371	6.53	<0.0001

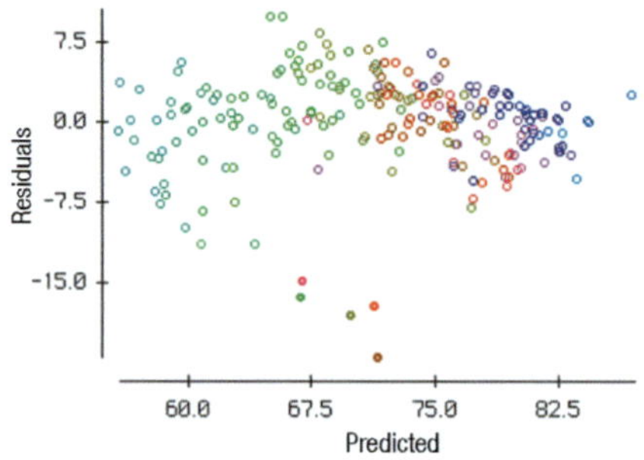

Comment on the model and the residuals. What do the coefficients say? Do you have any concerns about the analysis?

T **18. Another slice of pizza** A plot of Studentized residuals against predicted values for the regression model found in Exercise 16 now looks like this. It has been colored according to *Type* of pizza and separate regression lines fitted for each type:

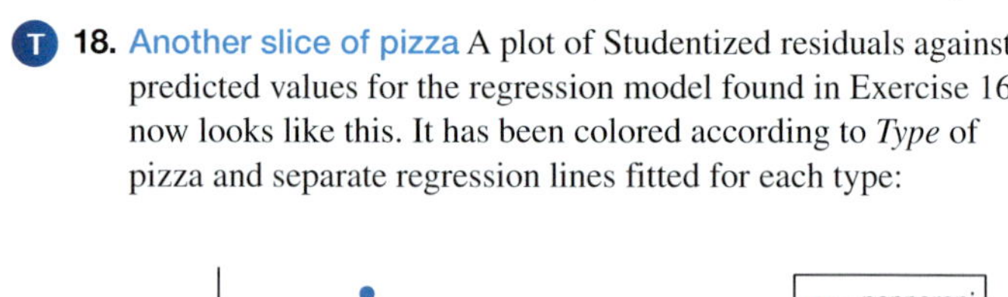

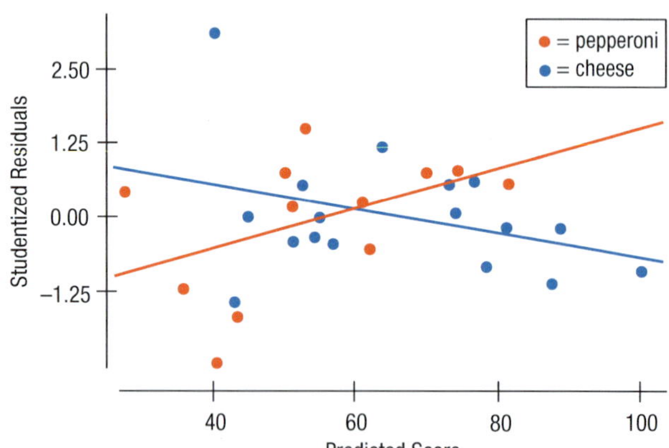

a) Comment on this diagnostic plot. What does it say about how the regression model deals with cheese and pepperoni pizzas?

Based on this plot, we constructed yet another variable consisting of the indicator *Type* multiplied by *Calories*:

Dependent variable is: Score
R-squared = 73.7%   R-squared (adjusted) = 66.5%
s = 12.82 with 29 − 7 = 22 degrees of freedom

Source	Sum of Squares	DF	Mean Square	F-ratio
Regression	10121.4	6	1686.90	10.3
Residual	3617.32	22	164.424	

Variable	Coefficient	SE(Coeff)	t-ratio	P-value
Intercept	−464.498	74.73	−6.22	<0.0001
Calories	1.92005	0.2842	6.76	<0.0001
Fat	−10.3847	1.779	−5.84	<0.0001
Type	183.634	59.99	3.06	0.0057
Type*Cals	−0.461496	0.1740	−2.65	0.0145
Reggio's	−64.4237	15.94	−4.04	0.0005
Michelina's	−51.4966	15.90	−3.24	0.0038

b) Interpret the coefficient of *Type*Cals* in this regression model.

c) Would you prefer this regression model to the model of Exercise 16? Explain.

19. World data diagnosis The residual plot of Exercise 17 calls out some countries that have particularly large negative residuals. They are Gabon, Swaziland, Botswana, Namibia, and South Africa.

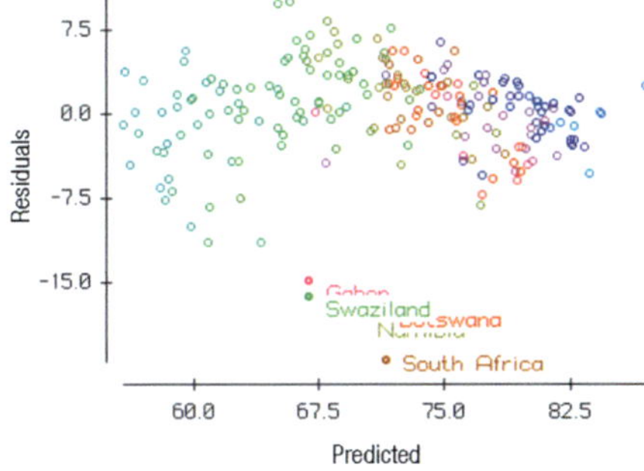

What do these countries have in common? (*Hint*: Consult a map.) What does it mean for a regression residual in this model to be so negative? Can you think of any reason these countries ought to have such large negative residuals?

20. The final slice Here's the residual plot corresponding to the regression model of Exercise 18:

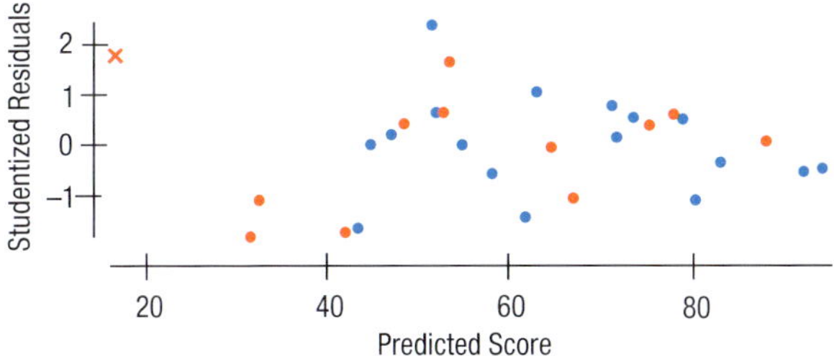

The extreme case this time is *Weight Watchers Pepperoni* (makes sense, doesn't it?). We can make one more indicator for *Weight Watchers*. Here's the model:

Dependent variable is: Score
R-squared = 77.1%   R-squared (adjusted) = 69.4%
s = 12.25 with 29 − 8 = 21 degrees of freedom

Source	Sum of Squares	DF	Mean Square	F-ratio
Regression	10586.8	7	1512.41	10.1
Residual	3151.85	21	150.088	

Variable	Coefficient	SE(Coeff)	t-ratio	P-value
Intercept	−525.063	79.25	−6.63	<0.0001
Calories	2.10223	0.2906	7.23	<0.0001
Fat	−10.8658	1.721	−6.31	<0.0001
Type	231.335	63.40	3.65	0.0015
Type*Cals	−0.586007	0.1806	−3.24	0.0039
Reggio's	−66.4706	15.27	−4.35	0.0003
Michelina's	−52.2137	15.20	−3.44	0.0025
Weight Wachr	28.3265	16.09	1.76	0.0928

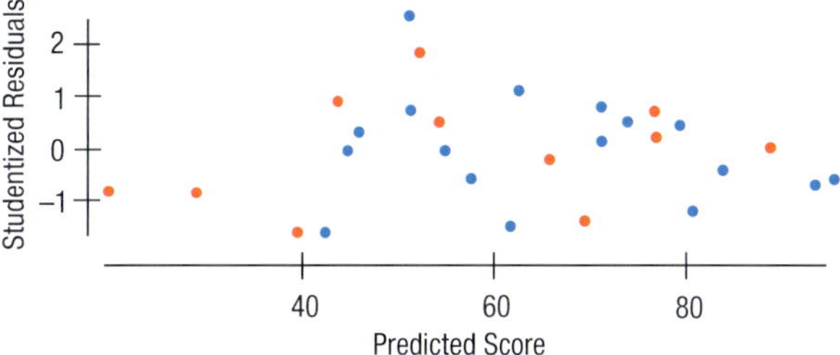

a) Compare this model with the others we've seen for these data. In what ways does this model seem better or worse than the others?

b) Do you think the indicator for *Weight Watchers* should be in the model? (Consider the effect that including it has had on the other coefficients also.)

c) What do the Consumers' Union tasters seem to think makes for a really good pizza?

T **21. World data with HIV** One ongoing health problem in the part of Africa encompassing the outlying countries for the regression model of Exercise 17 is HIV/AIDS. Could that explain these outliers? Here's another model, now with the logarithm of the HIV incidence included as a predictor.

Compare this model with the one found in Exercise 17. Use appropriate statistical terminology and choose what aspects of the model to compare.

Response variable is: Life expectancy
240 total cases of which 72 are missing
R squared = 82.8% R squared (adjusted) = 82.4%
$s = 3.870$ with $168 - 4 = 164$ degrees of freedom

Source	Sum of Squares	df	Mean Square	F-ratio
Regression	11793.4	3	3931.12	262
Residual	2456.55	164	14.9790	

Variable	Coefficient	SE(Coeff)	t-ratio	P-value
Intercept	66.0110	6.361	10.4	<0.0001
LogBirth	−15.4697	2.532	−6.11	<0.0001
LGDP/cap	5.63055	0.9102	6.19	<0.0001
LogHIV	−5.27471	0.5537	−9.53	<0.0010

JUST CHECKING

Answers

1. Histograms are used to examine the shapes of distributions of individual variables. We check especially for multiple modes, outliers, and skewness. They are also used to check the shape of the distribution of the residuals for the Nearly Normal Condition.
2. Scatterplots are used to check the Straight Enough Condition in plots of y *vs.* any of the x's. They are used to check plots of the residuals or Studentized residuals against the predicted values, against any of the predictors, or against *Time* to check for patterns. Scatterplots are also the display used in partial regression plots, where we check for influential points and unexpected subgroups.
3. The Normal model is needed only when we use inference; it isn't needed for computing a regression model. We check the Nearly Normal Condition on the residuals.

25

Analysis of Variance

WHO	Hand washings by four different methods, assigned randomly and replicated 8 times each
WHAT	Number of bacteria colonies
HOW	Sterile media plates incubated at 36°C for 2 days

Did you wash your hands with soap before eating? You've undoubtedly been asked that question a few times in your life. Mom knows that washing with soap eliminates most of the germs you've managed to collect on your hands. Or does it? A student decided to investigate just how effective washing with soap is in eliminating bacteria. To do this she tested four different methods—washing with water only, washing with regular soap, washing with antibacterial soap (ABS), and spraying hands with antibacterial spray (AS) (containing 65% ethanol as an active ingredient). Her experiment consisted of one experimental factor, the washing *Method*, at four levels.

She suspected that the number of bacteria on her hands before washing might vary considerably from day to day. To help even out the effects of those changes, she generated random numbers to determine the order of the four treatments. Each morning, she washed her hands according to the treatment randomly chosen. Then she placed her right hand on a sterile media plate designed to encourage bacteria growth. She incubated each plate for 2 days at 36°C, after which she counted the bacteria colonies. She replicated this procedure 8 times for each of the four treatments.

A side-by-side boxplot of the numbers of colonies seems to show some differences among the treatments:

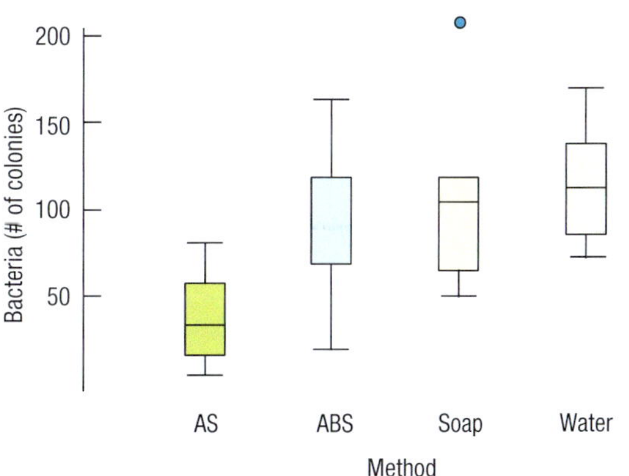

Figure 25.1
Boxplots of the bacteria colony counts for the four different washing methods suggest some differences between treatments. (Data in **Bacterial soap**)

When we first looked at a quantitative variable measured for each of several groups in Chapter 4, we displayed the data this way with side-by-side boxplots. And when we compared the boxes, we asked whether the centers seemed to differ, using the spreads of the boxes to judge the size of the differences. Now we want to make this more formal by testing a hypothesis. We'll make the same kind of comparison, comparing the variability among the means with the spreads of the boxes. It looks like the alcohol spray has lower bacteria counts, but as always, we're skeptical. Could it be that the four methods really have the same mean counts and we just *happened* to get a difference like this because of natural sampling variability?

What is the null hypothesis here? It seems natural to start with the hypothesis that *all the group means are equal*. That would say it doesn't matter what method you use to wash your hands because the mean bacteria count will be the same. We know that even if there were no differences at all in the *means* (for example, if someone replaced all the solutions with water) there would still be sample-to-sample differences. We want to see, statistically, whether differences as large as those observed in the experiment could naturally occur by chance in groups that have equal means. If we find that the differences in washing *Methods* are so large that they would occur only very infrequently in groups that actually have the same mean, then, as we've done with other hypothesis tests, we'll reject the null hypothesis and conclude that the washing *Methods* really have different means.[1]

EXAMPLE 25.1

Contrast baths are a treatment commonly used in hand clinics to reduce swelling and stiffness after surgery. Patients' hands are immersed alternately in warm and cool water. (That's the *contrast* in the name.) Sometimes, the treatment is combined with mild exercise. Although the treatment is widely used, it had never been verified that it would accomplish the stated outcome goal of reducing swelling.

Researchers[2] randomly assigned 59 patients who had received carpal tunnel release surgery to one of three treatments: contrast bath, contrast bath with exercise, and (as a control) exercise alone (Data in **Contrast baths**). Hand therapists who did not know how the subjects had been treated measured hand volumes before and after treatments in milliliters by measuring how much water the hand displaced when submerged. The change in hand volume (after treatment minus before) was reported as the outcome.

QUESTION: Specify the details of the experiment's design. Identify the subjects, the sample size, the experiment factor, the treatment levels, and the response. What is the null hypothesis? Was randomization employed? Was the experiment blinded? Was it double-blinded?

ANSWER: Subjects were patients who received carpal tunnel release surgery. Sample size is 59 patients. The factor was contrast bath treatment with three levels: contrast baths alone, contrast baths with exercise, and exercise alone. The response variable is the change in hand volume. The null hypothesis is that the mean changes in hand volume will be the same for the three treatment levels. Patients were randomly assigned to treatments. The study was single-blind because the evaluators were blind to the treatments. It was not (and could not be) double-blind because the patients had to be aware of their treatments.

[1]The alternative hypothesis is that "the means are not *all* equal." Be careful not to confuse that with "all the means are different." With 11 groups we could have 10 means equal to each other and 1 different. The null hypothesis would still be false.

[2]Janssen, Robert G., Schwartz, Deborah A., and Velleman, Paul F., "A Randomized Controlled Study of Contrast Baths on Patients with Carpal Tunnel Syndrome," *Journal of Hand Therapy*, 22:3, pp. 200–207. The data reported here differ slightly from those in the original paper because they include some additional subjects and exclude some outliers.

25.1 Testing Whether the Means of Several Groups Are Equal

We saw in Chapter 20 how to use a t-test to see whether two groups have equal means. We compared the difference in the means to a standard error estimated from all the data. And when we were willing to assume that the underlying group variances were equal, we pooled the data from the two groups to find the standard error.

Now we have more groups, so we can't just look at differences in the means.[3] But all is not lost. Even if the null hypothesis were true, and the means of the populations underlying the groups were equal, we'd still expect the sample means to vary a bit. We could measure that variation by finding the variance of the means. How much should they vary? Well, if we look at how much the data themselves vary, we can get a good idea of how much the means should vary. And if the underlying means are actually different, we'd expect that variation to be larger.

It turns out that we can build a hypothesis test to check whether the variation in the means is bigger than we'd expect it to be just from random fluctuations. We'll need a new sampling distribution model, called the F-model, but that's just a different table to look at (Table F can be found at the end of this chapter).

To get an idea of how it works, let's start by looking at the following two sets of boxplots:

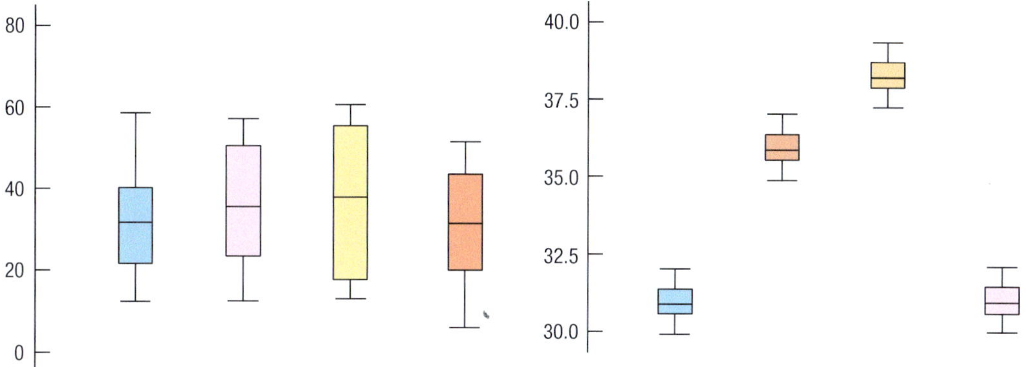

Figure 25.2
It's hard to see the difference in the means in these boxplots because the spreads are large relative to the differences in the means.

Figure 25.3
In contrast with Figure 25.2, the smaller variation makes it much easier to see the differences among the group means. (Notice also that the scale of the y-axis is considerably different from the plot on the left.)

We're trying to decide if the means are different enough for us to reject the null hypothesis. If they're close, we'll attribute the differences to natural sampling variability. What do you think? It's easy to see that the means in the second set differ. It's hard to imagine that the means could be that far apart just from natural sampling variability alone. How about the first set? It looks like these observations *could* have occurred from treatments with the same means.[4] This much variation among groups does seem consistent with equal group means.

Believe it or not, the two sets of treatment means in both figures are the same. (They are 31, 36, 38, and 31, respectively.) Then why do the figures look so different? In the second figure, the variation *within* each group is so small that the differences *between* the means stand out. This is what we looked for when we compared boxplots by eye back in Chapter 4. And it's the central idea of the F-test. We compare the differences *between* the means of the groups with the variation *within* the groups. When the differences between means are large compared with the variation within the groups, we reject the null hypothesis and conclude that the means are not equal. In the first figure, the differences among the means look as though they could have arisen just from natural sampling variability from groups with equal means, so there's not enough evidence to reject H_0.

---

[3] You might think of testing all pairs, but that method generates too many Type I errors. We'll see more about this later in the chapter.

[4] Of course, with a large enough sample, we can detect any differences that we like. For experiments with the same sample size, it's easier to detect the differences when the variation *within* each box is smaller.

How can we make this comparison more precise statistically? All the tests we've seen have compared differences of some kind with a ruler based on an estimate of variation. And we've always done that by looking at the ratio of the statistic to that variation estimate. Here, the differences among the means will show up in the numerator, and the ruler we compare them with will be based on the underlying standard deviation—that is, on the variability *within* the treatment groups.

EXAMPLE 25.2

RECAP: Fifty-nine postsurgery patients were randomly assigned to one of three treatment levels. Changes in hand volume were measured. Here are the boxplots. The recorded values are volume after treatment – volume before treatment, so positive values indicate swelling. Some swelling is to be expected. (Data in **Contrast baths**)

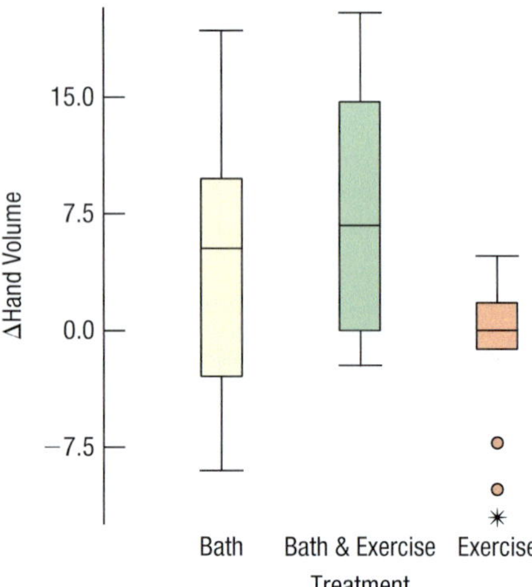

QUESTION: What do the boxplots say about the results?

ANSWER: There doesn't seem to be much difference between the two contrast bath treatments. The exercise only treatment may result in less swelling.

How Different Are They?

The challenge here is that we can't take a simple difference as we did when comparing two groups. In the hand-washing experiment, we have differences in mean bacteria counts across *four* treatments. How should we measure how different the four group means are? With only two groups, we naturally took the difference between their means as the numerator for the *t*-test. It's hard to imagine what else we could have done. How can we generalize that to more than two groups? When we've wanted to know how different many observations were, we measured how much they vary, and that's what we do here.

How much natural variation should we expect among the means if the null hypothesis were true? If the null hypothesis were *true*, then each of the treatment means would estimate the *same* underlying mean. If the washing methods are all the same, it's as if we're just estimating the mean bacteria count on hands that have been washed with plain water. And we have several (in our experiment, four) different, independent estimates of this mean. Here comes the clever part. We can treat these estimated means as if they were observations and simply calculate their (sample) variance. This variance is the measure we'll use to assess how different the group means are from each other. It's the generalization of the difference between means for only two groups.

WHY VARIANCES?

We've usually measured variability with standard deviations. Standard deviations have the advantage that they're in the same units as the data. Variances have the advantage that for independent variables, the variances add. Because we're talking about sums of variables, we'll stay with variances before we get back to standard deviations.

Level	n	Mean
Alcohol Spray	8	37.5
Antibacterial Soap	8	92.5
Soap	8	106.0
Water	8	117.0

The more the group means resemble each other, the smaller this variance will be. The more they differ (perhaps because the treatments actually have an effect), the larger this variance will be.

For the bacteria counts, the four means are listed in the table to the left. If you took those four values, treated them as observations, and found their sample variance, you'd get 1245.08. That's fine, but how can we tell whether it is a big value? Now we need a model, and the model is based on our null hypothesis that all the group means are equal. Here, the null hypothesis is that it doesn't matter what washing method you use; the mean bacteria count will be about the same:

$$H_0: \mu_1 = \mu_2 = \mu_3 = \mu_4 = \mu.$$

As always when testing a null hypothesis, we'll start by assuming that it is true. And if the group means are equal, then there's an overall mean, μ—the bacteria count you'd expect all the time after washing your hands in the morning. And each of the observed group means is just a sample-based estimate of that underlying mean.

We know how sample means vary. The variance of a sample mean is σ^2/n. With eight observations in a group, that would be $\sigma^2/8$. The estimate that we've just calculated, 1245.08, should estimate this quantity. If we want to get back to the variance of the *observations*, σ^2, we need to multiply it by 8. So $8 \times 1245.08 = 9960.64$ should estimate σ^2.

Is 9960.64 large for this variance? How can we tell? We'll need a hypothesis test. You won't be surprised to learn that there is just such a test. The details of the test, due to Sir Ronald Fisher in the early 20th century, are truly ingenious, and may be the most amazing statistical result of that century.

The Ruler Within

We need a suitable ruler for comparison—one based on the underlying variability in our measurements. That variability is due to the day-to-day differences in the bacteria count even when the same soap is used. Why would those counts be different? Maybe the experimenter's hands were not equally dirty, or she washed less well some days, or the plate incubation conditions varied. We randomized just so we could see past such things.

We need an independent estimate of σ^2, one that doesn't depend on the null hypothesis being true, one that won't change if the groups have different means. As in many quests, the secret is to look "within." We could look in *any* of the treatment groups and find its variance. But which one should we use? The answer is, *all* of them!

At the start of the experiment (when we randomly assigned experimental units to treatment groups), the units were drawn randomly from the same pool, so each treatment group had a sample variance that estimated the same σ^2. If the null hypothesis is true, then not much has happened to the experimental units—or at least, their means have not moved apart. It's not much of a stretch to believe that their variances haven't moved apart much either. (If the washing methods are equivalent, then the choice of method would not affect the mean *or* the variability.) So each group variance still estimates a common σ^2.

We're assuming that the null hypothesis is true. If the group variances are equal, then the common variance they all estimate is just what we've been looking for. Since all the group variances estimate the same σ^2, we can pool them to get an overall estimate of σ^2. Recall that we pooled to estimate variances when we tested the null hypothesis that two proportions were equal—and for the same reason. It's also exactly what we did in a pooled t-test. The variance estimate we get by pooling we'll denote, as before, by s_p^2.

For the bacteria counts, the standard deviations and variances are listed below.

Level	n	Mean	Std Dev	Variance
Alcohol Spray	8	37.5	25.56	705.43
Antibacterial Soap	8	92.5	41.96	1760.64
Soap	8	106.0	46.96	2205.24
Water	8	117.0	31.13	969.08

If we pool the four variances (here we can just average them because all the sample sizes are equal), we'd get $s_p^2 = 1410.10$. In the pooled variance, each variance is taken around its *own* treatment mean, so the pooled estimate doesn't depend on the treatment means being equal. But the estimate in which we took the four means as observations and took their variance does. That estimate gave 9960.64. That seems a lot bigger than 1410.10. Might this be evidence that the four means are not equal?

Let's see what we have. We have an estimate of σ^2 from the variation *within* groups of 1410.10. That's just the variance of the *residuals* pooled across all groups. Because it's a pooled variance, we could write it as s_p^2. Traditionally this quantity is also called the **Error Mean Square**, or sometimes the **Within Mean Square** and denoted by $\mathbf{MS_E}$. These names date back to the early 20th century when the methods were developed. If you think about it, the names do make sense—variances are means of squared differences.[5]

But we also have a *separate* estimate of σ^2 from the variation *between* the groups because we know how much means ought to vary. For the hand-washing data, when we took the variance of the four means and multiplied it by n we got 9960.64. We expect this to estimate σ^2 too, *as long as we assume the null hypothesis is true*. We call this quantity the **Treatment Mean Square** (or sometimes the **Between Mean Square**[6]) and denote by $\mathbf{MS_T}$.

The *F*-Statistic

Now we have two different estimates of the underlying variance. The first one, the MS_T, is based on the differences *between* the group means. If the group means are equal, as the null hypothesis asserts, it will estimate σ^2. But, if they are not, it will give some bigger value. The other estimate, the MS_E, is based only on the variation *within* the groups around each of their own means, and doesn't depend at all on the null hypothesis being true.

So, how do we test the null hypothesis? When the null hypothesis is true, the treatment means are equal, and both MS_E and MS_T estimate σ^2. Their ratio, then, should be close to 1.0. But, when the null hypothesis is false, the MS_T will be *larger* because the treatment means are not equal. The MS_E is a pooled estimate in which the variation within each group is found around its own group mean, so differing means won't inflate it. That makes the ratio MS_T/MS_E perfect for testing the null hypothesis. When the null hypothesis is true, the ratio should be near 1. If the treatment means really are different, the numerator will tend to be larger than the denominator, and the ratio will tend to be bigger than 1.

> **NOTATION ALERT**
>
> Capital F is used only for this distribution model and statistic. Fortunately, Fisher's name didn't start with a Z, a T, or an R.

Of course, even when the null hypothesis *is* true, the ratio will vary around 1 just due to natural sampling variability. How can we tell when it's big enough to reject the null hypothesis? To be able to tell, we need a sampling distribution model for the ratio. Sir Ronald Fisher found the sampling distribution model of the ratio in the early 20th century. In his honor, we call the distribution of MS_T/MS_E the ***F*-distribution**. And we call the ratio MS_T/MS_E the ***F*-statistic**. By comparing this statistic with the appropriate F-distribution we (or the computer) can get a P-value.

The resulting ***F*-test** is one-tailed because any differences in the means make the F-statistic larger. Larger differences in the treatments' effects lead to the means being more variable, making the MS_T bigger. That makes the F-ratio grow. So the test is significant if the F-ratio is big enough. In practice, we find a P-value, and big F-statistic values go with small P-values.

The entire analysis is called the Analysis of Variance, commonly abbreviated **ANOVA** (and pronounced uh-NŌ-va). You might think that it should be called the analysis of means, since it's the equality of the means we're testing. But we use the *variances* within and between the groups for the test.

[5]Well, actually, they're sums of squared differences divided by their degrees of freedom—$(n - 1)$ for the first variance we saw back in Chapter 3, and other degrees of freedom for each of the others we've seen. But even back in Chapter 3, we said this was a "kind of" mean, and indeed, it still is.

[6]Grammarians would probably insist on calling it the *Among Mean Square,* since the variation is among *all* the group means. Traditionally, though, it's called the *Between Mean Square* and we have to talk about the variation between all the groups (as bad as that sounds).

NOTATION ALERT

What, first little n and now big N? In an experiment, it's standard to use N for *all* the cases and n for the number in each treatment group.

Like Student's t-models, the F-models are a family. F-models depend on not one, but two, degrees of freedom parameters. The degrees of freedom come from the two variance estimates and are sometimes called the *numerator df* and the *denominator df*. The *Treatment Mean Square*, MS_T, is the sample variance of the observed treatment means. If we think of them as observations, then since there are k groups, this variance has $k - 1$ degrees of freedom. The *Error Mean Square*, MS_E, is the pooled estimate of the variance within the groups. If there are n observations in each group, then we get $n - 1$ degrees of freedom from each, for a total of $k(n - 1)$ degrees of freedom.

A simpler way of tracking the degrees of freedom is to start with all the cases. We'll call that N. Each group has its own mean, costing us a degree of freedom—k in all. So we have $N - k$ degrees of freedom for the error. When the groups all have equal sample size, that's the same as $k(n - 1)$, but this way works even if the group sizes differ.

We say that the F-statistic, MS_T/MS_E, has $k - 1$ and $N - k$ degrees of freedom.

Back to Bacteria

For the hand-washing experiment, the $MS_T = 9960.64$. The $MS_E = 1410.14$. If the treatment means were equal, the *Treatment Mean Square* should be about the same size as the *Error Mean Square*, about 1410. But it's 9960.64, which is 7.06 times bigger. In other words, $F = 7.06$. This F-statistic has $4 - 1 = 3$ and $32 - 4 = 28$ degrees of freedom.

An F-value of 7.06 is bigger than 1, but we can't tell for sure whether it's big enough to reject the null hypothesis until we check the $F_{3,28}$ model to find its P-value. (Usually, that's most easily done with technology, but we can use printed tables.) It turns out the P-value is 0.0011. In other words, if the treatment means were actually equal, we would expect the ratio MS_T/MS_E to be 7.06 or larger about 11 times out of 10,000, just from natural sampling variability. That's not very likely, so we reject the null hypothesis and conclude that the means are different. We have strong evidence that the four different methods of hand washing are not equally effective at eliminating germs.

25.2 The ANOVA Table

You'll often see the mean squares and other information put into a table called the **ANOVA table**. Here's the table for the washing experiment:

Analysis of Variance Table

Source	DF	Sum of Squares	Mean Square	F-Ratio	P-Value
Method	3	29882	9960.64	7.0636	0.0011
Error	28	39484	1410.14		
Total	31	69366			

CALCULATING THE ANOVA TABLE

This table has a long tradition stretching back to when ANOVA calculations were done by hand. Major research labs had rooms full of mechanical calculators operated by women. (Yes, always women; women were thought—by the men in charge, at least—to be more careful at such an exacting task.) Three women would perform each calculation, and if any two of them agreed on the answer, it was taken as the correct value.

The ANOVA table was originally designed to organize the calculations. With technology, we have much less use for that. We'll show how to calculate the sums of squares later in the chapter, but the most important quantities in the table are the F-statistic and its associated P-value. When the F-statistic is large, the Treatment (here *Method*) Mean Square is large compared to the Error Mean Square (MS_E), and provides evidence that in fact the means of the groups are not all equal.

You'll almost always see ANOVA results presented in a table like this. After nearly a century of writing the table this way, statisticians (and their technology) aren't going to change. Even though the table was designed to facilitate hand calculation, computer programs that compute ANOVAs still present the results in this form. Usually the P-value is found next to the F-ratio. The P-value column may be labeled with a title such as "Prob $>$ F," "sig," or "Prob." Don't let that confuse you; it's just the P-value.

You'll sometimes see the two mean squares referred to as the *Mean Square Between* and the *Mean Square Within*—especially when we test data from observational studies rather than experiments. ANOVA is often used for such observational data, and as long as certain conditions are satisfied, there's no problem with using it in that context.

EXAMPLE 25.3

RECAP: An experiment to determine the effect of contrast bath treatments on swelling in postsurgical patients recorded hand volume changes for patients who had been randomly assigned to one of three treatments.

Here is the Analysis of Variance for these data:

Analysis of Variance for Hand Volume Change

Source	df	Sum of Squares	Mean Square	F-Ratio	P-Value
Treatment	2	716.159	358.080	7.4148	0.0014
Error	56	2704.38	48.2926		
Total	58	3420.54			

QUESTION: What does the ANOVA say about the results of the experiment? Specifically, what does it say about the null hypothesis?

ANSWER: The F-ratio of 7.4148 has a P-value that is quite small. We can reject the null hypothesis that the mean change in hand volume is the same for all three treatments.

The *F*-Table

Usually, you'll get the P-value for the F-statistic from technology. Any software program performing an ANOVA will automatically "look up" the appropriate one-sided P-value for the F-statistic. If you want to do it yourself, you'll need an F-table. F-tables are usually printed only for a few values of α, often 0.05, 0.01, and 0.001. They give the critical value of the F-statistic with the appropriate number of degrees of freedom determined by your data, for the α-level that you select. If your F-statistic is greater than that value, you know that its P-value is less than that α-level. So, you'll be able to tell whether the P-value is greater or less than 0.05, 0.01, or 0.001, but to be more precise, you'll need technology (or an interactive table like the one in the *ActivStats* program online at www.pearsonglobaleditions.com/deveaux).

Here's an excerpt from an F-table for $\alpha = 0.05$:

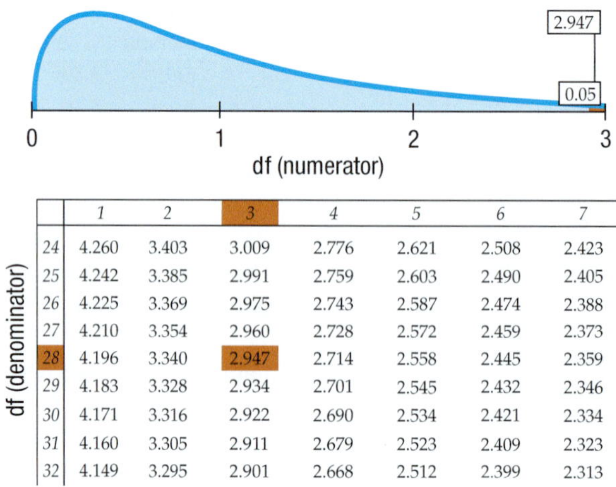

df (denominator)	1	2	3	4	5	6	7
24	4.260	3.403	3.009	2.776	2.621	2.508	2.423
25	4.242	3.385	2.991	2.759	2.603	2.490	2.405
26	4.225	3.369	2.975	2.743	2.587	2.474	2.388
27	4.210	3.354	2.960	2.728	2.572	2.459	2.373
28	4.196	3.340	2.947	2.714	2.558	2.445	2.359
29	4.183	3.328	2.934	2.701	2.545	2.432	2.346
30	4.171	3.316	2.922	2.690	2.534	2.421	2.334
31	4.160	3.305	2.911	2.679	2.523	2.409	2.323
32	4.149	3.295	2.901	2.668	2.512	2.399	2.313

Figure 25.4
Part of an F-table showing critical values for $\alpha = 0.05$ and highlighting the critical value, 2.947, for 3 and 28 degrees of freedom. We can see that only 5% of the values will be greater than 2.947 with this combination of degrees of freedom.

Notice that the critical value for 3 and 28 degrees of freedom at $\alpha = 0.05$ is 2.947. Since our F-statistic of 7.06 is larger than this critical value, we know that the P-value is less than 0.05. We could also look up the critical value for $\alpha = 0.01$ and find that it's 4.568 and the critical value for $\alpha = 0.001$ is 7.193. So our F-statistic sits between the two critical values 0.01 and 0.001, and our P-value is slightly *greater* than 0.001. Technology can find the value precisely. It turns out to be 0.0011.

JUST CHECKING

A student conducted an experiment to see which, if any, of four different paper airplane designs results in the longest flights (measured in inches). The boxplots look like this (with the overall mean shown in red): (Data in **Paper airplanes**)

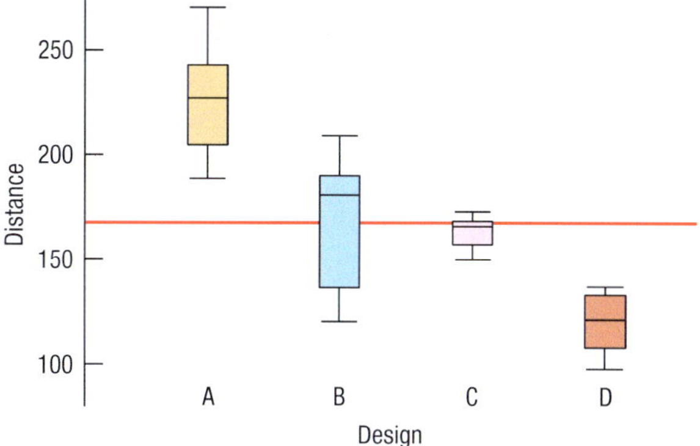

The ANOVA table shows:

Analysis of Variance

Source	DF	Sum of Squares	Mean Square	F-Ratio	Prob > F
Design	3	51991.778	17330.6	37.4255	<0.0001
Error	32	14818.222	463.1		
Total	35	66810.000			

1. What is the null hypothesis?
2. From the boxplots, do you think that there is evidence that the mean flight distances of the four designs differ?
3. Does the *F*-test in the ANOVA table support your preliminary conclusion in (2)?
4. The researcher concluded that "there is substantial evidence that all four of the designs result in different mean flight distances." Do you agree?

The ANOVA Model

To understand the ANOVA table, let's start by writing a model for what we observe. We start with the simplest interesting model: one that says that the only differences of interest among the groups are the differences in their means. This **one-way ANOVA model** characterizes each observation in terms of its mean and assumes that any variation around that mean is just random error:

$$y_{ij} = \mu_j + \varepsilon_{ij}.$$

That is, each observation is the sum of the mean for the treatment it received plus a random error. Our null hypothesis is that the treatments made no difference—that is, that the means are all equal:

$$H_0: \mu_1 = \mu_2 = \cdots = \mu_k.$$

It will help our discussion if we think of the overall mean of the experiment and consider the treatments as adding or subtracting an effect to this overall mean. Thinking in this way, we could write μ for the overall mean and τ_j for the deviation from this mean to get to the jth treatment mean—the *effect* of the treatment (if any) in moving that group away from the overall mean:

$$y_{ij} = \mu + \tau_j + \varepsilon_{ij}.$$

Thinking in terms of the effects, we could also write the null hypothesis in terms of these treatment *effects* instead of the means:

$$H_0: \tau_1 = \tau_2 = \cdots = \tau_k = 0.$$

We now have three different kinds of parameters: the overall mean, the treatment effects, and the errors. We'll want to estimate them from the data. Fortunately, we can do that in a straightforward way.

To estimate the overall mean, μ, we use the mean of all the observations: $\bar{\bar{y}}$ (called the "grand mean"[7]). To estimate each treatment effect, we find the difference between the mean of that particular treatment and the grand mean:

$$\hat{\tau}_j = \bar{y}_j - \bar{\bar{y}}.$$

There's an error, ε_{ij}, for each observation. We estimate those with the residuals from the treatment means: $e_{ij} = y_{ij} - \bar{y}_j$.

We can write each observation as the sum of three quantities that correspond to our model:

$$y_{ij} = \bar{\bar{y}} + (\bar{y}_j - \bar{\bar{y}}) + (y_{ij} - \bar{y}_j).$$

What this says is simply that we can write each observation as the sum of

- the grand mean,
- the effect of the treatment it received, and
- the residual.

Or:

$$\textit{Observations} = \textit{Grand mean} + \textit{Treatment effect} + \textit{Residual}.$$

If we look at the equivalent equation

$$y_{ij} = \bar{\bar{y}} + (\bar{y}_j - \bar{\bar{y}}) + (y_{ij} - \bar{y}_j)$$

closely, it doesn't really seem like we've done anything. In fact, collecting terms on the right-hand side will give back just the observation, y_{ij} again. But this decomposition is actually the secret of the Analysis of Variance. We've split each observation into "sources"—the grand mean, the treatment effect, and error.

Where does the residual term come from?

Think of the annual report from any Fortune 500 company. The company spends billions of dollars each year and at the end of the year, the accountants show where each penny goes. How do they do it? After accounting for salaries, bonuses, supplies, taxes, etc., etc., etc., what's the last line? It's always labeled "other" or miscellaneous. Using "other" as the difference between all the sources they know and the total they start with, they can always make it add up perfectly. The residual is just the statisticians' "other." It takes care of all the other sources we didn't think of or don't want to consider, and makes the decomposition work by adding (or subtracting) back in just what we need.

[7]The father of your father is your grandfather. The mean of the group means should probably be the grandmean, but we usually spell it as two words.

Let's see what this looks like for our hand-washing data. Here are the data again, displayed a little differently:

	Alcohol	AB Soap	Soap	Water
	51	70	84	74
	5	164	51	135
	19	88	110	102
	18	111	67	124
	58	73	119	105
	50	119	108	139
	82	20	207	170
	17	95	102	87
Treatment Means	37.5	92.5	106	117

The grand mean of all observations is 88.25. Let's put that into a similar table:

Alcohol	AB Soap	Soap	Water
88.25	88.25	88.25	88.25
88.25	88.25	88.25	88.25
88.25	88.25	88.25	88.25
88.25	88.25	88.25	88.25
88.25	88.25	88.25	88.25
88.25	88.25	88.25	88.25
88.25	88.25	88.25	88.25
88.25	88.25	88.25	88.25

The treatment *means* are 37.5, 92.5, 106, and 117, respectively, so the treatment *effects* are those minus the grand mean (88.25). Let's put the treatment effects into their table:

Alcohol	AB Soap	Soap	Water
−50.75	4.25	17.75	28.75
−50.75	4.25	17.75	28.75
−50.75	4.25	17.75	28.75
−50.75	4.25	17.75	28.75
−50.75	4.25	17.75	28.75
−50.75	4.25	17.75	28.75
−50.75	4.25	17.75	28.75
−50.75	4.25	17.75	28.75

Finally, we compute the residuals as the differences between each observation and its treatment mean:

Alcohol	AB Soap	Soap	Water
13.5	−22.5	−22	−43
−32.5	71.5	−55	18
−18.5	−4.5	4	−15
−19.5	18.5	−39	7
20.5	−19.5	13	−12
12.5	25.5	2	22
44.5	−72.5	101	53
−20.5	2.5	−4	−30

Now we have four tables for which

$$Observations = Grand\ Mean + Treatment\ Effect + Residual.$$

(You can verify, for example, that the first observation, $51 = 88.25 + (-50.75) + 13.5$).

Why do we want to think in this way? Think back to the boxplots in Figures 25.2 and 25.3. To *test* the hypothesis that the treatment effects are zero, we want to see whether the treatment effects are large *compared* to the errors. Our eye looks at the variation *between* the treatment means and compares it to the variation *within* each group.

The ANOVA separates those two quantities into the Treatment Effects and the Residuals. Sir Ronald Fisher's insight was how to turn those quantities into a statistical test. We want to see if the Treatment Effects are large compared with the Residuals. To do that, we first compute the Sums of Squares of each table. Fisher's insight was that dividing these sums of squares by their respective degrees of freedom lets us test their ratio by a distribution that he found (which was later named the F in his honor). When we divide a sum of squares by its degrees of freedom we get the associated *mean square*.

When the Treatment Mean Square is *large* compared to the Error Mean Square, this provides evidence that the treatment means are different. And we can use the F-distribution to see how large "large" is.

The sums of squares for each table are easy to calculate. Just take every value in the table, square it, and add them all up. For the *Methods*, the Treatment Sum of Squares, $SS_T = (-50.75)^2 + (-50.75)^2 + \cdots + (28.75)^2 = 29882$. There are four treatments, and so there are 3 degrees of freedom. So,

$$MS_T = SS_T/3 = 29882/3 = 9960.64.$$

In general, we could write the Treatment Sum of Squares as

$$SS_T = \sum\sum(\bar{y}_j - \bar{\bar{y}})^2.$$

Be careful to note that the summation is over the *whole* table, rows and columns. That's why there are two summation signs.

And,

$$MS_T = SS_T/(k - 1).$$

The table of residuals shows the variation that remains after we remove the overall mean and the treatment effects. These are what's left over after we account for what we're interested in—in this case the treatments. Their variance is the variance *within* each group that we see in the boxplots of the four groups. To find its value, we first compute the Error Sum of Squares, SS_E, by summing up the squares of every element in the residuals table. To get the Mean Square (the variance) we have to divide it by $N - k$ rather than by $N - 1$ because we found them by subtracting each of the k treatment means.

So,

$$SS_E = (13.5)^2 + (-32.5)^2 + \cdots + (-30)^2 = 39484$$

and

$$MS_E = SS_E/(32 - 4) = 1410.14.$$

As equations:

$$SS_E = \sum\sum(y_{ij} - \bar{y}_j)^2,$$

and

$$MS_E = SS_E/(N - k).$$

Now where are we? To test the null hypothesis that the treatment means are all equal we find the F-statistic:

$$F_{k-1,N-k} = MS_T/MS_E$$

and compare that value to the F-distribution with $k - 1$ and $N - k$ degrees of freedom. When the F-statistic is large enough (and its associated P-value small) we reject the null hypothesis and conclude that at least one mean is different.

There's another amazing result hiding in these tables. If we take each of these tables, square every observation, and add them up, the sums add as well!

$$SS_{Observations} = SS_{Grand\ Mean} + SS_T + SS_E$$

The $\text{SS}_{\text{Observations}}$ is usually very large compared to SS_T and SS_E, so when ANOVA was originally done by hand, or even by calculator, it was hard to check the calculations using this fact. The first sum of squares was just too big. So, usually the ANOVA table uses the "Corrected Total" sum of squares. If we write

$$Observations = Grand\ Mean + Treatment\ Effect + Residual,$$

we can naturally write

$$Observations - Grand\ Mean = Treatment\ Effect + Residual.$$

Mathematically, this is the same statement, but numerically this is more stable. What's amazing is that the sums of the squares still add up. That is, if you make the first table of observations with the grand mean subtracted from each, square those, and add them up, you'll have the SS_{Total} and

$$SS_{Total} = SS_T + SS_E.$$

That's what the ANOVA table shows. If you find this surprising, you must be following along. The tables add up, so sums of their elements must add up. But it is not at all obvious that the sums of the *squares* of their elements should add up, and this is another great insight of the Analysis of Variance.

Back to Standard Deviations

We've been using the variances because they're easier to work with. But when it's time to think about the data, we'd really rather have a standard deviation because it's in the units of the response variable. The natural standard deviation to think about is the standard deviation of the residuals.

The variance of the residuals is staring us in the face. It's the MS_E. All we have to do to get the **residual standard deviation** is take the square root of MS_E:

$$s_p = \sqrt{MS_E} = \sqrt{\frac{\sum e^2}{(N - k)}}.$$

The p subscript is to remind us that this is a *pooled* standard deviation, combining residuals across all k groups. The denominator in the fraction shows that finding a mean for each of the k groups costs us one degree of freedom for each.

This standard deviation should "feel" right. That is, it should reflect the kind of variation you expect to find in any of the experimental groups. For the hand-washing data, $s_p = \sqrt{1410.14} = 37.6$ bacteria colonies. Looking back at the boxplots of the groups, we see that 37.6 seems to be a reasonable compromise standard deviation for all four groups.

25.3 Assumptions and Conditions

Just as you would never find a linear regression without looking at the scatterplot of y vs. x, you should never embark on an ANOVA without first examining side-by-side boxplots of the data comparing the responses for all of the groups. You already know what to look for—we talked about that back in Chapter 4. Check for outliers within any of the groups and correct them if there are errors in the data. Get an idea of whether the groups have similar spreads (as we'll need) and whether the centers seem to be alike (as the null hypothesis claims) or different. If the spreads of the groups are very different—and especially if they seem to grow consistently as the means grow—consider re-expressing the response variable to make the spreads more nearly equal. Doing so is likely to make the analysis more powerful and more correct. Likewise, if the boxplots are skewed in the same direction, you may be able to make the distributions more symmetric with a re-expression.

Don't ever carry out an Analysis of Variance without looking at the side-by-side boxplots first. The chance of missing an important pattern or violation is just too great.

When we checked assumptions and conditions for regression we had to take care to perform our checks in order. Here we have a similar concern. For regression we found that displays of the residuals were often a good way to check the corresponding conditions. That's true for ANOVA as well.

Independence Assumptions The groups must be independent of each other. No test can verify this assumption. You have to think about how the data were collected. The assumption would be violated, for example, if we measured subjects' performance before some treatment, again in the middle of the treatment period, and then again at the end.[8]

The data *within* each treatment group must be independent as well. The data must be drawn independently and at random from a homogeneous population, or generated by a randomized comparative experiment.

We check the **Randomization Condition:** Were the data collected with suitable randomization? For surveys, are the data drawn from each group a representative random sample of that group? For experiments, were the treatments assigned to the experimental units at random?

We were told that the hand-washing experiment was randomized.

Equal Variance Assumption The ANOVA requires that the variances of the treatment groups be equal. After all, we need to find a pooled variance for the MS_E. To check this assumption, we can check that the groups have similar variances:

Similar Spread Condition: There are some ways to see whether the variation in the treatment groups seems roughly equal:

- Look at side-by-side boxplots of the groups to see whether they have roughly the same spread. It can be easier to compare spreads across groups when they have the same center, so consider making side-by-side boxplots of the residuals. If the groups have differing spreads, it can make the pooled variance—the MS_E—larger, reducing the F-statistic value and making it less likely that we can reject the null hypothesis. So the ANOVA will usually fail on the "safe side," rejecting H_0 less often than it should. Because of this, we usually require the spreads to be quite different from each other before we become concerned about the condition failing. If you've rejected the null hypothesis, this is especially true.
- Look at the original boxplots of the response values again. In general, do the spreads seem to change *systematically* with the centers? One common pattern is for the boxes with bigger centers to have bigger spreads. This kind of systematic trend in the variances is more of a problem than random differences in spread among the groups and should not be ignored. Fortunately, such systematic violations are often helped by re-expressing the data. (If, in addition to spreads that grow with the centers, the boxplots are skewed with the longer tail stretching off to the high end, then the data are pleading for a re-expression. Try taking logs of the dependent variable for a start. You'll likely end up with a much cleaner analysis.)
- Look at the residuals plotted against the predicted values. Often, larger predicted values lead to larger magnitude residuals. This is another sign that the condition is violated. (This may remind you of the **Does the Plot Thicken? Condition** of regression. And it should.) When the plot thickens (to one side or the other), it's usually a good idea to consider re-expressing the response variable. Such a systematic change in the

[8]There is a modification of ANOVA, called *repeated measures* ANOVA, that deals with such data. (If the design reminds you of a paired-t situation, you're on the right track, and the lack of independence is the same kind of issue we discussed in Chapter 21.)

spread is a more serious violation of the Equal Variance Assumption than slight variations of the spreads across groups.

Let's check the conditions for the hand-washing data. Here's a boxplot of residuals by group and a scatterplot of residuals by predicted value:

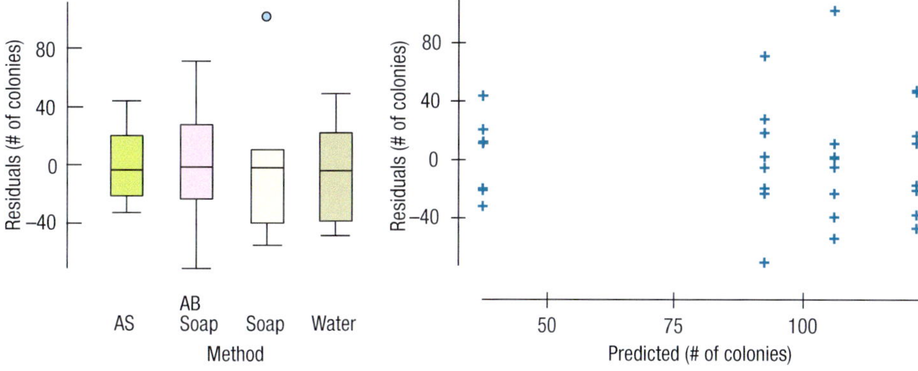

Figure 25.5
Boxplots of residuals for the four washing methods and a plot of residuals vs. predicted values. There's no evidence of a systematic change in variance from one group to the other or by predicted value.

Neither plot shows a violation of the condition. The IQRs (the box heights) are quite similar and the plot of residuals vs. predicted values does not show a pronounced widening to one end. The pooled estimate of 37.6 colonies for the error standard deviation seems reasonable for all four groups.

Normal Population Assumption Like Student's *t*-tests, the *F*-test requires the underlying errors to follow a Normal model. As before when we've faced this assumption, we'll check a corresponding **Nearly Normal Condition**.

Technically, we need to assume that the Normal model is reasonable for the populations underlying *each* treatment group. We can (and should) look at the side-by-side boxplots for indications of skewness. Certainly, if they are all (or mostly) skewed in the same direction, the Nearly Normal Condition fails (and re-expression is likely to help).

In experiments, we often work with fairly small groups for each treatment, and it's nearly impossible to assess whether the distribution of only six or eight numbers is Normal (though sometimes it's so skewed or has such an extreme outlier that we can see that it's not). Here we are saved by the Equal Variance Assumption (which we've already checked). The residuals have their group means subtracted, so the mean residual for each group is 0. If their variances are equal, we can group all the residuals together for the purpose of checking the Nearly Normal Condition.

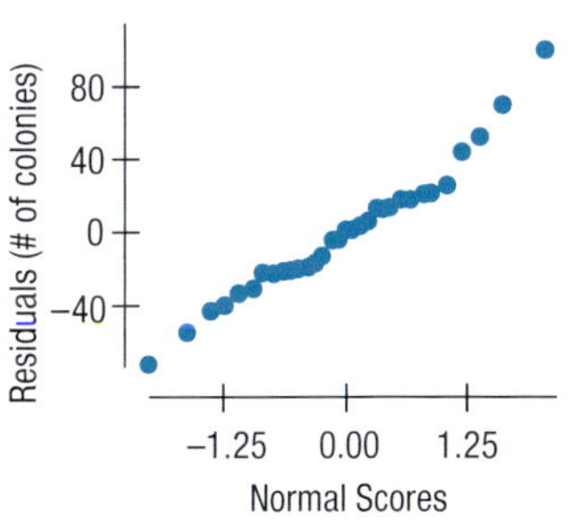

Figure 25.6
The hand washing residuals look nearly Normal in this Normal probability plot.

Check Normality with a histogram or a Normal probability plot of all the residuals together. The hand-washing residuals look nearly Normal in the Normal probability plot, although, as the boxplots showed, there's a possible outlier in the Soap group.

Because we really care about the Normal model within each group, the Normal Population Assumption is violated if there are outliers in any of the groups. Check for outliers in the boxplots of the values for each treatment group. The Soap group of the hand-washing data shows an outlier, so we might want to compute the analysis again without that observation. (For these data, it turns out to make little difference.)

One-way ANOVA *F*-test

We test the null hypothesis H_0: $\mu_1 = \mu_2 = \cdots = \mu_k$ against the alternative that the group means are not all equal. We test the hypothesis with the *F*-statistic, $F = \frac{MS_T}{MS_E}$, where MS_T is the Treatment Mean Square, found from the variance of the means of the treatment groups, and MS_E is the Error Mean Square, found by pooling the variances within each of the treatment groups. If the *F*-statistic is large enough, we reject the null hypothesis.

STEP-BY-STEP EXAMPLE

Analysis of Variance

In Chapter 4, we looked at side-by-side boxplots of four different containers for holding hot beverages. The experimenter wanted to know whether there was any difference among the different types of containers in their ability to keep his coffee hot. To test it, he heated water to a temperature of 180°F, placed it in the container, and then measured the temperature of the water again 30 minutes later. He randomized the order of the trials and tested each container 8 times. His response variable was the difference in temperature (in °F) between the initial water temperature and the temperature after 30 minutes. (Data in **Cups**)

QUESTION: Do the four containers maintain temperature equally well?

THINK

PLOT Plot the side-by-side boxplots of the data.

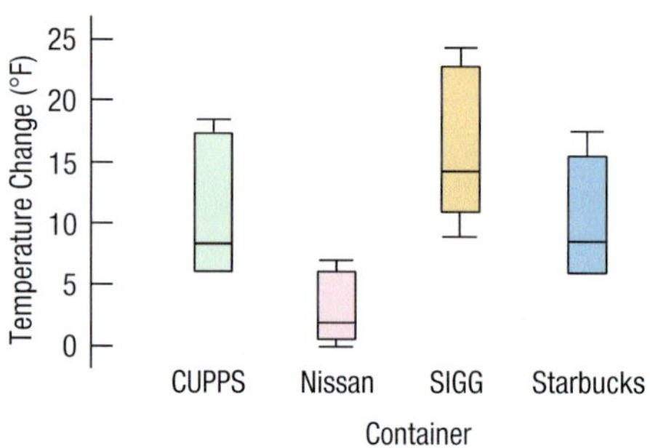

PLAN State what you want to know and the null hypothesis you wish to test. For ANOVA, the null hypothesis is that all the treatment groups have the same mean. The alternative is that at least one mean is different.

I want to test whether there is any difference among the four containers in their ability to maintain the temperature of a hot liquid for 30 minutes. I'll write μ_k for the mean temperature difference for container k, so the null hypothesis is that these means are all the same:

$$H_0: \mu_1 = \mu_2 = \mu_3 = \mu_4.$$

The alternative is that the group means are not all equal.

Think about the assumptions and check the conditions.

✓ **Randomization Condition:** The experimenter performed the trials in random order, so it's reasonable to assume that the performance of one tested cup is independent of other cups.

✓ **Similar Spread Condition:** The Nissan cup variation seems to be a bit smaller than the others. I'll look later at the plot of residuals vs. predicted values to see if the plot thickens.

SHOW

MECHANICS Fit the ANOVA model.

Analysis of Variance

Source	DF	Sum of Squares	Mean Square	F-Ratio	P-Value
Container	3	714.1875	238.063	10.713	<0.0001
Error	28	622.1875	22.221		
Total	31	1336.3750			

✓ **Nearly Normal Condition, Outlier Condition:** The Normal probability plot is not very straight, but there are no outliers.

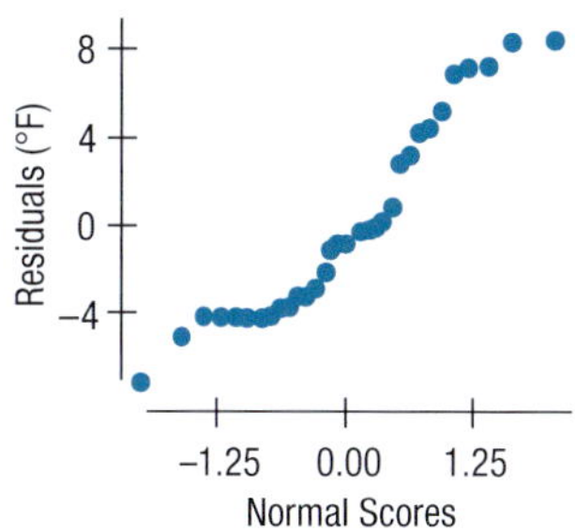

The histogram shows that the distribution of the residuals is skewed to the right:

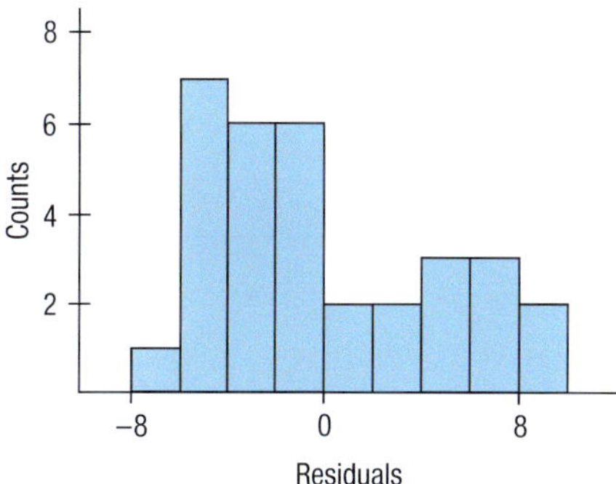

The table of means and SDs (below) shows that the standard deviations grow along with the means. Possibly a re-expression of the data would improve matters.

Under these circumstances, I cautiously find the P-value for the F-statistic from the F-model with 3 and 28 degrees of freedom.

The ratio of the mean squares gives an F-ratio of 10.7134 with a P-value of <0.0001.

SHOW Show the table of means.

From the ANOVA table, the Error Mean Square, MS_E, is 22.22, which means that the standard deviation of all the errors is estimated to be $\sqrt{22.22} = 4.71$ degrees F.

This seems like a reasonable value for the error standard deviation in the four treatments (with the possible exception of the Nissan cup).

Level	*n*	Mean	Std Dev
CUPPS	8	10.1875	5.20259
Nissan	8	2.7500	2.50713
SIGG	8	16.0625	5.90059
Starbucks	8	10.2500	4.55129

TELL	**INTERPRETATION** Tell what the *F*-test means.	An *F*-ratio this large would be very unlikely if the containers all had the same mean temperature difference.
THINK	State your conclusions. (You should be more worried about the changing variance if you fail to reject the null hypothesis.) More specific conclusions might require a re-expression of the data.	**Conclusions**: Even though some of the conditions are mildly violated, I still conclude that the means are not all equal and that the four cups do not maintain temperature equally well.

The Balancing Act

Two of the examples we've looked at so far share a special feature. Each treatment group has the same number of experimental units. For the hand-washing experiment, each washing method was tested 8 times. For the cups, there were also 8 trials for each cup. This feature (the equal numbers of cases in each group, not the number 8) is called **balance**, and experiments that have equal numbers of experimental units in each treatment are said to be balanced or to have balanced designs.

Balanced designs are a bit easier to analyze because the calculations are simpler, so we usually try for balance. But in the real world, we often encounter unbalanced data. Participants drop out or become unsuitable, plants die, or maybe we just can't find enough experimental units to fit a particular criterion.

Everything we've done so far works just fine for unbalanced designs except that the calculations get a bit more complicated. Where once we could write n for the number of experimental units in a treatment, now we have to write n_k and sum more carefully. Where once we could pool variances with a simple average, now we have to adjust for the different n's. Technology clears these hurdles easily, so you're safe thinking about the analysis in terms of the simpler balanced formulas and trusting that the technology will make the necessary adjustments.

EXAMPLE 25.4

RECAP: An ANOVA for the contrast baths experiment had a statistically significant *F*-value.

Here are summary statistics for the three treatment groups:

Group	Count	Mean	StdDev
Bath	22	4.54545	7.76271
Bath & Exercise	23	8	7.03885
Exercise	14	−1.07143	5.18080

QUESTION: What can you conclude about these results?

ANSWER: We can be confident that there is a difference. However, it is the exercise treatment that appears to reduce swelling and not the contrast bath treatments. We might conclude (as the researchers did) that contrast bath treatments are of limited value.

25.4 Comparing Means

When we reject H_0, it's natural to ask which means are different. No one would be happy with an experiment to test 10 cancer treatments that concluded only with "We can reject H_0—the treatments are different!" We'd like to know more, but the F-statistic doesn't offer that information.

What can we do? If we can't reject the null, we've got to stop. There's no point in further testing. If we've rejected the simple null hypothesis, however, we *can* do more. In particular, we can test whether any pairs or combinations of group means differ. For example, we might want to compare treatments against a control or a placebo, or against the current standard treatment.

In the hand-washing experiment, we could consider plain water to be a control. Nobody would be impressed with (or want to pay for) a soap that did no better than water alone. A test of whether the antibacterial soap (for example) was different from plain water would be a simple test of the difference between two group means. To be able to perform an ANOVA, we first check the Similar Variance Condition. If things look OK, we assume that the variances are equal. If the variances *are* equal then a pooled t-test is appropriate. Even better (this is the special part), we already have a pooled estimate of the standard deviation based on *all* of the tested washing methods. That's s_p, which, for the hand-washing experiment, was equal to 37.55 bacteria colonies.

The null hypothesis is that there is no difference between water and the antibacterial soap. As we did in Chapter 20, we'll write that as a hypothesis about the difference in the means:

$$H_0\colon \mu_W - \mu_{ABS} = 0. \text{ The alternative is}$$

$$H_0\colon \mu_W - \mu_{ABS} \neq 0.$$

The natural test statistic is $\bar{y}_W - \bar{y}_{ABS}$, and the (pooled) standard error is

$$SE(\mu_W - \mu_{ABS}) = s_p\sqrt{\frac{1}{n_W} + \frac{1}{n_{ABS}}}.$$

Level	n	Mean	Std Dev
Alcohol Spray	8	37.5	25.56
Antibacterial Soap	8	92.5	41.96
Soap	8	106.0	46.96
Water	8	117.0	31.13

The difference in the observed means is $117.0 - 92.5 = 24.5$ colonies. The standard error comes out to 18.775. The t-statistic, then, is $t = \frac{24.5}{18.775} = 1.31$. To find the P-value we consult the Student's t-distribution on $N - k = 32 - 4 = 28$ degrees of freedom. The P-value is about 0.2—not small enough to impress us. So we can't discern a significant difference between washing with the antibacterial soap and just using water.

Our t-test asks about a simple difference. We could also ask a more complicated question about groups of differences. Does the average of the two soaps differ from the average of three sprays, for example? Complex combinations like these are called *contrasts*. Finding the standard errors for contrasts is straightforward but beyond the scope of this text. We'll restrict our attention to the common question of comparing pairs of treatments after H_0 has been rejected.

Bonferroni Multiple Comparisons

Our hand-washing experimenter *was* pretty sure that alcohol would kill the germs even before she started the experiment. But alcohol dries the skin and leaves an unpleasant smell. She was hoping that one of the antibacterial soaps would work as well as alcohol so she could use that instead. That means she really wanted to compare *each* of the other treatments against the alcohol spray. We know how to compare two of the means with a t-test. But now we want to do several tests, and each test poses the risk of a Type I error. As we do more and more tests, the risk that we might make a Type I error grows bigger than the α-level of each individual test. With each additional test, the risk of making an error grows. If we do enough tests, we're almost sure to reject one of the null hypotheses by mistake—and we'll never know which one.

There is a defense against this problem. In fact, there are several defenses. As a class, they are called **methods for multiple comparisons**. All multiple comparisons methods

Carlo Bonferroni (1892–1960) was a mathematician who taught in Florence. He wrote two papers in 1935 and 1936 setting forth the mathematics behind the method that bears his name.

require that we first be able to reject the overall null hypothesis with the ANOVA's *F*-test. Once we've rejected the overall null, then we can think about comparing several—or even all—pairs of group means.

Let's look again at our test of the water treatment against the antibacterial soap treatment. This time we'll look at a confidence interval instead of the pooled *t*-test. We did a test at significance level $\alpha = 0.05$. The corresponding confidence level is $1 - \alpha = 95\%$. For *any* pair of means, a confidence interval for their difference is $(\bar{y}_1 - \bar{y}_2) \pm ME$, where the margin of error is

$$ME = t^* \times s_p\sqrt{\frac{1}{n_1} + \frac{1}{n_2}}.$$

As we did in the previous section, we get s_p as the pooled standard deviation found from *all* the groups in our analysis. Because s_p uses the information about the standard deviation from *all* the groups it's a better estimate than we would get by combining the standard deviation of just two of the groups. This uses the **Equal Variance Assumption** and "borrows strength" in estimating the common standard deviation of all the groups. We find the critical value t^* from the Student's *t*-model corresponding to the specified confidence level found with $N - k$ degrees of freedom, and the n_k's are the number of experimental units in each of the treatments.

To reject the null hypothesis that the two group means are equal, the difference between them must be larger than the ME. That way 0 won't be in the confidence interval for the difference. When we use it in this way, we call the margin of error the **least significant difference** (**LSD** for short). If two group means differ by more than this amount, then they are significantly different at level α *for each individual test.*

For our hand-washing experiment, each group has $n = 8$, $s_p = 37.55$, and $df = 32 - 4 = 28$. From technology or Table T, we can find that t^* with 28 df (for a 95% confidence interval) is 2.048. So

$$LSD = 2.048 \times 37.55 \times \sqrt{\frac{1}{8} + \frac{1}{8}} = 38.45 \text{ colonies},$$

and we could use this margin of error to make a 95% confidence interval for any difference between group means. Any two washing methods whose means differ by more than 38.45 colonies could be said to differ at $\alpha = 0.05$ by this method.

Of course, we're still just examining individual pairs. If we want to examine *many* pairs simultaneously, there are several methods that adjust the critical t^*-value so that the resulting confidence intervals provide appropriate tests for all the pairs. And, in spite of making *many* such intervals, the overall Type I error rate stays at (or below) α.

One such method is called the **Bonferroni method**. This method adjusts the LSD to allow for making many comparisons. The result is a wider margin of error called the **minimum significant difference,** or **MSD**. The MSD is found by replacing t^* with a slightly larger number. That makes the confidence intervals wider for each contrast and the corresponding Type I error rates lower for *each* test. And it keeps the *overall* Type I error rate at or below α.

The Bonferroni method distributes the error rate equally among the confidence intervals. It divides the error rate among *J* confidence intervals, finding each interval at confidence level $1 - \frac{\alpha}{J}$ instead of the original $1 - \alpha$. To signal this adjustment, we label the critical value t^{**} rather than t^*. For example, to make the three confidence intervals comparing the alcohol spray with the other three washing methods, and preserve our overall α risk at 5%, we'd construct each with a confidence level of

$$1 - \frac{0.05}{3} = 1 - 0.01667 = 0.98333.$$

The only problem with this is that *t*-tables don't have a column for 98.33% confidence (or, correspondingly, for $\alpha = 0.01667$). Fortunately, technology has no such constraints. For the hand-washing data, if we want to examine the three confidence intervals comparing each of the other methods with the alcohol spray, the t^{**}-value (on 28 degrees of freedom) turns out to be 2.546. That's somewhat larger than the individual t^*-value of 2.048 that we would have used for a single confidence interval. And the corresponding ME is 47.69 colonies (rather than 38.45 for a single comparison). The larger critical value along with correspondingly wider intervals is the price we pay for making multiple comparisons.

Many statistics packages assume that you'd like to compare all pairs of means. Some will display the result of these comparisons in a table like this:

Level	n	Mean	Groups
Alcohol Spray	8	37.5	A
Antibacterial Soap	8	92.5	B
Soap	8	106.0	B
Water	8	117.0	B

This table shows that the alcohol spray is in a class by itself and that the other three hand-washing methods are indistinguishable from one another.

25.5 ANOVA on Observational Data

So far we've applied ANOVA only to data from designed experiments. That's natural for several reasons. The primary one is that, as we saw in Chapter 12, randomized comparative experiments are specifically designed to compare the results for different treatments. The overall null hypothesis, and the subsequent tests on pairs of treatments in ANOVA, address such comparisons directly. In addition, as we discussed earlier, the **Equal Variance Assumption** (which we need for all of the ANOVA analyses) is often plausible in a randomized experiment because the treatment groups start out with sample variances that all estimate the same underlying variance of the collection of experimental units.

Sometimes, though, we just can't perform an experiment. When ANOVA is used to test equality of group means from observational data, there's no *a priori* reason to think the group variances might be equal at all. Even if the null hypothesis of equal means were true, the groups might easily have different variances. But if the side-by-side boxplots of responses for each group show roughly equal spreads and symmetric, outlier-free distributions, you can use ANOVA on observational data.

Observational data tend to be messier than experimental data. They are much more likely to be unbalanced. If you aren't assigning subjects to treatment groups, it's harder to guarantee the same number of subjects in each group. And because you are not controlling conditions as you would in an experiment, things tend to be, well, less controlled. The only way we know to avoid the effects of possible lurking variables is with control and randomized assignment to treatment groups, and for observational data, we have neither.

ANOVA is often applied to observational data when an experiment would be impossible or unethical. (We can't randomly break some subjects' legs, but we *can* compare pain perception among those with broken legs, those with sprained ankles, and those with stubbed toes by collecting data on subjects who have already suffered those injuries.) In such data, subjects are already in groups, but not by random assignment.

Be careful; if you have not assigned subjects to treatments randomly, you can't draw *causal* conclusions even when the F-test is significant. You have no way to control for lurking variables or confounding, so you can't be sure whether any differences you see among groups are due to the grouping variable or to some other unobserved variable that may be related to the grouping variable.

Because observational studies often are intended to estimate parameters, there is a temptation to use pooled confidence intervals for the group means for this purpose. Although these confidence intervals are statistically correct, be sure to think carefully about the population that the inference is about. The relatively few subjects that happen to be in a group may not be a Simple Random Sample of any interesting population, so their "true" mean may have only limited meaning.

STEP-BY-STEP EXAMPLE One More Example

Here's an example that exhibits many of the features we've been discussing. It gives a fair idea of the kinds of challenges often raised by real data.

A study run as a student project at a liberal arts college attempted to find out whether men and women watch the same amount of TV, on average, and whether it matters if students are varsity athletes or not. Student researchers asked 200 randomly selected students questions about their backgrounds and about their television-viewing habits and received 197 legitimate responses. The researchers found that men watch, on average, about 2.5 hours per week more TV than women, and that varsity athletes watch about 3.5 hours per week more than those who are not varsity athletes. But is this the whole story? To investigate further, they divided the students into four groups: male athletes (MA), male non-athletes (MNA), female athletes (FA), and female non-athletes (FNA). (Data in **TV watching**)

QUESTION: Is the mean amount of time spent watching TV about the same for these four groups of students?

THINK

VARIABLES Name the variables, report the W's, and specify the questions of interest.

I have the number of hours spent watching TV in a week for 197 randomly selected students. We know their sex and whether they are varsity athletes or not. I wonder whether TV watching differs according to sex and athletic status.

PLOT Always start an ANOVA with side-by-side boxplots of the responses in each of the groups. Always.

These data offer a good example why.

Here are the side-by-side boxplots of the data:

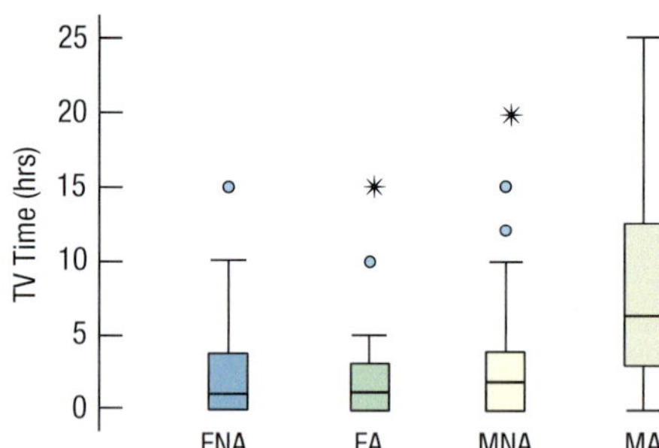

The responses are counts—numbers of TV hours. You may recall from Chapter 6 that a good re-expression to try first for counts is the square root.

This plot suggests problems with the data. Each box shows a distribution skewed to the high end, and outliers pepper the display, including some extreme outliers. The box with the highest center (MA) also has the largest spread. These data just don't pass our first screening for suitability. This sort of pattern calls for a re-expression.

Here are the boxplots for the square root of TV hours.

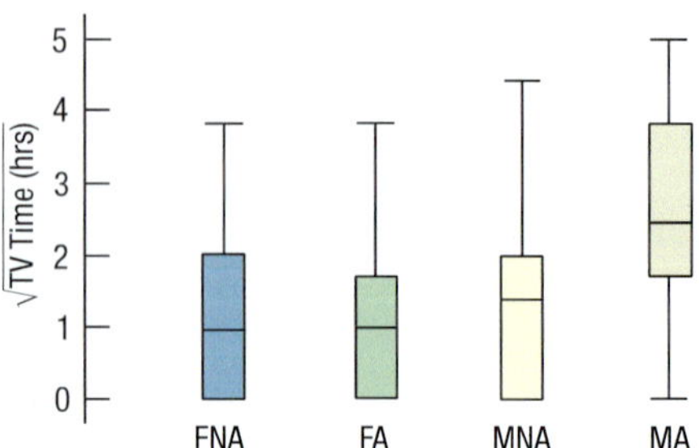

The spreads in the four groups are now more similar and the individual distributions more symmetric. And now there are no outliers.

Think about the assumptions and check the conditions.

✓ **Randomization Condition:** The data come from a random sample of students. The responses should be independent. It might be a good idea to see if the number of athletes and men are representative of the campus population.

✓ **Similar Spread Condition:** The boxplots show similar spreads. I may want to check the residuals later.

Fit the ANOVA model.

The ANOVA table looks like this:

Source	DF	Sum of Squares	Mean Square	F-Ratio	P-Value
Group	3	47.24733	15.7491	12.8111	<0.0001
Error	193	237.26114	1.2293		
Total	196	284.50847			

Nearly Normal Condition, Outlier Condition:
A histogram of the residuals looks reasonably Normal:

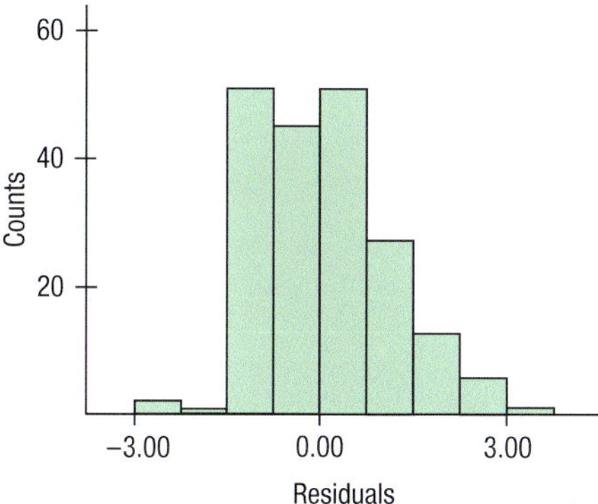

Interestingly, the few cases that seem to stick out on the low end are male athletes who watched no TV, making them different from all the other male athletes.

Under these conditions, it's appropriate to use Analysis of Variance.

TELL INTERPRETATION

The F-statistic is large and the corresponding P-value small. I conclude that the TV-watching behavior is not the same among these groups.

So Do Male Athletes Watch More TV?

Here's a Bonferroni comparison of all pairs of groups:

	Difference	Std. Err.	P-Value
FA–FNA	0.049	0.270	0.9999
MNA–FNA	0.205	0.182	0.8383
MNA–FA	0.156	0.268	0.9929
MA–FNA	1.497	0.250	<0.0001
MA–FA	1.449	0.318	<0.0001
MA–MNA	1.292	0.248	<0.0001

DIFFERING STANDARD ERRORS?

In case you were wondering . . . The standard errors are different because this isn't a balanced design. Differing numbers of experimental units in the groups generate differing standard errors.

Three of the differences are very significant. It seems that among women there's little difference in TV watching between varsity athletes and others. Among men, though, the corresponding difference is large. And among varsity athletes, men watch significantly more TV than women.

But wait. How far can we extend the inference that male athletes watch more TV than other groups? The data came from a random sample of students made during the week of March 21. If the students carried out the survey correctly using a Simple Random Sample, we should be able to make the inference that the generalization is true for the entire student body during that week.

Is it true for other colleges? Is it true throughout the year? The students conducting the survey followed up the survey by collecting anecdotal information about TV watching of male athletes. It turned out that during the week of the survey, the NCAA men's basketball tournament was televised. This could explain the increase in TV watching for the male athletes. It could be that the increase extends to other students at other times, but we don't know that. Always be cautious in drawing conclusions too broadly. Don't generalize from one population to another.

WHAT CAN GO WRONG?

- **Watch out for outliers.** One outlier in a group can change both the mean and the spread of that group. It will also inflate the Error Mean Square, which can influence the *F*-test. The good news is that ANOVA fails on the safe side by losing power when there are outliers. That is, you are less likely to reject the overall null hypothesis if you have (and leave) outliers in your data. But they are not likely to cause you to make a Type I error.
- **Watch out for changing variances.** The conclusions of the ANOVA depend crucially on the assumptions of independence and constant variance, and (somewhat less seriously as n increases) on Normality. If the conditions on the residuals are violated, it may be necessary to re-express the response variable to approximate these conditions more closely. ANOVA benefits so greatly from a judiciously chosen re-expression that the choice of a re-expression might be considered a standard part of the analysis.
- **Be wary of drawing conclusions about causality from observational studies.** ANOVA is often applied to data from randomized experiments for which causal conclusions are appropriate. If the data are not from a designed experiment, however, the Analysis of Variance provides no more evidence for causality than any other method we have studied. Don't get into the habit of assuming that ANOVA results have causal interpretations.
- **Be wary of generalizing** to situations other than the one at hand. Think hard about how the data were generated to understand the breadth of conclusions you are entitled to draw.

- **Watch for multiple comparisons.** When rejecting the null hypothesis, you can conclude that the means are not *all* equal. But you can't start comparing every pair of treatments in your study with a t-test. You'll run the risk of inflating your Type I error rate. Use a multiple comparisons method when you want to test many pairs.

CONNECTIONS

We first learned about side-by-side boxplots in Chapter 4. There we made general statements about the shape, center, and spread of each group. When we compared groups, we asked whether their centers looked different compared with how spread out the distributions were. Now we've made that kind of thinking precise. We've added confidence intervals for the difference and tests of whether the means are the same.

We pooled data to find a standard deviation when we tested the hypothesis of equal proportions. For that test, the assumption of equal variances was a consequence of the null hypothesis that the proportions were equal, so it didn't require an extra assumption. Means don't have a linkage with their corresponding variances, so to use pooled methods we must make the additional assumption of equal variances. In a randomized experiment, that's a plausible assumption.

Chapter 10 offered a variety of designs for randomized comparative experiments. Each of those designs can be analyzed with a variant or extension of the ANOVA methods discussed in this chapter. Entire books and courses deal with these extensions, but all follow the same fundamental ideas presented here.

ANOVA is closely related to regression analysis. (In fact, most statistics packages offer an ANOVA table as part of their regression output.) The assumptions are similar—and for good reason. The analyses are, in fact, related at a deep conceptual (and computational) level, as parts of a model for data called the General Linear Model, but those details are beyond the scope of this text.

The pooled two-sample t-test for means is a special case of the ANOVA F-test. If you perform an ANOVA comparing only two groups, you'll find that the P-value of the F-statistic is exactly the same as the P-value of the corresponding pooled t-statistic. That's because in this special case the F-statistic is just the square of the t-statistic. The F-test is more general. It can test the hypothesis that several group means are equal.

CHAPTER REVIEW

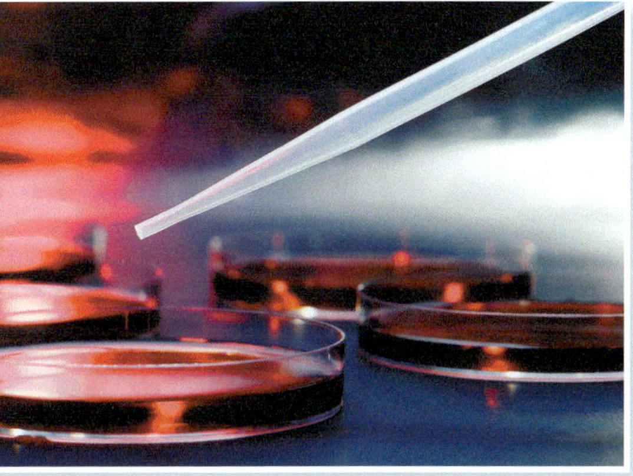

Recognize when to use an Analysis of Variance (ANOVA) to compare the means of several groups.

Know how to read an ANOVA table to locate the degrees of freedom, the Mean Squares, and the resulting F-statistic.

Know how to check the three conditions required for an ANOVA:

- Independence of the groups from each other and of the individual cases within each group.
- Equal variance of the groups.
- Normal error distribution.

Know how to create and interpret confidence intervals for the differences between each pair of group means, recognizing the need to adjust the confidence interval for the number of comparisons made.

REVIEW OF TERMS

Error (or Within) Mean Square (MS_E) The Error Mean Square (MS_E) is the estimate of the error variance obtained by *pooling* the variances of each treatment group. The square root of the (MS_E) is the estimate of the error standard deviation, s_p (p. 828).

Treatment (or Between) Mean Square (MS_T) The Treatment Mean Square (MS_T) is the estimate of the error variance under the assumption that the treatment means are all equal. If the (null) assumption is not true, the MS_T will be larger than the error variance (p. 828).

***F*-distribution** The *F*-distribution is the sampling distribution of the *F*-statistic when the null hypothesis that the treatment means are equal is true. It has two degrees of freedom parameters, one for the numerator, $(k - 1)$, and one for the denominator, $(N - k)$, where N is the total number of observations and k is the number of groups (p. 828).

***F*-statistic** The *F*-statistic is the ratio MS_T/MS_E. When the *F*-statistic is sufficiently large, we reject the null hypothesis that the group means are equal (p. 828).

***F*-test** The *F*-test tests the null hypothesis that all the group means are equal against the one-sided alternative that they are not all equal. We reject the hypothesis of equal means if the *F*-statistic exceeds the critical value from the *F*-distribution corresponding to the specified significance level and degrees of freedom (p. 828).

ANOVA An analysis method for testing equality of means across treatment groups (p. 828).

ANOVA table The ANOVA table is convenient for showing the degrees of freedom, the Treatment Mean Square, the Error Mean Square, their ratio, the *F*-statistic, and its P-value. There are usually other quantities of lesser interest included as well (p. 829).

One-way ANOVA model The model for a one-way (one response, one factor) ANOVA is

$$y_{ij} = \mu_j + \varepsilon_{ij}.$$

Estimating with $y_{ij} = \bar{y}_j + e_{ij}$ gives predicted values $\hat{y}_{ij} = \bar{y}_j$ and residuals $e_{ij} = y_{ij} - \bar{y}_j$ (p. 831).

Residual standard deviation The residual standard deviation,

$$s_p = \sqrt{MS_E} = \sqrt{\frac{\sum e^2}{N - k}},$$

gives an idea of the underlying variability of the response values (p. 835).

Balance An experiment's design is balanced if each treatment level has the same number of experimental units. Balanced designs make calculations simpler and are generally more powerful (p. 840).

Methods for multiple comparisons If we reject the null hypothesis of equal means, we often then want to investigate further and compare pairs of treatment group means to see if they differ. If we want to test several such pairs, we must adjust for performing several tests to keep the overall risk of a Type I error from growing too large. Such adjustments are called methods for multiple comparisons (p. 841).

Least significant difference (LSD) The standard margin of error in the confidence interval for the difference of two means is called the least significant difference. It has the correct Type I error rate for a single test, but not when performing more than one comparison (p. 842).

Bonferroni method One of many methods for adjusting the length of the margin of error when testing the differences between several group means (p. 842).

Minimum significant difference (MSD) The Bonferroni method's margin of error for the confidence interval for the difference of two group means is called the minimum significant difference. This can be used to test differences of several pairs of group means. If their difference exceeds the MSD, they are different at the overall rate (p. 842).

TECH SUPPORT

ANOVA

Most analyses of variance are found with computers. And all statistics packages present the results in an ANOVA table much like the one we discussed. Technology also makes it easy to examine the side-by-side boxplots and check the residuals for violations of the assumptions and conditions.

Statistics packages offer different choices among possible multiple comparisons methods (although Bonferroni is quite common). This is a specialized area. Get advice or read further if you need to choose a multiple comparisons method.

As we saw in Chapter 4, there are two ways to organize data recorded for several groups. We can put all the response values in a single variable and use a second, "factor," variable to hold the group identities. This is sometimes called *stacked format*. The alternative is to place the data for each group in its own column or variable. Then the variable identities become the group identifiers.

Most statistics packages expect the data to be in stacked format because this form also works for more complicated experimental designs. Some packages can work with either form, and some use one form for some things and the other for others. (Be careful, for example, when you make side-by-side boxplots; be sure to give the appropriate version of the command to correspond to the structure of your data.)

Most packages offer to save residuals and predicted values and make them available for further tests of conditions. In some packages you may have to request them specifically.

DATA DESK

- Select the response variable as Y and the factor variable as X.
- From the **Calc** menu, choose **ANOVA**.
- Data Desk displays the ANOVA table.
- Select plots of residuals from the ANOVA table's HyperView menu.

COMMENTS

Data Desk expects data in "stacked" format. The GLM (General Linear Model) command found in the toolbar offers a wide range of related analyses including ANOVA.

EXCEL

- In Excel 2003 and earlier, select **Data Analysis** from the Tools menu.
- In Excel 2007, select **Data Analysis** from the Analysis Group on the Data Tab.
- Select **ANOVA Single Factor** from the list of analysis tools.
- Click the **OK** button.
- Enter the data range in the box provided.
- Check the **Labels in First Row** box, if applicable.
- Enter an alpha level for the F-test in the box provided.
- Click the **OK** button.

COMMENTS

The data range should include two or more columns of data to compare. Unlike all other statistics packages, Excel expects each column of the data to represent a different level of the factor. However, it offers no way to label these levels. The columns need not have the same number of data values, but the selected cells must make up a rectangle large enough to hold the column with the most data values.

JMP

- From the **Analyze** menu select **Fit Y by X**.
- Select variables: a quantitative Y, Response variable, and a categorical X, Factor variable.
- JMP opens the **Oneway** window.
- Click on the red triangle beside the heading, select **Display Options**, and choose **Boxplots**.
- From the same menu choose the **Means/ANOVA.t-test** command.
- JMP opens the oneway ANOVA output.

COMMENTS

JMP expects data in "stacked" format with one response and one factor variable.

MINITAB

- Choose **ANOVA** from the Stat menu.
- Choose **One-way** . . . from the **ANOVA** submenu.
- In the One-way ANOVA dialog, assign a quantitative Y variable to the Response box and assign a categorical X variable to the Factor box.
- Check the **Store Residuals** check box.
- Click the **Graphs** button.
- In the ANOVA-Graphs dialog, select **Standardized residuals**, and check **Normal plot of residuals** and **Residuals versus fits**.
- Click the **OK** button to return to the ANOVA dialog.
- Click the **OK** button to compute the ANOVA.

COMMENTS

If your data are in unstacked format, with separate columns for each treatment level, choose **One-way (unstacked)** from the **ANOVA** submenu.

R

To perform an analysis of variance of a variable y on a categorical variable (factor) x:

- myaov = aov(y~x)
- summary(myaov) # gives the ANOVA table

To get confidence or prediction intervals use:

- predict(myaov,interval = "confidence")

or

- predict(myaov, interval = "prediction")

SPSS

- Choose **Compare Means** from the **Analyze** menu.
- Choose **One-way ANOVA** from the **Compare Means** submenu.
- In the One-Way ANOVA dialog, select the Y-variable and move it to the dependent target. Then move the X-variable to the independent target.
- Click the **OK** button.

COMMENTS

SPSS expects stacked format. The **Contrasts** and **Post Hoc** buttons offer ways to test contrasts and perform multiple comparisons. See your SPSS manual for details.

STATCRUNCH

To compute an ANOVA:

- Click on **Stat**.
- Choose **ANOVA** > **One Way**.
- Choose the **Columns** for all groups. (After the first one, you may hold down the ctrl or command key to choose more.)

OR

Choose the **single column** containing the data **(Responses)** and the column containing the **Factors**.

- Click on **Compute!**

EXERCISES

SECTION 25.1

1. Popcorn A student runs an experiment to test four different popcorn brands, recording the number of kernels left unpopped. She pops measured batches of each brand 4 times, using the same popcorn popper and randomizing the order of the brands. After collecting her data and analyzing the results, she reports that the F-ratio is 13.55.

a) What are the null and alternative hypotheses?
b) How many degrees of freedom does the Treatment Sum of Squares have? How about the Error Sum of Squares?
c) Assuming that the conditions required for ANOVA are satisfied, what is the P-value? What would you conclude?
d) What else about the data would you like to see in order to check the assumptions and conditions?

2. Skating A figure skater tried various approaches to her Salchow jump in a designed experiment using 3 different places for her focus (takeoff leg, free leg, and free). She tried each jump 7 times in random order, using two of her skating partners to judge the jumps on a scale from 0 to 6. After collecting the data and analyzing the results, she reports that the F-ratio is 6.46.

a) What are the null and alternative hypotheses?
b) How many degrees of freedom does the Treatment Sum of Squares have? How about the Error Sum of Squares?
c) Assuming that the conditions are satisfied, what is the P-value? What would you conclude?
d) What else about the data would you like to see in order to check the assumptions and conditions?

3. Gas mileage A student runs an experiment to study the effect of three different mufflers on gas mileage. He devises a system so that his car uses gasoline from a one-liter container. He tests each muffler 6 times, carefully recording the number of miles he can go in his car on one liter of gas. After analyzing his data, he reports that the F-ratio is 2.38 with a P-value of 0.0622.

a) What are the null and alternative hypotheses?
b) How many degrees of freedom does the Treatment Sum of Squares have? How about the Error Sum of Squares?
c) What would you conclude?
d) What else about the data would you like to see in order to check the assumptions and conditions?
e) If your conclusion in part c is wrong, what type of error have you made?

4. Darts A student interested in improving her dart-throwing technique designs an experiment to test 3 different stances to see whether they affect her accuracy. After warming up for several minutes, she randomizes the order of the 3 stances, throws a dart at a target using each stance, and measures the distance of the dart in centimeters from the center of the bull's-eye. She replicates this procedure 5 times. After analyzing the data, she reports that the F-ratio is 3.33.

a) What are the null and alternative hypotheses?
b) How many degrees of freedom does the Treatment Sum of Squares have? How about the Error Sum of Squares?
c) What would you conclude?
d) What else about the data would you like to see in order to check the assumptions and conditions?
e) If your conclusion in part c is wrong, what type of error have you made?

SECTION 25.2

T **5. Activating baking yeast** To shorten the time it takes him to make his favorite pizza, a student designed an experiment to test the effect of sugar and milk on the activation times for baking yeast. Specifically, he tested three different recipes and measured how many seconds it took for the same amount of dough to rise to the top of a bowl. He randomized the order of the recipes and replicated each treatment 4 times.

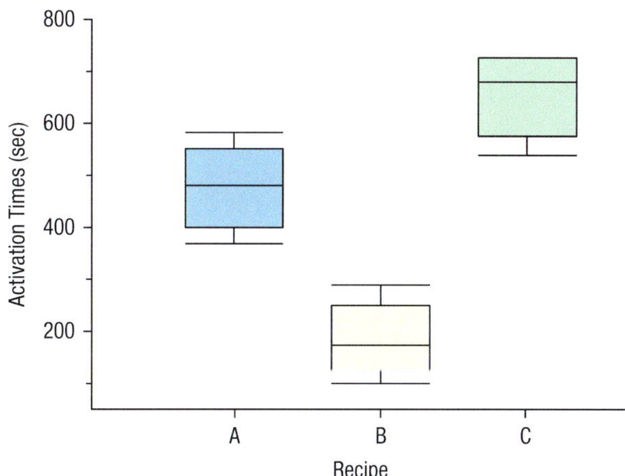

Analysis of variance

Source	DF	Sum of Squares	Mean Squares	F-Ratio	P-Value
Recipe	2	638915.29	319458	6.8316	0.0157
Error	9	420860.19	46762		
Total	11	1059775.48			

a) State the hypotheses about the recipes (both numerically and in words).
b) Assuming that the assumptions for inference are satisfied, perform the hypothesis test and state your conclusion. Be sure to state it in terms of activation times and recipes.
c) Would it be appropriate to follow up this study with multiple comparisons to see which recipes differ in their mean activation times? Explain.

T **6. Frisbee throws** A student performed an experiment with three different grips to see what effect it might have on the distance of a backhanded Frisbee throw. She tried it with her normal grip, with one finger out, and with the Frisbee inverted. She measured in paces how far her throws went. The boxplots and the ANOVA table for the three grips are shown below:

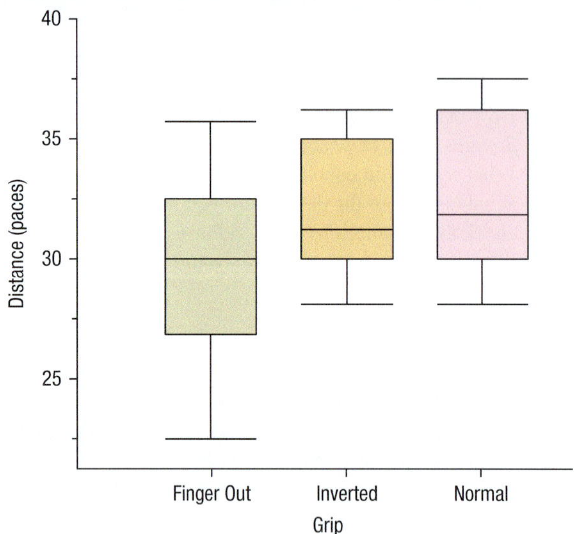

Analysis of variance

Source	DF	Sum of Squares	Mean Squares	F-Ratio	P-Value
Grip	2	58.04747	29.0237	2.0237	0.1560
Error	21	299.84323	14.2782		
Total	23	357.89070			

a) State the hypotheses about the grips.
b) Assuming that the assumptions for inference are satisfied, perform the hypothesis test and state your conclusion. Be sure to state it in terms of Frisbee grips and distance thrown.
c) Would it be appropriate to follow up this study with multiple comparisons to see which grips differ in their mean distance thrown? Explain.

SECTION 25.3

T **7. Cars** Here are boxplots that show the relationship between the number of cylinders a car's engine has and the car's fuel economy for a sample of cars.

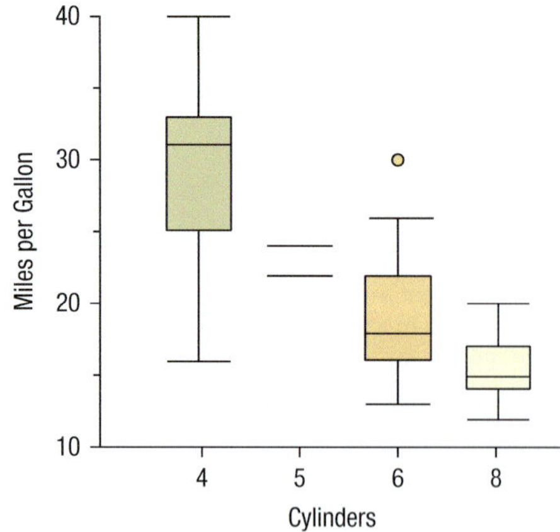

a) State the null and alternative hypotheses that you might consider for these data.
b) Do the conditions for an ANOVA seem to be met here? Why or why not?

T **8. Wine production** Here are case prices (in dollars) of wines produced by wineries in three regions of a state.

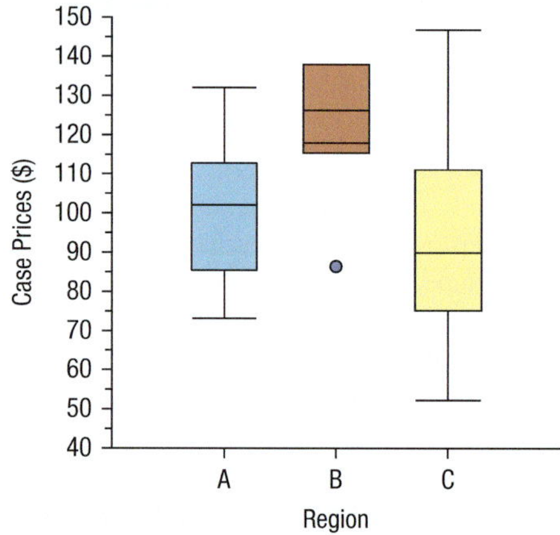

a) What null and alternative hypotheses would you test for these data? Talk about prices and location, not symbols.
b) Do the conditions for an ANOVA seem to be met here? Why or why not?

SECTION 25.4

T **9. Tellers** A bank is studying the time that it takes 6 of its tellers to serve an average customer. Customers line up in the queue and then go to the next available teller. Here is a boxplot of the last 140 customers and the times it took each teller:

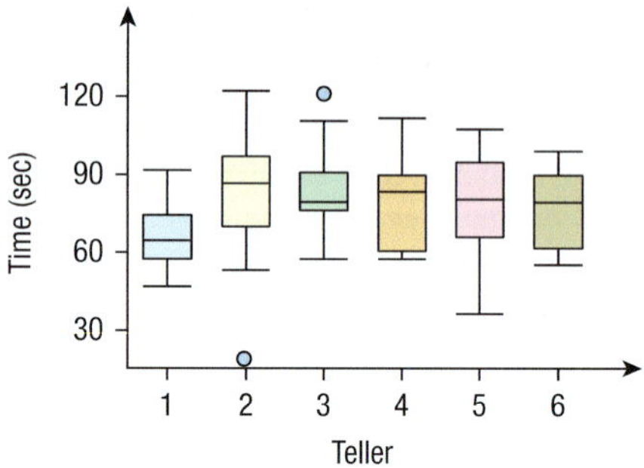

Source	DF	Sum of Squares	Mean Squares	F-Ratio	P-Value
Teller	5	7560.31	1512.061	3.439	0.0059
Error	134	58919.1	439.695		
Total	139	66479.4			

a) What are the null and alternative hypotheses?
b) What do you conclude?
c) Would it be appropriate to run a multiple comparisons test (for example, a Bonferroni test) to see which tellers differ from each other? Explain.

T **10. Hearing 4 lists** A researcher investigated four different word lists for use in hearing assessment. She wanted to know whether the lists were equally difficult to understand with a noisy background. To find out, she tested 96 subjects with normal hearing, randomly assigning 24 to each of the four word lists and measured the number of words perceived correctly in the presence of background noise. Here are the boxplots of the four lists:

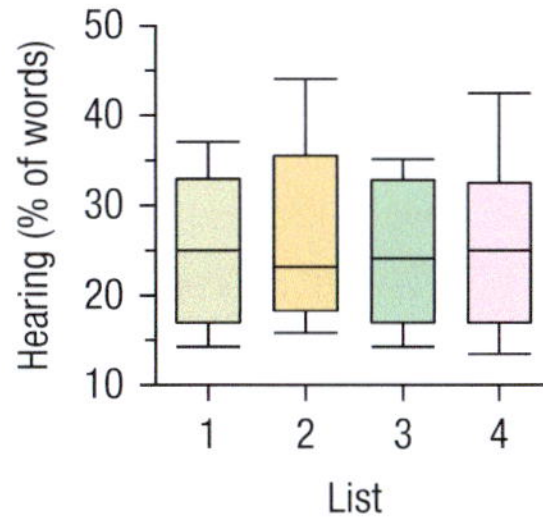

Source	DF	Sum of Squares	Mean Squares	F-Ratio	P-Value
List	3	374.523	124.841	1.7902	0.1545
Error	92	6415.620	69.735		
Total	95	6790.143			

a) What are the null and alternative hypotheses?
b) What do you conclude?
c) Would it be appropriate to run a multiple comparisons test (for example, a Bonferroni test) to see which lists differ from each other in terms of mean percent correct? Explain.

SECTION 25.5

T **11. Eye and hair color** A survey of 1744 school-age children was conducted by randomly selecting children from several elementary schools. Two of the questions concerned eye and hair color. In the survey, the following codes were used:

Hair Color	Eye Color
1 = Blond	1 = Grey
2 = Black	2 = Brown
3 = Brown	3 = Green
4 = Red	4 = Blue
5 = Other	5 = Other

The students analyzing the data were asked to study the relationship between eye and hair color. They produced this plot:

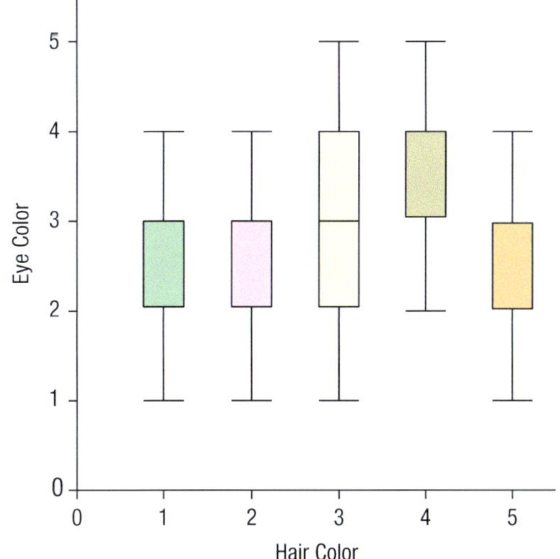

They then ran an Analysis of Variance with *Eye Color* as the response and *Hair Color* as the factor. The ANOVA table they produced follows:

Source	DF	Sum of Squares	Mean Squares	F-Ratio	P-Value
Hair Color	4	2.08399	0.520998	1.3529	0.2481
Error	1739	669.68662	0.385099		
Total	1743	671.770610			

What suggestions do you have for the statistics students? What alternative analysis might you suggest?

12. ZIP codes An intern from the marketing department at an online piercing salon has recently finished a study of the company's 400 customers. He wanted to know whether the mean ZIP code of customers purchasing different products varied according to the last product they bought. They have 13 different products, and the ANOVA table of ZIP code by product showed the following:

Source	DF	Sum of Squares	Mean Squares	F-Ratio	P-Value
Product	12	3.234e10	2.69500e9	4.1852	<0.0001
Error	378	2.43408e11	6.43936508		
Total	390	2.7578e11			

(Nine customers were not included because of missing ZIP code or product information.) What criticisms of the analysis might you make? What alternative analysis might you suggest?

CHAPTER EXERCISES

13. Yogurt An experiment to determine the effect of several methods of preparing cultures for use in commercial yogurt was conducted by a food science research group. Three batches of yogurt were prepared using each of three methods: traditional, ultrafiltration, and reverse osmosis. A trained expert then tasted each of the 9 samples, presented in random order, and judged them on a scale from 1 to 10. A partially completed Analysis of Variance table of the data follows.

Source	Sum of Squares	Degrees of Freedom	Mean Squares	F-Ratio
Treatment	20.507			
Residual	0.073			
Total	20.580			

a) Calculate the mean square of the treatments and the mean square of the error.
b) Form the F-statistic by dividing the two mean squares.
c) The P-value of this F-statistic turns out to be <0.00001. What does this say about the null hypothesis of equal means?
d) What assumptions have you made in order to answer part c?
e) What would you like to see in order to justify the conclusions of the F-test?
f) What is the average size of the error standard deviation in the judge's assessment?

14. Smokestack scrubbers Particulate matter is a serious form of air pollution often arising from industrial production. One way to reduce the pollution is to put a filter, or scrubber, at the end of the smokestack to trap the particulates. An experiment to determine which smokestack scrubber design is best was run by placing four scrubbers of different designs on an industrial stack in random order. Each scrubber was tested 5 times. For each run, the same material was produced, and the particulate emissions coming out of the scrubber were measured (in parts per billion). A partially completed Analysis of Variance table of the data looks like this:

Source	Sum of Squares	Degrees of Freedom	Mean Squares	F-Ratio
Treatment	88.8			
Residual	30.3			
Total	119.1			

a) Calculate the mean square of the treatments and the mean square of the error.
b) Form the F-statistic by dividing the two mean squares.
c) The P-value of this F-statistic turns out to be 0.0000515. What does this say about the null hypothesis of equal means?
d) What assumptions have you made in order to answer part c?
e) What would you like to see in order to justify the conclusions of the F-test?
f) What is the average size of the error standard deviation in particulate emissions?

15. Eggs A student wants to investigate the effects of real vs. substitute eggs on his favorite brownie recipe. He enlists the help of 10 friends and asks them to rank each of 8 batches on a scale from 1 to 10. Four of the batches were made with real eggs, four with substitute eggs. The judges tasted the brownies in random order. Here is a boxplot of the data. The mean score for the real eggs was 6.77 with a standard deviation of 0.651. The mean score for the substitute eggs was 4.73 with a standard deviation of 0.395.

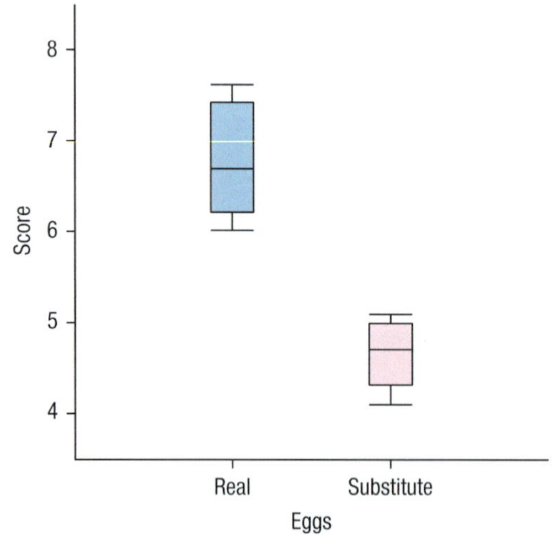

ANOVA (Analysis of Variance)

Source	DF	Sum of Squares	Mean Squares	F-Ratio	P-Value
Eggs	1	8.311680	8.31168	28.7095	0.0017
Error	6	1.737068	0.28951		
Total	7	10.048748			

a) What are the null and alternative hypotheses?
b) What do you conclude from the ANOVA table?
c) Do the assumptions for the test seem to be reasonable?
d) Perform a two-sample pooled t-test of the difference. What P-value do you get? Show that the square of the t-statistic is the same (to rounding error) as the F-ratio.

16. Auto noise filters In a statement to a Senate Public Works Committee, a senior executive of Texaco, Inc., cited a study on the effectiveness of auto filters on reducing noise. Because of concerns about performance, two types of filters were studied, a standard silencer and a new device developed by the Associated Octel Company. Here are the boxplots from the data on noise reduction (in decibels) of the two filters. Type 1 = standard; Type 2 = Octel.

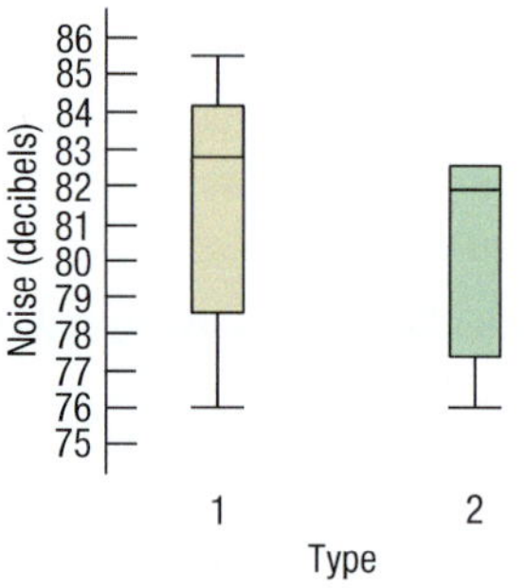

ANOVA Table

Source	DF	Sum of Squares	Mean Square	F-Ratio	P-Value
Type	1	6.31	6.31	0.7673	0.3874
Error	33	271.47	8.23		
Total	34	722.78			

Means and Std Deviations

Level	n	Mean	StdDev
Standard	18	81.5556	3.2166
Octel	17	80.7059	2.43708

a) What are the null and alternative hypotheses?
b) What do you conclude from the ANOVA table?
c) Do the assumptions for the test seem to be reasonable?
d) Perform a two-sample pooled t-test of the difference. What P-value do you get? Show that the square of the t-statistic is the same (to rounding error) as the F-ratio.

17. School system A school district superintendent wants to test a new method of teaching arithmetic in the fourth grade at his 10 schools. He plans to select 8 students from each school to take part in the experiment, but to make sure they are roughly of the same ability, he first gives a test to all 80 students. Here are the scores of the test by school:

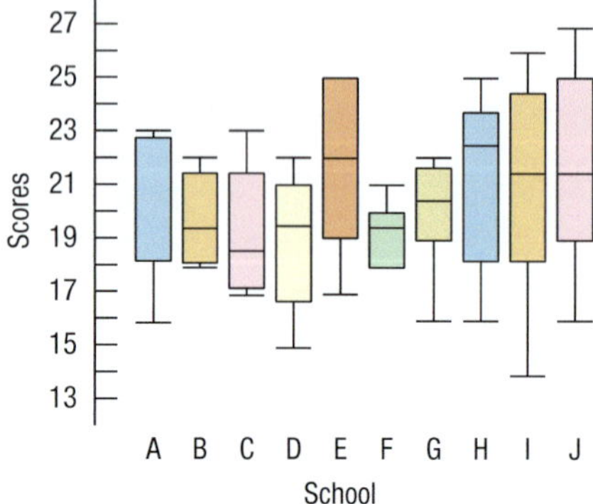

Source	DF	Sum of Squares	Mean Squares	F-Ratio	P-Value
School	9	82.236	9.137	0.962	0.4725
Error	70	664.926	9.499		
Total	79	747.16			

a) What are the null and alternative hypotheses?
b) What does the ANOVA table say about the null hypothesis? (Be sure to report this in terms of scores and schools.)
c) An intern reports that he has done t-tests of every school against every other school and finds that several of the schools seem to differ in mean score. Does this match your finding in part b? Give an explanation for the difference, if any, of the two results.

18. Fertilizers A biology student is studying the effect of 10 different fertilizers on the growth of mung bean sprouts. She sprouts 12 beans in each of 10 different petri dishes, and adds the same amount of fertilizer to each dish. After one week she measures the heights of the 120 sprouts in millimeters. Here are boxplots and an ANOVA table of the data:

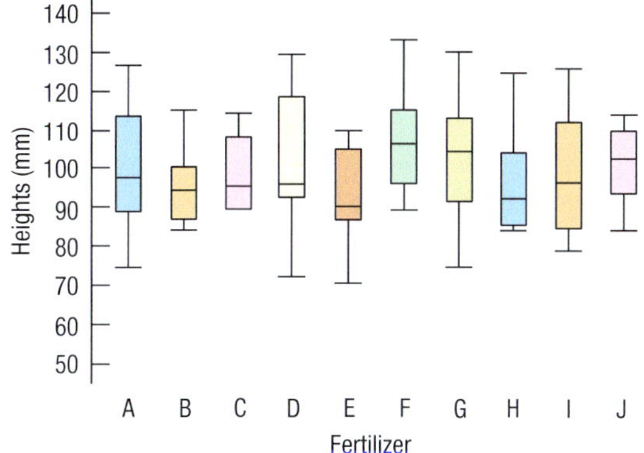

Source	DF	Sum of Squares	Mean Square	F-Ratio	P-Value
Fertilizer	9	2073.708	230.412	1.1882	0.3097
Error	110	21331.083	193.919		
Total	119	23404.791			

a) What are the null and alternative hypotheses?
b) What does the ANOVA table say about the null hypothesis? (Be sure to report this in terms of heights and fertilizers).
c) Her lab partner looks at the same data and says that he did t-tests of every fertilizer against every other fertilizer and finds that several of the fertilizers seem to have significantly higher mean heights. Does this match your finding in part b? Give an explanation for the difference, if any, between the two results.

19. Cereals Supermarkets often place similar types of cereal on the same supermarket shelf. We have data on the shelf as well as the sugar, sodium, and calorie content of 77 cereals. Does sugar content vary by shelf? At the top of the next column is a boxplot and an ANOVA table for the 77 cereals.

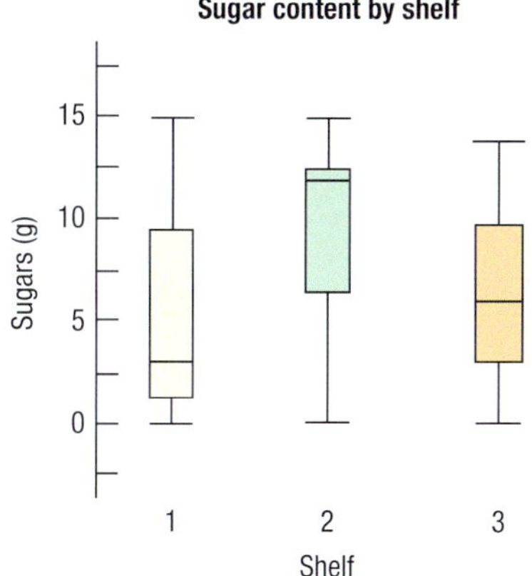

Source	DF	Sum of Squares	Mean Squares	F-Ratio	P-Value
Shelf	2	250.7721	125.386	6.6240	0.0023
Error	74	1400.7369	18.929		
Total	76	1651.5090			

Level	n	Mean	StdDev
1	20	4.81000	4.22727
2	21	9.55308	4.08734
3	36	6.20987	4.55783

a) What are the null and alternative hypotheses?
b) What does the ANOVA table say about the null hypothesis? (Be sure to report this in terms of *Sugars* and *Shelves*.)
c) Can we conclude that cereals on shelf 2 have a higher mean sugar content than cereals on shelf 3? Can we conclude that cereals on shelf 2 have a higher mean sugar content than cereals on shelf 1? What *can* we conclude?
d) To check for significant differences between the shelf means, we can use a Bonferroni test, whose results are shown below. For each pair of shelves, the difference is shown along with its standard error and significance level. What does it say about the questions in part c?

Dependant Variable: SUGARS

	(I) SHELF	(J) SHELF	Mean Difference (I-J)	Std. Error	P-Value	95% Confidence Interval	
Bonferroni						Lower Bound	Upper Bound
	1	2	−4.743	1.359	0.001	−8.073	−1.413
		3	−1.400	1.213	0.254	−4.372	1.572
	2	1	4.743	1.359	0.001	1.413	8.073
		3	3.343	1.195	0.007	0.416	6.270
	3	1	1.400	1.213	0.254	−1.572	4.372
		2	−3.343	1.195	0.007	−6.270	−0.416

T 20. Cereals protein by shelf We also have data on the protein content of the cereals in Exercise 19 by their shelf number. Here are the boxplot and ANOVA table:

Protein content by shelf

Source	DF	Sum of Squares	Mean Squares	F-Ratio	P-Value
Shelf	2	21.4719	10.736	9.0829	0.0003
Error	74	87.4437	1.182		
Total	76	108.9156			

Level	*n*	Mean	StdDev
1	20	2.74000	1.47751
2	21	1.83746	1.16045
3	36	3.10527	0.73743

a) What are the null and alternative hypotheses?
b) What does the ANOVA table say about the null hypothesis? (Be sure to report this in terms of protein content and shelves.)
c) Can we conclude that cereals on shelf 2 have a lower mean protein content than cereals on shelf 3? Can we conclude that cereals on shelf 2 have a lower mean protein content than cereals on shelf 1? What *can* we conclude?
d) To check for significant differences between the shelf means, we can use a Bonferroni test, whose results are shown below. For each pair of shelves, the difference is shown along with its standard error and significance level. What does it say about the questions in part c?

	(I) SHELF	(J) SHELF	Mean Difference (I–J)	Std. Error	P-Value	95% Confidence Level Lower Bound	Upper Bound
Bonferroni	1	2	0.90	0.340	0.011	0.07	1.73
		3	−0.37	0.303	0.233	−1.11	0.37
	2	1	−0.90	0.340	0.011	−1.73	−0.07
		3	−1.27	0.298	<0.011	−2.00	−0.54
	3	1	0.37	0.303	0.233	−0.37	1.11
		2	1.27	0.298	<0.001	0.54	2.00

T 21. Downloading To see how much of a difference time of day made on the speed at which he could download files, a college sophomore performed an experiment. He placed a file on a remote server and then proceeded to download it at three different time periods of the day. He downloaded the file 18 times in all, 6 times at each *Time of Day*, and recorded the *Time* in seconds that the download took.

Early (7 a.m)	Evening (5 p.m)	Late night (12 a.m)
71 sec	198 sec	210 sec
134 sec	309 sec	180 sec
88 sec	250 sec	177 sec
207 sec	315 sec	221 sec
120 sec	232 sec	190 sec
81 sec	208 sec	169 sec

a) State the null and alternative hypotheses, being careful to talk about download *Time* and *Time of Day* as well as parameters.
b) Perform an ANOVA on these data. What can you conclude?
c) Check the assumptions and conditions for an ANOVA. Do you have any concerns about the experimental design or the analysis?
d) (Optional) Perform a multiple comparisons test to determine which times of day differ in terms of mean download time.

T 22. Analgesics A pharmaceutical company tested three formulations of a pain relief medicine for migraine headache sufferers. For the experiment, 27 volunteers were selected and 9 were randomly assigned to one of three drug formulations. The subjects were instructed to take the drug during their next migraine headache episode and to report their pain on a scale of 1 = no pain to 10 = extreme pain 30 minutes after taking the drug.

Drug	Pain	Drug	Pain	Drug	Pain
A	4	B	6	C	6
A	5	B	8	C	7
A	4	B	4	C	6
A	3	B	5	C	6
A	2	B	4	C	7
A	4	B	6	C	5
A	3	B	5	C	6
A	4	B	8	C	5
A	4	B	6	C	5

a) State the null and alternative hypotheses, being careful to talk about *Drug* and *Pain* levels as well as parameters.
b) Perform an ANOVA on these data. What can you conclude?
c) Check the assumptions and conditions for an ANOVA. Do you have any concerns about the experimental design or the analysis?
d) (Optional) Perform a multiple comparisons test to determine which drugs differ in terms of mean pain level reported.

JUST CHECKING

Answers

1. The null hypothesis is that the mean flight distance for all four designs is the same.
2. Yes, it looks as if the variation between the means is greater than the variation within each boxplot.
3. Yes, the *F*-test rejects the null hypothesis with a P-value <0.0001.
4. No. The alternative hypothesis is that *at least* one mean is different from the other three. Rejecting the null hypothesis does not imply that all four means are different.

26

Multifactor Analysis of Variance

WHO	Attempts at hitting a bull's-eye in a dartboard
WHAT	Distance from the center of the target
UNITS	Inches
HOW	Two-factor experiment, randomized order of runs

How accurately can you throw a dart? It probably depends on which hand you use and on how far from the target you're standing. A student designed an experiment to test the effects of both factors. He used three levels of the factor *Distance: Near, Middle*, and *Far*—and the two natural levels *Left* and *Right* for *Hand.* (He was right-handed.) In random order, he threw six darts at a dartboard under each of the six treatment conditions, and measured the distance each dart landed from the bull's-eye in inches.[1] (Data in **Darts**)

It makes sense that dart throws would be more accurate from nearer distances, especially when using his regular throwing hand. Side-by-side boxplots of the distance from the bull's-eye by each factor seem to support our intuition:

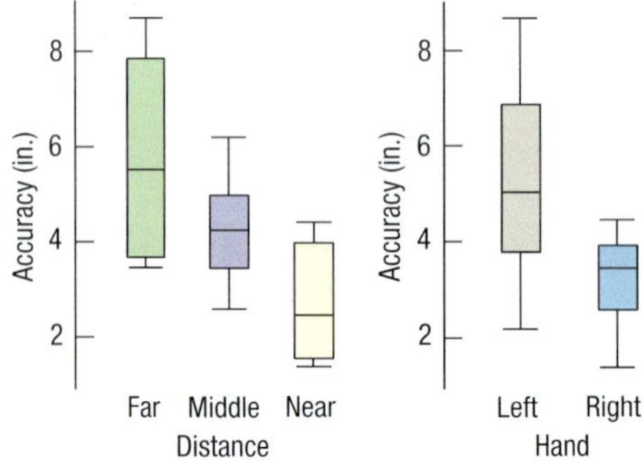

Figure 26.1
Boxplots of *Accuracy* (distance in inches from the bull's-eye) plotted for each experimental factor, *Distance* and *Hand*.

[1]He called the response variable *Accuracy,* although it measures inches *away* from the target. That means that bigger values correspond to *less* accuracy, not more.

26.1 A Two-Factor ANOVA Model

It appears that both the distance and the hand he used affected the accuracy of his throw, and by about the same amount. How can we measure the effect of each variable? And how can we test whether the observed differences are larger than we'd expect just from sampling variation?

Two Factors at Once?!

Unlike the experiments we've seen so far, this experiment has not one but two factors. You might expect this to confuse things, but the surprising fact is that it actually improves the experiment and its analysis. It also lets us see something we'd never be able to see if we ran two different one-factor experiments.

True, with two factors, we have two hypothesis tests. Each of those hypotheses asks whether the mean of the response variable is the same for each of the treatment levels. In comparing means, we know that any excess variation just gets in the way. Back in Chapter 11, we considered ways to remove or avoid extra variation in designing experiments.

A two-factor experiment does just that. If both factors do in fact have an effect on the response (as we'd expect hand choice and distance to do for dart accuracy), then removing the effects of one factor should make it easier to see and assess the effects of the other. We'll have removed variability due to an identifiable cause.

EXAMPLE 26.1

Students set up an experiment to investigate factors that determine how far a paper airplane will fly. Among the factors are whether the plane is constructed of copier paper or heavier construction paper, whether it is a "dart" or a glider, and whether the ends of the wings are folded up into "flaps." We'll combine the paper and shape factor into a single *Design* factor with four levels; *Dart/Copier, Dart/Construction, Glider/Copier,* and *Glider/Construction.* The second factor, *Flaps*, has two levels: *Flaps* and *None*. (Data in **Paper airplanes**)

Four planes were constructed of each of the four designs. For each design, two planes were given flaps and two were not. Each plane was flown 4 times and the distance of the flight carefully measured.

QUESTION: What are the details of the experiment design?

We have two factors: *Design* and *Flaps*. *Design* has 4 levels, Dart/Copier, Dart/Construction, Glider/Copier and Glider/Construction. *Flaps* has 2: Flaps and None. There are thus $4 \times 2 = 8$ treatment groups. Each treatment group has 2 planes and each plane was flown 4 times for a total of 8 observations. The response variable is the distance flown.

An ANOVA Model

The model for one-way ANOVA represented each observation as the sum of simple components. It broke each observation into the sum of three effects: the grand mean, the treatment effect, and an error:

$$y_{ij} = \mu + \tau_j + \varepsilon_{ij}.$$

Now we have two factors, *Hand* and *Distance*. So each treatment is a combination of a *Hand* assignment and a *Distance*. For example, throwing with the left hand from a near distance would be one treatment. Our model should reflect the effects of both factors.

We've already used τ for the treatment effects for the first factor, so for the second factor, we'll denote the effects by γ. Now, for each observation, we write:

$$y_{ijk} = \mu + \tau_j + \gamma_k + \varepsilon_{ijk}.$$

Here we see the *i*th observation at level *j* of factor A and level *k* of factor B. Why the *i* subscript? We're told that the student recorded six trials at each treatment, so the subscript *i* denotes the different observations at each combination of the two factors. The error has three subscripts because we associate a possibly different error with each observation. (If there were only one observation at each treatment, we wouldn't need the extra subscript.)

When you write out a model, you'll usually find it clearer to name the factors rather than using Greek letters. So the model will look like this:

$$y_{ijk} = \mu + \textit{Hand effect}_j + \textit{Distance effect}_k + \textit{Error}_{ijk}.$$

To estimate the effect of each treatment level for each factor, we take the *difference* between the mean response of that treatment and the grand mean, $\bar{\bar{y}}$. So, to find the effect of throwing with the left hand, we'd find the mean accuracy of *all* runs using the left hand and subtract the grand average from that.

In general, we'd like to know whether the mean dart accuracy changes with changes in the levels of either factor. The model helps us to think about this question. Our null hypothesis on each factor is that the effects of that treatment are all zero. So we can write two null hypotheses we'd like to test as

$$\text{H}_0\text{: } \tau_1 = \tau_2 \quad \text{and} \quad \text{H}_0\text{: } \gamma_1 = \gamma_2 = \gamma_3,$$

where the τ's (taus) denote the effects due to the two levels of *Hand* and the γ's (gammas) denote the effects due to the three levels of *Distance*.

The alternative hypotheses are that the treatment effects are not all equal. The details of the analysis are pretty much the same as for a one-factor ANOVA. We want to compare the differences across the treatment effects with the underlying variability within the treatment. But now with two factors, that underlying variability has the effect of both factors removed from it. So the error term holds the variability that's left after removing the effects of both factors. Here's the ANOVA table for the dart experiment:

Analysis of Variance for Accuracy

Source	DF	Sum of Squares	Mean Square	F-Ratio	Prob
Distance	2	51.0439	25.5219	28.561	<0.0001
Hand	1	39.6900	39.6900	44.416	<0.0001
Error	32	28.5950	0.893594		
Total	35	119.329			

This analysis splits up the total variation into three sources: the variation due to changes in *Distance*, the variation due to changes in *Hand*, and the error variation. The square root of the Error Mean Square, $\sqrt{0.894} = 0.946$ inches, estimates the standard deviation we'd expect if we made repeated observations at any treatment condition. Both F-ratios are large. Looking back at the boxplots, we shouldn't be surprised. It was clear that there were differences in accuracy among *near, middle,* and *far* even without accounting for the fact that each level contained both hands. The F-ratio for *Distance* is 28.56. The probability of an F-ratio that big occurring by chance is very small (<0.0001). So we can reject the hypothesis that the mean effects of *near, middle,* and *far* distances are equal.

The means of *left* and *right* don't look the same either. The F-ratio of 44.42 with a P-value of <0.0001 leads us to reject the null hypothesis about *Hand* effects as well.

Of course, there are assumptions and conditions to check. The good news is that most are the same as for one-way ANOVA.

26.2 Assumptions and Conditions

Even before checking any assumptions, you should have already plotted the data. Boxplots give a first look that draws our attention to the aspects tested by the ANOVA.

Plot the Data . . .

You could start by examining side-by-side boxplots of the data across levels of *each factor* as we did earlier, looking for outliers within any of the groups and correcting them if they are errors, or consider omitting them if we can't correct them.[2] (There don't seem to be any.) We'd like to get an idea of whether the groups have similar spreads (as we'll want) and whether the centers seem to be alike (as the null hypotheses claim) or different.

The problem of just looking at the boxplots is that the responses at each level of one factor contain *all levels of the other factor*. For example, the responses for the *left* hand were measured at all three distances. As a result, the boxplots show more variability than they need to. Our model will deal with both factors, so it will take care of this extra variability but the simple side-by-side boxplots can't do that.

On the other hand, if you do see a difference here, that's a difference *in spite of this extra variation*, so it's very likely that the F-test will reject the null hypothesis for that factor. Just be careful not to rule out the possibility of differences that aren't apparent.

A better alternative might be to make boxplots for each factor level *after removing the effects of the other factor.* How would you do that? You could compute a one-way ANOVA on one factor and find the residuals. Then make boxplots of those residuals for each level of the other factor. We might call such a display a **partial boxplot**.[3] By removing the variability due to the other factor, you should see differences in the responses among the levels of the factor you are graphing more readily. For example, here are the boxplots of dart toss *Accuracy* plotted for the two levels of *Hand* both in the original data and after removing the effects of *Distance*:

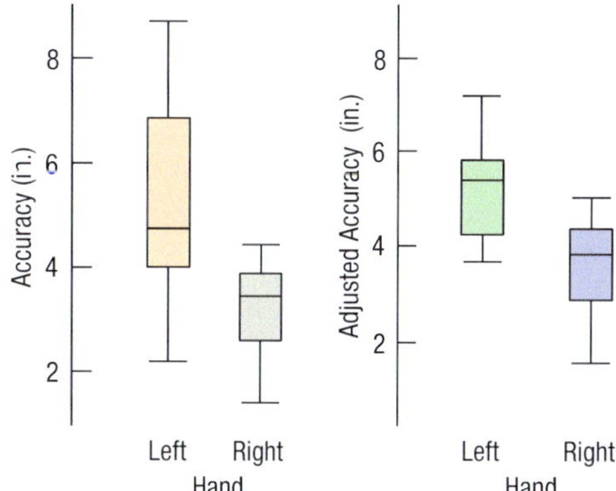

Figure 26.2
Boxplots of *Accuracy* by *Hand* (as in Figure 26.1) show the effect of changing *Hand* less clearly than the corresponding partial boxplots on the right, which show the effects of *Hand* after the effect of *Distance* has been removed. The effect of changing hands is much easier to see without the unwanted variation caused by changing the other factor, *Distance*.

Removing the variation due to *Distance* makes the differences between the two *Hands* in the partial boxplots much clearer. We'll want to do this for each factor, removing the effects of the other factor by computing the one-way ANOVA and finding the residuals.

[2]Unlike regression coefficients, which can be changed drastically by outliers, ANOVA tests tend to fail on the "safe side." Outliers generally inflate the Error Mean Square, which reduces the F-statistics and increases the P-values. That makes it harder to reject the null hypotheses, so we are more likely to make Type II errors. That's generally thought to be safer than making Type I errors. For example, we are not likely to be mistakenly led to approve a drug that isn't really effective. But it also means that we might fail to approve a drug that really is effective. Generally, it is best to remove outliers and report them separately.

[3]This is not a standard term. You are not likely to find it in a statistics package. But it is a natural term because we've removed *part* of the variability.

Here are the side-by-side and the partial boxplots for *Distance* removing the variability due to *Hand:*

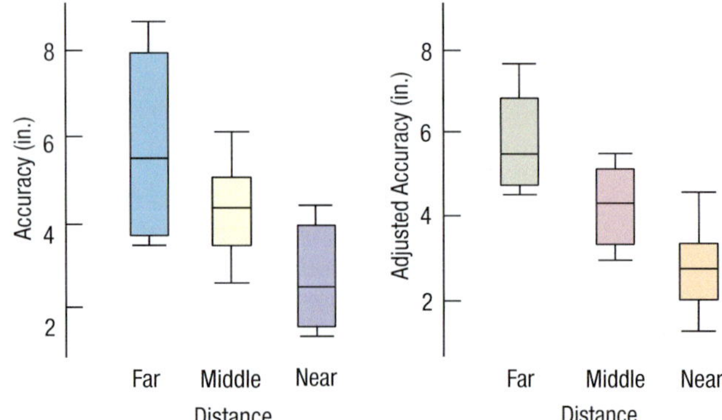

Figure 26.3
Partial boxplots after removing the effect of *Hand* (on the right) show the differences in *Accuracy* among the three distances more clearly than the original boxplots of *Accuracy* by *Distance* (on the left).

We'll use the partial boxplots for diagnostic purposes as we used the side-by-side boxplots in a one-way analysis. If the partial boxplots were all skewed in the same direction, or if their spreads changed systematically from level to level across a factor, we would have considered re-expressing the response variable to make them more symmetric. Re-expression would be likely to make the analysis more powerful and more correct.

Additivity

Each observation is assigned a treatment that combines two factor levels, one from each factor. Our model assumes that we can just *add* the effects of these two factor levels together:

$$y_{ijk} = \mu + \textit{Hand effect}_j + \textit{Distance effect}_k + \textit{Error}_{ijk}.$$

That's a nice simple model, but it may not be a good one for our data. Just like linearity in regression, **additivity** is an assumption. Like all assumptions, it helps us by making the model simpler, but it may be wrong for our data. Of course, we can't know for sure, but we can check the **Additive Enough Condition.** You won't be surprised that we check it with displays.

For the effects of *Hand* and *Distance* to be additive, changing hands must make the *same* difference in accuracy no matter what distance you throw from (or *vice versa*). Is this reasonable for this experiment? Alternatively, we could ask whether moving away from the target makes a greater difference to accuracy when throwing with the weaker hand. That's a *conditional* question, so we look at how *Accuracy* changes by *Distance* for each *Hand* separately.

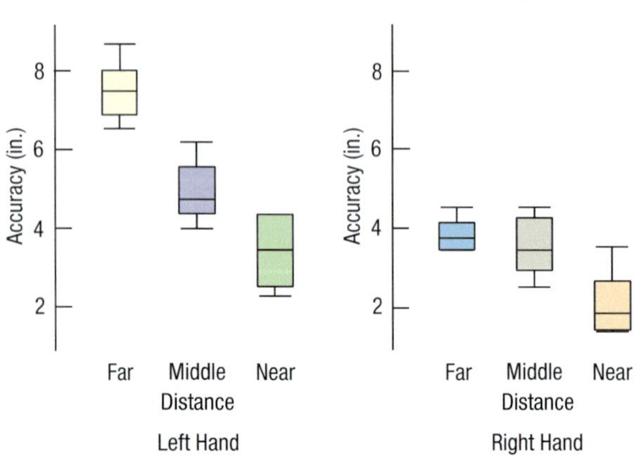

Figure 26.4
Boxplots of *Accuracy* by *Distance* conditional on each *Hand*. Changing the *Distance* seems to have a greater effect on left *Hand* accuracies.

The changes in accuracy due to *Distance* do look larger for the *left* hand, especially at the *far* distance.

When the effects of one factor *change* for different levels of another factor, we say that there's an **interaction**. To show the interaction, we often plot the treatment means in a single display, called an **interaction plot**. This plot shows the average of the observations at *each* level of one factor broken up by the levels of the other factor.

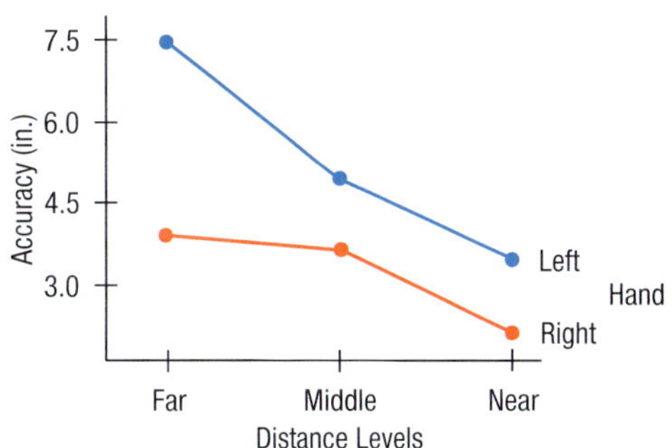

Figure 26.5
An interaction plot showing the mean accuracy at the six combinations of *Distance* and *Hand*. Connecting lines for right and left hands shows the greater effect of the far *Distance* on left-hand *Accuracy* than on right-hand *Accuracy*.

Here it's easier to see that throwing from the Far distance has more of an effect for the Left hand. If the effect of *Distance* were constant, the lines in this plot would be parallel. If the lines aren't parallel, that's evidence that the effects are not additive. But how parallel is enough? Are these lines sufficiently far from parallel for us to doubt whether our data satisfy the **Additive Enough Condition?** A graphical check like this is good enough to proceed. Later in the chapter, we'll see a way to include a term for non-additivity in our model and test for it.

Independence Assumptions

> "Without randomization, your inferences haven't got a chance."
>
> —George Cobb, famous statistician

The **Independence Assumption** for the two-factor model is the same as in one-way ANOVA. That is, the observations *within each treatment group* must be independent of each other. As usual, no test can verify this assumption. Ideally, the data should be generated by a randomized comparative experiment or at least be drawn independently and at random from a homogeneous population. Without randomization of some sort, our P-values are meaningless. We have no sampling distribution models and no valid inference.

Check the **Randomization Condition.** Were the data collected with suitable randomization? For surveys, are the data for each group a representative random sample of that group? For experiments, was the experiment randomized?

We were told that the order of treatment conditions in the dart experiment was randomized. This is especially crucial for an experiment like this because of the possibility of either learning or tiring. Your accuracy might increase after a few throws as you warm up, but after an hour or two, your concentration—and your aim—might start to wander.

If we are satisfied with the Additivity and Independence Assumptions, then we can look at the ANOVA table. For the dart data, we might have some doubts about the additivity, but we'll go ahead for now.

Analysis of Variance for Accuracy

Source	DF	Sum of Squares	Mean Square	F-Ratio	Prob
Distance	2	51.0439	25.5219	28.561	<0.0001
Hand	1	39.600	39.6900	44.416	<0.0001
Error	32	28.5950	0.893594		
Total	35	119.329			

These may look like small P-values, but inference requires additional assumptions. Before we believe the F-tests and use the P-values to make decisions, we have other assumptions to think about.

Equal Variance Assumption

Like the one-way ANOVA, the two-factor ANOVA requires that the variances of all treatment groups be equal. It's the residuals after fitting *both* effects that we'll pool for the Error Mean Square, so it's their variance whose stability we care about most. We check this assumption by checking the **Similar Variance Condition**.

In one-way ANOVA, we looked at the side-by-side boxplots of the groups to see whether they had roughly the same spread. But here, we need to check for equal spread across all treatment groups. The easiest way to check this is to plot the residuals from the two-way ANOVA model.

- Look at the residuals plotted against the predicted values. If the variance is constant, the plot should be patternless. If the plot thickens (to one side or the other), it's a sign that the variance is changing *systematically*. Then it's usually a good idea to consider re-expressing the response variable.

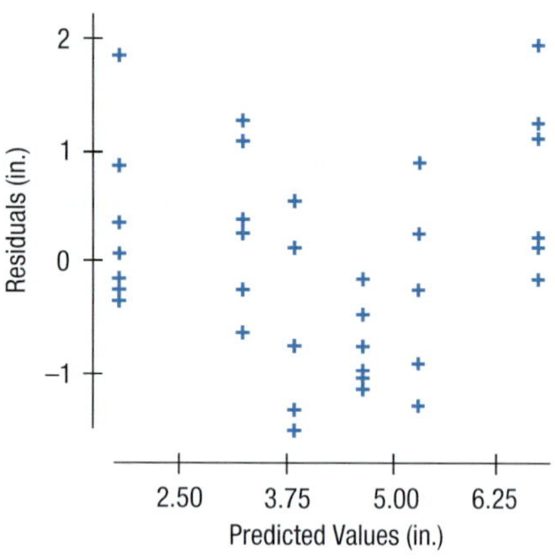

Figure 26.6
A scatterplot of residuals vs. predicted values from the two-way ANOVA model for *Accuracy* shows a U-shaped pattern. This suggests that a condition is violated.

This plot doesn't seem to show changing variance, but it's certainly not patternless. Residuals at both ends of the predicted values are higher than those in the middle. This could mean that something is missing from our model. Could that nagging suspicion we had about the additivity be resurfacing? Stay tuned.

- We can also plot the residuals grouped by each factor. Here's a boxplot of residuals by treatment level for each factor in the dart experiment:

Figure 26.7
Boxplots of the residuals from the two-way ANOVA of the dart accuracy experiment show roughly equal variability when plotted for each factor.

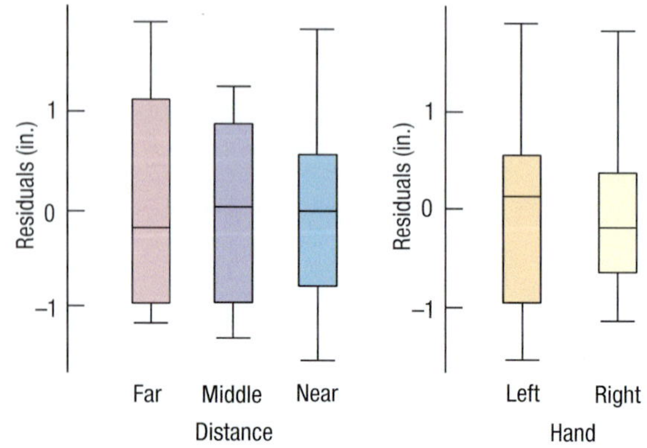

Remember, we can't use the original boxplots to check changing spread. From those it seemed that the spread of the *near* level was smaller than that of the *far* level. But much of that difference was due to changes in the way the factor *Hand* behaves at different distances rather than a change in the underlying variance. By contrast, these boxplots of residuals look fine.

Normal Error Assumption

As with one-way ANOVA, the *F*-tests require that the underlying errors follow a Normal model. We'll check a corresponding **Nearly Normal Condition** with a Normal probability plot or histogram of the residuals.

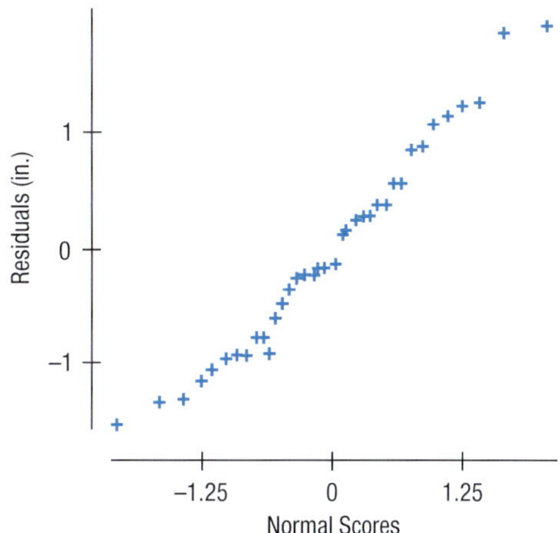

Figure 26.8
A Normal probability plot of the residuals from the two-way ANOVA model for dart accuracy seems reasonably straight.

This condition appears to be satisfied.

What Have We Learned So Far?

The ANOVA table confirmed what we saw in the boxplots. Both *Distance* and *Hand* affect accuracy of throwing a dart, at least at the levels chosen in this experiment. The small P-values give us confidence that the differences we see are due to the factors and not to chance. We did worry that the Additive Enough Condition might not be satisfied, and there seems to be something amiss in the plot of residuals versus predicted values. Before we examine that more, let's look at a different example with two factors, Step-By-Step.

STEP-BY-STEP EXAMPLE

Two-Factor Analysis of Variance

Another student, who prefers the great outdoors to damp pub basements, wonders whether leaving her tennis balls in the trunk of her car for several days after the can is opened affects their performance, especially in the winter when it can get quite cold. She also wonders if the more expensive brand might retain its bounce better. To investigate, she performed a two-factor experiment on *Brand* and *Temperature,* using two *Brands* and three levels of *Temperature*. She bounced three balls under each of the six treatment conditions by first randomly selecting a *Brand* and for that ball, randomly selecting whether to leave it at room temperature or to put it in the refrigerator or the freezer. After 8 hours she dropped the balls from a height of 1 meter to a concrete floor, recording the Height of the bounce in centimeters. (Data in **Tennis balls**)

WHO	Tennis balls of two *Brands* allocated to three *Temperatures*.
WHAT	*Height* of the ball after it bounces from an initial height of 100 centimeters
UNITS	Centimeters
HOW	Two-factor experiment, randomized order of runs

This is a completely randomized replicated design in two factors. The factor *Brand* has two levels: *Premium* and *Standard*. The factor *Temperature* has three levels: *Room, Fridge, and Freezer*. The null hypotheses are that *Brand* has no effect and that *Temperature* has no effect.

Brand	Temperature	Bounce Height
Standard	Freezer	37
Standard	Fridge	59
Standard	Room	59
Premium	Freezer	45
Premium	Fridge	60
Premium	Room	63
Standard	Freezer	37
Standard	Fridge	58
Standard	Room	60
Premium	Freezer	39
Premium	Fridge	64
Premium	Room	62
Standard	Freezer	41
Standard	Fridge	60
Standard	Room	61
Premium	Freezer	37
Premium	Fridge	63
Premium	Room	61

THINK State what we want to know and the null hypotheses we wish to test. For two-factor ANOVA, the null hypotheses are that all the treatment groups have the same mean for each factor. The alternatives are that the means are not all equal.

I want to know whether the storage *Temperature* affects the mean *Bounce Height* of tennis balls. I also want to know whether the two *Brands* have different mean *Bounce Heights*. Writing τ_j for the effect of Brand j, the null hypothesis is that the two brand effects are the same:

$$H_0: \tau_P = \tau_S.$$

Writing γ_k for the effect of Temperature level k, then the other null hypothesis is

$$H_0: \gamma_{room} = \gamma_{fridge} = \gamma_{freezer}.$$

The alternative for the first hypothesis is that the effects of the two *Brands* on the *Bounce Height* are different. The alternative for the second is that the effects of the *Temperature* levels are not all equal.

PLOT Examine the side-by-side partial boxplots of the data.

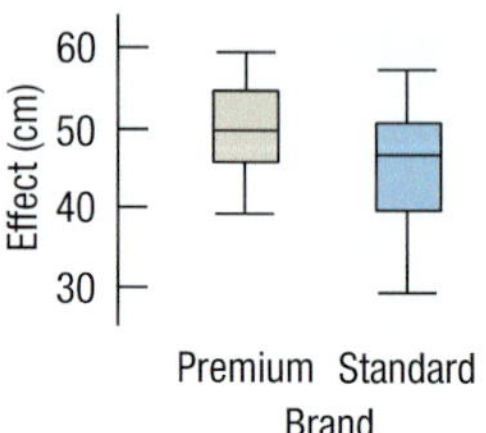

PLAN Think about the assumptions and check the appropriate conditions.

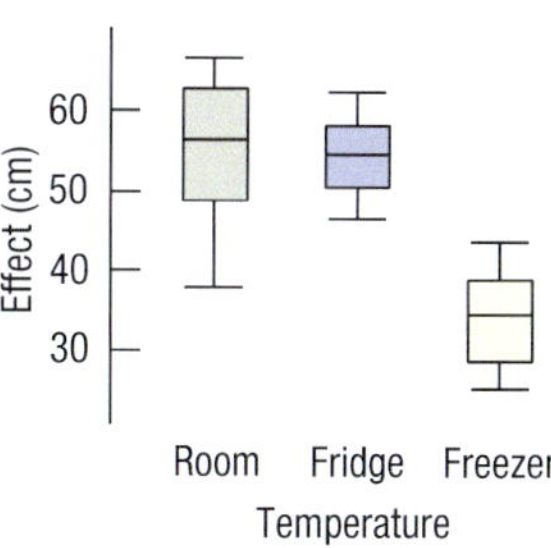

It's pretty clear that *Temperature* has an effect, but we can't be sure about the effect of Brand just from the plots.

✓ **Independence Assumption**: Because the order was randomized, I can assume the observations are independent. However, I might want to check to see if the balls of each brand came from the same box. It might be better to have selected balls of the same brand from different boxes.

✓ **Randomization Condition**: The experimenter performed the trials in random order.

✓ **Additive Enough Condition**: Here's the plot of the mean *Bounce Heights* at each treatment condition:

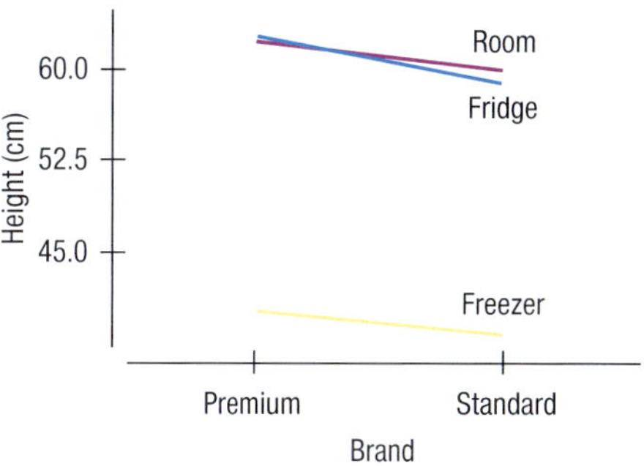

The lines look reasonably parallel. Now I can fit the additive model in two factors:

$$y_{ijk} = \mu + Brand\ effect_j + Temp\ effect_k + Error_{ijk}.$$

Under these conditions it's appropriate to fit the two-way Analysis of Variance model.

Show the ANOVA table.

Analysis of Variance for Height

Source	DF	Sum of Squares	Mean Square	F-ratio	P-value
Temp	2	1849.33	924.665	209.55	<0.0001
Brand	1	26.8889	26.8889	6.0935	0.0271
Error	14	61.7778	4.41270		
Total	17	1938.00			

The Error Mean Square is 4.41, which means that the common error standard deviation is estimated to be $\sqrt{4.41}$, a bit more than 2 centimeters. This seems like a reasonable accuracy for measuring *Bounce Heights* of tennis balls dropped from a height of 1 meter.

Before testing the hypotheses, I need to check some more conditions:

✔ **Similar Variance Condition:** A plot of residuals vs. predicted values shows some increased spread on the left, but with only 5 points it's hard to say.

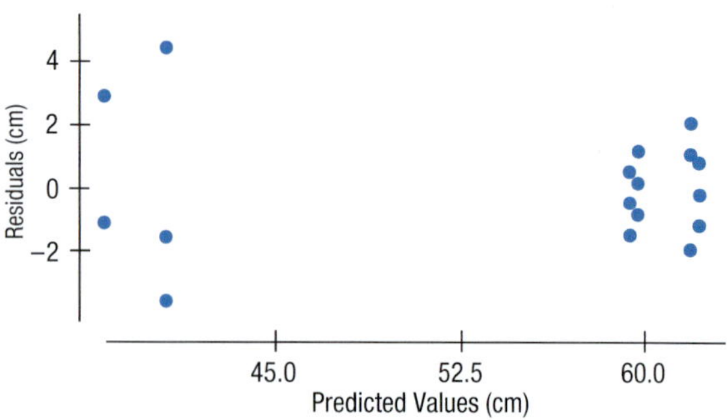

✔ **Nearly Normal Condition, Outlier Condition:** A Normal probability plot of the residuals looks OK too:

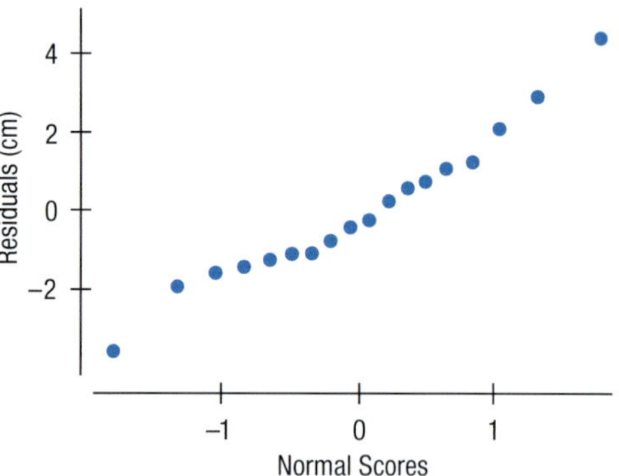

Under these conditions, it is appropriate to perform and interpret the *F*-test.

SHOW **MECHANICS** Discuss the tests.

The *F*-ratio for *Brand* is 6.09, with a P-value of 0.0271. The *F*-ratio for *Temperature* is 209.55, with a P-value less than 0.0001.

Display the effects for each level of the significant factors. In the next section, we'll show how they're calculated. Remember, significance does not guarantee that the differences are meaningful.

The overall mean *Bounce Height* is 53.67 cm.

The *Brand* effects are

Premium	1.22
Standard	−1.22

The *Temperature* effects are

Room	7.33
Fridge	7.00
Freezer	−14.33

TELL **INTERPRETATION** Tell what the *F*-tests mean.

F-ratios this large would be very unlikely if the factors had no effect. I conclude that the means of the two *Brand* levels are not equal.
I also conclude that the means of the three *Temperature* levels are not all equal.

Overall, the premium brand bounces about 2.44 cm higher than the standard brand, averaging across the temperature ranges. Generally, there seems to be little difference between the average *Bounce Height* for room and fridge temperatures, but tennis balls from the freezer bounce almost 22 cm less on average.

Conclusions: It appears that the premium brand outperforms the standard brand overall. It looks from the plots like our student needn't be too worried about playing tennis on a cold day, but if the temperature drops below freezing, she should try to warm the tennis balls first.[4]

EXAMPLE 26.2

RECAP: In the experiment to investigate the effects of paper airplane design on flight distance two factors were tested.

Here are boxplots of distance by each of the factors:

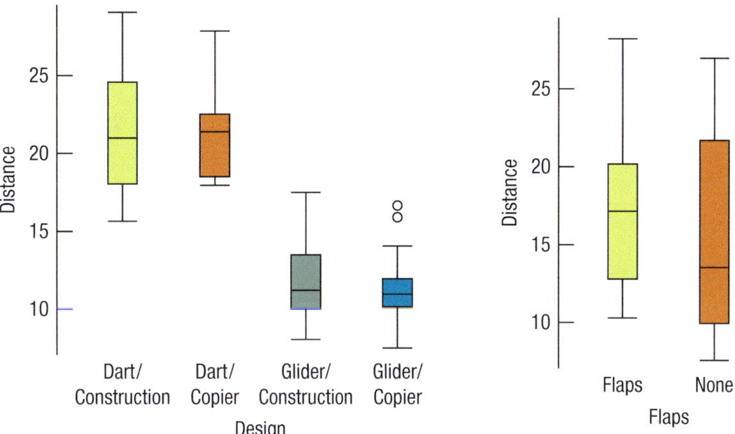

QUESTION: What can we learn from the boxplots?

It appears that the Darts travel farther than the Gliders. There does not appear to be an effect for flaps. The spreads of all the groups are comparable. There are two observations plotted as outliers for the Glider/Copier design, but they don't seem to be very extreme and are probably not a problem.

[4]It's quite obvious from the plots and the tables of effects that there's a difference between *freezer* and the other temperatures, but little difference between *room* and *fridge*. A Bonferroni test would confirm this.

*How ANOVA Works

You'll probably never carry out a two-way ANOVA by hand. But to understand the model it can help to see exactly how it decomposes each observation into components. Our tennis ball model says

$$y_{ijk} = \mu + \textit{Brand effect}_j + \textit{Temp effect}_k + \textit{Error}_{ijk}.$$

To apply this model to data, we need to *estimate* each of these quantities:

$$y_{ijk} = \bar{\bar{y}} + \widehat{\textit{Brand effect}}_j + \widehat{\textit{Temp effect}}_k + \textit{Residual}_{ijk}.$$

Here are the data again, displayed a little differently:

Brand		Temperature		
		Room	Fridge	Freezer
	Premium	63	60	45
		62	64	39
		61	63	37
	Standard	59	59	37
		60	58	37
		61	60	41

The first term in the model is μ, the overall mean. It's reasonable to estimate that with the grand mean, $\bar{\bar{y}}$, which is just the mean of all the observations. For the tennis balls, the average bounce was 53.67 cm. The grand mean has the same value for every observation, which we can lay out as a table like this:

Brand		Temperature		
		Room	Fridge	Freezer
	Premium	53.67	53.67	53.67
		53.67	53.67	53.67
		53.67	53.67	53.67
	Standard	53.67	53.67	53.67
		53.67	53.67	53.67
		53.67	53.67	53.67

Now we turn to the *Brand* effects. To estimate the effect of, say, the *premium Brand*, we find the mean of all *premium* balls and subtract the grand mean. That's 54.89 cm − 53.67 cm = 1.22 cm. In other words, the *premium Brand* balls bounced an average of 1.22 cm higher than the overall average. The *standard Brand* balls bounced an average of 52.444 cm, so their effect is 52.444 cm −53.667 cm = −1.22 cm. Of course, with only two levels, the fact that these two effects are equal and opposite is no coincidence. Effects *always sum to* 0.

We can put the *Brand* effects into a table like this:

Brand		Temperature		
		Room	Fridge	Freezer
	Premium	1.22	1.22	1.22
		1.22	1.22	1.22
		1.22	1.22	1.22
	Standard	−1.22	−1.22	−1.22
		−1.22	−1.22	−1.22
		−1.22	−1.22	−1.22

We can find a *Temperature* effect table in a similar way. We take the treatment average at each level of *Temperature* and subtract the grand mean. Here's the table:

Brand		Temperature		
		Room	Fridge	Freezer
	Premium	7.33	7.00	−14.33
		7.33	7.00	−14.33
		7.33	7.00	−14.33
	Standard	7.33	7.00	−14.33
		7.33	7.00	−14.33
		7.33	7.00	−14.33

We're missing only one table, the *Residuals*. Finding the residuals is easy, too. The additive model represents each observation as the sum of these three components. As always, the residuals are the differences between the observed values and those given by the model. Look at the first ball in the table. It was a *premium Brand* ball at *room Temperature*. From the effects, we'd predict it to bounce 53.67 (grand mean) +1.22 (from being premium) +7.33 (from being at room temperature) = 62.22 cm. Because it actually bounced 63 cm, its residual is 63 − 62.22 = 0.78 cm. If we compute all the residuals this way (or have the computer do it), we'd find the residual table to be

Brand		Temperature		
		Room	Fridge	Freezer
	Premium	0.78	−1.89	4.44
		−0.22	2.11	−1.56
		−1.22	1.11	−3.56
	Standard	−0.78	−0.44	−1.11
		0.22	−1.44	−1.11
		1.22	0.56	2.89

What we have done with all these tables is to decompose each original bounce height into the three estimated components plus a residual:

Bounce Height = *mean Bounce Height* + *Brand effect* + *Temperature effect* + *Residual*.

These tables actually display the workings of the ANOVA. Each term in the ANOVA model corresponds to one of these tables. Because the overall mean doesn't vary, we usually leave it out of the ANOVA table. Each Sum of Squares term in the ANOVA is simply the sum of the squares of all the values in its table.

The degrees of freedom are easy to find too. Look at the grand mean table. Because it has the same value for all observations, it has 1 degree of freedom. The *Brand* effect table seems to have 2 degrees of freedom, but we know that the two effects always sum to 0. So it really has only 1. In general, the degrees of freedom for a factor is one less than its number of levels. So *Temperature* has 2 degrees of freedom.

The original table has 18 observations and so 18 degrees of freedom. But what about the residuals? Remarkably enough, the degrees of freedom from the tables add. That means that we can find the degrees of freedom for the residuals by subtraction. They must have $18 - (1 + 1 + 2) = 14$ degrees of freedom.

The mean squares are just the sum of squares for each effect divided by the degrees of freedom. The F-ratios are the ratios of each Treatment Mean Square to the Error Mean Square. You can check the degrees of freedom and the mean squares in the step-by-step ANOVA table. The original table has 18 observations and so 18 degrees of freedom, but most programs choose not to display the overall mean and subtract out the degree of freedom associated with it to give what's sometimes called a "corrected" total or just "total." That's why our ANOVA table has only 17 degrees of freedom in "total."

Blocks

Back in Chapter 11, we thought about a *randomized block design*.[5] We imagined obtaining tomato plants from two stores and randomly assigning plants from each store to one of three fertilizer treatments. We can view the *Store* as a factor in this design. We now have a two-factor model, with *Store* and *Fertilizer* as factors. The ANOVA looks just the same, but now we're not likely to be interested in testing the *Store* effect. We *assumed* that the plants might be different from the different stores—that's why we used a blocked design.

The two-way ANOVA takes advantage of the blocking to remove the unwanted variation due to *Store*. As in a completely randomized two-factor design, removing the unwanted variation makes the effects of the other factor easier to see.

We wouldn't call a blocked design *completely randomized,* because we can't assign the plants at random to which store they come from. We must randomize *within* the blocks when we assign plants to levels of the *Fertilizer* factor.

26.3 Interactions

Up to now, we have assumed that whatever effects the two factors have on the response can be modeled by simply adding the separate effects together. What if that's not good enough? In the dart-throwing experiment, we worried that the experimenter's accuracy when throwing from farther away may deteriorate *even more* when he uses his nondominant hand. Looking at the interaction plot added evidence to that suspicion.

[5]You may have noticed in the tennis ball example that balls couldn't be randomly assigned to a *Brand*. *Brand* is actually a blocking factor, and the experiment is a randomized block design. But it doesn't change the analysis at all.

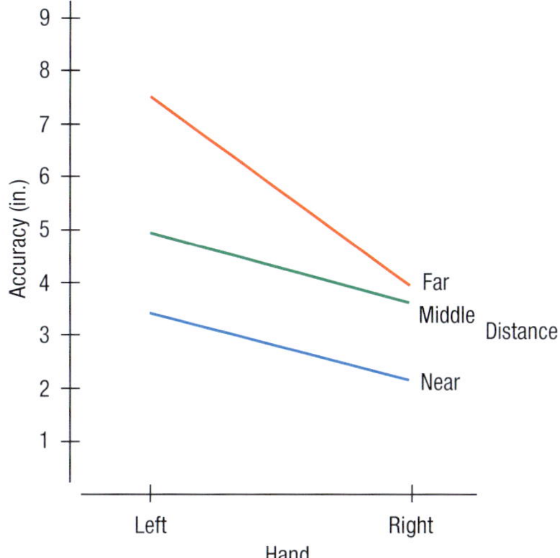

Figure 26.9
Interaction plot of *Accuracy* by *Hand* and *Distance*. The effect of *Distance* appears to be greater for the left hand.

It doesn't look like the effect of *Distance* is constant for the two hands. In particular, moving from *middle* to *far* has a much larger effect when using the *left Hand*. It's not unusual for the effect of a factor's level to depend on the level of the other factor. Unfortunately, our model doesn't take this into account. So, we need a new model.

When the effect of one factor *changes* depending on the levels of the other factor, the factors are said to *interact*. We can model this interaction by adding a term to our model and testing whether the adjustment is worthwhile with an *F*-test.

We include a new term to our model, adding an interaction effect at every *combination* of the levels of the other two factors. The model can now be written:

$$y_{ijk} = \mu + \tau_j + \gamma_k + \omega_{jk} + \varepsilon_{ijk}.$$

The new term, ω_{jk}, represents the *interaction* effect at level j of factor 1 and level k of factor 2.

We add a corresponding line to the ANOVA table. Of course, we'll have to spend some degrees of freedom on this new term. We'll have to take them from the error term where we parked all the leftover degrees of freedom after fitting the two factors. We'll need the *product* of each term's degrees of freedom. For the darts, this is only 2 degrees of freedom (1 from *Hand* and 2 from *Distance*), but for factors with more levels, this could get expensive.

Here's the ANOVA table for the dart experiment:

ANOVA Table for Accuracy

Source	DF	Sum of Squares	Mean Square	F-ratio	P-value
Distance	2	51.044	25.522	41.977	<0.0001
Hand	1	39.690	39.690	65.28	<0.0001
Distance × Hand	2	10.355	5.178	8.516	0.0012
Error	30	18.240	0.608		
Total	35	119.329			

What has changed? Nearly everything. (Of course, the Total Sum of Squares remains the same because it was calculated from the observations minus the overall mean, and the overall mean is still the same.) The interaction term is significant. Because we've now removed a significant source of variability, our Error Mean Square is smaller in spite of the loss of the 2 degrees of freedom devoted to the interaction term.

EXAMPLE 26.3

RECAP: An experiment on paper airplanes considered two factors; *Design* and *Flaps*. Here is the two-factor ANOVA:

Source	DF	Sum of Squares	Mean Square	F-ratio	P-value
Design	3	1549.50	516.499	56.610	<0.0001
Flap	1	27.0140	27.0140	2.9608	0.0905
Error	59	538.303	9.12378		
Total	63	2114.81			

State and test the null hypotheses. What can you conclude?

The first null hypothesis is that the mean distance flown will be the same for each design, after allowing for any effects of flaps. That hypothesis is rejected. As we saw in the boxplots, the darts fly farther.

The second null hypothesis is that the mean distance will be the same whether or not flaps are present, after allowing for the differences due to design. The P-value for that factor is too large to allow us to reject the null hypothesis.

Inference When Variables Are Related

The F-test for the interaction leads us to reject the null hypothesis of no interaction effect and indicates that the effect of *Distance* really does change depending on which *Hand* is used. (Or, equivalently the effect of *Hand* changes depending on the *Distance.*)

What about the main effects? When a significant interaction term is present, it's difficult to *Tell* about the main effect of each factor. How much does switching from *right* hand to *left* hand matter? It depends on the *Distance.* It no longer makes sense to talk about the main effects of each factor separately. The main effects are just averages of the effects over all the levels of the other factor. As we saw with Simpson's paradox (back in Chapter 3), it can be misleading to average over very different situations.

When a significant interaction is present, the best way to display the results is with an interaction plot.

Now that we have the interaction term in our model, we recalculate the residuals. They're the values that are left over after accounting for the overall mean, the effect of each factor, and the interaction effect. Here's a plot of residuals versus predicted values for the two-factor model with interaction:

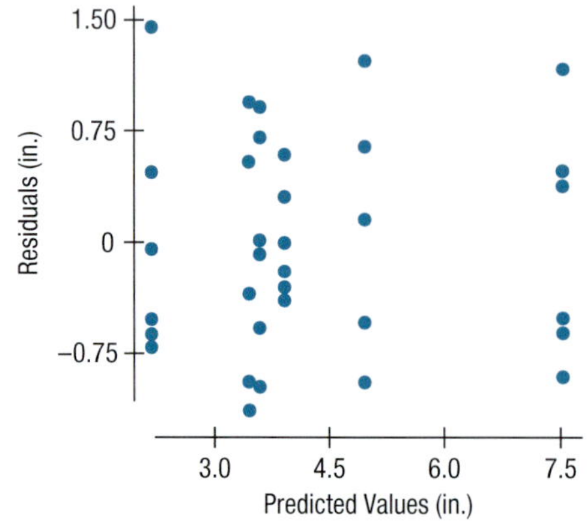

Figure 26.10
Residuals for the two-way ANOVA of dart accuracy with an interaction term included. Now, unlike Figure 26.6, there is no U-shaped pattern.

How has the interaction term helped? These residuals have less structure. Fitting the interaction term succeeded in removing structure from the error. This new model seems to satisfy the assumptions more successfully, and so our inferences are likely to be closer to the truth.

Why Not Always Start by Including an Interaction Term?

With two factors in the model, there's always the possibility that they interact. We checked the Additive Enough Condition with an interaction plot. Now we know we can include the interaction term and use the *F*-test for added information. If the interaction plot shows parallel lines and the *F*-test for the interaction term is not significant, you can comfortably proceed with a two-way ANOVA model without interaction. This will put back into the error term the degrees of freedom you used for the interaction.

On the other hand, if the interaction is significant, the model *must* contain more than just the main effects. In this case, the interpretation of the analysis depends crucially on the interaction term, and the interaction plot shows most of what to *Tell* about the data. When there's a significant interaction term, the effect of each factor depends on the level of the other factor, so it may not make sense to talk about the effects of a factor by itself. However, we may still be able to talk about the main effects. For example, we can see that the *right* hand is better *on average* than the *left* at any distance, and the closer you are to the target, the better the accuracy, no matter which hand you use. So both main effects are significant, and we can discuss them. A significant interaction effect tells us more. It says that *how much* better the performance is at near distances *depends* on which hand you use.

But if the lines in the interaction plot cross, you need to be careful. Here's an interaction plot from an experiment on fuels and alternative engine designs:

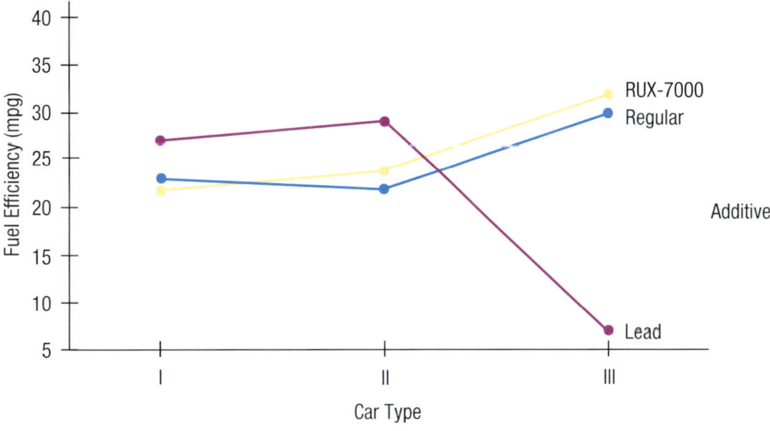

Figure 26.11 An interaction plot of gas mileage by *Car Type* and *Gas Additive*. If the lines cross, as they do here, be careful of interpreting the main effects.

It looks like the fuel with lead is great for two of the engine designs, but a disaster for the third. Suppose someone asked you to summarize the effect of adding lead. Nothing simpler than saying it *depends* on the engine type makes sense. *On average*, it lowers gas mileage, but is that relevant? The real message is that sometimes it improves mileage and sometimes it hurts. You won't want to *Tell anything* about the main effect. Its *F*-test is irrelevant.

Whether the lines cross or not, including an interaction term in our model is always a good way to start. But sometimes we just can't. Suppose we had run the dart experiment with only one dart at each treatment. A design like that is called an **unreplicated two-factor design**. Imagine that the throw at the combination of left hand and far distance missed the bull's-eye by a mile. Would this be evidence of an interaction between *Distance* and *Hand*, or was it just a bad throw? How could we tell? The obvious thing to do would be to *repeat* the treatment. If accuracy was consistently poor at this combination of factors, we'd want to model the interaction. If the other throws were no worse than we'd expect from the two main effects, then we might set the outlier aside. We can't distinguish an interaction effect from error unless we replicate the experiment. In fact, the design tells us that. Without replication, if we try to fit an interaction term, there are exactly 0 degrees of freedom left for the error. That makes any inference or testing impossible. If we are willing to *assume* that the additive model is adequate, then we can reduce the number of runs required and use an unreplicated design. In that case, although the residuals are indistinguishable from the interaction effects, the residual plot may help reveal whether the assumption made sense.

When you do replicate, it's best to replicate *all* treatment conditions equally. (Unbalanced and other more complicated designs can be analyzed, but they are beyond the scope of this text.)

JUST CHECKING

The student conducting the paper airplane experiment decides to do a follow-up experiment to investigate two other factors in more detail. Using his favorite design, he uses one of three different size weights and places the weight either on the nose or on the tail of the plane. At each combination, he performs four test flights (randomizing the order of the flights). An interaction plot looks like this:

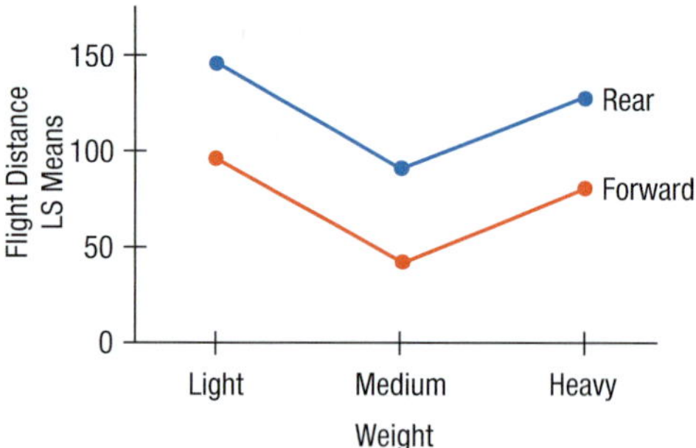

The ANOVA table shows:

Analysis of Variance

Source	DF	Sum of Squares	Mean Square	F-Ratio	P-value
Position	1	13371.9314	13371.9314	53.3868719	<0.0001
Weight	2	12625.5486	6312.7743	25.2034851	<0.0001
Position × Weight	2	10.6208	5.3104	0.02120155	0.9790
Error	18	4508.501	250.47		
Total	23	30516.602			

1. What are the null hypotheses?
2. From the interaction plot, do the effects appear to be additive?
3. Does the F-test in the ANOVA table support your conclusion in (2)?
4. Does the position of the weight seem to matter?
5. Does the amount of weight appear to affect flight distance?
6. What would you recommend to the student to increase flight distance?

STEP-BY-STEP EXAMPLE

Two-Factor Anova with Interaction

In Chapter 25 we looked at how much TV four groups of students watched on average. Let's look at their grade point averages (GPAs). Back in that chapter, we treated the four groups (male athletes, female athletes, male non-athletes, and female non-athletes) as four levels of the factor *Group*. Now we recognize that there are really two factors: the factor *Sex* with levels *male* and *female* and the factor *Varsity* with levels *yes* and *no*. Let's analyze the GPA data with a two-factor ANOVA. (Data in **GPA**)

As with many Social Science studies, this isn't an experiment. (Some fans might object if we assigned students at random to varsity teams.[6]) Instead, it is an application of ANOVA to an observational study.

THINK State what we want to know and the null hypotheses we wish to test. For two-factor ANOVA with interaction, the null hypotheses are that all the treatment groups have the same mean for each factor and that the interaction effect is 0. The alternatives are that at least one effect is not 0.

I want to know whether the mean *GPA* is the same for men and women, whether the mean *GPA* is the same for athletes and non-athletes, and whether the factors interact. Writing τ_j for the effect of *Sex* level *j*, the null hypothesis is that the effects are the same:

$$H_0: \tau_M = \tau_F$$

Writing γ_k for the effect of Varsity level k, the other null hypothesis is

$$H_0: \gamma_Y = \gamma_N$$

Finally, the null hypothesis for interaction is that the interaction terms are all 0.

The alternative for the first hypothesis is that the mean GPA is different for men and women. The alternative for the second is that the mean GPA is different for Varsity athletes than for other students. The alternative for the last is that there is an interaction effect.

PLOT Examine the side-by-side partial boxplots of the data.

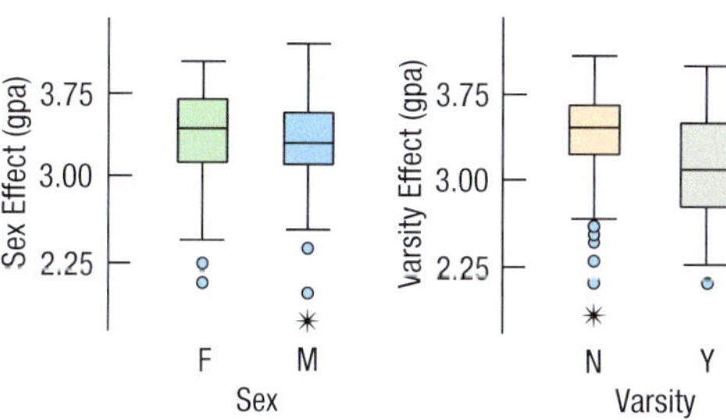

PLAN Think about the assumptions and check the conditions.

It appears that there is no difference in mean GPA between men and women (after accounting for varsity status), but there may be a difference between varsity athletes and non-athletes (after accounting for sex).

[6]And randomly assigning the *other* factor would be problematic.

The study is a random sample, so students' GPAs should be independent of each other.

✓ **Independence Assumption, Randomization Condition:** The study was based on a random sample of students.

✓ **Additive Enough Condition:** Here's the interaction plot of the means at each treatment condition:

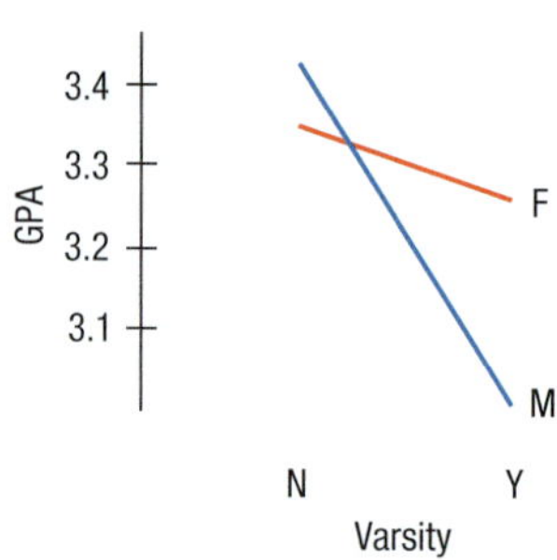

The lines cross, so I'll add an interaction term to the model and fit the two-way ANOVA with interaction:

$$y_{ijk} = \mu + \textit{Sex effect}_j + \textit{Varsity effect}_k + \textit{Interaction}_{jk} + \textit{Error}_{ijk}.$$

Show the ANOVA table.

Analysis of Variance for GPA

Source	DF	Sum of Squares	Mean Square	F-ratio	P-value
Sex	1	0.3040	0.3040	1.7681	0.1852
Varsity	1	2.4345	2.4345	14.1871	0.0002
Sex × Varsity	1	1.0678	1.0678	6.2226	0.0134
Error	196	33.6397	0.1716		
Total	199	37.4898			

✓ **Similar Variance Condition:** A plot of residuals vs. predicted shows no thickening or other patterns:

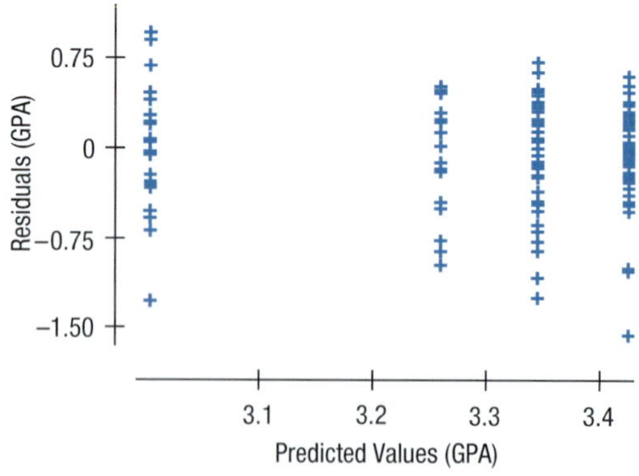

✓ **Nearly Normal Condition, Outlier Condition:** The Normal probability plot of the residuals is reasonable as well.

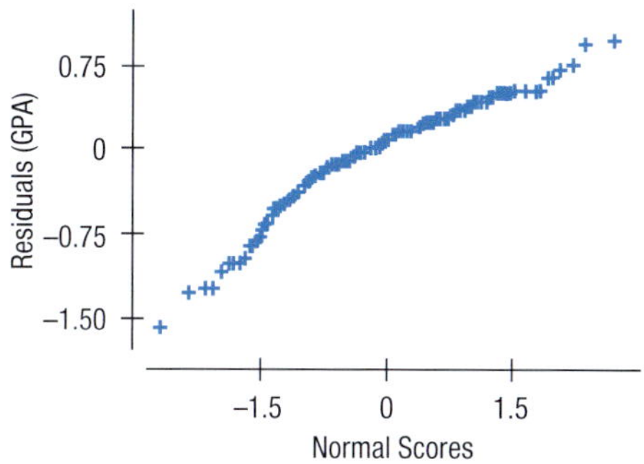

Under these conditions it is appropriate to interpret the F-tests and their P-values.

SHOW **MECHANICS**

The Error Mean Square is 0.172 points, which means that the common error standard deviation is estimated to be $\sqrt{0.172} = 0.415$ points (nearly half a letter grade).

The F-ratio for the interaction term is 6.223, with a P-value of 0.0134. Based on this, I reject the hypothesis of no interaction. Because there is interaction, I am cautious in testing the main effects, but will examine the means further to understand the factor effects.

Overall mean $GPA = 3.32$.

Level of Varsity	Effect
N	0.1281
Y	−0.1281

Show the table of means.

The main effect of Varsity is significant. Varsity athletes have lower GPA by about a quarter of a grade point averaged over all, but this effect is much greater for men than for women. Here are the four group mean GPAs:

Group	Mean GPA
Female, non-athlete	3.35
Female, varsity	3.26
Male, non-athlete	3.43
Male, varsity	3.00

TELL **INTERPRETATION** Tell what the *F*-test means.

The interaction term is significant, and it appears that most of the effect we see is due to a difference of 0.43 in mean GPAs between male athletes and non-athletes. It doesn't look like any of the other differences is large enough to be important. Probably none of them is statistically significant either.

EXAMPLE 26.4

RECAP: The ANOVA for paper airplane distance on two factors, *Design* and *Flaps,* showed a significant effect for *Design*, but not for *Flaps*. But maybe there is an interaction. Here is the ANOVA with the interaction term and an interaction plot.

Source	DF	Sum of Squares	Mean Square	F-ratio	P-value
Design	3	1549.50	516.499	81.065	<0.0001
Flaps	1	27.0140	27.0140	4.2399	0.0441
Design × Flaps	3	181.503	60.5009	9.4956	<0.0001
Error	56	356.800	6.37144		
Total	63	2114.81			

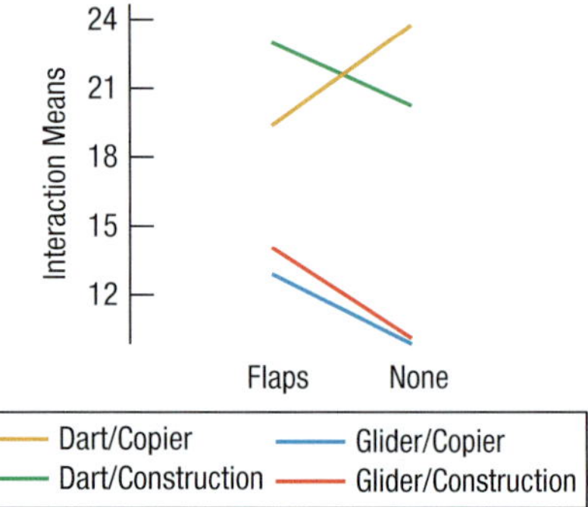

QUESTION: Interpret the interactions. What do they say about paper airplane design?

Including the interaction term reveals that the *Flaps* factor is significant. But it matters in different ways for different designs. Flaps help improve the distance of both kinds of gliders. But flaps hinder the performance of darts made of lighter copier paper, although they help darts made of construction paper.

What Next?

This chapter introduces some of the issues that arise in advanced experimental design and analysis. We have scratched the surface of this large and important area by showing the analysis of a balanced two-way design with interaction.

Of course, there's no reason to stop with two factors. And if we have more than two, there are many ways in which they may interact. As you can imagine, real-life experiments and observational studies may pose many interesting challenges. Anyone who is serious about analyzing or designing multifactor studies should consult a professional statistician. You now know the vocabulary and principles, so you are well equipped to do that.

WHAT CAN GO WRONG?

- **Beware of unreplicated designs unless you are sure there is no interaction.** Without replicating the experiment for each treatment combination, there is no way to distinguish the interaction terms from the residuals. If you are designing a two-factor experiment, you must be willing to assume that there is no interaction if you choose not to replicate. In such a case, you can fit an additive model only in the two factors. If there is an interaction, it will show up in the error term. You should examine a residual plot to help reveal a possible interaction effect. You must be prepared to defend the assumption of no interaction and the decision not to replicate.

- **Don't attempt to fit an interaction term to an unreplicated two-factor design.** If you have an unreplicated two-factor experiment or observational study, you'll find that if you try to fit an interaction term, you'll get a strange ANOVA table. The design exhausts the degrees of freedom for error, so fitting the interactions leaves no degrees of freedom for residuals. That wipes out the mean square errors, *F*-ratios, and P-values, which may appear in the computer output as dots, dashes, or some other indication of things gone wrong. Remove the interaction term from the model and try again.
- **Be sure to fit an interaction term when it exists.** When the design is replicated, it is always a good idea to fit an interaction term. If it turns out not to be statistically significant, you can then fit a simpler two-factor main effects model instead.
- **When the interaction effect is significant, don't interpret the main effects.** Main effects can be very misleading in the presence of interaction terms. Look at this interaction plot.

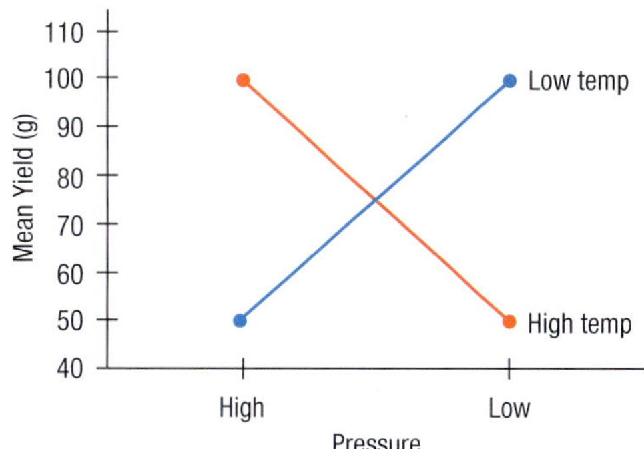

Figure 26.12
An interaction plot of *Yield* by *Temperature* and *Pressure*. The main effects are misleading. There is no (main) effect of *Pressure* because the average *Yield* at the two pressures is the same. That doesn't mean that *Pressure* has no effect on the *Yield*. In the presence of an interaction effect, be careful when interpreting the main effects.

- The experiment was run at two temperatures and two pressure levels. High amounts of material were produced at high pressure with high temperature and at low pressure with low temperature. What's the effect of *Temperature*? Of *Pressure*? Both main effects are 0, but it would be silly (and wrong) to say that neither *Temperature* nor *Pressure* was important. The real story is in the interaction.
- **Always check for outliers.** As in any analysis, outliers can distort your conclusions. An outlier can inflate the Error Mean Square so much that it may be hard to discern any effect, whether it exists or not. Use the partial boxplots to search for outliers. Consider setting outliers aside and re-analyzing the results. An outlier can make an interaction term appear significant. For example, a single male varsity athlete with a very low GPA could account for the results we saw.
- **Check for skewness.** If the underlying data distributions are skewed, you should consider a transformation to make them more symmetric.
- **Beware of unbalanced designs and designs with empty cells.** We've been assuming that the data are *balanced* over the design—that we have equal numbers of observations in each cell. There are methods that will easily compensate for small amounts of imbalance, but empty cells and other more serious violations of balance require different methods and additional assumptions.

CONNECTIONS

We first discussed designing experiments in Chapter 11. Now we know how to analyze a two-factor design. In Chapter 25, we saw that the one-way ANOVA generalized the pooled *t*-test to more than two levels of one factor. Now we've added a second factor.

Of course, we're still relying on boxplots, Normal probability plots, and scatterplots of residuals to help us check conditions and understand relationships. What we first said we'd look for in these displays back in Chapters 3, 4, 5, and 6 is still what we're concerned with here.

CHAPTER REVIEW

In Chapter 25, we learned that the Analysis of Variance is a natural way to extend the *t*-test for testing the means of two groups to compare several groups. Often, those groups are different levels of a single factor. Now we've learned that we can extend the Analysis of Variance to designs with more than one factor. We've learned that partial boxplots are a good way to examine the effect of each factor on the response. We've seen that separate tests of hypotheses can be carried out for each of the factors, testing whether the factor has any discernible effect on the response across any of the levels tested.

We've also learned that sometimes factors can interact with each other. When we have at least two observations at each combination of factor levels, we can add an interaction term to our model to account for the possible interaction.

Of course, we still need to check the appropriate assumptions and conditions as we did for simple ANOVA. All the methods we learned for displaying data in the early chapters are still important. And we've begun to see the deep connection between the design concepts we learned in Chapter 13, the exploratory methods we learned in the early chapters, and the inference methods of the latter part of the text.

REVIEW OF TERMS

Two-way ANOVA model (main effects model)
The model for a two-way (one response, two factors) ANOVA is

$$y_{ijk} = \mu + \tau_j + \gamma_k + \varepsilon_{ijk},$$

where τ_j represents the effect of (level *j*) of factor one and γ_k represents the effect of (level *k*) of factor two. The subscript *i* designates the *i*th replication of the treatment combination (p. 859).

Partial boxplot
Partial boxplots are found by plotting the response at each of the levels of one factor with the effects of the second factor removed. A simple way to generate partial boxplots of factor A is to first save the residuals from a one-way ANOVA of the response on factor B. Add the overall mean to the residuals and make boxplots of the resulting data against factor A (p. 861).

Additivity, additive model
A model in two factors is additive if it simply adds the effects of the two factors and doesn't include an interaction term (p. 862).

Interaction
When the effects of the levels of one factor change depending on the level of the other factor, the two factors are said to interact. When interaction terms are present, it is misleading to talk about the main effect of one factor because how large it is *depends* on the level of the other factor (p. 863).

Interaction plot
A plot that shows the means at each treatment combination, highlighting the factor effects and their behavior at all the combinations (p. 863).

Two-way ANOVA model with interaction (main effects model)

For a replicated two-way design, an interaction term can also be fit. The resulting model is

$$y_{ijk} = \mu + \tau_j + \gamma_k + \omega_{jk} + \varepsilon_{ijk},$$

where τ_j represents the effect of (level *j*) of factor 1 and γ_k represents the effect of (level *k*) of factor 2. Now ω_{jk} represents the effect of the interaction at levels *j* of factor one and *k* of factor two. The subscript *i* designates the *i*th replication of the treatment combination (p. 873).

TECH SUPPORT Two-Way ANOVA

Some statistics packages distinguish between models with one factor and those with two or more factors. You must be alert to these differences when analyzing a two-factor ANOVA. It's not unusual to find ANOVA models in several different places in the same package.

Usually, you must specify the interaction term yourself. That's because these features of statistics packages typically are designed for models with three or more factors where the number of interactions can explode unless the data analyst carefully selects the interactions to be included in the model.

DATA DESK

- Select the response variable as Y and the factor variable as X.
- From the **Calc** menu, choose **ANOVA** > **ANOVA** or **ANOVA** > **ANOVA with Interactions.**
- Data Desk displays the ANOVA table.
- Select plots of residuals from the ANOVA table's Hyper-View menu.

COMMENTS

Data Desk expects data in "stacked" format. The GLM (General Linear Model) command in the toolbar can also compute ANOVAs as well as a wide variety of related analyses, and make a variety of plots, including interaction plots. See the documentation for details.

EXCEL

- The Excel Data Analysis Add-in offers a two-way ANOVA "with and without replication."

COMMENTS

Excel requires that the data be in a special format and cannot deal with unbalanced data, estimate interactions, or make boxplots.

JMP

- From the **Analyze** menu select **Fit Model.**
- Select variables: and **Add** them to the **Construct Model Effects** box.
- To specify an interaction, select both factors and press the **Cross** button.
- Click **Run Model.**
- JMP opens a **Fit Least Squares** window.
- Click on the red triangle beside each effect to see the means plots for that factor. For the interaction term, this is the interaction plot.
- Consult the JMP documentation for information about other features.

COMMENTS

JMP expects data in "stacked" format with one continuous response and two nominal factor variables.

MINITAB

- Choose **ANOVA** from the **Stat** menu.
- Choose **Two-way . . .** from the **ANOVA** submenu.
- In the Two-way ANOVA dialog, assign a quantitative Y variable to the Response box and assign the categorical X factors to the Factor box.
- Specify interactions.
- Check the **Store Residuals** check box.
- Click the **Graphs** button.
- In the ANOVA-Graphs dialog, select **Standardized residuals,** and check **Normal plot of residuals** and **Residuals versus fits.**

COMMENTS

Minitab expects "stacked" format data for two-way designs.

R

To perform an Analysis of Variance of a variable *y* on factors x_1 and x_2:

- myaov = aov(y ~ x_1 + x_2)
- anova(myaov) # gives the ANOVA table

To include the interaction of x_1 and x_2:

- myaov = aov(y ~ x_1 + x_2 + x_1:x_2) or equivalently
- myaov = aov(y ~ x_1*x_2)
- summary(myaov) # shows the individual effects
- anova(myaov) # gives the ANOVA table

COMMENTS

To get confidence or prediction intervals use:

- predict(myaov, interval = "confidence")

or

- predict(myaov, interval = "prediction")

SPSS

- Choose **Analyze** > **General Linear Model** > **Univariate.**
- Assign the response variable to the **Dependent Variable** box.
- Assign the two factors to the **Fixed Factor(s)** box. This will fit the model with interactions by default.
- To omit interactions, click on **Model.** Select **Custom.** Highlight the factors. Select **Main Effects** under the **Build Terms** arrow and click the arrow.
- Click **Continue** and **OK** to compute the model.

STATCRUNCH

To find an ANOVA of response variable *y* on two factors, x_1 and x_2, from the Stat menu,

- choose ANOVA > Two Way
- In the dialog, specify the response variable.
- Specify the factors. It really doesn't matter which you specify as the "row factor" and which as the "column factor."

StatCrunch automatically includes an interaction term in your model. Choose the "Fit Additive Model" option to omit the interaction term.

A table of cell means is available.

TI-83/84 PLUS

COMMENTS

You need a special program to compute a two-way ANOVA on the TI-83.

EXERCISES

SECTION 26.1

1. Student TV time In the previous chapter we considered TV watching by male and female student athletes. In that example, we categorized the students into four groups, but now we have seen that these data could be analyzed with two factors, *Sex* and *Athlete*. Write the ANOVA model for the design in this form.

T **2. Cookies** A student performed an experiment to compare chocolate chip cookie recipes. He baked batches of cookies with different amounts of *Sugar*: 0.5, 0.375, and 0.25 cups, and with different kinds of *Chips*: milk, semisweet, and dark chocolate. Cookie quality was scored on a 0 to 10 scale by other students in the class, with the mean score being the result for each recipe. Write the ANOVA model for the design of this experiment.

SECTION 26.2

3. Student TV assumptions The TV watching study of Exercise 1 was collected as a survey of students at a small college. Do the assumptions of ANOVA appear to be met? Here are some displays to help in your decision:

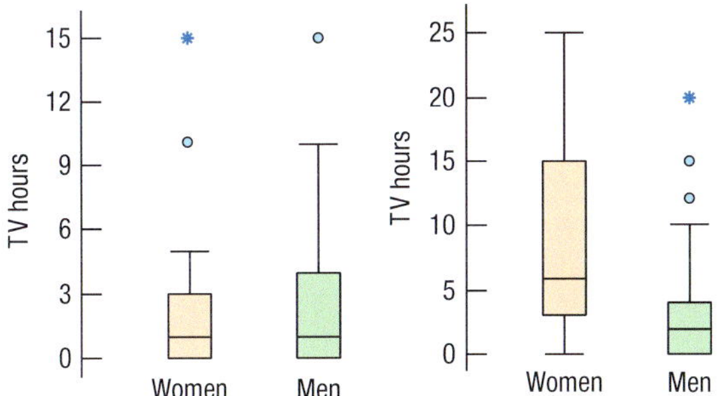

T **4. Cookie assumptions** The student performing the chocolate chip cookie experiment of Exercise 2 planned to analyze his results with an Analysis of Variance on two factors. Here are some displays. Do you think the assumptions for ANOVA are satisfied?

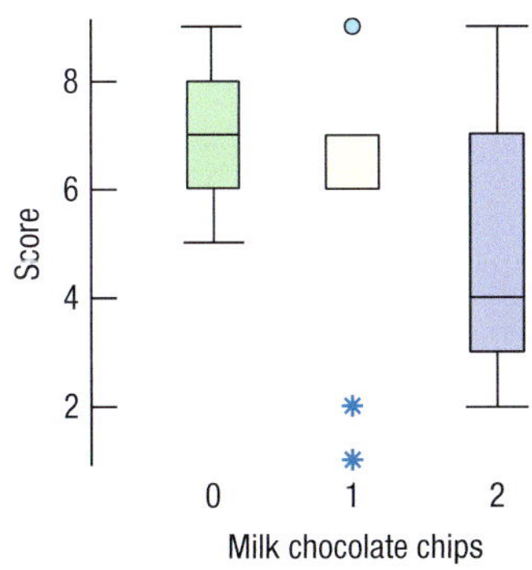

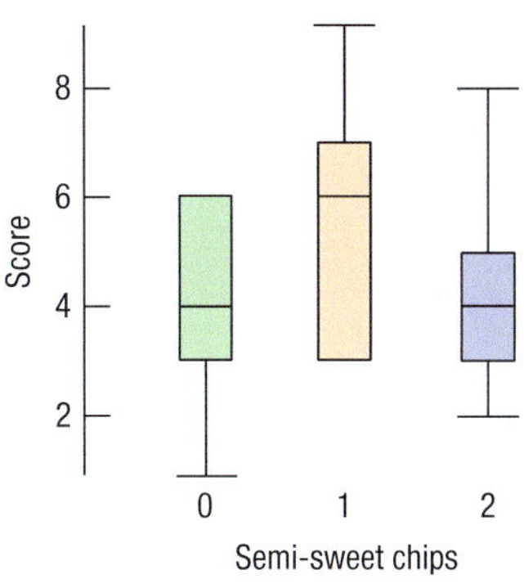

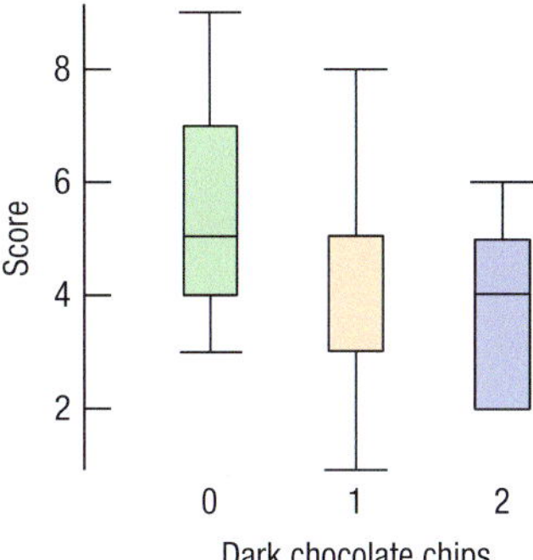

SECTION 26.3

5. Student TV interactions Here is an ANOVA table with an interaction term and the corresponding interaction plot for the TV watching data of Exercise 1. What does the interaction term mean here?

Analysis of Variance for No Selector **TVhours**

Source	DF	Sums of Squares	Mean Square	F-ratio	P-value
Intercept	1	2635.13	2635.13	134.51	<0.0001
Sex	1	464.839	464.839	23.728	<0.0001
Ath	1	374.634	374.634	19.123	<0.0001
Sex × Ath	1	312.877	312.877	15.971	<0.0001
Error	193	3780.97	19.5905		
Total	196	4765.12			

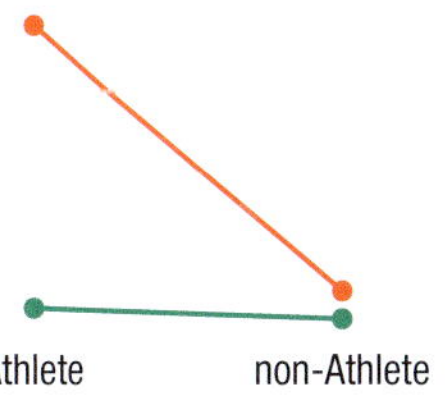

T **6. Have ANOVA cookie** Here is the ANOVA table for the cookie experiment of Exercise 2 along with an interaction plot. What does the interaction term say about the cookie recipes?

Analysis of Variance for Score

Source	DF	Sums of Squares	Mean Square	F-ratio	P-value
Intercept	1	2515.16	2515.16	677.74	<0.0001
Ch	2	27.8990	13.9495	3.7588	0.0270
Sgr	2	27.6566	13.8283	3.7262	0.0279
Ch × Sgr	4	38.2828	9.57071	2.5789	0.0426
Error	90	334.000	3.71111		
Total	98	426.838			

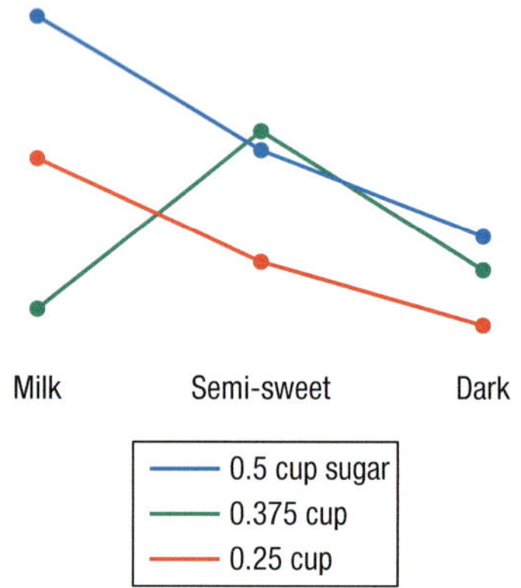

CHAPTER EXERCISES

7. Popcorn revisited In the previous chapter we saw a two-factor experiment to test how microwave power and temperature affect popping. She chooses 3 levels of *Power* (*low*, *medium*, and *high*) and 3 *Times* (*3 minutes, 4 minutes*, and *5 minutes*), running one bag at each condition. She counts the number of uncooked kernels as the response variable.

a) What are the null and alternative hypotheses for the main effects?
b) How many degrees of freedom does each factor sum of squares have? How about the Error Sum of Squares?
c) Should she consider fitting an interaction term to the model? Why or why not?

8. Gas mileage revisited In Chapter 25 we saw a student's experiment to study the effect of *Tire Pressure* and *Acceleration* on gas mileage. He devises a system so that his Jeep Wagoneer uses gasoline from a one-liter container. He uses 3 levels of *Tire Pressure* (*low, medium*, and *full*) and 2 levels of *Acceleration*, either holding the pedal *steady* or *pumping* it every few seconds. He randomizes the trials, performing 4 runs under each treatment condition, carefully recording the number of miles he can go in his Jeep Wagoneer on one liter of gas.

a) What are the null and alternative hypotheses for the main effects?
b) How many degrees of freedom does each Treatment Sum of Squares have? How about the Error Sum of Squares?
c) Should he consider fitting an interaction term to the model? Why might it be a good idea?
d) If he fits an interaction term, how many degrees of freedom would it have?

9. Popcorn again Refer to the experiment in Exercise 7. After collecting her data and analyzing the results, the student reports that the F-ratio for *Power* is 13.56 and the F-ratio for *Time* is 9.36.

a) What are the P-values?
b) What would you conclude?
c) What else about the data would you like to see in order to check the assumptions and conditions?

10. Gas mileage again Refer to the experiment in Exercise 8. After analyzing his data the student reports that the F-ratio for *Tire Pressure* is 4.29 with a P-value of 0.030, the F-ratio for *Acceleration* is 2.35 with a P-value of 0.143, and the F-ratio for the *Interaction effect* is 1.54 with a P-value of 0.241.

a) What would you conclude?
b) What else about the data would you like to see in order to check the assumptions and conditions?
c) If your conclusion about the *Acceleration* factor in part a is wrong, what type of error have you made?

11. Crash analysis In a crash test, stock automobiles are crashed into a wall at 35 mph with dummies in both the passenger and the driver's seats. A particular crash dummy model is capable of recording 134 channels of data on the impact of the crash at various sites on the dummy. In this test, 335 cars are crashed. The response variable is a measure of head injury. Researchers want to know if which seat the dummy is in affects head injury severity, as well as whether the type of car affects severity.

Here are partial boxplots for the 2 different *Seat* (*driver, passenger*) and the 6 different *Size* classifications (*compact, light, medium, mini, pickup, van*):

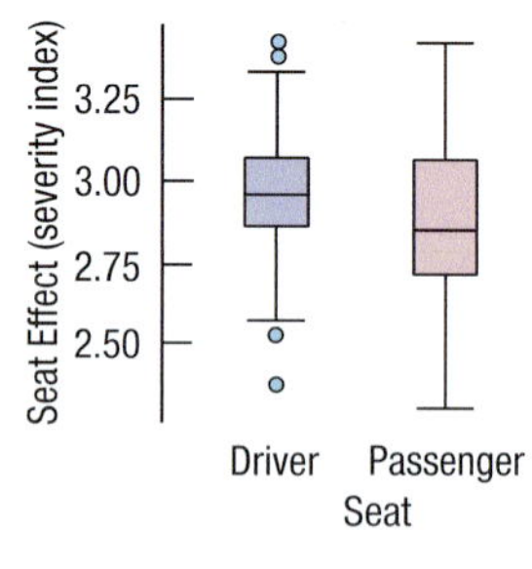

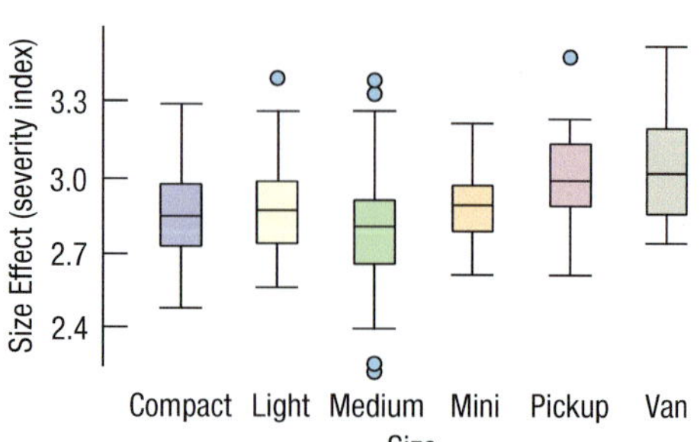

An interaction plot shows:

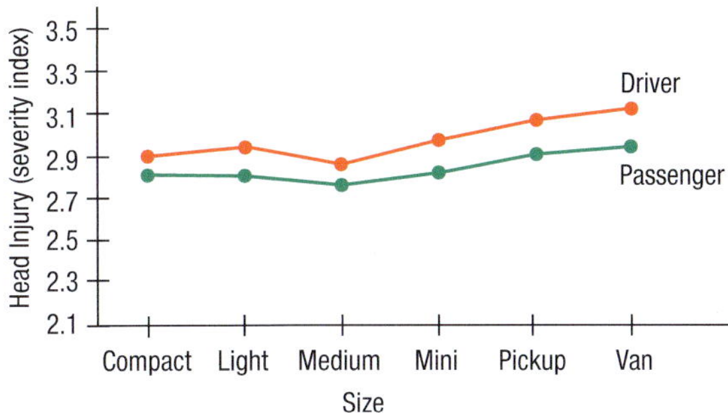

A scatterplot of residuals vs. predicted values shows:

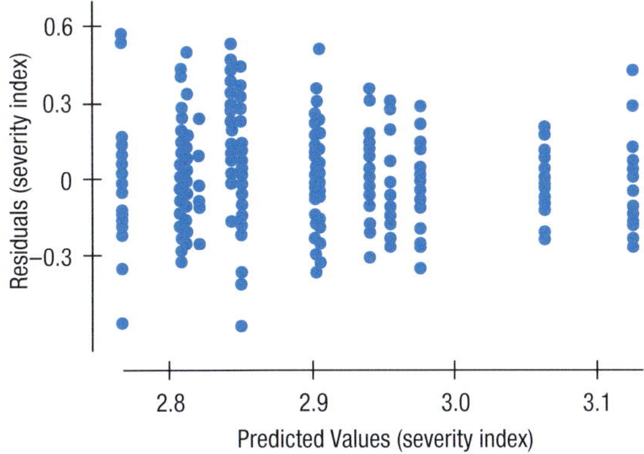

The ANOVA table follows:

Analysis of Variance for Head Severity: 293 cases

Source	DF	Sum of Squares	Mean Square	F-ratio	P-value
Seat	1	0.88441	0.88441	25.807	<0.0001
Size	5	1.33097	0.26619	7.767	<0.0001
Seat × Size	5	0.07205	0.01441	0.420	0.835
Error	289	9.90381	0.03427		
Total	300	12.19124			

a) State the hypotheses about the main effects (both numerically and in words).
b) Are the conditions for two-way ANOVA met?
c) If so, perform the hypothesis tests and state your conclusion. Be sure to state it in terms of head injury severity, seats, and vehicle types.

T **12. Sprouts** An experiment on mung beans was performed to investigate the environmental effects of salinity and water temperature on sprouting. Forty beans were randomly allocated to each of 36 petri dishes that were subject to one of four levels of *Salinity* (0, 4, 8, and 12 *ppm*) and one of three *Temperatures* (32°, 34°, or 36° C). After 48 hours, the biomass of the sprouts was measured.

At the top of the next column are partial boxplots of *Biomass* on *Salinity* and *Temperature:*

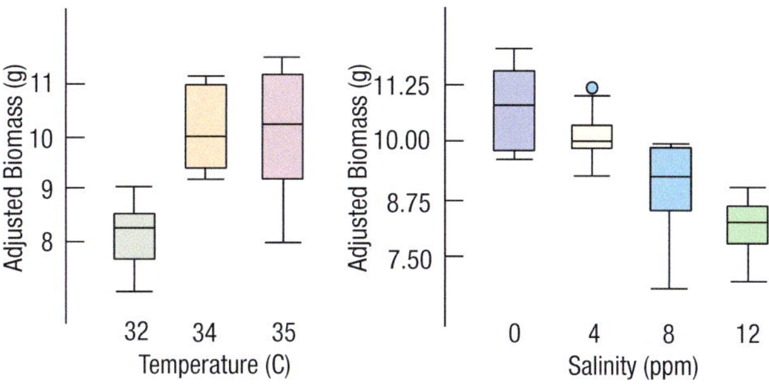

The interaction plot shows:

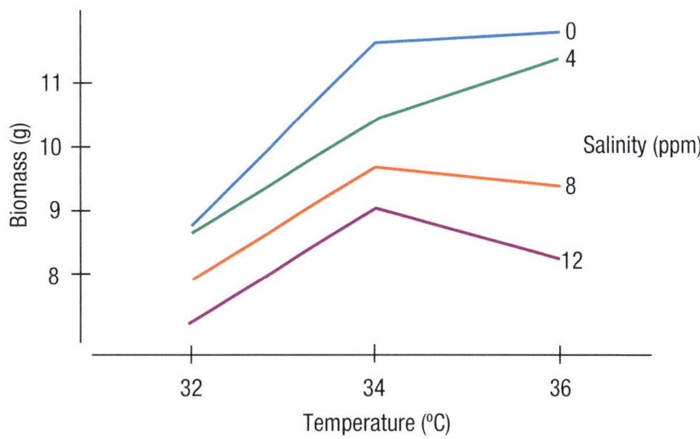

A two-way ANOVA model is fit, and the following ANOVA table results:

Analysis of Variance for Biomass (g)

Source	DF	Sum of Squares	Mean Square	F-ratio	P-value
Salinity	3	36.4701	12.1567	16.981	<0.0001
Temp	2	34.7168	17.3584	24.247	<0.0001
Salinity × Temp	6	5.2972	0.8829	1.233	0.3244
Error	24	17.1816	0.7159		
Total	35	93.6656			

The plot of residuals vs. predicted values shows:

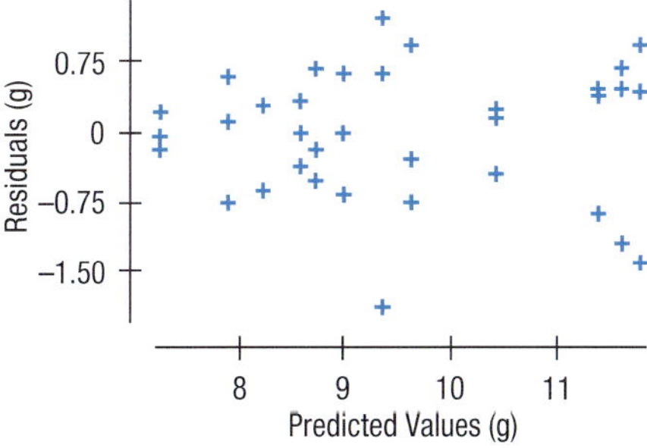

a) State the hypotheses about the factors (both numerically and in words).
b) Are the conditions for two-way ANOVA met?
c) Perform the hypothesis tests and state your conclusions. Be sure to state your conclusions in terms of biomass, salinity, and water temperature.

13. Baldness and heart disease A retrospective study examined the link between baldness and the incidence of heart disease. In the study, 1435 middle-aged men were selected at random and examined to see whether they showed signs of *Heart Disease* (or not) and what amount of *Baldness* they exhibited (*none, little, some*, or *much*).

Baldness	Heart Disease	Number of Men
None	Yes	251
None	No	331
Little	Yes	165
Little	No	221
Some	Yes	195
Some	No	185
Much	Yes	52
Much	No	35

A student runs a two-factor ANOVA on these data and finds the following ANOVA table:

Source	DF	Sum of Squares	Mean Square	F-Ratio	P-value
Baldness	3	62441.375	20813.792	17.956	0.020
Heart disease	1	1485.125	1485.125	1.281	0.340
Error	3	3477.375	1159.125		

a) Comment on her analysis. What problems, if any, do you find with the analysis?
b) What sort of analysis might you do instead?

14. Fish and prostate The Chapter 3 Step-By-Step looked at a Swedish study that asked 6272 men how much fish they ate and whether or not they had prostate cancer. (Data in **Fish diet**) Here are summary counts:

		Prostate Cancer?	
		No	Yes
Eat Fish?	Never/seldom	110	14
	Small part of diet	2420	201
	Moderate part	2769	209
	Large part	507	42

Armed with the methods of this chapter, a student performs a two-way ANOVA on the data. Here is her ANOVA table:

Source	DF	Sum of Squares	Mean Square	F-ratio	P-value
Fish	3	3110203.0	1036734.3	1.3599	0.4033
Prostate cancer	1	3564450.0	3564450.0	4.6756	0.1193
Error	3	2287051.0	762350.0		

a) Comment on her analysis. What problems, if any, do you find with the analysis?
b) What sort of analysis might you do instead?

15. Baldness and heart disease again Refer back to Exercise 13. Perform your own analysis of the data to see if baldness and heart disease are related. Do your conclusions support the claim that baldness is a cause of heart disease? Explain.

16. Fish and prostate Refer back to Exercise 14. Perform your own analysis of the data to see if eating fish and contracting prostate cancer are related.

17. Basketball shots A student performed an experiment to see if her favorite sneakers and the time of day might affect her free throw percentage. She tried shooting with and without her favorite sneakers and in the early morning and at night. For each treatment combination, she shot 50 baskets on 4 different occasions, recording the number of shots made each time. She randomized the treatment conditions by drawing numbers out of a hat. Here are her data:

Time of Day	Shoes	Shots Made
Morning	Others	22, 23, 33, 33
Morning	Favorite	32, 23, 35, 35
Night	Others	26, 27, 32, 31
Night	Favorite	28, 31, 34, 37

a) What are the null and alternative hypotheses?
b) Write a short report on your findings, being sure to include diagnostic analysis as well as practical conclusions.

18. Washing For his final project, Jonathan examined the effects of two factors on how well stains are removed when washing clothes. On each of 16 new white handkerchiefs, he spread a teaspoon of dirty motor oil (obtained from a local garage). He chose 4 *Temperature* settings (each of which is a combination of wash and rinse: *cold-cold, cold-warm, warm-hot,* and *hot-hot*) and 4 *Cycle* lengths (*short, med short, med long*, and *long*). After its washing, each handkerchief was dried in a dryer for 20 minutes and hung up. He rounded up 10 family members to judge the handkerchiefs for cleanliness on a scale of 1 to 10 and used the average score as his response. Here are the data:

Temp	Cycle	Score
Cold-cold	Med long	3.7
Warm-hot	Med long	6.5
Cold-warm	Med long	4.9
Hot-hot	Med long	6.5
Cold-cold	Long	4.6
Warm-hot	Long	8.3
Cold-warm	Long	4.7
Hot-hot	Long	9.1
Cold-cold	Short	3.4
Warm-hot	Short	5.6
Cold-warm	Short	3.8
Hot-hot	Short	7.1
Cold-cold	Med short	3.1
Warm-hot	Med short	6.3
Cold-warm	Med short	5
Hot-hot	Med short	6.1

You may assume, as Jonathan did, that interactions between *Temperature* and *Cycle* are negligible. Write a report showing what you found about washing factors and stain removal.

T **19. Sprouts sprouted** An experiment on bean sprouts was performed to investigate the environmental effects of salinity and water temperature on sprouting. Forty beans were randomly allocated to each of 36 petri dishes that were subject to one of four levels of salinity (0, 4, 8, and 12 ppm) and one of three temperatures (32° Celsius, 34° Celsius, or 36° Celsius). The students kept track of the number of beans sprouting (out of 40) for each of the 36 dishes, and constructed the ANOVA table and various plots given below.

An ANOVA table shows:

Source	DF	Sum of Squares	Mean Square	F-ratio	P-value
Salinity	3	2002.017	667.3	53.341	<0.0001
Temp	2	57.943	28.972	2.316	0.1203
Salinity × Temp	6	96	16	1.279	0.3039
Error	24	300.3	12.51		
Total	35	2456.3			

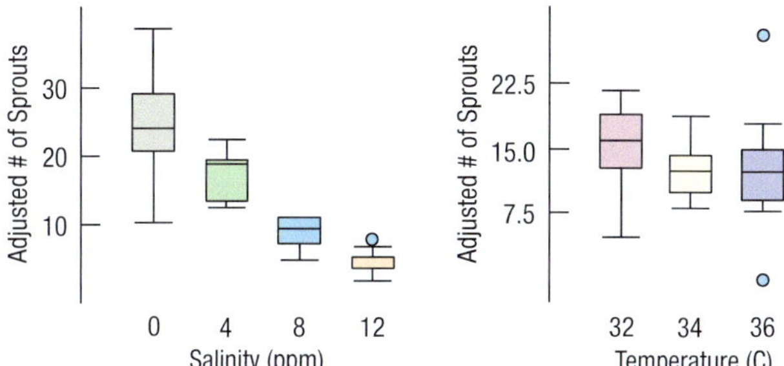

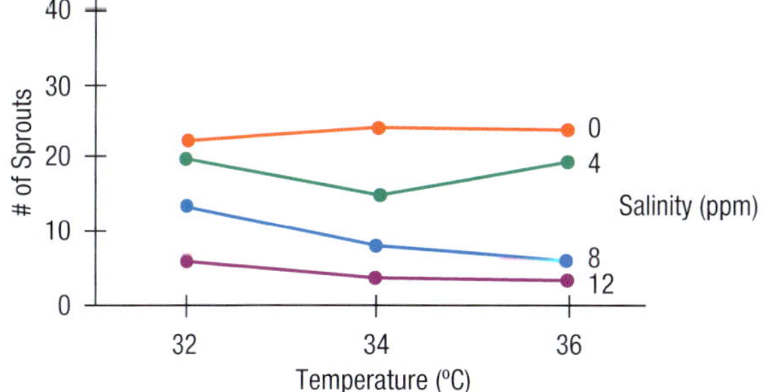

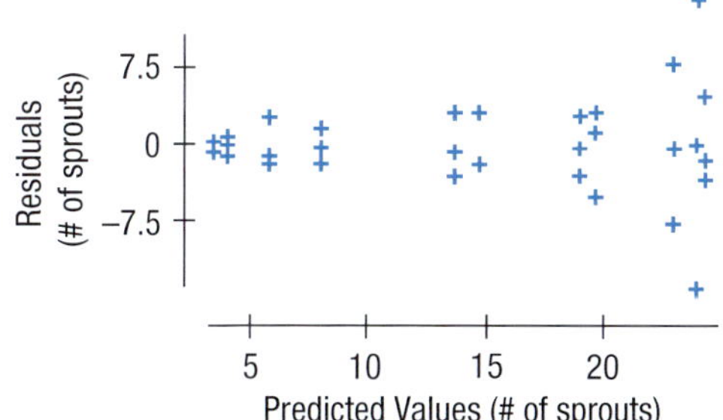

a) Comment on their analysis. What problems, if any, do you find with the analysis?

b) What sort of analysis might you do instead?

T **20. Containers revisited** Building on the cup experiment of the Chapter 4 Step-By-Step, a student selects one type of container and designs an experiment to see whether the type of *Liquid* stored and the outside *Environment* affect the ability of a cup to maintain temperature. He randomly chooses an experimental condition and runs each twice:

Liquid	Environment	Change in Temperature
Water	Room	13
Water	Room	14
Water	Outside	31
Water	Outside	31
Coffee	Room	11
Coffee	Room	11
Coffee	Outside	27
Coffee	Outside	29

After fitting a two-way ANOVA model, he obtains the following interaction plot, ANOVA table, and effects table:

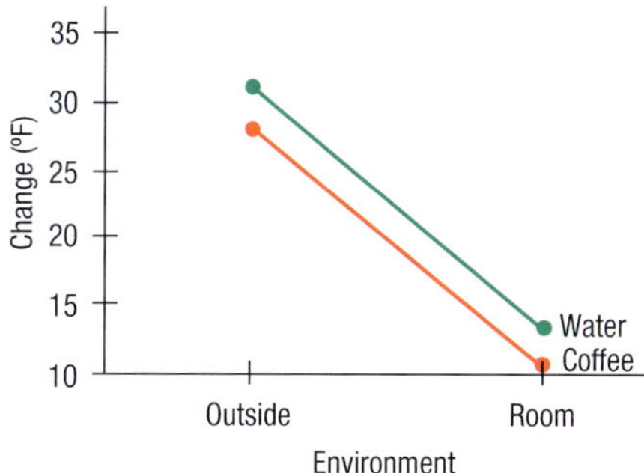

Source	DF	Sum of Squares	Mean Square	F-ratio	P-value
Liquid	1	15.125	15.125	24.2	0.0079
Environ	1	595.125	595.125	952.2	<0.0001
Interaction	1	0.125	0.125	0.2	0.6779
Error	4	2.500	0.625		
Total	7	612.875			

Term	Estimate
Overall mean	20.875
Liquid II [Coffee]	−1.375
Liquid II [Water]	1.375
Environment II [Outside]	8.625
Environment II [Room]	−8.625

a) State the null and alternative hypotheses.

b) Test the hypotheses at $\alpha = 0.05$.

c) Perform a residual analysis.

d) Summarize your findings.

T 21. Gas additives An experiment to test a new gasoline additive, Gasplus, was performed on three different cars: a sports car, a minivan, and a hybrid. Each car was tested with both *Gasplus* and regular gas on 10 different occasions and their gas mileage was recorded. Here are the partial boxplots:

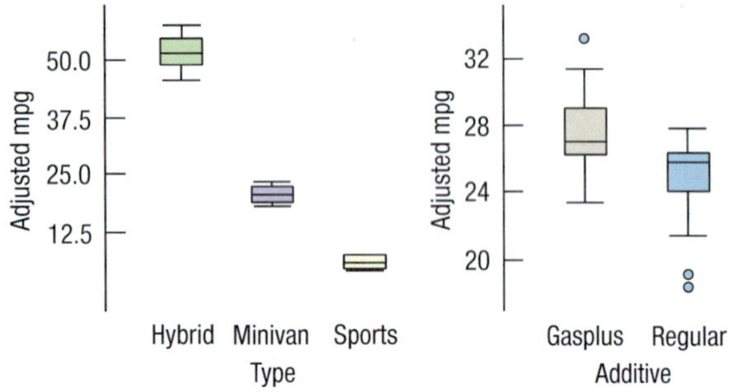

A two-way ANOVA with interaction model was run, and the following ANOVA table resulted:

Source	DF	Sum of Squares	Mean Square	F-ratio	P-value
Type	2	23157.4	11578.7	2739.54	<0.0001
Additive	1	98.7001	98.7001	23.35	<0.0001
Type × Additive	2	59.5954	29.7977	7.05	0.002
Error	54	228.232	4.22652		
Total	59	23543.9			

A plot of the residuals vs. predicted values showed:

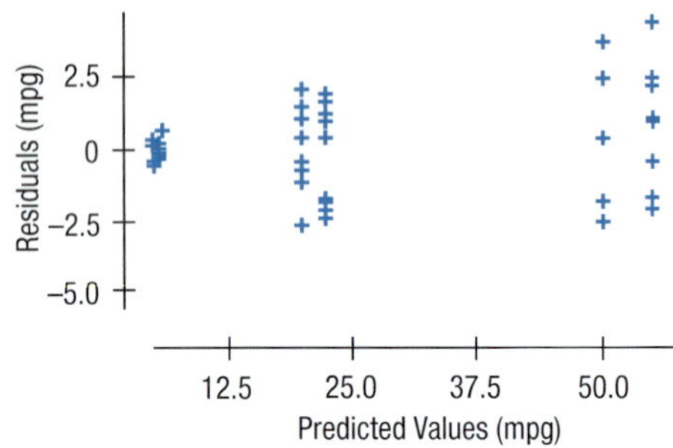

What conclusions about the additive and car types do you draw? Do you see any potential problems with the analysis?

T 22. Chromatography A gas chromatograph is an instrument that measures the amounts of various compounds in a sample by separating its constituents. Because different components are flushed through the system at different rates, chromatographers are able to both measure and distinguish the various constituents of the sample. A counter is placed somewhere along the instrument that records how much material is passing at various times. By looking at the counts at various times, the chemist is able to reconstruct the amounts of various compounds present. The total number of counts is proportional to the amount of the compound present.

An experiment was performed to see whether slowing down the flow rate would increase total counts. A mixture was produced with three different *Concentration* levels: *low, medium,* and *high*. The two *Flow Rates* used were *slow* and *fast*. Each mixture was run 5 times and the total counts recorded each time. Partial boxplots for *Concentration* and *Flow Rate* show:

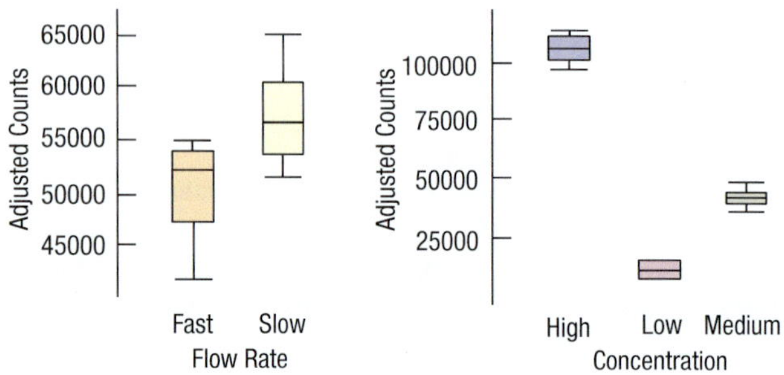

A two-way ANOVA with interaction model was run, and the following ANOVA table resulted:

Source	DF	Sum of Squares	Mean Square	F-ratio	P-value
Conc	2	483655E5	241828E5	1969.44	<0.0001
Flow rate	1	364008E3	364008E3	29.65	<0.0001
Interaction	2	203032E3	101516E3	8.27	0.0019
Error	24	294698E3	122791E3		
Total	29	492272E5			

A plot of residuals vs. predicted values showed:

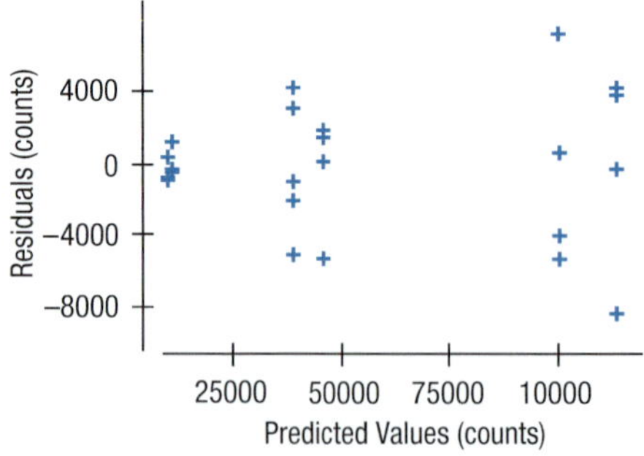

What conclusions about the effect of flow rate do you draw? Do you see any potential problems with the analysis?

23. Gas additives again Refer back to the experiment in Exercise 21. Instead of *mpg* redo the analysis using log (*mpg*) as the response. Do your conclusions change? How? Are the assumptions of the model better satisfied?

24. Chromatography again Refer back to the experiment in Exercise 22. Instead of *Total counts,* redo the analysis using log(*Total counts*) as the response. Do your conclusions change? How? Are the assumptions of the model better satisfied?

25. Batteries A student experiment was run to test the performance of 4 brands of batteries under 2 different Environments (room temperature and cold). For each of the 8 treatments, 2 batteries of a particular brand were put into a flashlight. The flashlight was then turned on and allowed to run until the light went out. The number of minutes the flashlight stayed on was recorded. Each treatment condition was run twice.

Partial boxplots showed:

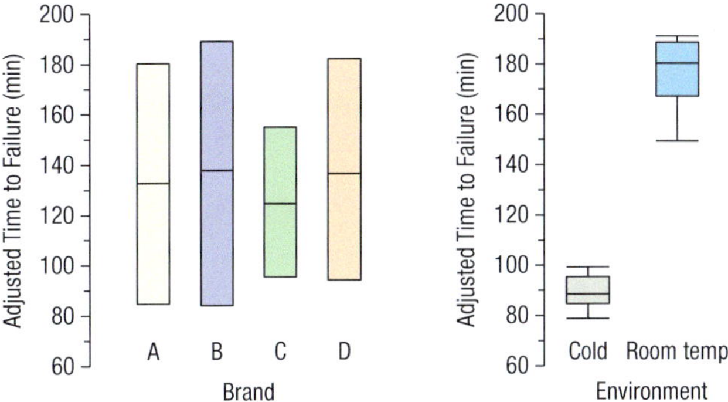

An interaction plot showed:

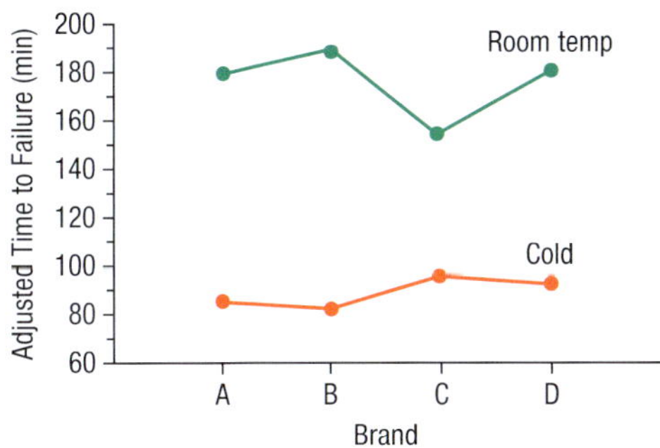

An ANOVA table showed:

Source	DF	Sum of Squares	Mean Square	F-ratio	P-value
Envir	1	30276.0	30276.0	706.15	<0.0001
Brand	3	374.750	124.917	2.9135	0.1007
Interaction	3	1190.00	396.667	9.252	0.0056
Error	8	343.0	42.8750		
Total	15	32183.8			

a) What are the main effect null and alternative hypotheses?
b) From the partial boxplots, do you think that the Brand has an effect on the time the batteries last? How about the condition?
c) Do the conclusions of the ANOVA table match your intuition?
d) What does the interaction plot say about the performance of the brands?
e) Why might you be uncomfortable with a recommendation to go with the cheapest battery (brand C)?

26. Peas In an experiment on growing sweet peas, a team of students selected 2 factors at 4 levels each and recorded *Weight, Stem Length,* and *Root Length* after $6\frac{1}{2}$ days of growth. They grew plants using various amounts of *Water* and *Quickgrow* solution, a fertilizer designed to help plants grow faster. Each factor was run at 4 levels: little, some, moderate, and full. They grew 2 plants under each of the 16 conditions.

An interaction plot of *Weight* in mg (*x*-axis is *Quickgrow*—levels are water) shows:

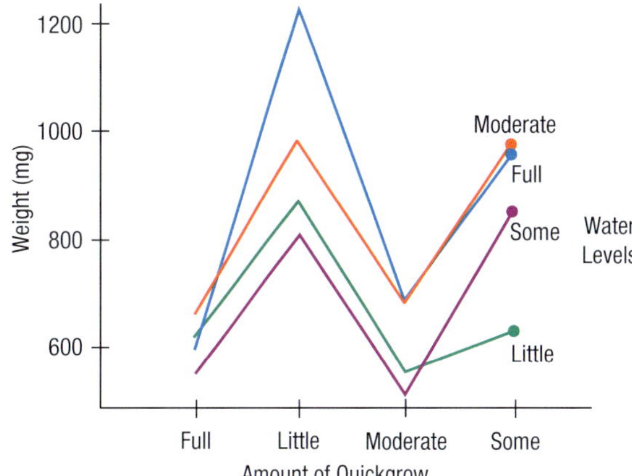

Because of this, a two-way ANOVA with interaction model was fit to the data, resulting in the following ANOVA table:

Analysis of Variance for Weight

Source	DF	Sum of Squares	Mean Square	F-ratio	P-value
Water	3	246255	82084.8	2.5195	0.0948
QS	3	827552	275851	8.4669	0.0013
Water × QS	8	176441	22055.1	0.6770	0.7130
Error	16	521275	32579.7		
Total	31	1771523			

Residuals plots showed no violations of the variance or Normality conditions. A table of effects for *Quickgrow* shows:

Level of Quickgrow	Effect
Little	213.8
Some	97.1
Moderate	−155.5
Full	−155.5

A table of effects for *Water* shows:

Level of Water	Effect
Little	−91.1
Some	−81.9
Moderate	66.4
Full	106.0

Write a report summarizing what the students have learned about the effects of *Water* and *Quickgrow* solution on the early stages of sweet pea growth as measured by *Weight*.

T **27. Batteries once more** Another student analyzed the battery data from Exercise 25, using a one-way ANOVA. He considered the experimental factor to be an 8-level factor consisting of the 8 possible combinations of *Brand* and *Environment*. Here are the boxplots for the 8 treatments and a one-way ANOVA:

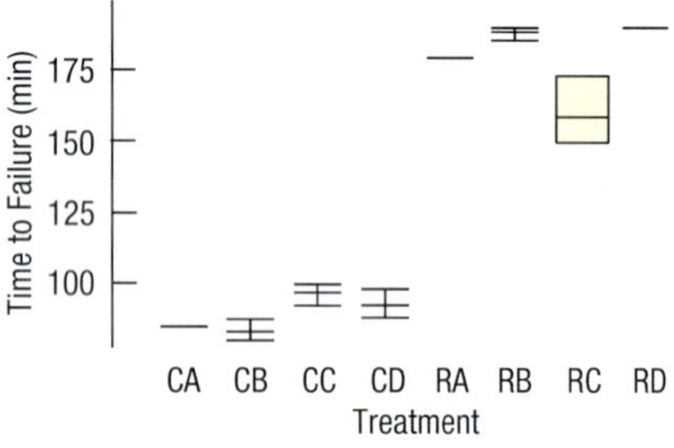

Source	DF	Sum of Squares	Mean Square	F-ratio	P-value
Treatment	7	31895.8	4556.54	93.189	<0.0001
Error	8	391.167	48.8958		
Total	15	32286.9			

Compare this analysis with the one performed in Exercise 25. Which one provides a better understanding of the data? Explain.

T **28. Containers one more time** Another student performs a one-way ANOVA on the container data of Exercise 20, using the 4 treatments *water room, water outside, coffee room,* and *coffee outside.* Perform this analysis and comment on the differences between this analysis and the one in Exercise 20.

JUST CHECKING

Answers

1. The null hypotheses are that the mean flight distance for all three *Weights* is the same, that the mean flight distance for both *Positions* is the same, and that the interaction effects are all 0.
2. Yes, the effects appear to be additive. The lines are nearly parallel.
3. Yes, the F-test provides no evidence to reject the null hypothesis with a P-value of >0.9.
4. Yes, we reject the null hypothesis with a P-value <0.0001.
5. Yes, we reject the null hypothesis with a P-value <0.0001.
6. Because both factors are significant, it appears that using the light *Weight* in the rear *Position* may result in the longest mean flight distance.

27

Introduction to Statistical Learning and Data Science

The Paralyzed Veterans of America (PVA) is a philanthropic and service organization chartered by the U.S. government to serve the needs of U.S. veterans who suffer from spinal cord injury or disease. Since 1946, PVA has raised money to support a variety of activities, including advocacy for veterans' health care, research and education in spinal cord injury and disease, and support for veterans' benefits and rights.

PVA raises money in a variety of ways, but the majority of their fund-raising comes from direct mail campaigns. They send free address labels or greeting cards to potential donors on their mailing list and request a donation for receiving these gifts.

PVA sends out regular solicitations to a list of more than four million donors. In 2016, PVA received about \$100 million in donations, but the effort cost them over \$40 million in postage, administrative, and gift expenses. Much of that money was spent on mailings to people who never responded. In fact, in any one solicitation, groups like PVA are lucky to get responses from more than a few percent of the people they contact. Response rates for commercial companies such as large credit card banks are so low that they often measure responses not in percentage points, but in basis points—hundredths of a percent.

If the PVA could avoid mailing to just half of the people who won't respond, they could save \$20 million a year and produce only half as much wasted paper. Can statistical methods help them decide who should get what mail?

The technological advances of the past few decades have enabled us to capture information at a rate that was unthinkable a generation ago. When you visit a social networking site, data about you, your likes, your photos, your friends, and your habits are uploaded to servers, waiting for analysis. Your smartphone sends information about the calls you make, the texts you send, and even your location to central servers. Every purchase you make online enables marketers to understand more and more about your likes and dislikes and changes the content of the ads, and perhaps the stories, that you see. Whether this information is being used in a prudent or ethical way by the government and industry is the subject of recent news and citizen concerns,[1,2] but the collection of data by business, science, and government is exploding and will continue to explode into the foreseeable future. And these data will be used to solve problems using techniques that are based on methods similar to those you've learned to master in this course.

You've learned to explore, visualize, and model data to help answer complex questions about the world. But the world of data continues to bring fresh challenges, like the one at PVA. In that dataset, there are 479 variables that can be recorded on each potential donor. How can we decide which variables to focus on? Multiple regression is powerful, but it can't model all the world's phenomena. New statistical models are being invented and deployed all the time. What you've learned in this text should help prepare you to learn about these new techniques and help answer questions like the PVA challenge. (For a source of interesting new data challenges see the website Kaggle.[3])

27.1 Data Science and Big Data

Data science is an umbrella term for the whole interdisciplinary field concerned with learning from data and communicating the results. What you've learned in this course is naturally a part of this endeavor, but typically data science involves the entire process of collecting, storing, and managing data as well as building models, making conclusions, and communicating the results. A child of both computer science and statistics, data science grew out the need to derive knowledge from increasingly large and complex datasets for which the existing tools were no longer adequate. This synergy between the disciplines has led to some notable successes. Genetic researchers are connecting genomic, behavioral, and medical information and are beginning to gain insights in the prediction and management of disease. Social scientists and business analysts are now able to sift through huge amounts of text to gain insights into everything from predicting future trends in society to the more workaday problem of predicting who is more likely to switch cell-phone providers. At the galactic scale, modern telescopes are now collecting terabytes of image data on literally billions of objects in the sky. The potential knowledge contained about the universe is vast, as are the challenges of cataloging and managing so much information.

Challenges arise from the fact that current methods in both computer science and statistics were not built to solve these problems. On the computer science side, it is challenging to store, access, and combine large datasets efficiently. On the statistics side, building models with thousands of potential predictors and their possible interactions remains a major area of research. Research in data science is growing, and demand for people who have data science skills is exploding. IBM predicts that "annual demand for the fast-growing new roles of data scientist, data developers, and data engineers will reach nearly 700,000 openings by 2020"[4]. Universities are scrambling to keep up with the demand. Twenty years ago you couldn't major in data science. Now there are hundreds of degree programs in data science all over the world. (For an interesting history of the term data science see "A Very Short History of Data Science."[5])

---

[1] "Momentum Builds Against N.S.A. Surveillance," *New York Times*, July 28, 2013.

[2] The Ethics Conversation We're Not Having about Data, *Harvard Business Review*, Nov 12, 2015. https://hbr.org/2015/11/ the-ethics-conversation-were-not-having-about-data

[3] www.kaggle.com

[4] www.forbes.com/sites/louiscolumbus/2017/05/13/ibm-predicts-demand-for-data-scientists-will-soar-28-by-2020/#18de2d877e3b

[5] www.forbes.com/sites/gilpress/2013/05/28/a-very-short-history-of-data-science/#1783a4a555cf

WHAT MAKES BIG DATA "BIG"? THE FOUR V'S

The size or "volume" of the dataset is just one of the aspects that make data "big," but complexity is equally important. Scientists at IBM have defined the 4 V's that make Big Data truly big: Volume, Variety, Velocity, and Veracity. So, not only is volume important, but also different types of data (variety), the speed and timeliness at which it is collected (velocity), and the quality (or lack thereof) of the data (veracity) make Big Data challenging to manage and analyze.

WHAT MAKES BIG DATA BIG? AN EXAMPLE

The U.S. Census Bureau collects information about nearly every person living in the United States. You might think that this would be Big Data. Although it is certainly voluminous, it lacks at least two crucial criteria: velocity and variety. Currently the Bureau collects demographic data only every 10 years on the entire population. It gathers economic data more frequently, but on smaller samples. According to Census Bureau Chief of Research and Methodology Tommy Wright, adding real-time information about individuals' purchases through credit and loyalty card transactions and linking those to the demographics would go a long way to helping provide a more accurate and instantaneous picture of the state and health of our economy.

The term **Big Data** is used to describe the challenges of large collections of data (see the Four V's sidebar). With that term has come a lot of hype about what all these data can do and solve. As you've learned throughout the course, the quality of the data and how the data were collected are crucial factors in determining whether any data can actually help answer a question correctly, be the basis for a reasonable model, or help make a sound decision. Unfortunately, in much of the literature about Big Data, these data issues tend to be downplayed and unrealistic promises about the kinds of things data can do are made and believed all the time.

From the name Big Data you may think that any dataset with a large number of cases is challenging. However, it's often the number and the variety of *variables* that makes analyzing a dataset difficult. If the number of cases in our dataset is too large for building a model, we know that sampling can lead to reasonable insights. But deciding which variables to use in a model is one of the most difficult tasks of this kind of analysis. It's not practical to sample from the variables as we can from the cases. In addition, how fast the dataset may be changing and the quality of the data (or lack thereof) provide more challenges. See the sidebar for an example of why analyzing the U.S. Census Bureau's huge database is not necessarily a Big Data problem.

The many terms associated with data science and Big Data can be confusing. The term **machine learning** is often used to describe models or algorithms that involve heavy computation and use data to predict new cases. Any statistical model, including multiple regression, uses computers to estimate the parameters and make predictions from data. Where the boundary between machine learning and computational statistics lies is unclear. Statisticians prefer the term **statistical learning** and use **data mining** to describe the process of discovering relationships in data using computationally intensive models. In the business world, the term **business analytics** (or sometimes **predictive analytics** or simply **analytics**) describes the use of these models to drive business decisions from large datasets. Recently, a particular type of model called an artificial neural network (see Section 27.3) outperformed other machine-learning algorithms in a contest involving image recognition, a skill that's extremely important for self-driving vehicles and other vision applications. Because these networks have more than one layer (described later), they were given the term **deep learning**. Like many of the terms in this realm, the name suggests something profound. It is part of the marketing of these methods to make them sound like they will deliver more than they actually do. While there are many success stories in Big Data, it is important to guard against the hype used by vendors who promise unrealistic results for these models.

In this chapter, we'll introduce some of the most commonly used methods in data mining and statistical learning and compare them to the tools you already know. Most of the methods won't be found in standard statistics packages or software, but open source versions of them are easily found, especially as implemented in the widely used R statistics program.[6,7] For simplicity, we'll use the term data mining to describe the process of exploring, modeling, and predicting from data using any of these statistical or machine learning models. As we'll see, some involve more automatic re-expressions and more flexible fitting than multiple regression, and most of the models are less interpretable than multiple regression, but the skills you've acquired by learning to fit models, diagnose them, and assess their limitations already prepare you to start entering the world of Big Data.

[6] cran.r-project.org

[7] www.nytimes.com/2009/01/07/technology/business-computing/07program.html

DATA MINING—SOME HISTORY

In the 1980s, credit card companies realized that with the incredible amount of data they routinely had to collect, there might be some extra value in the data other than simply recording all their transactions. They knew that customers had different "profiles." For example, some customers used their cards primarily for travel, while others used them for everyday purchases. This market segmentation was enabled by analyzing the transactional data.

By putting customers into different groups, credit card companies could more effectively choose the offers each group would receive.

The Goals of Data Mining

The broad purpose of statistical learning and data mining is to extract useful information hidden in large databases. With a database as large as a typical **data warehouse** (where the data are stored), that search can be like looking for a needle in a digital haystack. Analysts may start their search by asking specific questions about the data. Guided by their domain knowledge, they may try a query-driven approach, asking a series of specific questions to deduce patterns. Such an approach is **online analytical processing** or **OLAP**. An OLAP question for the PVA data might be: "What percentage of potential donors under the age of 65 with incomes between \$40,000 and \$60,000 in the Western region who have not donated in the previous two years gave more than \$25 to the most recent solicitation?" Although OLAP is efficient in answering such multivariable queries, it only produces answers for specific queries. Because it does not build a model, it can't generalize to other situations, or help develop a theory.

In contrast to an OLAP query, the outcome of a data mining analysis is a **predictive model**—a model that uses predictor variables to predict a response. For a quantitative response variable the model will predict the value of the response (as in linear regression). For a categorical response, the model estimates the most probable category of the response, often with the probability that the response variable is in that class (as in logistic regression). Both linear and logistic multiple regressions are common tools of a data miner. But data mining models, like all statistical models, are predictive and can be applied to other situations.

When the goal is to predict a quantitative response variable, the problem is generically called a **regression problem**, regardless of whether linear regression is used as one of the models. When the response variable is categorical, the problem is called a **classification problem** because the model will either guess the most likely category for each or assign a probability to each class for that code. For example, for the classification problem of predicting whether or not a particular donor will give to the next campaign, a model will produce either the most probable class (donate or not) or the probabilities of each of these. Predicting the amount of money a donor will donate is a regression problem.

For example, an analyst using an OLAP query might find that customers in a certain age group responded to a recent product promotion. But without building a model, the analyst can't understand the relationship between the customer's age and the success of the promotion, and thus may be unable to predict how the product will do in a more general setting. The goal of a data mining project is to increase understanding and **knowledge** by building a model to answer a specific set of questions raised at the beginning of the project that can generalize to future situations.

Both regression and classification problems are referred to as **supervised problems** because the response variable's outcome is known for the data on which the model is based. For example, for the PVA data we encounter, we *know* for at least one group of donors the values of the responses *TARGET-B* (did you give?) and *TARGET-D* (how much did you give?). We can use these values to construct the model. All the models you've seen so far in the text have been for supervised problems.

By contrast, there are problems for which there is no variable that measures what we are trying to predict. For example, an anthropologist might want to cluster pot shards from a site into groups with similar features to try to assess whether they came from a single culture or several different groups. That's a fine goal, but, unfortunately, pot shards don't come labeled this way. In such a **cluster analysis** there is no response variable—we just want to group together similar items. Whether the grouping we find is "correct" may be impossible to determine. Problems like these are known as **unsupervised problems** because there is no actual response variable.

Cluster analysis, grouping similar objects together, is the most common way of dealing with unsupervised problems. Many different algorithms are available to find clusters, but they all rely on a measure of how close (or, equivalently, how distant) the objects are to each other. (Further discussion of clustering and other unsupervised methods is beyond the scope of this text.)

Data Mining and Statistics

Data mining is similar to traditional statistical analysis in that it involves exploratory data analysis and modeling. However, several aspects of data mining distinguish it from more traditional statistical analysis. Although there is no consensus on exactly what data mining is and how it differs from statistics, some of the most important differences include:

- **The size of the databases.** Although no particular size is required for an analysis to be considered data mining, an analysis involving only a few hundred cases or only a handful of variables would not usually be considered data mining.
- **The exploratory nature of data mining.** Unlike a statistical analysis that might test a hypothesis or produce a confidence interval, the outcome of a data mining effort is typically a model used for prediction. Usually, the data miner is not interested in the values of the parameters of a specific model or in testing hypotheses. A classic example is automatic reading of ZIP codes from handwritten envelopes. The manager of the New York Post Office is unlikely to care what variables were included in the algorithm that reads the correct ZIP code automatically, as long as it can do so with a high level of accuracy.
- **The data are "happenstance."** In contrast to data arising from a designed experiment or survey, the typical data on which data mining is performed have not been collected in a systematic way. This is an important difference. Warnings not to confuse association with causation are especially pertinent for data mining. In addition, the sheer number of variables involved makes any search for relationships among variables prone to Type I errors.

Humans love to see signals and relationships among variables even when none exists. Resist the temptation to believe too strongly in these models. Although they can be useful for generating hypotheses to be tested, they are, by nature, exploratory. Be especially careful to check the "pedigree" of the data. The phrase "Garbage in, garbage out" was never more true than when applied to data mining.

- **The results of a data mining effort should be "actionable."** Before starting the analysis, there should be a consensus of what the problem of interest is and how the model will help solve the problem. There should be an action plan in place for a variety of possible outcomes of the model. Exploring large databases out of curiosity or just to see what they contain is not likely to be productive.
- **The modeling choices are automatic.** Typically the data miner will try several different types of models to see what each has to say, but doesn't want to spend a lot of time choosing which variables to include in the model or being careful in the kinds of choices he or she might make in a more traditional statistical analysis. Unlike the analyst who wants to understand the terms in a stepwise regression, the data miner is more concerned with the predictive ability of the model. As long as the resulting model can help make decisions, the data miner is likely to be satisfied.

Data Mining Challenges

One of the main challenges of data mining is, not surprisingly, data. In Part III we discussed the two main sources of data used to generalize to the world: sample surveys and designed experiments. Unfortunately, most of the data used in data mining analyses are neither of those. Companies collect data on transactions that their customers make. When you purchase a pair of earbuds on Amazon, that transaction is recorded and linked to several databases. It affects many aspects of Amazon's business: the inventory, items that need to be shipped, the revenue Amazon receives, and your purchase history. These data may be stored in different databases, or warehouses, often in different physical locations. These data are collected so that Amazon can run its business to know how many items they have in stock, how much money was received, and how and where to ship the item. Data

WHAT DOES A DATA SCIENTIST DO?

In response to this question on Quora (quora.com), James Madison (data scientist) answered:

1. Clean data.
2. Clean data.
3–12. Clean data.
13. Do some math.
14. Try to get everyone to understand my findings.
15–25. Try to get everyone to understand my findings.
26. Repeat 1-25.

like these are called **transactional data** because they are based on the transactions between the company and the customer.

An analyst may be asked to investigate how profitable a 10% coupon on a certain item might be if given to the right customers. But who are the right customers, and how much extra profit (or loss) is the coupon likely to generate? That's an important question, and a great way to answer it might be to perform an experiment, giving the coupon to a randomly selected group and comparing their purchases to the control group. Often, however, analysts attempt to build predictive models based on the data at hand — usually past transactional data. Whether those data contain the answer to the question is not clear, even if powerful data mining tools are brought to bear on the problems.

Another challenge in large databases is manipulating, or wrangling, the data themselves. In a large database, there may be many variables and cases that are irrelevant to the study. There may be missing data or inaccurate data. You may need to combine data from several sources. To perform operations on data efficiently you will probably need to write some computer code or to build a team of people to work with who have the expertise to do this. There are commercial software programs available to help with this, but often researchers write code in open source languages such as Python and R.

Most data mining projects will spend a significant amount of the total time devoted to the project cleaning, simplifying and transforming the data into a form that is suitable for analysis. In Chapter 24 we talked about some of the challenges of preparing a dataset for multiple regression. The larger datasets analyzed by data miners have even more data preparation challenges (see the sidebar for a humorous answer to the question of what a data scientist does).

Data Mining Myths

Data mining software usually contains a variety of exploratory and model-building tools and a graphical user interface designed to guide the user through the data mining process. Data mining can often *assist* the analyst to find meaningful patterns and help predict future customer behavior, but the more the analyst knows about the domain being studied, the more likely he or she will be successful using data mining. Data mining is not a magic wand that can overcome poor data quality or collection. A product may have tools for detecting outliers and may be able to impute (assign a value for) missing data values, but all the issues you learned about good statistical analysis are still relevant for data mining.

Here are some of the more common myths about what data mining can do.

- **Myth 1:** Find answers to unasked questions.

  Even though data mining can build a model to help answer a specific question, it does not have the ability to answer questions that haven't been asked. In fact, formulating a precise question is a key first step in any data mining project.

- **Myth 2:** Automatically monitor a database for interesting patterns.

  Data mining techniques build predictive models and can answer queries, but do not find interesting patterns on their own any more than regression models do.

- **Myth 3:** Eliminate the need to understand the context.

  In fact, the more an analyst understands his or her subject area, or domain, the more effective the data mining effort will be.

- **Myth 4:** Eliminate the need to collect good data.

  Good data are as important for a data mining model as for any other statistical model that you've encountered. While some data mining software contains tools to help with missing data and data transformations, there is no substitute for quality data.

- **Myth 5:** Eliminate the need for good data analysis skills.

  The better the data analysis skills of the miner—the skills you've learned in every chapter of this text—the better the analysis will be when using data mining tools. Data mining tools are more powerful and flexible than a statistical tool like regression but are similar in the way they work and in how they're implemented.

- **Myth 6:** Machine learning algorithms think like humans.

  Some of the hype about data mining and machine learning revolves around the belief that these algorithms, or models, think and learn. In fact, they "learn" about the relationship between the predictors (x) and response (y) no more, and no less, than linear regression learned about it. They simply fit a model to the data. Some do it in an automatic way, which may lead one to think they are actually learning, but this is simply a human interpretation of a powerful algorithm.

JUST CHECKING

A bank vice president brings in her analytic team to talk about several things she wants to know about a recent offer the bank mailed out to open an IRA (individual retirement account). For each of the following, identify if it's an appropriate data mining problem. If so, is it a supervised or an unsupervised problem? If supervised, is it a regression or a classification problem?

1. Why did some people not respond to the offer?
2. What percentage of those under 65 who have been bank customers for more than 5 years responded to the offer?
3. Using the bank products that customers currently use and other demographic information, group customers into different segments for the purposes of sending different retirement offers.
4. On the basis of past data, what is the chance that a given customer will accept a new offer that the bank sends out?
5. How many transactions is a given customer likely to make next month based on past behavior and other demographics?

27.2 The Data Mining Process

Data mining projects require a number of different skills. For that reason, a data mining project should be a team effort. No one person is likely to have the domain knowledge, computer and database management skills, software expertise, and statistical training needed for all the steps in the process. Because data mining projects tend to be complex, it's useful to map out the steps for a successful project. A group of data mining experts have shared their combined expertise in a project called the Cross Industry Standard Process for Data Mining (CRISP-DM).[8] A CRISP-DM schematic of the data mining cycle appears in Figure 27.1.

Let's use PVA as an example and follow the data mining cycle from beginning to end. The process starts with the *business understanding* phase, in which the problem to be addressed is carefully articulated. For any data exploration, it is best to have a specific problem before starting, but for data mining, with potentially very large databases, this is absolutely crucial. For PVA, the goal of "understanding how to best manage donors" might sound like a good idea, but is not precise enough for a data mining project. A better, more specific question might be to build a model to discover which donors are most likely to give enough in the next solicitation to justify sending them a free gift. It's important to have all the members of the data mining team involved at this stage, and the team should be representative of all parts of the business that may be affected by the resulting decisions. If key constituents aren't represented, the model is in danger of not answering the right question. It can't be overemphasized how important it is to have consensus on a precise, correctly formulated problem to be addressed before continuing with the data mining effort.

The next phase of *data understanding* is central to the entire data mining project. If you want to know which donors are likely to participate in the next mailing, you have to have data that can actually answer that question. This is where many data mining projects fail.

[8] In 2015, IBM announced an updated version of CRISP-DM called ASUM-DM, which stands for Analytics Solutions Unified Method for Data Mining/Predictive Analytics, but the basic principles remain the same.

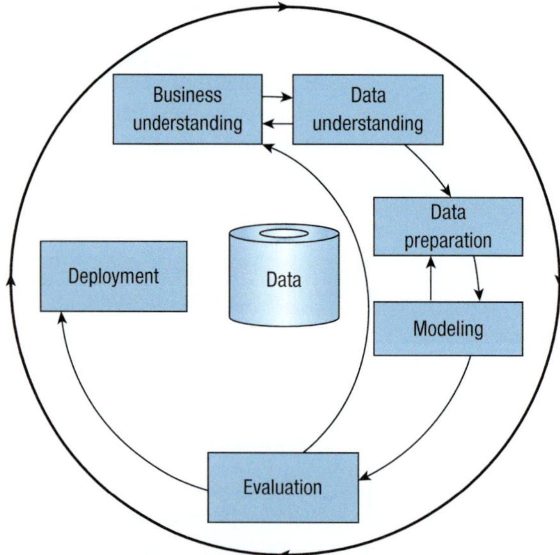

Figure 27.1
A picture of the CRISP-DM data mining process.

A traditional statistics approach might be to survey donors before the next mailing, but, typically, a data mining project uses only data the company has already collected. Much of that data is **transactional data**. Other data are **demographic**—information about the customer that might include the customer's income, number of children, education, and other customer-specific information. This information may have been supplied by the customer or may be purchased from other companies. Still other data are *historical*, based on past history between the customer and the company. These data sources are typically stored on different computers or servers, which means that they need to be merged before the analyst can access them. Usually, someone well versed in database architecture and data storage is needed to extract the relevant information in different databases and to merge them into a form that is useful for analysis. This requires teamwork and communication skills. And, unfortunately, all of this means that the answer to a specific question may not be found in these data, even when using a very powerful algorithm designed to search for it.

Customer-specific data can also be purchased from a variety of commercial organizations. For example, a credit card company may want to send out an offer of free flight insurance to customers who travel frequently. To help them know which customers those are, the company may buy information about customers' magazine subscriptions to identify customers who subscribe to travel or leisure magazines. The sharing and selling of individual information is controversial and raises privacy concerns, especially when the purchased data involve health records and personal information. In fact, concerns over the sharing of health data in the United States led to a set of strict guidelines known as HIPAA (Health Insurance Portability and Accountability Act) regulations. Restrictions on the collection and sharing of information about customers vary widely from country to country.

The PVA customer database[9] is a typical mix of variables. It contains 481 demographic and historical variables on each of nearly 100,000 donors. There are 479 potential predictor variables and two responses: *TARGET_B*, a 0/1 variable that indicates whether the donor contributed to the most recent campaign, and *TARGET_D*, which gives the dollar amount of the contribution. The demographic variables are both donor specific (e.g., age, income, number of children) and ZIP code specific (e.g., percent with college education, average income, percent minority). Both types of variables can be collected by the company itself or purchased from other organizations, both private and governmental (the U.S. Census

[9]The PVA made some of their data available for the 1998 Knowledge Discovery and Data Mining (KDD) contest. The object of the contest was to build a model to predict which donors should receive the next solicitation based on demographic and past giving information. The results were presented at the KDD conference (www.kdnuggets.com). The variables discussed in this chapter are some of the ones that PVA made available. We have used various subsets of the PVA dataset in previous chapters, sometimes changing the variable names for clarity. The entire dataset can be found at the University of California at Irvine Data Repository: kdd.ics.uci.edu/databases/kddcup98/kddcup98.html.

Bureau is a source of information). About 10% of the PVA variables describe past giving behavior collected by the PVA itself. More than half of the variables are regional data based on ZIP codes, most likely purchased from the Census Bureau. The rest of the variables are donor specific, either gathered by the PVA or purchased from other organizations. The historical data pertain to the past behavior of the donor and include whether and how much the individual donated to various campaigns in the past. Table 27.1 shows the first 18 records for a subset of the 481 variables found in the PVA data. We can guess what some of the variables mean from their names, but others are more mysterious.

Becoming familiar with a complex dataset can take many months. Information about the variables, including their definitions, how they are collected, the date of collection, and so forth, is collectively called **metadata**. The variables in the PVA dataset shown in Table 27.1 are typical of the types of data found in the customer records of many companies. We shouldn't be surprised to find out that *AGE* is the donor's age measured in years and that *ZIP* is the donor's postal ZIP code. But without the metadata, it would be hard to know that *TCODE* is the code used before the title on the address label (0 = blank, 1 = Mr., 2 = Mrs., 28 = Ms., and so on) or that *RFA_2A* is a summary of past giving.

The choice of the variables to explore depends on the subject matter knowledge of the team. As is clear from the PVA data, it is probably inefficient to spend a lot of time exploring each variable. This is one of the great challenges of a data mining project. At the beginning of the project, it's best to include as many variables as possible that might be useful, but keep in mind that having *too* many variables can make the model selection phase more difficult.

Once the variables have been selected and the response variable(s) agreed upon, it's time to begin the **data preparation** phase for modeling. As mentioned earlier, this can be a time-consuming part of the process and is also likely to be a team effort. Investigating missing values, correcting wrong and inconsistent entries, reconciling data definitions, and merging data sources are all challenging issues. Some of these can be handled automatically, while others require painstaking, detailed analysis. The team will have to

ODATEDW	OSOURCE	TCODE	STATE	ZIP	DOB	RFA_2A	AGE	OWN	INC	SEX	WEALTH	AVGGIFT	TARGET_B	TARGET_D
9401	L16	2	GA	30738	6501	F	33	U	5	F	2	11.66667	0	0
9001	L01	1	MI	49028	2201	F	76	H	1	M	2	8.777778	0	0
8601	DNA	1	TN	37079	0	E		U	1	M		8.619048	1	10
8601	AMB	1	WI	53719	3902	G	59			M		16.27273	0	0
8601	EPL	2	TX	79925	1705	E	81	H	2	F	6	10.15789	0	0
8701	LIS	1	IN	46771	0	F				M		8.871333	0	0
9201	GRI	1	IL	60016	1807	F	79	H	4	M	6	13.8	0	0
9401	HOS	0	KS	67218	5001	G	48	U	7	F	7	18.33333	0	0
8901	DUR	0	MI	48304	1402	F	84	H	7	M	9	12.90909	0	0
8601	AMB	0	FL	34746	1412	F	83	H	2	F	3	9.090909	0	0
9501	CWR	2	LA	70582	0	D		U	5	F		5.8	0	0
9501	ARG	0	MI	48312-	4401	E	54			F		8	0	0
8601	ASC	0	TX	75644	2401	G	74	U	7	F		13.20833	0	0
9501	DNA	28	CA	90059	2001	E	78	H	1	F		10	0	0
9201	SYN	0	FL	33167	1906	F	79	U	2	M	3	10.09091	0	0
9401	MBC	2	MO	63084	3201	F	66	H	5	F		10	0	0
9401	HHH	28	WI	54235	0	F		H	7	F		20	0	0
9101	L02	28	AL	36108	4006	F	58	H	5	F		10.66667	0	0

Table 27.1
Part of the donor records from the PVA dataset. Shown here are 15 of the 481 variables and 18 of the nearly 100,000 donor records used in the 1998 KDD (Knowledge Discovery and Data Mining) data mining competition.[10] The actual PVA customer database contains several million records.

[10]The KDD Cup is the leading data mining competition in the world and is organized by the SIGKDD interest group of the ACM (Association for Computing Machinery).

decide how much effort is reasonable to make the dataset as complete and reliable as possible, given the time and resource constraints of the project.

Part of the data preparation phase involves both transforming predictor variables (as we did in Chapter 24) and creating new variables from combinations of them. This is often referred to as **feature creation**. A new feature (variable) can be created *a priori* from the analyst's domain knowledge or sometimes more automatically by the data mining algorithms themselves. For example, a banker trying to model the chance of customer bankruptcy might see that customer *debt* and *income* have been measured and, from his or her experience, decide to create the *debt/income* ratio. For the Paralyzed Veteran's data, the numerous past giving variables were combined into three new variables that summarized when the donor last contributed (called *recency*), how often the donor gives (*frequency*), and how much the donor gives (*monetary*). This type of RFM (recency, frequency, monetary) analysis is widely used in marketing science.

Once the data have been prepared, the analysts begin the *modeling* phase by exploring and developing models. If the number of variables is very large, a preliminary, exploratory model (such as a **decision tree**, described in a later section) might be a good way to narrow the candidate predictor variables down to a reasonable number. However, if the number of predictors is reasonably small, modelers can use traditional graphical displays (histograms, bar charts) to explore each variable and then investigate the relationship between each predictor and the response with displays such as scatterplots, boxplots, or segmented bar charts. The more knowledge of the data and the variables that go into the model, the higher the chance of success for the project.

Once the collection of variables has been determined, the analysts typically fit several models for the response variable. At this point the *evaluation* phase can begin. Looking at the structure of each model, including which predictors are important for each model, should give the analysts information about the relationship between the predictors and the response that each model is built on. In the *evaluation* phase, the candidate models are tested by predicting the response values on the **test set** as we discussed in Chapter 24, Section 3. Various criteria are used to judge the models, depending on the type of problem. For example, for a regression problem with a quantitative response, the sum of squares of the residuals (or a version of R^2—discussed in the next section) on the test set might be compared across models. For a problem predicting a two-level categorical variable (with levels NO and YES), the two types of errors that arise (predicting YES when the true response is NO and vice versa) can be balanced in different ways. Placing different costs on the two types of errors might be warranted, and the total cost of misclassification should reflect that (see sidebar).

> **WHY NOT JUST REPORT THE TOTAL ERROR RATE WHEN JUDGING MODELS?**
>
> Fraud detection is a huge problem in the credit card industry and costs billions of dollars a year. Fortunately, it is fairly rare, affecting less than 1% of all transactions (wallethub.com/edu/credit-debit-cardfraud-statistics/25725/). So, simply reporting all transactions as fraud-free will result in an overall error rate of less than 1%. An accuracy rate of greater than 99% may sound good, but it's much better to look at the two types of errors (that we learned about in Chapter 19) separately, even weighing the different costs of each mistake.

At this point, the business question that motivated the project should be revisited. Does the model help to answer the question? If not, it might be necessary to go back to one of the previous steps to investigate why that happened. This kind of cycling through the various steps of the process is not uncommon. The reason that so many phases of the CRISP-DM diagram (Figure 27.1) have arrows both to and from them is that this process is not a straightforward movement through the phases—it is an iterative and interconnected process.

If a particular model (or average of several models) seems to give insight into the problem and predicts well enough to be useful, then it's time for the *deployment* phase. Usually, that means using the model to predict an outcome on a larger dataset than the original, or on more recent data than were used to build the model. Knowledge gained at any phase may trigger a re-examination of an earlier phase. Even the "final" phase, deployment, is not really final. In many business situations, the environment changes rapidly, so models can become stale quickly. Although a data mining project is complex and involves the efforts of many people, mapping out the different phases can help ensure that the project is as successful and reproducible as possible.

Successful Data Mining

The size of a typical data warehouse makes any analysis challenging. The ability to store data is growing faster than the ability to use it effectively. Commercial data warehouses often contain terabytes (TB)—more than 1,000,000,000,000 (1 trillion) bytes—of data

(1 TB is equivalent to about 260,000 digitized songs), and warehouses containing petabytes (1 PB = 1000 TB) are now common. According to *Gizmodo* magazine,[11] about 300 million photos and 500 TB of data are uploaded to Facebook every day. All the U.S. Census data from 1790 to 2000 would take about 600 TB. Data miners hope to uncover some important strategic information lying hidden within these massive collections of data.

As we have seen, to have a successful data mining outcome, the first step is to have a well-defined business problem. With 500 variables, there are over 100,000 possible two-way relationships between pairs of variables. The number of pairs that will be related just by chance is likely to be large. And it's human nature to find many of these relationships interesting and even to posit plausible reasons why two variables might be associated. Some of these variables may appear to provide a useful predictive model, when in fact they do not. A well-defined objective can help keep the analysts from going down a lot of blind paths.

As in painting a house, much of the effort of a data mining project lies in preparation and cleaning. A painter would like to get right to the more enjoyable part of painting, but unless enough time is spent prepping the wood, all that effort will ultmately be wasted. It is estimated that for any data mining project, 65% to 90% of the total project time is spent in such data preparation—investigating missing values, correcting wrong and inconsistent entries, reconciling data definitions, and possibly creating new variables from the original ones. The data may need to be extracted from several databases and combined. Errors must be corrected or eliminated, and outliers must be identified.

So, a successful data mining effort often requires a substantial amount of time devoted to basic preparation of the data before any modeling is performed. It will have a clear objective, agreed to by a team of people who will share the work and the responsibility of the data mining. It will have an action plan once the results of the data mining are known, whether those results are what the team expected or desired or not. Finally, the data mining should be accompanied by as much knowledge about both the data and the subject in question as possible. Blind searches for patterns in large databases are rarely fruitful and will waste valuable analytic resources.

EXAMPLE 27.1 Data Preparation

We've started looking at our (Paralyzed Veterans of America) database to try to understand who our current donors are. Just looking at the variable *AGE*, can we use it to predict who is most likely to give to our next campaign?

QUESTION: What corrections or adjustments would be appropriate?

ANSWER: Here are a histogram and boxplot for the variable *AGE* for all 94,649 records of the PVA dataset.

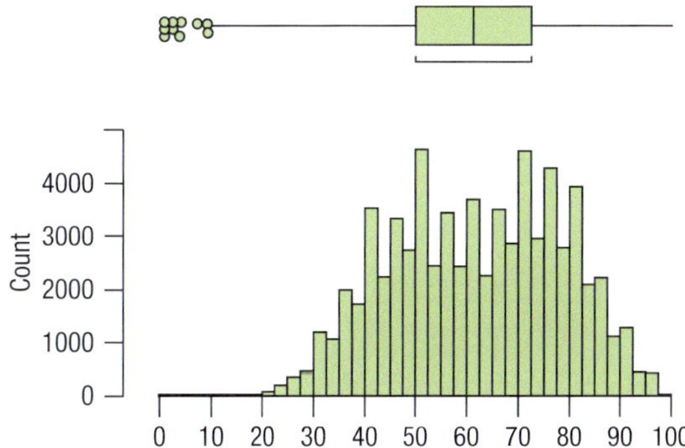

[11]gizmodo.com/5937143/what-facebook-deals-with-everyday-27-billion-likes-300-million-photos-uploaded-and-500-terabytes-of-data

	Quantiles	
100.0%	maximum	98.000
99.5%		95.000
97.5%		90.000
90.0%		83.000
75.0%	quartile	75.000
50.0%	median	62.000
25.0%	quartile	48.000
10.0%		39.000
2.5%		31.000
0.5%		26.000
0.0%	minimum	1.000
Mean		61.598019
Std Dev		16.666865
Std Err Mean		0.0624749
upper 95% Mean		61.720469
lower 95% Mean		61.475568
N		71170

There is a group of cases with ages below 20 that should immediately draw our attention. Are people that young likely to be donors? A closer examination reveals that there are even a handful of cases whose ages are below 15, of whom 17 are younger than 5 years old and 9 are 1 year old. There are also 23,479 missing values for the variable *AGE*. From what I know about our database, ages below 20 (and certainly those below 10) are likely to be errors. I could go back and investigate those individually, but I think given how few there are, simply deleting the values below 21 is probably the wisest course of action for now.[12]

JUST CHECKING

The bank vice president from the previous Just Checking wants to know some things about the data and the data mining process. Respond briefly to her assumptions and questions to a data mining consultant.

6. To me, it seems best and most efficient to build the team from one department, like my information technology group, rather than make a cross-disciplinary team for the project. What do you think?
7. I assume that once we identify the data sources we'll use, we can get right into modeling the issue, correct?
8. These data mining techniques are so powerful that I'm virtually guaranteed the predictions will be accurate, right?
9. Rather than look at one small problem, I think it might be a good idea to just explore the database and see what's in there. What do you think?
10. Once we test the model and it seems to work well, we'll be able to use it for the next few years at least.

27.3 Data Mining Algorithms: A Sample

Data mining models are often referred to as **algorithms**, a term that describes a sequence of programming steps with a specific purpose. You may hear a method (even one like linear regression) referred to as a model, an algorithm, a tool, or generically as a method. The

[12]Further analysis revealed that age was calculated from date of birth by a computer algorithm, but only the last two digits of the date of birth year were used. So, some people (perhaps no longer living) who were in the database had ages 100 years too young.

terms seem to be used interchangeably. Technically, the algorithm is the set of rules that the computer uses to compute the value of the model. The algorithm then uses the data to find the optimal values of the parameters, to fit the data most closely. As with linear regression, by most closely we typically mean minimizing the sum of squares of the differences between the data and the predicted values.

This section touches on only a few of the models used in data mining. Some of the most common models used for prediction are decision trees, neural networks and random forests (discussed below), and the regression methods discussed in Chapters 7-9, 23, and 24. It's a lively research field and new algorithms are appearing all the time.

Model Selection: Training the Model

Data mining algorithms are powerful. That's great, but with that power comes the ability to fit every pattern in the data, including noise. As mentioned in Chapter 24, to avoid overfitting, the data are typically split randomly into training and test sets. The **training** set is used to fit (or train) the model. Then the model predicts the response of the test data. The test set has not been used to choose the parameters or the variables in the model. We use this cross-validation to see how well the model predicts on data that it has never seen.

What Is a Data Mining Algorithm?

"A data mining algorithm is a well-defined* procedure that takes data as input and produces output in the form of models or patterns."

Principles of Data Mining, Hand, Mannila, and Smyth

**Well-defined* means it can be encoded into a set of commands into a computer.

A more thorough way to evaluate models is to split the dataset randomly into training and test repeatedly, cross-validating more than once as we did for regression in Chapter 24, Section 3. Typically, 90% of the data are used to fit and 10% used to evaluate in each iteration. When repeated test sets are used, they are often called **hold outs**. The model fitting is repeated on all the training sets, computing a performance measure on each hold out. The average performance on all the hold out sets will be the final measure of how well the model fits.

What do we mean by predicting well? For a continuous predictor, we might use the sum of squares of the deviations on the test set, or the R^2 on the test set defined as:

$$R^2 = 1 - SSE/SST.$$

This is the same definition we used for multiple regression, but for the new data, R^2 can actually go negative if the predictions for the test are farther away from the actual response than they are from their mean.

For categorical data there are a variety of performance measures all based on the number of misclassified cases in the test set. The simplest is to count up all the misclassifications. However, it's often more serious to misclassify one of the levels of the response than another, so weighting the misclassifications is also popular.

For example, it is far worse to miss finding cancer than it is to tell someone that he or she may have cancer, and then to discover that it is not there. Similarly, it may well be much worse to permit credit card fraud than to inconvenience people with a false alarm (when the company says that a transaction may be fraud, but, in fact, it was not).

Fortunately, many algorithms do some sort of cross-validation automatically and report back the best model. We'll see how this works in practice in the following examples.

Tree Models

Trees are simple models used for both classification and regression problems. To illustrate how a tree works, imagine that we want to predict whether someone will default on a mortgage. To build the tree, we use past data from a group of customers on whom we have measured some predictor variables and for whom we know whether they defaulted or not.

(So, this is a supervised classification problem.) For this simplified example, let's assume we'll predict *Default* only from the following variables:

- *Age* (years)
- *Household Income* ($)
- *On the Job* (years)
- *Debt* ($)
- *Homeowner* (Yes/No)

Perhaps the easiest way to understand what a tree is doing is to look at its output. Figure 27.2 shows a tree fit to our hypothetical default example. At each level of the tree, data are split into two parts based on the values of a predictor variable. The first split is on *Household Income* at the value $40,000. Those whose incomes are > $40,000 (indicated by "Yes") are on the right. Following down the tree, the tree next splits that group on household *Debt.* For people who have both *Income* > $40,000 and *Debt* > $10,000, the tree estimates a 0.05 probability of *Default.* For those with *Income* > $40,000 but *Debt* not > $10,000, the probability is only 0.01. Going back up to the top, for the left branch of the first split, the tree splits on years *On the Job* (and not *Debt*). For the customers with *Income* not > $40,000 and more than 5 years *On the Job,* the *Default* probability is estimated to be 0.06, and for those with *Income* not > $40,000 and 5 years or less *On the Job*, the probability is highest—at 0.11.

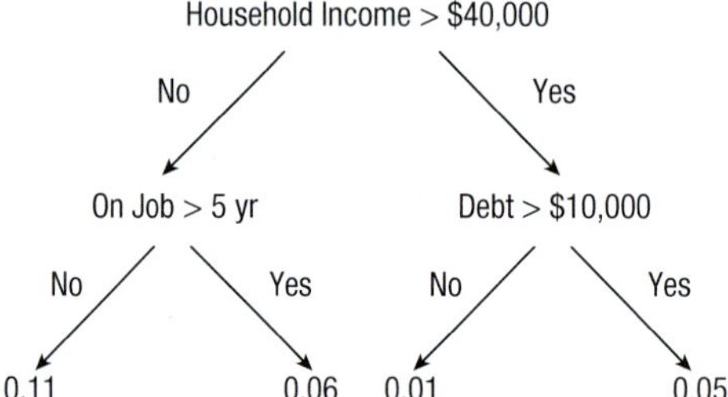

Figure 27.2
Part of a tree model on a hypothetical examination of mortgage default. The tree selects *Household Income* as the most important variable to split on and selects $40,000 as the split point. For those customers whose *Income* is more than $40,000, *Debt* is the next most important variable, while for those whose income is less than $40,000, how long they've been *On* their current *Job* is more important. For each of the four groups, the estimated probability of *Default* is shown at the bottom at the terminal nodes.

One of the benefits of the tree is how simple it is to interpret. But how does the tree choose the variables and the values to split on? The answer might be surprising. The tree essentially does it by brute force, trying every possible predictor variable and split point to see which combination produces the largest *difference* in default rates.[13]

The tree starts by considering all the data, and tries splitting the data into two groups based on each predictor variable, looking at the impact that makes on the response variable. It searches for the combination of variable and split that produces the largest difference in the response.[14] For a nonordered categorical predictor, the model considers every possible way of splitting the categories into two groups. For an ordered categorical variable, the modeler may choose to respect the order and restrict the possible splits (see sidebar). For a quantitative variable, the possible splits are at every possible value the variable can take (other then the min and the max).

After the algorithm finds the first split, it continues searching again on the two resulting groups, finding the next best variable and split point (possibly using the same variable as the previous split again). It continues in this way until one of several criteria is met (for example, number of customers is too small, not a large enough difference in rates is found) and then stops at what are called **terminal nodes**, where the model produces a prediction. The predictions at the terminal nodes are simply the average (if the response is quantitative) or the proportions of each category (for a classification problem) of the cases at that node.

[13]This is a simplification. Tree algorithms have shortcuts so they don't have to try every possible combination. In fact, they are computationally very efficient. Readers who want to know the details can read more advanced books on data mining or decision trees.

[14]There are several different criteria for measuring the difference in the response on the two sides of the split.

> **SPLITTING A CATEGORICAL VARIABLE**
>
> If our categorical predictor is *Region (East, North, South, West),* there are 7 possible splits:
>
> *East—North, South, West*
>
> *North—East, South, West*
>
> *South—East, North, West*
>
> *West—East, North, South*
>
> *East, North—South, West*
>
> *East, South—North, West*
>
> *East, West—North, South*
>
> If the categorical variable is ordered and the modeler wants to keep that order, then the splits are restricted. For example, if the variable is *Age (Child, Young Adult, Adult, Senior),* then the only possible splits are:
>
> *Child—Young Adult, Adult, Senior*
>
> *Child, Young Adult—Adult, Senior*
>
> *Child, Young Adult, Adult—Senior*

Tree models are very easy to implement and, in principle, are easy to interpret. They have the advantage of showing their logic clearly and are easy to explain to people who don't have a deep background in statistics. Even when they aren't used as a final model in a data mining project, they can be very useful for selecting a smaller subset of variables on which to do further analysis.

Unlike many other data mining algorithms, a tree can handle a great number of potential predictor variables, both categorical and quantitative. This makes it a great model for starting a data mining project. It also handles outliers and high influence values in the predictors well because it simply splits the predictor into two groups. Once the tree model is fit, the analyst can then choose whether to use this model as a final model for predicting the response, or simply to use the variables suggested by the tree as inputs to other models that are more computationally intensive and less able to handle very large numbers of predictors.

EXAMPLE 27.2 Gaining Insights from a Tree Model

QUESTION: How can a tree model help us understand who responded to our most recent solicitation?

ANSWER: The tree model starts at the root node with all 94,649 donors (listed under Count in the top box of the figure).

The first split of the tree is on the variable *RFA_4,* a variable that summarizes past giving. This is one of the RFM variables that marketing created to summarize past giving. The fact that the tree chose it first is a great credit to their efforts! The algorithm identifies this as the best single predictor of *TARGET_B,* splitting the levels of that variable into two groups, with 29,050 customers split to the left and the remaining 65,599 customers to the right. The variable *RFA_4* comprises codes that contain information on how recently the last gift solicitation was received, how frequently the donor has given, and how large the last gift was. (Examples of these codes are A3C, S4B, and so on, seen in parentheses after the variable name in the output display.) The percentages next to the levels 0 and 1 indicate the proportion of those who donated or not (*TARGET_B 1 or 0*, respectively) in each node. Notice that 7.34% of those in the left node contributed, which was almost twice the 4.06% in the right node who contributed.

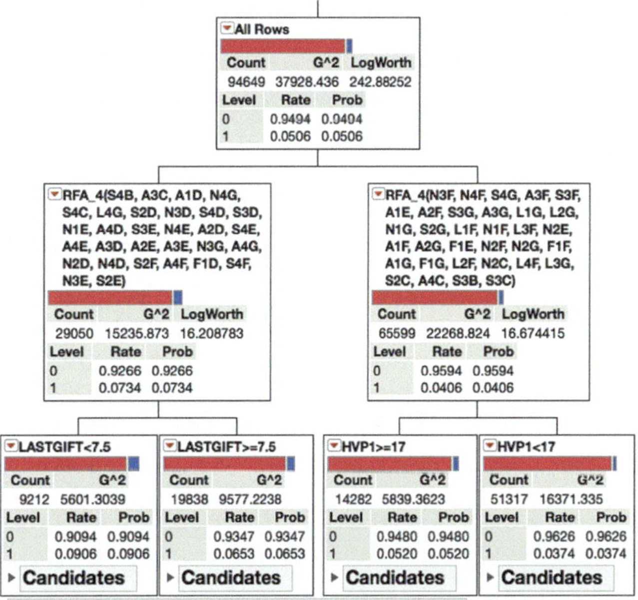

Figure 27.3
Part of a tree model run on the PVA data using JMP® software.

The tree continues to split each subpopulation. Notice that the variable chosen to split the left branch of the tree (*LASTGIFT*) is not the same as the variable chosen to split the right branch (*HVP1*). The terminal nodes are found in the bottom row. The terminal node at the bottom left of the figure shows that 9.06% of the 9212 potential donors in this subgroup donated, a substantial improvement over the average of all potential donors (5.06%) shown in the root node at the top. Even at this stage of the analysis, if we directed our solicitations to names on our list that satisfied the conditions found in the path through the tree to this node, we might reduce our costs and increase our yield substantially.

It's interesting to see that of the 479 potential predictor variables, 2 of the top 3 variables selected by the decision tree algorithm involve past giving history rather than demographics. *RFA_4* is a summary of past giving and *LASTGIFT* measures the amount of the last gift. *HVP1* is a demographic variable, measuring the percentage of residents in the donor's ZIP code whose homes are worth at least $200,000. By letting the tree grow further, I've increased the list of potential predictors to the top 20 (see Figure 27.4). Notice that most of the important variables involve past giving of the donors and a few key demographic variables about the donor's neighborhood. I suggest we look at the data only for those 20 variables and spend some time visualizing these variables and assessing data quality and dealing with missing data issues before proceeding with more model building.

Column Contributions

Term	Number of Splits	G^2	Portion
RFA_4	1	423.738142	0.4194
CARDGIFT	2	80.107907	0.0793
LASTGIFT	2	78.4402851	0.0776
LASTDATE	3	74.1487712	0.0734
HVP1	1	58.1273009	0.0575
CARDPROM	1	32.3495992	0.0320
EC2	1	29.4524333	0.0291
MC3	1	25.6105524	0.0253
HUPA3	1	22.7727912	0.0225
VC2	1	22.5842906	0.0224
ADATE_22	1	18.5629611	0.0184
HHN3	1	17.7936628	0.0176
HU5	1	17.4192077	0.0172
MINRDATE	1	17.0423069	0.0169
HHAS3	1	16.7292576	0.0166
HHN1	1	16.6986151	0.0165
ETHC1	1	15.8670199	0.0157
INCOME	1	15.549749	0.0154
AGEC6	1	14.9778014	0.0148
AGE	1	9.93632478	0.0098

Figure 27.4

The 20 most important predictor variables from a large tree model. Notice that the top 4 account for nearly two-thirds of the total variable importance and all involve past donor giving.

*How Big (or Deep) Should the Tree Be?

If we continue to split the data enough times, eventually each data point will be in its own cell and the tree will fit the data "perfectly," a classic example of overfitting. Splitting only once is likely to underfit the data. So, what's the right compromise? Although there are some theoretical criteria that attempt (like adjusted R^2) to choose the right size, cross-validation is usually a good way, in practice, to answer the question. Here's an example using 30% of the data as a hold out on the PVA data:

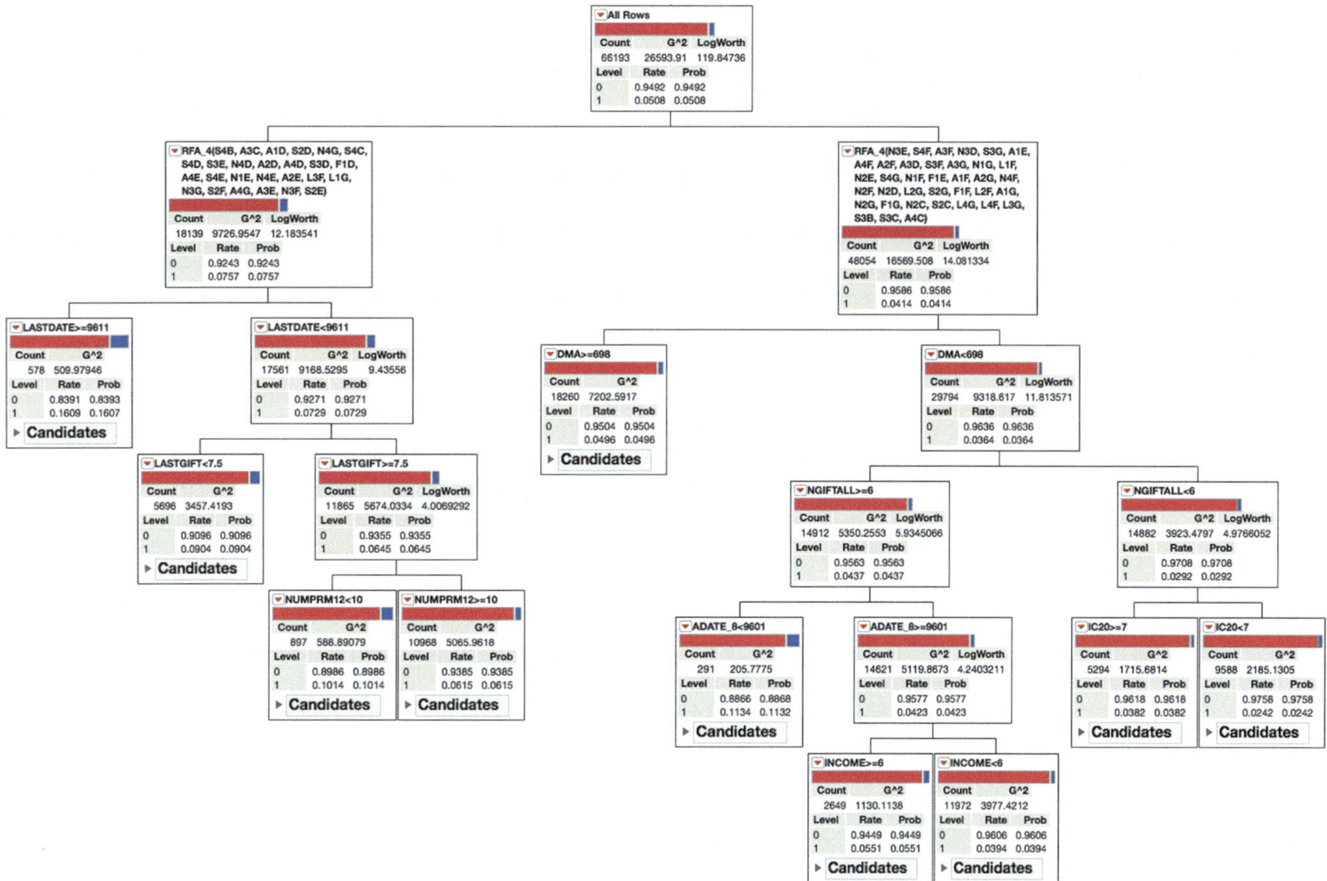

Figure 27.5

With a random 30% of the data as hold out, the tree algorithm found that this size tree, using 70% of the data as training to fit the model, had the best performance on the remaining 30% test set. The criterion used was based on the number of misclassifications—predicting that a person would respond to the solicitation when they didn't and vice versa.

Exploring the tree by splitting by hand (as we did in Example 27.2) can be a useful way to *explore* the data, but some sort of cross-validation should always be used to avoid overfitting.

JUST CHECKING

The vice president is interested in using a tree model to predict the likelihood that a customer will sign up for the IRA offer recently sent out by the bank. She has a few questions.

11. When we run the tree model, will it automatically find the best combination of variables to predict the sign-up probability?
12. Last year we used a logistic regression to predict a similar sign-up probability. Will the tree perform better?
13. We have a few customers with very high incomes that proved to be a problem for logistic regression because of their high impact. We had to take them out of the model and treat them separately. How will the tree deal with them?

Neural Networks

Another popular data mining tool is the multilayer perceptron or (artificial) **neural network**. This algorithm sounds more impressive than it really is. Although it was inspired by models that tried to mimic the function of the brain, it's really just an automatic, flexible, nonlinear regression model. The neural network, like multiple regression, models the response variable with a function of the predictor variables but, unlike multiple regression, the functions are not linear. They are more closely related to smoothing functions that you saw back in Sections 4.3 and 23.7. And unlike stepwise regression, the neural network retains all the predictors, even if some of them have little effect on the response.

Because the "network" is made up by adding together many small nonlinear functions, the equation that predicts the response is uninterpretable and results in what's called a **black box** model. That's a major disadvantage for many applications. In multiple regression, the fitted model shows how each predictor influences the response. In a tree, the splits show the role of each predictor as you move down the tree. Sometimes, however, a black box model is acceptable. For example, if the neural network can correctly distinguish the digits of a hand-written ZIP code on a random letter with high accuracy, the U.S. Postal Service might not really care how it's doing it. But, if a neural network determined that you are not qualified for a student loan, you'd probably want to know why—something that would be hard to tell from a black box model!

The basic idea of a neural network is to first create new functions (called **nodes** or **features**) from the original predictors. These nodes are just linear combinations (weighted averages) of the original predictors. The weights used are similar to coefficients in a linear regression. The collection of these new functions is called the **hidden layer**. The reason they are called hidden is that the response, y, is not used directly to estimate the weights. The fitting is done by a computationally intensive method, the most common of which is called **back propagation**. The hidden functions are then transformed by an S-shaped curve (like the one you saw in logistic regression) and are given the colorful name **sigmoidal activation functions** (which really just means S-shaped curve!). Finally, the response is fit by a regression on these transformed functions. If the response is quantitative, the regression is linear. If it's categorical, then another sigmoid is applied to model the probability of belonging to each class. The fact that the nodes are transformed using a nonlinear function gives the model its flexibility—and the ability to approximate a wide variety of phenomena.

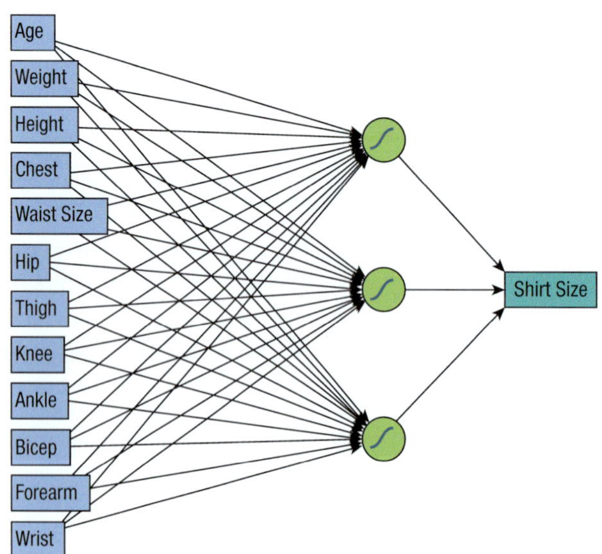

Figure 27.6
A neural network to predict *Shirt Size* from various other body measurements. The network has 3 nodes in the hidden layer. The S symbol indicates that the network is using sigmoidal activation functions.

A diagram of such a network with three nodes in the hidden layer is shown in Figure 27.6. In this example, a shirt manufacturer wants to predict a man's shirt size (to the nearest half inch) from his other measurements (data in **Shirt sizes full**). Various methods are employed to ensure that the model doesn't overfit the data, and the final model is chosen by cross-validation.

Although the model is, in general, uninterpretable, there are some implementations of neural networks that provide insight into the effect of the predictors by profiling the effect of each predictor using sensitivity curves and measuring the importance of the predictors. Figure 27.7 shows the profile of predictor influences on *Shirt Size* (measured in inches), with a listing of variable importance as produced by JMP® (Table 27.2).

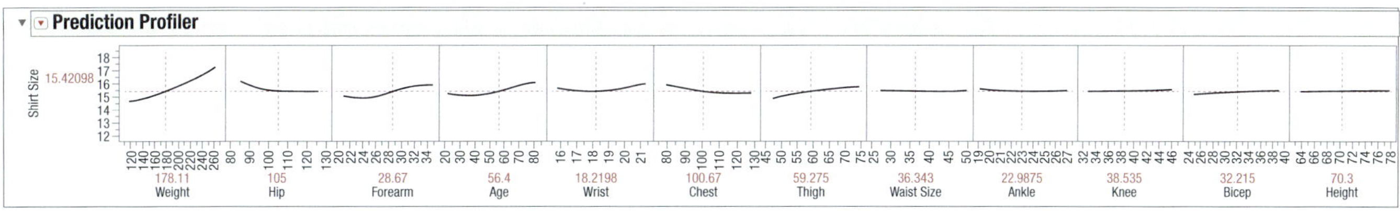

Figure 27.7

The profiler in JMP® shows the influence of each predictor variable on *Shirt Size*. The neural network model is nearly linear in *Weight*, and somewhat curved in the next few variables. The last few have relatively little effect on the response.

Table 27.2

The Summary Report shows numerical values for the relative importance of each predictor. *Weight* is by far the most important predictor of *Shirt Size*, accounting for more than half of the predictive power of the model. The model reports an R^2 of 0.69 on the training set and higher on the test set. (This increase is probably just an anomaly, but it at least indicates that the model is not overfit.)

Summary Report

Column	Main Effect	Total Effect
Weight	0.432	0.557
Hip	0.069	0.187
Forearm	0.049	0.157
Age	0.019	0.151
Wrist	0.033	0.132
Chest	0.011	0.086
Thigh	0.029	0.069
Waist Size	0.012	0.051
Ankle	0.013	0.043
Knee	0.008	0.03
Bicep	0.008	0.022
Height	0.003	0.01

Crossvalidation

Source	RSquare	RASE	Freq
Training Set	0.6911	0.48978	186
Validation Set	0.7641	0.50461	37
Test Set	0.8702	0.30148	25

This model was fit using 75% of the data for training and 10% for testing (out of 250 cases). Another 15% was set aside as a "tuning set." We specified 3 hidden nodes and used the tuning set to select the best model. Table 27.2 shows the performance on each set. On the training data, the model reports an R^2 of 0.69, and the model actually fit the test set best, but this is probably an anomaly.

When the R^2 on the test set is much lower than the R^2 on the training set, this usually indicates overfitting. When the performance is approximately the same, it doesn't necessarily indicate a powerful model, but at least it's evidence that the model is not overfitting. When the model actually performs best on the test set (data that it hasn't seen) it's probably just random noise. Models in general perform best on the data on which they are fit (or trained).

The language used to describe neural network models and the method of estimating their parameters is colorful and designed to make you think of the way the human brain works. Don't be fooled. The model is really just a flexible function fitter and the network doesn't "learn" any more (or less) than any other statistical model. If you applied these terms to multiple regression, the intercept and slopes would be called weights. The act of fitting the regression to a set of training data would be called training the network, or "learning." The resulting predictions would be called knowledge. A glossary of terms is given in Table 27.3.

Glossary	
Neural network	A nonlinear regression model using transformed linear combinations of the predictors to predict the response
Weight	A coefficient in the neural network model
Back propagation	A computational method for estimating all the weights in the model
Hidden node	A linear combination of the original predictor variables
Sigmoidal activation function	An S-shaped curve used to transform the hidden "nodes"
Training and learning	Colorful names for the process of estimating the coefficients of the model from a set of training data
Knowledge	The output from the model—really just the predictions
Deep learning	A neural network with more than one hidden layer

Table 27.3
A glossary of some of the most common terms used in neural networks.

In the 1980s some mathematical theorems were proved that showed that neural networks can approximate nearly any function (with enough data). They also showed that one hidden layer (with enough nodes) is all that is needed. In spite of this, some researchers have put a lot of effort into studying multi-layer networks. One of these research groups won an image recognition contest in 2012 with an algorithm that was twice as accurate as the runner-up. These multi-layer networks now go by the name **deep learning**, continuing the trend in making fanciful names that sound like they are doing more than they are.

JUST CHECKING

Our vice president is back with a couple of questions.

14. Because the neural network thinks like a human, it will be the best model that we can run, correct?
15. A customer who had a very high probability of taking our offer, as predicted by the neural network, didn't. Can the neural network tell us why?

27.4 Models Built from Combining Other Models

The president of the United States often consults the cabinet when making an important decision. Similarly, a CEO may ask the board of directors for advice. But it's not often obvious how to combine disparate opinions. After all, the decision maker has to make a final choice. Knowing that no single model is "correct," data miners often consider several models. They can then choose the "best" model — perhaps defined as predicting most accurately on a test set, but they, like the CEO, may want to combine the models, trying to combine the wisdom and insights of all of them. Finding the best ways to combine many different models is an active and exciting area of current research in data mining. We'll discuss a few of these ideas in this section.

The most common methods for combining models are called **bagging** and **boosting**. Each of these methods creates "committees" of models. In bagging (short for **b**ootstrap **agg**regat**ing**), you fit a model (say, a tree) to the training data. Then you take a bootstrap sample of the data (see Chapter 17) and fit another tree. Repeat this a large number of times (say, 1000). Now use the average prediction (in the case of a regression problem, or the most frequently predicted class for a classification problem) as the prediction for each case in the test set.

Boosting a model is more technical, but one way to think about it (for classification) is to fit a small model (for example, a tree with just 2 or 3 splits) to the training data. Now, give more weight to the cases that were misclassified, and create another tree. Repeat this over and over, each time giving more weight to the misclassified cases. The later trees will fit the harder cases better (and the easier cases perhaps less well). Now average all the trees by their overall performance. In principle, you can use bagging and boosting for nearly any model, but commonly trees are used.

Random Forests

Perhaps the most common version of bagged trees is the **random forest**. The random forest creates trees on many bootstrapped samples of the training data and then averages them, but with one important wrinkle. Every time each tree looks for a split, it does so on only a small random sample of the predictors (often 10%). This seems counterintuitive. For example, suppose that only 10 or so of the nearly 500 predictors in the PVA data are important. If we restrict each split to only 50 randomly chosen predictors each time, there's a good chance that none of those 10 will be in the sample and the tree will make a uninformative split. But it turns out that that doesn't really hurt performance. In fact, it provides a richer set of trees that seem to find more signal in the data than simply bagging lots of trees.

Continuing our *Shirt Size* example, here is the output of a random forest of 100 trees fit to the training data (Table 27.4) and the profiler, showing the effect of the predictors (Figure 27.8).

Table 27.4

The Column Contributions show the relative importance of each predictor. The order is quite different than the order produced by the neural network, although both models claim to fit the test set fairly well.

Column Contributions

Term	Number of Splits	SS	Portion
Chest	32	16.2302808	0.2282
Forearm	31	11.4753715	0.1613
Weight	29	10.4729397	0.1472
Wrist	31	7.70337005	0.1083
Waist Size	28	6.04566438	0.0850
Hip	25	5.81539012	0.0818
Bicep	26	5.36643865	0.0754
Knee	29	2.7386453	0.0385
Thigh	18	2.02304012	0.0284
Ankle	27	1.42607054	0.0200
Age	26	1.27974214	0.0180
Height	16	0.55646824	0.0078

	RSquare	RMSE	N
Training	0.811	0.3835496	186
Validation	0.686	0.5820115	37
Test	0.773	0.3985855	25

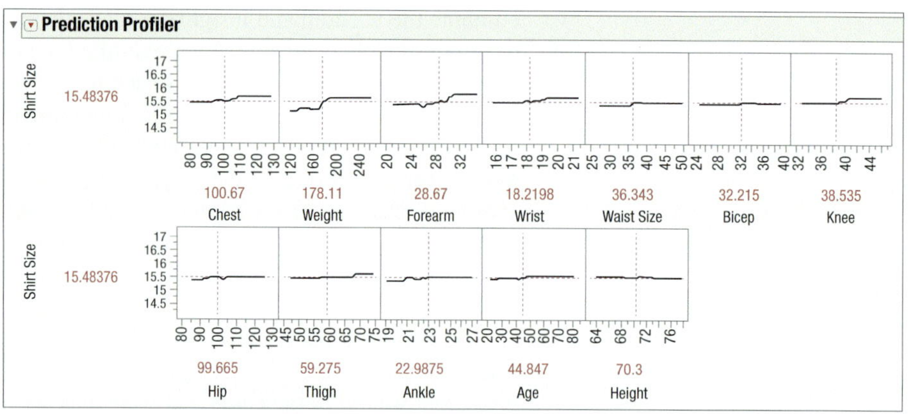

Figure 27.8
The profiler for the random forest shows a typical pattern for tree-based models, with sharp step-like increases for *Shirt Size* at various levels of the predictors.

The random forest fit an R^2 of 0.811 on the training set and then slightly lower (0.773) on the test set (Table 27.4). Although it is a somewhat automatic method, performance can often be improved by optimizing some of the parameters (number of trees fit, percentage of predictors chosen at each split, size of each tree, and so on). Notice also that the random forest used the predictors quite differently than the neural network did, placing more importance on *Chest* and *Forearm* size and less on *Waist Size*. It's important to remember that these models are sensitive to the training data that were selected as well as some random aspects of their fitting. Large differences between data mining models can provide insights into the phenomenon being studied, or they can be an indication that the models are fitting noise as well as signal. It is always wise to be cautious in putting too much faith in a model that is automatically fit, especially on a "found" set of data.

Boosted Trees

Trees can also be boosted, and the resulting ensemble (set) of trees is usually a more powerful predictor than a single tree. Of course, like any ensemble, it's much less interpretable, especially compared to the simple structure of a tree. The random forest creates an ensemble by fitting trees on different bootstrap samples of the dataset. The boosted tree does not resample the data, but fits a *sequence* of trees on the original sample. However, after the first tree, it gives more importance (weight) to the cases that were predicted badly. So the next tree focuses more on the hard-to-predict cases. At each stage, the misclassified cases are upweighted so that later trees focus on repeatedly misclassified cases. These later trees may not do as well overall on classifying the data, but they may perform better on the hard-to-predict cases. The final model averages the predictions from all the trees, with more weight given to the trees that do best overall. Boosting can be applied to a variety of models, and the resulting ensemble often outperforms a single model.

The boosted tree on the *Shirt Size* data gave the summary in Table 27.5 and the profiler in Figure 27.9. The model produces the typical step functions of a tree, as one can see in the profiler. In contrast to the random forest, however, the importance it gives to the variables is similar to that given by the neural network. Its performance on the test set ($R^2 = 0.792$) falls between the two previous models.

Table 27.5

The Column Contributions show that the boosted tree agrees with the neural network that *Weight* is the most important predictor. Its performance is quite consistent with an R^2 on the test set of 0.792, close to the R^2 of 0.834.

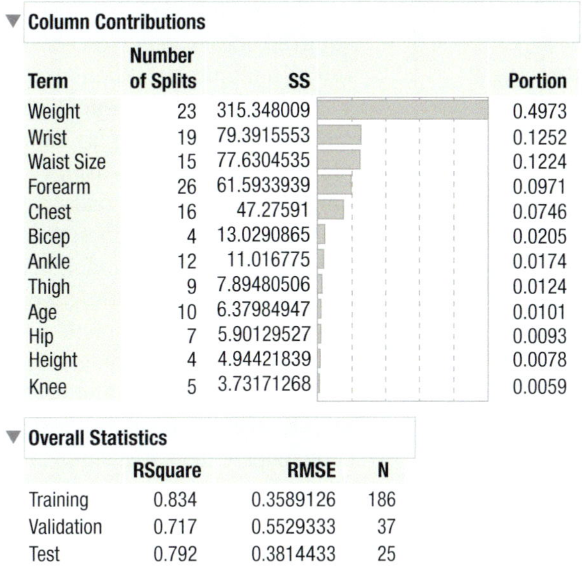

Column Contributions

Term	Number of Splits	SS	Portion
Weight	23	315.348009	0.4973
Wrist	19	79.3915553	0.1252
Waist Size	15	77.6304535	0.1224
Forearm	26	61.5933939	0.0971
Chest	16	47.27591	0.0746
Bicep	4	13.0290865	0.0205
Ankle	12	11.016775	0.0174
Thigh	9	7.89480506	0.0124
Age	10	6.37984947	0.0101
Hip	7	5.90129527	0.0093
Height	4	4.94421839	0.0078
Knee	5	3.73171268	0.0059

Overall Statistics

	RSquare	RMSE	N
Training	0.834	0.3589126	186
Validation	0.717	0.5529333	37
Test	0.792	0.3814433	25

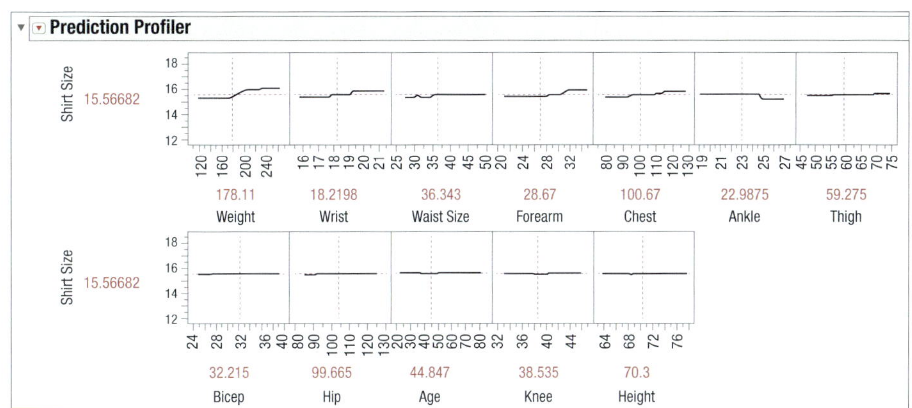

Figure 27.9

The profiler for the boosted tree shows the same typical pattern for tree-based models with sharp step-like increases for *Shirt Size* at various levels of the predictors. (*Source*: Created with JMP® 13 Pro software. Copyright 2016, SAS Institute, Cary, NC, USA. All Rights Reserved. Reproduced with permission of SAS Institute, Inc, Cary, NC.)

JUST CHECKING

And . . .

16. Which of the 1000 trees that the Random Forest uses is the best? Can we look at it?
17. When we classified people into "Likely to buy" and "Not likely to buy," using a Boosted Tree we got a 20% error rate on the training set, but a 50% error on the test set. What's the likely cause of that?

27.5 Comparing Models

Usually there is no way of knowing which of many models will predict best on future data. Choosing a model (or models) requires judgment. If several models perform about equally well, one might be preferred on the basis of scientific or other domain knowledge. Performance measures for quantitative responses include R^2 on the test set (defined earlier) and

MEASURING TEST SET PERFORMANCE FOR QUANTITATIVE RESPONSES:

$$R^2 = 1 - SSE/SST$$

$$\text{RASE} = \text{RMSE} = \sqrt{\frac{SSE}{n}}$$

the **RASE or RMSE (square root of averaged squared prediction error or root mean squared error)** on the test set, which is the same as the residual standard deviation that we saw in multiple regression (except that it's evaluated on the test set).

For categorical responses, there is always a trade-off between Type I and Type II errors. A graphical display called an ROC curve[15] can often help compare models over a range of choices. This display shows the trade-off between Power and Type I error over a range of possible error rates.

These measures (in addition to business knowledge) can help guide researchers in choosing the "best" model. Sometimes, however, there is no reason to prefer one over another and the analyst might be just as happy to average the predictions from all of them. This average can be a simple arithmetic average of each prediction, or a weighted average, or the predictions from various models can even be used as inputs to another neural network, which will then optimize a function of them based on other training data. In the next Step-by-Step Example, we will display all the models we have just discussed for *Shirt Size* and compare them to a simple multiple regression.

STEP-BY-STEP EXAMPLE

Model Comparisons

Let's return to the problem of estimating a man's shirt size (inches around the neck to the nearest half inch) from other body measurements.

THINK

PLAN Define the problem—state the objective of the study.

Identify the variables.

If reasonable, plot the variables and look for relationships among the variables, outliers, and other interesting features.

We want to build and compare several models to predict *Shirt Size* from other body measurements. The available variables on men from a scientific study include:

Variable	Unit
Age	*(yr)*
Weight	*(lb)*
Height	*(in.)*
Chest	*(cm)*
Waist	*(in.)*
Hip	*(cm)*
Thigh	*(cm)*
Knee	*(cm)*
Ankle	*(cm)*
Bicep	*(cm)*
Forearm	*(cm)*
Wrist	*(cm)*
Shirt Size	*(nearest half in.)*

[15]Originally called receiver operating characteristic (ROC) curves, they plot the true positive rate (TPR) against the false positive rate (FPR) for all values between 0 and 1. The TPR is equivalent to the Power (the probability of correctly classifying a positive) and the FRP is the probability of a Type I error (probability of misclassifying a negative).

If there are too many variables to display, running an exploratory model such as a tree can help select the most important variables, which can then be displayed.

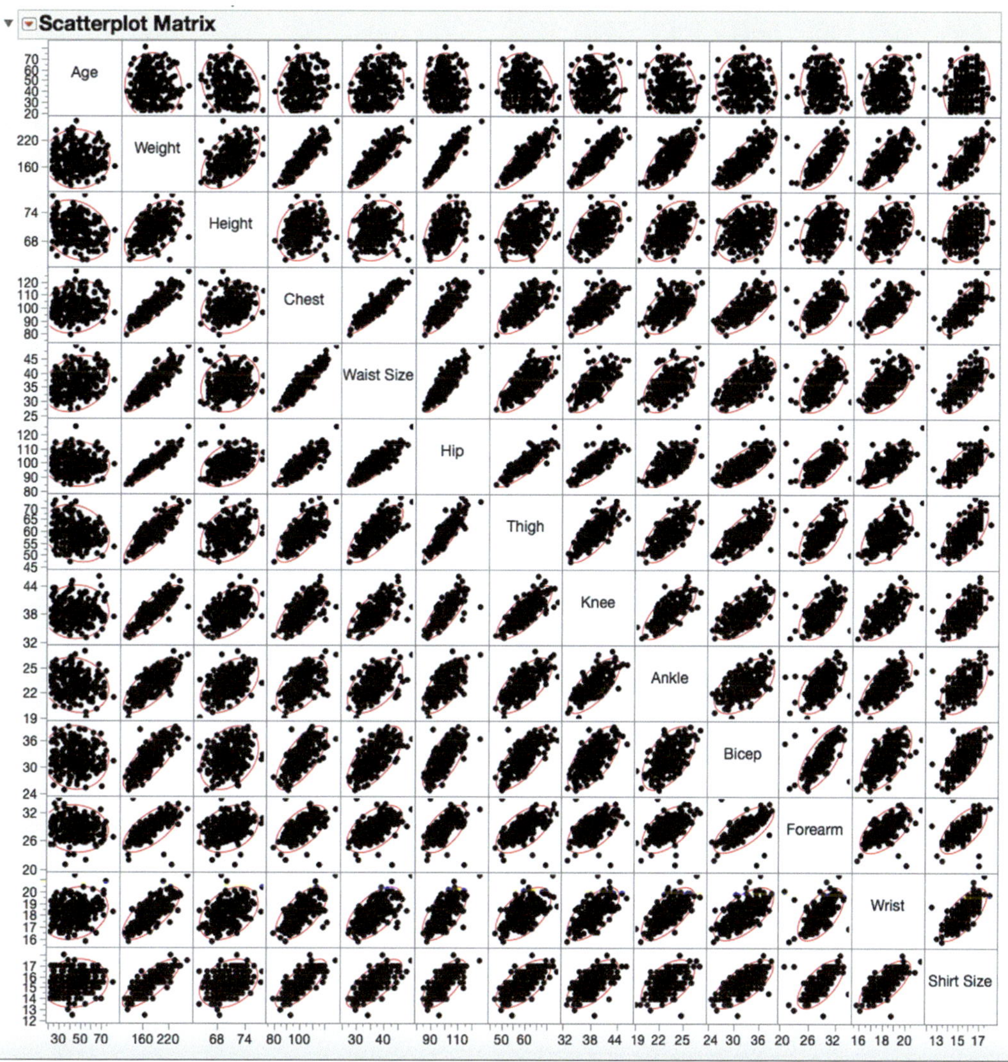

There are many strong, nearly linear relationships among the variables. In particular, most of the measurements (except *Age*) seem to have a positive linear relationship with *Shirt Size*, which makes sense. Men who are larger, heavier, and taller tend to have larger shirt sizes. There are no obvious outliers.

SHOW **MECHANICS** Fit a stepwise multiple regression model, starting with all the predictors and their interactions.

Here is the computer output for the stepwise multiple regression, starting with all 13 predictor variables and their interactions, and proceeding forward until no more candidates were nominated for exclusion.

Dependent variable is: Shirt Size
R-squared = 69.1% R-squared (adjusted) = 68.1%
s = 0.499 with 186 − 7 = 179 degrees of freedom

Term	Estimate	Std Error	t-Ratio	Prob > |t|
Intercept	13.34017106	2.02415021	6.59	<0.0001
Weight	0.032632385	0.005077533	6.43	<0.0001
Height	−0.04587073	0.017441901	−2.63	0.0093
Hip	−0.044188241	0.016614473	−2.66	0.0085
Ankle	−0.109697048	0.041553449	−2.64	0.009
Forearm	0.047588487	0.024155895	1.97	0.0504
Wrist	0.278018404	0.063206759	4.4	<0.0001

The estimated equation is:

$$\widehat{Shirt\ Size} = 13.34 + 0.033\ Weight - 0.046\ Height - 0.044\ Hip - 0.110\ Ankle + 0.048\ Forearm + 0.278\ Wrist$$

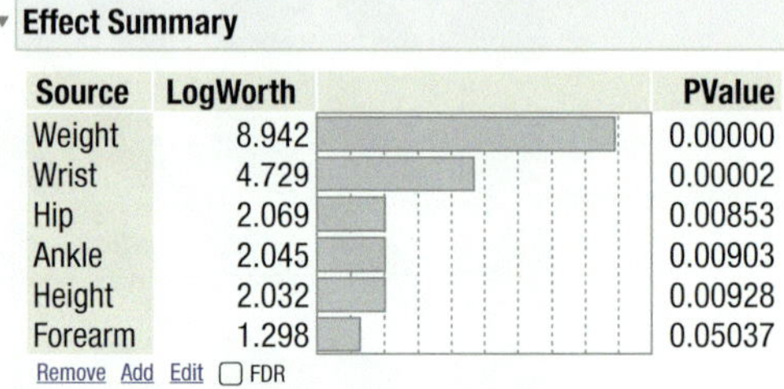

Effect Summary

Source	LogWorth	PValue
Weight	8.942	0.00000
Wrist	4.729	0.00002
Hip	2.069	0.00853
Ankle	2.045	0.00903
Height	2.032	0.00928
Forearm	1.298	0.05037

Remove Add Edit FDR

Crossvalidation

Source	RSquare	RASE	Freq
Training Set	0.6911	0.48978	186
Validation Set	0.7641	0.50461	37
Test Set	0.8702	0.30148	25

Variable importance shows that *Weight* and *Wrist* size are the most important predictors. The model gets an R^2 of 0.691 on the training set and a surprisingly high 0.870 on the test set.

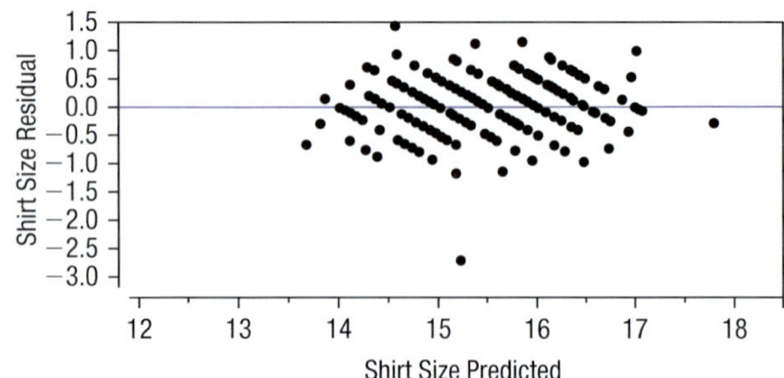

Fit a tree, neural network, random forest, and boosted tree.

The residual plot shows one case with a low residual, but without large influence on the model. Other than the striped pattern because the shirt sizes are discrete (every half inch), the plot is patternless.

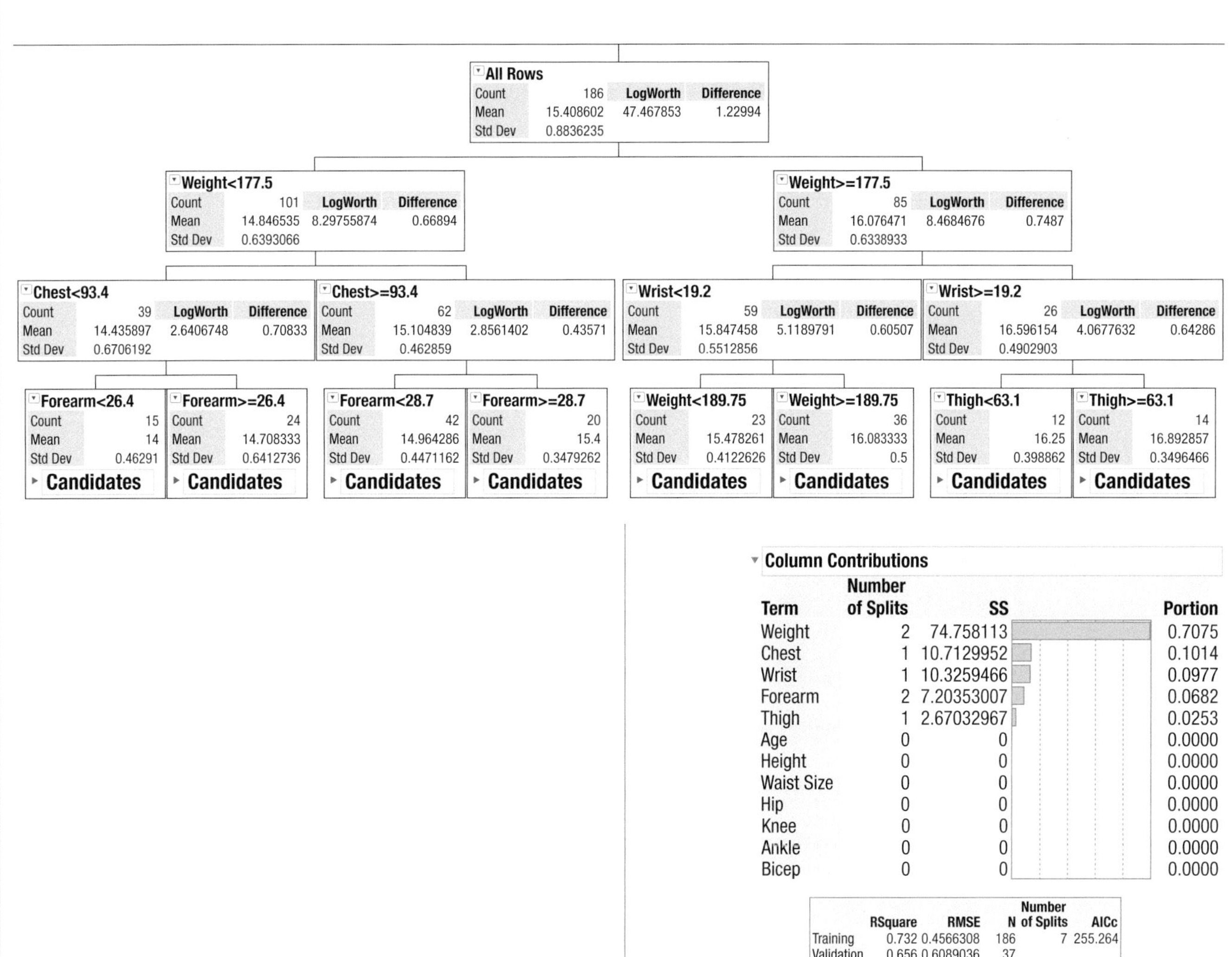

Column Contributions

Term	Number of Splits	SS	Portion
Weight	2	74.758113	0.7075
Chest	1	10.7129952	0.1014
Wrist	1	10.3259466	0.0977
Forearm	2	7.20353007	0.0682
Thigh	1	2.67032967	0.0253
Age	0	0	0.0000
Height	0	0	0.0000
Waist Size	0	0	0.0000
Hip	0	0	0.0000
Knee	0	0	0.0000
Ankle	0	0	0.0000
Bicep	0	0	0.0000

	RSquare	RMSE	N	Number of Splits	AICc
Training	0.732	0.4566308	186	7	255.264
Validation	0.656	0.6089036	37		
Test	0.699	0.4590413	25		

The tree fits less well on the test set than the multiple regression, with an R^2 of 0.699 on the test set. The most important variable is *Weight*, followed by *Chest* size. *Wrist*, second in the multiple regression, is third here.

Results for the neural network, random forest, and boosted tree are found in the previous section. Here is a summary of all 5 models and their performance on the test set:

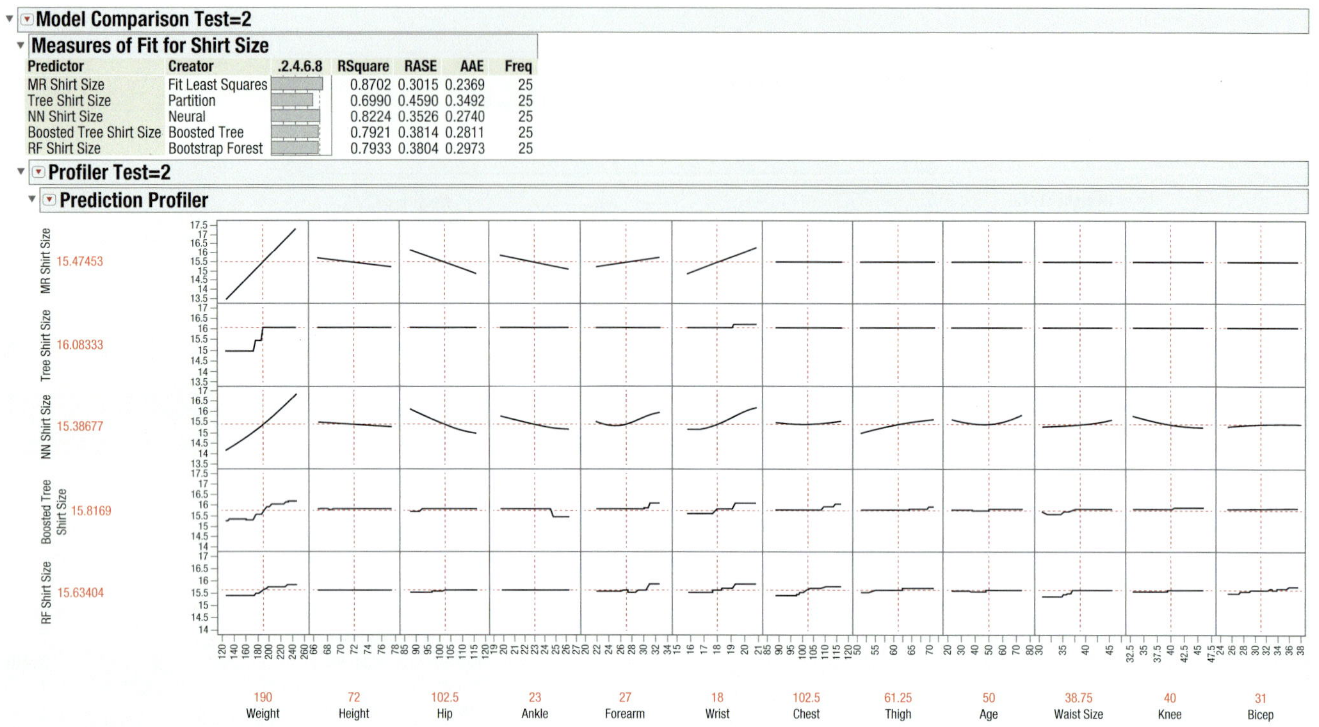

Model Comparison Test=2

Measures of Fit for Shirt Size

Predictor	Creator	.2.4.6.8	RSquare	RASE	AAE	Freq
MR Shirt Size	Fit Least Squares		0.8702	0.3015	0.2369	25
Tree Shirt Size	Partition		0.6990	0.4590	0.3492	25
NN Shirt Size	Neural		0.8224	0.3526	0.2740	25
Boosted Tree Shirt Size	Boosted Tree		0.7921	0.3814	0.2811	25
RF Shirt Size	Bootstrap Forest		0.7933	0.3804	0.2973	25

TELL **CONCLUSION** Summarize your results and state any limitations of your model in the context of your original objectives.

Five models for *shirt size* were fit to a training set and predicted on a small test set of 25 randomly chosen cases. The models included multiple regression, a decision tree, a neural network, a random forest, and a boosted tree.

The models were similar in finding that *Weight* and *Wrist* size are important predictors. *Chest*, *Hip*, and *Forearm* size also appeared important in some models to a lesser degree. *Weight* and *Waist* size are positively related to *Shirt Size*, but because of collinearity with those variables in the model, *Hip* size is negatively related. Most of the predictors are roughly linear with *Shirt Size*, which helps to explain why a multiple linear regression was competitive with more flexible models.

All of the models gave R^2 values on the test set between 0.70 and 0.87, with the multiple regression highest and the tree lowest. The RASE averaged about 0.38 inch, which means that, roughly, we can predict a typical men's shirt size to within about 0.75 inch (2*0.38), which is about a size and a half. This may not be accurate enough for remote fitting. We recommend that the man try on the shirt for best fit.

27.6 Summary

There are many similarities between the modeling process of data mining and the basic approach to modeling that you've been learning throughout this text. What makes data mining different really has to do with the large number of algorithms and types of models available to the data miner and with the size and complexity of the datasets. But most of what makes a data mining project successful would make any statistical analysis successful. The same principles of understanding and exploring variables and their relationships are key to both processes. A famous statistician, Jerry Friedman, was once asked if there was a difference between statistics and data mining. Before answering, he asked if the questioner wanted the long answer or the short answer. Because the response was "the short answer," Jerry said simply, "No." We never got to hear the long answer, but we suspect it may have contained some of the differences discussed in this chapter.

Having a good set of statistical and data analysis tools is a great beginning to becoming a successful data miner. Being willing to learn new techniques, whether they come from statistics, computer science, machine learning, or other disciplines, is essential. Learning to work with other people whose skills complement yours not only will make the task more pleasant but also is key to a successful data mining project. The need to understand the information contained in large databases will increase in the years ahead, so it will continue to be important to know about this rapidly growing field.

WHAT CAN GO WRONG?

- **Be sure that the question to be answered is specific.** Make sure that the business question to be addressed is specific enough so that a model can help to answer it. A goal as vague as "increasing profits" is not likely to lead to a successful data mining project.
- **Be sure that the data have the potential to answer the question.** Check the variables to see whether a model can reasonably be built to predict the response. For example, if you want to know which type of customers are going to a particular web page, make sure that data are being collected that link the web page to the customer visiting the site.
- **Beware of overfitting to the data.** Because data mining tools are powerful and because the datasets used to train them are usually large, it is easy to think you're fitting the data well. Make sure you validate the model on a test set—a dataset not used to fit the model.
- **Make sure that the data are ready to use in the data mining model.** Typically, data warehouses contain data from several different sources. It's important to confirm that variables with the same name actually measure the same thing in two different databases. Missing values, incorrect entries, and different time scales are all challenges to be overcome before the data can be used in the model-building phase.
- **Don't try it alone.** Data mining projects require a variety of skills and a lot of work. Assembling the right team of people to carry out the effort is crucial.

CHAPTER REVIEW

Understand the uses and value of data mining in business.

Recognize data mining approaches.

- OLAP approaches answer specific questions about data in a database that may involve many variables.
- Predictive model approaches build complex models that attempt to predict a response.

Be aware that data preparation is essential to sound data mining.

- Data must be cleaned of errors and impossible or implausible values.
- Missing values may be a serious issue.
- Data preparation may take the majority of time and effort devoted to a data mining project.

Be able to articulate the goal of a data mining project and understand that data mining efforts differ according to the kind of dependent variable.

- When the goal is to predict a quantitative response, the problem is called a regression problem.
- When the goal is to classify individuals according to a categorical variable, the problem is called a classification problem.

Be aware of the variety of models and algorithms that are used to build predictive models.

- Tree models build interpretable models that select predictor variables and split points to best predict the response in either a regression or classification problem.
- Neural networks build black box models that use all the predictor variables for both regression and classification problems.
- Many other data mining algorithms such as random forests, boosting, and ensemble methods that combine models are used as well, and new competitors are continually being developed.

Understand that successful data mining is a team effort that requires a variety of skills.

- Defining the problem and objective of interest is crucial for success.
- The CRISP data mining process provides a good outline for a data mining project.

REVIEW OF TERMS

Data science An umbrella term for the interdisciplinary field concerned with learning from data and communicating the results. (p. 894)

Big Data The collection and analysis of datasets so large and complex that traditional methods typically brought to bear on the problem would be overwhelmed. (p. 895)

Machine learning The name given to a collection of models that are heavily computational and often black box models. Neural networks and random forests are common machine-learning algorithms or models. (p. 895)

Statistical Learning The process of estimating the parameters of the model. (p. 895)

Data mining A process that uses a variety of data analysis tools to discover patterns and relationships in data that are useful for making predictions. (p. 895)

Business analytics (or analytics or predictive analytics) The process of using statistical analysis and modeling to make predictions and drive business decisions. (p. 895)

Deep learning A term often used to describe a neural network with more than one hidden layer. (p. 895)

Data warehouse A digital repository for several large databases. (p. 896)

Online analytical processing (OLAP) An approach for providing answers to queries that typically involve many variables simultaneously. (p. 896)

Predictive model A model that provides predictions for the response variable. (p. 896)

Regression problem A prediction problem that has a quantitative response variable. (See also classification problem.) (p. 896)

Classification problem A prediction problem that involves a categorical response variable. (See also regression problem.) (p. 896)

Knowledge The insights gained from the model and its predictions. (p. 896)

Supervised problem A classification or regression problem in which the analyst is given a set of data for which the response is known to use to build a model. (p. 896)

Cluster analysis A task of grouping similar objects (rows) based on their values of a set of variables.(p. 896)

Unsupervised problem A problem, unlike classification or regression problems, with no response variable. The goal of an unsupervised problem is typically to put similar cases together into homogeneous groups or clusters. (p. 896)

Transactional data Data describing an event involving a transaction, usually an exchange of money for goods or services. (p. 898)

Demographic variable A variable containing information about a customer's personal characteristics or the characteristics of the region in which the customer lives. Commonly used demographics include age, income, race, and education. (p. 900)

Metadata Information about the data including when and where the data were collected. (p. 901)

Data preparation The process of cleaning data and checking its accuracy prior to modeling. Data preparation includes investigating missing values, correcting wrong and inconsistent entries, and reconciling data definitions. (p. 901)

Feature (and feature creation) A variable in a machine learning model. Feature creation implies that a new variable has been created. (p. 902)

Decision tree (data mining version) A model that predicts either a categorical or quantitative response in which the branches represent variable splits and the terminal nodes provide the predicted value. (p. 902)

Algorithm A set of instructions used for calculation and data processing. The algorithm specifies how the model is built from the data. (p. 904)

Hold out (or test) set The dataset used in a supervised classification or regression problem not used to build the predictive model. This dataset is withheld (held out) from the model-building stage of the process, and the model's predictions on the test set are used to evaluate the model performance. (p. 905)

Training Fitting a model, usually from a training set. (p. 905)

Training set The dataset used in a supervised classification or regression problem to build the predictive model. (p. 905)

Terminal node The final leaves of a decision tree where the predictions for the response variable are found. (p. 906)

Neural network A model based on analogies with human brain processing that uses combinations of predictor variables and nonlinear regression to predict either a categorical or quantitative response variable. (p. 910)

Black box A model whose equation is either hidden or so complicated that it is impossible to interpret. Examples include neural networks and random forests. (p. 910)

Node The variables in a neural network model, including the variables in the "hidden" layer, are referred to as nodes. The variables in the first hidden layer are linear combinations of the original variables. (p. 910)

Hidden node (hidden layer) A linear combination of the original predictor variables used to predict the response (sometimes called a latent feature or variable). The ensemble of hidden nodes is called the hidden layer. (p. 910)

Back propagation A computational method for estimating the weights in the neural network model. (p. 910)

Sigmoidal activation function An S-shaped curve used to transform the hidden "nodes." The logistic function from logistic regression is one of the most common activation functions. (p. 910)

Bagging Abbreviation for **B**(ootstrap) **agg**(regat)**ing** a model. For any model, take a bootstrap sample and refit the model. Repeat this many times and use the average of all the models as the prediction. (p. 913)

Boosting A process of taking a simple model, fitting it to a set of data, and then using the results of that model to build a new model, putting more emphasis on the values that were predicted poorly. (p. 913)

Random forest A model built by bagging a set of trees, with the added restriction that only a small random subset of predictors are used at each split of each tree. (p. 913)

RMSE (or RASE) The square root of the average squared error of the predictions from the model. (p. 916)

EXERCISES

SECTION 27.1

1. Your supervisor is confused by Big Data. Explain what makes "big data" different than "data."

2. Your supervisor learned all about databases, and frequently makes queries such as: "What fraction of our customers who bought a product in the last six months are female and live within 5 miles of the store?" She says that she is data mining. Is she right? Explain.

SECTION 27.2

3. Your supervisor has read a couple of articles about data mining and artificial intelligence. He wants to use data mining to find the best customers. What would you suggest to structure this problem?

4. Is any one portion of the CRISP-DM more important than the others? Why?

SECTION 27.3

5. What are the advantages and disadvantages of using tree vs. neural network models?

6. Is it a good idea to use the total accurate classification rate (percent of all cases properly classified) as the metric to evaluate the division's models? Why or why not?

SECTION 27.4

7. What are the pros and cons of combining multiple models to produce a prediction? Should we always combine models?

8. Explain why boosting or other ways of averaging several models might be better than trying to find "the best."

SECTION 27.5

9. Your supervisor asked you to fit many models to predict which customers will buy this holiday season. The firm is going to pick only one. What criteria would you recommend that the firm use to evaluate the various models and to compare them?

JUST CHECKING

Answers

1. Not a data mining problem. Too vague.
2. Not data mining. OLAP / Database query.
3. Data mining, unsupervised clustering.
4. Data mining, supervised, classification because Accept is dichotomous.
5. Yes, supervised regression.
6. Bad idea to have everyone from one department. Good solutions need people with business knowledge, too.
7. No. Need cleaning.
8. No. Need good data, and data mining is not magic.
9. Bad idea. First, problem is too vague. Second, even if the problem were exploratory, there are too many relationships, and some will look significant, even if they are not.
10. Depends on the business environment. Probably not.
11. Ambiguous. Some software will cross-validate by default.
12. Maybe.
13. Tree will handle with aplomb, because it essentially functions like a median, splitting in two parts.
14. Neural nets do not "think" at all, nor do they "think" like humans. They may or may not be a good choice of model.
15. No, in general, neural networks are "black boxes" and do not provide information about the basis of classification or prediction.
16. Not really.
17. Tree is overfitting, and fitting the noise is not helpful.

Review of Part VII

INFERENCE WHEN VARIABLES ARE RELATED

Quick Review

This part introduces inference for the most widely used class of statistical models: linear models. You may have read one, two, or all of the chapters in this part. They have a consistent central theme, and it is one you'll likely see if you study more statistics.

- Linear models predict a single quantitative variable from one or more other variables. The predictor variables can be quantitative or categorical. Linear models require only that the effects of the predictor variables be simply added together (rather than, for example, being multiplied or exponential).
- When the predictors are categorical, we are finding an Analysis of Variance (ANOVA).
- When the predictors are quantitative, we are finding a regression model. With two or more predictors, we have a *multiple* regression model.
- The overall fit of a linear model is tested with an F-test. The null hypothesis for this test is that we'd do better just modeling y with its mean.
- If the overall F-test is significant, we can move on to consider the individual coefficients. For regression, this means considering the t-statistics. For ANOVA, it means first considering the F-statistics for individual factors. We may then move on to consider *contrasts* among different levels of significant factors.
- Models with two or more predictors can fail conditions in even more ways than those with only one, and some of these can be quite subtle. These models require care in checking assumptions and conditions.

Remember (have we said this often enough yet?): Never use any inference procedure without first checking the assumptions and conditions. We summarized those in the last unit; have another look.

Here come more opportunities to practice using these concepts and skills, but we've also sprinkled in some exercises that call for methods learned in earlier chapters, so keep alert.

Quick Guide to Inference

Think			Show				Tell?
Inference about?	One sample or two?	Procedure	Model	Parameter	Estimate	SE	Chapter
Proportions	One sample	1-Proportion *z*-Interval	z	p	$\hat{p}$	$\sqrt{\frac{\hat{p}\hat{q}}{n}}$	16
		1-Proportion *z*-Test				$\sqrt{\frac{p_0 q_0}{n}}$	18
	Two independent groups	2-Proportion *z*-Interval	z	$p_1 - p_2$	$\hat{p}_1 - \hat{p}_2$	$\sqrt{\frac{\hat{p}_1\hat{q}_1}{n_1} + \frac{\hat{p}_2\hat{q}_2}{n_2}}$	20
		2-Proportion *z*-Test				$\sqrt{\frac{\hat{p}\hat{q}}{n_1} + \frac{\hat{p}\hat{q}}{n_2}}, \hat{p} = \frac{y_1 + y_2}{n_1 + n_2}$	20
Means	One sample	*t*-Interval *t*-Test	t df $= n - 1$	μ	$\bar{y}$	$\frac{s}{\sqrt{n}}$	17, 18
	Two independent groups	2-Sample *t*-Test 2-Sample *t*-Interval	t df from technology	$\mu_1 - \mu_2$	$\bar{y}_1 - \bar{y}_2$	$\sqrt{\frac{s_1^2}{n_1} + \frac{s_2^2}{n_2}}$	20
	n Matched pairs	Paired *t*-Test Paired *t*-Interval	t df $= n - 1$	μ_d	$\bar{d}$	$\frac{s_d}{\sqrt{n}}$	21
Distributions (one categorical variable)	One sample	Goodness of fit	χ^2 df $= cells - 1$			$\sum\frac{(\text{obs} - \text{exp})^2}{\text{exp}}$	22
	Many independent samples	Homogeneity χ^2 Test	χ^2 df $= (r - 1)(c - 1)$				
Independence (two categorical variables)	One sample	Independence χ^2 Test					
Association (two quantitative variables)	One sample	Linear Regression *t*-Test or Confidence Interval for β	t df $= n - 2$	β_1	b_1	$\frac{s_e}{s_x\sqrt{n-1}}$ (compute with technology)	23
		Confidence Interval for μ_ν		μ_ν	$\hat{y}_\nu$	$\sqrt{SE^2(b_1)\cdot(x_\nu - \bar{x})^2 + \frac{s_e^2}{n}}$	23
		Prediction Interval for y_ν		y_ν	$\hat{y}_\nu$	$\sqrt{SE^2(b_1)\cdot(x_n - \bar{x})^2 + \frac{s_e^2}{n} + s_e^2}$	23
Inference about?	One sample or two?	Procedure	Model	Parameter	Estimate	SE	Chapter

REVIEW EXERCISES

T R7.1. Mutual fund returns 2017 We have a sample of municipal mutual funds that report their return (%) in the previous 3 years and the previous 5 years.

a) Create a 95% confidence interval for the difference in rate of return for the 3- and 5-year periods covered by these data. Clearly explain what your interval means.
b) It's common for advertisements to carry the disclaimer that "past returns may not be indicative of future performance," but do these data indicate that there was an association between 3- and 5-year rates of return?

Fund	3 year (%)	5 year (%)
Muni California Long	3.81	3.85
Muni National Long	3.41	3.12
Muni California Intermediate	2.63	2.71
Muni National Interm	2.64	2.55
Muni New York Long	3.35	2.84
Muni New Jersey	2.89	2.52
Muni Pennsylvania	3.1	2.68
Muni Massachusetts	2.69	2.3
Muni New York Intermediate	2.57	2.39
Muni Minnesota	2.66	2.52
Muni Ohio	2.64	2.38
Muni Single State Interm	2.36	2.16
Muni Single State Long	2.67	2.23
Muni Single State Short	1.24	1.16
Muni National Short	0.93	0.95

T R7.2. Polling How accurate are pollsters in predicting the outcomes of Congressional elections? The table shows the actual number of Democrat seats in the House of Representatives and the number predicted by the Gallup organization for non-presidential election years since World War II.

a) Is there a significant difference between the number of seats predicted for the Democrats and the number they actually held? Test an appropriate hypothesis and state your conclusions.
b) Is there a strong association between the pollsters' predictions and the outcomes of the elections? Test an appropriate hypothesis and state your conclusions.

Democratic Congressmen

Year	Predicted	Actual
1946	190	188
1950	235	234
1954	232	232
1958	272	283
1962	259	258
1966	247	248
1970	260	255
1974	292	291
1978	277	277
1982	275	269
1986	264	258
1990	260	267
1994	201	204
1998	211	211

R7.3. Football A student runs an experiment to test four different grips on his football throwing distance, recording the distance in yards that he can throw the football using each grip. He randomizes the grip used each time by drawing numbers out of a hat until each grip has been used 5 times. After collecting his data and analyzing the results, he reports that the P-value of his test is 0.0032.

a) What kind of test should he have performed?
b) What are the null and alternative hypotheses?
c) Assuming that the conditions required for the test are satisfied, what would you conclude?
d) What else about the data would you like to see in order to check the assumptions and conditions?
e) What might he want to test next?

R7.4. Golf A student runs an experiment to test four different clubs on her putting accuracy, recording the distance in centimeters from a small target that she places on the green. She randomizes the club used each time by drawing numbers out of a hat until each club has been used 6 times. After collecting her data and analyzing the results, she reports that the P-value of her test is 0.0245.

a) What kind of test should she have performed?
b) What are the null and alternative hypotheses?
c) Assuming that the conditions required for the test are satisfied, what would you conclude?
d) What else about the data would you like to see in order to check the assumptions and conditions?
e) What might she want to test next?

R7.5. Horses, a bit less wild In an attempt to control the growth of the herds of wild horses, managers sterilized some of the stallions. The variable *Sterilized* is coded 0 for herds and years in which no stallions were sterilized and 1 for herds and years in which some stallions were sterilized. The resulting regression looks like this:

Dependent variable is: Foals
R-squared = 84.8% R-squared (adjusted) = 83.9%
s = 4.814 with 38 − 3 = 35 degrees of freedom

Source	Sum of squares	DF	Mean square	F-ratio
Regression	4511.85	2	2255.92	97.3
Residual	811.229	35	23.1780	

Variable	Coefficient	SE(Coeff)	t-ratio	P-value
Intercept	6.23667	4.801	1.30	0.2024
Adults	0.112271	0.0268	4.18	0.0002
Sterilized	−6.43657	3.769	−1.71	0.0965

a) What is the name for the kind of variable used here to represent sterilizing?
b) What is the interpretation of the coefficient of *Sterilized?*
c) Does sterilizing appear to work? That is, do these data show a statistically significant effect of sterilizing stallionson the number of foals born?

T **R7.6. Hard water** The data on water hardness and mortality considered in Part VI Review Exercise 30 also included information on whether the town is north or south of Derby. When that information is coded in a variable as 1 for "north of Derby" and 0 for "south of Derby," and that variable is included in the regression, the resulting table looks like this:

Dependent variable is: Mortality
R-squared = 56.2% R-squared (adjusted) = 54.7%
s = 126.4 with 61 − 3 = 58 degrees of freedom

Variable	Coefficient	SE(Coeff)	t-ratio	P-value
Intercept	1537.50	42.02	36.6	≤0.0001
Calcium	−2.16011	0.4979	−4.34	≤0.0001
North of Derby	158.892	37.87	4.20	≤0.0001

a) What is the name for the kind of variable that the 0/1 variable is for being north of Derby?
b) What does the coefficient of that variable mean in this regression?
c) Would you prefer this regression to the one of Review VI. Explain why or why not.

R7.7. Lost baggage 2018 The Bureau of Transportation Statistics of the U.S. Department of Transportation reports statistics about airline performance. For September 2018 they report the following number of bags lost per 1000 passengers.

Airline	lost/1000
DELTA AIR LINES	1.83
JETBLUE AIRWAYS	1.96
SPIRIT AIRLINES	2.11
HAWAIIAN AIRLINES	2.43
UNITED AIRLINES	2.77
ALASKA AIRLINES	2.79
FRONTIER AIRLINES	3.1
SOUTHWEST AIRLINES	3.39
SKYWEST AIRLINES	4.32
AMERICAN AIRLINES	4.39
EXPRESSJET AIRLINES	5.46
ENVOY AIR	6.37

Are the airlines roughly equal in their baggage performance? Perform a chi-square goodness-of-fit test, or explain why that would not be appropriate.

R7.8. Bowling A student runs an experiment to test how different factors affect her bowling performance. She uses 3 levels for the weight of the ball (low, medium, and high) and 2 approaches (standing and walking), throwing 4 balls at each condition and choosing the conditions in random order. She counts the number of pins knocked down as the response variable.

a) What are the null and alternative hypotheses for the main effects?
b) How many degrees of freedom does each factor sum of squares have? How about the Error Sum of Squares?
c) Should she consider fitting an interaction term to the model? Why or why not? How many degrees of freedom would it take?

R7.9. Video racing A student runs an experiment to test how different factors affect his reaction time while playing video games. He uses a specified race course with random hazards and times how long it takes him (in seconds) to finish 4 laps under a variety of experimental conditions. As factors, he uses three different types of *Mouse* (cordless ergonomic, cordless regular, or corded) and keeps the *Lights* on or off. He measures his time once at each condition. To avoid a learning effect, he runs the race 5 times before selecting the conditions in random order.

a) What are the null and alternative hypotheses for the main effects?
b) How many degrees of freedom does each factor sum of squares have? How about the Error Sum of Squares?
c) Should he consider fitting an interaction term to the model? Why or why not? How many degrees of freedom would it take?

R7.10. Resume fraud In 2002 the Veritas Software company found out that its chief financial officer did not actually have the MBA he had listed on his resume. They fired him, and the value of the company's stock dropped 19%. Kroll, Inc., a firm that specializes in investigating such matters, said that they believe as many as 25 percent of background checks might reveal false information. How many such random checks would they have to do to estimate the true percentage of people who misrepresent their backgrounds to within ±5% with 98% confidence?

R7.11. Paper planes In preparation for a regional paper airplane competition, a student tried out her latest design. The distances her plane traveled (in feet) in 11 trial flights are given here. (The world record is an astounding 193.01 feet!) The data were 62, 52, 68, 23, 34, 45, 27, 42, 83, 56, and 40 feet.

Here are some summaries:

Count	11
Mean	48.3636
Median	45
StdDev	18.0846
StdErr	5.45273
IntQRange	525
25th %tile	35.5000
75th %tile	60.5000

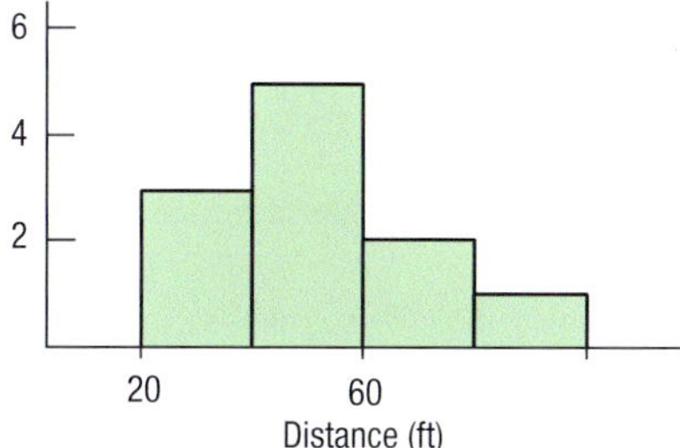

a) Construct a 95% confidence interval for the mean distance.
b) Based on your confidence interval, is it plausible that the mean distance is 40 ft? Explain.
c) How would a 99% confidence interval for the mean distance differ from your answer in part a? Explain briefly, without actually calculating a new interval.
d) How large a sample size would the student need to get a confidence interval half as wide as the one you got in part a, at the same confidence level?

R7.12. Nuclear plants Here are data on 32 light water nuclear power plants. The variables are:

Cost: In \$100,000, adjusted to 1976 base.

Date: Date that construction permit was issued in years after 1900. Thus, 68.58 is roughly halfway through 1968.

Mwatts: Power plant net capacity in megawatts.

We are interested in the *Cost* of the plants as a function of *Date* and *Mwatts*.

Cost	Mwatts	Date	Cost	Mwatts	Date
345.39	514	67.92	712.27	845	69.50
460.05	687	68.58	289.66	530	68.42
452.99	1065	67.33	881.24	1090	69.17
443.22	1065	67.33	490.88	1050	68.92
652.32	1065	68.00	567.79	913	68.75
642.23	1065	68.00	665.99	828	70.92
272.37	822	68.17	621.45	786	69.67
317.21	457	68.42	608.80	821	70.08
457.12	822	68.42	473.64	538	70.42
690.19	792	68.33	697.14	1130	71.08
350.63	560	68.58	207.51	745	67.25
402.59	790	68.75	288.48	821	67.17
412.18	530	68.42	284.88	886	67.83
495.58	1050	68.92	280.36	886	67.83
394.36	850	68.92	217.38	745	67.25
423.32	778	68.42	270.71	886	67.83

a) Examine the relationships between *Cost* and *Mwatts* and between *Cost* and *Date*. Make appropriate displays and interpret them with a sentence or two.
b) Find the regression of *Cost* on *Mwatts*. Write a sentence that explains the relationship as described by the regression.
c) Make a scatterplot of residuals vs. predicted values and discuss what it shows. Make a Normal probability plot or histogram of the residuals. Discuss the four assumptions needed for regression analysis and indicate whether you think they are satisfied here. Give your reasons.
d) State the standard null hypothesis for the slope coefficient and complete the t-test at the 5% level. State your conclusion.
e) Estimate the cost of a 1000-mwatt plant. Show your work.
f) Compute the residuals for this regression. Discuss the meaning of the R-squared in this regression. Plot the residuals against *Date*. Does it appear that *Date* can account for some of the remaining variability?
g) Compute the multiple regression of *Cost* on both *Mwatts* and *Date*. Compare the coefficient in this regression with those you have found for each of these predictors.
h) Would you expect *Mwatts* and *Date* to be correlated? Why or why not? Examine the relationship between *Mwatts* and *Date*. Make a scatterplot and find the correlation coefficient, for example. It's only because of the extraordinary nature of this relationship that the relationships you saw at earlier steps were this simple.

R7.13. Barbershop music At a barbershop music singing competition, choruses are judged on three scales: *Music* (quality of the arrangement, etc.), *Performance*, and *Singing*. The scales are supposed to be independent of each other, and each is scored by a different judge, but a friend claims that he can predict a chorus's singing score from the other two scores. He offers the following regression based on the scores of all 34 choruses in a recent competition:

Dependent variable is: Singing
R-squared = 90.9% R-squared (adjusted) = 90.3%
s = 6.483 with 34 − 3 = 31 degrees of freedom

Variable	Coefficient	SE(Coeff)	t-ratio	P-value
Intercept	2.08926	7.973	0.262	0.7950
Performance	0.793407	0.0976	8.13	≤0.0001
Music	0.219100	0.1196	1.83	0.0766

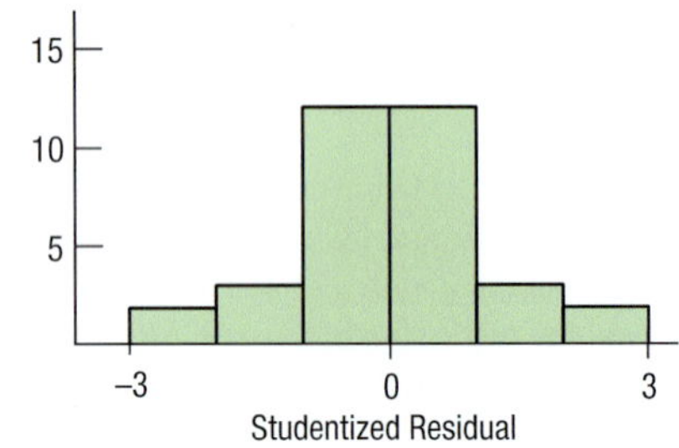

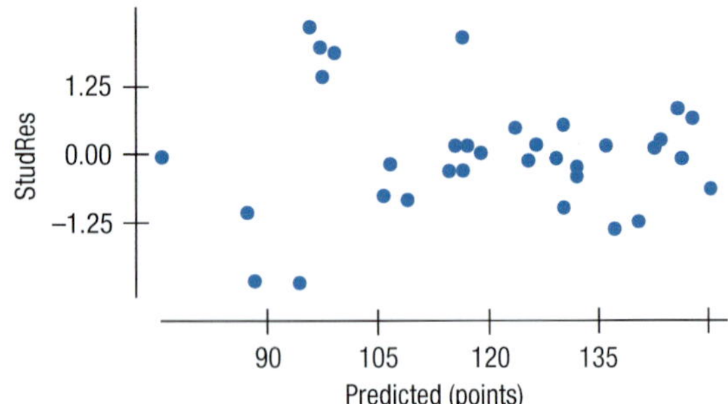

a) What do you think of your friend's claim? Can he predict singing scores? Explain.
b) State the standard null hypothesis for the coefficient of *Performance* and complete the t-test at the 5% level. State your conclusion.
c) Complete the analysis. Check assumptions and conditions to the extent you can with the information provided.

R7.14. Sleep Using a Simple Random Sample, a student group asked 450 students about their sleep and study habits and received about 200 responses. The group wanted to know if the average amount of sleep varied by sex (F or M) or by year (Freshman, Sophomore, Junior, or Senior) of the respondent. Partial boxplots of the amount of sleep last night by the two factors are shown below with the interaction plot:

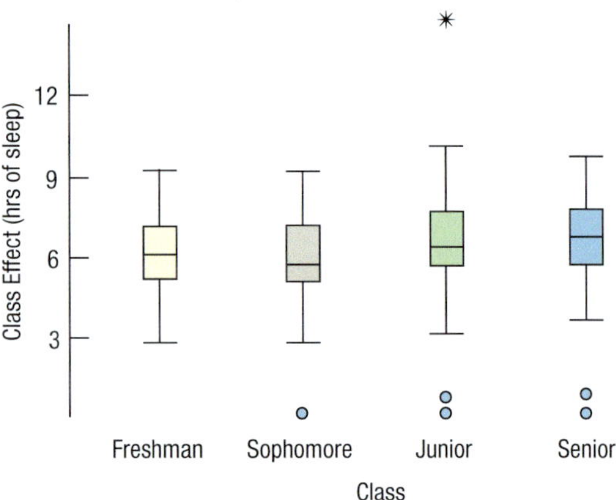

The ANOVA table shows:

Source	DF	Sum of squares	F-ratio	P-value
Sex	1	8.0658	2.2821	0.1325
Year	3	8.8990	0.8393	0.4739
Sex × Year	3	19.4075	1.8303	0.1431
Total	196	707.0180		

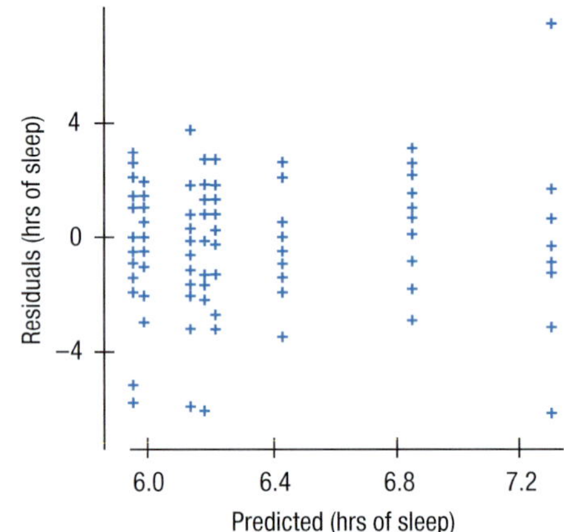

a) What are the null and alternative hypotheses for the main effects?
b) Should the group consider fitting an interaction term to the model? Why or why not?
c) What effects appear to be significant?
d) What, if any, reservations do you have about the conclusions?

R7.15. Study habits The survey in Exercise 14 also asked students about the number of hours they studied. Those doing the survey wanted to know if the average amount of studying varied by sex or by class of the respondent. Partial boxplots of the Hours *Studied* last night by the two factors are shown below:

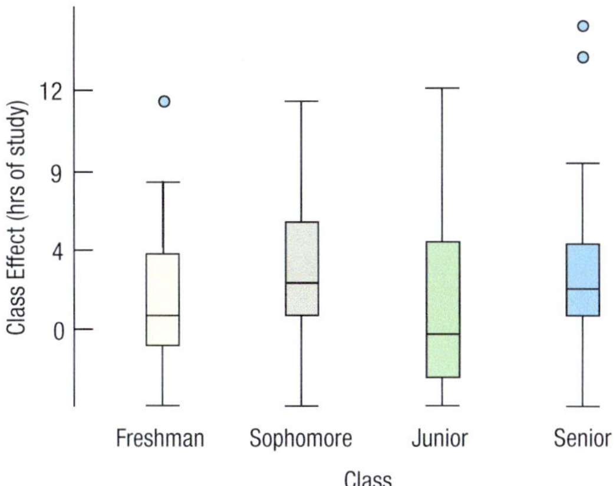

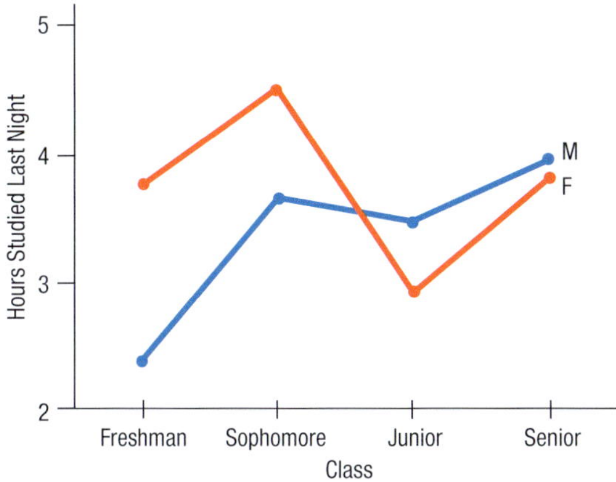

Source	DF	Sum of squares	F-ratio	P-value
Sex	1	7.1861	1.0744	0.3013
Year	3	34.9922	1.7440	0.1595
Sex × Year	3	27.8188	1.3865	0.2483
Total	196	1329.8020		

A plot of residuals vs. predicted shows:

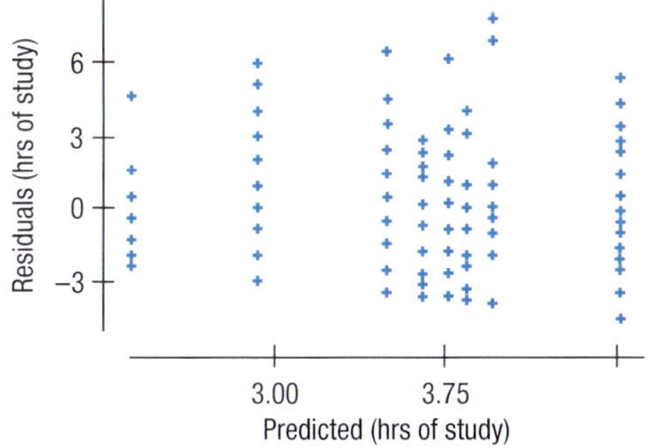

a) What are the null and alternative hypotheses for the main effects?
b) Should those doing the survey consider fitting an interaction term to the model? Why or why not?
c) What effects appear to be significant?
d) What, if any, reservations do you have about the conclusions?

R7.16. Is Old Faithful getting older? We have data on Old Faithful eruptions that includes another variable recording the *Day* on which the eruption occurred (where 1 is the first day and each successive day just counts one more). The correlation of *Interval* (minutes until the next eruption) with *Day* is −0.004. But when we include *Day* in a multiple regression along with the *Duration* (in minutes) of the previous eruption, we get the following model:

Dependent variable is: Interval
R-squared = 77.6% R-squared (adjusted) = 77.3%
s = 6.092 with 222 − 3 = 219 degreets of freedom

Variable	Coefficient	SE(Coeff)	t-ratio	P-value
Intercept	35.2463	1.509	23.4	≤0.0001
Duration	10.4348	0.3794	27.5	≤0.0001
Day	−0.126316	0.0523	−2.42	0.0166

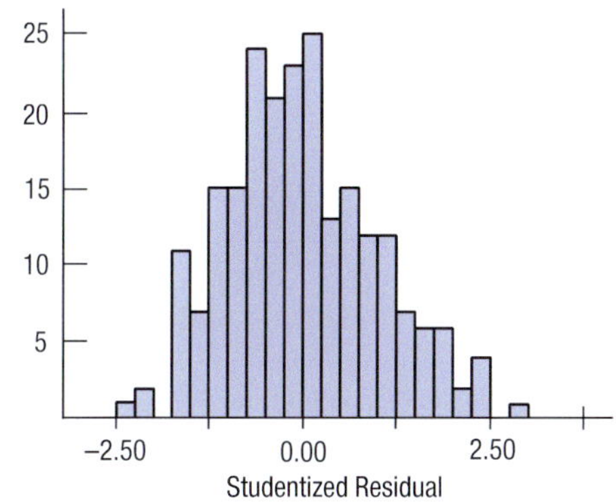

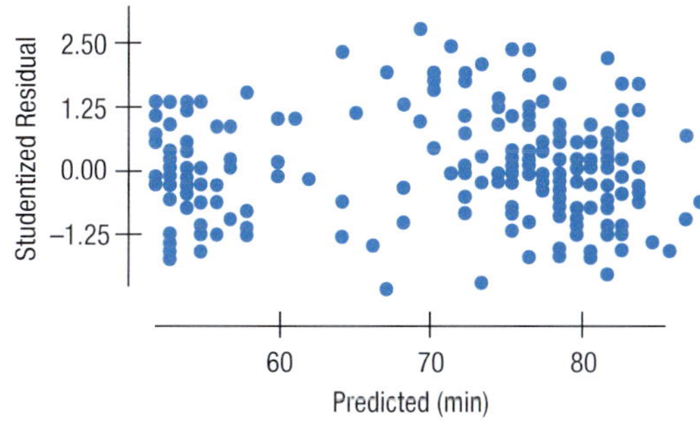

a) What is the model fit by this regression?
b) Is the *Interval* changing over time? Perform a formal test of the relevant hypothesis.
c) Doesn't the small P-value for *Day* contradict the correlation between *Interval* and Day being virtually zero? Explain.
d) Is the amount of change in *Interval* due to *Day* meaningful?

T **R7.17. Traffic fatalities 2013** The Insurance Institute for Highway Safety publishes data on a variety of traffic related risks. One report gives the numbers of people killed in highway accidents during each year from 1975 to 2013. Here is a regression predicting *Total traffic fatalities* by *Year*:

Response variable is: Total traffic fatalities
R-squared = 53.0% R-squared (adjusted) = 51.2%
s = 3013 with 27 − 2 = 25 degrees of freedom

Source	Sum of Squares	DF	Mean Square	F-ratio
Regression	256443883	1	256443883	28.2
Residual	227016875	25	9080675	

Variable	Coefficient	SE(Coeff)	t-ratio	P-value
Intercept	689752	122140	5.65	<0.0001
Year	−324.680	61.10	−5.31	<0.0001

a) Here's a scatterplot of residuals vs. predicted values for this regression. Discuss the assumptions and conditions of the regression.

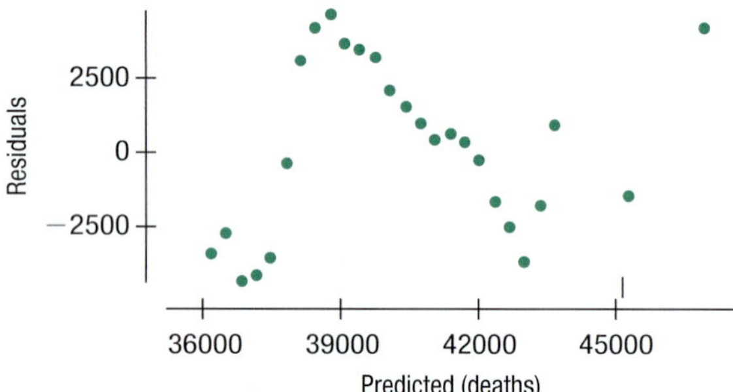

b) What is the meaning of the R^2 value of 53.0% for this regression?
c) What is the regression equation?
d) Give an interpretation of the coefficient of *Year* in this regression.

R7.18. Births The National Vital Statistics Report provides information on live *Births* (per 1000 women), according to the age of the woman (in 5-year brackets—*Age* used here is the midpoint of the bracket) and the *Year* from 1990 to 1999. The report isolates births to women younger than 20 as a separate category. Looking only at women 20 years old and older, we find the following regression:

Dependent variable is: Births
R-squared = 98.1% R-squared (adjusted) = 98.0%
s = 15.55 with 50 − 3 = 47 degrees of freedom

Variable	Coefficient	SE(Coeff)	t-ratio	P-value
Intercept	4422.98	1527	2.90	0.0057
Age	−15.1898	0.3110	−48.8	≤0.0001
Year	−1.89830	0.7656	−2.48	0.0168

a) Write out the regression model.
b) How would you interpret the coefficient of *year* in this regression? What happened to pregnancy rates during the decade of the 1990s?
Here's a scatterplot of the Studentized residuals against the predicted values.

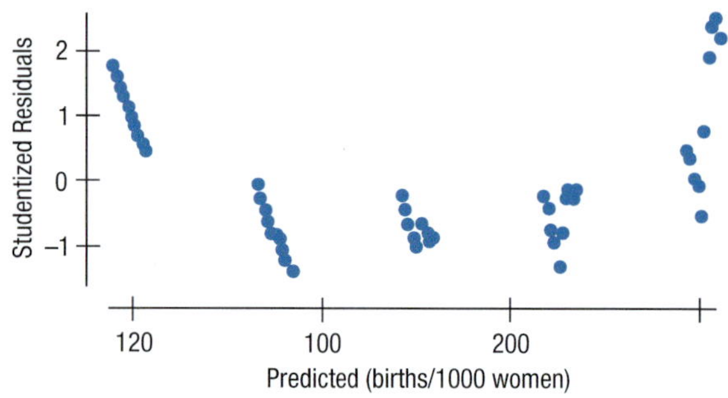

c) What might you do to improve the regression model?

R7.19. Typing For a class project, Nick M. designed and carried out an experiment to see if the room *Temperature* and the wearing of *Gloves* affected his typing speed. He ran each combination of hot and cold temperature and gloves on and off 8 times, recording the net number of words typed (words typed minus mistakes). Partial boxplots and interaction plots show:

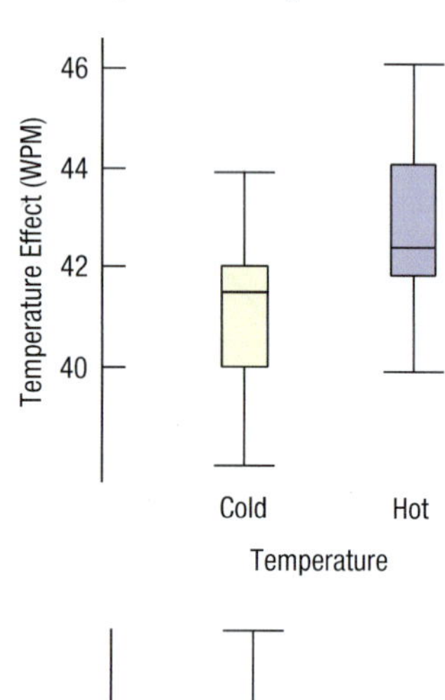

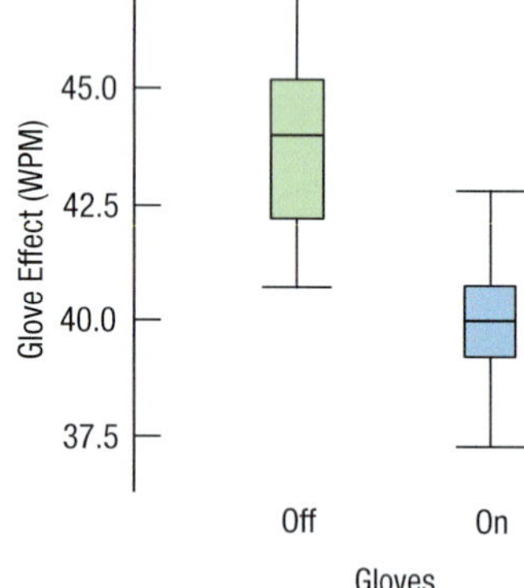

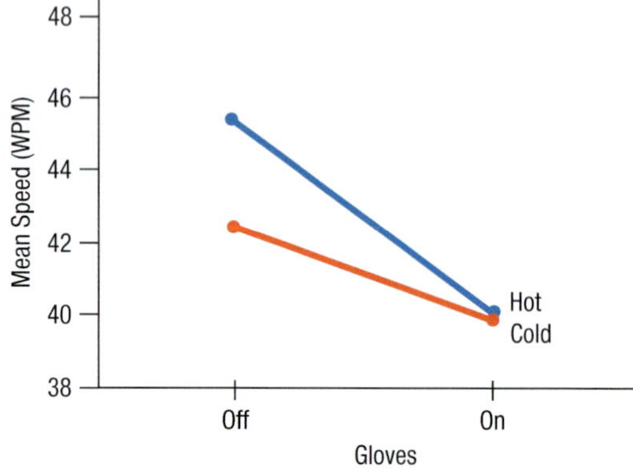

The ANOVA table shows:

Source	DF	Sum of squares	F-ratio	P-value
Gloves	1	120.1250	57.2511	<0.0001
Temperature	1	18.0000	8.5787	0.0067
Gloves*Temp	1	15.1250	7.2085	0.0121
Total	28	58.75		

A scatterplot of residuals vs. predicted values shows:

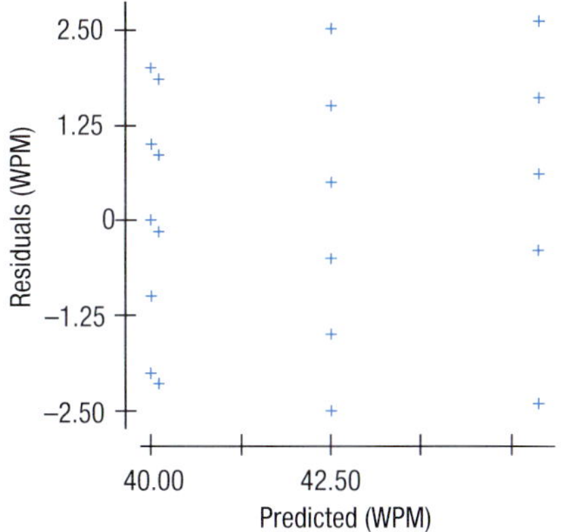

a) What are the null and alternative hypotheses for the main effects?
b) Given the partial boxplots, what do you suspect the *F*-test will say about the two main effects?
c) Should he consider fitting an interaction term to the model? Why or why not?
d) If he does fit an interaction term, do you suspect it will be significant on the basis of the interaction plot? Explain.
e) Which effects appear to be significant?
f) Describe the effects of the factors.
g) What is the size of the estimated standard deviation of the errors? Does this seem reasonable given the partial boxplots? Explain.
h) If Nick wants to increase his typing speed, what recommendations would you give him?
i) What reservations, if any, do you have about the conclusions?

R7.20. Typing again Nick (see Exercise 19) designed a follow-up experiment to see if having *Music* or *Television* on would affect his typing speed. In particular, he'd like to know if he can type just as effectively with the music and/or the TV on while he types. He ran each combination of *Music* and *Television* on or off 8 times, recording the net number of words typed (words typed minus mistakes). Partial boxplots and interaction plots show:

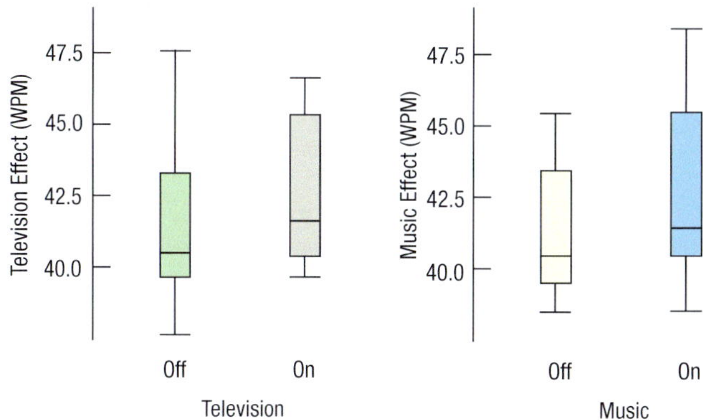

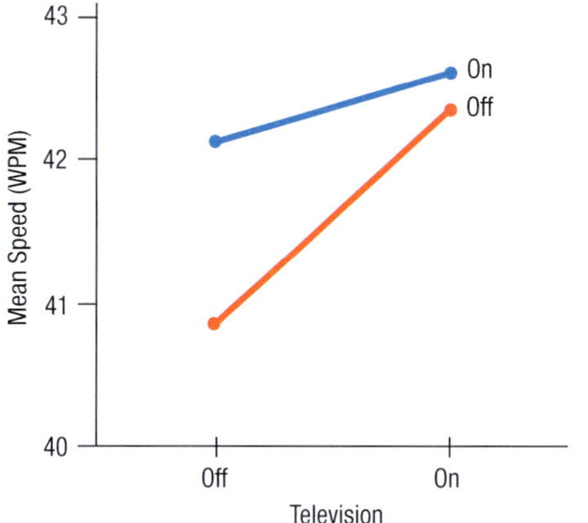

The ANOVA table shows:

Source	DF	Sum of squares	F-ratio	P-value
Music	1	4.5000	0.6380	0.4312
Television	1	8.0000	1.1342	0.2960
Interaction	1	2.0000	0.2835	0.5986
Total	28	197.5000		

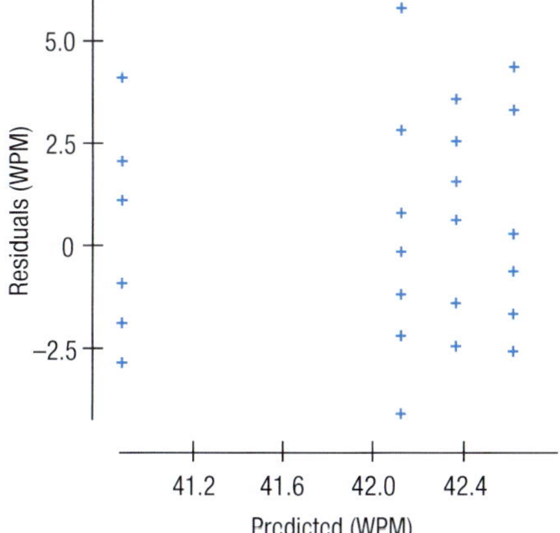

a) What are the null and alternative hypotheses for the main effects?
b) Given the partial boxplots, what do you suspect the F-test will say about the two main effects?
c) Should he consider fitting an interaction term to the model? Why or why not?
d) If he does fit an interaction term, do you suspect it will be significant on the basis of the interaction plot? Explain.
e) What effects appear to be significant?
f) Describe the effects of the factors.
g) What is the size of the estimated standard deviation of the errors? Does this seem reasonable given the partial boxplots? Explain.
h) If Nick wants to increase his typing speed, what recommendations would you give him?
i) What reservations, if any, do you have about the conclusions?

R7.21. NY Marathon The *New York Times* reported the results of the 2003 NY Marathon by listing time brackets and the number of racers who finished within that bracket. Because the brackets are of different sizes, we look at the number of racers finishing *per minute* against the time at the middle of the time bracket. The resulting regression looks like this:

Dependent variable is: #/minute
R-squared = 10.9% R-squared (adjusted) = 9.5%
s = 78.45 with 63 − 2 = 61 degrees of freedom

Variable	Coefficient	SE(Coeff)	t-ratio	P-value
Intercept	45.1407	51.74	0.873	0.3863
Mid time ...	0.519037	0.1899	2.73	0.0082

a) How would you interpret the coefficient of *Mid Time?*

Here's a scatterplot of the Studentized residuals against the predicted values for this regression:

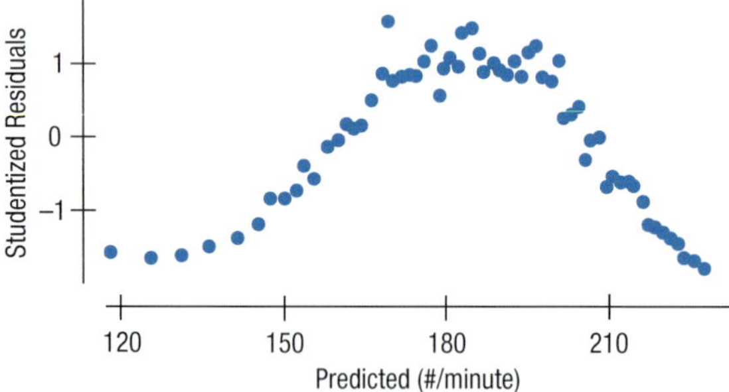

b) Comment on the regression in the light of this plot.

R7.22. Learning math Developers of a new math curriculum called "Accelerated Math" compared performances of students taught by their system with control groups of students in the same schools who were taught using traditional instructional methods and materials. Statistics about pretest and posttest scores are shown in the table. (J. Ysseldyke and S. Tardrew, *Differentiating Math Instruction*, Renaissance Learning, 2002)

a) Did the groups differ in average math score at the start of this study?
b) Did the group taught using the Accelerated Math program show a significant improvement in test scores?
c) Did the control group show significant improvement in test scores?
d) Were gains significantly higher for the Accelerated Math group than for the control group?

		Instructional Method	
		Acc. math	**Control**
Number of students		**231**	**245**
Pretest	**Mean**	560.01	549.65
	St. Dev	84.29	74.68
Posttest	**Mean**	637.55	588.76
	St. Dev	82.9	83.24
Individual gain	**Mean**	77.53	39.11
	St. Dev.	78.01	66.25

R7.23. Pesticides A study published in the journal *Environmental Health Perspectives* examined the sex ratios of children born to workers exposed to dioxin in Russian pesticide factories. The data covered the years 1961 to 1988 in the city of Ufa, Bashkortostan, Russia. Of 227 children born to workers exposed to dioxin, only 40% were male. Overall in the city of Ufa the proportion of males was 51.2%. Is this evidence that human exposure to dioxin may results in the birth of more girls? (An interesting note: It appeared that paternal exposure was most critical; 51% of babies born to mothers exposed to the chemical were boys.)

R7.24. Video pinball A student runs an experiment to test how different factors affect his score while playing video pinball. Here are the results of 16 runs of an experiment performed in random order. Factor *Eyes* has two levels: *both* open and *right eye* closed. Factor *Tilt* has two levels: tilt *on* and tilt *off*. The response is the score of one ball at a combination of factors. Each combination was repeated 4 times in random order.

a) What are the null and alternative hypotheses for the main effects?
b) Analyze the data and write a short report on your findings. Include appropriate graphics and diagnostic plots.

Eyes	Tilt	Score
Both	On	67059
Right eye	Off	21036
Both	On	59520
Right eye	Off	3100
Both	Off	61272
Right eye	On	55957
Both	Off	72472
Right eye	On	18460
Right eye	On	16556
Both	Off	89553
Right eye	On	37950
Both	Off	74336
Right eye	Off	700
Both	On	79037
Right eye	Off	36591
Both	On	74610

R7.25. Javelin Brianna, a member of the track team, runs an experiment to test how different factors affect her javelin throw. She wants to know if the more expensive (*premium*) javelin is worth the extra price and is curious to know how much warming up helps her distance. She tries all four combinations of the two *Javelins* (*standard* and *premium*) and *Preparation* (no warm-up—*cold*—and *warm-up*), repeating each combination twice in random order. She measures the distance of her throw to the nearest meter. Here are the results of the 8 runs.

Javelin	Preparation	Distance (meters)
Premium	Warm-up	46
Standard	Warm-up	39
Premium	Cold	37
Standard	Cold	30
Premium	Warm-up	45
Standard	Warm-up	40
Premium	Cold	35
Standard	Cold	32

a) What are the null and alternative hypotheses for the main effects?

b) Analyze the data and write a short report on your findings. Include appropriate graphics and diagnostic plots and a recommendation for Brianna to optimize her javelin distance.

T **R7.26.** LA rainfall 2018 The Los Angeles Almanac website reports recent annual rainfall (in inches), as shown in the table.

a) Create a 90% confidence interval for the mean annual rainfall in LA.

b) If you wanted to estimate the mean annual rainfall with a margin of error of only 2 inches, how many years' data would you need?

c) Do these data suggest any change in annual rainfall as time passes? Check for an association between rainfall and year.

d) Consider the residuals. If there are outliers, look them up ("Rainfall in Los Angeles in *####*"). Comment.

Year	Rain(in)	Year	Rain(in)
1991	11.47	2005	37.25
1992	21	2006	13.19
1993	27.36	2007	3.21
1994	8.11	2008	13.53
1995	24.35	2009	9.08
1996	12.46	2010	16.36
1997	12.4	2011	20.2
1998	31.01	2012	8.69
1999	9.09	2013	5.85
2000	11.57	2014	6.08
2001	17.94	2015	8.52
2002	4.42	2016	9.65
2003	16.49	2017	19
2004	9.24	2018	4.79

R7.27. TV and athletics Using a Simple Random Sample, a student asked 200 students questions about their study and workout habits, and received 124 responses. One of the questions asked, "Do you participate in intramural (IM) athletics, varsity athletics, or no athletics?" while the second asked, "How many hours of television did you watch last week?" The student wants to see if participation in athletics is associated with amount of TV watched. Here are the boxplots of *TV Watching* by *Athletic Participation:*

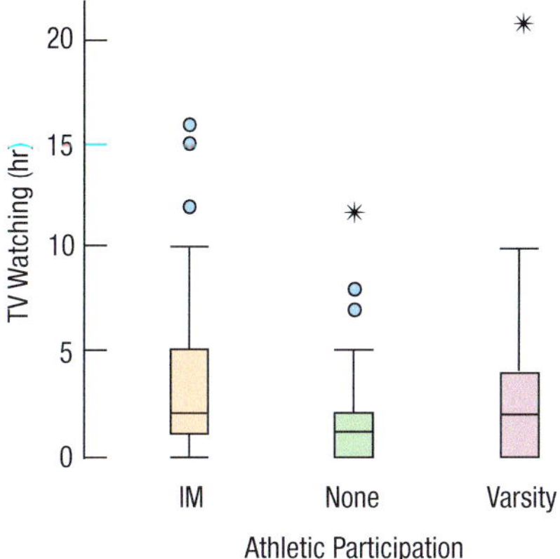

Analysis of Variance

Source	DF	Sum of squares	Mean square	F-ratio	P-value
Athl. Part.	2	128.271	64.1353	4.2343	0.0167
Error	121	1832.72	15.1465		
Total	123	1960.99			

a) State the hypothesis about the students (both numerically and in words).

b) Do the conditions for ANOVA appear to be satisfied? What concerns, if any, do you have?

c) Assuming that the assumptions for inference are satisfied, perform the hypothesis test and state your conclusion.

Be sure to state it in terms of television watching and athletic participation.

d) An analysis of the data with all the points highlighted as outliers removed was performed. The *F*-test showed a P-value of 0.0049. How does this affect your answer to part b?

R7.28. Weight and athletics Using the same survey as in Exercise 27, the student examined the relationship between *Athletic Participation* and *Weight*. Here are the boxplots of *Weight* by *Athletic Participation:*

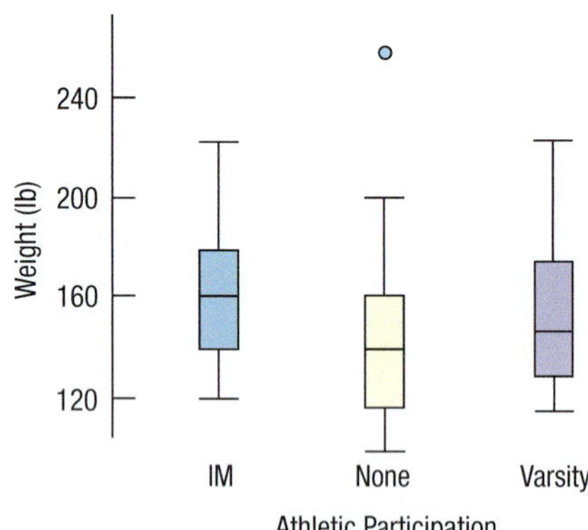

Analysis of Variance

Source	DF	Sum of squares	Mean square	F-ratio	P-value
Athl. Part.	2	9025.55	4512.78	5.7368	0.0042
Error	121	95183.2	786.638		
Total	123	104209			

a) State the null hypothesis about the students (both numerically and in words).

b) Do the conditions for ANOVA appear to be satisfied? What concerns, if any, do you have?

c) Assuming that the assumptions for inference are satisfied, perform the hypothesis test and state your conclusion. Be sure to state it in terms of *Weight* and *Athletic Participation.* What might explain the apparent relationship?

d) An analysis of the data with the points highlighted as an outlier removed was performed. The *F*-test showed a P-value of 0.0030. How does this affect your answer to part b?

R7.29. Weight loss A weight loss clinic advertises that its program of diet and exercise will allow clients to lose 10 pounds in one month. A local reporter investigating weight reduction gets permission to interview a randomly selected sample of clients who report the given weight losses during their first month in this program. Create a confidence interval to test the clinic's claim that typical weight loss is 10 pounds.

Pounds	Lost
9.5	9.5
13	9
9	8
10	7.5
11	10
9	7
5	8
9	10.5
12.5	10.5
6	9

T **R7.30. Cramming** Students in two basic Spanish classes were required to learn 50 new vocabulary words. One group of 45 students received the list on Monday and studied the words all week. Statistics summarizing this group's scores on Friday's quiz are given. The other group of 25 students did not get the vocabulary list until Thursday. They also took the quiz on Friday after "cramming" Thursday night. Then, when they returned to class the following Monday they were retested—without advance warning. Both sets of test scores for these students are shown.

Group 1

Fri.

Number of students = 45

Mean = 43.2 (of 50)

StDev = 3.4

Students passing (score ≥ 40) = 33%

a) Did the week-long study group have a mean score significantly higher than that of the overnight crammers?

b) Was there a significant difference in the percentages of students who passed the quiz on Friday?

c) Is there any evidence that when students cram for a test their "learning" does not last for 3 days?

d) Use a 95% confidence interval to estimate the mean number of words that might be forgotten by crammers.

e) Is there any evidence that how much students forget depends on how much they "learned" to begin with?

Group 2			
Fri.	**Mon.**	**Fri.**	**Mon.**
42	36	50	47
44	44	34	34
45	46	38	31
48	38	43	40
44	40	39	41
43	38	46	32
41	37	37	36
35	31	40	31
43	32	41	32
48	37	48	39
43	41	37	31
45	32	36	41
47	44		

Cumulative Review Exercises

T **1. Igf** The data set **Igf** contains measurements on people of various ages. The main variable of interest is the level of insulin-like growth factor (*igf*) (*J. Clin. Endocrinol. Metab.* 78(3): 744–752, March 1994). Each row in the data set corresponds to one individual. Data were collected on these variables:

age	quantitative (years, decimal)
sex	categorical (M = male; F = female)
igf	quantitative (insulin-like growth factor, μg/l)
tanner	ordered categorical (codes 1–5: stages of puberty, as defined by Tanner)
testvol	quantitative (testicular volume, ml)
menarche	categorical (Has menarche (menstruation) occurred? yes/no)
weight	quantitative (kg)
height	quantitative (cm)

a) Using appropriate graphics and summary statistics, describe the distribution of *igf*.
b) Describe the relationship between *igf* and *sex*.
c) Describe the relationship between *igf and age* using both graphics and summary statistics.
d) Is a linear regression appropriate for modeling the relationship between *igf* and *age*?

T **2. Igf13** The data set **Igf13** contains the data from **Igf** for children under 13 years old. Most of the data was collected from physical examinations in schools.

a) Fit a linear regression to *igf* using *age* as the predictor variable. Comment on the appropriateness of the model.
b) Use the model from part a to predict *igf* for a 20-year-old person. Do you think this prediction is accurate? Support your answer briefly.
c) If your model does not satisfy the assumptions and conditions of regression, find a suitable transformation of *igf* and refit the model. Are the assumptions and conditions met for the new model?
d) Would you use the model in either part a or part c to predict *igf* for a 30-year-old person? Why or why not?
e) Predict *igf* for a 20-year-old person using the model in part c. Do you think this prediction is better than the one you found in part a? Do you think it is accurate? Support your answer briefly.

T **3. More Igf13** Now consider the variables *igf* and *weight* in the data set **Igf13**.

a) Fit a regression model. If the data violate any assumptions, find a suitable re-expression of *igf*.
b) Add *sex* to the model in part a as a predictor. Interpret the coefficient of *sex*.
c) Use the model in part b to predict *igf* for a 100-pound female.
d) Add *height* to the model in part b. Are all the coefficients statistically significant? What happened? Explain.
e) Based on the model in part d, what model would you consider next?

T **4. Speed and density** The following data relate traffic *Density* (measured as number of cars per mile) and the average *Speed* of traffic (in mph) on city highways. The data were collected at the same location at 10 different times randomly selected within a span of 3 months.

	Density	Speed
	69	25.4
	56	32.5
	62	28.6
	119	11.5
	84	21.3
	74	22.1
	73	22.3
	90	18.5
	38	37.2
	22	44.6
Mean	**68.7**	**26.4**
SD	**27.07**	**9.64**

a) Describe the relationship between *Speed* and *Density*.
b) Fit a model to predict *Speed* from *Density*. Comment on the appropriateness of the model.
c) If the traffic density is 56 cars/mile and you observe a speed of 32.5 mph, what is the residual? What is the residual for an observed speed of 36.1 mph with a traffic density of 20 cars/mile?
d) Which of the two residuals in part c is more unusual? Explain.
e) A new data point has come in. For this observation, *Density* = 125 cars/mile and *Speed* = 55 mph. What would happen to the slope and the correlation if this point were included?
f) Predict the speed of traffic for a density of 200 cars/mile. Is your prediction reasonable? Explain briefly.
g) If you standardize both variables, what equation predicts the standardized speed from the standardized density?
h) If you standardize both variables, what equation predicts the standardized density from the standardized speed?

5. Hospital variables A random sample of 1000 patients was taken from a large medical center database. The six variables included are *Sex* (M/F), *Age* (years), *Insurance coverage* (yes/no), *Number of children* (0–10), *Marital status* (single, married, divorced, widowed), *Number of appointments that patient had in 2017*, and *Number of appointments that patient had in 2018*.

For each question below, tell what type of test or analysis should be used to make an inference about the large database from the data in the sample. When appropriate, specify the test, the statistic (e.g., *t*, *z*, etc.), whether one or two samples are used, and whether the test is one- or two-sided. (For example, your answer may be a one-sided, two-sample *t*-test for the difference of two means.)

a) Is the percentage of males 50%?
b) Is the percentage covered by insurance the same for both men and women?
c) Were there more appointments in 2018 than in 2017?
d) Is the mean age of women the same as the mean age of men?
e) Can you predict the number of appointments in 2018 from the number in 2017?

6. Lake source cooling Since 2000, Cornell University has used a lake source cooling plant that circulates water from Cayuga Lake through a heat transfer system to chill water for use on campus in refrigeration and air conditioning. Supporters of this method cite reduced carbon emissions and energy costs. Critics fear that higher levels of chlorophyll may damage the ecology of the lake. Cornell monitors the lake at seven test sites near the discharge pipe and at a control site farther away. A report found a statistically significant (at $\alpha = 0.05$) higher mean chlorophyll level relative to the control site at site #7—one of the monitoring sites.

a) A long-time critic of lake source cooling is quoted by the *Ithaca Journal* saying, "The finding shows that the lake source cooling is impacting the . . . lake." Which of these comments about this statement is most appropriate?
   i. He's correct: That's what a significant finding means.
   ii. He's wrong: The finding is probably due to chance.
   iii. He's correct: Comparing to a control site makes this an experiment.
   iv. He's wrong: This is really an observational study, not an experiment. We can conclude that there is a discernible difference, but we cannot infer a cause.
   v. He's correct: There are no other plausible explanations, so lake source cooling must be the cause.

b) Cornell appears to have performed a hypothesis test. What, specifically, was the null hypothesis of this test?
   i. H_0: Chlorophyll levels have not changed in the 8 years of lake source cooling.
   ii. H_0: $\mu_{TestSite\#7Chlorophyll} - \mu_{ControlSiteChlorophyll} = 0$
   iii. H_0: $\bar{y}_{TestSite\#7Chlorophyll} - \bar{y}_{ControlSiteChlorophyll} = 0$
   iv. H_0: $\mu_{TestSite\#7Chlorophyll} = 0$
   v. None of the above

c) The alternative hypothesis of Cornell's test was most likely
   i. one-sided, testing for higher chlorophyll levels at the test site.
   ii. one-sided, testing for lower chlorophyll levels at the test site.
   iii. two-sided.
   iv. rejected.
   v. irrelevant.

d) An associate dean of the university is quoted by the *Journal* pointing out that 28 tests were actually performed and that the one cited by the critic is the only one that showed a statistically significant result. He says, "Scientists expect that pure chance will result in a statistically significant result in every 1 in 20 comparisons." In proper statistical language, what is the dean implying?
   i. This result may have been a sampling error.
   ii. This result may have been due to response bias.
   iii. This result may have been a Type I error.
   iv. This result may have been a Type II error.
   v. This result may have been a Type III error.

e) You've been called in to consult as a statistics expert. You are asked to comment on whether the hypothesis test is appropriate. What would you like to know or see about the data?
   i. A timeplot showing the chlorophyll levels at the site for the past 8 years to check for linearity.
   ii. The correlation of chlorophyll levels at test site #7 and at the control site.
   iii. A histogram of the chlorophyll measurements at test site #7 and at the control site to check for outliers, skewness, or bimodality.
   iv. An independent assessment of the chlorophyll levels performed by New York State environmental officials.
   v. Boxplots of the chlorophyll levels at all seven of the test sites to check that the variation is approximately equal.

f) An environmental expert has noted that chlorophyll levels are ordinarily quite variable and has questioned whether the effect size of this finding is large enough to be of environmental concern. Statistically, what is he saying?
   i. The observed difference isn't really significant.
   ii. This was probably a Type I error.
   iii. The test isn't powerful enough.
   iv. Although the difference is significant, it may not be meaningful.
   v. Although the difference is profound, it may be too subtle.

7. Life expectancy and literacy Consider the relationship between the life expectancy (in years) and the illiteracy rate (per hundred people) in the 50 U.S. states plus Washington, DC. A linear model is run and the output is presented here:

Coefficients:

	Estimate	Std. Error	t value	P-value
Intercept	72.3949	0.3383	213.973	<0.0001
Illiteracy	−1.2960	0.2570	−5.043	<0.0001

Residual standard deviation: 1.097 on 48 degrees of freedom
Multiple R-squared: 0.3463,

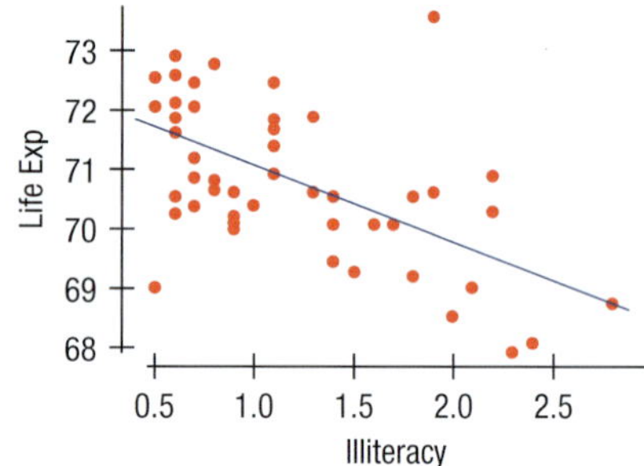

a) Colorado has an illiteracy rate of 0.70. What is its predicted life expectancy?

b) Based on the analysis, which of the following can you conclude about this relationship?
   i. Reducing illiteracy will increase life expectancy.
   ii. Reducing illiteracy will reduce life expectancy.
   iii. If you move to a state that spends less money on teachers, your life expectancy will go down, on average, due to lurking variables.
   iv. Higher levels of illiteracy are associated with generally lower life expectancies.
   v. States with lower life expectancies generally have lower illiteracy rates.
   vi. None of the above

c) What is the correlation between life expectancy and illiteracy?

d) Tennessee's illiteracy rate is about 1 SD above the mean for all states. What do you predict its life expectancy to be?
   i. About 1.296 SDs below the mean life expectancy.
   ii. About 1 SD below the mean life expectancy.
   iii. About 1 SD above the mean life expectancy.
   iv. About 0.59 SD below the mean life expectancy.
   v. None of the above

e) High school graduation rate has a correlation of 0.60 with life expectancy. A simple regression of *life expectancy* on high school *graduation rate* shows a positive slope with a very low P-value. If you add high school *graduation rate* as a predictor to the regression of *life expectancy* on *illiteracy*, and fit a multiple regression on *high school graduation* and *Illiteracy*, which of the following is true?
   i. The R^2 of this model is at least as high as the R^2 of either single predictor model.
   ii. The slope of the high school graduation rate is positive.
   iii. The slope of the high school graduation rate is negative.
   iv. The slope of the high school graduation rate is statistically significant.
   i. None of the above

Hotel maids A Harvard psychologist recruited 75 female hotel maids to participate in a study. She randomly selected 41 and informed them (truthfully) that the work they do satisfies the Surgeon General's recommendations for an active lifestyle, providing examples to show them that their work is good exercise. The other 34 maids were told nothing. Various characteristics of the maids, such as weight, body fat, body mass index, and blood pressure, were recorded at the start of the study and then again after four weeks. The researcher was interested in whether the information she provided would result in measurable physical changes. If there is a difference, it could challenge our understanding of the placebo effect (in which subjects who receive the null treatment are not informed) by showing that being informed about a treatment can make a difference. Complete Exercises 8–12 related to this study.

8. Which type of study is this?

a) A prospective observational study because it followed the maids for four weeks.
b) A retrospective observational study because at the end of four weeks, the researcher had to look back at the original measurements.
c) A survey because the maids were randomly selected.
d) A designed experiment because a randomly selected group was treated differently and the researcher was interested in the effect of that treatment.
e) A voluntary response study because the maids volunteered to participate.

9. Now let's consider only the maids who were *informed*. Here is a display of their body mass index (BMI), a measure of body fat, at the start (BMI) and at the end (BMI2) of the study.

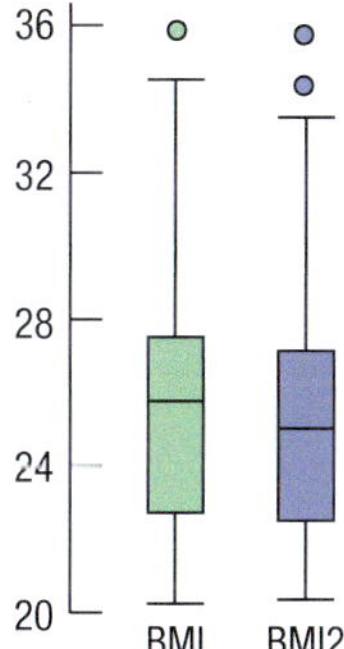

Which of these is the best comment to make about this display?

a) This is not an appropriate plot for this study.
b) This bar chart has three outliers.
c) These boxplots show that informing the maids made no difference in their BMI.
d) The maid with the highest BMI was the same at the start of the study and at the end.
e) There is a significant drop in the median BMI over the four-week study.

10. The researcher was concerned about whether each maid's BMI changed over the four-week study. Here are the results of a test performed using technology:

Mean of Paired Differences = 0.2068
t-Statistic = 3.853 w/72 df
P = 0.0003

Using the proper notation, how would you write the null hypothesis tested?

a) $H_0: \bar{y}_{Beginning} = \bar{y}_{End}$
b) $H_0: \bar{y}_{Beginning} - \bar{y}_{End} = 0$
c) $H_0: \bar{d}_{Beginning-End} = 0$
d) $H_0: \mu_{Beginning} - \mu_{End} = 0$
e) $H_0: \mu_d = 0$

11. What conclusion can you draw from the test in Exercise 10?

a) The maids lost weight, on average.
b) The mean BMI for the informed maids changed during the study.
c) The mean difference in BMI was not zero for the maids who were informed that their work qualifies as an active lifestyle.
d) The mean change in BMI was greater for the maids who were informed that their work qualifies as an active lifestyle than for those who were not informed.
e) Maids should concentrate on the exercise aspect of their jobs.

12. But maybe the BMIs of *all* the participants changed during the four-week period. We can calculate the change in BMI for each participant in the study, and then compare the mean change for the 41 maids in the *informed* group with the mean change for the 34 in the *uninformed* group. Here is a display (Informed = 1 means *informed*, Informed = 0 means *uninformed*):

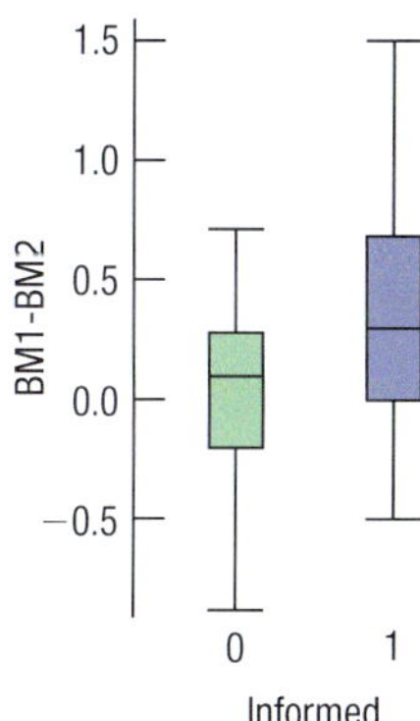

Which of these inference methods would you use to tell whether informing the maids made a difference in their BMIs?

a) See if the boxes in the boxplot overlap. The difference is significant if they don't.
b) Find a confidence interval for the paired changes in BMI for each group. If the confidence intervals don't overlap, then they are different.
c) Perform a two-sample t-test comparing whether the mean changes for the two groups are equal.
d) Use a χ^2 test of homogeneity.
e) Use a two-sample z-test for the proportions that gained weight in each group.

13. Inference Methods For each of the following descriptions, select the letter of the inference method you would use. (It is possible for a method to be used more than once or not at all.)

A **Student's *t* for a mean**
B **one-proportion *z***
C **χ^2 goodness-of-fit**
D **two-proportion *z*-test**
E **paired *t***

a) **Ithaca garbage bags** In Ithaca, New York, every garbage bag must be tagged and not exceed 20 lb in weight. The city wants to know whether residents comply with the rules. One day, a truck driver randomly selected 135 bags left on the curb and weighed them. The driver found a mean weight of 21.5 lb with a standard deviation of 4 lb. Are Ithacans complying with the rule?

b) **Google flu prediction** The Centers for Disease Control and Prevention (CDC) gathers data from physicians and publishes the number of flu cases two weeks later. Researchers at Google recorded the number of flu-related queries and constructed an estimate of flu incidence that was immediate because it didn't depend on reports from physicians. Google used its method to predict the number of flu cases for each week of the flu season and compared that prediction to the actual number of cases reported (two weeks later) by the CDC. Did Google's predictions match the actual counts?

c) **Gambling students** A sample of 1979 college students completed the South Oaks Gambling Screen and Interpersonal Guilt Questionnaire and answered questions about their substance use. Students who were identified on the Gambling Screen as pathological gamblers ($n = 145$) were matched to non–problem gamblers with respect to demographics and substance use to see whether there was any difference in guilt (as assessed by the Guilt Questionnaire). Pathological gamblers had significantly higher interpersonal guilt than their non–problem-gambling peers.

d) **Vitamin D and colds** Researchers randomly assigned 322 healthy adults in New Zealand to take either a placebo or a high dose of vitamin D. Researchers had hoped to show that vitamin D could prevent or reduce the symptoms of colds. But after 18 months (including two winter seasons), the proportion of participants suffering from an upper respiratory infection was no lower in the vitamin D takers than in those who took the placebo. (Source: *JAMA* 308:1333, 2012)

e) **Gallup Poll bias?** Polls of voters taken just before an election have the special feature that after the election, we learn the true population proportion (vote percentage for each candidate) that they were attempting to estimate. The Gallup Poll has been criticized for reporting values that were biased in favor of Republican candidates. Did their final prediction for the presidential election come close enough to the true value, or is there evidence that their methods are faulty?

T Olympic archery Olympic Archery is an event that doesn't get the attention it deserves. Both men and women start with a field of 64 qualifiers. Each archer shoots a round of 72 arrows (total possible score: 720) to establish a seeding position. Then they participate in a single-elimination contest. Thus, the seeding round is the only one that provides data for all archers (because some are eliminated at each step of the elimination rounds). Exercises 14–17 make use of the seeding round data. The data file **Archery.txt** contains the seeding round scores for both men and women.

An arrow that hits the yellow disk in the center of the archery target scores 10 points. But there is an even smaller circle within that disk. Arrows that hit the smaller circle score an additional X. The X's are used to break ties, if necessary.

14. Here are summary statistics on X hits for each of the 64 female archers in the 2008 Olympics:

Archer count	64
Arrow count	4608
Total X's	375
Proportion of X's	0.0814
Mean no. of X's	5.85938
StdDev in no. of #X's	2.86671

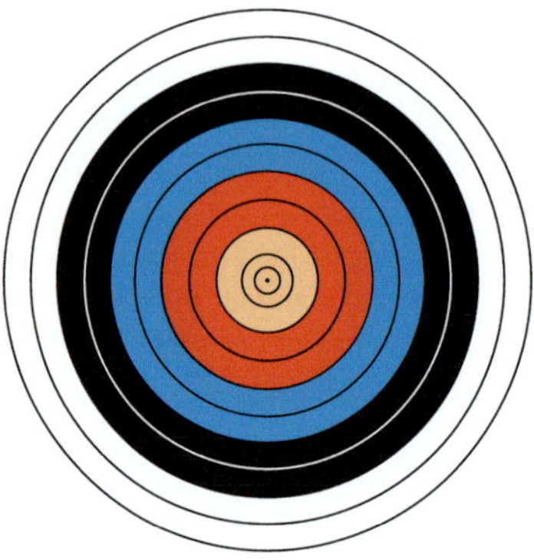

a) Construct a confidence interval for the proportion of arrows that hit the X circle. Write out your calculation; that is, don't just report a result from a statistics program. Do the calculation "by hand" (using a calculator or the computer, of course) and show all your work.

b) Interpret the resulting interval.

15. In the final round of competition, archers shoot only 12 arrows. Is it likely that an Olympic archer may make no X shots at all? Explain your reasoning. For seeding purposes (that is, to set up the final elimination round matches), the total score on the initial 72 arrows is used.

a) Using the data in the file **Archery.txt**, construct and interpret a confidence interval for the women's mean seeding score.

b) In the final round of the 2008 women's archery competition, the gold medal winner, Zhang Juanjuan, scored 9.1667 points per arrow—an equivalent of 660 for 72 shots. Do you consider her score to be extraordinary? Explain.

16. Although Zhang Juanjuan won the gold medal in 2008, she did not perform extraordinarily well in the seeding round. How consistent are archers? The seeding round for the women is reported in two halves ("1st half" and "2nd half" in the data file). Are the scores in the two halves consistent, or are they different? Perform and interpret an appropriate test.

17. Archery is one of the few Olympic sports in which men and women compete on an equal footing. Both contests start with 64 archers, and the competitors first launch 72 arrows at a distance of 70 m from the target to establish seeding for their elimination rounds. Thus we can compare the performances of men and women. Do men or women score higher on their seeding rounds, on average?

18. Olympic medals. Over the course of the modern Olympics, France, Italy, and Great Britain have participated in roughly the same number of Olympic games. But is the distribution of their medals the same, or do they differ? Here is a table summarizing their Summer Olympic performances:

	Gold	Silver	Bronze	Total
FRA	191	212	233	**636**
GBR	297	255	253	**805**
ITA	191	157	174	**522**
Total	**679**	**624**	**660**	**1963**

Chi-square = 9.505 with 4 df
P-value = 0.0497

a) State and test an appropriate hypothesis using the chi-square value reported here. Be clear about the type of test you are performing.
b) State your conclusion from that test.
c) Here are the standardized residuals from the chi-square analysis:

	Gold	Silver	Bronze
FRA	−1.95467	0.691188	1.31053
GBR	1.11173	−0.055889	−1.07328
ITA	0.776994	−0.693534	−0.113745

Discuss what the residuals show. What does this add to your answer in part b?

T **19.** Belmont Stakes The Belmont Stakes is the last and longest of the three horse races that make up the Triple Crown. Curiously, some of the Belmont races have been run clockwise around the track, and others have been run counterclockwise. Do the horses care? Here are boxplots and a confidence interval for the difference in mean speeds (in miles per hour). (Data in **Belmont Stakes 2015**)

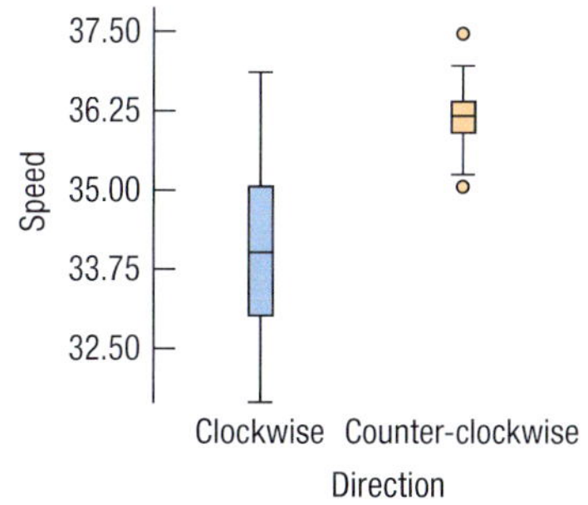

With 95.00% confidence,

$$-2.5355089 < \mu_{\text{Clockwise speed}} - \mu_{\text{Counterclockwise speed}} < -1.7628508$$

a) What conclusion can you draw?
b) But wait. Not only has the direction been inconsistent, but so has the length of the race (in miles).

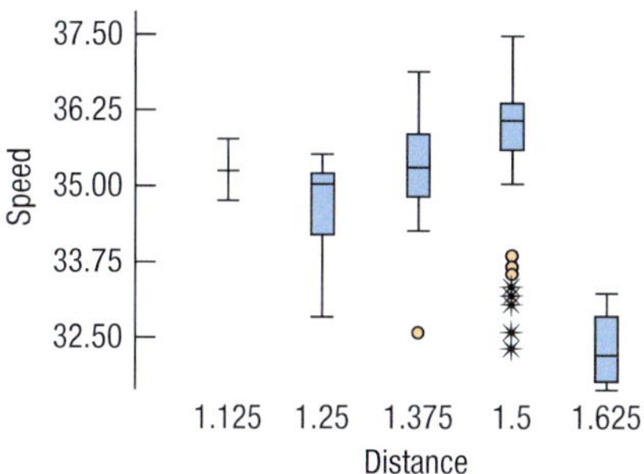

Comment on what you see in this plot.

c) The most common, and the current, length of the race is 1.5 miles. So let's restrict our attention to only the races of that length.

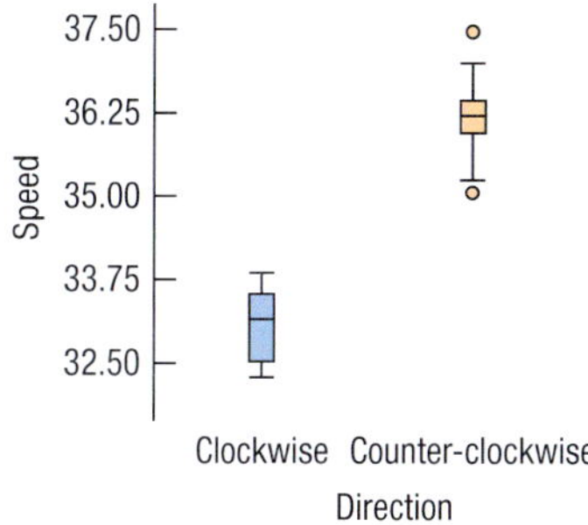

Do you expect a confidence interval for the difference in mean speeds based only on these races to be wider, narrower, or about the same as the interval found earlier? Explain.

d) There is, in fact, a lurking variable in these data. Using the data file **Belmont Stakes 2015**, identify the lurking variable and, if necessary, update your conclusion about the direction that racehorses prefer to run.

T **20.** Body fat In the text you have examined data in the file **Bodyfat**. In particular, you made regression models to predict *%Body Fat* from *Waist* size. But there are many other variables in the file. Using the statistics program of your choice, examine the ability of other variables in the file to predict *%Body Fat* either individually or in a model with several predictors.

21. PVA The Paralyzed Veterans of America (PVA) is a Congressionally chartered veterans' service organization that represents the interests of paralyzed veterans. The agency provides a range of services to veterans who have spinal cord injury or dysfunction. It derives most of its funding from contributions. The data set **PVA** contains a sample of the donors who recently gave money to the organization. We would like to use the sample to understand the donors in the entire database. The following variables are included:

a) Describe the distribution of the donors' ages, including any unusual features. Give a 95% confidence interval for the mean age of all donors in the database.

b) The average age at which people buy homes in the United States is about 34. So, in general, homeowners are older than non-homeowners. Test whether homeowners, on average, are older than non-homeowners in this data set. Explain the discrepancy.

c) Are men more likely than women to own homes? Use the data set to test a hypothesis and comment on the result.

d) Describe the distribution of *Current Gift* for these donors.

Variable Name	Units (if applicable)	Description	Remarks
Age	Years		
Own Home?	H = Yes; U = No or unknown		
Children	Counts		
Income		1 = Lowest; 7 = Highest	Based on national medians and percentiles
Sex	M = Male; F = Female		
Total Wealth		0 = Lowest; 9 = Highest	Based on national medians and percentiles
Gifts to Other Orgs	Counts	Number of Gifts (if known) to other philanthropic organizations in the same time period	
Number of Gifts	Counts	Number of Gifts to this organization in this time period	
Time Between Gifts	Months	Time between first and second gifts	
Smallest Gift	$	Smallest Gift (in $) in the time period	See also Sqrt(Smallest Gift)
Largest Gift	$	Largest Gift (in $) in the time period	See also Sqrt(Largest Gift)
Previous Gift	$	Gift (in $) (or previous campaign	See also Sqrt(Previous Gift)
Average Gift	$	Total amount donated divided by total number of gifts	See also Sqrt(Average Gift)
Current Gift	$	Gift (in $) to organization this campaign	See also Sqrt(Current Gift)
Sqrt(Smallest Gift)	Sqrt($)	Square Root of Smallest Gift in $	
Sqrt(Largest Gift)	Sqrt($)	Square Root of Largest Gift in $	
Sqrt(Previous Gift)	Sqrt($)	Square Root of Previous Gift in $	
Sqrt(Average Gift)	Sqrt($)	Square Root of Average Gift in $	
Sqrt(Current Gift)	Sqrt($)	Square Root of Current Gift in $	

APPENDIX

A Answers

Here are the "answers" to the exercises for the chapters and the unit reviews. The answers are outlines of the complete solution. Your solution should follow the model of the Step-by-Step examples, where appropriate. You should explain the context, show your reasoning and calculations, and draw conclusions. For some problems, what you decide to include in an argument may differ somewhat from the answers here. But, of course, the numerical part of your answer should match the numbers in the answers shown.

Chapter 1

1. Retailers, and suppliers to whom they sell the information, will use the information about what products consumers buy to target their advertisements to customers more likely to buy their products.

3. Owners can advertise about the availability of parking. They can also communicate with businesses about hours when more spots are available and when they should encourage more business.

5. The individual draws.

7. The sample is over 5000 people; the *Who* is the selected subjects; the *What* includes data on health, family life, marriage, divorce, pregnancy, infertility, and use of contraception.

9. a) Sample—A principal wants to know how many from each grade will be attending a performance of a school play.
b) Sample—A principal wants to see the trend in the amount of mathematics learned as students leave each grade.

11. Quantitative.

13. Categorical.

15. They might consider whether a person voted previously or whether he or she could name the candidates (indicating greater interest in the election).

17. Answers will vary.

19. *Who*—121 women and 84 men aged 20–82; *What*—Metabolic brain age of women vs men; *Population*—All women and men.

21. *Who*—2500 cars; *What*—Distance from car to bicycle; *Population*—All cars passing bicyclists.

23. *Who*—Coffee drinkers at a Newcastle University coffee station; *What*—Amount of money contributed; *Population*—All people in honor system payment situations.

25. *Who*—474 participants in the San Antonio Longitudinal Study of Aging; *What*—Diet soda consumption and waist size change; *Population*—All diet soda drinkers.

27. *Who*—54 bears; *Cases*—Each bear is a case; *What*—Weight, neck size, length, and sex; *When*—Not specified; *Where*—Not specified; *Why*—To estimate weight from easier-to-measure variables; *How*—Researchers collected data on 54 bears they were able to catch.
Variable—Weight; *Type*—Quantitative; *Units*—Not specified; *Variable*—Neck size; *Type*—Quantitative; *Units*—Not specified; *Variable*—Length; *Type*—Quantitative; *Units*—Not specified; *Variable*—Sex; *Type*—Categorical.

29. *Who*—Cocktails; *Cases*—Each cocktail is a case; *What*—Components, number of calories, and serving size; *When*—Not specified; *Where*—Cocktail stall; *Why*—To assess nutritional value of cocktails; *How*—Report by cocktail owner; *Variable*—Components; *Type*—Categorical (multiple response); *Variable*—Number of calories; *Type*—Quantitative; *Units*—Calories; *Variable*—Serving size; *Type*—Quantitative; *Units*—Millilitres.

31. *Who*—882 births; *Cases*—Each of the 882 births is a case; *What*—Mother's age, length of pregnancy, type of birth, level of prenatal care, birth weight of baby, sex of baby, and baby's health problems; *When*—1998–2000; *Where*—Large city hospital; *Why*—Researchers were investigating the impact of prenatal care on newborn health; *How*—Not specified exactly, but probably from hospital records; *Variable*—Mother's age; *Type*—Quantitative; *Units*—Not specified, probably years; *Variable*—Length of pregnancy; *Type*—Quantitative; *Variable*—Birth weight of baby; *Type*—Quantitative; Not specified, probably pounds and ounces; *Variable*—Type of birth; *Type*—Categorical; *Variable*—Level of prenatal care; *Type*—Categorical; *Variable*—Sex; *Type*—Categorical; *Variable*—Baby's health problems; *Type*—Categorical.

33. *Who*—Experiment subjects; *Cases*—Each subject is a case; *What*—Treatment (herbal cold remedy or sugar solution) and cold severity; *When*—Not specified; *Where*—Not specified; *Why*—To test efficacy of herbal remedy on common cold; *How*—The scientists set up an experiment; *Variable*—Treatment; *Type*—Categorical; *Variable*—Cold severity rating; *Type*—Quantitative (perhaps ordinal categorical); *Units*—Scale from 0 to 5; *Concerns*—The severity of a cold seems subjective and difficult to quantify. Scientists may feel pressure to report negative findings of herbal product.

35. *Who*—Streams; *Cases*—Each stream is a case; *What*—Name of stream, substrate of the stream, acidity of the water, temperature, BCI; *When*—Not specified; *Where*—Upstate New York; *Why*—To study ecology of streams; *How*—Not specified; *Variable*—Stream name; *Type*—Identifier; *Variable*—Substrate; *Type*—Categorical; *Variable*—Acidity of water; *Type*—Quantitative; *Units*—pH; *Variable*—Temperature; *Type*—Quantitative; *Units*—Degrees Celsius; *Variable*—BCI; *Type*—Quantitative; *Units*—Not specified

37. *Who*—353 refrigerator models; *Cases*—Each of the 353 refrigerator models is a case; *What*—Brand, cost, size, type, estimated annual energy cost, overall rating, and repair history; *When*—2013; *Where*—United States; *Why*—To provide information to the readers of *Consumer Reports*; *How*—Not specified; *Variable*—Brand; *Type*—Categorical; *Variable*—Cost; *Type*—Quantitative; *Units*—Not specified (dollars); *Variable*—Size; *Type*—Quantitative; *Units*—Cubic feet; *Variable*—Type; *Type*—Categorical; *Variable*—Estimated annual energy cost; *Type*—Quantitative; *Units*—Not specified (dollars); *Variable*—Overall rating; *Type*—Categorical (ordinal); *Variable*—Percent requiring repair in the past 5 years; *Type*—Quantitative; *Units*—Percent of refrigerators.

39. *Who*—Top 10 male 18–39 finishers of the London 2019 Marathon; *What*—Place, bib, name, official time, and country; *When*—April 28; *Where*—London; *Why*—Not specified (probably to see trends in racing); *How*—Official statistics collected at race; *Variable*—Place; *Type*—Categorical; *Variable*—Bib; *Type*—Identifier and Categorical; *Variable*—Name; *Type*—Identifier and Categorical; *Variable*—Official Time; *Type*—Quantitative; *Units*—Hours, minutes, and seconds; *Variable*—Country; *Type*—Categorical.

41. a) Nico Rosberg
b) Germany (5 wins)
c) 1 hour, 29 minutes and 30.259 seconds
d) Valtteri Bottas in 2019

Chapter 2

1.

Labour Status	Frequency
Employed	0.58107
Unemployed	0.02130
Inactive	0.39762

3. a) Yes; each movie is categorized in a single genre.
b) Other

5. i. C, ii. A, iii. D, iv. B

7. a) Data for the years 1995–2019 is available.
b) We have one reading for each year.
c) The distribution of active volcanoes is slightly skewed to the left and ranges between 52 and 81 active volcanoes per year. The distribution is bimodal, with a mode at 70 and another at 71 active volcanoes.

9. Based on the height of the tallest points, about 85 of these 250 men have biceps close to 13 inches around. Most are between 12 and 15 inches around. But there are two as small as 10 inches and several that are 16 inches.

11. The distribution is centered at around 7 calls, with many of the patients not calling during the entire month or calling just once. The distribution has no outliers, however there was one patient who phoned 20 times in a month.

13. The distribution is mound-shaped and roughly symmetric.

15. a) The distribution is skewed to the left.
b) One mode is at about 74 to 76 years. The fluctuations from bar to bar don't seem to rise to the level of defining additional modes, although opinions can differ.

17. a) The median. The distribution is skewed to the left, so the mean will be pulled by the tail in the lower direction.
b) The median is resistant to the skewed shape of the distribution, so it is a better choice for most summaries.

19. Because the distribution of bicep circumferences is unimodal and symmetric, the mean and the median should be very similar. The usual choice is to report the mean or to report both.

21. a) IQR
b) For a skewed distribution, is it more appropriate to report the IQR. The skewness inflates the standard deviation.

23. Standard deviation. The distribution is reasonably symmetric and mound-shaped. The standard deviation is generally more useful whenever it is appropriate.

25. Answers will vary.

27. Answers will vary.

29. Answers will vary.

31. Solution:
a) Who—Eurostat; What—average annual net earning salary of full-time single EU workers; When—2018; Where—European Union; How—gathering yearly data from member states; Why—to see trends in wages.
b) Euro.
c) Mean

33. a) Unimodal (near 0) and skewed to the right. Many seniors will have 0 or 1 speeding ticket. Some may have several, and a few may have more than that.
b) Probably unimodal and slightly skewed to the right. It is easier to score 15 strokes over the mean than 15 strokes under the mean.
c) Probably unimodal and symmetric. Weights may be equally likely to be over or under the average.
d) Probably bimodal. Men's and women's distributions may have different modes. The distribution may also be skewed to the right, since it is possible to have very long hair, but hair length can't be negative.

35. a) Thriller/Suspense
b) It is easy to tell from either chart; sometimes differences are easier to see on the bar chart because slices of the pie chart look too similar in size.

37. Solution: In 2011, the total number of cases was 2,744, out of which 26.31% were new, 26.02% were completed and the other 47.67% were pending.

39. a)

Cause	Percentage
Diseases of the circulatory system	36.42%
Cancer	24.92%
Diseases of the respiratory system	8.47%
Accidents and other external causes	4.41%
Diseases of the digestive system	4.20%
Mental and behavioral diseases	4.11%
Diseases of the nervous system	4.08%

b) Since the percentages listed add up to 86.61%, other causes must account for 13.39% of EU deaths in 2015.
c) A bar chart including the 'Other' category will do (or else a pie chart).

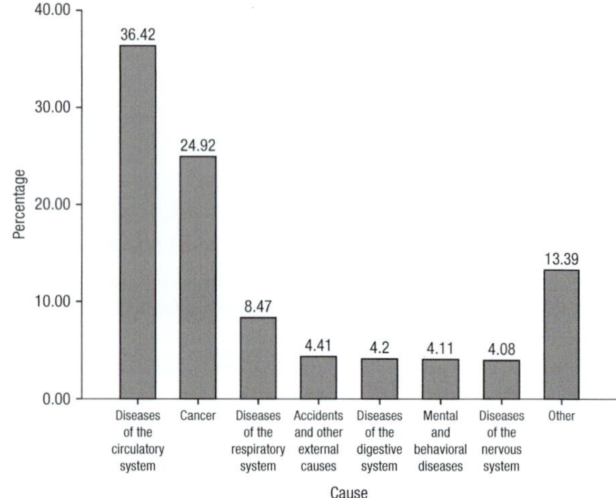

41. a) There are too many categories to make a meaningful bar chart or pie chart by genre.
b) They combined several smaller categories into the category "Other."

43. The percentages total 141%, and the three-dimensional display distorts the sizes of the regions, violating the area principle.

45. a) The distribution is bimodal. It looks like there may be two groups of cereals. The modes are near 13 gm and 22 gm. There are no outliers.
b) Corn Chex, Corn Flakes, Cream of Wheat (Quick), Crispix, Just Right Fruit & Nut, Kix, Nutri-Grain Almond-Raisin, Product 19, Rice Chex, Rice Krispies, Shredded Wheat 'n' Bran, Shredded Wheat Spoon Size, Total Corn Flakes, Triples

47. a) The distribution of the ages of the female purchasers is skewed to the left, so the mean is smaller than the median.
b) Unimodal and skewed to the left, with ages ranging from 20 to 75 years. The majority of the purchasers are between 50 and 60 years of age. There are relatively few purchasers who are younger than 30 or older than 70 years of age.
c) The median is appropriate. To summarize the spread of the date, the interquartile range would suffice.

49. a) 46 points
b) $Q1 = 37, Q3 = 55$
c)

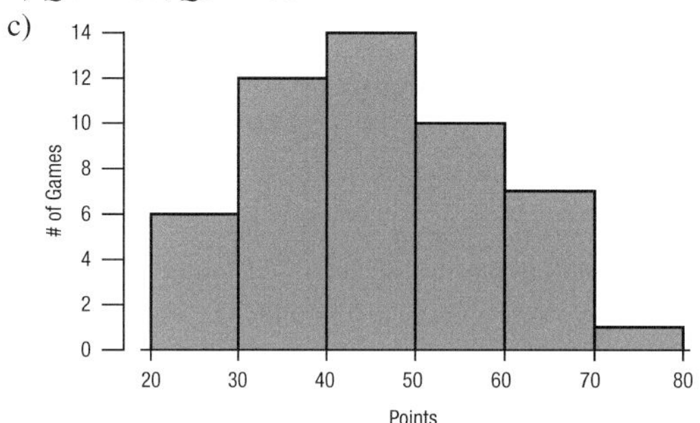

The distribution is fairly symmetric, but slightly right skewed because no games had fewer than 20 points. It is centered at 46 with an IQR of 18 points. The minimum is 21 and the maximum is 75.

51. a) The distribution is bimodal with one mode at about 62 and one at about 78. The higher mode might be math majors, and the lower mode might be non-math majors.
b) Because the distribution is bimodal, neither the mean nor the median tells much about a typical score.

53. a) The median will probably be unaffected; the mean will be higher.
b) The range and standard deviation will increase; the IQR will be unaffected.

55. a) 19647.42.
b) 11098, 12533, 13186, 15310, 15708, 18066, 18753, 20487, 22719, 25722, 29424, 32763. The median is 18409.5. The Q1 is 14779, the Q3 is 23469.75.
c) The range of the distribution of causalities is $32763 - 11098 = 21665$ while the IQR is 8690.75.

57. Slightly skewed to the left and bimodal. There is one mode at about 21,000 casualties and one at about 30,000 casualties.

59. The data appears to be fairly symmetric, so the mean and the standard deviation are good choices to describe the center and spread of the data.

61. a) The mean is around €400
b) The standard deviation is around €200

63. a) About 105 minutes
b) Yes, only 3 of these 150 movies, or 2% of them, run that long.
c) We cannot tell from the histogram. Overall, the distribution is skewed a bit to the higher values, but the outlier on the lower end would also affect the mean.

65. a) i. The middle 50% of movies ran between 98 and 116 minutes.
ii. Typically, movie lengths varied from the mean run time by 16.6 minutes.
b) We should be cautious in using the standard deviation because the distribution of run times is skewed to the right, and the outlier might inflate the value.

67. The publication is using the median; the watchdog group is using the mean, pulled higher by the several very successful movies in the long right tail.

69. a)

Stem	Leaf
31	1
31	5
32	1233
32	6678
33	
33	9
34	23
34	556

32|1 = $3.21/gallon

b) The distribution of gas prices is bimodal and skewed to the left (lower values) with peaks around $3.25 and $3.45. The lowest and highest prices were $3.11 and $3.46.
c) There is a rather large gap between the $3.28 and $3.39 prices.

71. a) The median is 4.5 million. The IQR is 5 million ($Q1 = 2$ million, $Q3 = 7$ million).
b) The distribution of populations is unimodal and skewed to the right. The median population is 4.5 million. One state is an outlier, with a population of 37 million. Another, at 25 million, could also be considered an outlier.

73. The distribution is reasonably symmetric except for a second mode at the low end (around 10 home runs).

75. a) This is not a histogram. The horizontal axis should split the number of home runs hit in each year into bins. The vertical axis should show the number of years in each bin.
b)

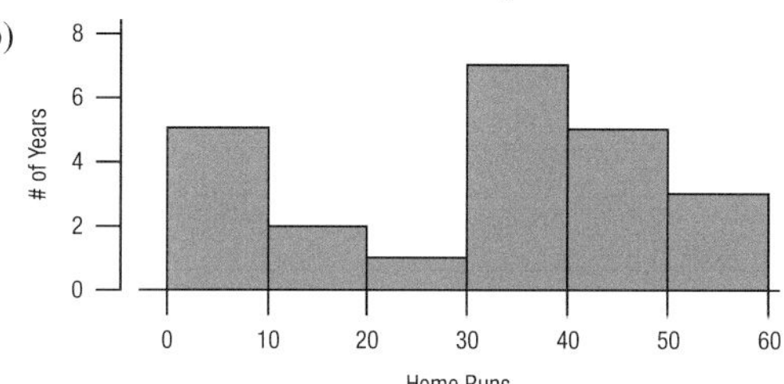

77. a) Skewed to the right, possibly bimodal with one fairly symmetric group near 4.4, another at 5.6. Two outliers in middle seem not to belong to either group.

Stem	Leaf
57	8
56	27
55	1
54	
53	
52	9
51	
50	8
49	
48	2
47	3
46	034
45	267
44	015
43	0199
42	669
41	22

41|2 = 4.12 pH

b) The cluster of high outliers contains many dates that were holidays in 1973. Traffic patterns would probably be different then, which might account for the difference.

79. The histogram bars are so wide that the distribution cannot be seen.

81. It is neither appropriate nor useful. ZIP codes are categorical data, not quantitative. But they do contain some information. The leading digit gives a rough east-to-west placement in the United States. So, we see that the company has almost no customers in the Northeast. A bar chart by leading digit would be more appropriate.

83. a) Median 285, IQR 9, mean 284.360, SD 6.845
b) Because it's skewed to the left, probably better to report median and IQR.
c) Skewed to the left. The center is around 284. The middle 50% of states scored between 280 and 289.

85. The histogram shows that the distribution of *Percent Change* is unimodal and skewed to the right. The states vary from a minimum of −0.6% (Michigan) to 35.1% (Nevada) growth in the decade. The median was 7.8% and half of the states had growth between 4.3% and 14.1%. Not surprisingly, the top three states in terms of growth were all in the West: Nevada (35.1%), Arizona (24.6%), and Utah (23.8%).

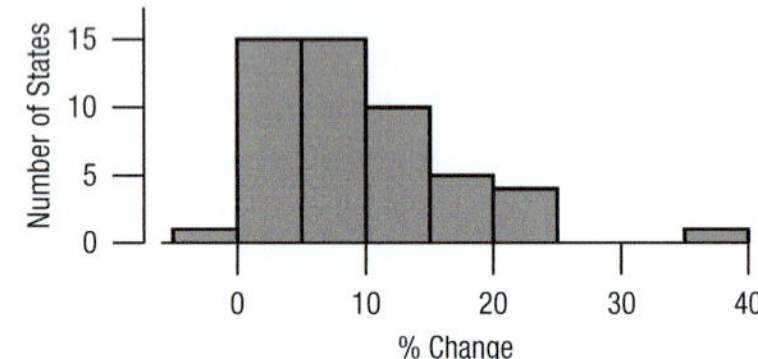

Chapter 3

1. a) 26.26%
b) 38.84%; 35.62%.
c) 26.26% of male students and 32.81% of female students strongly agree. The percent of male students who agree, 32.42%, is higher than that of female students, 27.19%. A higher percentage of female students (4.38%) are unsure than male students (2.47%). The percentages of male and female students who disagree are 20.79% and 20.76% respectively. A higher percentage of male students (18.05%) strongly disagree than female students (14.86%).
d) This sample percentage is an estimate of the percentage of all University students who feel the same way; thus, the corresponding population percentage should be probably close to, but not exactly, 30%.

3. a) 32.81% Strongly Agree, 27.19% Agree, 4.38% Unsure, 20.76% Disagree, 14.86% Strongly Disagree.
b) 58.68% Positive; 2.47% Neutral; 38.84% Negative.

5. a) Omnivores are more liberal and less conservative. Vegetarians are the most liberal. Other comments are appropriate.
b) The differences are very large. It does appear there is a strong association between diet preference and political alignment.

7. a) About 10%
b) There are more men who didn't have cancer who never/seldom ate fish.
c) The percentage of men who never/seldom ate fish is lower in the group with no cancer than in the group with cancer.

9. a) Men. For men this is 10.4%, but only 1.4% of women are conservative carnivores.
b) Women. Of the 17 liberal vegetarians more than 70% are women.

11. a) $452/1529 = 29.6\%$
b) $124/1529 = 8.1\%$
c) $124/452 = 27.4\%$
d) $124/312 = 39.7\%$

13. a) 1200.
b) 63.33%.
c) 19.58%.
d) 4.87%; 16.59%.

15. **Solution:** There does not seem to be an association between Instagram usage and gender.
There is a strong association between Instagram usage and age, since the majority of Instagram users seem to be aged between 18 and 34 years. Instagram users of age 55 or over are quite low in number.

17. a) 8.63%
b) It is difficult to make a good display because the numbers are of such different magnitudes. A display of the two largest values would not show the two smallest ones.

19. a) 42.1%
b) declined to 18.5%
c) declined in a similar way to 19.9%

21. a) All six
b) South Africa—18%, Ghana—20%, Senegal—21%, Nigeria—13%, Kenya—11%, Tanzania—5%.
c) The 2013 and 2017 percentages of smartphone ownership in Tanzania are 8% and 13% respectively. The increase of smartphone ownership is Tanzania is the lowest out of all six sub-Saharan countries, showing that smartphone ownership has remained rather low and stable in Tanzania.

23. More than 75% of both male and female teenagers stating that this is one of their most common ways of communicating. Female teenagers are more likely to text than male teenagers (84% for girls versus 75% for boys). Phone calls and social media sites are more popular among female teenagers. Male teenagers are much more likely to use gaming as a way of communicating to their close friend than female teenagers (39% for boys versus just 3% for girls).

25. a) Men. The male columns are wider than the female columns.
b) Yes, Females are more likely to be Liberal, and Males are more likely to be Conservative.
c) Yes, Liberals are more likely to be Vegetarians, and Conservatives are more likely to be Carnivores.
d) Yes, the differences in Vegetarians are more pronounced in Females than in Males. The differences in Carnivores are more pronounced in Males than in Females.

27. a) Row percentages. The two totals are 100%.
b) i. Can't tell ii. 57.9%
iii. Can't tell iv. 6.8%

29. a) 82.5% b) 12.9% c) 11.1%
d) 13.4% e) 85.7%

31. a) Column percents, because the column totals are 100%
b) G and PG movies are less common and PG-13 and R movies have become more common.
c) For Dramas, the percentages of PG and R have decreased while the percentage of PG-13 dramas has increased significantly. For Comedies, there has been a large increase in the percentage of R-rated films.

33. a) 73.9% 4-year college, 13.4% 2-year college, 1.5% military, 5.2% employment, 6.0% other
b) 77.2% 4-year college, 10.5% 2-year college, 1.8% military, 5.3% employment, 5.3% other
c) Many charts are possible. Here is a side-by-side bar chart.

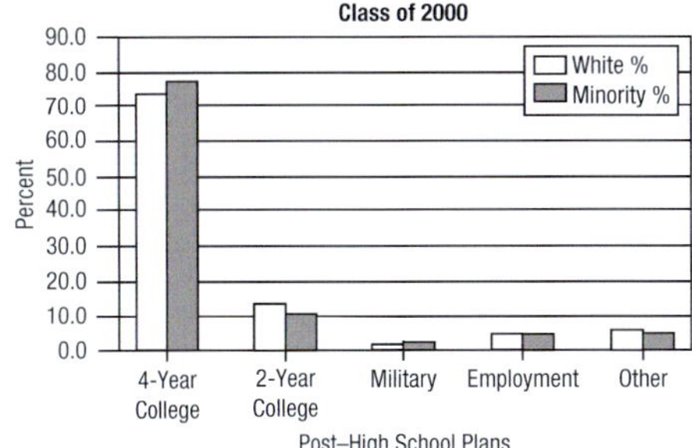

d) The white and minority students' plans are very similar. The small differences should be interpreted with caution because the total number of minority students is small. There is little evidence of an association between race and plans.

35. a) 28.86% b) 18.71%
c) 69.15% d) 29.48%

37. The distributions of commencements across sectors for the years 2018 and 2019 are similar to each other. For 2018, 37.69% of commencements pertained to higher education, 28.86% pertained to VET and 21.71% pertained to ELICOS. The remaining 8.36% and 3.39% were non-award and schools commencements respectively. For 2019, 40.22% of commencements pertained to higher education, 29.84% pertained to VET and 18.71% pertained to ELICOS. The remaining 7.60% and 3.63% were non-award and schools commencements respectively.

39. a) 52% b) 70% c) 18%
d) For questions with an actual Yes answer, the student answered correctly 18 out of 24 times (75%). For questions with an actual No answer, the student answered correctly 17 out of 26 times (65.38%). The discrepancy between these two percentages is significant, so there seems to be evidence of an association between the type of answer and the ability of the student to choose the correct answer.

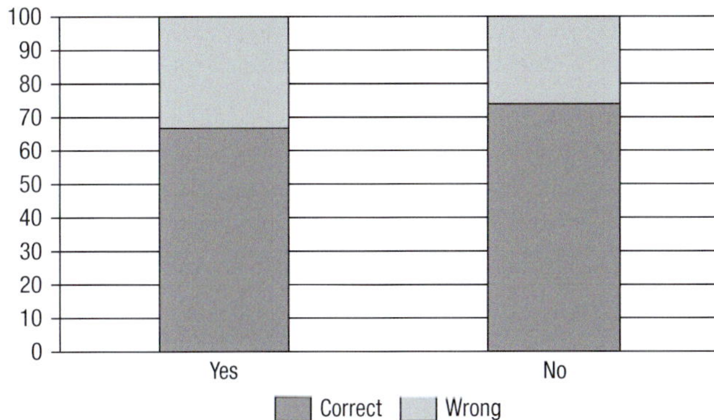

41. a) Low 20.0%, Normal 48.9%, High 31.0%.
b)

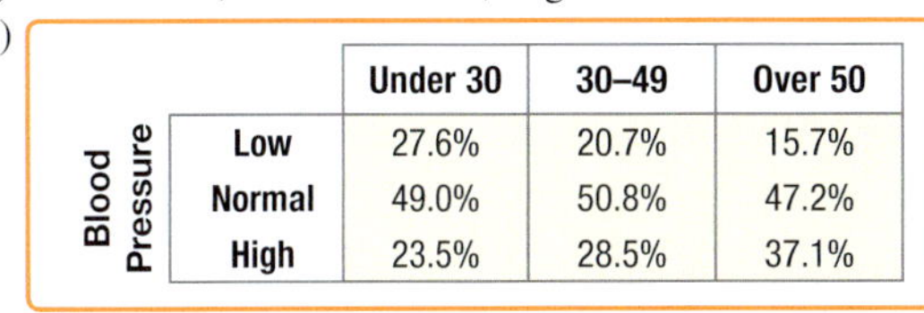

Blood Pressure	Under 30	30–49	Over 50
Low	27.6%	20.7%	15.7%
Normal	49.0%	50.8%	47.2%
High	23.5%	28.5%	37.1%

c)

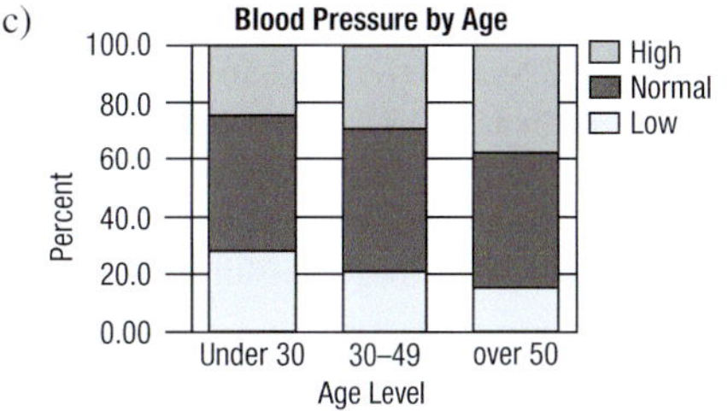

d) As age increases, the percent of adults with high blood pressure increases. By contrast, the percent of adults with low blood pressure decreases.
e) No, but it gives an indication that it might. There might be additional reasons that explain the differences in blood pressures.

43. No, there's no evidence that Prozac is effective. The relapse rates were nearly identical: 28.6% among the people treated with Prozac, compared to 27.3% among those who took the placebo.

45. a) 4%
b) 49.5%
c) A slightly higher percentage of younger drivers are male than in the overall population. This difference decreases as drivers are older. But in the oldest categories, the drivers are increasingly women.
d) No, the answer to part c describes a relationship that is not independent.

47. a) The upper right bar chart shows the original data.
b) It's clear that shuffled data did not produce associations as obvious as the original data. The randomly scrambled data look different from the original data, and this supports the belief that there is an association between the variables.

49. a) 160 of 1300, or 12.3%
b) Yes. Major surgery: 15.3% vs. minor surgery: 6.7%.
c) Large hospital: 13%; small hospital: 10%.
d) Large hospital: Major 15% vs. minor 5%.
Small hospital: Major 20% vs. minor 8%.
e) No. Smaller hospitals have a higher rate for both kinds of surgery, even though it's lower "overall."
f) The small hospital has a larger percentage of minor surgeries (83.3%) than the large hospital (20%). Minor surgeries have a lower delay rate, so the small hospital looks better "overall."

51. a) 42.6%
b) A higher percentage of males than females were admitted: Males: 47.2% to females: 30.9%.
c) Program 1: Males 61.9%, females 82.4%.
Program 2: Males 62.9%, females 68.0%.
Program 3: Males 33.7%, females 35.2%.
Program 4: Males 5.9%, females 7.0%.
d) The comparisons in part c show that males have a lower admittance rate in every program, even though the overall rate shows males with a higher rate of admittance. This is an example of Simpson's paradox.

Chapter 4

1. Both distributions are unimodal. The domestic distribution is slightly skewed to the left. The centers of the distributions (both mean and median) are very similar. The IQRs and standard deviations show that the domestic load factors vary a bit more.

3. The standard deviation will be much lower. (It actually becomes 4.6.) The IQR will be affected very little, if at all.

5. Load factors are generally highest and least variable in the summer months (June–August). They are lower and more variable in the winter and spring. Several months have low outliers, and April has a very low outlier that is the minimum of all the data. (The low value is for April of 2003. At that time, there was concern over the deadly SARS virus, which may have depressed international travel.)

7. We are aware of a possible explanation for the depressed air travel that has nothing to do with general monthly or overall annual patterns. If we want to understand these patterns, it might be best to set these months aside.

9. a) After a period of little change in 2000–2001, load factors have been increasing steadily.
b) No. These are percentages, so they can't exceed 100%. Moreover, we have no reason to believe this trend will continue.

11. Data with a distribution this skewed are difficult to summarize. The extremely large values will dominate any summary or description.

13. a) Who—a group of 25 friends, What—car values (in Euro), When—end of 2019, Where—unspecified, Why—for comparison purposes.
b) Boxplot
c) There exists at least one outlier with a very high car value.

15. a) On average, a cappuccino is the most expensive of the three. The first quartile of cappuccino prices is higher than all the water prices. The middle 50% of cappuccino prices is less variable than egg prices, and the top 25% is higher.
b) No, the minimum price of a cappuccino is lower than the minimum price of a carton of eggs. And some of the bottom 25% of cappuccino prices are lower than the egg prices. (But traveler beware the cheap cup of coffee!)

17. a) The distribution is essentially symmetric, very slightly skewed to the right with two high outliers at 36 and 48. Most victims are between the ages of 16 and 24.
b) The slight increase between ages 22 and 24 is apparent in the histogram but not in the boxplot. It may be a second mode.
c) The median would be the most appropriate measure of center because of the slight skew and the outliers.
d) The IQR would be the most appropriate measure of spread because of the slight skew and the outliers.

19. a) About 59%
b) Bimodal.
c) Some cereals are very sugary; others are healthier low-sugar brands.
d) Yes.
e) Although the ranges appear to be comparable for both groups (about 28%), the IQR is larger for the adult cereals, indicating that there's more variability in the sugar content of the middle 50% of adult cereals.

21. a)

Pop Growth: 30.0, 22.5, 15.0, 7.5, 0.0
NE/MW S/W
Region

b) Growth rates in NE/MW states are tightly clustered near 5%, but with two high outliers. S/W states are more variable, with a median near 13% and two outliers as well. The distributions are fairly symmetric.

23. a) The graphs should be put on the same scale.
b) Female distribution: skewed to the right, higher range (96–100.9 F)
Male distribution: skewed to the left, lower range (96.3–99.7 F)
c) No

25. a) Both girls have a median score of about 17 points per game, but Scyrine is much more consistent. Her IQR is about 2 points, while Alexandra's is over 10.
b) If the coach wants a consistent performer, she should take Scyrine. She'll almost certainly deliver somewhere between 15 and 20 points. But if she wants to take a chance and needs a "big game," she should take Alexandra. Alex scores over 24 points about a quarter of the time. (On the other hand, she scores under 11 points as often.)

27. Female distribution: skewed to right, one outlier below minimum, median ≈ €72, lower spread, higher center; Male distribution: skewed to right, no outliers, median ≈ €63, higher spread, lower center.

29. a) Time series data
b) Total employment—declined between 2010–2013 and increased between 2013–2017; Male employment—same pattern as total employment; Female employment—increased during each year, lower than male employment in all years.
c) 13% in 2010

31. The class A is 1, class B is 2, and class C is 3.

33. a) Probably slightly left skewed. The mean is slightly below the median, and the 25th percentile is farther from the median than the 75th percentile.
b) No, all data are within the fences.
c)

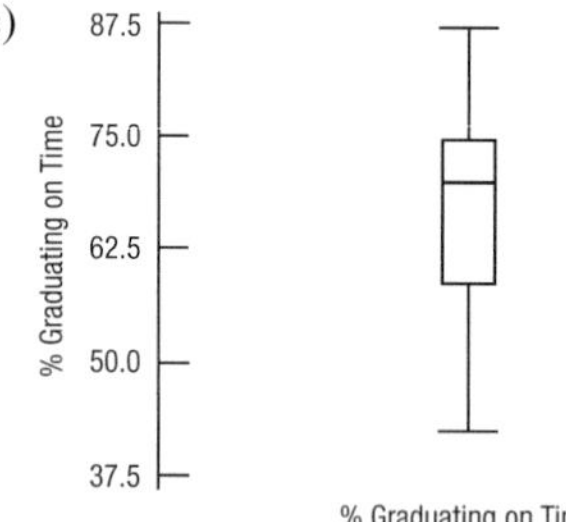

d) The 48 universities graduate, on average, about 68% of freshmen "on time," with percentages ranging from 43% to 87%. The middle 50% of these universities graduate between 59% and 75% of their freshmen in 4 years.

35. a) *Who*—Student volunteers; *What*—Memory test; *Where, when*—Not specified; *How*—Students took memory test 2 hours after drinking caffeine-free, half-dose caffeine, or high-caffeine soda; *Why*—To see if caffeine makes you more alert and aids memory retention.
b) Drink: categorical; Test score: quantitative.

c) 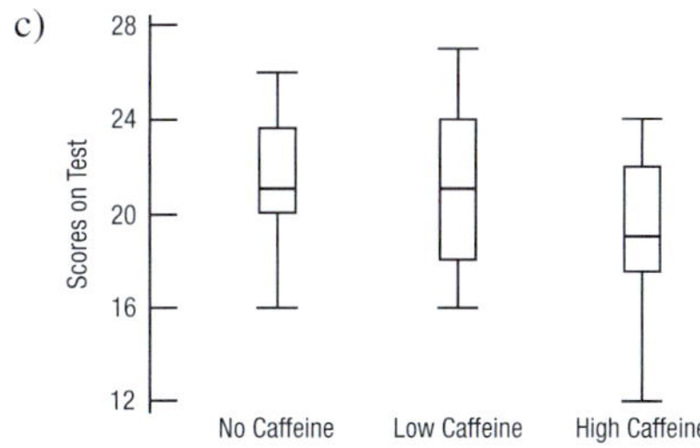

d) The participants scored about the same with no caffeine and low caffeine. The medians for both were 21 points, with slightly more variation for the low-caffeine group. The high-caffeine group generally scored lower than the other two groups on all measures of the 5-number summary: min, lower quartile, median, upper quartile, and max.

37. a) About 36 mph
b) Q1 about 35 mph and Q3 about 37 mph.
c) The range appears to be about 7 mph, from about 31 to 38 mph. The IQR is about 2 mph.
d) We can't know exactly, but the boxplot may look something like this:

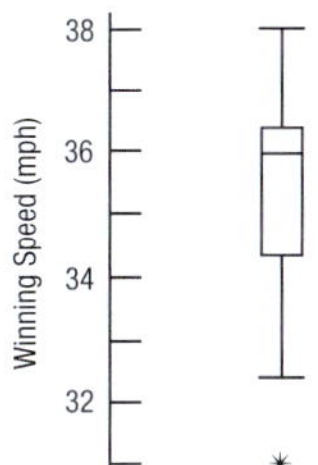

e) The median winning speed has been about 36 mph, with a max of about 38 and a min of about 31 mph. Half have run between about 35 and 37 mph, for an IQR of 2 mph.

39. a) Girls b) Girls c) Boys
d) Boys e) 81.86.

41. 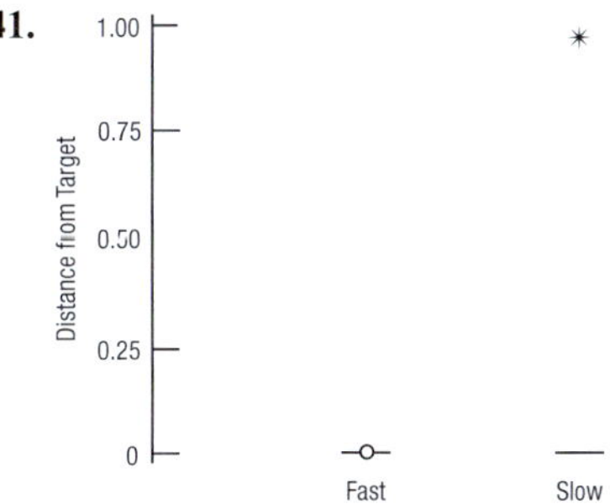

There appears to be an outlier! This point should be investigated. We'll proceed by redoing the plots with the outlier omitted:

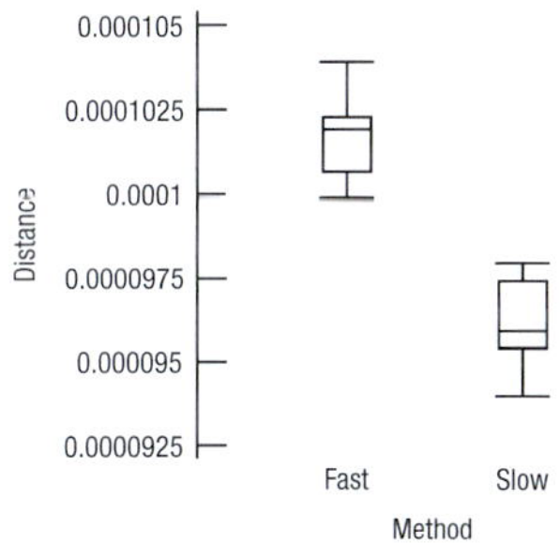

It appears that slow speed provides much greater accuracy. But the outlier should be investigated. It is possible that slow speed can induce an infrequent very large distance.

43. a) 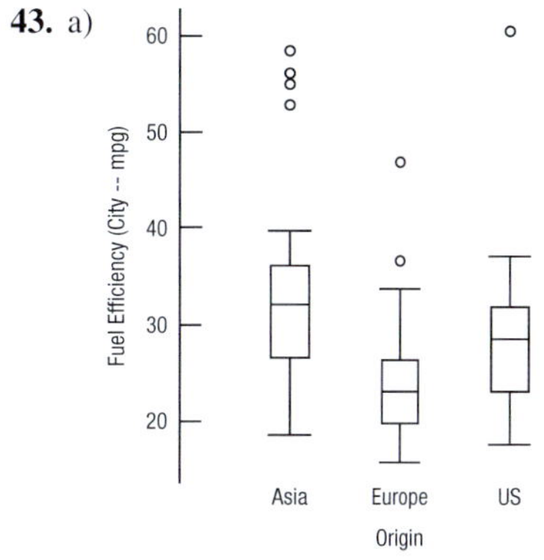

b) All three groups have high outliers, potentially hybrid or electric cars. Asian cars generally have the highest fuel efficiency, while European cars have the lowest. The Asian median is about 32 mpg, European cars around 23 mpg, and U.S. cars 28 mpg.

45. a) Most of the data are found in the far left of this histogram. The distribution is very skewed to the right.
b) Re-expressing the data by, for example, logs or square roots might help make the distribution more nearly symmetric.

47. a) The logarithm makes the histogram more symmetric. It is easy to see that the center is around 3.5 in log assets.
b) That has a value of around $2500 million.
c) That has a value of around $1000 million.

49. a) Fusion time and group.
b) Fusion time is quantitative (units = seconds). Group is categorical.
c) Both distributions are skewed to the right with high outliers. The boxplot indicates that visual information may reduce fusion time. The median for the Verbal/Visual group seems to be about the same as the lower quartile of the No/Verbal group.

51. a) The histogram would be centered at 0 if the true difference were 0.
b) (Answers may vary.) Because 0 never occurred in 1000 random samples, this provides strong evidence that the difference we saw was not due to chance. We can conclude that there is a real (if small) difference between the mean prices of the two items.

53. a) Prices appear to be both higher on average and more variable in Baltimore than in the other three cities. Prices in Chicago may be slightly higher than in Dallas and Denver, but the difference is very small.
b) There are outliers on the low end in Baltimore and Chicago and one high outlier in Dallas, but these do not affect the overall conclusions reached in part a.

Chapter 5

1. 65

3. In January, a high of 55° is not quite 2 standard deviations above the mean, whereas in July a high of 55° is more than 2 standard deviations lower than the mean. So it's less likely to happen in July.

5. a) 3.1 kg, 1.9 kg b) 6.834 lb, 4.189 lb

7. a) 11.12 in. b) 1.597 in.

9. a)

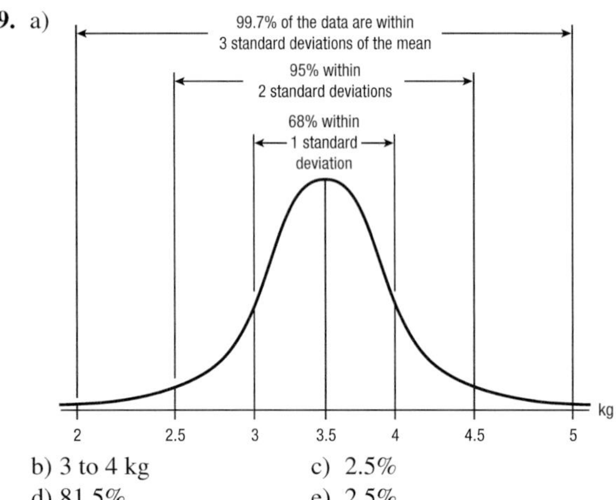

b) 3 to 4 kg c) 2.5%
d) 81.5% e) 2.5%

11. The boy's height is 1.88 standard deviations below the mean height of American children his age.

13. a) 40.13% b) 30.85% c) 38.02%

15. a) 2.8% b) 29.2% c) 68.0%

17. a) No. The plot is not straight.
b) It is skewed to the right.

19. a) Skewed to the right; the mean is higher than median.
b) \$350 and \$950
c) Minimum \$350. Mean \$750. Median \$550. Range \$1200. IQR \$600. Q1 \$400. SD \$400.
d) Minimum \$330. Mean \$770. Median \$550. Range \$1320. IQR \$660. Q1 \$385. SD \$440.

21. Lowest score = 910. Mean = 1230. SD = 120. Q3 = 1350. Median = 1270. IQR = 240.

23. 1.202

25. The z-scores, which account for the difference in the distributions of the two tests, are 1.5 and 0 for Derrick and 0.5 and 2 for Julie. Derrick's total is 1.5, which is less than Julie's 2.5.

27. a) Faye b) Pearl

29. a) About 0.125 standard deviations below the mean.
b) 1.71 m ($z = 0.275$) is more unusual than 1.53 m ($z = -0.175$).

31. a) Mean = 4.3 + 0.2 = 4.5 mm; SD is unchanged at 0.12 mm.
b) Mean = 9.95(4.5) = € 44.78; SD = 9.95(0.12) = € 1.19

33. Min = 9.95(4.1) − 17.56 = €23.24;
median = 9.95(4.42) − 17.56 = €26.42;
SD = 9.95(0.098) = €0.98; IQR = 9.95(1.34) = €13.33

35. Middle-aged individuals have age approximately between 35 and 59 years. A workable standard deviation would cover most of that range of values with ±3 standard deviations around the mean age of 48 years. If the standard deviation were 1 year, ±3 standard deviations would give a range of just 6 years around the mean, which is not reasonable as individuals aged 40, say, are not covered. If the standard deviation were 5 years, ±2 standard deviations would give a range of 20 years around the mean which is not reasonable as a person aged 60+ would be considered as a middle-aged individual. Hence, the SD must be 2 years.

37. A park visitor who has height more than 2 standard deviations less than the mean, i.e., a height less than 1.6 − 2(0.4) = 0.8 metres, is unusually low.

39. a)

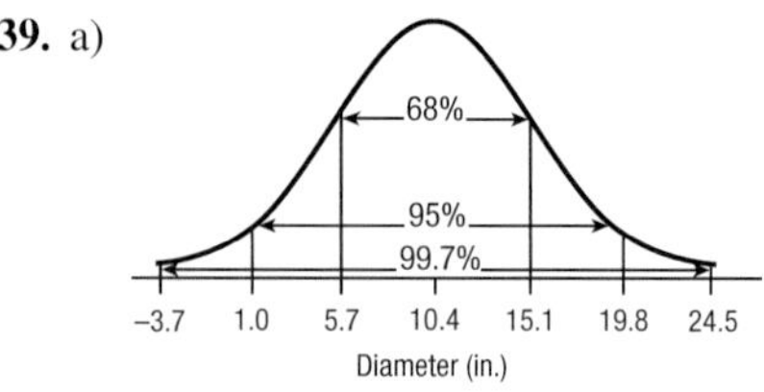

b) Between 1.0 and 19.8 inches c) 2.5%
d) 34% e) 16%

41. No. Since the histogram is not unimodal and symmetric, it is not wise to have faith in numbers from the Normal model.

43. a) 16% b) 13.2%
c) Because the Normal model doesn't fit the data distribution well.

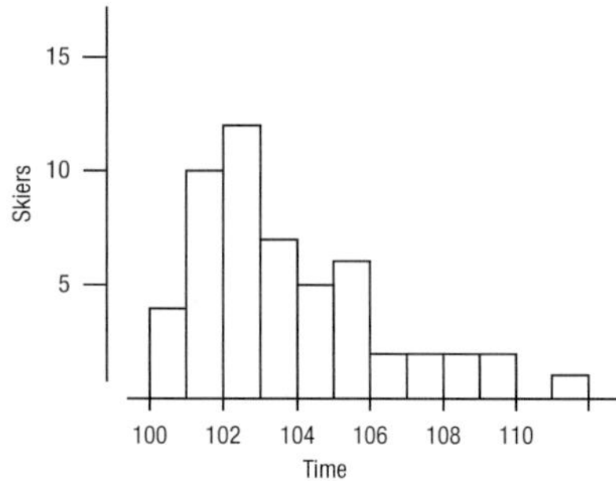

d) The distribution is skewed and thus not a good match to the Normal mode.

45. a) 2.5%
b) 2.5% of the 483 receivers, or about 12 of them, should gain more than 274.73 + 2 × 327.32 = 929.37 yards. (In fact, 31 receivers exceeded this value.)
c) The distribution is strongly skewed to the right, not symmetric.

47. a) 2.5%
b) No, the distribution is skewed.
c) No, the distribution is skewed.
d) The rule would work fairly well even though there is a slight asymmetry to the distribution.
e) Yes. Now the distribution of the means is nearly Normal. The 68–95–99.7 Rule works well.

49. a) 2.113 m b) 1.087 m c) 1.446 to 1.754 m

51. a) 1130.7 lb b) 1347.4 lb c) 113.3 lb

53. a)

68%
95%
99.7%
116 140 164 188 212 236 260
Cholesterol (mg/dL)

b) 30.85% c) 17.00% d) 32.4 mg/dL e) 212.9 mg/dL

55. a) 30.85% b) (72.18, 97.82) mmHg
c) 101.45 mmHg

57. a) 4.98 b) 2.40
c) This year's because SD is smaller

Part I Review

R1.1. a)

b) Median 49 cents, IQR 6 cents.

c) The distribution is unimodal and left skewed. The center is near 50 cents; values range from 42 cents to 53 cents.

R1.3. a) If enough sopranos have a height of 65 inches, this can happen.

b) The distribution of heights for each voice part is roughly symmetric. The basses are slightly taller than the tenors. The sopranos and altos have about the same median height. Heights of basses and sopranos are more consistent than those of altos and tenors.

c) Perhaps from low to high voice: bass, tenor, alto, soprano.

R1.5. a) It means their heights are also more variable.

b) The z-score for women to qualify is 2.40, compared with 1.75 for men, so it is harder for women to qualify.

R1.7. a) *Who*—People who live near State University; *What*—Age, attended college? Favorable opinion of State?; *When*—Not stated; *Where*—Region around State U; *Why*—To report to the university's directors; *How*—Sampled and phoned 850 local residents.

b) Age—Quantitative (years); attended college?—categorical; favorable opinion?—categorical.

c) The fact that the respondents know they are being interviewed by the university's staff may influence answers.

R1.9. a) These are categorical data, so mean and standard deviation are meaningless.

b) Not appropriate. Even if it fits well, the Normal model is meaningless for categorical data.

R1.11. a)

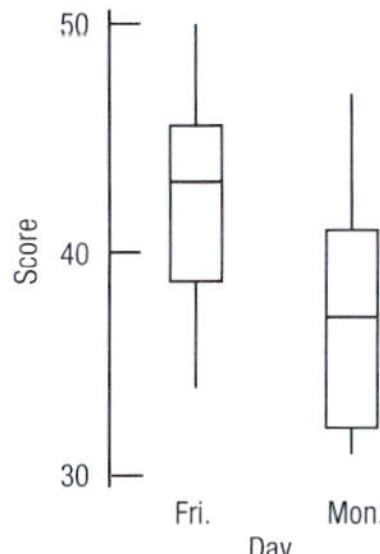

b) The scores on Friday were higher by about 5 points on average. This is a drop of more than 10% off the average score and shows that students fared worse on Monday after preparing for the test on Friday. The spreads are about the same, but the scores on Monday are a bit skewed to the right.

c)

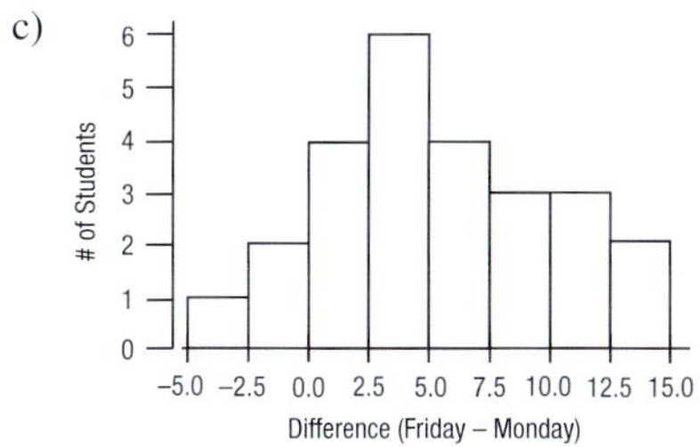

d) The changes (Friday−Monday) are unimodal and centered near 4 points, with a spread of about 5 (SD). They are fairly symmetric, but slightly skewed to the right. Only 3 students did better on Monday (had a negative difference).

R1.13. a) Categorical.

b) Go fish. All you need to do is match the denomination. The denominations are not ordered. (Answers will vary.)

c) Gin rummy. All cards are worth their value in points (face cards are 10 points). (Answers will vary.)

R1.15. a) Annual mortality rate for males (quantitative) in deaths per 100,000 and water hardness (quantitative) in parts per million.

b) Calcium is skewed right, possibly bimodal. There looks to be a mode down near 12 ppm that is the center of a fairly tight symmetric distribution and another mode near 62.5 ppm that is the center of a much more spread out, symmetric (almost uniform) distribution. Mortality, however, appears unimodal and symmetric with the mode near 1500 deaths per 100,000.

R1.17. a) They are on different scales.

b) January's values are lower and more spread out.

c) Roughly symmetric but slightly skewed to the left. There are more low outliers than high ones. Center is around 40 degrees with an IQR of around 7.5 degrees.

R1.19. a) Bimodal with modes near 2 and 4.5 minutes. Fairly symmetric around each mode.

b) Because there are two modes, which probably correspond to two different groups of eruptions, an average might not make sense.

c) The intervals between eruptions are longer for long eruptions. There is very little overlap. More than 75% of the short eruptions had intervals less than about an hour (62.5 minutes), while more than 75% of the long eruptions had intervals longer than about 75 minutes. Perhaps the interval could even be used to predict whether the next eruption will be long or short.

R1.21. a)

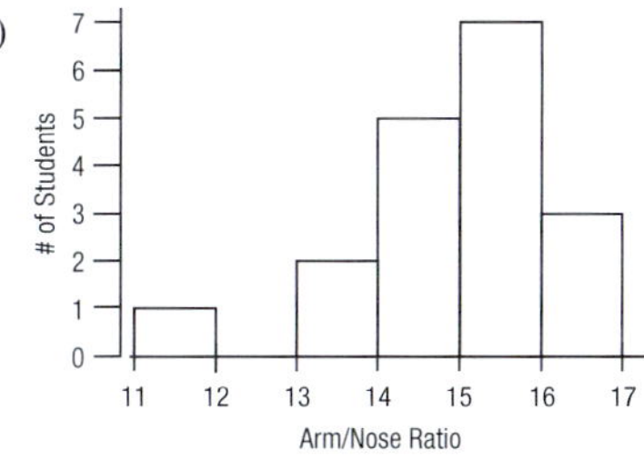

The distribution is left skewed with a center of about 15. It has an outlier between 11 and 12.

b) Even though the distribution is somewhat skewed, the mean and median are close. The mean is 15.0 and the SD is 1.25.
c) Yes. 11.8 is already an outlier. 9.3 is more than 4.5 SDs below the mean. It is a very low outlier.

R1.23. If we look only at the overall statistics, it appears that the follow-up group is insured at a much lower rate than those not traced (11.1% of the time compared with 16.6%). But most of the follow-up group were black, who have a lower rate of being insured. When broken down by race, the follow-up group actually has a higher rate of being insured for both blacks and whites. So the overall statistic is misleading and is attributable to the difference in race makeup of the two groups.

R1.25. a)

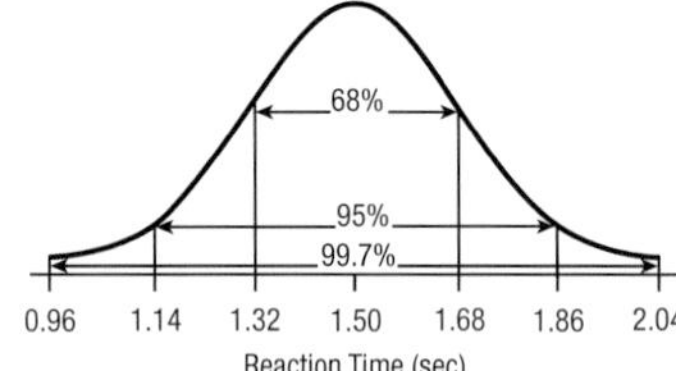

b) According to the model, reaction times are symmetric with center at 1.5 seconds. About 95% of all reaction times are between 1.14 and 1.86 seconds.
c) 8.2%
d) 24.1%
e) Quartiles are 1.38 and 1.62 seconds, so the IQR is 0.24 second.
f) The slowest 1/3 of all drivers have reaction times of 1.58 seconds or more.

R1.27. a)

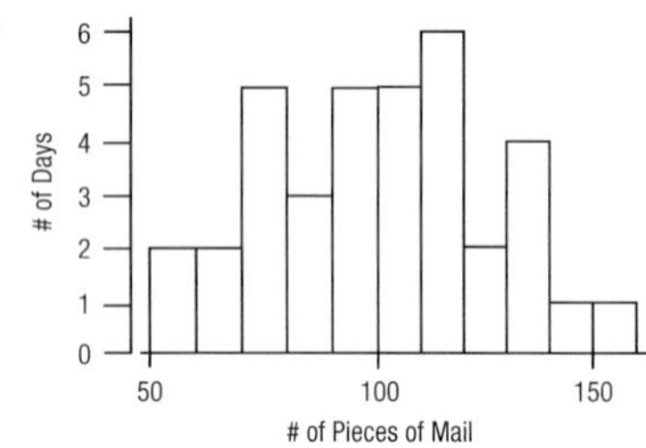

b) Mean 100.25, SD 25.54 pieces of mail.
c) The distribution is somewhat symmetric and unimodal, but the center is rather flat, almost uniform.
d) 64%. The Normal model seems to work reasonably well, since it predicts 68%.

R1.29. a) *Who*—100 health food store customers; *What*—Have you taken a cold remedy?, and Effectiveness (scale 1 to 10); *When*—Not stated; *Where*—Not stated; *Why*—Promotion of herbal medicine; *How*—In-person interviews.
b) Have you taken a cold remedy?—categorical. Effectiveness—categorical or ordinal.
c) No. Customers are not necessarily representative, and the Council had an interest in promoting the herbal remedy.

R1.31. a) 38 cars
b) Possibly because the distribution is skewed to the right.
c) Center—median is 148.5 cubic inches. Spread—IQR is 126 cubic inches.
d) No. It's bigger than average, but smaller than more than 25% of cars. The upper quartile is at 231 inches.
e) No. 1.5 IQR is 189, and 105 − 189 is negative, so there can't be any low outliers. 231 + 189 = 420. There aren't any cars with engines bigger than this, since the maximum has to be at most 105 (the lower quartile) + 275 (the range) = 380.
f) Because the distribution is skewed to the right, this is probably not a good approximation.
g) Mean, median, range, quartiles, IQR, and SD all get multiplied by 16.4.

R1.33. a) 44.0%
b) If this were a random sample of all voters, yes.
c) 38.4%
d) 0.87%
e) 9.96%
f) 8.96%

R1.35. a) Republican—3705, Democrat—3976, Neither—733; or Republican—44.0%, Democrat—47.3%, Neither—8.7%.
b)

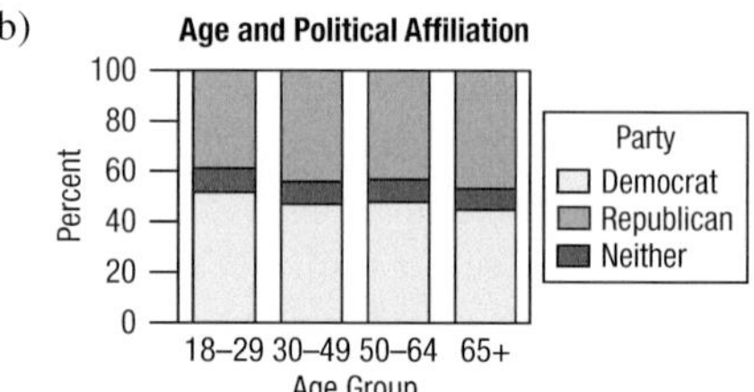

c) It appears that the older the voter, the less likely they are to lean Democratic and the more likely to lean Republican.
d) No. There is an association between age and affiliation. Younger voters tend to be more Democratic and less Republican.

R1.37. a) 0.43 hour
b) 1.4 hours
c) 0.89 hour (or 53.4 minutes)
d) Survey results vary, and the mean and the SD may have changed.

R1.39. a)

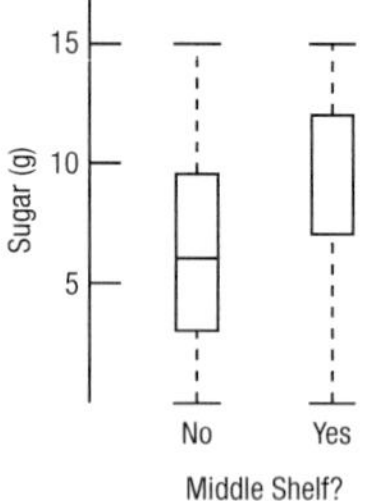

The 21 middle shelf cereals have a mean sugar content of 9.62 g/serving, compared with 5.93 g/serving for the others, for a difference of 3.69 g. (Medians are 12 and 6, respectively.)
b) Answers will vary slightly.

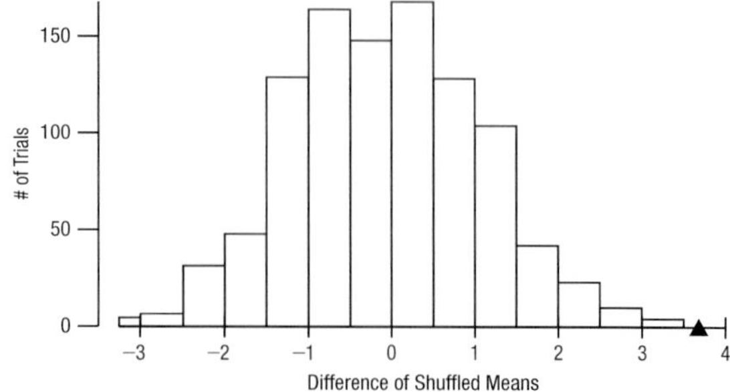

Because none of the 1000 shuffled differences were as large as the observed difference, we can say that a difference that large is unlikely to be produced by chance. There is evidence to suggest that the cereals on the middle shelf have a higher mean sugar content.

R1.41. Answers will vary.

R1.43. Answers will vary.

Chapter 6

1. a) Weight in ounces: explanatory; Weight in grams: response. (Could be other way around.) To predict the weight in grams based on ounces. Scatterplot: positive, straight, strong (perfectly linear relationship).
b) Circumference: explanatory. Weight: response. To predict the weight based on the circumference. Scatterplot: positive, possibly not straight, moderately strong.
c) Shoe size: explanatory; GPA: response. To try to predict GPA from shoe size. Scatterplot: no direction, no form, very weak.
d) Miles driven: explanatory; Gallons remaining: response. To predict the gallons remaining in the tank based on the miles driven since filling up. Scatterplot: negative, straight, moderate.

3. a)

b) Positive. c) Linear. d) Strong. e) No.

5. a) True.
b) False. It will not change the correlation.
c) False. Correlation has no units.

7. Correlation does not demonstrate causation. The analyst's argument is that sales staff cause sales. However, the data may reflect the store hiring more people as sales increase, so any causation would run the other way.

***9.** 4, 5.176 and 6.097. Yes, it will be straight.

11. a) Altitude: explanatory; Temperature: response. (Other way around possible as well.) To predict the temperature based on the altitude. Scatterplot: negative, possibly straight, weak to moderate.
b) Ice cream cone sales: explanatory. Air-conditioner sales: response—although the other direction would work as well. To predict one from the other. Scatterplot: positive, straight, moderate.
c) Age: explanatory; Grip strength: response. To predict the grip strength based on age. Scatterplot: curved down, moderate. Very young and elderly would have grip strength less than that of adults.
d) Reaction time: explanatory; Blood alcohol level: response. To predict blood alcohol level from reaction time test. (Other way around is possible.) Scatterplot: positive, nonlinear, moderately strong.

13. a) None. b) (1) and (4). c) All.
d) (1) and (3). e) (2) and (4).

15. There seems to be a very weak—or possibly no—relationship between brain size and performance IQ.

17. a)

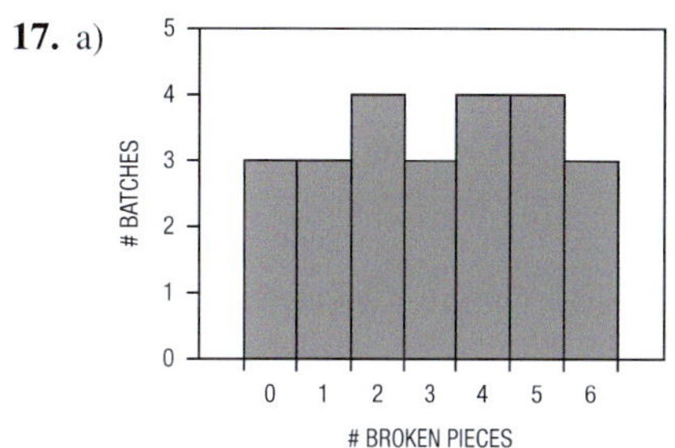

b) Uniform between 0 and 6 broken pieces. The uniformity.
c) The positive and somewhat linear relationship between batch number and number of broken pieces.

19. a) 0.006 b) 0.777 c) −0.977 d) −0.487

21. There may be an association, but not a correlation unless the variables are quantitative. There could be a correlation between average number of hours of TV watched per week per person and number of crimes committed per year. Even if there is a relationship, it doesn't mean one causes the other.

23. a) Yes, scatterplot shows a linear form with no outliers.
b) There is a strong, positive, linear association between *Drop* and *Speed*; the greater the initial drop, the higher the top speed.

25. The scatterplot is not linear; correlation is not appropriate.

27. The correlation may be near 0. We expect nighttime temperatures to be low in January, increase through spring and into the summer months, then decrease again in the fall and winter. The relationship is not linear.

29. The correlation coefficient won't change, because it's based on z-scores. The z-scores of the prediction errors are the same whether they are expressed in nautical miles or miles.

31. a) Assuming the relation is linear, a correlation of −0.772 shows a strong relation in a negative direction.
b) Continent is a categorical variable. Correlation does not apply.

33. a) Yes, taller children are generally better readers, but this does not imply causation.
b) Age is a lurking variable; older children are generally taller and have better reading skills.

35. a) No, this cannot be concluded from the correlation alone. The relationship may be nonlinear or may contain outliers.
b) No, the form of the relationship cannot be determined from the correlation.
c) No, the correlation does not provide evidence for or against the existence of outliers.
d) No, the correlation does not depend on the units used to measure the variables.

37. These are categorical data even though they are represented by numbers. The correlation is meaningless.

39. a) There is a positive, moderately strong, linear association between the *Housing Cost Index* and the *Median Family Income*.
b) The correlation would remain 0.66.

c) The correlation would remain 0.66.
d) The correlation would have the same sign and would decrease slightly since the region has a low HCI for its median family income.
e) No, we can only say that higher median family incomes are associated with higher housing cost index scores, but we do not know why. Other variables may be at play.
*f) Tau says that there is a positive association between median income and housing costs, but it makes no claims about the form of this association. Tau has no requirement that the relationship be linear. It appears that the plot "thickens" from left to right. That could affect the correlation, but not tau.

41. a)

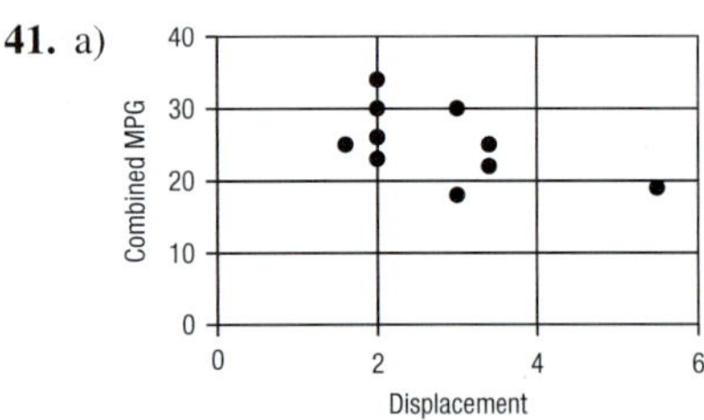

b) Negative, moderately strong, and linear.
c) −0.546
d) The relationship between displacement and combined MPG is linear and negative. Therefore, vehicles in the selected group with higher displacement have lower mileage.

43. a)

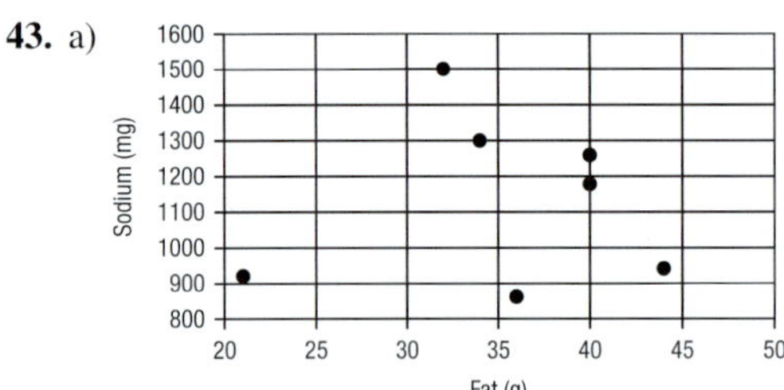

(Plot could have explanatory and predictor variables swapped). Correlation is 0.061 (including the outlier). There is no apparent association between fat and sodium content in burgers, especially with the low-sodium, low-fat burger included. Without this outlier, the correlation is −0.578, yielding a negative, moderately strong association.
*b) Spearman's rho is −0.126. It is slightly negative and less than the correlation coefficient because Spearman's rho uses ranks and thus minimizes the influence of the outlying point.

45. a) Yes, relationship between number of runs scored and attendance appears to be straight and without any outliers.
b) The association between attendance and runs scored is positive, straight, and moderately strong.
c) There is evidence of an association between attendance and runs scored, but this does not prove that more fans would come if the number of runs increased. Association does not imply causation.

47. The distribution for means is unimodal and symmetric, so we could use the 68–95–99.7 Rule for it. The distribution of medians is too skewed for the rule.

49. The association between length and duration is linear, positive, and moderately strong, with no outliers. Generally, rides on coasters with a greater length tend to last longer, with a correlation of 0.7.

51. a) Lower rank is better, so variables like speed and duration should have negative associations with rank. We would expect that as one variable (say length of ride) increases, the rank will improve, which means it will decrease.
b) Max vertical angle is the only variable with a negative correlation with rank–and that is due almost entirely to a single coaster, Nemesis.

53. a)

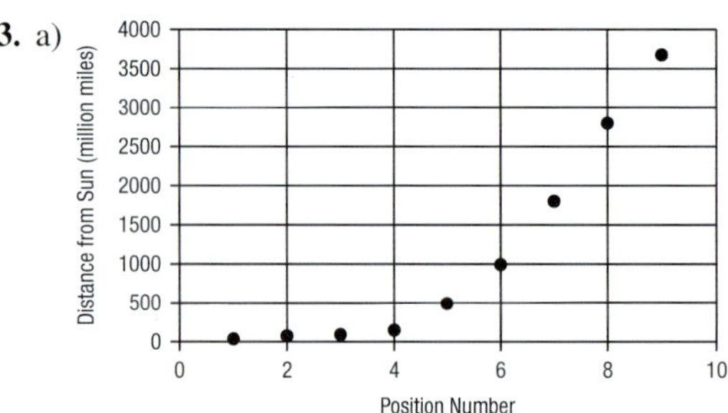

The relationship between *Position Number* and *Distance* is nonlinear, positive, and very strong.
b) The relationship is nonlinear.
c)

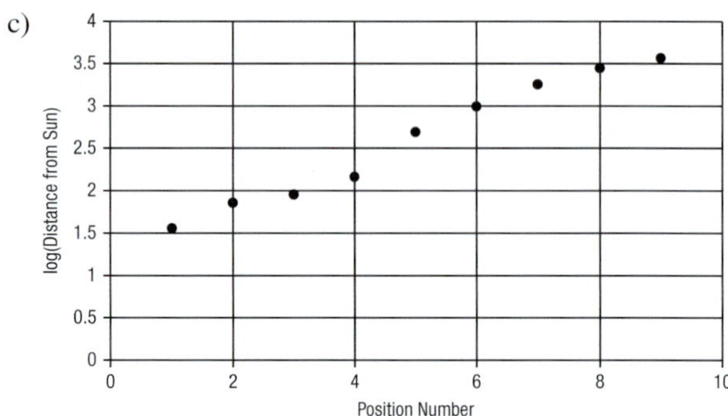

This scatterplot is better since the relationship between position number and log of distance appears to be roughly linear.

Chapter 7

The calculations in this chapter are particularly sensitive to the rounding. If your answers differ from those here even in the second place, they may still be correct in the sense that you calculated them by a correct method, but you may have used values that were rounded differently than those used to find these answers. In general, we use the original data and do no intermediate rounding. Don't be concerned about minor differences—what is important is your interpretations of the results.

1. a) False. The line usually touches none of the points. We minimize the sum of the squared errors.
b) True.
c) False. It is the sum of the squares of all the residuals that is minimized.

3. The weight of a newborn boy can be predicted as −5.36 kg plus 0.1675 kg per cm of length. No, this is a model fit to data, and no baby should be expected to fit this model exactly.

5. a) $b_1 = 0.914$ if found by hand. $b_1 = 0.913$ if found by technology. (Difference is due to rounding error.)
b) It means that an additional 0.914 ($1000) or $914 of sales is associated with each additional sales person working.
c) $b_0 = 8.10$
d) It would mean that, on average, we expect sales of 8.10 ($1000) or $8100 with 0 sales people working. Doesn't really make sense in this context.

e) $\widehat{Sales} = 8.10 + 0.914$ *Number of Sales People Working*
f) \$24.55 (\$1000) or \$24,550. (24,540 if using the technology solution.)
g) 0.45 (\$1000) or \$450. (\$460 with technology.)
h) Underestimated.

7. The winners may be suffering from regression to the mean. Perhaps they weren't really better than other rookie executives, but just happened to have a lucky year.

9. a) Thousands of dollars
b) 2.77 (the largest residual in magnitude)
c) 0.07 (the smallest residual in magnitude)

11. $R^2 = 93.2\%$. About 93% of the variance in *Sales* can be accounted for by the regression of *Sales* on *Number of Sales Workers*.

13. a) Linearity Condition
b) Equal Spread Condition
c) Outlier Condition

15. 319 milligrams.

17. The potassium content is actually lower than the model predicts for a cereal with that much fiber.

19. The model predicts that cereals will have approximately 27 more milligrams of potassium, on average, for every additional gram of fiber.

21. 81.5%

23. The true potassium contents of cereals vary from the predicted amounts with a standard deviation of 30.77 milligrams.

25.

	$\bar{x}$	s_x	$\bar{y}$	s_y	r	$\hat{y} = b_0 + b_1x$
a)	49	4.7	8	5.1	0.3	$\hat{y} = -8.17 + 0.33x$
b)	16	7.7	34	4.5	−0.8	$\hat{y} = 41.52 - 0.47x$
c)	20	9.2	41	92	−0.8	$\hat{y} = 201 - 8x$
d)	40.7	1.9	446.5	35.5	0.8	$\hat{y} = -164 + 15x$

27. a) Relationship is nonlinear, so linear model is not appropriate.
b) Relationship is nonlinear, so linear model is not appropriate.
c) Linear model is appropriate.

29. a) *Price* (in thousands of dollars) is y and *Size* (in square feet) is x.
b) Slope is thousands of \$ per square foot.
c) Positive. Larger homes should cost more.

31. 300 pounds/foot. If a "typical" car is 15 feet long, all of 3, 30, and 3000 would give ridiculous weights.

33. A linear model on *Size* accounts for 71.4% of the variation in home *Price*.

35. a) R^2 does not tell whether the model is appropriate. High R^2 could also be due to an outlier.
b) Predictions based on a regression line are for average values of y for a given x. The actual wingspan will vary around the prediction.

37. a) 0.845
b) Price should be 0.845 SDs above the mean in price.
c) Price should be 1.690 SDs below the mean in price.

39. a) Probably not. Your score is better than about 97.5% of people, assuming scores follow the Normal model. Your next score is likely to be closer to the mean.
b) The friend probably should retake the test. His score is better than only about 16% of people. His next score is likely to be closer to the mean.

41. a) *Price* increases by about \$0.061 per 1000 square feet or \$61.00 per square foot.
b) 230.82 thousand, or \$230,820.
c) \$115,020; \$6000 is the residual.

43. a) Maybe, there is some curvature to the residuals but not enough to completely disregard the linear model.
b) The linear model on tar content accounts for 92.4% of the variability in nicotine content.

45. a) $r = 0.902$
b) Nicotine should be 1.804 SDs below average.
c) Tar should be 0.902 SDs above average.

47. a) $\widehat{Nicotine} = 0.1483 + 0.06216$ *Tar*
b) 0.397 mg
c) Nicotine content increases by 0.062 mg of nicotine per milligram of tar.
d) We'd expect a cigarette with no tar to have 0.1483 mg of nicotine.
e) 0.533 mg

49. a) Yes. The relationship is straight enough, with a few outliers. The spread increases a bit for states with large median incomes, but we can still fit a regression line.
b) From summary statistics:
$\widehat{HCI} = -147.515 + 0.0106$ *Median income*
From technology: $\widehat{HCI} = -148.15 + 0.0106$ *Median income*
c) 329.4
d) 218.62
e) $\widehat{z_{HCI}} = 0.624\, z_{MFI}$
f) $\widehat{z_{MFI}} = 0.624\, z_{HCI}$

51. a) $\widehat{Total} = 545.6584 + 0.8472$ *Age*
b) Yes. Both variables are quantitative; there are no apparent outliers; no curved pattern in scatter plot; scatter plot does not thicken.
c) \$561.76; \$587.17
d) 0.08%
e) No, the model accounts for almost none of the variation in *Total Yearly Purchases* and the scatterplot shows no evidence of positive linear relationship.

53. a) Moderately linear and positive. One possible outlier (higher-than-expected math score).
b) The student with a verbal score of slightly more than 200 and a math score of 800.
c) Positive, moderate linear relationship. 33.18% of the variability in math score is explained by variability in verbal score.
d) $\widehat{Math} = 291.1879 + 0.5121$ *Verbal*.
e) Every point of verbal score adds 0.5121 points to the predicted average math score.
f) 512.4151 points.
g) 99.1321 points.

55. a) 0.685
b) $\widehat{Verbal} = 171.3 + 0.69$ *Math*
c) The observed verbal score is higher than predicted from the math score.
d) 518.5 points
e) 559.6 points

f) Regression to the mean. Someone whose math score is below average is predicted to have a verbal score below average, but not as far (in SDs). So if we use *that* verbal score to predict math, they will be even closer to the mean in predicted math score than their observed math score. If we kept cycling back and forth, eventually we would predict the mean of each and stay there.

57. a) The relationship is straight enough, but very weak, as shown by the large spread in the scatterplot.
b) The amount of land damaged by wildfires increases by about 0.144 million acres per year.
c) Yes, the intercept represents the amount of land damaged by wildfires in 1982.
d) The residuals have a standard deviation of 1.804 million acres. Compared to the observed values, most of which are between 3 and 8 million acres, this amount of error in the model's predictions is quite large, so the model isn't very effective for prediction.
e) 22.7% of the variation in the number of acres damaged by wildfires can be accounted for by the linear model on *Year since 1982*. This indicates that there is very little association between the two variables. Other variables should be considered for more accurate prediction.

59. a)

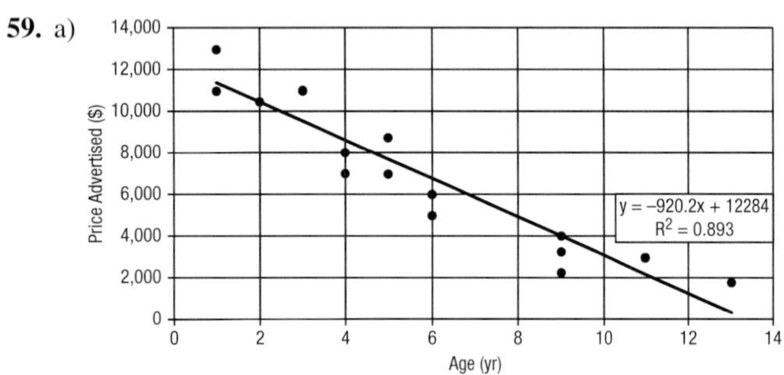

b) Linear, strong, negative.
c) Yes.
d) −0.94
e) 89.3% of the variability in price is accounted for by variability in age.
f) Other factors—condition, mileage, etc.—account for remaining variability.

61. a) $\widehat{Price} = 17{,}674 - 844.5\ Years$
b) Prices decline at the rate of about \$844.5 per year.
c) The average new Corolla costs \$17,674.
d) \$11,762.50
e) Negative residual. Its price is below the predicted value for its age.
f) −\$729
g) No. As of age 21, the model predicts negative prices. The relationship is no longer linear.

63. a)

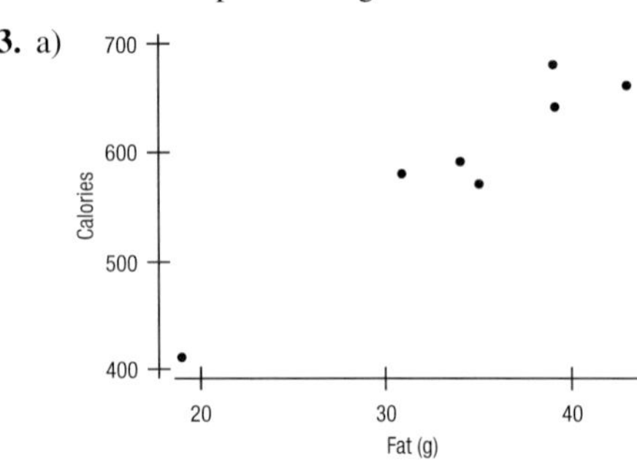

b) 92.3% of the variation in calories can be accounted for by the fat content.
c) $\widehat{Calories} = 211.0 + 11.06\ Fat$
d)

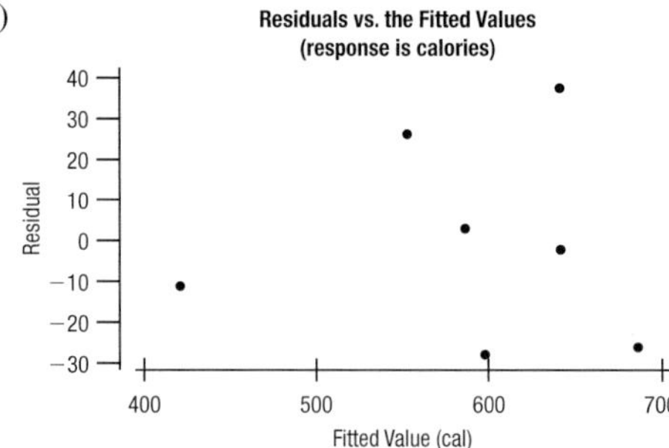

Residuals show no clear pattern, so the model seems appropriate.
e) Could say a fat-free burger still has 211.0 calories, but this is extrapolation (no data close to 0).
f) Every gram of fat adds 11.06 calories, on average.
g) 553.5 calories

65. a) The regression was for predicting calories from fat, not the other way around.
b) $\widehat{Fat} = -15.0 + 0.083\ Calories$. Predict 35.1 grams of fat.

67. a) The model predicts that all bridges are deficient. This is because the intercept is lower than 5 and the slope is negative.
b) The condition of bridges decreases by an average of 0.00481 per year—much less rapid than the decrease found by the earlier model.
c) The R^2 value is just 2.6%, so the model does not have much predictive power.

69. a) 0.5805.
b) 33.7% of the variability in mean air temperature is accounted for by variability in the concentration of CO_2.
c) $\widehat{Mean\ Temperature} = 15.308 + 0.004\ CO_2$.
d) An increase in CO_2 level of 1 ppm yields an increase of 0.004 degrees Celsius in mean annual air temperature.
e) The y-intercept does not have a meaningful interpretation in the context of this problem, so we treat it just as a base value for the prediction.
f) No.
g) 16.784 degrees Celsius.
h) No. The model does not give an exact prediction of what will happen under any circumstance. It just gives a general idea of what to expect.

71. a) $\widehat{\%\ Body\ Fat} = -26.925 + 0.2493\ Weight$
b) Residuals look randomly scattered around 0, so a linear model seems appropriate.
c) Percentage body fat increases, on average, by 0.2493% per pound of weight.
d) No. $R^2 = 42.87\%$, and the standard deviation of the residuals is around 7%, which is quite large.
e) −2.6661%.

73. a) $\widehat{LongJump} = 7.595 - 0.01\ (800mTime)$
b) 9.4%
c) Yes, the slope is negative. Faster runners tend to jump higher as well.
d) There is an extraordinary point (Akela Jones) that is influential. This model is not appropriate.

e) No. The residual standard deviation is 0.23 meters, almost exactly the same as the SD of all long jumps. The influential point dominates the model, making it unsafe to use.

75. a) Fairly strong, negative, linear relationship. Towns with harder water tended to have lower mortality rates.

b) $\widehat{Mortality} = 1579.734 - 2.811\ Calcium$

c) Mortality decreases, on average, by 2.811 deaths per 100,000 for every unit increase in calcium. For towns with no calcium, the predicted mortality rate is 1579.734 deaths per 100,000 on average.

d) One town had a mortality rate which was 86.1 deaths per 100,000 higher than the amount predicted by the model.

e) 1391.397 deaths per 100,000 on average.

f) 90.3% of the variability in the mortality rate is accounted for by the variability in the calcium concentration.

77. Least squares means that the sum of the squared residuals from the line is as small as possible. For example, the residual at the point $(10, 10)$ is $10 - (7.0 + 1.1 \times 10) = -8$, which contributes $(-8)^2$ or 64 to the sum of squared residuals.

79. a) The distribution of the slopes is unimodal and slightly skewed to the low end. 95% of the random slopes fell between -4.4 and -2.2.

b) The slope for the sample from Chapter 6 is just barely inside the interval that holds the middle 95% of the randomly generated slopes, so by that definition, it is not unusual.

c) Answers will vary.

Chapter 8

1. The different segments are not scattered at random throughout the scatterplot. Each segment may have a different relationship.

3. Yes, it is clear that the relationship between January and December spending is not the same for all five segments. Using one overall model to predict January spending would be very misleading.

5. Your friend is extrapolating. It is impossible to know if a trend like this will continue so far into the future.

7. a) $354,472

b) An extrapolation this far from the data is unreliable.

9. Yes, this observation was influential. After it was removed, the slope of the regression line increased by 142.86% and the R-squared value improved by 44.86%.

11. No. In warm weather, more children will go outside and play.

13. Individual student scores will vary greatly. The class averages will have much less variability and may disguise important trends.

15. a) No re-expression needed.

b) Re-express to straighten the relationship.

c) Re-express to equalize spread.

17. Improve homoscedasticity (more nearly equal variances between groups).

19. Yes. The square root doesn't make the spreads as nearly equal. The reciprocal clearly goes too far on the Ladder of Powers.

21. a) The trend appears to be somewhat linear up to about 1940, but from 1940 to about 1970 the trend appears to be nonlinear with a large drop from 1940 to 1950. From 1975 or so to the present, the trend appears to be linear.

b) Relatively strong for certain periods.

c) No, as a whole the graph is clearly nonlinear. Within certain periods (ex: 1975 to the present) the correlation is high.

d) Overall, no. You could fit a linear model to the period from 1975 to 2003, but why? You don't need to interpolate, since every year is reported, and extrapolation seems dangerous.

23. a) The relationship appears to be nonlinear.

b) The residuals will appear clustered above and below the horizontal axis.

25. a) No. We need to see the scatterplot first to see if the conditions are satisfied, and models are always wrong.

b) No, the linear model might not fit the data everywhere.

27. a) Millions of dollars per minute of run time.

b) Costs for dramas increase at approximately the same rate per minute of run time as do costs for other kinds of movies.

c) On average dramas cost about $20 million less for the same run time.

29. a) There are several features to comment on in this plot. There is a strong monthly pattern around the general trend. From 1997 to 2008, passengers increased fairly steadily with a notable exception of Sept. 2001, probably due to the attack on the twin towers. Then sometime in late 2008; departures dropped dramatically, possibly due to the economic crisis. Recently, they have been recovering, but not at the same rate as their previous increase.

b) The trend was fairly linear until late 2008; then passengers dropped suddenly.

c) The trend since 2009 has been linear (overall, ignoring monthly oscillations). If the increase continues to be linear, the predictions should be reasonable for the short term.

31. a) 1) High leverage, small residual.

2) No, not influential.

3) Correlation would become weaker because the point exhibits large values on both axes, thus increasing correlation.

4) The slope would remain more or less the same since the point is in line with the other points.

b) 1) Small leverage, large residual.

2) No, not influential.

3) Correlation would become slightly stronger since scatter would decrease.

4) The slope would change very little.

c) 1) Small leverage, large residual.

2) No, not influential.

3) Correlation would become slightly stronger since scatter would decrease.

4) The slope would change very little.

d) 1) High leverage, large residual.

2) Yes, influential.

3) Correlation would become stronger since scatter would decrease.

4) The slope would increase because the point is significantly above and to the left of the other points.

33. 1) a 2) e 3) b 4) c 5) d

35. Perhaps high blood pressure causes high body fat, high body fat causes high blood pressure, or both could be caused by a lurking variable such as a genetic or lifestyle issue.

37. a) The graph shows that, on average, students progress at about one reading level per year. This graph shows averages for each grade. The linear trend has been enhanced by using averages.
b) Very close to 1.
c) The individual data points would show much more scatter, and the correlation would be lower.
d) A slope of 1 would indicate that for each 1-year grade level increase, the average reading level is increasing by 1 year.

39. a) Model predicts a \$2.16 decrease in heating cost with every unit increase in Fahrenheit temperature.
b) Model predicts a heating cost of \$101 when the temperature is 0° Fahrenheit.
c) The predictions are generally accurate, because the residual is close to 0.
d) \$79.4.
e) About \$64.
f) No, because the residuals plot shows a curved pattern. The association is probably not linear.
g) No, the relationship does not depend on the units, and, hence, there would be no difference.

41. a) 0.8729.
b) On average, interest rates during this period increased at about 0.26% per year, starting from an interest rate of about 0.43% in 1950.
c) Substituting 50 in the model yields a predicted rate of 13.51%.
d) Not really. Extrapolating 20 years beyond the final year of the data would be dangerous and unlikely to be accurate.

43. a) Interest rates peaked around 1980 and decreased afterward. This regression model has a negative slope and a high intercept.
b) This model predicts −2.16%! Much lower than the prediction with the other model.
c) Even though we separated the data, there is no way of knowing if this trend will continue. And the rate cannot become negative, so we have clearly extrapolated far beyond what the data can support.
d) It is clear from the scatterplot that we can't count on TBill rates to change in a linear way over many years, so it would not be wise to use any regression model to predict rates.

45. a) Stronger. Both slope and correlation would increase.
b) Yes, restricting the study to nonhuman animals would justify it.
c) Very strong.
d) On average, for every one-year increase in life expectancy, the gestation period increases by about 12.97 days.
e) About 503.28 days.

47. a) Removing hippos would make the association stronger, since hippos are more of a departure from the pattern.
b) Increase.
c) No, there must be a good reason for removing data points.
d) Yes, removing it lowered the slope from 15.5 to 11.6 days per year.

49. a) Answers may vary. Using the data for 1955–2015 results in a scatterplot that is relatively linear with some curvature. You might use the data after 1955 only to predict 2025, but that would still call for extrapolation and would not be safe. The prediction is 26.13.
b) Not much, since the data are not truly linear and 2025 is 10 years from the last data point (extrapolating is risky).
c) No, that extrapolation of more than 50 years would be absurd. There's no reason to believe the trend from 1955 to 2015 will continue.
d) 28.97 years old
e) Not very much. We would have to assume that the current trend would continue until 2025. That's a strong assumption, and probably not true.
f) No. That would simply be an unreasonable extrapolation.

51. a) The residual plot is reasonably scattered with no evidence of nonlinearity, so it is appropriate to fit a regression model. However, care should be taken due to the outlier present in the data.
b) Yes, an outlier which has an unusually high life expectancy occurs at an approximate fertility rate of 7.75 births per woman.
c) 72.7% of the variation in *Life Expectancy* is explained by variability in *Fertility*.
d) No. Although there seems to be an association, there is no reason to expect causality. Lurking variables are likely to affect both *Fertility* and *Life Expectancy*.

53. a)

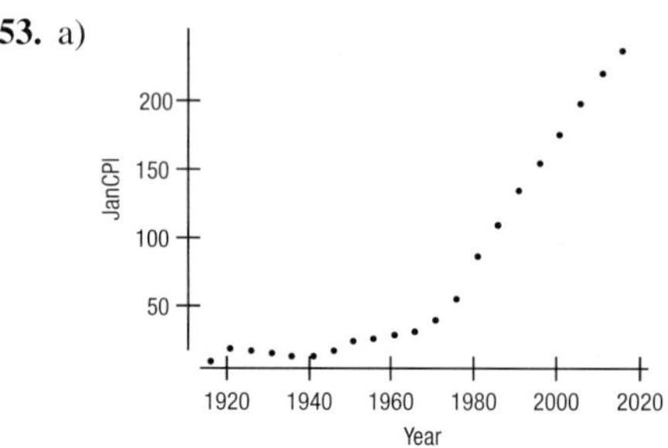

The scatterplot is clearly nonlinear; however, the last few years—say, from 1970 on—do appear to be linear.
b) Using the data from 1970 to 2011 gives $r = 0.999$ and
$\widehat{CPI} = -8797.47 + 4.4839\ Year$
Predicted CPI in 2025 = 282 (an extrapolation of doubtful accuracy).

55. a) There's an annual pattern in when people fly, so the residuals cycle up and down. There was also a sudden decrease in passenger traffic after 2008.
b) No, this kind of pattern can't be helped by re-expression.

57. a) Strong negative linear association.
b) Gas mileage decreases by an average of 8.20243 mpg for each 1000 lb of weight.
c) No, residuals show uneven, increasing spread and a fluctuating pattern.

59. a) Residuals are more randomly spread around 0, with some low outliers.
b) $\widehat{Fuel\ Consumption} = 0.625 + 1.178 \times Weight$.
c) For each additional 1000 pounds of *Weight*, an additional 1.178 gallons will be needed to drive 100 miles.
d) 21.06 miles per gallon.

61. a) No, although nearly all of the variation is accounted for by the model, it is still necessary to examine a residual plot to see if the regression is appropriate.
b) No, since the residuals show decreasing variation.

63. No. The residual pattern shows that this model is still not appropriate.

65. a) The scatterplot shows a strong, positive, linear association between speed and stopping distance.

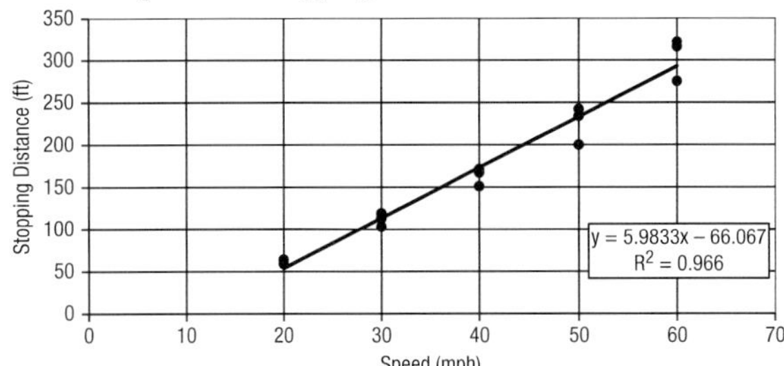

However, the residual plot shows a curved pattern, which indicates that the linear model is not appropriate.

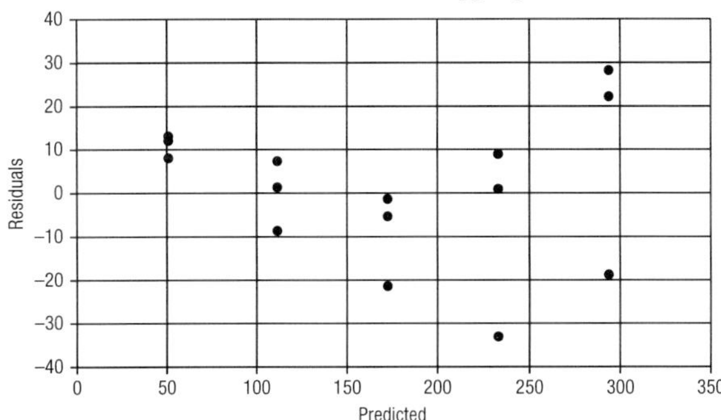

b) The square root transformation of the response variable is appropriate to straighten the scatterplot.

c) $\widehat{\sqrt{Distance}} = 3.3024 + 0.2356\,Speed$

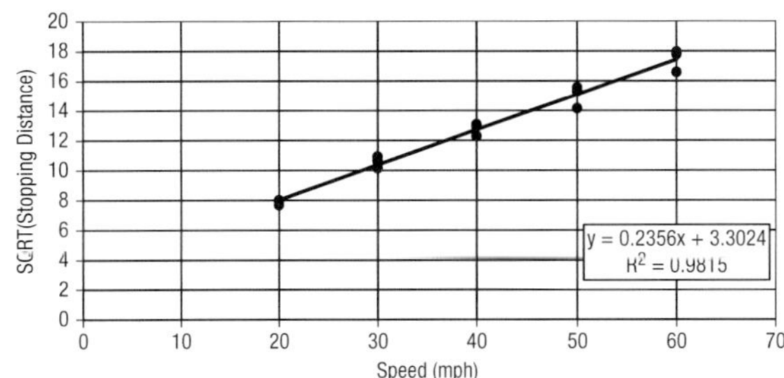

d) $13.9044^2 \approx 193.33$ ft

e) $19.7944^2 \approx 391.82$ ft

f) Quite high confidence, since R^2 is very high, and the residual plot is scattered. The prediction for a 55-mph speed should be accurate while that for a 70-mph speed is an extrapolation and is probably less reliable.

67. a) No, salaries didn't change much until the 1980s but then rose rapidly. The trend is not linear and the plot is not straight enough for a regression model.

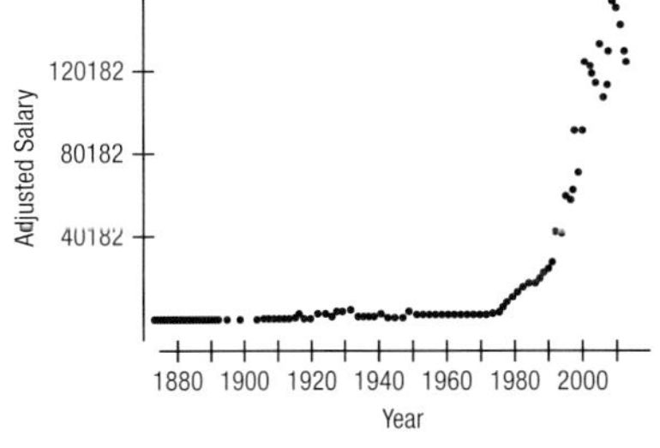

b) The logarithm works well:

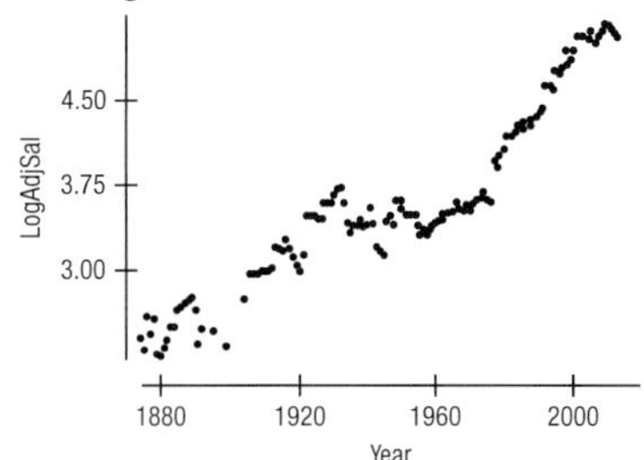

c) Response variable is: LogAdjSal
R squared = 87.7% s = 0.2788

Variable	Coefficient
Intercept	−32.3897
Year	0.018492

d) Salaries have been increasing on a log scale by about .018 logmillion \$/year. Transforming back to dollars, that's a bit more than a million dollars/year.

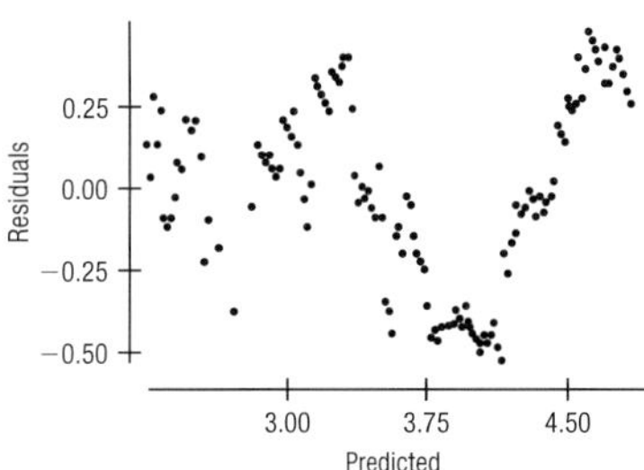

e) The residuals show a fluctuating pattern, which suggests that the linear model may not be the best description of the data.

69. a) $\widehat{\log(Distance)}$ against position works pretty well.

$\widehat{\log(Distance)} = 1.245 + 0.271 \times Position\ number.$

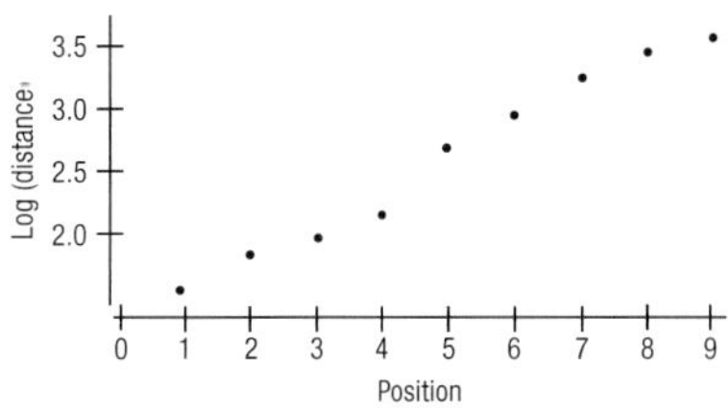

b) Pluto's residual is not especially large in the log scale. However, a model without Pluto predicts the 9th planet should be 5741 million miles. Pluto, at "only" 3684 million miles, doesn't fit very well, giving support to the argument that Pluto doesn't behave like a planet.

71. The predicted log (*Distance*) of Eris is 3.86, corresponding to a distance of 7379 million miles. That's more than the actual average distance of 6300 million miles.

73. a)

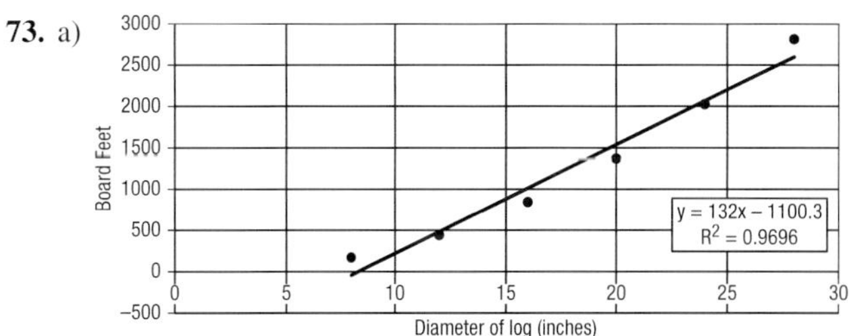

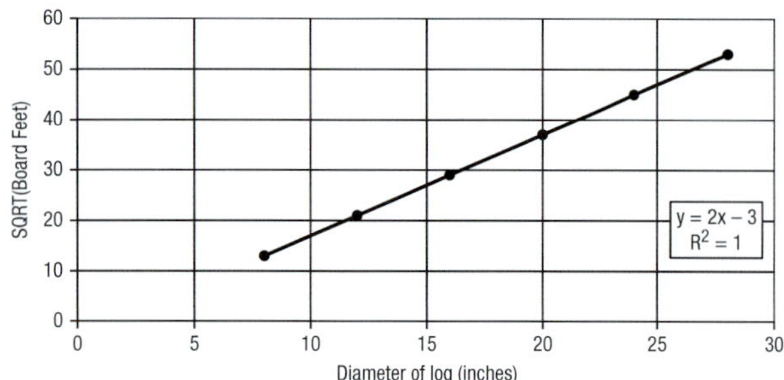

$\widehat{\sqrt{Board\ Feet}} = -3 + 2\ Diameter$
The model is exact.

b) 2,209 board feet of lumber.
c) 3,249 board feet of lumber.

75. $\widehat{log_{10}(Life\ Expectancy)} = 1.6469 + 0.214\ log_{10}(Decade)$.

77. The association is curved and cannot be straightened by use of a re-expression. A linear model does not fit the data well.

79. a) $\widehat{Life\ Expectancy} = 62.873 - 0.6619\ Age$.
No, the residuals plot shows a curved pattern.
b) $\widehat{\sqrt{Life\ Expectancy}} = 8.6013 - 0.0704\ Age$.
Predicted life expectancy is 50.74 years.
c) Yes, the residuals plot looks reasonably scattered and has very small residuals.

Chapter 9

1. a) $88,675.71.
b) $95,324.29.
c) The house sold for about $95.300 more than the model's estimate.

3. a) $\widehat{Gross} = -45.8877 + 0.9895\ Budget + 0.3855\ Run\ Time + 0.6308\ Critics\ Score$
b) After allowing for the effects of *Run Time* and *Critics Score*, each million dollars spent making a film yields about 0.9895 million dollars in gross revenue.

5. a) Linearity: The plot is reasonably linear with no bends.
b) Equal spread: The plot fails the Equal Spread Condition. It is much more spread out to the right than on the left.
c) Normality: A scatterplot of two of the variables doesn't tell us anything about the distribution of the residuals.

7. a) 0.9723
b) *Avatar* pulls the line toward it. Without that film, the slope would be smaller.

9. a) Use an indicator. Code Male = 0, Female = 1 (or the reverse).
b) Treat it as a quantitative predictor.
c) Use an indicator. Code Older than 65 = 1, Younger than 65 = 0.

11. a) Doesn't mention the other predictors.
b) This is correct.
c) Can't predict from response to predictor.
d) R^2 is about the fraction of variability accounted for by the regression model, not the fraction of data values.

13. a) $\widehat{Final} = -6.74 + 0.2570\ Test1 + 0.3917\ Test2 + 0.9011\ Test3$
b) 77.7% of the variation in final exam scores is accounted for by the regression model.
c) After allowing for the linear effects of the other predictors, each point on *Test3* is associated with an average increase of 0.9011 points on the final score.
d) Most probably, test scores are collinear. If the R^2 is almost as good, then *Test1* may not add much to the regression; however, it is probably associated with the final exam score.

15. a) *Won* and *Runs* are probably correlated. Including *Won* in the model is then very likely to change the coefficient of *Runs*, which now must be interpreted after allowing for the effects of *Won*.
b) The Indians' actual *Attendance* was less than we would predict for the number of *Runs* they scored.

17. a) $\widehat{Price} = -150765 + 9306\ Bathrooms + 141.74\ Living\ Area$
b) *Price* increases, on average, by about $141.74 for each additional square foot of living area, for homes with the same number of bathrooms.
c) The number of bathrooms is probably correlated with the living area of the house, even after taking into account the area of the bathroom itself. This correlation between the predictors may affect the coefficient of *Bathrooms*. In addition, the regression model is not capable of predicting what will happen if, for example, existing space is converted into a new bathroom.

19. The plot of residuals vs. predicted values looks bent, rising in the middle and falling at both ends. This violates the Straight Enough Condition. The Normal probability plot and the histogram of the residuals suggest that the highest five residuals (which we know are in the middle of the predicted value range) are extraordinarily high. These data may benefit from a re-expression.

21. a) $\widehat{Salary} = 9.876 + 0.111\ Service + 0.068\ Education + 0.098\ Score + 0.017\ Speed + 0.065\ Dictation$.
b) $34,653.
c) After removing the effects of months of service and years of education from *Age*, what is left is years not spent in education or service. These non-productive years may have a negative effect on *Salary*.

23. a) Each pound of weight is associated, on average, with an increase of 0.189 in *%Body Fat*.
b) Among men with the same *Waist* and *Height* measurements, being a pound heavier is associated with having a *%Body Fat* that is 0.1 lower.
c) I'd like to see the partial regression plot for *Weight*. It would have an *x*-axis that is *Weight* with the effects of *Waist* and *Height* removed, so I could understand that better.

25. a) $\widehat{Calories} = 83.0469 + 0.057211\ Sodium - 0.019328\ Potassium + 2.38757\ Sugars$
b) A scatterplot of residuals vs. predicted values and a Normal probability plot of the residuals.
c) No. A regression model doesn't predict what would happen if we change a value of a predictor. It only models the data as they are.

27. a) For children of the same *Age*, their *Dominant Hand* was faster on this task by about 0.304 seconds on average.
b) Yes. The relationship between *Speed* and *Age* is straight and the lines for dominant and non-dominant hands are very close to parallel.

29. a) The change in *Time* with *Distance* is different for men and women, so a different slope is needed in the model.
b) After accounting for *Distance*, the increase in race time with distance is greater for men than for women. Said another way, women's times increase less rapidly for longer races than do men's.

Part II Review

R2.1. % over 50, 0.69.
% under 20, −0.71.
% graduating on time, −0.51.
% full-time faculty, 0.09.

R2.3. a) There does not appear to be a linear relationship.
b) Nothing, there is no reason to believe that the results for the Finger Lakes region are representative of the vineyards of the world.
c) $\widehat{CasePrice} = 92.77 + 0.567 \times Years$.
d) Only 2.7% of the variation in case price is accounted for by the ages of vineyards. Most of that is due to two outliers. We are better off using the mean price rather than this model.

R2.5. a) $\widehat{TwinBirthsRate} = 17.77 + 0.551\ Years\ Since\ 1980$.
b) Twin births have increased, on average, by 0.55 births per year.
c) 36.5 births predicted; 33.22 observed in fact. Reasonably close for an extrapolation like that.
d) $\widehat{TwinBirths} = 18.21 + 0.508\ Years\ Since\ 1980$. The fit is very good. R^2 is 97%. However, the residual plot reveals a fluctuating pattern, so there may be more to understand about twin births than is available from these data.

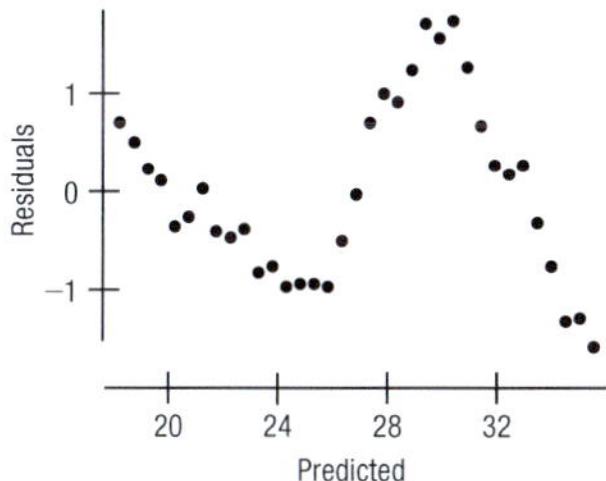

R2.7. a) −0.520
b) Negative, not strong, somewhat linear, but with more variation as *pH* increases.
c) The *BCI* would also be average.
d) The predicted *BCI* will be 1.56 SDs of *BCI* below the mean *BCI*.

R2.9. a) $\widehat{BCI} = 2733.37 - 197.694\ pH$.
b) *BCI* decreases by about 197.69 points per point of *pH*. BCI
c) $BCI = 1112.3$

R2.11. a) $\widehat{BCI} = 2342.95 - 137.833\ pH - 0.3372\ Hard$
b) Only slightly. R^2 and s improved only a small amount.
c) $BCI = 1144$
d) 165. The residual.
e) After allowing for the linear effect of the *Hardness* of the water, *BCI* decreases by about 137.8 per unit of *pH*.

R2.13. a) −0.984 b) 96.9%
c) 32.95 mph d) 1.66 mph
e) Slope will increase.
f) Correlation will weaken (become less negative).
g) Correlation is the same, regardless of units.

R2.15. a) Weight (but unable to verify linearity).
b) Heavier cars get, on average, lower gas mileage.
c) *Weight* accounts for 81.5% of the variation in *Fuel Efficiency*.

R2.17. a) $\widehat{Horsepower} = 3.50 + 34.314 \times Weight$.
b) Thousands. For the equation to have predicted values between 60 and 160, the *X*-values would have to be in thousands of pounds.
c) Yes. The residual plot does not show any pattern.
d) 115.0 horsepower.

R2.19. a) No. There is a curve in the plot.
b) Yes. The new residual plot now appears to satisfy the conditions for inference. There is one outlier to consider.

R2.21. a) The scatterplot shows a fairly strong linear relation in a positive direction. There seem to be two distinct clusters of data.
b) $\widehat{Interval} = 33.967 + 10.358 \times Duration$.
c) The time between eruptions increases by about 10.4 minutes per minute of *Duration* on average.
d) Since 77% of the variation in *Interval* is accounted for by *Duration* and the error standard deviation is 6.16 minutes, the prediction will be relatively accurate.
e) 75.4 minutes.
f) A residual is the observed value minus the predicted value. So the residual = 79 − 75.4 = 3.6 minutes, indicating that the model underestimated the interval in this case.

R2.23. a) $r = 0.888$. Although r is high, you must look at the scatterplot and verify that the relation is linear in form.
b)

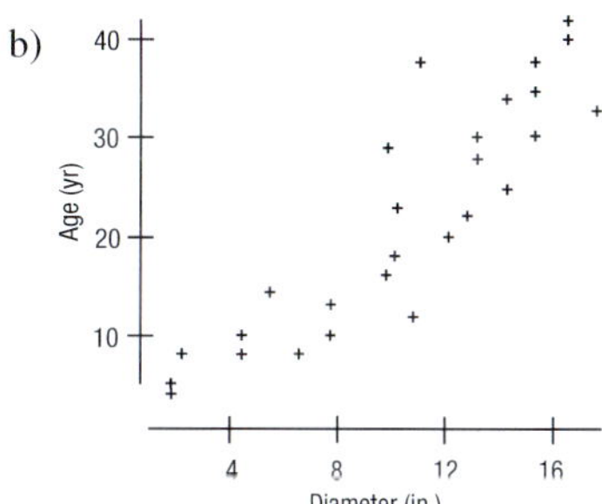

The association between diameter and age appears to be strong, somewhat linear, and positive.
c) $\widehat{Age} = -0.97 + 2.21 \times Diameter$.
d)

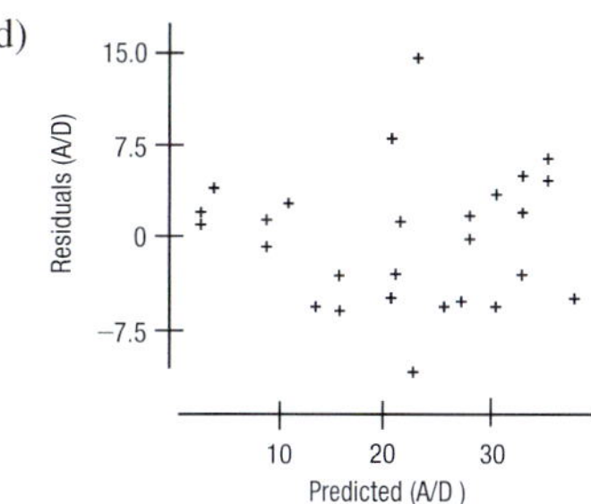

The residuals show a curved pattern (and two outliers).
e) The residuals for the larger predicted values tend to be positive, so it seems that larger trees may have their ages underestimated.

R2.25. TVs are 19 to 70 inches. A TV with a screen that was 10 inches larger would be predicted to cost $10(0.03) = \$30$, $10(0.3) = \$300$, $10(3) = \$3000$, or $10(30) = \$30{,}000$. Only \$300 is reasonable, so the slope must be 0.3.

R2.27. a) The model predicts *% Smoking* from *Year*, not the other way around.
b) $\widehat{Year} = 2027.17 - 1.9823 \times \%\ Smoking$.
c) The model predicts 2027, but an extrapolation to $x = 0$ is probably too far from the given data. The prediction is not very reliable in spite of the strong correlation.

R2.29. The relation shows a negative direction, with a somewhat linear form, but perhaps with some slight curvature. There are several model outliers.

R2.31. a) 71.9%
b) At higher latitudes the average January temperature is lower.
c) $\widehat{January\ Temperature} = 108.80 - 2.111 \times Latitude$.
d) Average January temperature is lower by 2.11 °F per degree of latitude.
e) The intercept predicts an average January temperature of 108.8° at the equator (where latitude $= 0$). This is an extrapolation and probably not trustworthy.
f) 24.4 degrees.
g) The equation underestimates the average January temperature.

R2.33. a) The scatterplot shows a strong, linear, positive association.
b) There is an association, but it isn't causal. Athletes who perform well in the high jump are likely to also perform well in the long jump.
c) Neither; the change in units does not affect the correlation.
d) I would predict the winning long jump to be 0.910 SDs above the mean long jump.

R2.35. a) No relation; the correlation would probably be close to 0.
b) The relation would have a positive direction and the correlation would be strong, assuming that students were studying French in each grade level. Otherwise, no correlation.
c) No relation; correlation close to 0.
d) The relation would have a positive direction and the correlation would be strong, since vocabulary would increase with each grade level.

R2.37. $\widehat{Calories} = 560.7 - 3.08 \times Time$.
Toddlers who spend an additional minute at the table consume, on average, 3.08 fewer calories.

R2.39. There seems to be a strong, positive, linear relationship with one high-leverage point (Northern Ireland) that makes the overall R^2 quite low. Without that point, the R^2 increases to 61.5%. Of course, these data are averaged across thousands of households, so the correlation appears to be higher than it would be for individuals. Any conclusions about individuals would be suspect.

R2.41. a) 3.842 b) 501.187 c) 4.0

R2.43. a) 30,818 pounds. b) 1302 pounds.
c) 31,187.6 pounds.
d) Negative residuals will be more of a problem, as the predicted weight would overestimate the weight of the truck; trucking companies might be inclined to take the ticket to court.

R2.45. The original data are nonlinear, with a significant curvature. Using reciprocal square root of diameter gave a scatterplot that is nearly linear:

$$\widehat{1/\sqrt{Drain\ Time}} = 0.0024 + 0.219\ Diameter.$$

R2.47. a) *LogAssets* increase on average by 0.868 Log\$ per Log\$ of sales for both banks and other companies.
b) Bank assets are, on average, 0.946 Log\$ higher than other companies, after allowing for the linear effect of *LogSales*.
c) Yes. The lines are roughly parallel.

R2.49. a) After accounting for the linear effect of *Living Area*, houses with another *Bathroom,* on average, have a *Price* that is \$75,020 higher.
b) 75,020.3

R2.51. a) After accounting for the linear effect of *Depth*, the *log(HeartRate)* decreases by 0.045 bpm per minute *Duration* of the dive.
b) Yes. There is a strong, negative, linear relationship. The longer the *Duration*, the slower the *HeartRate*.

Chapter 10

1. The university administrators should take a survey amongst a sample of the student body, selecting respondents with a randomization method. They should be sure to draw a sufficiently large sample.

3. The proportion in the sample is a statistic. The proportion of all students is the parameter of interest. The statistic estimates that parameter, but is not likely to be exactly the same.

5. This is not a simple random sample. Each student may have an equal chance to be in the survey; however, the selection is not independent since groups of friends who sit together will either all be in or out of the sample.

7. a) Cluster sample. b) Stratified sample.
c) Systematic sample.

9. Several terms are poorly defined. The survey needs to specify the meaning of "family" for this purpose and the meaning of "higher education." The term "seek" may also be poorly defined (for example, would applying to college but not being admitted qualify for seeking more education?).

11. a) This would suffer from voluntary response bias.
b) This would be a convenience sample.

13. a) No. It would be nearly impossible to get exactly 500 males and 500 females from every country by random chance.
b) A stratified sample, stratified by whether the respondent is male or female.

15. a) A cluster sample, with teams being the clusters.
b) Yes, an entire team can be sampled at once relatively easily, but a random group of all players cannot efficiently be sampled on the same day.

17. a) **Population:** Probably all adults in the country.
b) **Population Parameter of Interest:** Proportion who have used and benefited from alternative medicine.
c) **Sampling Frame:** All Consumer Union subscribers.
d) **Sample:** All subscribers who responded.
e) **Sampling Method:** Voluntary response sampling (no randomization)

f) **Left Out of Study:** Those who are not Consumer Union subscribers.
g) **Sources of Bias:** Voluntary response bias and undercoverage.

19. a) **Population:** All voters in the state.
b) **Population Parameter of Interest:** Most important issues to voters.
c) **Sampling Frame:** All state residents.
d) **Sample:** Stratified sample, with each district being represented by the four blocks.
e) **Sampling Method:** Multistage sample, using stratified sample with districts as strata and cluster sample within each district (no randomization is specified)
f) **Left Out of Study:** People not home during the survey.
g) **Sources of Bias:** Undercoverage and sampling within clusters is non-random.

21. a) **Population:** All cars.
b) **Population Parameter of Interest:** Proportion of cars with up-to-date registration, insurance, and safety inspection stickers.
c) **Sampling Frame:** Cars on the road where roadblock is set up.
d) **Sample:** Cars stopped by the roadblock.
e) **Sampling Method:** Cluster sampling, census with cluster (no randomization specified)
f) **Left Out of Study:** Cars not passing through that road during the time of the roadblock.
g) **Sources of Bias:** Undercoverage, and the location of the roadblock is probably non-random.

23. a) **Population:** All produce farms that produce oranges.
b) **Population Parameter of Interest:** Proportion passing inspection.
c) **Sampling Frame:** All oranges produced at all produce farms that day.
d) **Sample:** Unclear, but probably a random sample of farms and then random oranges within each farm.
e) **Sampling Method:** Multistage sampling (no randomization specified)
f) **Left Out of Study:** Nothing.
g) **Sources of Bias:** There is no clear indication of bias, as long as the day of inspection is chosen randomly. Potential bias from convenience sampling.

25. Sampling error. The description of the sampling method suggests that samples should be representative of the voting population. Nonetheless, random chance in selecting the individuals who were polled means that sample statistics will vary from the population parameter, perhaps by quite a bit.

27. a) Voluntary response. Only those who see the show *and* feel strongly will respond.
b) Possibly more representative than part a, but only strongly motivated parents go to PTA meetings (voluntary response).
c) Multistage sampling, with cluster sample within each school. Probably a good design if most of the parents respond.
d) Systematic sampling. Probably a reasonable design.

29. They will get responses only from people who come to the park to use the playground. Parents who are dissatisfied with the playground may not come.

31. The first sentence points out problems the respondents may not have noticed and might lead them to feel they should agree. However, the last phrase mentions higher fees, which could make people reject improvements to the playground.

33. a) This is a voluntary response survey. Even a large sample size can't make it representative.
b) The wording seems fair enough. It states the facts and gives voice to both sides of the issue.
c) The sampling frame is, at best, those who visit this particular site and even then depends on their volunteering to respond to the question.
d) This is a true statement.

35. a) Seems neutral.
b) Biased toward yes because of "great tradition." Better to ask, "Do you favor continued funding for the space program?"

37. Cell phones may be more likely among younger citizens, which could introduce a bias. As cell phones grow in use, this problem will be lessened.

39. a) **Sample:** Mean of the gas mileage for the last six fill-ups of my car.
b) **Population:** Mean of the gas mileage for my car.
c) Recent driving conditions may not be typical of the overall driving conditions.
d) **Population:** Mean of the gas mileage for all cars of that make, model, and year.

41. a) Undercoverage for project managers and response bias if a member of management asks its employees directly about employee discontent.
b) Assign a random number from 001 to 772 to each employee and use a list of random numbers or software to choose the employees.
c) Responses for each group may be different, so each group should be examined individually. There are relatively few project managers, and we want to make sure their opinions are noted.
d) Stratified sampling by job type and randomly choose a proportionate number of people from each group.
e) Answers will vary. Assign each person a number from 01 to 14 and use a list of random numbers or software to do the selection.

43. If discolored fish are not equally likely to be caught as normal fish, or fisherman are more disposed to bring discolored fish than normal fish, then the sample will be biased and resulting estimation will be biased.

45. a) If the sampling is performed from regular (line) listings, this is fair as long as all doctors are listed. If from ads, the sampling is unfair and not representative.
b) Not appropriate. The cluster sample will probably contain listings for only one or two types of businesses.

Chapter 11

1. This is an observational study because the sports writer is not randomly assigning players to take steroids or not take steroids; the writer is merely observing home run totals between two eras. It would be unwise to conclude steroids caused any increases in home runs because we need to consider other factors besides steroids—factors possibly leading to more home runs include better equipment, players training more in the off-season, smaller ballparks, better scouting techniques, etc.

3. Each of the 40 deliveries is an experimental unit. He has randomized the experiment by flipping a coin to decide whether or not to phone.

5. The factor is calling, and the levels are whether or not he calls the customer. The response variable is the tip percentage for each delivery.

7. By calling some customers but not others during the same run, the driver has controlled many variables, such as day of the week, season, and weather. The experiment was randomized because he flipped a coin to determine whether or not to phone and it was replicated because he did this for 40 deliveries.

9. Because customers don't know about the experiment, those that are called don't know that others are not, and vice versa. Thus, the customers are blind. That would make this a single-blind study. It can't be double-blind because the delivery driver must know whether or not be phones.

11. Yes. Driver is now a block. The experiment is randomized within each block. This is a good idea because some drivers might generally get higher tips than others, but the goal of the experiment is to study the effect of phone calls.

13. Answers will vary. The cost or size of a delivery may confound his results. Larger orders may generally tip a higher or lower percentage of the bill.

15. a) No. There are no manipulated factors. Observational study.
b) There may be lurking variables that are associated with both parental income and performance on the SAT.

17. a) Retrospective and observational study.
b) Yes, because the disease in the study is a rare disease.
c) Exchange students.
d) The level of vitamin D for the subject, and whether the subject developed the disease.

19. a) A placebo-controlled, randomized, comparative experiment.
b) Yes, because the goal was to determine the effects of black cohosh on hot flashes.
c) 393 women aged 45 to 55 who reported at least two hot flashes a day.
d) The treatment variables were a placebo, estrogen, a multi-herb supplement plus advice to consume more soy foods, black cohosh, and a multiherb supplement with black cohosh. The response variables were the women's symptoms.

21. a) Experiment.
b) 72 people with bipolar disorder.
c) 1 factor (omega-3 fats from fish oil) at 2 levels.
d) 2 treatments.
e) Improvement.
f) Unspecified design.
g) Single-blind (through the placebo), unknown if double-blind.
h) High dosage of omega-3 fats from fish oil causes improvements in patients with bipolar disorder.

23. a) Observational study.
b) Prospective.
c) Roughly 560 widows with moderately high and normal blood pressure, unknown selection process.
d) Memory and reaction time.
e) As there was no random assignment, there is no way to know that high blood pressure caused subjects to do worse on memory and reaction time tests. A cause-and-effect relationship cannot be established since a lurking variable, such as age, might have been the cause.

25. a) Experiment.
b) 25 women over the age of 57.
c) 1 factor (type of drink) at 2 levels (alcoholic and nonalcoholic). Blocking variable—supplemental estrogen.
d) 2 treatments.
e) Decrease in estrogen level.
f) Randomized block design.
g) No blinding was used.
h) This experiment indicates that drinking alcohol leads to decreased estrogen levels among those taking estrogen supplements.

27. a) Observational study.
b) Retrospective.
c) Women in Finland, unknown selection process with data coming from church records.
d) Difference in average lifespan between mothers of sons and mothers of daughters.
e) Without random assignment, there is no way to know that having sons or daughters causes a difference in the lifespan of mothers.

29. a) Observational study.
b) Prospective.
c) Women and men with and without depression, unknown selection process.
d) Frequency of crying in response to sad situations.
e) There is no apparent difference in crying response to sad movies for the depressed and nondepressed groups.

31. a) Experiment.
b) People experiencing sharp neck pain.
c) 2 factors (pain reliever and water temperature) at 2 levels each.
d) 4 treatments.
e) Level of pain relief.
f) Completely randomized over 2 factors.
g) This experiment was single-blind to the pain reliever factor. The subjects were not blinded to the water temperature factor as they knew whether they were drinking ice-cold or regular water.
h) The experiment may indicate whether pain reliever alone or in combination with ice-cold water relieves sharp neck pain.

33. a) Experiment.
b) Athletes who had suffered hamstring injuries.
c) 1 factor (type of exercise) at 2 levels.
d) 2 treatments.
e) Time before returning to sports.
f) Completely randomized.
g) No blinding was used in the experiment as the athletes were aware of which exercise program they were following.
h) Experiment can help determine which of the two exercise programs is more effective at rehabilitating hamstring injuries.

35. They need to compare omega-3 results to something. Perhaps bipolarity is seasonal and would have improved during the experiment anyway.

37. a) Subjects' responses might be related to many other factors (diet, exercise, genetics, etc). Randomization should equalize the two groups with respect to unknown factors.

b) More subjects would minimize the impact of individual variability in the responses, but the experiment would become more costly and time consuming.

39. People who engage in regular exercise might differ from others with respect to bipolar disorder, and that additional variability could obscure the effectiveness of this treatment.

41. Generate random values of 0 or 1 from an Internet site. Assign participants with a 1 to one program and those with a 0 to the other. Participants must follow the program to which they are assigned; they can't choose to switch programs.

43. a) First, they are using athletes who have a vested interest in the success of the shoe by virtue of their sponsorship. They should choose other athletes. Second, they should randomize the order of the runs, not run all the races with their shoes second. They should blind the athletes by disguising the shoes if possible, so athletes don't know which is which. The timers shouldn't know which athletes are running with which shoes, either. Finally, they should replicate several times, since times will vary under both shoe conditions.

b) Because of the problems in part a, the results they obtain may favor their shoes. In addition, the results obtained for Olympic athletes may not be the same as for the general runner.

45. a) Allowing athletes to self-select treatments could confound the results. Other issues such as severity of injury, diet, age, etc., could also affect time to heal; randomization should equalize the treatment groups with respect to any such variables.

b) A control group could have revealed whether either exercise program was better (or worse) than just letting the injury heal.

c) Doctors who evaluated the athletes to approve their return to sports should not know which treatment the subject had.

d) It's hard to tell. The difference of 15 days seems large, but the standard deviations indicate that there was a great deal of variability in the times.

47. a) The differences among the Mozart and quiet groups were more than would have been expected from sampling variation.

b)

c) The Mozart group seems to have the smallest median difference and thus the *least* improvement, but there does not appear to be a significant difference.

d) No, if anything, there is less improvement, but the difference does not seem significant compared with the usual variation.

49. a) Prospective, observational study.

b) The relation between health and wine consumption might be explained by the lurking variables of socioeconomic status and education.

c) None of these. While the variables have a relation, there is no indication of a cause-and-effect relationship.

51. a) Arrange the 28 containers in 28 separate locations. Number the container from 01 to 28 and use a random number generator to identify the 14 containers that should be filled with water.

b) The dowser is expected to be correct about 50% of the time just by guessing. A record of approximately 57% (16 out of 28) does not appear to be statistically significantly different.

c) Answers may vary. At least 90% success would be convincing; that is, more than 25 correct identifications.

53. Randomly assign half the reading teachers in the district to use each method. Students should be randomly assigned to teachers as well. Make sure to block both by school and grade (or control grade by using only one grade). Construct an appropriate reading test to be used at the end of the year, and compare scores.

55. a) The difference is greater than would be expected due to natural sampling variation.

b) Prospective, observational study.

c) No, this study cannot prove anything about causation. There are likely to be confounding variables that account for the higher death rate on the weekends.

d) Alcohol use, which may lead to traffic accidents and violence, might have something to do with the higher death rate on the weekends.

57. Answers may vary. This experiment has 1 factor (pesticide), at 3 levels (pesticide A, pesticide B, no pesticide), resulting in 3 treatments. The response variable is the number of beetle larvae found on each plant. Randomly select a third of the plots to be sprayed with pesticide A, a third with pesticide B, and a third with no pesticide (since the researcher also wants to know whether the pesticides even work at all). To control the experiment, the plots of land should be as similar as possible with regard to amount of sunlight, water, proximity to other plants, etc. If not, plots with similar characteristics should be blocked together. If possible, use some inert substance as a placebo pesticide on the control group, and do not tell the counters of the beetle larvae which plants have been treated with pesticides. After a given period of time, count the number of beetle larvae on each plant and compare the results.

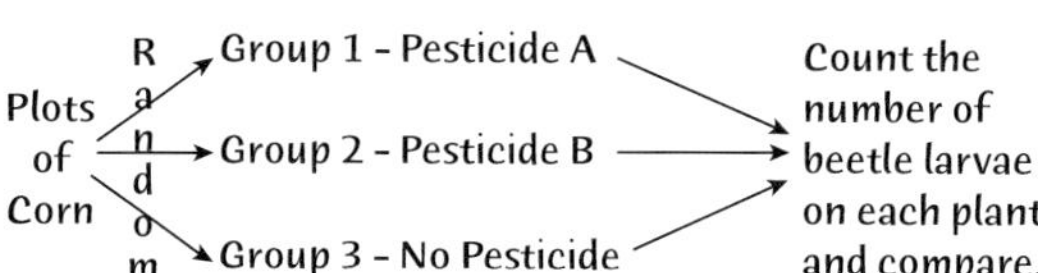

59. Answers may vary. Find a group of volunteers. Each volunteer will be required to shut off the machine with his or her left hand and right hand. Randomly assign the left or right hand to be used first. Complete the first attempt for the whole group. Now repeat the experiment with the alternate hand. Check the differences in time for the left and right hands.

61. a) Jumping with or without a parachute.

b) Volunteer skydivers (the dimwitted ones).

c) A parachute that looks real but doesn't work.

d) A good parachute and a placebo parachute.

e) Whether parachutist survives the jump (or extent of injuries).

f) All should jump from the same altitude in similar weather conditions and land on similar surfaces.

g) Randomly assign people the parachutes.

h) The skydivers (and the people involved in distributing the parachute packs) shouldn't know who got a working chute. And the people evaluating the subjects after the jumps should not be told who had a real parachute either!

Part III Review

R3.1. Observational prospective study. Indications of behavior differences can be seen in the two groups. May show a link between premature birth and behavior, but there may be lurking variables involved.

R3.3. Experiment, matched by gender and weight, randomization within blocks of two pups of same gender and weight. Factor: type of diet. Treatments: low-calorie diet and allowing the dog to eat all it wants. Response variable: length of life. Can conclude that, on average, dogs with a lower-calorie diet live longer.

R3.5. Completely randomized experiment, with the treatment being receiving folic acid or not (one factor, two levels). Treatments assigned randomly and the response variable is the number of/occurrence of additional precancerous growths. Neither blocking nor matching is mentioned, but in a study such as this one, it is likely that researchers and patients are blinded. Since treatments were randomized, it seems reasonable to generalize results to all people with precancerous polyps, though caution is warranted since these results contradict a previous study.

R3.7. Sampling. Probably a Simple Random Sample, although may be stratified by type of firework. Population is all fireworks produced each day. Parameter is proportion of duds. Can determine if the day's production is ready for sale.

R3.9. Observational retrospective study. Researcher can conclude that for anyone's lunch, even when packed with ice, food temperatures are rising to unsafe levels.

R3.11. Experiment, with a control group being the genetically engineered mice who received no antidepressant and the treatment group being the mice who received the drug. The response variable is the amount of plaque in their brains after one dose and after four months. There is no mention of blinding or matching. Conclusions can be drawn to the general population of mice and we should assume treatments were randomized. To conclude the same for humans would be risky, but researchers might propose an experiment on humans based on this study.

R3.13. Experiment. Factor is gene therapy. Hamsters were randomized to treatments. Treatments were gene therapy or not. Response variable is heart muscle condition. Can conclude that gene therapy is beneficial (at least in hamsters).

R3.15. Sampling. Population is all oranges on the truck. Parameter is proportion of unsuitable oranges. Procedure is probably stratified random sampling with regions inside the truck being the strata. Can conclude whether or not to accept the truckload.

R3.17. Observational retrospective study performed as a telephone-based randomized survey. Based on the excerpt, it seems reasonable to conclude that more education is associated with a higher Emotional Health Index score, but to insist on causality would be faulty reasoning.

R3.19. Answers will vary. The sample does show skewness to the right that is in the population. The sample suggests there may be a high outlier or two, but the population has a smooth tail and no outliers.

R3.21. a) 100 obese persons.
b) Restrictive diet, alternate fasting diet, control (no special diet).
c) Weight loss (reported as % of initial weight).
d) No, it could not be blinded because participants must know what they are eating.
e) No, it is not necessary that participants be a random sample from the population. All that is needed is that they be randomly assigned to treatment groups.

R3.23. a) Experiment. Actively manipulated candy giving, diners were randomly assigned treatments, control group was those with no candy, lots of dining parties.
b) It depends on when the decision was made. If early in the meal, the server may give better treatment to those who will receive candy—biasing the results.
c) A difference in response so large it cannot be attributed to natural sampling variability.

R3.25. There will be voluntary response bias, and results will mimic those only of the visitors to sodahead.com and not the general U.S. population. The question is leading responders to answer "yes" though many might understand that the president's timing for his vacation had nothing to do with the events of the week.

R3.27. a) Eyes (open or closed), Music (On or Off), Moving (sitting or moving).
b) It is a blocking variable.
c) Each of 4 subjects did 8 runs, so 32 in all.
d) Randomizing completely is better to reduce the possibility of confounding with factors that weren't controlled for.

R3.29. a) Yes.
b) No. Residences without phones are excluded. Residences with more than one phone number had a higher chance.
c) No. People who respond to the survey may be of age but not registered voters.
d) No. Households who answered the phone may be more likely to have someone at home when the phone call was generated. These may not be representative of all households.

R3.31. a) Does not prove it. There may be other confounding variables. Only way to prove this would be to do a controlled experiment.
b) Alzheimer's usually shows up late in life. Perhaps smokers have died of other causes before Alzheimer's is evident.
c) An experiment would be unethical. One could design a prospective study in which groups of smokers and nonsmokers are followed for many years and the incidence of Alzheimer's is tracked.

R3.33.

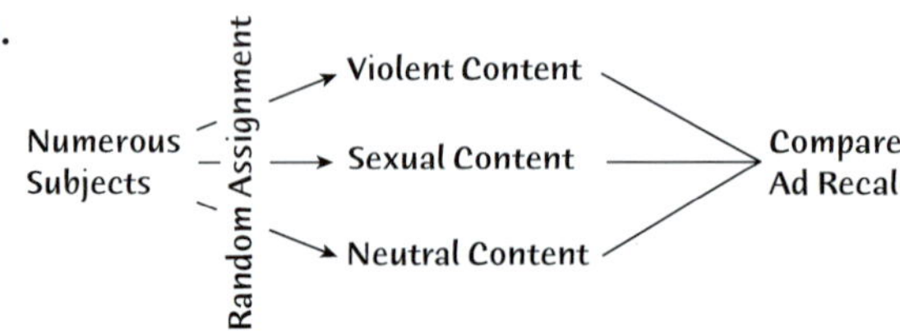

Numerous subjects will be randomly assigned to see shows with violent, sexual, or neutral content. They will see the same commercials. After the show, they will be interviewed for their recall of brand names in the commercials.

R3.35. a) May have been a Simple Random Sample, but given the relative equality in age group, it may have been stratified.
b) 38.2%

c) We don't know. If data were collected from voting precincts that are primarily Democratic or primarily Republican, that would bias the results. Because the survey was commissioned by NBC News, we can assume the data collected are probably OK.
d) Do party affiliations differ for different age groups?

R3.37. The factor in the experiment will be type of bird control. I will have three treatments: scarecrow, netting, and no control. I will randomly assign several different areas in the vineyard to one of the treatments, taking care that there is sufficient separation that the possible effect of the scarecrow will not be confounded. At the end of the season, the response variable will be the proportion of bird-damaged grapes.

R3.39. a) This was the control group. Subjects needed to have some sort of treatment to perceive they were getting help for their lower back pain.
b) Even though patients were volunteers, their treatments were randomized. To generalize the results, we would need to assume these volunteers have the same characteristics of the general population of those with lower back pain (probably reasonable).
c) It means that the success rates for the acupuncture groups were higher than could be attributed to chance when compared to the conventional treatment group. In other words, researchers concluded that both proper and "fake" acupuncture reduced back pain.

R3.41. a) Use stratified sampling to select 2 first-class passengers and 12 from coach.
b) Number passengers alphabetically, 01 = Bergman to 20 = Testut. Read in blocks of two, ignoring any numbers more than 20. This gives 65, 43, 67, 11 (selects Fontana), 27, 04 (selects Castillo).
c) Number passengers alphabetically from 001 to 120. Use the random-number table to find three-digit numbers in this range until 12 different values have been selected.

Chapter 12

1. In the long run, a fair coin will generate 50% heads and 50% tails, approximately. But for each flip we cannot predict the outcome.

3. There is no law of averages for the short run. The first five flips do not affect the sixth flip.

5. a) 0.30 b) 0.80

7. a) 0.2267 b) 0.7733 c) 0.9908

9. a) S = {HH, HT, TH, TT}, equally likely
b) S = {0, 1, 2, 3}, not equally likely
c) S = {H, TH, TTH, TTT}, not equally likely
d) S = {1, 2, 3, 4, 5, 6}, not equally likely

11. In this context, "truly random" should mean that every number is equally likely to occur and outcomes are mutually independent.

13. There is no "Law of Averages." She would be wrong to think they are "due" for a harsh winter.

15. There is no "Law of Averages." If the tennis player's games are independent, he could very well lose his next game.

17. a) There is some chance you would have to pay out much more than the $1500.
b) Many customers pay for insurance. The small risk for any one customer is spread among all.

19. a) Legitimate. b) Legitimate.
c) Not legitimate (sum more than 1). d) Legitimate.
e) Not legitimate (can't have negatives or values more than 1).

21. A family may own both a computer and an HD TV. The events are not disjoint, so the Addition Rule does not apply.

23. When cars are traveling close together, their speeds are not independent, so the Multiplication Rule does not apply.

25. a) He has multiplied the two probabilities.
b) He assumes that being accepted at the universities are independent events.
c) No. Universities use similar criteria for acceptance, so the decisions are not independent.

27. a) 0.73 b) 0.93 c) 0.07

29. a) 0.3890 b) 0.0197 c) 0.1794

31. a) Repair needs for two cars must be independent.
b) Maybe not. An owner may treat three cars similarly, taking good (or poor) care of all three. This may decrease (or increase) the likelihood that two or more cars may simultaneously need repairing.

33. a) 353/1028
b) 406/1028

35. a) 0.0243 b) 0.134
c) Responses are independent.
d) People were polled at random.

37. a) 0.5332 b) 0.9132
c) $(1 - 0.62) + 0.62(1 - 0.14)$

39. a) 1) 0.30 2) 0.30 3) 0.90 4) 0.0
b) 1) 0.027 2) 0.128 3) 0.512 4) 0.271

41. a) Disjoint (can't be both red and orange).
b) Independent (unless you're drawing from a small bag).
c) No. Once you know that one of a pair of disjoint events has occurred, the other is impossible.

43. a) 0.0046 b) 0.125 c) 0.296
d) 0.421 e) 0.995

45. a) 0.6628 b) 0.1247 c) 0.2707 d) 0.0005

47. a) 0.1664 b) 0.4219 c) 0.657

49. 0.266

51. a) For any day with a valid three-digit date, the chance is 0.001, or 1 in 1000. For many dates in October through December, the probability is 0. (No three digits will make 10/15, for example.)
b) There are 65 days when the chance to match is 0. (Oct. 10–31, Nov. 10–30, and Dec. 10–31.) The chance for no matches on the remaining 300 days is 0.741.
c) 0.259 d) 0.049

Chapter 13

1. 0.42 **3.** 0.675 **5.** 0.135

7. No $P(S) = 0.323$ but $P(S|FC) = 0.625$. These are not the same.

9. We have joint probabilities and marginal probabilities, not conditional probabilities, so a table is the better choice.

	Use Internet	Do Not Use Internet	Total
Asia	0.28	0.27	**0.55**
Rest of the World	0.29	0.16	**0.50**
Total	**0.57**	**0.43**	**1.00**

11. A tree is better because we have conditional and marginal probabilities. The joint probabilities are found at the end of the branches.

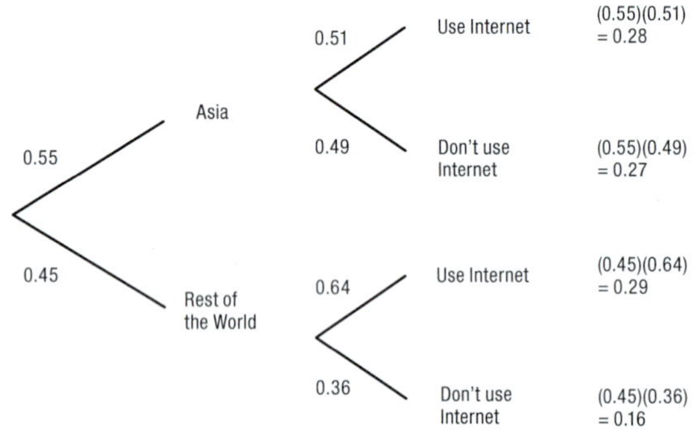

13. $$P(Asia \mid Uses\ Internet) = \frac{P(Asia\ and\ Uses\ Internet)}{P(Uses\ Internet)} = \frac{0.28}{0.28 + 0.29} = 0.49$$

It has decreased.

15. a) 0.97 b) 0.03 c) 0.77

17. a) 0.31 b) 0.48 c) 0.31

19. a) 0.2025 b) 0.7023 c) 0.2404 d) 0.0402

21. a) 0.50 b) 1.00 c) 0.077 d) 0.333

23. a) 0.66 b) 0.72 c) 0.834 d) 0.083

25. a) 0.011 b) 0.222 c) 0.054 d) 0.337 e) 0.436

27. 0.21

29. a) 0.145 b) 0.118 c) 0.414 d) 0.217

31. a) 0.318 b) 0.955 c) 0.071 d) 0.009

33. a) 32% b) 0.135
c) No. 7% of juniors have taken both.
d) No. The probability that a junior has taken a computer course is 0.23. The probability that a junior has taken a computer course *given* he or she has taken a statistics course is 0.135.

35. a) 0.525 b) 0.25
c) The events are independent.
d) The events mutually exclusive.

37. Yes. $P(\text{Ace}) = 1/13$. $P(\text{Ace} \mid \text{Spade}) = 1/13$. And the same is true for all four suits.

39. a) 0.33
b) No. 9% of the chickens had both contaminants.
c) No. $P(C \mid S) = 0.64 \neq P(C)$. If a chicken is contaminated with salmonella, it's more likely also to have campylobacter.

41. No. 28.8% of men with OK blood pressure have high cholesterol, but 40.7% of those with high blood pressure do.

43. a) 72.8%
b) Probably. 71.5% of people with cell phones had landlines, and 72.8% of all people did.

45. No. Only 34% of men were Democrats, but over 39% of all voters were.

47. a) Not independent as probability of enjoying sports programs depends on whether the household enjoys music programs. The probability of enjoying sports programs if the household also enjoys music programs is 0.71 and only 0.16 if not.
b) 0.6636

49. 0.5412

51. a) No. The probability of missing work for day-shift employees is 0.01. It is 0.02 for night-shift employees. The probability depends on whether they work day or night shift.
b) 1.4%

53. 57.1%

55. a) 0.20
b) 0.272
c) 0.353
d) 0.033

57. 0.563

59. 0.9995

Chapter 14

1. 1.55 pets

3. 1.07 pets

5. \$1700, \$141

7. 0.055

9. a) 19
b) 4.2

11. a)

Change in Value	−0.2	−0.1	0	0.2
Probability	0.2	0.2	0.4	0.2

b) −0.02
c) No, it's not worth it, because, on average, you lose

13. a)

Attempts	1	2	3
Probability	0.9	0.09	0.01

b) 1.11 c) 0.111

15. \$27,000

17. a) 7 b) 1.89

19. 0.1327

21. 0.3433

23. a) 3.1 b) 1.044

25. $\mu = 0.64$, $\sigma = 0.93$

27. a) \$50 b) \$100

29. a) No. The probability of winning the second depends on the outcome of the first.
b) 0.42 c) 0.08
d)

Games won	0	1	2
P(Games won)	0.42	0.50	0.08

e) $\mu = 0.66, \sigma = 0.62$

31. a)

Number good	0	1	2
P(Number good)	0.067	0.467	0.467

b) 1.40 c) 0.61

33. a) $\mu = 30, \sigma = 6$
b) $\mu = 26, \sigma = 5$
c) $\mu = 30, \sigma = 5.39$
d) $\mu = -10, \sigma = 5.39$
e) $\mu = 20, \sigma = 2.83$

35. a) $\mu = 240, \sigma = 12.80$
b) $\mu = 140, \sigma = 24$
c) $\mu = 720, \sigma = 34.18$
d) $\mu = 60, \sigma = 39.40$
e) $\mu = 600, \sigma = 22.63$

37. a) 1.8 b) 0.87
c) Cartons are independent of each other.

39. $\mu = 31, \sigma = 3.3015$
(This is only valid if reservations are independent of one another.)

41. a) $\mu = 23.4, \sigma = 2.97$
b) We assume each truck gets tickets independently.

43. a) There will be many gains of \$150 with a few large losses.
b) $\mu = \$300, \sigma = \$8485.28$
c) $\mu = \$1{,}500{,}000, \sigma = \$600{,}000$
d) Yes. \$0 is 2.5 SDs below the mean for 10,000 policies.
e) Losses are independent of each other. A major catastrophe with many policies in an area would violate the assumption.

45. a) 1 oz
b) 0.5 oz
c) 0.023
d) $\mu = 4$ oz, $\sigma = 0.5$ oz
e) 0.159
f) $\mu = 12.3$ oz, $\sigma = 0.54$ oz

47. a) 12.2 oz b) 0.51 oz
c) 0.058

49. a) $\mu = 200.57$ sec, $\sigma = 0.46$ sec
b) No, $z = \dfrac{199.48 - 200.57}{0.461} = -2.36$. There is only 0.009 probability of swimming that fast or faster.

51. a) A = price of a pound of apples; P = price of a pound of potatoes; Net Income = $100A + 50P - 2$
b) \$63.00 c) \$20.62
d) Mean—no; SD—yes (independent sales prices).

53. a) 0.0014
b) Not reasonable, because 500 is over 6 standard deviations above the mean.
c) 0.9452

Chapter 15

1. a) No. More than two outcomes are possible.
b) Yes, assuming the people are unrelated to each other.
c) No. The chance of a heart changes as cards are dealt.
d) No, 500 is more than 10% of 3000.
e) If packages in a case are independent of each other, yes; otherwise, no.

3. 0.0596

5. 0.011

7. 0.140

9. a) A Uniform; all numbers should be equally likely to be selected.
b) 0.02 c) 0.10

11. a) $\lambda = 1/3$. b) 0.811

13. a) Use single random digits. Let 0, 1 = Hope. Count the number of random numbers until a 0 or 1 occurs.
c) Results will vary.
d)

x	1	2	3	4	5	6	7	8	≥9
P(x)	0.2	0.16	0.128	0.102	0.082	0.066	0.052	0.042	0.168

15. a) Use single random digits. Let 0, 1 = Hope. Examine random digits in groups of five, counting the number of 0's and 1's.
c) Results will vary.
d)

x	0	1	2	3	4	5
P(x)	0.33	0.41	0.20	0.05	0.01	0.0

17. Departures from the same airport during a 2-hour interval may not be independent. All could be delayed by weather, for example.

19. a) 0.0729 b) 0.009 c) 0.01

21. 10 pings

23. 20 calls

25. a) 25 b) 0.185 c) 0.217 d) 0.693

27. a) 50 b) $E(X) = np = 100(0.5) = 50$.

29. a) 0.147 b) 0.8824 c) 0.357
d) 0.3241 e) 0.7443 f) 0.2557

31. a) 1.8 sales. b) 1.1225 sales. c) 3.3333

33. a) $E[Y] = 7$ and $SD[Y] = 1.45$ b) 0.9717
c) 0.0106 d) 0.1029 e) 0.8497

35. $\mu = 20.28, \sigma = 4.22$

37. a) 0.3487 b) 0.1937 c) 0.9298 d) 0.2639

39. a) $E[X] = 108$ passing inspection.
$SD[X] = 3.29$.
b) $np = 108$, $nq = 12$, greater than 10, hence normal model approximation holds.

c) Between 104.71 and 111.29 are expected to pass first inspection 68% of the time, between 101.42 and 114.58 are expected to pass first inspection 95% of the time, and between 98.13 and 117.87 are expected to pass first inspection 99.7% of the time.
d) Binomial, approx.: 0.9921, Normal, exact: 0.9925

41. a) Assuming apples fall and become blemished independently of each other, Binom(300, 0.06) is appropriate. Since $np \geq 10$ and $nq \geq 10$, N(18, 4.11) is also appropriate.
b) Normal, approx.: 0.072; Binomial, exact: 0.085
c) No, 50 is 7.8 SDs above the mean.

43. Binomial, approx: 0.0139, Normal, approx: 0.0121

45. The mean number of sales should be 24 with SD 4.60. Ten sales is more than 3.0 SDs below the mean. He was probably misled.

47. a) 0.0863 b) 0.0365

49. a) 4 cases b) 0.9817

51. a) 5 b) 0.066 c) 0.107
d) $\mu = 24, \sigma = 2.19$
e) Normal, approx.: 0.819; Binomial, exact: 0.848

53. $\mu = 20, \sigma = 4$. I'd want *at least* 32 (3 SDs above the mean). (Answers will vary.)

55. The probability that the ice hockey player scores 3 goals in a row with a 33% probability is $0.33^3 \approx 0.036$. We can expect this to happen 1 in every 28 times for every set of 3 shots he makes. This is not a very common occurrence and, so, is somewhat unusual.

57. Answers may vary. The player's performance seems to have increased. 30 is $\frac{30 - 19.8}{3.64} \approx 2.8$ standard deviations above the mean, quite an extraordinary feat, unless his score probability has increased. This does not necessarily mean that the blades are responsible for the increase in the score probability, but it is possible.

59. a) The Poission model b) 0.9502 c) 0.0025

61. a) The Exponential model
b) 1/3 minutes c) 0.0473

Part IV Review

R4.1. a) 0.34 b) 0.27 c) 0.069
d) No, 2% of cars have both types of defects.
e) Of all cars with cosmetic defects, 6.9% have functional defects. Overall, 7.0% of cars have functional defects. The probabilities here are estimates, so these are probably close enough to say the defects are independent.

R4.3. a) C = Price to China; F = Price to France; Total $= 3C + 5F$
b) $\mu = \$5500, \sigma = \$672.68$
c) $\mu = \$500, \sigma = \$180.28$
d) Means—no. Standard deviations—yes; ticket prices must be independent of each other for different countries, but all tickets to the same country are at the same price.

R4.5. a) $\mu = -\$0.20, \sigma = \$1.89$
b) $\mu = -\$0.40, \sigma = \$2.67$

R4.7. a) 0.106 b) 0.651 c) .442

R4.9. a) 0.590 b) 0.328 c) 0.00856

R4.11. a) $\mu = 15.2, \sigma = 3.70$
b) Yes, $np \geq 10$ and $nq \geq 10$
c) Normal, approx.: 0.080; Binomial, exact: 0.097

R4.13. a) 0.0173
b) 0.591
c) Left: 960; right: 120; both: 120
d) $\mu = 120, \sigma = 10.39$
e) About 68% chance of between 110 and 130; about 95% between 99 and 141; about 99.7% between 89 and 151.

R4.15. a) Men's heights are more variable than women's.
b) Men
c) M = Man's height; W = Woman's height; $M - W$ is how much taller the man is.
d) 5.1″
e) 3.75″
f) 0.913
g) If independent, it should be about 91.3%. We are told 92%. This difference seems small and may be due to natural sampling variability.

R4.17. a) The chance is 1.6×10^{-7}. b) 0.952
c) 0.063

R4.19. −$2080.00

R4.21. a) 0.717 b) 0.588

R4.23. a) $\mu = 100, \sigma = 8$
b) $\mu = 1000, \sigma = 60$
c) $\mu = 100, \sigma = 8.54$
d) $\mu = -50, \sigma = 10$
e) $\mu = 100, \sigma = 11.31$

R4.25. a) Many do both, so the two categories can total more than 100%.
b) No. They can't be disjoint. If they were, the total would be 100% or less.
c) No. Probabilities are different for boys and girls.
d) 0.0524

R4.27. a) 21 days b) 1649.73 som
c) 3300 som extra. About 157-som "cushion" each day.

R4.29. No, you'd expect 541.2 homeowners, with an SD of 13.56. 523 is 1.34 SDs below the mean; not unusual.

R4.31. a) 0.027 b) 0.240 c) 0.30

R4.33. a) 6 b) 15 c) 0.402

R4.35. a) 34% b)35% c) 31.4%
d) 31.4% of classes that used calculators used computer assignments, while in classes that didn't use calculators, 30.6% used computer assignments. These are close enough to think the choice is probably independent.

R4.37. a) A Poisson model
b) 0.1 failures
c) 0.090
d) 0.095

R4.39. a) 1/11 b) 7/22 c) 5/11
d) 0 e) 19/66

R4.41. a) Expected number of stars with planets.
b) Expected number of planets with intelligent life.
c) Probability of a planet with a suitable environment having intelligent life.

d) f_l: If a planet has a suitable environment, the probability that life develops.
f_i: If a planet develops life, the probability that the life evolves intelligence.
f_c: If a planet has intelligent life, the probability that it develops radio communication.

R4.43. 0.991

Chapter 16

1. a) Unimodal and symmetric (roughly Normal).
b) 0.36 c) 0.034

3. All the histograms are centered near 0.05. As n gets larger, the histograms approach the Normal shape, and the variability in the sample proportions decreases.

5. a) This means that 16% of the 1300 young adults aged 18–25 years in the sample said that they are addicted to the internet. This is our best estimate of p, the proportion of all UK young adults aged 18–25 years who would say they do so.
b) $SE(\hat{p}) = \sqrt{\frac{(0.16)(0.84)}{1300}} \approx 0.0102$
c) Because we don't know p, we use $\hat{p}$ to estimate the standard deviation of the sampling distribution. The standard error is our estimate of the amount of variation in the sample proportion we expect to see from sample to sample when we ask 1300 young adults aged 18–25 years whether they go online several times a day.

7. Check conditions: The data come from a random sample, so the randomization condition is met. We don't know the exact value of p, but if the sample is anywhere near correct, $np \approx 2705 \times 0.12 = 325$ and $nq \approx 2705 \times 0.88 = 2380$. So, there are well over 10 successes and 10 failures, meeting the Success/Failure Condition. Since there are more than $(10)(2705) = 27{,}050$ men in Japan, the 10% condition is met. A normal model is appropriate for the sampling distribution of the sample proportion.

9. a) The histogram is unimodal and symmetric for $n = 200$.
b) The Success/Failure Condition says that np and nq should both be at least 10, which is not satisfied until $n = 200$ for $p = 0.05$. The theory predicted my choice.

11. a) No. The sampling distribution of the maximum is skewed to the left.
b) No. The 95% rule is based on the Normal distribution.

13. a) We are 95% confident that, if we were to ask all UK young adults aged 18–25 years whether they go online several times per day, between 14% and 18% of them would say they do.
b) If we were to collect many random samples of 1300 young adults, about 95% of the confidence intervals we construct would contain the proportion of all UK young adults aged 18–25 years who say they go online several times per day.

15. a) $SE(\hat{p}) = \sqrt{\frac{\hat{p}\hat{q}}{n}} = \sqrt{\frac{(0.54)(0.46)}{2066}} = 0.0110$
$ME = 1.645 \times 0.0110 = 0.018$
b) We are 90% confident that the observed proportion responding "Wrong" is within 0.018 of the population proportion.

17. a) Yes. Random sample and sufficiently large sample.
b) Larger. Higher confidence requires a wider confidence interval.

19. a) 158 (using $\hat{p} = 0.30$)
b) 356 c) 632

21. She believes the true proportion is with 3% of her estimate, with some (probably 95%) degree of confidence.

23. a) Population—all cars; sample—those actually stopped at the checkpoint; p—proportion of all cars with safety problems; $\hat{p}$—proportion actually seen with safety problems (10.4%); if sample (a cluster sample) is representative, then the methods of this chapter will apply.
b) Population—general public; sample—those who logged onto the website; p—population proportion of those who favor prayer in school; $\hat{p}$—proportion of those who voted in the poll who favored prayer in school (81.1%); can't use methods of this chapter—sample is biased and nonrandom.
c) Population—parents of students at the school; sample—those who returned the questionnaire; p—proportion of all parents who favor uniforms; $\hat{p}$—proportion of respondents who favor uniforms (60%); should not use methods of this chapter, since not SRS. (Possible nonresponse bias.)
d) Population—students at the college; sample—the 1632 students who entered that year; p—proportion of all students who will graduate on time; $\hat{p}$—proportion of that year's students who graduate on time (85.0%); can use methods of this chapter if that year's students (a cluster sample) are viewed as a representative sample of all possible students at the school.

25. a) Not correct. This implies certainty.
b) Not correct. Different samples will give different results. Many fewer than 95% will have 88% on-time orders.
c) Not correct. The interval is about the population proportion, not the sample proportion in different samples.
d) Not correct. In this sample, we *know* 88% arrived on time.
e) Not correct. The interval is about the parameter, not the days.

27. a) False. b) True. c) True. d) False.

29. On the basis of this sample, we are 90% confident that the proportion of Japanese cars is between 29.9% and 47.0%.

31. a) (0.182, 0.318)
b) We are 95% confident that between 18.2% and 31.8% of all 'cod' or 'haddock' sold in Ireland are mislabeled.
c) The size of the population is irrelevant. If a random sample was used, 95% of intervals generated by studies like this will capture the true level of mislabeling.

33. a) 0.026
b) We're 90% confident that this poll's estimate is within $\pm 2.6\%$ of the true proportion of people who are baseball fans.
c) Larger. To be more certain, we must be less precise.
d) 0.040
e) Less confidence.

35. a) (0.0457, 0.0495). The assumptions and conditions for constructing a confidence interval are satisfied.
b) The confidence interval gives the set of plausible values with 95% confidence. Since 5.5% is above the interval, it seems to be a bit optimistic.

37. a) (12.7%, 18.6%)
b) We are 95% confident, based on this sample, that the proportion of all auto accidents that involve teenage drivers is between 12.7% and 18.6%.

c) About 95% of all random samples of this size will produce confidence intervals that contain the true population proportion.
d) Contradicts. The interval is completely below 20%.

39. Probably nothing. Those who bothered to fill out the survey may be a biased sample.

41. a) Response bias (wording).
b) (0.554, 0.615)
Answers may vary slightly, since counts are not given.
c) Smaller. The sample size was larger.

43. a) (18.2%, 21.8%)
b) We are 98% confident, based on the sample, that between 18.2% and 21.8% of English children are deficient in vitamin D.
c) About 98% of all random samples of this size will produce a confidence interval that contains the true proportion of English children deficient in vitamin D.

45. a) Wider. The sample size is probably about one fourth of the sample size for all adults, so we'd expect the confidence interval to be about twice as wide.
b) There are fewer "young" people than seniors, so the CI for them would be expected to be a bit wider.

47. a) (15.5%, 26.3%)
b) 612
c) The sample may not be random or representative. Deer that are legally hunted are mostly adult bucks, so does and fauns are not likely to be represented. There is no guarantee that the deer legally killed are a representative sample in terms of health, nutrition, age, or other variables.

49. 2017, using $\hat{p} = 0.30$.

51. 323 total, using $\hat{p} = 0.16$.

53. 90%

Chapter 17

1. a) *SD*s are 1 lb, 0.5 lb, and 0.2 lb, respectively.
b) The distribution of pallets. The CLT tells us that the Normal model is approached in the limit regardless of the underlying distribution. As samples get larger, the approximation gets better.

3. a) Can't use a Normal model to estimate probabilities. The distribution is skewed right—not Normal.
b) 5 is probably not a large enough sample to say the average follows the Normal model.
c) No. This is 3.30 *SD*s above the mean.

5. a) 0 Model is $N(8.50, 0.608)$
b) $386.09 or less

7. a) 2.65 b) 1.66

9. The shape is unimodal, symmetric, and bell-shaped. It becomes closer to Normal. The center is at 0. The standard deviation becomes smaller.

11. a) Houses are independent; randomly sampled. Should check Nearly Normal by making a histogram. Sample size: 36 should be big enough
b) ($9,052.50, $10,067.50)

13. We are 95% confident that the interval ($9052.50, $10,067.50) contains the true mean loss in home value. That is, 95% of all random samples will produce intervals that will contain the true mean.

15. a) The confidence interval is for the population mean, not the individual cows in the study.
b) The confidence interval is not for individual cows.
c) We *know* the average gain in this study was 56 pounds!
d) The average weight gain of all cows does not vary. It's what we're trying to estimate.
e) No. This interval is not a standard. There is a 95% chance that another sample will have its average weight gain within two standard errors of the true mean.

17. a) (232.434, 237.13)
b) We are 95% confident that the true mean cholesterol level is between 232.4 and 237.1 mg/dL.
c) That the individual values in the sample were independent. For example, it helps if the sample was random.

19. The interval is a range of possible values for the mean shoe size. The average is not a value that any individual in the population will have, but an average of all the individuals.

21. a) No. A confidence interval is not about individuals in the population.
b) No. It's not about individuals in the sample, either.
c) No. We know the mean cost for students in the sample was €750.
d) No. A confidence interval is not about other sample means.
e) Yes. A confidence interval estimates a population parameter.

23. a) Based on this sample, we can say, with 95% confidence, that the mean pulse rate of adults is between 70.9 and 74.5 beats per minute.
b) 1.8 beats per minute
c) Larger. More confidence requires a larger margin of error.

25. The assumptions and conditions for a *t*-interval are not met. The distribution is highly skewed to the right and there is a large outlier.

27. a) No. The CLT is about means and proportions, not individual observations.
b) Yes, if we assume these adults are a representative, random sample of French adults. With a sample size this large, even some skewness will not be a problem.
c)

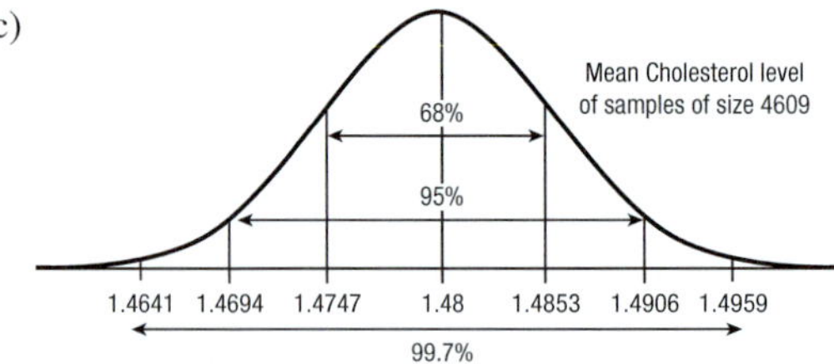

29. a) Yes. Randomly selected group, so they are independent of each other; the histogram is not unimodal and symmetric, but it is not highly skewed and there are no outliers, so with a sample size of 52, the CLT says $\bar{y}$ is approximately Normal.
b) (98.06, 98.51) degrees F
c) We are 98% confident, based on the data, that the average body temperature for an adult is between 98.06°F and 98.51°F.
d) 98% of all such random samples will produce intervals containing the true mean temperature.
e) These data suggest that the true normal temperature is somewhat less than 98.6°F.

31. a) Narrower. A smaller margin of error, so less confident.
b) Advantage: more chance of including the true value. Disadvantage: wider interval.
c) Narrower; due to the larger sample, the *SE* will be smaller.

33. a) (299,709.90, 299,802.54)
b) With 95% confidence, based on these data, the speed of light is between 299,709.9 and 299,802.5 km/sec.
c) Normal model for the distribution, independent measurements. These seem reasonable here, but it would be nice to see if the Nearly Normal Condition held for the data.

35. a) Given no time trend, the monthly on-time departure rates should be independent. Though not a random sample, these months should be representative. The histogram looks unimodal, but slightly left-skewed; not a concern with this large sample.
b) $77.60\% < \mu(\text{OT Departure\%}) < 78.60\%$
c) We can be 90% confident that the interval from 77.60% to 78.60% holds the true mean monthly percentage of on-time flight departures.

37. The 95% confidence interval lies entirely above the 0.08 ppm limit, evidence that mirex contamination is too high and consistent with rejecting the null. We used an upper-tail test, so the *P*-value should therefore be smaller than $\frac{1}{2}(1 - 0.95) = 0.025$, and it was.

39. Because even the lower bound of the confidence interval is above 220 mg/dL, we are quite confident that the mean cholesterol level for those who eat frozen pizza is a level that indicates a health risk.

41. a) We are 95% confident that the interval from 22.1 mpg to 26.6 mpg contains the true mean gas mileage for 2016 vehicles.
b) It appears that the data are a mix of small, mid-size, and large vehicles, but without knowing how the data were selected, we are cautious about generalizing to all 2016 vehicles.

43. a) The histogram is unimodal and slightly skewed to the right, centered at 36 inches with a standard deviation near 4 inches.
b) All the histograms are centered near 36 inches. As *n* gets larger, the histograms approach the Normal shape and the variability in the sample means decreases. The histograms are fairly Normal by the time the sample reaches size 5.

45. a)

n	Observed Mean	Theoretical Mean	Observed st. dev.	Theoretical st. dev.
2	36.314	36.33	2.855	2.842
5	36.314	36.33	1.805	1.797
10	36.341	36.33	1.276	1.271
20	36.339	36.33	0.895	0.899

b) They are all very close to what we would expect.
c) For samples as small as 5, the sampling distribution of sample means is unimodal and very symmetric.
d) The distribution of the original data is nearly unimodal and symmetric, so it doesn't take a very large sample size for the distribution of sample means to be approximately Normal.

47.

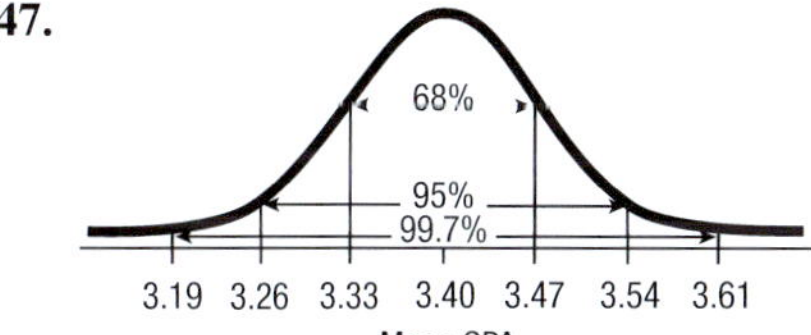

Normal, $\mu = 3.4, \sigma = 0.07$. We assume that the students are randomly assigned to the seminars and that individuals' GPAs are independent of one another.

49. a) As the CLT predicts, there is more variability in the smaller outlets.
b) If the lottery is random, all outlets are equally likely to sell winning tickets.

51. a) 21.1% b) 276.8 days or more
c) $N(266, 2.07)$ d) 0.002

53. a) There are more premature births than very long pregnancies. Modern practice of medicine stops pregnancies at about 2 weeks past normal due date.
b) Parts a and b—yes—we can't use Normal model if it's very skewed. Part c—no—CLT guarantees a Normal model for this large sample size.

55. a) Probably a representative sample; the Nearly Normal Condition seems reasonable. (Show a Normal probability plot or histogram.) The histogram is nearly uniform, with no outliers or skewness.
b) $\bar{y} = 28.78, s = 0.40$
c) (28.36, 29.21) grams
d) Based on this sample, we are 95% confident the average weight of the content of Ruffles bags is between 28.36 and 29.21 grams.
e) The company is erring on the safe side, as it appears that, on average, it is putting in slightly more chips than stated.
f) The sample is too small to provide a good basis for a bootstrap.

57. Yes. The 95% confidence interval is (3.73%, 9.82%). This is below the 10% desired goal.

59. a) Random sample; the Nearly Normal Condition seems reasonable from a Normal probability plot. The histogram is roughly unimodal and symmetric with no outliers.
b) (1187.9, 1288.4) chips
c) Based on this sample, the mean number of chips in an 18-ounce bag is between 1187.9 and 1288.4, with 95% confidence. The *mean* number of chips is likely to be greater than 1000. However, if the claim is about individual bags, then it's not necessarily true. If the mean is 1188 and the *SD* is near 94, then 2.28% of the bags will have fewer than 1000 chips, using the Normal model. If in fact the mean is 1288, the proportion below 1000 will be about 0.1%, but the claim is still false.

61. a) (46.03, 58.38) seconds
b) The Normal probability plot is relatively straight, with one outlier at 93.8 sec. Without the outlier, the conditions seem to be met. The histogram is roughly unimodal and symmetric, with no other outliers. (Show your plot.)
c) (45.49, 54.77) seconds
d) No. Both confidence intervals show that the maze can be solved in an average time of less than 1 minute.

63. a) 287.4 yards $< \mu_{(\text{Drive Distance})} <$ 290.0 yards
b) These data are not a random sample of golfers. The top professionals are (unfortunately) not representative and were not selected at random. We might consider these data to represent the population of all professional golfers, past, present, and future.
c) If these data were the means for each golfer rather than individual drives, then they would be less variable than individual drives and the inference would be invalid.

Chapter 18

1. The new drug is not more effective than aspirin, or H_0: $p = 0.42$, where p = rate of reduction in heart attacks.

3. a) Let p = probability of winning on a slot machine. H_0: $p = 0.02$ vs. H_A: $p \neq 0.02$
b) Let μ = mean spending per customer this year. H_0: $\mu = \$54.81$ vs. H_A: $\mu \neq \$54.81$
c) Let p = proportion of patients cured by the new drug. H_0: $p = 0.27$ vs. H_A: $p \neq 0.27$
d) Let p = proportion of clients now using the website. H_0: $p = 0.35$ vs. H_A: $p \neq 0.35$

5. a) One-sided. They are interested in discovering only if their drug is more effective than aspirin, not if it is less effective than aspirin.
b) There is convincing evidence to conclude that the new drug is better than aspirin.
c) There is no convincing evidence that the new drug is more effective than aspirin.

7. a) H_0: The proportion of people in Wealden district who voted to leave the UK is $p = 0.519$.

$$H_A: p > 0.519$$

b) This is a one-proportion z-test. The 1300 district residents were a random sample of all district residents. 1300 is almost certainly less than 10% of the population of the district. We expect $np_0 = (1300)(0.519) = 674.7$ successes and $nq_0 = (1300)(0.481) = 625.3$ failures, which are both more than 10. The conditions are satisfied, so it's ok to use a Normal model and perform a one-proportion z-test.
c) $\hat{p} = \dfrac{744}{1300} = 0.572;$

$$SD(\hat{p}) = \sqrt{\frac{p_0 q_0}{n}} = \sqrt{\frac{0.519 \times 0.481}{1300}} \approx 0.014$$

$$z = \frac{\hat{p} - p_0}{SD(\hat{p})} = \frac{0.572 - 0.519}{0.014} = 3.79;$$

$P\text{-}value = P(z > 3.79) < 0.01$

d) Because the P-value is extremely low, there is evidence that the proportion of people in favor of leaving the UK in Wealden district is higher than those in favor of the referendum.

9. a) An increase in the mean score would mean that the mean difference (After – Before) is positive. So H_0: $\mu_{diff} = 0$, H_A: $\mu_{diff} > 0$.
b) Because the P.value is large, there is not convincing evidence that the course works.
c) Because the P.value is greater than 0.5 and the alternative is one-sided (>0), we can conclude that the actual difference was less than 0. (If it had been positive, the probability to the right of that value would have to be less than 0.5.)

11. If, in fact, the mean cholesterol level of pizza eaters does not indicate a health risk, then only 7 of every 100 people sampled, on average, would have a mean cholesterol level as high as (or higher than) was observed in this sample.

13. a) The drug may not be approved for use. Then people will miss out on a beneficial product and the company will miss out on potential sales.
b) The drug will go into production and people will suffer the side effect.

15. a) H_0: $p = 0.30$; H_A: $p < 0.30$
b) H_0: $p = 0.50$; H_A : $p \neq 0.50$
c) H_0: $p = 0.20$; H_A : $p > 0.20$

17. Statement d is correct. It talks about the probability of seeing the data, not the probability of the hypotheses.

19. No. We can say only that there is a 27% chance of seeing the observed effectiveness just from natural sampling variation. There is insufficient evidence that the new formula is more effective, but we can't conclude that they are equally effective.

21. a) No. There's a 25% chance of losing twice in a row. That's not unusual.
b) 0.125
c) No. We expect that to happen 1 time in 8.
d) Answers may vary. The chance of 5 losses in a row is only 1 in 32, which seems unusual. On the other hand, I'm very skeptical of the claim and will need extraordinary evidence to be convinced (and to verify that he's not cheating!).

23. 1) Use p, not $\hat{p}$, in hypotheses.
2) The question was about failing to meet the goal, so H_A should be $p < 0.96$.
3) Did not check $0.04(200) = 8$. Since $nq < 10$, the Success/Failure Condition is violated. Didn't check 10% Condition.
4) $188/200 = 0.94$; $SD(\hat{p}) = \sqrt{\dfrac{(0.96)(0.04)}{200}} = 0.014$
5) z is incorrect; should be $z = \dfrac{0.94 - 0.96}{0.014} = -1.43$.
6) $P = P(z < -1.43) = 0.076$
7) There is only weak evidence that the new instructions do not work.

25. a) H_0: $p = 0.30$; H_A : $p > 0.30$
b) Possibly an SRS; we don't know if the sample is less than 10% of his customers, but it could be viewed as less than 10% of all possible customers; $(0.3)(80) \geq 10$ and $(0.7)(80) \geq 10$. Wells are independent only if customers don't have farms on the same underground springs.
c) $z = 0.73$; P-value $= 0.232$
d) If his dowsing is no different from standard methods, there is more than a 23% chance of seeing results as good as those of the dowser's, or better, by natural sampling variation.
e) These data provide insufficient evidence that the dowser's chance of finding water is any better than normal drilling.

27. a) H_0: $p = 0.35$ H_A: $p \neq 0.35$
b) Students were randomly sampled and should be independent. 35% and 65% of 7500 are greater than 10; 7500 students are less than 10% of student population in the region.
c) P-value is 0.0456.
d) The P-value provides weak evidence against null hypothesis.
e) No. A difference this small, even if statistically significant, is probably not meaningful. We might look at new data in a few years.

29. a) H_0: $p = 0.055$ vs. H_A: $p < 0.055$.
b) We assume the whole mailing list has over 500,000 names. This is a random sample, and we expect 2750 successes and 47,250 failures.
c) $z = -7.258$; P-value $= 1.96 \times 10^{-13}$, so we reject H_0; there is strong evidence that the donation rate would be below 5.5%.

31. H_0: $p = 0.20$; H_A: $p > 0.20$. SRS (not clear from information provided); 22 is more than 10% of the population of 150; $(0.20)(22) < 10$. Do not proceed with a test.

33. H_0: $p = 0.03$; $p \neq 0.03$. $\hat{p} = 0.015$. One mother having twins will not affect another, so observations are independent; not an SRS; sample is less than 10% of all births. However, the mothers at this hospital may not be representative of all teenagers; $(0.03)(469) = 14.07 \geq 10$; $(0.97)(469) \geq 10$. $z = -1.92$; P-value $= 0.055$. These data show some (although weak) evidence that the rate of twins born to teenage girls at this hospital may be less than the national rate of 3%. It is not clear whether this can be generalized to all teenagers.

35. H_0: $p = 0.25$; H_A: $p > 0.25$. SRS; sample is less than 10% of all potential subscribers; $(0.25)(500) \geq 10$; $(0.75)(500) \geq 10$. $z = 1.24$; P-value $= 0.1075$. The P-value is high, so do not reject H_0. These data do not provide sufficient evidence that more than 25% of current readers would subscribe; the company should not go ahead with the WebZine solely on the basis of these data.

37. H_0: $p = 0.40$; H_A: $p < 0.40$. Data are for all executives in this company and may not be able to be generalized to all companies; $(0.40)(43) \geq 10$; $(0.60)(43) \geq 10$. $z = -1.31$; P-value $= 0.096$. Because the P-value is high, we fail to reject H_0. These data do not show that the proportion of women executives is less than the 40% of women in the company in general.

39. H_0: $p = 0.015$; H_A: $p > 0.015$. $\hat{p} = 0.167$; $z = 1.545$; P-value $= 0.061$. Because the P-value is not very low, we fail to reject H_0. These data do not provide convincing evidence that the dropout rate is increasing.

41. H_0: $p = 0.90$; H_A: $p < 0.90$. $\hat{p} = 0.844$; $z = -2.05$; P-value $= 0.020$. Because the P value is so low, we reject H_0. There is strong evidence that the actual rate at which passengers with lost luggage are reunited with it within 24 hours is less than the 90% claimed by the airline.

43. a) Yes. Assuming this sample to be a typical group of people, P-value $= 0.0008$. This cancer rate is very unusual.
b) No, this group of people may be atypical for reasons that have nothing to do with the radiation.

45. a) H_0: $\mu = 98.6$°F; H_A: $\mu \neq 98.6$°F
b) The sample was randomly selected. $n = 52 > 30$. 52 is less than 10% of the population of adults. Conditions are met.
c) $t = -3.33$ with 51 df, P-value $= 0.0016$. Because the P-value is low, we reject the null hypothesis. We have convincing evidence that the mean body temperature of adults is not 98.6°F.

47. If the mean cholesterol level really does not exceed the value considered to indicate a health risk, there is a 7% probability that a random sample of this size would have a mean as high as (or higher than) that in this sample.

49. a) Probably a representative sample; the Nearly Normal Condition seems reasonable. (Show a Normal probability plot or histogram.) The histogram is nearly uniform, with no outliers or skewness.
b) $\bar{y} = 28.78$, $s = 0.40$
c) $t = 2.94$, $df = 5$, P-value $= 0.032$. Because the P-value is low, we reject H_0. We have convincing evidence that the mean weight of bags of Ruffles potato chips is not 28.3 grams.

51. Yes. These are a random sample of bags and the Nearly Normal Condition is met. (Show a Normal probability plot or histogram.) $t = -2.51$ with 7 df for a one-sided P-valueof 0.0203.

53. a) Random sample; the Nearly Normal Condition seems reasonable from a Normal probability plot. The histogram is roughly unimodal and symmetric with no outliers. This is definitely less than 10% of all bags of Chips Ahoy!
b) H_0: $\mu = 1000$, H_A: $\mu > 1000$, where μ is the mean number of chips per bag; $\bar{y} = 1238.2$, $s = 94.3$, $t = 10.1$, $df = 15$, P-value < 0.0001. Because the P-value is so low, we reject the null hypothesis. We have convincing evidence that the mean number of chips per bag is greater than 1000. However, their statement isn't about the mean. They claim that *all* bags have at least 1000 chips. So this test doesn't really answer the question.

55. a) The Normal probability plot is relatively straight, with one outlier at 93.8 sec. Without the outlier, the conditions seem to be met. The histogram is roughly unimodal and symmetric with no other outliers. (Show your plot.)
b) Because the question asks for "at most one minute," H_0: $\mu = 60$, H_A: $\mu < 60$, where μ is the mean number of seconds to complete the maze. With the outlier, $\bar{y} = 52.2$, $s = 13.6$, $t = -2.63$, $df = 20$, P-value $= 0.008$. Without the outlier, the evidence for the alternative hypothesis should be even stronger because the mean will be lower, as will the standard deviation. The numbers without the outlier are $\bar{y} = 50.13$, $s = 9.9$, $t = -4.46$, $df = 20$, P-value $= 0.0001$. Because the P-value is low both with and without the outlier, we reject the null hypothesis. We have convincing evidence that the mean number of seconds to complete the maze is less than 60. It appears that, even with the occasional slow mouse, this maze meets the requirement that the time be at most one minute. However, if the researcher wants the mean time to be one minute, we have evidence that this requirement is not met.

57. a) $(45.271, 59.939)$
b) Because of the outlier at 93.8 seconds.
c) It makes the interval not symmetric around the sample mean.
d) Because 60 seconds is not in the interval, it is not plausible that the mean time is 60 seconds. The mean time appears to be less than 60 seconds.
e) We want the proportion of cases farther than 7.79 seconds from 60 seconds. On the left, that is fewer than 0.05% of cases. On the right, it is fewer than 0.5% of cases. We might estimate the P-value at about 0.004.

59. a) 234.701 mg/dL
b) (Answers may vary slightly.) The 95% bootstrapped interval is $(232.35, 237.26)$.
c) The bootstrapped P-value is < 0.001 using 1000 samples.
d) Both the interval and the test indicate that a mean of 240 is not plausible. The mean cholesterol level of the Framingham participants is lower than 240 by an amount too large to be attributed to random variability.

Chapter 19

1. a) False. It provides evidence against it but does not show it is false.
b) False. The P-value is not the probability that the null hypothesis is true.
c) True
d) False. Whether a P-value provides enough evidence to reject the null hypothesis depends on the risk of a type I error that one is willing to assume (the α-level).

3. a) False. A high P-value shows that the data are consistent with the null hypothesis, but provides no evidence for rejecting the null hypothesis.
b) False. It results in rejecting the null hypothesis, but does not prove that it is false.
c) False. A high P-value shows that the data are consistent with the null hypothesis but does not prove that the null hypothesis is true.
d) False. Whether a P-value provides enough evidence to reject the null hypothesis depends on whether it is smaller than the limit you set.

5. a) True.
b) False. The alpha level is set independently and does not depend on the sample size.
c) False. The P-value would have to be less than 0.01 to reject the null hypothesis.
d) False. It simply means we do not have enough evidence at that alpha level to reject the null hypothesis.

7. a) $z^* = \pm 1.96$
b) $z^* = 1.645$
c) $t^* = \pm 2.03$
d) $z^* = 2.33$; n is not relevant for critical values for z.
e) $z^* = -2.33$

9. a) There is very little chance of seeing a sample proportion as low as 97.4% vaccinated by natural sampling variation in a sample as large as this one if 98% have really been vaccinated.
b) We conclude that p is below 0.98, but a 95% confidence interval would suggest that the true proportion is between (0.971, 0.977). Most likely, a decrease from 98% to 97.6% would not be considered important. On the other hand, with 1,000,000 children a year vaccinated, even 0.1% represents about 1000 kids—so this may very well be important.

11. a) Type I error. The actual value is not greater than 0.3 but they rejected the null hypothesis.
b) No error. The actual value is 0.50, which was not rejected.
c) Type II error. The actual value was 55.3 points, which is greater than 52.5.
d) Type II error. The null hypothesis was not rejected, but it was false. The true relief rate was greater than 0.25.

13. If there is no difference in effectiveness, the chance of seeing an observed difference this large or larger is 0.021% by natural sampling variation. Assuming he is testing with an alpha of 5%, his P-value of 2.1% is quite small.

15. $\alpha = 0.10$: Yes. The P-value is less than 0.05, so it's less than 0.10. But to reject H_0 at $\alpha = 0.01$, the P-value must be below 0.01, which isn't necessarily the case.

17. a) SRS so responses are independent, successes and failures both >10; (0.448; 0.476).
b) Because 43% is not in the interval, we have strong evidence that more than 43% of French adults go grocery shopping on Sundays.
c) $\alpha = 0.025$; it's an upper-tail test based on a 95% confidence interval.

19. a) (0.499, 0.561)
b) Since 42% is not in the interval, we can reject the hypothesis that $p = 42\%$.

21. a) The Success/Failure Condition is violated: only 5 pups had dysplasia.
b) No. The sample size is too small

23. a) Type II error.
b) Type I error.
c) By making it easier to get the loan, the bank has reduced the alpha level.
d) The risk of a Type I error is decreased and the risk of a Type II error is increased.

25. a) Power is the probability that the bank denies a loan that would not have been repaid.
b) Raise the cutoff score.
c) A larger number of trustworthy people would be denied credit, and the bank would miss the opportunity to collect interest on those loans.

27. a) The null is that the level of home ownership remains the same. The alternative is that it rises.
b) The city concludes that home ownership is on the rise, but, in fact, the tax breaks don't help.
c) The city abandons the tax breaks, but they were helping.
d) A Type I error causes the city to forgo tax revenue, while a Type II error withdraws help from those who might have otherwise been able to buy a home.
e) The power of the test is the city's ability to detect an actual increase in home ownership.

29. a) It is decided that the shop is not meeting standards when it is.
b) The shop is certified as meeting standards when it is not.
c) Type I.
d) Type II.

31. a) The probability of detecting a shop that is not meeting standards.
b) 40 cars. Larger n, which will reduce the standard error, making it easier to detect a failing shop.
c) 10%. Greater chance to reject H_0.
d) A lot. A larger effect size is easier to detect.

33. a) One-tailed. The company wouldn't be sued if "too many" women were hired.
b) Deciding the company is discriminating when it is not.
c) Deciding the company is not discriminating when it is.
d) The probability of correctly detecting actual discrimination.
e) Increases power.
f) Lower, since n is smaller.

35. a) One-tailed. The graded assignments are supposed to help increase the final exam scores.
b) H_0: $\mu = 67$; H_A: $\mu > 67$
c) He implements the graded assignments system when it doesn't help students improve.
d) He doesn't implement the graded assignments when it does help students improve.
e) The probability of correctly deciding that the graded assignments are helpful.

37. a) t = 7.06; P < 0.0001. The change is statistically significant. A 95% confidence interval is (71.2; 72.8). This is clearly higher than 67. Graded assignments do appear to improve exam scores.
b) Students' scores improved by approximately 4 to 6 points on the exam. This could be enough to be worth the effort.

39. a) Upper-tail. We want to show it will hold 500 pounds (or more) easily.
b) They will decide the stands are safe when they're not, which could be dangerous.
c) They will decide the stands are unsafe when they are in fact safe.

41. a) Decrease α. This means a smaller chance of declaring the stands safe if they are not.
b) The probability of correctly detecting that the stands are not capable of holding more than 500 pounds.
c) Decrease the standard deviation—probably costly. Increase the sample size—takes more time for testing and is costly. Increase α—more Type I errors. Increase the "design load" to be well above 500 pounds—again, costly. But the cost of a Type I error could be an expensive lawsuit.

43. a) H_A: The coin is the 30% heads coin.
b) Reject the null hypothesis if the coin comes up tails—otherwise fail to reject.
c) $P(\text{tails given the null hypothesis}) = 0.1 = \alpha$
d) $P(\text{tails given the alternative hypothesis}) = \text{power} = 0.70$
e) Spin the coin more than once and base the decision on the sample proportion of heads.

45. a) 0.0859
b) Type I.
c) 54.4%
d) Increase the number of shots. Or keep the number of shots at 10, but increase alpha by declaring that 8, 9, or 10 will be deemed as having improved.

47. P-value < 0.0001

Part V Review

R4.1. a) Because the data from Australia are census data, not a sample.
b) H_0: $\mu = 7.86$; H_A: $\mu \neq 7.86$; df = 111; P-value = 0.1487. With such a high P-value, we do not reject H_0. Assuming that Missouri babies fairly represent babies born in the United States, these data provide no evidence that mean weights of American babies are different from Australian babies.

R4.3. a) Yes, $0.08 \times 325 = 26$, so we expect more than 10 successes and more than 10 failures.
b) $\mu = 0.08$, $\sigma = 0.015$
c)

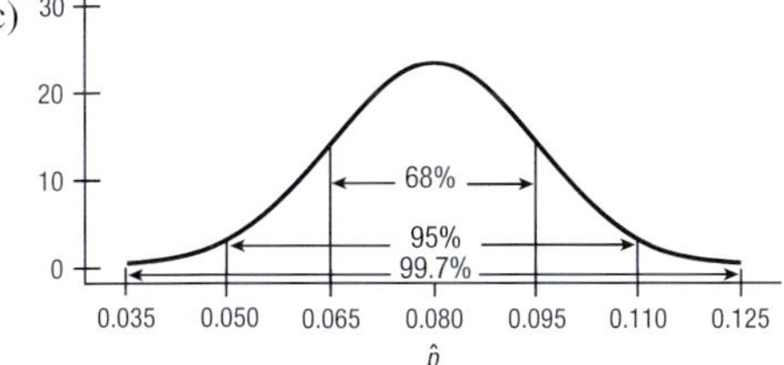

d) There is about a 68% chance of observing between 6.5% and 9.5% of color-blind males in a class of this size and a 95% chance of having between 5% and 11% color-blind males. Almost all classes of this size will have a percentage between 3.5% and 12.5% color-blind males.

R4.5. a) No. There is always natural sampling variation.
b) 42.8% and 49.0% (using 1.96 SDs).
c) A percentage of 1.06% is too close to 0 for the Normal model to be appropriate.
d) Less. Proportions farther from 0.5 have smaller standard deviations.

R4.7. a) The researcher believes that the true proportion of "A's" is within 10% of the estimated 54%—namely, between 44% and 64%.
b) The standard error of the sample is large. This is because the true proportion of A's is close to 50%, the most difficult proportion to predict.
c) No, 63% is contained in the interval.

R4.9. a) 60.1% to 63.9%
b) Since the interval is entirely above 50%, it is not plausible that only a minority of U.S. adults support legalizing gay marriage. There is strong evidence that legalizing gay marriage does have majority support.
c) About 3315 people

R4.11. a) 32.0% to 32.2%.
b) Based on this survey, we are 95% confident that the proportion of 18- to 34-year-olds who are living with parents is between 32.0% and 32.2%.
c) 95% of all such random samples will produce confidence intervals that contain the true proportion of 18- to 34-year-olds who are living with parents.

R4.13. About 1800.

R4.15. a) Random sample; less than 10% of all high school students; many more than 10 successes and 10 failures. The 90% confidence interval for proportion who cheat is 72.9% to 75.1%.
b) Based on this information, we are 90% confident that the proportion of all high school students who have cheated at least once on a test is between 72.9% and 75.1%.
c) 90% of all such random samples will produce confidence intervals that contain the true proportion of students who cheat.
d) Wider. More confidence means a larger margin of error.

R4.17. a) 22.4% to 23.2%
b) Based on this study, we are 95% confident that the proportion of American adults who are not affiliated with any organized religion or have no religious beliefs is between 22.4% and 23.2%.
c) 95% of all such random samples will produce confidence intervals that contain the true proportion of Americans who are not affiliated with any organized religion or have no religious beliefs.

R4.19. At least 423, assuming that p is near 50%.

R4.21. a) We are 95% confident that between 11.6% and 16.4% of Texas children wear helmets when biking, roller skating, or skateboarding, based on these data.
b) The data might not be a random sample.
c) About 408, using the previous 14% as $\hat{p}$.

R4.23. a) Increase. b) Decrease.

R4.25. The null hypothesis is that the percentage of students who attain a GPA of at least 3.5 remained 20% in 2012. The sample proportion of 25% is more than four standard deviations above the hypothesized rate, strong evidence that the results are not due to chance. This may be an indication of grade inflation.

R4.27. a) Although 27% of the people polled could identify her, her name recognition rate in the whole population could be less than the required 25%.
b) Type II.
c) Higher. A larger effect size leads to greater power.

R4.29. H_0: $p = 0.26$; H_A: $p \neq 0.26$. $z = 0.946$; P-value $= 0.3443$. Because the P-value is high, we do not reject H_0. These data do not provide evidence that the Denver area rate is different from the national rate in the proportion of businesses with women owners.

R4.31. a) 88.6% b) 82.2%
c) The petition would be certified when there are not enough valid signatures.
d) A valid petition is not certified.
e) H_0: $p = 0.822$; H_A: $p > 0.822$. $z = 7.48$; P-value $<$ 0.0001. With such a low P-value, we reject H_0. This sample provides sufficient evidence for certification of the petition.
f) Increase sample size.

R4.33. Based on the survey, we are 95% confident that the proportion of American adults who agree with the statement is between 56.7% and 63.1%.

R4.35. a) We are 95% confident that between 45.36% and 61.22% of patients with metastatic melanoma will have at least a partial response to vemurafenib, based on this study.
b) 266 or 267 patients (using 0.5 or 0.53 as the proportion)

R4.37. a) H_0: $\mu = 35.4$; H_A: $\mu < 35.4$
b) There is one high value, but it may or may not be an outlier. Otherwise the histogram appears roughly unimodal and symmetric (only 6 values).
c) $t = 0.726$ on 5 df; P-value $= 0.7497$. With such a large P-value, we do not reject H_0. The data do not show that the average *Weight* is less than claimed.
d) $t = -0.53$ on 4 df; P-value $= 0.3107$. Without the high outlier, the data still do not show that the average *Weight* is significantly less than claimed.
e) The stated weight seems legitimate based on these 6 bags.

R4.39. a) Different samples give different means; this is a fairly small sample. The difference may be due to natural sampling variation.
b) H_0: $\mu = 100$; H_A: $\mu < 100$
c) The batteries were independent of each other, were a representative sample of the company's batteries, and had lifetimes that were nearly Normal.
d) $t = -1.0$ on 15 df; P-value $= 0.166$; we do not reject the null hypothesis. We cannot conclude that the mean lifetime of these batteries is less than 100 hours.
e) Yes, a Type II error.

Chapter 20

1. 0.0161

3. We are 95% confident that, based on these data, the proportion of foreign-born Canadians is between 3.24% and 9.56% higher than the proportion of foreign-born Americans.

5. If we were to take repeated samples of these sizes of Canadians and Americans, and compute two-proportion confidence intervals, we would expect 95% of the intervals to contain the true difference in the proportions of foreign-born citizens.

7. We must assume the data were collected randomly and that the Americans selected are independent of the Canadians selected. Both assumptions should be met. Also, for both groups, we have at least 10 national-born and foreign-born citizens and the sample sizes are less than 10% of the population sizes. All conditions for inference are met.

9. a) 0.064
b) 3.964
c) P-value is < 0.001. Very strong evidence, so reject the null hypothesis that the proportions are the same in the two countries.

11. We must assume the samples were random or otherwise independent of each other. We also assume that the distributions are roughly normal, so it would be a good idea to check a histogram to make sure there isn't strong skewness or outliers.

13. We are 95% confident that the mean purchase amount at Sainsbury's is between £2.07 and £13.93 less than the mean purchase amount at Tesco.

15. The difference is −£8 with an SE of 3.005, so the t-stat is −2.662. With 157.525 df, the P-value is 0.0086, which is less than 0.05. Reject the null hypothesis that the means are equal.

17. The t-statistic is −2.655 using the pooled estimate of the standard deviation, 3.013. There are 158 df, so the P-value is still 0.0087. We reach the same conclusion as before. Because the sample standard deviations are nearly the same and the sample sizes are large, the pooled test is essentially the same as the two-sample t-test.

19. No. The two-sample test is almost always the safer choice. In this case, the variances are likely to be quite different. The purchase prices of prêt-à-porter items are much lower and may be less variable than the prices of fancy designer dresses. They should use the two-sample t-test.

21. It's very unlikely that samples would show an observed difference this large if in fact there is no real difference in the proportions of internet users worldwide who use the social media platform between the two time periods.

23. The responses are not from two independent groups, but are from the same individuals.

25. a) Stratified.
b) 6% higher among males.
c) 4%
d)

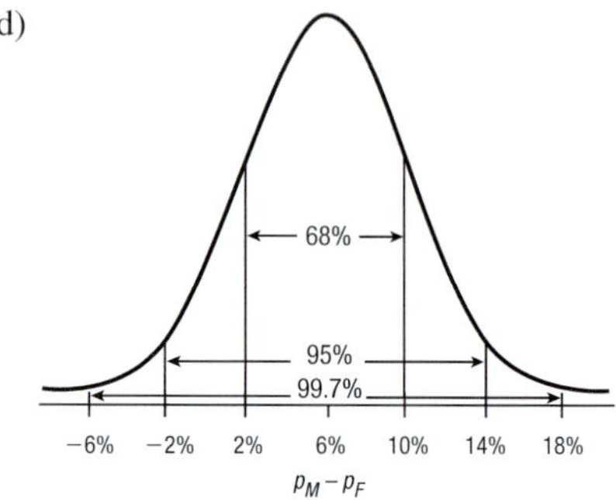

e) Yes. A poll result showing little difference is only 1–2 standard deviations below the expected outcome.

27. a) Yes. Random samples; less than 10% of the population; samples are independent; more than 10 successes and failures in each sample.
b) $(0.131, 0.217)$
c) We are 95% confident, based on these samples, that the proportion of Australian women aged 65 and older who suffer from arthritis is between 13.1% and 21.7% higher than the proportion of Australian men of the same age who suffer from arthritis.
d) Yes. The entire interval lies above 0.

29. a) 0.032
b) $(0.353, 0.478)$
c) We are 95% confident, based on these data, that the proportion of pets with a malignant lymphoma in homes where herbicides are used is between 35.3% and 47.8% higher than the proportion of pets with lymphoma in homes where no pesticides are used.

31. a) Experiment. Men were randomly assigned to have surgery or not.
b) $(0.006, 0.080)$
c) Since the entire interval lies above 0, there is evidence that surgery may be effective in preventing death from prostate cancer.

33. a) Yes. Subjects were randomly divided into independent groups, and more than 10 successes and failures were observed in each group.
b) $(4.7\%, 8.9\%)$
c) We're 95% confident that the rate of infection is 5–9 percentage points lower. That's a meaningful reduction, considering the 20% infection rate among the unvaccinated kids.

35. a) H_0: $p_V - p_{NV} = 0$; H_A: $p_V - p_{NV} < 0$
b) Because 0 is not in the confidence interval, reject the null. There's evidence that the vaccine reduces the rate of ear infections.
c) Type I.
d) Babies would be given ineffective vaccinations.

37. a) Prospective study.
b) H_0: $p_1 - p_2 = 0$; H_A: $p_1 - p_2 \neq 0$, where p_1 is the proportion of teens whose parents disapprove of smoking who become smokers and p_2 is the proportion whose parents are lenient about smoking who become smokers.
c) Yes. We assume the students were randomly selected; they are less than 10% of the population; samples are independent; at least 10 successes and failures in each sample.
d) $z = -1.17$, P-value $= 0.2844$. These samples do not show evidence that parental attitudes influence teens' decisions to smoke.
e) If there is no difference in the proportions, there is about a 28% chance of seeing the observed difference or larger by natural sampling variation.
f) Type II.
g) $(-0.221, 0.065)$
h) We are 95% confident that the proportion of teens whose parents disapprove of smoking who will become established smokers within two years is between 22.1% lower and 6.5% higher than for teens with parents who are lenient about smoking.
i) 95% of all random samples will produce intervals that contain the true difference.

39. a) H_0: $p_1 - p_2 = 0$; H_A: $p_1 - p_2 > 0$, where p_1 is the proportion of mothers exposed to soot and ash and p_2 is the proportion of mothers not exposed. $z = 1.949$, P-value $= 0.026$. With a P-value this low, we reject H_0 at $\alpha = 0.10$. This study shows a significantly higher rate of low-weight babies born to mothers exposed to high levels of air pollution during pregnancy.
b) We are 90% confident that the proportion of low-birthweight babies will be between 0.8% and 7.7% higher for mothers exposed to high levels of air pollution than those who are not.

41. a) H_0: $p_1 - p_2 = 0$; H_A: $p_1 - p_2 > 0$, where p_1 is the proportion of women who never had mammograms who die of breast cancer and p_2 is the proportion of women who undergo screening who die of breast cancer. $z = 2.17$, P-value $= 0.0148$. With a P-value this low, we reject H_0. These data do suggest that mammograms may reduce breast cancer deaths.
b) Type I

43. a) We are 95% confident, based on this study, that between 67.0% and 83.0% of patients with joint pain will find medication A effective.
b) We are 95% confident, based on this study, that between 51.9% and 70.3% of patients with joint pain will find medication B effective.
c) Yes, they overlap. This might indicate no difference in the effectiveness of the medications, although this is not a proper test.
d) We are 95% confident that the proportion of patients with joint pain who will find medication A effective is between 1.7% and 26.1% higher than the proportion who will find medication B effective.
e) No. There is evidence of a difference in the effectiveness of the medications.
f) To estimate the variability in the difference of proportions, we must add variances. The two one-sample intervals do not. The two-sample method is the correct approach.

45. A 95% confidence interval is $(0.081; 0.143)$; 0%—or no significant bounce—is not a plausible value.

47. The conditions are satisfied to test H_0: $p_{young} - p_{old} = 0$ against H_A: $p_{young} - p_{old} > 0$. $z = 1.56$. The one-sided P-value is 0.0602, so we may not reject the null hypothesis at $\alpha = 0.05$, but would at $\alpha = 0.10$. Although the evidence is not strong, *Time* may be justified in saying that younger men are more comfortable discussing personal problems.

49. H_0: $p_{urban} - p_{rural} = 0$ vs. H_A: $p_{urban} - p_{rural} \neq 0$. Yes. With a low P-value less than 0.001, reject the null hypothesis of no difference. There is strong evidence to suggest that the percentages are different for the two groups: People from urban areas are more likely to agree with the statement than those from rural areas.

51. Yes. The high P-value means that we lack evidence of a difference, so 0 is a plausible value for $\mu_{Meat} - \mu_{Beef}$.

53. a) Plausible values of $\mu_{Meat} - \mu_{Beef}$ are all negative, so the mean fat content is probably higher for beef hot dogs.
b) The difference in sample means is significant.
c) 0.10

55. a) False. The confidence interval is about means, not about individual hot dogs.
b) False. The confidence interval is about means, not about individual hot dogs.

c) True.
d) False. CIs based on other samples will also try to estimate the true difference in population means; there's no reason to expect other samples to conform to this result.
e) True.

57. a) 2.927
b) Larger.
c) Based on this sample, we are 95% confident that students who learn Math using the CPMP method will score, on average, between 5.57 and 11.43 points higher on a test solving applied Algebra problems with a calculator than students who learn by traditional methods.
d) Yes. 0 is not in the interval.

59. a) H_0: $\mu_C - \mu_T = 0$ vs. H_A: $\mu_C - \mu_T \neq 0$
b) Yes. Groups are independent, though we don't know if students were randomly assigned to the programs. Sample sizes are large, so CLT applies.
c) If the means for the two programs are really equal, there is less than a 1 in 10,000 chance of seeing a difference as large as or larger than the observed difference just from natural sampling variation.
d) On average, students who learn with the CPMP method do significantly worse on Algebra tests that do not allow them to use calculators than students who learn by traditional methods.

61. a) (1.36, 4.64)
b) No. 5 minutes is beyond the high end of the interval.

63. These were random samples, both less than 10% of properties sold. Prices of houses should be independent, and random sampling makes the two groups independent. The boxplots make the price distributions appear to be reasonably symmetric, and with the large sample sizes the few outliers don't affect the means much. Based on this sample, we're 95% confident that, in New York, having a waterfront is worth, on average, about \$59,121 to \$140,898 more in sale price.

65.

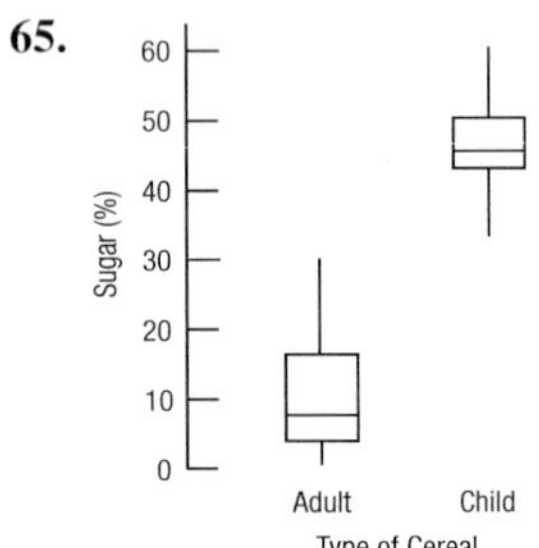

Random sample—questionable, but probably representative, independent samples, less than 10% of all cereals; boxplot shows no outliers—not exactly symmetric, but these are reasonable sample sizes. Based on these samples, with 95% confidence, children's cereals average between 32.49% and 40.80% more sugar content than adult's cereals.

67. H_0: $\mu_N - \mu_C = 0$ vs. H_A: $\mu_N - \mu_C > 0$; $t = 2.207$; P-value $= 0.034$; df $= 37.28$. Because of the small P-value, we reject H_0 (at $\alpha = 0.05$). These data do suggest that new activities are better. The mean reading comprehension score for the group with new activities is significantly (at $\alpha = 0.05$) higher than the mean score for the control group.

69. a)

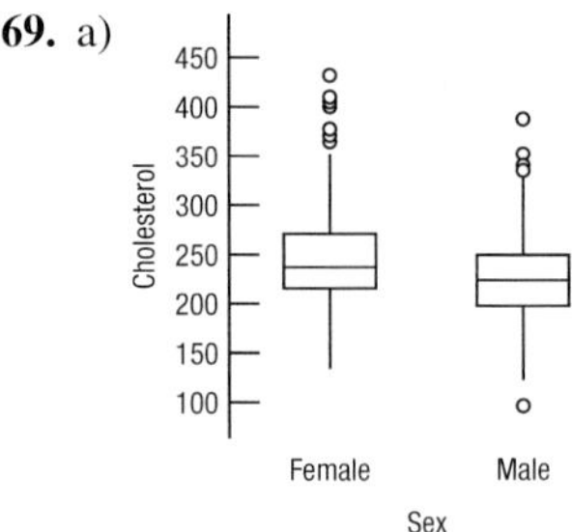

b) No. There is a great deal of overlap between the two genders.
c) The test statistic t is 6.97 with 1401 df, giving a P-value < 0.0001. We reject the null hypothesis because the P-value is so small. There is strong evidence that there is a difference in mean cholesterol level between men and women. The 95% confidence interval is (12.1, 21.6).
d) No. The evidence is stronger than it appeared from the boxplots.
e) No. Removing the outliers would reduce the standard error (which would actually make stronger evidence of a difference) and not change the means by much. The overall effect on the test statistic would probably be small.

71. a)

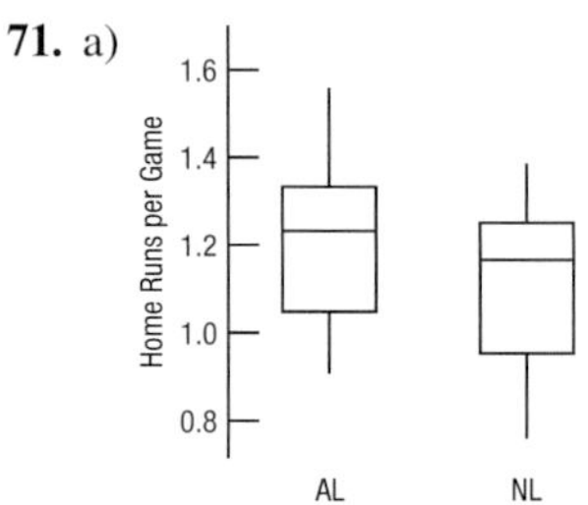

Both are reasonably symmetric with no outliers.
b) Based on these data, the average number of home runs hit per game in an American League stadium is between 1.12 and 1.31 with 95% confidence.
c) No. The boxplot indicates that there are no National League outliers (and Colorado is not even the highest).
d) We want to work directly with the average difference. The two separate confidence intervals do not answer questions about the difference. The difference has a different standard deviation, found by adding variances.
e) (−0.021, 0.263) (AL − NL)
f) Based on these data, with 95% confidence, American League stadiums average between 0.021 fewer and 0.263 more home runs per game than National League stadiums.
g) No, 0 is in the interval.

73. These are not two independent samples. These are before and after scores for the same individuals.

75. a) These data do not provide evidence of a difference in the mean number of items recalled by viewers of red and black cardboards.
b) H_0: $\mu_R - \mu_B = 0$ vs. H_A: $\mu_R - \mu_B \neq 0$; $t = 2.378$; df ≈ 198; P-value $= 0.018$. Because the P-value is low, we reject H_0. These data suggest that item recall between red and blue cardboards is different; those who saw red cardboards had higher average recall.

77. H_0: $\mu_{big} - \mu_{small} = 0$ vs. H_A: $\mu_{big} - \mu_{small} \neq 0$; bowl size was assigned randomly; amount scooped by individuals and by the two groups should be independent. With 34 df, $t = 2.104$ and P-value $= 0.0428$. The low P-value leads us to reject the null hypothesis (at $\alpha = 0.05$) and conclude that there is a difference in the average amount of ice cream that people scoop when given a bigger bowl.

79. a) The interval is $(-135.98, 207.58)$. If the assumptions and conditions are met, we can be 95% confident that the mean time for women to cross Lake Ontario is between 135.98 minutes less and 207.58 minutes more than the mean time for men to cross. Because the interval includes 0, we cannot be confident that there is any difference at all.
b) The P-value for a two-sided test is 0.676.
c) Independence Assumption: There is no reason to doubt that the swims are independent or to think that the two groups are not independent of each other (except for a few multiple swims by the same swimmer). Randomization Condition: The swimmers are not a random sample from any identifiable population. They may be representative of swimmers who tackle challenges such as this but we have no way of knowing.
Nearly Normal Condition: The boxplots show two high outliers for the men and some skewness for both. Removing the outliers may make the difference in times larger, but there is no justification for doing so. The histograms are unimodal, but somewhat skewed to the right. We are reluctant to draw any conclusions about the difference in mean times it takes men or women to swim the lake. The sample is not random, we have no way of knowing if it is representative, and the data are skewed with some outliers.

81. Independent Groups Assumption: The runners are different women, so the groups are independent. The Randomization Condition is satisfied, since the runners are selected at random for these heats.

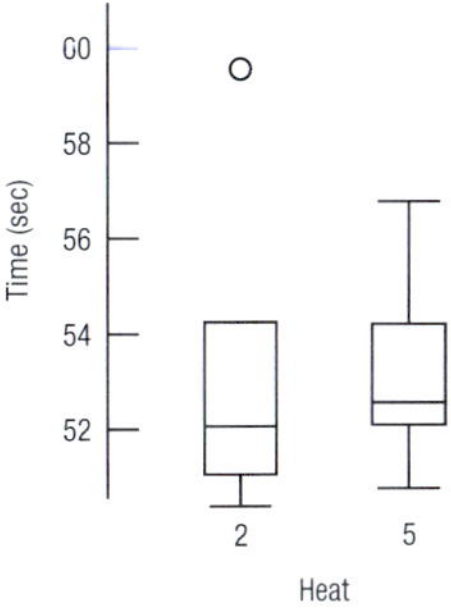

Nearly Normal Condition: The boxplots show an outlier, but we will proceed and then redo the analysis with the outlier deleted. When we include the outlier, $t = 0.0484$ with a two-sided P-value of 0.963. With the outlier deleted, $t = -1.10$ with $P = 0.300$. Either P-value is so large that we fail to reject the null hypothesis of equal means and conclude that there is no evidence of a difference in the mean times for randomly assigned heats.

*There are only two "exceedences," so Tukey's test does not find a significant difference. The Rank-sum test agrees, with a P-value of 0.70.*Using the randomization test, 500 simulated differences are plotted below. The actual difference of 0.07833 is almost exactly in the middle of the distribution of simulated differences. The simulated P-value is 0.952, which means that 95.2% of the simulated differences were at least 0.7833 in absolute value; there is no evidence of a difference in mean times between heats.

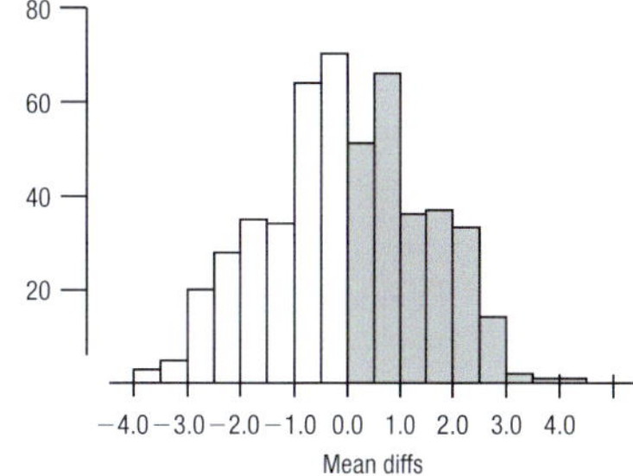

83. H_0: $\mu_{Regular} - \mu_{Stinger} = 0$ vs. H_A: $\mu_{Regular} - \mu_{Stinger} < 0$. With $t = -4.57$ and a very low P-value of 0.0013, we reject the null hypothesis of equal mean velocities. There is strong evidence that golf balls hit off Stinger tees have a higher mean initial velocity.

85. a) H_0: $\mu_M - \mu_R = 0$ vs. H_A: $\mu_M - \mu_R > 0$. $t = -0.70$; df $= 45.88$; P-value $= 0.7563$. Because the P-value is so large, we do not reject H_0 for any reasonable α-level. These data provide no evidence that listening to Mozart while studying is better than listening to rap. Now you might have noticed that subjects listening to Mozart recalled fewer items. And so, there is no need for a statistical test. Looking and thinking first can save you some calculation effort. If you did notice, and said there's no evidence of a benefit, without calculation—that's a good answer to this exercise.
b) With 90% confidence, the average difference in score is between 0.189 and 5.351 objects more for those who listen to no music while studying, based on these samples.

87. a)

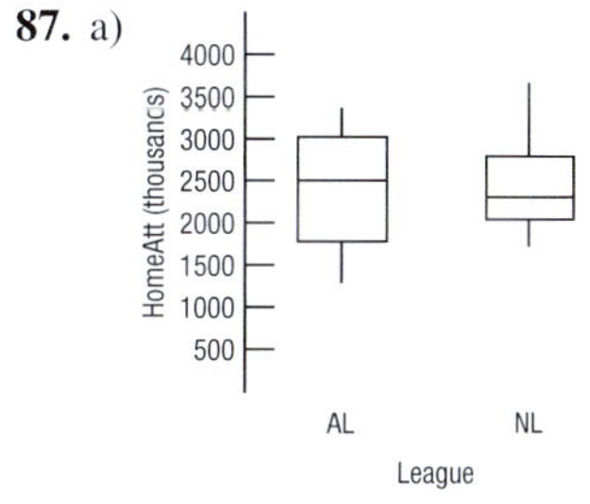

b) A difference of 60,000 is very near the center of the distribution of simulated differences. The vast majority of the simulated differences are more than 60,000 in absolute value, so the difference is not statistically significant.

89. Using a randomization test with 500 simulated differences, the actual difference of -16.8 is far from the simulated values. The P-value is 0. In fact, none of the simulated differences are even close to an absolute value of 16.8. This means that there is strong evidence of a difference in mean cholesterol levels between men and women.

Chapter 21

1. a) Paired b) Not paired c) Paired

3. a) No. The vehicles have no natural pairing.
b) Possibly. The data are quantitative and paired by vehicle.

c) The sample size is large, but there is at least one extreme outlier that should be investigated before applying these methods. And there are several values near 0 that should be investigated as well.

5. (7.10, 7.40). We are 95% confident that the interval 7.10 mpg to 7.40 mpg captures the true improvements in highway gas mileage compared to city gas mileage.

7. The difference between fuel efficiency of cars and that of trucks can be large but isn't relevant to the question asked about highway vs. city driving. Pairing places each vehicle in its own block to remove that variation from consideration.

9. a) Randomly assign 50 cows to each of the two kinds of feed. Compare production at the end of the month.
b) Give all 100 cows the new feed for 2 weeks and the old feed for 2 weeks, randomly selecting which feed the cows get first. Analyze the differences in production for all 100 cows.
c) Matched pairs. Because cows vary in milk production, the matched-pairs design will control for that.

11. a) Show the same people ads with and without celebrity images, and record how many products they remember in each group. Randomly decide which ads a person sees first. Examine the differences for each person.
b) Randomly divide volunteers into two groups. Show one group ads with celebrity images and the other group ads without them. Compare how many products each group remembers.

13. a) Matched pairs—same cities in different periods.
b) There is a significant difference (P-value $= 0.0244$) in the labor force participation rate for women in these cities; women's participation seems to have increased between 1968 and 1972.

15. a) Use the paired t-test because we have pairs of Fridays in 5 different months. Data from consecutive Fridays within a month may be more similar than data from randomly chosen Fridays.
b) We conclude that there is evidence (one-sided P-value $=$ 0.0004) that the mean number of cars found on the M25 motorway on Friday the 13th is lower than on the previous Friday.
c) We don't know if these Friday pairs were selected at random. If these are the Fridays with the largest differences, this will affect our conclusion. The Nearly Normal Condition appears to be met by the differences, but the sample size is small.

17. Adding variances requires that the variables be independent. These price quotes are for the same cars, so they are paired. Drivers quoted high insurance premiums by the local company will be likely to get a high rate from the online company, too.

19. a) The histogram—we care about differences in price.
b) Insurance cost is based on risk, so drivers are likely to see similar quotes from each company, making the differences relatively smaller.
c) The price quotes are paired; they were for a random sample of fewer than 10% of the agent's customers; the histogram of differences looks approximately Normal.

21. H_0: $\mu(Local - Online) = 0$ vs. H_A: $\mu(Local - Online) > 0$; with 9 df, $t = 0.76$. With a high P-value of 0.233, we don't reject the null hypothesis. These data don't provide evidence that online premiums are lower, on average.

23.

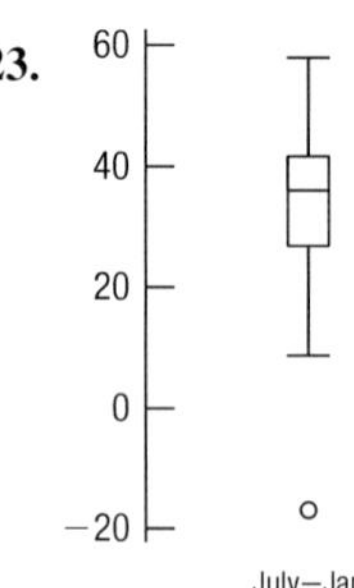

a) The boxplot shows a low outlier, which turns out to be Auckland, New Zealand. A southern hemisphere city would have seasons reversed, so it doesn't belong in these data.
b) A 95% CI for the remaining 34 cities is (30.94, 39.53) degrees. We can be 95% certain that this interval contains the true mean difference in temperatures between July and January in cities like these. Data are paired for each city; cities are independent of each other; boxplot shows the temperature differences are reasonably symmetric, with no outliers. This is probably not a random sample, so we might be wary of inferring that this difference applies to all European cities. Based on these data, we are 90% confident that the average temperature in European cities in July is between 30.9°F and 39.5°F higher than in January.
c) Answers will vary. Here is one possible answer: (28.9, 38.1).

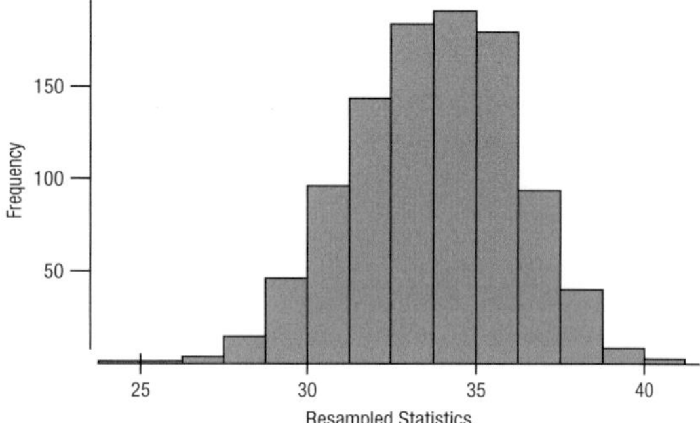

25. Based on these data, we are 90% confident that boys, on average, can do between 2.0 and 14.2 more push-ups than girls (independent samples—not paired).

27. a) Paired sample test. Data are before/after for the same workers; workers randomly selected; assume fewer than 10% of all this company's workers; boxplot of differences shows them to be symmetric, with no outliers.

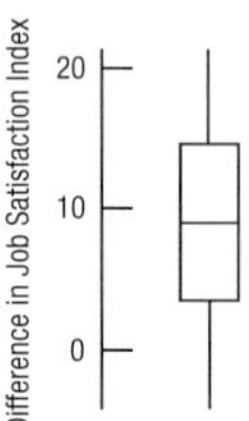

b) H_0: $\mu_D = 0$ vs. H_A: $\mu_D > 0$. $t = 3.60$; P-value $= 0.0029$. Because $P < 0.01$, reject H_0. These data show evidence that average job satisfaction has increased after implementation of the program.
c) Type I.

29. H_0: $\mu_D = 0$ vs. H_A: $\mu_D \neq 0$. Data are paired by brand; brands are independent of each other; fewer than 10% of all yogurts (questionable); boxplot of differences shows an outlier (100) for Great Value:

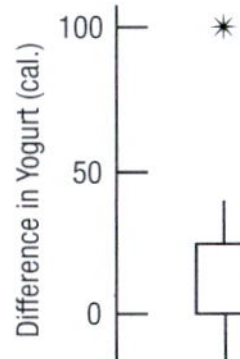

With the outlier included, the mean difference (Strawberry – Vanilla) is 12.5 calories with a t-stat of 1.332, with 11 df, for a P-value of 0.2098. Deleting the outlier, the difference is even smaller, 4.55 calories with a t-stat of only 0.833 and a P-value of 0.4241. With P-values so large, we do not reject H_0. We conclude that the data do not provide evidence of a difference in mean calories.

31. a) Cars were probably not a simple random sample but may be representative in terms of stopping distance; the boxplot does not show outliers, but does indicate right skewness. A 95% confidence interval for the mean stopping distance on dry pavement is (131.5, 148.9) feet.

b) Data are paired by car; cars were probably not randomly chosen, but representative; the boxplot shows an outlier (car 3) with a difference of 12. With deletion of that car, a Normal probability plot of the differences is relatively straight.

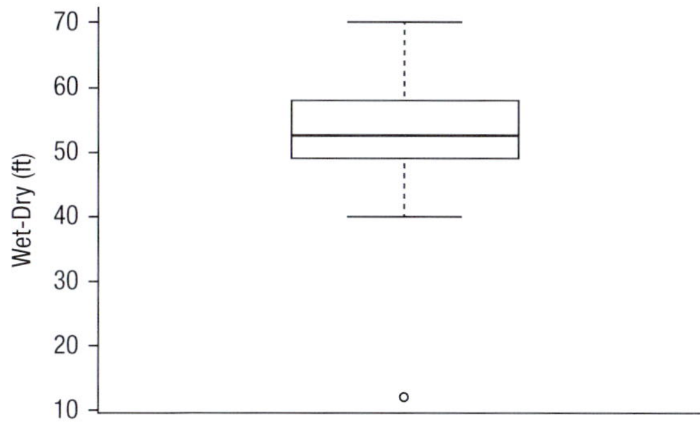

Retaining the outlier, we estimate with 95% confidence that the average braking distance is between 38.8 and 62.2 feet more on wet pavement than on dry, based on this sample. (Without the outlier, the confidence interval is 47.8 to 61.8 feet.)

33. a) Paired Data Condition: Data are paired by college. Randomization Condition: This was a random sample of public colleges and universities.

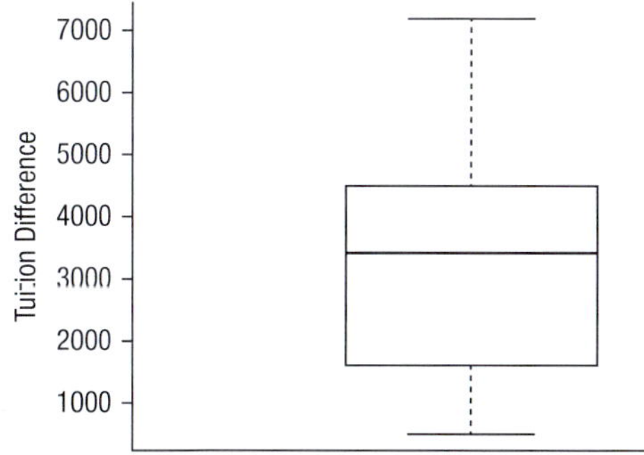

Normal population assumption: There are no outliers.

b) We are 90% confident, based on the data, that those from overseas pay, on average, between £2763 and £4045 more than residents.

c) The assertion is reasonable; without the outliers, £3500 is in the confidence interval.

35. a) 60% is 30 strikes; H_0: $\mu = 30$ vs. H_A: $\mu > 30$. $t = 6.07$; P-value $= 3.92 \times 10^{-6}$. With a very small P-value, we reject H_0. There is very strong evidence that players can throw more than 60% strikes after training, based on this sample.

b) H_0: $\mu_D = 0$ vs. H_A: $\mu_D > 0$. $t = 0.135$; P-value $= 0.4472$. With such a high P-value, we do not reject H_0. These data provide no evidence that the program has improved pitching in these Little League players.

c) Answers will vary. A 95% confidence interval is (32.2, 34.5). This interval is significantly above 30!

37. a) The data are clearly paired. Both races may have improved over time, but the pairwise differences are likely to be independent. We can only check the Nearly Normal Condition by using the computer files; however, with a sample size of 40 we are not very concerned.

b) With 95% confidence we can say the mean time difference is between -18.00 (women are faster) and 3.47 (men are faster).

c) The interval contains 0, so we would not reject the hypothesis of no mean difference at $\alpha = 0.05$. We can't discern a difference between the mean female wheelchair times and the male running times.

39. a) Same cows before and after injection; the cows should be representative of others of their breed; cows are independent of each other; less than 10% of all cows; don't know about Nearly Normal differences.

b) (18.69, 21.31)

c) Based on this sample, with 95% confidence, the average increase in milk production for a herd of 70 Ayrshire cows given BST is between 18.69 and 21.31 pounds per day.

d) $0.25(45) = 11.25$. The average increase is much more than this, so we would recommend he incur the extra expense.

Chapter 22

1. a) 30 births per season.

b) 2.8.

c) 3 df.

3. a) 3

b) No. It's smaller than the mean.

c) It would say that there is not enough evidence to reject the null hypothesis that births are distributed uniformly across the seasons.

d) 7.815

e) Do not reject the null hypothesis. As in part c, there is not enough evidence to suggest that births are not distributed uniformly across the seasons.

5. a) The age distributions of the customers are the same at the two branches.

b) χ^2 test of homogeneity.

c)

	Age			
	Less Than 30	**30–55**	**56 or Older**	**Total**
In-Town Branch	27.5	45	27.5	**100**
Mall Branch	27.5	45	27.5	**100**
Total	**55**	**90**	**55**	**200**

d) 30.8081.
e) 2 df.
f) P-value < 0.00001.
g) We reject the null hypothesis and conclude that the distribution of ages is not the same at the two branches.

7. a) $(-0.913, 0.913, 0.365, -0.365)$
b) No. They are quite small for z-values.
c) Because we did not reject the null hypothesis, we shouldn't expect any of the standardized residuals to be large.

9. a)

		Lethal?		
		Fatal	**Not fatal**	**Total**
Injury Site	**Body**	50.06	11.94	62
	Head/neck	35.53	8.47	44
	Limb	23.41	5.59	29
	Total	109	26	135

b) 51.6698. c) 2 df. d) P-value < 0.00001.
e) We reject the null hypothesis and conclude that there is strong evidence of an association between the site of injury and the lethality of the injury.

11. a) Chi-square test of independence. We have one sample and two variables. We want to see if the variable *Account Type* is independent of the variable *Trade Type*.
b) Other test. *Account Size* is quantitative, not counts.
c) Chi-square test of homogeneity. We want to see if the distribution of one variable, *Courses*, is the same for two groups (resident and nonresident students).

13. a) 16 times.
b) Goodness-of-fit.
c) H_0: The die is fair (All faces have the same probability of coming up), versus H_A: The die is not fair.
d) Count data; rolls are random and independent; expected frequencies are all bigger than 5.
e) 5 df.
f) $\chi^2 = 17.125$; P-value ≈ 0.0043.
g) We reject the null hypothesis and conclude that there is strong evidence that the die is unfair.

15. a) Weights are quantitative, not counts.
b) Count the number of each kind of nut, assuming the company's percentages are based on counts rather than weights.

17. H_0: The police force represents the population, versus H_A: The police force is not representative of the population. $\chi^2 \approx 19906.8881$; df $= 4$; P-value < 0.00001. We reject the null hypothesis and conclude that the police force is not representative of the population. In particular, there are too many white officers in relationship to their membership in the community.

19. a) $\chi^2 \approx 5.778$; df $= 3$; P-value $= 0.1229$. We fail to reject the null hypothesis and conclude that there is not sufficient evidence to support the claim that the ratio of traits is different from the theoretical ratio.
b) $\chi^2 \approx 11.556$; df $= 3$; P-value $= 0.0091$. We reject the null hypothesis and conclude that there is strong evidence that the ratio of traits is different from the theoretical ratio.
c) With smaller samples, more data sets will be consistent with the null hypothesis. With larger samples, smaller discrepancies will show evidence against the null hypothesis.

21. a) $96/16 = 6$
b) Goodness-of-fit.
c) H_0: The number of large hurricanes remains constant over decades. H_A: The number of large hurricanes has changed.
d) 15
e) P-value $= 0.63$
f) The very high P-value means these data offer no evidence that the number of large hurricanes has changed.

23. a) Independence.
b) H_0: Breastfeeding success is independent of having an epidural. H_A: There's an association between breastfeeding success and having an epidural.

25. a) 1 b) 159.34
c) Breastfeeding behavior should be independent for these babies. They are fewer than 10% of all babies; we assume they are representative. We have counts, and all the expected counts are at least 5.

27. a) 5.90 b) P-value < 0.005
c) The P-value is very low, so reject the null. There's evidence of an association between having an epidural and subsequent success in breastfeeding.

29. a) $\dfrac{(190 - 159.34)}{\sqrt{159.34}} = 2.43$
b) It appears that babies whose mothers had epidurals during childbirth are much less likely to be breastfeeding 6 months later.

31. These factors would not be mutually exclusive. There would be yes or no responses for every baby for each.

33. a) 40.26% b) 8.15% c) 62.04% d) 286.67
e) H_0: Survival was independent of status on the ship. H_A: Survival was not independent of the status.
f) 3
g) We reject the null hypothesis and conclude that survival and ticket class are not independent.

35. First-class passengers were most likely to survive, while third-class passengers and crew were underrepresented among the survivors.

37. a) Experiment—actively imposed treatments (different drinks).
b) Homogeneity.
c) H_0: The rate of urinary tract infection is the same for all three groups. H_A: The rate of urinary tract infection is different among the groups.
d) Count data; random assignment to treatments; all expected frequencies larger than 5.
e) 2
f) $\chi^2 = 7.776$; P-value $= 0.020$
g) With a P-value this low, we reject H_0 at $\alpha = 0.05$. These data provide reasonably strong evidence that there is a difference in urinary tract infection rates between cranberry juice drinkers, lactobacillus drinkers, and the control group.
h) The standardized residuals are

	Cranberry	**Lactobacillus**	**Control**
Infection	−1.87276	1.19176	0.68100
No Infection	1.24550	−0.79259	−0.45291

From the standardized residuals (and the sign of the residuals), it appears those who drank cranberry juice were less likely to develop urinary tract infections; those who drank lactobacillus were more likely to have infections.

39. a) Independence.
b) H_0: *Political Affiliation* is independent of *Sex*. H_A: There is a relationship between *Political Affiliation* and *Sex*.
c) Counted data; probably a random sample, but can't extend results to other states; all expected frequencies greater than 5.
d) $\chi^2 = 4.851$; df $= 2$; P-value $= 0.0884$
e) Because of the high P-value, we do not reject H_0 at $\alpha = 0.05$. These data do not provide evidence of a relationship between *Political Affiliation* and *Sex*.

41. H_0: *Political Affiliation* is independent of *Region*. H_A: There is a relationship between *Political Affiliation* and *Region*. $\chi^2 = 13.849$; df $= 4$; P-value $= 0.0078$. With a P-value this low, we reject H_0 at $\alpha = 0.01$. *Political Affiliation* and *Region* are related. Examination of the residuals shows that those in the West are more likely to be Democrat than Republican; those in the Northeast are more likely to be Republican than Democrat.

43. a) Homogeneity.
b) H_0: The grade distribution is the same for both professors. H_A: The grade distributions are different.
c)

	Prof. Alpha	Prof. Beta
A	6.667	5.333
B	12.778	10.222
C	12.222	9.778
D	6.111	4.889
F	2.222	1.778

Three cells have expected frequencies less than 5.

45. a)

	Prof. Alpha	Prof. Beta
A	6.667	5.333
B	12.778	10.222
C	12.222	9.778
Below C	8.333	6.667

All expected frequencies are now larger than 5.
b) Decreased from 4 to 3.
c) $\chi^2 = 9.306$; P-value $= 0.0255$. Because the P-value is so low, we reject H_0 at $\alpha = 0.05$. The grade distributions for the two professors are different. Dr. Alpha gives fewer A's and more grades below C than Dr. Beta.

47. $\chi^2 = 14.058$; df $= 1$; P-value $= 0.0002$. With a P-value this low, we reject H_0 at $\alpha = 0.001$. There is evidence of racial steering. Blacks are much less likely to rent in Section A than Section B.

49. $\chi^2 = 2.699$; df $= 3$; $P = 0.4404$. Because the P-value is > 0.05, these data do not show enough evidence of an association between the mother's age group and the outcome of the pregnancy.

Chapter 23

1. $\widehat{Earn} = 14318.69 + 27.12SAT$. On average, earnings increase by \$27.12/year per additional point in SAT score.

3. The residual plot has no structure, and there are not any striking outliers. The histogram of residuals is symmetric and bell-shaped. All conditions are satisfied to proceed with the regression analysis.

5. $s = \$5603$. This is the standard deviation of the residuals and thus indicates how much the data points vary about the linear regression model.

7. The standard error of the slope is the estimated standard deviation of the sampling distribution for the slope. It tells us how much the slope of the regression equation would vary from sample to sample. The P-value is essentially zero, which confirms that the slope is statistically significantly different than zero.

9. We can reject the null hypothesis and conclude that the slope of the relationship between *Earn* and *SAT* is not zero. It seems that those who score higher on their SAT tend to earn more.

11. We are 95% confident that the true slope relating *Earn* and *SAT* is between 24.23 and 30.3 \$/year per SAT point.

13. a) $\widehat{Earn} = 23{,}974 + 23.188(SAT) - 8501(\%need)$
b) On average, *Earn* increases by about \$23/year per *SAT* point after allowing for the effects of *%need*.

15. a) *ACT* and *SAT* are highly correlated with each other. After all, they are very similar measures. Thus *SAT*, after allowing for the effect of *ACT*, is not really a measure of test performance but rather a measure of how students who take the SAT may differ from those who take the ACT at the colleges in question.
b) (1.56, 18.66); the interval covers much smaller values as the plausible ones.
c) We are less confident that this coefficient is different than zero. The collinearity with *ACT* has inflated the variance of the coefficient of *SAT*, leading to a smaller *t*-ratio and larger P-value.

17. A prediction interval for an individual value will be too wide to be of much use.

19. No, regression models describe the data as they are. They cannot predict what would happen if a change were made. We cannot conclude that earning a higher SAT score will lead to higher earnings; that would suppose a causal relationship that clearly isn't true.

21. a) $\widehat{Error_24h} = 4205.28 - 2.06$ year; according to the model, the error made in predicting a hurricane's path has been declining at a rate of about 2.06 nautical miles per year.
b) $H_0: \beta_1 = 0$; there has been no change in prediction accuracy. $H_A: \beta_1 \neq 0$; there has been a change in prediction accuracy.
c) With a P-value < 0.0001, I reject the null hypothesis and conclude that prediction accuracies have in fact been changing during this period.
d) 77.5% of the variation in hurricane prediction accuracy is accounted for by this linear model on time.

23. a) $\widehat{Extent} = 97.3704 - 6.01784(MeanGlobalTemperature)$. We would expect 97.3704 km^2 of sea ice if the mean global temperature were 0°C. The extent of sea ice decreases by 6.01784 km^2 with each increase of 1°C in mean global temperature.
b) The scatter plot shows a moderate, linear relationship with no striking outliers. The plot of the residuals versus the predicted values has no apparent bend but has a slightly greater variability on the left than on the right. The Normal probability plot is reasonably straight.
c) The amounts by which sea ice extents differ from predictions vary with a standard deviation of 0.4437 km^2.

d) 0.3633.

e) The standard error is the estimated standard deviation of the sampling distribution of the slope coefficient. The slopes of regression lines for models based on different samples of years would vary with a standard deviation of about 0.3633 km^2 per °C.

f) No. We can see an association, but we cannot establish causation from this study.

25. We are 95% confident that the slope is between -5.6 and -3.24.

27. a) $H_0: \beta_1 = 0$; there is no linear association between calories and sodium content.
$H_A: \beta_1 \neq 0$: there is a linear association between calories and sodium content.

b) The P-value is 0.0011, so we reject the null hypothesis and conclude that there is evidence of a linear association between calories and sodium content. Since the slope is positive, hot dogs with more calories generally have higher sodium contents.

29. a) Among all-beef hot dogs with the same number of calories, the sodium content varies, with a standard deviation of about 60 mg.

b) 0.561 mg/cal

c) If we tested many other samples of all-beef hot dogs, the slopes of the resulting regression lines would be expected to vary, with a standard deviation of about 0.56 mg of sodium per calorie.

31. I'm 95% confident that for every additional calorie, all-beef hot dogs have, on average, between 1.07 and 3.53 mg more sodium.

33. a) H_0: Difference in age at first marriage has not been changing, $\beta_1 = 0$.
H_A: Difference at first marriage has been changing, $\beta_1 \neq 0$.

b) Plot of residuals vs. predicted values shows a "wave" shape, oscillating. The histogram of the residuals is bimodal. However, the Normal probability plot suggests that the residuals are nearly Normal. We should be cautious in interpreting the regression model.

c) $t = -13.3$, $P < 0.0001$. We can be confident that the slope is not zero. It looks like the gap in age between men and women at first marriage has been shrinking.

35. We are 95% confident that the slope of *Men-Women* vs *Year* is between -0.018 and -0.014.

37. a) $H_0: \beta_1 = 0$; there is no linear association between pH and BCI.
$H_A: \beta_1 \neq 0$: there is a linear association between pH and BCI.

b) $t \approx -1.2941$ with 17 df. P-value $= 0.213$.

c) Fail to reject the null hypothesis and conclude that there is no evidence of a linear association between pH and BCI.

39. a) H_0: After accounting for Alkali content, *BCI* and *pH* are not (linearly) related, $\beta_1 = 0$.
H_A: *BCI* changes with *pH*, $\beta_1 \neq 0$.

b) With a very large P-value of 0.47 we fail to reject H_0. We failed to find evidence that *pH* is related to *BCI* after allowing for the effects of *Alkali*.

c) *pH* is likely to be correlated with *Alkali*. After allowing for the effect of *Alkali*, there is no remaining effect of *pH*. The collinearity has inflated the SE of the coefficient of *pH*, reducing its *t*-ratio.

41. a) $\widehat{Global\ Temperature\ Anomaly} = -2.84396 + 0.00902(CO_2)$.

b) Yes, since the P-value for the slope is significant and the scatterplot is fairly straight.

c) Yes, since the standard deviation of the residuals is small and the scatterplot is straight.

d) No, the model is consistent with the claim that increasing CO_2 is causing global warming, but it does not by itself prove that this is the mechanism. There may be another lurking variable.

43. a) (0.0092, 0.011) degrees per CO_2 ppm.

b) We are 90% confident that *Temp* increases between 0.0092 and 0.011 degrees per ppm of CO_2.

45. a) $1.28 \pm 1.67 \times 0.0407 = (1.21, 1.35)$

b) $1.28 \pm 1.67 \times 0.0974 = (1.12, 1.44)$

c) Yes, it is within the interval so it is a plausible value.

47. a) H_0: No linear relationship between *Brain Size* and *IQ*, $\beta_1 = 0$.
H_A: There is evidence of a relationship, $\beta_1 \neq 0$. $t = 1.12$; this will not be significant.

b) With $R^2 = 6.5\%$, the relationship is very weak. There seem to be three students with large brains who also scored high. Without them, there seems to be no association at all.

49. The **Does the plot thicken?** and **Nearly Normal** conditions are violated. The residuals are more variable for lower predicted values than for higher ones and the histogram of the residuals is severely skewed to the right.

51. a) $H_0: \beta_1 = 0$; there is no linear association between ozone level and population.
$H_A: \beta_1 \neq 0$; there is a linear association between ozone level and population.
P-value $= 0.0021$. Since the P-value is so low, we reject the null hypothesis and conclude that there is evidence of a linear association between ozone level and population.

b) Yes, $s = 5.942$ parts per million is small and 80.7% of the variability in ozone level is accounted for by population.

53. a) Based on this regression, each additional million residents corresponds to an increase in average ozone level of between 3.29 and 10.01 ppm, with 90% confidence.

b) The mean *Ozone* level for cities with 600,000 people is between 18.69 and 27.06 ppm, with 90% confidence.

55. a) 69 tablet computers.

b) Yes. The scatter plot is roughly linear with lots of scatter. The plot of the residuals versus the predicted values looks random and shows consistent variability. The normal probability plot of residuals is reasonably straight.

c) $H_0: \beta_1 = 0$; there is no linear association between maximum brightness of the screen and battery life.
$H_A: \beta_1 \neq 0$; there is a linear association between maximum brightness of the screen and battery life.
P-value $= 0.0463$. We reject the null hypothesis and conclude that there is evidence of a linear association between maximum brightness of the screen and battery life.

d) No, R-squared $= 4.43\%$ and $s = 1.946$ hours.

e) $\widehat{BatteryLife} = 6.24114 + 0.0042(MaxBrightness)$.

f) $(-0.0012, 0.0100)$ hours per cd/m^2 units.

g) Battery life increases, on average, by an amount between -0.0012 and 0.0100 hours per increase in brightness of one cd/m^2, with 95% confidence.

57. a) R-squared = 30.9%
$s = 16.448$ with $24 - 2 = 22$ degrees of freedom

Variable	Coefficient	S.E.	t-ratio	Prob.
Intercept	11.753	18.179	0.646	0.525
Midterm1	0.716	0.228	3.139	0.005

The regression model is
$\widehat{Midterm2} = 11.753 + 0.716(Midterm1)$.

b) The scatter plot is roughly linear with lots of scatter, showing a weak, positive relationship. The plot of the residuals versus the predicted values looks random and shows consistent variability. The normal probability plot of residuals is reasonably straight, and the histogram of the residuals is symmetric. The regression model appears appropriate. The small P-value for the slope (0.005) shows that the slope is significantly different from zero.

c) No. $R^2 = 30.9\%$ and $s = 16.448$ points indicate that he would not be able to predict performance on *Midterm2* accurately.

59. H_0: Slope of *Effectiveness* vs. *Initial Ability* = 0.
H_A: Slope $\neq 0$

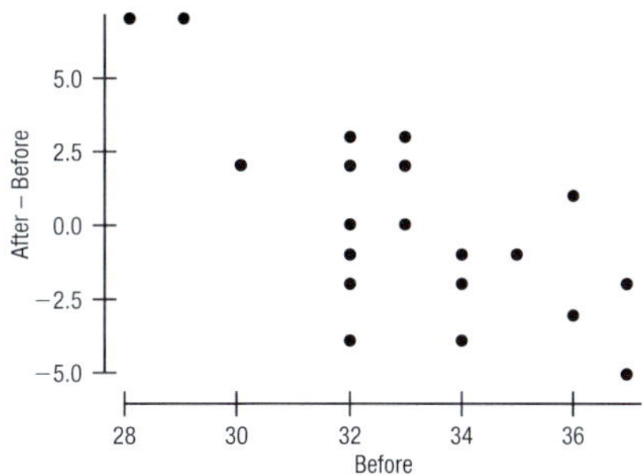

Scatterplot is straight enough. Regression conditions appear to be met. $t = -4.34$; df = 18; P-value = 0.0004. With a P-value this small, we reject the null hypothesis. There is strong evidence that the effectiveness of the video depends on the player's initial ability. The negative slope indicates that the method is more effective for those whose initial performance was poorest and less so for those whose initial performance was better. This looks like a case of regression to the mean. Those who were above average initially tended to be worse after training. Those who were below average initially tended to improve.

61. a) The scatter plot is straight enough. The plot of the residuals versus the predicted values looks random and shows consistent variability. The histogram of the residuals is unimodal and symmetric with no outliers.

b) $H_0: \beta_1 = 0$; there is no linear association between level of education and mortality rate.
$H_A: \beta_1 \neq 0$; there is a linear association between level of education and mortality rate.
P-value = 0.0002. We reject the null hypothesis and conclude that there is evidence of a linear association between level of education and mortality rate.

c) No. Data are on cities, not individuals. Moreover, the data are observational, so a causal relationship cannot be determined.

d) $(-55.7487, -20.1799)$ deaths per 100,000 people.

e) Mortality rate decreases, on average, by an amount between 20.1799 and 55.7487 deaths per 100,000 people for an increase of one year in average education level, with 95% confidence.

f) Based on the regression, the mean mortality rate for cities with an average of 12 years of schooling will be between 889.5761 and 939.4807 deaths per 100,000 people, with 95% confidence.

63. a) $\widehat{Logit\,(Embrace)} = 0.5651 - 0.0152\,Age$.

b) Yes, the P-value is 0.0015, which means there is strong evidence to suggest the association.

c) The coefficient of *Age* is negative, so an older person is less likely to respond "Embrace".

Part VI Review

R6.1. H_0: There is no difference in cancer rates, $p_1 - p_2 = 0$.
H_A: The cancer rate in those who use the herb is higher, $p_1 - p_2 > 0$.

R6.3. a) Experiment.

b) H_0: There is no difference, $p_1 - p_2 = 0$.
H_A: Patients with carbolic acid are more likely to live, $p_1 - p_2 > 0$. $z = 3.03$, P-value = 0.0012; with a P-value so low, we reject H_0. These data show that carbolic acid is effective in increasing the chances of surviving an amputation.

c) We are not told whether the patients were randomized to the treatments. We are not told whether the experiment was double-blind or even blinded at all. This could have biased the results toward a more favorable outcome.

R6.5. H_0: There is no difference, $p_1 - p_2 = 0$. H_A: Early births have increased, $p_1 - p_2 < 0$. $z = -0.729$, P-value = 0.2329. Because the P-value is so high, we do not reject H_0. These data do not show an increase in the incidence of early birth of twins.

R6.7. a) H_0: There is no difference, $p_1 - p_2 < 0$. H_A: Treatment prevents deaths from eclampsia, $p_1 - p_2 < 0$.

b) Samples are random and independent; less than 10% of all pregnancies (or eclampsia cases); more than 10 successes and failures in each group.

c) 0.8008

d) There is insufficient evidence to conclude that magnesium sulfate is effective in preventing eclampsia deaths.

e) Type II.

f) Increase the sample size, increase α.

g) Increasing sample size: decreases variation in the sampling distribution, is costly. Increasing α: Increases likelihood of rejecting H_0, increases chance of Type I error.

R6.9. a) Experiment.

b) A one-sided test, since they are interested only in an increase in fuel economy.

c) Deciding the additive increases fuel economy when there really is no difference.

d) Deciding the additive makes no difference when it really does increase fuel economy.

e) Type II.

f) Given that the two groups received roughly the same use and care, yes. They can't necessarily claim it will work for all cars, only cars similar to their fleet.

R6.11. a) H_0: $\mu_{Jan} - \mu_{Jul} = 0$; H_A: $\mu_{Jan} - \mu_{Jul} \neq 0$. $t = -1.94$, df = 43.68, P-value = 0.0590. This is mild evidence of a difference, but the P-value is slightly greater than the usual 0.05 cutoff.

b) H_0: $\mu_{Apr} - \mu_{Oct} = 0$; H_A: $\mu_{Apr} - \mu_{Oct} \neq 0$. $t = -0.92$; df = 59.40; P-value = 0.3610. Since $P > 0.10$, do not reject the null; these data do not show a significant difference between April and October with regard to the mean age at which crawling begins.

c) There is mild evidence of a possible difference in mean crawling age between January and July babies, but stronger evidence is needed to make a convincing argument. There is no evidence of a difference between mean crawling times for babies born in April and October.

R6.13. Based on these data, we are 95% confident that the mean difference in aluminum oxide content is between −3.37 and 1.65. The means in aluminum oxide content of the pottery made at the two sites could reasonably be the same.

R6.15. a) If there is no difference in the average fish sizes, the chance of seeing an observed difference this large just by natural sampling variation is less than 0.1%.

b) If cost justified, feed them a natural diet.

c) Type I.

R6.17. a) Assuming the conditions are met, from these data we are 95% confident that patients with cardiac disease average between 3.39 and 5.01 years older than those without cardiac disease.

b) Older patients are at greater risk from a variety of other health issues and perhaps more depressed.

R6.19. a) Based on the data, we are 95% confident that the difference in eating disorders between Muslim students and Christian students is between 4.6% and 22.7%.

b) Although caution in generalizing must be used since the study was restricted to the Spanish city of Ceuta, it appears there is a true difference in the prevalence of eating disorders. We can conclude this because the entire interval is above 0.

R6.21. H_0: $\mu_{Cert} - \mu_{UC} = 0$; H_A: $\mu_{Cert} - \mu_{UC} > 0$. $t = 1.57$; df = 86; P-value = 0.0598. The P-value is marginal for rejecting the null hypothesis at alpha = 0.05.

R6.23. There is a significant difference in Math performance; students of certified teachers do better. There is a significant difference in language performance at $\alpha = 0.05$ between students of certified teachers and those of uncertified teachers.

R6.25. a) H_0: $p_{High} - p_{Low} = 0$; H_A: $p_{High} - p_{Low} \neq 0$. $z = 3.83$; P-value = 0.00012. Because the P-value is so low, we reject H_0. These data show the IRS risk is different in the two groups; people who consume dairy products often have a lower risk, on average.

b) Doesn't prove it; this is not an experiment.

R6.27. a)

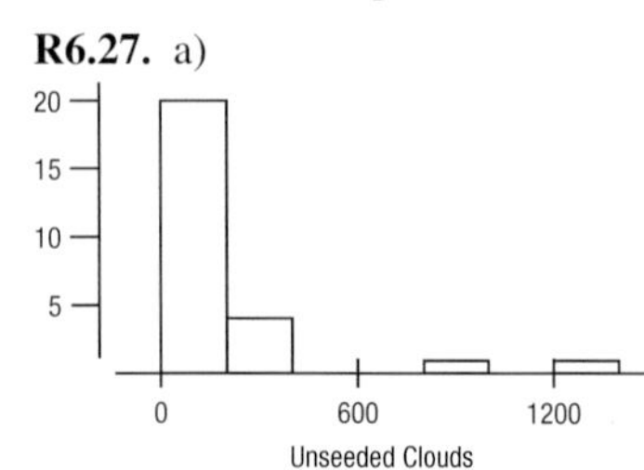

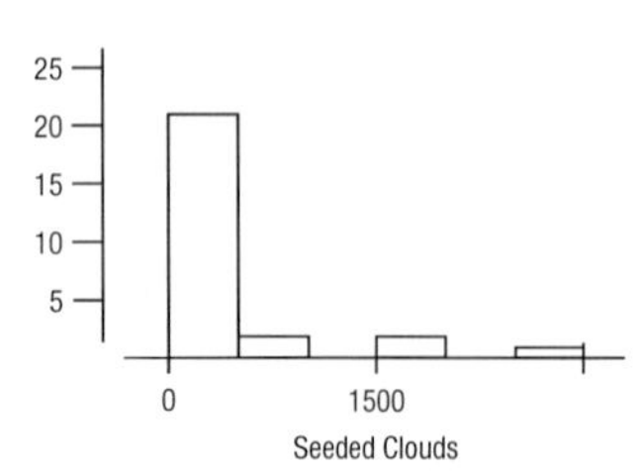

b) Logarithms make both distributions more symmetric.

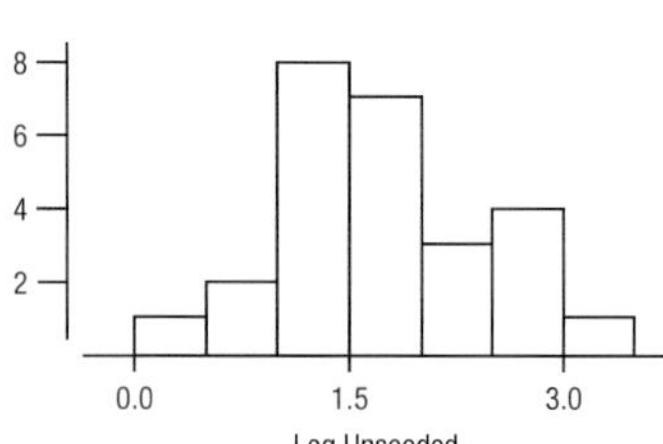

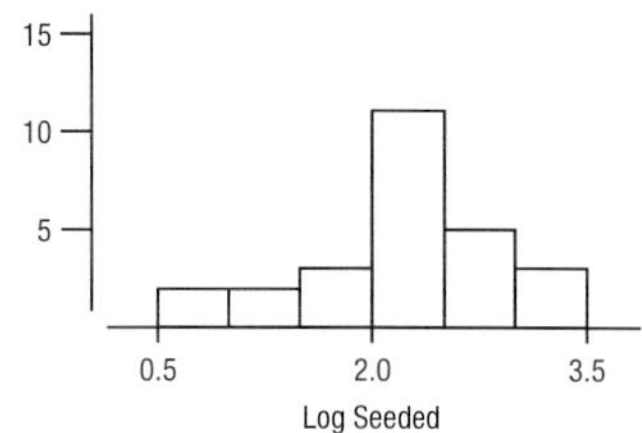

c) Both versions of the test now have much smaller P-values, and the randomization test agrees better with the two-sample *t*-test.

R6.29. a) 59 products

b) 84.5%

c) (2.21, 2.78) dollars per minute

d) Based on this regression, average price increases between $2.21 and $2.78 for each minute of polishing time, with 95% confidence.

R6.31. H_0: Income and opinion are independent. H_A: Income and opinion are not independent. $\chi^2 = 123.02$; P-value < 0.0001. With such a small P-value, we reject H_0. These data show evidence that income level and opinion about redistribution of the nation's wealth are not independent. Examination of the components shows that the low-income respondents are more likely to approve of redistribution compared to the high-income respondents.

R6.33. H_0: $p_L - p_R = 0$; H_A: $p_L - p_R \neq 0$. $z = 1.40$; P-value = 0.1628. Since $P > 0.05$, we do not reject H_0. These data do not provide evidence of a difference in musical abilities between right- and left-handed people.

R6.35. $\chi^2 = 6.14$; P-value = 0.1887. Since $P > 0.05$, we do not reject H_0. These data do not provide evidence of an association between duration of pregnancy and level of care.

R6.37. H_0: $p_{Men} - p_{Women} = 0$; H_A: $p_{Men} - p_{Women} < 0$. $z = -2.31$; P-value = 0.0103. With such a low P-value, we reject H_0. These data provide evidence that the proportion of all men who feel "a lot behind schedule" is significantly lower than the proportion of women when it comes to retirement planning.

R6.39. a) This is a linear regression that is meaningless—the data are categorical.

b) This is a two-way table that is appropriate. H_0: *Eye* and *Hair* color are independent. H_A: *Eye* and *Hair* color are not independent. However, four cells have expected counts less than 5, so the χ^2 analysis is technically not appropriate unless cells are merged. However, with a χ^2 value of 223.6 with 16 df and a P-value < 0.0001, the results are not likely to change if we merge appropriate eye colors.

R6.41. a) For the simple regression: Calories increase by about 1.26 calories per gram of carbohydrate. For the multiple regression: After allowing for the effects of *fiber* in these cereals, calories increase at the rate of about 0.83 calories per gram of carbohydrate.

b) After accounting for the effect of *fiber*, the effect of *carbo* is no longer statistically significant.

c) I would like to see a scatterplot showing the relationship of *carbo* and *fiber*. I suspect that they may be highly correlated, which would explain why after accounting for the effects of *fiber*, there is little left for *carbo* to model. It would be good to know the R^2 value between the two variables.

R6.43. These three cereals are creating a strong, negative collinearity between *carbo* and *fiber*.

R6.45. a) H_0: $\beta_1 = 0$; H_A: $\beta_1 \neq 0$. $t = 4.16$, P-value < 0.0001. These data show evidence of a positive relationship between number of meals eaten together and grades.
b) No. R^2 is small and $s = 0.66$ points. So we could predict only to within 1.32 grade points at best.
c) No. The slope is clearly not 0, but that doesn't mean the relationship is strong or the predictions are useful.

R6.47. a) We are 99% confident that the interval 1.64% to 20.8% contains the true difference in violent felony rates when comparing individual therapy to MST.
b) Since the entire interval is above 0, we can conclude that MST is successful in reducing the proportion of juvenile offenders who commit violent felonies. The population of interest is adolescents with mental health problems.

R6.49. $\chi^2 = 8.23$; P-value $= 0.0414$. There is evidence of an association between *cracker type* and *bloating*. Standardized residuals for the gum cracker are -1.32 and 1.58. Prospects for marketing this cracker are not good.

R6.51. Based on the data, we are 95% confident that the mean difference in words misunderstood is between -3.76 and 3.10. Because 0 is in the confidence interval, I cannot reject the null hypothesis that the two tapes are equivalent.

Chapter 24

1. a) filtering
b) merging

3. a) Likely high leverage.
b) Likely outlier.

5. No. The predictors are almost certainly collinear, which would affect the coefficient of *Age*.

7. a) CO_2 and *Year* are clearly collinear. That will make the t-ratios small even though the R^2 is large. It would not be valid to conclude that CO_2 has no effect on *Avg Global temp*.
b) The data were filtered to select these years. Without a good reason for selecting these particular years, we might be suspicious of the analysis.

9. a) The slope of a partial regression plot is the coefficient of the corresponding predictor, in this case -1.02047.
b) Quaker Oatmeal is influential on the slope. It makes the slope more strongly negative.
c) Omitting Quaker Oatmeal changes the coefficient of Fiber to be positive. This second regression model has a higher R^2, suggesting that it fits the data better. Without the influential point, the second regression is probably the better model.
d) The coefficient of Fiber is not significantly different from zero since its P-value is large. We have no evidence that it contributes significantly to Calories after allowing for the other predictors in the model.

11. a) After allowing for the effects of *Sodium* and *Sugars*, each gram of *Potassium* is associated with a 0.019-calorie decrease in *Calories*.
b) Those points pull the slope of the relationship down. Omitting them should increase the value of the coefficient of *Potassium*. It would likely become positive, since the remaining points show a positive slope in the partial regression plot.
c) These appear to be influential points. They have both high leverage and large residual, and the partial regression plot shows their influence.
d) If our goal is to understand the relationships among these variables, then it might be best to omit these cereals because they seem to behave as if they are from a separate subgroup.

13. (Answers may vary.)
a) The distributions of *Population, Birthrate, GDP/capita, Imports/Capita, Exports/capita,* and possibly *Airports/capita* are skewed to the right and might benefit from re-expression. The distributions of *Birthrate* and *Obesity%* may be bimodal.
b) The re-expressions have improved the symmetry of variables. After re-expression, Obesity% seems more clearly bimodal.

15. a) Most show reasonably straight relationships. *LogPopulation* and *Education* seem to be only weakly associated with other variables. (Answers may vary.)
b) The variables most closely (linearly) associated with *Life Expectancy* are *LogBirthrate, LogGDP/capita, LogImports/capita,* and $\sqrt{Airports/capita}$. The scatterplots in the plot matrix suggest that any of these is linearly associated with *Life Expectancy*, but several are also linearly associated with each other, so we must beware of collinearity if we choose several of them as predictors.

17. The regression seems like a good model. 73.2% of the variation in *Life Expectancy* is accounted for by the model and all t-ratios are significant. The model says that, after allowing for *LogGDP/capita*, *LogBirthrate* is negatively associated with life expectancy—higher birthrates go with lower life expectancy. After allowing for *LogBirthrate*, higher GPD/capita is associated with a higher life expectancy.

However, the residuals show 5 low outliers that deserve attention and explanation.

19. All five outlying countries are in southern Africa. We would like to find something they have in common that could depress life expectancy, after allowing for GDP/capita and birthrate.

21. This model has a higher R^2 and higher adjusted R^2. The coef ficients of the previously used predictors have not changed very much and both are still statistically significant. The coefficient of *LogHIV* says that after allowing for the effects of the other predictors, higher HIV incidence is associated with lower life expectancy—as we would expect.

Chapter 25

1. a) H_0: $\mu_1 = \mu_2 = \mu_3 = \mu_4$ versus H_A: Not all means are equal. Here, μ_k represents the mean numbers of unpopped kernels for popcorn brand k.
b) MS_T has 3 df, MS_E has 12 df.
c) $F = 13.55$; P-value $= 0.000369$. We reject the null hypothesis and conclude that there is strong evidence that the mean number of unpopped kernels is not the same for all four popcorn brands.
d) Look at the side-by-side boxplots of the treatment groups, the normal probability plot of the residuals, the histogram of the residuals, and the plot of the residuals versus the predicted values.

3. a) H_0: The mean gas mileage is the same for each muffler versus H_A: The mean gas mileages for the different mufflers are not all the same.
b) MS_T has 2 df, while MS_E has 15 df.
c) $F = 2.38$; P-value $= 0.126556$. We fail to reject the null hypothesis and conclude that there is no evidence to suggest that gas mileage associated with any single muffler is different than the others.
d) Look at the side-by-side boxplots of the treatment groups, the normal probability plot of the residuals, the histogram of the residuals, and the plot of the residuals versus the predicted values.
e) Type II error.

5. a) H_0: $\mu_1 = \mu_2 = \mu_3$ versus H_A: Not all means are equal. Here, μ_k represents the mean activation time for recipe k.
b) $F = 6.8316$; P-value $= 0.0157$. We reject the null hypothesis and conclude that there is strong evidence that the mean activation times for the three recipes are not all the same.
c) Yes, because the null hypothesis was rejected. Proceed with a multiple comparisons method to compare all the groups.

7. a) H_0: $\mu_4 = \mu_5 = \mu_6 = \mu_8$ versus H_A: Not all means are equal. Here, μ_k is the mean mileage for an engine with k cylinders.
b) No, the Similar Variance condition is not satisfied. The side-by-side boxplots show distributions with different spreads, so we cannot proceed with an ANOVA. A re-expression of the response variable may equalize the spreads.

9. a) H_0: $\mu_1 = \mu_2 = \mu_3 = \mu_4 = \mu_5 = \mu_6$ versus H_A: Not all means are equal. Here, μ_k is the mean time teller k takes to serve a customer.
b) $F = 3.439$; P-value $= 0.0059$. We reject the null hypothesis and conclude that there is strong evidence that the mean time to serve a customer is not the same for each teller.
c) Yes, because the null hypothesis was rejected. Proceed with a multiple comparisons test.

11. An ANOVA is not appropriate because *Eye Color* is a categorical variable. A chi-square test of independence between eye and hair color is appropriate.

13. a) $MS_T = 10.2535$; $MS_E \approx 0.0122$.
b) $F \approx 840.4508$.
c) We reject the null hypothesis and conclude that there is strong evidence that the mean taste test scores for each method of preparation are not all the same.
d) The variances of the treatment groups are similar, the residuals are nearly Normal, and the experimental runs were performed in random order.
e) A boxplot of the scores by method, a normal probability plot of the residuals, a histogram of the residuals, and a plot of residuals versus predicted values.
f) $s_p = \sqrt{0.0122} \approx 0.1105$ points.

15. a) H_0: The mean taste test score is the same for both real and substitute eggs, versus H_A: The mean taste test scores are different for real and substitute eggs.
b) $F = 28.7095$; P-value $= 0.0017$. We reject the null hypothesis and conclude that there is strong evidence that the mean taste test scores are different for real and substitute eggs. The real eggs have a higher mean taste score.
c) The Independence and Similar Variance conditions appear to be reasonable. It would be nice to see a plot of the residuals to check whether the Nearly Normal condition is reasonable.
d) The pooled estimate of the variance is equal to MS_E. The t-ratio is 5.3586 and $5.3586^2 \approx 28.7146$, which is approximately the F-ratio. The two-sided P-value for the t-value with 6 df is 0.00017, which agrees with the P-value for the F-statistic.

17. a) H_0: The mean math test scores are all the same at the 10 schools, versus H_A: The mean math test scores are not all the same at the 10 schools.
b) $F = 0.962$; P-value $= 0.4725$. We fail to reject the null hypothesis and conclude that there is not enough evidence that the mean math test scores are not all the same at the 10 schools.
c) No, the Type I error rate is higher for several t-tests than for an analysis of variance or a multiple comparisons method.

19. a) H_0: The mean sugar content is the same for each of the 3 shelves, versus H_A: The mean sugar contents are not all the same for each of the 3 shelves.
b) $F = 6.624$; P-value $= 0.0023$. We reject the null hypothesis and conclude that there is strong evidence to suggest that the mean sugar contents are not the same for each of the 3 shelves.
c) We cannot conclude that cereals on shelf 2 have a higher mean sugar content than cereals on shelf 3, or that cereals on shelf 2 have a higher mean sugar content than cereals on shelf 1. We can only conclude that the mean sugar contents are not all equal.
d) There is evidence to conclude that the cereals on shelf 2 have a higher mean sugar content than cereals on shelves 1 and 3.

21. a) H_0: The mean download time is the same for each of the three times of day, versus H_A: The mean download time is not the same for each of the three times of day.
b)

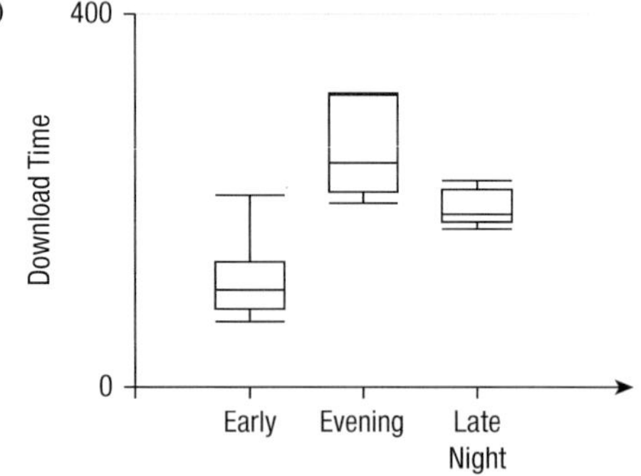

Analysis of Variance

Source	Sum of Squares	df	Mean Square	F-ratio	P-value
Time of Day	54992.333	2	27496.167	15.170	0.000249
Error	27187.667	15	1812.511		
Total	82180.000	17			

$F = 15.170$; P-value $= 0.000249$. We reject the null hypothesis and conclude that there is strong evidence to suggest that the mean download time is not the same for each of the three times of day.

c) The runs were not randomized because they were in specific time periods. It would be good to check a timeplot of residuals by run order to see if there is dependence from one run to the next. The Similar Variance condition is questionable, since the distribution of download times for the last category (Late Night) has less variation than the others. The normal probability plot of the residuals is reasonably straight, and the residuals plot shows no evidence of violating the Nearly Normal condition.

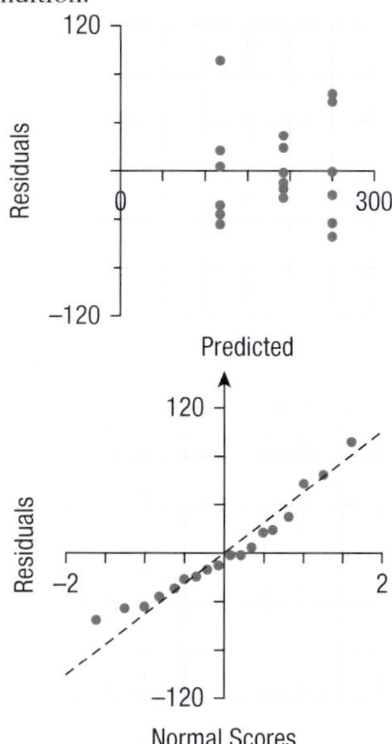

d) A Bonferroni test shows that pairs Early & Evening and Early & Late Night are significantly different from each other at $\alpha = 0.05$.

Chapter 26

1. $TV_{ijk} = \mu + Sex_j + Athlete_k + \varepsilon_{ijk}$

3. The boxplots show different distributions for Women and Men, so the Additivity Assumption is probably not satisfied. The plots show that, at least for men, the Equal Variance Assumption seems unlikely. And both plots show outliers, making the Normality Assumption questionable. If the survey was randomized, the responses may be mutually independent, but we don't have enough information to tell.

5. The interaction term reflects the fact that there is very little difference in TV watching among women between athletes and non-athletes, but there is a difference for men with athletes watching more TV.

7. a) H_0: $\gamma_1 = \gamma_2 = \gamma_3$ where γ represents the effect of the *Power* level vs. H_A: Not all of the *Power* levels have the same effect on the response.
   H_0: $\tau_1 = \tau_2 = \tau_3$ where τ represents the effect of *Time* level vs. H_A: Not all of the *Time* levels have the same effect on the response.
   b) The *Power* sum of squares has 2 df; the *Time* sum of squares has 2 df; the error sum of squares has 4 df.
   c) There are no degrees of freedom left for the interaction term. She must assume that the interaction effects are negligible.

9. a) The P-value for *Power* is 0.0165. The P-value for *Time* is 0.0310.
   b) We reject the null hypothesis that *Power* has no effect and conclude that the mean number of uncooked kernels is not equal across all 3 *Power* levels. We also reject the null hypothesis that *Time* has no effect and conclude that the mean number of uncooked kernels is not the same across all 3 *Times*.
   c) Partial boxplots, scatterplots of residuals vs. predicted values, and a Normal probability plot of the residuals to check the assumptions.

11. a) H_0: $\gamma_1 = \gamma_2$, where γ represents the effect of the *Seat* level vs. H_A: One of the Seat levels has a different effect. The null hypothesis states that seat choice has no effect on mean head injury sustained. The alternative states that it does.
   H_0: $\tau_1 = \tau_2 = \tau_3 = \tau_4 = \tau_5 = \tau_6$, where τ represents the effect of the *Size* level vs. H_A: The Size level effects are not all equal. The null hypothesis states that size of vehicle has no effect on mean head injury sustained. The alternative states that it does.
   b) The conditions appear to be met. The data is assumed to be collected independently, the effects are additive enough, the boxplots show that the variance is roughly constant, and there are no patterns in the scatterplot of residuals vs. predicted values.
   c) There is no significant interaction, since the P-value is greater than the level of significance. The P-values for both Seat and Size are <0.0001. Thus, we reject the null hypotheses and conclude that both *Seat* and *Size* affect the severity of head injury.

13. a) A two-factor ANOVA must have a quantitative response variable. Here the response is whether they exhibited baldness or not, which is a categorical variable. A two-factor ANOVA is not appropriate.
   b) We could use a chi-square analysis to test whether *Baldness* and *Heart Disease* are independent.

15. a) A chi-square test of independence gives a chi-square statistic of 14.510 with a P-value of 0.0023. We reject the hypothesis that *Baldness* and *Heart Disease* are independent.
   b) No, the fact that these are not independent does NOT mean that one causes the other. There could be a lurking variable (such as age) that influences both.

17. a) H_0: $\gamma_1 = \gamma_2$, where γ represents the effect of the *Time* level vs H_A: The Time levels have a different effect. The null hypothesis states that time of day has no effect on the number of shots made. The alternative states that it does.
   H_0: $\tau_1 = \tau_2$, where τ represents the effect of the *Shoe* level vs H_A: The Size levels have a different effect. The null hypothesis states that the type of shoe has no effect on the number of shots made. The alternative states that it does.
   H_0: The interaction term between *Time* and *Shoe* is 0 vs H_A: There is an interaction effect between *Time* and *Shoe*.
   b)

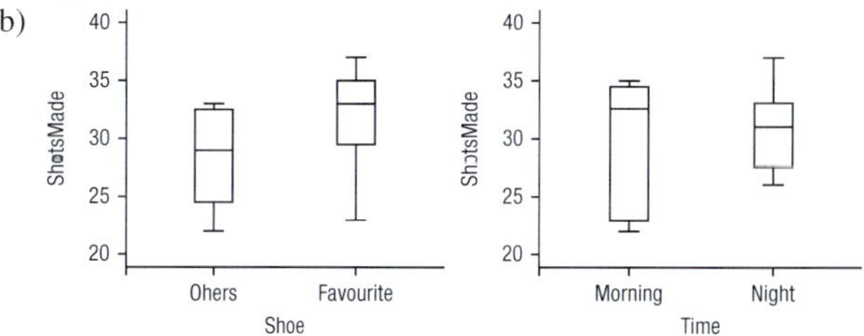

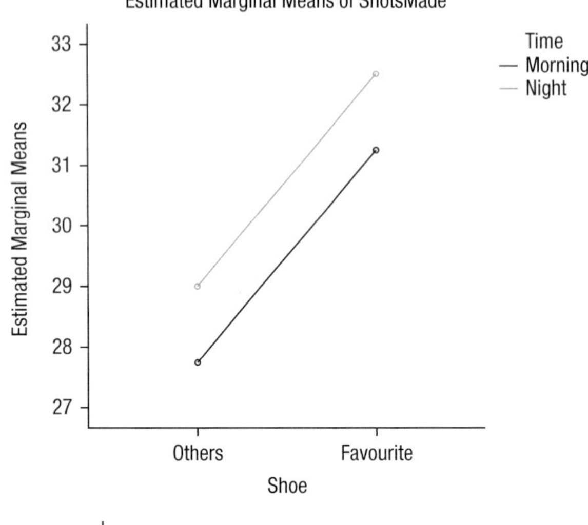

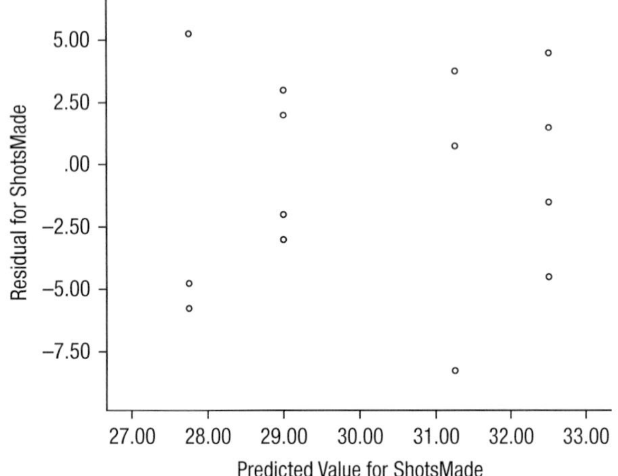

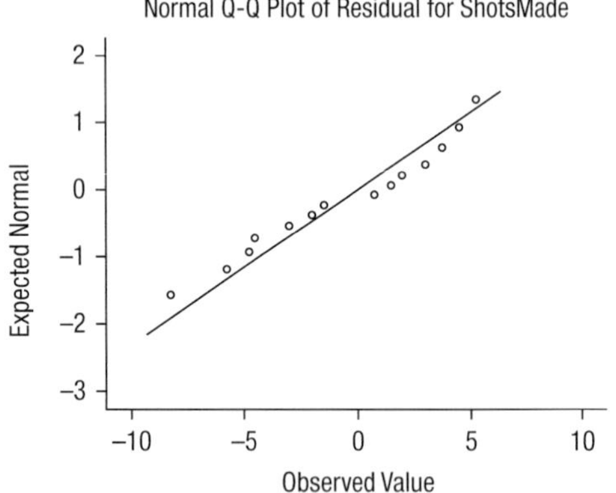

The partial boxplots show little effect of either *Shoe* or *Time* on the number of shots made.

Neither the scatterplot of the residuals vs. predicted values nor the Normal probability plot of residuals show any conditions that are not met. We assume that the number of shots made were independent from one treatment condition to another.

The interaction plot does not show a possible interaction effect. Indeed, the F-test shows that we fail to reject the null hypothesis that there is no interaction effect. Moreover, neither of the main effects appears to be significant. It looks as though she can conclude that neither *Shoe* nor *Time* affects the number of shots made.

19. a) Not all conditions of two-way ANOVA are met, so the results should be treated with caution. There appears to be more spread in the number of sprouts for the lower salinity levels and the residual plot has a fanned shape. The partial boxplots show that *Salinity* appears to have an effect on the number of bean sprouts, while *Temperature* does not. Moreover, it appears that neither the interaction effect nor *Temperature* have a significant effect on the number of bean sprouts.

b) We could use a chi-square analysis to test whether salinity levels and number of bean sprouts are independent.

21. Not all conditions of two-way ANOVA are met, so no general conclusion can be made. There are some differences in the spread of the boxplots and the residual plot shows a clear pattern. Moreover, the interaction effect between *Additive* and *Type* is significant, which makes the interpretation of the main effects problematic.

23. The ANOVA table now shows only main effects to be significant:

Source	DF	Sum of Squares	Mean Square	F-ratio	P-value
Type	2	10.1254	5.06268	5923.1	<0.0001
Additive	1	0.026092	0.0260915	30.526	<0.0001
Interaction	2	7.57E-05	3.78E-05	0.044265	0.9567
Error	54	0.046156	8.55E-04		
Total	59	10.1977			

The residuals now show *no* violation of the conditions:

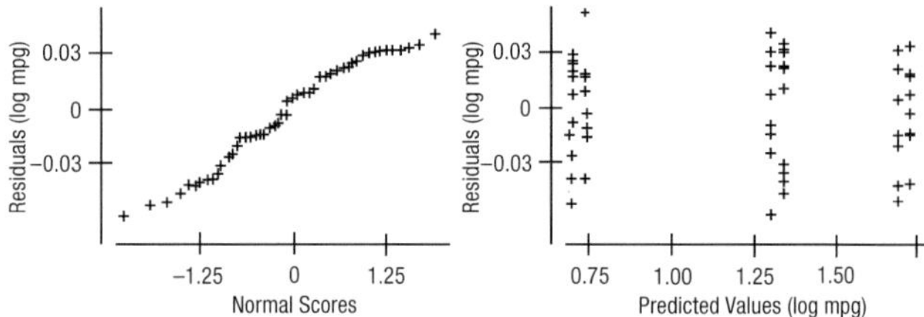

We can conclude that both the *Car Type* and *Additive* have an effect on mileage and that the effects are constant (in log10 *Mpg*) over the values of the various levels of the other factor.

25. a) Let γ represent the effect of the *Environment* level and τ represent the effect of the *Brand* level.

H_0: $\gamma_{RT} = \gamma_C$ vs H_A: The mean times are different under the two environments.

H_0: $\tau_A = \tau_B = \tau_C = \tau_D$ vs. H_A: Not all τ equal.

b) It is not clear whether *Brand* has an effect, since the boxes generally overlap. The condition clearly has an effect since the Cold boxplot is much lower than the Room temp boxplot.

c) Yes, *Brand* is non-significant with a P-value of 0.1007, and *Environment* is clearly significant with a P-value <0.0001.

d) There is also a significant interaction between *Brand* and *Environment*, which makes the statement about *Brand* problematic. Not all brands are affected by the environment in the same way. Brand C works best in a cold environment and performs worst at room temperature.

e) Since it performs worst at room temperature.

27. In this one-way ANOVA, we can see that the means vary across treatments. (However, boxplots with only 2 observations are not appropriate.) By looking closely, it seems obvious that the four flashlights at room temperature lasted much longer than the ones in the cold. But it is much harder to see whether the means of the four brands are different, or whether they differ by the same amounts across both environmental conditions. The two-way ANOVA with interaction makes these distinctions clear.

Part VII Review

R7.1. a) Based on these data, with 95% confidence 5-year yields are between 11.2% and 31.9% higher than 3-year yields on average (paired data).
b) Yes, for these data, the regression equation is 3-Year Yield = 0.134 + 1.03 × 5-Year yield. The P-value for the slope is less than 0.0001. (But we don't know that this was an SRS or even a representative sample of large cap funds.)

R7.3. a) An ANOVA and an F-test.
b) The null hypothesis is that the mean *Distance* thrown is the same for all 4 grips. The alternative is that it is not.
c) Conclude that the mean *Distance* thrown is not the same for all the grips.
d) Boxplots of *Distance* by *Grip*, residual vs. predicted value scatterplot, and a Normal probability plot of the residuals.
e) A multiple comparison test to see which grip is best.

R7.5. a) Indicator variable.
b) There were, on average, 6.4 fewer foals in herds in which some stallions were sterilized, after allowing for the number of adults in the herd.
c) The P-value for *Sterilized* is 0.096. That's not significant at the 0.05 level.

R7.7. These are not counts, so they fail the Counted Data Condition. We cannot use chi-square methods.

R7.9. a) The null hypotheses are that the mean *Time* is the same for all three *Mice* and for *Lights* on and off. The alternatives are that the means are not all the same for the three *Mice* or for the two lighting conditions.
b) *Mouse* has 2 df; *Light* has 1. The error has 1 df.
c) No, the interaction would use 2 df. There are none left, so the interaction can't be estimated.

R7.11. a) (36.21, 60.51) feet.
b) Yes, 40 is in the interval.
c) Wider; we'd need to have a wider interval to be more confident.
d) Roughly 4 times as big—44 flights.

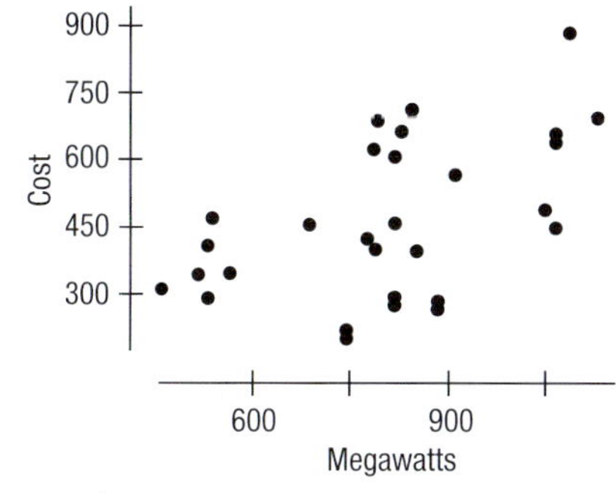

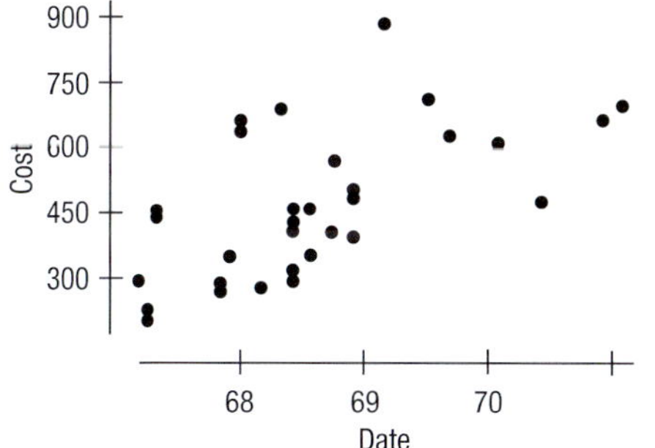

R7.13. a) With an R^2 of 90.9%, your friend is right about being able to predict.
b) H_0: $\beta_{PRS} = 0$, $t = 8.13$ on 31 df. P is <0.0001, so we reject the null hypothesis.
c) The plots show evidence that all 4 conditions are met.

R7.15. a) Null hypotheses are that mean hours studied are the same for both sexes and that the mean hours studied are the same for all four classes. Alternatives are that the mean hours studied are not the same for both sexes and that not all four classes have the same mean hours studied.
b) Yes, an interaction term should be fit because the Additive Enough Condition does not seem to be met.
c) None appear to be significant.
d) There are a few outliers. The Constant Variance Condition appears to be met, but we do not have Normal probability plots of residuals. The main concern is that we may not have enough power to detect differences between groups. We would need more data to increase the power.

R7.17. a) Linearity: The residuals show a time-related pattern of rising and falling. We should be cautious in interpreting the regression model.
Independence: These data are measured over time. We should plot the data and the residuals against time to look for failures of independence.
Constant variance: The scatterplot of the residuals shows no evidence of thickening.
Normality: Without a histogram or Normal probability plot of the Studentized residuals, we can't confirm whether the residuals are nearly Normal.
b) 53.0% of the variation in traffic deaths is accounted for by the linear trend over time.
c) Predicted *Deaths* = 689752 − 324.68 *Year*.
d) *Deaths* have been declining at the rate of –324.68 per *Year*. (With a standard deviation of the errors estimated to be 3013, this doesn't seem like a very meaningful decrease.)

R7.19. a) Null hypotheses are that the mean number of *Words Typed* is the same for both *Temperatures* and for both *Gloves* on and off. The alternatives are that the mean number of *Words Typed* is different for the two *Temperatures* and for *Gloves* on and off.
b) Based on the partial boxplots, both effects appear to be significant.
c) Yes, an interaction term should be fit because the Additive Enough Condition does not seem to be met.
d) The interaction term appears to be real. The temperature effect is only active with gloves off.
e) All three effects are significant.
f) *Gloves* decrease *Words Typed* at both *Temperatures*. Hot temperature increases words typed with gloves off, but has little effect with gloves on.
g) $s = \sqrt{\frac{58.75}{28}} = 1.45$ words per minute. Yes, this seems consistent with the size of the variation shown in the partial boxplots.
h) Tell him to type in a warm room with his gloves off.
i) Even though we haven't seen a histogram or Normal probability plot of residuals, the other conditions seem to be satisfied. For the levels of the factors that he used, it seems clear that a warm room and not wearing gloves are beneficial for his typing.

R7.21. a) The number of finishers per minute appears to grow at about 0.523 finisher per minute for each additional minute of race time.

b) Oops! Clearly, the Linearity Assumption is violated. We can't interpret this regression model at all, although there seems to be a strong—but not linear—association.

R7.23. H_0: $p = 0.512$; H_A: $p < 0.512$. $z = -3.35$; P-value $= 0.0004$. With such a small P-value, we reject H_0. These data provide evidence that exposure to dioxin is related to a reduced rate of male births.

R7.25. a) H_0: Both javelin *Brands* give the same average *Distance*.
H_A: There is a difference in average *Distance* between *Brands*.
H_0: Both *Preparation* conditions lead to the same average *Distance*.
H_A: There is a difference in average *Distance* between *Preparation* conditions.

b) Partial boxplots show that both factors seem to have effects:

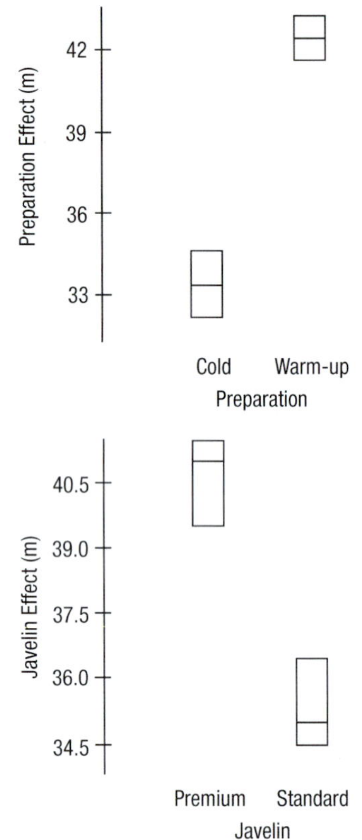

No interaction seems to be present. The Additive Enough Condition appears to be satisfied:

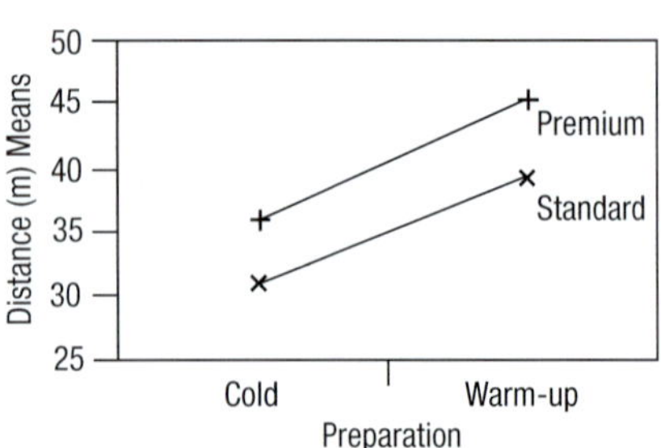

The conditions appear to be met to fit a two-way ANOVA without interaction:

Source	DF	Sum of squares	F-ratio	P-value
Javelin	1	60.58	55	0.0007
Preparation	1	162	147.27	≤0.0001
Error	5	5.5		

Both effects are significant. It appears that depending on the cost, the premium javelin may be worth it—it increases the distance about 5.5 meters on average. Warming up increases distance about 9 meters on average. She should always warm up and consider using the premium javelin.

R7.27. a) The null hypothesis is that mean *TV Watching* is the same for all three groups. The alternative is that it is not.

b) The variance for the None group appears to be slightly smaller, and there are outliers in all three groups. We don't have a Normal probability plot of the residuals, but we suspect that the data may not be Normal enough.

c) The F-test indicates that the mean amount of *TV Watching* is not the same for the three groups.

d) It seems that the differences are evident even when the outliers are removed. It now seems that the conclusion is valid.

R7.29. Based on these data, the average *Weight Loss* for the clinic is between 8.24 and 10.06 pounds, with 95% confidence. The clinic's claim is plausible.

Cumulative Review

1. a) Bimodal, skewed right:

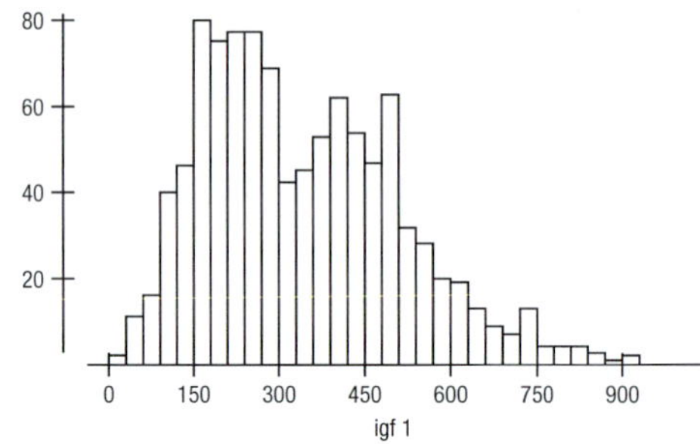

b) Bimodal, skewed right for both males and females, but shifted right (higher) for females.

Males:

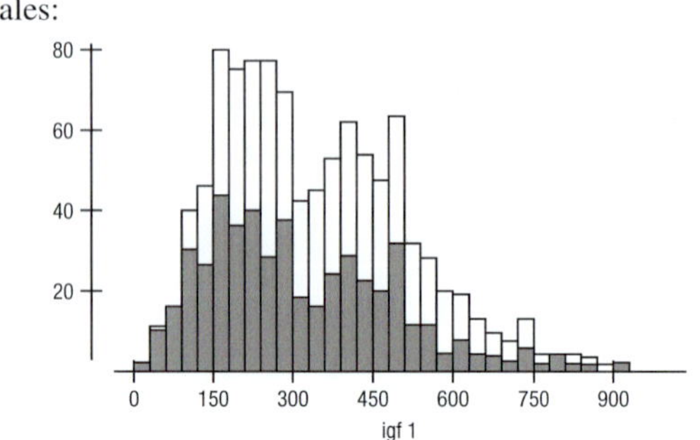

Mean 310.9, median 280.

Females:

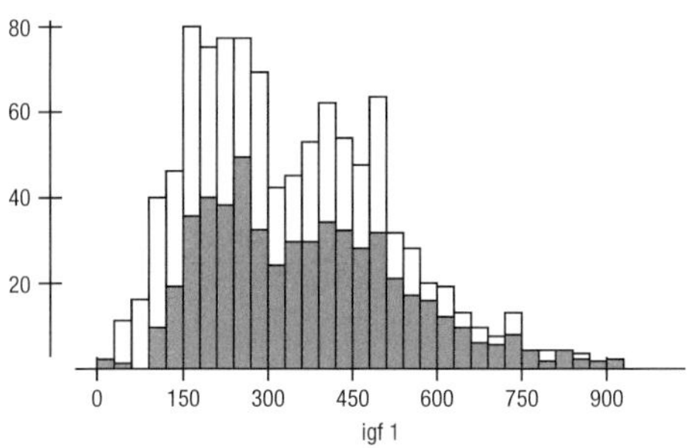

Mean 368.1, median 352.

Both groups show a few outliers on the high side. Because of the right skew, better to compare medians which is 72 $\mu g/l$ higher.

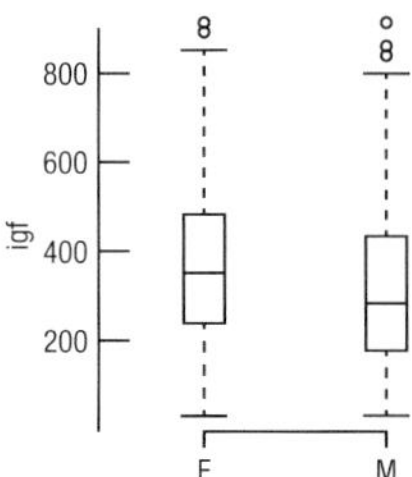

c) Below the age of 20 there is a strong, curved, positive increase of igf with age. After age 20, igf decreases gradually.

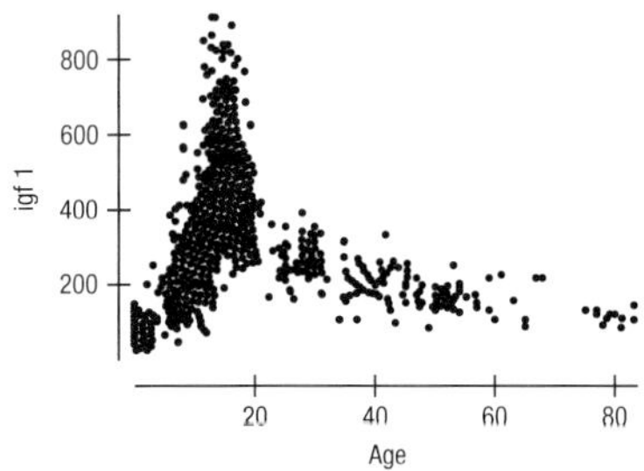

d) No; the relationship is not linear or consistent. It is not appropriate to summarize it with a regression.

3. a)

Response variable is: igf
R squared = 31.2%
s = 121.3 with 405 − 2 = 403 degrees of freedom

Variable	Coefficient	SE(Coeff)	t-ratio	P-value
Intercept	2.94202	21.81	0.135	0.8928
weight	8.06787	0.5972	13.5	< 0.0001

A scatterplot of the residuals does not have a constant variance:

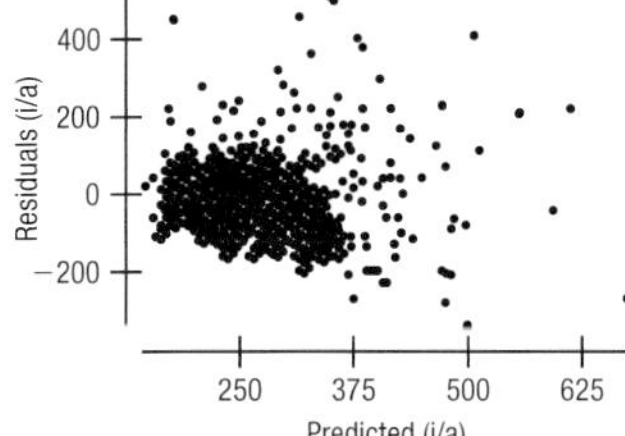

Log(igf) works better:

Response variable is: Log(igf)
R squared = 29.8%
s = 0.1705 with 405 − 2 = 403 degrees of freedom

Variable	Coefficient	SE(Coeff)	t-ratio	P-value
Intercept	2.02218	0.0307	66.0	< 0.0001
weight	0.010989	0.0008	13.1	< 0.0001

b)

Response variable is: Log(igf)
R squared = 34.4%
s = 0.1650 with 405 − 3 = 402 degrees of freedom

Variable	Coefficient	SE(Coeff)	t-ratio	P-value
Intercept	2.05963	0.0305	67.5	< 0.0001
weight	0.011009	0.0008	13.5	< 0.0001
sex#	−0.087745	0.0165	−5.30	< 0.0001

The interpretation of the coefficient is that boys have, on average, an igf level that is 61.6 units lower than girls after allowing for the effects of weight. However, a plot of the residuals shows that the slopes for boys and girls are not parallel for regression on log(igf):

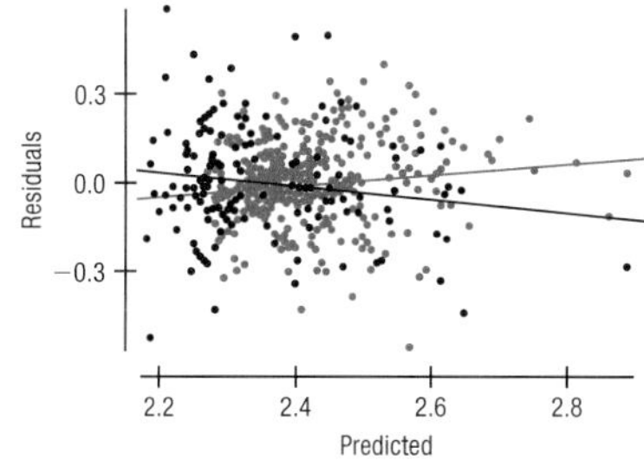

So, a careful student might conclude that this isn't an appropriate regression.

c) For a 100-pound female, $\log(igf) = 2.05963 + 0.011 \times 100 = 3.15963$. In natural units, $10^{3.15963} = 1444$.

d) No; $height$ and $weight$ are collinear (correlation 0.887).

e) Use age as a predictor. It is odd to leave that out when we have ages from infants to teens and are trying to use height and weight as predictors.

5. a) z-test; $H_0 : p = 0.5$
b) z-test of difference between proportions.
c) Two-sample t-test, one-sided alternative specified.
d) Two-sample t-test, two-sided alternative specified.
e) Regression of the number of appointments in 2017 predicted from the number of appointments in 2018.

7. a) 71.49 years
b) IV c) 0.588
d) IV e) I

9. a

11. c

13. a) A b) C c) E
d) D e) B

15. a) The 95% t-interval is (617.28, 631.53).
b) No, it is less than 2 SDs above the mean.

17. Men score higher. The difference is significant (with a two-sample t-test), $P < 0.0001$.

19. a) The clockwise races were slower than the counterclockwise races.
b) Oddly, horses appear to run faster in longer races, with the 1.625-mile races being an exception. But that doesn't make sense, so there may be a lurking variable.

c) I expect the confidence interval to be narrower. There is a bigger effect and less variation.
d) Year is the lurking variable. In fact, horses have gotten faster, so more recent races (run counterclockwise) were faster. Horses probably don't care which way they run.

21. a) The distribution is unimodal and symmetric, but it includes some questionable values, such as donors who are 4, 13, and 15 years old. Nevertheless, a 95% confidence interval is (61.186, 61.862).
b) $t = 2.412$, $P = 0.0159$, reject the null hypothesis of equal mean age.
c) Yes; 69% of men own homes but only 66.1% of women. Chi-square for a test of independence is 7.386 with a P-value of 0.0066, so I reject the hypothesis that men and women are equally likely to own homes.
d) The distribution is unimodal and skewed to the high end.

APPENDIX

B Credits

6: Paul F. Velleman; **6:** Richard D. De Veaux **6:** David E Bock

Chapter 1

27 Peterhowell/iStock/Getty Images **30** Makistock/Fotolia **31** Donatas1205/Fotolia **33** David A. Barnes/Alamy Stock Photo **34** Victoria/Fotolia **35** Peterhowell/iStock/Getty Images

Chapter 2

43 AF archive/Alamy Stock Photo **47** © 2013 Randall Munroe. Reprinted with permission. All rights reserved. **55** Olivier Juneau/Shutterstock **57** AF archive/Alamy Stock Photo **64** Car Collection/Alamy Stock Photo **71** AF archive/Alamy Stock Photo

Chapter 3

90 Syda Productions/Fotolia **97** Pieter De Pauw/Fotolia **99** Elnariz/Fotolia **108** Syda Productions/Fotolia

Chapter 4

121 Darren Green Photography/Alamy Stock Photo **125** David Wall/Alamy Stock Photo **126** 23d/Shutterstock **128** Earl Eliason/iStock/Getty Images **129** Graeme Dawes/Shutterstock **130** Erin Paul Donovan/Alamy Stock Photo **137** Ithaca Times **138** Darren Green Photography/Alamy Stock Photo

Chapter 5

152 Diego Azubel/Epa/Shutterstock **158** Fotolia **162** Skynesher/E+/Getty Images **166** David Gallaher/Shutterstock **167** WavebreakmediaMicro/Fotolia **168** Olga Nayashkova/Shutterstock **173** Diego Azubel/Epa/Shutterstock

Chapter 6

193 Nature and Science/Alamy Stock Photo **197** © 2013 Randall Munroe. Reprinted with permission. All rights reserved. **201** Shutterstock **205** Richards Frank/Pearson Education, Inc. **206** Pearson Education, Inc. **207** © 2013 Randall Munroe. Reprinted with permission. All rights reserved. **213 (top)** HO/AP Images **213 (bottom)** Nature and Science/Alamy Stock Photo

Chapter 7

226 Canadapanda/Shutterstock **229** NASA **232** US Navy Photo/Alamy Stock Photo **241** © 2013 Randall Munroe. Reprinted with permission. All rights reserved. **243** Mara Zemgaliete/Fotolia **247** Canadapanda/Shutterstock

Chapter 8

265 Vladsilver/Shutterstock **269** © 2013 Randall Munroe. Reprinted with permission. All rights reserved. **272 (top)** St Petersburg Times/Zumapress/Newscom **272 (bottom)** Jim Cole/AP Images **273** World History Archive/Alamy Stock Photo **276** Moodboard/Fotolia **283** TravelMediaProductions/Shutterstock **288** Vladsilver/Shutterstock

Chapter 9

308 Airet/iStock/Getty Images **311** Pearson Education, Inc. **316** Leung Cho Pan/123RF **328** Airet/iStock/Getty Images

Chapter 10

351 SDI Productions/E+/Getty Images **352** The Literary Digest **353** Pearson Education, Inc. **361** Freeograph/Shutterstock **367** SDI Productions/E+/Getty Images

Chapter 11

375 Hero Images/Getty Images **376** Shimmo/E+/Getty Images **377** Science History Images/Alamy Stock Photo **378** Houghton Library, Harvard University **379** Lurii Davydov/Shutterstock **380** Brian Chase/Shutterstock **387** Artem Kursin/Shutterstock **388** Joss/iStock/Getty Images **390** Hero Images/Getty Images

Chapter 12

405 Jessica Bethke/Shutterstock **407** Pearson Education, Inc. **408** SpxChrome/iStock/Getty Images **409** OJO Images Ltd/Alamy Stock Photo **410** Pearson Education, Inc. **414** Carolyn Jenkins/Alamy Stock Photo **417** Jessica Bethke/Shutterstock

Chapter 13

423 Worradirek/Shutterstock **424** U.S. Department of the Treasury **426** Aspen Photo/Shutterstock **432** Ventdusud/Shutterstock **436** Alexander Raths/Shutterstock **437** Pearson Education, Inc. **438** Photos.com/Getty Images **439** Worradirek/Shutterstock

Chapter 14

445 Digital Vision/Getty Images **454** Pearson Education, Inc. **458** N-Photos/Fotolia **461** Digital Vision/Getty Images

Chapter 15

468 (top) Olena Kachmar/123RF **468 (bottom)** Brad Smith/ISI/Shutterstock **469** Pearson Education, Inc. **471** Keith Brofsky/Photodisc/Getty Images **482** Lebrecht Music & Arts/Alamy Stock Photo **486** Olena Kachmar/123RF **498** Courtesy of Richard D. De Veaux

Chapter 16

501 B C morris/iStock/Getty Image **503** Pearson Education, Inc. **508** Pj Aun/Shutterstock **517** Rudall30/Fotolia **522** B C morris/iStock/Getty Image

Chapter 17

532 Shutt2016/Shutterstock **537** Pearson Education, Inc. **540** Pearson Education, Inc. **542** Samuel Borges Photography/Shutterstock **547** Juanmonino/ iStock Unreleased/Getty Images **549** Planet Observer/Universal Images Group North America LLC/Alamy Stock Photo **551** Shutt2016/Shutterstock

Chapter 18

563 Avatar_023/Shutterstock **567** Aroonsak thiranuth/123RF **570** © 2013 Randall Munroe. Reprinted with permission. All rights reserved. **571** Teamdaddy/Shutterstock **576** Moodboard/Fotolia **577** Science History Images/Alamy Stock Photo **580** sirtravelalot/Shutterstock **585** © 2013 Randall Munroe. Reprinted with permission. All rights reserved. **586** Avatar_023/Shutterstock

Chapter 19

598 Danlogan/iStock Editorial/Getty Images **600** AlcelVision/Fotolia **602** Paul Velleman **616** Danlogan/iStock Editorial/Getty Images

Chapter 20

630 Peathegee Inc/Blend Images/Getty Images **634** Hero/Fancy/AGE Fotostock **636** CandyBox Images/Shutterstock **637** Don Tremain/Photodisc/Getty Images **645** Westend61/Getty Images **648** Pearson Education, Inc **655** Peathegee Inc/Blend Images/Getty Images

Chapter 21

675 Dorit H Aaron **677** David E Bock **679** Vladislav Gajic/Shutterstock **683** Dorit H Aaron **688** Dorit H Aaron **699** Bruce Leighty/Sports Images/Alamy Stock Photo

Chapter 22

700 Baloncici/123RF **704** Rob/Fotolia **708** Sebastian Janicki/Shutterstock **710** Iofoto/Shutterstock **715** Semmick Photo/123RF **720** Baloncici/123RF

Chapter 23

732 Christian Mueller/Shutterstock **737** Stocktrek/Photodisc/Getty Images **738** Dean Drobot/Shutterstock **751** Edward S. Curtis/Library of Congress **757** Christian Mueller/Shutterstock **783** David E Bock

Chapter 24

788 Sorbis/Shutterstock **796** Mahout/Fotolia **799** Andrew Fox/Alamy Stock Photo **802** Andrey_Popov/Shutterstock **811** Sorbis/Shutterstock

Chapter 25

823 Olivier Le Queinec/Shutterstock **838** Sirtravelalot/Shutterstock **842** Chandler Studios/Pearson Education, Inc **844** Tomas Skopal/Shutterstock **847** Olivier Le Queinec/Shutterstock

Chapter 26

858 Dzimin/Fotolia **865** Sebastian Duda/Fotolia **882** Dzimin/Fotolia

Chapter 27

893 Robert Carner/123RF **896** ProStockStudio/Shutterstock **916** Cemark/Shutterstock

APPENDIX C

Index

Note: Page numbers in **boldface** indicate chapter-level topics; page numbers followed by n indicate footnotes.

D

APPENDIX D

Tables and Selected Formulas

Row	Table Of Random Digits									
1	96299	07196	98642	20639	23185	56282	69929	14125	38872	94168
2	71622	35940	81807	59225	18192	08710	80777	84395	69563	86280
3	03272	41230	81739	74797	70406	18564	69273	72532	78340	36699
4	46376	58596	14365	63685	56555	42974	72944	96463	63533	24152
5	47352	42853	42903	97504	56655	70355	88606	61406	38757	70657
6	20064	04266	74017	79319	70170	96572	08523	56025	89077	57678
7	73184	95907	05179	51002	83374	52297	07769	99792	78365	93487
8	72753	36216	07230	35793	71907	65571	66784	25548	91861	15725
9	03939	30763	06138	80062	02537	23561	93136	61260	77935	93159
10	75998	37203	07959	38264	78120	77525	86481	54986	33042	70648
11	94435	97441	90998	25104	49761	14967	70724	67030	53887	81293
12	04362	40989	69167	38894	00172	02999	97377	33305	60782	29810
13	89059	43528	10547	40115	82234	86902	04121	83889	76208	31076
14	87736	04666	75145	49175	76754	07884	92564	80793	22573	67902
15	76488	88899	15860	07370	13431	84041	69202	18912	83173	11983
16	36460	53772	66634	25045	79007	78518	73580	14191	50353	32064
17	13205	69237	21820	20952	16635	58867	97650	82983	64865	93298
18	51242	12215	90739	36812	00436	31609	80333	96606	30430	31803
19	67819	00354	91439	91073	49258	15992	41277	75111	67496	68430
20	09875	08990	27656	15871	23637	00952	97818	64234	50199	05715
21	18192	95308	72975	01191	29958	09275	89141	19558	50524	32041
22	02763	33701	66188	50226	35813	72951	11638	01876	93664	37001
23	13349	46320	01856	29935	80563	03742	49470	67749	08578	21956
24	69238	92878	80067	80807	45096	22936	64325	19265	37755	69794
25	92207	63527	59398	29818	24789	94309	88380	57000	50171	17891
26	66679	99100	37072	30593	29665	84286	44458	60180	81451	58273
27	31087	42430	60322	34765	15757	53300	97392	98035	05228	68970
28	84432	04916	52949	78533	31666	62350	20584	56367	19701	60584
29	72042	12287	21081	48426	44321	58765	41760	43304	13399	02043
30	94534	73559	82135	70260	87936	85162	11937	18263	54138	69564
31	63971	97198	40974	45301	60177	35604	21580	68107	25184	42810
32	11227	58474	17272	37619	69517	62964	67962	34510	12607	52255
33	28541	02029	08068	96656	17795	21484	57722	76511	27849	61738
34	11282	43632	49531	78981	81980	08530	08629	32279	29478	50228
35	42907	15137	21918	13248	39129	49559	94540	24070	88151	36782
36	47119	76651	21732	32364	58545	50277	57558	30390	18771	72703
37	11232	99884	05087	76839	65142	19994	91397	29350	83852	04905
38	64725	06719	86262	53356	57999	50193	79936	97230	52073	94467
39	77007	26962	55466	12521	48125	12280	54985	26239	76044	54398
40	18375	19310	59796	89832	59417	18553	17238	05474	33259	50595

Table *Z*
Areas under the standard Normal curve

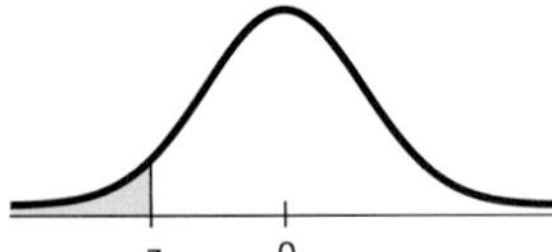

Second decimal place in *z*										
0.09	*0.08*	*0.07*	*0.06*	*0.05*	*0.04*	*0.03*	*0.02*	*0.01*	*0.00*	*z*
									0.0000†	−3.9
0.0001	0.0001	0.0001	0.0001	0.0001	0.0001	0.0001	0.0001	0.0001	0.0001	−3.8
0.0001	0.0001	0.0001	0.0001	0.0001	0.0001	0.0001	0.0001	0.0001	0.0001	−3.7
0.0001	0.0001	0.0001	0.0001	0.0001	0.0001	0.0001	0.0001	0.0002	0.0002	−3.6
0.0002	0.0002	0.0002	0.0002	0.0002	0.0002	0.0002	0.0002	0.0002	0.0002	−3.5
0.0002	0.0003	0.v0003	0.0003	0.0003	0.0003	0.0003	0.0003	0.0003	0.0003	−3.4
0.0003	0.0004	0.0004	0.0004	0.0004	0.0004	0.0004	0.0005	0.0005	0.0005	−3.3
0.0005	0.0005	0.0005	0.0006	0.0006	0.0006	0.0006	0.0006	0.0007	0.0007	−3.2
0.0007	0.0007	0.0008	0.0008	0.0008	0.0008	0.0009	0.0009	0.0009	0.0010	−3.1
0.0010	0.0010	0.0011	0.0011	0.0011	0.0012	0.0012	0.0013	0.0013	0.0013	−3.0
0.0014	0.0014	0.0015	0.0015	0.0016	0.0016	0.0017	0.0018	0.0018	0.0019	−2.9
0.0019	0.0020	0.0021	0.0021	0.0022	0.0023	0.0023	0.0024	0.0025	0.0026	−2.8
0.0026	0.0027	0.0028	0.0029	0.0030	0.0031	0.0032	0.0033	0.0034	0.0035	−2.7
0.0036	0.0037	0.0038	0.0039	0.0040	0.0041	0.0043	0.0044	0.0045	0.0047	−2.6
0.0048	0.0049	0.0051	0.0052	0.0054	0.0055	0.0057	0.0059	0.0060	0.0062	−2.5
0.0064	0.0066	0.0068	0.0069	0.0071	0.0073	0.0075	0.0078	0.0080	0.0082	−2.4
0.0084	0.0087	0.0089	0.0091	0.0094	0.0096	0.0099	0.0102	0.0104	0.0107	−2.3
0.0110	0.0113	0.0116	0.0119	0.0122	0.0125	0.0129	0.0132	0.0136	0.0139	−2.2
0.0143	0.0146	0.0150	0.0154	0.0158	0.0162	0.0166	0.0170	0.0174	0.0179	−2.1
0.0183	0.0188	0.0192	0.0197	0.0202	0.0207	0.0212	0.0217	0.0222	0.0228	−2.0
0.0233	0.0239	0.0244	0.0250	0.0256	0.0262	0.0268	0.0274	0.0281	0.0287	−1.9
0.0294	0.0301	0.0307	0.0314	0.0322	0.0329	0.0336	0.0344	0.0351	0.0359	−1.8
0.0367	0.0375	0.0384	0.0392	0.0401	0.0409	0.0418	0.0427	0.0436	0.0446	−1.7
0.0455	0.0465	0.0475	0.0485	0.0495	0.0505	0.0516	0.0526	0.0537	0.0548	−1.6
0.0559	0.0571	0.0582	0.0594	0.0606	0.0618	0.0630	0.0643	0.0655	0.0668	−1.5
0.0681	0.0694	0.0708	0.0721	0.0735	0.0749	0.0764	0.0778	0.0793	0.0808	−1.4
0.0823	0.0838	0.0853	0.0869	0.0885	0.0901	0.0918	0.0934	0.0951	0.0968	−1.3
0.0985	0.1003	0.1020	0.1038	0.1056	0.1075	0.1093	0.1112	0.1131	0.1151	−1.2
0.1170	0.1190	0.1210	0.1230	0.1251	0.1271	0.1292	0.1314	0.1335	0.1357	−1.1
0.1379	0.1401	0.1423	0.1446	0.1469	0.1492	0.1515	0.1539	0.1562	0.1587	−1.0
0.1611	0.1635	0.1660	0.1685	0.1711	0.1736	0.1762	0.1788	0.1814	0.1841	−0.9
0.1867	0.1894	0.1922	0.1949	0.1977	0.2005	0.2033	0.2061	0.2090	0.2119	−0.8
0.2148	0.2177	0.2206	0.2236	0.2266	0.2296	0.2327	0.2358	0.2389	0.2420	−0.7
0.2451	0.2483	0.2514	0.2546	0.2578	0.2611	0.2643	0.2676	0.2709	0.2743	−0.6
0.2776	0.2810	0.2843	0.2877	0.2912	0.2946	0.2981	0.3015	0.3050	0.3085	−0.5
0.3121	0.3156	0.3192	0.3228	0.3264	0.3300	0.3336	0.3372	0.3409	0.3446	−0.4
0.3483	0.3520	0.3557	0.3594	0.3632	0.3669	0.3707	0.3745	0.3783	0.3821	−0.3
0.3859	0.3897	0.3936	0.3974	0.4013	0.4052	0.4090	0.4129	0.4168	0.4207	−0.2
0.4247	0.4286	0.4325	0.4364	0.4404	0.4443	0.4483	0.4522	0.4562	0.4602	−0.1
0.4641	0.4681	0.4721	0.4761	0.4801	0.4840	0.4880	0.4920	0.4960	0.5000	−0.0

†For $z \leq -3.90$, the areas are 0.0000 to four decimal places.

Table *Z* (cont.)
Areas under the standard Normal curve

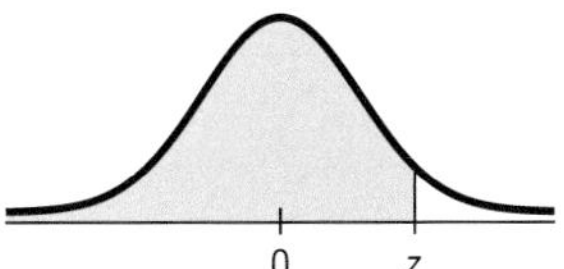

z	Second decimal place in *z*									
	0.00	*0.01*	*0.02*	*0.03*	*0.04*	*0.05*	*0.06*	*0.07*	*0.08*	*0.09*
0.0	0.5000	0.5040	0.5080	0.5120	0.5160	0.5199	0.5239	0.5279	0.5319	0.5359
0.1	0.5398	0.5438	0.5478	0.5517	0.5557	0.5596	0.5636	0.5675	0.5714	0.5753
0.2	0.5793	0.5832	0.5871	0.5910	0.5948	0.5987	0.6026	0.6064	0.6103	0.6141
0.3	0.6179	0.6217	0.6255	0.6293	0.6331	0.6368	0.6406	0.6443	0.6480	0.6517
0.4	0.6554	0.6591	0.6628	0.6664	0.6700	0.6736	0.6772	0.6808	0.6844	0.6879
0.5	0.6915	0.6950	0.6985	0.7019	0.7054	0.7088	0.7123	0.7157	0.7190	0.7224
0.6	0.7257	0.7291	0.7324	0.7357	0.7389	0.7422	0.7454	0.7486	0.7517	0.7549
0.7	0.7580	0.7611	0.7642	0.7673	0.7704	0.7734	0.7764	0.7794	0.7823	0.7852
0.8	0.7881	0.7910	0.7939	0.7967	0.7995	0.8023	0.8051	0.8078	0.8106	0.8133
0.9	0.8159	0.8186	0.8212	0.8238	0.8264	0.8289	0.8315	0.8340	0.8365	0.8389
1.0	0.8413	0.8438	0.8461	0.8485	0.8508	0.8531	0.8554	0.8577	0.8599	0.8621
1.1	0.8643	0.8665	0.8686	0.8708	0.8729	0.8749	0.8770	0.8790	0.8810	0.8830
1.2	0.8849	0.8869	0.8888	0.8907	0.8925	0.8944	0.8962	0.8980	0.8997	0.9015
1.3	0.9032	0.9049	0.9066	0.9082	0.9099	0.9115	0.9131	0.9147	0.9162	0.9177
1.4	0.9192	0.9207	0.9222	0.9236	0.9251	0.9265	0.9279	0.9292	0.9306	0.9319
1.5	0.9332	0.9345	0.9357	0.9370	0.9382	0.9394	0.9406	0.9418	0.9429	0.9441
1.6	0.9452	0.9463	0.9474	0.9484	0.9495	0.9505	0.9515	0.9525	0.9535	0.9545
1.7	0.9554	0.9564	0.9573	0.9582	0.9591	0.9599	0.9608	0.9616	0.9625	0.9633
1.8	0.9641	0.9649	0.9656	0.9664	0.9671	0.9678	0.9686	0.9693	0.9699	0.9706
1.9	0.9713	0.9719	0.9726	0.9732	0.9738	0.9744	0.9750	0.9756	0.9761	0.9767
2.0	0.9772	0.9778	0.9783	0.9788	0.9793	0.9798	0.9803	0.9808	0.9812	0.9817
2.1	0.9821	0.9826	0.9830	0.9834	0.9838	0.9842	0.9846	0.9850	0.9854	0.9857
2.2	0.9861	0.9864	0.9868	0.9871	0.9875	0.9878	0.9881	0.9884	0.9887	0.9890
2.3	0.9893	0.9896	0.9898	0.9901	0.9904	0.9906	0.9909	0.9911	0.9913	0.9916
2.4	0.9918	0.9920	0.9922	0.9925	0.9927	0.9929	0.9931	0.9932	0.9934	0.9936
2.5	0.9938	0.9940	0.9941	0.9943	0.9945	0.9946	0.9948	0.9949	0.9951	0.9952
2.6	0.9953	0.9955	0.9956	0.9957	0.9959	0.9960	0.9961	0.9962	0.9963	0.9964
2.7	0.9965	0.9966	0.9967	0.9968	0.9969	0.9970	0.9971	0.9972	0.9973	0.9974
2.8	0.9974	0.9975	0.9976	0.9977	0.9977	0.9978	0.9979	0.9979	0.9980	0.9981
2.9	0.9981	0.9982	0.9982	0.9983	0.9984	0.9984	0.9985	0.9985	0.9986	0.9986
3.0	0.9987	0.9987	0.9987	0.9988	0.9988	0.9989	0.9989	0.9989	0.9990	0.9990
3.1	0.9990	0.9991	0.9991	0.9991	0.9992	0.9992	0.9992	0.9992	0.9993	0.9993
3.2	0.9993	0.9993	0.9994	0.9994	0.9994	0.9994	0.9994	0.9995	0.9995	0.9995
3.3	0.9995	0.9995	0.9995	0.9996	0.9996	0.9996	0.9996	0.9996	0.9996	0.9997
3.4	0.9997	0.9997	0.9997	0.9997	0.9997	0.9997	0.9997	0.9997	0.9997	0.9998
3.5	0.9998	0.9998	0.9998	0.9998	0.9998	0.9998	0.9998	0.9998	0.9998	0.9998
3.6	0.9998	0.9998	0.9999	0.9999	0.9999	0.9999	0.9999	0.9999	0.9999	0.9999
3.7	0.9999	0.9999	0.9999	0.9999	0.9999	0.9999	0.9999	0.9999	0.9999	0.9999
3.8	0.9999	0.9999	0.9999	0.9999	0.9999	0.9999	0.9999	0.9999	0.9999	0.9999
3.9	1.0000†									

†For $z \geq 3.90$, the areas are 1.0000 to four decimal places.

Table T

Values of t_α

Two tails: $\frac{\alpha}{2}$, $-t_{\alpha/2}$, 0, $t_{\alpha/2}$, $\frac{\alpha}{2}$

One tail: 0, t_α, α

	Two-tail probability	0.20	0.10	0.05	0.02	0.01	
	One-tail probability	0.10	0.05	0.025	0.01	0.005	
	df						df
	1	3.078	6.314	12.706	31.821	63.657	1
	2	1.886	2.920	4.303	6.965	9.925	2
	3	1.638	2.353	3.182	4.541	5.841	3
	4	1.533	2.132	2.776	3.747	4.604	4
	5	1.476	2.015	2.571	3.365	4.032	5
	6	1.440	1.943	2.447	3.143	3.707	6
	7	1.415	1.895	2.365	2.998	3.499	7
	8	1.397	1.860	2.306	2.896	3.355	8
	9	1.383	1.833	2.262	2.821	3.250	9
	10	1.372	1.812	2.228	2.764	3.169	10
	11	1.363	1.796	2.201	2.718	3.106	11
	12	1.356	1.782	2.179	2.681	3.055	12
	13	1.350	1.771	2.160	2.650	3.012	13
	14	1.345	1.761	2.145	2.624	2.977	14
	15	1.341	1.753	2.131	2.602	2.947	15
	16	1.337	1.746	2.120	2.583	2.921	16
	17	1.333	1.740	2.110	2.567	2.898	17
	18	1.330	1.734	2.101	2.552	2.878	18
	19	1.328	1.729	2.093	2.539	2.861	19
	20	1.325	1.725	2.086	2.528	2.845	20
	21	1.323	1.721	2.080	2.518	2.831	21
	22	1.321	1.717	2.074	2.508	2.819	22
	23	1.319	1.714	2.069	2.500	2.807	23
	24	1.318	1.711	2.064	2.492	2.797	24
	25	1.316	1.708	2.060	2.485	2.787	25
	26	1.315	1.706	2.056	2.479	2.779	26
	27	1.314	1.703	2.052	2.473	2.771	27
	28	1.313	1.701	2.048	2.467	2.763	28
	29	1.311	1.699	2.045	2.462	2.756	29
	30	1.310	1.697	2.042	2.457	2.750	30
	32	1.309	1.694	2.037	2.449	2.738	32
	35	1.306	1.690	2.030	2.438	2.725	35
	40	1.303	1.684	2.021	2.423	2.704	40
	45	1.301	1.679	2.014	2.412	2.690	45
	50	1.299	1.676	2.009	2.403	2.678	50
	60	1.296	1.671	2.000	2.390	2.660	60
	75	1.293	1.665	1.992	2.377	2.643	75
	100	1.290	1.660	1.984	2.364	2.626	100
	120	1.289	1.658	1.980	2.358	2.617	120
	140	1.288	1.656	1.977	2.353	2.611	140
	180	1.286	1.653	1.973	2.347	2.603	180
	250	1.285	1.651	1.969	2.341	2.596	250
	400	1.284	1.649	1.966	2.336	2.588	400
	1000	1.282	1.646	1.962	2.330	2.581	1000
	∞	1.282	1.645	1.960	2.326	2.576	∞
	Confidence levels	80%	90%	95%	98%	99%	

Right-tail probability		0.10	0.05	0.025	0.01	0.005
Table X Values of χ^2_α	df					
	1	2.706	3.841	5.024	6.635	7.879
	2	4.605	5.991	7.378	9.210	10.597
	3	6.251	7.815	9.348	11.345	12.838
	4	7.779	9.488	11.143	13.277	14.860
	5	9.236	11.070	12.833	15.086	16.750
	6	10.645	12.592	14.449	16.812	18.548
	7	12.017	14.067	16.013	18.475	20.278
	8	13.362	15.507	17.535	20.090	21.955
	9	14.684	16.919	19.023	21.666	23.589
	10	15.987	18.307	20.483	23.209	25.188
	11	17.275	19.675	21.920	24.725	26.757
	12	18.549	21.026	23.337	26.217	28.300
	13	19.812	22.362	24.736	27.688	29.819
	14	21.064	23.685	26.119	29.141	31.319
	15	22.307	24.996	27.488	30.578	32.801
	16	23.542	26.296	28.845	32.000	34.267
	17	24.769	27.587	30.191	33.409	35.718
	18	25.989	28.869	31.526	34.805	37.156
	19	27.204	30.143	32.852	36.191	38.582
	20	28.412	31.410	34.170	37.566	39.997
	21	29.615	32.671	35.479	38.932	41.401
	22	30.813	33.924	36.781	40.290	42.796
	23	32.007	35.172	38.076	41.638	44.181
	24	33.196	36.415	39.364	42.980	45.559
	25	34.382	37.653	40.647	44.314	46.928
	26	35.563	38.885	41.923	45.642	48.290
	27	36.741	40.113	43.195	46.963	49.645
	28	37.916	41.337	44.461	48.278	50.994
	29	39.087	42.557	45.722	59.588	52.336
	30	40.256	43.773	46.979	50.892	53.672
	40	51.805	55.759	59.342	63.691	66.767
	50	63.167	67.505	71.420	76.154	79.490
	60	74.397	79.082	83.298	88.381	91.955
	70	85.527	90.531	95.023	100.424	104.213
	80	96.578	101.879	106.628	112.328	116.320
	90	107.565	113.145	118.135	124.115	128.296
	100	118.499	124.343	129.563	135.811	140.177

Table F

Denominator df	Numerator dF 1	2	3	4	5	6	7	8	9	10	11	12	13	14	15	16	17	18	19	20	21	22
1	4052.2	4999.3	5403.5	5624.3	5764.0	5859.0	5928.3	5981.0	6022.4	6055.9	6083.4	6106.7	6125.8	6143.0	6157.0	6170.0	6181.2	6191.4	6200.7	6208.7	6216.1	6223.1
2	98.50	99.00	99.16	99.25	99.30	99.33	99.36	99.38	99.39	99.40	99.41	99.42	99.42	99.43	99.43	99.44	99.44	99.44	99.45	99.45	99.45	99.46
3	34.12	30.82	29.46	28.71	28.24	27.91	27.67	27.49	27.34	27.23	27.13	27.05	26.98	26.92	26.87	26.83	26.79	26.75	26.72	26.69	26.66	26.64
4	21.20	18.00	16.69	15.98	15.52	15.21	14.98	14.80	14.66	14.55	14.45	14.37	14.31	14.25	14.20	14.15	14.11	14.08	14.05	14.02	13.99	13.97
5	16.26	13.27	12.06	11.39	10.97	10.67	10.46	10.29	10.16	10.05	9.96	9.89	9.82	9.77	9.72	9.68	9.64	9.61	9.58	9.55	9.53	9.51
6	13.75	10.92	9.78	9.15	8.75	8.47	8.26	8.10	7.98	7.87	7.79	7.72	7.66	7.60	7.56	7.52	7.48	7.45	7.42	7.40	7.37	7.35
7	12.25	9.55	8.45	7.85	7.46	7.19	6.99	6.84	6.72	6.62	6.54	6.47	6.41	6.36	6.31	6.28	6.24	6.21	6.18	6.16	6.13	6.11
8	11.26	8.65	7.59	7.01	6.63	6.37	6.18	6.03	5.91	5.81	5.73	5.67	5.61	5.56	5.52	5.48	5.44	5.41	5.38	5.36	5.34	5.32
9	10.56	8.02	6.99	6.42	6.06	5.80	5.61	5.47	5.35	5.26	5.18	5.11	5.05	5.01	4.96	4.92	4.89	4.86	4.83	4.81	4.79	4.77
10	10.04	7.56	6.55	5.99	5.64	5.39	5.20	5.06	4.94	4.85	4.77	4.71	4.65	4.60	4.56	4.52	4.49	4.46	4.43	4.41	4.38	4.36
11	9.65	7.21	6.22	5.67	5.32	5.07	4.89	4.74	4.63	4.54	4.46	4.40	4.34	4.29	4.25	4.21	4.18	4.15	4.12	4.10	4.08	4.06
12	9.33	6.93	5.95	5.41	5.06	4.82	4.64	4.50	4.39	4.30	4.22	4.16	4.10	4.05	4.01	3.97	3.94	3.91	3.88	3.86	3.84	3.82
13	9.07	6.70	5.74	5.21	4.86	4.62	4.44	4.30	4.19	4.10	4.02	3.96	3.91	3.86	3.82	3.78	3.75	3.72	3.69	3.66	3.64	3.62
14	8.86	6.51	5.56	5.04	4.69	4.46	4.28	4.14	4.03	3.94	3.86	3.80	3.75	3.70	3.66	3.62	3.59	3.56	3.53	3.51	3.48	3.46
15	8.68	6.36	5.42	4.89	4.56	4.32	4.14	4.00	3.89	3.80	3.73	3.67	3.61	3.56	3.52	3.49	3.45	3.42	3.40	3.37	3.35	3.33
16	8.53	6.23	5.29	4.77	4.44	4.20	4.03	3.89	3.78	3.69	3.62	3.55	3.50	3.45	3.41	3.37	3.34	3.31	3.28	3.26	3.24	3.22
17	8.40	6.11	5.19	4.67	4.34	4.10	3.93	3.79	3.68	3.59	3.52	3.46	3.40	3.35	3.31	3.27	3.24	3.21	3.19	3.16	3.14	3.12
18	8.29	6.01	5.09	4.58	4.25	4.01	3.84	3.71	3.60	3.51	3.43	3.37	3.32	3.27	3.23	3.19	3.16	3.13	3.10	3.08	3.05	3.03
19	8.18	5.93	5.01	4.50	4.17	3.94	3.77	3.63	3.52	3.43	3.36	3.30	3.24	3.19	3.15	3.12	3.08	3.05	3.03	3.00	2.98	2.96
20	8.10	5.85	4.94	4.43	4.10	3.87	3.70	3.56	3.46	3.37	3.29	3.23	3.18	3.13	3.09	3.05	3.02	2.99	2.96	2.94	2.92	2.90
21	8.02	5.78	4.87	4.37	4.04	3.81	3.64	3.51	3.40	3.31	3.24	3.17	3.12	3.07	3.03	2.99	2.96	2.93	2.90	2.88	2.86	2.84
22	7.95	5.72	4.82	4.31	3.99	3.76	3.59	3.45	3.35	3.26	3.18	3.12	3.07	3.02	2.98	2.94	2.91	2.88	2.85	2.83	2.81	2.78
23	7.88	5.66	4.76	4.26	3.94	3.71	3.54	3.41	3.30	3.21	3.14	3.07	3.02	2.97	2.93	2.89	2.86	2.83	2.80	2.78	2.76	2.74
24	7.82	5.61	4.72	4.22	3.90	3.67	3.50	3.36	3.26	3.17	3.09	3.03	2.98	2.93	2.89	2.85	2.82	2.79	2.76	2.74	2.72	2.70
25	7.77	5.57	4.68	4.18	3.85	3.63	3.46	3.32	3.22	3.13	3.06	2.99	2.94	2.89	2.85	2.81	2.78	2.75	2.72	2.70	2.68	2.66
26	7.72	5.53	4.64	4.14	3.82	3.59	3.42	3.29	3.18	3.09	3.02	2.96	2.90	2.86	2.81	2.78	2.75	2.72	2.69	2.66	2.64	2.62
27	7.68	5.49	4.60	4.11	3.78	3.56	3.39	3.26	3.15	3.06	2.99	2.93	2.87	2.82	2.78	2.75	2.71	2.68	2.66	2.63	2.61	2.59
28	7.64	5.45	4.57	4.07	3.75	3.53	3.36	3.23	3.12	3.03	2.96	2.90	2.84	2.79	2.75	2.72	2.68	2.65	2.63	2.60	2.58	2.56
29	7.60	5.42	4.54	4.04	3.73	3.50	3.33	3.20	3.09	3.00	2.93	2.87	2.81	2.77	2.73	2.69	2.66	2.63	2.60	2.57	2.55	2.53
30	7.56	5.39	4.51	4.02	3.70	3.47	3.30	3.17	3.07	2.98	2.91	2.84	2.79	2.74	2.70	2.66	2.63	2.60	2.57	2.55	2.53	2.51
32	7.50	5.34	4.46	3.97	3.65	3.43	3.26	3.13	3.02	2.93	2.86	2.80	2.74	2.70	2.65	2.62	2.58	2.55	2.53	2.50	2.48	2.46
35	7.42	5.27	4.40	3.91	3.59	3.37	3.20	3.07	2.96	2.88	2.80	2.74	2.69	2.64	2.60	2.56	2.53	2.50	2.47	2.44	2.42	2.40
40	7.31	5.18	4.31	3.83	3.51	3.29	3.12	2.99	2.89	2.80	2.73	2.66	2.61	2.56	2.52	2.48	2.45	2.42	2.39	2.37	2.35	2.33
45	7.23	5.11	4.25	3.77	3.45	3.23	3.07	2.94	2.83	2.74	2.67	2.61	2.55	2.51	2.46	2.43	2.39	2.36	2.34	2.31	2.29	2.27
50	7.17	5.06	4.20	3.72	3.41	3.19	3.02	2.89	2.78	2.70	2.63	2.56	2.51	2.46	2.42	2.38	2.35	2.32	2.29	2.27	2.24	2.22
60	7.08	4.98	4.13	3.65	3.34	3.12	2.95	2.82	2.72	2.63	2.56	2.50	2.44	2.39	2.35	2.31	2.28	2.25	2.22	2.20	2.17	2.15
75	6.99	4.90	4.05	3.58	3.27	3.05	2.89	2.76	2.65	2.57	2.49	2.43	2.38	2.33	2.29	2.25	2.22	2.18	2.16	2.13	2.11	2.09
100	6.90	4.82	3.98	3.51	3.21	2.99	2.82	2.69	2.59	2.50	2.43	2.37	2.31	2.27	2.22	2.19	2.15	2.12	2.09	2.07	2.04	2.02
120	6.85	4.79	3.95	3.48	3.17	2.96	2.79	2.66	2.56	2.47	2.40	2.34	2.28	2.23	2.19	2.15	2.12	2.09	2.06	2.03	2.01	1.99
140	6.82	4.76	3.92	3.46	3.15	2.93	2.77	2.64	2.54	2.45	2.38	2.31	2.26	2.21	2.17	2.13	2.10	2.07	2.04	2.01	1.99	1.97
180	6.78	4.73	3.89	3.43	3.12	2.90	2.74	2.61	2.51	2.42	2.35	2.28	2.23	2.18	2.14	2.10	2.07	2.04	2.01	1.98	1.96	1.94
250	6.74	4.69	3.86	3.40	3.09	2.87	2.71	2.58	2.48	2.39	2.32	2.26	2.20	2.15	2.11	2.07	2.04	2.01	1.98	1.95	1.93	1.91
400	6.70	4.66	3.83	3.37	3.06	2.85	2.68	2.56	2.45	2.37	2.29	2.23	2.17	2.13	2.08	2.05	2.01	1.98	1.95	1.92	1.90	1.88
1000	6.66	4.63	3.80	3.34	3.04	2.82	2.66	2.53	2.43	2.34	2.27	2.20	2.15	2.10	2.06	2.02	1.98	1.95	1.92	1.90	1.87	1.85

Table F(cont.)

Denominator df	Numerator dF																					
$\alpha = .01$	23	24	25	26	27	28	29	30	32	35	40	45	50	60	75	100	120	140	180	250	400	1000
1	6228.7	6234.3	6239.9	6244.5	6249.2	6252.9	6257.1	6260.4	6266.9	6275.3	6286.4	6295.7	6302.3	6313.0	6323.7	6333.9	6339.5	6343.2	6347.9	6353.5	6358.1	6362.8
2	99.46	99.46	99.46	99.46	99.46	99.46	99.46	99.47	99.47	99.47	99.48	99.48	99.48	99.48	99.48	99.49	99.49	99.49	99.49	99.50	99.50	99.50
3	26.62	26.60	26.58	26.56	26.55	26.53	26.52	26.50	26.48	26.45	26.41	26.38	26.35	26.32	26.28	26.24	26.22	26.21	26.19	26.17	26.15	26.14
4	13.95	13.93	13.91	13.89	13.88	13.86	13.85	13.84	13.81	13.79	13.75	13.71	13.69	13.65	13.61	13.58	13.56	13.54	13.53	13.51	13.49	13.47
5	9.49	9.47	9.45	9.43	9.42	9.40	9.39	9.38	9.36	9.33	9.29	9.26	9.24	9.20	9.17	9.13	9.11	9.10	9.08	9.06	9.05	9.03
6	7.33	7.31	7.30	7.28	7.27	7.25	7.24	7.23	7.21	7.18	7.14	7.11	7.09	7.06	7.02	6.99	6.97	6.96	6.94	6.92	6.91	6.89
7	6.09	6.07	6.06	6.04	6.03	6.02	6.00	5.99	5.97	5.94	5.91	5.88	5.86	5.82	5.79	5.75	5.74	5.72	5.71	5.69	5.68	5.66
8	5.30	5.28	5.26	5.25	5.23	5.22	5.21	5.20	5.18	5.15	5.12	5.09	5.07	5.03	5.00	4.96	4.95	4.93	4.92	4.90	4.89	4.87
9	4.75	4.73	4.71	4.70	4.68	4.67	4.66	4.65	4.63	4.60	4.57	4.54	4.52	4.48	4.45	4.41	4.40	4.39	4.37	4.35	4.34	4.32
10	4.34	4.33	4.31	4.30	4.28	4.27	4.26	4.25	4.23	4.20	4.17	4.14	4.12	4.08	4.05	4.01	4.00	3.98	3.97	3.95	3.94	3.92
11	4.04	4.02	4.01	3.99	3.98	3.96	3.95	3.94	3.92	3.89	3.86	3.83	3.81	3.78	3.74	3.71	3.69	3.68	3.66	3.64	3.63	3.61
12	3.80	3.78	3.76	3.75	3.74	3.72	3.71	3.70	3.68	3.65	3.62	3.59	3.57	3.54	3.50	3.47	3.45	3.44	3.42	3.40	3.39	3.37
13	3.60	3.59	3.57	3.56	3.54	3.53	3.52	3.51	3.49	3.46	3.43	3.40	3.38	3.34	3.31	3.27	3.25	3.24	3.23	3.21	3.19	3.18
14	3.44	3.43	3.41	3.40	3.38	3.37	3.36	3.35	3.33	3.30	3.27	3.24	3.22	3.18	3.15	3.11	3.09	3.08	3.06	3.05	3.03	3.02
15	3.31	3.29	3.28	3.26	3.25	3.24	3.23	3.21	3.19	3.17	3.13	3.10	3.08	3.05	3.01	2.98	2.96	2.95	2.93	2.91	2.90	2.88
16	3.20	3.18	3.16	3.15	3.14	3.12	3.11	3.10	3.08	3.05	3.02	2.99	2.97	2.93	2.90	2.86	2.84	2.83	2.81	2.80	2.78	2.76
17	3.10	3.08	3.07	3.05	3.04	3.03	3.01	3.00	2.98	2.96	2.92	2.89	2.87	2.83	2.80	2.76	2.75	2.73	2.72	2.70	2.68	2.66
18	3.02	3.00	2.98	2.97	2.95	2.94	2.93	2.92	2.90	2.87	2.84	2.81	2.78	2.75	2.71	2.68	2.66	2.65	2.63	2.61	2.59	2.58
19	2.94	2.92	2.91	2.89	2.88	2.87	2.86	2.84	2.82	2.80	2.76	2.73	2.71	2.67	2.64	2.60	2.58	2.57	2.55	2.54	2.52	2.50
20	2.88	2.86	2.84	2.83	2.81	2.80	2.79	2.78	2.76	2.73	2.69	2.67	2.64	2.61	2.57	2.54	2.52	2.50	2.49	2.47	2.45	2.43
21	2.82	2.80	2.79	2.77	2.76	2.74	2.73	2.72	2.70	2.67	2.64	2.61	2.58	2.55	2.51	2.48	2.46	2.44	2.43	2.41	2.39	2.37
22	2.77	2.75	2.73	2.72	2.70	2.69	2.68	2.67	2.65	2.62	2.58	2.55	2.53	2.50	2.46	2.42	2.40	2.39	2.37	2.35	2.34	2.32
23	2.72	2.70	2.69	2.67	2.66	2.64	2.63	2.62	2.60	2.57	2.54	2.51	2.48	2.45	2.41	2.37	2.35	2.34	2.32	2.30	2.29	2.27
24	2.68	2.66	2.64	2.63	2.61	2.60	2.59	2.58	2.56	2.53	2.49	2.46	2.44	2.40	2.37	2.33	2.31	2.30	2.28	2.26	2.24	2.22
25	2.64	2.62	2.60	2.59	2.58	2.56	2.55	2.54	2.52	2.49	2.45	2.42	2.40	2.36	2.33	2.29	2.27	2.26	2.24	2.22	2.20	2.18
26	2.60	2.58	2.57	2.55	2.54	2.53	2.51	2.50	2.48	2.45	2.42	2.39	2.36	2.33	2.29	2.25	2.23	2.22	2.20	2.18	2.16	2.14
27	2.57	2.55	2.54	2.52	2.51	2.49	2.48	2.47	2.45	2.42	2.38	2.35	2.33	2.29	2.26	2.22	2.20	2.18	2.17	2.15	2.13	2.11
28	2.54	2.52	2.51	2.49	2.48	2.46	2.45	2.44	2.42	2.39	2.35	2.32	2.30	2.26	2.23	2.19	2.17	2.15	2.13	2.11	2.10	2.08
29	2.51	2.49	2.48	2.46	2.45	2.44	2.42	2.41	2.39	2.36	2.33	2.30	2.27	2.23	2.20	2.16	2.14	2.12	2.10	2.08	2.07	2.05
30	2.49	2.47	2.45	2.44	2.42	2.41	2.40	2.39	2.36	2.34	2.30	2.27	2.25	2.21	2.17	2.13	2.11	2.10	2.08	2.06	2.04	2.02
32	2.44	2.42	2.41	2.39	2.38	2.36	2.35	2.34	2.32	2.29	2.25	2.22	2.20	2.16	2.12	2.08	2.06	2.05	2.03	2.01	1.99	1.97
35	2.38	2.36	2.35	2.33	2.32	2.30	2.29	2.28	2.26	2.23	2.19	2.16	2.14	2.10	2.06	2.02	2.00	1.98	1.96	1.94	1.92	1.90
40	2.31	2.29	2.27	2.26	2.24	2.23	2.22	2.20	2.18	2.15	2.11	2.08	2.06	2.02	1.98	1.94	1.92	1.90	1.88	1.86	1.84	1.82
45	2.25	2.23	2.21	2.20	2.18	2.17	2.16	2.14	2.12	2.09	2.05	2.02	2.00	1.96	1.92	1.88	1.85	1.84	1.82	1.79	1.77	1.75
50	2.20	2.18	2.17	2.15	2.14	2.12	2.11	2.10	2.08	2.05	2.01	1.97	1.95	1.91	1.87	1.82	1.80	1.79	1.76	1.74	1.72	1.70
60	2.13	2.12	2.10	2.08	2.07	2.05	2.04	2.03	2.01	1.98	1.94	1.90	1.88	1.84	1.79	1.75	1.73	1.71	1.69	1.66	1.64	1.62
75	2.07	2.05	2.03	2.02	2.00	1.99	1.97	1.96	1.94	1.91	1.87	1.83	1.81	1.76	1.72	1.67	1.65	1.63	1.61	1.58	1.56	1.53
100	2.00	1.98	1.97	1.95	1.93	1.92	1.91	1.89	1.87	1.84	1.80	1.76	1.74	1.69	1.65	1.60	1.57	1.55	1.53	1.50	1.47	1.45
120	1.97	1.95	1.93	1.92	1.90	1.89	1.87	1.86	1.84	1.81	1.76	1.73	1.70	1.66	1.61	1.56	1.53	1.51	1.49	1.46	1.43	1.40
140	1.95	1.93	1.91	1.89	1.88	1.86	1.85	1.84	1.81	1.78	1.74	1.70	1.67	1.63	1.58	1.53	1.50	1.48	1.46	1.43	1.40	1.37
180	1.92	1.90	1.88	1.86	1.85	1.83	1.82	1.81	1.78	1.75	1.71	1.67	1.64	1.60	1.55	1.49	1.47	1.45	1.42	1.39	1.35	1.32
250	1.89	1.87	1.85	1.83	1.82	1.80	1.79	1.77	1.75	1.72	1.67	1.64	1.61	1.56	1.51	1.46	1.43	1.41	1.38	1.34	1.31	1.27
400	1.86	1.84	1.82	1.80	1.79	1.77	1.76	1.75	1.72	1.69	1.64	1.61	1.58	1.53	1.48	1.42	1.39	1.37	1.33	1.30	1.26	1.22
1000	1.83	1.81	1.79	1.77	1.76	1.74	1.73	1.72	1.69	1.66	1.61	1.58	1.54	1.50	1.44	1.38	1.35	1.33	1.29	1.25	1.21	1.16

Table F(cont.)

Numerator dF; Denominator df

α = .05	1	2	3	4	5	6	7	8	9	10	11	12	13	14	15	16	17	18	19	20	21	22
1	161.4	199.5	215.7	224.6	230.2	234.0	236.8	238.9	240.5	241.9	243.0	243.9	244.7	245.4	245.9	246.5	246.9	247.3	247.7	248.0	248.3	248.6
2	18.51	19.00	19.16	19.25	19.30	19.33	19.35	19.37	19.38	19.40	19.40	19.41	19.42	19.42	19.43	19.43	19.44	19.44	19.44	19.45	19.45	19.45
3	10.13	9.55	9.28	9.12	9.01	8.94	8.89	8.85	8.81	8.79	8.76	8.74	8.73	8.71	8.70	8.69	8.68	8.67	8.67	8.66	8.65	8.65
4	7.71	6.94	6.59	6.39	6.26	6.16	6.09	6.04	6.00	5.96	5.94	5.91	5.89	5.87	5.86	5.84	5.83	5.82	5.81	5.80	5.79	5.79
5	6.61	5.79	5.41	5.19	5.05	4.95	4.88	4.82	4.77	4.74	4.70	4.68	4.66	4.64	4.62	4.60	4.59	4.58	4.57	4.56	4.55	4.54
6	5.99	5.14	4.76	4.53	4.39	4.28	4.21	4.15	4.10	4.06	4.03	4.00	3.98	3.96	3.94	3.92	3.91	3.90	3.88	3.87	3.86	3.86
7	5.59	4.74	4.35	4.12	3.97	3.87	3.79	3.73	3.68	3.64	3.60	3.57	3.55	3.53	3.51	3.49	3.48	3.47	3.46	3.44	3.43	3.43
8	5.32	4.46	4.07	3.84	3.69	3.58	3.50	3.44	3.39	3.35	3.31	3.28	3.26	3.24	3.22	3.20	3.19	3.17	3.16	3.15	3.14	3.13
9	5.12	4.26	3.86	3.63	3.48	3.37	3.29	3.23	3.18	3.14	3.10	3.07	3.05	3.03	3.01	2.99	2.97	2.96	2.95	2.94	2.93	2.92
10	4.96	4.10	3.71	3.48	3.33	3.22	3.14	3.07	3.02	2.98	2.94	2.91	2.89	2.86	2.85	2.83	2.81	2.80	2.79	2.77	2.76	2.75
11	4.84	3.98	3.59	3.36	3.20	3.09	3.01	2.95	2.90	2.85	2.82	2.79	2.76	2.74	2.72	2.70	2.69	2.67	2.66	2.65	2.64	2.63
12	4.75	3.89	3.49	3.26	3.11	3.00	2.91	2.85	2.80	2.75	2.72	2.69	2.66	2.64	2.62	2.60	2.58	2.57	2.56	2.54	2.53	2.52
13	4.67	3.81	3.41	3.18	3.03	2.92	2.83	2.77	2.71	2.67	2.63	2.60	2.58	2.55	2.53	2.51	2.50	2.48	2.47	2.46	2.45	2.44
14	4.60	3.74	3.34	3.11	2.96	2.85	2.76	2.70	2.65	2.60	2.57	2.53	2.51	2.48	2.46	2.44	2.43	2.41	2.40	2.39	2.38	2.37
15	4.54	3.68	3.29	3.06	2.90	2.79	2.71	2.64	2.59	2.54	2.51	2.48	2.45	2.42	2.40	2.38	2.37	2.35	2.34	2.33	2.32	2.31
16	4.49	3.63	3.24	3.01	2.85	2.74	2.66	2.59	2.54	2.49	2.46	2.42	2.40	2.37	2.35	2.33	2.32	2.30	2.29	2.28	2.26	2.25
17	4.45	3.59	3.20	2.96	2.81	2.70	2.61	2.55	2.49	2.45	2.41	2.38	2.35	2.33	2.31	2.29	2.27	2.26	2.24	2.23	2.22	2.21
18	4.41	3.55	3.16	2.93	2.77	2.66	2.58	2.51	2.46	2.41	2.37	2.34	2.31	2.29	2.27	2.25	2.23	2.22	2.20	2.19	2.18	2.17
19	4.38	3.52	3.13	2.90	2.74	2.63	2.54	2.48	2.42	2.38	2.34	2.31	2.28	2.26	2.23	2.21	2.20	2.18	2.17	2.16	2.14	2.13
20	4.35	3.49	3.10	2.87	2.71	2.60	2.51	2.45	2.39	2.35	2.31	2.28	2.25	2.22	2.20	2.18	2.17	2.15	2.14	2.12	2.11	2.10
21	4.32	3.47	3.07	2.84	2.68	2.57	2.49	2.42	2.37	2.32	2.28	2.25	2.22	2.20	2.18	2.16	2.14	2.12	2.11	2.10	2.08	2.07
22	4.30	3.44	3.05	2.82	2.66	2.55	2.46	2.40	2.34	2.30	2.26	2.23	2.20	2.17	2.15	2.13	2.11	2.10	2.08	2.07	2.06	2.05
23	4.28	3.42	3.03	2.80	2.64	2.53	2.44	2.37	2.32	2.27	2.24	2.20	2.18	2.15	2.13	2.11	2.09	2.08	2.06	2.05	2.04	2.02
24	4.26	3.40	3.01	2.78	2.62	2.51	2.42	2.36	2.30	2.25	2.22	2.18	2.15	2.13	2.11	2.09	2.07	2.05	2.04	2.03	2.01	2.00
25	4.24	3.39	2.99	2.76	2.60	2.49	2.40	2.34	2.28	2.24	2.20	2.16	2.14	2.11	2.09	2.07	2.05	2.04	2.02	2.01	2.00	1.98
26	4.23	3.37	2.98	2.74	2.59	2.47	2.39	2.32	2.27	2.22	2.18	2.15	2.12	2.09	2.07	2.05	2.03	2.02	2.00	1.99	1.98	1.97
27	4.21	3.35	2.96	2.73	2.57	2.46	2.37	2.31	2.25	2.20	2.17	2.13	2.10	2.08	2.06	2.04	2.02	2.00	1.99	1.97	1.96	1.95
28	4.20	3.34	2.95	2.71	2.56	2.45	2.36	2.29	2.24	2.19	2.15	2.12	2.09	2.06	2.04	2.02	2.00	1.99	1.97	1.96	1.95	1.93
29	4.18	3.33	2.93	2.70	2.55	2.43	2.35	2.28	2.22	2.18	2.14	2.10	2.08	2.05	2.03	2.01	1.99	1.97	1.96	1.94	1.93	1.92
30	4.17	3.32	2.92	2.69	2.53	2.42	2.33	2.27	2.21	2.16	2.13	2.09	2.06	2.04	2.01	1.99	1.98	1.96	1.95	1.93	1.92	1.91
32	4.15	3.29	2.90	2.67	2.51	2.40	2.31	2.24	2.19	2.14	2.10	2.07	2.04	2.01	1.99	1.97	1.95	1.94	1.92	1.91	1.90	1.88
35	4.12	3.27	2.87	2.64	2.49	2.37	2.29	2.22	2.16	2.11	2.07	2.04	2.01	1.99	1.96	1.94	1.92	1.91	1.89	1.88	1.87	1.85
40	4.08	3.23	2.84	2.61	2.45	2.34	2.25	2.18	2.12	2.08	2.04	2.00	1.97	1.95	1.92	1.90	1.89	1.87	1.85	1.84	1.83	1.81
45	4.06	3.20	2.81	2.58	2.42	2.31	2.22	2.15	2.10	2.05	2.01	1.97	1.94	1.92	1.89	1.87	1.86	1.84	1.82	1.81	1.80	1.78
50	4.03	3.18	2.79	2.56	2.40	2.29	2.20	2.13	2.07	2.03	1.99	1.95	1.92	1.89	1.87	1.85	1.83	1.81	1.80	1.78	1.77	1.76
60	4.00	3.15	2.76	2.53	2.37	2.25	2.17	2.10	2.04	1.99	1.95	1.92	1.89	1.86	1.84	1.82	1.80	1.78	1.76	1.75	1.73	1.72
75	3.97	3.12	2.73	2.49	2.34	2.22	2.13	2.06	2.01	1.96	1.92	1.88	1.85	1.83	1.80	1.78	1.76	1.74	1.73	1.71	1.70	1.69
100	3.94	3.09	2.70	2.46	2.31	2.19	2.10	2.03	1.97	1.93	1.89	1.85	1.82	1.79	1.77	1.75	1.73	1.71	1.69	1.68	1.66	1.65
120	3.92	3.07	2.68	2.45	2.29	2.18	2.09	2.02	1.96	1.91	1.87	1.83	1.80	1.78	1.75	1.73	1.71	1.69	1.67	1.66	1.64	1.63
140	3.91	3.06	2.67	2.44	2.28	2.16	2.08	2.01	1.95	1.90	1.86	1.82	1.79	1.76	1.74	1.72	1.70	1.68	1.66	1.65	1.63	1.62
180	3.89	3.05	2.65	2.42	2.26	2.15	2.06	1.99	1.93	1.88	1.84	1.81	1.77	1.75	1.72	1.70	1.68	1.66	1.64	1.63	1.61	1.60
250	3.88	3.03	2.64	2.41	2.25	2.13	2.05	1.98	1.92	1.87	1.83	1.79	1.76	1.73	1.71	1.68	1.66	1.65	1.63	1.61	1.60	1.58
400	3.86	3.02	2.63	2.39	2.24	2.12	2.03	1.96	1.90	1.85	1.81	1.78	1.74	1.72	1.69	1.67	1.65	1.63	1.61	1.60	1.58	1.57
1000	3.85	3.00	2.61	2.38	2.22	2.11	2.02	1.95	1.89	1.84	1.80	1.76	1.73	1.70	1.68	1.65	1.63	1.61	1.60	1.58	1.57	1.55

Table F(cont.)

	Denominator dF																					
α = .01	23	24	25	26	27	28	29	30	32	35	40	45	50	60	75	100	120	140	180	250	400	1000
1	248.8	249.1	249.3	249.5	249.6	249.8	250.0	250.1	250.4	250.7	251.1	251.5	251.8	252.2	252.6	253.0	253.3	253.4	253.6	253.8	254.0	254.2
2	19.45	19.45	19.46	19.46	19.46	19.46	19.46	19.46	19.46	19.47	19.47	19.47	19.48	19.48	19.48	19.49	19.49	19.49	19.49	19.49	19.49	19.49
3	8.64	8.64	8.63	8.63	8.63	8.62	8.62	8.62	8.61	8.60	8.59	8.59	8.58	8.57	8.56	8.55	8.55	8.55	8.54	8.54	8.53	8.53
4	5.78	5.77	5.77	5.76	5.76	5.75	5.75	5.75	5.74	5.73	5.72	5.71	5.70	5.69	5.68	5.66	5.66	5.65	5.65	5.64	5.64	5.63
5	4.53	4.53	4.52	4.52	4.51	4.50	4.50	4.50	4.49	4.48	4.46	4.45	4.44	4.43	4.42	4.41	4.40	4.39	4.39	4.38	4.38	4.37
6	3.85	3.84	3.83	3.83	3.82	3.82	3.81	3.81	3.80	3.79	3.77	3.76	3.75	3.74	3.73	3.71	3.70	3.70	3.69	3.69	3.68	3.67
7	3.42	3.41	3.40	3.40	3.39	3.39	3.38	3.38	3.37	3.36	3.34	3.33	3.32	3.30	3.29	3.27	3.27	3.26	3.25	3.25	3.24	3.23
8	3.12	3.12	3.11	3.10	3.10	3.09	3.08	3.08	3.07	3.06	3.04	3.03	3.02	3.01	2.99	2.97	2.97	2.96	2.95	2.95	2.94	2.93
9	2.91	2.90	2.89	2.89	2.88	2.87	2.87	2.86	2.85	2.84	2.83	2.81	2.80	2.79	2.77	2.76	2.75	2.74	2.73	2.73	2.72	2.71
10	2.75	2.74	2.73	2.72	2.72	2.71	2.70	2.70	2.69	2.68	2.66	2.65	2.64	2.62	2.60	2.59	2.58	2.57	2.57	2.56	2.55	2.54
11	2.62	2.61	2.60	2.59	2.59	2.58	2.58	2.57	2.56	2.55	2.53	2.52	2.51	2.49	2.47	2.46	2.45	2.44	2.43	2.43	2.42	2.41
12	2.51	2.51	2.50	2.49	2.48	2.48	2.47	2.47	2.46	2.44	2.43	2.41	2.40	2.38	2.37	2.35	2.34	2.33	2.33	2.32	2.31	2.30
13	2.43	2.42	2.41	2.41	2.40	2.39	2.39	2.38	2.37	2.36	2.34	2.33	2.31	2.30	2.28	2.26	2.25	2.25	2.24	2.23	2.22	2.21
14	2.36	2.35	2.34	2.33	2.33	2.32	2.31	2.31	2.30	2.28	2.27	2.25	2.24	2.22	2.21	2.19	2.18	2.17	2.16	2.15	2.15	2.14
15	2.30	2.29	2.28	2.27	2.27	2.26	2.25	2.25	2.24	2.22	2.20	2.19	2.18	2.16	2.14	2.12	2.11	2.11	2.10	2.09	2.08	2.07
16	2.24	2.24	2.23	2.22	2.21	2.21	2.20	2.19	2.18	2.17	2.15	2.14	2.12	2.11	2.09	2.07	2.06	2.05	2.04	2.03	2.02	2.02
17	2.20	2.19	2.18	2.17	2.17	2.16	2.15	2.15	2.14	2.12	2.10	2.09	2.08	2.06	2.04	2.02	2.01	2.00	1.99	1.98	1.98	1.97
18	2.16	2.15	2.14	2.13	2.13	2.12	2.11	2.11	2.10	2.08	2.06	2.05	2.04	2.02	2.00	1.98	1.97	1.96	1.95	1.94	1.93	1.92
19	2.12	2.11	2.11	2.10	2.09	2.08	2.08	2.07	2.06	2.05	2.03	2.01	2.00	1.98	1.96	1.94	1.93	1.92	1.91	1.90	1.89	1.88
20	2.09	2.08	2.07	2.07	2.06	2.05	2.05	2.04	2.03	2.01	1.99	1.98	1.97	1.95	1.93	1.91	1.90	1.89	1.88	1.87	1.86	1.85
21	2.06	2.05	2.05	2.04	2.03	2.02	2.02	2.01	2.00	1.98	1.96	1.95	1.94	1.92	1.90	1.88	1.87	1.86	1.85	1.84	1.83	1.82
22	2.04	2.03	2.02	2.01	2.00	2.00	1.99	1.98	1.97	1.96	1.94	1.92	1.91	1.89	1.87	1.85	1.84	1.83	1.82	1.81	1.80	1.79
23	2.01	2.01	2.00	1.99	1.98	1.97	1.97	1.96	1.95	1.93	1.91	1.90	1.88	1.86	1.84	1.82	1.81	1.81	1.79	1.78	1.77	1.76
24	1.99	1.98	1.97	1.97	1.96	1.95	1.95	1.94	1.93	1.91	1.89	1.88	1.86	1.84	1.82	1.80	1.79	1.78	1.77	1.76	1.75	1.74
25	1.97	1.96	1.96	1.95	1.94	1.93	1.93	1.92	1.91	1.89	1.87	1.86	1.84	1.82	1.80	1.78	1.77	1.76	1.75	1.74	1.73	1.72
26	1.96	1.95	1.94	1.93	1.92	1.91	1.91	1.90	1.89	1.87	1.85	1.84	1.82	1.80	1.78	1.76	1.75	1.74	1.73	1.72	1.71	1.70
27	1.94	1.93	1.92	1.91	1.90	1.30	1.89	1.88	1.87	1.86	1.84	1.82	1.81	1.79	1.76	1.74	1.73	1.72	1.71	1.70	1.69	1.68
28	1.92	1.91	1.91	1.90	1.89	1.38	1.88	1.87	1.86	1.84	1.82	1.80	1.79	1.77	1.75	1.73	1.71	1.71	1.69	1.68	1.67	1.66
29	1.91	1.90	1.89	1.88	1.88	1.37	1.86	1.85	1.84	1.83	1.81	1.79	1.77	1.75	1.73	1.71	1.70	1.69	1.68	1.67	1.66	1.65
30	1.90	1.89	1.88	1.87	1.86	1.85	1.85	1.84	1.83	1.81	1.79	1.77	1.76	1.74	1.72	1.70	1.68	1.68	1.66	1.65	1.64	1.63
32	1.87	1.86	1.85	1.85	1.84	1.83	1.82	1.82	1.80	1.79	1.77	1.75	1.74	1.71	1.69	1.67	1.66	1.65	1.64	1.63	1.61	1.60
35	1.84	1.83	1.82	1.82	1.81	1.80	1.79	1.79	1.77	1.76	1.74	1.72	1.70	1.68	1.66	1.63	1.62	1.61	1.60	1.59	1.58	1.57
40	1.80	1.79	1.78	1.77	1.77	1.76	1.75	1.74	1.73	1.72	1.69	1.67	1.66	1.64	1.61	1.59	1.58	1.57	1.55	1.54	1.53	1.52
45	1.77	1.76	1.75	1.74	1.73	1.73	1.72	1.71	1.70	1.68	1.66	1.64	1.63	1.60	1.58	1.55	1.54	1.53	1.52	1.51	1.49	1.48
50	1.75	1.74	1.73	1.72	1.71	1.70	1.69	1.69	1.67	1.66	1.63	1.61	1.60	1.58	1.55	1.52	1.51	1.50	1.49	1.47	1.46	1.45
60	1.71	1.70	1.69	1.68	1.67	1.66	1.66	1.65	1.64	1.62	1.59	1.57	1.56	1.53	1.51	1.48	1.47	1.46	1.44	1.43	1.41	1.40
75	1.67	1.66	1.65	1.64	1.63	1.63	1.62	1.61	1.60	1.58	1.55	1.53	1.52	1.49	1.47	1.44	1.42	1.41	1.40	1.38	1.37	1.35
100	1.64	1.63	1.62	1.61	1.60	1.59	1.58	1.57	1.56	1.54	1.52	1.49	1.48	1.45	1.42	1.39	1.38	1.36	1.35	1.33	1.31	1.30
120	1.62	1.61	1.60	1.59	1.58	1.57	1.56	1.55	1.54	1.52	1.50	1.47	1.46	1.43	1.40	1.37	1.35	1.34	1.32	1.30	1.29	1.27
140	1.61	1.60	1.58	1.57	1.57	1.56	1.55	1.54	1.53	1.51	1.48	1.46	1.44	1.41	1.38	1.35	1.33	1.32	1.30	1.29	1.27	1.25
180	1.59	1.58	1.57	1.56	1.55	1.54	1.53	1.52	1.51	1.49	1.46	1.44	1.42	1.39	1.36	1.33	1.31	1.30	1.28	1.26	1.24	1.22
250	1.57	1.56	1.55	1.54	1.53	1.52	1.51	1.50	1.49	1.47	1.44	1.42	1.40	1.37	1.34	1.31	1.29	1.27	1.25	1.23	1.21	1.18
400	1.56	1.54	1.53	1.52	1.51	1.50	1.50	1.49	1.47	1.45	1.42	1.40	1.38	1.35	1.32	1.28	1.26	1.25	1.23	1.20	1.18	1.15
1000	1.54	1.53	1.52	1.51	1.50	1.49	1.48	1.47	1.46	1.43	1.41	1.38	1.36	1.33	1.30	1.26	1.24	1.22	1.20	1.17	1.14	1.11

Denominator df

Table F(cont.)

	Denominator dF																					
$\alpha = .01$	1	2	3	4	5	6	7	8	9	10	11	12	13	14	15	16	17	18	19	20	21	22
1	39.9	49.5	53.6	55.8	57.2	58.2	58.9	59.4	59.9	60.2	60.5	60.7	60.9	61.1	61.2	61.3	61.5	61.6	61.7	61.7	61.8	61.9
2	8.53	9.00	9.16	9.24	9.29	9.33	9.35	9.37	9.38	9.39	9.40	9.41	9.41	9.42	9.42	9.43	9.43	9.44	9.44	9.44	9.44	9.45
3	5.54	5.46	5.39	5.34	5.31	5.28	5.27	5.25	5.24	5.23	5.22	5.22	5.21	5.20	5.20	5.20	5.19	5.19	5.19	5.18	5.18	5.18
4	4.54	4.32	4.19	4.11	4.05	4.01	3.98	3.95	3.94	3.92	3.91	3.90	3.89	3.88	3.87	3.86	3.86	3.85	3.85	3.84	3.84	3.84
5	4.06	3.78	3.62	3.52	3.45	3.40	3.37	3.34	3.32	3.30	3.28	3.27	3.26	3.25	3.24	3.23	3.22	3.22	3.21	3.21	3.20	3.20
6	3.78	3.46	3.29	3.18	3.11	3.05	3.01	2.98	2.96	2.94	2.92	2.90	2.89	2.88	2.87	2.86	2.85	2.85	2.84	2.84	2.83	2.83
7	3.59	3.26	3.07	2.96	2.88	2.83	2.78	2.75	2.72	2.70	2.68	2.67	2.65	2.64	2.63	2.62	2.61	2.61	2.60	2.59	2.59	2.58
8	3.46	3.11	2.92	2.81	2.73	2.67	2.62	2.59	2.56	2.54	2.52	2.50	2.49	2.48	2.46	2.45	2.45	2.44	2.43	2.42	2.42	2.41
9	3.36	3.01	2.81	2.69	2.61	2.55	2.51	2.47	2.44	2.42	2.40	2.38	2.36	2.35	2.34	2.33	2.32	2.31	2.30	2.30	2.29	2.29
10	3.29	2.92	2.73	2.61	2.52	2.46	2.41	2.38	2.35	2.32	2.30	2.28	2.27	2.26	2.24	2.23	2.22	2.22	2.21	2.20	2.19	2.19
11	3.23	2.86	2.66	2.54	2.45	2.39	2.34	2.30	2.27	2.25	2.23	2.21	2.19	2.18	2.17	2.16	2.15	2.14	2.13	2.12	2.12	2.11
12	3.18	2.81	2.61	2.48	2.39	2.33	2.28	2.24	2.21	2.19	2.17	2.15	2.13	2.12	2.10	2.09	2.08	2.08	2.07	2.06	2.05	2.05
13	3.14	2.76	2.56	2.43	2.35	2.28	2.23	2.20	2.16	2.14	2.12	2.10	2.08	2.07	2.05	2.04	2.03	2.02	2.01	2.01	2.00	1.99
14	3.10	2.73	2.52	2.39	2.31	2.24	2.19	2.15	2.12	2.10	2.07	2.05	2.04	2.02	2.01	2.00	1.99	1.98	1.97	1.96	1.96	1.95
15	3.07	2.70	2.49	2.36	2.27	2.21	2.16	2.12	2.09	2.06	2.04	2.02	2.00	1.99	1.97	1.96	1.95	1.94	1.93	1.92	1.92	1.91
16	3.05	2.67	2.46	2.33	2.24	2.18	2.13	2.09	2.06	2.03	2.01	1.99	1.97	1.95	1.94	1.93	1.92	1.91	1.90	1.89	1.88	1.88
17	3.03	2.64	2.44	2.31	2.22	2.15	2.10	2.06	2.03	2.00	1.98	1.96	1.94	1.93	1.91	1.90	1.89	1.88	1.87	1.86	1.86	1.85
18	3.01	2.62	2.42	2.29	2.20	2.13	2.08	2.04	2.00	1.98	1.95	1.93	1.92	1.90	1.89	1.87	1.86	1.85	1.84	1.84	1.83	1.82
19	2.99	2.61	2.40	2.27	2.18	2.11	2.06	2.02	1.98	1.96	1.93	1.91	1.89	1.88	1.86	1.85	1.84	1.83	1.82	1.81	1.81	1.80
20	2.97	2.59	2.38	2.25	2.16	2.09	2.04	2.00	1.96	1.94	1.91	1.89	1.87	1.86	1.84	1.83	1.82	1.81	1.80	1.79	1.79	1.78
21	2.96	2.57	2.36	2.23	2.14	2.08	2.02	1.98	1.95	1.92	1.90	1.87	1.86	1.84	1.83	1.81	1.80	1.79	1.78	1.78	1.77	1.76
22	2.95	2.56	2.35	2.22	2.13	2.06	2.01	1.97	1.93	1.90	1.88	1.86	1.84	1.83	1.81	1.80	1.79	1.78	1.77	1.76	1.75	1.74
23	2.94	2.55	2.34	2.21	2.11	2.05	1.99	1.95	1.92	1.89	1.87	1.84	1.83	1.81	1.80	1.78	1.77	1.76	1.75	1.74	1.74	1.73
24	2.93	2.54	2.33	2.19	2.10	2.04	1.98	1.94	1.91	1.88	1.85	1.83	1.81	1.80	1.78	1.77	1.76	1.75	1.74	1.73	1.72	1.71
25	2.92	2.53	2.32	2.18	2.09	2.02	1.97	1.93	1.89	1.87	1.84	1.82	1.80	1.79	1.77	1.76	1.75	1.74	1.73	1.72	1.71	1.70
26	2.91	2.52	2.31	2.17	2.08	2.01	1.96	1.92	1.88	1.86	1.83	1.81	1.79	1.77	1.76	1.75	1.73	1.72	1.71	1.71	1.70	1.69
27	2.90	2.51	2.30	2.17	2.07	2.00	1.95	1.91	1.87	1.85	1.82	1.80	1.78	1.76	1.75	1.74	1.72	1.71	1.70	1.70	1.69	1.68
28	2.89	2.50	2.29	2.16	2.06	2.00	1.94	1.90	1.87	1.84	1.81	1.79	1.77	1.75	1.74	1.73	1.71	1.70	1.69	1.69	1.68	1.67
29	2.89	2.50	2.28	2.15	2.06	1.99	1.93	1.89	1.86	1.83	1.80	1.78	1.76	1.75	1.73	1.72	1.71	1.69	1.68	1.68	1.67	1.66
30	2.88	2.49	2.28	2.14	2.05	1.98	1.93	1.88	1.85	1.82	1.79	1.77	1.75	1.74	1.72	1.71	1.70	1.69	1.68	1.67	1.66	1.65
32	2.87	2.48	2.26	2.13	2.04	1.97	1.91	1.87	1.83	1.81	1.78	1.76	1.74	1.72	1.71	1.69	1.68	1.67	1.66	1.65	1.64	1.64
35	2.85	2.46	2.25	2.11	2.02	1.95	1.90	1.85	1.82	1.79	1.76	1.74	1.72	1.70	1.69	1.67	1.66	1.65	1.64	1.63	1.62	1.62
40	2.84	2.44	2.23	2.09	2.00	1.93	1.87	1.83	1.79	1.76	1.74	1.71	1.70	1.68	1.66	1.65	1.64	1.62	1.61	1.61	1.60	1.59
45	2.82	2.42	2.21	2.07	1.98	1.91	1.85	1.81	1.77	1.74	1.72	1.70	1.68	1.66	1.64	1.63	1.62	1.60	1.59	1.58	1.58	1.57
50	2.81	2.41	2.20	2.06	1.97	1.90	1.84	1.80	1.76	1.73	1.70	1.68	1.66	1.64	1.63	1.61	1.60	1.59	1.58	1.57	1.56	1.55
60	2.79	2.39	2.18	2.04	1.95	1.87	1.82	1.77	1.74	1.71	1.68	1.66	1.64	1.62	1.60	1.59	1.58	1.56	1.55	1.54	1.53	1.53
75	2.77	2.37	2.16	2.02	1.93	1.85	1.80	1.75	1.72	1.69	1.66	1.63	1.61	1.60	1.58	1.57	1.55	1.54	1.53	1.52	1.51	1.50
100	2.76	2.36	2.14	2.00	1.91	1.83	1.78	1.73	1.69	1.66	1.64	1.61	1.59	1.57	1.56	1.54	1.53	1.52	1.50	1.49	1.48	1.48
120	2.75	2.35	2.13	1.99	1.90	1.82	1.77	1.72	1.68	1.65	1.63	1.60	1.58	1.56	1.55	1.53	1.52	1.50	1.49	1.48	1.47	1.46
140	2.74	2.34	2.12	1.99	1.89	1.82	1.76	1.71	1.68	1.64	1.62	1.59	1.57	1.55	1.54	1.52	1.51	1.50	1.48	1.47	1.46	1.45
180	2.73	2.33	2.11	1.98	1.88	1.81	1.75	1.70	1.67	1.63	1.61	1.58	1.56	1.54	1.53	1.51	1.50	1.48	1.47	1.46	1.45	1.44
250	2.73	2.32	2.11	1.97	1.87	1.80	1.74	1.69	1.66	1.62	1.60	1.57	1.55	1.53	1.51	1.50	1.49	1.47	1.46	1.45	1.44	1.43
400	2.72	2.32	2.10	1.96	1.86	1.79	1.73	1.69	1.65	1.61	1.59	1.56	1.54	1.52	1.50	1.49	1.47	1.46	1.45	1.44	1.43	1.42
1000	2.71	2.31	2.09	1.95	1.85	1.78	1.72	1.68	1.64	1.61	1.58	1.55	1.53	1.51	1.49	1.48	1.46	1.45	1.44	1.43	1.42	1.41

Denominator df

Table F(cont.)

Denominator dF

α = .01	23	24	25	26	27	28	29	30	32	35	40	45	50	60	75	100	120	140	180	250	400	1000
1	61.9	62.0	62.1	62.1	62.1	62.2	62.2	62.3	62.3	62.4	62.5	62.6	62.7	62.8	62.9	63.0	63.1	63.1	63.1	63.2	63.2	63.3
2	9.45	9.45	9.45	9.45	9.45	9.46	9.46	9.46	9.46	9.46	9.47	9.47	9.47	9.47	9.48	9.48	9.48	9.48	9.49	9.49	9.49	9.49
3	5.18	5.18	5.17	5.17	5.17	5.17	5.17	5.17	5.17	5.16	5.16	5.16	5.15	5.15	5.15	5.14	5.14	5.14	5.14	5.14	5.14	5.1
4	3.83	3.83	3.83	3.83	3.82	3.82	3.82	3.82	3.81	3.81	3.80	3.80	3.80	3.79	3.78	3.78	3.78	3.77	3.77	3.77	3.77	3.76
5	3.19	3.19	3.19	3.18	3.18	3.18	3.18	3.17	3.17	3.16	3.16	3.15	3.15	3.14	3.13	3.13	3.12	3.12	3.12	3.11	3.11	3.11
6	2.82	2.82	2.81	2.81	2.81	2.8[illegible]	2.80	2.80	2.80	2.79	2.78	2.77	2.77	2.76	2.75	2.75	2.74	2.74	2.74	2.73	2.73	2.72
7	2.58	2.58	2.57	2.57	2.56	2.56	2.56	2.56	2.55	2.54	2.54	2.53	2.52	2.51	2.51	2.50	2.49	2.49	2.49	2.48	2.48	2.47
8	2.41	2.40	2.40	2.40	2.39	2.39	2.39	2.38	2.38	2.37	2.36	2.35	2.35	2.34	2.33	2.32	2.32	2.31	2.31	2.30	2.30	2.30
9	2.28	2.28	2.27	2.27	2.26	2.26	2.26	2.25	2.25	2.24	2.23	2.22	2.22	2.21	2.20	2.19	2.18	2.18	2.18	2.17	2.17	2.16
10	2.18	2.18	2.17	2.17	2.17	2.16	2.16	2.16	2.15	2.14	2.13	2.12	2.12	2.11	2.10	2.09	2.08	2.08	2.07	2.07	2.06	2.06
11	2.11	2.10	2.10	2.09	2.09	2.08	2.08	2.08	2.07	2.06	2.05	2.04	2.04	2.03	2.02	2.01	2.00	2.00	1.99	1.99	1.98	1.98
12	2.04	2.04	2.03	2.03	2.02	2.02	2.01	2.01	2.01	2.00	1.99	1.98	1.97	1.96	1.95	1.94	1.93	1.93	1.92	1.92	1.91	1.91
13	1.99	1.98	1.98	1.97	1.97	1.96	1.96	1.96	1.95	1.94	1.93	1.92	1.92	1.90	1.89	1.88	1.88	1.87	1.87	1.86	1.86	1.85
14	1.94	1.94	1.93	1.93	1.92	1.92	1.92	1.91	1.91	1.90	1.89	1.88	1.87	1.86	1.85	1.83	1.83	1.82	1.82	1.81	1.81	1.80
15	1.90	1.90	1.89	1.89	1.88	1.88	1.88	1.87	1.87	1.86	1.85	1.84	1.83	1.82	1.80	1.79	1.79	1.78	1.78	1.77	1.76	1.76
16	1.87	1.87	1.86	1.86	1.85	1.85	1.84	1.84	1.83	1.82	1.81	1.80	1.79	1.78	1.77	1.76	1.75	1.75	1.74	1.73	1.73	1.72
17	1.84	1.84	1.83	1.83	1.82	1.82	1.81	1.81	1.80	1.79	1.78	1.77	1.76	1.75	1.74	1.73	1.72	1.71	1.71	1.70	1.70	1.69
18	1.82	1.81	1.80	1.80	1.80	1.79	1.79	1.78	1.78	1.77	1.75	1.74	1.74	1.72	1.71	1.70	1.69	1.69	1.68	1.67	1.67	1.66
19	1.79	1.79	1.78	1.78	1.77	1.77	1.76	1.76	1.75	1.74	1.73	1.72	1.71	1.70	1.69	1.67	1.67	1.66	1.65	1.65	1.64	1.64
20	1.77	1.77	1.76	1.76	1.75	1.75	1.74	1.74	1.73	1.72	1.71	1.70	1.69	1.68	1.66	1.65	1.64	1.64	1.63	1.62	1.62	1.61
21	1.75	1.75	1.74	1.74	1.73	1.73	1.72	1.72	1.71	1.70	1.69	1.68	1.67	1.66	1.64	1.63	1.62	1.62	1.61	1.60	1.60	1.59
22	1.74	1.73	1.73	1.72	1.72	1.7[illegible]	1.71	1.70	1.69	1.68	1.67	1.66	1.65	1.64	1.63	1.61	1.60	1.60	1.59	1.59	1.58	1.57
23	1.72	1.72	1.71	1.70	1.70	1.69	1.69	1.69	1.68	1.67	1.66	1.64	1.64	1.62	1.61	1.59	1.59	1.58	1.57	1.57	1.56	1.55
24	1.71	1.70	1.70	1.69	1.69	1.68	1.68	1.67	1.66	1.65	1.64	1.63	1.62	1.61	1.59	1.58	1.57	1.57	1.56	1.55	1.54	1.54
25	1.70	1.69	1.68	1.68	1.67	1.67	1.66	1.66	1.65	1.64	1.63	1.62	1.61	1.59	1.58	1.56	1.56	1.55	1.54	1.54	1.53	1.52
26	1.68	1.68	1.67	1.67	1.66	1.66	1.65	1.65	1.64	1.63	1.61	1.60	1.59	1.58	1.57	1.55	1.54	1.54	1.53	1.52	1.52	1.51
27	1.67	1.67	1.66	1.65	1.65	1.64	1.64	1.64	1.63	1.62	1.60	1.59	1.58	1.57	1.55	1.54	1.53	1.53	1.52	1.51	1.50	1.50
28	1.66	1.66	1.65	1.64	1.64	1.63	1.63	1.63	1.62	1.61	1.59	1.58	1.57	1.56	1.54	1.53	1.52	1.51	1.51	1.50	1.49	1.48
29	1.65	1.65	1.64	1.63	1.63	1.62	1.62	1.62	1.61	1.60	1.58	1.57	1.56	1.55	1.53	1.52	1.51	1.50	1.50	1.49	1.48	1.47
30	1.64	1.64	1.63	1.63	1.62	1.62	1.61	1.61	1.60	1.59	1.57	1.56	1.55	1.54	1.52	1.51	1.50	1.49	1.49	1.48	1.47	1.46
32	1.63	1.62	1.62	1.61	1.60	1.60	1.59	1.59	1.58	1.57	1.56	1.54	1.53	1.52	1.50	1.49	1.48	1.47	1.47	1.46	1.45	1.44
35	1.61	1.60	1.60	1.59	1.58	1.58	1.57	1.57	1.56	1.55	1.53	1.52	1.51	1.50	1.48	1.47	1.46	1.45	1.44	1.43	1.43	1.42
40	1.58	1.57	1.57	1.56	1.56	1.55	1.55	1.54	1.53	1.52	1.51	1.49	1.48	1.47	1.45	1.43	1.42	1.42	1.41	1.40	1.39	1.38
45	1.56	1.55	1.55	1.54	1.53	1.53	1.52	1.52	1.51	1.50	1.48	1.47	1.46	1.44	1.43	1.41	1.40	1.39	1.38	1.37	1.37	1.36
50	1.54	1.54	1.53	1.52	1.52	1.5[illegible]	1.51	1.50	1.49	1.48	1.46	1.45	1.44	1.42	1.41	1.39	1.38	1.37	1.36	1.35	1.34	1.33
60	1.52	1.51	1.50	1.50	1.49	1.49	1.48	1.48	1.47	1.45	1.44	1.42	1.41	1.40	1.38	1.36	1.35	1.34	1.33	1.32	1.31	1.30
75	1.49	1.49	1.48	1.47	1.47	1.46	1.45	1.45	1.44	1.43	1.41	1.40	1.38	1.37	1.35	1.33	1.32	1.31	1.30	1.29	1.27	1.26
100	1.47	1.46	1.45	1.45	1.44	1.43	1.43	1.42	1.41	1.40	1.38	1.37	1.35	1.34	1.32	1.29	1.28	1.27	1.26	1.25	1.24	1.22
120	1.46	1.45	1.44	1.43	1.43	1.42	1.41	1.41	1.40	1.39	1.37	1.35	1.34	1.32	1.30	1.28	1.26	1.26	1.24	1.23	1.22	1.20
140	1.45	1.44	1.43	1.42	1.42	1.4[illegible]	1.41	1.40	1.39	1.38	1.36	1.34	1.33	1.31	1.29	1.26	1.25	1.24	1.23	1.22	1.20	1.19
180	1.43	1.43	1.42	1.41	1.40	1.40	1.39	1.39	1.38	1.36	1.34	1.33	1.32	1.29	1.27	1.25	1.23	1.22	1.21	1.20	1.18	1.16
250	1.42	1.41	1.41	1.40	1.39	1.39	1.38	1.37	1.36	1.35	1.33	1.31	1.30	1.28	1.26	1.23	1.22	1.21	1.19	1.18	1.16	1.14
400	1.41	1.40	1.39	1.39	1.38	1.37	1.37	1.36	1.35	1.34	1.32	1.30	1.29	1.26	1.24	1.21	1.20	1.19	1.17	1.16	1.14	1.12
1000	1.40	1.39	1.38	1.38	1.37	1.36	1.36	1.35	1.34	1.32	1.30	1.29	1.27	1.25	1.23	1.20	1.18	1.17	1.15	1.13	1.11	1.08

Denominator df

Selected Formulas

$Range = Max - Min$

$IQR = Q3 - Q1$

Outlier Rule-of-Thumb: $y < Q1 - 1.5 \times IQR$ or $y > Q3 + 1.5 \times IQR$

$$\bar{y} = \frac{\sum y}{n}$$

$$s = \sqrt{\frac{\sum (y - \bar{y})^2}{n - 1}}$$

$$z = \frac{y - \mu}{\sigma} \text{ (model based)}$$

$$z = \frac{y - \bar{y}}{s} \text{ (data based)}$$

$$r = \frac{\sum z_x z_y}{n - 1}$$

$$\hat{y} = b_0 + b_1 x \qquad \text{where } b_1 = r\frac{s_y}{s_x} \text{ and } b_0 = \bar{y} - b_1\bar{x}$$

$$P(\mathbf{A}^C) = 1 - P(\mathbf{A})$$

$$P(\mathbf{A} \text{ } or \text{ } \mathbf{B}) = P(\mathbf{A}) + P(\mathbf{B}) - P(\mathbf{A} \text{ } and \text{ } \mathbf{B})$$

$$P(\mathbf{A} \text{ } and \text{ } \mathbf{B}) = P(\mathbf{A}) \times P(\mathbf{B} \mid \mathbf{A})$$

$$P(\mathbf{B} \mid \mathbf{A}) = \frac{P(\mathbf{A} \text{ } and \text{ } \mathbf{B})}{P(\mathbf{A})}$$

If **A** and **B** are independent, $P(\mathbf{B} \mid \mathbf{A}) = P(\mathbf{B})$

$E(X) = \mu = \Sigma x P(x)$	$Var(X) = \sigma^2 = \Sigma(x - \mu)^2 P(x)$
$E(X \pm c) = E(X) \pm c$	$Var(X \pm c) = Var(X)$
$E(aX) = aE(X)$	$Var(aX) = a^2 Var(X)$
$E(X \pm Y) = E(X) \pm E(Y)$	$Var(X \pm Y) = Var(X) + Var(Y)$ if X and Y are independent

Geometric: $P(x) = q^{x-1}p \qquad \mu = \frac{1}{p} \qquad \sigma = \sqrt{\frac{q}{p^2}}$

Binomial: $P(x) = {}_nC_x p^x q^{n-x} \qquad \mu = np \qquad \sigma = \sqrt{npq}$

$$\hat{p} = \frac{x}{n} \qquad \mu(\hat{p}) = p \qquad SD(\hat{p}) = \sqrt{\frac{pq}{n}}$$

Sampling distribution of $\bar{y}$:
(CLT) As n grows, the sampling distribution approaches the Normal model with

$$\mu(\bar{y}) = \mu_y \qquad SD(\bar{y}) = \frac{\sigma}{\sqrt{n}}$$

Inference:

Confidence interval for parameter = ***statistic*** $\pm$ ***critical value*** $\times$ ***SE(statistic)***

$$\text{Test statistic} = \frac{statistic - parameter}{SD(statistic)}$$

Parameter	Statistic	*SD*(statistic)	*SE*(statistic)
p	$\hat{p}$	$\sqrt{\frac{pq}{n}}$	$\sqrt{\frac{\hat{p}\hat{q}}{n}}$
$p_1 - p_2$	$\hat{p}_1 - \hat{p}_2$	$\sqrt{\frac{p_1q_1}{n_1} + \frac{p_2q_2}{n_2}}$	$\sqrt{\frac{\hat{p}_1\hat{q}_1}{n_1} + \frac{\hat{p}_2\hat{q}_2}{n_2}}$
μ	$\bar{y}$	$\frac{\sigma}{\sqrt{n}}$	$\frac{s}{\sqrt{n}}$
$\mu_1 - \mu_2$	$\bar{y}_1 - \bar{y}_2$	$\sqrt{\frac{\sigma_1^2}{n_1} + \frac{\sigma_2^2}{n_2}}$	$\sqrt{\frac{s_1^2}{n_1} + \frac{s_2^2}{n_2}}$
μ_d	$\bar{d}$	$\frac{\sigma_d}{\sqrt{n}}$	$\frac{s_d}{\sqrt{n}}$
σ_ε	$s_e = \sqrt{\frac{\sum(y - \hat{y})^2}{n - 2}}$	(dividing $n - k - 1$ in multiple regression)	
β_1	b_1	(in simple regression)	$\frac{s_e}{s_x\sqrt{n - 1}}$
μ_ν	$\hat{y}_\nu$	(in simple regression)	$\sqrt{SE^2(b_1) \cdot (x_\nu - \bar{x})^2 + \frac{s_e^2}{n}}$
y_ν	$\hat{y}_\nu$	(in simple regression)	$\sqrt{SE^2(b_1) \cdot (x_\nu - \bar{x})^2 + \frac{s_e^2}{n} + s_e^2}$

Pooling: For testing difference between proportions: $\hat{p}_{pooled} = \frac{y_1 + y_2}{n_1 + n_2}$

For testing difference between means: $s_p = \sqrt{\frac{(n_1 - 1)s_1^2 + (n_2 - 1)s_2^2}{n_1 + n_2 - 2}}$

Substitute these pooled estimates in the respective SE formulas for both groups when assumptions and conditions are met.

Chi-square: $\chi^2 = \sum \frac{(Obs - Exp)^2}{Exp}$

ANOVA: $SS_T = \sum\sum(\bar{y}_j - \bar{\bar{y}})^2; \quad MS_T = SS_T/(k - 1)$

$SS_E = \sum\sum(y_{ij} - \bar{y}_j)^2; \quad MS_E = SS_E/(N - k)$

$F_{k-1, N-k} = MS_T/MS_E$

Assumptions for Inference	And the Conditions That Support or Override Them
Proportions (z)	
◆ **One sample**	
1. Individuals are independent.	1. SRS and $n < 10\%$ of the population.
2. Sample is sufficiently large.	2. Successes and failures ≥ 10.
◆ **Two sample**	
1. Samples are independent.	1. (Think about how the data were collected.)
2. Data in each sample are independent.	2. Both are SRSs and $n < 10\%$ of populations OR random allocation.
3. Both samples are sufficiently large.	3. Successes and failures ≥ 10 for both.
Means (t)	
◆ **One sample** (df $= n - 1$)	
1. Individuals are independent.	1. SRS and $n < 10\%$ of the population.
2. Population has a Normal model.	2. Histogram is unimodal and symmetric.*
◆ **Two independent samples** (df from technology)	
1. Samples are independent.	1. (Think about the design.)
2. Data in each sample are independent.	2. SRSs and $n < 10\%$ OR random allocation.
3. Both populations are Normal.	3. Both histograms are unimodal and symmetric.*
◆ **Matched pairs** (df $= n - 1$)	
1. Data are matched; n pairs.	1. (Think about the design.)
2. Individuals are independent.	2. SRSs and $n < 10\%$ OR random allocation.
3. Population of differences is Normal.	3. Histogram of differences is unimodal and symmetric.
Distributions/Association (χ^2)	
◆ **Goodness-of-fit** [df $=$ # of cells -1; one variable, one sample compared with population model]	
1. Data are counts.	1. (Are they?)
2. Data in sample are independent.	2. SRS and $n < 10\%$ of the population.
3. Sample is sufficiently large.	3. All expected counts ≥ 5.
◆ **Homogeneity** [df $= (r - 1)(c - 1)$; samples from many populations compared on one variable]	
1. Data are counts.	1. (Are they?)
2. Data in samples are independent.	2. SRSs and $n < 10\%$ OR random allocation.
3. Groups are sufficiently large.	3. All expected counts ≥ 5.
◆ **Independence** [df $= (r - 1)(c - 1)$; sample from one population classified on two variables]	
1. Data are counts.	1. (Are they?)
2. Data are independent.	2. SRSs and $n < 10\%$ of the population.
3. Group is sufficiently large.	3. All expected counts ≥ 5.
◆ **Regression with k quantitative predictors** (t, $df = n - k - 1$)	
1. Form of relationship is linear.	1. Scatterplot of residuals against predicted values shows no special structure.
2. Errors are independent.	2. No apparent pattern in plot of residuals against predicted values.
3. Variability of errors is constant.	3. Plot of residuals against predicted values has constant spread, doesn't "thicken."
4. Errors follow a Normal model.	4. Histogram of residuals is approximately unimodal and symmetric, or Normal probability plot is reasonably straight.*
Analysis of Variance (F, df depends on number of factors and number of levels in each.)	
◆ **Equality** of the mean response across levels of categorical predictors	
1. Additive Model (if there are 2 factors with no interaction term).	1. Interaction plot shows parallel lines (otherwise include an interaction term if possible).
2. Independent errors.	2. Randomized experiment or other suitable randomization
3. Equal variance across treatment levels.	3. Plot of residuals against predicted values has constant spread, Box plots (partial boxplots for 2 factors) show similar spreads.
4. Error follow a Normal model.	4. Histogram of residuals is unimodal and symmetric, or Normal probability plot is resonably straight.

*Less critical as n increases